Gas Turbines

Gas Turbines:
A Handbook of Air, Land and Sea Applications

Second Edition

Claire Soares

ELSEVIER

AMSTERDAM • BOSTON • HEIDELBERG • LONDON • NEW YORK • OXFORD • PARIS
SAN DIEGO • SAN FRANCISCO • SINGAPORE • SYDNEY • TOKYO

Butterworth-Heinemann is an imprint of Elsevier

BH

Butterworth-Heinemann is an imprint of Elsevier
The Boulevard, Langford Lane, Kidlington, Oxford, OX5 1AJ, UK
225 Wyman Street, Waltham, MA 02451, USA.

ISBN: 978-0-12-410461-7

Library of Congress Cataloging-in-Publication Data
A catalog record for this book is available from the Library of Congress

British Library Cataloguing in Publication Data
A catalogue record for this book is available from the British Library

For all information on all Butterworth-Heinemann publications visit our website at http://store.elsevier.com/

Typeset by TNQ Books and Journals
www.tnq.co.in

Printed in the USA

Dedication

To all my engineering friends, including but not limited to: the late Jim Hartsel, Dave Wisler, Dick Greenland, Stan Nathanson, the late Jim Pugh, John Russell, Peter McDermott, Graham Reynolds, Bob Kobierski, John Allingham, Roy Cleaver, and Fred Geitner. You taught and encouraged me.

To the dozens of men who appeared on the annual ASME IGTI conference panel sessions (1985 through 2003) on "Engine Condition Monitoring Systems as They Relate to Life Extension of Gas Turbine Components." The vision I gained from working with all of you is the basis for much of this book.

To Dr. D. Brian Spalding who got my career started. A better start, no engineer could have.

To Heinz Bloch who asked me to coauthor my first book.

To my family and friends, for all your patience. Thank you for just being there.

To the gas turbine engines in my life, especially IAE's V2500 "My lady V."

"The measure of a man is the friends he keeps."

Anonymous.

Contents

3. Gas Turbine Configurations and Heat Cycles

4. Gas Turbine Major Components and Modules

5. Cooling and Load Bearing Systems

8. Accessory Systems

13. Installation

14. The Business of Gas Turbines

18. Future Trends in the Gas Turbine Industry

19. Basic Design Theory

20. Additional References and Appendix for Unit Conversion

Preface 2014

In the years since I assembled the first edition of this book, gas turbine hardware remains, in principle, as it was then. True, a few major manufacturers have developed and released their J-category technology and with this come the refinements in metallurgy and design that help the J machines achieve their rating. OEMs get cleverer with exploring waste fluids for fuel, substituting fluids in case of disasters like earthquakes (air instead of water for cooling), and a myriad of other adaptations. With aviation and marine applications, there are refinements and improvements as always, but the gas turbine itself stays the same in operating principles.

It's the world in which the gas turbines operate that has changed, not always for the better. I made scant if any mention of the two wars that the United States waged in the 2008 first edition preface, the consequences on the world of that activity and the abysmal economy that started during the 43rd president of the United States' tenure, thinking all that couldn't last much longer. Clearly, I'm an optimist. Clearly, I was wrong.

Even the major original equipment manufacturers (OEMs) that were smart enough to survive years of a bad economy and emerge relatively unscathed, had to lose valuable people and expertise in the process. Some OEMs are still reorganizing and acquiring divisions of smaller competitors. Many smaller players and even related industries have gone under. For instance, about ten years ago, the turboexpander was a small but thriving turbomachinery category. Due to the specialized metallurgy it had to develop to handle corrosive or cryogenic fluids, it could have contributed much to gas turbine systems technology. Besides, a turboexpander often used what would have otherwise been a wasted stream of by-product fluid to develop power. Today the few remaining players in the turboexpander business have had to amalgamate and/or downsize to stay alive.

The wars in the Middle East breathed new life into the excess and used machinery business. When restructuring Iraq, for instance, the U.S. development teams would buy out-of-date gas turbine packages that may have had layers of dust on them to put them to use. This was more expedient than ordering new systems.

The wars also shifted attention and development funds to the development of weaponry, surveillance, and other defense electronic systems. World Wars I and II had seen the emergence of the jet engine because of necessity. This then led to the establishment of the peacetime gas turbine business. No similar steep development curve for gas turbine engines came with the Iraq (2003 to 2011) and the Afghanistan (2001 to -) wars.

Since money has been short, gas turbine owners are keener to keep what they already own. This is responsible for some of this second edition's updates, but relatively little material has been removed from the first edition. The latter's information is all still relevant. Even in cases where there has been development (like with the J size gas turbines), it is still useful to note the development work that preceded earlier developments. An engine fleet's history is often a major clue in determining overhaul and fleet management strategy. Consider as well, that in the developing world, many countries still keep the archaic models that came with communism's or colonialism's original overlords.

Global evidence of climate change has increased dramatically in the last decade, causing the United States to step up its drive to promote renewables and increase efficiency in fossil-fueled power plants. It is clear that fossil fuels will be around for another half century at least. The "smart grid" that promotes the integration of renewables into the power mix is now being constructed in some locations and then the rest of the world will have to catch up. While still not bending a knee to emissions protocols that were accepted globally, the United States is clearly keen on reducing its emissions and is stepping up incentives and legislation in that vein. "Repowering" (replacing steam turbines with gas turbine combined cycles or even replacing steam turbine coal fuel with gas fuel) is on the increase. The increased availability of gas due to the fracking movement is in part responsible. Since fracking may sometimes be handled by small, ruthless contractors who cut corners and create environmental hazards, fracking has a bad name with the public. If properly regulated and using the technology as it was designed, fracking need not be a nuisance. Further mass increases in drilling will cause seismic unrest below ground that can have consequences of its own, outside of the fracking process.

As a result, gas turbine and gas turbine system developments that favor component life extension, repairability, increased efficiency, fuel flexibility, derating and uprating, working in a hybrid system, better grid distribution, and a host of other items that affect cost of ownership, feature prominently in this second edition. It is clearly a different world than that in which the first edition emerged less than a decade ago. We ought to consider adapting.

The good news is that the gas turbine itself, fuel variations (with different IGCC cycles and experiments with new fuels) notwithstanding, is now a well-developed and relatively predictable beast. When OEMs develop newer models (like the J machines and their CC packages), they do so around a reliable core, albeit with refinements in design, manufacturing, and sometimes performance testing. Some of my OEM friends chuckle at the reduced level of gas turbine expertise among end-users (although there are always exceptions), but quite apart from the growth of power-by-the-hour and similar contracts, the end-user engineer can often get by with knowing less than one had to, two or more decades ago. One friend who retired from a major OEM went back to teach young gas turbine design engineers there and commented that the new engineer workforce would never get to do work as challenging and interesting as those who started their careers anywhere between 1965 and 1985. When I teach courses in industry, I observe that he is frequently right. I also see a trend among engineers and their management wanting material presented in more visual, easier to assimilate teaching formats. My media hobby skills have grown into a set of media resources in case the client requests animations, and digital videos versus static displays, in courses they order.

Does the gas turbine still offer the potential for reduced costs per fired hour? Certainly. Fossil fuels are alive and well and will be in use (although the renewable mix will increase thanks to smart grids and other factors) past the life span of anyone alive today. Refinements with respect to fuel technologies, repair and overhaul, metallurgy, environmental strategy, and emissions economies shall all continue. Emissions taxation and credits will increasingly affect our lives. Coal as "syngas" will continue to be refined as a gas turbine fuel. Hence this second edition.

Preface 2008

The current models of aviation engines on transpacific flights develop about 90,000 pounds of thrust, power generation gas turbines have broken the 450 megawatt gas turbine barrier, and gas turbines are now being used on cruise ships. Gas turbines have come a long way since my first meeting with them. At that point over thirty years ago, we waited a day for the casing on an old 20 kilowatt Brown Boveri to cool sufficiently for us to be lowered by rope harness into the intake for an inspection.

Most end users do not part easily with their old workhorses, so many of them are still around. The Rolls Royce Avon fleet on the Alaska pipeline, the huge number of globally installed General Electric Frame 5s (the original version, not the newly introduced model), the myriad of Solar Centaurs and Saturns everywhere in the world that needed "just about 3,000 or 1,000 horses" for a pipeline or oil and gas application, the many models of Pratt and Whitney's JT8D that still make up one of the world's largest commercial aircraft engine fleet. They work reliably, if inefficiently by today's standards, surrounded by a work force that can often hear the slightest whimper of distress from their machine—often because they rarely hear one. To some extent, these turbines owe their longevity to the continual design development in the form of service bulletins, decreed "mandatory" or "optional" by their manufacturers. I use quotation marks, as sometimes there are manufacturers who have used the "mandatory" label as a means of upgrading end-user fleets for their own revenue extension. Rather than actual end-user power requirements, the OEM's motivation was to lower the number of configurations that required a stock of spares, and other profit-motivated objectives. And then other times, as with the JT8D, the bulletins developed took a generic 9,000 pound thrust engine, born in the 1950s to just under 20,000 pounds thrust by the 1980s with one of the most enviable safety records for a gas turbine fleet. Many land based gas turbines, like the old GE Frame 5 have a proven record of specific steam injection designs raising their power output by 20–25%.

There are many types of basic applications of gas turbines. There are land, sea, and air gas turbines. On land, there are power generation and mechanical drive gas turbines. In aviation, there are large commercial, high-performance military, mid-range commercial, small fixed-wing, and helicopter engines, maintained to commercial or military specifications. At sea, there are large vessel turbines and smaller ferry turbines. There are offshore applications that must incorporate the sturdiness associated with land use turbines, with the light weight associated with aviation applications, with the corrosion resistance associated with marine applications.

There are many types of engineers who are fortunate enough to work with gas turbines. There are end users and OEMs (original equipment manufacturers). Gas turbine specialists and turbomachinery specialists who work on all rotating machinery. Overall systems and project engineers. And manufacturer design specialists who will work on one major turbine component all of their working lives.

There is an indefinable quality about gas turbines that favors those that who somehow develop an instinct for them, regardless of working years spent or formal education accumulated. I have watched humble mechanics point the way for befuddled technical gurus. There are brilliant design engineers who can miss a misalignment source that a millwright can spot blindfolded. There are some engineers who can actually troubleshoot a practical problem, and others who can't.

With gas turbines, there are systems design and specification, commissioning, troubleshooting, failure analysis, retrofit and reengineering, training, technical writing, design development, repair and overhaul, fleet management, and regular operations functions. I have been singularly fortunate in that I have run that entire gauntlet back and forth in power generation, oil and gas, process, military aviation, and commercial aviation on three different continents. No credit to my astuteness: the state of the world kept moving me on (politics is a good thing sometimes).

One flash of discernment, however, did make it possible for me to hold all of that exposure together not just as a cohesive whole, but one where all sectors could gain from each other. Then in the Canadian Air Force, I was about to take on all the six helicopter engine fleets the Canadian military branches flew. The presentation before me at the 1984 annual American Society of Mechanical Engineers International Gas Turbine Institute meeting (ASME IGTI's

"TurboExpo") featured an offshore oil and gas man who was displaying the control panel that ruled a platform. In Canada, we had just "piggybacked" on the US F-18 fighter program with a few of the same. The F-18's F-404 General Electric engine had a condition monitoring system that was, to say the least, intriguing. It occurred to me that the panel displayed on the screen was very similar to the one on the HUD (head up display) of the F-18.

And so a "joint" ASME IGTI session that I ran annually from 1985 through 2003, patiently assisted by luminaries like Jim Hartsel (one of the General Electric turbine engineers most responsible for the superior performance of the F-404 and the T700), was born. The committee sponsors included the Aircraft Engine, Controls, Instrumentation & Diagnostics; Materials, Metallurgy & Manufacturing; Marine, and Electric Utilities committees. The idea is to get land, sea, and air people to learn from each others' experiences. It works. The panel has hosted some of the best brains from commercial and military aviation, power generation, oil and gas, manufacturing, process and petrochemical, performance analysis, marine applications, metallurgical development, and controls instrumentation and diagnostics. Those attendees who are fortunate enough to show up have benefited enormously. It has given aspects of my work a rare flair that is attributable to the company I have been blessed to keep.

This book represents much of the expertise in the gas turbine field available today. It is 80–90% adapted and edited from many brilliant sources and about 10% is original writing. That latter portion serves to give the reader a point of reference that they can measure the extent of their agreement—or disagreement—against. It's practice for when you have to make decisions that your underwriters, insurance company, and mechanics may challenge. The book avoids the just-one-application bias (say, just mechanical drive or just power generation or just aviation or just plain theory) that all other gas turbine books I know of adopt.

Gas turbine engineers in all sectors, disciplines, and specialties, who looked at the draft, have told me they found its contents useful. Just as importantly, they gleaned information from others' applications. So besides imparting applications and basic design knowledge, this book is meant to get readers to think across disciplines, across land, sea, and air to the heart of this demanding, powerful, and infinitely variable mistress—the gas turbine.

Introduction, 2nd Edition

In this second edition, the single largest factor that is responsible for added material is technology that accommodates a wider range of fuels and fuel parameters in gas turbines (GTs) and gas turbine (including steam turbine) cycles. GTs are now on the verge of adopting coal as a fuel, as syngas in an integrated gasification combustion cycle (IGCC). Novel fuel technology has been around for decades, sometimes out of sheer necessity (as far back as the RAF using bunker fuel instead of light diesel during the Falklands war), but it's the degree of expanding success in this vein that is noteworthy. Some GTs can handle biomass. Others have been modified to burn low BTU process gas or waste fuels.

The size of the GT powerplant has been steadily growing. A GTCC is not quite in the size league of the largest steam plants yet, but it's growing.

The world has been on a roller coaster economy for about two decades now. The aviation business has been hard hit. GT owners are keener to keep their GT components for longer than their stated lives and their GTs for even longer. Overhaul and repair, as well as performance optimization, continue to develop.

In the last decade, the gas industry forges ahead with fracking. This drops the price of gas dramatically, making the development of IGCC technologies less vital, but politics, labor markets, and low gas resources in many poorer countries insist that coal as a fuel (in GTs or otherwise) must continue to develop. Emissions concerns prompt the development of carbon capture and sequestration. Carbon emissions can be "turned around" as carbon credits and then traded or taxed.

Technology is now accessible enough that power can be made in ever smaller distributed packages: microturbines, with or without fuel cells, solar, wind, wave, tide generators, hybrids of all of them, some of which include small (micro) gas turbines. Ironically, even as small users develop options with respect to their own energy independence, large power continues to grow. Metallurgy improves to accommodate ultra supercritical steam, not just supercritical steam, helping coal rich countries with no gas, to avoid paying heavy premiums for imported oil and LNG: thus the steam turbine business continues to rival the gas turbine sector.

Support technologies like smart grids and smart materials play their part. New grid technology helps renewables become a contender, albeit smaller than fossil fuel plants. Smart materials will slowly transform manufacture.

At last, there is widespread recognition, even among stubborn politicians, that emissions need to be mitigated, so carbon-capture and sequestration updates are a larger part of this second edition. This is one of the strongest areas of international cooperation, within the EU, in the Americas and Asia and increasingly, across everyone's oceans. Reduced emissions generally also mean not just maximized fuel efficiency, but longer times between overhauls. That was not always a connection made by all.

Introduction, 1st Edition

In the gas turbine world, it is essential for all industry sectors to learn from each other. Despite how expensive reinventing the wheel might be, this does not happen enough or sufficiently.

The extent to which it does happen, however, is owed largely to the inception of the aeroderivative gas turbine engine. In part fostered by the offshore industry's need for a lighter-than-industrial-engine-frame, OEMs (original engine manufacturers) took specific aircraft engines and placed them on a light, strong, and flexible base. Some of industry's largest fleets are aeroderivative. The land-based Rolls Royce RB211, Trent, and Avon all had mothers who fly (or flew). A General Electric CF6-80C2 eventually "produced" an LM2500 on the ground. In fact the metallurgy of contemporary General Electric Frame 7 and 9 engines is quite similar to that used for the CF6 mature models. The Rolls Royce Olympus and Spey that are used so effectively in marine, offshore, and conventional land-based applications have aero roots.

The logic for the panel that I discussed in the Preface continues. When the General Electric first released their F404 engine triple redundancy architecture was relatively new to industrial users. It is now commonplace in modern power plants.

The concept of a cycle of gas turbine life used versus a calendar hour evolved from realizing that the leader of an aerobatics squadron might only develop 1/20 of the wear on his engines, as compared with the engines of his followers who have to "hunt and follow" a specific distance from his wings. Algorithm development uses the parameters of time, temperature, and speed essentially, to calculate cycles. Unless the engine is among specific models of Rolls Royce that can use just time and speed parameters for the most part. Additional cooling may cost in some ways but pays off in others. Land based users slowly caught on with experimental work on how much a stop and start to a conventional industrial machine (such as an original General Electric Frame 5) versus a much smaller workhorse (such as a Solar Centaur) was worth.

Profit margins are directly affected by how quickly different sectors can learn from each other. The sheer size of GE's aviation CF680C2 fleet and their land-based LM2500 fleet (the stimulus of GE's unparalleled financing program notwithstanding) can attest to that. As can the sizes of the Rolls Royce RB211 (land and air fleets). As can the success of the Alstom (once ABB) GT35 both on land and at sea.

Note also that as personnel and technology travel across national boundaries, proprietary technical innovations follow them. The wide chord fan blade is an acknowledged Rolls Royce "first." It was an enviable one as its performance attributes, both in terms of aerodynamic performance and bird (FOD or foreign object damage) chopping ability, verified. The latter rang the bell at 24 pounds of bird (a vulture over the Indian Air approach in Calcutta). In fairly short order, the wide chord fan blade has appeared on other OEM's models, albeit with different "internals."

Gas turbines and their development are plagued with the whims and dips of global finance and politics. Rarely do all sectors have the development money required for their progress at the same time. Military aviation engines may have large budgets at a time when funds in the land-based sector are scarce. In times of peace, certain military engine development programs may be totally suspended. In more recent wars, the priority with military engines may shift from performance to longevity.

The end-user parameter throws the OEM a variable and sometimes exasperating curve. Not all end users are as astute as they'd like to believe. Many of them "shoot from the hip" with unfortunate "brainwaves." Even when OEMs phase out a model by refusing to service it, some less affluent user generally salvages it, illegally or otherwise. The fact that they might not know the meaning of terms like "not under the stress endurance curve" or "hydrogen embrittlement" does not stop some from midnight raids on the scrap heaps of legitimate shops.

That brings us to the "buyer beware" issue. I have spent a little space discussing cases that books rarely touch, such as the case where 5 JT8Ds were sold to a hapless U.S. courier service by a trusted U.S. ally that claimed the engines had undergone "ESV2" (major overhaul complete status). The engines were missing significant components such as inner air seals on the low-pressure turbine. The diffuser cases (that have to hold significant pressure) had cracks long enough in them that they should have been scrapped. "Zero-timed" overhaul may thus be a function of who's making that claim.

One way the end user can get leverage with OEMs is by joining an end-user group: a lobby of sorts. This is discussed in some detail in Chapters 14 and 17. In the mechanical drive land-based world, one former group that did this was called the Gas Turbine Users Association. It was the strength of this forum that persuaded a major manufacturer of 3,000 and 1,000 horsepower turbines to back off obliging many end users who had no use for a specific service bulletin to "adopt it or void warranty." It was similar "users' strength in numbers" that had Rolls Royce developing cycle use algorithms for the major hot section components in their Spey and Olympus models, which gave the end user a hefty additional lease of life.

In the power generation world, each major model has its own end-user group. Alstom's (formerly ABB) GT11N group will meet separately from the Alstom GT24 group and so forth. Here the division between end user and OEM can get blurred. Many OEMs now own large shares of power stations. Or they may participate in "BOOT" (build, own, operate, transfer) contracts. Another notable advance is that oil and gas giants are now entering the independent power generation business. Exxon-Mobil, Chevron, and Shell are among those who have moved in a big way on the opportunity to build their own power plants, to whom their refineries and gas fields could then sell their own fuel. I say "in a big way." When dictated by demographics, oil companies have always made their own power. I cut my power generation teeth at the first Syncrude tarsands project that has always sold excess power back to the public grid.

In military aviation, there are CIP (component improvement meetings). In my military chopper days, I recall a number of CIPs with the U.S. Coastguard most in attendance. In terms of mandate and mission profile, the Canadian helicopter fleets (Army, Navy, and Air Force) most commonly resemble the U.S. Coast Guard for search and rescue and smuggling/drug enforcement patrols. In fact, some of those profiles might sometimes be more demanding than conventional military operations impose.

And so to get maximum benefit for this book, although you might scour the index for mention of "your" model, the more useful stuff you learn may be from applications that run 40,000 feet higher—or lower—than yours. Or a wartime pressed version of "your" aeroengine at sea that had to use heavier fuel than the manufacturer specified. Or a power generation version of your mechanical drive application whose end users happen to have a great deal more budget for financing new repair development or performance analysis system development.

OEM design development that also considers the results from these meetings may include optimized controls and diagnostics (Chapter 9), performance methods (Chapter 10), environmental strategy (Chapter 11), repair and overhaul (Chapter 12), improved testing and installation (Chapter 13), business methods (Chapter 14), and manufacture (Chapter 15). Chapter 19 deals with the design and calculation strategy used by OEMs. The chapter on education also deals with OEM and agency participation in gas turbine educational programs.

Hybrid systems (Chapter 16) are taking on a larger profile for energy conservation and other reasons: combined cycle power generation for instance. Fuel cells and microturbines are proving to have their place, independently and in combination with more conventional machinery.

The book contains detailed discussion of all gas turbine components (Chapters 1 and 2) and elements of a gas turbine system including instrumentation, monitors, filters, and other accessories (Chapters 3 through 8). Specific landmark and case histories on interesting contemporary applications in plants have also been included.

Acronyms

ABB Asea Brown Boveri
ABRT Accreditation Board of Engineering and Technology
AC ammonium carbonate
ACC active clearance control
ADC adjusted direct cooling
ADH advanced diffusion healing
AI Artificial intelligence
AIC adjusted indirect cooling
AMB active magnetic bearings
AOH Actual Operating Hours
APS Arizona Public Service's
APU Auxiliary Power Units
ARC Albany Research Center
ASME American Society of Mechanical Engineers
ASU air separation unit
ATF Altitude Test Facility
ATS Advanced Turbine System
AVA Aerodynamische Versuchs-Anstalt
AVT All-Volatile Equipment
BAT best available technology
BFO blended fuel oil
BOOT build, own, operate, transfer
BOP balance of plant
BOV blow-off valves
BP booster pump
BPST Back-Pressure Steam Turbine
BVM Blade Vibration Monitor
CAS close air support
CC Combined-Cycle Power Plan
CCPP combined cycle power plant
CDA controlled diffusion airfoils
CE Coulombic efficiency
CEC Community Environmental Council
CEM continuous emission monitoring
CF cycle fatigue
CFCC continuous fiber reinforced ceramic composites
CFD computational fluid dynamic
CHP Combined Heat and Power Plant
CMC ceramic matrix composite
CMM coordinate measurement machine
CMS condition monitoring system
CO carbon monoxide
COD chemical oxygen demand
COG Coke Oven Gas
CP Constant Pressure
CPP captive power plant
CPUC California Public Utilities Commission
CSP concentrating solar power
CV Constant Volume
DCS Distributed Control System
DDA direct drive alternator
DENR Department of Environment and Natural Resources
DLE Dry Low Emissions
DLN dry low NOx
DOE The US Department of Energy
DS directionally solidification
DSS Daily Start and Stop
EAM electronics assembly module
EC engineering criteria
ECMS Engine Condition Monitoring Systems
EEL Exxon Energy Limited
EGAT Electrical Generating Authority of Thailand
EGR Exhaust Gas Recirculation
EGT European Gas Turbine
EIA Energy Information Administration
EOH equivalent operating hours
EOH equivalent operating hours
EPA engine performance analysis
EPC Engineering, Procurement, and Construction
EPDC Electric Power Development Co.
EPEC Existing Plants, Emissions & Capture
ESR effective structural repair
ESS engine section stator
ETATH Thermal efficiency for shaft power engines
ETN European Turbine network
FADEC full-authority digital engine controls
FBH front bearing housing
FCCU fluid catalytic cracking units
FCFC full coverage film cooling
FGD flue gas desulfurization
FHV fuel heating value
FIC fixed indirect cooling
FOCM Fiber Optic Communication Module
FOG Finex Oven Gas
FOVM Fiber Optic Vibration Monitor
FPA fuel purchase agreement
FT Fischer-Tropsch
GE General Electric's
GEAE General Electric Aircraft Engine
GG gas generator
GM General Motors
GT GAS TURBINE
GTCU gas turbine change unit
GTUA Gas Turbine Users Association
HAP hazardous air pollutants
HAT humid air turbine
HCF high-cycle fatigue
HEFPP High Efficiency Fossil Power Plants
HFO Heavy Fuel Oil
HGP hot-gas-path
HHV higher heating value
HIP hot isostatic pressing
HMIP Her Majesty's Inspectorate of Pollution
HP High Pressure
HPC high-pressure compressor
HRATE Heat rate for shaft power cycles

HRSG heat recovery steam generator
HTDU high temperature demonstration unit
IAE International Aero Engines
IAEA International Atomic Energy Agency
IBA independent bearing assembly
IEEE Institute of Electronics and Electrical Engineers
IGCC Integrated Gasification Combined Cycle
IGV Inlet Guide Vane
IHPTET Integrated High Performance Turbine Engine Technology
IMO International Maritime Organisation
IOR improved oil recovery
IP Infrared Pyrometry
IP intermediate pressure
IPC Integrated Pollution Control
IPC intermediate pressure compressor
IPG International Power Generation
IPP independent power producers
IR infra-red
IR Ingersoll Rand
ISA International Standard Atmosphere
ITM ion transport membrane
JDF Japan, the DME forum
JSF Joint Strike Fighter
LCA life cycle analysis
LCA life cycle assessment
LCC Life Cycle Cost
LCF low-cycle fatigue
LD liquidated damages
LEED Leadership in Energy and Environmental Design
LHV lower heating value
LLNL Lawrence Livermore National Laboratory
LNCFS low NOx concentric firing system
LP Low Pressure
LPG Liquefied petroleum gas
LSB last stage blades
LTSA long-term service agreements
MAT moisture air turbine
MCA multiple circular arc
MCFC molten carbonate fuel cell
MCR maximum continuous rating
MFC Microbial fuel cells
MI Minor inspection
MITI Ministry of International Trade and Industry
MMI man-machine interface
MPI multi-passage injector
MPS Mitsubishi Power Systems
MSF multistage flash
MTU Motoren Turbinen Union
NG Natural Gas
NGTE National Gas Turbine Establishment
NIC newly industrialized country
NO nitrogen oxides
NPC National Power Corporation
NREC Northern Research and Engineering Company
OA overfire air
OEM Original engine manufacturers
OGV outlet guide vanes
OHSU Oregon Health and Science University
OMCR Oil Movements Control Room
ORNL Oak Ridge National Laboratory
OTM oxygen transport membrane
PA performance analysis
PA performance assessment
PAE Project application engineers
PAH Polycyclic aromatic hydrocarbons
PC pressure compressor
PCA parametric cycle analysis
PCC Post-combustion capture
PCFBC pressurized circulating fluidized-bed combustion
PCS Petrochemical Corporation of Singapore
PDI Product Design and Improvement
PDP Siemens Product Development Process
PEM polymer electrolyte membrane
PFBC pressurized fluidized-bed combustion
PHCR Power House Control Room
PI power island
PM Particulate matter
PMG Power Marketing Group
PR pressure ratio
PRDS Pressure Relieving Desuperheating Station
PT personal turbine
PTC Performance Test Code
PTET power turbine entry temperature
PV Paxman Valenta
QC quality control
RAM reliability-availability-maintainability
RBT Resistance Bulb Thermometers
RC Rankine Cycle
RCC Residue Catalytic Cracker
RCT refinery crude train
RFG Refinery Fuel Gas
RFM Radio Frequency Monitor
RFP request for proposal
RH relative humidity
RIC reactive ion coating
ROI return on investment
RPS Renewable Portfolio Standard
RPV Remotely Piloted Vehicle
RR Rolls Royce
SAE scale altitude effect
SAGBO stress assisted grain boundary oxidation
SB service bulletin
SB service bulletins
SCE Southern California Edison
SCR selective catalytic reduction
SEV sequential environmental
SFC specific fuel consumption
SFR steam fuel ratio
SGT Siemens Westinghouse's
SGT Small Gas Turbine
SIL sound intensity level
SMP standard maintenance procedures
SOFC solid oxide fuel cell
SOP standard operating procedures
SPL sound pressure level
SPP small power producers

SPW Specific power or thrust
SRB surface reaction braze
SRI sound reduction index
ST static temperature
ST Steam Turbine
STOL short takeoff and landing
STOVL short/vertical takeoff/landing
SWPC Siemens Westinghouse Power Corporation
T/C Thermocouple
TBA Test Bed Analysis
TBC Thermal Barrier Coating
TCM Test Centre Mongstad
TCP Topologically close-packed
TDS Total Dissolved Solids
TE Temperature Element
TF tangential firing
TFO treated fuel oil
TGA thermo gravimetric analysis
TGO thermally grown oxide
TI Turbulence Intensity
TIT turbo inlet temperature
TMF thermal-mechanical fatigue
TNB Tenaga Nasional Berhad
TOC Total Organic Carbon
TRU thrust reverser unit
TSC Turbine Stress Controller
TSTC Texas State Technical College
TT temperature transmitter
UAE United Arab Emirates
UCG underground coal gasification
USAFA US Air Force Academy
USC ST Ultra Supercritical Steam
VA vibration analysis
VAN variable area nozzles
VCRF Vertical Combustion Research Facility
VDU visual display unit
VGV variable guide vanes
VIS Viscosity
VM vibration monitoring
VOC volatile organic compounds
VPC variable production cost
VR variable reluctance
VTOL Vertical takeoff and landing
VWO valve wide open
WB World Bank
WFO Waste Fuel Oil
WI Wobbe Index
WW World War

Notes to the Reader

From the author:

1. When I assembled the second edition of this book, I found the material I included in the first edition more relevant today than ever. That's primarily because most of the updates I have added in the second edition platform on the technology I covered in the first edition. Also gas turbines require massive financial outlay, so no GT owner parts with his older models because there are new ones. The GT field is one where end-users hire consultants to engineer steam injection uprates to their older existing machines to equate to buying a new one. Therefore, for the most part, it made sense to leave the first edition material where it is.
2. For 19 years, I ran a panel session at ASME's IGTI annual meeting that resulted in attendees considering the gas turbine (GT) experience of end-users in other sectors. My own time in the gas turbine industry has gone from power generation to mechanical drive to aeroengine and back again, so it is second nature to be able to apply the theory, applications, and experience from one sector to another. As this book covers "land, sea, and air," my suggestion to the reader is that he or she use this book to do the same, if that's not already "standard."
3. Case studies in this book are generally adapted extracts from academic papers that are included with the authors' or the authors' parent companies' permission. If the reader decides to get the complete original paper, I suggest contacting the originating or originating OEM company or authors (if the authors are "independent"). The publisher of said work may be an academic society. As such, its technical work may be run by volunteers. Its core staff may be entirely non-technical and would not be as seasoned as the writer of the work or the OEM company involved. The latter may, for instance, point out that there have been more papers on that same subject, which they may have presented at another society's venue. Academic society or conference/publisher offices, however, can sell overall proceedings of a meeting, generally supplied on a CD Rom. They could also be a source of archived papers that the originating OEM, whether because of new technology and information attrition through joint ventures and acquisitions, has lost. Most OEMs, however, carefully safeguard their "old" material and one ought to consider contacting them first.
4. The global economy and consequential mergers and acquisitions have complicated the gas turbine sector technically, sometimes for the better. One may note that the cross-section of a Siemens W501 resembles the equivalent Mitsubishi (MHI) turbine. This is logical if one recalls that Siemens acquired Westinghouse (a U.S. company), which prior to its acquisition, had worked on the forerunner to the W501 in a joint venture with MHI. So in considering any engine, one needs to recall who did the original development engineering and of what the system or component most of interest, comprises. This is especially true in joint venture engines. Purchasers of IAE's V2500 were heard to sigh with relief when they heard that the oil system was a Rolls Royce design. Rolls Royce is known to not skimp on their cooling whether it be via internal air or lubrication.
5. The changing gas turbine world has also created logistical issues. Brown Boveri (BB) became Asea Brown Boveri (ABB), which still exists and very healthily too. ABB now deals with fans, control systems, and many other critical GT system features. The part of ABB that made gas turbines and GT CCs became ABB Alstom, and then later Alstom (Alstom Power). ABB originally had two main branches that made entirely different size ranges of turbines. ABB Stal was in Sweden and built the smaller GTs. ABB Switzerland built the larger machines. Siemens (after it had acquired Westinghouse) acquired parts of Alstom Power that essentially included ABB Stal. Soon thereafter, Siemens developed a standardized numbering system for all its GT models that essentially changed the designations of the acquired engine lines and Siemens existing gas turbine model numbers.

   After the content of this edition was final, Siemens bought part of Rolls Royce. So model number changes within engine lines, will continue.

6. Due to the complexities that come with item 4 above, I have done the following with respect to all case histories included here. If a paper was originally written by "ABB" for instance, I have left that name as is in the text, even if that engine line now is, for instance, part of the Siemens engine line. This does no disservice to Siemens, as the "change over in model numbers" key by Siemens is included both in this text* and on (potentially updated) the Siemens website. This way, the reader can assess the chronological state of the technology at the time the case was written.
7. Although items 4 and 5 give the reader some responsibility for checking the logistics of the equivalent current model numbers, this avoids the far more serious alternative of confusion on technical issues. If, for instance, in the future, the acquiring OEM were to institute a technical modification to a model that may then change the outcomes noted in the subject case, the reader then understands the chain of technical development custody. This is crucial when making decisions regarding, for instance, independent facility repairs or retrofit engineered systems. Most end-users (provided they are engineers and not accountants or lawyers) would rather wade through logistical records than acquire a technical problem because they did not understand the design development history of their engine.
8. The other reason I favor this approach is it gives credit for the original case study to the individual people who wrote it. The company they worked for at the time may still be their employer or they may have moved to another OEM, via an acquisition. Or they may work for yet another employer or have retired as "independent(s)." Gas turbine engineers would favor this approach, because they can then read the case about the engine they are interested in, as well as recognize the original authors by name, if they were to meet them at some future date.
9. In the interests of making this book affordable, several source illustrations that were originally in color were adapted to be readable in gray scale. The reader has a choice here; to make future reference easier, I get my students to use colored markers as a learning exercise and for future clarity. The reader can also purchase a copy of all the source documents in which these figures appear in color.

***Siemens Gas Turbines Industrial Applications**

Product Type	Previous	New
Siemens Gas Turbines	Typhoon	SGT-100
	Typhoon Single Shaft	SGT-100-1S
	Typhoon Twin Shaft	SGT-100-2S
	Tornado	SGT-200
	Tornado Single Shaft	SGT-200-1S
	Tornado Twin Shaft	SGT-200-2S
	Tempest	SGT-300
	Cyclone	SGT-400
	GT35	SGT-500
	GT10B	SGT-600
	GT10C	SGT-700
	GTX100	SGT-800
	W251	SGT-900
	V64.3A	SGT-1000F

Large-Scale Power Applications

Product Type	Hz	Previous	New
Siemens Gas Turbines			
Examples:	50		SGTS-PAC (GT Driver)
		V64.3A	SGT-1000F
		V94.2	SGT5-2000E
		V94.2A	SGT5-3000E
		V94.3A	SGT5-4000F
Examples:	60		SGT6-PAC (GT Driver)
		V64.3A	SGT-1000F
		V84.2	SGT6-2000E
		W501D5A	SGT6-3000E
		V84.3A	SGT6-4000F
		W501F	SGT6-5000F
		W501G	SGT6-6000G

* Siemens standardized model numbers (as per Siemens catalogs) are currently as in the tables below. Consult Siemens catalogues for additional model number details.

About the Author

A professional engineer registered in Alberta Canada, and a Fellow of the American Society of Mechanical Engineers (ASME), Claire Soares has worked on rotating machinery for over 20 years. Claire's extensive experience includes the specification of new turbomachinery systems, retrofit design, installation, commissioning, troubleshooting, operational optimization, and failure analysis of all types of turbomachinery used in power generation, oil and gas, petrochemical and process plants, and aviation. The turbines (gas, steam, or combined cycle; land or aero applications) in question were typically made by General Electric, Siemens Westinghouse, Rolls, Rolls Allison, Solar, Alstom Power, or the companies they formerly were, before some of them merged.

Her career experience also includes intensive training programs for engineers and technologists employed by heavy industrial clients. Her specialty areas include turbomachinery diagnostic systems, failure analysis and troubleshooting as well as subsequent retrofit and re-engineering.

In her years spent with large aircraft engine overhaul and aircraft engine fleet programs in the United States and Canada, Claire worked on turbine metallurgy and repair procedures, fleet asset management, and aeroengine crash investigation. She also was engineering manager for the first overhaul program in the United States for the V2500 engine (commissioned in 1991).

Gas turbines, land, air, and sea, are Soares' primary area within the turbomachinery field. Her perspective with respect to gas turbines is that of an operations troubleshooter with extensive design experience in gas turbine component retrofits and repair specification as well as retrofit system design development.

Claire has authored or co-authored six books for Butterworth-Heinemann and McGraw-Hill on rotating machinery. See the links below for book details. She also writes as a freelancer for various technical journals.

Claire has an MBA in International Business (University of Dallas, TX), and a B.Sc.Eng. (University of London, external). She is a commercial pilot. Her scuba diving certification and training were in high altitude conditions. She has lived and worked on four continents. Her "non-engineering" time is partly spent on cinematography and still photography, both of which have grown to complement her training courses and writing. She develops training packages presented in video or other audio-visual formats.

http://books.elsevier.com/bookscat/search/results.asp?country=United+States&ref=&community=listing&mscssid=0589M7ACKL658H5QPFMW2650RBQ26XGD

http://books.mcgraw-hill.com/search.php?keyword=claire+ soares&template=&subjectarea=113&search=Go

"History will be kind to me for I intend to write it."

Winston Churchill

Chapter 1

Gas Turbines: An Introduction and Applications

"The farther backwards you can look, the farther forward you are likely to see."

—Winston Churchill

Chapter Outline

The gas turbine is the most versatile item of turbomachinery today. It can be used in several different modes in critical industries such as power generation, oil and gas, process plants, aviation, as well domestic and smaller related industries.

A gas turbine essentially brings together air that it compresses in its compressor module, and fuel, which are then ignited. Resulting gases are expanded through a turbine. That turbine's shaft continues to rotate and drive the compressor, which is on the same shaft, and operation continues. A separate starter unit is used to provide the first rotor motion until the turbine's rotation is up to design speed and can keep the entire unit running. The relationship between pressure, volume and temperature is discussed later in this chapter. Note that this relationship is common to gas turbines regardless of the application.

The compressor module, combustor module, and turbine module connected by one or more shafts are collectively called the gas generator. Figure 1—1 illustrates a typical gas generator in schematic format.

The second half of this chapter will deal in more detail with different applications and their subdivisions. At this time, we will look at some typical examples of land, air, and sea use.

GAS TURBINES ON LAND

The gas turbine itself operates essentially in the same manner, regardless of whether it is on land, in the air, or at sea. However, the operating environment and criticality of the application in question may make design and system modifications necessary. For instance, the gas generator

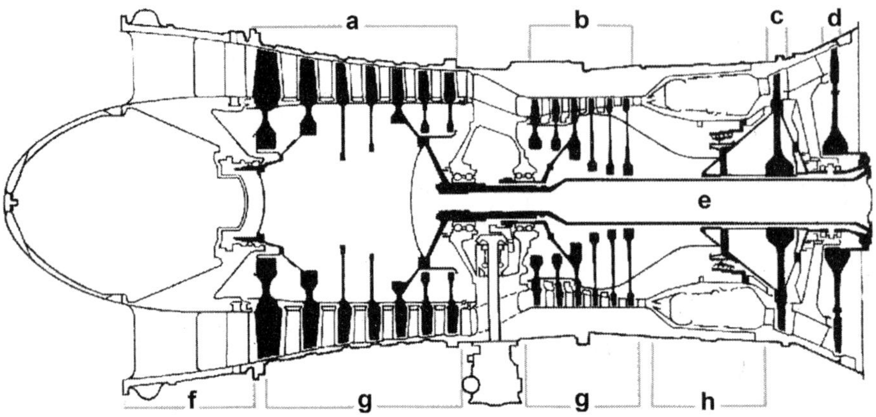

FIGURE 1–1 Schematic of gas turbine. *(Source: Bloch and Soares,* Process Plant Machinery, *Second Edition. Boston: Butterworth-Heinemann, 1998.)*

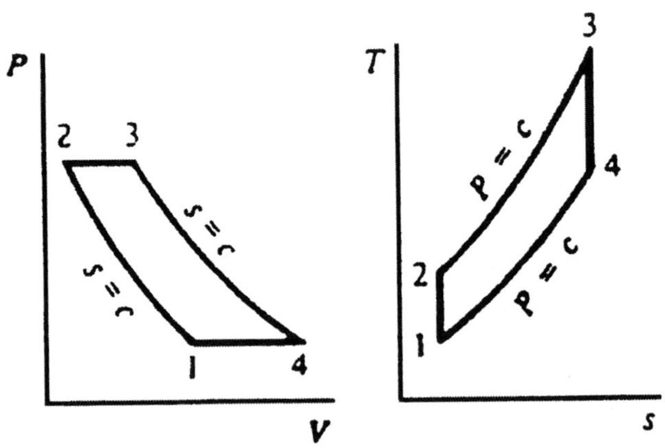

FIGURE 1–2 The basic gas turbine cycle. *(Source: Kiameh,* Power Generation Handbook. *New York: McGraw-Hill, 2003.)*

shown in Figure 1–1 may be operating in mechanical drive service to drive compressors that move gas down a pipeline. Essentially the same machine can be used to generate power. It can also be used as a power plant on an aircraft. However, the layout, the other turbomachinery supplied with the gas turbine, and optional systems will vary in each case.

Let us first look at the basic gas turbine cycle (see Figure 1–2).

A comparison can be drawn between the gas turbine's operating principle and a car engine's (see Figure 1–3). A car operates with a piston engine (reciprocating motion) and typically handles much smaller volumes than a conventional gas turbine.

Direct Drive and Mechanical Drive

With land-based industries, gas turbines can be used in either direct drive or mechanical drive application.

With power generation, the gas turbine shaft is coupled to the generator shaft, either directly or via a gearbox: "direct drive" application. A gearbox is necessary in applications where the manufacturer offers the package for both 60 and 50 cycle (Hertz, Hz) applications. The gearbox will use roughly 2% of the power developed by the turbine in these cases.

Power generation applications extend to offshore platform use. Minimizing weight is a major consideration for this service and the gas turbines used are generally "aeroderivatives" (derived from lighter gas turbines developed for aircraft use).

For mechanical drive applications, the turbine module arrangement is different. In these cases, the combination of compressor module, combustor module, and turbine module is termed the gas generator. Beyond the turbine end of the gas generator is a freely rotating turbine. It may be one or more stages. It is not mechanically connected to the gas generator, but instead is mechanically coupled, sometimes via a gearbox, to the equipment it is driving. Compressors and pumps are among the potential "driven" turbomachinery items (see Figure 1–4).

In power generation applications, a gas turbine's power/size is measured by the power it develops in a generator (units watts, kilowatts, megawatts). In mechanical drive applications, the gas turbine's power is measured in horsepower (HP), which is the torque developed multiplied by the turbine's rotational speed.

In aircraft engine applications, if the turbine is driving a rotor (helicopter) or propeller (turboprop aircraft), then its power is measured in horsepower. This means that the torque transmission from the gas turbine shaft is, in principle, a variation of mechanical drive application. If an aircraft gas turbine engine operates in turbothrust or ramjet mode (i.e., the gas turbine expels its exhaust gases and the thrust of that expulsion propels the aircraft forward), its power is measured in pounds of thrust. The following are examples of operational specifications for land-based gas turbines.

Applications Versatility with Land-Based Gas Turbines

The gas turbine's operational mode gives it unique size adaptation potential. The largest gas turbines today are over 200 MW (megawatts), which then places gas turbines in an

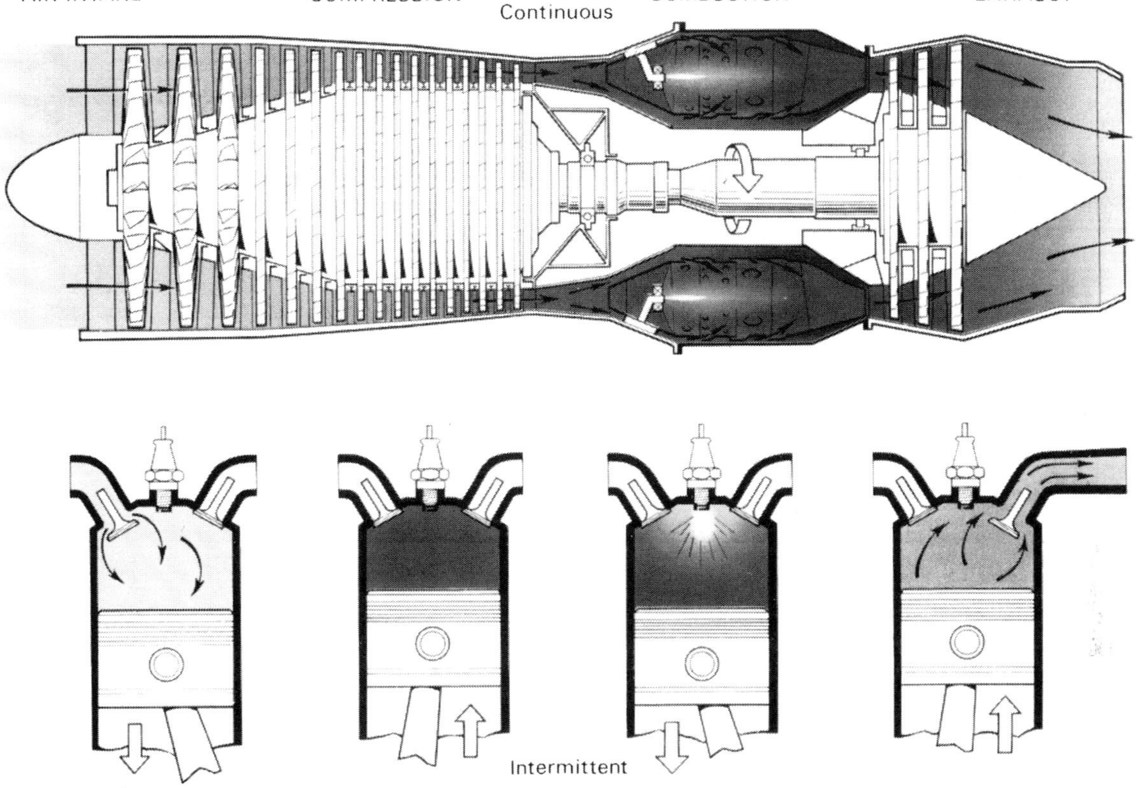

FIGURE 1–3 Comparison of the gas turbine and reciprocating engine cycles. *(Source: Rolls Royce,* The Jet Engine. *UK: Rolls Royce Plc, 1986.)*

applications category that until recently only steam turbines had owned.

The smallest gas turbines are microturbines. The smallest commercially available microturbines are frequently used in small power generation (distributed power) applications and can be as small as 50 kW (kilowatts). Work continues on developing microturbines that will be thumbnail size. The world of "personal turbines" where one might plug this turbine into a "drive slot" in their car, come home from work and plug it into a "household slot" for all one's household power is a discernible, if as yet unpredictable, target.

Understanding the gas turbine's historical origins and other applications gives the gas turbine community a better handle on optimized design, operation, and maintenance. Gas turbines came into their own in the Second World War. In peacetime, NASA took over the research that led to better alloys, components, and design techniques. This technology was then handed down to military aviation, then commercial aviation.

However, the same manufacturers generally also make gas turbines for land and marine use. So "aeroderivative" gas turbines were a natural offshoot of their flying forerunners.

Aeroderivative gas turbines are essentially aviation gas turbines that are installed on a light frame on a flat surface (ground-based, marine craft, or offshore platform). Aeroderivatives are commonly used in power generation service, particularly where a relatively light package is required, such as in offshore service.

The Rolls Royce Spey and Olympus engines, for instance, are both aeroengines but are also popular when packaged as aeroderivatives in land-based and offshore platform service.

Pratt and Whitney's (PW) JT-8D was once the largest aircraft engine family in existence. The engine first made its

Alstom's GT 24/GT 26 (188 MW 60 Hz, 281 MW 50 Hz). Both Used in Simple Cycle, Combined Cycle, and Other Cogen Applications

GT24 (ISO 2314;1989)	
Fuel	Natural gas
Frequency	60 Hz
Gross electrical output	187.7 MW*
Gross electrical efficiency	36.9%
Gross heat rate	9251 Btu/kWh
Turbine speed	3600 rpm
Compressor pressure ratio	32:1

(Continued)

Alstom's GT 24/GT 26 (188 MW 60 Hz, 281 MW 50 Hz). Both Used in Simple Cycle, Combined Cycle, and Other Cogen Applications—cont'd

Exhaust gas flow	445 kg/s
Exhaust gas temperature	612°C
NO_x emissions (corr. to 15% O2, dry)	<25 vppm
GT26 (ISO 2314;1989)	
Fuel	Natural gas
Frequency	50 Hz
Gross electrical output	281 MW*
Gross electrical efficiency	38.3%
Gross heat rate	8910 Btu/kWh
Turbine speed	3000 rpm
Compressor pressure ratio	32:1
Exhaust gas flow	632 kg/s
Exhaust gas temperature	615°C
NO_x emissions (corr. to 15% O2, dry)	<25 vppm

**In a combined cycle, approximately 12 MW (GT26) or 10 MW (GT24) is indirectly produced via the steam turbine through heat released in the gas turbine cooling air coolers into the water steam cycle.*
(Source: Alstom Power.)

Alstom's GT 11N2, Either 60 Hz or 50 Hz (with a gearbox). Used in Simple Cycle, Combined Cycle, and Other Cogeneration Applications

GT11N2 (50Hz)	
Fuel	Natural gas
Frequency	50 Hz
Gross electrical output	113.6 MW
Gross electrical efficiency	33.1%
Gross heat rate	10,305 Btu/kWh
Turbine speed	3600 rpm
Compressor pressure ratio	15.5:1
Exhaust gas flow	399 kg/s
Exhaust gas temperature	531°C
NO_x emissions (corr. to 15% O2, dry)	<25 vppm
GT11N2 (60Hz)	
Fuel	Natural gas
Frequency	60 Hz
Gross electrical output	115.4 MW

Alstom's GT 11N2, Either 60 Hz or 50 Hz (with a gearbox). Used in Simple Cycle, Combined Cycle, and Other Cogeneration Applications—cont'd

Gross electrical efficiency	33.6%
Gross heat rate	10,150 Btu/kWh
Turbine speed	3600 rpm
Compressor pressure ratio	15.5; 1
Exhaust gas flow	399 kg/s
Exhaust gas temperature	531°C
NO_x emissions (corr. to 15% O2, dry)	<25 vppm

(Source: Alstom Power.)

SGT-600 Industrial Gas Turbine—25 MW (Former Designation, Alstom's GT10)

Technical specifications:	
Dual fuel	Natural gas and liquid
Frequency	50/60 Hz
Electrical output	24.8 MW
Electrical efficiency	34.2%
Heat rate	10,535 kJ/kWh
Turbine speed	7700 rpm
Compressor pressure ratio	14.0:1
Exhaust gas flow	80.4 kg/s
Exhaust gas temperature	543°C
NO_x emissions (corr. to 15% O2, dry)	<25 vppm

(Source: Siemens Westinghouse.)

appearance in the 1950s and delivered about 10,000 pounds of thrust even then. Several variations on the basic core produced a version that delivered roughly 20,000 pounds of thrust about twenty years later. This incremental power development around the same basic design saves on development costs, spares stocking costs, and maintenance. PW's FT-8D is their aeroderivative equivalent used in both power generation and mechanical drive application.

Similarly, General Electric's (GE's) LM2500 and LM6000 family (aeroderivative) are essentially CF6-80C2 aeroengines that have been adapted for land-based use. What was ABB's GT35 (land-based), then Alstom's GT35 (change of corporate ownership), then Siemens Westinghouse's SGT500 (yet another corporate purchase) is another example of an aeroderivative. Most aeroderivatives

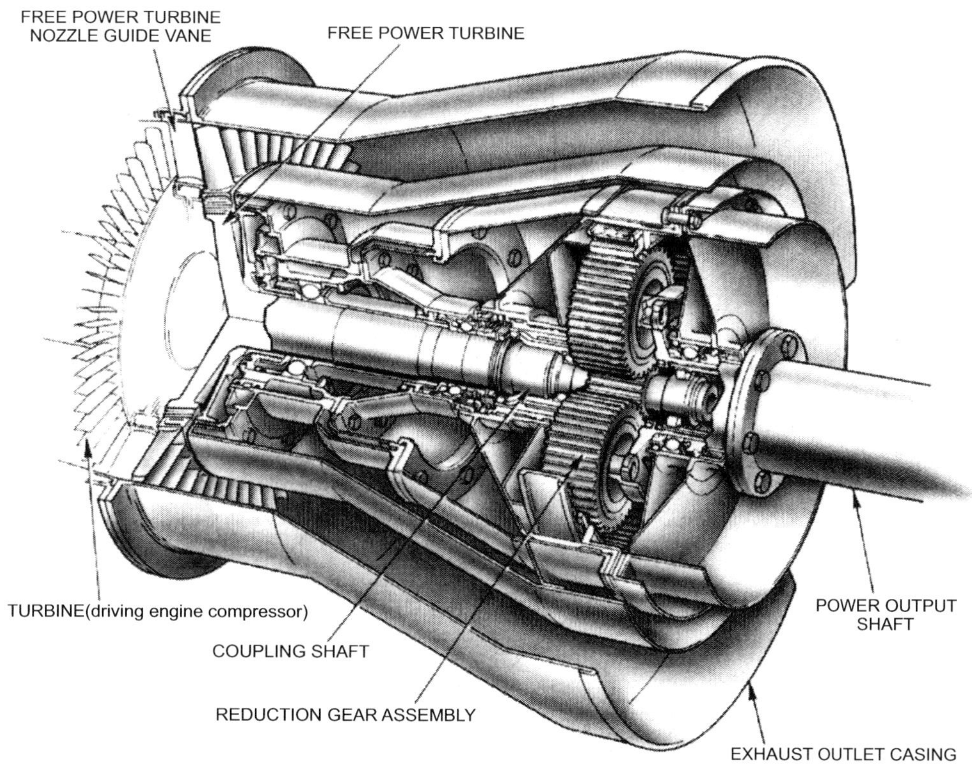

FIGURE 1–4 A typical free power turbine. *(Source: Bloch and Soares,* Process Plant Machinery, *Second Edition. Boston: Butterworth-Heinemann, 1998.)*

can also be used in marine (ferry, ship) applications. Some of them are also used on mobile land applications, such as in military tanks.

Aero- and aeroderivative gas turbine engines are likely to be built in modular construction. This means that one module of the gas turbine engine may be removed from service and the other modules left in place. A substitute module may be inserted in place of the removed module so the gas turbine can resume service. An industrial engine is more likely to be constructed in a non-modular format. If part of an industrial engine has serious problems, it is likely that the entire engine will be "down for maintenance."

The term "industrial" gas turbine implies a heavier frame and a gas turbine model that was not intended for service where the mass (weight) to power ratio (in other words, weight minimization for the power plant) was of paramount concern. That said, the metallurgical selections for contemporary industrials reflect the best developments in metallurgical selections. The gas turbine field is a highly competitive one, and the highest turbine inlet temperatures (TITs) that can be tolerated by the metallurgical and fuel selections are sought, as this optimizes the gas turbine's peak power rating. In other words, GE's industrial Frame 7s and 9s (be they "-F", "-G", or "-H" technology) may incorporate similar metallurgy to that used on their aircraft engines. The letters F, G, and H refer to temperature ceilings and therefore imply higher power (with "later" alphabet letters).

Some turbine model designations can appear confusing in terms of their corporate ownership. This is partly due to the fact that the OEM (original equipment manufacturer) gas turbine scene changes constantly with corporate mergers, partial mergers, buyouts of specific divisions, and joint ventures. This section and the one on combined cycles therefore have several notes about specific engines' model designation history and previous ownership. This has considerable relevance when it comes to noting the finer points of any gas turbine's design. This is critical to operators as they can then make better decisions regarding the overhaul, performance optimization, component updates, and retrofit systems on their turbine systems.

Any Application of a Gas Turbine could have a Great Deal to Offer End-Users in other Industrial Sectors

Power generation is often the least demanding application for a given gas turbine, unless it is used in variable load/peaking service. Mechanical drive units are more likely to experience load swings. One example would be turbines driving pumps that reinject (into the soil) varying volumes of seawater that accompany "mixed field" (oil, gas, and seawater deposits) oil and gas production.

Aircraft engine turbines may see varying stresses depending on their service. If, for instance, one considers an aerobatic squadron, one needs to be aware that the engines on the planes trying to stay a fixed distance from the wing tip of the formation's leader may accumulate life cycle losses of twenty times that of the formation leader's engines.

In other words, the variations in all parameters that pertain to a gas turbine's overall life, component lives, or time between overhauls (TBOs) offer insight to gas turbine operators regardless of whether that turbine operates in "their" industry or not. Lessons learned in one sector of industry on gas turbine metallurgy and operating systems, such as controls or condition monitoring, can be applied in some way to other gas turbine applications.

As we will see in Chapter 2, gas turbine historians can argue that the aircraft gas turbine engine was developed well before its land-based counterpart, or vice versa. Different schools of thought will be presented on that subject so the reader can draw his or her own conclusions. What is undisputed, however, is that the gas turbine is also a versatile propulsion mechanism in terms of aviation applications.

AEROENGINE GAS TURBINES

One way to subdivide aeroengines is by whether they have a centrifugal compressor or an axial compressor. In very general terms, the former type offers more in terms of simplicity and ruggedness. The axial compressor, however, is used in most high performance, more complex designs.

Another subdivision that can be made is whether the aeroengine drives a propeller (via the gas generator shaft or a free power turbine) and just basically pushes its exhaust gases out its exhaust section and thus pushes the plane forward (jet propulsion). This operational mode (turboshaft or turbojet) is independent of what type of compressor the gas turbine has, as we see in the family tree figure (see Figure 1—5). A turboshaft, which also has a large fan at the front (air intake) end, is called a turbofan engine.

Examples of turbofan engines are the Trent engine family (Rolls Royce), the CF-6 engine family (General Electric), and the JT-8D engine family (Pratt and Whitney) (Figures 1—6 to 1—10).

Relationships Between Pressure, Volume, and Temperature*

Note that these Relationships are Essentially the Same Regardless of Whether the Gas Turbine is Used in Land, Sea, or Air Applications.

During the working cycle of the turbine engine, the airflow or "working fluid" receives and gives up heat, so producing changes in its pressure, volume, and temperature. These changes as they occur are closely related, for they follow a common principle that is embodied in a combination of the laws of Boyle and Charles. Briefly, this means that the product of the pressure and the volume of the air at the various stages in the working cycle is proportional to the absolute temperature of the air at those stages. This relationship applies for whatever means are used to change the state of the air. For example, whether energy is added by combustion or by compression, or is extracted by the turbine, the heat change is directly proportional to the work added or taken from the gas.

There are three main conditions in the engine working cycle during which these changes occur. During compression, when work is done to increase the pressure and decrease the volume of the air, there is a corresponding rise in the temperature. During combustion, when fuel is added to the air and burned to increase the temperature, there is a corresponding increase in volume while the pressure remains almost constant. During expansion, when work is taken from the gas stream by the turbine assembly, there is a decrease in temperature and pressure with a corresponding increase in volume.

Changes in the temperature and pressure of the air can be traced through an engine by using the airflow diagram in Figure 1—11. With the airflow being continuous, volume changes are shown up as changes in velocity.

The efficiency with which these changes are made will determine to what extent the desired relations between the pressure, volume, and temperature are attained. For the more efficient the compressor, the higher the pressure generated for a given work input; that is, for a given temperature rise of the air. Conversely, the more efficiently the turbine uses the expanding gas, the greater the output of work for a given pressure drop in the gas.

When the air is compressed or expanded at 100% efficiency, the process is said to be adiabatic. Since such a change means there is no energy loss in the process, either by friction, conduction, or turbulence, it is obviously impossible to achieve in practice; 90% is a good adiabatic efficiency for the compressor and turbine.

Changes in Velocity and Pressure

During the passage of the air through the engine, aerodynamic and energy requirements demand changes in its velocity and pressure. For instance, during compression, a rise in the pressure of the air is required and not an increase in its velocity. After the air has been heated and its internal energy increased by combustion, an increase in the velocity of the gases is necessary to force the turbine to rotate. At the propelling nozzle a high exit velocity is required, for it is

* Source: Adapted, with permission, from Rolls Royce, *The Jet Engine*, 1986, Rolls Royce Plc: UK.

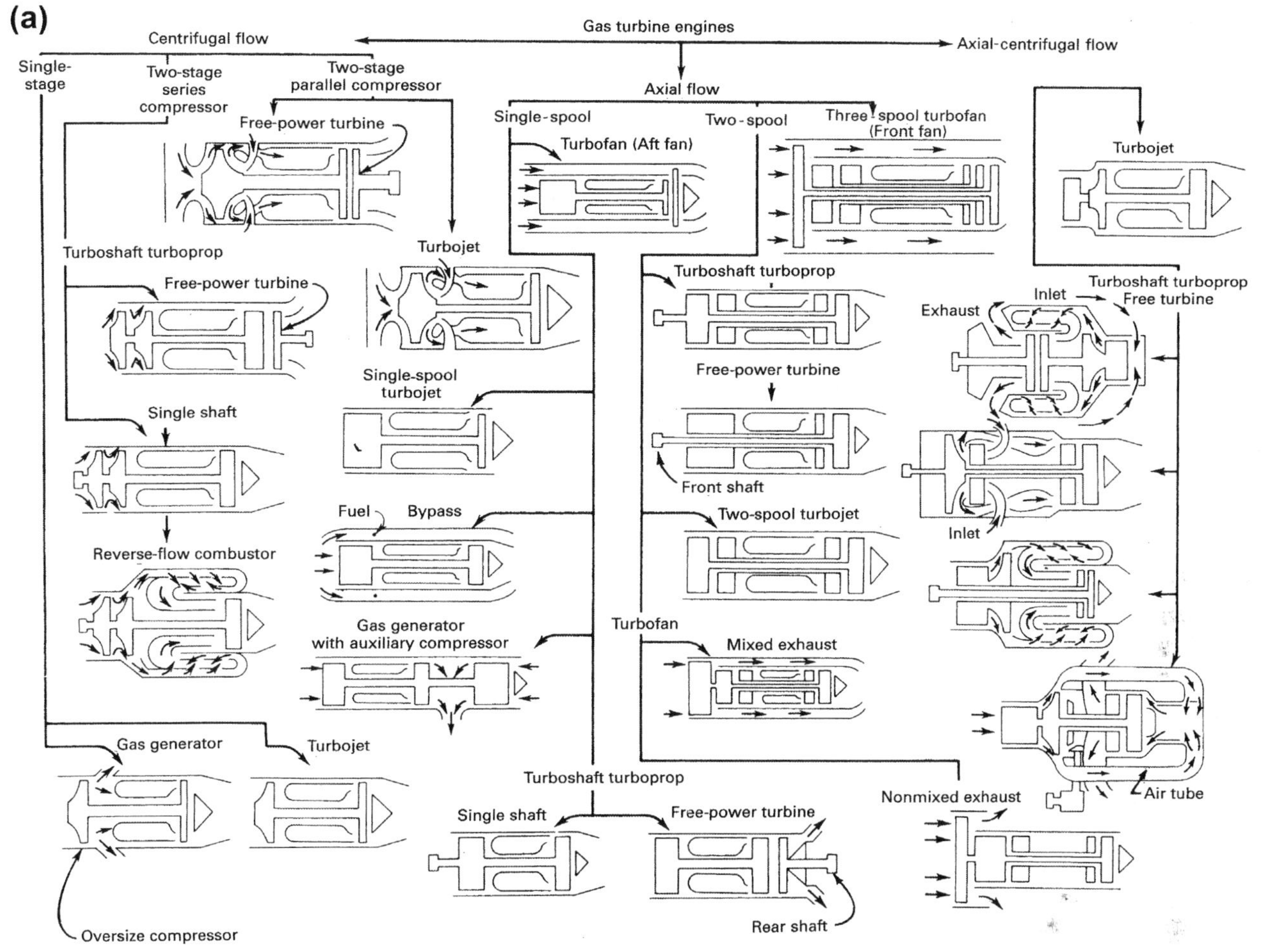

FIGURE 1–5 a. The family tree. b. Energy distribution of the turbojet, turboprop, and turbofan engines. *(Source: Treager,* Aircraft Gas Turbine Engine Technology, *Third Edition. New York: McGraw-Hill, 1996.)*

the change in the momentum of the air that provides the thrust on the aircraft. Local decelerations of airflow are also required, for instance, in the combustion chambers, to provide a low velocity zone for the flame to burn.

These various changes are effected by means of the size and shape of the ducts through which the air passes on its way through the engine. Where a conversion from velocity (kinetic) energy to pressure is required, the passages are divergent in shape. Conversely, where it is required to convert the energy stored in the combustion gases to velocity energy, a convergent passage or nozzle (Figure 1–12) is used. These shapes apply to the gas turbine engine where the airflow velocity is subsonic or sonic, i.e., at the local speed of sound. Where supersonic speeds are encountered, such as in the propelling nozzle of the rocket, athodyd, and some jet engines, a convergent-divergent nozzle or venturi (Figure 1–13) is used to obtain the maximum conversion of the energy in the combustion gases to kinetic energy.

The design of the passages and nozzles is of great importance, for upon their good design will depend the efficiency with which the energy changes are effected. Any interference with the smooth airflow creates a loss in efficiency and could result in component failure due to vibration caused by eddies or turbulence of the airflow.

Airflow

The path of the air through a gas turbine engine varies according to the design of the engine. A straight-through flow system (Figure 1–11) is the basic design, as it provides for an engine with a relatively small frontal area and is also suitable for use of the bypass principle. In contrast, the reverse flow system gives an engine with greater frontal area, but with a reduced overall length. The operation, however, of all engines is similar. The variations due to the different designs are described in the subsequent paragraphs.

The major difference of a turbo-propeller engine is the conversion of gas energy into mechanical power to drive the

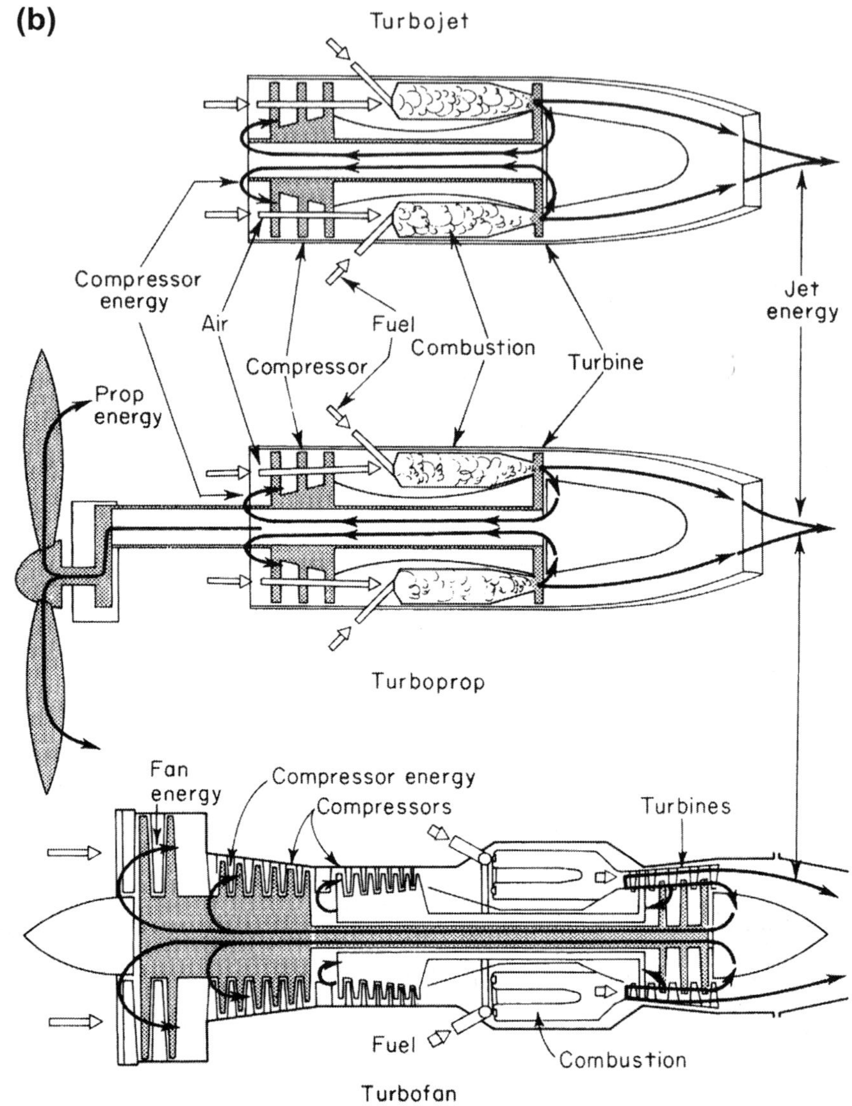

FIGURE 1–5 *(Cont'd).*

propeller. Only a small amount of "jet thrust" is available from the exhaust system. The majority of the energy in the gas stream is absorbed by additional turbine stages, which drive the propeller through internal shafts.

As can be seen in Figure 1–11, the bypass principle involves a division of the airflow. Conventionally, all the air taken in is given an initial low compression and a percentage is then ducted to bypass, the remainder being delivered to the combustion system in the usual manner. This principle is conducive to improved propulsive efficiency and specific fuel consumption.

An important design feature of the bypass engine is the bypass ratio; that is, the ratio of cool air bypassed through the duct to the flow of air passed through the high-pressure system. With low bypass ratios, i.e., in the order of 1:1, the two streams are usually mixed before being exhausted from the engine. The fan engine may be regarded as an extension of the bypass principle, and the requirement for high bypass ratios of up to 5:1 is largely met by using the front fan in a twin or triple-spool configuration (on which the fan is, in fact, the low-pressure compressor) both with and without mixing of the airflows. Very high bypass ratios, in the order of 15:1, are achieved using propfans. These are a variation on the turbo-propeller theme but with advanced technology propellers capable of operating with high efficiency at high aircraft speeds.

On some front fan engines, the bypass airstream is ducted overboard either directly behind the fan through short ducts or at the rear of the engine through longer ducts; hence the term "ducted fan." Another, though seldom used, variation is that of the aft fan.

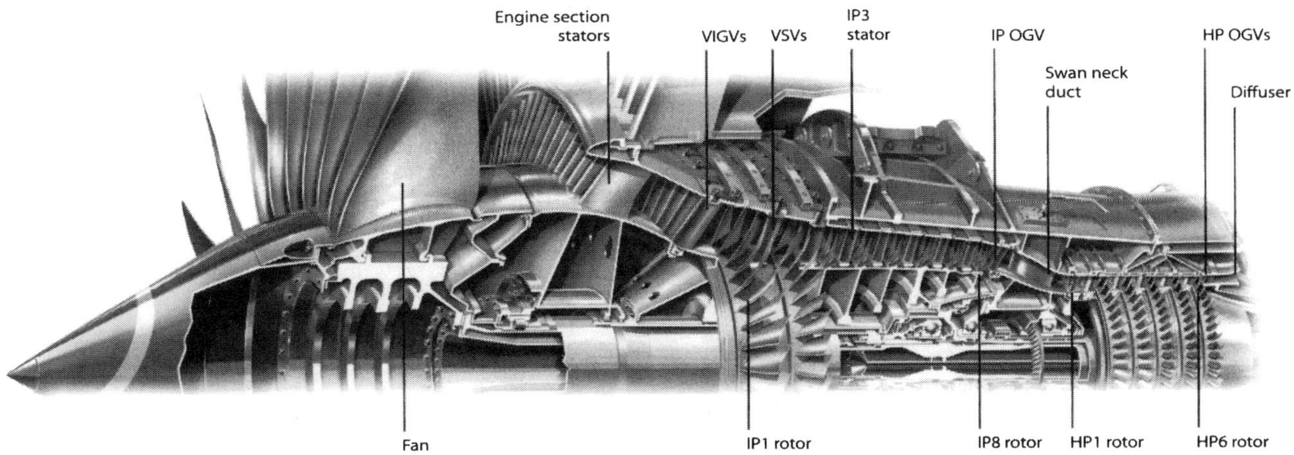

FIGURE 1–6 The fan and compressors form the compression system of the Trent 500, a high-bypass civil engine. *(Source: Rolls Royce.)*

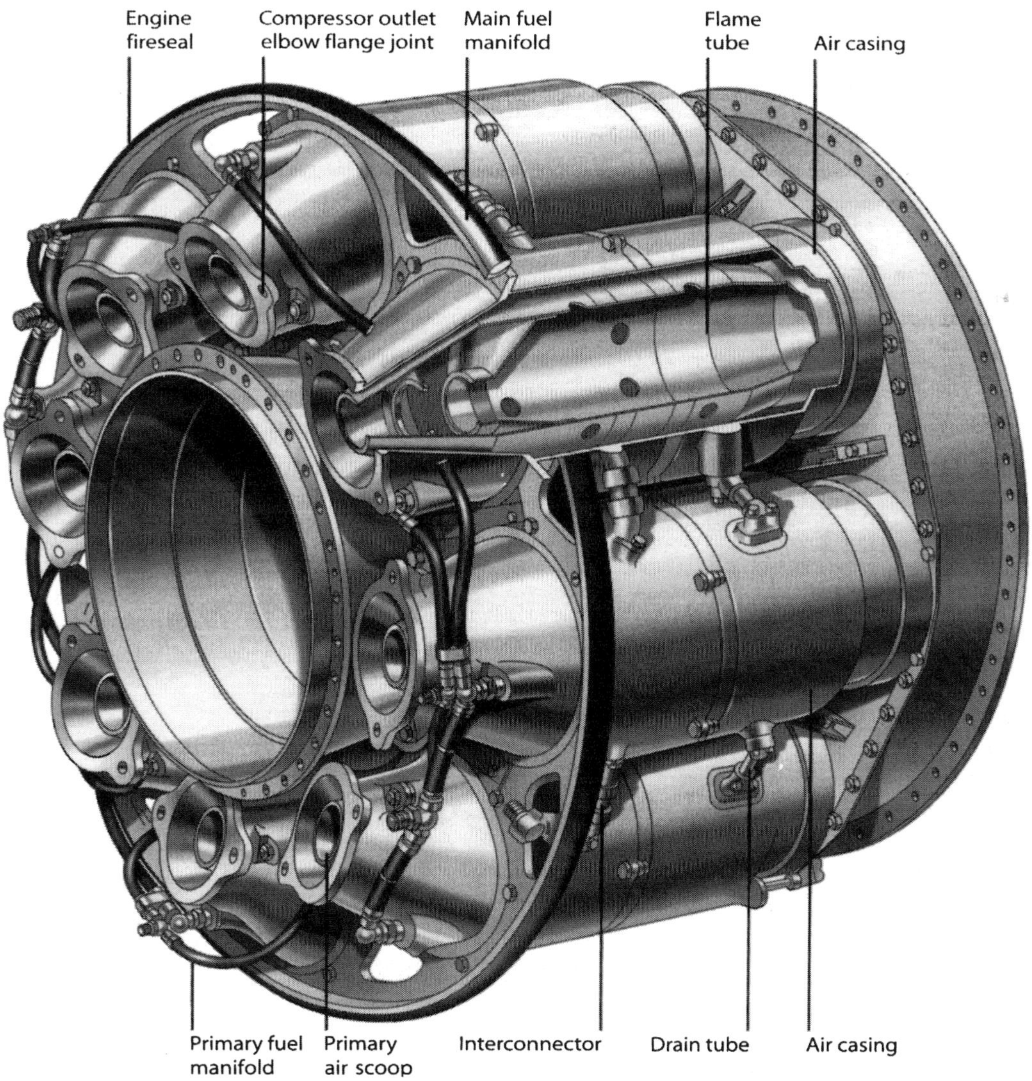

FIGURE 1–7 A multiple combustion chamber system with individual flame tubes and air casings. *(Source: Rolls Royce.)*

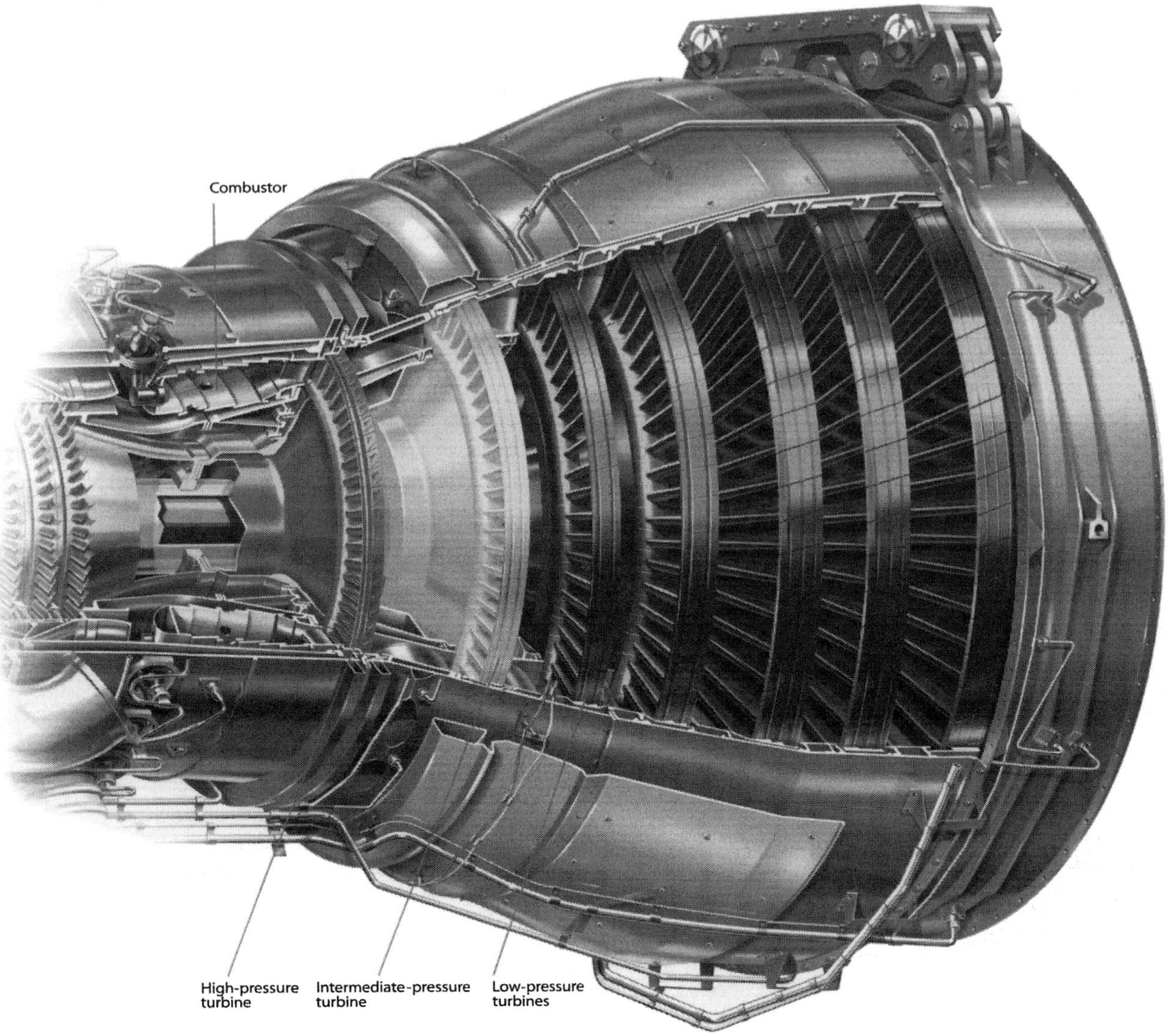

FIGURE 1—8 A three-shaft turbine system. *(Source: Rolls Royce.)*

GAS TURBINES AT SEA

In marine applications, the gas turbine is generally driving the ship's or ferry's propellers, via a gearbox, if it is providing the vessel's propulsion. However, it may also be used to generate onboard power, as in the example below.

The gas turbine used is essentially the "same" machine as would be used for land service. The aeroderivative GT35 (now the SGT-500 under Siemens ownership) shown in a marine application in Figure 1—14 is also used in land-based service.

In Figure 1—14, this vessel is an FPSO (floating production, storage, and offloading) vessel, and power on board is provided by two SGT-500 gas turbines. One WHRG (waste heat recovery generator) for each gas turbine heats process water. The SGT-500 is a lightweight, high-efficiency, heavy-duty industrial gas turbine. Its special design features are high reliability and fuel flexibility. It is also designed for single lift, which makes the unit suitable for all offshore applications. The modular, compact design of the GT35C facilitates on-site modular exchange.

GAS TURBINES: DETAILS OF INDIVIDUAL APPLICATIONS

In the previous section we saw some basic examples of the gas turbine's role in industry. In this section we will look at more detailed specifics of the individual applications in land, sea, and air usage.

Of all the industries today, power generation is by far the largest and also experiences the most growth.

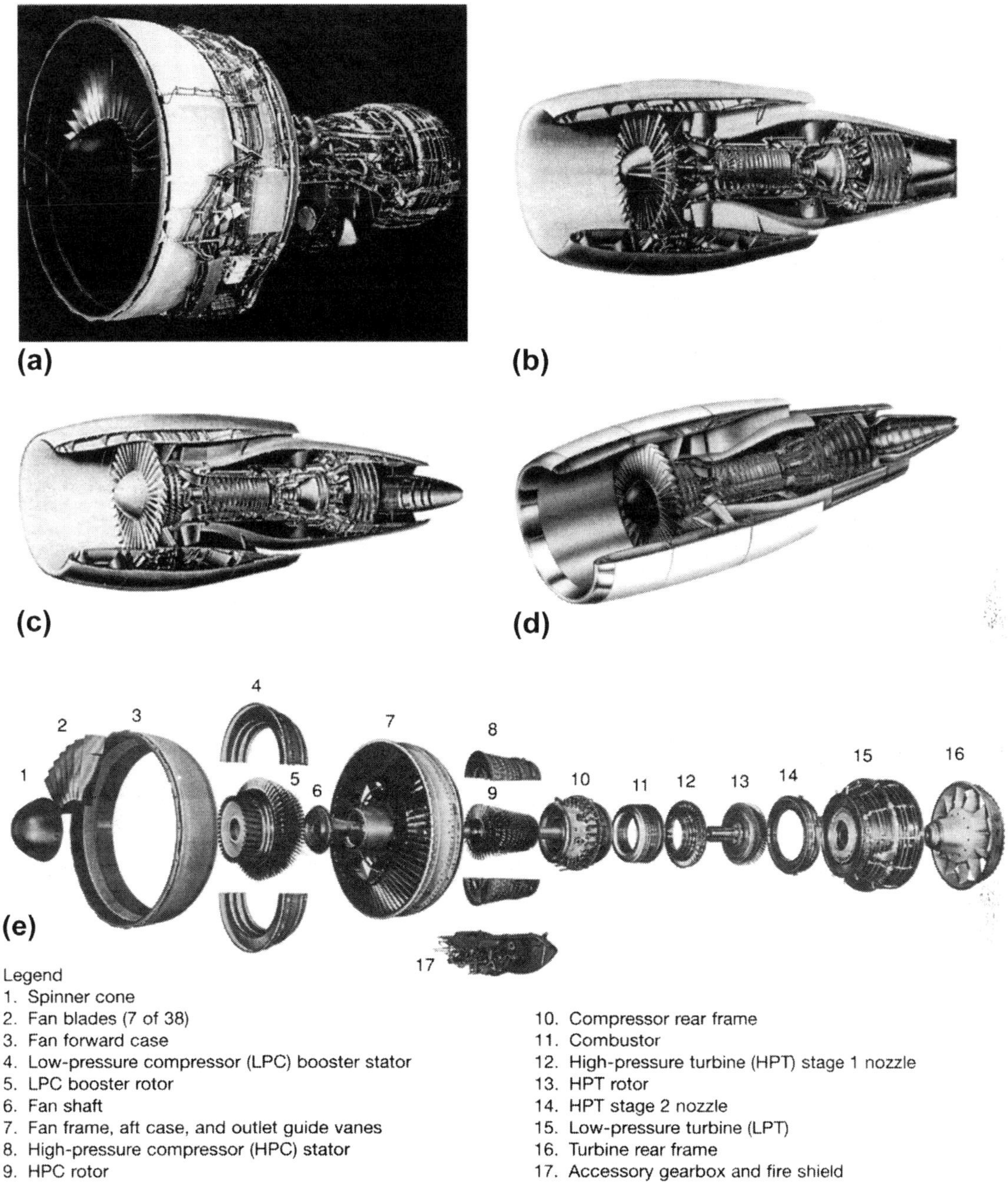

FIGURE 1–9 General Electric CF6 series engine was developed from the TF39 engine, which was used in the Lockheed Georgia C-5A Galaxy. a. General external view of the GE CF6-80C2 showing the very large diameter fan. b. Cutaway view of the GE CF6-6 high-bypass-ratio turbofan. c. Cutaway view of the GE CF6-50. d. Cutaway view of the GE CF6-80C2 high-bypass-ratio turbofan engine. e. Exploded view of the GE CF6-80C2 high-bypass turbofan engine. *(Source: Treager, Aircraft Gas Turbine Engine Technology, Third Edition. New York: McGraw-Hill, 1996.)*

Major Classes of Power Generation Application*

Table 1–1 summarizes the main classes of power generation application for which the gas turbine is a candidate, including examples of actual engines used. The descriptions that follow refer to items provided in Table 1–1.

Figure 1–15 presents thermal efficiency and utilization for these applications in graphical form, again using the item numbers from Table 1–1. Utilization is the number of hours per year the application is typically fired.

* Source: Courtesy Rolls Royce, *Gas Turbine Performance,* Walsh and Fletcher, Blackwell Science, 1998. Adapted with permission.

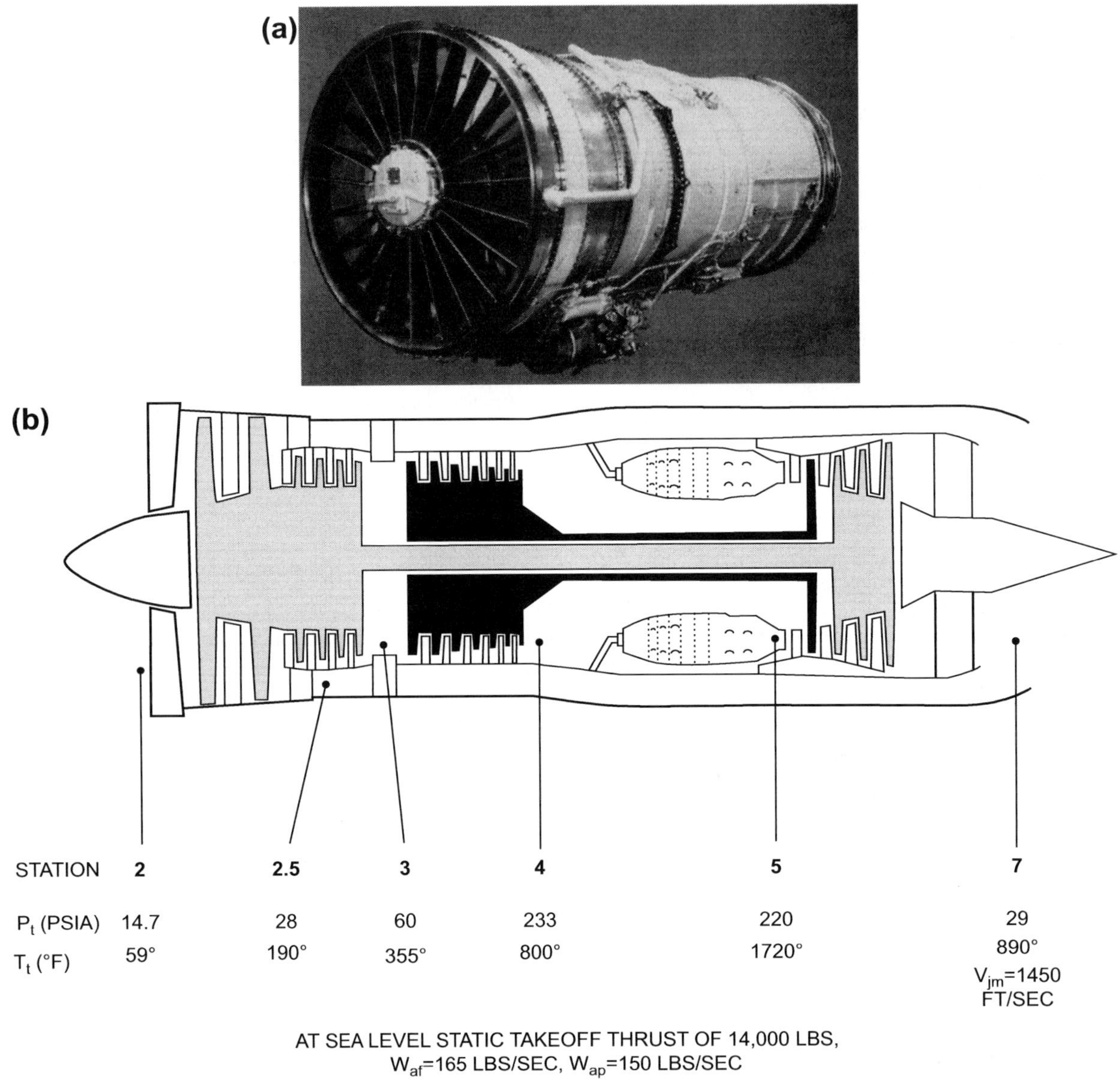

FIGURE 1–10 Pratt & Whitney JT8D Series Engine. a. External view of the Pratt & Whitney JT8D-200. b. A schematic view showing the arrangement of internals parts, pressures, and temperatures at various points of the JT8D engine. c. Cutaway view of an early version of the JT8D turbofan engine, naming some of the internal and external parts. d. Schematic view of the JT8D-200 series engine showing the single-stage fan at the front and the mixer at the rear. e. Cutaway view of the JTD8-200. *(Source: Treager, Aircraft Gas Turbine Engine Technology, Third Edition. New York: McGraw-Hill, 1996.)*

Grid System

Figure 1–16 illustrates a typical grid system for distribution of electrical power, showing voltages at key points. The shaft power system must operate at a constant *synchronous speed* to deliver electrical power at fixed frequency via an alternator. Frequencies are usually 50 Hz or 60 Hz depending upon the country.

Until recently the trend was for grids to be supplied by a few large power stations. More flexible *distributed power* systems are now becoming popular, however, due largely to the gas turbine's viability even at smaller sizes. Here electricity is generated locally to the consumers whether by a CHP (combined heat and power) plant, or to a lesser extent the *mid-merit* power stations described later. Excess power is exported to the grid.

To illustrate the levels of power that must be transported, a city of one million people may have a peak demand of up to 2 GW. The average and peak consumption in a home for a family of four is around 1 kW and 6 kW, respectively, excluding space heating.

Standby Generators (Items 1 and 2, Table 1–1)

Standby generators are employed for emergency use, where there may be a loss of main supply that cannot be

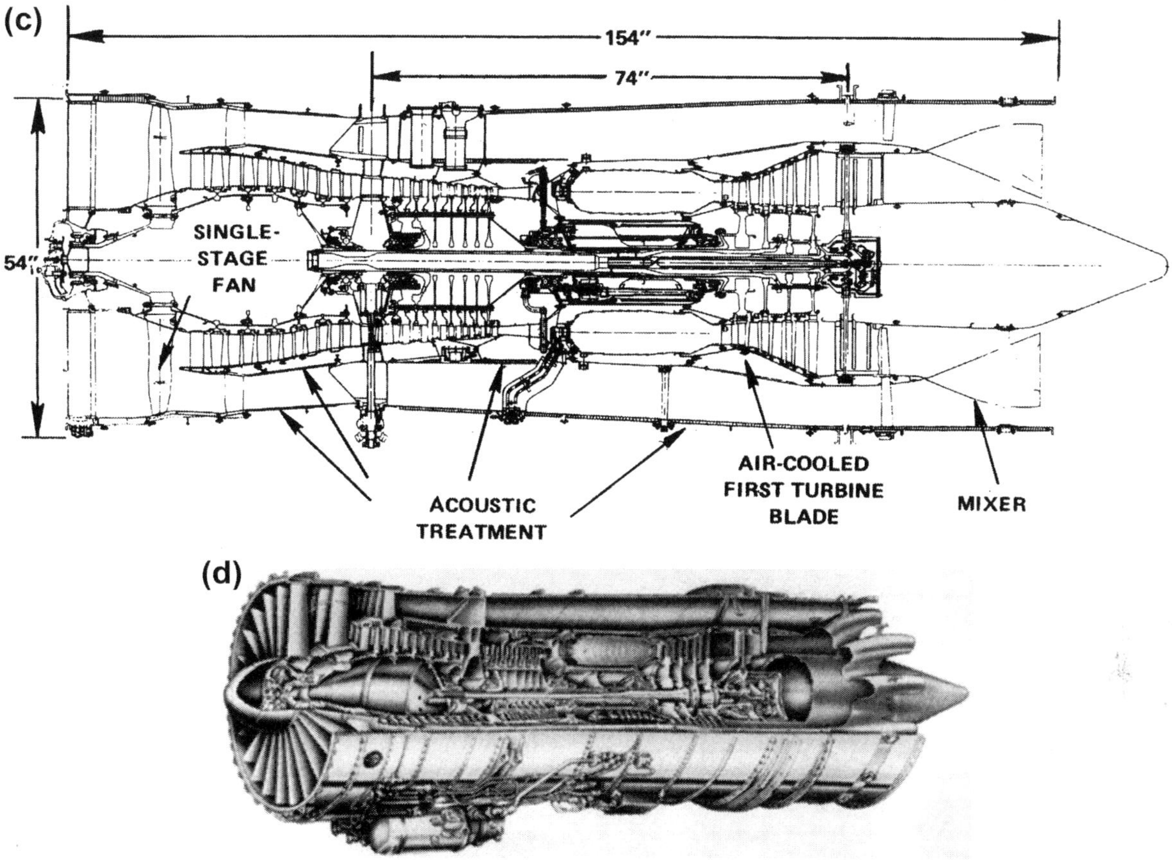

FIGURE 1–10 *(Cont'd).*

tolerated. Examples include hospitals and other public buildings in areas such as Japan, which may be prone to earthquakes. The power generated is used locally and the units are not connected to the grid system. Usually the fuel type is diesel. Key requirements are predominantly driven by the low utilization, and are outlined below in order of importance:

- Low unit cost
- Often low unit weight and volume are crucial (see below)
- Fast start and acceleration to rated power times may be very important
- Thermal efficiency and emissions levels are of secondary importance

The diesel engine is most popular, primarily due to the plethora of automotive and marine engines in the required power bracket, which reduces unit cost via high production volume. The gas turbine has made some inroads, particularly where weight and volume must be limited, for instance, if the standby generator is located on the roof of an office block with limited load-bearing capability.

Gas turbine engines in this sector are usually simple cycle, single spool rather than free power turbine engines. This is because the reduced number of components reduces unit cost and because part speed torque is unimportant. Following the start sequence the engine remains at synchronous speed to ensure constant electrical frequency. Centrifugal compressor systems with pressure ratios of 5:1–10:1 are employed to minimize unit cost, and because the efficiency of axial flow compressors at such low flow rates is poor. The turbine blades, and usually the nozzle guide vanes, are uncooled, leading to SOT (stator outlet temperature) levels of typically 1100–1250 K.

Major Shaft Power Producing Systems

These typically are:

- A gas turbine combined cycle plant, where gas turbine waste heat raises steam in a heat recovery steam generator (HRSG) to drive steam turbines.
- A coal-fired boiler also generates steam to drive steam turbines.

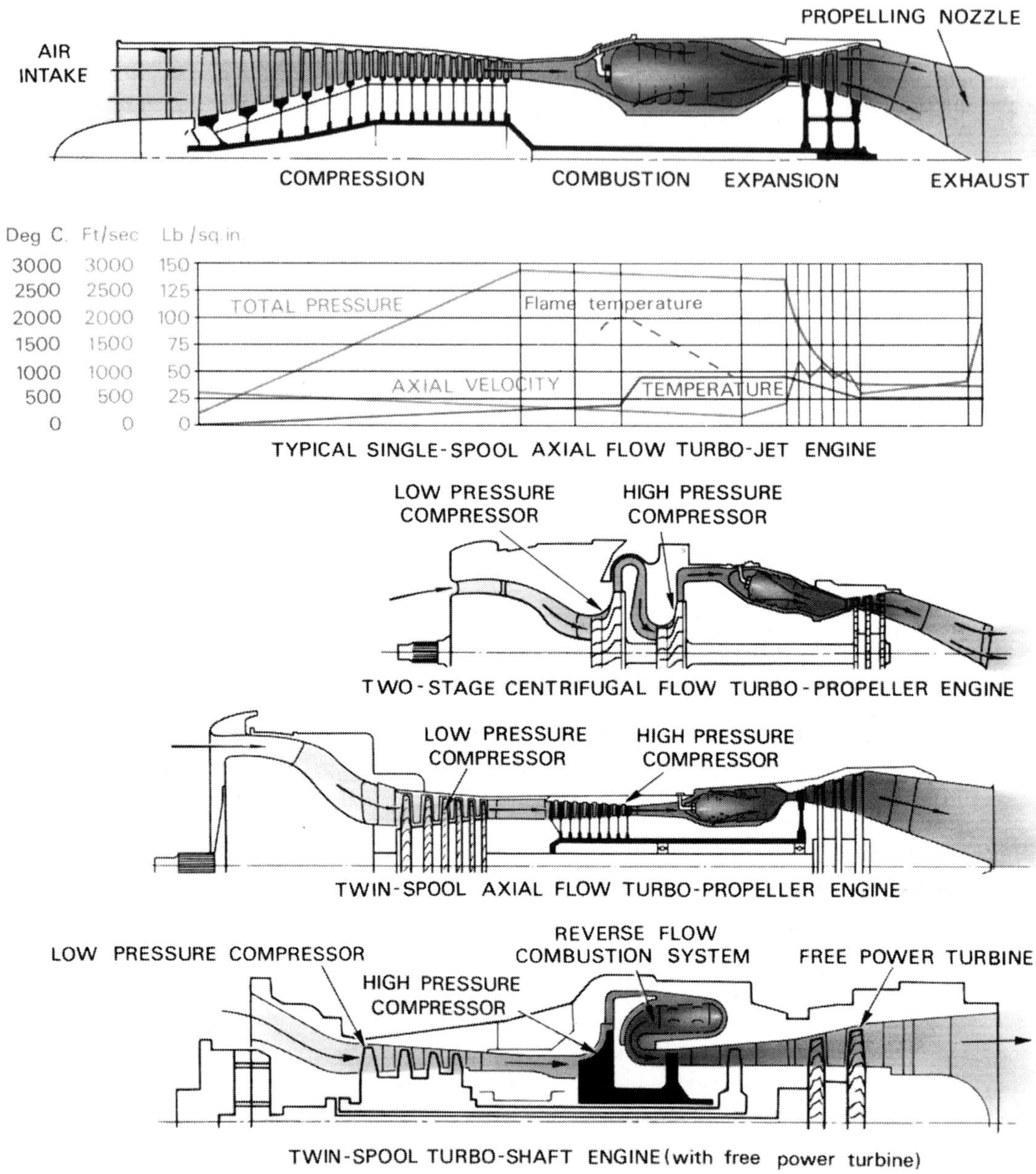

FIGURE 1–11 Airflow systems. *(Source: Rolls Royce.)*

- A nuclear powered boiler that raises steam to drive steam turbines.

Small-Scale Combined Heat and Power—CHP (Items 3 and 4, Table 1–1)

In this application the waste heat is typically utilized in an industrial process. The heat may be used directly in drying processes or more usually it is converted by an HRSG (heat recovery steam generator) into steam for other uses. Table 1–1 shows the presentation manufacturers usually employ to publish the steam raising capability of an engine. Most CHP systems burn natural gas fuel. The electricity generated is often used locally, and any excess exported to the grid. The key power plant selection criteria in order of importance are:

1. Thermal efficiency, for both CHP and simple cycle operation. The latter becomes more significant if for parts of the year there is no use for the full exhaust heat.
2. Heat to power ratio is important as electricity is a more valuable commodity than heat. Hence a low ratio is an advantage as the unit may be sized for the heat requirement and any excess electricity sold to the grid.

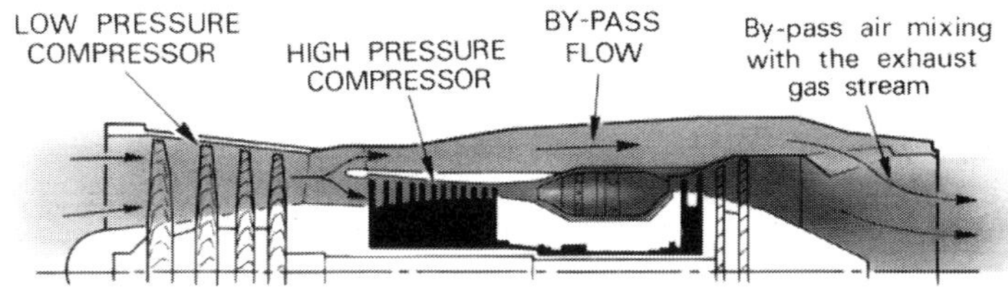

TWIN-SPOOL AXIAL FLOW BY-PASS TURBO JET ENGINE(low by-pass ratio)

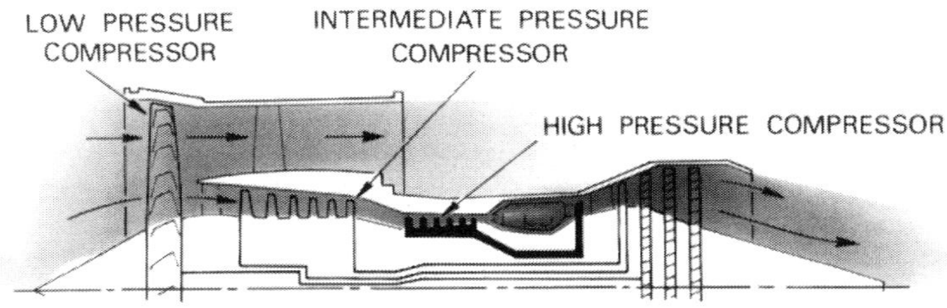

TRIPLE-SPOOL AXIAL FLOW FRONT FAN TURBO JET ENGINE(high by-pass ratio)

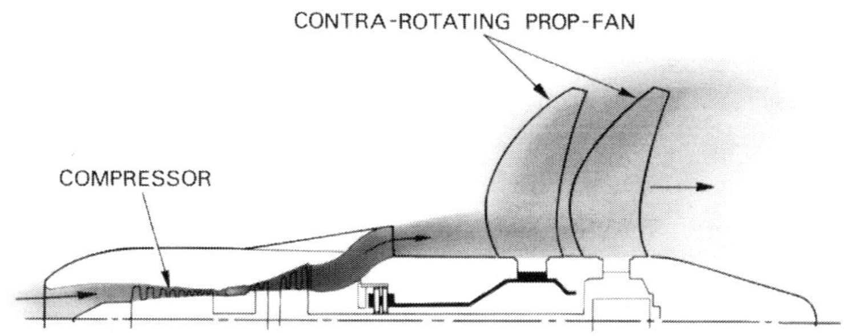

AXIAL FLOW CONTRA-ROTATING PROP-FAN(with free power turbine)

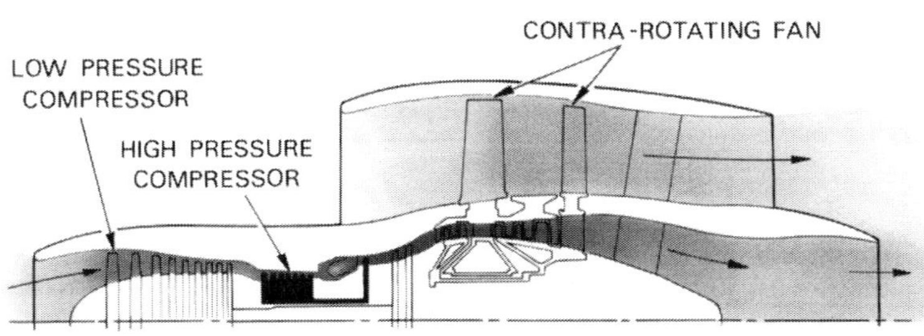

TWIN-SPOOL AXIAL FLOW CONTRA-ROTATING REAR FAN(with free power turbine)

FIGURE 1–11 *(Cont'd)*.

3. The grade (temperature) of the heat is very important in that the process usually demands a high temperature.
4. Owing to the high utilization, low unit cost, start, and acceleration times are all of secondary importance, as are weight, volume, and part speed torque.

The attributes of the gas turbine engine best meet the above criteria, and hence it is the market leader. The diesel engine still retains a strong presence, however, particularly for applications where substantial low-grade heat is acceptable, or where the importance of simple cycle *thermal* efficiency is paramount.

Gas turbines for this application are usually custom designed and tend to be single spool. Below 3 MW centrifugal compressors are used exclusively, with pressure ratios of between 8:1 and 15:1. This is a compromise between unit cost and simple cycle and CHP thermal efficiencies. At the lower end of the power bracket SOT tends to be 1300–1400 K, which requires only the first stage turbine nozzle guide vanes to be cooled. At the higher end of the power bracket the first stage rotor blades may also be cooled, allowing SOT levels of up to 1450 K. This becomes viable because of the increased size of the blades, and because the increased unit cost may be supported at the higher power.

Large-Scale CHP (Item 5, Table 1–1)

Here the waste heat is almost exclusively used to raise steam, which is then used in a large process application

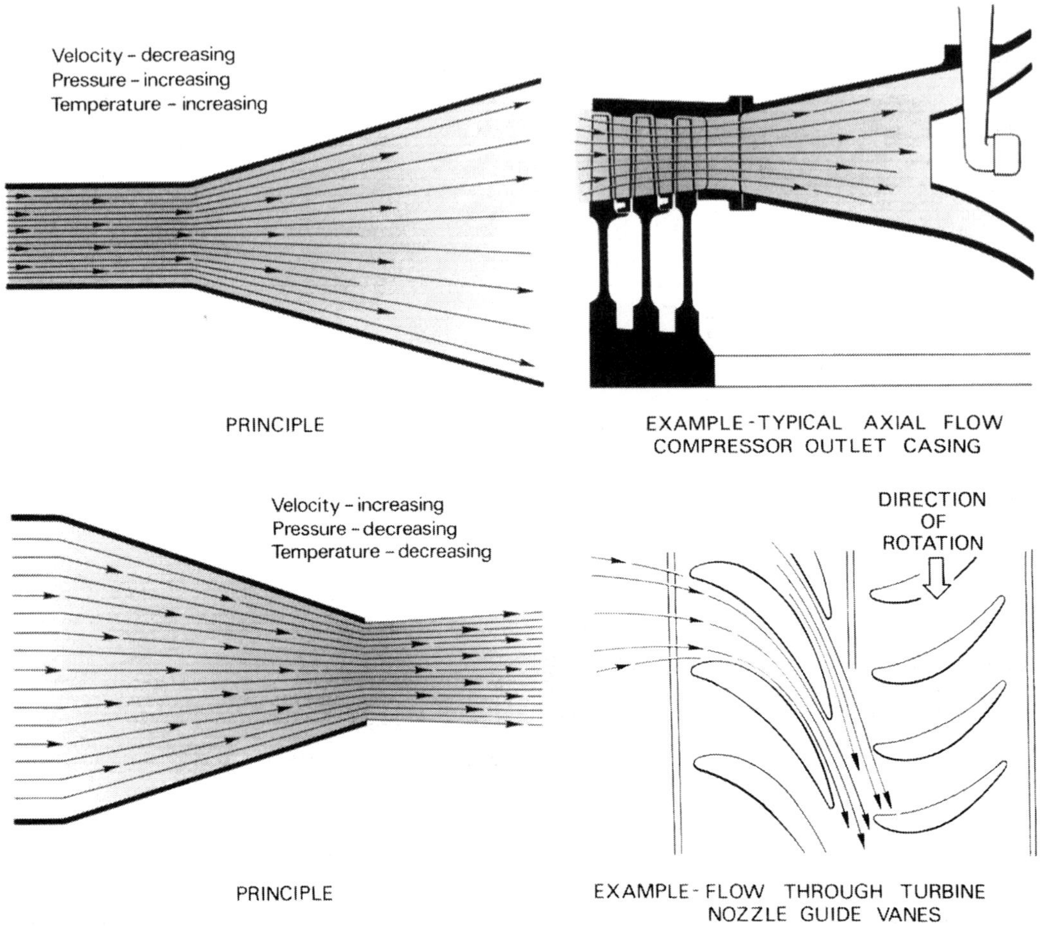

FIGURE 1–12 An airflow through divergent and convergent ducts. *(Source: Rolls Royce.)*

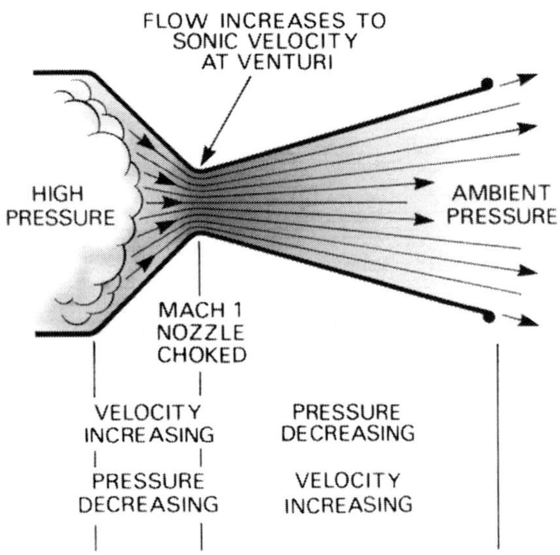

FIGURE 1–13 Supersonic airflow through a convergent-divergent nozzle or venturi. *(Source: Rolls Royce.)*

FIGURE 1–14 Gas turbines marine service: SGT-500 Industrial Gas Turbine—17 MW. Application: Two SGT-500 power packages for FPSO vessel in the Leadon Oilfields. (Note that the SGT-500 was Alstom's, formerly ABB's, GT-35; designation changed after Siemens acquisition.)

TABLE 1–1 Main Classes of Power Generation Application

Item	Plant Type	Examples of Applications	Examples of Engine	Power per Engine (MW)
1	Standby generator, simple cycle gas turbine	Office block Hospital	Yanmar AT36C, 60C, 180C Turbomeca Astazou	0.25–1.5
2	Standby generator, diesel engine	Office block Hospital	Caterpillar 352 V12 MTU 396	0.25–1.5
3	Small-scale CHP, gas turbine	Hospital Small process factory	EGT Hurricane NP PGT2 Allison 501 Solar mars EGT Tempest	0.5–10
4	Small-scale CHP, diesel, or natural gas-fired piston engine	Hospital Small process factory	Petter A MB 190	0.5–10
5	Large-scale CHP, gas turbine	Electricity and district heating for town of up to 25,000 people Large process factory, exporting electrical power	ABB STAL GT10 GE LM2500 Coberra 6000	10–60
6	Peak lopping units, simple cycle gas turbine	Supply to grid	ABB GT10 RR RB211	20–60
7	Mid-merit power station, simple cycle gas turbine	Supply to grid	GE LM6000 RR Trent	30–60
8	Base load power station, gas turbine in combined cycle with steam plant	Supply to grid	WEC 50W GE PG9331(FA)	50–450
9	Base load power station, coal-fired steam plant	Supply to grid		200–800
10	Base load power station, nuclear powered steam plant	Supply to grid		800–2000

MTU = Motoren Turbinen Union; EGT = European Gas Turbines; NP = Nuovo Pignone;
MB = Mirrlees Blackstone; ABB = Asea Brown Broveri; GE = General Electric; RR = Rolls Royce;
WEC = Westinghouse Electric Company (now part of Siemens); CHP = Combined heat and power.
To convert MW to hp, multiply by 1341.
(Source: Rolls Royce.)

such as a paper mill, or for district heating. Again, the electricity generated may be used locally or exported to the grid. The importance of performance criteria to engine selection are as for small-scale CHP, except that emissions legislation is more severe at the larger engine size (Figure 1–17).

Here gas turbines are used almost exclusively. High-grade heat is essential and the weight and volume of diesel engines prohibitive at these power outputs. Furthermore, the gas turbines used are often applicable to other markets, such as oil and gas and marine, which reduces unit cost. Aeroderivative gas turbines are the most common, though some *heavyweight engines* are used. Aeroderivatives usually employ the core from a large civil turbofan as a gas generator, with a custom designed free power turbine for industrial use. Heavyweight engines are designed specifically for industrial applications and as implied are far heavier than aeroderivatives, their low cost construction employing solid rotors, thick casings, etc.

The gas turbine configuration is usually a free power turbine. While this is not necessary for CHP applications, it is essential to also allow use in oil and gas and marine. Axial flow compressors are used exclusively with overall pressure ratios between 15:1 and 25:1. The aeroderivatives are at the top end of this range as this pressure ratio level results from a civil turbofan core. This pressure ratio is a compromise between that required for optimum CHP thermal efficiency of 20:1 and the 35:1 for optimum simple cycle efficiency. These values apply to the typical SOT of between 1450 K and 1550 K. Advanced cooling systems are employed for at least both the HP turbine first stage nozzle guide vanes and blades.

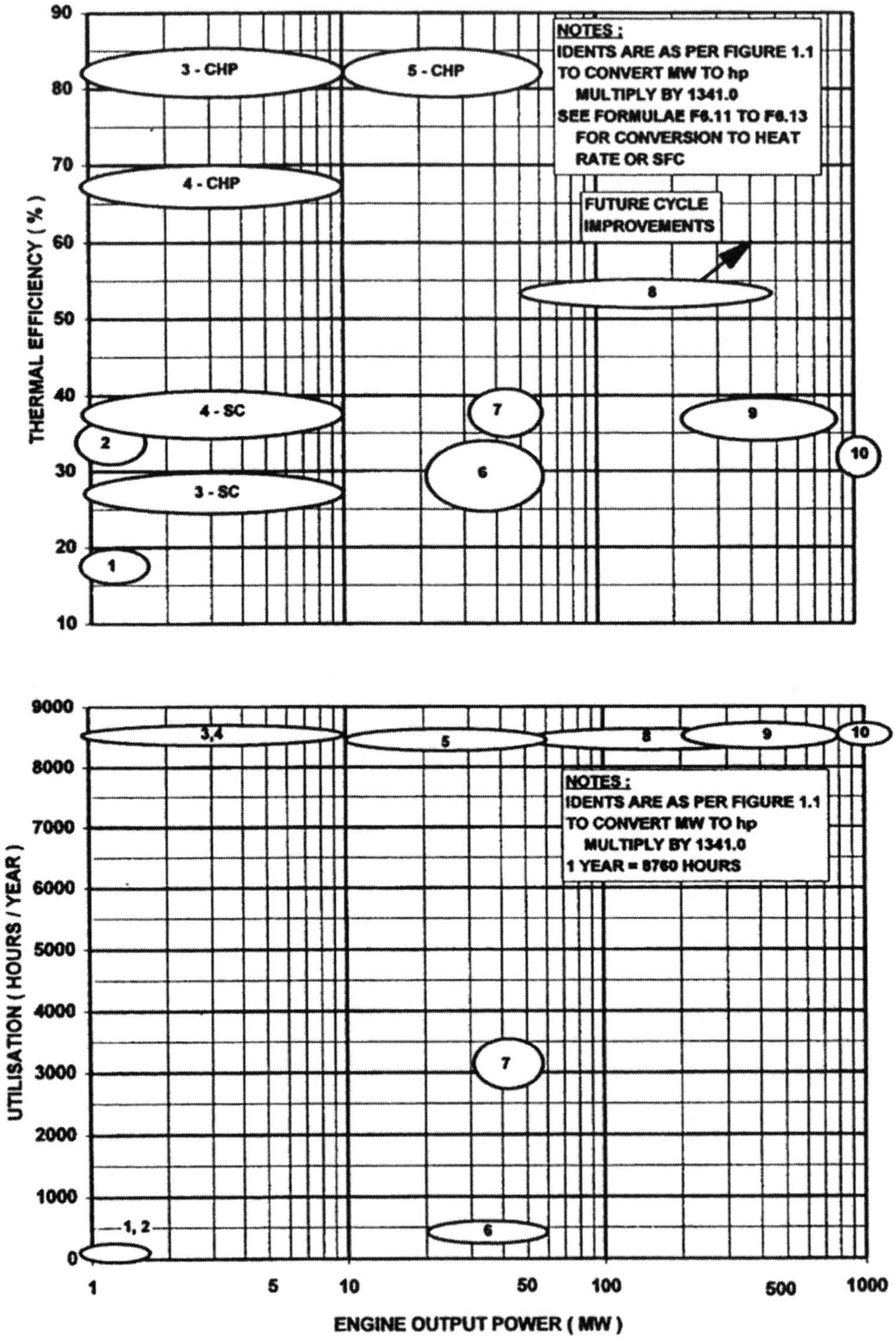

FIGURE 1–15 Characteristics of a power generation plant. *(Source: Rolls Royce.)*

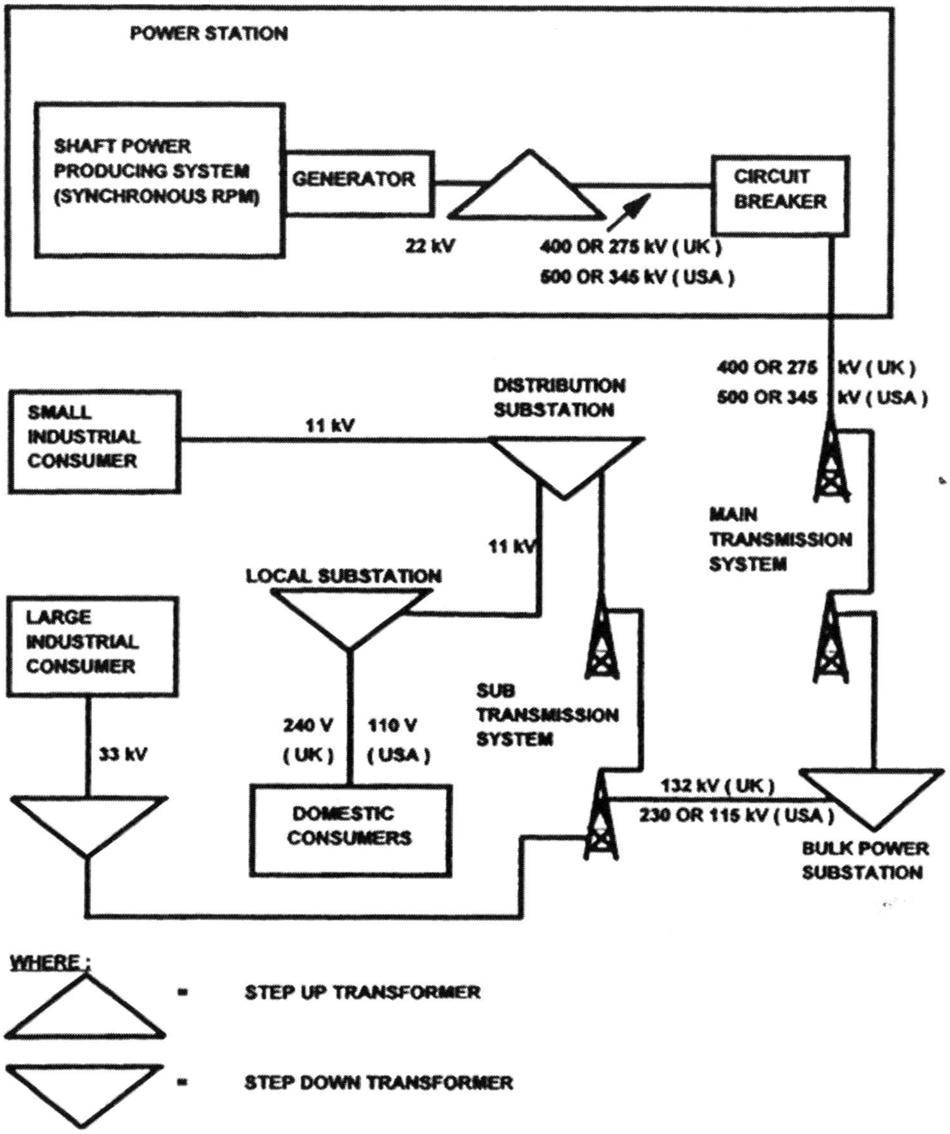

Major shaft power producing systems:
Gas turbine combined cycle plant, where gas turbine waste heat raises steam in a heat recovery steam generator (HSRG) to drive steam turbines
Coal fired boiler which raises steam to drive steam turbines
Nuclear powered boiler which raises steam to drive steam turbines

FIGURE 1–16 Grid for electrical power generation and distribution. *(Source: Rolls Royce.)*

Applications that Supply Solely to a Grid System (Items 6 to 10, Table 1–1)

Power plants supplying a grid fall into three categories:

1. *Peak lopping* engines have a low utilization, typically less than 10%. They are employed to satisfy the peak demand for electrical power that may occur on mid-weekday evenings as people return home and switch on a multitude of appliances.
2. *Base load power plants* achieve as near to 100% utilization as possible to supply the continuous need for electrical power.
3. *Mid-merit power plants* typically have 30–50% utilization. They serve the extra demand for electricity that is seasonal, such as the winter period in temperate climates where demand increases for domestic heating and lighting.

The considerations in selecting the type of power plant for a base load power station are as follows:

1. Thermal efficiency and availability are paramount.
2. Unit cost is of as high importance as the capital investment and period of time before the power station comes on line to generate a return on the investment are large.
3. Cost of electricity is a key factor in selecting the type of power plant and fuel price is a major contributor to this. Coal, nuclear, and oil-fired plants all compete with the gas turbine.

In all cases weight and volume are of secondary importance. Other specific comments are as follows:

- For base load plants, start and acceleration times are unimportant.
- For peak lopping power stations unit cost is crucial, time onto full load is very important, and thermal efficiency relatively unimportant.
- Mid-merit power stations are a compromise with some unit cost increase over and above peak loppers being acceptable in return for a moderate gain in thermal efficiency.

Peak loppers are mostly simple cycle gas turbines burning either diesel or natural gas, and some diesel engines are used at the lower power end. This is because the unit cost and time onto load are far lower than for other available alternatives, which involve steam plant. Both aeroderivative and heavyweight gas turbines are employed as peak loppers, either single spool or free power turbine, and with pressure ratios between 15:1 and 25:1. SOT may be as high as 1500 K, particularly where the unit is also sold for CHP and mechanical drive applications that demand high thermal efficiency.

For base load applications the gas turbine is used in combined cycle, to achieve the maximum possible thermal efficiency. It competes here with coal- and the nuclear-fired steam plants.

Historically, coal-fired plants have always had the biggest market share. In recent years the combined cycle as turbine has taken an increasing number of new power station orders due to the availability of natural gas leading to a competitive fuel price, higher thermal efficiency, and lower emissions, and the fact that the power stations may often be built with a lower capital investment. This has been supported by advances in gas turbine technology, which have increased both thermal efficiency and the feasible power output from a single engine. In particular, improvements in mechanical design have allowed SOT, and the last stage turbine stress level, to increase significantly. This particular stress is a limiting feature for large single spool engines in that as mass flow is increased at synchronous speed, so too must the turbine exit area to keep to an acceptable Mach number.

In some countries a relatively modern coal-fired plant has even "slid down the merit table" and is now only being used in mid-merit applications. However, in parts of the world where there is no natural gas and an abundance of coal, coal-fired steam plants will continue to be built for the foreseeable future. For nuclear power the case is complex, depending upon individual government policies and subsidies.

For base load applications above 50 MW the gas turbines are almost exclusively custom designed, single spool heavyweight configurations. The chosen pressure ratio is the optimum or combined cycle thermal efficiency at the given SOT, though this curve is relatively flat over a wide range of pressure ratios. Usually the higher pressure ratio on this flat portion is chosen to minimize steam plant entry temperature for mechanical design considerations. Engines currently in production are in the 1450–1550 K SOT range, and pressure ratios range from 13:1–16:1. For a number of concept engines SOT levels of 1700–1750 K are under consideration, with pressure ratios of 19:1–25:1. These employ advanced cycle features such as steam cooling of NGVs and blades. These engines are targeted at a combined cycle thermal efficiency of 60%. Aeroderivative gas turbines are currently limited to around 50 MW due to the size of the largest aeroengines. In this power bracket they are competitive in combined cycle, particularly at higher SOT levels.

Mid-merit power stations employ simple cycle gas turbines of a higher technology level than those used for peak lopping. The higher unit cost is justified by the higher thermal efficiency, given the higher utilization. Most engines are aeroderivative but at pressure ratios of the order of 25:1–35:1 for optimum simple cycle thermal efficiency. Corresponding SOT levels are 1500–1600 K.

Closed Cycles

Here the working fluid, often helium, is recirculated from turbine *exit* to compressor entry via pre-cooling heat exchangers. Advantages of a *closed*, as opposed to *open*, cycle include the following:

- No inlet filtration requirements or blade erosion problems.
- Reduced turbomachinery size, due to the working fluid being maintained at a high pressure and density. In addition, helium offers a high specific heat.
- The use of energy sources unsuited to combustion within an open gas turbine cycle, such as nuclear reactors or alternative fuels such as wood and coal. Helium offers a short half-life for use in radioactive environments.
- A flat SFC characteristic at part power as compressor entry pressure may be modulated, preserving cycle pressure ratio and SOT.

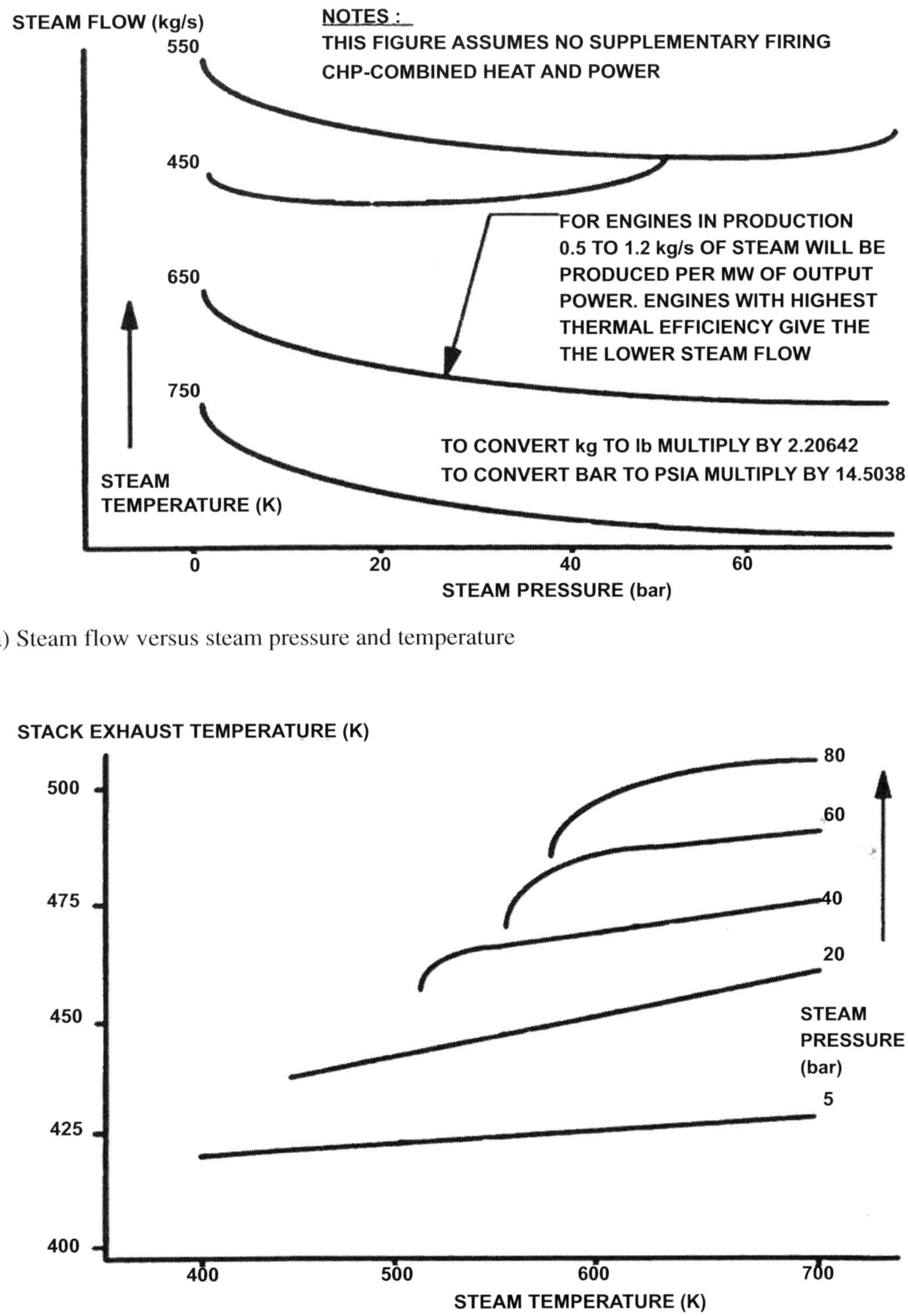

(a) Steam flow versus steam pressure and temperature

(b) Stack exhaust temperature versus steam temperature and pressure

FIGURE 1–17 Gas turbine CHP steam production capability. *(Source: Rolls Royce.)*

However, few closed cycle plants have been manufactured despite numerous studies for power generation and submarine propulsion. This is because the above advantages have been offset by high unit cost and modest thermal efficiency due to the SOT limit of around 1100 K dictated by nuclear reactor or heat exchanger mechanical integrity limits. The high unit cost results from the plant complexity and the implications of designing for very high pressures.

Industrial Mechanical Drive Applications

Here the engine is used to drive a pump or compressor. The most prolific example is the gas and oil industry that typically orders 1 GW per year of new engines. The majority of engines are installed onshore, although there is an offshore sector where engines are located on platforms.

This industry also has the need for some local primary power generation and emergency power generation. The requirements here are as per conventional power generation but the importance of low weight and volume is amplified.

The Gas and Oil Pipeline System

Figure 1–18 shows the configuration of a natural gas pipeline system, in which gas is pumped from a well head to industrial and domestic consumers. Pipelines have diameters of typically 915 mm (36 in) to 1420 mm (56 in), and are usually underground. A notable exception is in permafrost areas where they must be raised to avoid melting the permafrost. These systems may extend over thousands of kilometers, with compression stations approximately every 200 km. For example, the trans-Canada pipeline runs from the Alberta province in Canada to the east coast of the United States. The power plant burns natural gas tapped off the pipeline, and drives a centrifugal compressor. Figure 1–18 also shows the typical flow rate of natural gas versus pumping power, and the pipeline compressor pressure ratio. For comparison, a family of four may consume up to 10 standard cubic meters per day in the winter period. A further use for gas turbines is to pump water into depleted natural gas fields to increase gas extraction.

Oil pipelines are less complex. Oil is pumped from a well head to a refinery, and occasionally distillate fuels are then pumped to large industrial users. Extracting oil from the well may involve pumping gas down to raise pressure and to force oil up the extraction pipe by bubbling gas through it.

Engine Requirements

The major power blocks required are around 6–10 MW, 15 MW, and 25–30 MW. These power levels are generally beyond the practical size for a high-speed diesel engine given the requirements outlined below, and hence gas turbines are used almost exclusively. In order of importance these requirements are:

1. Low weight, as the engines often have to be transported to remote locations, where it also may be difficult and costly to build substantial foundations.
2. Good base load thermal efficiency, since utilization is as near 100% as possible.
3. Reasonable part power torque, to respond to load changes on the gas compressor. However, a fast start time is not essential, and a loading rate of 2 minutes from idle to full power is typical.

For offshore well heads the engine must be located on a platform; hence the importance of low weight is amplified and low volume essential. While the gas or oil may have a high pressure as it comes out of the ground it invariably needs further pressurization to pipe it back onshore. The engine may drive the compression unit mechanically as described above, or sometimes via an electric motor. In the latter instance a CHP arrangement supplies power for other needs, such as electricity for the drill and heat for uses such as natural gas processing or space heating.

For the middle and higher power bracket, simple cycle, free power turbine aeroderivatives best meet the above criteria, and are used almost exclusively. Pressure ratios of 20:1–25:1 and SOT levels of 1450–1550 K are typical, leading to thermal efficiency levels in the mid to high thirties. Engines include the Cooper Rolls Coberra 6000 and the GE LM2500, which are also utilized in power generation applications. For the lower power bracket, both custom designed industrial engines such as the Solar Mars, and aeroderivatives such as the Allison 501, are used with lower pressure ratio and SOT levels leading to thermal efficiencies in the low thirties.

Automotive Applications

The Gas Turbine Versus Reciprocating Engines

The first gas turbine propelled automotive vehicle was the Rover JETI produced in the U.K. in 1950, the design team being led by Maurice Wilks and Frank Bell. The engine had a free power turbine and produced 150 kW from a simple cycle; the vehicle fuel consumption was 5.4 km/liter (15.2 mpg). Over the ensuing decades significant effort has been spent on automotive programs; however, the diesel and gasoline engines have continued to dominate, with the gas turbine only achieving a presence in specialist applications. The contributory issues are explored in this section, but there are three key reasons:

1. The poor part load thermal efficiency of the gas turbine, despite the use of a recuperated cycle with variable area nozzle guide vanes. To improve thermal efficiency, ceramic turbine technology has been researched for decades, but progress towards a production standard has been frustratingly slow.
2. There is a relatively long acceleration time of the gas turbine gas generator spool from idle to full load.
3. Huge capital investment would be required in gas turbine manufacturing facilities.

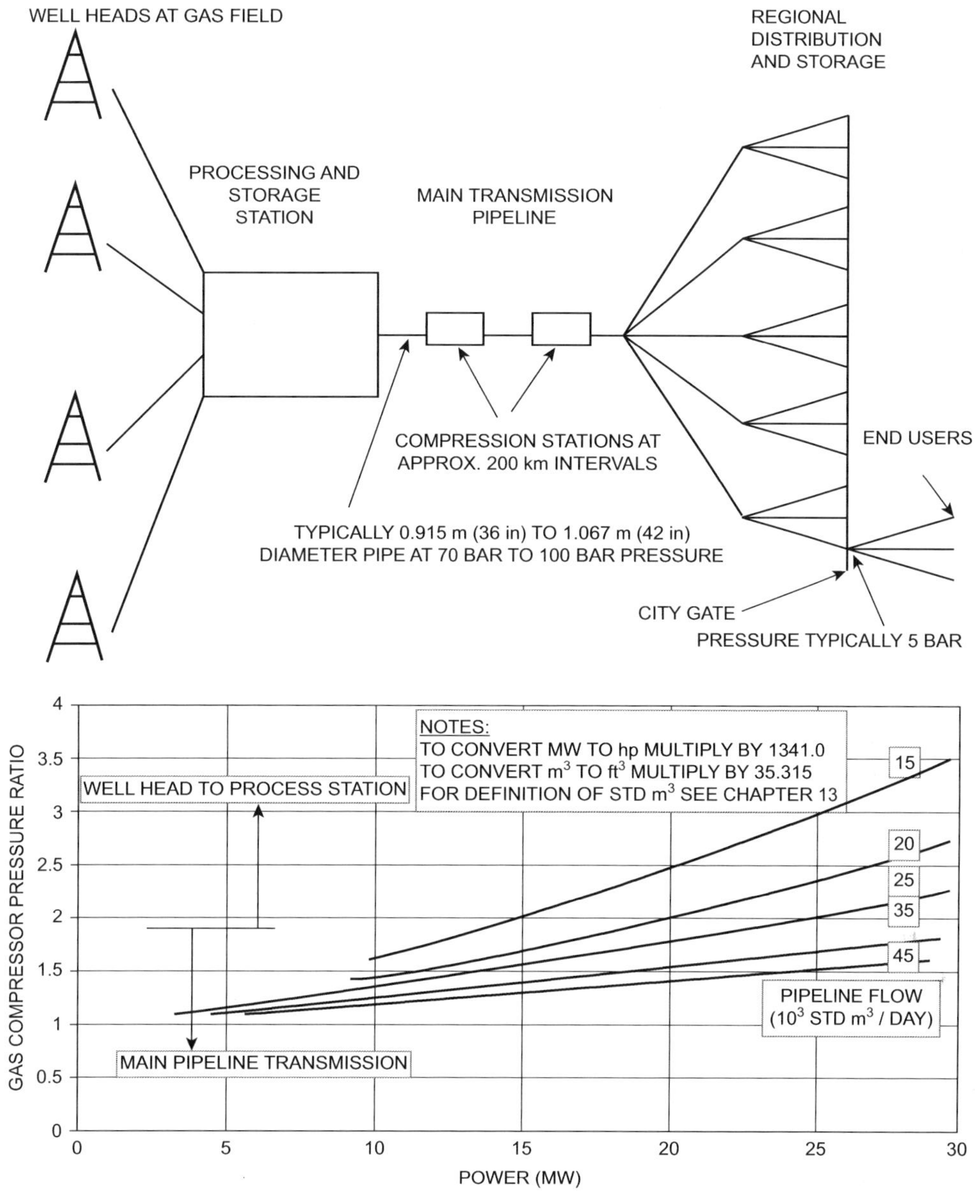

FIGURE 1–18 Natural gas transmission system and power requirements. *(Source: Rolls Royce.)*

These disadvantages have mostly outweighed the benefits of the gas turbine, which are:

- Better part speed torque capability as described in Figure 1–20, which reduces the need for varying gear ratios.
- Lower weight and volume per unit power.
- Potential for significantly lower emissions.

One other use for gas turbine engines has been in thrust-propelled vehicles for attempts on the world land speed record. In 1983 Richard Nobel's "Thrust 2" achieved 1019 km/h (633 mph) using a Rolls Royce Avon turbojet. In 1997 his "Thrust SSC" piloted by Andrew Green exceeded the speed of sound, and set a new world land speed record of 1220 km/h (763 mph), using two Rolls Royce Spey turbofans.

The Gasoline Engine Versus the Diesel Engine

Figure 1–19 includes curves of thermal efficiency and torque versus part load power for the gasoline engine. Overall it has worse SFC than the diesel engine because to avoid pre-ignition its compression ratio is lower, typically 10:1, contrasting with 15:1–20:1 for diesels. Whereas the Otto cycle in a gasoline engine has combustion at constant volume, producing increased pressure, in a diesel engine it is at constant pressure. Both engines can be turbocharged to increase power by raising inlet density and hence air mass flow. The weight and size saving can be significant, though with some expense in terms of response time, due to turbo lag as the turbocharger spool accelerates.

The main advantage of the gasoline engine is that it has lower weight and volume, which approach those for the gas turbine at the 50 kW required for a typical family sedan. Hence gasoline engines are used where fast vehicle acceleration is essential, space is at a premium, and worsening of SFC is acceptable. Diesel engines dominate for applications such as trucks where fuel consumption is paramount due to high utilization, engine weight and volume relative to the vehicle are low, and high vehicle acceleration is not a priority.

Table 1—2 lists the major categories of the automotive vehicle and the engines used in them.

The Hybrid Electric Vehicle

Stringent emissions legislation is creating a niche market in certain parts of the world, such as California. Pure electric vehicles have limited performance and range before battery recharging is required, even those using the most advanced battery systems such as sodium sulfur. To overcome this, a heat engine may also be fitted in one of two possible *hybrid* configurations, currently the subject of many development studies.

Figure 1—20 describes the various modes in which the two hybrid electric vehicle configurations may operate. For a *range extender* there are two modes of operation. Mode A is that of a conventional electric vehicle, where the battery provides traction power to the wheel motors via power electronics. In mode B the heat engine drives a generator that provides power to charge the battery. In this application when operative the gas turbine runs at maximum power, where response is unimportant and the SFC difference versus piston engines is lowest.

For a full hybrid power plant, in mode B the engine may also drive the motor.

Marine Applications

Marine propulsion uses diesel engines, gas turbines, or oil- or nuclear-fired steam plants. Diesel engines are split into two main types. The smaller *high speed* (1500 rpm) variety burn a highly refined light diesel fuel as per marine gas turbines. Larger *low speed* or *cathedral* diesels burn far heavier diesel oil, the low speed (120 rpm) and indirect injection not requiring rapid fuel vaporization for combustion. While most marine propulsion uses diesel engines the gas turbine is popular in certain applications.

The first instance of naval propulsion using gas turbines was in 1947 in the U.K. using a Metrovick "Gatric" engine in a modified gun boat. This was based on the F2 jet engine but with a free power turbine in the tail pipe and burning diesel. Sea trials lasted four years and convinced skeptics that operation of a simple cycle lightweight engine at sea was practical. Metopolitan Vickers was later taken over by Rolls Royce.

Another early development was the Rolls Royce RM60 double intercooled and recuperated engine, of 4.0 MW. This had a flat SFC curve and was intended as a single engine for small ships, and as a cruise engine for larger ones. It was fitted to *HMS Grey Goose* in 1953, which became the world's first solely gas turbine propelled ship and spent four years at sea. Though mostly technically successful, the engine did not see production, being too complex for the patrol boat role and inferior to diesels as a cruise engine.

The first operational case was the use of three Bristol Engine Company (again later taken over by Rolls Royce) Proteus engines in a fast patrol boat in 1958.

Marine propulsion system requirements differ significantly from land-based units. Owing to the large vessel inertia engine, acceleration time is generally not critical. Also the impact of current emissions legislation is negligible, particularly out at sea where pollution concentrations are low. The International Maritime Organization (IMO) is reluctant to introduce stringent legislation that a gas turbine could meet but diesel engines could not.

Major Classes of Marine Vessel

The major classes of marine vessel for which gas turbine engines are candidates are summarized in Table 1—3. Examples of actual vessels and the engines used are provided. The gas turbine competes with the diesel engine and nuclear power plant utilizing boilers and steam turbines. At the time of writing the oil-fired steam plant is becoming rare in new vessels, but remains in service.

Marine Vessel Propulsion Requirements

Figure 1—21 illustrates and quantifies the elements that comprise the total power requirement for vessel forward motion. A vessel moving through calm water creates two waveforms, one with a high water pressure at the bow, the other a reduced pressure at the stern. The energy to create this wave system is derived from the vessel via the wave making resistance. At high speed the wave resistance is dominant. Indeed for a given hull design a critical hull speed for wave making resistance is reached, where the vessel literally climbs a hill of water, with the propulsion thrust tilting upward, and it is uneconomical to go beyond this speed. The sinusoidal effect visible superimposed on the curve of wave making resistance is due to interactions of the bow and stern wave systems. The skin friction resistance, or friction form resistance, is also a major contributor to the total resistance. This is the friction between the hull and the water.

The pressure resistance, hydrodynamic drag, or form drag is due to flow separations of the water around the hull creating an adverse pressure field. Any resulting eddies or vortices are in addition to the waves created by the wave making resistance.

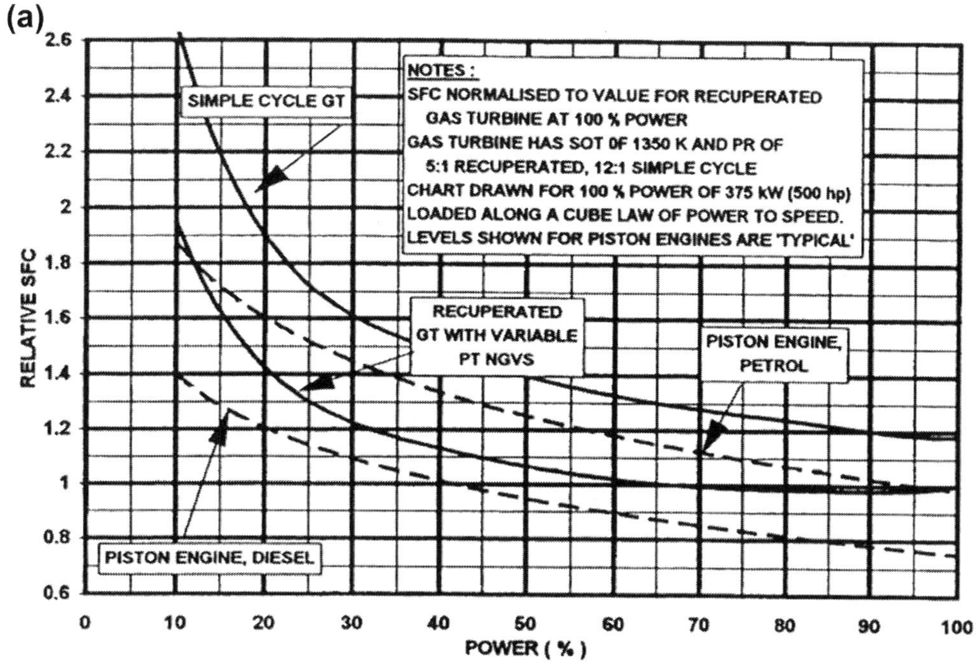

SFC versus power

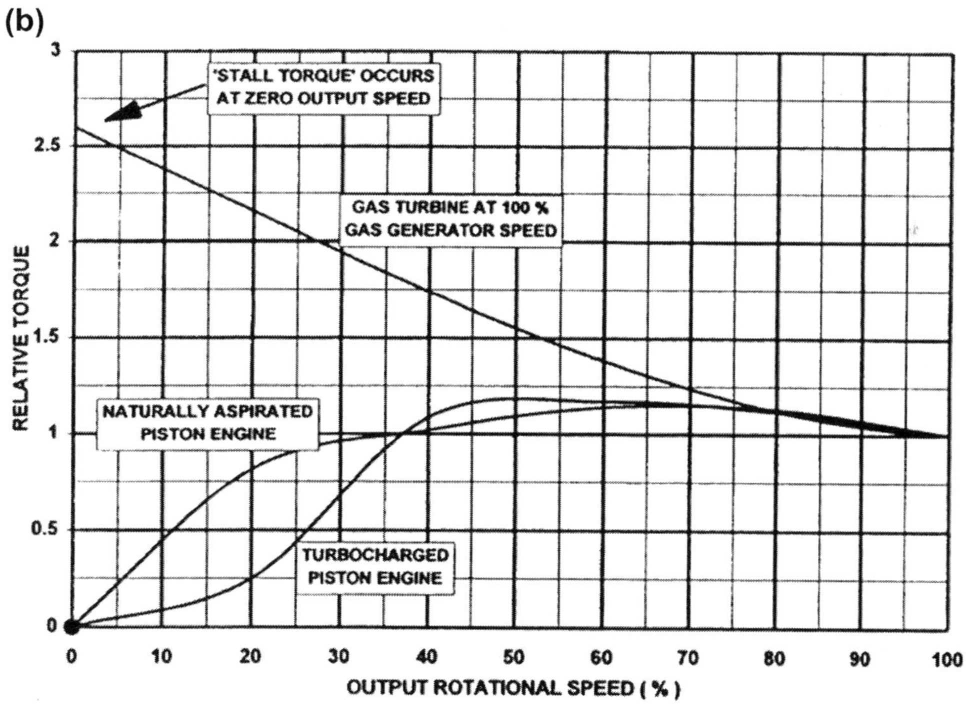

Torque versus engine output rotational speed

FIGURE 1–19 Performance of gas turbines compared with piston engines. a. SFC versus power. b. Torque versus engine output rotational speed. *(Source: Rolls Royce.)*

TABLE 1–2 Major Categories of Automotive Vehicle. Examples and Engine Types

Item	Vehicle Class	Examples of Vehicles	Engines Utilized	Power at ISO (kW)
1	Family sedan	GM Vauxhall Cavalier Volkswagen Jetta Pontiac Phoenix (experimental)	4 CYL, 1.3–1.8 liter PE 4 CYL, 1.3–2.5 liter PE Allison AGT 100 GT	40–100
2	Family sedan: hybrid electric vehicle	Volvo ECC (experimental)	Sodium sulfur battery and gas turbine	50–60
3	Family sedan: luxury	Jaguar XJ12 Mercedes Benz 320	12 CYL, 6 liter PE 6 CYL, 3.2 liter PE	190–220
4	Sports car	Porsche 911 turbo Ferrari Testarossa	6 CYL, 3.3 liter TC PE 12 CYL Flat 5.3 liter PE	180–350
5	Formula 1 racing car	Williams FW12 Benetton 8189	8 CYL, 3.5 liter PE 8 CYL, Ford HBV8 PE	500–550
6	Large truck	Scania 4 Series Ford Transcontinental H Series British Leyland Marathon T37 (experimental)	11.7 liter 6 CYL DSC12 DE Cummins NTC 355 DE Rover 2S/350R GT	300–450
7	Main battle tank	Royal Ordanance Challenger Chrysler M1 Abrams Bofors STRV 103 (experimental)	Caterpillar 12CYL DE TL AGT-1500 GT 2 × DE, 1 × GT (boost)	900–1150

PE = Petrol engine; TC = Turbocharged; RR = Rolls Royce; GM = General Motors; DE = Diesel engine; GT= Gas turbine; TL = Textron Lycoming; ISO = Standard ambient, at sea level; CYL = Cylinder.
To convert kW to hp, multiply by 1.341.
(Source: Rolls Royce.)

The air resistance due to the drag of the vessel above the water line contributes less than 5% of the total resistance. Often for low speed vessels little effort is spent in aerodynamic profiling.

The above four resistances comprise the naked resistance. The appendage resistance must then be added to evaluate the total resistance. This is the loss incurred by rudders, bilge keels, propellers, etc., and is less than 10% of the total resistance.

Traditionally these resistances for a vessel design are evaluated by model testing in a water tank and then using non-dimensional groups to scale up the resulting formulae and coefficients to the actual vessel size. This process is complex. The power requirement approximates to a cube law versus vessel velocity for displacement hulls, which support weight by simple buoyancy. Resistance is also dependent on vessel displacement (i.e., weight). Power approaches a square law for semi-planing hulls, which produce lift hydrodynamically.

Engine Load Characteristics

Ship engines drive either a conventional propeller or a waterjet via a gearbox. The latter consists of an enclosed pump that sends a jet of water rearwards. Since both devices pump incompressible water, power versus shaft speed adheres closely to a fluid dynamics cube law. For a propeller the vessel speed determines the shaft speed; in the absence of propeller blade slippage these are uniquely related. As the number of engines driving changes, engine operation moves between different possible cube laws, with slippage likely for fewer propellers driving. In the engine concept design phase cube laws are a reasonable assumption, but at the earliest opportunity the law(s) for the actual propeller or waterjet should be obtained from the manufacturer. Variable pitch propellers are often employed, which mainly affect the lowest speed characteristics.

Also shown are the resulting characteristics of engine speed versus ship speed. The cube laws for power versus ship speed, and power versus engine output speed combine to make engine output speed directly proportional to ship speed. With one engine driving, however, engine output speed is almost 20% higher than with two to achieve a given ship speed, hence propeller speed, due to slippage of the loaded propeller. These multiple load characteristics must be considered when designing a gas turbine for a multi-engine vessel.

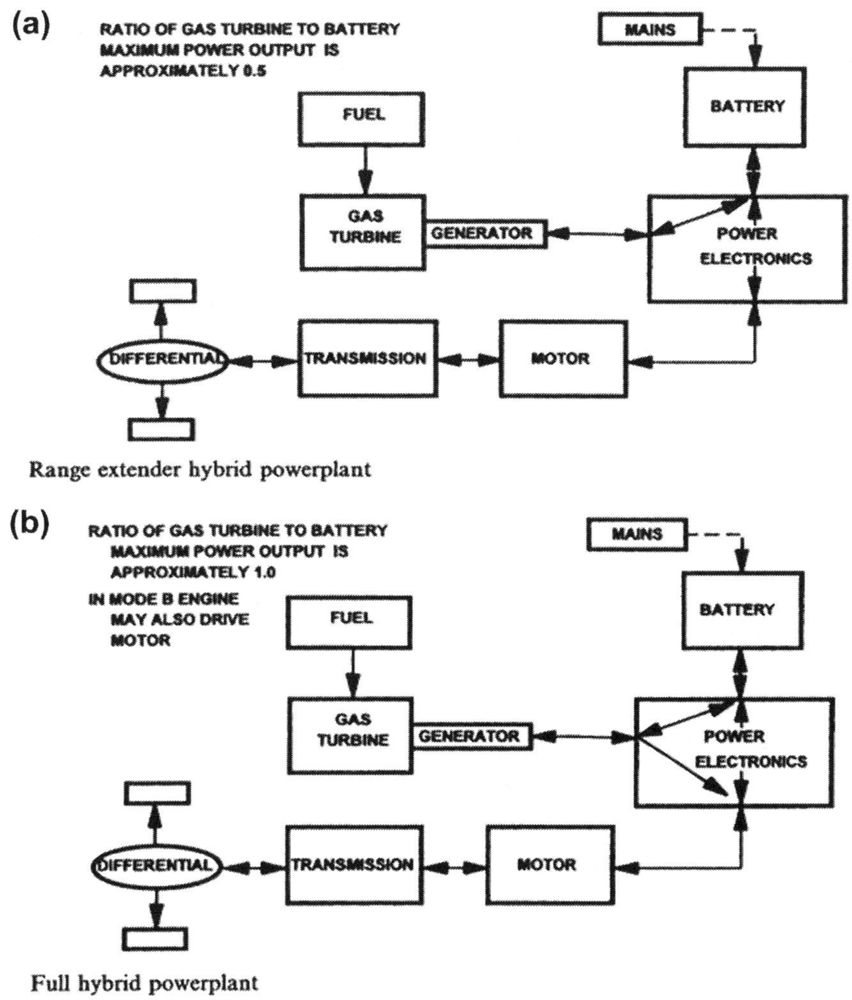

FIGURE 1–20 Hybrid electric vehicle power plant configurations. *(Source: Rolls Royce.)*

CODAG, CODOG, COGAG, and CODLAG Propulsion Systems

Figure 1–22 presents schematic diagrams of the above systems.

- In *CODAG* (COmbined Diesel And Gas turbine) systems a diesel engine provides propulsive power at low ship speed, but at high speeds the gas turbine is fired providing the relatively large additional power requirement dictated by the cube law.
- In *CODOG* (COmbined Diesel Or Gas turbine) systems the gearing is arranged such that only the diesel or gas turbine may drive the propeller or water jet at a given time.
- *COGAG* (COmbined Gas turbine And Gas turbine) and COGOG systems use a small gas turbine at low ship speed and/or a larger engine for high ship speeds.
- *CODLAG* (COmbined Diesel eLectric And Gas turbine) systems use an electric motor and diesel powered generator for low speeds. An important feature is low noise for anti-submarine work.
- *IED* (Integrated Electric Drive) or *FEP* (Full Electric Propulsion) is the subject of serious study for large naval vessels. Here the gas turbine drives a generator to provide electrical power for propulsion, ship services or, crucially, future weapons systems. A constant output speed of 3600 rpm is likely, though a gearbox or even new high power frequency converters may be employed. Such electric propulsion has long been employed on merchant vessels, but not to date on gas turbine warships.

Hovercraft (Items 1 and 2, Table 1–3)

Hovercraft use a fan to maintain pressure under side skirts to hover above the water surface and air propellers to provide thrust for propulsion. The ratio of power required for propulsion to that for hovering is between 5:1 and 10:1.

TABLE 1–3 Major Classes of Marine Vessels

Item	Vessel Type	Examples of Vessel	Engines Utilized	Total Power (MW)
1	Medium hovercraft	BHC AP1-88 Multi purpose Textron LACV-30 landing craft	4 × DEUTZ BF12L 12 CYL diesels 2 × PW ST6T GTs	1.5–3
2	Large hovercraft	BHC SR N4 passenger vessel Westamarin passenger vessel	4 × RR Proteus GTs 2 × MTU 396 diesels 2 × DEUTZ MWM diesels	3.5–10
3	Patrol boat	Souter Shipyard Wasp Bollinger Shipyard Island Glass	2 × GM 16V diesels 2 × PV diesels	2.5–4.5
4	Luxury yacht	Chritensen CXV Denison Marine Thunderbolt	2 × CAT 3412 diesels 2 × MTU 12U 396 diesels	0.5–3
5	Fast ferry	Yuet Hing Marine Catamaran Aquastrada Monohull	2 × TL TF40 GTs 1 × LM2500 GT, plus 2 × MTU 595 diesels in CODOG	6–30
6	Large merchant container	Hellenic Explorer Lloyd Nipponica	6 × diesels Boiler plus STs	20–40
7	Ultra-large tanker	Sumitomo King Opama Uddevalla Nanny	Boiler plus STs Boiler plus STs	30–40
8	Attack submarine	General Dynamics Sturgeon (USN) Vickers Fleet Class (RN)	1 × PWR plus STs 1 × PWR plus STs	10–20
9	Ballistic submarine	General Dynamics Ohio Class (USN) Vickers Vanguard Class (RN)	1 × PWR plus STs 1 × PWR plus STs	40–45
10	Frigate	Yarrow Shipyard Type 23 (RN) BIW Olivers Hazard Perry Class (USN)	2 × RR SM1C GTs plus 4 × PV diesels in CODLAG 2 × GE LM2500 GTs	30–40
11	Destroyer	BIW Arleigh Burke Class (USN) RN Type 22	4 × GE LM2500 GTs 2 × RR SM1C GTs plus 2 × RR Tyne GTs in COGAG	45–75
12	Light aircraft carrier	BIW Intrepid Class (USN) Vickers Invincible Class (RN)	Boilers plus 4 × STs 4 × RR Olympus GTs	100–120
13	Large aircraft carrier	Newport News Nimitz Class (USN) Newport News J F Kennedy Class (USN)	2 × PWRs plus STs Boilers plus 4 × STs	180–220

131-1C = British Hovercraft Co; RR = Rolls Royce; TL = Textron Lycoming; PW = Pratt & Whitney;
GM = General Motors; PV = Paxman Valenta; CAT = Caterpillar; PWR = Pressurized water reactors;
GTs = Gas turbines; STs = Steam Turbines; BIW = Bath Iron Works; RN = UK Royal Navy; USN = US Navy.
(Source: Rolls Royce.)

These vessels are designed for high speed. They are generally used in applications such as commercial passenger ferries or landing craft where the vessel spends most of its time at maximum speed. The key power plant criteria in order of importance are:

1. Low weight
2. Good high power SFC
3. Good part power SFC

The output power is below 10 MW, hence a high-speed diesel engine is practical. The diesel engine has the best high power SFC but the gas turbine the lowest weight. Consequently both power plant types are used. The gas turbine configuration is simple cycle for minimum weight, with a free power turbine. Typically SOT levels of 1250 K are employed with pressure ratios of around 7:1, the low values partly reflecting the predominance of older engine designs in current applications.

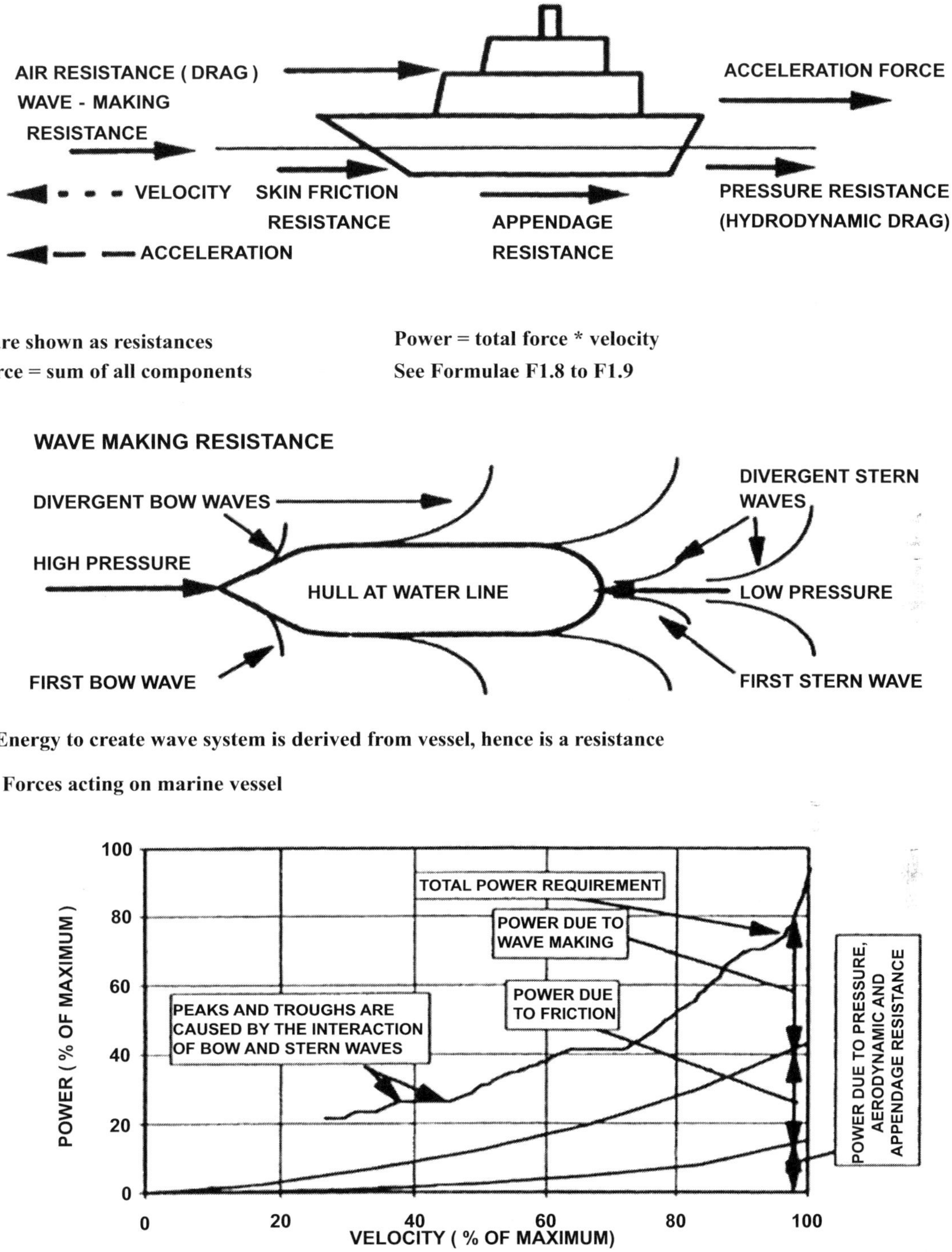

(b) Relative magnitudes of components of total power requirements (no acceleration)

FIGURE 1–21 Marine vessel power and force requirements. *(Source: Rolls Royce.)*

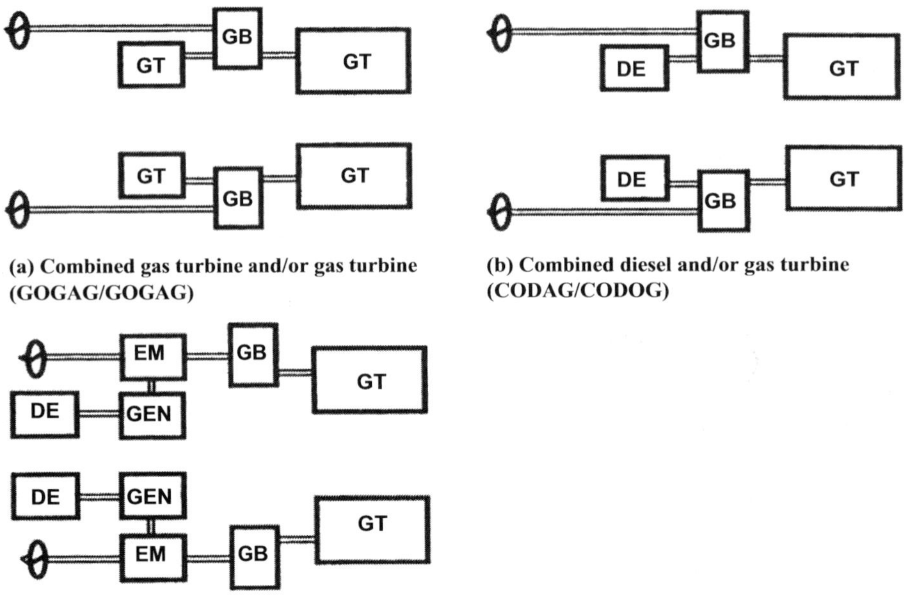

(a) Combined gas turbine and/or gas turbine (GOGAG/GOGAG)

(b) Combined diesel and/or gas turbine (CODAG/CODOG)

(c) Combined diesel -electric and gas turbine (CODLAG)

GT=Gas Turbine, DE=Diesel Engine, EM=Electric Motor, GEN=Generator

FIGURE 1–22 Marine power plant configurations using gas turbines. *(Source: Rolls Royce.)*

Monohull Patrol Boat and Luxury Yacht (Items 3 and 4, Table 1–3)

These vessels spend most of their time cruising at low speed. Because of the cube law relationship of output power to vessel speed most of the operational time is at a very low percentage output power, even allowing for twin engine vessels only cruising on one engine. The key requirement is good part load SFC; given this and the small size the diesel engine dominates.

Fast Ferry (Item 5, Table 1–3)

Fast ferries for commercial passenger transport represent a growing market. These often employ *catamaran* multiple hull configurations and *hydrofoil* operation where a lifting surface raises much of the hull above the water at high speeds. While the vessel speed is still below that of the hovercraft it has the advantage of being able to operate in rough seas. Like the hovercraft, most of its time is at high power, and key requirements are as for the hovercraft. Low weight is particularly important given the higher speed than for other classes. Power required increases with weight even for a hydrofoil, and the fuel weight for the range must be considered when operating at high speed.

For large fast ferries the power requirement is beyond that of a diesel engine and the gas turbine dominates. There is often a CODOG arrangement with diesel engines for harbor maneuvering. At the smaller end the gas turbine and diesel share the market.

Simple cycle gas turbines are employed, usually low weight aeroderivatives, with pressure ratios of 15:1–25:1. SOT is between 1450 and 1550 K with advanced cooled nozzle guide vanes and rotor blades.

Large Merchant Container and Ultra-Large Tanker (Items 6 and 7, Table 1–3)

Large merchant container ships are also propelled by either oil-fired boilers with steam turbines or nowadays diesel engines, giving top speeds of around 25 knots. The choice of top forward speed has varied, depending largely on fuel price. When fuel prices are low getting the cargo to market faster becomes dominant, and transatlantic carriers with speeds of up to 40 knots have been proposed using gas turbines. When fuel prices are high fuel cost considerations dictate lower speeds.

The ultra-large tanker (or supertanker) is the largest vessel class at sea. It operates mostly at its relatively low maximum speed of around 15 knots. Owing to the huge size, engine weight and volume are relatively unimportant, and there is free space available beneath the crew accommodation superstructure. These vessels are almost exclusively propelled by large, slow speed cathedral diesels, though older designs used oil-fired steam plant. The diesel power plant is extremely heavy and bulky but the fuel is less refined and of far lower cost per kilowatt than that used for high-speed diesels and gas turbines.

Attack and Ballistic Submarines (Items 8 and 9, Table 1–3)

Owing to the elimination of refueling and the ability to sustain full speed under water, nuclear reactors and steam turbines are used for most modern submarines. Some lower cost, smaller attack submarines are diesel–electric.

Frigate, Destroyer, and Light Aircraft Carrier (Items 10, 11, and 12, Table 1–3)

These vessels spend most of their time on station, at low vessel speed. However, substantially higher power levels are also required for sustained periods for transit to an operational zone. Hence the key powerplant criteria are:

1. Good part load SFC
2. Minimum weight, to be able to achieve high vessel speeds
3. Minimum volume, due to the need for many on-board systems and personnel
4. High availability, i.e., low maintenance and high reliability

Here CODOG, COGAG, and CODLAG systems predominate. The required output power at high speed would require a large number of diesels with unacceptable weight and volume, and so gas turbines are used for main engines. To achieve good SFC at cruise either diesels or smaller gas turbines are also employed. The gas turbine configuration utilized is as for fast ferries.

At the time of writing, one significant new marine gas turbine development program is the WR21, a 25 MW class intercooled and recuperated engine funded by the U.S., U.K., and French navies. The aim is to reduce fuel usage by 30% versus existing simple cycle engines, the heat exchangers and variable power turbine nozzle guide vanes providing a very flat SFC curve to suit naval operating profiles. Rotating components are closely based on the Rolls Royce RB211 and Trent aero turbofans.

Large Aircraft Carrier (Item 13, Table 1–3)

The "supercarrier" has the largest power requirement of any marine vessel type. The operational profile is akin to that of the other naval vessels described above.

At this power level, diesel or gas turbine installations are significantly larger than nuclear ones, especially considering the number of engines required and their fuel tanks and ducting. Hence a pressurized water nuclear reactor is employed with boilers and steam turbines. The size of the island (superstructure) is reduced without engine intake and exhaust ducts, allowing deckspace for around two more aircraft for the same vessel displacement. The resulting smaller ship profile also reduces radar cross-section, and the lack of exhaust smoke and heat further reduces signatures. The elimination of engine refueling is an advantage, though tanker support is still needed for the embarked aircraft.

Aircraft Applications—Propulsion Requirements

The concept of using a gas turbine for jet propulsion was first patented by Guillame in France in 1921. Prior to this René Lorin had obtained a patent for a ramjet as early as 1908. In January 1930 Sir Frank Whittle, unaware of the earlier French patents, also obtained a patent for a turbojet in the UK. Whittle's first engine, the world's first, ran on a test bed in April 1937. The world's first flight of a turbojet-propelled aircraft was the Heinkel He 178 in Germany, with Hans von Ohain's He S-3b engine, on 27 August 1939. This had been bench tested in early 1939; an earlier test in March 1937 had been hydrogen fueled and hence not a practical engine. Whittle, dogged by lack of investment, finally got his W.1A engine airborne propelling the Gloster E28139 on 15 May 1941. The first flight of a turboprop was on 20 September 1945, the Rolls Royce Trent powering a converted Meteor. The Trent was discontinued after five were built as Rolls Royce concentrated on the Dart, which became the first turboprop in airline service. It should be noted that Rolls Royce has used the name "Trent" again in the 1990s for its latest series of large civil turbofan engines.

The gas turbine has entirely replaced the piston engine for most aircraft applications. This is in marked contrast with the automotive market discussed earlier. The difference for aircraft propulsion was that the gas turbine could deliver something the piston engine is incapable of—practical high-speed aircraft, and much lower engine weight and size. For example, the thrust of the four turbofans on a modern Boeing 747 would require around one hundred World War II Merlin engines, which would then be far too heavy.

The Flight Mission and Aircraft Thrust Requirements

The major phases of a flight mission are takeoff, climb, cruise, descent, and landing. For military aircraft *combat* must also be considered, and all aircraft must turn, albeit briefly. By definition, the relationship between drag and equivalent airspeed is independent of altitude. At low and medium altitudes considerable excess thrust beyond aircraft drag is available.

During takeoff, high excess thrust is available for acceleration. Typical takeoff velocity and distance for a fighter are 140 kt (0.21 Mach number) and 1.2 km, respectively. Corresponding values for a civil aircraft are up to 180 kt (0.27 Mach number) and 31 km. Takeoff is a key flight condition for engine design, with usually the highest SOT.

In order to climb, additional upwards force is required. This is achieved by maintaining a high angle of attack to increase lift coefficient. The resulting increased drag is overcome by increasing thrust, excess being available beyond that required for steady flight. Also, a component of the thrust is directed vertically. The excess thrust available at low altitudes provides a high rate of climb, typically 500 m/min for a subsonic transport, and up to 8000 m/min for a fighter. Flight speed during climb is initially at a fixed level of equivalent air speed due to airframe structural considerations (maintaining constant dynamic pressure), and then at the limiting flight Mach number for airframe aerodynamics once achieved. At the top of climb the maximum engine thrust is just equal to the aircraft drag. This is a key sizing condition for the engine, with highest referred speeds and hence referred airflow. It is not the highest SOT, due to the lower ambient temperature.

Aircraft usually cruise at high altitude because here the true air speed achieved for the given level of equivalent air speed is significantly higher, and because engine fuel consumption is minimized by the correspondingly lower thrust requirement. The choice of cruise altitude is complex, and depends on engine size required to achieve the altitude, the true air speed, and range demanded by the market sector. The optimum altitude for cruise generally increases with the required level of flight Mach number. Over a long period at cruise required thrust may reduce by 20% of that at the top of climb, due to the reduction in aircraft fuel weight. Some aircraft therefore climb gradually as weight reduces, known as *cruise-climb*.

During descent the engines are throttled back to a *flight idle rating* and the aircraft angle of attack reduced. Both these effects reduce lift, and the flight direction is below horizontal. A component of the weight now acts in the direction of travel, supplementing the engine thrust to overcome drag. With zero engine thrust this would be *gliding*.

Turning requires centripetal force, provided by *banking* the aircraft to point the wings' lift radially inwards. To support the weight the overall lift must be increased, hence also the thrust as drag thereby increases.

The approach for landing is on a glide slope of approximately 3°, with a high angle of attack and flaps set to reduce aircraft speed as far as possible to give the required lift. Typically landing speeds are between 120 kt (0.18 Mach number) and 140 kt (0.21 Mach number). Landing distances are substantially less than those required for takeoff, as deceleration due to *reverse thrust* or brakes and spoilers is faster than the takeoff acceleration. Most turbofan propelled aircraft employ engines with a reverse thrust capability, where the bypass air is diverted forward using either louvers in the nacelle or rearward clamshell doors.

Afterburning or reheat is often employed for fighter aircraft and supersonic transport. Fighters generally use it only for short durations due to the high fuel consumption. These are specific maneuvers such as takeoff, transition to supersonic flight, combat, or at extreme corners of the operational envelope. Generally supersonic transports such as Concorde use it for takeoff and supersonic transition.

Engine Configuration Selection for a Required Flight Regime

The key parameters of interest are:

- SFC, especially at a reasonably high thrust or power level corresponding to cruise. Other levels such as climb and descent become more important for short ranges.
- Weight and frontal area (hence engine nacelle drag), particularly for high Mach number applications.
- Cost—this can increase with engine/aircraft size, but for expendable applications such as missiles it must be as low as practical.

The gas turbine engine achieves adequate acceleration times of around 5 seconds for civil engines and 4 seconds for military, so this does not give it or a piston engine any competitive advantage.

Range factor is the most commonly used parameter to assess the suitability of engine configurations for a required flight mission. It is the ratio of the weight of fuel and engine to the engine net thrust less pod drag for a range and flight speed. Clearly a low value of range factor is better. Figure 1−23 presents range factor versus flight Mach number for ranges of 1000 km and 8000 km, and a number of engine configurations including a piston engine.

It is immediately apparent why the gas turbine so readily replaced the piston engine for most aircraft propulsion: the latter is only in contention at low Mach numbers, below around 0.3. This is primarily because propulsion power requirements increase rapidly with Mach number. The weight and frontal area of a piston engine increase far more rapidly with output power than they do for the gas turbine. The immense importance of these factors at high flight speed is quantified by the range factor diagram.

Above 0.3 Mach number the weight and frontal area considerations mean the turboprop takes over from the piston engine as the optimum power plant. It has better fuel consumption than a turbojet or turbofan, due to a high propulsive efficiency, achieving thrust by a high mass flow of air from the propeller at low jet velocity. Above 0.6

Mach number the turboprop in turn becomes uncompetitive, due mainly to higher weight and frontal area. In addition, high propeller tip speeds required are a difficult mechanical design issue, and the high tip relative Mach numbers create extreme noise.

Above 0.6 Mach number the turbofan and turbojet compete, the optimum choice depending on the application. The turbofan has a better SFC than the turbojet, but at the expense of worse specific thrust and hence weight and frontal area. Increasing bypass ratio provides the following engine tradeoffs:

- SFC improves
- The capability for reverse thrust improves
- Weight per unit thrust increases
- Frontal area per unit thrust increases
- The number of LP turbine stages to drive the fan increases rapidly
- The cost per unit of thrust increases
- Auxiliary power and bleed offtake have a more powerful detrimental effect upon performance

The high bypass ratio engine is most competitive at flight Mach numbers of approximately 0.8, whereas at 2.2 Mach number the ideal bypass ratio is less than 1 and a turbojet becomes increasingly competitive.

Above about 2.0 Mach number the specific thrust of the ramjet becomes even better than that of a turbojet; however, it has poorer specific fuel consumption. The impact of this on range factor is shown on Figure 1—23. The low engine frontal area and weight resulting from the high specific thrust dominates at low range and high Mach number where the ramjet becomes the most competitive power plant. Also applications to date requiring this flight regime have been missiles and hence the lower unit cost of the ramjet is beneficial. For turbojet mechanical integrity the compressor delivery temperature must be kept to below approximately 950 K, hence above 2.5 flight Mach number there is very little room for compressor temperature rise.

The other possible power plant is a rocket, which is beyond the scope of this discussion.

Shaft-Powered Aircraft—Turboprops and Turboshafts

This section describes the requirements of shaft-powered aircraft, while the next section covers thrust propelled aircraft. The term turboprops usually refers to gas turbine engines that provide shaft power to drive a propeller for fixed wing aircraft propulsion. Those providing power for a rotary wing aircraft, or helicopter, are referred to as turboshafts.

Comparison of Propulsion Requirements of Shaft Power and Thrust Propelled Aircraft

The equivalent thrust and equivalent SFC of a turboprop may be calculated, allowing first cut comparisons of thrust and shaft power engines for a given application. The small amount of thrust available in a turboprop exhausts into an equivalent shaft power. This may be added to the delivered shaft power to get a total equivalent shaft power, and a corresponding SFC may be defined.

Major Classes of Shaft Powered Aircraft

Table 1—4 presents the major classes of shaft powered aircraft together with examples of actual aircraft and the engines utilized. Figure 1—25 presents key and interesting characteristics of these aircraft classes using the items from Table 1—4. Aircraft takeoff weight, range, maximum speed, and number of seats are plotted versus required power.

Fixed Wing Aircraft (Items 1, 2, and 3, Table 1—4)

Light aircraft are often privately owned, and used for short range transport or recreation. The business/executive turboprop is usually owned corporately to give flexibility in transporting executives. The commuter, or regional, transport turboprop is operated by commercial airlines on routes of moderate range, where the reduction in journey time offered by thrust aircraft would be of minimal benefit.

The piston engine now has only a few applications in the aircraft industry, one being for light aircraft with top speeds of less than 200 kt (0.30 Mach number). The piston engine is only competitive at such low flight speeds.

For the flight speeds and ranges demanded by business and commuter aircraft the range factor diagrams show the turboprop to be more competitive. Also at the engine powers above 250 kW the gas turbine is clear of the worst of the small-scale effects.

Engines are almost always of free power turbine configuration with a single spool or occasionally a two spool gas generator. Compressors are either centrifugal or axicentrifugal as this minimizes cost; efficiency is reasonably competitive with axial compressors at such low flow levels, and because frontal area is not critical at the moderate flight speeds involved. Pressure ratio is usually in the range 7:1—10:1. Axial flow turbine systems are employed with SOT levels of between 1250 and 1450 K. Above 1350 K, rotor blade cooling is employed. The choice of pressure ratio reflects a compromise between a lower value reducing the cost and weight of the compression system, and a higher value improving SFC and specific power if compressor efficiency is maintained.

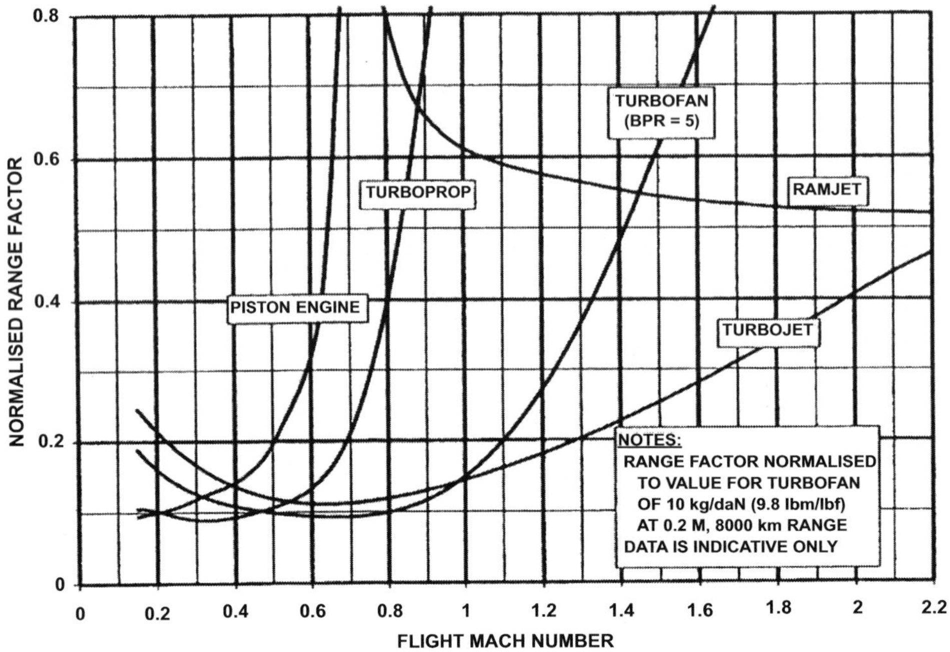

(a) Range of 1000 km

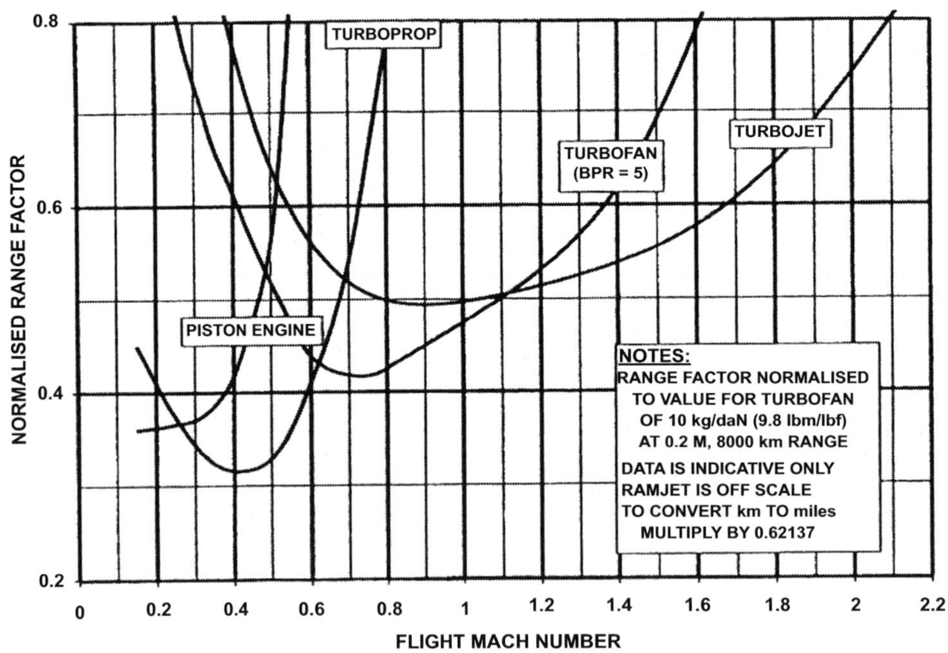

FIGURE 1–23 Aircraft range factor versus Mach number, for ranges of 1000–8000 km. *(Source: Rolls Royce.)*

Rotary Wing Aircraft (Items 4, 5, 6, and 7, Table 1–4)

Here the key criteria in order of importance are:

1. Engine weight
2. Part power SFC, as maximum power will either be sized for hot day operation, or for a multi-engine helicopter the engine failure case
3. Rated SFC
4. Engine frontal area is not particularly significant due to the low flight speeds and "buried" installation

To minimize weight some small turboshafts are single spool, which is possible because rotor pitch may be varied to change load at constant speed. For medium turboshaft helicopters the engine configuration is as per the turboprop engines described above. Levels of pressure ratio

TABLE 1–4 Major Categories of Turboprop/Turboshaft Aircraft

Item	Aircraft Type	Examples of Aircraft	Engines Utilized	Total Shaft Power (KW)
1	Light aircraft, piston engines	Piper Warrior II Beech Bonanza	1 × TL 0320-133G flat twin 1 × TC 10 520 BB flat 6	120–220
2	Business/executive Turboprop	Piper Cheyenne 400 Cessna Caravan Dornier 228-100	2 × Garrett TPE331 1 × PW PT6A-114 2 × Garrett TPE331	500–1200
3	Commuter/regional Transport turboprop	BAE Jetstream 41 Shorts 330 BAe ATP Fokker 50	2 × Garrett TPE331 2 × PW PT6A-45R 2 × PW 126A 2 × PW 125B	1800–4000
4	Light helicopter, piston engines	Robinson R22 Schweizer 300C	1 × TL 0-32-132C flat 4 1 × TL H10-360-DIA	120–170
5	Light helicopter, turboshaft engines	Bell-Jetranger III Bell 406	1 × Allison 250-C20J 1 × Allison 250-C30R	300–500
6	Multirole medium helicopter	Sikorsky S-70A (Black Hawk) Westland/Augusta EH101	2 × GE T700-700 3 × GE T700-401A, or 3 × RR/TM RTM322	2300–3500
7	Heavy lift helicopter	Sikorsky H53E Boeing Chinook CH-47	3 × GE T64-416 2 × TLT55-712	6500 10,000

TL =Textron Lycoming; GE = General Electric; TC = Teledyne Continental;
RR = Rolls Royce; PW = Pratt & Whitney; BAe = British Aerospace
To convert kW to hp multiply by 1.341.
(Source: Rolls Royce.)

and SOT are up to around 17:1 and 1500 K, respectively, the latter requiring turbine blade cooling. This pressure ratio is the optimum for specific power. At the largest engine size fully axial compressors are employed. Occasionally recuperated cycles have been considered for long-range helicopters to minimize fuel weight, though none have come to fruition. This is primarily due to the increased engine cost, weight and volume, and reliability concerns. Piston engines are used only at the lowest power levels. Figure 1–24.

Thrust Propelled Aircraft—Turbofans, Turbojets, and Ramjets

Major Classes of Thrust Propelled Aircraft

Table 1–5 presents the major classes of thrust propelled aircraft, together with examples of actual aircraft and the engines utilized. Figure 1–25 presents characteristics of these aircraft classes using the item numbers from Table 1–5.

Unmanned Vehicle Systems (Item 1, Table 1–5)

Unmanned vehicle systems include aircraft such as target and reconnaissance drones, decoys used by military aircraft to divert threats, and long range cruise missiles.

For expendable target drones and decoys the highest priority is minimum unit cost. A Mach number of at least 0.8 is usually required, with only a low range requirement. Single spool turbojets are usually used, often with centrifugal compressors because of their low cost and the low mass flow rates. Any increased weight and frontal area are accepted. Engine pressure ratios are usually between 4:1 and 8:1 as a compromise between low values favoring weight and frontal area, and high values favoring SFC and specific thrust. Low SOT levels of around 1250 K avoid the need for turbine cooling (and also give better SFC for a turbojet). Both axial and radial turbines are used.

The long range required by cruise missiles means that they fit the turbofan regime with SFC a key issue, though engine size and cost are also important as the vehicle must be transported and is expendable. Medium bypass ratio turbofans are employed, with centrifugal compressors. Indicative cycle parameters are 1.5:1 bypass ratio, 10:1 pressure ratio, and 1250 K SOT.

Subsonic Commercial Aircraft and Military Trainer (Items 2, 3, 4, and 6, Table 1–5)

Business/executive jets and civil subsonic transports all have range and flight Mach number requirements fitting the

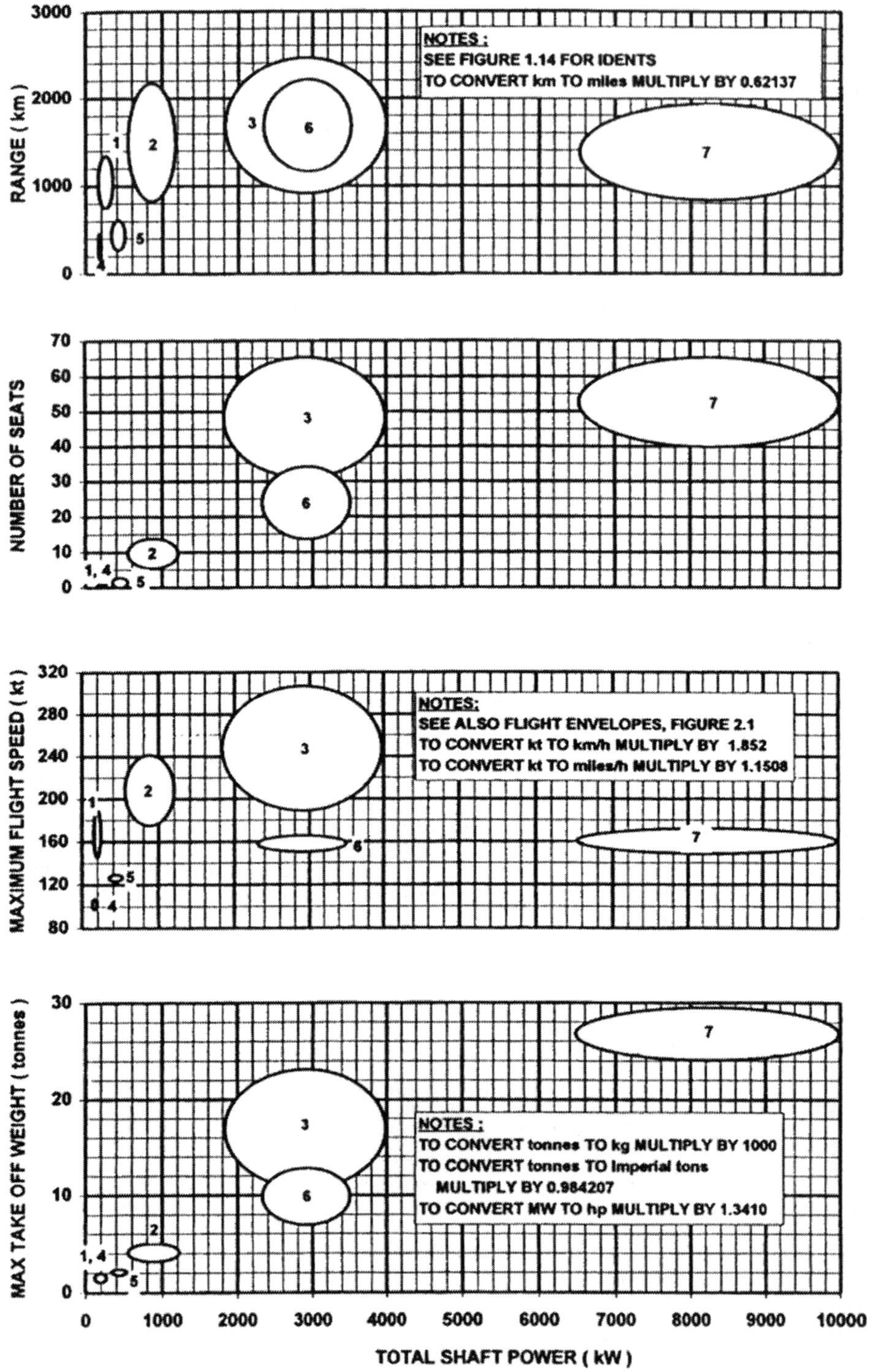

FIGURE 1−24 Turboprop/turboshaft aircraft: leading characteristics. *(Source: Rolls Royce.)*

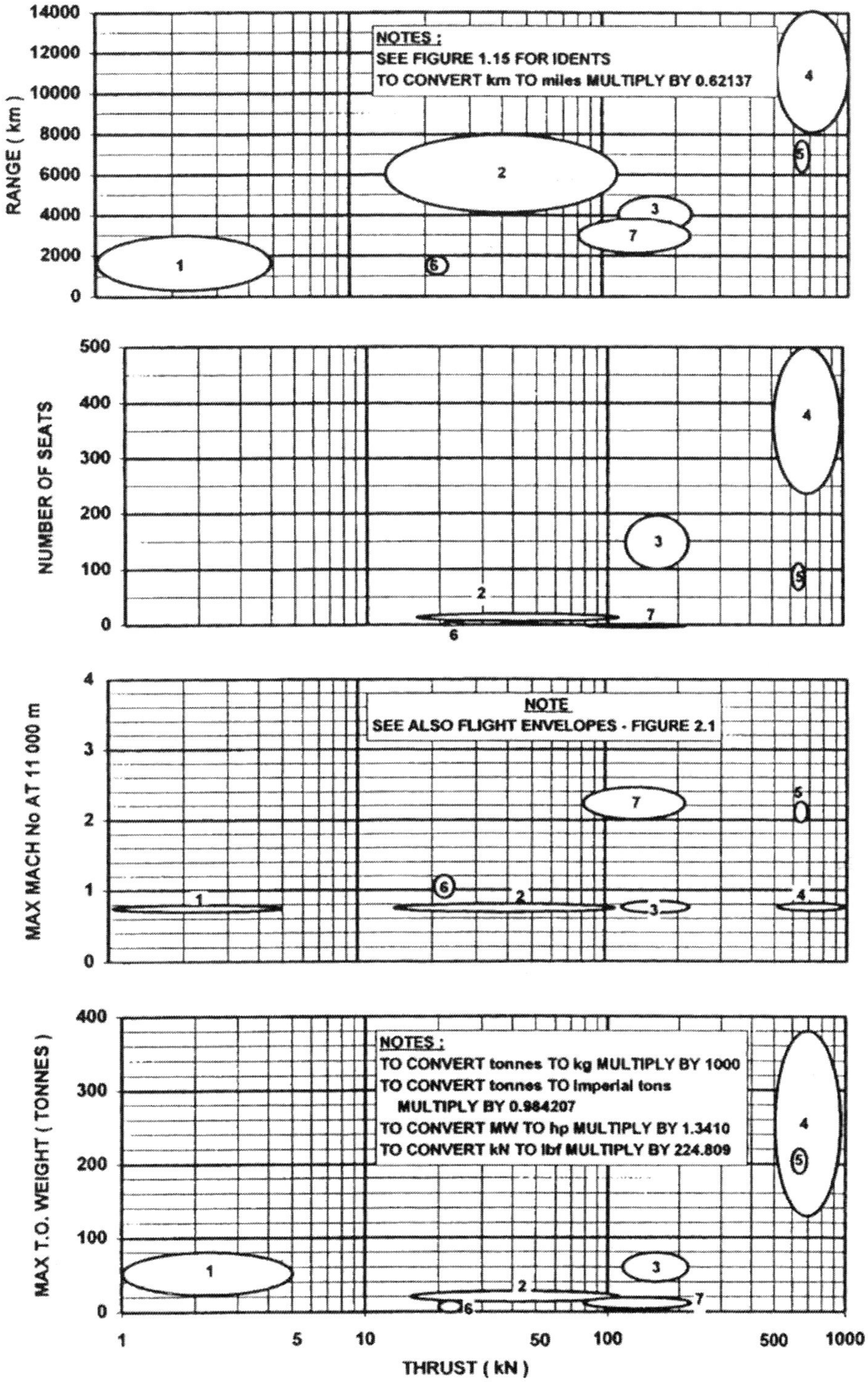

FIGURE 1–25 Thrust propelled aircraft: leading characteristics. *(Source: Rolls Royce.)*

TABLE 1–5 Major Classes of Thrust Propelled Aircraft

Item	Aircraft Type	Examples of Aircraft	Engines Utilized	Total Thrust ISA SLS TIO (kN)
1	Unmanned Vehicle Systems (UVS)	Beech MOM 107B Target Drone IMI Delilah Decoy GD BGM-109 Tomahawk long range cruise missile	1 × MT TR160-2-097 TJET 1 × NPT 151 TJET 1 × WI F107-WR-103 TFAN	1–5
2	Business/executive jet	Swearingen SJ30 Gulfstream IV BAe 125 Series 800	2 × WI/RR FJ44 TFANS 2 × RR TAY 611-8 TFANS 2 × GT TFE731-5R-1H	15–120
3	Short-medium range civil transport	Fokker 100 Boeing 737-400 Airbus A320	2 × RR TAY 620 TFANS 2 × CFM56-3B-2 TFANS 2 × IAE V2500 A1 TFANS or 2 × CFM56-5 TFANS	120–220
4	Long range civil transport	Airbus A310 Boeing 747 400	2 × PW4152 TFANS or 2 × GE CF6-80C2-A8 4 × PW4056 TFANS or 4 × GE CF6-80C2 TFANS or 4 × RR RB211-524H TFANS	500–1000
5	Supersonic civil transport	BAe/Aerospatiale Concorde	4 × SNECMA/RR Oympus 593 TJETS	600–700
6	Military trainer/light attack aircraft	BAe Hawk Aermachi MB-339C	1 × RR/TM Adour TFAN 1 × RR Viper TJET	20–25
7	Advanced military fighter	General Dynamics F16 Falcon Panavia Tornado (BAe, MBB, Aeritalia) McDonnell Douglas F15C	1 × GE F110-GE-100 TFAN or 1 × PW F100-PW-220 TFAN 2 × TU R8199-34R TFANS 2 × PW F100-PW220 TFANS	80–220
8	Ramjet propelled missiles	BAe Sea Dart (ship to air) BAe Bloodhound (gr. to air)	1 × RR TODINHOR 1 × RR	N/A N/A

WI = Williams International; RR = Rolls Royce; CFM = GE/SNECMA Joint Venture; MT = Microturbo;
NPT = Noel Penny Turbines; IAE = International Aero Engines; IMI = Israeli Military Industries;
BAe = British Aerospace; TU = Turbo Union; PW = Pratt & Whitney; GE = General Electric; GT = Garrett;
TM = Turbomeca.
To convert kN to lbf multiply by 224.809.
(Source: Rolls Royce.)

turbofan regime. They all use multi-spool gas generators with axial flow turbomachinery (except at the smallest sizes) and sophisticated turbine blade cooling for the best SFC. The pressure ratio is selected from cycle charts to give the best cruise SFC for the given SOT.

The highest bypass ratio for an engine in production at the time of writing is 8.5:1. At ISA SLS takeoff, advanced engines utilize a fan pressure ratio of around 1.8:1, and overall pressure ratio exceeds 40:1. The corresponding SOT is around 1650 K, rising to over 1750 K on a hot day. At ISA cruise overall pressure ratio is around 10% lower, and SOT around 1400 K. The highest overall pressure ratio in the flight envelope is around 45:1 at the top of climb. For lower technology engines bypass ratio is nearer to 4:1. At ISA SLS takeoff, fan pressure ratio is approximately 1.8:1 and overall pressure ratio 25:1, with SOT around 1525 K. At cruise, pressure ratio is around 10% lower, and SOT around 1350 K.

Military trainer aircraft are again in the subsonic regime, but range requirements are shorter and unit cost very important. Here turbojets and turbofans compete.

Supersonic Civil Transport and Advanced Military Fighter (Items 5 and 7, Table 1–5)

The only engines viable here are turbojets, or turbofans with a bypass ratio of less than 1:1. Multi-spool configurations with all axial turbomachinery are used for maximum efficiency and minimum frontal area. All engines have reheat systems that are employed at key points in the operational envelope.

For the limited civil applications to date, such as Concorde, and U.S. and Russian development programs, afterburning turbojets have been utilized. Takeoff SOT exceeds 1600 K, though higher values might be chosen for more modern engine designs. Pressure ratios of around 14:1 have been employed to minimize weight, and because higher values are not practical due to the high compressor delivery temperature at high flight Mach number. In addition, this value is the optimum for a pure turbojet specific thrust around a Mach number of 1.0, and also for reheated operation at 2.2 Mach number. At this flight speed the reheat fuel is burned at a high enough pressure that SFC is little worse than for a pure jet, though thrust is significantly higher. Studies for future applications encompass variable cycles, where higher bypass ratio minimizes noise and SFC during subsonic overland flight.

Advanced military fighters use low bypass ratio afterburning turbofans, with maximum SOT exceeding 1850 K and pressure ratio around 25:1. Combustor inlet temperature approaches 900 K. Future engine designs are considering SOT levels of 2000–2100 K, with combustor inlet temperatures nearer 1000 K, requiring ceramic materials. Again, engine designs with variable cycles have been proposed, to achieve higher bypass ratio to improve SFC at low flight Mach number.

One other military aircraft application is for short/vertical takeoff/landing (VTOL or STOVL), operational forms of which have utilized two main approaches. The U.K./U.S. Harrier has a fixed geometry turbofan (RR Pegasus) with four rotatable propelling nozzles, two for the core stream and two for the bypass. In contrast, Russian Yakalov aircraft have used separate vertically mounted lift jets. Future variable cycle engines could be beneficial for a Harrier type approach, providing additional bypass air for jet borne flight.

Ramjet Propelled Missiles (Item 8, Table 1–5)

At Mach numbers in excess of 2.5 the ramjet is the ideal powerplant. Combustion temperatures approach the *stoichiometric* value, where all oxygen is used. This ranges from 2300 to 2500 K, depending on inlet temperature and hence flight Mach number, and is feasible as there are no stressed turbine blades to consider. At these Mach numbers the only competitor engine is a rocket. Starting a ramjet requires a short duration booster rocket, to accelerate the vehicle to a Mach number where operation is possible.

Air to air missiles to date have been almost entirely rocket powered, as this better suits the requirement of high thrust for a short duration. However, experimental ramjet versions have been produced, particularly in France and the former USSR. Several current proposals involve ramjets, as air-to-air missile range requirements increase.

Surface to air missiles with ramjets have seen production, such as the U.K. "Bloodhound," as range requirements are more suitable. A typical mission would be launch, climb to around 20,000 m, followed by a loiter phase and then attack. The distance covered would be around 50 km.

Auxiliary Power Units (APUs)

Aircraft APUs have normally fulfilled several functions in an aircraft, namely:

- Main engine starting
- Supply of cooling air for aircraft secondary systems, particularly when at ground idle in hot climates
- Supply of electrical power when main engines are shut down, including for ground checkout of aircraft systems

These functions give an aircraft self sufficiency when on the ground. In addition an APU will be required to fire up at altitude in case of main engine flame out, to power electrical systems—vital for fly by wire aircraft—and if at low flight Mach number to provide crank assistance to help restart the engines.

Until recently, new developments have been rare, but APU sophistication is now increasing to match that of recent aircraft, where APU operation is becoming less intermittent. For civil applications APU requirements may now include operation in all regions of the flight envelope, and for military aircraft advanced systems with start times as low as a second. A current typical start time is around 6 seconds at 15,000 m. There are occasional studies on the benefits of permanent running power units that avoid compromising the design of the propulsion engines by power and bleed offtake.

Historically the main requirements for APUs have been:

- Low development and unit costs
- High reliability and maintainability
- Low volume and weight
- Good SFC

Gas Turbines Versus Piston Engines

APUs for aircraft are almost exclusively simple cycle gas turbines. Power density in terms of weight and volume per unit of shaft output power are vastly superior to a piston engine, around 4.4 kW/kg and 8 MW/m^3. This effectively makes a piston engine impractical, despite its lower unit cost. Fuel consumption becomes a secondary issue where operation is intermittent.

APU Power Requirements of Major Aircraft Classes

The output power range of APUs is between 10 kW and 300 kW, with bleed supplied requiring additional turbine

TABLE 1–6 Auxiliary Power Unit (APU) Examples and Applications

Model	Configuration	Application	Power (kW)
Turbomach T-62T-40-8	Single shaft: 1 Stage centrifugal compressor Reverse flow annular combustor 1 Stage radial turbine	Jet fuel starter General Dynamics F16 Fighter	190
Allied Signal 131-9(D)	Single shaft: 1 Stage centrifugal compressor 2 Stage axial turbine 1 Stage centrifugal load compressor	Bleed, E.G. engine start Electrical power ENV conditioning McDonnell Douglas MD90	300/100
Allied Signal 331-500B	Single shaft: 2 Stage centrifugal compressor Reverse flow annular combustor 2 Stage axial turbine 1 Stage centrifugal load compressor	Bleed, E.G. engine start Electrical power Boeing 777	850/170
APIC APS 3200	Single shaft: 1 Stage centrifugal compressor Reverse flow annular combustor 2 Stage axial turbine 1 Stage centrifugal load compressor	Electrical power Oil and fuel pumps Airbus A321	385/90

All data is indicative.
Where two powers are shown the higher figure includes the load compressor drive power.
To convert kW to hp multiply by 1.3410.
To convert kg to lb multiply by 2.2046.
APIC—Auxiliary Power International Company.
(Source: Rolls Royce.)

power. Table 1–6 presents specific examples of APUs employed in production aircraft.

APU Configurations

For all configurations centrifugal compressors are used exclusively and often radial inflow turbines, even occasionally combined as a monorotor to minimize cost. SOT levels are typically 1250 to 1260 K to minimize the need for turbine cooling. Pressure ratio is generally between 4:1 and 8:1, though the trend is towards higher levels.

The most common forms of APU provide high-pressure air to the main engine mounted air turbine starter. These are referred to as *pneumatic APUs.* Air must usually be supplied at around five or more times ambient pressure, with the APU sized to enable hot day main engine starting. The most common pneumatic APU is a single shaft gas turbine with integral bleed. Here the engine is of single spool configuration but with the pneumatic air supply bled off from compressor delivery. This is the simplest unit and hence has the lowest cost. Also generators or pumps may be driven off the spool to provide electrical or hydraulic power.

Single shaft gas turbines driving a centrifugal load compressor, as well as application pumps or generator are growing in popularity. This configuration has the highest power output per unit mass and volume, though it is of higher cost.

A small number of APUs apply torque directly to the main engine HP shaft via its gearbox and a clutch, rather than supplying high-pressure air. These are termed *jet fuel starters.* In this instance the APU is often of free power turbine configuration to provide an adequate partial speed torque characteristic.

Chapter 2

Historical Development of the Gas Turbine

"I only hope that we never lose sight of one thing—that it was all started by a mouse."

—Walt Disney

Chapter Outline

EARLY HISTORY OF THE GAS TURBINES

The development of the gas turbine took place in several countries. Several different schools of thought and contributory designs led up to Frank Whittle's 1941 gas turbine flight. Despite the fact that NASA's development budget now trickles down to feed the improvement of flight, land-based, and marine engines, the world's first jet engine owed much to early private aircraft engine pioneers and some lower profile land-based developments.

The development of the gas turbine is a source of great pride to many engineers worldwide and, in some cases, takes on either industry sector fervor (for instance, the aviation versus land-based groups) or claims that are tinged with pride in one's national roots. People from these various sectors and subsectors can therefore be selective in their reporting.

So for understanding the history of the gas turbine, one would have to read several different papers and select material written by personnel from the aviation and land-based sectors. At that point, one can "fill in the gaps."

What follows therefore are three different accounts of the gas turbine's development. Each version is correct; it is mainly a matter of perspective.

Land-Based Gas Turbine Development Perspective

The first timeline contains many of the relevant *land-based gas turbine design developments.** Note that this also contains some timeline references to aircraft engine development.

Switzerland (and Swiss abroad, in italics)

1921 J. Ackeret, high-speed aerodynamics scientist at ETH Zurich, arrives at L. Prandtl's AVA (Aerodynamische Versuchs-Anstalt), Gottingen; stays seven years.

* Source: ASME 2001—GT-0395, "Advanced Gas Turbine Technology—ABB/BBC Historical Firsts," by Eckardt, D., and Rufli, P., Alstom Power Ltd.

1925 CEM (G. Darrieus), a French subsidiary of BBC (Brown Boveri Company), produces a series of windmills, using airfoil design theory.

1926 BBC's four-stage axial test compressor designed, first with untwisted blades, later swirl adapted.

1932 BBC sold a number of 11-stage axial compressors, PR = 3.4, for the Mondeville project and high-speed windtunnels at ETH-Zurich and Rome.

1934 C. Keller, assistant to J. Ackeret at ETH Zurich, designed one of the windtunnel blowers (second blower for high-speed tunnel came from BBC).

1939 A. Meyer, BBC's technical director, presents a comprehensive paper on GT design achievements (including gas turbine usage for compact and lightweight ship/destroyer propulsion) at the Institute of Mechanical Engineering, London. First commercial industrial gas turbine (GT) from BBC is operational at Neuchâtel. BBC delivers first industrial GT to RAE, 1.6 MW, 20-stage axial compressor. In 1940, BBC delivers axial aircraft superchargers, 190 hp, PR = 2.5, to complete a Rolls Royce (RR) purchase order.

Germany (and Germans abroad)

1922 W. Bauersfeld suggests the use of airfoil theory for fluid machinery.

1935(?) At AVA Gottingen, a four-stage axial turbocharger, seven-stage compressor design undergoes development (Encke et al. design), PR = 3.8 [in production PR = 3.1].

1935 H. P. von Ohain gets a secret turbo-engine patent no. 317/38.

1937 H. P. von Ohain's test engine HeS313 runs.

1939 The first jet-powered flight He 178 aircraft with HeS313, on Sunday August 27, 1939. R. Friedrich, Junkers Magdeburg, designs the 14-stage axial compressor for the RTO engine (Riickstoss-Turbine ohne Leistungsabgabe with a propeller), for the helium aircraft S30 engine, based on a Gottingen airfoil design.

1942 Me 262 fighter aircraft enters service with two Jumo 004 engines, first test flight on August 18.

England (and English abroad, in italics)

1926 A. A. Griffith releases "An Aerodynamic Theory of Turbine Design," which discusses a GT as an aircraft's power plant.

1930 F. Whittle gets the first patent for a turbo aero-engine. Tizard, Gibson & Glauert committee denies that the gas turbine could be superior to the piston engine.

1937 F. Whittle, radial compression engine, has a test run on December 4.

1938 A delegation at BBC decides that "Exclusivity on the BBC (axial) compressor design would not be granted."

1941 The Whittle engine has its first flight.

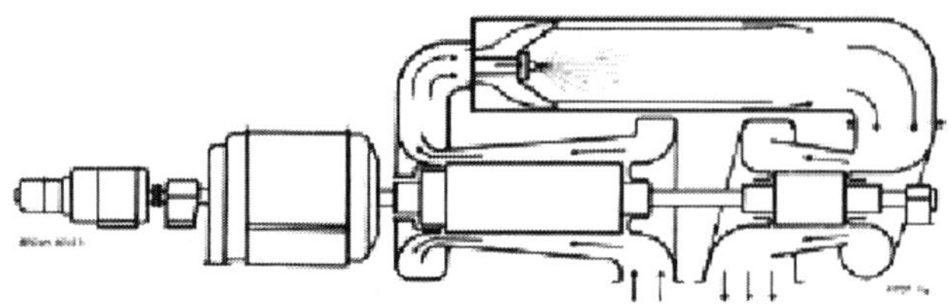

FIGURE 2–1 Cross-section of first gas turbine for public power generation, BBC, Neuchâtel, Switzerland, 1939. *(Courtesy: Alstom Power.)*

Note that, in the early 1930s, BBC designed components used for the Velox project boilers. It developed a turbine that had enough power to drive the compressor and could generate excess power through the inverse operation of the electric starter motor. Also, in 1936, BBC's 4 MW all-axial process gas turbine/blower train with PR = 4 was supplied to a US refinery. In July 1939, BBC commissioned the world's first utility gas turbine at Neuchâtel, Switzerland. The gas turbine had one 23-stage axial compressor, one single-can combustor, one 7-stage axial turbine, and a synchronously operated generator on the same shaft (Figure 2–1).

Aircraft Engine Development: A US Perspective

The second timeline presents an *aircraft engine development perspective (more a US perspective).***

Attempts to develop gas turbines were first undertaken in the early 1900s, with pioneering work done in Germany. The most successful early gas turbines were built by Holzwarth who developed a series of models between 1908 and 1933. The first industrial application of a gas turbine was installed in a steel works in Hamborn, Germany, in 1933. In 1939, a gas turbine was installed in a power plant in Neuchâtel.

1925 R. E. Lasley of Allis-Chalmers receives the first of several patents on gas turbines. Around 1930 he forms the Lasley Turbine Motor Company in Waukegan, Illinois, with a goal of producing a gas turbine for aircraft propulsion.

1929 Haynes Stellite develops Hastelloy alloy for turbine buckets, allowing operation up to gas temperatures of over 1800°F. This superior alloy was later crucial to the successful operation of the I-A and it gave US turbine manufacturers the ability to use uncooled designs rather than include the complexity of blade cooling.

1931 US Army awards GE a turbine-powered turbosupercharger development contract.

1934 US Army personnel from Wright Field visit Lasley's shop and inspect his hardware and the engine he had filmed in operation earlier that year. However, neither the Army nor Navy would fund Lasley.

** Reference: J. St. Peter, "The History of Aircraft Gas Turbine Development in the United States," published IGTI, American Society of Mechanical Engineers, 1999.

1935 US Army, Northrop, TWA, and GE combine to test fly a Northrop Gamma at 37,000 feet from Kansas City to Dayton. This led to a production contract for GE to build 230 units of the "Type B" supercharger and to establishment of the GE Supercharger Department in Lynn, Massachusetts (later the site of the I-A development based on the Whittle engine).

1938 Wright Aeronautical Corporation designs its own vaned superchargers for its own engines, although the superchargers were manufactured for Wright by GE.

1939 GE studies gas turbine aircraft propulsion options and concludes the turbojet is preferable to the turboprop. Note, however, that two years later the company changed its mind and proposed a turboprop to the Durand Committee.

1940 NACA joins with Wright, Allison, and P&W to standardize turbosupercharger testing techniques.

1941 GE Steam Turbine Division (Schenectady) participates in the Durand Special Committee on Jet Propulsion and proposes a turboprop, designated the TG-100 (later the T31), which ran successfully in May 1943 under Army sponsorship.

1941 GE Turbosupercharger Division (Lynn, Massachusetts) receives the Whittle W.1.X engine and drawings for the W.2.B improved version. A top-secret effort begins to build an improved version, known as the I-A, for flight test in the Bell P-59.

1941 The Durand Committee also awards Navy contracts to Allis-Chalmers and Westinghouse. The Westinghouse W19, a small booster turbojet, resulted from this but Allis-Chalmers dropped out of the "gas turbine race" in 1943.

1942 In April, the GE I-A runs for the first time in a Lynn test cell. In October, it powers the Bell P-59 on its first flight at Muroc Dry Lake, California.

By the latter part of 1942, the following "native" aircraft gas turbine efforts were proceeding. These projects included:

1. Northrop Turbodyne turboprop
2. P&W PT-1 turboprop
3. GE/Schenectady TG-100 turboprop
4. Allis-Chalmers turbine-driven ducted fan
5. NACA piston-driven ducted fan
6. Westinghouse 19A turbojets
7. Turbo Engineering Corporation's booster-sized turbojet

Ultimately, the history of gas turbines is a subject that has almost as many versions as there are engineer historians. Each person credits one development more or less than another and every account has some merit. Also of interest is the following perspective, in a more narrative style.

The development* of the gas turbine engine as an aircraft power plant has been so rapid that it is difficult to appreciate that, prior to the 1950s, very few people had heard of this method of aircraft propulsion. The possibility of using a reaction jet had interested aircraft designers for a long time, but initially the low speeds of early aircraft and the unsuitability of a piston engine for producing the large high velocity airflow necessary for the "jet" presented many obstacles.

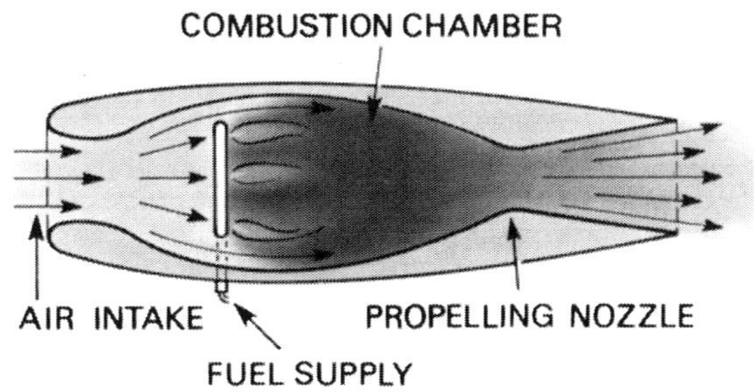

FIGURE 2–2 Lorin's jet engine. *(Source: Rolls Royce.)*

A French engineer, René Lorin, patented a jet propulsion engine (Figure 2–2) in 1913, but this was an athodyd and was at that period impossible to manufacture or use, since suitable heat resisting materials had not then been developed. In the second place, jet propulsion would have been extremely inefficient at the low speeds of the aircraft of those days. However, today the modern ramjet is very similar to Lorin's conception.

In 1930, Frank Whittle was granted his first patent for using a gas turbine to produce a propulsive jet, but it was 11 years before his engine completed its first flight. The Whittle engine formed the basis of the modern gas turbine engine and from it was developed the Rolls Royce Welland, Derwent, Nene, and Dart engines. The Derwent and Nene turbojet engines had worldwide military applications; the Dart turbopropeller engine became world famous as the power plant for the Vickers Viscount aircraft. Although other aircraft may be fitted with later engines termed *twin-spool*, *triple-spool*, *by-pass*, *ducted fan*, *unducted fan*, and *propfan*, these are inevitable developments of Whittle's early engine.

The jet engine (Figure 2–3), although appearing so different from the piston engine/propeller combination,

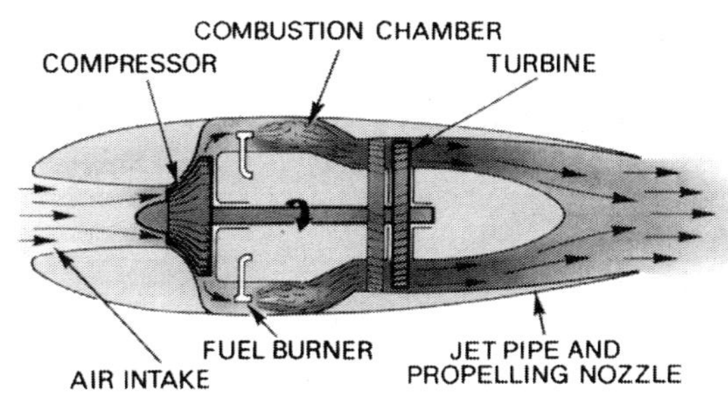

FIGURE 2–3 A Whittle-type turbojet engine. *(Source: Rolls Royce.)*

* Source: Adapted, with permission, from Rolls Royce, *The Jet Engine*, 1986, Rolls Royce Plc: UK.

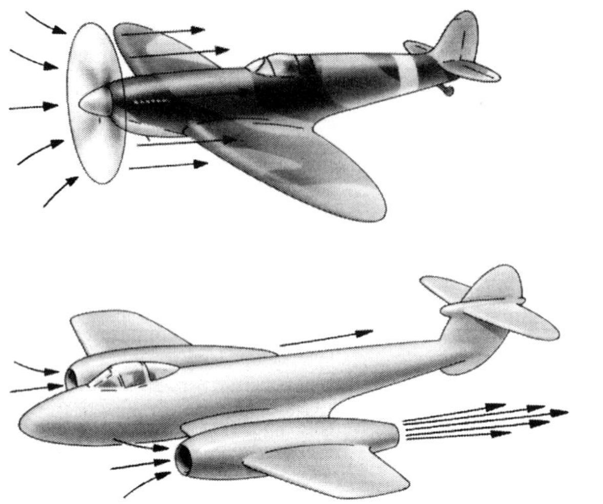

FIGURE 2–4 Propeller and jet propulsion. *(Source: Rolls Royce.)*

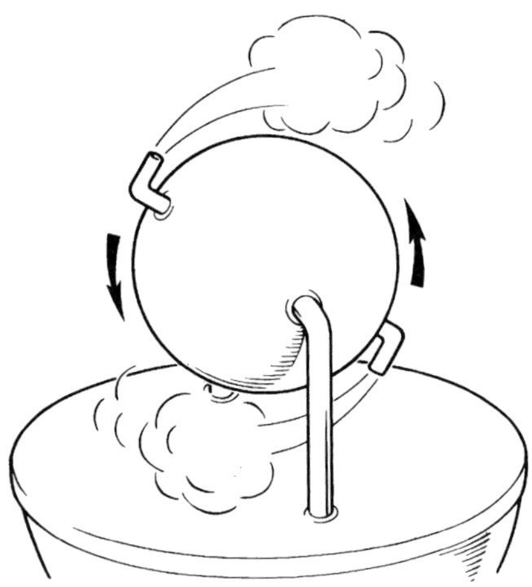

FIGURE 2–5 Hero's engine—probably the earliest form of jet reaction. *(Source: Rolls Royce.)*

applies the same basic principles to effect propulsion. As shown in Figure 2–5, both propel their aircraft solely by thrusting a large weight of air backwards.

Although today jet propulsion is popularly linked with the gas turbine engine, there are other types of jet-propelled engines, such as the ramjet, the pulse jet, the rocket, the turbo/ramjet, and the turborocket.

PRINCIPLES OF JET PROPULSION

Jet propulsion is a practical application of Sir Isaac Newton's third law of motion, which states that, "for every force acting on a body there is an opposite and equal reaction." For aircraft propulsion, the "body" is atmospheric air that is caused to accelerate as it passes through the engine. The force required to give this acceleration has an equal effect in the opposite direction acting on the apparatus producing the acceleration. A jet engine produces thrust in a similar way to the engine/propeller combination. Both propel the aircraft by thrusting a large weight of air backwards (Figure 2–4), one in the form of a large air slipstream at comparatively low speed and the other in the form of a jet of gas at very high speed.

This same principle of reaction occurs in all forms of movement and has been usefully applied in many ways. The earliest known example of jet reaction is that of Hero's engine (Figure 2–5) produced as a toy in 120 BC. This toy showed how the momentum of steam issuing from a number of jets could impart an equal and opposite reaction to the jets themselves, thus causing the engine to revolve.

The familiar whirling garden sprinkler (Figure 2–6) is a more practical example of this principle, for the mechanism rotates by virtue of the reaction to the water jets. The high-pressure jets of modern firefighting equipment are an example of "jet reaction," for often, due to the reaction of the water jet, the hose cannot be held or controlled by one firefighter. Perhaps the simplest illustration of this principle is afforded by the carnival balloon, which when the air or gas is released, rushes rapidly away in the direction opposite to the jet.

Jet reaction is definitely an internal phenomenon and does not, as is frequently assumed, result from the pressure of the jet on the atmosphere. In fact, the jet propulsion engine, whether rocket, athodyd, or turbojet, is a piece of apparatus designed to accelerate a stream of air or gas and to expel it at high velocity. There are, of course, a number of ways of doing this, as described next, but in all instances the resultant reaction or thrust exerted on the engine is proportional to the mass or weight of air expelled by the engine and the velocity change imparted to it. In other words, the same thrust can be provided either by giving a large mass of air a little extra velocity or a small mass of air a large extra velocity. In practice the former is preferred, since by lowering the jet velocity relative to the atmosphere, a higher propulsive efficiency is obtained.

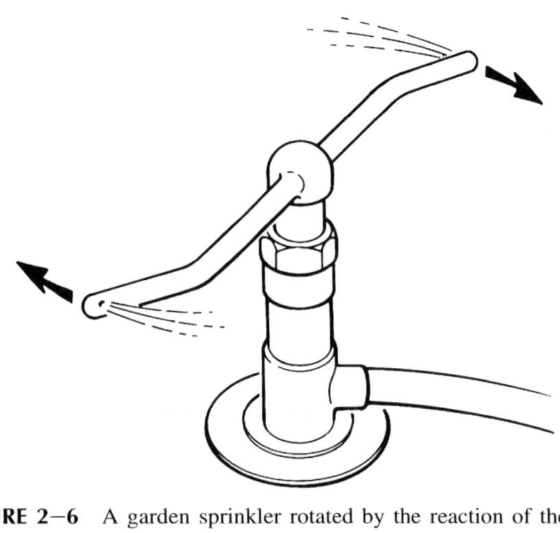

FIGURE 2–6 A garden sprinkler rotated by the reaction of the water jets. *(Source: Rolls Royce.)*

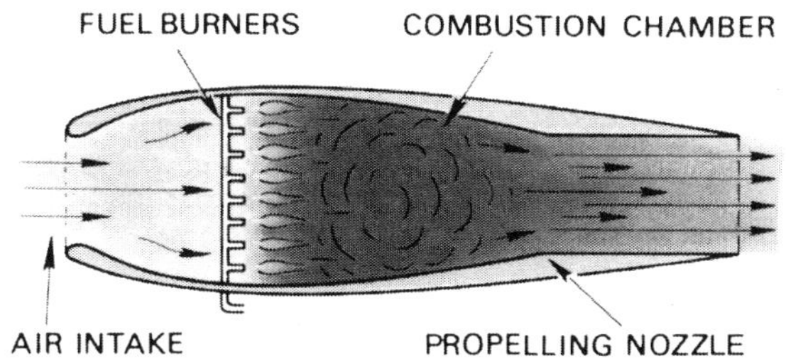

FIGURE 2–7 A ramjet engine. *(Source: Rolls Royce.)*

Methods of Jet Propulsion

The types of jet engine, whether ramjet, pulse jet, rocket, gas turbine, turbo/ramjet, or turborocket, differ only in the way in which the "thrust provider," or engine, supplies and converts the energy into power for flight.

The ramjet engine (Figure 2–7) is an athodyd, or aerothermodynamic-duct to give it its full name. It has no major rotating parts and consists of a duct with a divergent entry and a convergent or convergent/divergent exit. When forward motion is imparted to it from an external source, air is forced into the air intake, where it loses velocity or kinetic energy and increases its pressure energy as it passes through the diverging duct. The total energy is then increased by the combustion of fuel and the expanding gases accelerate to atmosphere through the outlet duct. A ramjet is often the power plant for missiles and target vehicles but is unsuitable as an aircraft power plant because it requires forward motion imparting to it before any thrust is produced.

The pulse jet engine (Figure 2–8) uses the principle of intermittent combustion, and unlike the ramjet, it can be run at a static condition. The engine is formed by an aerodynamic duct similar to the ramjet but, due to the higher pressures involved, it is of more robust construction. The duct inlet has a series of inlet "valves" that are spring-loaded into the open position. Air drawn through the open valves passes into the combustion chamber and is heated by the burning of fuel injected into the chamber. The resulting expansion causes a rise in pressure, forcing the valves to close, and the expanding gases are then ejected rearwards. A depression created by the exhausting gases allows the valves to open and repeat the cycle. Pulse jets have been designed for helicopter rotor propulsion and some dispense with inlet valves by careful design of the ducting to control the changing pressures of the resonating cycle. The pulse jet is unsuitable as an aircraft power plant because it has a high fuel consumption and is unable to equal the performance of the modern gas turbine engine.

Although a rocket engine (Figure 2–9) is a jet engine, it has one major difference in that it does not use atmospheric air as the propulsive fluid stream. Instead, it produces its own propelling fluid by the combustion of liquid or chemically decomposed fuel with oxygen, which it carries, thus enabling it to operate outside the earth's atmosphere. It is, therefore, suitable only for operation over short periods.

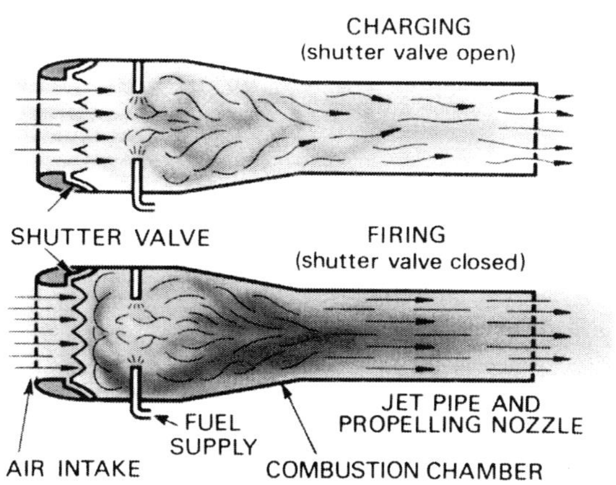

FIGURE 2–8 A pulse jet engine. *(Source: Rolls Royce.)*

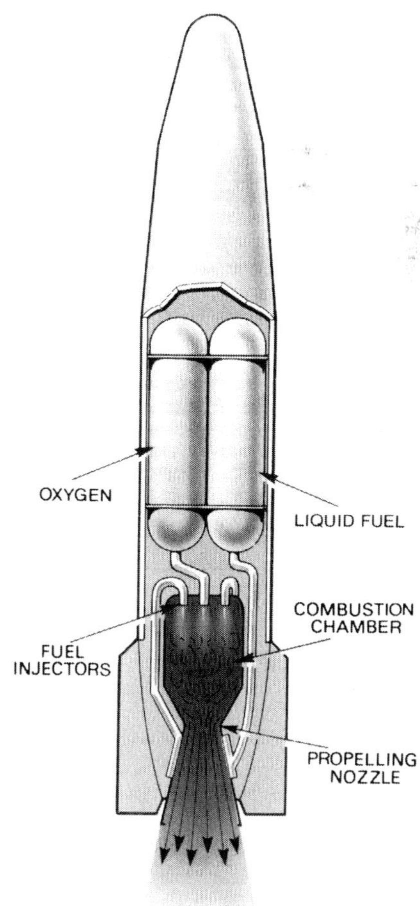

FIGURE 2–9 A rocket engine. *(Source: Rolls Royce.)*

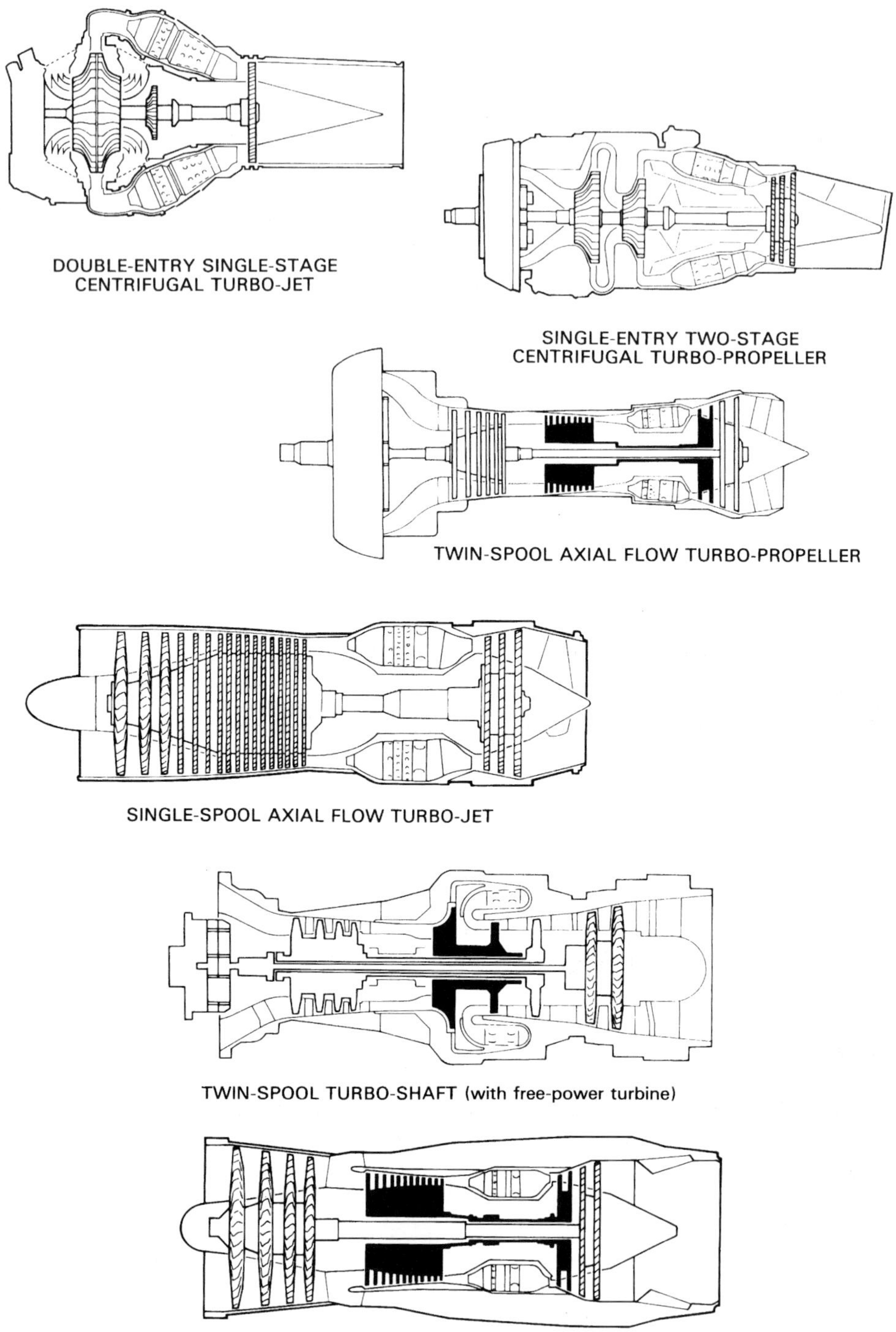

FIGURE 2–10 Mechanical arrangement of gas turbine engines. *(Source: Rolls Royce.)*

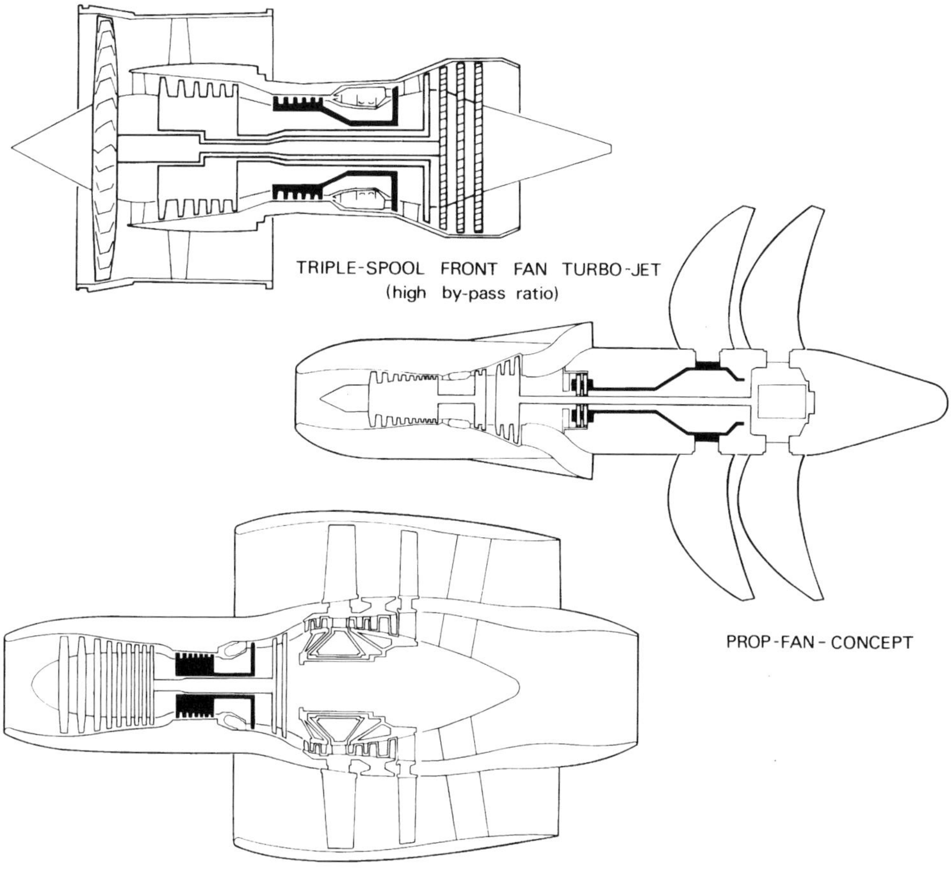

FIGURE 2–10 Cont'd

The application of the gas turbine to jet propulsion has avoided the inherent weakness of the rocket and the athodyd, for by the introduction of a turbine-driven compressor, a means of producing thrust at low speeds is provided. The turbo-jet engine operates on the "working cycle" as already described. It draws air from the atmosphere, and after compressing and heating it, a process that occurs in all heat engines, the energy and momentum given to the air forces it out of the propelling nozzle at a velocity of up to 2000 feet per second or about 1400 miles per hour. On its way through the engine, the air gives up some of its energy and momentum to drive the turbine that powers the compressor.

The mechanical arrangement of the gas turbine engine is simple, for it consists of only two main rotating parts, a compressor and a turbine, and one or a number of combustion chambers. The mechanical arrangement of various gas turbine engines is shown in Figure 2–10. This simplicity, however, does not apply to all aspects of the engine, for as described in subsequent sections, the thermo- and aerodynamic problems are somewhat complex. They result from the high operating temperatures of the combustion chamber and turbine, the effects of varying flows across the compressor and turbine blades, and the design of the exhaust system through which the gases are ejected to form the propulsive jet.

At aircraft speeds below approximately 450 miles per hour, the pure jet engine is less efficient than a propeller-type engine, since its propulsive efficiency depends largely on its forward speed; the pure turbojet engine is, therefore, most suitable for high forward speeds. The propeller efficiency does, however, decrease rapidly above 350 miles per hour due to the disturbance of the airflow caused by the high blade-tip speeds of the propeller. These characteristics have led to some departure from the use of pure turbojet propulsion, where aircraft operate at medium speeds by the introduction of a combination of propeller and gas turbine engine.

The advantages of the propeller/turbine combination have to some extent been offset by the introduction of the bypass, ducted fan, and propfan engines. These engines deal with larger comparative airflows and lower jet

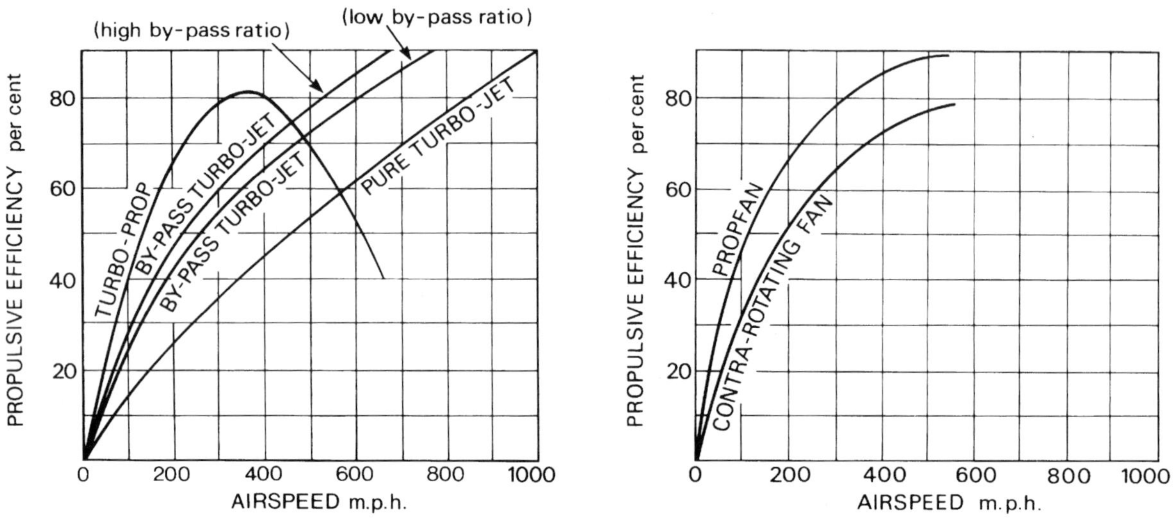

FIGURE 2–11 Comparative propulsive efficiencies. *(Source: Rolls Royce.)*

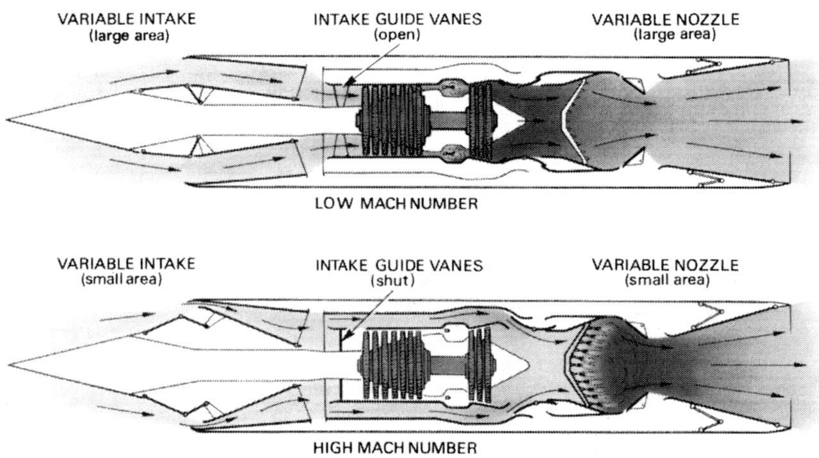

FIGURE 2–12 A turbo/ramjet engine. *(Source: Rolls Royce.)*

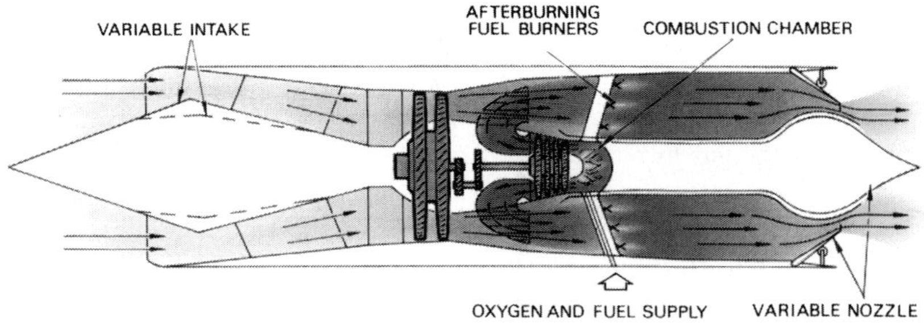

FIGURE 2–13 A turborocket engine. *(Source: Rolls Royce.)*

velocities than the pure jet engine, thus giving a propulsive efficiency that is comparable to that of the turboprop and exceeds that of the pure jet engine (Figure 2–11).

The turbo/ramjet engine (Figure 2–12) combines the turbojet engine (which is used for speeds up to Mach 3) with the ramjet engine, which has good performance at high Mach numbers.

The engine is surrounded by a duct that has a variable intake at the front and an afterburning jet pipe with a variable nozzle at the rear. During takeoff and acceleration, the engine functions as a conventional turbojet with the afterburner lit; at other flight conditions up to Mach 3, the afterburner is inoperative. As the aircraft accelerates through Mach 3, the turbojet is shut down and the intake air is diverted from the compressor, by guide vanes, and ducted straight into the afterburning jet pipe, which becomes a ramjet combustion chamber. This engine is suitable for an aircraft requiring high speed and sustained high Mach number cruise conditions where the engine operates in the ramjet mode.

The turborocket engine (Figure 2–13) could be considered as an alternative engine to the turbo/ramjet; however, it has one major difference in that it carries its own oxygen to provide combustion.

The engine has a low-pressure compressor driven by a multistage turbine; the power to drive the turbine is derived from combustion of kerosene and liquid oxygen in a rocket-type combustion chamber. Since the gas temperature will be on the order of 3500°C, additional fuel is sprayed into the combustion chamber for cooling purposes before the gas enters the turbine. This fuel-rich mixture (gas) is then diluted with air from the compressor and the surplus fuel burned in a conventional afterburning system.

Although the engine is smaller and lighter than the turbo/ramjet, it has higher fuel consumption. This tends to make it more suitable for an interceptor or space-launcher-type of aircraft that requires high speed, high altitude performance, and normally has a flight plan that is entirely accelerative and of short duration.

THE GAS TURBINE GLOBAL FLEET: MODEL DESIGNATION AND PRODUCTION PROGNOSIS AS OF 2013–2022

The following appendices indicate the current complexity of the gas turbine market in terms of different models, sizes, and applications. The appendices for the first edition have been left in, so that the reader can gauge the shift in the gas turbine market over the last several years.

Appendix 2A-1: Gas Turbine Engines Powering Aircraft: Aircraft Engine Production*

Aircraft Engine Production

Company	Description	Total Units 2013–2022
Aviadvigatel OJSC	TF	3
Aviadvigatel OJSC	TF	13
Aviadvigatel OJSC	TF	17
Aviadvigatel OJSC	TF	87
Aviadvigatel OJSC	TF	120
Avio SpA	TS	91
Avio SpA	TS	142
Avio SpA	TS	233
CFM International Inc	TF	156
CFM International Inc	TF	21
CFM International Inc	TF	253
CFM International Inc	TF	1728
CFM International Inc	TF	12
CFM International Inc	TF	19
CFM International Inc	TF	23
CFM International Inc	TF	50
CFM International Inc	TF	57
CFM International Inc	TF	246
CFM International Inc	TF	518
CFM International Inc	TF	921
CFM International Inc	TF	3513
CFM International Inc	TF	38
CFM International Inc	TF	513
CFM International Inc	TF	1035
CFM International Inc	TF	4135
CFM International Inc	TF	13238
China Aviation Gas Turbine Co Ltd	TJ	226

(Continued)

* **Source: Forecast International, CT, USA**. Please contact Forecast International for additional information regarding this data. In the interest of space, this data is a summary of what Forecast International has on file.

Aircraft Engine Production—cont'd

Company	Description	Total Units 2013–2022
China National South Aeroengine Co	TP	132
		358
Dongan Engine Manufacturing Company	TP	21
Dongan Engine Manufacturing Company	TP	33
Dongan Engine Manufacturing Company	TP	54
Eurojet Turbo GmbH	TF	85
Eurojet Turbo GmbH	TF	358
Eurojet Turbo GmbH	TF	443
Europrop International GmbH	TP	964
GE Aviation	TF	122
GE Aviation	TF	234
GE Aviation	TF	82
GE Aviation	TF	890
GE Aviation	TF	380
GE Aviation	TF	79
GE Aviation	TF	351
GE Aviation	TF	269
GE Aviation	TF	296
GE Aviation	TF	25
GE Aviation	TF	174
GE Aviation	TF	57
GE Aviation	TF	138
GE Aviation	TF	3
GE Aviation	TS	365
GE Aviation	TS	318
GE Aviation	TS	396
GE Aviation	TS	36
GE Aviation	TS	26
GE Aviation	TS	13
GE Aviation	TS	37
GE Aviation	TS	152
GE Aviation	TS	77
GE Aviation	TP	50

Aircraft Engine Production—cont'd

Company	Description	Total Units 2013–2022
GE Aviation	TP	127
GE Aviation	TF	12
GE Aviation	TF	202
GE Aviation	TF	31
GE Aviation	TF	85
GE Aviation	TF	117
GE Aviation	TF	38
GE Aviation	TF	52
GE Aviation	TF	81
GE Aviation	TF	193
GE Aviation	TF	128
GE Aviation	TS	323
GE Aviation	TF	6
GE Aviation	TF	47
GE Aviation	TF	370
GE Aviation	TF	1279
GE Aviation	TF	90
GE Aviation	TF	580
GE Aviation	TF	1099
GE Aviation	TF	396
GE Aviation	TF	311
GE Aviation	TS	189
GE Aviation	TS	357
GE Aviation	TS	4
GE Aviation	TS	9
GE Aviation	TS	107
GE Aviation	TS	167
GE Aviation	TS	274
GE Aviation	TS	4
GE Aviation	TS	40
GE Aviation	TS	530
GE Aviation	TS	33
GE Aviation	TS	82
GE Aviation	TS	469
GE Aviation	TS	1416
GE Aviation	TF/TP/TS	13818

Aircraft Engine Production—cont'd

Company	Description	Total Units 2013–2022
GE Aviation Czech s.r.o.	TP	58
GE Aviation Czech s.r.o.	TP	260
GE Aviation Czech s.r.o.	TP	318
GE Honda Aero Engines LLC	TF	973
GE-P&W Engine Alliance LLC	TF	713
Hindustan Aeronautics Ltd—Engine Division	TJ	226
Honeywell Aerospace	TF	331
Honeywell Aerospace	TF	194
Honeywell Aerospace	TF	431
Honeywell Aerospace	TF	397
Honeywell Aerospace	TF	478
Honeywell Aerospace	TS	19
Honeywell Aerospace	TS	36
Honeywell Aerospace	TS	541
Honeywell International Inc	TS	33
Honeywell International Inc	TF	54
Honeywell International Inc	TF	12
Honeywell International Inc	TF	45
Honeywell International Inc	TF	247
Honeywell International Inc	TF	307
Honeywell International Inc	TF	362
Honeywell International Inc	TF	303
Honeywell International Inc	TP	11
Honeywell International Inc	TP	3
Honeywell International Inc	TP	26
Honeywell International Inc	TP	62
Honeywell International Inc	TP	198
Honeywell International Inc	TP	20
Honeywell International Inc	TP	50
Honeywell	TF/TP/TS	4160
IHI Aerospace Co Ltd	TF	131
IHI Corporation	TS	36
IHI Corporation	TS	70
IHI Corporation	TS	16
IHI Corporation	TS	8

Aircraft Engine Production—cont'd

Company	Description	Total Units 2013–2022
IHI Corporation	TF/TS	261
International Aero Engines AG (IAE)	TF	1412
International Aero Engines AG (IAE)	TF	133
International Aero Engines AG (IAE)	TF	194
International Aero Engines AG (IAE)	TF	580
International Aero Engines AG (IAE)	TF	2319
Kawasaki Heavy Industries (KHI)—Gas Turbine Division	TS	16
Klimov Corporation	TF	28
Klimov Corporation	TF	218
Klimov Corporation	TF	192
Klimov Corporation	TS	375
Klimov Scientific Industrial Enterprise	TS	200
Klimov Scientific Industrial Enterprise	TS	91
Klimov Scientific Industrial Enterprise	TS	186
Klimov Scientific Industrial Enterprise	TS	1953
Klimov Scientific Industrial Enterprise	TS	2617
Klimov Scientific Industrial Enterprise	TS	433
Klimov	TF/TS	6293
Light Helicopter Turbine Engine Co (LHTEC)	TS	12
Light Helicopter Turbine Engine Co (LHTEC)	TS	114
Light Helicopter Turbine Engine Co (LHTEC)	TS	210
Light Helicopter Turbine Engine Co (LHTEC)	TS	336
Microturbo Inc	TJ	100
Microturbo Inc	TJ	365
Microturbo Inc	TJ	303
Microturbo Ltd	TJ	151

(Continued)

Aircraft Engine Production—cont'd

Company	Description	Total Units 2013–2022
Microturbo SA	TJ	378
Microturbo SA	TJ	598
Microturbo SA	TJ	217
Microturbo SA	TJ	443
Microturbo SA	TJ	2555
Mitsubishi Heavy Industries Ltd (MHI)	TJ	33
Mitsubishi Heavy Industries Ltd (MHI)	TJ	15
Mitsubishi Heavy Industries Ltd (MHI)	TJ	44
Mitsubishi Heavy Industries Ltd (MHI)	TJ	92
Motor Sich JSC/Progress (ZMKB)	TF	167
Motor Sich JSC/Progress (ZMKB)	TF	290
Motor Sich JSC/Progress (ZMKB)	TP	130
Motor Sich JSC/Progress (ZMKB)	TS	77
Motor Sich JSC/Progress (ZMKB)	TP	9
Motor Sich JSC/Progress (ZMKB)	TF	133
Motor Sich JSC/Progress (ZMKB)	TF/TP	806
MTU Turbomeca Rolls-Royce (MTR) GmbH	TS	38
MTU Turbomeca Rolls-Royce (MTR) GmbH	TS	59
MTU Turbomeca Rolls-Royce (MTR) GmbH	TS	92
MTU Turbomeca Rolls-Royce (MTR) GmbH	TS	93
MTU Turbomeca Rolls-Royce (MTR) GmbH	TS	282
Nationalist Chinese Arsenals	TJ	400
NPO Saturn OJSC	TF	802
NPO Saturn OJSC	TF	290
NPO Saturn OJSC	TF	130

Aircraft Engine Production—cont'd

Company	Description	Total Units 2013–2022
NPO Saturn OJSC	TF	1222
PowerJet JSC	TF	458
Pratt & Whitney	TF	14
Pratt & Whitney	TF	50
Pratt & Whitney	TF	4
Pratt & Whitney	TF	17
Pratt & Whitney	TF	66
Pratt & Whitney	TF	150
Pratt & Whitney	TF	230
Pratt & Whitney	TF	921
Pratt & Whitney	TF	737
Pratt & Whitney	TF	265
Pratt & Whitney	TF	329
Pratt & Whitney	TF	2776
Pratt & Whitney	TF	798
Pratt & Whitney	TF	17
Pratt & Whitney	TF	19
Pratt & Whitney	TF	278
Pratt & Whitney	TF	8
Pratt & Whitney	TF	23
Pratt & Whitney	TF	70
Pratt & Whitney AeroPower	TJ	3594
Pratt & Whitney	TJ/TF	10366
Pratt & Whitney Canada	TP	148
Pratt & Whitney Canada	TP	92
Pratt & Whitney Canada	TP	50
Pratt & Whitney Canada	TP	36
Pratt & Whitney Canada	TP	119
Pratt & Whitney Canada	TP	72
Pratt & Whitney Canada	TP	879
Pratt & Whitney Canada	TP	365
Pratt & Whitney Canada	TP	124
Pratt & Whitney Canada	TP	71
Pratt & Whitney Canada	TP	126
Pratt & Whitney Canada	TP	184
Pratt & Whitney Canada	TP	357

Aircraft Engine Production—cont'd

Company	Description	Total Units 2013–2022
Pratt & Whitney Canada	TP	56
Pratt & Whitney Canada	TP	186
Pratt & Whitney Canada	TP	417
Pratt & Whitney Canada	TP	324
Pratt & Whitney Canada	TP	745
Pratt & Whitney Canada	TP	984
Pratt & Whitney Canada	TP	33
Pratt & Whitney Canada	TP	30
Pratt & Whitney Canada	TP	22
Pratt & Whitney Canada	TP	65
Pratt & Whitney Canada	TP	424
Pratt & Whitney Canada	TP	412
Pratt & Whitney Canada	TP	115
Pratt & Whitney Canada	TP	151
Pratt & Whitney Canada	TP	782
Pratt & Whitney Canada	TP	100
Pratt & Whitney Canada	TP	118
Pratt & Whitney Canada	TP	140
Pratt & Whitney Canada	TP	198
Pratt & Whitney Canada	TS	243
Pratt & Whitney Canada	TS	568
Pratt & Whitney Canada	TS	196
Pratt & Whitney Canada	TS	1374
Pratt & Whitney Canada	TS	55
Pratt & Whitney Canada	TP	66
Pratt & Whitney Canada	TP	16
Pratt & Whitney Canada	TP	153
Pratt & Whitney Canada	TP	162
Pratt & Whitney Canada	TP	260
Pratt & Whitney Canada	TP	17
Pratt & Whitney Canada	TP	109
Pratt & Whitney Canada	TP	1344
Pratt & Whitney Canada	TP	875
Pratt & Whitney Canada	TS	62
Pratt & Whitney Canada	TS	304
Pratt & Whitney Canada	TS	1097

Aircraft Engine Production—cont'd

Company	Description	Total Units 2013–2022
Pratt & Whitney Canada	TS	71
Pratt & Whitney Canada	TS	73
Pratt & Whitney Canada	TS	134
Pratt & Whitney Canada	TS	927
Pratt & Whitney Canada	TS	134
Pratt & Whitney Canada	TS	315
Pratt & Whitney Canada	TS	809
Pratt & Whitney Canada	TF	9
Pratt & Whitney Canada	TF	8
Pratt & Whitney Canada	TF	425
Pratt & Whitney Canada	TF	1317
Pratt & Whitney Canada	TF	549
Pratt & Whitney Canada	TF	9
Pratt & Whitney Canada	TF	29
Pratt & Whitney Canada	TF	103
Pratt & Whitney Canada	TF	1028
Pratt & Whitney Canada	TF	474
Pratt & Whitney Canada	TF	1297
Pratt & Whitney Canada	TF	1165
Pratt & Whitney Canada	TF/TP/TS	23702
Rolls-Royce Corp	TP	33
Rolls-Royce Corp	TP	12
Rolls-Royce Corp	TP	50
Rolls-Royce Corp	TS	66
Rolls-Royce Corp	TS	54
Rolls-Royce Corp	TS	11
Rolls-Royce Corp	TS	25
Rolls-Royce Corp	TS	26
Rolls-Royce Corp	TS	10
Rolls-Royce Corp	TS	22
Rolls-Royce Corp	TS	35
Rolls-Royce Corp	TS	38
Rolls-Royce Corp	TS	128
Rolls-Royce Corp	TS	101
Rolls-Royce Corp	TS	115
Rolls-Royce Corp	TS	49

(*Continued*)

Aircraft Engine Production—cont'd

Company	Description	Total Units 2013–2022
Rolls-Royce Corp	TS	825
Rolls-Royce Corp	TS	7
Rolls-Royce Corp	TP	28
Rolls-Royce Corp	TP	39
Rolls-Royce Corp	TP	155
Rolls-Royce Corp	TP	508
Rolls-Royce Corp	TP	512
Rolls-Royce Corp	TP	30
Rolls-Royce Corp	TF	2
Rolls-Royce Corp	TF	3
Rolls-Royce Corp	TF	3
Rolls-Royce Corp	TF	4
Rolls-Royce Corp	TF	5
Rolls-Royce Corp	TF	6
Rolls-Royce Corp	TF	6
Rolls-Royce Corp	TF	33
Rolls-Royce Corp	TF	41
Rolls-Royce Corp	TF	530
Rolls-Royce Corp	TF	13
Rolls-Royce Corp	TS	16
Rolls-Royce Corp	TS	41
Rolls-Royce Corp	TS	365
Rolls-Royce Corp	TS	2782
Rolls-Royce Corp	TP	140
Rolls-Royce plc	TF	91
Rolls-Royce plc	TF	560
Rolls-Royce plc	TF	695
Rolls-Royce plc	TF	71
Rolls-Royce plc	TF	212
Rolls-Royce plc	TF	519
Rolls-Royce plc	TF	11
Rolls-Royce plc	TF	16
Rolls-Royce plc	TF	575
Rolls-Royce plc	TF	332
Rolls-Royce plc	TF	473
Rolls-Royce plc	TF	1419

Aircraft Engine Production—cont'd

Company	Description	Total Units 2013–2022
Rolls-Royce	TF/TP/TS	11843
Rolls-Royce Deutschland Ltd & Co KG	TF	323
Rolls-Royce Deutschland Ltd & Co KG	TF	396
Rolls-Royce Deutschland Ltd & Co KG	TF	448
Rolls-Royce Deutschland Ltd & Co KG	TF	1037
Rolls-Royce Deutschland Ltd & Co KG	TF	444
Rolls-Royce Deutschland Ltd & Co KG	TF	2648
Rolls-Royce Turbomeca Ltd	TF	80
Rolls-Royce Turbomeca Ltd	TF	27
Rolls-Royce Turbomeca Ltd	TS	5
Rolls-Royce Turbomeca Ltd	TS	8
Rolls-Royce Turbomeca Ltd	TS	15
Rolls-Royce Turbomeca Ltd	TS	17
Rolls-Royce Turbomeca Ltd	TS	32
Rolls-Royce Turbomeca Ltd	TS	61
Rolls-Royce Turbomeca Ltd	TS	73
Rolls-Royce Turbomeca Ltd	TS	134
Rolls-Royce Turbomeca Ltd	TS	239
Rolls-Royce Turbomeca Ltd	TS	12
Rolls-Royce Turbomeca Ltd	TF/TS	703
Samsung Techwin Co Ltd	TS	568
Shanghai Aviation Industry Group Corp (SAIC)	TS	15
Shanghai Aviation Industry Group Corp (SAIC)	TS	29
Shanghai Aviation Industry Group Corp (SAIC)	TS	40
Shanghai Aviation Industry Group Corp (SAIC)	TS	204
Shanghai Aviation Industry Group Corp (SAIC)	TS	288
Snecma	TF	46
Snecma	TF	54
Snecma	TF	7

Aircraft Engine Production—cont'd

Company	Description	Total Units 2013–2022
Snecma	TF	57
Snecma	TF	265
Snecma	TF	429
Teledyne Turbine Engines	TJ	3456
Teledyne Turbine Engines	TF	629
	TF/TJ	4085
Turbomeca SA	TJ	150
Turbomeca SA	TS	549
Turbomeca SA	TS	21
Turbomeca SA	TS	505
Turbomeca SA	TS	107
Turbomeca SA	TS	198
Turbomeca SA	TS	251
Turbomeca SA	TS	3
Turbomeca SA	TS	6
Turbomeca SA	TS	4
Turbomeca SA	TS	109
Turbomeca SA	TS	115
Turbomeca SA	TS	23
Turbomeca SA	TS	535
Turbomeca SA	TS	1737
Turbomeca SA	TS	560
Turbomeca SA	TS	32
Turbomeca SA	TS	160
Turbomeca SA	TS	189
Turbomeca SA	TS	23
Turbomeca SA	TS	54
Turbomeca SA	TS	869
Turbomeca SA	TS	17
Turbomeca SA	TS	247
Turbomeca SA	TS	198
Turbomeca SA	TS	142
Turbomeca SA	TS	109
Turbomeca SA	TS	39
Turbomeca SA	TS	48
Turbomeca SA	TS	17
Turbomeca SA	TS	262
Turbomeca SA	TS	503

Aircraft Engine Production—cont'd

Company	Description	Total Units 2013–2022
Turbomeca SA	TS	107
Turbomeca SA	TJ/TS	7889
Volvo Aero Corp	TF	2
Williams International	TJ	191
Williams International	TF	1544
Williams International	TF	438
Williams International	TF	56
Williams International	TF	731
Williams International	TF	515
Williams International	TF	929
Williams International	TJ	20
Williams International	TF/TJ	4424
WSK PZL-Rzeszow SA	TS	49
WSK PZL-Rzeszow SA	TS	60
WSK PZL-Rzeszow SA	TS	109

Appendix 2A-2: Electrical Generation/Mechanical Drive/Marine Power Turbine Engines

Electrical Generation/Mechanical Drive/Marine Power Turbine Engines

Company	Total Units 2013–2022
Alstom	495
Bharat Heavy Electricals Ltd	25
Daihatsu Diesel Manufacturing Company Ltd	151
Dresser-Rand Co	16
Ebara Corp	25
GE Energy	3888
Hitachi Ltd	169
Ingersoll-Rand Energy Systems	318
Ingersoll-Rand Energy Systems	1418
Ingersoll-Rand Energy Systems	1905
Kawasaki Heavy Industries (KHI)—Gas Turbine Division	1336
MAN Group	179

(Continued)

Electrical Generation/Mechanical Drive/Marine Power Turbine Engines—cont'd

Company	Total Units 2013–2022
Mitsubishi Heavy Industries Ltd (MHI)	395
Mitsui Engineering & Shipbuilding Company Ltd	329
OPRA Optimal Radial Turbine BV	365
Pratt & Whitney Canada	86
Pratt & Whitney Power Systems	572
Rolls-Royce	1048
Siemens	1592
Solar Turbines Inc	5899
Turbomeca	14
Vericor Power Systems Inc	101

Appendix 2A-3: APU Engine Production

APU Engine Production

Company	Description	Total Units 2013–2022
Honeywell Aerospace	TS	14603
Light Helicopter Turbine Engine Co (LHTEC)	TS	2
Microturbo SA	TS	1192
Mitsubishi Heavy Industries Ltd (MHI)	TS	10
Pratt & Whitney AeroPower	TS	7022
Pratt & Whitney Canada	TS	387

Appendix 2A-4: Missile/Drone/RPV Engine Production

Missile/Drone/RPV Engine Production

Company	Description	Total Units 2013–2022
China Aviation Gas Turbine Co Ltd	TJ	226
Hindustan Aeronautics Ltd—Engine Division	TJ	226
Honeywell International Inc	TP	286
Microturbo SA	TJ	2555
Mitsubishi Heavy Industries Ltd (MHI)	TJ	92
Nationalist Chinese Arsenals	TJ	400
Pratt & Whitney AeroPower	TJ	3594
Pratt & Whitney Canada	TS	62
Rolls-Royce Corp	TF/TS	97
Teledyne Turbine Engines	TF/TJ	4085
Turbomeca SA	TJ	150
Williams International	TF/TJ	1755

Appendix 2A-5: Microturbines for Electrical Generation

Microturbines for Electrical Generation

Company	Total Units 2013–2022
Capstone Turbine Corp	8794
Elliott Turbomachinery Company Inc	4102
Ingersoll-Rand Energy Systems	2755
Turbec SpA	1104

GAS TURBINE GLOBAL FLEET: MODEL DESIGNATION AND PRODUCTION AS OF MAY 2006†

The following appendices indicate the anticipated complexity of the gas turbine market in terms of different models, sizes, and applications.

Appendix 2B-1: Gas Turbine Engines Powering Aircraft*

† For any updates or additional information on these data, please contact the Source, Forecast International in Connecticut, USA.

* Source: Forecast International. In the interest of space, this data is a summary of what Forecast International has on file.

Engine Production Matrix by Manufacturer Years to Print (2006–2015) (10 Years)

Record Number	Engine	Application	Engine Type	06	07	08	09	10	11	12	13	14	15	Total 06–15
Aviadvigatel JSC														
Aviadvigatel JSC														
102192	PS-90A	IL-96-300	TF	7	14	14	10	5	0	0	0	0	0	50
105119	PS-90A	IL-96-400	TF	6	10	9	11	15	17	14	14	14	12	122
102193	PS-90A	TU-204	TF	17	17	13	12	11	9	9	9	8	9	114
Aviadvigatel JSC				30	41	36	33	31	26	23	23	22	21	286
Avio SPA														
Avio SPA (Licensee)														
105443	PT6T-3B/3D	AB412/412SP (CIVIL)	TS	11	9	8	3	0	0	0	0	0	0	31
9696	PT6T-3B/3D	AB412/412SP (MIL)	TS	12	9	7	0	0	0	0	0	0	-	28
Avio SPA (Co-Product)														
1462	T700/T6A	EH101 MILITARY UTILITY	TS	0	0	0	0	1	3	4	7	8	9	32
1456	T700/T6A	EH101 NAVAL	TS	7	7	5	0	0	2	7	6	5	7	46
102722	T700/T6E	NH90/NFH	TS	2	7	8	9	10	12	12	12	12	14	98
102724	T700/T6E	NH90/TTH	TS	22	20	16	17	20	20	22	25	24	22	208
Avio SPA				54	52	44	29	31	37	45	50	49	52	443
CFE Company														
CFE Company (Consortium)														
11138	CFE738	FALCON 2000	TF	10	6	0	0	0	0	0	0	0	0	16
CFE Company				10	6	0	0	0	0	0	0	0	0	16
CFM International SA														
CFM International SA (Consortium)														
104466	CFM56-5B	A318	TF	23	20	15	18	19	21	22	20	18	17	193
9786	CFM56-5B	A320-100/-200	TF	218	204	164	151	114	95	102	98	98	66	1310

(Continued)

Engine Production Matrix by Manufacturer Years to Print (2006–2015) (10 Years)—cont'd

Record Number	Engine	Application	Engine Type	Engine Units 06	07	08	09	10	11	12	13	14	15	Total 06–15
100911	CFM56-5B	A321-200	TF	32	36	26	25	31	28	26	24	26	16	270
102594	CFM56-5B/-5A	A319	TF	190	160	99	89	91	84	74	77	85	57	1006
102571	CFM56-5C4	A340-300X	TF	11	10	11	9	6	0	0	0	0	0	47
105048	CFM56-7	737 AEW&C	TF	5	6	9	6	5	7	9	8	3	0	58
103995	CFM56-7	737 BBJ	TF	10	13	12	10	8	8	7	8	7	6	89
104859	CFM56-7	737 BBJ2	TF	6	6	6	4	2	3	3	0	0	0	30
103399	CFM56-7	737-600	TF	12	12	13	11	9	10	11	9	9	6	102
102586	CFM56-7	737-700	TF	289	257	232	219	194	179	163	150	135	93	1911
103074	CFM56-7	737-800	TF	377	423	384	343	328	315	294	270	230	152	3116
104111	CFM56-7	737-900	TF	14	16	20	28	35	38	34	32	29	18	264
105040	CFM56-7	P-8A	TF	0	1	3	3	1	7	17	29	36	27	124
CFM International SA				1184	1164	994	916	843	795	762	725	676	458	8520
Dongan Engine MFG Company														
Dongan Engine MFG Company (Licensee) 102517	PT6A-27	Y-12-II (CIVIL)	TP	5	6	6	2	0	0	0	0	0	0	19
105430	PT6A-27	Y-12-II (MIL.)	TP	0	2	3	1	0	0	0	0	0	0	6
Dongan Engine MFG Company				5	8	9	3	0	0	0	0	0	0	25
Eurojet														
Eurojet (Consortium) 1757	EJ200	TYPHOON	ATF	74	95	108	108	108	108	108	108	108	75	1000
105382	EJ200	TYPHOON (AUSTRIA)	ATF	3	12	17	9	0	0	0	0	0	0	41
106077	EJ200	TYPHOON (SAUDI ARABIA)	ATF	0	0	4	13	22	27	27	16	0	0	109
Eurojet				77	107	129	130	130	135	135	124	108	75	1150

Europrop International														
Europrop International (Consortium)														
105222	TP400-D6	A400M	TP	0	4	18	45	91	112	112	112	112	112	718
Europrop International				0	4	18	45	91	112	112	112	112	112	718
GE Aircraft Engines														
GE Aircraft Engines														
105721	CF34-10A	ARJ21	TF	1	3	4	8	24	43	52	55	62	65	317
104436	CF34-10E	190/195	TF	136	144	124	109	104	100	97	100	104	107	1125
101525	CF34-3A1/3B	CRJ 100/200	TF	26	35	30	25	18	18	21	22	20	17	232
105466	CF34-3B	CHALLENGER 604 (CIVIL)	TF	57	38	22	8	0	0	0	0	0	0	125
103058	CF34-3B	CHALLENGER 604 (MIL)	TF	2	2	2	0	2	0	2	0	0	0	10
106131	CF34-3B	CHALLENGER 605	TF	0	0	0	0	0	0	0	0	0	0	0
101826	CF34-8C	CRJ 700	TF	109	96	91	96	96	101	101	95	96	67	948
104396	CF34-8E	170/175	TF	112	108	100	98	99	103	103	97	97	97	1014
10408	CF6-80C	767-300	TF	14	10	8	5	0	0	0	0	0	0	37
102792	CF6-80C	767-300F	TF	7	9	7	5	0	0	0	0	0	0	28
103843	CF6-80C	767-400ER	TF	2	0	0	0	0	0	0	0	0	0	2
10376	CF6-80C2	747-400	TF	36	25	22	21	17	16	12	7	0	0	156
105689	CF6-80C2	767 T/T	TF	6	7	6	2	0	0	0	0	0	0	21
10398	CF6-80C2	767-200	TF	4	7	0	0	0	0	0	0	0	0	11
105690	CF6-80C2	C-X	TF	1	2	1	0	5	10	11	13	14	13	70
104848	CF6-80C2	C-5M (MOD)	TF	6	7	6	13	27	36	42	54	54	54	299
105623	CF6-80C2	E-10A MC2A	TF	0	0	0	1	1	0	1	3	3	0	9
105978	CF6-80E1	KC-30 MRTT (RAAF)	TF	0	3	6	6	0	0	0	0	0	0	15
102435	CT7-6	EH101 CIVIL	TS	0	0	1	4	7	6	4	6	4	7	39
103177	CT7-8A	S-92	TS	35	29	22	24	26	22	24	28	24	0	234
105653	CT7-8B5	MH-60M (USSOCOM MOD)	TS	0	2	0	8	16	20	34	42	0	0	122
103452	CT7-8C	H-92	TS	2	12	27	27	26	22	24	27	28	24	219

(*Continued*)

Engine Production Matrix by Manufacturer Years to Print (2006–2015) (10 Years)—cont'd

Record Number	Engine	Application	Engine Type	Engine Units 06	07	08	09	10	11	12	13	14	15	Total 06–15
105896	CT7-8E	US101	TS	2	6	9	16	12	15	15	14	8	0	97
101970	CT7-9C3	CN-235M	TP	12	13	15	18	16	14	14	12	14	10	138
101971	CT7-9C3	CN-235M	TP	5	7	10	9	9	9	5	0	0	0	54
104778	F110-GE-129	F-15K (ROK)	ATF	27	27	27	27	27	18	0	0	0	0	153
106070	F110-GE-129	F-15SG (SINGAPORE)	ATF	0	1	7	14	14	6	0	0	0	0	42
104793	F110-GE-129	F-16C/D (CHILE)	ATF	8	0	0	0	0	0	0	0	0	0	8
105105	F110-GE-129	F-16C/D (OMAN)	ATF	8	0	0	0	0	0	0	0	0	0	8
103818	F110-GE-132	F-16E/F (UAE)	ATF	28	20	0	0	0	0	0	0	0	0	48
105386	F136	F-35 (PRODUCTION)	ATF	0	0	0	0	0	4	13	22	41	50	130
105810	F136	F-35 (SDD)	ATF	0	3	6	8	5	1	0	0	0	0	23
105289	F404-GE-F2J3	LCA INITIAL PRODUCTION	ATF	9	8	0	0	0	0	0	0	0	0	17
105099	F414-GE-400	EA-18	ATF	2	10	21	31	40	40	37	20	0	0	201
101771	F414-GE-400	F/A-18E/F (US NAVY)	ATF	90	88	80	60	54	53	39	7	0	0	471
100917	GE90	777-200/200ER	TF	18	19	22	22	21	18	16	18	17	15	186
103926	GE90-110	777-200LR	TF	3	9	22	29	30	33	32	31	32	31	252
104830	GE90-115	777-300ER	TF	87	100	79	78	81	81	84	84	81	80	835
105835	GENX	A350	TF	0	1	3	9	26	47	57	66	66	44	319
106075	GENX	747-8	TF	0	0	5	22	40	57	76	95	101	99	495
105819	GENX	787	TF	1	10	41	73	101	119	119	119	119	89	791
103471	T700-GE-401	AH-1Z (MOD)	TS	4	12	26	42	51	56	71	71	53	0	386
106012	T700-GE-401	UH-1Y	TS	0	3	23	34	34	34	34	29	0	0	191
104067	T700-GE-401	UH-1Y (USMC MOD)	TS	6	9	7	0	0	0	0	0	0	0	22
104000	T700-GE-401	SH-2G (EXPORT MOD)	TS	2	5	7	7	2	0	0	0	0	0	23
105008	T700-GE-401C	MH-60R	TS	9	27	44	54	54	54	54	54	54	40	444
103984	T700-GE-401C	MH-60S	TS	40	40	44	54	54	52	34	0	0	0	318

105986	T700-GE-401C	MH-60S (SINGAPORE)	TS	0	1	4	4	3	0	1	2	1	0	16
105987	T700-GE-401C	MH-60S (TURKEY)	TS	0	1	7	12	6	0	3	9	9	6	53
2197	T700-GE-700/701C/D	UH-60/S-70 EXPORT	TS	24	26	28	29	34	38	38	38	35	34	324
2198	T700-GE-700/701C/D	UH-60A/L/M (U.S. ARMY)	TS	80	82	118	169	170	170	170	175	192	202	1528
101300	T700-GE-701C	AH-64D (EXPORT)	TS	33	18	20	19	22	24	24	24	25	24	233
105991	T700-GE-701C	AH-64D (U.S. ARMY)	TS	10	19	0	0	0	0	0	0	0	0	29
105834	T700-GE-701D (MOD)	AH-64D BL 3 (US ARMY MOD)	TS	0	0	0	12	55	101	161	161	161	161	812
104755	T700-GE-701D (MOD)	UH-60M (U.S. ARMY MOD)	TS	3	0	0	0	0	0	0	0	0	0	3
GE Aircraft Engines (Associate)														
104564	CF34-8C5	CRJ 900	TF	34	34	38	44	50	55	59	64	67	50	495
GE Aircraft Engines				1101	1138	1192	1356	1479	1596	1684	1664	1582	1386	14,178
GE-P&W Engine Alliance														
GE-P&W Engine Alliance														
104772	GP7200	A380	TF	15	34	63	86	102	111	122	112	107	77	829
GE-P&W Engine Alliance				15	34	63	86	102	111	122	112	107	77	829
H A L, Engine Division														
H A L, Engine Division														
1636	ADOUR MK 811/871	JAGUAR INTERNATIONAL	ATF	23	17	0	0	0	0	0	0	0	0	40
1637	GTX-35VS	LCA	ATF	0	0	0	0	2	4	9	16	20	20	71
H A L, Engine Division (Licensee)														
1626	TPE 331-5	228-200/-201 (MIL)	TP	10	12	0	0	0	0	0	0	0	0	22
H A L, Engine Division				33	29	0	0	2	4	9	16	20	20	133
Honeywell														
Honeywell														
104346	HTF7000	CHALLENGER 300	TF	106	101	84	72	67	69	68	72	48	0	687
106005	HTS900	407X	TS	2	9	24	40	52	63	64	68	73	70	465
106007	HTS900	407X (MIL)	TS	1	5	10	9	11	13	13	14	13	13	102

(Continued)

Engine Production Matrix by Manufacturer Years to Print (2006–2015) (10 Years)—cont'd

Record Number	Engine	Application	Engine Type	Engine Units 06	07	08	09	10	11	12	13	14	15	Total 06–15
106006	HTS900	USA ARH/407X	TS	0	2	11	41	65	76	76	72	44	0	387
105330	T53-L-703 (MOD)	HUEY II (MOD)	TS	12	13	14	12	12	10	10	10	8	0	101
103720	T55-GA-714A (MOD)	CH-47F (MOD)	TS	10	34	30	46	50	64	66	66	66	68	500
105023	T55-GA-714A (MOD)	MH-47G (MOD)	TS	24	12	12	12	12	0	0	0	0	0	72
1283	T55-L-712/714	CH-47D (MOD)	TS	6	0	0	0	0	0	0	0	0	0	6
101298	T55-L-712/714/714A	CH-47D/SD/F	TS	17	19	16	15	14	17	19	14	11	10	152
101731	TFE 731-2-2A	K-8	TF	12	15	17	17	14	13	13	13	12	10	136
105094	TFE 731-2-2C	AT-63	TF	3	7	4	0	0	0	0	0	0	0	14
106054	TFE 731-2-2C	AT-63 (MOD)	TF	11	0	0	0	0	0	0	0	0	0	11
105315	TFE 731-20	LEARJET 40	TF	45	41	41	42	42	45	45	48	51	50	450
104583	TFE 731-40	FALCON 50 (MOD)	TF	9	9	9	9	6	6	0	0	0	0	48
103411	TFE 731-40	FALCON 50EX	TF	15	12	10	10	10	10	10	9	6	7	99
105057	TFE 731-40	G100	TF	4	0	0	0	0	0	0	0	0	0	4
105316	TFE 731-40	G150	TF	30	47	40	36	37	39	40	39	38	35	381
103400	TFE 731-5BR	HAWKER 800XP (CIVIL)	TF	13	0	0	0	0	0	0	0	0	0	13
105581	TFE 731-5BR	HAWKER 800XP/U-125A (MIL)	TF	0	0	0	0	0	0	0	0	0	0	0
106163	TFE 731-5BR	HAWKER 850XP (CIVIL)	TF	0	0	0	0	0	0	0	0	0	0	0
106164	TFE 731-5BR	HAWKER 850XP (MIL.)	TF	0	0	0	0	0	0	0	0	0	0	0
106133	TFE 731-60	FALCON 900DX	TF	20	20	17	18	15	13	13	13	12	9	150
103389	TFE 731-60	FALCON 900EX	TF	59	55	50	48	48	47	45	47	44	41	484
103706	TPE 331-12	C-212-400 (CIVIL)	TP	2	2	2	3	4	1	1	2	2	1	20
105419	TPE 331-12	C-212-400 (MIL.)	TP	3	4	5	6	4	4	4	4	2	1	37
102945	TPE 331-14	AN-38-100	TP	4	5	6	6	5	4	4	3	2	3	42
Honeywell (Licensee)														
106055	F124	M-346	TF	0	1	6	13	13	15	25	25	25	25	148

Honeywell (Consortium)														
103318	TFE 1042–70	L-159	TF	2	0	0	0	0	0	0	0	0	0	2
Honeywell (Associate)														
102753	TFE 731-20	LEARJET 45	TF	58	62	63	63	69	72	70	65	61	59	642
Honeywell				468	475	471	518	550	581	586	584	518	402	5153
International Aero Engines AG														
International Aero Engines AG (Consortium)														
9791	V2500-A1	A320-100/-200	TF	185	200	196	190	164	143	133	137	148	129	1625
103131	V2500-A5	A319	TF	119	110	102	96	87	83	84	80	77	45	883
103412	V2533-A5	A321-200	TF	29	37	35	17	13	18	24	30	33	18	254
International Aero Engines AG				333	347	333	303	264	244	241	247	258	192	2762
Ishikawajima-Harima, Aero Engn														
Ishikawajima-Harima, Aero Engn														
100153	F3	DEV & TEST	TF	2	0	0	0	0	0	0	0	0	0	2
105691	XF7-10	P-X	TF	2	4	3	0	4	12	19	26	31	31	132
Ishikawajima-Harima, Aero Engn (Licensee)														
102210	F110-GE-129	F-2	ATF	10	9	8	8	8	7	4	0	0	0	54
104650	T700	UH-60J	TS	5	3	0	0	0	0	0	0	0	0	8
104651	T700	UH-60JA	TS	2	3	4	5	7	7	7	7	7	10	59
105820	T700-GE-401C	SH-60K	TS	11	18	18	18	18	18	18	18	18	13	168
105031	T700-GE-701C	AH-64D (JAPAN)	TS	5	5	5	5	4	2	2	4	10	12	54
Ishikawajima-Harima, Aero Engn				37	42	38	36	41	46	50	55	66	66	477
Kawasaki Heavy Ind. Ltd (JET)														
Kawasaki Heavy Ind. Ltd (JET) (Licensee)														
103316	T53-K-703	UH-1J	TS	2	0	0	0	0	0	0	0	0	0	2
1694	T55-L-712	CH-47J	TS	4	3	5	5	7	6	2	0	0	0	32
Kawasaki Heavy Ind. Ltd (JET)				6	3	5	5	7	6	2	0	0	0	34
Klimov Corp (Klimov/Motor-SICH)														
Klimov Corp (Klimov/Motor-SICH)														
101788	RD-33	MIG-29	ATF	2	8	14	8	6	8	10	10	8	8	82
101789	RD-33	MIG-29UB	ATF	2	6	0	0	4	4	0	0	4	0	20

(Continued)

Engine Production Matrix by Manufacturer Years to Print (2006–2015) (10 Years)—cont'd

Record Number	Engine	Application	Engine Type	Engine Units 06	07	08	09	10	11	12	13	14	15	Total 06–15
102348	TV2/TV3/VK-2500	MI-8/17 (CIVIL)	TS	31	28	26	24	24	24	23	21	20	19	240
105468	TV2/TV3/VK-2500	MI-8/17 (MIL.)	TS	110	98	88	81	78	79	75	70	67	67	813
102343	TV3-117	KA-50/52	TS	0	0	2	7	10	11	10	12	13	14	79
102349	TV3-117	MI-24/25/35	TS	13	8	7	6	3	0	0	0	0	0	37
102780	TV3/VK-2500	KA-32	TS	13	11	12	10	9	8	7	5	4	2	81
101548	TV7-117	IL-114	TP	6	14	12	12	14	10	10	8	8	8	102
102350	TV7-117V	MI-28	TS	2	4	13	14	16	16	16	17	19	21	138
104200	VK-1500	KA-60	TS	1	2	2	6	8	11	13	13	15	15	86
102346	VK-1500	KA-62	TS	0	1	3	3	7	12	16	18	18	17	95
Klimov Corp (Klimov/Motor-SICH) (Licensee)														
105329	LARZAC 04 R20	MIG-AT	TF	19	22	22	24	25	22	22	22	22	22	222
103040	RD-35	YAK-130	TF	16	14	18	20	23	23	23	25	28	28	218
Klimov Corp (klimov/Motor-SICH) (Consortium)														
103555	RD-93	FC-1	ATF	4	12	23	30	35	35	35	35	35	35	279
Klimov Corp (Klimov/Motor-SICH)				219	228	242	245	262	263	260	256	261	256	2492
LHTEC														
LHTEC (Consortium)														
105270	CTS800	FUTURE LYNX	TS	0	1	3	6	13	30	40	40	30	0	163
104221	CTS800	SUPER LYNX 200/ 300	TS	11	14	18	16	12	6	0	0	0	0	77
LHTEC				11	15	21	22	25	36	40	40	30	0	240
Manufacturer Varies														
Manufacturer Varies (Consortium)														
104348	VARIOUS	ACJ	TF	13	11	11	13	14	13	13	13	13	14	128
Manufacturer Varies				13	11	11	13	14	13	13	13	13	14	128
Mitsubishi Heavy Ind Ltd														
Mitsubishi Heavy Ind Ltd														
101677	TS1-10	OH-1	TS	9	9	9	9	9	7	8	0	0	0	60

Mitsubishi Heavy Ind Ltd				9	9	9	9	9	7	8	0	0	0	60
Motor Sich/Progress (ZMKB)														
Motor Sich/Progress (ZMKB)														
105447	D-18T	AN-124 (CIVIL)	TF	0	2	6	12	18	18	21	27	25	13	142
104594	D-27	AN-70	PF	5	7	7	0	0	0	0	0	0		19
102254	D-36	YAK-42/142	TF	0	2	6	9	7	6	7	5	3	0	45
102256	D-436T	TU-334	TF	2	8	12	14	16	20	21	19	19	17	148
Motor sich/progress (ZMKB)				7	19	31	35	41	44	49	51	47	30	354
Motorlet Aero-Engines														
Motorlet Aero-Engines														
102901	M 601Z	PZL-106BT	TP	1	1	2	1	1	2	1	2	1	0	12
Motorlet Aero-Engines				1	1	2	1	1	2	1	2	1	0	12
MTR GMBH														
MTR GMBH (Consortium)														
105064	MTR390	TIGER (EXPORT)	TS	0	0	0	0	0	0	3	11	16	17	47
12507	MTR390	TIGER (FRANCE)	TS	18	22	19	16	24	27	27	27	27	25	232
12508	MTR390	TIGER ARH (AUSTRALIA)	TS	12	11	5	0	0	0	0	0	0	0	28
105667	MTR390	TIGER HAD (SPAIN)	TS	0	0	0	2	7	7	9	9	6	0	40
106011	MTR390	TIGER HAP (SPAIN)	TS	5	0	0	0	0	0	0	0	0	0	5
101672	MTR390	TIGER UHT (GERMANY)	TS	22	22	22	22	22	21	12	0	0	0	143
MTR GMBH				57	55	46	40	53	55	51	47	49	42	495
Not Selected														
Not Selected														
105842	NOT SELECTED	MAPL	TS	0	2	2	2	0	0	0	0	0	0	6
106139	NOT SELECTED	CITATION XX	TF	0	0	0	0	0	0	0	0	0	0	0
105106	NOT SELECTED	F-16C/D (EXPORT)	ATF	0	0	1	7	16	22	25	20	14	0	105
105885	NOT SELECTED	F-16C/D (PAKISTAN)	ATF	0	0	3	10	15	15	15	5	0	0	63
105387	NOT SELECTED	AEJPT	TF	0	0	0	0	0	0	0	0	3	18	21

(Continued)

Engine Production Matrix by Manufacturer Years to Print (2006–2015) (10 Years)—cont'd

Record Number	Engine	Application	Engine Type	Engine Units 06	07	08	09	10	11	12	13	14	15	Total 06–15
105046	NOT SELECTED	E-X (ROK)	TF	0	1	6	7	4	3	0	0	0	0	21
105047	NOT SELECTED	TANKER/ TRANSPORT (MOD)	TF	4	6	9	9	9	10	13	13	15	13	101
105989	NOT SELECTED	NH90 (SPAIN)	TS	0	0	0	1	7	14	15	15	15	16	83
105844	NOT SELECTED	US ARMY LUH	TS	21	72	108	108	108	108	108	78	12	0	723
105992	NOT SELECTED	USAF PRV	TS	0	0	1	3	4	9	26	43	52	52	190
104757	NOT SELECTED	CH-53 HLR	TS	0	0	0	0	0	0	1	4	5	0	10
Not Selected (Consortium)														
106106	NOT SELECTED	A320NSR	TF	0	0	0	0	0	0	2	4	8	104	118
106074	NOT SELECTED	737X	TF	0	0	0	0	0	1	4	2	40	211	258
Not Selected				25	81	130	147	163	182	209	184	164	414	1699
OMSK Engine Engine Bureau OJSC														
OMSK Engine Engine Bureau OJSC														
105236	TVD-20	AN-38-200	TP	5	7	10	11	10	10	10	8	7	8	86
OMSK Engine Engine Bureau OJSC				5	7	10	11	10	10	10	8	7	8	86
Pratt & Whitney														
Pratt & Whitney														
104360	F100-PW-229	F-16C/D (ISRAEL)	ATF	27	19	3	0	0	0	0	0	0	0	49
105379	F100-PW-229	F-16C/D (POLAND)	ATF	10	19	17	5	0	0	0	0	0	0	51
103039	F100-PW-229	F-16C/D (SINGAPORE)	ATF	1	0	0	0	0	0	0	0	0	0	1
103174	F100-PW-229	F-16G (GREECE)	ATF	0	0	3	26	12	0	0	0	0	0	41
106117	F117-PW-100	C-17 (AUSTRALIA)	TF	1	7	10	0	0	0	0	0	0	0	18
12721	F117-PW-100	C-17 (USAF)	TF	71	56	17	0	0	0	0	0	0	0	144
1749	F119-PW-100	F-22	ATF	60	59	58	56	31	0	0	0	0	0	264
104804	F135	F-35 (PRODUCTION)	ATF	0	0	3	13	28	48	52	54	57	57	312
104803	F135	F-35 (SDD)	ATF	1	2	6	6	0	0	0	0	0	0	15

10381	PW4000	747-400	TF	12	10	13	13	13	12	9	6	0	0	88
10413	PW4000	767-300	TF	3	1	2	2	0	0	0	0	0	0	8
100918	PW4000	777-200/200ER	TF	27	18	20	19	18	20	19	15	15	15	186
103854	PW4090/4098	777-300	TF	1	2	3	5	7	7	7	8	7	7	54
9743	PW4158	A300-600	TF	13	10	0	0	0	0	0	0	0	0	23
101802	PW4168	A330-200	TF	19	17	16	13	12	7	0	0	0	0	84
9780	PW4168	A330-300	TF	8	7	8	7	6	3	0	0	0	0	39
104392	PW6000	A318	TF	9	18	15	10	14	15	16	16	15	16	144
rec_type='A' ENGTYPE='TF'.OR.ENGTYPE='ATF'.OR.ENGTYPE='TJ'.OR.ENGTYPE='ATJ'.OR.ENGTYPE='TP'.OR.ENGTYPE='TS'.OR.ENGTYPE='PF'. OR.ENGTYPE='VC'														
Pratt & Whitney				263	245	194	175	141	112	103	99	94	95	1521
Pratt & Whitney Canada														
Pratt & Whitney Canada														
101866	JT15D-5	HAWKER 400XP	TF	111	103	68	56	45	36	39	32	34	32	556
105234	PT6A (VARIOUS)	AT-602	TP	4	5	4	6	7	7	6	5	6	4	54
13396	PT6A-112	F406	TP	6	4	1	0	0	0	0	0	0	0	11
104165	PT6A-114A	CARAVAN 675	TP	15	13	12	13	13	15	15	13	13	10	132
101259	PT6A-114A	CARAVAN I 208B	TP	79	76	75	70	74	70	73	74	72	57	720
106159	PT6A-135A	C90GT	TP	0	0	0	0	0	0	0	0	0	0	0
10001	PT6A-135A	KING AIR C90	TP	0	0	0	0	0	0	0	0	0	0	0
105213	PT6A-135A	F406 MARK II	TP	2	3	5	6	5	6	7	8	6	4	52
13206	PT6A-27	PC-6	TP	3	3	3	3	3	3	3	2	2	1	26
9991	PT6A-42	KING AIR B200	TP	63	52	43	43	38	35	34	30	26	22	386
102878	PT6A-45R	AT-502A	TP	13	15	18	16	16	17	17	19	17	13	161
102880	PT6A-45R	AT-802/A	TP	10	11	11	10	11	11	10	12	10	0	96
105817	PT6A-60/65/67AG	660 TURBO-THRUSH	TP	7	9	11	12	9	11	11	11	12	8	101
103449	PT6A-62	KT-1	TP	7	9	8	7	8	7	7	9	6	0	68
100827	PT6A-64	TBM 700 (CIVIL)	TP	0	0	0	0	0	0	0	0	0	0	0
102904	PT6A-64	TBM 700 (MIL)	TP	0	0	0	0	0	0	0	0	0	0	0
105974	PT6A-65/67	THRUSH 660	TP	11	12	12	11	14	12	12	10	13	12	119

(Continued)

Engine Production Matrix by Manufacturer Years to Print (2006–2015) (10 Years)—cont'd

Record Number	Engine	Application	Engine Type	Engine Units 06	07	08	09	10	11	12	13	14	15	Total 06–15
102421	PT6A-65B	M28 SKYTRUCK (CIVIL)	TP	6	8	6	4	5	6	4	2	3	4	48
105635	PT6A-65B	M28 SKYTRUCK (MIL.)	TP	11	11	10	8	6	5	7	6	4	4	72
105461	PT6A-66	P180 (CIVIL)	TP	55	50	47	45	45	45	43	42	41	40	453
11716	PT6A-66	P180 (MIL.)	TP	0	0	1	4	1	1	4	2	1	1	15
106160	PT6A-66	TBM 850 (CIVIL)	TP	0	0	0	0	0	0	0	0	0	0	0
106161	PT6A-66	TBM 850 (MIL)	TP	0	0	0	0	0	0	0	0	0	0	0
105662	PT6A-66A	AE270	TP	21	20	24	24	21	20	22	21	26	20	219
102922	PT6A-67B	PC-12 (COMMUTER)	TP	11	11	12	13	14	14	15	15	15	14	134
101559	PT6A-67B	PC-12 (CORP./ UTILITY)	TP	79	67	60	56	54	53	53	55	54	53	584
102381	PT6A-68	SUPER TUCANO	TP	0	1	5	6	8	8	9	8	6	6	57
105242	PT6A-68	PC-21	TP	8	11	11	10	9	10	10	10	12	9	100
103444	PT6A-68	T-6A (EXPORT)	TP	10	12	13	11	11	11	12	15	16	9	120
103442	PT6A-68	T-6A (USAF)	TP	59	61	53	39	9	0	0	0	0	0	221
103443	PT6A-68	T-6A (USN)	TP	0	3	15	36	54	36	0	0	0	0	144
103484	PT6B-37	A119	TS	21	21	21	22	23	23	23	25	25	15	219
104022	PT6C-67A	BA609	TS	2	4	24	55	60	48	49	50	53	57	402
105460	PT6C-67C	AB139 (CIVIL)	TS	71	85	87	78	71	70	79	81	82	64	768
102721	PT6C-67C	AB139 (MIL)	TS	4	10	19	23	25	25	29	28	31	21	215
100870	PT6T-3B/3D	412	TS	48	33	25	24	25	18	12	6	0	0	191
10176	PT6T-3B/3D	412/412SP/412HP (MIL)	TS	13	12	0	0	0	0	0	0	0		25
8384	PW123	Q200	TP	2	0	1	2	0	0	0	0	0	0	5
1446	PW123	Q300	TP	18	17	12	9	8	7	8	7	5	0	91
1329	PW123AF	415	TP	1	3	4	4	3	4	2	2	0	0	23
103271	PW127	ATR-42-500	TP	12	13	13	11	10	9	8	7	6	0	89
104579	PW127	ATR-72-500	TP	46	62	71	66	58	54	47	45	42	26	517

104424	PW127	C-295	TP	22	23	26	25	23	23	26	25	25	25	243
102382	PW127	MA60	TP	13	18	20	22	22	25	27	24	23	22	216
103386	PW150	Q400	TP	52	52	51	46	42	38	36	36	39	29	421
102307	PW206B/B2	EC 135 (CIVIL)	TS	64	65	63	61	62	64	65	65	65	63	637
105479	PW206B2	EC 635	TS	0	0	1	3	4	2	2	3	4	3	22
103469	PW206C	A109 POWER (CIVIL)	TS	33	26	19	12	11	11	12	13	13	9	159
105459	PW207D	427 (CIVIL)	TS	12	12	7	0	0	0	0	0	0	0	31
105837	PW207D	429	TS	9	30	62	95	111	99	89	80	83	59	717
104998	PW207E	COMBAT EXPLORER	TS	0	1	3	6	9	10	13	13	13	10	78
103795	PW207E	MD902 (CIVIL)	TS	19	29	38	47	54	54	61	64	60	38	464
105840	PW207K	A109 GRAND	TS	20	29	36	40	40	36	29	34	31	25	320
106010	PW210S	S-76D	TS	0	8	36	57	54	51	48	51	49	45	399
102073	PW305A	LEARJET 60	TF	39	37	39	40	38	34	29	27	25	21	329
105058	PW306A	G200	TF	49	51	50	49	52	53	52	50	49	51	506
104298	PW306C	CITATION SOVEREIGN	TF	109	92	82	75	74	70	76	79	77	59	793
104875	PW307A	FALCON 7X	TF	13	56	103	106	108	107	105	106	107	103	914
103797	PW308A	HAWKER 4000	TF	25	41	54	65	68	62	63	59	60	46	543
106134	PW308C	FALCON 2000DX	TF	1	6	10	12	11	11	11	10	8	8	88
104784	PW308C	FALCON 2000EX	TF	57	57	50	49	49	46	44	42	41	44	479
103375	PW530A	CITATION BRAVO	TF	35	29	20	15	8	0	0	0	0	0	107
104297	PW535A	CITATION ENCORE	TF	38	53	58	50	43	43	39	41	38	27	430
106063	PW535E	PHENOM 300	TF	0	0	4	30	42	44	50	56	68	64	358
105735	PW545B	CITATION EXCEL XLS	TF	110	86	75	73	74	72	69	67	67	76	769
105053	PW610F	ECLIPSE 500	TF	60	130	209	331	446	572	706	793	827	788	4862
105280	PW615F	MUSTANG	TF	35	93	165	234	278	294	299	315	312	300	2325
106060	PW617F	PHENOM 100	TF	0	4	28	120	156	170	200	210	230	230	1348
Pratt & Whitney Canada (Consortium)														
102002	PT6A-60A	SUPER KING AIR 350	TP	67	43	31	30	22	21	17	16	10	0	257

(Continued)

Engine Production Matrix by Manufacturer Years to Print (2006–2015) (10 Years)—cont'd

Record Number	Engine	Application	Engine Type	Engine Units 06	07	08	09	10	11	12	13	14	15	Total 06–15
Pratt & Whitney Canada (Associate)														
104169	PT6A-42	MALIBU MERIDIAN	TP	34	36	39	38	39	39	38	38	36	34	371
103702	PT6A-68	ALX (BRAZIL)	TP	20	20	19	15	8	0	0	0	0	0	82
Pratt & Whitney Canada				1765	1957	2203	2549	2711	2748	2876	2969	2999	2685	25,462
Progress (ZMKB Progress)														
Progress (ZMKB Progress)														
102352	D-136	MI-26	TS	7	6	5	4	1	0	0	0	0	0	23
106049	D-436	AN-148	TF	8	17	30	39	39	39	38	36	38	37	321
106105	SM 146	RUSSIAN REGIONAL JET	TF	0	0	0	0	0	0	0	0	0	0	0
Progress (ZMKB Progress)				15	23	35	43	40	39	38	36	38	37	344
Rolls Royce DE														
Rolls Royce DE														
103722	BR710	NIMROD MRA4	TF	3	6	0	0	0	0	0	0	0	0	9
105076	BR710	GLOBAL 5000	TF	45	52	55	57	52	52	49	51	34	0	447
101528	BR710	GLOBAL EXPRESS	TF	43	52	49	47	48	41	37	37	42	28	424
104710	BR710	GLOBAL EXPRESS ASTOR	TF	3	0	3	0	3	3	0	3	0	3	18
105317	BR710	G500/G550 (CIVIL)	TF	70	71	64	62	64	66	68	69	71	74	679
105752	BR710	G500/G550 (MIL.)	TF	5	6	4	4	2	2	3	4	1	1	32
105212	BR710	GULFSTREAM V ELINT/SIGINT	TF	3	1	4	5	6	4	3	3	0	0	29
103194	BR715	717-200	TF	6	0	0	0	0	0	0	0	0	0	6
Rolls Royce DE				178	188	179	175	175	168	160	167	148	106	1644
Rolls Royce North America														
Rolls Royce North America														
1790	250-B17B/C/D/E/F	CONVERSIONS	TP	6	6	5	5	6	4	5	0	0	0	37
13656	250-B17C	SF. 260TP	TP	7	6	3	0	0	0	0	0	0	0	16

102340	250-B17C	M/MT/MX/MXT-7-420	TP	1	2	1	0	1	2	2	1	1	0	11
101975	250-B17C/F	BN-2T SPECIAL MISSION	TP	6	7	9	10	9	13	16	13	13	12	108
105240	250-B17C/F	TURBINE ISLANDER BN-2T	TP	4	6	9	9	9	12	13	14	12	0	88
104294	250-B17F	T-7	TP	10	5	2	1	0	0	0	0	0	0	18
104339	250-C20/28/30	CONVERSIONS	TS	10	9	6	0	0	0	0	0	0	0	25
1674	250-C20B	NBO-105 (MIL.)	TS	0	0	0	0	0	0	0	0	0	0	0
1919	250-C20B	500E (CIVIL)	TS	2	1	2	2	3	3	2	3	4	0	22
1208	250-C20J	206B III (CIVIL)	TS	3	2	0	0	0	0	0	0	0	0	5
101622	250-C20R	520N (CIVIL)	TS	4	6	8	9	9	11	6	0	0	0	53
102466	250-C20R	520N (MIL)	TS	1	1	1	2	2	1	0	0	0	0	8
102061	250-C20W	480 (CIVIL)	TS	24	22	21	23	24	24	26	25	26	17	232
102062	250-C20W	TH-28 (MIL)	TS	1	2	2	2	0	2	1	3	2	2	17
104703	250-C20W	333 (CIVIL)	TS	12	13	14	16	17	16	16	15	13	0	132
105671	250-C20W	333 (MIL)	TS	3	2	1	3	3	4	4	4	3	0	27
1210	250-C30P	206L III/IV (CIVIL)	TS	7	7	5	0	0	0	0	0	0	0	19
102057	250-C40B	430 CORPORATE	TS	19	22	23	24	19	19	22	20	17	10	195
105458	250-C47B	407 (CIVIL)	TS	45	41	28	10	0	0	0	0	0	0	124
103009	250-C47B	407 (MIL)	TS	5	5	5	6	3	0	0	0	0	0	24
103475	250-C47M	600N	TS	13	19	21	20	22	23	29	31	29	19	226
102436	AE 1107C	CV-22 (USAF)	TS	6	4	4	6	12	13	11	11	12	13	92
102438	AE 1107C	MV-22 (U.S. NAVY)	TS	0	0	0	0	0	0	1	4	5	7	17
103021	AE 1107C	MV-22 (USMC)	TS	22	21	24	31	44	70	78	78	78	78	524
104421	AE 2100	C-27J	TP	16	29	38	36	38	36	38	33	35	26	325
103325	AE 2100	C-130J	TP	56	59	48	46	52	61	67	67	69	73	598
103782	AE 2100	US-1AKAI	TP	0	2	6	13	19	13	0	0	0	0	53
101896	AE 3007	CITATION X	TF	35	38	37	34	35	37	34	33	30	24	337
104865	AE 3007	ERJ 140	TF	0	4	12	12	12	8	0	0	0	0	48
103137	AE 3007	ERJ 145 (CIVIL)	TF	20	13	10	14	21	22	22	19	17	15	173
104566	AE 3007	ERJ 145 (MIL)	TF	0	1	3	4	5	6	4	4	3	4	34

(Continued)

Engine Production Matrix by Manufacturer Years to Print (2006–2015) (10 Years)—cont'd

Record Number	Engine	Application	Engine Type	Engine Units 06	07	08	09	10	11	12	13	14	15	Total 06–15
104897	AE 3007	LEGACY 600	TF	51	45	40	38	39	41	38	37	38	36	403
104024	AE 3007A	ERJ 135	TF	0	0	2	5	4	3	4	3	3	0	24
1596	T56-A-427	E-2C (USN)	TP	4	4	4	3	0	0	0	0	0	0	15
106031	T56-A-427	E-2D	TP	0	0	0	0	0	0	0	0	0	0	0
ROLLS ROYCE NORTH AMERICA				393	404	394	384	408	444	439	418	410	336	4030
Rolls Royce Plc														
Rolls Royce Plc 106086	NOT SELECTED	A340X	TF	0	0	0	0	0	0	0	0	0	0	0
104607	PEGASUS 11-61 (MOD)	HARRIER GR7A/9A (MOD)	TF	8	0	0	0	0	0	0	0	0	0	8
100938	RB211-524G/H/T	747-400	TF	8	8	4	4	0	5	4	4	0	5	42
102072	RB211-535E4/F5	TU-204	TF	5	6	5	7	10	10	8	7	9	9	76
105749	TAY 611-8C	G350	TF	26	28	32	33	35	36	34	31	26	23	304
104895	TAY 611-8C	G450	TF	56	58	52	50	53	56	57	54	51	52	539
105390	TRENT 1000	787	TF	1	13	46	79	107	124	124	124	124	83	825
106093	TRENT 1700	A350	TF	0	0	1	3	3	15	31	48	66	67	234
103592	TRENT 500	A340-500	TF	9	10	17	30	40	39	34	25	18	13	235
104810	TRENT 500	A340-600	TF	69	70	74	90	87	78	69	60	43	13	653
101803	TRENT 700	A330-200	TF	14	9	4	4	4	3	0	0	0	0	38
100915	TRENT 700	A330-300	TF	16	13	10	9	8	5	0	0	0	0	61
105215	TRENT 700	KC-30 FSTA	TF	0	0	1	6	10	11	10	6	0	0	44
100945	TRENT 800	777-200/200ER	TF	23	21	23	23	23	22	21	19	17	17	209
103855	TRENT 800	777-300	TF	2	4	7	8	7	9	8	7	7	7	66
104771	TRENT 900	A380	TF	11	41	74	79	65	59	54	59	65	46	553
Rolls Royce Plc				248	281	350	425	452	472	454	444	426	335	3887

Rolls Royce Turbomeca Ltd														
Rolls Royce Turbomeca Ltd (Consortium)														
1304	ADOUR MK 871	HAWK 100	TF	7	10	12	11	5	0	0	0	0	0	45
1305	ADOUR MK 871	HAWK 200	TF	0	0	1	4	5	4	3	4	0	0	21
105692	ADOUR MK 871	HAWK 100	TF	0	0	5	11	16	16	8	0	0	0	56
104608	ADOUR MK 900 (MOD)	HAWK 100	TF	10	10	10	4	0	0	0	0	0	0	34
105693	ADOUR MK 951	HAWK 120/128 LIFT	TF	15	8	11	13	12	15	17	18	17	16	142
1995	F405-RR-400/401	T-45A	TF	14	9	8	3	0	0	8	0	0	0	34
102676	RTM322	EH101 MILITARY UTILITY	TS	28	7	7	7	8	12	13	12	14	14	122
11256	RTM322	EH101 NAVAL	TS	0	0	0	0	0	1	5	7	6	4	23
105985	RTM322	KHI-01	TS	0	0	0	0	0	0	0	0	0	0	0
105643	RTM322	KHI-01	TS	1	3	3	5	7	7	7	7	5	0	45
105851	RTM322	MRH90 (AUSTRALIA)	TS	3	10	10	3	0	0	0	0	0	0	26
105647	RTM322	NH90 (GREECE)	TS	12	13	14	15	10	3	0	0	0	0	67
105841	RTM322	NH90 (OMAN)	TS	0	5	16	16	8	0	0	0	0	0	45
105101	RTM322	NH90 (SCANDINAVIA)	TS	0	0	0	0	0	0	0	0	0	0	0
12496	RTM322	NH90/NFH	TS	16	18	16	18	21	19	17	15	12	12	164
102723	RTM322	NH90/TTH	TS	27	25	20	21	28	40	46	49	49	48	353
105068	RTM322	NH90 (SCANDINAVIA)	TS	16	16	16	18	18	17	12	0	0	0	113
Rolls Royce Turbomeca Ltd				149	134	149	149	138	134	128	112	103	94	1290
Samsung Techwin Co. Ltd.														
Samsung Techwin Co. Ltd. (Consortium)														
103315	F404-GE-102	T/A/F-50	TF	10	13	13	15	20	22	27	27	26	23	196
Samsung Techwin Co. Ltd.				10	13	13	15	20	22	27	27	26	23	196
Saturn (NPO Saturn)														
Saturn (NPO Saturn)														
101782	AL-31F	SU-27 (ALL)	ATF	96	113	116	113	99	86	86	84	83	84	960
105883	AL-31FN	J-10	ATF	26	31	31	31	31	31	31	31	31	16	290
104462	AL-55F/I	HJT-36	TJ	0	2	5	5	2	0	0	0	0	0	14

(Continued)

Engine Production Matrix by Manufacturer Years to Print (2006–2015) (10 Years)—cont'd

Record Number	Engine	Application	Engine Type	Engine Units 06	07	08	09	10	11	12	13	14	15	Total 06–15
Saturn (NPO Saturn)				122	146	152	149	132	117	117	115	114	100	1264
Shenyang Liming Aero Engine GP														
Shenyang Liming Aero Engine GP (Licensee)														
105431	WZ-8	Z-9 (CIVIL)	TS	4	4	3	2	1	0	0	0	0	0	14
12041	WZ-8	Z-9 (MIL.)	TS	2	2	2	2	1	0	0	0	0	0	9
Shenyang Liming Aero Engine GP				6	6	5	4	2	0	0	0	0	0	23
SNECMA														
SNECMA														
105380	M53-P2	MIRAGE 2000-5 (INDIA)	ATF	0	1	6	8	0	0	0	0	0	0	15
105381	M53-P2	MIRAGE 2000-5 (INDIA)	ATF	0	0	0	2	9	17	24	27	18	4	101
102558	M88-2	RAFALE B	ATF	9	9	9	9	9	11	14	14	11	0	95
102777	M88-2	RAFALE C	ATF	16	16	16	18	21	21	21	23	27	19	198
101742	M88-2	RAFALE M	ATF	3	8	8	8	11	17	17	15	7	0	94
SNECMA (Co-Product)														
101801	CF6-80E	A330-200	TF	38	34	31	25	19	9	0	0	0	0	156
9779	CF6-80E	A330-300	TF	18	18	16	12	8	5	0	0	0	0	77
SNECMA (Consortium)														
9736	CF6-80C2	A300-600	TF	5	0	0	0	0	0	0	0	0	0	5
SNECMA				89	86	86	82	77	80	76	79	63	23	741
Turbomeca														
Turbomeca														
105104	ARDIDEN	DHRUV ALH (MIL.)	TS	5	27	55	45	41	42	41	43	43	44	386
102028	ARRIEL 1B/1D1	AS 350B1/B2 (CORPORATE)	TS	4	4	4	3	2	3	2	2	3	2	29
105433	ARRIEL 1B/1D1	HB 350 (CIVIL)	TS	4	3	3	3	3	3	3	2	2	2	28
102027	ARRIEL 1B/1D1	HB 350/550 (MIL.)	TS	3	3	2	2	2	2	2	2	2	2	22
9631	ARRIEL 1C1/1C2/2C	AS 365N (CIVIL)	TS	29	27	25	23	22	21	22	20	18	15	222

101658	ARRIEL 1D1	AS 350B2 (CIVIL)	TS	30	20	19	17	17	14	12	12	11	10	162
102499	ARRIEL 1E2	BK 117C-1	TS	0	0	0	0	0	0	0	0	0	0	0
104708	ARRIEL 1E2	EC 145	TS	28	25	26	24	22	21	20	21	19	17	223
104707	ARRIEL 1E2	BK 117C-2	TS	5	7	6	5	7	7	6	4	4	4	55
9616	ARRIEL 1M/1M1/2C	AS 565 (MIL)	TS	4	7	8	7	7	6	4	4	2	2	51
103481	ARRIEL 2B	AS 350B3 (CIVIL)	TS	51	55	52	52	51	48	43	40	36	33	461
104016	ARRIEL 2B	AS 550U/C/A3 (MIL.)	TS	3	3	4	4	4	4	4	3	3	3	35
104203	ARRIEL 2B1	EC 130B4 (CIVIL)	TS	33	32	31	29	26	25	24	24	22	21	267
104207	ARRIEL 2C	EC 155 (CIVIL)	TS	24	23	23	25	26	26	25	26	24	23	245
105845	ARRIEL 2C	HH-65C (MOD)	TS	86	25	0	0	0	0	0	0	0	0	111
105267	ARRIEL 2C	H410A/H425 (CIVIL)	TS	7	7	8	6	4	4	4	4	2	1	47
106009	ARRIEL 2S2	S-76C++	TS	80	55	24	0	0	0	0	0	0	0	159
102508	ARRIUS 1	AS 355N (CIVIL)	TS	17	14	12	12	11	10	10	11	11	10	118
101008	ARRIUS 1	AS 555U/S/M/C/A (MIL.)	TS	5	6	5	5	6	4	3	5	5	4	48
100876	ARRIUS 2B	EC 135 (CIVIL)	TS	81	72	67	65	65	65	65	64	65	63	672
105480	ARRIUS 2B	EC 135 (MIL.)	TS	1	3	1	2	1	0	0	0	0	0	8
105478	ARRIUS 2B	EC 635	TS	0	4	11	11	11	11	10	8	7	7	80
101395	ARRIUS 2F	EC 120 (CIVIL)	TS	53	60	64	72	77	82	86	87	86	85	752
102036	ARRIUS 2F	EC 120 (MIL)	TS	0	1	4	5	5	5	5	6	7	6	44
104352	ARRIUS 2K1	A109 POWER (CIVIL)	TS	26	21	18	16	17	17	17	15	15	10	172
104787	ARRIUS 2K2	A109 POWER (MIL)	TS	30	30	27	25	24	25	24	27	25	15	252
105469	MAKILA 1A1	AS 332C/L (CIVIL)	TS	6	5	5	5	5	5	5	4	3	4	47
9511	MAKILA 1A1	AS 532/332C/L (MIL.)	TS	9	12	10	7	6	5	4	2	3	4	62
13126	MAKILA 1A1/1A2	NAS-332 (MIL.)	TS	8	6	0	0	0	0	0	0	0	0	14
9507	MAKILA 1A2/2A	AS 332 MK2/EC 225 (CIV)	TS	19	15	13	12	12	11	10	11	8	7	118
9508	MAKILA 1A2/2A	AS 532 MK2/EC 725 (MIL)	TS	5	5	9	12	12	11	10	12	10	7	93
105660	TM333	DHRUV ALH (CIVIL)	TS	8	8	9	10	12	12	12	14	14	14	113
11996	TM333	DHRUV ALH (MIL.)	TS	49	37	0	0	0	0	0	0	0	0	86

(Continued)

Engine Production Matrix by Manufacturer Years to Print (2006–2015) (10 Years)—cont'd

Record Number	Engine	Application	Engine Type	Engine Units 06	07	08	09	10	11	12	13	14	15	Total 06–15
Turbomeca				713	622	545	504	498	489	473	473	450	415	5182
Volvo/GE														
Volvo/GE (Consortium)														
101730	RM12	JAS 39C/D/EXPORT	ATF	14	11	9	9	11	11	10	11	11	10	107
Volvo/GE				14	11	9	9	11	11	10	11	11	10	107
Williams International														
Williams International														
105878	FJ33	A700	TF	2	9	27	81	121	139	128	123	112	108	850
106065	FJ44-3A	SPN UTILITY JET	TF	9	34	44	50	53	60	63	62	61	60	496
Williams International				11	43	71	131	174	199	191	185	173	168	1346
Williams Rolls Inc														
Williams Rolls Inc (Consortium)														
104190	FJ44-1AP	CITATIONJET CJ1/ CJ1+	TF	33	31	25	12	0	0	0	0	0	0	101
104593	FJ44-2	EAGLE2 CITATION 500 (MOD)	TF	6	0	0	0	0	0	0	0	0	0	6
103510	FJ44-2	PREMIER I	TF	69	74	73	66	60	65	68	69	75	57	676
101870	FJ44-2	SJ30-2	TF	51	84	87	86	82	81	80	7	73	69	770
105142	FJ44-3A	CITATIONJET CJ3	TF	153	161	150	137	130	124	122	128	118	83	1306
104299	FJ44-3A-24	CITATIONJET CJ2/ CJ2+	TF	61	65	66	60	61	65	70	71	67	49	635
Williams Rolls Inc				373	415	401	361	333	335	340	345	333	258	3494
Total				8052	8450	8620	9138	9461	9675	9844	9793	9478	8310	90,821

Appendix 2B-2: Gas Turbines for Electrical Generation, Mechanical Drive, and Marine Power*

Record Number	Engine	Application	Engine Units 06	07	08	09	10	11	12	13	14	15	Total 06–15
ALSTOM													
ALSTOM													
103115	GT 24	GENERATION	3	4	7	9	9	9	9	9	9	8	76
103116	GT 26	GENERATION	7	7	12	12	12	12	12	12	12	11	109
102596	TYPE 11N/N2	GENERATION	9	12	14	14	14	14	13	13	14	14	131
100974	TYPE 13E/E2	GENERATION	9	11	13	15	15	15	15	15	14	14	136
106050	TYPE BC2	GENERATION	3	5	7	7	8	8	8	8	7	7	68
ALSTOM			31	39	53	57	58	58	57	57	56	54	520
Ansaldo Energia S.P.A.													
Ansaldo Energia S.P.A. (Licensee)													
105918	SGT5-2000E (V94.2)	GENERATION	6	6	6	5	5	5	4	5	4	4	50
105919	SGT5-3000E (V94.2A)	GENERATION	0	1	1	2	2	2	1	0	2	0	11
105567	SGT5-4000F (V94.3A)	GENERATION	1	2	0	2	1	2	1	1	2	3	15
Ansaldo Energia S.P.A.			7	9	7	9	8	9	6	6	8	7	76
Bharat Heavy Electricals Ltd													
Bharat Heavy Electricals Ltd (Licensee)													
105569	SGT5-2000E (V94.2)	GENERATION	2	2	1	0	1	1	0	0	0	0	7
Bharat Heavy Electricals Ltd (Co-Product)													
102229	MODEL 5000	GENERATION	3	2	3	3	4	4	3	3	2	3	30
Bharat heavy electricals ltd			5	4	4	3	5	5	3	3	2	3	37
Daihatsu Diesel MFG Co Ltd													
Daihatsu Diesel MFG Co Ltd													
104884	DT SERIES												
	(3 MW & LARGER)	GENERATION	3	3	2	3	2	3	3	3	4	3	29
105751	DT SERIES												
	(3000 SHP & UP)	MECHANICAL DRIVE	2	2	2	2	2	1	2	1	1	0	15
104883	DT SERIES												
	(TO 3 MW)	GENERATION	3	4	5	5	5	4	5	4	4	4	43
105750	DT SERIES												
	(TO 3000 SHP)	MECHANICAL DRIVE	3	2	2	3	4	4	4	3	3	4	32
Daihatsu Diesel MFG Co Ltd			11	11	11	13	13	12	14	11	12	11	119
Dresser-Rand													
Dresser-Rand													
3074	KG2	GENERATION	1	2	2	1	0	0	0	0	0	0	6
Dresser-Rand			1	2	2	1	0	0	0	0	0	0	6

(*Continued*)

—cont'd

Record Number	Engine	Application	Engine Units 06	07	08	09	10	11	12	13	14	15	Total 06–15
Ebara Corp.													
Ebara Corp. (Licensee)													
104113	FT8	GENERATION	0	2	1	1	0	1	1	2	2	2	12
Ebara Corp.			0	2	1	1	0	1	1	2	2	2	12
GE Energy													
GE Energy													
29190	LM 1600	GENERATION	1	1	0	0	0	0	0	0	0	0	2
29895	LM 1600	MECHANICAL DRIVE	0	0	0	0	0	0	0	0	0	0	0
102235	LM 6000	GENERATION	40	38	40	40	38	40	40	40	38	37	391
29195	LM2500/LM2500+	GENERATION	30	42	46	56	60	60	60	60	48	48	510
29800	LM2500/LM2500+	MARINE POWER	26	26	26	26	24	24	24	24	24	23	247
30075	LM2500/LM2500+	MECHANICAL DRIVE	18	22	24	24	24	26	26	26	24	24	238
105727	LMS100	GENERATION	3	5	8	12	18	18	18	18	18	17	135
105926	MODEL 6000B/C (MS6001B/C)	GENERATION	13	11	12	12	11	11	12	14	14	14	124
102408	MODEL 6000B/C (MS6001B/C)	MECHANICAL DRIVE	2	2	4	5	5	5	4	5	5	4	41
103089	MODEL 6000FA (MS6001FA)	GENERATION	8	9	12	15	14	15	14	14	15	14	130
105687	MODEL 7001EA	GENERATION	10	12	12	14	12	12	14	12	12	11	121
105864	MODEL 7001EA	MECHANICAL DRIVE	3	3	3	4	2	2	2	3	3	2	27
29321	MODEL 7001F/FA	GENERATION	30	42	42	44	46	46	46	43	42	42	423
105075	MODEL 7001FB	GENERATION	5	12	17	22	25	25	23	21	25	25	200
103619	MODEL 7001H	GENERATION	3	9	16	18	18	18	18	22	22	21	165
106090	MODEL 9001E	GENERATION	6	18	12	12	12	12	14	14	13	13	126
105891	MODEL 9001E	MECHANICAL DRIVE	3	1	2	2	0	2	1	2	1	0	14
103899	MODEL 9001F/FA	GENERATION	12	8	11	12	16	18	18	16	17	18	146
105260	MODEL 9001FB	GENERATION	4	2	3	6	6	6	6	6	6	6	51
103901	MODEL 9001H	GENERATION	0	1	2	4	4	5	5	5	4	3	33
GE energy			217	264	292	328	335	345	345	345	331	322	3,124
GE Licensees/Affiliated Firms													
GE Licensees/Affiliated Firms													
106021	MODEL 7001A-E	GENERATION	1	1	1	1	1	0	0	0	0	0	5
106022	MODEL 7001EA	GENERATION	0	1	2	1	0	2	3	3	2	3	17
106023	MODEL 7001EA	MECHANICAL DRIVE	1	2	1	2	3	4	4	4	3	3	27
106024	MODEL 7001F/FA	GENERATION	3	5	5	5	5	5	4	4	5	5	46

—cont'd

Record Number	Engine	Application	Engine Units 06	07	08	09	10	11	12	13	14	15	Total 06–15
GE Licensees/Affiliated Firms (Associate)													
101815	LM2500/LM2500+	GENERATION	4	6	6	6	6	6	4	4	6	6	54
102595	LM2500/LM2500+	MARINE POWER	4	4	3	3	4	3	3	4	4	4	36
101816	LM2500/LM2500+	MECHANICAL DRIVE	4	4	6	6	8	8	8	7	7	7	65
105781	MODEL 6000B/C (MS6001B/C)	GENERATION	3	5	5	4	5	5	5	4	5	4	45
105782	MODEL 6000B/C (MS6001B/C)	MECHANICAL DRIVE	0	2	1	3	2	3	5	5	5	4	30
105783	MODEL 6000FA (MS6001FA)	GENERATION	2	5	4	6	6	6	5	6	5	5	50
106087	MODEL 9001E	GENERATION	3	2	2	2	1	3	3	4	4	3	27
106092	MODEL 9001E	MECHANICAL DRIVE	3	6	9	6	4	4	3	6	8	8	57
100964	MODEL 9001F/FA	GENERATION	0	2	2	1	2	2	1	2	2	3	17
104844	MODEL 9001H	GENERATION	3	0	2	3	4	4	4	4	3	3	30
GE Licensees/Affiliated Firms			31	45	49	49	51	55	52	57	59	58	506
GE Oil & gas (Nuovo Pignone)													
GE Oil & Gas (Nuovo Pignone)													
100795	GE-10/1	GENERATION	4	5	5	5	5	5	4	4	4	3	44
100794	GE-10/2	MECHANICAL DRIVE	10	12	15	15	15	14	13	12	12	12	130
105263	GE-5/1	GENERATION	0	2	1	1	2	1	2	3	1	1	14
105262	GE-5/2	MECHANICAL DRIVE	2	1	2	3	3	3	2	3	3	2	24
30455	MODEL 5000	GENERATION	10	8	8	9	9	8	8	7	7	6	80
30335	MODEL 5000	MECHANICAL DRIVE	35	32	32	32	30	28	28	30	28	26	301
GE Oil & Gas (Nuovo Pignone) (Licensee)													
105372	LM 1600	GENERATION	2	1	1	0	0	0	0	0	0	0	4
105376	LM 1600	MECHANICAL DRIVE	1	0	1	0	0	0	0	0	0	0	2
GE Oil & Gas (Nuovo Pignone)			64	61	65	65	64	59	57	59	55	50	599
Hitachi Ltd													
Hitachi Ltd													
101805	H-25	GENERATION	8	9	10	10	10	12	12	12	12	11	106
Hitachi ltd			8	9	10	10	10	12	12	12	12	11	106
Interturbo (Zao Interturbo)													
Interturbo (Zao Interturbo) (Licensee)													
105571	SGT5-2000E (V94.2)	GENERATION	0	1	1	0	2	0	1	1	2	2	10
Interturbo (Zao Interturbo)			0	1	1	0	2	0	1	1	2	2	10

(Continued)

—cont'd

Record Number	Engine	Application	06	07	08	09	10	11	12	13	14	15	Total 06–15
Kawasaki Heavy Industries													
Kawasaki Heavy Industries													
105139	L20A	GENERATION	3	5	6	6	5	6	8	8	8	8	63
3093	M1A-03/-06/-13/-13CC	GENERATION	26	24	24	24	24	30	30	28	28	27	265
102274	M1A-23	GENERATION	18	20	23	22	22	22	20	18	18	17	200
101827	M1T-03/-06/-13	GENERATION	24	26	24	22	22	22	20	20	20	19	219
105812	M1T-23	GENERATION	24	26	26	28	28	28	26	24	24	24	258
104866	M7A-01	GENERATION	1	0	0	0	0	0	0	0	0	0	1
104867	M7A-02	GENERATION	2	3	5	5	5	5	5	4	4	4	42
3152	S2A-01	GENERATION	25	30	40	40	40	40	35	28	27	25	330
Kawasaki Heavy Industries (Licensee)													
101746	INDUSTRIAL SPEY	MARINE POWER	2	3	2	0	0	2	0	0	2	0	11
Kawasaki Heavy Industries			125	137	150	147	146	155	144	130	131	124	1,389
Man Group Machines													
Man Group Machines													
3178	THM 1304												
	(ALL MODELS)	GENERATION	1	2	1	0	0	0	0	0	0	0	4
3027	THM 1304												
	(ALL MODELS)	MECHANICAL DRIVE	3	4	2	2	0	0	0	0	0	0	11
Man Group Machines (Co-Product)													
103326	FT8	GENERATION	4	2	3	4	4	4	4	3	4	4	36
103078	FT8	MECHANICAL DRIVE	1	2	0	2	2	2	0	2	1	2	14
Man group machines			9	10	6	8	6	6	4	5	5	6	65
Manufacturer Varies													
Manufacturer Varies													
105797	ENGINE VARIES	GENERATION: 125–180 MW	0	0	0	0	0	0	0	42	60	60	162
105796	ENGINE VARIES	GENERATION: 50–125 MW	9	48	63	68	68	68	68	63	63	63	581
105826	ENGINE VARIES	MARINE: 10,000–19,999 SHP	0	0	0	2	3	4	6	6	6	6	33
105813	ENGINE VARIES	MD: 10,000–19,999.99 HP	0	0	0	3	3	10	15	20	20	18	89
105814	ENGINE VARIES	MD: 20,000 HP & LARGER	0	0	0	3	7	13	15	15	20	18	91
Manufacturer varies			9	48	63	76	81	95	104	146	169	165	956
Mitsubishi Heavy Ind Ltd													
Mitsubishi Heavy Ind Ltd													
106081	MF-111B	GENERATION	1	1	2	0	1	2	2	0	1	0	10
29361	MODEL 501F	GENERATION	2	3	4	5	6	6	6	6	5	5	48

—cont'd

Record Number	Engine	Application	06	07	08	09	10	11	12	13	14	15	Total 06–15
104145	MODEL 501G	GENERATION	3	6	7	7	6	6	7	7	6	6	61
105928	MODEL 501H	GENERATION	0	1	2	2	4	5	5	6	8	8	41
101791	MODEL 701F	GENERATION	10	10	10	9	12	12	12	12	12	12	111
104146	MODEL 701G	GENERATION	2	2	4	6	7	8	8	8	8	8	61
Mitsubishi Heavy Ind Ltd (Co-Product)													
103077	FT8	MARINE POWER	1	2	0	2	2	0	4	2	2	4	19
Mitsubishi Heavy Ind Ltd			19	25	29	31	38	39	44	41	42	43	351
Mitsui Engineering & Shipbldng													
Mitsui Engineering & Shipbldng													
101521	SB5	GENERATION	10	15	20	27	27	27	27	20	18	17	208
Mitsui Engineering & Shipbldng			10	15	20	27	27	27	27	20	18	17	208
OPRA Optimal Radial Turbine BV													
OPRA Optimal Radial Turbine BV													
105944	OP16	GENERATION	7	12	22	40	40	40	35	35	35	34	300
OPRA Optimal Radial Turbine BV			7	12	22	40	40	40	35	35	35	34	300
Rolls Royce (Indust. & Marine)													
Rolls Royce (Indust. & Marine)													
3225	501-K (MARINE SERIES)	MARINE POWER	4	6	6	6	5	5	4	2	2	4	44
3037	501-KB5	GENERATION	9	9	11	11	11	11	8	8	8	7	93
103628	501-KB7	GENERATION	18	20	22	22	20	20	16	12	12	12	174
3314	501-KC5	MECHANICAL DRIVE	9	9	9	9	9	9	9	8	8	9	88
103629	501-KH5	GENERATION	0	2	1	2	0	2	3	4	4	3	21
29045	INDUSTRIAL AVON	GENERATION	1	1	0	0	0	0	0	0	0	0	2
29940	INDUSTRIAL AVON	MECHANICAL DRIVE	0	2	1	0	0	0	0	0	0	0	3
29400	INDUSTRIAL RB211	GENERATION	5	6	9	9	9	8	8	9	9	9	81
29850	INDUSTRIAL SPEY	MARINE POWER	0	2	0	0	2	0	0	2	0	2	8
103123	INDUSTRIAL TRENT	GENERATION	2	3	5	7	8	8	8	8	7	7	63
105730	INDUSTRIAL TRENT	MECHANICAL DRIVE	6	2	3	3	4	4	4	4	4	3	37
105332	INDUSTRIAL TRENT MT30	MARINE POWER	4	7	7	9	9	10	12	14	14	14	100
Rolls Royce (Indust. & Marine)			58	69	74	78	77	77	72	71	68	70	714
Rolls Royce (U.K.)													
Rolls Royce (U.K.)													
30135	INDUSTRIAL RB211	MECHANICAL DRIVE	10	9	10	12	12	12	12	11	11	10	109
105127	INDUSTRIAL RB211 (WR21)	MARINE POWER	3	4	6	6	8	8	8	8	8	8	67
Rolls Royce (U.K.)			13	13	16	18	20	20	20	19	19	18	176

(Continued)

—cont'd

Record Number	Engine	Application	Engine Units 06	07	08	09	10	11	12	13	14	15	Total 06–15
Siemens AG, Power Gen Group													
Siemens AG, Power Gen Group 3147	SGT-100 (TYPHOON)	GENERATION	4	5	7	8	8	10	10	10	10	9	81
3021	SGT-100 (TYPHOON)	MECHANICAL DRIVE	3	6	8	8	7	7	8	8	7	7	69
3008	SGT-200 (TORNADO)	GENERATION	3	5	7	7	7	6	6	6	6	6	59
3309	SGT-200 (TORNADO)	MECHANICAL DRIVE	6	8	10	13	13	12	12	12	11	11	108
104288	SGT-300 (TEMPEST)	GENERATION	6	10	12	12	12	12	12	12	11	11	110
104375	SGT-400 (CYCLONE)	GENERATION	1	2	2	3	3	2	2	2	2	2	21
104988	SGT-400 (CYCLONE)	MECHANICAL DRIVE	6	9	12	12	11	11	10	10	11	10	102
29110	SGT-500 (GT 35)	GENERATION	2	1	1	0	0	0	0	0	0	0	4
30045	SGT-500 (GT 35)	MECHANICAL DRIVE	3	3	3	2	2	0	0	0	0	0	13
104224	SGT-600/-700 (GT 10)	GENERATION	7	7	8	9	10	10	10	10	10	9	90
29500	SGT-600/-700 (GT 10)	MECHANICAL DRIVE	13	15	16	16	16	18	18	17	17	17	163
104225	SGT-800 (GTX100)	GENERATION	7	10	13	15	15	15	15	15	14	14	133
105916	SGT5-2000E (V94.2)	GENERATION	3	4	6	6	6	6	6	6	6	6	55
105917	SGT5-3000E (V94.2A)	GENERATION	2	1	2	0	2	1	2	2	0	0	12
103908	SGT5-4000F (V94.3A)	GENERATION	12	13	15	18	18	18	17	17	18	17	163
Siemens AG, Power Gen Group			78	99	122	129	130	128	128	127	123	119	1,183
Siemens Westinghouse Power													
Siemens Westinghouse Power 105927	SGT6-3000E (W501D5A)	GENERATION	1	1	0	1	0	0	0	0	0	0	3
102252	SGT6-5000F (W501F)	GENERATION	5	6	10	10	12	12	12	12	12	12	103
103396	SGT6-6000G (W501G)	GENERATION	4	4	10	14	15	15	14	14	15	14	119
Siemens Westinghouse Power			10	11	20	25	27	27	26	26	27	26	225
Solar Turbines Inc													
Solar Turbines Inc 3002	CENTAUR 40/50	GENERATION	33	33	42	42	33	27	33	27	24	21	315
3017	CENTAUR 40/50	MECHANICAL DRIVE	39	39	27	27	27	21	21	21	18	18	258
3004	MARS 90/100	GENERATION	63	78	78	78	78	78	72	70	70	68	733
3019	MARS 90/100	MECHANICAL DRIVE	57	78	78	78	78	75	72	75	72	72	735

—cont'd

Record Number	Engine	Application	Engine Units 06	07	08	09	10	11	12	13	14	15	Total 06–15
103407	TAURUS 60/65/70	GENERATION	200	241	302	302	302	302	289	256	243	240	2,677
103408	TAURUS 60/70	MECHANICAL DRIVE	96	144	160	160	160	160	144	144	135	135	1,438
105044	TITAN 130	GENERATION	56	72	80	80	80	80	72	70	66	66	722
105045	TITAN 130	MECHANICAL DRIVE	32	40	48	56	56	48	48	56	48	45	477
Solar Turbines Inc			576	725	815	823	814	791	751	719	676	665	7,355
Turbomeca													
Turbomeca 101515	MAKILA TI	GENERATION	2	2	2	0	0	0	0	0	0	0	6
101516	MAKILA TI	MECHANICAL DRIVE	2	1	0	0	0	0	0	0	0	0	3
Turbomeca			4	3	2	0	0	0	0	0	0	0	9
UTC PWPS													
UTC PWPS 30012	FT8	GENERATION	18	16	20	20	20	20	18	20	20	18	190
101830	ST18	GENERATION	0	1	1	0	1	1	0	1	1	1	7
3007	ST18	MARINE POWER	6	5	2	0	4	5	2	0	4	5	33
101533	ST18	MECHANICAL DRIVE	0	1	0	1	0	0	0	0	0	0	2
104896	ST40	GENERATION	1	1	2	3	3	2	3	2	3	3	23
105753	ST40	MARINE POWER	6	5	6	8	8	8	8	8	6	8	71
105754	ST40	MECHANICAL DRIVE	1	0	2	2	0	2	0	0	0	0	7
105588	ST6	GENERATION	1	2	1	3	0	2	0	0	0	0	9
3251	ST6	MARINE POWER	0	1	1	0	0	0	0	0	0	0	2
3297	ST6	MECHANICAL DRIVE	7	9	12	13	13	13	12	12	11	10	112
UTC PWPS			40	41	47	50	49	53	43	43	45	45	456
Vericor Power Systems													
Vericor Power Systems 104126	ASE40	GENERATION	2	0	1	0	0	0	0	0	0	0	3
105756	ASE40	MECHANICAL DRIVE	0	2	1	0	0	0	0	0	0	0	3
104127	ASE50	GENERATION	1	2	3	2	3	4	3	2	2	3	25
3197	ASE8	GENERATION	2	1	1	0	0	0	0	0	0	0	4
105931	ETF 40B	MARINE POWER	32	40	40	0	0	0	0	0	0	0	112
3016	TF25/35/40/50	MARINE POWER	1	2	2	1	2	0	0	0	0	0	8
Vericor Power Systems			38	47	48	3	5	4	3	2	2	3	155
Total			1371	1702	1929	1991	2006	2018	1949	1937	1899	1855	18,657

rec_type='IM'
ENGTYPE='TS'
.NOT. (SOURCE='GA')
Source: Forecast International.

Appendix 2B-3: Gas Turbines for APU/GPU Units*

Record Number	Engine	Application	Engine Type	Engine Units 06	07	08	09	10	11	12	13	14	15	Total 06–15
APIC														
APIC 103222	APS 1000	400	TS	24	24	24	22	20	18	17	16	18	18	201
105246	APS 1000	MA60	TS	4	4	4	5	5	6	6	5	5	5	49
103225	APS 2000	737-300/400/500												
		(RETROFIT)	TS	24	24	24	18	16	16	14	7	7	7	157
103730	APS 2000	USN GPU	TS	115	33	0	0	0	0	0	0	0	0	148
103467	APS 2100	717-200/300	TS	5	0	0	0	0	0	0	0	0		5
104632	APS 2300	ERJ-170/190	TS	95	97	100	99	98	100	102	103	103	103	1,000
102700	APS 3200	A318/319/320/321	TS	132	120	106	112	120	124	110	110	110	110	1,154
104966	APS 3400	C-5M (MOD)	TS	14	24	24	24	8	8	0	0	0		102
102708	APS 500	200	TS	8	7	5	5	5	5	4	4	4	4	51
104222	APS 500	ERJ-135/140/145	TS	58	58	55	52	63	66	70	72	70	70	634
105821	APS 500	LEGACY 600	TS	11	15	15	15	15	15	17	17	17	17	154
105831	NOT SELECTED	787	TS	2	4	27	62	79	97	114	105	112	110	712
APIC				492	410	384	414	429	455	454	439	446	444	4,367
Hamilton Sundstrand														
Hamilton Sundstrand 102496	TITAN T-62T	V-22	TS	15	13	13	18	23	38	42	45	46	46	299
7034	TITAN T-62T	C-135/E-3/E-8 APU	TS	6	6	6	0	0	0	0	0	0	0	18
7036	TITAN T-62T	CH-47D/SD/F	TS	5	8	8	9	8	8	8	0	0	0	54
49320	TITAN T-62T	F-16	TS	73	78	36	28	22	26	22	20	17	17	339
7103	TITAN T-62T	APU/GPU AFTER-MARKET SPARE	TS	75	65	60	60	60	60	65	60	55	55	615
7134	TITAN T-62T	H-53	TS	0	0	0	0	1	2	0	0	0	0	3
7135	TITAN T-62T	H-60/S-70	TS	72	76	102	152	158	157	159	155	134	134	1,299
Hamilton Sundstrand				246	246	225	267	272	291	296	280	252	252	2,627
Honeywell														
Honeywell 102495	G250	F-22	TS	27	30	30	34	34	34	35	35	35	35	329
7121	JFS 190-1	F-15	TS	13	13	15	13	13	15	6	6	6	6	106
103731	MODEL 131-9[A]	A318/319/320/321	TS	117	116	116	127	128	121	114	114	114	114	1,181
103226	MODEL 131-9[B]	737-600/700/800/900	TS	224	208	227	222	209	201	206	202	202	202	2,103

—cont'd

Record Number	Engine	Application	Engine Type	Engine Units										Total
				06	07	08	09	10	11	12	13	14	15	06–15
101621	MODEL 165	NEW SPARES	TS	0	0	0	0	0	0	0	0	0	0	0
105202	MODEL 331	TANKER/ TRANSPORT (MOD)	TS	5	7	6	9	7	7	5	5	5	5	61
7087	MODEL 331	FORCAST SPARES	TS	14	14	14	14	16	16	16	18	18	18	158
7028	MODEL 331-200/400	767	TS	11	9	9	9	0	0	0	0	0		38
7016	MODEL 331-250	A300	TS	14	11	14	12	11	11	11	8	8	8	108
7119	MODEL 331-250	C-17A	TS	18	18	20	18	5	0	0	0	0	0	79
101321	MODEL 331-350	A330	TS	52	46	41	39	37	26	18	9	7	7	282
101322	MODEL 331-350/600	A340	TS	31	29	28	33	31	29	29	27	27	27	291
102204	MODEL 331-500	777	TS	64	74	83	84	84	83	84	81	82	82	801
7019	MODEL 36	A318/319/320/321	TS	18	19	16	17	14	15	12	12	12	12	147
7122	MODEL 36	F/A-18	TS	52	52	50	54	56	56	54	34	34	34	476
104212	MODEL 36	AH-64D (EXPORT)	TS	15	20	20	21	24	22	20	20	20	20	202
103227	MODEL 36	AH-64D LONGBOW	TS	39	26	27	29	32	32	25	24	24	24	282
104470	MODEL 36	CHALLENGER 300	TS	55	44	40	35	34	33	36	39	37	37	390
104925	MODEL 36	CRJ 100/200	TS	55	46	40	42	39	42	40	35	35	35	409
7045	MODEL 36	CHALLENGER	TS	19	17	19	18	17	19	18	17	15	12	171
102686	MODEL 36	CITATION X	TS	26	29	26	28	26	31	28	29	28	27	278
102205	MODEL 36	FALCON 2000	TS	0	0	0	0	0	0	0	0	0	0	0
105637	MODEL 36	FALCON 2000EX	TS	37	36	35	33	32	32	31	29	29	29	323
7056	MODEL 36	FALCON 50/900/ 50EX/900EX	TS	25	23	23	22	20	20	20	18	18	18	207
105081	MODEL 36	FALCON 7X	TS	10	20	24	28	29	29	30	30	30	30	260
103732	MODEL 36	DHC-8	TS	10	9	9	9	8	7	6	6	6	6	76
101644	MODEL 36	328JET/ENVOY 3	TS	12	14	12	10	12	14	12	12	12	12	122
105828	MODEL 36	G350	TS	16	18	21	20	22	20	19	20	20	20	196
105638	MODEL 36	G450	TS	8	29	31	32	31	31	30	30	30	30	282
101647	MODEL 36	ASTRA/SP/SPX/ G100	TS	5	0	0	0	0	0	0	0	0		5
105640	MODEL 36	G150	TS	7	14	15	17	16	17	16	15	15	15	147
104471	MODEL 36	G200	TS	20	23	25	25	23	24	24	24	25	25	238
7102	MODEL 36	APU/GPU SPARES (ALL)	TS	62	62	58	56	56	55	55	53	53	53	563
7041	MODEL 36	HAWKER 800/XP	TS	33	33	29	30	33	31	28	26	26	26	295
104214	MODEL 36	HORIZON	TS	25	24	25	22	24	28	25	28	26	26	253

(Continued)

—**cont'd**

Record Number	Engine	Application	Engine Type	Engine Units 06	07	08	09	10	11	12	13	14	15	Total 06–15
104215	MODEL 36	S-92	TS	8	12	10	10	12	12	10	12	11	11	108
7074	MODEL 85	C-130/L-100	TS	16	16	16	18	18	18	18	18	18	18	174
7082	MODEL 85	APU/GPU SPARES (ALL)	TS	50	50	50	50	50	50	50	50	50	50	500
103725	RE100	CITATION EXCEL	TS	38	31	28	28	28	26	26	26	26	26	283
105248	RE100	CITATION SOVEREIGN	TS	43	50	41	35	32	32	32	32	32	32	361
105981	RE100	G150	TS	16	23	17	17	17	17	18	19	18	18	180
103508	RE100	LEARJET 45	TS	33	33	33	33	36	36	36	35	35	35	345
104011	RE220	CRJ 700	TS	48	42	40	44	41	46	44	40	40	40	425
104639	RE220	CRJ 900	TS	17	17	17	18	22	26	26	26	26	26	221
105091	RE220	GLOBAL 5000	TS	18	27	27	27	27	25	25	25	25	25	251
103238	RE220	GLOBAL EXPRESS	TS	22	25	25	25	24	20	20	20	20	20	221
103239	RE220	G500/G550	TS	30	30	32	32	32	34	34	34	34	34	326
Honeywell				1478	1489	1484	1499	1462	1443	1392	1343	1334	1330	14,254
Kawasaki Heavy Industries														
Kawasaki Heavy Industries (Co-Product)														
105823	RE220	US-1AKAI	TS	0	0	2	3	5	5	0	0	0	0	15
Kawasaki Heavy Industries				0	0	2	3	5	5	0	0	0	0	15
LHTEC														
LHTEC (Consortium)														
105320	T800-4K	US-1AKAI	TS	0	0	3	5	5	0	0	0	0		13
LHTEC				0	0	3	5	5	0	0	0	0	0	13
Lucas Aerospace Ltd														
Lucas Aerospace Ltd														
7125	CT 2106 MK IV	AV/TAV-8B HARRIER II	TS	6	6	6	6	6	6	0	0	0	0	36
Lucas Aerospace Ltd				6	6	6	6	6	6	0	0	0	0	36
Microturbo Inc														
Microturbo Inc (Subsidiary)														
104992	HAWK 047/096 GTS	T-45A	TS	14	8	0	0	0	0	0	0	0	0	22
Microturbo Inc				14	8	0	0	0	0	0	0	0	0	22
Microturbo Ltd														
Microturbo Ltd (Subsidiary)														
104993	HAWK 047/096 GTS	HAWK 100	TS	27	18	19	22	20	18	20	22	16	16	198
104989	SAPHIR 10	HAWK 200	TS	4	4	0	0	6	4	5	4	0	0	27
Microturbo Ltd				31	22	19	22	26	22	25	26	16	16	225

—cont'd

Record Number	Engine	Application	Engine Type	Engine Units 06	07	08	09	10	11	12	13	14	15	Total 06–15
Microturbo SA														
Microturbo SA														
101643	RUBIS	RAFALE APU (ALL)	TS	18	20	20	25	27	27	27	22	20	20	226
7091	RUBIS	GCAT 200	TS	0	0	3	0	0	3	0	0	3		9
104990	SAPHIR 100	NH 90	TS	42	53	54	54	55	59	59	54	53	53	536
101637	SAPHIR 20	AS 332 MK2/EC 225 (CIV)	TS	9	6	6	7	6	5	4	4	4	4	55
101638	SAPHIR 20	AS 532 MK2/EC 725 (MIL)	TS	13	13	12	13	11	9	9	8	8	8	104
103752	SAPHIR 20	415	TS	4	4	4	4	3	0	0	0	0	0	19
104991	SAPHIR 4	FALCON 20 (MOD)	TS	2	0	0	0	0	0	0	0	0	0	2
105990	TGA 15	GROUND START CARTS	TS	0	2	0	2	0	2	0	0	2	0	8
7131	TGA 15	JAS 39A/B/C	TS	16	12	10	10	10	12	12	10	10	10	112
Microturbo SA				104	110	109	115	112	117	111	98	100	95	1,071
Mitsubishi Heavy Ind Ltd														
Mitsubishi Heavy Ind Ltd (Licensee)														
105838	TITAN T-62T	CH-47J	TS	3	3	2	0	0	0	0	0	0	0	8
102709	TITAN T-62T	SH-60J	TS	6	6	6	8	8	10	8	8	8	8	76
Mitsubishi Heavy Ind Ltd				9	9	8	8	8	10	8	8	8	8	84
Pratt & Whitney Canada														
Pratt & Whitney Canada														
7031	PW901A	747-400	TS	14	9	9	12	12	9	9	12	9	12	107
104811	PW980A	A380	TS	5	10	20	28	36	36	38	38	38	38	287
Pratt & Whitney Canada				19	19	29	40	48	45	47	50	47	50	394
Prvni Brnenska Strojirna														
Prvni Brnenska Strojirna (Licensee)														
103755	SAPHIR 5	L-159	TS	4	0	0	0	0	0	0	0	0	0	4
Prvni Brnenska Strojirna				4	0	0	0	0	0	0	0	0	0	4
Rolls Royce DE														
Rolls Royce DE														
7128	T 312	TORNADO	TS	8	0	0	6	0	0	6	0	0		20
Rolls Royce DE				8	0	0	6	0	0	6	0	0	0	20
Total				2411	2319	2269	2385	2373	2394	2339	2244	2203	2195	23,132

rec_type='PU'
ENGTYPE='TS'
[+] Additional Fields Printed
ENGTYPE as Engine, Type
**Source: Forecast International.*

Appendix 2B-4: Gas Turbines Powering Missiles/Drones/RPV/UAV*

Record Number	Engine	Application	Engine Type	Engine Units 06	07	08	09	10	11	12	13	14	15	Total 06–15
Chinese State Factories														
Chinese State Factories														
103641	UNSPECIFIED	C802	TJ	121	158	158	158	137	0	0	0	0	0	732
105140	UNSPECIFIED	CHINESE CRUISE MISSILE	TJ	0	0	0	1	12	13	15	16	24	24	105
103643	UNSPECIFIED	HY-4	TJ	32	27	0	0	0	0	0	0	0	0	59
Chinese State Factories				153	185	158	159	149	13	15	16	24	24	896
DENEL														
DENEL														
102841	UNKNOWN	HTD-1	TJ	17	22	23	23	23	23	21	21	23	17	213
DENEL				17	22	23	23	23	23	21	21	23	17	213
H A L, Engine Division														
H A L, Engine Division														
101291	PTAE-7	PTA-11	TJ	8	11	11	12	20	30	46	44	44	44	270
H A L, Engine Division				8	11	11	12	20	30	46	44	44	44	270
Hamilton Sundstrand														
Hamilton Sundstrand														
105257	TJ-120	NEW MALD	TJ	0	42	195	299	367	415	490	449	309	242	2,808
Hamilton Sundstrand				0	42	195	299	367	415	490	449	309	242	2,808
Honeywell														
Honeywell														
105171	TPE 331-10	MQ-9 PREDATOR B	TP	4	5	7	9	9	9	9	10	9	7	78
Honeywell				4	5	7	9	9	9	9	10	9	7	78
Manufacturer Unknown														
Manufacturer Unknown														
105085	UNSPECIFIED	S-100	TJ	0	2	53	80	123	127	131	153	165	165	999
102543	WP-11	CHANG HONG 1	TJ	21	19	14	8	0	0	0	0	0		62
102742	WP-6	CHANG KONG 1C	TJ	25	26	23	21	26	26	26	24	0	0	197
Manufacturer Unknown				46	47	90	109	149	153	157	177	165	165	1,258
Microturbo Inc														
Microturbo Inc (Subsidiary)														
105120	TRI 60	BQM-167A SKEETER	TJ	37	41	41	41	43	43	41	40	37	23	387
105173	TRI 60-30	RBS15M MK 2[†]	TJ	0	0	0	0	0	0	0	0	0	0	0

—cont'd

Record Number	Engine	Application	Engine Type	Engine Units 06	07	08	09	10	11	12	13	14	15	Total 06–15
103126	TRI 60-30	RBS15M MK 3	TJ	49	52	53	51	51	50	45	42	41	41	475
5024	TRI 60-5	MQM-107D	TJ	6	0	6	0	0	0	0	0	0	0	12
Microturbo Inc				92	93	100	92	94	93	86	82	78	64	874
Microturbo Ltd														
Microturbo Ltd (Subsidiary)														
102776	TRI 60	STORM SHADOW	TJ	92	94	91	84	77	67	53	51	41	0	650
Microturbo Ltd				92	94	91	84	77	67	53	51	41	0	650
Microturbo SA														
Microturbo SA														
101246	TRI 40	NSM	TJ	21	29	33	37	37	40	41	41	41	31	351
103111	TRI 40	MM.40 BLOCK 3	TJ	12	16	27	48	60	61	61	61	61	61	468
104350	TRI 60	FRENCH AERIAL TARGET	TJ	18	13	14	11	15	17	18	16	15	12	149
103562	TRI 60-30	SCALP NAVALE	TJ	0	0	0	1	16	32	46	53	53	53	254
103561	TRI 60-30	SCALP-EG	TJ	59	59	59	59	59	59	59	59	58	52	582
5197	TRS 18-1	MIRACH 100	TJ	31	13	14	12	15	18	15	0	0	0	118
Microturbo SA				141	130	147	168	202	227	240	230	228	209	1,922
Mitsubishi Heavy Ind Ltd														
Mitsubishi Heavy Ind Ltd														
103056	NOT SELECTED	XJ/AQM-5	TJ	0	0	1	11	21	0	0	0	0	0	33
101501	TJM2	TYPE 90 SSM-1B	TJ	19	0	0	0	0	0	0	0	0	0	19
5211	TJM2	TYPE 93 ASM-2	TJ	28	0	0	0	0	0	0	0	0	0	28
103125	TJM2	XSSM-2	TJ	2	14	26	35	42	43	44	45	42	42	335
5095	TJM3	J/AQM-1	TJ	22	21	20	17	0	0	0	0	0	0	80
101432	TJM4	J/AQM-3	TJ	0	0	0	0	0	0	0	0	0	0	0
105745	TJM4	XJ/AQM-6	TJ	0	0	0	5	6	10	10	10	9	0	50
Mitsubishi Heavy Ind Ltd				71	35	47	68	69	53	54	55	51	42	545
Nationalist Chinese Arsenals														
Nationalist Chinese Arsenals														
104399	UNSPECIFIED	HSIUNG FENG IV	TJ	0	0	0	0	2	7	10	15	18	15	67
Nationalist Chinese Arsenals (Licensee)														
104398	ARBIZON IV (TAIWAN)	HSIUNG FENG IIA	TJ	36	35	29	20	17	0	0	0	0	0	137
Nationalist Chinese Arsenals				36	35	29	20	19	7	10	15	18	15	204

(Continued)

—cont'd

Record Number	Engine	Application	Engine Type	Engine Units 06	07	08	09	10	11	12	13	14	15	Total 06–15
Not Selected														
Not Selected														
105059	NOT SELECTED	UCAV	TF	0	0	0	0	0	1	2	2	3	1	9
105103	NOT SELECTED	UCAV-N	TF	0	0	0	0	0	0	0	0	1	0	1
5202	NOT SELECTED	MIRACH NG	TJ	0	0	0	3	14	22	24	21	21	21	126
103377	NOT SELECTED	ISRAEL ADV. ASH	TJ	19	24	40	60	70	70	70	70	70	63	556
103552	NOT SELECTED	JASSM BLOCK 2	TJ	0	49	147	220	331	353	367	366	355	258	2,446
105612	NOT SELECTED	LAM	TF	0	0	0	0	2	51	85	128	166	175	607
104025	NOT SELECTED	CANADIAN NAVAL UAV	TS	0	1	4	7	5	3	0	0	0	0	20
104387	NOT SELECTED	CL-289 FOLLOW-ON	TJ	0	1	18	22	25	28	32	31	29	0	186
103020	NOT SELECTED	GERMAN NAVAL UAV	TS	0	0	0	0	4	8	0	0	0	0	12
103128	NOT SELECTED	ITALIAN ADV. ASH	TJ	0	0	0	0	0	1	18	43	66	65	193
105528	NOT SELECTED	ITALIAN ENDURANCE UAV	TJ	0	0	3	5	5	5	5	5	0	0	28
103060	NOT SELECTED	LRCM	TJ						0	0	4	33	65	102
104330	NOT SELECTED	TARGET 21	TJ	3	66	103	119	126	134	134	134	134	129	1,082
104568	NOT SELECTED	US MR-E	TF	6	10	11	11	9	4	0	0	0	0	51
104775	UNSPECIFIED	ERCM	TJ	0	0	0	0	0	0	0	0	0	0	0
Not Selected (Consortium)														
105789	NOT SELECTED	DARKSTAR FOLLOW-ON	TF	0	0	0	0	0	2	2	4	4	4	16
Not selected				28	151	326	447	591	682	739	808	882	781	5,435
Pratt & Whitney Canada														
Pratt & Whitney Canada														
105584	PW200/55	EAGLE EYE	TS	8	5	4	5	6	8	9	9	8	5	67
Pratt & Whitney Canada				8	5	4	5	6	8	9	9	8	5	67
Rolls Royce North America														
Rolls Royce North America														
104393	250-C20W	MQ-8 FIRE SCOUT	TS	1	3	3	1	4	7	8	8	8	6	49
105945	AE 3007	RQ-4B GLOBAL HAWK	TF	8	11	12	11	12	12	9	6	9	11	101
Rolls Royce North America				9	14	15	12	16	19	17	14	17	17	150
Teledyne CAE														
Teledyne CAE														
103554	J402-CA-100-2/B9	JASSM BLOCK 1	TJ	258	262	216	163	150	144	143	133	102	77	1,648
102819	J402-CA-400	HARPOON BLOCK II	TF	81	77	69	70	81	90	87	75	69	50	749

—cont'd

Record Number	Engine	Application	Engine Type	Engine Units										Total
				06	07	08	09	10	11	12	13	14	15	06–15
5030	J402-CA-702	MQM-107D/E	TJ	20	0	0	0	0	0	0	0	0	0	20
Teledyne CAE				359	339	285	233	231	234	230	208	171	127	2,417
Turbomeca														
Turbomeca														
101552	ARBIZON IIID	MILAS	TJ	31	26	20	0	0	0	0	0	0	0	77
104888	ARBIZON IIID	OTOMAT MK.4	TJ	49	53	55	56	57	54	59	61	48		492
Turbomeca				80	79	75	56	57	54	59	61	48	0	569
Unspecified														
Unspecified														
103661	NOT SELECTED	TU-300	TJ	0	0	2	13	18	18	20	20	20	16	127
103660	TR-3–117	REIS	TJ	11	8	10	0	0	0	0	0	0		29
105331	UNSPECIFIED	DT-25/DT-35	TJ	25	27	30	38	39	49	55	55	55	55	428
105334	UNSPECIFIED	NISHAN-MK2	TJ	4	3	2	2	0	0	0	0	0	0	11
104460	UNSPECIFIED	SAEQEH	TJ	29	28	24	21	16	0	0	0	0	0	118
105856	UNSPECIFIED	PAKISTAN CRUISE MISSILE	TJ	0	0	15	19	25	27	29	28	30	31	204
104084	UNSPECIFIED	AG-1	TJ	32	32	32	32	30	0	0	0	0		158
104507	UNSPECIFIED	ASAT/FALCONET FOLLOW-ON	TJ	15	21	23	27	26	25	17	19	20	14	207
103055	UNSPECIFIED	UK SSGW	TJ	0	47	51	60	61	64	63	62	60	0	468
104589	UNSPECIFIED	SS.N-27 KLUB	TF	17	26	26	20	16	17	27	32	32	28	241
104774	UNSPECIFIED	DELILAH-GL	TJ	14	14	14	17	17	18	20	21	20	19	174
105930	UNSPECIFIED	AS-18 KAZOO	TF	12	14	15	15	12	11	15	15	12	10	131
104588	UNSPECIFIED	AS-20 KAYAK	TF	37	37	39	64	65	80	80	77	42	29	550
Unspecified				196	257	283	328	325	309	326	329	291	202	2,846
Williams International														
Williams International														
104083	F122	KEPD 150	TJ	0	6	29	48	50	43	19	0	0	0	195
103810	F122	KEPD 350	TJ	106	108	108	89	46	42	31	31	31	23	615
103036	F415	TACTICAL TOMAHAWK	TF	452	452	456	402	350	304	219	198	161	118	3,112
103564	WR24-8	BQM-74C	TJ	10	9	0	0	0	0	0	0	0	0	19
101718	WR24-8	BQM-74E	TJ	60	52	0	0	0	0	0	0	0	0	112
105714	WR24-8	BQM-74F	TJ	61	64	94	118	125	133	133	132	125	123	1,108
Williams International				689	691	687	657	571	522	402	361	317	264	5,161
Total				2029	2235	2573	2781	2975	2918	2963	2940	2724	2225	26,363

rec_type='MR'
ENGTYPE='TF'.OR.ENGTYPE='ATF'.OR.ENGTYPE='TJ'.OR.ENGTYPE='ATJ'.OR.ENGTYPE='TP'.OR.ENGTYPE='TS'.OR.ENGTYPE='PF'. OR.ENGTYPE='VC'
ENGTYPE as Engine, Type
Source: Forecast International.
†*Additional Fields Printed.*

Appendix 2B-5: Gas Turbines Powering On-Land/Surface Vehicles*

			Engine Units										Total
Record Number	**Engine**	**Application**	**06**	**07**	**08**	**09**	**10**	**11**	**12**	**13**	**14**	**15**	**06–15**
Honeywell													
Honeywell 101439	AGT 1500 (ALL)	M1 SERIES (ALL)	54	72	72	20	0	0	0	0	0	0	218
Honeywell			54	72	72	20	0	0	0	0	0	0	218
Total			54	72	72	20	0	0	0	0	0	0	218

rec_type='V' ENGTYPE='TS'

Appendix 2B-6: Gas Turbines Powering Microturbines for Electrical Generation*

			Engine Units										Total
Record Number	**Engine**	**Application**	**06**	**07**	**08**	**09**	**10**	**11**	**12**	**13**	**14**	**15**	**06–15**
Capstone Turbine Corp													
Capstone Turbine Corp 104917	C30	GENERATION: MICROS/MINIS	425	450	530	540	580	590	590	590	590	590	5,475
104924	C65	GENERATION: MICROS/MINIS	70	80	100	120	130	130	130	130	130	130	1,150
Capstone Turbine Corp			495	530	630	660	710	720	720	720	720	720	6,625
Elliott Energy Systems Inc.													
Elliott Energy Systems Inc. 105184	TA-100	GENERATION: MICROS/MINIS	90	125	185	235	265	275	275	280	280	280	2,290
Elliott Energy Systems Inc.			90	125	185	235	265	275	275	280	280	280	2,290
Ingersoll-Rand Energy Systems													
Ingersoll-Rand Energy Systems 105795	250 SERIES (MT250)	GENERATION: MICROS/MINIS	10	15	20	25	30	30	30	35	35	35	265
104919	70L SERIES	GENERATION: MICROS/MINIS	60	75	100	110	125	130	130	140	140	140	1,150
Ingersoll-Rand Energy Systems			70	90	120	135	155	160	160	175	175	175	1,415
Turbec S.P.A.													
Turbec S.P.A. 104906	T100 CHP SYSTEM	GENERATION: MICROS/MINIS	45	55	65	70	70	70	70	75	75	75	670
Turbec S.P.A.			45	55	65	70	70	70	70	75	75	75	670
Total			700	8001	000	100	1200	1225	1225	1250	1250	1250	11,000

rec_type='IM'
ENGTYPE='TS'
SOURCE='GA'
Source: Forecast International.

Chapter 3

Gas Turbine Configurations and Heat Cycles

"Every child is an artist. The problem is how to remain an artist once he grows up."

—Pablo Picasso

Chapter Outline

Gas Turbines

When defining heat cycles, one needs to specify the reference engine stations or the positions used for the pressure and temperature measurements that will detail the heat cycle. One also needs to define the flow of that cycle in its simplest terms: In other words, which fluid (air, fuel, hot gases, injected fluid, heat exchanger flows) is flowing where and how. This information helps define the gas turbine's configuration and efficiency.

GAS TURBINE CONFIGURATIONS

The following description* of main configurations for gas turbine engines does not mention "land, sea, or air." Some of them, for instance "conventional turbojet," obviously refer to an aeroengine application. However, a "simple-cycle, free-power turbine engine" can be land (mechanical drive), sea (propulsion), or air (turboprop, helicopter) based.

Figure 3−1(a) shows a conventional single-spool turbojet above the center line and one with the addition of an afterburner, convergent-divergent (con-di) intake, and con-di nozzle.

Ambient air passes from free stream to the flight intake leading edge. The air accelerates from the free stream if the engine is static, whereas at high flight Mach numbers, it diffuses from the free stream, ram conditions. Usually, it then diffuses in the flight intake before passing through the engine intake to the compressor face, resulting in a small loss in total pressure.

The compressor then increases both the pressure and temperature of the gas. Work input is required to achieve the pressure ratio; the associated temperature rise depends on efficiency level. Depending on the complexity, the turbojet compressor pressure ratio ranges from 4:1 up to 25:1.

The compressor exit diffuser passes the air to the combustor. Here, fuel is injected and burned to raise exit gas temperature to between approximately 1200 K and 2000 K, depending on the engine technology level. Both the diffuser and combustor impose a small total pressure loss.

The hot, high-pressure gas then is expanded through the turbine, where work is extracted to produce shaft power; both temperature and pressure are reduced. The shaft power is required to drive the compressor and any engine and "customer" auxiliaries and to overcome engine mechanical losses such as disc windage and bearing friction. The turbine nozzle guide vanes and blades often are cooled to ensure acceptable metal temperatures at elevated gas temperatures. This utilizes relatively cool air from the compression system, which bypasses the combustor via *air system* flow paths that feed highly complex internal cooling passages within the vanes and blades.

On leaving the turbine, the gas still is at a pressure typically at least twice that of ambient. This results from the higher inlet temperature to the turbine and the fundamental form of the temperature−entropy (T−S) diagram.

Downstream of the turbine, the gas diffuses in the jet pipe. This short duct transforms the flow path from annular to a full circle at entry to the propelling nozzle. The jet pipe imposes a small total pressure loss. The propelling nozzle is a convergent duct that accelerates the

* Source: [3-1] Courtesy of Rolls Royce, *Gas Turbine Performance*, Walsh and Fletcher, Blackwell Science, 1998. Adapted with permission.

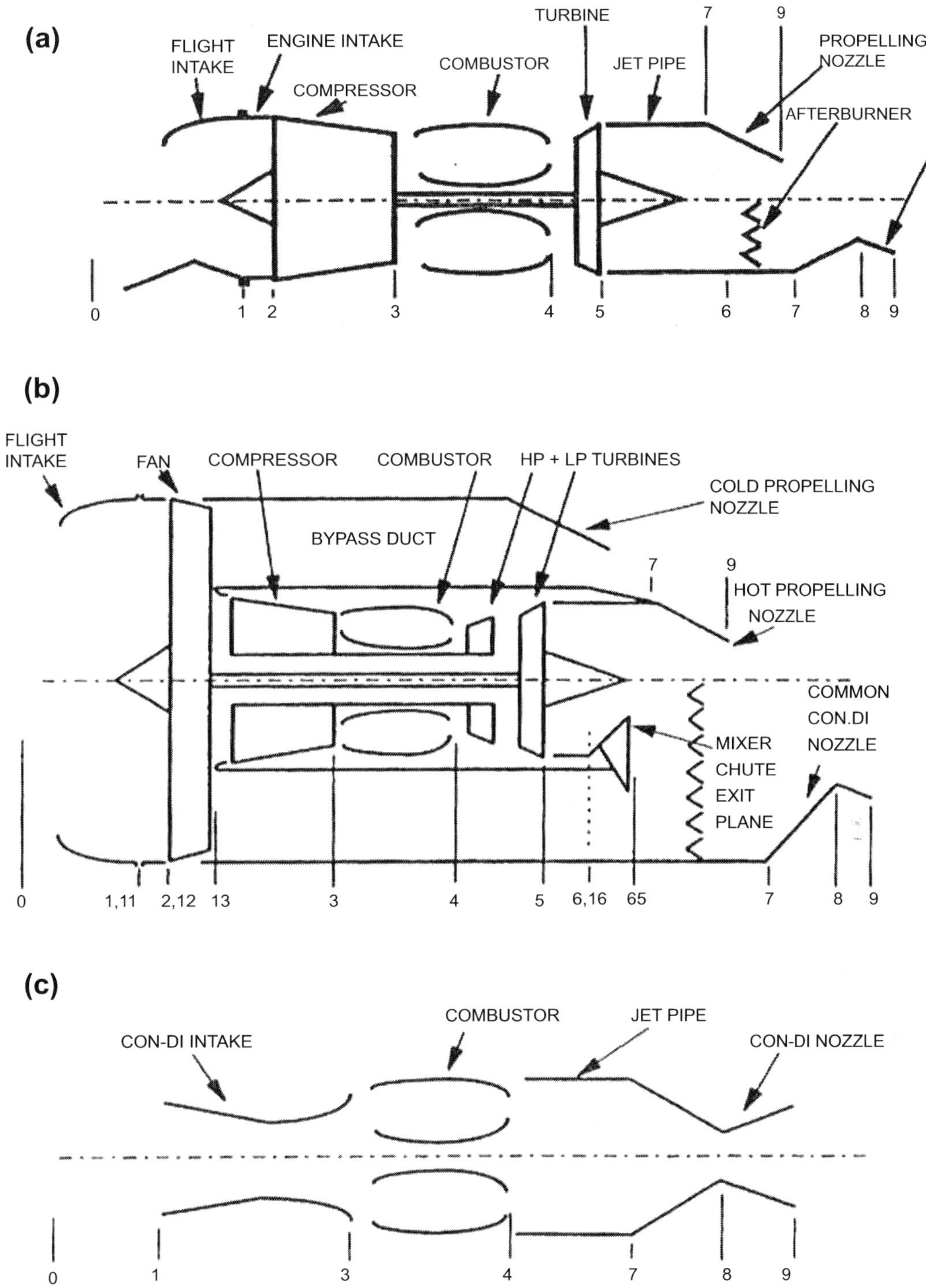

FIGURE 3–1 Thrust engine configurations and station numbering: (a) Conventional turbojet and afterburning turbojet with con-di intake and nozzle. (b) Separate jets, turbofan and mixed, afterburning turbofan with con-di nozzle. (c) Ramjet with con-di intake and nozzle [3-1].

flow to provide the high-velocity jet to create the thrust. If the available expansion ratio is less than the choking value, the static pressure in the exit plane of the nozzle will be ambient. If it is greater than the choking value, the Mach number at the nozzle will be unity (i.e., sonic conditions), the static pressure will be greater than ambient, and shock waves will occur downstream. In the latter instance, the higher static pressure at nozzle exit plane relative to the intake creates thrust additional to that of the jet momentum.

In a two-spool engine there are both low-pressure (LP) and high-pressure (HP) compressors driven by LP and HP turbines. Each spool has a different rotational speed, with the LP shaft inside of and concentric to that of the HP spool. If the spool gas paths are at different radii, this arrangement necessitates short intercompressor and interturbine ducts, which incur small total pressure losses.

Turbojet with Afterburner and Convergent-Divergent Nozzle (Figure 3–1[a])

For high-flight Mach number applications, an afterburner is often employed, which offers higher thrust from the same turbomachinery. This, also called reheat, involves burning fuel in an additional combustor downstream of the jet pipe. The greatly increased exhaust temperature provides a far higher jet velocity, and the ratios of engine thrust to weight and thrust to unit frontal area are greatly increased.

To enable the jet efflux to be supersonic, and hence achieve the full benefit of the afterburner, a convergent-divergent nozzle may be employed. A nozzle downstream of an afterburner must be of variable area to avoid compressor surge problems due to the increased back pressure on the engine when the afterburner is lit. Usually, for engines utilized in this high flight Mach number regime, a convergent-divergent intake also is employed. This enables efficient diffusion of the ram air from supersonic flight Mach numbers to subsonic flow to suit the compressor ci fan. This is achieved via a series of oblique shock waves, which impose a lower total pressure loss than a normal shock wave.

Separate Jets Turbofan (Figure 3–1[b])

A schematic diagram of a two-spool separate jets turbofan is presented above the center line in Figure 3–1(b). Here, the first compressor, termed a fan, supplies flow to a bypass as well as a core stream. The core stream is akin to a turbojet and provides the hot thrust; however, the core turbines also provide power to compress the fan bypass stream.

The bypass stream bypasses the core components via the bypass duct, incurring a small total pressure loss. It then enters the cold nozzle. The total thrust is the sum of those from both the hot and cold nozzles. The purpose of the bypass stream is to generate additional thrust with a high mass flow rate, but low jet velocity, which improves specific fuel consumption (SFC) relative to a pure turbojet. However, this results in lower ratios of engine thrust to frontal area and weight.

Some turbofans have three spools, with an intermediate pressure (IP) spool as well as the HP and LP spools.

Mixed Turbofan with Afterburner (Figure 3–1[b])

This configuration is shown below the center line in Figure 3–1(b). Here, the two streams are combined in a mixer upstream of a common jet pipe with an afterburner and convergent-divergent nozzle to provide high jet velocities for supersonic flight. It often also is beneficial to mix the two streams for turbofans without afterburners.

Ramjet (Figure 3–1[c])

The ramjet is the simplest thrust engine configuration, employing no rotating turbomachinery. The ram air is diffused in a convergent-divergent intake then passed directly to the combustor. It is accelerated to supersonic jet velocity using a convergent-divergent nozzle. The ramjet is practical only for high supersonic flight regimes.

Simple-Cycle Single-Spool Shaft-Power Engine (Figure 3–2[a])

This engine configuration appears similar to a turbojet, apart from the intake and the exhaust. The main difference is that all the available pressure at entry to the turbine is expanded to ambient to produce shaft power, apart from a small total pressure loss in the exhaust. After diffusion in the exhaust duct, the gas exit velocity is negligible. This results in turbine power substantially greater than that required to drive the compressor; hence, excess power drives the load, such as a propeller (turboprop) or electrical generator (turboshaft). The gas temperature at the exhaust exit plane is typically 250–350°C hotter than ambient, which represents considerable waste heat for an industrial application.

The style of the intake and exhaust varies greatly depending on the application, although fundamentally, the exhaust normally is a diverging, diffusing system as opposed to the jet pipe and nozzle employed by the turbojet for flow acceleration.

The term simple cycle is used to distinguish this configuration from the complex cycles described later, which utilize additional components such as heat exchangers or steam boilers.

Simple-Cycle Free-Power Turbine Engine (Figure 3–2[b])

Here, the load is driven by a free-power turbine separate from that driving the engine compressor. This has significant impact on off-design performance, allowing far greater flexibility in output speed at power.

Gas Generator

The term gas generator describes either the compressor and turbine combination providing the hot, high-pressure gas that enters the jet pipe and propelling nozzle for a turbojet or the free-power turbine for a turboshaft. It is

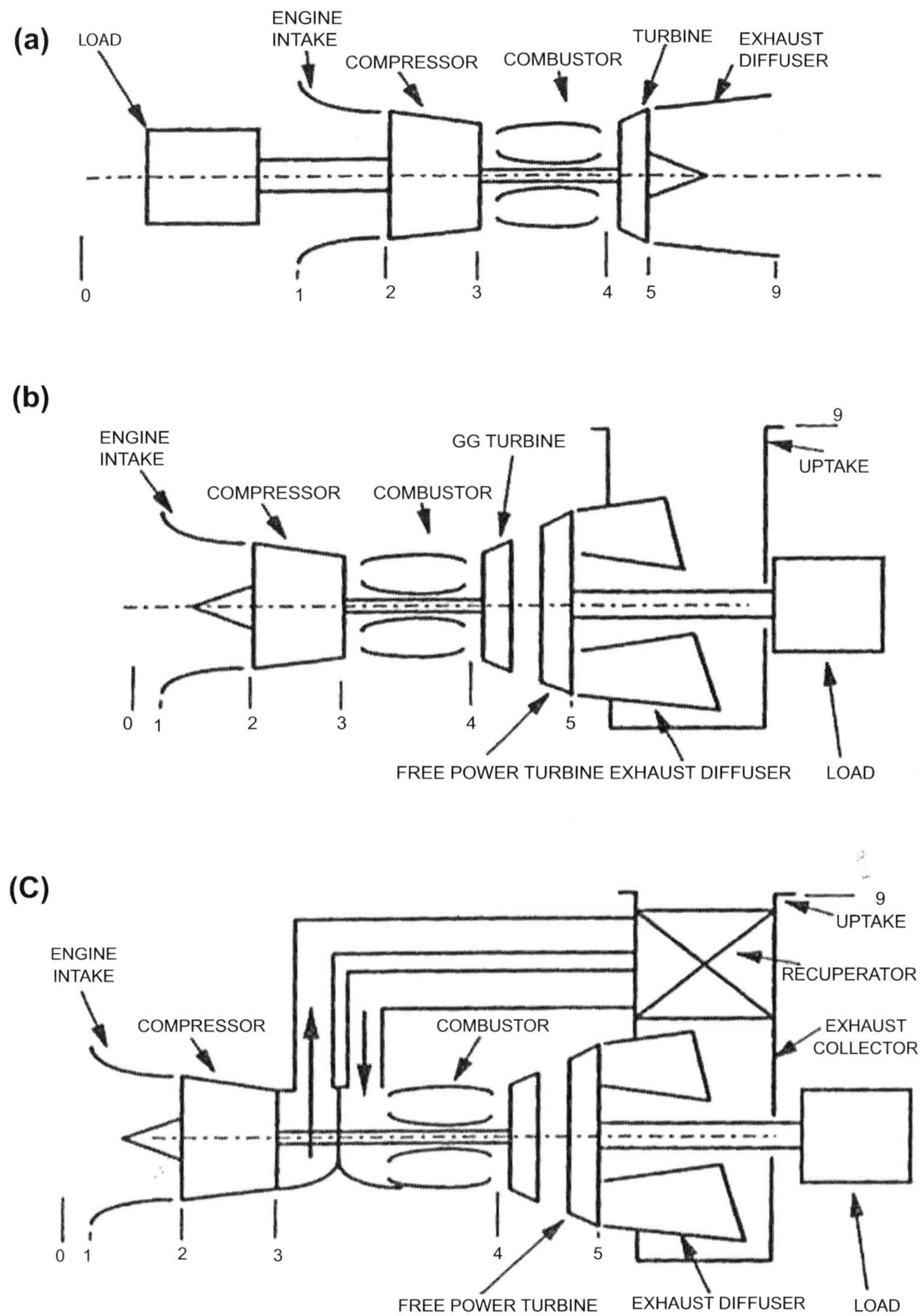

FIGURE 3–2 Shaft power engine configurations and station numbering: (a) Single-spool shaft-power engine—shown with cold end drive. (b) Free-power turbine engine—shown with hot end drive. (c) Recuperated free-power turbine engine—hot end drive shown [3-1].

common practice to use a given gas generator design for both a turbojet (or turbofan) and an aeroderivative free-power turbine engine. Here, the jet pipe and propelling nozzle are replaced by a power turbine and exhaust system; for turbofans, the fan and bypass duct are removed.

Recuperated Engine (Figure 3–2[c])

Here, some of the heat that would be lost in the exhaust of a simple cycle is returned to the engine. The heat exchanger used is either a recuperator or regenerator, depending on its configuration.

The compressor delivery air is ducted to the air side of the heat exchanger, where it receives heat from the exhaust gas passing through the gas side. The heated air then is ducted back to the combustor, where now less fuel is required to achieve the same turbine entry temperature, which improves specific fuel consumption. Pressure losses occur in the heat exchanger air and gas sides and the transfer ducts.

Intercooled Shaft-Power Engine (Figure 3—3[a])

Here, heat is extracted by an intercooler between the first and second compressors. As might be expected, rejecting heat normally worsens SFC, since more fuel must be burned to raise cooler compressor delivery air to any given turbine entry temperature. However, intercooling improves engine power output and, potentially, even SFC at high-pressure ratios via reduced power absorption in the second compressor. This is due to the lower inlet temperature reducing the work required for a given pressure ratio.

The intercooler rejects heat to an external medium such as seawater. The air side of the intercooler and any ducting impose total pressure losses.

Intercooled Recuperator Shaft-Power Engine (Figure 3—3[b])

Here, both an intercooler and recuperator are employed. The increase in power from intercooling is accompanied by an SFC improvement, as the heat extraction also results in increased heat recovery in the recuperator, due to the lower compressor delivery temperature.

Closed Cycle (Figure 3—3[b])

All the engine configurations just described are open cycle, in that air is drawn from the atmosphere and passes through the engine only once. In a closed cycle configuration, the working fluid is continuously recirculated. It may be air or another gas, such as helium.

Usually, the gas turbine is of an intercooled recuperated configuration, as shown in Figure 3—3(b). However, the combustor is replaced by a heat exchanger, as fuel cannot be burned directly. The heat source for the cycle may be a separate combustor burning normally unsuitable fuels, such as coal, or a nuclear reactor.

On leaving the recuperator, the working fluid must pass through a precooler, where heat is ejected to an external medium such as seawater to return it to the fixed inlet temperature, usually between 15°C and 30°C. The pressure at inlet to the gas turbine is maintained against leakage from the system by an auxiliary compressor supplying a large storage tank, called an accumulator. The high density of the working fluid at engine entry enables very high power output for a given size of plant, which is the main benefit of the closed cycle. Pressure at inlet to the gas turbine typically is around 20 times atmospheric. In addition, varying the pressure level allows power regulation without changing SFC.

Combined Cycle (Figure 3—4[a])

Figure 3—4(a) shows the simplest combined-cycle configuration. The gas turbine otherwise is of simple-cycle configuration, but with a significant portion of the waste heat recovered in an HRSG (heat recovery steam generator). This is a heat exchanger with the gas turbine exhaust on the hot side and pumped high-pressure water, which forms steam, on the cold side. The first part of the HRSG is the economizer, where the water is heated at constant pressure until it reaches its saturation temperature then vaporizes. Once the steam is fully vaporized, its temperature is increased further in the superheater.

The high-pressure, high-temperature steam is then expanded across a steam turbine, which provides up to an extra 45% power in addition to that from the gas turbine. On leaving the steam turbine, the steam wetness fraction typically is 10%. The rest of the steam is then condensed in one of several possible ways. The most common method uses cooling towers, where heat is exchanged to cold water, usually pumped from a local source, such as a river. When all the steam is condensed, the water passes back to the pumps, ready to be circulated again. Hence, the steam plant is also a "closed cycle."

Figure 3—4(a) presents a single-pressure steam cycle configuration. The most complex form of steam cycle used is the triple-pressure reheat, where steam expands through three turbines in series. In between successive turbines, it is returned to the HRSG and the temperature is raised again, usually to the same level as at entry to the first turbine. This cycle has the highest efficiency and specific power.

In the combined-cycle plant, the gas turbine often is referred to as the topping cycle, being the hotter one, and the steam plant as the bottoming cycle.

Combined Heat and Power (Figure 3—4[b])

Several forms of combined heat and power, or CHP (cogeneration), plants are described next, in order of increasing complexity.

In the simplest arrangements, the gas turbine waste heat is used directly in an industrial process, such as for drying in a paper mill or cement works.

Adding an HRSG downstream of the gas turbine allows conversion of the waste heat to steam, giving greater flexibility in the process for which it may be used, such as chemical manufacture or space heating in a hospital or factory.

Finally, Figure 3—4(b) shows the most complex CHP configuration, which employs supplementary firing. Here,

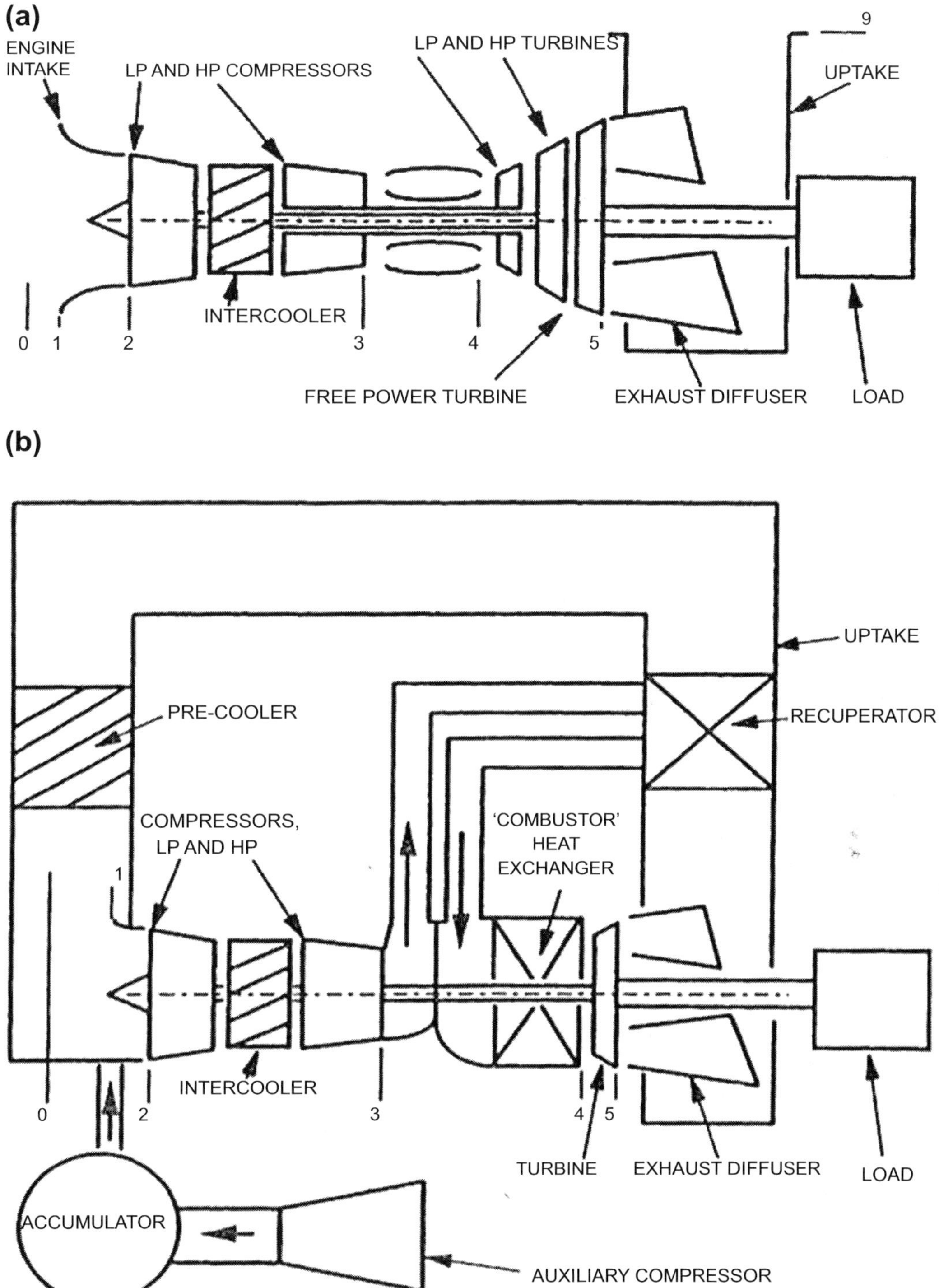

FIGURE 3–3 Intercooled engine configurations and station numbering: (a) intercooled free power turbine engine—hot end drive shown, (b) closed-cycle, single-spool, intercooled, recuperated shaft-power engine [3-1].

the simple-cycle gas turbine waste heat again is used to raise steam in an HRSG, which then passes to a boiler, where fuel is burned in the vitiated air to raise additional steam. The boiler provides flexibility in the ratio of heat to electrical power. Once the steam has lost all its useful heat, it passes to a condenser and pumps for recirculation.

Aeroderivative and Heavyweight Gas Turbines

Outside aero applications, gas turbines for producing shaft power fall into two main categories: aeroderivative and heavyweight (or industrial). As implied, the former are

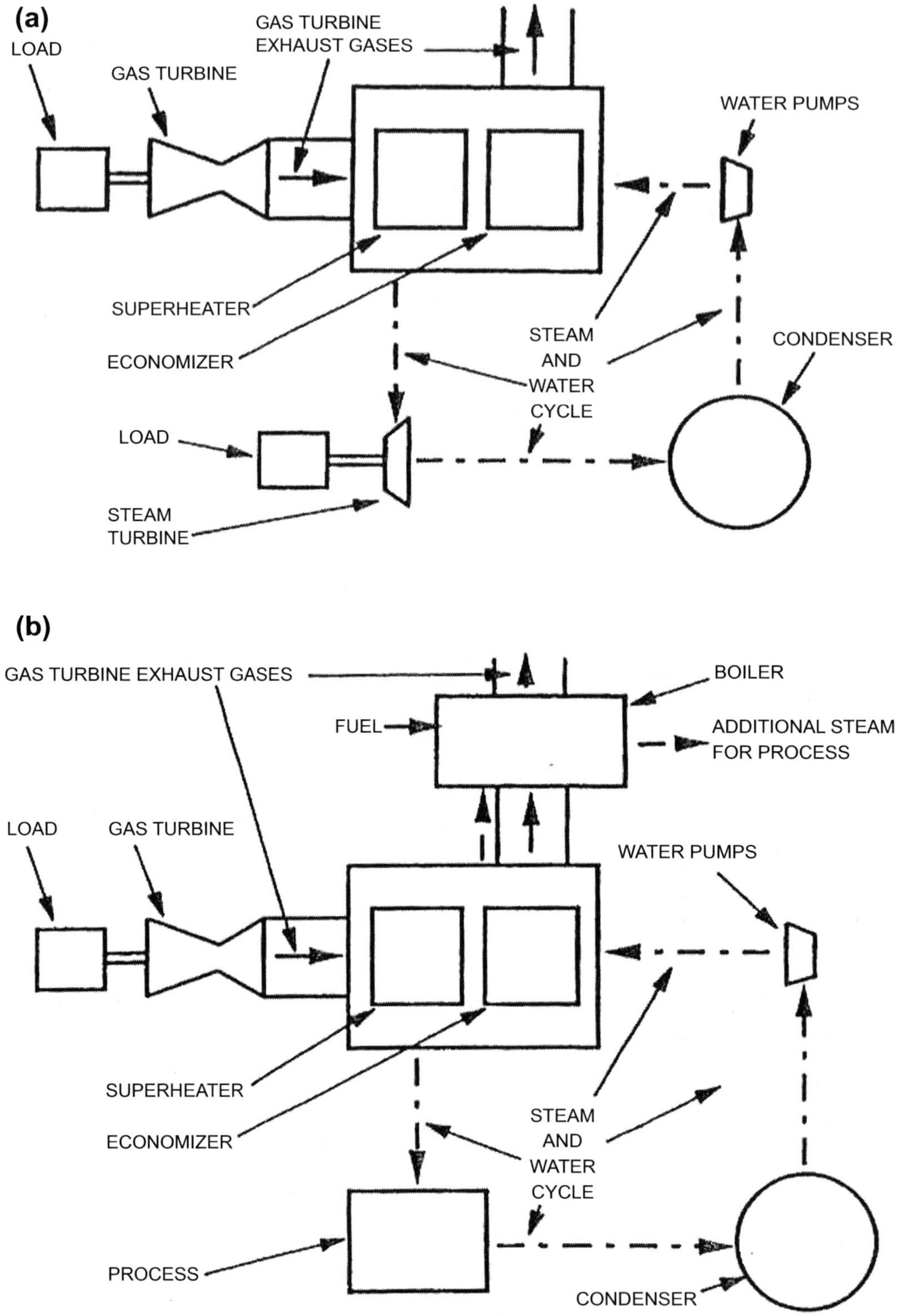

FIGURE 3–4 Combined-cycle configurations: (a). single-pressure combined cycle, (b) combined heat and power (CHP) with supplementary firing [3-1].

direct adaptations of aeroengines, with many common parts. The latter are designed with emphasis on low cost rather than low weight and, hence, may employ such features as solid rotors and thick casings.

GAS TURBINE CYCLES: SUMMARIZED THEORY AND ECONOMICS*,**

As we saw in the first chapter, the basic gas turbine cycle consists of four main processes. Optimizing that cycle frequently involves reducing fuel consumption. This reduction requires changes in the basic cycle that allow the same fuel to provide more power.

The power generation industry is the largest industry sector in the world and has a higher growth rate than any other industry. So the following pages cover cycle variations to the basic gas turbine cycle mainly for power generation applications. The variations in other industrial sectors, however, use much the same logic, and the technology may be essentially similar. For instance, in land-based power generation, the gas turbine's waste gases are used to generate steam that then drives a steam turbine. This configuration is called a *combined cycle*. Another variation for this cycle is when the gas turbine's exhaust gases (which are still rich in oxygen) have fuel added to their mix and the overall mixture is ignited. This is exactly the principle behind the aircraft engine afterburner.

Similarly, "active clearance controls" (ACC), a method for increasing the cooling to hot turbine blades, thus allowing the turbine to increase fuel flow (and therefore final delivered power), is used both in land-based and aircraft engine technology. Original engine manufacturers (OEMs) sometimes use the same terms, such as ACC, to describe systems that may differ considerably.

With ACC in land-based gas turbines in the combined cycle mode, the OEM has the opportunity to use steam as the cooling medium. Steam provides more cooling for a given application than air. Further, cooling steam can be injected back into the gas turbine for a boost in overall power generated due to steam injection.

In aircraft engine applications, however, normally no steam is available. The cooling medium is air.

Aircraft engines, however, may have larger fans than are found on land-based engines, allowing them to "swallow" more air, use more fuel to combine with it, and therefore increase power delivered.

So, as mentioned previously, we now look at basic cycle modifications for land-based gas turbines. Typically, these modifications occur more in power generation applications, although they also serve mechanical drive and marine (including offshore) applications. Ultimately, the decision to use any of these cycle modifications depends on weight allowance (much lower in aircraft engine and offshore platform applications), space available, and whether steam is available.

The success of these cycle modifications frequently is governed by the fuel used in the application. Gas turbine designers get increasingly better at burning "low-Btu" fuels and waste fuels (such as paper liquor and steel mill flue gas) of all kinds. Fuels are discussed in further detail in Chapter 7.

Additional cycle modifications tailored to specific applications are discussed at various points throughout this book, for instance, Chapter 10 on performance optimization, so the material in this chapter merely lays the groundwork.

As we saw earlier, the primary applications of gas turbines are:

- Aircraft engines—large turbo fan (ramjet), helicopter (power via transmission to rotor)
- Direct drive, land-based—power generation
- Mechanical drive—driving compressors, pumps, blowers (generally via gearbox)
- Marine service—power generation on board or for propulsion

In land and marine applications, the main advantages of gas turbine plants versus steam turbine plants include:

- Small in size, mass, and purchase cost per unit of power
- Quick installation; delivery time may be better
- Quick starting (start to full speed is ~10 sec); can be started remotely
- Smoother running
- Capacity factor (running time at full power) of 96–98%
- Can run on a wide variety of fuels, including gasified coal, flue gas from steel mills, liquor from paper waste, biomass, residual
- Environmental considerations and restrictions can be more economical to manage (consider precipitators and flue gas desulfurization)

Power Generation Gas Turbine, Simple and Combined Cycles

Figures 3–5, 3–6, and 3–7 describe the gas turbine simple cycle. The gas turbine simple cycle consists of the following:

- Air intake is at atmospheric pressure.
- The compressor pressure ratio (PR) typically is 15–25 for most land-based gas turbines (higher in the case of some aeroengines); the temperature increases to typically 750–850°F (400–465°C).

* Source: [3-2] Course notes, Claire M. Soares. "Basic Operations and Theory of Steam and Gas Turbines, Cogeneration and Combined Cycle Plants," 2005.

** See Appendix 3A for ST power plant theory as applicable to CC and STs operating "solo", in completion with GTs or GT CCs.

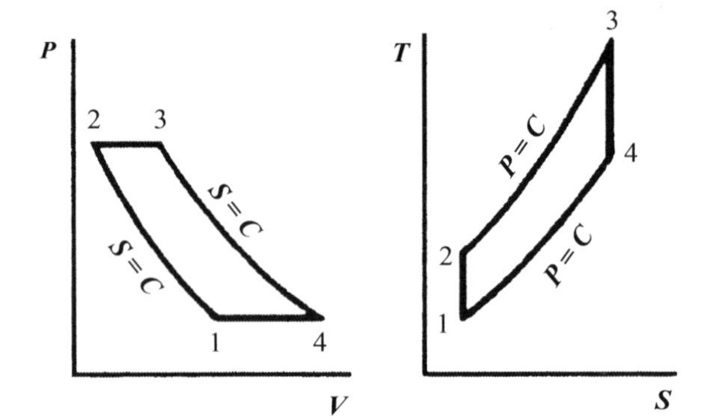

FIGURE 3–5 P-v and T-s diagrams of an ideal Brayton cycle [3-3].

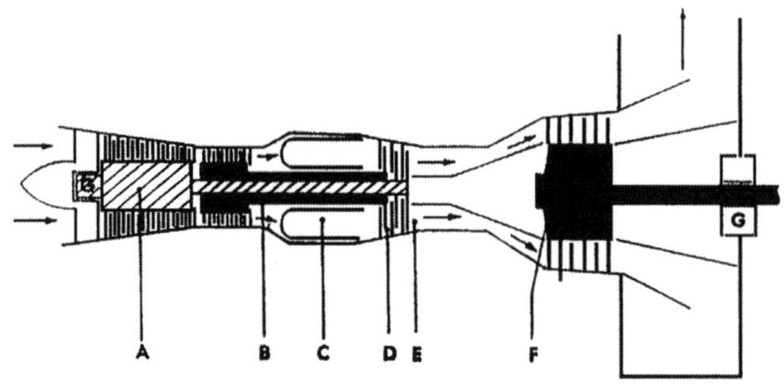

FIGURE 3–7 Dual-shaft gas turbine. A: LP compressor; B: HP compressor; C: combustors; D: HP compressor turbine; E: LP compressor turbine; F: power turbine; G: bearings [3-3].

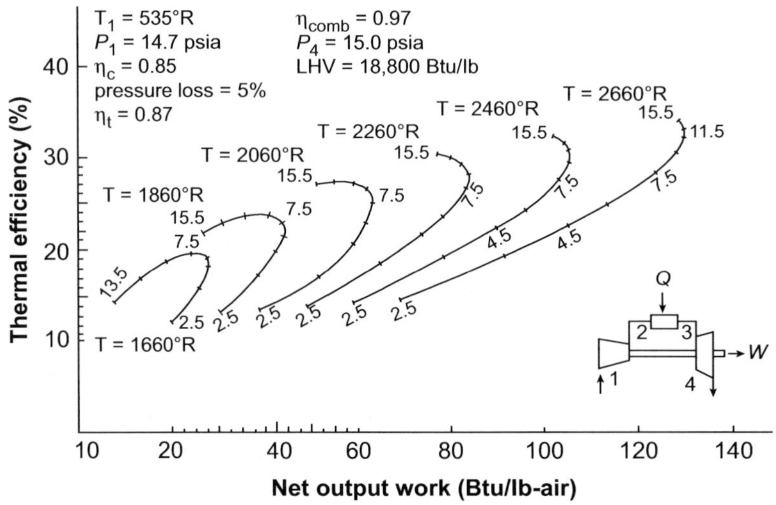

FIGURE 3–6 Performance map of a simple cycle [3-3].

- The combustion zone typically is 2500–2600°F (1370–1427°C) for land-based gas turbines. Flame temperatures may exceed 2000°C in some aircraft engines.
- First-stage turbine inlet nozzles (1st stage NGVs) convert high-enthalpy gas to high-velocity gas.
- High-velocity gas turns the turbine rotor blades and buckets.
- The turbine exhaust temperature is typically 900–1180°F (482–638°C). For every 100°F (56°C) increase in the turbo inlet temperature (TIT), typically a 1.5% efficiency increase can result. Work output can increase by 10%.

All of these parameters modify with cycle adaptations. Regeneration lowers the heat rate (involves the heat recovery of turbine exhaust gases and adds between 15 and 20% efficiency; see Figure 3–8). The optimum PR for a regenerative cycle is about 7; for a simple cycle, it is about 18 (see Figures 3–6 and 3–8).

As steam turbines form such a beneficial partnership with gas turbines in combined cycle applications, we consider at a brief summary of steam turbine theory as it applies to gas turbine combined cycle operation, and steam turbine (solo) operation, both primarily in power generation service. This material on steam turbines is covered in Appendix 3A. (For this section, see Table 3–1 for the symbols used.)* We now continue the discussion of combined cycle and other gas turbine cycle modifications.

In the first edition, the steam turbine theory was not in an appendix. The content of the original steam turbine material remains unchanged, however. Also, consider that the Benson boiler design that is prevalent on gas turbine combined cycles is also used in steam turbine installations. The boiler size may vary, but the working principle does not.**

* Source: [3-1]

** See Appendix 3A for ST power plant theory as applicable to CC and STs operating "solo", in completion with GTs or GT CCs.

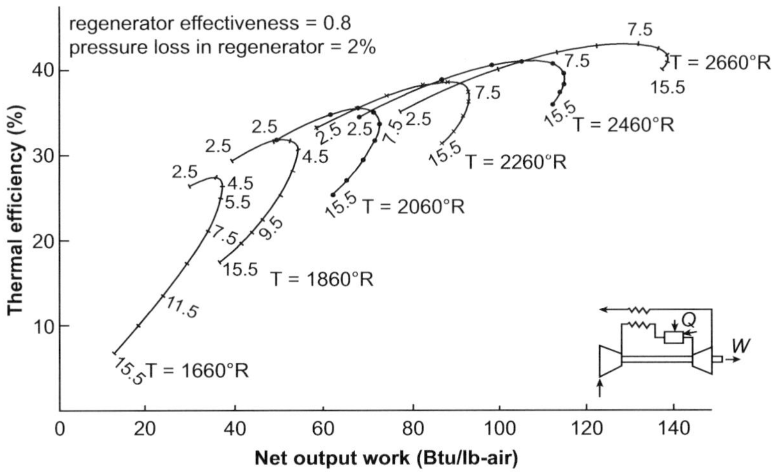

FIGURE 3–8 Performance map of a regenerative cycle [3-3].

TABLE 3–1 Thermodynamic Symbols
c_p = specific heat at constant pressure, Btu/(lb_m · °F) [J/(kg · K)]
c_v = specific heat at constant volume, Btu/(lb_m · °F) [J/(kg · K)]
h = specific enthalpy, Btu/lb_m (J/kg)
H = total enthalpy, Btu (J)
J = energy conversion factor = 778.16 ft · lb_f/Btu (1.0 N m/J)
M = molecular mass, lb_m/lb · mol or kg/kg · mol
n = polytropic exponent, dimensionless
P = absolute pressure (gauge pressure + barometric pressure), lb_f/ft^2; unit may be lb_f/in^2 (commonly written psia, or Pa)
Q = heat transferred to or from system, Btu or J, or Btu/cycle or J/cycle
R = gas constant, lb_f · ft/(lb_m · °R) or J/(kg · K) = $\overline{R}/M$
R = universal gas constant = 1.545.33, lb_f · ft/(lb · mol · °R) or 8.31434 × 10^3 J/(kg · mol · K)
s = specific entropy, Btu/(lb_m · °R) or J/(kg · K)
S = total entropy, Btu/°R or J/kg
t = temperature, °F or °C
T = temperature on absolute scale, °R or K
u = specific internal energy, Btu/lb_m or J/kg
U = total internal energy, Btu or J
v = specific volume, ft^3/lb_m or m^3/kg
V = total volume, ft^3 or m^3
W = work done by or on system, lb_f · ft or J, or Btu/cycle or J/cycle
x = quality of a two-phase mixture = mass of vapor divided by total mass, dimensionless
k = ratio of specific heats, c_p/c_v dimensionless
η = efficiency, as dimensionless fraction or percent
Subscripts Used in Vapor Tables
f refers to saturated liquid
g refers to saturated vapor
fg refers to change in property because of change from saturated liquid to saturated vapor

Combined Cycles and Other GT Cycle Modifications

A Brayton cycle, P-v and T-s diagrams are illustrated in Figure 3–9 (= 1234).

Compressor polytropic efficiency, $\eta_c = \dfrac{\text{ideal work}}{\text{actual work}}$

$$= \frac{h_{2s} - h_1}{h_2 - h_1}$$

Assume constant specific heats

$$\eta_c = \frac{T_{2s} - T_1}{T_2 - T_1}$$

And

Turbine polytropic efficiency, $\eta_T = \dfrac{\text{actual work}}{\text{ideal work}}$

$$= \frac{h_3 - h_4}{h_3 - h_{4s}}$$

Assume constant specific heats

$$\eta_T = \frac{T_3 - T_4}{T_3 - T_{4s}}$$

The net power of the cycle:

$\dot{W}_n$ = power of turbine − power of compressor

For constant specific heats:

$$\dot{W}_n = \dot{m}c_p[(T_3 - T_4) - (T_2 - T_1)]$$

Or

$$\dot{W}_n = \dot{m}_{c_p}\left[(T_3 - T_{4s})_{\eta_T} - \frac{T_{2s} - T_1}{\eta_c}\right]$$

In terms of

$T_1,\ T_3 = \dot{W}_n$

$$= \dot{m}c_pT_1\left[\left(\eta_T\,\frac{(T_3)}{T_1} - \frac{r_p^{(k-1)/k}}{\eta_c}\right)\left(1 - \frac{1}{r_p^{(k-1)/k}}\right)\right]$$

The second bracketed term is ideal cycle efficiency.

Specific power of the non-ideal cycle is:

Heat added in the cycle:

$$\dot{Q}_A = \dot{m}c_p\,T_3 - T_2$$

$$= \dot{m}c_p\left[(T_3 - T_1) - \left(T_1\frac{r_p^{(k-1)/k} - 1}{\eta_c}\right)\right]$$

Note many losses/irreversibilities have not been allowed for, including piping losses, pressure drop between 2 and 3, and pressure difference between 4 and 1, to name a few.

See Figure 3–10 for an illustration of a simple cycle and a regenerative cycle (dashed lines).

Data used for the simple cycle was:

T_i = 15°C = 59°F = constant

P_1 = 1.013 bar = 1 atm = constant

η_c = 90% and η_T = 87%

Mechanical losses = 1%

Combustion chamber losses = 2%

Air bypass = 3%

Pressure losses: Air inlet 1%

In combustion chamber 3%

At outlet 2%

In regeneration 4%

Efficiency and specific work depend heavily on the max. temperature reached ("TIT" = turbine inlet temperature).

Cycle Modifications

The most common cycle modifications are:

- Regeneration
- Compressor intercooling
- Turbine reheat
- Water injection

Regeneration ⇒ internal exchange of heat within the cycle. Note the regenerator or recuperator in Figure 3–11. Regenerator effectiveness or efficiency is less than 100%. If

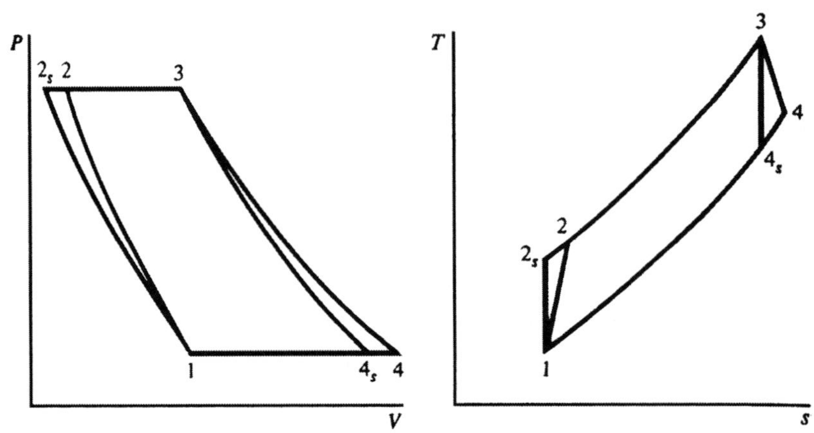

FIGURE 3–9 P-v and T-s diagrams of ideal and non-ideal Brayton cycle [3-3].

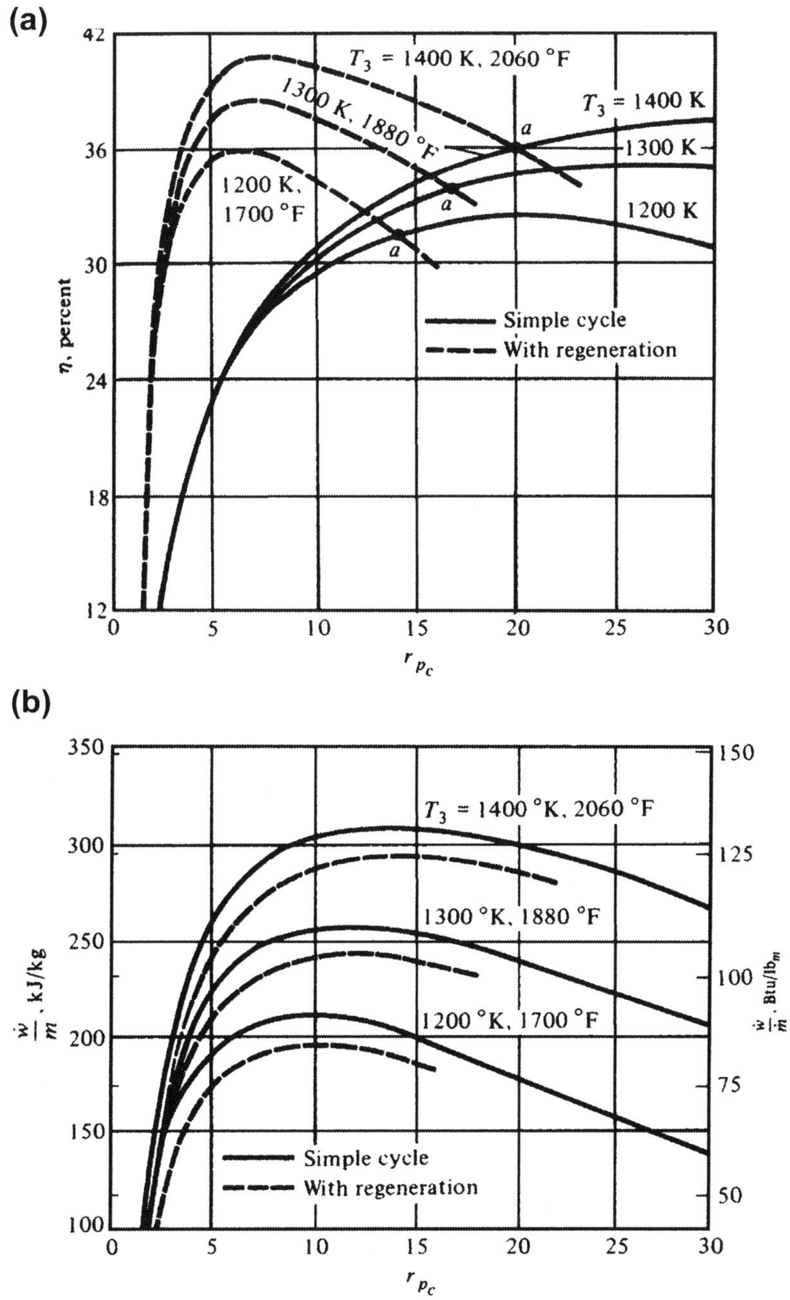

FIGURE 3—10 (a) Efficiency versus compressor pressure ratio of a non-ideal Brayton cycle, showing effects of maximum temperature and regeneration. (b) Specific power versus compressor pressure ratio of a non-ideal Brayton cycle, showing effects of maximum temperature and regeneration [3-3].

the compressor outlet is heated to $T_2{}'$, then regenerator effectiveness is:

$$\varepsilon_R = \frac{T_{2'} - T_4}{T_4 - T_2}$$

Let's say that = 75%.

The rise in efficiency is notes in Figure 3—10. However, the optimum pressure ratio for an efficiency value drops,

$$\because PR\downarrow \ \Rightarrow (T_4 - T_2)\uparrow$$

Cycle heat input drops further.

At some point, the temperature of exhaust gases < post-compression air, or the regenerator is ineffective.

Intercooling involves taking compressed air at a given stage of compression, cooling it (which increases its density and therefore the mass of air that can be compressed) and leading the cooled air back to the next stage of

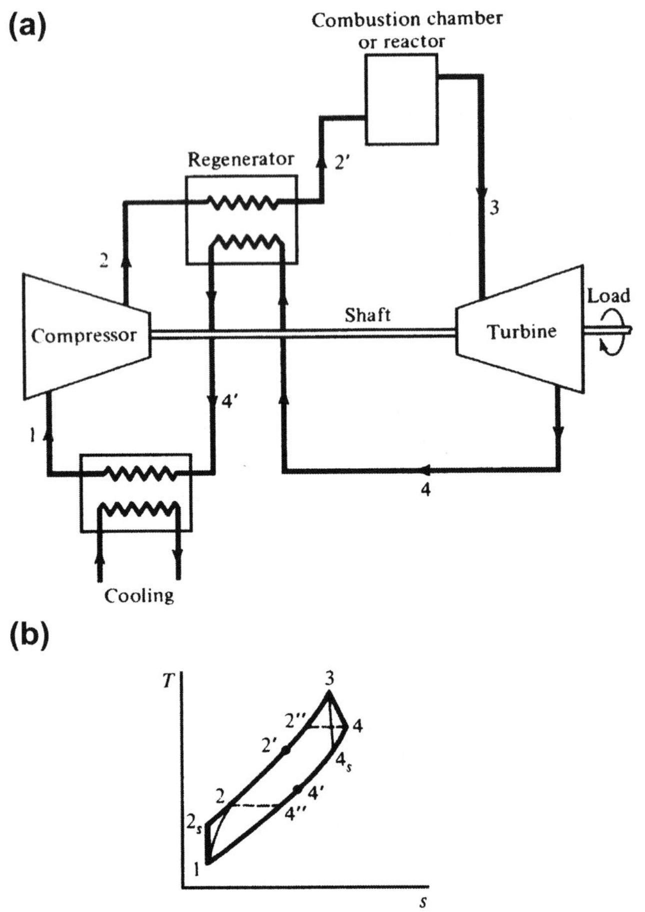

FIGURE 3–11 (a) Flow and (b) T-s diagrams of a closed, non-ideal Brayton cycle with regeneration [3-3].

compression. There may be more than one stage of intercooling (see Figure 3–12).

In Figure 3–12, the increase of work provided by intercooling is 21′2′1′2′ × 2. More heat had to be added to get this work increase. Per unit mass that additional heat is:

$$h_x - h_{2''}$$

However, a net efficiency increases results.

Reheat involves taking gas from an interim stage in the turbine, adding heat to the gas and then leading it back to the turbine (see Figure 3–12). One stage of reheat is shown.

- The increase in cycle work is 43′4′ *y*.
- Additional heat required to get this is: $H_{3'} - H_4$
- However, net work and efficiency rise.
- Efficiency rises with the number of stages of intercooling and reheat, but so does plant capital cost.

Water injection increases power output significantly and efficiency slightly. Water is injected into the compressor and vaporizes as the air temperature increases. The heat of vaporization reduces the compressed air temperature. This lowers compressor work (see Figure 3–13).

- 124579′1 is the cycle with no water injection.
- 9′ is exhaust gas from the regenerator.
- 3 represents the evaporating water.
- With water injection, compressor air cools from 2 to 3.
- The regenerator preheats the air from 3 to 4. The heat required to raise temperature from 3 to 2 comes from exhaust gas from 9′ to 9.
- T_3 is saturation temperature for air.
- Further increase in water vapor increases net work required and reduces efficiency.

Combined Cycles

The most common types are:

- HRSG or HRB (boilers)
- STAG (steam and gas) CC
- CC with multipressure steam

Figure 3–14 illustrates the first kind (HRB).

$$\begin{aligned}\text{HRB} &= \text{EC (economizer)} + \text{B (boiler)}\\ &\quad + \text{SU (superheater)}\end{aligned}$$

To increase output through limited duration peaks, SF (supplementary fuel) is used.

STAG CC

In Figure 3–15, 4 GE Frame 7s exhaust to air or produce steam in 4 HRBs that send steam to one ST. The combination can provide 600, 800, 1000, or 1250 psi steam.

CC with Multipressure Steam

In Figure 3–16, the HRB has two steam circuits:

- A high-pressure circuit, feeding the ST inlet
- A low-pressure circuit, feeding an intermediate stage of the ST

In Figure 3–17,

- 10–11 = feedwater heating in the economizer
- 11–12 = evaporation
- 12–13 = superheat
- Water is pumped by a booster pump (BP) from the low-pressure steam drum at 11 to 14.
- Note the single HP circuit is 10′15 – 16 – 17, with gas leaving the stack at 6′.
- With the low circuit, the gas leaves at 6.
- As 6 is lower than 6′, this ⇒ that more energy has been extracted, so efficiency is higher.

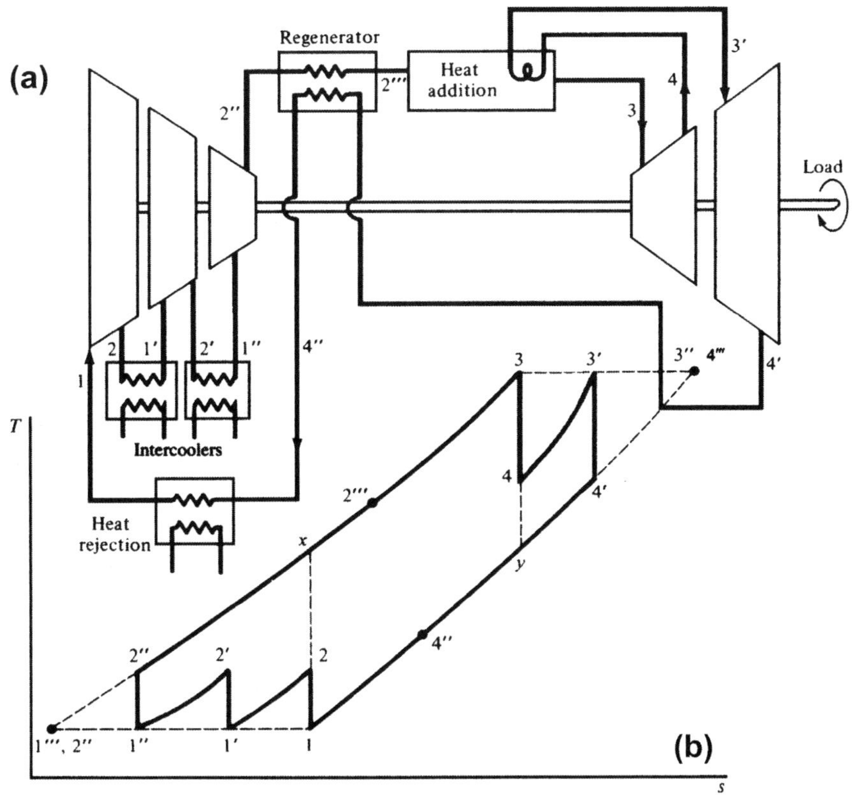

FIGURE 3–12 (a) Flow and (b) T-s diagrams of a closed, ideal Brayton cycle with two stages of intercooling, one stage of reheat, and regeneration [3-3].

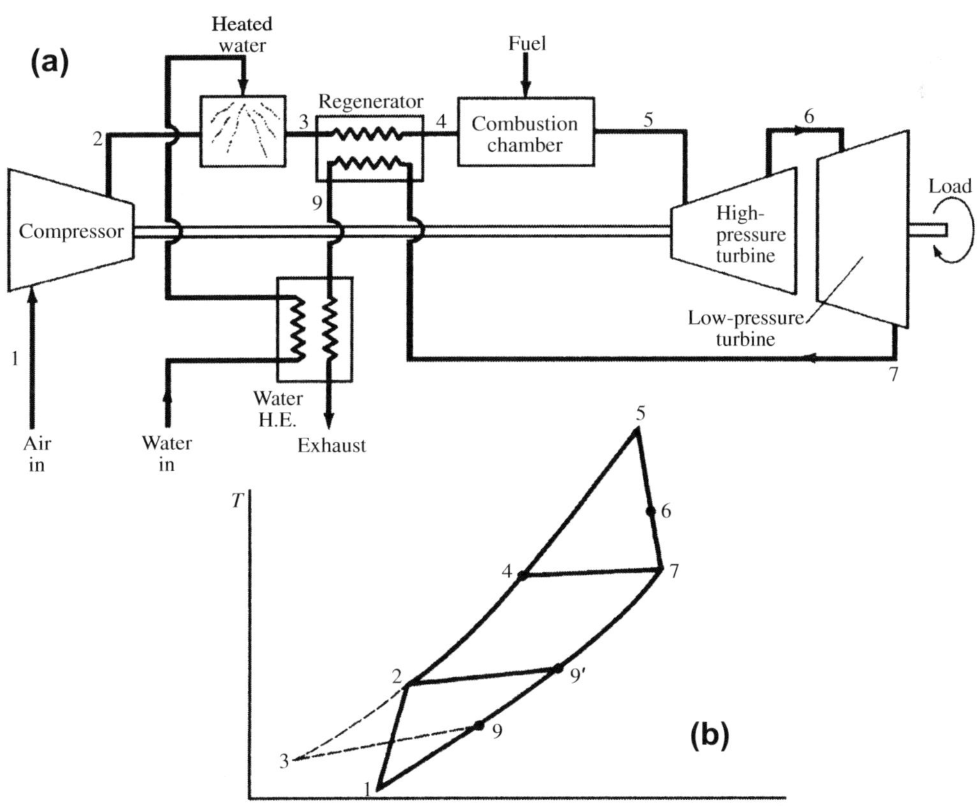

FIGURE 3–13 (a) Flow and (b) T-s diagrams of a two-shaft gas turbine cycle with water injection and regeneration [3-3].

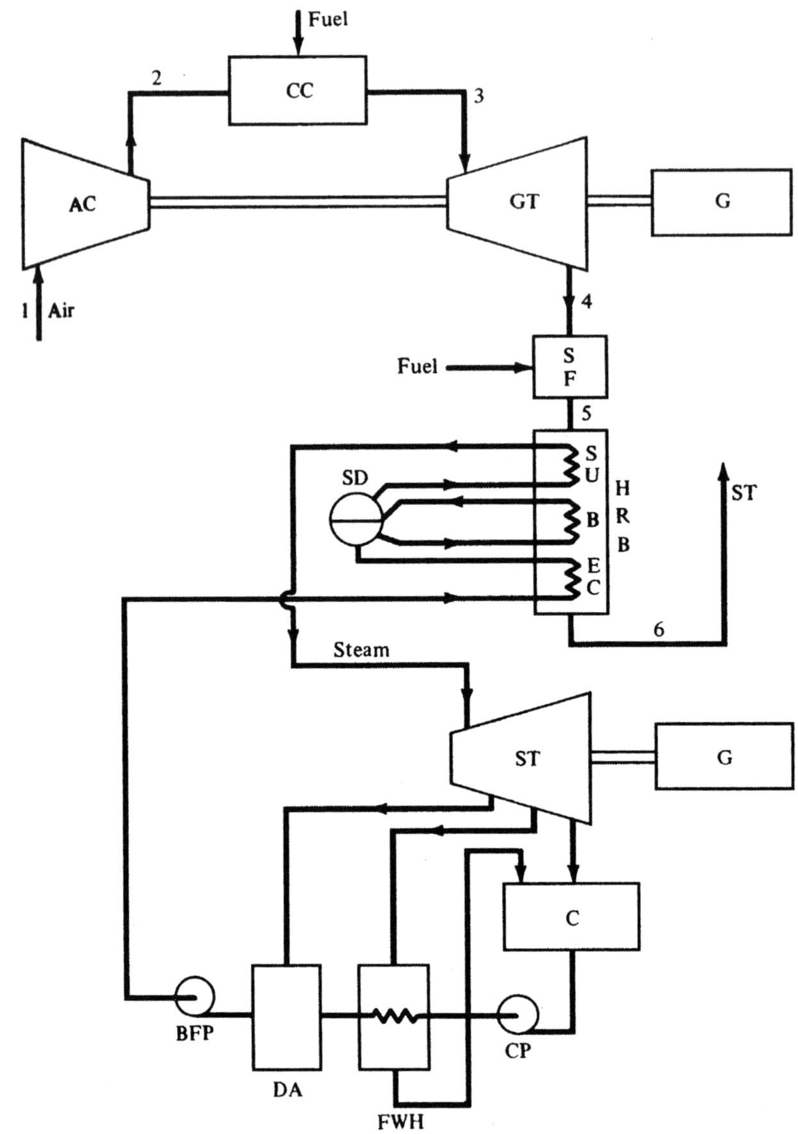

FIGURE 3–14 Schematic flow diagram of a combined cycle with a heat recovery boiler (HRB) [3-3].

Combined-Cycle Economics*

Within the field of power generation, the combined cycle is seen as a major step towards fuel economy and lessening GHG (greenhouse gas) emissions. Combined-cycle plant efficiencies now typically are up to between 58 and 60%.

Gas flapper valves allow the gas turbine exhaust to bypass the heat recovery boiler (HRSG), allowing the gas turbine to operate if the steam unit is down for maintenance. In earlier designs, supplementary oil or gas firing was included to permit steam unit operation with the gas turbine down. This is not generally included in contemporary combined-cycle designs, as it adds to capital cost, complicates the control system, and reduces efficiency.

Sometimes as many as four (but most frequently two) gas turbines, each with individual boilers, may be associated with a single steam turbine. As stated previously, the gas turbine, steam turbine, and generator may be arranged as a single-shaft design. A multi-shaft arrangement can also be used: Each gas turbine drives a generator and has its own HRSG and steam turbine, which in turn may also add power to the generator.

In areas such as Scandinavia, additional criteria such as cogeneration in combined heat and power (CHP) plants or district heating, as well as demanding conditions (e.g., available space, emissions, noise level, architecture, environmental

* Source: [3-2]

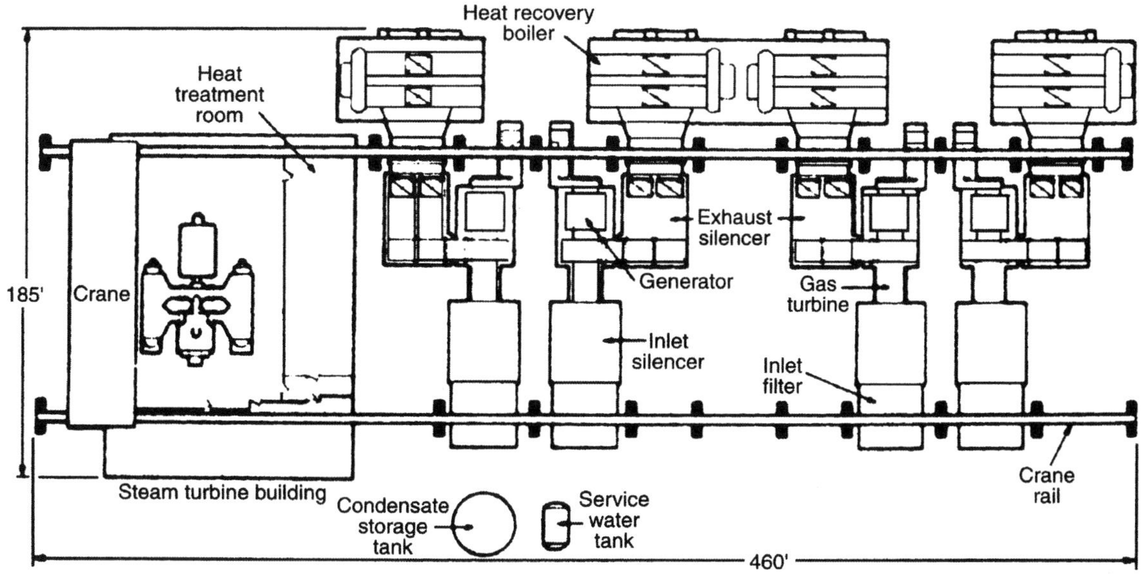

FIGURE 3–15 Layout of the STAG combined-cycle power plant [3-3].

permits) associated with existing sites and available infrastructure, must be considered. A customer's preferences regarding fuel selection, personnel training level required, and service requirements also must be accommodated.

A major portion of the economic considerations that make up an end-user's selection of gas turbine packages has to do with the candidate OEMs' "state of proven technology." This means that as OEMs continue to develop newer gas turbine technologies, they need to test customized packages ordered by the end-user. At some point this needs to be a full load test and depending on which OEM is the supplier, this full load test may be done in the OEM's test facility or on-site. In other words, the end-user can contract with the OEM for financial consideration if the gas turbine package does not perform as specified in the sales contract. This chapter ends with a case study from an EPC (engineering, procurement, and construction) contractor that has dealt with gas turbine packages involving the OEM's latest technology and notes the various delays that may arise when one is commissioning models that are not yet mature in the market.

Combined-Cycle Module Flexibility*

With combined cycles, capacity can be installed in modules or module stages. Gas turbines (GTs) can be commissioned initially (one- to two-year project construction) and then the HRSG(s) and steam turbine(s) (ST) (an additional six months to one year).

For instance, an Alstom 13E2 CC module can consist of two 13E2 gas turbines with heat recovery steam generation and one steam turbine, as in their Kuala Langat, Malaysia, plant. In this way, combined-cycle capacity can be installed in segments. This further assists generation dispatching, as each gas turbine can be operated with or without the steam turbine. This then provides better efficiency at partial load than operating one large machine with the total capacity equal to the gas turbine(s) and the steam turbine.

Another case** illustrating application of the two GT and one ST module is Alstom's contract for the Sohar Aluminum Company to do the turnkey construction of a 1000 MW gas-fired combined-cycle power plant in Oman. The power plant, which will supply electricity to power a new aluminum smelter, will include four 13E2s, four heat recovery steam generators, two steam turbines, and six generators. The size of the modules then provides the option for Sohar to add an additional 500 MW of capacity in the future (two GT13E2 gas turbines, two heat recovery steam generators, a steam turbine, and three generators).

Gas turbine or combined-cycle (CC) construction cost per kilowatt does not increase much for smaller turbines. With steam turbines, it would to a far greater extent, because of the high additional construction work that comes with a steam turbine plant. A CC unit typically can be installed in two to three years; and a steam plant often takes four to five years, with no incremental power

* Source [3-4] Courtesy of Siemens. Compiled from I. Paul, J. Karg, and D. O'Leary Sr., *Tailor-made Off the Shelf: Reducing the Cost and Construction Time of Thermal Power Plants*; R. Taub, J. Karg, and D. O'Leary Sr., *Gas Turbine Power Plants: A Technology of Growing Importance for Developing Countries*. Alstom Power press releases, December 9 and 14, 2005.

** Source [3-5] Courtesy of Alstom Power. Alstom Power press release, December 14, 2005.

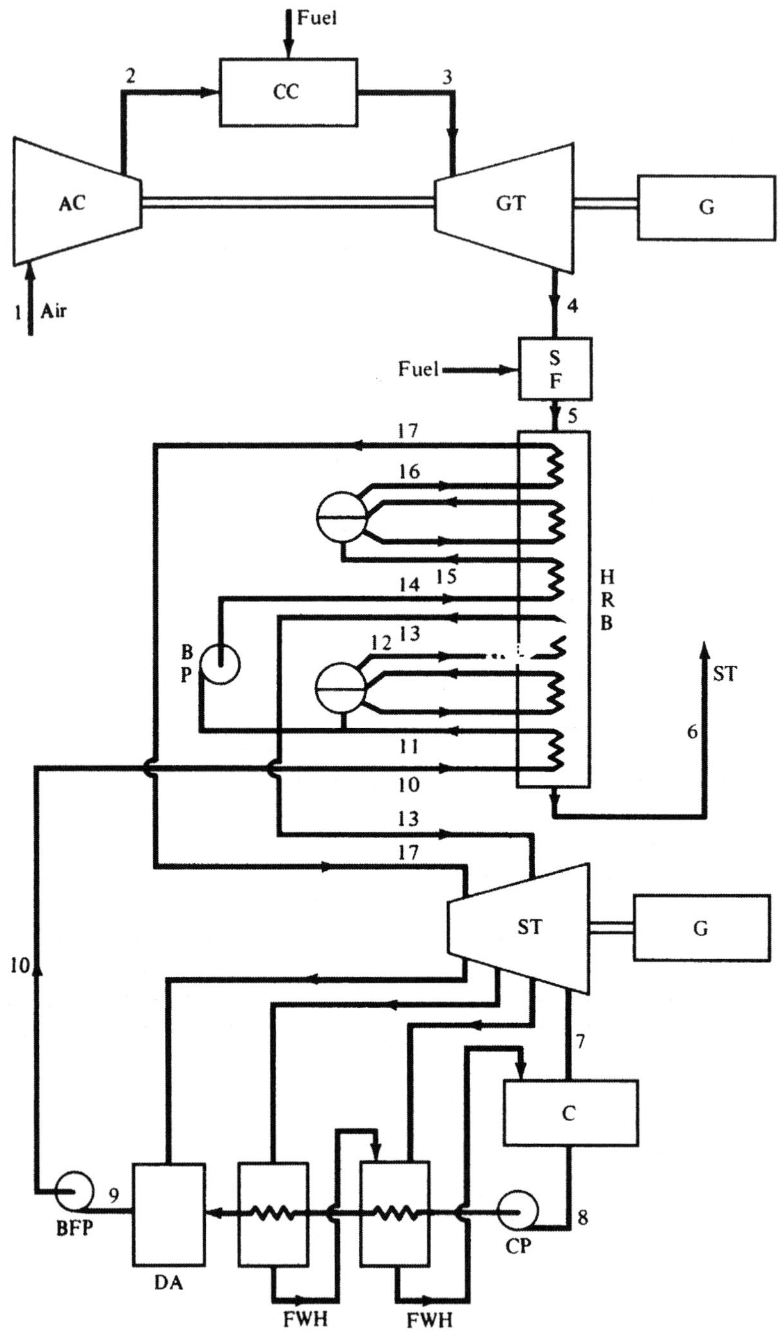

FIGURE 3–16 A schematic diagram for a dual-pressure combined cycle [3-3].

available until the complete plant is commissioned. An application case that illustrates the availability of power in increments is an Alstom recent project† from an Australian energy company, Alinta Ltd, to supply two 172 MW GT13E2 gas turbines for the first stage of a major cogeneration facility at Alcoa's Wagerup alumina refinery in Australia. That power plant also will provide reserve capacity to the new wholesale electricity market in the state of Western Australia. The Alstom turbines initially will operate in open cycle (Wagerup Stage 1). At a later stage (Wagerup Stage 2), the turbines will be part of a cogeneration plant, operating as a baseload power station providing both steam and electricity.

A project database (developed by Siemens KWU) was used to analyze all combined, open cycle, and steam power plants globally with respect to capacity (MW), fuel requirements, power system frequency, and regional location. The database lists projected orders through 2005.

† Alstom Power press release, December 9, 2005.

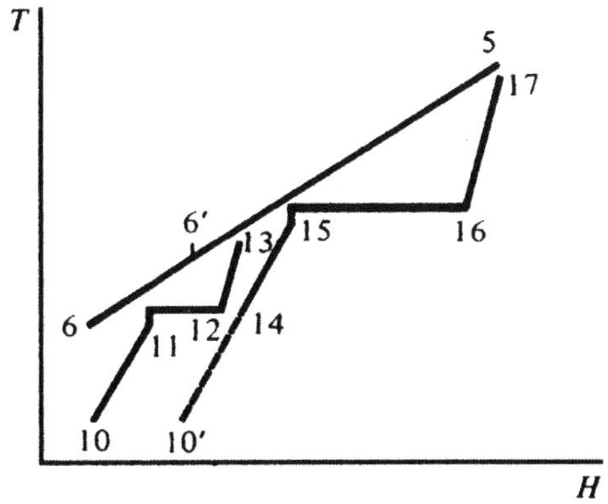

FIGURE 3—17 Temperature-enthalpy (T-H) diagram of the heat recovery boiler of the dual-pressure combined cycle shown in Figure 3—23 [3-3].

In terms of overall plant size, 300—600 MW combined-cycle plants are the most favored plant size in both the 50 and 60 Hz markets (Figure 3—18). A combination of more than one block improves economics, and 300—600 MW fits well with the demand curve of most power grids in well-developed countries. Financiers also are familiar with these economies of scale.

Countries with large grids and high power demand growth prefer combined-cycle plants in the range of 600—2500 MW. For this combination, two to six parallel units (single-shaft or multi-shaft) suffice.

Power systems in countries with relatively small generating capacity, which require smaller capacity additions, need combined-cycle power plants in the range of 100—300 MW. A large gas turbine and a steam turbine located on a single shaft can deliver this range.

Countries with smaller or specialized grids buy multi-shaft combined-cycle plants with several smaller gas turbines and one or more steam turbines. Dirty fuels, for instance, or residuals, promote requests for stolid, highly reliable trains that may run derated, over higher-efficiency turbines.

For peaking power or power systems with very low-cost fuels, gas turbines in an open cycle system serve the power range between 50 and 300 MW.

New order forecasts show the market evenly divided between 50 Hz and 60 Hz customers. Rising gas and oil prices everywhere, including the United States, mean renewed strength in technologies that use alternative fuels, such as pulverized coal, paper liquor waste, and steel mill flue gas.

Steam-Only (Coal-Fired) Power Plants

The forecast projects 10% of the new orders will be steam power plants in 60 Hz market from 1999 to 2003 (Figure 3—19). In the 50 Hz market, the key ranges are 300—500 MW and 500—700 MW. Above 700 MW, supercritical technology represents a small but growing market share.

OEM Modular Strategy

As previously discussed, to save on costs to both OEMs and end-users, OEMs developed modular plants. Siemens has 12 basic power plant combinations (Table 3—2): four for open cycle gas turbine plants, six for combined-cycle plants, and two for coal-fired steam power plants (with sub- and supercritical technology). Each combination covers a specific power range, efficiency, and fuel specification, with allowance for cogeneration system additions.

For design flexibility, options to the reference version for each major functional unit (Figure 3—20) are provided.

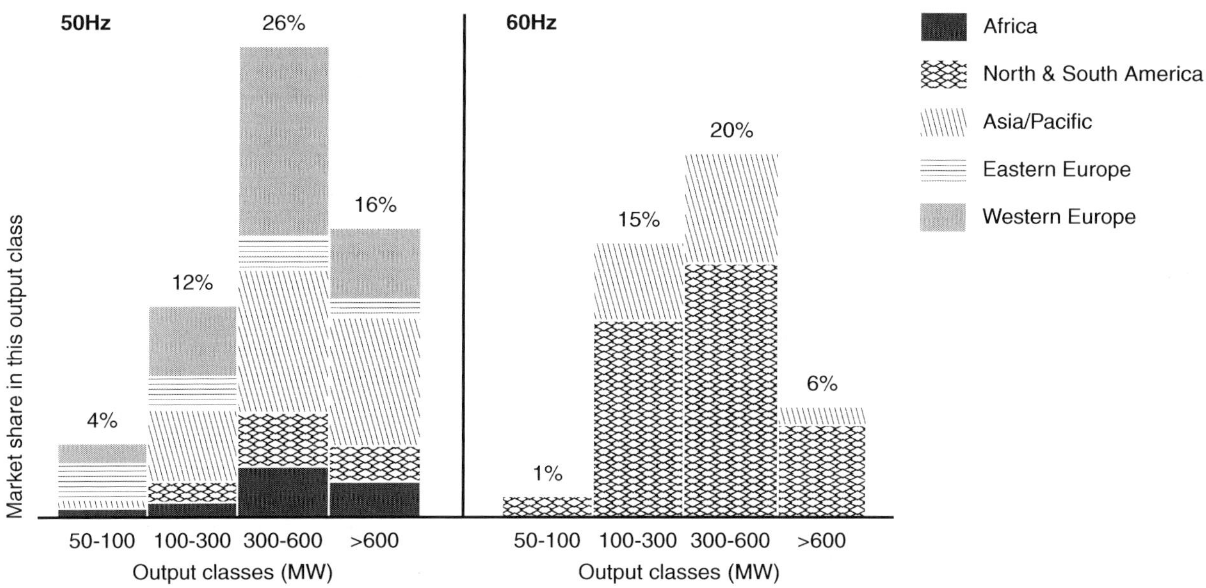

FIGURE 3—18 Markets for gas turbines, 1995—2005. Greater than 300 MW combined-cycle dominates in both the 50 and 60 Hz markets [3-4]. *(Source: Adapted for reproduction in black and white.)*

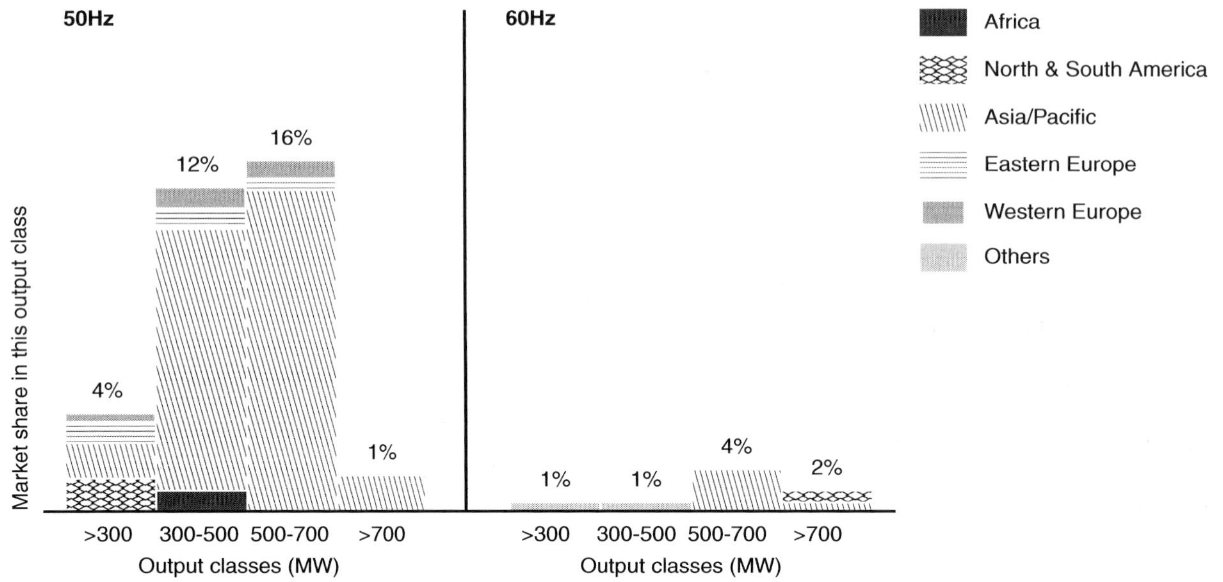

FIGURE 3–19 Markets for steam turbines, 1999–2005. 300–450 MW and 500–700 MW dominate in the 50 Hz market. The 60 Hz market is very small [3-4]. *(Source: Adapted for reproduction in black and white.)*

TABLE 3–2 Reference Power Plant Data [3-4]

Ref. Plant Type	Frequency	Output	Efficiency (%)	Fuel
Gas Turbine Power Plants				
GT PP 2.84.2	60 Hz	200–220 MW	33.5–34.0	Gas/Oil
GT PP 2.94.2	50 Hz	300–320 MW	34.0–34.5	Gas/Oil
GT PP 2.84.3A	60 Hz	340–360 MW	38.0–38.5	Gas/Oil
GT PP 2.94.3A	50 Hz	480–510 MW	38.0–38.5	Gas/Oil
Combined-Cycle Power Plants				
GUD1S.64.3A	50/60 Hz	100–105 MW	53.7–54.0	Gas/Oil
GUD1S.84.3A	60 Hz	250–260 MW	57.8–58.0	Gas/Oil
GUD1S.94.3A	50 Hz	360–380 MW	57.8–58.0	Gas/Oil
GUD 2.84.3A	60 Hz	500–520 MW	57.8–58.0	Gas/Oil
GUD 2.94.3A	50 Hz	700–760 MW	57.8–58.0	Gas/Oil
GUD 2.94.2	50 Hz	470–480 MW	52.2–52.3	Gas/Oil
Steam Turbine Power Plants				
ST PP 300/450	50 Hz	2×300–2×450 MW	38.0–39.0	Coal
ST PP 500/700	50 Hz	2×500–2×700 MW	40.2–41.6	Coal

For example, via ship is the reference for the functional unit coal supply with delivery via rail as an option. Flexible design requires breaking down the power plant into functional units, each of which directly affects only one or two other modules. For a combined-cycle plant, the functional units are arranged around the gas turbine and steam turbine. With the gas turbine, as we saw earlier, OEMs strive to maintain core feature commonalities.

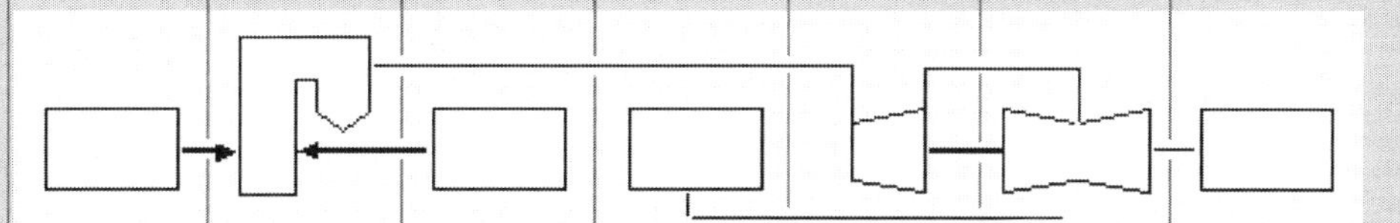

	Coal delivery	Boiler	Feed pumps	Main condensate pumps	High Pressure turbine	Low Pressure turbine	Generator
Reference	by ship	Supercritical once-through 250 bar 540/560°C	2× 50 % electric motor drive	2× 50 % electric motor drive	H 30-100	N 4 ×8 m^2	THDF 115/67
Options	by rail	Drum 167 bar 540/540°C	3× 50 % electric motor drive 2× 50 % steam turbine drive 1× 30 % electric motor drive	3× 50 % electric motor drive	H 30-63	N 4 ×10 m^2 N 2 ×10 m^2	THDF 115/53

FIGURE 3–20 A 2 × 700 MW steam reference power plant [3-4].

Fuels for Combined Cycles

Gas turbine operators prefer to burn natural gas and light oil (diesel, No. 2). As we saw previously, crude oil, residuals, and "bunker" fuel contain corrosive components. They require fuel treatment equipment. Also, ash deposits from these fuels can result in gas turbine derating of up to 15%. As we also saw previously (in the case of the Shunde plant in south China), they still may be economically attractive fuels, particularly in combined-cycle plants.

Sodium and potassium are removed from residual, crude, and heavy distillates by a water-washing procedure. A simpler and less-expensive purification system does the same job for light crude and light distillates. A magnesium additive system reduces vanadium.

Note that reduced availability results, due to water cleaning shutdowns to remove blade deposits, as online washing, even at reduced speeds, is not effective. A shutdown with a crank soak every 100–120 hours is required. Reduced component life due to hot gas path corrosion caused by vanadium deposits and other corrosion is another factor to consider.

Table 3–3 provides a sample of naphtha- and heavy oil-fired power plants in operation and in the planning stage. As this table shows, some plants (e.g., Kot Addu and Valladolid) accumulated 30–60,000 hrs of successful operation over their first five years plus.

Design and operation of these plants requires more attention than natural gas-fired plants, particularly in relation to fuel variables such as calorific content, density, composition, concentration of contaminants and emissions, as well as different burning behaviors (e.g., ignitability, flame velocity, and stability).

To overcome these difficult fuel properties, technological adaptation, additional equipment, and operational requirements are necessary. These include GT layout (compressor, turbine) for the changed mass flows, different burner technology (burner design, burner nozzles), and additional startup/shutdown fuel system and safety measures. Performance, availability, and operation and maintenance (O&M) expenses can be affected. To illustrate this, Table 3–4 shows some key non-standard fuels and their effect on a standard fuel system.

An example of a gas turbine combined-cycle plant burning a nonconventional fuel is the 220 MW Valladolid plant in Mexico. This plant, commissioned in 1994, burns heavily contaminated fuel oil, containing 4.2% sodium and up to 300 ppm vanadium. Fuel impurities (sodium, potassium, and vanadium) tend to form ash particles in the combustion process, form deposits, and corrode the gas turbine blades. In the case of the Valladolid plant, "Epsom salts," consisting mainly of magnesium sulfate ($MgSO_4$ $7H_2O$), are dissolved in water injected into the gas turbine combustor through special orifices. This converts the vanadium into a stable water-soluble product (magnesium vanadate), which is deposited downstream of the combustor on the gas turbine blades, and causes only minor blade corrosion. To prevent major performance loss with salt buildup (as with the Shunde, China, plant that we read about previously), washing every 150 hours was necessary to restore aerodynamic performance and plant efficiency. Good manhole access was a critical success factor for this project, as servicing and maintenance during turbine washing shutdowns are simplified. (The plan is to eventually convert the Valladolid plant to natural gas operation.)

TABLE 3–3 Operation Experience with Naphtha- and Heavy Oil-Fired CC Plants

"Naphtha Class" Projects

Plant Name	Location	Fuel	Gas Turbine Model	Rating	Operating Hours
Paguthan	India	Naphtha, Distillate, NG	3 × V94.2 1 GUD 3.94.2	630 MW	19,000 hours
Santa Rita	Philippines	Naphtha, Condensate, Distillate, NG	4 × V84.3A 4 GUD 1S.84.3A	950 MW	Erection phase
Faridabad	India	NG, Naphtha, HSD	2 × V94.2 1 GUD 2.94.2	440 MW	Engineering phase

Note: 9 Gas turbines ordered for fuels of "Naphtha class," "Naphtha class" means light, low boiling liquid.

Latest Plants Ash-Forming Fuels

Plant Name	Location	Fuel	Gas Turbine Model	Rating	OperatingHours
Valladolid	Mexico	Residual oil	2 × V84.2 1 GUD 2.84.2	220 MW	24,000 hrs
Kot Addu	Pakistan	Furnace oil (Heavy oil)	4 × V94.2 2 GUD 2.94.2	820 MW	60,000 hrs
Rousch	Pakistan	Heavy oil	2 × V94.2 1 GUD 2.94.2	390 MW	Commissioning phase

Note: 34 Gas turbines delivered for ash-forming fuels in total accumulated operating hours (approx. 140,000 hours).

Factors that Affect Costs per Fired Hour

Fuel type and mode of operation (steady load/partial load) determine the maintenance intervals and work items required. Some estimate that burning residual or crude oil increases maintenance costs by a factor of 3 (assuming a base of 1 for natural gas and by a factor of 1.5 for distillate fuel) and those costs are three times higher for the same number of fired hours if the unit is started every fired hour, instead of starting once very 1000 fired hours. "Peaking" at a 110% rating increases maintenance costs by a factor of 3 relative to base-load operation at rated capacity, for any given period.

The control system on combined-cycle units is automatic. When an operator starts the unit, it accelerates, synchronizes, and loads "by itself." Fewer operators are required than in a steam plant.

Trends in Global Combined-Cycle Installations

A few hundred power generation plants are ordered from about a dozen OEMs every year. This means the market is exceptionally competitive. Given that most of the new plants are going into newly developing countries, the biggest factor in determining the winner of each project (bid on by several OEMs or not) is the financial deal the OEM can put together for the end-user.

As one might expect, maintenance costs are higher for any type of plant in countries that have not had as much exposure to the OEMs' technology. As a significant extension of their revenue, OEMs offer overall "power by the hour" maintenance contracts. These costs vary, even for the same basic modular configuration and mechanical design, depending on the location's demographics. So then will the actual and contractually set "cost per fired hour" figures. There is a significant difference between what actual operational costs are for the same OEM's CC block in a well-developed area of the United States and a remote area in Azerbaijan, for instance. Demographics also alter construction costs. (As an illustration, in 1990s figures, costs varied from $592/kW for a new 1080 MW combined-cycle plant in Egypt to $875/kW for a steam addition to convert four gas turbines in Pakistan to a combined-cycle plant, according to World Bank data.) OEMs are aware that end-users compare cost data at various meetings and forums and that price variations are a sore and much negotiated point. Therefore, OEMs continually strive to optimize designs and assembly methods to minimize the steepness of new operators' learning curve.

"Modularization" (for instance, the Siemens Westinghouse GUD block, which is two V94.3 gas turbines, their HRSG boiler capacity, and a steam turbine) reduces construction costs. Compared with the customized design and construction, modularization can reduce project costs of detailed engineering, material price contingencies, and financial loan interest during construction.

Downsizing power delivery (to the grid) requirements changes overall operational cost figures. "Repowering" changes operational statistics significantly. *Repowering* is a term used to define the reconfiguration of a power station.

TABLE 3–4 Gas Turbines for Non-Standard Fuels, Critical Fuel Properties [3-4]

Critical Fuel Properties	Fuels	Effect on Standard Fuel System
• Low viscosity (reduced ludricity)	Naphtha, Kerosene	Effect on fuel supply system (e.g., pump design)
• Low density (high volume flow)	Condensates,Naphtha, Condensates	Limits for fuel supply system and bumer nozzles
• Low flash point • Low boiling point (high vapour pressure)	Naphtha, Kerosene, Liquefied Petroleum Gases (LPG) High-Speed Diesel (HSD), Condensates	Increased explosion protection effort Increased ventilation effort
• Low auto ignition point • Contaminants (high temperature corrosion, ash deposits)	Naphtha, Condensates, Contaminated fuel oils, Crude oils, Heavy oils, Heavy residues	Effect on premix capability Increased explosion protection effort. Increased ventilation effort Effect on GT blading and hot gas path. Counter-measures: -Temperature reduction (reduced performance) -Fuel treatment (washing inhibitor dosing)
• Low heating value (high volume flow)	Process and synthesis gases (low calorific gases), Low BTU natural gas	Effect on layout of GT compressor, bumer nozzles, fuel supply system
• High H content	Process and synthesis gases (low calorific gases)	High flame velocity, effect on premix capability
• High dewpoint	Gases with high boiling components	Droplets causes erosion and non-constant heat flow

It may mean replacing a steam turbine with a gas turbine or combined-cycle unit. One example of a repowering option offered by an OEM is Alstom combining its 181 MW GT24 gas turbine with a dual pressure reheat cycle consisting of a 70 MW LP/IP steam turbine and a 20 MW HP steam turbine, to generate a total of 270 MW.

The most common configuration (Figure 3–21) is called *parallel powering*, where the gas turbine exhausts are used in the existing steam cycle. This is achieved by feeding the exhausts into a heat-recovery steam generator, which provides additional steam to the existing steam turbine. Typically, parallel powering requires the addition of a

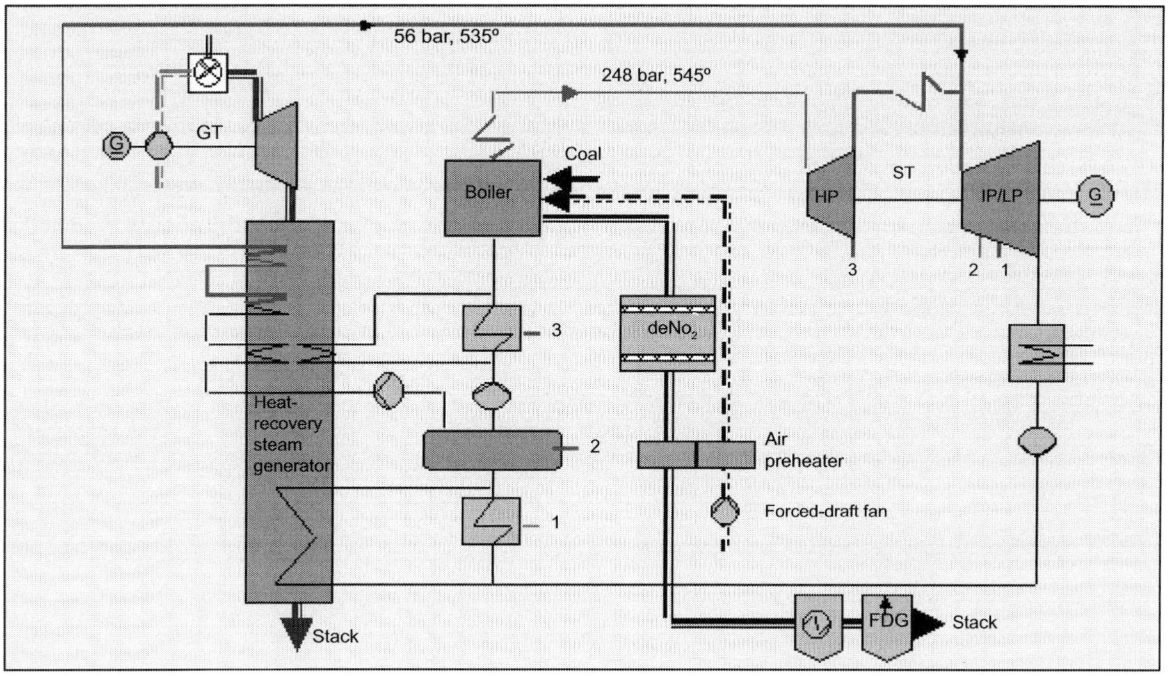

FIGURE 3–21 Schematic diagram of a parallel-powered combined-cycle block with full flue gas cleaning [3-4].

FIGURE 3–22 Driving forces in power plant construction. *(Source: Adapted from [3-4].)*

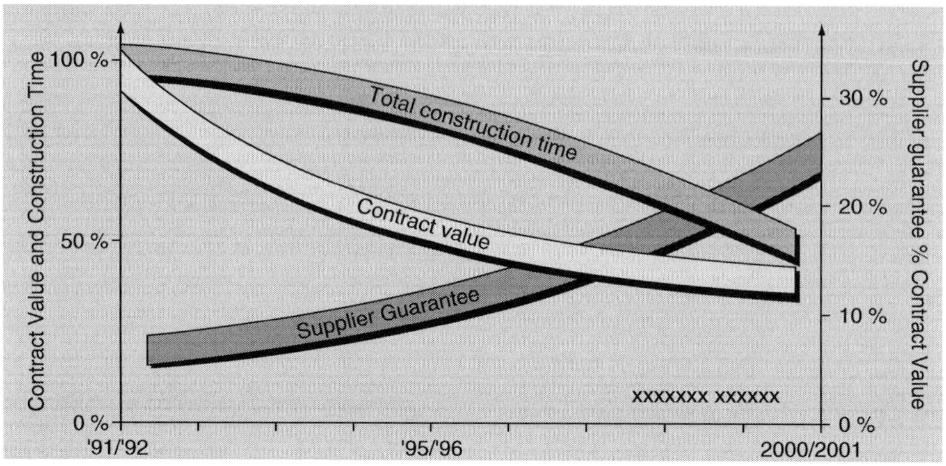

gas turbine, associated electrical and instrumentation and control equipment, civil engineering, HRSG, additional piping, and pumps, as well upgrading the steam turbine. Generally, parallel powering can be undertaken separately from the existing part of the plant, with a final integration phase and a plant downtime of 1.5–2 months. The typical cost range is $300–500/kW.

In some cases, national or international markets alter a power plant's budget by changing available fuels. An example would be the United Kingdom's temporary moratorium on their indigenous natural gas (which promoted coal for that period). When the decision was made to allow North Sea petrochemical liquid deposits to vaporize and be delivered as gas instead, that move created operational ripples in all industries that used petrochemical fuel, including power generation.

Since the late 1980s, market growth in plant additions and optimization technology retrofits has shifted, in part, from Europe, North America, and Japan to newly industrializing countries in Asia and Latin America. Financial means keep many of the end-users in these regions from using newer technologies that would extend their power generation capacity and reduce their costs per fired hour. Nevertheless, they are becoming increasingly aware of these design developments and seek to incorporate them where and when possible.

For the OEM, the main challenges are minimizing project cost, construction time, and risk guarantees (Figure 3–22). Between the 1980s and 2000, project cost and construction time of coal- and gas-fired units dropped by 50%. However, to compete, OEMs must offer better warranty packages. So the standardization of core design to minimize spare costs and make factory assembly methods and repair and overhaul methods "foolproof" increases in importance.

Table 3–5 shows the cost breakdown for combined-cycle plants (350–700 MW capacity) based on Siemens experience in the following categories: integrated services (project management or subcontracting; plant and project engineering or project management software, plant erection, commissioning, and training; transport and insurance) and lots (civil works, gas- and steam-turbine and generator sets, balance of plant, electrical systems, instrumentation and control systems, and the boiler island).

Trends in the Power Generation Market

From 1994 to 1999, power plant contract awards for fossil fuel-fired power plants (above 50 MW) averaged 63 GW

TABLE 3–5 Cost Breakdown for CC Power Plants

Integrated Services	4%	Project Management/Subcontracting					
	2%	Plant and Project Engineering/Software					
	8%	Plant Erection/Commissions/Training					
	1%	Transport, Insurance					
	Σ15%						
Lots		Civil Works	Gas- and Steam Turbine Set	Balance of Plants	Electrical Systems	Instrumental and Control	Boiler Island
	Σ85%	15%	32%	16%	7%	4%	11%

Basis: 350/700 MW CC Plant with a V94.3A Gas Turbine.
(Adapted from [3-4].)

per year. In 1999, sales were forecasted to average about 67 GW per year over the period 1999–2004. The market in the Asia Pacific Basin was declining, while showing a moderate growth in Europe and (starting from a low level) strong growth in North America.

In comparison with coal-fueled power plants, open and closed cycle power plants are characterized by lower investment costs. However, US fuel-related costs (i.e., fuel price and plant efficiency) have changed with the rise in oil and gas prices in the United States, which was precipitated by the Iraq War and Hurricane Katrina. At the turn of the century, gas prices ranged from about US$2.0/GJ to US$4.5/GJ, with the North American prices being at the lower end of the range.

The only fact that anyone can attest to, in terms of oil and gas prices in 2006, is that they will go up. One Canadian forecast agency suggests that gas prices in 2006 in Canada will stay at about C$8/GJ. This then means that the fierce inter-OEM rivalry with respect to fuel efficiency will escalate.

Three major infrastructure changes continue to drastically alter the face of the power generation industry and directly or indirectly promote technological innovation. They are:

- *Deregulation.* This means that independent power producers (IPPs), and some of them small power producers (SPPs), help make large plant new construction or expansion unnecessary. Consider the earlier examples of the PCS company's use of what were Alstom Power GT10s (this model was part of the Siemens' acquisition of Alstom's smaller engine divisions) in combined-cycle operation. The waste hydrocarbon fluids they used as fuel helped further develop the low-Btu fuel technology experience. Many SPPs can sell their excess power back to the utility grid.
- *Oil companies as IPPs.* Shell in the United Kingdom is a good example of a growing trend. As IPPs, oil companies can be their own customer for their oil and gas. This then short circuits much of the fuel purchase agreement contractual formalities that other IPPs have to negotiate.

Integrated Gasification Combined-Cycle (IGCC) Plants
Integrated gasification combined-cycle (IGCC) plants consist of three main sections:

- The "gas island" for conversion of coal or refinery residues (such as heavy fuel oil, vacuum residues, or petroleum coke). This includes gasification and downstream gas purification (removal of sulfur and heavy metal compounds in accord with required emissions levels).
- The air separation unit.
- The combined-cycle plant. The modular design (gas generation, gas turbine system, HRSG, and the steam turbine system) allows phased construction as well as retrofitting of the CC plant with a gasification plant. This replaces the "standard" gas turbine fuels (natural gas or fuel oil) by syngas produced from coal or refinery residues.

IGCC is a combination of two proven technologies; however, proper integration depends on using the lessons learned from several demonstration projects in Europe and the United States. Currently, more than 350 gasifiers are operating commercially worldwide and there are at least seven technology suppliers. About 100 CC unit plants are ordered per year, but there is limited experience in IGCC commercial operation. Currently, we refer to operating experience at five IGCC plants: the 261 MW Wabash River plant, the 248.5 MW Tampa plant, the 253 MW Buggenum plant, the 99.7 MW Pinon Pine plant, and the 318 to 300 MW Puertollano plant. IGCC will see commercial application in developed countries, such as Italy, for residual refinery fuels and gasified coal. However, great care needs to be taken in implementing a commercialization strategy for developing countries.

Through 2015, the potential for refinery-based integrated coal gasification combined-cycle plants is estimated to be 135 GW. Currently, over 6 GW of coal- and refinery residue-based IGCC projects are either under construction or planned.

Technology, Performance, Environment, and Demand Trends

Figure 3–23 compares the supply flows, emissions, and by-products of different 600 MW-class plants. With environmental emissions (including greenhouse gases, GHG), IGCC plants compare well with pulverized coal-fired steam power plants.

Depending on the degree of integration between the gas turbine and the air separation unit (ASU), either standard gas turbine or compressor configurations can be applied. If not, the mismatch between turbine and compressor mass flows, which results from the application of gases with low heating values, requires limited modifications to compensate.

Three options are available. The selection of the appropriate air and nitrogen integration concept depends on a number of factors, to be considered on a case-by-case basis. A summary of the important criteria is provided in Figures 3–24 and 3–25.

Figure 3–23 sets out the principal criteria for selection of the different IGCC integration concepts. The "fully integrated approach" (selected for the European coal-based demonstration plants) results in the highest efficiency potential, but it can prove more difficult to operate. Nevertheless, after some initial operational problems, the Buggenum IGCC facility demonstrated that the design can provide good availability.

The nonintegrated concept with a completely independent ASU is simpler in terms of plant operation and possibly in achievable availability. However, the loss in overall IGCC net plant efficiency compared with the

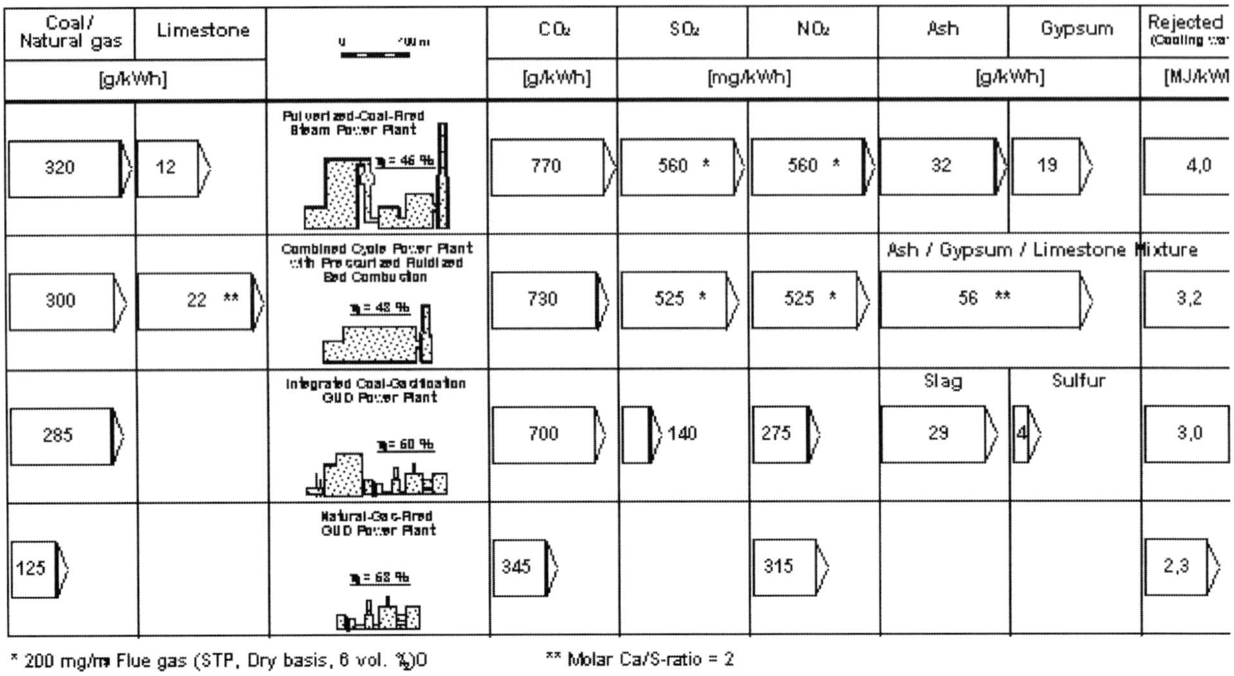

FIGURE 3–23 Comparison of supply flows, emissions, and by-products of different 600 MW-class plants [3-4].

Integration Options

1. Non-integrated (independent) ASU
2. Partially integrated ASU
3. Fully integrated ASU

Criteria for Selection of the Integration Concept

- Gasification process and waste heat recovery (syngas cooler, quench, cooler/saturator cycle)
- ASU process
- Fuel gas analysis (with or without nitrogen return)
- Limits for NOx emissions
- Overall plant efficiency
- Investment costs
- Operational aspects
- Site-specific aspects
- Necessary modifications of standard gas turbines
- Available fuel flow

FIGURE 3–24 Main criteria for the selection of the IGCC integration concept. *(Source: Adapted from [3-4].)*

fully integrated concept is 1.5–2.5%. So this concept is of interest for applications where efficiency is not the key factor (e.g., for the gasification of refinery residues).

The concept with partial air-side integration is a compromise, with only a moderate loss in efficiency but improved plant flexibility, when compared with the fully integrated concept.

What follows are some examples of gas turbines that can be incorporated into combined-cycle packages. The Siemens SGT6-5000F (198 MW, 60 Hz) operates under simple cycle, combined-cycle, and other cogeneration applications. The SGT6-5000F gas turbine has more than 1 million hours of fleet operation. It can be used in either simple cycle or heat recovery applications, including cogeneration, combined cycle, and repowering.

The SGT6-5000F provides online generation for peaking duty, intermediate operation, or continuous service.

The SGT6 5000F's technical features include:

- A 16 can-type combustors in a circular array
- A 16-stage axial-flow compressor with advanced 3D design technology
- Variable inlet guide vanes
- Four-stage turbine
- Advanced cooling technology
- Optional multiple fuels capability
- Low-NO_x combustion
- Cooled, filtered rotor cooling air
- Steam power augmentation (optional)

The Siemens SGT5-2000E (163 MW, 50 Hz) operates under simple or combined cycle and other cogeneration applications. The SGT5-2000E is used for simple or combined-cycle processes with or without combined heat and power and for all load ranges, particularly peak-load operation. For integrated coal gasification combined-cycle applications, Siemens provides the SGT5-2000E (LCG) machine, a 2-type machine with modified compressor. The SGT5-2000E has more than 120 units in operation, accounting for approximately 70,000 starts and more than 4 million operating hours.

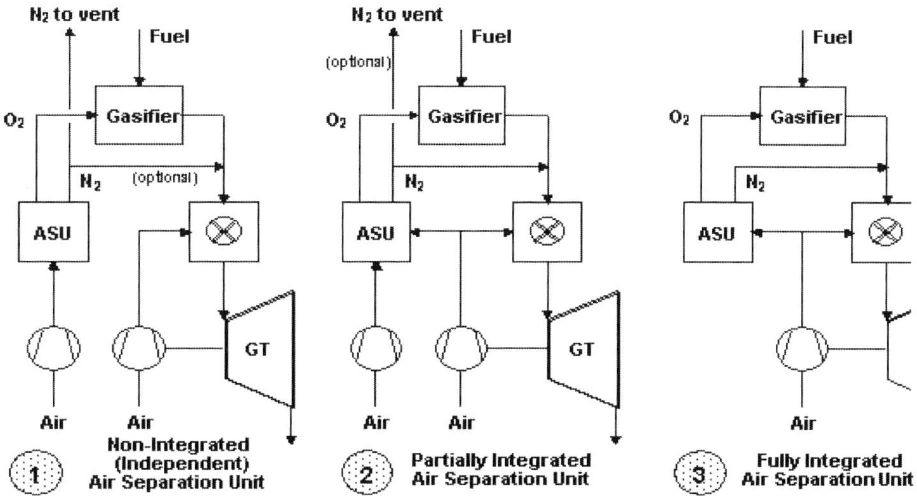

FIGURE 3–25 Integration options for IGCC power plants [3-4].

The SGT5 2000E's technical features include:

- Horizontally split casing
- Two walk-in combustion chambers for hot-gas-path inspection without cover lift
- Combustion chambers lined with individually replaceable ceramic tiles
- A 16-stage axial-flow compressor with variable-pitch inlet guide vanes
- Optional fast inlet guide vanes for peak-load operation and frequency stabilization
- A 4-stage turbine
- All blades removable with rotor in place
- Hybrid burners for premix and diffusion mode operation with natural gas, fuel oil, and special fuels, such as heavy oil and refinery residue
- Low-NO_x combustion

CASE STUDY 1: AN END-USER/EPC CONTRACTOR'S EXPERIENCE WITH SOME OF THE OEMS' LATEST GAS TURBINE MODELS IN POWER GENERATION SERVICE*

The writers cover EPC experience gained using "FX" (a generic term for advanced "F" technology) and "G" class GTs. This equipment was provided by major GT suppliers, such as General Electric (GE) (7FA, 7FB, 9FA), Alstom (GT26), and Siemens-Westinghouse (501FD, 501G, V94.3A2). This case illustrates the potential economic implications of using gas turbine technology that is still relatively new.

* Reference: [3-6] Courtesy of J. Zachary, the material includes extracts from J. Zachary and R. Narula, "Next Generation Gas Turbines, an EPC Contractor's Experience."

Revolutionary ("Grassroots" New) versus Evolutionary (Based on a Known Core) GT Changes

Brand New Models

In recent years, the heavy-duty GT industry made important technical strides to cater to the all-important combined-cycle power plants market. The introduction of the "G" and "H" technologies, offered currently by manufacturers, creates an inseparable thermodynamic and physical link between the primary and secondary power generation systems by using steam (in lieu of air) in a closed loop to perform most, if not all, GT cooling.

These new GT designs rely on an evolutionary process using a proven existing design base and manufacturers' accumulated expertise. However, to achieve substantial performance improvements and meet stringent low-emissions requirements, major design changes or even new designs must be implemented.

Note the "cross pollination" between stationary and aeroengine designs. The aeroengine manufacturers, such as Pratt & Whitney and Rolls Royce, entered into commercial agreements with heavy-duty GT manufacturers, such as Siemens-Westinghouse and Alstom. The GE R&D group and Aero-Engine Division likewise played an important role in GE Power Systems' heavy-duty GT development. Compressor and turbine aero-design, air and steam cooling, combustors, and intake and exhaust systems are just a few examples of this type of technology transfer.

The development process requires individual component testing for all major GT features, including the compressor, combustion system, and turbine. Despite extensive multiphase validation and integration programs, all new GTs have exhibited many "first of a kind" technical challenges in the field. Table 3–6 summarizes some of the issues publicly acknowledged by the manufacturers.

TABLE 3–6 Partial List of Gas Turbine Problems Experienced at Introduction Phase [3-6]

Component	Problem	Equipment Type
Compressor	Tip rubbing	MHI-M501G
	Second stage stator—high cycle fatigue cracks	SWPC-W501G
	Vibrations	SWPC-501G
Combustor	High metal temperature—transition piece inlet	MHI-M501G
	Flashback	MHI-M501G, Siemens V84.3A, Alstom GT24/26
	Cracks in transition panel or transition piece outlet	MHI M501G, SWPC W501G
	Combustion oscillations	SWPC W501G, Siemens V94.3A2, GE 7FA, Alstom GT24/GT26
	Liner overheating	Alstom GT24/GT26
Turbine	Shorter useful life of nozzles and blades	All
	TBC problems	All
	LPT first row cooling	GT24/GT26
	LPT second row shroud	GT24/GT26

OEMs, EPC contractors, and insurance carriers all paid a hefty price to correct these problems in the field.

Manufacturers (including Alstom, Siemens, and MHI) established their own in-house (load) test facilities to evaluate equipment performance in a completely controlled environment and identify potential problems earlier. Others selected the alternative approach of using power plants as validation sites, where plant owner(s) allow the OEM to conduct tests using additional instrumentation and to implement changes in exchange for special commercial arrangement. Table 3–7 lists test facilities for major manufacturers.

The testing of the equipment at these facilities is useful in identifying short-term issues for validating compatibility of various components and optimizing performance. Many solutions for the dry low-NO_X (DLN) combustor problems were worked out at these sites, where complete combustion systems could be tested.

Testing of the complete GT (compressor, combustor, and turbine) is necessary to validate the combustion system design, since rig tests alone are not sufficient to determine the interaction between individual components. Note that the test facility cannot replicate the actual operating condition(s) (fuel variability, load following requirements, and

TABLE 3–7 Testing Facilities for Advanced Gas Turbines [3-6]

Manufacturer	Location	Type of Gas Turbine	Remarks
Alstom	Birr, Switzerland	GT26	In-house test facility with generator
Siemens-Westinghouse	Berlin, Germany	V84.3A2 501 FD	In-house test facility with water brake
Siemens-Westinghouse	Lakeland, FL US	501G	First installation—validation site
Siemens	Cottam, United Kingdom	V94.3A2	Demonstration site
MHI	Takasago, Japan	M501G, M501H	In-house verification plant with generator
GE	Baglan Bay, United Kingdom	109H	"H" system—validation site

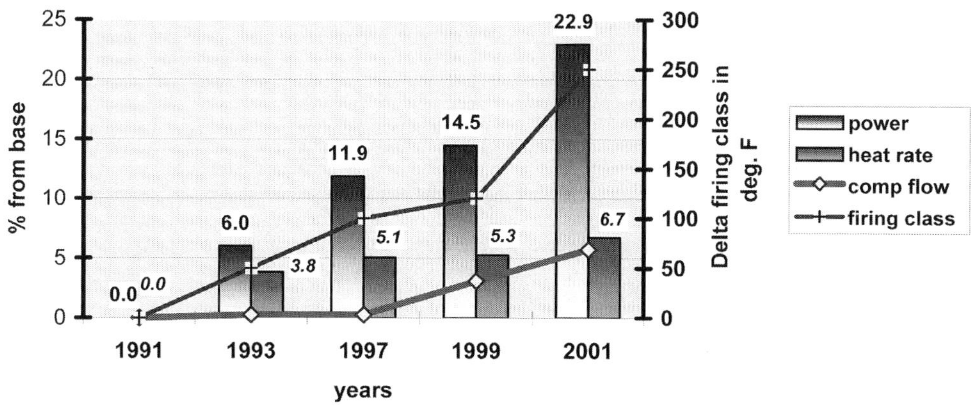

FIGURE 3–26 Evolutionary process of "F" technology for power output and heat rate [3-6].

extreme ambience) of other locations and the machine's long-term behavior.

Existing Model Upgrades

It is typical of OEMs to upgrade and improve the thermal or mechanical performance of their existing models. This, however, poses a very interesting question—should these upgrades be treated as "tweaks" or as brand new models? A good example involves the GE 7FB and Siemens-Westinghouse 501 FD models, which provide thermal performance far superior to that of the original "F" models.

The evolutionary process for the GE "F" technology class is illustrated in Figure 3–26. The power output increased from 150 MW to 174 MW and the heat rate improved by about 7% over a period of 10 years. The major changes include increased airflow and pressure through the compressor and higher firing temperature resulting from use of better materials for the turbine blades and nozzles (single crystal).

The two tracks of performance betterment—brand new models and existing model upgrades—interact. Many of the improvements from the "H" class of GT are flowing back to the "F" and "G" class equipment.

The most recent examples include the GE 9FB's implementation of many GE "H" system features, and the MHI M701G2's superior performance over that of the original M701G GT, based on M501H technological enhancements.

When GT development work had to be carried out in the field, the manufacturers stood behind their products and provided the needed support until the issues were solved. Performance shortfalls and schedule were dealt with through the contractual commercial agreements.

Role of the EPC Contractor

Equipment Selection

An experienced EPC contractor remains technology neutral but proactive in identifying the best available technology meeting owner requirements. Therefore, before selecting the equipment, the EPC contractor must understand the owner's pro forma objectives for plant output, heat rate, reliability, availability, and emissions requirements.

The process entails a comprehensive technology assessment with the supplier and other previous users of this technology. The performance offered by the OEMs for a specific project must be normalized and correlated with the performance of the same GT, achieved at other executed projects. The analysis should also include a comparison with other OEMs' competitive products.

The EPC contractor creates an in-house database for field test data. Figure 3–27 summarizes test data from 29 units recently completed by Bechtel and includes all major manufacturers.

In addition to the power output and heat rate, the major issue related to evaluation of GT performance in combined-cycle applications is quantification of the GT exhaust flow and temperature. All four of these parameters for a GT are heavily interdependent. For example, a GT that exceeds the guaranteed power output usually has a better-than-expected component efficiency, particularly for the turbine section. As a result, the exhaust temperature is lower. A lower turbine exhaust temperature has a negative impact on the heat recovery steam generator's steam production and steam turbine output. Conversely, higher GT exhaust energy has a positive impact on ST output.

No GT meets all four guarantees concurrently. When establishing the HRSG and ST design conditions, it is good engineering practice to allow for some variability of GT exhaust flow and temperature. This way, the design could accommodate either shortfalls or better-than-guaranteed exhaust energy (flow and temperature) of the GT.

Figure 3–28 quantifies the effect on the combined-cycle heat rate, due to a 5°F lower exhaust temperature and a 1% better turbine section efficiency. Figure 3–28 also shows that, for the same firing temperature, a 1% improvement in GT performance overcomes the performance loss due to a 5°F shortfall in exhaust temperature and still results in 11 Btu/kWh (0.16%) improvement in combined cycle heat rate. These numbers may vary for different GT makes and models.

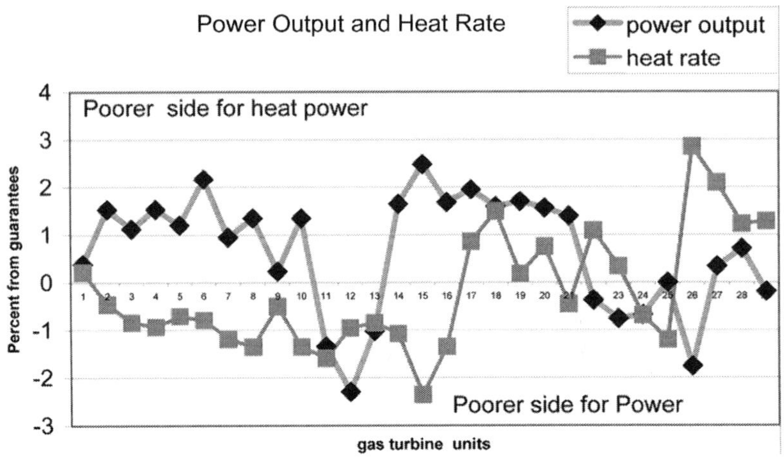

FIGURE 3–27 Comparison between guarantees and test results for power output and GT heat rate [3-6].

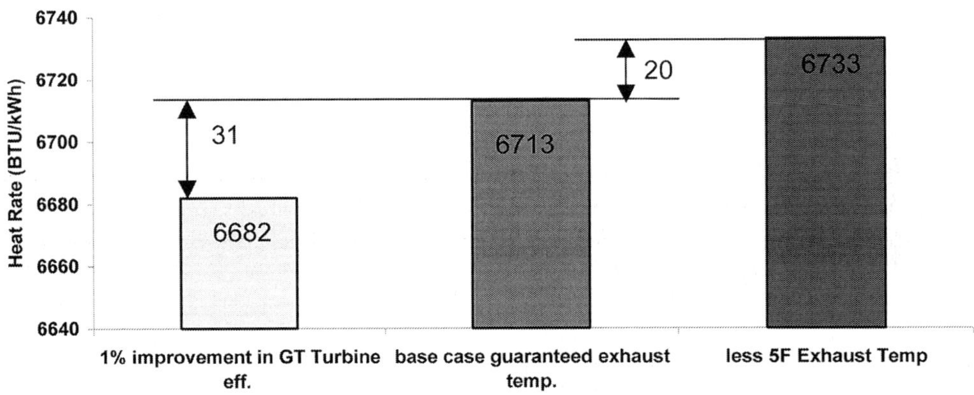

FIGURE 3–28 Influence of the improved efficiency of the turbine section and the penalties associated with lower exhaust temperature for a 1 × 1 combined-cycle configuration [3-6].

To properly assess the impact, owners or operators need to either hire an EPC contractor experienced in purchase, have a reservation agreement for a GT or power island (PI), or have the appropriate skills secured from a different source. The EPC contractor can verify:

- That terms and conditions essential to managing project design, construction, and commissioning are adequately covered in these agreements.
- That the scope is complete and all interfaces are well defined.

They can also assist by:

- Assuring adequate coverage of performance test tolerances and measurement uncertainty.
- Addressing the impact of GT performance offsets (between power output and exhaust energy) on HRSG and ST sizing.
- Ensuring consistency in pollutant levels and units included in the plant permits and those guaranteed by GT suppliers.

Table 3–8 lists technologies and locations for various projects around the globe involving GTs from three major suppliers.

TABLE 3–8 List of Projects with Three Major OEMs and Bechtel Participation [3-6]

Number of GTs	Type of GT	Manufacturer	Country
15	7FA	GE	US, Mexico
13	9FA	GE	UK, Turkey
4	GT26	Alstom	UK
2	V94.3A	Siemens-Westinghouse	Netherlands
4	W501G	Siemens-Westinghouse	US
9	W501F	Siemens-Westinghouse	Brazil, Mexico, US

Power Plant Configuration

The most common configurations (as noted in Chapter 10 on performance) include:

- 1 × 1 (one GT, one HRSG, and one ST)
- 2 × 1 (two GTs, two HRSGs, and one ST)
- 3 × 1 (three GTs, three HRSGs, and one ST)

Some of the advantages of the 1 × 1 configuration are pointed out in the chapter on performance as well and include:

- Greater flexibility
- Shorter construction duration of the first unit
- Several units in a 1 × 1 layout and using supplemental duct firing can respond more effectively and economically to grid requirements
- Plant redundancy
- Spare parts inventory can be optimized

Design Challenges

Emissions

GT designers have to balance two contradictory requirements: better performance and lower plant emissions. On the one hand, to achieve higher output and better heat rate, OEMs have to increase the firing temperature or the airflow through the machine, thus leading to higher amounts of pollutants. On the other hand, regulatory agencies in the United States require ultra-low emissions of less than 2.5 ppm for NO_x, CO, and ammonia slip.

Power plants based on either "F" or "G" technology cannot meet the ultra-low emission level without post-combustion emissions control methods. Figure 3—29 depicts the history of tightening NO_x emissions requirements at the stack in the United States.

Post-combustion devices, such as selective catalytic reduction (SCR) for NO_x and CO oxidizer for CO and volatile organic compounds (VOCs), are commonly used. They add complexity and cost and lengthen the startup schedule. In some "G" class machines, the effectiveness of the SCR must be over 90% to meet the 2.5 ppm NO_x stack emissions limit.

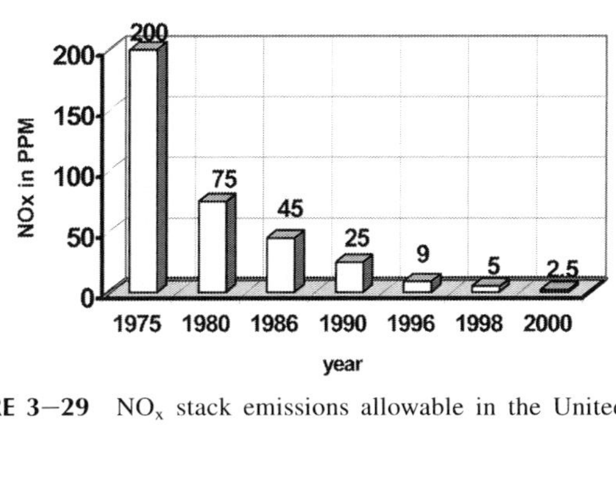

FIGURE 3—29 NO_x stack emissions allowable in the United States [3-6].

Fuel

Due to the higher cycle pressure ratio, the "G" GT needs higher fuel pressure at the GT interface flange. Because the gas pipeline pressure typically is lower, a suitable fuel compression system with associated interfacing controls is required. A filtration and purification system also is necessary to avoid entraining droplets of liquid lubrication oil (heavy hydrocarbons) in the gas. Premix combustion systems are very sensitive to the presence of heavy hydrocarbons in the fuel, causing premature ignition "flashback." From a reliability viewpoint, fuel system redundancy is an important consideration in selecting the fuel gas compressors.

In combined-cycle applications, it is common to preheat the fuel with hot water or steam to improve the plant heat rate. However, the fuel preheating system has been an issue for some new GTs, mainly due to combustion oscillations.

Cooling Steam Quality and Blowdown Design

In "G" technology, intermediate pressure steam cools combustion transition ducts. The use of steam as a cooling medium calls for special attention to achieving and maintaining the specified steam quality. While the steam quality requirements (silica, sodium, chlorides, and total organic compounds, or TOCs) for advanced GTs using steam cooling are no more stringent than the normal steam quality requirements of a combined cycle, deviations or excursions from the requirements can lead to serious consequences.

To avoid such situations, state-of-the-art reverse osmosis or electrodialysis ion exchange demineralizer water treatment systems have been used. Special attention should be given to the transient conditions during startup and power ramping. In these modes of operation, drum water characteristics, especially pH, are difficult to control.

As a result, passivity films can be removed, corrosion products can be displaced, internal treatment chemicals can be over- or underfed, and finally, steam purity can be compromised. Exceeding equipment manufacturers' limits poses a significant risk to the owner-operator.

Balance of Plant Issues

Proper system integration is vital to achieving guaranteed performance. System integration via balance of plant (BOP) equipment and interconnecting or support systems is the key that enables the GT, HRSG, and ST to operate smoothly and efficiently at all design loads and configurations.

While the BOP requirements for the new generation of GTs are no different than earlier GT designs, the larger size of BOP systems (amount of circulating water, steam, air, and fuel) warrants careful attention.

TABLE 3–9 Types of Heat Sink Used for Various Projects [3-6]

Number of GTs	GT Type	Heat Sink Type
1	9FA	Once-through seawater
2	9FA	Cooling tower
5	9FA	Indirect dry cooling—Heller type
2	9FA	Air-cooled condenser
4	GT26	Air-cooled condenser
1	W501G	Cooling tower
3	W501G	Air-cooled condenser
7	W501FD	Cooling tower

In several GT models, the cooling air circuits include external heat exchangers. This process increases the overall efficiency of the plant by utilizing the rejected heat in the steam cycle rather than dissipating it in the atmosphere. However, the mechanical layout design and control of the associated piping for air and steam are challenging tasks.

A majority of the recently completed plants using advanced GTs are merchant plants (MPPs) and dispatch only when competitive. Accordingly, the BOP design should respond to cycling operation requirements.

The selection of an optimum heat sink is site specific, since it is based on many variables, such as site configuration, water availability, water disposal, ambient conditions, level of target power output, anticipated hours of operation, the power purchase agreement structure, and the owner's economic evaluation factors. Table 3–9 lists the heat sinks selected by Bechtel for some recent projects.

Construction and Field Problems

Startup

EPC contractors generally are responsible for plant startup. Significant penalties are assigned if timely operation is not achieved.

Dual-Fuel Operation

Typically, advanced GTs use natural gas as the primary fuel. Duel-fuel capability, using distillate oil, is attractive to many customers with interruptible gas supply contracts. However, dual-fuel capability adds complexity to already very complicated combustion systems and controls. A difficult challenge is to achieve switchover from one fuel to another at a reasonably high GT power level.

Implementation of Modifications in the Field

One of the most critical challenges to erection of advanced GTs has been implementation of field modifications. On many occasions, the OEMs used unscheduled outages, not only to correct a problem but also to implement a number of design changes based on lessons learned on other sites. This process created a "ripple effect" requiring additional changes in the plant control software and start sequences.

Combustion System Commissioning and Tuning

To meet strict emissions requirements, all advanced GT combustion systems operate with DLN combustion systems. The combustion process between 70% and 100% load takes place in premix mode with a lean equivalence ratio, which creates a lower localized flame temperature and therefore lower NO_x.

The premix combustion operation, in which fuel and air are mixed in advance of the combustion process, is less stable than diffusion flame. It is common practice to use a pilot operating in diffusion mode to provide stability and inhibit the excessive combustion-related pressure fluctuations inherent in the lean premix operation.

The combustion system operation from diffusion mode at low loads to full premix mode at base load takes place in several complicated steps and stages, requiring very close control of fuel flow and exhaust temperature. The process is sensitive to ambient conditions, combustion-associated instabilities, and even physical combustor dimensions, due to manufacturing or assembly tolerances.

Currently, each GT is individually adjusted to meet the performance guarantees and emissions requirements without combustion oscillations. This practice has become a standard feature of the GT commissioning. However, this activity has an adverse impact on the EPC contractor.

The execution schedule is extended to perform a water wash of the compressor prior to the tuning process and to install and remove temporary instrumentation for full-blown GT performance testing. Because emissions limits must be met at all ambient conditions, adjustments made in the field might modify the performance correction curves for ambient temperature.

Low Load Flexibility

Gas turbines tend to run best at peak load. Frequently, however, operating conditions dictate that load varies. Examples include gas turbines used in gas and oil production facilities. Gas turbines driving pumps that extract seawater (from an underground mix of oil, gas and seawater) may not always have the same amount of seawater to move. Similarly, gas turbine driven compressors delivering production gas from a mixed field may not always have the same amount of gas supplied at their inlet and the gas may not always be the same molecular weight either. The implications for load and efficiency of driver and driven units is obvious.

To some degree, loads will also vary on gas turbines in marine service and aeroengines, especially those that drive a propeller. In today's power generation market, load conditions may also vary.

Due to its broad potential applicability to the global gas turbine fleet, the engineering and testing described in the following case study must be thoroughly validated in the event of wanting to confirm any gas turbine will run at low load.

CASE STUDY 2: AN OEM'S DEVELOPMENT OF A GAS TURBINE THE SGT6-5000F* (FORMERLY KNOWN AS W501F) ENGINE

Introduction

The SGT6-5000F (formerly known as W501F) engine has demonstrated an exceptional operational record over the 16-year, 5.7 million fleet hour operational history. Since its introduction in 1993, this F-class gas turbine has undergone continuous development to improve performance, reliability, and operational flexibility and to reduce emissions and life cycle costs. This gas turbine, which was designed for both simple-cycle and combined-cycle (CC) power generation in utility and industrial applications, represented the next model in the successful W501 family. Its design was based on fundamental, time-proven design concepts as well as new concepts and technologies incorporated to increase efficiency, reduce NO_x and CO emissions, and enhance reliability. It was designed to operate on all conventional fuels, as well as coal-derived low Btu gas. New technologies were validated for engine application by extensive rig and two full-load engine shop tests, as well as field tests in the initial installation.

Figure 3—30 shows the latest evolution of the SGT6-5000F(4) engine. The 13-stage compressor is connected to the 4-stage turbine by a single tie bolt, and the Ultra Low NOx combustion system employs 16 can-annular baskets. The 213 SGT6-5000F engines currently in service are employed in peaking, intermediate, and continuous duty operation. The fleet has amassed more than 5.7 million operating hours and has demonstrated excellent reliability, availability, and starting reliability.

Due to design enhancements, development efforts and technology cross-flow from other Siemens' advanced gas turbines, the rated simple cycle output has increased from 150 MW to 208 MW and its rated efficiency from 35% to 38+%. In 1×1 CC applications, the rated net plant output and efficiency can now be more than 300 MW and 57.5%, respectively.

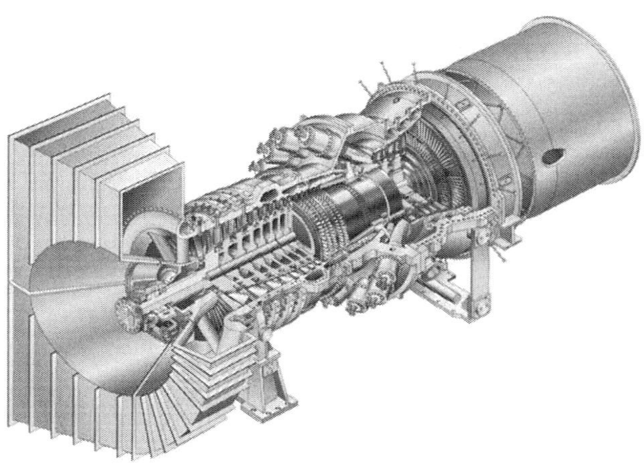

FIGURE 3—30 Siemens SGT6-5000F(4) gas turbine.

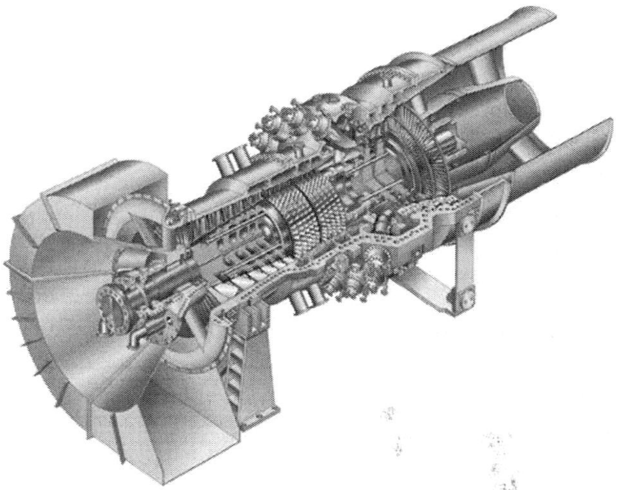

FIGURE 3—31 Siemens W501G gas turbine.

The W501G (also known as SGT6-6000G) gas turbine (Figure 3—31) design concept was driven by the changing power market. Between 1993 and 1995, the power market was moving toward deregulation and replacement of aging base load plants, such as coal-based power plants. The market was demanding clean and highly efficient combined cycle plants. Expectations of deregulation in the North American electricity market caused prospective plant buyers to turn to cleaner plants with shorter installation and commissioning times compared to traditional coal plants which had a six year (or more) lead time to permit and build. The belief, expressed by many of them, at that time was that the new high-efficiency, low emissions, clean fuel plants would be the economic and environmental choice and displace coal plants. Because the W501G had a high rated efficiency, it was intended to be operated primarily at base load.

In the late 1990s there was an increased demand for electric power and a relatively low price for natural gas, at

* Source: Courtesy of Siemens. Extracts from "Low Load Operational Flexibility for Siemens F- & G-Class Gas Turbines" GT2010-22055.

one point about \$2.50/MMBtu, causing an increased demand for gas turbines for simple- and combined-cycle operation. By 2002, the demand for power was subsiding and some areas were over capacity. Natural gas had increased to above \$6/MMBtu and the price for electricity had decreased causing gas turbine combined cycle plants to be operated on the average (based on reported information) at only 30% capacity (higher utilization was reported for advanced frames including the Siemens F & G fleet). The increased natural gas price caused some combined-cycle plants to move lower in the dispatch order causing them to operate in cycling duty mode. In this context, the amount of time some merchant plants can operate profitably may be significantly reduced. In response to the change in market demand for more cyclic operation, design changes and improvements are regularly being incorporated into the SGT6-5000F and the W501G engines.

Market Drivers for Low Load Operational Flexibility

Since the beginning of the power market's liberalization in the mid-1990s, the power plant business has been changing. Today, many power plant operators can find themselves in a more challenging market situation with the presence of strong competition, higher fluctuation of fuel prices, and in some cases without long-term power purchase agreements. Despite these new challenges, the market liberalization also presents new business opportunities such as the utilization of market price fluctuations for operation and maintenance optimization, participation in ancillary service markets and short-term trading. All of these opportunities can contribute to significantly improve operating margins. By knowing how to approach these opportunities, an operator can in some cases achieve higher profits when compared to operation under a long-term power purchase agreement.

The changed market conditions can have an influence on the operating profile of virtually every power plant. Combined-cycle power plants often do not operate in a base load regime running 8000 hours per year. Many units are operating in a daily start-stop regime with some units starting up twice a day. In this market environment, an economic model that incorporates only a certain amount of base load hours with fixed power revenues will not describe the full picture. Additional earnings from the above-mentioned market opportunities would not be considered. To be more accurate, an extended approach for evaluating a cycling plant with high flexibility is necessary. Key parameters for operational flexibility are, for example, startup time, standby operation, and shutdown time.

From the mid-1990s to 2000, published reports indicated that there was a steady reduction in the US electric power reserve margins. It was believed by some that deregulation and cleaner, more efficient combined-cycle plants would replace aging base load generation stations, such as nuclear and coal-based plants. Increasing demand for electricity and high electricity prices caused a surge in new orders for both simple-cycle and combined-cycle plants. The result was an increase in total generated electricity capacity and reserve margins in all US regions, as well as a decrease in CC capacity factor (defined as the ratio of actual generation to the total possible generation over a time period). The capacity factor reduction led operators to operate in peaking and intermediate modes rather than in the planned base load, thus increasing demand for cyclic operation capability.

Demand growth, economic dispatchability, and operational flexibility are key factors that influence the electricity-generating plant's ability to improve its dispatch rate (i.e., the order in which it is dispatched as demand for electric power increases during the day). Due to the reported current overcapacity and the increase in reserve margins, the units that excel in economic dispatchability and operational flexibility will dispatch before other competing units. The dispatch order is determined by the unit's variable production cost (VPC). Fuel cost and variable operation and maintenance (O&M) cost are used to calculate VPC. Small changes in VPC can significantly affect the unit's dispatch ranking. Fuel cost is directly impacted by the gas turbine's efficiency, thus increased efficiency improves not only the revenue per megawatt hour but also the unit's total dispatch hours. Reduced O&M costs will also lower VPC, improve dispatchability, and increase net cash flow. Units that are operationally flexible and can load follow, cycle on and off more economically, which will allow improved dispatchability and a competitive advantage in the current market. Design improvements made in the Siemens fleet of SGT6-5000Fs and W501Gs are intended to help enhance the units' dispatchability, increase efficiency and lower life cycle costs (hence reduced VPC), and improve operational flexibility.

Design and Implementation

A key issue in operating at lower loads is an increase in carbon monoxide (CO) emissions. When the engines are base loaded, the combustion system operates at a higher firing temperature and most of the CO is oxidized to carbon dioxide (CO_2). However, at part loads, the firing temperature is lower and the CO-to-CO_2 oxidation reaction is quenched by the cool regions near the walls of the combustion liner. This can result in increased CO emissions at low loads. Hence, these engines were likely operating between 70% and 100% of GT base load. To address this, Siemens has developed a modification to the engine designed to allow the gas turbine to continue to operate at lower loads while maintaining emissions at set levels.

The "low load turndown" design is based on a combination of reducing the compressor airflow and bypassing air around the combustor to achieve higher firing temperature at part loads. Compressed air is bypassed around the combustor into the turbine to increase the fuel-to-air ratio inside the combustor. This along with a combination of inlet guide vane (IGV) and secondary air system controls allows the combustor to operate at higher primary zone temperatures (as compared to non-LLCO configuration) at lower loads thus resulting in improved CO-to-CO_2 conversion.

Figure 3—32a is a conceptual schematic for the additional LLCO-specific bleed air bypass system for the G-class engine. It illustrates how—during low load operation—extra bleed air is bypassed from the combustor shell and compressor bleeds into the turbine stages. The approach for the F-class design is also similar with some modifications (additional bleed into the exhaust) as needed for the operating conditions for the F-class engines (Figure 3—32b). Some of the instrumentation (thermocouples, annubars, etc.) that was used during the test to verify operation is also illustrated in Figure 3—32a and 3—32b.

This low load turndown design (also referred to as LLCO for brevity due to improved low load carbon monoxide emissions) was first installed in one of the operating engines of the Siemens G-class gas turbine fleet in October 2007. Subsequently, a slight variation of the design (tuned for frame operating conditions) was installed on an SGT6-5000F operating engine in April 2008.

The installations included modification to the cooling air piping and installation of additional valves (Figure 3—33). Changes were made to the control logic to accommodate the changes to the operation of the gas turbine. Several sections of the gas turbine were instrumented to record the impact of the LLCO product. The same test was conducted on the unit before and after LLCO was implemented in order to obtain a direct comparison of the engine response between the baseline and LLCO operation. A combined cycle emissions and performance test was also conducted in order to measure the differences in emissions and performance between the baseline and LLCO operation.

Results

In close cooperation with the customer, Siemens was able to test this design while the engine was in commercial operation. The focus of the test was to turn down the engine to the lowest load possible while maintaining CO emissions at or below 10 parts per million (ppm) (corrected to 15% O_2). All the instrumentation was continuously monitored for deviation from pre-set limits.

A maximum turndown of approximately 40% GT load was obtained while maintaining 10 ppm CO out of the engine and the turndown was up to 28% GT load with 10 ppm CO out of the exhaust stack using the CO catalyst. There may be potential methods of further improving the turndown capability at lower ambient temperatures such as combining LLCO with an inlet heating upgrade.

For the W501G engines, LLCO control settings are activated approximately around 28% GT load. At these very low loads, this control methodology was calculated to provide a significant CO reduction during startup and shutdown. Figure 3—34 shows that the CO emissions out of the gas turbine are calculated to drop from about 2400 ppm (corrected to 15% O_2) at about 30% load with normal control to about 500 ppm (average) when LLCO is installed. It is also significant to note that the relatively high CO emissions period appears to last for a much shorter load transient. With the normal control (black curve), the high CO emissions are experienced between 30% and 70% GT load. With LLCO implemented, the relatively higher CO emissions transient period is only between 30% and 40% load. This can be particularly beneficial for plants with permit limits capped by total annual CO emissions. Site-specific plant configurations need to be evaluated for use of this product at all startup conditions.

Figure 3—34 shows measured CO emissions data (corrected to 15% O_2) after the LLCO upgrade was installed at four different sites (engines). It can be noted that there is some variation between the different sites in terms of CO emissions reduction. It is evident that the CO emission curve is almost exponential to %load at lower load regimes (30—45%). Hence relatively small differences in the engine performance parameters can have a relatively larger impact on the CO emissions.

The steam turbine also operated reliably during these combined-cycle tests. The primary goal was to maintain enough heat input to the Heat Recovery Steam Generator (HRSG) to generate enough steam to keep the steam turbine operational, which was achieved. In the case of the G-class gas turbines, the transition section of the combustion system is cooled with steam. It was also observed that transition steam temperatures were within design operating limits. In certain cases, some control settings or capacity changes may be needed for attemperators to optimize turndown capability.

The impact to the HRSG due to LLCO operation was also evaluated during the test. No apparent failure modes were observed. Furthermore, appropriate steam cycle data from the LLCO tests was evaluated by the HRSG vendors (both from the F- and G-engine tests) using their analytical models. The results of the evaluation were that all "as tested" conditions were within the design criteria and no detrimental impact was reported by the HRSG vendors in their analyses. In fact, one vendor reported that it was their view that LLCO operation could conceivably be better in comparison to daily cycling due to the reduced thermal cycling fatigue.

Validation and Field-Follow

Since the LLCO upgrade was installed it has been extensively used as is demonstrated in Figure 3—35.

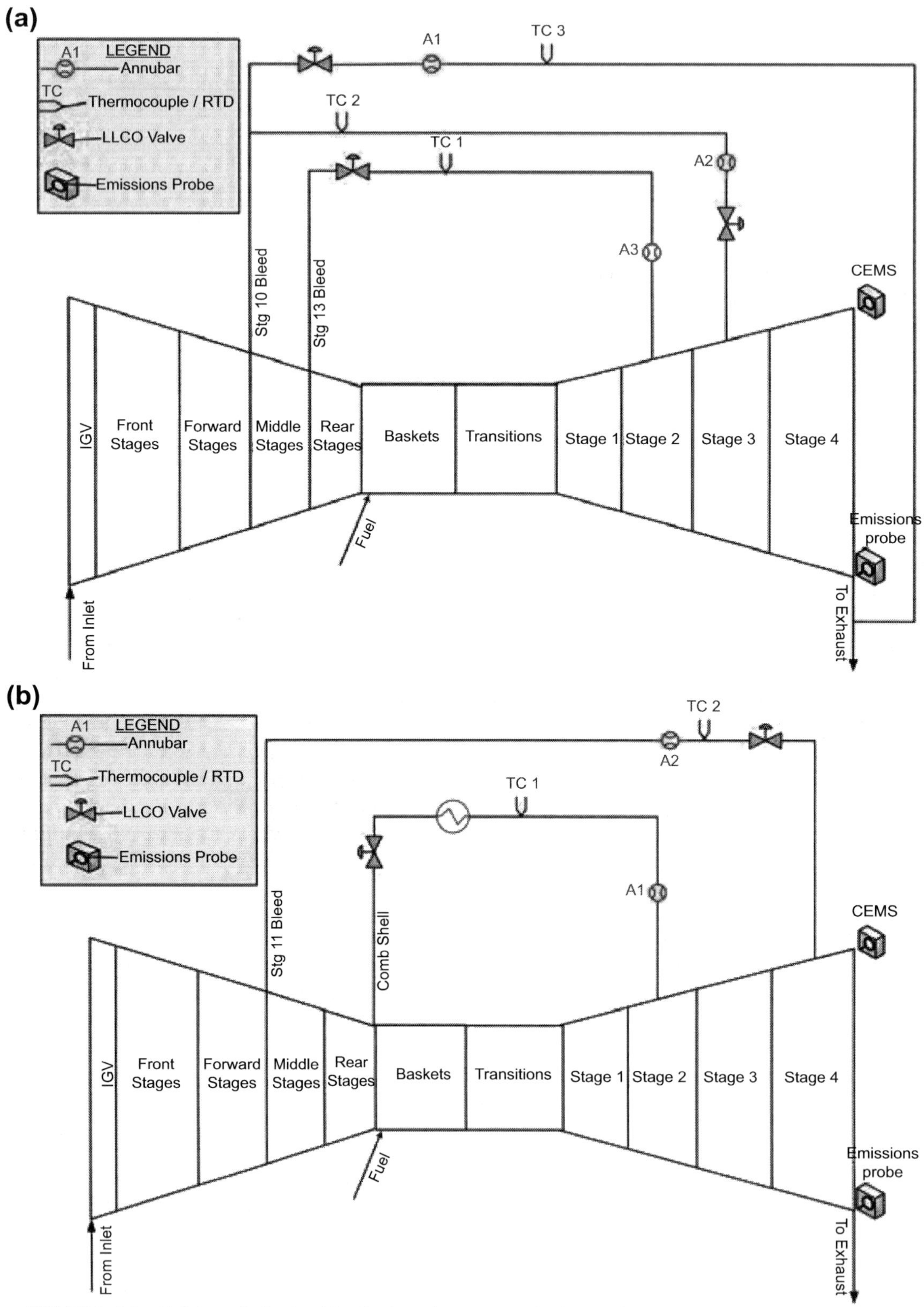

FIGURE 3–32 (a) Schematic for the G low load turndown system. (b) Schematic for the F low load turndown system.

FIGURE 3—33 LLCO upgrade being implemented on a site.

Before the LLCO upgrade, this unit was reportedly limited to operation between 100% and 70% load and hence was reportedly often shut down during low power demand periods. Now, following the installation of the LLCO upgrade, it has been reportedly turned down to minimum load (~50% GT load) and left there while the demand is low. When demand for electricity is higher, the GT has been brought to higher loads as needed in a relatively shorter period of time.

The F-class engines are also accumulating operational hours with this upgrade and continue to take advantage of the turndown flexibility, which allows them to follow the varying demands of the power market.

In the work described by the OEM Siemens (above), that OEM worked closely with the end-user. This dialogue is always beneficial.

It is appropriate now to provide a perspective from an EPC contractor, who may also handle commissioning and operations services for a client. This work is more general in that it covers experience with a number of different GT systems, GT sizes, and therefore varying "N2" (which in this case refers to the minimum number of GTs required to meet the power demand). By "large" this author refers primarily to "G" and "H" sized GTs.

CASE STUDY 3: OPERATIONAL EXPERIENCE WITH LARGE ADVANCED GAS TURBINES IN VARIABLE LOAD CONDITIONS*

Nomenclature

AVT = All-Volatile Equipment
BOP = Balance of Plant
BPST = Back-Pressure Steam Turbine
CC = Combined-Cycle Power Plant
CPP = Captive Power Plant
DCS = Distributed Control System
EPC = Engineering, Procurement, and Construction
GT = Gas Turbine
HP = High Pressure
HRSG = Heat Recovery Steam Generator
IGV = Inlet Guide Vane
IP = Intermediate Pressure
LP = Low Pressure
NG = Natural Gas
PRDS = Pressure Relieving Desuperheating Station
RCT = Refinery Crude Train
RFG = Refinery Fuel Gas
ST = Steam Turbine
TDS = Total Dissolved Solids
TE = Temperature Element
TOC = Total Organic Carbon
TT = Temperature Transmitter
WFO = Waste Fuel Oil

Introduction

The unique challenges involved in designing a power generating station to support the special needs of an industrial processing facility are discussed. A conventional combined-cycle (CC) power plant has one basic goal: to produce electricity at a minimal cost for sale to the grid. In industrial facilities (e.g., oil refineries, smelters, and chemical and desalinization plants), electricity and steam produced are for internal use rather than export. Since power and/or steam supply interruptions might have catastrophic effects on facility processes, the paramount requirements for this type of dedicated plant or captive power plant (CPP) are availability and reliability rather than high thermal performance.

In recent years, the advanced H- and G-class GTs, which use steam to perform GT cooling duty, have successfully demonstrated close to 60% efficiency in CC applications and accumulated thousands of operating hours. Many improvements in the G and H class were integrated into the FB and FD class of aircooled GTs. However, industrial applications commonly use older D- and E-class medium GTs as the prime mover. Given that the fundamental purpose of a CPP is different than a traditional power station, this case covers CPP application requirements and the means to optimize these requirements in key areas to support required industrial processes. To illustrate the design methodology, three cases (a refinery, a cogeneration facility, and a smelter) are presented, describing an engineering, procurement, and construction (EPC) contractor's perspective on the unique industrial application constraints and proposed solutions to meet owner requirements. It also identifies the specific applications in which use of advanced classes of GTs might be suitable.

* [3-7] Courtesy of J. Zachary. Extracts from "Are Advanced Large Gas Turbine Ready for Use in Captive Applications?" GT2008-50720.

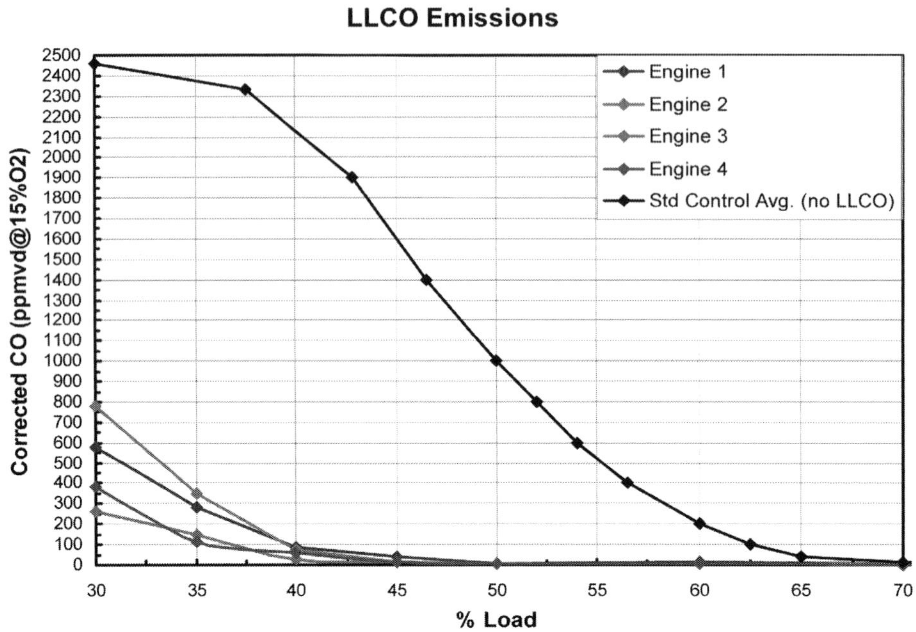

FIGURE 3–34 CO emissions results in G-class operating sites.

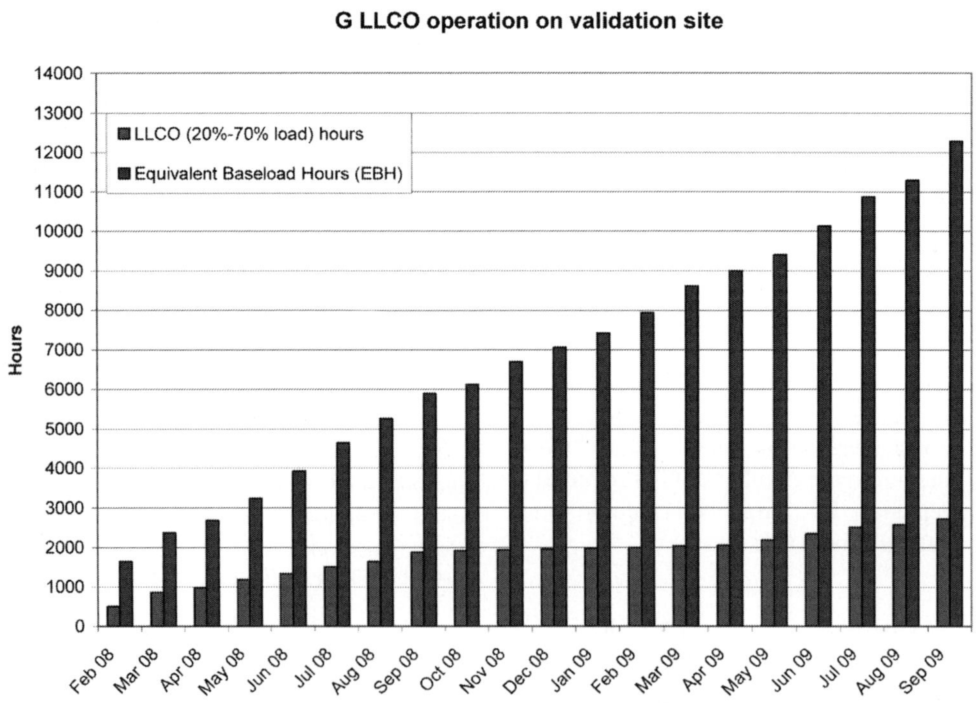

FIGURE 3–35 Operational data from validation unit.

Refinery Case

Captive Power Plant Concept

A CPP uses same major components as a traditional CC power plant; however, it is designed to achieve a different objective: the production of power and steam is "captured" for industrial use and is only sold to the grid when applicable. The differences between the two types of plants in an oil refinery environment are described below.

Assume a CPP is being constructed to support a new refinery crude train (RCT), which will produce 1 million barrels of oil/day. The oil refining process will require a constant supply of high pressure (HP) steam and electricity. If the steam or power supplies are interrupted or

TABLE 3–10 Typical Refinery Requirements [3-7]

Max Design Conditions	Unit	Quantity
Site Ambient Temperature	°C	40
Site Relative Humidity	%	90
Refinery Power Demand	MW	650
Refinery HP Steam Demand	Tons/hr	2500
Refinery IP Steam Demand	Tons/hr	1500
Refinery LP Steam Demand	Tons/hr	1000
Refinery Crude Train Production	Barrels/day	1,000,000
Normal Design Conditions		
Refinery Power Demand	MW	500
Refinery HP Steam Demand	Tons/hr	600
Refinery IP Steam Demand	Tons/hr	800
Refinery LP Steam Demand	Tons/hr	350

fall outside the narrow band of specified supply conditions, the associated processes could be ruined. As a result, a batch of products will be disposed of, leading to heavy financial penalties. Refer to Table 3–10 for typical CPP requirements. In this case, in addition to using GTs as prime movers, the critical demand for steam will require that additional conventional boilers be incorporated.

Key CPP Design Criteria

Major design areas to be analyzed include:

- Cycle selection (combined cycle versus boilers)
- Reliability and redundancy
- Fuel selection
- Auxiliary equipment design

Cycle Selection

Optimal CPP arrangements must ultimately be based on specific site conditions, owner's requirements, and equipment available to meet project schedule. When considering power demand, steam demand, redundancy, reliability, cost, and schedule, the analysis results may vary for different combinations of technologies and operating conditions. In this case, the following design criteria for a typical RCT are assumed.

Power Demand

Early in the evaluation process, it was determined that the refinery electricity requirements could be met by CC components only (GTs and back-pressure steam turbines [BPSTs]). Several combinations were considered to meet the 650 MW power demand. Various GT and BPST candidates and their MW ratings are listed in Table 3–11. The selection of larger frame GTs can decrease equipment quantity, but do not meet the N-2 criteria (minimum number of operating GTs necessary to meet the power demand).

Due to the large oil production rate (1 million barrels/day) and associated power requirements (650 MW) of the RCT, as shown in Table 3–10, six GT Model 3 units were selected for this application. Ultimately, their relatively large unit power output and operational track record made them the most suitable choice for this application. It is worth noting that GT Models 4 and 5 would also be good candidates. These six Frame E class GTs can produce a maximum of 690 MW of power. An additional 60 MW can be generated by two BPSTs using HP to intermediate pressure (IP) letdown steam. A maximum of 750 MW can be generated, which is far above the required 450 MW power demand for normal operation.

Steam Demand

The high efficiency CC designs incorporate heat recovery steam generators (HRSGs) with either two or three pressure and reheat levels. However, in this case, only single-pressure HRSGs were selected to be coupled with the six GTs for steam production. Each HRSG produces approximately 300 tons per hour (tph) of HP steam at GT base load. Additionally, moderate supplemental firing can increase steam production and add system controllability. If there are minor fluctuations in RCT steam conditions, duct firing can rapidly be adjusted to match the demand.

However, the six GT/HRSG trains do not provide enough steam to meet all RCT demands. Four single-pressure conventional boilers were selected to augment steam production. Each of the four boilers is rated at approximately 300 tph of HP steam. The boilers were sized

TABLE 3–11 Typical GT and ST Power Ratings [3-7]

	Technology Class	Rating MW (at site conditions)
GT Model 1	B	35
GT Model 2	FA	60
GT Model 3	E	115
GT Model 4	E	135
GT Model 5	E	130
ST Model 1	Large	100
ST Model 2	Small	30

to match the steam output of a GT/HRSG train. When a GT/HRSG train is down because of a failure or a planned maintenance outage, a boiler can be ramped up to take its place. By producing HP steam at the same pressure and temperature, the HRSG and boilers could be fed into a single common header. HP steam is the primary supply stream to the refinery, but additional low pressure (LP) steam is also needed. To maximize use of the HP stream, HP steam is let down to an IP steam header through two BPSTs, as described above. HP steam is also used for other auxiliary power users (e.g., fan drivers, steam tracing). The steam is then discharged to an LP steam header that is shared with the refinery. See Figure 3–36 for a simplified CPP layout diagram. The use of steam in BPSTs and for other local balance of plant (BOP) customers is more efficient than using pressure relieving desuperheating stations (PRDSs) to let down the pressure.

Combining CC and boilers provides the client with the system flexibility necessary to further optimize CPP operation during all possible processing scenarios demanded by the refinery.

Reliability and Redundancy

Reliability is a refinery owner's highest priority. If the CPP trips, the RCT that it supports shuts down. In this case, the RCT produces 1 million barrels of oil/day, which amounts to substantial financial losses for every day that the unit is not operating. To meet reliability requirements, it is recommended that the CPP be designed for normal operation and have two spares (N-2 concept). This N-2 concept allows for normal operation if two components generating steam or power are down for maintenance or fail simultaneously.

Under normal operating conditions, the power load is 450 MW. Four GTs operating at base load produce 460 MW, which is more than needed. Therefore, the other two GTs can be considered as the spares in the N-2 philosophy. One of the most critical system requirements is the capability to respond rapidly to changes in RCT electric load and steam demand. For this reason, all six GTs run at 60% part load instead of running four at base load and having two shut down. In this scenario, each GT can be quickly ramped up or down to meet the process controllability and responsiveness requirements. Similarly, during normal operation all six HRSGs operate at 90% load and all four boilers operate at 30% load.

Fuel System Design

The fuel system must be able to combust an assortment of liquid and gaseous fuels available in the refinery: natural gas (NG), diesel, distillate fuel oil (e.g., naphtha, light slurry oil, kerosene, etc.), heavy fuel oil, waste fuel oil (WFO) (a bottom distillation column fuel), and refinery fuel gas (RFG). Coal is not included due to economics and

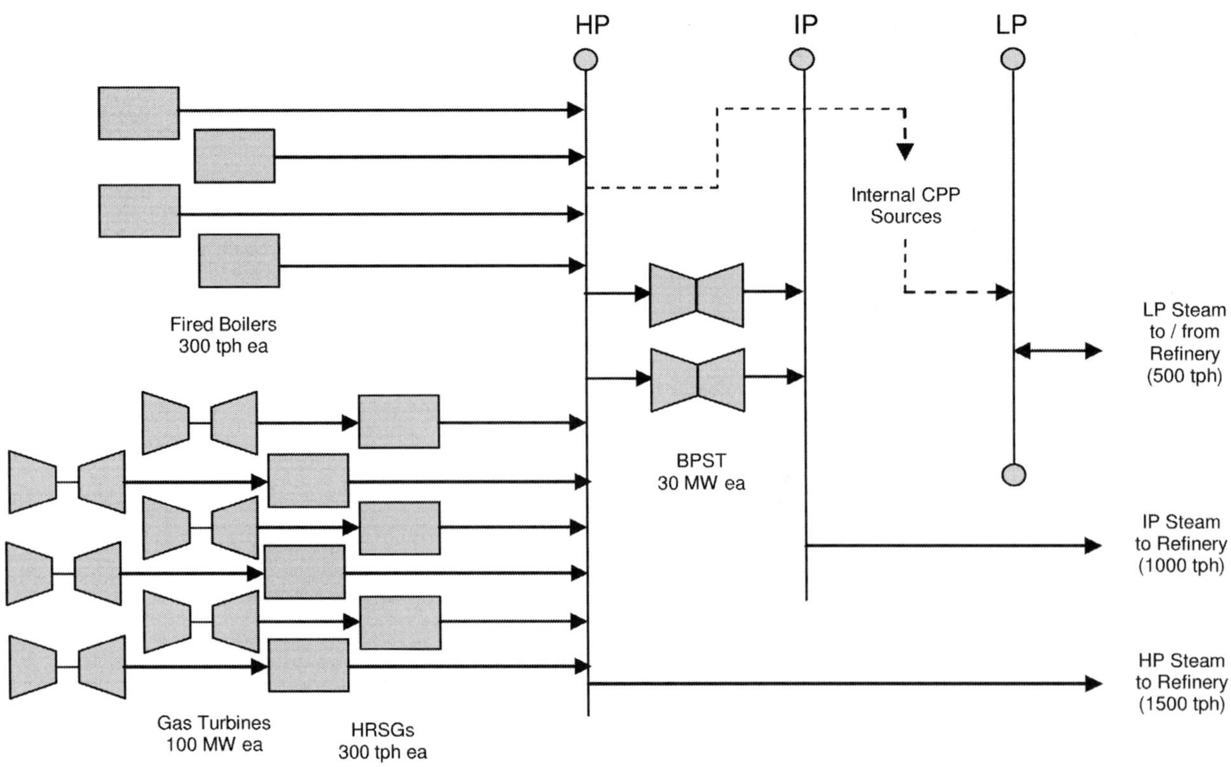

FIGURE 3–36 CPP layout diagram. [3-7]

TABLE 3–12 Possible CPP Fuels and Their Selected Users [3-7]

		GT		HRSG		Boiler	
Fuel Source	Cost at LHV	Primary Source	Secondary Source	Primary Source	Secondary Source	Primary Source	Secondary Source
NG		X		X		X	
RFG			X		X		X
Diesel			X		X		
Distillate			X		X		
WFO							X

availability constraints. Designing the CPP to use fuel oil and RFG decreases operational costs. However, refinery production changes constantly and is not a reliable fuel source. For this reason, fuel supplied to the CPP is stored in large fuel tanks. Once enough fuel has accumulated, it is burned and the tanks are refilled. See Table 3–12 for a list of possible fuels and their selected CPP users.

The GT/HRSG trains are designed to fire a different secondary fuel than the boilers. If NG (primary fuel) is unavailable, the equipment switches to secondary fuels. This approach is tied to the N-2 redundancy concept and reliability design requirements.

Auxiliary Equipment Design Considerations

All auxiliary equipment such as fans, pumps, and deaerators is designed in accordance with the N-2 concept. The reliability requirements often drive the CPP components to be redundant (compared to traditional power plants) so that the CPP never experiences total failure. As a result, the cost to build a CPP could be considerably higher than that of a conventional power plant. The additional capital cost associated with the N-2 concept is justified when compared with the penalties related to a refinery shutdown. To minimize the price of the plant, experienced designers must use simulation tools to conduct detailed risk assessments for each component and its integration with other systems. For process optimization, the CPP is designed to use steam, which is an abundant and cost-effective alternative to electricity. Because of the N-2 requirement, a minimum of three pumps or fans is provided for every application. Whenever possible, at least one of these pumps or fans is built with a steam turbine (ST) driver instead of a motor driver. If, for example, there are five feedwater pumps with three normally in operation (N-2 philosophy), then three pumps should be designed with ST drivers and two with motor drivers. By using both streams of energy, the system is controllable, reliable, and does not fully depend on either steam or electricity.

Off-Design Case Operation

The CPP must be able to adapt to (and continuously operate at) 15 off-design cases. For the application defined in this section, the minimum power requirement is 300 MW (corresponding to 1500 tph of steam) and the maximum power requirement is 650 MW (corresponding to 3000 tph of steam).

Unique Control Systems

The CPP DCS must be linked to the refinery DCS. It is imperative that these two control systems be fully integrated and closely coordinated. Instrumentation redundancy is also important. For example, a high temperature element (TE)/temperature transmitter (TT) reading can trigger major CPP operation changes, which could disrupt refinery production. To avoid errors, two-out-of-three logic is used extensively to discriminate between real upsets and false signals.

Transient Behavior

The transient behavior of the entire system, including the steam host and the CPP, must be modeled in an early plant design stage.

Cogeneration Case

Cogeneration facilities pose significant up-front design challenges. The design must balance the requirements of two different customers with dissimilar sets of priorities—one entity demanding electricity and the other needing steam. For the steam host, steam supply reliability at specific flow, temperature, and pressure conditions is of paramount importance. Therefore, process steam requirements need to be identified early, not only for one condition, but for the entire allowable range of operation (nominal, maximum, and minimum). Since the steam production will affect the amount of available electrical power, it is also appropriate to define the number of short- and long-term steam demand changes.

To further complicate the matter, in many cases, the steam demand has more than one pressure level. In the design optimization process, all power and steam requirements must be superimposed to ensure that any predefined steam demand will be met at all operating conditions.

Supplementary Firing

Similar to industrial applications, a popular cogeneration feature is the use of supplementary firing (duct firing) within the HRSG. The reduction in electric power that occurs when process steam demand is increased can be easily overcome by duct firing in the HRSG to generate more steam. Hence, the plant can meet the increased steam demand without reducing the power output. The availability of an additional source of unconventional off-gas fuel (often used by chemical plants for duct firing) would be an ideal application.

Low-Pressure Turbine Sizing

The design team and the owner should size the ST and associated equipment based on the "zero export steam" case. An ST passing all the steam generated will have a larger LP section, larger electrical equipment (generator, transformer, etc.), and a higher electrical output. Economic and thermal performance considerations must be part of the decision-making process.

Reliability and Availability

Since the export steam in many cases is used as the heat source for a chemical or industrial process, its reliability is a primary consideration in equipment selection and plant design. This is a common requirement for all industrial applications. A steam outage could have a devastating effect on the host facility and cause heavy financial losses.

Plant Configuration

During transients, the host imposes other design constraints, which are dictated by the narrow allowable variability range for process steam flow, pressure, and temperature. These constraints and individual component reliability values define the plant configuration options. The first option comprises multiple identical trains (1 × 1 × 1) operating independently. The second option involves a single ST, with common steam headers fed from two or three HRSGs and their respective GTs (2 × 2 × 1 and 3 × 3 × 1).

Boiler feedwater purity and steam quality issues need to be carefully addressed throughout the range of planned operation.

Aluminum Smelter Case

Since aluminum smelting is energy intensive, most of the world's smelters are located in areas that have access to abundant power resources (hydro-electric, natural gas, coal, or nuclear). It takes 15 MWh of electric power to produce one ton of aluminum. Many locations are remote and the electricity is generated specifically for the aluminum plant by a CPP. A smelter comprises a large number of pots (steel containers with carbon lining). The product of the chemical reaction of the aluminum oxide under high temperature conditions is molten aluminum. Given that the smelting process is continuous, a smelter cannot be easily stopped and restarted. If production is interrupted by a power supply failure that lasts more than 4 hours, the metal in the pots solidifies, often requiring an expensive rebuilding process. From time to time, individual pot linings reach the end of their useful life, and the pots are taken out of service and relined. In order to avoid the grave consequences of pot solidification, redundant power must be considered.

Many of the design features for an oil refinery are also applicable to smelter applications (common steam header and supplementary firing [to compensate for power loss due to ambient conditions]) and have been addressed earlier in this section.

N-2 Criteria

Before selecting equipment from different suppliers, a thorough investigation is necessary to ensure that the special requirements of the process are met for power output, operational flexibility, reliability, and availability based on N-2 criteria, etc. The normal reliability of GTs for this application (including scheduled maintenance) is close to 96%.

The calculated reliability of multiple units indicates that a 4 × 50% configuration will yield a reliability of 0.99998%, whereas a 2 × 50% configuration will yield a reliability of only 0.97%. The high reliability does not come cheap. In order to justify backup power, many projects opt for an N-1 configuration during time periods when scheduled maintenance is not required. Given the importance of N-2 criteria, special consideration is given to selecting GT size and CC configuration. An example of the procedure is given in Figure 3–37. The Y axis represents the percentage of excess power above the target value for a given plant configuration. The target value is the minimum power requirement for the plant. The X axis represents different configurations. It can be seen that a configuration of two blocks of 2 × 1 (two GTs and one ST) plus two GTs in simple cycle will have a margin of 11.3% above minimum power in an N-2 operational scenario. The selection process must also be based on economic factors such as

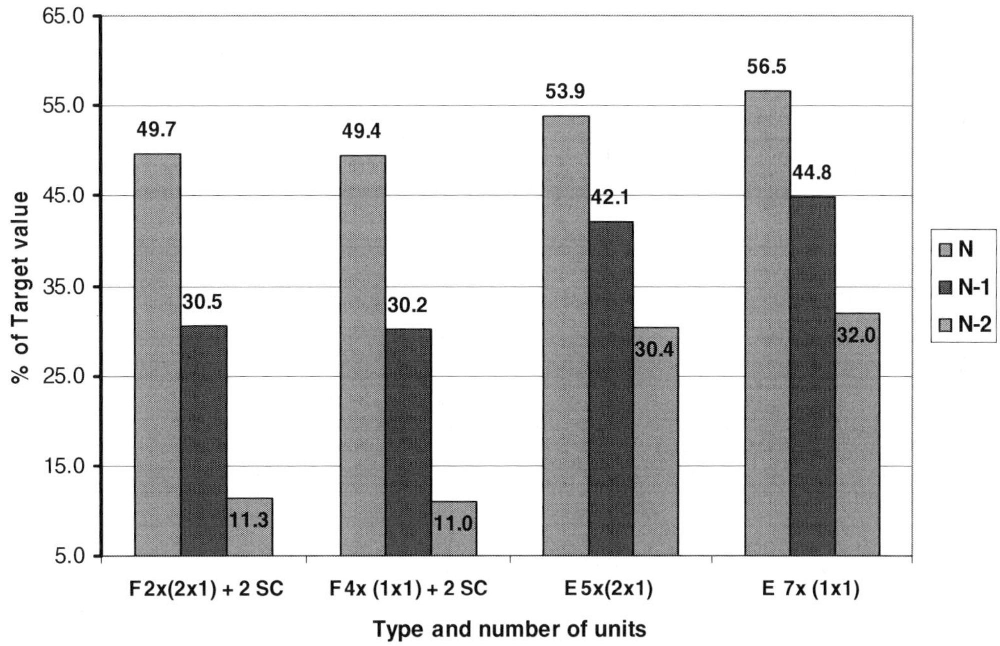

FIGURE 3–37 N-2 selection criteria for a smelter. [3-7]

capital and operating and maintenance cost. While it is very difficult to generalize, the tendency is to use medium-size turbines for this type of application. In case of a trip, the loss of power is less than for a larger unit. It should be mentioned that in selecting the associated STs, the choice must account for the cycle pressures and locations of the steam extractions limited by the commercial availability of STs that can meet the specified conditions. Another decision is needed on whether to design the ST and associated steam cycle equipment for the case where no steam is required for the process.

Load Variations

Smelters present some unique load variation demands. For example, sudden changes occur due to anode effects on the pot line—gas bubbles emerge and disappear in the aluminum production process. The GT must follow with rapid changes of 10–20 MW while maintaining grid frequency stability. The selected equipment should be able to demonstrate ability to follow load accordingly. Original equipment manufacturers have incorporated special attributes such as variable inlet guide vanes (IGV) rapid adjustments, fuel system control, and other features to achieve the desired response time. The use of simple cycle GTs can provide rapid startup, load ramping, and unloading.

Equipment Selection Criteria

Early selection of major plant equipment is an important ingredient for a successful plant. The process must identify the owner's objectives for plant output, heat rate, and reliability, availability, and emissions requirements. In addition to power output and heat rate, the major issue related to evaluating GT performance in captive applications is quantifying the GT exhaust flow and temperature as major contributors to steam production. All four parameters are heavily interdependent. For example, a GT that exceeds the guaranteed power output usually has better-than-expected component efficiency, particularly for the turbine section. As a result, the exhaust temperature is lower. A lower turbine exhaust temperature has a negative impact on HRSG steam production.

To properly assess these impacts, owners/operators are well advised to carefully evaluate the purchase or reservation agreement for the major equipment. While such evaluation is always important, it is crucial for projects involving special industrial applications. Direct experience with the equipment being considered can be brought to bear to ensure that it complies with the specific requirements of the project scope and that all interfaces are well defined. Other considerations include:

- Impact of GT performance offsets (between power output and exhaust energy) on HRSG and ST sizing
- Consistency in pollutant levels and units included in the plant permits and those guaranteed by GT suppliers. In developing CC projects for captive applications, an important aspect of equipment selection is the plant configuration. Usually, suppliers offer standardized CC plant design configurations:
    - 1 × 1 (one GT, one HRSG, and one ST)

- 2 × 1 (two GTs, two HRSGs, and one ST)
- 3 × 1 (three GTs, three HRSGs, and one ST)

This approach allows the customer to choose the configuration that best meets its objectives. For multiple units or trains, a 1 × 1 configuration offers many advantages and may be more desirable in certain applications. Some of the advantages of a 1 × 1 configuration are:

- A phased construction allows greater flexibility for the future addition of units to respond to a potential plant expansion.
- Output can be more closely matched with load demand. A site with several units in a 1 × 1 layout and using supplemental duct firing can respond more effectively and economically to steam and power requirements.
- Plant redundancy (N-2 criteria) can be achieved because the units are redundant.
- Spare parts inventory can be optimized, since each component is identical for all trains.

APPENDIX 3A: STEAM TURBINE POWER PLANT THEORY APPLICABLE TO COMBINED CYCLE AND 'SOLO' (AS COMPETITION TO GAS TURBINE CYCLE) OPERATION*

Rankine Cycle

The Rankine cycle is the most common cycle in power generation. Figure 3A–1 is a simplified Rankine cycle.

Figure 3A–2 shows the ideal Rankine cycle. Cycle 1234B1 is a saturated Rankine cycle (saturated vapor enters the turbine). Cycle 1′2′34B1′ is a superheated Rankine cycle.

The following assumptions are made:

- The cycles shown are internally irreversible.
- Processes through the turbine and pump are adiabatic irreversible (see the vertical on the T-s diagram).

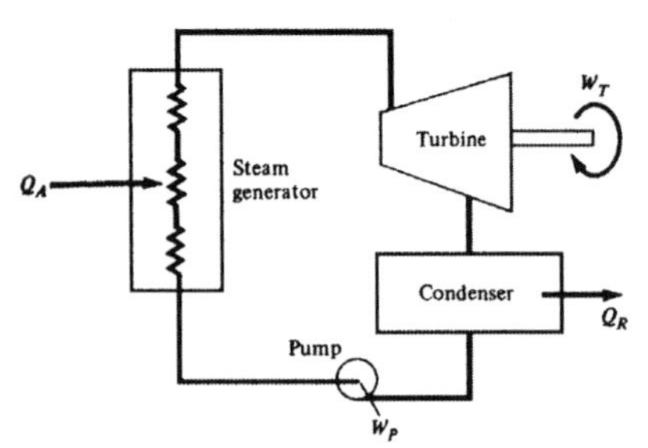

FIGURE 3A–1 Schematic flow diagram of a Rankine cycle [3-3].

* Reference: [3-3] Courtesy of McGraw-Hill. Adapted from Kiameh, Power Generation Handbook, New York: McGraw-Hill, 2003.

- No piping losses.
- 4B11′ is a constant pressure line.
- 1-2 or 1′-2′ is an adiabatic reversible expansion (exhaust is in two phases).
- 2-3 or 2′-3′ is a constant-temperature, two-phase mixture with constant-pressure heat rejection in the condenser.
- 3-4 is adiabatic reversible compression by the pump of saturated liquid at condenser pressure 3 to subcooled liquid at steam generator pressure 4. Note: 3-4 is vertical on both the P-V and T-S diagrams (incompressibility and the pump is adiabatic reversible).
- 4-1 or 4-1′ is a constant-pressure heat addition in the steam generator. 4-B-1-1′ is a constant pressure line on both diagrams.
- 4-B is subcooled liquid to saturated liquid at B at constant temperature and pressure (two-phase mixture).
- B-1 in the steam generator is the boiler or evaporator.
- 1-1′ indicates heating saturated vapor at 1 to 1′ (superheat).
- Per unit mass in the cycle,

Heat added: $q_A = h_1 - h_4$ Btu/lbm (or J/kg)
Turbine work: $w_T = h_1 - h_2$ Btu/lbm (or J/kg)
Heat rejected: $|q_R| = h_2 - h_3$ Btu/lbm (or J/kg)
Pump work: $|w_p| = h_4 - h_3$

Net work: $\Delta w_{net} = \overbrace{(h_1\ h_2)}^{\text{Turbine work}}\ \overbrace{(h_4\ h_3)}^{\text{Pump work}}$ Btu/lbm (or J/kg)

Thermal efficiency: $\eta_{th} = \frac{\Delta w_{net}}{q_A} = \frac{(h_1-h_2)-(h_4-h_3)}{(h_1-h_4)}$

For small units: $P_4 \approx P_3$, $h_3\ h_3 \approx h_4$

Therefore, pump work is small compared with turbine work,

$$\eta_{th} = \frac{h_1 - h_2}{h_1 - h_3}$$

In modern plants, this assumption is not true. The term P_4 may be 1000 psi (70 bar) and P_3 about 1 psi (0.07 bar).

To calculate pump work from h_3 (saturated liquid at P_3),

Pump work ≈ Change in flow work, or

$$W_p = V_3(P_4 - P_3)$$

Calculate h_4 from subcooled liquid tables at T_4 and P_4 (assume $T_3 = T_4$).

Reheaters

Reheating is illustrated in Figures 3A–3 and 3A–4.

We make the following assumptions:

- Processes through the turbine and pump are reversible and adiabatic.
- No pressure drop (losses) in the cycle.

Note: Line *ab* = primary coolant in a counterflow steam generator (primary heat source = combustion gases from the steam generator).

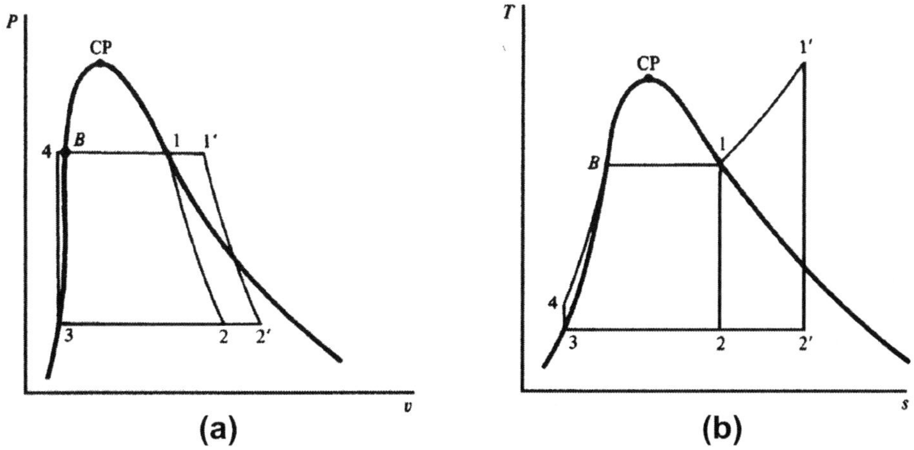

FIGURE 3A–2 Ideal Rankine cycles of the (a) P-V and (b) T-S diagrams. Line 1-2-3-4-B-1 = saturated cycle. Line 1′-2′-3-4-B-1′ = superheated cycle. CP = critical point [3-3].

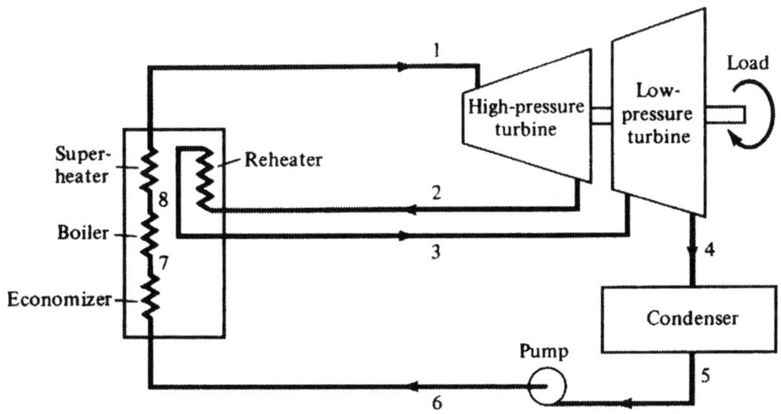

FIGURE 3A–3 Schematic of a Rankine cycle with superheat and reheat [3-3].

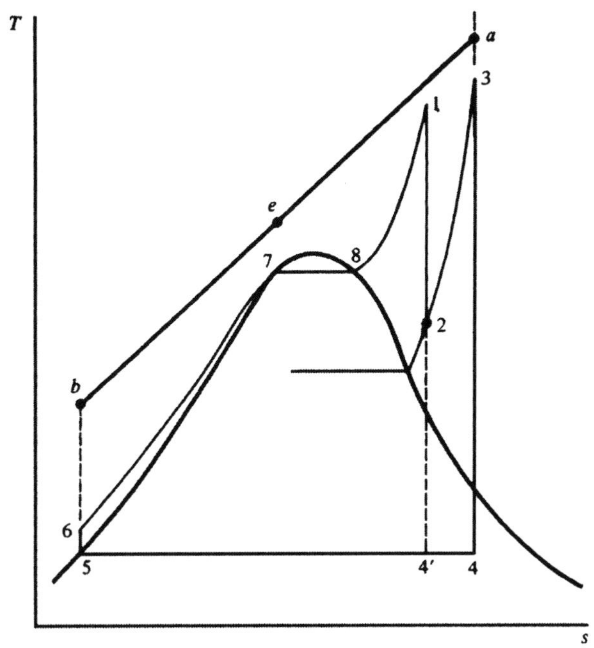

FIGURE 3A–4 T-s diagram of Rankine cycle of Figure 3–10 [3-3].

One stage of reheat works economically. Two stages are not a good return on investment.

The steam may be reheated in the steam generator or a separate heat exchanger reheater.

The reheat cycle contains two turbine work terms and two heat addition terms.

Figure 3A–5 shows cycle efficiency versus the ratio of P_2/P_1 (reheat pressure/initial pressure).

If the reheating pressure is too close to the initial pressure, the increase in cycle efficiency is minimal, since only a small portion of heat is added at high pressure. For optimum reheating, P_2/P_1 is 20–25%.

Lowering reheat pressure causes further efficiency decrease.

Exhaust steam is drier in a reheat cycle. A superheat-reheat plant is designated $P_1/T_1/T_3$ (e.g., 2500 psi/°F/1000°F), see Table 3A–1. Note efficiency rises with reheating and drops with non-ideal fluids.

Regeneration

Figure 3A–6 illustrates the Rankine cycle. The working fluid is represented by 4B1234, where

FIGURE 3A–5 Effect of reheat-to-initial-pressure ratio on efficiency, high-pressure turbine exit temperature, and low-pressure turbine exit quality. Data for cycle with initial steam at 2500 psia and 1000°F and steam reheated to 1000°F (2500/1000/1000) [3-3].

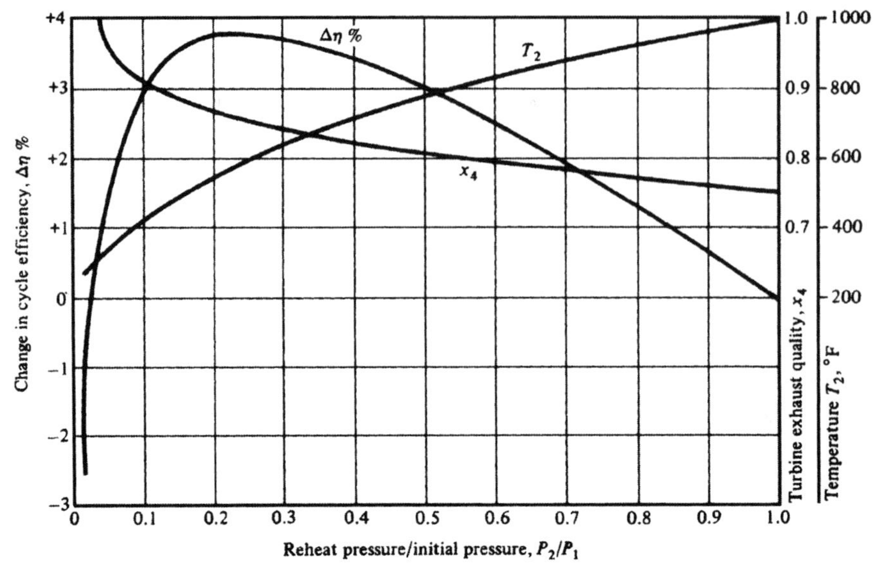

TABLE 3A–1 Steam Power Plant Performance Comparison [3-3]

	Cycle				
Data	(A) Superheat 2500/1000	(B) 2500 Saturated	(C) Superheat 1000/668.11	(D) 2500/1000/ 1000	(E) 2500/1000 Non-ideal
Turbine inlet pressure (P_1), psia	2500	2500	1000	2500	2500
Turbine inlet temperature (T_1), °F	1000	668.11	668.11	1000	1000
Condenser pressure, psia	1	1	1	1	1
Inlet steam enthalpy, Btu/lbm	1457.5	1093.3	1303.1	1457.5	1457.5
Exhaust steam enthalpy, Btu/lbm	852.52	688.36	834.44	970.5	913.02
Turbine work, Btu/lbm	604.98	404.94	468.66	741.8	544.48
Pump work, Btu/lbm	7.46	7.46	2.98	7.46	11.52
Net work, Btu/lbm	597.52	397.48	465.68	734.34	532.96
Heat added, Btu/lbm	1380.31	1016.11	1230.39	1635.10	1376.25
Exhaust steam quality	0.7555	0.5971	0.7381	0.8694	0.8139
Cycle efficiency, %	43.29	39.12	37.85	44.91	38.73

a-b = primary coolant in a counterflow steam generator (combustion gases).

c-d = heat sink fluid in a counterflow heat exchanger (condenser cooling water or air).

Line *a-b* too close to 4-*B*-1 implies that the *T* between the primary coolant and working fluid is small, which implies that irreversibility is small but the steam generator has to be large; expensive.

Line *a-b* to far from 4-*B*-1 implies that irreversibility is large, efficiency is low, and the steam generator is small.

Maximum irreversibility occurs at the maximum distance between *a-b* and 4-*B*. This distance is minimum at point *B*. So irreversibility is minimized if the fluid is added to the steam generator at point *B* not 4. Regeneration helps this minimization by adding heat from the

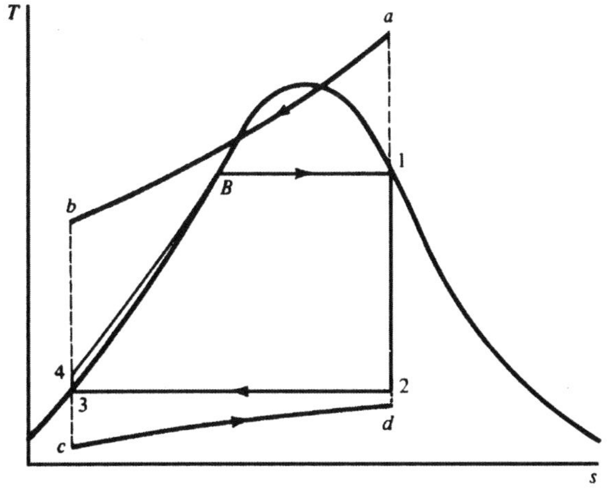

FIGURE 3A–6 External irreversibility with the Rankine cycle [3-3].

expanding fluid in the turbine to the compressed fluid before heat addition.

Feedwater Heating

With feedwater heating at point 4, heat is added in several heat exchangers by steam bled from the turbine at selected stages. Power plants generally use between five and eight feedwater heating stages.

Efficiency and Heat Rate of Power Plants

Gross efficiency is calculated using the power (in MW) produced before supplying power to the plant's internal equipment (pumps, compressors, and so forth).

Net efficiency is calculated based on the net power of the plant (the gross power minus the power needed for internal plant equipment).

Cogeneration

The term cogeneration means that electricity and steam (heat) are generated at the same time. Process plants and small municipalities often use cogeneration, as they can usually get better return on investment (ROI) on the process than a utility.

Cogeneration is useful only if it saves money over separate generation of electricity and steam.

Cogeneration efficiency: $\eta_{co} = \frac{E+\Delta H_s}{Q_A}$

where

E = the electric energy generated.
ΔH_S = the heat energy from the process steam.

Enthalpy of steam entering the process—enthalpy of process condensate.

Q_A = the heat added to the plant (from fuel).

For separate electricity and steam generation, the heat added per unit of total energy is

$$\frac{e}{\eta_e} + \frac{(1-e)}{\eta_h}$$

where

e = electrical fraction of total energy output = $[E/(E + \Delta H_S)]$.
η_e = electric plant efficiency.
η_h = steam (or heat) generator efficiency.

The equation for combined efficiency for separate generation (from previous equations) is

$$\eta_c = \frac{1}{(e/\eta_e) + [(1-e)\eta_h]}$$

If $\eta_{co} > \eta_c$, then cogeneration offers a better ROI.

Types of Cogeneration

There are two main kinds of cogeneration cycle: a topping cycle and a bottoming cycle.

The topping cycle really is the only economic cycle of the two. The main heat source generates high-enthalpy steam and electricity. Low-enthalpy steam is taken from an intermediate turbine stage or the turbine exhaust for process requirements. If the steam is taken from the turbine exhaust, this is called a back-pressure turbine.

The range of pressures for process steam varies and generally is on the order of 0.5 bar to 40 bar.

The common subdivisions for types of topping cycle plant are

- Steam electric plant with steam extraction from a condensing turbine (note Figure 3–15).
- Steam electric power plant with a back-pressure turbine.
- Gas turbine power plant with waste heat recovery and a steam generator (boiler).
- Combined-cycle gas turbine plus steam turbine plant (steam turbine can be back-pressure or condensing type).

The bottoming cycle is not very efficient. High-enthalpy heat is used directly for process needs (e.g., cement manufacture), and low-enthalpy waste heat is used to generate electricity.

Steam Turbine Basic Components and Main Systems

In a steam turbine, high-pressure, high-temperature steam expands to increase the steam's kinetic energy at the expense of its original pressure energy. That kinetic energy is used to turn the rotor and produce mechanical energy as torque and rotor speed increase.

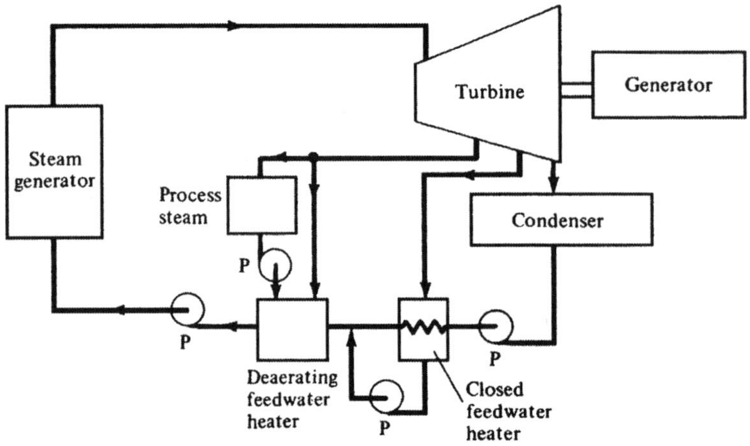

FIGURE 3A—7 Schematic of basic cogeneration plant with extraction-condensing turbine [3-3].

There are two main kinds of steam turbine: impulse and reaction. The bulk of steam turbines today are reaction turbines. In impulse turbines, the flow momentum changes as a steam jet pushes the rotor blades forward. In reaction turbines, a reaction force is felt by the moving blades as steam flow accelerates through a decreasing cross-sectional area.

Figures 3A—7 and 3A—8 illustrates a turbine with impulse blading. It has one velocity-compounded stage where the velocity is reduced in two steps through two rows of moving blades, as the pressure stays constant.

Figure 3A—9 illustrates a reaction turbine with one velocity compounded impulse stage, where a large pressure drop occurs.

Steam Turbine Components

The block diagram Figure 3A—10 illustrates the main mechanical components in a basic steam turbine. Steam turbines that deliver between 40 and 60 MW usually are single-cylinder machines (see Figure 3A—11). For larger steam turbines, more cylinders are used to extract the steam's energy.

Single-Cylinder Turbines

The two main subdivisions here are condensing and back-pressure (noncondensing) turbines. Automatic extraction turbines allow part of the steam to be drawn off at intermediate stage(s), while the rest of the steam is exhausted to a condenser. These turbines use controls and valves to get constant pressure of extraction steam during conditions of varying load and extraction demand. Uncontrolled extraction (pressure varies at extraction points with load) steam is used to add heat in feedwater heaters.

Some plants add high-pressure noncondensing turbines to increase capacity. The added turbines are called topping or superposed units.

Nozzles*

Nozzles, together with blades, play an important role in achieving the high efficiency of a turbine. Nozzles are designed to attain the highest thermal efficiency. They are made of 12% or 18% chrome and 8% nickel stainless steel and have excellent mechanical strength and erosion resistance. The diaphragm in which the nozzle is built is split at the horizontal centerline and the spring-backed labyrinth packing is fitted to minimize steam leakage. Four types of construction are standard, as shown in Figure 3A—12, each of which has its own application range based on specific operating conditions.

Blades

Blades are designed based on accumulated aerodynamic data from various tests and research. The blades are subjected to a large centrifugal force and an exciting force. The blades are made of 12% or 17% chrome and 4% nickel stainless steel and have excellent mechanical strength and damping properties. If necessary, erosion shields made of Stellite are used in the last stages of the condensing turbines to protect the leading edges of the blades from moisture erosion.

Compound Turbines

Compound turbines have more than one cylinder: one high pressure and one low pressure. The low-pressure cylinder generally is of the double-flow type to handle a large volume of low-pressure steam. The turbine blades are limited in length by design and the double-flow design adds potential for steam flow. Large plants have more than two cylinders: an intermediate-pressure cylinder and four low-pressure cylinders are not uncommon.

---

* Source: [3-4] Courtesy of MPS (Mitsubishi Power Systems).

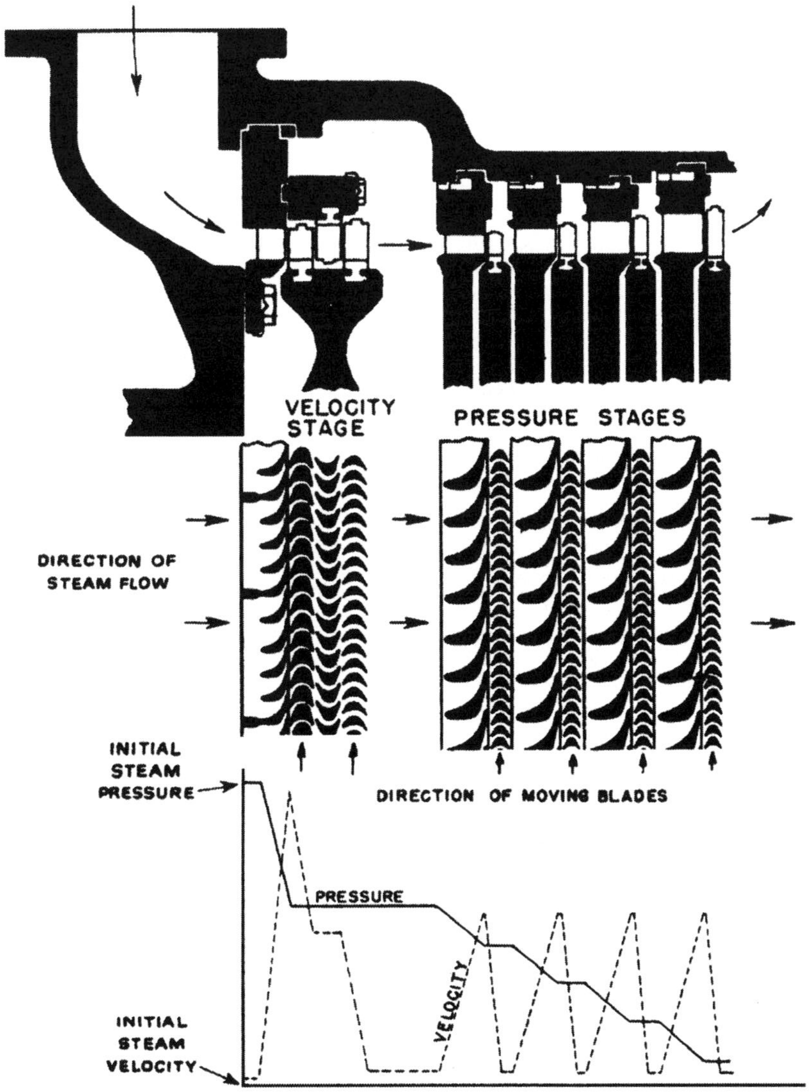

FIGURE 3A–8 Turbine with impulse blading. Velocity compounding is accomplished in the first two stages by two rows of moving blades between which is placed a row of stationary blades that reverses the direction of steam flow as it passes from the first to the second row of moving blades. Other ways of accomplishing velocity compounding involve redirecting the steam jets so that they strike the same row of blades several times with progressively decreasing velocity [3-3].

If the cylinders are on one shaft, the arrangement is termed a tandem compound. If they are on more than one shaft, the arrangement is termed a cross compound (see Figure 3A–13).

Condensing Turbines

Straight-condensing turbines (Figure 3A–14) are advantageous, especially when large quantities of a reliable power source are required or an inexpensive fuel, such as a process by-product gas, is readily available. To improve plant thermal efficiency, steam usually is extracted from the intermediate stage of the turbine for feedwater heating.

Extraction-Condensing Turbines

Extraction-condensing turbines (Figure 3A–15) generate both process steam and stable electric power. Process steam, at one or more fixed pressures, can be automatically extracted as needed. This type of turbine has the flexibility to satisfy wide variations of process steam at a constant pressure and meet electric power demands.

Back-Pressure Turbines

Back-pressure turbines (Figure 3A–16) can be used when a large quantity of process steam is required. The turbine exhaust steam is supplied to the process, and the electric output depends on the demand for the process steam. These

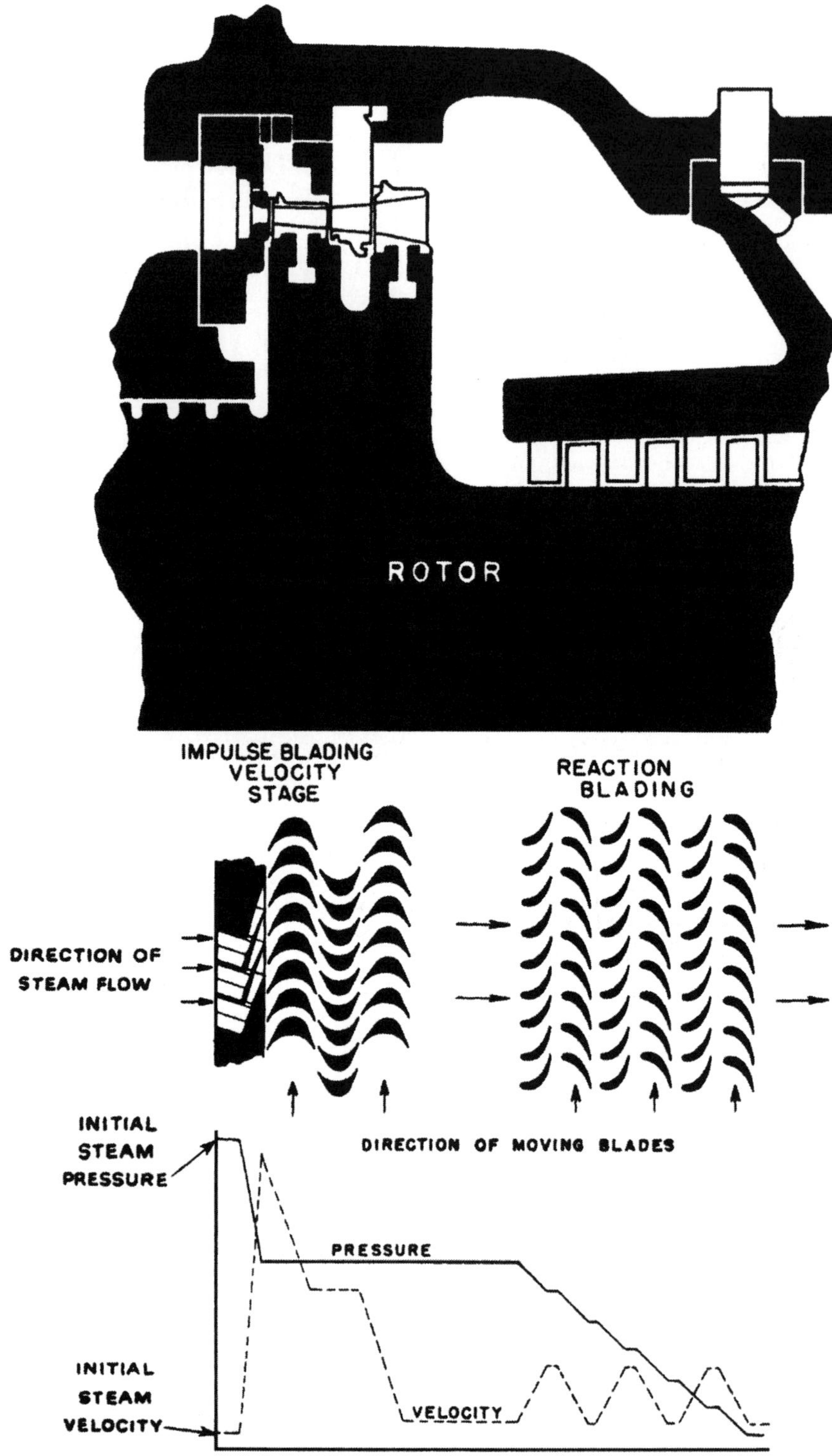

FIGURE 3A–9 Reaction turbine with one velocity-compounded impulse stage. The first stage of this turbine is similar to the first velocity-compounded stage of Figure 3–14. However, in the reaction blading of this turbine, both pressure and velocity decrease as the steam flows through the blades. The graph at the bottom shows the changes in pressure and velocity through the various stages [3-3].

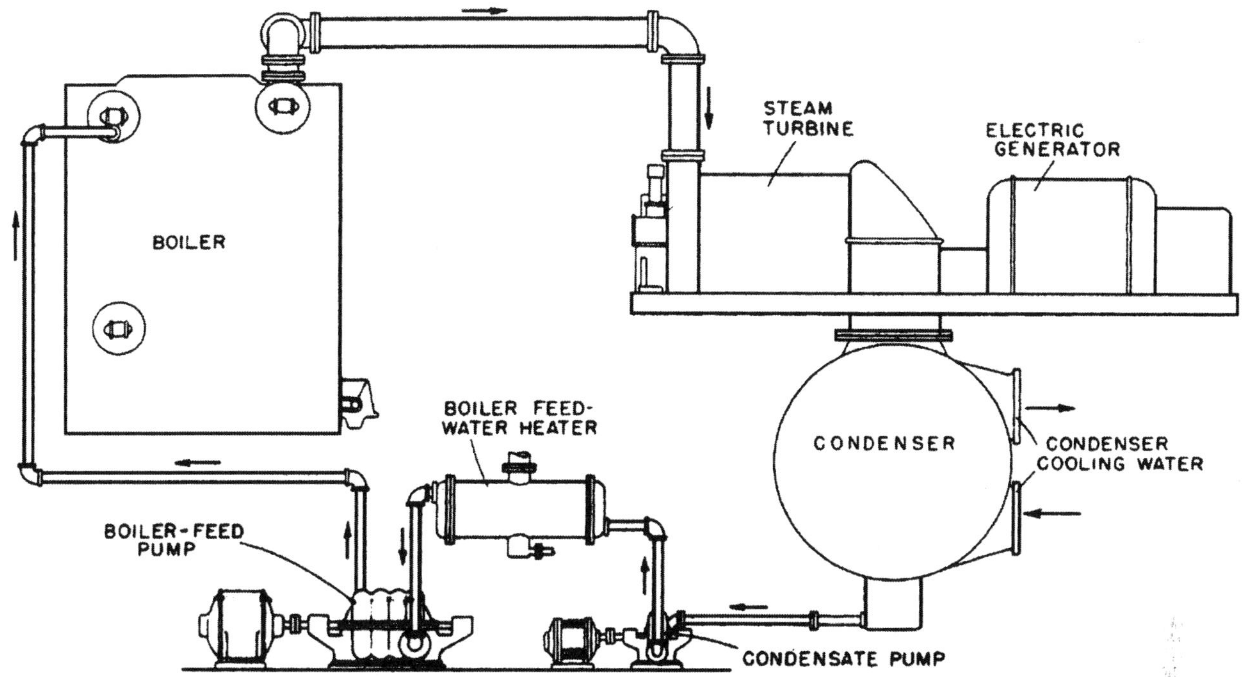

FIGURE 3A–10 Simple power plant cycle. This diagram shows that the working fluid, steam and water, travels a closed loop in the typical power plant cycle [3-3].

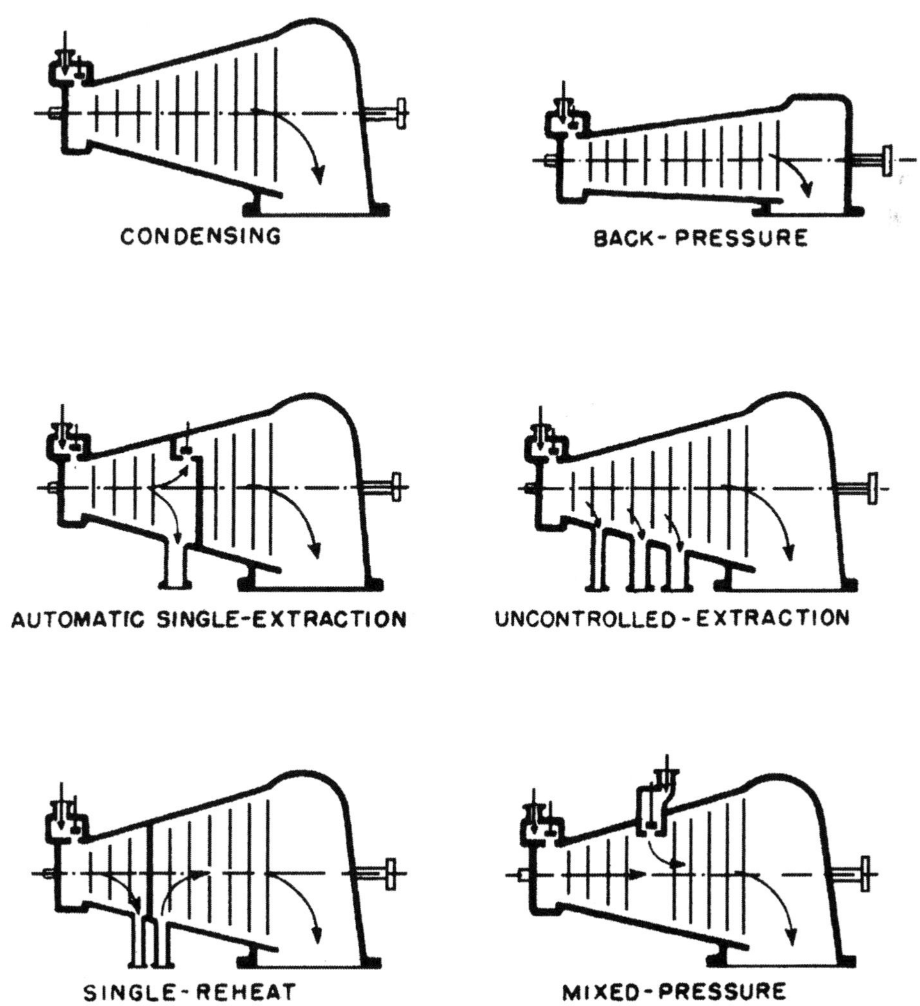

FIGURE 3A–11 Typical single-cylinder turbine types. As shown, condensing turbines, as compared to back-pressure turbines, must increase more in size toward the exhaust end to handle the larger volume of low-pressure steam [3-3].

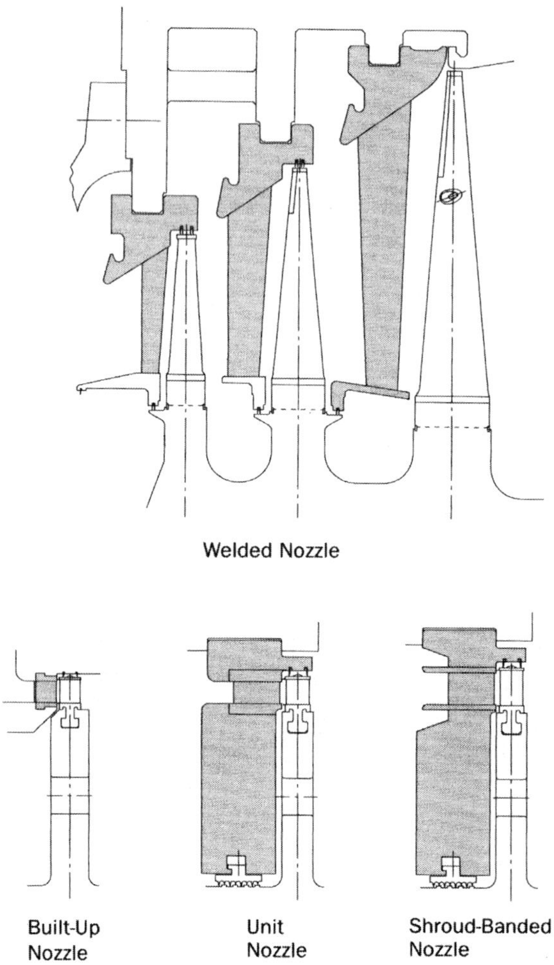

FIGURE 3A—12 Standard nozzles [3-4].

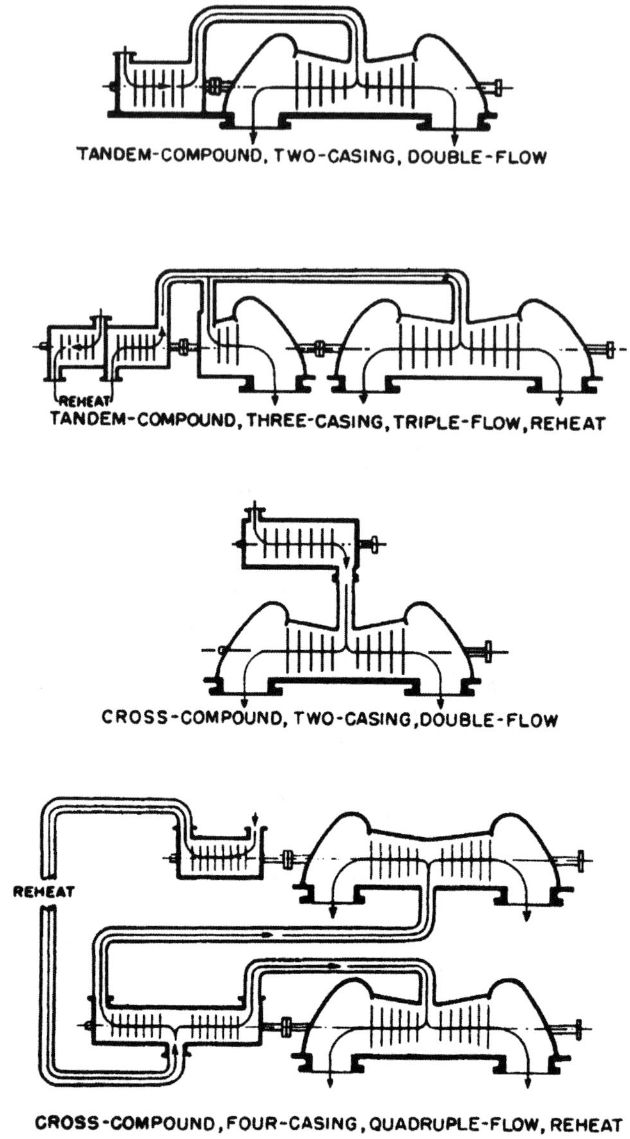

FIGURE 3A—13 Some arrangements of compound turbines. While many arrangements are used, these diagrams illustrate some of the more common ones [3-3].

turbines also can be used as top turbines to supply exhaust steam to existing units; this improves the entire plant's thermal efficiency.

Extraction Back-Pressure Turbines

Extraction back-pressure turbines (Figure 3A—17) can be used when two or more kinds of process steam are required. High-pressure steam is supplied through the extraction openings, and low-pressure steam is supplied as the turbine exhaust. Electric output depends on the demand for process steam.

Mixed-Pressure Turbines

Mixed-pressure turbines (Figure 3A—18) are driven by two or more kinds of steam, admitted independently to the turbine. In applying dual heat sources, the optimum steam condition for each source can be selected. This type of turbine also can be used to combine an existing boiler and a new boiler, which makes it an effective means of improving plant thermal efficiency.

Geared Turbines

Geared turbines (Figure 3A—19) can be applied to smaller power generation units of up to around 40 MW. Compared with direct-coupled turbines, geared turbines have many advantages:

- Higher efficiency
- Easier operation and maintenance
- Smaller initial investment
- Smaller space requirement
- Shorter delivery time

The maximum capacity of geared turbines is increasing every year.

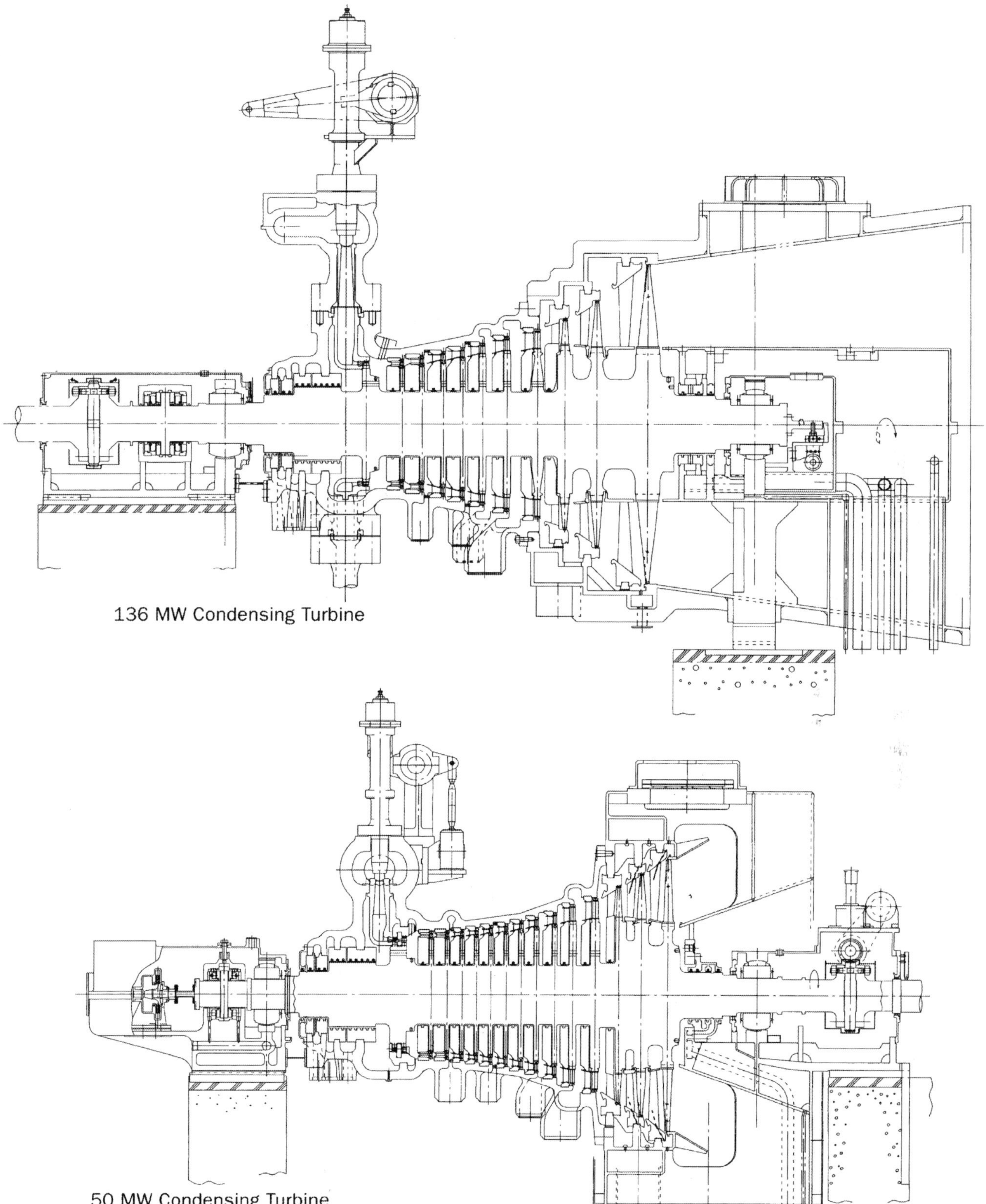

FIGURE 3A–14 Condensing turbines [3-4].

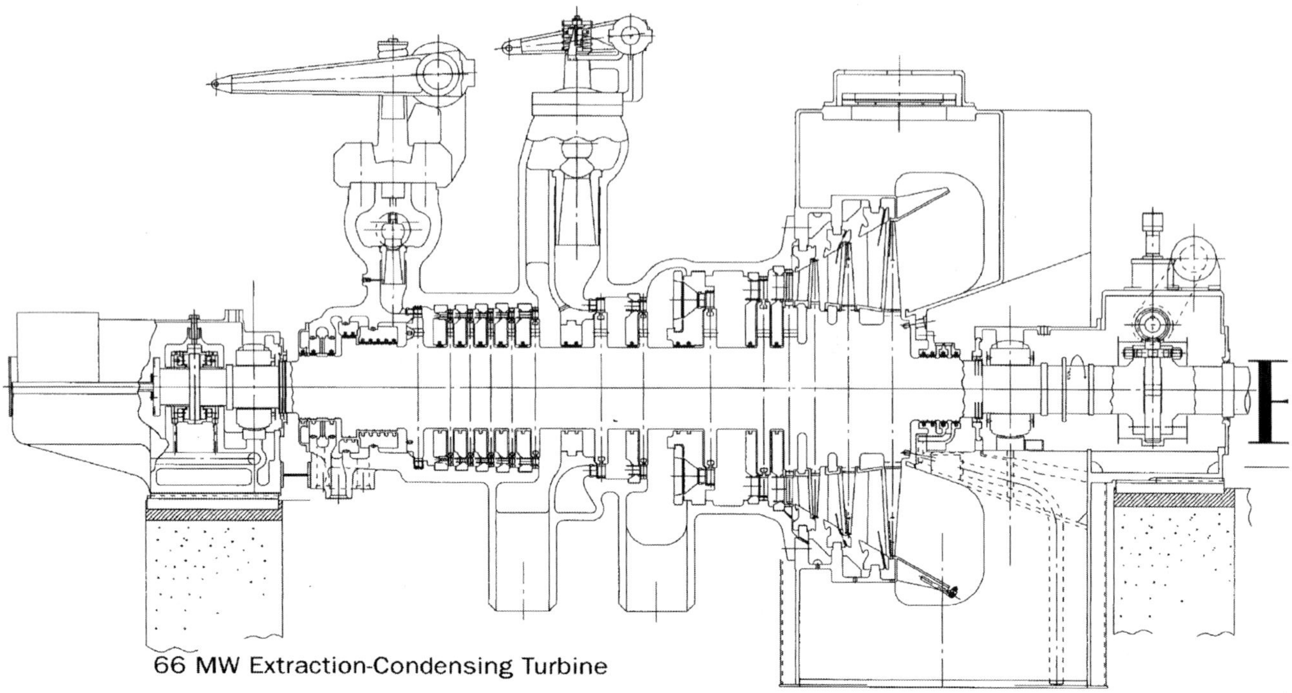

66 MW Extraction-Condensing Turbine

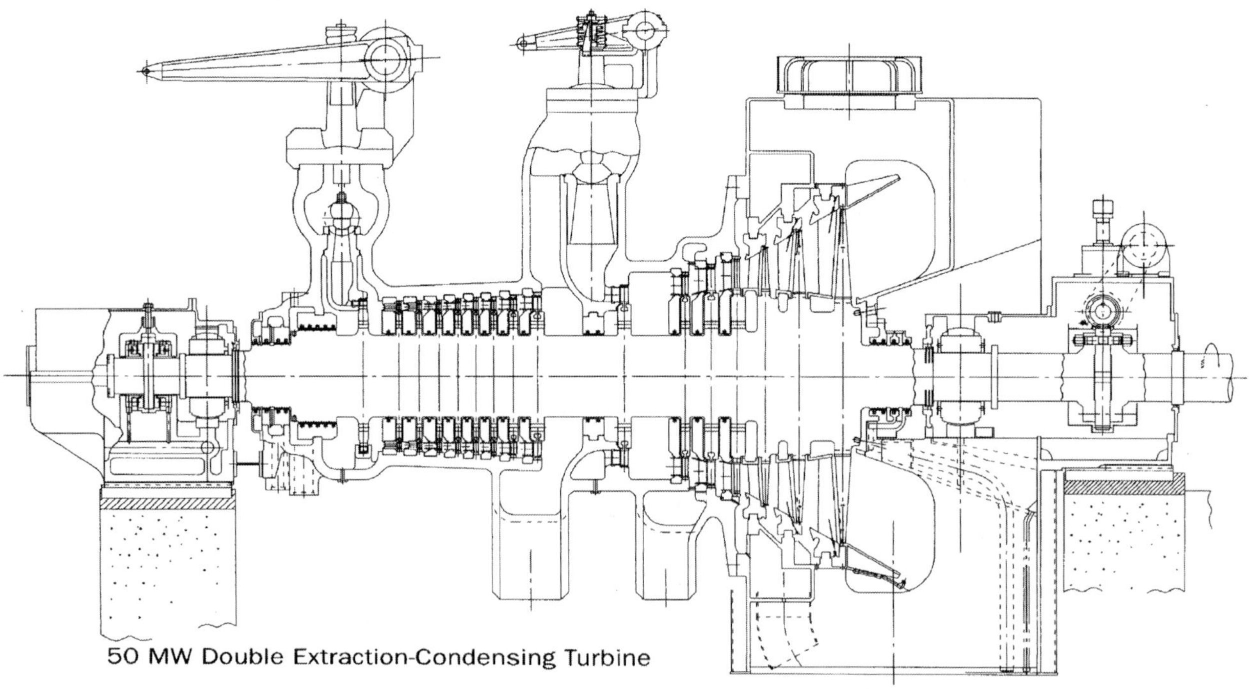

50 MW Double Extraction-Condensing Turbine

FIGURE 3A–15 Extraction-condensing turbines [3-4].

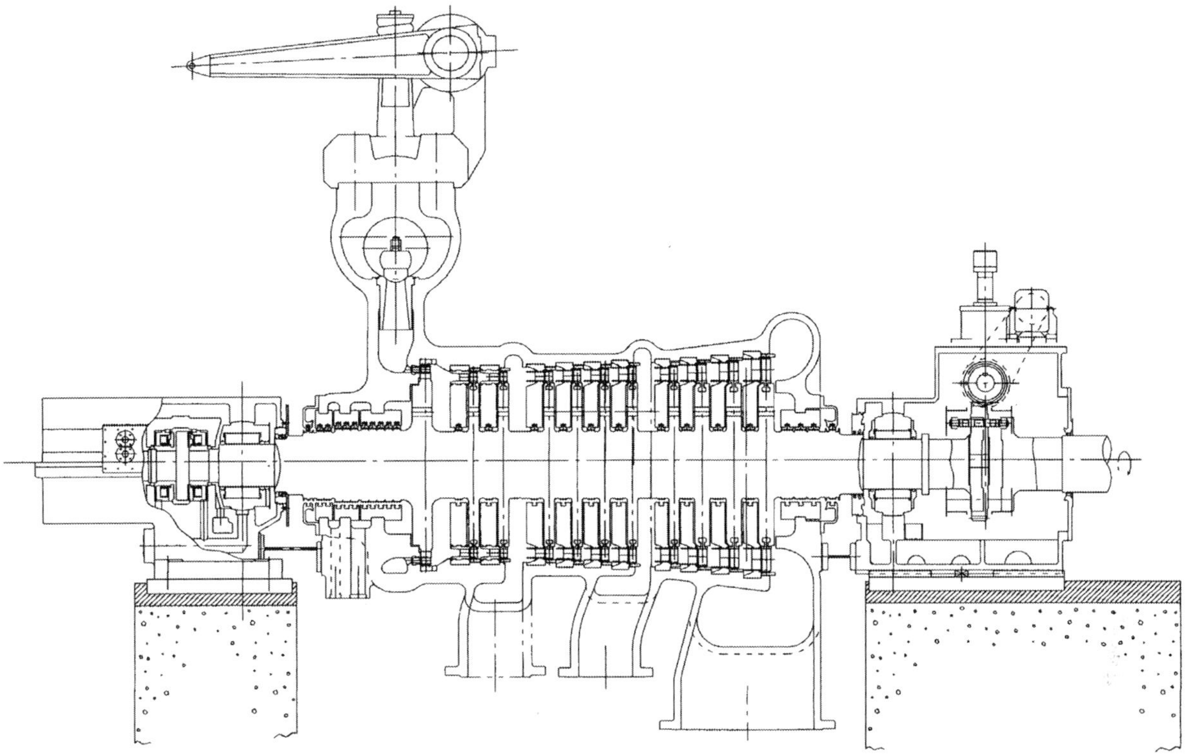

54 MW Back-Pressure Turbine

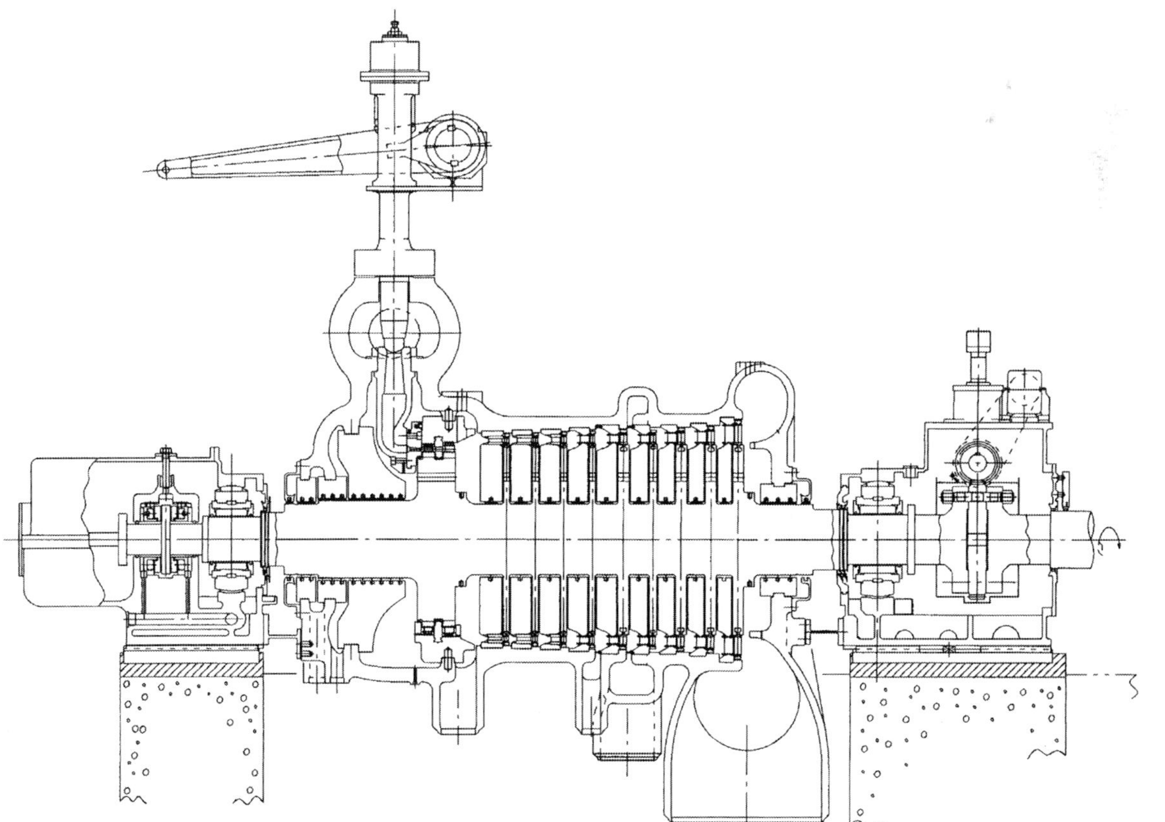

55 MW Back-Pressure Turbine

FIGURE 3A–16 Back-pressure turbines [3-4].

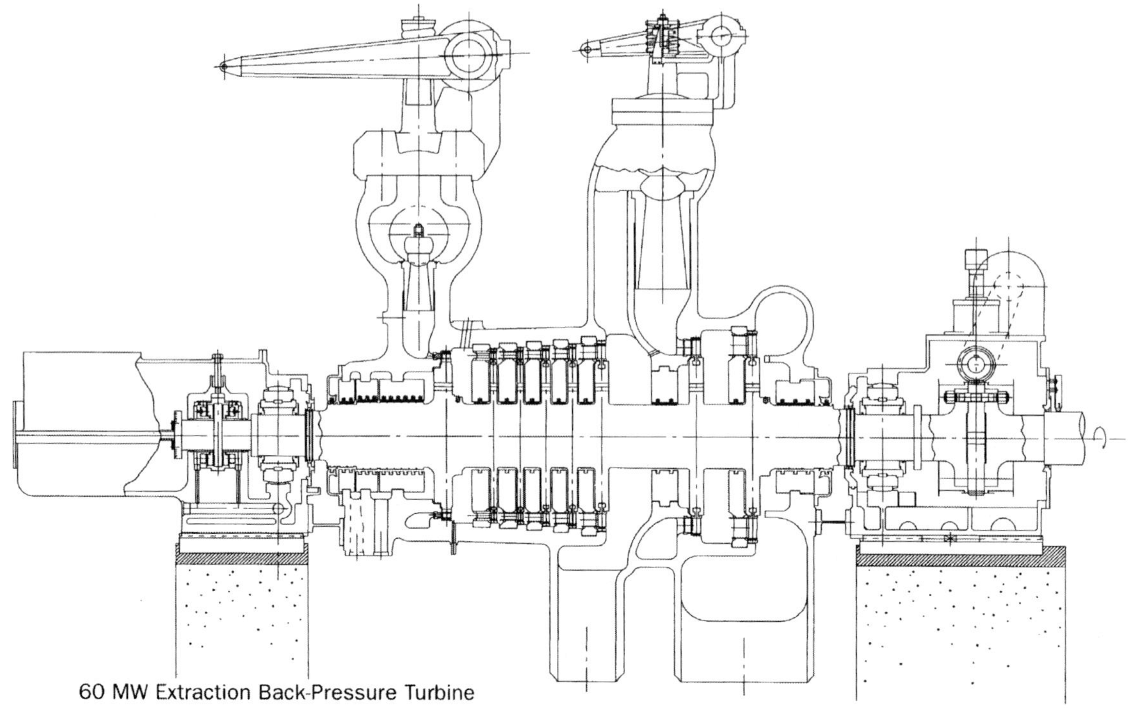

60 MW Extraction Back-Pressure Turbine

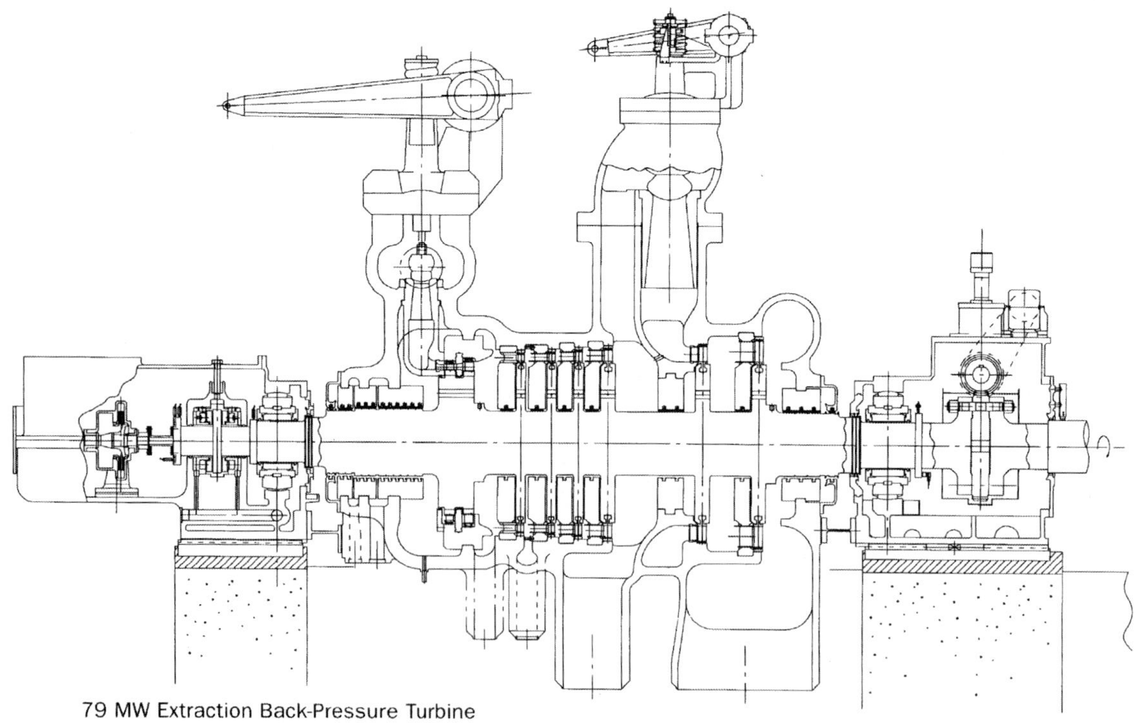

79 MW Extraction Back-Pressure Turbine

FIGURE 3A–17 Extraction back-pressure turbines [3-4].

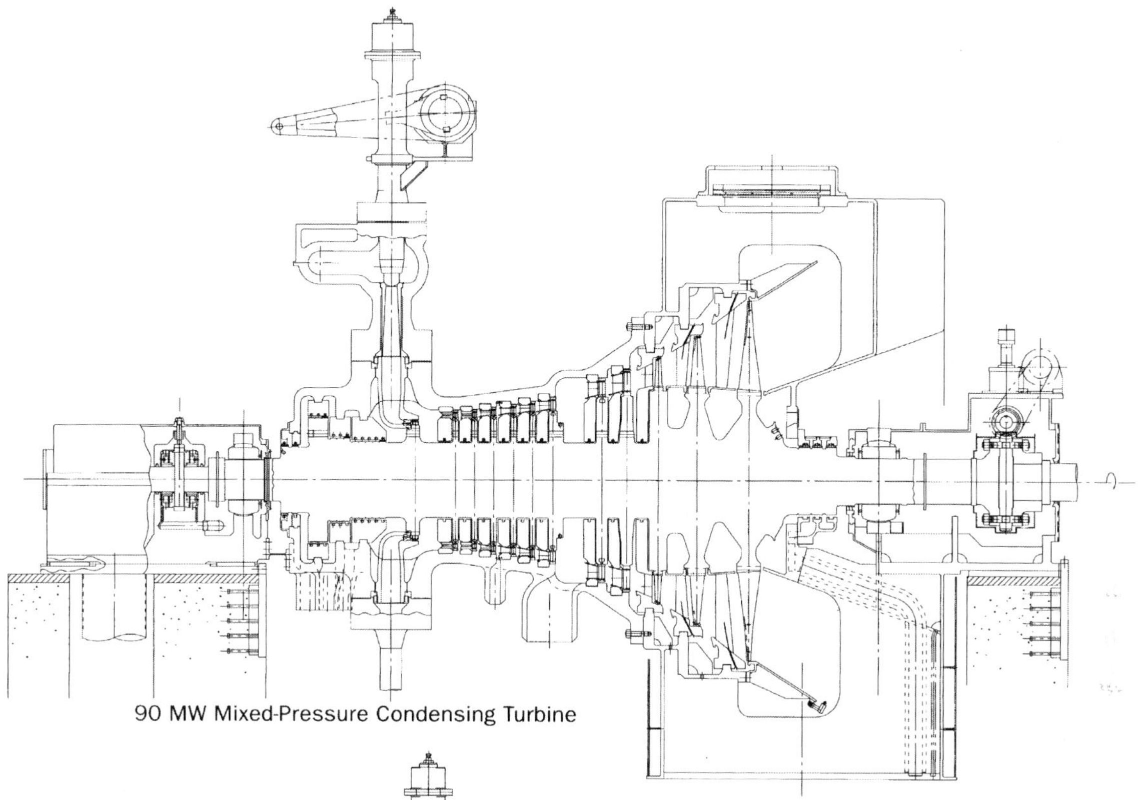

90 MW Mixed-Pressure Condensing Turbine

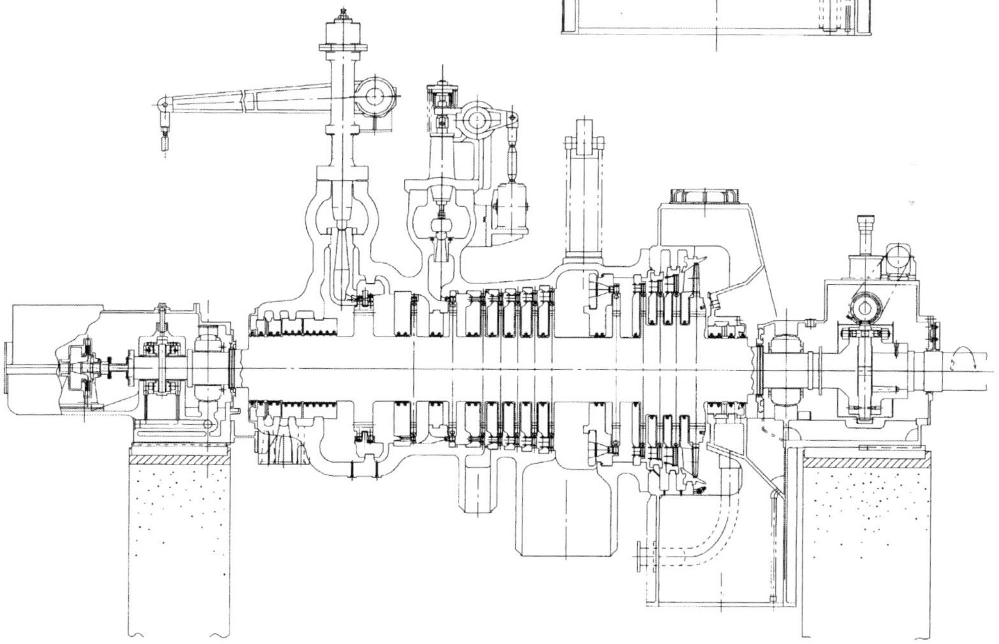

35 MW Mixed-Pressure Extraction-Condensing Turbine

FIGURE 3A—18 Mixed-pressure turbines [3-4].

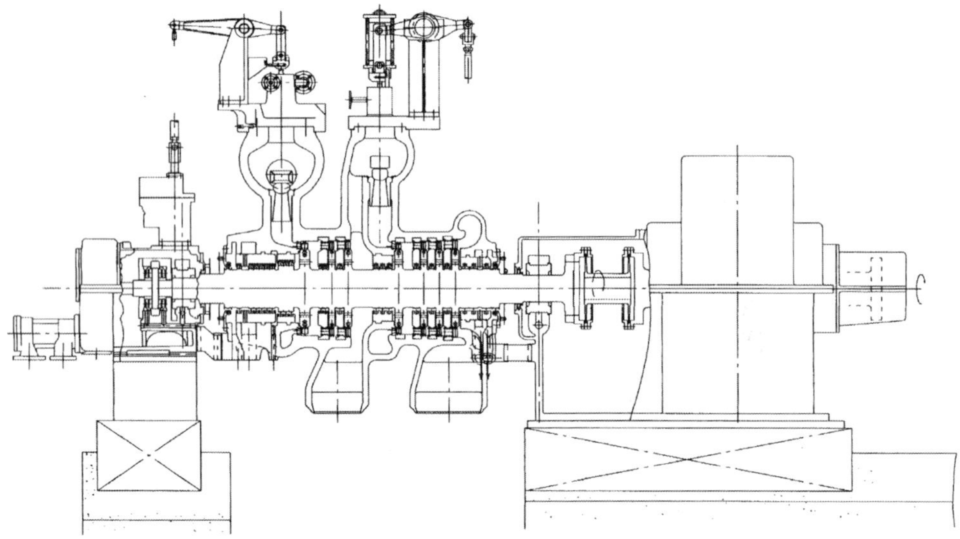

24 MW Geared-Extraction Back-Pressure Turbine

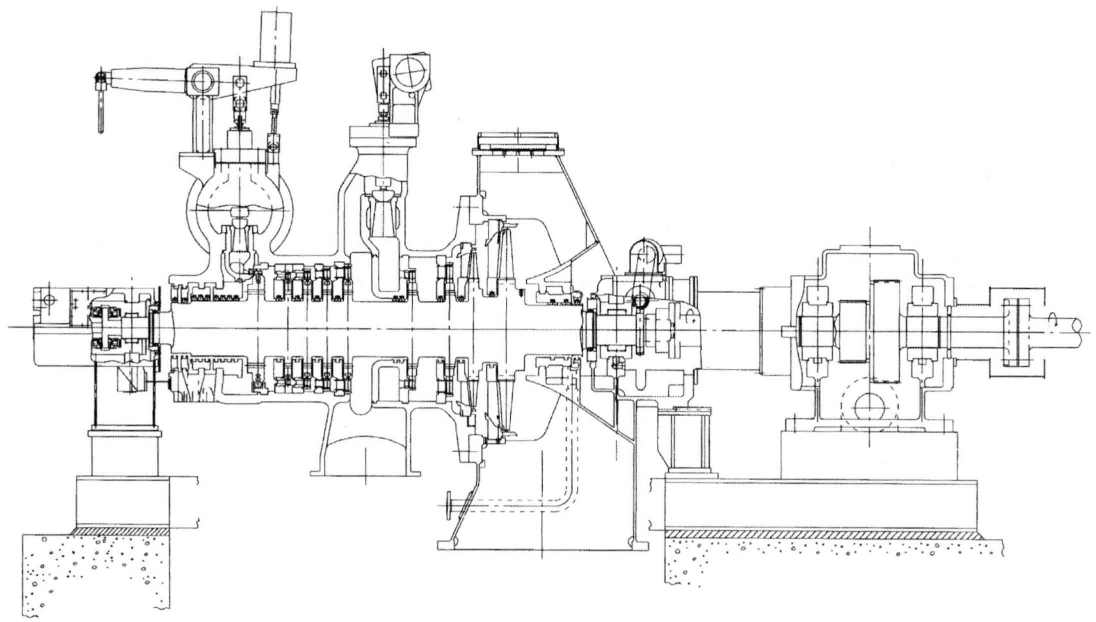

40 MW Geared-Extraction Condensing Turbine

FIGURE 3A–19 Geared turbines [3-4].

Single-Cylinder Reheat Turbines

Traditionally, nonheat turbines have been used for industrial applications. Recent demands for higher efficiency and larger unit capacity, however, call for reheat turbines in this field. Taking these demands into consideration, single-cylinder reheat turbines have been developed that are applicable in the 75 MW to 200 MW range. Single-cylinder reheat turbines (Figure 3A–20) offer:

- Smaller space requirements
- Shorter construction and erection periods
- Easier operation and maintenance
- Shorter overhaul periods
- Smaller initial investments

Single-cylinder reheat turbines have been used at conventional boiler-turbine plants and combined cycle plants instead of two-cylinder reheat turbines.

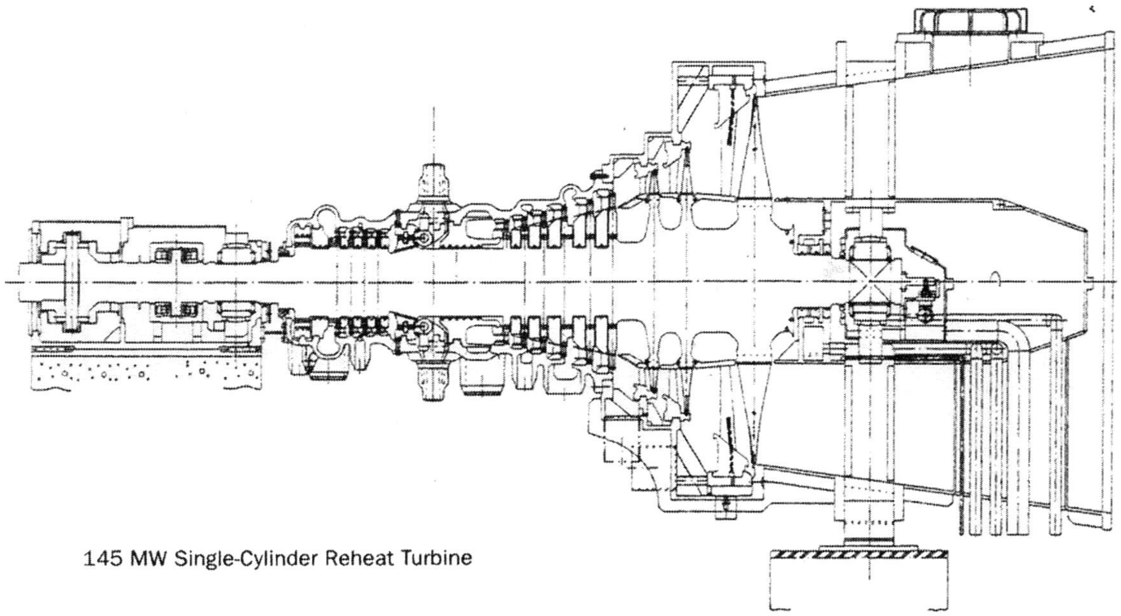

145 MW Single-Cylinder Reheat Turbine

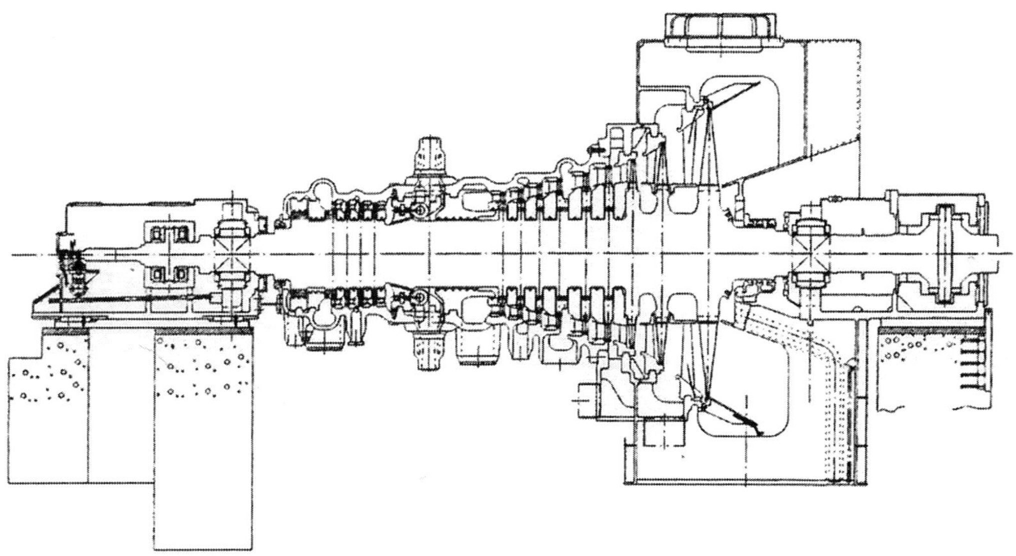

149 MW Single-Cylinder Reheat Turbine

FIGURE 3A–20 Single-cylinder reheat turbines [3-4].

Two-Cylinder Reheat Turbines

Two-cylinder reheat turbines (Figure 3A–21) can be used when a very high efficiency is required for steam turbines larger than 75 MW.

Rotors

The rotor is one of the most important rotating parts of a turbine. Special attention must be paid to the design, materials, and workmanship that go into its construction and the inspections that ensure the highest-quality product. The rotor and the discs, of which critical speed is below the rated speed, are machined completely from a solid forging and are of the flexible type.

Casings

The turbine casing is split at the horizontal centerline. Symmetrical design and simple construction minimize the thermal stress created by temperature changes, and the casing is allowed to expand and contract concentrically with the rotor. The casing is made of cast steel or steel plate, depending on its operating steam conditions.

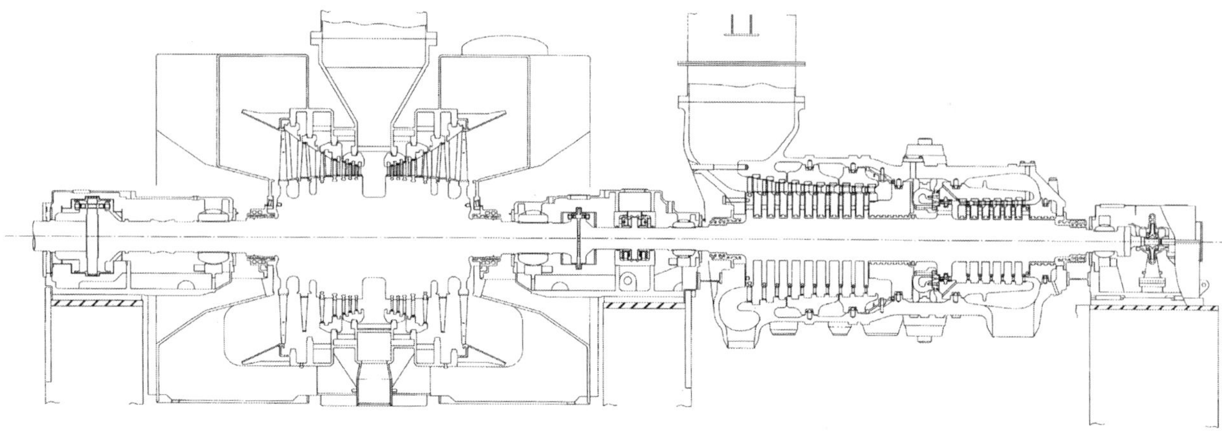

125 MW Two-Cylinder Reheat Turbine
Double-Flow Type

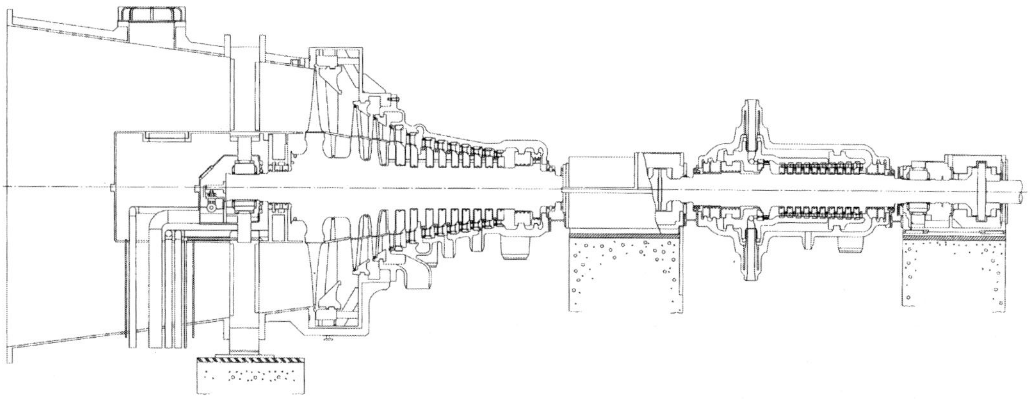

150 MW Two-Cylinder Reheat Turbine
Single-Flow Type

FIGURE 3A–21 Two-cylinder reheat turbines [3-4].

Axial-Exhaust Steam Turbines

Axial-exhaust steam turbines, with the condensers located at the back of the turbines, are effective in reducing the height of the turbine/generator house. This results in a lower initial investment. Customers commonly request that steam turbine suppliers deliver factory-assembled units to reduce the period for and cost of installation.

Steam Turbine Control Systems

Control systems have a minimum of two independent governors. One (the emergency trip) shuts off the steam supply if the turbine overspeeds. The other throttles steam flow to maintain constant speed if the unit is not synchronized to the grid or varies the load in units that are synchronized to the grid.

For extraction, mixed-pressure, and back-pressure turbines, complex governor(s) control steam flow for a range of variable speeds and pressures.

Speed Governor Systems

Speed governor systems consist of:

- A speed-sensitive element
- A linkage or force-amplifying mechanism that transmits motion from the governor to the steam control valves
- Steam control valves (governing valves)

The simplest governors (usually on small, simple turbines) are centrifugal or flyball governors. Opposite weights revolving on a spindle move outward with centrifugal force acting against a spring (with increasing centrifugal force). This changes the position of the steam admission valve with either:

- Mechanical linkage
- Operation of a hydraulic system pilot valve that releases fluid to opposite sides of a power piston or one side of a spring-loaded piston that in turn opens or closes the steam valve

Systems for large turbines are complex and involve booster relays to reduce response time. Intercept valves are installed upstream of the intermediate turbine and fulfill the function served by an emergency trip on a simple turbine.

Pressure Governors

Pressure governors generally are used in back-pressure or extraction turbines to maintain constant extraction or exhaust pressure at all loads.

Main Stop Valves

Using a positive spring closing force, the main stop valve (Figure 3A–22) closes rapidly in an emergency. It is standard to provide a built-in pilot valve to equalize pressure between inlet and outlet at startup and an integral steam strainer to protect the turbine from particle damage.

Lubrication System

Bearings

Sleeve (hydrodynamic journal) bearings support the turbine rotor and generator. Jacking oil (oil lift) is provided during startup and shutdown so the weight of the rotor will not cause the shaft to contact the bearing surfaces. The jacking oil also reduces the starting load on the turning gear. This oil lifts the shaft and floats it on an oil film until the shaft speed is high enough to create a hydrodynamic film between the shaft and the babbitt.

Oil whirl or whip (vibration when the center of journal starts to rotate about the stable position, which is the center of the shaft) is eliminated by use of a pressure pad, tilting pad, or three-lobe designs. (See Figures 3A–23 through 3A–25.)

Thrust bearings absorb the axial thrust that occurs because of the pressure difference across each row of blades. Small turbines use babbitt-faced ends on journal bearings or rolling-element thrust bearings. Larger turbines use tilting-pad or tapered-land bearings. (See Figures 3A–26 and 3A–27.)

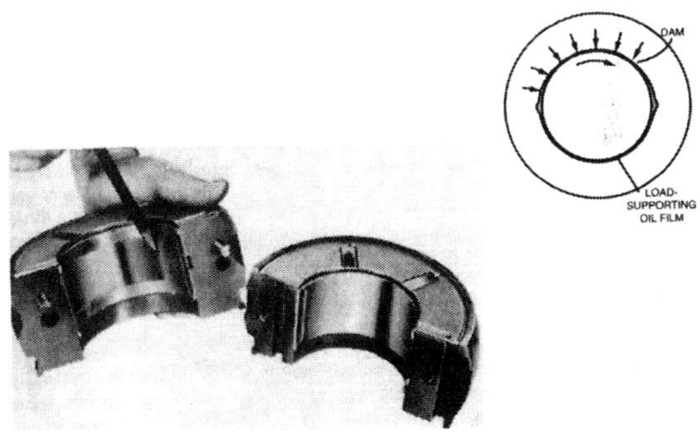

FIGURE 3A–23 Pressure bearing. The wide groove in the upper half ends in a sharp dam at the point indicated. As shown in the insert, this causes a downward pressure that forces the journal into a more eccentric position, which is more resistant to oil whirl [3-3].

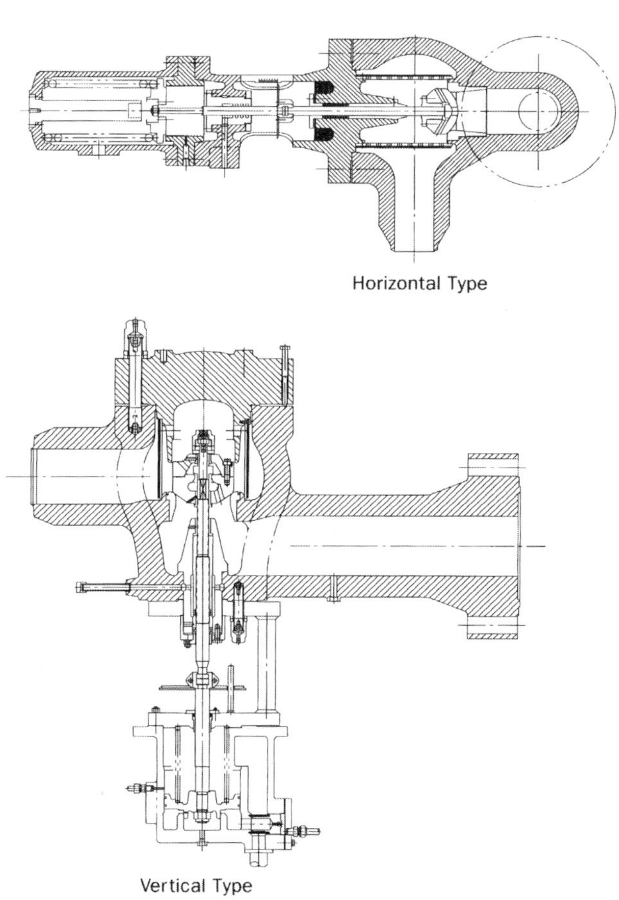

FIGURE 3A–22 Main stop valves [3-4].

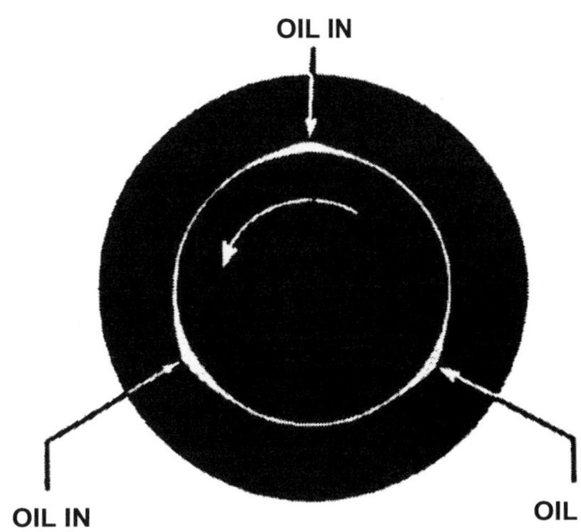

FIGURE 3A–24 Three-lobe bearing. The shape of the bearing is formed by three arcs of radius somewhat greater than the radius of the journal. This has the effect of creating a separate hydrodynamic film in each lobe, and the pressures in these films tend to keep the journal in a stable position [3-3].

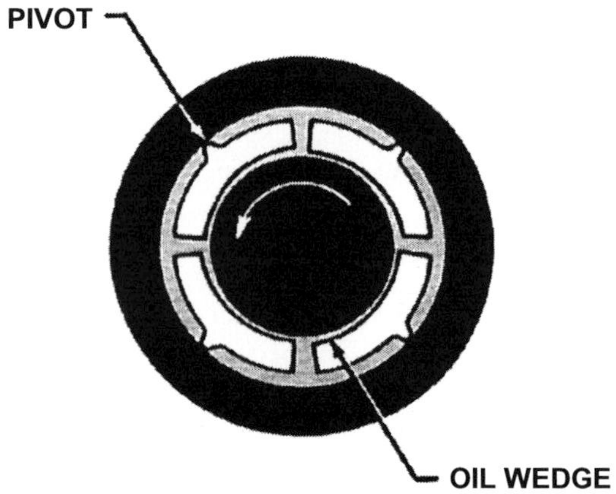

FIGURE 3A–25 Tilting-pad antiwhip bearing. As in the three-lobe bearing, the multiple oil films formed tend to keep the journal in a stable position [3-3].

FIGURE 3A–26 Combined journal and tilting-pad thrust bearing. A rigid collar on the shaft is held centered between the stationary thrust ring and a second stationary thrust ring (not shown) by two rows of tilting pads [3-3].

Oil

Oil leaving the bearings is about 70°C. The oil system, like the hydraulic system, must be kept free of dirt and small particles.

The factors that affect lubricating oil are:

- Overheating and exposure to air result in oxidation. Fine metal particles from wear accelerate oxidation. This changes the oil composition and viscosity. Sludge may result.
- Water contamination from:
  - Leaking turbine and pump seals.
  - Condensation in humid air.
  - Water leaks in heat exchangers. Emulsion forms when oil and water mix. Oxidation increases emulsification.

FIGURE 3A–27 Tapered-land thrust bearing and plain journal bearing. The thrust bearing consists of a collar on the shaft and two stationary bearing rings, one on each side of the collar. The babbitted thrust faces of the bearing rings are cut into sectors by radial grooves. About 80% of each sector is beveled to the leading radial groove, to permit the formation of wedge oil films. The unbeveled portions of the sectors absorb the thrust load when speed is too low to form hydrodynamic films [3-3].

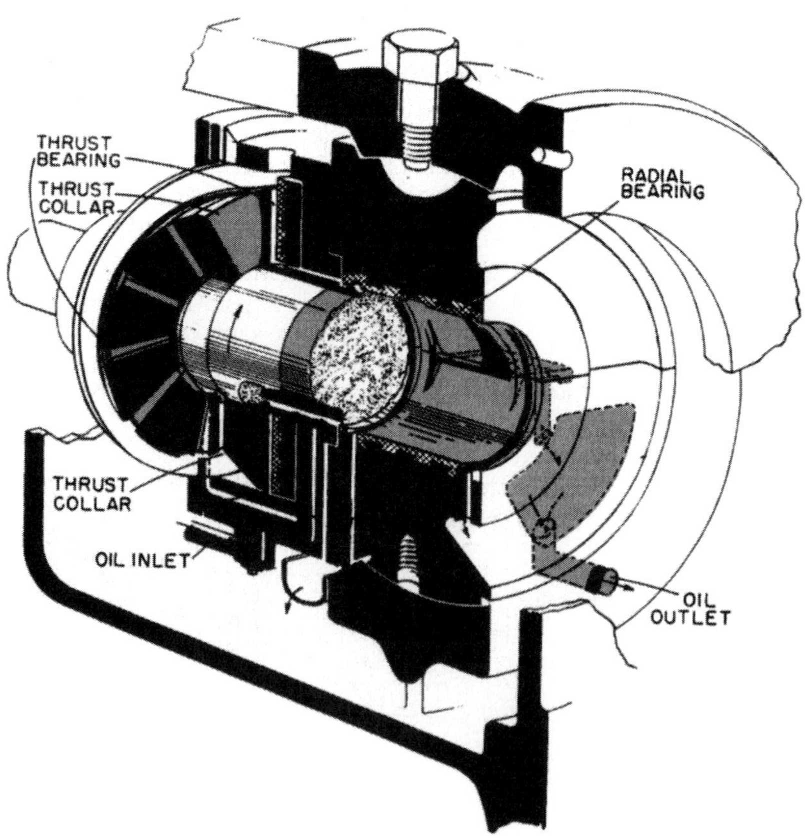

- Rust contamination. Rust particles.
- Catalyze the rate of oil oxidation.
- Scratch journals.
- Block small clearances in the governing system
- Air contamination. Air bubbles cause sponginess in hydraulic controls. They reduce load capacity of oil films. Air accelerates oxidation. It can generate foaming.

Oil, for the turbine model and application in question, needs the correct:

- Viscosity
- Load carrying ability
- Oxidation stability
- Protection against rusting
- Water separating ability
- Foam resistance
- Entrained air release
- Fire resistance

Steam Turbine Governing System

Major Components

How a basic governor works and its major components were covered previously. See Figure 3A–28 for a simple mechanical governor design.

The main functions of a governing system are:

1. To limit the speed rise to specific boundaries when the load is rejected (unit disconnected from the load). This is also covered by the emergency trip.
2. To control the power developed by the steam turbine (by controlling the steam governing valve position).
3. To control the turbine speed during run-up and synchronization.
4. To match the power generated to the power required by the load by responding to frequency changes (when the generator is not connected to the grid, also termed *islanding*).

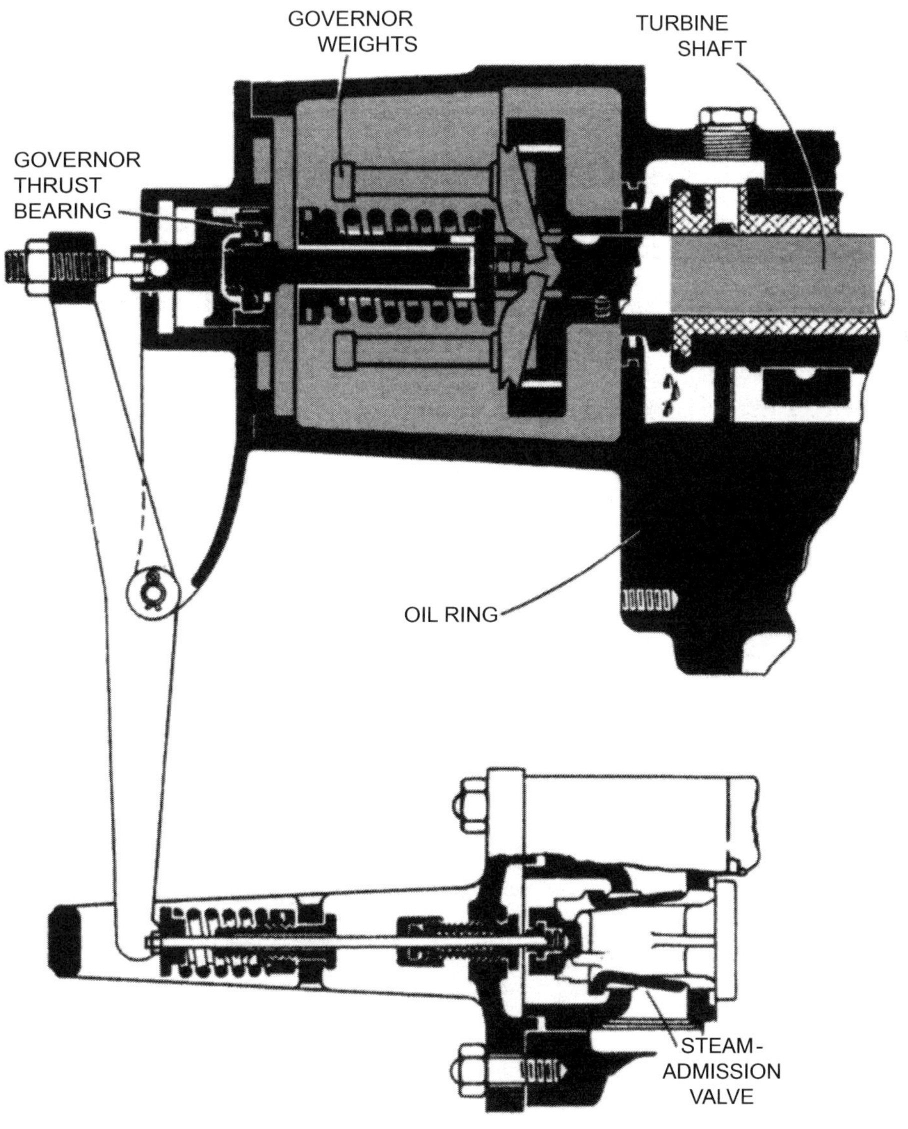

FIGURE 3A–28 Mechanical speed governor. A simple arrangement such as this, using a flyball governor, is suitable for many small turbines [3-3].

Most grid problems last less than an hour, so the governor needs to keep the unit at running speed, so that resynchronization is possible after the fault condition is over and to ensure continuity of the unit's power from the generator through the transformer. After a load rejection, there will always be a transient overspeed, due to the response time of the governing valve and the stored energy of the steam in the turbine and rest of the system.

Steam Turbine Operation

A basic governing system block diagram is shown in Figure 3A–29. Its operation varies with each set of application parameters. In general, however, it:

- Controls the speed versus load curve when the machine is synchronized to the grid
- Determines the position of the governing valve and intermediate valve(s)
- Limits the maximum speed of the generator
- Limits the power output
- Allows testing of the system

Steam Chests and Governing Valves

Numerous standard types of steam chests and governing valves cover a wide range of inlet steam conditions. Steam chests (Figure 3A–30) can be:

- Integral-cast type
- Top-mounted type
- Side-mounted type

Governing valves can be:

- Inner-bar type
- Semi-individual type
- Individual type

The governing valves have a multivalve arrangement and effective diffusers to minimize inlet energy loss. The valves are operated sequentially to control the inlet steam flow into the turbine with minimum inlet pressure loss. Thus, governing valves enable the turbine to achieve higher efficiency, even at partial loads.

Extraction Control Valves

There are two types of automatic extraction control valves (Figure 3A–31):

- Diffuser type, the standard extraction control valve. Its construction is the same as that of governing valves.
- Grid type, the type of valve commonly used when the extraction pressure is low. It consists of a stationary diaphragm installed in the turbine casing and a rotating port ring actuated by a servo system. The extraction steam flow is controlled by sequential port operation.

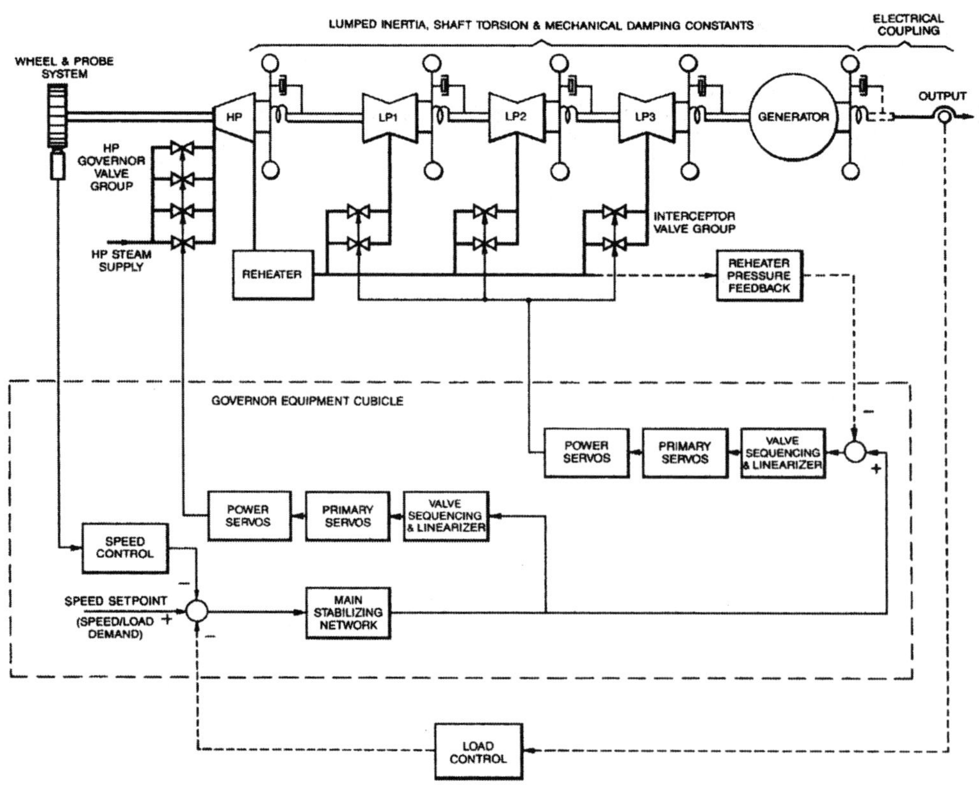

FIGURE 3A–29 Electrical governing system applied to a wet-steam turbine [3-3].

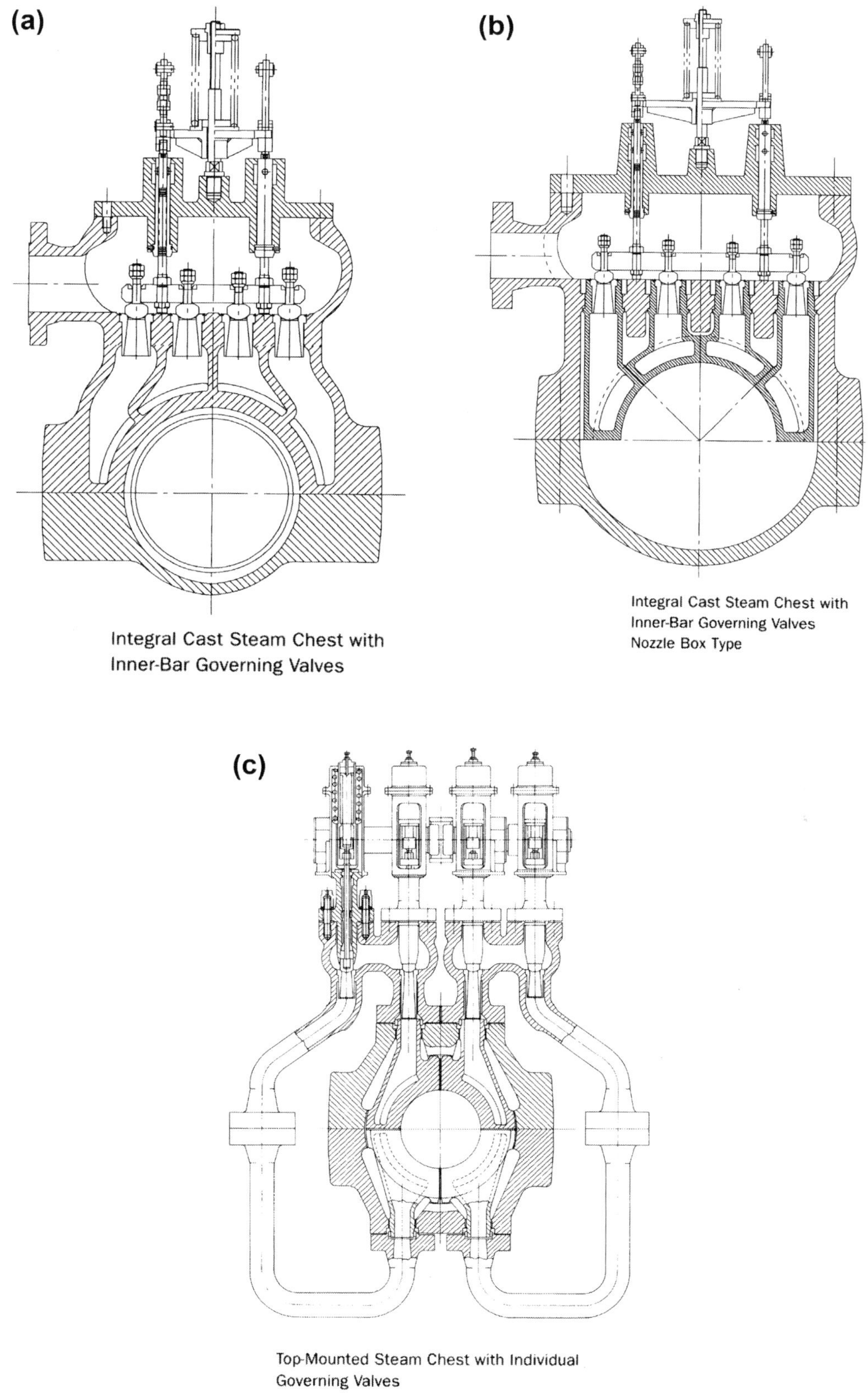

FIGURE 3A–30 (a) Integral cast steam chest with inner-bar governing valves. (b) Integral cast steam chest with inner-bar governing valves, nozzle box type. (c) Top-mounted steam chest with individual governing valves. (d) Side-mounted steam chest with individual governing valves [3-4].

(d)

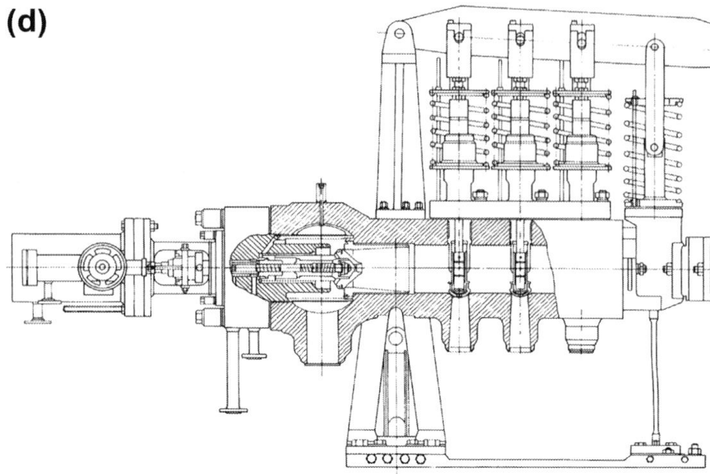

Side-Mounted Steam Chest with Individual Governing Valves

FIGURE 3A–30 (*Cont'd*).

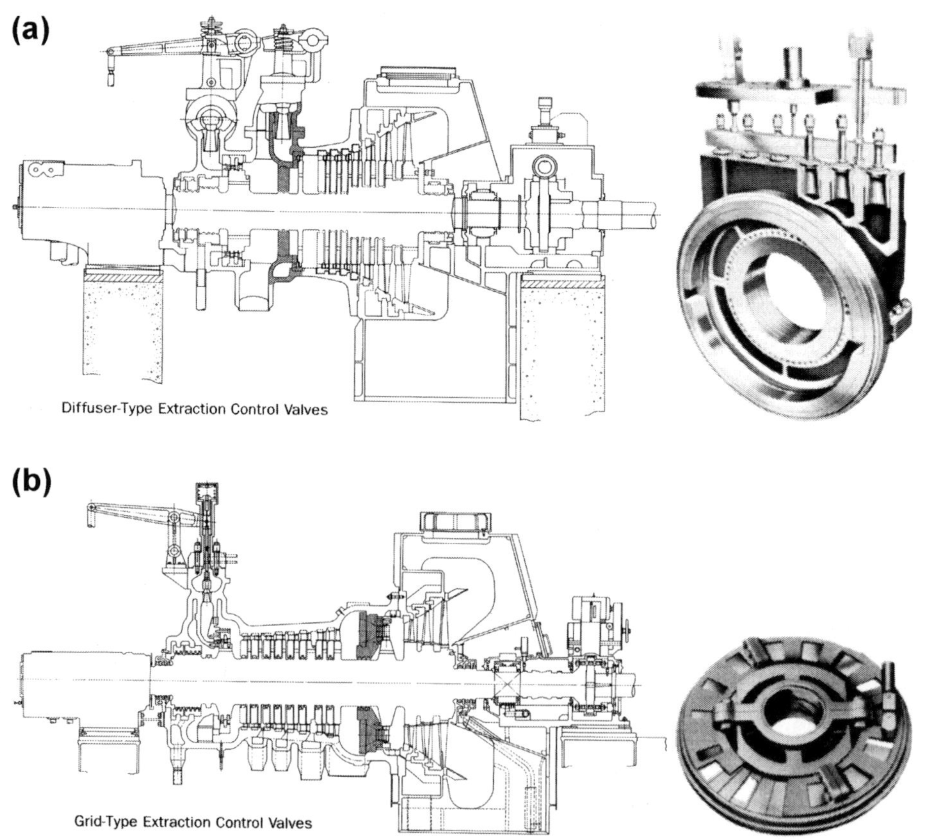

FIGURE 3A–31 Extraction control valves: (a) diffuser type, (b) grid type [3-4].

Glands

A typical gland sealing system is shown in Figure 3A–32.

To minimize steam leakage from the turbine gland, spring-backed labyrinth packings are used at both ends of the rotor. The labyrinth packings are divided into several segments to facilitate maintenance. The automatic steam regulator is provided to maintain the seal pressure constant at any turbine load.

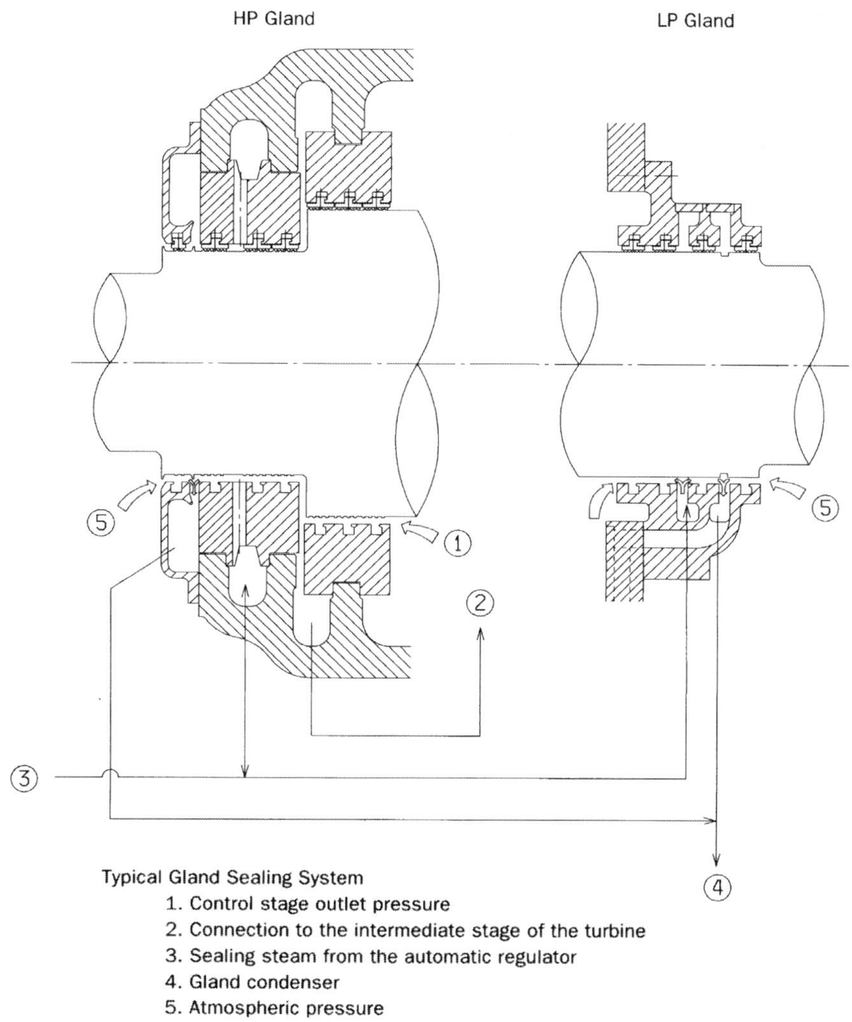

FIGURE 3A–32 Typical gland sealing system: (1) control stage outlet pressure, (2) connection to the intermediate stage of the turbine, (3) sealing steam from the automatic regulator, (4) gland condenser, and (5) atmospheric pressure [3-4].

Turbine Run-Up

As run-up progresses, the steam flow increases to full load steam flow. At the start of the run-up, 2–3% of full load power is required to overcome mechanical friction and windage losses. A toothed wheel coupled to the main turbine shaft and a speed-sensing device provide speed readings.

For constant steam inlet conditions, power versus steam flow is linear (see Figure 3A–33(a)). It would be desirable to have the steam demand input to the valve vary linearly with load (steam flow). However, Figure 3A–33(b) shows this is not the case. Note that the steam flow varies directly with the boiler pressure (as well as the load).

The valve lift, in turn, varies with the valve flow area, see Figure 3A–33(c).

Linearizing circuitry results in the characteristic shown in Figure 3A–33(d).

Turbine Trip (o/s)

See the discussion in the preceding section. For load rejection (unloading), three channels are used to conduct steam demand signals to each valve controller (see Figure 3A–34). The controller selects an operating demand point based on "majority votes." The valve position transducer provides feedback on its position to the controller (also see Figure 3A–35).

Hydraulic Fluid

Hydraulic fluid is used to conduct large forces to control the valves. The fluid needs to flow around and through some small clearances. For steam turbines below 500 MW, the hydraulic fluid needs to develop 35 bar or less. For larger turbines, the hydraulic fluid needs to conduct between 70 and 150 bar.

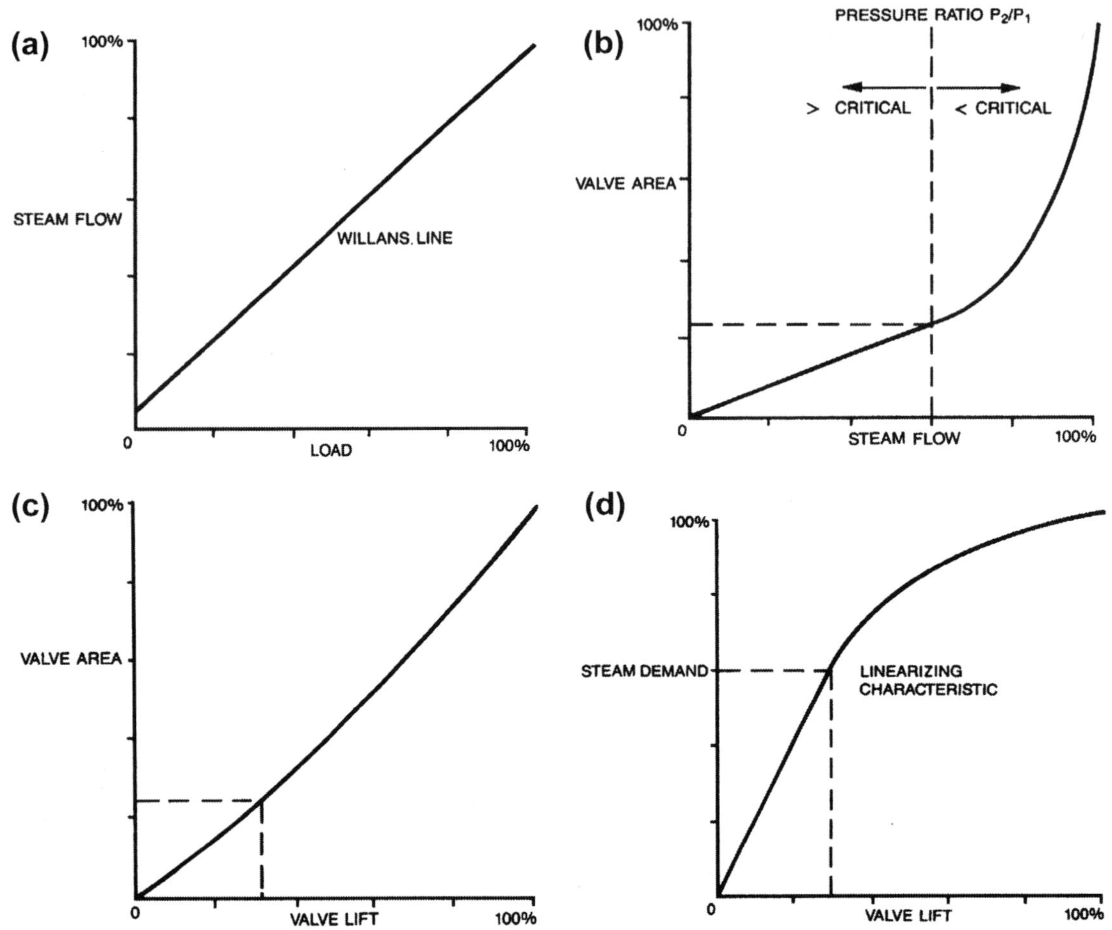

FIGURE 3A–33 Typical steam load/valve characteristics: (a) variation of steam flow with the load, (b) variation of valve area with steam flow, (c) variation of valve area with valve lift, and (d) variation of steam demand with valve lift (known as a linearizing characteristic) [3-3].

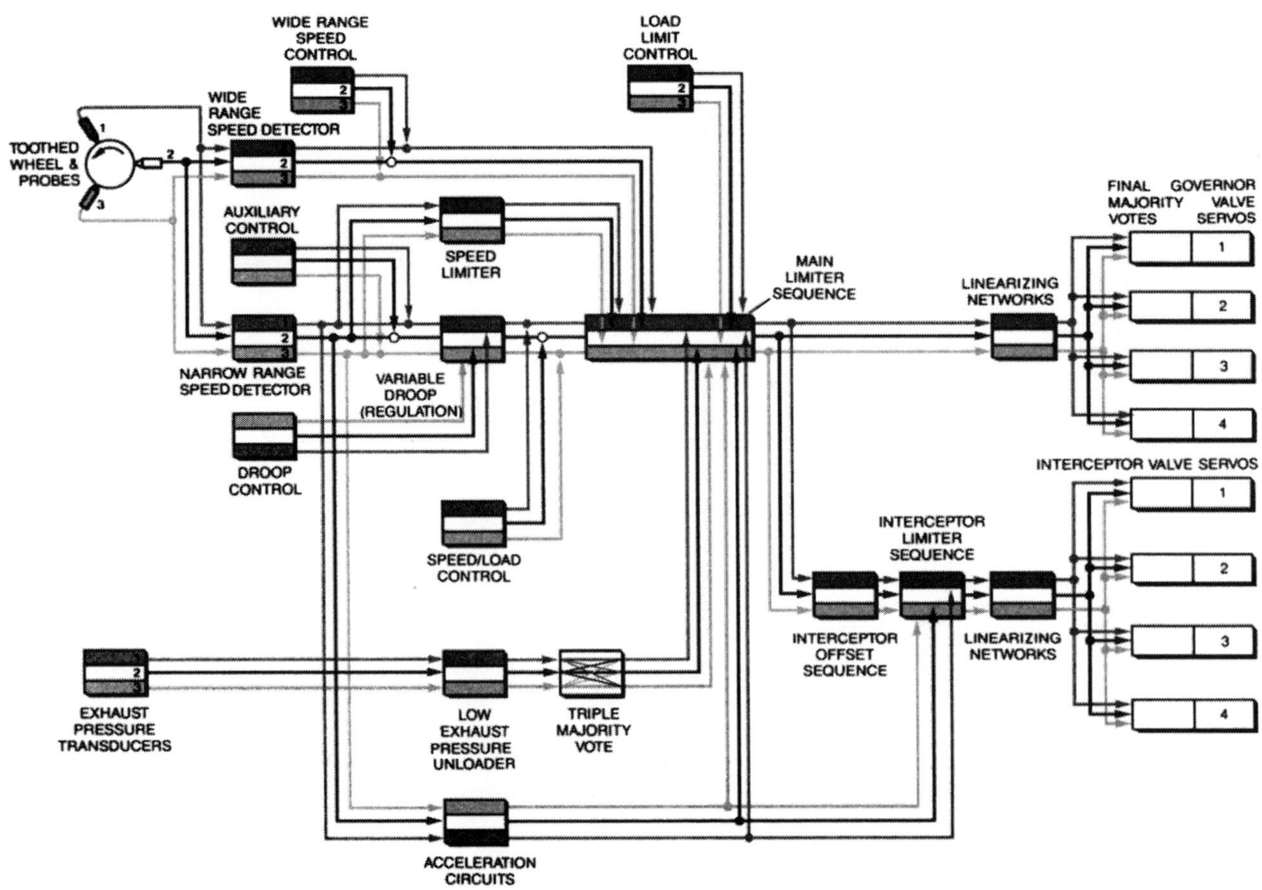

FIGURE 3A–34 Block diagram of a three-channel governing system [3-3].

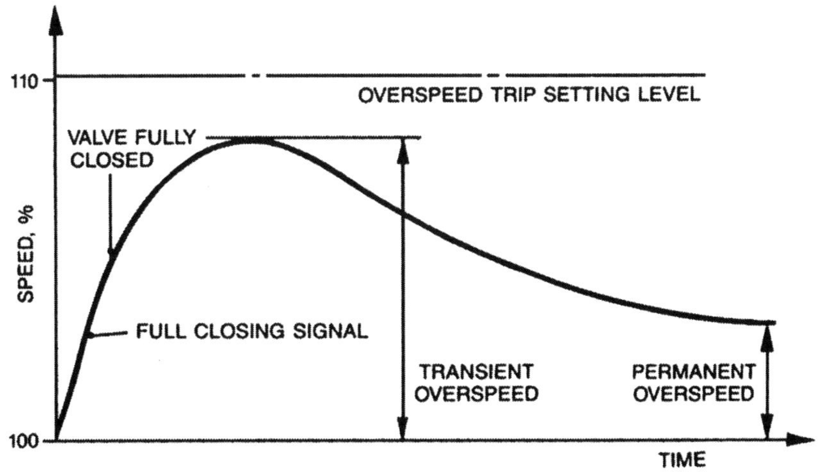

FIGURE 3A—35 Turbine response to load rejection [3-3].

As the fluid pressures are high, a small leak causes a potentially flammable spray. Normal oil ignites on contact with hot pipes. However, phosphate esters are used. They are not flammable and have a long service life. The viscosity of hydraulic fluid needs to be monitored diligently. A change in viscosity may mean slow valve reaction with disastrous pressure build-up or seal failure.

See Figure 3A—36 for a schematic of a hydraulic system. A redundant pumping line also capable of conducting hydraulic fluid adds system reliability.

The system pumps are screw or piston type with a coarse inlet filter. A separate centrifugal pump is used to boost fluid volume. The system attempts to maintain 40°C fluid temperature for optimum viscosity. See Figure 3A—37, which depicts a hydraulic system skid.

Tripping Signals

In addition to the main overspeed trip, other relief valves are used in the system to protect individual system components and systems. For instance, reheater valves prevent reheater overpressurization. There may be additional release valves to vent steam under certain conditions, thus avoiding damage to the turbine blades.

These release valves operate hydraulically or pneumatically. In most modern turbine systems, the governor system is electronic and the valves are opened by solenoid valves. They open for:

- Loss of tripping pressure
- High shaft acceleration
- Closure of all interceptor steam relays. This includes all turbine trips or load rejection with no trip.

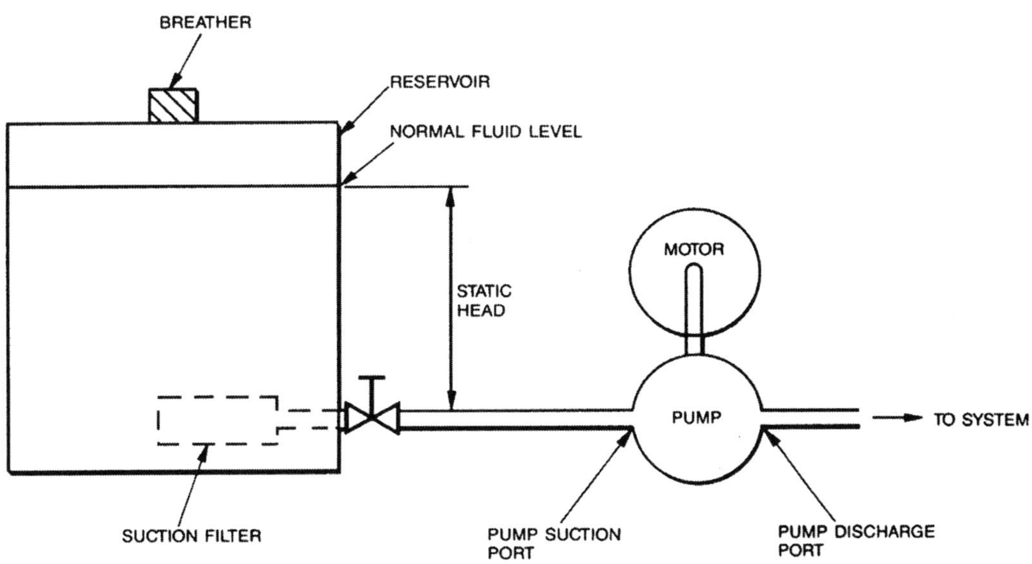

FIGURE 3A—36 Pump suction arrangement for a separately mounted pump [3-3].

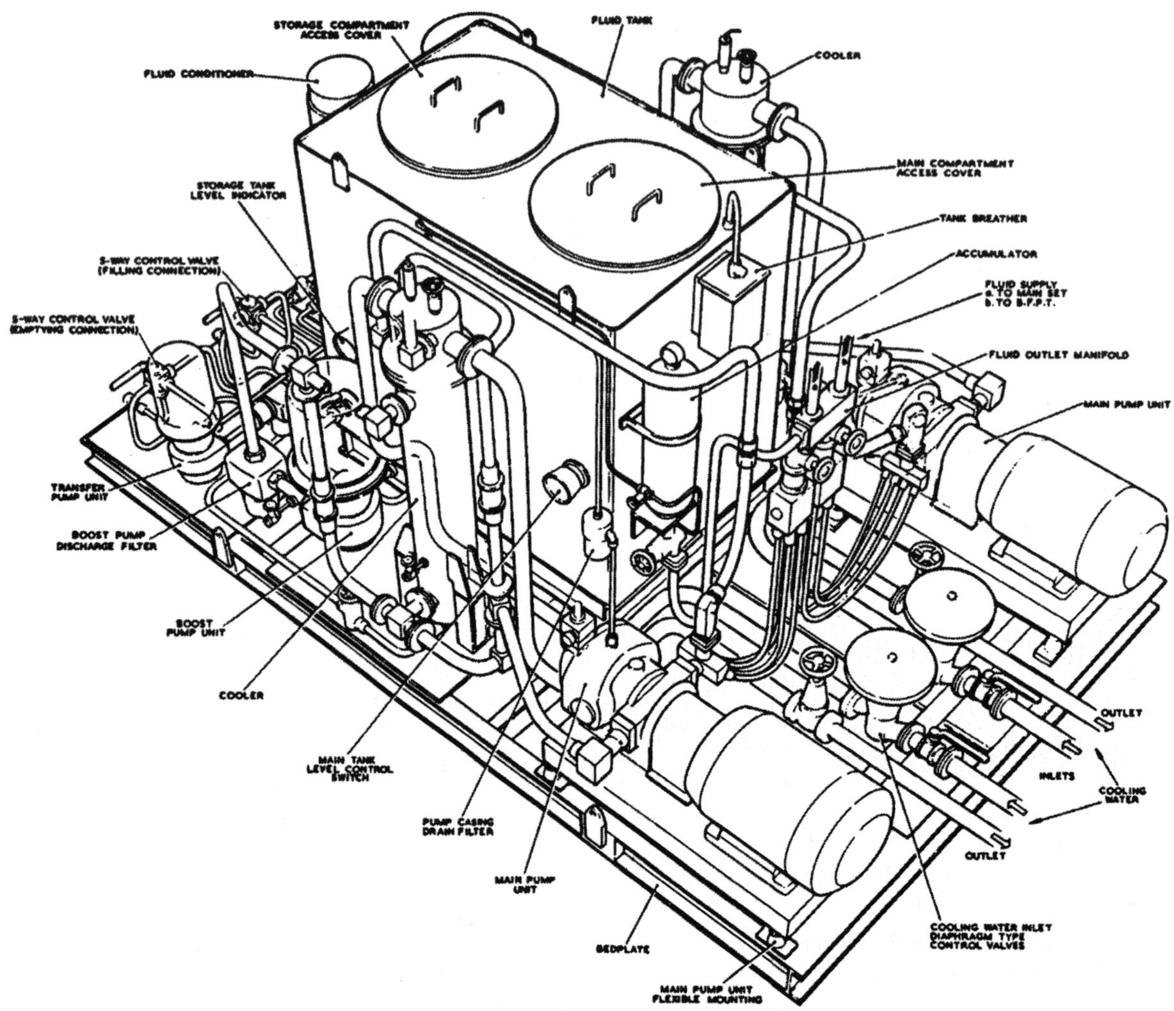

FIGURE 3A–37 Fire-resistant fluid packaged unit [3-3].

This does not protect against potential steam churning during run-down after a trip or as interceptor valves close after load rejection.

Supercritical Systems: Targeting 700+ °C Steam Temperature

Supercritical steam conditions of 250 bar, 600°C HP, and 610°C reheat are now described as state of the art in supercritical steam cycle systems. The prospect of further development towards steam temperatures higher than 700°C has been under active investigation since the early 1990s in Europe, Japan, Russia, and the United States. Discussions with key experts at Siemens* in Mülheim indicate that while the technology looks feasible, further investigation is needed to demonstrate the economic advantages. Efficiency increases from turbine blade improvements may lower life cycle costs.

Deregulation in the power supply industries around the world does little to advance the cause of supercritical coal fired generating plant systems development, even in countries with a continuing commitment to coal burning.

Countries with limited indigenous fuel resources and a high commitment to environmental protection such as Japan and Denmark, which appear to be dependent on imported coal for the foreseeable future, do have a continuing interest in increasing generating plant efficiency while reducing generating costs.

Australia, Germany, Russia, South Africa, the UK, and to some extent the United States are economically motivated to burn coal from indigenous resources with the highest possible efficiency without sacrificing availability, but the key economic criterion in these days of increasing

* Source: Courtesy Siemens from their work on steam turbines including http://www.siemens.com/innovation/en/publikationen/publications_pof/pof_spring_2008/energy/effiziente_kraftwerke.htm (extracted June 2014) and http://www.energy.siemens.com/hq/pool/hq/power-generation/power-plants/steam-power-plant-solutions/benson%20boiler/Steam_Generators_for_the_Next_Generation_of_Power_Plants.pdf (extracted June 2014)

competition in utility services continues to be the minimization of lifetime operating costs.

Under these conditions the trade-offs between initial design criteria and materials selection against operation and maintenance costs may be conflicting. It is interesting, on the other hand, that China is increasingly interested in installing supercritical coal-fired power plants to gain maximum thermal efficiency.

Current supercritical power plants reach thermal efficiencies of just over 40%, with a few more recent highly supercritical power plants reaching 45%. The leading developers of advanced coal burning power plant technology claim that thermal efficiency as high as 50%, notwithstanding Carnot, is now within reach. But the main competition is from new gas turbine combined-cycle plants which are now aspiring to an overall efficiency of 60%, making a huge difference to generating costs and life cycle costs.

On the other hand, the newest gas turbines that are about to be introduced to the market will exhaust into waste heat recovery steam generators at temperatures above 600°C, necessitating the use of the same high chromium steel alloys such as ASTM-A335 P91 (X10CrMoVNb9-1) in the boiler tube bundles as is used in the more highly supercritical coal-fired conventional generating plants now being built.

While increasing steam temperature is generally agreed to be the most rewarding route to increasing efficiency in coal fired power plant, there are recognized thresholds of investment costs above which the returns are not just diminishing, but increasingly negative.

This has led companies such as Siemens KWU, which is currently assimilating the technology of Westinghouse Electric in the United States, and Parsons in the UK, to concentrate more on the development of advanced steam turbine blading and shaft design than on the use of the newer austenitic steels and nickel-based alloys. Nonetheless, Siemens claims it is able to offer plants designed to work at steam conditions of 300/600/610 and is working towards steam conditions in the order of 300/700/720.

The development of highly supercritical power plants in Germany has been substantially set back since the period of peak interest in 1994/95 with the shelving of the Hessler, Lübeck, and now RWE's newest Niederaussem power plant unit (250/580/600). But if the EU Energy Charter's aspirations towards cost transparency ever materialize, operating experience from recently built supercritical coal fired plants like the four big VEAG lignite stations in former eastern Germany—Boxberg, Lippendorf, Schwarze Pumpe (268/547/565), and Jänschwalde—and also the IPP Schkopau, none of which exceeds 40% efficiency by much, will do a good deal to settle the outstanding issues.

Perhaps the most significant reference will be Siemens KWU's latest supercritical plant contract, the 600 MWe Isogo unit being built for Tokyo Electric Power for operation in 2001, which is rated at 250/600/610, since blades of nickel-based alloys are generally thought to be necessary for reheat temperatures of 610°C and higher. The steam turbines for this will be supplied directly from KWU's Mülheim works rather than from the local licensee Fuji Electric.

This represents a substantial increase in steam temperature over Chubu Electric's 700 MWe Kawagoe double reheat systems which run on steam conditions of 311 bar, 566/566/566°C. The first two of these units have been running since June 1989 at efficiencies of 45%, an efficiency gain of 5% over previous conventional plants.

These make strategic use of 12% chrome cast steel in the Toshiba turbines and blades and 9% chrome in the pipework.

This was the first use of such alloys in Japan since the standard single reheat steam conditions of 242 bar, 538/566°C were established for large fossil fueled steam turbines. Since the Kawagoe units demand the high levels of availability that has been essential for large fossil fired middle load power plants in recent years, operating experience from this station is of inestimable importance to advanced supercritical system development.

International Development Programs

By 1995 research and development anticipating 50% net plant efficiency within five to seven years was proceeding worldwide with the Brite Euram and COST in the European Community, the VKR Hessler and Preussenelektra Lübeck projects in Germany, the ELSAM "Convoy" double reheat power projects in Denmark, the KEMA work in the Netherlands, EPRI 1403—50 in the United States, and CRIEPI in Japan.

Much of the attention under these programs was directed at the metallurgical problems of the steam generators and the tendency for thinning of the superheater tubes in service using the ferritic/martensitic steels in common use. Some parties believe that in practice the tube thinning will not happen, and the new plants now beginning to operate in Germany, Japan, and Denmark will give valuable insight and design validation for future projects.

The Schwarze Pumpe 2 × 800 MWe power plant recently commissioned near Dresden in the Lausitz lignite mining area is a good example. These are thought to be largest lignite boilers ever built, with physical dimensions of 161 m in height, 24 m × 24 m in cross section. They are once-through, single pass boilers designed by EVT as 900 MWe units but with a high process steam output of up to 800 t/h to various process plants in the region. Then plants have steam parameters of 268 bar/547°C/565°C for the maximum steam generating capacity of 2420 t/h.

This is said to be the first time that P91 has been used for a power plant steam generator of this size. Used for the live steam and hot reheat piping, this material was selected as having higher creep rupture stress than the X20CrMoV121 normally used, which leads to lower tube wall thicknesses and hence reduction in thermal stresses. The weight reduction in turn reduces stress levels at the boiler connections and at the turbine connections as well as on the structural steelwork.

A seven stage regenerative feed heating system driven by a turbine driven feed pump delivers feedwater at a final temperature of 270°C. The supercritical steam output supplies a single reheat turbine system in sliding pressure mode with main steam at 638 kg/s, 253 bar, 544°C, and reheat at 52 bar, 562°C.

Feedwater temperature is becoming an increasingly critical parameter in recent years with the added emphasis on emissions control. There are economic benefits in raising steam pressure to 300 bar since, because of the lower thermal energy transfer required in the evaporator, the heat exchange area required is reduced with a corresponding saving in costly high chromium alloys.

On the other hand, the maximum furnace temperatures have to be retained, even to the point of possibly removing the economizer, but this will then not cater for reheat inlet temperatures above 310°C because the limitations on maximum SCR temperature will be exceeded.

Since with the high process heat output at Schwarze Pumpe a fuel utilization efficiency of some 55% is claimed, it may be considered that cogeneration of heat and steam could be a more economic approach to beating Carnot, but there are not that many instances of adequately large heat loads existing close to the mine mouth sites adjacent to the coal resources.

With such constraints and the economic limitations inherent in using P91, P92, austenitic steel alloys, and nickel-based alloys for which good creep and manufacturability data are only available from the nuclear industry, Siemens have placed increasing reliance on steam turbine design advances in their quest to break the magic 50% thermal efficiency barrier.

Boiler Design

Siemens has been promoting the once-through boiler concept since they acquired the Mark Benson patent in the 1920s because it brought the capacity for high-pressure boiler potential into the field of power generation for the first time. They claim responsibility for the spiral configuration of the evaporator tubes, and they have since further developed this vertical tube design with a view to reducing investment costs while at the same time introducing boilers operating at supercritical pressures in which the drum type boiler has been the dominant design.

By using rifled tubes with high heat transfer capacities, they were able to reduce mass velocity in the tubes as much as 2500 kg/m^2s to less than 1200 kg/m^2s. The popular reference for this design is the 550 MW Staudinger Unit 5. This operates at 260 bar, 545/562°C, still using low-cost martensitic steels for the final stages of the HP and reheater heating surfaces.

Bent Blades

Fullly three-dimensional, variable reaction blading with compound lean, a Siemens development designated 3DV—the already well publicized curved blades—is the main ingredient in optimizing HP and IP turbine efficiencies. Relevant here is a dispute among turbine engineers that goes back more than 100 years over the choice of stage reaction, i.e., the split of pressure drop and velocity increase between stationary blades and moving blades. Defined more scientifically, stage reaction is the ratio of the enthalpy drop in the rotating row to the enthalpy drop of a whole stage.

A reaction turbine is characterized by a stage reaction of 50%, and the enthalpy drop is equally divided between across stator and rotor rows. The symmetry in enthalpy drop entails a symmetry in flow relative to stator and rotor flows and allows the same profile to be used for both blade rows. In this case, flow velocities and flow directions along the blade path are moderate and profile and secondary losses are quite low. Since half of the enthalpy drop occurs in the rotor row, the pressure differential across the rotor blades is rather high and exerts a large axial thrust on the rotor.

In order to compensate for this axial thrust, a dummy balance piston is required in single flow designs. The leakage flow across this dummy piston reduces the turbine cylinder efficiency.

In impulse turbines with zero stage reaction, the total stage enthalpy is converted into kinetic energy in the stator row, while the rotor row merely deflects the steam without further acceleration. In this case different profiles must be used for stator and rotor blades because their flows are asymmetrical.

Flow velocities and flow deflection along the blade path are significantly larger than the equivalent reaction stages, incurring higher profile and secondary losses. Since the pressure differential across the rotor is much less than that for reaction turbines, a much smaller dummy balance piston with lower losses can be used.

Three-dimensional blades have the potential to exploit the best of both worlds, and Siemens maintain that by the use of computer controlled five axis milling machines, any three-dimensional blade that the design engineer can invent can now be manufactured.

For the first stages of HP and IP turbines, a special three-dimensional blade with compound lean has been developed by Siemens.

In these stages, the secondary losses are significant due to the low volume flow rates and short blade lengths. Compound lean and twist both serve to reduce the secondary losses at the root and the tip of blade. The reaction of each stage is set individually and may vary between 10 and 60%.

Extensive measurements were performed on a four-stage test turbine to confirm efficiency improvements of up to 2%—a major gain—compared with conventional cylindrical blading.

Another feature of the Siemens KWU supercritical IP turbine is the vortex cooling principle. For rotor stress

TABLE 3A–2 Main Specifications and Features of Recently Designed Steam Turbines for Conventional Plants [3-5]

			Steam Condition					
	Unit	Capacity (MW)	Pressure (MPa)	Temp. (°C)	Vacuum (kPa)	Revolution (rpm)	Year in Operation	Features
A	Hirono #5	600	24.6	600/600	5.1	3000	2004	• 600 MW, two casings (HP/IP + LP) • 3000 rpm 48 inch steel blades • 600°C high temperature turbine
B	Kobe #1	700	24.2	538/566	5.1	3600	2002	• 700 MW, three casings (HP/IP+2LP) • 3600 rpm 40 inch steel blades
C	Ratchaburi #1, 2	840	24.2	538/566	8.0/10.0	3000	2000	• Tandem 840 MW, four casings (HP + IP + 2LP)
D	Tachibanawan #2	1050	25.1	600/610	5.1	3600	2000	• 1000 MW class, world's highest steam condition
							1800	• World's highest efficiency • Cross-compound type

reduction in the steam admission zone, the rotor is built without an axial through bore.

There is a double "T" root configuration for the two first stages with tilted first row stationary blades mounted on a heat shield ring with a cooling passage beneath the root. Cooling steam flows through tangential bores to form a vortex that cools the rotor since the steam temperature is reduced by its expansion to a high level of kinetic energy.

The rotor surface temperature distribution shows a temperature reduction of up to 16°C at the critical rotor section, which is enough to greatly improve the stress characteristics of the design.

Economic Advantage

Every 1% of efficiency increase gained by the use of more highly supercritical steam conditions is reckoned to increase the output of a 660 MW turbine system in Germany by some 3 MW at normal input conditions.

The higher efficiency saves about 5600 t of coal per year, which in turn saves about 12,500 t of CO_2 emissions. On top of this, life cycle costs in Germany are reduced by about $6 million.

Case Study 3A-1: Advanced Design of Mitsubishi Large Steam Turbines*

Mitsubishi has developed modern high temperature steam turbines based on advanced technologies, including advanced steam path design through fully 3-D computational fluid dynamic (CFD) analysis methodologies, high loading reaction blades, high efficiency seals due to active clearance control, rotor cooling structure, enhanced LP section, and state-of-the-art material technologies, which are now market ready. The LP enhancement includes the world longest LP last blades such as 3600 rpm 40 inch and 3000 rpm 48 inch steel blade.

The main specifications and features for recently designed steam turbines of different power output ranges are shown in Table 3A–2. The 600 MW high-temperature tandem-compound turbine design (A) of Hirono No. 5 unit, Tokyo Electric Power Co., is comprised of one combined high/intermediate pressure (HP/IP) turbine and one double flow low-pressure (LP) turbine with 3000 rpm 48 inch steel LP last stage blades. The 700 MW (B) Kobe # 1 unit started commercial operation in 2002 and it is composed of one combined HP/IP casing and two LP casings with 3600 rpm 40 inch last stage blade. The 840 MW tandem-compound four casings design (C) of Thailand Ratchaburi units No. 1/No. 2 started commercial operation in 2000. This machine is composed of one double flow HP, IP turbine and two double flow LP turbines.

The 1050 MW Tachibana-wan power plant unit 2 (D), Electric Power Development Co., entered commercial operation in December 2000. This turbine is a typical cross-compound machine designed with the industry highest steam temperatures and achieving the highest gross efficiency in the world.

As for 50 Hz application, the optimum LP last blade design for 50 Hz will be selected in accordance with the required power output and the available power train combination for the 50 Hz application. Table 3A–4 presents two examples of the main features of modern MHI steam turbines for combined cycle application.

* [3-5] Courtesy of Mitsubishi Power Systems. Extracts from Y. Tanaka et al., "Advanced Design of Mitsubishi Large Steam Turbines."

TABLE 3A–3 Example of Recent MHI Steam Turbine Designs for Combined Cycle Applications [3-5]

	Unit	Capacity (MW)	Steam Condition Press. (MPa)	Temp. (°C)	Vacuum (kPa)	Revolution (rpm)	Year in Operation	Features
E	Alabama	178	11.2	566/566	10.9	3600	2001	• 3600 rpm 40 inch steel blades • Large capacity SRT for F 2 on 1
F	ILIJAN #1, 2	243	14.7	538/566	6.7	3600	2002	• Two casings (HP/IP + LP) • 3600 rpm 40 inch steel blades

Units E and F in Table 3A–3 are typical steam turbine frames for 2 on 1 combined cycles. The E unit is designed with single casing (HP/IP and LP combined type: single reheat turbine). Both E and F units apply 3600 rpm 40 inch steel LP last blades.

The main design features of MHI large capacity steam turbines are summarized as follows:

- Compact HP/IP combined turbine with high efficiency reaction blades
- New advanced LP last stage blades
- High temperature design

This section describes these design features and operating experience of large steam turbines recently designed for fossil-fuel and combined-cycle power plants.

Product Line of MHI's New Advanced LP Last Stage Blade

A new standard series of these LP last stage blades was completed for 50 Hz and 60 Hz unit application including 3600 rpm 45 inch titanium blade, 3000 rpm 48 inch steel blade, and 1500/1800 rpm 54 inch steel blade, which have the largest exhaust annulus area among the same class blades in the industry.

High Temperature Design

Tachibana-wan No. 2 unit in Shikoku, Japan, for Electric Power Development Co. (EPDC) is a 1000 MW cross-compound plant with maximum temperature of 600/610°C (see Table 3A–4). The HP and IP turbines are positioned on the high-speed shaft (3600 rpm) and two double-exhaust LP turbines on the low-speed shaft (1800 rpm). This unit has been operating successfully since December 2000 and the acceptance test demonstrated that the gross efficiency figures achieved a new benchmark (about 49%) for 1000 MW class steam turbine.

Ferritic heat-resistant steel materials are widely used for 600°C class high temperature turbines, such as new 12 Cr forged, 12 Cr cast, and 9 Cr forged steels. In high temperature rotating blades, austenitic refractory alloys are used, while 12 Cr forged steel having sufficient creep rupture strength for operation at 600°C class temperatures is used as the rotor material. In the journal and thrust collar sections of the 12 Cr rotor, a low Cr overlay material is used in order to reduce wear of bearing materials. As for the materials for stationary parts, 12 Cr cast steel (MJC-12), which has excellent creep rupture strength, is used in the nozzle chamber and inner casing. A 9 Cr forged steel is used in the main stop valve, HP turbine inlet, governing valve, and connection piping between the valves and casing (see Table 3A–5).

The upper limit for steam temperatures initially applied for new 12 Cr steel rotors designated TMK-1/2 was 600°C class. However, this limit has been raised after development of advanced 12 Cr (with Co) steel for recent projects on Mitsubishi advanced steam turbines of large capacity. This material is superior to new 12 Cr or 9 Cr steel in terms of high temperature strength.

Ultra Supercritical Steam (USC ST) Turbines

Super-critical steam has been around since the 1950s. The countries that took advantage of this technology included countries like China, which are relatively poor in oil and gas resources. Even with the onset of fracking technology, the increase in energy demands globally are such that coal will continue in its use as a fuel for the forseeable future. The development of renewables continues, but that curve is not as steep as environmentalists would have it. Also new gas resources discovered by fracking still will not totally fill the energy need, even in the USA with its own domestic reserves of conventional oil and gas (supplemented by synthetic crude imports from Canada and oil imports from the rest of the world). Since coal is a reality that will not go away, research work, with organizations such as the US DOE, continue to develop metallurgy that handles USCS well. Several plants around the world (a case study follows here) prove that this is attainable.

TABLE 3A–4 Major Specifications of 600°C Class 1000 MW Steam Turbines [3-5]

Item	Specifications		
Unit name	Matsuura No. 2 Unit, Electric Power Development Co.	Misumi No. 1 Unit, Chugoku Electric Power Co., Inc.	Tachibanawan No. 2 Unit Electric Power Development Co.
Type	Cross compound, 4 exhaust flows, reheat, regenerative condensing type		
Output (rated)	1,000,000 kW		1,050,000 kW
Steam Condition			
(Main steam press.)	24.2 MPa	24.6 MPa	25.1 MPa
(Main steam temp.)	593°C	600°C	600°C
(Reheat steam temp.)	593°C	600°C	610°C
Rotating speed	Primary shaft	3600 rpm	
	Secondary shaft	1800 rpm	
Vacuum	5.06 kPa (722 mmHg)		
End blade length feed	1170 mm (46 inches)		
Water heater	8 stages		

TABLE 3A–5 Materials for HP, IP Turbines [3-5]

	EPDC Matsuura No. 2 1000 MW (538/566°C)	Cyuubu Elec. Co., Inc. Heklmar No. 3 700 MW (538/593°C)	EPDC Tachbana-wan No. 2 1050 MW (600/610°C)
HP Turbine			
Rotor	Cr MoV forging	CrMoV forging	New 12 Cr forging
Inner Casing	2¼ CrMoV cast steel	2¼ CrMoV cast Steel	12 Cr cast steel
MSV/CV	2¼ CrMoV forging	2¼ CrMoV forging	9 Cr forging
Main Steam Inlet	2¼ CrMoV forging	2¼ CrMoV forging	9 Cr forging
Pipe			
IP Turbine			
Rotor	MoV forging	1 C forging	New 12 Cr forging
Inner Casing	2¼ CrMoV cast steel	12 Cr cast steel	12 Cr cast steel
RSS/VC	2¼ CrMoV forging	9 Cr forging	9 Cr forging
Reheat Steam Inlet	2¼ CrMoV forging	9 Cr forging	9 Cr forging

MHI Development Material Rotor: TMK-1,2, Casing MJC-12.
US Development Material Super 9 Cr.

Metallurgists, however, push still harder to attain steam temperatures of over 750°C. A brief extract of US DOE work in this vein follows the case study.

Consider also that GT systems operators, especially in power generation, are becoming more astute with cogeneration applications, the use of alternative fuels, and CHP systems. Also, as transmission technology improves, plant operators are more adept at integrating GT systems with ST systems, as well as renewable energies (better handled now by smart grid technology).

Item	Not Exceed Limit	Requirement
Engine oil tank	Flight termination	Check oil level. Replenish as necessary. Record amount taken
Cowls	Transit	Check the pod cowls for damage and external evidence of fuel and oil leaks
Caps and access panels	Transit	Check secure
Engine intake	Transit	Check clear. Free from damage and loose objects
Turbine and exhaust collector	Transit	Visually inspect for signs of damage and metal deposits
Engine intake	25 hours	Visually inspect front of engine through air intake for signs of damage paying particular attention to intake guide vanes and leading stage rotor blades
Turbine and exhaust collector	25 hours	Visually inspect L.P.2 turbine blades, nozzle guide vanes and mixer unit for cracking and damage by viewing from rear using a strong spotlight
Fuel filter	125 hours	Drain sample and check for water contamination
Magnetic chip detector	200 hours	Remove and inspect
Igniter plugs	200 hours	Audibly check operation
Oil pressure filter	600 hours	Check and clean/renew filter element
Fuel filter	800 hours	Remove filter element, check and renew

Case Study 3A-2: One OEM's Supercritical ST Product Range with Specific Reference to the 800 MW Trianel Power Project in Lünen*

This case presents Siemens products and solutions for ultra supercritical steam power plants and their application in the 800 MW Trianel Power project in Lünen. Several German and European municipal utility companies hold a share in Trianel Kohlekraftwerk Lünen GmbH & Co. KG. Advanced steam parameters (280 bar/600°C/610°C), a net efficiency above 45% (LHV basis, hard coal), and specific CO_2 emissions well below 800 g/kWh are characteristic features of this project.

The Lünen power plant will meet the most stringent environmental protection requirements of the German authorities and will have among the lowest environmental emissions of any coal-fired power plant in Germany.

The flue gas cleaning system includes equipment for removing nitrogen oxides (SCR reactor), particulates (electrostatic precipitators) and sulfurous components (flue gas desulfurization unit). Typical emissions limits are as follows:

- SO_x 200 mg/Nm^3
- NO_x 200 mg/Nm^3
- CO $<$200 mg/Nm^3
- Particulates 20 mg/Nm^3

Completely enclosed conveyor belts will supply the fuel from the ship unloading station to the closed coal silos and next to the coal bunkers of the steam generator. Such a complex coal-handling system avoids emissions of respirable dust to a large extent. Clean flue gas is supplied to the natural-draft wet cooling tower. Large amounts of moisturized air in the cooling tower ensure that the emissions are highly diluted before they are rejected to the environment.

Noise emissions measured above background levels at a distance of 0.5 km from the power plant site will be no greater than 60 dB(A) during the day and 45 db(A) during the night, whereby the proportionate noise rating level generated exclusively by the power plant has to be 10 dB (A) lower at these immission points.

Due to the high efficiency of the overall plant, specific CO_2 emissions are well below 800 g/kWh. Technologies for removing CO_2 from the exhaust gas of coal-fired steam power plants are still under development. Several small-scale postcombustion CO_2 capture pilot plants based on different CO_2 solvents are in operation (e.g., the Castor project in Esbjerg) or have been announced. Significant efficiency penalties, tougher requirements for flue gas desulfurization, and solvent degradation in oxygen-rich exhaust gases are some of the difficulties associated with postcombustion CO_2 capture that need to be addressed (see Table 3A–6).

* Courtesy Siemens: Adapted with permission. "Lünen–State-of-the Art Ultra Supercritical Steam Power Plant" (released at PowerGen Europe 2009).

TABLE 3A–6 Key Technical Features

Gross power output	813 MW (rated output; 50 Hz); single unit
Net efficiency (LHV basic)	~45.6% (@ design point)
Steam generator	Tower-type once-through boiler with vertical evaporator tubing
Gas cleaning	Selective catalytic reactor ($DeNO_X$), electrostatic precipitators (particulate matters), and wet limestone flue gas desulfurization (SO_X)
Steam parameters	280 bar/600°C/610°C steam parameters at boiler output
Steam turbine	SST5-6000 with single reheat and two double-flow LP turbines (4×12.5 m^2 exhaust annular area)
Generator	SGen5-3000 W, water/hydrogen-cooled
Feedwater preheating	9-stages: 3 high-pressure FWPH (header-type) with one external desuperheater, 5 low-pressure FWPH (plate-type); feedwater heaters A1 and A2 are located in the condenser neck as a duplex heater
Final feedwater temperature	308°C
Feedwater pump concept	2 × 50% electric motor-driven feedwater pumps
Condenser	Dual-pressure serial condenser operating at 30 and 45 mbar respectively
Flue gas discharge	Via the natural-draft wet cooling tower
Distributed control system	SPPA-T3000 power plant automation system

Turbine-Generator

For the 50 Hz market, Siemens offers full speed tandem compound turbo-sets for steam power plants (SPP) with ultra supercritical steam parameters in the gross power output range of 600–1200 MW per unit. The steam turbine set SST5-6000 used in Lünen is a four-casing design with separate HP, IP, and two LP turbines (Figure 3A–38).

High parameter values for the main steam (270 bar, ~600°C) and reheat steam (60 bar, ~610°C) at the turbine inlet pose special requirements on both the design and the materials. For example, the HP cylinder is designed as a barrel-type turbine and has an inner casing. Ultra supercritical steam conditions usually require the use of thick-walled components. The rate of heat transfer into these components is often the limiting factor for the duration of

FIGURE 3A–38 SST5-6000 Steam turbine.

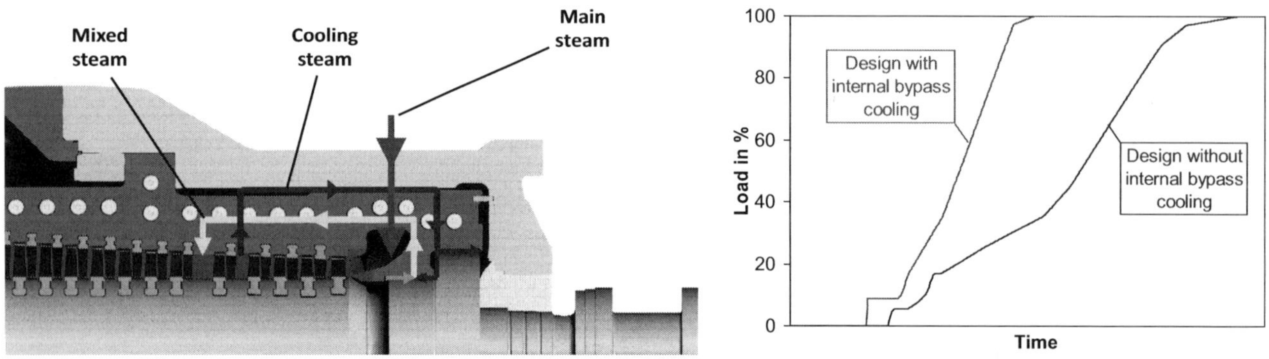

FIGURE 3A–39 HP turbine: Internal bypass cooling system and reduced startup time.

the startup process. In order to remove this restriction, a special feature has been developed for HP turbine modules: An internal bypass cooling system that allows for a more flexible operation (startup/load changes). In a nutshell, a small amount of cooling steam passes through radial bores into the small annulus between the inner and outer HP casing (Figure 3A–39).

This approach effectively protects the inner surface of the outer casing (which would be exposed to main steam temperature without the internal bypass cooling). As a consequence it was possible to reduce the wall-thickness of the outer casing and thus enable faster heat-up of the casing. An improved starting performance is the main customer benefit of this innovative concept. Components exposed to high temperatures such as the HP inlet barrel as well as HP/IP rotors and inner casings are made of 9–12% CrMoV steel. Siemens advanced 3DV™ technology (three-dimensional design with variable reaction levels) for HP/IP blades is used in Lünens' steam turbine generator set. With 3DV™ blades the stage reaction and stage loading for each row is optimized to gain highest HP and IP efficiencies. Stage reaction describes the split of pressure drop and velocity increase between stationary and moving blades, and is defined by the ratio of the enthalpy drop through the moving blade row to the enthalpy drop through the whole stage. Both low-pressure turbines are double-flow designs. Free-standing 1150 mm steel last stage blades (LSB) provide an annular area of 12.5 m^2 per flow.

US DOE Work on USC ST Materials**

As part of a larger collaborative effort, the Albany Research Center (ARC) is examining the steam-side oxidation behavior for ultrasupercritical (USC) steam turbine applications. Initial tests are being done on six alloys identified as candidates for USC steam boiler applications: ferritic alloy SAVE12, austenitic alloy Super 304H, the high Cr-high Ni alloy HR6W, and the nickel-base superalloys Inconel 617, Haynes 230, and Inconel 740. Each of these alloys has very high strength for its alloy type.

The Advanced Research (AR) program has two efforts in ultra-supercritical (USC) steam—one on USC boilers and one on USC turbines. The overall steam temperature and pressure goal is 760°C (1400°F) and 38.5 MPa (5586 psi) by 2020. There are intermediate goals for demonstrating the use of materials with steam temperatures of 650°C (1202°F) by 2010 and 760°C (1400°F) by 2015.

The Albany Research Center (ARC) is taking part in the steamside oxidation research in the USC turbine effort. A major part of the USC turbine effort is the selection or development of candidate alloys suitable for use in the USC turbine. Until candidate alloys are selected, initial tests at ARC will be done using the six candidates from the USC boiler project that have already been identified. These are ferritic alloy SAVE12, austenitic alloy Super 304H, the high Cr-high Ni alloy HR6W, and the nickel-base superalloys Inconel 617, Haynes 230, and Inconel 740. Each of these alloys has very high strength for its alloy type. Three types of oxidation experiments are planned: cyclic oxidation in air plus steam at atmospheric pressure, thermogravimetric analysis (TGA) in steam at atmospheric pressure, and exposure tests in supercritical steam up to 650°C (1202°F) and 34.5 MPa (5000 psi).

Alloys

Initial oxidation experiments are planned for the six alloys selected as candidates for the USC boiler effort. The primary aim of these initial experiments is to ensure that the three types of experiments are working properly and will be ready when the USC turbine alloys are selected. However, these results should also benefit the USC boiler research and overall USC goals. The six alloys are ferritic alloy SAVE12, austenitic alloy Super 304H, the high Cr-high Ni

** Extracts from "Ultrasupercritical Steam Corrosion." Work performed at the Albany Research Center in Oregon, USA, led by G.R. Holcomb.

TABLE 3A–7 Composition of Ferritic Stainless Steel SAVE12

	C	Si	Mn	Cr	W	Co	V	Nb	N	Ta	Nd
SAVE12	0.01	0.3	0.20	11.0	3.0	3.0	0.20	0.07	0.04	0.07	0.04

TABLE 3A–8 Composition of Austenitic Stainless Steel SUPER304H

	C	Si	Mn	Ni	Cr	Nb	Cu	N
SUPER304H	0.10	0.2	0.8	9.0	18.0	0.40	3.0	0.10

TABLE 3A–9 Composition of High Cr-high Ni Alloy HR6W

	C	Si	Mn	Ni	Cr	W	Nb	Ti	B
HR6W	0.08	0.4	1.2	43.0	23.0	6.0	0.08	0.08	0.003

alloy HR6W, and the nickel-base superalloys Inconel 617, Haynes 230, and Inconel 740.

One efficiency, as used here, is based on higher heating value (HHV), or gross calorific value. Ferritic stainless steels with 9–12% Cr are currently used with steam temperatures of about 600°C (1112°F). Most estimates of the upper temperature limit are about 650°C (1202°F), with high temperature strength being the limiting factor. Two very similar alloys, SAVE12 (Sumitomo Metal Industries) and NF12 (Nippon Steel Co.), have 10^5 h creep rupture strengths of 180 MPa at 600°C (1112°F). The composition of SAVE12 is given in Table 3A–7.

Austenitic stainless steels maintain their strength at higher temperatures than ferritic alloys, and so were used in the early USC plants in the 1950s and 1960s. However, severe thermal fatigue problems prevented their continued use at the original design temperatures and pressures. Because thermal fatigue becomes more of an issue in thicker component sections, austenitic alloys may still find use in certain thinner components. One alloy was selected as a candidate USC boiler alloy: SUPER304H. Its composition is shown in Table 3A–8. It has a 10^5 h creep rupture strength of 168 MPa at 600°C (1112°F). It is currently used in boiler sections of advanced Japanese USC power stations (including Matsuura 2, Haramachi 2, Misumi 1, Tsuruga 2, Tachibanawan 2, and Isogo 1) with steam parameters as high as 28.0 MPa/605°C/613°C (Isogo 1).

The high Cr/high Ni alloy HR6W is also a candidate USC boiler alloy. Its composition is shown in Table 3A–9.

Three nickel-base superalloys are candidates: Inconel 617, Haynes 230, and Inconel 740. Extensive prior testing for gas turbine applications should reduce their development time and cost as compared to new superalloys. Their compositions are shown in Table 3A–10.

TABLE 3A–10 Composition of Nickel-Base Superalloys

		Cr	Ni	Co	Mo	W	Al	Fe	C	B	Other
Inconel	617	22.0	55.0	12.5	9.0		1.0		0.07		
Haynes	230	22.0	Bal	5.0 max	2.0	14.0	0.35	3.0 max	0.10	0.015 max	0.02 La
Inconel	740	25.0	48.3	20.0	0.5		0.9	0.7	0.03		1.8 Ti 2.0 Nb 0.30 Mn

Chapter 4

Gas Turbine Major Components and Modules

"We are what we repeatedly do. Excellence then, is not an act, but a habit."

—Aristotle

Chapter Outline

ECONOMICS DICTATES DESIGN*

Although this is not intended to be a chapter on the business economics of gas turbines, it would be unfair to not point out to the reader how intimately financial considerations literally shape a gas turbine's components. A basic explanation of the primary hardware is followed by some detailed design cases, but these first pages provide the economic frame of mind to fully appreciate those cases.

Specification of a gas turbine and gas turbine system that then fits its working role well requires:

- Understanding of all hardware components technically and economically (including life cycle usage, spares costs, and repair costs)
- Knowledge of case histories that point out potential benefits and pitfalls for one's own application
- The economics of the above two factors and how they form the overall "cost per fired hour" figures for a given application

In the working lifetime of anyone alive today, gas turbines will be a major and growing player in 80% of the global economy (power generation, process industries, energy, and fuels). They will increase their currently small role in the remaining 20% (domestic sector) of the economy. Factors that currently limit their more widespread use are:

- The global abundance of coal
- The global glut of residual fuel that only a handful of gas turbine models can burn successfully
- Other favored forms of power generation such as hydro (in countries rich in this resource such as Sweden and

* Reference: [4-1] Personal working notes, various projects, Claire Soares, 1979 through 2006.

Canada) and nuclear power (in fuel-poor countries such as Japan)
- The slow rate of technology development with using pulverized coal or some form of coal gas "mined" from coal seams as a fuel for gas turbines

The factor that determines the best gas turbine OEM (original equipment manufacturers) "sellers" is the strength of their financial services offered. In financial services, GE is the undisputed leader and the size of its fleets, particularly in newly developing countries, prove this. Other OEMs have formed successful joint ventures with foreign national power companies, countries, and banks. The bottom line is that if the customer does not have sufficient financial resources, an OEM's

- Deep pockets,
- Willingness to countertrade, as well as
- Political muscle in the United States and international terms

weigh much heavier in the machinery selection process than they would with an end user who can easily afford to pay for his gas turbine systems. These end users' list of highest consideration factors generally includes the gas turbine system's cost per fired hour.

There are many factors that make up the value of cost per fired hour. They include:

- Fuel costs, including fuel efficiency considerations
- Design features, such as thrust bearing selection, aerodynamic losses, that may account for considerable power loss
- The signing of any "power by the hour" contracts with the OEM
- Warranty package costs (often hidden in the initial capital cost of the turbine package)
- Personnel training (also often hidden in the initial capital cost)
- Type of initial contract (BOOT and other options, see Chapter 15 in this book) with the OEM
- Fuel purchase contract and considerations such as whether the fuel supplier is also the turbine package owner and operator)
- Turbine system availability
- Turbine system reliability
- Regular O&M, including considerations for ease of operation and operator skill level required
- Repair and overhaul costs
- Costs of spares, including spares in stock availability, potential for repair of components
- Regulatory costs (including lost efficiency with anti-pollution devices)
- Costs incurred with OEM model consolidation (for example, discontinuing a model to reduce stocking spares and "duplicating" the old model's place by replacement with an existing model and a gearbox, obliging the user to absorb the 2–3% power loss)
- Taxes and other costs of doing business

Typically, although OEMs consider all of the above factors, they are biased towards the first three, the first two of those because that involves their technical expertise and product pride. The third (power by the hour) is where an OEM can turn a healthy profit. These contracts are on the increase because many gas turbine operators, particularly in the power patch and fuels business, are starting not to maintain high levels of in-house turbine expertise. If it is to an OEM's benefit to install new spares instead of concentrate on repair development, he can do so more easily with a power by the hour contract.

How all these considerations turn out depends to some extent on what the turbine, and more specifically the turbine components, offers in terms of design performance. Design and performance is what OEMs market; and the other, often more valid, business considerations are hammered out in negotiation phases. And so gas turbine designers strive for every small improvement with basic hardware. Most design specialists spend entire lifetimes with just one component, so their individual targets may include increased pressure ratio on the compressor, higher TITs (turbine inlet temperatures), more heat-resistant first stage nozzle guide vane metallurgy, higher thrust to weight, or power/weight ratios.

Eventually, every design feature is a compromise. Higher TITs will require more cooling features. The extent to which that happens (or not) will mean higher capital cost and fired costs per unit of power. More cooling air than the absolute minimum dictated by conventional design then reduces efficiency. The OEM's marketing staff, depending on which OEM's design philosophy we are talking about, may not appreciate that. So the end user may not always agree with the OEM on the priorities exercised at that decision point. Those few decimal points or more of efficiency may be a price the end users pay in blood, if a combination of operational circumstances should need more cooling air than the OEM calculated. Additional cooling air can make the difference between a pilot with a demanding and unpredictable mission profile (like a military fighter pilot) getting home safely or not.

And so the numbers of knowledgeable end users diminish as increases occur in the number of power by the hour contracts and OEMs who are also power plant owners or non-OEM asset managers of some sort. The few that are left battle for their share of input into OEMs' final designs, with their respective priorities. Those may, for instance, include:

- Reliability and availability (a few efficiency points squeezed out of any model "in trade" mean little to an

end user if he's a remote arctic operator who needs an uninterruptible power supply or a marine operator who hopes to not have to dock for several months)
- The ability to derate for lower TITs (he may need longer lives for his hot section components and a cooler operating turbine if he's in a distant and isolated desert)

Priorities vary. An end user may be thrilled to gain 0.1% efficiency on the turbine module of his Frame 7, if it means he "gains" $700,000 in fuel costs per turbine, or he may not care if he's an oil sheik with bottomless oil wells. The sheik may opt for derated, less efficient operation, if he still gets what his personal refinery needs. If he cannot get any additional profit by selling power excess to the national grid, derating may mean his turbine assets then need fewer trips to the nearest overhaul shop in Scotland.

A customer in a Southeast Asian country whose laws dictate "BAT" (best available technology) as opposed to specific NO_x emission volumes in ppm would be happy to settle for an OEM that offers a single-burner silo combustor over one with multiple burners that offers better temperature distribution and lower NO_x production. If and when environmental regulations in that country tighten (Korea, China, and Taiwan have incrementally tightened their environmental standards over the past decade), the single-burner OEM then has the opportunity to sell a low NO_x multiburner retrofit or a water or steam injection system. The end user then has to deal with the requirement for boiler feedwater quality for the injection system, but if he is a combined cycle operator, he already has that at this disposal. As did a chemical plant in Thailand (which had a steam source from another process) looking at the potential for getting another 20,000 horsepower out of its five older model GE Frame 5s in a physical location where it did not have enough real estate to place another machine. When properly designed, steam injection can result in 20–30% additional horsepower. Or, depending on the design, only NO_x reduction may result.

Sometimes, gas turbines owe the size of the global fleet of any one model to the year of their market entry. The world has a large fleet of Solar Saturn and Centaur gas turbines and gas turbine-compressor trains. When these models made their debut some decades ago, their current competitors, such as the smaller models made by what was part of Alstom (formerly European Gas Turbine, formerly Ruston), were not yet a financially backed glimmer in any designer's eye. Solar also helped its market share with the feature of several readily available (if expensive) rental units in the event of required removals.

This commanding market share fostered a tendency to pass on design development costs to the end user to an extent that many considered unfair. The earliest model of the Centaur delivered a nominal 3300 hp. When Solar designed uprate packages to deliver 3500 and 3800 hp, the customer base that depended on Solar to do its overhauls were obliged to buy the higher-hp kits. Customers complied for the most part, faced with nonwarranty of any overhaul and the fact that Solar did not wish to support the earlier components any longer than they could be changed out during overhaul. End-user pressure started to make itself felt at about the start of the 1980s and Solar evolved to a far more end-user-responsive OEM. Solar then extended its product range with larger and more efficient models in land-based power generation and mechanical drive application, to rival those of other OEMs. Meanwhile, the old Centaur and Saturn fleets continue operation to this day. Not models of optimized efficiency, they are rugged enough to have lasted all these decades, and many of their owners are faced with declining production in the fields the machinery was meant to tap.

Sometimes, a global gas turbine fleet owes its size in part to the financial genius and deep pockets of the OEM in question. One OEM, to secure a large sale to an aviation customer in China, paid the customer a 4% fuel efficiency shortfall (compared with the nearest OEM competitor) over the stated life of the engines (20,000–25,000 pound thrust size). In other words, that OEM gave away the initial capital expenditure to the end user. It was a shrewd and calculated move. The winning OEM's inlet geometry had much lower FOD resistance. A 2-pound pigeon had been observed to "corncob" one of their engines as far back as the final turbine stages. The unsuccessful rival had a wide chord fan blade and bypass annulus features that had, on one occasion, ingested a 24-pound vulture on landing approach, broken two fan blades (and essentially little else), and kept on running.

All of this serves to underline a few salient action items (if the end user does not already practice them):

1. Even if it does not directly affect daily operations, regard all detailed technical knowledge gained (such as from the cases in this chapter) as the potential basis for future negotiation in after-sales business with an OEM.
2. Join end-user groups, participate, and help sponsor to any extent possible their annual meetings. The end-user groups are varied and several. A few examples follow. (If "your" group is not mentioned below, the Internet and other end users will be able to tell you.)
   - The GTUA (Gas Turbine Users Association) is a land-based applications end-user group. Its applications are predominantly mechanical drive and pipeline related. Every second year, it has a separate meeting for OEMs to provide feedback. With GTUA, there are separate meeting rooms for each discussing each OEM's turbines.
   - Power-generation users have a vastly more specific agenda. They have separate groups per model of any specific OEM. The membership, particularly in the United States (which is where most of the meetings are held), consists of large and powerful utilities. All

end-user groups theoretically exclude OEMs from "rap" sessions, where they are tabling complaints about them. However, the twist to this, given the acquisitions climate fostered during the 1990s, is that OEMs frequently today own, or partially own, and operate power plants. They are often their own end user.

- Similarly, large oil companies are becoming IPPs (independent power producers) at an increasing rate, so they can buy their own fuel in a fuel purchase agreement (FPA) that they orchestrate. About 20 years ago, the main reason for a large oil company producing its own power was because of potential required power shortfall in remote locations. Oil companies and refineries today can also get tax rebates (in countries such as Thailand that encourages any IPPs, small and large) for supplying their excess power to the grid. So the line between the two preceding groups of end users is blurring.
- The militaries of the Western world run their own formalized CIPs (component improvement) meetings for aircraft engines. Adjacent nations may attend the same meetings depending on the relative size of the military branch. For instance, the relatively small Canadian Forces helicopter fleets that serve much the same purpose as the U.S. Coast Guard do in the United States, join them for their meetings.

3. Scrutinize (if you are an end user) OEM service bulletins (SBs) whether mandatory, optional, or "campaign," for applicability to your operation.
4. Consider item 3 in the light of your extended business plans. When Shell decided on the best harvesting options for its Brent field, it decided to let the deposits of light ends depressurize, so that they could be extracted as gas. At that decision point, it opted to retrofit more sophisticated instrumentation packages (non-OEM supplied, but OEM qualified source) on existing turbine packages.
5. Shell also gained considerable sophistication with life cycle assessment (LCA) of major gas turbine components in the process of doing 4, and the ability to extend certain component lives by as much as 100% or more. In other words, it pays to know your package accessories and instrumentation as well as your gas turbine components and view them all as an integral system.
6. Learn the design priorities of individual OEMs, and therefore their inherent weaknesses and strengths: the "raise efficiency by raising TITs and be as stingy with the cooling air as possible" group, the "same tooling function for a smaller premium" (and when this works, one saves a great deal of money) types, the "excess cooling air and rugged FOD resistant design" school, the guys whose oil system always works. This gets particularly interesting with joint venture products, where different

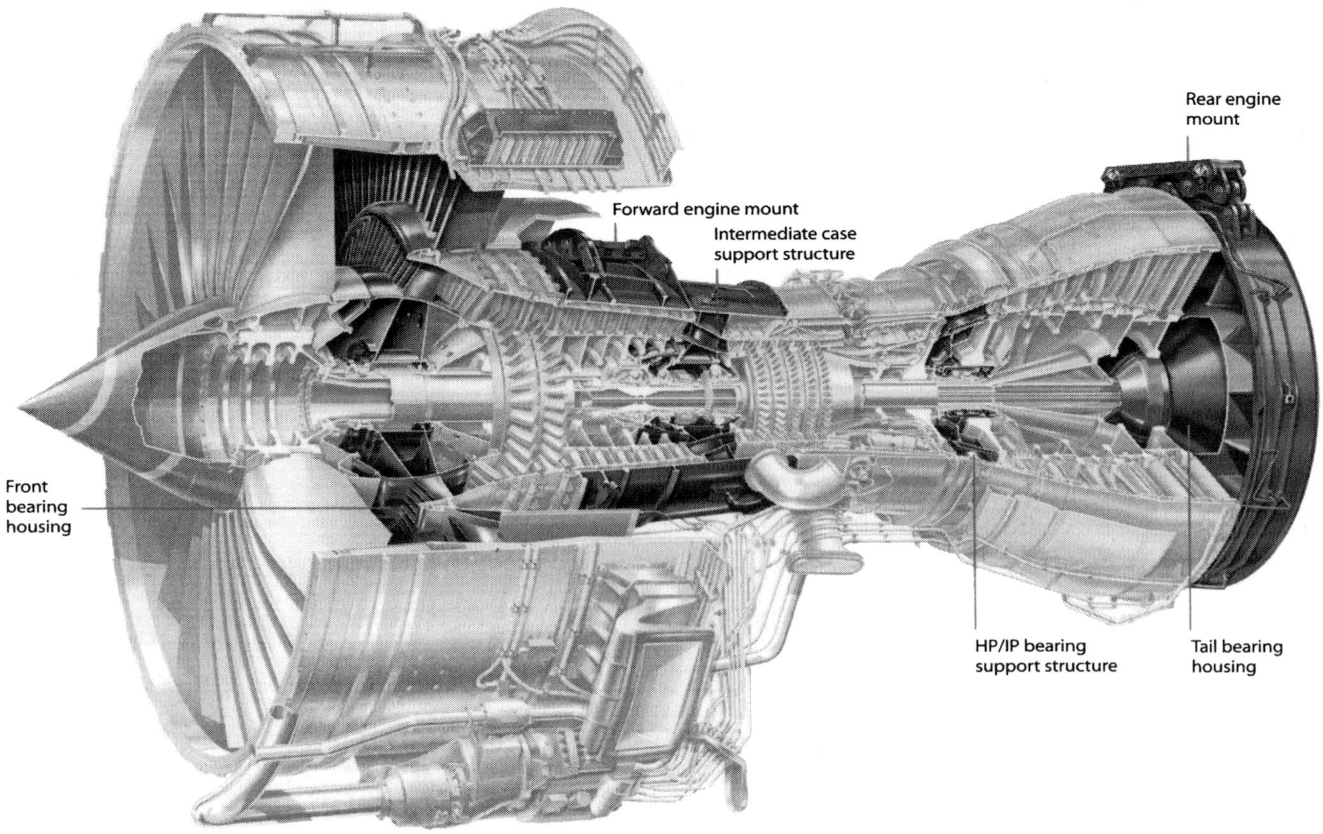

FIGURE 4–1 The main support structures and engine mounts on a three-shaft engine. *(Source: Rolls Royce (2005).)*

OEMs design different modules, interfaces, and common-to-all-module systems. Sometimes this division of responsibility works in the endusers' favor, in that it contributes to reduced costs per fired hour.

7. Expect that reliable gas turbine workhorses will still be around several decades hence, albeit heavily tweaked and individually modified. One good example of this is the old GE Frame 5. GE recently released a "new version" of this old and beloved model. However, owners of the old version will not give up their machines. The "tweaked and individually modified" occurs when existing machines do not, for any reason, meet fleet standards. Take the French power facility that had its Frame 5 fleet developing premature cracks in their turbine discs. An external consultant designed the steam injection that took the discs out of danger of cracking (and gave the old machines a 20% additional power bonus). Also the metallurgy of these old machines makes them relatively easy to repair (by welding or heat treating).
8. Be aggressive in learning from experience in different applications of "your" engine. For instance: The GE LM2500 fleet is a vast one, in power generation, mechanical drive, and marine service. The land and marine end-user base benefited from the fact that the LM2500 is essentially a GE CF6 80C2 "sitting" on land or the sea. The Rolls Royce RB211, Trent, and Avon fleets benefit similarly. The additional number of operating hours in a different environment with demands of its own may increase the overall expertise the end users have to draw upon. For instance: The question posed by land users of LM2500's additional required systems for high vanadium content in fuel was preceded by NATO studies done on the same question on behalf of the naval military LM2500 fleet. Another example: The recent "revolutionary" development of injecting fuel into already fuel-rich exhaust gases of a power generation system and reigniting the mix for power augmentation is essentially a "grounded" afterburner.

The source of the basic detailed component information that follows is Rolls Royce (RR). This section is based on one of their aviation engine models, which is appropriate, given the sophisticated nature of aeroderivatives in land and marine use as well as new industrials. The newer models are "officially" called by the same designation with RR only because they are just as sophisticated as a flight engine in terms of metallurgy or instrumentation systems.

GAS TURBINE ENGINE MODULES

Rotor Support Structures

Fundamentally, the engine outer structure is a pressure vessel that contains hot, flowing air. The rotor support structures extend inside the pressure vessel to support the rotating components of the engine while allowing air to pass through from front to rear. They are generally circular, with a number of struts or vanes joining the inner and outer rings and a bearing housing located in the middle. Inside the bearing housings, the bearings allow free rotation, yet precise centering, of the rotors. On the outside, support structures may provide mounting lugs as attachment points for external engine components or the engine-to-aircraft mounting. Some lugs also transmit engine thrust loads to restrain forward and reverse motion. The support structures are joined together by compressor or turbine casings to form a complete support frame for the engine.

Each engine rotor requires two or even three rotor support structures; however, a single support structure like the intercase can be used for up to three rotors. Because of this, the RB211 and Trent engines need only four structures to support the three rotor systems.

The engine rotors transmit loads generated by the rotors to the stationary engine structure through the rotor support structures. The outer engine structure collects the loads and transfers them to the aircraft at the engine mounts. When an aircraft performs maneuvers, the structures maintain the centering of the engine rotors. In the event of a component failure, the structures ensure that the engine will not create a hazard to the aircraft, although the engine may stop operating.

The rotor loads enter the support structures at the bearings, which are inside an annulus of flowing air. Struts or vanes transmit the loads through the gaspath to the outer stationary structure of the engine. Both struts and vanes minimize disruption and pressure loss in the gaspath, but vanes are more sophisticated, and are used to significantly redirect the air/gas. The struts or vanes also provide a path for lubricating oil to be provided to, and returned from, the bearing chambers.

Each structure must withstand a wide range of extreme conditions to ensure the engine's safe and reliable operation. In an aircraft engine, weight must be stringently controlled. Jet engines for aircraft propulsion use the lightest possible materials. In cooler locations, light alloys such as aluminum or magnesium perform well. For moderate temperatures, titanium, though expensive, provides the necessary qualities of high strength, low weight, and temperature. In the very hottest locations, heavy materials such as nickel alloys provide sufficient temperature resistance. To ensure the most structurally efficient configurations, engineers use extensive Finite Element Analysis to evaluate the ability of the structures to withstand engine and aircraft loads. (See figures 4–2 and 4–3).

The innermost part of the support structure is the bearing chamber, which provides a favorable environment for the bearings. Inside, oil nozzles distribute lubrication to the bearings and gears. Around the shafts, labyrinth seals prevent the oil from leaking out and limit the amount of hot air entering the chamber. Buffer air at higher pressure

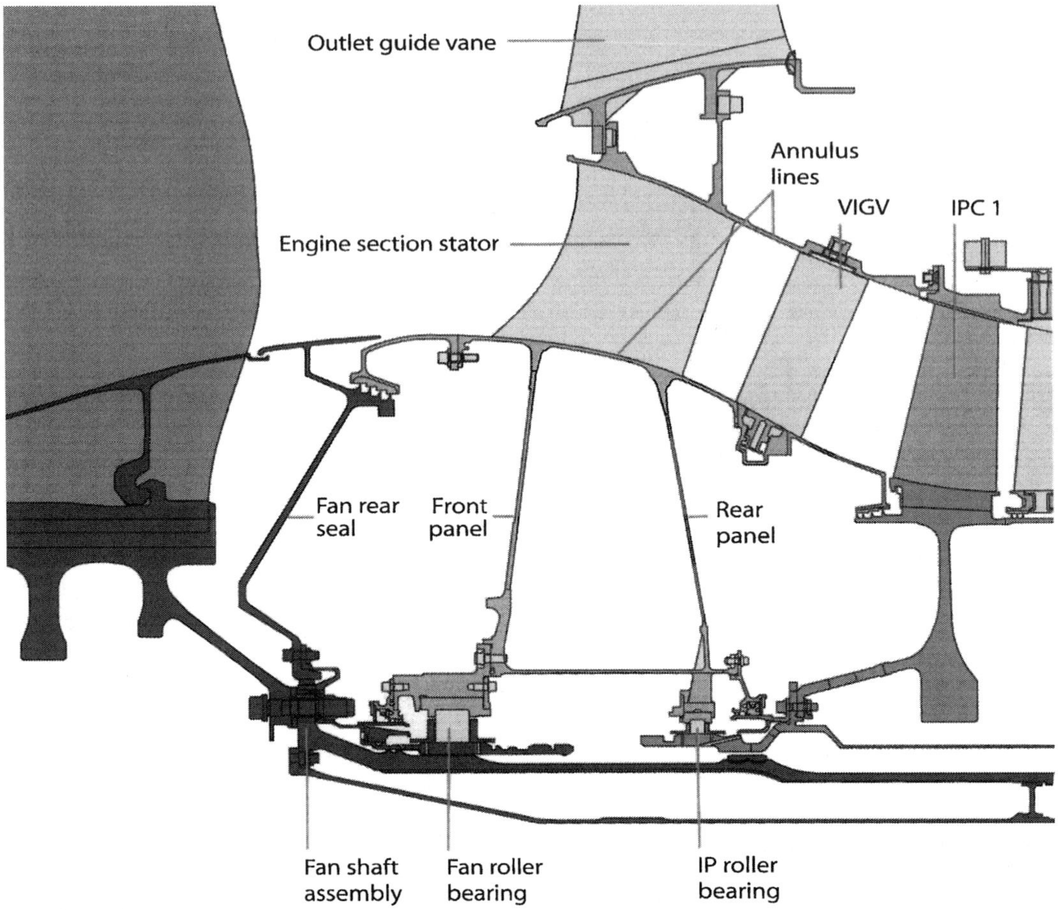

FIGURE 4–2 Front bearing housing on a three-shaft engine locating the LP and IP front roller bearings. The configuration and functions of the FBH are highly dependent on the engine architecture. *(Source: Rolls Royce.)*

FIGURE 4–3 The intercase: on a three-shaft engine, it locates the thrust bearings for all three rotors. *(Source: Rolls Royce.)*

surrounds the bearing chambers outside the labyrinth seals so that air flows inward through the seals. This inward flow of buffer air prevents oil from migrating out of the labyrinth seal.

Several engine families use a four-rotor support structure, which includes:

- the front bearing housing
- the intermediate case
- the HP/IP structure
- the tail bearing housing

Front Bearing Housing

The front bearing housing (FBH) provides support near the front of the fan rotor (also known as the low-pressure or LP rotor) and at the front of the intermediate-pressure (IP) rotor. The bearing chamber on the inside contains the forward LP and IP roller bearings. The engine section stator (ESS) vanes direct a portion of the fan airflow into the core of the engine and carry structural loads to the splitter area. Two conical panels attach the ESS vanes to the bearing chamber. The ESS structure is manufactured as either a machined cast ring of vanes, or built up from individual forgings that are welded together to form a ring. In addition to structure, the ESS ring provides:

- aerodynamic functionality to feed the IP compressor, delivered by the ESS aerofoil shape
- routing for services, which can include oil feed, oil scavenge and oil and air vent, and speed probe wires.

Intermediate Case

Also called the intercase, this is a structure between the HP and IP compressor cases, which houses the main shaft thrust bearings and carries the rotor gas loads through struts to the engine casing and thrust mounts. It also houses the internal gearbox, which incorporates a bevel-gear drive shaft linking the HP rotor to the external gearbox. The intercase provides support for all three rotor systems. Thrust bearings contained in the intercase bearing chamber, away from the hot end of the engine, provide mid-rotor support for the LP and IP rotors, and forward support for the HP rotor. These bearings transmit all of the axial forces of the rotors to the engine structure. On Trent and RB211 engines, lugs on the intercase transmit the engine thrust to the nacelle structure. Therefore, unlike the other rotor support structures, the intercase must be strong in the axial direction as well as the radial direction. The intercase, therefore, literally pulls the aircraft through the air.

Because of the location of the intercase at the forward end of the HP rotor, it is called upon for an additional and unique function. It provides an internal gearbox in the bearing chamber to transmit power to and from the HP rotor. This is necessary for engine starting and to drive mechanical units such as oil pumps and generators that are mounted on the engine. The internal "gearbox" includes a pair of bevel gears mounted within the intercase. One gear is mounted on the HP rotor; its mating gear is connected to a small shaft that runs through a strut. This small shaft is the radial drive, which is part of the system that transmits mechanical power to and from the external gearbox. (See figure 4–4).

As in all rotor support structures, the intercase contains a passage for the engine's core airflow. This passage is known as a "swan neck" duct, because it sweeps from the larger radius of the IP compressor exit to the smaller radius of the HP compressor inlet, giving the appearance of a swan neck on drawings. Struts to carry structural loads across the flowpath span the swan neck. The hollow struts in the swan neck duct allow oil services to, and venting from, the bearing chamber, as well as a place for the radial drive. The structure is usually cast titanium with some amount of welding necessary due to the complexity of the structure.

HP/IP Structure

The HP/IP turbine bearing support structure is located between the HP and IP turbine discs to provide support to the aft end of the HP and IP rotors. The bearing chamber houses the HP and aft IP roller bearings. The structure transmits radial bearing loads through the hub and into the outer casing. This structure operates in a very challenging environment. It is surrounded by very hot engine parts and must carry load through the HP turbine exit airflow, which

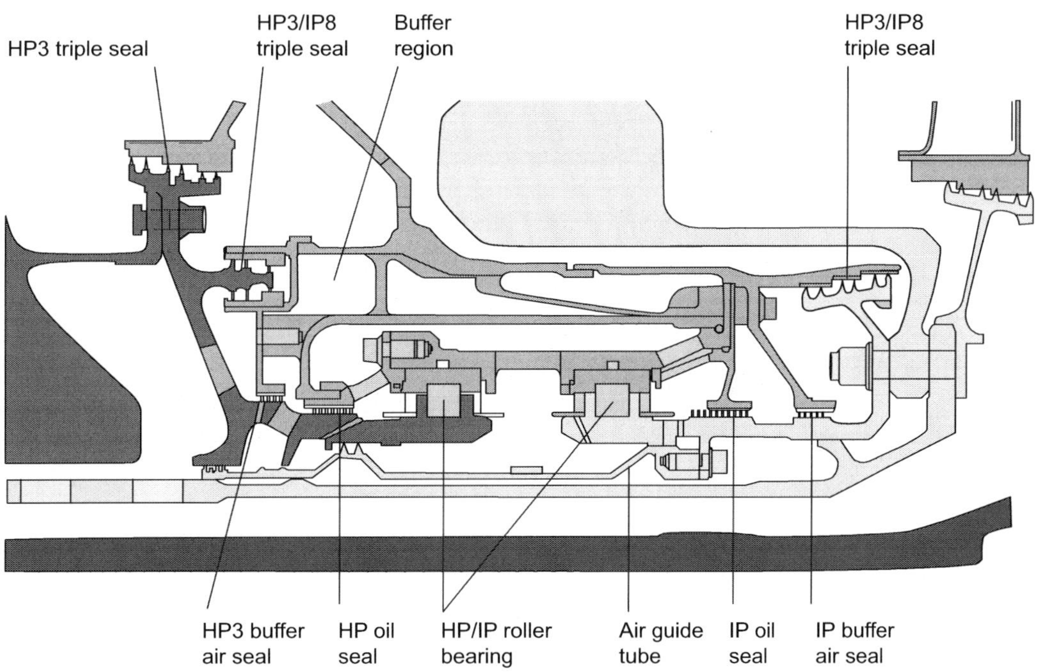

FIGURE 4–4 HP/IP hub structure on a three-shaft engine. This structure houses the HP and rear IP roller bearings. *(Source: Rolls Royce.)*

is one of the hottest parts of the engine. Struts connect the inner structure to the turbine case while allowing air to pass. In this environment, the only fluid available for cooling is the oil supply, which must pass through the hot flowpath along with the struts. Due to their interaction with the outer casings, the bearing support structure has a major influence on the control of blade tip clearances and shaft dynamics. The bearing support structure, therefore, must have sufficient stiffness to withstand extreme maneuvers, while maintaining an adequate fatigue life.

The blade tip clearances are further influenced by the use of an oil squeeze film damper. The damper consists of a narrow oil-filled gap between the bearing outer race and the bearing chamber structure. The rotor system, though precisely balanced, will still have unbalance present. The damper provides fluid support for the bearing race in a way that allows the rotor system to rotate about its true mass center. In addition, the fluid film reduces the unbalance forces transmitted to the structure. The bearing chamber components operate near the maximum permissible limits for bearings and engine oil. Typically, nickel alloys are the materials used to make the housings, shafts, support structures, and air seals for high bypass three-shaft engines.

Tail Bearing Housing

The tail bearing housing is the bearing chamber that supports the aft end of the LP rotor and contains the rear engine mounts. Exit guide vanes provide structural support of the bearing chamber and provide the pathways for oil, air, and instrumentation cables. The exit vane shape is simpler than the FBH ESS vanes because they need to provide less turning of the airflow.

The rear-bearing chamber provides a protected environment, housing the LP shaft rear roller bearing and LP turbine over-speed probe. Roller bearings transfer radial loads into the structure. Oil transfer routes through the exit vanes provide oil lubrication, scavenging, and an oil film damper.

Although the tail bearing housing must operate in the environment of the LP turbine exhaust, it is not as severe as that endured by the HP/IP support. Due to the prevailing high temperatures from the turbine, the structure material is a nickel alloy. In the quest for lower production costs, manufacturing methods have varied between a fully cast structure to a fabricated structure, but both methods have proven comparable. The bearing chamber housing is traditionally manufactured from cast steel alloy, but has also been made from cast nickel alloy.

Fan Module Basics

The primary modules in a gas turbine are the:

- Compressor module
- Combustion module
- Turbine module

A gas turbine also has an inlet section/module and an exhaust section/module (see Figure 4–5).

Most advanced and large gas turbines have compressors that are of the axial design type. Some of the earlier, smaller, or deliberately compact gas turbines have centrifugal compressors (see Figures 4–6, 4–7, and 4–8).

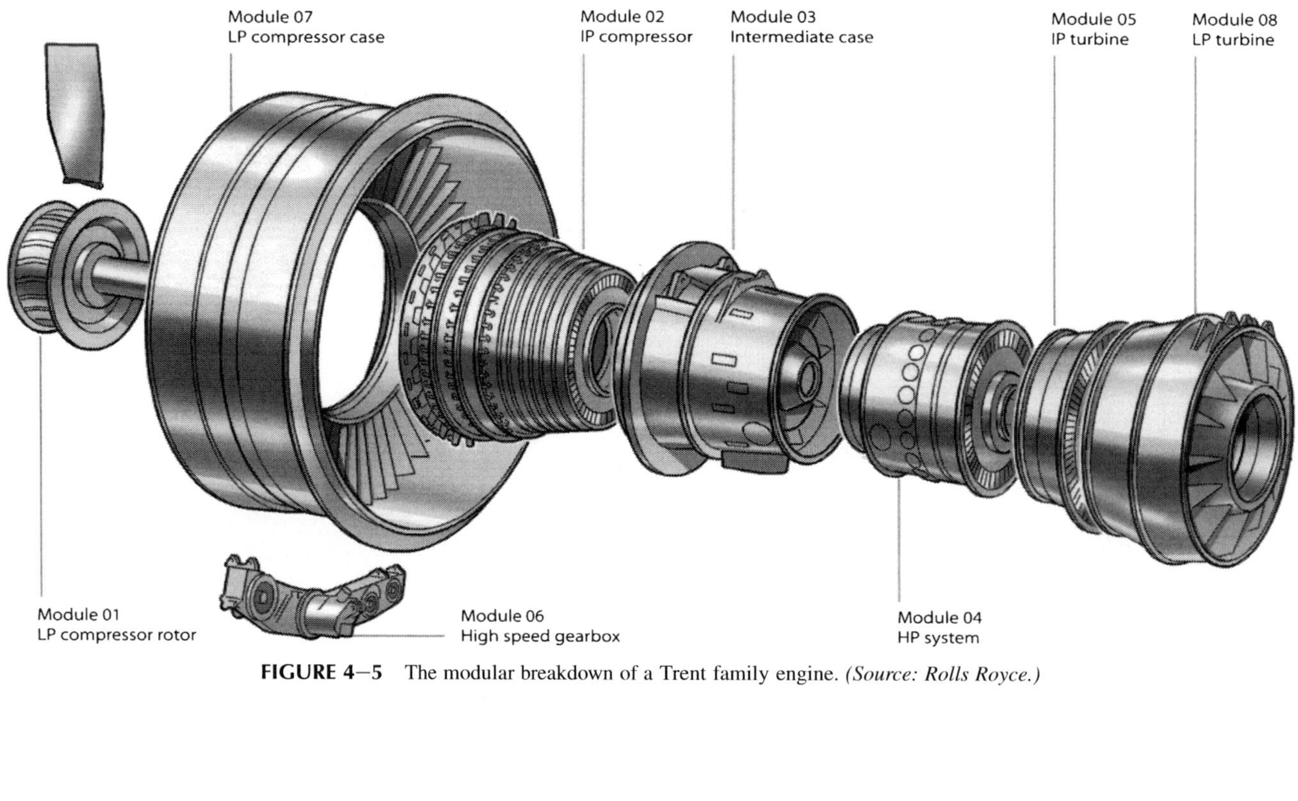

FIGURE 4–5 The modular breakdown of a Trent family engine. *(Source: Rolls Royce.)*

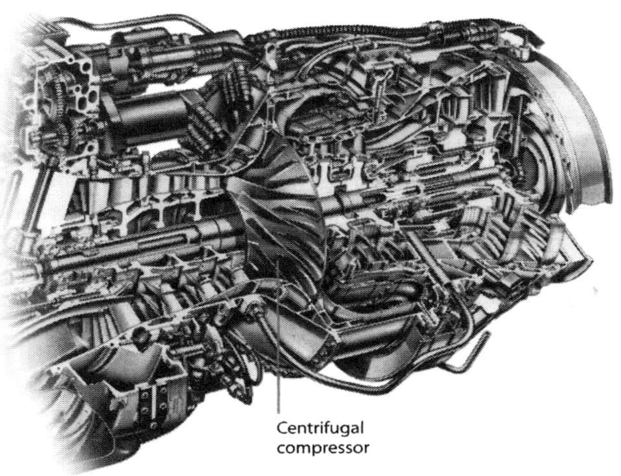

FIGURE 4–6 The Turbomeca centrifugal compressor in the RTM 322. *(Source: Rolls Royce.)*

Each compressor stage provides an opportunity for stepping up the overall compressor pressure ratio (PR), so although an axial stage may not offer as much of a PR as a centrifugal stage of the same diameter, a multi-stage axial compressor offers far higher PR (and therefore mass flow rates and resultant power) than a centrifugal design (see Figures 4–7, 4–8, and 4–9).

Therefore, most gas turbine designs incorporate axial compressors. In newer designs, the compressor and turbine modules may be split into further submodules to lessen the stress on individual components and achieve better efficiencies.

So a compressor may have a low-pressure (LP) module, or LPC, and a high-pressure module (HPC). In this case, there will be equivalent high- and low-pressure turbine modules (LPT and HPT). The LPC and LPT will operate on one long shaft at the same speed. The HPC and HPT will operate on a shorter shaft that fits around, and is concentric to, the low-pressure shaft and is at a higher speed than the low-pressure module (see Figure 4–5).

Some contemporary gas turbines have three modules in their compressor and turbine sections, designated low, intermediate, and high pressure, each with their own shaft. This modular concept allows for module replacement or exchange, if maintenance to a module is required, without taking the entire gas turbine out of service.

The Nose Cone

The nose cone provides the inner annulus profile in front of the fan for smooth airflow into the fan blades' roots and must withstand bird impact, erosion, and the build-up of ice. To meet this functionality, the cone is made from glass fiber, laid and curved so that the cone achieves maximum strength. The thickness of the cone is based on the experience of bird impact tests from previous engines; the angle of the cone is optimized for both bird impact and ice shedding behavior, while delivering satisfactory airflow into the fan. The nose cone also has a rubber tip at the front to dislodge any ice accretion. There is also a seal bonded to fit beneath the fairing and a polyurethane coating for protection against erosion. The nose cone is generally secured to the fan module by a bolted flange and internal spigot arrangement.

The cone thickness on more recent engines has been modified to meet the heavier bird certification requirements and a double-ply lay-up technique is now widely used. Manufacture is based on the autoclave molding technique. The layers of fibers are pre-impregnated with resin, placed in the cone tooling, and covered by a pressure bag.

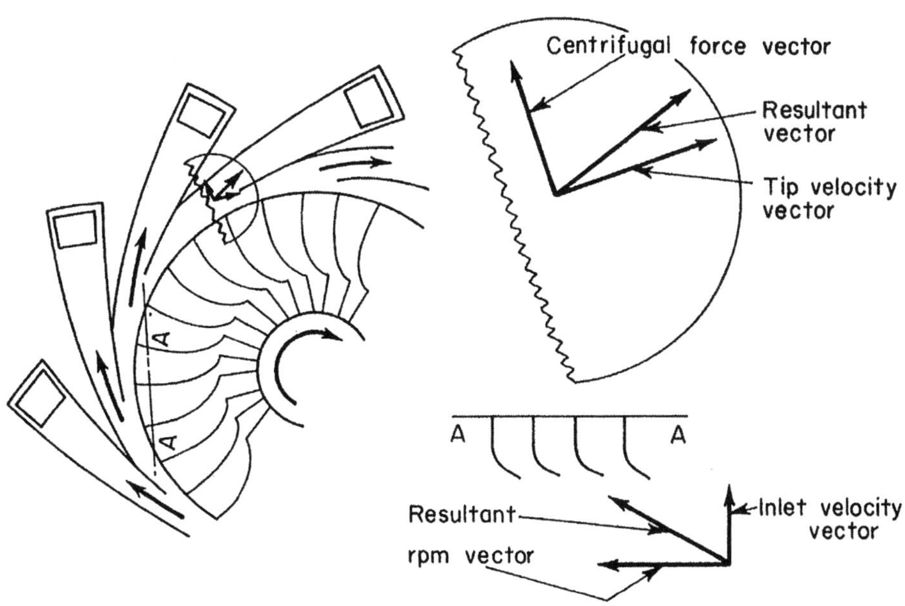

FIGURE 4–7 Centrifugal-flow compressor (plan view). *(Source: Rolls Royce.)*

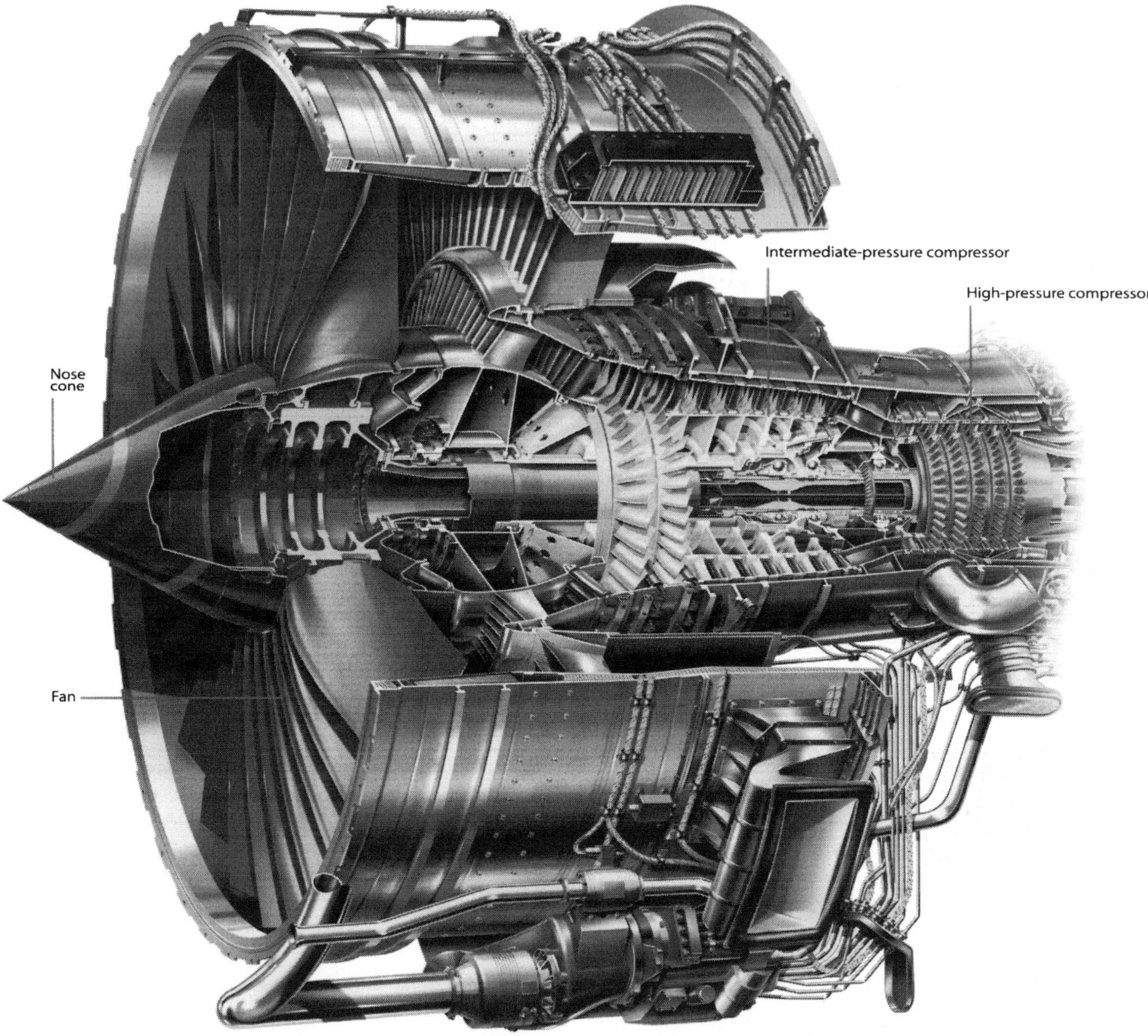

FIGURE 4–8 The fan and compressors on the Trent 500. *(Source: Rolls Royce.)*

The assembly is then placed in the autoclave with sufficient heat and gas pressure to cure the resin and consolidate the layers. Once set, the cone is machined around the flange (removing material from sacrificial layers) and the holes are drilled. The cone is then painted, coated with polyurethane, and the rubber parts bonded to it. (See figure 4–10).

Fans

The fan system has two primary functions:

- compress the bypass air
- feed supercharged air into the core

In a turbofan, a proportion of the air from the low-pressure compressor passes into the core compression system—the remainder of the air, the bypass flow, is ducted around the core compression system. Both flows eventually pass through separate or integrated propelling nozzles at the rear of the engine to generate thrust. The civil, high bypass ratio fan has a pressure ratio approaching 2:1. This bypass air expands through the exhaust nozzle and contributes around 75% of the engine thrust. A low bypass ratio military fan has a pressure ratio typically in the range 3:1–4:1. This air passes down the bypass duct, and is then mixed with the core airflow from the turbines, and expanded through the exhaust nozzle. The bypass air is also used for afterburning and to cool the reheat and nozzle system. (See figure 4–11).

The fan system of the Pegasus V/STOL engine in the Harrier is an exception; here, the bypass air is passed

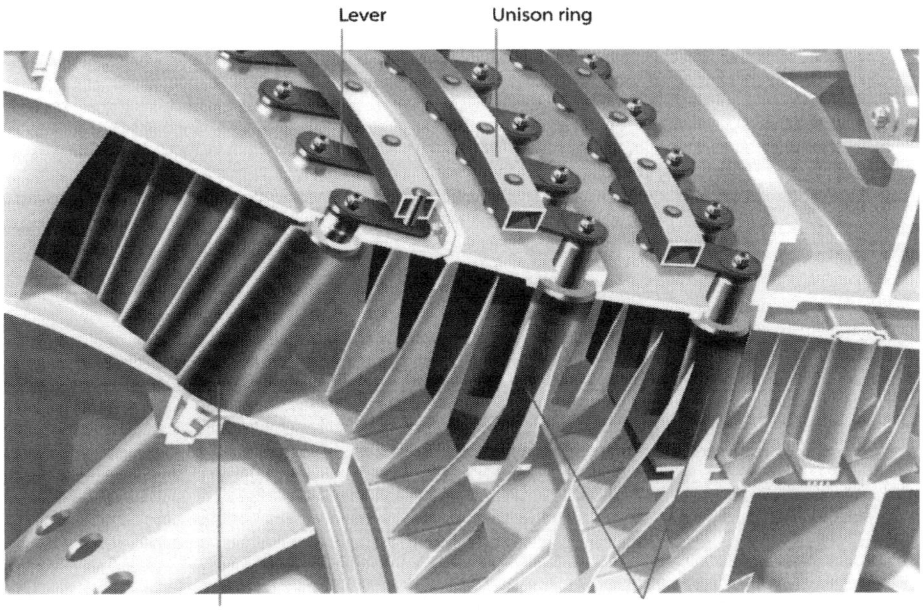

FIGURE 4–9 The VIGVs and two VSV stages of the IP compressor of the Trent 500. *(Source: Rolls Royce.)*

FIGURE 4–10 The nose cone with rubber tip to prevent and remove ice build-up. *(Source: Rolls Royce.)*

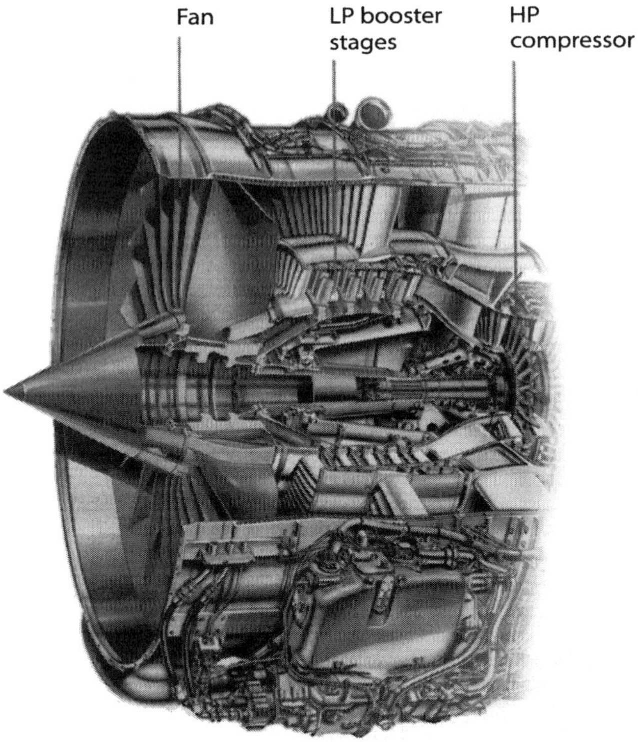

FIGURE 4—11 Booster stages in the core section on the two-shaft V2500. *(Source: Rolls Royce.)*

FIGURE 4—12 Wide-chord, swept fan blades on a high bypass turbofan. *(Source: Rolls Royce.)*

directly to the front nozzles of the lift system to generate thrust. This functionality needs to be achieved at a high level of aerodynamic efficiency, at a low life-cycle cost, weight, and diameter, and at a low level of noise (civil rather than military). The system must also have an adequate stability margin and be able to cope with harsh operating environments.

The system has to pass rigorous certification tests: rain, hail, icing, operability, bird strike requirements, fan-blade-off, any distortion of inlet airflow resulting from aircraft maneuvers or cross-wind, altitude, and compatibility with intake and thrust reverser. Achievement of noise targets is also of crucial importance.

The fan system must be designed to cope with impact from a range of bird sizes at various positions on the fan face; the size of the bird is a function of intake diameter, so the larger the diameter of the fan intake, the larger the weight of bird that must be accepted. The system has to be able to demonstrate integrity for all types of bird specified in the certification requirements.

Distortion of inlet airflow is a significant issue for military fans, given the comparatively extreme maneuvers of military aircraft and the often more complex air intake system. Because of the need for a higher pressure ratio, military fans tend to be multi-stage, and may be configured with VIGVs. The splitter on military fans is usually downstream of the fan bypass vanes, or outlet guide vanes (OGVs). (See figures 4—12 and 4—13).

FIGURE 4—13 The three-stage military fan on the RB199. *(Source: Rolls Royce.)*

Civil Aero Engine Fans

The modern civil aero engine has a very high bypass ratio turbofan configuration. In this configuration, the intake air undergoes only one stage of compression in the fan before being split between the core (or gas generator system) and the bypass stream. For modern engines, the bypass ratio can be as high as ten. This results in the optimum configuration for passenger and transport aircraft flying at just below the speed of sound. For large engines, a three-shaft configuration is preferred, with an intermediate-pressure (IP) compressor and high-pressure (HP) compressor in the core section.

The major components of the civil fan system are the fan blades, fan disc, containment casing, and the FBH structure containing the bypass vanes and engine section stators.

To reduce the fan diameter, and therefore weight and drag, the inlet hub-tip ratio is minimized, subject to meeting mechanical criteria for the hub design. The fan blade comprises an aerofoil with a root attachment that secures the blade into the fan disc. The rotor is attached to the fan shaft, which is connected to and driven by the LP turbine. The whole fan rotor assembly is supported by the FBH. The flow leaving the OGVs is axial. The flow leaving the engine section stators may be axial or swirling, depending on the engine configuration.

Fan Blade

The hollow, wide-chord fan blade allows higher flow, higher efficiency, and is quieter than its predecessor, the snubbered blade. A snubbered blade consists of a solid aerofoil, which has two appendages, or snubbers, attached at right angles to the aerofoil span at about three quarters of the blade height. These are also known as clappers. When the blade is assembled, these snubbers form a support, which resists twisting of the aerofoil when subjected to cyclic loading caused by aerodynamic distortion and wakes. They also raise the natural frequencies of the blade and provide a course of damping. Simply making the blade snubberless results in a design that is too flexible (its natural frequencies are too low) and removes the mechanism for damping any aerofoil vibration. To overcome this, the blade chord is increased, stiffening the blades and allowing a reduction in the number of aerofoils.

One of the principal reasons why civil designs adopted the wide-chord fan is efficiency. Snubbers introduce a significant amount of aerodynamic loss, resulting in a very inefficient design; they also present a blockage to the airflow, requiring the frontal area to be increased. To avoid excessive fan module weight, the aerofoil is hollow; this not only lightens the individual fan blade but also lightens the whole system (disc, front structure, containment casing).

A hollow blade has a cavity within the aerofoil and is formed from three sheets of titanium: two outer sheets and one inner sheet—a very thin membrane. These blades are produced using diffusion bonding and super-plastic forming processes. Hollow blades behave in a very similar manner to solid blades and there is no detriment in stiffness of bird strike capability. The larger the blade, the greater the benefit from hollow blade technology as more weight can be saved. As blades reduce in size they can no longer be hollow because the panels would become too thin. (See figure 4–14 and 4–15).

Fan Disc

The fan disc is one of the most critical components in the engine and has four main functions:

- react to centrifugal loads from the fan blades—both during normal running and in the event of a fan-blade-off
- provide attachment from the LP shaft to drive the fan and to retain the fan blades

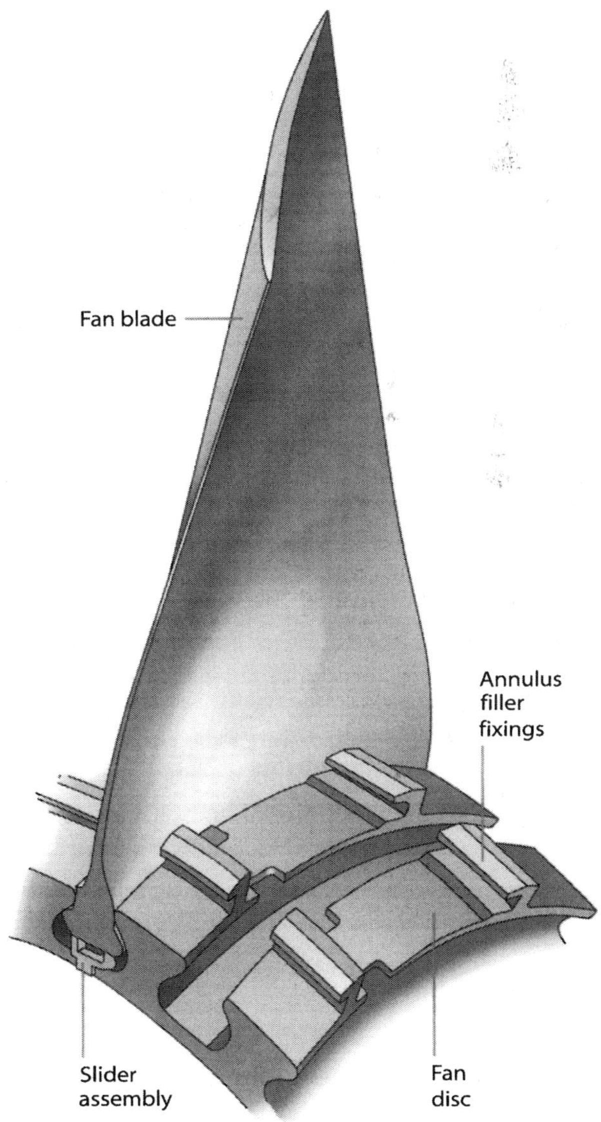

FIGURE 4–14 Fan fixing arrangement of a typical wide-chord civil fan. *(Source: Rolls Royce.)*

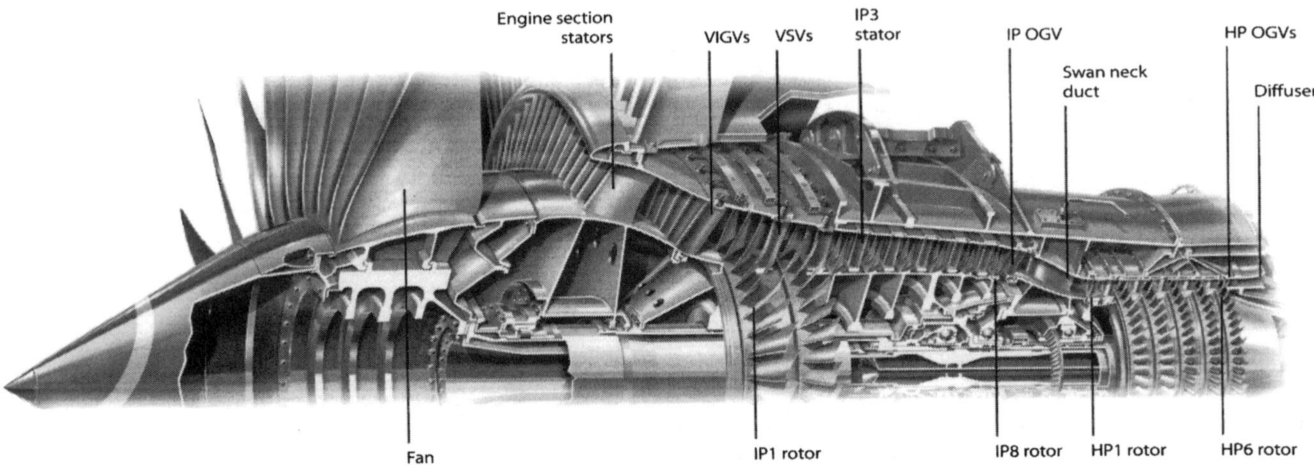

FIGURE 4–15 The fan and compressors form the compression system of the Trent 500, a high-bypass civil engine. *(Source: Rolls Royce.)*

- absorb impact loads
- provide attachment for the nose cone and other peripheral components

As disc failure is hazardous to the aircraft, this component is classified as a critical part. The disc contains a number of slots into which the fan blades are fitted and there is a front drive arm, which provides attachment to the nose cone assembly. The disc material is usually forged titanium. The mechanical design of the disc is one of the key design areas, because it is a critical part, and it is an extremely heavy component of the fan system. The role of the disc is to ensure that the blades continue to travel in a circular path and resist their high centrifugal loads—about one hundred tons, equivalent to ten double-decker buses hanging from each fan blade.

The total disc stress is a combination of inertia stresses of the disc itself and the stresses imposed by centrifugal force on the blades. Two key issues govern the amount of stress the disc is designed to withstand. First, the burst criteria state that, if the assembly overspeeds, the disc will not burst and compromise the integrity of the engine; this provides the minimum cross-sectional area for the disc. The second major issue is the life of the disc; this sets the maximum stress in the disc. If it is unable to meet the life criteria, then various strategies can be employed:

- increase the size of the disc, so that the stress in the disc reduces to acceptable levels. Any extra material added is put at the bore, the most weight efficient location
- decrease or eliminate any stress concentrations in the disc, such as small holes or tight radii
- increase the capability of the material
- if the material properties exceed the life requirement, then the disc size can be reduced until it reaches the minimum size and weight, as specified by the burst overspeed margin. (See figures 4–16 to 4–18).

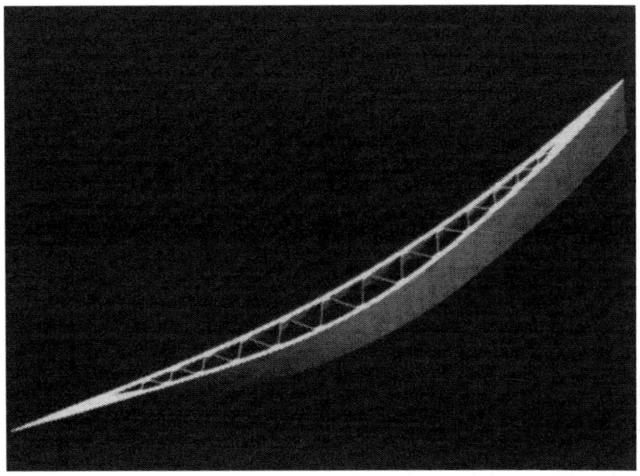

FIGURE 4–16 Section through a hollow fan blade model. *(Source: Rolls Royce.)*

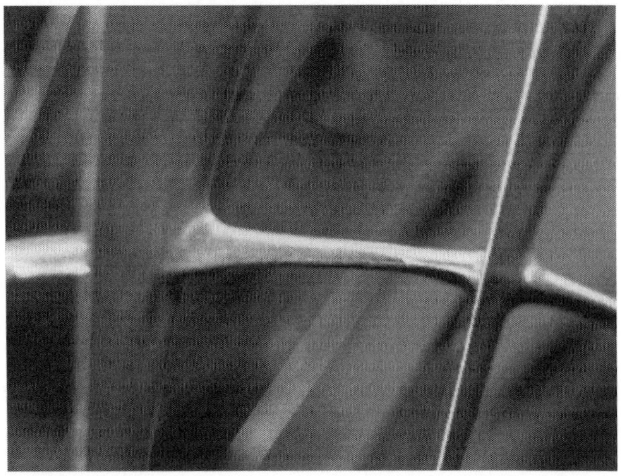

FIGURE 4–17 First stage snubbered compressor blades. *(Source: Rolls Royce.)*

FIGURE 4–18 A Trent fan disc: the force locating a Trent fan blade is powerful enough to throw a car 30 meters into the air. *(Source: Rolls Royce.)*

Fan Casing

The primary functions of the fan case are to form the outer gas path, and contain a fan blade should it disintegrate during flight. The casing must be capable of absorbing the energy of a complete fan blade, without releasing blade or case fragments, and maintain the integrity of the engine. The energy of a released fan blade is equivalent to a family saloon car at 100 kmh (60 mph). The casing therefore needs to have high strength and high ductility.

In some engines, the fan case is part of the engine mounting system and thus transmits thrust from the core engine to the aircraft. It interfaces at its front flange with the nacelle and at its rear flange with the rear case (typically military) or with an OGV ring (typically high bypass civil). The fan case also provides mounts for the gearbox, ground support equipment, and other accessories mounted on the accessories flange. The casing assembly also contains acoustic liners to attenuate noise generated by the fan. The panels are made from a honeycomb structure of composite construction. The fan case inner profile, when fully assembled with the in-fill panels, fan track liner, acoustic panels, and ice impact panel, forms the outer annulus line. Containment system weight is a function of the fan diameter cubed so high bypass ratio engines with a large fan blade tip-to-tip diameter have much heavier containment systems.

Military Aero Engine Fans

The major components of the military fan module are the rotor drum, casing and other statics, fan blades, and support structures. Military fans can be configured with either a rear support structure (an overhung rotor), or a front and rear support structure (a "straddle-mounted" rotor). Where a VIGV stage is present, this is incorporated into the front bearing structure.

To minimize the frontal area, and hence weight, drag, and ultimately airframe visibility, the inlet hub-tip ratio is kept as low as possible. The need for low frontal area and low weight means that, in modern military engines, the rotor assembly often has blisks, where the blades and disc are integrated into a single component. The stators are usually of shrouded construction and are mounted into the casing. The flow leaving the OGV is axial. The casing may also have a casing treatment to improve the fan's stall and surge characteristics.

Fan Rotor

The fan rotor configuration has traditionally consisted of two or three discs, each with a set of rotor blades of aerofoil cross-section. The discs and, in more modern engines, blisks, can be bolted or welded together. The blisk is a challenging component to manufacture. There are two very different methods:

- machining from a single, solid piece of metal
- linear friction welding—this allows the engineer both to optimize the properties of the aerofoil and disc, and also to use hollow aerofoil technology so that the blades and disc can be even lighter.

In the fan system, blades, discs, and blisks are usually made of titanium.

Fan Casing and Statics

The military fan casing has various functions:

- form the outer gas path, and provide close control of tip clearance for the rotors
- support the statics (stator vanes), and also the front bearing structure
- provide a mount for the VIGV actuation system where present
- provide a containment system for the rotor blades
- mount engine accessories

The casing needs to have high strength and ductility to achieve these requirements. The inside surface of the casing incorporates an abradable lining material similar to those used in civil casings. The abradable material is axially aligned with the rotor tips and helps to maintain tight tip clearances, which are critical for performance and stability.

The casing is normally split horizontally in two halves, with the vanes secured into the casing via a dovetail fixing. The vanes are shrouded and are fitted with an inner shroud ring to provide integrity. The inside

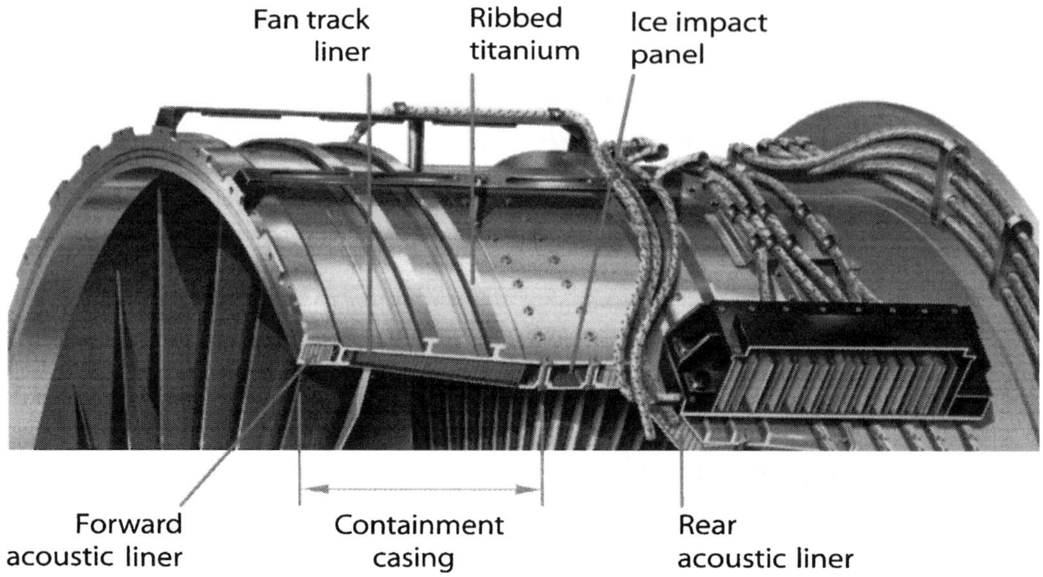

FIGURE 4–19 Section through a Trent 500 fan casing. This casing mounts the engine accessories, minimizes fan noise impact, and must contain a fan blade during a blade-off event. *(Source: Rolls Royce.)*

diameter of the shroud ring incorporates abradable material, which provides a sealing face against the rotor labyrinth seal thereby preventing the leakage of air from stator exit to stator inlet.

The casing and vanes can also be of integral construction: the vanes are built up into rings, like cartwheels, and then assembled to form the casing. For part-speed operation,VIGVs and VSVs can be used and casing treatments can also be applied. The most common form of casing treatment is circumferential grooves, but slotted style treatments are also used. (see figure 4–19).

Main Modules in a Gas Turbine (Air, Land, or Sea Applications)

*Compressors**

In the gas turbine engine, compression of the air before expansion through the turbine is effected by one of two basic types of compressor, one giving centrifugal flow and the other axial flow. Both types are driven by the engine turbine and are usually coupled direct to the turbine shaft.

The centrifugal flow compressor (Figure 4–20) is a single- or two-stage unit employing an impeller to accelerate the air and a diffuser to produce the required pressure rise. The axial flow compressor (shown later in Figures 4–26 and 4–27) is a multi-stage unit employing alternate rows of rotating (rotor) blades and stationary (stator) vanes, to accelerate and diffuse the air until the required pressure rise is obtained. In some cases, particularly on small engines, an axial compressor is used to boost the inlet pressure to the centrifugal compressor.

With regard to the advantages and disadvantages of the two types, the centrifugal compressor is usually more robust than the axial compressor and is also easier to develop and manufacture. The axial compressor, however, consumes far more air than a centrifugal compressor of the same frontal area and can be designed to attain much higher pressure ratios. Since the airflow is an important factor in determining the amount of thrust, this means the axial compressor engine will also give more thrust for the same frontal area. This, plus the ability to increase the pressure ratio by addition of Odra stages, has led to the adoption of axial compressors in most engine designs. However, the centrifugal compressor is still favored for smaller engines, where its simplicity and ruggedness outweigh any other disadvantages.

The trend to high-pressure ratios that has favored the adoption of axial compressors is because of the improved efficiency that results, which in turn leads to improved specific fuel consumption for a given thrust, Figure 4–24.

Centrifugal Flow Compressor

Centrifugal flow compressors have a single- or double-sided impeller and occasionally a two-stage, single-sided impeller is used, as on the Rolls Royce Dart. The impeller is supported in a casing that also contains a ring of diffuser vanes. If a double-entry impeller is used, the airflow to the rear side is reversed in direction and a plenum chamber is required.

* Source: [4-1] Adapted, with permission, from Rolls Royce, *The Jet Engine* 1986, Rolls Royce Plc: UK.

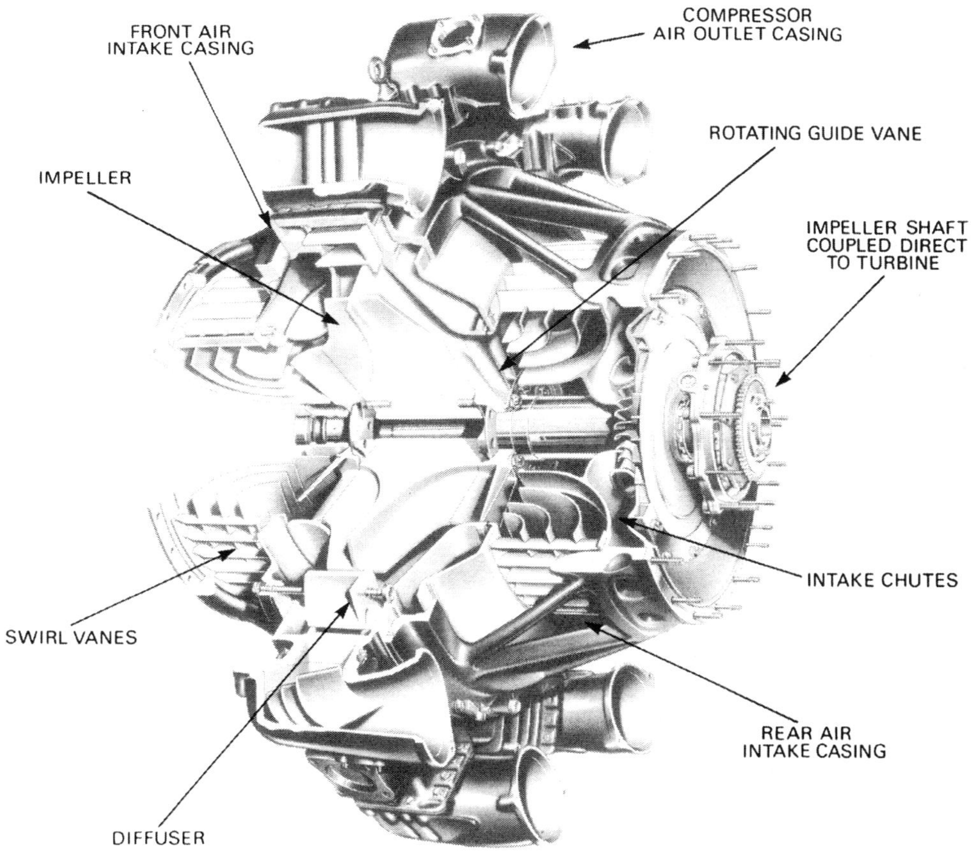

FIGURE 4–20 A typical centrifugal flow compressor. *(Source: Rolls Royce.)*

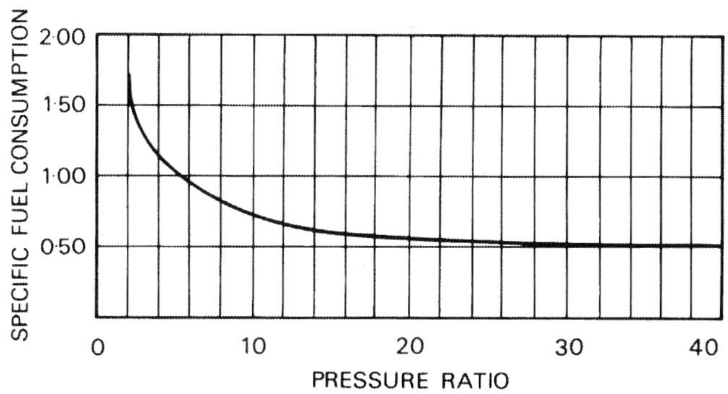

FIGURE 4–21 Specific fuel consumption and pressure ratio. *(Source: Rolls Royce.)*

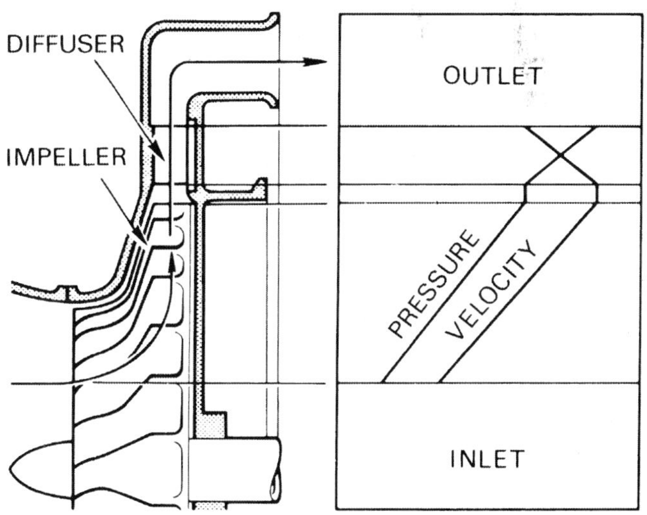

FIGURE 4–22 Pressure and velocity changes through a centrifugal compressor. *(Source: Rolls Royce.)*

Principles of Operation The impeller is rotated at high speed by the turbine and air is continuously induced into the center of the impeller. Centrifugal action causes it to flow radially outwards along the vanes to the impeller tip, thus accelerating the air and also causing a rise in pressure to occur. The engine intake duct may contain vanes that provide an initial swirl to the air entering the compressor.

The air, on leaving the impeller, passes into the diffuser section where the passages form divergent nozzles that convert most of the kinetic energy into pressure energy, as illustrated in Figure 4–25. In practice, it is usual to design the compressor so that about half of the pressure rise occurs in the impeller and half in the diffuser.

To maximize the airflow and pressure rise through the compressor requires the impeller to be rotated at high

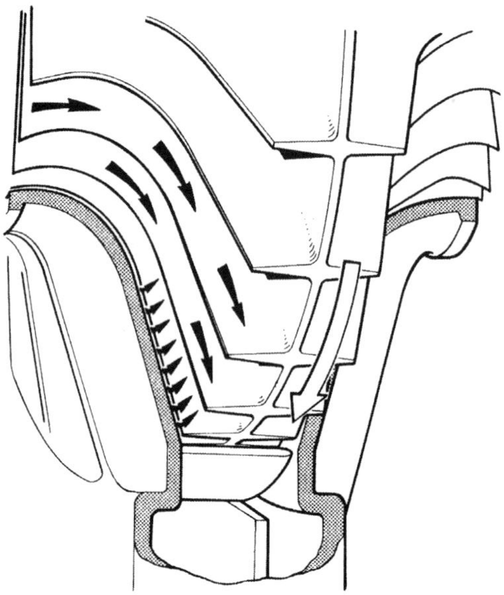

FIGURE 4–23 Impeller working clearance and air leakage. *(Source: Rolls Royce.)*

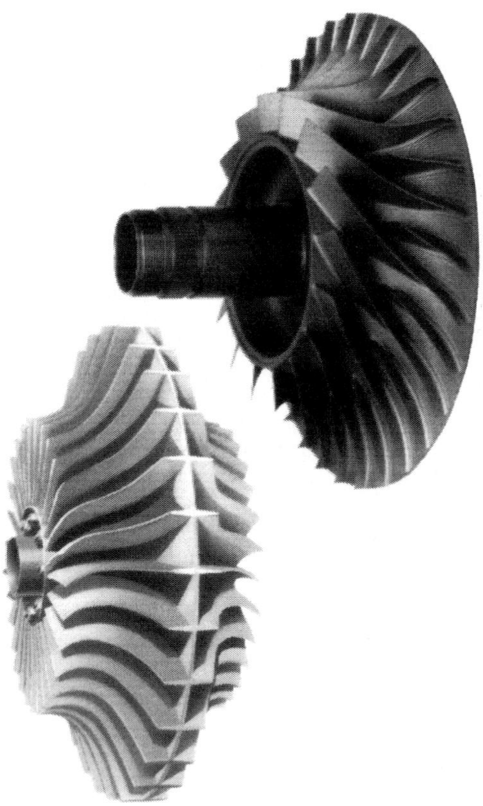

FIGURE 4–24 Typical impellers for centrifugal compressors. *(Source: Rolls Royce.)*

speed, therefore impellers are designed to operate at tip speeds of up to 1600 ft per sec. By operating at such high tip speeds, the air velocity from the impeller is increased so that greater energy is available for conversion to pressure.

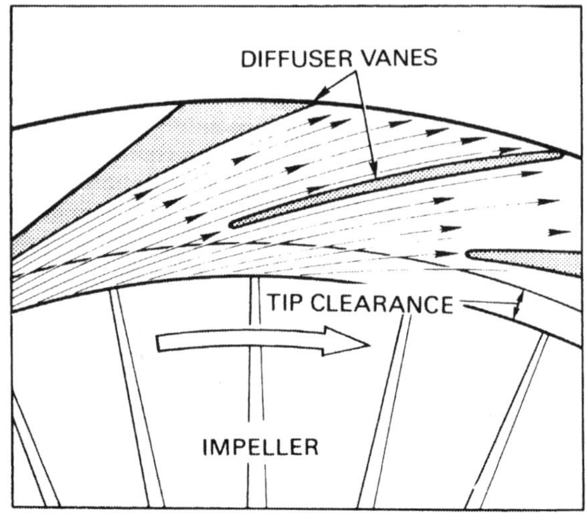

FIGURE 4–25 Airflow at entry to diffuser. *(Source: Rolls Royce.)*

To maintain the efficiency of the compressor, it is necessary to prevent excessive air leakage between the impeller and the casing: this is achieved by keeping their clearances as small as possible (Figure 4–26).

Construction The construction of the compressor centers around the impeller, diffuser, and air intake system. The impeller shaft rotates in ball and roller bearings and is either common to the turbine shaft or split in the center and connected by a coupling, which is usually designed for ease of detachment.

Impellers The impeller consists of a forged disc with integral, radially disposed vanes on one or both sides (Figure 4–27) forming convergent passages in conjunction with the compressor casing. The vanes may be swept back, but for ease of manufacture, straight radial vanes are usually employed. To ease the air from axial flow in the entry duct on to the rotating impeller, the vanes in the center of the impeller are curved in the direction of rotation. The curved sections may be integral with the radial vanes or formed separately for easier and more accurate manufacture.

Diffusers The diffuser assembly may be an integral part of the compressor casing or a separately attached assembly. In each instance it consists of a number of vanes formed tangential to the impeller. The vane passages are divergent to convert the kinetic energy into pressure energy and the inner edges of the vanes are in line with the direction of the resultant airflow from the impeller (Figure 4–28a-c). The clearance between the impeller and the diffuser is an important factor, as too small a clearance will set up aerodynamic buffeting impulses that could be transferred to the impeller and create an unsteady airflow and vibration.

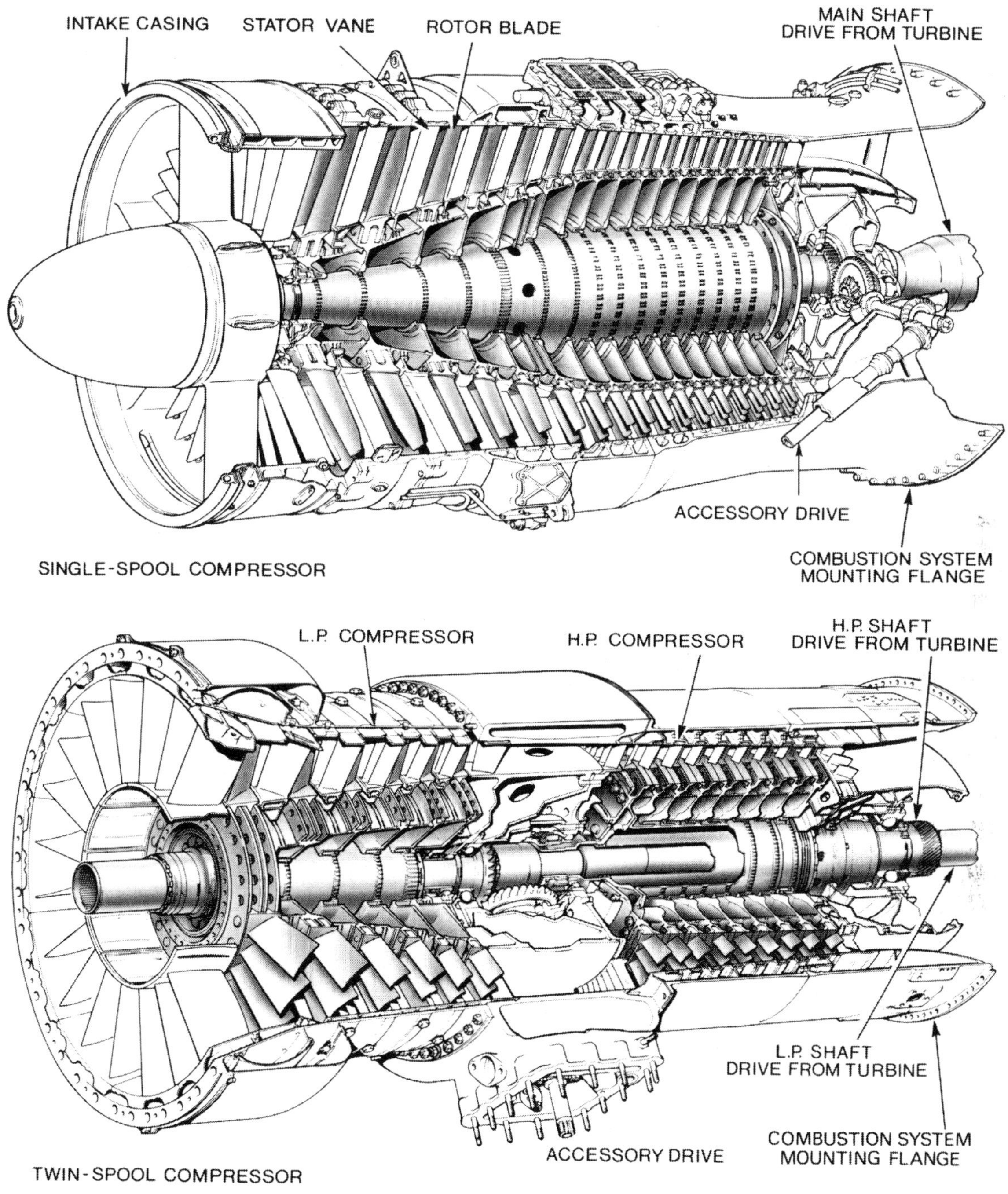

FIGURE 4–26 Typical axial flow compressors. *(Source: Rolls Royce.)*

Axial Flow Compressor

An axial flow compressor (Figures 4–26 and 4–27) consists of one or more rotor assemblies that carry blades of airfoil section. These assemblies are mounted between bearings in the casings that incorporate the stator vanes. The compressor is a multi-stage unit as the amount of pressure increase by each stage is small; a stage consists of a row of rotating blades followed by a row of stator vanes. Where several stages of compression operate in series on one shaft, it becomes necessary to vary the stator vane angle to enable the compressor to operate effectively at speeds below the design condition. As the pressure ratio is increased the incorporation of variable stator vanes ensures that the airflow is directed onto the succeeding stage of rotor blades at an acceptable angle; see the section "Airflow Control."

From the front to the rear of the compressor, i.e., from the low- to the high-pressure end, there is a gradual reduction of the air annulus area between the rotor shaft and

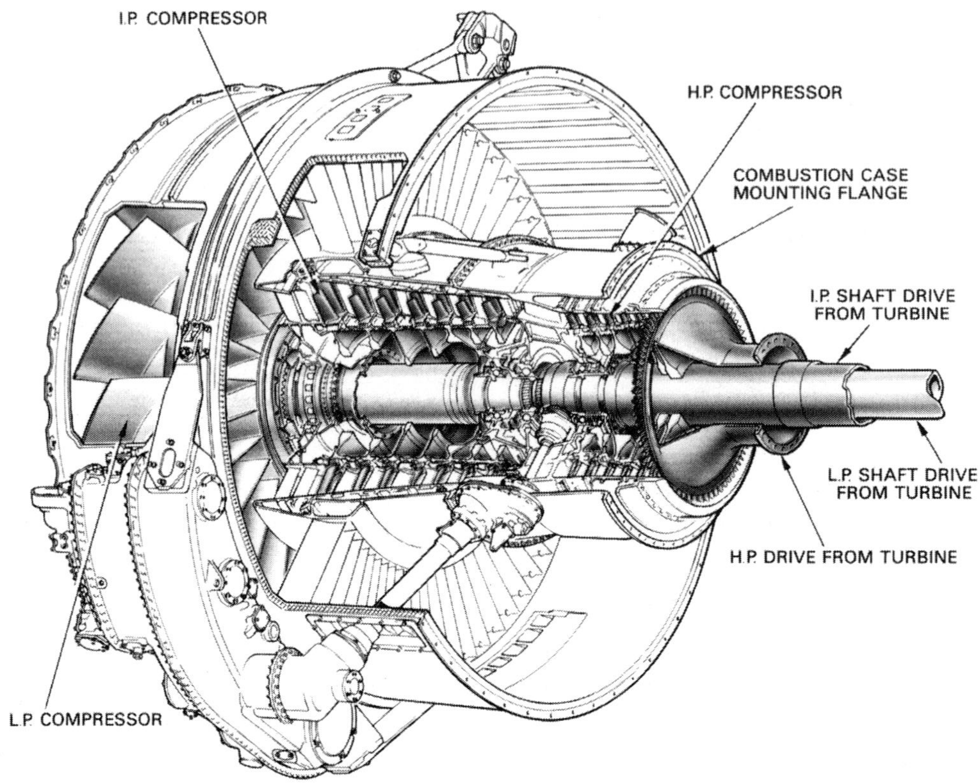

FIGURE 4–27 Typical triple spool compressor. *(Source: Rolls Royce.)*

the stator casing. This is necessary to maintain a near constant air axial velocity as the density increases through the length of the compressor. The convergence of the air annulus is achieved by the tapering of the casing or rotor. A combination of both is also possible, with the arrangement being influenced by manufacturing problems and other mechanical design factors.

A single-spool compressor (Figure 4–26) consists of one rotor assembly and stators with as many stages as necessary to achieve the desired pressure ratio, and all the airflow from the intake passes through the compressor.

The multispool compressor consists of two or more rotor assemblies, each driven by their own turbine at an optimum speed to achieve higher pressure ratios and to give greater operating flexibility.

Although a twin-spool compressor (Figure 4–26) can be used for a pure jet engine, it is most suitable for the bypass type of engine, where the front or low-pressure compressor is designed to handle a larger airflow than the high-pressure compressor. Only a percentage of the air from the low-pressure compressor passes into the high-pressure compressor; the remainder of the air, the bypass flow, is ducted around the high-pressure compressor. Both flows mix in the exhaust system before passing to the propelling nozzle. This arrangement matches the velocity of the jet nearer to the optimum requirements of the aircraft and results in higher propulsive efficiency, hence lower fuel consumption. For this reason the pure jet engine where all the airflow passes through the full compression cycle is now obsolete for all but the highest speed aircraft.

With the high bypass ratio turbofan this trend is taken a stage further. The intake air undergoes only one stage of compression in the fan before being split between the core or gas generator system and the bypass duct in the ratio of approximately 1 to 5 (Figure 4–27). This results in the optimum arrangement for passenger and/or transport aircraft flying at just below the speed of sound. The fan may be coupled to the front of a number of core compression stages (two shaft engine) or a separate shaft driven by its own turbine (three shaft engine).

Principles of Operation During operation the rotor is turned at high speed by the turbine so that air is continuously induced into the compressor, which is then accelerated by the rotating blades and swept rearwards onto the adjacent row of stator vanes. The pressure rise results from the energy imparted to the air in the rotor, which increases the air velocity. The air is then decelerated (diffused) in the following stator passage and the kinetic energy translated into pressure. Stator vanes also serve to correct the deflection given to the air by the rotor blades and to present the air at the correct angle to the next stage of rotor blades. The last row of stator vanes usually act as air straighteners to remove swirl from the air prior to entry into the

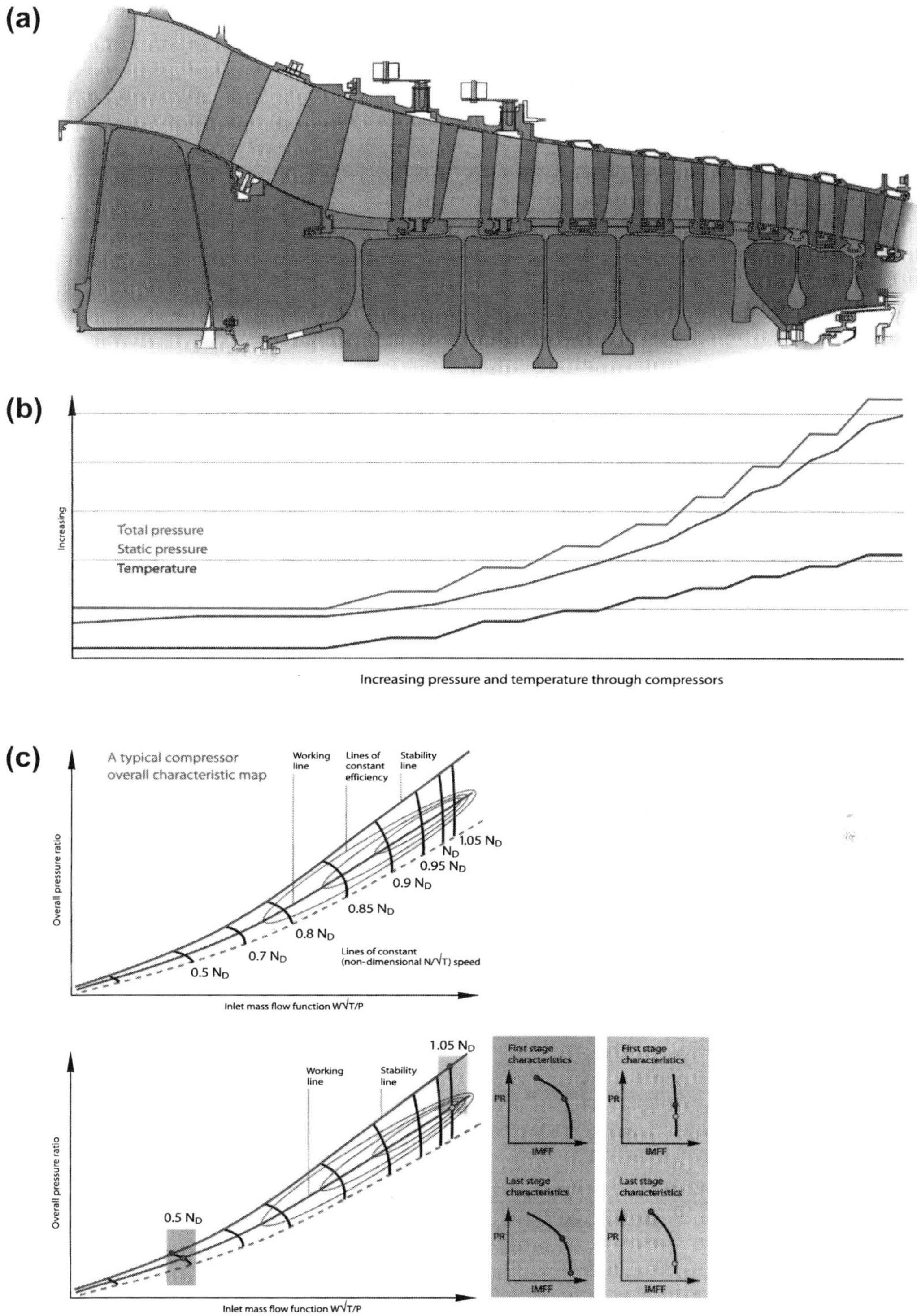

FIGURE 4–28a-c Increasing pressure and temperature through compressors; inlet mass flow function. *(Source: Rolls Royce.)*

combustion system at a reasonably uniform axial velocity. Changes in pressure and velocity that occur in the airflow through the compressor are shown diagrammatically in Figure 4–28a-c. The changes are accompanied by a progressive increase in air temperature as the pressure increases.

Across each stage the ratio of total pressures of outgoing air and inlet air is quite small, being between 1:1 and 1:2. The reason for the small pressure increase through each stage is that the rate of diffusion and the deflection angle of the blades must be limited if losses due to air breakaway at the blades and subsequent blade stall are to be avoided. Although the pressure ratio of each stage is small, every stage increases the exit pressure of the stage that precedes it. So while this first stage of a compressor may only increase the pressure by 3–4 lb per sq. in., at the rear of a 30:1 compression system the stage pressure rise can be up to 80 lb per sq. in. The ability to design multi-stage axial compressors with controlled air velocities and straight through flow minimizes losses and results in a high efficiency and hence low fuel consumption. This gives it a further advantage over the centrifugal compressor, where these conditions are fundamentally not so easily achieved.

The more the pressure ratio of a compressor is increased the more difficult it becomes to ensure that it will operate efficiently over the full-speed range. This is because the requirement for the ratio of inlet area to exit area, at the high-speed case, results in an inlet area that becomes progressively too large relative to the exit area as the compressor speed and hence pressure ratio is reduced. The axial velocity of the inlet air in the front stages thus becomes low relative to the blade speed, this changes the incidence of the air onto the blades and a condition is reached where the flow separates and the compressor flow breaks down. Where high-pressure ratios are required from a single compressor this problem can be overcome by introducing variable stator vanes in the front stages of the system. This corrects the incidence of air onto the rotor blades to angles that they can tolerate. An alternative is the incorporation of interstage bleeds, where a proportion of air after entering the compressor is removed at an intermediate stage and dumped into the bypass flow. While this method corrects the axial velocity through the preceding stages, energy is wasted and incorporation of variable stators is preferred.

The fan of the high bypass ratio turbofan is an example of an axial compressor that has been optimized to meet the specific requirements of this cycle. While similar in principle to the core compressor stage, the proportions of design are such that the inner gas path is similar to that of the core compressor that follows it, while the tip diameter is considerably larger. The mass flow passed by the fan is typically six times that required by the core, the remaining five sixths bypass the core and is expanded through its own coaxial nozzle or may be mixed with the flow at exit from the core in a common nozzle. To optimize the cycle the bypass flow has to be raised to a pressure of approximately 1.6 times the inlet pressure. This is achieved in the fan by utilizing very high tip speeds (1500 ft per sec) and airflow such that the bypass section of the blades operates with a supersonic inlet air velocity of up to Mach 1.5 at the tip. The pressure that results is graded from a high value at the tip, where relative velocities are highest, to the more normal values of 1.3–1.4 at the inner radius, which supercharges the core where aerodynamic design is more akin to that of a conventional compressor stage. The capability of this type of compressor stage achieves the cycle requirement of high flow per unit of frontal area, high efficiency, and high-pressure ratio in a single rotating blade row without inlet guide vanes within an acceptable engine diameter, thus keeping weight and mechanical complexity at an acceptable level.

Construction The construction of the compressor centers around the rotor assembly and casings. The rotor shaft is supported in ball and roller bearings and coupled to the turbine shaft in a manner that allows for any slight variation of alignment. The cylindrical casing assembly may consist of a number of cylindrical casings with a bolted axial joint between each stage or the casing may be in two halves with a bolted center line joint. One or other of these construction methods is required in order that the casing can be assembled around the rotor.

Rotors In compressor designs (Figure 4–29) the rotational speed is such that a disc is required to support the centrifugal blade load. Where a number of discs are fitted onto one shaft they may be coupled and secured together by a mechanical fixing but generally the discs are assembled and welded together, close to their periphery, thus forming an integral drum.

Typical methods of securing rotor blades to the disc are shown in Figure 4–30; fixing may be circumferential or axial to suit special requirements of the stage. In general the aim is to design a securing feature that imparts the lightest possible load on the supporting disc thus minimizing disc weight. While most compressor designs have separate blades for manufacturing and maintainability requirements, it becomes more difficult on the smallest engines to design a practical fixing. However, this may be overcome by producing blades integral with the disc: the so-called blisk.

Rotor Blades The rotor blades are of airfoil section (Figure 4–31) and usually designed to give a pressure gradient along their length to ensure that the air maintains a reasonably uniform axial velocity. The higher pressure towards the tip balances out the centrifugal action of the rotor on the airstream. To obtain these conditions, it is necessary to "twist" the blade from root to tip to give the correct angle of incidence at each point. Air flowing through a compressor creates two boundary layers of slow to stagnant air on the inner and outer walls. In order to compensate for the slow air in the boundary layer a localized increase in blade camber both at the blade tip and root has been introduced. The blade extremities appear as if formed by bending over each corner, hence the term "end-bend."

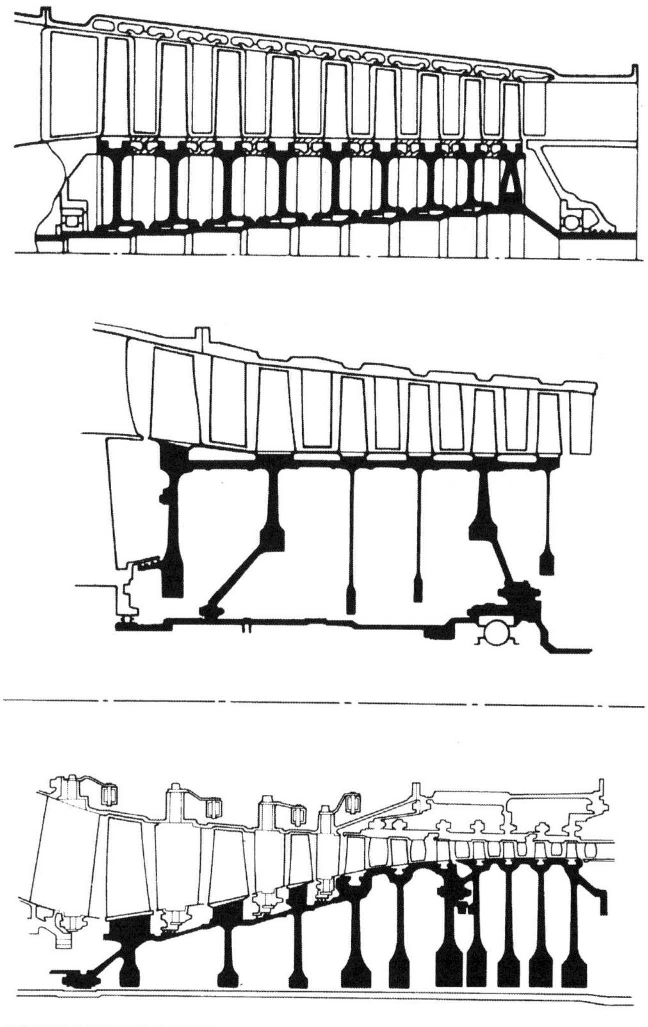

FIGURE 4–29 Rotors of drum and disc construction. *(Source: Rolls Royce.)*

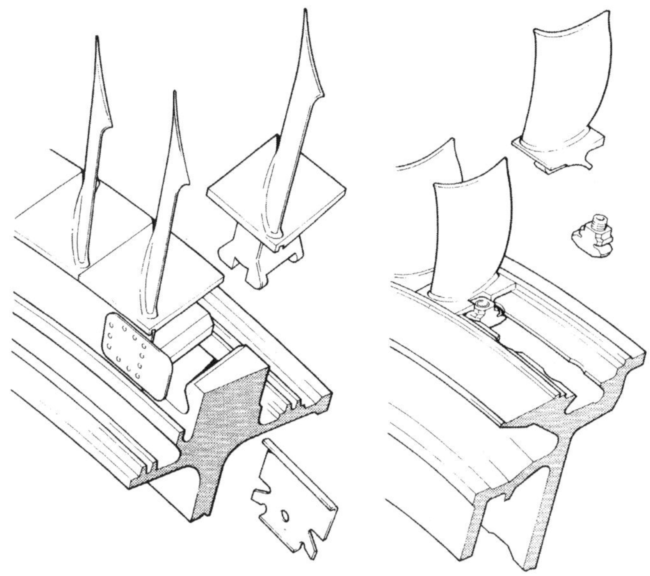

FIGURE 4–30 Methods of securing blades to disc. *(Source: Rolls Royce.)*

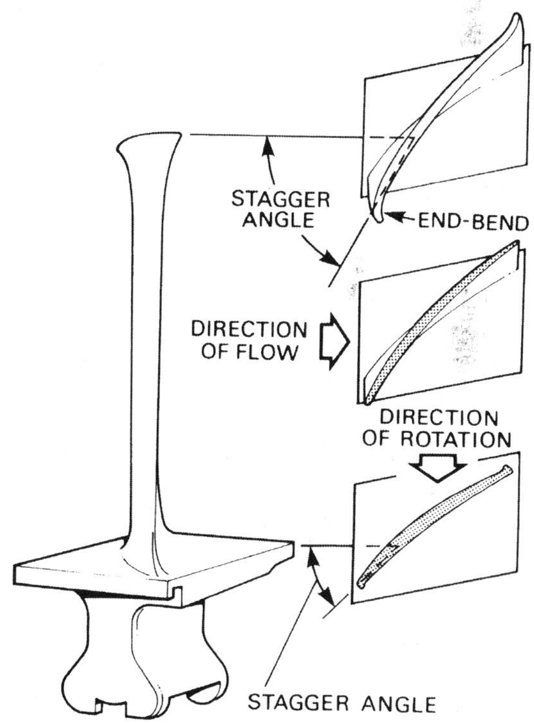

FIGURE 4–31 A typical rotor blade showing twisted contour. *(Source: Rolls Royce.)*

Stator Vanes The stator vanes are again of airfoil section and are secured into the compressor casing or into stator vane retaining rings, which are themselves secured to the casing (Figure 4–32). The vanes are often assembled in segments in the front stages and may be shrouded at their inner ends to minimize the vibrational effect of flow variations on the longer vanes. It is also necessary to lock the stator vanes in such a manner that they will not rotate around the casing.

Operating Conditions

Each stage of a multi-stage compressor possesses certain airflow characteristics that are dissimilar from those of its neighbor; thus to design a workable and efficient compressor, the characteristics of each stage must be carefully matched. This is a relatively simple process to implement for one set of conditions (design mass flow, pressure ratio, and rotational speed), but is much more difficult when reasonable matching is to be retained with the compressor operating over a wide range of conditions such as an aircraft engine encounters.

If the operating conditions imposed upon the compressor blade depart too far from the design intention, breakdown of airflow and/or aerodynamically induced vibration will occur. These phenomena may take one of two

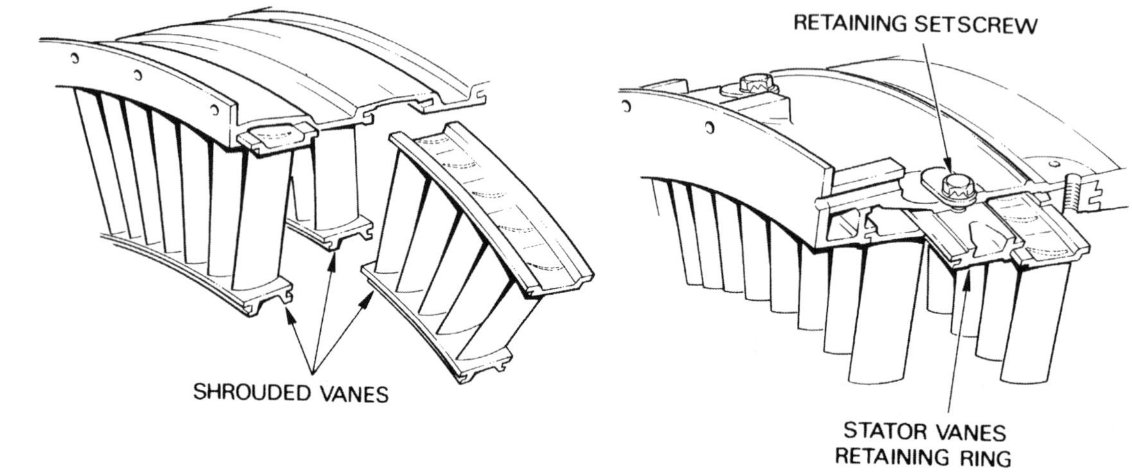

FIGURE 4—32 Methods of securing vanes to compressor casing. *(Source: Rolls Royce.)*

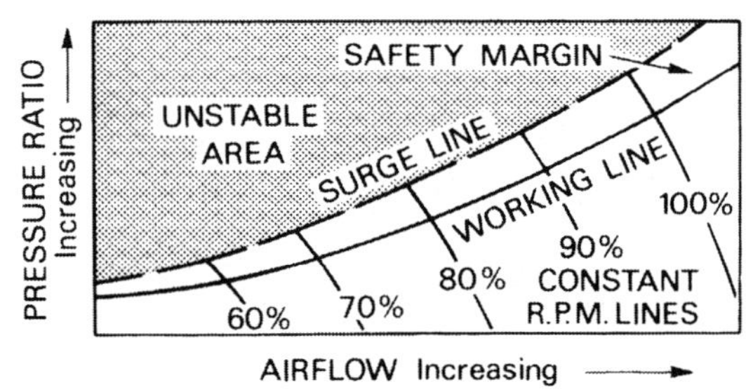

FIGURE 4—33 Limits of stable airflow. *(Source: Rolls Royce.)*

forms: the blades may stall because the angle of incidence of the air relative to the blade is too high (positive incidence stall) or too low (negative incidence stall). The former is a front stage problem at low speeds and the latter usually affects the rear stages at high speed; either can lead to blade vibration that can induce rapid destruction. If the engine demands a pressure rise from the compressor that is higher than the blading can sustain, "surge" occurs. In this case there is an instantaneous breakdown of flow through the machine and the high-pressure air in the combustion system is expelled forward through the compressor with a loud "bang" and a resultant loss of engine thrust. Compressors are designed with adequate margin to ensure that this area of instability (Figure 4—33) is avoided.

Airflow Control

Where high-pressure ratios on a single shaft are required it becomes necessary to introduce airflow control into the compressor design. This may take the form of variable inlet guide vanes for the first stage plus a number of stages incorporating variable stator vanes for the succeeding stages as the shaft pressure ratio is increased (Figure 4—34). As the compressor speed is reduced from its design value these static vanes are progressively closed in order to maintain an acceptable air angle value onto the following rotor blades. Additionally interstage bleed may be provided but its use in design is now usually limited to the provision of extra margin while the engine is being accelerated, because use at steady operating conditions is inefficient and wasteful of fuel. Three types of air bleed systems are illustrated as follows: Figure 4—35, hydraulic; Figure 4—36, pneumatic; and Figure 4—37, electronic.

Materials

Materials are chosen to achieve the most cost-effective design for the components in question, in practice for aeroengine design this need is usually best satisfied by the lightest design that technology allows for the given loads and temperatures prevailing.

For casing designs the need is for a light but rigid construction enabling blade tip clearances to be accurately maintained, ensuring the highest possible efficiency. These needs are achieved by using aluminum at the front of the compression system followed by alloy steel as compression temperature increases, while for the final stages of the compression system, where temperature requirements possibly exceed the capability of the best steel, nickel-based alloys may be required. The use of titanium in preference to aluminum and steel is now more common; particularly in military engines where its high rigidity to density ratio can result in significant weight reduction. With the development of new manufacturing methods, component costs can now be maintained at a more acceptable level in spite of high initial material costs.

Stator vanes are normally produced from steel or nickel-based alloys, a prime requirement being high fatigue strength when "notched" by ingestion damage. Earlier designs specified aluminum alloys but because of its inferior ability to withstand damage its use has declined. Titanium may be used for stator vanes in the low-pressure area but is unsuitable for the smaller stator vanes further rearwards in

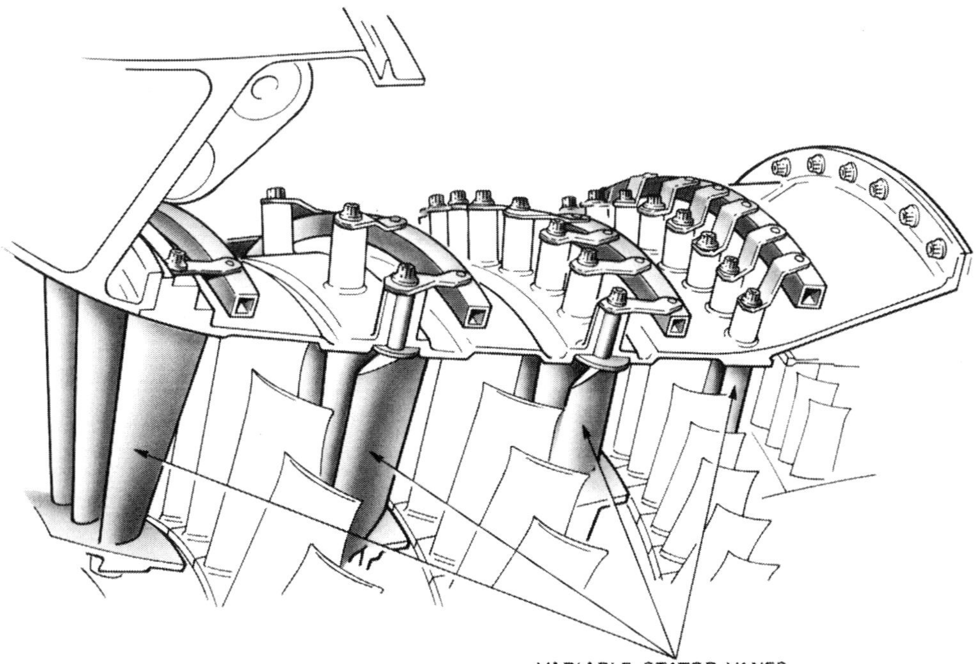

FIGURE 4–34 Typical variable stator vanes. *(Source: Rolls Royce.)*

the compression system because of the higher pressures and temperatures encountered. Any excessive rub that may occur between rotating and static components as a result of other mechanical failures can generate sufficient heat from friction to ignite the titanium. This in turn can lead to expensive repair costs and a possible airworthiness hazard.

In the design of rotor discs, drums, and blades, centrifugal forces dominate and the requirement is for metal with the highest ratio of strength to density. This results in the lightest possible rotor assembly, which in turn reduces the forces on the engine structure enabling a further reduction in weight to be obtained. For this reason, titanium even with its high initial cost is the preferred material and has replaced the steel alloys that were favored in earlier designs. As higher temperature titanium alloys are developed and produced, they are progressively displacing the nickel alloys for the disc and blades at the rear of the system.

The high bypass ratio fan blade (Figure 4–38) only became a design possibility with the availability of titanium, conventional designs being machined from solid forgings. A low weight fan blade is necessary because the front structure of the engine must be able to withstand the large out of balance forces that would result from a fan blade failure. To achieve a sufficiently light solid fan blade, even with titanium, requires a short axial length (or chord). However, with this design, the special feature of a mid-span support ("snubber" or "clapper") is required to prevent aerodynamic instability. This design concept has the disadvantage of the snubber being situated in the supersonic flow where pressure losses are greatest, resulting in inefficiency and a reduction in airflow. This disadvantage has been overcome with the introduction of the Rolls Royce designed wide chord fan blade; stability is provided by the increased chord of the blade thus avoiding the need for snubbers. The weight is maintained at a low level by fabricating the blade from skins of titanium incorporating a honeycomb core.

Centrifugal impeller material requirements are similar to those for the axial compressor rotors. Titanium is thus normally specified though aluminum may still be employed on the largest low-pressure ratio designs where robust sections give adequate ingestion capability and temperatures are acceptably low.

Balancing

The balancing of a compressor rotor or impeller is an extremely important operation in its manufacture. In view of the high rotational speeds and the mass of materials, any unbalance would affect the rotating assembly bearings and engine operation. Balancing on these parts is effected on a special balancing machine.

*Combustors**

Combustion Chambers

The combustion chamber (Figure 4–39) has the difficult task of burning large quantities of fuel, supplied through the

* Source: [4-1] Adapted, with permission, from Rolls Royce, *The Jet Engine* 1986, Rolls Royce Plc: UK.

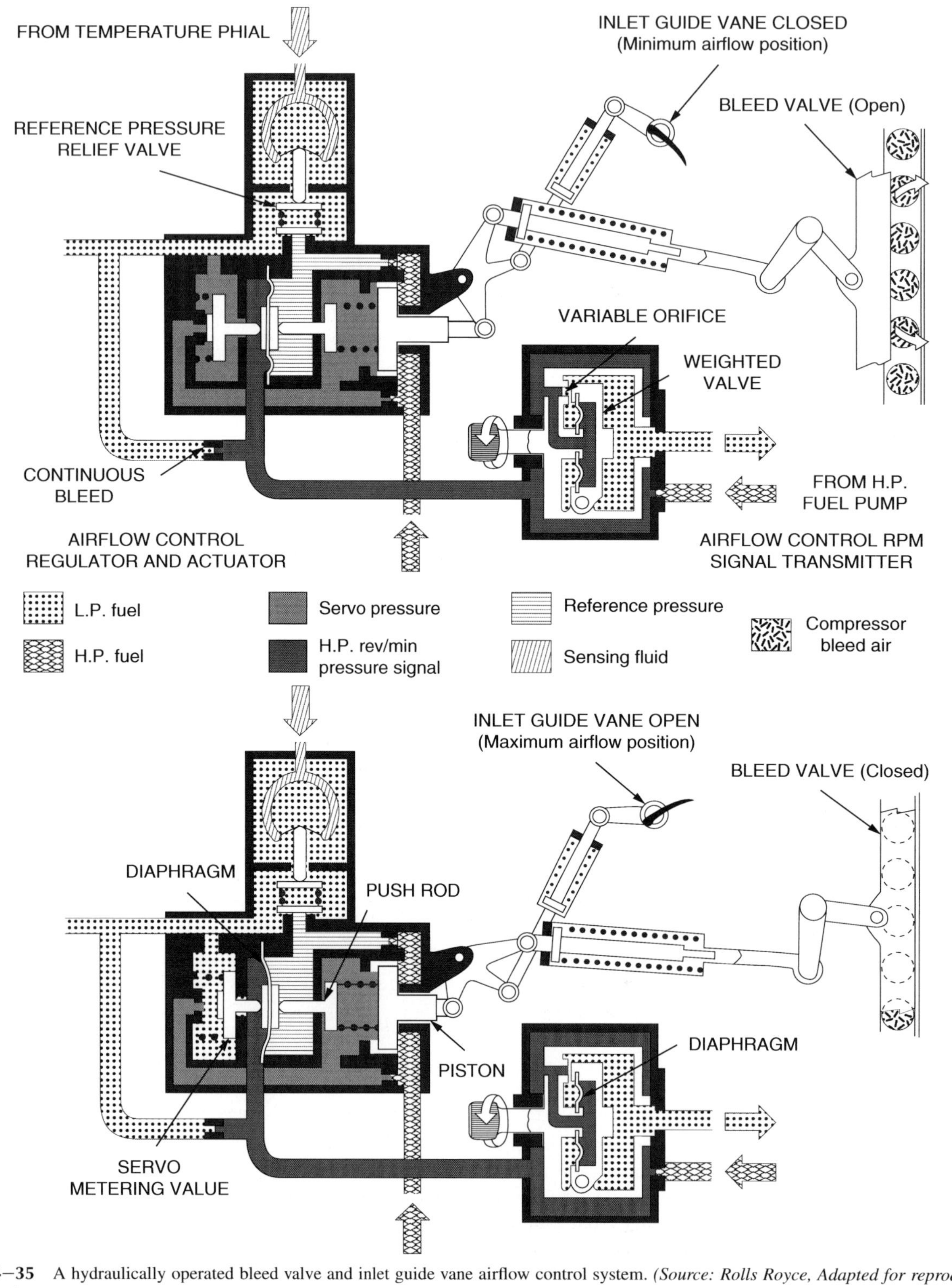

FIGURE 4–35 A hydraulically operated bleed valve and inlet guide vane airflow control system. *(Source: Rolls Royce, Adapted for reproduction in black and white.)*

fuel spray nozzles, with extensive volumes of air, supplied by the compressor and releasing the heat in such a manner that the air is expanded and accelerated to give a smooth stream of uniformly heated gas at all conditions required by the turbine. This task must be accomplished with the minimum loss in pressure and with the maximum heat release for the limited space available.

The amount of fuel added to the air will depend upon the temperature rise required. However, the maximum temperature is limited to within the range of 850–1700°C

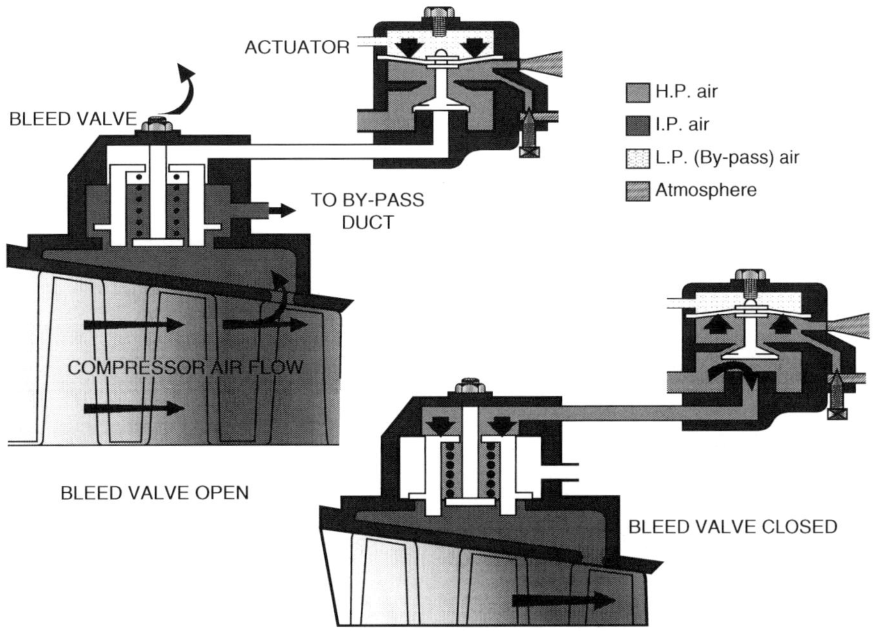

FIGURE 4–36 A pneumatically operated bleed valve system. *(Source: Rolls Royce, Adapted for reproduction in black and white.)*

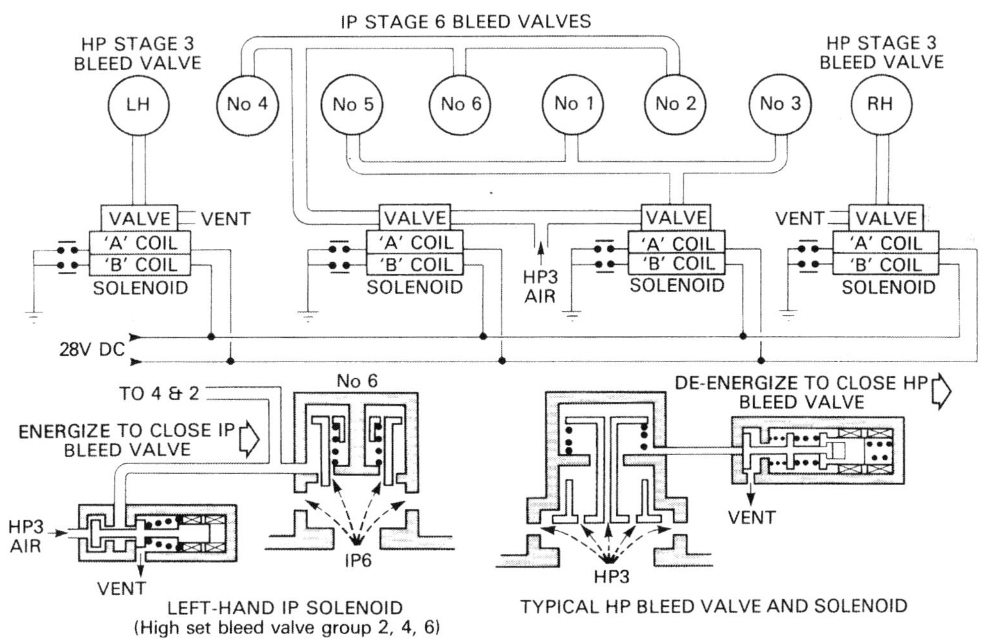

FIGURE 4–37 An electronically operated bleed valve system. *(Source: Rolls Royce.)*

by the materials from which the turbine blades and nozzles are made. The air has already been heated to between 200 and 550°C by the work done during compression, giving a temperature rise requirement of 650–1150°C from the combustion process. Since the gas temperature required at the turbine varies with engine thrust, and in the case of the turbopropeller engine upon the power required, the combustion chamber must also be capable of maintaining stable and efficient combustion over a wide range of engine operating conditions.

Efficient combustion has become increasingly important because of the rapid rise in commercial aircraft traffic and the consequent increase in atmospheric pollution, which is seen by the general public as exhaust smoke.

Combustion Process

Air from the engine compressor enters the combustion chamber at a velocity up to 500 ft per sec, but because at this velocity the air speed is far too high for combustion, the

FIGURE 4—38 Typical types of fan blades. *(Source: Rolls Royce.)*

first thing that the chamber must do is to diffuse it, i.e., decelerate it and raise its static pressure. Since the speed of burning kerosene at normal mixture ratios is only a few feet per second, any fuel lit even in the diffused air stream, which now has a velocity of about 80 ft per sec, would be blown away. A region of low axial velocity has therefore to be created in the chamber, so that the flame will remain alight throughout the range of engine operating conditions.

In normal operation, the overall air/fuel ratio of a combustion chamber can vary between 45:1 and 130:1. However, kerosene will only burn efficiently at, or close to, a ratio of 15:1, so the fuel must be burned with only part of the air entering the chamber, in what is called a primary combustion zone. This is achieved by means of a flame tube (combustion liner) that has various devices for metering the airflow distribution along the chamber.

Approximately 20% of the air mass flow is taken in by the snout or entry section (Figure 4—40). Immediately downstream of the snout are swirl vanes and a perforated flare, through which air passes into the primary combustion zone. The swirling air induces a flow upstream of the center of the flame tube and promotes the desired recirculation. The air not picked up by the snout flows into the annular space between the flame tube and the air casing.

Through the wall of the flame tube body, adjacent to the combustion zone, are a selected number of secondary holes through which a further 20% of the main flow of air passes into the primary zone. The air from the swirl vanes and that from the secondary air holes interact and create a region of low velocity recirculation. This takes the form of a toroidal vortex, similar to a smoke ring, which has the effect of stabilizing and anchoring the flame (Figure 4—41). The recirculating gases hasten the burning of freshly injected fuel droplets by rapidly bringing them to ignition temperature.

It is arranged that the conical fuel spray from the nozzle intersects the recirculation vortex at its center. This action, together with the general turbulence in the primary zone,

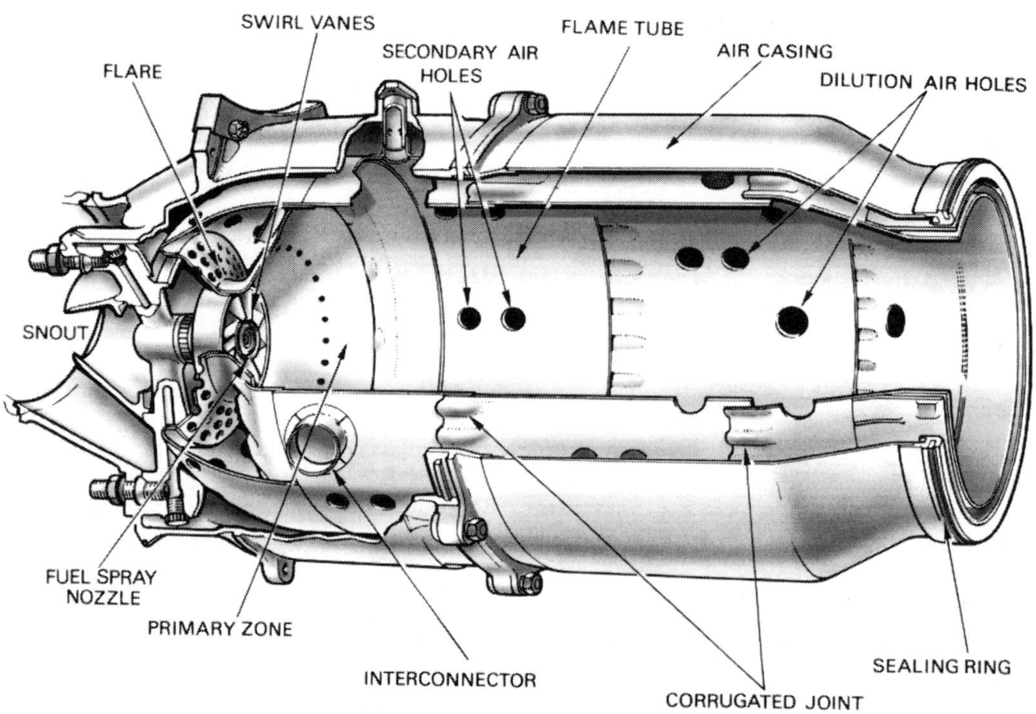

FIGURE 4—39 An early combustion chamber. *(Source: Rolls Royce.)*

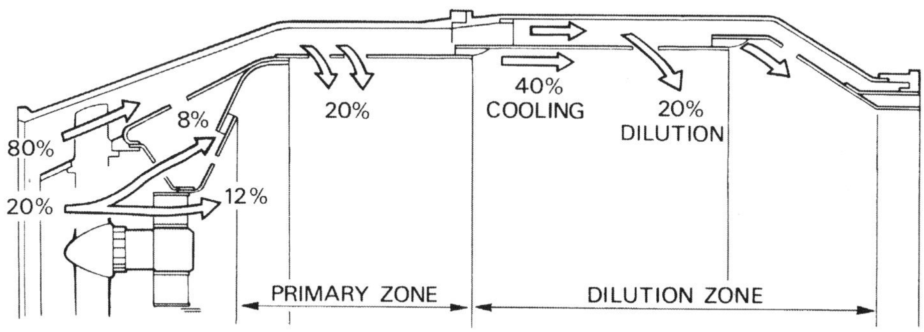

FIGURE 4–40 Apportioning the airflow. *(Source: Rolls Royce.)*

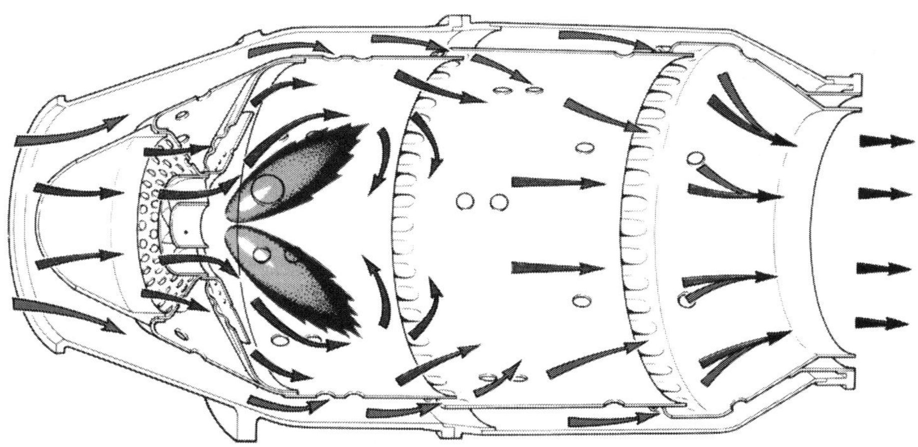

FIGURE 4–41 Flame stabilizing and general airflow pattern. *(Source: Rolls Royce.)*

greatly assists in breaking up the fuel and mixing it with the incoming air.

The temperature of the gases released by combustion is about 1800–2000°C, which is far too hot for entry to the nozzle guide vanes of the turbine. The air not used for combustion, which amounts to about 60% of the total airflow, is therefore introduced progressively into the flame tube. Approximately a third of this is used to lower the gas temperature in the dilution zone before it enters the turbine and the remainder is used for cooling the walls of the flame tube. This is achieved by a film of cooling air flowing along the inside surface of the flame tube wall, insulating it from the hot combustion gases (Figure 4–42). A recent development allows cooling air to enter a network of passages within the flame tube wall before exiting to form an insulating film of air; this can reduce the required wall cooling airflow by up to 50%. Combustion should be completed before the dilution air enters the flame tube, otherwise the incoming air will cool the flame and incomplete combustion will result.

An electric spark from an igniter plug initiates combustion and the flame is then self-sustained.

The design of a combustion chamber and the method of adding the fuel may vary considerably, but the airflow distribution used to effect and maintain combustion is always very similar to that described.

Fuel Injectors

The fuel has to be delivered to the combustion chamber where it is thoroughly mixed with air before combustion. For liquid fuels, there are two distinct methods of doing this: vaporizers and fuel spray nozzles, the latter comprising the two main types of pressure-jets and airspray injectors.

Vaporizers

Vaporizers are comparatively simple, cheap, and lightweight structures that serve to mix the fuel and air. Fuel is injected through a fuel-feed tube or sprayer into an L- or T-shaped tube that turns the fuel/air mixture through 180 degrees. The corners of the vaporizer are typically sharp and are intended to create vortices and promote mixing. These may be supplemented by weirs inside the vaporizer, which also encourage turbulence and mixing. Although the fuel/air mixture is heated inside the vaporizer, most of the mixture leaves the vaporizer and impinges on the combustor baseplate as a series of droplets that receive heat and are vaporized by the high temperatures in the primary zone of

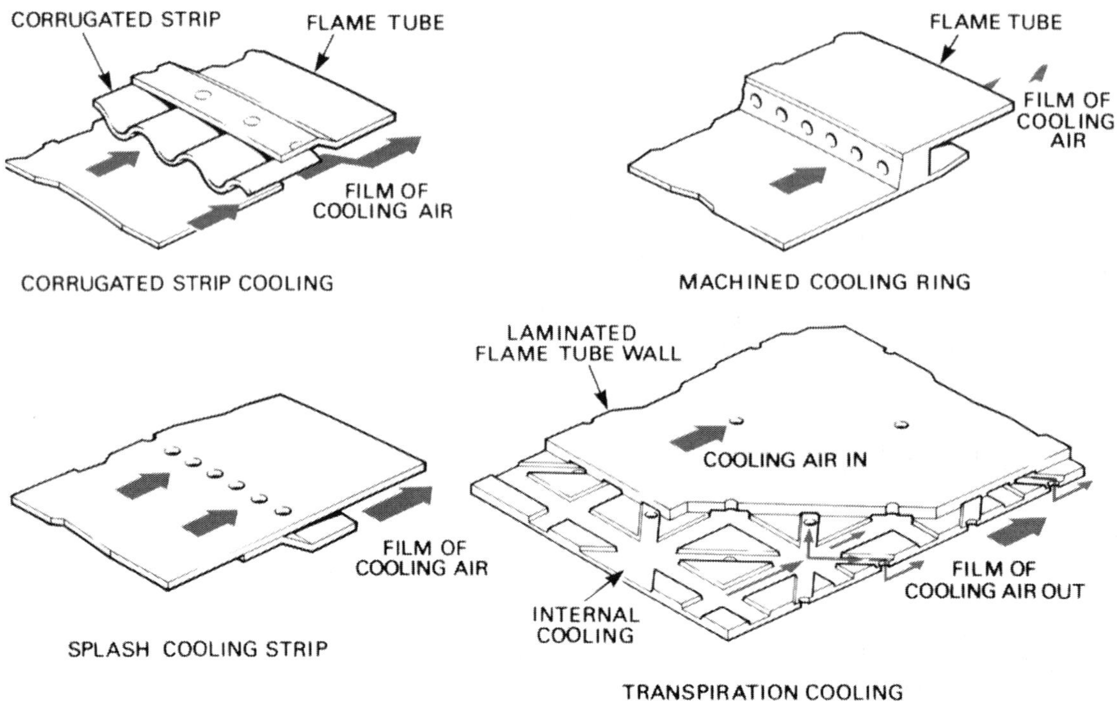

FIGURE 4–42 Flame tube cooling methods. *(Source: Rolls Royce.)*

the combustor. Some combustor designs require the addition of specialized air feed features such as "blown rings" to blow fuel away from the walls to improve efficiency. Engines with vaporizers additionally require primers, which are pressure-jet fuel injectors, to improve ignition characteristics by delivering atomized fuel near the igniters.

The vaporizer is fuel-cooled and has a tendency to overheat when the engine decelerates because the combustion gases in the primary zone are still radiating and conducting heat, but there is little fuel to cool the vaporizer. Because it is fuel-cooled, the vaporizer is also susceptible to overheating caused by blockage of the fuel feed tube.

Vaporizers have been predominant in applications requiring simple, cheap, and lightweight fuel injectors, particularly military aero engines like the Pegasus and RB199, and the RTM322 and Gem helicopter engines. They were also used in the Olympus 593 that powered Concorde. They have not been favored on large civil aero engines because of durability and emissions requirements.

While vaporizers are able to offer high efficiencies and can give low smoke at reasonably high pressures, they are unable to produce satisfactorily low smoke at the *very* high temperatures and pressures seen in the latest generation of civil and military high-thrust aero engines.

Fuel Spray Nozzles

The fuel spray nozzles atomize the fuel to ensure its rapid evaporation and burning when mixed with air. This combustion is a difficult process for two reasons: the velocity of the air stream from the compressor creates a hostile environment for the flame, while the short length of the combustion system means there is little time for burning to occur.

Pressure-Jet Injectors

One technique of atomizing the fuel is to pass it through a swirl chamber where tangential holes or slots impart swirl to the fuel. The fuel is then passed through the discharge orifice, where the fuel is atomized to form a cone-shaped spray. This is called pressure-jet atomization. The rate of swirl and pressure of the fuel at the fuel spray nozzle are important factors in good atomization. The shape of the spray is an indication of the degree of atomization: at low fuel pressures, a continuous film of fuel is formed, known as a "bubble"; at intermediate fuel pressures, the film breaks up at the edges to form a "tulip"; at high fuel pressures, the tulip shortens towards the orifice and forms a finely atomized spray.

The simplex spray nozzle is a pressure-jet atomizer with a single fuel manifold. Used on early jet engines, it consists of a chamber that induces a swirl into the fuel and a fixed-area atomizing orifice. This nozzle gave good atomization at the higher fuel flows (at high fuel pressures) but was very unsatisfactory at the low pressures required at low engine speeds and especially at high altitude. The simplex is, by the nature of its design, a "square law" spray nozzle; that is, the flow through the nozzle is proportional to the square of the pressure drop across it. This meant that if the

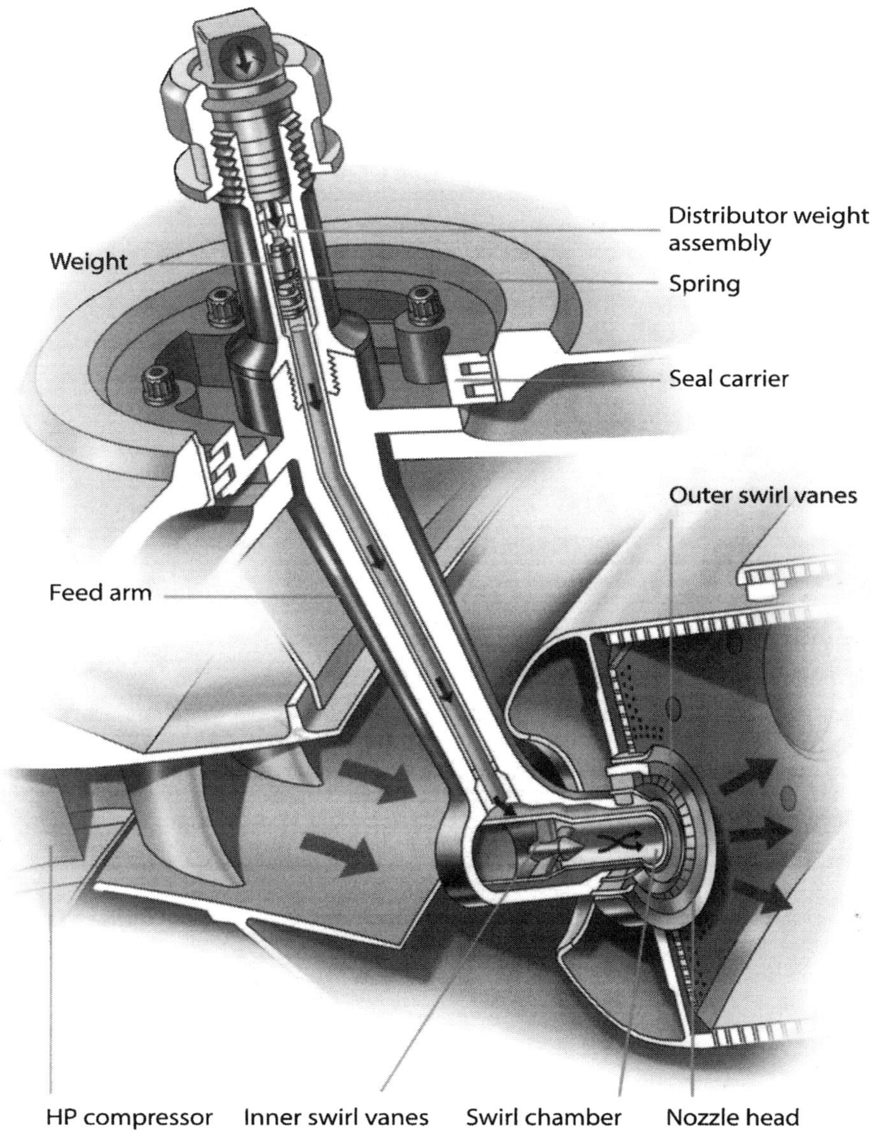

FIGURE 4–43 Section through an airspray nozzle. *(Source: Rolls Royce.)*

minimum pressure for effective atomization were 200 kPa, the pressure needed to give maximum flow would be about 40,000 kPa. The fuel pumps available at that time were unable to cope with such high pressures.

The duplex and duple fuel spray nozzles require a primary and a main fuel manifold and have two independent orifices, one much smaller than the other. The smaller orifice handles the lower flows; the larger deals with the higher flows as the pressure increases.

A pressurizing valve may be employed with this type of spray nozzle to apportion fuel to the two manifolds. As the fuel flow and pressure increase, the pressurizing valve moves to admit fuel progressively into the main manifold and the main orifices. This combined flow down both manifolds allows the duplex and duple fuel spray nozzles to give effective atomization over a wider flow range than the simplex spray nozzle for the same fuel pressure. The duple has two fuel chambers and two orifices, whereas the duplex has one fuel chamber and two orifices.

Airspray Nozzles

The airspray nozzle uses compressor discharge air to create a finely atomized fuel spray. By aerating the spray, the local fuel-rich concentrations produced by other types of spray nozzle are avoided, giving a reduction in both carbon deposition and exhaust smoke, The airspray fuel spray nozzle will typically have two or three air swirler circuits: an inner, an outer, and a dome. An annular fuel passage between the inner and outer air circuits feeds air onto a prefilming lip. This forms a sheet of fuel that breaks down into ligaments. These ligaments are then broken up into

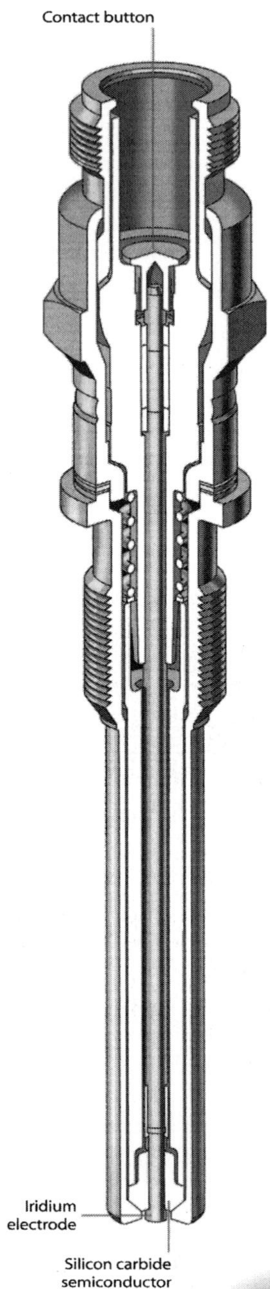

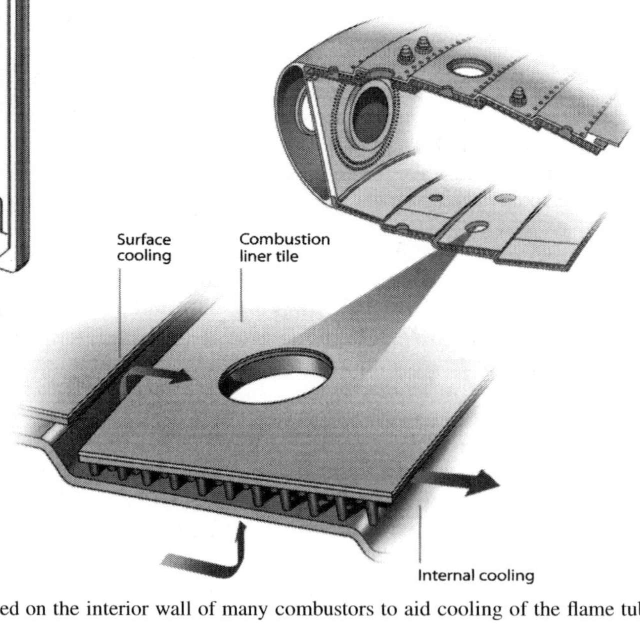

FIGURE 4–44 Ceramic-coated tiles are used on the interior wall of many combustors to aid cooling of the flame tube. *(Source: Rolls Royce.)*

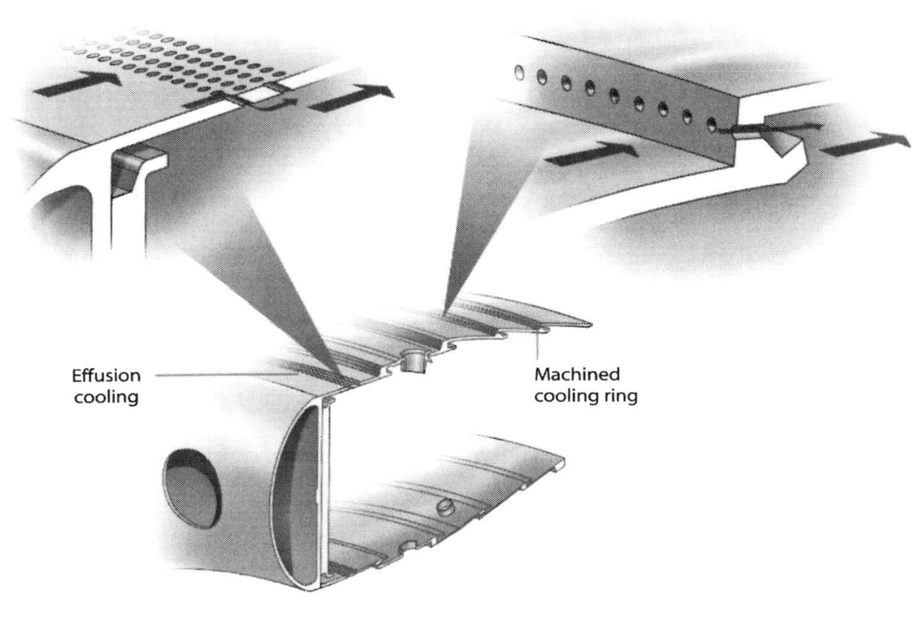

FIGURE 4–45 Machined cooling ring and diffusion cooling holes. *(Source: Rolls Royce.)*

droplets within the shear layers of the surrounding highly swirling air.

The fuel spray nozzle designer not only has to consider optimizing the atomization of fuel, but also where the fuel droplets are directed. These characteristics can be fine-tuned by altering the quantities of air that pass through each air circuit and the amount of swirl that is imparted. An additional advantage is that the low fuel pressure required for atomization permits the use of the comparatively light gear-type pump.

Fuel Distribution

For larger diameter combustion chambers, a flow distributor valve is often required to compensate for the gravity head across the manifold at low fuel pressures to make sure that all the spray nozzles pass an equal quantity of fuel especially at ignition conditions.

This ensures that all sectors of the combustor operate in the same way, giving repeatability in the temperature distribution seen by the high-pressure (HP) turbine. Small diameter combustion chambers, such as those used on military engines, do not have flow distributor valves, but may nevertheless have to cope with an irregular distribution of fuel pressure caused by high-g maneuvers.

Industrial and Marine Fuel Injectors

Industrial engines have an additional complication in that they may be required to run on both liquid and gaseous fuels. This is approached in different ways, depending upon how quickly the change-over is required: "dual fuel" combustion systems have a single set of fuel injectors and can switch between fuels while running; "double fuel" combustion systems require the swapping of fuel injectors when fuels are changed. Dual fuel nozzles are evolved from aero liquid-fuel spray nozzles; gas-only fuel injectors operate at lower pressures, and some may use a series of plane orifices to impart swirl to the fuel flow.

Igniters

There are two basic types of igniter plug; the constricted or constrained air gap type and the shunted surface discharge type. The air gap type is similar in operation to the conventional reciprocating engine spark plug, but has a larger air gap between the electrode and igniter body for the spark to cross. A potential difference of approximately 25,000 volts is required to ionize the gap before a spark will occur. This high voltage requires very good insulation throughout the circuit. The surface discharge igniter plug has the end of the insulator formed by a semi-conducting pellet, which permits an electrical leakage from the central high-tension electrode to the body. This ionizes the surface of the pellet to provide a low resistance path for the energy stored in the capacitor. The discharge takes the form of a high intensity flashover from the electrode to the body and only requires a potential difference of approximately 2000 volts for operation.

The normal spark rate of a typical ignition system is between 60 and 100 sparks per minute. Periodic replacement of the igniter plug is necessary due to the progressive erosion of the igniter electrodes caused by each discharge.

The igniter tip has a range of immersions into the combustor flame tube of plus or minus one millimeter depending on flame tube design and wall cooling technology. During operation, the spark penetrates a further 20 mm. The fuel mixture is ignited in the relatively stable boundary layer; the flame then propagates throughout the combustion

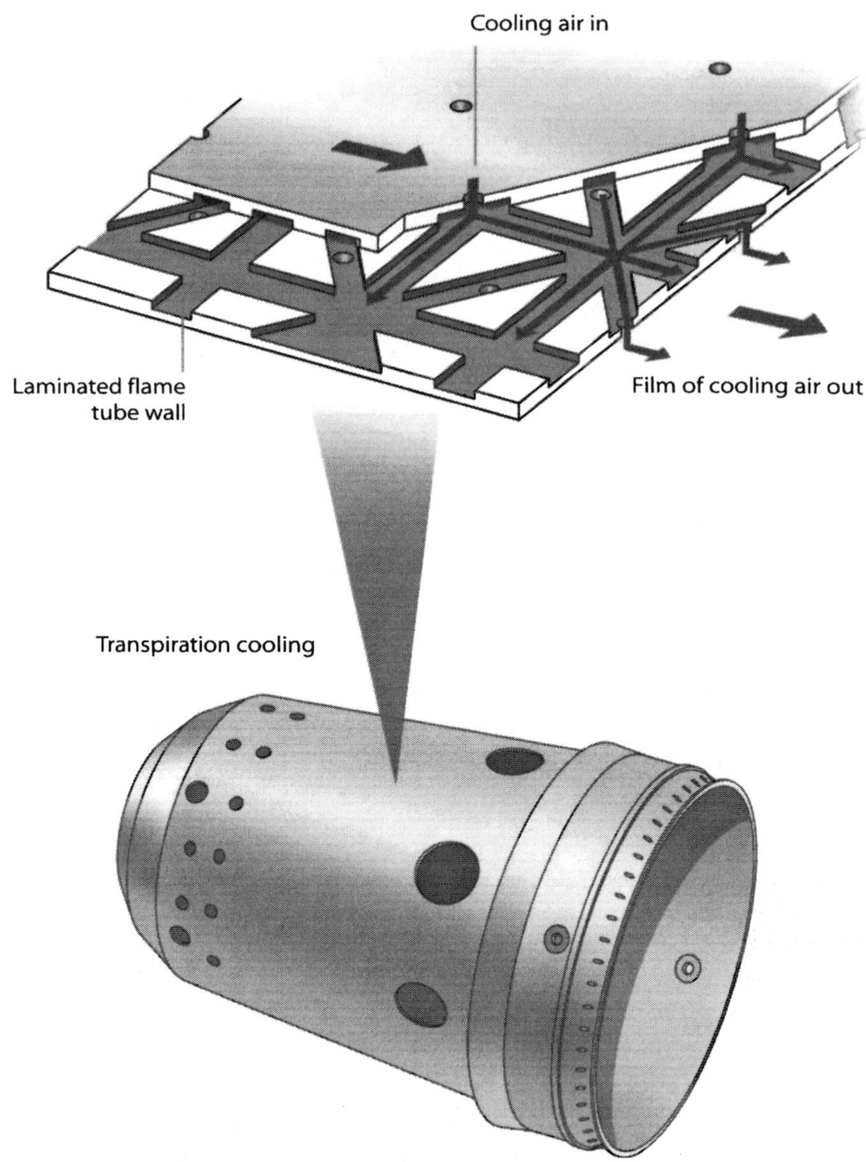

FIGURE 4–46 Transpiration cooling uses laminated materials with a network of internal air passages. *(Source: Rolls Royce.)*

system. A modern annular combustion system usually has two igniters on opposite sides of the annulus.

Cooling

The temperature of the gases released by the combustion process may peak above 2100°C and average 1500°C; this is much higher than the melting point of the combustion chamber and turbine materials. The designer must ensure all of the metal surfaces that are exposed to the hot gas are adequately cooled—quite a challenge when the "cold" air used for cooling may itself be at a temperature approaching 700°C. Furthermore, the amount of air used for cooling must be minimized in order to maximize the air available for emissions control.

A commonly employed technique for cooling the combustor wall is to introduce a cooling film at several locations along the wall. The way this film is introduced varies with the manufacturing method of the combustor wall. For example, a combustor manufactured from sheet metal may use a splash cooling strip or a machined cooling ring, whereas a forged or cast wall could accommodate a Z-ring. This may be supplemented by the use of local effusion cooling (holes) and a ceramic thermal barrier coating on the combustor wall.

Many combustors employ ceramic-coated tiles to line the combustor wall. The individual tiles are attached to a cold "skin," and cooling air passes through holes in the combustor wall and impinges on the tile. The air then moves through a series of pedestals designed to improve the convective heat transfer coefficient, before exiting the front and rear of the tile to form an insulating film. The tiles are designed to be removable for maintenance.

An alternative cooling technique, called transpiration, is to use laminated materials that allow cooling air to enter a network of passages within the flame tube wall before exiting to form an insulating film of air.

The thermal management of fuel-wetted surfaces within the fuel injector is a particular concern. If fuel is exposed to excessive temperatures within the fuel injector, it will decompose to form lacquers and carbon deposits that may block fuel passages or cause distortion. For this reason, the fuel injectors feature complex heat shielding and are carefully designed to prevent regions of stagnant fuel from occurring.

This issue can be more of a problem for industrial and marine applications, where the liquid diesel fuels have lower thermal stability. Subtle combustor cooling changes may also be necessary for industrial and marine applications due to the increased radiation caused by diesel fuel properties.

Predictive Modeling

The modeling of metal temperatures is necessary to determine the displacement, thermal stresses, and life of a component. This modeling is done using finite element analysis. In order to calculate metal temperatures, it is necessary to input material property data, engine performance data, air system data, and heat transfer coefficients. These heat transfer coefficients may be validated by computational fluid dynamics (CFD) analysis and/or rig or engine thermocouple measurements. CFD can also allow the designer to model, first, the flow of air in, through, and out of the combustor, second, the complicated air/fuel mixing, and third, the chemistry behind the combustion process.

Types of Combustion Chamber

There are three main types of combustion chamber in use for gas turbine engines. These are the multiple chamber, the tubo-annular chamber, and the annular chamber.

Multiple Combustion Chamber This type of combustion chamber is used on centrifugal compressor engines and the earlier types of axial flow compressor engines. It is a direct development of the early type of Whittle combustion chamber. The major difference is that the Whittle chamber had a reverse flow as illustrated in Figure 4—47 but, as this created a considerable pressure loss, the straight-through multiple chamber was developed by Joseph Lucas Limited.

The chambers are disposed around the engine (Figure 4—48) and compressor delivery air is directed by ducts to pass into the individual chambers. Each chamber has an inner flame tube around which there is an air casing. The air passes through the flame tube snout and also between the tube and the outer casing as already described.

The separate flame tubes are all interconnected. This allows each tube to operate at the same pressure and also allows combustion to propagate around the flame tubes during engine starting.

Tubo-annular Combustion Chamber The tubo-annular combustion chamber bridges the evolutionary gap between the multiple and annular types. A number of flame tubes are fitted inside a common air casing (Figure 4—49). The airflow is similar to that already described. This arrangement combines the ease of overhaul and testing of the multiple system with the compactness of the annular system.

Annular Combustion Chamber This type of combustion chamber consists of a single flame tube, completely annular in form, which is contained in an inner and outer casing

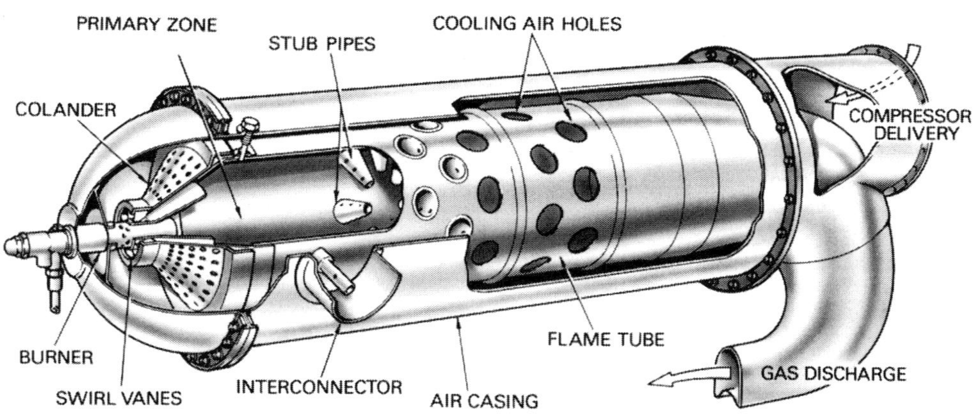

FIGURE 4—47 An early Whittle combustion chamber. *(Source: Rolls Royce.)*

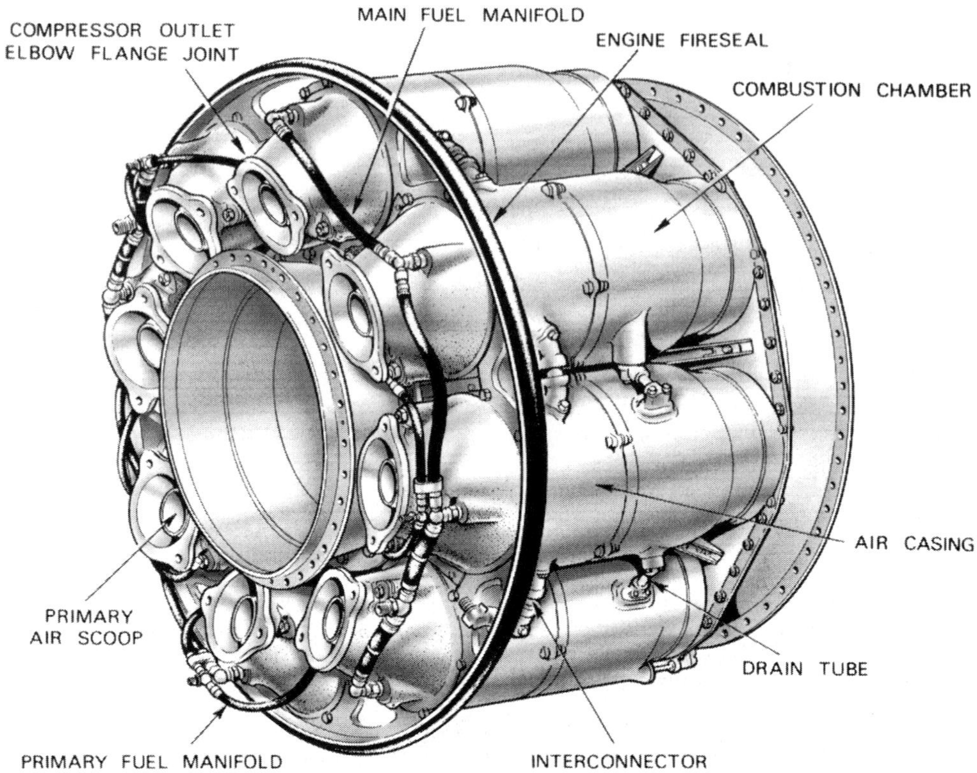

FIGURE 4–48 Multiple combustion chambers. *(Source: Rolls Royce.)*

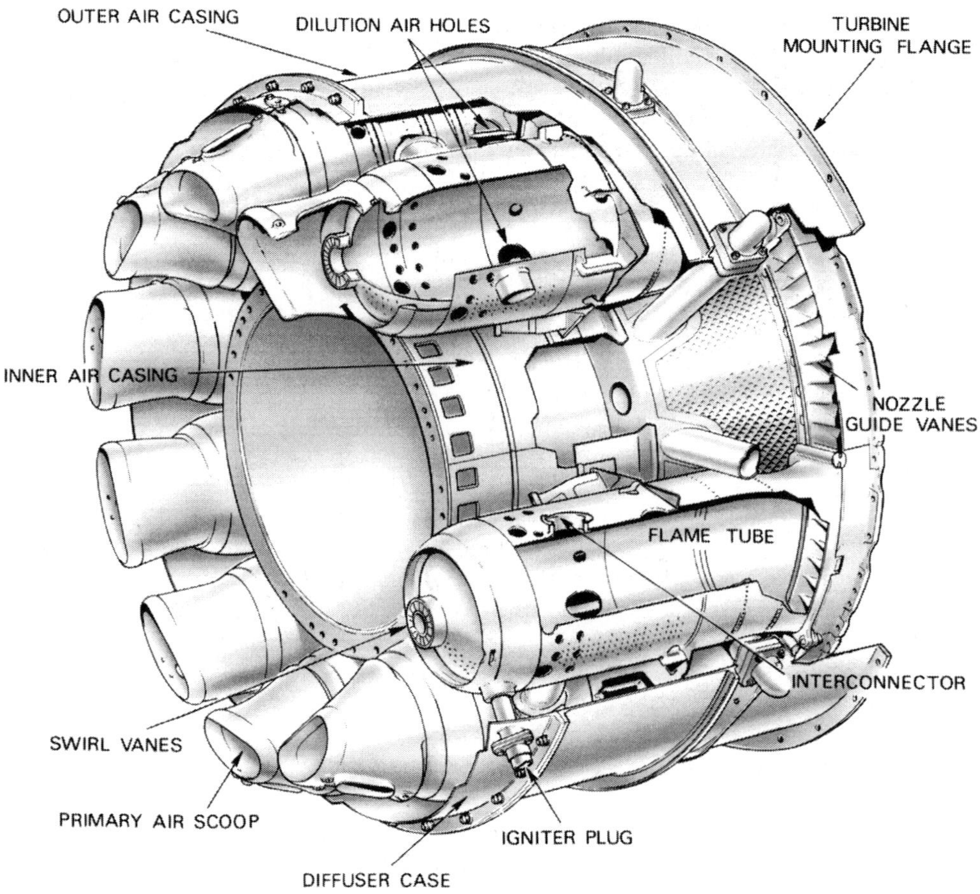

FIGURE 4–49 Tubo-annular combustion chamber. *(Source: Rolls Royce.)*

(Figure 4—50). The airflow through the flame tube is similar to that already described, the chamber being open at the front to the compressor and at the rear to the turbine nozzles.

The main advantage of the annular chamber is that, for the same power output, the length of the chamber is only 75% of that of a tubo-annular system of the same diameter, resulting in considerable saving of weight and production cost. Another advantage is the elimination of combustion propagation problems from chamber to chamber.

In comparison with a tubo-annular combustion system, the wall area of a comparable annular chamber is much less; consequently the amount of cooling air required to prevent the burning of the flame tube wall is less, by approximately 15%. This reduction in cooling air raises the combustion efficiency to virtually eliminate unburned fuel and oxidizes the carbon monoxide to nontoxic carbon dioxide, thus reducing air pollution.

The introduction of the air spray type fuel spray nozzle to this type of combustion chamber also greatly improves the preparation of fuel for combustion by aerating the over-rich pockets of fuel vapors close to the spray nozzle; this results in a large reduction in initial carbon formation.

Combustion Chamber Performance

A combustion chamber must be capable of allowing fuel to burn efficiently over a wide range of operating conditions without incurring a large pressure loss. In addition, if flame extinction occurs, then it must be possible to relight. In performing these functions, the flame tube and spray nozzle atomizer components must be mechanically reliable.

The gas turbine engine operates on a constant pressure cycle, therefore any loss of pressure during the process of combustion must be kept to a minimum. In providing adequate turbulence and mixing, a total pressure loss varying from about 3—8% of the air pressure at entry to the chamber is incurred.

Combustion Intensity The heat released by a combustion chamber or any other heat generating unit is dependent on the volume of the combustion area. Thus, to obtain the required high power output, a comparatively small and compact gas turbine combustion chamber must release heat at exceptionally high rates.

For example, at takeoff conditions a Rolls Royce RB211-524 engine will consume 20,635 lb of fuel per hour. The fuel has a calorific value of approximately 18,550 British thermal units per lb, therefore the combustion chamber releases nearly 106,300 Btus per second. Expressed in another way, this is an expenditure of potential heat at a rate equivalent to approximately 150,000 hp.

Combustion Efficiency The combustion efficiency of most gas turbine engines at sea-level takeoff conditions is almost 100%, reducing to 98% at altitude cruise conditions, as shown in Figure 4—53.

Combustion Stability Combustion stability means smooth burning and the ability of the flame to remain alight over a wide operating range.

For any particular type of combustion chamber there is both a rich and weak limit to the air/fuel ratio, beyond which the flame is extinguished. An extinction is most likely to occur in flight during a glide or dive with the engine idling, when there is a high airflow and only a small fuel flow, i.e., a very weak mixture strength.

The range of air/fuel ratio between the rich and weak limits is reduced with an increase of air velocity, and if the air mass flow is increased beyond a certain value, flame extinction occurs. A typical stability loop is illustrated in Figure 4—54. The operating range defined by the stability loop must obviously cover the air/fuel ratios and mass flow of the combustion chamber.

The ignition process has weak and rich limits similar to those shown for stability in Figure 4—54. The ignition loop, however, lies within the stability loop since it is more difficult to establish combustion under "cold" conditions than to maintain normal burning.

Fission The unwanted pollutants that are found in the exhaust gases are created within the combustion chamber. There are four main pollutants that are legislatively controlled; unburned hydrocarbons (unburned fuel), smoke (carbon particles), carbon monoxide, and oxides of nitrogen. The principal conditions for the formation of pollutants are pressure, temperature, and time.

In the fuel-rich regions of the primary zone, the hydrocarbons are converted into carbon monoxide and smoke. Fresh dilution air can be used to oxidize the carbon monoxide and smoke into nontoxic carbon dioxide within the dilution zone. Unburned hydrocarbons can also be reduced in this zone by continuing the combustion process to ensure complete combustion.

Oxides of nitrogen are formed under the same conditions as those required for the suppression of the other pollutants. Therefore it is desirable to cool the flame as quickly as possible and to reduce the time available for combustion. This conflict of conditions requires a compromise to be made, but continuing improvements in combustor design and performance have led to a substantially "cleaner" combustion process.

Materials

The containing walls and internal parts of the combustion chamber must be capable of resisting the very high gas

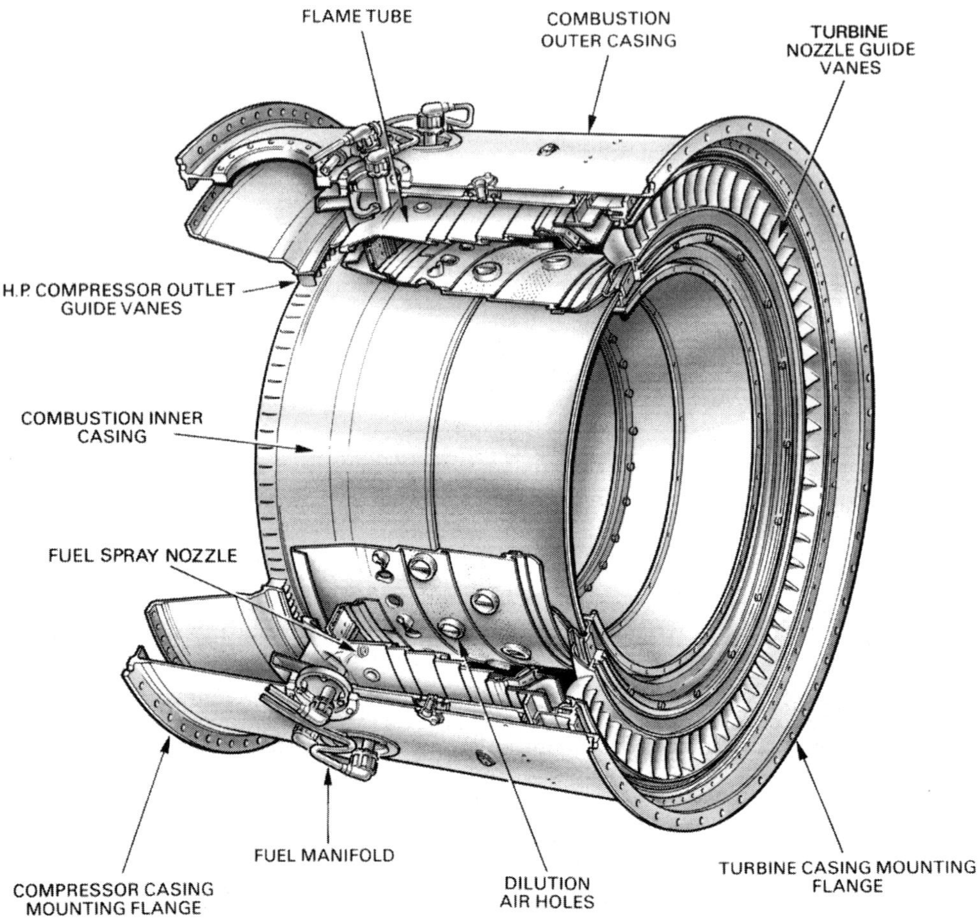

FIGURE 4–50 Annular combustion chamber. *(Source: Rolls Royce.)*

temperature in the primary zone. In practice, this is achieved by using the best heat-resisting materials available, the use of high heat resistant coatings, and by cooling the inner wall of the flame tube as an insulation from the flame.

The combustion chamber must also withstand corrosion due to the products of the combustion, creep failure due to temperature gradients, and fatigue due to vibrational stresses.

Low NO_x Combustors*

From the basic gas turbine cycle discussed in the previous section, one sees that raising the temperature at which the combustion gases enter the turbine (temperature just past the first stage inlet guide vanes or turbine inlet temperature) will also raise the efficiency of the gas turbine cycle. This method of increase in gas turbine efficiency is quite common with certain manufacturers. Care has to be taken, however, that an increase in TITs does not cause other operational problems, such as overheating of turbine components and turbine lubrication oil. If the TIT increase is not accompanied with sufficient additional cooling, this could happen.

Also, one needs to consider that the amount of oxides of nitrogen (NO_x) produced by a combustor increases with the value of the flame temperature in the combustor and the corresponding value of TIT. NO_x contributes to acid rain and legislation against NO_x production has become increasingly stringent. Hence lower TITs, to the extent permitted by optimized efficiency, are desirable. This fact needs to be kept in focus when selecting and/or specifying gas turbines for particular applications and specific demographics (i.e., country or state concerned and their particular legislation).

The NO_x products of combustion can be "cleaned up," or mitigated at any rate, by an external process, such as SCR (selective catalytic reduction), which occurs after combustion.

Low NO_x combustors are designed, optimized, and promoted extensively for both performance- and profit-based

* Reference: [4-2] Proceedings of the panel sessions ASME IGTI annual conferences 1985 through 2003 on "Using ECMS (Engine Condition Monitoring Systems) to Extend Component Lives in Gas Turbine Engines," chairman Claire Soares.

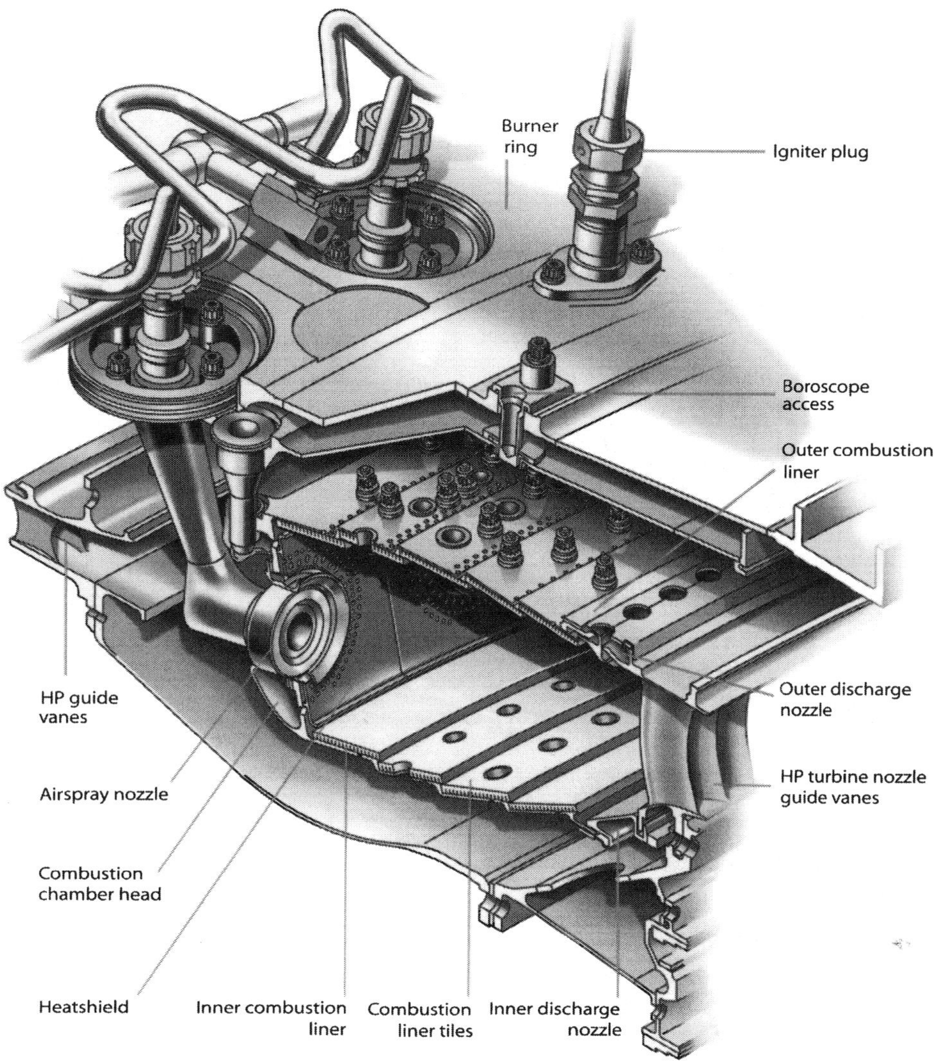

FIGURE 4–51 Section through an annular combustion chamber showing an airspray nozzle. *(Source: Rolls Royce.)*

reasons. The extent of the profit they represent varies with the demographics of the location in question. Specifically,

1. In the United States, as flameless combustor (see later this section) designers are quick to point out, their ultralow single-digit NO_x designs succeed in getting their operators legally permitted (to commence power production) in some cases a few months ahead of their rivals, who may have quite respectable NO_x levels ranging from 9–15 ppm. This represents a considerable amount of revenue.
2. Both the United States and Canada deal with emissions trading. Regardless of any opinion on the technical wisdom of such measures with respect to the overall atmospheric load, low NO_x abilities represent revenue to an operator, who can then sell his or her "spare" credits.
3. In Scandinavian countries, operators pay taxes per unit weight of NO_x and SO_x emissions. This source of revenue method is spreading through the Western world.
4. Low NO_x means that other emissions such as CO and CO_2 are also lowered. CO_2 taxes may soon be reality in global, particularly Western world, terms. In this aspect, once again Scandinavian countries point the way for other operators.
5. End users may also note that reduced NO_x generally means lower TITs, hence reduced wear on hot section components and therefore reduced costs per fired hour.

To study low NO_x combustor design by the major OEMs, case studies that are extracts of design and development work by the OEMs follow. The reader may note that:

- The design strategy of different OEMs is quite different.
- The design strategy of different model teams within each OEM is different.

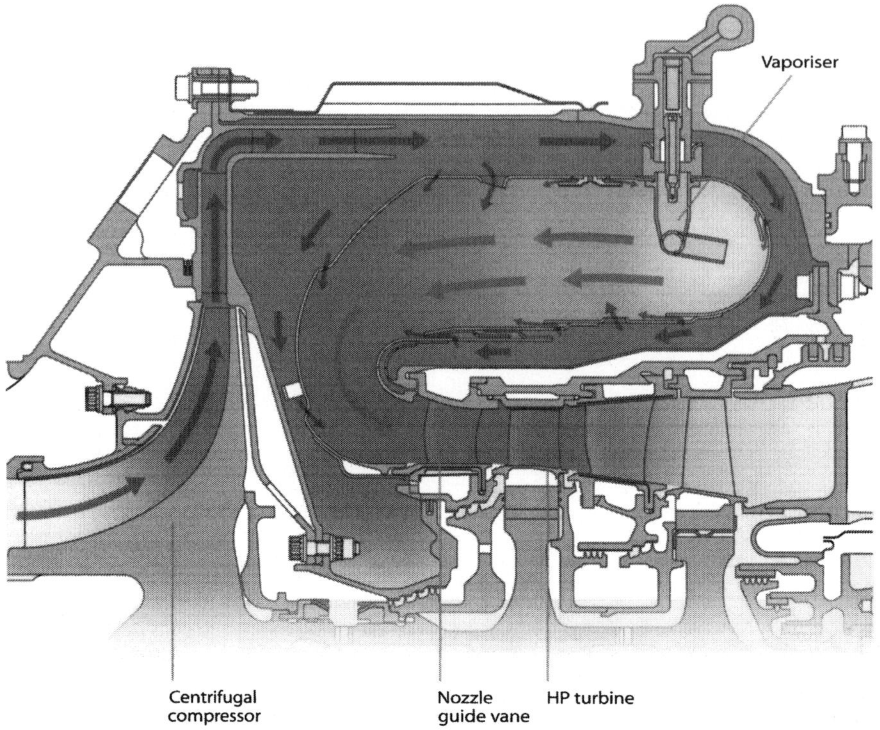

FIGURE 4–52 Airflow through the reverse-flow annular combustion chamber of the RTM 322. *(Source: Rolls Royce.)*

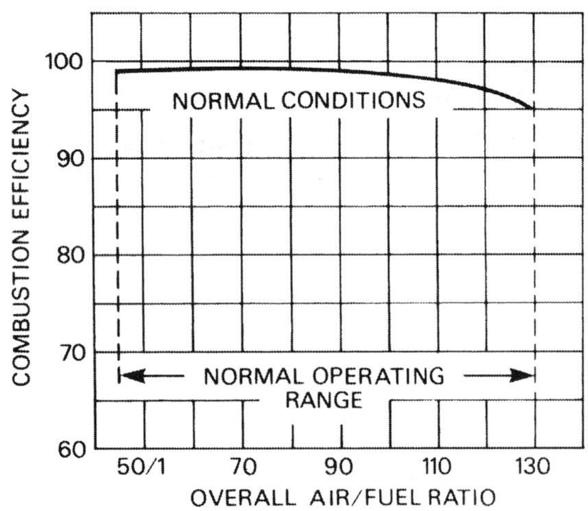

FIGURE 4–53 Combustion efficiency and air/fuel ratio. *(Source: Rolls Royce.)*

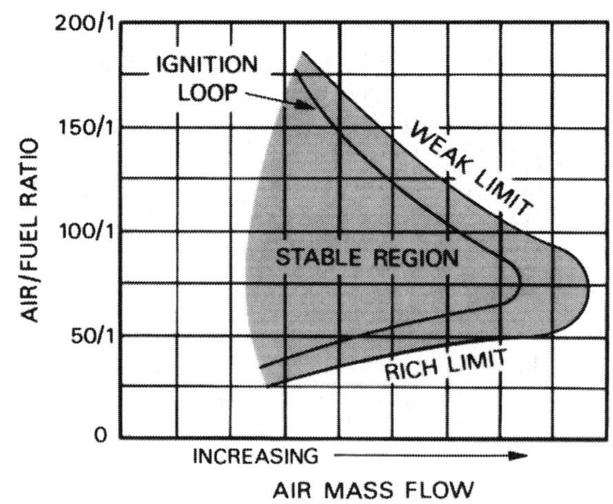

FIGURE 4–54 Combustion stability limits. *(Source: Rolls Royce.)*

- The peak temperatures that the combustor in question must withstand radically affects the design.
- The testing strategy among OEMs also differs.
- Fuel strategies among OEMs may vary depending on global location and individual customer requests. OEMs are keen to accommodate customer needs that may involve low BTU fuel (like gas from a steel furnace or a waste liquid hydrocarbon stream from a petrochemical plant), especially in power hungry areas.

Flameless (Catalytic) Combustors In the first edition of this book, material on Xonon (trademark) flameless combustors made by a company called Catalytica was featured. It would appear that Catalytica have not survived to this point. Nevertheless, research on flameless combustors continues. Progress is currently at the experimental stage.

These extracts from the abstract of a recent conference paper indicate a facet of the current ongoing level of research.

FIGURE 4–55 The SGT-600 dry, low-emission (DLE) combustion system (Siemens Westinghouse). The SGT-600 fleet has clocked up 1.5 million operating hours with its DLE combustion system, which significantly reduces environmental impact. The DLE combustion system was developed for the SGT-600 in 1990 (original design developed by ABB, later Alstom Power, then acquired by Siemens Westinghouse). The SGT-600 burner lowers NO_x by reducing the flame temperature in its combustion chamber. The SGT-600 annular combustor has a total of 18 burners. Each burner consists of a cone split in two halves, which are slightly offset to form two slots for the combustion air to enter (original Alstom designation was the EV burner). The main gas supply also enters through these slots, via tubes fitted along them. Primary fuel is injected at the tip of the cone. This results in a richer fuel mixture, enabling a control feature to stabilize the flame over a range of load conditions. Further combustion control can be provided by means of an optional bypass system that allows the amount of dilution air to be varied.

The current design achieves NO_x emission levels of less than 25 ppmv (at 15% O_2), operating on natural gas in 50–100% load range. Single-digit NO_x levels have been measured in some plants. The DLE system for the SGT-600 has been operating successfully in a variety of applications, including mechanical drives for pipeline and gas storage compressors; cogeneration for industrial duty as well as municipal district-heating systems; and power generation, in both combined-cycle and simple-cycle operation. Installations cover a range of environments, including offshore, from arctic to tropical, at altitudes of up to 1500 m.

Although DLE technology is suitable for dual-fuel combustion, water injection is required to reduce NO_x emissions when burning liquid fuel. Emission levels for operation on liquid fuel are below 42 ppmv, at full load, with a modest water-to-fuel ratio of 0.8. (Source: www.powergeneration.siemens.com)

This study* investigates the performance and the conditions under which flameless oxidation can be achieved for a given annular adiabatic combustor. Numerical modeling of velocity, temperature, and species fields are performed for different flow configurations of air and methane streams injected into a proposed design of a gas-turbine combustor. Parametric analysis was performed by systematically varying several parameters: radius of a recirculation zone, radius of the combustor, location of air and fuel ports, air and fuel velocity magnitudes, and injection angles. The analysis was performed initially using a three-step global chemistry model to identify a design (geometry and operating conditions) that yields flameless combustion regime.

Overall, similar qualitative flow, temperature, and species patterns were predicted by both kinetics models; however, the detailed mechanism provides quantitatively more realistic predictions. An optimal flow configuration was achieved with exhaust NO_X emissions of <7.5 ppm, CO <35 ppm, and a pressure-drop <5%, hence meeting the design criteria for gas turbine engines. This study demonstrates the feasibility of achieving ultra-low NO_X and CO emissions utilizing a flameless oxidation regime.

* Extracted from "Design and Performance Analysis of a Gas Turbine Flameless Combustor Using CFD Simulations," by Y. Levy, F. C. Christo, I. Gaissinski, V. Erenburg, and V. Sherbaum. http://proceedings.asmedigitalcollection.asme.org/proceeding.aspx?articleid=1694001 (Accessed on July 19, 2013).

Legislative Trends Legislative trends help foster the development of this combustor type. The issue taken with all low NO_X combustors at some point is the width of their stable operating range. Experimental work with different kinds of fuel supply techniques continues.

The case studies presented are as follows:

1. Case 1, extracts from IGTI 2001-GT-0024, "Industrial Trent Dry Low Emissions Gas Fuel Control System." This case discusses low NO_X combustors on the 50 MW Trent and outlines the Rolls Royce approach to this gas turbine feature for gas fuel. Note the insights into fuel delivery challenges presented.

Note that the next cases offer perspective on how technology for one model (smaller in this case) may be adapted for larger power models:

2. Case 2, extracts from IGTI 2000-GT-112 "Dual Fuel DLE Typhoon Commercial Operating Experience and Improvement Upgrades." This case was written by the end user in concert with the OEM (the then ABB Alstom, although the design has, like the previous two cases, EGT roots) on operation of Typhoon 4.9 MW gas turbines. The Typhoons were taking over from two aging Ruston (EGT evolved from Ruston) TA1750s and a TB5000 that operated at pollution levels that the operator was obliged to change, in the face of current UK legislative directives. The case discusses how burner development was coincident with end-user changing requirements with benefits for both the end user and OEM.
3. Case 3, extracts from IGTI 2001-GT-0076, "Tempest Dual Fuel Development and Commercial Operating

Experience and Ultra Low NO_x Operation." In this case, dual fuel actual field operation is discussed on a 7.7 MW unit. The effect of specific auxiliary system(s) such as variable guide vanes (VGV) is discussed.

4. Case 4, extracts from "Update on design and operating experience of low calorie gas firing gas turbine for steel works" (presented by Mitsubishi Heavy Industries at PowerGen 2007). This case deals with three different GT sizes operating with this low BTU gas (30 MW, 90 MW, and 180 MW).

Case Study 1: Application of a Dry Low Emmissions (DLE) gas turbine fuel control system (Model Designation Trent)* Regulations increasingly require that emissions from land-based gas turbines be tightly controlled. To limit emissions, the industrial Trent turbine controls flame temperature through a premixed multistage combustor. This requires a multipath fuel delivery system that provides accurate and repeatable metering of the fuel. There are also requirements that the fuel system quickly stop the flow of fuel and vent appropriate lines in the case of a load rejection or an emergency shutdown.

The gas system is a four-path fuel metering system that incorporates shutoff and vent capabilities in addition to high-accuracy multipath flow metering. One of the fuel paths is dedicated to the ignition torch and is significantly smaller and flow accuracies are less demanding than for the three main metering legs.

The concentration of work was dedicated to providing highly accurate and repeatable flow metering for the three main fuel legs. There was also a drive to minimize cost and provide commonality of parts between each of the three legs. Current fuel schedules require metering through a 20:1 turndown ratio.

Flow through each metering leg is measured in the same manner as flow measurement performed across an orifice flow meter. The metering valve is a primary element for which the flow characteristics are well known under a large number of valve positions and flow conditions. Analysis and testing has been conducted to define flow accuracy and repeatability. Further work has been conducted to minimize pressure drop across the system and to reduce the number of sensors required.

The industrial Trent is derived from the aero Trent 892, which was certified at 92,000 lbf in 1995. The industrial Trent is designed to provide 50 MW of power with a thermal efficiency of 42%.

The engine incorporates a dry, low-emissions (DLE) cannular-type combustor, a cross-section of which is shown in Figure 4—56. Fuel and air are mixed in three separate manifolds, referred to as the primary, secondary, and tertiary premixers. The combustor is a staged design, where only the primary is self-stabilized; that is, it can operate alone. The primary therefore provides the combustion stability required during instances such as fast transient maneuvers as it ensures that the combustor remains lit. The secondary and tertiary are each ignited by their respective upstream stage.

This allows the secondary and tertiary to be operated at a much lower flame temperatures than is normally required for flame stabilization. Consequently, a large turndown ratio is possible for the secondary and tertiary fuel/air ratios.

The combustor design also gives a high degree of flexibility in fuel scheduling, as the primary and secondary temperatures can be controlled independently with the tertiary taking the balance of the total fuel. The relationship between combustor inlet and outlet temperatures can vary considerably by daily condition. This means that the primary, secondary, and tertiary fuel schedules, which are set to give the best possible performance for each day's condition, can vary considerably as well.

One of the main challenges in DLE design is to meet the required emissions targets while at the same time avoiding thermoacoustic resonance. As shown in Figure 4—57, as the flame temperature increases, the NO_x level increases, as the flame temperature decreases, the level of CO increases. The result of this relationship is that the fuel must be controlled in such a fashion that the flame temperature is

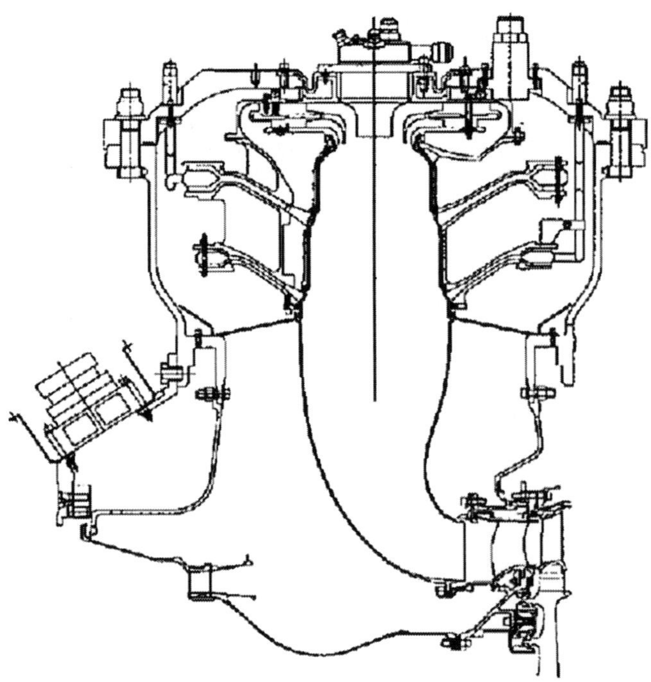

FIGURE 4—56 Industrial Trent combustor [4-3].

* Source: [4-3] Adapted extracts from 2001-GT-0024, J. Gessaman and A. Goetz. "Industrial Trent Dry Low Emissions Gas Fuel Control System."

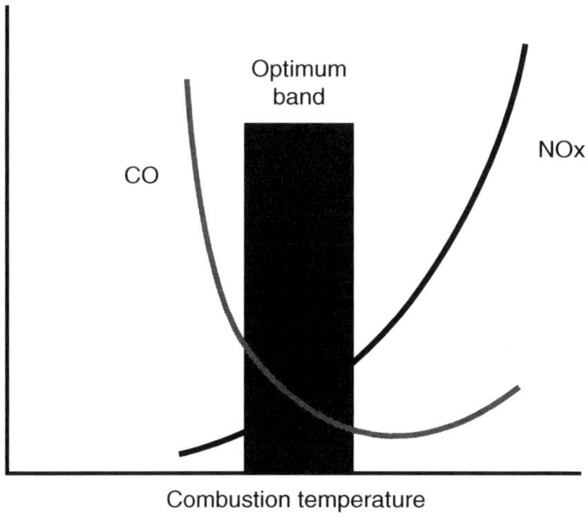

FIGURE 4–57 Relationship between emissions and combustion temperature [4-3].

controlled within an optimum band, which is generally agreed to be around 1850 K.

As with most other DLE-type combustors, the industrial Trent fuel system must have sufficient fuel metering flexibility and accuracy to manage a variety of requirements (including thermoacoustic resonance, emissions, combustion hardware thermal loading, etc.).

In addition to metering the fuel accurately, the delivery system must also meet stringent repeatability requirements. This is due to the need to reliably deliver the required amount of fuel given the same set of input conditions. If the fuel delivery system did not possess a high degree of repeatability, the scenario could arise where the fuel schedules would have to be frequently adjusted to compensate for changing input conditions, such as changes in ambient temperature.

The fuel delivery system must also provide capability to vent the fuel manifolds during fast transient maneuvers in order to prevent undesired fuel flow to the engine. Minimal pressure loss across the system is also a design requirement, as the higher the pressure loss across the delivery system, the higher is the pressure that the gas supply system must provide.

Fuel Flow Metering. Combustor flame temperature is controlled by accurately metering fuel flow through the three fuel legs. The method of flow measurement chosen for the Trent gas fuel system (Figure 4–58) is based on the principle of the differential producer (pressure drop across a restriction). In this application the primary element (restriction) is the fuel-metering valve.

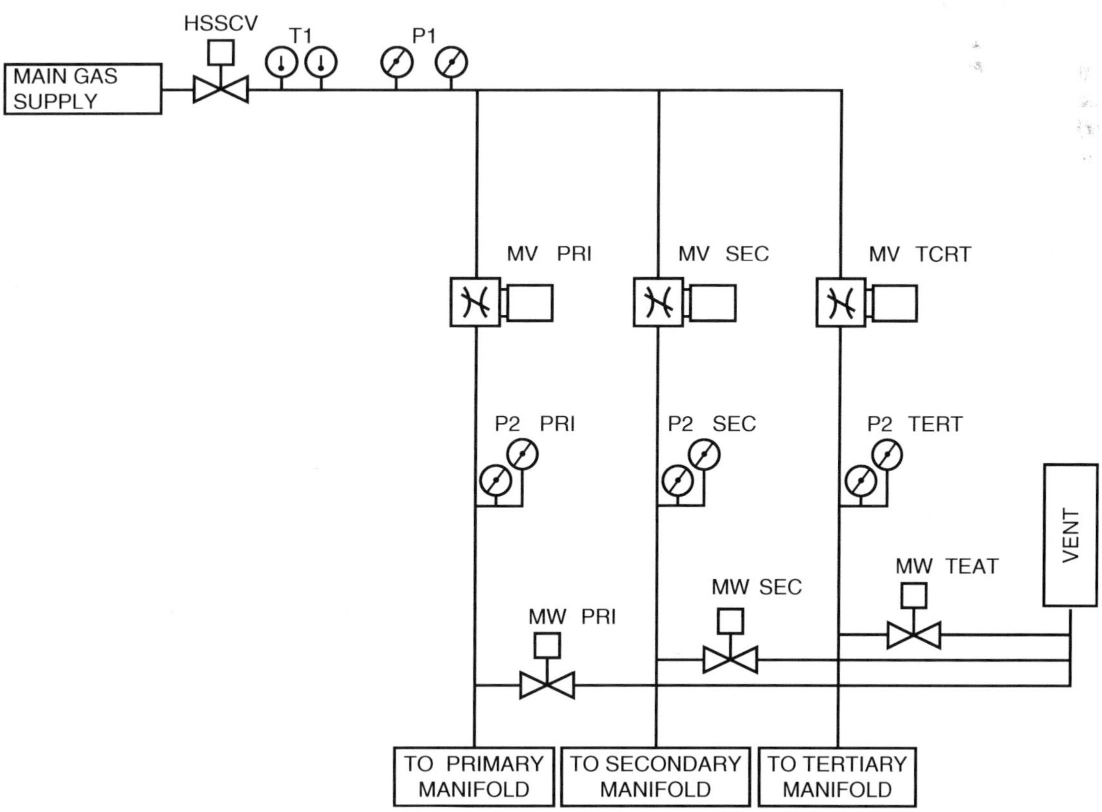

FIGURE 4–58 Schematic of the Trent gas fuel system [4-3].

Other methods of fuel metering were investigated and discarded due to specific application weaknesses. Coriolis meters are known as highly accurate tools for direct mass flow measurement; however, signal noise, vibration sensitivity, and cost of redundancy were the factors that prevented their use. Thermal mass flow meters were not used because the high time constant of the device would not allow adequate dynamic control of the fuel system. Measurement of flow across a venturi was also studied, but the lack of flexibility in the case of fuel schedule changes prevents the use of this method.

Upstream pressure, upstream temperature, downstream pressure, fuel composition, and valve Cd must be precisely known to provide accurate fuel flow metering. Pressure measurements are made using transducers that were originally developed for the aerospace full-authority digital engine controls (FADEC). The transducers have an accuracy of better than ± 0.5 psi from 5 to 1000 psia. Fuel temperature is measured using a high-accuracy platinum RTD and fuel composition is measured using a spectral gas analyzer.

Using the valve as the primary element requires that valve positioning be very accurate and that the valve discharge coefficient be precisely known for a wide range of flow conditions. The metering valves are positioned with a brushless servomotor and position feedback is provided by dual revolvers. Closed-loop position control is better than $\pm 0.2\%$ of valve stroke.

Fuel Shutoff and Venting. Upon a load rejection the fuel metering valves can slew from open to closed in under 125 milliseconds if required. In addition, manifold vent valves quickly open to rapidly drop pressure in the fuel manifold, stopping fuel flow into the combustor. Rapid and precise venting of the fuel, combined with fast and accurate fuel flow control using the fuel metering valves, is an important factor in allowing the Trent to be smoothly brought to synchronous idle speed after a high power load rejection.

In the case of an emergency shutdown, the high-speed shutoff valve (HSSOV) will quickly block off the main fuel supply. Simultaneously the fuel metering valves will close and the manifold vent valves will open. This manages the deceleration control requirements.

The manifold vent valves and HSSOV must have a high flow capacity with a very low-pressure drop. They are also required to shift to the safe state in less than 100 milliseconds.

Fuel Metering Optimization. For all turbine applications it is important to minimize pressure drop across the fuel delivery system. Pressure drop across the metering system represents losses in energy due to higher supply pressure requirements. This can also increase the initial investment in the form of larger fuel-compression hardware. The high compressor ratio of the Trent turbine (40:1) makes minimizing pressure losses in the fuel delivery system essential.

Measuring flow across a differential producer requires that there is a pressure drop across a restriction and that this pressure drop is measured accurately. To accomplish this while minimizing fuel system pressure drop, accuracy of the pressure measurements is critical. This drives the use of very-high-accuracy pressure transducers and an understanding of well-established standards for measuring flow across a differential producer.

During development of the fuel system, minimizing cost was a key objective of the design team. Flow measurement standards define the location for pressure measurement relative to the restriction. This requires pressure measurement at the inlet and outlet of each metering valve. Redundancy requirements drive the use of two transducers at each pressure measurement location. Two pressure transducers at each measurement point on three metering legs means that 12 pressure transducers would be required in total. Twelve very-high-accuracy pressure transducers would represent a high cost element of the gas fuel system.

To reduce cost for the system, P1 measurement for each leg is made at a common point on the inlet fuel header. This lowered the number of pressure transducers from 12 to 8. Measuring pressure away from the valve inlet has the potential of reducing flow-metering accuracy, so analysis was conducted to minimize the pressure difference between the header pressure port and that at the inlet of each valve. Analysis was conducted for all ambient conditions under worst-case fuel conditions. Several iterations were made to find that a 3 in. header feeding the 2 in. fuel metering legs provided the smallest error. This arrangement also provides very low-pressure losses.

The manifold is designed to minimize P1 measurement error; however, the flow accuracy requirements demand tighter P1 measurement accuracy. This is further complicated because the production metering valves are calibrated with P1 pressure measurements at the inlet of the valve. To achieve flow accuracy while measuring pressure at a common header port, an algorithm was devised to correct header pressure measurement to the pressure that would be measured at each of the legs. This algorithm is implemented in the control and is an integral part of the flow control loop.

To meet the flow accuracy metering requirements and to provide good dynamic control, the minimum pressure drop across the metering valves is 50 psid. Additional sources for pressure drop are the HSSOV, the inlet manifold, and the outlet spool pieces. The design is such that the pressure ratio across the fuel system is minimized as much as possible.

Summary

The industrial Trent gas fuel system provides highly accurate and repeatable fuel metering for a wide range of flow conditions. The system meets aggressive targets for pressure loss and cost and offers high parts commonality despite having to meet a wide variety of flow requirements. Reliability was not sacrificed at the expense of cost, as the fuel system incorporates a high degree of redundancy.

Testing on the Woodward flow rig allowed a high degree of confidence to be achieved in the system as it gave the opportunity to test the valves under conditions representative of engine operation. Use of the flow rig provided a method of mapping the system to provide extremely accurate and repeatable flow control.

Case Study 2: Dry Low Emmissions (DLE) Combustor System Application (Thyphon)* The first Typhoon generator set fitted with ABB Alstom Power's dual fuel dry, low-emissions system entered service in June 1998 at the Boots Company headquarters in Nottingham, England. Running on distillate fuel NO_x emissions of 30 ppmvd at 15% O_2, they have been commercially demonstrated without water or steam injection. This section describes the operating experience gained from the gas turbine manufacturer's and operating customer's points of view. The gas and liquid fuel burner configurations along with the operating concept are described. Discussion of the initial site issues encountered, the solutions introduced, and the effect on operational impact are then explained in detail.

End-User Profile. Today the main Boots site is located at Beeston to the west of Nottingham. The site covers 300 acres and contains over 100 individual buildings. In the region of 7000 people are employed on the site. A number of different activities are carried out on site, including:

- Head office functions
- Consumer and health-care product manufacture and development
- Warehouse and goods distribution
- Computing and data processing
- Chemical manufacture

The site has had its own power station since the development of the site in 1928 when the powerhouse was constructed. Steam and electricity were provided initially from two coal-fired boilers supplying steam via a 1 MW back pressure turboalternator. As time progressed the powerhouse was developed as site activities expanded and its energy requirements increased.

By 1996 what had evolved was a highly complex powerhouse that could raise steam by coal, gas, gas oil, or heavy fuel oil and generate electricity by gas turbine or steam turbine using either back pressure or condensing turbines. The gas turbine plant comprised two Ruston TA1750s and a TB5000.

This combination of plant led to an operation that had become uneconomical, with high maintenance and operating costs, and some plant was coming to the end of its useful life. The powerhouse operation was also regulated by Her Majesty's Inspectorate of Pollution (HMIP) under the Environmental Protection Act (EPA) of 1990 and required an Integrated Pollution Control (IPC) authorization to operate. The station's authorization to operate contained a number of improvement notices, one of which was to investigate NO_x reductions on gas turbine and boiler plant for improvements by April 1997.

The requirement to improve the environmental impact of the powerhouse initiated a feasibility study that led to the design, construction, and commissioning of a new combined heat and power (CHP) energy center. The salient parts of the new energy center comprised:

- Three 4.9 MWe Typhoon gas turbines
- Three 10 tonne/h waste heat recovery boilers
- Four 20 tonne/h package boilers
- Gas compressor
- Water treatment plant and station auxiliaries
- Distributed control system

Gas Turbine Procurement. Procurement of the gas turbines started in the summer of 1995. At the time dry, low-NO_x reduction on circa 5 MW turbines was not as well advanced as in larger engines. However, Boots was aware that dry, low-NO_x techniques represented the best available technology (BAT) solution and as such were keen to pursue this route. NO_x reduction by steam injection was an option, but in this case was not a practical solution since the steam raised within the CHP was not of sufficient pressure or quality.

Subsequent gas turbine tender evaluation revealed that dual-fuel DLE was commercially available in only one circa 5 MW turbine and in our opinion the UK support for this engine was not satisfactory.

Discussions with ABB Alstom Power (then European Gas Turbines) revealed that it was undertaking a development program to provide dual-fuel DLE capability for the

* Source: [4-4] Extracts from 2000-GT-112, P. Taylor, R. McMillan, and D. Baker, "Dual Fuel DLE Typhoon Commercial Operating Experience and Improvement Upgrades." At the time of the paper's release in 2001, the Tempest engine (initially European Gas Turbine) was part of ABB Alstom's (later ABB Alstom became Alstom) product line. It is now part of the Siemens engine product line. The table at the front of this book, as well as the Siemens Power Generation website, provides a cross-reference for model numbers.

Typhoon. At this stage ABB Alstom Power was anticipating availability by summer of 1998 with guaranteed NO_x emissions of 25 ppmvd on gas and 85 ppmvd on gas oil (at 15% O_2). With this in mind Boots approached HMIP and gained dispensation to run uncontrolled engines until September 1, 1998, by which time the turbines must be retrofitted with DLE technology. This requirement was subsequently incorporated as an improvement notice into the energy center's IPC authorization to operate.

A contract for three 4.9 MWe Typhoons was awarded to ABB Alstom Power, which included the retrofitting of DLE technology when commercially available. In view of this ABB Alstom Power brought forward the development program in order to meet the agreed time scales.

The three Typhoons were installed during the summer of 1996 and commissioned in the autumn with the new energy center fully operational by the end of the year. Each of these has since accumulated over 20,000 operating hours.

Dual DLE Burner ABB Alstom Power has successfully completed over 750,000 operating hours with its G30 dry, low-emission combustion technology on Typhoon 4.35 MWe, 4.7 MWe, 5.05 MWe, and 5.25 MWe gas turbine engines. The single-shaft generator sets have guaranteed NO_x and CO emissions of less than 25 ppmvd when operating on natural gas fuels and less than 50 ppmvd when operating on distillate 2 fuels (at 15% O_2). The DLE combustion system was designed to retrofit into existing diffusion flame Typhoon engines in minimum time and with minimum operating impact. The units have been designed to offer a dual-fuel starting capability for greater customer flexibility, particularly on offshore installations where initial commissioning requires power without gas fuel supplies. CO emission (below 50 ppmvd) turndown to 70% engine load on gas fuel and 65% on liquid is achieved using modulation of the compressor variable guide vanes.

Three main combustor parameters are measured to monitor the engine and combustor health. These are:

- Pilot tip temperatures. Each pilot burner assembly has a thermocouple positioned behind the pilot burner front face; this is used for ignition recognition and ensures that the face temperature is acceptable.
- Combustor dynamics. Engine flame front pressure fluctuations are monitored using a PCB transducer connected to the combustor pressure casing.
- Exhaust gas temperature profile. There are a number of thermocouples situated downstream of the turbine exit to monitor the gas temperature for deviation from normal.

Burner Concept Design. The premix burner design consists of 12 radial inlet slots configured to allow a predetermined amount of air into the main head of the combustor can. The air swirler inlet vanes are of the same dimensions and are positioned in order to create a swirling vortex. A considerable amount of analytical design (including CFD) and test evaluation was carried out in order to determine the correct number of swirler vanes, their thickness, the size of air passage way, and the elimination of any recirculatory zones or wakes from the trailing edge of the "cheese slice" shapes. Substantial testing of this design was carried out in order to optimize fuel placement while avoiding undesirable factors like fuel impingement on combustor walls and burner slot faces. The optimization of these parameters led to a design that is inherently safe against flashback, and since the main fuel supply is introduced into these "mixing slots" with their high velocities, excellent homogeneous air fuel mixtures exist at the operating condition. Production of a cone-shaped flame that is anchored onto the pilot face therefore produces a simple while robust flame. Anchoring of the flame on the pilot face allows for acceptable levels of combustion noise; low levels of combustion noise when operating on either gas or liquid fuel produce an advantageous environment for combustion hardware life and lead to long-term durable turbine usage.

Dual-Fuel DLE Burner Configuration. The dual-fuel DLE burner is composed of a two-stream gas and two-stream liquid fuel system, both systems being operated by independently controlled electrically actuated valves. The premixed main fuel flame is supported during starting and transients with a pilot (gas) or primary (liquid) fuel input, which can be separately configured at full load for various ambient temperatures or local emission requirements. Cross-light tubes have been removed from the DLE combustion cans in order to limit emissions; this gives a requirement for individual ignitors mounted in the pilot body which emerge onto the pilot face. Purging systems assist in keeping non-operational pilot fuel injector ports free from blockage. The main fuel injector ports, due to their position within the burner swirler, do not require any purging.

A single air-assisted primary nozzle is used in each combustor to initiate liquid fuel ignition and to support the main premixed fuel during running.

The primary or pilot flame assists the premixed zone stability during engine transients and can be adjusted using the control panel in order to adjust emissions for the local requirements when running. This task is carried out while monitoring pilot tip temperatures, as pilot split variations can have a marked effect on the component temperature. Configurable maps allow pilot percentage changes to be made easily and quickly. This is normally carried out during final commissioning with continuous emission monitoring (CEM) and combustor dynamic monitoring (FFT) equipment.

DLE Retrofit During the dual-fuel DLE program Boots was kept informed of the development progress and attended regular update meetings with ABB Alstom Power. Originally Boots had hoped to embark on the retrofit program only when the technology was commercially available and proven in the field. However, due to the September 1, 1998 deadline it became necessary for Boots to commit to being the first site to have dual-fuel DLE installed on a Typhoon. This decision followed extensive talks with ABB Alstom Power and relied on the successful demonstration of dual-fuel operation. However, at this stage work was still progressing on fuel changeover at the ABB Alstom Power site in Lincoln.

While a DLE retrofit can take place without core removal, the retrofit at Boots was planned to coincide with the planned two-year service; therefore a rolling core change-out approach was taken. Each change-out was planned to take eight weeks although it was accepted at the time this was very optimistic as the retrofit involved significant work for both ABB Alstom Power and Boots.

The ABB Alstom Power scope of work included:

- Rolling change-out of the engine core with a core fitted with the DLE combustion equipment
- Replacement of the existing fuel systems with the new DLE fuel system
- Control hardware and software upgraded to DLE standard, including additional I/O

The Boots scope of work included:

- Installation of additional interface cabling between the skid and turbine control module
- Modifications to the fuel supply lines
- Installation of an extensive vents and drains system
- Provision of additional compressed air supplies for purging
- MCC modification associated with change from dc to ac ignitors
- Re-engineering of the MODBUS interface to the station distributed control system and the associated turbine graphical displays

The first engine (GT1) was shut down on March 27, 1998 to facilitate the conversion. During this time electricity had to be imported due to the lost generation and additional steam had to be raised on the package boilers. This had a significant financial impact on the business, and as such, it was essential that the work was completed in minimal time.

The retrofit work on GT1 was completed on May 22, 1998, and commissioning started with the first start on gas fuel occurring towards the end of May. Subsequent work demonstrated turbine operation on both gas and liquid fuel including load shed and fuel changeover. However, at this stage ABB Alstom Power were unable to demonstrate liquid starting successfully due to liquid-air-assist supply-side contaminants. Emissions performance was demonstrated successfully using mobile continuous monitoring equipment set up on site by ABB Alstom Power, the results of which were compared with Boots-installed continuous monitoring equipment. On June 10, 1998, Boots accepted GT1 despite the liquid fuel starting restriction for a seven-day reliability run. Following successful completion of the reliability run, GT2 and GT3 retrofits followed.

The final turbine was shut down on August 28, 1998, just prior to the September 1 deadline after which Boots IPC authorization did not permit operation of uncontrolled engines. The work had taken longer than planned, particularly on GT1 due to a number of reasons, including:

- Delivery of essential new DLE parts
- Physical limitations of technicians working within the turbine skid
- Additional vents and drains system
- Turbine exhaust gas leakage from waste-heat boiler ductwork. This was not attributed to the turbine but delayed commissioning of GT1.

During the retrofit program it became apparent that there was also a reliability problem with starting on gas fuel. ABB Alstom Power were able to assure Boots that the gas starting reliability would be resolved by the implementation of a software change that had been proved on the test bed and was undergoing final software validation. In the meantime, starting could be achieved but required the engine to be closely monitored by a field technician. This never affected operations but resulted in ABB Alstom Power maintaining a daytime site presence to ensure high turbine availability.

Commercial Operations Post Retrofit

Gas Starting Reliability. The gas starting reliability was resolved by the installation and commissioning of revised software modules incorporating a new starting philosophy and an improved method of detecting "flame on."

Restriction of Liquid Operation. Problems with liquid running meant that operation was restricted to gas fuel in November 1998. Boots has an interruptible gas contract and as such is susceptible to gas interruptions during the winter months, particularly January and February. It was therefore essential to have liquid fuel operation available for this period. Careful planning between Boots and ABB Alstom Power allowed the remedial work to be carried out during weekends in December and during the Christmas holiday period. This ensured liquid fuel operation was available for January while minimizing the financial implications of the turbine outages. Liquid starting was still not available, but reliable fuel changes could be easily achieved.

Liquid Starting. Initial observed starting problems were traced to contaminant blockage of the air-assist system. Following successful testing and validation of an improved purge system, retrofit kits were installed to the three turbines during September 1999 with the work on each turbine being completed over successive weekends. Currently during summer weekends only two turbines are required for optimal economic operation, the work was therefore completed without effecting commercial operation of the energy center.

Availability and Reliability. As has been discussed a number of issues were encountered under normal site operating conditions with the dual-fuel DLE systems that resulted in lost generation while problems were investigated and subsequently resolved. In the majority of instances Boots was able to accommodate the required outages for the remedial work during weekends or holiday periods when the commercial effects could be minimized. Putting aside the outages required to accommodate the remedial work availability has been high and running reliability has been excellent with no notable shutdowns due to the DLE technology.

Emissions. The reduction in total NO_x mass emissions from the three Typhoons at the Boots energy center is shown in Figure 4–59. The graph shows clearly the significant environment improvement that the retrofit has achieved. An additional important benefit of the retrofit is the absence of any visible plume from the chimney on both gas and liquid fuel.

Solutions to Initial Operating Issues. Initial running of the dual-fuel units in the field under normal operating schedules highlighted several combustion shortcomings resulting in undesirable fuel leakages of the main and pilot burners. The following is a description of the findings and the work carried out to overcome the operational difficulties.

Main Air Swirler—Issue 1. The main burner with its integral gas and liquid fuel galleries is a multicomponent part. The various assembly bolts have been carefully positioned to hold the combustor can and retain the fuel gallery integrity. During production works testing the cores demonstrated repeatable and stable fuel changeovers in both directions with no deterioration in emissions. During extended site running it was noticed that the emissions were exhibiting a step change in the NO_x after consecutive fuel changeovers. Inspection of the main fuel burner highlighted that the gas gallery seal was leaking due to distortion of the swirler plate base. This led to premature gas injection upstream of the air swirlers and undesirable misplaced fuel positioning. NO_x emissions would gradually increase with successive fuel changeovers and pilot tip temperatures would rise. Operational limitations were due to high tip temperatures and the units were restricted to running on gas fuel.

Solution to Issue 1. Revised manufacturing heat treatment methods and a more comprehensive leak check

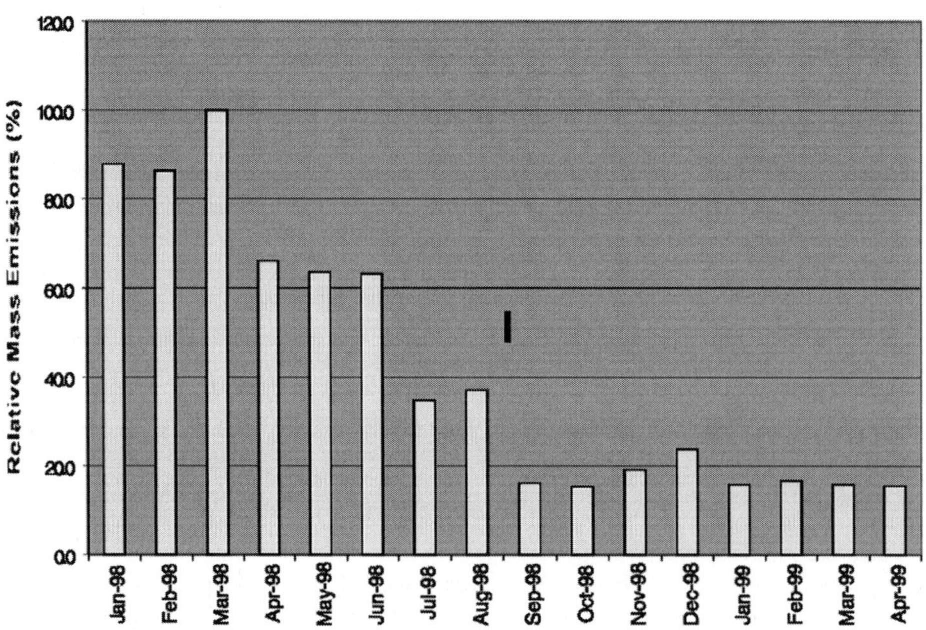

FIGURE 4–59 Total NO_x emissions for the three units before, during, and after the retrofits from January 1998 to April 1999. See text for retrofit dates [4-4].

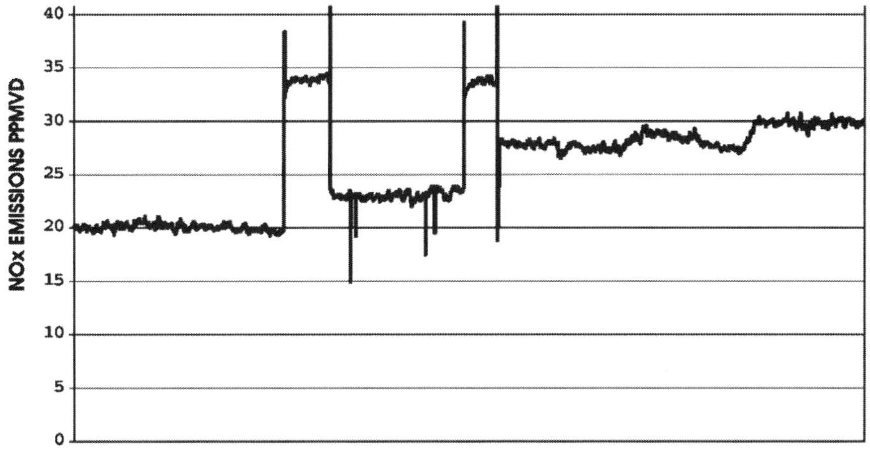

FIGURE 4–60 Several fuel changeovers showing the deteriorating NO_x emissions due to swirler plate warping and poor fuel placement over a time of four hours. Gas operation initially producing 20 ppmvd gradually increasing to 30 ppmvd after two liquid runs (~34 ppmvd) [4-4].

process were implemented; successful and repeatable emissions have now been demonstrated over a long period of operation, as shown in Figure 4–60.

Pilot Liquid Fuel Lances—Issue 2. The pilot burner houses both gas and liquid fuel galleries for each combustion chamber. Gas is channeled to the outer diameter of the pilot while the liquid is transported through lances to the pilot face. Initial design consideration for the lances suggested a screwed connection, which would simplify exchange and overhaul; however, repeated problems were encountered with leakage at the base of the lances (Figures 4–61 and 4–62). Subsequent contamination of the pilot tip thermocouple with liquid fuel would lead to very high indicated tip temperatures due to boiling off of the fuel just beneath the pilot face. Liquid operation would be possible up until the point where the lances would leak; this problem occurred randomly.

Solution to Issue 2. In order to overcome the leakage the lances were brazed in during the manufacturing process, instead of relying on a screwed connection. A sealing compound was also applied between the manifold block and the pilot cap to stop the hydraulic ingress of liquid fuel into the tip thermocouple hole.

Air-Assist Supply—Issue 3. Initial commissioning of the liquid fuel system showed reliable ignition and successful liquid starting capability. Continuous operation with intermittent water washing over a period of several months would see a reduction in the reliability of the liquid ignition performance, while fuel changeovers were operational and unaffected. Subsequent examination of the burners revealed that the air-assist porting just prior to the liquid entry into the combustor chamber (Figure 4–63) had accumulated debris from the instrument air or compressor

FIGURE 4–61 Gas burner manifold leakage during calibration and pressure test [4-4].

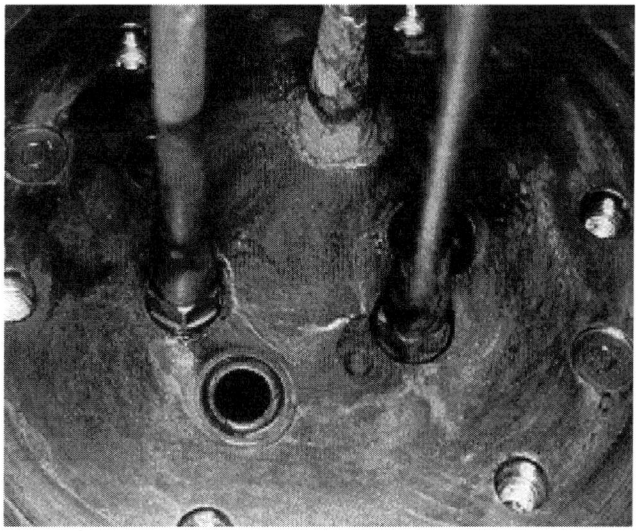

FIGURE 4–62 Liquid fuel leakage from the base of the lances led to contamination of the pilot tip thermocouple and coking or blockage of the air-assist-supply porting [4-4].

FIGURE 4–63 Site analysis using visual inspection of the pilot ignition lance, with operational air assist and water, would indicate the condition of the air supply porting and hence atomization performance [4-4].

purge system. Subsequent analysis showed compressor dirt and pipe manufacturing debris would travel along the purge system after continuous running and deposit itself in the critical swirl entry ports seriously impairing the atomization performance.

Solution to Issue 3. The development work carried out to overcome this difficulty centered on an improved purge system using finned cooling pipe lengths, which offered a reduced pressure drop and used an inline main filter and individual last-chance burner filters. In stopping the debris from reaching the atomization entry ports, it was found that liquid ignition reliability did not deteriorate with running time. Implementation of this improvement has now been completed on all three units.

Starting Improvements—Issue 4. Original ignition philosophies of increasing engine crank speed while inputting a constant fuel flow showed that, during adverse ambient temperature changes, the unit would miss the ignition window. This was easily corrected by fuel light flow adjustments; however, it did require a field support visit and missed valuable running time for the customer.

Solution to Issue 4. A new starting philosophy was developed using a fixed crank speed and a variable fuel flow ramp, the range of fuel flows was set large enough to counter the air density changes encountered at adverse weather sites. Can "flame-on" monitoring using the pilot tip thermocouples showed a marked increase in light-up reliability; previous software attempted to interpret power turbine exit temperatures and to equate this rather transient condition to can lit indications. Ignitor durability improvements and revised gas valve drive couplings and calibration methods also improved startup repeatability. These improvements have now been installed and have led to increased start reliability.

Combustor Dynamics Measurements. Acceptable combustion dynamics or flame front pressure oscillations are critical if mechanical integrity of the combustion hardware is to have a long life. Mechanical life is closely related to the amount of energy produced by a combustion system, by using a relatively small and robustly constructed combustion can with unique design features, acceptable levels of combustion noise have been demonstrated.

Adjustment of the amount of pilot or primary fuel inputs assist in stabilizing the flame, adjusting the pilot face temperatures, and changing the level of emissions. Optimum settings for all three parameters can be achieved during final site commissioning. When operating on gas fuel the pilot is set up in order to just meet emission obligations, with typical pilot splits around 7–10% of total fuel flow. The split of the pilot is set at these relatively high levels to ensure all year round flame stability and good transient capabilities. When operating on liquid fuel the primary fuel split is typically set around 4–6% of total fuel flow. Pilot tip temperatures when operating on liquid exhibit a more sensitive characteristic than when operating on gas with respect to pilot/primary adjustments.

Figures 4–64 and 4–65 show typical plots of the readings obtained while running on gas and liquid fuel at full load. Dynamics measurements while running at full load with gas fuel generally exhibit sub-100 mpsi peaks while with liquid fuel sub-300 mpsi peaks are evident.

Operating Milestones. At 12,000 hours commercial operation (December 1999) on the lead engine with ABB Alstom Power's dual-fuel DLE system found the following results:

- Demonstrated NO_x emissions of 20 and 30 ppmvd at 15% O_2 on gaseous and liquid fuels, respectively. This allows for emission guarantees for NO_x of 25 ppmvd at 15% O_2 when operating on gas and 50 ppmvd at 15% O_2 when operating on liquid fuel.
- Operation on both fuels over total load range with acceptable pilot tip temperatures and combustor dynamics.
- Establishment of combustion hardware life and life extension programs.
- Establishment of commissioning procedures for future DLE dual-fuel units.

In summary, ABB Alstom Power's G30 dry, low-emission Typhoon is a dual-fuel system that is capable of delivering around 30 ppmvd NO_x while running on liquid fuel. Reliable fuel changeovers and liquid starting capability have now been allied to a proven low-emission

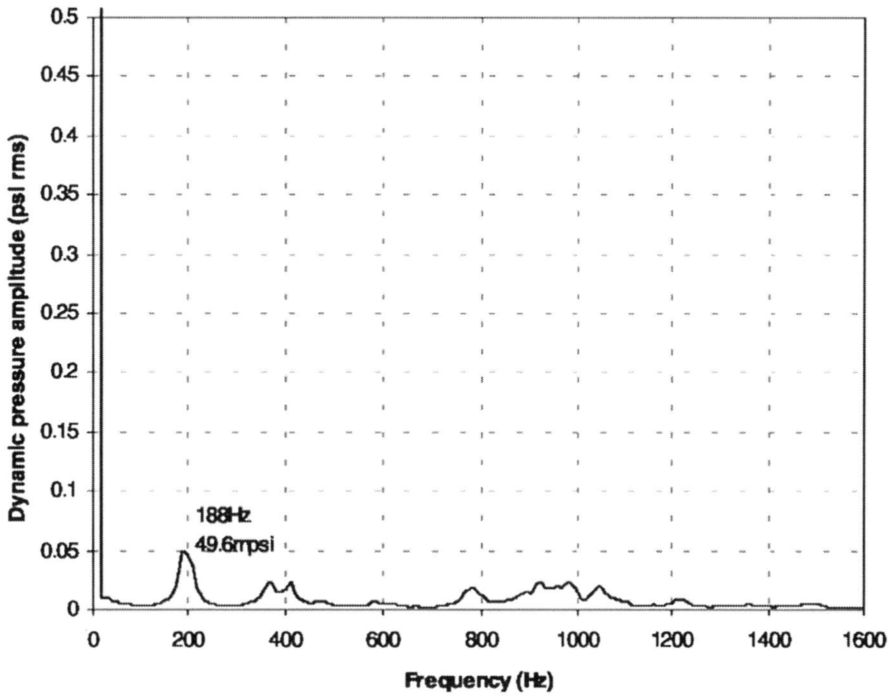

FIGURE 4–64 Flame front pressure oscillations at full-load conditions with gas fuel measured at the engine casing [4-4].

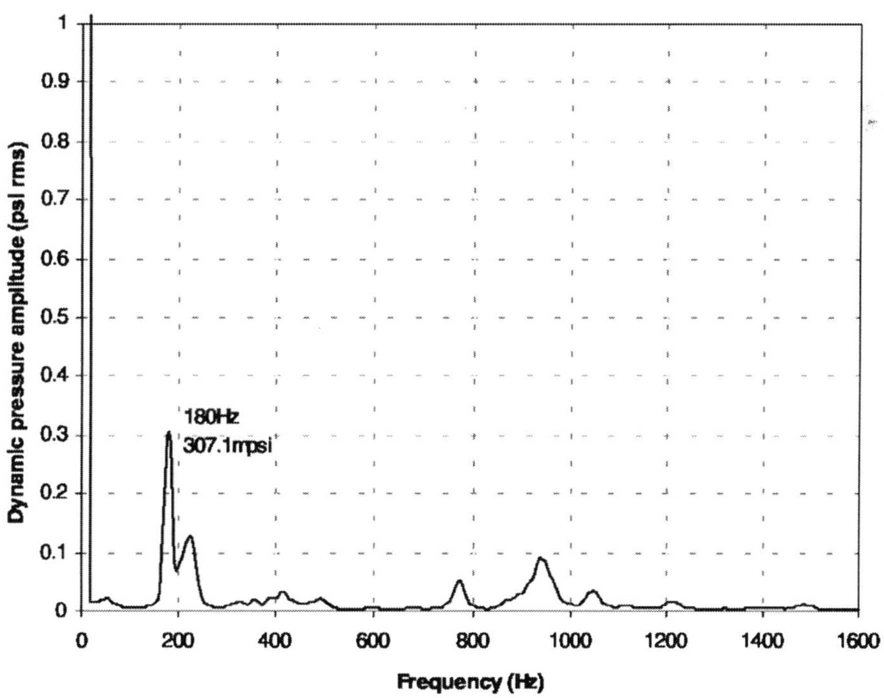

FIGURE 4–65 Full-load liquid fuel operation casing dynamics measurement showing characteristic flame frequencies [4-4].

gas system in a compact power generation unit. The DLE retrofit has demonstrated the performance of the DLE technology and given substantial reduction in NO_X mass emissions at the Boots site. As can be expected with the introduction of such technology for the first time, a number of issues arose. These presented ABB Alstom Power with a number of technically interesting challenges, solutions for which have been validated on the

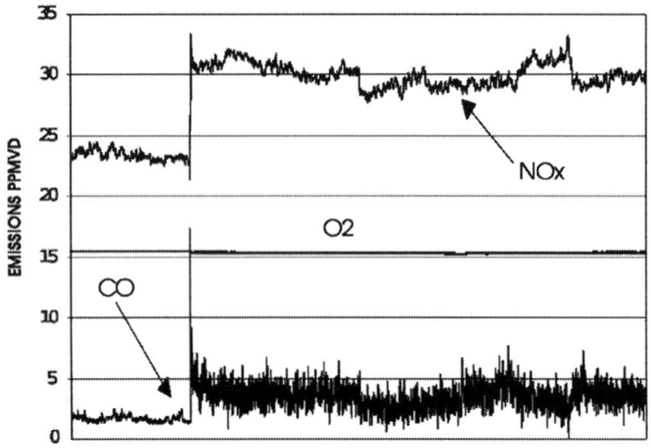

FIGURE 4–66 Gas-to-liquid fuel changeover and subsequent full-day operation, NO_x, and CO emissions [4-4].

core engineering's development facilities. The major achievement of this work has been to demonstrate that very low emissions operating on distillate fuel without water or steam injection is a commercial reality (Figure 4–66).

Case Study 3: Another DLE Application (Former Model Designation* Tempest) * Following the experience gained on the Typhoon generator set, Alstom Power's generic dual-fuel dry, low-emissions combustion system has been successfully developed for operation on the Tempest generator set. The first engine fitted with the full DLE dual-fuel system entered service in October 2000.

NO_x emissions levels of less than 30 vppm at 15% O_2 have been demonstrated on distillate fuel. This section describes the development of the dual-fuel DLE system on the Tempest as well as the commercial operating experience gained to date on both the full dual-fuel DLE and liquid standby DLE systems. The commercial operating experience will cover areas such as operating concept, liquid starting reliability, and site issues encountered. The lead Tempest gas-only DLE units have successfully completed 22,000 hours of operation.

Emission guarantee levels are set at less than 25 ppmvd on NO_x down to 65% load when operating on gas fuel and 50 ppmvd on NO_x down to 60% load when operating on liquid fuel. CO turndown guarantee levels are 50 ppmvd down to 65% load on gas fuel and down to 60% on liquid fuel.

Development of Dual DLE Tempest. The Tempest gas turbine engine was launched in 1995. The engine is a single-shaft configuration producing 7.7 MWe for the electrical power generation market. Initially offered with either conventional (diffusion flame) gas-only/dual-fuel or gas-only DLE(3), applications for dual fuel operation with DLE configuration were met with gas DLE and a liquid capability, but without any emissions compliance, referred to as gas DLE + LSB (liquid standby). Table 4–1 shows the Tempest's operating conditions.

High-Pressure Rig Test Work. By the end of 2000 the Tempest fleet consisted of 38 engines, 20 of these configured for DLE operation. To recognize the need for liquid fuel capability, the requirement for a dual-fuel DLE variant was confirmed.

The Tempest dual-fuel DLE combustion system consists of six reverse flow canular combustion chambers that are connected to the center casing by means of six individual transition ducts. The dual-fuel version uses the same combustion cans and transition ducts as the gas-only version. Figure 4–67 illustrates the pilot burner and main burner including liquid core comprising the main liquid nozzles, and combustion chamber.

As with the gas-only version, the dual-fuel combustion system uses a lean premix burner to control the flame temperature at levels that limit thermal NO_x formation. A central pilot burner is used for ignition and flame stability.

Prior to engine test development of the combustion system, a series of tests were performed on a single combustor high-pressure test facility. The main objective of the high-pressure rig work was to establish the optimum size and configuration of the liquid fuel nozzles as well as establish an operating regime with respect to pilot split and combustion dynamics.

TABLE 4–1 Tempest Operating Conditions [4-5]

Fuel	Natural Gas or Diesel
Electrical output	7.7 MW
Turbine speed	14,010 rpm
Compressor pressure ratio	13.7:1
Exhaust gas mass flow	29.5 kg/s
Exhaust gas temperature	540°C
Turbine entry temperature	1142°C

* Source: [4-5] Extracts from 2001-GT-0076, Cramb and R. McMillan, "Tempest Dual Fuel DLE Development and Commercial Operating Experience and Ultra Low NO_x Gas Operation." At the time of the paper's release, in 2001, the Tempest engine (initially European Gas Turbine) was part of Alstom's product line. It is now part of the Siemens engine product line. The table at the front of this book, as well as the Siemens Power Generation website, provide a cross-reference for model numbers.

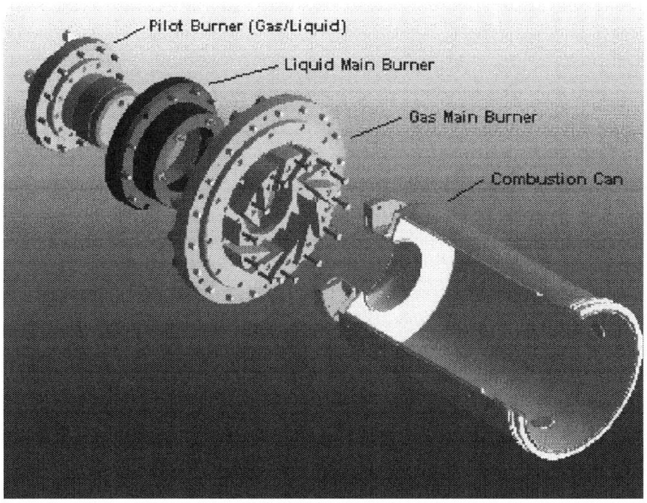

FIGURE 4–67 Tempest dual-fuel DLE combustor assembly [4-5].

The main liquid fuel nozzles were investigated first. The nozzle size chosen was scaled from the Typhoon dual-fuel combustor. This gave the minimum flow number, maximum fuel jet penetration possible without compromising the fuel nozzle passage size. This nozzle size equated to a six main fuel nozzles per burner configuration as with the Typhoon combustor. Using this main fuel nozzle size a variety of tests were conducted using from 3 main fuel nozzles per burner up to 12 nozzles. From this it was concluded that using less than six main nozzles compromised the stability of flame particularly at lower loads, while more than six nozzles gave poor fuel jet penetration which in turn was shown by the high pilot burner tip temperatures and eventual damage of the pilot burner face. Therefore the final configuration of six main fuel nozzles was selected.

The circumferential position of the liquid main nozzles was also investigated by testing two configurations. The first was scaled from the Typhoon combustor and the second to give a similar circumferential position to that of the proposed Cyclone dual-fuel combustion system. The test proved that the circumferential position of the main fuel nozzle in a Tempest-size combustor had no effect on the stability, emissions, or burner component temperatures.

The liquid standby version of the Tempest has five liquid fuel nozzles in the pilot burner. All nozzles have the same flow number with one being an ignition nozzle with air assist. For the starting point for the dual-fuel testing the liquid standby ignition nozzle was used. This nozzle had already been proven successful in engine operation for ignition and following the rig tests proved satisfactory for flame stabilization in conjunction with a six main nozzle per burner configuration. A lower flow number ignition nozzle (35% lower flow number) was also tested with the view to improving smoke emissions and light-up reliability by having improved atomization with the higher-pressure, lower-flow-number nozzle. Light-up reliability and smoke emissions at pull-away and low-load operating conditions are not measured on the high-pressure (HP) test rig due to the fact that the engine operating conditions cannot be simulated close enough to give reliable information. This proved unacceptable for engine operation as the amount of pilot required at reduced load to ensure flame stability would not be achievable due to a limit on skid-edge fuel pressure. Therefore the final configuration of one ignition nozzle (standard nozzle) per pilot burner with the same flow number as the liquid standby ignition nozzle was selected.

Engine Development Test Work. A Tempest core was fitted with a set of dual-fuel pilot and main burners, based on the best configuration proved on the HP rig. No alterations were required for the fuel system to accommodate the dual-fuel testing, as the existing liquid standby system was compatible, which gives the option for fairly easy conversion of liquid standby systems to full DLE dual fuel.

The initial testing was performed on gas fuel in order to compare the performance of the dual-fuel version with the gas-only version. Using the standard settings for light-up, pull-away, and load running from the gas-only and liquid standby engines, the full dual-fuel DLE hardware gave good startup and pull-away reliability. Emissions, burner component temperatures, and combustion dynamics were then measured throughout the load range, with all parameters giving similar levels to those expected. NO_x emissions at full load with a standard pilot split was 22 ppmvd @15% O_2.

Completion of the initial gas fuel operation was followed by light-up performance optimization on liquid fuel. As the ignition nozzle being used was identical to that used in the liquid standby burners, the liquid light-up settings, for example fuel pressure, air-assist pressure, fuel ramp rates, and pilot split, were all known. It was found that by increasing the air-assist pressure the ignition reliability was reduced. One reason for this reduction in reliability on ignition with increasing air-assist pressure was noted when visually checking the spray quality on the calibration rig. When the air-assist pressure was increased the spray cone narrowed and lifted further away from the area where the ignition source was, thus reducing the chances of successful light-up. One alteration to the pull-away on the full dual-fuel version was the need to increase the amount of pilot split from the corresponding value on the standby engines. This was to allow for good flame stability and acceleration of the engine to idle.

At idle, with the VGVs in their normal open position, the pilot tip temperatures reached mid-800°C, but by closing the VGV on the compressor the pilot tip temperatures reduced by approximately 100°C. This also resulted in improving the CO and smoke emissions levels. The smoke levels recorded indicated that above 50% load UK Bacharach smoke number was 1 and below 50% load was between 3 and 4. Variations in pilot fuel split at idle conditions were completed to optimize emissions. The engine load was increased in stages; at each point the pilot split altered to allow the optimization of emissions, combustion dynamics, and burner component temperatures. This test allowed the VGV position to be modulated manually, keeping them closed until a certain operating temperature was reached then opening them up to maintain this temperature until fully open.

The exhaust emissions achieved at full load showed NO_x levels were 34 ppmvd at 15% O_2 (Figure 4–68), with CO levels remaining below 50 ppmvd at 15% O_2 down to 60% load (Figure 4–69). These are compared in Figure 4–70.

Operational Experience, Gas Only. To date the Tempest has accumulated over 150,000 hours operating the DLE combustion system. The lead engines have both to date completed over 22,000 hours. Regular inspection of the combustion system has indicated the design life of the components has been met. Main burners, combustion

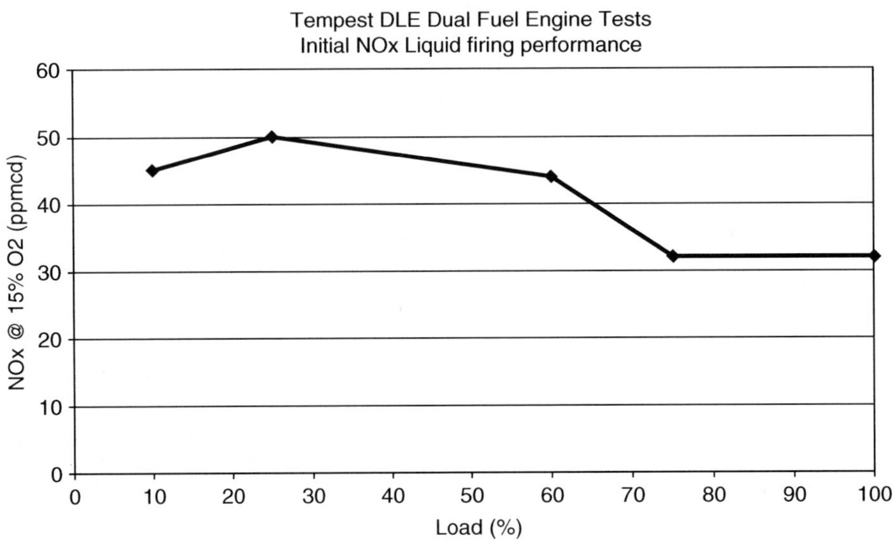

FIGURE 4–68 NO_x emissions across load range when firing liquid fuel on the development engine [4-5].

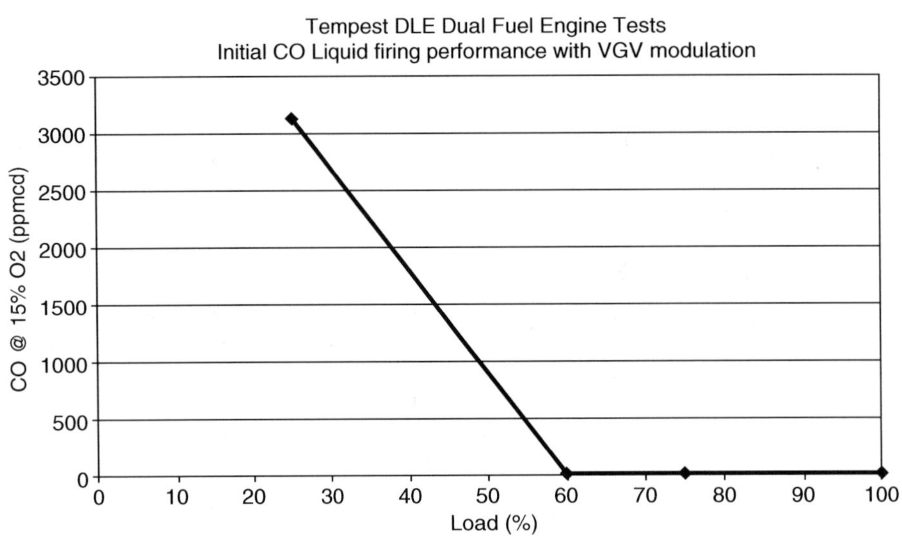

FIGURE 4–69 CO emissions across load range with VGV modulation when firing liquid fuel on the development engine [4-5].

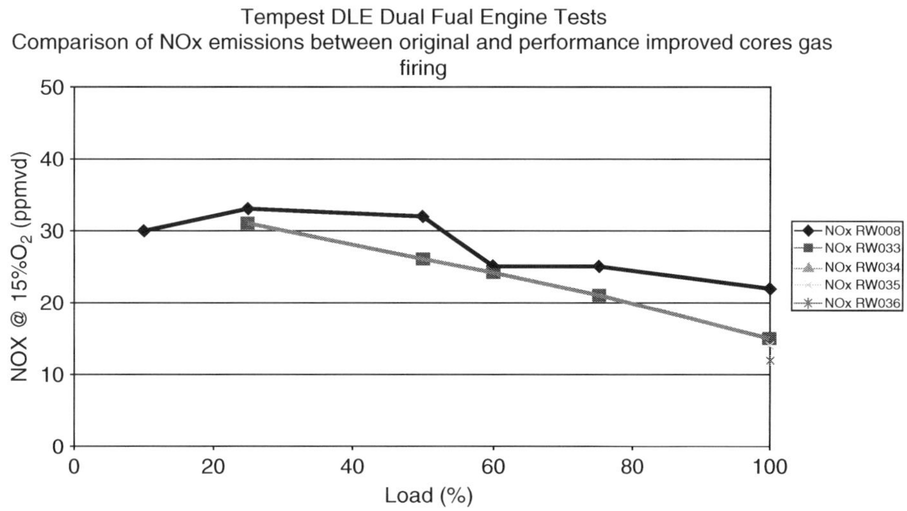

FIGURE 4–70 Comparison of NO_x emissions performance between the two standards of engine configuration firing liquid fuel [4-5].

cans, and transition ducts are in excellent condition and expected to easily achieve design life. Pilot burner life on these engines has also been extremely encouraging, with 16,000 hours being achieved without component distress being identified. Emissions performance on these engines has also been consistent over the engine life, with guarantee levels still being met comfortably after 22,000 hours, based on there being no change from commissioning through to independent measurements completed at 16,000 hours.

Liquid Standby. Four gas DLE liquid-standby Tempest engines in commercial operation have accumulated over 50,000 hours of operation. The lead engines have both to date completed in excess of 18,000 hours. Like the gas-only engines regular inspections of the combustion system have indicated that all components are in excellent condition and are expected to reach the design life. Estimation of pilot and main burner life has though been effected by oil carryover issues on the lead engines. Both engines have gas supplied via an on-site gas compressor. During the initial period of operation problems were encountered with start reliability and load running. On inspection of the combustion and fuel system it was found that the burners had become contaminated with lube oil, which had been carried through the system from the gas compressors. After installation of coalescers downstream of the gas compressors and replacement of the main and pilot burners, the engines have operated without interruption. Startup reliability on both fuels as well as fuel changeover capability has been available since the installation of these coalescers. Both these units are to be retrofitted to full dual-fuel DLE following the 24,000 hour inspection which is planned for summer 2001.

Dual DLE. In addition to the completion of the standard works test on the first production core with dual-fuel DLE combustion equipment fitted, the core was installed and tested as a complete package before being shipped to the site (Figure 4–71). The combustion system performed as expected with good startup reliability on both fuels, load running on both fuels, fuel changeovers, load accept and rejects on both fuels all being established while operating within the normal package layout against the generator. Following installation at the site the engine package was successfully commissioned during September and performed its acceptance run during October. Emissions measurements taken on-site indicated that the unit was running with a NO_x level of around 12 ppmvd on natural gas with a pilot split of 7%. During the first half of 2001 the full load emissions will be optimized to obtain continuous sub-10 ppmvd NO_x operation in conjunction with Alstom's ultra-low-emissions development program. Emissions levels of around 20 ppmvd were measured when running on diesel. Engine light-up reliability on both fuels has so far been met comfortably, with the contractual availability being achieved. At the time of writing this section the engine had accumulated over 2000 hours operation.

FIGURE 4–71 Dual-fuel DLE combustion system installed into Tempest package prior to on-site installation [4-5].

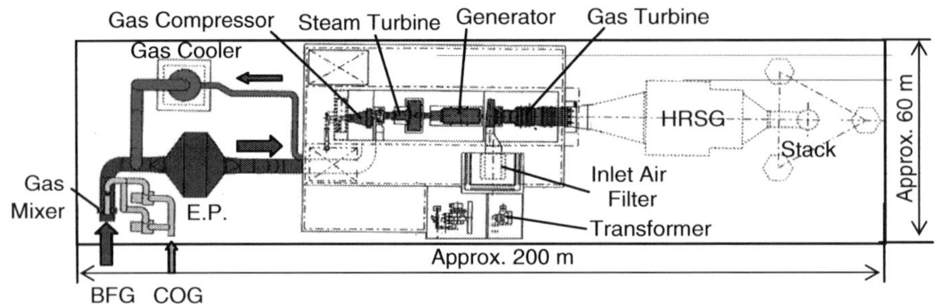

FIGURE 4–72 Overall equipment layout.

In summary, the G30 dual-fuel dry, low emissions on the Alstom Typhoon gas turbine has been successfully applied and developed on the Alstom Tempest gas turbine engine.

This dual-fuel DLE combustion system has been shown to deliver, reliably, less than 30 ppmvd NO_x on liquid fuel accompanied with a NO_x emissions performance on gas fuel of less than 10 ppmvd.

Reliable starting on both fuels as well as fuel change-over capabilities are all proven features of this simple yet durable combustion system.

The dual-fuel Tempest has been introduced into commercial operation using the new modularized package.

Case Study 4: DLE Applications in Gas Turbines (30 MW to 180 MW)* Depending on the amount of by-product gas available and power generation needs, there are three configurations that have been used as of 2007:

- Small gas turbine size (30 MW Class M251)
- Medium gas turbine size (90 MW Class M501DA or M701DA)
- Large gas turbine size (180 MW Class M701F)

Commercial operation of the world's first low-Btu BFG gas turbine (MW-151S) with turbine inlet temperature of 1000°C (1832°F) started in 1982, and was followed by a 145 MW large frame M701D gas turbine operating with BFG commissioned in 1987. After accumulating long-term practical operation both in middle size BFG firing gas turbines and large frame D class gas turbines and combined cycles. The first BFG firing gas turbine combined cycle plant export unit was delivered to UNA in the Netherlands in 1997. MHI further up-rated the low calorie technology to F class gas turbines for BFG firing. The first F class firing BFG with 1300°C (2372°F) turbine inlet temperature started commercial operation for a domestic 300 MW power plant in 2004. This combined cycle includes impressive and unprecedented features such as a 56 meter long single shaft configuration that includes a 87.3 MW centrifugal BFG compressor driven through a step-up gearbox (refer to Figure 4.72).

MHI has manufactured numerous plants burning unconventional by-product gases not limited to blast furnace gases. Other gases included Coke Oven Gas (COG), Finex Oven Gas (FOG), hydrogen-rich refinery gases as well as different types of IGCC.

The result of the long experience and the reliable operation results of these high efficiency units have made MHI a leading manufacturer for gas turbines firing low calorie steel mills by-product gases. Table 4–2 lists MHI low calorie experience, including 15 new low-calorie by-product projects that have been recently commissioned or are under construction for China, Ukraine, and steel companies in other countries. This includes the development of a combined cycle in Korea with a gas turbine burning a mixture of different furnace by-product gases.

Combustor Design Considerations for Low BTU Fuel Application. Compared to burning natural gas firing, the efficient combustion of low calorie fuel presents several difficult challenges:

- Reduced flammability limits → fuel air ratio in the combustion area is critical.
- Reduced flame speed → the flow speed in the combustion area must be reduced.
- Higher combustion air requirements → potentially high combustor metal temperature due to less cooling air.

On the other hand, the thermal NO_x level is low, because of the lower adiabatic flame temperature associated to the combustion of low calorie by-product gases. Therefore water or steam injection for NO_x reduction, or the use of premixed gas firing are not required. During the early low calorie gas burning applications, MHI used a single large silo type combustor in order to achieve suitable combustion in small or mid-size gas turbines.

* Source: Courtesy of Mitsubishi Heavy Industries. Extracts from "Update on design and operating experience of low calorie gas firing gas turbine for steel works," presented by Mitsubishi Heavy Industries at PowerGen 2007.

TABLE 4–2 Experience list of low heating value gas firing

Customer	Application	Start-Up	Model	GT Unit Rating (Incl.G.C.)	Combined Plant Rating	Fuel Main	Combuster Stand-By	
NSC (Nippon Steel Co.) Yahata works	Power supply	1958	–	850 kw	–	BFG (3.2MJ/Nm^3)	–	Single-can
NSC Yahata works	Power supply	1964	–	4000 kw	–	BFG (3.2MJ/Nm^3)	–	Single-can
Sumitomo Metal Co. Wakayama works	Co-generation (WHB,G-M,BLOWER)	1965	MW171	15,000 kw	–	BFG (3.1MJ/Nm^3)	–	Single-can
Shikoku Elec. Pwr Co. Sakaide PS	Combined cycle	1970	MW301	34,000 kw	225,000 kw	COG (3.1MJ/Nm^3)		Multi-can
Mitsubishi Coal Mining Co. Minami Oyubari Plant	Co-generation (with air-preheater)	1970	MW101	9000 kw	–	Coal mine (20MJ/Nm^3)	OIL	Multi-can
NSC Kamaishi works	Combined cycle with existing STs	1982	M151	16,000 kw	23,000 kw	BFG (2.8MJ/Nm^3)	OIL	Single-can
JFC Steel Co. Chiba works	Combined cycle (single shaft)	1987	M701	87,400 kw	145,000 kw	BFG/COG (4.2MJ/Nm^3)	–	Multi-can with BP-V
Mitsubishi Gas-Chemical Co. Mizushima-factory	Co-generation (WHB)	1988	MF111	16,250 kw	–	BFG/COG (10MJ/Nm^3)	OIL	Multi-can
NSC Hirohata works	Co-generation (WHB)	1989	M251	30,200 kw	–	LDG (7.6MJ/Nm^3)	OIL	Multi-can
Nissin Steel Co. Kure works	Combined cycle with existing STs	1989	M251	32,000 kw	50,000 kw	BFG (2.9MJ/Nm^3)	–	Multi-can with BP-V
Nakayama Steel Co. Funamachi works	Combined cycle (single shaft)	1991	M151	15,000 kw	37,000 kw	BFG/LDG (4.2MJ/Nm^3)	–	Multi-can with BP-V
Mizhushima Joint Thermal Power Co.	Combined cycle (single shaft)	1994	M501	86,250 kw	145,000 kw	BFG/M (4.0MJ/Nm^3)	–	Multi-can with BP-V
Fukuyana Joint Thermal Power Co.	Combined cycle (single shaft)	1995	M501	86,250 kw	145,000 kw	BFG/M (4.0MJ/Nm^3)	–	Multi-can with BP-V
NUON Netherlands	Combined cycle (single shaft)	1997	M701	87,400 kw	145,000 kw	BFG/COG (4.2MJ/Nm^3)	–	Multi-can with BP-V
NSC Ooita Works	Combined cycle	2001	M251	31,000 kw	65,000 kw	BFG (2.9MJ/Nm^3)	–	Multi-can with BP-V
Nippon Petroleum Co. Ltd. Negishi Refinery	Combined cycle (single shaft)	2003	M701F	301,000 kw	431,000 kw	Syn gas+N2 (5.9MJ/Nm^3)	OIL	Multi-can

(Continued)

TABLE 4–2 Experience list of low heating value gas firing—cont'd

Customer	Application	Start-Up	Model	GT Unit Rating (Incl.G.C.)	Combined Plant Rating	Fuel Main	Combuster Stand-By	
Kimitsu Cooperative Thermal Power Co.	Combined cycle (single shaft)	2004	M701F	180,700 kw	300,000 kw	BFG/COG (4.4MJ/Nm^3)	–	Multi-can with BP-V
Zhangjiagang Hongchang Plate Co., LTD., China	Combined cycle (multi shaft)	2006	M251	30,000 kw	–	BFG/COG (3.1MJ/Nm^3)	–	Multi-can with BP-V
Aanshan Iron & Steel Co., Ltd., China	Combined cycle (single shaft)	2006	M701F	183,000 kw	300,000 kw	BFG/COG (4.4MJ/Nm^3)	–	Multi-can with BP-V
Maanshan Iron & Steel Co., Ltd., China	Combined cycle (single shaft)	2006	M701DA	91,300 kw	153,000 kw	BFG/COG (4.4MJ/Nm^3)	–	Multi-can with BP-V
Handan Iron & Steel Group Co., LTD., China	Combined cycle (multi shaft)	2006	M251	30,000 kw	–	BFG/COG (3.1MJ/Nm^3)	–	Multi-can with BP-V
Clean Coal Power R&D Co., Ltd. Nakoso	Combined cycle (single shaft)	2007	M701DA	129,200 kw	250,000 kw	Syngas (4.2MJ/Nm^3)	–	Multi-can
Posco, Korea	Combined cycle (single shaft)	2007	M501DA	88,100 kw	146,000 kw	FOG (–)	–	Multi-can with BP-V
Lianyuan Iron & Steel Group CO.LTD	Combined cycle (multi shaft)	2007	M251	30,000 kw	–	BFG (3.1MJ/Nm^3)	–	Multi-can with BP-V
Baotou Iron & Steel Co.	Combined cycle (single shaft)	2007/ 2008	M701DA×2	82,500 kw	1380,00 kw×2	BFG/COG (4.4MJ/Nm^3)	–	Multi-can with BP-V
Anshan Iron & Steel Group Co., Ltd	Combined cycle (single shaft)	2008	M701DA	89,7000 kw	151,000 kw	BFG/COG (4.4MJ/Nm^3)	–	Multi-can with BP-V
Ukraine/OJSC Alchevsk Iron & Steelworks	Combined cycle (single shaft)	2009	M701DA×2	89,700 kw	151,000 kw×2	BFG/COG (4.4MJ/Nm^3)	–	Multi-can with BP-V
Taiyuan Steel Co. Ltd.	Combined cycle (multi shaft)	2008	M251	30,000 kw	–	BFG (3.1MJ/Nm^3)	–	Multi-can with BP-V
Confidential	Combined cycle (single shaft)	2009	M701DA	89,700 kw	151,000 kw	BFG/COG (4.4MJ/Nm^3)	–	Multi-can with BP-V

BP-V; Bypass value.

Multi-cannular was later applied for large capacity gas turbines, as shown in Figure 4—73. This design is favorable because of the smaller combustor surface area compared with the large single can silo type combustor design.

Several design improvements were introduced together with the multi-cannular combustor to achieve suitable combustion, as follows.

Bypass Valve. A bypass valve is equipped on the transition piece, as shown in Figure 4.73. This valve allows regulation of the airflow to the combustion area by adjusting the valve opening improving the combustion characteristic during the gas turbine part load operation. Combustion efficiency in excess of 99% can be achieved from ignition to full load as shown in Figure 4.74. A suitable combustion maintained at load rejection is also achieved by using the combustor by-pass valve.

Flame Speed in Combustion Area. Combustion of low calorie gas exhibits lower burning velocity and lower adiabatic flame temperature. Therefore, longer residence time is needed in the combustion area to achieve flame stability and high combustion efficiency. Compared with combustors designed for natural gas, low calorie combustors are designed with larger cross-sectional area to achieve longer residence time. However, careful consideration of the combustor metal temperature is required. The bypass valve contributes to maintain stable combustion and avoid high metal temperatures in the combustor.

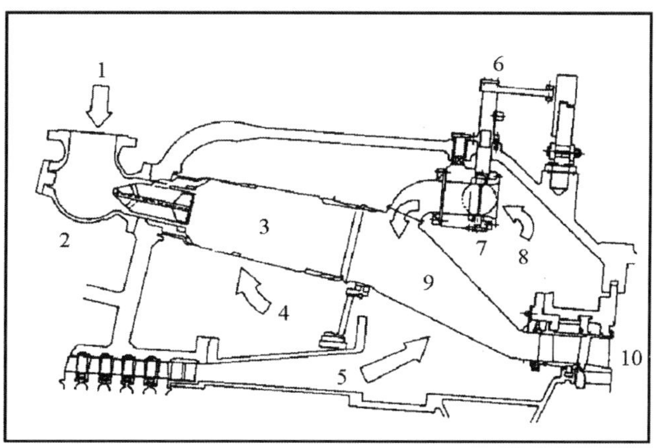

1. BFG+COG Fuel gas
2. Spherical elbow
3. Combustor
4. Air for combustion
5. Compressor discharge air
6. Variable ring
7. Bypass valve
8. Bypass air
9. Transition piece
10. Turbine

FIGURE 4—73 Combustor configuration for BFG firing.

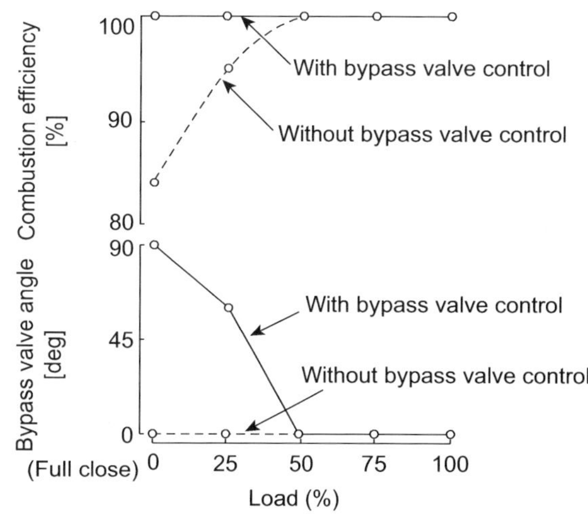

FIGURE 4—74 Combustor efficiency improvement using bypass valve.

Configurations and Field Experience with Low Calorie Fuels. Depending on the amount of by-product gas available and the power generation needs, Mitsubishi applies one of three configurations:

- Small gas turbine size (30 MW Class M251)
- Medium gas turbine size (90 MW Class M501DA or M701DA)
- Large gas turbine size (180 MW Class M701F)

This section describes features of each configuration and presents relevant operational data of examples of each configuration.

Small Gas Turbine Size (30 MW Class M251). As shown in the table above, several small-size low calorie plants have been built in Japan with different class units including MW171, MW, 301, MF111, and M251. The M251 has become the standard Mitsubishi gas turbine for small-size applications.

M251 Series Gas Turbine Combustor with BFG and Pilot COG. An interesting application of an M251 gas turbine burning BFG with COG as pilot fuel started in commercial operation in 2006. This combination was implemented to meet the end-user required operation without fuel oil. Figure 4—75 shows the combustor configuration for this gas turbine. A COG pilot nozzle is implemented at the center of the BFG nozzle in order to maintain stable combustion with the low calorie BFG flame.

Mitsubishi has applied this type of combustor on five units within a BFG calorie range of 700 kcal/Nm^3 ~ 800 kcal/Nm^3.

Field Experience. Special instrumentation was installed during the commissioning period in order to take measurements for confirming performance and reliability. The instrumentation included combustor and transition

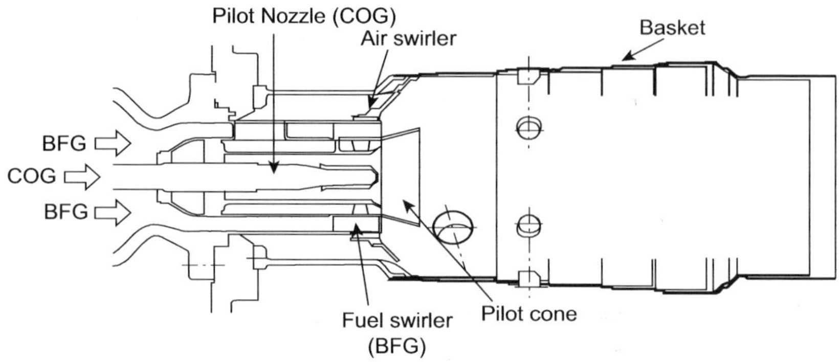

FIGURE 4–75 Configuration of nozzle and basket for BFG and COG pilot.

piece metal temperatures as well as combustor pressure fluctuations measurements. Data confirmed operation below the allowable levels. NO_x levels at base load condition were confirmed to be lower than 25ppm (15%O_2). CO also achieved low level throughout the entire load range.

Medium Gas Turbine Size (90 MW Class M501DA or M701DA). M501DA and M701DA are being used for mid-size low calorie combined cycles. In the last three years, as many as eight units were booked, two of them recently commissioned and the rest are under manufacturing or commissioning.

M501DA Series Gas Combustor with FOG. A M501DA gas turbine burning a non-conventional by-product gas called Finex Oven Gas (FOG) started commercial operation this year. Figure 4.76 shows the configuration of the gas fuel nozzle, fuel-swirler, air-swirler and basket for this plant. In this application, the FOG is mixed with N_2 or COG in order to control the fuel gas calorie. It can achieve suitable combustion from ignition to full load condition using only gas fuel.

Field Experience. Combustor basket and transition piece metal temperature were measured throughout the entire load range. Both temperatures were confirmed within allowable levels. The combustor pressure fluctuations were confirmed to be lower than the allowable level. The emission levels were also monitored. Both NO_x and CO levels were low and within the contractual limits.

Large Gas Turbine Size (180 MW Class M701F). A high efficiency combined cycle based on the M701F gas turbine firing BFG with an inlet temperature exceeding 1300°C was developed and commissioned in 2004 for a Japanese steel factory (Figure 4–77). A similar plant was later built and commissioned in China.

These combined cycle plants include impressive and unprecedented features:

- 50% (LHV base) plant thermal efficiency.
- A multi-can-annular combustor with air bypass valve to allow low caloric gas operation over the entire operating range (from turbine startup to full load operation).

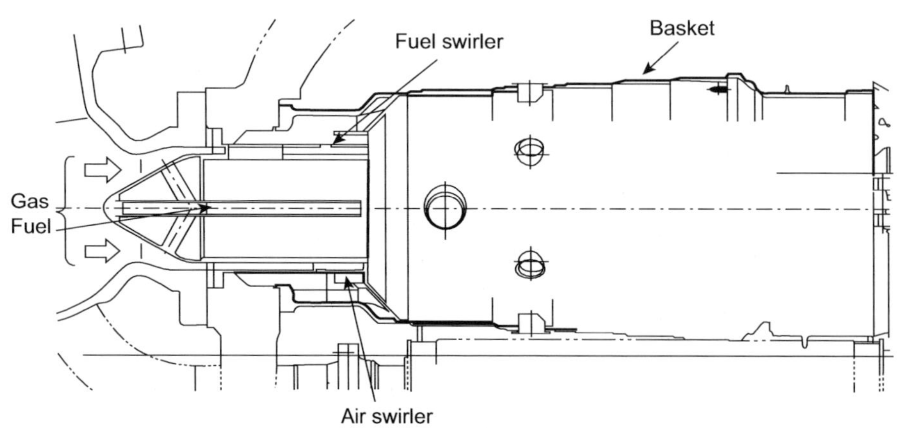

FIGURE 4–76 Configuration of nozzle and basket for FOG.

FIGURE 4–77 Overall plant view.

- The gas turbine, generator, steam turbine, and gas compressor are coupled on a single 56 meter long shaft.
- An 87 MW gas compressor is connected to the single shaft arrangement through a step-up gear.
- In order to prevent mutual interference resulting from longitudinal thermal expansion of the shaft, both ends of the steam turbine shaft are fitted with flexible couplings.
- The plant is started up by the main steam turbine using steam from an existing boiler.

This eliminates the need for startup device.

- The multi-can-annular low NO_x combustor with air bypass valve allows low NO_x operation without steam or water injection.
- A gas decompression device with direct water cooling system, developed by Mitsubishi, is installed to enable recirculation of the high temperature and high-pressure gas
- discharged from the gas compressor outlet to the gas supply line in case of emergency
- or during normal shutdown.
- A full automatic control system makes it possible for one or two operators to control and monitor the plant from a central control room.

*Turbines**

The turbine has the task of providing the power to drive the compressor and accessories and, in the case of engines that do not make use solely of a jet for propulsion, of providing shaft power for a propeller or rotor. It does this by extracting energy from the hot gases released from the combustion system and expanding them to a lower pressure and temperature. High stresses are involved in this process, and for efficient operation, the turbine blade tips may rotate at speeds over 1500 ft per sec. The continuous flow of gas to which the turbine is exposed may have an entry temperature between 850 and 1700°C and may reach a velocity of over 2500 ft per sec in parts of the turbine.

To produce the driving torque, the turbine may consist of several stages each employing one row of stationary nozzle guide vanes and one row of moving blades (Figure 4–78). The number of stages depends upon the relationship between the power required from the gas flow, the rotational speed at which it must be produced, and the diameter of turbine permitted.

The number of shafts, and therefore turbines, varies with the type of engine; high compression ratio engines usually have two shafts, driving high and low-pressure compressors (Figure 4–79). On high bypass ratio fan engines that feature an intermediate pressure system, another turbine may be interposed between the high- and low-pressure turbines, thus forming a triple-spool system (Figure 4–80). On some engines, driving torque is derived from a free-power turbine (Figure 4–81). This method allows the turbine to run at its optimum speed because it is mechanically independent of other turbine and compressor shafts.

The mean blade speed of a turbine has considerable effect on the maximum efficiency possible for a given stage output. For a given output the gas velocities, deflections, and hence losses are reduced in proportion to the square of higher mean blade speeds. Stress in the turbine disc increases as the square of the speed, therefore to maintain the same stress level at higher speed, the sectional thickness, hence the weight, must be increased disproportionately. For this reason, the final design is a compromise between efficiency and weight. Engines operating at higher turbine inlet temperatures are thermally more efficient and have an

* Source: [4-1] Adapted, with permission, from Rolls Royce, *The Jet Engine*, 1986, Rolls Royce plc: UK.

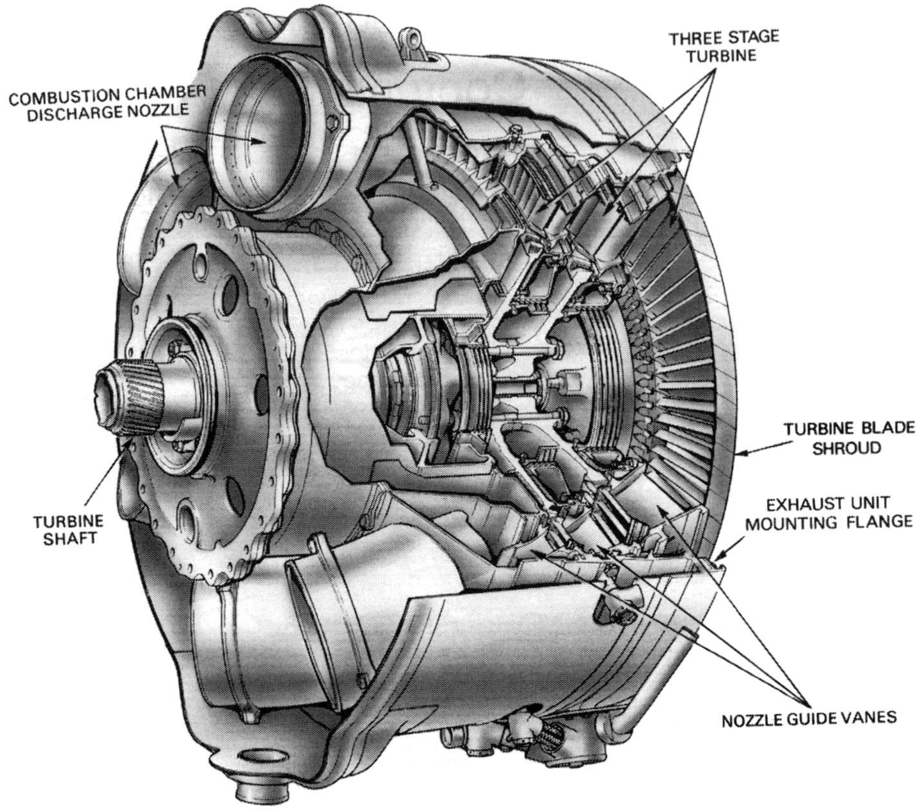

FIGURE 4–78 A triple-stage turbine with a single-shaft system. *(Source: Rolls Royce.)*

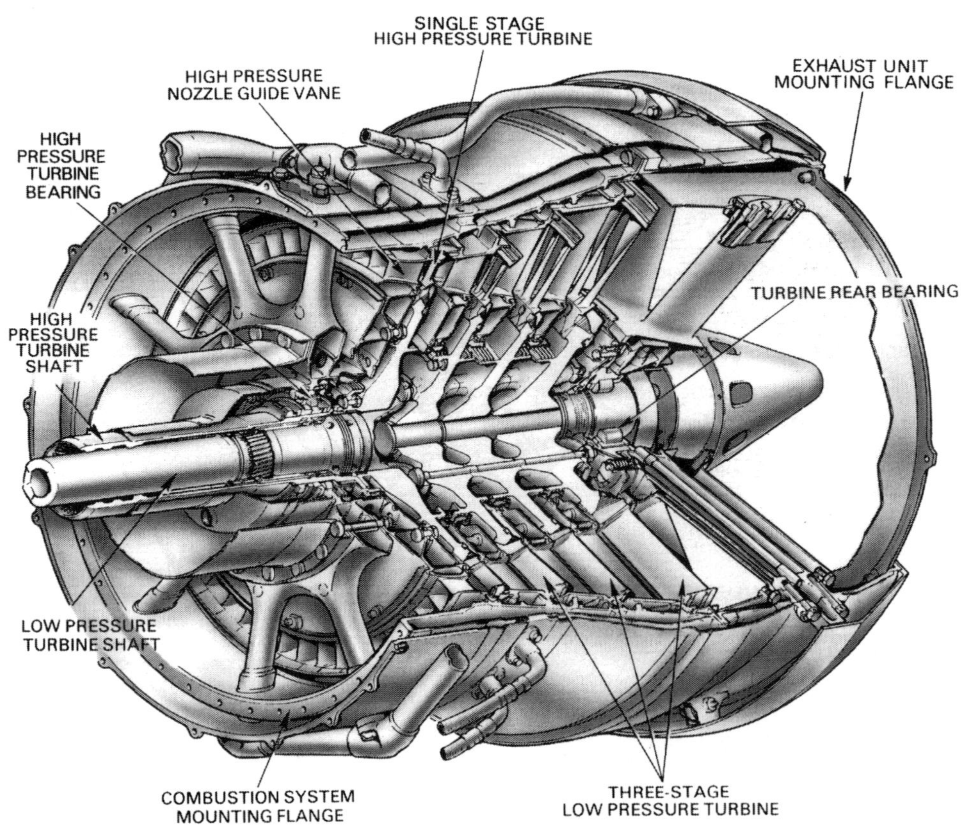

FIGURE 4–79 A twin turbine and shaft arrangement. *(Source: Rolls Royce.)*

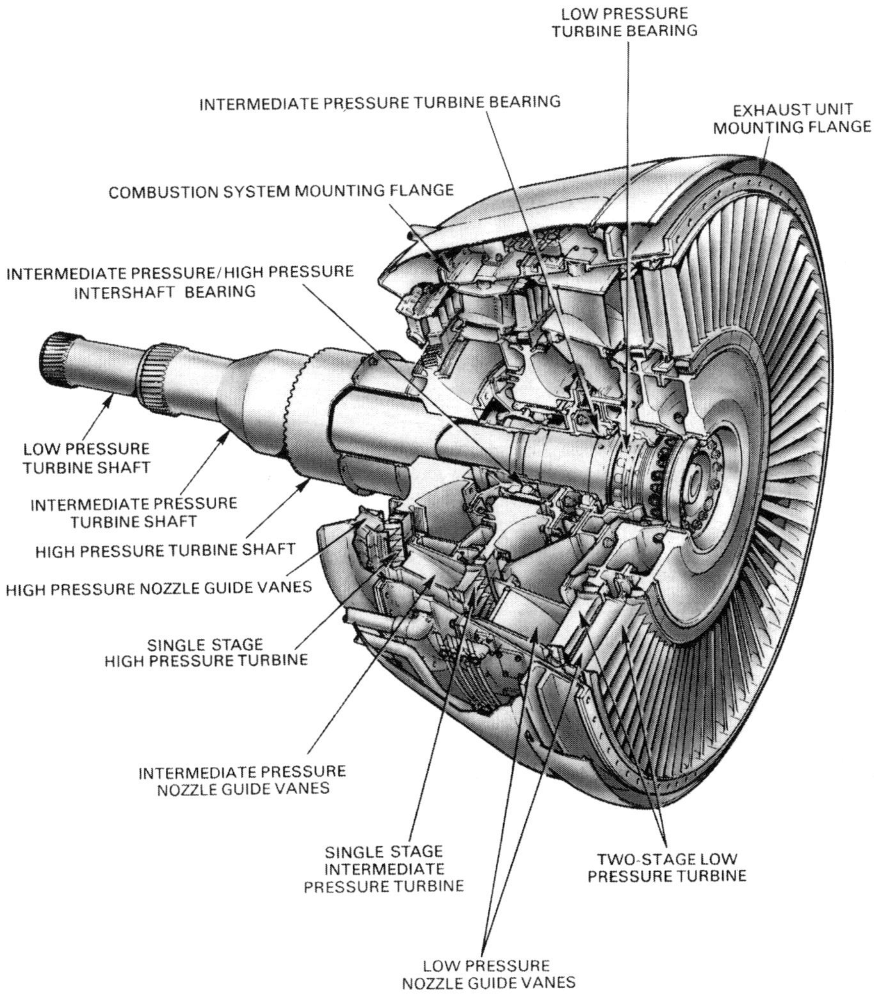

FIGURE 4–80 A triple turbine and shaft arrangement. *(Source: Rolls Royce.)*

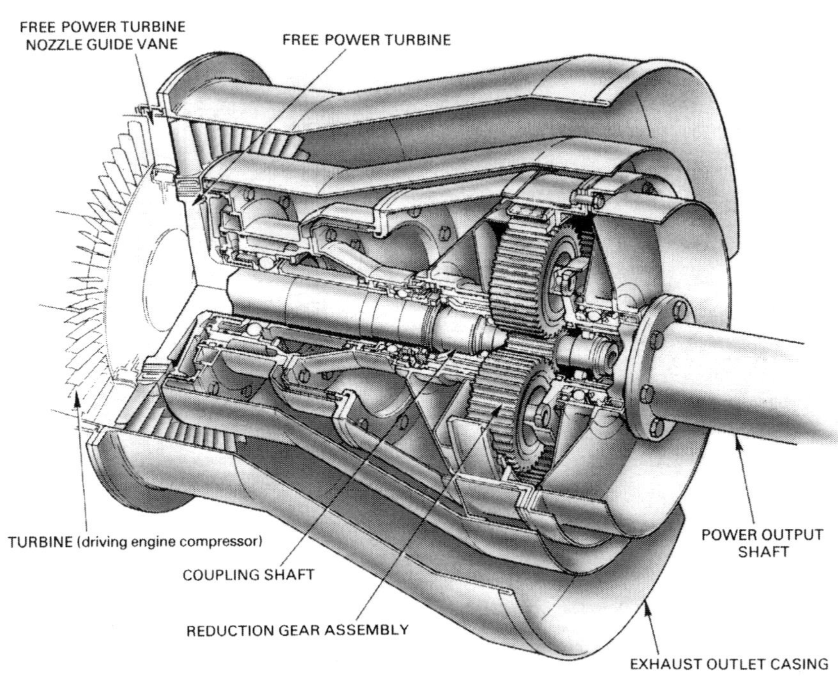

FIGURE 4–81 A typical free-power turbine. *(Source: Rolls Royce.)*

improved power to weight ratio. By-pass engines have a better propulsive efficiency and thus can have a smaller turbine for a given thrust.

The design of the nozzle guide vane and turbine blade passages is based broadly on aerodynamic considerations, and to obtain optimum efficiency, compatible with compressor and combustion design, the nozzle guide vanes and turbine blades are of basic aerofoil shape. There are three types of turbine: impulse, reaction, and a combination of the two known as impulse/reaction. In the impulse type the total pressure drop across each stage occurs in the fixed nozzle guide vanes, which, because of their convergent shape, increase the gas velocity while reducing the pressure. The gas is directed onto the turbine blades which experience an impulse force caused by the impact of the gas on the blades. In the reaction type the fixed nozzle guide vanes are designed to alter the gas flow direction without changing the pressure. The converging blade passages experience a reaction force resulting from the expansion and acceleration of the gas. Normally gas turbine engines do not use pure impulse or pure reaction turbine blades but the impulse/reaction combination (Figure 4–82). The proportion of each principle incorporated in the design of a turbine is largely dependent on the type of engine in which the turbine is to operate, but in general it is about 50% impulse and 50% reaction. Impulse-type turbines are used for cartridge and air starters.

Energy Transfer from Gas Flow to Turbine

From the previous description, it will be seen that the turbine depends for its operation on the transfer of energy between the combustion gases and the turbine. This transfer is never 100% because of thermodynamic and mechanical losses.

When the gas is expanded by the combustion process, it forces its way into the discharge nozzles of the turbine, where, because of their convergent shape, it is accelerated to about the speed of sound, which, at the gas temperature, is about 2500 ft per sec. At the same time the gas flow is given a "spin" or "whirl" in the direction of rotation of the turbine blades by the nozzle guide vanes. On impact with the blades and during the subsequent reaction through the blades, energy is absorbed, causing the turbine to rotate at high speed and so provide the power for driving the turbine shaft and compressor.

The torque or turning power applied to the turbine is governed by the rate of gas flow and the energy change of the gas between the inlet and the outlet of the turbine blades. The design of the turbine is such that the whirl will be removed from the gas stream so that the flow at exit from the turbine will be substantially "straightened out" to give an axial flow into the exhaust system. Excessive residual whirl reduces the efficiency of the exhaust system and also tends to produce jet pipe vibration, which has a detrimental effect on the exhaust cone supports and struts.

It will be seen that the nozzle guide vanes and blades of the turbine are "twisted," the blades having a stagger angle that is greater at the tip than at the root (Figure 4–83). The reason for the twist is to make the gas flow from the combustion system do equal work at all positions along the length of the blade and to ensure that the flow enters the exhaust system with a uniform axial velocity. This results in certain changes in velocity, pressure, and temperature

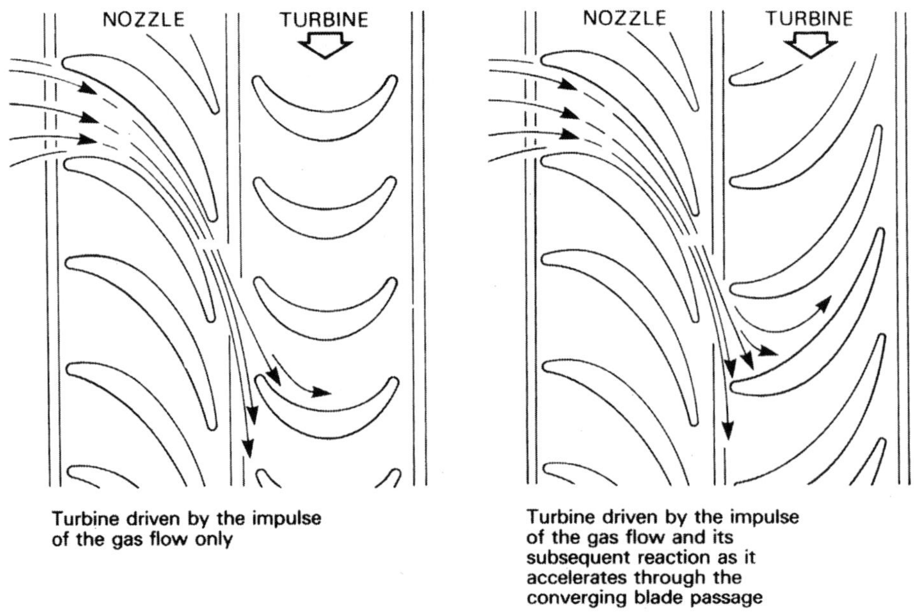

FIGURE 4–82 Comparison between a pure impulse turbine and an impulse/reaction turbine. *(Source: Rolls Royce.)*

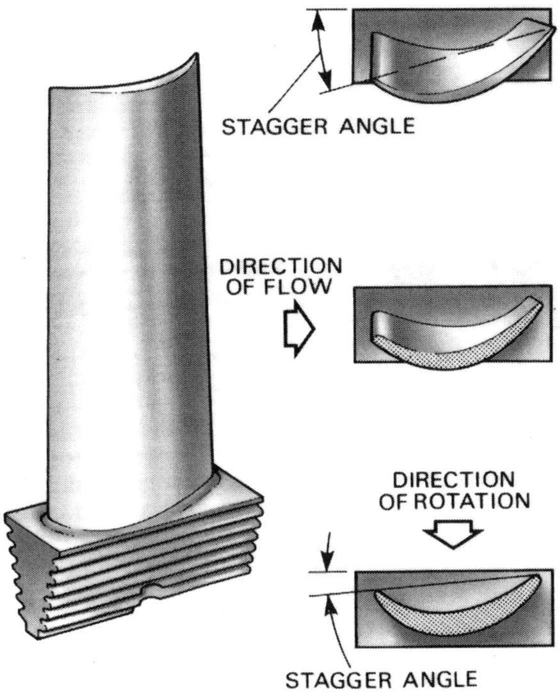

FIGURE 4–83 A typical turbine blade showing twisted contour. *(Source: Rolls Royce.)*

occurring through the turbine, as shown diagrammatically in Figure 4–84.

The "degree of reaction" varies from root to tip, being least at the root and highest at the tip, with the mean section having the chosen value of about 50%.

The losses that prevent the turbine from being 100% efficient are due to a number of reasons. A typical uncooled three-stage turbine would suffer a 3.5% loss because of aerodynamic losses in the turbine blades. A further 4.5% loss would be incurred by aerodynamic losses in the nozzle guide vanes, gas leakage over the turbine blade tips, and exhaust system losses; these losses are of approximately equal proportions. The total losses result in an overall efficiency of approximately 92%.

Construction

The basic components of the turbine are the combustion discharge nozzles, the nozzle guide vanes, the turbine discs, and the turbine blades. The rotating assembly is carried on bearings mounted in the turbine casing and the turbine shaft may be common to the compressor shaft or connected to it by a self-aligning coupling.

Nozzle Guide Vanes The nozzle guide vanes are of an aerofoil shape with the passage between adjacent vanes forming a convergent duct. The vanes are located (Figure 4–85) in the turbine casing in a manner that allows for expansion.

The nozzle guide vanes are usually of hollow form and may be cooled by passing compressor delivery air through them to reduce the effects of high thermal stresses and gas loads.

Turbine Discs Turbine discs are usually manufactured from a machined forging with an integral shaft or with a flange onto which the shaft may be bolted. The disc also has, around its perimeter, provision for the attachment of the turbine blades.

To limit the effect of heat conduction from the turbine blades to the disc, a flow of cooling air is passed across both sides of each disc.

Turbine Blades The turbine blades are of an aerofoil shape, designed to provide passages between adjacent blades that give a steady acceleration of the flow up to the "throat," where the area is smallest and the velocity reaches that required at exit to produce the required degree of reaction.

The actual area of each blade cross-section is fixed by the permitted stress in the material used and by the size of any holes that may be required for cooling purposes. High efficiency demands thin trailing edges to the sections, but a compromise has to be made so as to prevent the blades

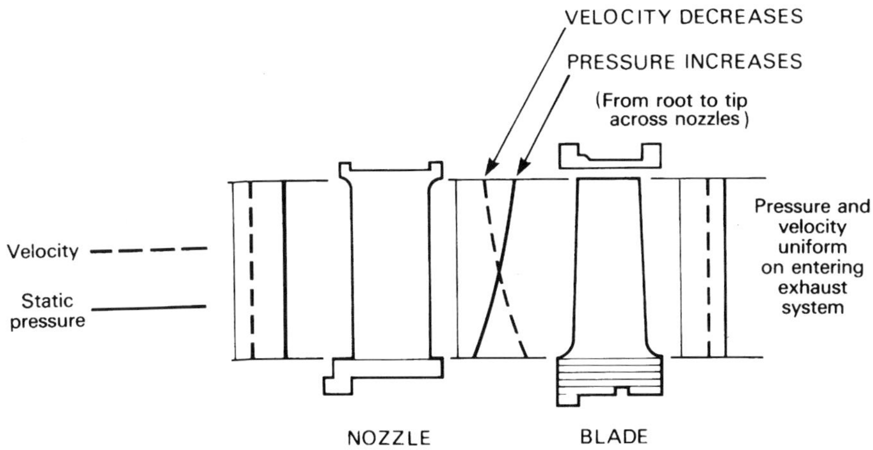

FIGURE 4–84 Gas flow pattern through nozzle and blade. *(Source: Rolls Royce.)*

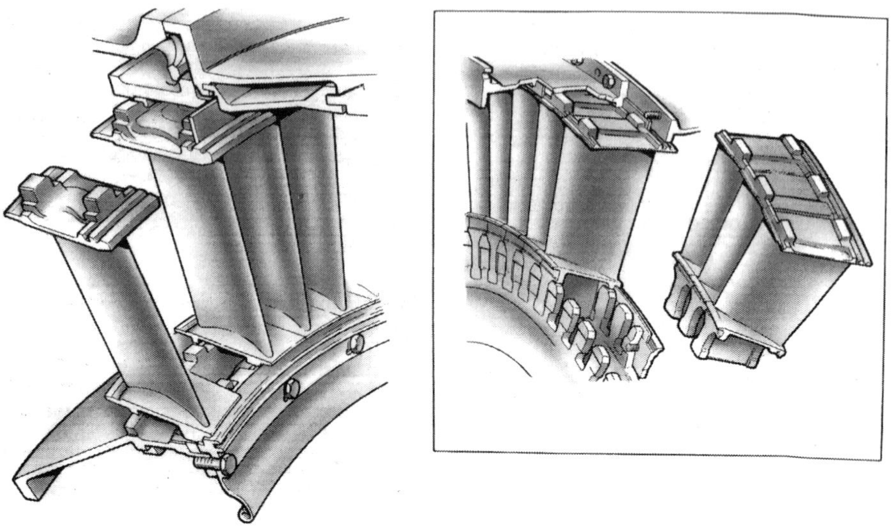

FIGURE 4–85 Typical nozzle guide vanes showing their shape and location. *(Source: Rolls Royce.)*

cracking due to the temperature changes during engine operation.

The method of attaching the blades to the turbine disc is of considerable importance, since the stress in the disc around the fixing or in the blade root has an important bearing on the limiting rim speed. The blades on the early Whittle engine were attached by the de Laval bulb root fixing, but this design was soon superseded by the "fir-tree" fixing that is now used in the majority of gas turbine engines. This type of fixing involves very accurate machining to ensure that the loading is shared by all the serrations. The blade is free in the serrations when the turbine is stationary and is stiffened in the root by centrifugal loading when the turbine is rotating. Various methods of blade attachment are shown in Figure 4–86; however, the B.M.W. hollow blade and the de Laval bulb root types are not now generally used on gas turbine engines.

A gap exists between the blade tips and casing, which varies in size due to the different rates of expansion and contraction. To reduce the loss of efficiency through gas leakage across the blade tips, a shroud is often fitted as was shown in Figure 4–86. This is made up by a small segment at the tip of each blade, which forms a peripheral ring around the blade tips. An abradable lining in the casing may also be used to reduce gas leakage. Active clearance control (A.C.C.) is a more effective method of maintaining minimum tip clearance throughout the flight cycle. Air from the compressor is used to cool the turbine casing and, when used with shroudless turbine blades, enables higher temperatures and speeds to be used.

Contrarotating Turbine Figure 4–87 shows a 12-stage contrarotating free-power turbine driving a contra-rotating rear fan. This design has only one row of static

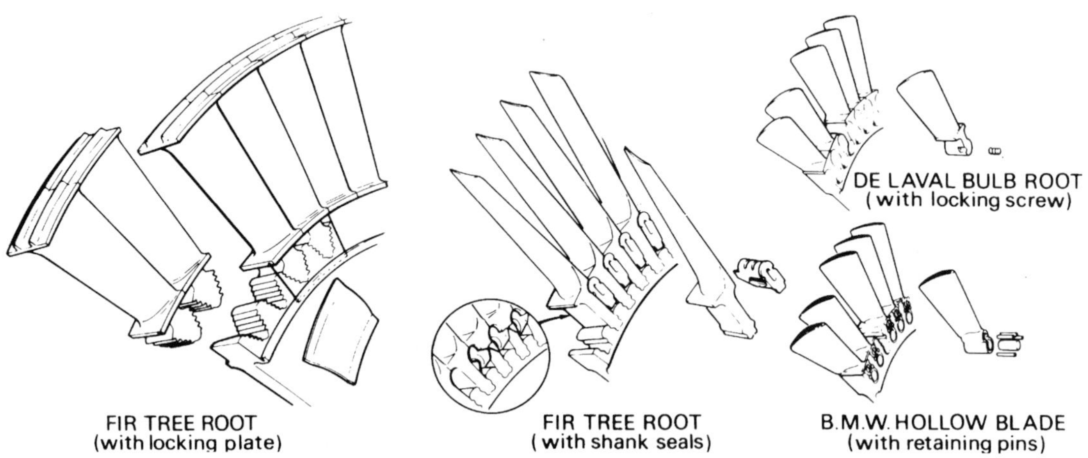

FIGURE 4–86 Various methods of attaching blades to turbine discs. *(Source: Rolls Royce.)*

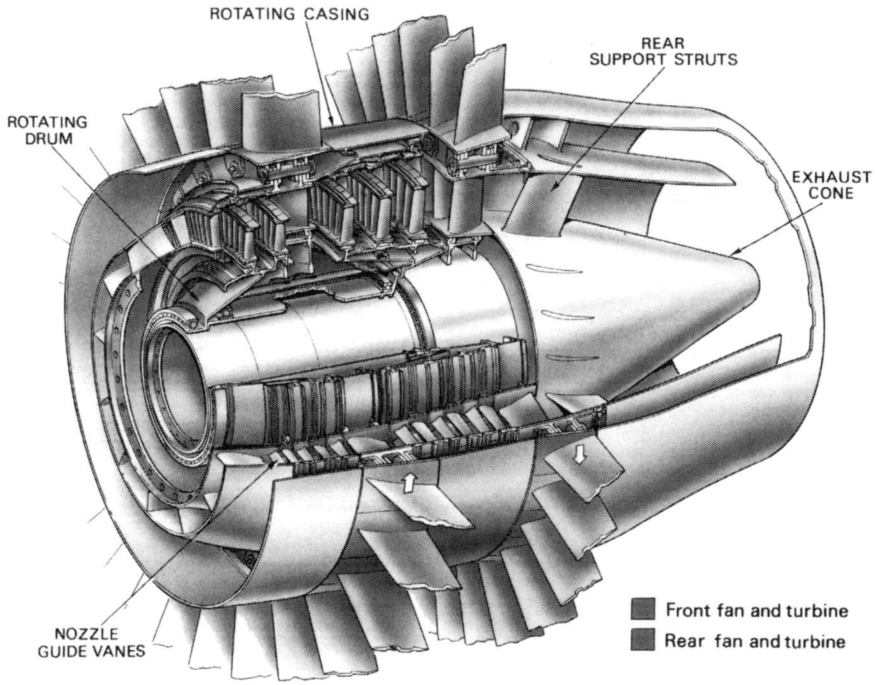

FIGURE 4–87 Free-power contrarotating turbine. *(Source: Rolls Royce.)*

nozzle guide vanes. The remaining nozzle guide vanes are, in effect, turbine blades attached to a rotating casing, which revolves in the opposite direction to a rotating drum. Since all but one aerofoil row extracts energy from the gas stream, contrarotating turbines are capable of operating at much higher stage loadings than conventional turbines, making them attractive for direct drive applications.

Dual-Alloy Discs Very high stresses are imposed on the blade root fixing of high work rate turbines, which make conventional methods of blade attachment impractical. A dual alloy disc, or "blisk," as shown in Figure 4–88, has a ring of cast turbine blades bonded to the disc. This type of turbine is suitable for small high-power helicopter engines.

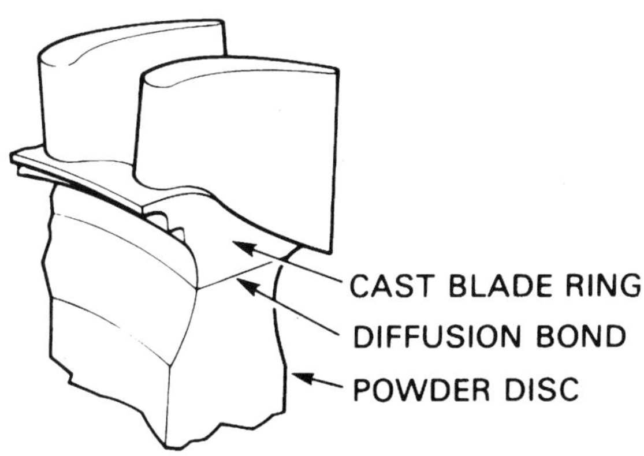

FIGURE 4–88 Section through a dual-alloy disc. *(Source: Rolls Royce.)*

Compressor–Turbine Matching The flow characteristics of the turbine must be very carefully matched with those of the compressor to obtain the maximum efficiency and performance of the engine. If, for example, the nozzle guide vanes allowed too low a maximum flow, then a back pressure would build up causing the compressor to surge; too high a flow would cause the compressor to choke. In either condition a loss of efficiency would very rapidly occur.

Materials

Among the obstacles in the way of using higher turbine entry temperatures have always been the effects of these temperatures on the nozzle guide vanes and turbine blades. The high speed of rotation, which imparts tensile stress to the turbine disc and blades, is also a limiting factor.

Nozzle Guide Vanes Due to their static condition, the nozzle guide vanes do not endure the same rotational stresses as the turbine blades. Therefore, heat resistance is the property most required. Nickel alloys are used, although cooling is required to prevent melting. Ceramic coatings can enhance the heat resisting properties and, for the same set of conditions, reduce the amount of cooling air required, thus improving engine efficiency.

Turbine Discs A turbine disc has to rotate at high speed in a relatively cool environment and is subjected to large rotational stresses. The limiting factor that affects the useful disc life is its resistance to fatigue cracking.

In the past, turbine discs have been made in ferritic and austenitic steels but nickel-based alloys are currently used. Increasing the alloying elements in nickel extends the life limits of a disc by increasing fatigue resistance. Alternatively, expensive powder metallurgy discs, which offer an additional 10% in strength, allow faster rotational speeds to be achieved.

Turbine Blades A brief mention of some of the points to be considered in connection with turbine blade design will give an idea of the importance of the correct choice of blade material. The blades, while glowing red-hot, must be strong enough to carry the centrifugal loads due to rotation at high speed. A small turbine blade weighing only 2 ounces may exert a load of over 2 tons at top speed and it must withstand the high bending loads applied by the gas to produce the many thousands of

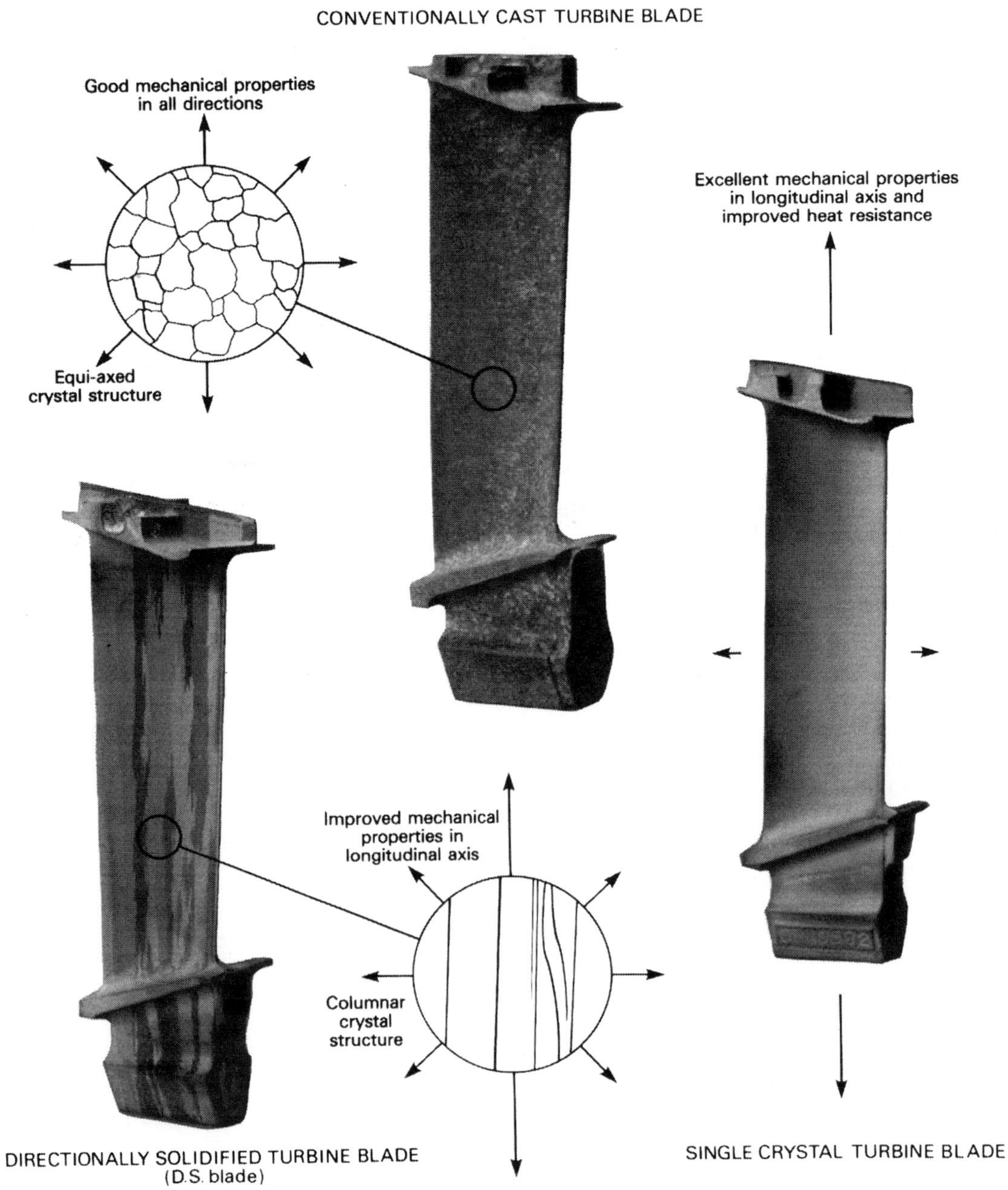

FIGURE 4–89 Various turbine blade crystal structures. *(Source: Rolls Royce.)*

turbine horsepower necessary to drive the compressor. Turbine blades must also be resistant to fatigue and thermal shock, so that they will not fail under the influence of high frequency fluctuations in the gas conditions, and they must also be resistant to corrosion and oxidization. In spite of all these demands, the blades must be made in a material that can be accurately formed and machined by current manufacturing methods.

From the foregoing, it follows that for a particular blade material and an acceptable safe life there is an associated maximum permissible turbine entry temperature and a corresponding maximum engine power. It is not surprising, therefore, that metallurgists and designers are constantly searching for better turbine blade materials and improved methods of blade cooling.

Over a period of operational time the turbine blades slowly grow in length. This phenomenon is known as "creep" and there is a finite useful life limit before failure occurs.

The early materials used were high temperature steel forgings, but these were rapidly replaced by cast nickel base alloys that give better creep and fatigue properties.

Close examination of a conventional turbine blade reveals a myriad of crystals that lie in all directions (equiaxed). Improved service life can be obtained by aligning the crystals to form columns along the blade length, produced by a method known as "directional solidification." A further advance of this technique is to make the blade out of a single crystal. Examples of these structures are shown in Figure 4–89. Each method extends the useful creep life of the blade (Figure 4–90) and in the case of the single crystal blade, the operating temperature can be substantially increased.

A non-metal-based turbine blade can be manufactured from reinforced ceramics. Their initial production application is likely to be for small high-speed turbines that have very high turbine entry temperatures. An example of a ceramic blade is shown in Figure 4–91.

The following three cases* illustrate some of the issues involved in uprating the power and temperature ceilings of a gas turbine. The cases are drawn from Mitsubishi's work on "J" and "H" technology development. As with other cases in this book, the reader is asked to note the year of publication for the H technology paper is 2001 and for the J papers is 2012. This serves to illustrate how involved design development is, even when based on a similar core design. I have left in the earlier paper as it helps illustrate the careful increments with which OEMs must proceed when enlarging their core design.

Certain details, such as actual load testing, will be significant to endusers who have had issues with gas turbines that have not had benefit of load test at the OEM's facility. It is appropriate when buying a new turbine package or even a fleet of them to consider load testing, particularly full load testing, as one of the primary turbine selection factors when comparing OEMs.

* Source: Courtesy Mitsubishi Power Systems.

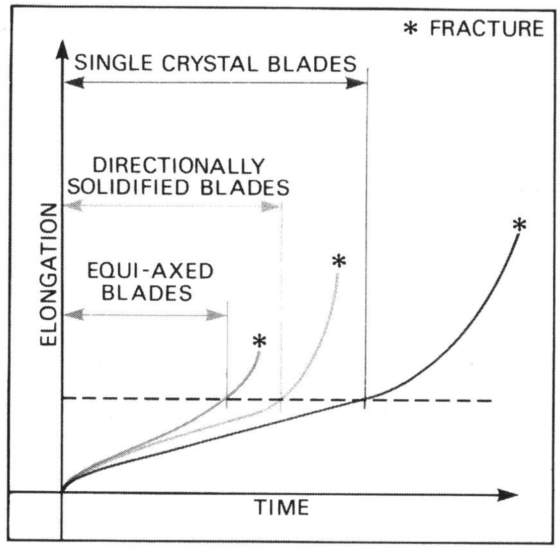

FIGURE 4–90 Comparison of turbine-blade life properties. *(Source: Rolls Royce.)*

FIGURE 4–91 Ceramic turbine blades. *(Source: Rolls Royce.)*

In Mitsubishi's case, their work was being supported by a Japanese national grant. The overall objective was to attain a TIT of 1600 degrees Celsius and raise efficiency for the GT CC package by 5–6% thus taking it past the 60% efficiency point to 62% and higher.

Case Study 5: Development of a 60cps "J" Technology Turbine*

High-efficiency, natural gas-fired combined cycles generate considerably lower CO_2 emissions, especially compared with conventional power generation based on coal or liquid hydrocarbons.

Since 2004, MHI started participation in a Japanese government-funded project called the Japanese National Project. This project is still ongoing and targets combined cycle efficiencies in the 62 to 65% range through the development of innovative technology. The gas turbine for this National Project is expected to operate at a Turbine Inlet Temperature of 1700°C. The first phase of this project was completed in 2007 and involved considerable R&D efforts. It steered the following developments:

- Enhanced turbine aerodynamics
- Higher turbine cooling efficiency
- Advanced thermal barrier coating
- Higher pressure ratio compressor
- Advanced heat resistant turbine materials
- Exhaust gas recirculation combustor for lower emissions

The M501J operates at an intermediate TIT between the 1700°C goal of the Japanese National Project and the current 1500°C technology. This approach allows early application of the technology developed for the 1700°C TIT class gas turbine and provides initial validation of the technology at a lower temperature exposure.

Features of the M501J/M701J Several component technologies developed for the National Project were used to design the J series (refer to Figure 4–92). High-performance cooling schemes with reduced-cooling air requirements and new thermal barrier coating materials facilitated the increase of TIT to the record-breaking 1600°C, while maintaining the turbine reliability and operating life.

The new thermal barrier coating materials provided substantial reduction in thermal conductivity compared with conventional YSZ type without compromising durability.

By combining a 1600°C TIT and a pressure ratio of 23:1, the J-series gas turbine provides large power generation capacity and high thermal efficiency, as shown in Table 4–3. These parameters were revised upward based on better-than-expected performance during the validation period.

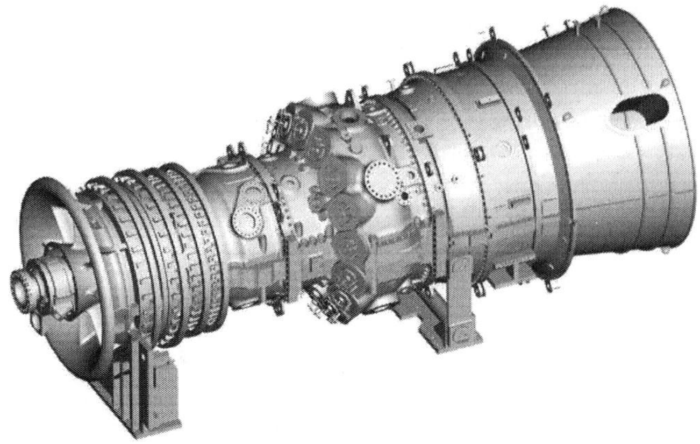

FIGURE 4–92 Isometric representation of the M501J gas turbine.

The Economy of Large Frame Gas Turbine Validation The design and testing of new gas turbines is expensive and time-consuming. During the initial development stage, there are large R&D expenditures associated with the design and individual component testing. This effort is typically followed by shop tests where the new engine is carefully monitored through numerous temporary sensors. These tests typically begin with turning operation and are followed by first ignition, no load operation, first synchronizing, and operation at different loads. Shop tests are typically run off-grid and the maximum load and duration will depend on the capacity of the hydraulic brakes or other energy dissipating devices, as well as the validation budgets to cover the spent fuel.

A study conducted by EPRI determined that fuel costs constitute 68% of a large-frame combined cycle plant's total cash flow (net present value basis). The power generated off-grid does not provide any revenue and cannot be economically sustained for extended periods of time.

TABLE 4–3 J-series Gas Turbine Performance

Model	M501J	M701J
Frequency (Hz)	60	50
Gas turbine power output (MW)	327	470
Combined cycle output (MW)	470	680
Combined cycle efficiency (%)	61.5	61.7

* Courtesy of Mitsubishi Heavy Industries. Extracts from "Mitsubishi 1,600 TIT J Series Gas Turbine" (presented at PowerGen 2012).

FIGURE 4–93 T-Point Validation Power Plant.

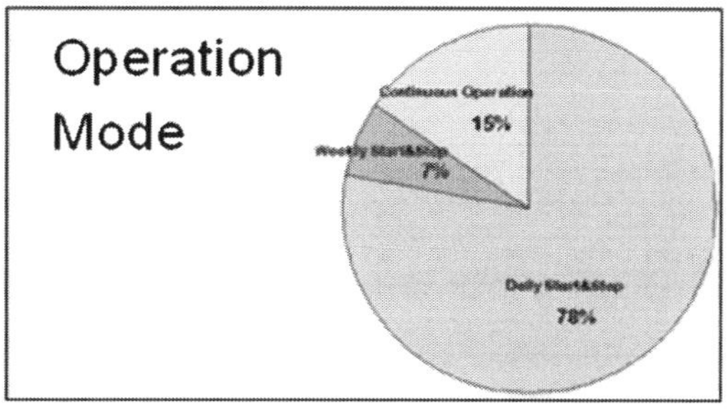

FIGURE 4–94 T-Point historical operating modes.

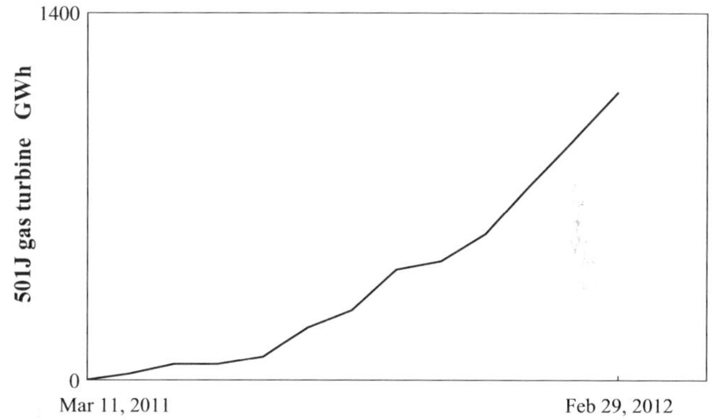

FIGURE 4–95 M501J gas turbine GWh at T-Point.

Testing while connected to the grid imposes a rotating speed that is dictated by the grid frequency. However, if the power generated is sold, it allows sustained validation for extended time. Testing a prototype off the grid allows adjustments of the operating speed to confirm "under" and "over" frequency characteristics of the new engine, as well as compressor surge margins.

MHI's gas turbine manufacturing facility in Japan has the energy dissipation capabilities (water break) required to perform off-grid testing. However, the final long-term "load dynamic" validation takes place at the demonstration plant on the premises of Takasago Machinery Works (Figure 4–93). This approach expedites the maturity process while avoiding clients' exposure to prototype units.

FIGURE 4–96 M501J in transit to the nearby T-Point Validation Power Plant.

Status of M501J Validation T-Point Power Station started commercial operation in 1997. Prior to the installation of the M501J gas turbine this plant facilitated the deployment of the M501G gas turbine, which became the largest steam-cooled fleet of gas turbines in the market. Different upgrades of the G series were validated in the plant accumulating close to 40,000 actual hours and more than 2100 starts. T-Point operating hours, as well as the loading profiles, are determined by Kansai Electric who purchases the output from MHI. The pie chart in Figure 4–94 shows the historical distribution of operating modes (78% daily start/stop or DSS, 7% weekly start/stops or WSS, and 15% continuous operation).

The T-Point service factor has been increased after the March 11, 2011 earthquake and tsunami to alleviate the power shortage. Figure 4–95 shows the GWh generated since that tragic event.

Figure 4–96 shows the first M501J in transit from Takasago Machinery Works' assembly building to T-Point, where it was installed in November 2010. The first fire took place in February 2011 and the record-breaking 1600°C TIT was reached on February 23, 2011.

Figures 4–97 to 4–100 show measurement results recorded during the commissioning period with all parameters well below the allowable limits.

Figure 4–97 includes transition piece metal temperature data as well as combustion dynamics measured under 1600°C operation. It also shows the condition of the hardware during the removal of temporary instrumentation.

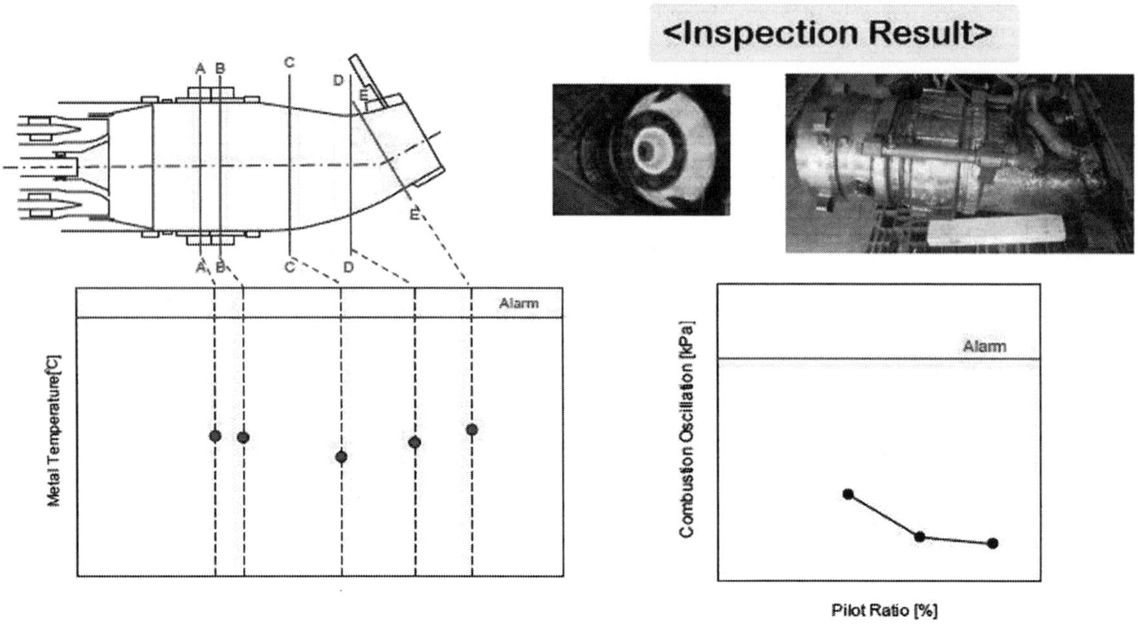

FIGURE 4–97 Combustion system metal temperatures and pressure fluctuations.

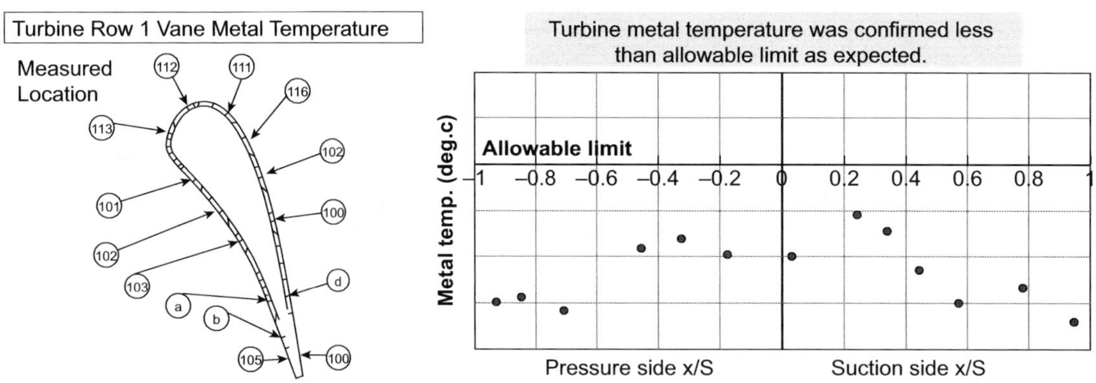

FIGURE 4–98 Turbine row 1 vane metal temperatures.

Figure 4–98 shows the Turbine Row 1 Vane metal temperature measured during 1600°C operation. Turbine Row 1 Blade surface temperature was also measured by the use of pyrometer and the results are shown in Figure 4–99.

The M501J applies the same Active Clearance Control technology successfully used in the G Fleet, with demonstrated improvement of efficiency under load operating conditions. Figure 4–100 shows the measured clearance for Turbine Row 2 Blade.

At the end of the initial trial run, the unit was inspected and all parts were found in good condition.

The commissioning period of this unit was completed in June 2011 and commercial operation run started in July. At the time of submitting this article, the unit had accumulated more than 5300 actual operating hours and 62 starts.

With a combined cycle efficiency of 61.5%, the J-series GTCC power generation will considerably reduce the CO_2

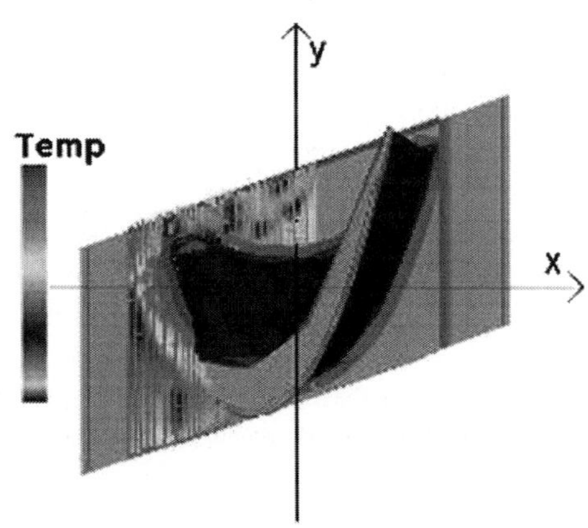

FIGURE 4–99 Turbine row 1 blade surface temperature by the use of a pyrometer.

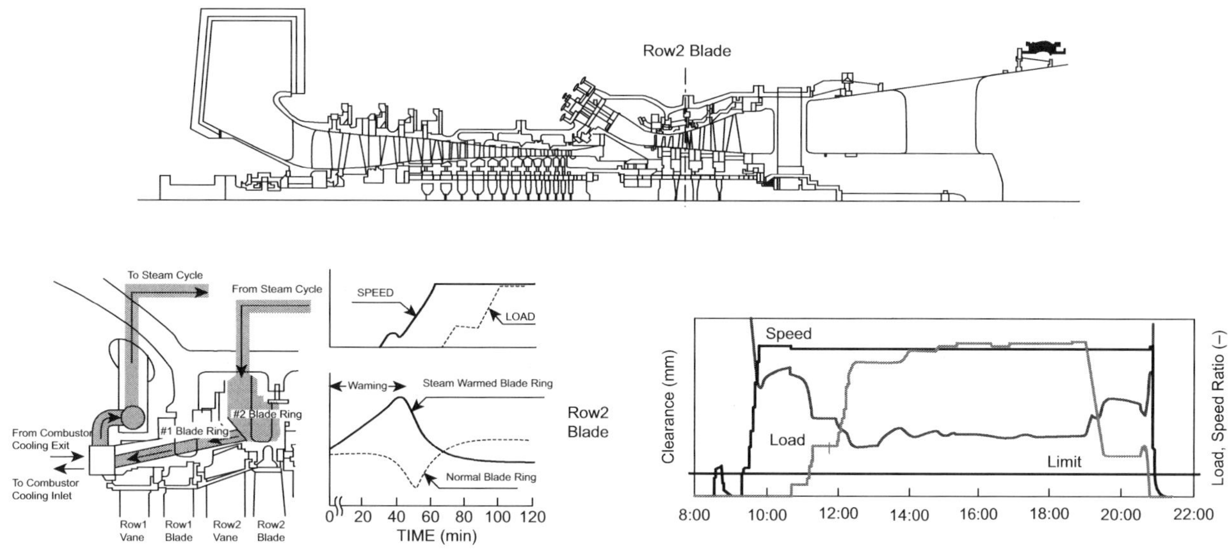

FIGURE 4–100 Effect of the Active Clearance Control (ACC).

emissions released by the existing technology and therefore will contribute to reducing global warming.

Case Study 6: Development and Operation Experience of Mitsubishi 50 Hz Large-Frame (M701F) Gas Turbines*

This case illustrates how successive models of gas turbine build on the previous update. Similar to the J Series, the M701F5 targets increased combined cycle efficiency and the associated reduction in fuel consumption and CO_2 emissions.

M701F4 Features and Fleet Update The M701F4 is the upgraded version of the M701F3 which was developed in 1999. It incorporates a 6% increase of the compressor inlet flow to raise the power output. It also includes an advanced air-cooled combustor with 50°F higher firing temperature (compared with the preceding M701F3). The cooling technology applied to this machine and the last stage turbine blades were retrofitted from the G series gas turbines (refer to Figure 4–101). It further illustrates how adjustments can be made to the gas turbine's basic systems to accommodate shortage of resources (such as cooling water) to adjust for disasters such as Japan's Sendai earthquake in 2011. Reminder: the Japanese power grid operates both 50 and 60 cps systems.

The first M701F4 unit went commercial in July 2010 at the Sendai combined cycle Power Station in Japan (refer to Figure 4–102). The construction of this plant was handled as a turnkey project by MHI and the commissioning proceeded as scheduled. This is a single-shaft plant with a generation capacity of 446 MW. It features a Mitsubishi tandem compound double-flow TC2F-40.5 ISB steam turbine with 40.5-inch last stage blades (refer to Figure 4–103). This plant operated as the fleet leader unit until it was temporarily taken out of production by the Sendai earthquake and tsunami. Until then, it had been operated on a daily start/stop (DSS) approach accumulating 4643 Actual Operating Hours (AOH) and 166 starts without any trips.

The plant was restored and brought back into operation in early 2012, increasing the number of hours to 6200 and 175 starts.

Another 2 on 1 combined cycle plant was built in Turkey and started commercial operation in December 2010 (refer to Figure 4–104). The plant is rated 919 MW, powered by two M701F4 gas turbines and a tandem compound four-flow steam turbine with 40.5-inch last row blades (Mitsubishi TC4F-40.5 ISB). It achieved a net efficiency level of more than 59%, less than 15 ppm of NO_x and 10 ppm of CO emissions.

These units surpassed the Sendai unit and became fleet leaders with the #11GT accumulating 10,059 AOH with 112 starts, while the #12GT accumulated 9591 AOH with 115 starts.

Temporary Measurements of the First M701F4 A comprehensive set of temporary measurements was used during the commissioning of the first M701F4 to confirm its performance and reliability. More than 500 sensors were mounted on the gas turbine (refer to Figure 4–105). All

* Courtesy of MHI. Extracts of "Development and Operation Experience of Mitsubishi 50Hz Large-Frame Gas Turbines" presented at PowerGen Europe 2012.

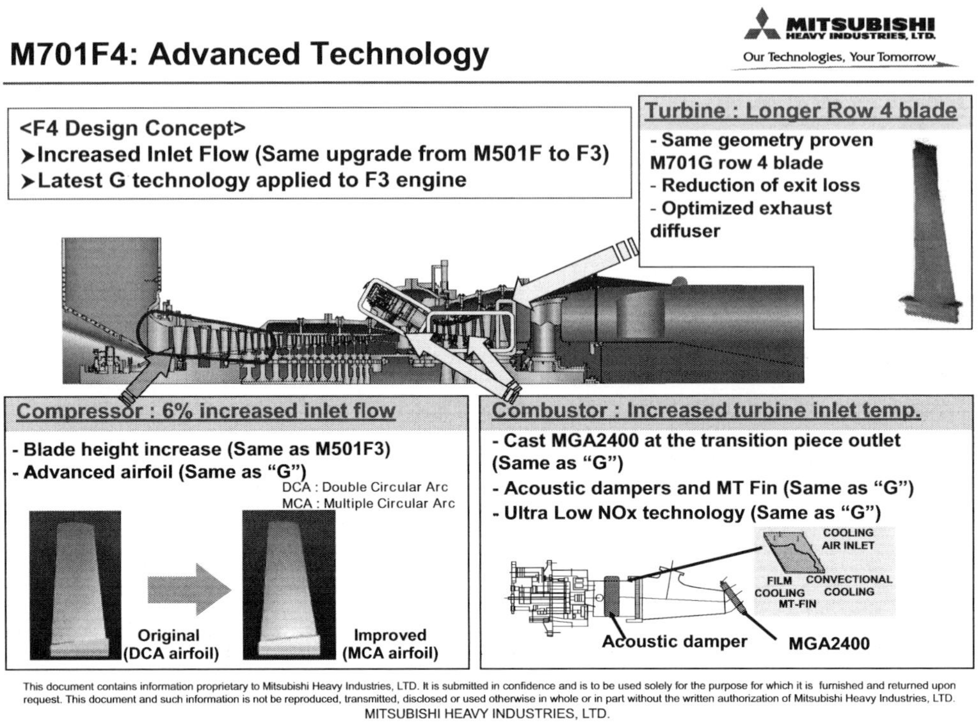

FIGURE 4–101 Main features of the M701F4.

FIGURE 4–102 Facade of Sendai Power Station.

FIGURE 4–103 Sendai Power Station single-shaft configuration.

FIGURE 4–104 Turkey's Bandirma Combined Cycle plant.

measurement points displayed values within expected levels. Figure 4–106 shows the metal temperature measured on the Turbine Row 1 Vane.

Source of New Technology MHI has been engaged in a Japanese government-funded project called the "National Project" since 2004. The main target of this project is the development of a 1700°C (3092°F) class gas turbine in order to achieve 62–65% (LHV) combined cycle efficiency. This represents a 5–10% increase compared with the current available technology.

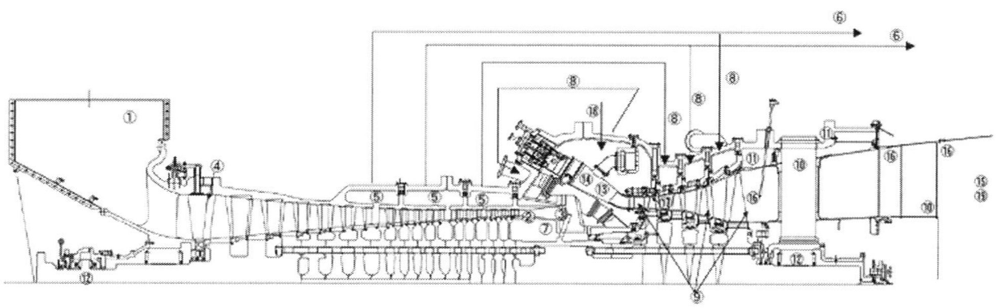

MESUREMENT ITEM

1 AIR FLOW
2 COMPRESSOR DIFFUSER PRESSURE
3 COMPRESSOR OUTLET TEMP. & PRESS.
4 COMPRESSOR INLET PRESSURE DYNAMICS
5 COMPRESSOR BLEED AIR PRESSURE
6 COMPRESSOR BLEED AIR FLOW
7 ROTOR CAVITY TEMP.
8 COOLING AIR FLOW
9 TURBINE ROW 1 VANE METAL TEMP.
TURBINE ROW 4 BLADE INLET/OUTLET GAS TEMP. & PRESS.
10 EXHAUST STRUT METAL TEMP.

EXHAUST CASING TEMP.
BEARING CAVITY TEMP.
COMBUSTOR METAL TEMP.
COMBUSTOR PRESSURE DYNAMICS
EXHAUST EMISSIONS (NOx, CO, UHC, O2)
EXHAUST DIFFUSER PRESS.
#1 RING SEGMENT METAL TEMP.
CASING COOLING AIR FLOW
EXHAUST TEMP.

FIGURE 4–105 Temporary measurement points and items.

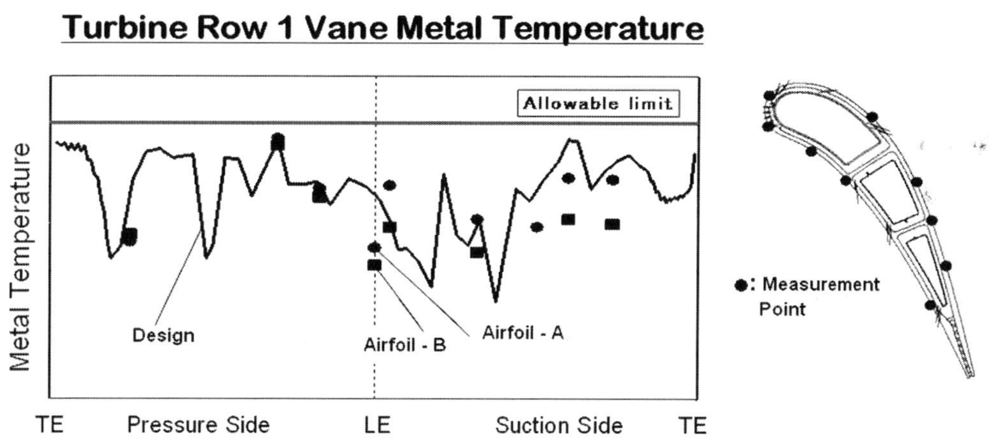

FIGURE 4–106 Metal temperature measurement results on Turbine Row 1 Vane.

Several technological improvements derived from the first stage of the National Project include:

- Exhaust Gas Recirculation (EGR) combustor
- Higher cooling efficiency
- Lower thermal conductivity TBC
- Higher loading turbine aerodynamics
- Higher pressure ratio compressor and
- Advanced turbine material

Mitsubishi has retrofitted several of the developed technologies into existing as well as new gas turbine developments.

Conversion of Steam-Cooled 50 Hz G Units into GAC The Japanese power grid has experienced severe shortages after the 2011 earthquake and tsunami. Motivated to alleviate the situation, Tokyo Electric and MHI engaged in a quick installation of three 50 Hz steam-cooled engines that were stored at Takasago Machinery Works for a delayed project. In order to quickly bring these units into simple cycle operation, they were converted to air-cooled units by replacing the steam-cooled liners with air-cooled hardware.

Two of the three M701GAC units were brought into commercial operation by the summer of 2011 (refer to Figure 4–107), barely 5–6 months after the earthquake, helping alleviate the difficult power crises caused by the earthquake and tsunami.

Development of the M701J Gas Turbine Several articles listed in the reference section present a description of the 60 Hz M501J. The first engine of this revolutionary

FIGURE 4–107 Two M701GAC units operating a simple cycle in Chiba.

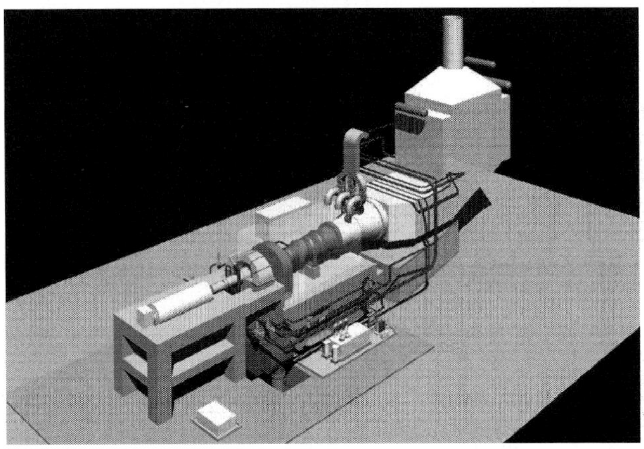

FIGURE 4–108 Bird's eye view of M501J combined cycle (1:1).

technology has been in commercial operation since June 2011 and has accumulated more than 5300 AOH and 62 starts.

MHI has already delivered the first of six M501J gas turbines to Kansai Electric Power Company (KEPCO) in Japan. The six-unit single-shaft combined cycle plant called Himeji is being built by MHI on a turn-key basis with the first unit scheduled to go commercial by 2014. Ten additional units will follow in Korea.

The 50 Hz counterpart was designed as a scaled-up version of the M501J. It will operate at the same turbine inlet temperature (TIT) of 1600°C and pressure ratio of 23:1 to achieve a combined cycle thermal efficiency of 61.7%.

This engine was designed based on the proven steam cooling approach developed and extensively used in the G series, while incorporating state-of-the-art technologies developed by Japanese National Project. The incorporated technologies include advanced turbine cooling, improved turbine aerodynamics, and advanced TBC.

The resulting higher exhaust gas temperature will be effectively used in combined cycle application (refer to Figure 4–108).

Figure 4–109 shows the evolution of TIT and its effect on efficiency and power output. The J-series combined cycle is expected to achieve over 61% efficiency (LHV), while its power generation capacity will be about 20% higher than that of the G-series.

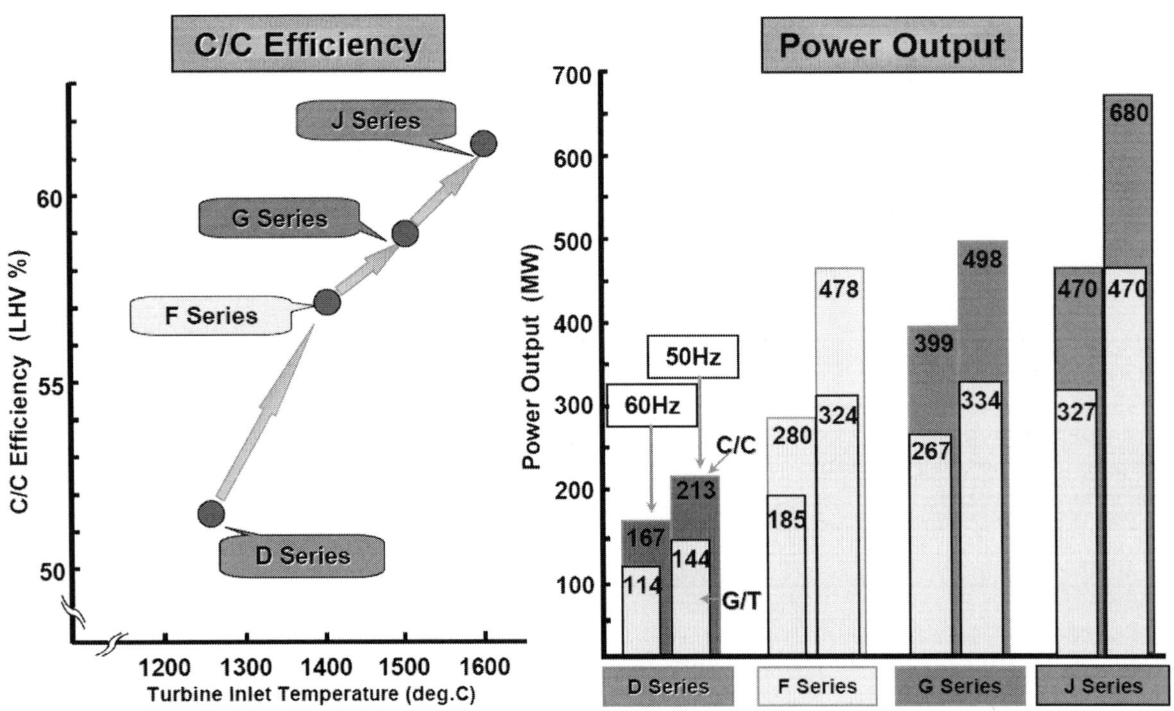

FIGURE 4–109 MHI large gas turbine performance.

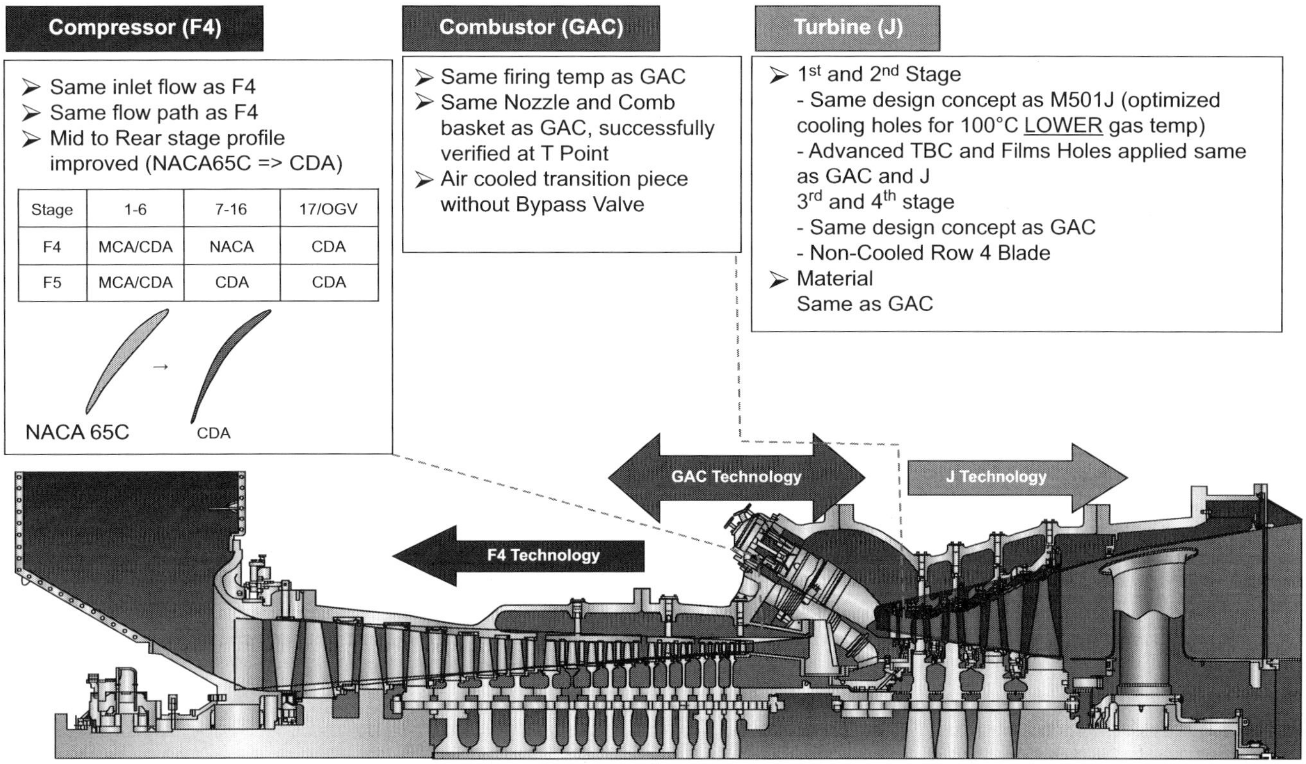

FIGURE 4–110 Outline of M701F5 gas turbine.

Table 4–3 shows the J-series gas turbine rated power output for 60 Hz and 50 Hz in 1:1 single-shaft combined cycle configuration.

The first unit of the M701J is scheduled to start commercial operation in 2016.

Development of the M701F5 Gas Turbine (50 Hz) By applying the state-of-the-art technologies developed for the J and the air-cooled GAC engines (M501GAC, M701GAC), MHI is developing the M701F5 gas turbine, which is the upgraded version of the proven M701F4 (refer to Figure 4–110).

The M701F5 consists of the same F4 compressor, air-cooled combustors with TIT of 1500°C proven by GAC engines and the J turbine components. The M701F5 gas turbine is expected to maintain high reliability based on the use of proven components. A 1:1 combined cycle with the M701F5 will generate the power output of 525 MW with an efficiency of 61% (refer to Figure 4.111).

Case Study 7: Development of an "H" Technology Turbine*

Mitsubishi Heavy Industries Ltd. (MHI) developed the M501H gas turbine to improve the efficiency and power of the G series using closed steam cooling of stage 1 and 2 blades and vanes following the firm's experience in the construction of the D, F, and G series. Development included elementary tests for preverification and trial operation using a combined plant verification plant to reach 220 MW in May 1999 and to verify the stability of the steam cooling rotor, cooling properties of blades and vanes, and the efficiency of major components.

Mitsubishi Heavy Industries Ltd. developed the 1150°C class large capacity gas turbine, M701D, in 1981 and subsequently verified its high plant total thermal efficiency, high reliability, and low pollution at Tohoku Electric Power Co. Inc., Higashi Niigata Thermal Power Plant #3 Power Train. In 1985, MHI started development of the 1350°C class "F" series gas turbine (M501F/M701F), and an actual loading shop test was conducted on the initial unit in 1989. MHI has already received orders for 90 "F" series gas turbines as the main engines for combined cycle power plants in both domestic and overseas locations, and 41 out of 90 units have been accumulating experience with successful operation.

MHI started development of the 1500°C class "G" series gas turbine (M501G/M701G) in 1993. A trial operation of the M501G gas turbine for the 60 Hz market started in February 1997 at the Takasago in-house verification plant for combined cycle power plant ("T"-Point), and a verification operation started in June 1997 as a long-term

* Extracts from 2001-GT-0050, A. Maekawa, K. Uematsu, E. Ito, K. Hirokawa, and Y. Fukuizumi, "Development of H Series Gas Turbine."

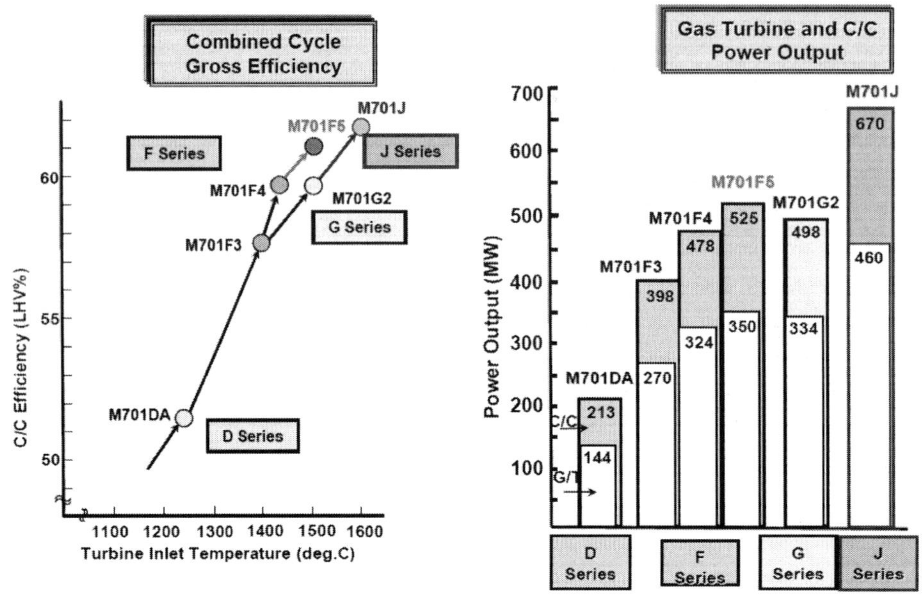

FIGURE 4—111 Performance of MHI 50 Hz gas turbines.

reliability verification unit. On the other hand, the M701G1 gas turbine for 50 Hz market was started with a trial operation at Tohoku Electric Power Co. Inc., Higashi Niigata Thermal Power Plant #4 Power Train on October 15, 1998, and was put into commercial operation in July 1999. Currently, MHI has received orders for 45 "G" series gas turbines, and three out of 45 units are now being successfully operated. Based on abundant experience of electric utility gas turbines such as the "D," "F," and "G" series, MHI developed the "H" series gas turbine. The "H" series is a 1500°C class gas turbine, which is the same as the "G" series gas turbine; however, the "D," "F," and "G" series gas turbines adopted air cooling for turbine blades and vanes, while the "H" series gas turbine adopted steam cooling, which is a main feature of this type of gas turbine. Thanks to steam cooling, high plant total thermal efficiency could be obtained.

This section describes the results of trial operation of the M501H gas turbine for the 60 Hz market at "T"-Point conducted in May 1999 in addition to the features of 1500°C class "H" series gas turbine.

Features of the "H" Series Gas Turbine The "H" series gas turbine is a high performance gas turbine with 1500°C turbine inlet temperature. Compared with the existing gas turbines, the "H" series gas turbine is higher in plant total thermal efficiency by approximately 2% (absolute value), and larger in output by approximately 20%.

Overall Structure The basic structure of the "H" series gas turbine is modeled after the "F" and "G" series gas turbines and inherits their proven technologies as follows:

- For the rotor, the two bearing support is adopted, with the support from both the compressor side bearing and turbine side bearing.
  1. For the connection with the generator shaft, the compressor shaft end drive system, of which the amount of thermal effect such as thermal expansion is small and the flexible coupling is not required, is adopted.
  2. The exhaust system is an axial flow exhaust structure, which is optimum for the layout of a combined cycle power plant.
  3. As for the bearing support structure, the compressor side bearing is supported by eight radial struts, and the turbine side bearing is supported by tangential struts, which can easily absorb differential thermal expansion by keeping the shaft at the center.
- For the connecting structure, discs with torque pins are connected with bolts on the compressor rotor side, while discs with curvic couplings are connected with bolts on the turbine rotor side. Both of them have a structure of delivering torque without fail.

Steam Cooling Structure of Blades and Vanes The "H" series gas turbine has a main feature that the rows 1 and 2 blades and vanes are cooled with steam being generated at the bottoming, while the existing ones are cooled with compressed high-pressure air discharged from the compressor.

Thanks to the application of a steam cooling system, the amount of cooling air can be reduced to approximately one half of the existing one, the mixture loss of cooling air can be also reduced, and the plant total thermal efficiency can be improved. In addition, combustion gas flow increases by the amount of reduced cooling air, and efficiency at the turbine section is increased, so these factors can improve the output.

Furthermore, the exhaust gas temperature increases by the reduced amount of low-temperature cooling air mixing with combustion gas at the same pressure ratio. So, the exhaust gas temperature keeps the existing level by increasing pressure ratio. For the compressor, the latest type blades and vanes were applied, and a high efficiency compressor having higher pressure ratio and reduced number of stages was developed. This compressor was adopted based on thorough verification using a 0.29 scale compressor.

Development of the "H" Series Gas Turbine As shown in the Table 4–4, development of the "H" series gas turbine was started in 1996, and a trial operation was started in February 1999. The development for such a short period was supported by reliable verification through various component tests conducted abreast. The improvement required was reflected in the design in detail without losing an opportunity.

Confirmation Test of Steam-Sealing Characteristics. For the connections at the supply and recovery passage to/from steam cooled blades and vanes, a special seal which had not been used for the existing gas turbines was adopted. The "H" series gas turbine applies steam for cooling; however, if the same amount of leakage as the existing cooling air is allowed, a large amount of steam leakage will occur. Steam leakage reduces the plant total thermal efficiency and causes the increase of makeup water at the bottoming. Therefore, this seal is of critical importance.

In order to solve these issues in advance, sealing ability of the selected seal structure was confirmed through various component tests. First of all, a component test for checking sealing ability under the stationary condition was conducted on the components installed in both stationary system and rotating system to select optimum structure and dimension. Second, the components selected for rotating system were tested using the model rotor, which could simulate an actual rotating condition in order to check that sealing ability could be kept as designed.

Confirmation Test of Steam-Cooled Blades and Vanes. Several pieces of turbine row 1 vanes of the M701F gas turbine unit installed at MHI Yokohama Machinery Works ("K"-Point) were modified to steam cooled vanes. These modified vanes were cooled with steam under actual operation, and the expected data of cooling characteristics were obtained.

Prior to a trial operation of the "H" series gas turbine, a high temperature cascade test simulating the actual engine was conducted using the actual turbine row 1 blades cooled with steam as a medium of cooling. Through the test, the data of cooling characteristics were obtained in advance.

TABLE 4–4 Development of the "H" series Gas Turbine

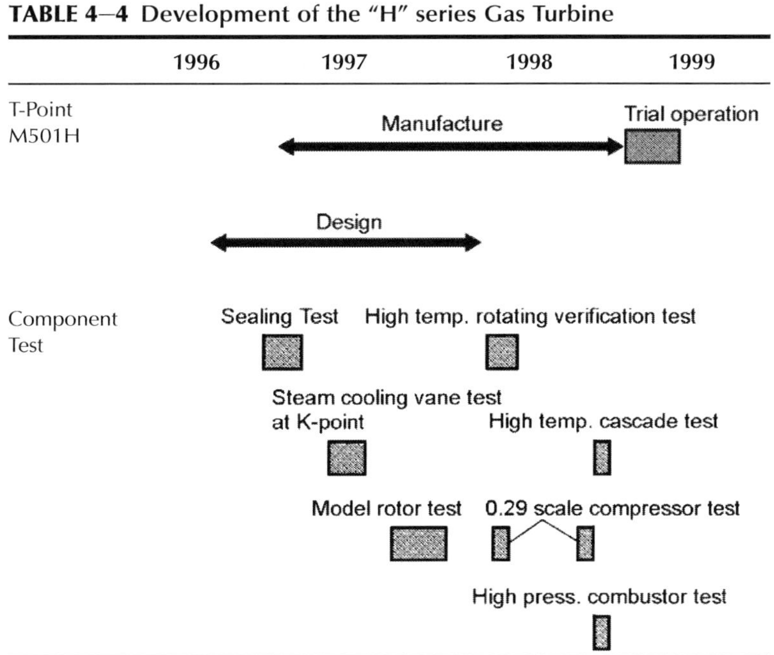

(Source: MHI.)

Confirmation Test of Compressor. A 0.29 scale compressor was manufactured for the "H" series gas turbine. This compressor was driven by a two-shaft gas turbine of M252 and equipped with a control valve for increasing pressure at the outlet.

Under such a condition, a model compressor test was conducted simulating various operating conditions. This compressor was confirmed to have the same level characteristics and performance as design from the data each of various starting up characteristics and compressor performance which could be obtained by changing pressure and rotating speed freely.

A high-pressure combustion test was conducted using the above-mentioned compressor, of which a high-pressure combustion test rig was equipped on the downstream side. A high-temperature rotation verification test was also performed.

This test rig is equipped with a combustor to which high-pressure air is supplied from the above-mentioned compressor, and natural gas is also supplied from a pipe line for combustion. A 1500°C combustion gas through the combustor is introduced to a 0.6 scale single-stage turbine for rotation. The rotating load is controlled by absorbing with a dynamometer to keep the rotating speed constant at 6000 rpm.

For development of the "H" series gas turbine, six blades of a scale design of "G" series gas turbine were modified to the steam cooling structure, and the steam passage was made on the bladed rotor for manufacturing a new type of gas turbine having the structure which can supply and recover steam from the shaft end.

As in the "H" series gas turbine, this gas turbine was operated under the condition where steam was supplied to the rotor from the shaft end for cooling steam cooled blades and then recovered from the shaft end through the rotor again. During operation, data from each of the cooling characteristics of blades, temperature, and vibration at each portion were obtained and confirmed to be at the same level as designed.

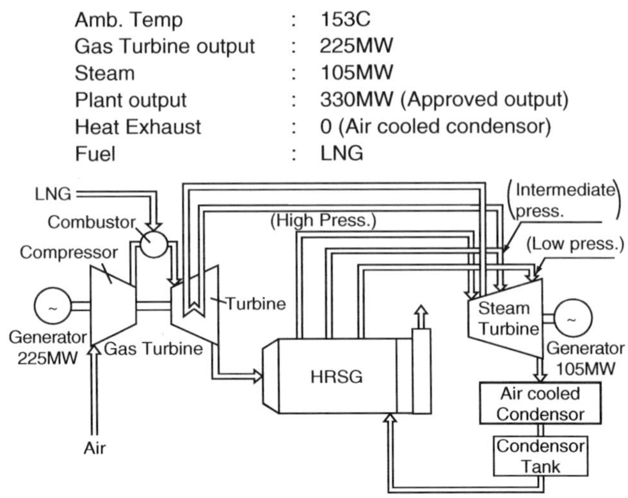

FIGURE 4–112 Main line of H series test. *(Source: MHI.)*

Results of "H" Series Gas Turbine Trial Operation

Outline of Trial Operation. A trial operation of the M501H gas turbine was conducted using the in-house combined cycle verification plant of MHI Takasago Machinery Works. Generally, this in-house verification plant is used for verifying long-term reliability of the M501G gas turbine; however, this plant was tentatively replaced with the M501H gas turbine for a trial operation. As shown in Figure 4–112, high-pressure steam generated at the exhaust heat recovery steam generator was introduced to the high-pressure steam turbine from which outlet steam was introduced to the steam cooled blades and vanes for cooling. The steam after cooling was recovered to the inlet of the middle pressure steam turbine.

Results of Trial Operation. The trial operation was started on February 8, 1999, and 220 MW (GT output: 60 MW, ST output: 60 MW) output was achieved through various confirmations of actuation. From this trial operation, the first step targeting the verification of high-pressure compressor characteristics and steam system (cooling characteristics of steam cooled blades) was successfully completed. On the next step, the operation up to 1500°C will be planned to be carried out (Figure 4–113).

From the measurement results, it could be confirmed that cooling steam flowed through the turbine rotor, blades, blade rings, and vanes and proved effective cooling performance as designed. In addition, the condition after various tests was good.

Effects of Trial Operation. In this trial operation, a special measurement was carried out at 1500 points. The effects of this trial operation are shown below.

Confirmation Test of Steam-Sealing Characteristics. Steam is supplied from the rotor shaft end to the steam cooled blades through the inside of the rotor and recovered to the rotor shaft end through inside of the rotor again.

As shown in Figure 4–114, stable vibration characteristics were obtained in this trial operation. This trial operation was the first run for the "H" series gas turbine, so the operation was manually done as a general rule so that timely actions could be taken for various operating conditions.

As mentioned previously, the characteristics of steam cooled blades and vanes had been confirmed in advance. In the results of the trial operation, expected cooling

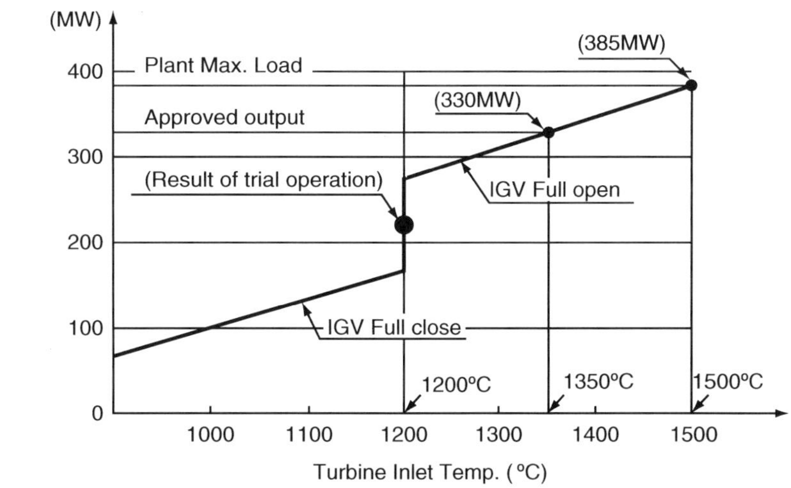

FIGURE 4–113 Load test curve. *(Source: MHI.)*

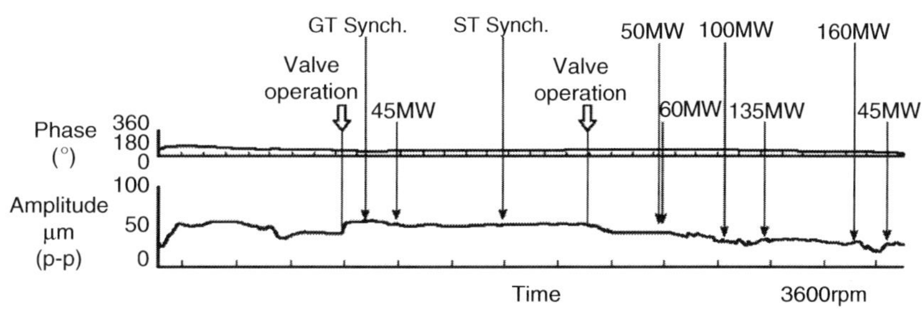

FIGURE 4–114 Stability of steam cooling rotor. *(Source: MHI.)*

performance could be obtained. The output and the turbine inlet temperature were up to 220 MW and 1200°C in this trial operation; however, it was verified that the metal temperature was within the allowable limit by extrapolating the measurement data theoretically into the condition of 385 MW/T, 1500°C.

Confirmation of Performance of Compressor and Turbine. For the compressor, the data from each of the starting up characteristics and compressor performance were obtained in advance by a performance test using a 0.29 scale compressor as mentioned previously. In the results of the trial operation, the characteristics were quite similar to those obtained from the performance test. Therefore, starting up characteristics and compressor adiabatic efficiency were confirmed to be the same as designed.

Turbine aerodynamic efficiency was confirmed to have achieved the design efficiency from calculation with correction of steam leakage. The plant total thermal efficiency was confirmed to almost satisfy the design value.

In summary, the "H" series gas turbine adopted a steam cooling system from the bottoming for turbine blades and vanes, which had not been applied to the existing gas turbines. Thanks to the steam cooling system, the plant total thermal efficiency and output are both improved compared with other gas turbines of the same scale.

The above-mentioned new development elements have been verified by various component tests in advance. The basic structure inherited those of the existing gas turbines with high reliability. From the start of development, the trial operation was carried out in a short period, only 30 months, and the first step of operation was successfully completed with 220 MW output achieved.

In this trial operation, confirmation of each of stable operation of steam-cooled rotor, cooling characteristics of steam-cooled blades and vanes, and performance of compressor and turbine was successfully finished.

As shown above, this trial operation could verify that the "H" series gas turbine could achieve higher plant total thermal efficiency than the existing gas turbines.

This type of combined cycle power plant with LNG firing such as the "H" series gas turbine is now the focus of the world's attention as a trump card for the countermeasure of global warmth because of its low CO_2 emission, and it is quickly gaining in popularity in European and American countries.

In the domestic and Asian countries, spread of such a type of gas turbine is slow due to reduced growth of power demand caused by the global business recession; however, it will spread steadily accompanied by recovery of the business situation.

Chapter 5

Cooling and Load Bearing Systems

"Learn from the mistakes of others. You'll not live long enough to make them all yourself."

—Old Pilot's Saying

Chapter Outline

Seals* are found within the gas turbine's internal air system. In summary, the seals serve to separate hot (working) gases from the cooling air required to ensure that the lubricating oil and bearings stay cool. Fully understanding their function in any one engine involves knowing the internal air system thoroughly.

Some manufacturers are known for their superior cooling design. Better cooling adds cost to the overall engine both in terms of manufacture and performance efficiency. With manufacture, items such as laser-drilled internal passages on blades and vanes would add to the overall cost of cooling seen in the cost of manufacture. With performance, any air used for cooling is not available to be part of the working gases (which turn the turbine and develop the power). This then reduces the efficiency possible for any given air mass intake at the compressor inlet. (Note also that non-working air is also used for subsidiary systems other than cooling. In the case of aircraft engines, for instance, the air may be used for, among other uses, the cabin air-conditioning system.)

The priorities that different manufacturers set on this item are well known among seasoned gas turbine engineers who have worked with every major manufacturer's engines. They are also well known to overhaul and repair shops that see the results of less cooling. It is the operators who wear the final results. As a shrewd design engineer once told his design team (who were anxious to squeeze every last fraction of a percent of efficiency out of a new fighter engine model): he would "rather give the pilot a better chance to get home after a mission than myself a chance to meet his angry widow" (see Chapter 14 in this book).

* Reference: Proceedings of the panel sessions ASME IGTI annual conferences on "Using ECMS (Engine Condition Monitoring Systems) to Extend Component Lives in Gas Turbine Engines," chairman Claire Soares.

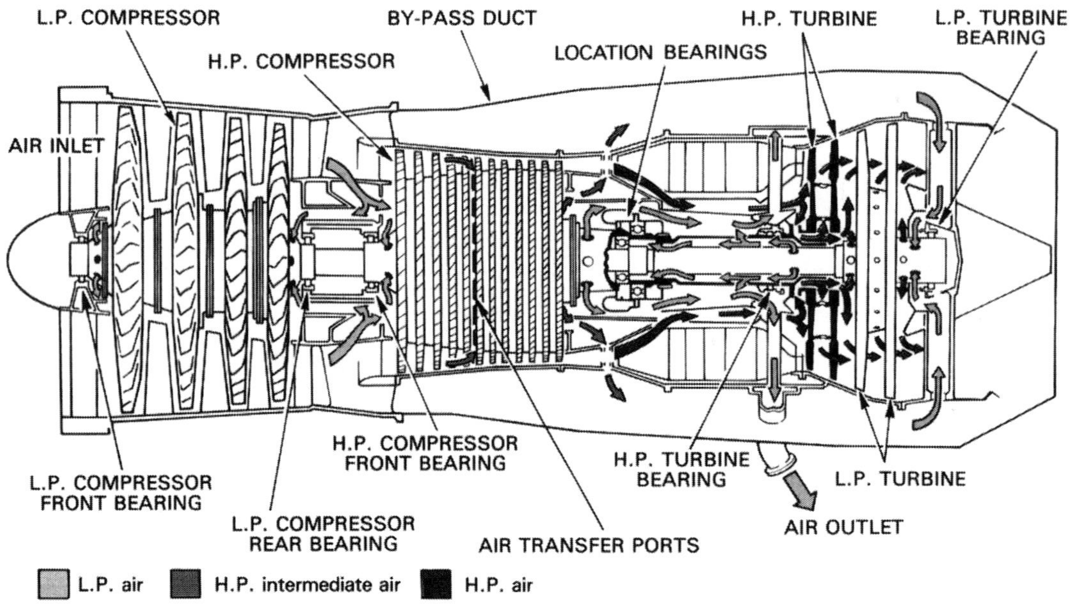

FIGURE 5–1 General internal airflow pattern. *(Source: Rolls Royce.)*

There is another way of keeping the engine cooler: just reduce the turbine inlet temperature. This then extends the framework of basic operation in other ways. As is detailed in Chapter 16, it can help make the use of heavier fuels possible. Residual fuel is used in some models, where customers would like to use cheap residual fuel. At the same operating temperatures that some other manufacturers use for the equivalent power in their models, even with fuel washing and pretreatment, the residual fuel would cause deposits that cake hard on the turbine blades (instead of coming off in the regular washes that use of this fuel requires). For now, let us return to the internal air system.

INTERNAL AIR SYSTEM**

The engine internal air system is defined as those airflows that do not directly contribute to the engine thrust. The system has several important functions to perform for the safe and efficient operation of the engine. These functions include internal engine and accessory unit cooling, bearing chamber sealing, prevention of hot gas ingestion into the turbine disc cavities, control of bearing axial loads, control of turbine blade tip clearances, and engine anti-icing. The system also supplies air for the aircraft services. Up to one-fifth of the total engine core mass airflow may be used for these various functions.

** Source: Adapted, with permission, from Rolls Royce, *The Jet Engine*, 1986, Rolls Royce Plc: UK.

An increasing amount of work is done on the air, as it progresses through the compressor, to raise its pressure and temperature. Therefore, to reduce engine performance losses, the air is taken as early as possible from the compressor commensurate with the requirement of each particular function. The cooling air is expelled overboard via a vent system or into the engine main gas stream, at the highest possible pressure, where a small performance recovery is achieved.

Cooling

An important consideration at the design stage of a gas turbine engine is the need to ensure that certain parts of the engine, and in some instances certain accessories, do not absorb heat to the extent that is detrimental to their safe operation. The principal areas that require air cooling are the combustor and turbine.

Cooling air is used to control the temperature of the compressor shafts and discs by either cooling or heating them. This ensures an even temperature distribution and therefore improves engine efficiency by controlling thermal growth and thus maintaining minimum blade tip and seal clearances. Typical cooling and sealing airflows are shown in Figure 5–1.

Turbine Cooling

High thermal efficiency is dependent upon high turbine entry temperature, which is limited by the turbine blade and nozzle guide vane materials. Continuous cooling of these components allows their environmental operating temperature to exceed the material's melting point without affecting the blade and vane integrity. Heat conduction from the turbine blades to the turbine disc requires the discs to be cooled and thus prevent thermal fatigue and uncontrolled expansion and contraction rates.

An air-cooled high-pressure nozzle guide vane and turbine blade arrangement illustrating the cooling airflow is shown in Figure 5–2. Turbine vane and turbine blade life depends not only on their form but also on the method of cooling; therefore the flow design of the internal passages is important. There have been numerous methods of turbine vane and turbine blade cooling, which have been used throughout the history of gas turbines. Generally, single pass internal (convection) cooling was of great practical benefit but development has led to multi-pass internal cooling of blades, and impingement cooling of vanes with external air film cooling of both vanes and blades; these are shown in Figures 5–3 and 5–4.

The "preswirl nozzles" (Figure 5–2) reduce the temperature and pressure of the cooling air fed to the disc for blade cooling. The nozzles also impart a substantial whirl velocity to assist efficient entry of the air into the rotating cooling passages. Cooling air for the

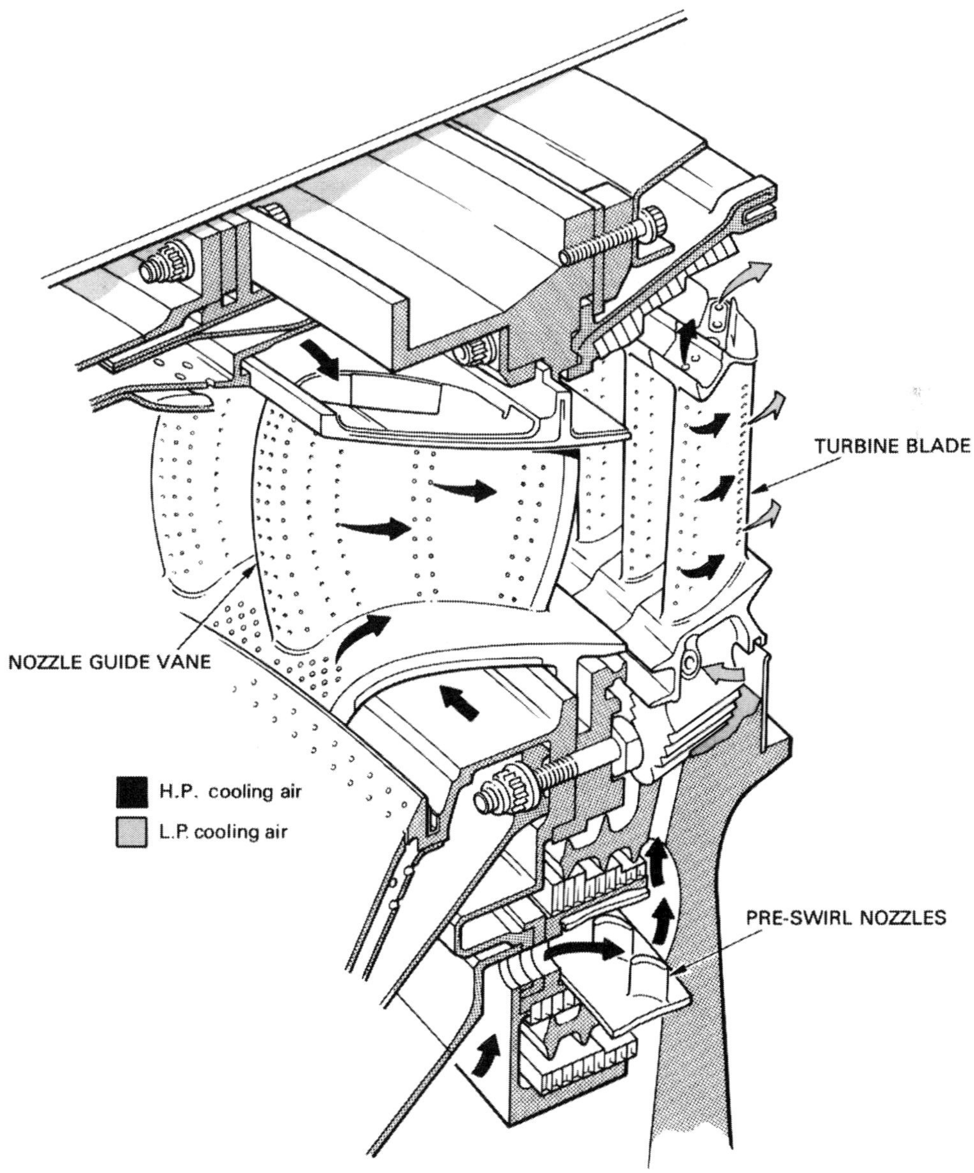

FIGURE 5–2 Nozzle guide vane and turbine blade cooling arrangement. *(Source: Rolls Royce.)*

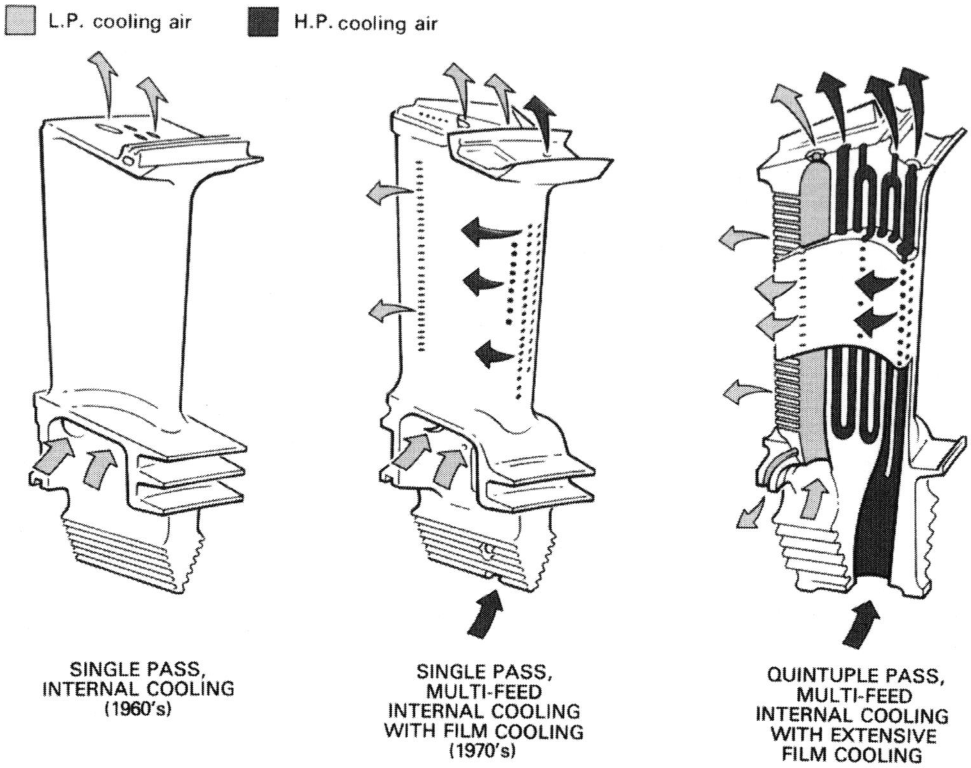

FIGURE 5–3 Development of high-pressure turbine blade cooling. *(Source: Rolls Royce.)*

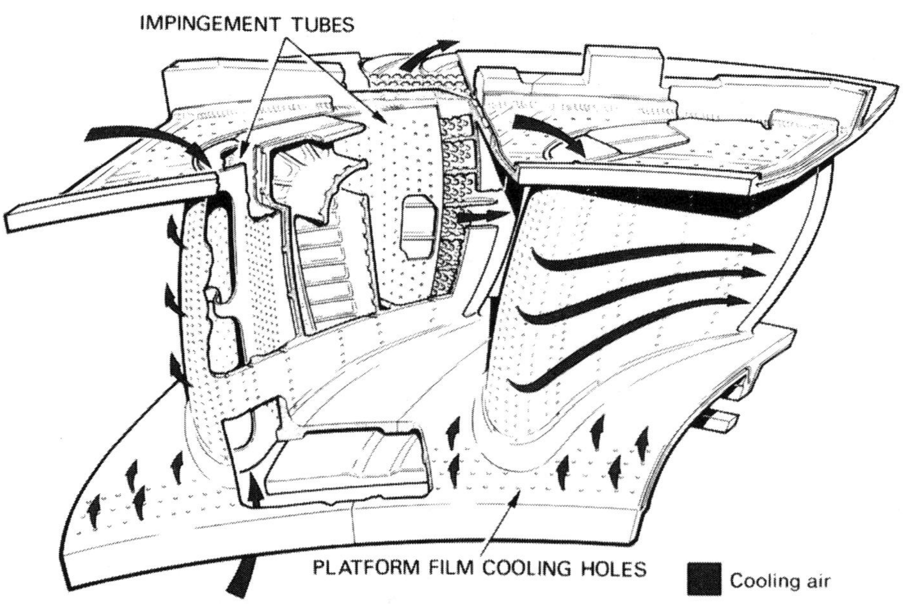

FIGURE 5–4 High-pressure nozzle guide vane construction and cooling. *(Source: Rolls Royce.)*

turbine discs enters the annular spaces between the discs and flows outwards over the disc faces. Flow is controlled by interstage seals and, on completion of the cooling function, the air is expelled into the main gas stream (Figure 5–5): see the discussion on hot gas ingestion.

Bearing Chamber Cooling

Air cooling of the engine bearing chambers is not normally necessary since the lubrication system is adequate for cooling purposes. Additionally, bearing chambers are located, where possible, in the cooler regions of the engine.

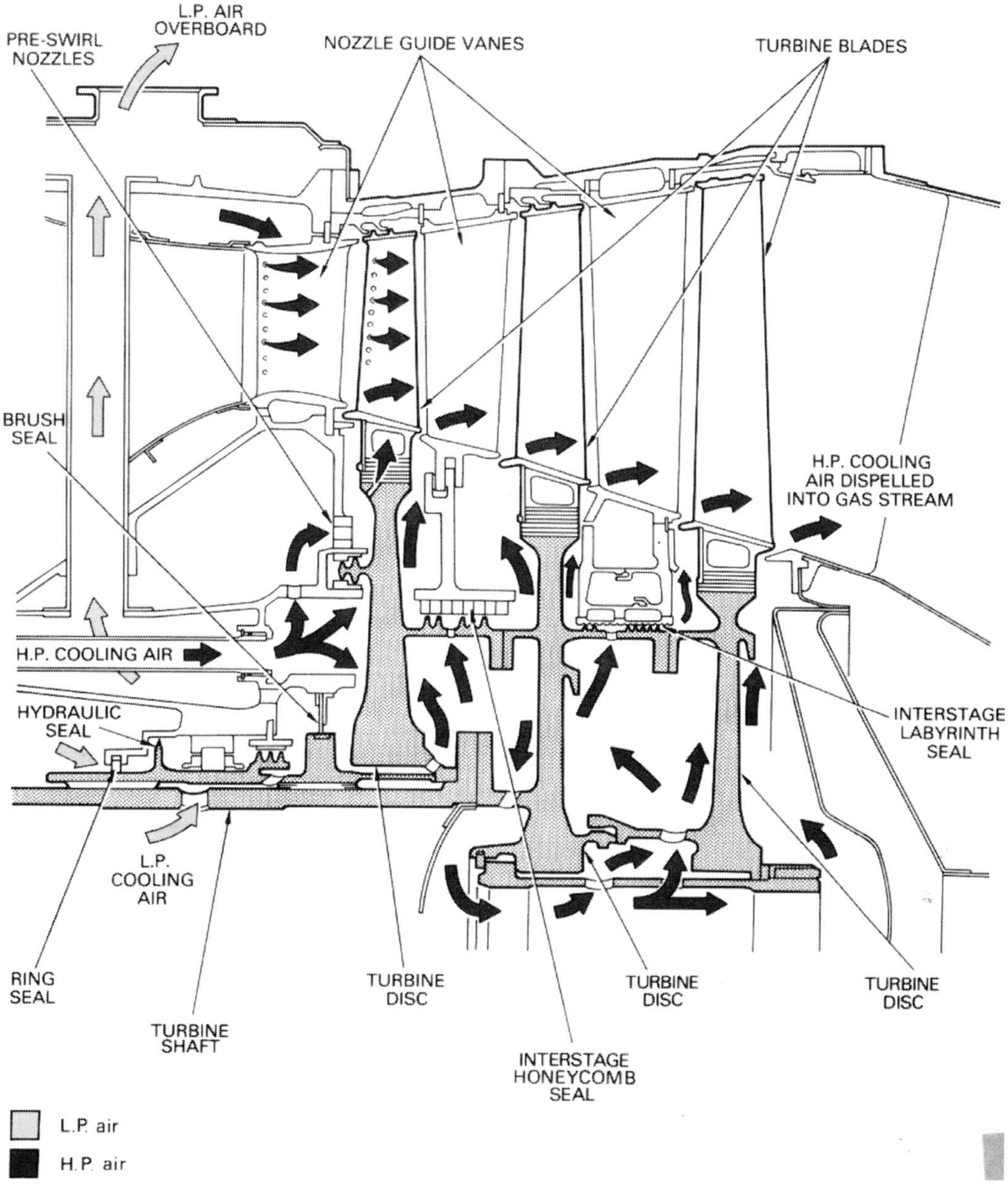

FIGURE 5–5 A hypothetical turbine cooling and sealing arrangement. *(Source: Rolls Royce.)*

In instances where additional cooling is required, it is good practice to have a double skinned bearing housing with cooling air fed into the intermediate space.

Accessory Cooling

A considerable amount of heat is produced by some of the engine accessories, of which the electrical generator is an example, and these may often require their own cooling circuit. When air is used for cooling, the source may be the compressor or atmospheric air ducted from intake louvers in the engine cowlings.

When an accessory unit is cooled during flight by atmospheric air, it is usually necessary to provide an induced circuit for use during static ground running when there would be no external airflow. This is achieved by allowing compressor delivery air to pass through nozzles situated in the cooling air outlet duct of the accessory. The air velocity through the nozzles creates a low-pressure area that forms an ejector, so inducing a flow of atmospheric air through the intake louvers. To ensure that the ejector system only operates during ground running, the flow of air from the compressor is controlled by a valve. A generator cooling system with an ejector is shown in Figure 5–6.

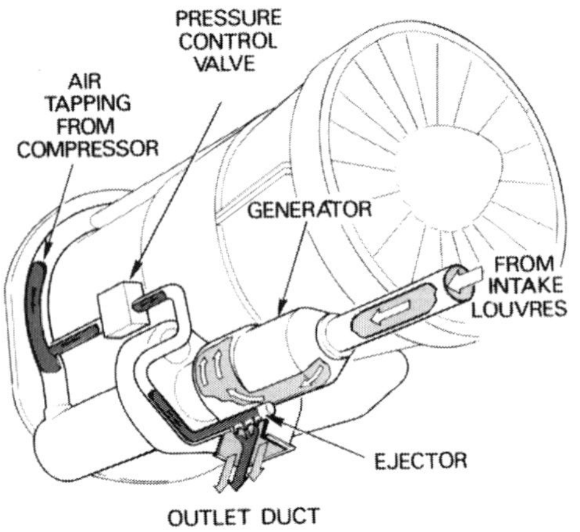

FIGURE 5–6 A generator cooling system. *(Source: Rolls Royce.)*

Sealing

Seals are used to prevent oil leakage from the engine bearing chambers, to control cooling airflows, and to prevent ingress of the mainstream gas into the turbine disc cavities.

Various sealing methods are used on gas turbine engines. The choice of which method is dependent upon the surrounding temperature and pressure, wearability, heat generation, weight, space available, ease of manufacture, and ease of installation and removal. Some of the sealing methods are described in the following paragraphs. A hypothetical turbine showing the usage of these seals is shown in Figure 5–5.

Labyrinth Seals

This type of seal is widely used to retain oil in bearing chambers and as a metering device to control internal airflows. Several variations of labyrinth seal design are shown in Figure 5–7.

A labyrinth seal comprises a finned rotating member with a static bore that is lined with a soft abradable material or a high temperature honeycomb structure. On initial running of the engine the fins lightly rub against the lining, cutting into it to give a minimum clearance. The clearance varies throughout the flight cycle, dependent upon the thermal growth of the parts and the natural flexing of the rotating members. Across each seal fin there is a pressure drop that results in a restricted flow of sealing air from one side of the seal to the other. When this seal is used for bearing chamber sealing, it prevents oil leakage by allowing the air to flow from the outside to the inside of the chamber. This flow also induces a positive pressure, which assists the oil return system.

Seals between two rotating shafts are more likely to be subject to rubs between the fins and abradable material due to the two shafts deflecting simultaneously. This will create excessive heat, which may result in shaft failure. To prevent this, a non-heat producing seal is used where the abradable lining is replaced by a rotating annulus of oil. When the shafts deflect, the fins enter the oil and maintain the seal without generating heat (Figure 5–7).

Ring Seals

A ring seal (Figure 5–7) comprises a metal ring that is housed in a close fitting groove in the static housing. The normal running clearance between the ring and rotating shaft is smaller than that which can be obtained with the labyrinth seal. This is because the ring is allowed to move in its housing whenever the shaft comes into contact with it.

Ring seals are used for bearing chamber sealing, except in the hot areas where oil degradation due to heat would lead to ring seizure within its housing.

Hydraulic Seals

This method of sealing is often used between two rotating members to seal a bearing chamber. Unlike the labyrinth or ring seal, it does not allow a controlled flow of air to traverse across the seal.

Hydraulic seals (Figure 5–7) are formed by a seal fin immersed in an annulus of oil, which has been created by centrifugal forces. Any difference in air pressure inside and outside of the bearing chamber is compensated by a difference in oil level either side of the fin.

Carbon Seals

Carbon seals (Figure 5–7) consist of a static ring of carbon that constantly rubs against a collar on a rotating shaft. Several springs are used to maintain contact between the carbon and the collar. This type of seal relies upon a high degree of contact and does not allow oil or air leakage across it. The heat caused by friction is dissipated by the oil system.

Brush Seals

Brush seals (Figure 5–7) comprise a static ring of fine wire bristles. They are in continuous contact with a rotating shaft, rubbing against a hard ceramic coating. This type of seal has the advantage of withstanding radial rubs without increasing leakage.

Hot Gas Ingestion

It is important to prevent the ingestion of hot mainstream gas into the turbine disc cavities as this would cause overheating and result in unwanted thermal expansion and fatigue. The pressure in the turbine annulus forces the hot

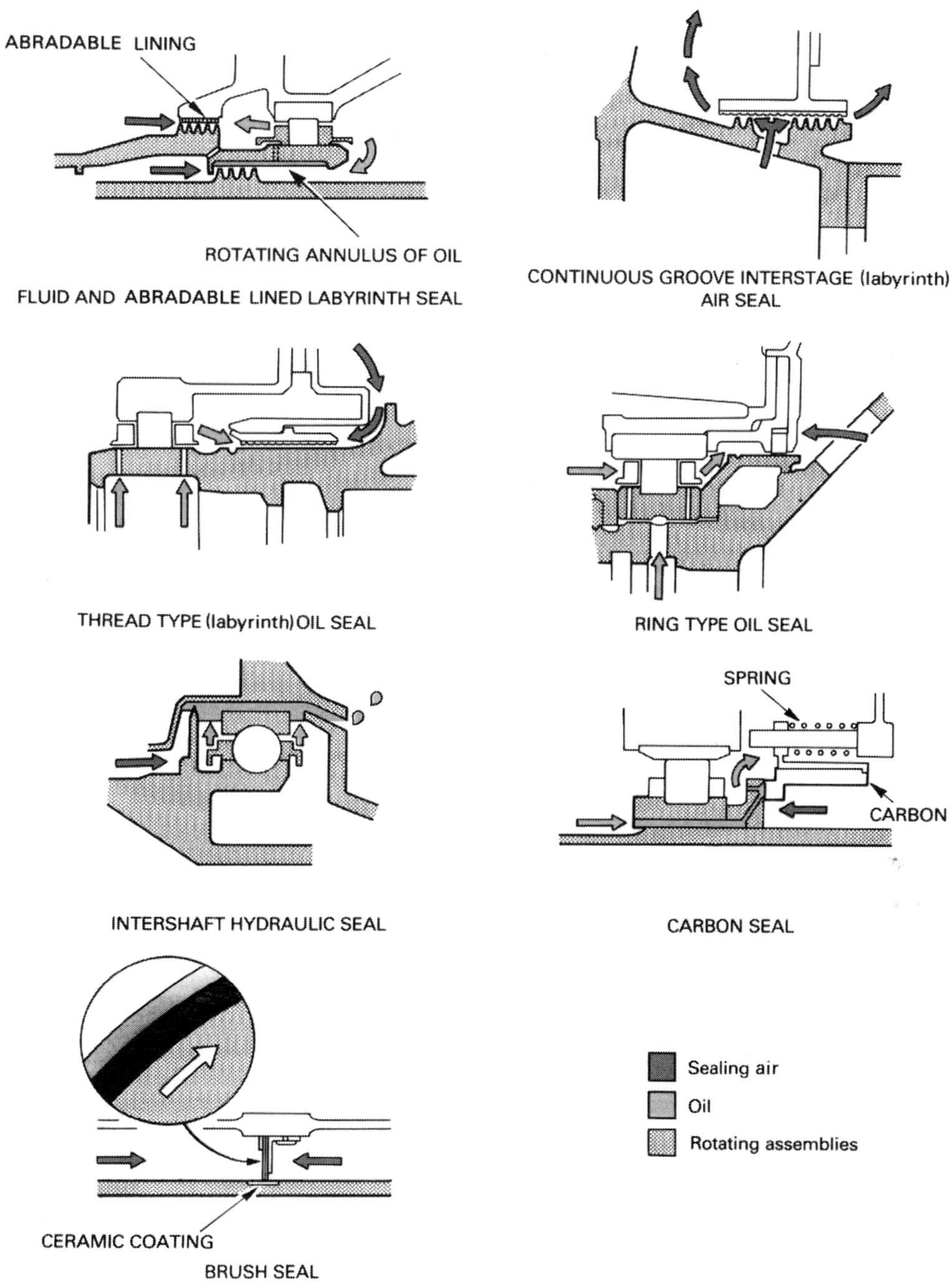

FIGURE 5–7 Typical seals. *(Source: Rolls Royce.)*

gas between the rotating discs and the adjacent static parts into the turbine disc rim spaces. In addition, air near the face of the rotating discs is accelerated by friction causing it to be pumped outwards. This induces a complementary inward flow of hot gas.

Prevention of hot gas ingestion is achieved by continuously supplying the required quantity of cooling and sealing air into the disc cavities to oppose the inward flow of hot gas. The flow and pressure of the cooling and sealing air is controlled by interstage seals (see Figure 5–5).

Control of Bearing Loads

Engine shafts experience varying axial gas loads, which act in a forward direction on the compressor and in a rearward direction on the turbine. The shaft between them is therefore always under tension and the difference between the loads is carried by the location bearing, which is fixed in a static casing (Figure 5–8). The internal air pressure acts upon a fixed diameter pressure balance seal to ensure the location bearing is adequately loaded throughout the engine thrust range.

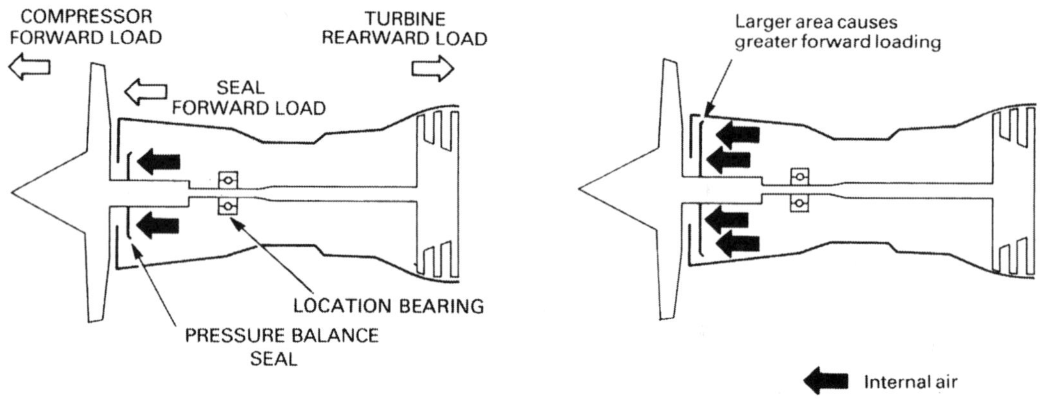

FIGURE 5–8 Control of axial bearing load. *(Source: Rolls Royce.)*

Aircraft Services (for Aircraft Engines)

To provide cabin pressurization, airframe anti-icing, and cabin heat, substantial quantities of air are bled from the compressor. It is desirable to bleed the air as early as possible from the compressor to minimize the effect on engine performance. However, during some phases of the flight cycle it may be necessary to switch the bleed source to a later compressor stage to maintain adequate pressure and temperature.

The other* primary critical component in the cooling system is the load bearing system for the rotor, in other words, the bearings. All turbomachinery bearings depend on the lubrication system to be able to function. Further, they need an oil supply to remove dirt particles and wear debris and cool their working metal surfaces. The oil flow rate to each bearing is dictated by the orifice (supply) plate that designers place upstream of the bearing cavity. The flow varies according to the service temperature of the bearing in question. Bearings close to the hottest areas of the gas turbine get the most oil flow.

"All oil systems are not created equal." Just as there are some OEMs that err on the side of "more cooling air at the cost of efficiency" there are OEMs that prefer "too much rather than just enough or too little cooling oil." Just these two facts of life can affect the reception a group of operations-savvy end users give a new engine or a joint-venture design new engine (see Chapter 14).

LUBRICATION**

The lubrication system is required to provide lubrication and cooling for all gears, bearings, and splines. It must also be capable of collecting foreign matter, which, if left in a bearing housing or gearbox, can cause rapid failure. Additionally, the oil must protect the lubricated components that are manufactured from non-corrosion resistant materials. The oil must accomplish these tasks without significant deterioration.

Lubricating Systems

The requirements of a turbopropeller engine are somewhat different to any other type of aero-gas turbine. This is due to the additional lubrication of the heavily loaded propeller reduction gears and the need for a high-pressure oil supply to operate the propeller pitch control mechanism.

Most gas turbine engines use a self-contained recirculatory lubrication system in which the oil is distributed around the engine and returned to the oil tank by pumps. However, some engines use a system known as the total loss or expendable system in which the oil is spilled overboard after the engine has been lubricated.

There are two basic recirculatory systems, known as the "pressure relief valve" system and the "full-flow" system. The major difference between them is in the control of the oil flow to the bearings. In both systems the temperature and pressure of the oil are critical to the correct and safe running of the engine. Provision is therefore made for these parameters to be indicated in the cockpit.

Pressure Relief Valve System

In the pressure relief valve system the oil flow to the bearing chambers is controlled by limiting the pressure in the feed line to a given design value. This is accomplished by the use of a spring-loaded valve that allows oil to be directly returned from the pressure pump outlet to the oil tank or pressure pump inlet when the design value is exceeded. The valve opens at a pressure that corresponds to the idling speed of the engine, thus giving a constant feed pressure over normal engine operating speeds. However, increasing engine speed causes the bearing chamber pressure to rise sharply. This reduces the pressure difference between the

* Reference: Proceedings of the panel sessions ASME IGTI annual conferences on "Using ECMS (Engine Condition Monitoring Systems) to Extend Component Lives in Gas Turbine Engines," chairman Claire Soares.
** Source: Adapted, with permission, from Rolls Royce, *The Jet Engine*, 1986, Rolls Royce Plc: UK.

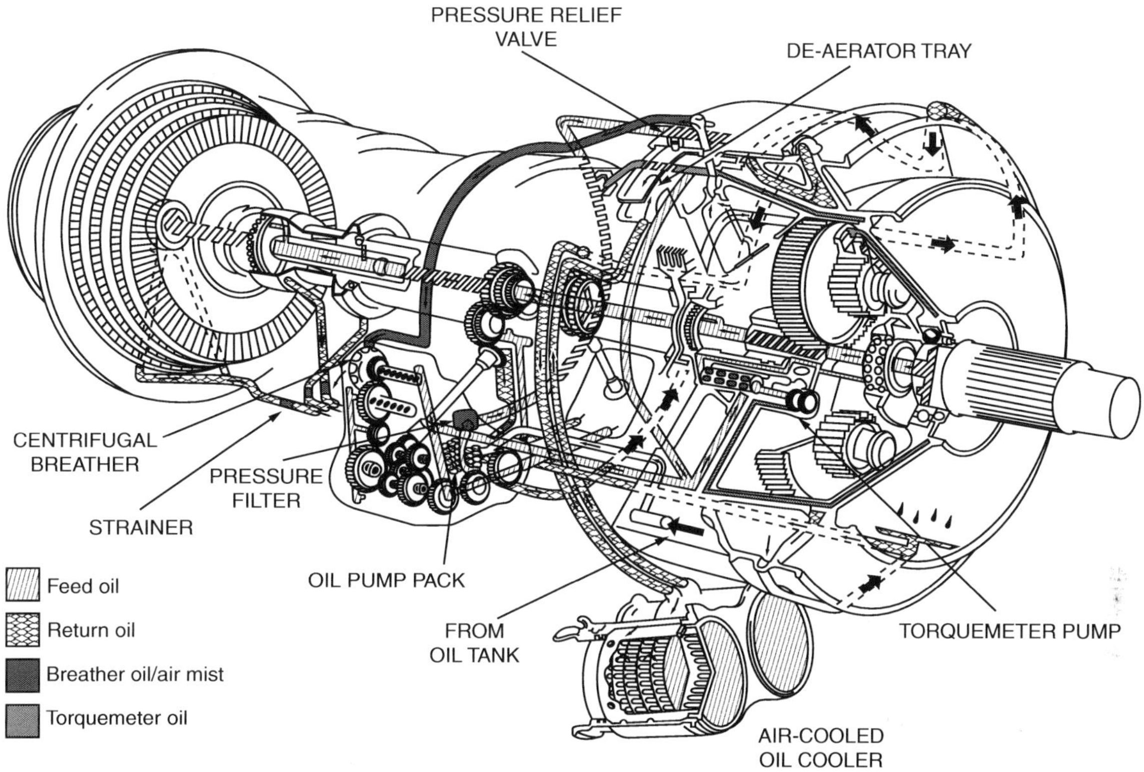

FIGURE 5–9 A pressure relief valve type oil system. *(Source: Rolls Royce. Adapted for reproduction in black and white.)*

bearing chamber and feed jet, thus decreasing the oil flow rate to the bearings as engine speed increases. To alleviate this problem, some pressure relief valve systems use the increasing bearing chamber pressure to augment the relief valve spring load. This maintains a constant flow rate at the higher engine speeds by increasing the pressure in the feed line as the bearing chamber pressure increases.

Figure 5–9 shows the pressure relief valve system for a turbopropeller engine and indicates the basic components that comprise an engine lubrication system. The oil pressure pump draws oil from the tank through a strainer, which protects the pump gears from debris that may have entered the tank. Oil is then delivered through a pressure filter to the pressure relief valve that maintains a constant oil delivery pressure to the feed jets in the bearing chambers. Some engines may have an additional relief valve (pressure limiting valve) fitted at the oil pressure pump outlet. This valve is set to open at a much higher value than the pressure relief valve to return the oil to the inlet side of the oil pressure pump in the event of the system becoming blocked. A similar valve may also be fitted across the pressure filter to prevent oil starvation of the bearing chambers should the filter become partially blocked or the oil having a high viscosity under cold starting conditions, preventing sufficient flow through the filter. Provision is also made to supply oil to the propeller pitch control system, reduction gear, and torquemeter system. Scavenge pumps return the oil to the tank via the oil cooler. On entering the tank, the oil is deaerated, ready for recirculation.

Full Flow System

Although the pressure relief valve system operates satisfactorily for engines that have a low bearing chamber pressure, which does not unduly increase with engine speed, it becomes an undesirable system for engines that have high chamber pressures. For example, if a bearing chamber has a maximum pressure of 90 lb per sq in., it would require a pressure relief valve setting of 130 lb per sq in. to produce a pressure drop of 40 lb per sq in. at the oil feed jet. This results in the need for large pumps and difficulty in matching the required oil flow at slower speeds.

The full flow system achieves the desired oil flow rates throughout the complete engine speed range by dispensing with the pressure relief valve and allowing the pressure pump delivery pressure to supply directly the oil feed jets. Figure 5–10 shows an example of this system, which can be found on a turbofan engine. The pressure pump size is determined by the flow required at maximum engine speed. The use of this system allows smaller pressure and scavenge pumps to be used since the large volume of oil that is spilled by the pressure relief valve system at maximum engine speed is obviated.

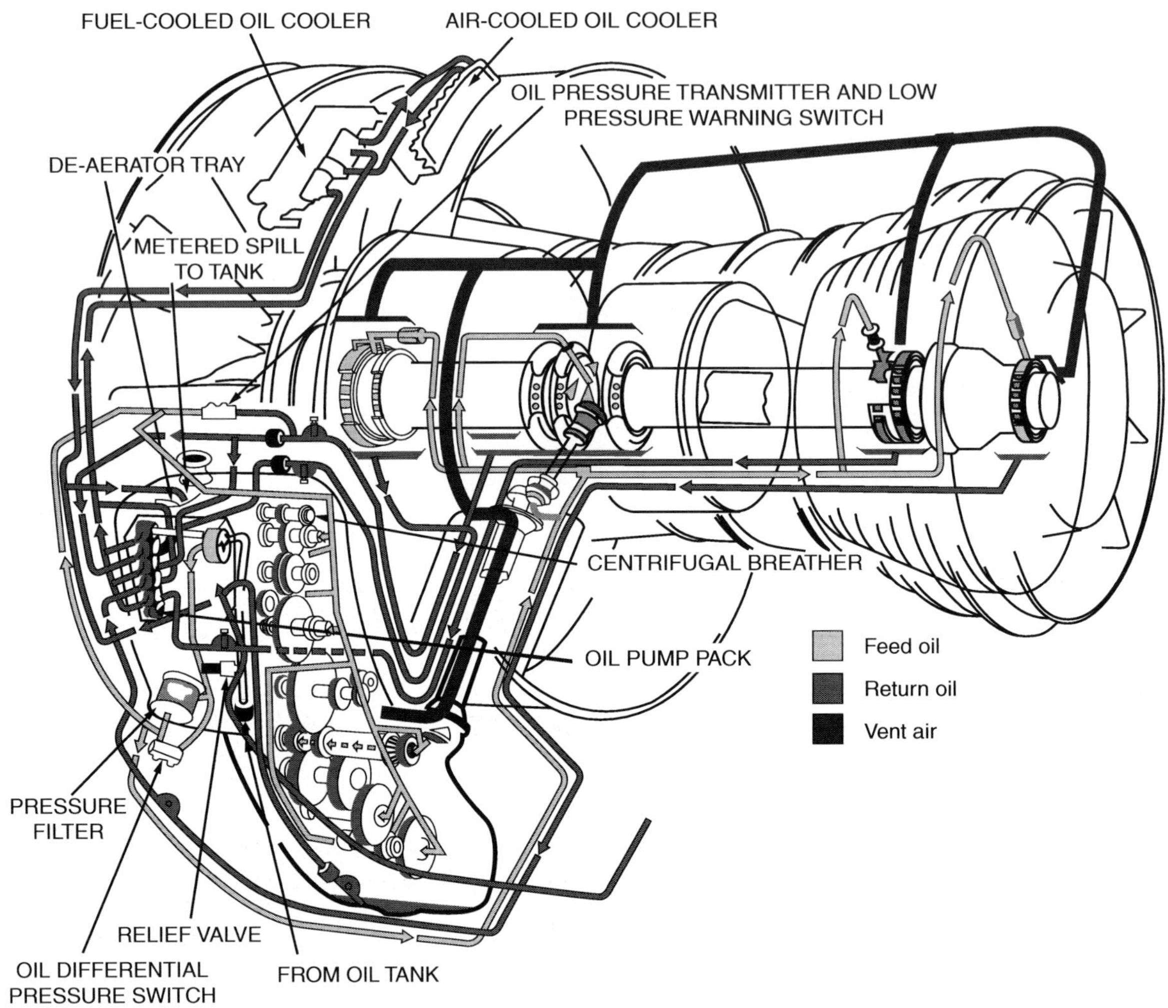

FIGURE 5–10 A full-flow-type oil system. *(Source: Rolls Royce. Adapted for reproduction in black and white.)*

To prevent high oil pressures from damaging filters or coolers, pressure-limiting valves are fitted to bypass these units. These valves normally only operate under cold starting conditions or in the event of a blockage. Advance warning of a blocked filter may be indicated in the cockpit by a differential pressure switch, which senses an increase in the pressure difference between the inlet and outlet of the filter.

Total Loss (Expendable) System

For engines that run for periods of short duration, such as booster and vertical lift engines, the total loss oil system is generally used. The system is simple and incurs low weight penalties because it requires no oil cooler, scavenge pump, or filters. On some engines oil is delivered in a continuous flow to the bearings by a plunger-type pump, indirectly driven from the compressor shaft; on others it is delivered by a piston-type pump operated by fuel pressure (Figure 5–11). In the latter, the oil supply is automatically selected by the high-pressure fuel shutoff valve (cock) during engine starting and is delivered as a single shot to the front and rear bearings. On some engines provision is made for a second shot to be delivered to the rear bearing only, after a predetermined period.

After lubricating the fuel unit and front bearings, the oil from the front bearing drains into a collector tray and is then ejected into the main gas stream through an ejector nozzle. The oil that has passed through the rear bearing drains into a reservoir at the rear of the bearing where it is retained by centrifugal force until the engine is shut down. This oil then drains overboard through a central tube in the exhaust unit inner cone.

Oil System Components

The oil tank (Figure 5–12) is usually mounted on the engine and is normally a separate unit, although it may also be an integral part of the external gearbox. It must have provision to allow the lubrication system to be drained and replenished. A sight glass or a dipstick must also be

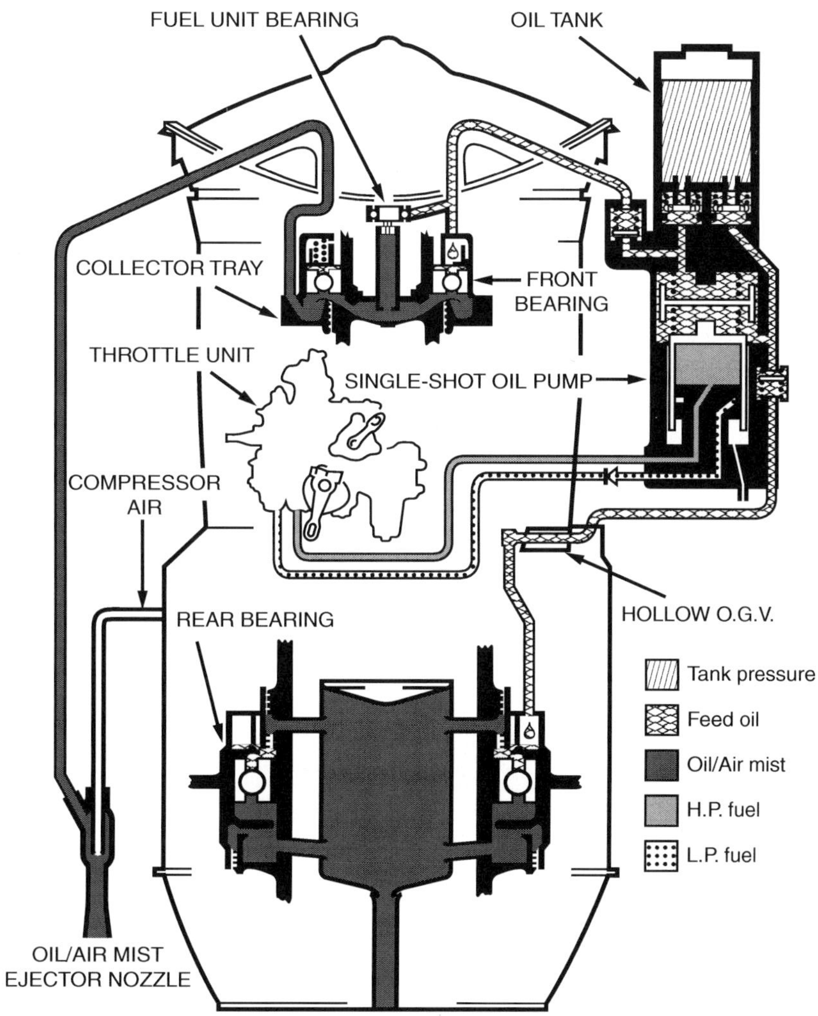

FIGURE 5–11 A total loss (expendable) oil system. *(Source: Rolls Royce. Adapted for reproduction in black and white.)*

incorporated to allow the oil system contents to be checked. The filler can be either the gravity or pressure filling type; on some engines both types are fitted. Provision is also made for a continuous supply of oil to be made available in aircraft that are designed to operate during inverted flight conditions. Since air is mixed with the oil in the bearing chambers, a deaerating device is incorporated within the oil tank, which removes the air from the returning oil.

The oil pumps are vital to the efficient operation of the engine. Failure of the pumps will necessitate a rapid shutdown of the engine. For this reason, the oil pump driveshafts do not incorporate a weak shear-neck because they must continue to supply oil for as long as possible, regardless of damage.

As the feed oil is distributed to all the lubricated parts of the engine, a substantial amount of sealing air mixes with it and increases its volume. Additionally, the bearing chambers operate under differing pressures. Therefore, to prevent flooding it is usually necessary to have a scavenge pump for each chamber.

Gear type pumps are normally used in recirculatory oil systems but vane and gerotor pumps are employed in some engines. The simplicity of single-shot pumps make them ideal for engines that run for a short duration and use the total loss type of oil system.

Gear pumps (Figure 5–13) consist of a pair of intermeshing steel gears that are housed in a close-fitting aluminum casing. When the gears are rotated, oil is drawn into the pump, carried around between the teeth and casing, and delivered at the outlet.

Since a small quantity of incompressible oil becomes trapped in the gear mesh, which can cause a hydraulic lock and possible pump damage, a relief slot is machined into the end faces of the casing to provide an escape route for the oil.

Gear pumps are used both as pressure (feed) pumps and scavenge (return) pumps and are incorporated within a

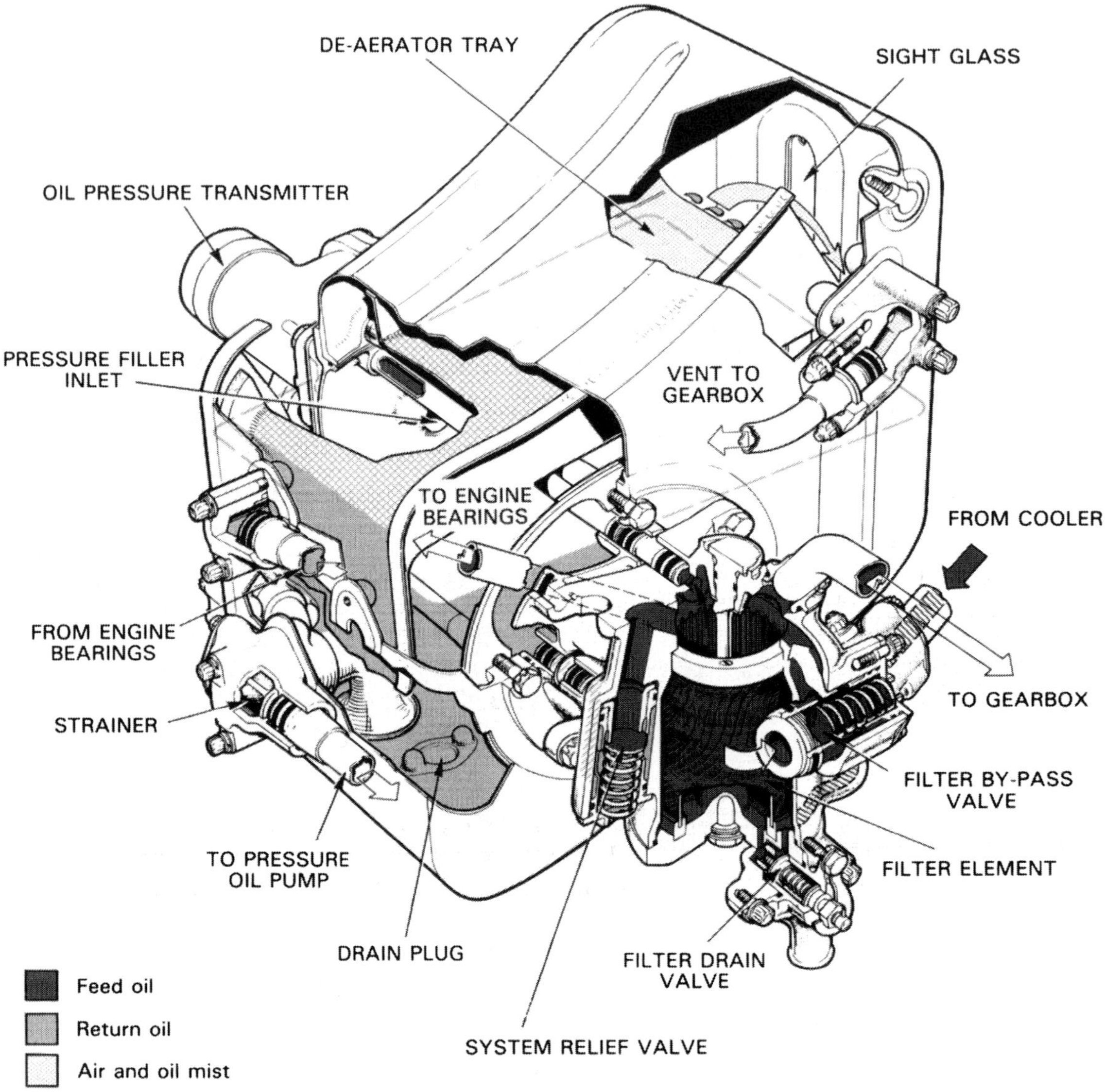

FIGURE 5—12 An oil tank. *(Source: Rolls Royce.)*

common casing. The oil pumps pack is driven by the accessory drive system.

Single-shot pumps (Figure 5—14) have a quantity of oil contained within a cylinder. When the piston is forced up the cylinder bore, under the control of the throttle unit, the oil forces the outlet valves to open allowing a flow of oil to the parts required to be lubricated. When the piston reaches the top of the cylinder bore the outlet valves close due to the reduced oil pressure. Recharging of the oil pump cylinder is achieved by a spring forcing the piston to its original position. This reduces the pressure between the cylinder and the oil tank, which allows the oil replenishing valves to open until the cylinder is recharged.

The most common type of oil distribution device is a simple orifice that directs a metered amount of oil onto its target. These jet orifices are positioned as close to the target area as possible to overcome the possibility of the local turbulent environment deflecting the jet of oil. The smallest diameter of a jet orifice is 0.04 in., which allows a flow of 12 gallons per hour when operating at a pressure of 40 lb per sq in. The use of restrictors upstream can reduce the flow rate if required.

All engines transfer heat to the oil by friction, churning, and windage within a bearing chamber or gearbox. It is therefore common practice to fit an oil cooler in recirculatory oil systems. The cooling medium may be fuel or air and, in some instances, both fuel-cooled and air-cooled coolers are used.

Some engines that utilize both types of cooler may incorporate an electronic monitoring system that switches in the air-cooled cooler only when it is necessary. This

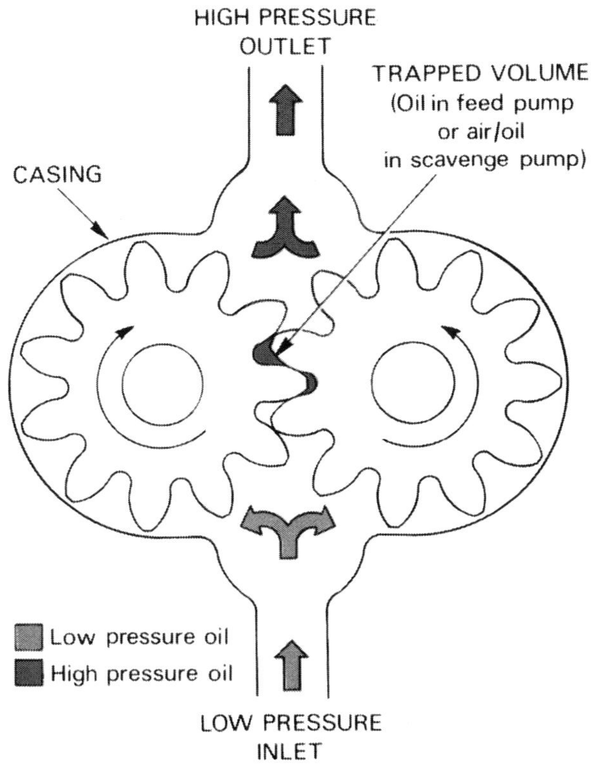

FIGURE 5–13 Principle of a gear pump. *(Source: Rolls Royce.)*

maintains the ideal oil temperature and improves the overall thermal efficiency.

The fuel-cooled oil cooler (Figure 5–15) has a matrix divided into sections by baffle plates. A large number of tubes convey the fuel through the matrix, the oil being directed by the baffle plates in a series of passes across the tubes. Heat is transferred from the oil to the fuel, thus lowering the oil temperature.

The fuel-cooled oil cooler incorporates a bypass valve fitted across the oil inlet and outlet. The valve operates at a preset pressure difference across the cooler and thus prevents engine oil starvation in the event of a blockage. A pressure maintaining valve is usually located in the feed line of the cooler, which ensures that the oil pressure is always higher than the fuel pressure. In the event of a cooler internal fault developing, the oil will leak into the fuel system rather than the potentially dangerous leakage of fuel into the oil system.

The air-cooled oil cooler is similar to the fuel-cooled type in both construction and operation; the main difference is that air is used as the cooling medium.

Magnetic plugs, or chip detectors (Figure 5–16), are fitted on the scavenge (return) side to collect ferritic debris from each bearing chamber. They are basically permanent magnets inserted in the oil flow and are retained in self-sealing valve housings. Safety features incorporated in the design ensure correct retention within the housing. Upon examination they can provide a warning of impending failure without having to remove and inspect the filters. They are designed to be removed during maintenance inspection, for condition, monitoring purposes, without oil loss occurring. Additionally, they may be connected to a cockpit warning system to give an in-flight indication.

Bearings

Bearings provide a means of accurately locating the rotors while transmitting high forces with very little rotational resistance. Jet engines tend to use rolling element bearings, but occasional application of plain bearings can be found. There are two types of bearing used in a gas turbine: ball bearings and roller bearings. Ball bearings use balls as the rolling elements, which, because of their shape, can withstand both radial and axial forces. This makes ball bearings suitable for transmitting thrust. Roller bearings use cylinders as the rolling elements. The rollers can transmit radial load across their diameters, but allow the shaft to slide lengthwise. Using a single ball bearing for thrust and one or more roller bearings to support a rotor allows positioning at the thrust bearing, but freedom for growth at the roller bearings.

Bearings can be used between rotating and fixed structures, or can be used between rotating components. For example, the LP shaft thrust bearing on three-shaft engines is mounted between the LP and IP rotors. All rotating shafts in the engine, including the drive shafts from the internal gearbox to the accessory gearbox and the gear shafts within the accessory gearbox, are mounted on rolling element bearings.

All rolling element bearings consist of an inner and outer race, a cage, and the rolling elements themselves. One or both of the races have a raceway formed within it to guide the rolling elements. The cage is used to maintain spacing of the rolling elements, which are trapped inside pockets. The cage has a clearance with respect to both the inner and outer races, but is primarily located by one or the other, depending upon the requirements of the bearing application. To ensure that the cage runs concentrically, the clearance between the locating lands on the race and the cage is small and well lubricated so that it operates without appreciable wear. The cage may also have features to assist in catching and directing lubrication to the rolling elements (Figures 5–17 to 5–19).

Ball Bearings

Ball bearings provide axial location for a rotating shaft, but will usually carry a substantial radial load. A rotating shaft is supported by at least two bearings: normally, one is a ball bearing and the other a roller bearing.

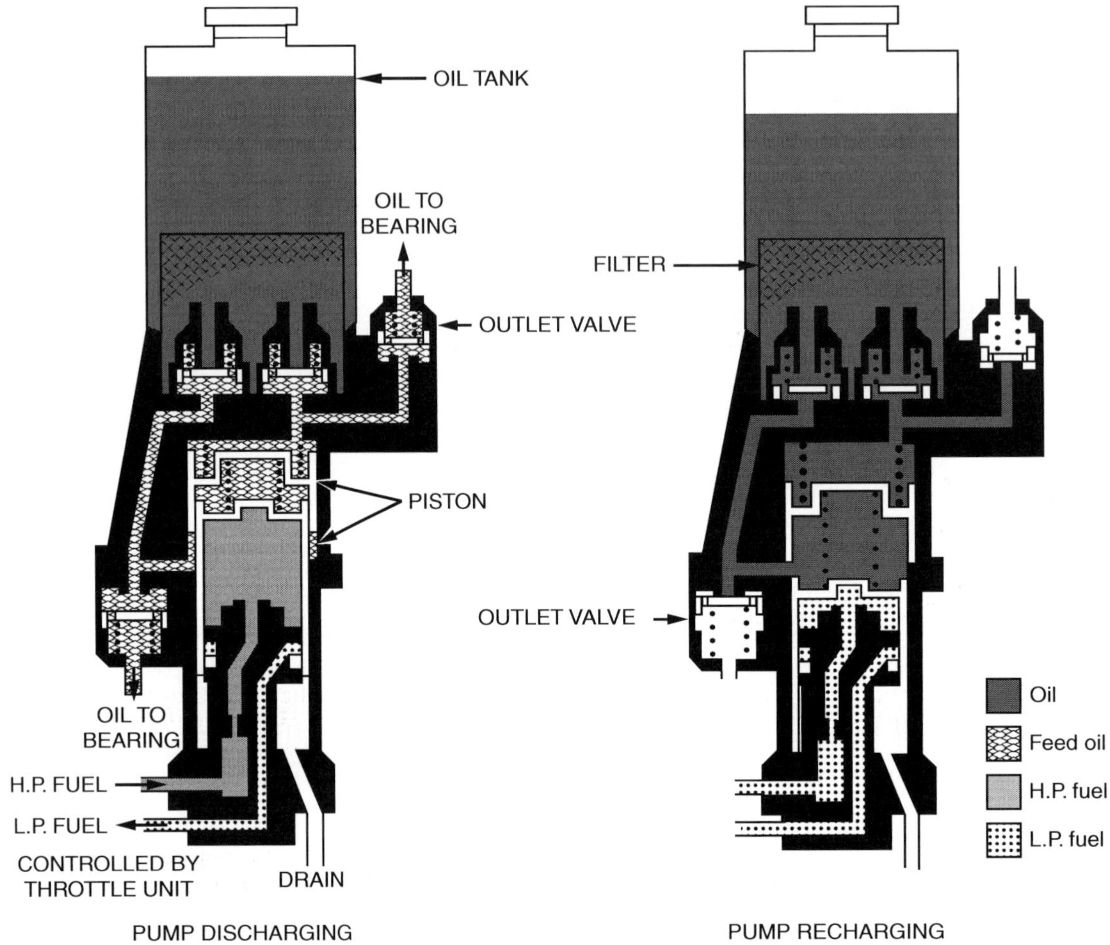

FIGURE 5–14 A single-shot oil pump. *(Source: Rolls Royce. Adapted for reproduction in black and white.)*

Main shaft location bearings are situated in the internal gearbox on three-shaft engines and on many two-shaft engines. Putting these highly-loaded bearings in a relatively cool part of the engine greatly simplifies design of the load paths through the engine structures. Accurate axial location provided by the ball bearings is essential for close control of compressor tip clearances.

Deep Groove

Deep-groove ball bearings have single-piece inner and outer rings. The cage is made, therefore, from two pieces to allow the bearing to be assembled. The inner and outer track forms are both derived from a single radius, and so the balls can only make single-point contact with each race. They are often used for applications with moderate radial loads and light axial loads.

Two-Piece Raceway Type

This bearing commonly has a single-piece outer race and two-piece inner race, although it is possible to have a two-piece outer and single-piece inner. Splitting one of the races allows the bearing to be assembled and to have a single-piece cage. The raceway in each race is formed from two radii (one for each half of the raceway) struck from different centers so that the form of the track is a gothic arch. Since one of the races must be split, the thrust load must be maintained at a high level during operation to prevent the balls from contacting the split. Therefore these bearings are used more in applications that require high thrust-carrying capacity.

The gothic arch form allows oil to be fed into the center of the inner track without the risk of damage that might result from the balls running over the edges of the oil feed holes. Supplying oil to the center of the inner race gives good lubrication at the ball contacts. This configuration is most commonly used for main shaft location bearings, as used on the Trent LP, IP, and HP main shafts (Figures 5–20 and 5–21).

Roller Bearings

Roller bearings are used in all main shaft and auxiliary drive shaft applications to support pure radial load, and

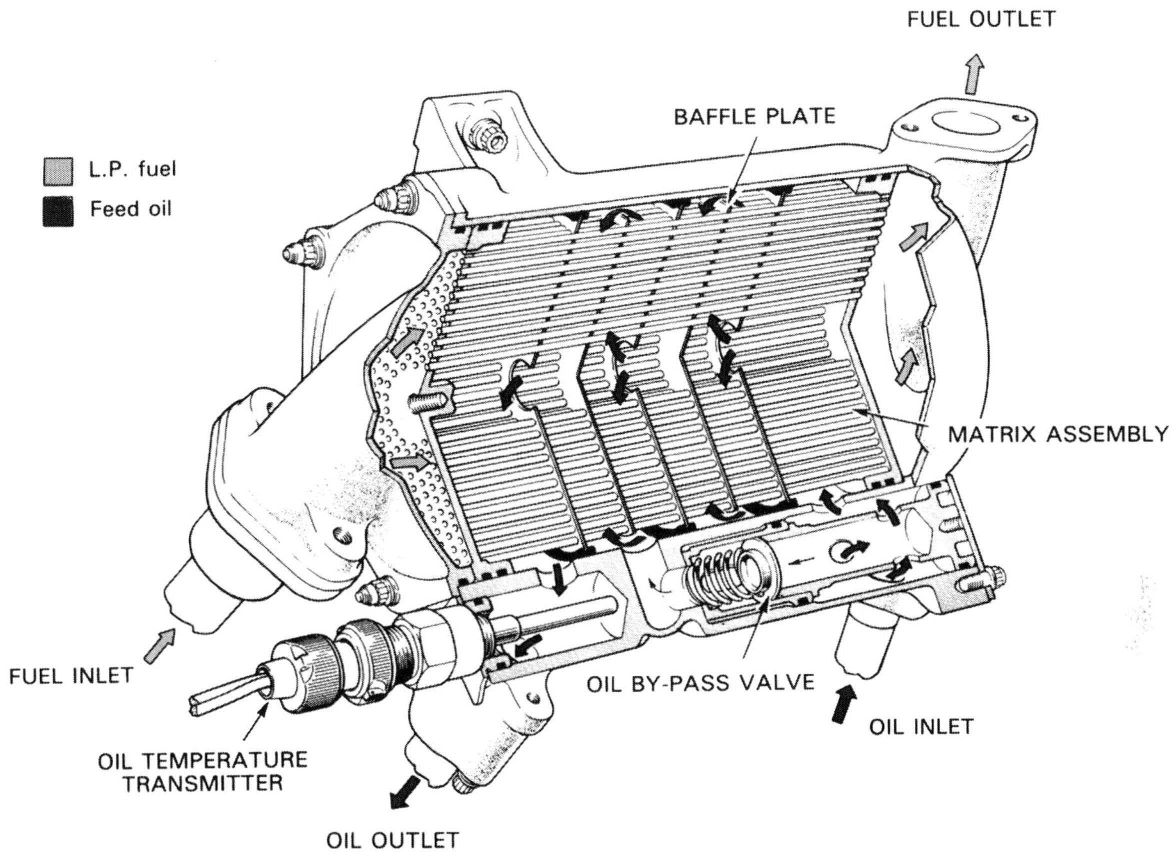

FIGURE 5–15 A low-pressure fuel-cooled oil cooler. *(Source: Rolls Royce.)*

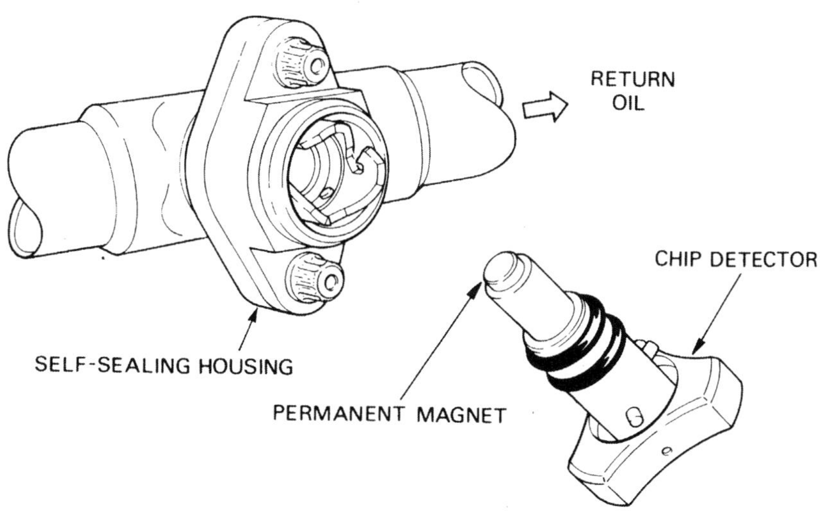

FIGURE 5–16 A magnetic chip detector. *(Source: Rolls Royce.)*

allow for axial shaft elongation due to temperature changes with no additional load effect on the bearing. They are usually located at the ends of the turbine and compressor shafts and are often mounted in a housing, but separated from it by a layer of pressurized oil known as a squeeze film damper.

In many cases, instead of having a separate inner race for roller bearings, the "inner race" is an integral part of the shaft or stub shaft. This reduces complexity, weight, and build-up of concentricity tolerances. Overall, this is cost effective, but the cost of replacement or repair is likely to be higher than for separate inner races.

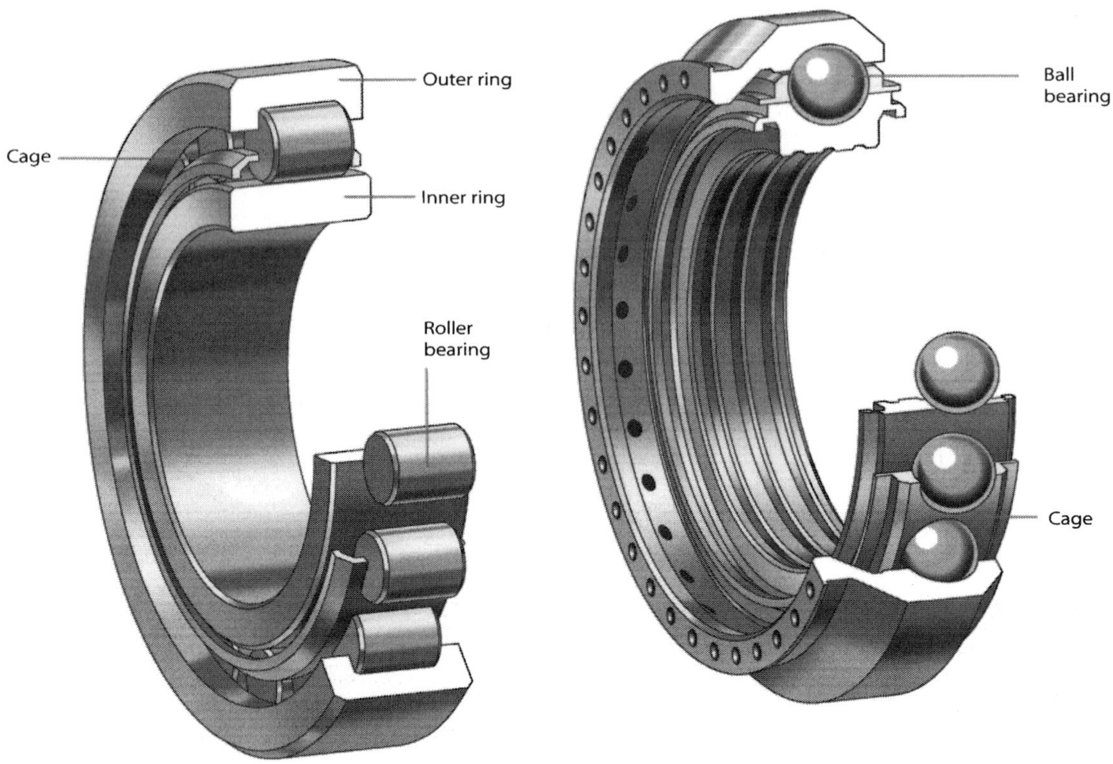

FIGURE 5–17 Ball and roller bearing components. *(Source: Rolls Royce.)*

FIGURE 5–18 View of an assembled ball bearing, also called a thrust or location bearing. *(Source: Rolls Royce.)*

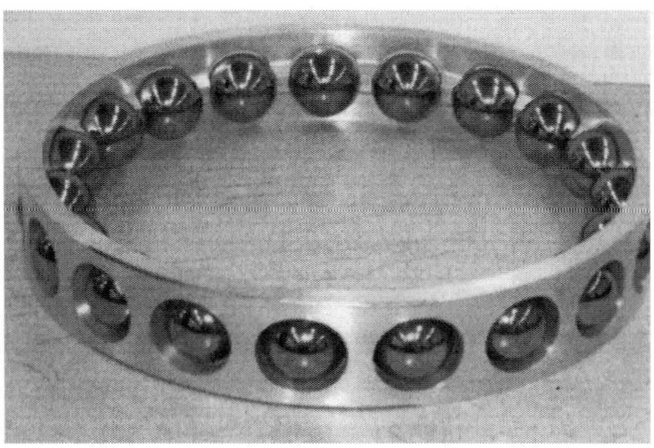

FIGURE 5–19 The balls and cage removed from the inner and outer race. *(Source: Rolls Royce.)*

Bearing Internal Clearance

Bearing diametral clearance is the total free movement between the inner and outer races in the radial direction. For ball bearings, there must be some positive diametral clearance under all operating conditions. Roller bearings and ball bearings that are mainly radially loaded benefit from low diametral clearance. This maximizes the number of loaded elements and reduces rolling element-to-race stress levels. For roller bearings, a low diametral clearance also helps to reduce the risk of roller skidding.

In some engines, to minimize the effect of the dynamic loads transmitted from the rotating assemblies to the bearing housings, a "squeeze film" type of bearing is used (Figure 5–22). They have a small clearance between the outer race of the bearing and housing with the clearance being filled with oil. The oil film dampens the radial motion of the rotating assembly and the dynamic loads transmitted to the bearing housing thus reducing the vibration level of the engine and the possibility of damage by fatigue.

Bearing Materials

Bearings are currently manufactured from steels that may be either case-hardened or through-hardened to suit the

FIGURE 5—20 An LP fan roller bearing from a three-shaft engine. *(Source: Rolls Royce.)*

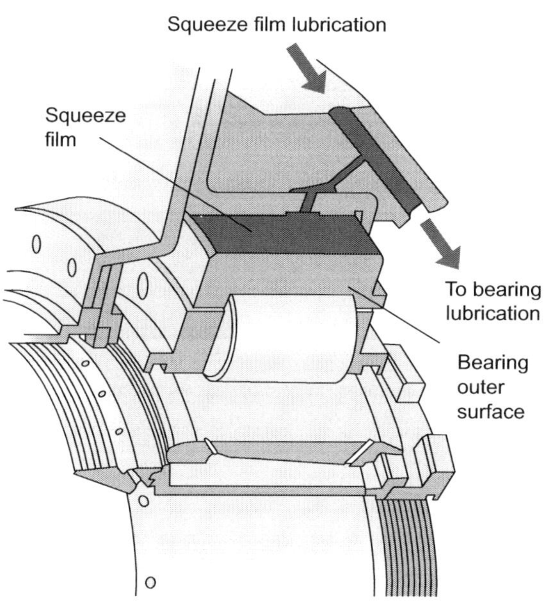

FIGURE 5—21 The squeeze film damper reduces vibration transmission to the engine structure and reduces loads on the shaft due to flexing of the engine casings. *(Source: Rolls Royce.)*

application. Rolling element bearings operate with high local stress levels at the contacts between the rolling elements and the races. This means that the material used must have a very high resistance to rolling contact fatigue. Other requirements of the material are a high

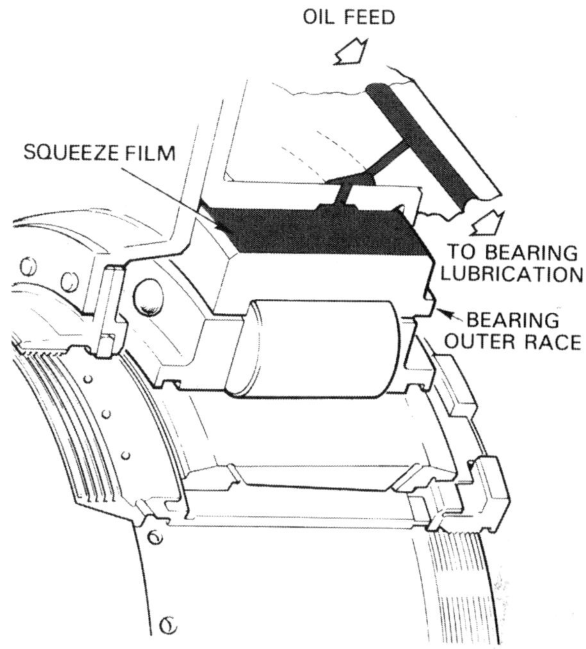

FIGURE 5—22 A squeeze film bearing. *(Source: Rolls Royce.)*

level of hardness at the surface, high temperature and wear resistance, and often a tough core. The effects of rotation and installation fits can further increase these stress levels. Surface-hardened materials have an additional attribute: a surface that is usually in compression. This is beneficial to a surface in tension and tends to cancel out the effects of rotation and fit. Corrosion resistance and damage tolerance may be other important attributes in some applications.

Most bearings employ high-quality steels for the cage material. However, lower duty bearings may use phosphor bronze or brass cages. Silver plating and phosphate coating enhance friction, lubrication, and wear properties on steel cages.

Bearing Developments

The demands for future gas turbine bearings will be longer life, higher speeds, higher load capacity, smaller diameters, and (for aero engines) less weight. Steel processing continues to improve and is delivering cleaner, inclusion-free materials, leading to higher fatigue resistance.

Current technology goes some way to meeting these needs. However, alternative materials such as ceramics, polymers, and composites will play a future role in aerospace bearing technology, particularly in high-speed applications. They offer high strength for low weight and work well in high temperatures and poor lubrication conditions. Specialist surface treatments are also being developed that will enhance bearing performance.

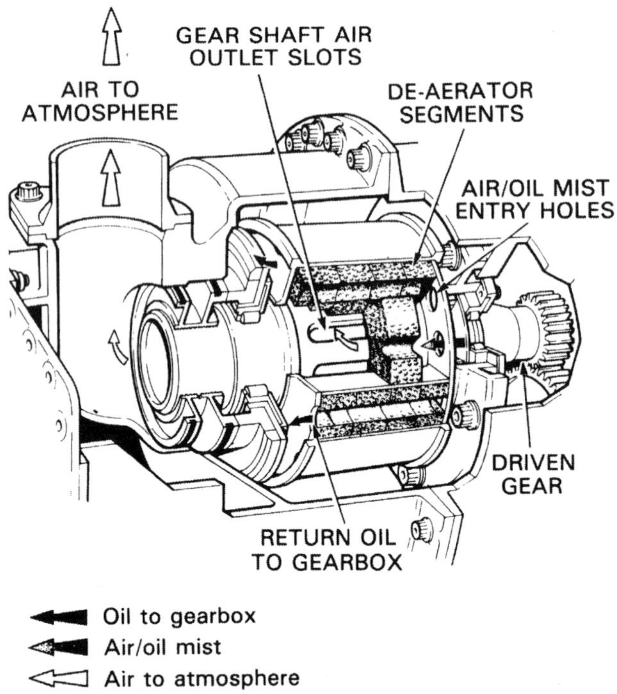

FIGURE 5–23 A centrifugal breather. *(Source: Rolls Royce.)*

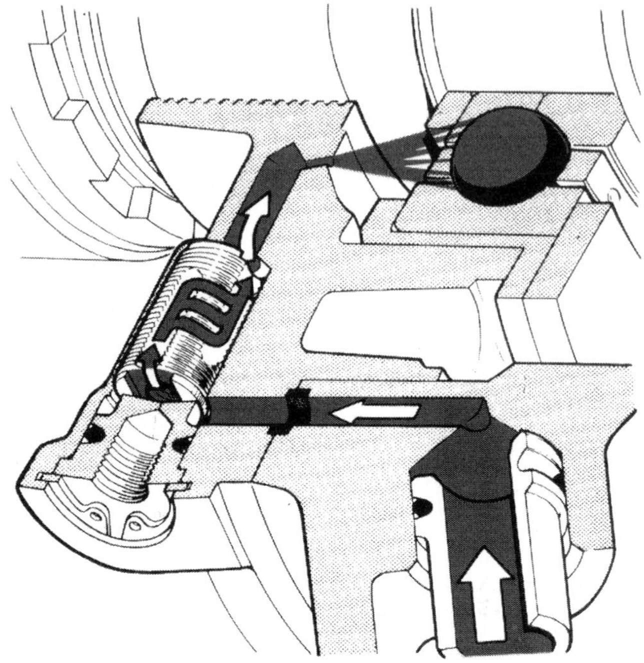

FIGURE 5–24 A thread-type oil filter. *(Source: Rolls Royce.)*

To prevent excessive air pressure within the oil tank, gearboxes, and bearing chambers, a vent to atmosphere is incorporated within the lubrication system. Any droplets in the air are separated out by a centrifugal breather prior to the air being vented overboard. Some breathers may incorporate a porous media, forming de-aerator segments, which improves the efficiency of the oil separation (Figure 5–23).

To prevent foreign matter from continuously circulating around the lubricating system, a number of filters and strainers are positioned within the system. Coarse strainers are usually fitted at the outlet of the oil tank or immediately prior to the inlet of the oil pumps to prevent debris from damaging the pumps. A fine pressure filter is fitted at the pressure pump outlet, which retains any small particles that could block the oil feed jets. Thread-type filters (Figure 5–24) are often fitted as a "last chance" filter immediately upstream of the oil jets. Sometimes perforated plates or gauze filters are used for this application. Scavenge filters are fitted in each oil return line to collect any debris from the lubricated components. An example of a pressure and scavenge filter is shown in Figure 5–25. They are invariably of tubular construction with a pleated woven wire cloth or a resin impregnated with fibers, as the filtering medium. Some filters comprise one or more wire wound elements but these tend to be insufficient for fine filtration. A "pop-up indicator" may be fitted to the filter housing to give a visual warning of a partially blocked filter.

Oil System Differences for Marine Applications

A typical marine gas turbine installation may consist of a gas turbine change unit (GTCU), a power turbine, and the associated installation module. The GTCU oil system may share oil with a hydraulic system pressurized by a GTCU-driven pump. The hydraulic system provides power to the airflow control regulator.

Module-Mounted Components

Some oil system components, such as the oil tank, may be located in the installation module for ease of access. Other components, which do not require regular attention (for example, the engine oil-pumping unit), may be located on the high-speed gearbox on the GTCU, as on an aero application.

The main filtration is carried out in the scavenge side of the system and the filter may be a duplex unit, located in the installation module. This allows the oil flow to be switched between two identical filter elements, allowing one to be replaced without stopping the engine or losing filtration. If a filter becomes blocked a bypass valve will open, allowing unfiltered oil to flow, permitting the engine to continue running with no loss of oil pressure. A visual indication of the filter condition (pressure drop) is provided to ensure that the filter is changed before bypass occurs. Instead of using the fuel (or air) supply to cool the oil, it is usual to pass seawater through a module-mounted heat exchanger.

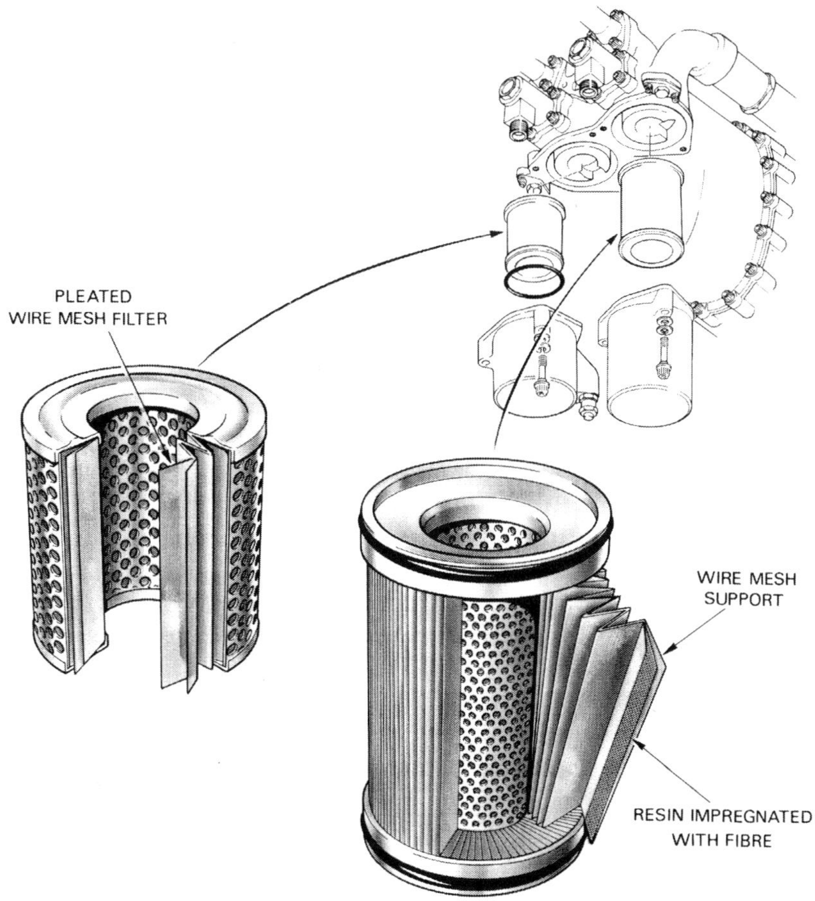

FIGURE 5–25 A typical pressure and scavenger filter. *(Source: Rolls Royce.)*

Power Turbine (Mechanical Drive) Oil System

The power turbine oil supply system is independent of the GTCU system. Oil is delivered from the ship's supply system, through an adjustable orifice valve and distribution block, to the power turbine bearings; the oil is returned to the same supply system.

Lubricating Oils

Early gas turbines used thinner oils than those used in piston engines but were produced from the same mineral crude oil. As gas turbines were developed to operate at higher speeds and temperatures these mineral oils oxidized and blocked the filters and oilways. The development of low viscosity (thin) synthetic oils overcame the major problems encountered with the early mineral oils.

The choice of a lubricating oil is initially decided by the need to start the engine at very low temperatures, when the viscosity of the oil is high, while being able to survive in an engine environment which exhibits very high temperatures. Having met these fundamental requirements, the need to provide improved lubrication characteristics using additives must also be investigated. Special laboratory and engine tests are done to prove the suitability of a particular oil for a specific type of engine. Assessments are made as the extent to which it deteriorates and the corrosive effects it may have on the engine.

Most gas turbines use low viscosity oil due to the absence of reciprocating parts and heavy-duty gearing. This reduces the power required for starting, particularly at low temperatures. In fact normal starts can be made in temperatures as low as −40°C without having to preheat the oil.

Turbopropeller engines use a slightly higher viscosity oil due to the additional requirements of the reduction gear and propeller pitch change mechanism.

The field of lubrication is a massive one, and there are deservedly books dedicated exclusively to this topic. We will not attempt to cover all that technical territory, but there is a list of several excellent references in Chapter 20. However, in this chapter, an operator's perspective on his turbine oil (both gas turbine and steam turbine [with steam turbines in combined cycle applications or solo] and their systems) is presented next. The original author is an oil company turbomachinery end user and reflects quite comprehensively the issues this author dealt with as an end user in the oil and gas and process (including refining industries).

AN OPERATOR'S PERSPECTIVE ON TURBINE OIL SELECTION*

The answer to "how long will this turbine oil last?" varies depending on the application. Turbine oil suppliers can give fairly wide-ranging estimates, say 5–15 years, for gas turbines. An exact answer depends on many variables: water, heat, contamination, operating hours, and maintenance practices will impact turbine oil longevity. Properly tested and maintained, higher-quality turbine oils will provide longer life than poorly tested and maintained, lower-quality products. New turbine oil performance characteristics will promote longer, trouble-free service.

In turbines, what is potentially more than 100 tons of rotating steel is supported by plain bearings on a cushion of oil that is thinner than a human hair. In power plants around the world, the same fluid dynamics take place day-in and day-out without much notice. Lost revenue at seasonal peaks can be counted in millions of dollars. Poor selection and maintenance of turbine oil can result in production losses exceeding $500,000 per day.

When selecting turbine oil for steam, gas, hydro, and aeroderivative turbines, oil supplier service abilities ought to be evaluated as part of the selection process. Turbine oils differ from most other lubricating oils and one needs to understand the basics of how and why.

Steam, gas, and hydro turbines oils are R&O (rust and oxidation inhibited) oils. Turbine equipment geometry, operating cycles, maintenance practices, operating temperatures, and potential for system contamination need to be considered.

Utility steam and gas turbine sump capacities range in size from 1000 to 20,000 gallons and up. Low turbine oil makeup rates (approximately 5% per year) are easier with high-quality, long-life lubricants. Without significant oil contamination issues, turbine oil life is primarily dictated by oxidation stability. Oxidation stability is adversely affected by heat, water aeration, and particulate contamination. Antioxidants, rust inhibitors, and demulsibility additives are blended with premium-quality base stock oil to extend oil life. Lube oil coolers, water removal systems, and filters are installed in turbine lubrication systems for the same purpose.

Unlike most gasoline and diesel engine oil applications, turbine oil is formulated to shed water and allow solid particles to settle where they can be removed through sump drains or kidney loop filtration systems periodically. To aid in contaminant separation, most turbine oils are not "doped" with high levels of detergents or dispersants that clean and carry away contaminants. Turbine oils are not exposed to fuel or soot and so do not need to be drained and replaced on a frequent basis.

Characteristics of Turbine Oil and Applications

Steam Turbines

A well-maintained steam turbine oil with moderate makeup rates should last 20–30 years. When steam turbine oil fails early through oxidation, it is often due to water contamination. Water reduces oxidation stability and supports rust formation, which among other negative effects acts as an oxidation catalyst.

Varying amounts of water will constantly be introduced to the steam turbine lubrication systems through gland seal leakage. Because the turbine shaft passes through the turbine casing, low-pressure steam seals minimize steam leakage or air ingress leakage to the vacuum condenser. Water or condensed steam is channeled away from the lubrication system but some water will penetrate the casing and enter the lube oil system. Gland seal condition, gland-sealing steam pressure, and the condition of the gland seal exhauster will affect the amount of water introduced to the lubrication system. Typically, vapor extraction systems and high-velocity downward-flowing oil create a vacuum that can draw steam past shaft seals into the bearing and oil system. Water can also enter via lube oil cooler failures, improper powerhouse cleaning practices, water contamination of makeup oil, and condensed ambient moisture.

In many cases, the impact of poor oil/water separation can be offset with the right additives including antioxidants, rust inhibitors, and demulsibility improvers.

Excess water may also be removed continuously with water traps, centrifuges, coalescers, tank headspace dehydrators, or vacuum dehydrators. If turbine oil demulsibility has failed, exposure to water-related lube oil oxidation then depends on water separation systems.

Heat will also cause reduced turbine oil life through increased oxidation. In utility steam turbine applications, bearing temperatures of 120–160°F (49–71°C) and lube oil sump temperatures of 120°F (49°C) are typical. Heat doubles the oxidation rate for every 18° above 140°F (10° above 60°C).

A conventional mineral oil rapidly oxidizes at temperatures above 180°F (82°C). Most tin-babbited journal bearings will begin to fail at 250°F (121°C). That is well above the temperature limit of conventional turbine oils. High-quality antioxidants can delay thermal oxidation but excess heat and water must be minimized to gain long turbine oil life.

* Reference: J.B. Hannon. "How to Select and Service Turbine Oils," *Machinery Lubrication*, July/August 2001.

Gas Turbines

For most large gas turbine frame units, high operating temperature is the leading cause of premature turbine oil failure. Higher turbine efficiencies and firing temperatures in gas turbines have been the main incentive for more thermally robust turbine oils. Today's land-based large-frame units operate with bearing temperatures in the range of 160–250°F (71–121°C). OEMs have increased their suggested limits on RPVOT ASTM D2272 (rotation pressure vessel oxidation test) and TOST ASTM D943 (turbine oil oxidation stability) performance to meet higher operating temperatures.

Changes in operating cycles are also introducing new lubrication hurdles. Lubrication issues specific to gas turbines in cyclic service started to appear in the mid-1990s. Higher bearing temperatures and cyclic operation foul system hydraulics. This delays equipment startup. Properly formulated hydrocracked turbine oils were developed to remedy this problem and to extend gas turbine oil drain intervals. Products such as Exxon Teresstic GTC and Mobil DTE 832 have demonstrated excellent performance for almost five years of service life in cyclically operated gas turbines. Conventional mineral oils had failed in one to two years.

Aeroderivative Gas Turbines

Aeroderivative gas turbines require oils with much higher oxidation stability. Lube oil in aeroderivative turbines is in direct contact with metal surfaces ranging from 400–600°F (204–316°C) typically and sump lube oil temperatures can range from 160–250°F (71–121°C). These turbines' cyclical operation imparts significant thermal and oxidative stress on the lubricating oil. This then dictates the use of high purity synthetic lubricating oils. Average lube oil makeup rates of 0.15 gallons per hour help rejuvenate the turbo oil under these conditions.

Aeroderivative turbines operate with much smaller (than industrial) lube oil sumps, typically 50 gallons or less.

Synthetic turbo oils are formulated for military aircraft gas turbo engines identified in military (MIL) specification format. These MIL specifications ensure that similar quality and fully compatible oils are available throughout the world and as referenced in OEM lubrication specifications.

Type II turbo oils were commercialized in the early 1960s to meet US Navy demands for improved performance, which created MIL-L (PRF)-23699. The majority of aeroderivatives in power generation today deploy these Type II, MIL-L (PRF)-23699, polyol ester-base stock, synthetic turbo oils. These Type II oils offer significant performance advantages over the earlier Type I diester-based synthetic turbo oils.

Enhanced Type II turbo oils were commercialized in the early 1980s for the US Navy's demand for better high-temperature stability. This led to the new specification MIL-L (PRF)-23699 HTS. In 1993, Mobil JetOil 291 was commercialized as the first fourth-generation turbo oil for advanced high-temperature and high-load conditions of jet oils.

Generator bearings typically use an ISO 32 R&O or hydraulic oil. The lower pour points of a hydraulic vs. an R&O oil may dictate the use of a hydraulic oil in cold environments.

Turbine Oil Procurement Standard

Steam, gas, and hydro turbine oils blend highly refined or hydroprocessed petroleum-based oils, usually ISO VG 32 and 46 or 68. Lubricant suppliers have developed turbine oils specifically for propulsion and power generation, based on turbine OEM specifications. Many turbine OEMs have moved away from specific turbine oil brand-name approvals due to improvements in specific turbine oils. OEMs recommend lube oil performance test criteria and typically ask that an oil, successful in the field, still be used even if all recommended values have not been satisfied. Industry standard lube oil bench tests reveal the performance and life expectancy of turbine oils. However, turbine OEMs and oil suppliers agree that past performance of an oil under similar conditions is the best overall recommendation.

Regardless of service type, the quality of the base stocks and additive chemistry will affect longevity. High-quality base stocks have higher-percentage saturates, lower-percentage aromatics, and lower sulfur and nitrogen levels. Additives must be extensively tested. They must also be blended in a tightly controlled process.

The key to good turbine oil is property retention. Some turbine oils have good lab test data but can experience premature oxidation because of additive dropout and base stock oxidation. Again, lube oil laboratory analysis can determine turbine oil longevity, but field experience should take precedence. Note: turbine oil suppliers offer typical lube oil analysis data to help assess predicted performance. Typical data are used due to variations in base stock variations.

Utility steam and gas turbine oils can be conventional mineral-based (Group 1) or hydroprocessed (Group 2). Conventional mineral-based oils have performed well in both steam and gas turbine service for more than 30 years. The trend toward higher efficiency, cyclically operated gas turbines has spurred the development of hydroprocessed, Group 2, turbine oils.

Most hydroprocessed turbine oils have better initial RPVOT and TOST performance than conventional turbine oils. This oxidation stability performance advantage helps heavy-duty gas turbine applications.

The oxidation performance advantages of a hydroprocessed turbine oil may be unnecessary in less demanding steam and gas turbine applications. Conventional mineral-based oils have better solvency than hydroprocessed oils. This, in turn, can provide better additive package retention and increased ability to dissolve oxidation products that could otherwise potentially lead to varnish and sludge.

Compatibility testing between turbine oil brands should also be addressed with a turbine oil specification for systems not available for a complete drain and flush. Clashing additive chemistries or poor in-service oil quality may prohibit mixing different and incompatible turbine oils. Oil suppliers can provide compatibility testing to confirm suitability for service. This testing assesses the condition of the in-service oil compared to various possible blends with the proposed new oil. The in-service oil should first be tested for suitability for continued service. Then a 50/50 blend should be tested for oxidation stability (RPVOT ASTM D2272), demulsibility (ASTM D1401), foam (ASTM D892, Sequence 2), and the absence of additive package dropout as witnessed in a seven-day storage compatibility test.

Lube Oil System Flushing

Turbine lube oil system flushing and initial filtration should be addressed when selecting turbine oil. Lubrication system flushing may be either a displacement flush after a drain and fill or a high-velocity flush for initial turbine oil fills. A displacement flush is performed concurrently during turbine oil replacement and a high-velocity flush is designed to remove contaminants entering from transport and during the commissioning of a new turbine.

Displacement flushes using a separate flush oil remove residual oil oxidation product that is not removed by draining or vacuum. A displacement flush utilizes lubrication system circulation pumps without any modification to normal oil circulation flow paths, except for potential kidney loop filtration. This flush is typically done based on a time interval vs. cleanliness (particle levels) to allow removal of soluble and insoluble contaminants that would not typically be removed by system filters.

Most turbine OEMs provide high-velocity flushing and filtering guidelines. Some contractors and oil suppliers also give flushing and filtering guidelines. Often during turbine commissioning, these guidelines are scaled back to reduce cost and time. There are common elements of a high-velocity flush that are generally supported by interested parties. There are also some procedural concerns that may differ and are addressed on a risk vs. return on investment basis.

Common elements of mutual agreement in high-velocity flushing are as follows:

- Supply and storage tanks should be clean, dry, and odor free. Diesel flushing is not acceptable.
- Two to three times normal fluid velocity achieved with external high-volume pumps or by sequential segmentation flushing through bearing jumpers.
- Removal of oil after flush is completed to inspect and manually clean (lint-free rags) turbine lube oil system internal surfaces.
- High-efficiency bypass system hydraulics eliminates the risk of fine particle damage.

Possible supplemental or alternative elements of a high-velocity flush are as follows:

- Use of a separate flush oil to remove oil soluble contaminants that can impact foam, demulsibility, and oxidation stability.
- Need to filter the initial oil charge at a level consistent with the filtration specification.
- Thermal cycling of oil during the flush.
- Pipe line vibrators and the use of rubber mallets at pipe elbows.
- Special cleanliness test strainers and sampling ports.
- Cleanliness criteria for flush buy-off.
- Lab ISO 17/16/14 to 16/14/11 acceptable particulate range.
- Use of on-site optical particle counters.
- 100-mesh strainer, no particles detectable by naked eye.
- Millipore patch test.

Prior planning and meetings with construction, startup, oil supplier, and the end user should be scheduled in advance to build consensus on these flushing procedures.

A good practice for turbine oil performance documentation is to take a 1-gallon sample from the supply tank and then a second gallon sample from the turbine reservoir after 24 hours of operation. The recommended testing is consistent with turbine oil condition assessment testing:

- Suitability for Continued Use (Annual)
- Viscosity ASTM D445
- RPVOT ASTM D2272
- Water by Karl Fischer Titration ASTM D1744
- Acid Number ASTM D664
- ISO Cleanliness Code 4406
- Rust ASTM D665 A
- Demulsibility ASTM D1401
- Foam ASTM D892 Sequence 2
- ICP Metals

Development Case study: Oil-Free Bearings

Another newer innovative design feature (still in development) that addresses the need for cooling, and as it turns out also lowers fuel consumption, is the oil-free bearing. This design feature is not yet commercially available on commercial gas turbines, but it does need to be considered in the future, particularly for aircraft engine applications. The work from which the extracts on this topic are drawn was published by NASA in 2010***. Since then, the budget to continue with this work by this source has been reassigned elsewhere. This is the kind of development that will be taken up again in better economic times because of the fuel savings realized. There is also the matter of weight savings and particularly in aeroengine applications, this is important.

The term oil-free turbomachinery has been used to describe a rotor support system for high-speed turbomachinery that does not require oil for lubrication, damping, or cooling. The foundation technology for oil-free turbomachinery is the compliant foil bearing. This technology can replace the conventional rolling element bearings found in current engines. Two major benefits are realized with this technology. The primary benefit is the elimination of the oil lubrication system, accessory gearbox, tower shaft, and one turbine frame. These components account for 8–13% of the turbofan engine weight. The second benefit that compliant foil bearings offer to turbofan engines is the capability to operate at higher rotational speeds and shaft diameters.

While traditional rolling element bearings have diminished life, reliability, and load capacity with increasing speeds, the foil bearing has a load capacity proportional to speed. The traditional applications for foil bearings have been in small, lightweight machines. However, recent advancements in the design and manufacturing of foil bearings have increased their potential size. An analysis, grounded in experimentally proven operation, was performed to assess the scalability of the modern foil bearing. This analysis was coupled to the requirements of civilian turbofan engines.

The application of the foil bearing to larger, high bypass ratio engines nominally at the 120 kN (~25,000 pound) thrust class has been examined. The application of this advanced technology to this system was found to reduce mission fuel burn by 3.05%. The present study is grounded within the NASA Fundamental Aeronautics Program – Subsonic Fixed Wing Project and its focus is limited to what is termed "N+1" propulsion systems. These systems are representative of the next generation of civilian aircraft. In terms of the propulsion system, this implies two spool, direct drive turbofan engines having bypass ratios in the range of 8–12.

Future studies will consider advanced propulsion systems in which a new design space is enabled by gas foil bearings. By considering clean sheet propulsion system designs that are not limited to current engineering design practices, enhanced benefits attributed to this bearing technology are expected in future system studies. Typically, gas foil bearings are used in much smaller turbomachines where the rotors are lighter, rotate faster, and where system auxiliary weights (such as the lubrication system) are a significant fraction of the overall system weight. However, advances in load capacity and scalability of the gas foil bearing warrant further study into larger systems. The gas foil bearing is the key technology in what has become known as "oil-free turbomachinery."

Oil-free turbomachinery, within the context of this study, is defined as highspeed rotating equipment operating without oil lubricated rotor supports, such as bearings, dampers, seals, and other components required in conventional turbo machines. Through the elimination of the lubricating oil, rolling element bearings, and the requisite supporting subsystems it is envisioned that revolutionary improvements in performance, efficiency, and reliability of turbomachinery propulsion systems can be realized. The enabling technology of oil-free turbomachinery is the modern development of high speed, high temperature, and high load capacity foil bearings.

Among the benefits of oil-free bearings are reduced machine weight due to the elimination of the oil systems, removal of the traditional diameter – rotational speed limit (DN) imposed by rolling element bearings, and a synergistic use of working fluid as a lubricant enabling a contaminant-free working fluid. However, in order to realize these benefits, foil bearings must be able to support the required machine load and operate reliably and without burden to machine performance.

A generic foil journal bearing is shown in Figure 5–26. It consists of a rigid, stationary shell that contains a compliant foundation. This compliant foundation can take different forms, but is usually portrayed as a thin layer of corrugated metal referred to as the bump foil. A smooth foil, known as the top foil, is then fastened to one edge of the bump foil. A rotating shaft fits inside the top foil. Foil

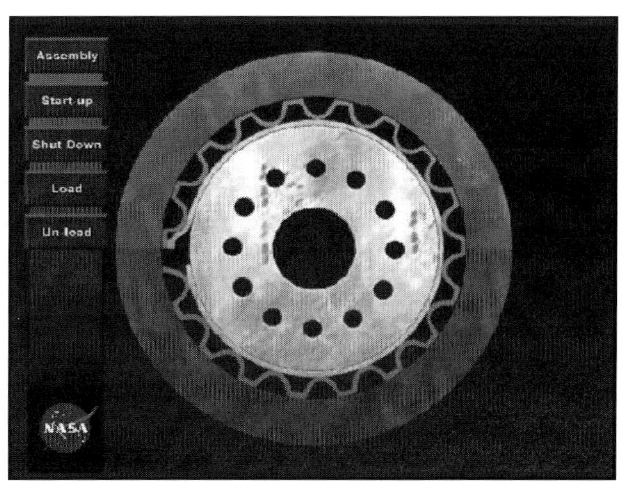

FIGURE 5–26 A typical bump foil style compliant foil bearing.

*** Adapted with permission from NASA's self-published work on oil-free foil bearings, also published as GT2010-22118 (R.J. Bruckner, NASA Glenn Research Center).

bearings typically operate without static clearance between the stationary top foil and rotating shaft. In fact there is often a preload force between the two parts. This static preload is unique to foil bearings and presents a technical challenge for both machine startup as well as analytic modeling. At a certain speed known as the lift off speed, the hydrodynamic gas film force overcomes the preload force and complete fluid film lubrication is established.

Chapter 6

Inlets, Exhausts, and Noise Suppression

"The course of true love never did run smooth."

—William Shakespeare

Chapter Outline

*The inlet to a gas turbine is designed to promote optimum air swallowing capability. Smooth geometry is essential, so its surface contour generally is kept as simple as possible. Its smooth surface may conceal pressure probes and heating boots that are part of an anti-icing system (depending on where the turbine is located geographically).

Land- and marine-based engines frequently are equipped with inlet air filtration systems to protect the gas turbine against ingestion of contaminants potentially including, but not limited to, salt sea spray, insects, sand, and oil stack fumes. These contaminants could coat themselves on the airfoils in the compressor or further downstream and cause erosive or corrosive damage to gas turbine components. They also may just adhere temporarily to the compressor blade surfaces and cause incremental performance loss.

However, the filter system must cause as small a pressure (power loss) drop as possible. The compromise of these two requirements results in a filter system that allows some dirt through which eventually clogs enough of the compressor air passages that a compressor wash is required to restore performance.

Most gas turbines, therefore, are fitted with wash systems (see Chapter 8, Accessory Systems, for details on wash systems). In summary, these systems include nozzles through which cleaning fluid may be conducted into the gas turbine compressor. The wash is conducted at a variety of speeds, from slow roll to "while online." Wash parameters depend on the operating conditions of the gas turbine. For instance, as we see in the chapter on fuels, gas turbines on

* Reference: [6-1] Claire Soares, Working case notes, 1975 through 2003 and Proceedings of ASME IGTI panel sessions, "ECMS as They Apply to Life Extension of Gas Turbine Engine Components," 1985 through 2003.

residual fuel develop deposits from fuel-treatment-caused compounds. To remove these from the turbine blades requires a crank soak every 100–150 hours.

Engine condition monitoring systems (ECMS; see Chapter 9, Controls, Instrumentation, and Diagnostics) generally have a performance assessment (PA) module. PA (see Chapter 10, Performance, Performance Testing, and Performance Optimization) measures compressor differential pressure versus compressor flow (and frequently the same with the turbine module) to confirm the onset of potential flow problems and attempt to diagnose these.

Due to differences in their design, all models of gas turbines require a slightly different wash system for different applications for optimum results. Attempts to standardize are made by the OEMs, which ask their subcontract wash system hardware suppliers to aim at a design that will fit as many of their models as possible.

The main aspects of the gas turbine inlet system, depending on where it is located and its application (land, sea or air), include:

- Inlet filtration (upstream of the gas turbine bell mouth).
- Inlet cooling and fogging (downstream of the gas turbine bell mouth). This topic is dealt with in Chapter 3, Gas Turbine Configurations and Heat Cycles, and Chapter 10, Performance, Performance Testing, and Performance Optimization.
- Inlet cooling (downstream of the gas turbine bell mouth) is a means of performance enhancement. Basically, if the inlet air is cooled, it is denser and occupies less volume for a given mass, so more mass is swallowed by the compressor. Based on CDP, more fuel (see section on controls) is supplied to be burned with the air, so more horsepower is developed for any given set of atmospheric conditions. This topic is dealt with in Chapter 3, Gas Turbine Configurations and Heat Cycles, and Chapter 10, Performance, Performance Testing, and Performance Optimization.

GAS TURBINE INLET AIR FILTRATION**

With gas turbine inlet air filtration, the filter size, design type, and filter media must be carefully matched to the end user's application.

** Source: Courtesy McGraw-Hill, *Process Engineering Equipment Handbook* (Soares, C. New York: McGraw-Hill, 2001); original source: Altair Filters International Limited, UK. Adapted with permission. Note that this section has been considerably condensed from this source. Note also that the original source of this material was Altair Filters International, UK., presented by R. Cleaver at the IGTI panel on ECMS (chair/organizer C. Soares) in Cologne, Germany, 1992.

Purposes for installing gas turbine air-inlet filtration include:

- Prevention or protection against icing
- Reduction or elimination of ingestion of insects, sand, oil fumes, and other atmospheric pollutants

Potential ice-ingestion problems can be avoided with a pulse-jet-type filter, commonly called a *huff and puff* design. Ice builds up on individual filter elements that are part of the overall filter. When a predetermined pressure drop is reached across the elements, a charge of air is directed through the filter elements and against the gas turbine intake flow direction. The ice (or dust "cake") then falls off and starts to build up once more.

Ice ingestion has caused disastrous failures of gas turbines. A few companies also make instruments that detect incipient ice formation by measurement of physical parameters at the turbine air inlet. Sometimes, if the pulse-type filter is retrofitted, the anti-icing detection instrument may already be there. The pulse filter, however, provides a preventive "cure" that will work regardless of whether the icing-detection instrument is accurate, as the cleaning pulse is triggered by a signal that depends on differential pressure drop across the filter elements. If a pulse filter is used, icing-detection instruments, which normally are necessary in a system that directs hot compressor air (bleed air) into the inlet airstream, are not required.

Many filtration applications are examples of retrofit engineering or reengineering because the original application may have been designed and commissioned without filters or the original choice of filters or filter elements was inappropriate. For tropical applications, filter media that swells or degrades (also rain must not be allowed to enter the filter system) cannot be used because of the intense humidity. This excludes cellulose media. Tropical installations present among the most severe applications.

Inlet Air Filters for the Tropical Environment

The factors that determine design include the following:

Rainfall

The tropics extend for 23°28′ either side of the equator, stretching from the Tropic of Cancer in the north to the Tropic of Capricorn in the south, and represent the tilt of the earth's axis relative to the path around the sun. The sun will pass overhead twice in a year, passing the equator on June 21 on its travel north and September 23 on its travel south. The sun's rays will pass perpendicular to the earth's atmosphere and so will have the least amount of filtration, giving high levels of ultraviolet rays.

The area has little seasonal variation; however, the main characteristic of the area is the pronounced periods of rainfall. Typhoons and cyclones are common to certain parts of this area.

It is not surprising that the records for the highest rainfall ever recorded are all within the tropics. Intense rainfall is difficult to measure since its maximum intensity only lasts for a few minutes. Rainfall can be expressed in many ways, either as the precipitation that has fallen within 1 hour (in millimeters per) or over shorter or longer periods but all relating back to that same unit of measurement. Since gas turbines experience problems due to rainfall within a few minutes, it is important to take account of the values of "instantaneous rainfall" that can occur.

The most intense rainfall ever recorded was in Barst, Guadeloupe (latitude 16°N), on November 26, 1970, when 38.1 mm fell in just 1 min.

Another important feature regarding rainfall is the effect of wind speed. "Horizontal rain" is often described, but in practice is unlikely to occur. However, wind speeds can give rain droplets significant horizontal components. The impact of this can be very important, particularly with small droplet sizes. Even droplets 5 mm in diameter will cause a vertical surface to be almost four times wetter in wind speeds of 37 m/s than on the horizontal surface (to which the rainfall rates relate).

The effect of rainfall in tropical environments on the operation of gas turbines has been very much underrated.

The humid environment also ensures that relative humidities are generally high, with the lowest humidities being experienced during the hottest part of the day and the highest occurring at night. During the rainy season, the humidity tends to remain constant throughout the day. The effect of humidity is important where airborne salt is concerned since salt can become dry if the humidity is below 70%.

Dust

Dust levels in tropical environments in Southeast Asia are generally low.

There are, of course, always specific exceptions to this, for example, near new construction sites or by unpaved roads. But, in general, dust is not a significant problem. Mother nature ensures this by casting her seeds on the fertile soil and quickly turning any unused open space into a mass of overgrown vegetation very quickly, thereby suppressing the dust in the most natural way.

Insects and Moths

In the tropics, the hot, humid environment is a natural encouragement to growth of all kinds. It is often said that if a walking stick is stuck into the rich fertile soil of the area and left for 3 months, it will sprout leaves and grow. Certainly the insect population reflects this both in size and quantity. Large moths are common to the area and tend to occur in quantity during specific breeding periods. These can quickly cover intake grills, obstructing airflow and even causing large gas turbines to trip. Some of the largest moths are found in the tropics. A common moth in India and Southeast Asia is the Swift moth (Hepralidae), which can have a wing span of some 15 cm and is said to lay up to 1200 eggs in one night. Another moth is the Homoprera shown in Figure 6–1, which has a similar wing span. Moths are attracted by the lights that often surround the turbine installations, as well as the airflow, which acts as a great vacuum cleaner.

On one installation in Sumatra, large gas turbines have been known to trip out after only 8 hours of operation due to blockage of the air filters with moths.

Fortunately, moths tend to confine themselves to within a few miles of land and so offshore installations do not tend to suffer these problems.

Design Problems Experienced

Many feel that standardization is the key to reducing costs and boosting profits. It is not surprising, therefore, that gas turbine air-filter systems were designed with this in mind.

FIGURE 6–1 Homoprera (insect type). *(Source: AFI Ltd.)*

Dust was important to system designers, and so filter systems may be chosen to be able to deal with prodigious amounts of it, whereas, in practice, dust is only normally a problem next to unpaved roads or construction sites.

Despite this, most gas turbines were fitted with elaborate and expensive solutions to overcome a problem that hardly existed, or at least only in a relatively small percentage of installations. Many of these systems employ bleed fans that need additional electrical energy and a constant maintenance requirement. A typical system is shown in Figure 6–2. This system employs spin tubes that swirl dust to the outside of the tube where a bleed slot extracts the dust while allowing the cleaner air to pass through the main core of the tube. Since the air is rotated, the peripheral speeds need to be high, which, in turn, results in a relatively high pressure-loss coefficient. The efficiency of the system is very reliant on the bleed air, which is provided by auxiliary fans.

Another bleed extract system uses a series of convergent vanes that funnel the air toward a central slot through which bleed air is extracted. The heavier dust particles are guided toward the bleed extract, while the main air passes between the vanes at almost 180° to the general direction of airflow. Again, the efficiency of the system is reliant on the provision of bleed air.

Protection against rain was elementary. On many systems that had dust-extract systems, no further provision was made. On others a coarse weather louver, often of plastic, was provided. Sometimes a partial weatherhood was provided, sometimes not.

The emphasis on the designer was to provide a "three-stage" filter system, without worrying too much about the suitability of those stages.

There was recognition of the high humidities that exist, and so most systems incorporated a coalescer, whose function was to coalesce small aerosol droplets into larger ones, which could then be drained away. These coalescer panels varied between pieces of knitted mesh wound between bars within the filter housing to separate panels with their own framing. Loose glass fiber pads were used as an inexpensive solution and also served as a prefilter pad.

FIGURE 6–2 A typical spin-tube inertia filter. *(Source: AFI Ltd.)*

The final stage in almost all of these systems was the high-efficiency filter element, either as a cartridge or as a bag. The high-efficiency cartridge was typically a deep-pleated glass-fiber paper sealed in its own frame and with a seal on its rear face. Other high-efficiency filters employed glass-fiber pockets that fitted into permanent wire baskets within the filter house.

Almost all of the filter systems were enclosed in housings constructed in carbon steel, finished with a variety of paint finishes. Other materials, such as stainless steel, were not common since their initial cost was thought to be excessive.

Protection against complete filter blockage was often provided by means of a bypass door. This normally was a counterbalanced door in which a weight held the door closed. When the pressure drop across the filter was high, this overcame the force exerted by the balance weight and the door opened, thereby bypassing the filter system with unfiltered air.

Offshore Environment Original Design Data

In Europe in the late 1960s, the only data generally available on the marine environment was generated from that found on ships. Since at that time there was considerable interest in using gas turbines as warship propulsion systems, several attempts were made to define the environment at sea, with particular respect to warships.

Not only was it found difficult to produce consistent data, but other factors such as ship speed, hull design, and height above water level had major effects. It became apparent that predicting salt in air levels was as difficult as predicting weather itself.

Since the gas turbine manufacturers had defined a total limit of the amount of the contaminants that the turbines could tolerate, some definition of the environment was essential to design filter systems that could meet these limits.

Many papers and conferences were held with little agreement, as can be seen in Figure 6–3. However, since the gas turbine industry is a conservative one, it adopted the most pessimistic values as its standard, namely the National Gas Turbine Establishment (NGTE) 30-knot aerosol (Table 6–1). It was treated more as a test standard rather than what its name implied. In the absence of any other data, this was used to define the environment on offshore platforms, despite the fact that they were much higher out of the water, and did not move around at 40 knots!

This then defined the salt in air concentration, but did not address any other particulates. In hindsight, it now

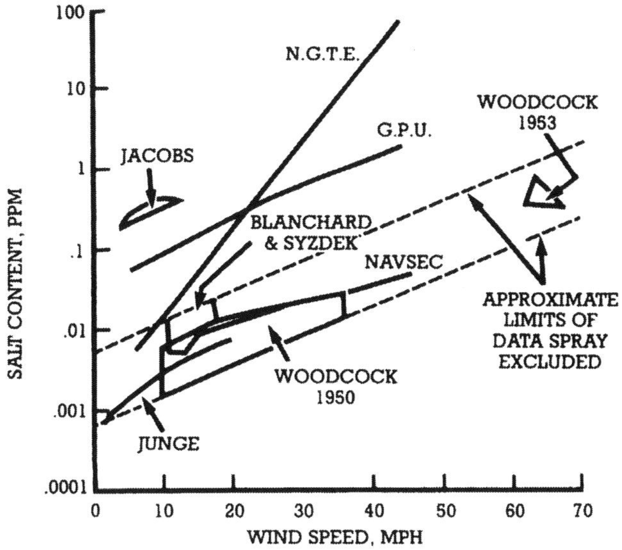

FIGURE 6–3 Airborne salt comparisons. *(Source: AFI Ltd.)*

TABLE 6–1 NGTE 30-Knot Aerosol

Microns	Salt Content, ppm
<2	0.0038
2–4	0.0212
4–6	0.1404
6–8	0.3060
8–10	0.4320
10–13	0.6480
>13	2.0486
Total	3.600

(Source: AFI Ltd.)

seems naive that the offshore environments were originally considered to be clean with no other significant problems than salt. Many equipment specifications were written at that time saying the environment was "dust free."

In the early 1970s there was also a lively debate as to whether the salt in the air was wet or dry. One argument was put forward that if the salt was wet it would require a further stage of vane separators as the final stage to prevent droplet reentrainment from the filters. The opposing argument maintained that vane separators were unnecessary and that a lower humidity resulted in evaporation of the droplet, giving a smaller salt particle that required a higher degree of filtration. Snow and insect swarms were largely ignored as a problem.

Initial Filter Designs

The types of systems that were used on the first phase of the developments in the North Sea fell into two categories: high-velocity systems where the design face velocity was normally around 6 m/s, and lower velocity systems that operated at 2.5 m/s.

The low-velocity system was a similar system used on land-based installations, and usually comprised a weather louver, followed by a prefilter and a high-efficiency bag or cartridge filter. Sometimes a demister stage was added and occasionally bleed extract inertials were supplied as a first stage.

The high-velocity system was attractive to packagers since it was lighter and occupied a much smaller space. It was the system derived from shipboard use and comprised a vane, coalescer, and vane system.

In general, both systems were housed in mild steel housings with a variety of paint finishes. The weather louvers and vanes were normally constructed from a marine grade aluminum alloy. The filter elements often had stainless steel or galvanized frames.

Bypass doors were used to protect the engine against filter blockage.

The emphasis by package designers was to include a provision for a "three or four stage system" often without regard for what those stages should comprise.

Actual Offshore Environment

The actual offshore environment is in many ways different from that originally envisaged.

Salt in air is present, although it is only a problem when the filtration system leaks or is poorly designed. Horizontal rain can be a severe problem although sea spray does not generally reach the deck levels even in severe storms.

Flare carbon and mud burning can be a significant problem if the flare stack is badly positioned or if the wind changes direction (see Figure 6–4). Not only do the filters block more quickly, but greasy deposits can cover the entire filter system, making the washing of cleanable filters more difficult.

The relative humidity offshore was found to be almost always high enough to ensure that salt was in its wet form. Some splendid work by Tatge, Gordon, and Conkey concluded that salt would stay as supersaturated droplets unless the relative humidity dropped below 45%. Further analysis of offshore humidities in the North Sea showed that this is unlikely to happen (Table 6–2).

Initially it was thought that the platforms were dust free, but this is far from the case.

Drilling cement, barytes, and many other dusts are blown around the rig as they are used or moved. But the main problem has resulted from grit blasting. As the

FIGURE 6–4 Flare carbon can cause problems. *(Source: AFI Ltd.)*

TABLE 6–2 Monthly Average Relative Humidity, North Sea

January	91	July	85
February	86	August	83
March	87	September	83
April	84	October	80
May	86	November	81
June	84	December	80

Source: ASME Report 80-GT 174.

platforms got older, repainting was found to be an accelerating requirement with grit blasting a necessary prerequisite.

In order to be effective, grit is sharp and abrasive by design and can be devastating if ingested into a gas turbine (see Figure 6–5). The quantities used can seem enormous. On one platform it was found that over a 12-month period, 700 tons of grit blast had been used!

FIGURE 6–5 Typical turbine damage. *(Source: AFI Ltd.)*

Offshore Application Design Problems Encountered

In general, problems were slow to appear, typically taking three to five years after start-up, but since a lot of equipment had been installed at about the same time, the problems manifested themselves like an epidemic.

These problems could be categorized as follows:

1. Erosion of compressor blading
2. Short intervals between compressor cleaning
3. Frequent filter change-out
4. Turbine corrosion
5. Corrosion of the filters and housing

By far the most serious of these problems was the erosion of compressor blading that was experienced almost simultaneously on many platforms. This occurred about three to five years after start-up, as this was the time that repainting programs were initiated. Grit blast found its way into the turbine intakes either through leaking intakes, bypass doors, or through the media itself (see Figure 6–6). Since the airborne levels were high, the air filters quickly blocked up, allowing the bypass doors to open. As filter maintenance is not a high priority on production platforms, considerable periods were spent with grit passing straight into the turbine through open bypass doors. Even where maintenance standards were more attentive, there were usually enough leaks in the intake housing and ducting to ensure delivery of the grit to the turbine.

It often seemed contradictory that the system designers would spend a lot of time specifying the filter system, but would pay little attention to ensuring the airtightness of the ducting downstream.

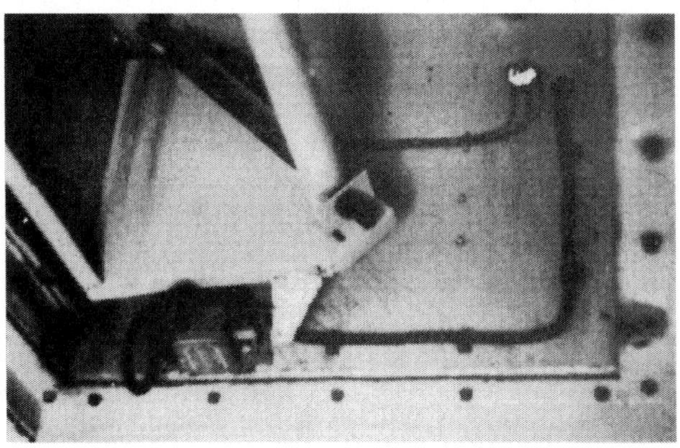

FIGURE 6–6 Typical leak caused by a missing cable gland. *(Source: AFI Ltd.)*

Since the grit was sharp, it sometimes damaged the filter media itself, reducing the system efficiency dramatically.

Bypass doors were a major problem. Early designs failed to take account of the environment or the movement in the large structures of the filter housings. Very few of those initial designs were airtight when shut, and it was not uncommon for them to be blown open by the wind.

Turbine corrosion could almost always be traced to leaky ducting or operation with open bypass doors. Very few systems gave turbine corrosion problems if the ducting was airtight. The few installations that did give problems were usually the result of low-velocity systems operating with poor aerodynamics, so that local high velocities reentrained salt water droplets into the airstream and onward to the engine.

Rapid compressor fouling was usually the prelude to more serious problems later, since it was usually caused by the combined problems of filter bypass.

Compressor cleaning almost once a week was fairly standard for systems with those problems.

As time progressed the marine environment took its toll on the carbon steel and severe corrosion was experienced on the intake housing and ducting. In some cases, corrosion debris was ingested into the turbine causing turbine failure. This again was accelerated by poor design, which allowed dissimilar metals to be put into contact, leading to galvanic corrosion.

How Good does your Filter Need to be for the above Applications

The end-user is the best judge of this. Clearly, in the clean air of northern Canada, the gas turbine engine might not be expected to ingest oil fumes (as in offshore applications) or insects (as in the tropics), but icing (ice ingestion) may be a factor. In the heavily industrialized areas of Europe however, dirty particulates are prevalent. The GT system may allow for online washing but some of these particulates bake onto the airfoil surfaces and destroy aerodynamic performance. Additional filtration capacity may be in order, but at what point does the additional capital cost justify itself? What is the GT power that has to be sacrificed for more efficienct filtration and can the end-user tolerate it? These are, once again, end-user questions.

The case study that follows does present some perspective on the testing and financial analysis on three = phase filtration systems. Again, the end-user ought to pay attention to the model numbers of GT involved in this case study and their respective power ratings. Sometimes accessories that work well with one power size bracket of gas turbine may not "scale up" well with larger power brackets due to system geometries, other accessory systems on the gas turbine and various other factors. The study does present the key parameters to be potentially studied were a similar study to be undertaken elsewhere.

CASE STUDY 1: A TEST CASE OF TWO- VS. THREE-PHASE FILTRATION*

These filtration systems were installed on two Mitsubishi M 701F, two Alstom GT13E2, two GE Frame 9FA, two GE LM 6000, four Solar engines packaged by TurboMach, an Alstom Tempest and two Ruston Typhoons, with corresponding ratings between 4.5 MW and 232 MW, operated at facilities in Singapore, Indonesia, France, Austria, the Netherlands, the UK and Germany. Despite a wide range of different temperatures, humidity levels, dust concentrations, and other local conditions, none of the machines required an online or offline washing routine within an operating period of one year.

High-quality air filters are able to reduce the fouling on the blades and enable stable power output P and efficiency h. Continuous development of filter media and filter construction have improved the filter performance in the past. Will there be further steps towards even better air filtration for gas turbines or has the development reached a plateau? The answer to this question can be found in an economic analysis taking into account a reduction of fouling due to better combustion air quality on the one hand and higher investment costs and higher pressure drop in the air intake on the other hand.

The evaluations and calculations are accomplished at so-called static filter systems where air filters with depth loading characteristics are used. With static filter systems several steps of filtration are arranged in a sequence and therefore the examined upgrades of the filtration efficiencies can be implemented relatively easily so that the effect could be studied. A comparison of two- and three-stage filter systems for intake air filtration at gas turbines produces important findings regarding the most operationally cost-efficient filter sequence overall. The upgrade of the air intake system on the one hand resulted in an increased static pressure loss in the air intake causing reduced turbine efficiency and less power output. On the other hand at the same time the soiling of the blades mainly in the compressor section (compressor fouling) is lowered and as a consequence efficiency and power output are enhanced. The effects of a higher pressure drop entailed by three-stage filtration are compared with those arising from reduced soiling on the blades. In the cases examined, including case studies from actual operation, definite advantages are found for three-stage filtration. Even the modification costs for

* Source: [6-2] From: Freudenberg Filtration Technologies SE & Co KG, Adapted, with permission, from GT2008-50280 "Economical Benefits of Highly Efficient Three-Stage Intake Air Filtration for Gas Turbines." Note that this manufacturer uses their in-house model filter designations to clarify size and type of filter used. The exact filter size specifications can be found on their website.

installing another filter stage can be amortized in what will sometimes be significantly less than two years.

The purity of the combustion air for gas turbines is increasingly a major focus for the manufacturers and operators, since these machines, due to continuous technical design enhancements, are reacting with progressively rising sensitivity to soiling on their blades. Properly functioning filters in the intake air system will reduce soiling on the blades, thus increasing the power output and efficiency of the gas turbine involved. Repeated advances in the fields of filter design and filter media are enhancing the performance capabilities of the air filters being used.

Will the efficiency of the air filters used continue to rise over the years ahead, or has a plateau been reached, and will the future more probably see stagnation at a high level of filtering efficiency? The answer to this question can be supplied only by a feasibility study, in which the advantages of even purer combustion air, and the concomitantly reduced level of blade soiling, are compared to the disadvantages entailed by higher capital investment costs for the intake air system and a higher pressure drop of the filters.

Nomenclature

P: gas turbine power output
h: gas turbine efficiency
t: time
Cp: pressure drop power reduction coefficient

An air filter system is required to significantly reduce the penetration of solid and liquid particles into the turbine under temporally fluctuating environmental conditions. The size of these particles lies within a bandwidth ranging from approx. 0.01 micrometer (μm) to approx. 3 millimeter (mm), thus according to the size ratio between a tennis ball and a 10,000-foot-high mountain. This analogy illustrates why filter systems are often designed in a multi-stage configuration, so as to capture large particles in the pre-filter and small particles in the fine filter.

At a highly polluted location—neglecting extreme situations like sandstorms—the air can be expected to contain average mass concentrations of up to 0.2 mg/mÑ or number concentrations per mÑ of up to 30 million particles of 0.5 μm and larger in size. A gas turbine with an intake volume flow of 1.5 million cubic meters per hour will ingest up to 30 trillion particles of 0.5 μm and larger in size per hour. If a fraction of this dust penetrates the filters, it will either deposit on the blades or cause erosion, leading to reduced power output and impaired efficiency for the gas turbine. The major effects for the deterioration include increased tip clearances, changes in airfoil geometry and increasing surface roughness. In the case of locations near the coast, salt particles being transported in the air may reach the blades, where they cause serious corrosion and consequently shorten the blades' lifetime quite considerably. The good water solubility of many salts in any case poses tough requirements for the filter system, since water passing through may wash the salts out of the filters again, and transport them into the turbine (see Table 6-3).

TABLE 6–3 Pollution Levels of Ambient Air

Typical concentration (by mass) of airborne particles in ambient air	0.01–0.2 mg/m^3
Typical concentration (by number) of airborne particles in ambient air	10^6–$3 \cdot 10^7$ particles/m^3
A gas turbine rated at approx. 150 MW ingests approx. 20,000,000,000,000 particles per hour	

Good air filtration significantly reduces the amount of caked-on dust deposits (fouling) on the blades, thus delaying a temporal fall in the gas turbine's power output and efficiency. If particles that cause erosion and corrosion are kept away from the gas turbine's compressor, the blade lifetime will be prolonged. Offline washing for removing caked-on dust particles is one of the substantial cost factors. The washing procedure causes downtimes lasting several hours, during which no power or steam can be generated. In the case of a 150-MW gas turbine, an offline washing routine can mean lost revenues totaling EUR 30,000 to 40,000.

Efficient air filtration is designed to significantly increase the time intervals between two offline washing routines, and thus enable the machine to be run for longer, with a concomitant increase in revenues. The lower costs for any washing agents used, or for fully demineralized washing water for online washing routines should also be mentioned in this context. The overview in Table 6–4 shows the most important beneficial and detrimental influences that combustion air filtration exerts on the operating behavior of gas turbines.

Besides the static filter systems explained in detail below, so-called cleanable filters are also used for filtering the combustion air of gas turbines. The advantages and disadvantages of these filter elements, which operate on the principle of surface filtration and are designed to have the filtered dust removed with a pressure pulse (pulse-jet or pulseclean), have already been discussed elsewhere. Cleanable systems are excluded from the analysis below, and will not be included in the basic analysis (Figure 6.7).

A Comparison of Two- and Three-Stage Filter Systems

From a purely technical point of view, it is definitely possible to reduce the amount of pollutants in the intake air entering a gas turbine, by installing a three-stage filter system with a high-efficiency final filter instead of the two-stage system commonly used today. This would reduce the amount of fouling on the blades; however, the average

TABLE 6–4 Influences of Intake Air Filtration on Operating Behavior of Gas Turbines

- Increase of power output and turbine efficiency thanks to less fouling on the blades.
- Reduction of maintenance costs by prolongation of the turbine's lifetime because of protection of the blades (e.g., less corrosion / fouling).
- Increase of power output by extension of on- and offline washing intervals.
- Reduction of the power output and turbine efficiency due to the additional static pressure drop in the air intake system.

FIGURE 6–7 Typical deposits caused by ambient air particles [6-2].

pressure drop of the filters would rise. In the analysis below, only two of the influencing factors explained above will initially be covered. The first of these is the increase in power output thanks to reduced fouling on the blades, and the second is the countervailing reduced power output due to an increased pressure drop. The criteria under scrutiny, adduced for examining the cost-efficiency of an improved filtration process, are listed below:

1. Increase of the average power output and turbine efficiency due to less fouling on the blades when using a three-stage system.
2. The thereto contrary reduced power output due to the increased pressure drop in the air intake system when using a three-stage system.

Figures 6–8 and 6–9 illustrate the two- and three-stage filter systems being compared, with filter classes defined under EN779: 2003 (coarse and fine dust filters) and EN 1822 (HEPA/ULPA filters) serving as the starting point for assessing the air filter collection efficiency capabilities. For the conventional two-stage filter system, filter classes F6 (pocket filters) and F8 (cassette filters) were selected. This filter combination has proved its worth in many field applications, with useful lifetimes of more than 2 years for both the first and the second filter stages.

The improvement in filtration is to be effected using the filter sequence F6-F9-H11. The first filter stage, featuring an F6 pocket filter, remains the same compared to the two-stage system, whereas for the second and third filter stages cassette filters of classes F9 and H11 are used. The upgrade job thus

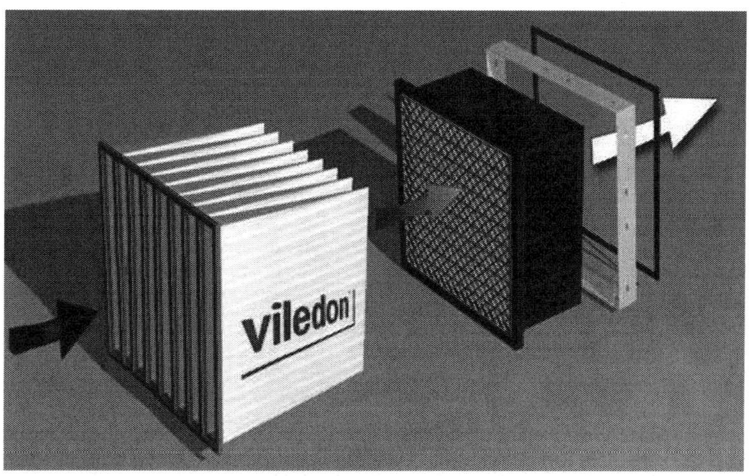

FIGURE 6–8 Two-stage filter system with pocket filters F6 and cassette filters F8 [6-2].

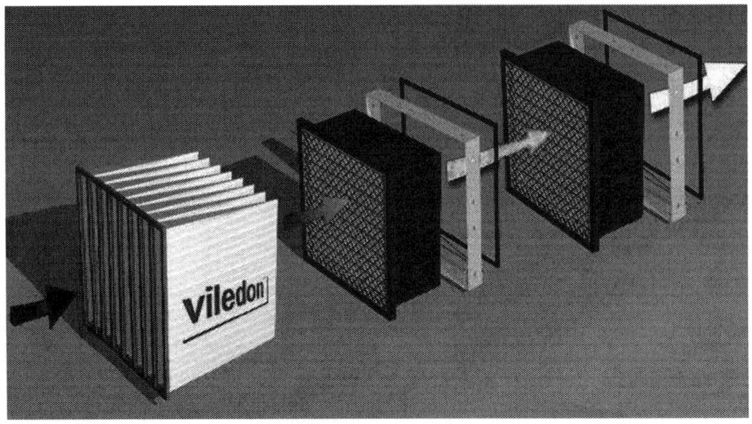

FIGURE 6–9 High efficient, three-stage filter system with pocket filters F6 and casssette filters F9 and H11 [6-2].

involves increasing the number of filter stages from two to three. The first task is to mathematically quantify the adverse effect of the increased pressure drop involved. The causal relationship between pressure drop in the intake system and power output of a gas turbine, as shown in Figure 6–10, is adduced for this purpose; i.e., for each 50 Pa of increased pressure drop the turbine's power output decreases by 0.1%.

With most gas turbines, the actual value is somewhat smaller than 0.1%, so that the effect of a reduced power output level due to the increased pressure drop tends to be weighted rather more heavily in the calculations below than it actually is. The diagrams in Figures 6–11 and 6–12 show a typical graph for the pressure drops of the above-described two- and three-stage filter systems over the course of 9000 operating hours (approx. 1 year in base-load operation). They are based on the evaluation of monitored pressure drop curves for different machines with two-stage systems and the first installation of a three-stage filtration system.

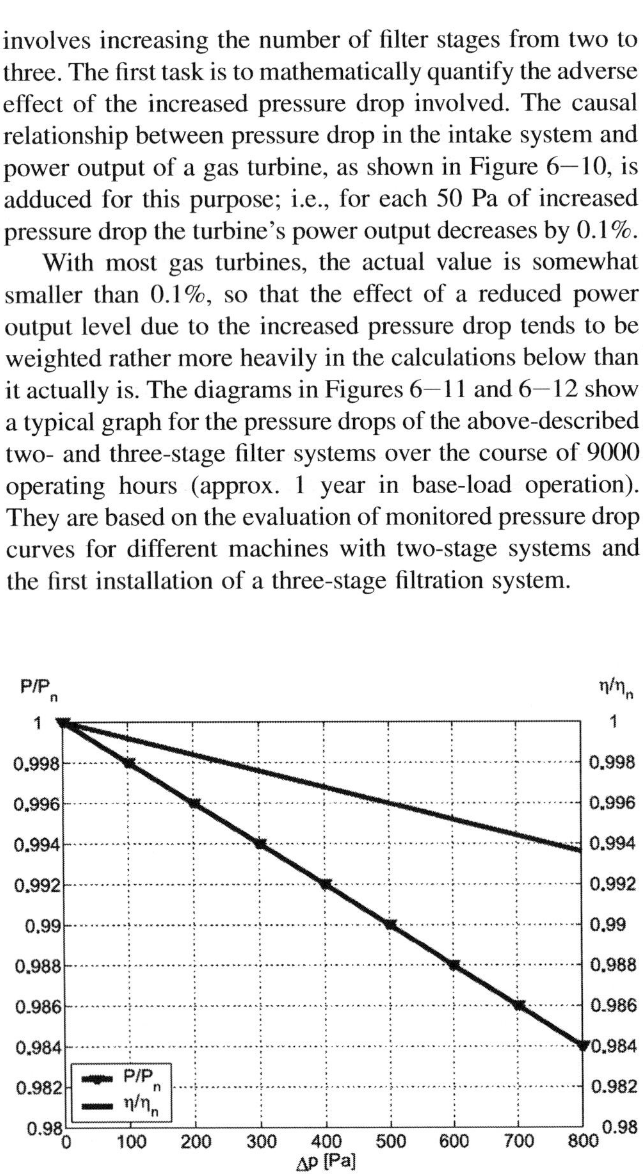

FIGURE 6–10 Output and efficiency losses of gas turbines due to the pressure drop in the air intake system [6-2].

The pressure drop in the air intake system of a gas turbine depends on many factors like, for example, relative humidity, temporarily increased dust concentrations and fluctuating volume flow rate of the turbine. Therefore real life monitored pressure drop curves do not show such a smooth behavior. The average pressure drop curves shown in Figures 6–11 and 6–12 result from flattening out short term fluctuations. The curves marked with "two-stage system" or "three-stage system," respectively, represent the total pressure drop of the two-stage or three-stage systems. The significantly lower pressure drop in the two-stage filter system can clearly be seen, with the average pressure drop over the entire year being approximately 300 Pa less compared to the three-stage system.

In order to understand the reasons behind the reduced fouling on the blades, the comparison provided in Table 6–5 is helpful. The diagram compares the particle concentrations in the clean air for the two filter sequences F6-F8 and F6-F9-H11, in each case for three particle-size fractions in the as-new condition of the filters. In the F6-F8 two-stage filter system, for example, approx. 7.2 million particles of

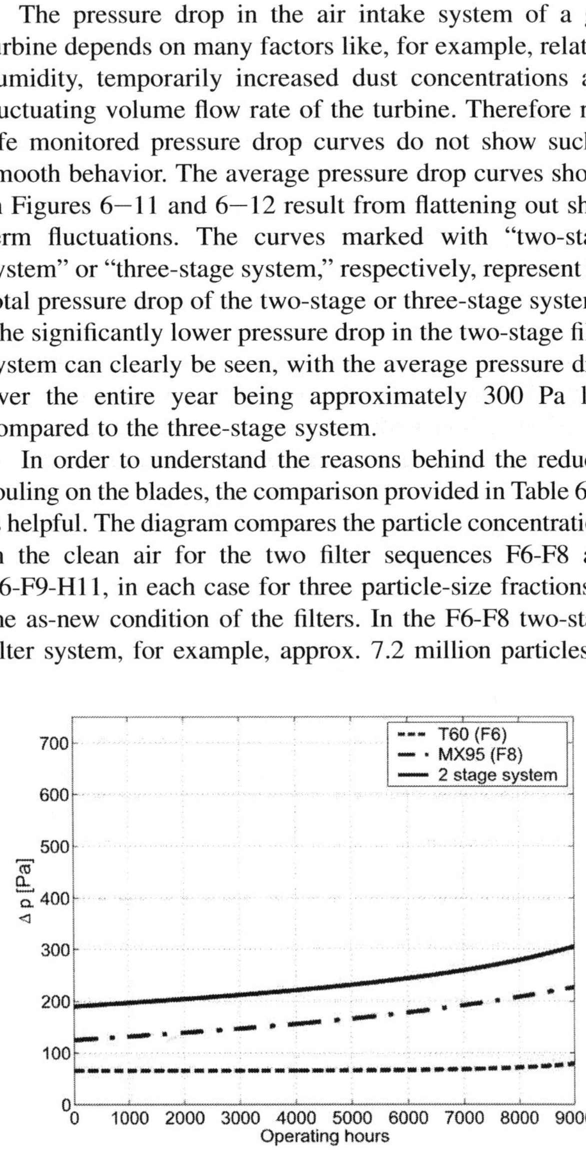

FIGURE 6–11 Pressure drop curve of a two-stage filter system at 4250 m^3/h volume flow rate per filter element [6-2].

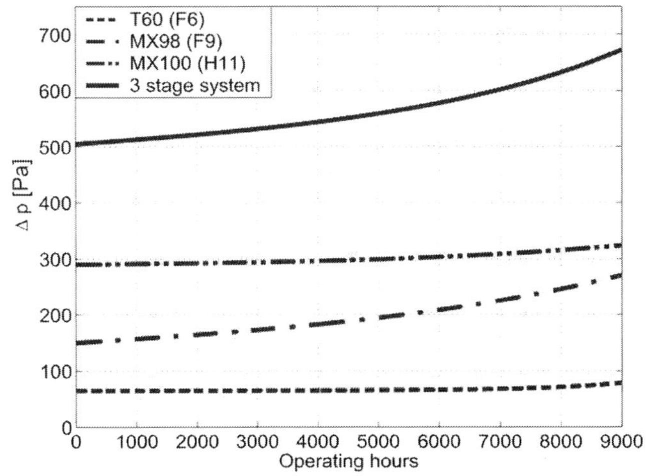

FIGURE 6–12 Pressure drop curve of a three-stage filter system at 4250 m^3/h volume flow rate per filter element [6-2].

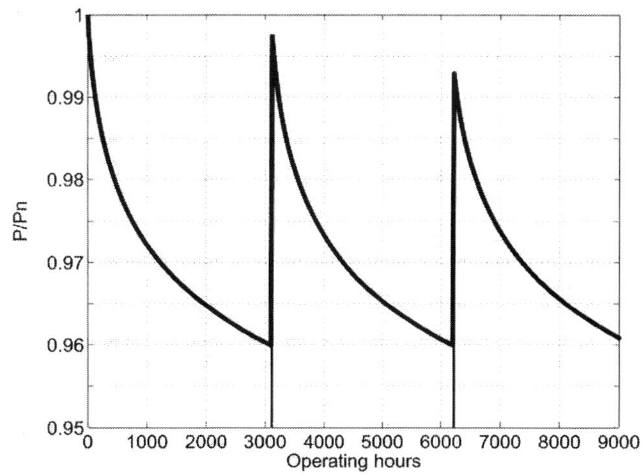

FIGURE 6–13 Output losses due to compressor fouling and recovery from "offline washing" in a two-stage system [6-2].

TABLE 6–5 Comparison of the Filter Collection Efficiencies of Two-stage and Three-stage Filter Systems [6-2]

Particle Size μm	Particles in Atmosphere Number/m^3	Initial Efficiency of Filtration	Particle Penetration Number/m^3	
0.3–0.5	20,000,000	≈64%	7,200,000	2-stage filtration
0.5–1.0	4,000,000	≈80%	800,000	
1.0–2.0	300,000	≈95%	15,000	
0.3–0.5	20,000,000	≈98.9%	220,000	3-stage filtration
0.5–1.0	4,000,000	≈99.9%	4000	
1.0–2.0	300,000	≈99.999%	3	

0.3–0.5 μm in size still reach the clean-air side of the final filter stage, whereas with the three-stage filter system F6-F9-H11 the figure falls to approx. 0.22 million particles. The filter's efficiency is approx. 64% in the two-stage configuration compared to 98.9% for the three-stage system; in other words, in comparison to a two-stage system the three-stage filtration holds back approximately 97% of the particles reaching the clean-air side of the final filter stage. This reduced penetration through the filters in a three-stage system is the reason for significantly reduced fouling.

Figure 6–13 shows the typical logarithmic graph for the power output of a gas turbine operated with a two-stage filter system. The graph is based on the evaluation of measured power output data on different machines and the experience of different gas turbine operators concerning the recoverable power output after an offline wash.[2] It was assumed that two offline washes per year are performed, so that after approx. 3100 h, the first offline washing routine is carried out, and power output capability is increased. The power output decreases again logarithmically, until after approx. 6200 h another offline washing is performed, and so on. In the case of three-stage filtration with the filter sequence already explained, the decrease in power output is significantly smaller, and the time interval before any washing routine becomes longer (see Figure 6–14).

In the example depicted here, taken from actual operation, an offline washing routine was not required within the 9000 operating hours examined. The measured data were approximated by the same logarithmic function as for the two-stage system, but with a new set of coefficients fitting the three-stage system power output (Figures 6–15 to 17).

The difference in power output of the gas turbines with two-stage or three-stage systems is calculated each time.[3]

2. The authors draw on empirical work for certain estimated values in this study. Consult the source for further details.

3. The integral equations used in this study are available from the source.

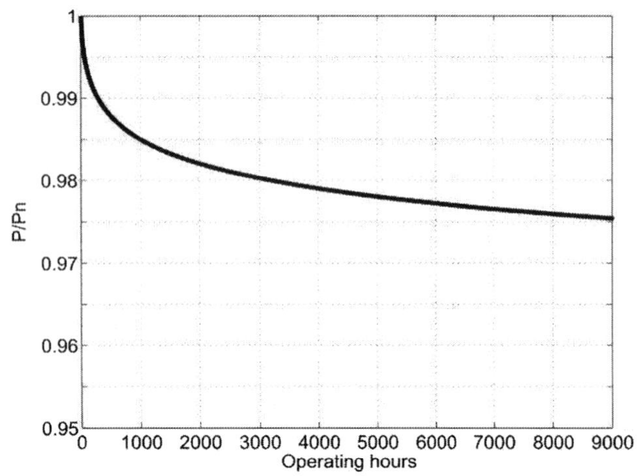

FIGURE 6–14 Lower output losses and longer "washing" intervals in a three-stage system [6-2].

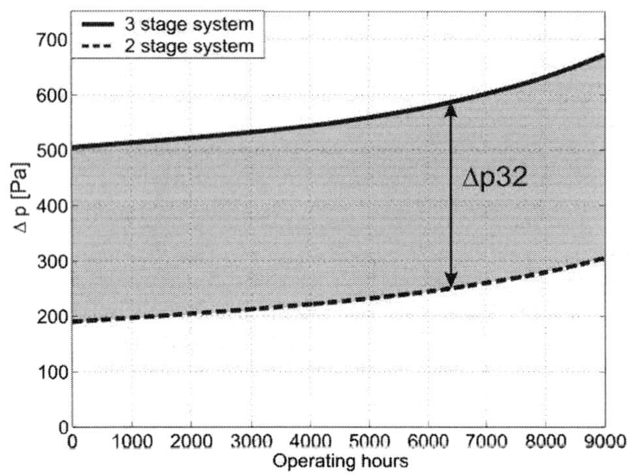

FIGURE 6–15 Comparison of two- and three-stage filtration [6-2].

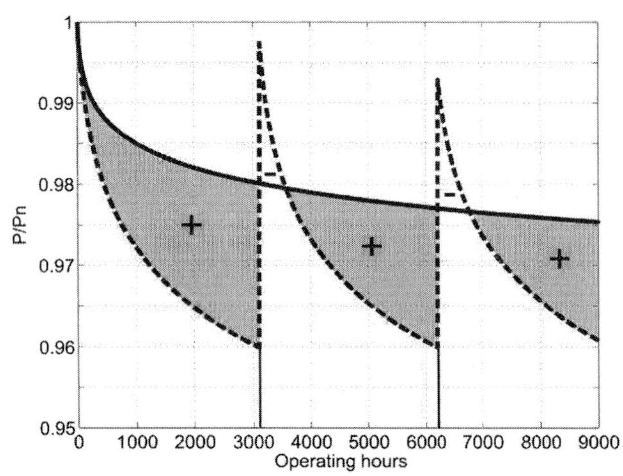

FIGURE 6–16 Comparison of two- and three-stage filtration: more mechanical energy produced due to less fouling [6-2].

For the same gas turbine, rated at 165 MW, this means additional energy output of 16,000 MWh over one year. To conclude the analysis, Table 6–6 shows the results of a comparison based on an electricity-generating power plant. For the 165-MW gas turbine, the additional annual revenues are calculated at EUR 189,000. This calculation must be regarded as on the conservative side, given the selling price assumed for electricity, since the revenue for a megawatt-hour of electricity will usually be higher. The decrease of power output for all systems was less than 2%, referenced to the power output in an absolutely clean condition, i.e., the operating behavior observed was more favorable than that depicted in Figure 6–19.

The useful filter lifetimes for the various filter stages involved, including the Class H11 cassette filters in the third stage, were in all systems longer than one year and reached up to three years. In offline washing routines carried out for trial purposes, the washing water remained clean, and no caked-on deposits of dirt were discernible when the machines were opened up. The above-discussed effects of an improved filtration process have thus all materialized at the 15 test systems, and the calculations can accordingly be verified.

In summary, a comparison of two- and three-stage filter systems for gas turbine intake air filtration reveals important arguments with regard to the most operationally cost-efficient filter sequence overall. The effects of a higher pressure drop entailed by three-stage filtration are compared with those arising from reduced soiling on the blades. In the examined systems, including case studies from actual operation, definite advantages are found for three-stage filtration with the filter sequence F6-F9-H11 in conformity with EN 779 and EN 1822. Even the modification costs for installing another filter stage can be amortized in what will sometimes be significantly less than two years.

GAS TURBINE EXHAUSTS†

The gas turbine's exhaust system conducts the products of combustion—any unburned fuel, excess air, and the heat carried by all these gases—away from the turbine.

In contemporary land-based applications, energy conservation policies, environmental law, and growing awareness of the exhaust steam's potential resulted in a variety of energy conservation designs that are added to the tail end of the exhaust system. These include mainly (also see Chapters 3 and 10 on gas turbine configurations and heat cycles and efficiency optimization):

† [6-1].

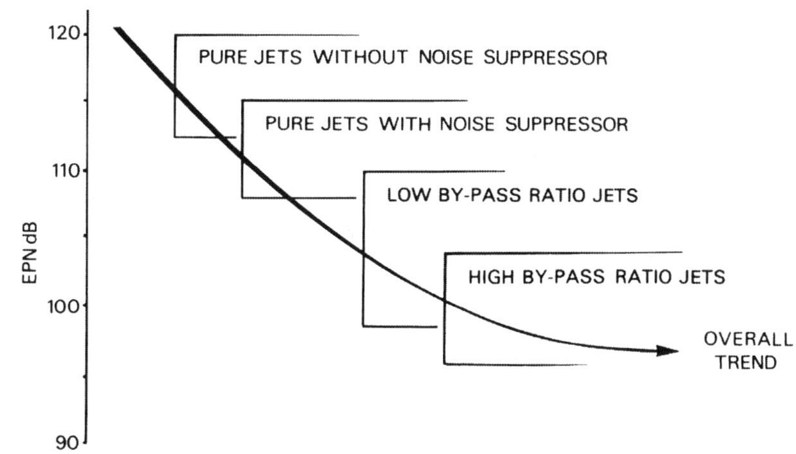

FIGURE 6–17 Comparative noise levels of various aircraft engine types. *(Source: Rolls Royce.)*

TABLE 6–6 Compilation of the Calculation Results [6-2]

Less power output due to higher pressure drop	−9700 MWh
Higher power output due to less fouling	+16,000 MWh
Total gain in power output	+6300 MWh
Cost savings @ 30 EUR/MWh @ 40 USD/MWh	198,000 EUR per annum 252,000 USD per annum

- Waste heat steam generation boilers for use in a combined-cycle unit
- Waste heat use for the civilian population's heating needs (CHPs, combined heat and power plants)
- Waste heat usage for greenhouses that grow vegetables and flowers

By far the most demanding application for exhaust systems is in aircraft engine turbojet applications. In turboprop applications, the bulk of the gas turbine's power is used to drive the propeller.

This also is the case with land-based power generation where the turbine's power is being used in part to drive the generator or in land-based mechanical drive where the power turbine uses most of the power generated by the gas generator (the term given to the gas turbine minus the power turbine in mechanical drive application). Much of the turbine module's power in all applications is used to drive the compressor.

The turbojet application, then, is where the exhaust duct has to withstand the bulk of the gas turbine's thrust. This is because the exhaust gases exiting the exhaust duct propel the aircraft and its entire payload forward.

The following extract deals with aeroengine exhaust systems. The main difference between aeroengines and nonaeroengines in this respect is the geometry of that section. The aeroengine's geometry is not configured for transporting exhaust gases to a stack or HRSG. The nonaeroengine's ducting that conducts the exhaust gases is relatively simple. However, the expansion joints require the most care and cause the most trouble in that area of the gas turbine system. Chapter 8, Accessory Systems, deals with expansion joints.

Aircraft* gas turbine engines have an exhaust system that passes the turbine discharge gases to atmosphere at a velocity, and in the required direction, to provide the resultant thrust. The velocity and pressure of the exhaust gases create the thrust in the turbo-jet engine but in the turbo-propeller engine only a small amount of thrust is contributed by the exhaust gases, because most of the energy has been absorbed by the turbine for driving the propeller. The design of the exhaust system therefore exerts a considerable influence on the performance of the engine. The areas of the jet pipe and propelling or outlet nozzle affect the turbine entry temperature, the mass airflow and the velocity and pressure of the exhaust jet.

The temperature of the gas entering the exhaust system is between 550 and 850°C according to the type of engine and with the use of afterburning can be 1500°C or higher. Therefore, it is necessary to use materials and a form of construction that will resist distortion and cracking, and prevent heat conduction to the aircraft structure.

A basic exhaust system is shown in Figure 6–18. The use of a thrust reverser noise suppressor and a two position propelling nozzle entails a more complicated system as shown in Figure 6–19. The low bypass engine may also

* Source: Courtesy Rolls Royce. Adapted, with permission. *The Jet Engine*, 1986, Rolls Royce Plc: UK.

FIGURE 6–18 A basic exhaust system. *(Source: Rolls Royce.)*

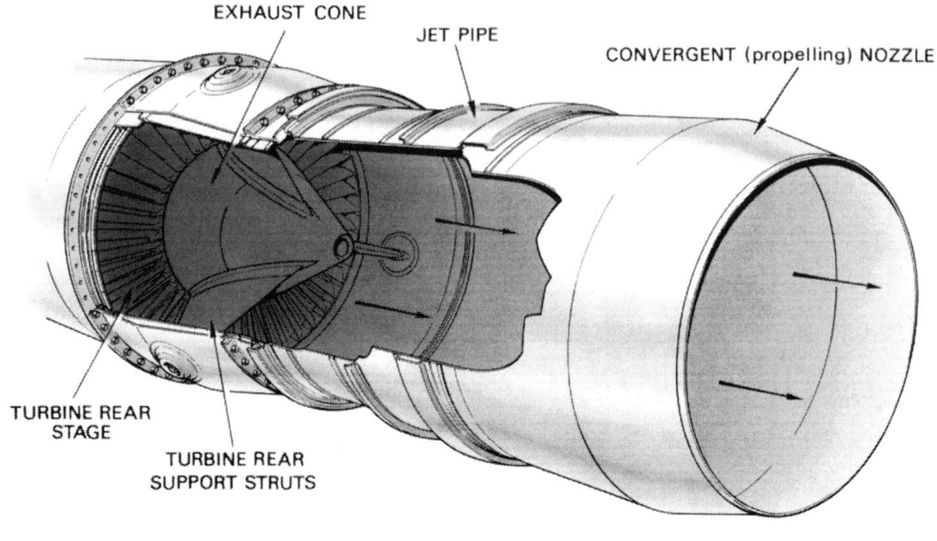

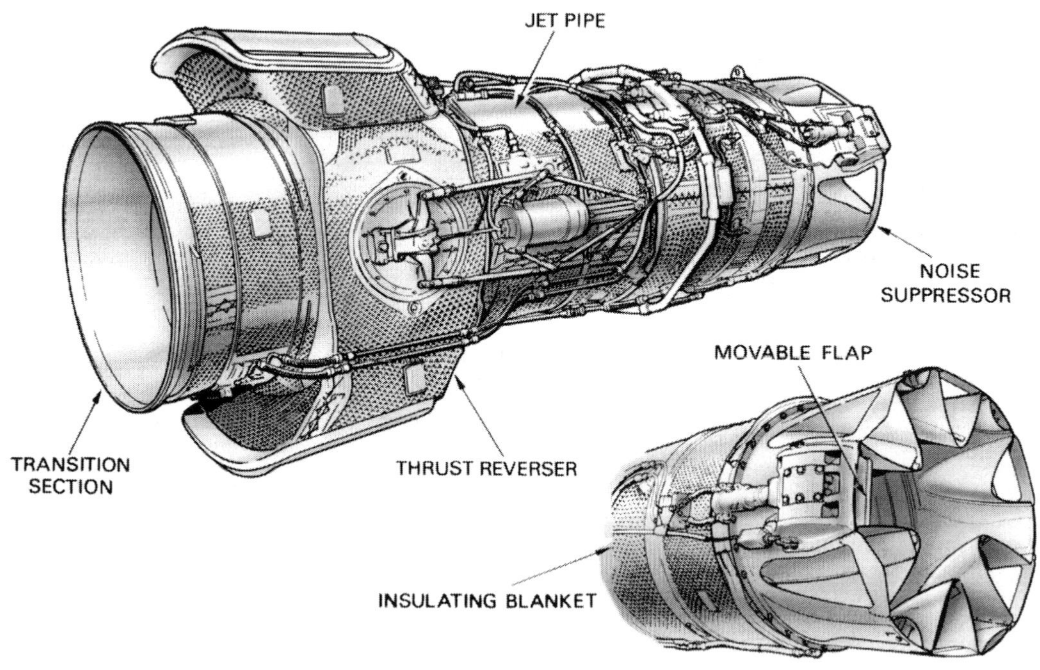

FIGURE 6–19 Exhaust system with thrust reverser, noise suppressor, and two position propelling nozzle. *(Source: Rolls Royce.)*

include a mixer unit (Figure 6–20) to encourage a thorough mixing of the hot and cold gas streams.

Exhaust Gas Flow

Gas from the engine turbine enters the exhaust system at velocities from 750–1200 feet per second, but, because velocities of this order produce high friction losses, the speed of flow is decreased by diffusion. This is accomplished by having an increasing passage area between the exhaust cone and the outer wall as shown in Figure 6–18. The cone also prevents the exhaust gases from flowing across the rear face of the turbine disc. It is usual to hold the velocity at the exhaust unit outlet to a Mach number of about 0.5, i.e.,

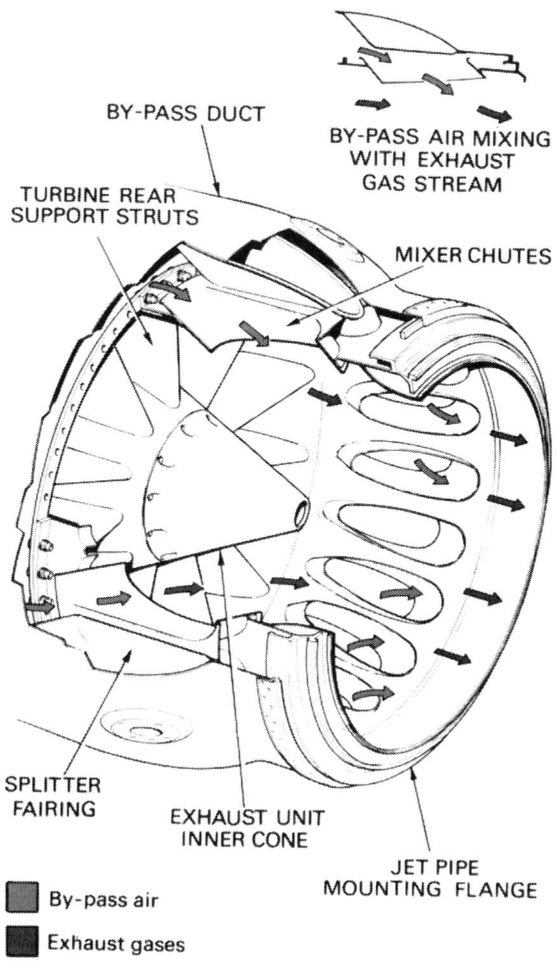

FIGURE 6–20 A low air bypass mixer unit. *(Source: Rolls Royce.)*

approximately 950 feet per second. Additional losses occur due to the residual whirl velocity in the gas stream from the turbine. To reduce these losses, the turbine rear struts in the exhaust unit are designed to straighten out the flow before the gases pass into the jet pipe.

The exhaust gases pass to atmosphere through the propelling nozzle, which is a convergent duct, thus increasing the gas velocity. In a turbo-jet engine, the exit velocity of the exhaust gases is subsonic at low thrust conditions only. During most operating conditions, the exit velocity reaches the speed of sound in relation to the exhaust gas temperature and the propelling nozzle is then said to be "choked;" that is, no further increase in velocity can be obtained unless the temperature is increased. As the upstream total pressure is increased above the value at which the propelling nozzle becomes "choked," the static pressure of the gases at exit increases above atmospheric pressure. This pressure difference across the propelling nozzle gives what is known as "pressure thrust" and is effective over the nozzle exit area. This is additional thrust to that obtained due to the momentum change of the gas stream.

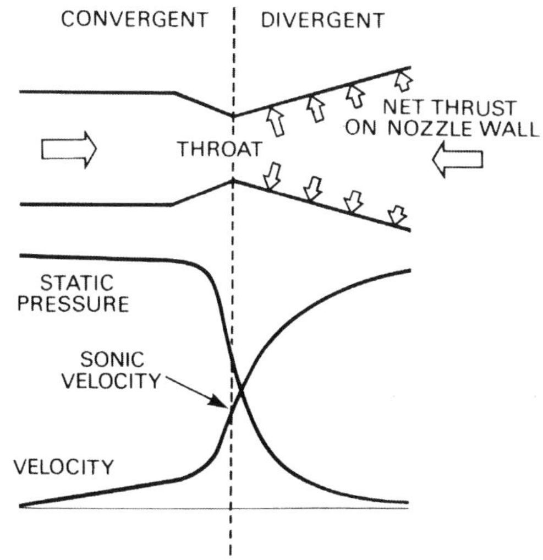

FIGURE 6–21 Gas flow through a convergent-divergent nozzle. *(Source: Rolls Royce.)*

With the convergent type of nozzle wastage of energy occurs, since the gases leaving the exit do not expand rapidly enough to immediately achieve outside air pressure. Depending on the aircraft flight plan, some high-pressure ratio engines can with advantage use a convergent-divergent nozzle to recover some of the wasted energy. This nozzle utilizes the pressure energy to obtain a further increase in gas velocity and, consequently, an increase in thrust.

From the illustration (Figure 6–21), it will be seen that the convergent section exit now becomes the throat with the exit proper now being at the end of the flared divergent section. When the gas enters the convergent section of the nozzle, the gas velocity increases with a corresponding fall in static pressure. The gas velocity at the throat corresponds to the local sonic velocity. As the gas leaves the restriction of the throat and flows into the divergent section, it progressively increases in velocity towards the exit. The reaction to this further increase in momentum is a pressure force acting on the inner wall of the nozzle. A component of this force acting parallel to the longitudinal axis of the nozzle produces the further increase in thrust.

The propelling nozzle size is extremely important and must be designed to obtain the correct balance of pressure, temperature and thrust. With a small nozzle these values increase, but there is a possibility of the engine surging, whereas with a large nozzle the values obtained are too low.

A fixed-area propelling nozzle is only efficient over a narrow range of engine operating conditions. To increase this range, a variable area nozzle may be used. This type of nozzle is usually automatically controlled and is designed to maintain the correct balance of pressure and temperature at all operating conditions. In practice, this system is seldom used as the performance gain is offset by the increase in weight. However, with afterburning a variable area nozzle is necessary.

The bypass engine has two gas streams to eject to atmosphere, the cool bypass airflow and the hot turbine discharge gases.

In a low bypass ratio engine, the two flows are combined by a mixer unit (Figure 6—20) which allows the bypass air to flow into the turbine exhaust gas flow in a manner that ensures thorough mixing of the two streams.

In high bypass ratio engines, the two streams are usually exhausted separately. The hot and cold nozzles are coaxial and the area of each nozzle is designed to obtain maximum efficiency. However, an improvement can be made by combining the two gas flows within a common, or integrated, nozzle assembly. This partially mixes the gas flows prior to ejection to atmosphere. An example of both types of high bypass exhaust system is shown in Figure 6—22.

Construction and Materials

The exhaust system must be capable of withstanding the high gas temperatures and is therefore manufactured from nickel or titanium. It is also necessary to prevent any heat being transferred to the surrounding aircraft structure. This is achieved by passing ventilating air around the jet pipe, or by lagging the section of the exhaust system with an insulating blanket (Figure 6—23). Each blanket has an inner layer of fibrous insulating material contained by an outer skin of thin stainless steel, which is dimpled to increase its strength. In addition, acoustically absorbent materials are sometimes applied to the exhaust system to reduce engine noise.

When the gas temperature is very high (for example, when afterburning is employed), the complete jet pipe is usually of double-wall construction with an annular space between the two walls. The hot gases leaving the propelling nozzle induce, by ejector action, a flow of air through the annular space of the engine nacelle. This flow of air cools the inner wall of the jet pipe and acts as an insulating blanket by reducing the transfer of heat from the inner to the outer wall.

The cone and streamline fairings in the exhaust unit are subjected to the pressure of the exhaust gases; therefore, to prevent any distortion, vent holes are provided to obtain a pressure balance.

The mixer unit used in low bypass ratio engines consists of a number of chutes through which the bypass air flows into the exhaust gases. A bonded honeycomb structure is

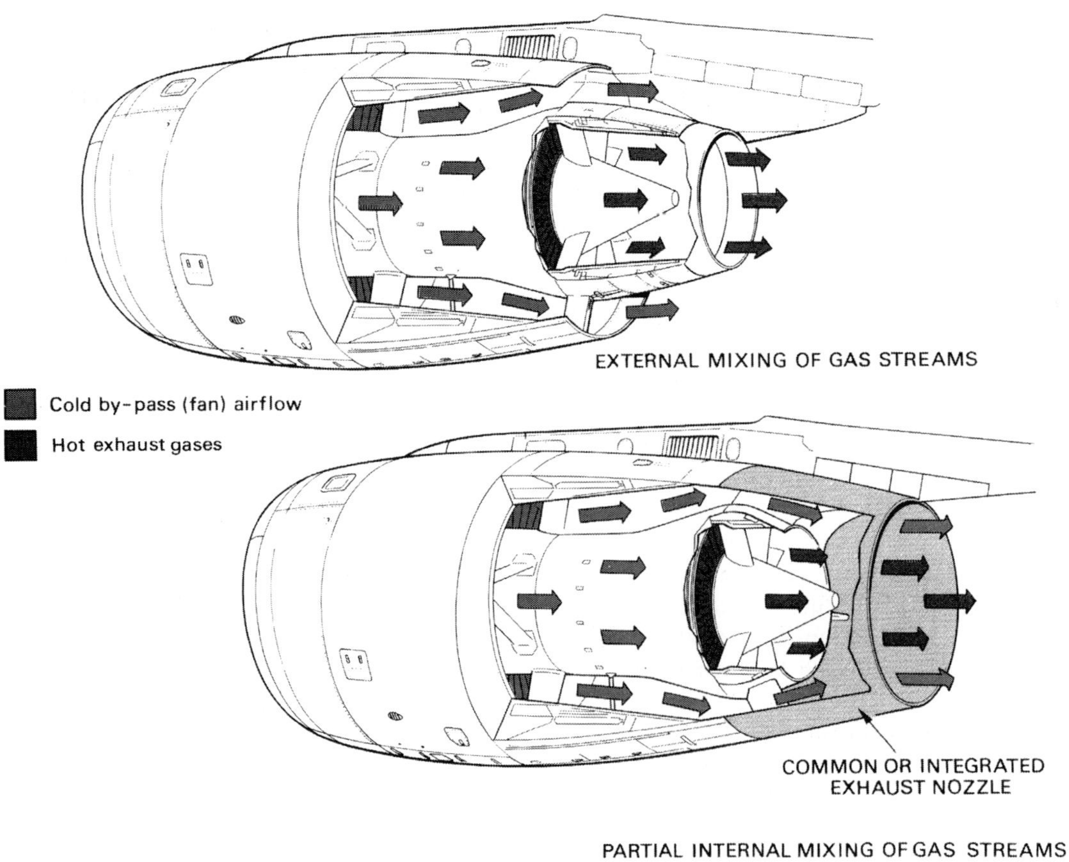

FIGURE 6—22 High bypass ratio engine exhaust systems. *(Source: Rolls Royce.)*

FIGURE 6–23 An insulating blanket. *(Source: Rolls Royce.)*

used for the integrated nozzle assembly of high bypass ratio engines to give lightweight strength to this large component.

Due to the wide variations of temperature to which the exhaust system is subjected, it must be mounted and have its sections joined together in such a manner as to allow for expansion and contraction without distortion or damage.

GAS TURBINE NOISE SUPPRESSION*

With land-based and marine applications, the nature of the application permits the designers to add "houses" around the gas turbine package to provide a measure of noise protection for the personnel working in the area. In applications such as offshore and some naval applications, where weight per delivered unit of horsepower is at a premium, the "house" may be lighter or nonexistent and personnel permanently equipped with hearing protection in the vicinity of the turbine. The design of these protective houses (which also mitigate the risk of explosion) is covered next.

By far the most demanding application for noise mitigation, then, is with aircraft gas turbines. These turbines expose the entire surrounding population on the ground to their noise during takeoff and landing. Tolerant during the Second World War, which accelerated the progress of aircraft gas turbine engines considerably, the public now is extremely demanding, expecting relief from aircraft engine noise.

In all gas turbine applications, the noise one produces is a function of both its aerodynamic design (gas velocity and so forth) and its overall system, which includes its potential for resonance and additional vibration that would not be noted during test cell runs. This would depend on the vibration potential of the gas turbine with:

- Its base mounting in the case of land-based and naval applications
- Its nacelle with aircraft applications
- The entire surrounding mechanical structure, which would include the plant concrete floor, piping, and all other accessories somehow "connected" to the gas turbine via a physical medium in land-based and marine applications
- The entire aircraft in aircraft engine applications

One of the best examples to illustrate this in the case of aeroengines was the following observation (made in the late 1980s) on the early rivalry between the General Electric/Snecma CFM-56 and IAE's (International Aeroengines) V2500. The first aircraft fitted with the V2500 was the Airbus A321 and the early noise trial records at John Wayne Airport revealed that the V2500 fitted aircraft was not detected by several of the airport's noise sensors on approach. The same aircraft fitted with CFM-56 engines was observed to "rattle the passenger's teeth" in a certain range of seat rows. However, all this can change in design development, as designers sort out which component or hardware combination causes the overall result to "rattle" and change it enough to get that configuration to run smoothly.

What follows now is a section on mitigating noise in aircraft engine applications. The reader should note that, for many aircraft engines developed in the days before protecting the public from aircraft engine noise was as high profile as it is today, hush kits with retrofit potential have been developed. A good example is the Pratt and Whitney JT-8D fleet. From the 1950s to the 1980s, this engine's growth took it from about 9000 pounds of thrust to 19,000 pounds. In the process, it became the world's largest commercial engine fleet. It no longer is the high-profile engine it once was; however, the number of these engines still out there and operating are legion, particularly in countries that need to buy "bargain used aircraft" from wealthier countries. Retrofit hush kits for these engines (so that they could fly into airports with more stringent noise rules) became popular in the late 1980s.

Airport* regulations and aircraft noise certification requirements, all of which govern the maximum noise level

* [6-1].

* Source: Courtesy Rolls Royce. Adapted, with permission. *The Jet Engine*, 1986, Rolls Royce Plc: UK.

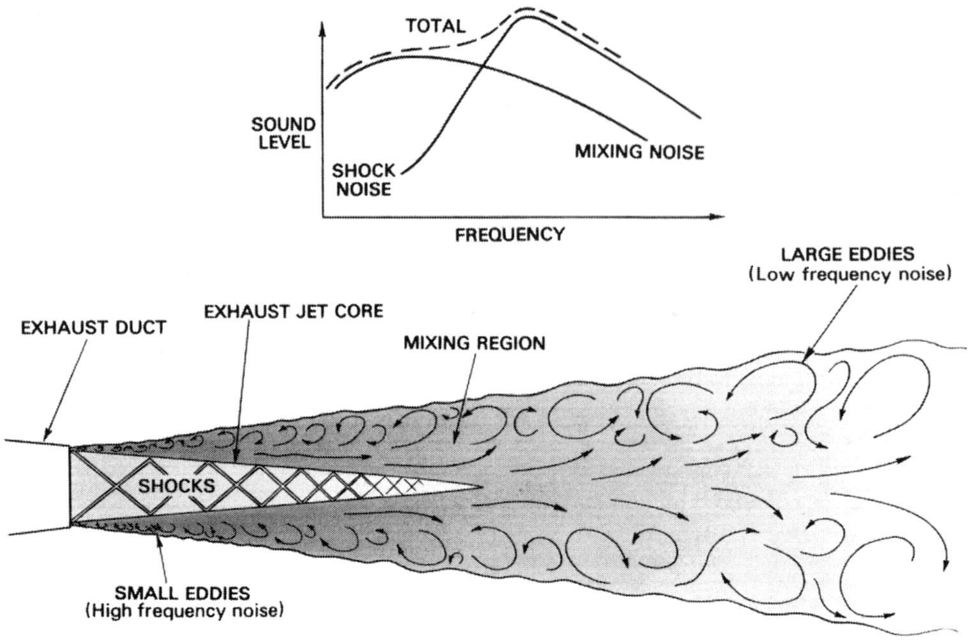

FIGURE 6–24 Exhaust mixing and shock structure. *(Source: Rolls Royce.)*

aircraft are permitted to produce, have made jet engine noise suppression one of the most important fields of research.

The unit that is commonly used to express noise annoyance is the Effective Perceived Noise deciBel (EPNdB). It takes into account the pitch as well as the sound pressure (decibel) and makes allowance for the duration of an aircraft flyover. Figure 6–24 compares the noise levels of various jet engine types.

Airframe self-generated noise is a factor in an aircraft's overall noise signature, but the principal noise source is the engine.

To understand the problem of engine noise suppression, it is necessary to have a working knowledge of the noise sources and their relative importance. The significant sources originate in the fan or compressor, the turbine and the exhaust jet or jets. These noise sources obey different laws and mechanisms of generation, but all increase, to a varying degree, with greater relative airflow velocity. Exhaust jet noise varies by a larger factor than the compressor or turbine noise, therefore a reduction in exhaust jet velocity has a stronger influence than an equivalent reduction in compressor and turbine blade speeds.

Jet exhaust noise is caused by the violent and hence extremely turbulent mixing of the exhaust gases with the atmosphere and is influenced by the shearing action caused by the relative speed between the exhaust jet and the atmosphere. The small eddies created near the exhaust duct cause high frequency noise but downstream of the exhaust jet the larger eddies create low frequency noise. Additionally, when the exhaust jet velocity exceeds the local speed of sound, a regular shock pattern is formed within the exhaust jet core. This produces a discrete (single frequency) tone and selective amplification of the mixing noise, as shown in Figure 6–24. A reduction in noise level occurs if the mixing rate is accelerated or if the velocity of the exhaust jet relative to the atmosphere is reduced. This can be achieved by changing the pattern of the exhaust jet as shown in Figure 6–25.

Compressor and turbine noise results from the interaction of pressure fields and turbulent wakes from rotating blades and stationary vanes, and can be defined as two distinct types of noise; discrete tone (single frequency) and broadband (a wide range of frequencies). Discrete tones are produced by the regular passage of blade wakes over the stages downstream causing a series of tones and harmonics from each stage. The wake intensity is largely dependent upon the distance between the rows of blades and vanes. If the distance is short then there is an intense pressure field interaction, which results in a strong tone being generated. With the high bypass engine, the low pressure compressor (fan) blade wakes passing over downstream vanes produce such tones, but of a lower intensity due to lower velocities and larger blade/vane separations. Broadband noise is produced by the reaction of each blade to the passage of air over its surface, even with a smooth airstream. Turbulence in the airstream passing over the blades increases the intensity of the broadband noise and can also induce tones.

With the pure jet engine the exhaust jet noise is of such a high level that the turbine and compressor noise is

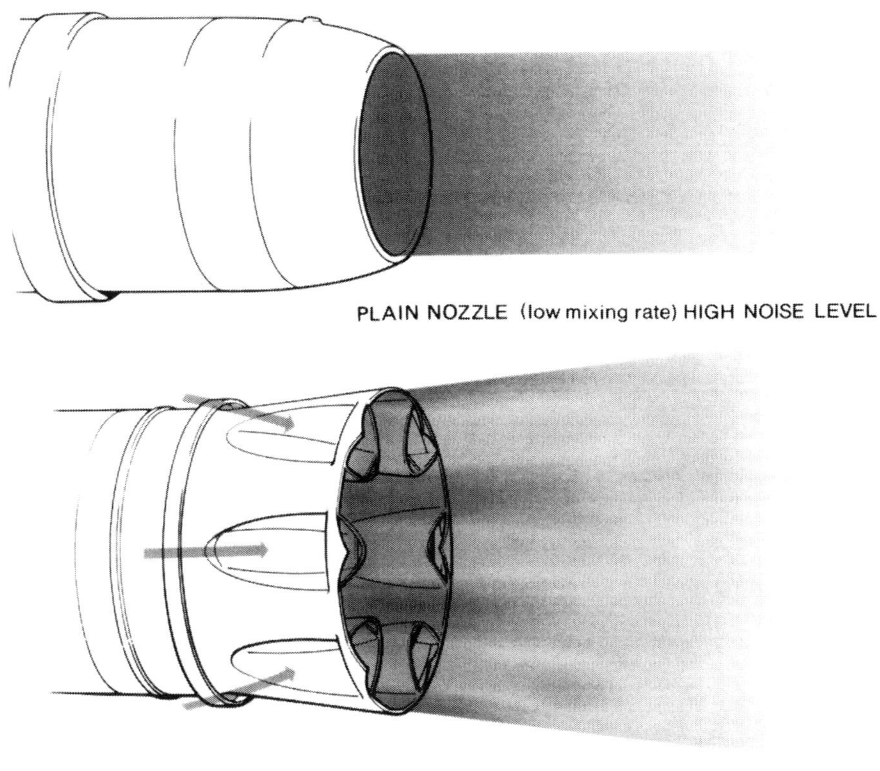

FIGURE 6–25 Change of exhaust jet pattern to reduce noise level. *(Source: Rolls Royce.)*

insignificant at all operating conditions, except low landing-approach thrusts. With the bypass principle, the exhaust jet noise drops as the velocity of the exhaust is reduced but the low pressure compressor and turbine noise increases due to the greater internal power handling. The introduction of a single-stage low-pressure compressor (fan) significantly reduces the compressor noise because the overall turbulence and interaction levels are diminished. When the bypass ratio is in excess of approximately 5 to 1, the jet exhaust noise has reduced to such a level that the increased internal noise source is predominant. A comparison between low and high bypass engine noise sources is shown in Figure 6–26.

Listed amongst the several other sources of noise within the engine is the combustion chamber. It is a significant but not a predominant source, due in part to the fact that it is "buried" in the core of the engine. Nevertheless it contributes to the broadband noise, as a result of the violent activities that occur within the combustion chamber.

Methods of Suppressing Noise

Noise suppression of internal sources is approached in two ways; by basic design to minimize noise originating within or propagating from the engine, and by the use of acoustically absorbent linings. Noise can be minimized by reducing airflow disruption, which causes turbulence. This is achieved by using minimal rotational and airflow velocities and reducing the wake intensity by appropriate spacing between the blades and vanes. The ratio between the number of rotating blades and stationary vanes can also be advantageously employed to contain noise within the engine.

As previously described, the major source of noise on the pure jet engine and low bypass engine is the exhaust jet, and this can be reduced by inducing a rapid or shorter mixing region. This reduces the low frequency noise but may increase the high frequency level. Fortunately, high frequencies are quickly absorbed in the atmosphere and some of the noise that does propagate to the listener is beyond the audible range, thus giving the perception of a quieter engine. This is achieved by increasing the contact area of the atmosphere with the exhaust gas stream by using a propelling nozzle incorporating a corrugated or lobe-type noise suppressor (Figure 6–27).

In the corrugated nozzle, free stream atmospheric air flows down the outside corrugations and into the exhaust jet to promote rapid mixing. In the lobe-type nozzle, the exhaust gases are divided to flow through the lobes and a

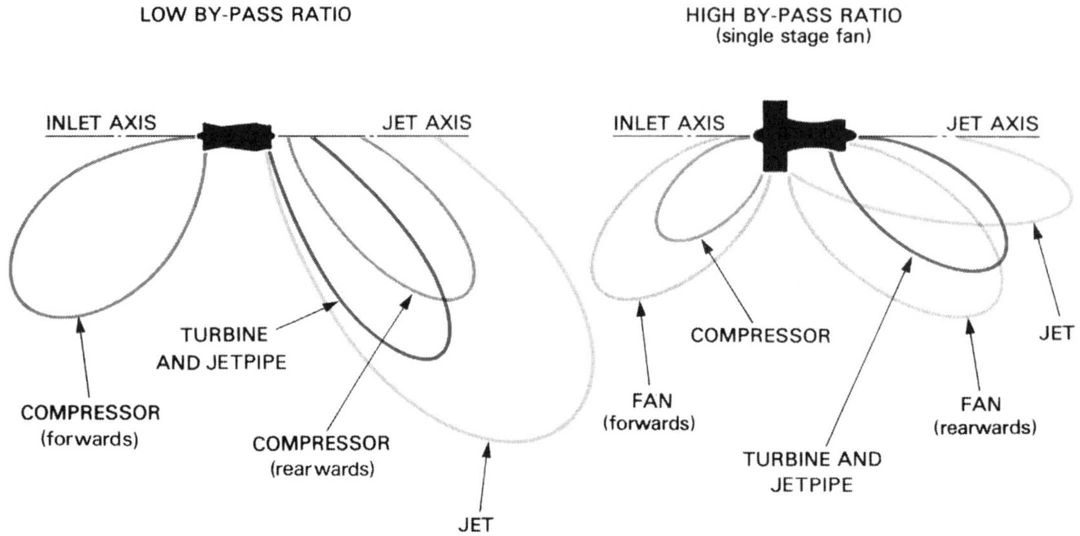

FIGURE 6–26 Comparative noise sources of low and high bypass engines. *(Source: Rolls Royce.)*

small central nozzle. This forms a number of separate exhaust jets that rapidly mix with the air entrained by the suppressor lobes. This principle can be extended by the use of a series of tubes to give the same overall area as the basic circular nozzle.

Deep corrugations, lobes, or multi-tubes give the largest noise reductions, but the performance penalties incurred limit the depth of the corrugations or lobes and the number of tubes. For instance, to achieve the required nozzle area, the overall diameter of the suppressor may have to be increased by so much that excessive drag and weight results. A compromise that gives a noticeable reduction in noise level with the least sacrifice of engine thrust, fuel consumption, or addition of weight is therefore the designer's aim. The high bypass engine has two exhaust streams to eject to atmosphere. However, the principle of jet exhaust noise reduction is the same as for the pure or low bypass engine, i.e., minimize the exhaust jet velocity within overall performance objectives. High bypass engines inherently have a lower exhaust jet velocity than any other type of gas turbine, thus leading to a quieter engine, but further noise reduction is often desirable. The most successful method used on bypass engines is to mix the hot and cold exhaust streams within the confines of the engine (Figure 6–27) and expel the lower velocity exhaust gas flow through a single nozzle.

In the high bypass ratio engine the predominant sources governing the overall noise level are the fan and turbine. Research has produced a good understanding of the mechanisms of noise generation and comprehensive noise design rules exist. As previously indicated, these are founded on the need to minimize turbulence levels in the airflow, reduce the strength of interactions between rotating

FIGURE 6–27 Types of noise suppressor. *(Source: Rolls Royce.)*

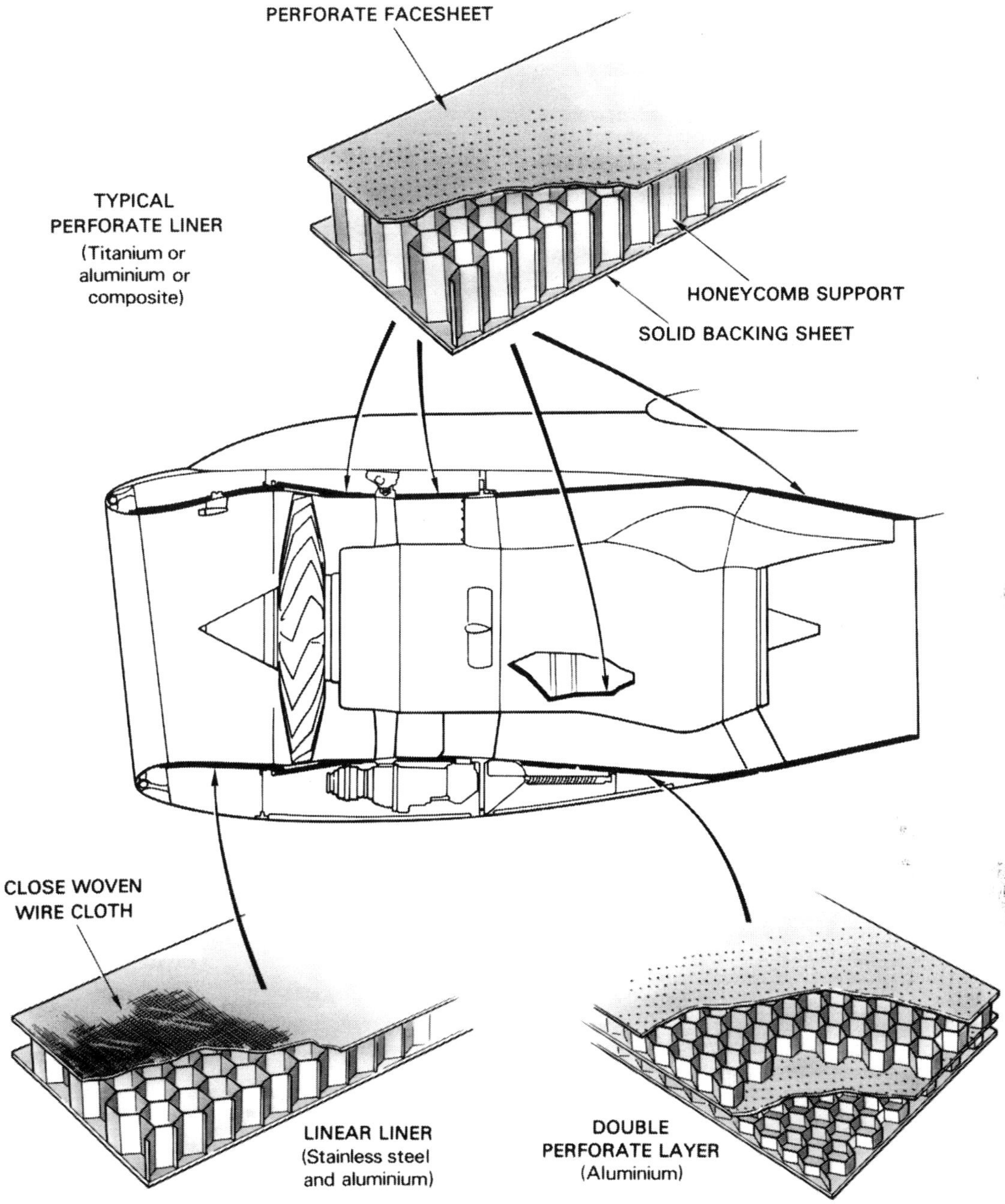

FIGURE 6–28 Noise absorbing materials and location. *(Source: Rolls Royce.)*

blades and stationary vanes, and the optimum use of acoustically absorbent linings.

Noise absorbing "lining" material converts acoustic energy into heat. The absorbent linings (Figure 6–28) normally consist of a porous skin supported by a honeycomb backing, to provide the required separation between the facesheet and the solid engine duct. The acoustic properties of the skin and the liner depth are carefully matched to the character of the noise, for optimum suppression. The disadvantage of liners is the slight increase in weight and skin friction and hence a slight increase in fuel consumption. They do, however, provide a very powerful suppression technique.

Construction and Materials

The corrugated or lobe-type noise suppressor forms the exhaust propelling nozzle and is usually a separate assembly bolted to the jet pipe. Provision is usually made to adjust the nozzle area so that it can be accurately calibrated. Guide vanes are fitted to the lobe-type suppressor to prevent excessive losses by guiding the exhaust gas smoothly through the lobes to atmosphere. The suppressor is a fabricated welded structure and is manufactured from heat-resistant alloys.

Various noise absorbing lining materials are used on jet engines. They fall mainly within two categories, lightweight composite materials that are used in the lower temperature regions and fibrous-metallic materials that are used in the higher temperature regions. The noise absorbing material consists of a perforate metal or composite facing skin, supported by a honeycomb structure on a solid backing skin, which is bonded to the parent metal of the duct or casing.

Nonaeroengine (Land-Based) Gas Turbine Noise Suppression*

Nonaeroengines generally have a "turbine house" around their main modules and systems.

Enclosures around noisy rotating machinery, particularly gas turbines, provide protection against noise and help contain risk from situations such as small gas leaks. The design of acoustic gas turbine enclosures is summarized in this section in text extracts from two papers on the subject. These extracts give the reader all the salient points to watch for if specifying or buying an acoustic enclosure. It also provides the reader with basic knowledge of how acoustic sound intensity measurements are conducted generally (i.e., for any kind of equipment). Note also that these designs were built for offshore applications where weight has to be minimized.

Applications of Sound Intensity Measurements to Gas Turbine Engineering

Nomenclature

A	=	total absorption in receiving room
I, I_0	=	intensity, pW/m^2
L_1, L_2	=	sound pressure levels, dB
LI	=	sound intensity level, dB
LK	=	reactivity index
P, P_0	=	pressure, Pa
P_a, P_b	=	pressure, Pa
S	=	surface area of the test panel, m^2
W, W_0	=	power, pW
$r, \Delta r$	=	distance, m
t	=	time, s
u	=	particle velocity, m/s
ρ	=	density, kg/m^2

* Source: Courtesy McGraw-Hill, *Process Engineering Equipment Handbook* (Soares, C. New York: McGraw-Hill, 2001), original extract provided by Altair Filters International (AFI), Limited UK: it includes adapted extracts from an Altair International paper published in *ASME Journal of Engineering for Gas Turbines and Power* 113 (October 1991).

Gas turbines that are supplied to the oil and power industries are usually given extensive acoustic treatment to reduce the inherent high noise levels to acceptable limits. The cost of this treatment may be a significant proportion of the total cost of the gas turbine installation. In the past it has been difficult to determine if the acoustic treatment is achieving the required noise limits because of a number of operational problems. These problems include: the presence of other nearby, noisy equipment, the influence of the environment, and instrumentation limitations. Traditionally, sound measurements have been taken using a sound level meter that responds to the total sound pressure at the microphone irrespective of the origin of the sound. So, the enforcement of noise limits has been difficult because of uncertainties concerning the origin of the noise.

Recent advances in signal processing techniques have led to the development of sound intensity meters that can determine the direction, as well as the magnitude, of the sound, without the need for expensive test facilities. These instruments enable the engineer to determine if large equipment, such as gas turbine packages, meet the required noise specification even when tested in the factory or on site where other noise sources are present.

There are, of course, limitations in the use of sound intensity meters, and there are some differences of opinion on measurement techniques. Nevertheless, the acoustic engineer's ability to measure and identify the noise from specific noise sources has been greatly enhanced.

In this section, the differences between sound pressure, sound intensity, and sound power are explained. Measurement techniques are discussed with particular references to the various guidance documents that have been issued. Some case histories of the use of sound intensity meters are presented that include field and laboratory studies relating to gas turbines and other branches of industry.

SOUND FUNDAMENTAL CONCEPTS

Sound Pressure, Sound Intensity, and Sound Power

Any item of equipment that generates noise radiates acoustic energy. The total amount of acoustic energy it radiates is the *sound power.* This is, generally, independent of the environment. What the listener perceives is the *sound pressure* acting on his or her eardrums and it is this parameter that determines the damaging potential of the sound. Unlike the sound power, the sound pressure is very dependent on the environment and the distance from the noise source to the listener.

Traditional acoustic instrumentation, such as sound level meters, detects the sound pressure using a single microphone that responds to the pressure fluctuations incident upon the microphone. Since pressure is a scalar quantity, there is no simple and accurate way that such instrumentation can determine the amount of sound energy radiated by a large source unless the source is tested in a specially built room, such as an echoic or reverberation room, or in the open air away from sound reflecting surfaces. This imposes severe limitations on the usefulness of sound pressure level measurements taken near large equipment that cannot be moved to special acoustic rooms.

Sound intensity is the amount of sound energy radiated per second through a unit area. If a hypothetical surface, or envelope, is fitted around the noise source, then the sound intensity is the number of acoustic watts of energy passing through 1 m of this envelope (see Figure 6–29). The sound intensity, I, normal to the spherical envelope of radius, r, centered on a sound source of acoustic power, W, is given by

$$I = \frac{W}{4\pi r^2}$$

Clearly, the total sound power is the product of the sound intensity and the total area of the envelope if the sound source radiates uniformly in all directions. Since the intensity is inversely proportional to the distance of the envelope from the noise source, the intensity diminishes as the radius of the envelope increases. But as this distance increases, the total area of the envelope increases also, so the product of the intensity and the surface area (equal to the sound power) remains constant.

When a particle of air is displaced from its mean position by a sound wave that is moving through the air there is a temporary increase in pressure. The fact that the air particle has been displaced means that it has velocity. The product of the pressure and the particle velocity is the sound intensity. Since velocity is a vector quantity, so is sound intensity. This means that sound intensity has both direction and magnitude.

It is important to realize that sound intensity is the time-averaged rate of energy flow per unit area. If equal amounts of acoustic energy flow in opposite directions through a hypothetical surface at the same time, then the net intensity at that surface is zero.

Reference Levels

Most parameters used in acoustics are expressed in decibels because of the enormous range of absolute levels normally considered. The range of sound pressures that the ear can tolerate is from 2×10^{-5} Pa to 200 Pa. This range is reduced to a manageable size by expressing it in decibels, and is equal to 140 dB.

The sound pressure level (SPL) is defined as

$$\text{SPL} = 20 \log_{10}\left(\frac{P}{P_0}\right) \text{dB}\left(\text{re.}2 \times 10^{-6}\text{Pa}\right)$$

Likewise, sound intensity level (SIL) and sound power level (PWL) are normally expressed in decibels. In this case,

$$\text{SIL} = 10 \log_{10}\left(\frac{I}{I_0}\right) \text{dB}\left(\text{re.}1\text{pW/m}^2\right)$$

$$\text{PWL} = 10 \log_{10}\left(\frac{W}{W_0}\right) \text{dB (re.1 pW)}$$

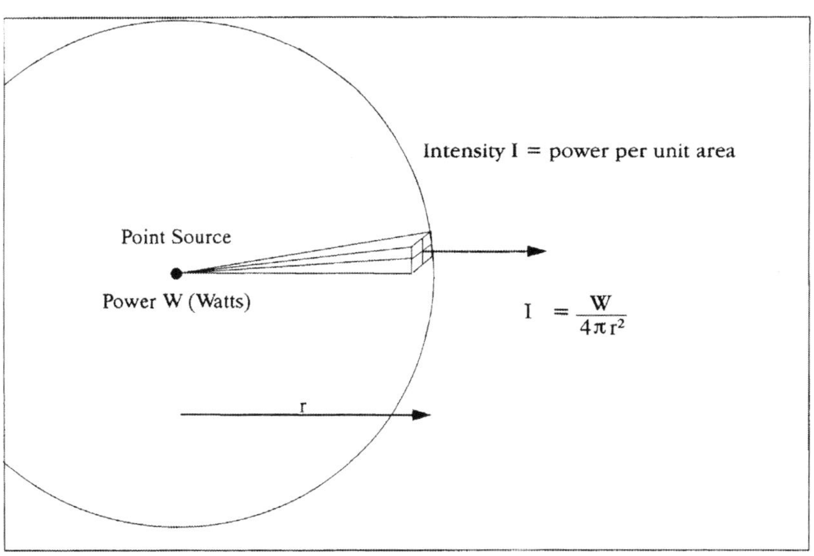

FIGURE 6–29 The intensity level from a point sound source. *(Source: AFI Ltd.)*

The Relationship between Sound Pressure Level and Sound Intensity Level

When the sound intensity level is measured in a free field in air, then the sound pressure level and sound intensity level in the direction of propagation are numerically the same. In practice most measurements of the sound intensity are not carried out in a free field, in which case there will be a difference between the sound pressure and intensity levels. This difference is an important quantity and is known by several terms, such as reactivity index, pressure-intensity index, *P-I* index, phase index, or *LK* value. This index is used as a "field indicator" to assess the integrity of a measurement in terms of grades of accuracy or confidence limits. This will be considered in more detail later in this section.

Instrumentation

Sound Intensity Meters

A sound intensity meter comprises a probe and an analyzer. The analyzer may be of the analog, digital, or FFT (fast Fourier transform) type. The analog type has many practical disadvantages that make it suitable only for surveys and not precision work.

Digital analyzers normally display the results in octave or $^1/_3$ octave frequency bands. They are well suited to detailed investigations of noise sources in the laboratory or on site. Early models tended to be large and heavy and require electrical main supplies, but the latest models are much more suited to site investigations.

FFT analyzers generate spectral lines on a screen. This can make the display very difficult to interpret during survey sweeps because of the amount of detail presented. Another disadvantage of FFT-based systems is that their resolution is generally inadequate for the synthesis of $^1/_3$ octave band spectra.

Sound Intensity Probes

There are several probe designs that employ either a number of pressure microphones in various configurations or a combination of a pressure microphone and a particle velocity detector. The first type of probe uses nominally identical pressure transducers that are placed close together. Various arrangements have been used with the microphones either side by side, face to face, or back to back. Each configuration has its own advantages and disadvantages.

If the output signals of two microphones are given by P_a and P_b, then the average pressure, P, between the two microphones is

$$P = \frac{1}{2}(P_a + P_b) \tag{6.2}$$

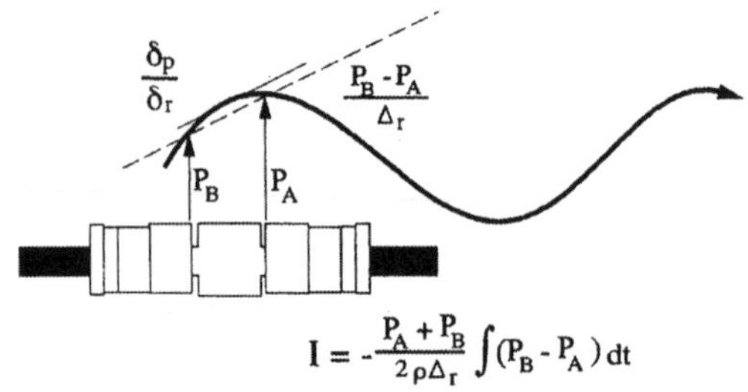

FIGURE 6–30 The finite difference approximation of sound intensity for a two-microphone configuration. *(Source: AFI Ltd.)*

The particle velocity, u, is derived from the pressure gradient between the two microphones by the relationship

$$\begin{aligned} u &= -\int \frac{1}{\rho}\cdot\frac{\partial p}{\partial r}\cdot dt \\ u &= -\frac{1}{\rho}\int\frac{(P_b - P_a)}{\Delta r}\cdot dt \end{aligned} \tag{6.3}$$

Since sound intensity, I, is the product of the pressure and particle velocity, combining equations (6.2) and (6.3) gives the intensity as

$$I = \left(\frac{P_a + P_b}{2\rho\Delta r}\right)\int(P_a - P_b)\cdot dt \tag{6.4}$$

Figure 6–30 shows a two-microphone probe, with a face-to-face arrangement, aligned parallel to a sound field. In this orientation the pressure difference is maximized, and so is the intensity. If the probe is aligned so that the axis of the two microphones is normal to the direction of propagation of the sound wave, then the outputs of the two

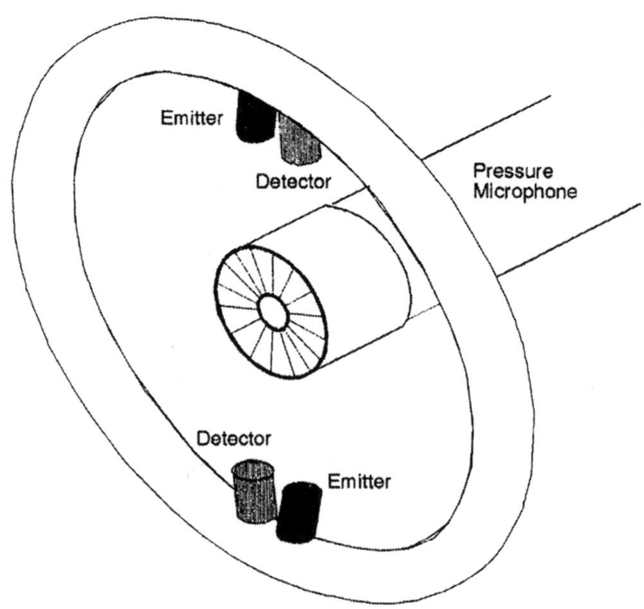

FIGURE 6–31 Schematic representation of a pressure/velocity probe. *(Source: AFI Ltd.)*

microphones would be identical in magnitude and phase. Since the particle velocity is related to the difference between the two pressures, P_a and P_b, then the intensity would be zero.

The second type of probe combines a microphone, to measure the pressure, and an ultrasonic particle velocity transducer. Two parallel ultrasonic beams are sent in opposite directions as shown in Figure 6–31. The oscillatory motion of the air caused by audio-frequency sound waves produces a phase difference between the two ultrasonic waves at their respective detectors. This phase difference is related to the particle velocity component in the direction of the beams. This measure of particle velocity is multiplied directly by the pressure to give the sound intensity.

TABLE 6–7 Uncertainty of the Determination of Sound Power Level (ISO/DP 9614)

Octave Band Center Frequencies, Hz	1/3 Octave Band Center Frequencies, Hz	Standard Deviations, dB Class 1	Class 2	Class 3
63–125	50–160	2	3	4
250–500	200–630	1.5	2	4
1000–4000	800–5000	1	1.5	4
	6300	2	2.5	4
A Weighted 160–6300 Hz		1	1.5	4

[1]Class 1 = Precision Grade, Class 2 = Engineering Grade, Class 3 = Survey Grade.
[2]The width of the 95% confidence intervals corresponds approximately to four times the dB values in this table.
(Source: AFI Ltd.)

Guidelines and Standards in Sound Intensity Measurements and Measurement Technique

Guidelines and Standards

Work began in 1983 on the development of an international standard on the use of sound intensity and the final document is about to be issued. Further standards are expected dealing specifically with instrumentation.

In the absence of a full standard the only guidance available was the draft ISO standard (ISO/DP 9614) and a proposed Scandinavian standard (DS F88/146).

The ISO document ISO/DP 9614 gives specific methods for determining the sound power levels of noise sources within specific ranges of uncertainty.

The proposed test conditions are less restrictive than those required by the International Standards series ISO 3740–3747, which are based on sound pressure measurements. The proposed standard is based on the sampling of the intensity normal to a measurement surface at discrete points on this surface. The method can be applied to most noise sources that emit noise that is stationary in time and it does not require special purpose test environments.

The draft document defines three grades of accuracy with specified levels of uncertainty for each grade. Since the level of uncertainty in the measurements is related to the source noise field, the background noise field, and the sampling and measurement procedures, initial procedures are proposed that determine the accuracy of the measurements. These procedures evaluate the "field indicators" that indicate the quality of the sound power measurements. These field indicators consider, among other things:

- The pressure–intensity index (or reactivity index)
- The variation of the normal sound intensities over the range of the measurement points
- The temporal variation of the pressure level at certain monitoring points

The three grades of measurement accuracy specified in ISO/DP 9614, and the associated levels of uncertainty, are given in Table 6–7.

The Scandinavian proposed standard DSF 88/146 was developed for the determination of the sound power of a sound source under its normal operating conditions and in situ. The method uses the scanning technique whereby the intensity probe is moved slowly over a defined surface while the signal analyzer time-averages the measured quantity during the scanning period.

The results of a series of field trials by several Scandinavian organizations suggested that the accuracy of this proposed standard is compatible with the "Engineering Grade," as defined in the ISO 3740 series.

The equipment under test is divided into a convenient number of subareas that are selected to enable a well-controlled probe sweep over the subarea. Guidance is given on the sweep rate and the line density. Measurement accuracy is graded according to the global pressure-intensity index, *LK*. This is the numerical difference between the sound intensity level and the sound pressure level. If this field indicator is less than or equal to 10 dB then the results are considered to meet the engineering grade of measurement accuracy. As this field indicator increases in value, the level of uncertainty in the intensity measurement increases. When the *LK* value lies between 10 and 15 dB the measurement accuracy meets the "Survey" grade.

Measurement Techniques

The precise measurement technique adopted in a particular situation depends on the objectives of the investigation and the level of measurement uncertainty that is required.

Subareas

It was mentioned earlier that the total sound power is the product of the intensity and the surface area of the measurement envelope around the noise source. In practice, most noise sources do not radiate energy uniformly in all directions so it is good practice to divide the sound source envelope into several subareas. Each subarea is then assessed separately, taking into account its area and the corresponding intensity level. The subarea sound powers can then be combined to give the total sound power of the source.

The number, shape, and size of each subarea are normally dictated by two considerations: the physical shape of the source and the variations in intensity over the complete envelope. Subareas are normally selected to conform to components of the whole source such that the intensity over the subarea is reasonably constant. It is important that the subareas are contiguous and the measurement envelope totally encloses the source under investigation.

Sweep or Point Measurement

Should one measure the intensity levels at discrete positions, with the probe stationary, or should the probe be swept over the subarea? This controversy has occupied much discussion time among practicing acousticians. For precision grade measurements, discrete points are used, but for lower grade work, sweeping is acceptable.

If discrete points are used then the number and distribution of the measurement points must be considered in relation to the field indicators.

In surroundings that are not highly reverberant and where extraneous noise levels are lower than the levels from the source under investigation, relatively few discrete points may be used, distributed uniformly over the surface. The distance from the source may be as great as 1 m.

As the extraneous noise levels increase and/or the environment becomes more reverberant, measurements must be made progressively closer to the source in order to maintain an acceptable level of uncertainty in the measurements. This also requires more measurement points to be used because of the increase in the spatial variation of the intensity distribution.

If sweeping is used then other factors must also be considered. The speed with which the probe is swept across the subarea must be uniform, at about 300 mm/sec, and the area should be covered by a whole number of sweeps with an equal separation between sweep lines. Care must be taken that excessive dwell time does not occur at the edges of the subarea when the probe's direction of sweep is reversed. The operator must also be careful that his or her body does not influence the measurements by obscuring sound entering the measurement area as he or she sweeps.

Distance between Source and Probe

Generally, the greater the extraneous noise and the more reverberant the environment then the closer should be the probe to the source. In extreme cases the probe may be only a few centimeters from the source surface in order to improve the signal-to-noise ratio. This is normally frowned upon when using conventional sound level meters because measurements of sound pressure, taken close to a surface, may bear little relation to the pressures occurring further away from the surface. This discrepancy is not due simply to the attenuation with distance that normally occurs in acoustics.

The region very close to a surface is called the "near field." In this region the local variations in sound pressure may be very complex because some of the sound energy may circulate within this near field and not escape to the "far field." This recirculating energy is known as the *reactive* sound field. The sound energy that does propagate away from the surface is called the *active* sound field because this is the component that is responsible for the acoustic energy in the far field.

Since sound intensity meters can differentiate between the active and reactive sound fields, measurements of intensity taken close to noise sources can faithfully indicate the radiated sound energy. However, using a conventional sound level meter near to a noise source may indicate higher sound power levels than occur in the far field because these instruments cannot differentiate between active and reactive fields.

Some Advantages and Limitations in Sound Intensity Measurements

Background Noise

One of the main advantages of the sound intensity method of measurement is that accurate assessments of sound power can be made even in relatively high levels of background noise. But this is only true if the background noise is steady (i.e., not time varying). Using conventional sound pressure level methods the background noise level should be 10 dB below the signal level of interest.

Using sound intensity techniques the sound power of a source can be measured to an accuracy of 1 dB even when the background noise is 10 dB *higher* than the source noise of interest. Figure 6—32(a) shows a noisy machine enclosed by a measurement surface. If the background noise is steady, and there is no sound absorption within the measurement surface, then the total sound power emitted by the machine will pass through the measurement surface, as shown.

If, however, the noisy machine is outside the measurement surface, as shown in Figure 6—32(b), then the sound

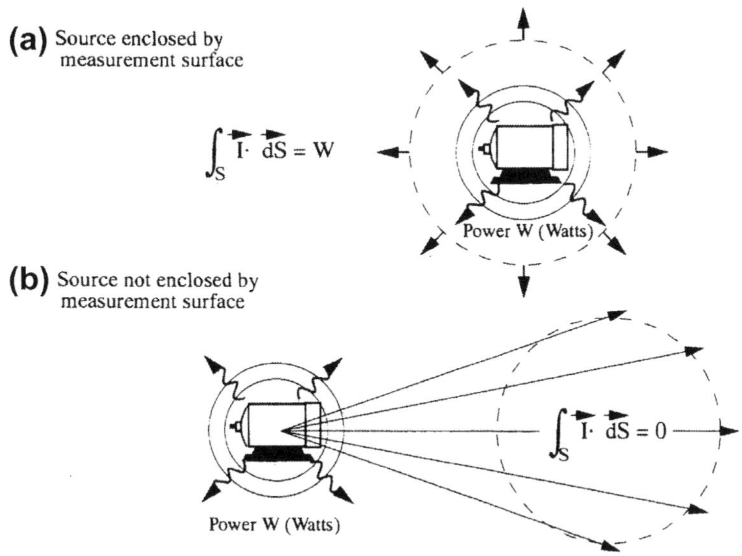

FIGURE 6–32 The effect of sound sources inside and outside the measurement surface. *(Source: AFI Ltd.)*

energy flowing into the surface on the left hand side will be emitted from the right hand side of the measurement surface. When the sound intensity is assessed over the whole measurement surface the net sound power radiated from the total surface will be zero.

Effects of the Environment

When the sound power of a noise source is evaluated in the field using sound pressure level techniques, it is necessary to apply a correction to the measured levels to account for the effects of the environment. This environmental correction accounts for the influence of undesired sound reflections from room boundaries and nearby objects.

Since a sound intensity survey sums the energy over a closed measurement surface centered on the source of interest, the effects of the environment are cancelled out in the summation process in the same way that background noise is eliminated. This means that, within reasonable limits, sound power measurements can be made in the normal operating environment even when the machine under investigation is surrounded by similar machines that are also operating.

Sound Source Location

Since a sound intensity probe has strong directional characteristics there is a plane at 90° to the axis of the probe in which the probe is very insensitive. A sound source just forward of this plane will indicate positive intensity, whereas if it is just behind this plane the intensity will be negative (Figure 6–33).

This property of the probe can be used to identify noise sources in many practical situations. The normal procedure is to perform an initial survey of the noise source to determine its total sound power. The probe is pointed toward the source system to identify areas of high sound intensity. Then the probe is reoriented to lie parallel to the measurement surface and the scan is repeated. As the probe moves across a dominant source the intensity vector will flip to the opposite direction.

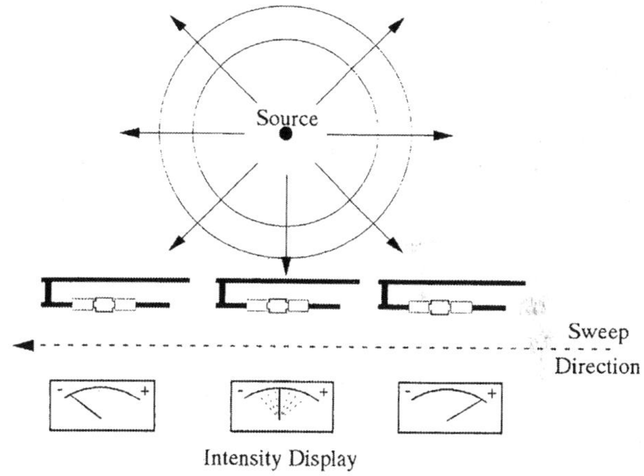

FIGURE 6–33 Sound source location using the intensity probe. *(Source: AFI Ltd.)*

Testing of Panels

The traditional procedure for measuring the transmission loss, or sound reduction index, of building components is described in the series of standards ISO 140. The test method requires the panel under investigation to be placed in an opening between two independent, structurally isolated reverberation rooms, as shown in Figure 6–34.

FIGURE 6–34 Comparison of the pressure and intensity methods for measuring the sound reduction index of panels. *(Source: AFI Ltd.)*

Sound is generated in the left hand room and the sound pressure levels in the two rooms are measured. Assuming that the sound energy in the right hand room comes through the panel then the sound reduction index (SRI) of the panel is given by

$$\text{SRI} = L_1 - L_2 + 10 * \log_{10}(S/A)\ \text{dB} \qquad (6.5)$$

For this method to give accurate results, flanking transmission (sound bypassing the test panel) must be minimal and both rooms must be highly reverberant.

If the measurements in the right hand room are carried out using sound intensity, then it is necessary to reduce the amount of reverberation in this room. Since the probe can measure the sound intensity coming through the panel then flanking transmission is no longer a limitation.

For these reasons one can dispense with the second reverberation room altogether. The sound reduction index is then given by

$$\text{SRI} = L_1 - LI - 6\ \text{dB} \qquad (6.6)$$

If the panel contains a weak area, such as a window, the sound reduction index of the window can be assessed separately. But this will only work if the panel is a greater sound insulator than the window.

MEASURING TONAL NOISE SOURCES

Measuring the sound power of tonal noise sources presents difficulties using traditional techniques (ISO 3740, 1980). Unfortunately, using sound intensity techniques on such sources is also fraught with problems. This is because the spatial distribution of the intensity is very sensitive to small alterations in source position and the presence of nearby sound reflective objects.

CASE STUDY 2: THE USE OF SOUND INTENSITY MEASUREMENT

Gas Turbine Package Witness Testing

Gas turbine packages are normally assembled in large factory buildings or in the open air between factory buildings. In either case, the environment is totally unsuitable for reliable acoustic tests to be carried out using sound level meters alone. Sound intensity techniques are especially relevant in these situations because of the location in which the tests are to be carried out and because some components, such as the compressor test loop, may not be contract items. By surveying each component with a sound intensity meter

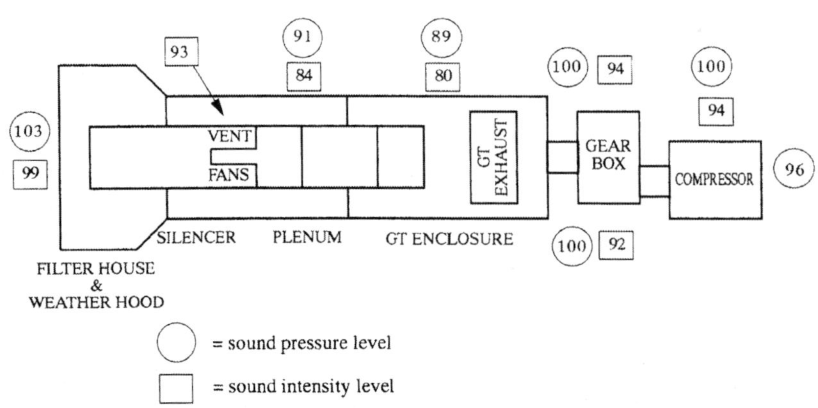

FIGURE 6–35 Sound pressure and sound intensity levels for a gas turbine package during witness testing. *(Source: AFI Ltd.)*

the sound power for each component can be determined separately.

Figure 6–35 shows a typical gas turbine driving a compressor. The figures on the drawing indicate the sound pressure and sound intensity levels that were measured during a particular witness test on an RB211 gas turbine package. The sound pressure levels in close proximity to the package were between 89 and 103 dB(A), with the higher levels dominating. Even so, reliable values of the intensity levels were obtained from which the sound power levels were determined. These values are given in Table 6–8. Since the sound intensity level is numerically equal to the sound pressure level in free field, the average sound intensity over a given surface area of a gas turbine package provides a direct indication of the average sound pressure level from that surface in free field conditions.

Referring again to Figure 6–35, the sound intensity level measured by the casing of the ventilation fan was 94 dB(A). It would not normally be possible to measure the output from this fan accurately, using a sound level meter, in this situation because of the relatively high sound pressure level in this area due to other sources.

During the testing of another package, the sound power levels from the ventilation fan casing and the fan motor were measured separately. The fan motor was found to be noisier than the manufacturer's stated levels. Discussions with the motor manufacturer revealed that the wrong cooling fans had been fitted to the motors, which accounted for this increase in noise level. The correct cooling fans were subsequently fitted.

This example clearly illustrates the benefits of sound intensity measurements to check compliance with noise specifications when the test items are very large and are cited in acoustically undesirable areas.

TABLE 6–8 Rank Ordering of Components in Terms of the A-Weighted Sound Power Levels

Description of Measured Item	Measured Sound Power Levels, dB(A)
Combustion air intake	110
Composition air plenum and silencer	104
Turbine comp. vent, air breakout	104
Turbine enclosure	103
Compressor casing	103
Gearbox	103
Breakout from temporary exhaust	96

(Source: AFI Ltd.)

CASE STUDY 3: COMPARISON OF NOISE ON TWO NOMINALLY IDENTICAL PRODUCTION MACHINES

This example is taken from an extensive survey of a production department that had many, relatively small, machines close together in a highly reverberant factory room. The two machines were nominally identical roller mills, as used in many production lines in the paint, flour, and confectionary industries. The drive motors were situated on the top of the machines. A routine sound intensity survey was carried out on each machine during normal production because it was not possible to run the machines in isolation. The two machines are identified as machine A and machine B.

Table 6–9 gives the overall, A-weighted sound pressure levels and sound power levels for each machine, and the sound power levels of the motors. The total sound power levels of the machines agreed very well with the values obtained by the manufacturer using sound pressure measurements to derive the sound power levels (ISO 3740, 1980). This technique can only give the total sound power of a machine; it cannot obtain the sound power levels of parts of a machine.

At the time of the survey two machines were each in areas of high noise levels and their total sound power levels were 96 and 94 dB(A). But comparing the motor sound power levels revealed a difference of 8 dB(A) in their respective levels even though both motors were classified as the "low noise" type and cost more than the standard motors. This is just one example where a significant degree of noise control might be achievable by selecting the correct one of two nominally identical electric motors. However, using the traditional method of sound pressure level measurement would not reveal any difference between the motors.

Acoustical Properties of Flexible Connectors

Heavy-duty flexible connectors are used to join separate components of a gas turbine package. When high-performance acoustic hardware is used it is imperative

TABLE 6–9 Comparison of the Sound Levels from Two Roller Mills

Description	Machine A, dB(A)	Machine B, dB(A)
Total sound power for machine	96	94
Sound pressure level by machine	92	89
Sound power level of the motor	93	85

(Source: AFI Ltd.)

that these flexible connectors do not compromise the total acoustic performance of the package. This is particularly so in the gas turbine exhaust system where multilayered flexible connectors are exposed to high temperatures and severe buffeting from exhaust gases. In some exhaust systems overlapping metal plates are inserted inside the flexible connectors to reduce the buffeting of the flexible material. If additional sound attenuation is required, a heavy, fibrous mat, or "bolster," is inserted between the plates and the flexible connector.

For the acoustics engineer, these flexible connectors are a problem because there are very few data on their acoustical performance, and the designs do not lend themselves to simple theoretical prediction. A brief laboratory investigation was carried out to compare the performances of seven types of flexible connectors with and without plates and bolsters. The tests were carried out in Altair's acoustical laboratory, which was designed to test materials using sound intensity techniques. The samples were physically quite small so the low-frequency performances were probably distorted by the small size of the samples. Nevertheless, the exercise yielded much valuable information and led to a simple engineering method of predicting a flexible connector's acoustic performance from knowledge of its basic parameters.

Figure 6–36 compares the sound reduction indices of a typical, multilayered flexible connector tested alone, with plates and with bolster and plates. All of the test results showed a sharp increase in performance at 250 Hz due to the plates. No satisfactory explanation can be offered at this stage for this effect, which may be related to the small size of the samples. But in the middle and high frequencies the results were generally as expected with the plate giving an additional attenuation of about 5 dB compared to the compensator alone. The combination of the plate and bolster gave an additional attenuation of between 10 and 15 dB compared to the flexible connector alone.

Sound Source Location

During two noise surveys the sound power levels from two different designs of lube oil console were measured using the sound intensity meter. The overall sound power levels were 93 dB(A) and 109 dB(A). The two consoles had electrically driven pumps and both emitted strong tonal noise. In the first case the tonal noise was centered on 8 kHz; using the sound intensity probe it was possible to "home in" on the pump section pipe as a major noise source. This was confirmed by vibration measurements.

In the second survey the pump outlet pipe gave the highest sound intensity reading. Since this was a relatively long pipe, which was rigidly attached to the frame of the console, it was a dominant source both in its own right and because it was "exciting" the framework. In these cases the sound intensity meter was a useful tool in identifying dominant sources among a number of small, closely packed noise radiators.

This section has discussed the concepts of sound intensity, sound power, and sound pressure. It has shown how sound intensity meters have given the acoustic engineer a very powerful diagnostic tool. Noise specifications for large, complex machinery can now be checked without the need for special acoustics rooms.

The advantages, and limitations, of sound intensity have been discussed in some detail and several applications have been illustrated by case histories taken from surveys carried out in the process and gas turbine industries.

The superiority of sound intensity meters over sound level meters is clearly apparent. Certain types of laboratory studies can also be carried out more cost effectively using intensity techniques.

Although there are some limitations in the use of sound intensity instrumentation, when used intelligently, it can yield valuable information on dominant noise sources, which, in turn, should provide more cost-effective solutions to noise control in the oil and power industries.

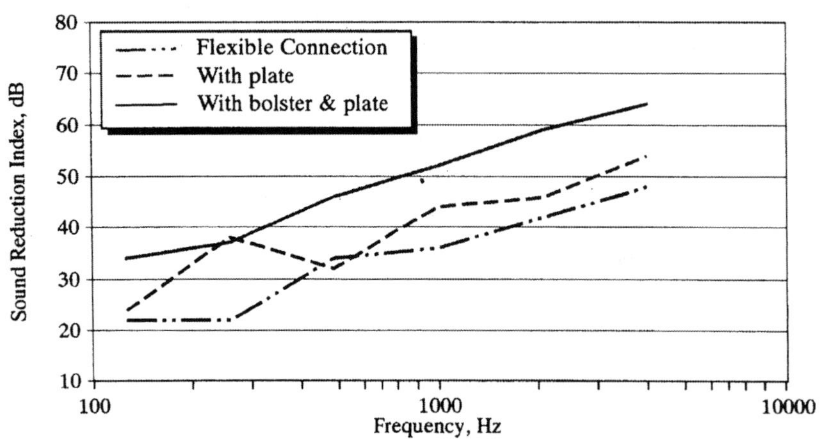

FIGURE 6–36 Sound reduction index of a flexible connection. *(Source: AFI Ltd.)*

The superiority of sound intensity meters over sound level meters is clearly apparent. Certain types of laboratory studies can also be carried out more cost effectively using intensity techniques.

Although there are some limitations in the use of sound intensity instrumentation, when used intelligently, it can yield valuable information on dominant noise sources, which, in turn, should provide more cost-effective solutions to noise control in the oil and power industries.

ACOUSTIC DESIGN OF LIGHTWEIGHT GAS TURBINE ENCLOSURES

Nomenclature

a, b = panel dimensions, m
B, Bx, By, Bxy = bending and torsional stiffnesses
c = speed of sound in air, m/s
d = fiber diameter, mm
f = frequency, Hz
$f1$ = first panel resonance, Hz
f_c, f_{cx}, f_{cy} = coincidence frequency, Hz
l = thickness of absorptive layer, m
ln = natural logarithm
m = mass per unit area, kg/m^2
$R1$ flow resistivity index or transmission loss, dB
R = sound reduction
S = stiffness
α = attenuation constant for the material, dB/m
η = damping
λ_m = wavelength of sound in the absorptive layer, m
υ = Poisson ratio
ρ_0 = density of air
ρ_m = density of absorptive layer
τ = transmission coefficient
ω = angular frequency

Gas turbines are used extensively in onshore and offshore environments for power generation, but their use introduces a number of potential hazards. To reduce the risks caused by fire and high noise levels, enclosures, with intake and exhaust silencers, are fitted around the turbines. These enclosures and silencers must be capable of withstanding large static loads produced by equipment sited on top of them and large dynamic loads due to wind.

Traditionally these enclosures are heavy and expensive, especially when stainless steel or aluminum is required for offshore use. This has led to a consideration of more cost-effective designs that still comply with the stringent demands of the oil and gas industry.

One approach that is proving successful is the use of a corrugated enclosure design, which employs a thinner steel wall than its flat panel counterpart, without compromising the structural and fire protection requirements. However, corrugated designs are intrinsically less effective as sound insulators than flat panels. These weaknesses must be understood so that multilayered panels based on the corrugated design can compensate for the deficiencies of unlined, corrugated panels.

This section presents the results of theoretical predictions and measurements on flat and corrugated panels, which were tested in the unlined condition and then with a sound absorbent lining. The effects of varying the profile of the corrugations are also considered.

This section considers first the behavior of flat, unlined panels, then describes the physical reasons why corrugated, unlined panels have a different acoustic response to flat panels. The effects of sound-absorbent linings on flat and corrugated panels are then considered.

Sound Transmission through Unlined Flat Panels

When a sound wave is incident on a wall or partition, some of the sound energy is transmitted through the wall. The fraction of incident energy that is transmitted is called the transmission coefficient. The accepted index of sound transmission is the sound reduction index, which is sometimes called the transmission loss. This is related to the transmission coefficient by the equation

$$R = 10\log_{10}(1/\tau),\ \text{dB} \tag{6.7}$$

The behavior of flat panels has been described extensively in the literature, so only the outline of their theoretical performance is given here. The general behavior of a single skin, isotropic panel is shown in Figure 6–37. This characteristic behavior is valid for a wide range of materials, including steel and aluminum.

The propagation of audio-frequency waves through panels and walls is primarily due to the excitation of bending waves, which are a combination of shear and compressional waves. When a panel is of finite extent then a number of resonances are set up in the panel that are

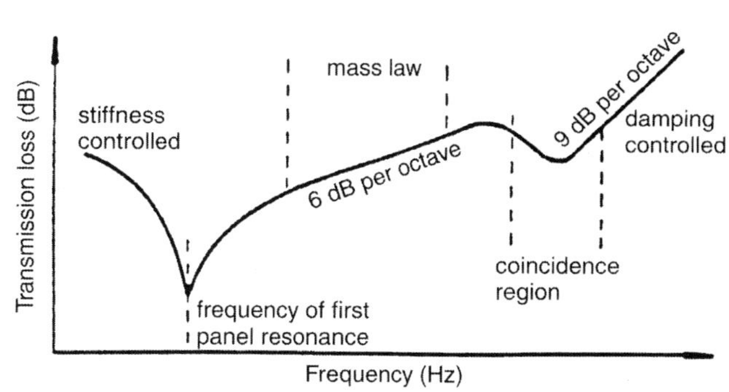

FIGURE 6–37 Sound reduction index for a typical flat, unlined panel. *(Source: AFI Ltd.)*

TABLE 6—10 Comparison of Predicted and Measured Sound Reduction Indices of Flat, Unlined Panels

		Sound Reduction Index, dB							
Material	**Measured or Predicted**	**63**	**125**	**250**	**500**	**1000**	**2000**	**4000**	**8000**
3-mm steel	Predicted	16	22	28	34	40	43	35	44
	Measured	16	21	27	33	38	39	33	—
6-mm steel	Predicted	22	28	34	40	43	35	44	53
	Measured	22	27	35	39	44	37	42	—
6-mm aluminum	Predicted	13	19	25	31	34	28	35	44
	Measured	13	19	25	30	36	30	32	—

dependent on the bending stiffness of the panel. The frequency of the first panel resonance, $f1$, is given by

$$f1 = (\pi/2)\sqrt{(B/m)(L/a^2 + 1/b^2)},\ \text{Hz} \tag{6.8}$$

The second resonance occurs at the "coincidence" frequency, fc, when the projected wavelength of the incident sound coincides with the wavelength of the bending wave in the panel. The coincidence frequency is given by

$$f = (c^2/2\pi)\sqrt{(m/B)},\ \text{Hz} \tag{6.9}$$

The amount of sound transmitted by a panel is dependent on the surface weight, the damping of the panel, and the frequency of the sound. For a finite panel exposed to a random noise field, the acoustic behavior is specified mathematically as follows:

$$R = 20\log_{10}S - 20\log_{10}f - 20\log_{10}(4\pi\rho, c), dB, f < f1 \tag{6.10}$$

$$R = 20\log_{10}(mf) - 47, \text{dB}, f1 < f < f_c \tag{6.10a}$$

$$R = 20\log_{10}(\pi fm/\rho_0 c) + 10\log_{10}(2\eta f/\pi f_c), \text{dB}, f > f_c \tag{6.10b}$$

In most practical situations the lowest resonance frequency is below the audio range. Above this frequency a broad frequency range occurs in which the transmission loss is controlled by the surface weight and increases with frequency at the rate of 6 dB per octave. In the coincidence region the transmission loss is limited by the damping of the panel. Above the coincidence frequency the transmission loss increases by 9 dB per octave and is determined by the surface weight and the damping.

Clearly for a high value of sound reduction index over the majority of the audio-frequency range (the mass-controlled region) it is better to have a high surface weight, a low bending stiffness, and a high internal damping.

The acoustic performances of some materials have been predicted and are compared to measured values in Table 6—10. The data for 3-mm steel and 6-mm aluminum are shown in Figures 6—38 and 6—39. Each of the examples shown has the characteristic shape described above and the agreement between the measured and predicted values of transmission loss is good.

Sound Transmission through Unlined Corrugated Panels

In corrugated panels the characteristics of the panel are not the same in all directions. The moment of inertia across the corrugations differs from that parallel to the corrugations; thus the bending stiffness varies with direction. This affects both the first panel resonance and the coincidence frequency so that the sound transmission characteristics for

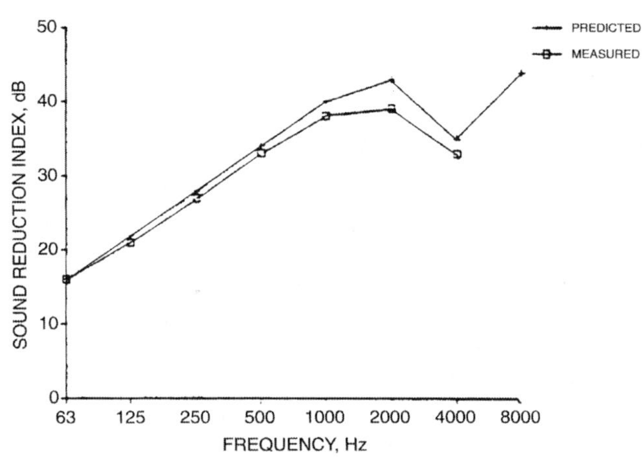

FIGURE 6—38 Predicted and measured sound reduction indices for 3-mm steel. *(Source: AFI Ltd.)*

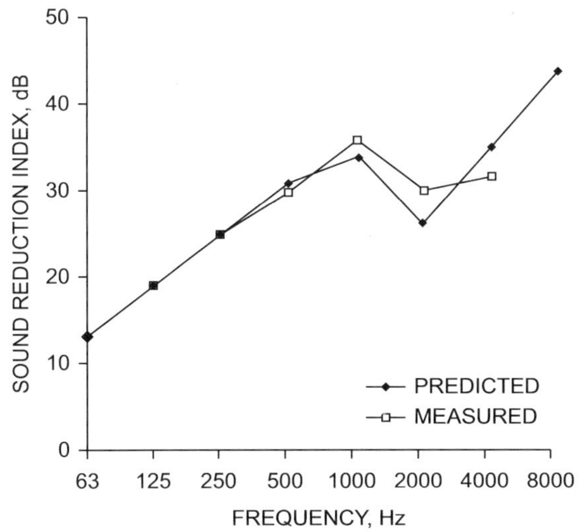

FIGURE 6–39 Predicted and measured sound reduction indices for 6-mm aluminum. *(Source: AFI Ltd.)*

these panels differ from flat panels with the same thickness. The first panel resonance is now given by

$$f1 = \left(\pi/2m^{0.5}\right) \times \sqrt{(Bx/a^{1}(1-\upsilon^{2}) + By/b^{4}(1-\upsilon^{2}) + Bxy/a^{2}b^{2})},\ \text{Hz} \quad (6.11)$$

where Bx and By are the bending stiffnesses in the two principal planes of the panel and Bxy accounts for the torsional rigidity of the plate.

For real panels, measuring several meters in length and height, the frequency of this first resonance may still be in the subaudio range but can be several octaves higher than the first resonance of a flat panel of the same dimensions.

Since the coincidence frequency is determined by the bending stiffness, the presence of two bending stiffnesses gives rise to two critical frequencies, f_{cx} and f_{cy}, where

$$f_{cx} = \left(c^{2}/2\pi\right)\sqrt{(m/Bx)},\ f_{cy} = \left(c^{2}/2\pi\right)\sqrt{(m/By)}$$

If the ratio of the bending stiffnesses, Bx and By, is less than 1.4, then the effects on the transmission loss of the panel will be small, but in typical panels the ratio of the bending stiffnesses is usually much greater than 1.4. This gives rise to a plateau in the transmission loss curve, which is illustrated in Figure 6–40. The plateau may extend over several decades for common corrugated or ribbed panels.

The sound transmission through orthotropic (corrugated) panels has been investigated by Heckl, who derived the following relationships for the diffuse field sound transmission:

$$\tau = (\rho_0 c f_{cx}/\pi\omega m f)[\ln(4f/f_{cx})]^{2},\ f_{cx} < f < f_{cy} \quad (6.12)$$

$$\tau = (\pi\rho_0 c/\omega m f)\left(f_{cx} f_{cy}\right)^{0.5},\ f > f_{cy} \quad (6.13)$$

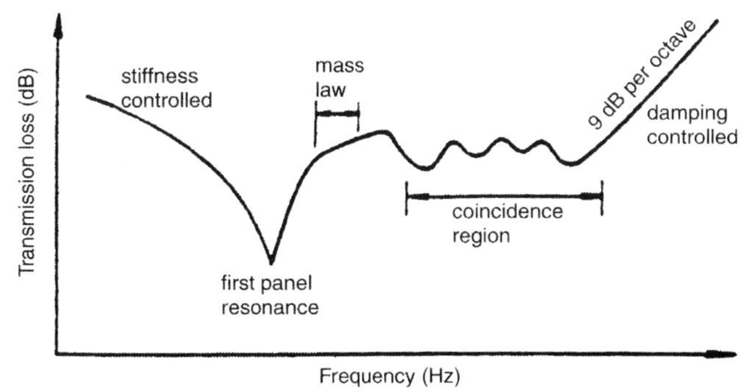

FIGURE 6–40 Sound reduction index for a typical corrugated, unlined panel. *(Source: AFI Ltd.)*

where f_{cx} and f_{cy} are the two coincidence frequencies and where f_{cx} is the lower of the two values.

The performances of two designs of corrugated panels have been predicted for panels made of 2.5-mm-thick steel with the designs shown in Figure 6–41. The predicted transmission losses are given in Table 6–11 and Figure 6–42.

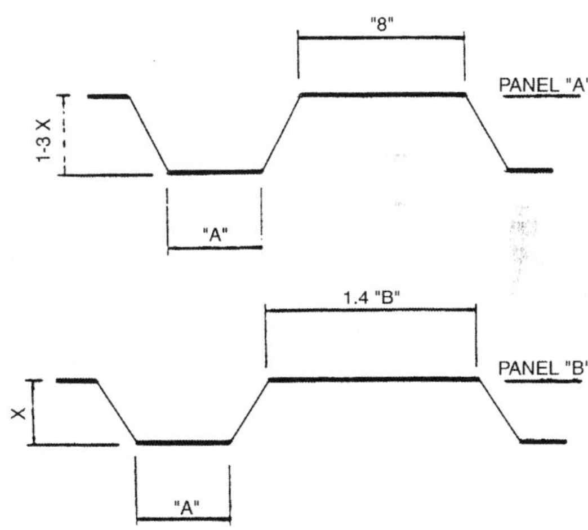

FIGURE 6–41 Two designs of corrugated panels. *(Source: AFI Ltd.)*

TABLE 6–11 Comparison of Predicted Sound Reduction Indices of Two Unlined Corrugated Panels

	Sound Reduction Index, dB							
	63	125	250	500	1000	2000	4000	8000
Panel A	15	18	21	24	28	32	37	38
Panel B	15	21	24	25	29	33	37	38

(Source: AFI Ltd.)

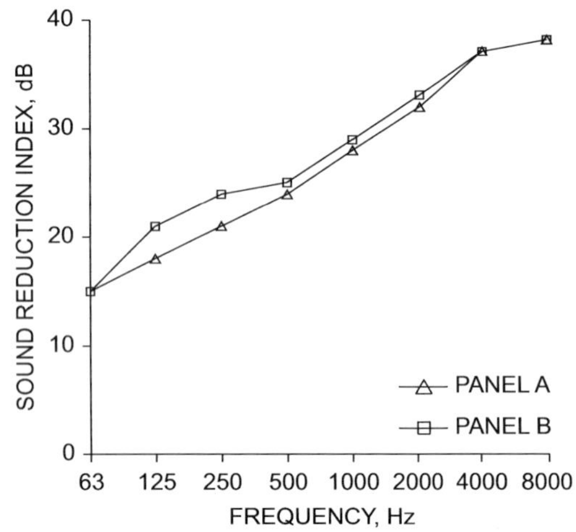

FIGURE 6–42 Predicted transmission loss for two designs of corrugated panels. *(Source: AFI Ltd.)*

The predicted panel bending stiffnesses and resonant frequencies for the two panels are tabulated in Table 6–12 for a simply supported panel 4 m wide by 2.5 m high. The values for a flat panel of the same overall dimensions have also been included. The bending stiffness of panel B, in the direction parallel to the corrugations, is slightly less than that for panel A. This causes a shift in the lower critical frequency of about half an octave, which gives rise to a slightly higher transmission loss at lower frequencies.

Transmission loss measurements have been carried out on a partition with the design of panel A. The test was carried out in accordance with the standard ISO 140 (1978). Table 6–13 and Figure 6–43 compare the predicted and measured performances. Both sets of curves show a plateau effect, which is more evident in the measured values. The predicted and measured values are within 4 dB of each other up to 4 kHz, above which the curves differ by about 7 dB. Although the agreement is not as close as for flat panels, it is encouraging. Of more interest is the comparison between the transmission losses of unlined, flat, and corrugated panels with the same thickness. Figure 6–44 and Table 6–14 show that the corrugated panel is substantially less effective in reducing sound transmission in the mid-frequencies compared to the flat panel equivalent.

TABLE 6–13 Predicted and Measured Sound Reduction Indices for Panel A

	Sound Reduction Index, dB							
	63	**125**	**250**	**500**	**1000**	**2000**	**4000**	**8000**
Predicted	15	18	21	24	28	32	37	38
Measured	20	23	25	23	23	28	29	29

(Source: AFI Ltd.)

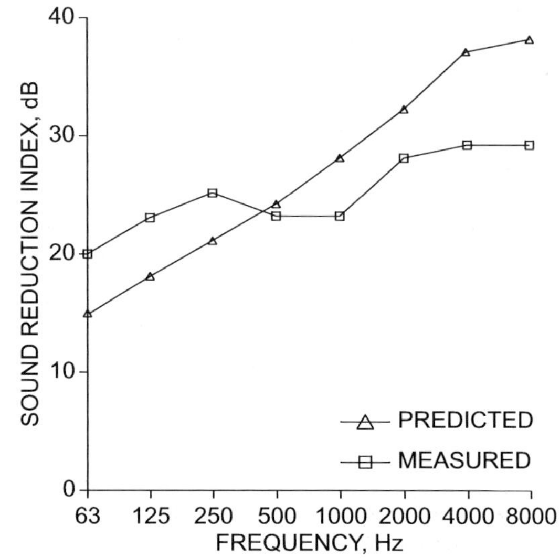

FIGURE 6–43 Predicted and measured sound reduction indices for panel A. *(Source: AFI Ltd.)*

Clearly, if noise control is a primary consideration then unlined corrugated panels are not recommended, unless other engineering considerations dictate their use. Corrugated panels offer a considerable increase in bending stiffness, compared to flat panels, which reduces the amount of additional stiffening that would be required for a flat panel. But the volume of material required to produce a corrugated panel is significantly greater than that for a flat

TABLE 6–12 Predicted Parameters for Three Panels (Panels Made of Steel Measuring 4 m by 2.5 m by 2.5 mm Thick)

	Bending Stiffness, N•m	First Panel Resonance, Hz	Coincidence Frequency, Hz
Panel A (corrugated)	trun -1240,000:241	30	163: 5115
Panel B (corrugated)	90,000:263	18	264: 4897
Flat panel	296	2.9	4825

(Source: AFI Ltd.)

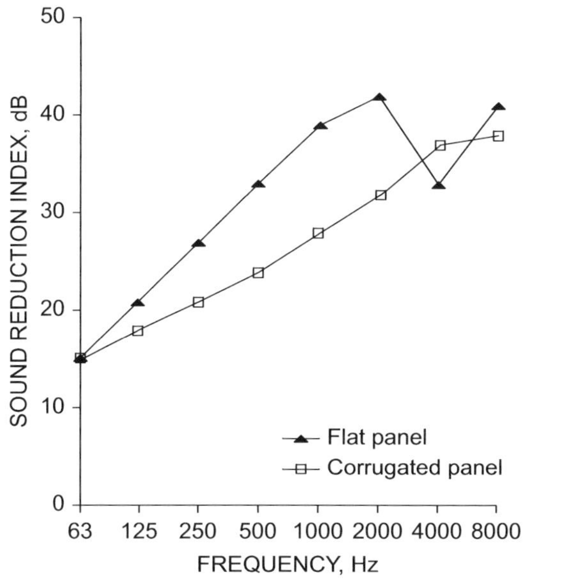

FIGURE 6–44 Predicted performance of flat and corrugated panels. *(Source: AFI Ltd.)*

TABLE 6–14 Predicted Sound Reduction Indices for Flat Corrugated Panels (Panel Thickness of 2.5 mm)

	Sound Reduction Index, dB							
	63	125	250	500	1000	2000	4000	8000
Flat panel	15	21	27	33	39	42	33	41
Corrugated panel	15	18	21	24	28	32	37	38

(Source: AFI Ltd.)

panel. For the two corrugated panel designs considered above, panel A contains 23% more material than the flat panel, and panel B contains 13% more material. Depending on the structural requirements, the additional stiffening required for a flat panel may result in a material savings when a corrugated design is chosen, especially if the flat panel is much thicker to compensate for its inherent low bending stiffness.

Attenuation of Sound by Absorptive Linings

Sound-absorbent linings are frequently fitted in acoustic enclosures to reduce the buildup of reverberant noise inside the enclosure. Typical reductions in the reverberant noise level may be between 3 and 10 dB depending on the application. An additional benefit is the increase in transmission loss of the enclosure panel, which further reduces the noise level outside the enclosure.

The increase in panel transmission loss arises because of several mechanisms. First, if the absorbent lining is sufficiently heavy and the panel is relatively thin, then the added layer may give sufficient additional weight to affect the "mass-law" performance and increase the damping of the panel. At high frequencies the absorbent may be relatively thick in comparison to the wavelength of the sound. The high-frequency sound may be attenuated not only because of the impedance mismatch between the air and the absorbent, but as the sound wave passes through the added layer a significant amount of acoustic energy is converted into heat by viscous losses in the interstices. In practice, the amount of heat generated is minute.

It is possible to distinguish between three frequency regions in which different attenuating mechanisms are predominant. For convenience these are described as regions A, C, and B, where A is the low-frequency region, C is the high-frequency region, and B is the transition region. The boundaries between these three regions are defined by the physical characteristics of the absorptive material in terms of the flow resistivity and the material thickness.

The flow resistivity of fibrous absorptive materials is dependent upon the bulk density and fiber diameter by the approximate relationship.

The frequency limits of the three regions, A, B, and C, are defined by

$$R1 = (3.18 \times 10^2)(\rho_\alpha^{15}/d^2)$$

Region A: $101 < \lambda_m$
Region B: $101 > \lambda_m$, $\alpha l < 9$ dB
Region C: $\alpha l > 9$ dB

The values of λ_m, the wavelength of the sound inside the absorptive layer, and α, the attenuation constant for the material, can be measured or predicted for semirigid materials:

$$\alpha = (\omega/c)\left[0.189(\rho_\alpha f/RI)^{0.280}\right] \tag{6.14}$$

$$\lambda_m = (c/f)\left[1 + 0.0978(\rho_\alpha f/R1)^{-0.7}\right]^{-1} \tag{6.15}$$

For an absorptive layer of known thickness and flow resistivity, the attenuation predicted from the equations given above is additive to that produced by the unlined panel.

The predicted and measured acoustic performances of two flat panels and one corrugated panel, each with an absorptive lining, are shown in Table 6–15 and Figures 6–45 to 6–47.

The predicted performances of the two flat panels are in good agreement with the measured performances over the majority of the frequency range. The largest discrepancies occur at 63 Hz and 8 kHz.

The agreement between the theoretical and measured performances of the corrugated panel is not as good as for

TABLE 6—15 Comparison of Predicted and Measured Sound Reduction Indices of Three Lined Panels

		Sound Reduction Index, dB							
	Measured or Predicted	63	125	250	500	1000	2000	4000	8000
Panel 1 (flat)	Predicted	14	22	31	39	48	47	65	63
	Measured	20	21	27	38	48	58	67	66
Panel 2 (flat)	Predicted	20	30	40	46	52	60	63	79
	Measured	31	34	35	44	54	63	62	68
Panel 3 (flat)	Predicted	16	19	22	27	36	44	52	56
(corrugated)	Measured	22	24	28	32	38	48	52	52

Panel 1: 1.6 mm flat steel lined with 100 mm thick glass fiber, 49 kg/m^3 density.
Panel 2: 5 mm flat steel lined with 100 mm thick glass fiber, 48 kg/m^3 density.
Panel 3: 2.5 mm corrugated steel lined with 50 mm thick mineral wool, 64 kg/m^3 density.
(Source: AFI Ltd.)

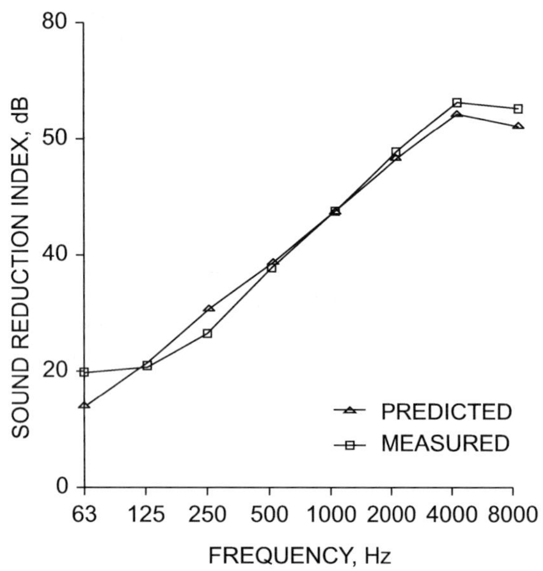

FIGURE 6—45 Predicted and measured sound reduction index of panel 1. *(Source: AFI Ltd.)*

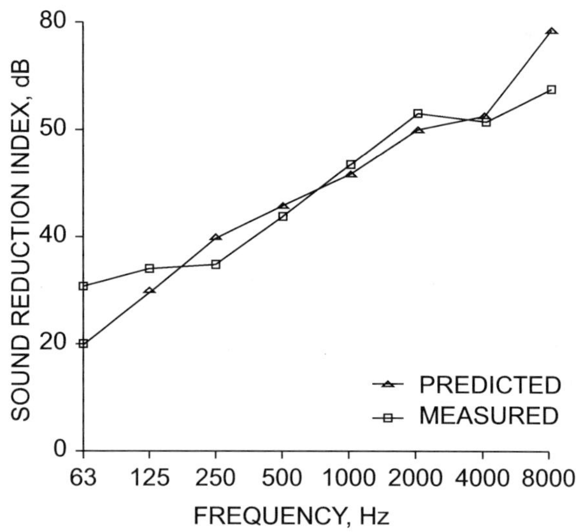

FIGURE 6—46 Predicted and measured sound reduction index of panel 2. *(Source: AFI Ltd.)*

the flat panel. The largest discrepancies occur at the lower frequencies with better agreement occurring at high frequencies. This follows the low-frequency trend shown in Figure 6—44 where the corrugated panel was unlined and the predicted performance was less than the measured performance by 5 dB. Nevertheless, the agreement is sufficiently close to support the theoretical model.

Experience Thus Far

It has been shown that the acoustic performance of lined and unlined panels can be predicted with reasonable accuracy for flat and corrugated panels. It has also been shown that, where noise control is important, unlined corrugated panels are not recommended unless other engineering considerations dictate their use, because corrugated panels are intrinsically less effective as sound insulators than flat panels of the same thickness.

A lining of sound absorptive material can substantially increase the sound reduction index of panels and the additional attenuation depends on the density, fiber diameter, and thickness of the lining. By careful selection of these parameters, the acoustic disadvantages of corrugated panels can be considerably reduced so that corrugated panels can be used confidently in situations where noise control is a primary requirement. The additional bending stiffness of corrugated panels permits a thinner outer skin to be employed and reduces the amount of additional bracing

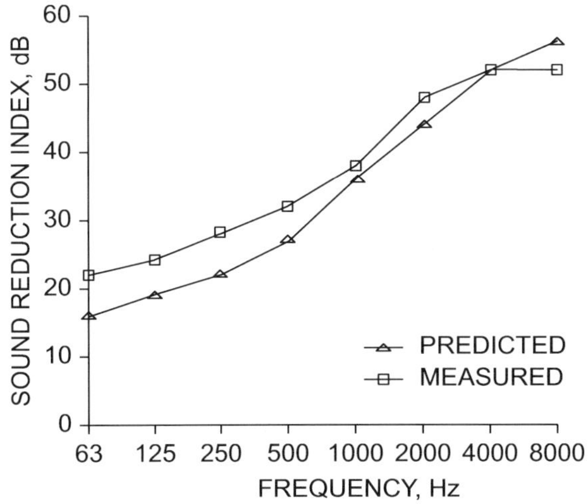

FIGURE 6–47 Predicted and measured sound reduction index of panel 3. *(Source: AFI Ltd.)*

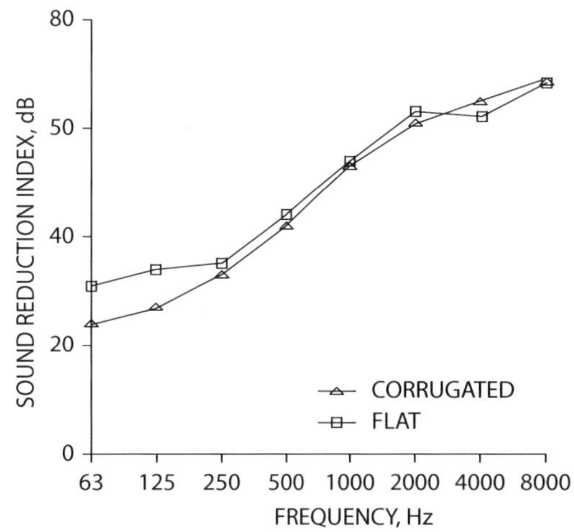

FIGURE 6–48 Predicted sound reduction indices of two high-performance panels. *(Source: AFI Ltd.)*

required to provide the structural integrity necessary in the demanding environment offshore. This reduction in overall weight compensates for the additional material used in forming the corrugations.

By careful design of the panel, a corrugation profile can be selected, which provides the most cost-effective solution when structural integrity, weight cost, ease of manufacture, and acoustic performance are considered. When expensive materials, such as stainless steel and aluminum, are employed, the reduction in cost by using a thinner-walled corrugated panel can be considerable.

A further consideration is the fire rating of lined corrugated panels. The normal requirement for bulkheads and decks offshore is the "A-60" class division. Corrugated panel designs of the type described here have been submitted to, and approved by, the appropriate authorities.

In some situations where a particularly high acoustic performance is called for, the corrugated design lends itself well to a multilayer construction employing an additional inner layer of heavy impervious material. Cheaper materials are used for the additional septum rather than for the outer skin. The acoustic attenuation of these multilayer designs is comparable to the performance of flat panels employing outer skins of twice the thickness of the corrugated outer skin. Figure 6–48 compares the measured performances of a traditional 5-mm-thick flat panel design with a 100-mm-thick absorptive lining and a multilayered panel based on a 2.5-mm-thick corrugated panel lined with a 50-mm absorptive layer.

The nominal surface weights of the two designs are 50 kg/m^2 and 40 kg/m^2 for the flat and corrugated panels, respectively. Except at 63 and 125 Hz, the performance of the two panels is very similar.

In Summary

The acoustic performance of corrugated and flat steel panels can be predicted. The acoustic behavior of corrugated panels is very different from that of flat panels. This means that if corrugated panels are required, careful consideration must be given to the design, since unlined corrugated panels are unsuitable on their own for noise control applications.

However, the greater bending stiffness of corrugated panels offers many financial and structural advantages in the demanding environment that exists offshore, especially for gas turbines.

By lining the interior of a corrugated panel with a material whose physical parameters have been carefully chosen, the inherent acoustic weaknesses can be overcome. Thus a more cost-effective approach to gas turbine enclosure design can be adopted, which considers the structural integrity, weight, cost, ease of manufacture, and acoustic performance. The resultant designs employ less bracing and thinner outer skins to achieve the same acoustic performance as flat-walled constructions weighing typically 25% more than the equivalent corrugated design.

Chapter 7

Gas Turbine Fuel Systems and Fuels

"All truths are easy to understand once they are discovered, the point is to discover them."

—Plato

Chapter Outline

The basics* of a gas turbine fuel system are similar for all turbines. The most common fuels are natural gas, LNG (liquid natural gas), and light diesel. With appropriate design changes, the gas turbine has proved to be capable of handling residual oil, pulverized coal, syngas from coal and various low BTU fluids, both liquid and gas, that may be waste streams of petrochemical processes or for instance gas from a steel (or other industry) blast furnace. Handling low BTU fuel can be a tricky operation, requiring long test periods and a willingness to trade the savings in fuel costs with the loss of turbine availability during initial prototype full load tests.

This chapter includes a case study (case 5) on blast furnace gas in a combined cycle power plant (CCPP). Chapter 4 has a case study that describes some of the combustion chamber component design changes required to handle low BTU waste gas.

As the world continues to industrialize, the applications that use waste fluids as low BTU fuel are likely to increase. Residence time in the combustion chamber is one factor that helps facilitate this economy of resources.

In general, a gas turbine fuel system consists of:

- A fuel delivery system that includes pump(s), valves, piping, and other accessories
- Fuel nozzles, simplex (one jet for one fuel type or phase, usually gas), duplex (gas or liquid), dual fuel (gas and liquid at the same time), or triplex (three fuel types)

If the fuel in question is nonconventional, for instance, heavy residual fuel, the fuel system also embodies

- Fuel additives (to deal with vanadium)
- Fuel washing (to deal with sodium and potassium salts)
- Modifications to the fuel delivery system (such as preheating to move more viscous fuels more easily) and fuel nozzles
- Any other accessories

If the fuel is pulverized coal, biomass (agricultural or otherwise), peat or black liquor from pulp and paper manufacture, waste hydrocarbons from plastics manufacture, flue gas from steel production, or a myriad of expanding options, the individual system has design refinements to suit those fuels. These design modifications may incorporate higher residence times in the combustion chamber (to allow for lower BTU values) as well as hardware changes.

Fuel efficiency is a key ingredient of gas turbine and gas turbine systems marketing, especially if the fuel is an optimum choice fuel, such as natural gas, which burns cleaner than other fossil fuels. All other aspects of gas turbine design, including aerodynamic efficiency and losses (aerodynamic or otherwise) affect fuel efficiency. As is indicated in the chapter on calculations, designers struggle for every fraction of an efficiency point (see also Chapter 14, The Business of Gas Turbines, and Chapter 18, Future Trends in the Gas Turbine Industry). How much they gain depends heavily on the type(s) of fuel in use and their delivery system.

What follows is the description of the basic fuel system of a gas turbine. The specific fuel system referred to belongs to a Rolls Royce aeroengine (the Gem 60). However, the basic components and fuel flow depicted are similar in most gas turbine engines, although design nuances and refinements vary greatly among OEMs. Note also that a turboprop application is similar to a land-based mechanical drive application.

BASIC GAS TURBINE FUEL SYSTEM**

The functions of the fuel system are to provide the engine with fuel in a form suitable for combustion and to control the flow to the required quantity necessary for easy starting, acceleration and stable running, at all engine operating conditions. To do this, one or more fuel pumps are used to deliver the fuel to the fuel spray nozzles, which inject it into the combustion system in the form of an atomized spray. Because the flow rate must vary according to the amount of air passing through the engine to maintain a constant selected engine speed or pressure ratio, the controlling devices are fully automatic with the exception of engine power selection, which is achieved by a manual throttle or power lever. A fuel shutoff valve (cock) control lever is also used to stop the engine, although in some instances these two manual controls are combined for single-lever operation.

It is also necessary to have automatic safety controls that prevent the engine gas temperature, compressor delivery pressure, and the rotating assembly speed from exceeding their maximum limitations.

With the turbo-propeller engine, changes in propeller speed and pitch have to be taken into account due to their effect on the power output of the engine. Thus, it is usual to interconnect the throttle lever and propeller controller unit, for by so doing the correct relationship between fuel flow and airflow is maintained at all engine speeds and the pilot is given single-lever control of the engine. Although the maximum speed of the engine is normally determined by the propeller speed controller, overspeeding is ultimately prevented by a governor in the fuel system.

* Reference: Claire Soares, Course Notes, Power Generation Gas Turbines, 2006, 2008, and 2013.

** Source: Adapted, with permission, from Rolls Royce, *The Jet Engine*, 1986, Rolls Royce Plc: UK.

The fuel system often provides for ancillary functions, such as oil cooling and the hydraulic control of various engine control systems; for example, compressor airflow control.

Manual and Automatic Control

The control of power or thrust of the gas turbine engine is effected by regulating the quantity of fuel injected into the combustion system. When a higher thrust is required, the throttle is opened and the pressure to the fuel spray nozzles increases due to the greater fuel flow. This has the effect of increasing the gas temperature, which in turn increases the acceleration of the gases through the turbine to give a higher engine speed and a correspondingly greater airflow, consequently producing an increase in engine thrust.

This relationship between the airflow induced through the engine and the fuel supplied is, however, complicated by changes in altitude, air temperature and aircraft speed. These variables change the density of the air at the engine intake and consequently the mass of air induced through the engine. A typical change of airflow with altitude is shown in Figure 7–1. To meet this change in airflow a similar change in fuel flow (Figure 7–2) must occur, otherwise the ratio of airflow to fuel flow will change and will increase or decrease the engine speed from that originally selected by the throttle lever position.

Described here are five representative systems of automatic fuel control; these are the pressure control and flow control systems, which are hydro-mechanical, and the acceleration and speed control and pressure ratio control systems, which are mechanical. With the exception of the pressure ratio control system, which uses a gear-type pump, all the systems use a variable-stroke, multi-plunger type fuel pump to supply the fuel to the spray nozzles.

Some engines are fitted with an electronic system of control and this generally involves the use of electronic circuits to measure and translate changing engine conditions to automatically adjust the fuel pump output. On helicopters powered by gas turbine engines using the free-power turbine principle, additional manual and automatic controls on the engine govern the free-power turbine and, consequently, aircraft rotor speed.

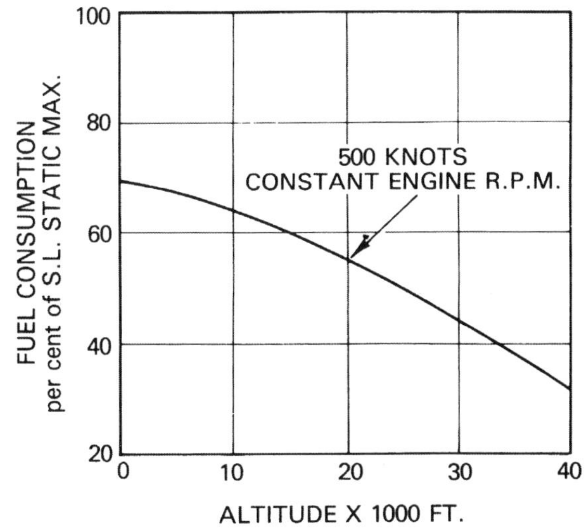

FIGURE 7–2 Fuel flow changing with altitude. *(Source: Rolls Royce.)*

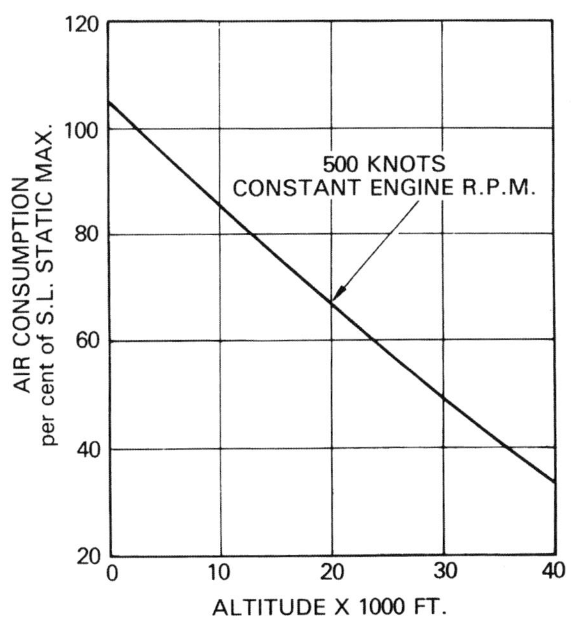

FIGURE 7–1 Airflow changing with altitude. *(Source: Rolls Royce.)*

Fuel Control Systems

Typical high-pressure (H.P.) fuel control systems for a turbo-propeller engine and a turbo-jet engine are shown in simplified form in Figure 7–3, each basically consisting of an H.P. pump, a throttle control, and a number of fuel spray nozzles. In addition, certain sensing devices are incorporated to provide automatic control of the fuel flow in response to engine requirements. On the turbo-propeller engine, the fuel and propeller systems are coordinated to produce the appropriate fuel/rpm combination.

The usual method of varying the fuel flow to the spray nozzles is by adjusting the output of the H.P. fuel pump. This is effected through a servo system in response to some or all of the following:

1. Throttle movement
2. Air temperature and pressure
3. Rapid acceleration and deceleration
4. Signals of engine speed, engine gas temperature, and compressor delivery pressure

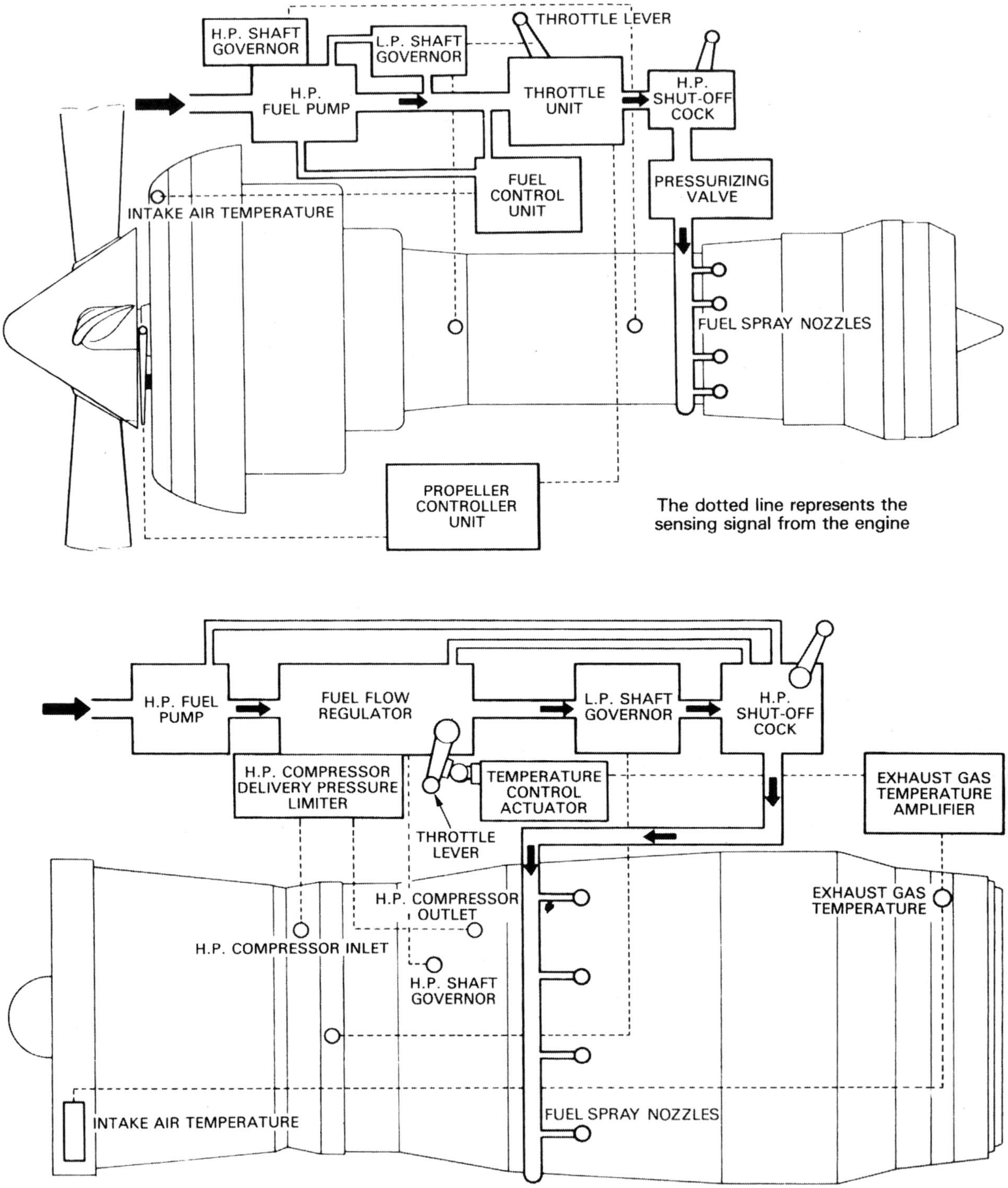

FIGURE 7–3 Simplified fuel systems for turbo-propeller and turbo-jet engines. *(Source: Rolls Royce.)*

Pressure Control (Turbo-Propeller Engine)

The pressure control system (Figure 7–4) is a typical system as fitted to a turbo-propeller engine where the rate of engine acceleration is restricted by a propeller speed controller. The fuel pump output is automatically controlled by spill valves in the flow control unit (F.C.U.) and the engine speed governor. These valves, by varying the fuel pump servo pressure, adjust the pump stroke to give the correct fuel flow to the engine.

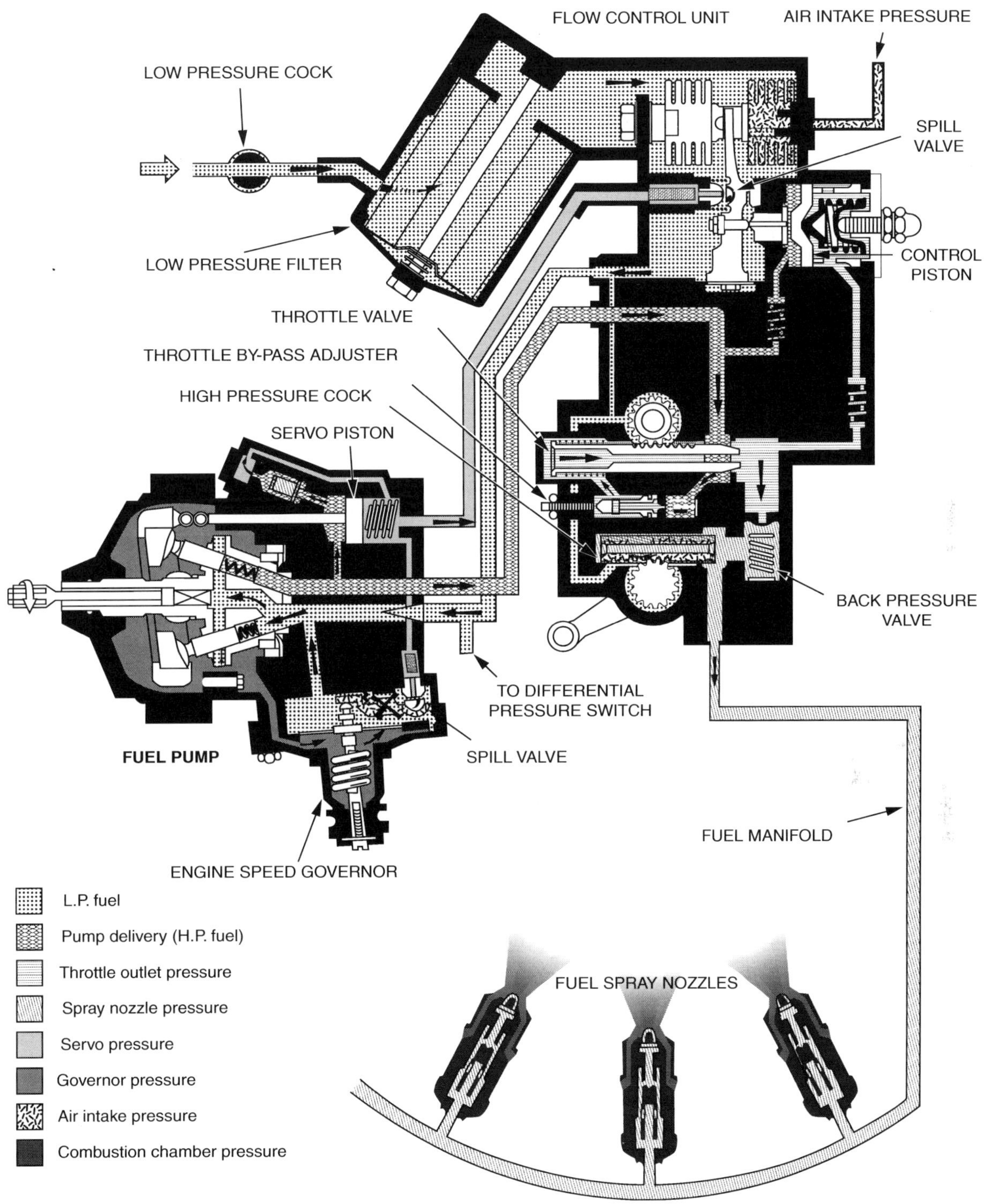

FIGURE 7–4 A pressure control system (turbo-propeller engine). *(Source: Rolls Royce. Adapted for reproduction in black and white.)*

At steady running conditions, at a given air intake pressure and below governed speed, the spill valve in the F.C.U. is in a sensitive position, creating a balance of forces across the fuel pump servo piston and ensuring a steady pressure to the throttle valve.

When the throttle is slowly opened, the pressure to the throttle valve falls and allows the F.C.U. spill valve to close, so increasing the servo pressure and pump delivery. As the pressure to the throttle is restored, the spill valve returns to its sensitive or controlling position, and the fuel pump stabilizes its output to give the engine speed for the selected throttle position. The reverse sequence occurs as the throttle is closed.

A reduction of air intake pressure, due to a reduction of aircraft forward speed or increase in altitude, causes the F.C.U. capsule to expand, thus increasing the bleed from the F.C.U. spill valve. This reduces fuel pump delivery until the fuel flow matches the airflow and the reduced H.P. pump delivery (throttle inlet pressure), allows the spill valve to return to its sensitive position. Conversely, an increase in air intake pressure reduces the bleed from the spill valve and increases the fuel flow. The compensation for changes in air intake pressure is such that fuel flow cannot be increased beyond the predetermined maximum permissible for static International Standard Atmosphere (I.S.A.) sea-level conditions.

The engine speed governor prevents the engine from exceeding its maximum speed limitation. With increasing engine speed, the centrifugal pressure from the fuel pump rotor radial drillings increases and this is sensed by the engine speed governor diaphragm. When the engine reaches its speed limitation, the diaphragm is deflected to open the governor spill valve, thus overriding the F.C.U., and preventing any further increase in fuel flow. Some pressure control systems employ a hydro-mechanical governor.

The governor spill valve also acts as a safety relief valve. If the fuel pump delivery pressure exceeds its maximum controlling value, the servo pressure acting on the orifice area of the spill valve forces the valve open regardless of the engine speed, so preventing any further increase in fuel delivery pressure.

Pressure Control (Turbo-Jet Engine)

In the pressure control system illustrated in Figure 7—5, the rate of engine acceleration is controlled by a dashpot throttle unit. The unit forms part of the fuel control unit and consists of a servo-operated throttle, which moves in a ported sleeve, and a control valve. The control valve slides freely within the bore of the throttle valve and is linked to the pilot's throttle by a rack and pinion mechanism. Movement of the throttle lever causes the throttle valve to progressively uncover ports in the sleeve and thus increase the fuel flow. Figure 7—6 shows the throttle valve and control valve in their various controlling positions.

At steady running conditions, the dashpot throttle valve is held in equilibrium by throttle servo pressure opposed by throttle control pressure plus spring force. The pressures across the pressure drop control diaphragm are in balance and the pump servo pressure adjusts the fuel pump to give a constant fuel flow.

When the throttle is opened, the control valve closes the low-pressure (L.P.) fuel port in the sleeve and the throttle servo pressure increases. The throttle valve moves towards the selected throttle position until the L.P. port opens and the pressure balance across the throttle valve is restored. The decreasing fuel pressure difference across the throttle valve is sensed by the pressure drop control diaphragm, which closes the spill valve to increase the pump servo pressure and therefore the pump output. The spill valve moves into the sensitive position, controlling the pump servo mechanism so that the correct fuel flow is maintained for the selected throttle position.

During initial acceleration, fuel control is as described previously; however, at a predetermined throttle position the engine can accept more fuel and at this point the throttle valve uncovers an annulus, so introducing extra fuel at a higher pressure (pump delivery through one restrictor). This extra fuel further increases the throttle servo pressure, which increases the speed of throttle valve travel and the rate of fuel supply to the spray nozzle.

On deceleration, movement of the control valve acts directly on the throttle valve through the servo spring. Control valve movement opens the flow ports through the control valve and throttle valve, to bleed servo fuel through the L.P. port. Throttle control pressure then moves the throttle valve towards the closed position, thus reducing the fuel flow to the spray nozzles.

Changes in air intake pressure, due to a change in aircraft altitude or forward speed, are sensed by the capsule assembly in the fuel control unit. With increased altitude and a corresponding decrease in air intake pressure, the evacuated capsule opens the spill valve, so causing a reduction in pump stroke until the fuel flow matches the airflow. Conversely, an increase in air intake pressure closes the spill valve to increase the fuel flow.

H.P. compressor shaft rpm is governed by a hydro-mechanical governor that uses hydraulic pressure proportional to engine speed as its controlling parameter. A rotating spill valve senses the engine speed and the controlling pressure is used to limit the pump stroke and so prevent overspeeding of the H.P. shaft rotating assembly. The controlling pressure is unaffected by changes in fuel specific gravity.

At low H.P. shaft speeds, the rotating spill valve is held open, but as engine speed increases, centrifugal loading moves the valve towards the closed position against the diaphragm loads. This restricts the bleed of fuel to the L.P. side of the valve until, at governed speed, the governor pressure deflects the servo control diaphragm and opens the servo spill valve to control the fuel flow and thereby the H.P. shaft speed.

If the engine gas temperature attempts to exceed the maximum limitation, the current in the L.P. speed limiter and temperature control solenoid is reduced. This opens the spill valve to reduce the pressure on the pressure drop

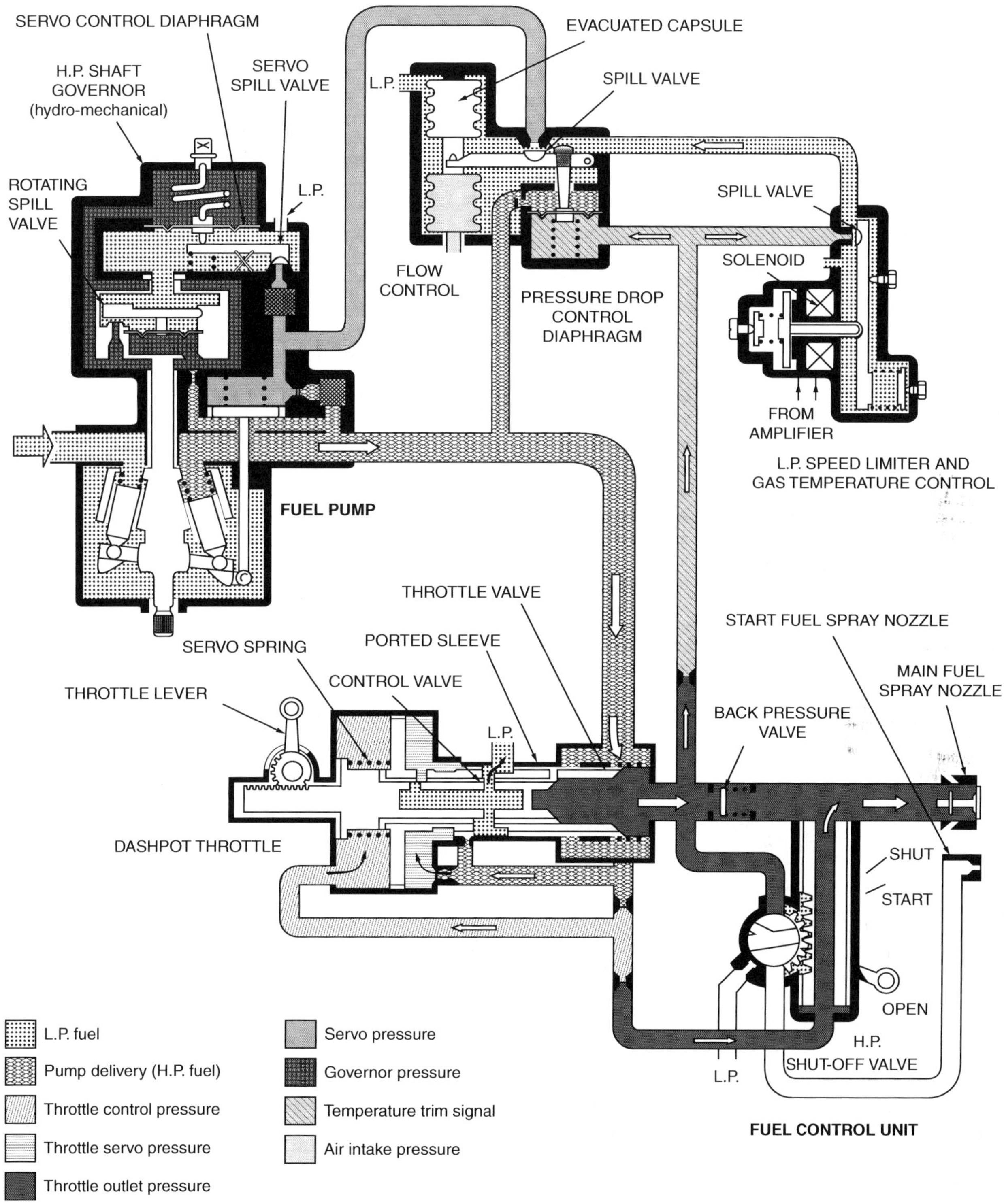

FIGURE 7–5 A pressure control system (turbo-jet engine). *(Source: Rolls Royce. Adapted for reproduction in black and white.)*

control diaphragm. The flow control spill valve then opens to reduce the pump servo pressure and fuel pump output.

To prevent the L.P. compressor from overspeeding, multi-spool engines usually have an L.P. compressor shaft speed governor. A signal of L.P. shaft speed and intake temperature is fed to an amplifier and solenoid valve, the valve limiting the fuel flow in the same way as the gas temperature control.

The system described uses main and starting spray nozzles under the control of an H.P. shutoff valve. Two starting nozzles are fitted in the combustion chamber, each

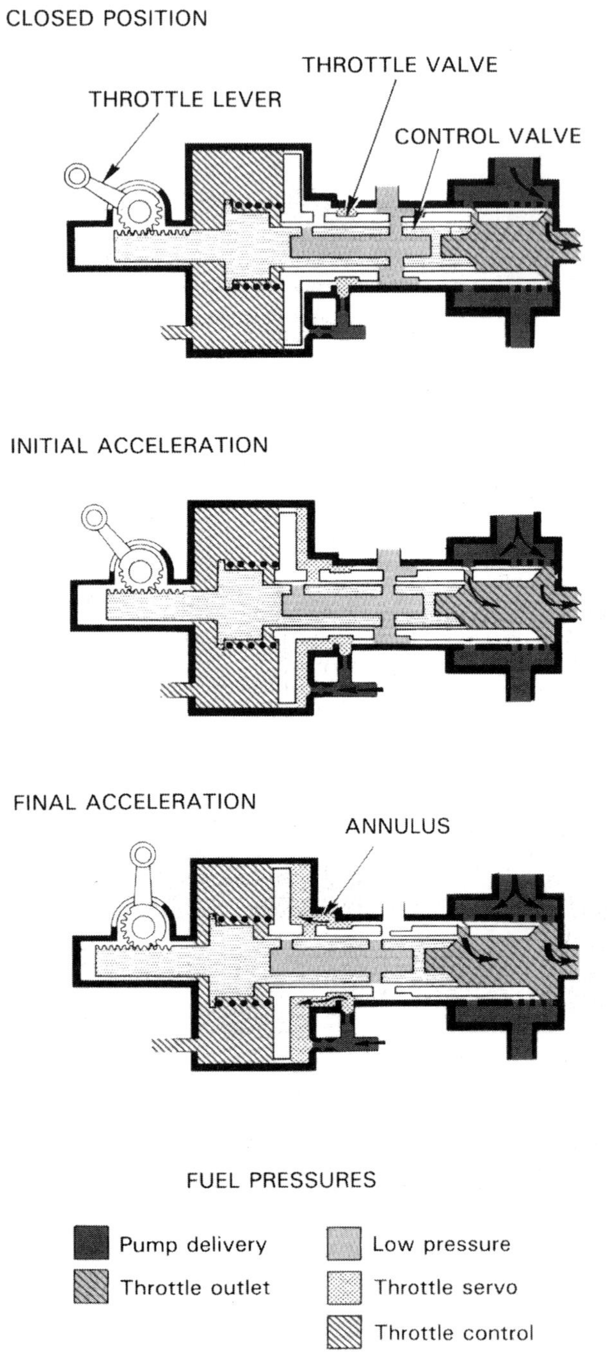

FIGURE 7–6 Acceleration control by dashpot throttle. *(Source: Rolls Royce.)*

being forward of an igniter plug. When the engine has started, the fuel flow to these nozzles is cut off by the H.P. shutoff valve.

To ensure that a satisfactory fuel pressure to the spray nozzles is maintained at high altitudes, a back pressure valve, located downstream of the throttle valve, raises the pressure levels sufficiently to ensure satisfactory operation of the fuel pump servo system.

Flow Control

A flow control fuel system is generally more compact than a pressure control system and is not sensitive to flow effect of variations downstream of the throttle. The fuel pump delivery pressure is related to engine speed; thus, at low engine speeds pump delivery pressure is quite low. The fuel pump output is controlled to give a constant pressure difference across the throttle valve at a constant air intake condition. Various devices are also used to adjust the fuel flow for air intake pressure variations, idling and acceleration control, gas temperature and compressor delivery pressure control.

A variation of the flow control system is the proportional flow control system (Figure 7–7), which is more suitable for engines requiring large fuel flows and which also enables the fuel trimming devices to adjust the fuel flow more accurately. A small controlling flow is created that has the same characteristics as the main flow, and this controlling or proportional flow is used to adjust the main flow.

A different type of spill valve, referred to as a kinetic valve, is used in this system. This valve consists of two opposing jets, one subjected to pump delivery pressure and the other to pump servo pressure, and an interrupter blade that can be moved between the jets (Figure 7–8). When the blade is clear of the jets, the kinetic force of the H.P. fuel jet causes the servo pressure to rise (spill valve closed) and the fuel pump moves to maximum stroke to increase the fuel flow. When the blade is lowered between the jets, the pressure jet is deflected and the servo pressure falls, so reducing the pump stroke and the fuel flow. When the engine is steadily running, the blade is in an intermediate position allowing a slow bleed from servo and thus balancing the fuel pump output.

All the controlling devices, except for the engine speed governor, are contained in one combined fuel control unit. The main parts of the control unit are the altitude sensing unit (A.S.U.), the acceleration control unit (A.C.U.), the throttle and pressurizing valve unit, and the proportioning valve unit.

At any steady running condition below governed speed, the fuel pump delivery is controlled to a fixed value by the A.S.U. The spill valve in this unit is held in the controlling position by a balance of forces, spring force and the piston force. The piston is sensitive to the pressure difference across the sensing valve, the pressure difference being created by fuel flowing from the proportioning valve back to the fuel pump inlet.

The proportioning valve diaphragm is held open in a balanced condition allowing fuel to pass to the A.S.U. This means that the restrictor outlet pressure is equal to the throttle outlet pressure and, as their inlet pressures are equal, it follows that the pressure difference across the

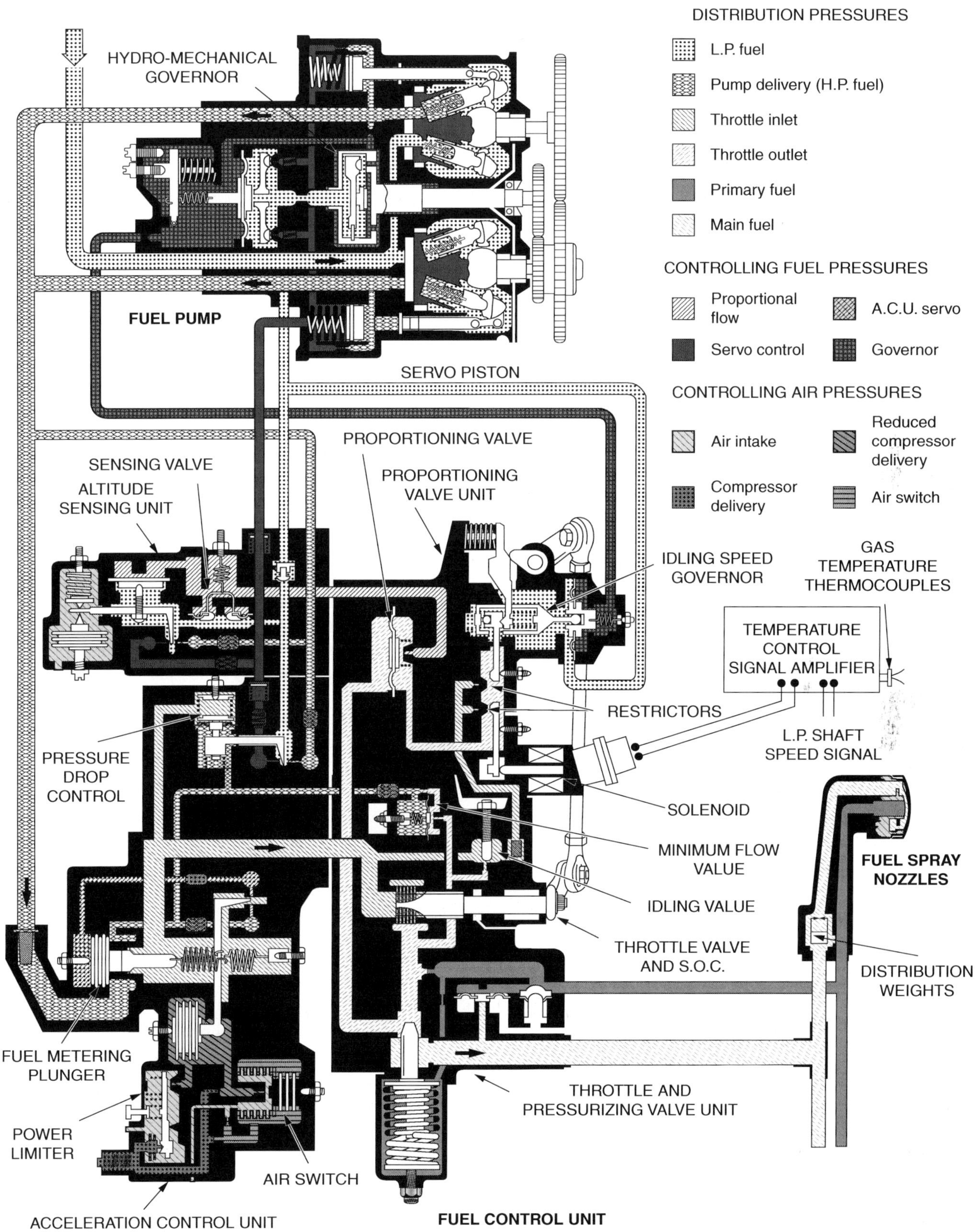

FIGURE 7–7 A proportional flow control system. *(Source: Rolls Royce. Adapted for reproduction in black and white.)*

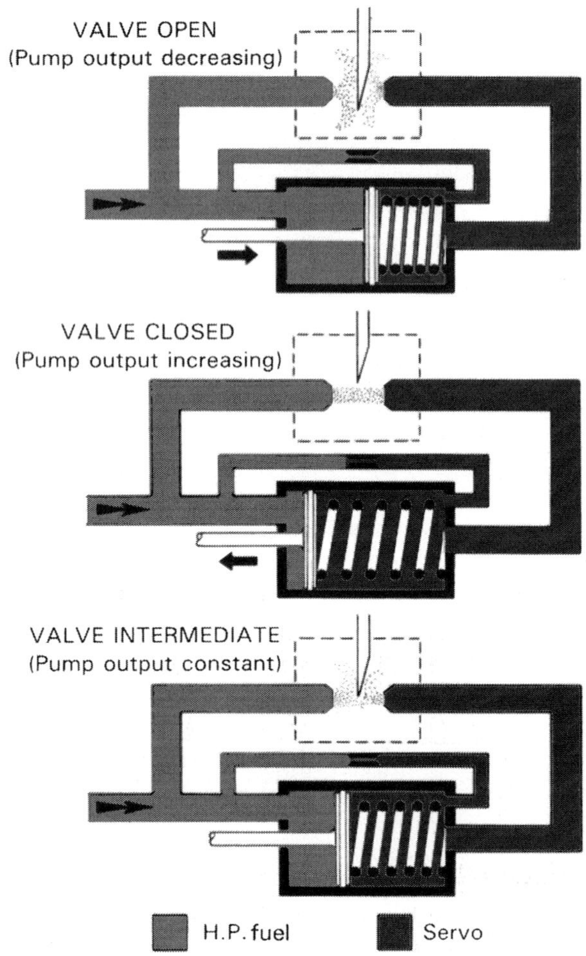

FIGURE 7–8 Servo pressure control by kinetic valve. *(Source: Rolls Royce.)*

restrictors and the throttle are equal; therefore, a constant fuel flow is obtained.

When the throttle is slowly opened, the pressure difference across the throttle valve and the proportioning flow restrictors decreases and the proportioning valve diaphragm adjusts its position. This reduces the proportional flow, which closes the A.S.U. spill valve and increases the servo pressure. The fuel pump increases its delivery and this restores the pressure difference across the throttle valve and equalizes the pressure difference across the restrictors. The proportional flow is restored to its original value and the balance of forces in the A.S.U. returns the spill valve to the controlling position.

A variation of air intake pressure, due to a change of aircraft forward speed or altitude, is sensed by the capsule in the A.S.U. A pressure reduction causes the A.S.U. capsule to expand, thus increasing the bleed from the spill valve. This reduces fuel pump delivery until the fuel flow matches the airflow and results in a lower pressure difference across the throttle valve and the proportioning valve restrictors. The reduced proportional flow restores the balance in the A.S.U. that returns the spill valve to its controlling position. Conversely, an increase in aircraft forward speed increases the air intake pressure, which reduces the bleed from the spill valve and increases the fuel flow.

During a rapid acceleration, the sudden decrease in throttle pressure difference is sensed by the A.S.U., causing the spill valve to close. Such a rapid increase in fuel supply would, however, create an excessive gas temperature and also cause the compressor to surge. This occurs because the inertia of the rotating assembly results in an appreciable time lag in the rate of airflow increase. It is essential, therefore, to have an acceleration control to override the A.S.U. to give a corresponding lag in the rate of fuel flow increase.

The rapid initial increase of fuel flow causes a rise in the pressure difference across the fuel metering plunger and this is sensed by a diaphragm in the pressure drop control section. At a fixed value of overfueling, the pressure drop control diaphragm opens its servo spill valve to override the A.S.U. and maintains a constant pressure difference across the metering plunger.

The increased fuel supply causes the engine to accelerate and the fuel metering plunger gives the maximum permissible fuel flow to match the increasing compressor delivery pressure. This it achieves through the A.C.U. servo system, which is under the control of a spill valve operated by compressor delivery air pressure acting on a capsule.

As the compressor delivery pressure continues to rise, the capsule is compressed to open the spill valve and to bleed pressure from above the metering plunger. Pump delivery pressure acting underneath the plunger causes it to lift, this increases the area of the main fuel flow passage.

The pressure drop control spill valve closes to increase the fuel pump delivery and maintains the controlling pressure difference across the plunger. The fuel flow, therefore, progressively rises as airflow through the compressor increases. The degree of overfueling can be automatically changed by the air switch, which increases the pressure signal on to the capsule. The full value of compressor delivery pressure is now passed on to the A.C.U. capsule assembly, thus increasing the opening rate of the metering plunger.

As the controlled overfueling continues, the pressure difference across the throttle valve increases. When it reaches the controlling value, the A.S.U. takes over due to the increasing proportional flow and again gives a steady fuel flow to the spray nozzles.

The engine speed governor can be of the pressure control type described previously, or a hydro-mechanical governor also described previously.

The control of servo pressure by the hydro-mechanical governor is very similar to that of the pressure control governor, except that the governor pressure is obtained from pump delivery fuel passing through a restrictor and the restricted pressure is controlled by a rotating spill valve; this type of governor is unaffected by changes in fuel specific gravity.

At low engine speeds, the rotating spill valve is held open; however, as engine speed increases, centrifugal loading moves the valve towards the closed position against the diaphragm loads. This restricts the bleed of H.P. fuel to the L.P. side of the drum until, at governed speed, the governor pressure deflects the diaphragm and opens the fuel pump servo pressure spill valve to control the maximum fuel flow and engine speed.

If the engine gas temperature exceeds its maximum limitation, the solenoid on the proportioning valve unit is progressively energized. This causes a movement of the rocker arm to increase the effective flow area of one restrictor, thus increasing the proportional flow and opening the A.S.U. spill valve to reduce servo pressure. The fuel flow is thus reduced and any further increase of gas temperature is prevented.

To prevent the L.P. compressor from overspeeding, some twin-spool engines have an L.P. shaft rpm governor. A signal of L.P. shaft speed is fed to an amplifier and solenoid valve, which limits the fuel output in the same way as the gas temperature control.

An idling speed governor is often fitted to ensure that the idling rpm does not vary with changing engine loads. A variation of idling rpm causes the rocker arm to move and alter the proportional flow, and the A.S.U. adjusts the pump delivery until the correct idling rpm is restored.

On some engines, a power limiter is used to prevent overstressing of the engine. To achieve this, compressor delivery pressure acts on the power limiter capsule. Excess pressure opens the power limiter atmospheric bleed to limit the pressure on the A.C.U. capsule and this controls the fuel flow through the metering plunger.

To enable the engine to be relit and to prevent flame-out at altitude, the engine idling rpm is made to increase with altitude. To achieve this, some engines incorporate a minimum flow valve that adds a constant minimum fuel flow to that passing through the throttle valve.

Combined Acceleration and Speed Control

The combined acceleration and speed control system (Figure 7–9) is a mechanical system without small restrictors or spill valves. It is also an all-speed governor system and therefore needs no separate governor unit for controlling the maximum rpm. The controlling mechanism is contained in one unit, usually referred to as the fuel flow regulator (F.F.R.). An H.P. fuel pump is used and the fuel pump servo piston is operated by H.P. fuel on one side and main spray nozzle (servo) pressure on the spring side.

The F.F.R. is driven by the engine through a gear train and has two centrifugal governors, known as the speed control governor and the pressure drop control governor. Two sliding valves are also rotated by the gear train. One valve, known as the variable metering sleeve, has a triangular orifice, known as the variable metering orifice (V.M.O.), and this sleeve is given axial movement by a capsule assembly. The V.M.O. sleeve moves inside a non-rotating governor sleeve that is moved axially by the speed control governor. The other valve, known as the pressure drop control valve, is provided with axial movement by the pressure drop control governor and has a triangular orifice, known as the pressure drop control orifice, and a fixed-area rectangular orifice. The speed control governor is set by the throttle lever through a cam, a spring, and a stirrup arm inside the regulator.

At any steady running condition, the engine speed is governed by the regulator controlling the fuel flow. The fuel pump delivery is fixed at a constant value by applying the system pressure difference to the fuel pump servo piston. This is arranged to balance the servo piston spring forces.

When the air intake pressure is at a constant value, the rotating V.M.O. sleeve is held in a fixed axial position by the capsule loading. The fixed throttle setting maintains a set load on the speed control governor and, as the rpm is constant, the governor sleeve is held in a fixed position.

The fuel pump delivery is passed to the annulus surrounding the V.M.O.; the annulus area is controlled by the governor sleeve, and the exposed area of the orifice is set by the axial position of the V.M.O. sleeve. Consequently, fuel passes to the inside of the sleeve at a constant flow and therefore at a constant pressure difference.

The pressure drop control valve, which also forms a piston, senses the pressure difference across the V.M.O. and maintains the fuel flow at a fixed value in relation to a function of engine speed, by controlling the exposed area of the pressure drop control orifice.

When the throttle is slowly opened, the load on the speed control governor is increased, so moving the governor sleeve to increase the V.M.O. annulus area. The effect of opening the V.M.O. is to reduce the pressure difference and this is sensed by the pressure drop control governor, which opens the pressure drop valve. The reduced system pressure difference is immediately sensed by the fuel pump servo piston, which increases the pump stroke and consequently the fuel output. The increased compressor delivery pressure acts on the capsule assembly that gradually opens the V.M.O. so that the fuel flow and engine speed continue to increase. At the speed selected, centrifugal forces acting on the speed control governor move the governor sleeve to reduce the V.M.O. annulus area. The resultant increased pressure difference is sensed by the pressure drop control governor, which adjusts the pressure drop valve to a point at which the pump servo system gives an output to match the engine requirements.

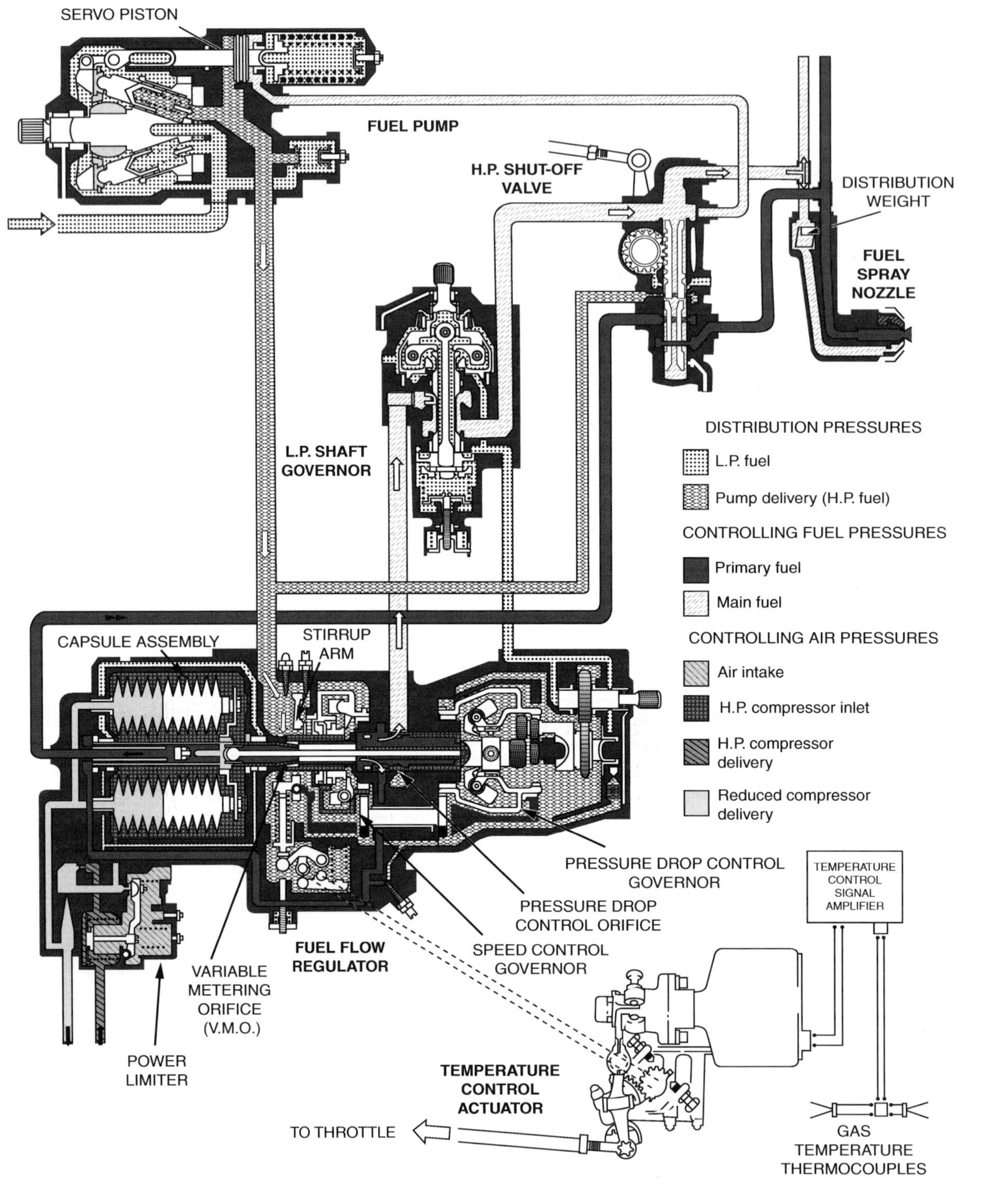

FIGURE 7–9 A combined acceleration and speed control system. *(Source: Rolls Royce. Adapted for reproduction in black and white.)*

The function of the governors and the control of the fuel flow is shown diagrammatically in Figure 7–10.

During a rapid acceleration, the initial degree of overfueling is mechanically controlled by a stop that limits the opening movement of the speed control governor sleeve. A similar stop also prevents the fuel supply from being completely cut off by the governor sleeve during a rapid deceleration.

Changes in altitude or forward speed of the aircraft vary the fuel flow required to maintain a constant engine speed. To provide this control, the capsule assembly senses changes in H.P. compressor inlet and delivery pressures and

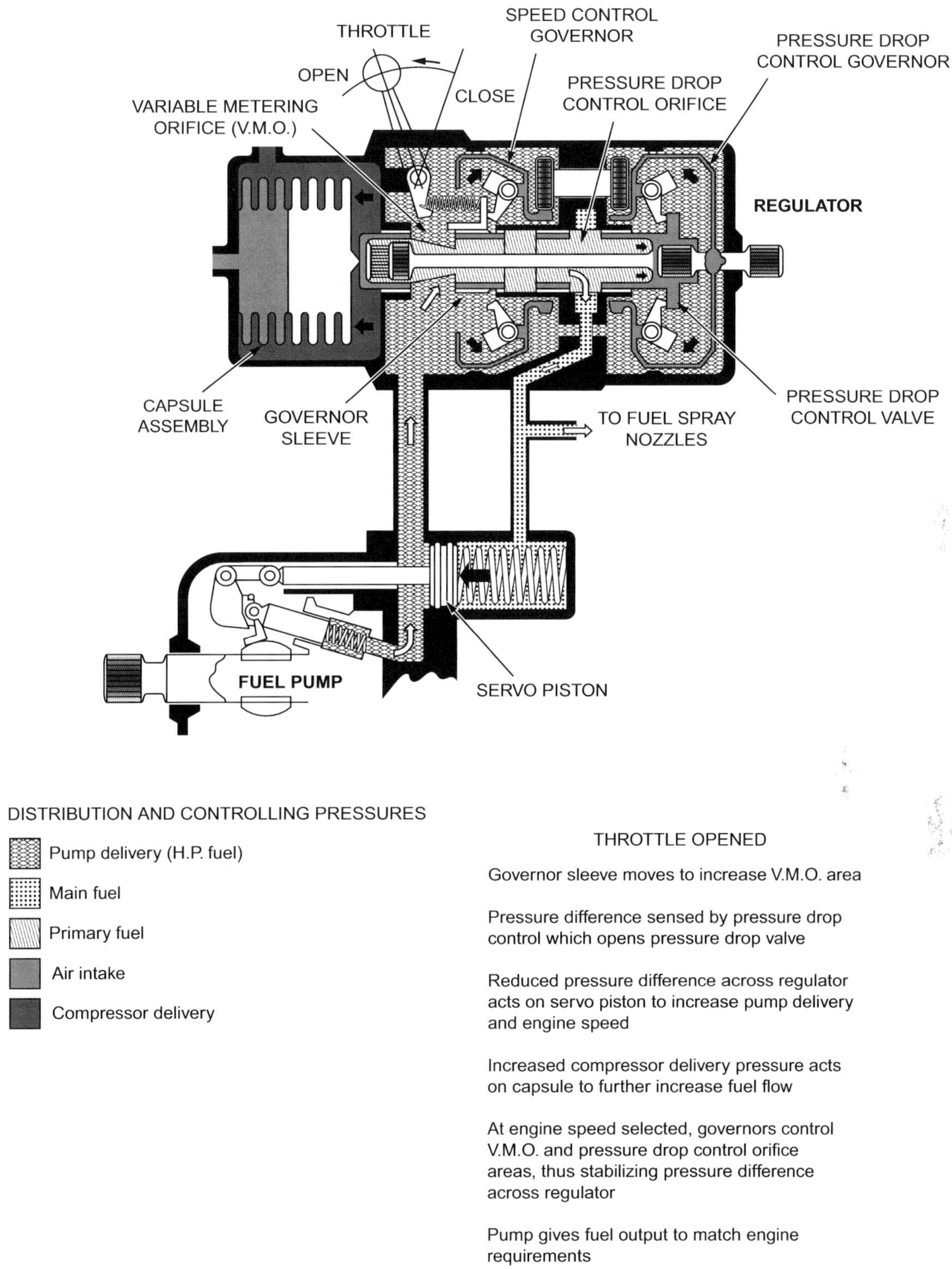

FIGURE 7–10 Governor movement and fuel flow control. *(Source: Rolls Royce. Adapted for reproduction in black and white.)*

adjusts the V.M.O. accordingly. For instance, as the aircraft altitude increases, the compressor delivery pressure falls and the capsule assembly expands to reduce the V.M.O. The increased system pressure drop is sensed by the fuel pump servo piston, which adjusts the pump output to match the reduced airflow and so maintain a constant engine speed. Conversely, an increase in aircraft forward speed causes the capsule assembly to be compressed and increase the V.M.O. The reduced system pressure drop causes the fuel pump to increase its output to match the increased airflow.

To prevent the maximum gas temperature from being exceeded, fuel flow is reduced in response to signals from thermocouples sensing the temperature. When the maximum temperature is reached, the signals are amplified and passed to a rotary actuator, which adjusts the throttle mechanism. This movement has the same effect on fuel flow as manual operation of the throttle.

To ensure that the engine is not overstressed, the H.P. compressor delivery pressure is controlled to a predetermined value. At this value, a pressure-limiting device, known as a power limiter, reduces the pressure in the capsule chamber, thus allowing the capsule assembly to expand and reduce the V.M.O. so preventing any further increase in fuel flow.

A governor prevents the L.P. compressor shaft from exceeding its operating limitations and also acts as a maximum speed governor in an event of a failure of the F.F.R. The governor provides a variable restrictor between the regulator and the main fuel spray nozzle manifold. Should the L.P. compressor reach its speed limitation, flyweights in the governor move a sleeve valve to reduce the flow area. The increased system pressure drop is sensed by the fuel pump servo piston, which reduces the fuel flow to the spray nozzles.

This fuel system has no pressurizing valve to divide the flow from the fuel pump into main and primary fuel flows. Primary fuel pressure is taken from the fixed-area orifice of the pressure drop control valve. This pressure is always higher than the main fuel pressure and it is not shut off by the pressure drop control piston. It therefore gives a satisfactory idling fuel flow at all altitudes.

On engines featuring water injection, a reset device (Figure 7—11), operated by a piston and reset cam, increases the loading on the throttle control spring and stirrup arm, thus selecting a higher engine speed during water injection. To prevent the power limiter (Figure 7—9) from canceling the effect of water injection, a capsule in the limiter is subjected to water pressure to raise the compressor delivery pressure at which the power limiter operates.

Pressure Ratio Control

The pressure ratio control (Figure 7—12) is a mechanical system similar to the combined acceleration and speed control system, but uses the ratio of H.R. compressor delivery pressure to air intake pressure $(1^{3}/_{4}/P_1)$ as the main controlling parameter. It needs no separate governor unit for controlling the maximum rpm The controlling mechanism is contained in one unit, which is usually referred to as a fuel flow regulator (F.F.R.). A gear-type pump is used, as described later, and the pump output to the F.F.R. is controlled by a pressure drop spill valve.

The F.F.R. is driven by the engine through a gear train and has two rotating valves. One valve, known as a variable metering sleeve, has a triangular orifice, known as the variable metering orifice (V.M.O.), and this sleeve is given axial movement by a capsule assembly. The other valve, known as the pressure drop control valve, is provided with axial movement by a centrifugal governor, known as a pressure drop control governor. Both valves form variable restrictors which control the fuel flow to the spray nozzles.

Control of the V.M.O. area is a function of a pressure ratio control unit housed in the F.F.R. A pressure ratio control valve, subjected to P_4 and P_1 pressures, regulates the movement of the F.F.R. capsule and thus controls the V.M.O. area to produce the pressure ratio dictated by the throttle or power lever.

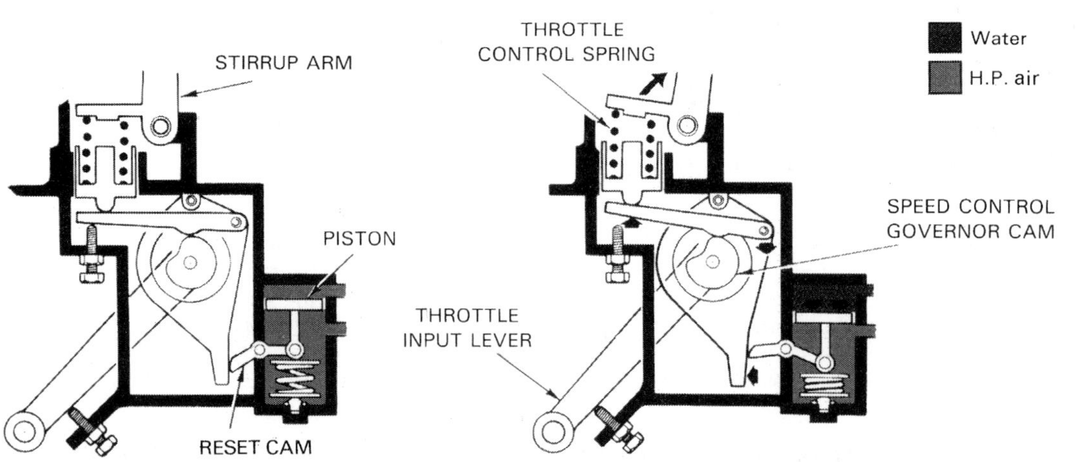

SPEED CONTROL GOVERNOR RESET DEVICE WITH WATER INJECTION NOT SELECTED

SPEED CONTROL GOVERNOR RESET DEVICE WITH WATER INJECTION SELECTED

FIGURE 7—11 Effect of water reset on speed control governor. *(Source: Rolls Royce.)*

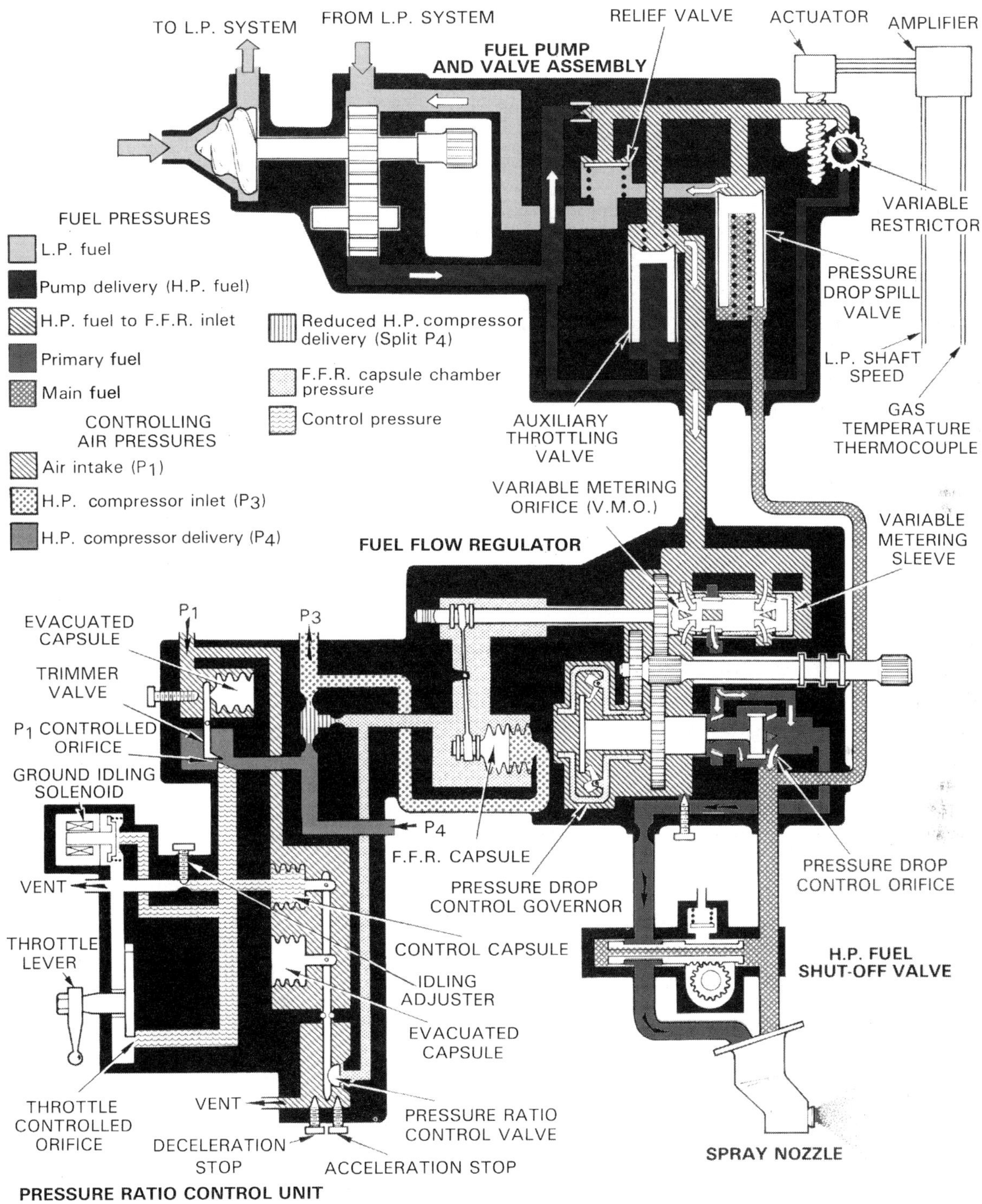

FIGURE 7–12 A pressure ratio control system. *(Source: Rolls Royce.)*

At any steady running condition, the output of the fuel pump is greater than the engine requirement. The pressure drop spill valve is open to allow surplus fuel to return to the inlet side of the pump. This action controls the fuel delivery to that demanded by the F.F.R.

When the throttle is slowly opened, the throttle-controlled orifice is increased and the control pressure falls, thus allowing the pressure ratio control valve to move towards the closed position (acceleration stop). F.F.R. capsule chamber pressure increases and the capsule moves the metering sleeve to increase the V.M.O. area. The effect of opening the V.M.O. is to reduce the pressure difference and this is sensed by the pressure drop governor, which opens the pressure drop control orifice. The reduced system pressure difference is immediately sensed by the pressure drop spill valve, which moves towards the closed position and consequently increases the fuel output. The increased fuel flow accelerates the engine with a subsequent increase in pressure ratio (P_4/P_1). When the required pressure ratio is reached, the pressure ratio control valve opens and the F.F.R. capsule chamber pressure reduces. The capsule assembly expands, moving the V.M.O. sleeve to reduce the orifice area. The resultant increased pressure difference is sensed by the pressure drop control governor, which adjusts the pressure drop control orifice to a point at which the pressure drop spill valve gives a fuel output consistent with steady running requirements.

During a rapid acceleration, the degree of overfueling is mechanically controlled by the acceleration stop, which limits the movement of the pressure ratio control valve. A similar stop prevents the fuel supply from being completely cut off during a rapid deceleration.

When accelerating to a higher P_4/P_1 ratio, the throttle control orifice is increased. The reduced pressure allows the pressure ratio control capsule to contract so that the valve contacts the acceleration stop. F.F.R. capsule chamber pressure increases and the capsule moves to increase the V.M.O. area. This action continues until the required P_4/P_1 ratio is reached. The increased P_4 pressure allows the pressure ratio control capsule to re-expand and the valve to return to the steady running position.

A change in altitude of the aircraft requires a variation in fuel flow to match the engine thrust and aircraft climb requirement. The normal effect of an altitude increase is to decrease the P_1 and P_4 pressures, thus opening the pressure ratio control valve and allowing the F.F.R. capsule to expand to reduce the V.M.O. area and, in consequence, the fuel flow. However, to match the engine thrust and aircraft climb requirement it is necessary to increase the P_4/P_1 ratio with increasing altitude. This is done by a trimmer valve and a capsule that is subjected to P_1 pressure. As P_1 pressure decreases, the trimmer valve moves across the P_1 controlled orifice to reduce the control pressure. This is sensed by the control capsule, which, by acting on the pressure ratio control valve, slows the closure of the V.M.O. as altitude is increased. This maintains the thrust requirement with the throttle at a fixed position.

To prevent the maximum L.P. compressor rpm and engine gas temperature from being exceeded, a valve known as the auxiliary throttling valve is fitted in the outlet from the fuel pump. Under steady running conditions, the valve is held open by spring force. When limiting conditions are reached, the fuel flow is reduced in response to speed and temperature signals from the engine. The signals are amplified and passed to a rotary actuator that reduces the area of a variable restrictor. The effect of this is to increase the fuel pressure, which partially closes the throttling valve. H.P. fuel pressure acting on the face of the pressure drop spill valve is increased and the spill valve opens to reduce the fuel flow to the spray nozzles.

H.P. shaft speed is also governed by the auxiliary throttling valve. Should other controlling devices fail and pump speed increase, the fuel pressure closes the throttling valve and opens the pressure drop spill valve to reduce the fuel flow.

With the throttle closed, idling condition is determined by controlling the amount of air being vented through the idling adjuster and the ground idling solenoid valve. With both bleeds in operation, satisfactory flight idling for the air off-takes is ensured. By closing the solenoid valve a lower power condition for ground idling is obtained.

This fuel system, like the combined acceleration and speed control system, has no pressurizing valve to divide the flow from the fuel pump into main and primary flows.

Electronic Engine Control

As stated previously, some engines utilize a system of electronic control to monitor engine performance and make necessary control inputs to maintain certain engine parameters within predetermined limits. The main areas of control are engine shaft speeds and exhaust gas temperature (E.G.T.) which are continuously monitored during engine operation. Some types of electronic control function as a limiter only, that is, should engine shaft speed or E.G.T. approach the limits of safe operation, then an input is made to the fuel flow regulator (F.F.R.) to reduce the fuel flow thus maintaining shaft speed or E.G.T. at a safe level. Supervisory control systems may contain a limiter function but, basically, by using aircraft generated data, the system enables a more appropriate thrust setting to be selected quickly and accurately by the pilot. The control system then makes small control adjustments to maintain engine thrust consistent with that pre-set by the pilot, regardless of changing atmospheric conditions. Full authority digital engine control (F.A.D.E.C.) takes over virtually all of the steady state and transient control intelligence and replaces most of the hydro-mechanical and pneumatic elements of the fuel system. The fuel system is thus reduced to a pump and control valve, an independent shutoff cock and a minimum of additional features necessary to keep the engine safe in the event of extensive electronic failure.

Full authority fuel control (F.A.F.C.) provides full electronic control of the engine fuel system in the same way as F.A.D.E.C. but has none of the transient control intelligence capability used to control the compressor airflow system as the existing engine control system is used for these.

Speed and Temperature Control Amplifiers

The speed and temperature control amplifier receives signals from thermocouples measuring E.G.T. and from speed probes sensing L.P. and in some cases I.P. shaft speeds (N_1 and N_2). The amplifier basically comprises speed and temperature channels that monitor the signals sensed. If either N_1, N_2, or E.G.T. exceeds pre-set datums, the amplifier output stage is triggered to connect an electrical supply to a solenoid valve or a variable restrictor that overrides the F.F.R. and causes a reduction in fuel flow. The limiter will only relinquish control back to the F.F.R. if the input conditions are altered (altitude, speed, ambient temperature, or throttle lever position). The limiter system is designed to protect against parameters exceeding their design values under normal operation and basic fuel system failures.

Engine Supervisory Control

The engine supervisory control (E.S.C.) system performs a supervisory function by trimming the fuel flow scheduled by the fuel flow governor (F.F.G.) to match the actual engine power with a calculated engine power for a given throttle angle. The E.S.C. provides supervisory and limiting functions by means of a single control output signal to a torque motor in the F.F.G. In order to perform its supervisory function the E.S.C. monitors inputs of throttle angle, engine bleed state, engine pressure ratio (E.P.R.), and air data computer information (altitude, Mach number, and temperatures). From this data the supervisory channel predicts the value of N_1 required to achieve the command E.P.R. calculated for the throttle angle set by the pilot. Simultaneously a comparison is made between the command E.P.R. and the actual E.P.R. and the difference is compared with a programmed datum.

During acceleration the comparitor connects the predicted value of N_1 to the limiter channel until the difference between the command and actual E.P.R. is approximately 0.03 E.P.R. At this point the predicted L.P. shaft speed is disconnected and the E.P.R. difference signal is connected to the limiter channel.

The final output from the supervisory channel, in the form of an error signal, is supplied to a "lowest wins" circuit along with the error signals from the limiter channel. While the three error signals remain positive (N_1 and E.G.T. below datum level and actual E.P.R. below command E.P.R.) no output is signaled to the torque motor. If, however, the output stage of the E.S.C. predicts that E.G.T. will exceed datum or that N_1 will either exceed its datum or the predicted level for the command E.P.R., then a signal is passed to the torque motor to trim the fuel flow.

Low-Pressure Fuel System

An L.P. system (Figure 7–13) must be provided to supply the fuel to the engine at a suitable pressure, rate of flow and temperature to ensure satisfactory engine operation. This system may include an L.P. pump to prevent vapor locking and cavitation of the fuel, and a fuel heater to prevent ice crystals forming. A fuel filter is always used in the system and in some instances the flow passes through an oil cooler. Transmitters may also be used to signal fuel pressure, flow, and temperature.

Fuel Pumps

There are two basic types of fuel pump, the plunger-type pump and the constant-delivery gear-type pump; both of these are positive displacement pumps. Where low pressures are required at the fuel spray nozzles, the gear-type pump is preferred because of its lightness.

Plunger-Type Fuel Pump

The pump shown in Figure 7–14 is of the single-unit, variable-stroke, plunger type; similar pumps may be used as double units depending upon the engine fuel flow requirements.

The fuel pump is driven by the engine gear train and its output depends upon its rotational speed and the stroke of the plungers. A single-unit fuel pump can deliver fuel at the rate of 100–2000 gallons per hour at a maximum pressure of about 2000 lb. per square inch. To drive this pump, as much as 60 horsepower may be required.

The fuel pump consists of a rotor assembly fitted with several plungers, the ends of which project from their bores and bear on to a non-rotating camplate. Due to the inclination of the camplate, movement of the rotor imparts a reciprocating motion to the plungers, thus producing a pumping action. The stroke of the plungers is determined by the angle of inclination of the camplate. The degree of inclination is varied by the movement of a servo piston that is mechanically linked to the camplate and is biased by springs to give the full stroke position of the plungers. The piston is subjected to servo pressure on the spring side and on the other side to pump delivery pressure; thus variations in the pressure difference across the servo piston cause it to move with corresponding variations of the camplate angle and, therefore, pump stroke.

Gear-Type Fuel Pump

The gear-type fuel pump (Figure 7–12) is driven from the engine and its output is directly proportional to its speed. The fuel flow to the spray nozzles is controlled by recirculating excess fuel delivery back to inlet. A spill valve, sensitive to the pressure drop across the controlling units in

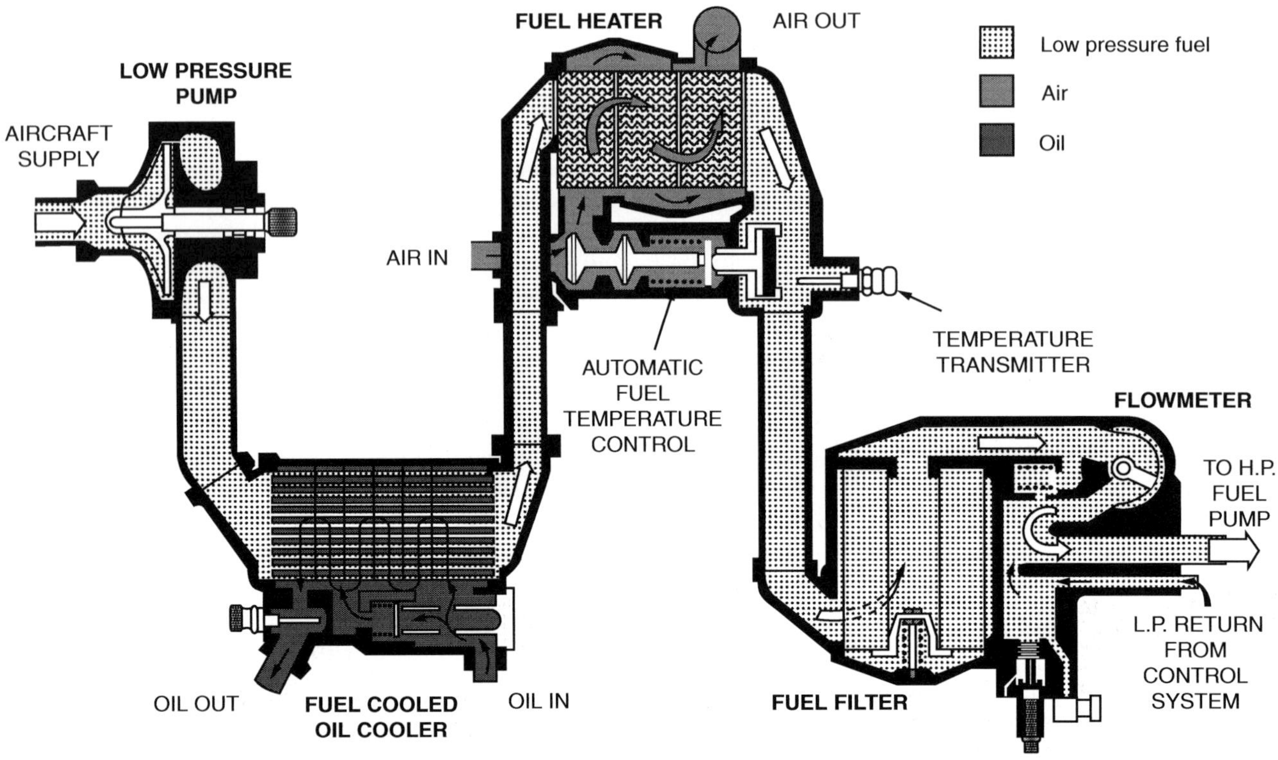

FIGURE 7–13 A low-pressure system. *(Source: Rolls Royce. Adapted for reproduction in black and white.)*

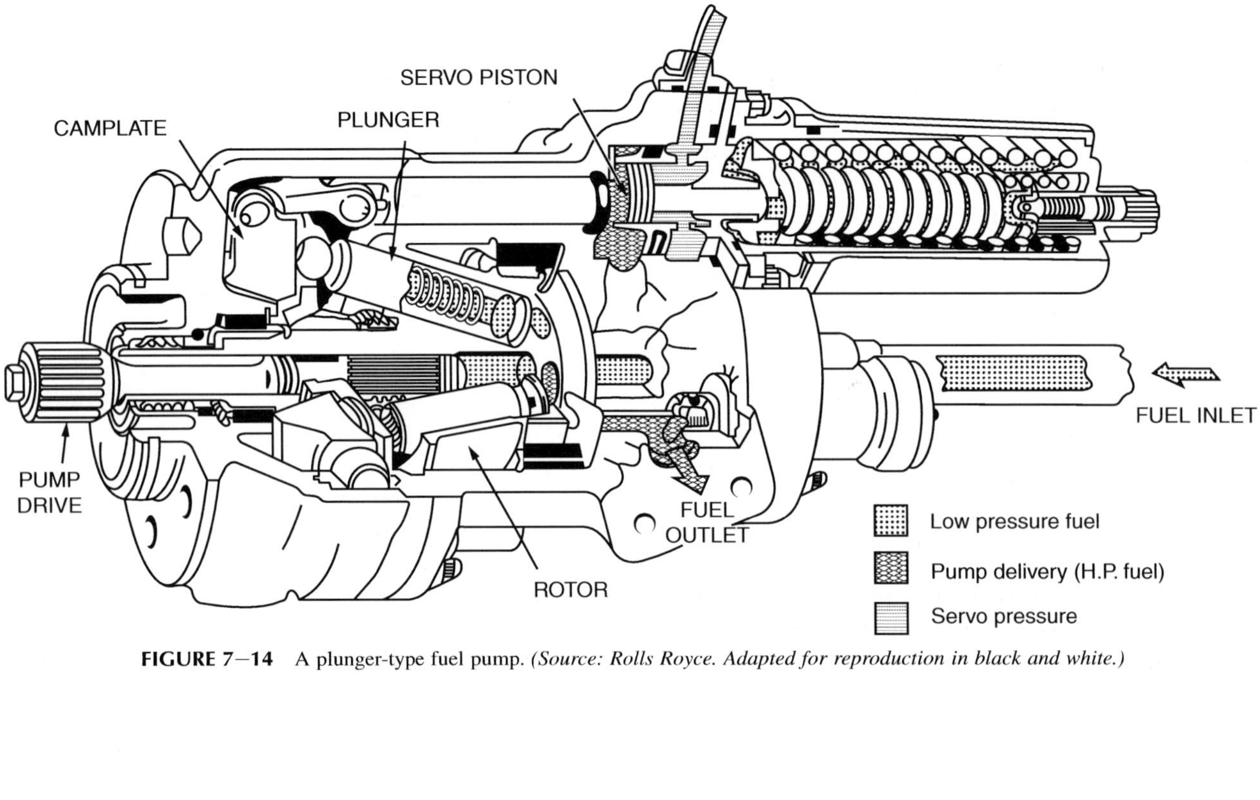

FIGURE 7–14 A plunger-type fuel pump. *(Source: Rolls Royce. Adapted for reproduction in black and white.)*

the system, opens and closes as necessary to increase or decrease the spill.

Fuel Spray Nozzles

The final components of the fuel system are the fuel spray nozzles, which have as their essential function the task of atomizing or vaporizing the fuel to ensure its rapid burning. The difficulties involved in this process can be readily appreciated when one considers the velocity of the air stream from the compressor and the short length of combustion system in which the burning must be completed.

An early method of atomizing the fuel is to pass it through a swirl chamber where tangentially disposed holes or slots imparted swirl to the fuel by converting its pressure energy to kinetic energy. In this state, the fuel is passed through the discharge orifice that removes the swirl motion as the fuel is atomized to form a cone-shaped spray. This is called "pressure jet atomization." The rate of swirl and pressure of the fuel at the fuel spray nozzle are important factors in good atomization. The shape of the spray is an indication of the degree of atomization as shown in Figure 7–15. Later fuel spray nozzles utilize the airspray principle that employs high velocity air instead of high velocity fuel to cause atomization. This method allows atomization at low fuel flow rates (provided sufficient air velocity exists) thus providing an advantage over the pressure jet atomizer by allowing fuel pumps of a lighter construction to be used.

The atomizing spray nozzle, as distinct from the vaporizing burner, has been developed in five fairly distinct types, the Simplex, the variable port (Lubbock), the Duplex or Duple, the spill type, and the airspray nozzle.

The Simplex spray nozzle shown in Figure 7–16 was first used on early jet engines. It consists of a chamber, which induces a swirl into the fuel, and a fixed-area atomizing orifice. This fuel spray nozzle gave good atomization at the higher fuel flows, that is, at the higher fuel pressures, but was very unsatisfactory at the low pressures required at low engine speeds and especially at high altitudes. The reason for this is that the Simplex was, by the nature of its design, a "square law" spray nozzle; that is, the flow through the nozzle is proportional to the square root of the pressure drop across it. This meant that if the minimum pressure for effective atomization was 30 lb. per square inch, the pressure needed to give maximum flow would be about 3000 lb. per square inch. The fuel pumps available at that time were unable to cope with such high pressures so the variable port spray nozzle was developed in an effort to overcome the square law effect.

Although now only of historical value, the variable port or Lubbock fuel spray nozzle (Figure 7–17) made use of a spring-loaded piston to control the area of the inlet ports to the swirl chamber. At low fuel flows, the ports were partly uncovered by the movement of the piston; at high flows, they were fully open. By this method, the square law pressure relationship was mainly overcome and good atomization was maintained over a wide range of

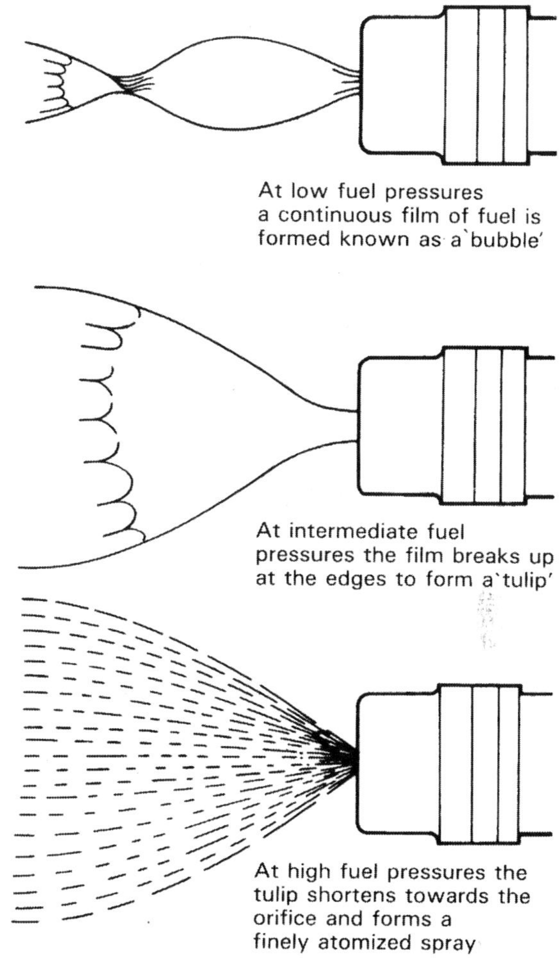

FIGURE 7–15 Various stages of fuel atomization. *(Source: Rolls Royce.)*

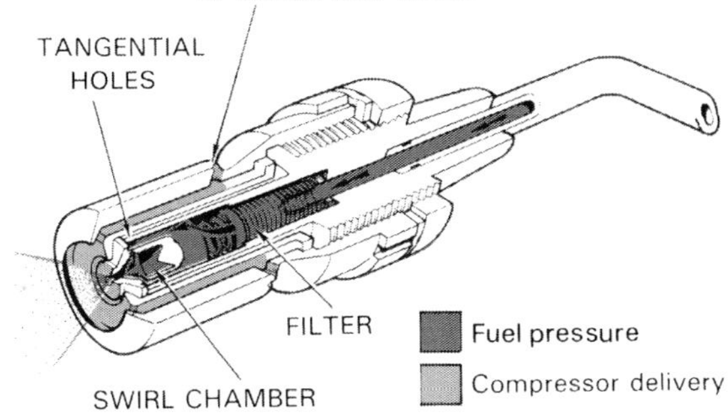

FIGURE 7–16 A Simplex fuel spray nozzle. *(Source: Rolls Royce.)*

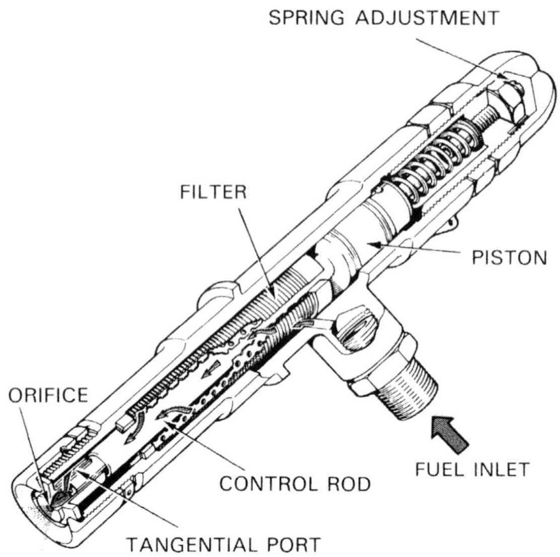

FIGURE 7—17 A variable port or Lubbock fuel spray nozzle. *(Source: Rolls Royce.)*

fuel flows. The matching of sets of spray nozzles and the sticking of the sliding piston due to dirt particles were, however, difficulties inherent in the design, and this type was eventually superseded by the Duplex and the Duple fuel spray nozzles.

The Duplex and the Duple spray nozzles require a primary and a main fuel manifold and have two independent orifices, one much smaller than the other. The smaller orifice handles the lower flows and the larger orifice deals with the higher flows as the fuel pressure increases. A pressurizing valve may be employed with this type of spray nozzle to apportion the fuel to the manifolds (Figure 7—18). As the fuel flow and pressure increase, the pressurizing valve moves to progressively admit fuel to the main manifold and the main orifices. This gives a combined flow down both manifolds. In this way, the Duplex and Duple nozzles are able to give effective atomization over a wider flow range than the Simplex spray nozzle for the same maximum fuel pressure. Also, efficient atomization is obtained at the low flows that may be required at high altitude. In the combined acceleration and speed control system, the fuel flow to the spray nozzles is apportioned in the F.F.R.

The spill-type fuel spray nozzle can be described as being a Simplex spray nozzle with a passage from the swirl chamber for spilling fuel away. With this arrangement it is possible to supply fuel to the swirl chamber at a high pressure all the time. As the fuel demand decreases with altitude or reduction in engine speed, more fuel is spilled away from the swirl chamber, leaving less to pass through the atomizing orifice. The spill spray nozzles' constant use of a relatively high pressure means that even at the extremely low fuel flows that occur at high altitude there is

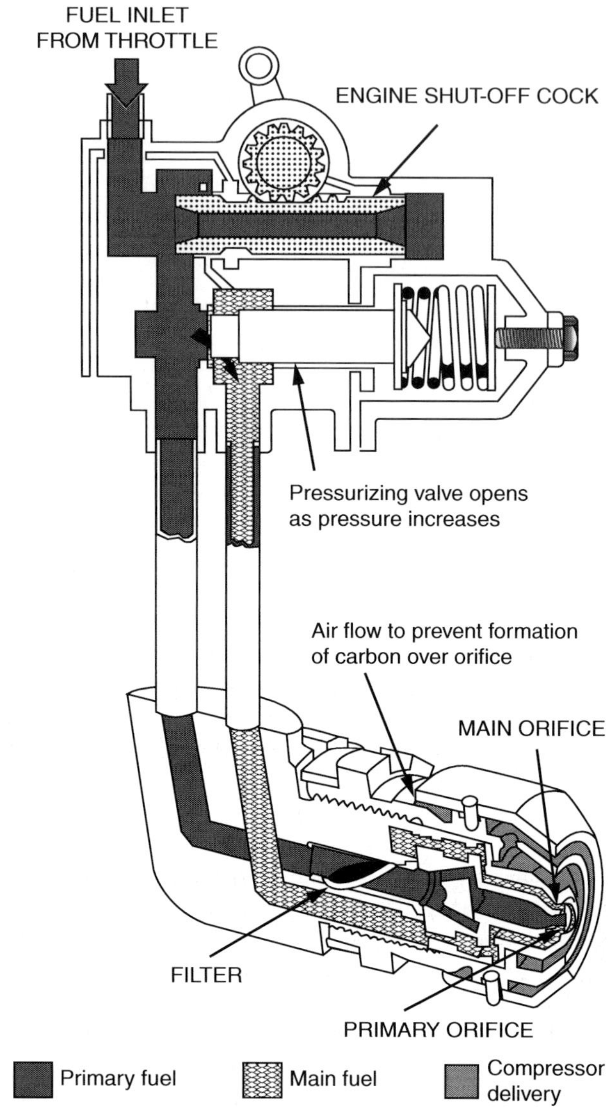

FIGURE 7—18 A Duple fuel spray nozzle and pressurizing valve. *(Source: Rolls Royce. Adapted for reproduction in black and white.)*

adequate swirl to provide constant and efficient atomization of the fuel.

The spill spray nozzle system, however, involves a somewhat modified type of fuel supply and control system from that used with the previous types. A means has to be provided for removing the spill and also for controlling the amount of spill flow at various engine operating conditions. A disadvantage of this system is that excess heat may be generated when a large volume of fuel is being recirculated to inlet. Such heat may eventually lead to a deterioration of the fuel.

The airspray nozzle (Figure 7—19) carries a proportion of the primary combustion air with the injected fuel. By aerating the spray, the local fuel-rich concentrations produced by other types of spray nozzle are avoided, thus

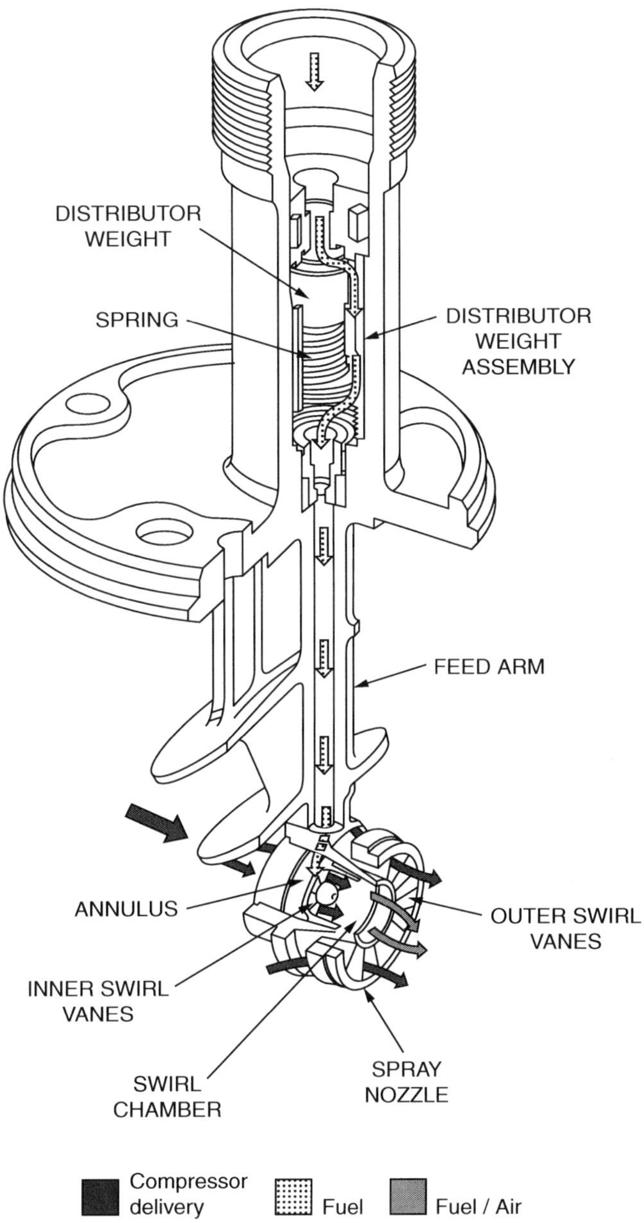

FIGURE 7—19 An airspray nozzle. *(Source: Rolls Royce. Adapted for reproduction in black and white.)*

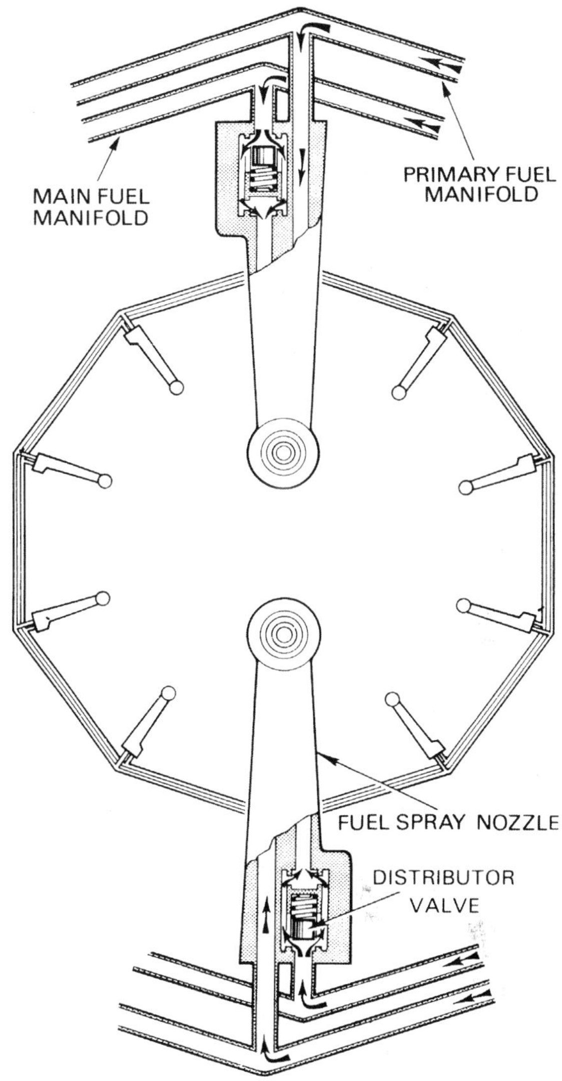

FIGURE 7—20 Fuel flow distributor. *(Source: Rolls Royce.)*

giving a reduction in both carbon formation and exhaust smoke. An additional advantage of the airspray nozzle is that the low pressures required for atomization of the fuel permit the use of the comparatively lighter gear-type pump.

A flow distributor (Figure 7—20) is often required to compensate for the gravity head across the manifold at low fuel pressures to ensure that all spray nozzles pass equal quantities of fuel.

Some combustion systems vaporize the fuel as it enters the combustion zone.

Fuel Heating

On many engines, a fuel-cooled oil cooler is located between the L.P. fuel pump and the inlet to the fuel filter (Figure 7—13), and advantage is taken of this to transfer the heat from the oil to the fuel and thus prevent blockage of the filter element by ice particles. When heat transference by this means is insufficient, the fuel is passed through a second heat exchanger where it absorbs heat from a thermostatically controlled airflow taken from the compressor.

Effect of a Change of Fuel

The main effect on the engine of a change from one grade of fuel to another arises from the variation of specific gravity and the number of heat units obtainable from a gallon of fuel. As the number of heat units per pound is practically the same for all fuels approved for gas turbine engines, a comparison of heat values per gallon can be obtained by comparing specific gravities.

Changes in specific gravity have a definite effect on the centrifugal pressure type of engine speed governor, for with an increase in specific gravity the centrifugal pressure acting on the governor diaphragm is greater. Thus the speed at which the governor controls is reduced, and in consequence the governor must be reset.

With a decrease in specific gravity, the centrifugal pressure on the diaphragm is less and the speed at which the governor controls is increased; in consequence, the pilot must control the maximum rpm by manual operation of the throttle to prevent overspeeding the engine until the governor can be reset. The hydro-mechanical governor is less sensitive to changes of specific gravity than the centrifugal governor and is therefore preferred on many fuel systems.

The pressure drop governor in the combined acceleration and speed control system is density compensated, by using a buoyant material for the governor weights, resulting in fuel being metered on mass flow rather than volume flow.

Changes to a lower grade of fuel can lead to production of carbon, giving a greater flame luminosity and temperature, leading to higher combustor metal temperatures and reduced combustor and turbine life.

GAS TURBINE FUELS

Fuels for aircraft gas turbine engines must conform to strict requirements to give optimum engine performance, economy, safety, and overhaul life. Fuels are classed under two headings, kerosene-type fuel and wide-cut gasoline-type fuel.

Fuel Requirements

In general, a gas turbine fuel should have the following qualities:

1. Be "pumpable" and flow easily under all operating conditions
2. Permit engine starting at all ground conditions and give satisfactory flight relighting characteristics
3. Give efficient combustion at all conditions
4. Have as high a calorific value as possible
5. Produce minimal harmful effects on the combustion system or the turbine blades
6. Produce minimal corrosive effects on the fuel system components
7. Provide adequate lubrication for the moving parts of the fuel system
8. Reduce fire hazards to a minimum

The pumping qualities of the fuel depend upon its viscosity or thickness, which is related to fuel temperature. Fuel must be satisfactory down to approximately −50°C. As the fuel temperature falls, ice crystals may form to cause blockage of the fuel filter or the orifices in the fuel system. Fuel heating and anti-icing additives are available to alleviate this problem.

For easy starting, the gas turbine engine depends upon the satisfactory ignition of the atomized spray of fuel from the fuel spray nozzles, assuming that the engine is being motored at the required speed. Satisfactory ignition depends upon the quality of fuel in two ways:

1. The volatility of the fuel; that is, its ability to vaporize easily, especially at low temperatures.
2. The degree of atomization, which depends upon the viscosity of the fuel, the fuel pressure applied, and the design of the atomizer.

The calorific value (Figure 7—21) of a fuel is an expression of the heat or energy content per pound or gallon that is released during combustion. This value, which is usually expressed in British thermal units, influences the range of an aircraft. Where the limiting factor is the capacity of the aircraft tanks, the calorific value per unit volume should be as high as possible, thus enabling more energy, and hence more aircraft range, to be obtained from a given volume of fuel. When the useful payload is the limiting factor, the calorific value per unit of weight should be as high as possible, because more energy can then be obtained from a minimum weight of fuel. Other factors that affect the choice of heat per unit of volume or weight must also be taken into consideration; these include the type of aircraft, the duration of flight, and the required balance between fuel weight and payload.

Turbine fuels tend to corrode the components of the fuel and combustion systems mainly as a result of the sulfur and water content of the fuel. Sulfur, when burned in air, forms sulfur dioxide; when mixed with water this

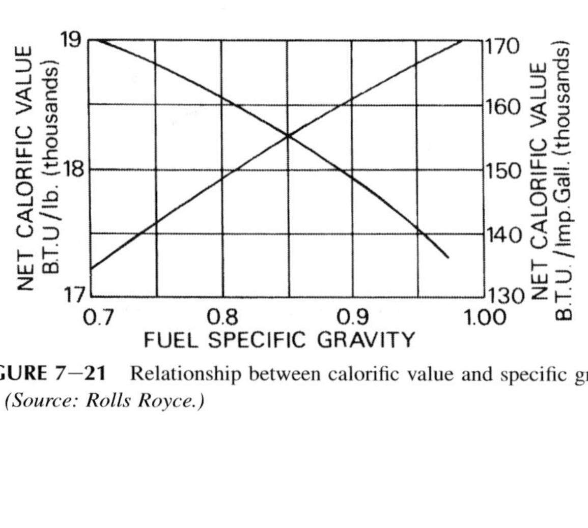

FIGURE 7—21 Relationship between calorific value and specific gravity. *(Source: Rolls Royce.)*

forms sulfurous acid and is very corrosive, particularly on copper and lead. Because it is impracticable to completely remove the sulfur content, it is essential that the sulfur be kept to a controlled minimum. Although free water is removed prior to use, dissolved water, i.e., water in solution, cannot be effectively removed, as a fuel would reabsorb moisture from the atmosphere when stored in a vented aircraft or storage tank.

All gas turbine fuels are potentially dangerous; handling storage precautions should be strictly observed.

Vapor Locking and Boiling

The main physical difference between kerosene and wide-cut fuels is their degree of volatility, the latter type of fuel having a higher volatility, thus increasing the problem of vapor locking boiling. With kerosene-type fuels, the volatility is controlled by distillation and flash point, but with the wide-cut fuels it is controlled by distillation Reid Vapor Pressure (R.V.P.) test. In this test, the absolute pressure of the fuel is recorded by special apparatus with the fuel temperature at 37.8°C (100°F).

Kerosene has a low vapor pressure and will boil only at extremely high altitudes or high temperatures, whereas a wide-cut fuel will boil at a much lower altitude.

The fuel temperature during flight depends upon altitude, rate of climb, duration at altitude, and kinetic heating due to forward speed. When boiling does occur, the vapor loss can be very high, especially with wide-cut fuels, and this may cause vapor locking with consequent malfunctions of the engine fuel system and fuel metering equipment.

To obviate or reduce the risk of boiling, it is usual to pressurize the fuel tanks. This involves maintaining an absolute pressure above the fuel in excess of its vapor pressure at any specific temperature. This may be accomplished by using an inert gas or by using the fuel vapor pressure with a controlled venting system.

For sustained supersonic flight, some measure of tank insulation is necessary to reduce kinetic heating effects, even when lower volatility fuels are used.

Fuel Contamination Control

Fuel can be maintained in good condition by well-planned storage and by making routine aircraft tank drain checks. The use of suitable filters, fuel/water separators, and selected additives will restrict the contamination level, e.g., free water and solid matter, to a practical minimum. Keeping the fuel free of undissolved water will prevent serious icing problems, reduce the microbiological growth, and minimize corrosion. Reducing the solid matter will prevent undue wear in the fuel pumps, reduce corrosion, and lessen the possibility of blockage occurring within the fuel system.

FUEL AND FUEL OIL PROPERTIES*

Fuel and fuel oil properties are required for engine performance calculations including design point, off design, windmilling, starting, transient and test data analysis. This section describes the parameters of relevance and provides a database to cover all needs for such calculations.

The more general topic of properties of fuels and oils in relation to all gas turbine engine design disciplines is exhaustive and cannot be covered here.

Combustion Process, Fuel Properties, and Gas Turbine Fuel Types

Combustion Process

The combustion process primarily entails the exothermic reaction of a hydrocarbon fuel with oxygen, to produce carbon dioxide and water. The number of moles of each is dictated by the fuel composition. For illustration the reaction for one of the hydrocarbons present in kerosene is described below.

$C_{10}H_{20}$	$+ 15O_2$	$=$	$10CO_2$	$+$	$10H_2O$	Chemical equation
1	15		10		10	By mole
10 × 12 + 20 × 1	15 × 32		10 × (12 + 32)		10 × (2 + 16)	By mass
140	480		440		180	By mass

The *stoichiometric* fuel air ratio, where all the atmospheric oxygen would he consumed, is then easily derived. The required number of moles of air is evaluated, and then converted to mass. This is shown below using the molar oxygen content of air, and then the molecular weight of air:

15 × (1/0.2095) = 71.6 Number of moles of air
71.6 × 28.964 = 2073.8 Mass of air
140/2073.8 = 0.0675 FAR by mass

In fact the chemical reaction is far more complex with many reactions taking place simultaneously. This includes the above for the whole range of hydrocarbons present, as well as formation and consumption of carbon monoxide and oxides of nitrogen (NO_x). The last is due to dissociation of atmospheric nitrogen.

Direct Firing and Indirect Firing

The vast majority of gas turbine engines employ *direct firing* where the fuel is injected into the engine combustion

* Source: Courtesy Rolls Royce, *Gas Turbine Performance*, Walsh and Fletcher, Blackwell Science, 1998. Adapted with permission.

chamber and then burned. This is the case for the three primary fuels: kerosene, diesel, and natural gas, and also for a number of less common fuels.

However, for certain obscure fuels considered for industrial engines direct firing may not be practical due to their corrosive or erosive nature. The only manner in which such fuel may be utilized is via *indirect firing* where it is burned external to the gas turbine, with a heat exchanger to transfer heat to the compressor delivery air. Another indirect firing case of interest is a *nuclear* powered closed cycle.

For indirect firing the following changes must be made to performance calculations:

- No mass of fuel is added in the combustor calculations.
- CP downstream of the combustor is that for air at the given temperature, as opposed to that of combustion products.

The impact of both of the above is to lower power output and thermal efficiency since the mass flow and CP in the turbines are both lower. The exact power loss depends upon the engine cycle, and is between 4 and 8%.

Kerosene

Kerosene is a faction of crude oil primarily comprising a band of hydrocarbons with an average composition of $C_{12}H_{23.5}$ and molecular weight of 167.7. There are a number of commercial grades available such as JP4, JP5, Jet Al, and AVTUR that are refined to a tight specification. The proprietary high cost JP10 is a high-density fuel used for certain military applications such as missiles. The term kerosene is usually also taken to encompass the tighter faction aviation gasoline, or AVGAS.

Aeroengines almost exclusively use kerosene as its high calorific value minimizes fuel weight, and because it is free of corrosive elements such as sulfur, which is essential for such high cost engines. The premium in fuel price is warranted by these gains.

Diesel

Diesel fuel is a heavier faction of crude oil than kerosene, again comprising a band of hydrocarbons, with an average composition of $C_{12.9}H_{23.9}$ and molecular weight of 178.6. It is less refined and hence has a lower cost, but therefore also contains small percentages of other elements such as the corrosive agent sulfur. The terms *diesel* and *fuel oil* are commonly used interchangeably. There are a number of *grades* spanning *Number 1 fuel oil* to *Number 6 fuel oil*, the tolerance relative to the nominal specification for each grade is significantly wider than for the kerosenes. The lower *Numbers* are closer to kerosene, have lower amounts of sulfur (typically up to 1.5% by weight being acceptable), and hence usually only Numbers 1 and 2 are used for gas turbine engines.

For cost considerations, diesel is almost exclusively used for marine engines, and in military applications it presents a lower risk of explosion. Its corrosive sulfur content is less of an issue, since marine engines must be designed to withstand a corrosive atmosphere anyway due to the seawater environment.

Though land-based power generation engines usually burn natural gas, a back-up liquid fuel capability is often required to guard against an interrupted gas supply. In addition some niche applications use only liquid fuel. Again, due to cost, fuel oil Numbers 1 or 2 are employed, but life is shorter than for natural gas.

Natural Gas

Natural gas comprises over 80% methane with minor amounts of ethane, propane, butane, and heavier hydrocarbons. It may also include carbon dioxide, nitrogen, and hydrogen. There are a plethora of blends of natural gas available worldwide. The composition on a molar basis of a common blend, used as a "typical" natural gas for deriving gas properties of combustion products, is listed below:

Methane:	95%	I-Pentane:	0.1%
Ethane:	1.9%	N-Pentane:	0.1%
Propane:	0.5%	Hexane	0.1%
I-Butane:	0.5%	Nitrogen:	1.5%
N-Butane:	0.1%	Carbon dioxide:	0.2%

As stated, industrial engines for power generation predominantly use natural gas. This is because of its abundance, competitive cost, and negligible content of corrosive elements such as sulfur. The benefits include lower emissions of carbon dioxide and good engine life, and the large volume required is not a logistics issue as it is for aircraft or marine propulsion. The lower emissions of carbon dioxide are because the lighter hydrocarbons contain more hydrogen and produce more water and less carbon dioxide when burned. As described later the higher water content of the combustion products leads to a higher power output and thermal efficiency for a given engine and SOT level. Engines for natural gas pumping burn solely the gas tapped off the pipeline.

Liquefied petroleum gas (LPG) may be burned in gas turbines, injected either as a liquid or as a gas. Refinery tail gases are also of interest, and are produced as a by-product of oil refining. Other gas fuels occur naturally in lower quantities, such as *condensates* and *naphtha*. Their composition does not comprise the high percentage of methane present in "natural gas" and their properties provide significant difficulties for satisfactory fuel injection. For instance, condensates condense into a liquid at ambient temperature under the action of pressure.

Other Fuels

For industrial engines other fuels have been considered for specific applications/projects, but none have managed to take a significant share of the market.

Extensive efforts over many decades have been spent on developing coal as a fuel for gas turbine engines; however, to date success has been limited. Coal gasification involves pyrolysis of coal in an atmosphere of steam and usually oxygen (as opposed to air). The resultant fuel comprises primarily carbon monoxide and hydrogen, with minor amounts of carbon dioxide, methane, and sulfur compounds. Other attempts have also been made to coarse mill the coal to around 1 mm particles and then chemically clean it to remove the corrosive and erosive elements such as sodium, potassium, vanadium, and ash (silicon-based compounds). It is fine milled to sub-10 μm particle size such that it has characteristics of a gas for direct firing.

Hydrogen has been considered on a research basis for aircraft engines due to its high calorific value by weight. However, the volume of fuel required is prohibitive, requiring complex tank arrangements. Hydrogen is present in many blends of natural gas; too high a content leads to problems with fuel injector flow number.

Biomass is a gas fuel produced from natural vegetation. While it has been considered more recently as a *renewable* fuel it is not widely used due to the large area of cultivated land required to fuel one engine continuously.

Database of Key Fuel Properties for Performance Calculations

This section presents descriptions of the key fuel properties for performance calculations, together with a database. There are many other properties essential to gas turbine design, such as volatility and cloud point, which are outside the scope of this book.

Calorific or Heating Value on a Mass or Volumetric Basis

Calorific value is the heat energy released per unit mass of fuel burned as shown by Formula F7.1. *Gross calorific value,* or *higher heating value* (*HHV*), is the total heat released making no allowance for the latent heat required to vaporize the liquid water produced by the combustion process. Hence it is a theoretical parameter and is rarely used. The *net calorific value*, or *lower heating value* (*LHV*), does account for the latent heat of vaporization and hence is the parameter most commonly used for gas turbine performance calculations. Strictly LHV is quoted as *the heat released under pressure in a constant volume when the combustion products are cooled to the initial temperature of 25°C.* Hence for exhaustively rigorous combustion calculations the enthalpy balance should include:

- Heat release/absorption to raise fuel to 25°C
- Heat release due to reducing compressor delivery temperature to 25°C
- Heat release from fuel
- The resultant heat from the above is that available to raise combustion products from 25°C to combustor outlet temperature

Occasionally for gas fuels calorific value is quoted on a volumetric basis (Formula F7.2) having units of either kJ/m^3 or kl/scm. The scm (standard cubic meter) is the amount of gas that would occupy 1 m^3 at the ISO pressure and temperature of 101.325 kPa and 288.15 K respectively. Formula F7.3 shows how the number of scm present in a volume of gas fuel at any given pressure and temperature may be calculated.

LHV is used in most performance calculations such as design point, off design and test data analysis. Nominal values for the three primary gas turbine fuels are presented below, followed by methodologies for deriving values for a specific batch of fuel used during an engine test.

- Kerosene: 43,124 kJ/kg
- Diesel: 42,600 kJ/kg
- Natural gas: 38,000–50,000 kJ/kg
- Sample natural gas: 48,120 kJ/kg

Liquid Fuels

The most accurate method is to measure LHV by testing a sample in a bomb calorimeter, and is normal practice for performance testing.

Natural Gas

For key performance tests using natural gas the fuel composition must first be determined using a gas chromatograph. The formula for then calculating LHV is complex, being a function of the wide range of constituents. Natural gas LHV can also be measured using a bubble calorimeter; however, this is only undertaken for "guarantee" performance tests as a further check on the calculated value.

Density and Specific Gravity

Fuel density, the mass of fuel in a given volume (Formula F7.4), is important during engine performance testing. Invariably volumetric flow rate is measured and density must be derived such that heat release may be calculated. It is not of interest for other gas turbine performance topics.

Specific gravity is commonly used for liquid fuels and is the ratio of the fuel density to that of water at 4°C as shown by Formula F7.5. Figure 7–22 presents SG versus fuel temperature for commonly used kerosenes and diesels.

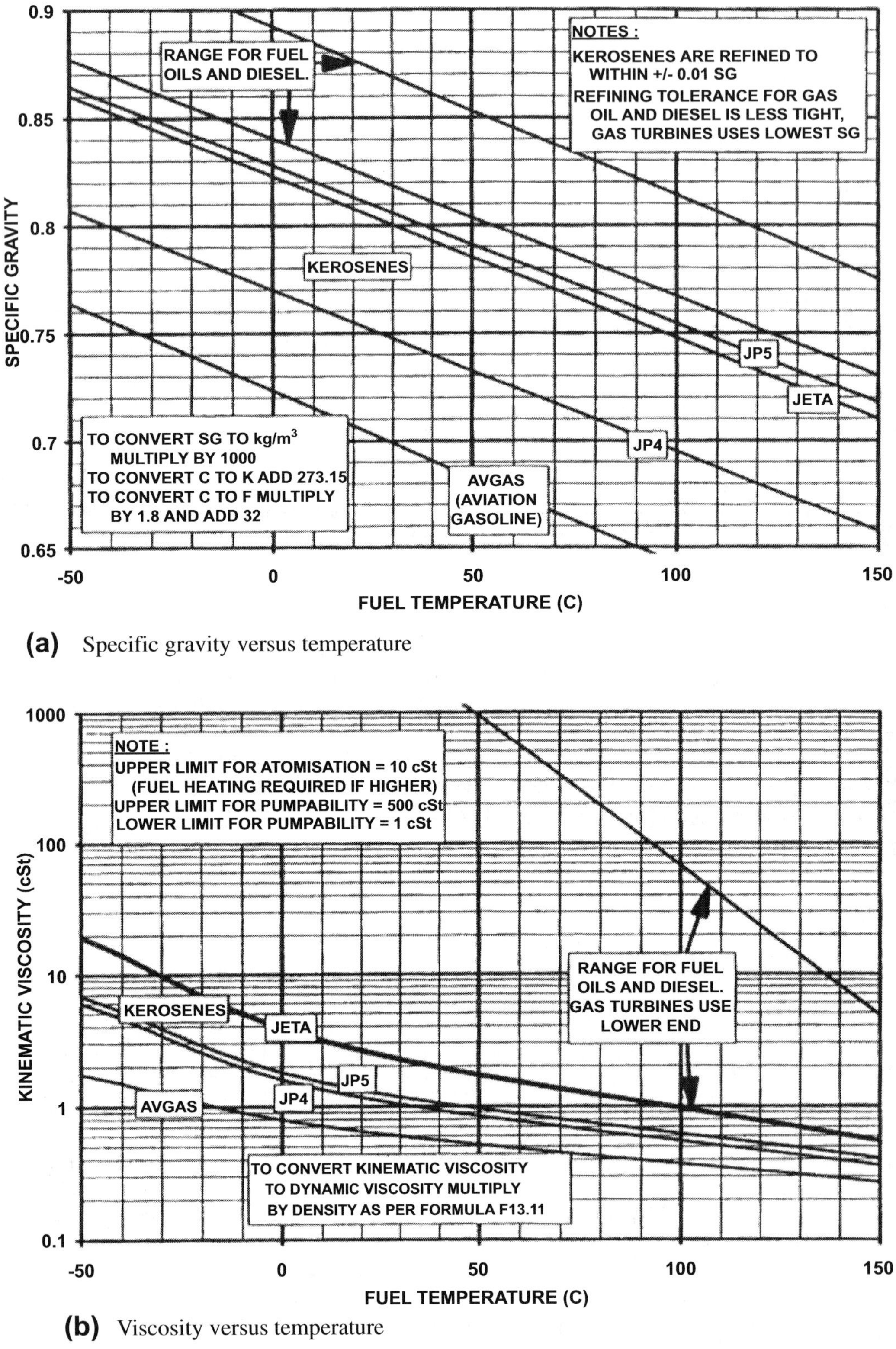

FIGURE 7–22 Fuel specific gravity and viscosity versus temperature: (a) specific gravity versus temperature. (b) Viscosity versus temperature. *(Source: Rolls Royce.)*

At 15°C the SG for AVGAS is 0.704, the rest of the kerosenes are between 0.76 and 0.818, and diesels range from 0.82–0.88. For a given kerosene type the tolerance is around ±0.1 and for diesel it is even wider. Hence Figure 7–22 can only be used as indicative of SG level and for key engine performance tests it must be measured. This is accomplished using a fuel sample in a laboratory with a calibrated hydrometer. The fuel sample temperature must be measured simultaneously and Formula F7.6 is used to derive SG at the fuel temperature measured during the engine performance test.

Gas fuel density may be calculated from Formula F7.7 where the "compressibility" term z is used to modify the perfect gas law such that it may be applied to non-perfect gases. Gas fuel pressure and temperature must be measured in the vicinity of the turbine flow meter. ASME (1993) *Performance Test Code PTC–22* and ASME (1969) *Gaseous Fuels PTC 3.3* show how R and z may be derived once the fuel gas composition is known. The calculation process for the latter is complex; however, if it is not followed then significant errors will result.

Viscosity, Dynamic and Kinematic

Dynamic viscosity is the resistance to movement of one layer of a fluid over another and is defined by Formula F7.8. *Kinematic viscosity* is dynamic viscosity divided by density (Formula F7.9) and is the ratio of viscous forces to inertia forces. Dynamic viscosity for liquid fuels is occasionally required for performance calculations in determining fuel pump power requirements such as during start modeling. Kinematic viscosity is a second-order variable in the calibration of bulk meters and turbine flow meters for measuring liquid fuel volumetric flow rate.

Figure 7–23 presents kinematic viscosity for gas turbine liquid fuels. It is common practice to arrange fuel dynamic viscosity as an input value to a performance computer program. Hence suitable values may be taken from Figures 7–22 and 7–23. If necessary the reader may "fit" a polynomial to the data presented in these charts to embed the properties within the computer program.

Liquid fuels are not "pumpable" with a kinematic viscosity of less than 1 cSt, and atomization will be unsatisfactory above 10 cSt. Figure 7–22 shows that kerosenes generally have a viscosity of less than 10 cSt even at −50°C and hence are always acceptable for atomization. However, diesels do exceed the limit and hence fuel heating is required if its temperature is allowed to fall below the threshold level.

None of the above effects are of importance for gas fuel as its viscosity is an order of magnitude lower; hence a database is not provided here.

CP of Fuels

For fully rigorous modeling fuel CP is required to calculate enthalpy input of the fuel prior to combustion. Approximate levels of CP for the primary gas turbine fuels are:

- Kerosene: 2.0 kJ/kg K
- Diesel: 1.9 kJ/kg K
- Natural gas: 2.0–2.2 kJ/kg
- Sample natural gas: 2.1 kJ/kg

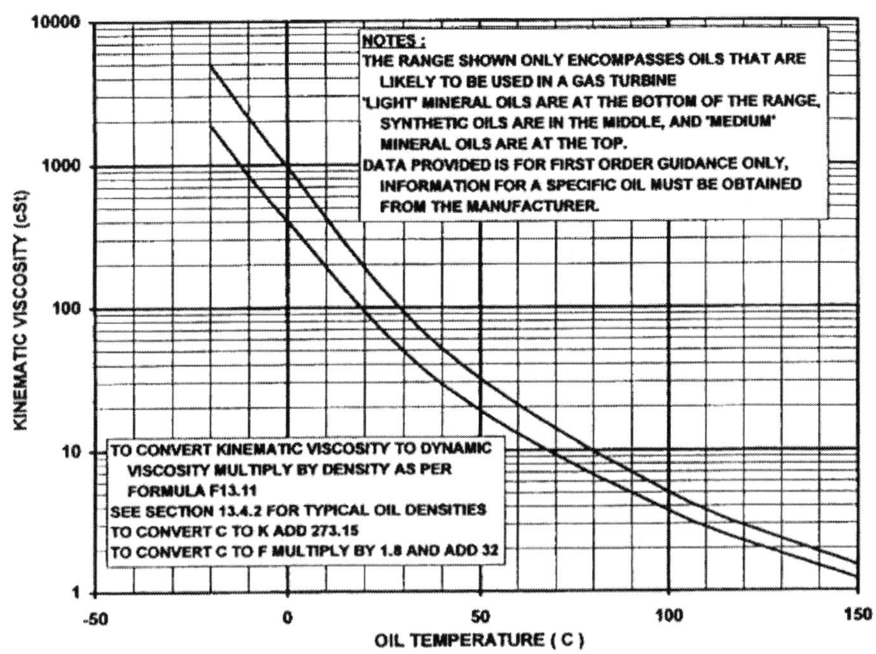

FIGURE 7–23 Oil viscosity versus temperature. *(Source: Rolls Royce.)*

CP of Combustion Products

CP of the products of combustion is required for all performance calculations involving combustion including design point, off design, and engine testing.

However, this is not so for natural gas that may come in a huge variety of blends, often with more than 10 constituents. Where second-order loss of accuracy is unacceptable then the CP of the products of natural gas combustion must be evaluated rigorously for the given natural gas composition.

Synthesis Exchange Rates for Primary Fuel Types

Industrial engines may often operate on more than one fuel type; in fact many engines are capable of switching between natural gas and diesel fuel in the field. This allows operators to pay a lower tariff for the gas fuel due to an *interruptible supply*, but not lose any availability. Also, engines may often undergo a production pass off test on kerosene at the manufacturer's facility, even though they will primarily operate on natural gas in the field.

Hence the second-order effect on engine performance of the different fuel types is often of great interest. The major contributor to performance change is the change in gas properties downstream of the combustor. Any difference in fuel mass flow resulting from different calorific values, and change in referred parameter turbine maps (full non-dimensional maps are unchanged), produces only tertiary effects.

Kerosene to Diesel

The gas properties of the combustion products of kerosene and diesel, as well as their lower heating values are very similar. Hence there is negligible change in performance when burning either of these fuels.

Liquid Fuel to Natural Gas

The CP of the combustion products of the natural gas is around 2% higher than for liquid fuels. This has a noticeable impact upon performance increasing both the fuel energy flow required for a given combustor temperature rise, and the shaft power output. The net effect is an improvement in SFC. The exact change in performance depends upon the engine cycle but at constant SOT the changes when burning gas relative to a liquid fuel are in the range:

- Power output: +4 to +6%
- Rotational speed: 0.5 to +1.5%
- Inlet mass flow: +2 to +3%
- HPC delivery pressure: +2 to +3%
- Fuel energy flow: +3 to +4%
- SFC: −1 to −2%

Oil Types and Database of Key Properties

Oil Types

Oils used for gas turbine engines fall into two major categories: mineral oils and synthetic oils. Mineral oils result from the refining of crude oil. Usually anti-oxidant and anti-corrosive additives are employed for gas turbine use. Synthetic oils are ester based and their cost is approximately 10 times that of mineral oils.

Synthetic oils are used almost exclusively in gas generators due to their vastly higher auto-ignition temperature. Mineral oils are usually used for industrial power turbines where bearing chamber temperatures are sufficiently low. This is particularly true where journal as opposed to ball bearings are used, because the former requires approximately 20 times the oil flow to dissipate the heat resulting from significantly higher friction, hence the cost of using synthetic oil is prohibitive. Journal bearings are often required in large power turbines as they are capable of reacting far higher thrust loads.

Density

Formula F7.10 enables oil density for a typical synthetic oil with a density of 1100 kg/m^3 at 15°C to be calculated as a function of oil temperature. The same formula may be factored to give the typical density of a mineral oil at 15°C, and then used for other oil temperatures. Typical values for oil density at 15°C are:

- Light mineral: 850 kg/m^3
- Medium mineral: 860 kg/m^3
- Synthetic: 1100 kg/m^3

Viscosity, Kinematic and Dynamic

Kinematic and dynamic viscosity were defined earlier. Oil viscosity is important for calculating bearing and gearbox losses.

Figure 7–23 presents kinematic viscosity versus temperature for oils typically used in gas turbines. Light mineral oils are at the lower end of the band, synthetic oils in the middle and medium mineral oils at the top.

The formulae presented provide sufficient accuracy for all performance calculations.

Formulae

1. Lower heating value is also commonly called calorific value.
2. "fn" is an abbreviation for "a function of."

F7.1 Fuel lower heating value (kJ/kg) = fn(heat release (kJ), fuel mass (kg))

$$\text{LHV} = \text{Q/Mass}$$

F7.2 Fuel lower heating value by volume (kJ/m^3) = fn(heat release (kJ), number of standard cubic meters of gas fuel (m^3))

$$\text{LHY} = \text{Q/SCM}$$

F7.3 Standard cubic meters (m^3) = fn(volume (m^3), gas fuel pressure, and temperature (kPa, K))

$$\text{SCM} = \text{Volume} \times (288.15 \times \text{P})/(101.325 \times \text{T})$$

This calculates the number of standard cubic meters that are equivalent to the amount of gas occupying the current volume at the prevailing P and T.

F7.4 Density (kg/m^3) = fn(mass (kg), volume (m^3))

$$\text{RHO} = \text{Mass/Volume}$$

F7.5 Fuel specific gravity = fn(actual density (kg/m^3), standard density (kg/m^3))

$$\text{FSG} = \text{RHO}/1000$$

The standard density of 1000 kg/m^3 is that of water at 4°C.

F7.6 Fuel specific gravity = fn(fuel sample SG, fuel sample temperature (°C), fuel temperature (°C))

$$\text{FSG} = \text{FSGsample} - (\text{Tfuel} - \text{Tsample}) \times 0.00074$$

"Sample" refers to a fuel sample for which the SG and temperature have been measured in a laboratory.

To convert from FSG to actual density multiply by 1000 kg/m3.

F7.7 Gas fuel density = fn(static pressure (kPa), gas constant (J/kgK), temperature (K), compressibility)

$$\text{RHO} = \text{P/R} \times \text{T} \times \text{Z}$$

F7.8 Dynamic viscosity (Ns/m^2) = fn(shear stress (N/m^2), velocity gradient (m/sm), static temperature (K))

$$\text{VIS} = \text{Fshear}/(\text{dV/dy})$$

Fshear is the shear stress in the fluid. V is the velocity in the direction of the shear stress. dV/dy is the velocity gradient perpendicular to the shear stress.

F7.9 Kinematic viscosity (cSt) = fn(dynamic viscosity (Ns/m^2), density (kg/m^3))

$$\text{VISkinematic} = 1{,}000{,}000 \times \text{VIS/RHO}$$

F7.10 Oil density (kg/m^3) = fn (oil temperature (K))

$$\text{RHOoil} = 1405.2 - 1.0592 \times \text{T}$$

This is for a typical synthetic oil.

F7.11 Average molecular weight of co[illegible] products = fn(molecular weights of c[illegible] number of moles of constituents)

$$\text{MWav} = \sum(\text{Moles} \times \text{MW})/[illegible]$$

UNCONVENTIONAL FUELS

Unconventional means different things to different people. Today, end users, with the OEMs' cooperation, are burning biowaste, paper production waste liquor, pulverized coal gas, low BTU waste (from plastics manufacture) hydrocarbons, and a variety of other "emerging" fuels.

What follows are extracts of recent studies conducted by an EPC contractor* and a gas turbine OEM,** both of which discuss different aspects of the use of unconventional fuels and contrast that with use of natural gas. The equipment supplied by an OEM of fuel treatment systems (Westfalia) is also discussed.† Further, the authors of these works represent engineering (the EPC contractor), economics and logistics/ global project finance (the OEM in concert with the World Bank), and accessory system design (Westfalia Separator makes treatment systems for fuel and systems that support the fuel system). The reader needs to note the different emphases placed on the same system with the same fuel, by these different technical and corporate functions.

Emerging Fuel Trends/New Fuel Technologies

The fracking movement has generated much controversy in the USA. The natural gas it yields may vary somewhat in its extent of impurities and geological "additives," but it is still natural gas. The process has opened up natural gas reserves that were previously unattainable. Other countries and continents are sure to follow. The potential problem is that with fracking, oil and gas companies may opt to use contractors. Contractors do not always use the same safety standards to which a major oil company may be held. Therefore "results" with respect to work items such as getting rid of associated waste water (and associated impurities) varies.

With fracking, the oil and gas companies saw a potential for subcontracting work. Employing contractors to drill

* Reference: [7-1]. Papers by J. Zachary, some with R. Narula, venues for content include PowerGen conferences, ASME papers, ISO committee work, including but not limited to "Unconventional Fuels for Combined Cycle Applications—An Attractive Alternative" (PowerGen Asia, 2004).
** Source: [7-2]. Courtesy of Siemens. Extracts from papers by R. Taud (head of Product Marketing in the Product Center of the Gas Turbine/ Combined Cycle Power Plant Division of Siemens Power Marketing Group (KWU) in Erlangen, Germany), J. Karg (manager of IGCC Power Plants and Gasification Technology for KWU), D. O'Leary (senior power engineer of the World Bank/Siemens/World Bank Staff Exchange Program).
† Source: [7-3]. Courtesy of Westfalia Separator product information.

into natural gas-rich formations, then create the "minor" explosions that break up the rock to the extent that the ground releases its entrained natural gas, then inject the fluid mix that moves the gas to the wellhead at the surface, is cheaper than the oil company doing it themselves. Their legal liability for what corners these contractors may cut may be less than had they done it themselves.

Fracking technology is reasonably well thought out. A casing sheath is supposed to protect surrounding rock and soil formations, as well as ground water sources, from being contaminated with the potentially poisonous mix of drilling fluid (water plus chemical additives). Sheath ruptures and punctures happen. This book shall not attempt to debate how often.

The risk, even were the technology perfect, is evident and its chances of increasing with smaller contractors who may feel they have to cut corners must be acknowledged. Will fracking be more competently done by seasoned oil and gas companies than contractors? Probably yes, they have the resources, the experience, the safety standards. Also their liability exposure is greater and they have much to lose if they can be found to have cut corners.

All this said, the bottom line is that gas prices dropped as a result of the increased production from fracking. In 1997, the known global gas reserves stood at 70 years' worth; coal at 200. A decade later, known gas reserves had increased, thanks to new exploration and improved deep sea drilling methods that attained depths that were formerly unknown.

However, even with a world giddy with the thought of more natural gas, coal refuses to go away. It won't disappear, even in the gas turbine world, and even in countries that are more heavily legislated with respect to emissions. The reason is advancing technologies with respect to coal as a gas turbine fuel, the capture of carbon-in-fuel or CO_2 resulting from coal combustion and the storage of CO_2. These technologies either permit or promise the potential of reducing the GHG emissions. The emphasis on this element of running turbines has grown considerably in terms of importance placed on it by OEMs and legislators.

This is interesting in places like the USA where some plant owners have not just converted their steam turbines to combined cycle operation that burns natural gas, but also converted their steam turbines to burn natural gas, without any modifications to the steam turbine that may also raise its effieiency. Conventional wisdom dictates that operators take up base load with a steam turbine, then use gas turbine(s) for peaking, but there have been cases of operators doing the reverse, probably hoping to reduce their on-record emissions.

Coal has been used as a gas turbine fuel in liquefied form and also in pulverized form. More recently, "syngas" from coal (consisted of CO and H_2) has been explored. If one neglects impurities, then syngas can burn like natural gas (1 unit of CO_2 per unit of syngas by weight).

Once a gas turbine has been modified to burn coal in any form, it can then spread its wings still further. Some OEMs have done work on buring biomass with DLN combustors.

Consider also that gas turbines, in combined cycle or otherwise, can be modified to burn other unconventional fuels. Flue gas from steel mills and other industries, waste liquids from petrochemical and plastics plants, black liquor from producing paper have all been used with varying degrees of success. Residual fuel is a popular fuel in some gas-poor countries as well, and this chapter deals considerably with the equipment and treatment required to successfully do this.

Returning to the subject of coal as a potential turbine fuel: many operators globally have operated steam turbines to supply the bulk of their power and gas turbines run on LNG (if they have no gas resources, like Japan) to fill in peaking demands. What this created in some countries, like China and Japan, is the drive to work with supercritical, ultra-supercritical (USC) and now advanced ultra-supercritical steam. The EU and the USA now have research programs of their own. Siemens won the EU grant to work on advanced steam turbine metallurgy research, but that program has run (maybe temporarily) out of money. The US DOE continues its work to explore boiler materials that will operate with 700°C temperature and over steam.

There have been some interesting studies of systems that integrate the heating and cooling streams from gas and steam turbines (see Chapters 10 and 12). Consider also that the Benson boiler design common in most GT CCs is essentially the same boiler design (neglecting scale) as is used with steam turbine cycles. The potential for cycle adaptation for fossil fueled turbines abounds.

IGCC is one of the most studied systems for using coal (PC) fuel in GTs. This chapter has more material on that technology. Oxy-combustion and hydrogen fuel are two other fuel technique/fuel systems being actively explored.

Oxy-combustion has now replaced IGCC as being the intended combustion technique to be used in Illinois' FutureGen project. Discussion on both oxy-combustion and hydrogen as a fuel has been addressed in Chapter 18, the Future of Gas Turbines, as these are fuel fields still in their relative youth.

All coal-fueled power plants are either using or planning carbon removal, then carbon sequestration and storage (CSS) systems. These are discussed in Chapter 11 (on emissions).

Economic Conditions that Affect Fuel Strategy

In the 2007 economic climate of rocketing natural gas prices, nonconventional fuels are more important than ever. Even when global politics stabilize, because fossil fuel deposits are finite, nonconventional fuels will tend to transition into "conventional." Let us first consider the market conditions relevant to fuel selection in gas turbines and their operating environment and how these may affect the various sectors of the gas turbine environment.

Market Demand[††]

Contract awards for the last five years for fossil fuel-fired power plants (above 50 MW) averaged 63 GW per year, sales are forecasted to average about 67 GW per year over the period 1999–2004. From a regional viewpoint, the market in the Asia Pacific Basin is declining, while showing a moderate growth in Europe (starting from a low level) and strong growth in North America. This is reflected in the growth of the 60 Hz market. In the past, the proportion of power plants with gas turbines, that is, open cycle and closed cycle (combining gas and steam turbines), was around 54% of total fossil-fuel-fired power plants. For the next five years, this proportion is expected to rise to 63%. For open- and closed-cycle power plants, this means that sales are expected to grow from 34 GW per annum to 42 GW per annum. This growth is predominantly due to market developments in North America.

In comparison with coal-fired power plants, open- and closed-cycle power plants are characterized by lower investment costs. On the other hand, fuel costs (i.e., fuel price and plant efficiency) play a relatively more important role. Therefore, gas pricing is a very important issue, which ranges presently from about $2.0/GJ to $4.5/GJ, with the North American prices being at the lower end of the range. Thus, particularly in North America, open- and closed-cycle power plants can be particularly attractive investments.

Technology, Performance, Environment, and Design Trends

Over the last 50 years, the efficiencies of gas turbines went up from 33–38% (LHV) and the efficiencies of combined-cycle plants from 48–58%. At present, there are a rather small number of global manufacturers and suppliers offering GT products covering a wide range of power output. The products of the different suppliers can be grouped in clusters by comparable output. A rough classification is small (S), that is, up to 50 MW; medium (M), above 50 MW; and large (L), which is the class of the largest, high-performing heavy-duty gas turbines in the market with a power output above 200 MW. For all major manufacturers, the small and medium classes include proven existing GT models with lower efficiencies and operating temperatures but with robust and reliable performance; the large classes are confined to high-performance machines.

The latest gas turbine technology also reflects the expertise of aero-jet engine manufacturers. As a result, using pressure ratios of around 17, performance features have been achieved for heat rates in the range of 9400–9800 kJ/kWh and for efficiencies up to 38%. These performance improvements have been achieved through improved compressor design; higher turbine inlet temperature (approximately 1250°C ISO); advanced blading (design, materials, coating); optimal design of the turbine blade cooling system, with or without the need for external coolers; improved cooling of the combustion system sections (chamber, canes, transitions); and less pressure drop in the turbine exhaust.

Less than 25-ppm NO_x emissions can be expected from these machines. This compares with the maximum allowable levels of 60 ppm for GTs fired with natural gas as set out in the World Bank's "Thermal Power-Guidelines for New Plants." Because of better materials (alloys and coatings), with a higher resistance to high temperature oxidation and corrosion, combined with better cooling techniques, gas turbines are expected to reach a unit capacity of around 250–350 MW and GT efficiencies of 40% over the next years. In conjunction with optimized main steam parameters and improved plant component efficiencies, this in turn will increase the CC plant capacity into the 375–500 MW range, with CC efficiencies around 60%. (See also Figure 7–24.)

From a plant aspect, plant construction time has been shortened by the concept of predesigning, introduction of modules, and advanced project management and logistics. Inter alia, this was effected by simplification of the electrical and I&C systems as well as the foundation and overall civil works construction requirements. In addition, plant operation has been simplified, for example, by allowing full 100% steam bypass operation in the event of steam turbine malfunctions.

Cost Trends

Typical combined-cycle blocks are roughly in the range of 50–500 MW and costs range from $500–750/kW.

Impacts on Design and Manufacturing in Developing Countries

Table 7–1 shows the cost breakdown for combined-cycle plants (350–700 MW capacity) based on Siemens experience into the following categories: integrated services (project management/subcontracting; plant and project engineering/project management software, plant erection/commissioning/training; transport/insurance) and lots (civil works; gas- and steam-turbine and generator sets;

[††] Source: [7-4]. Courtesy of Siemens. Extracts from papers by R. Taud (head of Product Marketing in the Product Center of the Gas Turbine/Combined Cycle Power Plant Division of Siements Power Marketing Grop (KWU) in Erlangen, Germany), J. Karge (manager of IGCC Power and Gasification Technology for KWU, D. O'Leary Program).

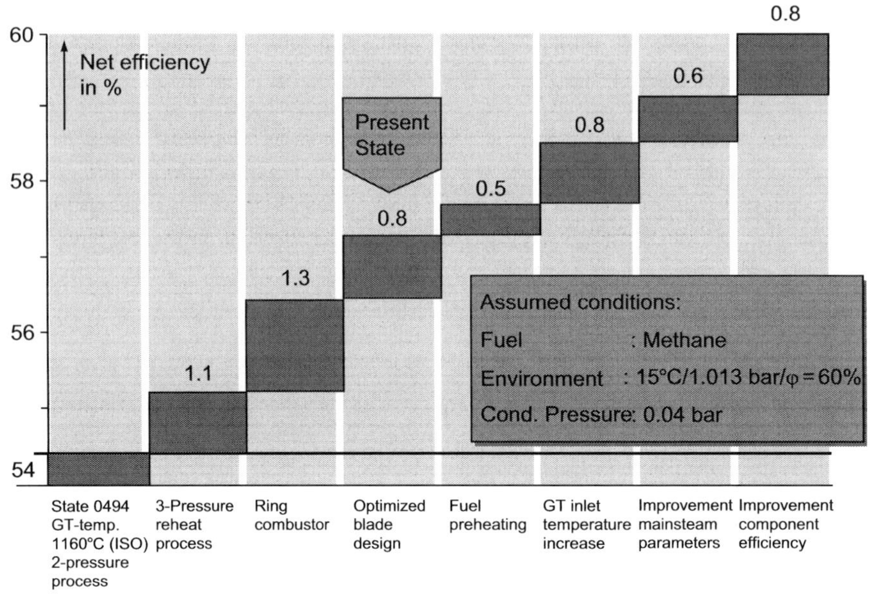

FIGURE 7–24 Improvement potential of combined-cycle efficiency. *(Source: Siemens.)*

balance of plant; electrical systems; instrumentation and control systems; and the boiler island). Up to approximately 25–35% of the costs of CC plants can be provided locally in countries such as Argentina, Brazil, China, India, Malaysia, the Philippines, South Africa, and Thailand.

These include some of the engineering services (project engineering/project management as well as plant erection/commissioning) as well as some or all of the following lots: civil works, balance of plant, electrical systems, and the boiler island.

Upgrading Fossil-Fueled Power Plant

The preceding descriptions of technology developments have been confined to "Greenfield" plant installations. However, in many cases the power plant operator looks to improve plant performance beyond what can be achieved through routine maintenance; in particular, the operator desires higher output, additional heat extraction, improved pollution control, and improved plant control. A common approach to this problem is to incorporate a gas turbine into an existing fossil-fired steam generation

TABLE 7–1 Cost Breakdown for CC Power Plants

Integrated Services	4%	Project Management/Subcontracting
	2%	Plant and Project Engineering/Software
	8%	Plant Erection/Commissions/Training
	1%	Transport, Insurance
	Σ15%	
Lots	15%	Civil Works
	32%	Gas- and Steam Turbine Set
	16%	Balance of Plants
	7%	Electrical Systems
	4%	Instrumental and Control
	11%	Boiler Island
	Σ85%	

Basis: 350/700 MW CC Plant with a V94. 3A Gas Turbine
(Source: Siemens.)

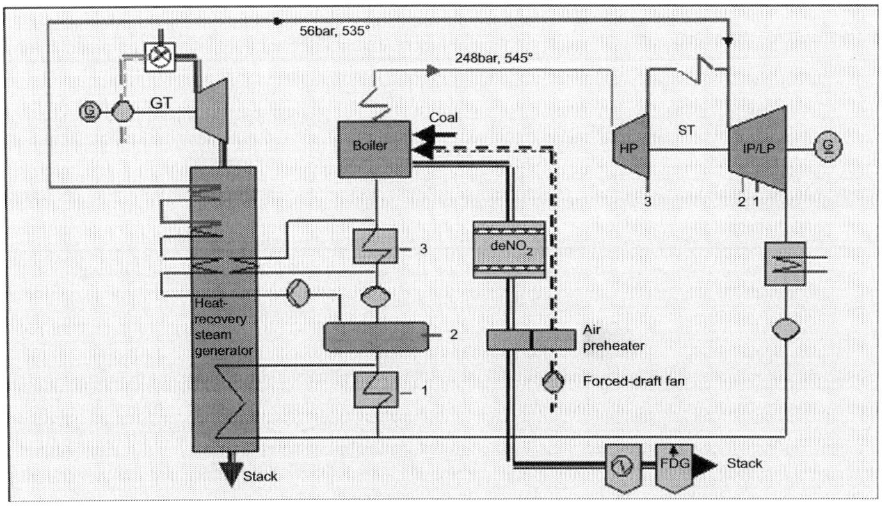

FIGURE 7–25 Schematic diagram of a parallel-powered combined-cycle block with full flue gas cleaning. *(Source: Siemens.)*

system. The most common configuration (Figure 7–25) is called parallel powering, where the gas turbine exhausts are used in the existing steam cycle. This is achieved by feeding the exhausts into a heat-recovery steam generator (HRSG), which provides additional steam to the existing steam turbine. Typically, parallel powering requires the addition of a gas turbine, associated electrical and instrumentation, and control equipment, civil engineering, HRSG, additional piping and pumps as well upgrading the steam turbine. Generally, parallel powering can be undertaken fairly separately from the existing part of the plant, with a final integration phase and a plant downtime of 1.5–2 months. The typical cost range is $300–500/kW.

Physical Properties that Affect Fuel Selection and System Design

Gas* fuels are more suitable than liquid fuels for continuous base-load operation. Gas turbine combustion system design reflects this fact. Burning natural gas in a premix mode is the most efficient fossil fuel technology with the lowest emissions.

A wide range of other gaseous fuels can also be used in gas turbines. Table 7–2 shows other gaseous fuel types and their impact on gas turbine design.

Fuels Rich in Hydrogen

Hydrogen (H_2) characteristics differ substantially from those of natural gas. H_2 exhibits low ignition energy, low molecular weight and density, and high flame temperature and propagation speed. Since H_2 has the ability to diffuse through and embrittle materials, the gas turbine itself and ancillary systems must allow for this fact.

For H_2-rich fuels, the combustors are particularly susceptible to premature ignition. Their design must avoid flashback by accelerating the fuel flow through burner passages while avoiding nozzle erosion caused by H_2's high-speed flame. The diffusion combustion must use substantial amounts of steam cooling diluents for NO_x reduction because of the high flame temperature. Therefore, the combustion products have a high moisture content and a heightened heat transfer capability, which affect the life of hot-gas-path components. So the use of gaseous fuel with high H_2 content (above 50% volume) is currently limited to firing temperature typically found in older E-class gas turbine technology.

However, several combustion testing facilities are working on studying the behavior of full-scale combustor operation for F-class technology at firing temperatures and at actual operating pressures and flows. In several IGCC plants, fuel with more moderate H_2 content (between 20 and 30%) is combusted in diffusion mode at F-class firing temperature. Development of dry low NO_x combustors operating with higher than 10% H_2 content (the present maximum allowable concentration) will be a major breakthrough.

From both the end users' and the EPC contractors' perspective, the use of H_2-rich fuels presents many challenges, such as providing a different startup fuel capability, ensuring adequate purging systems to cope with much larger fuel handling systems volume (low BTU value), and dealing with stringent safety issues for leakage detection, explosion-proofing, ventilation, and so on. Special consideration should be given to materials of construction for pipes and valves as well as welded connections and sealing materials.

* Source: [7-1].

TABLE 7–2 Other Gaseous Fuels [7-1]

Fuel Types	Leading Property	Lower Heating Value (BTU/scf)	Impact on Fuel System Designs
Natural gas	Basic fuel	800–1200	Basic design
Process gases	Low heating value (large volume flow)		
		100–150	Increased compressor flow
Syngas		200–400	Configuration of fuel supply system combustors
Low BTU natural gas		400–800	
Liquefied petroleum gases Other gaseous fuels	High heating value (low volume flow)	2300–3200	Configuration of fuel supply systems, combustors. Special constraints for startup and part-load operation
Process gases Syngas	High hydrogen content	300–1000	High flame velocity, special material, safety issues
Gases with heavy hydrocarbon content	High dew point		Formation of droplets, erosion, nonhomogeneous composition, variable heat flow

One EPC contractor's (Bechtel) experience with medium H_2 content fuel is at Tampa Electric Company's Polk Power Station Unit 1, an IGCC plant with a syngas fuel content of approximately 37% H_2 by volume at a lower heating value (LHV) of 200 BTU/ft^3 (20% of the natural gas LHV).

Liquefied Natural Gas

To supply natural gas from the extraction fields to end users over large distances or oceans, the most common method is to liquefy the natural gas by refrigeration at −160°C (256°F) and then transport it in spherical tankers. At the receiving terminal, the liquefied natural gas (LNG) is unloaded and stored in insulated tanks. From the storage tank, the LNG is pumped at the desired pressure and vaporized for use in gas turbines. The combustion of vaporized LNG in a gas turbine is not much different than that of regular pipeline-quality natural gas in gaseous form. The use of LNG can be noted in many Asia-Pacific countries, where local natural gas resources are scarce.

It is expensive and therefore is used in countries such as Japan, which also has higher emissions standards than those in the rest of SE Asia.

Based on at least nine major projects, one major EPC contractor's experience shows that integration of the LNG unloading, storage, and regasification facilities with the associated combined-cycle power plant yields substantial advantages to the owner. Sizable cost savings can be achieved by developing common auxiliary systems for cooling water and fire protection, by optimizing the site layout, and by integrating the unloading pier with power plant water intake and discharge structures.

A special feature of integration of the two facilities (LNG terminal and power plant) is the recovery of LNG cold energy from regasifying the LNG. In the past, the regasification was done by heating with seawater using open rack vaporizers without the heat. For the integrated system, one patented process uses a glycol-water mixture as the working fluid, which is cooled to 1°C in a closed-loop LNG vaporizer heat exchanger. The cold glycol water is then sent to the gas turbine to cool the gas turbine inlet air to 7°C. A fin-tube-type air-cooling coil is located downstream of the air filters. The return glycol water is pumped back to the LNG terminal, where it passes through a seawater heat exchanger to add supplemental heat as needed. Following the return of the warm glycol-water working fluid, the regasification process starts over again. Figure 7–26 illustrates the benefits of inlet cooling.

Liquefied Petroleum Gas

Liquefied petroleum gas (LPG) typically consists of butane and/or propane. LPG is a natural derivative of both natural gas and crude oil. Since LPG is composed mainly of heavier hydrocarbons, the gas turbine combustion system operates quite differently than for LNG. As the price for butane and propane varies seasonally, the receiving terminal should be designed to receive LPG in any combination of the two hydrocarbons, from 100% propane to 100% butane.

LPG is not widely used as a fuel for gas turbines because of its limited availability and the complexity of the

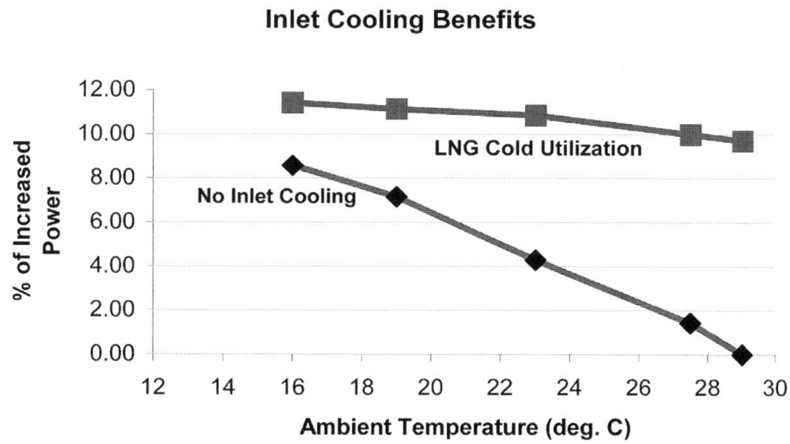

FIGURE 7–26 Benefits of inlet cooling for integrated LNG [7-1].

fuel handling equipment. In terms of phase change properties at atmospheric pressure, propane is liquid at −42°C (44°F) and butane at 0°C (32°F). For use in gas turbines, the LPG is supplied at 90 bar (1300 psi); at this pressure, it can easily change phase between liquid and gas. One result of LPG's volatility is that gas turbines can operate in only two modes, either "on" at full load or "off" at full speed, no load. With LPG supplied as 100% liquid at a lower pressure and with careful temperature control, gas turbine operation across the entire range of loads is possible. The limited market for LPG has kept manufacturers from developing dry low NO_x combustion systems for this fuel. The use of conventional diffusion burners necessitates the use of diluents to meet emissions limits, which affect the equipment life and increase the cost of the water treatment plant.

LPG vaporization can be also used to cool the air inlet to the gas turbine in the same way as LNG cold utilization. The economics of the inlet chilling systems, however, are quite different due to the diverse cryogenic properties of pentane and butane.

Today dimethyl-ether (DME) manufacture is commercialized. DME, which is interchangeable with LPG, can be manufactured from multifeedstocks such as coal, natural gas, and biomass in the same way as methanol. Table 7–3 lists properties of DME in comparison with butane and propane.

This multipurpose fuel is very attractive for developing countries (China, India) in permanent need of portable (bottled) or easily transportable fuels. OEMs (GE in particular) are looking at commercial offerings of DME-fired E- or F-class heavy-duty turbines. In Japan, the DME forum (JDF), formed in 2000, with 37 Japanese companies and two universities as members, works to ensure large-scale DME technology development. With low emissions and good thermal efficiency, DME could play an important role in specific power generation markets.

Fire detection and protection systems must be designed to allow for the fact that LPG and DME are heavier than air. Further, the fuel treatment system requires (1) fuel preheating to ensure that the fuel temperature is maintained above its dew point at all times and conditions and (2) proper handling and storage of the fuel.

Syngas and Other Low-Btu Gases

Gasification of coal and other low-grade fuels to produce syngas enables more efficient use of these fuels through IGCC technology. Over 40 projects worldwide (existing or in development) use heavy oil, petroleum coke, coal, biomass, and waste materials as fuel. The capital cost continues to drop as a result of technological advances and incorporation of lessons learned from existing facilities.

TABLE 7–3 Selected Dimethyl-Ether Properties [7-1]

Property	DME	Propane	Butane
Boiling point (°C)	−24.9	−42.1	−0.5
Vapor pressure at 20°C (bar)	5.1	8.4	2.1
Liquid density at 20°C (kg/m^3)	668	501	610
Specific density (gas)	1.59	1.52	2.01
Low heating value (MJ/kg)	28.4	46.4	45.7
Autoignition temperature at 1 bar (°C)	235–350	470	365
Explosion/flammability limit in air (volume %)	3.4–17	2.1–9.4	1.9–8.4

TABLE 7–4 Some Low BTU Gases [7-1]

Gas Source	H_2 Content	LHV (BTU/scf)	Remarks
Oxygen-blown gasification	Above 30% volume	200–400	Steam injection for NO_x control
Air-blown gasification	8–20% volume	100–150	Very large fuel systems
Blast furnace		Less than 100	Must be mixed with other fuels, usually natural gas
Coke oven	Very high value above 40% volume		Requires "scrubbing" before use in a gas turbine

Under the generic name of low-BTU gases, these fuels can vary significantly from one application to another and are highly dependent on the type of feedstock and gasification process. Table 7–4 summarizes some of the low-BTU gases based on hydrogen content and the range of LHV.

Conventional gas turbine combustion systems and associated fuel pretreatment apparatus cannot be used unmodified for these fuels. Since each fuel has different properties, analytical (mainly computational fluid dynamics; CFD) and experimental work is conducted before field implementation. Low-BTU gases on the combustion system necessitate full characterization of the combustor performance in test facilities. However, there is no substitute for actual field operational experience with each type of fuel. As a result, extended startup time is required to fine-tune the gas turbine fuel supply system. Table 7–4 illustrates the wide variety of low-BTU gases used in gas turbines.

The Wobbe Index (WI), a measurement of volumetric energy (calculated using the LHV and specific gravity of the fuel), indicates a combustor's capability to burn different types of fuels using a unique fuel supply system. Usually, the maximum allowable WI variability is around ±5% of its value. Beyond this limit, multiple gases can be combusted in the same gas turbine only if supplied by independent fuel trains with different control valves, manifolds, and fuel nozzles, which add complexity and cost. Table 7–5 depicts the composition of low-BTU gases in comparison with natural gas. The LHV and WI are calculated as a percent of their absolute values for natural gas.

As shown in the Table 7–5, the LHV of syngas is lower than that of natural gas, requiring a considerable fuel mass flow increase to match the necessary heat input. When fully fired on typical syngas compositions, the gas turbine produces up to 25% more power than a conventional machine operating on natural gas, due to this increased mass flow.

In addition to the complexity of the combustion process related to the fuel composition, a major concern for gas turbines firing syngas is the significant amount of contaminants usually found in this type of fuel.

Deposits and Degradation

Certain deposits, especially from trace metals, affect gas turbine integrity and performance. In one particular application, Ni deposits concentrated on the tip of the burner forced a redesign of the burner. In many cases, the deposits plug the blades' cooling holes and cause overheating as well as cracks in the material. Special coatings on the blades often are used to minimize the corrosion.

TABLE 7–5 Typical Syngas Fuel Composition [7-1]

Syngases	H_2	CO	H_2O	N_2	CO_2	CH_4	C_{2+}	LHV as % of Natural Gas LHV	Wobbe as % of Natural Gas WI
1	31.25	29.78	34.59	0.98	3.27	0.13	0	22.62	16.96
2	9.03	16.25	5.46	40.56	13.61	14.6	0.53	16.39	17.06
3	9.56	9.61	29.55	22.42	7.48	20.73	0.54	22.91	21.62
4	0	0	32.6	11.72	8.38	44.64	2.65	32.32	37.84
5	0	0	0	5	10	84.5	0.5	76.92	73.86
Natural gas	0	0	0	1.2	0.2	94.8	3.8	100	100

Performance Degradation

Another negative effect of deposits is the accelerated thermal performance degradation. Limits on the allowable degradation, recoverable and nonrecoverable from base load conditions can be provided by the OEMs. Operating below these limits will increase the risk of compressor surge. In the development phase of a project, the impact of the recoverable and nonrecoverable degradation, scheduled outages, and so on needs to be analyzed.

Global Experience with IGCC

As of 2004, 21 global integrated gasification combined-cycle (IGCC) power plants producing 5810 MW are in operation. GE heavy-duty gas turbines are used in or planned for some of these IGCC projects. These projects include various levels of integration within the gasification plant as well as a variety of fuel compositions. Siemens and Ansaldo (a Siemens licensee) are also active participants in this technology. Several units with silo combustors (V94.2) have been in operation for at least five years. MHI has entered the IGCC arena, using asphalt as the main feedstock and combusting the syngas in a modified M701F gas turbine. Table 7–6 lists gas turbines from various manufacturers currently being employed in IGCC plants.

Liquid Fuels

Light Distillates in Heavy-Duty Gas Turbines

Heavy-duty gas turbines are capable of burning a variety of liquid fuels ranging from light petroleum distillates to heavy residuals. The combustion process of light distillates such as naphtha, kerosene, and No. 2 fuel oil is well proven through millions of hours of operation. The real challenges for gas turbine design and operation come from the use of heavier ash-forming fuels. The combustion system capable must efficiently burn the fuel while maintaining emission limits and hardware integrity.

TABLE 7–6 Gas Turbine Models in IGCC Applications [7-1]

Manufacturer	Gas Turbine Model
Alstom	13E2, 11N2, 8B/C
Ansaldo	V94.2 K
GE	6B, 7EA, 9E, 6FA, 7FA, 9FA
MHI	M701F
Siemens Westinghouse	V94.2, V94.3

Residual Fuel

There is a glut of cheap residual fuel on the market. Countries like China are keen to use it. Even after the cost of transportation via tankers from the Middle East is added, it is still an attractive option for fast developing countries. Power generation is the largest industrial sector in China, which will grow to be the largest power producer in the world. A look at China's environmental standards reveals that the Chinese are tightening their limits, but they still do not address CO_2 greenhouse gas emissions, other than aiming for maximized fuel efficiency. For all other emissions, like NO_x, if OEMs keep to best available technology (BAT), the environmental regulation targets are easily met.

The OEMs' response to their newly industrialized country (NIC) clients is predictable. When presented with a technical challenge, residual fuel or otherwise, they find a way to work with it. So, overall, residual fuel, with its resultant CO_2 emissions that exceed those formed with natural gas combustion, will continue to grow in terms of usage. Not just gas- and oil-poor countries use it. One of Siemens's most successful residual fuel plants is in Mexico. Some OEMs are doing interesting work on CO_2 mitigation generally by postcombustion absorption, reinjection into rock formations (which they expect will hold the gas for some period of time), and similar initiatives. To this point, there has been no real breakthrough on this front.

*Residual or "Bunker" Fuel Combustion**

Traditionally, older E-class technology gas turbines have been used for combusting crude oils, residue oils, and other heavy oils, since their firing temperatures are lower. Because the fuel is cheap, high efficiency is not an issue. However, the combustion of these fuels can prove extremely detrimental to gas turbine life due to corrosion, erosion, and fouling of the blades. Hot corrosion due to trace elements such as sodium (Na), potassium (K), lead (Pb), and particularly vanadium (V) in the fuel has a major impact on gas turbine performance and service life of hot gas path components.

Small amounts of V, naturally found in liquid fuels, create low-temperature-melting (1250°F) vanadium pentoxide (V_2O_5) ash. At typical gas turbine operation temperatures, these V ash deposits are in molten form and accelerate oxidation of turbine blades and vanes. The presence of other alkali metal contaminants in the fuel, Na and K salts, leads to formation of sodium sulfates (Na_2SO_4) (the sulfidation attack) through reaction with sulfur (S) in the fuel, resulting in pitting of the hot path components.

* Source: [7-1].

TABLE 7—7 Limits of Major Contaminants in Liquid Fuels [7-1]

Element	Units	Liquid Fuel Limits	Remarks
Vanadium (V)	ppm (wt)	Less than 0.5	
Lead (Pb)	ppm (wt)	Less than 1.0	No violation allowed
Sodium (Na) and Potassium (Ka)	ppm (wt)	Less than 0.5	For coastal areas, even less. Also function of other additives
Ash	ppm (wt)	Less than 100	Additives increase the amount of ash
Water	% volume	Less than 1.0	Corrosion of fuel system components

Fuel Specifications

Crude composition varies widely in terms of the amount of trace metal contaminants, ash, sulfur, and wax as well as physical properties such as viscosity, gravity, and distillation range.

OEMs have specifications for allowable levels of these contaminants in the fuel as well as the appropriate test methodology to determine them. Their concentrations might be corrected based on the actual LHV of the fuel.

Typical values are presented in Table 7—7.

For crude fuels with a low flash point, the fuel system must meet explosion-proof design specifications.

Fuel Treatment Summary

Prior to combustion in the gas turbine, the fuel has to be pretreated to reduce the concentration of contaminants to meet the stringent requirements set in Table 7—7. The pretreatment process involves chemical reactions and mechanical cleaning systems.

Additives for Residual Fuel

Na and K salts are soluble in water and can be removed by on-site fuel washing. Other methods of fuel treatment include the use of high-temperature corrosion additives. Effective additives are tailored to a specific fuel composition. The most common additive is magnesium (Mg). Additives containing Mg are used primarily to control vanadium oxidation by increasing the ash melting point, allowing the ash to remain in a solid state on gas turbine blades and vanes. The melting temperature of the new ash compound, which contains Mg, V, and oxygen (O), is much higher (around 1238°C or 2260°F) and dictates the maximum allowable firing temperature of the gas turbine. Other additives, based on chromium (Cr) and silicon (Si), are used mainly to combat Na- and K-based contaminants.

Mg-based additives are generically of three types: water soluble, oil soluble, and oil dispersible. The water-soluble type is the cheapest; however, it requires extensive mechanical handling. Oil-soluble additives are the easiest to handle, since they are self-dispersing on line and off line in fuel storage tanks. However, they are five times more expensive than the water-based additive. Magnesium sulfate ($MgSO_4$) is the most common water-soluble product using a 1:1 ratio for Mg:V. An example of oil-soluble product is KI 150 from Petrolite using a 3:1 ratio for Mg:V.

Mechanical System Functions

To reduce the amount of expensive additives, a preliminary step for improving fuel quality entails use of commercially available mechanical fuel treatment skids composed of centrifugal separators, pumps, and sludge tanks. The devices can reduce the concentration of Na, K, S, and V. A typical two-stage system can reduce the Na content from 150 ppm to 1 ppm for heavy salts. An alternative is the use of an electrostatic desalting system, where the water is coalesced to large droplets to enhance settling. However, electrostatic systems require longer startup time and more filtration than a centrifugal system.

Operation and Maintenance Issues with Residual and Other Heavy Fuels

Many unique requirements are associated with the use of heavy oils as the main or backup fuel. Contaminants, formation of ash, and low flammability need to be considered.

Startup and Shutdown

To ensure that the entire gas turbine fuel system is free from waxy fuel buildup or plugging, startup and shutdown must be done using No. 2 fuel oil. Transfer to crude can occur after the unit has been synchronized. The process should be carried out slowly to ensure heat-up of the piping, since crude needs to be preheated to 275°F to ensure that no frozen blocks are formed. On shutdown, No. 2 fuel oil is required to purge the system from any residues of crude.

Normal Operation

During normal operation, it is necessary to remove the ash formation on first-stage nozzles. The rate of deposits depends on type of fuel and ambient temperature. Because the amount of air moving through the turbine is larger at low ambient temperatures, the allowable plugging should not exceed 5–7%. At higher ambient temperatures, such as in the summer, the allowable plugging can exceed 10–12%. To reduce the amount of off-line water wash, some on-line wash could be done using an on-line turbine abrasive system and injecting particles at the first-stage nozzle.

Off-Line Cleaning

This process takes 10–12 hours from shutdown to start and comprises the steps shown in Table 7–8. Table 7–9 describes a gas turbine typical cleaning cycle. The concentration of contaminants determines the actual cleaning schedule.

Inspection Intervals

Inspection intervals are specific to each installation, determined by the condition of the hot-gas-path parts. In addition to contaminant corrosion, fuel containing heavier hydrocarbons has a higher H_2 content, which releases a higher amount of thermal energy during combustion. Higher combustion temperature leads to poorer hardware condition and shorter intervals between inspections. The number of equivalent operating hours (EOH), on which maintenance intervals are based, increases more rapidly based on fuel contaminant concentration. This can mean a factor of 1.5 for each operating hour.

A typical expected maintenance schedule follows in Table 7–10.

TABLE 7–8 Off-Line Cleaning Process [7-1]

Procedure	Time	Remarks
Shutdown	9 hr.	Wheel space temperature to reach 300°F
Off-line compressor wash	5 min.	Injected in the compressor, gas turbine in turning gear
Soaking	20 min.	Unit is shut down
Water injection in the hot path	20 min.	Via atomizing air piping for all combustors, gas turbine in turning gear

TABLE 7–9 Gas Turbine Typical Cleaning Cycle [7-1]

Hours of Operation	Power Output (%)	Activity	Remarks
0	100	Start of operation	
150	87	On-line abrasive cleaning	Restored power to 92%
200	85	Unit shut down for off-line cleaning	Restored power to 96%
300	82	On-line abrasive cleaning	Restored power to 90%
350	85	Unit shut down for off-line cleaning	Restored power to 92%

TABLE 7–10 Typical Maintenance Schedule [7-5]

Inspection	Hours	Starts
Combustion system	3000	400
Hot gas path	6000–12,000	1200
Major	18,000–24,000	2400

According to GE, several older gas turbines (7EA, Frame 5, and Frame 6) have accumulated several millions of hours of operation using "light Arabian crude" with V concentration of 10 to 30 ppm. The fuel is supplied directly from the refinery and then washed on site. An Mg additive is added at a ratio of 3:1 Mg:V. The units start and stop on crude. GE reports that after 1 million hours of operation, no significant hot path damage was found. Note that the older machines operate at much lower firing temperature (949°C or 1740°F) than modern gas turbines with a firing temperature of 1343°C or 2450°F or higher.

*Success with Nonconventional Fuels**

Siemens' experience in this arena demonstrates additional success with nonconventional fuel. Table 7–11 provides a sample of naphtha- and heavy oil-fired power plants in operation and in the planning stage. As this figure shows, some plants (e.g., Kot Addu and Valladolid) accumulated 30–60,000 hrs of successful operation over five years or more.

As a result of the impacts of deregulation, environmental standards as well as usage as secondary fuels, up to 25% of the gas turbine output (in MWs) in the next 10 years could involve non-natural gas fuels. Naphtha is

* Source: [7-2].

TABLE 7—11 Operational Experience with Naphtha and Heavy-oil-fired CC Plants [7-1]

"Naphtha Class" projects*:					
Plant Name	*Location*	*Fuel*	*Gas Turbine Model*	*Rating*	*Operating Hours*
Paguthan	India	Naphtha, Distillate, NG	3 × V94.2 1 GUD 3.94.2	630 MW	19,000 hours
Santa Rita	Philippines	Naphtha, Condensate, Distillate, NG	4 × V84.3A 4 GUD 1S.84.3A	950 MW	Erection phase
Faridabad	India	NG, Naphtha, HSD	2 × V94.2 1 GUD 2.94.2	440 MW	Engineering phase
Latest plants with ash-forming fuels:**					
Plant Name	*Location*	*Fuel*	*Gas Turbine Model*	*Rating*	*Operating Hours*
Valladolid	Mexico	Residual oil	2 × V84.2 1 GUD 2.84.2	220 MW	24,000 hrs
Kot Addu	Pakistan	Furnace oil (Heavy oil)	4 × V94.2 2 GUD 2.94.2	820 MW	60,000 hrs
Rousch	Pakistan	Heavy oil	2 × V94.2 1 GUD 2.94.2	390 MW	Commissioning phase

**Note: 9 Gas Turbines ordered for fuels of "Naphtha Class". "Naphtha Class" means light, low boiling liquid fuels.*
***Note: 34 Gas Turbines delivered for ash-formlng fuels In total accumulated operating hours approx. 140,000 hours.*

expected to be the dominant fuel in this category with a few thousand MW of new capacity being added every year.

Technology, Performance, and Design Trends

Design and operation of these plants requires more attention than natural-gas-fired plants particularly in relation to fuel variables such as calorific content, density, composition, concentration of contaminants, and emissions, as well as different burning behaviors (e.g., ignitability, flame velocity and stability).

To overcome these difficult fuel properties, technological adaptation, additional equipment, and operational requirements are necessary. These include GT-layout (compressor, turbine) for the changed mass flows, different burner technology (burner design, burner nozzles), additional startup/shutdown fuel system, and safety measures. Performance, availability and operation and maintenance (O&M) expenses can be affected. To illustrate this, Table 7—12 shows important nonstandard fuels and their effect on a standard fuel system.

A Siemens example of gas turbine combined cycle plant burning a nonconventional fuel is the 220 MW Valladolid plant in Mexico. This plant has been operating since 1994, burning heavily contaminated fuel oil, containing 4.2% sodium and up to 300 ppm vanadium. Associated with this fuel are its impurities (sodium, potassium, and vanadium), which tend to form ash particles in the combustion process, form deposits, and corrode the gas turbine blades. In the case of the Valladolid plant, "Epsom salts," consisting mainly of magnesium sulfate ($MgSO_4$ $7H_2O$), was dissolved in water injected into the gas turbine combustor through special orifices. This converts the vanadium into a stable water-soluble product (magnesium vanadates). While this is deposited downstream of the combustor on the gas turbine blades, it causes only minor blade corrosion and has to be removed through blade washing. At Valladolid, it was determined that washing every 150 hours was necessary to keep the aerodynamic performance of the blades, the plant efficiency, and reliability near to design levels. Good manhole access was a critical success factor for this project as servicing and maintenance during turbine washing shutdowns are simplified. It is planned to convert the Valladolid plant to natural gas operation in the near future.

OEM Design Requirements as They Relate to Integrated Gasification Combined-Cycle Plants

IGCC plants consist of three main sections: the "gas island" for conversion of coal and refinery residues (such as heavy fuel oil, vacuum residues, or petroleum coke) including gasification and downstream gas purification (removal of sulfur and heavy metal compounds in accord with required emissions levels), the air separation unit, and the combined-cycle plant. The modular design (gas generation, gas turbine system, HRSG, and the steam turbine system) offers the possibility of phased construction as well as retrofitting of CC plants with a gasification plant, thus replacing the "standard" gas turbine fuels (natural gas or fuel oil) by syngas produced from coal or refinery residues.

In principle, IGCC is a combination of two mature technologies; however, proper integration is the key to the

TABLE 7—12 Gas Turbines for Nonstandard Fuels, Critical Fuel Properties [7-2]

Critical Fuel Properties	Fuels	Effect on Standard Fuel System
• Low viscosity (reduced lubricity) • Low density (high volume flow)	Naphtha, Kerosene, Condensates Naphtha, Condensates	Effect on fuel supply system (e.g., pump design) Limits for fuel supply system and bumer nozzles
• Low flash point • Low boiling point (high vapor pressure)	Naphtha, Kerosene, Liquefied Petroleum Gases (LPG), High Speed Diesel (HSD), Condensates	Increased explosion protection effort Increased ventilation effort
• Low auto ignition point	Naphtha, Condensates	Effect on premix capability. Increased explosion protection effort. Increased ventilation effort
• Contaminants (high temperature corrosion, ash deposits)	Contaminated fuel oils, Crude oils, Heavy oils, Heavy residues	Effect on GT blading and hot gas path. Counter-measures: – Temperature reduction (reduced performance) – Fuel treatment (washing, inhibitor dosing)
• Low heating value (high volume flow)	Process and synthesis gases (low calorific gases), Low BTV natural gas	Effect on layout of GT compressor burner nozzles, fuel supply system
• High heating value (low volume flow)	LPG, gaseous and liquid	Effect on layout of burners, fuel supply system. Limited startup and part-load capability
• High H_2 content	Process and synthesis gases (low calorific gases)	High flame velocity. Effect on premix capability
• High dewpoint	Gases with high boiling components	Droplets causes erosion and nonconstant heat flow

success of the IGCC projects from lessons learned from several demonstration projects in Europe and the United States. Although more than 350 gasifiers are operating commercially worldwide and at least seven technology suppliers, and orders of more than 100 CC units per year are being handled by the major manufacturers, there is only limited experience of commercial operation as an integrated IGCC plant. IGCC can be expected to be a field of commercial application in developed countries, such as Italy, for residual refinery fuels and gasified coal.

Through 2015, the potential for refinery-based integrated coal gasification combined-cycle plants is estimated to be 135 GW. Currently over 6 GW of coal- and refinery-residue-based IGCC projects are either under construction or planned. Figure 7—27 shows some of the IGCC plants planned or under construction.

Technology, Performance, Environment, and Demand Trends

Figure 7—28 compares the supply flows, emissions, and by-products of different 600 MW-class plants. With reference to environmental emissions (including GHG), it may be seen that IGCC plants compare favorably with pulverized-coal-fired steam power plants.

Depending on the degree of integration between the gas turbine and the air separation unit (ASU), either standard gas turbine/compressor configurations can be applied, or limited modifications are required, to compensate for the mismatch between turbine and compressor mass flows, which results from the application of gases with low heating values. Three principal options are available. The selection of the appropriate air and nitrogen integration concept depends on a number of factors to be considered on a case-by-case basis. A summary of the important criteria is provided in Figure 7—29.

Figure 7—30 sets out the principal criteria for selection of the different IGCC integration concepts. The denominated "fully integrated approach," which was selected for the European coal-based demonstration plants, resulted in the highest efficiency potential, but it is to be considered as highly sophisticated from the operational point of view. Nevertheless, after some initial operational problems, the Buggenum IGCC facility has demonstrated that this type of

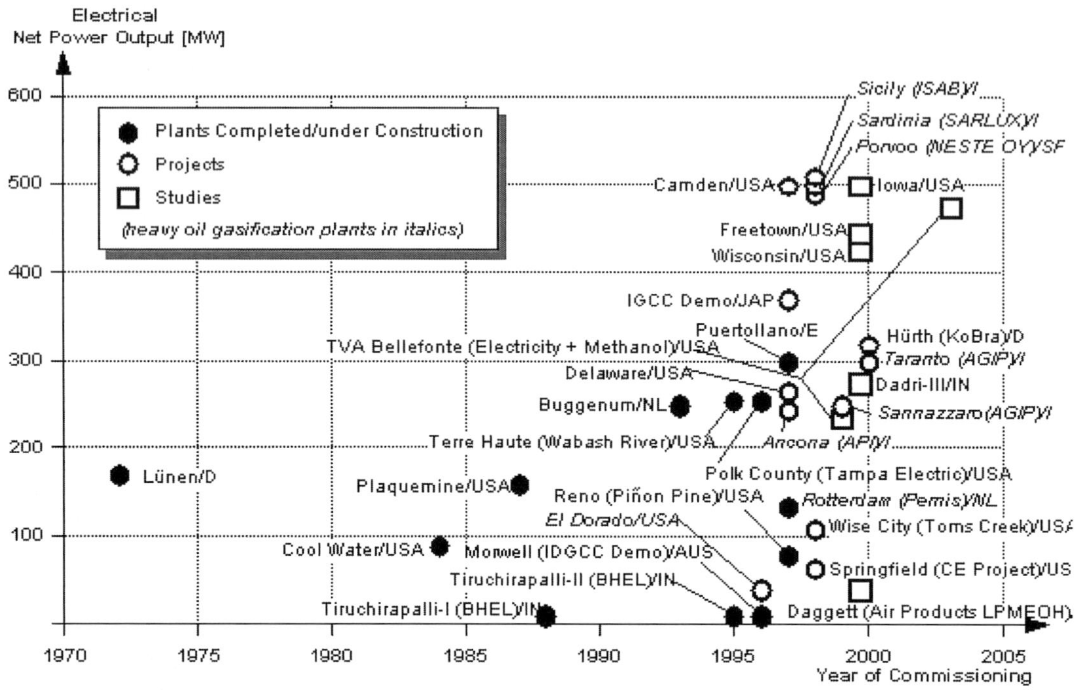

FIGURE 7–27 Combined-cycle power plants with gasification, worldwide activities [7-2].

plant can be operated successfully and with the required availability.

The nonintegrated concept with a completely independent ASU is expected to have advantages concerning simplification of plant operation and possibly in achievable availability. On the other hand, the loss in overall IGCC net plant efficiency compared with the fully integrated concept amounts to 1.5–2.5%. This concept therefore is primarily of interest for application in cases where efficiency is not the key factor (e.g., for the gasification of refinery residues).

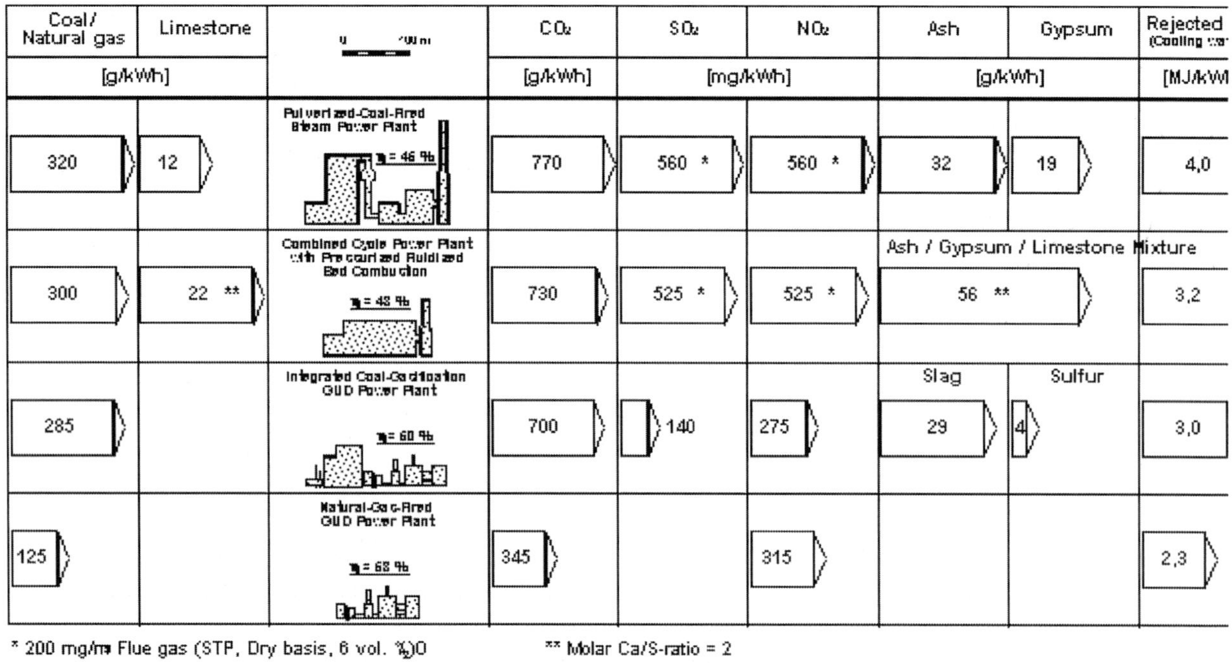

FIGURE 7–28 Comparison of supply flows, emissions, and by-products of different 600 MW-class power plants [7-2].

Main Criteria for Selection of the IGCC Integration Concept

Integration options

(1) Non-integrated (independent) ASU
(2) Partially integrated ASU
(3) Fully integrated ASU

Criteria for Selection of the integration Concept

- Gasification process and waste heat recovery (syngas cooler, quench, cooler/saturator cycle)
- ASU process
- Fuel gas analysis (with or without nitrogen return)
- Limits for NO_x emissions
- Overall plant efficiency
- Investment costs
- Operational aspects
- Site-specific aspects
- Necessary modifications of standard gas turbines
- Available fuel flow

IGCC Technology

FIGURE 7–29 Main criteria for selection of the IGCC integration concept [7-2].

The concept with partial air-side integration is an interesting compromise solution, with an only moderate loss in efficiency but improved plant flexibility compared with the fully integrated concept.

In summary, this OEM notes that

1. Industrial gas turbines, fired by natural gas, are a well-established technology. Ongoing development and the short-term introduction of the most advanced products improve GT efficiency to around 40%, with the largest GT capacities being in the range of 250–350 MW. This, in turn, improves the advanced CC efficiency to 60%, with corresponding plant capacities being in the range of 375–500 MW.
2. Using low cost salts, such as Epsom, as inhibitors, the GT components of CC plants can be successfully fired with refinery residues with high sodium and vanadium content. However, great care is needed in handling such difficult fuels.
3. IGCC can be expected to become a field of commercial application for residual refinery fuels (particularly in developed countries such as Italy) and gasified coal. However, great care needs to be taken in implementing an IGCC commercialization strategy for developing countries. Particular attention needs to paid to addressing integration issues.

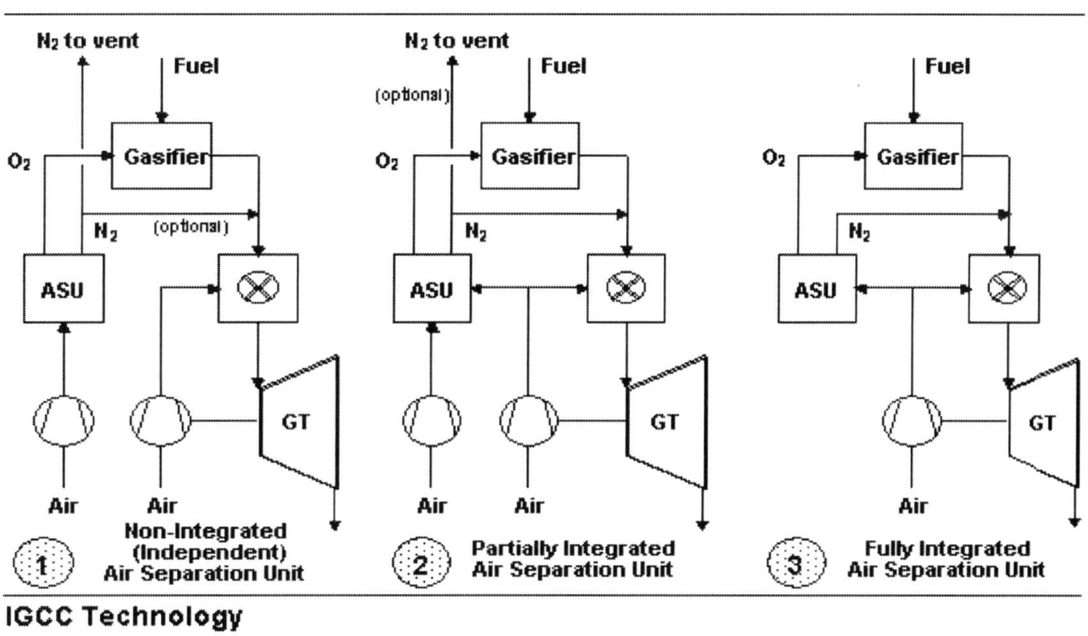

FIGURE 7–30 Integration options for IGCC power plants [7-2].

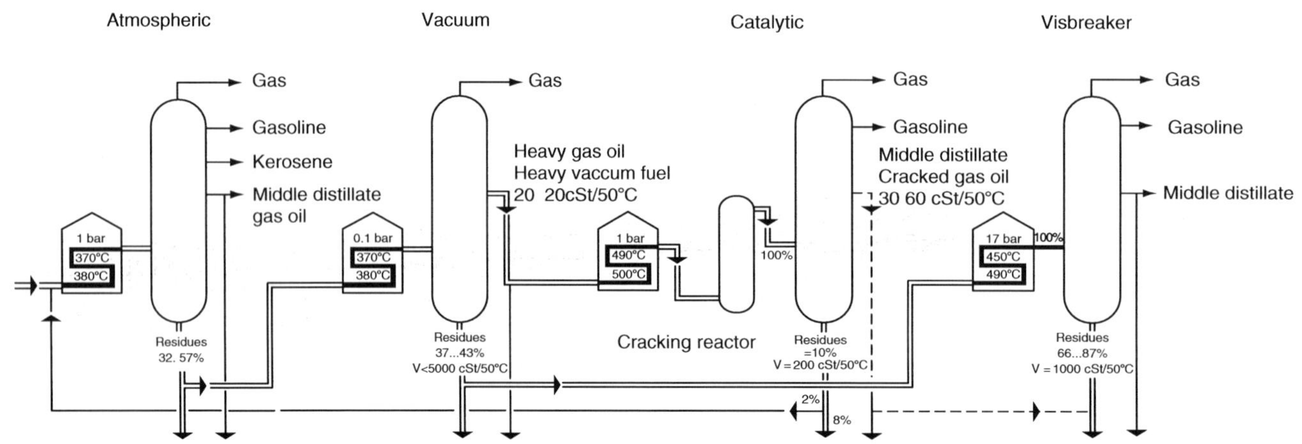

FIGURE 7—31 Refinery process for fuel production. *(Source: Westfalia.)*

FUEL TREATMENT HARDWARE*

Fuel Properties that Affect Fuel Treatment

Liquid Fuel

Liquid petroleum fuels fall into two basic classifications: true distillate and ash-forming oils. True distillate oils are most frequently the lighter oils such as naphtha, kerosene, and Number 2 distillate oil. This class of fuels is clean burning and behaves in a manner similar to natural gas as far as the internal portion of a gas turbine is concerned.

Naphtha, NGL, kerosene, and Light Number 1 distillate oil have little or no lubricity. Special handling equipment will be required in order to prevent damage to the peripheral equipment such as the main fuel pump, fuel nozzle plunger, flow dividers, etc.

Ash-forming oils are normally the heavier oils, such as blends, crude, and residual oils. Since they usually contain trace metal contaminants, fuel treatment is required to remove or modify the effects of harmful constituents prior to combustion in the power plant.

The degree of equipment sophistication, plant equipment investment and operating cost required for this treatment are tied directly to key physical and chemical properties of the fuel. Physically, the parameters of interest include specific gravity, viscosity, flash point, pour point, and wax content. Chemically, the prime interests are the amounts of corrosive trace metals present, the overall fuel washability, and compatibility of one fuel with another. The latter is particularly significant for users who obtain fuel from multiple sources (Figure 7—31).

* Source: [7-3].

Trace Metals

Trace metal contaminants of concern include sodium, potassium, calcium, lead, vanadium, and magnesium. Vanadium, sodium, potassium, and lead are corrosive to turbine components (buckets and nozzles) at elevated temperatures, particularly when present in amounts above specification limits. All of these materials plus calcium can also form hard deposits that are difficult to remove. These deposits can cause plugging of the turbine nozzle, form deposits on the compressor blades, and reduce turbine output.

Table 7—13 lists the trace metals of interest, their effect on the turbines, and the means of treatment to reduce harmful effects along with typical permissible limits. Dealing with these metal contaminants will be discussed in the fuel treatment section of this book.

TABLE 7—13 Trace Metal Effects

Trace Metal	Effects on Turbine	Type of Treatment
Sodium (Na)	High temperature corrosion	Fuel washing
Potassium (K)		
Calcium (Ca)	Fouling deposits	Fuel washing to a limited extent
Lead (Pb)	High temperature corrosion	None, limit spec.
Vanadium* (V)	High temperature corrosion	Inhibited by magnesium
Magnesium (Mg)	Inhibits vanadium, forms deposits	Used to inhibit vanadium

**Vanadium levels less than 0.5 ppm do not require inhibition. Maximum vanadium levels may be dictated by local codes regarding stack particulate emissions and the user's acceptable costs to inhibit.*
(Source: Westfalia.)

Density

The density of a fuel oil is the ratio of the weight of a given volume of the oil at 15°C to the weight of an equal volume of distilled water at the same temperature.

The closer the density of the fuel is to that of the water, the more difficult it is to treat the fuel. Westfalia's experience has shown that fuel with a high density of 0.991 can easily be processed by centrifugal separation.

Viscosity

The viscosity of a fluid is a measure of its resistance to flow and is expressed in various units. The most common viscosity units are kinematic viscosity in Centistokes (cSt), Saybolt Universal Seconds (SSU), and Saybolt Furol Seconds (SSF).

Westfalia's experience has shown that viscosities of 1000 cSt at 50°C can easily be processed. Greater viscosities can be handled but they are evaluated on an individual basis.

Flash Point

The flash point is the lowest temperature at which application of a flame causes the vapor above a sample to ignite. This value indicates the potential fire hazard associated with handling fuels. Consideration should be given to explosion proofing if the flash point of any fuel that is used in the system is below 60°C.

Pour Point

The pour point is the lowest temperature at which the oil will flow under standard test conditions.

High pour points are characteristic for paraffin-based oils and dictate the need for increased storage temperature, higher capacity suction heaters, heat tracing of lines, as well as certain equipment and recirculation circuits to establish and maintain the operating temperature.

Wax Content

Wax can be present in fuel oil, particularly crude oil and heavy distillates. The percentage of wax and its melting point must be determined and suitable heating and heat tracing must be provided in order to ensure that the wax in the fuel is dissolved at all times.

Stability of Fuels

Future fuels will be of considerably poorer quality in the years to come. The major factor of such oils may be their incompatibility when oils from different refineries are used.

Incompatibility

At power plant installations, sludge of an asphaltic nature may be produced if fuels from different refineries are mixed. When this occurs, the fuels are incompatible. Such incompatibility first becomes evident in the form of excessive amounts of sludge separating out from the fuel. Asphalt sludge may also clog the pipes, heaters, and filters and collect in the tank's bottom. Changing oil temperature has virtually no effect on the formation of sludge. The reason is that the more intensive refining process, particularly the use of the viscosity breakers (visbreakers), produces a higher content of asphaltenes. If a heavy oil of the above kind is mixed with a fuel that has a lower capacity for dissolving asphaltenes, the asphaltenes precipitate out of the fuel. Instability of this nature exists even when the mixed component is of a paraffin-base distillate.

Stability

Fuel oils are composed of many hydrocarbons of which four types are present in the largest volume. These are, in order of increasing instability: paraffin, naphthene, aromatics, and olefin being the most unstable. The stability of these hydrocarbons is governed by their resistance to oxidation or the ability to pick up oxygen and their reaction to heat and/or high temperature.

Sometimes an oil's instability may go undetected. The following items are indicative of instability of fuel oils:

1. Excessive amount of sludge in storage tanks. There will be a normal amount found in most tanks but it should not accumulate rapidly.
2. Large carbon formation in preheaters, due to coking of some excessively heavy carbonaceous materials.
3. Large formation of soot caused by the incomplete combustion of the heavy precipitated insoluble compounds.
4. Thick sludge, waxy and gum-like substances found on screens and strainers.

Classical Heavy Fuel Oil (HFO)

The classical HFO is a mixture of residues from atmospheric and/or vacuum distillation processes and light distillate.

Atmospheric Distillation Desalted and dehydrated crude oil is heated in the distillation tower to 370–380°C and is split up into compounds of differing molecular structure (fractions) in a distillate tower at atmospheric pressure. The temperature is relatively low, otherwise the mineral oil products would begin to crack (decompose).

Fractions with a high hydrogen content are drawn off from the crude oil. The remaining residue contains molecules with a fairly low hydrogen and a fairly high carbon content. The combustion characteristics of this residue are fairly good.

Depending upon the source of the crude oil, the amounts of residue collected are between 32% and 57% relative to the initial quantity of crude oil.

Crude oils from the Middle East are generally of the paraffin base. The residual from this crude burns well.

Vacuum Distillation Vacuum distillation is a follow-up distillation process to the atmospheric distillation. Residual oil from the atmospheric distillation process can be fractionated to a further degree in a vacuum distillation process. The distillation takes place under a vacuum at a pressure of 100 mm of mercury. By means of this process, the residual from the atmospheric distillation can be reduced to 37—43% with a viscosity of 6000 cSt at 50°C. At the same time, a vacuum heavy oil with an approximate viscosity of 20—200 cSt at 50°C is obtained.

The residual fuel from vacuum distillation is extremely viscous (2000—6000 cSt at 50°C) and cannot be further broken down by distillation. It can be processed into bitumen for road building.

At the present time, the vacuum residue is still being blended with inexpensive distillate to bring it to the viscosity of heavy fuel oil. However, it is increasingly becoming the practice to reduce its viscosity in the viscosity breaker (visbreaker) in order to save on distillate fuel.

Another possibility for further processing is to coke the vacuum residue into petroleum coke by splitting off light components.

The other refining method is called cracking. This method is divided into two distinct types: catalytic cracking and thermal cracking (viscosity breaking—visbreaking). The catalytic cracking method is the more modern and efficient method.

Low-Grade Heavy Fuel Oil (HFO)

The low grade HFO will contain residues from the catalytic cracker and visbreaker in addition to the classical HFO.

Catalytic Cracker The cracking processes involved in these cases are used to obtain distillates with a desired boiling point from vacuum heavy oil by molecular cleavage. In this distillation process the existing molecules are segregated into fractions. Large high-viscosity molecules are converted into small fluid low-viscosity hydrocarbon molecules. Light distillates are thus obtained but these distillates have a hydrogen deficiency.

There are many catalytic cracking processes. Heavy vacuum distillate is cracked to a further degree by the use of a powdered aluminum silicate catalyst. This catalyst is heated to approximately 700°C and mixed with the vacuum heavy oil, then preheated to approximately 420°C, before being fed into the reaction chamber. In the reaction chamber, the catalyst causes the molecules of the vacuum heavy oil to split at a process temperature of approximately 500°C and a pressure of 1 bar.

Cracked residual is particularly low in hydrogen and rich in carbon and it contains a large proportion of unsaturated hydrocarbons. This means increased sludge formation and accompanied in some cases by abrasive residues in powder form from the catalyst. The quantities of this catalyst may be as high as 200—300 ppm. Therefore, the fuel must be cleaned (remover of the catalytic fines) before it is used in diesel engines and gas turbines.

From the heavy vacuum fuel about 90% will be converted to middle distillate in a catalytic cracker. This distillate fuel will have a viscosity of 30—50 cSt at 50°C, of which 10% will be residue with a viscosity of 200 cSt at 50°C. From this, 2% will be routed back to the atmospheric distillation tower.

Viscosity Breaking (Visbreaking—Thermal Cracking)

Visbreaking is a simple process in which the very viscous residues from the vacuum distillation or even the atmospheric residues are subjected to thermal treatment at 450—490°C and 17 bars pressure. This results in the formation of low molecular weight, low-viscosity hydrocarbons that considerably reduce the residual viscosity.

Because the obtained residual is occasionally too viscous, it is diluted with cracked gas oil. Cracked gas oil is used as a diluent in order to reduce the tendency to sediment formation (the separating out of asphaltenes). Because of the high asphaltene content, the question of compatibility with other heavy fuel oils is particularly important (formation of excessive sludge in the centrifuge).

From the 100% stock at 450—490°C and 17 bars that enters the visbreaker, 66—87% of it is a residual fuel with a viscosity of up to 1000 cSt at 50°C.

Note: Based on the new distillation processes where sludge and catalytic fines are produced, no electrostatic desalter is recommended as a fuel treatment system for gas turbine fuels.

Fuel Treatment

Liquid Fuel Treatment

In the case of a true light distillate oil, treatment is required primarily to capture dirt and shipment contaminants that have not been removed by settling in the fuel storage tanks. Undetected and untreated, a single shipment of contaminated fuel can cause substantial damage to a power plant. Filters are the most commonly used device although centrifuges and sometimes other desalting methods may also be required to remove sodium if present as a result of seawater contamination.

Treatment and conditioning systems for ash-forming fuels are more complex because of the processing steps required to remove or control their characteristic trace metal contaminants. Particular attention must be paid to sodium, potassium, lead, and vanadium. Complete fuel treatment includes:

1. Washing the fuel to remove the water-soluble trace metals such as sodium, potassium, and certain calcium compounds. Centrifugal washing also removes much of the inorganic particulate material that is normally forwarded to the filtering system. Washing involves the addition of water to the fuel and subsequent removal of the contaminant laden water. Under special circumstances it may be possible to remove the water that is in the raw fuel without adding water and still be able to obtain adequate contaminant removal. This special process is called "fuel purification" and is frequently applied to light crude oils and distillate. Fuel purification is discussed in a separate technical paper.
2. Filtering the fuel to remove solids, oxides, silicates, and other related compounds that are not adequately removed prior to forwarding of the fuel to power plants. These particles can clog fuel pumps, fuel nozzle plungers, flow dividers, and fuel nozzles.
3. Inhibiting the vanadium in the fuel with magnesium compounds in a ratio of three parts of magnesium, by weight, to each part of vanadium. This form of treatment inhibits the corrosive characteristics of vanadium by forming high-melting temperature ash composed of magnesium sulfate, magnesium oxide, and magnesium vanadates.

Lead is also corrosive to power plant components (turbines) and should be avoided. Since there is no economical technique for removing or providing an inhibitor, it is necessary to only purchase fuel with extremely low amounts of lead (less than 1 ppm).

Fuel Washing

Since heavy fuel oils are almost universally contaminated with salts and salt water, they normally require water washing to remove the water-soluble salts of sodium and potassium. Fuel washing involves mixing heated fuel with 5–10% of potable water along with a small amount (0.02%) of an emulsion-breaking fluid to aid in separation. The washing process is a two-stage extraction procedure with the extraction water flowing counter-current to the fuel. Each stage has a mixer to initiate contact of the washwater with the fuel, followed by a mechanism to separate the salt-laden water from the extracted fuel.

The number of extraction stages required depends on the salt level in the fuel to be processed and on certain properties of the fuel such as specific gravity and viscosity. Since the successful operation of the fuel-washing system depends on a working difference between the specific gravities of fuel and that of water, fuels with specific gravities above 0.991 may not be washable unless they are first blended with a compatible liquid that has a lower specific gravity such as distillate fuel.

Chemicals referred to as demulsifiers are added to the fuel to assist in breaking up difficult oil/water emulsions.

Water is added to the fuel oil and is subjected to controlled mixing in order to distribute the sodium-seeking water throughout the raw fuel. The mixture of water and oil flows to a separator that is a mechanical centrifuge. In the centrifuge, the oil and water are separated under centrifugal force, the cleaned oil flows into one circuit and the latent wash water with a high concentration of salts flows out from the centrifuge through another circuit.

The effluent water may require a treatment system to remove the entrained oil to comply with local codes.

For example, a customer specification of "no visible sheen" which is approximately 15 ppm of free oil in the water. Reduction of oil content below these values may require a complicated system.

Elements of the Fuel Treatment System for Ash-Forming Fuels

Figure 7–32 presents a schematic arrangement of a heavy-duty gas turbine power station system developed after years of experience with ash-forming fuels.

The system is configured to provide simplified plant operation. It has the flexibility to incorporate future improvements.

Methods of Separation

Separation by Electrostatic

Figure 7–33 gives a schematic arrangement of a typical two-stage electrostatic system.

The principle behind any electrostatic system is that the oil/water separation process is induced through the application of an electric field. The electric field causes finely dispersed water droplets to coalesce, thereby increasing the droplet diameter. This phenomenon increases settling rates according to Stokes' law, which states that "the velocity of a water droplet falling through oil is proportional to the square of the diameter of the droplet."

There are two types of electrical separators presently used: direct current and alternating current. Direct-current electrostatic systems are devices employed with light refined fuels of low conductivity. Alternating-current electrostatic separators are applicable for heavier fuels having higher conductivity. Electrostatic separators are appealing because there are no high-speed rotating parts within the vessel.

Sludge removal is accomplished by the transport of solids in the effluent water from the vessel. This can be

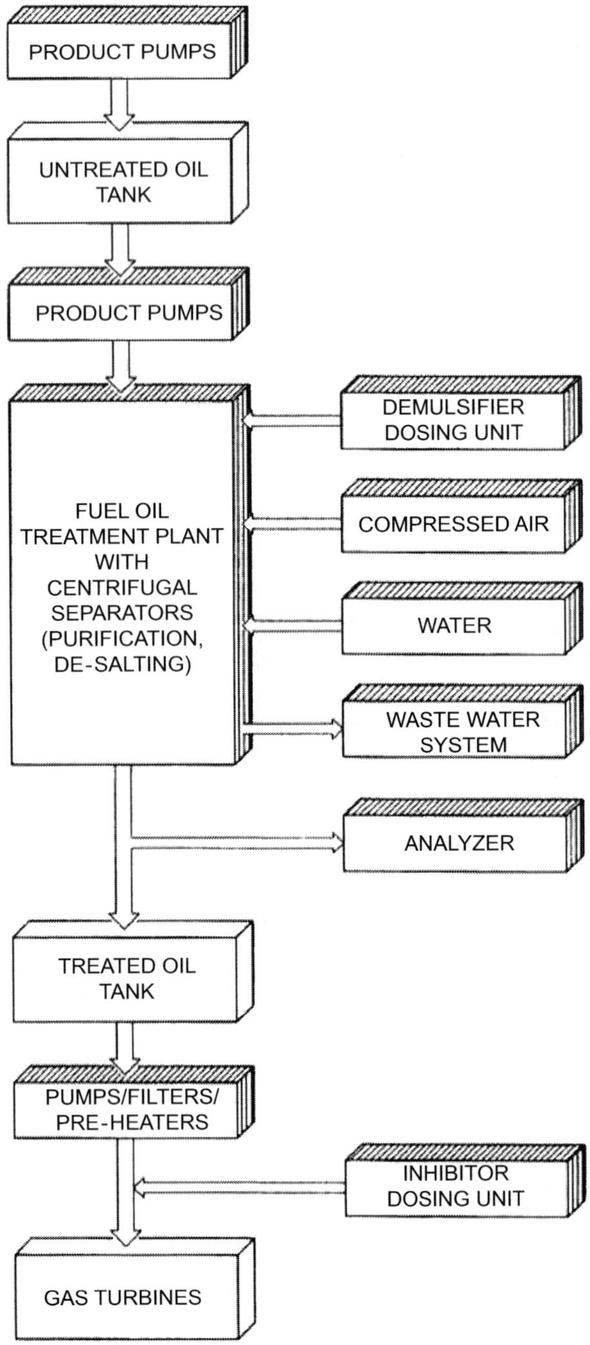

FIGURE 7–32 Fuel oil treatment system for gas turbine power plants. *(Source: Westfalia.)*

accelerated by a hydraulic system within the vessel that causes recirculation of the water.

The key parameter in sizing such a system is the selection of the number of stages. Generally, one can expect 85–90% sodium removal per stage. Three-stage systems are usually considered for initial sodium plus potassium inputs up to 100 ppm and two-stage systems are considered for initial sodium plus potassium inputs of up to 50 ppm. Although system startup times using electrostatic precipitators are somewhat long, usually taking three to four hours per stage, the operator is capable of detecting small changes in operating effectiveness of the system and can take corrective action with the system on line.

Electrostatic washing designed to gas turbine company specifications will remove sodium plus potassium concentrations to less than 1 ppm in fuels having specific gravities of 0.96 or less.

For efficient separation the temperature in electrostatic systems must be raised to 149°C (300°F), this requires a pressure vessel to prevent the water from boiling. This in turn can cause reduction of fuel viscosity and furthermore carbonizes the fuel oil.

Separation by Centrifugal Force

The mechanical separation process of washing fuel is shown schematically in Figure 7–34. This arrangement represents a typical two-stage centrifugal system. The principle behind this centrifugal process is that the oil and water are separated as a result of the centrifugal forces developed in the centrifuge. In contrast to electrostatic precipitators, which rely on increased particle diameters in a high voltage gradient field and an IG gravity field, the centrifuge employs a high gravity field of many thousands of Gs to accomplish the separation without the need of an electrostatic field. A variety of centrifuge models are currently available, having capacity ranges from 1.1–110 m/h.

Centrifugal fuel washing systems have been successfully applied to all gas turbines by various manufacturers using fuels with high density and high viscosity.

With correct system parameters, a typical two-stage centrifugal washing system can reduce sodium plus potassium from 150 ppm to a combined level of 1.0 or 0.5 part per million as specified by the turbine manufacturer.

Centrifuges are also effective in removing inorganic particles and dirt that are forced out with the effluent wash water. This added benefit minimizes the load on the downstream fuel filtration system.

Fuel Purification-Simplified Washing

The purification process is shown schematically in Figure 7–35.

Fuel purification is a one-stage extraction procedure that employs centrifuges to treat distillates and light crude oils without adding water to the fuel. This process relies on the fact that the salts are already dissolved in the entrained water; thus by removing the water the salts are removed as well.

Purification is practical with distillate fuel and light crude oils having a minimum 0.5% water in the fuel, with a salt concentration not exceeding 20 ppm of sodium plus potassium. Since the purification process eliminates the

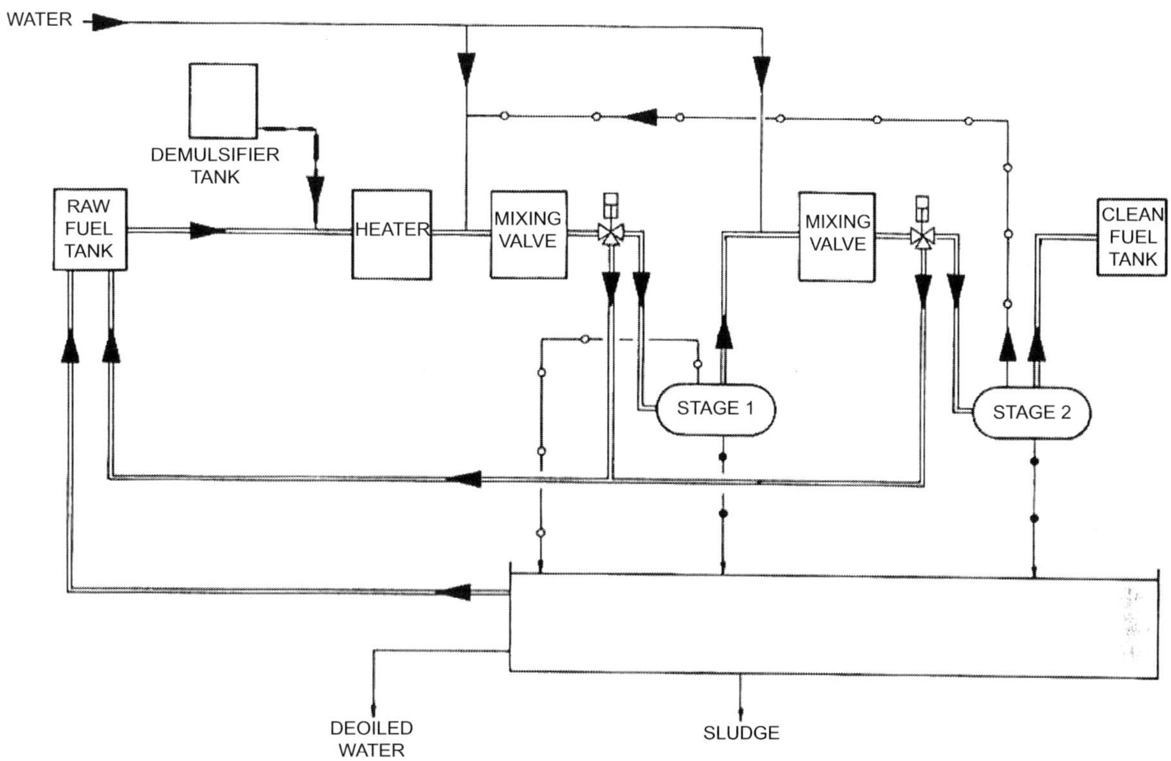

FIGURE 7–33 Heavy fuel treatment system—electrostatic. *(Source: Westfalia.)*

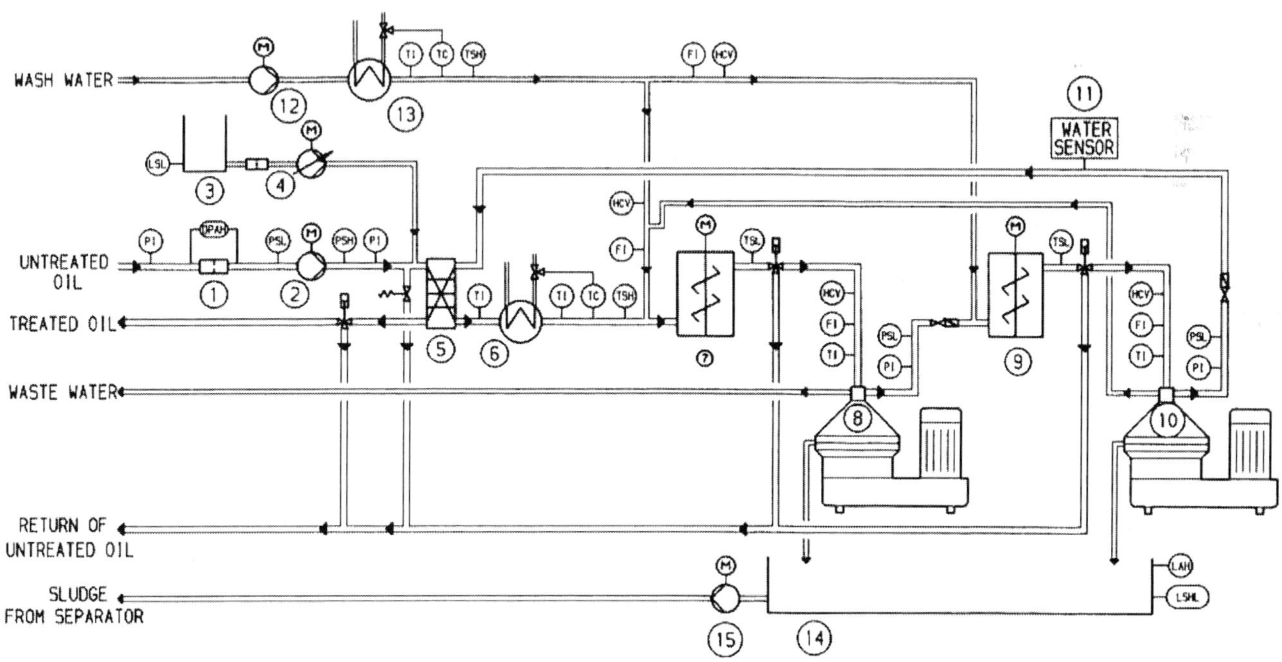

(1) PRE-STRAINER
(2) PRODUCT PUMP
(3) TANK FOR DEMULSIFIER
(4) DEMULSIFIER DOSING PUMP
(5) HEAT EX-CHANGER
(6) PRE-HEATER FOR OIL
(7) MULTI-STAGE MIXER STAGE 1
(8) CENTRIFUGAL SEPARATOR STAGE 1
(9) MULTI-STAGE MIXER STAGE 2
(10) CENTRIFUGAL SEPARATOR STAGE 2
(11) WATER SENSOR
(12) WASH WATER PUMP
(13) PRE-HEATER FOR WATER
(14) SLUDGE CHANNEL
(15) SLUDGE PUMP

FIGURE 7–34 Heavy fuel washing system. *(Source: Westfalia.)*

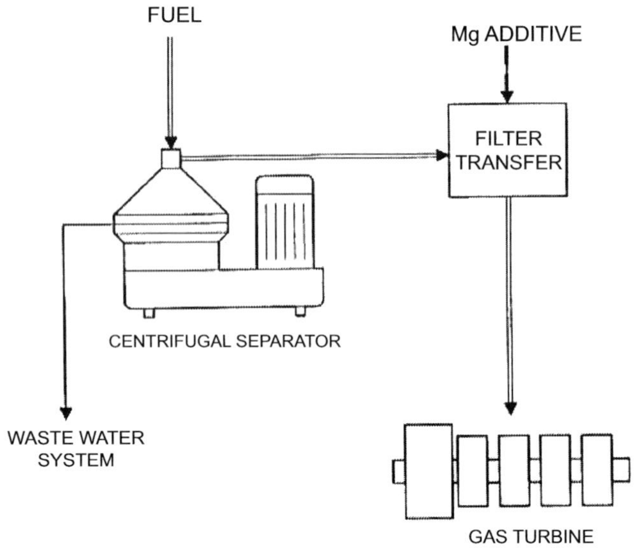

FIGURE 7–35 Fuel purification. *(Source: Westfalia.)*

need for added water and demulsifiers, it also simplifies the overall waste disposal process.

Selection of Fuel Washing Equipment

Comparison of Equipment

The selection of whether centrifugal or electrostatic precipitators are used to reduce the soluble trace metals to acceptable limits is usually dictated by the customer or the turbine manufacturer; for example where space is at a premium or in case of marine use, the natural choice is centrifuges. For systems with continuous flows over 100 m^3/hr, where space is not a major constraint, where the customer has at least 25 storage tanks with a capacity of 19,000 m^3 each, and it can sit in the tanks for a minimum of two weeks, electrostatic desalters are acceptable.

The following gives a brief comparison of the two systems: centrifugal desalting system versus electrostatic desalting system. First, let us have a look into the technical part: mixing.

As was already pointed out, the centrifugal desalting system is using a multi-stage mixer, Figure 7–36, with multi-stages; the mixer is a low shear mixer. It is a similar process to hand mixing dough when making bread at home, a slow and gentle mix in comparison with the electrostatic desalting system that uses a high shear mixing valve (Figure 7–37), causing a very violent mixing method.

An electrostatic desalting system company in east Texas tested the two mixing methods (Figure 7–38) and concluded that if a multi-stage mixer is used in the electrostatic desalting system the salt in the finished oil is about 5 ppm. After five hours they switched to the mixing valve and the salt content in the finished oil increased within two hours to 20 ppm. And, when after six hours the system was switched

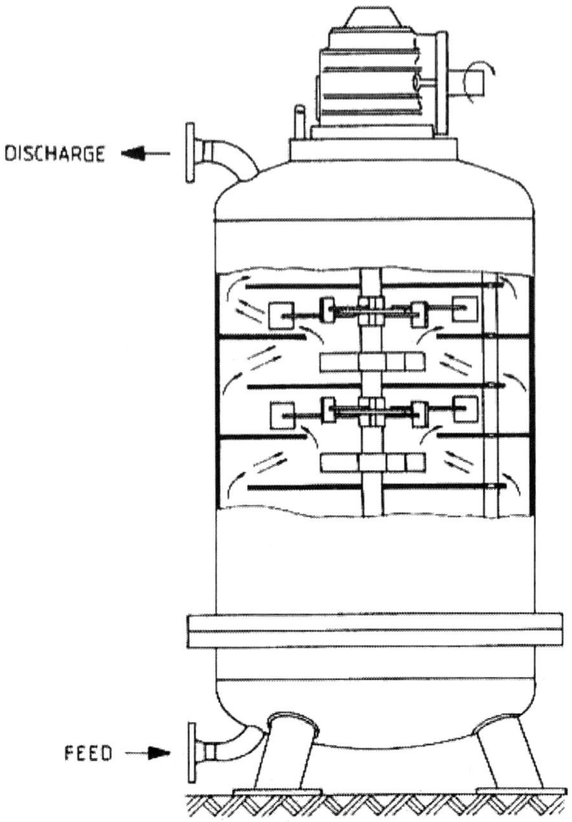

FIGURE 7–36 Multi-stage mixer. *(Source: Westfalia.)*

back to the multi-stage mixer it dropped back to 5 ppm within two hours. The electrostatic desalting system company published this study and their conclusion was that they should supply multi-stage mixer with their electrostatic desalting systems and also supply additional filtration.

In addition to Figure 7–38, Table 7–14 gives a comparison of 12 items within the electrostatic and centrifugal fuel treatment systems.

System Considerations

There is no one, basic, fuel-treatment system suitable for all applications, but a design can be tailored to the specific needs and requirements of each customer. The optimum

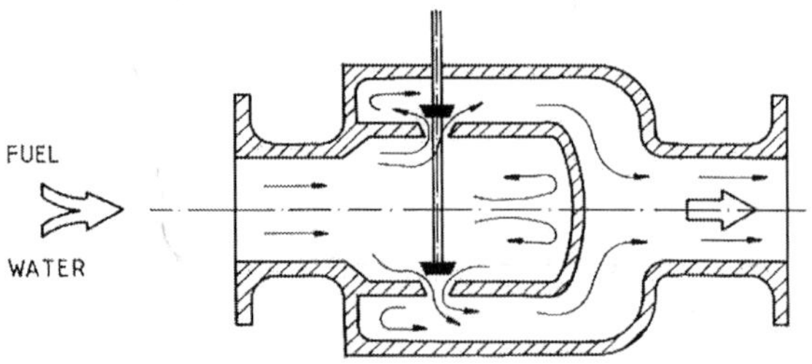

FIGURE 7–37 Mixing valve. *(Source: Westfalia.)*

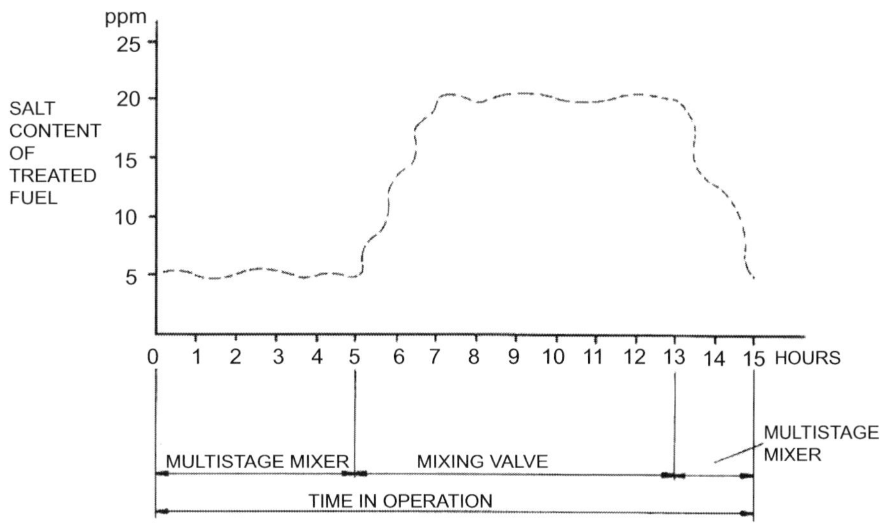

FIGURE 7–38 Comparison of mixing methods. *(Source: Westfalia.)*

TABLE 7–14 Comparison of Centrifuge/Electrostatic Features

Parameter	Electrostatic	Centrifuge
Time to reach startup stability	4 hours (per stage); 12 hours total	20 minutes total
Reliability	Good	Good
Maintenance	Low	Low
Desludging	By water effluent but can accumulate in vessels	By water effluent and by automatic bowl opening
	Approximately once a year vessels needs to be cleaned	
Performance in sodium + potassium removal	Meets spec. if fuel is within system design spec.	Exceeds spec. even if raw fuel does not completely meets design spec.
Oil content in water effluent	500–2000 ppm typical, 50,000 ppm max	50 ppm typical, 500 ppm max
Switch positions		
Response time to correct fuel out of spec. washed fuel	12 to 24 hours	20 to 40 minutes
Solid removal	Fair additional filtration needed	Excellent
Amount of heating required for fuel treatment	127–137°C (260–278°F)	98°C (208°F)
Fuel required to change system (cannot be used)	Approximately 1000–2000 m^3	1–2 m^3
Flexibility—servicing while operating	System cannot be partially shut down	Centrifuges can be removed from service while continuing to operate
Space required	Large area	Limited area

(Source: Westfalia.)

approach results from an evaluation of operational requirements for the gas turbine, the expected fuel characteristics, the desired maintenance philosophy, the choice of equipment, the choice of additive, and the desired system flexibility. Generally, there is a tendency to request "worst case" design without giving due thought to the many constraints that may exist, which may result in an over-designed and expensive installation.

Fuel-Washing Equipment Design

Once the choice of a centrifuge or electrostatic desalter is made (see Figures 7—39 and 7—40), a support-equipment design approach must be selected commensurate with the system objectives. This equipment, which includes mixers, demulsifier injection pumps, heat exchangers, economizers, etc., dictates the actual performance of the system, the resulting maintenance philosophy, and the actual equipment cost. For example, units that are periodically shut down, such as peaking mid-range gas turbines, can have maintenance performed on a planned downtime basis. Continuously running systems require on-line redundancy of critical components such as heat exchangers and economizers. This permits scheduled cleaning without incurring downtime. Further, since the equipment can be maintained while operating, the fuel-washing size can be kept smaller, reducing initial costs. Likewise, redundancy of control air, critical pumps, and motors is recommended.

Fuel Oil Treatment

With two-stage counter-current washing systems (see Figure 7—41) the degree of purity specified by the gas turbine manufacturers is attained, and the trace elements are reduced to the specified limits.

The residual oil is first fed to a prestrainer where coarse impurities are separated out to protect the downstream pump.

After adding demulsifier to facilitate separation of the wash water in the purifier, the oil is conveyed to a heat exchanger, where heat recovery takes place. The residual oil is heated to the required separation temperatures in the downstream preheater. The separation temperature is determined by the viscosity.

The wash water separated from the oil in the second stage is added to the oil upstream of the first stage mixer. The residual oil is thoroughly mixed with the water in a multistage mixer. Salts in solution with the oil will be extracted into the water. The oil/water mixture flows to the purifier in the first washing stage. The solids are spun out due to the high centrifugal force; the dissolved salts, together with the wash water, are simultaneously separated out.

The salt content in the oil is substantially reduced and contaminants such as sand, rust, and cat fines are removed from the oil. Water is again added to the purified oil upstream of the second stage mixer. After intensive mixing, the oil/water mixture is sent to the purifier in the second washing stage, where further purification and desalting takes place.

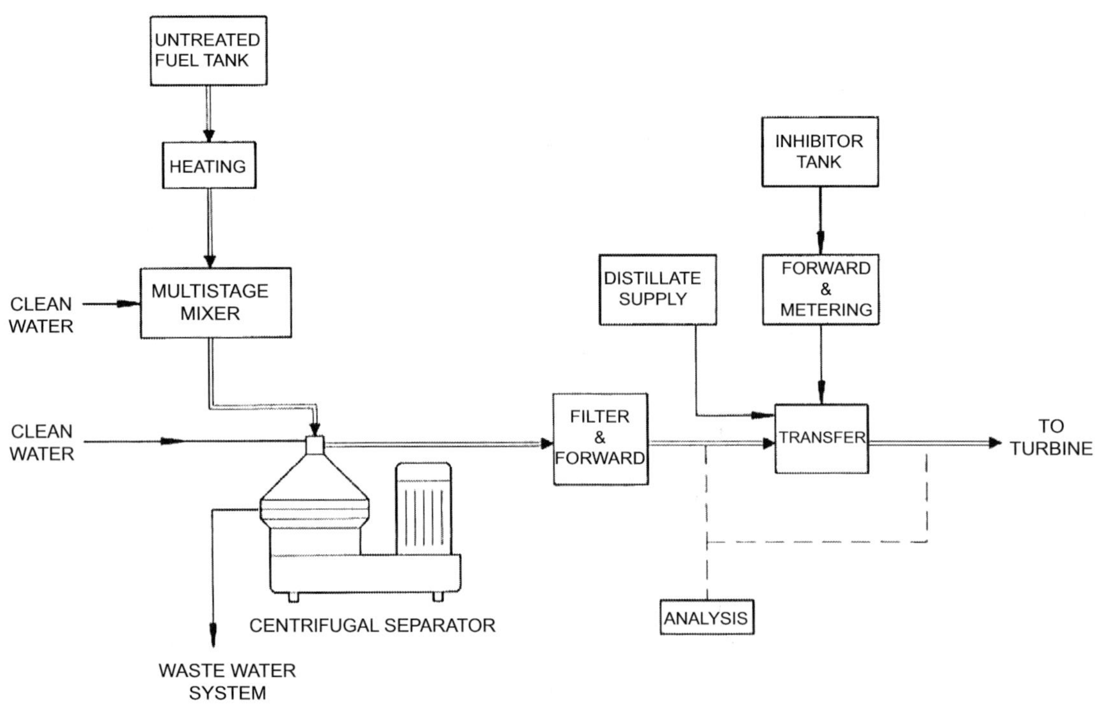

FIGURE 7—39 Centrifugal washing system. *(Source: Westfalia.)*

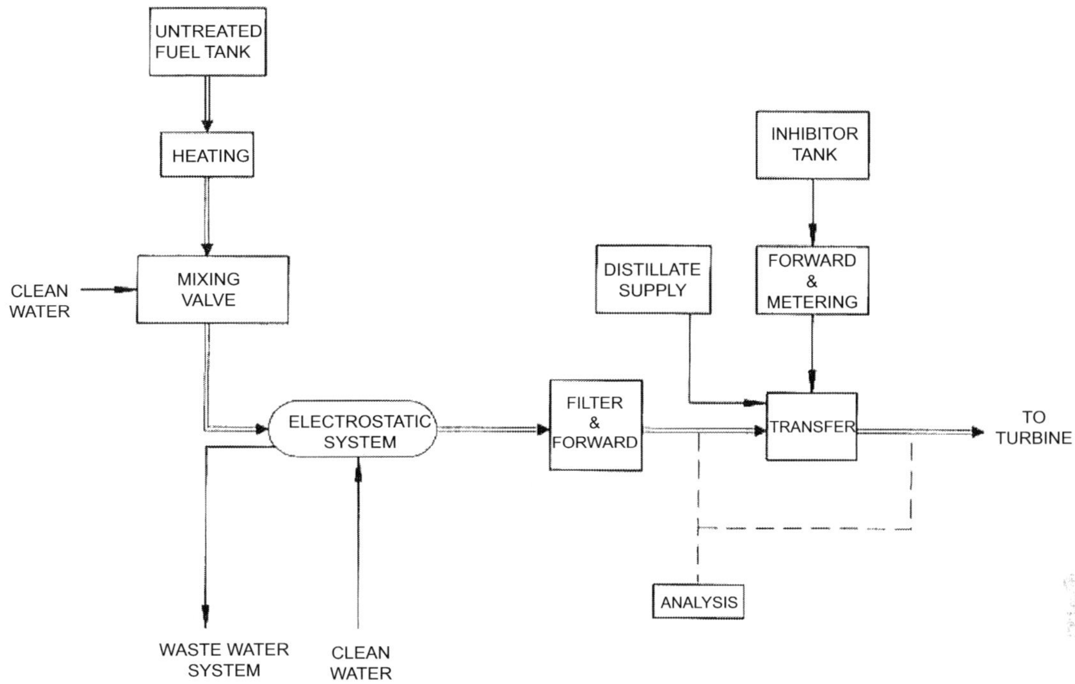

FIGURE 7–40 Electrostatic washing system. *(Source: Westfalia.)*

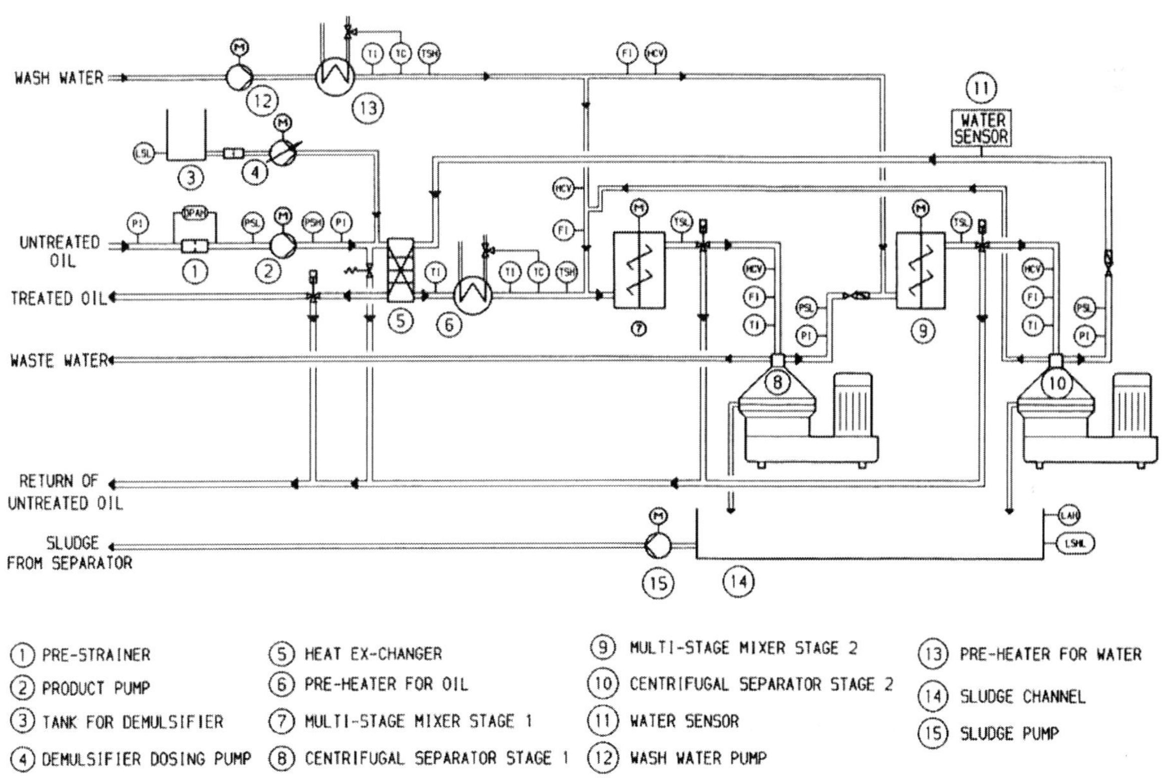

FIGURE 7–41 Fuel oil washing system. *(Source: Westfalia.)*

The required wash water volume depends on:

- The quality of the residual oil
- The sodium and potassium content in the dirty oil (ppm)
- The specified sodium and potassium content in the clean oil (ppm)
- The water quality

The clean oil discharges from the separator in the second washing stage into the clean oil tank via a heat exchanger.

If required, analyzers can be fitted in the clean oil discharge. They are used for continuous monitoring of the water and salt content. If the specified limits are exceeded, the oil is fed back into the dirty oil tank via a three-way valve. As part of a research project, optimum purification efficiency was attained using treatment systems from Westfalia Separator, even when processing high-viscosity oils (1650 mm^2/sec at 50°C).

When designing a system the following residual oil parameters are decisive for the capacity (see Figure 7—41):

- Viscosity (mm^2/sec)
- Density (g/ml)
- Sodium and potassium content in the dirty oil (ppm)
- Specific sodium and potassium content in the clean oil (ppm)

Filtration

Solids, oxides, silicates, and related compounds that are not adequately removed by the fuel washing systems must be removed by the low- and high-pressure filters prior to entering the fuel system's flow divider. The particles that are not removed may clog the flow divider and fuel nozzles.

Incorporating a motor-driven, self-cleaning filter minimizes the load on the downstream high-efficiency filters. The use of a duplex filtration system permits filter maintenance without turbine shutdown.

Inhibition

Crude oils and heavy residual fuels usually contain organic vanadium, which is extremely corrosive to the hot gas path parts of the turbine. The vanadium is not present in liquid fuels in a water-soluble form; hence it cannot be removed through the washing process. Instead, its effects are inhibited by the addition of magnesium to the fuel in the ratio of three parts of magnesium by weight to each part vanadium. At this level of inhibition, the corrosive vanadium pentoxide that is formed during combustion reacts to form magnesium orthovanadate. This has a melting point sufficiently high to allow its passage downstream through the turbine without deposition on the hot gas path parts and resultant corrosion.

A variety of magnesium-containing inhibition agents are available, and their choice is determined largely by the economics of the system and the overall amount of vanadium present in the fuel. Three additives frequently employed are magnesium sulfonate, which is oil soluble and delivered ready for use, magnesium oxide, which is an oil-based slurry, and magnesium sulfate, commonly known as epsom salts, which must be mixed with water on the site. On-line inhibition systems are the preferred method of applying the magnesium agent. The inhibitor is injected into the fuel immediately ahead of its introduction to the gas turbine (Figure 7—42). The system consists of a small in-line metering pump and a flow switch that provides an alarm signal in the event flow of inhibition stops. Either constant flow systems or systems that provide a flow varying with the actual fuel use rate are available. Certification tanks have been employed in some installations. In this approach, batches of the magnesium inhibition agent are mixed directly with the fuel. Use of this system is not currently recommended, since there is a strong possibility that the additive will settle out.

Both oil soluble (magnesium sulfonate) and water-soluble inhibitors (magnesium sulfate) are used with an equal degree of success. The mixing ratio is three parts magnesium to one part vanadium and should not exceed 3.5:1.

For reasons of economy, the oil soluble inhibitor is primarily used for low vanadium concentrations (up to 20 ppm) and the water-soluble inhibitor for high vanadium concentrations (above 20 ppm). Figure 7—43 shows a water-soluble inhibitor.

Dosing Unit for Inhibitors

In the dosing units from Westfalia Separator, dosing and conveyance of the inhibitor is by means of pumps. The pump output must be variable. This is possible by

- Stroke adjustment
- Speed adjustment
- Partial stream regulation

It is frequently the case that combinations of the different control systems are required. The pump can be adjusted as follows:

- As a function of fuel oil consumption by electrical speed adjustment
- As a function of the vanadium content by electrical stroke adjustment

Mixing Unit for Water-Soluble Inhibitor

A screw conveyor transports a certain amount of magnesium sulfate ($MgSO + 7\ H_2O$) into the mixing tank. Prior to adding magnesium sulfate, the tank is filled to a preset level with demineralized water. An agitator ensures uniform distribution.

A conductivity meter in the circulation line monitors the concentration. Once the required concentration is reached,

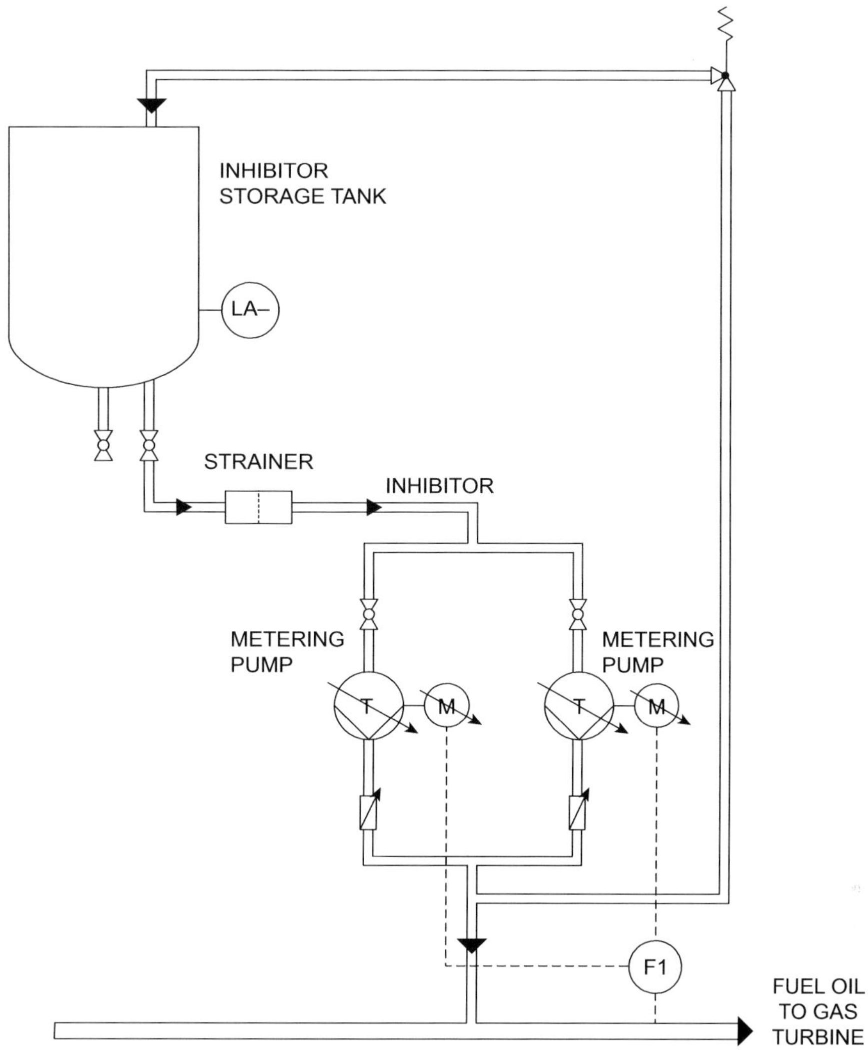

FIGURE 7–42 Fuel inhibition system (oil soluble). *(Source: Westfalia.)*

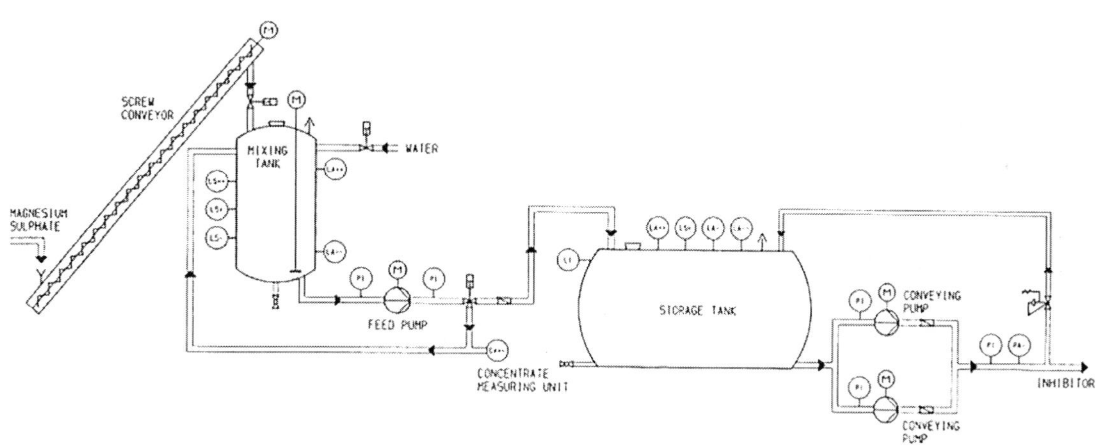

FIGURE 7–43 Water-soluble inhibition system. *(Source: Westfalia.)*

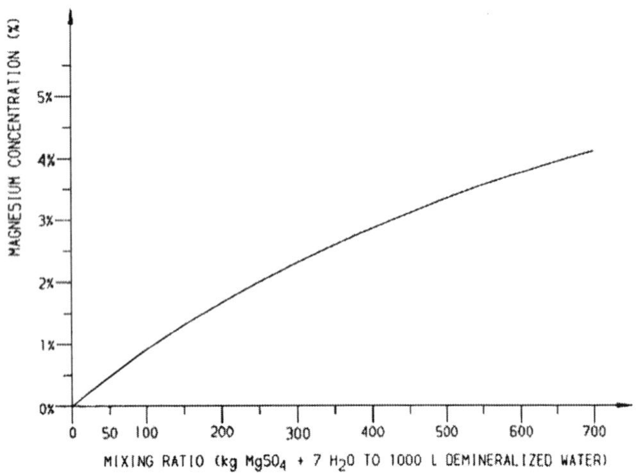

FIGURE 7–44 Preparation of water-soluble inhibitor as a function of the magnesium concentration. *(Source: Westfalia.)*

the solution is pumped into the storage tank. The entire process is automatically controlled and monitored.

Figure 7–44 shows the mixing ratio as a function of the magnesium concentration.

Fuel Selection

Residual fuels cover a wide range of properties including vanadium content, specific gravity, and viscosity. Depending largely on economic trade-offs, the user may or may not decide to burn all of the fuels available at this location. The maximum allowable vanadium content may be controlled by local regulations governing total stack particulate emissions. Lower vanadium level fuels, where available, can substantially reduce fuel treatment additive costs.

Users requiring a capability for washing fuels with a specific gravity and/or viscosity higher than specified for the fuel-washing system can accommodate these fuels by blending them with a compatible distillate fuel before washing.

Fuel Handling and Storage

Heavy fuels do not, generally, have consistent chemical and physical properties; purchased supplies can vary widely in their composition. Observation of the following precautions for storage and handling of fuel prior to use can enhance fuel washing and assure greater consistency in the fuel delivered to the gas turbine:

1. The use of three tanks sized to provide a 24-hour settling time will permit one tank to be in use while the second tank is being filled and the third is settling after being filled.
2. Tanks should be covered and provided for easy drainage of water and sediment from the bottom, which should be done on a daily basis.
3. Fuel should not be pumped directly from the bottom of the tank.
4. All fuel delivered should be filtered to remove any large particles, and piping should be arranged to minimize agitation of sediment at the bottom of the tank.
5. Storage tanks for high viscosity fuels such as residuals should be heated to keep the viscosity low enough so that the fuel can be pumped.
6. Oil storage temperature should be as low as possible consistent with the operation. Washed oil must be delivered to the day tank below the flash point.
7. Cadmium, zinc, and copper catalyze the decomposition of hydrocarbons. These elements and their alloys, therefore, should not be used in the construction of storage tanks and related items.

Fuel-Wash Capacity

The fuel-use profile is significant in establishing the size of the fuel-washing equipment and treated fuel storage tank.

Fuel demand is a function of the site ambient temperature and number of hours of use by each turbine, which determines the required fuel treatment capacity to support the site operation. Final sizing depends on the margin desired by the customer and the maintenance philosophy adopted to keep the washing equipment on line.

Fuel Analysis and System Controls

A modern, heavy-fuel gas turbine installation requires controls for the fuel treatment equipment as part of total system control. Actual philosophy is dictated by customer preference. Generally, each functional component is equipped with its own control system and has manual set points. Once the set points are preselected, the system can be initiated on-site (or at a remote area) after which the entire process is self-regulating. Monitoring of critical functions can be incorporated if desired.

Fail-safe and permissive features are provided as part of the control system. When a malfunction occurs in the fuel wash equipment, the fuel is automatically diverted to the raw fuel storage tank, rather than to the day tank, until the malfunction is cleared.

Additionally, the fuel quality must be monitored at various steps of the process: sodium and vanadium level after washing and total sodium and magnesium after inhibition.

For light fuels not requiring inhibition (vanadium less than 0.5 ppm) only the sodium level is monitored. This is most readily accomplished with a flame photometer, which is available as a manual instrument for spot checks.

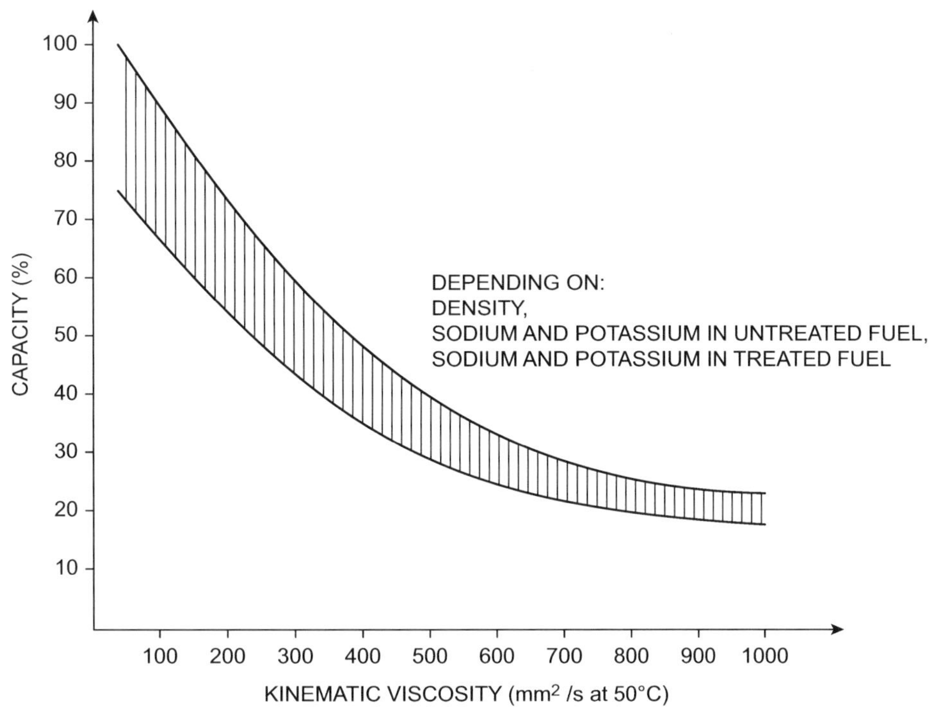

FIGURE 7–45 Effective plant capacity in residual oil treatment. *(Source: Westfalia.)*

Fuels requiring both washing and vanadium inhibition by a magnesium additive should be monitored with a direct-reading, automatic, emission spectrophotometer (shown in Figure 7–45). While this is a complex instrument, its operation is very simple and amenable to use by on-site labor. The operator collects samples, inserts them one-at-a-time in the excitation chamber, presses a button, and one minute later reads the analysis printed by a typewriter. About once every two weeks the instrument calibration is checked.

Treatment of Light Crude Oil

Operation on Refinery Crude

Light crude oil with a typical analysis shown in Table 7–15 is delivered by a pipeline to a power station located in central Saudi Arabia to operate six gas turbines. Sodium plus potassium level in "as-received" fuel runs from 15 to 4 ppm, only rarely does it come close to the value shown in Table 7–16 (21 ppm). Operation of this 54 m^3/h fuel purification system (Figure 7–46) began at the end of 1981.

The centrifuges used at this location are of the automatic desludging type. Water and solids accumulating at the bowl perimeter are discharged at intervals of about four hours. A batch analyzer provides a spectrographic flame analysis of the trace metal in the purified fuel. Since this was the first purification system, it was equipped with the ability to add small amounts of water to the fuel if it was found to be necessary. This system was also equipped with demulsifier injection capability to prevent emulsions.

During initial startup the water injection system was inadvertently activated, giving rise to erroneous indications that the light crude oil was being washed rather than purified. The interesting fact is that there was an improved performance of the system when the water injection was terminated. Water quality there is quite good; it is received from a local water desalination plant.

Purification tests have shown very easy formation of stable emulsions which can be troublesome due to their sedimentation at the oil/water interface and which cause a deleterious effect on centrifugal purification. The small amount of water used in the system (about 0.075 m^3/h) is utilized in the operation of the centrifuge bowl.

Operational difficulties have been observed as fuel temperature dropped below 54°C, which appears to be the approximate "wax point" or the temperature at which waxes begin to precipitate from solution. Waxes are interfacial in nature, and on several occasions have completely filled the centrifuge bowl with this semi-solid. This in turn negates the separation effect of the microscopic water droplets containing must of the trace metals so damaging to the hot-gas-path components. Since crude oil temperature has to be maintained above the wax point prior to the purification system, the crude pour point (wax separation point) is therefore one of the most important criteria of system design.

TABLE 7—15 Typical Crude Oil Analysis

	Unit	First Source (Refinery) Max.	First Source (Refinery) Min.	Second Source (Oil Field) Max.	Second Source (Oil Field) Min.
Gravity					
at 15.6°C	API	34	39	31	33.5
at 15.6/15.6°C	Sp. gr.	0.855	0.830	0.871	0.858
Viscosity					
at 37.8°C	cSt	11.0	5.5	10.0	6.0
at 50.0°C	cSt	8.5	4.0	6.5	4.5
Water content	Vol. %	1.0	0.01	1.0	0.1
BS&W	Vol. %	0.7	0.01	1.5	0.2
Flash point	°C	−6.7	−6.7	−6.7	−6.7
Pour point	°C	—	−30.0	−6.7	−17.8
Sulfur	Wt. %	1.2	1.0	3.5	3.3
H_2S	ppm	150	33	145	35
Wax	Wt. %	5	2	30	14
Ash	Wt. %	0.1	0.01	0.128	0.001
Sodium (Na)	ppm	20	1.5	50	5
Potassium (K)	ppm	1	0.1	3	0.4
Vanadium (V)	ppm	15	6	8	4

(Source: Westfalia.)

It is interesting to know that none of the problems that sometimes plague the high firing temperature gas turbines have occurred: failure of flow dividers, fuel nozzle plungers, high-pressure pumps, fuel nozzle coking, etc., due to crude purification and turbine operation cycle. The first turbine water wash was after 3000 hours of turbine operation.

The simultaneous elimination of water and demulsifier injection had increased filter life.

Filter life is influenced by many factors: moisture, asphaltic sludge, inorganic particulates, bacteria (algae), wax out of solution, as well as chemical additives (demulsifier and vanadium inhibitor).

Increase in filter life was at first thought to be coincidence, and little investigation was given to their interrelation. The later lab studies on the oil field crude demonstrated conclusively that one oil-soluble magnesium inhibitor can have a major effect on filter life in the presence of only trace amounts of moisture. And that these chemicals also can cause fuel nozzle coking is proven by the extremely high magnesium content of the deposits.

Operation on Oil Field Crude

During 1981 and 1982 centrifugal lab investigations were conducted on an oil field crude shown in Table 7—15. Experience dictates any such investigation should be conducted on-site as exportation of drum quantities of light crude to a remote laboratory has proven to be a futile effort because of numerous factors. Exporting large samples has resulted in highly questionable results.

Test runs on oil field crude were made on a quantity of oil taken only hours prior to testing. Tests were run on the light crude from many points in its processing train, and a great deal of data were obtained concerning desalting as well as separation rates under various processing conditions. Figure 7—47 shows the actual physical nature of the oil field crude.

The variations are considerably higher than those seen in the refinery crude oil. During the peak periods of January to May 1981, washed crude from an electrostatic system was observed with 4.5 ppm of sodium plus potassium in the day tank. Passage through the test centrifuge gave a further

TABLE 7–16 Data from Centrifugal Lab Test on Oilfield Crude without Water Washing

Ref. No.	Sample Point	Viscosity at 50°C cSt	Sp. Gravity at 15.6/15.6°C	Water Vol. %	Test Temp. °C	Trace Metal Tent (ppm) Na	K	Ca	V	Pb	Mg	Fe	Ni	Cu	Sn	Remarks
15/1	Feed	10.8	0.865	1.30	70	4	1.4	0.8	3	0.2	12	3.8	5	1.0	0	
	Discharge			>0.005		0.2	0.0	0.2	3	0	1	1.4	5	0.6	0	
15/2	Feed	9.5	0.875	1.40	60	4.5	1.7	2.1	4.2	0.8	0.6	12.5	7	0.4	0	
	Discharge			>0.005		0.6	0.1	0.5	4.7	0.7	0.1	4.6	6	0.0	0	
11/1*	Feed	9.8	0.865	6.00	52	104	34	32	6	0.8	21	62	11	0.8	6	Sample from tank bottom
	Discharge			>0.005		7	0.4	3	4	0	3	14	10	0.7	2	
11/1	Feed	9.8	0.865	6.00	52	104	0.4	3	4	0	3	14	10	0.7	2	1% water added to above sample
	Discharge			>0.005		<1.0	<0.4	1	4	0	<1.0	<10	10	<0.7	<2	
12/1*	Feed	10.5	0.873	14.00	77	143	36	31	7	1.0	11	54	9	0.4	2	Sample from tank bottom
	Discharge			>0.005		9	0.8	2	7	0.2	0	14	7	0.2	1	
12/1	Feed	10.5	0.873	14.00	77	143	26	31	7	1.0	11	54	9	0.4	2	1% water added to above sample
	Discharge			>0.005		<1.0	<0.8	<1.0	7	0.2	0	<5	7	0.2	<1	

*= *Two stage purification, no water added to the fuel.*
(Source: Westfalia.)

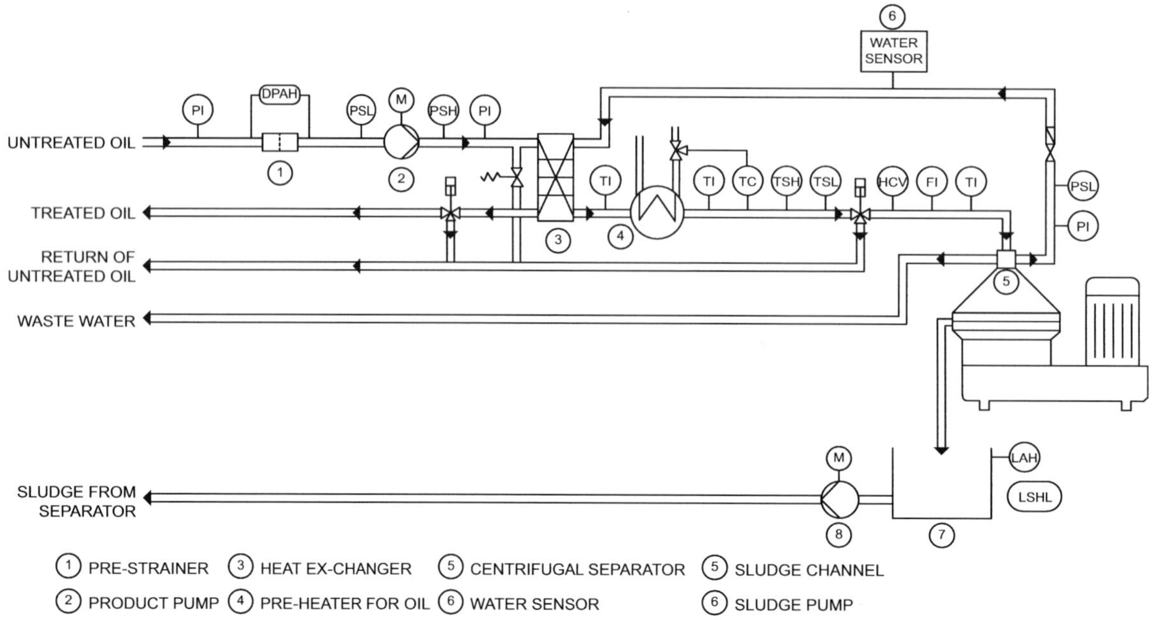

FIGURE 7–46 Crude oil purification system. *(Source: Westfalia.)*

reduction of sodium plus potassium to very acceptable levels as plotted in Figure 7–48. Note: Field experience showed a higher centrifuge efficiency.

Table 7–16 and Figure 7–49 show the extent to which sodium and potassium can be reduced in a centrifugal purification process on the oil field crude. Reduction of other trace metals is also given. Note: Field experience showed a higher centrifuge efficiency.

The above test also provided some answers to questions of current interest to gas turbine operators contemplating burning light crude oil. The conclusions reached and data presented are considered valid for the stations where studies have been made on refinery and oil field crude. Other crudes are under investigation, it is expected that most crudes are purifiable if their physical properties are similar.

Comparison of Centrifuge and Site Data

During 1981 a direct feed crude oil purification system was installed at the Gizan power station in order to purify crude oil for GE's Fram 6 gas turbines. The system operated very successfully. At a later date the customer switched over to contaminated distillate and utilized the same purification system without any modifications. Westfalia Separator's purification system can operate on crude oil or distillate fuel and obtain identical results.

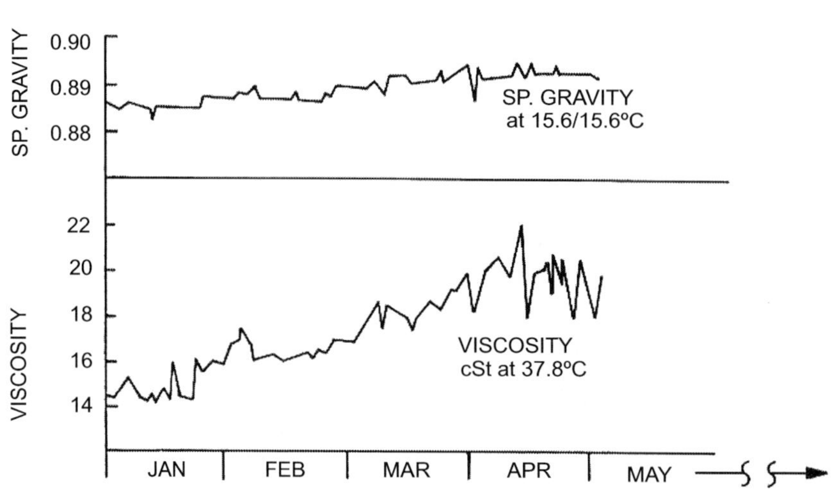

FIGURE 7–47 Gravity and viscosity of second source crude oil, 1981. *(Source: Westfalia.)*

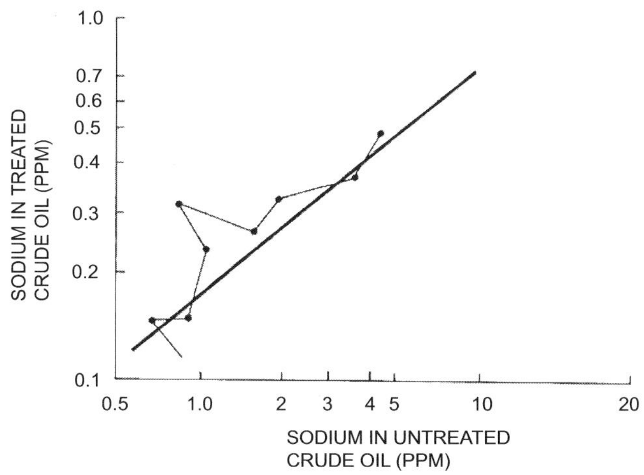

FIGURE 7–48 Performance of test centrifuge on crude oil. *(Source: Westfalia.)*

Westfalia's system reduced the sodium plus potassium level from an average of 2.92 ppm down to an average of 0.015 ppm (site data dated Sept. 26, 1989).

At a later date (1987) an additional system was installed (not Westfalia) for GE's Fram 7. This system could not supply purified fuel directly to the gas turbine (it has been modified at present time) and its sodium plus potassium reduction is only 88%.

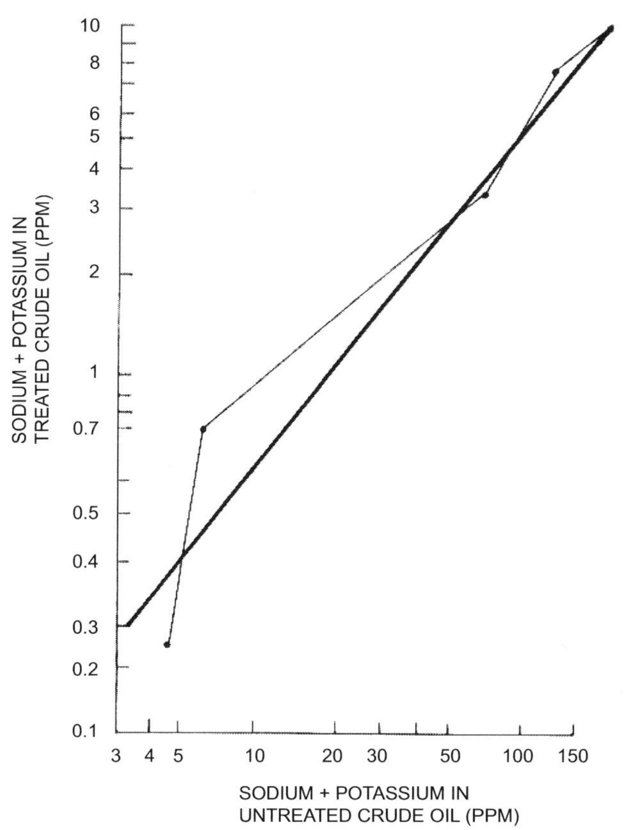

FIGURE 7–49 Centrifugal purification studies on oilfield crude. *(Source: Westfalia.)*

If we look into the purification system efficiency from the filter plugging data point of view, it can be noted (from site date) that the turbines that are being supplied with fuel that was purified in the Westfalia system have a filter life span of 1078–4103 hours on the low-pressure and high-pressure filters whereas the turbines that are being supplied with fuel that was purified by the other purification system have a filter life span of 95–1009 hours on the low-pressure filters and between 1578–1716 hours on the high-pressure filters.

From the above comparison of the two purification systems (summarized in Table 7–17) it is clearly noted that the Westfalia purification system performed as designed.

In summary, the data from centrifuge lab studies as well as from operations at various power stations, confirmed the viability of the process. A level of approximately 20 ppm of sodium plus potassium can be reduced to <0.5 ppm while operating at commercially acceptable flow rates for the standard centrifuges models available.

It has been found that purification operations on crude oil should be at 55–60°C to ensure that wax separation does not influence the operation. This appears to be true regardless of wax concentration, as proven by test runs on tank bottoms containing extremely high wax content (about 40% by volume).

Centrifugal purification of light crude oil can extend filter life on 5–10 μm paper cartridge filters to 1000 hours, as have been shown by both commercial and pilot plant operation. The operator must, however, be very careful in selection of additives to be used, as well as their injection point, as discussed previously.

Operating water on a typical light crude system of 54 m^3/h has a total consumption of about 0.075 m^3/h. This water is needed for the bowl hydraulic. For the same size plant (54 m^3/h), a washing operation having a moderate 2% by volume water would consume about 1.0 m^3/h of water.

Extremely high saltwater contamination level can be tolerated in fuel as received providing proper storage facilities are available and are rigorously maintained. Tests show sodium plus potassium level exceeding 1000 ppm can be reduced in 24 hours of quiet storage to about 40 ppm. Floating suctions and a floating pan with a fixed roof are also highly recommended to maintain clean fuel.

Special caution should be taken when planning for the successful desalting of certain crude oils by the centrifugal purification process without the addition of water to the fuel. The success or failure of this process is highly dependent upon both the design and configuration of the centrifuge and the processing systems.

TABLE 7–17 Field Experience with Crude Oil

		Tabuk		Gizan	
	Unit	Untreated Fuel	Treated Fuel	Untreated Fuel	Treated Fuel
Gravity at 15.6/15.6°C	Sp. Gr.	0.825	0.825	0.830	0.830
		0.876	0.876	0.855	0.855
Viscosity at 37.8°C	cSt	3.40	3.40	5.50	5.50
		7.30	7.30	11.00	11.00
Sulfur	Wt %	1.8	1.5	1.0	0.80
		1.2	1.7	1.2	1.0
Sodium & potassium	ppm	15	≤1.0	21	≤1.0
Vandium	ppm	23	23	15	15
Water & sediments	Vol. %	4.0	0.01	0.1	0.01
Ash	Wt %	Unknown	Unknown	0.1	0.01
Total flow	m^3/hr	61.32	61.32	54.5	54.5

(Source: Westfalia.)

Treatment of Residual Fuel

Tung Hsiao Power Plant (OEM: Alstom)

An OEM, such as Alstom for a model such as their IIN, might ask for a fuel treatment system that can handle 158.4 m^3/h (44 l/sec.) of residual fuel with a specific gravity of up to 0.986 at 16°C and a viscosity of up to 440 cSt at 50°C. Incoming fuel will have a sodium plus potassium level of 50 ppm and must be reduced to ≤0.5 ppm.

During the acceptance test the sodium plus potassium was reduced from 30 ppm to 0.08–0.10 ppm. It should be noted that this is the first (by any company) to achieve a Na + K reduction to 0.08–0.10 ppm. This fuel treatment system is the largest operating system in the world with the largest centrifuge in the world.

The fuel washing system consists of:

1. Fuel forwarding from tanks to centrifuge
2. Centrifuge section

Fuel Forwarding

Two 100% flow (158.4 m^3/h) speed controlled pumps are moving the 440 cSt at 50°C fuel from the raw fuel storage via a duplex strainer to the centrifuge section. Fuel pump speed had to vary from 1800 ppm to 360 ppm in order to support one or more fuel treatment lines (operation of a line is controlled by fuel demand number of turbine in operation).

Downstream of the pump a demulsifier is introduced to the fuel—200 ppm of Betz Petromeen EB-911—it should be noted that the demulsifier injection pump as well as the fuel forwarding pump, have a variable speed drive to compensate for the quantity of demulsifier required as a ratio to fuel flow.

After that the fuel flow powered five independent two-stage washing systems. Each system includes a steam heater, two multi-stage mixers and two SA-160/220 centrifuges. This arrangement permits the customer to save on power and chemicals.

When testing this system, each fuel washing line reduced the sodium plus potassium from a nominal average of 50 ppm to ≤0.5 ppm (see Table 7–18).

Naga Power Plant, the Philippines (Gas Turbine OEM: Alstom)

The OEM requested a fuel treatment system that can handle 13.8 m^3/h of residual fuel with a specific gravity of up to 0.952 at 16°C and a viscosity of up to 220 cSt at 50°C.

Incoming fuel will have a sodium plus potassium level of 50 ppm and must be reduced to ≤1.0 ppm. During the acceptance test the sodium plus potassium was reduced from 70 ppm to less than 1 ppm.

The fuel washing system consists of:

1. Pump-heating section
2. Centrifuge section
3. Waste water section

Fuel Forwarding

Two 100% flow pumps are moving the fuel from the raw fuel storage via a duplex strainer to the centrifuge section.

TABLE 7–18 TPC System Run Results

Date	Flow		Temperature		Demulsifier	Sodium and Potassium (ppm)					
							Out				
M^3/hr	Fuel M^3/hr	Water °C	Fuel °C	Water	ppm	In	Linei	Line 2	Line 3	Line 4	Line 5
09.04.91	95.0	4.2	99	94	300	20	0.14	—	0.30	0.20	—
10.04.91	126.7	5.6	99	94	300	20	0.18	—	0.19	0.12	0.39
19.04.91	126.7	5.6	99	94	300	10	0.09	—	0.14	0.12	0.09
30.04.91	126.7	9.6	99	94	300	10	0.09	—	0.09	0.09	0.09
09.05.91	158.4	7.0	99	94	300	10	0.09	0.08	0.09	0.08	0.09
10.05.91	126.7	5.6	99	94	300	10	0.08	0.09	0.09	0.09	—
20,09.91	158.4	7.0	99	94	300	10	0.08	0.09	0.09	0.09	0.09
30.05.91	158.4	7.0	99	94	300	10	0.08	0.09	0.09	0.09	0.09
09.06.91	158.4	7.0	99	94	300	7.8	0.09	0.09	0.09	0.09	0.09
10.06.91	158.4	7.0	99	94	300	6.5	0.09	0.11	0.09	0.09	0.09
19.06.91	158.4	7.0	99	94	300	8.8	0.20	0.13	0.10	0.11	0.10
29.06.91	126.7	5.6	99	94	300	20	0.19	0.28	—	0.30	0.32
08.07.91*	158.4	10.0	99	94	200	30	0.08	0.10	0.09	0.09	0.08

Acceptance run.
(Source: Westfalia.)

Downstream of the pump a demulsifier is introduced to the fuel.

After that the fuel flow powered a two stage washing system. The system includes the thermal heater, three multi-stage mixers, and three OSB 35/40 centrifuges—two in operation and one standby. This arrangement permits the customer to save on power.

When testing this system, the fuel washing line reduced the sodium plus potassium from a nominal average of 50 ppm to $\leq$1.0 ppm (Table 7–19).

The cases in this chapter add perspective to the subject of mature OEM-developed technology with specific and/or unconventional gas turbine fuels.

Case Study 1 is a residual "bunker" fuel case study* featuring the Limay Bataan plant in the Philippines (2001-GT-2013).

Case Study 2 is based on extracts of C. J. Goy, A. J. Moran, and G. O. Thomas, "Autoignition Characteristics of Gaseous Fuels at Representative Gas Turbine Conditions," IGTI 2001-GT-0051. In this case, optimization of the pre-mixer (of fuel and air) for minimized emissions by OEM (Rolls Royce) designers is discussed. The case is interesting in that it provides a basis to compare this OEM's design philosophy versus the others whose cases are included here.

In Case Study 3, the tri-fuel injector for LPG and naphtha applications made by Alstom for the Tornado gas turbine is explored. This case is based on extracts from A. Newman et al., "From Concept to Commercial Operation—the Tri-Fuel Injector Used for LPG and Naphtha Applications," IGTI 2001-GT-0079. The design philosophy discussed here has EGT (European Gas Turbine) roots and therefore is part of EGT's original strategy to cater to the requirements of the Asian and east European market (among others), which could access LPG and naphtha more easily than natural gas, LNG, and diesel. The design strategy taken parallels the end user's requirements and therefore presents a particularly interesting case.

Case 4 discusses the Siemens approach to the applications issue addressed by Alstom in Case 2. It is based on extracts from E. Deuker et al., "Multi Fuel Concept of the Siemens 3A-Gas Turbine Series," IGTI 2001-GT-0078.

Case 5 deals with the use of blast furnace gas to fuel a 300 MW CC plant. The steel plant the power plant serves is in China and makes its own power, as well as its CC fuel. The CC package is Mitsubishi's (MHI).

Case 6 discusses experimental work done on using biomass fuel in DLN combustors by Siemens.

TABLE 7–19 Naga System Run Results

Date	Flow		Temperature		Demulsifier	Sodium and Potassium, ppm	
	Fuel m^3/hr	Water m^3/hr	Fuel °C	Water °C	ppm	In	Out
14.02.92	8.0	0.4	98	98	250	≤50	≤0.5
	8.0	0.8	98	98	250	≤50	≤0.5
	10.0	0.5	98	98	250	≤50	≤1.0
to	10.0	1.0	98	98	250	V50	≤0.5
	12.0	0.6	98	98	250	≤50	≤1.0
	12.0	1.2	98	98	250	≤50	≤1.0
20.02.92	13.8	1.4	98	98	250	≤50	≤1.0

(Source: Westfalia.)

CASE STUDY 1: A RESIDUAL "BUNKER" FUEL CASE STUDY (METRO MANILA, LIMAY BATAAN COMBINED CYCLE)*

Alstom Power is among those OEM that have had success in the residual fuel arena.

Nomenclature

CCPP = combined cycle power plant
NPC = National Power Corporation
HRSG = heat recovery steam
BFO = generator blended fuel oil
TFO IFO = treated fuel oil, intermediate fuel oil
DFO = diesel fuel oil
HP/IP/LP = high/intermediate/low pressure
FOTP = fuel oil treatment plant
OEM = original equipment manufacturer

The use of ash-bearing oil fractions in gas turbines seems a rather exotic application. Of all the installed MW running on ash-bearing fuels, 27% have been installed by Alstom Power. The Limay Bataan CCPP installation in the Philippines is one of them, burning the heavy residual oil produced in the PETRON Bataan refinery near Limay.

Owned by National Power Corporation (NPC), this power station is fully operated and maintained by Alstom Power under an O&M contract for 15 years. This section highlights the operational experiences gained in this facility. In this section, operational aspects of using heavy oil in gas turbines are discussed using the experiences made at the Limay Combined Cycle power station with an output of roughly 600 MW. Having been in commercial operation for eight years on this specific fuel type, it can be said that this is one of the most efficient ways of burning this type of fuel.

The first combined cycle power plant contract in the Philippines was awarded to a consortium of ABB/KHI/Marubeni on a turnkey basis for the financing, design, build, and turn over to NPC. Among many requirements for the project the following were the most important ones:

- It must be possible to structure the financing scheme within a very short time.
- The GT should be equipped and capable of operating on different liquid fuels.
- The plant should be able to generate electricity within one year.

With the main components for Limay coming from various countries, such as the USA, Japan, Germany, and Switzerland, it was possible to structure a multi-national export loan facility with the corresponding governmental agencies, US-Exim, Japan-Exim, Hermes, and ERG.

Initial Power for Metro Manila in Record Time

To meet the very important third criteria and reduce the daily eight-hour brownouts, the project was split into two phases. The phase one gas turbines were completed in simple cycle mode by shipping and erecting preassembled blocks. It was possible to commence commercial operation only 8.5 months after the notice to proceed from National Power Corporation. The first block of three gas turbines was available after 13 months, the second block after 18 months.

* Source: [7-5]. Courtesy Alstom Power. Extracts from B. Basler and D. Marx, "Heavy Fuel Operation at Limay Bataan Power Station," 2001-GT-0213.

FIGURE 7–50 Limay Bataan power plant [7-5].

To allow the energy to be provided to Metro Manila, the 230 kV HV lines to the nearest substation some 50 km away from the plant had to be built within the same short time period. This task was managed by NPC on time.

With the restricted choice of fuels the location was chosen adjacent to Petron Bataan Refinery to take advantage of the available heavy residual fuel oil.

Limay Bataan Combined Cycle and Extension with Block "B"

The plant is depicted in Figure 7–50.

Phase two of completion, in which a 114 MW steam turbine (ST) was added using the steam produced in the heat recovery steam generators (HRSG) and thus completing the combined cycle, was achieved 12 months later, with the commissioning of the ST of Block "A," on 25 November 1994.

Financial constraints led the plant to be ordered in two separate blocks, to limit engineering efforts. Block "B" was to be identical to "A." Block "B" was handed over on 25 February 1995.

The combined cycle process is depicted in Figure 7–51. Each combined cycle block comprises three Alstom Power GT 1 1N gas turbines, three unfired natural circulation heat recovery steam generators, the steam turbine, main transformers, and high voltage switchyards. Overall control for each block from a central control room is operated by use of Procontrol P system. Each gas turbine is rated at 70 MW and is fired on treated fuel oil (TFO) with startup on diesel fuel oil. Water injection is available to augment power output. The steam turbine is rated at 114 MW. Total site

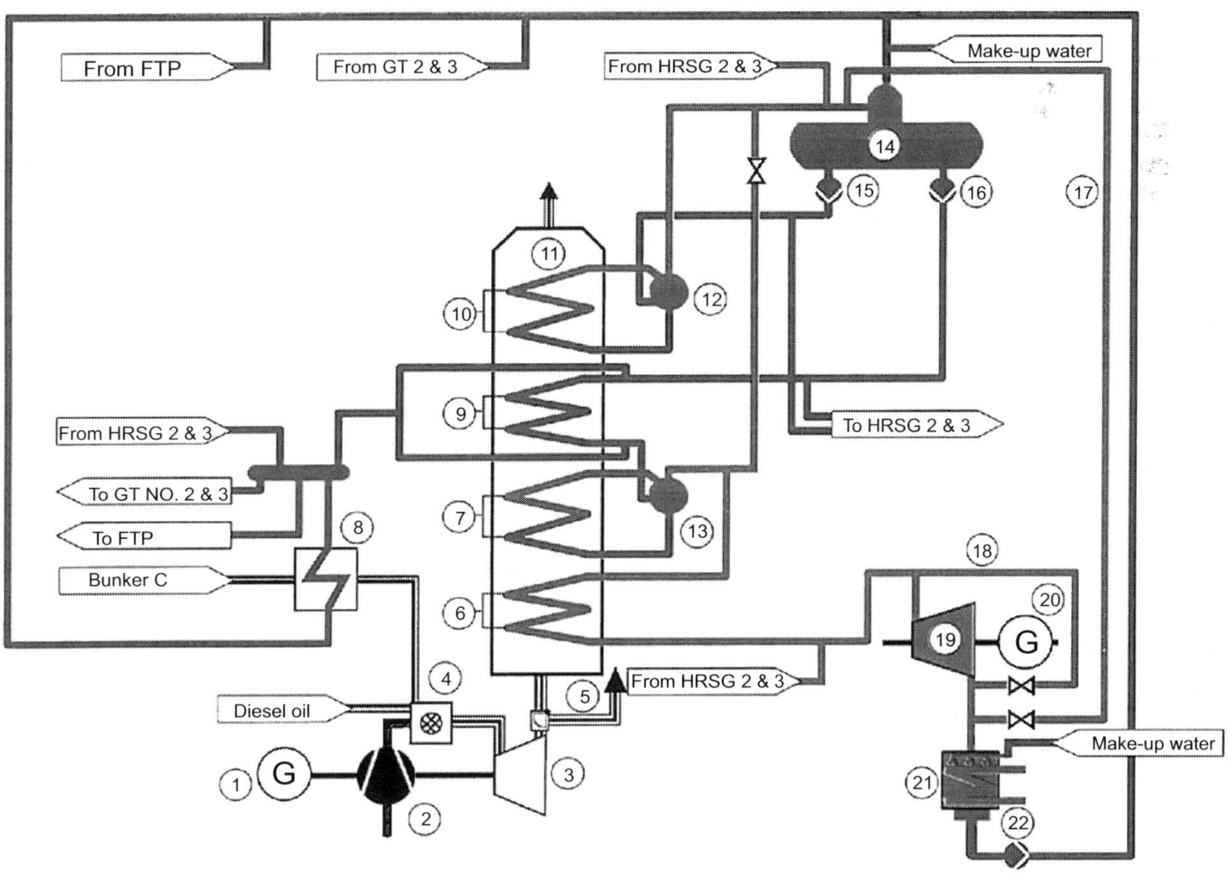

FIGURE 7–51 Plant scheme process [7-5].

KEY OPERATIONAL DATA October 2000								
	GT11A	GT12A	GT13A	ST14A	GT11B	GT12B	GT13B	ST14B
Prov. Acceptance Certificate	1993-05-28	1993-05-21	1993-04-20	1994-09-26	1993-11-04	1993-11-19	1993-11-25	1994-12-19
Availability (%) Since PAC	92.6	89.2	89.9	97.4	90.3	93	92.3	97.4
EOH (Hrs)	52'852	51'315	51'580	0	48'406	48'529	46777	0
Operating Hours	43'152	41'352	43'298	46'654	39'387	40'159	38'289	45'558
1st TFO Firing	1994-10-21	1994-10-06	1994-10-30	N/A	1994-11-07	1994-11-11	1994-11-15	N/A
Turbine Washes	107	92	97		101	94	96	
Compressor Washes	24	23	23		25	14	15	
Trips	89	93	42	92	69	35	51	67
Total Mwhrs Since PAC	2'129'546	2'125'875	2'155'804	2767'740	1'985'402	2'067'284	1'886'955	2'696'887
Average Load MW	49.3	51.4	49.8	59.3	50.4	51.5	49.3	59.2
No of Starts Since PAC	437	442	382	178	389	352	387	143
Running hrs/start	98.75	93.56	113.35	262.10	101.25	114.09	98.94	318.59

FIGURE 7—52 Limay Bataan statistics [7-5].

guaranteed output is 615.2 MW (dry) and 651.2 MW (with power augmentation) at an ambient temperature of 32°C. Sea water from Manila Bay is used for the main cooling system.

The heat recovery steam generators are supplied by KHI under a licensing arrangement. Each supplies 35.2 kg/s of steam at 50 bar/493°C to the steam turbine and 8.7 kg/s saturated steam for ancillary services in the plant.

The steam turbine has an integrated HP/IP casing and a single flow LP casing exhausting to a conventional titanium tube sea water-cooled condenser. For startup or ST trips each HRSG can bypass steam directly to the main condenser. Each power unit is coupled to an air-cooled generator producing power at 13.8 kV to be stepped up to 230 kV en route for the switchyards and then into the North Luzon Grid to feed Metro Manila.

The operating mode of the plant is normally in combined cycle. However, each gas turbine can be operated anytime on simple cycle as gas bypass stacks and diverter dampers are fitted before the HRSG. Figure 7—52 provides operational data.

The efficiency of the plant is—based on heavy fuel oil operation—at 46.6%.

Operating Experiences

Fuel and Fuel System

The residual fuel produced in the refinery is a very heavy product with varying quality, as can be seen in Figures 7—53 (specification) and 7—54 (actual supplies, demonstrated on the Na + K and heavy metals contents). The product would not meet the Philippines environmental standards with respect to SO_x emissions. Hence this "IFO" (intermediate fuel oil) must be blended with a low sulfur diesel oil (DFO) at the site to achieve a maximum level of 2.75% sulfur in the blended fuel oil (BFO). During the last year, due to some changes in the refining process, the refinery was able to provide fuel that met the requirements.

PROPERTY	UNIT	VALUE BFO	VALUE IFO
Viscosity	50°C, cSt	210	205
Density	15°C, g/cm3	0.97	0.95
Water + Sediment	%	1.0	0.12
Sulfur Content	%	2.75	3.2
FBN	%	0.245	0..1
Ash	525°C, ppm	550	190
Vanadium	ppm	100	54
Na+K	ppm	50	15
Pb+Zn	ppm	10	0.5
Ca	ppm	2	0.3
Ni	ppm	30	10
Sum heavy metals	ppm	100	67
Thermal stability	ASTM 4740	2	2
pH	--	7-8	

FIGURE 7—53 Typical properties of BFO and IFO [7-5].

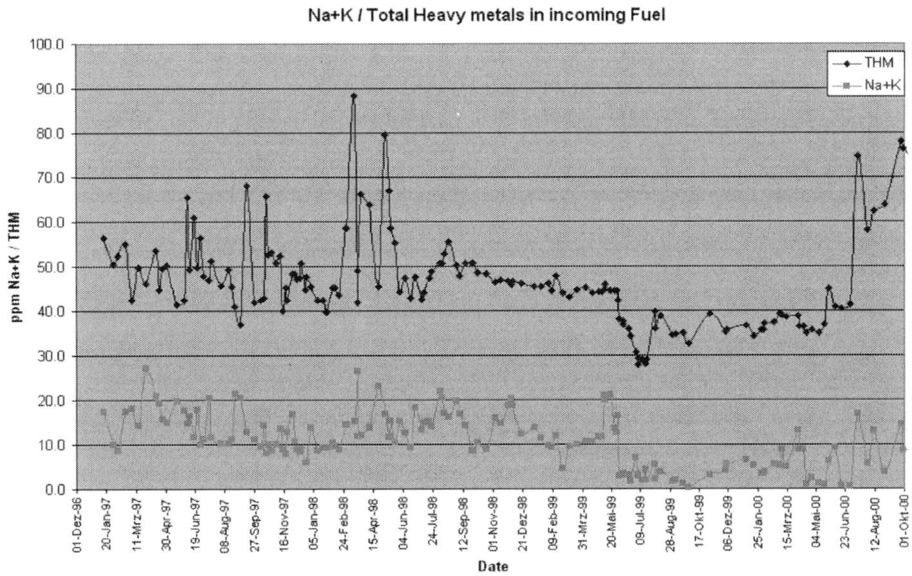

FIGURE 7—54 Actual fuel analysis as supplied (BFO) [7-5].

This situation, however, was only temporary, and by July 2000, the existing fuel mixing system was again in operation.

The fuel system is therefore quite complex and requires attention as it is not only a multi-step process, but also has to feed the two blocks simultaneously and continuously with fuel that meets the gas turbine fuel specification and can achieve the environmental standards set by the Department of Environment and Natural Resources (DENR). The system is described in Figure 7—55. The raw fuel, IFO, is stored in two large 15,000 m^3 tanks and is blended with diesel oil to meet the sulfur content criteria. As the sodium content is also quite high, fuel washing is required.

Each block therefore has its own fuel treatment plant (Figure 7—56) based on a two-stage washing process with electrostatic separation of the two phases, water and oil. This process adds water to the fuel to extract soluble salts and hydrophilic matter. This water is then separated by applying an electrostatic field that causes the water droplets to coagulate and separate out by gravity force. The separated oil must meet a sodium plus potassium content of <1 ppm and should therefore contain an absolute minimum of residual water. The extraction water must be virtually free of oily matter as it may only contain 10 ppm of residual oil. The fuel is then stored in four tanks of 5000 m^3 at approx. 60°C and then preheated to approx. 110—130°C for the combustion process in the GT combustor. In order to minimize high temperature corrosion caused by vanadium pentoxide, an Mg-based additive is added immediately before combustion to the fuel.

Experience with the fuel handling has shown various problems that occurred as a function of time. One occurred after three years of operation when suddenly the separation of the water was extremely difficult to handle and multiple recirculations of the oil became necessary to meet the requirements. Investigation of the problem revealed that part of the water/oil mixture had partially become a quasi-stable emulsion that did take longer to separate, although different demulsifiers were added.

A detailed fuel analysis finally showed that the pH of the fuel, due to processes at the refinery, was in the higher alkaline range and varied between 8 and 9.5 instead of 7—8. In this pH range the alkalinity is causing ideal conditions for stable emulsions. The reason for this alkalinity was investigated and found in the refining process.

When processing crude oils, the normal practice is to reduce chloride salts to very low concentrations via electrostatic desalting techniques. Primarily this is done to control chlorides carry-over into the distillation unit and thereby minimize the formation of corrosive hydrochloric acid. In cases where efficient desalting is not employed and where unacceptable levels of chloride salts may enter the distillation column, many refineries adopt the less desirable use of sodium hydroxide (caustic soda, NaOH) for corrosion control.

The main purpose of the sodium hydroxide is to convert the more volatile calcium and magnesium chlorides in the crude oil to the more stable sodium salt. Under high temperature distillation conditions, sodium chloride hydrolyzes less than calcium and magnesium chlorides, and therefore the rate of dissociation to form corrosive hydrogen chloride (HCl) is reduced.

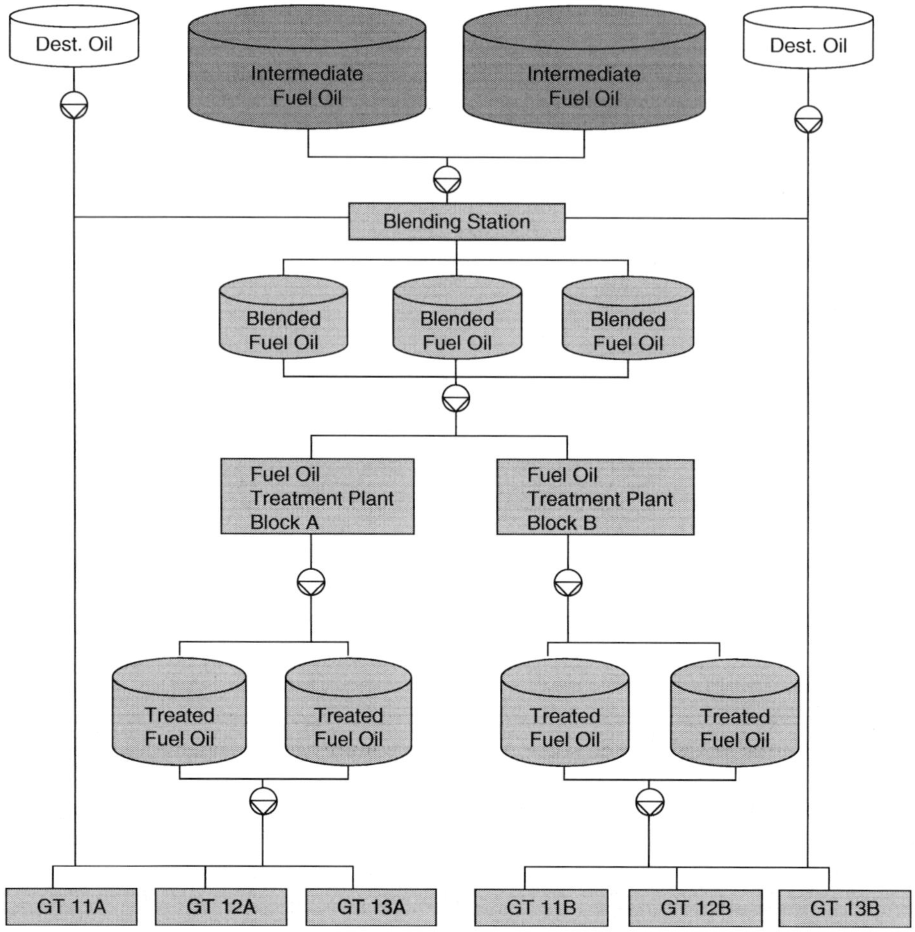

FIGURE 7–55 Fuel tank system drawing [7-5].

Also, naphthenic acids can be corrosive under certain high temperature conditions, and sodium hydroxide is sometimes utilized to control this type of corrosion in refinery equipment. Obviously, this may solve these refinery problems, but without looking at the effects in the power plant, where the presence of sodium is highly critical and treatment is carried out to remove sodium from the fuel.

FIGURE 7–56 Fuel treatment plant picture [7-5].

Essentially, the problems in the power plant result from the fact that it is difficult for refineries to add exactly the required concentration of sodium hydroxide for corrosion control, and the common tendency is to overtreat with an excess of NaOH. This results in high pH alkaline conditions in refinery effluent water streams, and the carry-over of "spent" sodium hydroxide into the residual oil fractions of the crude oil distillation process.

These high pH conditions, particularly if associated with oils containing naphthenic acids, result in the stabilization of oil-in-water emulsions and may lead to the generation of unacceptable oily effluent water streams. Experience shows that for pH of the effluent water higher than approximately 9.2, an effective effluent water treatment becomes difficult.

As the fuel spec in NPC's Fuel Supply Agreement and in the GT fuel specification had pH requirements, the solution had to be found together with the refinery and the OEM of the fuel treatment plant. After thorough testing of

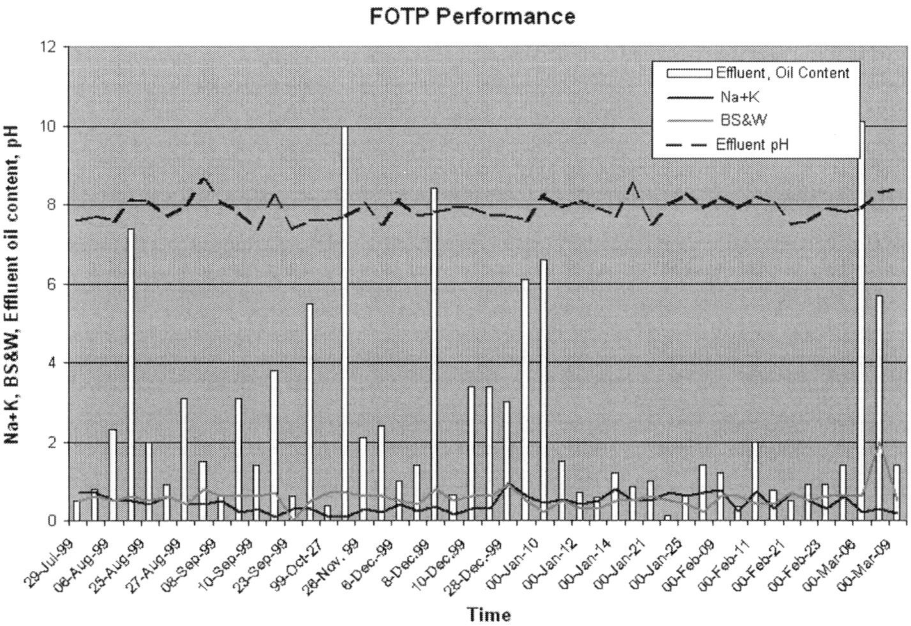

FIGURE 7–57 FOTP performance [7-5].

various demulsifiers with the fuel/water emulsion in the laboratory and in the field, and the introduction of a more strict quality assurance program on the refinery side, the fuel treatment plant is now operating with higher tolerance on pH. Excursions on the pH are limited as a consequence of stricter quality control of the fuel in the refinery.

The pH problem also caused a consequent problem in the treated fuel oil tanks. Usually the fuel arrives with 0.1–0.3% of bottoms, sediment, and water (BS&W), which would partially settle out. The above-mentioned fuel instability caused formation of a very high amount of sludge that accumulated at the bottom of the tanks. It was noticed that the fuel quality started to deteriorate at the tank outlet and an investigation was carried out by intensive sampling of the tank. This investigation revealed that approximately 250 m^3 of sludge type material was present in one tank; hence cleaning was the only way to restore the full tank capacity. The cleaning process was executed by a specialized company that separated fuel from the sludge such that approximately 90% could be recycled as fuel and the rest, cumulating in 66 drums of combustible material, was processed under control of the Department of Environment and Natural Resources (DENR). Tank bottom surveillance and regular tank sump drainage is now part of the daily operation program.

It can be summarized from this eight year experience in treating various fuels in Limay with the electrostatic equipment that this equipment is indeed maintenance friendly, as there is hardly any maintenance required and the process is very stable and runs reliably. Except for the case above and the wide range of basic fuels that was used, the electrostatic fuel oil treatment plant needed no further adjustment to achieve the requested fuel quality. The performance of the plant is depicted in Figure 7–57. Starting and stopping the plant has not been a problem. A restart of the process is easy and can be executed without difficulties. Within 4–5 hours after initiating the process, the plant is producing quality fuel at full throughput.

The fuel quality needs to be closely monitored for the following reasons: On one hand, the content of trace metals such as vanadium and sodium are critical for avoiding high temperature corrosion and on the other hand, physical and chemical parameters determine emissions and combustion quality. Therefore, a complete laboratory equipped for water and fuel analysis as well as emission monitoring is also part of the site.

Gas Turbine Operation

Burning residual oil in gas turbines involves skills and special precautions, because it is a low grade, ash-bearing fuel. This ash is corrosive and causes slag type deposits.

Sodium (Na) and vanadium (V) represent the main challenge. During combustion they form salts and oxides that are highly corrosive and lead to plugging and fouling. Alstom Power has developed the skills to apply ash-bearing crude and heavy oils successfully in the 50 years since it pioneered the burning of "low grade" fuels in its heavy-duty industrial gas turbines.

The two important steps to this are to (a) wash the fuel, i.e., to remove the alkali metals and (b) inject magnesium (Mg) additive to inhibit the vanadium by forming high melting Mg-vanadates. The resulting magnesium ash deposits and buildup on the hot turbine parts (particularly the blades) lead to deterioration of performance. Fortunately these deposits can be removed by shutting down the machine and water washing. Most of the deposits have a ceramic nature and spall off with the temperature differences and after cooling the remaining deposit is washed off with water, injected via nozzles strategically positioned within the turbine. After such "offline" washing the machine performance is restored.

Experience has shown that the critical issue in the first stages is to keep the deposits washable, as they tend to form oxides instead of the washable sulfates and hence they may be more difficult to remove by washing (Figure 7–58).

The deposit is thermally structured, i.e., the oxides usually prevail at the surface whereas sulfates are still present in the lower layers of the deposits. With decreasing temperature in the machine, the deposits are mainly sulfate and vanadate, with increasing V_2O_5 being present at the last stage of the blades as a yellowish powder, together with $MgSO_4$.

During the offline washing, the diverter dampers are closed and exhaust gases are led to the bypass stack. The deposits that are washed off are collected in the diverter damper sump area. This process is typically run until the wash water is clear. An immediate restart of the unit after the washing further removes deposits by the physical action of water in the deposits that evaporates and spalls deposits off. In general, the cleaning effect is very good and power is regained completely. A wash cycle, including cool-down and restart, has a typical duration of 12 h.

FIGURE 7–58 Turbine vanes and blades after washing [7-5].

Figure 7–58 shows the effect of the deposition of ash and its removal by washing. The pictures show the blading as it typically can be seen after the washing process. The washing is quite effective and restores most of the power and heat rate losses in the turbine.

The plant also performs regular compressor washing as a standard procedure, usually when the turbine is washed. In order to regain the complete mass flow, the compressor is only washed offline. As can be seen in Figure 7–59, the combination of compressor and turbine cleaning enhances the washing effect repeatedly and considerably. The deposits on the compressor side contain soot and hydrocarbonaceous material. Therefore a biodegradable compressor cleaner is added to the water.

HRSG Operation

Deposition in the HRSG is mainly consisting of $MgSO_4$ and V_2O_5, which has not deposited on the blading and which is the major portion of the ash. As the temperature is already quite low, sintering is no longer occurring; hence the deposits are light and easily removed (Figure 7–60).

Cleaning of the boiler is performed by soot blowing with a substantial effect on the heat transfer regain. Typically soot-blowing is carried out daily and results in a constant regain of power output. As the back pressure is also reduced, the gas turbine improves in output again (Figure 7–61).

Soot blowing is quite effective in cleaning the boiler tubes. This method may even gain focus in terms of the acidity of the deposits. In the lower parts of the cold end, any remaining MgO would of course attract free SO_x and neutralize it. These deposits can also absorb free SO_3 and form acidic deposits. It is therefore not surprising that washing water is rather acidic and acts as a sulfuric acid at the outlets of the wash water. Sufficient corrosion allowance and replacement when needed, combined with regular monitoring for safety reasons, are the main measures taken against acid corrosion up to now. There may, however, be an additional improvement by introducing a modified washing of the HRSG in order to reduce the effects of the sulfuric acid. The same study that is dealing with soot blowing will also investigate these possibilities, which will be reported on at a later date.

Heavy Oil Operation vs. Diesel Oil Operation

It is often questioned whether the operation on a heavy fuel oil is attractive in light of the additional measures that have

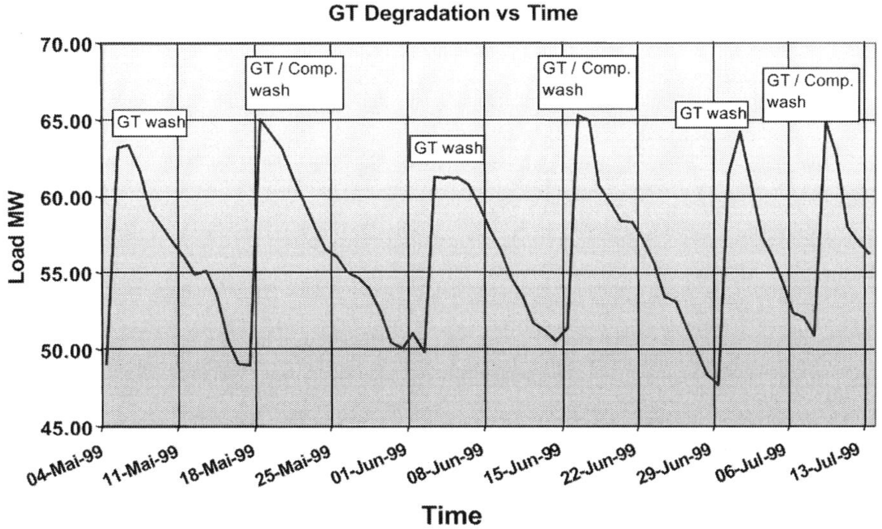

FIGURE 7–59 Effect on power output [7-5].

to be taken in order to reduce the immediate effects of the combustion of this fuel. Additional efforts are the fuel treatment itself and the additive dosing. Due to the corrosiveness of the environment hot gas path parts show higher wear and tear than comparative parts in clean fuel gas turbines. Comparing the effort on an annual basis, the additional cost for treatment, including all chemicals, maintenance, and operation of the FOTP is <5% of the annual fuel costs. Comparing the cost with the additional cost running on a fuel such as diesel oil, the savings are related to the annual diesel costs, in the range of approximately 10%. This takes into account that the diesel to heavy oil index has changed from >2 in 1996 to about 1.4 in 1999, which makes the difference smaller.

Taking into account the side effects on the turbine output and efficiency, these numbers are reduced to a benefit of

FIGURE 7–60 Pictures of HRSG deposits. [7-5] may change; however, the heavy fraction would still be produced and would need a user.

roughly 7% for heavy oil operation. The fact that this fuel is available from the refinery next door and must be used in a cost-effective way makes the scenario quite attractive.

The inspection concept of A, B, and C inspections has proven to give enough information to be able to plan the major (C-) inspections in detail and hence be efficient. Nevertheless the gas turbine is the most delicate part of the plant. Incidents in the early days were fortunately limited to a second stage blade failure caused by high cycle fatigue, in combination with a resonance excitation from grid instabilities. The blade was modified for higher tolerance and has operated satisfactorily ever since. In fact, part of the blading is now reaching longer lifetimes than actually expected under these conditions.

In order to be able to carry out the major inspections quickly and efficiently, the individual steps of the inspection were critically analyzed. This analysis was then transferred into a work plan. At the same time it was also decided to invest in additional hardware on site, which can be preassembled to a higher degree and hence reduces the actual outage time. All these actions reduced the original outage time from roughly four weeks to only 12–15 days.

As an example, one of the decisions in O&M was to have fully bladed vane carriers on site that could be prepared for exchange while the unit was still in operation and then be exchanged as a whole. The blading and deblading process of the stationary blading is therefore outside the inspection period and also the parts removed from the machine can be moved out of the space where the turbine is being inspected. There is more room, which improves overview and planning. Similar solutions have been found for the combustion chamber, which is now overhauled with the same principle. Using a second combustor to replace the one going for inspection has also given a boost to

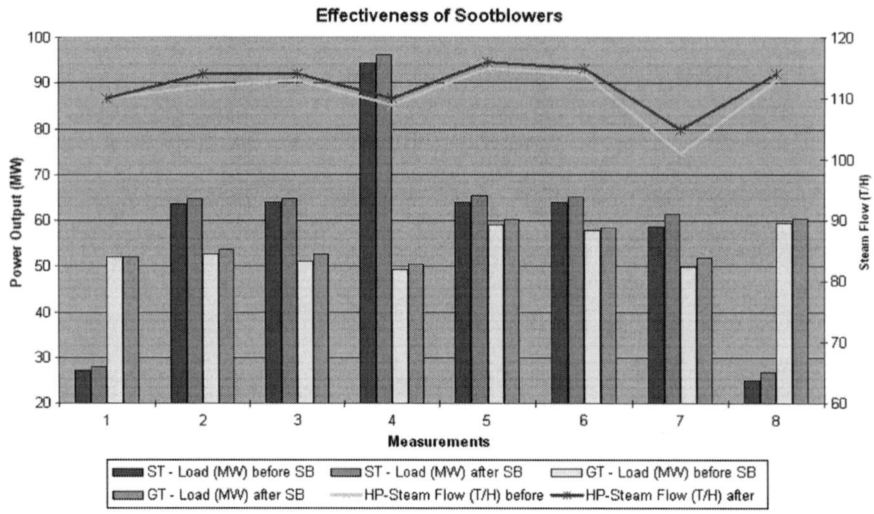

FIGURE 7–61 Effectiveness of soot blowers [7-5].

quality in a sense that during the pre-assembly, there is enough time for quality assurance during assembly and training on the actual material.

As there are six units that need to be inspected, the local team has developed excellent skills in performing the inspections. The Limay Bataan team will now only require a few specialists, and the major portion of the work can be organized locally, in close cooperation with Swiss headquarters. At the time of writing this case study, the first two gas turbines of Block "A" were again inspected after roughly 50,000 equivalent operating hours. Both units were back in operation after 15 days.

Plant Performance and Availability

Availability is defined as the period of time when the plant or part thereof is available for operation and is evaluated as an equivalent availability, i.e., if the load must be reduced for reasons that the operator has caused, the duration of this part load time is expressed as equivalent full load time loss.

Availability can be kept at a reasonably high level by proper operation and planning, considering the necessities of the specific fuel. The availability of the plant has been carefully monitored and actions were taken to minimize outage times, as described earlier.

Forced outage events were mainly related to the gas turbines, as the fuel effects nevertheless impacted the availability. Washing of the turbine is a necessary outage, and counts as a non-availability period. The high levels of availability are also influenced by the fact that the units are not actually dispatched by the National Grid Company to run at station base load for economic reasons. This allows longer intervals between washings, and flexible unit operation, providing the required power at all times.

CASE STUDY 2: AUTOIGNITION CHARACTERISTICS OF GASEOUS FUELS AT REPRESENTATIVE GAS TURBINE CONDITIONS*

The autoignition properties of gas turbine fuels have been studied for many years and by numerous researchers. The advent of ultra low emission industrial gas turbines using lean premixed technologies has given rise to premixer designs with longer residence times. This, in conjunction with the ever-increasing pressure ratios of aeroderivative machines, leads to the potential for autoignition within premix ducts, and has therefore renewed the interest in this field. Although much has been published, data in the region of interest to high-pressure ratio gas turbines are extremely sparse. Similarly, modeled autoignition delay times are not very accurate, as most reaction mechanisms were not generated to cover this range of conditions. Hence the uncertainties of autoignition delay times at gas turbine conditions are significant, thereby either imposing over-stringent design limitations or introducing risks of ignition occurrence in the early design process.

A series of experiments has been carried out for methane and simulated natural gas fuels in the region of interest, using shock tubes as the test vehicle. The experimental technique was chosen to isolate only the chemical kinetic component of the autoignition delay time, without any additional delays due to mixing and heating of the test gases. Predictive correlations and a chemical kinetic model (the

* Source: [7-6] Courtesy of Rolls Royce. Adapted extracts from C. J. Goy, A. J. Moran, and G. O. Thomas, "Autoignition Characteristics of Gaseous Fuels at Representative Gas Turbine Conditions," 2001-GT-0051.

GRI mechanism) have also been used to predict autoignition delay times at the same conditions. The correlation between experiment and prediction has been shown to be poor at representative temperatures. This case discusses some of the possible explanations for this poor agreement.

As worldwide emissions legislation becomes ever more stringent, there is a requirement for combustion engineers to design gas turbine combustors with the capability to produce extremely low levels of NO_x and CO in the exhaust. Such low levels of pollutant emissions can only be achieved by extending our understanding of current premixers, to maximize the mixing quality of fuel and air prior to entry into the combustion process. However, with the elevated inlet temperatures and pressures characteristic of high-pressure ratio aeroderivative machines, a limit is reached where the time required to fully premix the fuel and air streams becomes comparable with the autoignition delay time for the combustible mixture. A compromise is therefore sought between optimum mixing quality and freedom from autoignition. During the design process, this compromise is currently achieved by experiment. This approach is costly and time-consuming, as it involves the manufacture and testing of many design iterations. If validated predictive chemical kinetic schemes were available, and incorporated into computational fluid dynamics (CFD) codes, then the combustion engineer could have access to a predictive tool, for the optimization of future designs at minimum cost and in shorter timescales.

There are currently very little autoignition data available in the public domain covering the range of initial temperatures and pressures relevant to high-pressure ratio gas turbines. Many researchers have studied the autoignition characteristics of natural gases in shock tubes at high temperatures, typically above 1200 K, which is significantly above the inlet temperature range applicable to aero-derivative gas turbines. Much work has also been performed in flowing rigs at representative gas turbine temperatures, but to only moderate pressures. A gap exists in our knowledge over the range of temperature from 800–1000 K, and pressure from 10–40 atm. It is crucial that data are gathered in this region to enable the validation of chemical kinetic reaction mechanisms for use in combustion design tools.

This case presents the results of a series of shock tube experiments, aimed at determining the autoignition delay time of methane and natural gas mixtures at representative gas turbine combustor inlet conditions. The main objective is to determine the major factors influencing autoignition delay time in gas turbine combustion systems. Firstly, the autoignition characteristics of methane were determined over a range of temperatures and pressures. Then, to determine the sensitivity of autoignition delay time to variations in natural gas fuel composition, a series of tests were carried out with both ethane and propane added to the methane test gas. The effects on autoignition due to variations in stoichiometry and humidity levels were also investigated. The validity of the current experimental data is discussed, particularly in the lowest temperature range, where the measured autoignition delay times approach the observation time limit of the shock tube.

The resulting experimental data are compared against predictions from the GRI chemical kinetic mechanism and correlations from previous researchers (consult complete source work for reference list), to assess their suitability in predicting autoignition delay times in the gas turbine combustor. Finally, several suggestions are offered to explain why the agreement is poor at lower temperatures.

Nomenclature

τ = Autoignition delay time, in seconds
T = Temperature, in kelvin
P = Pressure, in atmospheres
$[O_2]$ = Oxygen concentration
$[CH_4]$ = Methane concentration
$[HC]$ = Non-methane hydrocarbon concentration (all in molecules/cubic centimeter)

Measurements of Autoignition Delay Time

All of the experimental data were generated in a shock tube of length 6.75 m and internal diameter 64 mm, shown schematically in the upper part of Figure 7–62. The driver and test section lengths were 3 m and 3.75 m, respectively, and a double mylar diaphragm separated the two sections prior to initiation of the test. Helium gas was used in the driver section to minimize boundary layer effects in the tube.

For the majority of the work presented here, the test gas simulated a methane/air mixture with an equivalence ratio of 0.5. Experiments were also performed on a number of other test gases, to investigate the effect on autoignition delay time of variable fuel composition, stoichiometry, and humidity. Each test gas mixture was produced in a vessel by the method of partial pressures, and allowed to mix thoroughly before its introduction into the shock tube. Before filling, the tube was brushed clean of debris from the previous test, and then evacuated. The length of the shock tube enabled a maximum test observation time in the reflected shock region of approximately 6 milliseconds, as shown schematically by the progression of the shock fronts with time in the lower part of Figure 7–62. The long observation times required for lower temperature autoignition measurements can cause the experiment to be compromised, due to low shock velocities in a long tube enabling boundary layer build-up and hence bifurcated shocks within the test section. This will be discussed later.

For all of the experimental work presented here, the onset of pressure rise at the test section end wall was used as the indicator of an autoignition event. This was chosen because

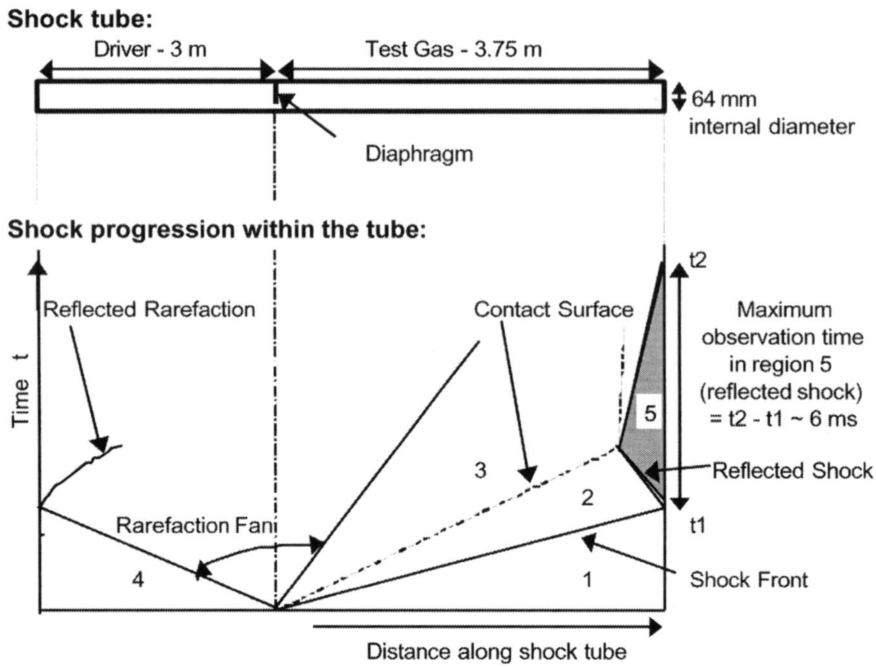

FIGURE 7—62 Schematic of shock tube. *(Source: Rolls Royce.)*

an increase in pressure is a definite indicator of exothermic reaction. As the ignition occurs at the end wall of the shock tube, the autoignition delay time is defined as the time interval between the shock front reaching the tube end wall and the pressure rise due to combustion. Light emission was recorded near the end wall of the shock tube at two wavelengths, corresponding to chemiluminescence from OH (308 nm) and CH (431 nm), and this gave good agreement with pressure rise data where all were measured simultaneously. The experimental results cover a range of initial temperatures from 900—1700 K, and initial pressures from 5—20 atm. The results are described for each test gas composition in turn.

During the course of the current experimental program, a series of control tests were performed to determine the characteristics of the shock tube without any fuel in it. The test gas mixture used in this case was bottled dry air. The results in Figure 7—63 show that, over the entire temperature range, light emission was observed at 431 nm in the unfueled tests at delay times of the same order as the recorded pressure rise during equivalent methane-fueled tests. At temperatures below 1000 K, the light emission was seen after shorter delay times than were typically expected for the autoignition of methane. This indicates that there may be some other source of ignition within the tube, which may affect the validity of the lower temperature data. However, no corresponding pressure rise was observed during the control tests, indicating that there was no significant exothermic reaction in the tube. As the current fueled tests used pressure rise as the indicator of autoignition, the presence of light emission in the control tests is not likely to affect the results presented here. This will be discussed later.

The autoignition delay time for methane is presented as a function of temperature and pressure in Figure 7—64. Predictions made using the GRI mechanism are also plotted; these will be discussed in the next section. The experimental data in Figure 7—64 show that the delay time is strongly dependent upon initial temperature, and is affected by pressure to a lesser extent; autoignition delay time reduces when either temperature or pressure is increased.

At the upper temperature limit of the present data, it is shown to be in good agreement with previous (1971) works. No other shock tube data were readily available for comparison with the lower temperature results, but flow rig data from a 1986 study, indicated on Figure 7—64, suggest a marked difference between the two experimental techniques at these conditions. Significantly longer delay times were observed in the flowing tests, possibly due to the influence of fuel/air mixing time on the overall time taken for autoignition. The current data appear to show a change in the activation energy at an initial temperature of approximately 1100 K at 20 atm. Similarly, the 10 atm data undergo a change in activation energy around 1300 K.

Having characterized the autoignition behavior of methane, the effect of adding 15% ethane (by volume) to the test fuel was investigated to simulate a possible extreme of natural gas composition. The results are shown in Figure 7—65, where an initial pressure of 10 atm and overall equivalence ratio of 0.5 were maintained for all tests. The data indicate that the autoignition delay time is

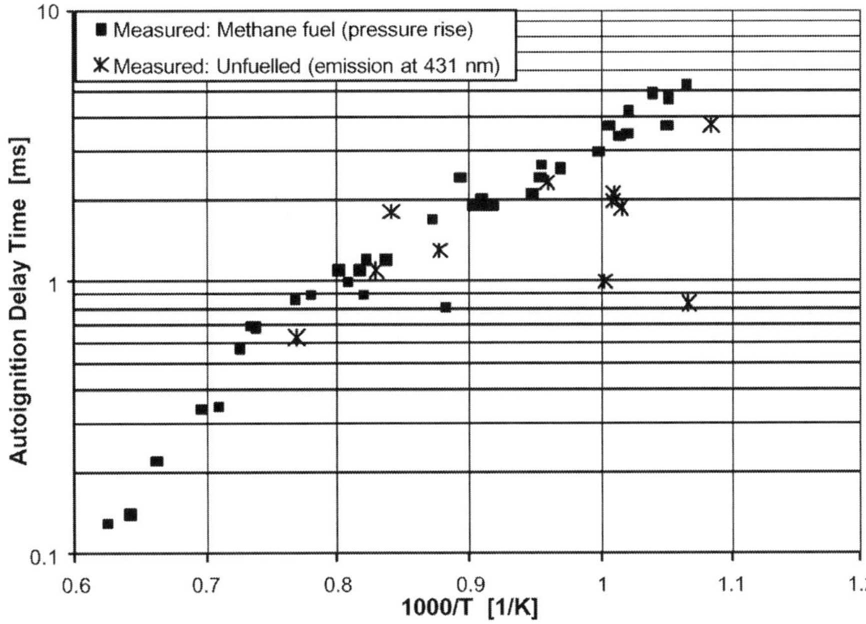

FIGURE 7–63 Emission from unfueled shock tube tests (initial pressure: 10 atm). *(Source: Rolls Royce)*

significantly shortened by the addition of ethane for initial temperatures greater than 1200 K. Below this temperature, the added ethane appears to have no effect on the autoignition characteristics of methane.

To further assess the impact of natural gas constituents on autoignition, a similar series of tests was performed with 15% propane added to methane. This data series does not extend to a sufficiently high temperature to fully determine its behavior relative to ethane. However, for all three test gas mixtures, shown graphically in Figure 7–65, the delay times begin to tend almost asymptotically towards the observation time limit of the shock tube at temperatures below 1200 K. Under these lower temperature conditions, the results collapse onto one curve, such that the effect of adding this amount of higher hydrocarbons to methane is negligible below 1100 K.

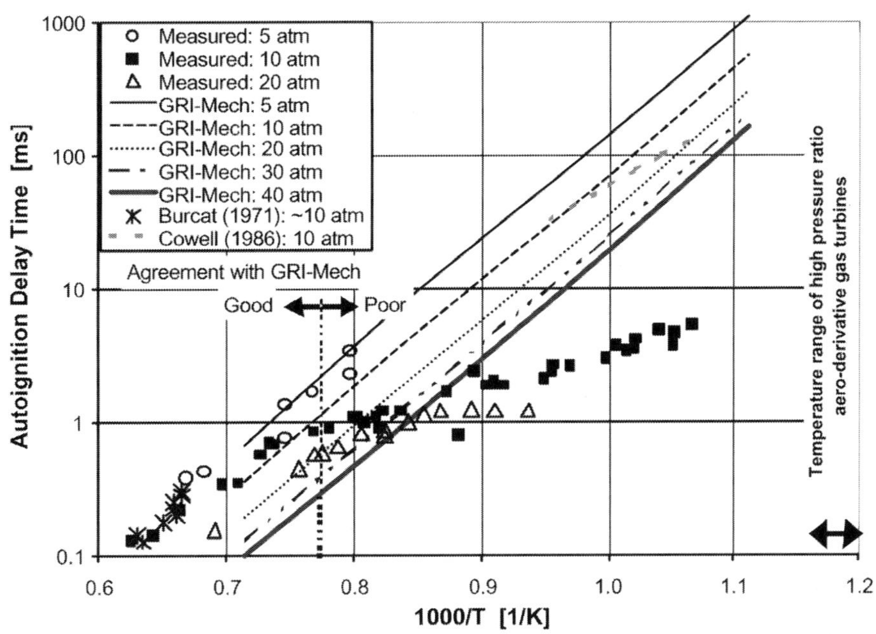

FIGURE 7–64 Methane autoignition delay time as a function of temperature and pressure. *(Source: Rolls Royce.)*

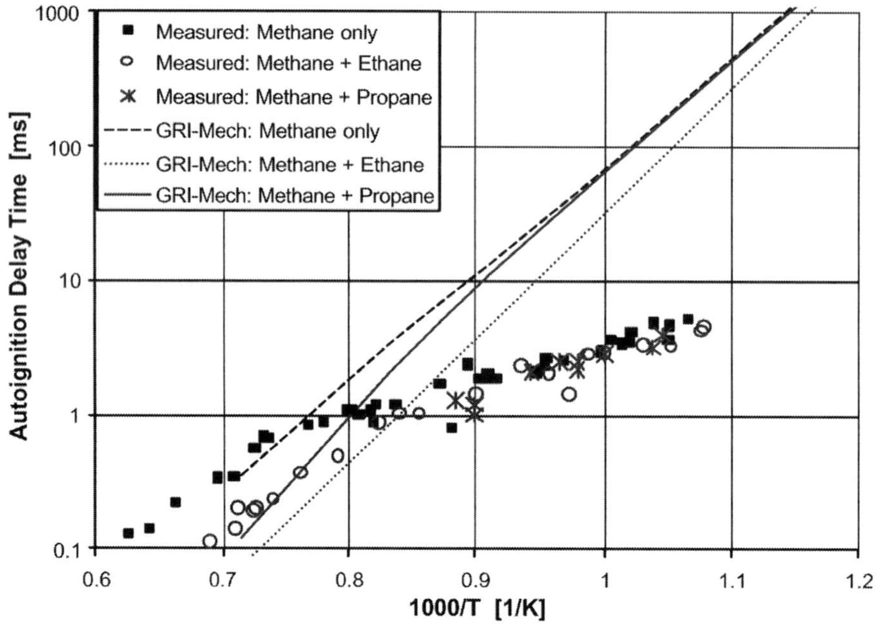

FIGURE 7–65 Effect of non-methane natural gas fuel. *(Source: Rolls Royce.)*

Lean premixed gas turbine combustion systems tend to be optimized to operate at equivalence ratios of around 0.5 at full load conditions, but this may vary widely during engine maneuvers. It is therefore important to understand the effect of variations in equivalence ratio on autoignition delay time, to prevent ignition events occurring during transient operation. Much of the data in the public domain suggest that the equivalence ratio has only a limited effect on the autoignition delay time. The current data, presented in Figure 7–66, confirms that varying the stoichiometry between 0.5 and 1 at 10 atm has only a small effect on the measured delay time. At temperatures above 1200 K the effect was negligible, but at lower temperatures the higher equivalence ratio tended to give rise to slightly shorter autoignition delay times.

In the final series of autoignition tests, varying amounts of water were added to the test gas to simulate the effects of high air humidity on autoignition delay time. A simulated methane/dry air mixture was used as the baseline test gas, with all experiments carried out at an initial pressure of 10 atm.

Although only a limited number of experiments were carried out with added water, the results in Figure 7–67 indicate that its effect was to slightly increase the autoignition delay time relative to the baseline test. It is also interesting to note that, for the test series with 8% water added, a sharp cut-off temperature was defined (at approximately 1250 K), below which no autoignition occurred.

Discussion

Where other published shock tube data are available, the agreement with the current data is good. However, it is acknowledged that there is a shortage of data in the region of interest to gas turbine combustion. The data tend to be at high temperature (greater than 1200 K), and pressures that are either too low (below 10 atm) or too high (over 50 atm).

The experimental data suggest a change in activation energy at low temperature (below ~1200–1300 K, depending on the initial pressure conditions). This is depicted by a change in the slope of the data series. The change in slope could be due to other factors than a change in activation energy. Several possible explanations are briefly discussed below:

- *A change in the chemistry important to autoignition (i.e., change in activation energy).* It is possible that the dominant chemistry at lower temperatures differs from that at higher temperatures. This has also been reported by Petersen et al. (1999) and Li and Williams (2000). The lower temperature chemistry may be dominated by simpler reactions, such as oxygen abstraction, the result of this being that the effects of temperature, pressure and added higher hydrocarbons become small, as was observed experimentally in Figures 7–64 to 7–67.
- *A physical limitation of the shock tube.* As the delay time approaches the shock tube's observation time limit, the measurements may be affected. Autoignition events may be spuriously recorded after artificially short delay times.

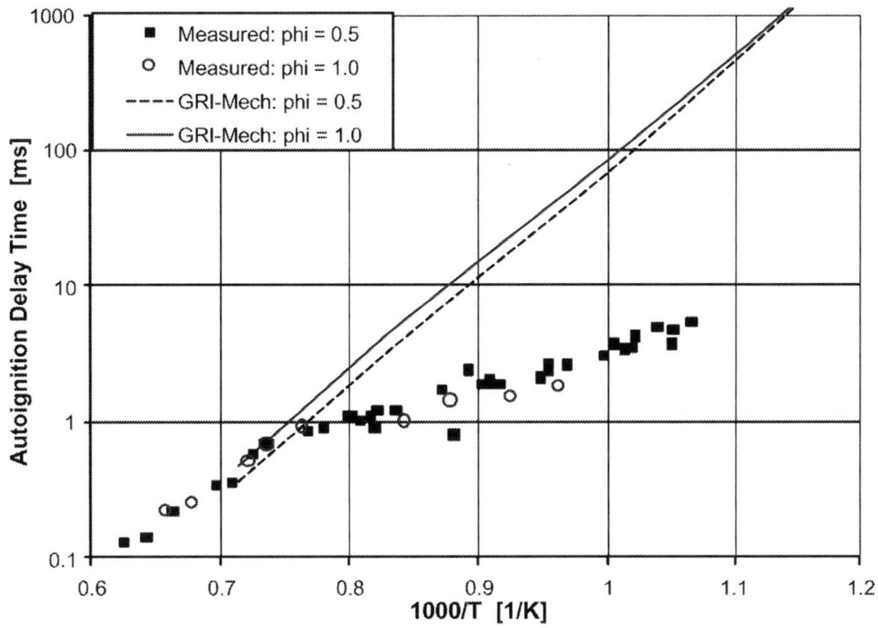

FIGURE 7–66 Effect of stoichiometry on the autoignition delay time of methane. *(Source: Rolls Royce.)*

The validity of this statement could be determined by the study of similar tests in different lengths of shock tube.

- *Ignition of the mylar diaphragm initiates reactions in the test gas.* The series of unfueled shock tube tests revealed light emission from the shocked gases, and a possible source of this emission is combustion of fragments of the mylar diaphragm. At high initial temperatures, where the autoignition delay time of the test gas is shorter than the delay time to emission from the mylar, the data are representative of the test gas mixture. However, at lower temperatures, where the delay time of the test gas becomes longer than that suspected of the mylar, the resulting data may be affected by the diaphragm material. As no pressure rise was observed, it is unlikely that sufficient heat would be generated to accelerate the methane reactions, but it is possible that it could act as a source of ignition. A study of the effect of alternative diaphragm materials would assess this risk.

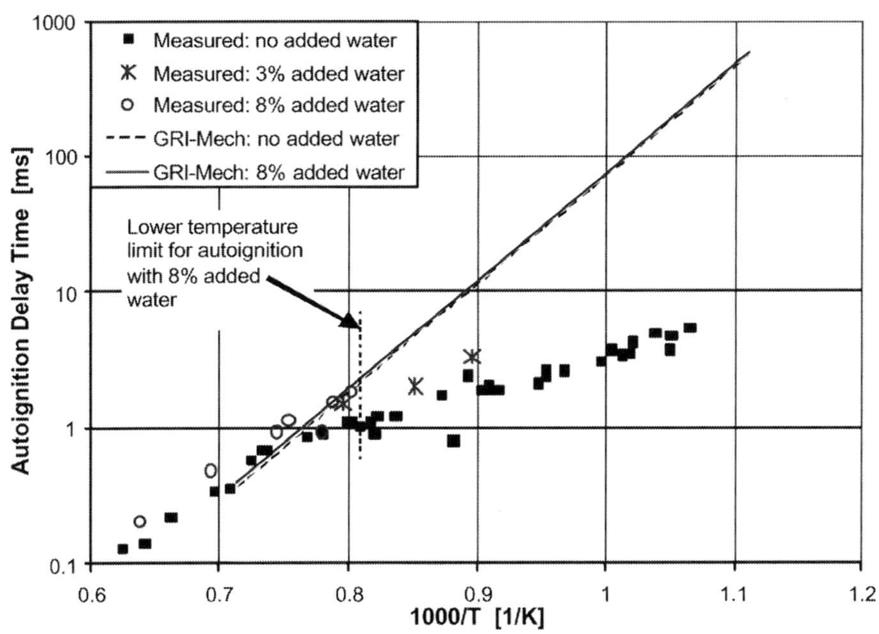

FIGURE 7–67 Effect of added water on the autoignition delay time of methane. *(Source: Rolls Royce.)*

- *Initial heating of the test gas by the incident shock.* The current work used the reflected shock technique for the determination of autoignition delay time. As the incident shock passes through the test gas section, it will heat and pressurize the gas, which may cause the initiation of some reactions. This may result in the measurement of artificially low delay times.
- *Non-ideal conditions in the shock tube.* When studying longer observation times, the likelihood of boundary layers forming in the test section increases. Replacing the nitrogen in the test gas mixture with a monatomic gas such as argon will minimize this effect, but preliminary comparison tests at 10 atm showed this to have a negligible effect on the resulting autoignition delay times.

If the integrity of the lower temperature data is shown to be good, several significant implications arise. Firstly, the suggestion that the actual composition of a natural gas fuel has no impact on the autoignition delay time for temperatures below 1100 K would, if confirmed, offer the potential to greatly simplify the predictive schemes. Also, for practical systems, the ability to elevate the minimum autoignition temperature by the addition of water could signify additional benefits for humid air cycles.

Without exception, all of the predictive methods assessed here show good agreement with the high temperature experimental data.

CASE STUDY 3: FROM CONCEPT TO COMMERCIAL OPERATION—TRI-FUEL INJECTOR USED FOR LPG AND NAPHTHA APPLICATIONS*

The tri-fuel injector is an extension of the standard dual fuel multi-passage injector to cover additional fuels such as liquid petroleum gas (LPG) and naphtha at medium pumping pressure (less than 40 Bara). There is an additional passage designed for metering the LPG or naphtha in liquid phase. The fuel system uses a non-contracting rotary pump with modern inverter technology for the pumping of these fuels.

Both tri-fuel injector and fuel system are now in commercial operation on two Tornado engines using LPG, natural gas and diesel fuels. The LPG system is operating just below 30 Bara pumping pressure, and has accumulated about 1029 operating hours on the lead engine. Commercial operation of the Typhoon on naphtha at medium pumping pressure (<40 Bara), with an improved tri-fuel injector, will commence in the near future.

* Source: [7-7] Courtesy of Alstom Power. Adapted extracts from A. Newman, P. S. Nixon, R. I. Wilms, and D. M. Taylor, "From Concept to Commercial Operation—the Tri-Fuel Injector Used for LPG and Naphtha Applications," 2001-GT-0079.

Nomenclature

CO = Carbon monoxide
FBP = Final boiling point
FSNL = Full speed no load
IBP = Initial boiling point
MPI = Multi-passage injector
NO_x = Oxides of nitrogen
OTDF = Overall temperature distribution factor
ppm = Parts per million
RMS = Root mean squared
RTDF = Radial temperature distribution factor
UHC = Unburned hydrocarbons

Alstom Power's industrial gas turbine range was originally designed for operation using standard pipeline quality natural gas, diesel, or both. However, recently there has been increasing worldwide customer interest to operate the turbines on the most economically available top industry and fuels such as liquid petroleum gas (LPG) and naphtha are cheaper alternatives to diesel.

Liquid petroleum gas (LPG) is a by-product of the natural gas treating process or an incidental gas recovered during the oil extraction process. LPG is generally composed of propane and butane, but it can contain lighter and/or heavier hydrocarbons that could be in unsaturated or isomeric forms. LPG fuel quality is closely controlled by international standards.

Table 7—20 shows the comparison of the main LPG fuel properties.

Naphtha is a generic, loosely defined term that covers a wide variety of light distillates. Naphtha is processed from crude oil through distillation towers in petroleum refineries and it is a primary ingredient in gasoline. It also has wide applications in the organic synthesis and polymer industries. Naphtha fuels are not recognized by standards in the same way as other liquid fuels, although compositions are reported in the literature. Table 7—21 compares fuel properties for gasoline and No. 2 diesel fuel samples with those for a typical naphtha fuel range.

This case describes the extension of combustion and fuel system technology already proven on natural gas and diesel to cover LPG and naphtha fuels. This technology is being applied to Alstom Power's Tornado and Typhoon engine range.

Design Objectives

Alstom's goal was to design a turbine fuel system that could use LPG and naphtha in liquid form, without the need to vaporize the fuel outside of the engine. Significant safety and cost advantages result if the external vaporizer is eliminated if this can be done without the need of a high-pressure fuel system.

TABLE 7–20 Main LPG Fuel Properties: Propane, Butane

Properties	Units	Butane	Propane
C_2 content	% Vol	NIL	5% max
C_3 content	% Vol	15% max	90% min
C_4 content	% Vol	85% min	10% max
C_5 content	% Vol	2% max	2% max
Olefins	% Vol	60% max	25% max
Vapor pressure at 40°C	Barg	5.05 max	15.6 max
Dynamic viscosity at 20°C	MPa.s	0.25 max	0.2 max
Specific gravity at 15°C		0.585 max	0.52 max
Heating value (nett)	MJ/kg	45.82	46.29
Autoignition temp (normal)	°C	360	450
Flammability limits	% Vol	2% to 9%	2% to 10%

(Source: Alstom Power.)

There are two engineering problems associated with the handling of liquid LPG and naphtha fuels. Firstly, the fuels must remain in liquid phase while flowing through the fuel system. Two-phase flows consisting of liquid and vapor will cause severe fuel maldistributions to the fuel injectors leading to unacceptable combustor exit temperature distribution and control system instability. Secondly, the fuel system components must be capable of handling low lubricity/high vapor pressure fuels. Lower pumping pressures allow the use of specialist LPG pumps as seen

TABLE 7–21 Fuel Properties: Naphtha, Gasoline, and Diesel

Properties	Units	Gasoline	No. 2 Diesel	Naphtha Range
Saturates	% vol	62	68	100 max
Aromatics	% vol	35	32	25 max
Olefins	% vol	4	<0.3	5 max
Distillation				
IBP	°C	33	200	30 min
50%	°C	96	297	
90%	°C	151	353	
FBP	°C	192	379	200 max
Vap. pressure 37.8°C	Bara	0.63	0.003	0.3 to 0.9
Kin. viscosity 37.8 °C	mm^2ls	0.7	3.8	0.6 to 0.9
Specific gravity		0.75	0.86	0.65 to 0.75
Heating value (nett)	MJ/kg	42.7	42.7	43 to 46
Autoignition temp (normal)	°C	260	245	220 to 290
Flammability limits	% vol	0.6 to 8		0.8 to 6

(Source: Alstom Power.)

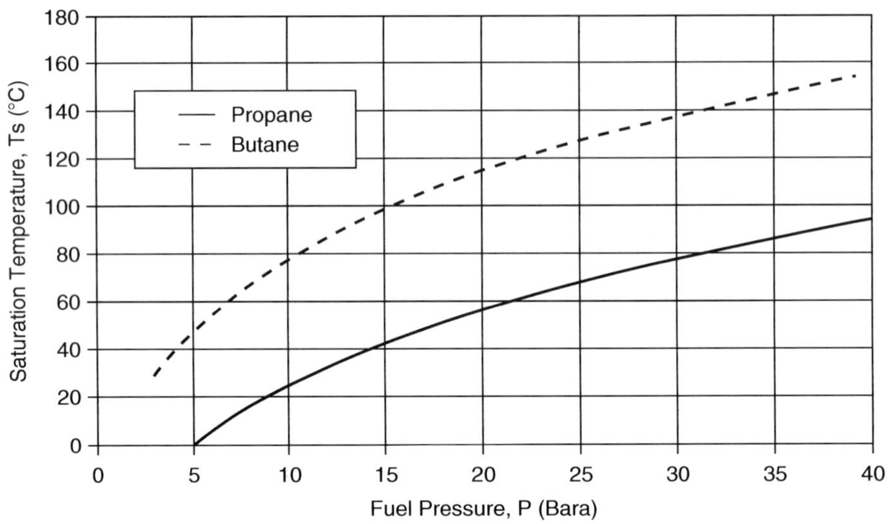

FIGURE 7–68 Vapor pressure curves for propane and butane. *(Source: Alstom Power.)*

later. The worst design case fuel is LPG propane due to its lowest viscosity and highest vapor pressure at a given temperature.

Figure 7–68 shows plots of saturation vapor pressure versus temperature for propane and butane LPG fuels. To maintain liquid phase the fuel pressure must exceed the saturation vapor pressure at the given supply temperature. Both naphtha and diesel have lower vapor pressures (Tables 7–20 and 7–21).

Alstom's goal was to achieve a maximum of 40 Bara fuel pressure by operating close to the fuel's liquid/vapor saturation line. This was achieved by using the thermodynamic properties of the fuel to refrigerate critical components of the fuel injector, this being the hottest part of the fuel system. It is understood that other systems for this type of application directly inject liquid LPG and naphtha at higher pressures between 80 and 90 Bara to ensure that the fuel is well above its saturation pressure.

System Overview

A liquid LPG fuel system was initially designed in order to meet a commercial requirement to supply tri-fuel capability (natural gas, diesel, and LPG) on two Tornado single shaft turbines. This system was designed to achieve maximum LPG pumping pressures of 30 Bara. The naphtha development is discussed later.

In order to meet tight delivery schedules set by the customer, the Tornados were delivered to site as standard dual fuel and the LPG was added later as a retrofit. This delivery matched the customer requirements to run mainly gas but with LPG standby during winter.

Burner and Combustor Installation

Figure 7–69 shows a typical Alstom combustion system, showing the injector to combustor interface. The Tornado system is comprised of eight reverse flow "diffusion flame" tubular combustors. The Typhoon has six similar combustors. There is always one injector per combustion chamber.

In a tri-fuel LPG system, the fuel system, like the fuel injector, allows the use of fuels close to their vapor point. The system comprises a metering and shut-off valve train inside the turbine enclosure and a separate pumping skid outside of the enclosure.

Due to the widely different thermodynamics properties between diesel and LPG, it was considered less risky to add a separate LPG pump stream to a proven dual fuel system, than to risk the consequences of modifying a single diesel pump stream to cater for LPG. There are also a number of advantages in using separate diesel and LPG pump streams, instead of a single pump stream. These are (a) no risk of cross-contamination between LPG and diesel tanks, (b) able to deliver turbine as standard dual fuel and add LPG later, (c) no requirement to purge the complete LPG system before restarting the turbine on either diesel or natural gas fuels.

Tri-Fuel Injector Design

The tri-fuel injector is an extension of the dual fuel multipassage injector (MPI) developed jointly with Delavan. The tri-fuel injector (Figure 7–70), like the MPI, has separate passages for No. 2 diesel fuel (pilot and main), gas fuel, compressor discharge air, water, and steam (for

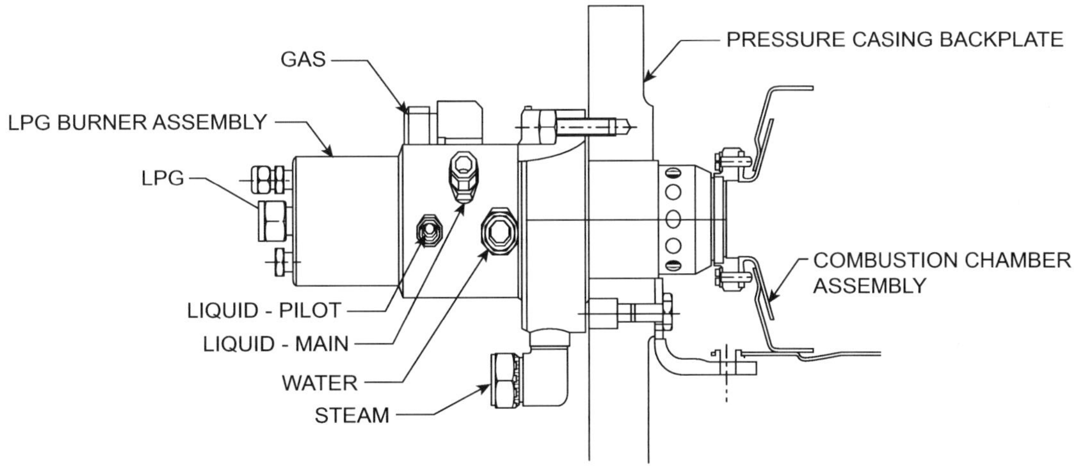

FIGURE 7–69 Typical combustion assembly. *(Source: Alstom Power.)*

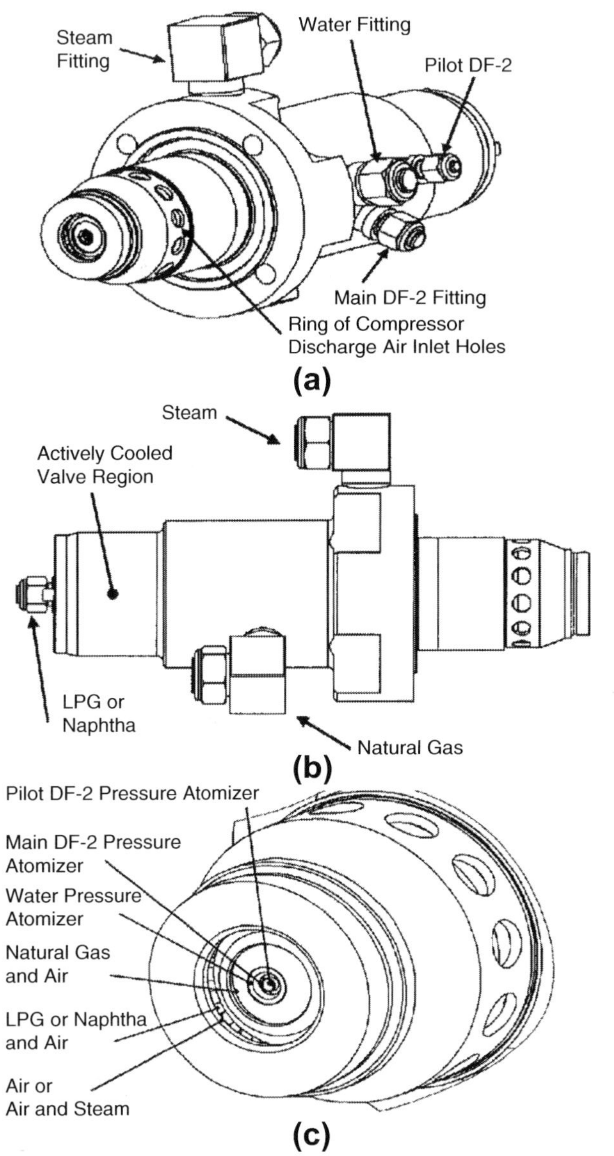

FIGURE 7–70 The tri-fuel injector—general. *(Source: Alstom Power.)*

emissions control). However, there is now an additional circuit (referred also as tertiary) for both LPG and naphtha.

The additional LPG/naphtha circuit comprises a fuel metering device (a flow metering valve and three fixed area metering orifices) and fuel transfer circuits. The valve and fixed area orifices were placed in the back of the injector to limit heat conduction to the valve. Fuel transfer circuits transfer the fuel to the tip where the fuel is vaporized in the LPG version and atomized in the naphtha version.

Fuel Metering Device

The injector's fuel metering device comprises a metering valve surrounded by cooling circuitry. This is designed to ensure metering of liquid LPG fuel during all turbine-operating conditions. The valve's piston operates under the action of differential pressure between fuel supply and combustion chamber pressures. Three metering ports are progressively opened as the valve piston moves. The valve crack point is set to give a minimum fuel system pressure of 20 Bara to prevent pure propane from phase changing in the fuel pump, manifold and injector.

As the LPG and diesel supplies are separate, it is necessary to pre-cool the tertiary valve before introducing LPG. For example, if the injector is started on diesel pilot and main and run at part load until steady state, the valve reaches temperatures of 100°C. Before valve crack, expanding a small fraction of the fuel into the separate cooling circuit at the lower combustor pressure actively cools the valve port. The latent heat of vaporization that results from the expansion and phase change, of this small fraction, is utilized to cool the injector valve, keeping the main flow of fuel in liquid phase.

The valve body will always be cooled to a temperature below the saturation temperature of the fuel at the fuel supply temperature. This is the result of the pressure differential and phase change. The lower absolute pressure cooling flow will always be lower in temperature than the saturation temperature of the fuel at supply pressure. This is always true for any composition of LPG even for extreme compositions containing large amounts of ethane, pentane and hexane.

Fuel Transfer Circuits

The fuel is then transferred to the injector tip via three fuel transfer circuits. The cooling fuel flow is kept separate from the main fuel until the fuel is vaporized at the injector tip. The low mass flow, but intermittent high volumetric flow rate of the cooling flow, would result in slug flow if the vapor cooling flow were to be mixed with the main fuel. Slug flow resulting in pulsations of liquid and vapor would be very detrimental to turbine operation.

The fuel transfer passages for the cooling flow have conflicting design requirements. The passages must be sufficiently large in flow area to accommodate a wide variation in volumetric flow rate while being sufficiently small to maintain sufficient high fuel velocity to maintain a dispersed two-phase flow. The volumetric fuel flow rate of the cooling flow varies significantly as the heat load varies. Initially when the LPG flow is initiated the LPG valve can be 100°C or greater. At these conditions the LPG cooling flow would be mostly vapor at the low fuel pressure at which this injector is designed to operate (20—30 Bara). Under prolonged steady state operations on LPG the cooling flow will be mostly liquid, as the steady state heat load becomes small. The ratio of vapor to liquid in the cooling flow will also vary as a function of fuel temperature, fuel composition, engine combustor pressure, and fuel supply pressure.

The coolant passages were designed to have a relatively large flow area in order to accommodate high volumetric flow, when the cooling flow is mostly vapor, while also having relatively small distances between opposing walls in order to limit the size of the bubbles that can be formed. Thin passages with high volumes will only form small bubbles, compared to other passages such as tubes that have minimized wall surface area to flow area. Using thin flow passages for the cooling flow allows large cooling flow areas to accommodate a wide range in volumetric flow rates without resulting in the formation of large bubbles that would cause unstable combustion.

The bubbles that do form in the fuel injector are after the metering device and are limited to the separate cooling flow, which is less than 10% of the fuel flow. Engines in the field have been shown to accept this amount of two-phase flow in the fuel system without resulting in unstable combustion.

Fuel Vaporizer

The active cooling concept of the fuel metering device combined with the design of the fuel transfer passages has provided controlled fuel flow of the mostly liquid main flow and at times mostly vapor cooling flow to the injector tip near to the combustion zone. The fuel must now be injected in a spatially controlled manner into the combustion chamber.

Most conventional fuel injectors would use a pressure atomization or air spray device to accomplish this task.

However, two-phase flow makes the use of any device that creates a pressure drop after the fuel is metered problematic. Only one metering device can be used per single stream of flow and the inlet of this device must have liquid only in order to have consistent predictable flow rates. Others have attempted to address the above problems with liquid LPG injection in two ways.

One way is to use a pressure atomizer at the injector tip with no other fuel-metering device such as a valve. However, this approach requires high minimum fuel pressure to maintain liquid phase (typically 100 Bara). For example, well-insulated fuel injector circuits would require a minimum pressure for propane to be 18—20 Bara (reference Figure 7—68). The difference between the minimum fuel pressure and the lowest operational combustor pressure could be 8 Bara for typical industrial gas turbine engines. This combined with these engines needing a 4:1 fuel range between maximum load and load reject results in very high maximum fuel pressures. The pressure atomizer would therefore require about a 16:1 pressure ratio to accommodate a 4:1 mass flow ratio. For engines that have at least an 8 Bar difference between minimum combustor pressure and the saturation line pressure, the fuel pressure could therefore exceed 100 Bara.

Another way is by placing a variable area metering device, such as a valve, near an air spray atomization device at the tip. This reduces the maximum fuel pressure. The close proximity of these components also prevents two-phase flow from separating into liquid and vapor components by minimizing the time and volume available for separation. However, these devices are problematic because of the difficulties of placing tightly toleranced moving components such as a spool valve in regions of high thermal gradients.

The tri-fuel injector developed jointly by Alstom Power and Delavan has been designed to overcome the reported difficulties with LPG injection, described below.

The fuel-metering device is placed in the back of the injector (see Figure 7—70) to limit heat conduction to the valve, thus minimizing mechanical problems. The injector's fuel transfer circuits, designed to keep the main and coolant flows separate, also allows the injector's fuel vaporizer to be positioned close to the injector tip where there is sufficient heat and air for both vaporization and atomization processes.

For the fuel vaporizer, the initial conceptual design was to use an air blast type of prefilming atomizer. The three cooling circuit flows and the three main fuel flows are introduced into an annular vaporizer in an evenly spaced symmetrical pattern. The idea was to use air to spin the LPG fuel into a film where it could be atomized by additional air outside of the injector.

Using air to add momentum to the fuel and spin the LPG fuel into a thin film had two advantages: it induces no pressure drop into the fuel circuit and the heat from the hot air partly expands the fuel thus increasing its volume. Also keeping the existing dual fuel design of the pilot, main, water, and gas circuits meant that the LPG filmer had to be at a larger diameter (see Figure 7–70). This would make the production of a continuous film difficult in this application without somehow increasing the volumetric flow rate of the fuel. The idea of adding more hot air to increase the volume of the fuel in order to create a film therefore developed. The philosophy was therefore just to add hot air and vaporize as much of the fuel as possible and then treat LPG as a gas when injecting it into the combustion chamber.

Carbon formation in the vaporizer is not an issue as long as the temperature of vaporization at the combustion pressure is less than the temperature at which carbon will form. For most industrial gas turbine engines this stays true at least until the fuel contains mostly C_8 and higher hydrocarbons. For this reason most naphthas can also be mixed internal to the fuel injector with hot air. As long as the fuel stays partly wet the temperature will stay low. This prevents auto-ignition and carbon formation from small fractions of C_{16} and higher hydrocarbons in naphtha.

Operating Philosophy

Turbine starting is on either diesel or natural gas fuels. Changeover from the starting fuel to LPG is conducted at part load condition. No direct LPG starting is permitted. During standard dual fuel running, the fuel injector tip components are continuously purged with turbine compressor discharge air.

During changeover to LPG, a thermocouple in intimate contact with the injector body in the valve region measures the local temperature and inhibits, via the turbine control system, the main flow of LPG until the temperature is below 45°C to ensure liquid flow.

After changing over from LPG to gas or diesel fuels, the LPG fuel lines from the shut-off valve to the injectors are nitrogen purged for about 20 seconds to remove any residual fuel.

Commercial LPG Operation

General Operation

The described tri-fuel system entered commercial operation in March 1999 on two Tornado gas turbines (units 1 and 2), both 6.45 MWe rating. Prototype 2 tri-fuel injector became the production standard. The lead engine has accumulated about 1029 hours, while the second has about 361 hours on LPG. Most of the above LPG running has occurred during winter, between November 1999 and March 2000. Including natural gas and some diesel operation total running has now exceeded about 8000 hours.

Fuel Pressures

The reported winter LPG running presented the coldest ambient conditions and therefore the highest power conditions. This operation resulted in the highest LPG fuel pumping pressures. Table 7–22 shows that fuel pumping pressures are within the original design intent (30 Bara).

Combustor Exit Temperature Spread

Figure 7–71 shows comparisons of combustor exit temperature spread between natural gas and LPG fuels. This is measured by 16 off thermocouples positioned circumferentially in the inter duct between compressor and power turbines. This spread level has shown no deterioration for LPG running, confirming that all eight injector LPG valves are operating consistently.

Transient Response

The tri-fuel engines can change over to and from LPG from a reasonable low load, say 1 MW, up to full load with complete confidence, although changeovers to LPG at FSNL have been locked out because of uncontrolled speed rises during the LPG cooling cycle. Changeovers from LPG

TABLE 7–22 LPG Fuel Pumping Pressures

Parameters		Date: Dec-99	Date: Jan-00
Turbine inlet temp	°C	18	−9
LPG system pressures			
Pump inlet	Bara	8.7	7.9
Burners manifold	Bara	25.6	27.3
LPG density		0.54	0.57
LPG manifold temp	°C	13.4	7.7

(Source: Alstom Power.)

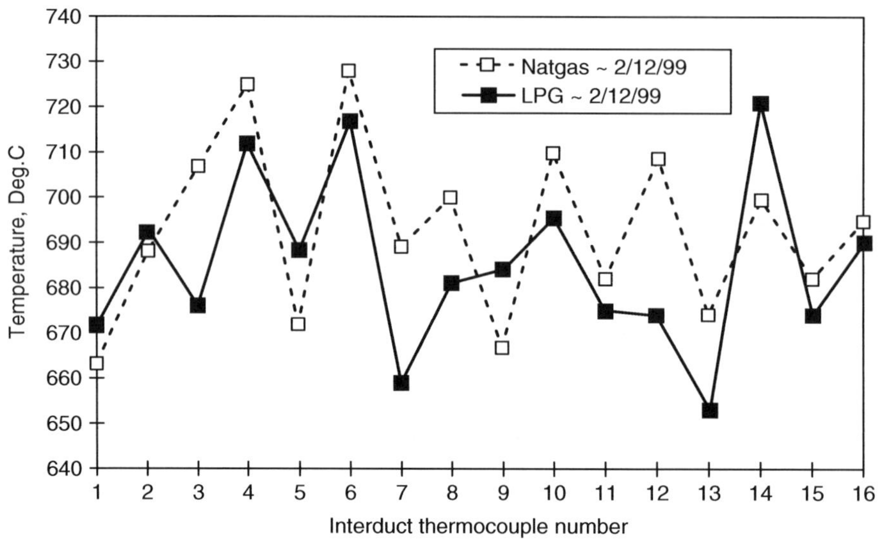

FIGURE 7–71 Combustor exit temperature spread. *(Source: Alstom Power.)*

at FSNL are possible but an uncontrolled speed rise (approximately 250 rpm) does occur during the 20-second nitrogen purge of the LPG manifold. No further work was conducted to optimize low load changeover, as it was not essential. Unit 2 was proven to LPG load shed from a variety of loads including full load; however, Unit 1 could only load shed from 2 MWe. The load-shed abilities are therefore not consistent and further work is required. It may be necessary to further optimize the LPG valves port shape to improve load shed. There are no issues with standard dual fuel operation.

Combustion Dynamics

During initial commissioning in March 1999, a combustion system dynamics survey was conducted for all fuels. An increased level of combustor dynamics, with a center frequency of about 100 Hz, was measured for LPG operation relative to both natural gas and diesel running. The amplitude of this dynamic increased in magnitude as the load increased; although maximum levels never exceeded 300 mpsi RMS. As these dynamics were within levels of experience (< 500 mpsi RMS), no further action was taken to reduce their levels.

Later in November 1999, a low frequency audible rumble in the exhaust became apparent on both turbines when running on LPG. There was no audible exhaust rumble for diesel and natural gas fuels. Combustion system dynamics showed no increase since the initial commissioning. The exhaust noise was reported to change with ambient conditions. It is probable that the exhaust noise is generated by the higher dynamic levels measured for LPG running, which only becomes audible at certain ambient conditions.

Naphtha Development Testing

The tri-fuel injector design has just recently been refined to reduce the dynamics seen during LPG operation, and to make the injector more suitable for naphtha operation. The latter is to support commercial operation of the Typhoon engine on naphtha.

The following modifications have been made to the fuel vaporizer in order to make it more like an air spray device. More air has been added to the injector's LPG vaporizer circuit. The internal air swirler geometry has been further optimized. The maximum fuel pressure has been increased from 30–40 Bara to account for increased pressure requirements of the proposed naphtha fuel system.

High-pressure rig testing of a naphtha prototype injector has recently been conducted on unleaded gasoline to simulate naphtha. An extreme over test has also been conducted using No. 2 diesel. LPG testing will be conducted in the near future.

Figure 7–72 shows a comparison of smoke emission for unleaded gasoline and No. 2 Diesel through the injector's LPG/naphtha circuit (tertiary). The results for unleaded gasoline (1–2 Bacharach) are slightly lower than those necessary to further optimize the LPG valves port shape to improve load shed. Table 7–23 summarizes corresponding full and no load emissions levels as well as OTDF measurements. There are no issues with standard dual fuel operation.

Similar full load NO_x emissions were measured for tertiary diesel and gasoline operation. This suggests similar mixing degrees for both fuels. Also full load measurements of UHC and CO are negligible indicating high levels of full load combustion efficiency for both fuels. At no load, gasoline results in lower part load CO and UHC emissions relative to No. 2 diesel. This indicates higher part load combustion efficiencies for the gasoline case.

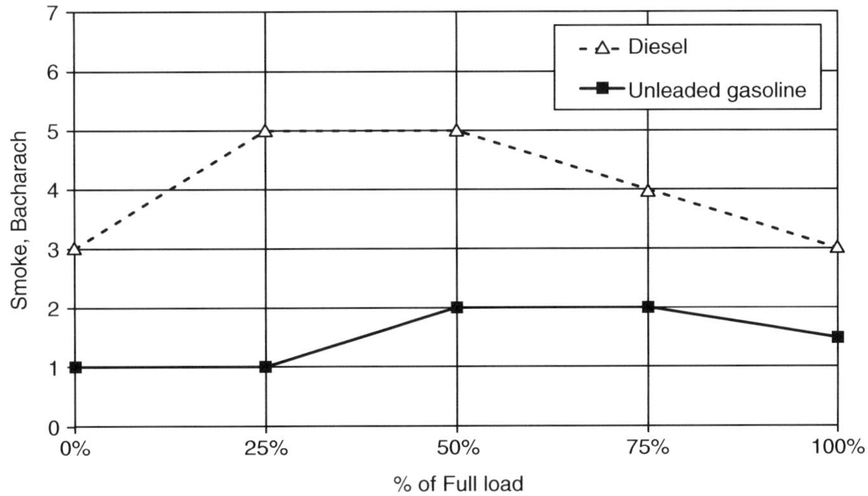

FIGURE 7–72 Naphtha prototype injector—smoke emissions. *(Source: Alstom Power.)*

TABLE 7–23 Naphtha Prototype Injector—Emissions Summary

	NO_x, ppmv	CO, ppmv	UHC, ppmv	OTDF, %
Tertiary (gasoline)	316	1.2 (96)	1.4 (4.1)	19.5
Tertiary (diesel)	312	0.72 (571)	2.5 (22)	15.5
P&M (diesel)	320	0.6 (368)	0.6 (15)	14.9
Natural gas	192	0.6 (405)	0.7 (61)	15.1

Notes: P&M denote pilot and main.
(Source: Alstom Power.)

Preliminary engine testing at Lincoln has identified a probable cure to the dynamics problem seen on LPG during commercial operation. Distillate fuel testing through the injector's LPG circuit has generated a similar dynamics phenomenon to that seen during LPG operation. Similar engine testing of the naphtha prototype injector, with the air spray device replacing vaporizer, has shown negligible dynamics levels. It is therefore probable that the revised injector standard used for naphtha will reduce the dynamics seen on LPG.

CASE STUDY 4: MULTI-FUEL CONCEPT OF THE SIEMENS 3A-GAS TURBINE SERIES*

To provide maximum fuel flexibility, Siemens has developed a liquid-fuel supply system for its 3A-Series gas turbines (Table 7–24), which in addition to customary distillate can also accommodate special fuels such as naphtha and condensates, and which also enables fuel changeover during gas turbine operation. The high vapor pressure and correspondingly low flash point of naphtha and condensates necessitate a complex fuel system safety concept as well as special flushing procedures when the gas turbine is started up and shut down. The addition of water to the fuel has been shown to be an efficient method of reducing NO_x emissions. This addition of water, coupled with the highly variable fuel density, places increased demands on the gas turbine control system. Moreover, it is possible to switch from natural-gas operation to fuel-oil operation and back again when the turbine is operated in the output range near base load.

* Source: [7-8] Courtesy of Siemens. Extracts from E. Deuker, R. Waldinger, H. U. Rauh, and F. Schade, "Multi Fuel Concept of the Siemens 3A-Gas Turbine Series," F 2001-GT-0078.

Liquid Multi-Fuel System

In the following, particular attention is given to the most recent innovations concerning the special liquid fuels naphtha and condensates, which are becoming a more attractive regional solution (e.g., in Southeast Asia and India) as a result of the increased cost of commercially available fuel oil on the world market.

Fuels

A large portion of the commercially available naphtha and fuel oil is derived from crude oil by atmospheric distillation. In contrast to naphtha and fuel oil, the term "condensates" does not apply to a specific fuel, but is rather used as a synonym for hydrocarbons that exist in a liquid state at standard temperature and pressure, and that occur as significant byproducts of the natural gas extraction process. Specific properties (Table 7–25) and their validation are of central importance in characterizing liquid fuels. Characteristic of condensates and naphtha are a high

TABLE 7–24 Siemens V-Series Gas Turbines with Multi-Fuel Systems

Reference	Paguthan	Faridabad	Santa Rita	San Lorenzo
Configuration	One CC	One CC	Four CC	Two CC
	3.V94.2	2.V94.2	1S.V84.3A	1S.V84.3A
Commissioning	9/97	6/99	12/99	06/02
Hours of operation	>18,000	>2,000	>11,500	–
	(June 00)	(June 00)	(Jan. 01)	
Natural gas	X	X	(X)	X
Naphtha	X	X	X	X
Condensates			X	X
Extra-light fuel oil	A	A	X, A	A

CC: Combined-cycle.
X: primary fuel; (X): planned for mid-2002; A: Fuel used for startup/shutdown.
(Source: Siemens.)

calorific value in excess of 42 MJ/kg, low density, low viscosity, a low flash point, and a comparatively high vapor pressure.

System Description

Fuel Delivery System at the Santa Rita Plant

The primary fuel (condensates, naphtha, and fuel oil) is delivered by tanker to the jetty. Following verification of fuel quality, the fuel is pumped into the three discharge tanks. The fuel treatment system draws fuel from these tanks, processes it, and feeds it into one of the two-day tanks. Two lines, each with redundant pumps, run from the day tanks to the two combined-cycle blocks (two combined-cycle units = one block). The system design allows a different fuel to be used for each block. Both blocks are joined via a common return line to the day tanks.

TABLE 7–25 Specific Properties of Naphtha, Condensates, and Fuel Oil

Fuel Type	Naphtha	Condensates	Distillate Fuel Oil No. 2-GT
LHV in MJ/kg*	>42	>42	>42
	43.980	43.590	42
Density in kg/m^3	650–775	650–850	820–880
	720	770	830
Flash point in °C	***	>55	
	<0	n.a.	>55
Vapor pressure in bar at 37.7°C	<0.9	<0.9 **	<0.003
Mercaptanes in ppm	—	4	—
Sulfur in % by wt.	<0.2	<0.5	<0.2
	<0.15	0.05	<0.2

Note: The auto-ignition temperature is approx. 232–277°C, for comparison: appr. 600°C for methane.
**Lower heating value, in the upper line the specified value and in the bottom line the tested value is listed.*
***Always to be taken as lower than ambient or operating temperature.*
****Specified by the fuel vendor (Santa Rita mercaptans (RSH) 350 ppm).*
(Source: Siemens.)

The startup fuel (distillate or fuel oil) is trucked to the site, and is treated the same way as the primary fuel. The startup system incorporates only one line with redundant pumps, which means that four gas turbines must be started up sequentially.

Gas-Turbine Liquid Fuel System

Whereas the fuel delivery system is designed around those characteristics specific to the power plant, for the gas turbine standard modular skids are used for natural gas and liquid fuels, so that use of a particular fuel is independent of the existing infrastructure.

The annular combustion chamber of Siemens 3A-Series gas turbines is equipped with 24 identical hybrid burners for natural gas and liquid fuels. Each hybrid burner in turn has a diffusion burner and a premix burner for liquid-fuel operation, as well as a diffusion burner, a premix burner and a pilot burner for operation using natural gas. In addition, a separate lance can be used to inject water directly into the flames to reduce NO_x formation.

The three natural-gas burners and the premix burner for liquid fuels have controlled feeds, i.e., the entire quantity of fuel supplied by the feed control valve in each case is injected into the combustion chamber. The liquid-fuel diffusion burner has a controlled feed and return, which allows control of not only the quantity of injected fuel but also the feed quantity and injection pressure, and thus permits control of the directly linked atomization quality. The liquid-fuel diffusion burner can operate in the load range from startup to base load (approx. 12 kg/s fuel), without the addition of water. In power operation the return mass flow from the diffusion burners is directly fed to the injection pump again (recirculation).

Compared to water injection via the lances, the direct mixing of water with the fuel and the operation of the premix burner achieve a greater reduction in NO_x emissions using the same quantity of water. In the first development phase, as represented by the Santa Rita plant, water is admixed via a jet mixer upstream of the injection pump (Figure 7—73). Moderate water pressure is required, allowing use of the water skid originally designed to be used with lance injection. The jet mixer increases the feed pressure of the liquid fuel/water mixture upstream of the injection pump by up to 10 bar, which also allows correspondingly greater injection pressures to be attained at the burners.

The density and viscosity of the liquid fuel/water mixture vary according to the mixture ratio and the composition of the fuel, which places heavy demands on the injection pump. The mixture ratio is defined as the ratio of the water mass flow to the fuel mass flow. The gas turbines at the Santa Rita plant have been successfully tested with mixture ratio over 100%. A major milestone in the development of the multi-fuel gas turbine was therefore the replacement of screw pumps with centrifugal pumps, as the latter are well suited to low-viscosity liquids (<1 mm^2/s for naphtha and condensates) and also have a characteristic favorable to the combined supply of diffusion and premix burners. The zero delivery head of the centrifugal pumps is approx. 1600 m.

Because the water is added prior to branching the mixture to the diffusion and premix feed lines (see Figure 7—73), both have already been implemented for the newest gas turbines via a standard dual-fuel system, water is added separately to the diffusion and premix feed lines, so that the mixture ratio can be varied independently for the two burners.

In gas operation, a switch off/switch on of burners, e.g., at changeover from diffusion to premix operation and back, can be implemented rather easily. No cooling or purging procedures are required. In liquid-fuel operation special precautions must be taken during startup and shutdown of the gas turbine, at changeover to liquid fuel premix operation and when changing over from gas to liquid fuel operation and back. If a liquid-fuel burner is not in operation, there is a risk of circulation flow among the individual burners as a result of slight local pressure differences in the combustion chamber, which would soon lead to overheating or the formation of coke at the nozzles. To prevent the ingress of hot gas when the liquid fuel burners are shut down, water is used to flush the residual fuel from the lines and nozzles, which are then supplied

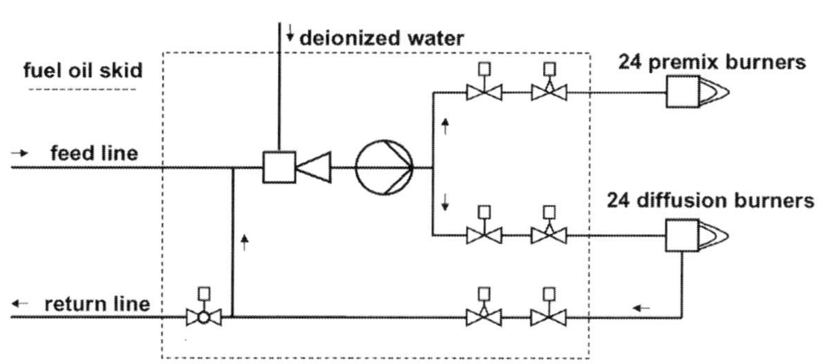

FIGURE 7—73 Gas turbine liquid-fuel system with addition of water via a jet mixer. *(Source: Siemens.)*

with cooled compressor outlet air, so-called seal air. Before a liquid fuel burner starts to operate, the empty sections of the line are filled with water. This not only allows the atomization pressure to build up rapidly, but also cools the nozzles so that the fuel cannot ignite prematurely.

The safety concept has given rise to the liquid multi-fuel requirement that all liquid fuel lines which are pressurized during startup of the gas turbine (including startup using natural gas) must not contain hazardous fuels such as naphtha and condensates, as these fuels would pose a significant risk to the gas turbine, such as in the event of a faulted start. For this reason Siemens multi-fuel gas turbines generally use fuel oil for startup and shutdown. Moreover, use of fuel oil means that in the event of a leak, which in many cases is not evident until there is pressure in the line, the escaping fuel is initially the much less hazardous startup fuel. Changeover to the primary operating fuel (and back) following gas turbine synchronization takes place within the fuel delivery system, which means that no noticeable changes occur in the operation of the gas turbine. In the event of a gas turbine trip, the lines downstream of the stop valves are completely evacuated and partially refilled with water. The liquid-fuel skid is then flushed.

Operating Experience

The gas turbines at the Santa Rita plant have been successfully tested using naphtha, condensates and fuel oil. When water is added to the fuel in a ratio of 1:1, NO_x emissions are well under the World Bank standard of 72 ppm.

In the part-load range the characteristics of the different fuels (primarily the density) led to variable turbine operation, which made it necessary to modify the structure of the gas turbine controller such that the operating characteristic of the gas turbine can be adapted to the fuel in use.

Base load is reached without restriction regardless of the fuel used.

Acceptance measurements were taken in Block 1 in May 2000, using condensates as the fuel, and in August 2000 in Block 2, with distillate oil. It should be emphasized that the expected gas turbine output and efficiency values were either met or exceeded.

Gas Turbine Controller

The gas turbine controller must meet unusually demanding requirements. It must be able to accommodate rapid load changes, e.g., when participating in frequency stabilization or in the event of load rejection to the auxiliary power level, and must also keep output fluctuations as low as possible when cutting a burner type in or out. To meet both of these requirements, it must be possible to quickly and very precisely vary the flow of fuel to each burner.

The flow of fuel to a particular burner is varied by changing the position of the associated control valve. For this reason it was particularly important to map the relationship between valve lift and fuel quantity in the controller:

- The valve characteristic (CV value vs. valve lift) of all control valves is stored in the controller, and checked and modified if necessary during the commissioning process.
- In the natural-gas system the pressure and temperature upstream of the control valves are measured, and used as input variables by the controller in calculating the correct valve lift. If there should be a sudden drop in the natural-gas supply pressure, for example, the natural-gas control valves are directly opened by an amount that is proportional to the change in pressure, thereby preventing a sudden drop in output.
- Similarly, the pressure upstream of the control valves in the liquid-fuel system is measured and evaluated by the controller, making on-line pump changeover possible over the entire operating range of the turbine in the case of systems with redundant injection pumps, for example.
- Because the density of the liquid fuel is highly variable, an algorithm was developed specifically for use in the multi-fuel system which determines the fuel density at any given time based on the pump characteristic and measurement of the pump outlet pressure.
- In adding water directly into the liquid fuel, the accumulator effect of the piping was taken into consideration. To increase volumetric flow of the water while holding turbine output constant, for example, the volumetric flow of the fuel must be decreased by the same amount until the new mixture is actually present at the burner nozzles. In this regard a path model consisting of dead time and delay modules was developed, which is optimized during commissioning.
- A fuel balancing system that acts on all active burners and is subordinate to the load controller ensures that when a burner is cut in or out, a corresponding quantity of fuel is added or subtracted at the main burner in each case. Appropriate dead time and delay modules take the effect of filling fuel lines into account.

Fuel Changeover

Quick fuel changeover with minimum gas-turbine load reduction is particularly important for operators who wish

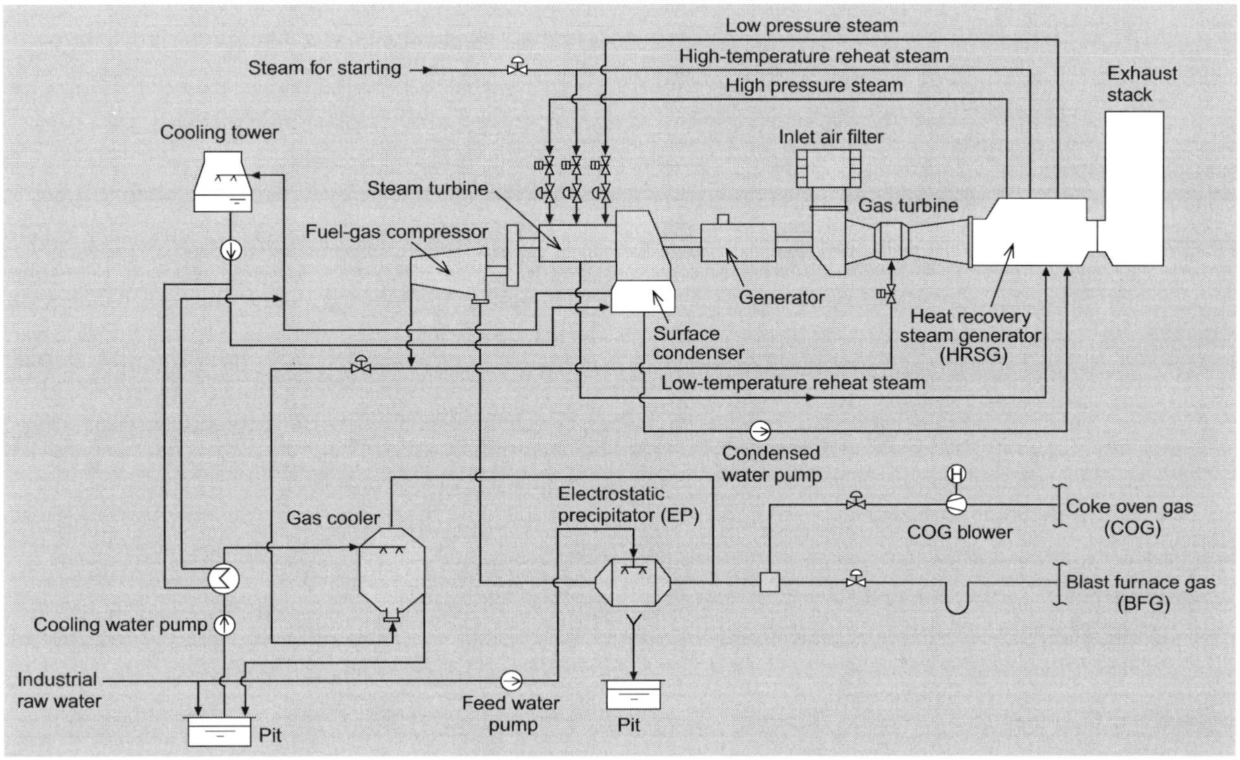

FIGURE 7–74 Scope of supply.

to use liquid fuels primarily as a backup source of energy if the supply of natural gas should fail. The on-line fuel changeover system, which switches between natural-gas premix operation and fuel-oil diffusion operation, was developed for use with 3A-Series gas turbines in the upper load range to satisfy this requirement.

For safety reasons, in the multi-fuel system startup fuel (fuel oil) is always used when changing over from natural-gas to liquid-fuel operation. The process is thus identical to fuel changeover in the standard dual-fuel system (natural gas and fuel oil).

On the natural-gas side, the fuel changeover process is in principle the same as the changeover from diffusion operation to premix operation, with the exception that the diffusion flame is a fuel-oil flame. On the liquid-fuel side, the fuel-oil diffusion burner is first operated as a feed burner during the ignition process (when changing over from natural gas to fuel oil), which avoids the danger posed by hot gas entering the initially unpressurized return line. When the fuel-oil burner is switched off (when changing over from fuel oil to natural gas), the fuel oil remaining in the lines and the burner is flushed out using water, and the water and fuel oil are removed via the return line. Thus the system can be effectively flushed using a high mass flow without resulting in significant load fluctuations.

Fuel changeover takes approximately two minutes, and has been proven successful at the Berlin test facility using a V84.3A gas turbine.

CASE STUDY 5: USE OF BLAST FURNACE GAS TO FUEL 300 MW CC PLANT*

The project has the following features:

1. It is the world's largest class for a BFG firing CCPP and was the first order in China.
2. The first design cooperation for power generation plant with the design and research institute of an iron and steel company in China.
3. The first installation of a power generation plant in an extremely cold area (the minimum monthly average temperature is less than −15°C).
4. Collaborative plant adopted Chinese-made products as related equipment for the bottoming cycle

Figure 7–74 shows the scope of supply of this plant.

* Source: [7-9] Courtesy of MHI (Mitsubishi Heavy Industries) Ltd. Extracts from "Anshan Iron & Steel Group Corporation, China, Construction and operation experience of 300 MW blast furnace gas firing combined cycle power plant."

FIGURE 7–75 Turbine floor layout.

Features of the Plant

The features of the plant are as follows:

1. Adoption of combined cycle power plant equipped with the most advanced, high efficiency blast furnace gas firing gas turbine model M701S (F).
2. Adoption of a gas turbine inlet air heater, considering the startup and load operation during extremely low atmospheric temperatures.
3. Realization of total integrated plant control including customer supplied equipment such as HRSG.
4. Adoption of monitoring and control graphic screens written in Chinese for the plant control system.

Figures 7–75 and 7–76 show the turbine floor layout and a longitudinal sectional drawing of the total assembly.

Main Specifications of Major Equipment

Table 7–26 shows the main specifications of the major equipment including the customer supplied HRSG. See Figure 7–77 for the plant output characteristics that show that the power plant is able to generate a maximum output of 300 MW at an ambient temperature of 15°C or below.

Measures against Low Ambient Temperatures

A gas turbine consumes a great deal of air as the driving source. While a lower ambient temperature helps increase the power output, too low a temperature causes the following problems:

1. Decreasing the compressor surge margin due to increasing thrust force and pressure ratio limits the maximum power output.
2. Combustion pressure fluctuations create an unstable combustion zone during startup and acceleration.

To cope with these problems, an inlet air heater was installed to heat up the gas turbine inlet air in cold weather. As the steam source for heating, auxiliary steam is used during startup and acceleration while low-pressure steam is extracted from the HRSG during load operations. The used steam is reclaimed by the surface condenser as a condensate to reduce the amount of waste water as an environmental measure.

Figure 7–78 is schematic of the inlet air heating system.

Automatic Control System

Figure 7–79 shows the control system configuration. MHI supplied the total plant control system which has the functions of automatic control, operation, and monitoring of all the equipment directly related to power generation including the customer supplied equipment such as HRSG and boiler feed water pump.

(1) Startup Characteristics

This power plant adopts a steam turbine starting system. After reaching the rated speed, it switches from the existing steam to HRSG steam once the conditions of the HRSG steam become acceptable for the steam turbine. The plant then increases its load automatically up to the minimum operational load of 150 MW. Once reaching 150 MW, the load can be changed between 150 and 300 MW depending on the power demand. The power plant also shuts down automatically. During load down, the cooling down process is designed to make the mismatch between the steam and the steam turbine metal temperature small so that the time to the next hot start is shortened. The expected time for starting (from ignition up to full-load operation) is

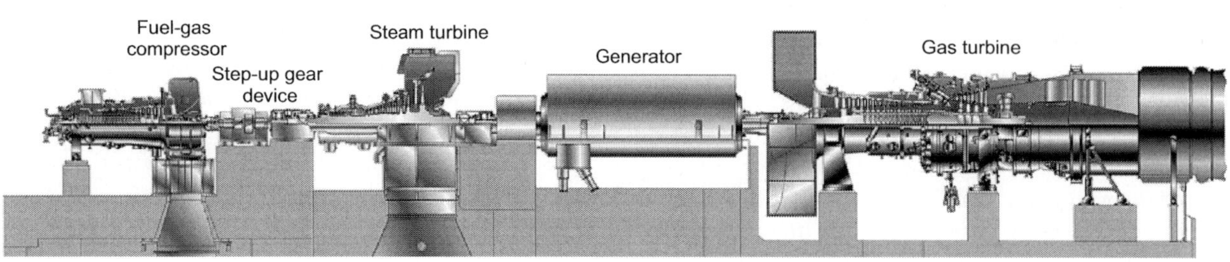

FIGURE 7–76 Longitudinal sectional drawing of complete assembly.

TABLE 7–26 Main Specifications of Major Equipment

Major Equipment	Particular	Specifications
Power generation facility	Model output	Single-shaft type combined cycle power plant 300 MW
Gas turbine	Model output Turbine inlet gas temperature Rotational speed	Open cycle single-shaft type M701S(F) (MHI) 183.0 MW 1300°C class 3000 rpm
Heat recovery steam generator (HRSG)	Model steam flow rate	Triple-pressure natural-circulation reheat boiler (China Hangzhou boiler) 586.4 tons/hr Details: High pressure 239.9 tons/hr Medium pressure 282.2 tons/hr Low pressure 46.3 tons/hr
Steam turbine	Model Output steam pressure/temperature Rotational speed	Single-casing reheat mixed-pressure condensing type SRT-30.5″ (MHI) 117.0 MW High pressure 10.2 MPa (a)/538°C Medium pressure 2.95 MPa (a)/538°C 245°C/245°C 3000 rpm
Generator	Model Capacity Cooling system Rotational speed	Horizontal-shaft tubular rotary field type (Mitsubishi Electric Corporation) 340,000 kVA Hydrogen gas cooled 3000 rpm
Fuel-gas compressor	Model Rotational speed	Single-casing axial flow type (MHI) 5025 rpm

approximately 90 minutes for a hot start, 150 minutes for a warm start and 200 minutes for a cold start.

(2) Plant Performance

Table 7–27 shows the actually measured performance. The plant output and plant thermal efficiency after compensation to the standard conditions (ambient temperature at 15°C and LHV base) sufficiently satisfy the planned values. The plant thermal efficiency reaches 51.2% compared to the planned 48.5%, and achieves the highest level in the world for BFG firing CCPPs.

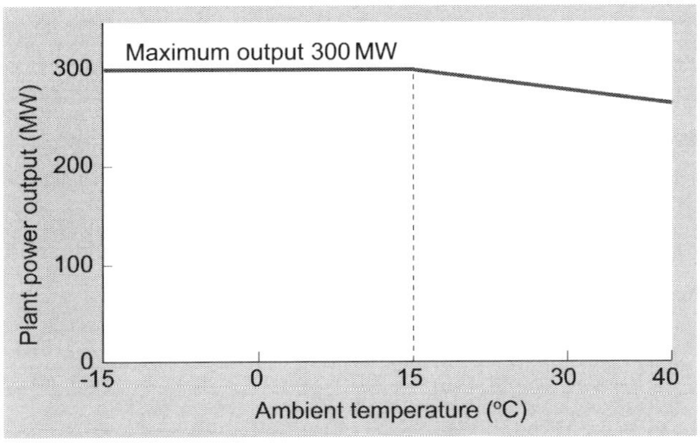

FIGURE 7–77 Plant output characteristics.

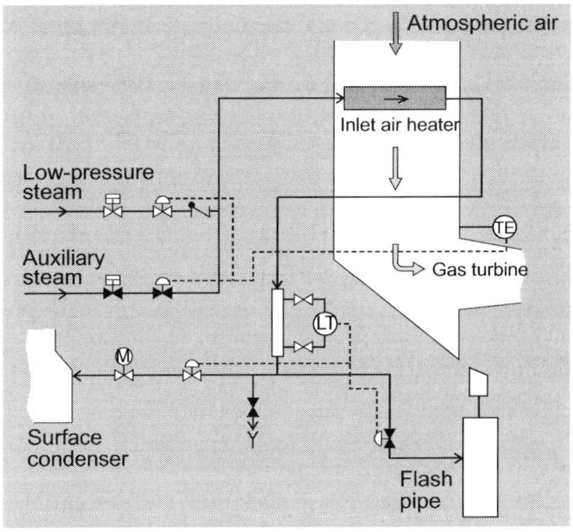

FIGURE 7–78 Outline schematic of inlet air heating system.

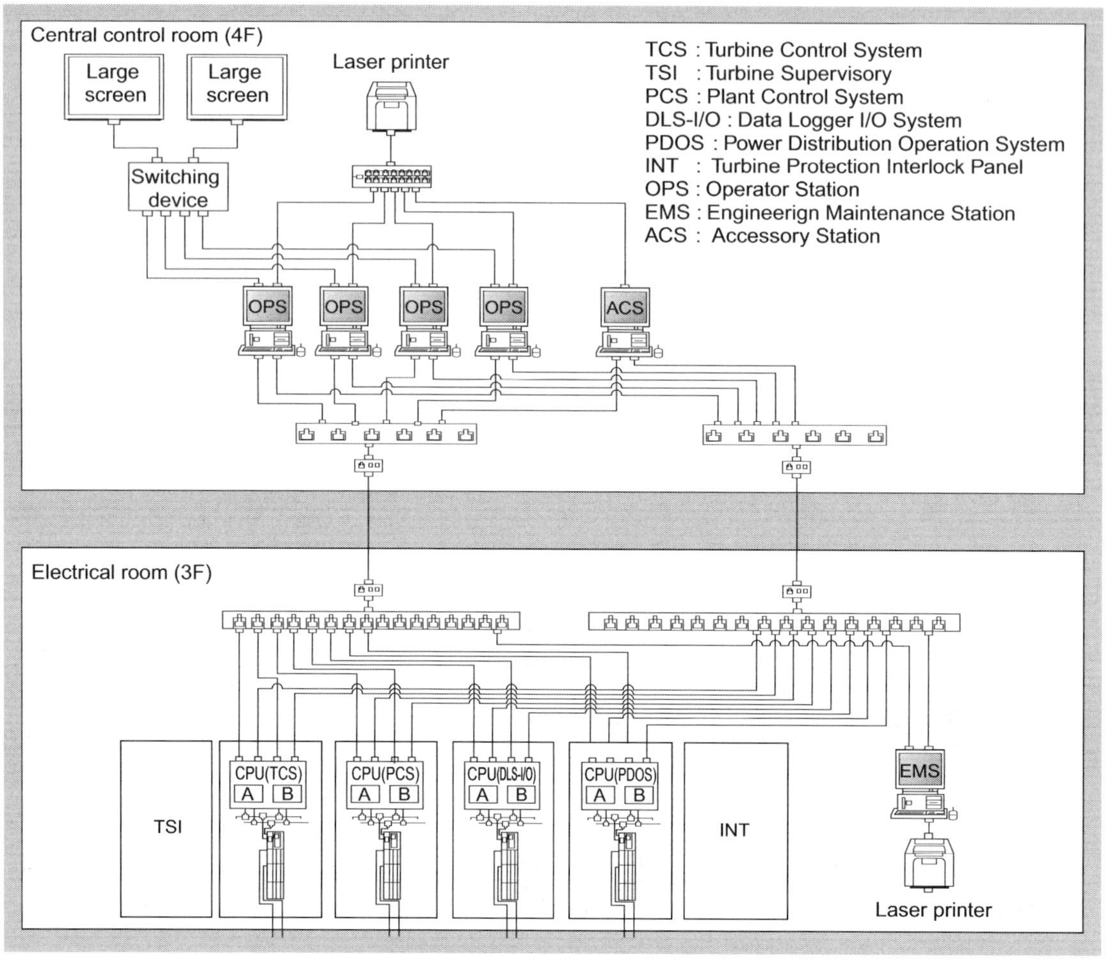

FIGURE 7–79 Control system configuration.

(3) Environmental Characteristics

By effectively transducing the low calorific blast furnace gas to electric power, this power plant contributes greatly to environmental preservation as follows. Firstly, this is a combined cycle power plant that effectively utilizes the blast furnace gas instead of the coal fired thermal power plants popular in China. It is estimated that the plant emits approximately 1,900,000 tons less CO_2 a year (assumed operation of 7000 hours per year) than that emitted by burning coal in a coal-fired thermal power plant of the same output. Secondly, although the environmental characteristics depend significantly on the composition of the blast furnace gas, as it contains nitride and sulfur compounds, its combustion produces less thermal NO_x than burning natural gas because of its low calorific value. Thus, this plant indicates the significantly low NO_x value of 15 ppm (15% O_2 base, without a de-NO_x system).

Operation Records

Table 7–28 shows the availability of the plant for about five months from May 2007, when the plant was taken over, to the end of September. While the plant availability depended on the scheduled shutdown due to maintenance of the blast furnace shortly after taking over the plant we experienced the degradation of plant availability because of initial failures and other incidents. After that, the plant reliability improved gradually, finally reaching 100% in September 2007.

TABLE 7–27 Actually Measured Performance

Guarantee Item	Planned Value	Measurement
Plant output	300 MW	313.9 MW
Plant thermal efficiency	48.5%	51.2%

TABLE 7–28 Plant Availability

Month and Year	Availability (%)
May 2007	82.3
June 2007	93.0
July 2007	91.9
August 2007	95.2
September 2007	100.0

CASE STUDY 6: BIODIESEL AS AN ALTERNATIVE FUEL IN SIEMENS DLE COMBUSTORS - ATMOSPHERIC AND HIGH PRESSURE RIG TESTING*

Atmospheric and high-pressure rig tests were conducted to investigate the feasibility of using biodiesel as an alternative fuel to power industrial gas turbines in one of the world's leading Dry Low Emissions (DLE) combustion systems, the SGT-100. At the same conditions, tests were also carried out for mineral diesel to provide reference information to evaluate biodiesel as an alternative fuel.

Siemens Industrial Turbomachinery Ltd. (Lincoln, UK) was involved in the recently finished European project AFTUR (**Al**ternative **F**uels for Industrial Gas **Tur**bines) to investigate the feasibility of using biodiesel as an alternative fuel to power industrial gas turbines. Siemens conducted both atmospheric and high-pressure rig tests in the SGT-100 using one of the world's leading DLE combustion systems. Reference data were obtained by carrying out the same tests with mineral diesel. In an atmospheric pressure rig, extensive ignition tests were conducted to investigate engine start reliability, lean ignition and extinction limits to establish biodiesel's operability in gas turbines. In high-pressure rig testing, emissions and combustion dynamics were investigated. The biodiesel selected for this study was supplied by C Zero Energy Ltd., produced from recycled cooking oil and was consequently derived from a mixture of feedstocks. All tests were with the standard DLE SGT-100 burner, combustor and transition duct and 100% biodiesel (BD100) was used. The reference mineral diesel was BS 2869 Class A2/D.

* Source: [7-10] Courtesy of Siemens Industrial Turbomachinery Ltd. Extracts from "Biodiesel as an Alternative Fuel in Siemens DLE Combustors: Atmospheric and High-Pressure Rig Testing" (GT2009-59065)

Siemens DLE Combustion System

Siemens Industrial Turbomachinery Ltd. (SIT Lincoln, UK) is the manufacturer of small industrial gas turbines from the SGT-100 at ca. 5 MW to the SGT-400 at 13.4 MW, which have now accumulated more than 3 million hours of experience. The SGT-300 and SGT-400 are offered commercially for less than 10 ppmv NO_x (@15% O_2) emission for certain markets, for example the SGT-300 operating at the University of New Hampshire, USA. The dual fuel DLE combustion system is designed as a reverse-flow tubo-annular type as shown in Figure 7–80. The combustor consists of three main sections: (1) the pilot burner, which houses the pilot fuel galleries and injectors for both gaseous and liquid fuel; (2) the main burner, which houses the main air swirler and main gas and liquid fuel injection systems; and (3) the combustor, which includes the prechamber and is of a double skin construction and is cooled through impingement cooling. Downstream of the combustor, a transition duct is used to condition the flow from the circular combustor exit to a sector of the turbine entry annulus. In order to cover all the power range, rather than having a single combustor design and applying different numbers of combustors to each engine type, the combustors have been scaled. Figure 7–81 shows the DLE combustor family. The SGT-100, 300, and 400 have 6 combustors per engine and the SGT-200 has 8.

A schematic of the combustion concept is shown in Figure 7–82. The main combustion air enters through a single radial swirler at the head of the combustor. The flow then turns through a right angle into the prechamber followed by a sudden expansion into the combustion chamber.

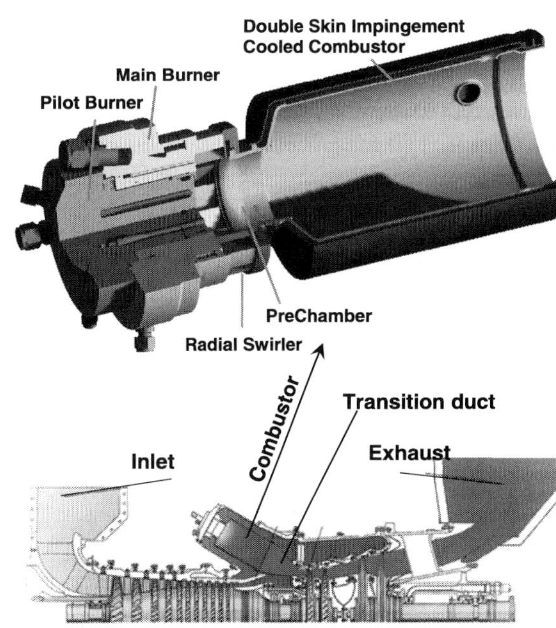

FIGURE 7–80 Siemens DLE combustor construction.

FIGURE 7–81 Siemens DLE combustor family.

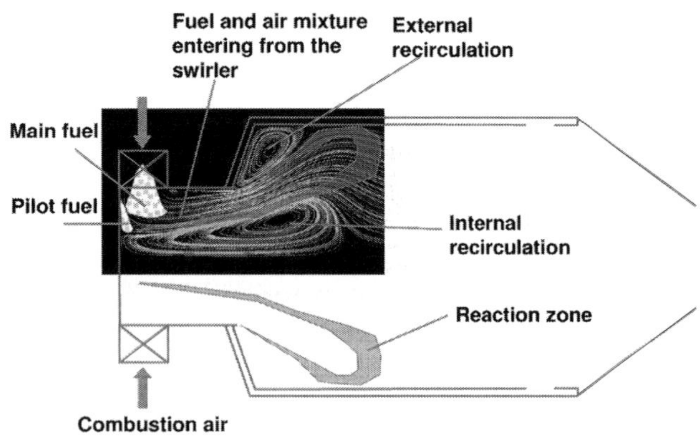

FIGURE 7–82 DLE combustion concept.

The swirl number is sufficiently high to induce a vortex breakdown reverse flow zone along the axis. This is termed the internal reverse flow zone. In the concept, this reverse flow zone remains attached to the surface of the pilot burner, thereby establishing a firm aerodynamic base for flame stabilization. In the wake of the sudden expansion, an external reverse flow zone is established. The flame is stabilized in the shear layers around the internal and external reverse flow zones.

The fuel, either gas or liquid, is introduced in two stages as shown in Figure 7–83: the main, which results in a high

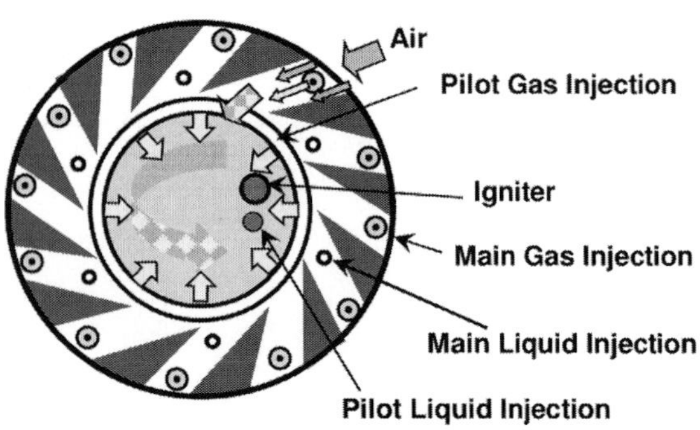

FIGURE 7–83 Conceptual details of fuel injectors and air paths.

degree of premixedness and hence low NO_x emissions, and the pilot, which is steadily increased as the load demand decreases in order to ensure flame stability. The pilot is arranged such that as the pilot fuel split increases, the fuel is biased towards the axis of the combustor.

Capability of the DLE System for Fuel Flexibility

Siemens has a continuous policy of expanding its fuels capability and as well as the above-mentioned DLE combustion system, Siemens also has conventional combustion systems which can burn various fuels in the Wobbe index range from 4–70 MJ/m^3.

Conclusions

Extensive ignition, emissions and combustion dynamics testing of biodiesel on standard DLE SGT-100 combustion hardware has been conducted to investigate the feasibility of using biodiesel as a fuel to power industrial gas turbines.

In ignition testing, two series of tests were performed to identify key properties of the proposed fuel. The first set of tests identified the likelihood of the engine lighting based on the measured probability of the ignition of a single combustor.

This demonstrated that a high likelihood of ignition would be possible depending on the choice of air assist pressure. The second series of tests investigated the lean ignition and extinction limits at various air temperatures. This showed that the ignition limit was different to that measured for mineral diesel, with the biodiesel being easier to ignite at high air mass flow rates and harder at lower flow rates. It was also demonstrated that air temperature did not have significant impact on the lean ignition limit. It was further shown that the ignition window for the SGT-100 lay well within the predicted curves for the lean ignition limit and therefore that good ignition should be possible. The lean extinction limits were found to be very similar to the ignition limits.

Emissions and combustion dynamics were measured in a high-pressure test rig. Performance of the burner on biodiesel was mapped by variation in operating conditions such as the air inlet pressure, temperature, the combustor pressure drop and flame temperature. The results were compared with those of diesel at the same conditions. For all the operating conditions tested, biodiesel had lower NO_x emission than diesel. For biodiesel at an air inlet pressure of 14 bar, there was a strong link between NO_x and the combustor pressure drop, where the NO_x level was reduced as the combustor pressure drop was increased. It was not the case for biodiesel at 9 bar and for mineral diesel at all pressures investigated.

However, after FBN correction, comparable NO_x emissions were obtained, although diesel showed slightly higher NOx emission than biodiesel at 9 bar inlet pressure.

Analysis of the high-pressure rig testing results showed that for biodiesel at low flame temperature and low air inlet pressure, a high degree of premixing was achieved, so NO_x level was independent of inlet pressure and combustor pressure drop. At high pressure and high flame temperature, the optimal level of mixing could not be achieved due to the amount of fuel supplied. Where mixing is not optimal the thermal NO_x formation is higher and dependent on inlet pressure and combustor pressure drop. However, for mineral diesel, good premixing was achieved at all pressures investigated and little dependence on inlet pressure and combustor pressure drop was observed.

At lower flame temperature, biodiesel had a significantly higher level of CO than diesel, particularly at a lower inlet pressure of 9 bar. For both fuels, the combustor pressure drop had apparent influence on CO at lower flame temperature with CO level decreasing with the combustor pressure drop decrease, and this influence diminished as flame temperature was increased. Both fuels had comparable levels of UHC emission at higher inlet pressure and biodiesel had slightly higher UHC emissions at lower inlet pressure.

For combustion dynamics, both fuels had comparable frequencies, but biodiesel had lower rms dynamic pressures at lower flame temperatures and higher when flame temperatures exceeded full load value.

Fuel droplet sizes were calculated and biodiesel had bigger droplet size, particularly at lower pressure, lower flame temperature and lower combustor pressure drop. The dominant contribution of big droplet size for biodiesel came from its much higher viscosity.

Chapter 8

Accessory Systems

"The shortest and surest way to live with honor in this world is to be in reality what we would appear to be."

—Socrates

Chapter Outline

Accessory systems in a gas turbine is every system that occurs outside of or additional to, the bare gas turbine hardware. In other words, in additional to everything listed in the table of contents for this chapter, it would also include systems such as inlet air fogging systems, steam injection (for cooling or power augmentation) systems, life cycle assessment systems. The latter systems are commonly retrofitted on gas turbines for optimal performance.

Systems such as one for online compressor washing would also be an accessory system; however, it is usually supplied with the gas turbine system when it is new. Commonly, it is made by another smaller OEM, although the GT OEM may have his nameplate on it. Compressor washing is considered in the chapter on performance optimization (Chapter 10), as are the previously mentioned performance enhancement systems. In some cases, certain cases on items such as steam injection to extend component lives or the TBO of the gas turbine appear in Chapter 12 on Maintenance, Repair and Overhaul.

This chapter is devoted primarily to accessory systems that are essential to the gas turbine basic system and will always be supplied with it, at initial purchase stage.

ACCESSORY DRIVES*

Accessory drive units provide the power for hydraulic, pneumatic, and electrical systems in addition to providing various pumps and control systems for efficient engine operation. The high level of dependence upon these units requires an extremely reliable drive system.

The drive for the accessory units is typically taken from a rotating engine shaft, via an internal gearbox, to an external gearbox that provides a mount for the accessories and distributes the appropriate geared drive to each accessory unit. A starter may also be fitted to provide an input torque to the engine. An accessory drive system on a high by-pass engine takes between 400 and 500 horsepower from the engine.

Gearboxes and Drives

Internal Gearbox

The location of the internal gearbox within the core of an engine is dictated by the difficulties of bringing the driveshaft radially outwards and the space available within the engine core.

Thermal fatigue and a reduction in engine performance, due to the radial drive shaft disturbing the gas flow, creates greater problems within the turbine area than the compressor area. For any given engine, which incorporates an axial-flow compressor, the turbine area is smaller than that containing the compressor and therefore makes it physically easier to mount the gearbox within the compressor section. Centrifugal compressor engines can have limited available space, so the internal gearbox is located within a static nose cone or, in the case of a turbo-propeller engine, behind the propeller reduction gear as shown in Figure 8—1.

On multi-shaft engines, the choice of which compressor shaft is used to drive the internal gearbox is primarily dependent upon the ease of engine starting. This is achieved by rotating the compressor shaft, usually via an input torque from the external gearbox. In practice the high pressure system is invariably rotated in order to generate an airflow through the engine and the high pressure compressor shaft is therefore coupled to the internal gearbox.

To minimize unwanted movement between the compressor shaft bevel gear and radial driveshaft bevel gear, caused by axial movement of the compressor shaft, the drive is taken by one of three basic methods (Figure 8—2).

* Source: Adapted, with permission, from Rolls Royce, *The Jet Engine*, 1986, Rolls Royce Plc: UK.

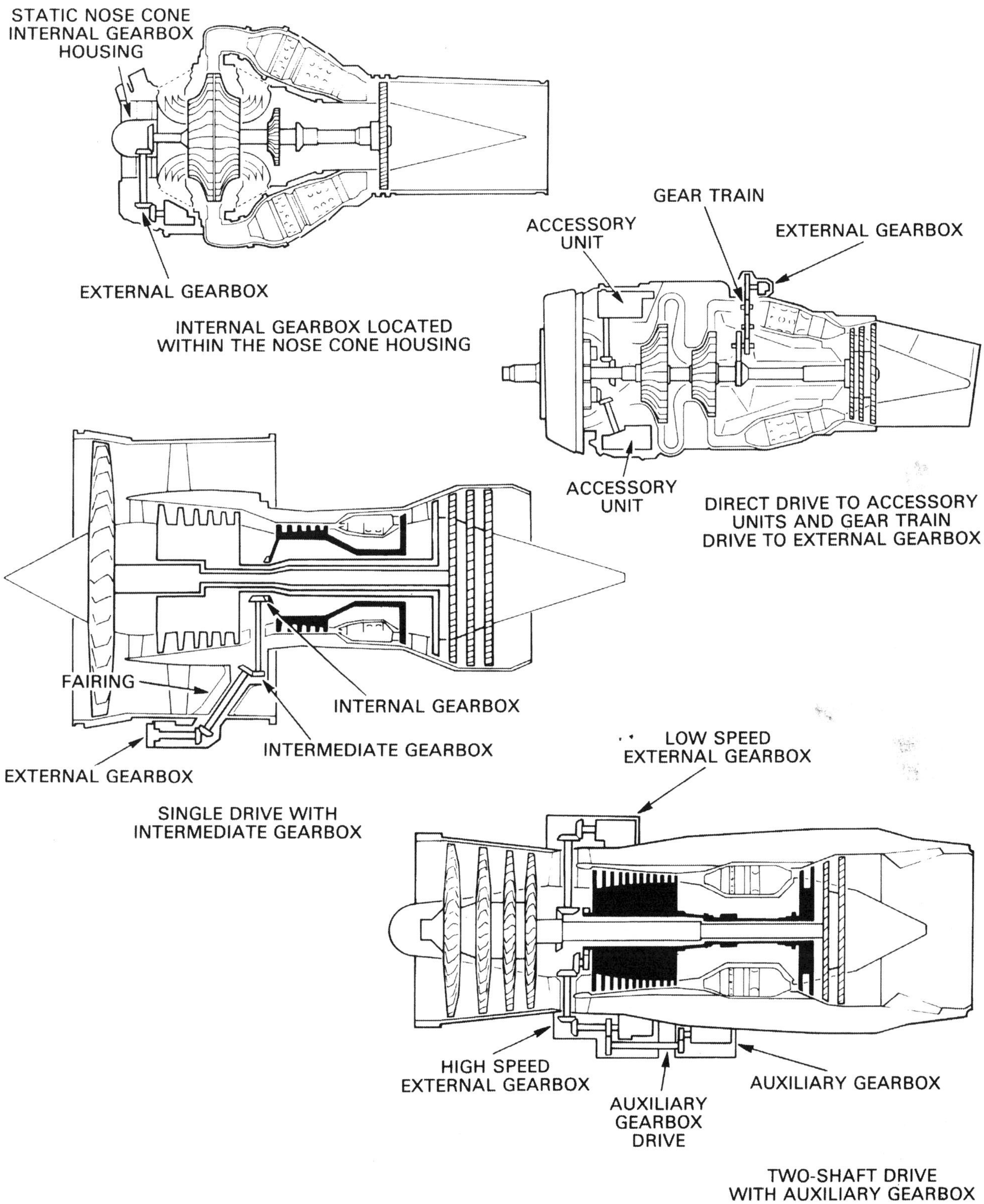

FIGURE 8–1 Mechanical arrangement of accessory drives. *(Source: Rolls Royce.)*

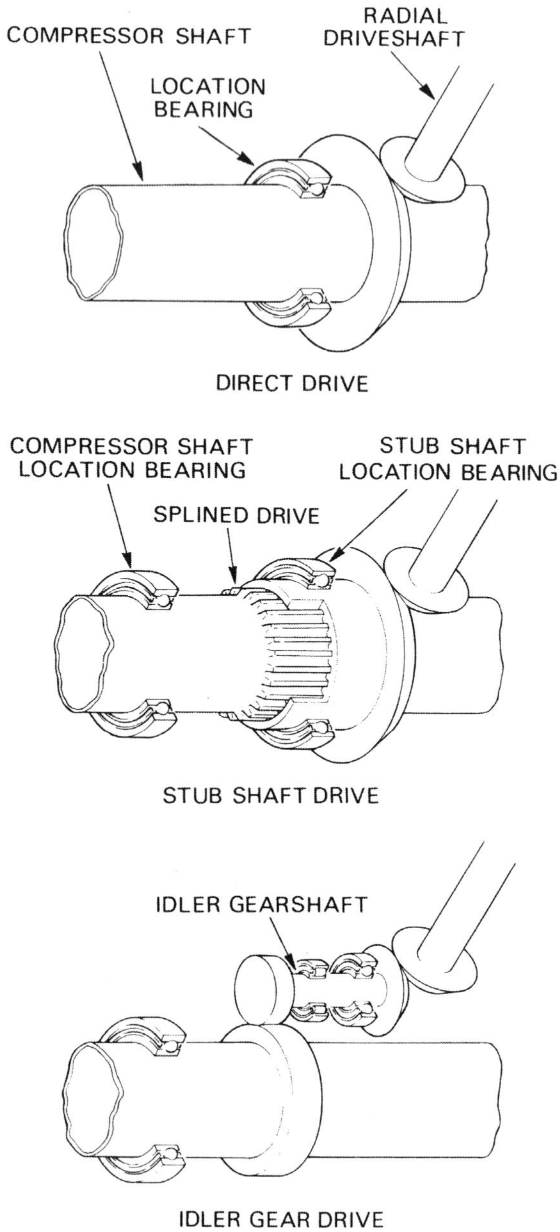

FIGURE 8–2 Mechanical arrangement of internal gearboxes. *(Source: Rolls Royce.)*

The least number of components is used when the compressor shaft bevel gear is mounted as close to the compressor shaft location bearing as possible, but a small amount of movement has to be accommodated within the meshing of the bevel gears. Alternatively, the compressor shaft bevel gear may be mounted on a stub shaft that has its own location bearing. The stub shaft is splined onto the compressor shaft that allows axial movement without affecting the bevel gear mesh. A more complex system utilizes an idler gear that meshes with the compressor shaft via straight spur gears, accommodating the axial movement, and drives the radial driveshaft via a bevel gear arrangement.

The latter method was widely employed on early engines to overcome gear engagement difficulties at high speed.

To spread the load of driving accessory units, some engines take a second drive from the slower rotating low pressure shaft to a second external gearbox (Figure 8–1). This also has the advantage of locating the accessory units in two groups, thus overcoming the possibility of limited external space on the engine. When this method is used, an attempt is made to group the accessory units specific to the engine onto the high pressure system, since that is the first shaft to rotate, and the aircraft accessory units are driven by the low pressure system. A typical internal gearbox showing how both drives are taken is shown in Figure 8–3.

Radial Driveshaft

The purpose of a radial driveshaft is to transmit the drive from the internal gearbox to an accessory unit or the external gearbox. It also serves to transmit the high torque from the starter to rotate the high-pressure system for engine starting purposes. The driveshaft may be direct drive or via an intermediate gearbox.

To minimize the effect of the driveshaft passing through the compressor duct and disrupting the airflow, it is housed within the compressor support structure. On by-pass engines, the driveshaft is either housed in the outlet guide vanes or in a hollow streamlined radial fairing across the low-pressure compressor duct.

To reduce airflow disruption it is desirable to have the smallest driveshaft diameter as possible. The smaller the diameter, the faster the shaft must rotate to provide the same power. However, this raises the internal stress and gives greater dynamic problems, which result in vibration. A long radial driveshaft usually requires a roller bearing situated halfway along its length to give smooth running. This allows a rotational speed of approximately 25,000 rpm to be achieved with a shaft diameter of less than 1.5 inch without encountering serious vibration problems.

Direct Drive

In some early engines, a radial driveshaft was used to drive each, or in some instances a pair of, accessory units. Although this allowed each accessory unit to be located in any desirable location around the engine and decreased the power transmitted through individual gears, it necessitated a large internal gearbox. Additionally, numerous radial driveshafts had to be incorporated within the design. This led to an excessive amount of time required for disassembly and assembly of the engine for maintenance purposes.

In some instances the direct drive method may be used in conjunction with the external gearbox system when it is impractical to take a drive from a particular area of the engine to the external gearbox. For example, Figure 8–1

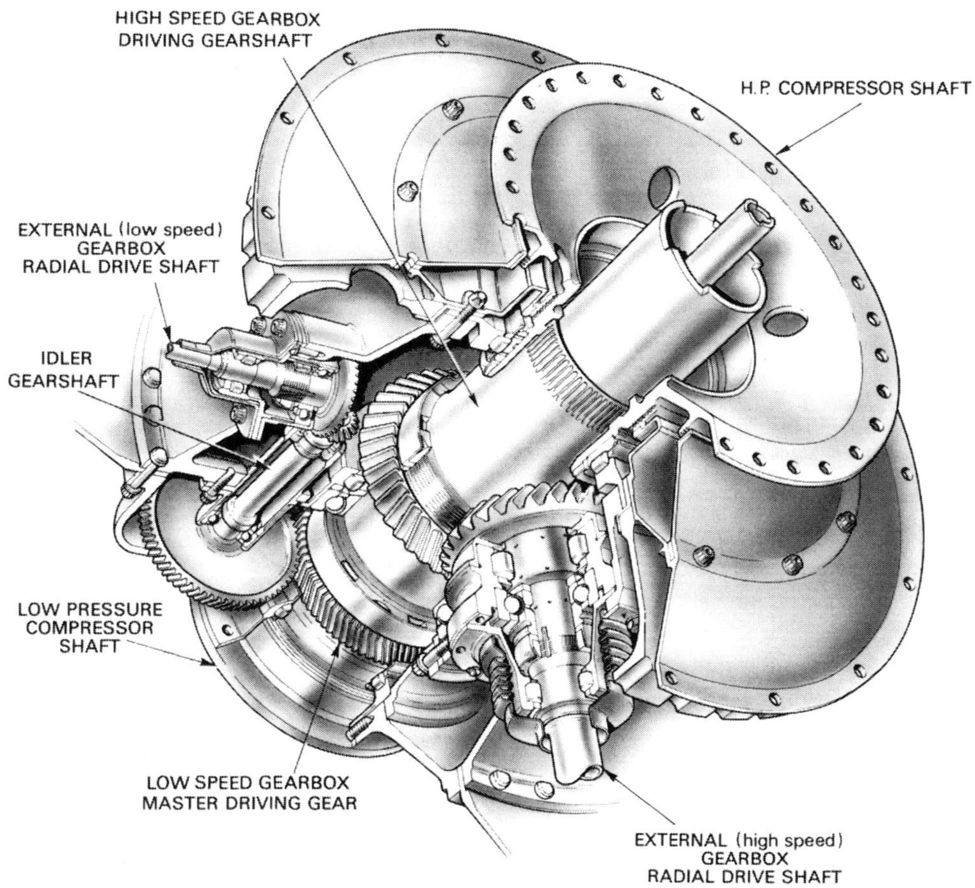

FIGURE 8–3 An internal gearbox. *(Source: Rolls Royce.)*

shows a turbo-propeller engine that requires accessories specific to the propeller reduction drive, but has the external gearbox located away from this area to receive the drive from the compressor shaft.

Gear Train Drive

When space permits, the drive may be taken to the external gearbox via a gear train (Figure 8–1). This involves the use of spur gears, sometimes incorporating a centrifugal breather. However, it is rare to find this type of drive system in current use.

Intermediate Gearbox

Intermediate gearboxes are employed when it is not possible to directly align the radial driveshaft with the external gearbox. To overcome this problem an intermediate gearbox is mounted on the high-pressure compressor case and redirects the drive, through bevel gears, to the external gearbox. An example of this layout is shown in Figure 8–1.

External Gearbox

The external gearbox contains the drives for the accessories, the drive from the starter, and provides a mounting face for each accessory unit. Provision is also made for hand turning the engine, via the gearbox, for maintenance purposes. Figure 8–4 shows the accessory units that are typically found on an external gearbox.

The overall layout of an external gearbox is dictated by a number of factors. To reduce drag while the aircraft is flying it is important to present a low frontal area to the airflow. Therefore the gearbox is "wrapped" around the engine and may look, from the front, similar to a banana in shape. For maintenance purposes the gearbox is generally located on the underside of the engine to allow ground crew to gain access. However, helicopter installation design usually requires the gearbox to be located on the top of the engine for ease of access.

The starter/driven gearshaft (Figure 8–4) roughly divides the external gearbox into two sections. One section provides the drive for the accessories that require low power while the other drives the high power accessories. This allows the small and large gears to be grouped together independently and is an efficient method of distributing the drive for the minimum weight.

If any accessory unit fails, and is prevented from rotating, it could cause further failure in the external gearbox by shearing the teeth of the gear train. To prevent secondary failure from occurring, a weak section is

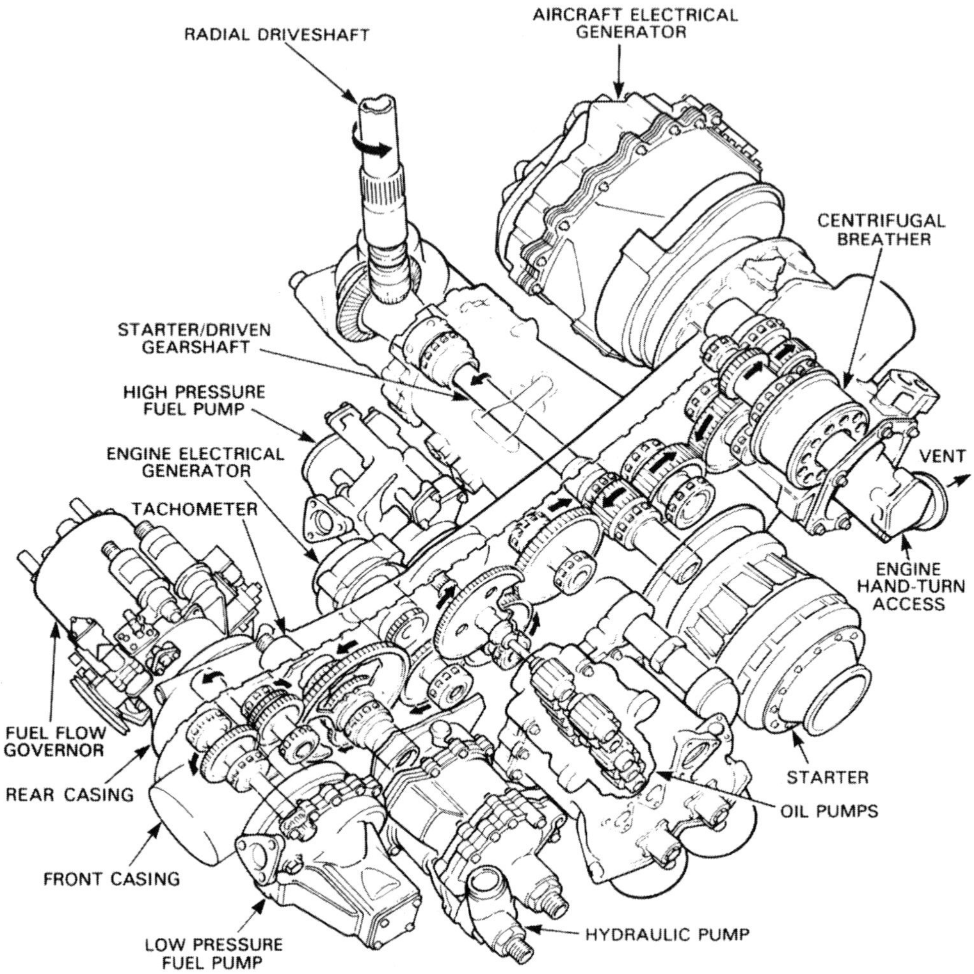

FIGURE 8–4 An external gearbox and accessory units. *(Source: Rolls Royce.)*

machined into the driveshafts, known as a "shear-neck," which is designed to fail and thus protect the other drives. This feature is not included for primary engine accessory units, such as the oil pumps, because these units are vital to the running of the engine and any failure would necessitate immediate shutdown of the engine.

Since the starter provides the highest torque that the drive system encounters, it is the basis of design. The starter is usually positioned to give the shortest drive line to the engine core. This eliminates the necessity of strengthening the entire gear train, which would increase the gearbox weight. However, when an auxiliary gearbox is fitted the starter is moved along the gear train to allow the heavily loaded auxiliary gearbox drive to pass through the external gearbox. This requires the spur gears between the starter and starter/driven gearshaft to have a larger face width to carry the load applied by the starter (Figure 8–5).

When a drive is taken from two compressor shafts, as discussed previously, two separate gearboxes are required. These are mounted on either side of the compressor case and are generally known as the "low speed" and "high speed" external gearboxes.

Auxiliary Gearbox

An auxiliary gearbox is a convenient method of providing additional accessory drives when the configuration of an engine and airframe does not allow enough space to mount all of the accessory units on a single external gearbox.

A drive is taken from the external gearbox (Figure 8–5) to power the auxiliary gearbox, which distributes the appropriate gear ratio drive to the accessories in the same manner as the external gearbox.

Construction and Materials

Gears

The spur gears of the external or auxiliary gearbox gear train (Figures 8–4 and 8–5) are mounted between bearings supported by the front and rear casings that are bolted

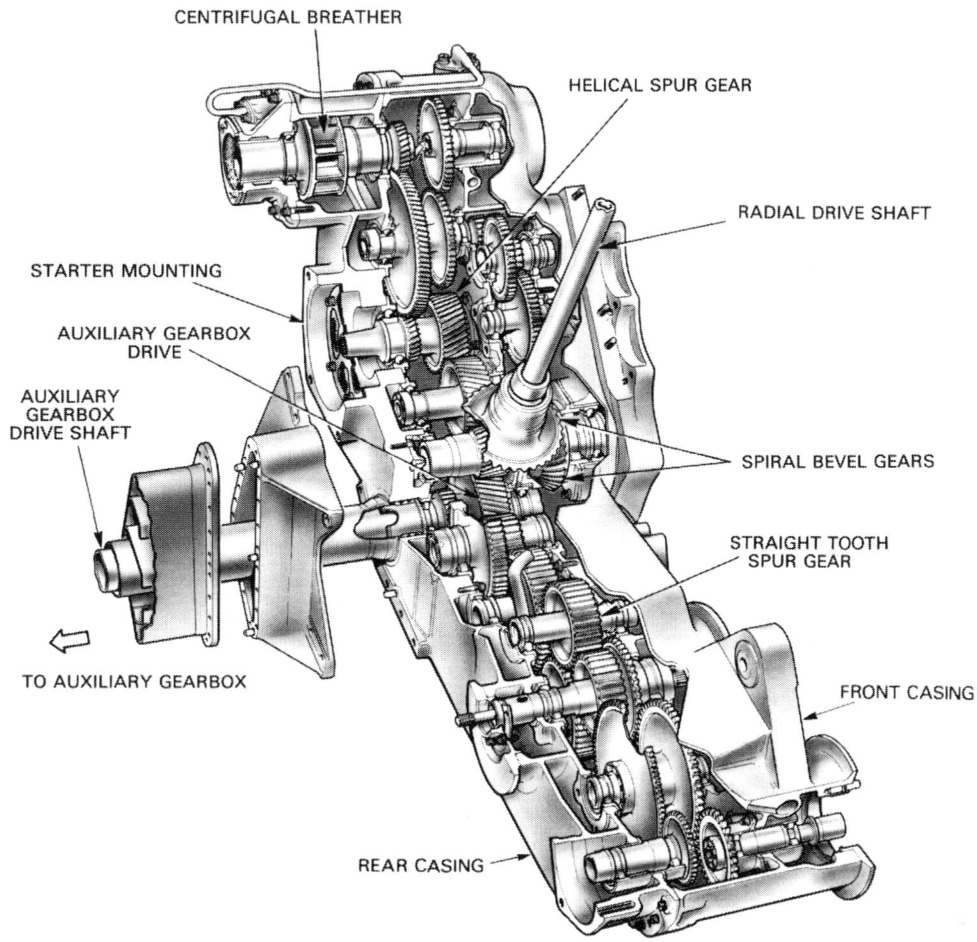

FIGURE 8–5 An external gearbox with auxiliary gearbox drive. *(Source: Rolls Royce.)*

together. They transmit the drive to each accessory unit, which is normally between 5000 and 6000 rpm for the accessory units and approximately 20,000 rpm for the centrifugal breather.

All gear meshes are designed with "hunting tooth" ratios that ensure that each tooth of a gear does not engage between the same set of opposing teeth on each revolution. This spreads any wear evenly across all teeth.

Spiral bevel gears are used for the connection of shafts whose axes are at an angle to one another but in the same plane. The majority of gears within a gear train are of the straight spur gear type; those with the widest face carry the greatest loads. For smoother running, helical gears are used but the resultant end thrust caused by this gear tooth pattern must be catered for within the mounting of the gear.

Gearbox Sealing

Sealing of the accessory drive system is primarily concerned with preventing oil loss. The internal gearbox has labyrinth seals where the static casing mates with the rotating compressor shaft. For some of the accessories mounted on the external gearbox, an air-blown pressurized labyrinth seal is employed. This prevents oil from the gearbox entering the accessory unit and also prevents contamination of the gearbox, and hence engine, in the event of an accessory failure. The use of an air-blown seal results in a gearbox pressure of about 3 lbs. per sq. in. above atmospheric pressure. To supplement a labyrinth seal, an "oil thrower ring" may be used. This involves the leakage oil running down the driving shaft and being flung outwards by a flange on the rotating shaft. The oil is then collected and returned to the gearbox.

Materials

To reduce weight, the lightest materials possible are used. The internal gearbox casing is cast from aluminum, but the low environmental temperatures that an external gearbox is subjected to allow the use of magnesium castings, which are lighter still. The gears are manufactured from non-corrosion resistant steels for strength and toughness. They are case hardened to give a very hard wear resistant skin and feature accurately ground teeth for smooth gear meshing.

STARTING AND IGNITION SYSTEMS*

Two separate systems are required to ensure that a gas turbine engine will start satisfactorily. Firstly, provision must be made for the compressor and turbine to be rotated up to a speed at which adequate air passes into the combustion system to mix with fuel from the fuel spray nozzles. Secondly, provision must be made for ignition of the air/fuel mixture in the combustion system. During engine starting the two systems must operate simultaneously, yet it must also be possible to motor the engine over without ignition for maintenance checks and to operate only the ignition system for relighting during flight.

The functioning of both systems is coordinated during a starting cycle and their operation is automatically controlled after the initiation of the cycle by an electrical circuit. A typical sequence of events during the start of a turbo-jet engine is shown in Figure 8–6.

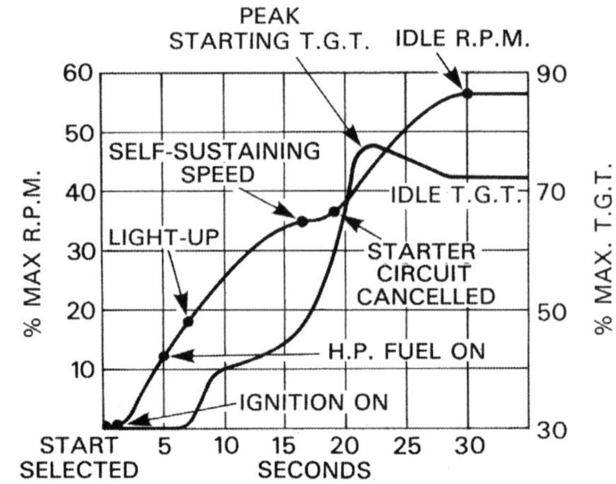

FIGURE 8–6 A typical starting sequence of a turbo-jet engine. *(Source: Rolls Royce.)*

Methods of Starting

The starting procedure for all jet engines is basically the same, but can be achieved by various methods. The type and power source for the starter varies in accordance with engine and aircraft requirements. Some use electrical power; others use gas, air, or hydraulic pressure, and each has its own merits. For example, a military aircraft requires the engine to be started in the minimum time and, when possible, to be completely independent of external equipment. A commercial aircraft, however, requires the engine to be started with the minimum disturbance to the passengers and by the most economical means. Whichever system is used, reliability is of prime importance.

The starter motor must produce a high torque and transmit it to the engine rotating assembly in a manner that provides smooth acceleration from rest up to a speed at which the gas flow through the engine provides sufficient power for the engine turbine to take over.

Electric

The electric starter is usually a direct current (D.C.) electric motor coupled to the engine through a reduction gear and ratchet mechanism, or clutch, which automatically disengages after the engine has reached a self-sustaining speed (Figure 8–7).

The electrical supply may be of a high or low voltage and is passed through a system of relays and resistances to allow the full voltage to be progressively built up as the starter gains speed. It also provides the power for the operation of the ignition system. The electrical supply is automatically cancelled when the starter load is reduced after the engine has satisfactorily started or when the time cycle is completed. A typical electrical starting system is shown in Figure 8–8.

Cartridge

Cartridge starting is sometimes used on military engines and provides a quick independent method of starting. The starter motor is basically a small impulse-type turbine that is driven by high velocity gases from a burning cartridge. The power output of the turbine is passed through a reduction gear and an automatic disconnect mechanism to rotate the engine. An electrically fired detonator initiates the burning of the cartridge charge. As a cordite charge provides the power supply for this type of starter, the size of the charge required may well limit the use of the cartridge starters. A triple-breech starter is illustrated in Figure 8–9.

Iso-Propyl-Nitrate

This type of starter provides a high power output and gives rapid starting characteristics. It has a turbine that transmits power through a reduction gear to the engine. In this instance, the turbine is rotated by high-pressure gases resulting from the combustion of iso-propyl-nitrate. This fuel is sprayed into a combustion chamber, which forms part of the starter, where it is electrically ignited by a high-energy ignition system. A pump supplies the fuel to the combustion chamber from a storage tank and an air pump scavenges the starter combustion chamber of fumes before each start. Operation of the fuel and air pumps, ignition systems, and cycle cancellation is electrically controlled by relays and time switches. An iso-propyl-nitrate starting system is shown in Figure 8–10.

* Source: Adapted, with permission, from Rolls Royce, *The Jet Engine*, 1986, Rolls Royce Plc: UK.

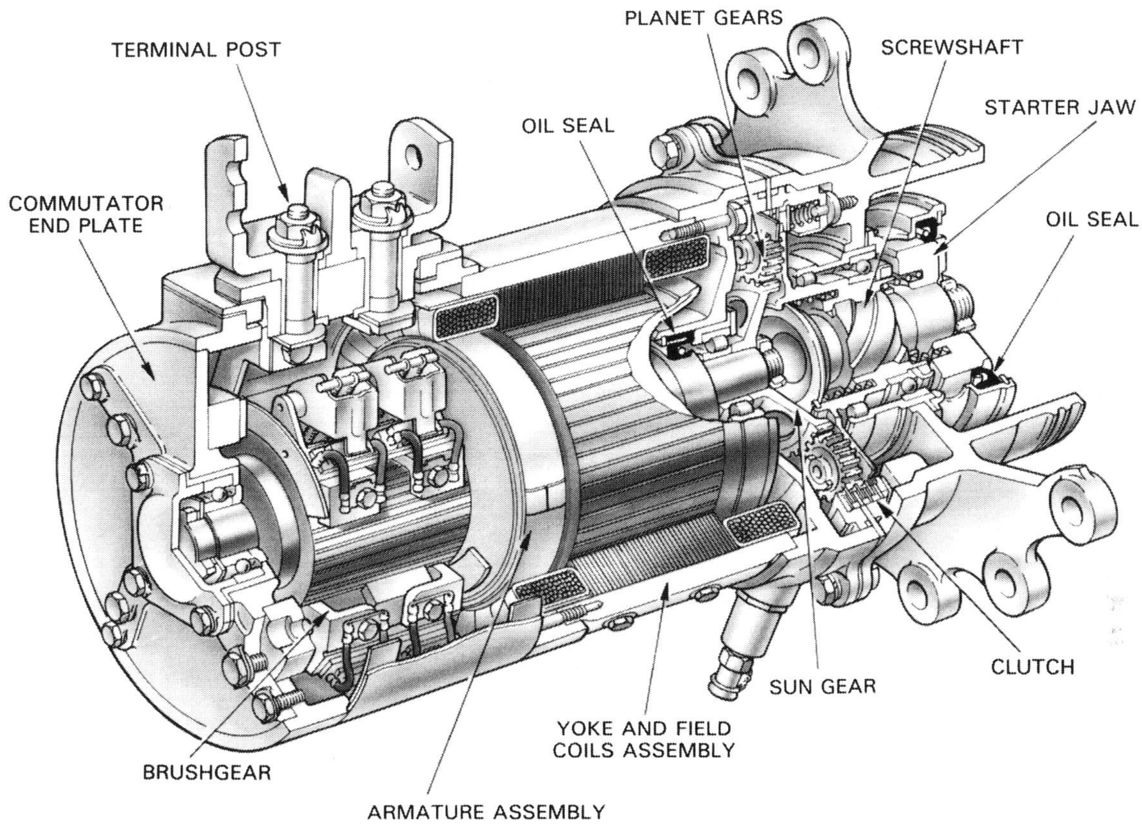

FIGURE 8–7 An electric starter. *(Source: Rolls Royce.)*

Air

Air starting is used on most commercial and some military jet engines. It has many advantages over other starting systems, and is comparatively light, simple, and economical to operate.

An air starter motor transmits power through a reduction gear and clutch to the starter output shaft, which is connected to the engine. A typical air starter motor is shown in Figure 8–11.

The starter turbine is rotated by air taken from an external ground supply, an auxiliary power unit (A.P.U.) or as a cross-feed from a running engine. The air supply to the starter is controlled by an electrically operated control and pressure-reducing valve that is opened when an engine start is selected and is automatically closed at a predetermined starter speed. The clutch also automatically disengages as the engine accelerates up to idling rpm and the rotation of the starter ceases. A typical air starting system is shown in Figure 8–12.

A combustor starter is sometimes fitted to an engine incorporating an air starter and is used to supply power to the starter when an external supply of air is not available. The starter unit has a small combustion chamber into which high pressure air, from an aircraft-mounted storage bottle, and fuel, from the engine fuel system, are introduced. Control valves regulate the air supply that pressurizes a fuel accumulator to give sufficient fuel pressure for atomization and also activates the continuous ignition system. The fuel/air mixture is ignited in the combustion chamber and the resultant gas is directed onto the turbine of the air starter. An electrical circuit is provided to shut off the air supply, which in turn terminates the fuel and ignition systems on completion of the starting cycle.

Some turbo-jet engines are not fitted with starter motors, but use air impingement onto the turbine blades as a means of rotating the engine. The air is obtained from an external source, or from an engine that is running, and is directed through non-return valves and nozzles onto the turbine blades. A typical method of air impingement starting is shown in Figure 8–13.

Gas Turbine

A gas turbine starter is used for some jet engines and is completely self-contained. It has its own fuel and ignition system, starting system (usually electric or hydraulic), and self-contained oil system. This type of starter is economical to operate and provides a high power output for a comparatively low weight.

The starter consists of a small, compact gas turbine engine, usually featuring a turbine-driven centrifugal compressor, a reverse flow combustion system, and a mechanically independent free-power turbine. The free-power turbine is connected to the main engine via a

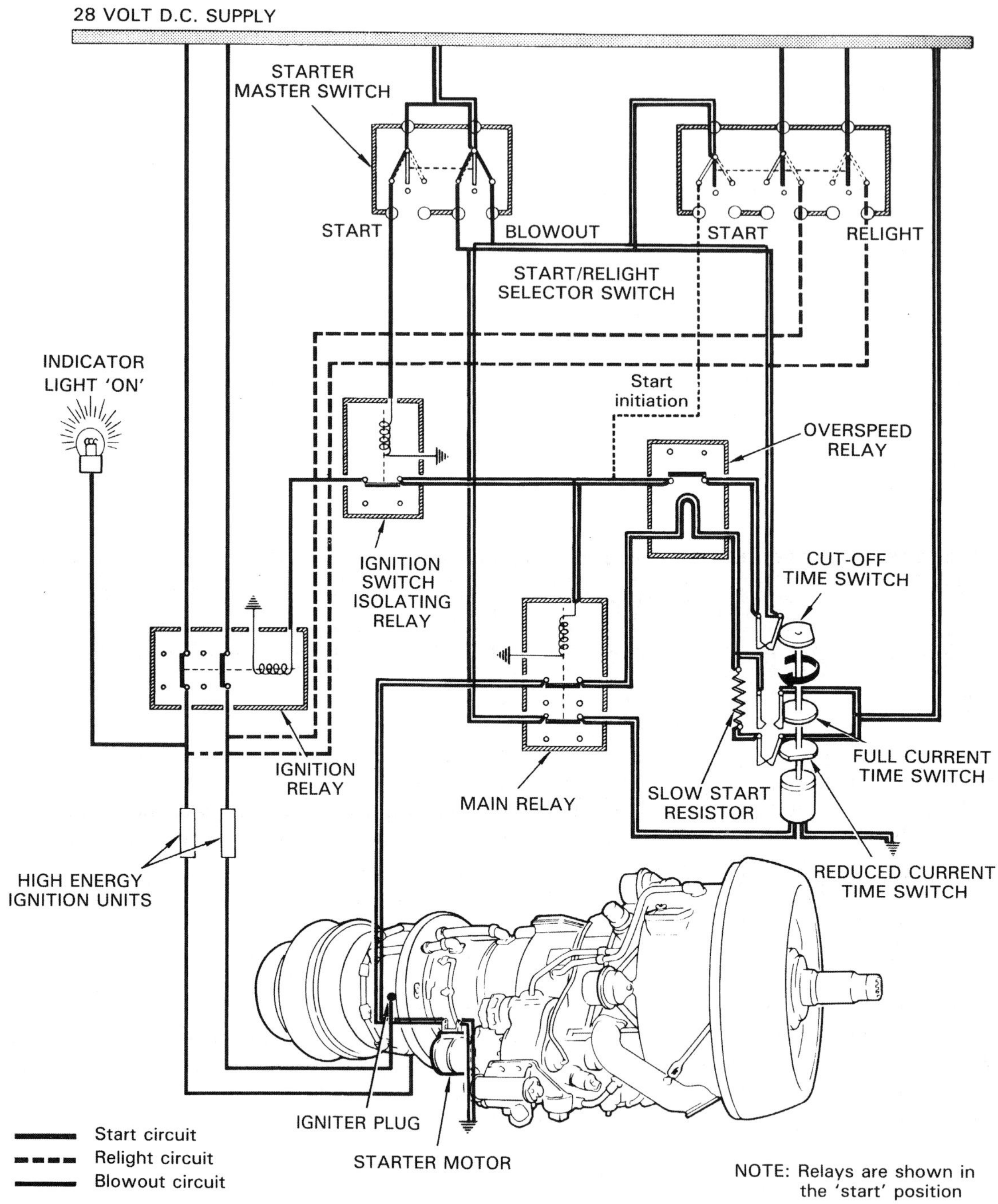

FIGURE 8–8 A low voltage electrical starting system. *(Source: Rolls Royce.)*

two-stage epicyclic reduction gear, automatic clutch and output shaft. A typical gas turbine starter is shown in Figure 8–14.

On initiation of the starting cycle, the gas turbine starter is rotated by its own starter motor until it reaches self-sustaining speed, when the starting and ignition systems are automatically switched off. Acceleration then continues up to a controlled speed of approximately 60,000 rpm. At the same time as the gas turbine starter engine is accelerating, the exhaust gas is being directed, via nozzle guide vanes, onto the free-power turbine to provide the drive to the main engine. Once the main engine reaches self-sustaining speed, a cut-out switch operates and shuts down the gas turbine starter. As the starter runs down, the clutch automatically disengages from the output shaft and the main engine accelerates up to idling rpm under its own power.

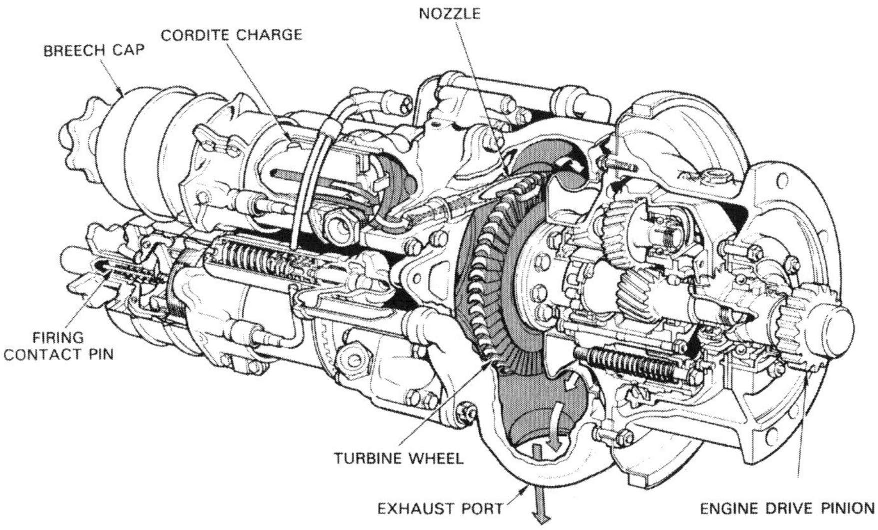

FIGURE 8–9 A triple-breech cartridge starter. *(Source: Rolls Royce.)*

Hydraulic

Hydraulic starting is used for starting some small jet engines. In most applications, one of the engine-mounted hydraulic pumps is utilized and is known as a pump/starter, although other applications may use a separate hydraulic motor. Methods of transmitting the torque to the engine may vary, but a typical system would include a reduction gear and clutch assembly. Power to rotate the pump/starter is provided by hydraulic pressure from a ground supply unit and is transmitted to the engine through the reduction gear and clutch. The starting system is controlled by an electrical circuit that also operates hydraulic valves so that on completion of the starting cycle the pump/starter functions as a normal hydraulic pump.

Ignition

High-energy (H.E.) ignition is used for starting all jet engines and a dual system is always fitted. Each system has an ignition unit connected to its own igniter plug, the two plugs being situated in different positions in the combustion system.

Each H.E. ignition unit receives a low voltage supply, controlled by the starting system electrical circuit, from the aircraft electrical system. The electrical energy is stored in the unit until, at a predetermined value, the energy is dissipated as a high voltage, high amperage discharge across the igniter plug.

Ignition units are rated in joules (one joule equals one watt per second). They are designed to give outputs that may vary according to requirements. A high value output (e.g., 12 joule) is necessary to ensure that the engine will obtain a satisfactory relight at high altitudes and is sometimes necessary for starting. However, under certain flight conditions, such as icing or takeoff in heavy rain or snow, it may be necessary to have the ignition system continuously operating to give an automatic relight should flame extinction occur. For this condition, a low value output (e.g., 3–6 joule) is preferred because it results in a longer life of the igniter plug and ignition unit. Consequently, to suit all engine operating conditions, a combined system giving a high and low value output is favored. Such a system would consist of one unit emitting a high output to one igniter plug, and a second unit giving a low output to a second igniter plug. However, some ignition units are capable of supplying both high and low outputs, the value being pre-selected as required.

An ignition unit may be supplied with direct current (D.C.) and operated by a trembler mechanism or a transistor chopper circuit, or supplied with alternating current (A.C.) and operated by a transformer. The operation of each type of unit is described in the subsequent paragraphs.

The ignition unit shown in Figure 8–15 is a typical D.C. trembler-operated unit. An induction coil, operated by the trembler mechanism, charges the reservoir capacitor (condenser) through a high voltage rectifier. When the voltage in the capacitor is equal to the breakdown value of a sealed discharge gap, the energy is discharged across the face of the igniter plug. A choke is fitted to extend the duration of the discharge and a discharge resistor is fitted to ensure that any residual stored energy in the capacitor is dissipated within 1 minute of the system being switched off. A safety resistor is fitted to enable the unit to operate safely, even when the high tension lead is disconnected and isolated.

Operation of the transistorized ignition unit is similar to that of the D.C. trembler-operated unit, except that the trembler-unit is replaced by a transistor chopper circuit. A typical transistorized unit is shown in Figure 8–16: such a unit has many advantages over the trembler-operated unit because it has no moving parts and gives a much

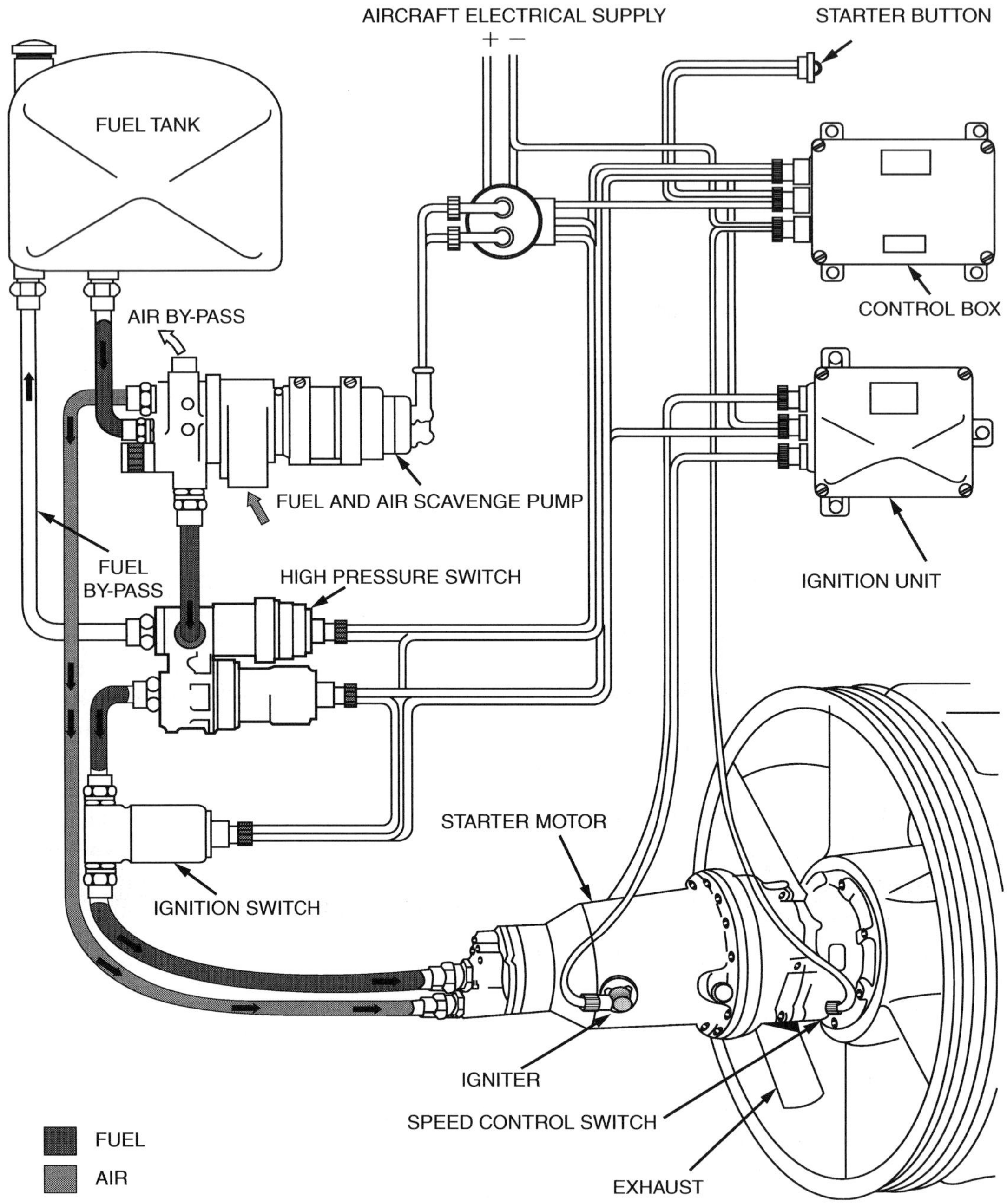

FIGURE 8–10 An iso-propyl-nitrate starting system. *(Source: Rolls Royce. Adapted for reproduction in black and white.)*

longer operating life. The size of the transistorized unit is reduced and its weight is less than that of the trembler-operated unit.

The A.C. ignition unit, shown in Figure 8–17, receives an alternating current that is passed through a transformer and rectifier to charge a capacitor. When the voltage in the capacitor is equal to the breakdown value of a sealed discharge gap, the capacitor discharges the energy across the face of the igniter plug. Safety and discharge resistors are fitted as in the trembler-operated unit.

There are two basic types of igniter plug; the constricted or constrained air gap type and the shunted surface discharge type. The air gap type is similar in operation to the conventional reciprocating engine spark plug, but has a larger air gap between the electrode and body for the spark to cross. A potential difference of approximately 25,000 volts is required to ionize the gap before a spark will occur. This high voltage requires very good insulation throughout the circuit. The surface discharge igniter plug (Figure 8–18) has the end of the insulator formed by a semi-conducting pellet that permits an electrical leakage

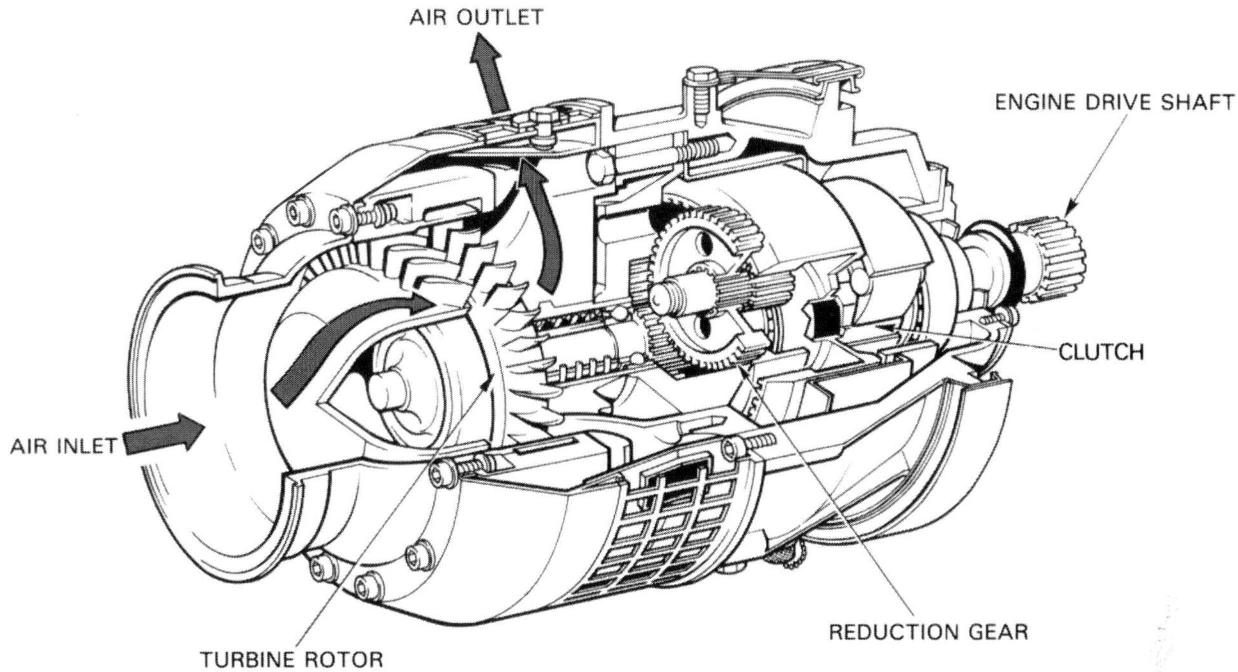

FIGURE 8–11 An air starter motor. *(Source: Rolls Royce.)*

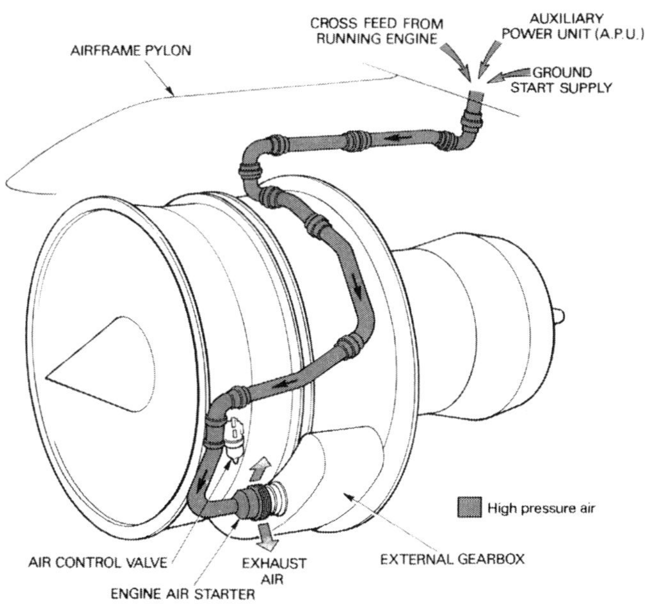

FIGURE 8–12 An air starting system. *(Source: Rolls Royce.)*

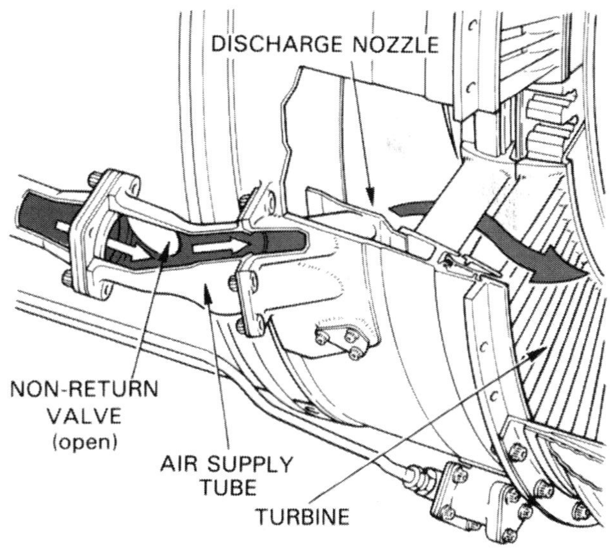

FIGURE 8–13 Air impingement starting. *(Source: Rolls Royce.)*

from the central high-tension electrode to the body. This ionizes the surface of the pellet to provide a low resistance path for the energy stored in the capacitor. The discharge takes the form of a high intensity flashover from the electrode to the body and only requires a potential difference of approximately 2000 volts for operation.

The normal spark rate of a typical ignition system is between 60 and 100 sparks per minute. Periodic replacement of the igniter plug is necessary due to the progressive erosion of the igniter electrodes caused by each discharge.

The igniter plug tip protrudes approximately 0.1 inch into the flame tube. During operation the spark penetrates a further 0.75 inch. The fuel mixture is ignited in the relatively stable boundary layer, which then propagates throughout the combustion system.

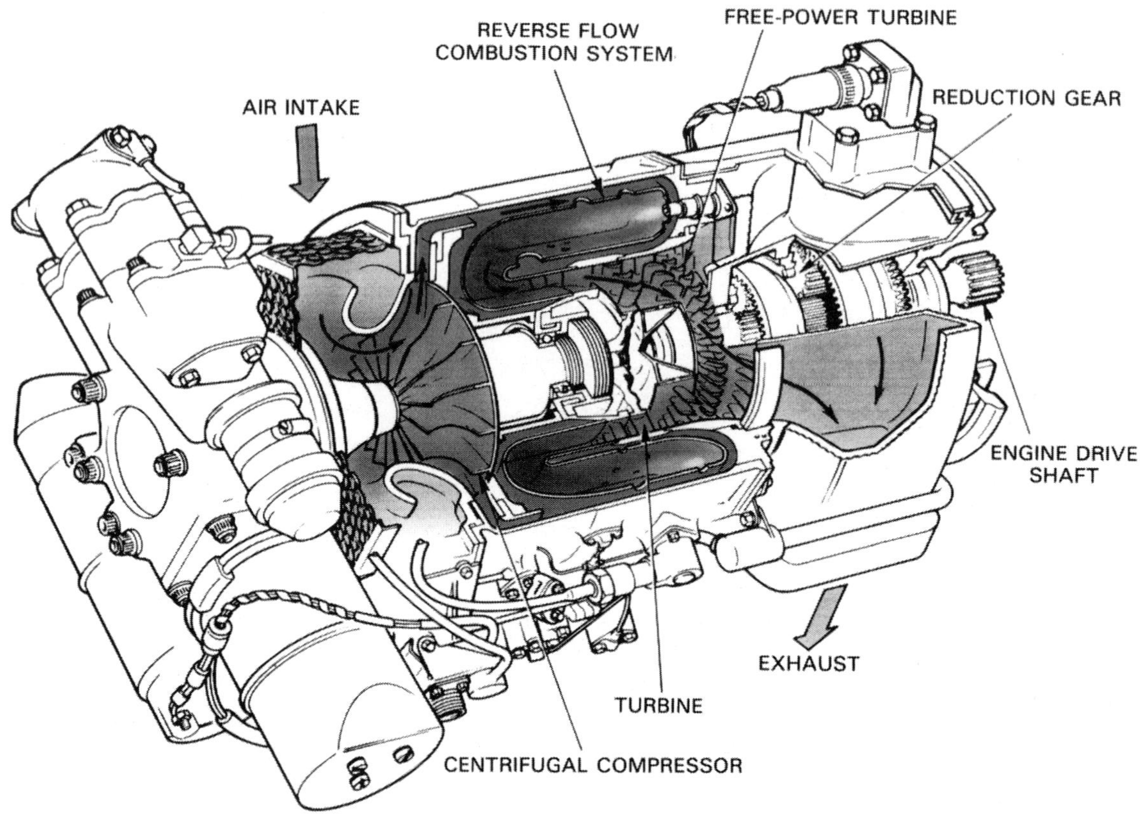

FIGURE 8–14 A gas turbine starter. *(Source: Rolls Royce.)*

Relighting

The jet engine requires facilities for relighting should the flame in the combustion system be extinguished during flight. However, the ability of the engine to relight will vary according to the altitude and forward speed of the aircraft. A typical relight envelope, showing the flight conditions under which an engine will obtain a satisfactory relight, is shown in Figure 8–19. Within the limits of the envelope, the airflow through the engine will rotate the compressor at a speed satisfactory for relighting; all that is required therefore, provided that a fuel supply is available, is the operation of the ignition system. This is provided for by a separate switch that operates only the ignition system.

ICE PROTECTION SYSTEMS*

With aeroengines, icing of the engine and the leading edges of the intake duct can occur during flight through clouds containing supercooled water droplets or during ground operation in freezing fog. Protection against ice formation may be required since icing of these regions can restrict considerably the airflow through the engine, causing a loss in performance and possible malfunction of the engine.

Additionally, damage may result from ice breaking away and being ingested into the engine or hitting the acoustic material lining the intake duct.

Given† the right conditions, icing can occur with land-based engines as well. With land-based engines, inlet duct heating may be used (exercising care that the melted ice does not cause damage to downstream airfoils). Inlet air filters (the "huff & puff" type that drop ice formed on filter cartridges with a blast of air) can also provide anti-icing protection.

An* ice protection system must effectively prevent ice formation within the operational requirements of the particular application. The system must be reliable, easy to maintain, present no excessive weight penalty, and cause no serious loss in engine performance when in operation.

Analyses are carried out to determine whether ice protection is required and, if so, the heat input required to limit ice build-up to acceptable levels. Figure 8–20 illustrates

* Source: Adapted, with permission, from Rolls Royce, *The Jet Engine*, 1986, Rolls Royce Plc: UK.

† Source: Working case notes, Claire Soares 1979.

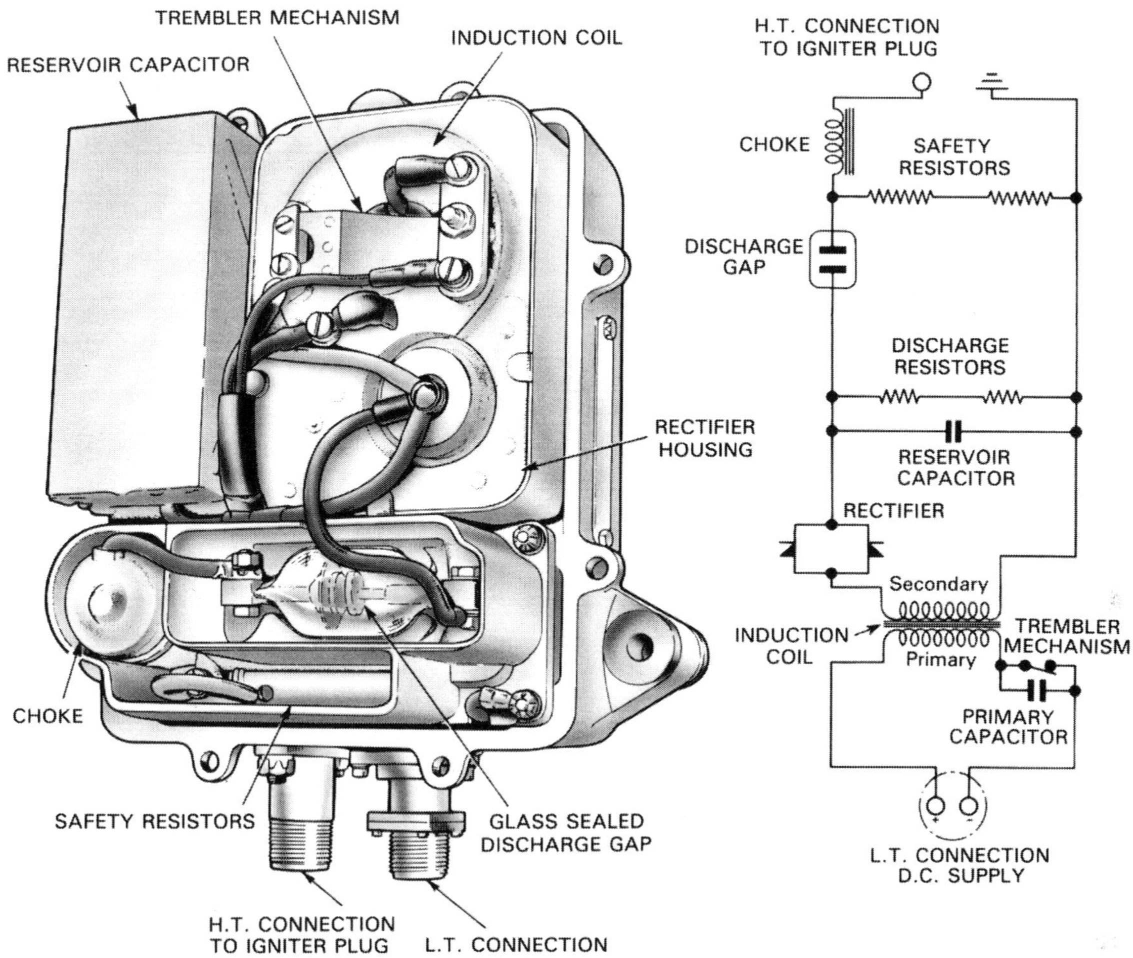

FIGURE 8–15 A D.C. trembler-operated ignition unit. *(Source: Rolls Royce.)*

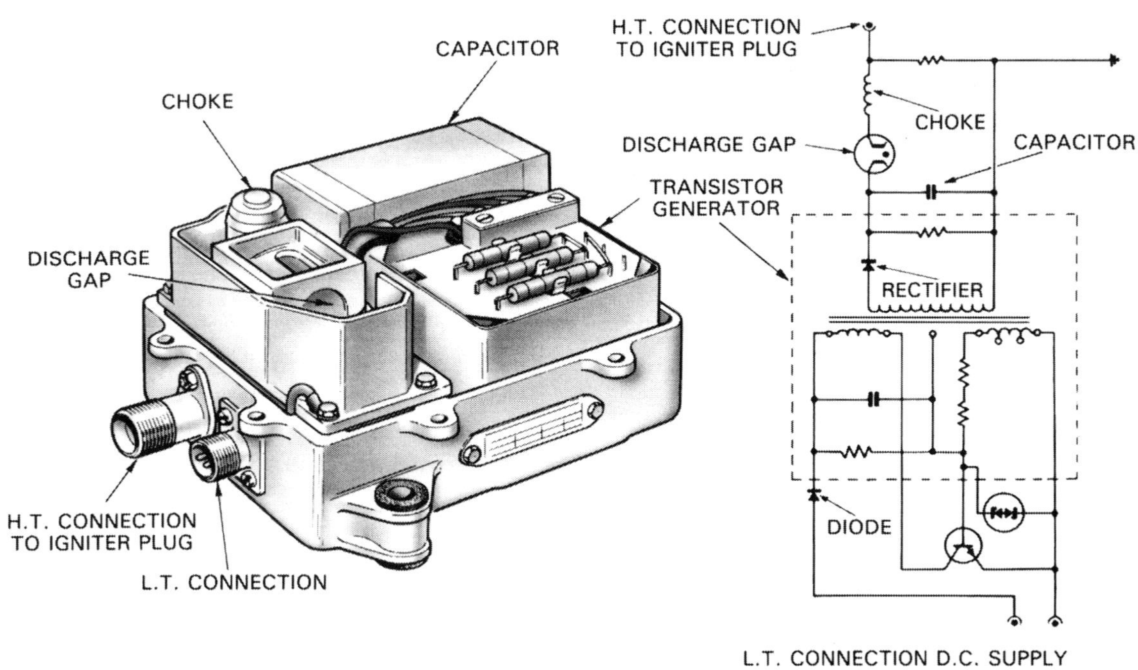

FIGURE 8–16 A transistorized ignition unit. *(Source: Rolls Royce.)*

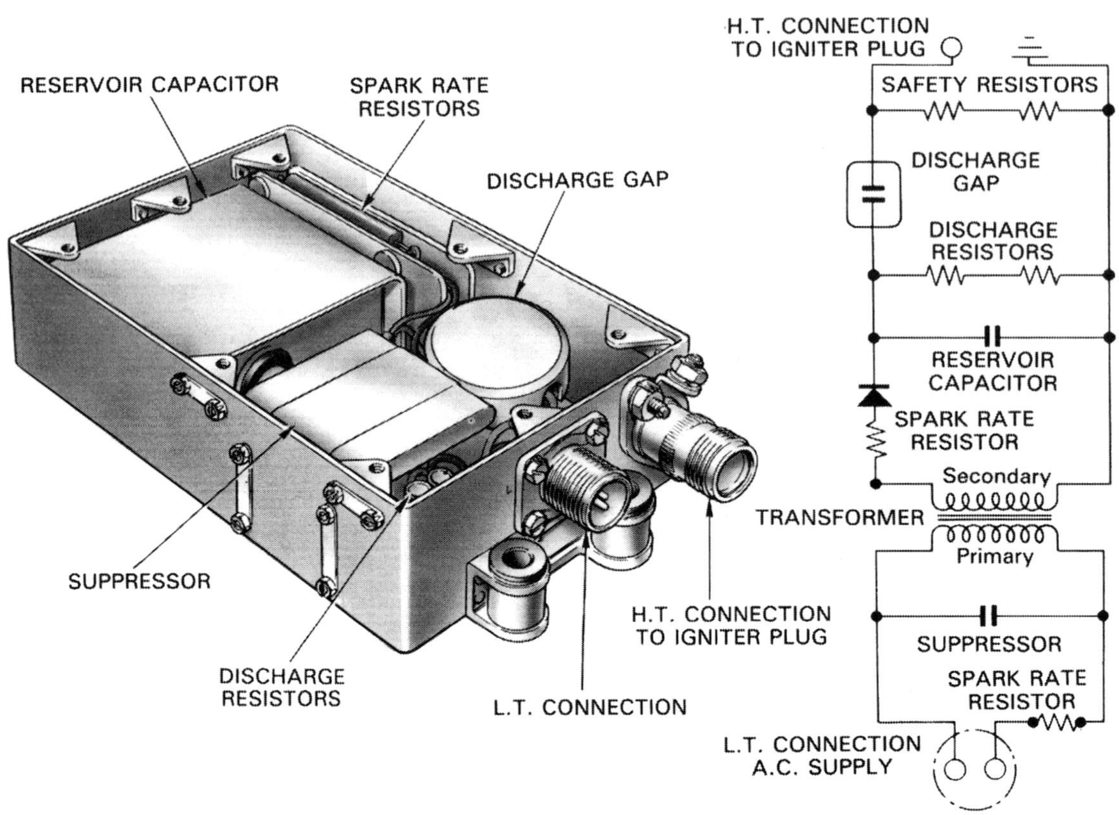

FIGURE 8–17 An A.C. ignition unit. *(Source: Rolls Royce.)*

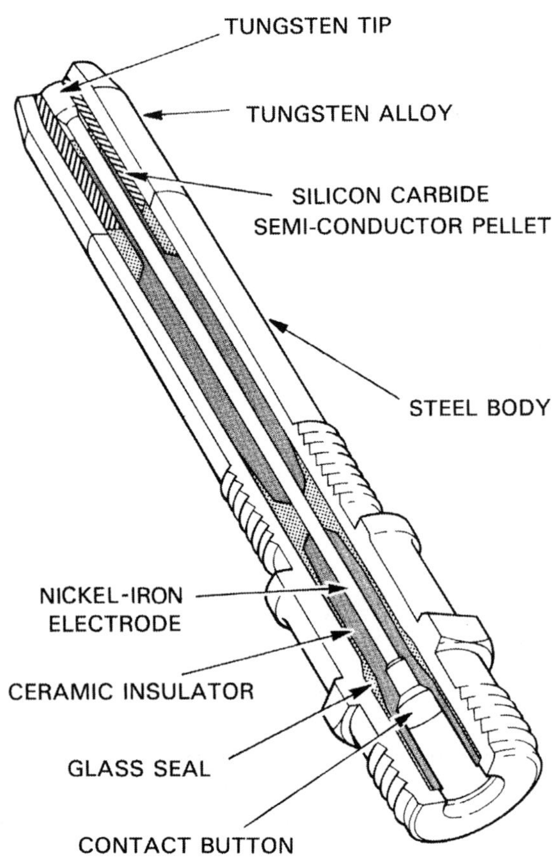

FIGURE 8–18 An igniter plug. *(Source: Rolls Royce.)*

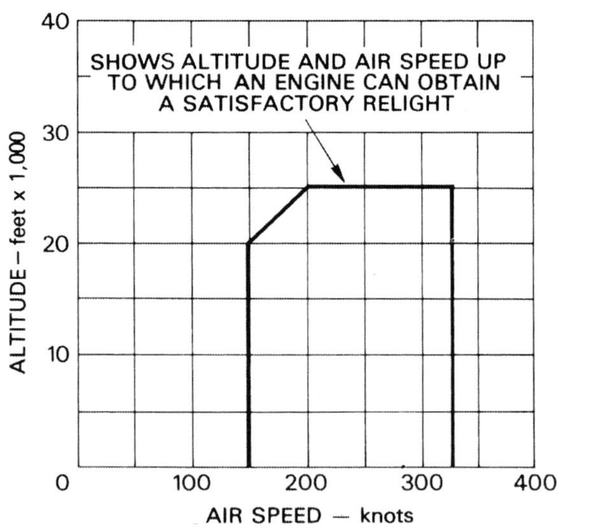

FIGURE 8–19 A typical flight relight envelope. *(Source: Rolls Royce.)*

the areas of a turbo-fan engine typically considered for ice protection.

There are two basic systems of ice protection; turbo-jet engines generally use a hot air supply (Figure 8–21), and turbo-propeller engines use electrical power or a combination of electrical power and hot air.

Protection may be supplemented by the circulation of hot oil around the air intake as shown in Figure 8–22. The hot air system is generally used to prevent the formation of

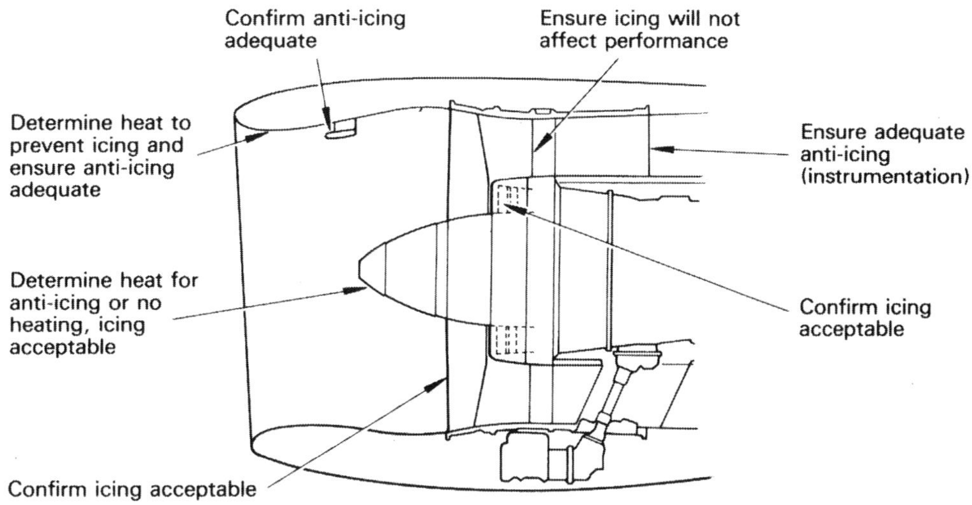

FIGURE 8–20 Areas typically considered for ice protection. *(Source: Rolls Royce.)*

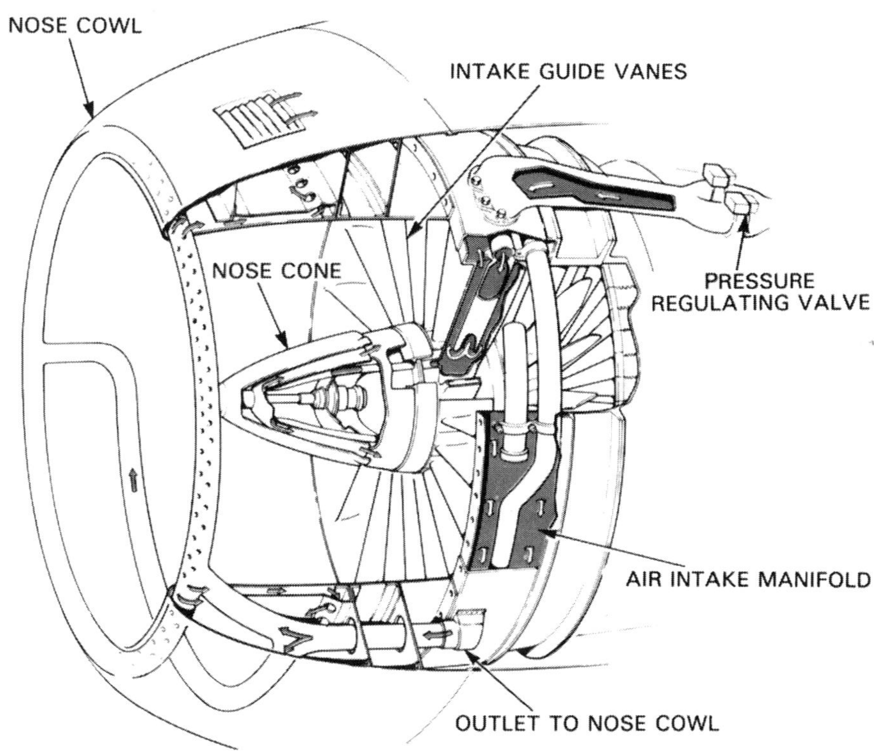

FIGURE 8–21 Hot air ice protection. *(Source: Rolls Royce.)*

ice and is known as an anti-icing system. The electrical power system is used to break up ice that has formed on surfaces and is known as a de-icing system.

HOT AIR SYSTEM

The hot air system provides surface heating of the engine and/or power plant where ice is likely to form. The protection of rotor blades is rarely necessary, because any ice accretions are dispersed by centrifugal action. If stators are fitted upstream of the first rotating compressor stage these may require protection. If the nose cone rotates it may not need anti-icing if its shape, construction, and rotational characteristics are such that likely icing is acceptable.

The hot air for the anti-icing system is usually taken from the high-pressure compressor stages. It is ducted through pressure regulating valves to the parts requiring anti-icing. Spent air from the nose cowl anti-icing system may be exhausted into the compressor intake or vented overboard.

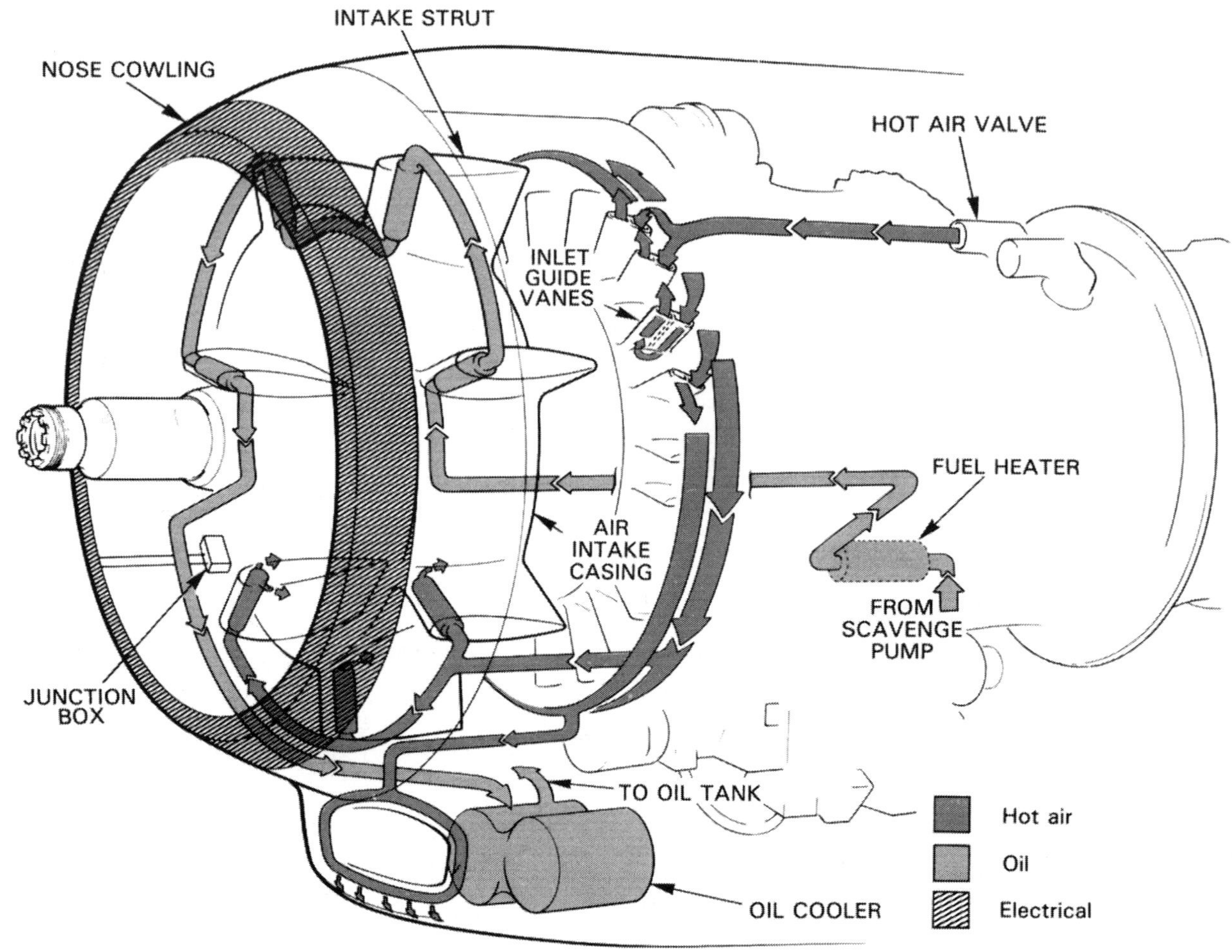

FIGURE 8–22 Combination of hot air, oil, and electrical ice protection. *(Source: Rolls Royce.)*

If the nose cone is anti-iced its hot air supply may be independent or integral with that of the nose cowl and compressor stators. For an independent system, the nose cone is usually anti-iced by a continuous unregulated supply of hot air via internal ducting from the compressor.

The pressure regulating valves are electrically actuated by manual selection, or automatically by signals from the aircraft ice detection system. The valves prevent excessive pressures being developed in the system, and act also as an economy device at the higher engine speeds by limiting the air offtake from the compressor, thus preventing an excessive loss in performance. The main valve may be manually locked in a pre-selected position prior to takeoff in the event of a valve malfunction, prior to replacement.

Electrical System

The electrical system of ice protection is generally used for turbo-propeller engine installations, as this form of protection is necessary for the propellers. The surfaces that require electrical heating are the air intake cowling of the engine, the propeller blades and spinner, and, when applicable, the oil cooler air intake cowling.

Electrical heating pads are bonded to the outer skin of the cowlings. They consist of strip conductors sandwiched between layers of neoprene, or glass cloth impregnated with epoxy resin. To protect the pads against rain erosion, they are coated with a special, polyurethane-based paint. When the de-icing system is operating, some of the areas are continuously heated to prevent an ice cap forming on the leading edges and also to limit the size of the ice that forms on the areas that are intermittently heated (Figure 8–23).

Electrical power is supplied by a generator and, to keep the size and weight of the generator to a minimum, the de-icing electrical loads are cycled between the engine, propeller, and, sometimes, the airframe.

When the ice protection system is in operation, the continuously heated areas prevent any ice forming, but the intermittently heated areas allow ice to form, during their "heat-off" period. During the "heat-on" period, adhesion of the ice is broken and it is then removed by aerodynamic forces.

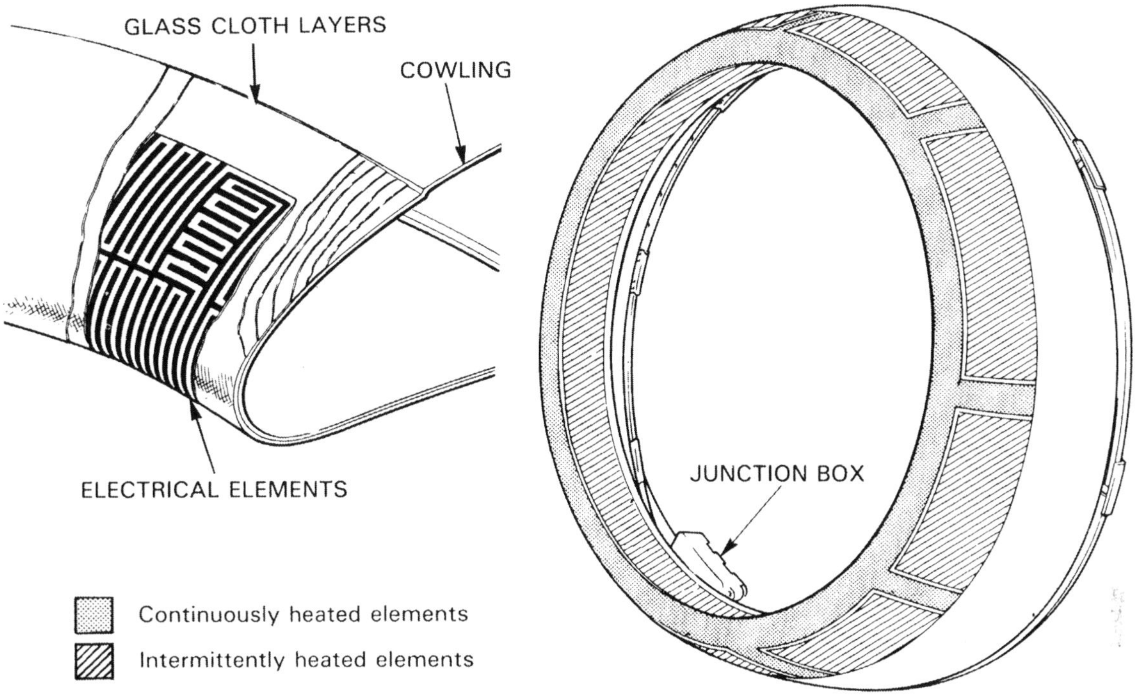

FIGURE 8–23 Electrical ice protection. *(Source: Rolls Royce.)*

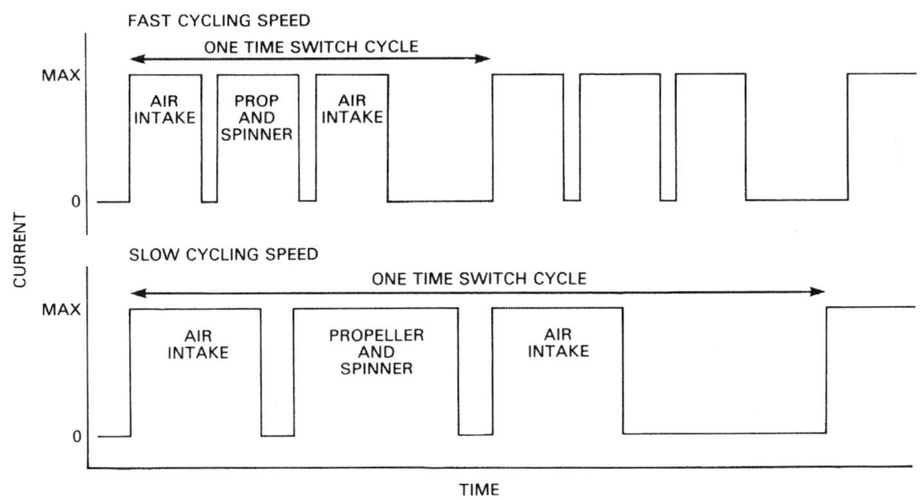

FIGURE 8–24 Typical ice protection cyclic sequence. *(Source: Rolls Royce.)*

The cycling time of the intermittently heated elements is arranged to ensure that the engine can accept the amount of ice that collects during the "heat-off" period and yet ensure that the "heat-on" period is long enough to give adequate shedding, without causing any run-back icing to occur behind the heated areas.

A two-speed cycling system is often used to accommodate the propeller and spinner requirements; a "fast" cycle at the high air temperatures when the water concentration is usually greater and a "slow" cycle in the lower temperature range. A typical cycling sequence chart is shown in Figure 8–24.

FIRE PROTECTION SYSTEMS*

All gas turbine engines and their associated installation systems incorporate features that minimize the possibility of an engine fire. It is essential, however, that if a failure does take place and results in a fire, there is provision for the immediate detection and rapid extinction of the fire, and for the prevention of it spreading. The detection and

* Source: Adapted, with permission, from Rolls Royce, *The Jet Engine*, 1986, Rolls Royce Plc: UK.

extinguishing systems must add as little weight to the installation as possible.

Prevention of Engine Fire Ignition

An engine/powerplant is designed to ensure that the prevention of engine fire ignition is achieved as far as possible. In most instances a dual failure is necessary before a fire can occur.

Most of the potential sources of flammable fluids are isolated from the "hot end" of the engine. External fuel and oil system components and their associated pipes are usually located around the compressor casings, in a "cool" zone, and are separated by a fireproof bulkhead from the combustion, turbine and jet pipe area, or "hot" zone. The zones may be ventilated, as described later, to prevent the accumulation of flammable vapors.

All pipes that carry fuel, oil, or hydraulic fluid are made fire resistant/proof to comply with fire regulations, and all electrical components and connections are made explosion-proof. Sparking caused by discharge of static electricity is prevented by bonding all aircraft and engine components. This gives electrical continuity between all the components and makes them incapable of igniting flammable vapor.

On some engines, tubes carrying flammable fluids in "hot areas" of the engine are constructed with a double skin. Should a fracture of the main fluid carrying tube occur the outer skin will contain any leakage, so preventing any possible fire ignition.

The powerplant cowlings are provided with an adequate drainage system to remove flammable fluids from the nacelle, bay, or pod, and all seal leakages from components are drained overboard at a position such that fluid cannot re-enter the pod and create a fire hazard.

Spontaneous ignition can be minimized on aircraft flying at high Mach numbers by ducting boundary layer bleed air around the engine. However, if ignition should occur, this high velocity air stream may have to be shut off; otherwise it would increase the flame intensity and reduce the effectiveness of the extinguishing system by rapid dispersal of the extinguishant.

External Cooling and Ventilation

The engine bay or pod is usually cooled and ventilated by atmospheric air being passed around the engine and then vented overboard (Figure 8–25). Convection cooling during ground running may be provided by using an internal cooling outlet vent as an ejector system. An important function of the airflow is to purge any flammable vapors from the engine compartment. By keeping the airflow minimal, the power plant drag is minimized and, as the required quantity of fire extinguishant is in proportion to the zonal airflow, any fire outbreak would be of low intensity.

On some engines a fireproof bulkhead is also provided to separate the "cool" area or zone of the engine, which contains the fuel, oil, hydraulic, and electrical systems from the "hot" area surrounding the combustion, turbine and exhaust sections of the engine. Differential pressures can be created in the two zones by calibration of the inlet and outlet apertures to prevent the spread of fire from the hot zone.

Figure 8–26 shows a more complex cooling and ventilation system used on a turbo-fan engine. Air is induced from the intake duct and also delivered from the fan to provide multi-zone cooling, each zone having its own calibrated cooling flow.

Fire Detection

The rapid detection of a fire is essential to minimize the fire period before engine shutdown drill and release of extinguishant is effected. It is also extremely important that a fire detection system will not give a false fire warning resulting from short circuiting caused by chafing or the ingress of moisture in the case of electrically operated systems and

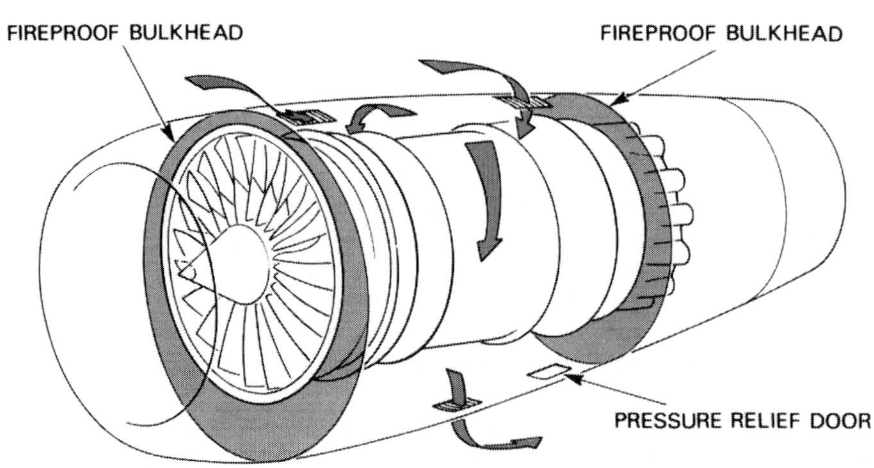

FIGURE 8–25 A typical cooling and ventilation system. *(Source: Rolls Royce.)*

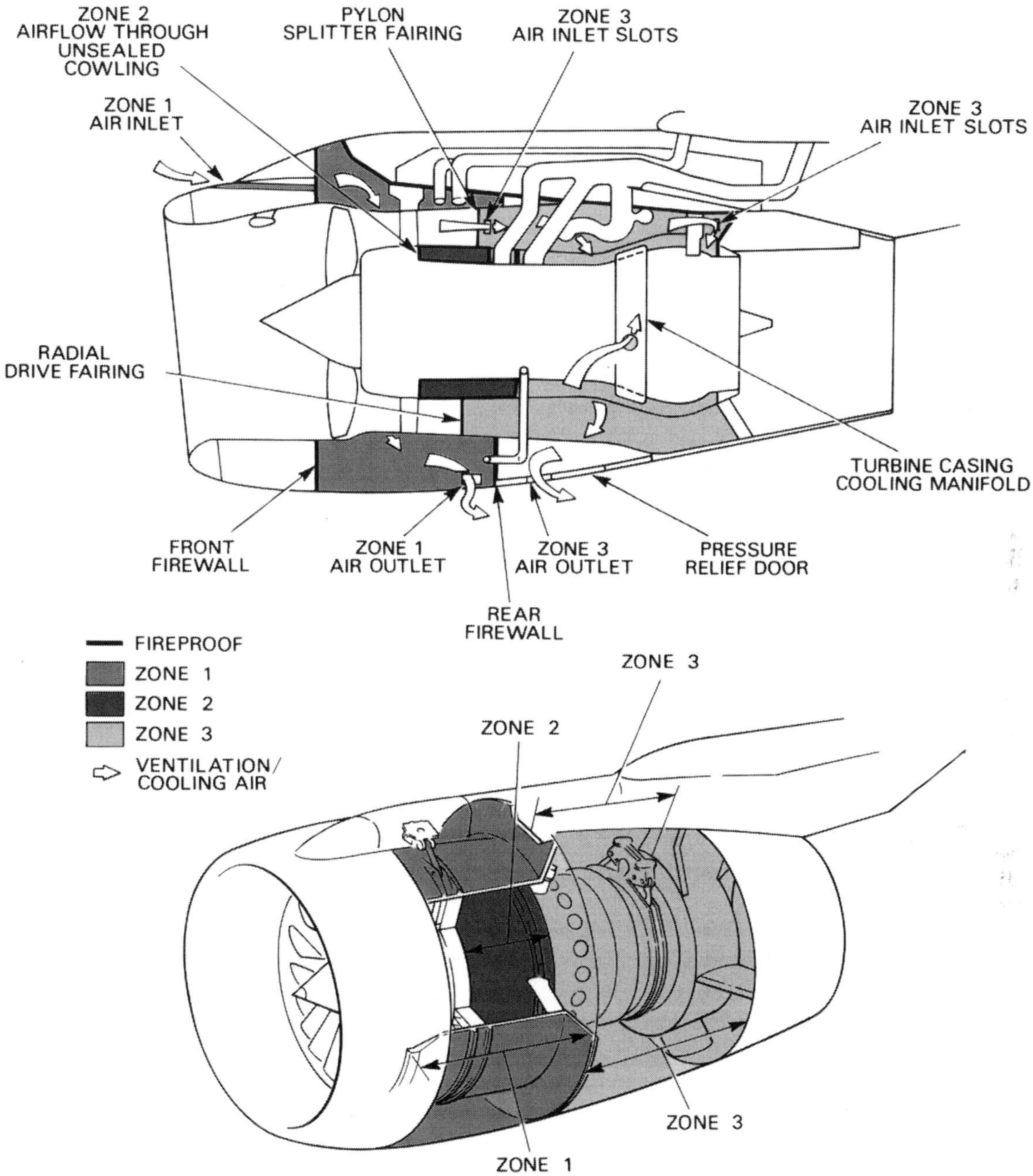

FIGURE 8–26 Cooling and ventilation—turbo-fan engine. *(Source: Rolls Royce.)*

chafes of the capillary resulting in loss of the contained gas in the case of the gas filled continuous element sensing type.

A detection system may consist of a number of strategically located detector units, or be of the continuous element (gas-filled or electrical) sensing type that can be shaped and attached to pre-formed tubes. The sensing element can be routed across outlet orifices, such as a zone extractor ventilation duct, to give early detection of a fire (Figure 8–27).

In the case of electrical systems the presence of a fire is signaled by a change in the electrical characteristics of the detector circuit, according to the type of detector, be it thermistor, thermocouple, or electrical continuous element. In these cases the change in temperature creates the signal that, through an amplifier, operates the warning indicator.

Both the thermocouple and thermistor detectors have properties making them ideally suited to this application. The thermocouple comprises two dissimilar metals that are joined together to form two junctions. As the temperature difference between the two junctions increases an E.M.F. is produced in the circuit and it is this E.M.F. that triggers the fire warning displays. The thermistor consists of a semiconductor material whose resistance changes as temperature

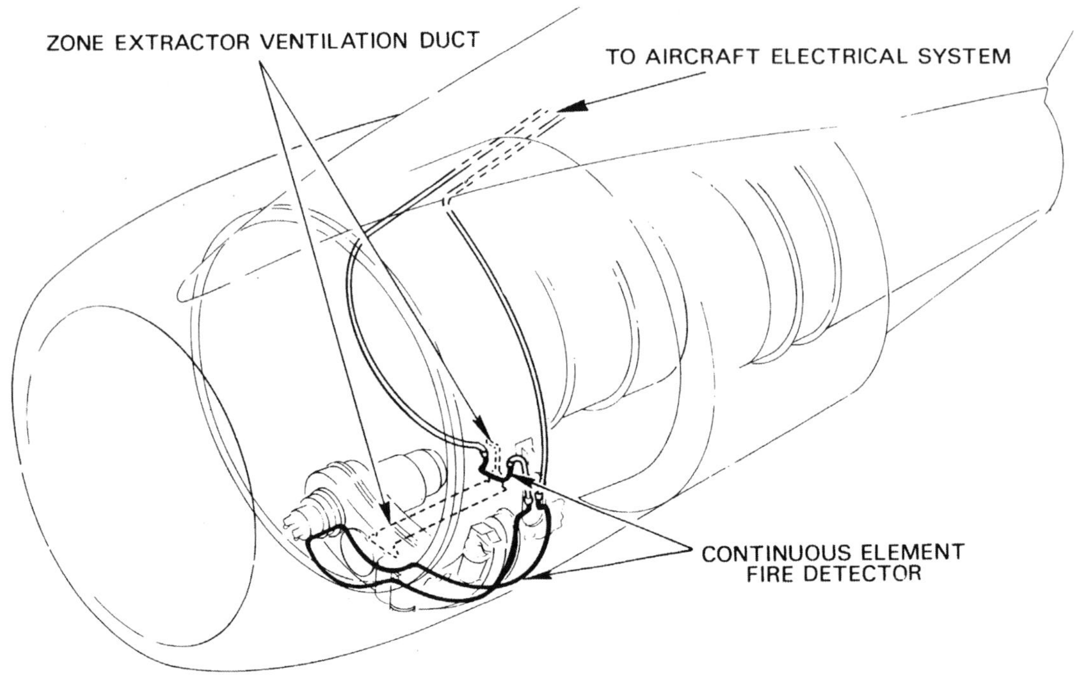

FIGURE 8–27 A continuous element fire detecting system. *(Source: Rolls Royce.)*

increases, with a corresponding change in the current flowing in the circuit. It is this change in the current that operates the warning indicators. A thermistor may be used as a single point detector or as a continuous element sensor.

Another form of continuous element sensor takes the form of a capacitor consisting of a tube containing a dielectric material with a conductor running through the center. A voltage difference is applied between the tube and the center conductor. As the temperature increases then the properties of the dielectric change with a corresponding change in the value of capacitance. This change of capacitance is displayed as a fire warning.

The gas filled detector consists of stainless steel tubing filled with gas absorbent material and in the event of a fire or overheat condition the temperature rise will cause the core of the sensing loop to expel the absorbed active gas into the sealed tube causing a rapid increase in pressure. This build-up of pressure is sensed by the detector alarm switch. Should the sensing loop become damaged causing a loss of the pressurized gas, an integrity switch will indicate a detection loop fault on the appropriate engine. Fire indication is given by a warning light and bell.

At high Mach numbers, the considerably higher temperature levels may be such as to render the thermistor or thermocouple fire detection system unsatisfactory. Thermal detectors that sense either a temperature rise, or a rate of temperature rise, may therefore prove most suitable.

Alternatives to the above types are surveillance detectors that respond to light radiation from a fire. These may be made so sensitive that they respond only to the ultraviolet and infrared rays emitted from a kerosene fire.

Fire Containment

An engine fire must be contained within the powerplant and not be allowed to spread to other parts of the aircraft. The cowlings that surround the engine are usually made of aluminum alloys, which would be unable to contain a fire when the aircraft is static. During flight, however, the airflow around the cowlings provides sufficient cooling to render them fireproof. Fireproof bulkheads and any cowlings that are not affected by a cooling airflow and sections of cowlings around certain outlets that may act as "flameholders" are usually manufactured from steel or titanium.

Fire Extinguishing

Before a fire extinguishing system is operated, the engine must be stopped to reduce the discharge of flammable fluids and air into the fire area. Any valves, such as the low-pressure fuel cock, that control the flow of flammable fluid must be situated outside the "hot" zone to prevent fire damage rendering them inoperative.

After a fire has been extinguished, no attempt must be made to start the engine again as this would probably reestablish the fluid leak and the ignition source that were the original causes of the fire. Furthermore, the extinguishing system may be exhausted.

The extinguishant that is used for engine fires is usually one of the Freon compounds. Pressurized containers are provided for the extinguishant and these are located outside the fire risk zone. When the relevant electrical circuit is manually operated, the extinguishant is discharged from the containers through a series of perforated spray pipes

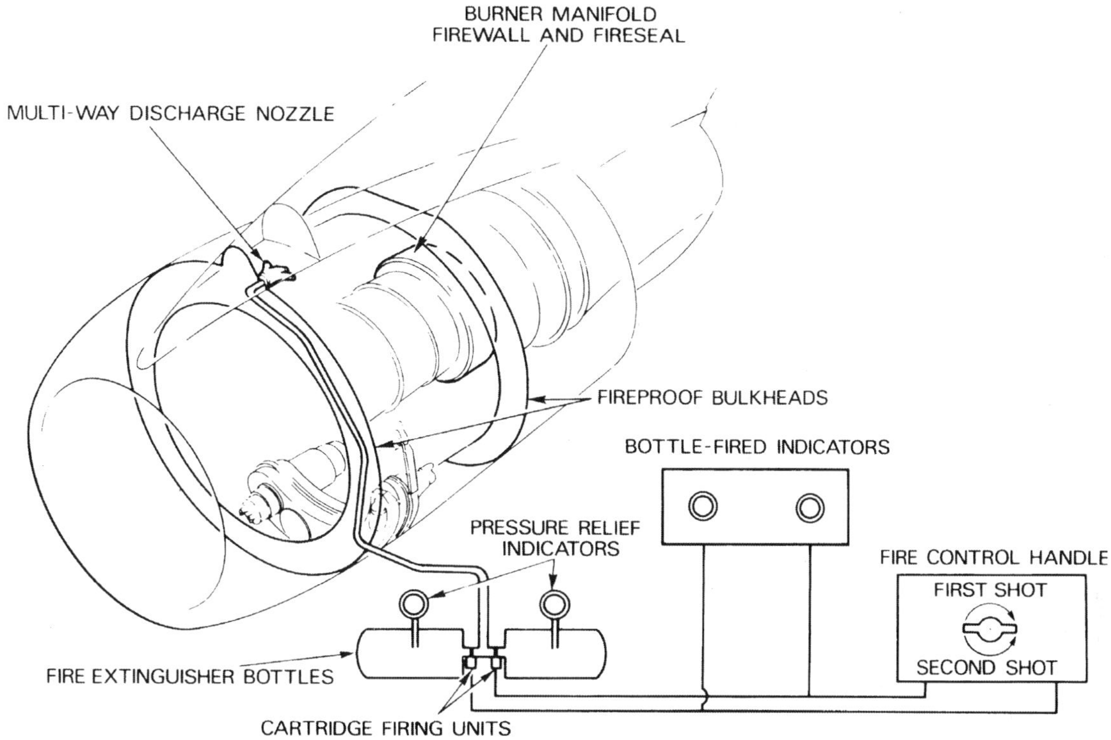

FIGURE 8–28 A typical fire extinguishing system. *(Source: Rolls Royce.)*

or nozzles into the fire (Figure 8–28). The discharge must be sufficient to give a predetermined concentration of extinguishant for a period that may vary between 0.5 seconds and 2 seconds. The system is generally one that enables two separate discharges to be made.

Engine Overheat Detection

Turbine overheat does not constitute a serious fire risk. Detection of an overheat condition, however, is essential to enable the pilot to stop the engine before mechanical or material damage results.

A warning system of a similar type to the fire detection system, or thermocouples suitably positioned in the cooling airflow, may be used to detect excessive temperatures. Thermal switches positioned in the engine overboard air vents, such as the cooling air outlets, may also be included to give an additional warning.

WATER INJECTION SYSTEMS*

The maximum power output of a gas turbine engine depends to a large extent upon the density or weight of the airflow passing through the engine. There is, therefore, a reduction in thrust or shaft horsepower as the atmospheric pressure decreases with altitude, and/or the ambient air temperature increases. Under these conditions, the power output can be restored or, in some instances, boosted for takeoff by cooling the airflow with water or water/methanol mixture (coolant). When methanol is added to the water it gives anti-freezing properties and also provides an additional source of fuel. A typical turbo-jet engine thrust restoration curve is shown in Figure 8–29 and a turbo-propeller engine power restoration and boost curve is shown in Figure 8–30.

There are two basic methods of injecting the coolant into the airflow. Some engines have the coolant sprayed directly into the compressor inlet, but the injection of coolant into the combustion chamber inlet is usually more suitable for axial flow compressor engines. This is because a more even distribution can be obtained and a greater quantity of coolant can be satisfactorily injected.

When water/methanol mixture is sprayed into the compressor inlet, the temperature of the compressor inlet air is reduced and consequently the air density and thrust are increased. If water only was injected, it would reduce the turbine inlet temperature, but with the addition of methanol the turbine inlet temperature is restored by the burning of methanol in the combustion chamber. Thus the power is restored without having to adjust the fuel flow.

The injection of coolant into the combustion chamber inlet increases the mass flow through the turbine, relative to that through the compressor. The pressure and temperature

* Source: Adapted, with permission, from Rolls Royce, *The Jet Engine*, 1986, Rolls Royce Plc: UK.

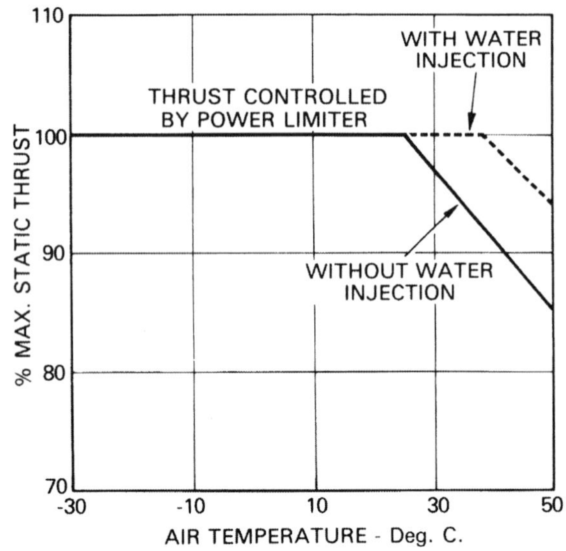

FIGURE 8–29 Turbo-jet thrust restoration. *(Source: Rolls Royce.)*

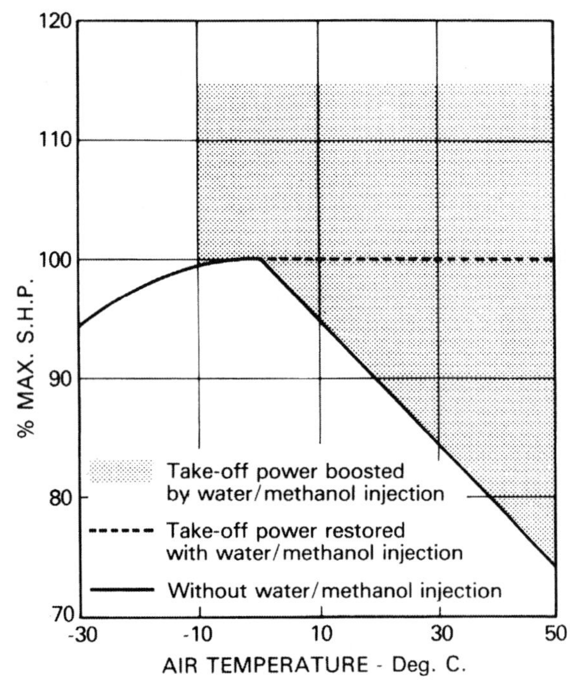

FIGURE 8–30 Turbo-propeller power boost. *(Source: Rolls Royce.)*

drop across the turbine is thus reduced and this results in an increased jet pipe pressure, which in turn gives additional thrust. The consequent reduction in turbine inlet temperature, due to water injection, enables the fuel system to schedule an increase of fuel flow to a value that gives an increase in the maximum rotational speed of the engine, thus providing further additional thrust. Where methanol is used with the water, the turbine inlet temperature is restored, or partially restored, by the burning of the methanol in the combustion chamber.

Compressor Inlet Injection

The compressor inlet injection system shown in Figure 8–31 is a typical system for a turbo-propeller engine. When the injection system is switched on, water/methanol mixture is pumped from an aircraft-mounted tank to a control unit. The control unit meters the flow of mixture to the compressor inlet through a metering valve that is operated by a servo piston. The servo system uses engine oil as an operating medium, and a servo valve regulates the supply of oil. The degree of servo valve opening is set by a control system that is sensitive to propeller shaft torque oil pressure and to atmospheric air pressure acting on a capsule assembly.

The control unit high-pressure oil cock control lever is interconnected to the throttle control system in such a manner that, until the throttle is moved towards the takeoff position, the oil cock remains closed, and thus the metering valve remains closed, preventing any mixture flowing to the compressor inlet. Movement of the throttle control to the takeoff position opens the oil cock, and the oil pressure passes through the servo valve to open the metering valve by means of the servo piston.

Combustion Chamber Injection

The combustion chamber injection system shown in Figure 8–32 is a typical system for a turbo-jet engine. The coolant flows from an aircraft-mounted tank to an air-driven turbine pump that delivers it to a water flow sensing unit. The water passes from the sensing unit to each fuel spray nozzle and is sprayed from two jets onto the flame tube swirl vanes, thus cooling the air passing into the combustion zone. The water pressure between the sensing unit and the discharge jets is sensed by the fuel control system, which automatically resets the engine speed governor to give a higher maximum engine speed.

The water flow sensing unit opens only when the correct pressure difference is obtained between compressor delivery air pressure and water pressure. The system is brought into operation when the engine throttle lever is moved to the takeoff position, causing microswitches to operate and select the air supply for the turbine pump.

The sensing unit also forms a non-return valve to prevent air pressure feeding back from the discharge jets and provides for the operation of an indicator light to show when water is flowing.

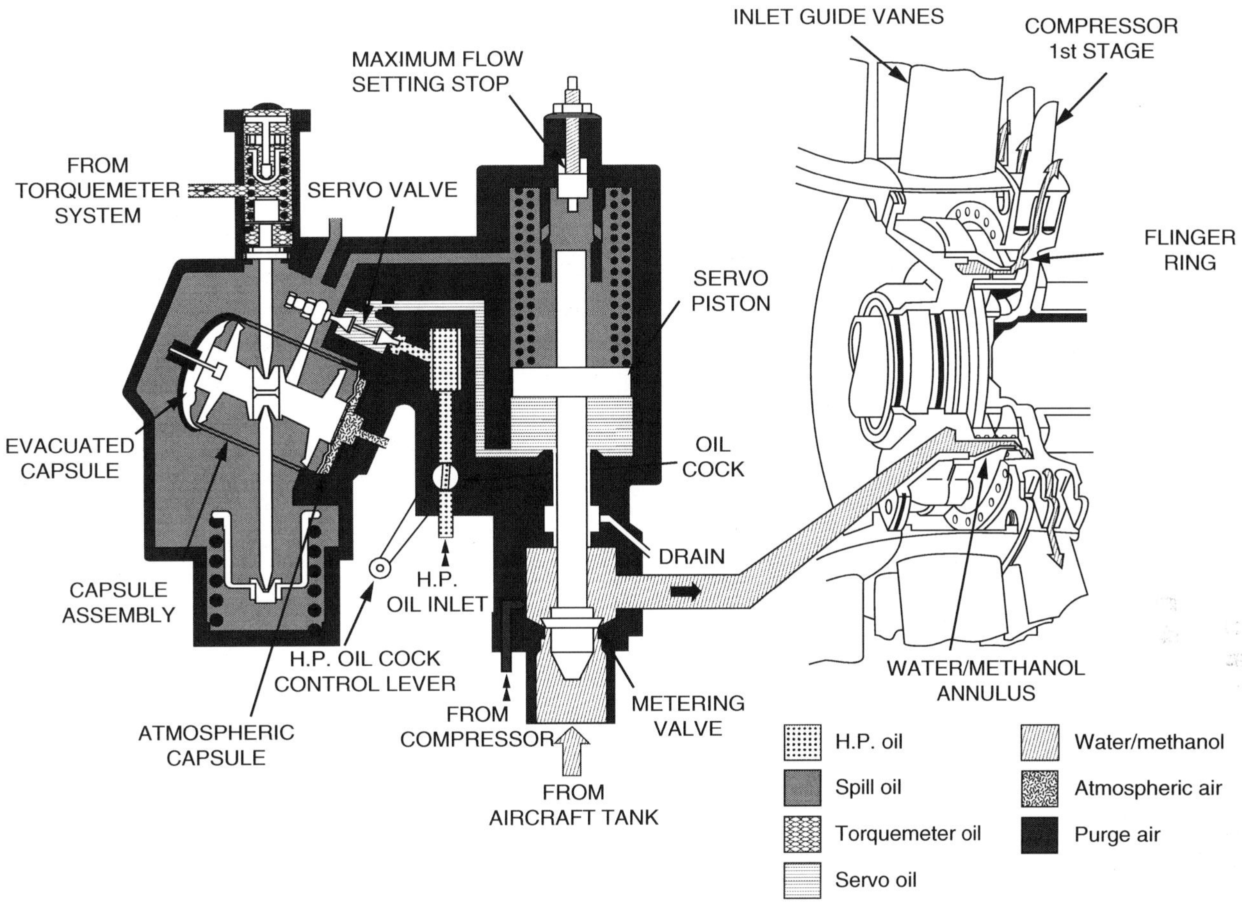

FIGURE 8–31 A typical compressor inlet injection system. *(Source: Rolls Royce. Adapted for reproduction in black and white.)*

SYSTEMS UNIQUE TO AIRCRAFT ENGINE APPLICATIONS

Thrust Reversal*

Modern aircraft brakes are very efficient but on wet, icy, or snow covered runways this efficiency may be reduced by the loss of adhesion between the aircraft tire and the runway thus creating a need for an additional method of bringing the aircraft to rest within the required distance.

A simple and effective way to reduce the aircraft landing run on both dry and slippery runways is to reverse the direction of the exhaust gas stream, thus using engine power as a deceleration force. Thrust reversal has been used to reduce airspeed in flight but it is not commonly used on modern aircraft. The difference in landing distances between an aircraft without reverse thrust and one using reverse thrust is illustrated in Figure 8–33.

On high by-pass ratio (fan) engines, reverse thrust action is achieved by reversing the fan (cold stream) airflow. It is not necessary to reverse the exhaust gas flow (hot stream) as the majority of the engine thrust is derived from the fan.

On propeller-powered aircraft, reverse thrust action is obtained by changing the pitch of the propeller blades. This is usually achieved by a hydro-mechanical system, which changes the blade angle to give the braking action under the response of the power or throttle lever in the aircraft.

Ideally, the gas should be directed in a completely forward direction. It is not possible, however, to achieve this, mainly for aerodynamic reasons, and a discharge angle of approximately 45° is chosen. Therefore, the effective power in reverse thrust is proportionately less than the power in forward thrust for the same throttle angle.

Principles of Operation

There are several methods of obtaining reverse thrust on turbo-jet engines; three of these are shown in Figure 8–34 and explained in the following paragraphs.

* Source: Adapted, with permission, from Rolls Royce, *The Jet Engine*, 1986, Rolls Royce Plc: UK.

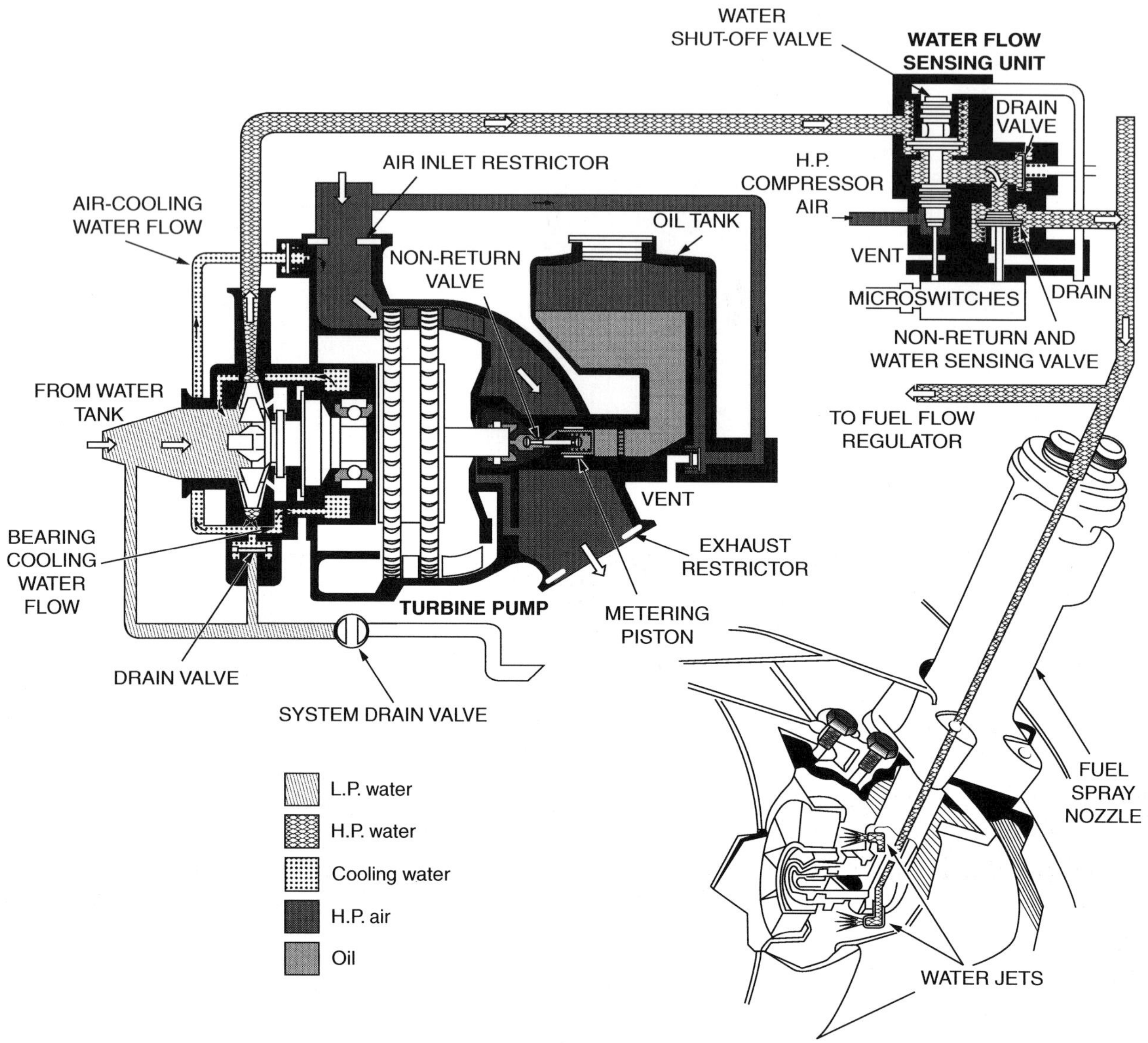

FIGURE 8–32 A typical combustion chamber injection system. *(Source: Rolls Royce. Adapted for reproduction in black and white.)*

One method uses clamshell-type deflector doors to reverse the exhaust gas stream and a second uses a target system with external type doors to do the same thing. The third method used on fan engines utilizes blocker doors to reverse the cold stream airflow.

Methods of reverse thrust selection and the safety features incorporated in each system described are basically the same. A reverse thrust lever in the crew compartment is used to select reverse thrust; the lever cannot be moved to the reverse thrust position unless the engine is running at a low power setting, and the engine cannot be opened up to a high power setting if the reverser fails to move into the full reverse thrust position. Should the operating pressure fall or fail, a mechanical lock holds the reverser in the forward thrust position; this lock cannot be removed until the pressure is restored. Operation of the thrust reverser system is indicated in the crew compartment by a series of lights.

Clamshell Door System

The clamshell door system is a pneumatically operated system, shown in detail in Figure 8–35. Normal engine operation is not affected by the system, because the ducts through which the exhaust gases are deflected remain closed by the doors until reverse thrust is selected by the pilot.

On the selection of reverse thrust, the doors rotate to uncover the ducts and close the normal gas stream exit. Cascade vanes then direct the gas stream in a forward direction so that the jet thrust opposes the aircraft motion.

The clamshell doors are operated by pneumatic rams through levers that give the maximum load to the doors in

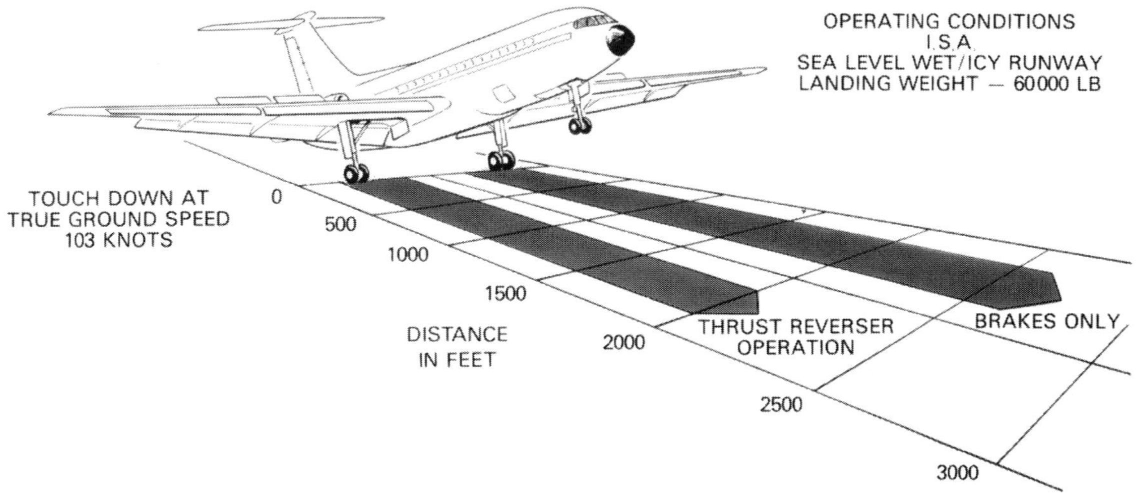

FIGURE 8–33 Comparative landing runs with and without thrust reversal. *(Source: Rolls Royce.)*

the forward thrust position; this ensures effective sealing at the door edges, so preventing gas leakage. The door bearings and operating linkage operate without lubrication at temperatures of up to 600°C.

Bucket Target System

The bucket target system is hydraulically actuated and uses bucket-type doors to reverse the hot gas stream. The thrust reverser doors are actuated by means of a conventional pushrod system. A single hydraulic powered actuator is connected to a drive idler, actuating the doors through a pair of pushrods (one for each door).

The reverser doors are kept in synchronization through the drive idler. The hydraulic actuator incorporates a mechanical lock in the stowed (actuator extended) position.

In the forward thrust mode (stowed) the thrust reverser doors form the convergent-divergent final nozzle for the engine.

Cold Stream Reverser System

The cold stream reverser system (Figure 8–36) can be actuated by an air motor, the output of which is converted to mechanical movement by a series of flexible drives, gearboxes and screw jacks, or by a system incorporating hydraulic rams.

When the engine is operating in forward thrust, the cold stream final nozzle is "open" because the cascade vanes are internally covered by the blocker doors (flaps) and externally by the movable (translating) cowl; the latter item also serves to reduce drag.

On selection of reverse thrust, the actuation system moves the translating cowl rearwards and at the same time folds the blocker doors to blank off the cold stream final nozzle, thus diverting the airflow through the cascade vanes.

Turbo-Propeller Reverse Pitch System

As mentioned earlier, reverse thrust action is affected on turbo-propeller powered aircraft by changing the pitch of the propeller blades through a hydro-mechanical pitch control system (Figure 8–37). Movement of the throttle or power control lever directs oil from the control system to the propeller mechanism to reduce the blade angle to zero, and then through to negative (reverse) pitch. During throttle lever movement, the fuel to the engine is trimmed by the throttle valve, which is interconnected to the pitch control unit, so that engine power and blade angle are coordinated to obtain the desired amount of reverse thrust. Reverse thrust action may also be used to maneuver a turbo-propeller aircraft backwards after it has been brought to rest.

Several safety factors are incorporated in the propeller control system for use in the event of propeller malfunction, and these devices are usually hydro-mechanical pitch locking devices or stops.

Construction and Materials

The clamshell and bucket target doors (Figure 8–38) described earlier form part of the jet pipe. The reverser casing is connected to the aircraft structure or directly to the engine. The casing supports the two reverser doors, the operating mechanism and, in the case of the clamshell door system, the outlet ducts that contain the cascade vanes. The angle and area of the gas stream are controlled by the number of vanes in each outlet duct.

The clamshell and bucket target doors lie flush with the casing during forward thrust operation and are hinged along

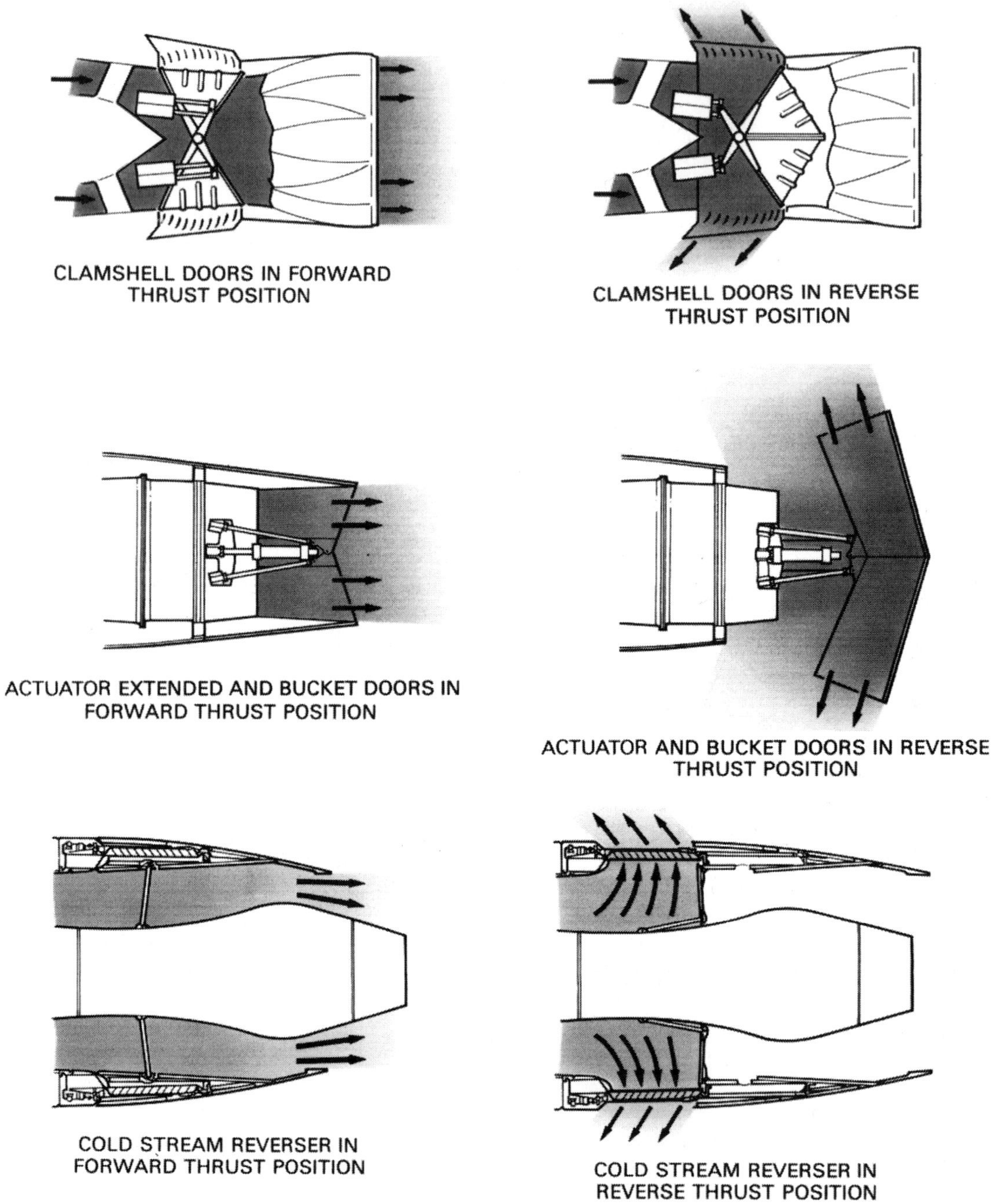

FIGURE 8–34 Methods of thrust reversal. *(Source: Rolls Royce.)*

the center line of the jet pipe. They are, therefore, in line with the main gas load and this ensures that the minimum force is required to move the doors.

Both the clamshell door system and the bucket target system are subjected to high temperatures and to high gas loads. The components of both systems, especially the doors, are therefore constructed from heat-resisting materials and are of particularly robust construction.

The cold stream thrust reverser casing (Figure 8–39) is fitted between the low-pressure compressor casing and the cold stream final nozzle. Cascade vane assemblies are arranged in segments around the circumference of the thrust reverser casing.

Blocker doors are internally mounted and are connected by linkages to the external movable (translating) cowl, which is mounted on rollers and tracks. Because the thrust reverser is not subjected to high temperatures, the casing, blocker doors, and cowl are constructed mainly of aluminum alloys or composite materials. The cowl is double-skinned, with the space between the skins containing noise absorbent material.

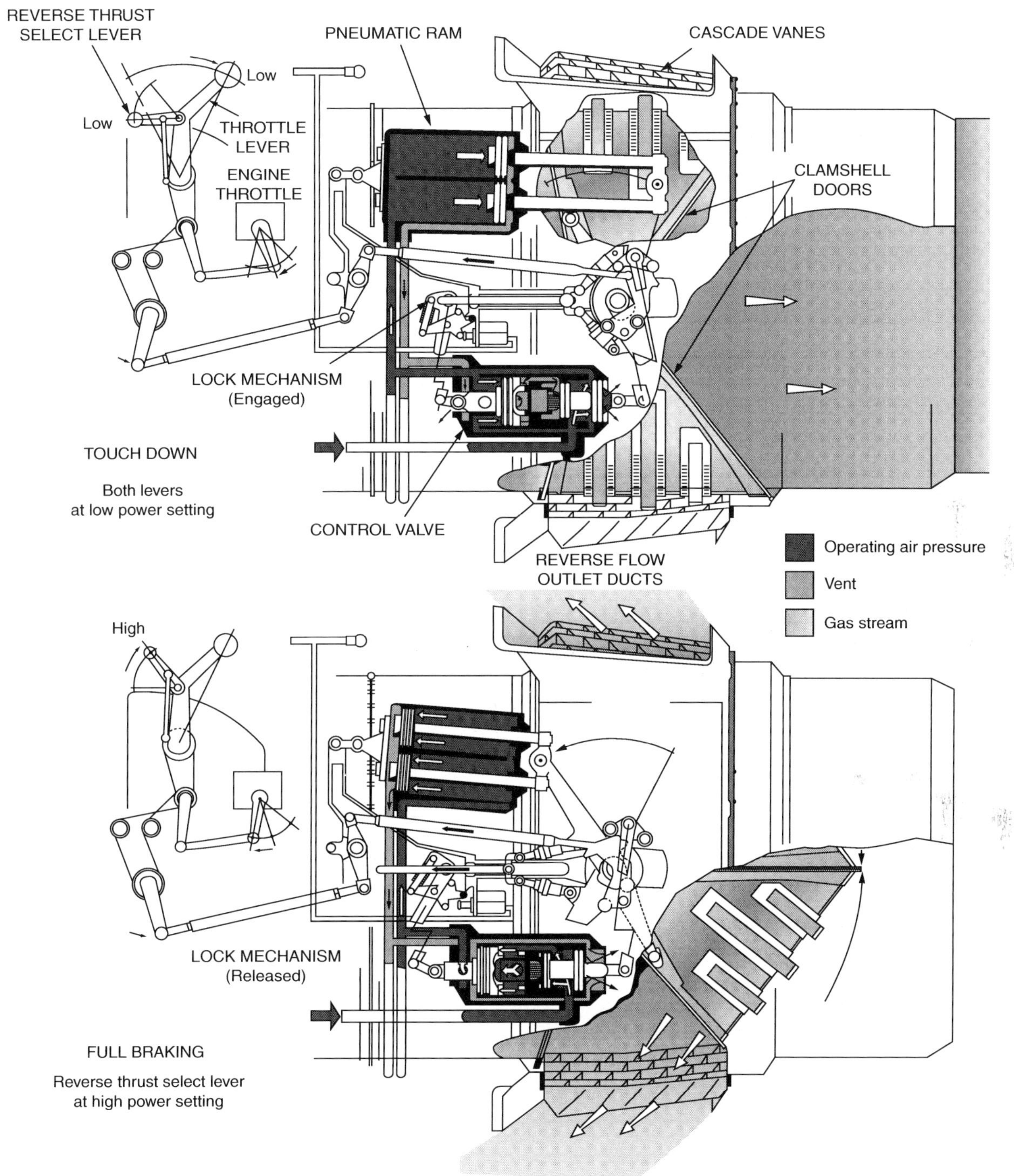

FIGURE 8–35 A typical thrust reverser system using clamshell doors. *(Source: Rolls Royce. Adapted for reproduction in black and white.)*

Afterburning*

Afterburning (or reheat) is a method of augmenting the basic thrust of an engine to improve the aircraft takeoff, climb, and (for military aircraft) combat performance. The increased power could be obtained by the use of a larger engine, but as this would increase the weight, frontal area, and overall fuel consumption, afterburning provides the best method of thrust augmentation for short periods.

Afterburning consists of the introduction and burning of fuel between the engine turbine and the jet pipe

* Source: Adapted, with permission, from Rolls Royce, *The Jet Engine*, 1986, Rolls Royce Plc: UK.

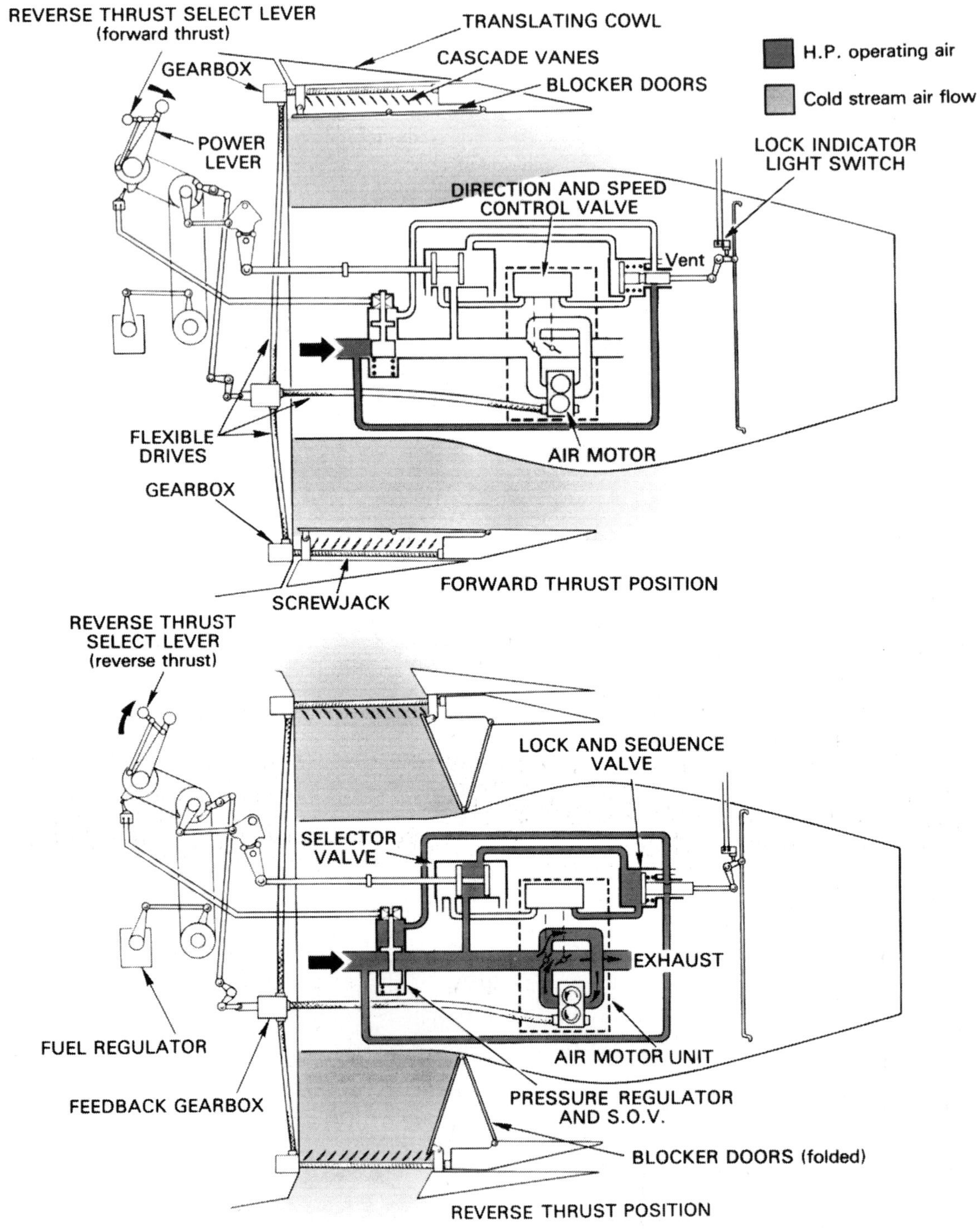

FIGURE 8–36 A typical fan cold stream thrust reversal system. *(Source: Rolls Royce.)*

propelling nozzle, utilizing the unburned oxygen in the exhaust gas to support combustion (Figure 8–40). The resultant increase in the temperature of the exhaust gas gives an increased velocity of the jet leaving the propelling nozzle and therefore increases the engine thrust.

As the temperature of the afterburner flame can be in excess of 1700°C, the burners are usually arranged so that the flame is concentrated around the axis of the jet pipe. This allows a proportion of the turbine discharge gas to flow along the wall of the jet pipe and thus maintain the wall temperature at a safe value.

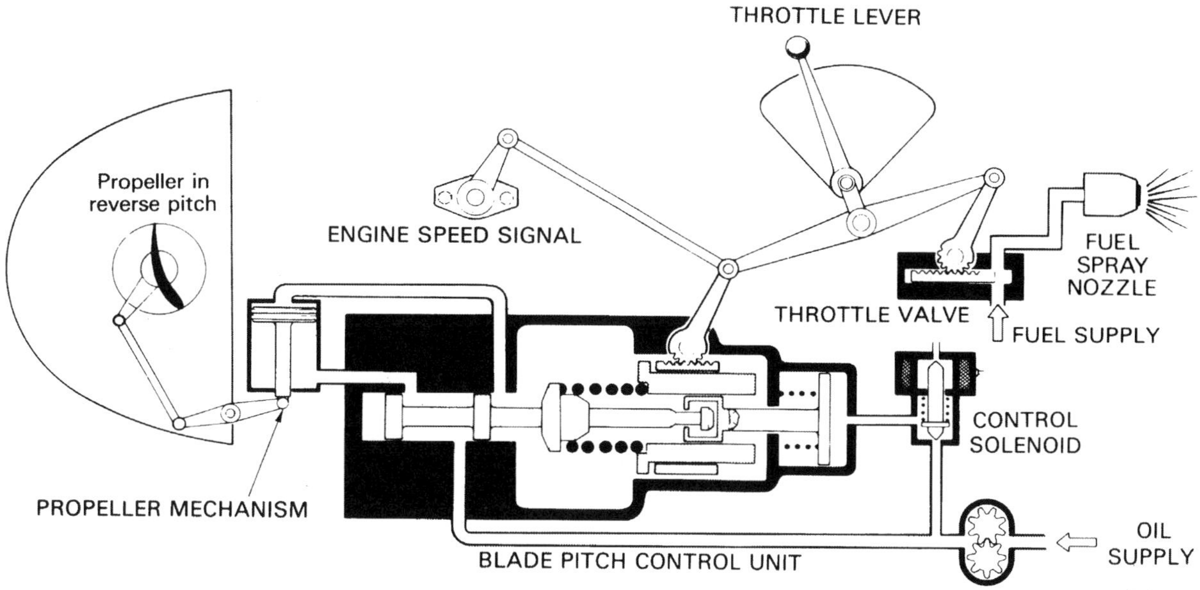

FIGURE 8–37 A propeller pitch control system. *(Source: Rolls Royce.)*

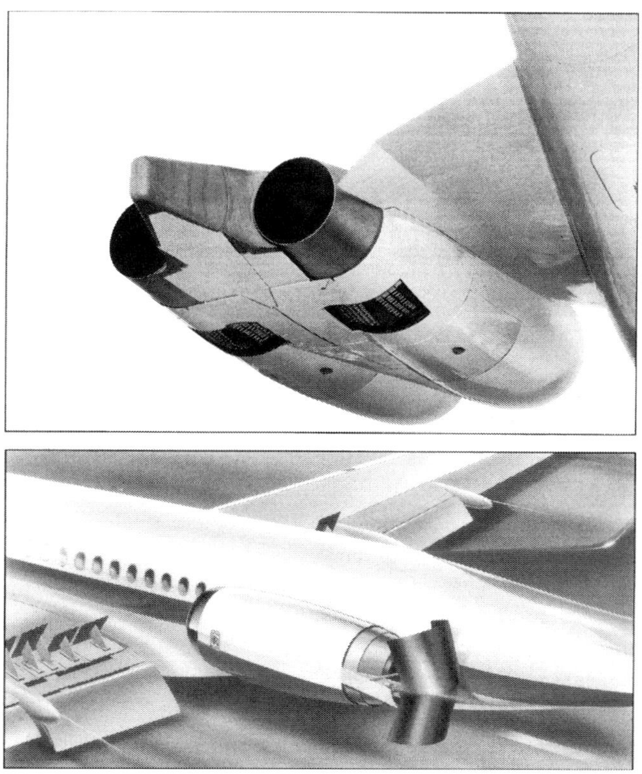

FIGURE 8–38 Hot stream thrust reverser installations. *(Source: Rolls Royce.)*

FIGURE 8–39 A cold stream thrust reverser installation. *(Source: Rolls Royce.)*

The area of the afterburning jet pipe is larger than a normal jet pipe would be for the same engine to obtain a reduced velocity gas stream. To provide for operation under all conditions, an afterburning jet pipe is fitted with either a two-position or a variable-area propelling nozzle (Figure 8–41). The nozzle is closed during non-afterburning operation, but when afterburning is selected the gas temperature increases and the nozzle opens to give an exit area suitable for the resultant increase in the volume of the gas stream. This prevents any increase in pressure occurring in the jet pipe that would affect the functioning of the engine and enables afterburning to be used over a wide range of engine speeds.

The thrust of an afterburning engine, without afterburning in operation, is slightly less than that of a similar engine not fitted with afterburning equipment; this is due to the added restrictions in the jet pipe. The overall weight of the powerplant is also increased because of the heavier jet pipe and afterburning equipment.

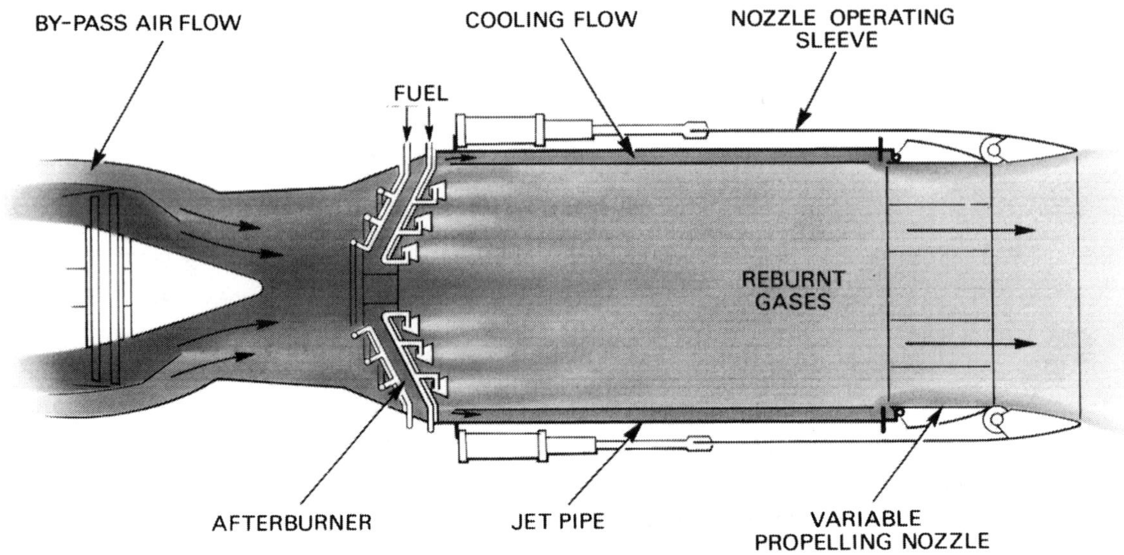

FIGURE 8–40 Principle of afterburning. *(Source: Rolls Royce.)*

Afterburning is achieved on low by-pass engines by mixing the by-pass and turbine streams before the afterburner fuel injection and stabilizer system is reached so that the combustion takes place in the mixed exhaust stream. An alternative method is to inject the fuel and stabilize the flame in the individual by-pass and turbine streams, burning the available gases up to a common exit temperature at the final nozzle. In this method, the fuel injection is scheduled separately to the individual streams and it is normal to provide some form of interconnection between the flame stabilizers in the hot and cold streams to assist the combustion processes in the cold by-pass air.

Operation of Afterburning

The gas stream from the engine turbine enters the jet pipe at a velocity of 750–1200 feet per second, but as this velocity is far too high for a stable flame to be maintained, the flow is diffused before it enters the afterburner combustion zone, i.e., the flow velocity is reduced and the pressure is increased. However, as the speed of burning kerosene at normal mixture ratios is only a few feet per second, any fuel lit even in the diffused air stream would be blown away. A form of flame stabilizer (vapor gutter) is, therefore, located downstream of the fuel burners to provide a region in which turbulent eddies are formed to assist combustion and where the local gas velocity is further reduced to a figure at which flame stabilization occurs while combustion is in operation.

An atomized fuel spray is fed into the jet pipe through a number of burners, which are so arranged as to distribute the fuel evenly over the flame area. Combustion is then initiated by a catalytic igniter, which creates a flame as a result of the chemical reaction of the fuel/air mixture being sprayed on to a platinum-based element, by an igniter plug adjacent to the burner, or by a hot streak of flame that originates in the engine combustion chamber (Figure 8–42); this latter method is known as "hot-shot" ignition. Once combustion is initiated, the gas temperature increases and the expanding gases accelerate through the enlarged area propelling nozzle to provide the additional thrust.

In view of the high temperature of the gases entering the jet pipe from the turbine, it might be assumed that the mixture would ignite spontaneously. This is not so, for although cool flames form at temperatures up to 700°C, combustion will not take place below 800°C. If, however, the conditions were such that spontaneous ignition could be effected at sea level it is unlikely that it could be effected at altitude where the atmospheric pressure is low. The spark or flame that initiates combustion must be of such intensity that a light-up can be obtained at considerable altitudes.

For smooth functioning of the system, a stable flame that will burn steadily over a wide range of mixture strengths and gas flows is required. The mixture must also be easy to ignite under all conditions of flight and combustion must be maintained with the minimum loss of pressure.

Construction

Burners

The burner system consists of several circular concentric fuel manifolds supported by struts inside the jet pipe. Fuel is supplied to the manifolds by feed pipes in the support struts and sprayed into the flame area, between the flame stabilizers, from holes in the downstream edge of the manifolds. The flame stabilizers are blunt nosed V-section annular rings located downstream of the fuel burners.

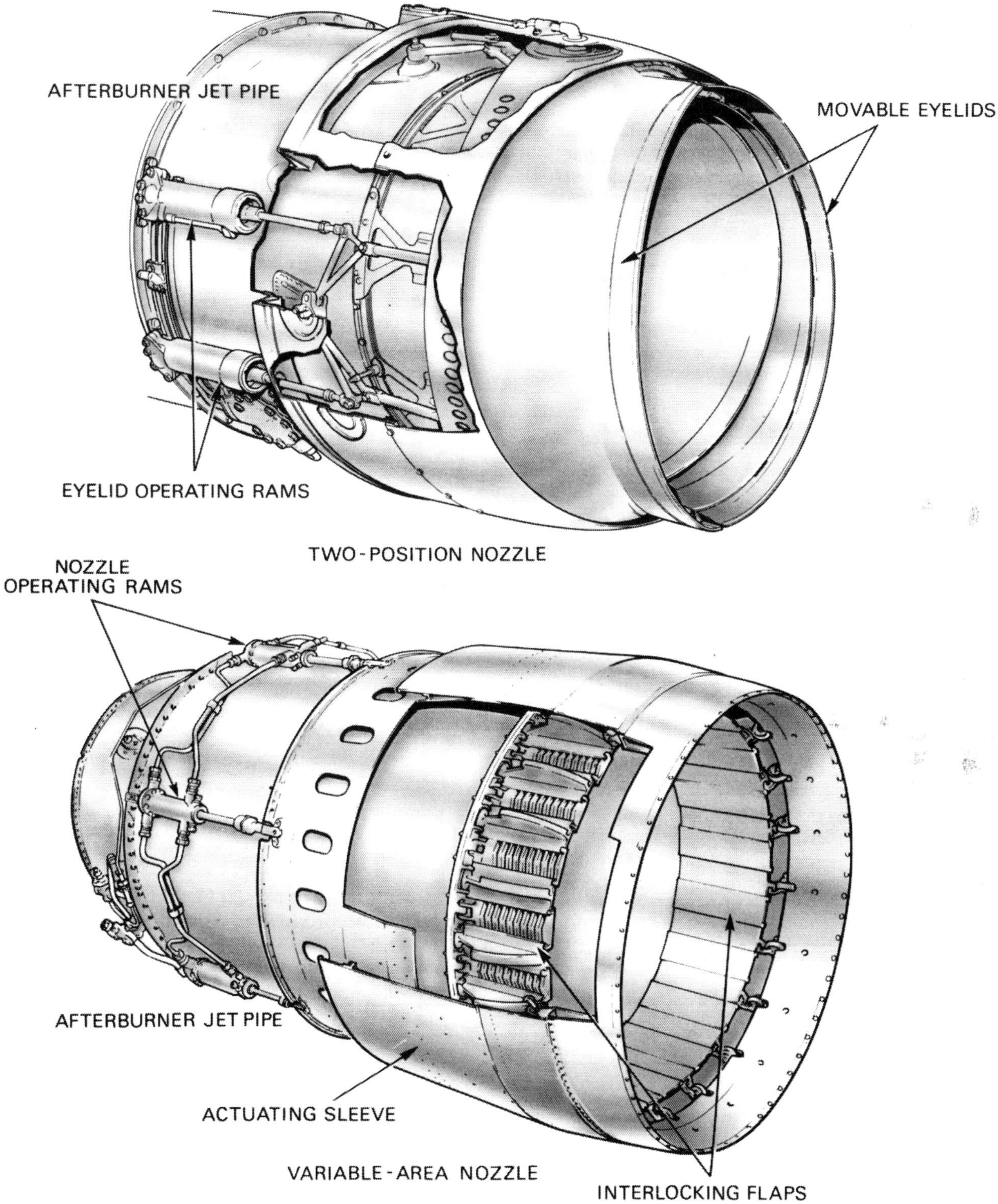

FIGURE 8–41 Examples of afterburning jet pipes and propelling nozzles. *(Source: Rolls Royce.)*

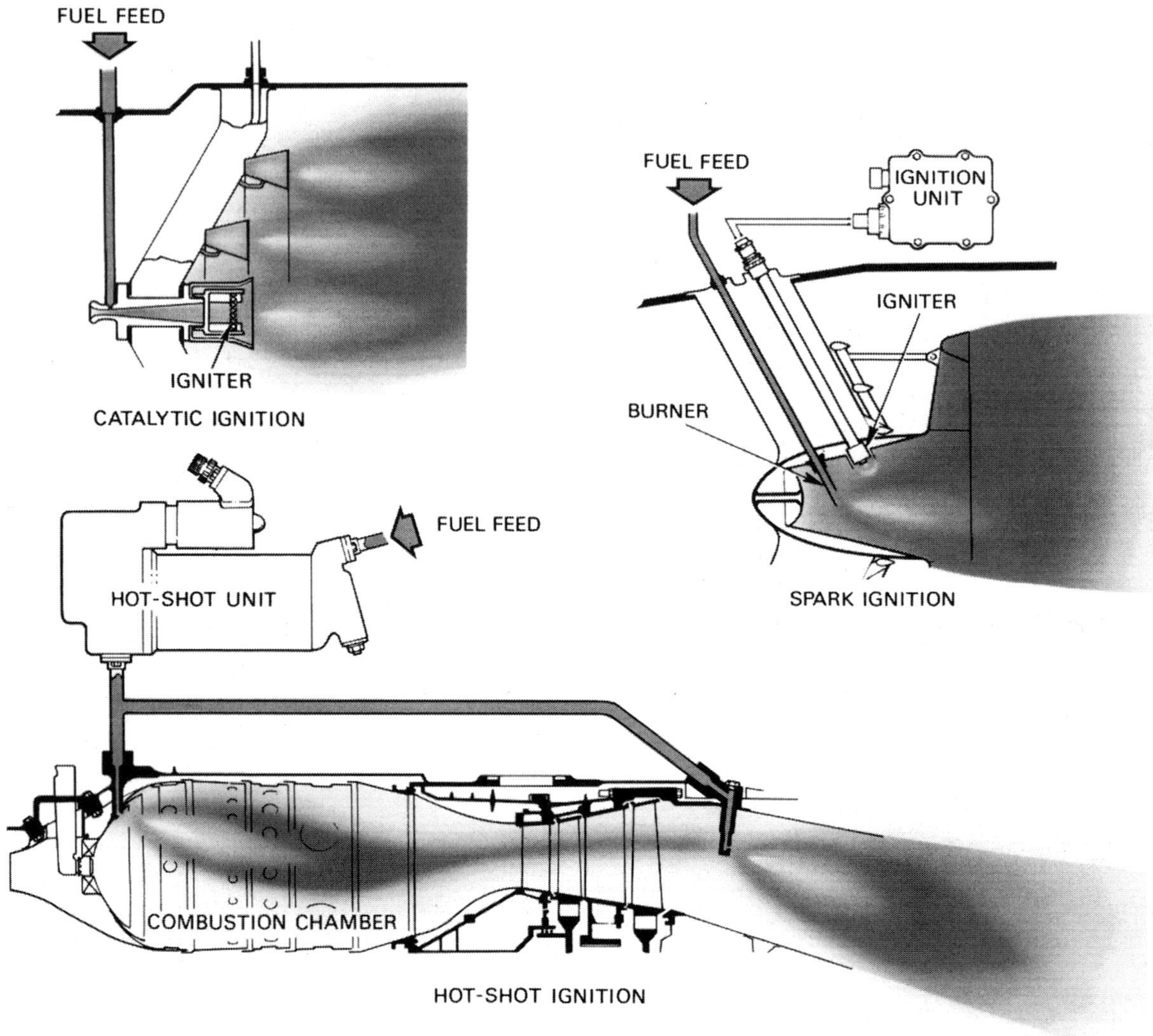

FIGURE 8–42 Methods of afterburning ignition. *(Source: Rolls Royce.)*

An alternative system includes an additional segmented fuel manifold mounted within the flame stabilizers. The typical burner and flame stabilizer shown in Figure 8–43 is based on the latter system.

Jet Pipe

The afterburning jet pipe is made from a heat-resistant nickel alloy and requires more insulation than the normal jet pipe to prevent the heat of combustion being transferred to the aircraft structure. The jet pipe may be of a double skin construction with the outer skin carrying the flight loads and the inner skin the thermal stresses; a flow of cooling air is often induced between the inner and outer skins. Provision is also made to accommodate expansion and contraction, and to prevent gas leaks at the jet pipe joints.

A circular heatshield of similar material to the jet pipe is often fitted to the inner wall of the jet pipe to improve cooling at the rear of the burner section. The heatshield comprises a number of bands, linked by cooling corrugations, to form a single skin. The rear of the heatshield is a series of overlapping "tiles" riveted to the surrounding skin (Figure 8–43). The shield also prevents combustion instability from creating excessive noise and vibration, which in turn would cause rapid physical deterioration of the afterburner equipment.

Propelling Nozzle

The propelling nozzle is of similar material and construction as the jet pipe, to which it is secured as a separate assembly. A two-position propelling nozzle has two movable eyelids that are operated by actuators, or pneumatic rams, to give an open or closed position. A variable-area propelling nozzle

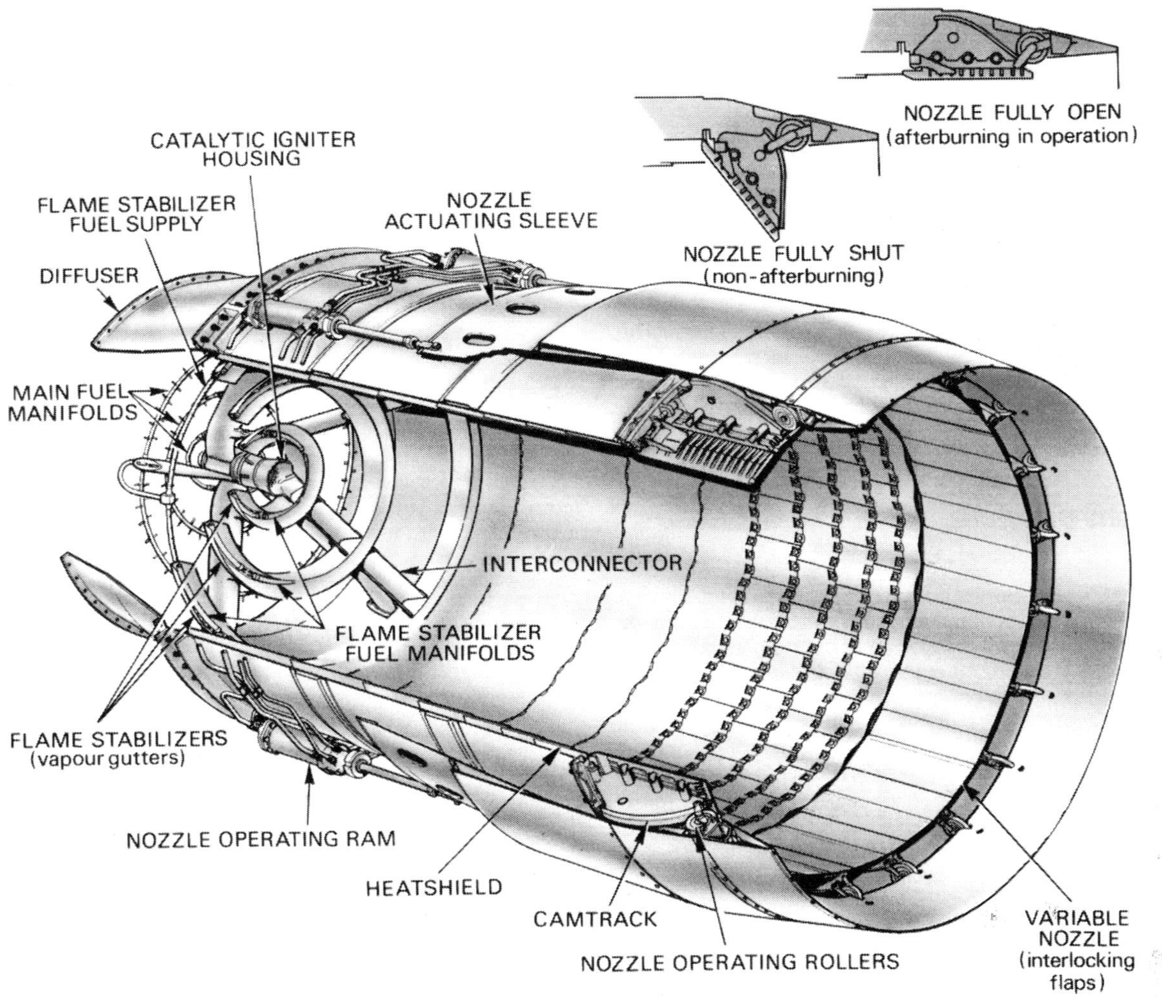

FIGURE 8–43 Typical afterburning jet pipe equipment. *(Source: Rolls Royce.)*

has a ring of interlocking flaps that are hinged to the outer casing and may be enclosed by an outer shroud. The flaps are actuated by powered rams to the closed position and by gas loads to the intermediate or the open positions; control of the flap position is by a control unit and a pump provides the power to the rams.

Control System

It is apparent that two functions, fuel flow and propelling nozzle area, must be coordinated for satisfactory operation of the afterburner system. These functions are related by making the nozzle area dependent upon the fuel flow at the burners or vice-versa. The pilot controls the afterburner fuel flow or the nozzle area in conjunction with a compressor delivery/jet pipe pressure sensing device (a pressure ratio control unit). When the afterburner fuel flow is increased, the nozzle area increases; when the afterburner fuel flow decreases, the nozzle area is reduced. The pressure ratio control unit ensures the pressure ratio across the turbine remains unchanged and that the engine is unaffected by the operation of afterburning, regardless of the nozzle area and fuel flow.

Since large fuel flows are required for afterburning, an additional fuel pump is used. This pump is usually of the centrifugal flow or gear type and is energized automatically when afterburning is selected. The system is fully automatic and incorporates "fail safe" features in the event of an afterburner malfunction. The interconnection between the control system and afterburner jet pipe is shown diagrammatically in Figure 8–44.

When afterburning is selected, a signal is relayed to the afterburner fuel control unit. The unit determines the total fuel delivery of the pump and controls the distribution of fuel flow to the burner assembly. Fuel from the burners is ignited, resulting in an increase in jet pipe pressure (P_6). This alters the pressure ratio across the turbine (P_3/P_6), and the exit area of the jet pipe nozzle is automatically increased until the correct P_3/P_6 ratio has been restored. With a further increase in the degree of afterburning, the

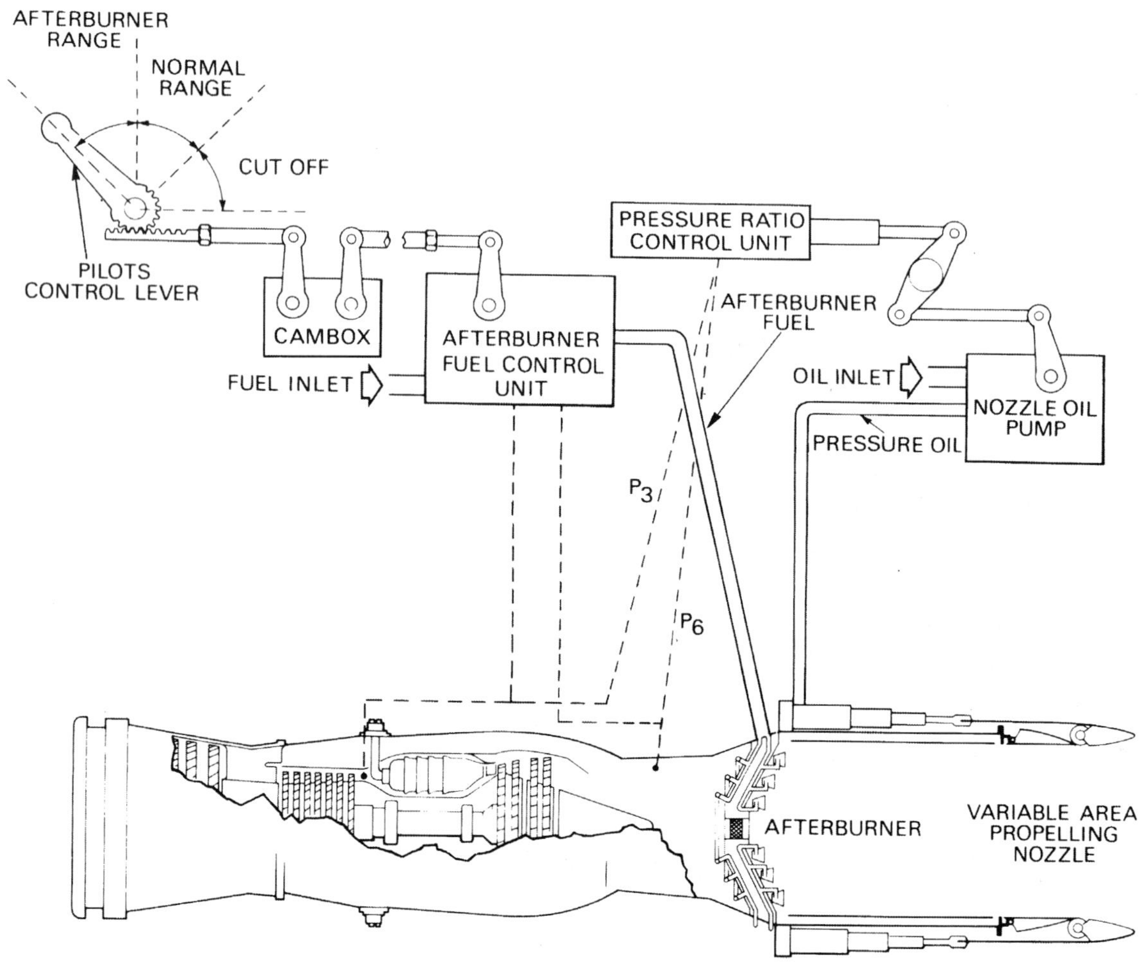

FIGURE 8–44 Simplified control system. *(Source: Rolls Royce.)*

nozzle area is progressively increased to maintain a satisfactory P_3/P_6 ratio. Figure 8–45 illustrates a typical afterburner fuel control system.

To operate the propelling nozzle against the large "drag" loads imposed by the gas stream, a pump and either hydraulically or pneumatically operated rams are incorporated in the control system. The system shown in Figure 8–46 uses oil as the hydraulic medium, but some systems use fuel. Nozzle movement is achieved by the hydraulic operating rams that are pressurized by an oil pump, pump output being controlled by a linkage from the pressure ratio control unit. When an increase in afterburning is selected, the afterburner fuel control unit schedules an increase in fuel pump output. The jet pipe pressure (P_6) increases, altering the pressure ratio across the turbine (P_3/P_6). The pressure ratio control unit alters oil pump output, causing an out-of-balance condition between the hydraulic ram load and the gas load on the nozzle flaps. The gas load opens the nozzle to increase its exit area and, as the nozzle opens, the increase in nozzle area restores the P_3/P_6 ratio and the pressure ratio control unit alters oil pump output until balance is restored between the hydraulic rams and the gas loading on the nozzle flaps.

Thrust Increase

The increase in thrust due to afterburning depends solely upon the ratio of the absolute jet pipe temperatures before and after the extra fuel is burned. For example, neglecting small losses due to the afterburner equipment and gas flow momentum changes, the thrust increase may be calculated as follows.

Assuming a gas temperature before afterburning of 640°C (913°K) and with afterburning of 1269°C (1542°K), then the temperature ratio = 1542/913 = 1.69. The velocity of the jet stream increases as the square root of the temperature ratio. Therefore, the jet velocity = $\sqrt{1.69} = 1.3$. Thus, the jet stream velocity is increased by 30%, and the increase in static thrust, in this instance, is also 30% (Figure 8–47).

Static thrust increases of up to 70% are obtainable from low by-pass engines fitted with afterburning equipment

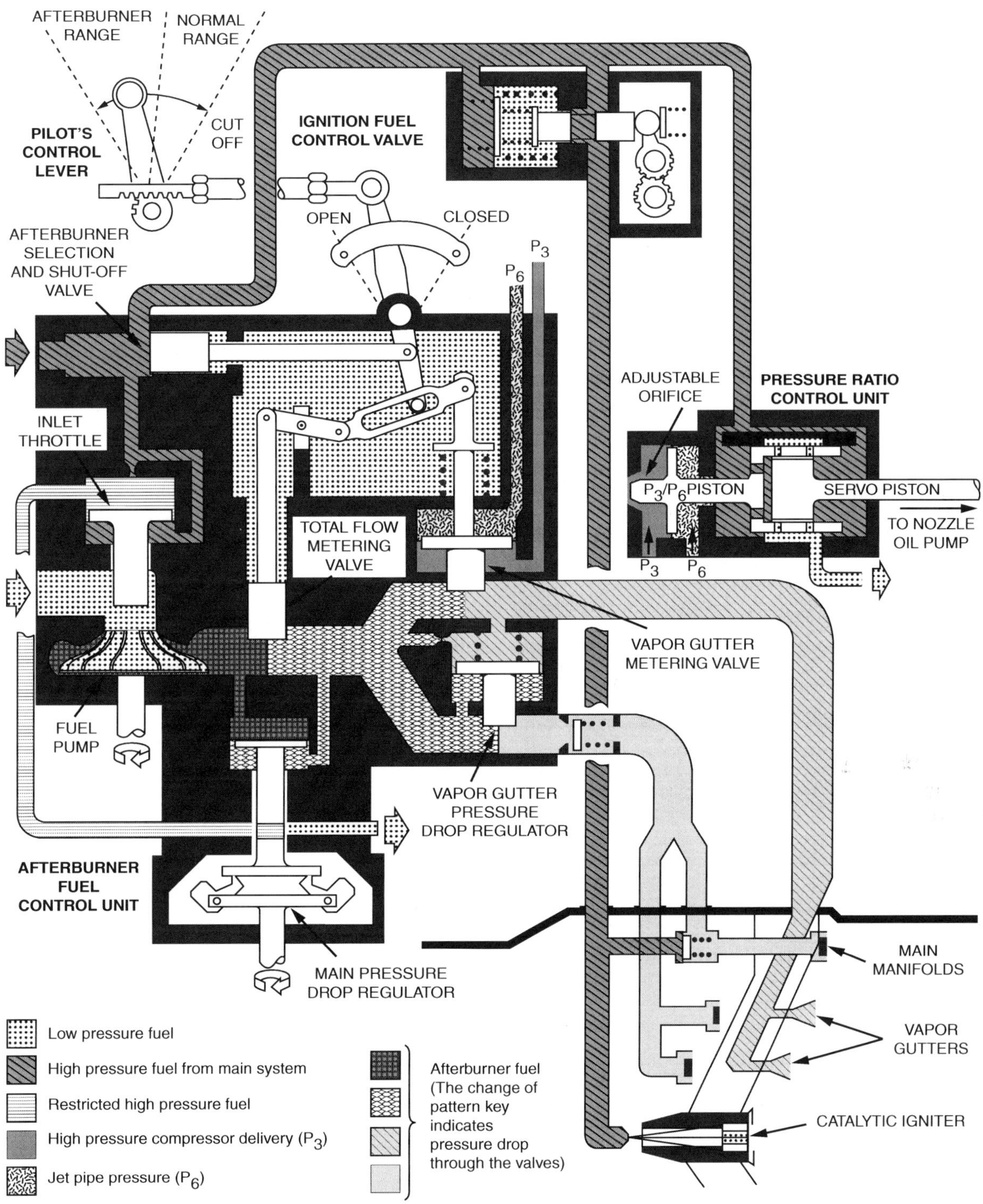

FIGURE 8–45 A simplified typical afterburner fuel control system. *(Source: Rolls Royce. Adapted for reproduction in black and white.)*

and, at high forward speeds, several times this amount of thrust boost can be obtained. High thrust boosts can be achieved on low by-pass engines because of the large amount of oxygen in the exhaust gas stream and the low initial temperature of the exhaust gases.

It is not possible to go on increasing the amount of fuel that is burned in the jet pipe so that all the available oxygen is used because the jet pipe would not withstand the high temperatures that would be incurred and complete combustion cannot be assured.

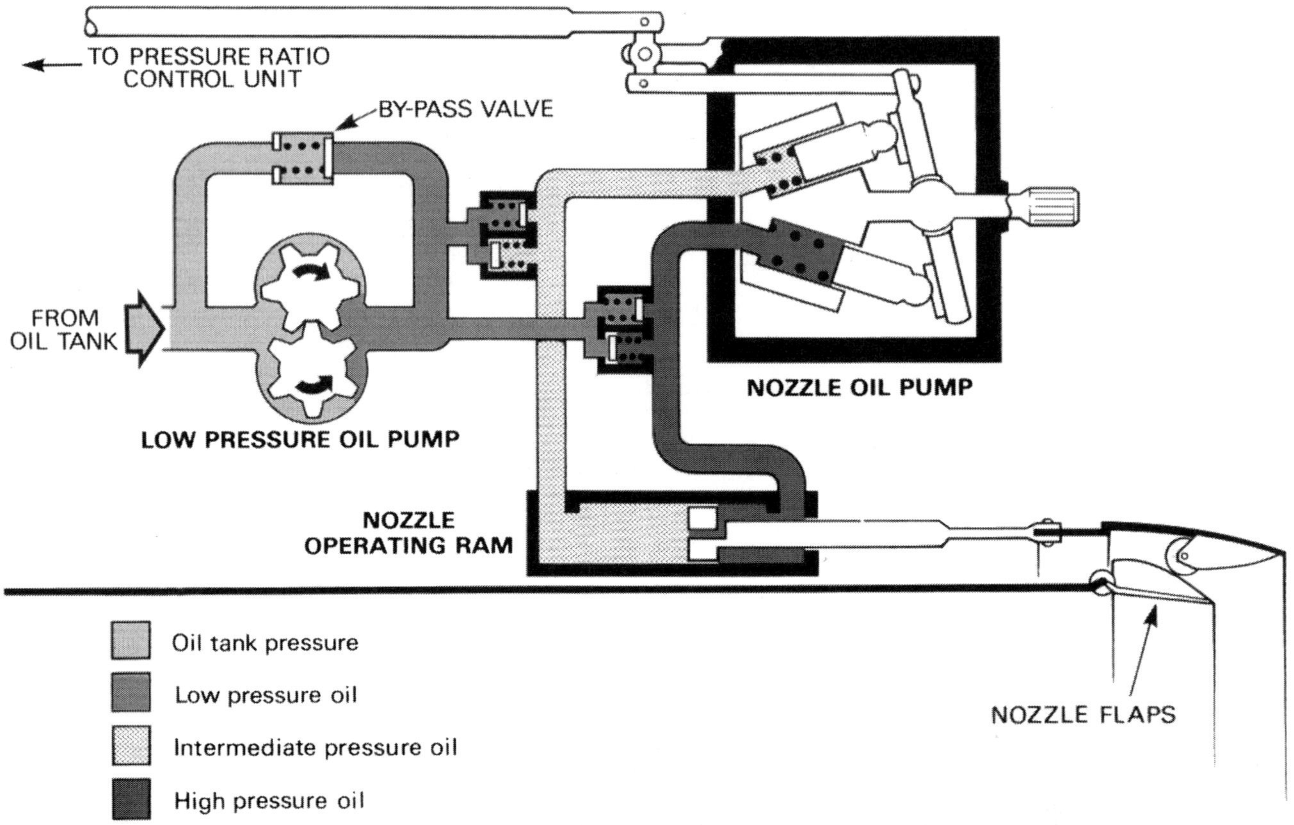

FIGURE 8—46 A simplified typical afterburner nozzle control system. *(Source: Rolls Royce.)*

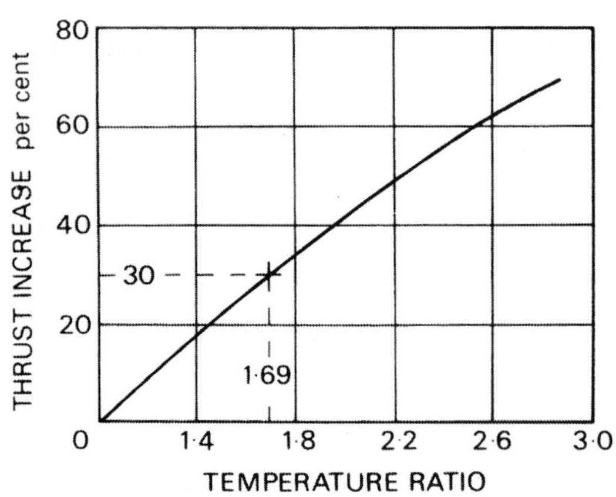

FIGURE 8—47 Thrust increase and temperature ratio. *(Source: Rolls Royce.)*

Fuel Consumption

Afterburning always incurs an increase in specific fuel consumption and is, therefore, generally limited to periods of short duration. Additional fuel must be added to the gas stream to obtain the required temperature ratio. Since the temperature rise does not occur at the peak of compression, the fuel is not burned as efficiently as in the engine combustion chamber and a higher specific fuel consumption must result. For example, assuming a specific fuel consumption without afterburning of 1.15 lb./hr./lb. thrust at sea level and a speed of Mach 0.9 as shown in Figure 8—48,

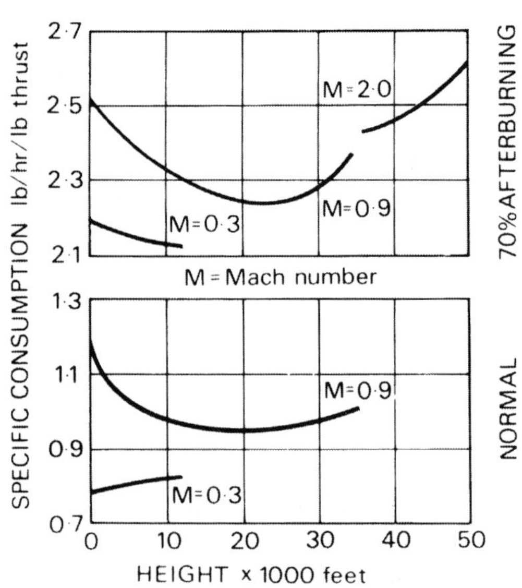

FIGURE 8—48 Specific fuel consumption. *(Source: Rolls Royce.)*

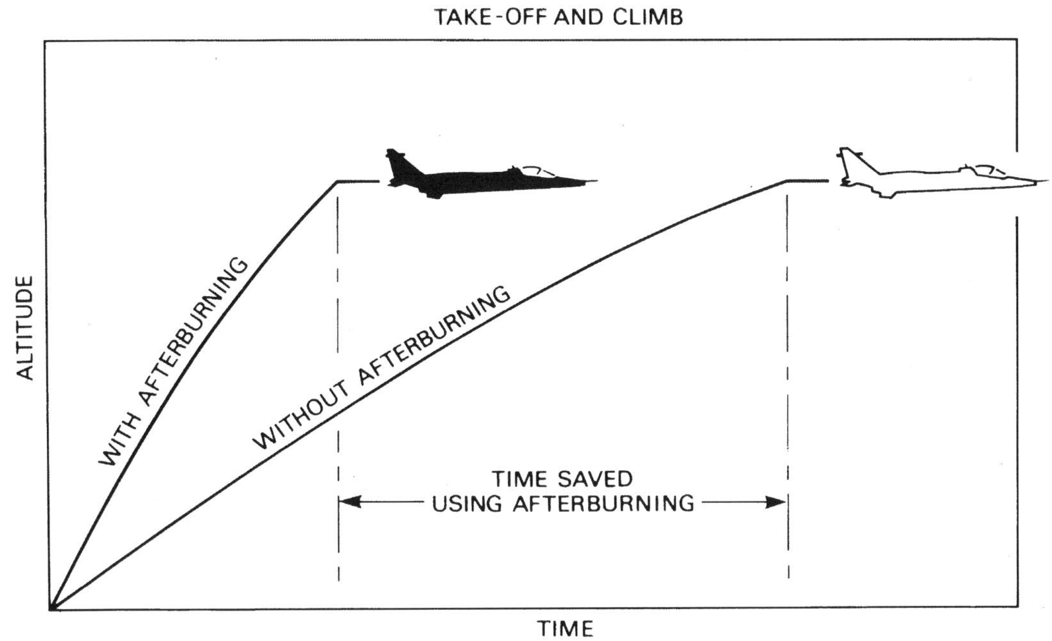

FIGURE 8–49 Afterburning and its effect on the rate of climb. *(Source: Rolls Royce.)*

then with 70% afterburning under the same conditions of flight, the consumption will be increased to approximately 2.53 lb./hr./lb. thrust. With an increase in height to 35,000 feet this latter figure of 2.53 lb./hr./lb. thrust will fall slightly to about 2.34 lb./hr./lb. thrust due to the reduced intake temperature. When this additional fuel consumption is combined with the improved rate of takeoff and climb (Figure 8–49), it is found that the amount of fuel required to reduce the time taken to reach operation height is not excessive.

Vertical/Short Takeoff and Landing*

Vertical takeoff and landing (VTOL) or short takeoff and landing (STOL) are desirable characteristics for any type of aircraft, provided that the normal flight performance characteristics, including payload/range, are not unreasonably impaired. Until the introduction of the gas turbine engine, with its high power/weight ratio, the only powered lift system capable of VTOL was the low disc loading rotor, as on the helicopter.

Early in 1941, the late Dr A.A. Griffiths, the then Chief Scientist at Rolls Royce, envisaged the use of the jet engine as a powered lift system. However, it was not until 1947 that a lightweight jet engine, designed by Rolls Royce for missile propulsion, existed and had a high enough thrust/weight ratio for the first pure lift-jet engine to be developed from it.

In 1956 the Bristol Aero-Engine Company was approached by Monsieur Michel Wibault with a proposal to use a turboshaft engine and a reduction gearbox to drive four centrifugal compressors that would be situated two on each side of the aircraft. The casing of these compressors could be rotated to change direction of the thrust (Figure 8–50). The concept incorporated two original ideas, i.e., the ability to deflect the thrust over the complete range of angles from the position for normal flight to that for vertical lift and a system where the resultant thrust always acted near to the center of gravity of the aircraft.

The principle proposed by M. Wibault was developed by using a pure jet engine with a free power turbine to drive an axial flow fan that exhausted into a pair of swiveling nozzles, one on each side of the aircraft. A further development was to use the fan to supercharge the engine, exhausting the by-pass air through one pair of swiveling nozzles and adding a second pair of swiveling nozzles to the exhaust system from the engine turbine. In this way the first ducted fan lift/propulsion engine (the Pegasus) evolved (Figure 8–51).

Subsequent experience with the Pegasus engine in the Harrier V/STOL fighter aircraft (Figure 8–52) led to the development of the short takeoff and vertical landing (STOVL) operational technique. In this way the additional lift generated by the aircraft wing, even after a short takeoff run, provided a large increase in the payload/range capability of the aircraft compared to a pure vertical takeoff. Vertical landing had several operational advantages compared to a short landing and so was maintained.

* Source: Adapted, with permission, from Rolls Royce, *The Jet Engine*, 1986, Rolls Royce Plc: UK.

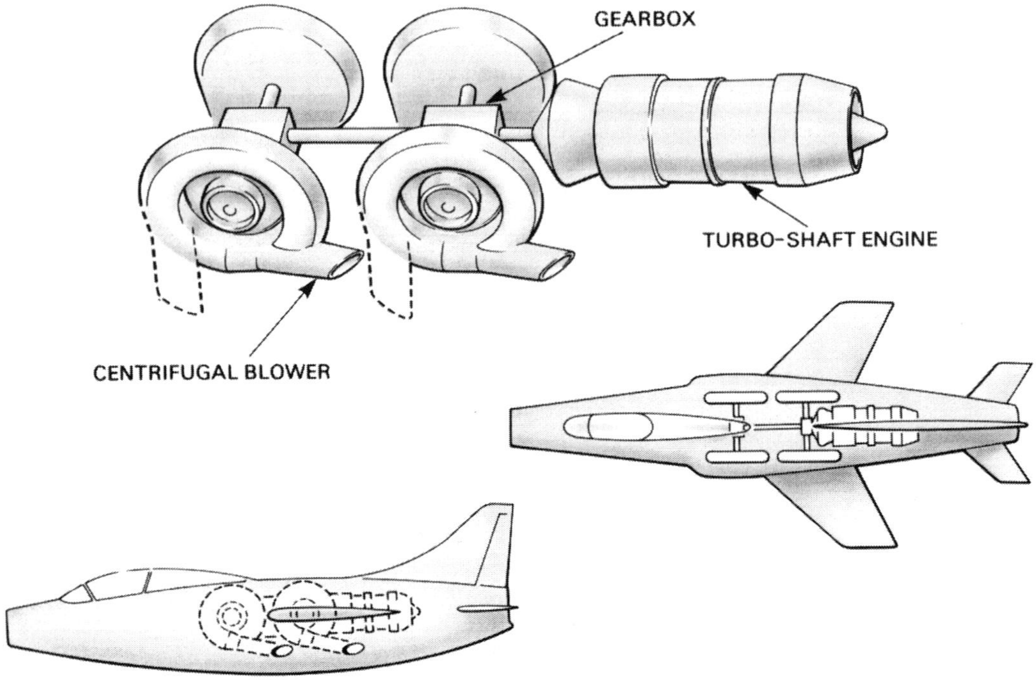

FIGURE 8–50 Michel Wibault's ground attack gyropter (concept), 1956. *(Source: Rolls Royce.)*

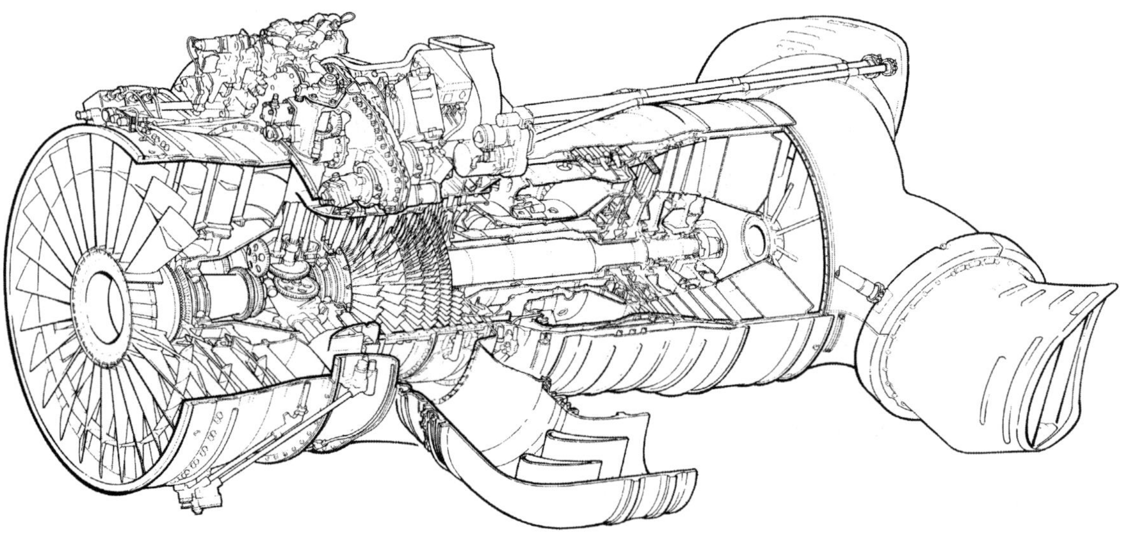

FIGURE 8–51 Lift/propulsion engine. *(Source: Rolls Royce.)*

FIGURE 8–52 V/STOL fighter aircraft. *(Source: Rolls Royce.)*

Methods of Providing Powered Lift

Although the Pegasus engine is the only V/STOL engine in operational service in the Western world there are several possible methods of providing powered lift, such as:

1. Deflecting (or vectoring) the exhaust gases and hence the thrust of the engine
2. Using specially designed engines for lift only
3. Driving a lift system, which is remote from the engine, either from the engine or by a separate power unit
4. Swiveling the engines
5. For STOL aircraft, using bleed air from the engines to increase circulation around the wing and hence increase lift

In several of the projected V/STOL aircraft a combination of two or more of these methods has been used.

Lift/Propulsion Engines

The lift/propulsion engine is capable of providing thrust for both normal wing-borne flight and for lift. This is achieved by changing the direction of the thrust either by a deflector system consisting of one, two, or four swiveling nozzles or by a device known as a switch-in deflector that redirects the exhaust gases from a rearward facing propulsion nozzle to one or two downward facing lift nozzles (Figure 8—53).

Thrust deflection on a single nozzle is accomplished by connecting together sections of the jet pipe, the joint faces of which are so angled that, when the sections are counter-rotated, the nozzle moves from the horizontal to the vertical position (Figure 8—54). To avoid either a side component of thrust or a thrust line offset from the engine axis during the movement of the nozzle it is necessary that the first joint face is perpendicular to the axis of the jet pipe. If it is desired that the nozzle does not rotate, as may be the case if it is a variable area nozzle, a third joint face that is perpendicular to the axis of the nozzle is required.

The two and four nozzle deflector systems use side mounted nozzles (Figure 8—55) that can rotate on simple bearings through an angle of well over 90° so that reverse thrust can be provided if required. A simple drive system, for example, a sprocket and chain, can be used and by mechanical connections all the nozzles can be made to deflect simultaneously. For forward flight, to avoid a high performance loss and consequent increase in fuel

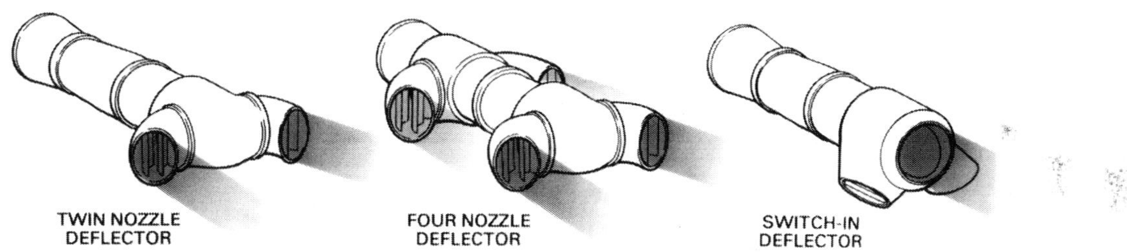

FIGURE 8—53 Thrust deflector systems. *(Source: Rolls Royce.)*

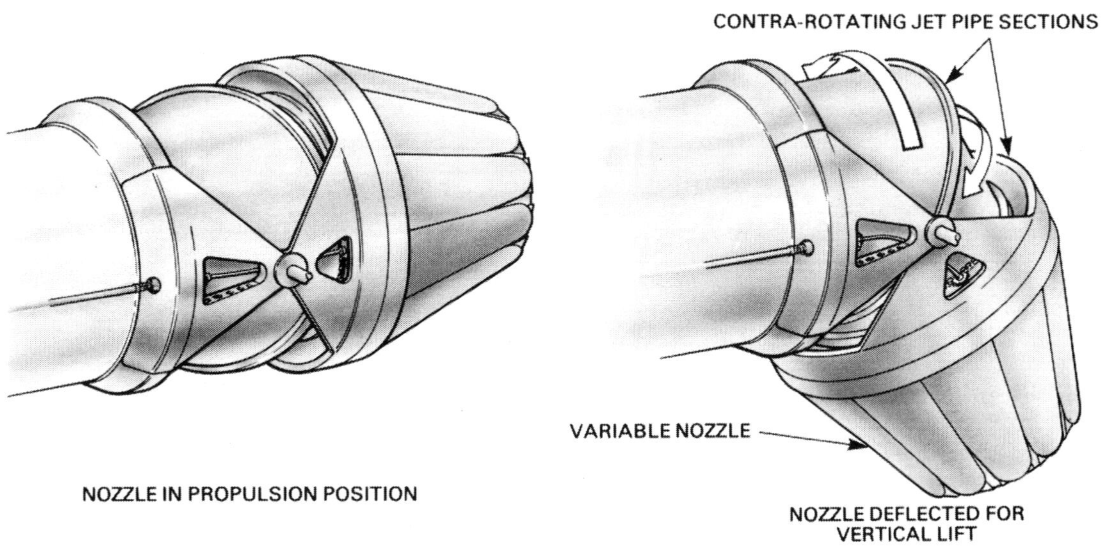

FIGURE 8—54 Deflector nozzle. *(Source: Rolls Royce.)*

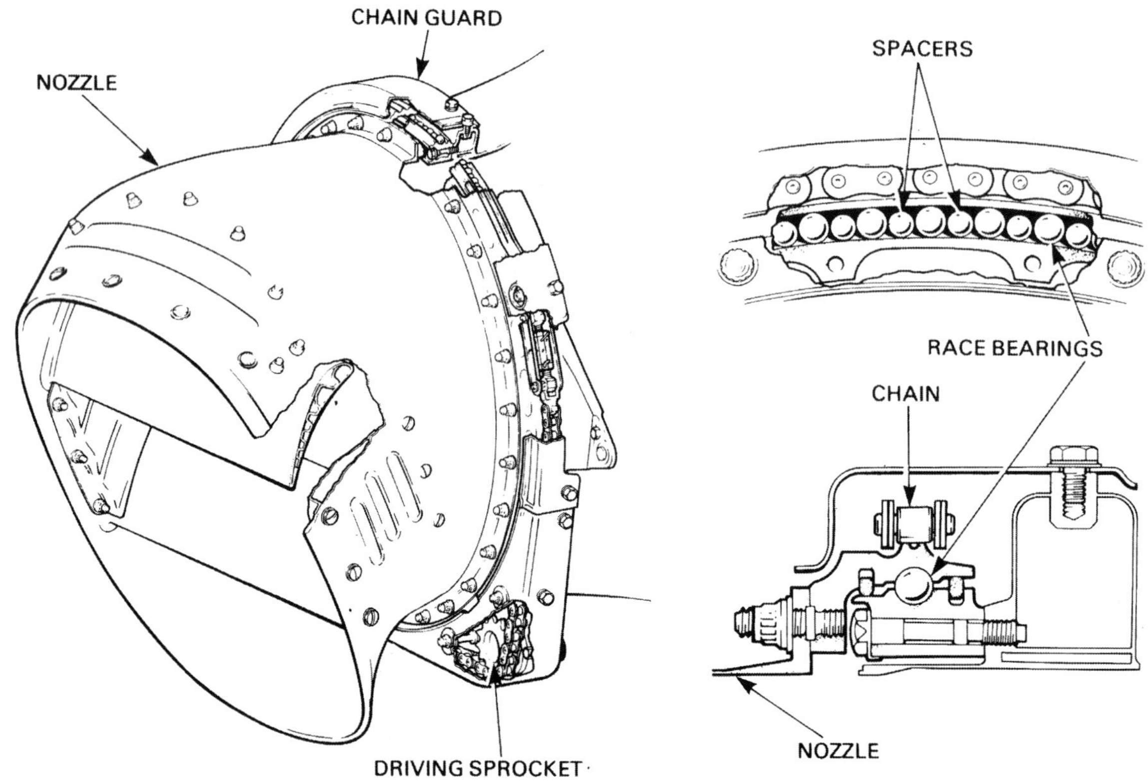

FIGURE 8–55 Side mounted swiveling nozzle. *(Source: Rolls Royce.)*

consumption, careful design of the exhaust unit and nozzle aerodynamic passages are essential to minimize the pressure losses due to turning the exhaust flow through two close coupled bends (Figure 8–56).

The switch-in deflector consists of one or a pair of heavily reinforced doors, which form part of the jet pipe wall when the engine is operating in the forward thrust condition. To select lift thrust, the doors are moved to blank off the conventional propelling nozzle and direct the exhaust flow into a lift nozzle (Figure 8–57). The lift nozzles may be designed so that they can be mechanically rotated to vary the angle of the thrust and permit intermediate lift/thrust positions to be selected.

A second type of switch-in deflector system is used on the tandem fan or hybrid fan vectored thrust engine (Figure 8–58). In this case the deflector system is situated between the stages of the fan of a mixed flow turbo-fan engine. In normal flight the valve is positioned so that the engine operates in the same manner as a mixed flow turbofan and for lift thrust the valve is switched so that the exhaust flow from the front part of the fan exhausts through downward facing lift nozzles and a secondary inlet is opened to provide the required airflow to the rear part of the fan and the main engine. On a purely subsonic V/STOL aircraft where fuel consumption is important the valve may

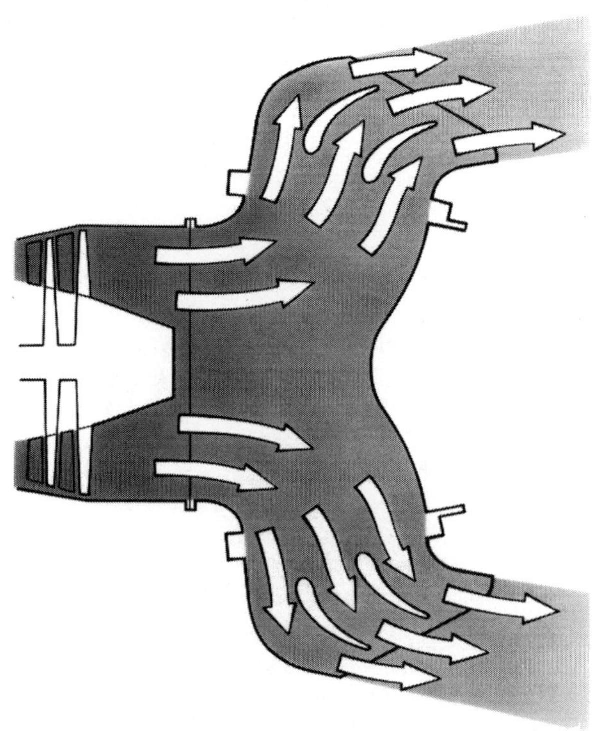

FIGURE 8–56 Nozzle duct configuration. *(Source: Rolls Royce.)*

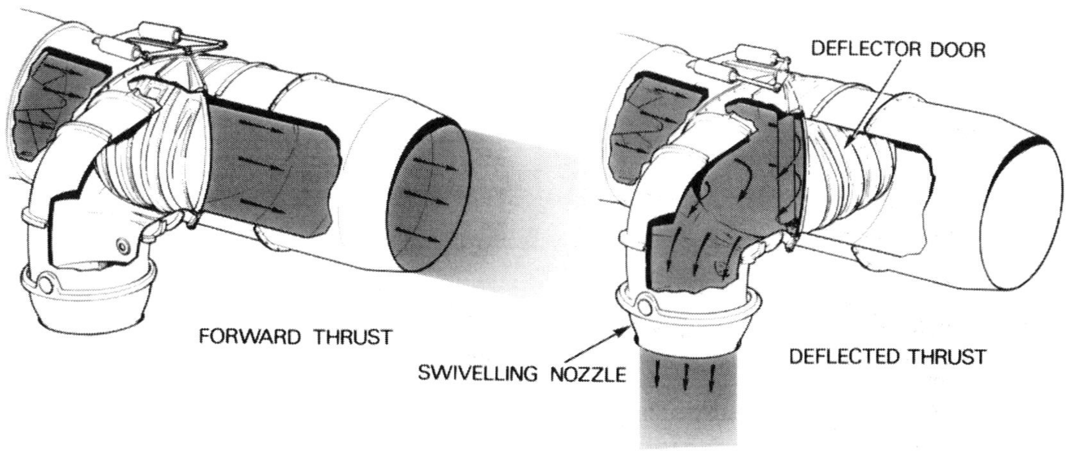

FIGURE 8–57 Switch-in deflector system. *(Source: Rolls Royce.)*

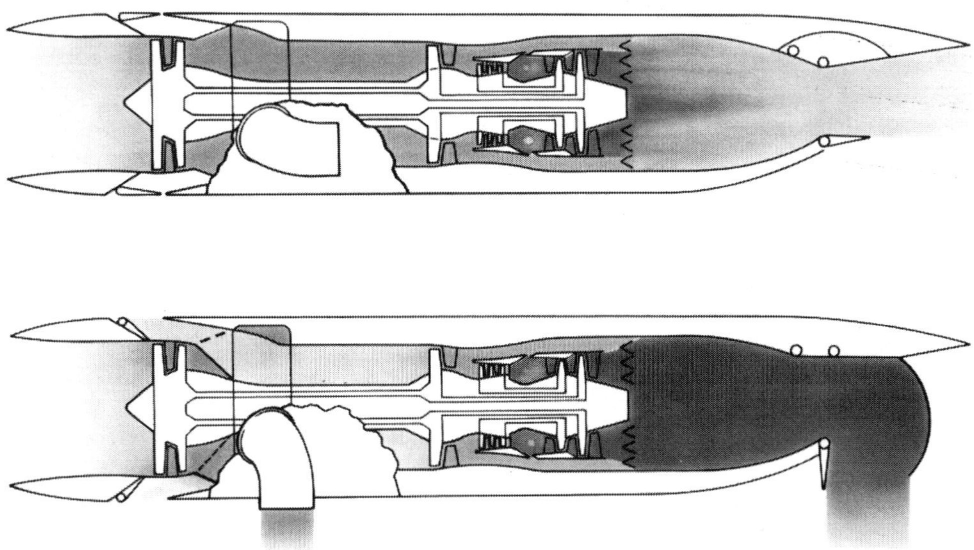

FIGURE 8–58 Vectored thrust engine. *(Source: Rolls Royce.)*

be dispensed with and the engine operated permanently in the latter high by-pass mode described above.

Thrust deflecting nozzles will create an upstream pressure distortion that may excite vibration of the fan or low-pressure turbine blades if the nozzle system is close to these components. Snubbers may be used on the fan blades to resist vibration. On the low-pressure turbine, shrouds at the blade tips or wire lacing may be used to achieve the same result.

Lift Engines

The lift engine is designed to produce vertical thrust during the takeoff and landing phases of V/STOL aircraft. Because the engine is not used in normal flight it must be light and have a small volume to avoid causing a large penalty on the aircraft. The lift engine may be a turbo-jet, which for a given thrust gives the lowest weight and volume. Should a low jet velocity be necessary a lift-fan may be employed.

Pure lift-jet engines have been developed with thrust/weight ratios of about 20:1 and still higher values are projected for the future. Weight is reduced by keeping the engine design simple and also by extensive use of composite materials (Figure 8–59). Because the engine is operated for only limited periods during specific flight conditions, i.e., during takeoff and landing, the fuel system can be simplified and a total loss oil system, in which the used lubricating oil is ejected overboard, can be used.

Lift engines can be designed to operate in the vertical or horizontal position and a thrust-deflecting nozzle fitted to provide some of the advantages of thrust vectoring. Alternatively, the engine may be mounted so that it can swivel through a large angle to provide thrust vectoring. The lift-jet engine will have an extremely hot, high velocity jet

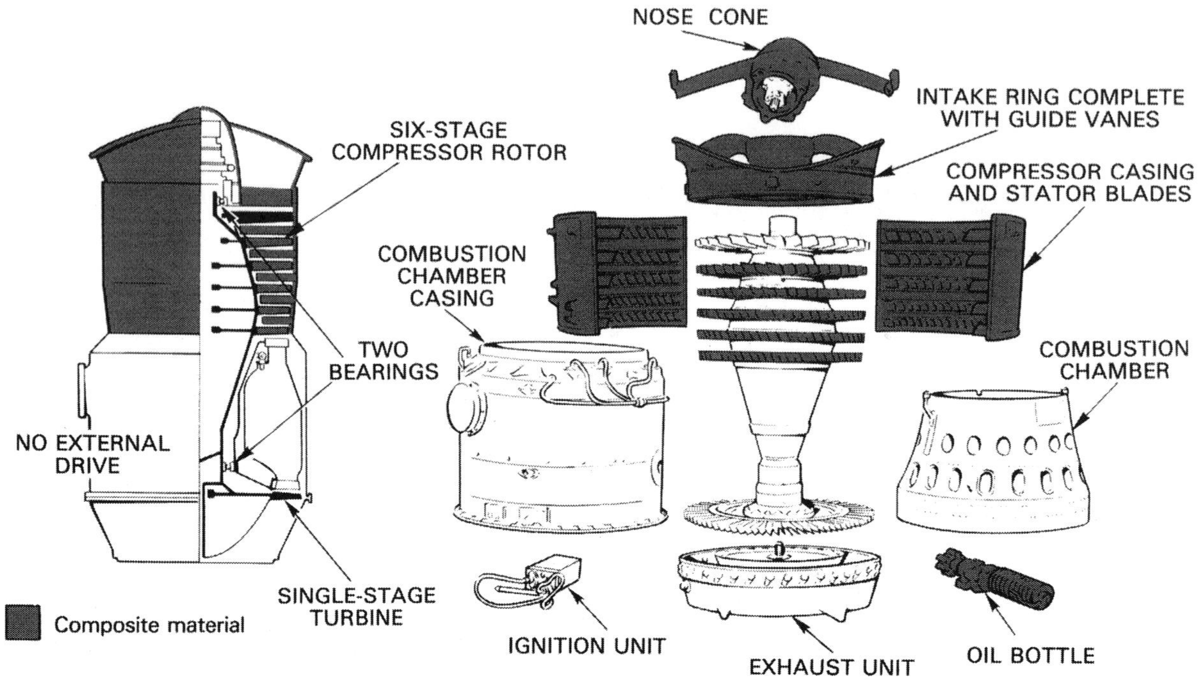

FIGURE 8–59 A lift-jet engine. *(Source: Rolls Royce.)*

exhaust and, to reduce ground erosion by the jet, the normal exhaust nozzle may be replaced by a multi-lobe nozzle to increase the rate of mixing with the surrounding air.

The lift-fan engine is designed to reduce the jet exhaust velocity, to reduce ground erosion, and allow operation from unprepared ground surfaces. It also reduces the jet noise significantly. A range of design options have been considered for this type of engine and some are shown in Figure 8–60.

Remote Lift Systems

Direct lift remote systems duct the by-pass air or engine exhaust air to downward facing lift nozzles remote from the engine. These nozzles may be in the front fuselage of the aircraft or in the wings. The engine duct is blocked by means of a diverter similar to that described earlier.

The remote lift-fan (Figure 8–61) is mounted in the aircraft wing or fuselage, and is driven mechanically or by air or gas ducted into a tip turbine. The drive system is provided by the main propulsion power plant or by a separate engine.

The advantage of the remote lift system is that it gives some freedom to the aircraft to position the propulsion system to the best advantage while still maintaining the resultant thrust near the aircraft center of gravity in the jet lift mode. This freedom is achieved at a cost of increased volume, particularly with the gas driven systems, due to

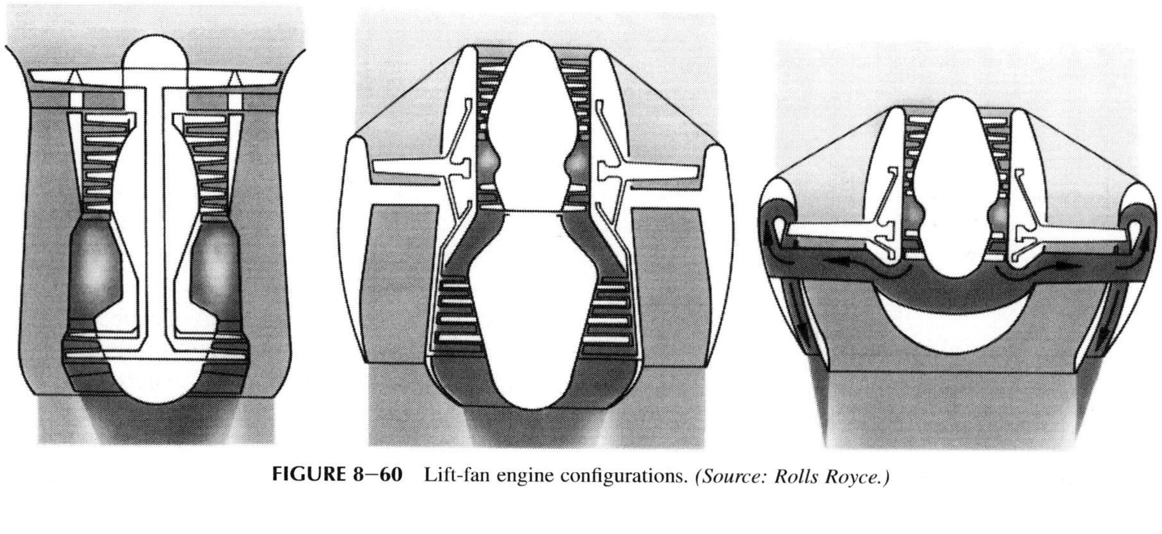

FIGURE 8–60 Lift-fan engine configurations. *(Source: Rolls Royce.)*

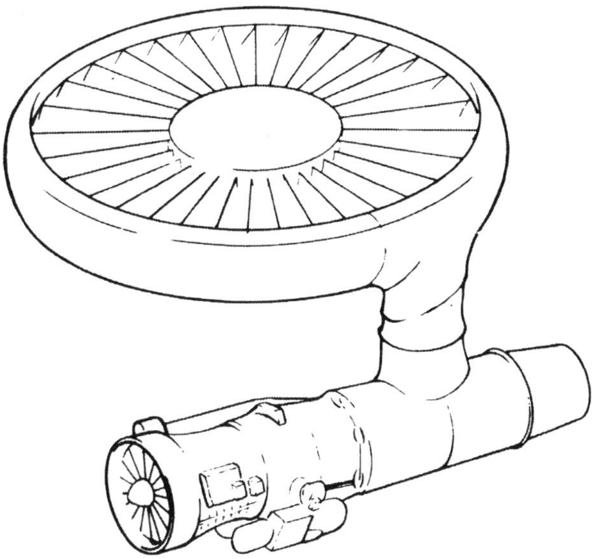

FIGURE 8–61 Remote lift-fan. *(Source: Rolls Royce.)*

the size of the ducts to feed the gas to the remote lift system. Although the mechanically driven remote lift-fan eliminates the need for these large gas ducts, it is done at the expense of long shafts and high power gearboxes and clutch systems.

Swiveling Engines

This method consists of having propulsion engines that can be mechanically swiveled through at least 90° to provide thrust vectoring (Figure 8–62). In addition to these propulsion engines, one or more lift engines may be installed

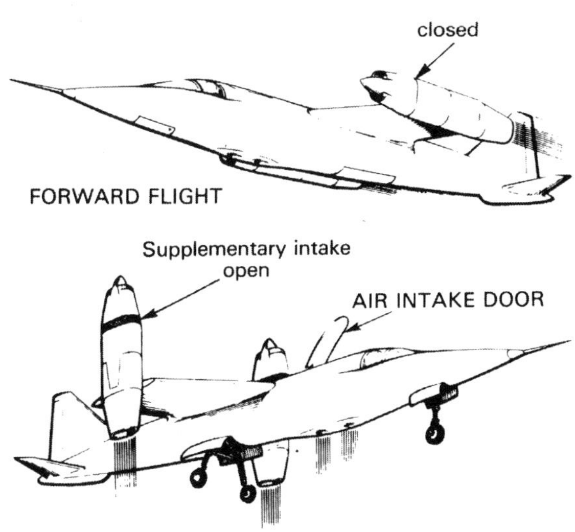

FIGURE 8–62 Jet lift with swiveling nozzles. *(Source: Rolls Royce.)*

to provide supplementary lift during the takeoff and landing phase of flight.

The swiveling engine system can only be used with two or more engines. This then introduces the problem of safety in the event of an engine failure. So, although there is only a small weight penalty and no increase in fuel consumption, safety considerations tend to offset these advantages compared to some of the other powered lift systems. The normal method of providing aircraft control at low speeds is by differential throttling and vectoring of the engines that simplifies the basic engine design but makes the control system more complex.

Bleed Air for STOL

Figure 8–63 shows one method how STOL can be achieved with a form of "flap blowing" The turbo-fan engine has a geared variable pitch fan and an oversized low pressure (L.P.) compressor from the exit of which air is bled and ducted to the flap system in the wing trailing edge. The variable pitch fan enables high L.P. compressor speed and thus high bleed pressure to be maintained over a wide range of thrusts. This gives excellent control at greatly different aircraft flight conditions.

Lift Thrust Augmentation

In many cases on V/STOL aircraft augmentation of the lift thrust is necessary to avoid an engine that is oversized for normal flight with the consequent effects of higher engine weight and fuel consumption than would be the case for a conventional aircraft. This lift thrust augmentation can be achieved in a number of different ways:

1. Using special engine ratings
2. Burning in the lift nozzle gas flow
3. By means of an ejector system

Special Engine Ratings

Experience has shown that an engine rating structure can be devised that provides high thrust levels for short periods of time without reducing engine life. Operation in ground effect and the takeoff and landing maneuvers require maximum thrust for less than 15 seconds so that use of a short lift rating for that time is feasible. Figure 8–64 shows an example of thrust permissible with a 15 second short lift rating compared to that with a 2.5 minute normal lift rating.

At high ambient temperatures, the engine may run into a turbine temperature limit before reaching its maximum rpm and suffer a thrust loss as a result. Restoration of the thrust can be achieved by means of water injection into the combustion chamber, which allows operation at a higher turbine gas temperature for a given turbine blade

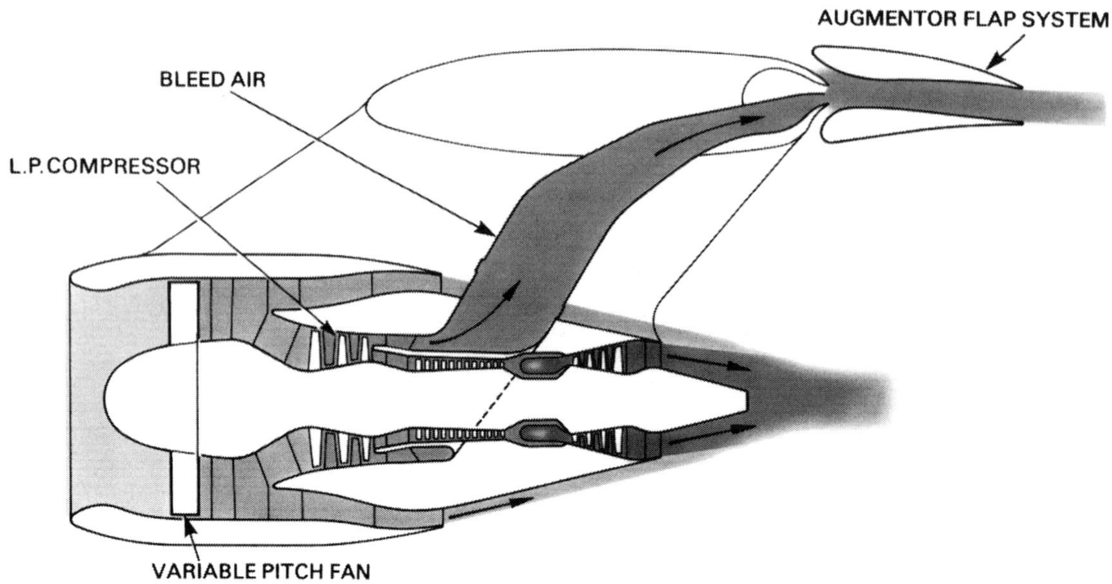

FIGURE 8–63 Flap blowing engine. *(Source: Rolls Royce.)*

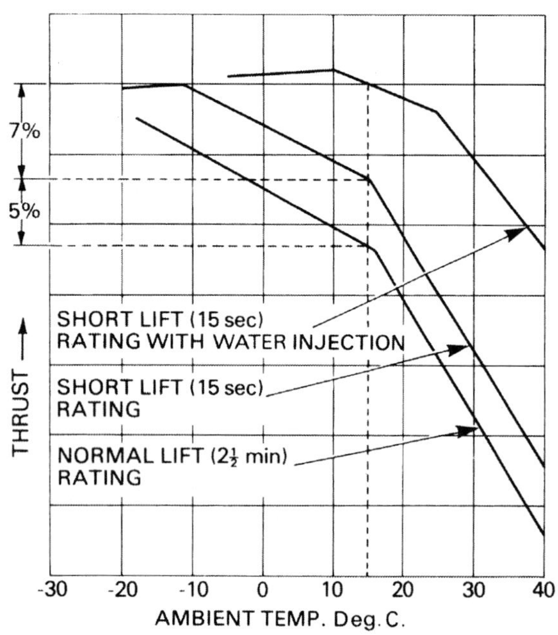

FIGURE 8–64 Thrust increases with short lift ratings. *(Source: Rolls Royce.)*

temperature. If desired, water injection can also be used to increase the thrust at low ambient temperatures.

Lift Burning Systems

The thrust of the four-nozzle lift/propulsion engine may be boosted by burning fuel in the by-pass flow in the duct or plenum chamber supplying the front nozzles. This is called plenum chamber burning (P.C.B.) (Figure 8–65) and thrust of the by-pass air may be doubled by this process. This thrust capability is available for normal flight as well as takeoff and landing and so can be used to increase maneuverability and give supersonic flight.

The thrust of a remote lift jet can also be augmented by burning fuel in a combustion chamber just upstream of the lift nozzle (Figure 8–66). This system is commonly known as a remote augmented lift system (R.A.L.S.). The thrust boost available from the burner reduces the amount of airflow to be supplied to it and therefore reduces the size of the ducting needed to direct the air from the engine to the remote lift nozzle.

Ejectors

The principle of the ejector is that a small, high energy jet entrains large quantities of ambient air by viscous mixing and an increase in thrust over that of the high energy jet results. A number of projected V/STOL aircraft have incorporated this concept using either all the engine exhaust air or just the by-pass flow.

Aircraft Control

The low forward speeds of V/STOL aircraft during takeoff and transition do not permit the generation of adequate aerodynamic forces from the normal flight control surfaces, it is therefore necessary to provide one or more of the following additional methods of controlling pitch, roll, and yaw.

Reaction Controls

This system bleeds air from the engine and ducts it through nozzles at the four extremities of the aircraft (Figure 8–67). The air supply to the nozzles is automatically cut off when

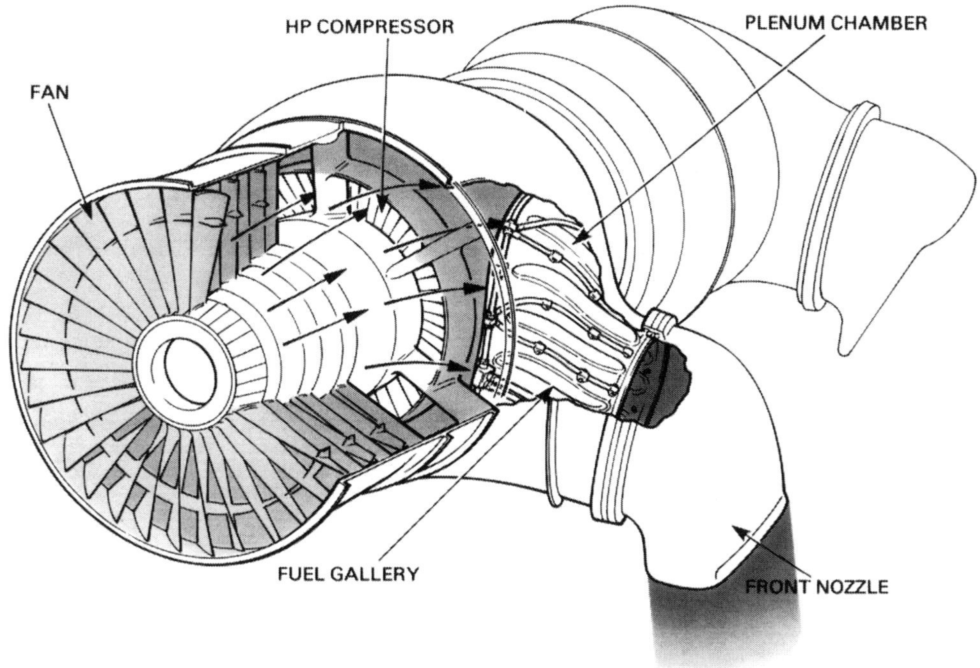

FIGURE 8–65 Plenum chamber burning. *(Source: Rolls Royce.)*

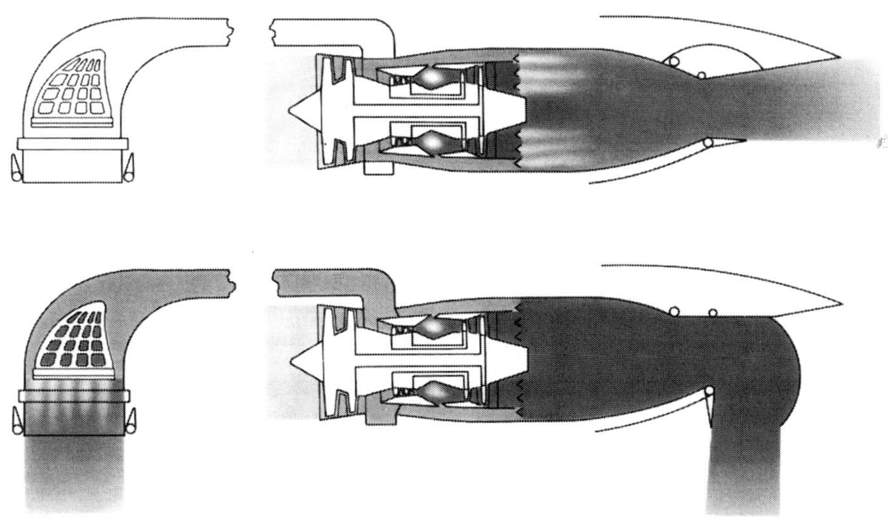

FIGURE 8–66 Remote augmented lift system. *(Source: Rolls Royce.)*

the main engine swiveling propulsion nozzles are turned for normal flight or when the lift engines are shut down. The thrust of the control nozzles is varied by changing their area that varies the amount of airflow passed.

Differential Engine Throttling

This method of control is used on multi-engined aircraft with the engines positioned in a suitable configuration. A rapid response rate is essential to enable the engines to be used for aircraft stability and control. It is usually necessary to combine differential throttling with differential thrust vectoring to give aircraft control in all areas.

Automatic Control Systems

Although it is possible for the pilot to control a V/STOL aircraft manually, some form of automation can be of benefit and in particular will reduce the pilot workload. The pilot's control column is electronically connected to a

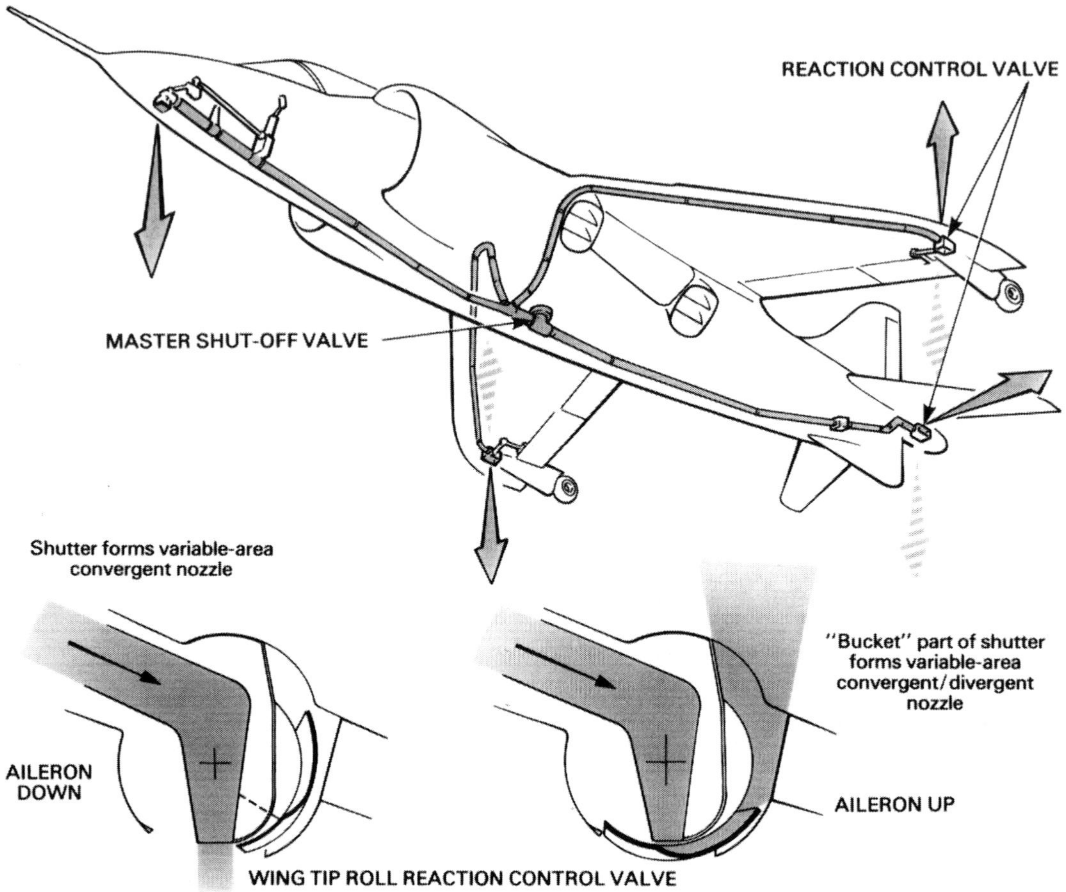

FIGURE 8–67 Reaction control system. *(Source: Rolls Royce.)*

computer or stabilizer that receives signals from the control column, compares them with signals from the sensors that measure the attitude of the aircraft, and automatically adjusts the reaction controls, differential throttling, or thrust vectoring controls to maintain stability.

Thrust Distribution*

Although the principles of jet propulsion will be familiar to the reader, the distribution of the thrust forces within the engine may appear somewhat obscure. These forces are in effect gas loads resulting from the pressure and momentum changes of the gas stream reacting on the engine structure and on the rotating components. They are in some locations forward propelling forces and in others opposing or rearward forces. The amount that the sum of the forward forces exceeds the sum of the rearward forces is normally known as the rated thrust of the engine.

* Source: Adapted, with permission, from Rolls Royce, *The Jet Engine*, 1986, Rolls Royce Plc: UK.

Distribution of the Thrust Forces

The diagram in Figure 8–68 is of a typical single-spool axial flow turbo-jet engine and illustrates where the main forward and rearward forces act. The origin of these forces is explained by following the engine working cycle shown.

At the start of the cycle, air is induced into the engine and is compressed. The rearward accelerations through the compressor stages and the resultant pressure rise produce a large reactive force in a forward direction. On the next stage of its journey the air passes through the diffuser where it exerts a small reactive force, also in a forward direction.

From the diffuser, the air passes into the combustion chambers where it is heated, and in the consequent expansion and acceleration of the gas, large forward forces are exerted on the chamber walls.

When the expanding gases leave the combustion chambers and flow through the nozzle guide vanes they are accelerated and deflected on to the blades of the turbine. Due to the acceleration and deflection, together with the

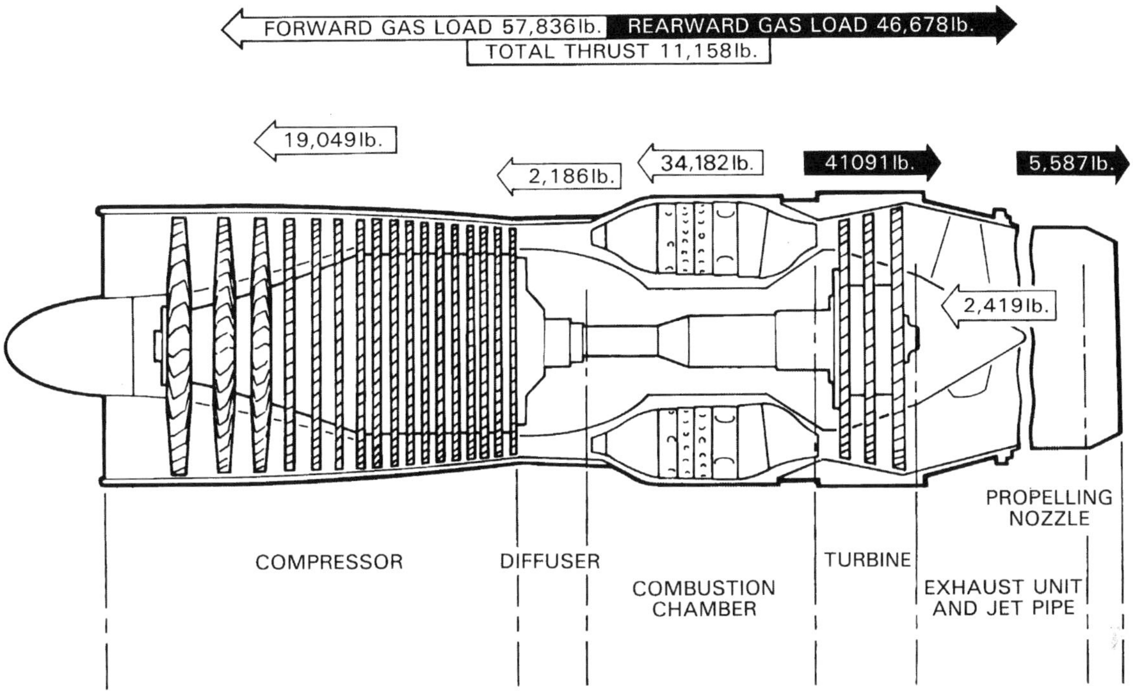

FIGURE 8–68 Thrust distribution of a typical single-spool axial flow engine. *(Source: Rolls Royce.)*

subsequent straightening of the gas flow as it enters the jet pipe, considerable "drag" results; thus the vanes and blades are subjected to large rearward forces, the magnitude of which may be seen in the diagram. As the gas flow passes through the exhaust system, small forward forces may act on the inner cone or bullet, but generally only rearward forces are produced and these are due to the "drag" of the gas flow at the propelling nozzle.

It will be seen that during the passage of the air through the engine, changes in its velocity and pressure occur. For instance, where a conversion from velocity (kinetic) energy to pressure energy is required the passages are divergent in shape, similar to that used in the compressor diffuser. Conversely, where it is required to convert the energy stored in the combustion gases to velocity, a convergent passage or nozzle, similar to that used in the turbine, is employed. Where the conversion is to velocity energy, "drag" loads or rearward forces are produced; where the conversion is to pressure energy, forward forces are produced.

Method of Calculating the Thrust Forces

The thrust forces or gas loads can be calculated for the engine, or for any flow section of the engine, provided that the areas, pressures, velocities and mass flow are known for both the inlet and outlet of the particular flow section.

The distribution of thrust forces shown in Figure 8–68 can be calculated by considering each component in turn and applying some simple calculations. The thrust produced by the engine is mainly the product of the mass of air passing through the engine and the velocity increase imparted to it (i.e., Newton's Second Law of Motion); however, the pressure difference between the inlet to and the outlet from the particular flow section will have an effect on the overall thrust of the engine and must be included in the calculation.

To calculate the resultant thrust for a particular flow section it is necessary to calculate the total thrust at both inlet and outlet, the resultant thrust being the difference between the two values obtained.

Calculation of the thrust is achieved using the following formula:

$$\text{Thrust} = (A \times P) + Wv_j/g$$

where

A = area of flow section in sq. in

P = Pressure in lb. per sec.

W = Mass flow in lb. per sec.

v_j = Velocity of flow in feet per sec.

g = Gravitational constant 32.2 feet per sec. per sec.

Calculating the Thrust of the Engine

When applying the above method to calculate the individual thrust loads on the various components it is assumed that the engine is static. The effect of aircraft forward speed on the engine thrust will be dealt with later. In the following calculations g is taken to be 32 for convenience. To assist in these calculations the locations concerned are illustrated by a number of small diagrams.

Thrust on Compressor Casing

To obtain the thrust on the compressor casing it is necessary to calculate the conditions at the inlet to the compressor and the conditions at the outlet from the compressor. Since the pressure and the velocity at the inlet to the compressor are zero, it is only necessary to consider the force at the outlet from the compressor. Therefore, given that the compressor

OUTLET Area (A) = 182 sq. in.
Pressure (P) = 94 lb. per sq. in. (gauge)
Velocity (v_j) = 406 ft. per sec.
Mass flow (W) = 153 lb. per sec.
The thrust
$= (A \times P) + Wv_j/g - 0$
$= (182 \times 94) + (153 \times 406)/32 - 0$
= 19,049 lb. of thrust in a forward direction.

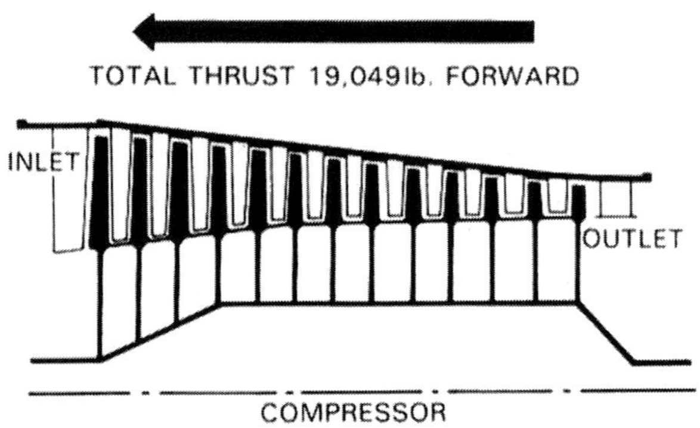

Compressor Casing Diagram. *(Source: Rolls Royce.)*

Diffuser Duct The conditions at the diffuser duct inlet are the same as the conditions at the compressor outlet, i.e., 19,049 lb.

Therefore, given that the diffuser

OUTLET Area (A) = 205 sq. in.
Pressure (P) = 95 lb. per sq. in. (gauge)
Velocity (v_j) = 368 ft. per sec.
Mass flow (W) = 153 lb. per sec.
The thrust
$= (205 \times 95) + (153 \times 368)/32 - 19{,}049$
$= 21{,}235 - 19{,}049$
= 2186 lb. of thrust in a forward direction.

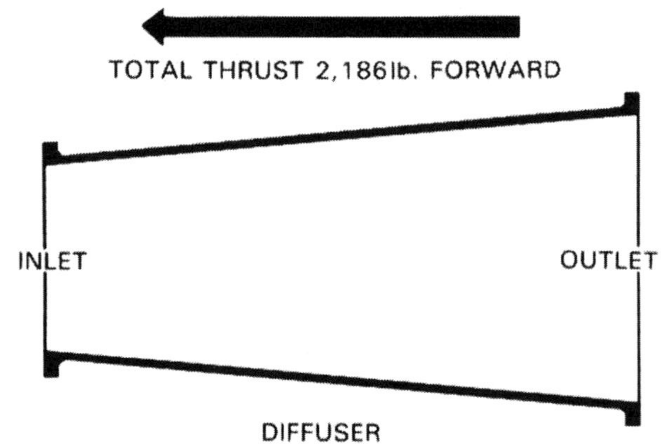

Diffuser Duct Diagram. *(Source: Rolls Royce.)*

Combustion Chambers The conditions at the combustion chamber inlet are the same as the conditions at the diffuser outlet, i.e., 21,235 lb. Therefore, given that the combustion chamber

OUTLET Area (A) = 580 sq. in.
Pressure (P) = 93 lb. per sq. in. (gauge)
Velocity (v_j) = 309 ft. per sec.
Mass flow (W) = 153 lb. per sec.
The thrust
$= (A \times P) + Wv_j/g - 21{,}235$ lb.
$= (580 \times 93) + (153 \times 309)/32 - 21{,}235$ lb.
$= 55{,}417 - 21{,}235$
= 34,182 lb. of thrust in a forward direction.

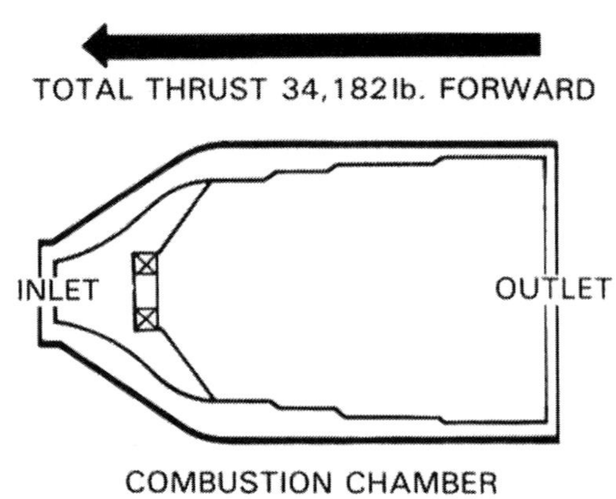

Combustion Chamber Diagram. *(Source: Rolls Royce.)*

Turbine Assembly The conditions at the turbine inlet are the same as the conditions at the combustion chamber outlet, i.e., 55,417 lb.

Therefore given that the turbine

OUTLET Area (A) = 480 sq. in.
Pressure (P) = 21 lb. per sq. in. (gauge)
Velocity (v_j) = 888 ft. per sec.
Mass flow (W) = 153 lb. per sec.
The thrust
$= (A \times P) + Wv_j/g - 55{,}417$
$= (480 \times 213) + (153 \times 888)/32 - 55{,}417$
$= 14{,}326 - 55{,}417$
$= -41{,}091.$

This negative value means a force acting in a rearward direction.

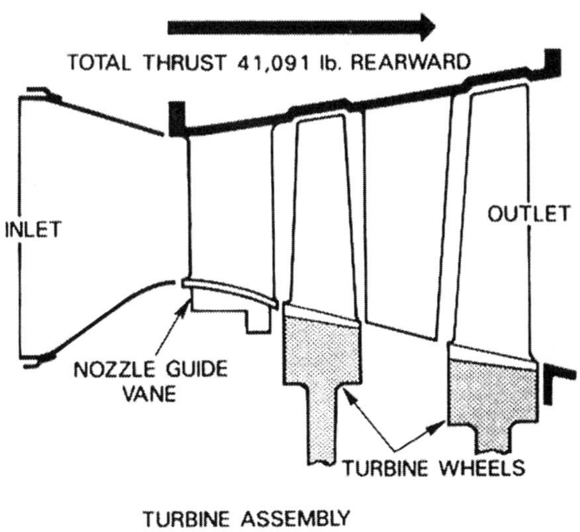

Turbine Assembly Diagram. *(Source: Rolls Royce.)*

Exhaust Unit and Jet Pipe The conditions at the inlet to the exhaust unit are the same as the conditions at the turbine outlet, i.e., ,14,326 lb.

Therefore, given that the jet pipe

OUTLET Area (A) = 651 sq. in.
Pressure (P) = 21 lb. per sq. in. (gauge)
Velocity (v_j) = 643 ft. per sec.
Mass flow (W) = 153 lb. per sec.
The thrust
$= (A \times P) + Wv_j/g - 14{,}326$ lb.
$= (651 \times 21) + (153 \times 643)/32 - 14{,}326$
$= 16{,}745 - 14{,}326$
= 2419 lb. of thrust in a forward direction.

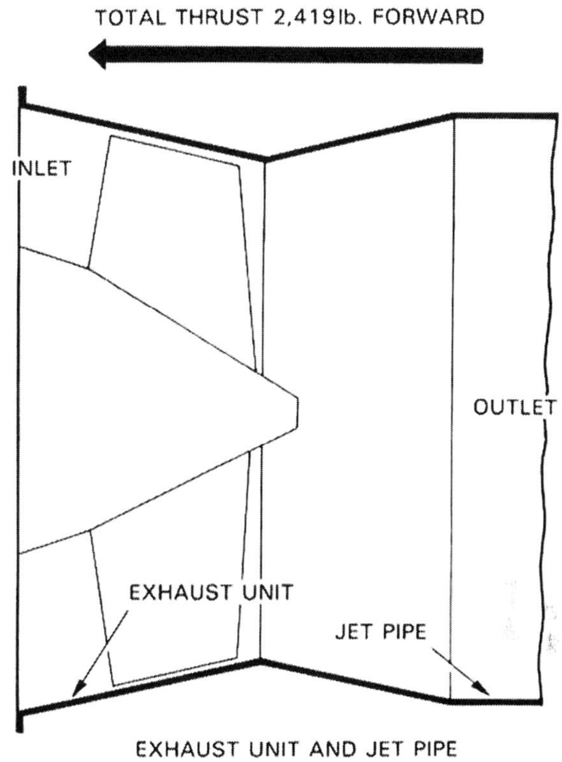

Exhaust Unit and Jet Pipe Diagram. *(Source: Rolls Royce.)*

Propelling Nozzle The conditions at the inlet to the propelling nozzle are the same as the conditions at the jet pipe outlet, i.e., 16,745 lb.

Therefore, given that the propelling nozzle

OUTLET Area (A) = 332 sq. in.
Pressure (P) = 6 lb. per sq. in. (gauge)
Velocity (v_j) = 1917 ft. per sec.
Mass flow (W) = 153 lb. per sec.
The thrust
$= (A \times P) + Wv_j/g - 16{,}745$ lb.
$= (332 \times 6) + (153 \times 1917)/32 - 16{,}745$
$= 11{,}158 - 16{,}745$
= 5587 lb, acting in a rearward direction.

It is emphasized that these are basic calculations and such factors as the effect of air offtakes have been ignored.

Based on the individual calculations, the sum of the forward or positive loads is 57,836 lb., and the sum of the

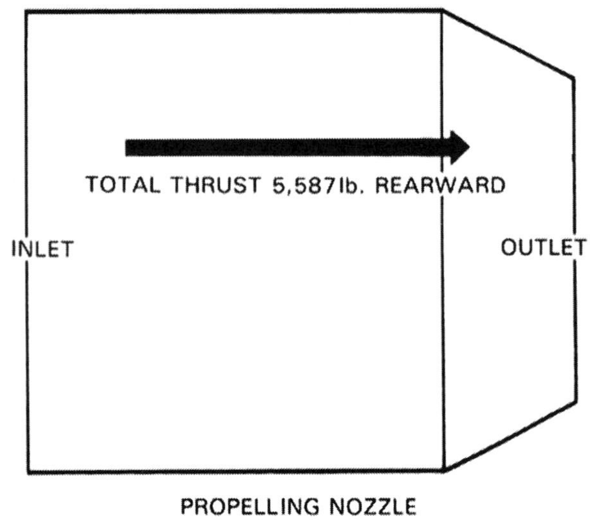

Propelling Nozzle Diagram. *(Source: Rolls Royce.)*

rearward or negative loads is 46,678 lb. Thus, the resultant (gross or total) thrust is 11,158 lb.

Engine It will be of interest to calculate the thrust of the engine by considering the engine as a whole, as the resultant thrust should be equal to the sum of the individual gas loads previously calculated.

Although the momentum change of the gas stream produces most of the thrust developed by the engine (momentum thrust $= Wv_j/g$), an additional thrust is produced when the engine operates with the propelling nozzle in a "choked" condition. This thrust results from the aerodynamic forces that are created by the gas stream and exert a pressure across the exit area of the propelling nozzle (pressure thrust). Algebraically, this force is expressed as $(P - P_0)A$.
where

A = Area of propelling nozzle in sq. in.
P = Pressure in lb. per sq. in.
P_0 = Atmospheric pressure in lb. per sq. in.

Therefore, assuming values of mass flow, pressure, and area to be the same as in the previous calculations, i.e.,

Area of propelling nozzle (A) = 332 sq. in.
Pressure (P) = 6 lb. per sq. in. (gauge)
Atmospheric pressure (P_0) = 0 lb. per sq. in. (gauge)
Mass flow (W) = 153 lb. per sec.
Velocity (v_j) = 1917 ft. per sec.
The thrust $= (P - P_0)A + Wv_j\ 0$
$= (6 - 0)\ 332 + (153 \times 1917)/32 - 0$
$= 1992 + 9166$
= 11,158 lb., the same as previously calculated by combining the gas loads on the individual engine locations.

On engines that operate with a non-choked nozzle, the $(P - P_0)A$ function does not apply and the thrust results only from the gas stream momentum change.

Inclined Combustion Chambers In the previous example the flow through the combustion chamber is axial; however, if the combustion chamber is inclined towards the axis of the engine, then the axial thrust will be less than for an axial flow chamber. This thrust can be obtained by multiplying the sum of the outlet thrust by the cosine of the angle (see Figure 8–69). The cosine = base/hypotenuse and for a given angle is obtained by consulting a table of cosines. It should be emphasized that if the inlet and outlet are at different angles to the engine axis, it is necessary to

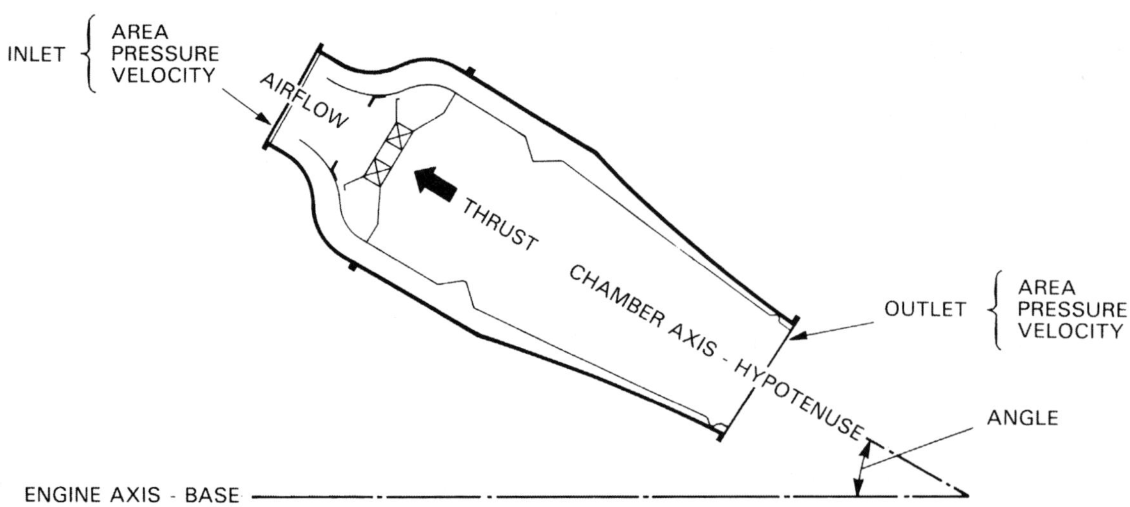

FIGURE 8–69 A hypothetical combustion chamber showing values required for calculating thrust. *(Source: Rolls Royce.)*

multiply the inlet and outlet thrusts separately by the cosine of their respective angles.

Thrust due to Afterburning

When the engine is fitted with an afterburner, the gases passing through the exhaust system are reheated to provide additional thrust. The effect of afterburning is to increase the volume of the exhaust gases, thus producing a higher exit velocity at the propelling nozzle.

Assuming that an afterburner jet pipe and propelling nozzle are fitted to the engine used in the previous calculations, and the new conditions at the propelling nozzle are as follows:

OUTLET Area (A) = 455 sq. in.
Pressure (P) = 5 lb. per sq. in. (gauge)
Velocity (v_j) = 2404 ft. per sec.
Mass flow (W) = 157 lb. per sec.
The thrust
$= (A \times P) + Wvj/g - 16{,}745$ lb.
$= (455 \times 5) + (153 \times 2404)/32 - 16{,}745$
$= 14{,}069 - 16{,}745$
$= 2676$ lb. acting in a rearward direction.

Therefore, compared with the previous calculation for a propelling nozzle, it will be seen that the negative thrust is reduced from −5587 lb. to −2676 lb.; the overall positive thrust is thus increased by 2911 lb., which is equivalent to a thrust increase of more than 25%.

To arrive at the total thrust of the engine with afterburning the calculations should use the above figures.

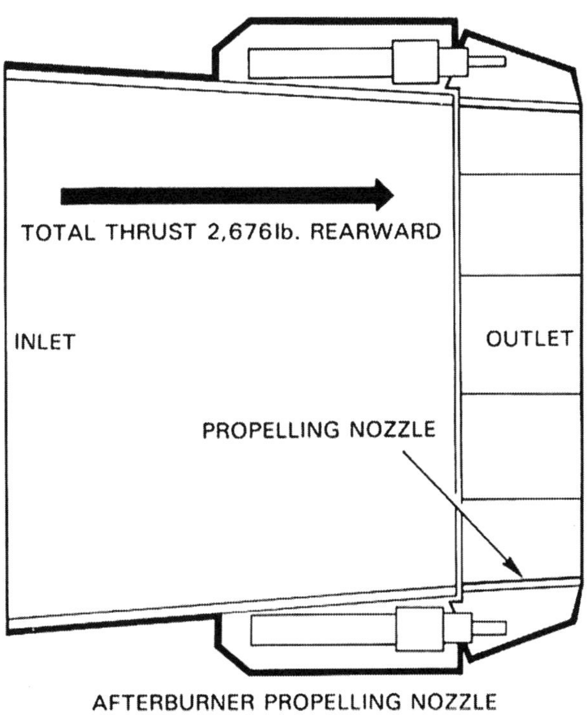

Afterburning Diagram. *(Rolls Royce.)*

SYSTEMS UNIQUE TO LAND OR MARINE APPLICATIONS

Generators*

Standard Design

The modular design of the turbogenerator permits the selection of a standard version with either an open or closed cooling system and static or brushless excitation systems.

The generators satisfy the requirement of IEC-34 and other relevant standards such as SEN, ANSI, NEMA, CSA, etc. This means that the generators can be used in countries in which these or comparable standards apply. Generators can be custom built to satisfy other standards.

Configuration

The turbogenerators are two-pole, air-cooled synchronous generators with cylindrical rotor and direct air-cooled rotor winding. They are intended for both basic and peak-load operations and designed to withstand high short-circuiting stresses. The configuration with journal bearings in the end shields permits delivery of the machines as a single unit, which considerably simplifies installation and commissioning (see Figure 8–70).

FIGURE 8–70 An installed generator package. *(Source: Alstom Power.)*

* Source: Courtesy of McGraw-Hill, from C. Soares, *Process Engineers Equipment Handbook* (New York: McGraw Hill, 2001). Note that the original information source was Alstom Power.

Degrees of Protection and Methods of Cooling

The design of the generators provides for the following degrees of protection and methods of cooling in accordance with IEC 34-5 and 34-6, 1983.

Degrees of Protection	Descriptions
IP 21	Protection against contact with live or moving parts inside the machine and drip-proof.
IP 54	Dust-proof, splash-proof.
Methods of Cooling	**Descriptions**
IC 01	The cooling air is supplied to the machine through a screen-protected opening for different types of air filters on the long sides of the stator and is exhausted through a flanged opening. This is located at the middle of the generator housing and can be connected to a hood for the exhaust air. The method of cooling is then IC 21.
IC 71	The cooling air circulates in a closed circuit and is cooled in turn by a water-cooled heat exchanger.

These combinations of degrees of protection and methods of cooling are standard for this information source.

The different cooling methods require two fans mounted on the generator rotor.

The use of the IC 01 and IC 21 methods of cooling presupposes that the cooling air supplied is cleaned by thorough filtration. For this reason, IC 01 and IC 21 should be avoided when the cooling air available contains corrosive gases or large quantities of other pollution.

Configuration/Arrangement Forms

The machines can be supplied in the following versions:

Arrangement Forms	Descriptions
IM 1005	Horizontal shaft, two bearings mounted in the bearing end shields, one free shaft extension end with coupling flange.
IM 1007	Horizontal shaft, two bearings mounted in the bearing end shields, two free shaft extensions with one coupling flange. This version permits powering of the generator from two directions. The stator is, in both cases, intended for foot mounting. If required, for example with an elevated turbine centerline, the above machines can be delivered with a base frame.

Excitation System

The generators can be delivered with one of two alternative excitation systems:

1. *Rotary brushless excitation system.* The winding of the generator rotor is supplied via a rectifier mounted on the shaft, from a directly connected three-phase AC exciter. The system includes a pilot AC exciter.
2. *Static excitation system.* This system consists of a static thyristor rectifier unit supplied from an external power source. The excitation current is connected to the rotor windings via conventional sliprings. The static thyristor rectifier unit can be supplied on request.

Accessories

The following accessories can be supplied on request:

- Fire extinguishing system
- Base frame
- Lubricating oil system
- Sound insulating enclosure
- Current transformer
- Voltage regulation system

The following information and requirements should be provided by the end user to the OEM with any request for a bid:

- Relevant standards and recommendations
- Rated power
- Temperature of the cooling medium
- Power factor with rated power
- Rated voltage and voltage range
- Rated main power supply frequency
- Rated rotational speed and overspeed
- Rotation direction (as seen from the exciter end)
- Generator power in relation to the maximum and minimum ambient temperatures
- Degree of protection and method of cooling
- Arrangement form
- Excitation system
- Application
- Cooling air quality (for methods of cooling IC 01 and IC 21)
- Cooling water quality (for method of cooling IC 71)
- Special requirements, e.g., thrust bearings
- Extra testing and documentation

Technical Data on Typical Available Generators

- Power range, 20–200 MVA (40°C cooling air, temperature rise in accordance with class B)

- Insulation class F (155°C)
- Power factor 0.8 for 50 Hz, 0.85 for 60 Hz
- Standard voltage up to 80 MVA is 11 kV for 50 Hz and 13.8 kV for 60 Hz. Equipment for other voltages can be provided on request.

Design Description

Stator Frame

The stator frame is of welded steel construction (see Figure 8–71).

The side plates are dimensioned to bear the weight of the complete generator during lifting. Openings are provided in the top or sides for cooling air supply and exhaust. Longitudinal foot plates are provided at the bottom of the long sides of the stator frame for fixing the stator to the foundation.

Stator Core

The stator core is built up of segments stamped from thin silicon-steel sheets that give high permeability and small losses (see Figure 8–71).

The segments are varnished on both sides with a heat-resistant varnish to form an effective and permanent insulation between the sheets.

The segments are stacked to form a number of axial packages. Radial cooling air channels are formed between the packages by means of support plate segments provided with spacers.

The sheet segments are guided into place by axial guide bars. The stator core is pressed with the upper pressure ring installed to give the sheet pressure specified. The pressure ring is then locked.

FIGURE 8–71 Generator stator. *(Source: Alstom Power.)*

The pressure rings and the superimposed pressure fingers are sprung into place during the pressing to maintain the necessary pressure in the stator core.

Stator Winding

The stator winding is a diamond winding installed in open slots with two coil sides per slot. The coils are manufactured in two halves that are brazed together.

The strands are transposed alternately through roebling in the slot section of the coil or by transposition, group by group, in the connections between the coils. The strands are insulated with impregnated glass fiber yarn.

The conductor insulation consists of MICAFOLD®.

The insulation of the coil sides is built up as continuous tape isolation, i.e., both the straight sections of the coil and the core ends are wound with tape. The insulation consists of epoxy-impregnated mica glass-tape. Both the pre-impregnated MICAREX® system and the vacuum-pressure impregnated MICAPACT® system satisfy the requirements for temperature class F (155°C). The insulation systems used are described in more detail in separate brochures.

The straight sections of the coils are fixed in the winding slots in the stator core by means of contra-wedging. Spring elements are inserted in the slots to hold the coils in the winding slots, also after long service. These exert uniform pressure on the complete length of the straight section of the coil.

The ends of the stator coils are supported against the pressure rings in the stator core by radial support plates of insulating material. The coil ends are anchored to the support plates and braced against each other. The bracing system is dimensioned to withstand the stresses that can develop during normal operations and following a sudden short-circuiting of the generator.

Stator Terminals

The connections to the busbars are located outside the generator casing (see Figure 8–72). The terminal bushings are connected to the terminal connections in the stator winding with clamps with generously dimensioned contact areas. The clamps are easily accessible for removing the stator terminals for, e.g., transport.

Fan Covers

The covers are manufactured of heat-resistant, glass-fiber-reinforced polyester. Each cover is divided into removable segments to simplify inspection of the coil ends.

The fan covers are mounted against the stator frame over the coil end area and lead the incoming cooling air to the axial fans on the rotor shaft.

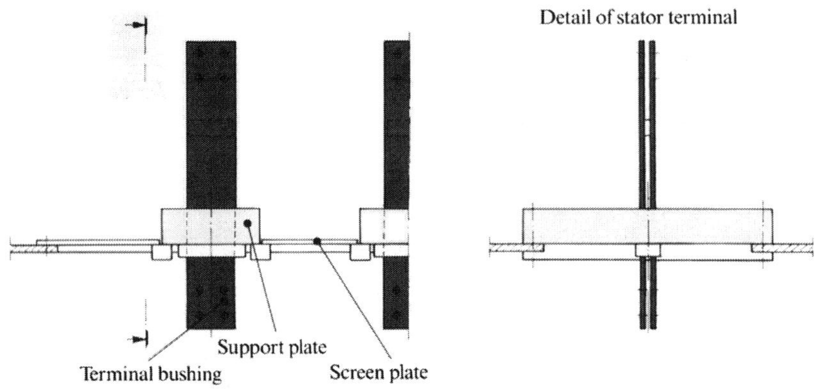

FIGURE 8–72 Stator terminal. *(Source: Alstom Power.)*

FIGURE 8–73 Bearing end shield. *(Source: Alstom Power.)*

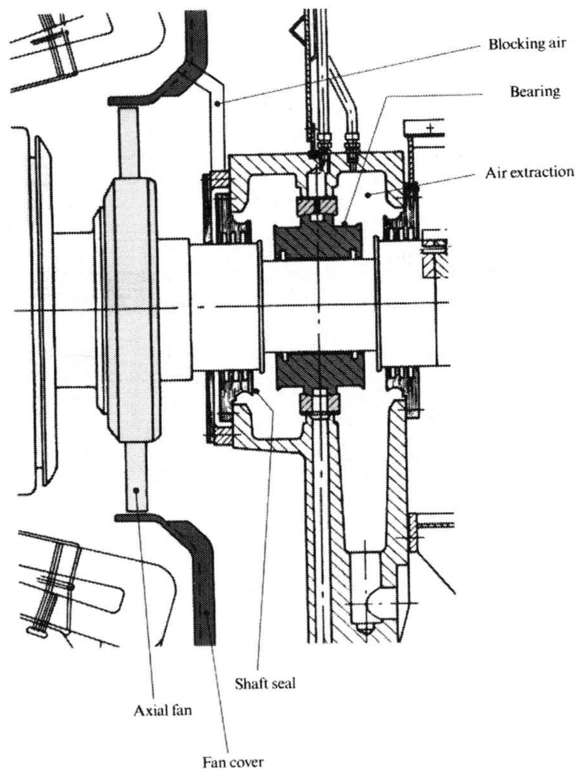

FIGURE 8–74 Fan bearing assembly. *(Source: Alstom Power.)*

Bearing End Shields

The bearing end shields consist of a very stiff lower half with bearing housing (see Figure 8–73). The upper half of the bearing housing is easily removed to facilitate inspection of the bearing. Holes for supply and drainage of oil are drilled in the lower half. The bearing housing is provided with connections for air extraction.

The upper half of the bearing end shield consists of a sound insulating screen plate divided into three parts to simplify removal for inspection.

Bearings

Each bearing housing contains a radial bearing (see Figure 8–74) for the generator rotor. The radial bearings are pressure-lubricated slide bearings with white-metal linings divided into staggered upper and lower segments. Lubricating oil is supplied to the bearing through drilled channels.

Where the bearings must take up axial forces, one is provided with axial thrust-bearing pads. The thrust bearing is double-acting, i.e., it takes up forces in both directions. One of the bearings is insulated from the bearing end shield to prevent the development of damaging bearing currents.

When the generator is to be driven from both ends, both bearings are insulated. The support bearing is always insulated. The shaft seals prevent oil leakage from the bearing housing and consist of a divided seal of oil-resistant insulation material. An air extraction system is connected to the bearing space to prevent oil leakage through the external seals. The internal shaft seals are provided with sealing caps and the intermediate space is connected to blocking air from the pressure side of the fan.

FIGURE 8–75 Assembling the rotor. *(Source: Alstom Power.)*

Rotor

The rotor body (see Figure 8–75) is manufactured from a cylindrical forging of high-alloy steel with suitable magnetic properties. A hole is drilled into the shaft extension for terminal conductors. The terminal conductors carry excitation current from the exciter to the rotor winding.

A number of axial winding slots for the excitation windings of the rotor are milled on both sides of the pole sections.

Rotor Terminals

The rotor terminals (see Figure 8–76) are located in the shaft extension and consist of two copper conductors, enclosed in a tube of insulating material and insulated from each other. The connections from the terminal conductor and radially outward to the winding consist of contact screws. The design of the contact screws eliminates the risk of fatigue failure in the connection details caused by the differential movement of the rotor winding and the shaft.

All joints are provided with gold-plated spring-contact elements to ensure effective current conduction between the contact surfaces.

Rotor Winding

The conductors in the rotor winding consist of silver-alloyed copper. Each conductor consists of two strands. The conductors are brazed to the rotor coils with silver solder.

The rotor coils are installed in the winding slots in the rotor body with conductor insulation of epoxy resin-impregnated glass fiber fabric between each conductor layer. The spaces at the side of the rotor coils function as ventilation channels.

The straight sections of the winding are supported tangentially with a number of bracing pins mounted in holes in the conductors and the coil insulation. The straight sections are fixed radially by means of pressure bars filled with synthetic resin under high pressure. The windings are thereby fixed radially and all play in the winding slots is eliminated.

The rotor coil ends are supported with blocks of glass fiber fabric laminate.

Impregnated glass fiber fabric is used as insulation between the retaining rings and rotor coil ends of the winding. The coil end insulation is provided with cooling channels after hardening.

The insulation system satisfies the requirements for temperature class F (155°C).

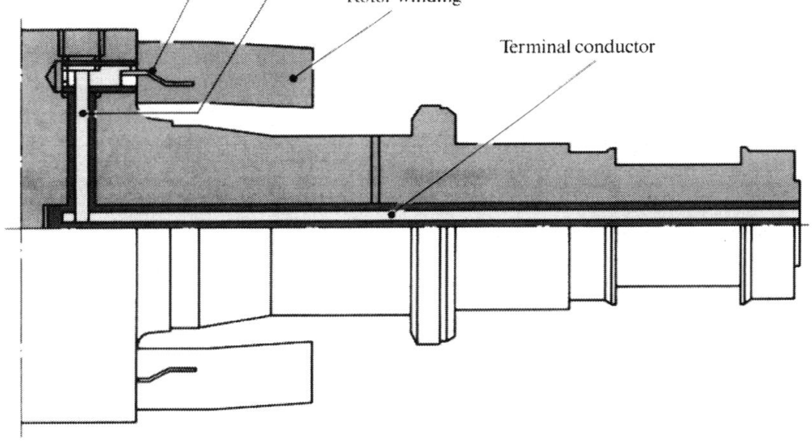

FIGURE 8–76 Rotor terminal. *(Source: Alstom Power.)*

Rotor Retaining and Support Rings

Rotor retaining rings of high alloy non-magnetic steel are shrunk on the rotor body to provide radial support of the rotor coil ends against centrifugal forces during rotation.

The retaining ring material is not affected by stress corrosion. To ensure that the retaining rings are sufficiently stiff to remain circular, a support ring is shrunk in the outer end of the retaining ring, free from the rotor shaft.

The retaining rings and the other parts of the bracing system are dimensioned to withstand the stresses that develop with short-circuiting and overspeeding.

Axial Fan

A number of aluminum fan blades are mounted on a fan hub, shrunk on each shaft extension on the rotor. The fan blades are easily removed and replaced when the rotor is to be installed.

Balancing and Overspeed Running

The rotor is balanced when the rotor winding, insulation, and retaining rings are in place. It is then heated to a high temperature and test-run at overspeed. This results in the winding adopting its final form before the rotor is finally balanced to satisfy the relevant requirements. The normal overspeed is 120% of the rated speed. Certain of the balancing planes remain accessible when the rotor is installed in the stator.

Rotor Cooling

The rotor winding (see Figure 8—77) is directly air-cooled by ventilation channels in direct contact with the winding slots. Cooling air is drawn in between the support ring and the shaft extension by the centrifugal fan effect of the rotor.

The air cooling the coil ends passes out radially through holes in the coil end insulation to the retaining ring and axially in channels under the retaining ring through holes in the support ring.

The cooling air in the winding slots is drawn in under the coil ends and into the axial cooling air channels between the sides of the slots and the conductor package. The rotor slot wedges and the beams between the winding slots are provided with a number of exhaust openings, connected in parallel, for the cooling air.

Cooling Systems

The generator is provided with two axial fans. The main task of the fans is to ventilate the stator (see Figure 8—78); the rotor itself functions as a centrifugal fan. The generator can be provided with an open or closed cooling system. The air circulation in the different cooling systems is shown in Figures 8—78 through 8—80.

In closed cooling systems (see Figure 8—79), the cooler housings are mounted, with the air/water coolers, on the long sides of the generator. The cooler housings are welded steel constructions that lead cooling air from the coolers to the axial fans. The air/water coolers are of the lamellar tube type. The materials chosen for the tubes and water chamber are dependent on the quality of the cooling water available.

Compensation air is drawn in through a filter to replace air leakage from the machine.

With an open cooling system (see Figure 8—80), the incoming air is to be filtered.

The choice of filter is determined by the site conditions. Recommendations for filter selection are based on the information regarding the environment provided with the request to tender.

Figure 8—81 illustrates a generator assembly that has not been installed.

FIGURE 8—77 Rotor winding. *(Source: Alstom Power.)*

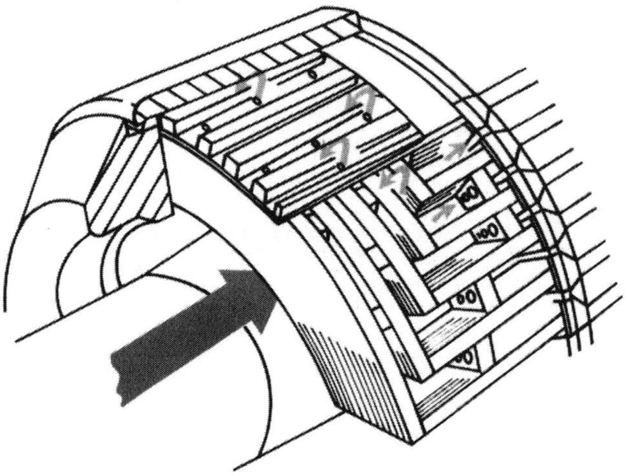

FIGURE 8—78 Cooling air direction. *(Source: Alstom Power.)*

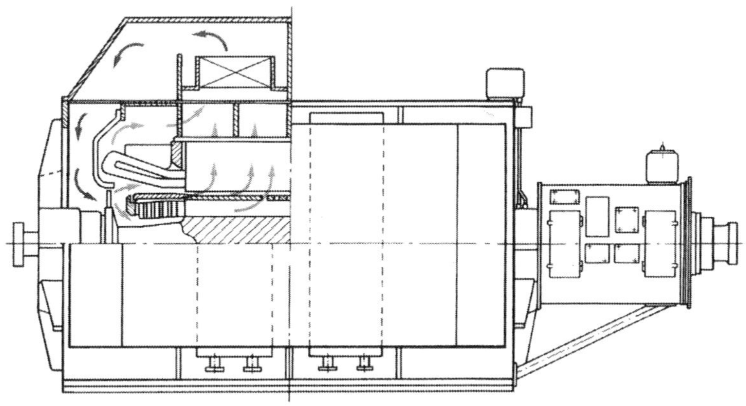

FIGURE 8–79 Closed cooling system. *(Source: Alstom Power.)*

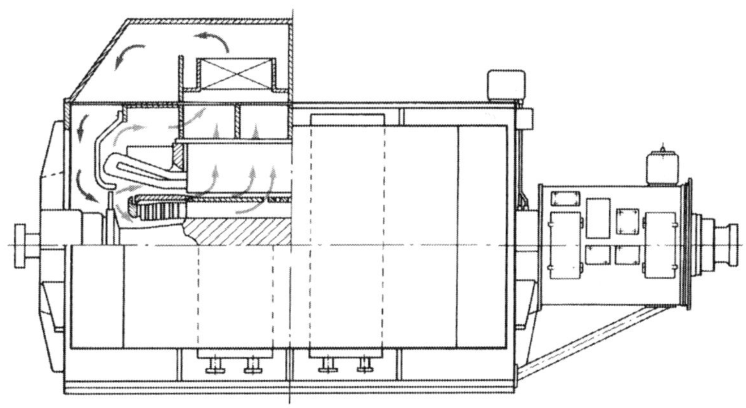

FIGURE 8–80 Open cooling system. *(Source: Alstom Power.)*

FIGURE 8–81 Generator assembly. *(Source: Alstom Power.)*

Brushless Excitation

Main Exciter

The main exciter (see Figure 8–82) is a multiple synchronous generator with salient poles in the stator.

The stator and rotor winding insulation satisfies the requirements for temperature class F (155°C). The windings are impregnated for protection against damp and vibration.

Rotating Rectifier

The purpose of the rectifier is to rectify the A.C. current from the main exciter and provide the rotor winding of the turbogenerator with D.C. current via connectors in the center of the shaft.

The electrical equipment in the rectifier consists of silicon diodes and RC protection.

The rectifier is built of two steel rings that are shrunk on the shaft of the exciter rotor with intermediate mica insulation. The steel rings are connected to the terminal conductors in the center of the shaft.

All contact surfaces are specially processed to guarantee a high degree of rectifier reliability and stability during operations.

Pilot Exciter

The pilot exciter is a synchronous generator with permanent magnets on the rotor.

The rotor magnets are enclosed in a short-circuited aluminum ring that prevents demagnetization of the poles because of short-circuiting in the stator winding.

The stator winding insulation satisfies the requirements for temperature class F (155°C).

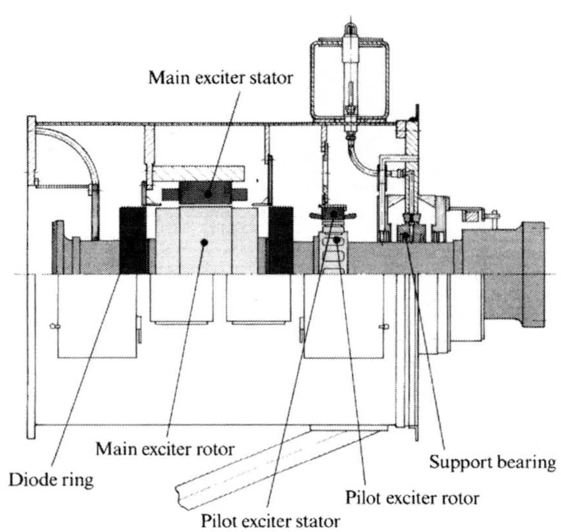

FIGURE 8–82 Main exciter. *(Source: Alstom Power.)*

Exciter Housing

Openings are provided in the side walls of the housing for cooling air intake and exhaust and for service activities.

Support Bearing

The support bearing consists of shield, bearing insulation, bearing, and shaft seals.

The bearing, which consists of a bearing body with white metal lining, is insulated from the shield. The shaft seals against oil leakage from the bearing housing consist of a split seal of oil-resistant insulation material. An air extractor is connected to the bearing space to prevent oil leakage through the external seals. Internal shaft seals are provided with sealing caps and the intermediate space is connected to the blocking air from the pressure side of the fan.

Cooling

The exciter can be provided with either an open or a closed cooling system.

With an open cooling system, a housing with filter cassettes is mounted on one side of the exciter housing for the incoming cooling air. The cooling air exit is directed downward under the exciter housing.

With a closed cooling system, the filter housing is replaced with supply and exhaust air channels connected to the cooler housing of the generator.

Surface Treatment

In its standard version, the generator is painted with a lacquer of two-component type based on ethoxylized chlorine polymer. The generator is primed inside and outside and then finished externally in a neutral blue color. The paint is resistant to corrosive, tropical, and other aggressive atmospheres.

Static Excitation

When the rotor winding of the generator is supplied with current from a static rectifier unit, the rotor is provided with a slipring shaft and supplied via brush gear.

Slipring Shaft

The slipring shaft (see Figures 8–83 and 8–84) consists of shaft extension, insulations, sliprings, contact screws, and terminal conductors.

The shaft extension consists of a steel forging with flange for connection to the shaft of the generator rotor. The center of the shaft extension is drilled for the terminal conductors.

FIGURE 8–83 Slipring shaft end. *(Source: Alstom Power.)*

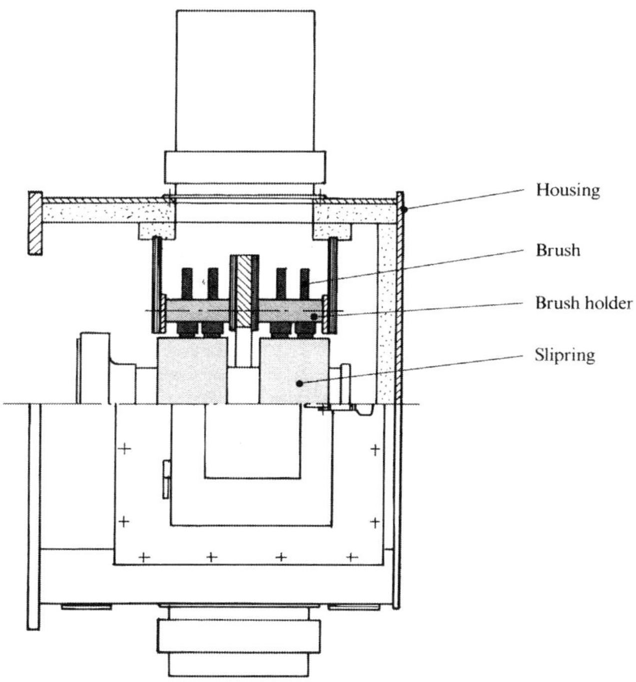

FIGURE 8–84 Slipring assembly detail. *(Source: Alstom Power.)*

The sliprings are manufactured of steel and have generously dimensioned contact surfaces for the carbon brushes. The spiral-machined contact surface is carefully ground and polished. This prevents current concentrations and reduces brush and slipring wear. The sliprings are shrunk on the shaft extension on a cylinder of insulating material. The radial connections from the sliprings to the terminal conductors consist of insulated contact screws through holes in the shaft extension.

The terminal conductors in the slipring shaft and the rotor shaft are connected with contact screws.

Brush Gear

The brush gear consists of a frame, insulation, brackets with brush holder, and carbon brushes.

The brackets are insulated from the frame with rings of insulation material and are connected to a supply ring to which cables for the excitation current are connected. Each bracket is provided with a brush holder pocket for connection of a handle to hold two carbon brushes.

The carbon brushes in the handle parts are mounted in holders of coil-spring type that give a constant brush pressure during the service life of the brush. The handle parts are insulated and the brush holders can be removed from the brush holder pockets by hand when the brushes are to be replaced. Brush replacement is thus possible during operations.

Slipring Housing

Openings are formed in the side walls of the housing for service. These are covered with hatches provided with air filters. The opening in the end wall of the housing, towards the generator, through which the slipring shaft passes, is provided with a seal.

Inspection and Testing

General basic inspection and testing points performed during the fabrication of the generators are included in a check plan. Each manufacturing operation is subject to extensive checking.

Final Tests

Tests in accordance with Table 8–1 constitute part of the normal testing before delivery. A type test is performed on the first machine in a manufactured series and its result is used as a reference for the subsequent machines of the same type. A more extensive test can be offered separately.

Control and Protection

Temperature Monitoring

A number of platinum wire resistance elements installed in different parts of the machine are used for continuous monitoring of the temperature of the parts. The connection cables of the elements are routed to junction boxes on the outside of the stator housing. The number and location of the elements are shown in the following list:

As a standard, the resistance elements have a resistance of 100 ohms at 0°C.

TABLE 8–1 Normal Testing

Test	Type Test	Routine Test
Overspeed test	X	X
Measurement of winding resistances	X	X
Generator characteristics	X	
Measurement of generator losses (through run-down test)	X	
Bearing vibration measurement	X	
Loading point with cos $\theta = 0$ overexcited	X	
Heat run	X	
Measurement of the voltage curve form under no-load conditions	X	
Measurement of reactances	X	
Voltage test	X	X
Measurement of insulation resistance	X	X
Sound measurement	X	
Measurement of bearing insulation resistance	X	X

(Source: Alstom Power.)

Number	Location
6	In stator winding, between coil sides (2/phase)
2	Cooling air supply
2	Cooling air exhaust
1	Cooling air exhaust from exciter

Bearing Vibration Measurement

Vibration transducers of seismic type for bearing vibration measurement can be delivered mounted on the bearing shields of the generator.

Heating Elements

Heating elements are installed in both the generator and exciter unit to prevent condensation during standstill of the generator at lower temperatures.

Current Transformers

Current transformers can be mounted on the stator terminals outside the generator casing. The transformers can be delivered in accordance with the purchaser's requirements.

Protective Equipment

The original equipment manufacturer recommends that the generator, as a minimum, be equipped with the following protective equipment:

- Overcurrent protection
- Overvoltage protection
- Differential protection
- Negative phase-sequence current protection
- Stator earth fault protection
- Rotor earth fault protection
- Underexcitation protection and/or underexcitation limiter
- Reverse power protection (depending on the drive machine type)
- Overexcitation and/or overexcitation limiter
- Loss-of-excitation protection; in installations where there is a risk of high overvoltages, a surge diverter is to be installed and, in certain cases, protective capacitors

Operating Characteristics

Operations with Constant Winding Temperature

With gas turbine operations, the principle described as follows is applied. This provides an optimum relation between the permitted power output of the generator and the power available from the turbine at varying cooling medium temperatures.

In accordance with international standards, particularly for gas turbine-powered generators, the generator can be loaded so that the maximum winding temperature permitted remains the same with a cooling air supply temperature other than 40°C. The winding temperature rises permit increase or decrease as much as the temperature of the cooling medium falls below, or exceeds, respectively, the values given previously.

Synchronous Compensator Operation

The generators are particularly suitable for synchronous compensator operation. To permit such operation, however, mechanical disconnection of turbine and generator is usually required and one of the main bearings must be provided with thrust bearings.

Operation at Low Ambient Temperatures

With very low temperatures the generator can be provided with a recirculation arrangement for cooling air or water.

Noise Reduction

When there are special acoustic requirements, the generator can be installed in a sound-absorbing enclosure consisting of a steel frame with panels of perforated steel sheets with sound-absorbing mineral wool in-fill.

The sealing against water leakage between the panels and the supporting structure consists of a self-adhesive rubber strip and silicon-rubber caulking.

The roof and walls of the enclosure are provided with service openings.

Base Frame

The stator frame of the generator is self-supporting and therefore requires no base frame to provide stiffness. If the center height is required to be higher than standard, the generator can be provided with a separate, welded, steel base frame.

Online Cleaning Systems*

Methods of Cleaning Fouled Gas Turbine Compressors

Offline Cleaning Methods

1. Crank soak chemical washing using in-built chemical injection water rinsing systems.
2. Crank soak chemical cleaning using handheld hose or lance.
3. Partial hand cleaning (e.g., struts, IGVs, first stage rotor and stator blading) using chemicals, rags, brushes, and water rinse.
4. Full hand cleaning with compressor covers removed using chemicals, various types of abrasives, or even light shotblasting techniques.
5. Steam cleaning.

Online Cleaning Methods

1. Injection of abrasives (e.g., crushed nutshells) into the compressor air stream to displace blade deposits by high velocity impingement.
2. Injection of plain water to remove water-soluble deposits.
3. Injection of special chemical solutions (solvent and aqueous based) to chemically dissolve and remove surface deposits from the blades.

However, with the popular resurgence of the gas turbine as an industrial prime mover over the past decades, serious interest in the problem of lost performance and increased fuel consumption caused by fouling has led to the development of so-called "online" or "fired-wash" compressor cleaning systems. The objective of these systems being to chemically clean the compressor while the engine remains in operation at up to full speed and load in order to extend the output for longer and avoid increase in heat rate and subsequent increases in fuel consumption.

In reality, the number of companies and individuals that have been seriously involved in the development of fired-wash systems over the years are few and far between. However, since the process has, of late, gained the official blessing of some major gas turbine manufacturers, there has been a sudden proliferation of system suppliers and even more running-wash chemical suppliers who, in many cases, may have scant knowledge or experience of the fired-wash process and gas turbines to which it is being applied.

If done properly, fired-washing can be a very safe and successful method of keeping gas turbines running more efficiently. The process is being constantly improved and perfected and is, without doubt, here to stay as more and more gas turbine manufacturers offer running-wash systems as a standard fit or recommended option. However, it can also be a dangerous process if injection systems or chemicals are incorrectly designed, fitted, or used and operators should be cautious when selecting any online cleaning system.

Questions to Ask When Selecting an Online Cleaning System

About the Injection System

- How long has the vendor been in business?
- Is there an installation reference list?
- Is the system known to the engine manufacturers? Do they approve it or have no objection to its installation and use?
- Is the system a recommended option or installed as standard in new gas turbines by any manufacturers/ packagers?
- Does the vendor design, manufacture, install, service and guarantee the system himself? (If not, why not?)
- Are the materials of the system of good quality?
- How long does the vendor say it should take to install the system? Some can take a few hours to install, others can take weeks to install if it involves drilling thick casting, etc.?
- Is the design safe? Could it possibly damage the engine or injure those using it?
- Does the vendor have sufficient liability insurance?

Since online washing has rather suddenly come into vogue even though deep routed knowledge and experience of the process is known to relatively few system suppliers and operators it is hoped that this section will be of sound

* Source: Courtesy of Peter McDermott, Rochem, UK.

practical help to those operators who would like to adopt online chemical washing procedures but who are unfamiliar with the concept.

Questions to Ask When Selecting an Online Cleaning System about the Cleaning Chemical(s)

- Does the vendor also manufacture special chemicals for use with the system and are they tried, tested, and approved?
- Was the chemical on offer solely developed for online cleaning or was it originally developed for some other application not connected with gas turbines?
- Does the vendor offer a choice of chemicals (i.e., solvent-based and water-based) to suit particular fouling and/or environmental requirements?
- Is the chemical supplied as a concentrate to save storage and transportation costs? Paying for water in ready-to-use chemical solutions can be very expensive and unnecessary.
- Can the chemical offered also be used safely and effectively for offline compressor washing if need be?
- Does the vendor offer ex-warehouse availability of chemicals?
- If the vendor only supplies chemicals, are you sure it is safe to use them in your injection system?

Recommendation: Be very wary of using any chemical especially for online fired-washing, unless it has been properly tested and approved and has a good long-term safety record behind it.

Can Online Washing Really Be Cost Effective and Is It Really Necessary—or Just Another Fad?

Due to the current Middle East crisis the cost of fuel has increased by roughly 70% over the past 3 months. For example, No. 2 diesel was, as of October 1, 1990, being quoted on the Rotterdam market at \$300/ton whereas 3 months prior it was priced in the region of \$180/ton.

Even if the fuel drops back to pre-crisis levels, in due time the fuel costs for operating any gas turbine are still substantial even when the machine is kept in perfect operating condition and at peak efficiency. A fouled compressor can easily increase fuel consumption by 5% or more and, in real terms, a 5% increase for the operator of a typical 25 MW heavy industrial unit running base load for say 8000 hours per year would add over \$0.5 million/year to the fuel bill at pre-Gulf crisis fuel prices and close to \$1 million extra at current prices.

Advantages and Disadvantages of Offline Compressor Washing

Advantages

- When carried out correctly can effectively clean compressor and restore majority of lost performance.

Disadvantages

- Gas turbine must be shut down completely
- Time-consuming process
- Labor-intensive process
- Costly problems in disposing of waste chemical and rinse water
- Lost power output cannot be recovered
- Increased fuel consumption cannot be recovered
- Shutdown/startup thermal cycles for offline wash are damaging
- Extra wear and tear on starting system during offline wash
- Can wash salt and corrosives into inaccessible parts of the engine
- Only a short-time cure and not a prevention for compressor fouling

This should be more than sufficient incentive for any turbine operator to at least investigate the potential benefits of online washing to control compressor fouling when the machine is running because the alternative of offline crank-soak washing can only be a temporary cure and not a prevention of the fouling and performance loss that only takes place when the machine is running. Lost performance is precisely that—lost!

Increased fuel consumption is, of course, only part of the story of compressor fouling, and loss of power output can be an even greater cost penalty if available power is directly related to production; such as in the case of an offshore oil production facility. In that instance loss of power can equate directly with loss of production and revenue. Similarly, co-generation facilities that sell excess energy to guarantee profitable survival can find themselves with major problems if projected amounts of saleable excess power are reduced due to performance loss and/or total shutdown to perform offline washing to restore lost performance.

The actual cleaning efficiency of the system is wholly dependent on both the method of injection and the cleaning solution but at the least one should expect to reduce by 50% the degradation that would otherwise be caused by compressor fouling.

If that were extrapolated out in the case of our typical 25 MW/8000 hour/year base load machine the additional extra fuel bill for every 1% increase in fuel burn would be about \$110,000 (or \$180,000 at present crisis price level). If the online washing process prevented even a minimal 50% of this extra fuel burn (a $55,000 or $90,000 saving) then the annual cost of cleaning chemical ($8000 to $16,000) would be well covered in addition to the capital cost of the system also being recovered from savings within the first 6 months.

Why Not Try to Avoid Fouling Altogether Instead of Spending Time and Money on Compressor Washing Techniques?

The simple answer—at least at this time—is that it is virtually impossible even with the finest multi-stage, high efficiency filtration systems to completely avoid compressor fouling.

There are many sources of compressor fouling but the worst and the most common is the ingestion of oily vapors that can readily pass through filtration systems.

Typical Causes of Gas Turbine Compressor Fouling

- Passage of oily vapors through filtration system
- Ingestion of oily vapors through breaches in air inlet plenum
- Leakage of oil from oil-bath type filtration systems
- Passage of very fine particulate matter through breaches in plenum
- Reingestion of exhaust gases through filtration system and breaches in air inlet plenum casing
- Ingestion of saturated salt droplets or salt crystals through filters and/or breaches in plenum casings
- Ingestion of seasonal tree and plant gums
- Ingestion of wide variety of chemicals and other pollutants generated at the site of gas turbine operation

Typically these oily vapors are the catalyst to compressor fouling in most cases. After restarting the gas turbine with a clean, dry compressor, a film of oily/greasy deposit can build up rapidly on the compressor airfoils (roughly in the first third of the compressor) and this forms a perfect "flytrap" to catch and absorb dry particulate matter that would otherwise have passed harmlessly through the compressor.

In more than 10 years of designing and developing compressor cleaning systems the writer has observed that classic fouling—particularly in polluted industrial environments—is basically a three-phase affair.

Firstly, there is the laying down of the oily film and the rapid absorption and entrapment of dry particulates. This can occur within days or even hours after restarting a cleaned engine and there can be a substantial reduction in output in a relatively short period of time.

Next, there is a slowing down of performance loss as the surface deposit gradually dries and the rate of fouling decreases.

Finally, there is a relatively long phase of slow decline over a period of weeks or even months followed by a further sharp decline as the aerodynamic tolerance of the compressor decreases and/or the surface deposit once more becomes sticky with more oily residues.

A fingernail scratch into the apparent dry, carbonaceous surface deposit on the bell mouth struts of many a compressor is almost sure to reveal the initial oily substrate that was laid down shortly after the cleaned engine was put back into service.

In general, one could say with confidence that compressor fouling would be much diminished if all traces of oil and grease could be prevented from entering the compressor and in this regard close attention to the complete integrity of the air inlet plenum after the filters is strongly recommended since even small breaches can allow inordinately large quantities of unfiltered vapors and particulates to pass freely into the compressor.

Because it is almost impossible to prevent some degree of compressor fouling the logical approach is to try to find a way of controlling and reducing the rate of fouling and ultimately to control it to a point that it does not affect performance at all.

This means regular compressor cleaning on a sufficiently frequent basis to prevent any meaningful amounts of deposit from building up between each wash and that could mean once a day or once per week depending on operating circumstances.

Advantages and Disadvantages of Online Chemical Cleaning of Compressor

Disadvantages

- Can have limited effect if chemical is not properly formulated
- Can have limited effect if chemical is not injected correctly
- Some online cleaning chemicals are quite expensive and results may not justify the costs in some cases
- Some online cleaning systems are very expensive and are priced against what the customer may save in operation costs rather than what it actually costs to develop and manufacture the equipment

Advantages

- No engine shutdown required
- Can arrest or slow down the rate of compressor fouling while the gas turbine remains in operation thus

maintaining heat rate and saving large amounts of money by avoiding additional fuel burn
- Helps maintain power output to avoid loss of production revenue and unnecessary use of standby machines
- There is no interruption to production or operation routines as washing can be carried out at full speed and load
- The cleaning process usually takes only a matter of minutes
- There is no waste chemical or water to deal with after the wash
- Assuming the chemical is properly formulated and injected it should not contribute to any additional atmospheric pollution
- By avoiding shutdowns for offline washing, thermal cycles are also considerably reduced thus extending life of the engine.
- No erosion, thermal shock, bearing damage, or cooling system blockage (if properly designed system is used)

It would be highly impractical and very costly to do this on a shutdown, crank-wash basis so the only practical solution is online or fired washing since this can be readily done without any interference to the normal operation of the engine.

Other Online Compressor Cleaning Methods

The question is, is online chemical washing the only online cleaning method?

The answer is no but of the other available choices it may be the most efficient, safe and practical online procedure.

The first alternative—and probably as old as the gas turbine itself—is abrasive cleaning using a variety of crushed nutshells or carbon-based pellets but in general this method is impractical for most modern gas turbines because, among other problems and drawbacks, the abrasive medium can damage coatings and plug turbine cooling holes with possible catastrophic results.

Advantages and Disadvantages of Abrasive Compressor Cleaning

Possible Advantages

- Can be used when gas turbine is operating
- Simple to use
- Relatively inexpensive
- Can remove certain deposits and restore performance

Possible Disadvantages

- Blade erosion
- Bearing damage
- Blockage of blade cooling systems leading to turbine failure
- Damage to blade coatings
- Blockage of instruments and/or control bleed air posts
- Jamming of variable guide vanes and/or stator vanes
- Removal of oily/greasy material to rear of compressor and/or into combustion system causing further performance loss
- Transfer of salt contamination from compressor to hot end and subsequent corrosion
- Ineffective against lacquered or hard carbonaceous deposits

The second alternative to online chemical cleaning is online plain water washing, which has been advocated by some as a cheap solution with the apparent simplistic argument that water must be inexpensive and chemical must be expensive.

This would, of course, be the perfect solution if all compressor fouling was water-soluble and the water used was virtually free of all dissolved and suspended solids so as to avoid corroding the engine in the process of trying to clean it.

Unfortunately, compressor fouling is rarely, if ever, of a water-soluble nature and good quality pure water is not always easy to come by. Many a gas turbine has been badly corroded by poor attention to water quality particularly when applied to a hot engine.

Water washing should not be confused with the use of so-called water-based or aqueous chemicals. Generally, water-based chemicals are formulated from various types of surfactants (surface acting agents) and corrosion inhibitors with water being in the majority (between 80 and 98% depending on the brand) carrying agent.

Some of the new formula water-based materials (some using very pure food-grade surfactants) are quite effective at breaking down stubborn oily/greasy deposits and are proving quite successful even for fired-washing but are particularly attractive for offline washing because of their high biodegradability factor, which makes the disposal of the waste chemical a lot easier.

Advantages and Disadvantages of Cleaning Compressors with Plain Water

Advantages

- Can be used when gas turbine is operating (if injected correctly)
- Simple to use
- Relatively inexpensive
- Can have some cleaning effects if fouling is purely water soluble

Disadvantages

- Not effective against oily/greasy contamination
- Not effective against lacquered or hard carbonaceous deposits

- May move deposits to hot section of engine to cause corrosion
- May directly cause corrosion if not inhibited
- May cause thermal shock/blade stress if not injected correctly
- Not effective for offline cleaning (unless deposits are water soluble)
- Must be high quality water with very low TDS and suspended solids

Expansion Joints*

An expansion joint is a device used to allow movement in a piping system while containing pressure and the medium running through it.

Frequently, thermal growth, equipment movement, vibration, or pressure pulsation can cause movement in a piping system. When flexibility for this movement cannot be designed into the piping system itself, an expansion joint is the ideal solution.

There are four basic movements that can be applied to a bellows. These are axial, lateral, angular, and torsional as illustrated below. Bellows behave like springs in a piping system. When they are compressed, they resist the movement the same as a spring would. The spring rate of a bellows is entirely dependent on bellows geometry and material properties.

- *Axial* movement is the change in dimensional length of the bellows from its free length in a direction parallel to its longitudinal axis.

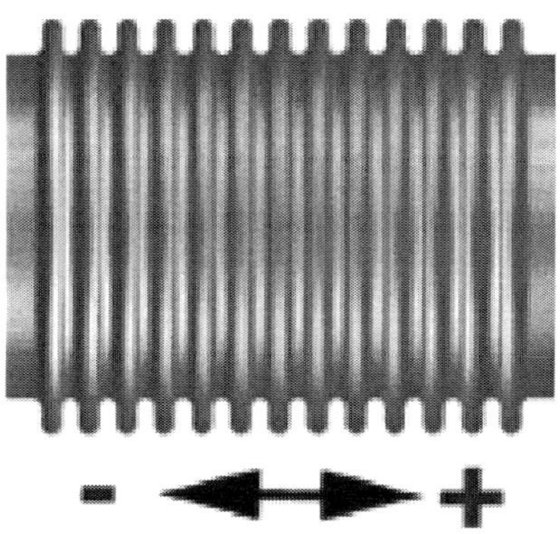

AXIAL MOVEMENT

- *Angular* movement is the rotational displacement of the longitudinal axis of the bellows toward a point of rotation.

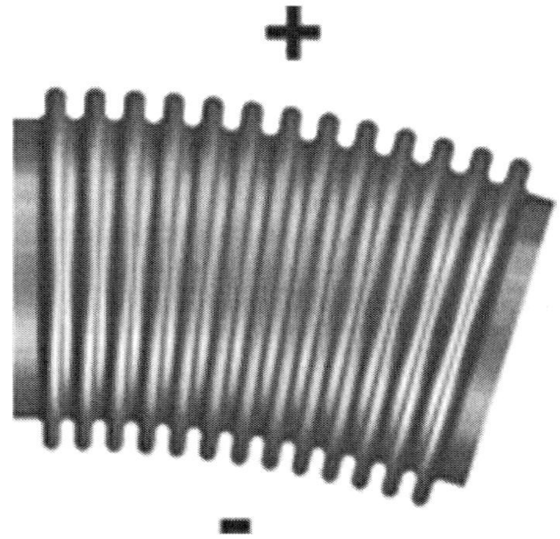

ANGULAR MOVEMENT

- *Lateral* movement is the relative displacement of one end of the bellows to the other end in a direction perpendicular to its longitudinal axis (shear).

LATERAL MOVEMENT

- *Torsional* movement is the rotation about the axis through the center of a bellows (twisting). *EJS discourages any torsional rotation of metal bellows expansion joints.* Torsion destabilizes an expansion joint, reducing its ability to contain pressure and absorb movement.

TORSIONAL MOVEMENT

* Source: EJS Expansion Joints USA. Also see a much larger section on expansion joints in C. Soares, *Process Engineers Equipment Handbook* (New York: McGraw-Hill, 2001).

End Connections

Flanges include special flanges, slip-on, or angle flanges.

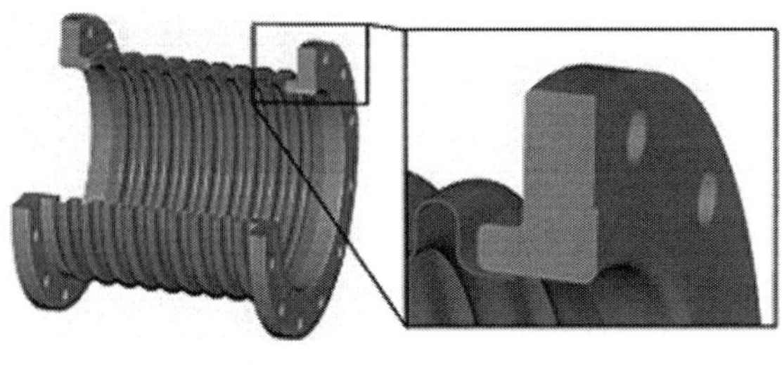

- *Vanstone* ends are modified flanged ends with the added flexibility of resolving bolt-hole misalignment.

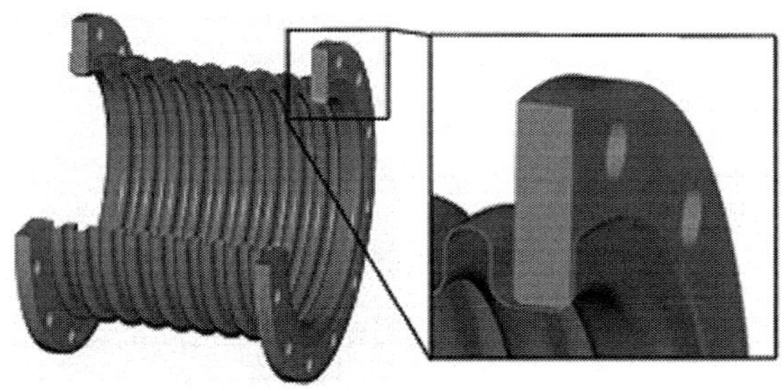

- *Weld* ends allow any pipe or duct to be attached to a bellows for welding into a system.

Types of Expansion Joints

Unrestrained Assemblies

- *Single expansion joint assemblies* are the simplest type of expansion joint consisting of a single bellows element welded to end fittings, either flange or pipe ends.

SINGLE EXPANSION JOINT

- *Universal expansion joint assemblies* consist of two bellows connected by a center spool piece with flange or pipe ends. The universal arrangement allows greater axial, lateral, and angular movements than a single bellows assembly.

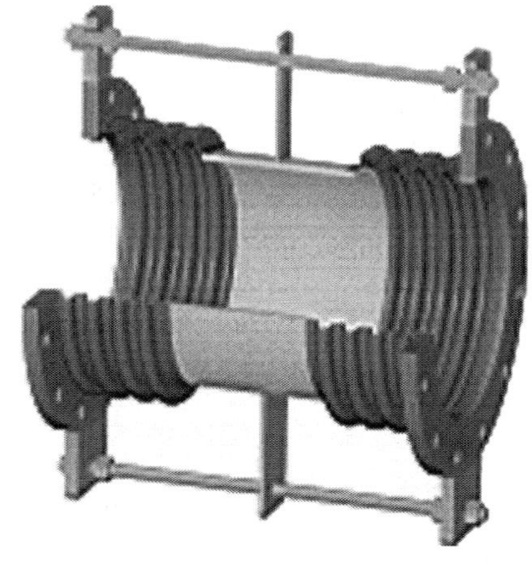

UNIVERSAL EXPANSION JOINT ASSEMBLIES

- In *externally pressurized expansion joints*, line pressure acts externally on the bellows by means of a pressure chamber. This allows a greater number of convolutions to be used for large axial movements, without fear of bellows instability. Externally pressurized expansion joints have the added benefit of self-draining convolutions if standing media are a concern. Anchors and guides are an essential part of a good installation.

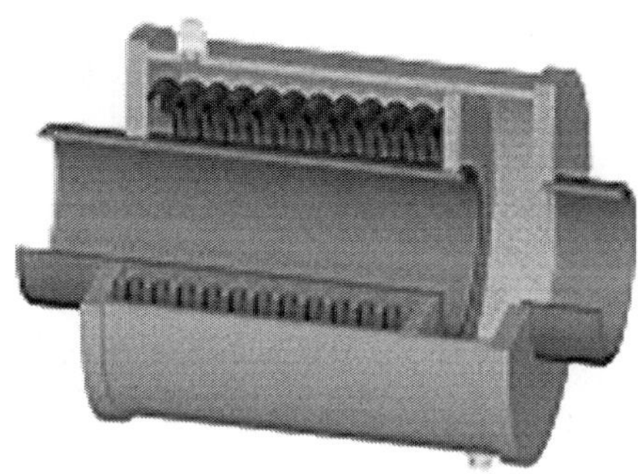

Externally Pressurized Expansion Joints

Restrained Assemblies

- *Tied single bellows assemblies* add tied rods to a single bellows assembly to increase design flexibility in a piping system. The tie rods are attached to the pipe or flange with lugs that carry the pressure thrust of the system, eliminating the need for main anchors.

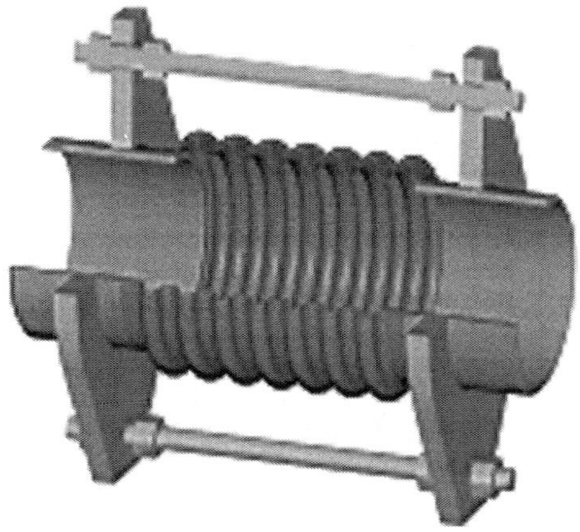

TIED SINGLE BELLOWS ASSEMBLIES

- *Tied universal assemblies* are similar in construction to a universal assembly except that tie rods absorb pressure thrust and limit movements to lateral offset and angulation only.

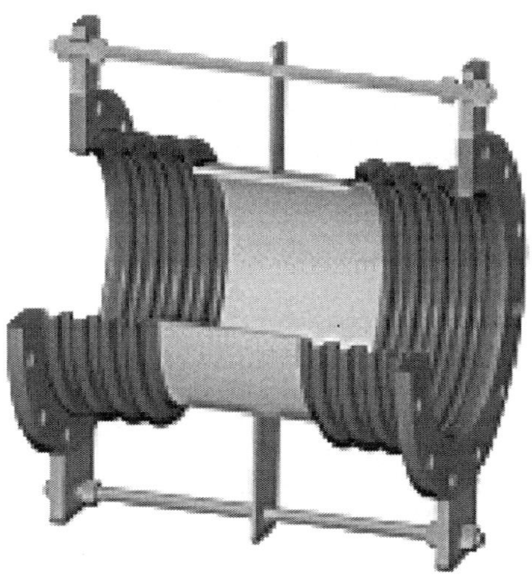

TIED UNIVERSAL ASSEMBLIES

- *Hinged bellows assemblies* limit movement to angulation in one plane. Hinged assemblies are normally used in sets of two or three to absorb large amounts of expansion in high pressure piping systems.

HINGED BELLOWS ASSEMBLIES

- *Gimbal bellows assemblies* are designed to absorb system pressure thrust and torsional twist while allowing angulation in any plane. Gimbal assemblies, when used in pairs or with a single hinged unit, have the advantage of absorbing movements in multi-planer piping systems.

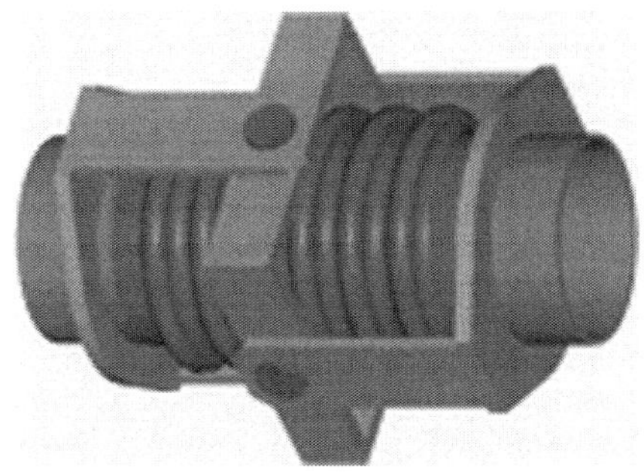

GIMBAL BELLOWS ASSEMBLIES

Accessories for Special Service

- *Tie rods* are devices, usually in the form of bars or rods, attached to the expansion joint assembly and are designed to absorb pressure loads and other extraneous forces like dead weight.

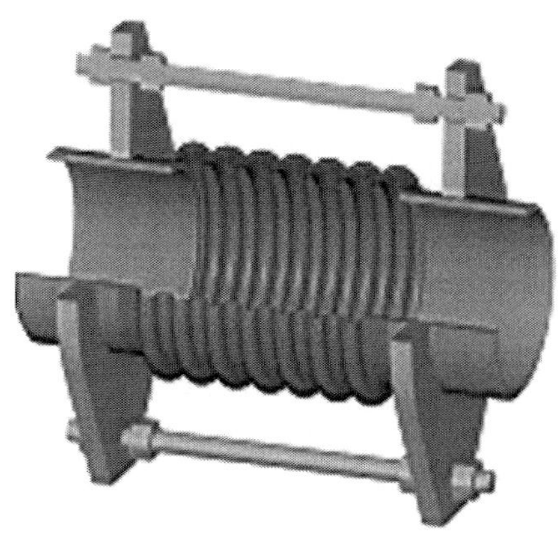

TIE RODS

- *Limit rods* are used to protect the bellows from movements in excess of design that occasionally occurs due to plant malfunction or the failure of an anchor. *Limit rods do not contain the pressure thrust during normal operation.*

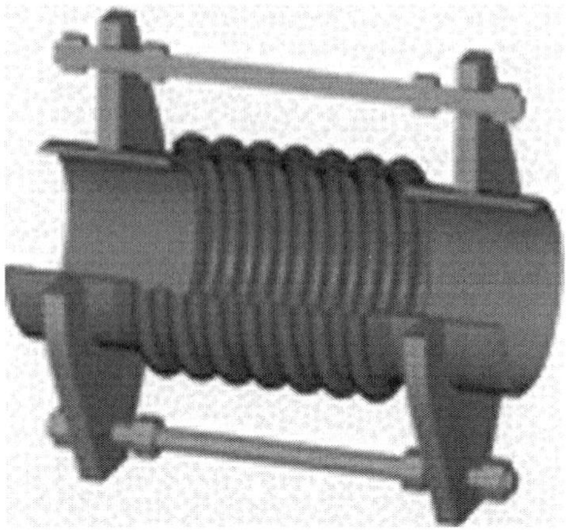

LIMIT RODS

- *Purge connections* are used in conjunction with internal liners to lower the skin temperature of the bellows in high temperature applications such as catalytic cracker bellows.

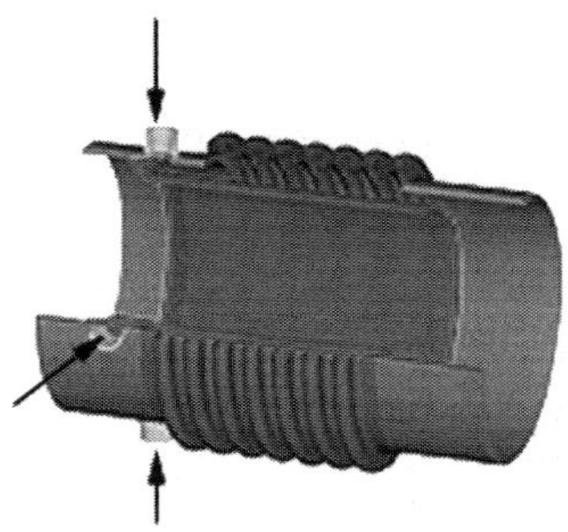

PURGE CONNECTIONS

- *Liners (Internal Sleeves)* are used to protect the bellows internally.

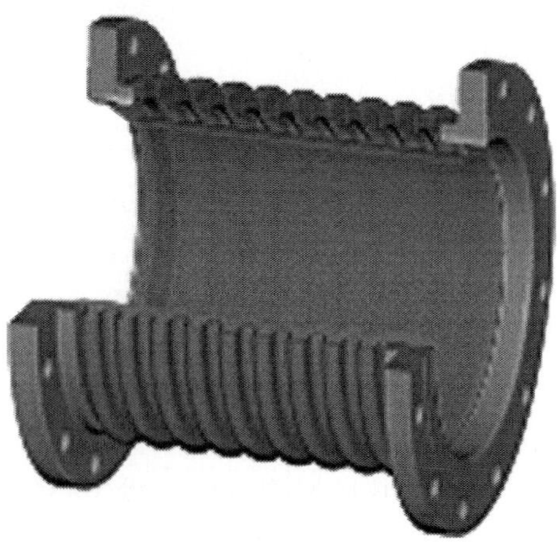

LINERS

- *Covers (Shrouds)* are used to protect the bellows externally.

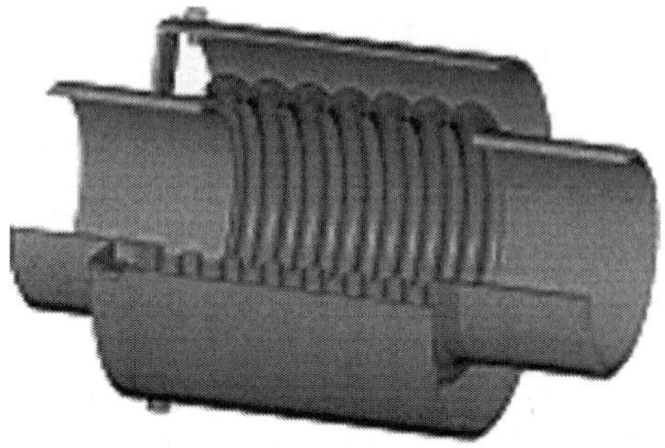

COVERS

Bellows Manufacture

The bellows element is the most important component of an expansion joint. This thin-walled, corrugated membrane allows flexibility in a piping system while containing the pressure and media. When complete, each bellows has a unique working pressure, spring rate, and cycle life that are entirely dependent on its geometry and material.

Bellows are manufactured using an expanding mandrel (punch forming) method followed by a finish rolling. A rectangular sheet is sheared and rolled into a tube. The tube is welded using an auto flat-bed welder with no filler metal added. The longitudinal seam weld is then "plannished" back down to the parent material thickness. Any dye penetrant, X-ray, or air testing is performed at this time. When testing is complete, the convolutions are punched individually drawing material from the top and bottom of the tube. This drawing process eliminates any thinning in the bellows material.

At this point, the convolutions have more of a "V" shape than the final "U" shape that is required. This is easily solved with the final re-rolling of the bellows between a series of aluminum bronze rolls. After trimming the bellows attachment ends (skirts), the bellows is complete and ready to have attachment ends installed.

Shipping

Every expansion joint that leaves the factory is provided with installation instructions. These instructions describe the simple, straightforward requirements that must be followed to ensure a trouble-free installation.

Shipping Bars

These are temporary attachments that "hold" the expansion joint at its correct installed length during shipping and installation. Angle iron or channel section is used and is always painted bright yellow. Shipping bars must never be removed until after the unit has been correctly welded or bolted into the piping system. Caution: tie rods or limit rods are sometimes mistaken for shipping bars.

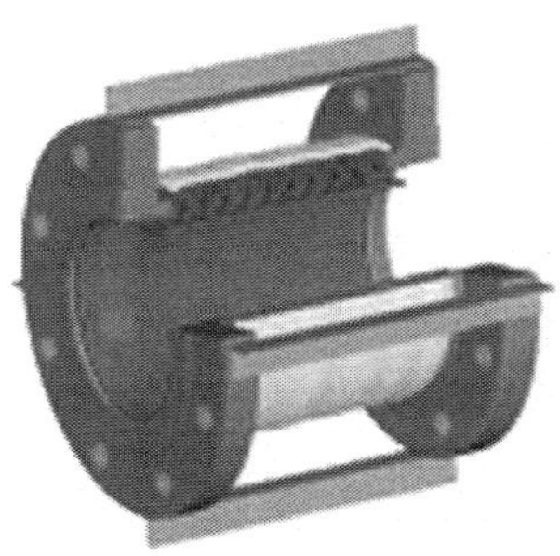

Never Tamper with These Attachments

Note: Great care must be taken when removing the shipping bars. If a welding or burning torch is used, *always* protect the bellows element from burn splatter with a flame-retardant cloth or other shielding material.

Installation Guidelines

Unrestrained Expansion Joints

Unrestrained expansion joints are not provided with attachments such as tie rods or hinges to restrain pressure thrust. Therefore, they can be used only in a piping system that incorporates correctly designed anchors and pipe alignment guides. These components prevent the bellows from overextension and damage due to distortion under operating conditions.

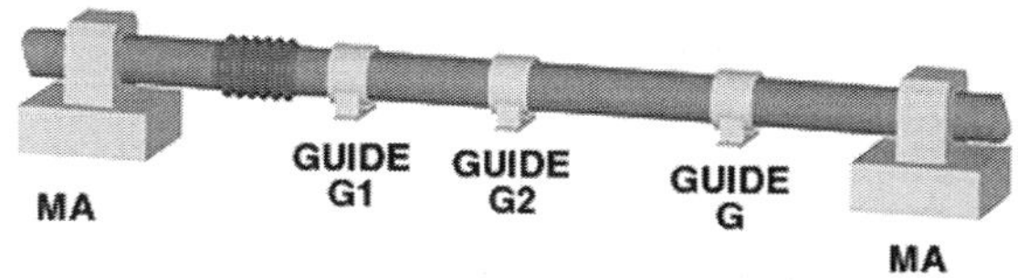

Restrained Expansion Joints

Tied expansion joints can be of the single or universal type provided with restraints such as tie rods, hinges or gimbals. Tie rods and gimbals allow the expansion joint to move in all planes. Hinges allow movement in a single plane only. These restraints are designed to absorb the pressure thrust and other external loads like pipe dead weight. For restraints to remain effective, the expansion joint can absorb only lateral offset or angulation in directional changes in the piping system, such as "Z" bends, "U" bends, or "S" bends. Tied units are used where the equipment or adjacent structures cannot accommodate pressure thrust.

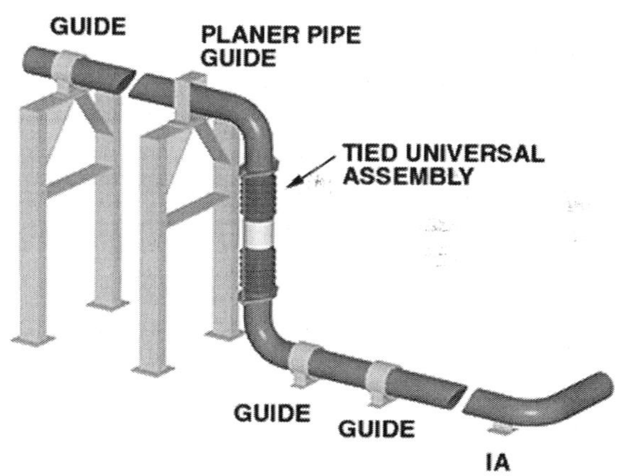

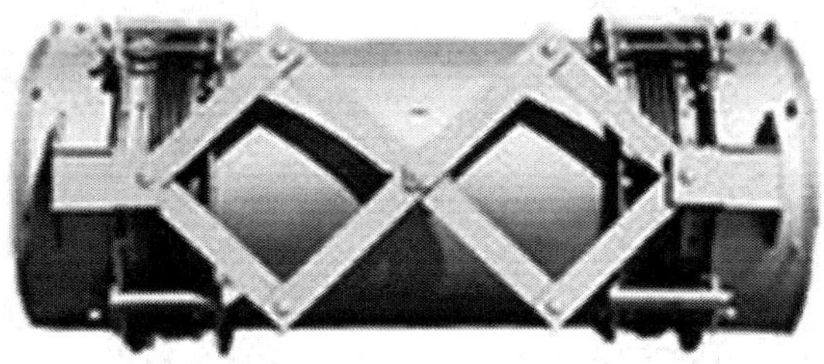

Specific Applications

Special product lines include a variety of rectangular joints, PenSeals (boiler seals), silencer bellows seals, expansion joints for fluid catalytic cracking units (FCCU), gas turbine exhaust (GTX) systems, aerospace, heat exchangers, and many other heavy industrial applications.

Expansion joints are made in both metal and fabric designs from 3′ nominal diameter to 200′ with process temperatures from −325−2500°F. Pressures for these designs range from full vacuum to 2000 psig. Single-ply, multi-ply, root ring, equalizer ring, and toroid bellows designs are also available.

Chapter 9

Controls, Instrumentation, and Diagnostics (CID)*

"Enjoy the little things in life, for one day you may look back and realize they were big things."

—Antonio Smith

Chapter Outline

* Sources (except where specified):
Notes on the proceedings of the panel session for ECMS as they relate to gas turbine component life extension, 1985 through 2003, C. Soares, chair and organizer.
Personal operations case studies/notes: turbomachinery in nonconventional energy (tarsands), conventional oil and gas, aeroengine fleet management, aeroengine repair and overhaul, Soares C.
Short course notes, C. Soares, "Vibration Analysis Theory and Applications," 2000.
Short course notes, C. Soares, "Engine Condition Monitoring Systems," 1998.

SYSTEM SCOPE AND SELECTION FOR GAS TURBINES

Controls and instrumentation on contemporary gas turbine engines serve several purposes. They include but are not limited to:

1. Providing the operator with indicators in terms of pressure, temperature, vibration, strain, and other parameters that give the operator a snapshot of how the gas turbine and gas turbine system is performing.
2. Making life easier for the operator in terms of providing a sometimes automated means to control a system with less manual labor and, in some cases, less "guessing." However, with complex systems, the easier the system is to operate, the higher the level of operator knowledge required.
3. Providing owners and operators with a means of controlling a turbine system remotely.
4. Coupled with PLCs, Pentiums, and other software, they provide an operator with an ability to diagnose the problem on a system.

With the advent of complex software architecture and other technology, such as fuzzy logic, potential diagnostics on a system have reached a new level. "Artificial intelligence" is now possible, allowing software to "select" a "solution" from a matrix of parameters, potential problems, and solutions. In most cases, artificial intelligence works better on highly monitored test systems, such as some of the cutting-edge research being done by the US Air Force laboratories. The "intelligence" can work with only the data it is given, however. "Real-world" systems are prone to mishaps, such as operators knocking into things (like sensitive proximity vibration probes), systems undergoing sudden surges, or excursions in temperature.

Most software monitors today can retain a "snapshot" of the condition at the time of its occurrence and over a predesignated time interval before and after it occurs. However, the actual prevention of such occurrences depends on overall system sophistication to a point. For example, if a gas turbine system is running with a crude fuel of variable quality or is designed for only gaseous fuel (and prone to slugs of liquid in the gaseous fuel), or tends to surge when it takes in a "mouthful" of sooty exhaust (when close to the discharge of a wing-mounted weapon system), that system will always be prone to spikes in operating parameters.

These spikes can often confuse an intelligent system if they were not part of the original programming. In other words, the old paradigm still holds: "Garbage in, garbage out" with respect to all software programs. Also there is not much use in providing a system with relatively crude controls (or instrumentation) with a high degree of sophistication in diagnostic potential. The mismatch will result in errors, few results of accurate significance, and worst of all, a false sense of security.

WHICH PARAMETERS ON WHAT APPLICATIONS*

Observations made on CMS or ECMS (engine condition monitoring systems) proposals and purchases in large plants include a case where a highly sophisticated system was being recommended as part of the retrofitting for a thermal power plant. The thermal plant had a large number of turbines, conservative in terms of peak operating temperatures. The turbines were steam turbines, but the following illustration might apply equally to a mature gas turbine fleet. The fleet had been operating relatively trouble free for about 20 years, as had the associated boiler feed pumps. The original OEM-supplied CMS would not have managed the parameter accuracies a newer turbomachinery system might require, but it was adequate for the limited temperature ranges and vibration those conservatively designed steam turbines and feed pumps would ever see.

One recommended system for power generation units would have been appropriate for the test cells of the latest in 90,000 pound plus thrust development aeroengines. Its specification described an expert system that also included instruments for remeasurement of basic parameters already measured on the existing system. For the turbine generator(s):

- Displacement probes vibration monitoring (VM)
- Velocity probe VM
- Eccentricity monitoring
- Dual thrust position monitoring
- Dual case expansion monitoring
- Differential expansion monitoring
- Dual valve position monitoring
- Rotor speed indicator (all of these with panel indicators)
- Phase angle transducer

For the boiler feed water pumps:

- Displacement probes VM
- Accelerometer monitoring for two positions on the hydraulic coupling
- Speed monitor
- Dual thrust position monitor
- Phase angle transducer

* Source: Courtesy McGraw-Hill, *Process Engineering Equipment Handbook* (Soares, C. New York: McGraw-Hill, 2001), adapted with permission.

Such a system might measure basic parameters to greater accuracy than the existing system, but greater accuracy for these parameters in this application was not necessary. The "expert system" then went into specific, in some instances overconservative, specifications with respect to:

- Operating temperature ranges of transducers
- Temperature sensitivity of probe and cable
- Differential pressure withstanding (of gear oil) potential of transducers
- Double braiding of shields on cables
- Transducer protection
- Electrical isolators
- Environmental specifications
- Cable pull strengths
- Probe mounting potential
- Transducer frequency response range
- Relay configuration and contact ratings
- Visual displays
- Calibration reprogramming or disabling of any monitor
- Online continuous monitoring
- Expert system diagnostics and a whole host of other factors that could effectively eliminate vendors that otherwise would be competent to supply CMS for the turbines and pumps in question

What might have been more to the point for turbines already advanced in total operating hours (with excellent availability history) was continuous monitoring for creep fatigue degradation. The need for this expense, too, would need to be assessed in terms of where the turbomachinery components lay on their stress endurance curves for normal operation and abnormal cyclic operations. Existing earlier methods would have used measurements taken from thermocouples and pressure transducers to perform offline calculations and used conservative design codes. The online method system supplied by some vendors has the potential for entering data from inspections at shutdown, modeling future potential temperature excursions, storing data for trending, and visual, real-time displays, some of which can be incorporated into LCA (life cycle assessment) counters.

In summary, the effectiveness of controls and instrumentation in performing their functions (which includes diagnostics) depends on their successful integration into the overall gas turbine system. In Chapter 10 on performance optimization, there are cases where the retrofit CID packages for PA, LCA, VA, or ECMS purposes or as part of power augmentation or metallurgical component life extension projects are far newer than the basic turbine, both in terms of actual service age or technology level. In these cases, the retrofit can succeed, but only if its appropriateness for the entire system is carefully considered during engineering retrofit design.

BASIC CONTROLS AND INSTRUMENTATION (C&I) ON GT SYSTEMS

At the 1984 ASME IGTI annual meeting, I noted that the control display portrayed in a presentation on offshore gas turbine systems closely resembled that of a HUD (head up display) of the gas turbine engines on a fighter aircraft. The annual IGTI session on "ECMS as They Affect Gas Turbine TBOs (time between overhauls) and Component Lives" was thus born.

Attendees at this panel session were, for the most part, operators who are prepared to learn from land, sea, and air users, regardless of where their turbines operate. The technology gap between land, sea, and air applications used to be considerable, with the most sophisticated technology deployed on military aeroengine applications. Large commercial aircraft engines, commuter aircraft engines, land-based aeroderivative engines, and land-based industrial engines followed, more or less in that order. When there was this distinction in terms of level of sophistication, very few gas turbines were used in marine applications. The contemporary stakes being what they are in terms of fuel efficiency, risk insurance, and environmental legislation and taxes, the sophistication boundaries first blurred and now are all but nonexistent for new installations.

The controls and instrumentation basic description that follows is for an aeroengine gas turbine; however, the reader can find system commonality with its features in land or marine applications. The analog dial gauges illustrated in this subsection are quite common on numerous plants that were supplied and commissioned before the digital age. Newer systems have digital gauges and readings. Later in this chapter, C, I, and D features of non-aeroengines also are discussed.

The gas turbine engine has many different applications, each with different requirements, and therefore each with its own control equipment and system architecture. However, the basic principles and functions of a gas turbine control system are essentially the same for all applications.

PRINCIPLES AND FUNCTIONS OF A CONTROL SYSTEM

After initial checks, the control system is required to start the engine, and accelerate it safely to a point where the gas turbine can sustain its speed without starter power and is stable (idle speed). Thereafter, the pilot or operator will require various levels of power output, depending on the operation required. The control system accelerates or decelerates the engine by changing the fuel flow and manipulating compressor variables (and others) to ensure the maneuvers are smooth and surge-free. During deceleration, care must be taken not to reduce fuel flow below the point at which combustion would be extinguished. When the pilot or operator shuts down the engine, the controller sets fuel flow to zero, and the engine decelerates to a stop. In some applications, further tasks are carried out to ensure that maintenance on the engine can be carried out safely and the engine is prepared for the next start. Before, during, and after the operation of the engine, data are transmitted by the control system for display to the operator.

Expressed in these terms, the control system's task is simple, but there are some additional complexities. For example, determining the engine power required by the pilot or operator involves a rating calculation, which, in an aero-engine application, involves flight condition (altitude and Mach number) and takes into account the non-propulsive power being extracted for aircraft services. Thus, for a given nominal power demand from the aircraft during climb, actual power will be varying continually.

The control system also has to perform self-checks; it ensures it is operating without failures and it must not be working with incorrect data—either situation would result in erroneous control decisions or incorrect data being sent to the pilot or operator. The rigor of the design and analysis of the control system reflects the safety, economic, and other consequences of such an error.

Above all, the control system must ensure that the engine is operating safely within its defined limits, even if the engine or control system fails. In some circumstances, the control system has no alternative but to shut down the engine—for instance, if there is a danger of rotor overspeed because the electronics can no longer control the flow of fuel to the engine. There is nothing the electronics can do in these circumstances and the rate of change of fuel flow may be too rapid to expect the operator to intervene. For this reason, all systems have independent means of measuring a limited set of data (typically rotor speeds) and commanding an immediate engine shutdown, or some other safe state, if set limits are exceeded. The control system contains many features designed to provide this safety protection, and the design and testing of these features is a major part of the designer's task.

Control Laws

Each manufacturer has different control strategies, and each engine type has detailed differences in its control laws. However, aerospace applications place certain common requirements on control:

- An engine must be able to accelerate from low power to high power in a fixed time so that an aircraft can abort a landing and achieve max take-off thrust—for example, to avoid a runway obstruction. The control laws may use a closed loop acceleration algorithm, where rate of change of speed is a function of current speed, to ensure that at a given condition acceleration time is always the same.
- As an engine wears during its life, the thrust it provides at a given condition must remain above a certain level if the aircraft is to achieve its take-off performance. Thus, a parameter must be chosen which provides a close measurement of thrust, and any inaccuracy in the measurement compensated by providing additional power. The control system must then control to that parameter very accurately.
- An engine must accelerate from stationary to idle in a reasonable time in order that the aircraft can taxi under its own power. The starting algorithms must accelerate the engine at a rapid rate, avoiding any stall or stagnation regions.
- The pilot must always be able to shut the engine down—the system's hardware must provide a separate mechanism to allow the pilot to override the control system if required.
- Above all, an engine must always be operated within its safe limits. The control system, therefore, must be programmed with data on all the relevant limitations and the action to be taken if such a limitation is approached.

This is necessarily only a small subset of the engine control requirements and consequences on the system (see Figures 9–1 to 9–6).

COMPONENTS OF A CONTROL SYSTEM

The complex functions described above are performed most effectively by digital electronics. All modern engines feature this form of control, and many older engine designs have been modified to include it. However, there are some purely mechanical control systems in service.

Control systems for aerospace (and some marine) applications often use bespoke electronic and mechanical equipment because these applications have limited space for their systems, which must also be low in weight. Energy and other marine applications do not have the same restrictions so their control systems can be implemented using equipment closer to industrial standards.

A typical engine control system has many constituents:

- An electronic controller that computes and commands the control functions; it contains one or more microprocessors and other circuitry, which read data from sensors, and control actuators and valves.

Engine parameter sensors, including pilot power demand and feedback signals from actuators.

A means of metering fuel being delivered to the engine, and of shutting the fuel *off*.

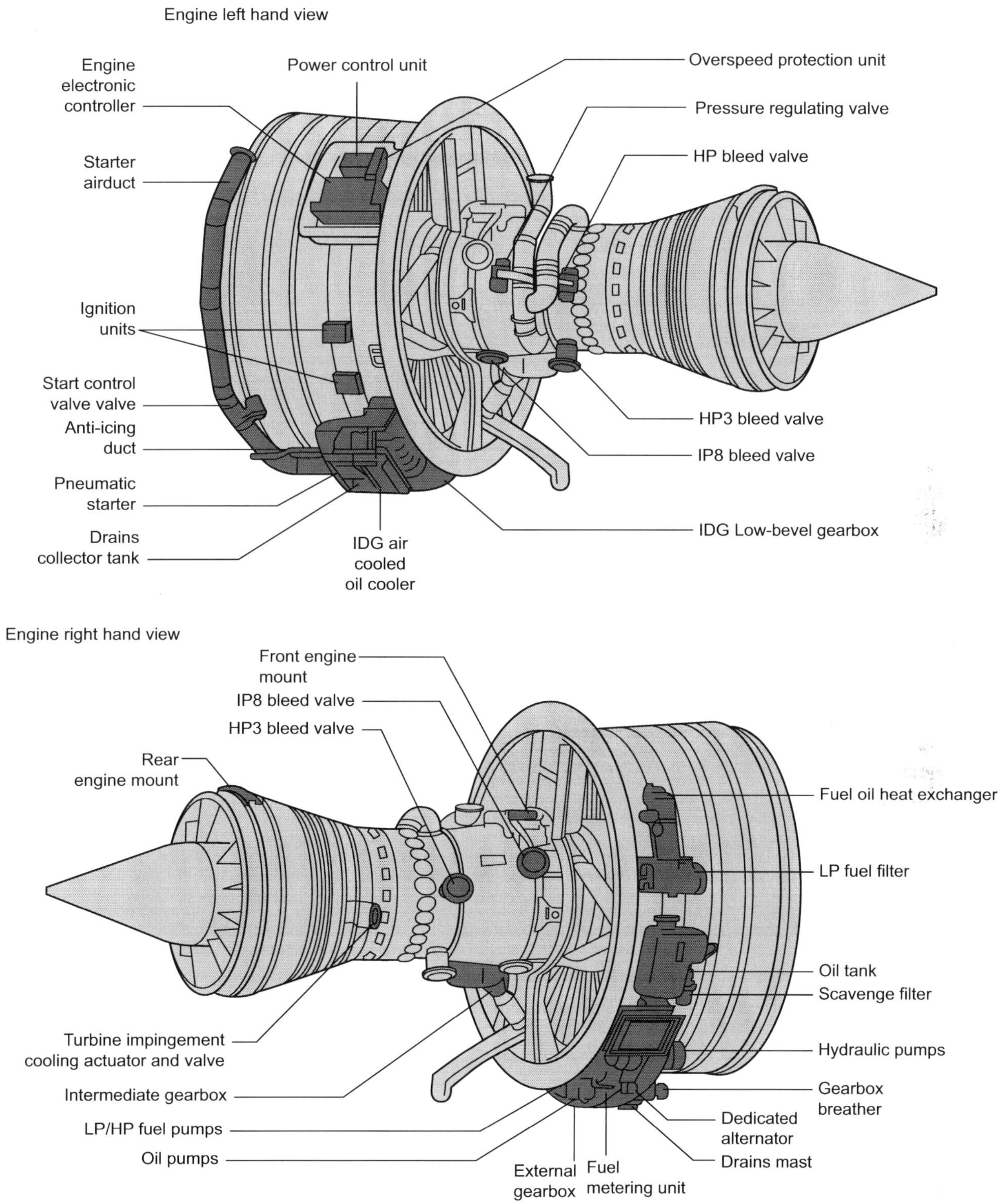

FIGURE 9–1 Typical control system components and other engine ancillaries. *(Source: Rolls Royce.)*

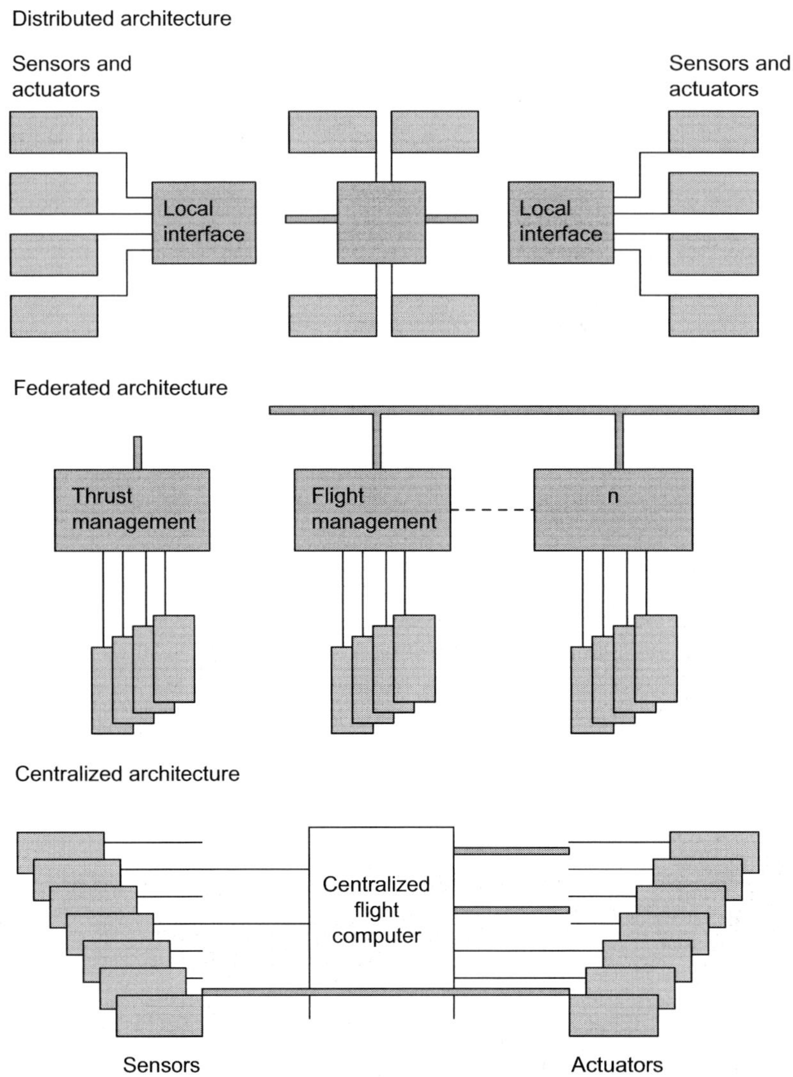

FIGURE 9–2 Three types of system architecture: federated, centralized, and distributed. *(Source: Rolls Royce.)*

- Actuator systems to provide variable geometry control and/or modulation of secondary systems (for example, bleed valves, variable compressor stator vanes and tip clearance control actuators).
- An electronic ignition system to generate high voltage sparks at the surface of an igniter plug in the combustion chamber, required to initiate combustion, which is then self-sustaining.

The most common form of starter is an air turbine system connected to the accessory gearbox. High-pressure air is used to rotate the HP turbine.

- A means of communication with the vehicle or plant systems. Today, this is usually with an electronic serial databus using an industry standard appropriate to the application, bandwidth, and integrity requirements.
- Separate systems dedicated to ensuring that control system failures cannot result in a dangerous condition.
- The other "component" of the system is the software in the microprocessor, which has to implement the complex functionality required. There are different standards for the development of this software in different industries.

AEROENGINE CONTROL SYSTEMS

Civil Aircraft Engine Controls

Controllers for modern engines are based on digital electronics. For historical reasons, the collection of control system elements in an aero engine is often referred to as the Full Authority Digital Electronic Controller (FADEC).

The components of a FADEC system are similar to those described in the general system above, with typically the following additions:

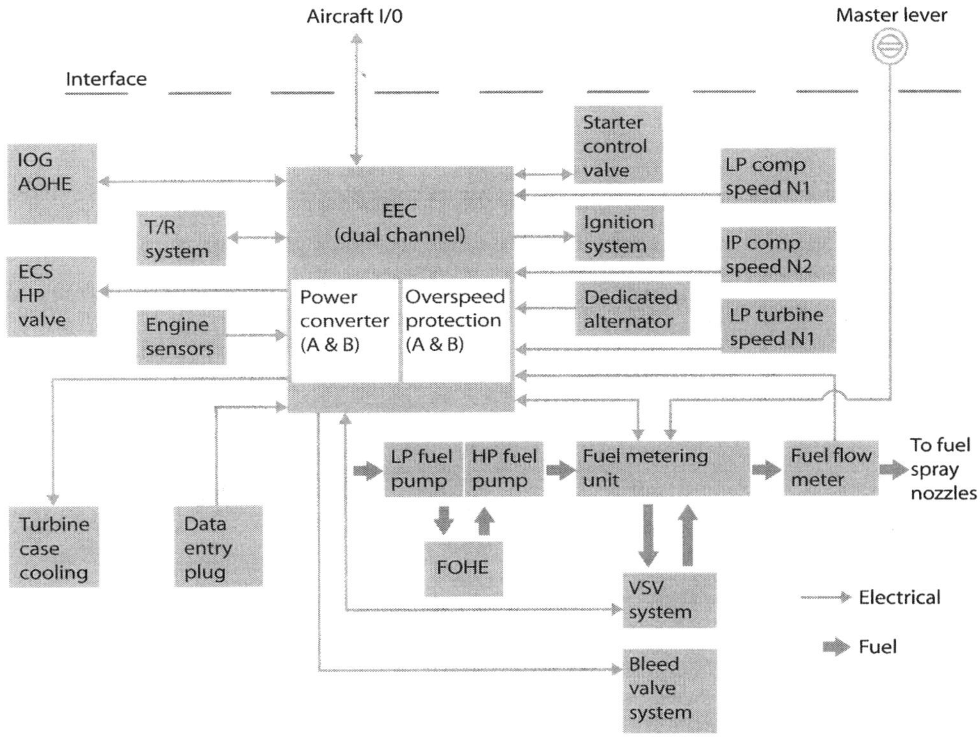

FIGURE 9–3 A typical FADEC structure. *(Source: Rolls Royce.)*

FIGURE 9–4 A FADEC circuit board. *(Source: Rolls Royce.)*

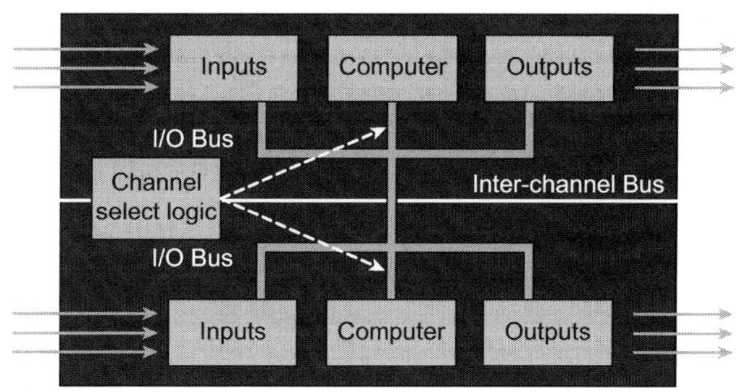

FIGURE 9–5 A typical dual channel EEC arrangement. *(Source: Rolls Royce.)*

- an engine-driven generator, dedicated to power the FADEC system
- means to control a thrust reverser, if fitted.

Engine Electronic Controller

At the center of the FADEC system is the engine electronic controller (EEC). The stringent requirements for safety and availability of an aero engine cannot be met with simplex electronics. For this reason, FADEC designs generally include two channels of electronics, sensors, wiring harnesses, and duplicated electrical parts of actuators. So that the system is fully operational following a single electrical or electronic failure. The two channels within the EEC include features that enable them to exchange data, which are used to detect failures in the system and to allow continued operation. However, the channels must be designed so that a fault in one channel cannot propagate to the other.

In some cases, the two electronic channels are housed in separate enclosures, but more usually, they are contained in a single unit. The EEC may be installed in the airframe, particularly in military aircraft or in civil fuselage-mounted engine installations. For wing-mounted engine applications, typical of large civil turbofans, the EEC is mounted

FIGURE 9–6 Cockpit of a modern aircraft with multi-function displays. *(Source: Rolls Royce.)*

on the engine. This installation places particularly harsh environmental requirements on the electronics, while further emphasizing the need for low weight and volume—a need reflected in the components used, the construction techniques, and mounting arrangements. One environmental threat, particular to electronic systems, is electro-magnetic radiation from, for example, lightning (both on the ground and in the air) and airport radar. The substantial connector housings used are in part designed to help alleviate these threats.

The EEC reads data from the sensors, other information from the aircraft avionic systems, and the pilot's inputs to calculate the new required position of the actuators, and uses its drive circuits to move them, often by means of secondary servos in the actuators. It also transmits data relating to the engine condition back to the aircraft, along industry standard serial data busses. The aircraft manufacturer is responsible for deciding which data are displayed to the pilot, subject to certification rules and the engine manufacturer's installation manual.

The EEC gathers information on any failures it has diagnosed within the electronics, the remainder of the FA DEC system, or in some cases in the gas turbine itself. This information is transmitted to the aircraft systems, but if the system considers itself to be in a safe configuration and no action is required in flight, the information is often not displayed in the cockpit—it is available to the pilot if required, but is intended for use by maintenance personnel on the ground.

This fault information may also be stored within the EEC itself for retrieval by the ground crew and may include more detail than is transmitted to the aircraft. Should the EEC be removed as a result of a suspected failure, these data are also used to assist in the diagnosis of the fault at the repair base.

Fuel Metering Unit (FMU)

In a FADEC system, a single unit is dedicated to accepting fuel from the piiDlping system and uses inputs from the

EEC to meter the flow of fuel to the engine. A proportion of the high-pressure fuel supply is used, after appropriate filtering, to power a hydraulic servo system, which operates valves within the unit. One of these valves maintains a constant pressure drop across a port in the sleeve of a second valve.

A two-stage servo uses the electrical current from the EEC to position a piston within this sleeve, which opens or covers the port. In this way the current is related to flow by the shape of the port in the sleeve. A feedback device measures the position of the piston and the reading used by the EEC to assist in control, and to ensure that the position and hence flow control is operating correctly.

The servo supply is also used to power other hydraulic circuits within the unit, for example, the fuel shut-off valve. In response to electrical signals from the EEC, it can also power the actuators controlling, for example, variable stator vanes in the compressor.

Actuation

Actuators can use various power sources, in addition to high-pressure fuel:

- Pneumatic systems are simple and rugged, but heavy and relatively slow in response.
- Hydraulic systems offer high levels of power and response, at a low weight, but require complex ancillary equipment.
- Low-pressure fuel systems have relatively low power but are sufficient to move components such as inlet guide vanes.
- Direct electro-mechanical systems, which offered low response times and were relatively heavy, can now be replaced with more modern technology, ensuring that they are lighter, less bulky, and operate at significantly higher speeds.

Fuel Pumps

The pumping system has to be able to deliver sufficient fuel flow to the engine under all conditions and at pressures high enough to overcome the gas pressure in the fuel spray nozzles generated by the engine compression system. Flow is also required to power the servo systems. The pumps are driven from the engine accessory gearbox.

The input pressure to the pumping system depends on the aircraft fuel system; in some cases, the engine pumps are required to continue to feed the engine with fuel if the aircraft fuel pumps fail—the engine pumps would then have to deliver fuel from the aircraft tanks. Generally, it is not possible to provide the suction and fuel pressure required with a single pump, and so two pumps are often used in series: a low-pressure centrifugal pump provides suction and delivers fuel to a high-pressure pump (most commonly a positive displacement gear pump), which in turn delivers fuel to the FMU.

In all cases, the pumping system must deliver enough fuel for the engine. The nature of the positive displacement pump sets a unique relationship between speed of rotation and flow delivered, resulting in more flow from the HP pump than is required by the engine. To accommodate this, the FMU returns unwanted fuel to the inlet of the HP pump, which then recirculates it. A consequence of this can be excessive heating of the fuel as it is repeatedly compressed and decompressed. A cooler is often required to dissipate this heat.

Bleed Valves

Many engines include bleed valves in the compression system, which allow intercompressor flows to be matched at low speeds. The position (open or closed) of these valves is signaled by the EEC, and fuel, electrical, or pneumatic servo systems are used to actuate them.

Electrical Power Supply

The reliability and availability of the electrical and electronic parts of the system are dependent on the quality and reliability of the electrical power supply. Each channel of the FADEC system requires independent power and, for aircraft, this is typically provided by a dedicated, gearbox-mounted generator. Aircraft power is also supplied for starting, and to provide a backup should the dedicated generator fail. In some military aircraft, only aircraft power is used, mainly because of the dependability of the aircraft supplies.

Software

The software embedded in the EEC defines the system behavior. The performance of this software is therefore vital to the operation of the engine. The software is generated from the requirements using disciplined processes and extensive testing. These processes are defined in industry standard documents and guidelines. Software developed to these standards is expensive to generate and can take a considerable time, particularly due to the effort required in testing and qualification. Software tools and techniques are becoming available to reduce this effort but these are far from mature.

Indication Systems

In modern systems, data from the control system and other sources is displayed on one or more display units mounted in the instrument panel; multi-function screens, which

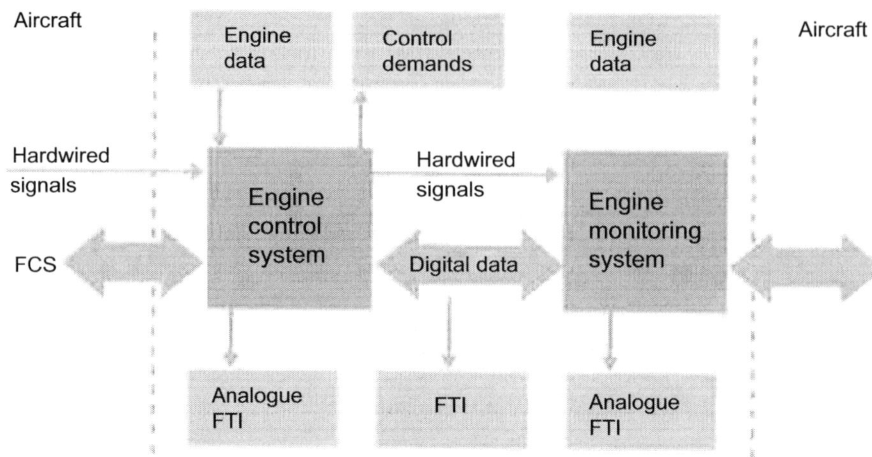

FIGURE 9–7 EEC and EHM integration. *(Source: Rolls Royce.)*

display basic engine data such as rotor speeds, turbine temperature, and power, have replaced the multitude of dials and individual instruments found in older aircraft. The multi-function screens are programmed to reconfigure themselves to display other data in response to abnormal circumstances, or as required by the operator. The information is displayed on the screen in the form of virtual dials with digital readouts and warnings; cautions and advisory messages are shown as text. A mimic diagram representing the physical layout of the equipment may be provided to assist in locating a problem. The displays are color-coded and, when necessary, linked to audible warning systems so that the operator is aware of the severity of any problem.

In military aircraft, these data may be displayed using a "head up display" (HUD). The HUD system projects information and instrument images onto the screen in front of the pilots.

Using this technology means that pilots do not have to divert their attention from the view around them. On some applications, information can also be shown on the visor as part of the headgear worn by the pilot.

Engine Health Monitoring

It is in the interests of all customers to minimize the cost of operation of the gas turbine and its associated equipment.

The costs of operation include fuel, scheduled maintenance, and unforeseen events that result in the engine not being available when required. Monitoring systems can help to reduce all of these costs. Scheduling of major engine maintenance (for example, to restore performance after many hours of operation) is a complex economic decision for which monitoring systems can provide important supporting data.

Although it is not strictly part of the control system, the EHM electronics are often housed within the control system enclosure, and the two systems are to some extent integrated.

It is important to note, however, that the safety requirements of the two systems are different and the design of each, and their integration, must reflect this. A function in the monitoring system cannot be adopted for use in the control system without considering the reliability of its implementation.

Data from the EHM systems are not generally available to the flight crew. Large amounts of data are stored although data reduction and analysis algorithms are used to make storage requirements more reasonable. Aircraft systems are used to transmit the data to a ground station, which in turn will forward the data to a center where further analysis can be carried out in order to inform maintenance logistics (Figure 9–7).

Sensors

Whatever the particular application, a series of parameters needs to be measured at various locations around the engine system in order to control the engine and provide useful indication of performance to the operator. Typically, these are temperature, pressure, and speed measurements. The transducers used to take these measurements are chosen on accuracy, response time, and durability requirements.

Temperature Sensors

Thermocouples

Thermocouples are used to measure high temperatures, typically at HP compressor exit, and in and around the turbines. The temperature of the gas in the turbine is measured at several radial and circumferential positions in

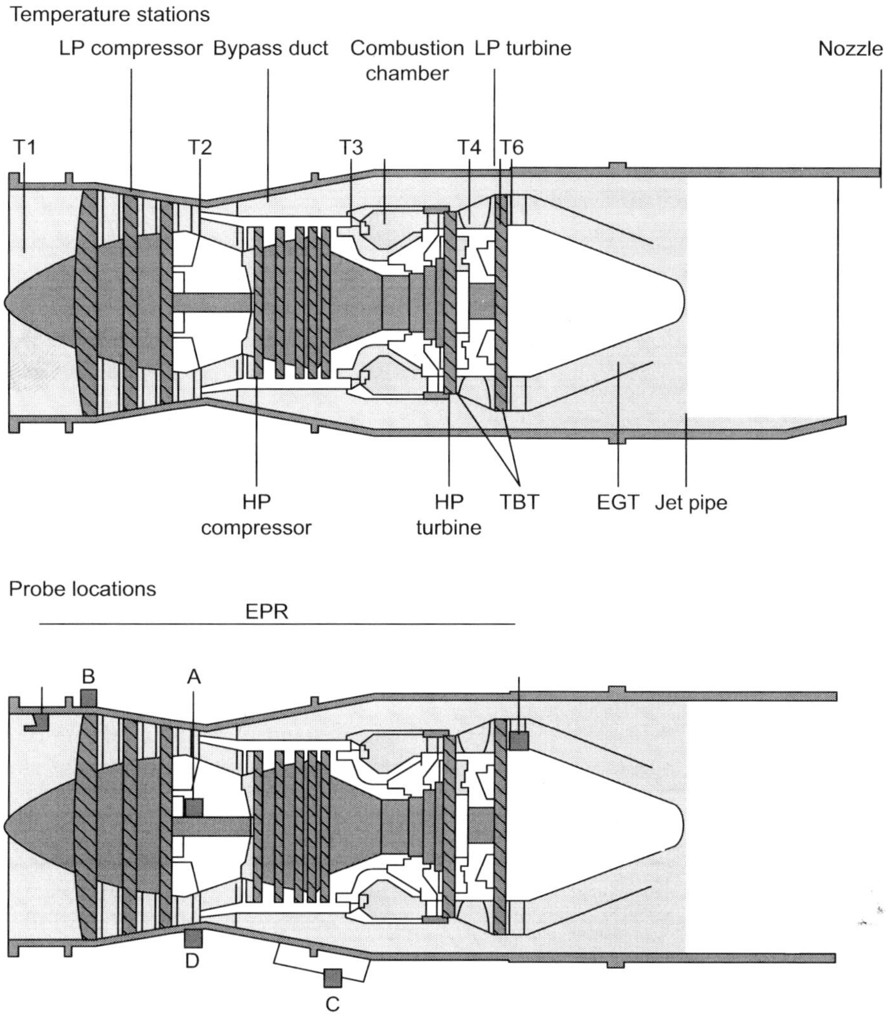

FIGURE 9–8 Typical parameter measurement points. *(Source: Rolls Royce.)*

order to even out any local temperature variations due to turbine entry temperature traverse effects. Thermocouples have the advantage of being very reliable, small, and cheap; they also have a relatively quick response time over a large temperature range, and generate their own output—and so do not require an external power supply. However, thermocouples are easily damaged, and can lose accuracy through oxidation (Figure 9–8).

Resistance Temperature Devices

These devices are most often used to monitor engine intake air temperature. They consist of a platinum coil, exposed to the airflow, which changes its electrical resistance with temperature. Often, the device will consist of a single probe with a twin output and a heated body for anti-icing. The advantages of these sensors are that they are resistant to damage (when in a housing) and give very accurate outputs with long-term stability. However, they have a slow response time when in a housing, need a constant current source to operate, and are relatively expensive.

Pyrometer

In this method, an optical device is used to view, for example, the turbine blades, connected to an infrared (IR) detector by a fibre-optic cable. This method enables rapid, accurate measurement of temperatures. However, a compressed air supply is needed to keep the lens clean, and the output needs sophisticated signal processing.

Pressure Sensors

Pressure sensors broadly divide into those required to provide high accuracy, and those that focus on transient response. Accurate measurement is required when pressure ratio is used to measure engine thrust. Transducers based on a variety of technologies are used for this purpose, but they generally need electronics to support their operation or to provide calibration information, and therefore the assembly is usually housed within the EEC, which can involve quite long pipe runs. If high bandwidth is required, simpler transducers, often based on strain gauge technology, are used and may be mounted close to the engine to avoid pipe delays.

Rotor Speed Sensors

Types of speed sensor include tachogenerators and magnetic variable reluctance (VR) probes. A tachogenerator is a shaft-driven, electrical generator with a variable frequency output, which is related to speed. These devices are very rugged, but produce a relatively low output signal. If a VR speed probe is used, it is positioned on the compressor casing in line with a small disc, which has accurately machined notches on its circumference and is mounted concentrically on the shaft. Rotation of the shaft results in a current being induced in the probe with a frequency content proportional to engine speed (Figures 9–9 and 9–10).

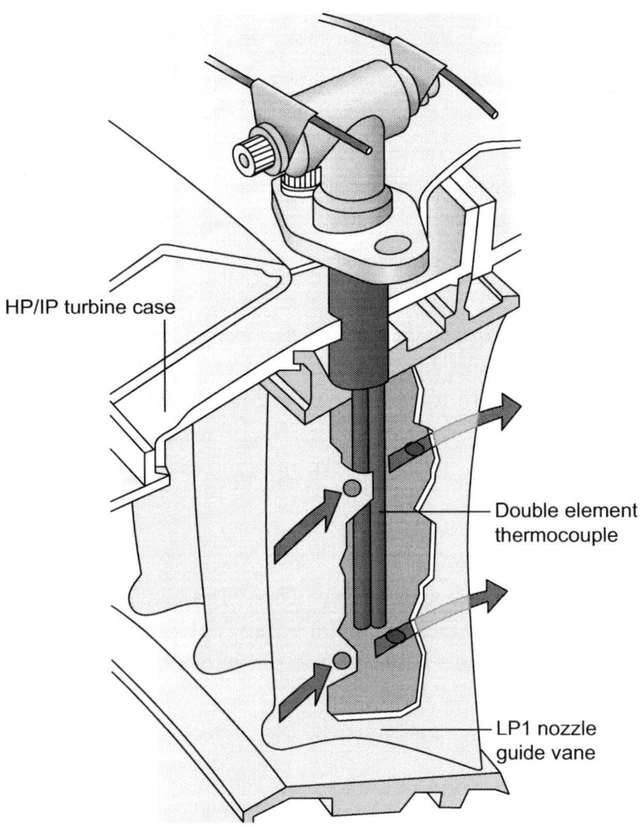

FIGURE 9–9 Typical thermocouple arrangement. *(Source: Rolls Royce.)*

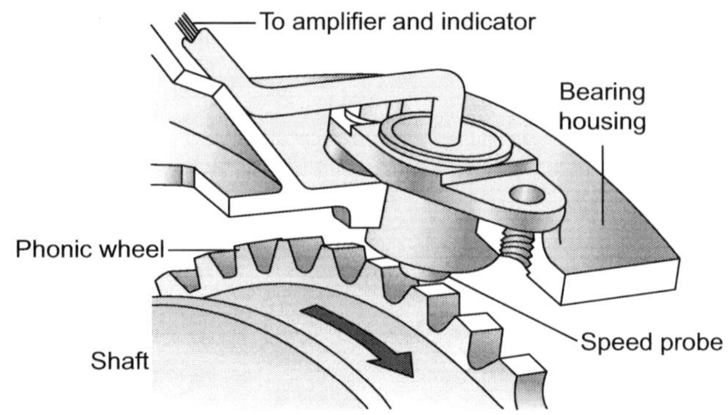

FIGURE 9–10 VR speed probe arrangement. *(Source: Rolls Royce.)*

Position Measurement

Position measurement is used to confirm that actuators are operating correctly and to assist in closed loop control. There are three main types of device used: the LVDT (linear variable differential transformer), RVDT (rotational variable differential transformer), and the resolver.

An LVDT consists of three adjacent coils of wire wound around a hollow form through which a core of permeable material (such as steel) can slide freely. The middle winding is known as the primary coil, and is excited by a relatively high frequency AC voltage. This sets up a magnetic flux, which is then coupled through the core to the other two, secondary windings, inducing a voltage in them. When the moving core is centered between the secondary coils, the voltage induced in them is equal and opposite. If the core is displaced, then an imbalance is set up, creating a voltage that can be read and calibrated to give a position.

RVDTs and resolvers are based on similar principles, but are used to measure rotational angles.

Vibration

Many engines are fitted with sensors that continuously monitor the vibration level of the engine. Indication of excessive vibration is shown on the control display unit using signals from engine-mounted transducers. There are three main types of vibration sensor:

- Piezoelectric accelerometers produce a very low value charge signal through deformation of a crystal lattice, and require the vibration signal to be processed using a charge amplifier and sophisticated cabling.
- Piezoresitive accelerometers change their resistance relative to an applied stress, and are easy to use and install, but require a separate power supply.
- Velocity pickups produce a voltage signal from a magnet moving in a coil, are easy to install, and require simple processing.

Safety and Availability

Safety is the most important design consideration in any gas turbine or installation; another high priority is availability—the loss of power from an engine, although not necessarily a safety hazard, can cause severe operational disruption. The duplication of the electrical elements of the system is evidence of this concern. Rigorous analyses and testing are necessary to ensure that faults in the system are correctly accommodated to allow for continued engine operation.

Just as it is safe to complete a flight during which a failure has occurred in the duplicated part of the system, it can also be shown by analysis that the aircraft can continue to operate for subsequent flights for a defined period before a fault is repaired. A "time limited dispatch" analysis is carried out to establish which faults can be treated in this way and for how long. This is of considerable benefit to the aircraft operator, who can continue to operate the aircraft normally and repair the fault at a convenient time, for example, when the aircraft next returns to the operator's main base.

Other safety features may also be required, implemented either in the software or in dedicated hardware to address the effects of adverse operating conditions, or of particular engine or control system failures, which could represent a threat to the aircraft if not accommodated.

Defense Applications

Much of the control system technology used in military applications of gas turbines is similar to civil aerospace engines. However, engine requirements can differ markedly between different military applications, depending on whether the aircraft is a single-engine trainer, a large, twin-engine fighter with full afterburner capability, or a propeller-driven military transport.

Afterburning, also known as wet thrust or reheat, requires additional fuel handling equipment such as pumps and metering valves. They employ similar technology to that described above. Afterburning also requires a variable area exhaust nozzle in order to control the LP system working line. In some applications, the final nozzle is not only variable in area, but the thrust can be vectored by limited angular displacement of the nozzle. This enables a significant increase in aircraft agility without the use of large control surfaces and their associated drag. A variable area nozzle is controlled with actuator rams, typically powered by HP fuel pressure, and an appropriate servo system signaled from the EEC.

The application of the aircraft may also involve vertical or short takeoff, hover, and vertical landing. There are a number of aircraft configurations used to deliver this functionality. Each requires different levels of power to be extracted from the engine to provide vertical thrust, and to stabilize and maneuver the aircraft until sufficient forward speed is attained that the conventional flight control surfaces become operational. The control system must ensure the engine remains stable during these maneuvers and can respond to the very rapid changes in power that are required.

A single-engine aircraft often warrants additional system provisions. This might take the form of a mechanical system which can be invoked by the pilot should all electronic means of control fail. However, as the functions required of such a controller increase, for example, in a complex STOVL application, a mechanical solution is not possible, and so the safety case must be justified based on the electronic system's reliability and built-in redundancy.

Helicopter Systems

In many respects, helicopter control systems function in much the same way as those of fixed-wing aircraft—sensors monitor engine parameters, which are communicated back to an engine controller. However, the nature of a helicopter and its engine configuration mean that there are different control system requirements.

The engine controller must closely control the engine in order to provide a stable power turbine shaft speed. This then allows a constant helicopter rotor speed, while the pitch of the rotor blades controls lift and horizontal helicopter speed.

Traditional control systems operated with a throttle, using the collective pitch lever as the main load demand, with a twist grip for the pilot to trim the demand and keep the rotor speed within defined limits. Modern engines do not have a conventional throttle; they operate on a governing system whereby the pilot demands a load and the control system and inherent control laws and schedules will control the engines to maintain the correct rotor speed. In such a system, the power turbine speed and torque are monitored and the fuel flow is modulated accordingly.

One of the key aspects of helicopter engine control is matching the torque provided by the engines on multi-engine aircraft. Torque mismatches can provide significant aircraft performance penalties. Engine parameter matching, through communication of data between the engines, can be used in order to enable isochronous control, and maintain an even loading of torque.

Although vibration absorbers can be used in some cases, helicopters experience significant levels of vibration; the engines are mechanically coupled through the drive train to the rotors and consequently there is very little vibration damping. It is therefore important to monitor vibration levels.

Typical Aircraft Engine C&I System Hardware*

The controls of the gas turbine engine are designed to remove, as far as possible, work load from the pilot while still allowing him ultimate control of the engine. To achieve this, the fuel flow is automatically controlled after the pilot has made the initial power selection.

All engine parameters require monitoring and instrumentation is provided to inform the pilot of the correct functioning of the various engine systems and to warn of any impending failure. Should any of the automatic governors fail, the engine can be manually controlled by the pilot selecting the desired thrust setting and monitoring the instruments to maintain the engine within the relevant operating limitations.

Main Control Functions

The control of a gas turbine engine generally requires the use of only one control lever and the monitoring of certain indicators located on the pilot's instrument panel (Figure 9—11). Operation of the control (throttle/power) lever selects a thrust level, which is then maintained automatically by the fuel system.

On engines fitted with afterburning, single lever control is maintained, although a further fuel system is required to supply and control the fuel to the afterburner.

On a turbo-propeller engine, the throttle lever is interconnected with the propeller control unit (P.C.U.), thus maintaining single lever operation of the engine. Similarly, the throttle control lever of a helicopter is interconnected with the collective pitch lever, so ensuring that an increase in pitch is accompanied by an increase in engine power.

The fuel system incorporates a high-pressure fuel shut-off cock to provide a means of stopping the engine. This may be operated by a separate lever, interconnected with the throttle lever, or electrically actuated and controlled by a switch on the pilot's instrument panel.

A turbo-jet engine fitted with a thrust reverser usually has an additional control lever that allows reverse thrust to be selected. On a turbo-propeller engine, a separate control lever is not required because the interconnected throttle and P.C.U. lever is operated to reverse the pitch of the propeller.

Basic Instrumentation Hardware

The performance of the engine and the operation of the engine systems are shown on gauges or by the operation of flag or dolls-eye type indicators. A diagrammatic arrangement of the control and instrumentation for a turbo-jet engine is shown in Figure 9—12.

Engine Thrust

The thrust of an engine is shown on a thrust-meter, which will be one of two basic types; the first measures turbine discharge or jet pipe pressure, and the second, known as an engine pressure ratio (E.P.R.) gauge, measures the ratio of two or three parameters. When E.P.R. is measured, the ratio is usually that of jet pipe pressure to compressor inlet pressure. However, on a fan engine the ratio may be that of integrated turbine discharge and fan outlet pressures to compressor inlet pressure.

In each instance, an indication of thrust output is given, although when only the turbine discharge pressure is measured, correction is necessary for variation of inlet pressure; however, both types may require correction for variation of ambient air temperature. To compensate for ambient atmospheric conditions, it is possible to set a correction figure to a sub-scale on the gauge; thus the minimum thrust output can be checked under all operating conditions.

Suitably positioned pitot tubes sense the pressure or pressures appropriate to the type of indication being taken from the engine. The pitot tubes are either directly connected to the indicator or to a pressure transmitter that is electrically connected to the indicator.

An indicator that shows only the turbine discharge pressure is basically a gauge, the dial of which may be marked in pounds per square inch (p.s.i.), inches of mercury (in. Hg.), or a percentage of the maximum thrust.

E.P.R. can be indicated by either electromechanical or electronic transmitters. In both cases the inputs to the transmitter are engine inlet pressure (P_1) and an integrated pressure (P_{INT}) comprised of fan outlet and turbine exhaust pressures. In some cases either fan outlet pressure or turbine exhaust pressure is used alone in place of P_{INT}.

The electro-mechanical system indicates a change in pressure by using transducer capsules (Figure 9—13) to deflect the center shaft of the pressure transducer causing the yoke to pivot about the axis A.A. This movement is sensed by the linear variable differential transformer (L.V.D.T.) and converted to an A.C. electrical signal, which is amplified and applied to the control winding of the servo motor.

The servo motor, through the gears, alters the potentiometer output voltage signal to the E.P.R. indicator and simultaneously drives the gimbal in the same direction as the initial yoke movement until the L.V.D.T. signal to the motor is canceled and the system stabilizes at the new setting.

* Source: Adapted, with permission, from Rolls Royce, *The Jet Engine*, 1986, Rolls Royce Plc: UK.

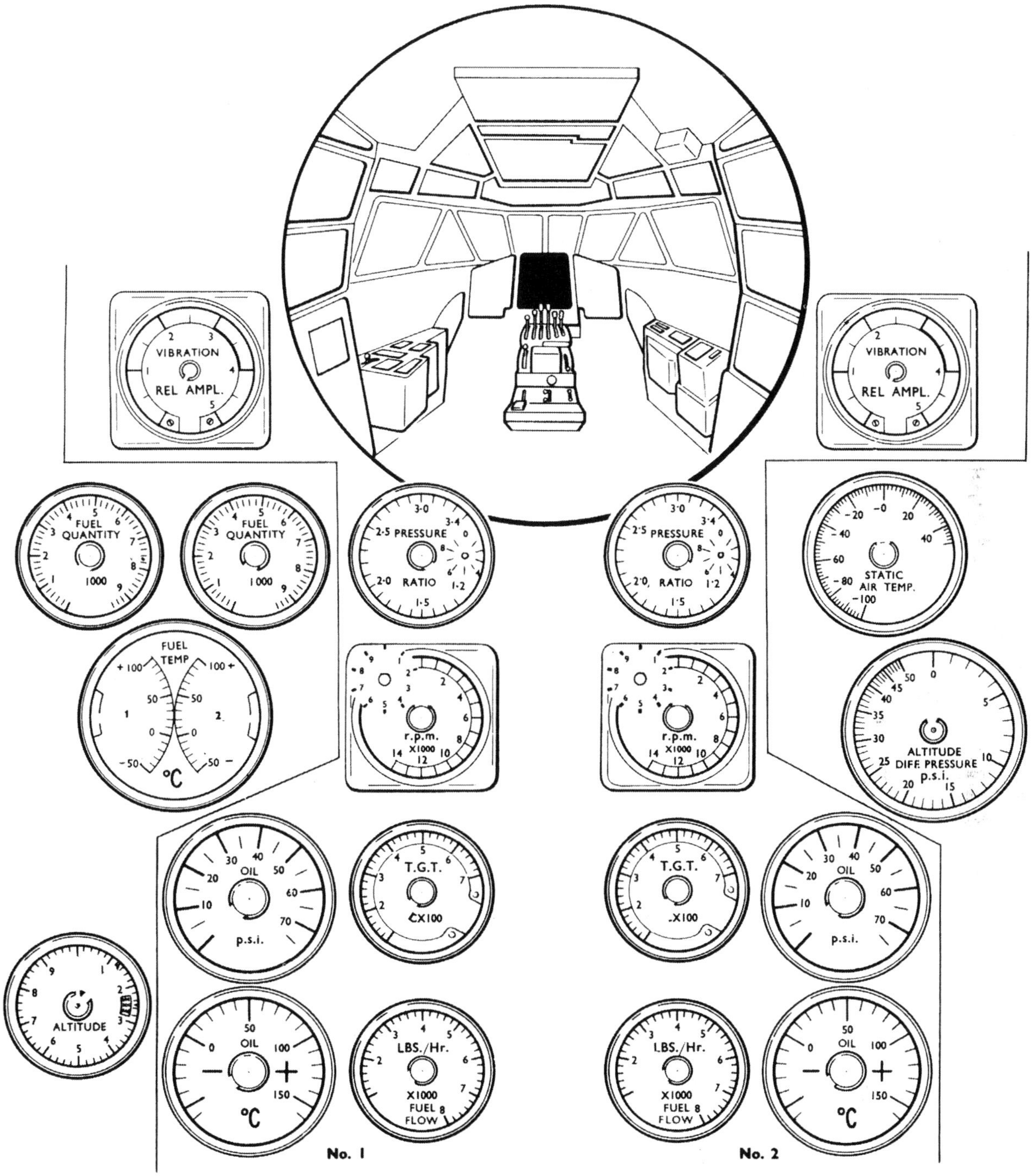

FIGURE 9–11 Pilot's instrument panel—turbo-jet engines. *(Source: Rolls Royce.)*

The electronic E.P.R. system utilizes two vibrating cylinder pressure transducers that sense the engine air pressures and vibrate at frequencies relative to these pressures. From these vibration frequencies electrical signals of E.P.R. are computed and are supplied to the E.P.R. gauge and electronic engine control system.

Engine Torque

Engine torque is used to indicate the power that is developed by a turbo-propeller engine, and the indicator is known as a torquemeter. The engine torque or turning moment is proportional to the horsepower and is transmitted through the propeller reduction gear.

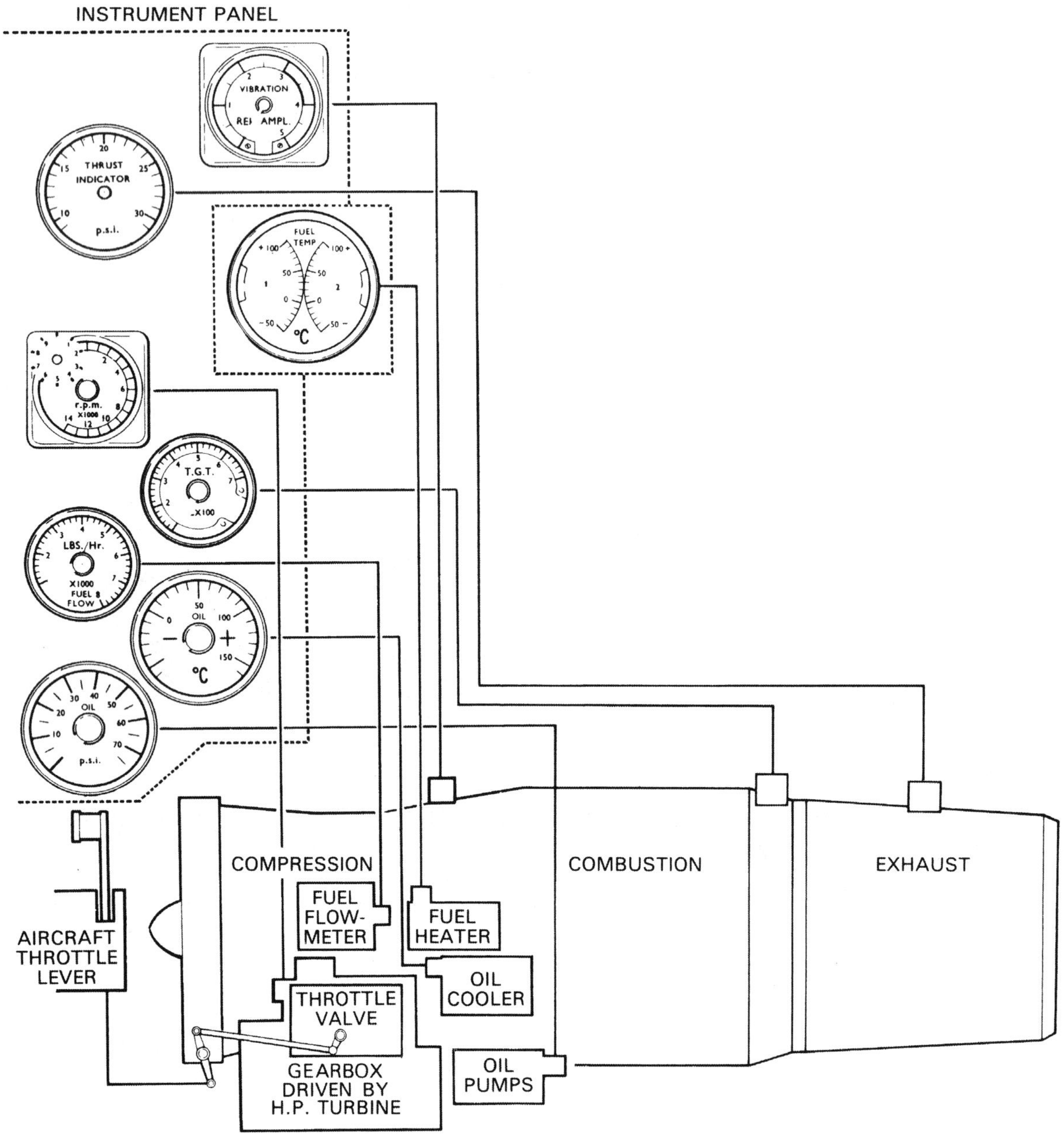

FIGURE 9–12 Diagrammatic arrangement of engine control and instrumentation. *(Source: Rolls Royce.)*

A torquemeter system is shown in Figure 9–14. In this system, the axial thrust produced by the helical gears is opposed by oil pressure acting on a number of pistons; the pressure required to resist the axial thrust is transmitted to the indicator.

In addition to providing an indication of engine power, the torquemeter system may also be used to automatically operate the propeller feathering system if the torquemeter oil pressure falls due to a power failure. It is also used, on some installations, to assist in the automatic operation of the water injection system to restore or boost the takeoff power at high ambient temperatures or at high altitude airports.

Engine Speed

All engines have their rotational speed (rpm) indicated. On a twin or triple-spool engine, the high-pressure assembly speed is always indicated; in most instances, additional indicators show the speed of the low pressure and intermediate pressure assemblies.

Engine speed indication is electrically transmitted from a small generator, driven by the engine, to an indicator that shows the actual revolutions per minute (rpm), or a percentage of the maximum engine speed (Figure 9–15). The engine speed is often used to assess engine thrust, but it does not give an absolute indication of the thrust being produced because inlet temperature and pressure conditions affect the thrust at a given engine speed.

The engine speed generator supplies a three-phase alternating current, the frequency of which is dependent

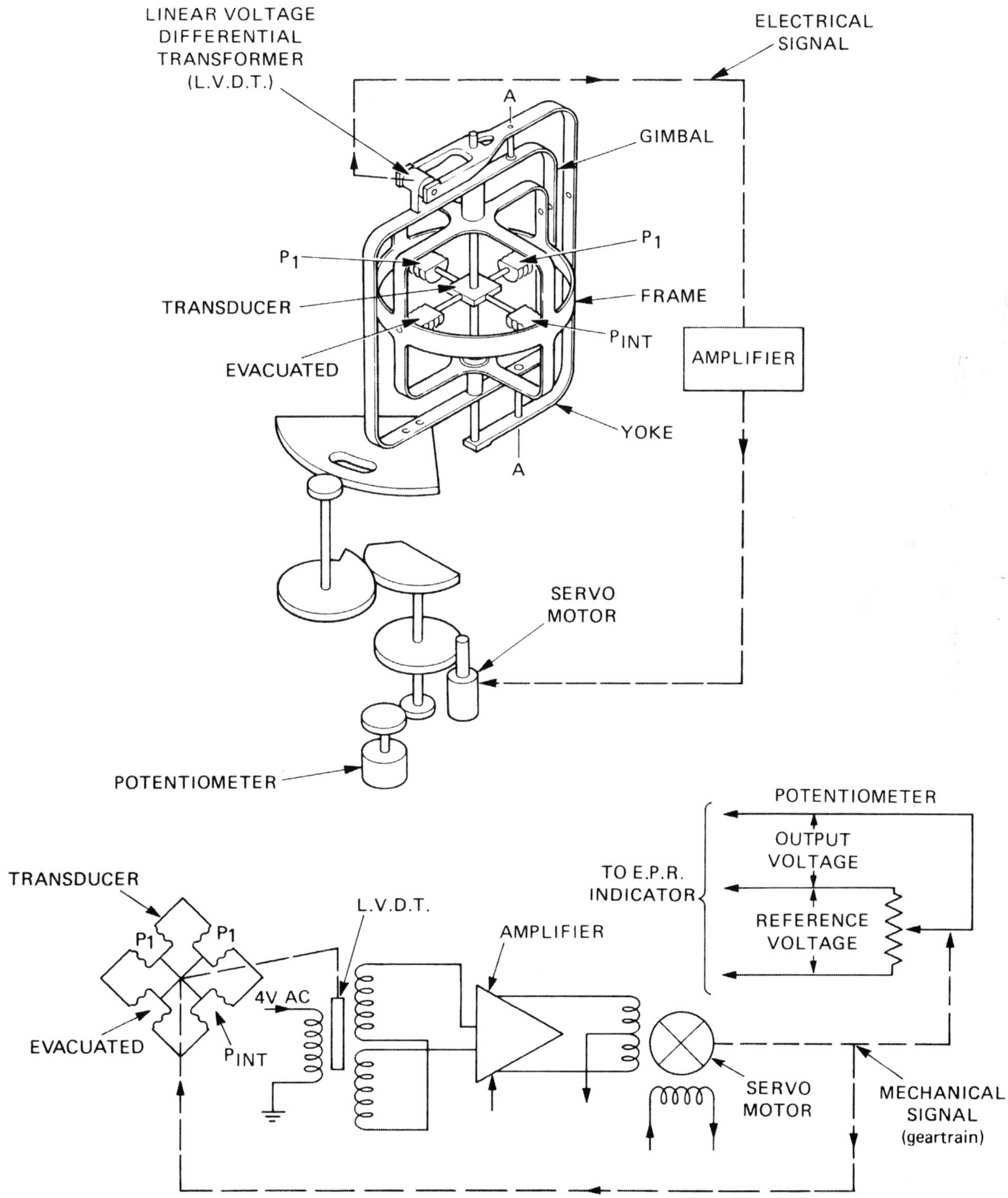

FIGURE 9–13 Electro-mechanical E.P.R. transmitter. *(Source: Rolls Royce.)*

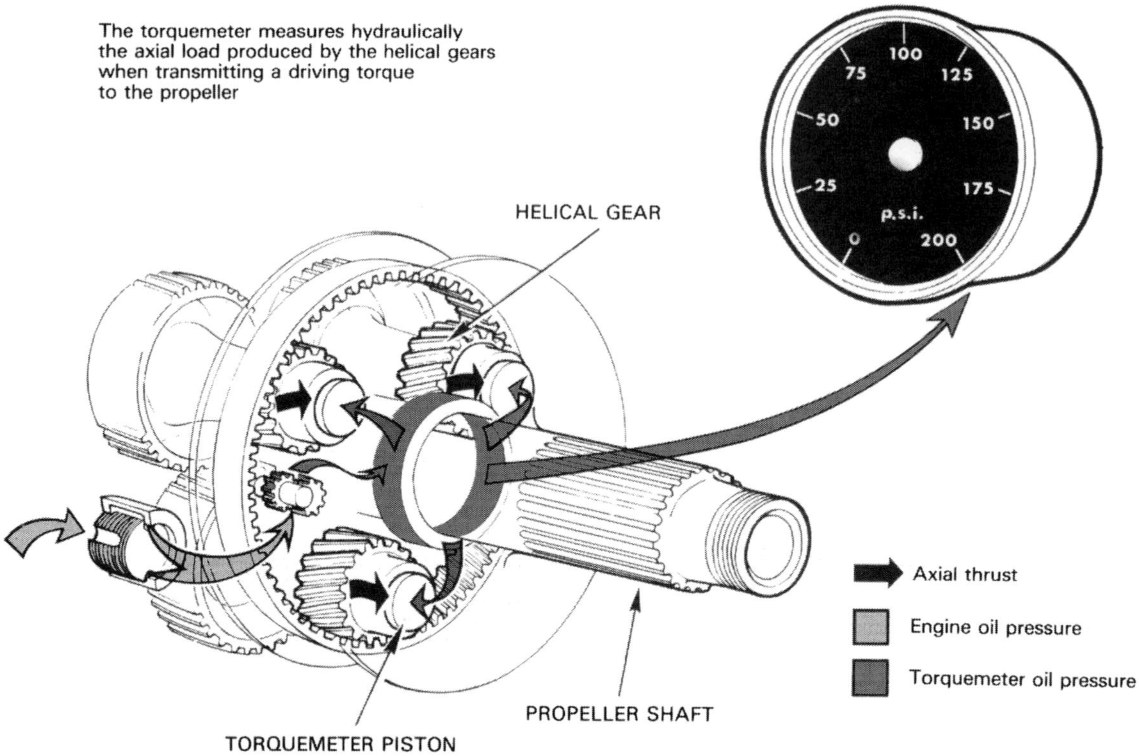

FIGURE 9–14 A simple torquemeter system. *(Source: Rolls Royce.)*

upon engine speed. The generator output frequency controls the speed of a synchronous motor in the indicator, and rotation of a magnet assembly housed in a drum or drag cup induces movement of the drum and consequent movement of the indicator pointer.

Where there is no provision for driving a generator, a variable-reluctance speed probe, in conjunction with a phonic wheel, may be used to induce an electric current that is amplified and then transmitted to an indicator (Figure 9–16).

This method can be used to provide an indication of rpm without the need for a separately driven generator, with its associated drives, thus reducing the number of components and moving parts in the engine.

The speed probe is positioned on the compressor casing in line with the phonic wheel, which is a machined part of the compressor shaft. The teeth on the periphery of the wheel pass the probe once each revolution and induce an electric current by varying the magnetic flux across a coil in the probe. The magnitude of the current is governed by the rate of change of the magnetic flux and is thus directly related to engine speed.

In addition to providing an indication of rotor speed, the current induced at the speed probe can be used to illuminate a warning lamp on the instrument panel to indicate to the pilot that a rotor assembly is turning. This is particularly important at engine start, because it informs the pilot when to open the fuel cock to allow fuel to the engine. The lamp is connected into the starting circuit and is only illuminated during the starting cycle.

Turbine Gas Temperature

The temperature of the exhaust gases is always indicated to ensure that the temperature of the turbine assembly can be checked at any specific operating condition. In addition, an automatic gas temperature control system is usually provided, to ensure that the maximum gas temperature is not exceeded.

Turbine gas temperature (T.G.T.) sometimes referred to as exhaust gas temperature (E.G.T.) or jet pipe temperature (J.P.T.), is a critical variable of engine operation and it is essential to provide an indication of this temperature. Ideally, turbine entry temperature (T.E.T.) should be measured; however, because of the high temperatures involved this is not practical, but, as the temperature drop across the turbine varies in a known manner, the temperature at the outlet from the turbine is usually measured by suitably positioned thermocouples. The temperature may alternatively be measured at an intermediate stage of the turbine assembly, as shown in Figure 9–17.

The thermocouple probes used to transmit the temperature signal to the indicator consist of two wires of dissimilar metals that are joined together inside a metal guard tube. Transfer holes in the tube allow the exhaust gas to flow across the junction. The materials from which the

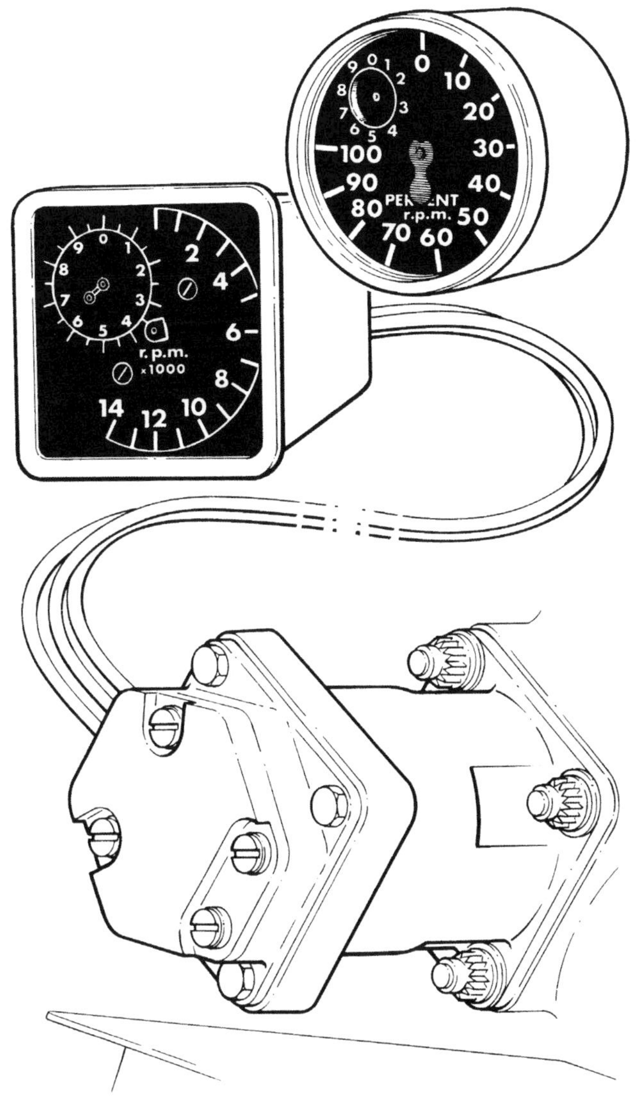

FIGURE 9–15 Engine speed indicators and generator. *(Source: Rolls Royce.)*

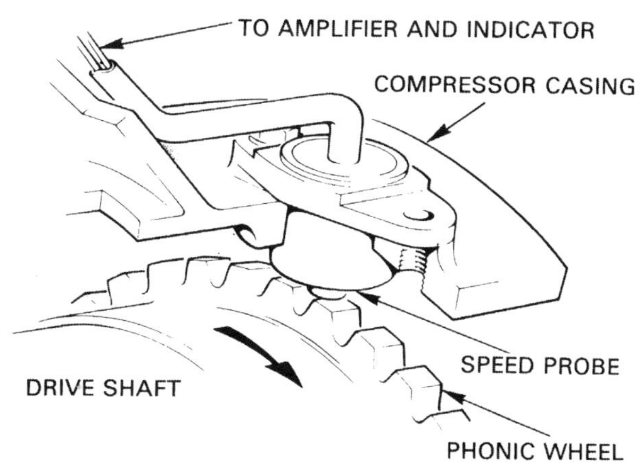

FIGURE 9–16 Variable-reluctance speed probe and phonic wheel. *(Source: Rolls Royce.)*

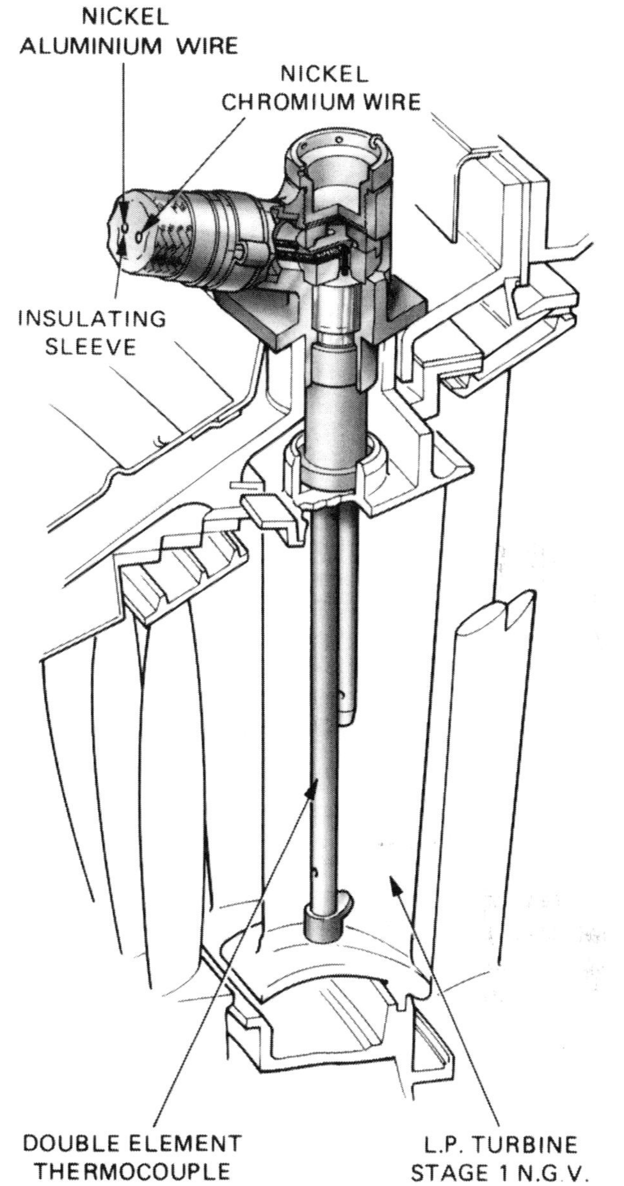

FIGURE 9–17 Turbine thermocouple installation. *(Source: Rolls Royce.)*

thermocouple's wires are made are usually nickel-chromium and nickel-aluminum alloys.

The probes are positioned in the gas stream so as to obtain a good average temperature reading and are normally connected to form a parallel circuit. An indicator, which is basically a millivoltmeter calibrated to read in degrees centigrade, is connected into the circuit (Figure 9–18).

The junction of the two wires at the thermocouple probe is known as the "hot" or "measuring" junction and that at the indicator as the "cold" or "reference" junction. If the cold junction is at a constant temperature and the hot junction is sensing the exhaust gas temperature, an

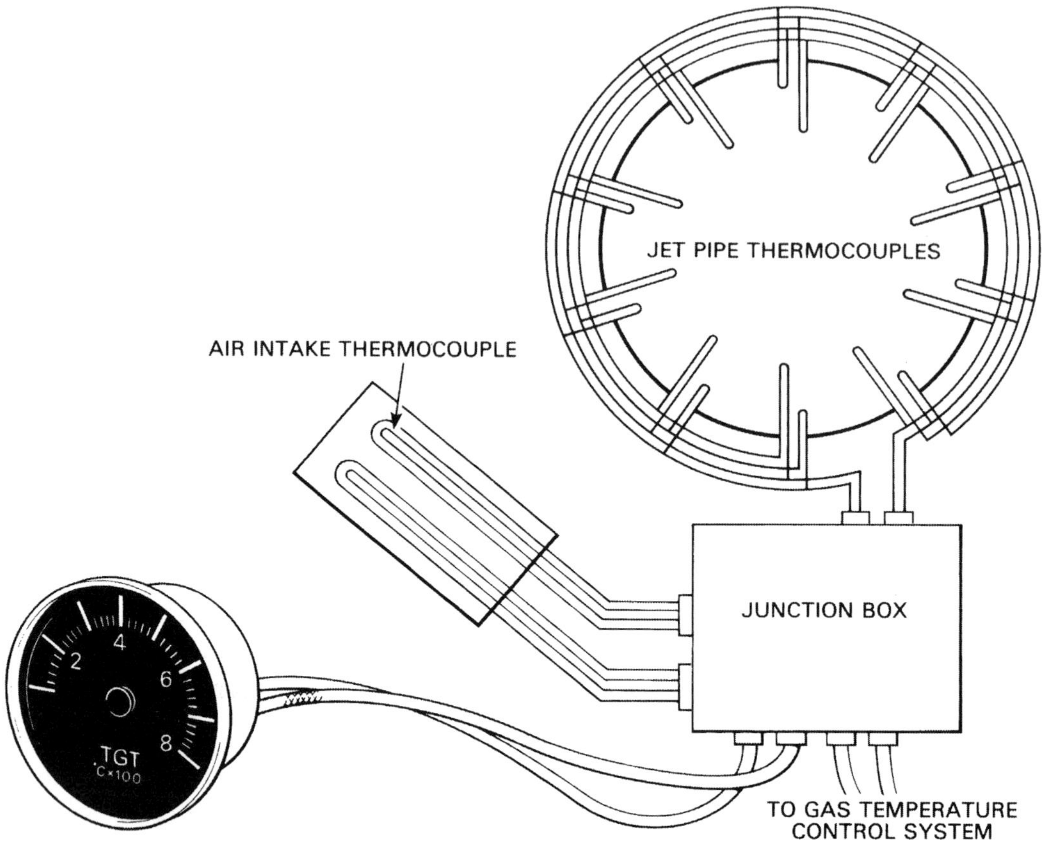

FIGURE 9–18 A typical double element thermocouple system. *(Source: Rolls Royce.)*

electromotive force (E.M.F.), proportional to the temperature difference of the two junctions, is created in the circuit and this causes the indicator pointer to move. To prevent variations of cold junction temperature affecting the indicated temperature, an automatic temperature compensating device is incorporated in the indicator or in the circuit

The thermocouple probes may be of single, double, or triple element construction. Where multiple probes are used they are of differing lengths in order to obtain a temperature reading from different points in the gas stream to provide a better average reading than can be obtained from a single probe (Figure 9–17).

The output to the temperature control system can also be used to provide a signal, in the form of short pulses, which, when coupled to an indicator, will digitally record the life of the engine. During engine operation in the higher temperature ranges, the pulse frequency increases progressively causing the cyclic-type indicator to record at a higher rate, thus relating engine or unit life directly to operating temperatures.

Thermocouples may also be positioned to transmit a signal of air intake temperature into the exhaust gas temperature indicating and control systems, thus giving a reading of gas temperature that is compensated for variations of intake temperature. A typical double-element thermocouple system with air intake probes is shown in Figure 9–18.

Oil Temperature and Pressure

It is essential for correct and safe operation of the engine that accurate indication is obtained of both the temperature and pressure of the oil. Temperature and pressure transmitters and indicators are illustrated in Figure 9–19.

Oil temperature is sensed by a temperature-sensitive element fitted in the oil system. A change in temperature causes a change in the resistance value and, consequently, a corresponding change in the current flow at the indicator. The indicator pointer is deflected by an amount equivalent to the temperature change and this is recorded on the gauge in degrees centigrade.

Oil pressure is electrically transmitted to an indicator on the instrument panel. Some installations use a flag-type indicator, which indicates if the pressure is high, normal or low; others use a dial-type gauge calibrated in pounds per square inch (psi).

Electrical operation of each type is similar; oil pressure, acting on the transmitter, causes a change in the electric

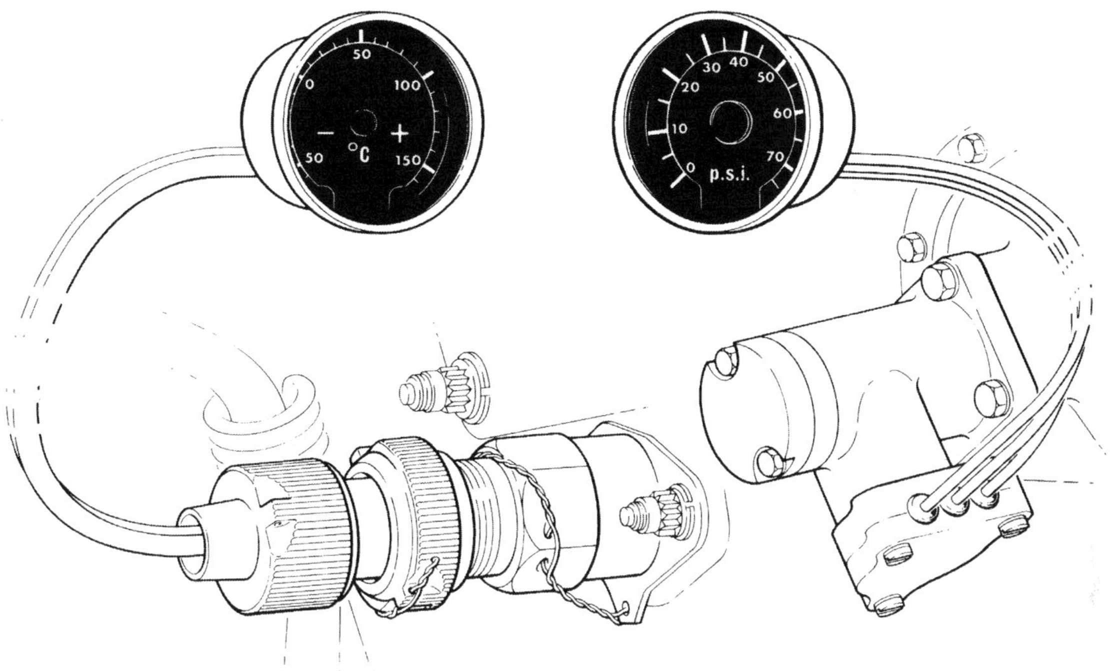

FIGURE 9–19 Oil temperature and pressure transmitters and indicators. *(Source: Rolls Royce.)*

current supplied to the indicator. The amount of change is proportional to the pressure applied at the transmitter.

The transmitter may be of either the direct or the differential pressure type. The latter senses the pressure difference between engine feed and return oil pressures, the return oil being pressurized by cooling and sealing air from the bearings.

In addition to a pressure gauge operated by a transmitter, an oil low-pressure warning switch may be provided to indicate that a minimum pressure is available for continued safe running of the engine. The switch is connected to a warning lamp in the flight compartment and the lamp illuminates if the pressure falls below an acceptable minimum.

Fuel Temperature and Pressure

The temperature and pressure of the low pressure fuel supply are electrically transmitted to their respective indicators and these show if the low pressure system is providing an adequate supply of fuel without cavitation and at a temperature to suit the operating conditions. The fuel temperature and pressure indicators are similar to those for oil temperature and pressure indication.

On some engines, a fuel differential pressure switch, fitted to the low-pressure fuel filter, senses the pressure difference across the filter element. The switch is connected to a warning lamp that provides indication of partial filter blockage, with the possibility of fuel starvation.

Fuel Flow

Although the amount of fuel consumed during a given flight may vary slightly between engines of the same type, fuel flow does provide a useful indication of the satisfactory operation of the engine and of the amount of fuel being consumed during the flight. A typical system consists of a fuel flow transmitter, which is fitted into the low-pressure fuel system, and an indicator, which shows the rate of fuel flow and the total fuel used in gallons, pounds or kilograms per hour (Figure 9–20). The transmitter measures the fuel flow electrically and an associated electronic unit gives a signal to the indicator proportional to the fuel flow.

Vibration

A turbo-jet engine has an extremely low vibration level and a change of vibration, due to an impending or partial failure, may pass without being noticed. Many engines are therefore fitted with vibration indicators that continually monitor the vibration level of the engine. The indicator is usually a milliammeter that receives signals through an amplifier from engine-mounted transmitters (Figure 9–21).

A vibration transmitter is mounted on the engine casing and electrically connected to the amplifier and indicator. The vibration-sensing element is usually an electro-magnetic transducer that converts the rate of vibration into electrical signals and these cause the indicator pointer to move proportional to the vibration level. A warning lamp on the instrument panel is incorporated in

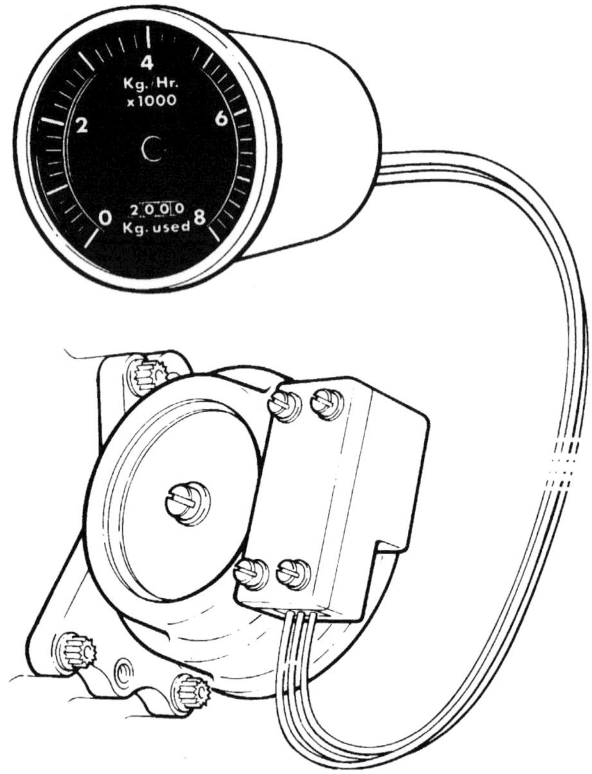

FIGURE 9–20 Fuel flow transmitter and indicator. *(Source: Rolls Royce.)*

the system to warn the pilot if an unacceptable level of vibration is approached, enabling the engine to be shut down and so reduce the risk of damage.

The vibration level recorded on the gauge is the sum total of vibration felt at the pick-up. A more accurate method differentiates between the frequency ranges of each rotating assembly and so enables the source of vibration to be isolated. This is particularly important on multi-spool engines.

A crystal-type vibration transmitter, giving a more reliable indication of vibration, has been developed for use on multi-spool engines. A system of filters in the electrical circuit to the gauge makes it possible to compare the vibration obtained against a known frequency range and so locate the vibration source. A multiple-selector switch enables the pilot to select a specific area to obtain a reading of the level of vibration.

Warning Systems

Warning systems are provided to give an indication of a possible failure or the existence of a dangerous condition, so that action can be taken to safeguard the engine or aircraft. Although the various systems of an aircraft engine are designed wherever possible to "fail safe," additional safety devices are sometimes fitted. Automatic propeller feathering should a power loss occur, and automatic closing of the high-pressure fuel shut-off cock should a turbine shaft failure occur, are but two examples. On some engine types, the fuel system is fitted with a control to enable the

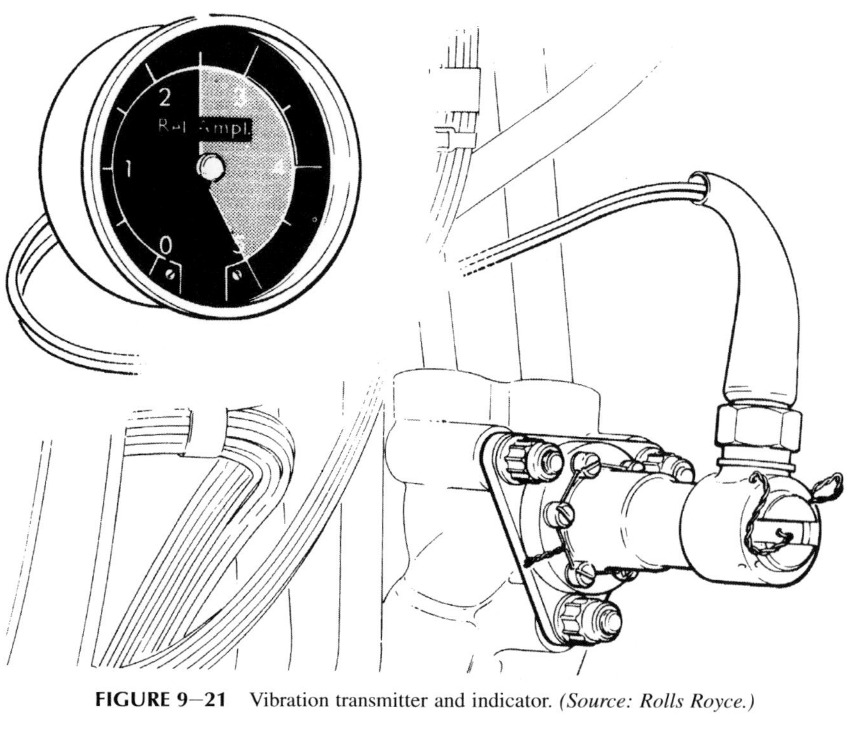

FIGURE 9–21 Vibration transmitter and indicator. *(Source: Rolls Royce.)*

engine to be operated by manual throttling should a main fuel system failure occur.

In addition to a fire warning system, a number of other audible or visual warning systems can be fitted to a gas turbine engine. These may be for low oil or fuel pressure, excessive vibration or overheating. Indication of these may be by warning light, bell, or horn. A flashing light is used to attract the pilot's attention to a central warning panel (C.W.P.) where the actual fault is indicated.

Other instruments and lights warn the pilot of the selected position of the thrust reverser, the fan reverser, or the afterburner variable nozzle, when applicable. Gauges also inform the pilot of such things as hydraulic pressure and flow and generator output, which are vital to the correct operation of the aircraft systems.

Aircraft Integrated Data System

The aircraft integrated data system (A.I.D.S.) is an extension of the "black box" aircraft accident data recorder. By monitoring and recording various engine parameters, either manually or automatically, it is possible to detect an incipient failure and thus prevent a hazardous situation arising.

Selected performance parameters may be recorded for trend analysis or fault detection. Existing instruments are used, wherever possible, to provide the signals to a magnetic tape. Further instrumentation, recording air pressure from points throughout the engine, oil contamination, tank contents and scavenge oil temperature, may be provided as required for flight recording.

After each flight the magnetic tape is processed by computer and the results are analyzed. Any deviation from the normal condition will enable a fault to be identified and the necessary remedial action to be taken.

Electronic Indicating Systems

Electronic indicating systems consolidate engine indications, systems monitoring, and crew alerting functions onto one or more cathode ray tubes (C.R.T.s) mounted in the instrument panel. The information is displayed on the screen in the form of dials with digital readout and warnings, cautions and advisory messages shown as text.

Only those parameters required by the crew to set and monitor engine thrust are permanently displayed on the screen. The system monitors the remaining parameters and displays them only if one or more exceed safe limitations. The pilot can, however, override the system and elect to have all main parameters in view at any time (Figure 9—22).

Warnings, cautions, and advisory messages are displayed only when necessary and are color coded to communicate the urgency of the fault to the flight crew. Provision is made to record any event or out of tolerance parameter in a nonvolatile memory for later evaluation by ground maintenance crews.

Electronic indicating systems offer improved flight operations by reducing the pilot workload through automatic monitoring of engine operation and a centralized caution and warning system. Reduced flight deck clutter is

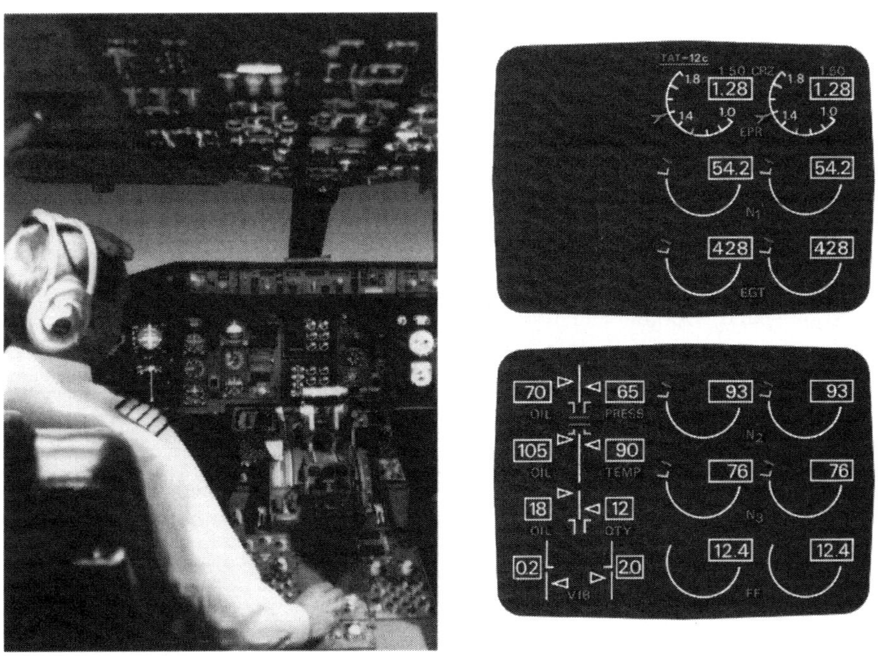

FIGURE 9—22 Typical electronic indicating display. *(Source: Rolls Royce.)*

another feature as the multitude of instruments traditionally present are replaced by the C.R.T.s.

Synchronizing and Synchrophasing

Synchronizing and synchrophasing systems are sometimes used on turbo-propeller engined aircraft to achieve a reduction of noise during flight.

On a multi-engined aircraft, a synchronizing system ensures the propeller speeds are all the same. This is achieved by an electrical system that compares speed signals from engine-mounted generators. Out-of-balance signals, using one engine as a master signal, are automatically corrected by electrically trimming the engine speeds until all signals are equal.

A synchrophasing system ensures that any given blade of an engine propeller is in the same relative position as the corresponding blade of the propeller on the master engine. This again is automatically achieved by very fine trimming of engine speeds resulting from phase signals from the synchrophasing generators.

On turbo-jet engines, synchronization can be achieved in a similar manner to that used for a turbo-propeller engine. On multi-spool engines, only one spool is synchronized. Manual trimming of engine or shaft speed can be done with the assistance of a synchroscope. This visually indicates, in comparison with a master engine, if the other engines are running at exactly the same speed; the normal engine speed indicator is, of course, not sufficiently sensitive to use for synchronizing.

We note that in the aircraft engine system just described, a switcher on the vibration monitors can choose which area of the engine the operator wishes to look at. With aircraft engines, the tendency is to use one or two probes set in the engine cowling near the location of atmospheric air intake. These probes generally are accelerometers with a large flat frequency response range that can detect vibration signals throughout the entire gas turbine (and transmission gearing if the aircraft engine in question is on a helicopter).

In land-based applications, it is intrinsically more practical to have a variety of probes (displacement, velocity, and acceleration) as space, weight, and footprint are not as big an issue as they are in aviation. What follows is a summary of control and instrumentation features on a gas turbine in power generation service. From this summary and the preceding aircraft engine C&I system description, we note that there may be a wide variation in the details of all C&I systems. Basically, however, all measure temperature, pressure, vibration, and physical position to varying extents and for different purposes.

Gas turbines in power generation service onboard ships and offshore platforms have very similar systems. Gas turbines in mechanical drive service do not differ greatly in this regard either.

MARINE C&I SYSTEMS

The environment around a marine gas turbine is somewhat more benign than most aircraft applications; with fewer constraints on weight and space, there is more scope for installation options and therefore the use of less rugged equipment. Even so, it is common for these products to use existing aerospace components, particularly in the fuel system. These may be mounted on the engine or assembled onto a fuel "skid" mounted in the engine enclosure. Similarly, electronics can be engine-mounted within the enclosure or outside in conventional equipment racks, or may be split between a number of these locations, communicating by means of digital data busses.

The marine gas turbine may mechanically drive a propeller, water jet, or other propulsion system, or may drive an electric generator providing power to electric motors, which in turn provide the motive power. The engine controller is therefore required to interface with the control system of this additional equipment in order to optimize overall performance.

In addition, there is usually a separate system providing the human-machine interface the means by which the ship's crew provide inputs to the propulsion system and monitors its performance. The system may need to be operated from several different locations on the ship: the engine room, the captain's chair in the center of the bridge, or from bridge "wings" during close harbor maneuvering.

TYPICAL C&I SYSTEM, LAND-BASED (POWER GENERATION)

Instrumentation*

Instrumentation on contemporary gas turbines, steam turbines, or GTCC can get quite involved. The basic instrumentation includes pressure and temperature readings. With the use of orifice plates, flow can also be measured.

The simplest of C&I systems may have analog indicators for these parameters. Modern plants have:

- Digital readouts of all parameters measured on a C&I system
- Automatic alarm and trip levels on several parameters
- Integration of the C&I with diagnostics
- "Artificial intelligence" (AI), in some of the more advanced engines (if my turbine's CDP is higher by a certain amount than usual, and the CDT has also risen

* Source: C. Soares, course notes, Engine Condition Monitoring and "Basic Operations and Theory of Gas and Steam Turbines, Cogeneration and Combined Cycle Plants," 2005.

by a certain amount with x and y conditions, for instance, then the wash cycle on my turbine needs to be automatically initiated)

Artificial intelligence tends to be popular with researchers and designers, particularly those who have security clearances and can get funding from specific militaries. The USAF (US Air Force) has specific firms under contract to work on nothing but AI for gas turbine engines.

Operations engineers, however, tend to be thoroughly skeptical of "AI that initiates control functions." Most of us know of many instances where the simplest parameter to read is "read wrong" at a critical moment. So AI that delivers options and leaves the decision making to human operators tends to be more popular. In any case, most alarm and trip levels have a manual override. Since human operators like to feel "in control," digital readouts of all parameters that affect the engine's health are provided within easy viewing distance of the on-duty operator's chair.

Other critical parameters include the ones that measure the turbine's vibration level, where measurement uses:

- Accelerometers
- Velocity probes
- Displacement (proximity) probes

Accelerometers are better at detecting high-frequency response problems, such as blade tip rub and gear tooth wear.

Velocity probes detect "intermediate" frequency problems (such as shaft misalignment). They have more directional sensitivity than accelerometers.

Proximity probes measure shaft movement in their own local "vision" and no further. They detect low-frequency problems, such as unbalance and oil whirl (whip). They may detect shaft misalignment.

Other items that may be measured include monitoring nonintrusive wear and oil debris.

Nonintrusive Wear Monitoring

A few users have tried nonintrusive wear monitors (in applications where bearing or other component deterioration might be an issue). This technology uses neutron bombardment and has not grown in popularity over the past decade although it aroused considerable interest initially.

Oil Debris Monitoring

Oil debris monitoring is better suited to reciprocating machines, because of the lead time involved in analyzing the readings. The technique may be used online but is a long way from being done in real time. Reciprocating machines turn slower, so a sample may be taken and analyzed in time to indicate a trend. This is rarely the case with rotating machinery. Once metal particulates are detectable in an oil sample on a rotating machine, progression to failure is generally swifter than the analysis cycle (see Figure 9–23).

Transducers

Pressure measurement is conducted by monitoring the deflection of a diaphragm with strain gauges.

Pressure transducers have a given range where output signal versus pressure change is a linear function. They are temperature limited to about 177°C. So, with GTs, probes installed on the casing direct air to the transducer. See Figure 9–24, which refers to a "typical" GT. This may be typical for older models. Contemporary GTs generally monitor far more pressure (and corresponding temperature) readings.

Temperature measurement is conducted using thermocouples and RTDs (resistive thermal detectors, normally platinum).

The primary locations for installing temperature measurement are as follows:

1. Turbine blade exhaust temperature. These readings measure blade path temperature for the blade stage they are next to and frequently employ a redundant sensor for increased reliability.
2. GT exhaust. Generally thermocouples are used in this location.
3. Redundant RTDs measure babbitt face bearing (oil) temperature.
4. Compressor inlet and discharge temperatures.

Typical (Nonaeroengine) Controls

The overall functions of a turbine system are to:

- Provide speed and temperature control
- Control the unit during operation, primarily normal operation
- Protect the gas turbine system
- Start up and shut down the system

Typically, magnetic transducers measure speed of a shaft that holds a toothed wheel (which engages with the main shaft). This is speed of the shaft at any given instant, which is compared to the desired rotational speed, and the error in shaft speed is computed. This reading is taken into account by a PID (proportional integral derivative) algorithm, and the fuel supply is trimmed to eliminate the error in minimal response time.

Typically, the temperature readings at the exhaust of the gas turbine are averaged and compared to a desired set point (for the relevant running conditions). The difference is called the *temperature error.* This is then

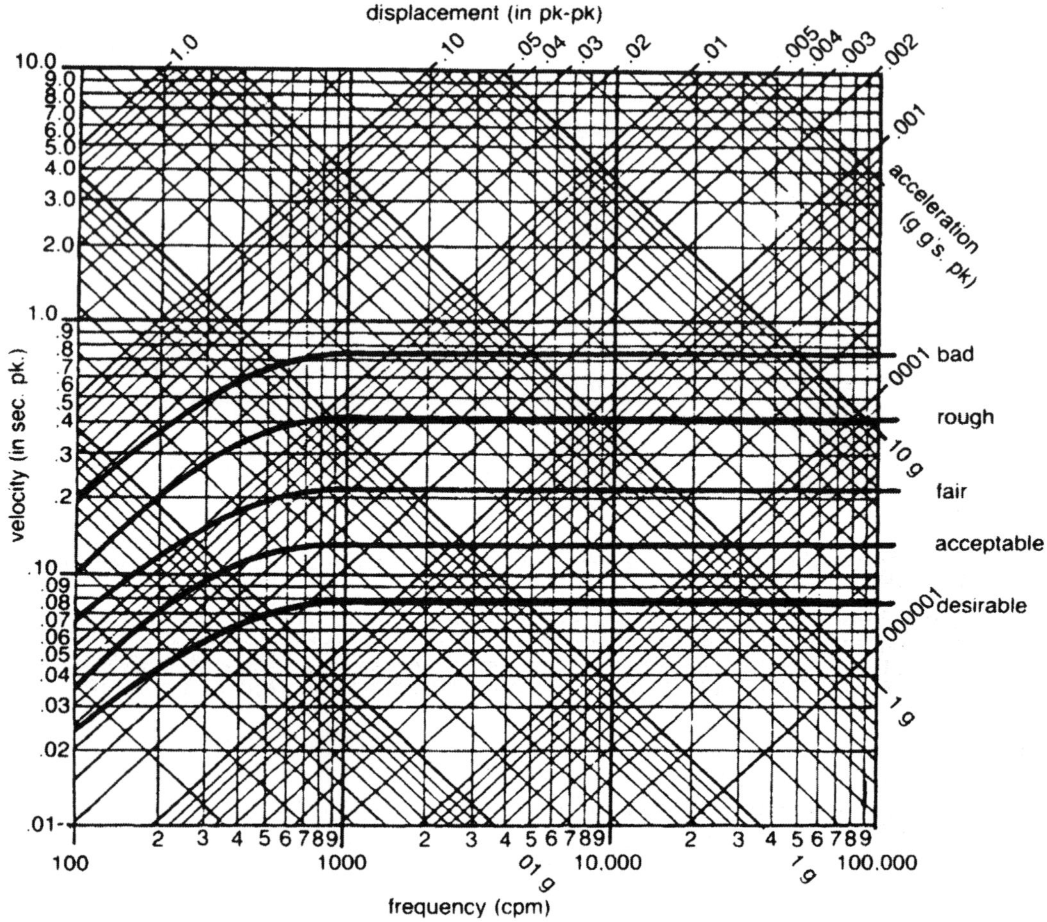

FIGURE 9–23 Vibration nomograph and severity chart. *(Source: Kiameh,* Power Generation Handbook. *New York: McGraw-Hill, 2003.)*

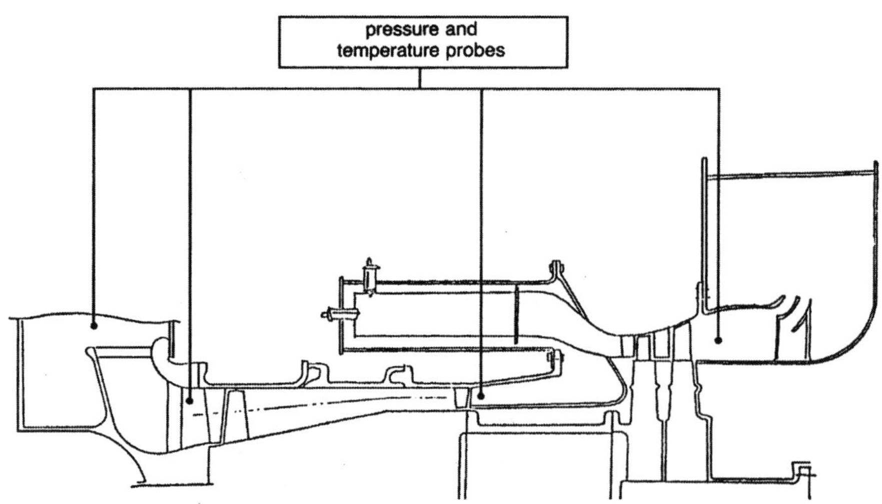

FIGURE 9–24 Locations of pressure and temperature probes on a typical gas turbine. *(Source: Kiameh,* Power Generation Handbook. *New York: McGraw-Hill, 2003.)*

adjusted for, according to a PID function, by fuel valve adjustment.

The protective systems listed earlier in this section protect the turbine system during the following events:

- Overspeed. A transducer mounted on the accessory gear shaft trips the unit at about 110% operating speed.
- Overtemperature. The system relies on thermocouples similar to those used for temperature control.
- Vibration. Sensors are set at alarm and trip levels.
- Flame out. Two ultraviolet flame detectors monitor flame status. In a can-annular design, flame detectors are mounted in cans that do not also hold the igniters. Two detectors have to indicate flame out to trip the machine.
- Loss of lubrication.
- Electrical open circuit, ground, or short circuit faults. The system warns the operator (signal light, sound, and/or reading, depending on system design).

Protective Systems

Generally, the GT I&C has the following protective measures:

- Low lube oil pressure. The turbine trips when oil pressure in the bearing header drops.
- High vibration. The turbine trips on high vibration. Sensor choice, technology, and placement vary considerably with different GT models and applications.
- Turbine overspeed. The turbine is protected against overspeed, generally by a triple-redundancy system: magnetic pickups inside the bearing cavities (near the thrust and journal bearings) monitor the speed. They trip the unit on overspeed. Overspeed bolts also may be used.
- High lube oil temperature. RTDs embedded in the babbitt monitor the temperature and, therefore viscosity, of the lube oil.
- Exhaust temperature. Thermocouples (numbers vary, 16 is a common number) measure exhaust temperature and trip the turbine if required.
- Blade path temperature. A "safe" (temperature limiting) location is chosen for turbine temperature readings. The turbine inlet temperature generally is too high for most thermocouples on the market. A number of thermocouples (16 is a typical number) measure temperature at a given turbine station. Allowing for combustor geometry, this then gives the state of individual combustors in a can-annular design or segments of the combustor in an annular design. Some designs average the readings, others take adjacent thermocouple readings. The latter also may set alarm levels, unload, and shutdown levels. A typical alarm level is 50°C, 67°C is a typical unload level, and 72°C is a typical shutdown level.
- High acceleration. Detectors trip the turbine on high acceleration, which may occur during startup, at incipient surge, or upon load rejection.
- High thrust bearing pad temperature. A thrust bearing, generally a tilting pad type, keeps the shaft from moving axially. Two of the pads generally support the thrust collar on either side. When they wear, the temperature inside the bearing cavity rises. RTDs are set to trip the turbine on high temperature. In some applications, proximity probes are set to serve the same function.
- Low or high GT inlet vacuum. A damaged or clogged filter can drop or raise, respectively, the GT inlet vacuum pressure. A pressure switch can be set to shut off the unit at predesignated values.
- High turbine exhaust pressure. A trip on high exhaust pressure can be set to shut off the unit in case of downstream (e.g., boiler) problems.

Permissives (Interlocks)

Permissives are conditions that must be met before a startup or operation can continue. Typically, they are as follows:

- Combustor outfire. The ultraviolets must confirm combustion within a specific time (generally a few seconds) or the turbine shuts down.
- Low compressor discharge pressure. This value must be high enough to sustain safe combustion. If not, the startup sequence stops.
- Low lube oil temperature. If this value is too low, startup stops.
- Low lube oil pressure. The turning gear stops if this value is too low.
- Too high or too low gas supply. Starting is prevented if these values occur.

Liquid fuel permissives:

- Low fuel pump suction pressure. The pump trips if this occurs.
- High or low differential pressure (pressure between the liquid fuel manifold and the compressor discharge) across the fuel manifold. The unit trips if either value is reached.
- Fuel transfer (switching from gas to liquid fuel) failure. The unit trips if this occurs.

Startup Sequence, GT

This proceeds as follows:

1. All auxiliaries in the starting position.
2. All gear and oil pumps operating.
   - A turbine system may have as many as three spare lube oil pumps, two AC and one DC (battery backup if AC fails) in addition to the main engine pump,

which is driven off the main GT shaft. The AC pumps keep the machine safe before the rotor has developed its speed on startup and on shutdown.
- When oil delivery pressure from the auxiliary lube pump reaches a certain value, the turbine turning gear is started. This must occur within a certain time, generally about 30 sec or the unit shuts down.

3. Fuel system ready.
4. Cranking phase:
   - Energize starting motor. Note: When the turbine gear is operating well, the starting motor is activated if the lube oil pressure is high enough. At about 15% speed, the turning gear can be de-energized.
   - Accelerate rotor to ignition speed (generally about 1000 rpm). Note: At ignition speed, the turbine o/s trip solenoid and vent solenoid are reset. When oil pressure is high enough, the o/s trip bolts (which trip at about 12.5% o/s) reset.
   - Energize ignitors. This occurs when the o/s trip bolts have reset. The energized ignition circuit in turn energizes the ignition transformers and ignition timer (allowed 30 sec to establish the flame on both detectors).
   - The preceding step, in turn, opens the fuel valves and starts atomizing air.
5. Acceleration phase. Note: Following fuel injection and ignition confirmation, the no-load set point (or speed reference) is increased. The fuel valve now opens further to increase speed. Shaft acceleration rate is limited by blade design and turbine EGT (exhaust gas temperature).
   - Start ramp controller accelerates the rotor to about 89% of operating speed over 20 min.
   - Start motor stops at about 50–66% of the operating speed. This value is called *self-sustaining speed.*
   - Speed controller accelerates rotor from 89–100% operating speed. Note: If desired acceleration is not maintained, the unit is tripped, as with high acceleration, compressor surge could occur and the rotor could heat faster than the stator, leading to tip rub.
6. Generator synchronization. Attempt only if
   - The generator frequency is slightly higher (say by 0.05 Hz) than the grid frequency.
   - The generator voltage matches grid voltage.
   - Generator phase voltage matches grid phase voltage.
   - Note: If commissioning, ensure that phase sequences in the generator and grid are matched.
   - Synchronous accepter relay must confirm independently that synchronization conditions are met (after an automatic synchronizer changes generator speed and voltage to match the grid) before the CB can be closed.

   Then the generator CB closes automatically or manually.
7. Loading: Load is increased by opening the fuel valve until required load is reached.
8. Operation for an exhaust temperature control system.

This system compares CDP with EGT and signals the fuel control valve to make changes as required. As CDP rises, the controller changes the set point of the EGT; it must drop. The controller signals the fuel control valve to close to the extent required. Power output stays at 100% while the change (less fuel flow) is made. This indicates the turbine runs more efficiently when ambient air temperature drops. However, this does not imply that the turbine is controlled by ambient air values. This is because CDP is also affected by:

- Compressor fouling
- Compressor blade and general module condition
- Ambient pressure
- Ambient humidity

Note: Other control mechanisms include:

- Inlet guide vanes (IGVs). These are partially closed when the unit is still developing power. When the GT develops more load, the IGVs open. For a combined cycle, when the GT is at part load, the IGVs close, which raises GT EGT and increases the heat recovery steam generator's efficiency.
- Compressor bleed valves. These valves vent compressor air during startup to prevent compressor surge. They close when the GT reaches about 92% speed. Note that compressor bleed air is used in various accessory systems.

Transmitters for pressure and temperature have no control function. They provide a 4–20 mA signal over the range specified. Their function is to indicate parameter values only.

Shutdown Sequence, GT

1. A local or remote shutdown (SD) request is made.
2. Fuel is reduced to zero load.
3. The main CB to the grid and the field CB (auxiliary loads) are opened and fuel valves tripped.
4. Note: For emergency SD, the CB and fuel valves are tripped before load reduction.
5. GT speed and oil pressure from the main pump drop. The DC auxiliary pump starts.
6. At 15% speed, the turning gear motor starts. When at 5 rpm (turning gear speed), the turning gear clutch is engaged.
7. The unit is purged of fuel with air (about five times volume of GT casing).
8. The turning gear is run for up to 60 hr to prevent shaft bow.
9. The turning gear and aux lube pump are shut down.

Starting Systems

The two energy sources for starting a gas turbine are

- Active energy, such as grid power, supplied to the starting motor or generator that acts as a startup motor during startup or an ICE (internal combustion engine).
- Stored energy, which may mean batteries, compressed air, compressed gas, or hydraulic oil. The stored energy method has obvious GT size limitations.

Engine starting systems can also be used to

- Rotate the compressor for offline washing.
- Supply air to cool the GT after an emergency trip before restart is attempted.

If the latter use is not employed, note that the rotor must be turned at slow speed (about 6 rpm for 4 hours) after emergency trips to prevent shaft sag.

SIGNIFICANT ADVANCES IN CONTROLS INSTRUMENTATION AND DIAGNOSTICS TECHNOLOGY

Advances in CID are varied and several. Field operations provide a proving ground for everyone from software engineers to gas turbine performance engineers hungry to squeeze every decimal point worth of efficiency out of their machines. There is no way that a book of any kind can cover everything of relevance to CID with GTs in one chapter or, for that matter, one volume. However, a few subfields within this specialty point the way to future concentrations of effort by both OEMs and end users. The case studies and program descriptions that follow provide details of current and future technologies in these areas. They include but are not limited to:

- Optical pyrometry
- Digital telemetry (data transfer)
- Pulsation measurements for diagnostic purposes (described with a case study)
- Performance and C + I system verification with modeling (described with a case study)

Many other new technologies would be of interest to a gas turbine engineer. In the interests of available space, I include some Internet addresses in Chapter 20. Again, the reader is cautioned about the interrelated nature of gas turbine technology, as this may mean that a link to an interesting diagnostic technology may appear in, for instance, Chapter 10, Gas Turbine Performance, Performance Testing, and Optimization, or Chapter 12, Maintenance, Repair, and Overhaul.

Performance verification with modeling is not a new field. However, when an OEM chooses to reveal some significant aspects of its methodology in this endeavor, it is worth a careful look. So there is a case study on a specific OEM's (Rolls Royce's) strategy for performance verification with modeling.

Optical pyrometry is used for accurate measurement of higher temperatures. Some OEMs, by design philosophy, expect to raise peak turbo inlet temperatures (TITs) to raise overall GT efficiency. Although the empirical equations confirm this to be possible, there are limitations based on metallurgy, cooling air flow available, hot spots in the hot section caused by unforeseen circumstances, quality control manufacturing or overhaul errors and a variety of other factors. This notwithstanding, accurate measurement of TITs, or a temperature close to the IGVs that can be extrapolated to provide TIT values, is essential for proving thermal efficiency values. This is particularly critical with the increase in "power by the hour" contract trends granted to OEMs, as hot section (generally the most expensive) component lives are particularly dependent on peak operating temperatures and uniformity of those temperatures.

TIT is extrapolated back from EGT readings taken by a series of thermocouples placed in the exhaust duct. Frequently, in older engines, all the thermocouple readings were averaged and no trouble was taken to determine temperature differentials between adjacent thermocouples to get indications of possible hot spots in the hot section further upstream. Newer engines have this feature.

Optical Pyrometry

Today, optical pyrometry is being optimized to provide accurate temperature readings of the TIT zone, as are other nonintrusive temperature measurement techniques. This is a highly specialized field, and very few players invest in developing this technology. Currently, it is used mainly in development work.

The optical pyrometry players include Rotadata in the UK and Siemens (sometimes in concert with the Technical University of Berlin). To help illustrate this new field, the following information is included in the sections that follow:

- Two (trademarked) products developed by Rotadata and their function (contact the company for more details regarding accuracies and tolerances)
- A case study of Siemens' work with the University of Berlin

The latter item is included for its interest value illustrating grassroots development in this new field. As always, once a technology is mature, much of the "hows and whys" are no longer evident in newer literature.

*Turbine Blade Optical Pyrometry Mapping (Rotamap 1™)**

Application

Rotamap (trade name) is a traversing infrared pyrometer system that is designed and programmed to scan the surface of rotating turbine blades in order to measure their surface temperatures at full engine operating speed. This approach to IR (infrared) pyrometry is to perform a radiance measurement of the turbine using small spot sizes and high bandwidth performance coupled with very high digital data transmission to ensure data integrity. These measurements can then be turned into very high quality temperature maps using various correctional techniques applied either by customers (mostly OEMs) or the manufacturer of this technology or jointly.

Principle of Operation

An air-cooled probe is located in a sealed traverse actuator, mounted onto the engine. The probe is inserted into the turbine annulus in front of or behind the blade row and is traversed radially into the engine. By correct orientation, radiance from almost any part of the rotating blade surface—from root to tip and on suction or pressure surface—can be scanned.

The pyrometer optics are set to a specific focal length, dictated by the engine installation. A periscope mirror at the base of the probe reflects the radiance from a small (2–3 mm diameter) "spot" on the blade surface. This is captured by the detector in the head of the pyrometer and fed via high-speed digital link to a groundstation. Here these data are stored and processed, prior to presentation in whatever manner is required.

Clear signals and significant "oversampling" through the use of a digital high speed acquisition system permits the filtering and removal of radiation spikes caused by combustion flares, reflection or particles.

By linking the captured data to a "once per revolution" signal from the engine, and the knowledge of the radial position of the scan, a map of radiance for each blade may be generated. When this is corrected for emissivity factors (blade incidence angle, material, surface finish or blade coating, etc.), blade temperatures are revealed.

*Turbine Blade Optical Pyrometry Mapping (Rotamap 2™)**

Application

Rotamap 2 is a self-contained infrared pyrometer system that is designed to operate in the hottest of gas turbines and is programmed to scan the surface of rotating turbine blades in order to measure their surface temperatures at full engine operating speed.

This approach to IR pyrometry is to perform a radiance measurement of the turbine using small spot sizes and high bandwidth performance coupled with very high digital data transmission to ensure data integrity. These measurements can then be turned into very high quality temperature maps using various correctional techniques applied either by customers or the manufacturer of this technology or jointly.

Principle of Operation

A special air-cooled probe and actuator mechanism is mounted on the engine casing. The probe is located just outside the hot gaspath. When blade temperature measurements are required, the actuator mechanism inserts the probe a short distance into the hot gas path (10–15 mm). Then a separate actuator system oscillates a special high temperature mirror at the base of the probe to reflect the radiance from a small (2–3 mm diameter) "spot" on the blade surface as it scans from the blade tips to their roots.

The pyrometer optics are set to an appropriate focal length to ensure the desired spot size at the targeted blades. The blade radiance is captured by the detector in the head of the pyrometer and fed via high-speed digital signal to a groundstation. Here these data are stored and processed, prior to presentation in whatever manner is required.

Clear signals and significant "oversampling" through the use of a digital high-speed acquisition system permit the filtering and removal of radiation spikes caused by combustion flares, reflection or particles.

By linking the captured data to a "once per revolution" signal from the engine, and the knowledge of the radial position of the scan, a map of radiance for each blade may be generated. When this is corrected for emissivity factors (blade incidence angle, material, surface finish or blade coating, etc.), blade temperatures may be mapped.

*Use of Optical Pyrometry with the Siemens Vx4.3 Gas Turbines***

This section will discuss two pyrometer systems that have been built that meet the design requirements for use in heavy-duty gas turbines. They are of the "test bed" standard probe design that consists of a 2000 mm long cylindrical probe with an outer diameter of 27 mm that is reduced to 22 mm outer diameter for the part that goes into the hot gas path (Figure 9–25). These versatile probes are designed for use in turbine stages one, two, and three. Thus, the

* Source: Courtesy of Rotadata, UK.

** Courtesy Siemens, adapted extracts from T.O. Holt, H. Eisenlohr, and D. Raake, "Application of a High Resolution Turbine Pyrometer to Heavy Duty Gas Turbines," 2001-GT-0577.

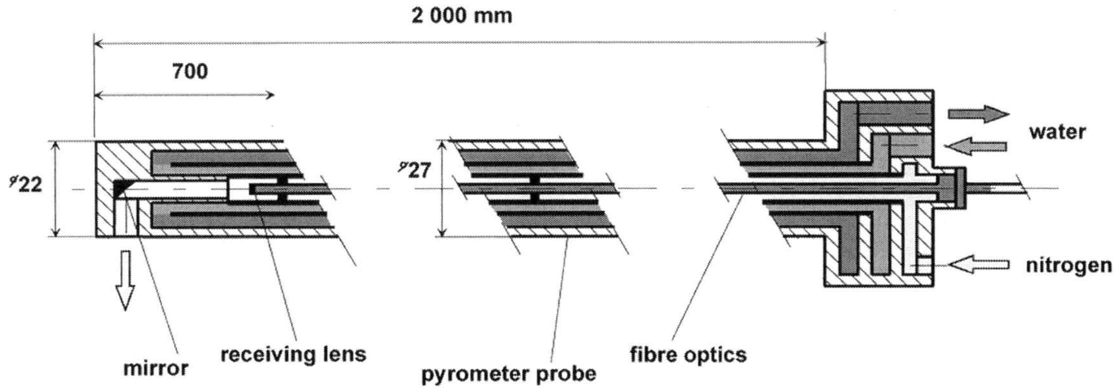

FIGURE 9–25 Schematic of the pyrometer probe. *(Source: Siemens.)*

maximum insertion depth ranges from 80 mm in the first stage of the V64.3A gas turbine (70 MW) to 400 mm in the third stage of the V84.3A (180 MW).

In the first stages the thermal stress is dominant with peak values of several hundred MPa, whereas in the second and third stage the bending stress becomes more and more important (up to 650 MPa). It is evident that only superior chrome nickel alloys and state-of-the-art manufacturing techniques such as electron beam welding have to be applied in order to achieve probes that are suitable for this wide field of application.

With regard to the high temperatures, the probe is water cooled by a stainless steel 24 stage high-pressure impeller pump. Furthermore, the optical parts are purged and cooled by high-pressure nitrogen, a portion of which is discharged through tiny slots in the mirror, in order to cool the very tip of the probe. A closed loop control allows a predetermined nitrogen overpressure to be kept constant at all operating conditions, e.g., $\Delta p = 1$ to 5 bar. Even though the actual pyrometer is miniaturized a 22 mm outer diameter of the hot gas part is required, due to space for the sturdy sight tube, the cooling water and the nitrogen channels.

Application to Heavy-Duty Gas Turbines

The main purpose of temperature measurements on turbine blading is the experimental evaluation of numerical calculations that have been applied during the design process. Figure 9–26 shows a schematic of traverse units for linear and angular movements of the probes mounted on a gas turbine. Usually at least two pyrometer systems are applied at the same time, such that the leading edge (pressure side) and the trailing edge (suction side) of one blade stage can be investigated simultaneously. In addition, new vanes and blades can be investigated, using the test bed gas turbine as component carrier. As a side effect, quality assurance of all rotating blades may be performed by optical pyrometry. Even though thermocouples and thermal paint are largely applied for temperature measurements, optical pyrometry is widely used by several gas turbine manufacturers for turbine blade and vane testing. The systems applied are based on commercially available sensors, e.g., by LAND.

Since the required technique for the test bed was not readily available at the time, a high-resolution pyrometer has been developed in a joint effort between Siemens Power Generation and the Technical University of Berlin, where the basic research work was carried out at the Institute of Internal Combustion Engines.

A visual impression of the high-resolution prototype pyrometer probe and the data acquisition unit is given in Figure 9–27. The optical access at the tip of the probe and the flexible light guide, which may be as long as 15 m, are clearly visible. During operation, the sturdy data acquisition unit, containing both the electronic receiver and a robust industrial PC, is mounted adjacent to the gas turbine.

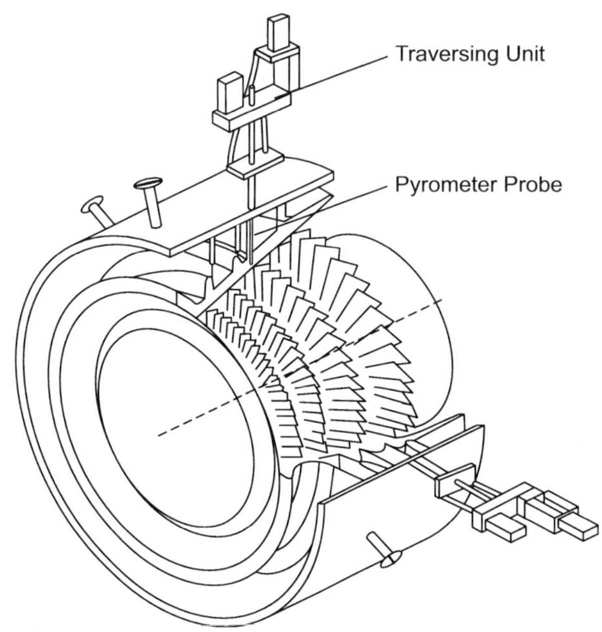

FIGURE 9–26 Pyrometer system mounted on a heavy-duty gas turbine. *(Source: Siemens.)*

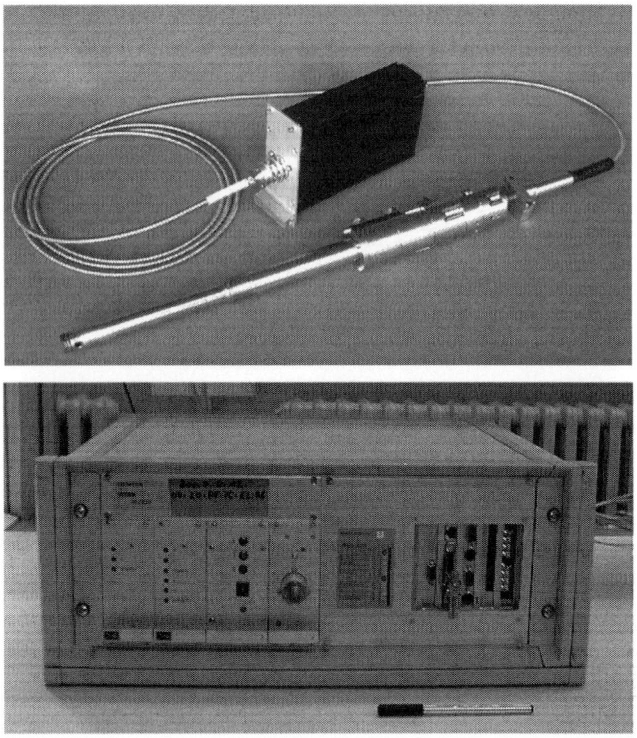

FIGURE 9–27 Pyrometer probe (prototype) and data acquisition unit. *(Source: Siemens.)*

Hence, measurement data is transferred via a local area network to the host PC in the remotely located control room. The pyrometer system is characterized by a temporal resolution of 1 μs, a minimum field of view of 1 mm, and a measurement range from 600 to about 1500°C. The cylindrical probe is of the traverse type with a modular design. Thus, the minimum field of view can be adjusted over a wide range and all components can be exchanged easily, e.g., in order to use up- or down-looking mirrors or to use optical components that are suitable for long or short infrared wavelength measurements.

Figure 9–28 displays the probes and traverse units mounted on a model V84.3A 180 MW gas turbine ready for testing. At the rear, the connections and flexible hoses for cooling water and nitrogen can be seen. Different traverse units with extension distances ranging from 500 to 3000 mm for probe diameters from 6 to 30 mm with maximum speeds of 1200 m/s are available for a large variety of investigations in the test bed.

Since the pyrometers are traversed into the hot gas path, it has to be demonstrated that the probes have no influence on the temperature distribution to be measured. This has been thoroughly investigated, making use of different traverse velocities and especially of different probe positions as shown in Figure 9–29(a).

The first stage blades can be measured directly from pyrometer 1 positioned at the first stage vane outlet or from

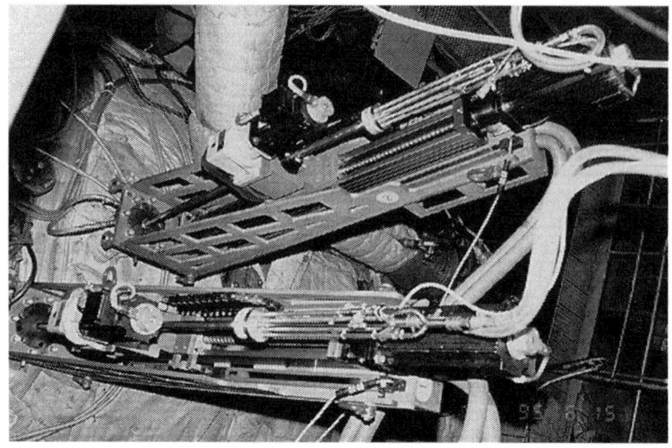

FIGURE 9–28 Pyrometer probes for surface temperature mapping on a model V84.3A gas turbine. *(Source: Siemens.)*

the combustor outlet (pyrometer 2). The result of the measurement is independent of the probe position for identical test conditions. Slight deviations are due only to decreased spatial resolution when the pyrometer is in position 2. Thus, unlike for flow probes, the 22 mm outer diameter size of the pyrometers has no influence on the results of measurement.

Figure 9–29(b) shows a schematic of the cylindrical pyrometer probe partly inserted into the hot gas path at the row 2 vane outlet in order to investigate the leading edge and pressure side of the row 2 blades. The blades are rotating past and the pyrometer receives temperature data from a certain radius. Additionally, a key phasor gives a once per revolution signal in order to determine the beginning of a new measurement.

During evaluation the 3D geometry software determines the intersection of the pyrometer beam and the rotating blade. Hence, each individual measurement point can be correlated to its origin on any blade.

The temperature vs. time plots of both systems for 10 blades are shown in Figure 9–30. The temperature values have been normalized by their mean value respectively. Both first stage blade measurements were accomplished during liquid fuel operation and are qualitatively in close agreement with one another.

However, maximum and minimum values are, as expected, considerably higher resp. lower for the high-resolution pyrometer. Thus, the state-of-the-art pyrometer provides a more realistic thermal image than the conventional system.

In order to evaluate the measurement results for the first stage, additional measurements are necessary, in particular for the stationary vanes. The radial hot gas temperature profile is determined by thermocouple probes that are traversed slowly into the flow channel, alternately to the pyrometer probes. Therefore, one can account for inhomogeneous temperature distributions at the outlet of the ring

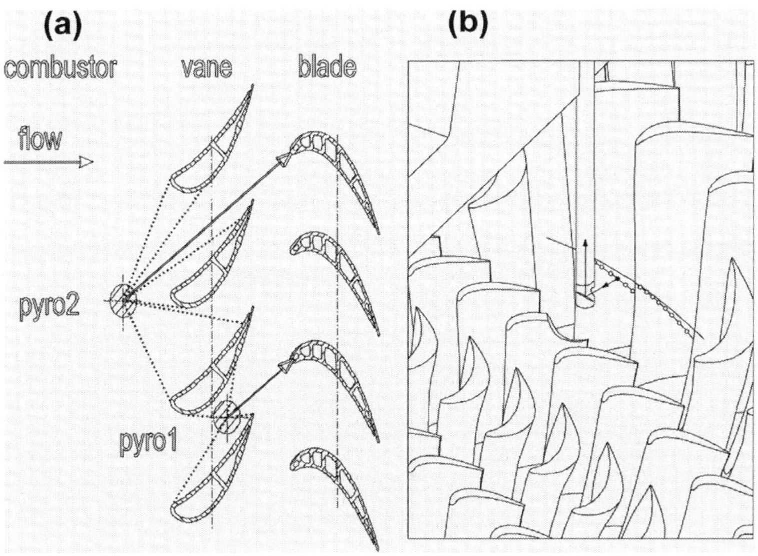

FIGURE 9–29 Schematic of the pyrometer probe in the GT: (a) Top view of the first stage vanes and blades with two positions for pyrometer and thermocouple probes. (b) Measuring the pressure side of the second stage blade. *(Source: Siemens.)*

combustor. Furthermore, the brightness temperature of the combustor outlet is measured by the pyrometer itself for natural gas operation at base load in premix mode. The results of these measurements are normalized by the maximum hot gas temperature and plotted against the normalized radius (Figure 9–31).

The hot gas temperature reaches its maximum relatively close to the hub, whereas the brightness temperature of the combustor shows maximum values close to the tip. The latter can be used for analytical corrections of the influence of combustor radiation. Please note that the brightness temperature of the combustor for liquid fuel operation is about 200 K higher at the same load conditions due to the bright yellow oil flames.

Thus, the error of the vane measurements can easily be in the magnitude of 100 K. Hence, the electronic receiver has been equipped with optical high pass filters that limit the effective wavelength from 0.75–1.10 μm, i.e., the visible portion is excluded. However, for vane measurements under natural gas premix operation (light blue flames, no soot) the combustor influence can be estimated as follows: a "true" vane temperature of 1000°C, a measured vane emissivity of 0.9 (±0.05) and a measured combustor brightness temperature of 1200°C yields a

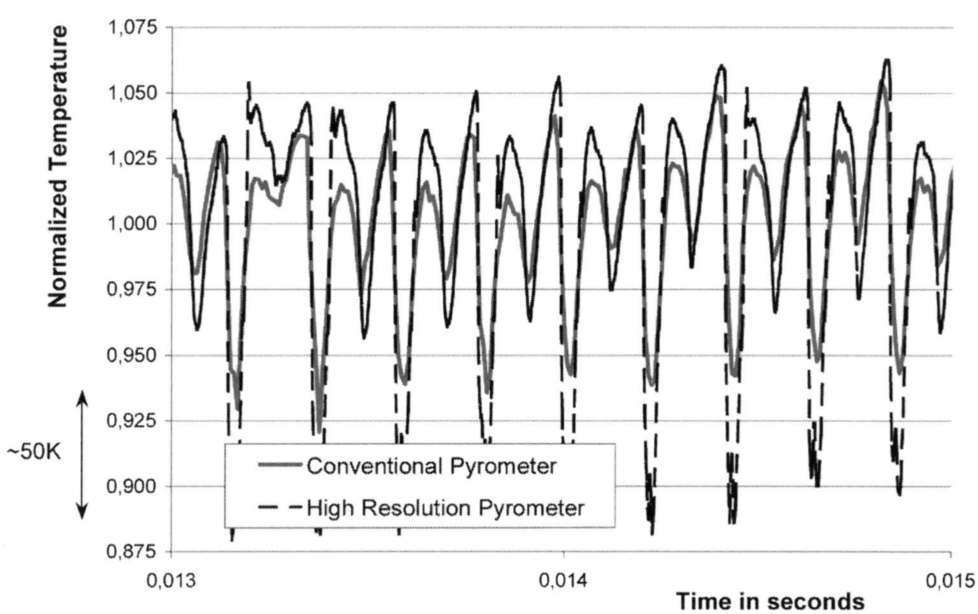

FIGURE 9–30 Comparison between normalized first stage blade temperature measurements with conventional and high-resolution pyrometers at midspan during fuel oil operation. *(Source: Siemens.)*

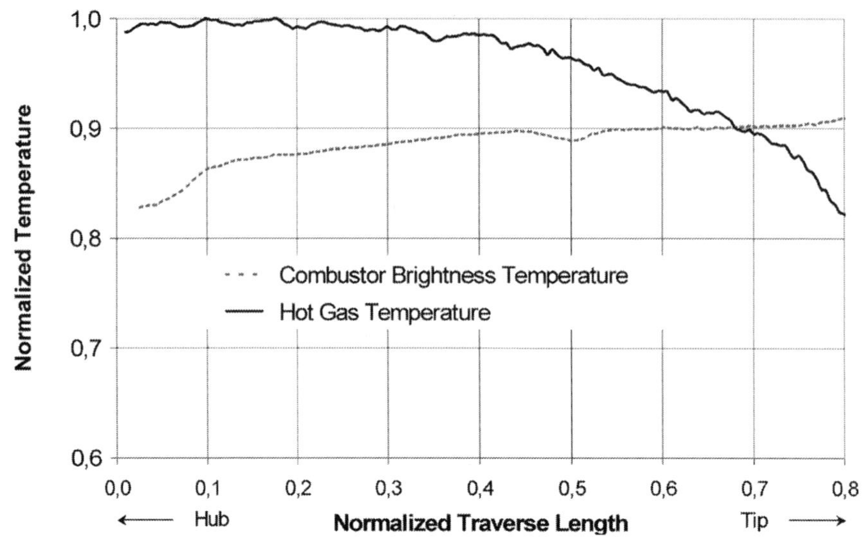

FIGURE 9–31 Normalized radial profiles of hot gas temperature and combustor brightness temperature at the combustor outlet (pyrometer position 2) for natural gas operation.

maximum error of about +40 K at the vane's leading edge; further downstream it is considerably lower.

In order to get temperature information from most of the blading surface, the pyrometer probes are operated in a scanning mode. For vane measurements, a certain viewing angle is chosen and the pyrometer traversed quickly (1200 mm/s) from the tip to the hub and subsequently withdrawn. Only the data acquired during the inward motion is evaluated, in order to process only undisturbed temperature values. After a waiting period of four thermal time constants of the vane a steady thermodynamic state was established again.

Pyrometer System*

Radiation from a body is governed by Planck's equation as described below.

$$q(T,\lambda) = \frac{\varepsilon \cdot C_1 \cdot \lambda^{-5}}{\exp\left(\frac{C_2}{\lambda \cdot T}\right) - 1} \quad (1)$$

where

C_1: the first Planck's constant

C_2: the second Planck's constant

q: radiance

T: target temperature in K

λ: wavelength

ε: emissivity

A pyrometer is an optical device that utilizes this correlation. It collects radiance emitted from an object and calculates the temperature.

A high-speed pyrometer system developed and supplied by Rotadata Ltd was used for L30A. Figure 9–32 shows the installation of the pyrometer. The probe head is fixed to the engine casing and the probe is inserted through the outer endwall of a vane. The vane is provided with a seal assembly that can reduce the loss of vane cooling air and absorb the relative displacement between the probe and the vane during engine operation. The probe usually stays outside the gas passage. During the measurement the probe is inserted into the gas passage and collects radiance. The measurement takes only about 20 seconds. After measurement, the probe is extracted from the gas passage again.

A small mirror is put inside the tip of the probe. The radiance emitted from a blade is reflected at the mirror and collected by the detector in the probe head. The data are

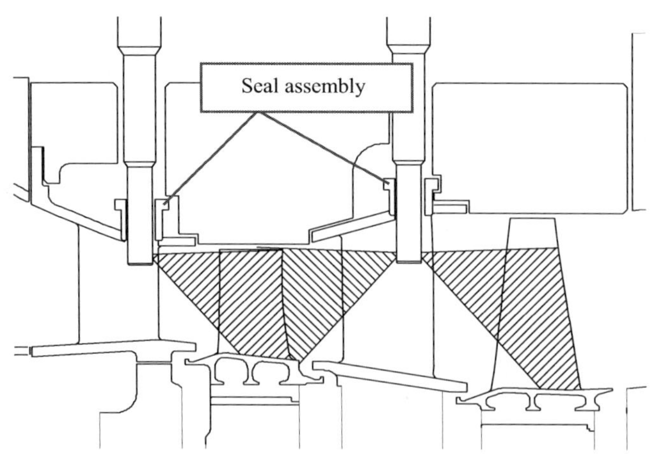

FIGURE 9–32 Pyrometer installation. *Proceedings of ASME Turbo Expo 2012: GT 2012 June 11–15, 2012, Copenhagen, Denmark. GT2012-68679. Application of an Optical Polymer to Newly Developed Industrial Gas Turbine, Taniguchi & Moritz et al., copyright 2012, courtesy of ASME.*

* Copyright: [9-1] Proceedings of ASME Turbo Expo 2012: GT 2012 June 11-15, 2012, Copenhagen, Denmark. GT2012-68679. Application of an Optical Polymer to Newly Developed Industrial Gas Turbine, Taniguchi & Moritz et al, copyright 2012, courtesy of ASME

transmitted via cable to the control computer. The mirror can continuously change angles to the probe axis. As the mirror oscillates, the radiance emitted from different radial positions can be collected. By synchronizing the data with once per rev signal of the rotor, each blade can be identified. The processing is automatically carried out in the system.

In the production engine, the vane with the seal assembly is assembled and plugged with special parts. When it becomes necessary to measure the blade temperature, the parts can be easily replaced with the pyrometer. And also, the seal assembly can be used as bore scope inspection ports and inspection inside the gas passage can be done through the ports at any time.

KHI have used Rotadata's pyrometer system for M7A and L20A gas turbines before and made most of the system. The probe head for them contains a Silicon PIN diode, which can measure temperature down to 700°C. And therefore, 1st and/or 2nd stage turbine blades are the main target of the measurement. But the new detector module for L30A is upgraded to the InGaAs detector. The new detector can give an output signal with less electrical noise on it and nominal operating temperature can be lowered to 600°C. KHI further made a calibration test and concluded that the new pyrometer can measure temperature down to 550°C with acceptable noise level. So the pyrometer system can be used for all stages of blades in L30A (see Table 9–1).

Pyrometer Calibration

The pyrometer was calibrated using a black body furnace before and after each engine testing. The calibration data were fitted to the following equation:

$$R = \frac{\varepsilon \cdot K \cdot C_1 \cdot \lambda^{-5}}{\exp\left(\frac{C_2}{\lambda \cdot T}\right) - 1} \tag{2}$$

T is the target temperature and R is the radiance measured and normalized by the pyrometer. ε is the emissivity of the target and is the value of 1 for a black body furnace. λ and K are the effective wavelength and the efficiency of the pyrometer respectively and these are determined using a minimum least squares error method.

TABLE 9–1 Major Specifications of Pyrometer System [9-1]

Measurement Range	600–1200 (nominal)
Resolution	+/– 1°C
Accuracy spot size	+/– 0.5% RD 2 mm diameter at 75 mm focal length
Detector	InGaAs
Sampling rate	1MHz

Figure 9–33 shows a calibration curve obtained before and after a certain engine testing. Slight degradation of sensitivity was found after engine testing. During the engine testing, high-pressure air supplied from a commercial compressor was used as purge air. Although a micro mist filter was installed between the compressor and the pyrometer, the optical system in the pyrometer was probably contaminated by the pollutant in the compressed air. Before next engine testing, the pyrometer was disassembled and the optical system was cleaned.

Repeatability of the pyrometer was also examined. Calibration was repeated six times and a calibration curve was obtained for each case. Figure 9–34 shows the temperature calculated with each calibration curve at a fixed and normalized radiance of 800. The deviation is within +/– 2°C, indicating that the pyrometer has sufficient repeatability.

In the calibration, the effective wavelength of the pyrometer is determined to be approximately 1.6 μm. Although gaseous medium like water vapor and CO_2 in the main gas stream also absorbs and emits radiation, these absorptions and emissions do not occur in the range between 1.5–1.75 μm.*

So in the engine testing, radiance measured by the pyrometer is not influenced by gas absorption and emission.

The short wavelength pyrometer which utilizes wavelength near 1 μm is not suitable for blades coated with TBC.**

Although TBC is applied to the production type of GGT-1B, several blades without TBC were assembled together.

Engine Installation

The pyrometer setup is shown in Figures 9–35 to 9–37. The pyrometer can be attached to several locations on the engine casing depending on the measurement target. At each position, it can be fixed at different circumferential angles. Thus, the detector can collect radiance emitted from a broad area on the blade surface.

As the probe is exposed to high-pressure and high-temperature main stream gas, it needs to be air-cooled during the engine testing. High-pressure air from a commercial compressor continued to be supplied to the probe at a maximum flow rate of 0.1 kg/s, which is far less than the main gas flow rate of 86.5 kg/s. In the engine testing of M7A, we confirmed that the cooling air of the pyrometer had little influence on the blade temperature. For L30A, the

* From: Schwerman, J. and Allen, G., 2003, "Special Topic Review Temperature Measurement Techniques; Full View Infrared Pyrometry," GE Global Research.

** From: Rooth, R.A. and Hiemstra, W., 2003, "Dual wavelength temperature monitoring of TBC coated Alstom 13E2 turbine blades," ASME paper GT2003-38814.

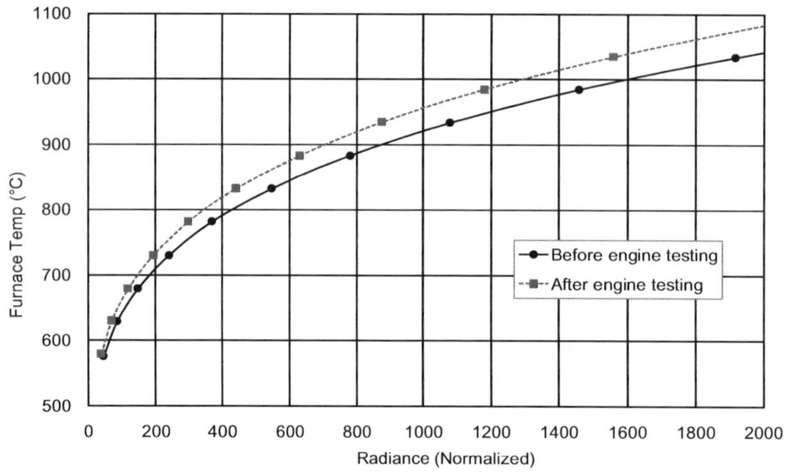

FIGURE 9–33 Pyrometer calibration curve. [9-1]

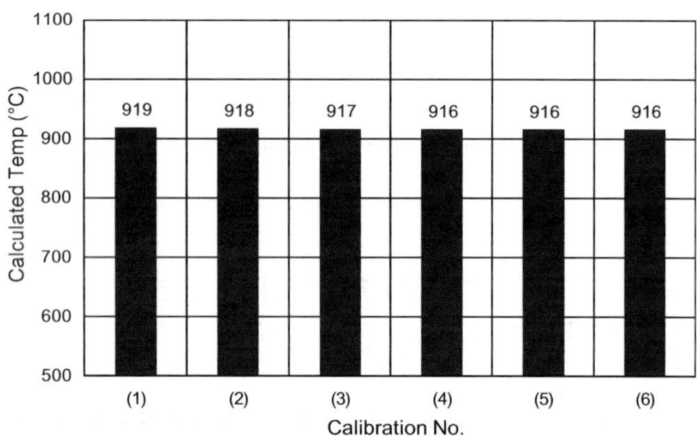

FIGURE 9–34 Calculated temperature at R=800. [9-1]

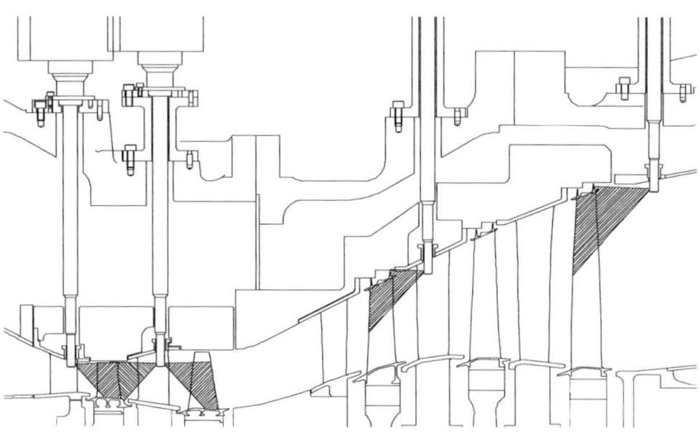

FIGURE 9–35 Pyrometer installation. [9-1]

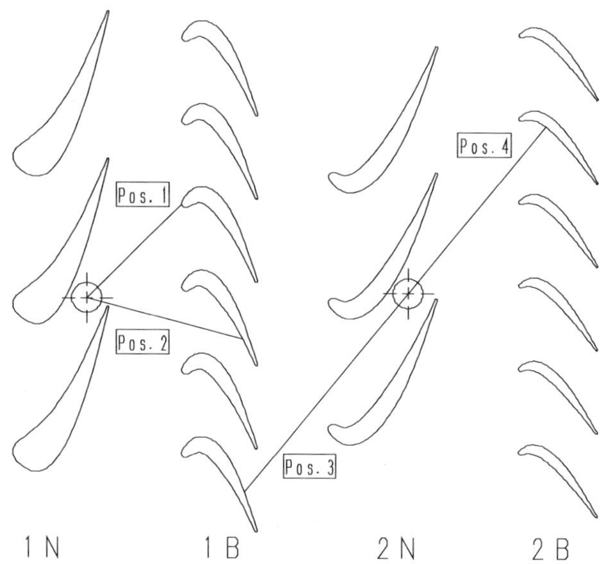

FIGURE 9–36 GGT pyrometer installation. [9-1]

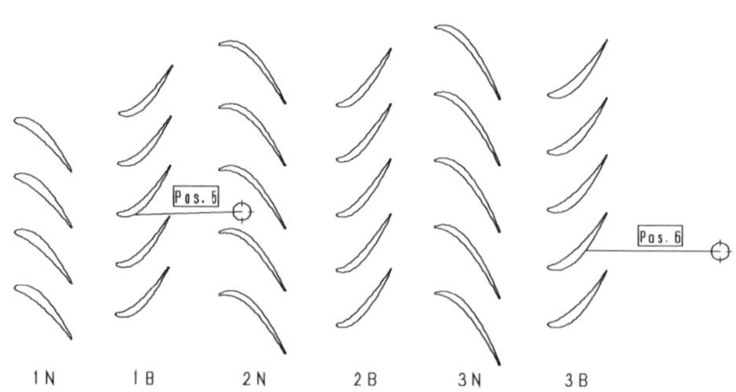

FIGURE 9–37 PT pyrometer installation. [9-1]

influence is even smaller since the ratio of cooling airflow rate to main gas flow rate is much smaller than that of M7A.

Digital Telemetry

Digital telemetry offers many "space age" possibilities. It is what enables a diagnosis team on the ground, a test pilot, or the diagnostic interface of an engine in flight to read and transmit data that, when interpreted by human or automated means, can save millions of revenue dollars. For instance, the data sent by telemetry (air to ground) could mean, for instance, that a module on a specific engine needs to be removed when the plane lands and a spare rotor used to replace it, before the engine can take off again.

Similarly, telemetry has great uses in offshore platform or remote plant and field operation. It offers the same potential to marine end users, too; however, ship and ferry engineers tend to be a rather self-sufficient crowd by necessity and culture. Their engine data can certainly be transmitted to on-shore offices; however, the engineer on board often has taken action before getting a call from a "land" office.

The two hardware units described next reflect the current state of digital telemetry instrumentation. The reader may contact the source for further operating data, as well as internet search for manufacturers of similar products.

Rotatel™ Telemetry Unit

Application

This compact, robust, digital telemetry unit is designed to provide clear signals of data transfer for up to 178 points of pressure, temperature and strain gauge measurements from rotating assemblies. Rotatel is an air-cooled unit, intended as a replacement for brush contact high-speed slip-rings, which have a short life and require CFC loaded freon as a coolant.

Principle of Operation

Rotatel consists of a series of rotating transmitter modules (and overhead modules), built into cassettes and mounted in a rotating assembly. These are hard wired to engine or rig mounted sensors in a self-contained unit complete with air-cooled bearings and power coils.

For simplicity of installation, this part is configured in two sections.

An independent bearing assembly (IBA) is mounted onto the rig or engine by a flange connection to ensure shaft concentricity, with a positive drive provided by a flexible hexagonal insert, through which the instrumentation cables enter the telemetry slip-ring unit. These wires are positively soldered to an integral circular PCP, which is set into the IBA.

A separate electronics assembly module (EAM) houses all the telemetry modules. The EAM is connected via a spigoted flange connection to the IBA. Instrumentation wiring to the modules is arranged via a separate, matching circular PCP to that in the IBA. Precision pins align the two PCBs and ensure a positive drive between the units, while a multi-pin connector provides a positive electrical bridge between the two.

A high-frequency supply is fed to the slip-ring telemetry unit from the telemetry groundstation. This powers the coils of the telemetry system, so that even when the unit is stationary output readings from the instrumentation may be recorded. An optical data link relays clean, noise-free signals of dynamic strain, pressure, and temperature, at rotational speeds up to 25,000 rpm.

The five equi-spaced rotating cassettes containing the telemetry modules are housed in a rotor, which is mounted on sealed, grease lubricated high-speed bearings. The air-tooled assembly is totally self-contained and includes the integral power transfer coils and an optical data link.

The module architecture can be varied to suit the variety of parameters being measured. Of the five cassettes, one is used for the power supply and to digitize the output signals. The remaining four can transmit either dynamic strain, temperature or pressure, in a configuration to suit the application.

This unit is supplied fully tested, fully balanced, spin tested and set up for immediate coupling to the test vehicle. It operates at speeds up to 25,000 rpm in a compact, air-cooled unit.

Minitel™ Telemetry Unit

Application

This compact digital telemetry unit is designed to provide clear signals of data transfer for up to 50 points of pressure, temperature, and strain gauge measurements from rotating assemblies. Minitel is an air-cooled unit, intended as a replacement for brush contact high speed slip-rings, which have a short life and require environmentally damaging coolant.

Principle of Operation

Minitel consists of a series of transmitter and overhead modules mounted in a rotating assembly. These are hard wired to engine mounted sensors in a self-contained unit complete with air-cooled bearings and power coils.

An optical data link transmits clean, noise-free signals of dynamic strain and temperature in a variety of combinations at speeds up to 35,000 rpm.

Specialized Diagnostic Systems

Today, different diagnostic techniques are used individually or in complementary application, as the following case studies illustrate.

CASE STUDY 1: A SURVEY OF NEW TECHNOLOGIES USED BY SIEMENS ENERGY FOR THE MONITORING AND DIAGNOSIS OF A GLOBAL FLEET OF POWER GENERATION SYSTEMS*

This case outlines one OEM's preferred methods in GT system fault diagnosis with their *Power Plant Automated Diagnostics Systems, Blade Vibration Monitor (BVM), Fiber Optic Vibration Monitor (FOVM), and the Radio Frequency Monitor (RFM).

WIN_TS is a passive, PC-based system that interfaces with the I&C system to acquire data. It is not designed to interface with any I&C plant operation functions or plant operating setpoints.

Data can be monitored in real time if certain events are detected that could be a potential issue to generating system operation. A watchdog function is programmed to automatically alert technicians in the Power Diagnostics® center when a defined plant operating anomaly is reported.

The analysis goals of acquired data can be broadly segregated into three categories: Monitoring, Diagnosis, and Prognosis. The goal of monitoring acquired data is the assessment of the quality of the data (i.e., sensor health, active data acquisition) and the assessment of the operational state of the monitored site (i.e., starting, stopped, producing power, operating within limits).

On a monitored site the Power Diagnostics® Services Automated Expert System registered a shift in engine parameters weeks before a maintenance outage was to be performed. A Siemens turbine specialist assessed the severity of the trend. Possible causes and effects were reported to the Siemens program manager. The most probable cause was identified as debris in the fuel nozzles (Figure 9–38) and the customer was made aware of the issue. Turbine hardware and operational impacts, such as potential complications during load changes, shutdown or startup were assessed. The customer and Siemens worked together to monitor the parameters. The plant remained in operation until maintenance could be scheduled at a time convenient to the customer's needs. Because Power Diagnostics® Services were able to pinpoint the location of the issue, when the outage was performed the clogged nozzle was clearly identified, the debris removed and the outage duration minimized.

FIGURE 9–38 Power diagnostics can detect a clean nozzle from one with debris.

* Source: Courtesy of Siemens. Adapted with permission, GT2009-59967 "A Survey of New Technologies Used by Siemens Energy for the Monitoring and Diagnosis of a Global Fleet of Power Generation Systems."

Fiber Optic Vibration Monitor

The Fiber Optic Vibration Monitor, or FOVM Model 2a, is used to monitor stator end winding vibration levels at high electric potential during generator operation. The system is designed to provide for the identification of changes within the generator and the opportunity to take early corrective action before extensive stator winding damages occur. In most cases, the monitor permits the utility to hold stator winding vibration below destructive levels by load, power factor, and cooling gas temperature adjustment.

The monitoring of stator winding vibration levels during operation has been made possible by the development of fiber optic vibration sensors. Fiber optic vibration sensors permit the transmission of stator vibration data from sensors mounted directly on the stator end windings to a multi-channel monitor external to the generator. The analysis of online vibration data provides a tool to assist in the determination of the present condition of components of the stator winding and to predict maintenance intervals based on vibration changes.

The accuracy of forced-vibration sensors, like those used in the FOVM Model 2a, is sensitive to changes in the sensor natural resonance frequency. The natural resonance frequency of a sensor can change due to changes in operating temperature and, to a lesser degree, sensor aging. To greatly reduce the effect of these changes on the monitor's vibration measurement accuracy, the FOVM Model 2a uses a patented algorithm that measures all sensors' natural frequencies automatically and online, and adjusts the reported vibration measurement for any detected natural frequency drift. In addition, the algorithm removes inherent variations in sensors introduced during manufacture. These capabilities are collectively referred to as autocalibration. Autocalibration provides a significant enhancement to vibration measurement accuracy, resolution, and the detection of failed sensors.

FOVM Sensor

The fiber optic sensor consists of a reed constrained on one end, with the other end free to vibrate. On the vibrating end of the reed is a grid, placed in between two optical fiber ends (see Figure 9–39).

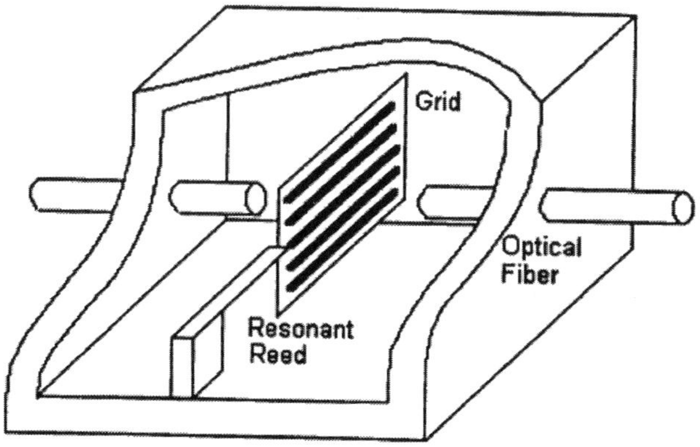

FIGURE 9—39 FOVM sensor schematic.

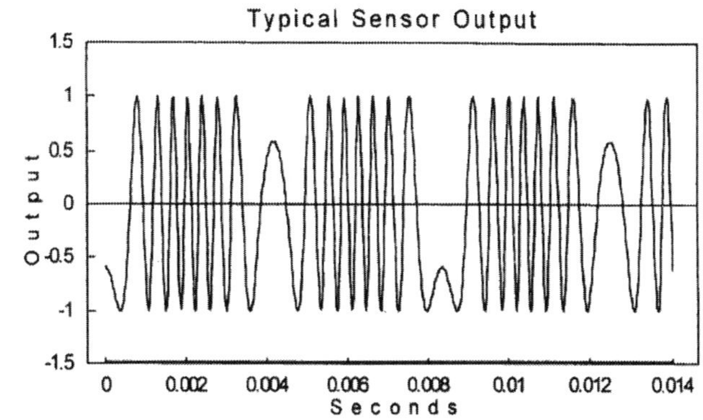

FIGURE 9—40 Typical fiber optic vibration sensor waveform.

A beam of light is sent to the sensor grid by the transmission fibers. The light is then chopped by the grid, and the return fiber brings the modulated light signal back to the detector. The grid, containing a series of closely spaced slits on its surface, is placed directly in the light path. As the reed vibrates, the grid alternately blocks and passes the light, creating a pattern of light pulses, characteristic of the vibration amplitude of the grid. The light pulses containing the vibration information leave the sensor through the optical fiber, pass through the pressure seal, and continue through to the detector. The frequency of light pulses generated in this manner is proportional to the reed amplitude. The signal waveform produced by the sensor is shown in Figure 9—40.

Blade Vibration Monitoring System

The Siemens BVM4 (Blade Monitoring System) is unique in many respects, the most notable being the sensing requirements. Most blade vibration systems require the installation of multiple sensors for each monitored row of blades, but the BVM4 system requires only a single sensor. Additional sensors can be installed to provide redundancy, but are not required for full system functionality. With only one sensor, the BVM4 offers greater accuracy than many competing monitoring systems.

The BVM4 has been installed successfully in many units in many locations. The primary use has been for steam turbines but the system can be installed in gas turbines as well. The BVM4 has a sophisticated analysis software package, which includes a user-friendly interface. The goal of the BVM4 is to allow the user to constantly monitor the behavior of installed blades. The monitor's software allows for a precise and sophisticated analysis of all of the blades in a monitored row.

The BVM4 system configuration consists of two electronics packages, the Fiber Optic Communication Module (FOCM) and BVM4 Main Chassis, and a probe set for each turbine row. The BVM4 system is provided as a four or eight row non-synchronous turbine blade vibration monitor.

Radio Frequency Monitor

Stator winding failures associated with deterioration of interstrand and groundwall insulation, and the mechanical failure of copper strands within the coil, can lead to an unscheduled unit outage. The cost of parts, labor, and replacement power associated with such events may represent a major loss to the utility. The Radio Frequency Monitor (RFM) has been designed to provide a warning of electrical malfunctions in and around the generator that may be associated with an arcing condition. The RFM is designed to detect the RF level that exists at the neutral lead of the generator at a point just above the neutral transformer. This activity represents a single variable, and by itself, may not contain sufficient information to diagnose the root cause of an RFM alarm. Trending, correlation with other generator variables and additional tests may be required to isolate the source of the RF energy.

The RF level and associated alarms are based solely on the average RF pulse height and duration of activity detected at the neutral lead. The sources of this RF signal may be divided into three groups:

1. Group 1. Signals from within the generator such as the stator windings and winding connections.
2. Group 2. The signal derived from generator subsystems such as the exciter, shaft and rotor winding ground detection brushes, rubbing seals, and buses.
3. Group 3. All sources external to the generator and generator subsystem such as motors, switches, welders, lightning, and communication signals.

The RFM detects that portion of the pulse having frequency components within the detection window. The amplitude of the detected RF is a symptom of an electrical malfunction in the system, but not a measure of its severity. Nor is it an indication of the progression of degradation. The amplitude of the detected RF is determined by such factors as the physical nature (type) of the RF source; the site of the malfunction within the generator; the attenuation of the RF signal as it travels to the detection site and the impedance at the detection site. A rapid fall in RF below alarm level may be indicative of progressive physical system degradation causing the arc to extinguish, instead of indicating that conditions have improved. Further high RF level can be expected as physical conditions at the fault change, and again favor an arcing condition.

CASE STUDY 2: PULSATION ANALYSIS: NEW TECHNIQUES AND THEIR LIMITATIONS*

This section describes the use of spectrum analysis on combustor pulsation measurements to diagnose hot section faults. The gas turbines are Alstom GT 11Ns (power generation application).

The emissions requirements imposed by the EPA, the trend towards higher gas turbine efficiency, lower cost per installed kW, etc., has caused gas turbines to operate much closer to flame stability limits and other design limits as was previously the case. This in combination with the need to maintain high availability and reliability requires OEMs to address the issues of monitoring systems and predictive maintenance in a new way.

The incident occurred at a power plant that was built and commissioned in the late 1980s by ABB may serve as an example to illustrate this need. The plant consists of 12 ABB type GT11N gas turbines that had been operation for approximately seven years. During the fall of 1996 some unusual and fluctuating bearing vibrations had been noted on one unit. The owner notified ABB, but by the time an investigation was initiated, one hot gas casing (transition piece between combustor liner and turbine inlet) had already failed causing a catastrophic failure. Within a short period of time it was found that combustor pulsations (pressure fluctuations of the combustor discharge pressure) were causing the problem. Not only were they high, but the dominant frequency was very close to the natural frequency of the hot gas casing. All units were inspected and another four damaged hot gas casings were found. Unfortunately, the available operating data trends were neither sufficient to indicate when the problem had begun nor what has caused it.

* Courtesy of Alstom Power, extracts from H. Anderson, "Early Detection of Combustor Pulsations and Optimized Operation through On-Line Monitoring Systems," 2000-GT-0180.

Prevention of Catastrophic Failures

From an OEM point of view, the early detection of component failures that could cause a catastrophic turbine failure clearly has a high priority. The analysis of the incident described in the previous sections makes it clear that the monitoring and trending of a few operating parameters would have given a clear indication of the deteriorating performance. In addition it shows that monitoring of the combustor pulsations is very important and cannot be replaced by monitoring of other parameters.

Increased Reliability, Availability, and Equipment Performance—User Requirements

From a user's point of view, the early detection of component failures is important as it impacts the reliability of the plant, but in addition the user has some additional requirements such as:

- Optimize equipment performance
- Provide the basis for or assistance in predictive and preventive maintenance programs
- If a component failure has been recognized, the system should allow, if possible, the safe operation at reduced power output until an outage can be scheduled with the objective to minimize the loss of revenue.

In today's competitive world, the above goals must be achievable in a timely manner without requiring especially skilled personnel.

Development of the Monitoring System

Selection of Relevant Operating Parameters

Taking the requirements of the previous sections into account (time, skill of personnel) and the fact that most operating parameters vary with varying ambient conditions, it becomes obvious that monitoring and displaying only directly measured parameters is not sufficient.

In order to allow an easy interpretation and financial evaluation in a timely manner, the following parameters must be provided by the system:

- Gas turbine power output corrected to ISO and/or guarantee conditions
- Gas turbine efficiency corrected to ISO and/or guarantee conditions
- Turbine efficiency corrected to ISO and/or guarantee conditions
- Compressor efficiency corrected to ISO and/or guarantee conditions
- Turbine firing temperature (based on a heat balance calculation; this temperature may differ from the temperature indicated by the control system)

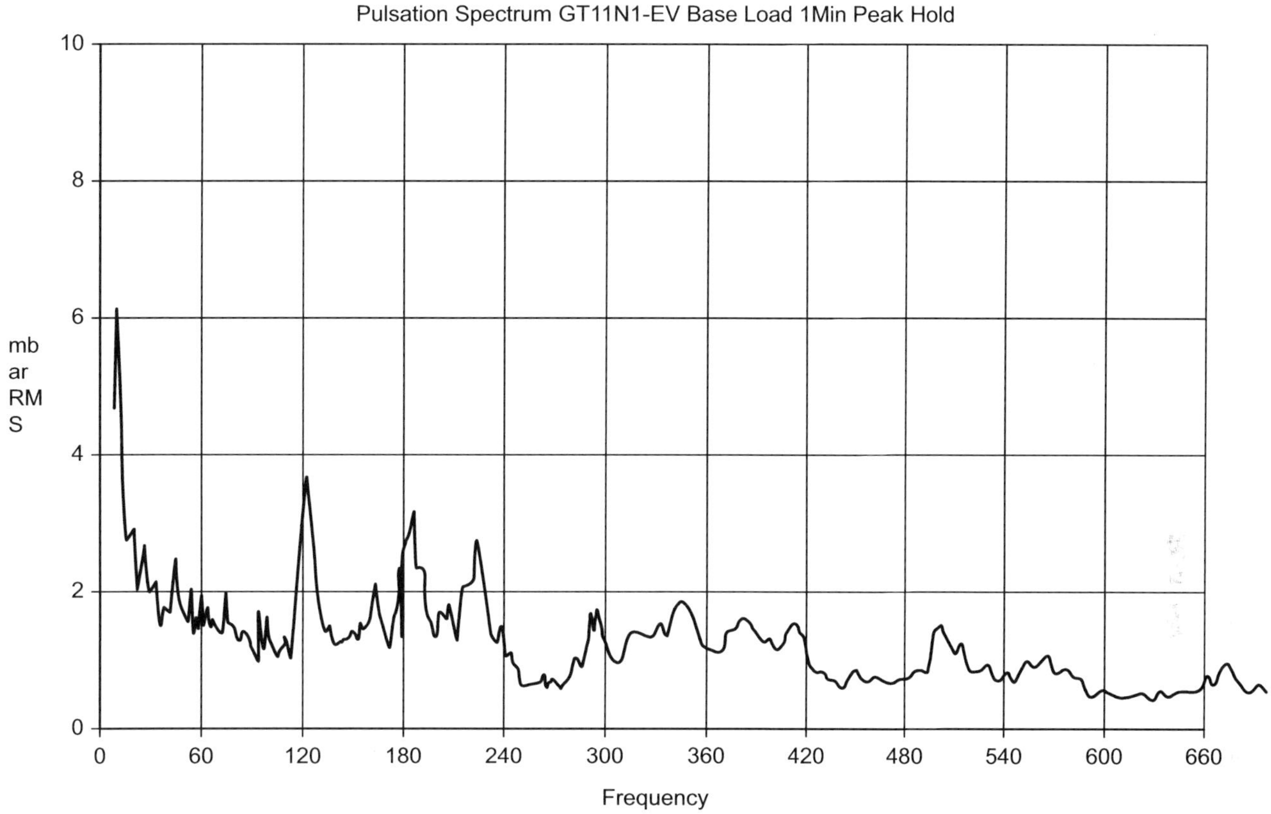

FIGURE 9–41 Typical pulsation spectrum of a GT11N1-EV on base load with natural gas as fuel. *(Source: Alstom Power.)*

And last but not least, the combustor pulsation must be monitored. This parameter requires a little more attention than the previously mentioned parameters. The evaluation of the failures that were caused by elevated pulsation shows the importance of the frequency spectrum of the pulsation. Figure 9–41 shows a typical pulsation spectrum obtained from an ABB GT11N-EV gas turbine. Obviously a well-defined peak at a given frequency is more damaging to the equipment than is a "white noise" type of pulsation, even though the overall RMS level of the two signals may be the same.

Thus the system must provide the following values:

- RMS level of the combustor pulsation
- Frequency spectrum of the combustor pulsation

In addition, the system must be capable of monitoring user-defined operating parameters.

Data Archiving

The system must provide a data archive that allows the following tasks:

- Trending operating parameters over the entire operating period of the plant
- Trending the more recent history with a higher time resolution
- Data are stored in ASCII files and can thus be read by almost by any software

The system is designed to provide the following:

- Hourly averaged operating parameters for the entire operating period of the plant
- Minute averaged operating parameters for the past 60 operating hours
- Non-averaged data (typical sample interval is 5 seconds) for the past hour.

Modular Program Structure

The monitoring system is a PC-based tool that reads process data from up to three different sources simultaneously (see Figure 9–42).

The modular structure was chosen, among other reasons, so that:

- Interfacing with different control systems does not require any changes in the main program
- Data can be read from multiple sources as not all process data may be available in the gas turbine control system
- Display and evaluation modules can be added/removed according to the users needs

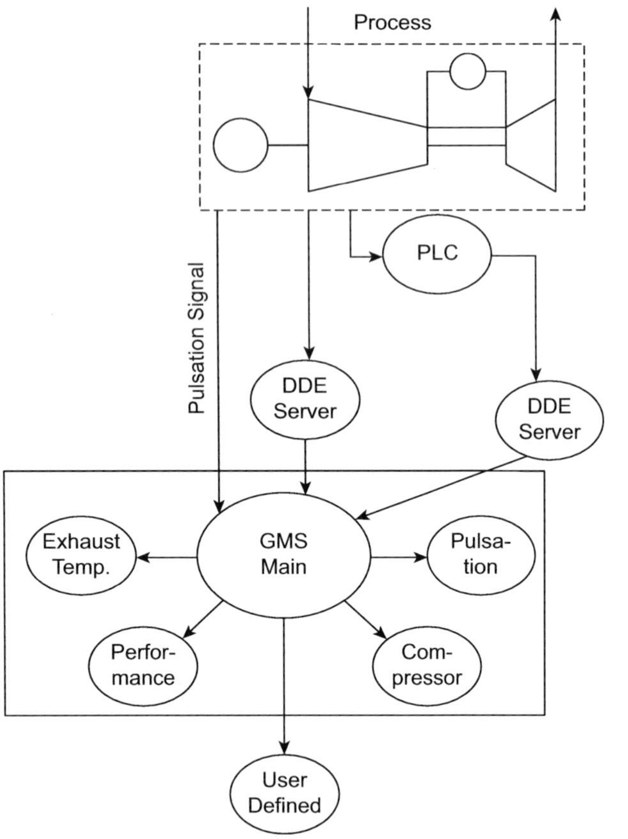

FIGURE 9–42 Program structure. Double-lined arrows: Dynamic Data Exchange link (DDE); can be accessed by the user. Single-lined arrows: Program internal communication link; cannot be accessed by the user. *(Source: Alstom Power.)*

Instrumentation

For the measurement of the combustor pulsations a heat resistant piezoelectric pressure transducer is used. The transducers have been used in several gas turbines for several years without failing. The transducer must be installed as close to the burner as possible. By placing the transducer close to the source for the pulsations, the likelihood of measuring in the vicinity of the pressure node is relatively small.

Installation of the Initial Systems

The monitoring software was first installed on two ABB type GT11N1-EV gas turbines in the summer of 1997. The main features of this gas turbines model are:

- 80 MW power output
- Top mounted silo combustor
- 37 dry low NO_x EV burners

On this type of gas turbine the pulsation probe is located in the transition piece between the silo combustor and the turbine inlet.

The system was monitoring combustor pulsations including the complete frequency spectrum and approximately 100 operating parameters. In the fall of 1998 the system was expanded to include exhaust gas (TAT) multipoint thermocouples and additional software to perform an on-line heat balance (including calculation of the true turbine inlet temperature) and compressor surge monitoring.

The past three years have shown that the following signals are the most valuable when trying to judge the condition of the hot gas path components:

- Combustor pulsation spectrum
- Temperature after turbine distribution (TAT profile)
- NO_x emissions

The following sections describe the conclusions that can be made based on these signals.

Pulsations

The combustor/burner specific spectrum is best seen when using a one-minute peak hold spectrum. A typical example for this turbine is shown in Figure 9–42.

This type of burner arrangement exhibits two characteristic frequencies.

The amplitude of the one in the 124 Hz range can be influenced by the overall fuel/air ratio and by the flow pattern downstream of the burners. Increased air leakage between the combustor inner liner and the burners will impact the amplitude of this frequency.

Large amplitudes in the range of 10–12 Hz indicate that some burners are starved of fuel. If the fuel and air distribution is correct, this frequency should not be dominant.

A crack in the general vicinity of the transmitter or a shift in the position of the casing, in which the transmitter is supposed, can falsify the signal significantly. An example is shown in Figure 9–43.

This is a particular good example of why the frequency spectrum is important. Would the above signal only be measured and displayed as an RMS value, one would only observe a huge increase in the signal level. If this were the only information available, the above signal would require the unit to be shut down and inspected. The spectrum, however, permits us to make the following observations:

- The increase in the RMS value is caused by an increase in the amplitudes across the entire spectrum
- The amplitudes of the characteristic frequencies are within the expected and permitted magnitudes

If this phenomenon is observed and the spectrum is available, the unit can be kept operational until an inspection can be scheduled. Naturally the unit needs to be monitored closely to watch for other indications of component deterioration.

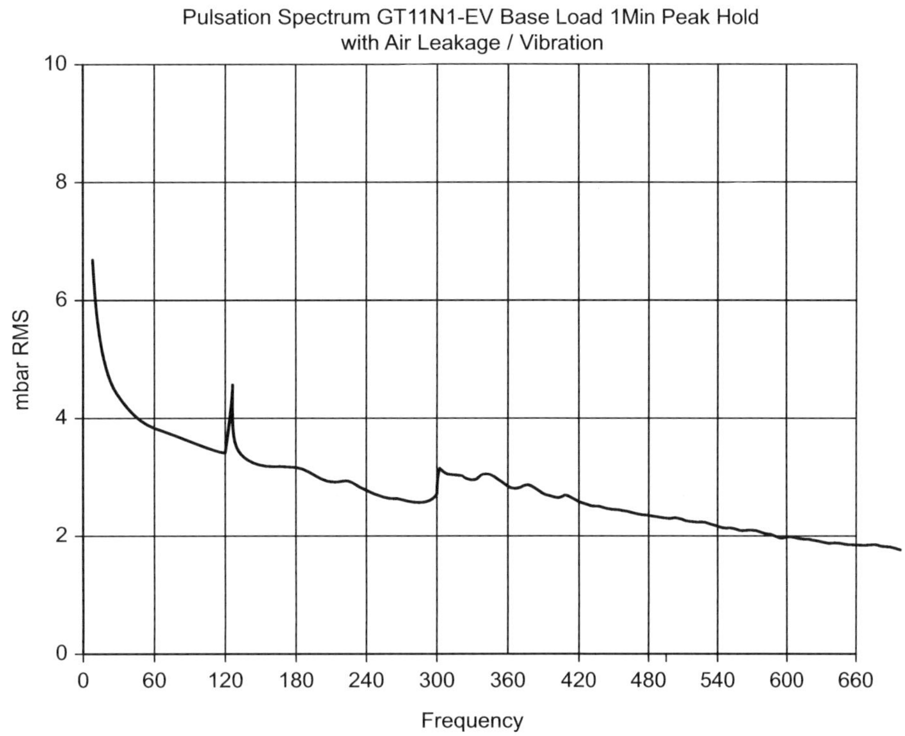

FIGURE 9–43 Pulsation spectrum of a GT11N1-EV on base load with natural gas as fuel where casing vibrations or leakage flows influence the signal. *(Source: Alstom Power.)*

Summarizing the above, the following statements can be made:

- The one-minute peak hold spectrum shows the characteristics of the combustion pulsation spectrum more clearly than a real-time spectrum
- The amplitude in the 124 Hz range is a function of the overall fuel/air ratio and the flow pattern downstream for the burners
- The amplitude in the 10 Hz range is an indication of burners being starved of fuel

Typical magnitudes of the characteristic frequencies but otherwise increased amplitudes across the spectrum indicate possible shifting casing positions and/or leakage airflow around the transmitter.

NO_X Emissions

On a GT11N1-EV the NO_x emissions are mainly a function of the fuel/air ratio. The NO_x emissions are very sensitive to the local fuel/air ratio. Thus if changing emissions are observed for a given operating condition, they can be caused by any of the following conditions:

1. Cracks in the combustor liner or hot gas casing or increasing clearances allowing air to bypass the burners
2. Burner fouling
3. Incorrect fuel distribution caused by manifold piping or valve positions
4. Change in fuel properties

Item 2 through 4 can usually be ruled out fairly easy by checking the condition of the gas filter, gas valve positions and the gas quality.

Thus if a change in the NO_x emissions is detected and items 2–4 can be eliminated as a root cause, one has a strong indication that one or more hot gas path components are deteriorating.

Turbine Exhaust Temperature Profile

Unlike the NO_x emissions that for a given operating point are pretty constant the temperature profile after the turbine is affected by ambient conditions and the period of time that the gas turbine has been operating at the given load. The latter is only significant during the first 4 hours after a start. On a unit that starts daily and only operates for 10–14 hours per day, this effect can make the detection of changes in the profile difficult as the ambient conditions typically change during operating period as well.

An inspection of the unit revealed that the combustor inner liner indeed had a crack that most likely was created during operation on fuel oil with high pulsations on January 8, 1999. Thus, the TAT spread is a good indicator for cracks

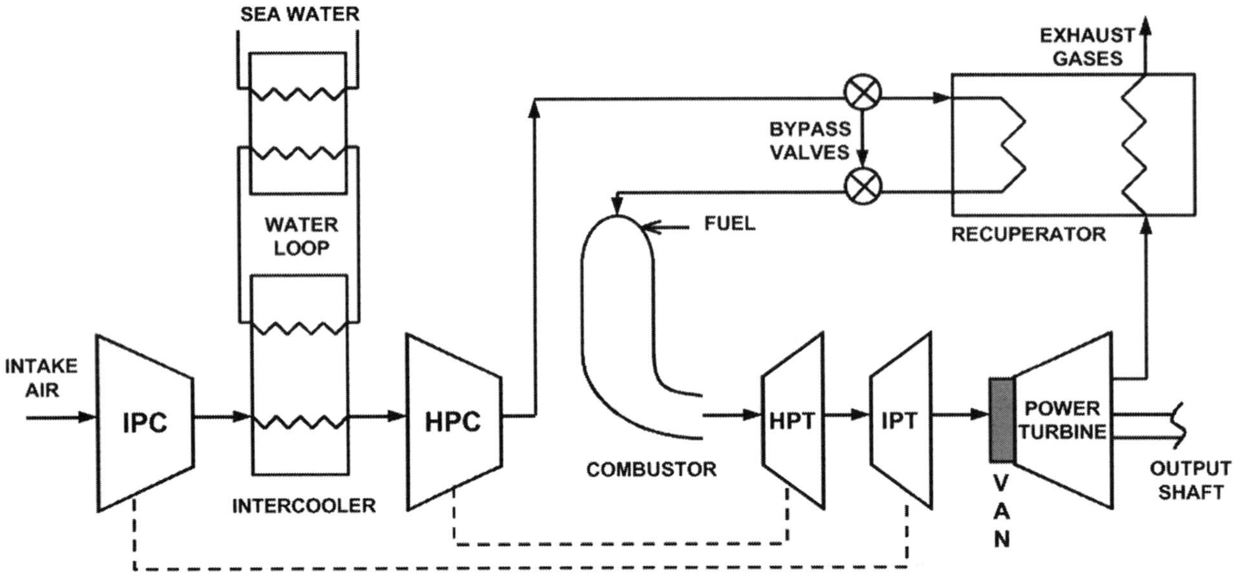

FIGURE 9—44 The WR-21 engine cycle. *(Source: Rolls Royce.)*

in the hot section of the turbine, but on daily cycling units the change in the profile may be difficult to detect.

CASE STUDY 3: PERFORMANCE AND C&I SYSTEM VERIFICATION WITH MODELING*

This section describes an OEM's (Rolls Royce) modeling strategy. The gas turbine in question is an intercooled, recuperated WR-21 (naval application). Due to the complexities of the operating cycle, getting the engine to a point that it can compete in three different markets requires careful analysis. The full paper is considerably more comprehensive than the illustrative extracts included here.

WR-21 Engine Cycle

Figure 9—44 presents the intercooled and recuperated WR-21 engine cycle.

Air is admitted through the intake ducting system and the radial air intake into the engine. The air is first compressed by the six-stage intermediate pressure compressor (IPC), before being fed to the on-engine intercooler. This heat exchanger uses a closed loop flow of freshwater and glycol, which is itself cooled by a seawater off-engine intercooler.

The aim of the intercooler is to reduce the following six-stage high-pressure compressor (HPC) air entry temperature, hence lowering the specific work required to achieve a given HPC pressure ratio. This enables more specific power to be extracted from the engine, and provides a lower recuperator air-side inlet temperature permitting maximum heat recovery in the recuperator. Intercooling alone is a fundamentally inefficient process, as heat is actually removed from the overall cycle, but its association with recuperation allows for the recovery of engine thermal efficiency, while preserving the power boost due to the intercooler.

The air coming out of the HPC is directed to the recuperator that is used to recover heat from the exhaust gases in order to preheat the compressor delivery air before its entry into the combustor. By returning this heat to the cycle, less temperature rise (and therefore fuel) is required in the combustors to achieve a given turbine entry temperature. After passing through the HP and IP turbines (HPT and IPT), driving their associated compressors, the combustion products enter the power turbine through first-stage variable area nozzles (VAN). The VAN are used to optimize the recuperator performance by controlling the power turbine capacity, which in turn maintains a high recuperator gas side inlet temperature. This enables optimized control over the air-side heat transfer, and hence high recuperator effectiveness across the power range.

The cycle gives significant specific fuel consumption (SFC) savings compared to a simple cycle gas turbine across the power range as presented in Figure 9—45.

Rolls Royce's Corporate Analysis Method

Representative simulation methods are required to accurately model gas turbine engines as complex as the WR-21. A predictive model is required, and this model needs to be

* Extracts from B. Biraud, A. Despierre, and S. Gayraud, "Simulation of the WR-21 Advanced Cycle Engine," 2001-GT-0020.

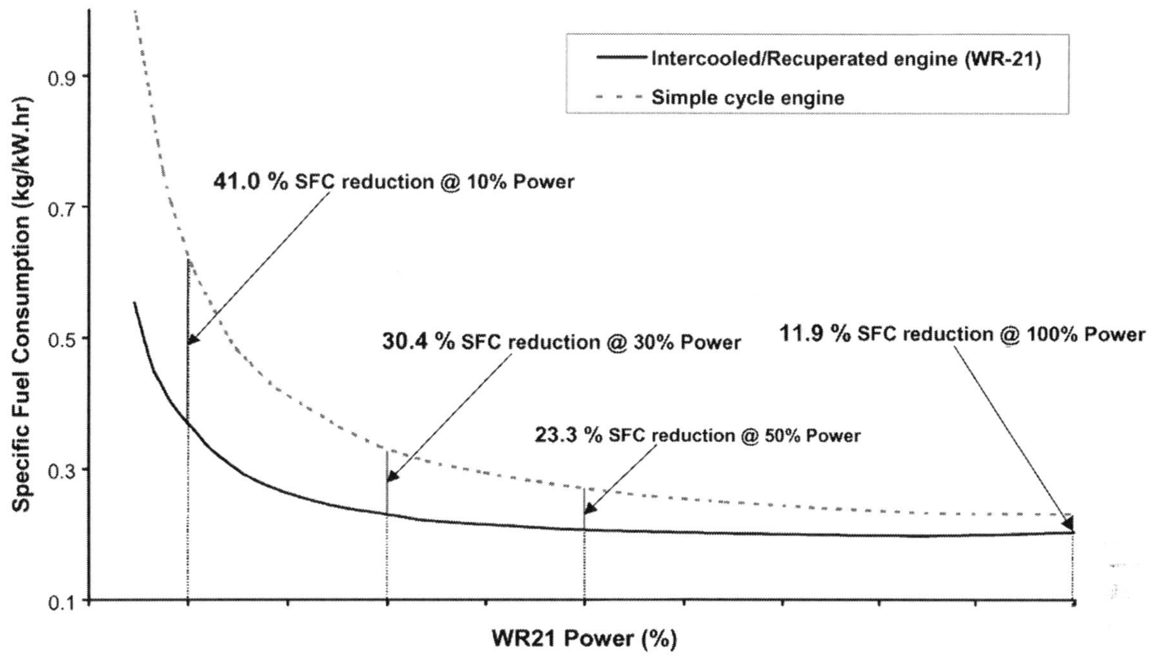

FIGURE 9–45 ICR vs. simple cycle SFC comparison. *(Source: Rolls Royce.)*

aligned to actual engine test data to produce a fully representative model. The Rolls Royce corporate method for aligning predictive models to test data is called ANSYN.

ANSYN is a shortcut for ANalysis-SYNthesis. An analysis program calculates the individual performance of each of the engine components (isentropic efficiency, pressure drops, effectiveness, etc.) from numerous measurements at different stations in the engine (air mass flow, fuel flow, shaft speeds, total pressures, and temperatures).

A synthesis program works in the opposite direction. It uses predictions (or assumptions) of component performance (compressor/turbine characteristics, pressure drops, air mass flows, etc.) to synthesize the value of thermodynamic parameters like air mass flow, shaft speeds, station temperatures, and pressures at each engine station. It is merely a predictive simulation program.

Both the synthesis and analysis models use numerical methods to simulate steady-state gas turbine performance. Model matching takes the usual form of a damped quasi-Newton method to solve the thermodynamics of the model through matching iterations. The program starts with guesses of the compressor working points, heat exchanger air-side exit conditions, combustion fuel/air ratio, power turbine nozzle area, intake mass flow, and output power. Using flow continuity and equality of powers and speeds, the turbine capacities, recuperator exit conditions, compressor inlet flows, and engine exit/inlet pressure are calculated. These are compared with the required values to derive a set of errors.

An iterative procedure is used to refine the guessed values until the errors are smaller than a specified tolerance.

The calculations that the paper describes with respect to gas and heat transfer dynamics, as well as dynamic controls simulation, will not be repeated here; however, the OEM may be contacted for the full paper and more recent work in this vein. A few figures of "operational events" and their description are included below, however.

Figure 9–46 shows a series of load rejections modeled with the WR-21 transient model in an electric drive configuration.

The load rejection is triggered instantaneously by simulating generator breakers opening. The engine detects that the generator load has been rejected by seeing a sudden power turbine acceleration. The load rejection control logic is activated that minimizes the power turbine overspeed, whilst ensuring constant engine stability, to bring the engine back to idle in the shortest time possible.

The amount of dynamic detail required in a model depends upon its application. Shaft dynamics may be sufficient for general engine operability and preliminary control assessment studies, but in the case of the WR-21 engine, where the model is also used for hardware validation and safety analyses, more comprehensive dynamic modeling is required. (See original Rolls Royce work for all details.)

Model data is featured in Figure 9–47. An emergency stop is the fastest deceleration that a gas turbine can experience in a mechanical drive configuration. The sequenced opening of the IP compressor blow-off valves (BOV) to maintain an

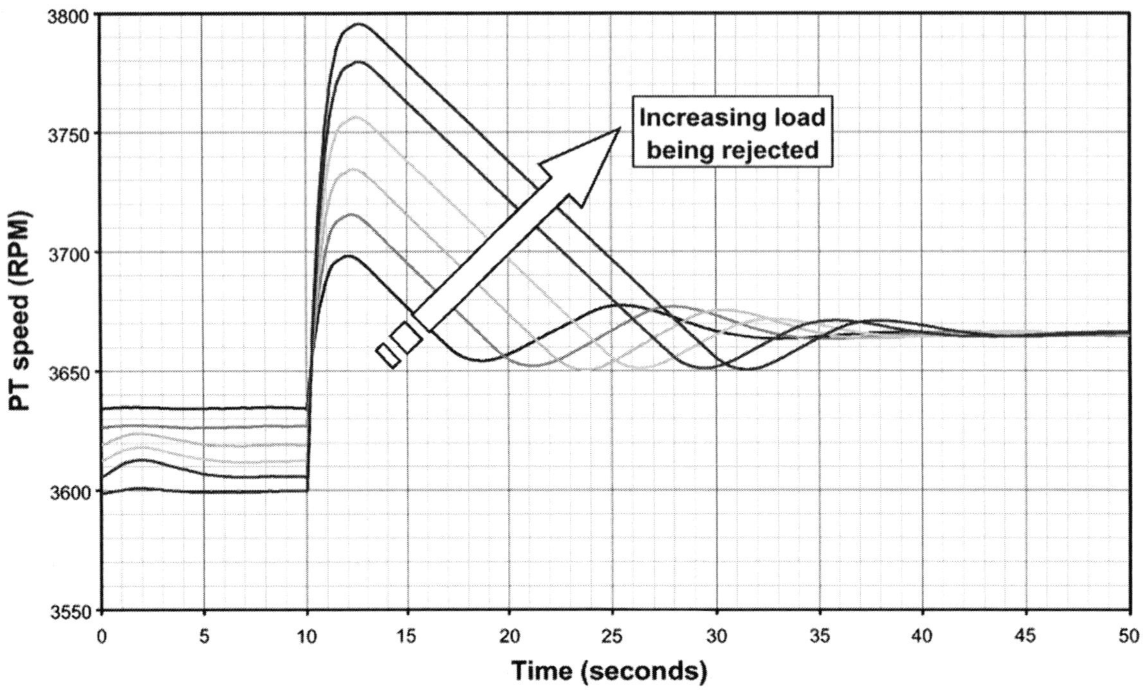

FIGURE 9–46 Power turbine speed trace after a load rejection. *(Source: Rolls Royce.)*

FIGURE 9–47 IP compressor map prediction during an emergency stop. *(Source: Rolls Royce.)*

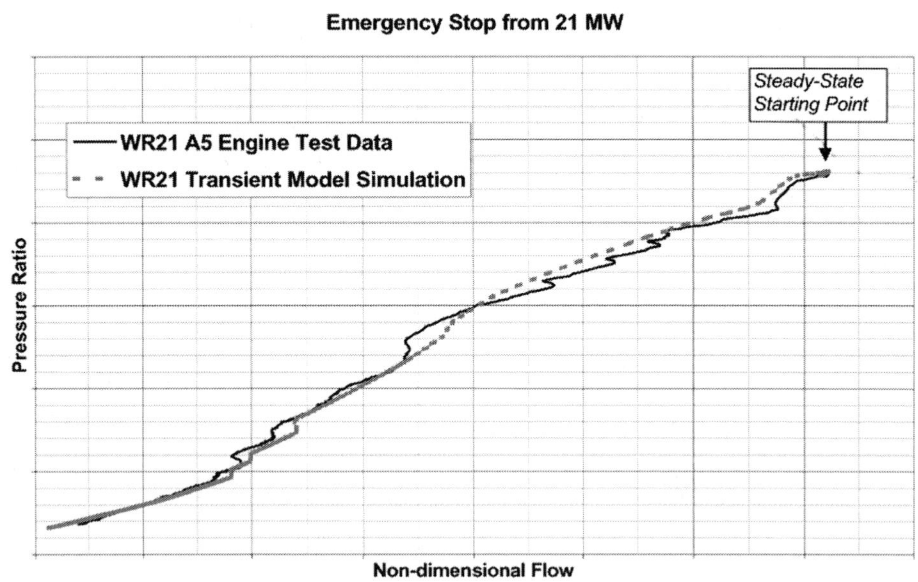

appropriate stability margin all along the transient maneuver explains the bigger "steps" seen on the running line.

The drivers for an elaborate control logic are the presence of the heat exchangers, the large volume of the recuperator and the presence of the VAN. The WR-21 has to be optimized for three modes of operations based on the heat exchangers' state.

The recuperator active and intercooler operative mode is the normal operation mode. It is also the mode where the specific fuel consumption benefits gained from the innovative thermodynamic cycle have to be optimized.

The engine must also be able to achieve the same power in the recuperator bypassed and intercooler operative mode as in the previous mode of operation.

Finally the recuperator bypassed and intercooler inoperative mode is considered a "failure mode" since the engine is supposed to experience it only when the intercooler is at fault, or during maintenance.

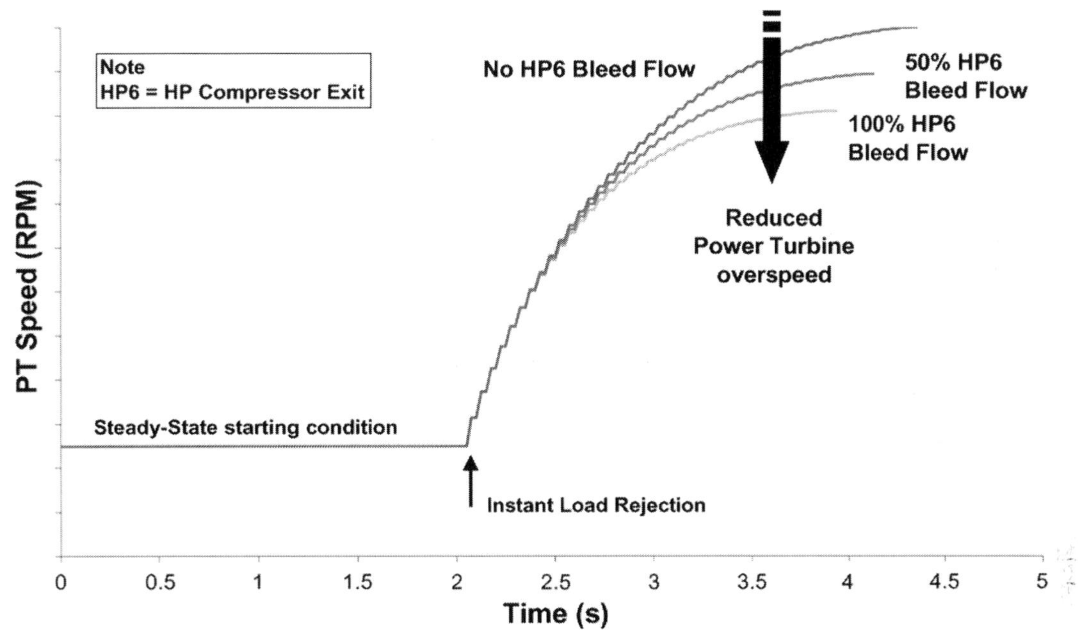

FIGURE 9–48 Effect of bleed extraction on a load rejection. *(Source: Rolls Royce.)*

Not only do the control laws have to withstand the three modes of operation, but they also must ensure that transitions between those modes can be performed at any power while the engine is running.

To compete in all markets the WR-21 engine must operate in a mechanical drive version (as developed originally) as well as in an electric drive version. The entire electric drive development has been performed with the transient model so far.

Since 1998, the growing interest from the Navies for an electric drive version of the WR-21 gas turbine engine encouraged the team to develop a specific logic to cope with the specific demands of this operation mode, especially during load sheds as shown in Figure 9–46. Using the transient model, the load shed detection, fuel flow drop, and bleed switching logics have been designed in order to limit the top speed and recover quickly to idle.

Figure 9–48 shows the difference in top speed achieved for different bleed configurations. This highlights the necessity of having an HP6 bleed (downstream of the HP compressor), and the benefits it brings in limiting the top speed. Extracting air from HP6 removes a significant amount of the air flow from the gas path. This in turn allows a larger drop in fuel flow without the risk of flameout since the fuel-to-air ratio remains high enough at any time. The presence of HP6 brings further benefits in permitting to open the remaining BOV earlier in the load rejection sequence.

In summary, since the beginning of the WR-21 program in December 1991, the Performance and Controls team has had to overcome significant technical challenges to successfully design and develop this intercooled/recuperated gas turbine engine.

Chapter 10

Performance, Performance Testing, and Performance Optimization*

"Our doubts are traitors and make us lose the good we oft might win by fearing to attempt."

—William Shakespeare

Chapter Outline

Gas turbine performance, performance verification, and maintenance are irrevocably linked in an end user's world. In this book, however, maintenance is dealt with in Chapter 12.

In summary, gas turbine performance verification (testing) is done at several levels. The testing of new GTs, done at the OEM's facility, may be witnessed by end-user reps. If the OEM is supplying a GT package (for instance, the GT and another OEM's associated equipment), the OEM may elect to do a different battery of tests. Sometimes functional, no-load tests are conducted. Load testing is far more expensive than functional testing but critical in cases where the application or engine model is in some way different from previous applications. One small component change or instrument system change could change what was a successful gas turbine system without that change.

In the manufacture of GTs, OEMs sometimes have licensee manufacturers for the GT itself or portions of the system. GE is one such manufacturer.

All OEMs have vendors who supply accessory packages of some kind to the GT system, for instance a compressor washing skid. In testing new GTs or GT systems, these external vendor systems may be "slaved." This means the OEM's test facility has a duplicate model of the same system that is used for their tests. After the test, the slave is removed and kept for the next test and the gas turbine package gets a new unit or system design duplicate of the item that was slaved. In some cases, the actual units supplied with the package may be on the skid and tested along with the rest of the gas turbine system.

* Sources (except where specified): Notes on the proceedings of the annual panel sessions for ECMS as they relate to gas turbine component life extension, 1985 through 2003, chair and organizer, C. Soares; operations case studies and notes, turbomachinery in nonconventional energy (tarsands), conventional oil and gas, aeroengine fleet management, aeroengine repair and overhaul; C. Soares, course notes "Basic Operations and Theory of Gas and Steam Turbines, Cogeneration and Combined Cycle Plants," 2005.

OEMs may conduct tests to check a modified component. Frequently, they negotiate with an existing end user to check a new development in an existing plant application.

Repair and overhaul is done by OEMs and independent shops. OEMs and independents may have licensees who are permitted to work on their machinery or specific components or processes (for instance, heat treatment, EDM, and powder metallurgy) within specific turbine models.

When an existing system needs to be optimized, the best designer for the project may or may not be the OEM: this is the end user's choice. An independent (non-OEM) firm may work with the OEM's knowledge and assistance or not. Some former-OEM-employee founded companies can be useful to end users for reverse engineering (for instance, making a component from studying an existing component), retrofit (adding to the existing GT package), or reengineering (redesigning a component or system that doesn't work) projects.

In an operational context, a GT system needs to be monitored for performance degradation. Frequently, the cause of performance degradation is flow blockage, either in the compressor (potentially due to fouling) or turbine (potentially due to component deterioration or failure).

Generally, unless components are damaged permanently, compressor performance can be recovered with washing the compressor. The washing can be online or offline, at operating speed or reduced speed, and done with water plus additives, cleaning fluid, rice husks, or other mildly abrasive solid material (which still will not damage metal airfoils).

Certain values of CDP (compressor differential pressure) and compressor mass flow rate (this may be computed with other readings including compressor outlet temperature) have set points that trigger a wash when required. The wash may be done "online" at anywhere from slow roll to operational speed. It may be done "offline" with the rotor being slow rolled. It may also be done with a "crank soak," where the wash fluid in inserted and left to sit while it dissolves and removes the contaminants.

The diagnostics required to pinpoint turbine module problems require similar readings for flow, pressure, and temperature. The preceding paragraphs in this section describe what is termed a performance analysis (PA) system.

The best PA systems work by comparing real live data (diagnostic) versus design (intended or "predicted") data. The reference points for such a system must be worked out for each turbine model type, a process that requires cooperation with the OEM (to the extent that it requires the design curves and design data for a specific model). In certain customized projects (such as one where old GE* Frame 5s with retrofitted steam injection took turbine model discs out of premature crack mode and provided augmented power), the original design data may be hard to find. In this case, reverse engineering may be required.

PA is extremely valuable, with a quick ROI (return on investment), especially in countries where fuel is expensive, fuel conservation is valued, or emissions taxed. Even if one is in a "fuel is cheap" location, PA can provide early warning signs of turbine component malfunction or failure. A couple of examples of a PA system "at work" follow.

Detecting turbine component deterioration and avoiding disastrous failure long before the turbine was scheduled for major overhaul. The turbine would have failed if regular TBOs had been observed.*

Gaining 0.1% efficiency on the turbine module of a GE Frame 7. For the reference US gas prices, that efficiency loss represented about $800,000 per GT annually.*

In terms of regular maintenance (covered in Chapter 12, Maintenance, Repair, and Overhaul), each GT has its own maintenance schedule, which is detailed in the O&M manuals. In general, there is:

Ongoing and regular maintenance (which may include oil sampling, borescope inspection, and items that do not require appreciable teardown).

Hot section inspections (HSIs) that check out the hot section (combustor including fuel nozzles and turbine).

Regular inspections (the extracts of a paper in this section give typical examples of those intervals).

Shutdown/turnaround/postfailure audits and other nonregular inspections occur when some major overhaul or parts replacement is made necessary by a machinery failure or major modification) or plant shutdown for other reasons, such as an expansion.

Note that maintenance is differentiated from GT engine condition monitoring systems (ECMS), which may include:

1. Vibration analysis (VA)
2. Performance analysis or PA (a form of performance verification)
3. Life cycle assessment (LCA)

ECMS, as we saw in the previous chapter, is a vast topic. While VA and to some extent PA were covered in the previous chapter, some aspects of LCA and PA are covered in this one.

PERFORMANCE

A good starting point for considering theoretical performance and performance optimization is theoretical cycle diagrams (see Chapter 3, Gas Turbine Configurations and Heat Cycles). From that chapter, it was noted that:

1. Intercooling during compression,
2. Adding (exhaust waste) heat to the compressed gases prior to combustion (recuperation),
3. Adding heat at an interim stage of the turbine (reheat),

* Reference: Liburdi Engineering, Canada.

all add to the efficiency of the gas turbine cycle.
Further cycle efficiency improvements can be made by:

1. Cooling inlet air (cooler air is denser and provides more power per unit volume of inlet air).
2. Other waste heat recovery and cogeneration, such as waste heat being used in a HRSG (heat recovery steam generator) to run a steam turbine with the gas turbine (a combined cycle).
3. A variation in principle of an aircraft engine afterburner in land-based applications injects a fuel source downstream of the gas turbine exhaust and ignites the exhaust air/supplementary fuel mix for additional power.

In addition to these basic cycle adaptations and variations on a theme thereof, gas turbine users now also use inlet air fogging (for cooling inlet air) and steam and/or water injection for power augmentation. The latter means, sometimes used for NO_X reduction or cooling (for instance, to take turbine discs out of heat damage range, such as a premature hot section component crack zone), can result in 20–25% additional power developed.

There are a myriad of other performance retention and optimization technologies. They include the science of performance recovery with (online or otherwise) compressor washing or cleaning with liquid or solid cleaning media.

Performance optimization can also be described as improved TBOs and increased MTBFs (mean time between failures). For optimization of TBOs, LCA is a valuable tool. The algorithms used by OEMs to calculate a cycle of life used per engine type vary among OEMs and different models for the same OEM. However, all generally result in extended service lives for major hot section components (over what used to be stated lives in hours of operation that had a large safety factor applied).

The author's course notes on a basic course on PA and LCA are included here. These notes do not give away any proprietary algorithms but do indicate how they may be derived, in principle, with specific parameters. Also covered are the applications of some of the hardware/software packages used to conduct LCA and potential for these packages to be retrofitted.

Note also that several performance-related basic accessory system descriptions were covered in Chapter 8, Accessory Systems. These include:

1. Thrust distribution and specific features that can direct thrust for different (vertical/short takeoff and landing systems, afterburners, and so forth) applications
2. Water injection

Before the case studies for this chapter, the following theory will be useful. The source is Rolls Royce's *The Jet Engine*; however, land-based users are advised again that all aircraft engine technology is directly applicable to *all* gas turbines at some point. This is true whether the aircraft engines are in helicopters (power stated in shaft horsepower) or jet service (power stated in pounds of thrust).

Performance Theory Summary*

The performance requirements of an engine are obviously dictated to a large extent by the type of operation for which the engine is designed. The power of the turbojet engine is measured in thrust, produced at the propelling nozzle or nozzles, and that of the turbo-propeller engine is measured in shaft horse-power (s.h.p.) produced at the propeller shaft. However, both types are in the main assessed on the amount of thrust or s.h.p. they develop for a given weight, fuel consumption and frontal area.

Since the thrust or s.h.p. developed is dependent on the mass of air entering the engine and the acceleration imparted to it during the engine cycle, it is obviously influenced, as subsequently described, by such variables as the forward speed of the aircraft, altitude, and climatic conditions. These variables influence the efficiency of the air intake, the compressor, the turbine, and the jet pipe; consequently, the gas energy available for the production of thrust or s.h.p. also varies.

In the interest of fuel economy and aircraft range, the ratio of fuel consumption to thrust or s.h.p. should be as low as possible. This ratio, known as the specific fuel consumption (s.f.c.), is expressed in pounds of fuel per hour per pound of net thrust or s.h.p. and is determined by the thermal and propulsive efficiency of the engine. In recent years considerable progress has been made in reducing s.f.c. and weight. These factors are further explained later.

Whereas the thermal efficiency is often referred to as the internal efficiency of the engine, the propulsive efficiency is referred to as the external efficiency. This latter efficiency, described later, explains why the pure jet engine is less efficient than the turbo-propeller engine at lower aircraft speeds leading to development of the bypass principle and, more recently, the propfan designs.

The thermal and the propulsive efficiency also influence, to a large extent, the size of the compressor and turbine, thus determining the weight and diameter of the engine for a given output.

These and other factors are presented in curves and graphs, calculated from the basic gas laws, and are proved in practice by bench and flight testing, or by simulating flight conditions in a high altitude test cell. To make these calculations, specific symbols are used to denote the

* Source: Adapted, with permission, from Rolls Royce *The Jet Engine* 1986, Rolls Royce Plc: UK.

pressures and temperatures at various locations through the engine; for instance, using the symbols shown in Figure 10–1 the overall compressor pressure ratio is P_3/P_1. These symbols vary slightly for different types of engine; for instance, with high bypass ratio engines, and also when afterburning is incorporated, additional symbols are used.

To enable the performance of similar engines to be compared, it is necessary to standardize in some conventional form the variations of air temperature and pressure that occur with altitude and climatic conditions. There are in use several different definitions of standard atmospheres, the one in most common use being the International Standard Atmosphere (I.S.A.). This is based on a temperature lapse rate of approximately 1.98K per 1000 ft., resulting in a fall from 288.15K. (15°C) at sea level to 216.65K (−56.5°C) at 36,089 ft. (the tropopause). Above this altitude the temperature is constant up to 65,617 ft. The I.S.A. temperature is constant up to 65,617 ft. The I.S.A. standard pressure at sea level is 14.69 pounds per square inch falling to 3.28 pounds per square inch at the tropopause (refer to I.S.A. Table 10–1, at the end of this section).

Engine Thrust in Test

The thrust of the turbo-jet engine on the test bench differs somewhat from that during flight. Modern test facilities are available to simulate atmospheric conditions at high altitudes thus providing a means of assessing some of the performance capability of a turbo-jet engine in flight without the engine ever leaving the ground. This is important as the changes in ambient temperature and pressure encountered at high altitudes considerably influence the thrust of the engine.

Considering the formula derived later for engines operating under "choked" nozzle conditions,

$$\text{Thrust} = (P - P_0)A + W_{v_j}/g$$

it can be seen that the thrust can be further affected by a change in the mass flow rate of air through the engine and by a change in jet velocity. An increase in mass airflow may be obtained by using water injection and increases in jet velocity by using afterburning.

As previously mentioned, changes in ambient pressure and temperature considerably influence the thrust of the engine. This is because of the way they affect the air density and hence the mass of air entering the engine for a given engine rotational speed. To enable the performance of similar engines to be compared when operating under different climatic conditions, or at different altitudes, correction factors must be applied to the calculations to return the observed values to those that would be found under I.S.A.

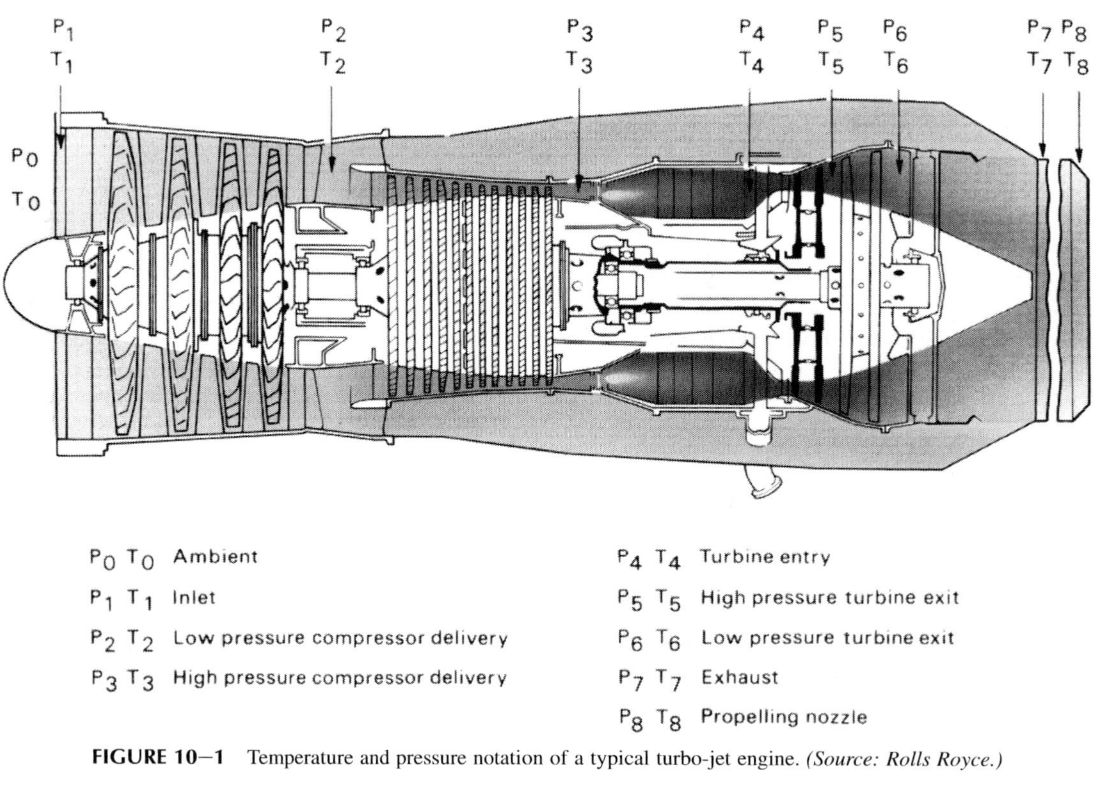

FIGURE 10–1 Temperature and pressure notation of a typical turbo-jet engine. *(Source: Rolls Royce.)*

TABLE 10–1 The International Standard Atmosphere (I.S.A.)

The International Standard Atmosphere (I.S.A.)									
Altitude (h)		Ambient Temperature (T_0)			Ambient Pressure (P_0)		Speed of Sound (A_0)		
Feet	Meters	Deg. K.	Deg. C.	Deg. F.	lb./sq. in.	Millibars	ft./sec.	knots	n./sec.
−1000	−304.8	290.13	+16.98	62.6	15.24	1050.4	1120.3	663.3	341.5
0	0	288.15	15.00	59.0	14.69	1013.2	1116.6	661.1	340.3
+1000	+304.8	286.17	13.02	55.4	14.17	977.1	1112.6	658.8	339.1
2000	609.6	284.19	11.04	51.9	13.66	942.1	1108.7	656.5	337.9
3000	914.4	282.21	9.06	48.3	13.17	908.1	1104.9	654.2	336.8
4000	1219.2	280.23	7.08	44.7	12.69	875.1	1100.9	651.9	335.6
5000	1524.0	278.24	5.09	41.2	12.23	843.0	1097.1	649.6	334.4
6000	1828.8	276.26	3.11	37.6	11.78	811.9	1093.2	647.8	333.2
7000	2133.6	274.28	1.13	34.0	11.34	781.8	1089.3	644.9	332.0
8000	2438.4	272.30	−0.85	30.5	10.92	752.6	1085.3	642.6	330.8
9000	2743.2	270.32	−2.83	26.9	10.51	724.3	1081.4	640.3	329.6
10,000	3048.0	268.34	−4.81	23.3	10.11	696.8	1077.4	637.9	328.4
11,000	3352.8	266.36	−6.79	19.8	9.72	670.2	1073.4	635.6	327.2
12,000	3657.6	264.38	−8.77	16.2	9.35	644.4	1069.4	633.2	325.9
13,000	3962.4	262.39	−10.76	12.6	8.98	619.4	1065.4	630.8	324.7
14,000	4267.2	260.41	−12.74	9.1	8.63	595.2	1061.4	628.4	323.5
15,000	4572.0	258.43	−14.72	5.5	8.29	571.7	1057.3	626.0	322.3
16,000	4876.8	256.45	−16.70	1.9	7.97	549.1	1053.3	623.6	321.1
17,000	5181.6	254.47	−18.68	−1.6	7.65	527.2	1049.2	621.2	319.8
18,000	5486.4	252.49	−20.66	−5.2	7.34	505.9	1045.1	618.8	318.5
19,000	5791.2	250.51	−22.64	−8.8	7.04	485.6	1040.9	616.4	317.3
20,000	6096.0	248.53	−24.62	−12.3	6.75	465.6	1036.9	613.9	316.1
21,000	6400.8	246.54	−26.61	−15.9	6.48	446.4	1032.7	611.5	314.8
22,000	6705.6	244.56	−28.59	−19.5	6.21	427.9	1028.6	609.0	313.5
23,000	7010.4	242.58	−30.57	−23.0	5.95	409.9	1024.4	606.5	312.2
24,000	7315.2	240.60	−32.55	−26.6	5.69	392.7	1020.2	604.1	310.9
25,000	7620.0	238.62	−34.53	−30.2	5.45	375.9	1015.9	601.6	309.7
26,000	7924.8	236.64	−36.51	−33.7	5.22	359.9	1011.8	599.1	308.4
27,000	8229.6	234.66	−38.49	−37.3	4.99	344.3	1007.5	596.6	307.1
28,000	8534.4	232.68	−40.47	−40.9	4.78	329.3	1003.2	594.0	305.8
29,000	8839.2	230.69	−42.46	−44.4	4.57	314.8	998.9	591.5	304.5
30,000	9144.0	228.71	−44.44	−48.0	4.36	300.9	994.7	588.9	303.2
31,000	9448.8	226.73	−46.42	−51.6	4.17	287.4	990.3	586.4	301.9
32,000	9753.6	224.75	−48.40	−55.1	3.98	274.5	986.0	583.8	300.5

(Continued)

TABLE 10–1 The International Standard Atmosphere (I.S.A.)—cont'd

The International Standard Atmosphere (I.S.A.)

Altitude (h)		Ambient Temperature (T_0)			Ambient Pressure (P_0)		Speed of Sound (A_0)		
Feet	Meters	Deg. K.	Deg. C.	Deg. F.	lb./sq. in.	Millibars	ft./sec.	knots	n./sec.
33,000	10058.4	222.77	−50.38	−58.7	3.80	261.9	981.7	581.2	299.2
34,000	10363.2	220.79	−52.36	−62.3	3.63	249.9	977.3	578.7	297.9
35,000	10668.0	218.81	−54.34	−65.8	3.46	238.4	972.9	576.1	296.5
36,000	10972.8	216.83	−56.32	−69.4	3.29	227.3	968.5	573.4	295.2
36,089	11000.0	216.65	−56.50	−69.7	3.28	226.3	968.1	573.2	295.1
37,000	11277.6	Ambient temperature remains constant from this point up to 65,617 ft.			3.14	216.6	Speed of sound remains constant from this point up to 65,617 ft.		
38,000	11582.4				2.99	206.5			
39,000	11887.2				2.85	196.8			
40,000	12192.0				2.72	187.5			
45,000	13716.0				2.14	147.5			
50,000	15240.0				1.68	115.9			
55,000	16764.0				1.32	91.2			
60,000	18288.0				1.04	71.7			
65,000	19812.0				0.82	56.4			

(Source: *Rolls Royce*)

conditions. For example, the thrust correction for a turbo-jet engine is:

$$\text{Thrust(lb.)(corrected)} = \text{Thrust(lb.)(observed)} \times 30/P_0$$

where

P_0 = atmospheric pressure in inches of mercury (in. Hg.) (observed)
30 = I.S.A. standard sea level pressure (in. Hg.)

The observed performance of the turbo-propeller engine is also corrected to I.S.A. conditions, but due to the rating being in s.h.p. and not in pounds of thrust the factors are different. For example, the correction for s.h.p. is:

$$\text{s.h.p. (corrected)} = \text{s.h.p. (observed)} \times 30/P_0 \times \sqrt{(273+15)/(273+T_0)}$$

where

P_0 = atmospheric pressure (in. Hg.) (observed)
T_0 = atmospheric temperature in °C (observed)
30 = I.S.A. standard sea level pressure (in. Hg.)
273 + 15 = I.S.A. standard sea level temperature in K
273 + T_0 = atmospheric temperature in K

In practice there is always a certain amount of jet thrust in the total output of the turbo-propeller engine and this must be added to the s.h.p. The correction for jet thrust is the same as that stated earlier.

To distinguish between these two aspects of the power output, it is usual to refer to them as s.h.p. and thrust horsepower (t.h.p.). The total equivalent horsepower is denoted by t.e.h.p. (sometimes e.h.p.) and is the s.h.p. plus the s.h.p. equivalent to the net jet thrust. For estimation purposes it is taken that, under sea-level static conditions, one s.h.p. is equivalent to approximately 2.6 lb. of jet thrust. Therefore:

$$\text{t.e.h.p.} = \text{s.h.p.} + (\text{jet thrust lb.})/2.6$$

The ratio of jet thrust to shaft power is influenced by many factors. For instance, the higher the aircraft operating speed the larger may be the required proportion of total output in the form of jet thrust. Alternatively, an extra turbine stage may be required if more than a certain proportion of the total power is to be provided at the shaft. In general, turbo-propeller aircraft provide one pound of thrust for every 3.5 h.p. to 5 h.p.

Comparison between Thrust and Horse-Power

Because the turbo-jet engine is rated in thrust and the turbo-propeller engine in s.h.p., no direct comparison between the two can be made without a power conversion factor. However, since the turbo-propeller engine receives its thrust mainly from the propeller, a comparison can be made by converting the horse-power developed by the engine to thrust or the thrust developed by the turbo-jet engine to t.h.p.; that is, by converting work to force or force to work. For this purpose, it is necessary to take into account the speed of the aircraft.

The t.h.p. is expressed as FV/(550 ft. per sec.) where

$$F = \text{lb. of thrust}$$
$$V = \text{aircraft speed (ft. per sec.)}$$

Since 1 horsepower is equal to 550 ft.lb. per sec. and 550 ft. per sec. is equivalent to 375 miles per hour, it can be seen from the above formula that 1 lb. of thrust equals 1 t.h.p. at 375 m.p.h. It is also common to quote the speed in knots (nautical miles per hour); one knot is equal to 1.1515 m.p.h. or 1 pound of thrust is equal to one t.h.p. at 325 knots.

Thus if a turbo-jet engine produces 5000 lb. of net thrust at an aircraft speed of 600 m.p.h. the t.h.p would be

$$(5000 \times 600)/375 = 8000$$

However, if the same thrust was being produced by a turbo-propeller engine with a propeller efficiency of 55% at the same flight speed of 600 m.p.h., then the t.h.p. would be

$$(8000 \times 100)/55 = 14,545$$

Thus at 600 m.p.h. 1 lb. of thrust is the equivalent of about 3 t.h.p.

Engine Thrust in Flight

Since reference will be made to gross thrust, momentum drag, and net thrust, it will be helpful to define these terms:

- Gross or total thrust is the product of the mass of air passing through the engine and the jet velocity at the propelling nozzle, expressed as:

$$(P - P_0)A + Wv_j/g$$

- The momentum drag is the drag due to the momentum of the air passing into the engine relative to the aircraft velocity is expressed as WV/g where

  W = Mass flow in lb. per sec.
  V = Velocity of aircraft in feet per sec.
  g = Gravitational constant 32.2 ft. per sec.

- The net thrust or resultant force acting on the aircraft in flight is the difference between the gross thrust and the momentum drag.

From the definitions and formulae just stated; under flight conditions, the net thrust of the engine, simplifying can be expressed as:

$$(P - P_0)A + W(Vj - V)/g$$

Figure 10–2 provides a diagrammatic explanation.

Effect of Forward Speed

Since reference will be made to "ram ratio" and Mach number, these terms are defined as follows:

- Ram ratio is the ratio of the total air pressure at the engine compressor entry to the static air pressure at the air intake entry.
- Mach number is an additional means of measuring speed and is defined as the ratio of the speed of a body to the local speed of sound. Mach 1.0 therefore represents a speed equal to the local speed of sound.

From the thrust equation, it is apparent that if the jet velocity remains constant, independent of aircraft speed, then as the aircraft speed increases the thrust would decrease in direct proportion. However, due to the "ram ratio" effect from the aircraft forward speed, extra air is taken into the engine so that the mass airflow and also the jet velocity increase with aircraft speed. The effect of this tends to offset the extra intake momentum drag due to the forward speed so that the resultant loss of net thrust is partially recovered as the aircraft speed increases. A typical curve illustrating this point is shown in Figure 10–3. Obviously, the "ram ratio" effect, or the return obtained in terms of pressure rise at entry to the compressor in exchange for the unavoidable intake drag, is of considerable importance to the turbo-jet engine, especially at high speeds. Above speeds of Mach 1.0, as a result of the formation of shock waves at the air intake, this rate of pressure rise will rapidly decrease unless a suitably designed air intake is provided; an efficient air intake is necessary to obtain maximum benefit from the ram ratio effect.

As aircraft speeds increase into the supersonic region, the ram air temperature rises rapidly consistent with the basic gas laws. This temperature rise affects the compressor delivery air temperature proportionately and, in consequence, to maintain the required thrust, the engine must be subjected to higher turbine entry temperatures. Since the maximum permissible turbine entry temperature is determined by the temperature limitations of the turbine assembly, the choice of turbine materials and the design of blades and stators to permit cooling are very important.

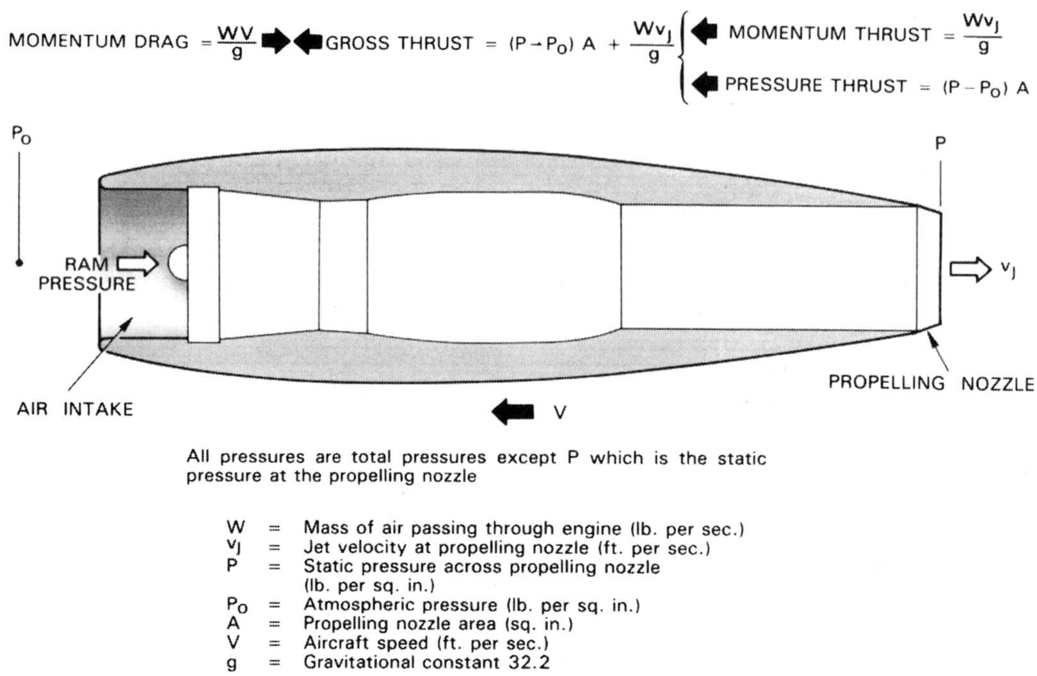

FIGURE 10–2 The balance of forces and expression for thrust and momentum drag. *(Source: Rolls Royce.)*

With an increase in forward speed, the increased mass airflow due to the "ram ratio" effect must be matched by the fuel flow and the result is an increase in fuel consumption. Because the net thrust tends to decrease with forward speed the end result is an increase in specific fuel consumption (s.f.c.), as shown by the curves for a typical turbo-jet engine in Figure 10–4.

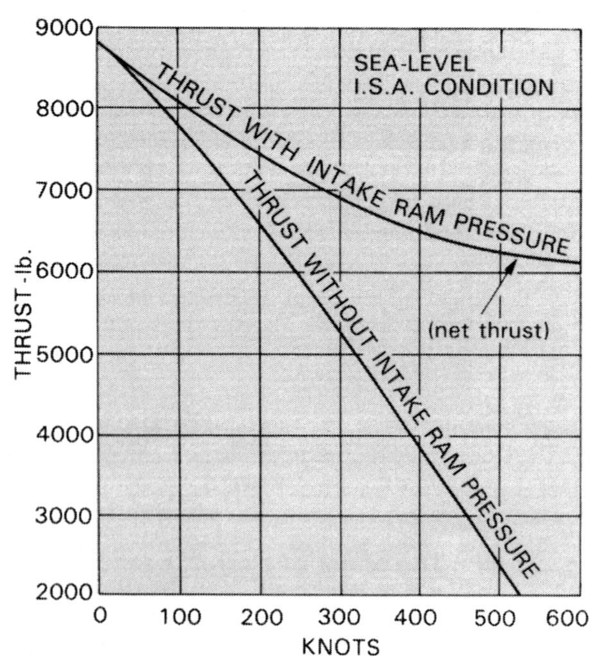

FIGURE 10–3 Thrust recovery with aircraft speed. *(Source: Rolls Royce)*

At high forward speeds at low altitudes the "ram ratio" effect causes very high stresses on the engine and, to prevent overstressing, the fuel flow is automatically reduced to limit the engine speed and airflow.

The effect of forward speed on a typical turbo-propeller engine is shown by the trend curves in Figure 10–5. Although net jet thrust decreases, s.h.p. increases due to the "ram ratio" effect of increased mass flow and matching fuel flow. Because it is standard practice to express the s.f.c. of a turbo-propeller engine relative to s.h.p., an improved s.f.c. is exhibited. However, this does not provide a true comparison with the curves shown in Figure 10–4, for a typical turbo-jet engine, as s.h.p. is absorbed by the propeller and converted into thrust and, irrespective of an increase in s.h.p., propeller efficiency and therefore net thrust deteriorates at high subsonic forward speeds. In consequence, the turbo-propeller engine s.f.c. relative to net thrust would, in general comparison with the turbo-jet engine, show an improvement at low forward speeds but a rapid deterioration at high speeds.

Effect of Afterburning on Engine Thrust

At takeoff conditions, the momentum drag of the airflow through the engine is negligible, so that the gross thrust can be considered to be equal to the net thrust. If afterburning is

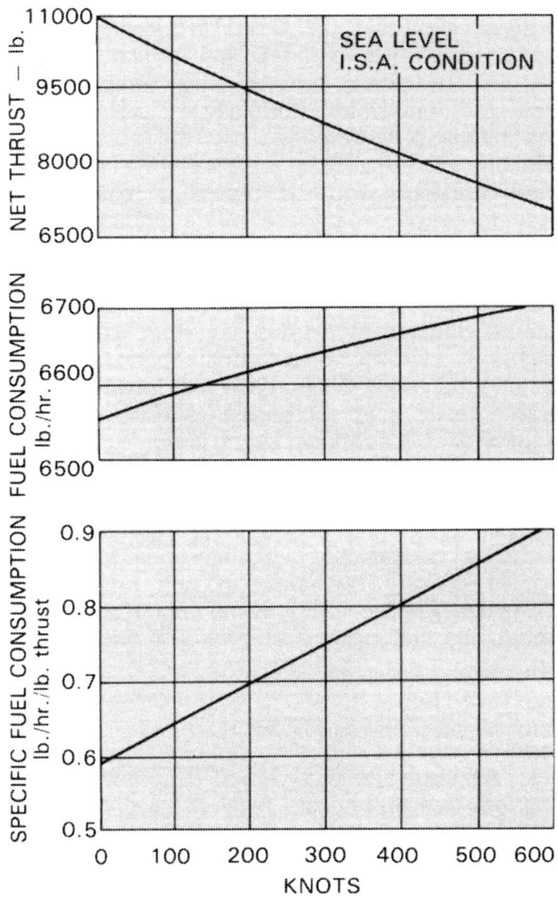

FIGURE 10–4 The effect of aircraft speed on thrust and fuel consumption. *(Source: Rolls Royce.)*

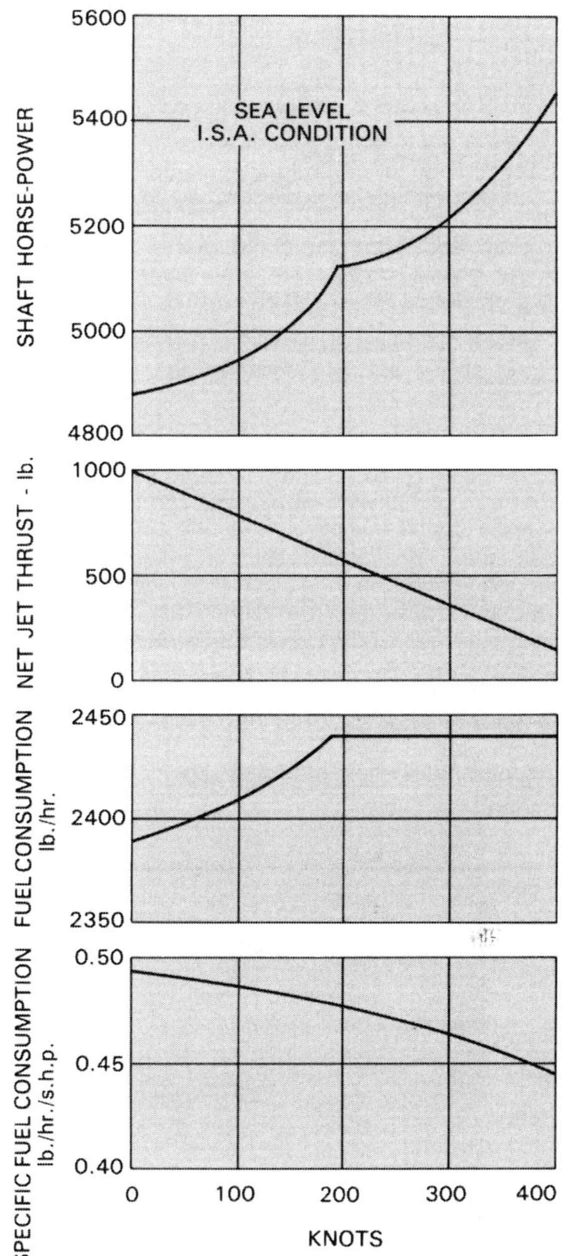

FIGURE 10–5 The effect of aircraft speed on s.h.p. and fuel consumption. *(Source: Rolls Royce)*

selected, an increase in takeoff thrust in the order of 30% is possible with the pure jet engine and considerably more with the bypass engine. This augmentation of basic thrust is of greater advantage for certain specific operating requirements.

Under flight conditions, however, this advantage is even greater, since the momentum drag is the same with or without afterburning and, due to the ram effect, better utilization is made of every pound of air flowing through the engine. The following example, illustrates why afterburning thrust improves under flight conditions.

Assuming an aircraft speed of 600 m.p.h. (880 ft. per sec.), then momentum drag is: 880/32.2 = 27.5 (approximately). This means that every pound of air per second flowing through the engine and accelerated up to the speed of the aircraft causes a drag of about 27.5 lb.

Suppose each pound of air passed through the engine gives a gross thrust of 77.5 lb. Then the net thrust given by the engine per lb. of air per second is 77.5 − 27.5 = 50 lb.

When afterburning is selected, assuming the 30% increase in static thrust given previously, the gross thrust will be 1.3 × 77.5 100.75 lb. Thus, under flight condition of 600 m.p.h., the net thrust per pound of air per second will be 100.75 − 27.5 = 73.25 lb. Therefore, the ratio of net thrust due to afterburning is 73.25/50 = 1.465. In other words a 30% increase in thrust under static conditions becomes a 46.5% increase in thrust at 600 m.p.h.

This larger increase in thrust is invaluable for obtaining higher speeds and higher altitude performances. The total and specific fuel consumptions are high, but not unduly so for such an increase in performance.

The limit to the obtainable thrust is determined by the afterburning temperature and the remaining usable oxygen in the exhaust gas stream. Because no previous combustion heating takes place in the duct of a bypass engine, these engines with their large residual oxygen surplus are particularly suited to afterburning and static thrust increases of up to 70% are obtainable. At high forward speeds several times this amount are achieved.

Effect of Altitude

With increasing altitude the ambient air pressure and temperature are reduced. This affects the engine in two interrelated ways:

- The fall of pressure reduces the air density and hence the mass airflow into the engine for a given engine speed. This causes the thrust or s.h.p. to fall. The fuel control system adjusts the fuel pump output to match the reduced mass airflow, so maintaining a constant engine speed.
- The fall in air temperature increases the density of the air, so that the mass of air entering the compressor for a given engine speed is greater. This causes the mass airflow to reduce at a lower rate and so compensates to some extent for the loss of thrust due to the fall in atmospheric pressure. At altitudes above 36,089 feet and up to 65,617 feet, however, the temperature remains constant, and the thrust or s.h.p. is affected by pressure only.

Graphs showing the typical effect of altitude on thrust, s.h.p., and fuel consumption are illustrated in Figures 10–6 and 10–7.

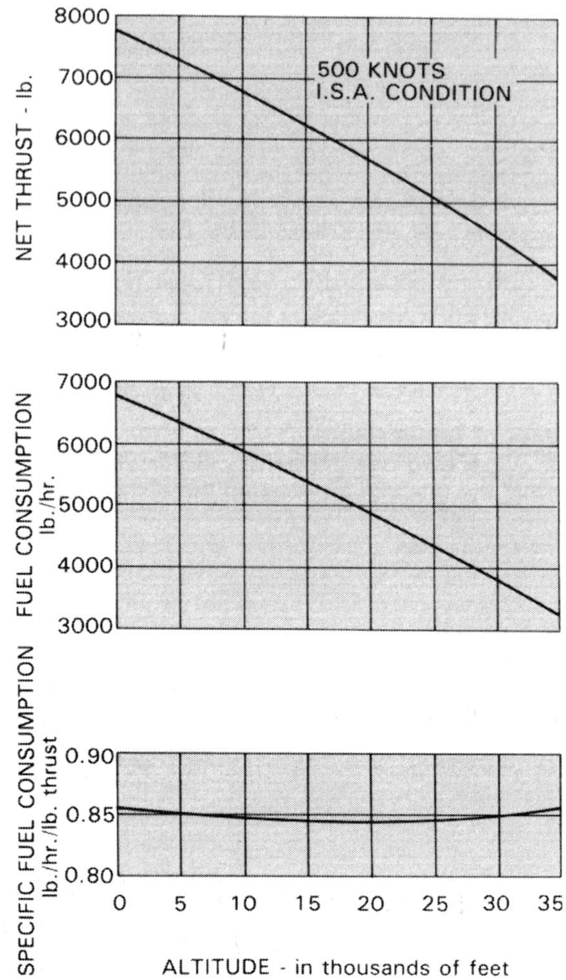

FIGURE 10–6 The effects of altitude on thrust and fuel consumption. *(Source: Rolls Royce.)*

Effect of Temperature

On a cold day the density of the air increases so that the mass of air entering the compressor for a given engine speed is greater, hence the thrust or s.h.p. is higher. The denser air does, however, increase the power required to drive the compressor or compressors: thus the engine will require more fuel to maintain the same engine speed or will run at a reduced engine speed if no increase in fuel is available.

On a hot day the density of the air decreases, thus reducing the mass of air entering the compressor and, consequently, the thrust of the engine for a given rpm. Because less power will be required to drive the compressor, the fuel control system reduces the fuel flow to maintain a constant engine rotational speed or turbine entry temperature, as appropriate; however, because of the decrease in air density, the thrust will be lower. At a temperature of 45°C, depending on the type of engine, a thrust loss of up to 20% may be experienced. This means that some sort of thrust augmentation, such as water injection, may be required.

The fuel control system controls the fuel flow so that the maximum fuel supply is held practically constant at low air temperature conditions, whereupon the engine speed falls but, because of the increased mass airflow as a result of the increase in air density, the thrust remains the same. For example, the combined acceleration and speed control fuel system schedules fuel flow to maintain a constant engine rpm, hence thrust increases as air temperature decreases until, at a predetermined compressor delivery pressure, the fuel flow is automatically controlled to maintain a constant compressor delivery pressure and, therefore, thrust.

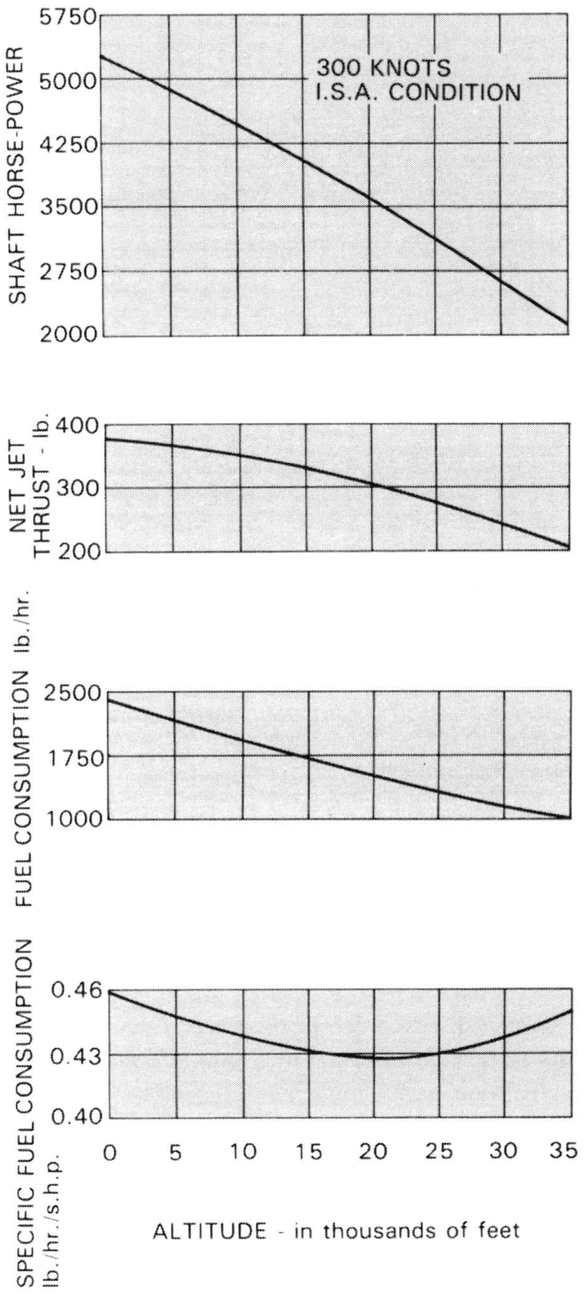

FIGURE 10–7 The effect of altitude on s.h.p. *(Source: Rolls Royce)*

Figure 10–8 illustrates this for a twin-spool engine where the controlled engine rpm is high-pressure compressor speed and the compressor delivery pressure is expressed as P_3. It will also be apparent from this graph that the low-pressure compressor speed is always less than its limiting maximum and that the difference in the two speeds is reduced by a decrease in ambient air temperature. To prevent the L.P. compressor overspeeding, fuel flow is also controlled by an L.P. governor which, in this case, takes a passive role.

The pressure ratio control fuel system schedules fuel flow to maintain a constant engine pressure ratio and, therefore, thrust below a predetermined ambient air temperature. Above this temperature the fuel flow is automatically controlled to prevent turbine entry temperature limitations from being exceeded, thus resulting in reduced thrust and, overall, similar curve characteristics to those shown in Figure 10–8. In the instance of a triple-spool engine the pressure ratio is expressed as P_4/P_1, i.e., H.P. compressor delivery pressure/engine inlet pressure.

Propulsive Efficiency

Performance of the jet engine is not only concerned with the thrust produced, but also with the efficient conversion of the heat energy of the fuel into kinetic energy, as represented by the jet velocity, and the best use of this velocity to propel the aircraft forward, i.e., the efficiency of the propulsive system.

The efficiency of conversion of fuel energy to kinetic energy is termed thermal or internal efficiency and, like all heat engines, is controlled by the cycle pressure ratio and combustion temperature. Unfortunately, this temperature is limited by the thermal and mechanical stresses that can be tolerated by the turbine. The development of new materials and techniques to minimize these limitations is continually being pursued.

The efficiency of conversion of kinetic energy to propulsive work is termed the propulsive or external efficiency and this is affected by the amount of kinetic energy wasted by the propelling mechanism. Waste energy dissipated in the jet wake, which represents a loss, can be expressed as $[W(v_J - V)^2]/2g$ where $(v_J - V)$ is the waste velocity. It is therefore apparent that at the aircraft lower speed range the pure jet stream wastes considerably more energy than a propeller system and consequently is less efficient over this range. However, this factor changes as aircraft speed increases, because although the jet stream continues to issue at a high velocity from the engine its velocity relative to the surrounding atmosphere is reduced and, in consequence, the waste energy loss is reduced.

Briefly, propulsive efficiency may be expressed as:

$$\frac{\text{Work done on the aircraft}}{\text{Energy imparted to engine airflow}}$$

or simply

$$\frac{\text{Work done}}{\text{Work done} + \text{work wasted in exhaust}}$$

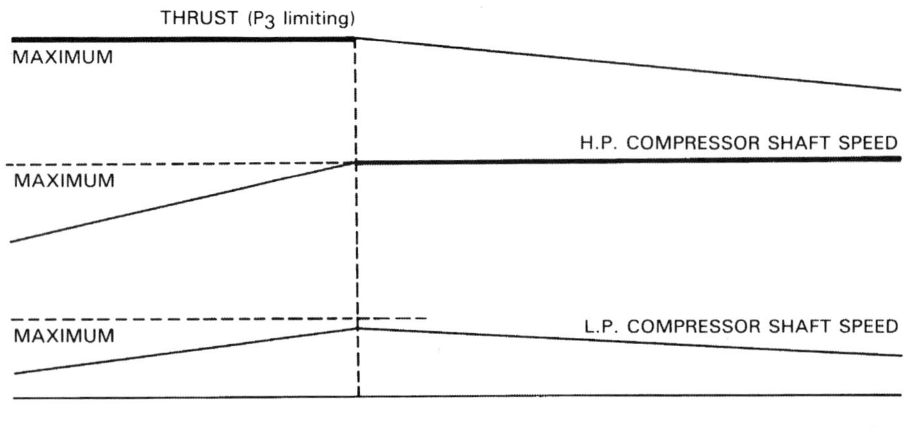

FIGURE 10–8 The effect of air temperature on a typical twin-spool engine. *(Source: Rolls Royce.)*

Work done is the net thrust multiplied by the aircraft speed. Therefore, progressing from the net thrust equation given earlier, the following equation is arrived at:

Propulsive efficiency

$$= \frac{V\left[(P - P_0)A + \frac{W(v_J - V)}{g}\right]}{V\left[(P - P_0)A + \frac{W(v_J - V)}{g}\right] + \frac{W(v_J - V)^2}{2g}}$$

In the instance of an engine operating with a non-choked nozzle, the equation becomes:

$$\frac{WV(v_J - V)}{WV(v_J - V) + \frac{1}{2}W(v_J - V)^2}$$

Simplified to: $2V/(V + v_j)$

This latter equation can also be used for the choked nozzle condition by using V_J to represent the jet velocity when fully expanded to atmospheric pressure, thereby dispensing with the nozzle pressure term $(P - P_0)A$.

Assuming an aircraft speed (V) of 375 m.p.h. and a jet velocity (v_J) of 1230 m.p.h., the efficiency of a turbo-jet is:

$$\frac{2 \times 375}{600 + 1230} = \text{approx. } 47\%$$

On the other hand, at an aircraft speed of 600 m.p.h. the efficiency is:

$$\frac{2 \times 600}{600 + 1230} = \text{approx. } 66\%$$

Propeller efficiency at these values of V is approximately 82 and 55%, respectively, and from reference to Figure 10–9 it can be seen that for aircraft designed to operate at sea level speeds below approximately 400 m.p.h. it is more effective to absorb the power developed in the jet engine by gearing it to a propeller instead of using it directly in the form of a pure jet stream. The disadvantage of the propeller at the higher aircraft

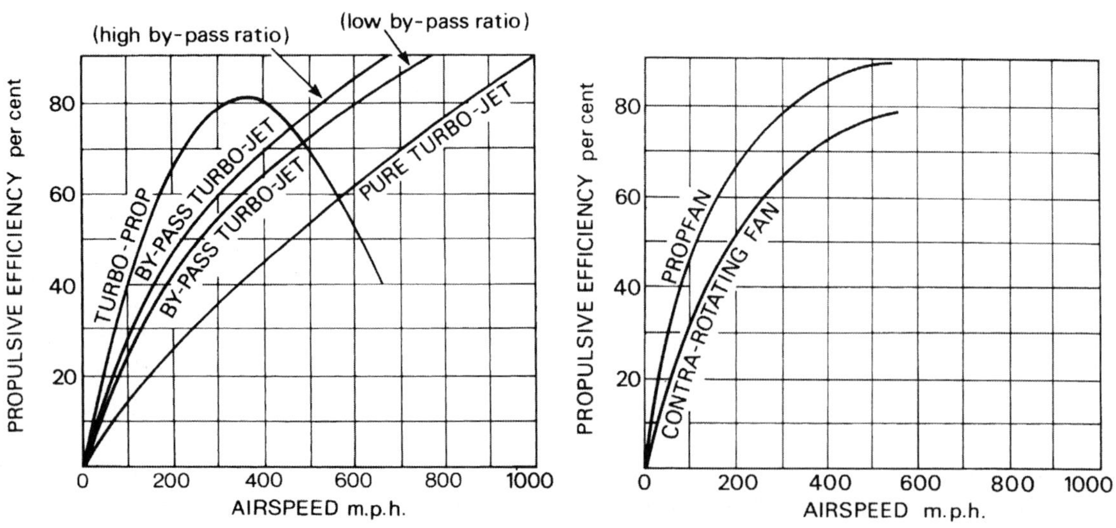

FIGURE 10–9 Propulsive efficiencies and aircraft speed. *(Source: Rolls Royce.)*

speeds is its rapid fall off in efficiency, due to shock waves created around the propeller as the blade tip speed approaches Mach 1.0. Advanced propeller technology, however, has produced a multi-bladed, swept back design capable of turning with tip speeds in excess of Mach 1.0 without loss of propeller efficiency. By using this design of propeller in a contra-rotating configuration, thereby reducing swirl losses, a "prop-fan" engine, with very good propulsive efficiency capable of operating efficiently at aircraft speeds in excess of 500 m.p.h. at sea level, can be produced.

To obtain good propulsive efficiencies without the use of a complex propeller system, the bypass principle is used in various forms. With this principle, some part of the total output is provided by a jet stream other than that which passes through the engine cycle and this is energized by a fan or a varying number of L.P. compressor stages. This bypass air is used to lower the mean jet temperature and velocity either by exhausting through a separate propelling nozzle, or by mixing with the turbine stream to exhaust through a common nozzle.

The propulsive efficiency equation for a high bypass ratio engine exhausting through separate nozzles is given below, where W_1 and v_{J_1} relate to the bypass function and W_2 and v_{J_2} to the engine main function.

Propulsive efficiency =

$$\frac{W_1V(v_{J_1} - V) + W_2V(v_{J_2} - V)}{W_1V(v_{J_1} - V) + W_2V(v_{J_2} - V) + ½W_1(v_{J_1} - V)^2 + ½W_2(v_{J_2} - V)^2}$$

By calculation, substituting the following values, which will be typical of a high bypass ratio engine of triple-spool configuration, it will be observed that a propulsive efficiency of approximately 85% results.

$V = 583$ m.p.h.
$W_1 = 492$ lb. per sec.
$W_2 = 100$ lb. per sec.
$V_{J_1} = 781$ m.p.h.
$V_{J_2} = 812$ m.p.h.

Propulsive efficiency can be further improved by using the rear mounted contra-rotating fan configuration of the bypass principle. This gives very high bypass ratios in the order of 15:1, and reduced "drag" results due to the engine core being "washed" by the low velocity aircraft slipstream and not the relatively high velocity fan efflux.

The improved propulsive efficiency of the bypass system bridges the efficiency gap between the turbo-propeller engine and the pure turbo-jet engine. A graph illustrating the various propulsive efficiencies with aircraft speed is shown in Figures 10–9.

Fuel Consumption and Power-to-Weight Relationship

Primary engine design considerations, particularly for commercial transport duty, are those of low specific fuel consumption and weight. Considerable improvement has been achieved by use of the bypass principle, and by advanced mechanical and aerodynamic features, and the use of improved materials. With the trend towards higher bypass ratios, in the range of 15:1, the triple-spool and contra-rotating rear fan engines allow the pressure and bypass ratios to be achieved with short rotors, using fewer compressor stages, resulting in a lighter and more compact engine.

S.f.c. is directly related to the thermal and propulsive efficiencies; that is, the overall efficiency of the engine. Theoretically, high thermal efficiency requires high pressures, which in practice also means high turbine entry temperatures. In a pure turbo-jet engine this high temperature would result in a high jet velocity and consequently lower the propulsive efficiency. However, by using the bypass principle, high thermal and propulsive efficiencies can be effectively combined by bypassing a proportion of the L.P. compressor or fan delivery air to lower the mean jet temperature and velocity as referred to previously. With advanced technology engines of high bypass and overall pressure ratios, a further pronounced improvement in s.f.c. is obtained.

The turbines of pure jet engines are heavy because they deal with the total airflow, whereas the turbines of bypass engines deal only with part of the flow; thus the H.P. compressor, combustion chambers, and turbines, can be scaled down. The increased power per lb. of air at the turbines, to take advantage of their full capacity, is obtained by the increase in pressure ratio and turbine entry temperature. It is clear that the bypass engine is lighter, because not only has the diameter of the high-pressure rotating assemblies been reduced but the engine is shorter for a given power output. With a low bypass ratio engine, the weight reduction compared with a pure jet engine is in the order of 20% for the same air mass flow.

With a high bypass ratio engine of the triple-spool configuration, a further significant improvement in specific weight is obtained. This is derived mainly from advanced mechanical and aerodynamic design, which in addition to permitting a significant reduction in the total number of parts, enables rotating assemblies to be more effectively matched and to work closer to optimum conditions, thus minimizing the number of compressor and turbine stages for a given duty. The use of higher strength light-weight materials is also a contributory factor.

For a given mass flow less thrust is produced by the bypass engine due to the lower exit velocity. Thus, to obtain the same thrust, the bypass engine must be scaled to pass a larger total mass airflow than the pure turbo-jet engine. The weight of the engine, however, is still less because of the reduced size of the H.P. section of the engine. Therefore, in addition to the reduced specific fuel consumption, an improvement in the power-to-weight ratio is obtained.

Performance Testing New Gas Turbine Engines: Parameters and Calculations*

Parameters

Typically the detailed design and development program for an engine takes 3–7 years from inception to service entry. Designing "right first time" is not practical for such a high technology product. Development comprises individual component tests followed by hundreds of hours of engine testing, based on which many design modifications are introduced. The resulting production engine standard will then comply as closely as possible with the original specification.

After service entry, *production acceptance* or *production pass-off* testing of each individual production engine is common practice, ensuring that it meets key *acceptance criteria.* This test is the final check on component manufacture and engine build quality prior to delivery to the customer.

This section outlines various types of engine test, and provides details of test beds, instrumentation, and analysis methods. Engineers from all disciplines involved in engine development or production acceptance must understand the fundamentals of performance testing technology, as it is central to both processes.

Types of Engine Test Bed

This section describes the configuration of engine test beds, and provides design to guidelines to ensure that:

- Engine inlet flow is uniform. Any distortion will affect compressor performance and will give an erroneous mass flow measurement. At worst, vortices may be shed from the floor or walls causing high cycle fatigue failure of compressor blades.
- There is no reingestion of hot exhaust gas. Should this occur the resulting distorted inlet temperature profile will again affect compressor performance. It will also prevent accurate measurement of inlet temperature, which is essential for referral of measured engine performance.
- For thrust engines, that the static pressure field around the engine is as close as practical to that of free stream conditions. As described below the thrust reading must be corrected for this effect: the smaller the effect the less scope for error in the correction.
- The static pressure distribution at the propelling nozzle exit plane allows accurate determination of a mean value. Otherwise the thrust measurement will be in error, and the engine performance may be affected.

Many other parameters are measured besides those mentioned above. For these, accuracy depends on good measurement practice rather than overall test bed design.

Outdoor Sea Level Thrust Test Bed

This is illustrated in Figure 10–10, and consists basically of an open air stand supporting an engine and providing thrust measurements. The effects of cross wind on entry conditions are negated by a large mesh screen fitted around the engine inlet. The immediate test bed area is free of obstructions to the airflow, to ensure the validity of the thrust and airflow readings. This is the most definitive thrust test bed, as for indoor test beds the thrust and airflow measurements are corrupted by the flow field generated by the side walls. Outdoor test beds are sited in remote areas, to minimize the environmental disturbance of the noise produced. Because of the resultant logistic difficulties and the impact of adverse weather conditions, indoor testing is preferred in most countries, with measurements *calibrated* versus outdoor facilities.

Indoor Sea Level Thrust Test Bed

Here a similar engine arrangement to that of Figure 10–10 is mounted indoors, as shown in Figure 10–11. The air flowpath to the engine is crucial, as flow disturbance must be minimized. The engine nozzle efflux enters a *detuner*, which exhausts hot gases and provides sound attenuation.

For a given engine the measured thrust may be up to 10% less than the value that would be recorded on an outdoor test bed. This is due to unrepresentative static pressure forces acting on the engine and cradle, caused by the velocity of air within the cell passing around the engine. This air is *entrained* into the detuner by the *ejector* effect of the engine jet; it prevents hot gas reingestion and also cools

* Source: Courtesy of Rolls Royce, *Gas Turbine Performance*, Walsh and Fletcher, Blackwell Science, 1998. Adapted with permission.

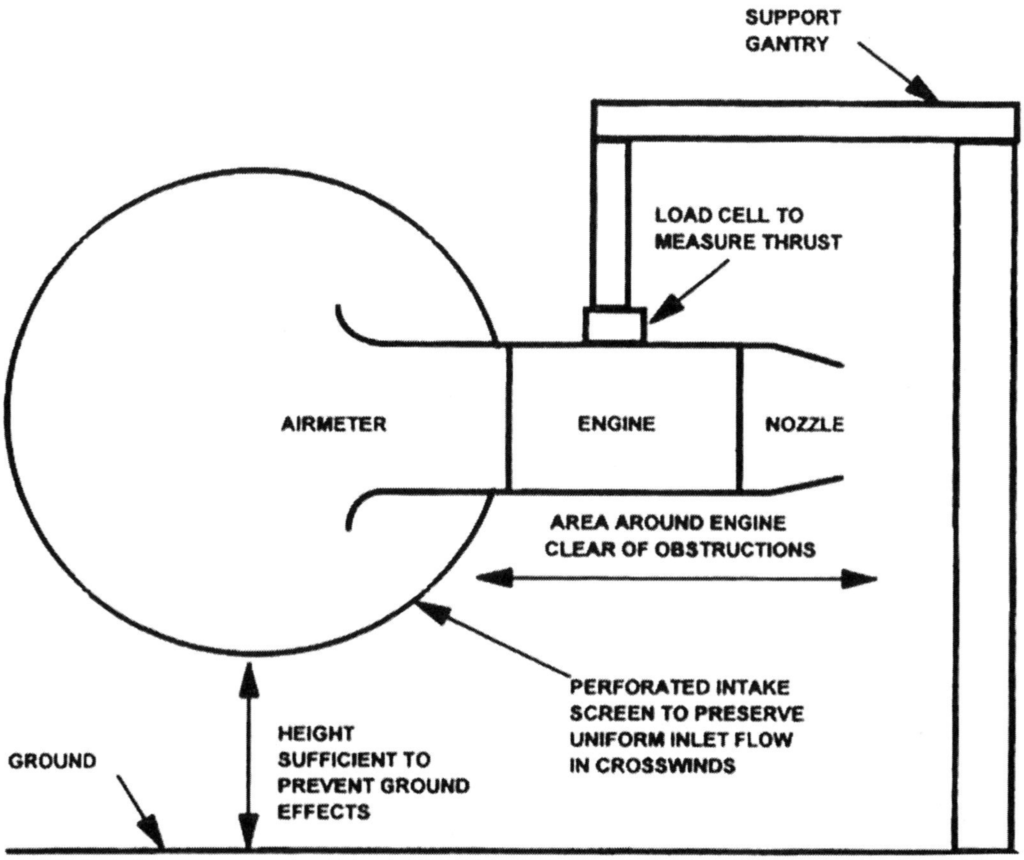

FIGURE 10–10 Outdoor sea level thrust test bed. *(Source: Rolls Royce.)*
Notes: To minimize noise disturbance outdoor test beds are normally sited in remote areas. Climatic conditions also influence choice of location.

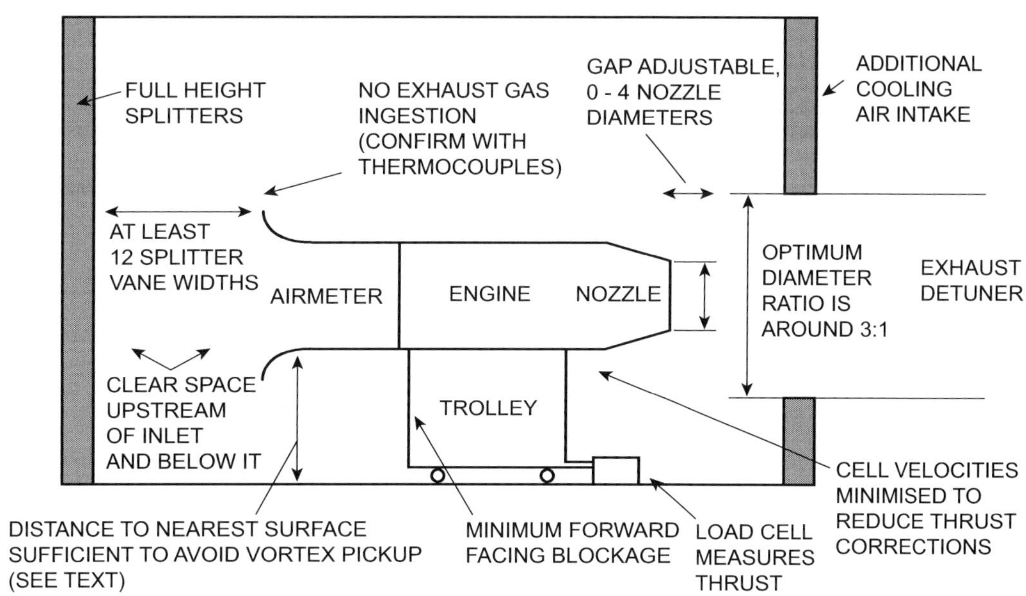

FIGURE 10–11 Indoor sea level thrust test bed. *(Source: Rolls Royce.)*
Note: Sea level gas generator test bed is similar but without thrust measurement.

the detuner. Furthermore, if the engine final nozzle is unchoked the test bed configuration can cause a rematch due to the local static pressure distribution at the nozzle exit. An indoor test bed gives useful all-weather availability, but unless the test bed is purely functional it should be calibrated against an outdoor test bed, as discussed later, to determine the effects of the static pressure field and rematch.

Key guidelines for designing an indoor test bed are presented below. These minimize the measured thrust deficiency described above, and also prevent extreme flow fields such as vortices from entering the airmeter. Owing to the complex flows within a test bed, some adjustability and subsequent experimentation may be required.

- Air should enter the test bed via a full height intake at the front, with *splitters* for flow straightening and noise attenuation. Additional side intakes should only be used if size constraints make them absolutely necessary.
- The distance from the inlet splitters to engine airmeter inlet should exceed a dozen splitter widths.
- Rear intakes, passing up to 10% of the main intake flow, may be employed to provide additional detuner cooling air while minimizing flow velocities past the engine.
- To avoid vortex pickup the height from the airmeter centerline to the floor or walls should exceed five airmeter throat diameters for an approach velocity of 0.01 times the throat velocity, and two throat diameters for 0.1 times the throat velocity. This may influence the choice of cell height and width.
- The space upstream of and around the airmeter should be unobstructed, to avoid flow disturbances that would impair the flow measurement.
- The air velocity within the cell flowing past the engine casing should not exceed around 10 m/s. This may also influence the cell dimensions, though it is heavily dependent on the entrained flow, which is discussed later.
- The engine cradle and slave equipment should present minimum forward facing area, to reduce thrust corrections.
- Recirculation of hot exhaust gas to the engine intake must be avoided; during test bed commissioning this should be confirmed using suitably located thermocouples.
- If the detuner turns flow vertically upwards it should incorporate a cascaded bend to minimize pressure loss.
- The detuner inlet diameter should be around three engine nozzle diameters. Increasing this increases the entrained flow.
- The axial gap between the detuner and the engine nozzle should be adjustable; values of at least two and ideally three or more engine nozzle diameters are recommended. Increasing this does not affect the entrained flow, but reduces thrust loss and rematching by avoiding pressure disturbances around the nozzle exit plane.
- The *entrainment ratio*, which is the ratio of entrained airflow to engine propelling nozzle flow, may be calculated from measured temperatures and a simple enthalpy balance. A good design target is 3:1 for turbojets, and lower for turbofans as engine flow is higher and discharge temperature lower.
- During test bed commissioning flow visualization using smoke should be employed as a further check on the quality of the test bed aerodynamics. This should confirm the absence of hot gas recirculation or vortex pickup, and determine suitable positions for cell static pressure measurement.

Indoor Sea Level Jet Bed or Turboshaft Gas Generator Tests

This test bed is employed when a turboshaft engine gas generator is tested with a final nozzle and exhaust detuner instead of its power turbine. The theoretical exhaust gas power is calculated based on gas generator exit flow, pressure, and temperature, and is the power that would be available from a power turbine of 100% efficiency with no duct pressure losses. Such testing is necessary when the power turbine either remains with the installation, or is supplied by another company. As this applies mainly to land-based engines, an altitude chamber would not normally be required. Key points are as follows:

- The accurate measurement of air mass flow is crucial to the power calculation. As for indoor thrust beds, flow disturbances near the airmeter must be avoided as well as vortex pickup.
- Hot gas recirculation to the intake must be avoided, as for thrust engines.
- The effect of the exhaust detuner is less critical, as the slave nozzle exit area may be "trimmed" to ensure capacity adequately matches that of the power turbine.

Indoor Sea Level Shaft Power Bed

This is used for turboprop or turboshaft engines, with output power measured directly. Figure 10–12 shows the key features of such test beds. The main differences from an indoor thrust bed are as follows.

- Air may be ducted directly to the engine from ambient rather than it flowing though the test cell, with flow measured outside the test bed at entry to the ducting. The test bed configuration does not affect measured airflow and hence measured performance; the building is only there to provide protection from adverse weather conditions.

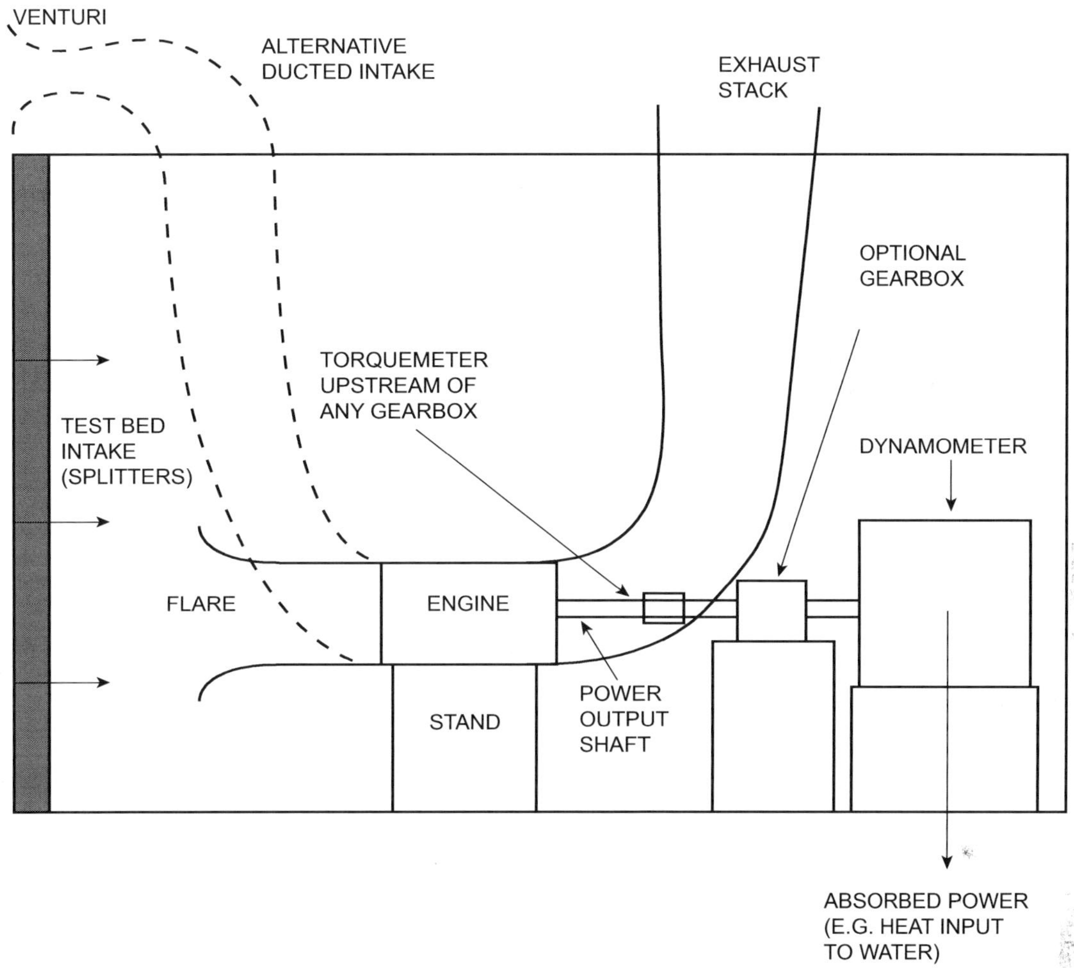

FIGURE 10–12 Indoor sea level shaft power test bed. *(Source: Rolls Royce.)*
Notes: If flare is used for mass flow measurement, positioning guidelines are per Figure 10–11. Shaft power test bed has almost no entrained airflow. If a gearbox is used to match power turbine speed to dynamometer then torque/power measurement must exclude gearbox losses.

- Alternatively air may enter the test bed through splitters and then into the engine as per a thrust bed; here test bed configuration does affect the measured airflow.
- The exhaust flow has a low velocity and is ducted directly to atmosphere with no detuner required. Entrainment effects therefore need not be considered.

On shaft power test beds some device must absorb the engine output power, providing suitable characteristics of load versus speed. There are several possibilities:

- For a turboprop, an aircraft propeller may be fitted on the test stand.
- An alternator may be used to generate electrical power, to be either dissipated in electrical resistance banks or passed to a grid system. The latter is appealing environmentally, but usually impractical during an engine development program. Setup costs are high, rotational speed is tied to grid frequency, and intermittent operation may be unacceptable to a grid operator.
- A *dynamometer* absorbs power over a range of power and speed combinations, and often also measures torque. In the hydraulic type a vaned rotor and stator arrangement pumps water through the vanes. The power absorbed heats the water, which must either be cooled or a fresh supply provided. Valves control the water level within the dynamometer, which changes the power absorbed at any given speed and allows for various power/speed laws. Torque measurement utilizes a load arm and weighing system on the external casing, which is freely mounted on hearings. The input torque is transmitted via the water and any bearing friction.

Altitude Test Facility (ATF)

Thrust or shaft power test beds may be housed within an *altitude test facility* (ATF), which reproduces the inlet conditions resulting from altitude and flight Mach number. Figure 10–13 shows the key features of an ATF. Unlike a sea level test bed the plant must provide a continuous

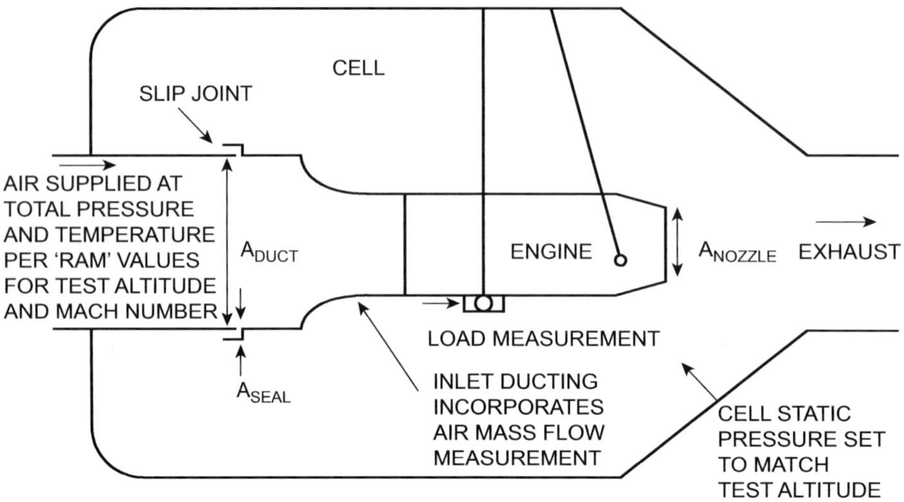

FIGURE 10—13 Altitude test facility (ATF). *(Source: Rolls Royce.)*
Notes: Gross thrust = Load + $A_{seal} \times (PS_{seal} - PS_{cell}) + A_{duct} \times (PS_{duct} - PS_{cell}) + W_{duct} \times V_{duct}$
Nett thrust = Gross thrust − $W_{duct} \times V_{duct}$
W_{duct}, load, PT_{duct}, T_{duct}, PS_{cell} and PS_{seal} are measured indirectly.
V_{duct} and PS_{duct} are calculated from the measurements using Q curves.

airflow even without the engine operating, to maintain reduced pressure and temperature.

Figure 10—14 illustrates two possible layouts of ATF plant. To simulate both ambient conditions and flight Mach number engine inlet total pressure and temperature must be controlled to the *ram* (free stream total) values for the altitude and Mach number. Also, the static pressure at the nozzle exit plane must be set to that of the test altitude. These parameters are mostly sub-ambient at altitude, hence common features of the various types of ATF are substantial pressure reduction, chilling and drying capabilities, and recompression of discharge air back to ambient. Accurate measurement of thrust is complex.

The other possibility for testing at flight conditions is *a flying test bed* as described below, where the engine is mounted on an aircraft. The main advantages of the ATF are:

- A full range of ambient and flight conditions may be tested in one geographical location
- Better instrumentation, including direct measurement of air mass flow and thrust
- High availability, independent of weather conditions

Flying Test Bed

A flying test bed is also often used for major aeroengine programs. Typically a four engined aircraft is modified to mount a single, new development engine at one berth. Compared with an ATF the advantages are:

- Better simulation of functional effects such as carcase loads and inlet distortion
- Lower capital cost

However, as mentioned there are no direct measurements of thrust and mass flow. These must be calculated as follows.

- Propelling nozzle thrust coefficient and capacity are obtained from rig and engine tests, ideally in an ATF.
- Nozzle entry total pressure and temperature are measured directly, with sufficient coverage to obtain valid average data.
- Nozzle mass flow may now be calculated, along with exit velocity.
- Any air offtake and nacelle ejector flows are estimated from design data.
- Fuel flow is measured directly.
- Inlet airflow may now be calculated.
- Nett thrust is the difference between the total exit and inlet momentum, and if the nozzle is choked any pressure thrust must be added.

Measurements and Instrumentation

Engine tests use differing amounts and sophistication of instrumentation, depending on their purpose. Many development tests require detailed performance investigation, hence pressures and temperatures are measured at virtually every station, as well as power or thrust, shaft speeds, fuel, and airflow, etc. At the other extreme, for production pass-off or endurance testing only a minimum of measurements are taken beyond those of the production control system, such as ambient conditions, power or thrust level, and fuel flow.

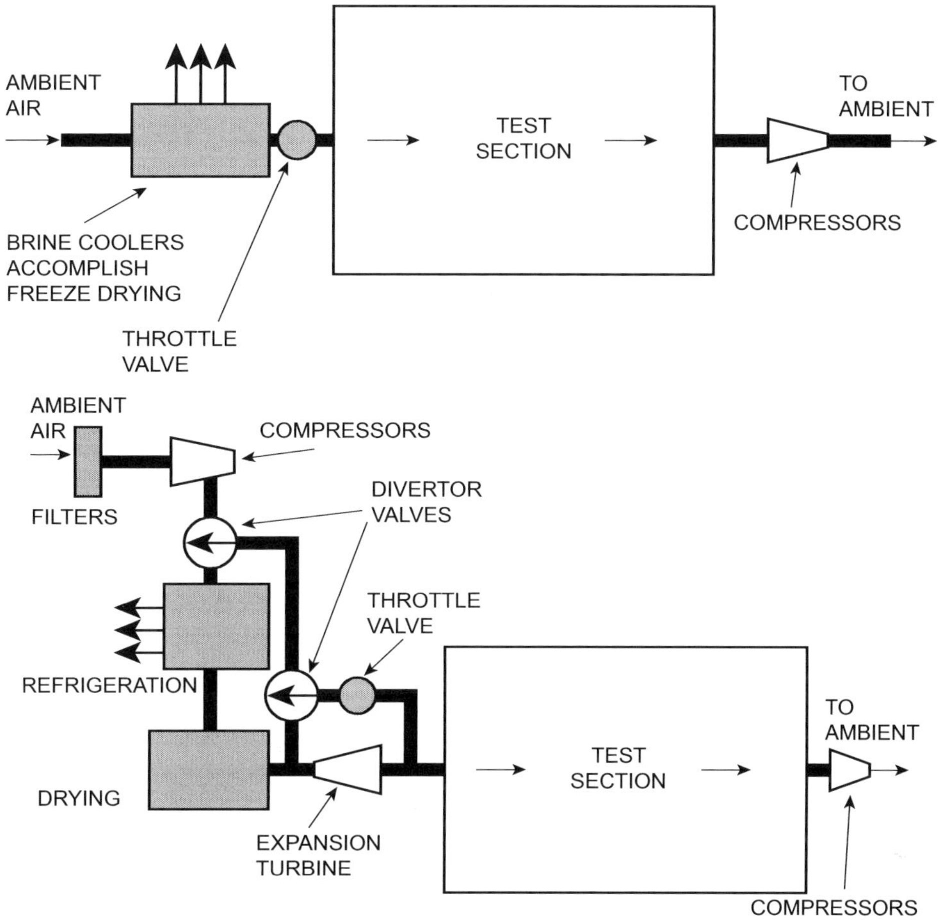

FIGURE 10–14 Altitude test facility (ATF): possible plant layouts.
Note: Different valve settings are used depending on the pressure and temperature levels required. *(Source: Rolls Royce)*

This section provides background to the instrumentation used for each possible measured parameter. Indicative accuracies and coverage requirements are provided, along with a summary of how the instruments work.

Engine testing is very expensive; hence to ensure good quality data are obtained the importance of the following cannot be overemphasized.

- The test bed and all instrumentation must be properly calibrated.
- Test planning should include careful specification of instrumentation requirements.
- Key measurements must be repeatedly checked during the testing, to ensure data are valid.
- Engine removal from the test bed must be delayed, and testing repeated, if necessary.

For all of the above an understanding of likely accuracy levels is required.

Pressures

Pressures are measured for a number of reasons:

- Determination of overall engine performance requires ambient or cell pressure so that parameters can be referred back to standard conditions.
- Engine station pressures help define component performance, e.g., pressure ratios, surge margins, and flow capacities.
- Mass flow measurement is based on the local difference between total and static pressure levels.

Figure 10–15 illustrates the main elements of a typical pressure measuring system. Local small holes called *tappings* allow the engine gas stream pressure to reach a measuring device outside the engine, via fine *capillary tubes* of 1–2 mm diameter. Generally at least three circumferential locations are used at a station, to increase coverage and to allow error detection by comparison of readings. A leaking line usually reads low, though thought

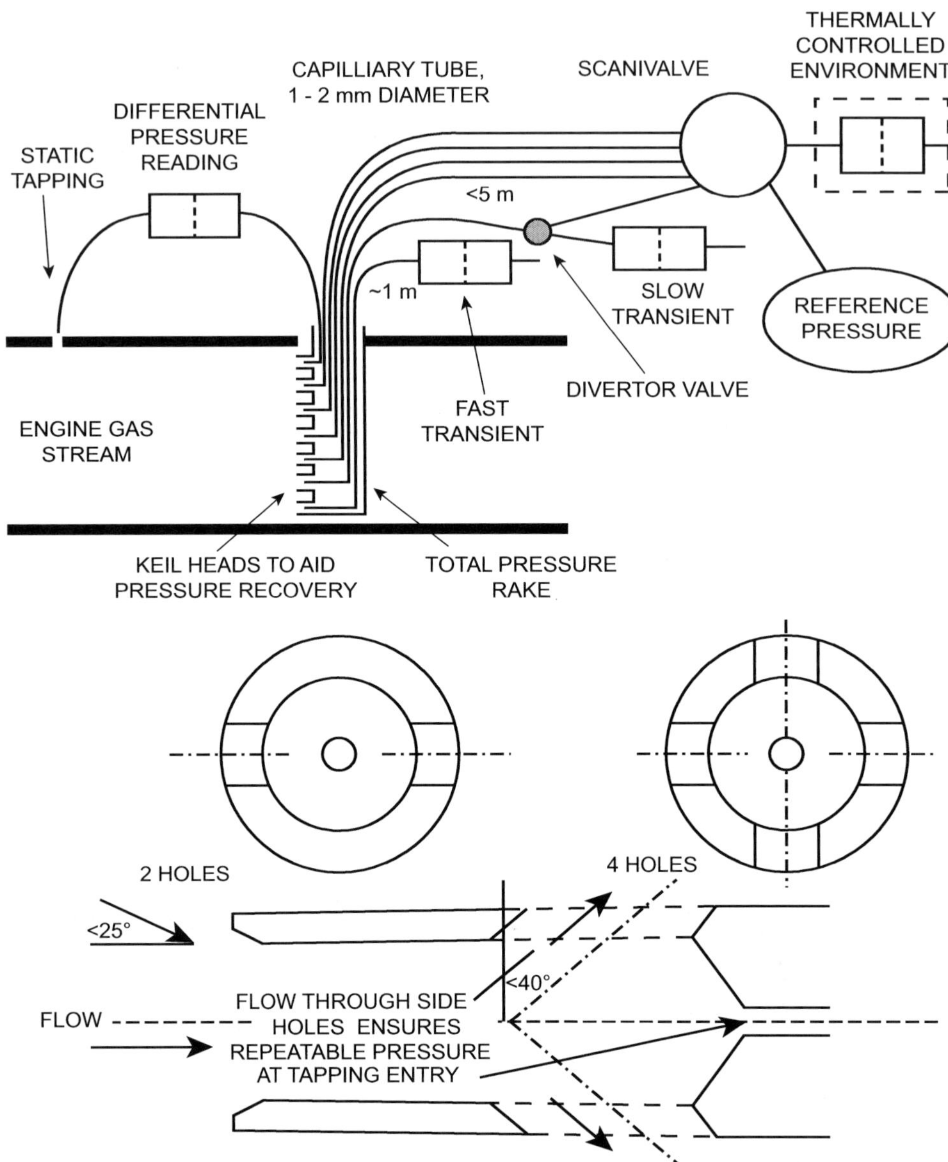

FIGURE 10–15 Pressure measuring system. *(Source: Rolls Royce.)*

should be given as to whether lines pass through higher pressure regions. An alternative is to *gang* the lines from several tappings together into a manifold and then read the manifold pressure as a *pneumatic average*. Though this method is relatively inexpensive a leak in any one pressure line is difficult to detect. The measuring devices used are described here.

Manometers

For pressures below around 2 bar, older test beds have used water or mercury *manometers*, where the height of a column of liquid in a glass tube is read visually. Small corrections are applied for the temperature of the liquid column. For a well-designed system accuracy is around 0.25%. As automatic data recording has become more prevalent manometers have virtually disappeared.

Transducers

Modern test beds use a *transducer*, where a pressure difference causes movement of a diaphragm, which is converted to an electrical signal. The other side of the diaphragm may be at ambient pressure or a vacuum, giving *gauge* and *absolute* readings respectively. The diaphragm movement is converted to a voltage, which is read by the data logging system. For many transducers the conversion uses an *energizing voltage* and a resistive straingauge on the diaphragm. Alternatives are piezoelectric, which generate their own voltage, or inductive. *Calibration curves* relate the electrical signal to pressure levels, and are obtained by either a *dead weight tester* that applies a known force and hence air pressure, or comparison with other calibrated transducers. Transducer designs are optimized for various pressure ranges; they should ensure operation is in the most linear part of the range, typically 10–90% of full scale. Transducer temperature should be controlled as this affects the straingauge resistance; even *compensating circuitry* does not fully eliminate the effect. Accuracies quoted herein are for a controlled transducer temperature.

Steady state, typical accuracies are around 0.1% of full scale for the basic transducer; however, a good overall accuracy is 0.5% for engine pressures. This figure allows for calibration drift, hysteresis, engine stability and pressure profiles, transducer nonlinearity, and drift in the voltage supply. Cell static pressure is less prone to engine effects, hence an accuracy of 0.25% is obtainable.

For steady state testing, many pressure tappings are read in turn via *scanivalves*, where rotating valvery connects a transducer to each tapping in turn. For best accuracy the transducer is kept in a temperature controlled environment, and known *reference pressures* are also read one or more times per scan giving continual, automatic update of the transducer calibration. With best practice overall accuracy can be around 0.25% as most transducer error effects are eliminated, leaving those due to engine pressure profiles and stability during the time taken to complete a scan. Such a system is unsuitable for transient use due to intermittent reading and volume packing of the pressure lines. The latter effect means around 4 minutes' stabilization is required before an accurate pressure reading can be obtained.

Comments on pressure measurements at the various engine stations are presented below.

Ambient Pressure—Barometers

Barometers are used to measure ambient pressure, with an accuracy of around 0.1%. They fall into two main categories:

1. An *aneroid* barometer consists of a dial and pointer controlled by an evacuated metal cylinder with corrugated sides and a spring action. Changes in ambient pressure cause movement which is read on the dial.
2. A mercury in glass or *Fortin* barometer uses a mercury column in a closed glass tube, evacuated at the closed, upper end. The height of the column is read visually.

Test Cell Static Pressure

This is required for all engines where the intake is inside a test bed, and is measured in at least two places of low cell velocity. Usual locations are on the side walls in the plane of the nozzle exit, at the same height as engine centerline. The instrumentation comprises the open end of a 1–2 mm diameter capillary tube surrounded by a perforated "pepper pot," which removes the effects of incident velocity. The tube is then connected to either a transducer or water manometer.

Engine Static Pressures

If the axis of a tapping is perpendicular to the flow direction then it will read static pressure, as no dynamic head will be recovered. Wall tappings give a more accurate reading than side or rearward facing tappings on immersed probes, as the presence of the probe disturbs the flow. Static pressure tappings may be used in place of total pressure readings if a calibration has already been obtained versus the total pressure reading. Such calibrations become tenuous, however, if there is swirl angle variation, as this changes the local Mach number.

Engine Total Pressures

To achieve full recovery of the stream dynamic head, and hence measure total pressure, the pressure tapping is mounted in a probe that points its axis towards the direction of gas flow. If the flow angle varies by more than ±5°, a *Kiel head* should be employed, as illustrated in Figure 10–15. This uses a chamfered entry to recover effectively the stream dynamic head for incidences of up to ±25°. Above gas temperatures of around 1300 K total pressure probes are not normally viable as they will require cooling and hence become so large that associated pressure drops are prohibitive.

Coverage requirements depend on how well understood the pressure uniformity is at a station, and should be agreed with the relevant component designer. Many radial and circumferential locations may be addressed via either multiple heads on vane leading edges, or several multihead *rakes* inserted into the gas stream. The heads are often placed on centers of equal flow area to assist in data averaging. Calibration of rig versus engine instrumentation standards may also be undertaken. Coverage in the cold end (compressors) should be at least three off rakes with one to five heads, depending on engine size. If large radial or circumferential nonuniformities are likely then more coverage may be employed, such as downstream of an aeroengine fan where up to 10 heads are used. For the hot end there is often significant swirl hence greater coverage may be needed. Around six rakes, or preferably instrumented vane leading edges, are suggested with two to five heads depending on engine size.

Total pressure rakes and wall static tappings may be used in combination, set in the same plane. Static tappings complete definition of the total pressure profile, as static and total pressures are equal at the walls. In addition, if the static pressure is reasonably uniform, which requires low swirl angle, the difference between total and static pressure indicates flow velocity.

Placement of total pressure rakes should consider obvious sources of error such as wakes downstream of struts. Good practice is to derive calibration factors between fitted instrumentation and that giving fuller coverage as listed above.

Differential Pressures

Normally pressure is measured as the difference between the gas stream and ambient, and an absolute pressure level obtained by addition of ambient pressure to the gauge reading. To read a pressure *difference* between two points, both sides of a transducer may be connected to tappings at the engine stations in question. This allows use of a more precise, lower range transducer, and avoids large inaccuracies due to the subtraction of similar numbers. One disadvantage is that recalibrating such a transducer is not possible without disconnecting the instrumentation, unlike for scanivalve systems.

Transient Pressures

For transient testing, dedicated pressure transducers, designed to optimize transient response, are required for each tapping. This provides a continuous reading, unlike a scanivalve system. The transducers are mounted local to the engine to minimize line volumes, and hence allow fast response to pressure changes; they may be water jacketed to enhance thermal stability. Line length limits are around 5 m for ordinary handling and 1 m for faster transients such as fuel spiking. In the former case a divertor valve may be employed to allow the same tapping to be read by the steady state scanivalve; for the shorter line length space does not permit this. Typical scan rates range from 10 to 500 scans per second. Absolute accuracies are lower for dedicated transient transducers than for a scanivalve system, around 1.5% of full range, and the transducers are more subject to drift.

A calibration curve should be run at the start of each working day to provide a comparison with the steady state instrumentation. In addition a transient maneuver should be performed to check for lag due to divertor valve faults.

Dynamic Measurements of Pressures

Dynamic measurements of pressure address high frequency pressure perturbations, rather than "dynamic pressure" as in the "velocity head." These measurements are employed to detect flow instabilities such as rotating stall or rumble that can occur in the compression and combustion systems. The accuracy in determining amplitude is low, around 10% of range, as the design is optimized for response and the thermal environment is normally uncontrolled. Probes such as the *kistler* (resistive) and *kulite* (piezo) varieties are utilized, which incorporate pressure transducers—the low volume allows response to the very high frequencies involved. The kistler probe is more vulnerable to vibration but can tolerate temperatures up to 350°C, which is 80°C higher than the kulite. The signal may be available in the control room on an oscilloscope and is normally recorded in analog form on tape. Later examination of amplitudes and frequencies helps pinpoint the cause or at least onset of instability. High frequency phenomena may also be shown qualitatively by noise on ordinary transient pressure signals.

Temperatures

Measurement of temperatures provides the following information on engine and component performance.

- Determination of overall engine performance requires inlet temperature so that parameters can be referred back to standard conditions.
- The temperatures local to a component are required to define its performance, i.e., efficiency and flow capacity.
- Temperatures are required to ensure the engine is not operated beyond limits stipulated for mechanical integrity.
- Mass flow measurement utilizes temperature levels.

Temperature measurement is complex. It is vital to follow the good design and working practices outlined herein, otherwise significant inaccuracies may result.

Temperature readings more closely reflect total rather than static conditions, as a rake in the gas stream brings the gas to rest on its surface. In fact it is impossible to measure purely static temperature. The fraction of the dynamic temperature recovered is termed the *recovery factor.* As for pressure measurement, temperature rakes employ Kiel heads where swirl angle variation is greater than 5°, to ensure the dynamic temperature is recovered over a wide range of incident flow angles. Little error is normally incurred if rake recovery factors are taken as a constant value of 0.94, assuming well-designed probes.

Figure 10–16 illustrates the main elements of a typical temperature measuring system. The following sections describe the instruments used.

Resistance Bulb Thermometers (RBT)

Here temperature is measured via changes in the resistance of a heated material. Platinum is frequently used, hence the

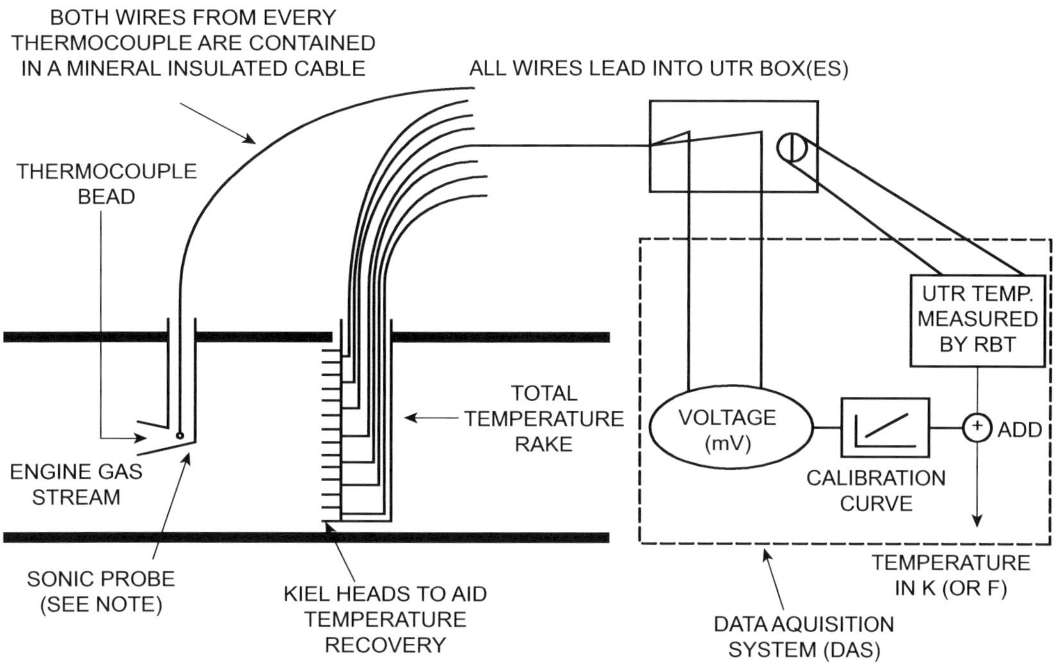

FIGURE 10–16 Temperature measuring system. *(Source: Rolls Royce.)*
Notes: Test bed inlet temperature is measured by multiple RBTs, or on older beds a "snake." It is not possible to measure static temperature, as most of the dynamic temperature is always recovered.
UTR = universal temperature reference
RBT = resistance bulb thermometer
Sonic probes increase heat transfer to the thermocouple by accelerating flow past the dead and to overboard. Aspirated probes use an auxiliary air ejector to accelerate flow past probe. Thermocouples may also be mounted on vane leading edges.

common alternative expression *PRT* (platinum resistance thermometer). In theory resistance thermometers are suitable for temperatures up to around 1000 K, and may give a high accuracy of potentially around 0.1 K if carefully calibrated. They are, however, comparatively delicate, and rarely used actually within an engine; the most common uses are for air inlet temperature and to measure *reference temperature* in thermocouple systems described later in this section.

Snakes

Snakes are resistance thermometers many meters long, sometimes used to measure average inlet temperature, and may for example be strung out over the inlet debris guard or splitter. One disadvantage is that the large physical size makes accurate calibration impossible, hence a preferred alternative is multiple resistance bulbs. Indicative overall accuracy for a snake is 1–2 K.

Thermocouples

If two dissimilar metal wires are connected at a *junction*, and the loose ends maintained at some *reference temperature*, a voltage is generated dependent on the temperature difference between the junction and the reference. Typically the junction is a welded head of up to 1.1 times the wire diameter. Thermocouples are less accurate than RBTs, but more robust. The loose ends' temperature is maintained by either a *UTR* (uniform temperature reference) box, whose own temperature is measured by RBTs, or an *Icell* (ice cell). Single pieces of wire should be used between the hot and cold ends, otherwise measurement uncertainties increase by around 2 K per extra junction. With single pieces *batch wire calibration* is applicable, where a calibration is obtained of a number of thermocouples made from a particular batch of cable. Providing the results agree, this calibration is applicable to other thermocouples made from that same batch.

Thermocouples that employ different wire materials produce different voltage curves. Standard curves are defined for different thermocouple types, and different *classes* of thermocouple lie within different tolerances of these curves. *Class 1* thermocouples will lie within the greater of 1.5 K or 0.4% of the standard curve, and *class 2* within 2.5 K or 0.75%. At higher temperatures drift and hysteresis together contribute a further 3 K inaccuracy, despite heat treatment to improve thermoelectric stability. In addition typically 1 K error will be contributed by both circuitry and conduction effects around the hot junction. Combining all these effects by root sum squaring gives overall system errors at 1000 K of 5.6 K and 8.2 K for class 1 and 2 respectively. *Type K* thermocouples use chromel-alumel (Ni–Cr/Ni–Al) wire. *Type N* thermocouples (Nicrosil/Nisil, Ni–Cr–Si/Ni–Si) were introduced around 1990 to provide longer life and improved stability over type K. Overall system errors at 1000 K are reduced to 4.2 K and 7.6 K.

In sitting thermocouples radiation from adjacent surfaces must be avoided, otherwise the temperature measured is not that of the gas stream. For locations where radiation may be severe, *shielded* thermocouples are employed, which use up to four concentric thin tubes surrounding the thermocouple bead.

The application of the above devices to measuring temperatures at key engine stations is described below.

Air Inlet Temperature

Recommended coverage is at least three RBTs mounted on the intake debris guard, and more if non-uniform inlet temperature profiles are suspected. The test bed layout should be adjusted to ensure that the difference between the readings is less than 1 K, otherwise it is difficult to be sure that the true average temperature is being measured, and the temperature profile may fundamentally affect engine performance. Snakes or even thermocouples may also be used; however, this results in lower accuracy as described earlier.

Cold End (Compressor) Temperatures

Usually thermocouples are employed, and accuracy is as described earlier. Coverage should be at least three points circumferentially, with rakes having one to five heads radially depending on engine size and the expected radial temperature profile. The heads are usually placed on centers of equal area to provide uniform coverage and assist in data averaging. An aeroengine fan is a special case; due to the temperature profiles and relatively low temperature levels rakes with up to 10 heads are employed.

Hot End (Turbine) Temperatures

Temperature measurement is significantly more difficult for turbines, for two main reasons:

1. Above temperatures of around 1300 K, the mechanical integrity of a probe becomes an issue, requiring bulky, cooled designs that are highly intrusive. Such measurements are rarely attempted.
2. At combustor exit, and to a decreasing extent rearwards through a turbine system, there is severe temperature "patternation" causing both circumferential and radial profiles. This is due to having discrete fuel injection points within the combustion system and cooling air influx downstream. To obtain a thermodynamically valid average temperature from a finite number of readings may be impractical.

For both reasons the temperature at combustor exit cannot be measured, and measurements are rarely possible at exit from any first HP turbine stage. For measurement stations further downstream, the minimum coverage required is *at least* eight locations circumferentially, and three to five thermocouple heads radially, depending on engine size. Patternation introduces a further error beyond the thermocouple inaccuracies described above. Rather than using centers of equal area, head placement is often biased towards the walls, where the temperature gradient is steepest.

Transient Temperatures

One further important thermocouple property is response time. Physically large thermocouples take time to respond to temperature changes, due to thermal inertia, and are unsuitable for transient development testing. Response is governed mainly by the time constant of the junction itself. For development testing physically small junctions are employed, mounted to minimize conduction and radiation.

For control system instruments, large robust "production" devices are required. Here the time constant of the thermocouple can be allowed for in the control algorithms, or heat transfer increased by increasing the flow past the thermocouple junction. Where the pressure ratio to ambient exceeds around 1.2, a *sonic probe* is used as shown in Figure 10–16. A small flow is extracted from the gas path through a venturi surrounding the thermocouple bead and then diffuses before being dumped overboard. Otherwise aspiration is employed, where higher pressure air is injected to draw flow past the bead by an ejector effect.

Liquid Fuel Energy Flow

Measuring fuel flow fulfills two essential purposes:

1. It is vital for calculating thermal efficiency and SFC.
2. Calculating temperature levels, hence lives, of the combustor and HP turbine requires fuel flow, as these temperatures cannot be measured.

In all cases the method involves the measurement of *volumetric* fuel flow, conversion to fuel *mass* flow based on the actual fuel density, and finally deriving energy flow using the fuel heating value (FHV). To obtain the density and FHV periodic laboratory analysis of fuel samples is required. This must be done at least once per fuel batch delivery, and even each day during key performance testing. Fuel energy flow is calculated from the volumetric flow, FHV and specific gravity.

Volumetric Flow

For liquid fuels three main instruments may be used:

1. A *bulkmeter* measures volumetric flow over a time period, using pistons connected to a rotating crankshaft. Volumetric flow rate is directly proportional to rotational speed as this determines the rate at which the piston volumes are filled and emptied. Bulkmeters also require calibration for fuel viscosity that has a second-order effect of up to 1%. Hence fuel temperature and fuel type such as diesel or kerosene must be allowed for. Bulkmeters are almost mandatory for steady state testing and are very accurate if engine operation is stable, around 0.25% or less. They are unaffected by inlet flow profile or swirl and hence no upstream flow conditioning is required.
2. A *turbine flow meter* indicates instantaneous volumetric fuel flow rate, via a calibration versus rotational speed. Again fuel viscosity must be allowed for. Initial accuracy is good, around 0.5%, but drifts by up to 1%, which is more than other instrument types, due to wear on vane tips and in the bearings. Turbine flow meters are affected by inlet flow distortion, hence a minimum settling length of 10 pipe diameters must be allowed upstream, with a flow straightener such as a colander plate at the inlet to this settling length.
3. A *glass bottle and stopwatch* may be used for steady state testing of small engines. This involves timing the visible consumption of a known fuel volume, and gives an accuracy of around 1%. The method is not suitable for automatic data recording.

Because fuel flow is so vitally important at least two of these devices should be used in series. Figure 10–17 shows the main elements of a fuel flow measuring system, including typical installation requirements.

Density

To measure density of a fuel sample a *hydrometer* is used, where a graduated scale on a float is read visually. The "units" are usually *relative density* or *specific gravity*,

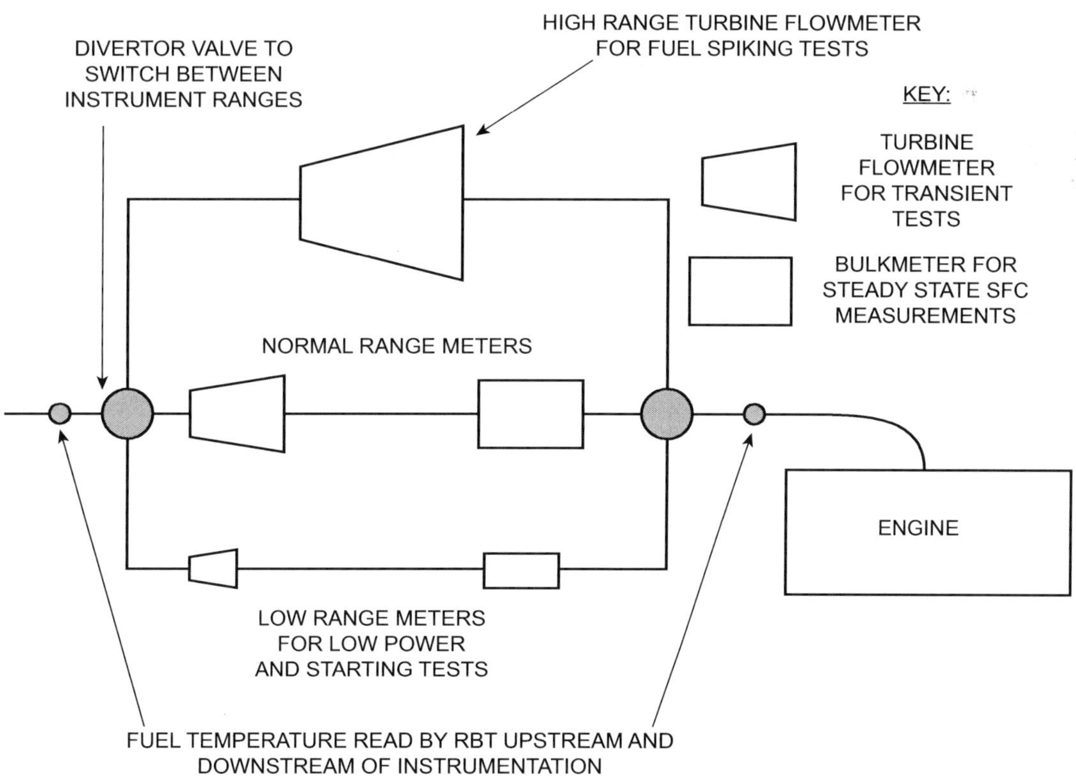

FIGURE 10–17 Fuel flow measuring system. *(Source: Rolls Royce.)*

Notes: Distances between meters and pipe bends should be at least 10 pipe diameters to prevent flow disturbances from corrupting the measurements. Steady state at least two meters of some kind should be used in series to confirm the measurement accuracy. Fuel samples should be taken for analysis at least once per fuel batch and daily during key steady state testing.

RBT = resistance bulb thermometer

Turbine flow meters require upstream flow straightening, as described in the text.

which is simply the ratio of the actual density (mass per unit volume) to a standard value of 1000 kg/m^3. The accuracy of measurement is around 0.1%.

Actual density varies with temperature, hence temperature readings are required for both the laboratory sample test and the fuel supply to the engine. For the latter, RTBs are placed in the fuel pipes, ideally ten pipe diameters away from the flow meter to prevent possible flow disturbances corrupting the flow measurement.

Fuel Heating Value

FHV is also obtained via laboratory analysis of samples. The main methods are:

- Bomb calorimeter, in which a fuel sample is burned and the heat release measured directly from the temperature rise.
- Analysis of composition, via spectroscopy. Here the percentages of carbon and hydrogen are used to calculate the energy that would be released in combustion.

Accuracy of the above methods is around 0.1%. If no laboratory analysis is available, an FHV for kerosene only may be calculated from the density.

Transient Fuel Flow Measurement

For measuring volumetric fuel flow transiently only a turbine flow meter is suitable as its speed gives an instantaneous flow rate. The usable range is around 10–100% of full scale, so different instruments are required for starting and fuel spiking, with ranges of around 20% and 300% of that for normal running. Fuel density and lower heating value are simply the same numbers as for steady state. Both a bulkmeter and flow bottle are unsuitable as they rely on fuel flow being steady.

One backup measurement that may be particularly useful transiently is via the combustion fuel injector *flow number*. Given the upstream fuel pressure measurement, flow is simply the flow number times the square root of the pressure drop. This is useful during fuel spiking where fuel flow changes so quickly that even a turbine flow meter reading may lag due to inertia. The method may be checked against the main instruments based on the previous steady state position.

Gas Fuel Energy Flow

The measurement process is similar to that for liquid fuels, with fuel energy flow evaluated from volumetric flow, density, and lower heating value.

Volumetric Flow

A bulkmeter cannot be used to measure gas fuel flow as it is *compressible*, i.e., its density changes with pressure, and the positive displacement principle upon which a bulkmeter relies is only applicable to incompressible liquid fuels. One of two techniques is employed as outlined below.

As for liquid fuel, a turbine flow meter may be utilized, with similar installation rules and measurement accuracies. It is not necessary to calibrate for viscosity.

An orifice plate may also be utilized, and an accuracy of ±1% can be achieved. The installation criteria include a flow straightener at the measuring section inlet, and typically upstream and downstream straight lengths of pipe of 10 and 5 diameters respectively. However, orifice plates have a number of significant disadvantages relative to a turbine flow meter:

- The accuracy is only maintained for a fuel flow *turndown ratio* of around 4:1, whereas for a turbine flow meter over 10:1 is common.
- The pressure drop of an orifice plate is higher, which must be allowed for in the fuel supply system.
- It is not possible to achieve the required manufacturing tolerances for engines of less than around 5 MW.

Calibration may utilize a series of *sonic nozzles in* parallel. This is an extremely sophisticated system and would normally be used in a laboratory rather than an engine test bed.

Density

The fact that gas fuel is compressible makes evaluation of its density complicated. The most accurate method of determining gas fuel density is from its pressure, temperature, gas constant, and *compressibility*. Static pressure should be the average of values measured around five diameters upstream and downstream of the volumetric flow measurement device; this upstream location is purposely after the flow straightener pressure losses. Temperature should be the average of measurements taken upstream of the flow straightener and approximately five diameters downstream of the volumetric measuring device using RTBs. For the gas constant and compressibility term (z) it is essential to analyze a sample using a gas chromatograph and then utilize formulae. The last measure is necessary as in fact very few gas supplies do not have some variability in the gas composition, and relatively small changes can significantly affect the gas constant and compressibility. Failure to do this can lead to inaccuracies of up to ±5%.

Taking all of the above measures will lead to accuracies for gas fuel mass flow of around ±1%. This is the best accuracy achievable, and is not vastly changed whether combining these effects is via root sum square (for 95% confidence) or arithmetic addition (for 100% confidence). The main inaccuracy is in volumetric flow, at best around ±0.75%, while pressure is only accurate to around ±0.25−±0.5%. Temperature has little effect, being only ±0.1 K if an RBT is used, while the compressibility and gas constant should be accurate if the formulae are used correctly.

Fuel Heating Value

The FHV can be evaluated to within 0.1% from the gas chromatograph output.

Air Mass Flow

Measurement of engine inlet mass flow is vital for various reasons:

- For thrust engines inlet mass flow determines momentum drag, and hence net thrust and SFC.
- For any engine temperature, levels in the combustor and HP turbine can only be determined by calculation from air mass flow, combustor inlet temperature, and fuel energy flow.
- Determining compressor surge margins and component flow capacities requires mass flow to be known.
- For shaft power engines tested without a free power turbine, air mass flow is required in the calculation of exhaust gas power.

Total and static pressures are read in a duct of known area, usually where contraction occurs and increases the dynamic head, along with total temperature. Measurements may also be taken at other stations, and where direct measurement is not possible calculations are performed based on flow continuity using design air system assumptions.

Airmeters

As shown in Figure 10−18, two similar instruments exist for measuring engine inlet flow:

1. A *flare* is a short duct with an entry bellmouth, fitted immediately in front of the engine. This imposes only a low-pressure drop on the engine.
2. A *venturi* is a contraction in a longer upstream duct, followed by a diffusing section. It is placed well upstream of any bow wave effects caused by the

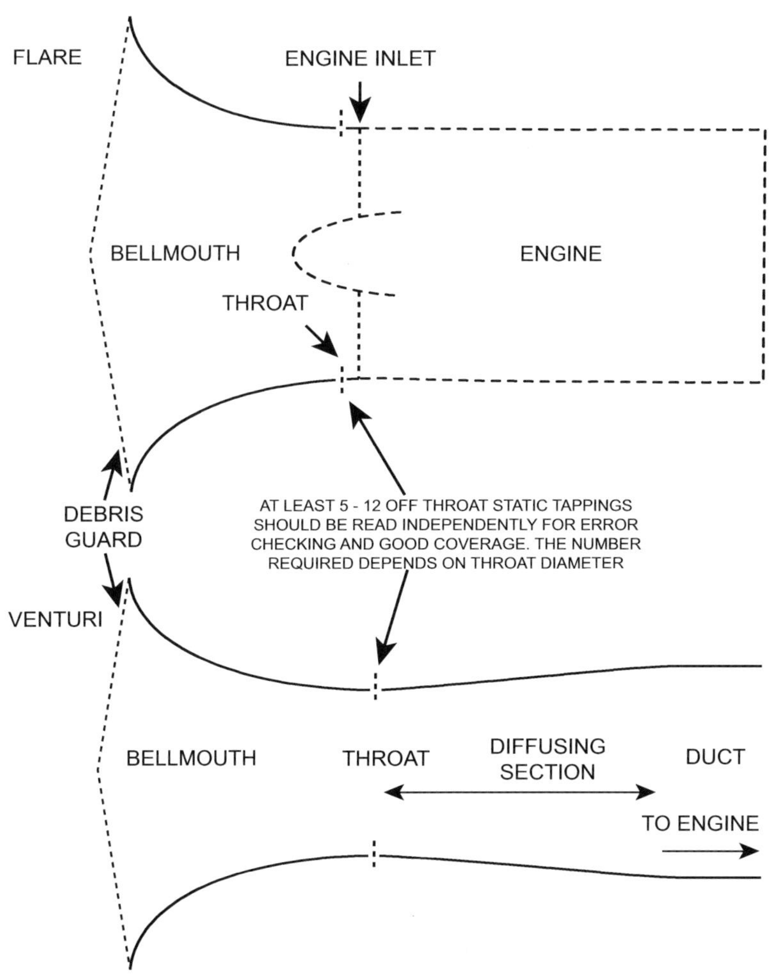

FIGURE 10−18 Engine inlet air mass flow measurement. *(Source: Rolls Royce.)*
Notes: Upstream temperature measurement normally uses multiple RBTs mounted on test cell inlet splitters or debris guard. Upstream pressure measurement is either that for cell static pressure, ambient (barometric), or total probes mounted on the debris guard.
RBT = resistance bulb thermometer
Airmeter throat should be sized such that Mach number does not exceed 0.7.

engine. Geometry dictates that the venturi throat generally gives a larger *depression*—the difference between total and static pressure—than a flare. This depression may then be more accurately measured down to low flow rates.

The term *airmeter* is applied to either device. Both require the use of a discharge coefficient (CD), which is the ratio of effective flow area to geometric area. This is less than unity because first, velocity is less than the ideal value, due to skin friction, i.e., total pressure is lost between the upstream measured value and the throat; and second, the flow does not fill the whole of the throat due to boundary layer growth. If the engine face is very close, such as with a flare, bow wave effects may add to this. The discharge coefficient may he determined by several possible methods:

- Cross calibration with an existing airmeter that has been calibrated on the same engine type and test bed
- Evaluation of throat conditions via a high coverage rake, with many total pressure heads
- Use of a standard airmeter geometry and CD calculation
- Theoretical predictions of flow behavior using *CFD* (computational fluid dynamics) methods

At least 6–12 throat static tappings should be used and read individually to allow error detection; the number depends on size. For a flare, total pressure should be that in the cell. For a venturi there may be some pressure loss upstream of the throat, hence total pressure must also be measured or the loss allowed for by calibration. Total temperature measurement ideally uses RBTs. Overall, a typical value for accuracy is around 0.5% of airflow, providing the tappings remain clean.

Turbofan Core Flow

For a turbofan the mass flow actually entering the core is of greatest interest, and may be determined in various ways.

Air System

Once a value has been established for first compressor inlet mass flow, the air system design data for cooling and leakage flows are initially assumed to obtain values for mass flows at other stations. The methods discussed under "Test Bed Analysis," or direct measurements of individual air system flows, are used to check this assumption.

Other Measurements

For all engine stations, including entry, measurement of total and static pressures, and total temperature may provide an indicative mass flow measurement. This may be calibrated against some true measurement, either directly or using air system data depending on the location.

Transient Air Mass Flow

For transient testing a *flow probe* is often fitted, especially at HP compressor entry. This reads a difference between total and static pressures via a dedicated differential transducer, with short length *balanced* pipework to remove aerodynamic noise. The mass flow is calculated using an effective area and calibration versus steady state data based on inlet flow readings from the airmeter. If the airmeter is a close coupled flare then it may be read transiently, using dedicated transducers with 1 m or 5 m line lengths.

Injected Steam Flow

Steam may be injected into an engine to reduce emissions and increase power. The supplied steam flow must be measured directly, as least for engine development. This is normally accomplished using a *vortex shedding flow meter.* Here a bluff body within the flow vibrates due to shedding of wakes (just as does a telephone wire in the wind). The frequency of vibration is roughly proportional to the volumetric flow. Such a device is calibrated to determine the exact dependency, versus a choked (*critical*) nozzle.

Thrust

For aeroengines, measuring thrust is essential for three main reasons:

1. This is a vitally important, fundamental design goal and will determine whether or not the engine/airframe combination can meet the desired mission.
2. It is required to calculate SFC.
3. Any inferred component filing information requires a thrust level to be meaningful.

Figures 10–10, 10–11 and 10–13 showed the basic layout of sea level and altitude thrust test beds. An engine is mounted in a cradle and restrained axially via *load cells*, which measure the axial force required. These contain springs and are often preloaded and mounted in opposition to maintain stiffness. Standard practice is to calibrate before and after a test run, often by hanging weights, and use the mean as valid during the test. An indicative accuracy level for the load cell is within 0.25% reading; to convert this to thrust requires test bed calibration, which is discussed later. Basically, for indoor test beds the impact of the static pressure field on the engine relative to that in an infinite atmosphere must also be evaluated. This leads to an overall thrust accuracy of around 1%.

In an altitude chamber additional effects contribute to measured thrust as shown in Figure 10–13. The inlet ducting incorporates a slip joint to prevent axial force transmission to the engine; this joint has a seal area larger than the upstream duct. The actual gross thrust produced by the engine is the measured load cell force plus inlet momentum drag and pressure loads acting on the duct and seal areas. Typical accuracy is around 1.5%. For both test beds, measurement of mass flow is also important, to corroborate predictions of momentum drag at flight conditions.

Shaft Speeds

Shaft speed measurements provide valuable component information:

- Anticipated turbomachinery performance is strongly influenced by speed, especially for compressors.
- Turbine lives and shaft critical speeds are crucially dependent on speed, and for aeroengines shaft speed levels are a certification issue.
- For shaft power engines output speed is of vital importance, and is also used in the calculation of output power.

A *phonic wheel* and pick up are normally used. This method is highly accurate and reliable, and hence available even as a production measurement. The passing of teeth cut in a wheel, integral with the shaft, is sensed by an electromagnetic coil. The pulses generated are converted to a shaft speed, knowing the number of pulses per revolution. Two main methods may be used for this:

1. A *clock and pulse counter* suits steady state operation, and as implied counts the number of pulses over a timed period. Accuracy is around 0.1%, assuming engine operation is stable.
2. A *frequency to DC converter* suits transient operation, and gives a readout of instantaneous shaft speed. Again accuracy is around 0.1%.

Engine Output Shaft Torque and Power

These are highly interrelated as power is simply the product of torque and rotational speed.

A *torquemeter* forms part of a load carrying shaft, and measures torque by sensing the relative rotation of two ends of a length of shaft. Shaft torque must be measured at an appropriate point along the output shafting, as shown in Figure 10–12, to exclude any losses that are not part of the engine supply. Figure 10–19 illustrates one method where phonic wheels, which also measure speed, are attached to both the shaft and an outer unloaded tube. By this means the teeth are in the same plane, and changes in the waveform picked up indicate angular displacement, and hence torque. Overall torquemeter accuracy is around 0.5 to 1%, depending on engine stability. An alternative method utilizes strain gauges to measure shaft twist; however, such devices are delicate and have not been widely successful.

A hydraulic dynamometer may also measure torque. Input torque is transmitted to the casing by the water and any bearing friction; the casing is freely mounted externally on bearings to allow torque measurement via a load arm and a weighing system.

Humidity

Humidity must be measured so that test data can be referred to either dry or ISO conditions, for evaluation on a

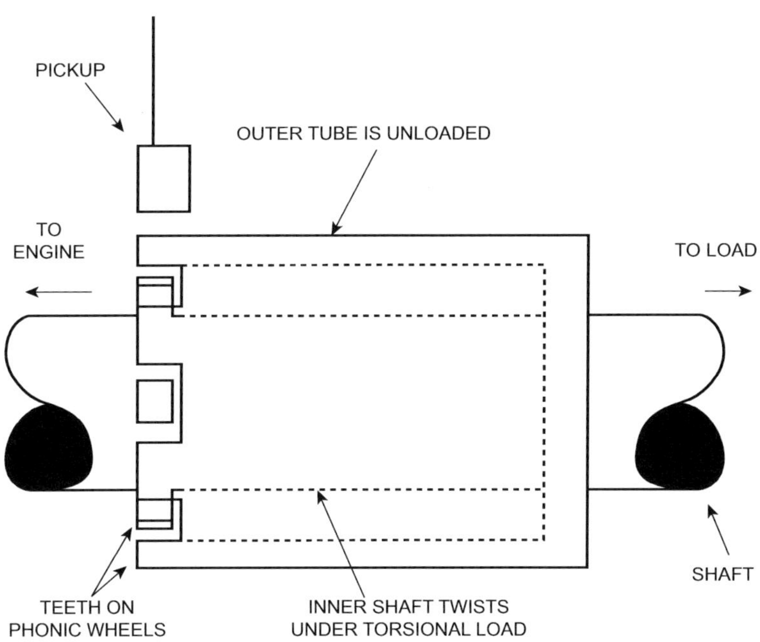

FIGURE 10–19 Shaft torque measurement. *(Source: Rolls Royce.)*
Notes: Twist of inner shaft changes angular relationship of phonic wheels. Changes in pickup waveform allow computation of shaft torque. The type illustrated is known as a "phase displacement meter."

defined basis. There are three main methods of humidity measurement:

1. *Capacitance sensors* utilize changes in the dielectric constant of a water absorbing material.
2. *Chilled mirrors* utilize changes in reflected light levels as different relative humidity results in different amounts of condensation.
3. *Wet and dry bulb thermometers* use an air draught to evaporate water from a wet lint on one of the bulbs, and reduce that bulb's temperature reading. Humidity is found from the temperature difference between the wet and dry bulbs, normally via tables. The traditional method was simple glass thermometers mounted on a "football rattle" frame, though ducted fans and other instruments are now used.

For all these methods an indicative accuracy level is the greater of 5% relative humidity or 0.2% specific humidity.

Geometric Parameters

Variable stator vanes are common on axial compressors, and recuperated engines often have variable turbine nozzles. To measure vane rotation directly, around three off *RVDTs* should be used. An RVDT is a "rotary variable displacement transformer," in which the relative movement of electric coils changes the mutual inductance. Accuracy is good, around 0.5°, assuming all vanes are held at the same angle. If alternatively the *linear* displacement of the actuator moving the ring is measured, via an LVDT, then other effects are relevant, such as ring tolerances and any slack in the total mechanism. The accuracy of measured vane position is then entirely dependent on the system design.

Turbine throat areas are subject to manufacturing variation and are normally measured geometrically on build, with a repeatability of around 0.1%; method (and hence definition) adjustments may give apparent "step" changes, however. The conversion factor between area and capacity is derived from the turbine aerodynamic design data.

For thrust engines, propelling nozzle areas are required. Defining and measuring certain areas may not be trivial, for example, on a turbofan the bypass nozzle plane may contain struts, or if a mixer is employed *its* area is required.

Test Bed Calibration

Engine performance measured on indoor test beds will differ from that in an "infinite atmosphere." *Test bed calibration* enables data from an indoor test bed to be adjusted to reflect infinite atmosphere conditions. In addition, engines of a given mark may be tested on more than one test bed, each with a different configuration and hence different *calibration factors.*

For shaft power engines where air is ducted from ambient as per typical installations, and the measurement made at duct entry, the test bed configuration does not normally need to be calibrated for. All that remains is the usual requirement for all instrumentation to be functioning satisfactorily.

Test Bed Definitions

Test bed *approval* is a formal process, involving regulation by the appropriate airworthiness authority for aircraft engines, and high customer involvement for other engine types. The following basic definitions apply, though other similar terms are also used:

- "Gold" standard test bed. This is the datum for *cross calibration* of all other beds for that engine type. Such calibration may be performed directly or via calibration of an intermediate "silver" standard test bed.
- "Silver" test bed. This has been directly cross calibrated against the gold standard test bed. It may be used for cross calibration of other beds for that engine type.
- "Bronze" test bed. This has been calibrated against a silver test bed. It may be used for performance testing, but not for calibrating other test beds.
- Functional test bed. This test bed has not been calibrated for performance purposes, and provides results that are indicative only. Such a test bed may be useful for endurance testing or demonstrating the functionality of a production engine, but not its performance.

A cross calibration exercise normally involves comparing results obtained running the same engine on two test beds, without any transportation of test bed slave equipment. The benefits of cross calibration are dubious for engine configurations where transportation between test beds would require significant rebuild.

Calibration of Gold Standard Test Bed

For thrust engines, the gold standard bed for the first engine of a type is declared by calibration versus an outdoor test bed, i.e., infinite atmosphere. For shaft power engines direct calibration of the mass flow measurement is required.

Steps in Test Bed Cross Calibration

The object of cross calibrating a test bed is to reproduce the performance measurements obtained on the gold standard test bed, after referral to standard conditions, such as ISA sea level static.

The steps involved are:

- All test bed instrumentation is calibrated.
- An *uncertainty analysis* is performed to determine the likely errors in all measurements and derived parameters accumulating from all sources. Clearly much is specific to a particular test bed.

- Calibration curves (i.e., sequences of throttle settings) are run on both test beds, consisting of around a dozen stabilized points from idle to maximum rating.
- Ideally a "1−2−1" sequence is used with the gold or silver bed as "2;" here any faults may be detected based on history. For example, the curve of inlet depression versus mass flow should be well known, and any deviation during the calibration run should be detectable. For the bed to be calibrated there may be no such history, hence having two runs proves repeatability. Other test sequences may be employed where high confidence levels make the cross calibration more of a confirmatory exercise.
- The physical configuration of the bed being calibrated is adjusted during the last test if necessary, with respect to how it affects airflow and hence thrust measurement. This includes inlet and exhaust details, engine position, and auxiliary equipment in the cell.
- Finally calibration factors are derived and are tabulated, versus speed say, for use in all future testing. These are the differences in thrust, airflow, and station temperatures and pressures.

Test Bed Audit

Test bed calibration exercises are expensive due to the engine running involved. Once a test bed has been calibrated the process is rarely repeated unless bed configuration is changed, or there is an inexplicable change in production engine performance. However, to maintain test bed approved status it should be audited, typically biannually. Important features to review in an audit are as follows:

- Trend plots of engine and overall parameters
- Instrumentation calibration status
- Original test bed calibration against a gold or silver test bed and back to infinite atmosphere
- Test bed configuration status against that formally recorded during the original calibration. Such items as physical layout, and instrumentation types and serial numbers should he examined
- Quality controls for data reduction software
- Actual working practices versus those in the engine test schedules
- Any significant events or changes noted since the previous audit
- Completion status of any "actions" from previous audits, or recorded since then

Steady State Development Testing

Specific Performance Tests

By the end of development the achieved performance standard must be demonstrated. Contractual implications mean both the customer and the engine manufacturer will be keenly interested. During development the standard is achieved via specific performance tests, which have the following aims:

- To validate the steady state performance model, or else establish which components do not achieve the predicted or rig-based bid level and hence require development effort.
- To determine the effect of component modifications implemented during a development program. Ideally these are tested singly, in a "back to back" sequence; however' this will not be the case if time is short or modifications are *known* to be of benefit.
- To derive control schedules for ratings and to optimize part load performance. For example, having both blow-off valves and variable vanes may allow many possible operating settings at part power for a given surge margin level.

A typical test is a performance curve consisting of around 5 to 10 stabilized points. These cover the range from the minimum power or thrust level important for the application (usually idle) to maximum rating. Stabilization times before taking readings range from around 4−20 minutes, the latter if large heat exchangers are utilized. Shorter times do not allow scanivalve pressure readings to stabilize.

For aircraft engines similar performance tests are conducted in an altitude test facility or flying test bed, to simulate the conditions encountered at cruise accurately. Owing to the time spent, altitude cruise is the most important condition for SFC.

Windmilling

For aircraft engines drag, auxiliary power capability and combustor inlet conditions are measured over a range of Mach numbers and altitudes in either an altitude test facility or flying test bed. Windmill testing is not normally conducted for other engine types.

Support of Engine Component Development

During engine development various tests are necessary where evaluating performance is not the goal but defining its level is vital to meet the primary test objective. For interest, examples of such tests are described next.

Ingestion Tests for Aircraft Engines

Ingestion events cannot be avoided for aircraft thrust engines since adequate filtration is impractical. Testing is a regulatory requirement to ensure satisfactory engine behavior, as outlined below.

- For *bird ingestion* tests, a number of dead birds are fired simultaneously into the engine when operating at maximum rating. The airworthiness requirements are either to maintain 75% thrust for a certain time after the impact, or shut down safely, depending on the size of bird.

- For *water ingestion*, the test requires static ingestion of up to 4% water, to simulate rain. However, to address the reduced fan centrifuge effect at flight speeds additional tests are normally carried out, which spray the water either at high pressure or even downstream of the fan.
- For *hail ingestion*, tests involve firing 25 mm and 50 mm hailstones at the engine in a specified pattern.
- At low airflows ice may build up in front of and even on the fan, and be shed and ingested; for *ice ingestion* the test is to ingest the ice that would build up if the anti-icing system were not switched on for 2 minutes after entering icing conditions. Ice has similar thermodynamic effects to water ingestion, but raises additional concerns about mechanical integrity.

Thermal Paint

This records the metal temperatures of key "hot end" components such as combustors and turbine blades. Special paint is applied, which changes to various colors depending on the temperature reached; several paint grades are available to suit different temperature ranges. The engine is first run at low power to heat the carcass to ensure representative seal clearances, hence cooling flows, and then held for between 3 and 10 minutes at the desired high power condition. The paint normally takes 3 minutes to cure; longer times are utilized when engine stabilization times are long, for example with a recuperated engine, though this must balance the risk of paint degradation. The paint is examined on engine strip after the test, or occasionally via borescope inspection. The indicated component surface temperatures are then related to the temperatures of the gas path and cooling air, assessed from the performance measurements and analysis.

Component Temperatures

In addition to thermal paint tests, key component areas such as disc faces may be fitted with thermocouples to measure metal temperatures. This allows the assessment of many operating conditions during a build, and provides safety monitoring during high power running or, in case of cooling flow inadequacy, in early development.

Air System

Pressure and temperature readings may be taken in key air system chambers, to validate design data for cooling flows, cooling temperatures, and pressure margins.

Straingauging

To assess the susceptibility of gas path components to resonant vibration, *straingauges* are cemented in key locations. These consist of fine wires embedded in a mounting plate; vibration causes the wires to stretch, which changes their electrical resistance. The test normally consists of slow acceleration and deceleration (around 4 minutes each) with various strain gauge selections being recorded. The frequency and amplitude of the measured signal characterizes blade response, identifying potentially harmful resonances. Testing explores the full range of mechanical and referred rotational speeds that the engine type will encounter. Mechanical speed determines the frequency of excitation forces, referred speed produces flow regimes that may cause excitation, and pressure level determines amplitude.

Achieving the highest speeds is often a challenge with limited test hardware, though in-service ambient variability [illegible]essed via an altitude chamber. Very high cycle temperatures may be reached, often requiring builds with special sizes of turbine and propelling nozzles to provide alleviation.

Bearing Loads

For this test special strain gauged equipment is built in to measure the axial loads on shaft bearings. The engine is run over a representative operating range to allow assessment of bearing lives; stabilization periods may be long (20 minutes and more) as air system pressure levels depend on seal clearances and hence disc growths.

Emissions

Exhaust gases are sampled, usually via a cruciform rake in the exhaust, and analyzed using a *spectrometer* for levels of pollutants such as unburned hydrocarbons, carbon monoxide, and oxides of nitrogen. Such tests are increasingly important for contractual purposes, as legislation becomes more stringent.

Endurance and Type Testing

All engines complete some form of arduous endurance test before service entry. This may take the following forms:

- Aviation authorities such as the FAA stipulate a *type test* comprising 150 hours' cyclic operation, including SOT levels that are the highest that then may be permitted for any engine in service. The engine manufacturer decides this temperature level, based on hottest day/highest altitude takeoff, with allowances for engine variation, deterioration, etc. The last item in particular is flexible, but if the type test temperature gives low margins relative to new or overhauled engines, the costs of frequent maintenance will be incurred.
- For marine engines one such test has been the US Navy's *qualification test*. This typically consists of 3000 hours' cyclic operation, 40% of which has the HP turbine temperature at the level required at 37.8°C (100°F) ambient temperature to achieve the power to be cleared.

This level must be continually adjusted to address any deterioration, and is normally achieved during the test via "customer" bleed extraction.

- For industrial engines, testing is normally specified by the engine manufacturer, and typically comprises 100–300 hours of cyclic testing with maximum SOT 10–20 K above that at base load.

Transient Development Testing

Transient performance testing aims to confirm that the engine can perform accels and decels (known as *handling*) within specified times. Potential operability problems include surge and weak extinction. Before a new engine variant is tested, transient performance work will normally be based on a transient model of the engine. This is validated using transient test data and can then be used to support testing and prove transient capability at extreme conditions that are impractical to test.

Handling

The maneuvers tested exceed the most severe to be met in normal service, in order to give high confidence that engines will behave satisfactorily. Examples include:

- Fast accel, which due to overfueling gives a high HP working line.
- Slow accels and decels, defined to just avoid tripping the transient bias on handling bleeds. This gives high working lines.
- Bodie or reslam, where the engine is decelerated and then almost immediately reaccelerated. Heat soakage raises the working line beyond that of a normal accel, and also within the HP compressor drops the surge line.
- Accel following startup from cold soak conditions. This gives the highest compressor tip clearances and hence low surge lines.

During these maneuvers key parameters are measured that allow definition of speed versus time, transient working line excursions, weak extinction margins, etc. A suitable scan rate is 10–50 scans per second. For aircraft engines most work is done on sea level test beds; however, some altitude chamber or flying test bed testing is essential to verify effects at the precise conditions to be encountered.

Surge Line Measurement

Surge lines are effected by phenomena such as engine structural loads creating asymmetric tip clearances, inlet distortion, or heat soakage within the HP compressor which lowers the surge line after a decel. It may be necessary to measure actual surge lines in an engine, or at least define a surge-free region, rather than relying on rig or predicted data. Key parameters are measured so that the actual trajectory to surge can be defined. Methods of initiating surge are as follows.

For an HP compressor the most usual method is *fuel spiking*. Here the working line is raised by momentarily injecting excess fuel, between 100% and 400% extra over around 200 ms. A suitable scan rate is 100–500 scans per second. Either a slave rig is used, or the standard control system reprogrammed.

For LP compressors or fans the most usual method is to build a development engine with a reduced capacity LP turbine or bypass nozzle respectively, to raise the working line. Bleed extraction is used to enable stable operation at high power; the engine is then slowly decelerated until it surges. This process is repeated with various levels of bleed to map out the surge line.

Other less common techniques include *in bleeding air* downstream of the subject compressor, for example on small engines, or if applicable utilizing heat exchanger bypass capabilities. The first surge on a build may increase compressor tip clearance, which raises operating temperatures and lowers the surge line. These effects should be quantified by further testing, and compressors examined by borescope to check for damage. The deterioration from subsequent surges is much less.

Controller and Engine Operability Tests

Transient performance and control are inseparable. The initial control strategy and algorithms are further developed during extensive sea level and altitude testing. One particular test of interest is surge recovery, where the controller must ensure that should an engine surge for any reason in service it will recover safely. This is demonstrated during development testing via a surge induced by one of the methods described.

Starting Tests

Specific tests addressing the light up region are:

- For an air starter, per aeroengine practice, measuring crank speed versus starter supply pressure, including that available in service such as from an aircraft APU.
- Starter performance verification. With zero fuel flow, the starter is cut from steady state cranking conditions; knowing the HP shaft inertia the initial deceleration rate indicates the cranking torque.
- Exploration of a matrix of crank speeds and light up flows, identifying the light off "window" and rotating stall boundaries.
- Sensitivity to variable stator vane positions and bleed flows.

Between light up and idle, tests are required to determine the upper and lower limits of acceptable fueling. Too much fuel results in compressor stall, which is affected by the bleed and whirl levels. Too little fuel results in hang, where at low speed some limitation is reached, such as over temperature, surge or rundown. This may be explored by reducing fuel flow at various starter power levels.

For industrial, marine and automotive engines start testing is usually conducted at the prevailing ambient conditions for the sea level test bed. Problems that may arise in service on cold or hot days are then dealt with as they arise. For aircraft engines cold and hot day start tests must be included in the development program. Methods include taking the engine to a cold or hot climatic region, chilling the engine in a cold room, or using large electrical heaters at engine inlet. Furthermore altitude restart capability compliant with the windmill and starter assist envelopes must be proven by testing in an altitude test facility or flying test bed.

A further test for all engine types is a hot restart, performed shortly after shutdown with the carcass still at high temperature. As well as the effect on seal clearances, heat in the gas path lowers surge lines and raises working lines.

Engine Failure Investigation

Engine mechanical failures during development invariably have a transient dimension. In addition to any transient performance data logging, key mechanical parameters such as vibration, speeds, and pressures must be recorded either analog or at least at 1000 scans per second to cater for such an event. To avoid storing immense amounts of data it may be overwritten, say every 30 minutes.

In the event of a failure this log may be interrogated to distinguish cause and effects. One example is to determine whether a surge occurred before (and hence caused) or after (and hence resulted from) mechanical damage to a compressor. If the surge came first, the amplitude of noise or vibration with a frequency corresponding to the spool speed will increase *after* the step decrease in compressor delivery pressure and step increase in broad band vibration caused by the surge.

Application Testing

Further development testing is often conducted with a new engine installed in its application, prior to production release. This enables the following key issues to be addressed.

- Overall performance levels are measured in the application, as opposed to a test bed where calibration factors are required. This is particularly important for aircraft thrust engines, where test bed effects are most significant. Such testing usually has high commercial importance, as it is the final determination of whether or not the engine/airframe combination meets its performance guarantees.
- Use of "real" hardware for accessories and installation ducting, rather than any slave test bed items.
- Engine mechanical integrity faced with representative structural loads, vibration levels, ambient conditions, and operating profiles.

Production Pass Off

Production Pass Off Test

It is common practice that each engine about to be delivered to a customer, whether new production, overhauled or repair, undergoes a *pass off* test. The inevitable variation in manufacturing and build dimensions, even within permitted tolerances, results in variation of component and overall engine performance. A typical test sequence is as follows:

- Starting: While an unavoidable test this is nonetheless vital.
- Running in: Here rotor seals are made to cut gradually into the static linings, by say a series of increasingly rapid accelerations, extended dry cranking, or dwells at increasing power or thrust levels.
- Performance curve: Here assessments are made of SFC, thrust or power versus temperature, working lines, shaft speeds, etc.
- Handling tests: Accels and decels are carried out to prove the engine meets its specification requirements.
- Control system setup: Any required control systems stops, trims and limiters are set. Trims are adjustments to measurement signals to address engine-to-engine variations.

For practicality an exception is made for heavyweight gas turbines, however.

Acceptance Criteria

There are various limiting levels that engine performance parameters must meet for pass off. These usually include demonstration of guaranteed power or thrust without exceeding set temperature and speed limits, and achieving guaranteed SFC at this power or thrust. The acceptance criteria are usually stated at some standard conditions, to which measured data is referred. Engines may initially fail their pass off test due to faults with instrumentation or less usually with the build.

In addition to performance parameters certain mechanical integrity issues are also addressed, such as fuel leaks, vibration, oil consumption, measured temperature spread, etc.

Engine Performance Trends

Changes in component manufacture or build practice may cause increasing numbers of engines to fail the pass off test. Identifying these changes early among the general scatter is challenging. Though some special instrumentation may be fitted for the pass off test, more modern practice is to rely on that of the production control system. Instrumentation is therefore often too sparse or inaccurate to suit the conventional analysis methods. Though these may be employed additional methods are necessary, which emphasize comparison with other engines. Figure 10—20 shows an example where pass off data from a production run are plotted for LP turbine inlet temperature:

- Mass plots: These show pass off performance parameters of interest plotted versus engine number over a time period. They truly define mean engine performance and scatter, and are useful for detecting measurement errors and "rogue" engines.
- CUSUM plots: These plot the cumulative sum of the difference between each engine's temperature (say) and an arbitrary value within its scatter band, versus engine number. Changes of slope of the mean line indicate changes of trend that may be almost impossible to detect from the mass plot.

Engine no. (–)	SOT (K)	SOT-1200 (K)	CUSUM (K)	Engine no. (–)	SOT (K)	SOT-1200 (K)	CUSUM (K)
1	1206	6	6	17	1203	3	21
2	1201	1	7	18	1199	–1	20
3	1197	–3	4	19	1205	5	25
4	1214	14	18	20	1212	12	37
5	1204	4	22	21	1195	–5	32
6	1208	8	30	22	1211	11	43
7	1190	–10	20	23	1202	2	45
8	1196	–4	16	24	1207	7	52
9	1905	5	21	25	1214	14	66
10	1192	–8	13	26	1190	–10	56
11	1207	7	20	27	1201	1	57
12	1202	2	22	28	1210	10	67
13	1201	1	23	29	1207	7	74
14	1196	–4	19	30	1213	13	87
15	1193	–7	12	31	1196	–4	83
16	1206	6	18	32	1211	11	94

(a) Tabular test data and CUSUM

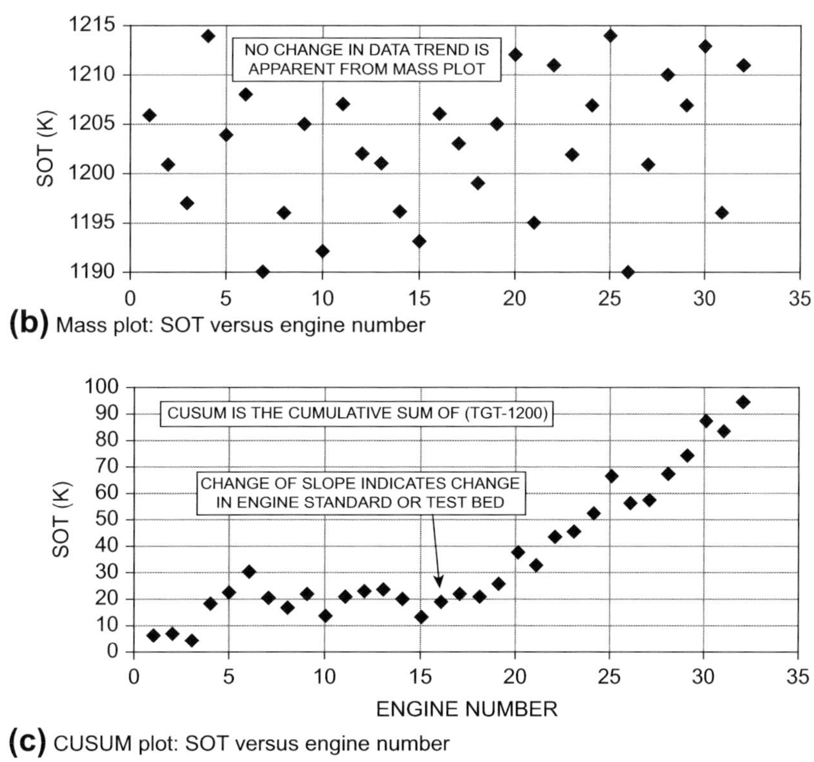

(c) CUSUM plot: SOT versus engine number

FIGURE 10—20 Test data trending: "mass" and CUSUM plots. *(Source: Rolls Royce.)*

Test Data Analysis

Steps in Test Data Analysis

Test data requires considerable processing before engineering conclusions may be drawn:

1. The instrument signals must be converted from say millivolts (mV) to engineering units of K, kPa, etc.
2. Erroneous readings must be detected.
3. Where there are a number of readings at a station they must be averaged.
4. The implied performance of both the engine and individual components must be computed. An error band analysis is essential for key parameters.
5. The evaluated component and engine performance levels must be compared with expectation.
6. The reasons for any differences must be ascertained.

These steps do not always occur in this ideal order. Measurement errors are not necessarily easily detectable, for example they may be systematic and apply to all readings at one station. The analysis process may take a long time to "finish" as new facts alter the understanding. Furthermore, certain methods involving cycle matching combine steps 4 and 5.

Initial Error Detection and Data Averaging

It is essential before using measured data to carry out error detection, for which some techniques are listed below.

- Compare rake head readings at the same radial immersion. Low-pressure values usually indicate leaks, and any outlying temperature readings are probably in error.
- Plot readings on each rake versus immersion. This gives a finer check on validity as the "rake profiles" should show a repeatable pattern, relative both to each other and to historical data, especially for compressor rakes.
- Check that any total to static pressure ratios are sensible.
- Compare cell and barometric pressures, especially for altitude chamber tests.
- Plot airmeter static tappings versus circumferential position and compare with historical data.
- Compare fuel flow calculated from the different volumetric devices arranged in series.
- Instrumentation readings should be consistent with other readings near that station, such as from the controller or air system instrumentation.

Rake data require averaging to form a single thermodynamic mean value at each instrumented station, as represented by the performance model. Such an average is usually either *area weighted* or *mass weighted.* Mass weighting is the ideal, but as it requires an accurate knowledge of the static pressure profile, which is difficult when swirl is present, area weighting is more usual. Either way, for the average to be meaningful any faulty heads must be detected and eliminated.

Test Bed Analysis (TBA) Calculations

These compute engine and component performance levels, e.g., thrust, SFC, isentropic efficiencies, directly from the measurements. Calculations proceed sequentially through the engine, evaluating almost all parameters at each station. This requires assumptions for bleeds, pressure losses, and power offtakes. Turbine exit conditions are assessed based on work, measured in the compressors and any power offtakes.

TBA calculations may be performed either at the tested engine condition or at standard conditions such as ISA sea level or ISO. In the latter case the measurements are referred using theta (θ), delta (δ), and the appropriate non-dimensional groups. To account for variation in gas properties due to changing temperature and fuel air ratio, theta is raised to exponents other than the standard 1.0 or 0.5, known as *theta exponents.* The actual values are obtained by running the engine synthesis model. For illustration, Table 10–2 presents a typical range.

At this stage, instrumentation error detection is again essential, using the derived parameters. Errors may be detected based on considerations such as the following:

- For a fixed geometry engine with fixed bleed configuration the plot of any parameter against another should be smooth. If not, plotting both against any third parameter should indicate which reading is in error.

TABLE 10–2 Theta Exponents for Performance Data Referral

Parameter	Theoretical Exponent	Typical Exponent
Temperatures	1	0.97–1.03
Pressures	0	0.0–0.06
Fuel flow	0.5	0.64–0.76
Shaft speeds	0.5	0.48–0.50
Air mass flow	0.5	0.47–0.52

Notes:
Theta is defined as T1/288.15 K (where T1 is engine inlet temperature).
To refer measured data to standard conditions raise theta to the powers shown.
Differences between theoretical and typical values reflect changes in gas properties due to fuel air ratio and temperature changes.
Theta exponents for a particular engine type are obtained by running a thermodynamic model.
Uses of theta exponents: To convert engine performance parameters to what would be obtained at other inlet temperatures. For example, test data may be referred to standard conditions for comparison with expectation.
(Source: *Rolls Royce.*)

- Calculated component efficiencies should not exhibit high scatter, and should of course lie below 100%. Genuine hardware defects may give values below rig levels.
- Calculated component flow capacities should match any available measured values, such as based on measured throat areas.
- For a thrust engine, propelling nozzle coefficients should be uniform versus thrust level and again should resemble expectation levels.
- If an airmeter is fitted the non-dimensional flow should fall on a unique plot versus depression, be it that of the cell, engine inlet, or throat.
- Apparent "reciprocal" component changes are usually due to a single measurement error, for example the pressure or temperature between two compressors.

One special case regarding TBA calculations is a turbofan, where the core mass flow is not measured directly but must be calculated. Several methods exist, and for convenience the two most commonly used are termed methods "A" and "B" herein. Method "A" uses a core heat balance (see Figure 10–21) and method "B" employs a knowledge of HP turbine capacity. An alternative approach is a transient mass flow probe, which gives indicative rather than absolutely accurate flow measurements.

Uncertainty Analysis

For key parameters produced by the TBA calculations it is essential to evaluate the likely error band resulting from the measurement uncertainty. The first step is to evaluate the potential errors in a calculated parameter resulting from that in each contributory measurement individually; the second step is to combine the effects of these errors. The most common practice is a root-sum-square addition, for 95% confidence. For 99% confidence, simple arithmetic addition is used. The process is also called *error band analysis.*

Comparison of Test Results with Expectation

It is usual to produce a *pre-test* matching model of an engine going to test. For development tests the effects of measured throat areas and any other specific hardware are normally included, whereas for a production engine the expectation—and indeed the requirement—will remain constant. It may be simply expressed via charts relating the measured parameters to thrust or power at standard conditions.

Valid comparison of the test results with expectation requires that both are based on identical conditions of inlet pressure and temperature, installation losses, etc. In

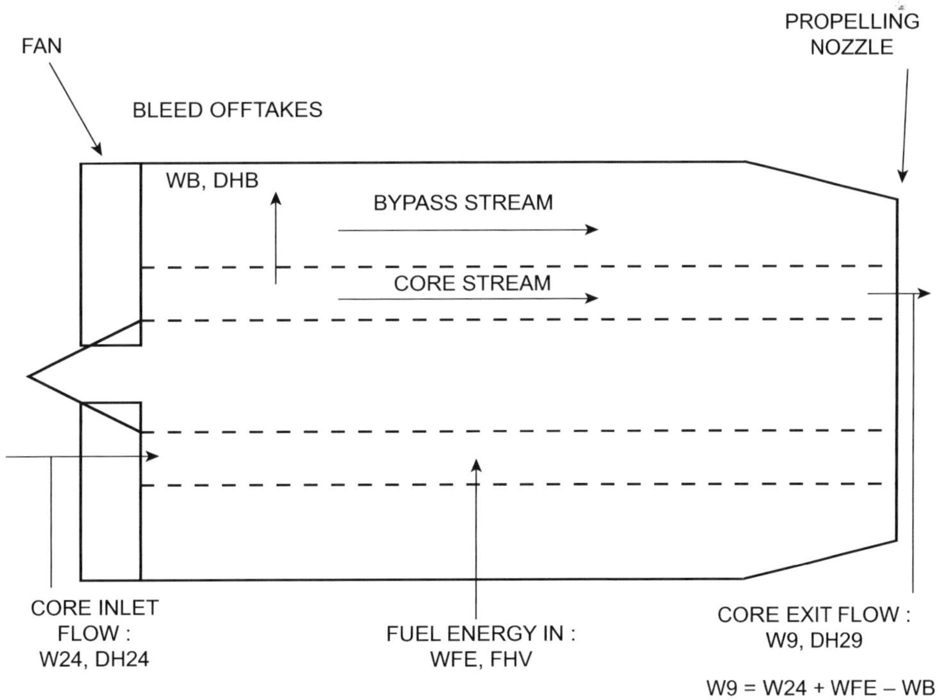

FIGURE 10–21 Turbofan mass flow calculation: method "A" core heat balance. *(Source: Rolls Royce.)*
Notes: Calculate specific enthalpy from temperature and fuel air ratio. Units are kJ/kg and are typically converted to be relative to a datum of 0°C if a mean specific heat is utilized. The following parameters are measured directly: Bleed offtake flow, WB (kg/s). Fuel flow, WFE (kg/s). Core inlet and exit temperatures, T24 and T9 (K). Heat (enthalpy) balance: $W24 \times DH24 + WFE \times FHV = WB \times DHB + (W24 + WFE - WB) \times DH9$
Hence $W24 = [WFE \times (DHB\ DH9)]/(DH24\ DH9)$

addition, for contractual purposes engine performance is normally required at some standard conditions, as stated. To achieve these aims there are two main approaches:

1. Referral of measurements to standard conditions. This provides engine performance data directly usable for contractual purposes.
2. Running a synthesis model at the tested power or thrust and ambient conditions. This is more rigorous, and is mandatory for engines that do not behave non-dimensionally. In this case the model is adjusted to match the test results (see below) and run again to predict performance levels at standard operating conditions.

For both approaches the model and test parameters are plotted versus some power setting parameter. For referred data it must be in a non-dimensional form.

Evaluation of Engine Component Performance

Various methods exist for attempting to explain anomalies, which will be due to either unidentified measurement error or engine component performance differences from prediction. Test data analysis is notoriously difficult, as once dealing with "real" data the possibilities for error are vast. The various methods are described below, and should preferably be used in combination.

Traditional Methods

Engineering judgment and logical thought have long been used for test data analysis. Knowledge of what *could* be suspect is invaluable, and may be based on known potential causes of component deficiency, the state of engine hardware on strip, other knowledge of the test such as what checks were performed for leakage, etc. Comparison of predicted and TBA component efficiencies and capacities gives one view of component performance levels, though being vulnerable to measurement error and rematch effects it is not definitive. This traditional approach together with some trial and error may be used to determine changes to the component performance in the pre-test prediction model which allow it to reproduce all test measurements deemed valid across the power range.

Achieving alignment between a revised model and the measurements is assisted by tables of exchange rates.

Synthesis exchange rates show the change in engine performance parameters if a given component efficiency is in error by 1%. *Analysis exchange rates* show the error in a parameter from the TBA calculations due to a 1% error in each contributory measurement. The advantages of traditional methods are that the result of such an analysis carries high credibility, as the process is easily understood, and no additional setting up is required. The disadvantage is that unlike the more automated methods described below no immediate answer is available.

Cycle Matching

Here an extended performance matching model is run at the tested conditions. It is used both to compare test data with expectation and to explain the results by changes in component performance. For test analysis the matching is extended with measurement values as additional matching constraints, and component scale factors as additional matching guesses. The model thereby changes component performance to replicate the input measurement values.

The main advantage is that analysis is performed automatically, and at the tested condition; this is more accurate than referring test data. In addition engine station data may be produced with however few measurements being taken, parameters at other engine stations being calculated based on the predicted component performance. There are some disadvantages, however. The method can only ascribe differences to component factors that are active in the matching, and the number of these is limited to the number of measurements; if some other component is at fault the answers are wrong. The matching iteration may not always work, or be ill-conditioned such that measurement errors even down to the repeatability level produce large, mutually cancelling changes in several components. Backup calculations are therefore required in online applications.

Matching methods have been successfully used by manufacturers of large turbofans and by some, though not all, airliner manufacturers. Use for industrial engines is not yet widespread.

Methods Based on Probability

Here a computer program "adjusts" both the measurement values and component performance levels so that calculated and measured station parameters match, using tables of exchange rates. By definition there is an infinite number of solutions. That given is the most likely solution, based on input values of the probability of error in components and measurements. Faced with no formal solution this gives a likely answer, and may produce useful suggestions. The solution is not unique, however, and carries little or no weight without engineering substantiation.

Transient Data

Here considerable processing is required to align instrument readings with the equivalent steady state parameters, given the reduced instrument coverage and transducer drift. Once this has been achieved, transient traces are compared with transient performance model predictions, and the model is then adjusted to achieve alignment. The model must already have a good alignment with engine *steady state* parameters, and adjustments are now made mainly in the modeling of heat soakage for transient operation, where real thermal masses and heat transfer coefficients are not normally known to great accuracy anyway.

Analysis/Calculations on the Effects of Water as a Vapor, Liquid, or Solid*

This section describes the effects of water in all forms—liquid, vapor, and ice—on gas turbine engine performance. When only water vapor is present, as opposed to air as well, the gas is referred to as *steam*. There are several reasons why water may be present in the air used by a gas turbine engine:

- Ambient humidity
- Water or ice ingestion
- Water or steam injection

The above are additional to the water vapor produced by combustion. There the amount of water and carbon dioxide produced are based on fuel properties and fuel air ratio, accounted in performance calculations (see Chapter 15). The presence of any additional water beyond that produced by combustion impacts engine performance via several different effects:

- Changes in gas properties due to the presence of water vapor.
- For liquid water or ice ingestion, solid or liquid water absorbs power and affects the compressor aerodynamics, which may also lower the surge line.
- Changes in temperature due to latent heat absorption or release in ice melting or local evaporation or condensation of water.
- Increases in mass flow—for example water or steam injection into the combustor provides additional mass flow through the turbines relative to the compressors.

Performance modeling of phenomena associated with *all* forms of water ingestion or injection, including engine test data correction for humidity, are covered herein and relevant tables and charts are provided. For most gas turbine cycles the water vapor concentration is sufficiently low that the water/air mixture remains essentially a perfect gas. For the rare occasions where the water concentration is higher than 10% this is not so, and steam tables must be used.

Gas Properties

Three fundamental gas properties need to be considered:

1. Specific heat at constant pressure, CP
2. Gas constant, R
3. Ratio of specific heats, gamma

* Source: Courtesy of Rolls Royce, *Gas Turbine Performance*, Walsh and Fletcher, Blackwell Science, 1998. Adapted with permission.

Changes in Gas Properties

The presence of water vapor changes the values of these gas properties, which can have a significant effect on the thermodynamic processes throughout the engine. Both CP and R increase significantly, whilst gamma diminishes more slowly; Figure 10–22 shows the variation of these parameters versus water vapor concentration. The molecular weight of water is lower than that of dry air. Formulae 10.1–10.3 at the end of this section give the variation of CP, R, and gamma with water content.

Effects on Component Performance

Engine components behave non-dimensionally and performance is mapped via various forms of parameter groups. Where the full dimensionless groups are used these maps are *unique even with varying amounts of water vapor present*. Quasi-dimensionless groups are also often used, where the gas properties are omitted and components mapped based on the gas properties of dry air. This form of the groups is useful as it relates the most easily observable engine parameters of flow, speed, pressure and temperature. Changes occur in these quasi-dimensionless maps due to the changes in gas properties described above. The use of such maps when water vapor is present is described in this section.

For a compressor, the presence of water vapor lowers $W\sqrt{T}/P$ at constant $N/\sqrt{T}$ and PR.

This is because the dimensionless group for speed, $N/\sqrt{(\gamma RT)}$, is reduced relative to dry air due to the increased R which outweighs the small reduction in γ. At this lower dimensionless speed the compressor passes a lower dimensionless flow, $W\sqrt{(RT)}/P\sqrt{\gamma}$. In addition, at this lower dimensionless flow the *quasi-dimensionless* flow $W\sqrt{T}/P$ is further reduced, again by the increased R.

For a turbine the presence of water vapor produces little or no change in $W\sqrt{T}/P$ at constant $N/\sqrt{T}$ and $\Delta H/T$. This is because, unlike a compressor, reduced speed normally either *raises* capacity or has no effect. One further effect of humidity is to increase the specific heat in the turbines, reducing the temperature drop for a given expansion ratio. The impact of these effects on engine performance depends on where through the engine the water is introduced.

Humidity

Description

Humidity is water vapor naturally present in the atmosphere:

- Specific humidity is the ratio of water vapor to dry air by mass
- Relative humidity is specific humidity divided by the saturated value

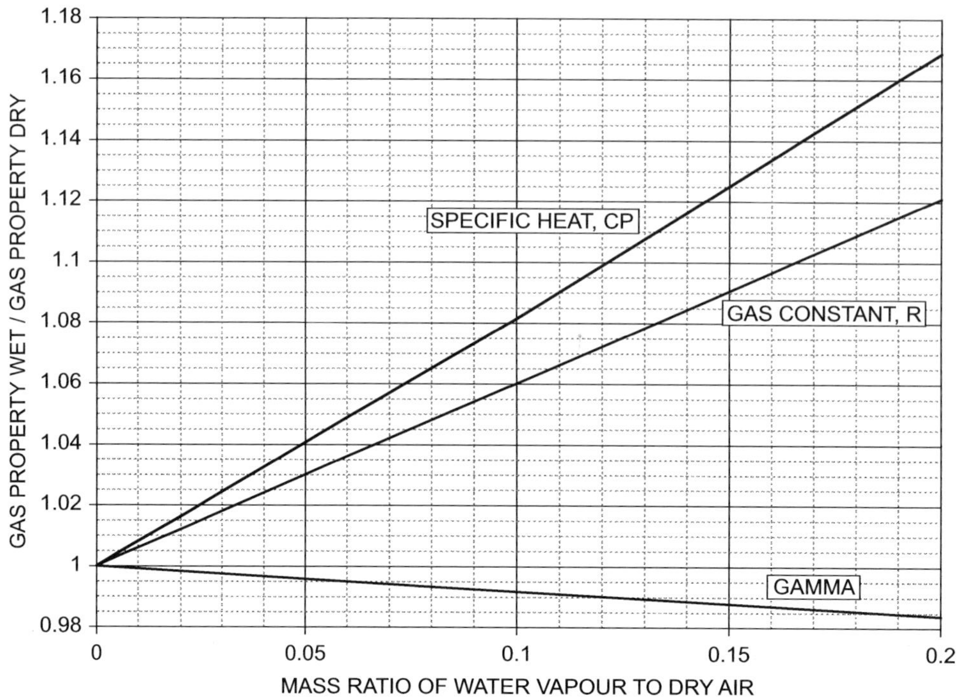

FIGURE 10–22 Variation of gas properties with water vapor content. *(Source: Rolls Royce.)*

Effects on Engine Performance

Table 10–3 shows qualitatively the effect of humidity on leading engine parameters. Figure 10–23 presents generic exchange rates that may be utilized for first-order accuracy to predict the impact of humidity on key performance parameters. These have been derived by considering all the full non-dimensional parameter groups to be constant at a given operating condition. The impact of humidity is greater than is immediately apparent, as *all* parameters must be factored. For example, at a specific humidity of 0.04 (100% relative humidity on a 36°C day) shaft power increases by about 1% and LP turbine inlet temperature falls by about 1%. Hence if governing to fixed LP turbine inlet temperature the increase in power would be around three to four times this. The precise impact depends on the engine cycle.

TABLE 10–3 Humidity: Effect on Leading Performance Parameters

	Changes At		
Parameter	**Fixed LP Speed**	**Fixed Last Turbine Inlet Temperature**	**Fixed Power or Thrust**
LP speed	None	Increase	Increase
HP speed	Negligible	Increase	Increase
Mass flow	Decrease	Decrease	Decrease
Fuel flow	Decrease	Increase	Increase
Output power	Decrease	Increase	None
Nett thrust	Decrease	Increase	None
SFC	Decrease	Increase	Increase
Temperatures	Decrease	See note 1	Decrease
Pressures	Decrease	Increase	Decrease

Notes: (1) Temperatures upstream of the last turbine inlet have negligible change, those downstream decrease. (2) Changes shown are for physical, not referred parameters.
(Source: *Rolls Royce.*)

Figure 10–23 may also be used for a first-order correction of engine test data from any humidity level to that for ISO conditions:

- Divide all parameters by the relevant factor from Figure 10–23 to obtain "dry" performance
- Multiply all parameters by the factor for a specific humidity of 0.0064 (60% relative humidity at 288.15 K and 101.325 kPa)

Water Injection

Description

Water injection has been employed for both aero- and industrial engines to improve performance, utilizing one of two injection points along the gas path:

- At first compressor entry—to boost power or thrust with minimal impact on SFC. Here a relatively small amount of water is used to lower the inlet temperature by evaporation.

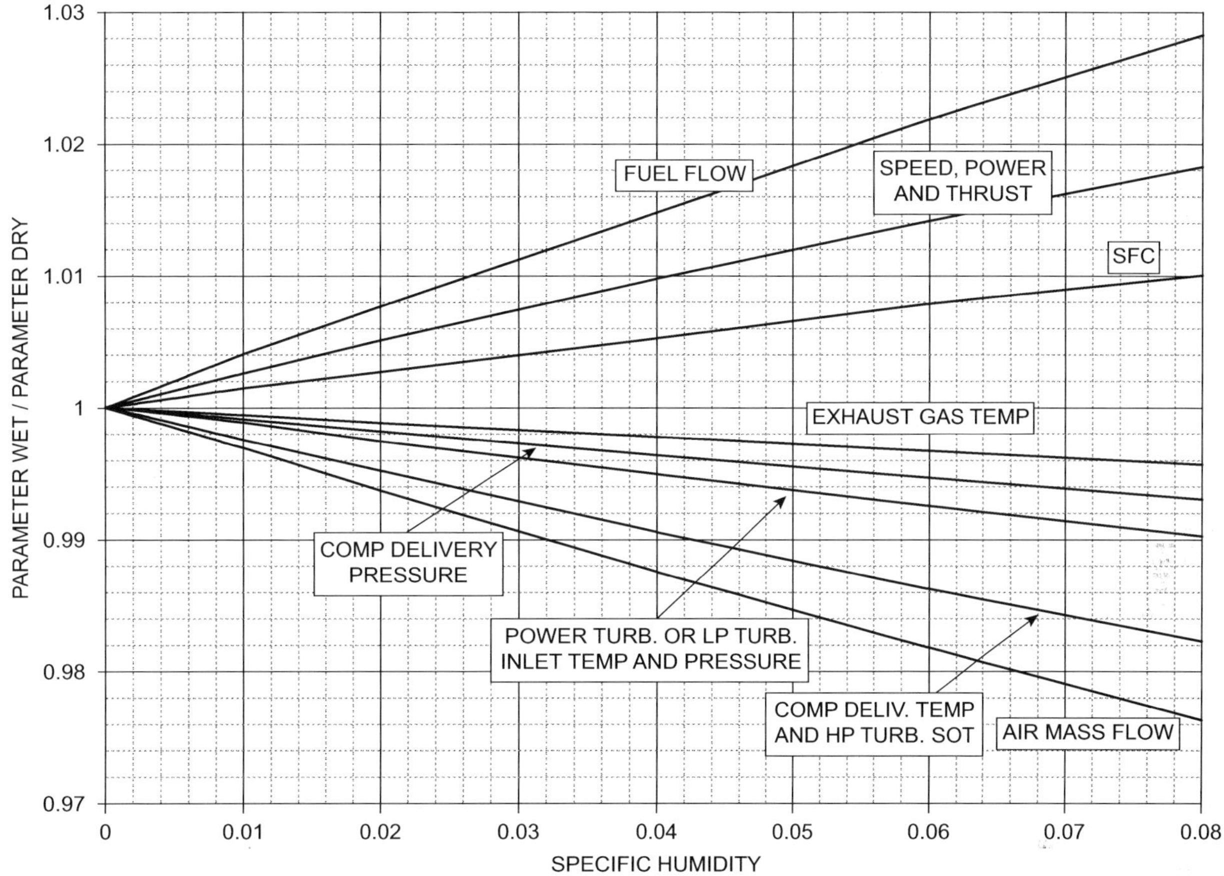

FIGURE 10–23 Effects of humidity on engine performance: generic exchange rates. *(Source: Rolls Royce.)*

- Into the combustor—primarily to reduce emissions for industrial engines. This also boosts power, but makes SFC worse.

Both practices are now largely superseded by other technologies. These are increased temperature capabilities and higher bypass ratios for aeroengines, and *dry low emissions* combustion technology for industrial engines, where low pollutant levels are achieved without the need for water injection.

Compressor Entry Injection—Aeroengines

The takeoff condition, especially for hot days or high airfields, was particularly important for older engines with zero or low bypass ratio. Water was injected using spray nozzles, and prevented from freezing by adding methanol. The power or thrust gain primarily came from the lower inlet temperature and intercooling within the compressor rather than the additional mass flow of water. If the amount of water injected exceeded the amount that could evaporate within the compressor, further power boost was achieved by similar effects as produced by water injection into the combustor. The weight of tanks, pumps, etc., was a disadvantage, as was the need to supply purified water.

Compressor Entry Injection—Industrial Engines

Industrial engines operating in hot, arid conditions often employ an evaporative cooler upstream of the intake. This consists of a screen with water flowing over and the air flowing through it. The engine experiences a colder, more humid day, and hence can run to higher power and/or lower operating temperatures. Compared with aeroengines water consumption is lower, as only that which immediately evaporates is used. Such equipment would clearly have been impractical for aeroengines.

Combustor Injection—Industrial Engines

Water injection into the combustor has been employed to reduce NO_x emissions, by lowering the temperature peaks in the combustor primary zone and hence the tendency for atmospheric nitrogen to dissociate. Side effects include increased power, but often increased emissions of carbon monoxide (CO) due to lower temperatures, especially for low-pressure ratios and short residence times. In addition combustion becomes more uneven, which may induce mechanical problems due to temperature patternation and noise. Water to fuel ratios (WFR) between 1 and 2 have been employed; as combustor temperatures are very high saturation is not a problem. The large amount of water and associated plant is a substantial disadvantage, the water being purified and pumped to the injection points via pipe work and manifolding. In addition, burners must incorporate water nozzles, and appropriate control capability is essential. As the water supply pipes etc. can be protected from low temperature no methanol is employed. SFC is worse due to the significant amount of heat required to evaporate the water.

Steam injection, described below, supplanted water injection for installations where steam plant could be used as it overcame the disadvantages associated with the chilling effect of liquid water. As stated, more modern industrial, and even marine, engines are moving instead to DLE combustion.

Effects on Engine Performance

For water injection, pumping power is not an engine performance concern, as the liquid water is incompressible and hence the power requirement is small. Logistic and weight issues of the associated plant are significant, however.

Compressor Entry

Table 10–4 shows the main effects on engine performance of water injection at compressor entry. As outlined, compressor inlet temperature is reduced by water evaporation, producing effectively a colder day and 100% relative humidity. The maximum overall performance boost at fixed SOT on a hot day is:

- 20% power or thrust increase
- 5–10% shaft power SFC improvement

The SFC improvement is because in non-dimensional terms the engine is suffering less from part load SFC deterioration. By coincidence the methanol prevents any increase in fuel requirement to maintain SOT (though unless liquid water persisted through to the combustor this effect would anyway be small). Saturation of the inlet air determines the feasible lowering of the inlet temperature, hence the initial humidity level has a strong limiting effect.

Figure 10–24 shows the effect on the compressor working lines, which are similar for both LP and HP compressors. At fixed mechanical speed reduced inlet temperature increases referred speed, causing operation to move further up the *same* compressor working line. At fixed SOT compressor referred speed increases further, again along the same working line.

TABLE 10–4 Compressor Entry Water Injection: Effects on Leading Performance Parameters

	Changes At		
Parameter	**Fixed LP Speed**	**Fixed Last Turbine Inlet Temperature**	**Fixed Power or Thrust**
LP speed	None	Increase	Decrease
HP speed	Decrease	Negligible	Decrease
Mass flow	Increase	Increase	Increase
Fuel flow	Increase	Increase	Negligible
Output power	Increase	Increase	None
Nett thrust	Increase	Increase	None
SFC	Decrease	Decrease	Negligible
Temperatures	Decrease	See note 1	Decrease
Pressures	Increase	Increase	Increase

Notes: (1) Temperatures upstream of the last turbine inlet have negligible change, those downstream decrease. (2) Changes shown are for physical, not referred parameters. (3) The use of methanol as an antifreeze reduces SFC. If no liquid water persists to the combustor then this effect is slight. (Source: *Rolls Royce.*)

Combustor

Table 10–5 lists the effects on key engine parameters of water injection into the combustor. Typically at fixed SOT an engine with a combustor design not optimized for emissions would achieve 85% NO_x reduction but with up to a three-fold CO increase. For a 1:1 WFR performance effects are:

- 10–20% power increase
- 6–8% worse SFC

The increase in power output is due both to the flow in the turbines exceeding that in the compressors, and the CP of the flow in the turbines being increased by the water. The latent heat of evaporation of water requires additional fuel flow, hence the worse SFC. At fixed SOT the compressor speeds are increased along with engine inlet mass flow.

Figure 10–24 shows that in both cases the HP compressor working line is *higher* as the HP turbine must pass the injected water vapor flow, raising compressor pressure ratio at a referred speed to maintain the required HP turbine capacity. Because the temperature drop in the turbines is reduced, different levels of power boost are achieved depending on which turbine inlet temperature is considered constant.

Unlike steam injection, described below, water injection into turbines downstream of the H turbine is not

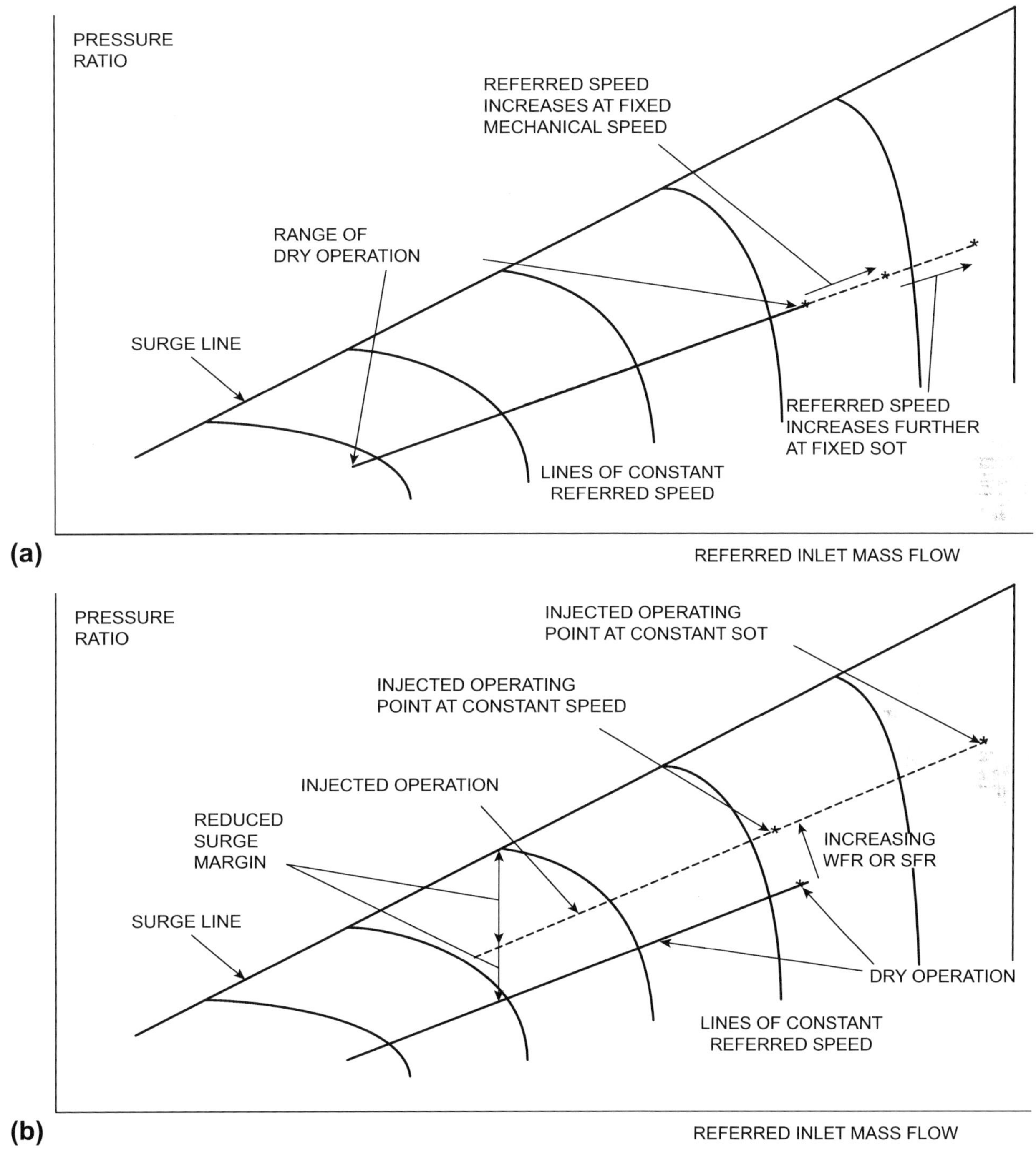

FIGURE 10–24 Water and steam injection: compressor working lines. a. Water injection upstream of compressor. b. Water or steam injection into combustor. *(Source: Rolls Royce.)*
Notes: For a single shaft engine with combustion injection, speed would not increase.
SFR, WFR steam/water to fuel ratio.

beneficial. The large heat absorption in evaporation drastically reduces that turbine's inlet temperature, and hence also its achievable specific power. The difference for water injection into the combustor is that extra fuel is added *simultaneously*, which offsets the temperature reduction due to the water. To add extra fuel to offset the temperature reduction for injection into other turbines would require impractically high temperatures in the HP turbine, or the complexity of additional combustors between the turbines.

TABLE 10–5 Combustor Water Injection: Effects on Leading Performance Parameters

Parameter	Changes At: Fixed LP Speed	Fixed Last Turbine Inlet Temperature	Fixed Power or Thrust
LP speed	None	Increase	Negligible
HP speed	Negligible	Increase	Negligible
Mass flow	Decrease	Increase	Decrease
Fuel flow	Increase	Increase	Increase
Output power	Increase	Increase	None
Nett thrust	Increase	Increase	None
SFC	Increase	Increase	Increase
Temperatures	Decrease	See note 1	Decrease
Pressures	Increase	Increase	Increase

Notes: (1) Turbine temperatures decrease apart from the last turbine. (2) Changes shown are for physical, not referred parameters. (3) The effects shown assume no use of methanol for antifreeze.
(Source: *Rolls Royce*.)

Operability and Control Philosophy

Compressor Entry

Water injection at entry to the compressor raises few issues as it has no impact on compressor running lines, only resulting in a higher referred speed for a given SOT. The aim is to increase power while keeping within the same operating limits; compressor *referred* speed is increased only on occasions where it would otherwise be well below its limit. The normal control strategies of temperature and speed limits remain applicable.

Combustor

Steady state the baseline control strategy is invariably constant water fuel ratio at all power levels, and the rated power governed to the same SOT as the dry engine. However on hot and cold days the compressor mechanical or referred speed limits are reached first and the power boost is limited. In practice very cold days often already have referred speed as the limiting parameter even for the dry engine, and no power boost is possible.

Transiently, however, there are two major consequences of water injection that must be addressed by control strategies:

1. Flame out avoidance: During engine deceleration water flow is normally reduced or shut off completely. This is because the additional heat required to evaporate the water would otherwise increase the tendency for weak extinction.
2. Surge avoidance: During engine acceleration water flow is again reduced or shut off completely. Otherwise the loss in HP compressor surge margin due to the overfueling for acceleration would be too great, given the surge margin reduction due to water injection. Indeed, only by this means can the lower steady state surge margin be accepted.

Specific Engine Design for High Injection Rates into the Combustor

Water injection into the combustor beyond around 2:1 WFR would usually require reselection of turbine capacities, to prevent unacceptable surge margin or overspeed as described above. Such modification is rarely if ever undertaken, since most engines must also operate "dry." However, both "wet" and "dry" cases may be considered in the design phase. Another item to address is increased power turbine entry temperature (PTET) at fixed SOT, due to the reduction in temperature drop in the turbine(s) driving the compressor(s).

The need for turbine capacity changes with high water injection rates arises because for the same compressor operating points compressor work is fixed, but turbine mass flow and specific heat are higher. For example, the requirements for a two shaft gas generator plus power turbine are illustrated below:

- HP turbine capacity must be increased. This accommodates the extra injected flow, preventing the rise in the HP compressor working line. It also reduces the HP turbine expansion ratio, and hence power output, to prevent an increase in HP compressor speed.
- LP turbine capacity must be decreased despite the higher mass flow, by a relatively modest amount. This forces a higher LP turbine inlet pressure, hence reducing the HP turbine expansion ratio further.
- Power turbine capacity must also be decreased, to reduce LP turbine expansion ratio, and prevent LP compressor overspeed. The lower HP and LP turbine expansion ratios are manifested as a higher power turbine inlet pressure and temperature, which give the increased engine power output.

By restagger of turbine aerofoils within casting allowances power boosts of up to 30% may be achieved for 4 WFR. As an example, at 6 WFR to retain acceptable compressor working lines and speeds the engine changes would be approximately:

- Power increase: 70%
- SFC deterioration: 10–15%
- HP turbine: 20% capacity increase, and 20% reduction in expansion ratio

- LP turbine: 5% capacity reduction, and 20% reduction in expansion ratio
- Power turbine: 20% capacity reduction, and 50% increase in expansion ratio

Steam Injection

Description

Steam injection is employed on industrial engines to reduce NO_x emissions and boost power. SFC also improves, as steam is usually raised using the engine exhaust heat, in a *heat recovery steam generator* (HRSG). The chilling effect of steam in the combustor is vastly less and more uniform than that of liquid water, hence there is little or no increase in CO emissions or combustion noise.

Steam injection into other turbines gives around half the SFC improvement and 1/5 of the power boost relative to the combustor, though it costs less in terms of compressor operability. Current systems have various proprietary names, such as *STIG* (steam injected gas turbine) or the *Cheng cycle. Supplementary firing* could also be employed, to increase steam production using a *boiler* with its own dedicated combustion.

Besides the need for steam plant and injection hardware, one significant issue for steam injection, like water injection, is the substantial need for purified water. Higher flow rates are considered than for water injection, as steam injection improves SFC. Owing to the cost and complexity of cooling towers etc. almost all systems are *total loss*, with the used steam simply passing up the exhaust stack. For all these reasons steam injection is employed almost solely on industrial engines rather than in mobile applications.

TABLE 10–6 Steam Injection: Effects on Leading Performance Parameters

Parameter	Changes At: Fixed LP Speed	Fixed Last Turbine Inlet Temperature	Fixed Power or Thrust
LP speed	None	Increase	Negligible
HP speed	Negligible	Increase	Negligible
Mass flow	Decrease	Increase	Decrease
Fuel flow	Decrease	Increase	Decrease
Output power	Increase	Increase	None
Nett thrust	Increase	Increase	None
SFC	Decrease	Decrease	Decrease
Temperatures	Decrease	See note 1	Decrease
Pressures	Increase	Increase	Increase

Notes: (1) Turbine temperatures decrease apart from the last turbine.
(2) Changes shown are for physical, not referred parameters.
(Source: *Rolls Royce.*)

Effects on Engine Performance

Table 10–6 summarizes the effects of steam injection on an engine cycle. Typically an aeroderivative industrial engine at constant SOT with 2:1 steam fuel ratio (SFR) would achieve:

- 15–20% power boost
- 10% SFC improvement

Raising steam at the highest possible temperature, dictated by the engine exhaust temperature and HRSG design, gives the best SFC; this reflects a maximum efficiency of energy recovery from the gas turbine exhaust.

In terms of component rematching the effects of steam and water injection are similar, additional power being available from the turbines due to their increased gas mass flow and specific heat. The HP turbine must pass extra flow, as for water injection, resulting in a higher HP compressor working line as per Figure 10–24. At fixed geometry the highest power boost occurs with the highest SFR that can be accepted within surge margin limits, which is normally not higher than 2.

Operability and Control Philosophy

Comments here almost exactly match those for water injection though the tendency to flame out during a decel is less for steam injection as without the need for evaporation it produces less chilling in the combustor.

Specific Engine Design for High Injection Rates into the Combustor

Almost all the same comments apply as for water injection, except that SFC improves rather than deteriorates. For a full turbine redesign, for 6:1 SFR the engine changes would be:

- Turbine capacity changes as per the 6:1 WFR water injection example above
- Power increase: 70%, again as for 6:1 WFR
- SFC improvement: 20%

The one difference relative to water injection is that due to the improved SFC going to high SFR is more attractive; however, as with water injection actually doing this is almost unknown in practice due to the compromise to "dry" performance.

In terms of engine cycle selection there are two other notable effects that impact the ability to produce steam:

1. Increasing SFR raises engine exit temperature for constant SOT and pressure ratio, as the temperature drop within the turbines reduces due to increased flow and specific heat.
2. Higher pressure ratio cycles allow less steam to be produced at constant SOT, as engine exit temperature is lower.

Condensation

Description

Condensation is confined to the "cold," compressor end of the engine and is due only to ambient humidity. The resultant water simply re-evaporates further along the gas path. Condensation may occur once flow conditions cause the local static temperature to be below the saturation value for the local water vapor partial pressure. This is determined by local static pressure and water vapor concentration; saturation corresponds to 100% relative humidity. The actual rate of condensation is determined by residence time, and by minute particles in the air that act as nuclei for water droplets. (See the discussion on the fundamental properties of the phases of water.)

In the cold end there are two regions where water vapor may condense:

1. Intake: Flow acceleration upstream of the first compressor reduces the static temperature, increasing the tendency for condensation. Figure 10–25 shows the range of humidity and local Mach number at which condensation may occur, at ambient pressure. At 40% ambient relative humidity (RH) the threshold Mach number is 0.5, and at 80% ambient RH it is 0.24.
2. Intercooler: The pressure increase in the first compressor raises the water vapor's partial pressure. This may exceed the saturated value once the intercooler reduces the temperature. The flow velocities are low within the intercooler heat exchanger matrix to achieve the required heat transfer; this gives adequate residence time for condensation, which occurs on the cold passage surfaces. If this produces water droplets, erosion of the downstream compressor blades may be a serious problem. The important difference from water injection or rain ingestion is that intercooler condensation may continue almost indefinitely in some geographical locations, unless the degree of intercooling is reduced.

Effects on Engine Performance

Intake

Intake condensation can cause a significant change in engine performance on a test bed. The apparent loss in power or thrust and increase in SFC at fixed fuel flow may be up to 2% for hot ambient conditions, and nearer 0.5% for temperate climates. The reason for this loss in performance may not be immediately apparent, being due to pressure and temperature changes occurring downstream of test bed inlet pressure and temperature measurements. Were it possible to measure accurately the pressure and temperature actually entering the

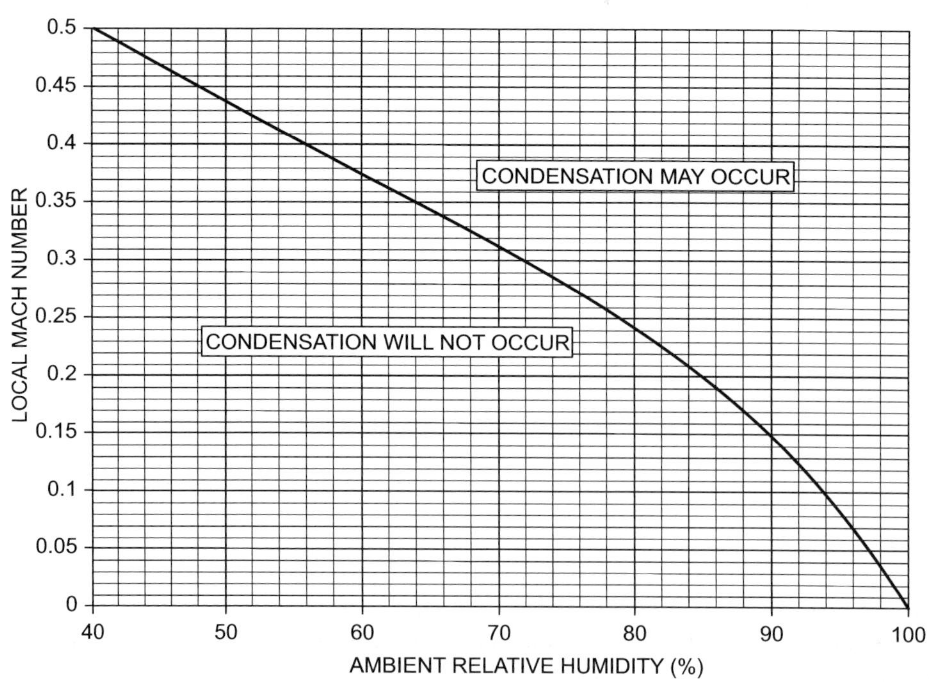

FIGURE 10–25 Intake condensation: threshold Mach number versus ambient relative humidity. *(Source: Rolls Royce.)*

compressor then there would be negligible impact on engine referred performance relative to that without condensation. In the absence of perfect measurements or correction methods, performance testing should not be undertaken during conditions of high humidity or precipitation.

The effects of condensation are an increase in air inlet temperature and a reduction in total pressure, as well as reduced specific humidity in the initial compressor stages. The increased temperature results from the latent heat release while the pressure loss reflects the momentum change in flow acceleration due to reduced density. Once within the engine the condensed water evaporates shortly downstream.

Any engine on the test bed (or even in service) will experience a real difference in performance as referred to the engine intake lip.

Intercooler

For an intercooler system some condensation control algorithm is often employed to prevent condensation causing blade erosion downstream. The coolest air temperature must be maintained above the calculated saturation value by control of the degree of cooling. For example, for a system with an intermediate transport loop between the cold sink and the engine this may involve a partial bypass of the cold sink. The intercooler exit temperature is thereby increased, with two main consequences for engine operation:

1. Reduced LP compressor surge margin due to decreased mass flow: The reduced intercooling means the HP compressor referred speed and hence referred flow are reduced by the higher inlet temperature. In addition for a given referred HP compressor flow the actual mass flow is reduced even further.
2. Increased turbine temperatures and actual HP shaft speed: The power required to drive the HP compressor is directly proportional to its absolute inlet temperature. If this power increases, the engine must run hotter and faster to produce the same output power.

Rain and Ice Ingestion

An aircraft will encounter large amounts of rain and hail when flying through storms. Furthermore in cold conditions ice may form on the intake surfaces, occasionally breaking off to pass into the engines. Significant ingestion is normally confined to turbofans and turbojets, which must pass ingestion tests for certification as described earlier. For aero turboprop and turboshaft engines all these phenomena may be reduced by *inertial separators* in the intake ducts. Here airflow is drawn in around a tight bend; heavier particles fail to take this bend, passing straight on and then overboard. Industrial, automotive, and marine engines are normally protected by intake filters.

Description—Rain Ingestion

The highest rain ingestion occurs when flying through monsoon conditions. For a turbofan the fan *centrifuges* a significant proportion of the liquid water into the bypass duct, i.e., the rotation of the blades forces particles radially outwards. This effect is dependent on residence-time and hence is increased by wide chord fan blades and reduced by high forward speed. However, some water always passes into the core compressors.

In extreme circumstances, should liquid water persist into the combustor the heat it absorbs in evaporation has a major chilling effect. This can be a serious problem, and on the CFM56 turbofan, rain ingestion did result in several rundowns and at least one forced landing.

Description—Ice and Hail Ingestion

Icing occurs when an aircraft encounters cold cloud conditions corresponding to "freezing fog," where air initially containing water vapor has dropped below 0°C. This causes ice to build up on the wings and engine intakes or nacelles. The ice may break off naturally when either it reaches some critical thickness, warmer air is encountered, or an *anti-icing* system is switched on. For engines, such a system consists of electrical or bleed air heating of the intake surfaces. The fan spinner is normally made of rubber to ensure flexibility which results in ice-being shed before it has reached a dangerous thickness.

Hail is ingested simply by flying through hailstorms, the worst often being produced by a thundercloud. In either case the main issues are of mechanical integrity.

Effects on Engine Performance

Mechanical issues aside, the effects of water and ice ingestion are basically similar. Control strategies must increase fuel flow to avoid engine rundown or even flameout. Though ice has a greater chilling effect than liquid water, in practice the overall concentration is less than that of liquid water and hence this is not a major concern.

Once fuel flow is increased the thermodynamics are closer to those for combustor water injection, described above, than to compressor entry water injection. However, within the core compressors there are additional phenomena that cannot be precisely predicted:

- Water progressively evaporates through the compressors, affecting gas properties, temperature levels, and gas mass flow. It also produces an intercooling effect that is specific to the compressor type. Evaporation begins when the partial pressure and temperature cross the saturated liquid line shown at the left-hand side of Figure 10–26. It continues until the saturated vapor line is reached on the right-hand side. The rate of

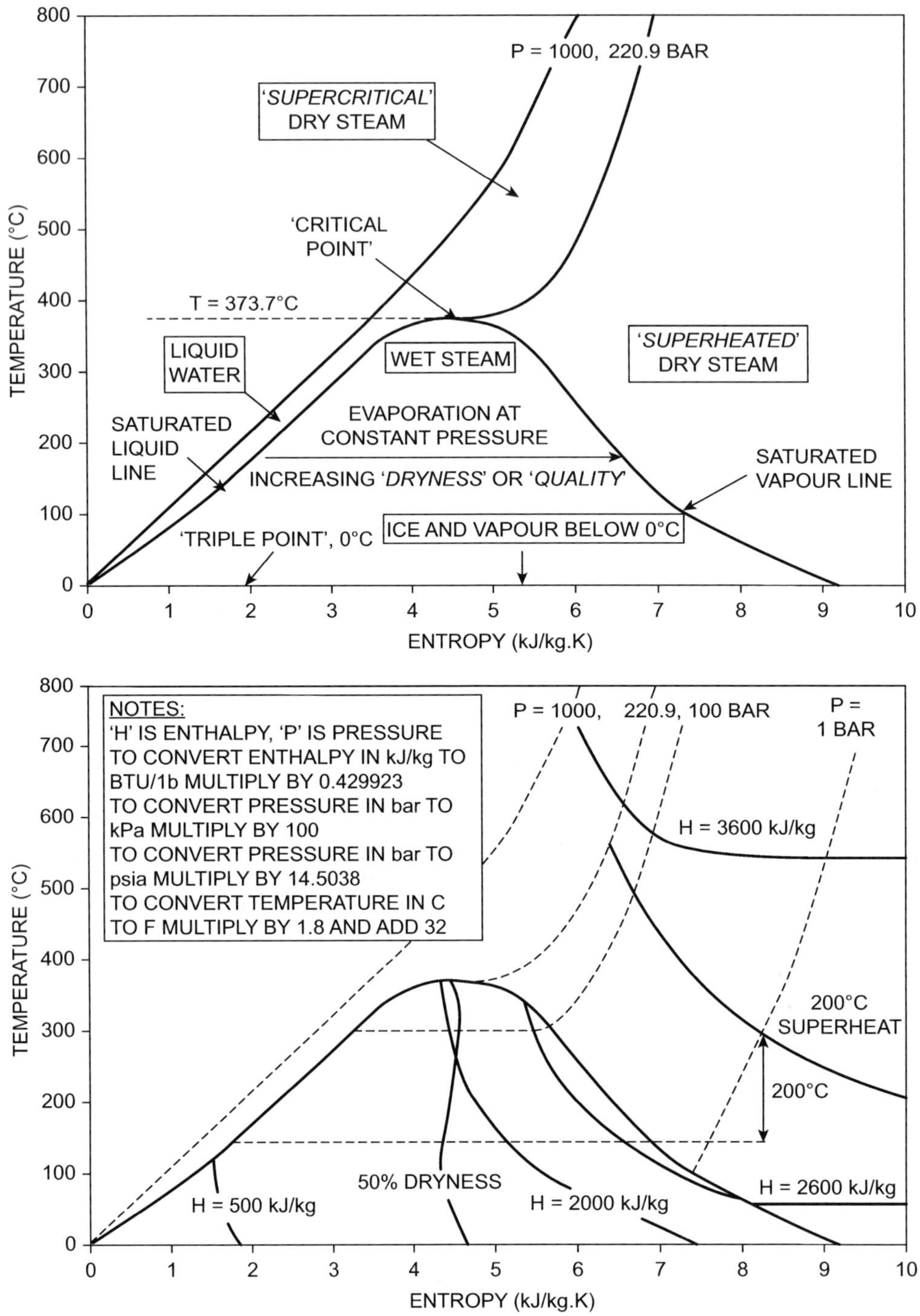

FIGURE 10–26 Temperature–entropy (T–s) diagram for water. *(Source: Rolls Royce.)*

evaporation is also specific to a compressor type, being also strongly dependent on residence time to transfer heat to the water, and hence can only be determined empirically.

- The surge line may be lowered, due to liquid water being centrifuged outwards which affects aerodynamics at the tip section.
- Mechanical power is absorbed by the work done in accelerating liquid water through a change in swirl velocity in each compressor stage.

Because of the various effects, rain normally results in aeroengine testing being suspended if an important performance curve is planned. The difference between rain injection and water injection into compressors is that there the intention is to drop the inlet temperature by evaporation, and not normally to pass large quantities of liquid water through the compressors.

The Thermodynamics of Water

As is well known, water may exist in both liquid and vapor phases over a wide temperature range. At a pressure of 1 bar and temperature of 100°C, water evaporates to steam if energy is supplied. This energy requirement is *the latent heat of evaporation*, and the process is known as *boiling*. At these conditions the volume increases by a factor of 1600. The boiling temperature rises as pressure increases, since a different change in volume then occurs. The inability to make good tea with water boiled on a mountain, which is then too cold, is well known. Conversely, if water is heated in a closed vessel the pressure and temperature both increase once boiling has begun.

Figure 10–26 shows schematically the properties of water in its three phases on a traditional temperature–entropy (T–S) diagram; for water it is known as the *Mollier diagram.* The characteristic "dome" shape defines the region through which evaporation may occur; inside this water exists as *wet steam*, a mixture of liquid and vapor. To the left of the dome only liquid water can exist, and to the right and above, only *dry steam.* Below 0°C liquid water cannot exist, only ice and an almost negligible vapor pressure due to *sublimation.* A temperature of 0°C is known as the *triple point*, where water may exist as ice, liquid, and steam. A pressure of 220.9 bar at 373.7°C is the apex of the "dome," known as the *critical point.*

Figure 10–26 includes several lines of constant pressure. At the left-hand end of each line there is the *saturated liquid line*, onto which lines collapse for a substantial pressure range; pressure changes have only a small effect, as liquid water is almost entirely incompressible.

At fixed pressure, heat input to liquid water increases the temperature until it crosses the saturated liquid line, when the boiling temperature is reached. Further heat input causes evaporation at constant temperature, and hence movement to the right within the dome; the degree of evaporation is described by the *dryness fraction* or *quality*, which is the ratio of vapor to total by mass. The right-hand edge of the dome is the *saturated vapor line*, where evaporation is complete and the steam is dry. Beyond the dome still further heat input results in a temperature increase, and the dry steam becomes *superheated. Superheating* is applied as a term in various forms to the temperature difference from the saturation line at that pressure. If the pressure is raised above the critical point (22.9 bar) it is *supercritical.* Here evaporation from liquid water to steam cannot be distinguished by any step change in density, which instead decreases uniformly as temperature increases.

Steam is not a perfect gas in that enthalpy is dependent on pressure as well as temperature, due to attraction forces between the polarized water molecules. This effect is least at low density and high temperature.

For illustration, Figure 10–26 also includes the following:

- A line of 50% dryness
- Lines of constant enthalpy, showing the large energy input required for evaporation
- A line of constant superheating

Use of steam in power cycles utilizes the pressure increase when heating liquid water in an enclosed volume. There is no need for large work input, unlike for gas turbine cycles, as pressurizing incompressible feed water requires very little power. It is mainly for this reason that steam cycles were developed much earlier than gas turbines. Injected steam in a gas turbine remains in the superheated region all through the engine, whereas in some purely steam cycles it condenses. Examples include the back stages of a steam turbine running "wet," and some reciprocating cycles utilizing the pressure reduction when steam is chilled.

There are different steam plant arrangements downstream of a gas turbine, including combined cycle, where a steam turbine produces additional shaft power.

The following are relevant:

- Formula 10.4 gives the enthalpy of saturated steam as a function of temperature.
- Formula 10.5 gives the evaporation temperature as a function of pressure.
- Formula 10.6 gives the partial pressure of water vapor as a function of its concentration.
- Formula 10.7 gives the enthalpy of liquid water as a function of temperature.
- Formula 10.10 gives the enthalpy of superheated and supercritical steam as a function of temperature and pressure.

Gas Turbine Performance Modeling and Test Data Analysis

The following sections present methods to model the individual effects occurring when water is encountered. These techniques are suitable both for predictions and test data analysis. Figure 10–27 presents a flowchart of how they interact; clearly many effects are common to several situations.

Ice and Melting

Accurate theoretical modeling is not practical; however, empiricism and approximation are sufficient to establish the impact on compressor efficiency and surge lines.

FIGURE 10–27 Performance modeling: main elements and sequence. *(Source: Rolls Royce.)*

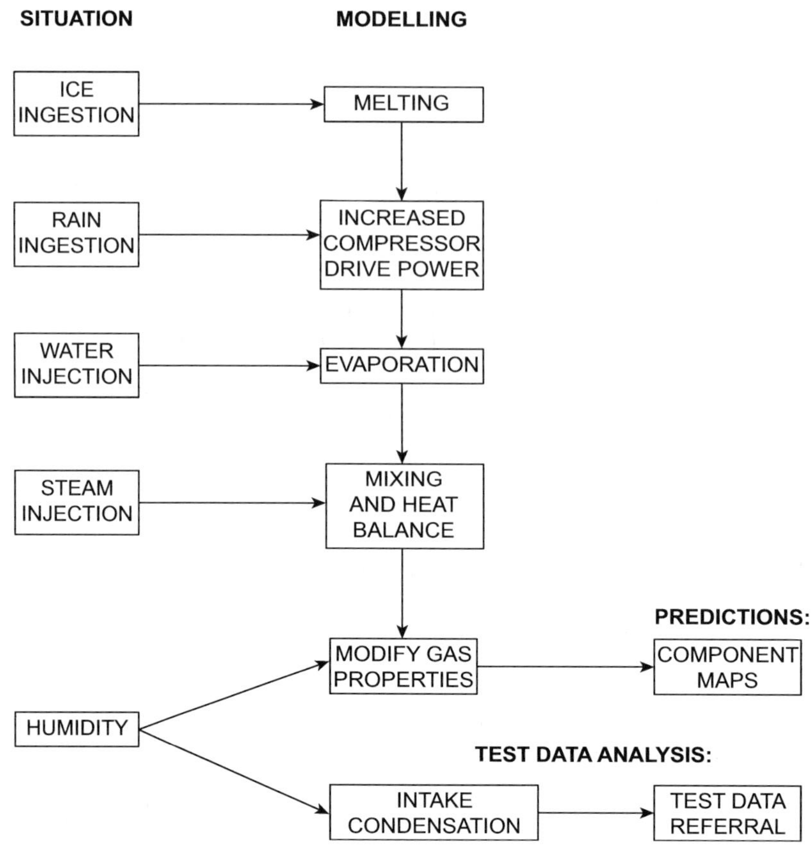

Whether ice melts in the combustor or the compressor depends on residence time and particle size; extremes may be established by assuming each in turn; the difference in terms of the overall cycle may be small. Once ice has melted it becomes water droplets, which are discussed below. Formula 10.8 enables the heat absorption in melting to be calculated, from which the resulting air temperature reduction may then be simply evaluated.

Liquid Water and Evaporation

For the compressor the existing characteristic may still be used for the airflow, a parallel water stream must also be considered, and the compressor drive power increased accordingly.

Accelerating liquid water to some swirl velocity in each stage absorbs power, around 40–60% of the work that would be required if it reached the full mean blade speed; this power absorption may be approximated by applying Formula 10.9 for each stage before evaporation has occurred. Some evaporation may occur in the compressor, though for large flows much will pass through to the combustor; judgment or empiricism must be utilized regarding the relative quantities, as discussed earlier.

During evaporation in either the compressor or the combustor the heat absorbed must be calculated from the difference between Formula 10.7 and 10.10 with the total temperature and partial pressure. The gas path temperature reduction due to this must then be calculated via a mixing and heat balance calculation as described. The sample calculation illustrates the approach.

Steam Injection

The mixing and heat balance calculation is described below. For higher steam flows giving more than 10% WAR (the ratio of water to dry air, by mass) the turbine calculations should employ steam formulae. Modeling the HRSG must consider several key points:

- Steam formulae must be used on the steam side, including the evaporation of the liquid water supplied.
- Steam formulae may also be required on the gas side, depending on the steam concentration.
- Enthalpy changes on the steam and gas sides must be equal.
- Design minimum levels of the main temperature differences must be maintained.

- Steam delivery pressure must be some specified amount above that at the engine injection point.

To meet the last two criteria the steam delivery temperature may vary as matching guesses in a matching model, and/or the rate of steam production.

Mixing and Heat Balance

Prior to evaporation the ice/water and air streams are considered in parallel. Afterwards, the latent heat absorption has been accounted for, the streams must be combined via a mixing and heat balance calculation:

- The changes in enthalpy for each stream must be equal and opposite. The sample calculation illustrates this process.
- Total pressure is normally assumed not to change during mixing, as the steam is normally a small fraction of the total flow.
- Gas properties downstream of the mixing plane should be adjusted by one of the methods described next.

Such calculations may require iteration, if so they become simply an extension of a "matching" model

Gas Properties

Three methods for adjusting gas properties to account the presence of water vapor for thermodynamic calculations are shown below. They are listed in order of increasing accuracy:

1. Using Formulae 10.1–10.3, which represent the factors to be applied to gas properties versus water vapor concentration as per Table 10–3. This method assumes that pressure does not change enthalpy (i.e., that water vapor is a perfect gas), and that the gas property ratios are not affected by temperature; it is adequate for first-order calculations.
2. Using the various formulae for specific heats given in Chapter 19. In all cases the overall specific heat should be obtained by averaging the gas constituents and additional water vapor on a molar basis, and the ratio of specific heats gamma is then easily derived. This method still assumes that water vapor is a perfect gas; however, where vapor content is less than 10% the error incurred is negligible; this method is the most widely used.
3. Evaluating CP for the additional water from Formula 10.11, which does not assume a perfect gas and accounts for the effect of pressure on steam properties. CP calculated in this fashion will be as per steam tables. The gas constant and ratio of specific heats are then derived as per method 2. This is the most rigorous method, and should be used for water content exceeding 10%.

Component Maps

Where there is water vapor but no liquid or ice present, component maps are normally read via the simplified quasi-non-dimensional groups. These groups relate flow, speed, pressure, and temperature and use implicit "dry air" values of the fundamental gas properties, gamma and R. As described earlier, water vapor changes these properties, hence with water vapor present the full forms of the non-dimensional groups must be recognized.

When water or ice are present, the use of the maps must incorporate gamma and R corresponding to 100% RH, or more often their ratios to the values for dry air. This is in addition to the calculations described in this section, where the water/ice is considered as a separate parallel stream. The changes in gas properties account for the effect on both the independent parameters against which the maps are tabulated, and on the dependent parameters returned from the map read. Evaporation within compressors produces an intercooling effect which is specific to the compressor design. Given the uncertainty of evaporation location, it is usual to simply retain the same compressor map.

Test Data

To understand engine behavior on test the effects of humidity must be accounted. These are due to the variation in gas properties, and potentially also to condensation. The first step is accurate humidity measurement at test conditions.

Non-Dimensional Corrections

Engine test data are referred to standard conditions via non-dimensional groups, to allow valid comparison of prediction and other test data. For humid conditions revised gas properties CP, gamma and R are evaluated as described above.

Rigorous Modeling

No attempt is made to correct test data to standard conditions. Instead it is compared with expectation via a rigorous thermodynamic prediction model, run at the tested conditions. The modeling accounts for effects of humidity on gas properties and hence on component maps, thermodynamic processes, etc. as described above. This applies equally to analysis of component performance changes via matching methods.

Intake Condensation

As stated the concentration and properties (hydrophobic, charged, etc.) of particles that may cause intake condensation cannot be determined. A theoretical, maximum rate at which it could occur may be calculated from relative humidity and intake Mach number. Formula 10.12 provides an approximate method. There are then two possible approaches, listed in order of increasing accuracy:

1. Assume condensation occurs at half the theoretical rate.
2. Correlate measured engine performance versus theoretical maximum rate. This would be done over a series of different engines, which also have build and manufacturing variabilities. Coefficients would be determined for the effects on main measured parameters, as the mean slopes of plots versus the theoretical temperature rise for measured ambient conditions. Comparison of engines would involve correction back to the same theoretical temperature rise using these coefficients.

Directly measuring the actual rate of condensation during a test would clearly be preferable; however, at the time of writing no absolutely successful technique has been demonstrated.

Intercooler Condensation

The control algorithms should adjust the degree of intercooling to prevent this. The requirement is therefore accurate modeling of the control action, which depends on the specific system. There should be no need to make adjustments to test data.

Formulae

Formula 10.1. Specific Heats Factor for Moist Air = fn(Water Air Ratio)

$$\text{CPfac} = [\text{WAR.molar} \times \text{CPW} + (1 - \text{WAR.molar}) \times \text{CPA}]/\text{CPA}$$

$$\text{WAR.molar} = \text{WAR} \times 28.96/18.015$$

(i) CPfac is the ratio of CP for moist air to that for dry air.
(ii) WAR.molar is the ratio of water to dry air, by number of moles.
(iii) CPW, CPA are the specific heats of water and dry air.
(iv) WAR is the ratio of water to dry air, by mass.
(v) 18.015, 28.96 are the molecular weights of water and dry air respectively.

Formula 10.2. Gas Constant Factor for Moist Air = fn(Water Air Ratio)

$$\text{Rfac} = \text{Ro}/(\text{MW} \times \text{RA})$$

$$\text{MW} = 1/\{(\text{WAR}/18.015) + [(1 - \text{WAR})/28.96]\}$$

(i) Rfac is the ratio of R for moist air to that for dry air.
(ii) Ro is the universal gas constant, 8.31 kJ/mole K.
(iii) MW is the molecular weight of moist air.
(iv) RA is the gas constant of dry air, 0.28705 kJ/kg K.
(v) WAR is the ratio of water to dry air, by mass.
(vi) 18.015, 28.96 are the molecular weights of water and dry air respectively.

Formula 10.3. Gamma Factor for Moist Air = fn(Water Air Ratio)

$$\text{GAMMAfac} = (\text{WAR.molar} \times \text{GAMMA W}) + (1 - \text{WAR.molar}) \times (\text{GAMMAA})/\text{GAMMAA}$$

$$\text{WAR.molar} = \text{WAR} \times 28.96.18.015$$

(i) GAMMAfac is the ratio of GAMMA for moist air to that for dry air.
(ii) WAR.molar is the ratio of water to dry air, by number of moles.
(iii) GAMMAW, GAMMAA are the ratio of specific heats for water and dry air.
(iv) WAR is the ratio of water to dry air, by mass.
(v) 18.015, 28.96 are the molecular weights of water and dry air respectively.

Formula 10.4. Enthalpy of Saturated Steam (kJ/kg) = fn [Temperature (°C)]

$$\text{H} = -7.352^{-06} \times \text{T}^3 - 2.333^{-03} \times \text{T}^2 + 2.437 \times \text{T} + 2492\,6949/(\text{T} - 387.5)$$

Accuracy is within 0.6%, range is 0–370°C.

Formula 10.5. Temperature Level of Evaporation (°C) = fn [Pressure(bar)]
For P = 1 to 25 bar:

$$\text{T} = -18.11^{-05} \times \text{P}^6 + 0.0014006 \times \text{P}^5 - 0.043 \times \text{P}^4 + 0.67482 \times \text{P}^3 - 5.9135 \times \text{P}^2 + 33.2486 \times \text{P} + 72.1585$$

For P = 25 to 210.5 bar:

$$\text{T} = -1.11726^{-11} \times \text{P}^6 + 8.97543^{-09} \times \text{P}^5 - 2.9476^{-06} \times \text{P}^4 + 0.00051476 \times \text{P}^3 - 0.05329436 \times \text{P}^2 + 3.933136 \times \text{P} + 152.0676$$

Accuracy is within 0.5°C.

Formula 10.6. Partial Pressure of Water Vapor in Moist Air (Bar) = fn[Specific Humidity, Pressure (Bar)]

$$\text{Pw} = \text{P}/[(0.622/\text{SH}) + 1]$$

(i) Pw is water vapor partial pressure.
(ii) SH is specific humidity, kg water vapor per kg dry air.
(iii) 0.622 is the ratio of molecular weights of water and dry air.

Formula 10.7. Enthalpy of Liquid Water (kJ/kg) = fn [Temperature (°C)]

$$H = 3.1566^{-12} \times T^6 - 2.9348^{-09} \times T^5 + 1.0407^{-06} \times T^4 - 0.16703^{-03} \times T^3 + 0.0120915 \times T^2 + 3.87675 \times T + 0.74591$$

(i) Average accuracy is within 0.7%.
(ii) Range is 0–370°C.

Formula 10.8. Latent Heat of Melting Ice (kJ/kg)

$$DHif = 333.5$$

Formula 10.9. Work on Liquid Water in Compressor (W) = fn[Water Flow (kg/s), Mean Blade Speed (m/s)]

$$DPW = 0.5 \times Wwater \times U^2$$

For an axial compressor this work is done in each stage.

Formula 10.10. Enthalpy of Dry Steam (kJ/kg) = fn[Temperature (°C), Pressure (Bar)]

Superheated:

$$H = 2.98^{-04} \times T^2 + 1.83 \times T + 2500 - 5.14207^{08} - P(T + 276)^3 - \left(1.03342^{37} \times P^3 - 6.42613^{31} \times P^5\right)/(T + 276)^{17.787}$$

(i) Average accuracy is 0.2%.
(ii) Range is 0.01–210 bar, and saturation to 800°C.

Supercritical:

$$H = P \times \left[T^3 \times (1 + 0.001634 \times P) + T^2 \times (1094.941 - 1.663087 \times P) - 8169907\right]/5.37^7 + 0.004505 \times T \times (738.0074 + 31,929.78/P - 0.30077 \times T) + 611.736 + 39.63429 \times (9.551098 - 0.002642 \times P) \times \arctan[(T + 0.005184 \times P)]$$

(i) Average accuracy is 0.5%.
(ii) Range is 220–400 bar, and 100–600°C.

Formula 10.11. Specific Heat of Steam (kJ/kg K) = fn [Temperature (°C), Enthalpy (kJ/kg)]

$$CP = dH/dT = [H(T + DT) - H(T - DT)]/(2 \times DT)$$

(i) H is enthalpy, evaluated via temperatures shown in Formula 10.10; T is temperature, °C.
(ii) DT is a small temperature increment, e.g., 5°C.

Formula 10.12. Theoretical Maximum Rate of Condensation = fn(Temperature, Pressure, Mach Number, Humidity)

The following outlines an approximate method:

- Calculate initial total enthalpy for air and water vapor using T0 and SH.
- For the case of no condensation, calculate local static temperature after flow acceleration using T0 and M.
- Calculate increased local static temperature TS after condensation as above + TRISE.
- Calculate local static pressure after flow acceleration using P0 and M.
- Calculate partial pressure of water vapor from Formula 10.11.
- Calculate saturated value of water vapor partial pressure. This also gives specific humidity at saturation, SHsat.
- Calculate amount of condensed water to reduce the partial pressure of water vapor to the saturated value.
- Calculate enthalpy of air and remaining water vapor using T0 + TRISE and SHsat.
- Calculate enthalpy of condensed liquid water from Formula 10.7.
- Iterate on TRISE until enthalpies balance before and after condensation.

The main approximation is to neglect the loss in total pressure that occurs due to the momentum change when heat release causes an increase in flow velocity.

Sample Calculations

Ingested rain evaporates within a compressor. Determine (i) the possible range of power absorption due to varying amounts of work done on the liquid water, (ii) gas conditions downstream and (iii) the variation of exit temperature resulting from changing the assumptions about how many stages the liquid water absorbs work in.

Conditions are as below:

Blade speed (m/s)	700
Compressor PR	5
Compressor isentropic efficiency	0.86
Number of compressor stages	6

	Water	Air
Temperature (K)	293	293
Pressure (kPa)	100	100
Mass flow (kg/s)	1.0	100

(i) Work Done on Liquid Water

From Formula 10.9 the work done in each compressor stage is DPW = 0.5 × Wwater × U^2:

$$\text{DPW} = 0.5 \times 1 \times 700^2$$

$$\text{DPW} = 254 \text{ kW per stage}$$

Since evaporation requires some temperature increase, work will be done in one stage at the very least. The maximum number of stages would be the full six:

$$\text{DPW} = 245 \text{ kW minimum, and } 6 \times 245 = 1470 \text{ kW maximum}$$

(ii) Gas Conditions at Compressor Exit

Compressor temperature rise for dry air. Ignore changes in compressor performance due to intercooling.

$$\text{T3} - \text{T2} = \text{T2} \times \left(\text{P3Q2}^{(\gamma-1)/\gamma} - 1\right)/\text{ETA2}$$

$$\text{T3} - \text{T2} = 293 \times \left(5^{2/7} - 1\right)/0.86$$

$$\text{T3} = 492 \text{ K} = 219^{\circ}\text{C}$$

Evaporation Calculation

Calculate enthalpy of ingested water using Formula 10.7:

$$\begin{aligned}\text{Hwater} = {} & 3.1566^{-12} \times 20^6 - 2.9348^{-09} \times 20^5 \\ & + 1.0407^{-06} \times 20^4 - 0.16703^{-03} \times 20^3 \\ & + 0.0120915 \times 20^2 + 3.87675 \times 20 \\ & + 0.74591\end{aligned}$$

$$\text{Hwater} = 81.94 \text{ kJ/kg}$$

Calculate partial pressure of steam at compressor exit using Formula 10.6:

$$\text{Pw} = \text{P}/[(0.622/\text{SH}) + 1]$$

$$\text{Pw} = 5/[(0.622/0.01) + 1]$$

$$\text{Pw} = 0.0791 \text{ bar}$$

Calculate enthalpy of steam using Formula 10.10 for superheated steam. Add work done on liquid water in first compressor stage.

$$\begin{aligned}\text{Hsteam} = {} & 2.98^{-04} \times \text{T}^2 + 183 \times \text{T}\,2500 - 5.14207^{08} \\ & \times \text{P}/(\text{T}+276)^3 - \left(1.03342^{37} \times \text{P}^3\right. \\ & \left.- 6.42613^{31} \times \text{P}^5\right)/(\text{T}+276)^{14.787}\end{aligned}$$

$$\text{QUwater} = (\text{Hsteam} - 81.94) \times 1.0 - 245$$

Now calculate change in air enthalpy due to heat absorbed by the water:

$$\text{QUair} = 100 \times 1.005 \times (219 - \text{Tmix})$$

Mixing Sum Iterate to make QUair = QUwater, using either inbuilt spreadsheet functions or "manual" updates. Converged solution gives Tmix = 193.5°C.

(iii) Effect of Variation in Mechanical Power Absorption

Recall work done on liquid water per stage = 245 kW.

Repeat above iteration with QUair = QUwater + 245:

$$\text{Tmix} = 195.9^{\circ}\text{C}$$

Hence a difference of one stage changes the mixed temperature by 2.4 K; five stages would change it by 12 K.

Note: For this small water concentration the presence of the water could have been neglected and the temperature change found by considering the air alone.

At this point, it is useful to consider a case study of testing/verification of a high performance gas turbine. In this case, the US DOE is funding work that raises the efficiency level of gas turbines made by OEMs in this US DOE program. Key objectives that stem from increased efficiency are fuel conservation and emissions reduction.

CASE STUDY 1: THE W501G TESTING AND VALIDATION IN THE SIEMENS WESTINGHOUSE ADVANCED TURBINE SYSTEMS PROGRAM*

Note to readers: *As per the footnote **, readers can note that this paper was written when Siemens was Siemens Westinghouse. As always in this book, even if an OEM has*

* Courtesy of Siemens Westinghouse**: Extracts from G. R. Gaul and T. S. Diakunchak, "The W501G Testing and Validation in the Siemens Westinghouse Advanced Turbine Systems Program," 2001-GT-0399.

** In 2001, what was to later be Siemens, was called Siemens Westinghouse. Note that this paper was written when Siemens was Siemens Westinghouse. As always in this book, even if an OEM has altered their name or ownership, if a technology or case study is still relevant today, then I will include the original name under which the work or paper was released. To those who study the nuances of design, this can have relevance in that the Siemens Westinghouse partnership brought with it some technology that had originated with Westinghouse. Further Westinghouse had had earlier associations with MHI (Mitsubishi) and some of its components bore evidence of same.

altered their name or ownership, if a technology or case study is still relevant today, then I will include the original name under which the work or paper was released. To those who study the nuances of design, this can have relevance in that the Siemens Westinghouse partnership brought with it, some technology that had originated with Westinghouse. Further Westinghouse had earlier associations with MHI (Mitsubishi) and some of its components bore evidence of same.

The Siemens Westinghouse Advanced Turbine System (ATS) has the ultimate goal of achieving greater than 60% LHV-based net plant thermal efficiency, less than 10 parts per million NO_X emissions, a 10% reduction in cost of electricity, and reliability-availability-maintainability (RAM) equivalent to modern advanced power generation systems. The ATS program, which is supported by the US DOE, introduces advanced technologies in three evolutionary steps to minimize risks and to increase the net benefits of the program. The W501Q, the first step in the ATS engine introduction incorporates many ATS technologies such as closed-loop steam cooling, advanced compressor design, and high temperature materials. The lead unit has completed full-load testing at the City of Lakeland McIntosh #5 site in Lakeland, FL, and has produced power and revenue for Lakeland Electric since May 2000. Results from the testing are presented and future developments are discussed. Building on the current W501Q advancements will include steam-cooled turbine vanes and leakage enhancements. Continuing this low risk step-wise introduction of new technology, the W501ATS engine adds further advanced designs that achieve the program objectives. Siemens Westinghouse is also infusing ATS technologies into its mature frames in both new units and service upgrades to maximize the benefit of the program.

The Advanced Turbine Systems Program (ATS) funded by the US Department of Energy, Office of Fossil Energy, is an ambitious multi-year effort whose goal is to develop technologies necessary for achieving significant increase in natural gas-fired power generation plant efficiency, a decrease in cost of electricity, and a reduction in harmful emissions, while maintaining the current state-of-the-art reliability, availability, and maintainability (RAM) levels. This three-phase technology development and demonstration program was started in 1992 and was completed in 2001.

To achieve the ATS Program goals for performance, emissions, electricity cost, and mechanical reliability, significant advancements were required in key technologies applied in gas turbine design. Successful developments were carried out in technologies relating to aerodynamics, combustion, cooling, sealing, materials, and coatings.

The W501ATS engine incorporates new technologies, as well as proven design features developed over the last 50 years and employed successfully in the W501 series of heavy-duty industrial and utility engines. These proven design features include single-shaft, two-bearing rotor; cold-end generator drive; compressor blade rings; low-alloy-steel rotor discs; curvic-clutched turbine rotor; four-stage turbine; cooled and filtered rotor cooling air; single first-stage turbine vane segments; tangential exhaust struts; and individual combustor baskets. The W501ATS engine is the latest in successful designs evolving from proven predecessors such as the 186 MW W501F and the 253 MW W501.

Evolutionary Approach

Siemens Westinghouse solicits input from an industry advisory panel comprised of members from major US and international utilities and independent power producers. Based on the input from this panel and market analyses, Siemens Westinghouse is pursuing an evolutionary introduction of the ATS, which incorporates ATS technology in stages culminating in an engine that meets or exceeds all of the program objectives. This approach has two main advantages. Firstly, the evolutionary approach mitigates the risk associated with introducing multiple, advanced technologies simultaneously. Secondly, the early introduction of ATS technology expands and accelerates the benefit of the program, as compared with limiting the technologies to only the ATS engine.

The evolutionary approach is shown schematically in Figure 10–28. Firstly, the introduction of the ATS frame begins with the W50IG. Many ATS technologies are incorporated in the W501G and are discussed later. Secondly, from the initial W501Q future enhancement include steam-cooled turbine vanes, leakage improvements, and increased burner temperature. Thirdly, the W501ATS engine evolves from the W501Q which reduces development risks through early demonstration of many critical technologies.

Siemens Westinghouse is further expanding the benefits of the ATS program by introducing ATS-developed

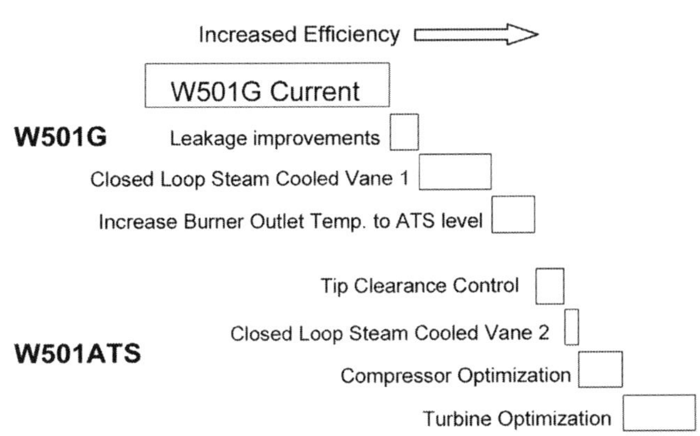

FIGURE 10–28 Evolutionary approach leading to the ATS engine. *(Source: Siemens.)*

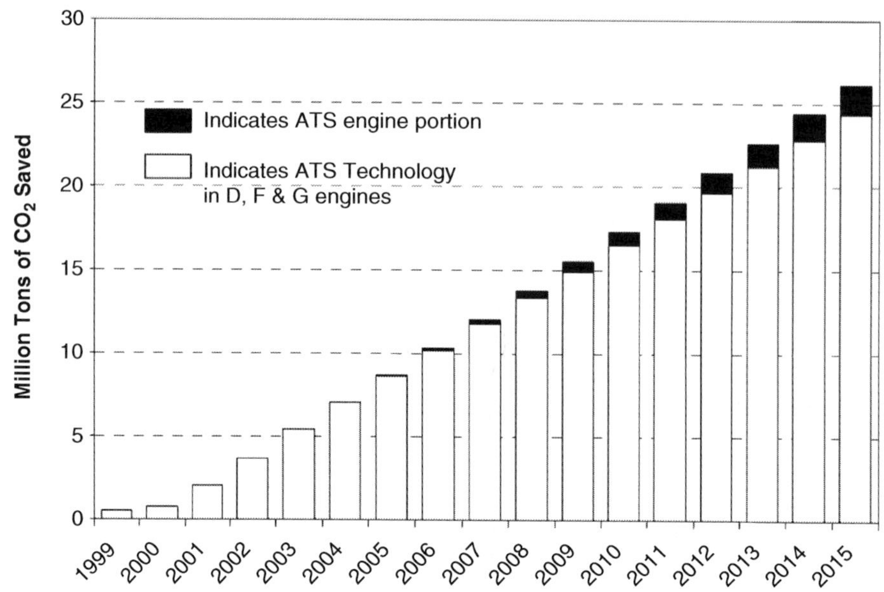

FIGURE 10–29 CO_2 net reductions from ATS technology. *(Source: Siemens.)*

technologies into its mature product lines. For example, the latest W501F incorporates ATS-developed brush seals, coatings, and compressor technology. Furthermore, many of these technologies can be retrofitted into operating units.

Because the F frame accounts for a majority of current new unit sales, this infusion of technology yields significant savings in fuel and emissions. Figure 10–29 shows the total impact of Siemens Westinghouse ATS technology on CO_2 emissions. Note that much of the net benefit is the result of Siemens Westinghouse's approach of expediting and expanding ATS technology through evolutionary introduction and infusion into mature frames.

ATS Technology in Operation

The W501G is the first major introduction of the ATS technology. The W501G incorporates the following ATS engine features:

- ATS advanced 3D compressor
- Advanced brush seals and abradable coatings
- Closed-loop steam cooling
- High temperature thermal barrier coatings
- ATS Row 4 turbine blade.

ATS Advanced 3D Compressor

The W501G incorporates the first 16 stages of the 27:1 pressure ratio, 19-stage, ATS compressor with slight modification of the last 3 stages, and with vanes 1 and 2 fixed instead of variable. As a result, the W501G operates the ATS mass flow of 558 kg/sec (1230 lbs./sec), but at a pressure ratio of 19:1—optimized for the G cycle performance.

The design is based on three-dimensional inviscid flow analyses and on custom-designed, controlled-diffusion airfoil shapes. Controlled-diffusion airfoil design technology has been successfully applied in the aircraft industry for many years. The mechanical integrity of each stationary and rotating airfoil was verified by finite element analyses to satisfy steady stress and endurance strength criteria. Each airfoil was tuned to avoid potentially harmful resonant frequencies.

To verify the aerodynamic performance and mechanical integrity of the new high-pressure ratio design, the full-scale W501ATS compressor was manufactured and tested in 1997 at a specially-designed facility at the US Navy Base in Philadelphia. To reduce the required power to the 25 MW available at the test facility, the compressor test was carried out at subatmospheric inlet conditions.

The ATS-developed compressor technology has also been retrofitted into the W501F product line. Using the analytical techniques developed and proven in the ATS program, the W501F compressor was upgraded in the latest improvement to this successful frame. This advanced compressor is utilized on new W501F units. In addition, the redesigned compressor can be retrofitted to any 42 W501Fs that were built with the original W501F compressor. Applying this ATS technology to the W501F expands the benefit of the ATS program since the W501F compromises more than 70% of future units that are sold or on order at Siemens Westinghouse.

Brush Seals and Abradable Coatings

To minimize air leakage, as well as hot gas ingestion into turbine disc cavities, brush seals were incorporated into the

W501ATS engine design at several locations: under the compressor diaphragms, at the turbine disc front, under turbine rims, and at the turbine interstages. Tests were carried out on test rigs for the different brush seal locations to develop effective, rugged, reliable, and long life brush seal systems. At the Philadelphia US Navy Base, full-scale brush seals were tested as part of the ATS compressor test, which verified the brush seal low leakage and wear characteristics.

To date, ATS-developed brush seals have been successfully incorporated and operated in W501G and later W501F product lines. Pre- and post-upgrade tests have demonstrated performance improvement in retrofit applications.

Considerable performance benefits can be obtained by reducing compressor and turbine blade tip clearances. Abradable coatings permit tip clearances to be minimized without fear of damaging hardware, and they provide more uniform tip clearances circumferentially. Abradable coatings, identified for compressor and turbine applications, were tested to determine abradability, tip-to-seal wear rate, and erosion characteristics. These ATS-developed abradable coatings have been incorporated into the W501F and W501G compressor and front turbine stages (1 and 2). The later turbine stages (3 and 4) employ shrouded blades with honeycomb seals.

Closed-Loop Steam Cooling

Using closed-loop steam cooling on transitions and turbine stationary components has two advantages. First, more compressor delivery air is available for premixing with the fuel gas in the combustor hot end. This allows very lean premixed combustion and makes possible the restriction of NO_x emissions to single digits. Second, closed-loop steam cooling significantly improves cycle efficiency by reducing the amount of chargeable air used for cooling and sealing.

The ATS transitions, which duct the hot combustor exit gases to the turbine inlet, are closed-loop steam cooled with air as an alternate coolant at part load. Steam enters the engine through four external connections and is routed to each transition supply manifold through internal piping. The supply manifold feeds the steam to an internal wall cooling circuit. After cooling the transition walls, the steam is collected in an exhaust manifold and ducted out of the engine. The W501G employs the ATS transition.

High-Temperature Thermal Barrier Coatings

Thermal barrier coatings (TBC) are an integral part of the W501ATS engine design. A development program is in progress to develop an advanced bond coat/TBC system with a projected service life of more than 24,000 hours. Different bond coats and ceramic materials were evaluated under accelerated oxidation test conditions and down selected. An advanced bond coat/TBC system mechanical integrity and durability was demonstrated in more than 24,000 hours of cyclic testing at 1010°C (1850°F). This advanced bond/coat TBC system has been incorporated on the W501G Row 1 and 2 blades. This coating will improve both the life and durability of these parts, and it can potentially improve future engine performance by reducing the amount of cooling air required.

ATS Row 4 Turbine Blade

The 25% increase in engine mass flow, compared with the baseline F class machines, necessitated an advanced design Row 4 turbine blade to avoid increasing turbine exhaust losses. The ATS Row 4 blade is an uncooled, interlocked, Z-shrouded, cast airfoil. Because the W501G employs the ATS compressor and associated mass flow, this blade was first introduced on the W501G. On the first W501G at the City of Lakeland McIntosh #5 site, the blade was instrumented with vibration monitors and strain gauge telemetry. In testing to date the blade has performed as predicted in both aerodynamic performance and mechanical strength.

Engine Test Results

The first W501G was ignited in April 1999, at the City of Lakeland MacIntosh #5 site. The unit has undergone extensive testing and verification. Since March 2000, the customer has dispatched the unit based on power demands. This lead W501G engine has accumulated over 239 starts and 850 fired hours as of November 2000. Currently, the unit operates in a simple cycle mode with a once-through steam generator for cooling steam production. Construction of the combined-cycle plant is under way and will complete in 2002. The W501G Design Plant Performance is shown in Table 10–7.

The test program included over 3000 sensors and measured parameters. An engine schematic showing the various sensors is shown in Figure 10–30. The test

TABLE 10–7 W501G Design Plant Performance

Power, net MW	253
Heat Rate	
kJ/kW-hr	6206
BTU/kW-hr	5884
Airflow	
kg/sec	558
lbs/sec	1230
Pressure ratio	19.5:1

(Source: Siemens.)

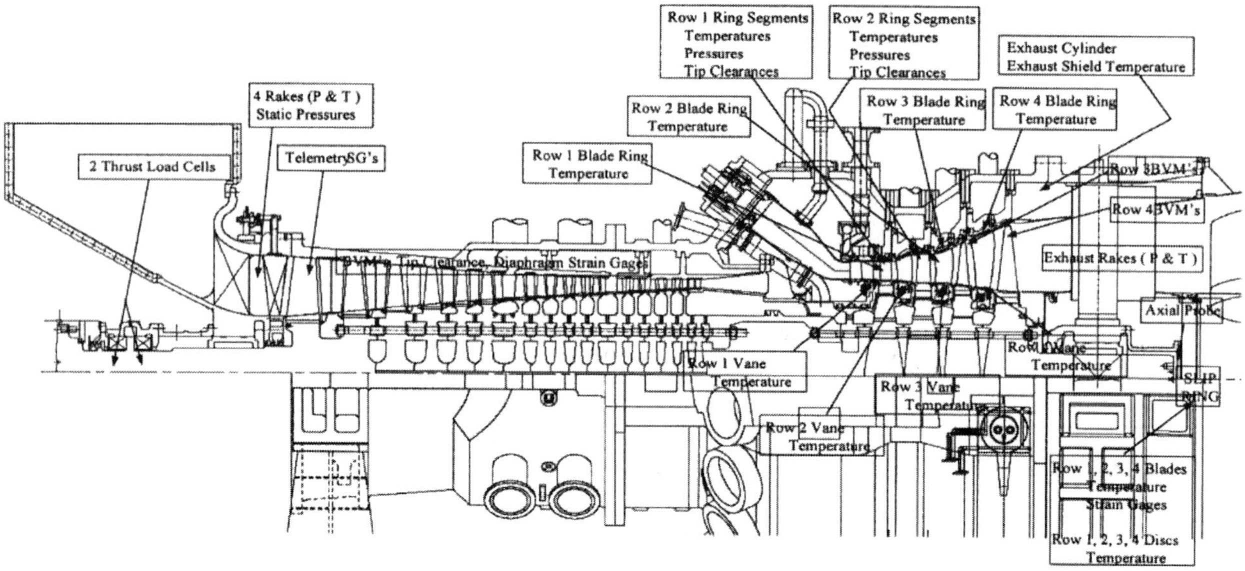

FIGURE 10–30 W501G instrumentation at City of Lakeland McIntosh #5 site in Lakeland, FL. *(Source: Siemens.)*

program consisted of two phases—emissions/performance mapping and thermal paint testing. The emissions/performance mapping phase tested targeted combustion system variables and provided engine performance mapping for different operating conditions such as IGV position and exhaust temperature.

Following the initial testing, turbine flowpath components and combustion components were prepared with thermally reactive paint and installed. The thermally reactive paint changes color based on exposed temperature. This method is used extensively in aero engine validation since it provides a complete and accurate temperature map of the components at operating conditions. To react the thermal paint, the engine was ramped up to full load, run for approximately 5 minutes at full load and then shut down. The thermal paint test was conducted in two phases. In July 2000, the transitions and Row 1 vanes were painted and tested. These components are removable without a major cover lift. In October 2000, a full paint test was conducted which included all turbine blades and vanes and areas of the rotor. Figure 10–31 shows the scope of the painted components. In addition to the base design, several components were installed with different cooling schemes. The different schemes were tested to evaluate possible cooling flow reductions and component durability enhancements. Both tests were conducted successfully and results are being evaluated.

The testing validated the closed-loop steam cooling in a commercial application. The steam turbine temperatures were measured under both air cooling at part load before steam quality was achieved and under full-load closed-loop steam cooling. The testing at Lakeland has confirmed the ability to switch between steam and alternate air cooling.

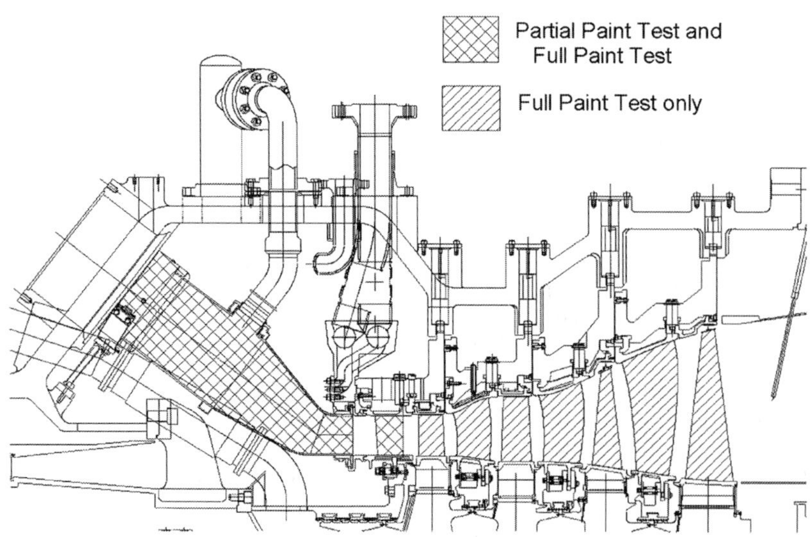

FIGURE 10–31 W501G paint test. *(Source: Siemens.)*

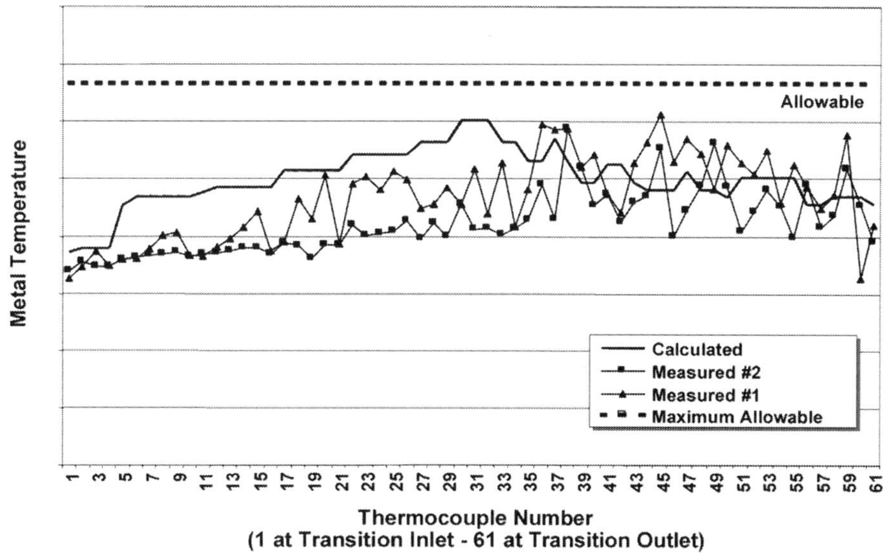

FIGURE 10–32 W501G transition temperatures. *(Source: Siemens.)*

As anticipated, actual measured metal temperatures are lower in the new closed-loop steam-cooled design than in existing open-loop air-cooled design. The temperatures were measured along the length of the transition and are shown in Figure 10–32.

Operating Experience

During testing and validation, two significant issues were encountered and resolved: vibration on front compressor diaphragms and combustor high frequency dynamics (see Figures 10–33 and 10–34).

The front compressor diaphragms on stages 1, 2, and 3 exhibited signs of distress after approximately 200 hours of operation. A root cause investigation identified high cycle fatigue due to high excitation of airfoils combined with residual stresses and high stress concentration factors. A minor design modification was initially applied and validated in the W501G at the City of Lakeland McIntosh #5 site. Additional dynamic strain gauges were applied for monitoring dynamic stresses. The redesign has operated successfully for a total of over 1000 hours in W501G engines. A full cover lift and NDE inspection at the City of Lakeland McIntosh #5 site was completed and no operating restrictions are in effect.

During testing, high combustion dynamics were observed at approximately 2200 Hz causing distress to combustion system components. A root cause investigation

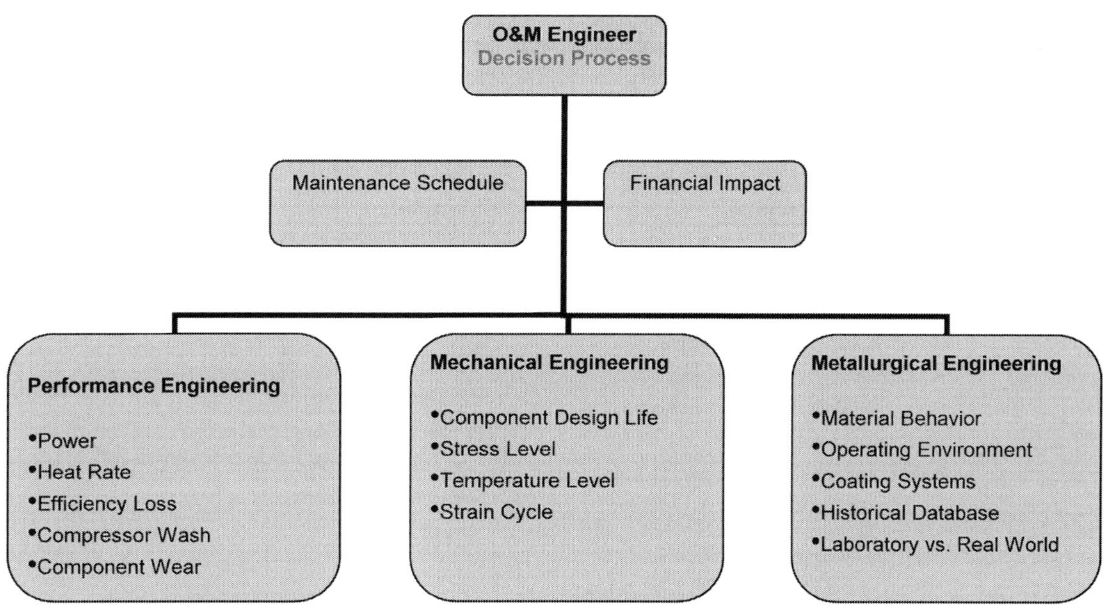

FIGURE 10–33 Engineering roles in operations and maintenance. *(Source: Siemens.)*

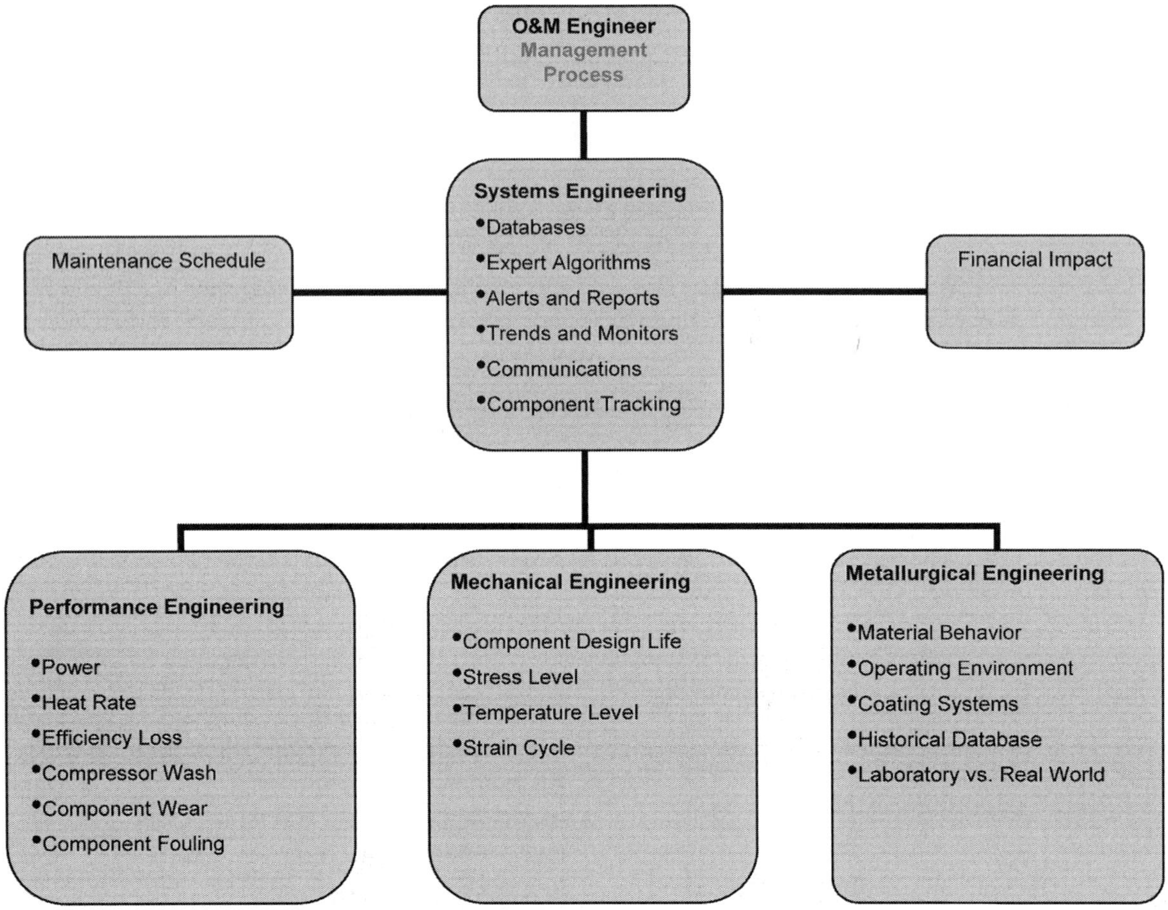

FIGURE 10–34 System engineering approach to O&M. *(Source: Siemens.)*

identified insufficient aerodynamic damping as result of closed-loop steam cooling of the transition. To eliminate the dynamics, resonators tuned for 2200 Hz dynamics were added to the transitions. The resonator design was first tested and validated at the Siemens Westinghouse high-pressure combustor test facility at Arnold Engineering Development Center in Tullahoma, TN. Subsequently, the modified transitions were validated at the city of Lakeland and successfully dampened the combustor dynamics. As an added precaution, a combustor dynamics monitor has been added as standard supervisory instrumentation as part of the digital control system. In the event that combustor dynamics occur, the system will automatically adjust the engine operation to eliminate the dynamics and avoid distress on the components. Validation continues with alternate fuel sources.

Development Activities

Steam-Cooled Vane

Development activities are focused on extending the W501G frame to ATS efficiencies through the introduction of additional technology advancements. The next major step will add a thin-walled, closed-loop, steam-cooled Row 1 turbine vane to the W501G. The steam-cooled vane will extend the benefits of the steam-cooled transition by eliminating most cooling air from the Row 1 vane. The result will be a combination of increased rotor inlet temperature and decreased burner outlet temperature. The benefit will be improved efficiency and reduced NO_x. Single-crystal casting trials have been successfully completed at PCC Airfoils, Inc. in Mentor, Ohio.

The ATS steam-cooled vane will be first tested in an engine sector rig. The test rig consists of a full-scale combustor basket and transition and a 1/16th-sector vessel, which will operate up to full ATS pressures and temperatures. The rig will be located at Arnold Engineering Development Center at the Arnold Air Force Base in Tennessee. The vane will be instrumented to verify analytical predictions of metal temperatures, heat transfer coefficients, and stress.

After validation in the 1/16th-sector rig, the vane will be retrofitted into a W501G. A comprehensive test program will verify vane performance and improved plant performance. Test parameters will include vane metal temperature, stress, and steam temperatures.

Coatings

Thermal barrier coatings (TBC) are an integral part of the W501ATS engine design. A development program is in progress to develop an advanced bond coat/TBC system with a projected service life of more than 24,000 hours. Different bond coats and ceramic materials were evaluated under accelerated oxidation test conditions and down selected. The mechanical integrity and durability of an advanced bond coat/TBC system was demonstrated in more than 24,000 hours of cyclic testing at 1010°C (1850°F).

Under a related program, DOE-Oak Ridge National Laboratory (Contract DE-ACO5-95OR22242), new ceramic compositions and TBC concepts were identified which have a sintering resistance and phase stability superiorted to that of 8YSZ TBC. These compositions and concepts are being further optimized and will be transferred to components for an 8000 hr engine demonstration under a DOE-Chicago contract DE-FCO2-000H11048.

Performance Optimization Case Histories and Discussion

In this book's introduction, the author points out that maintenance, repair, and overhaul (MR&O) works best in planned synergy with engine condition monitoring (ECMS), performance analysis (PA, in some cases managed as a component of ECMS), and performance (recovery or otherwise) testing. One may regard ECMS and MR&O as "chicken and egg" in that either may happen first in a plant's life, depending on the designer's grasp of maintenance philosophies. In Chapter 12, Maintenance, Repair, and Overhaul, the author discusses maintenance philosophies. To reiterate briefly those are

- Reactive (do nothing until failure occurs).
- Preventive (do only what the OEM suggests in his manual, and if it does not work, you can blame the OEM, maybe).
- Predictive, also called *on condition* (keep an eye on the system via a comprehensive ECMS).

Yet another option, of which this author is not fond, considering the power unleashed in the case of catastrophic failure, such a turbomachinery case rupture, is "run to failure."

Nevertheless this has been done by operators such as those in refinery service who intend that their plant be shut down and abandoned when they run out of product to process or incur a failure in a critical plant location, whichever comes first.

In larger organizations today, the ECMS and MR&O functions are typically done by different people. Further, within MR&O, there are maintenance engineers, R&O engineers, and metallurgists. It then becomes critical that these various individuals communicate well. Since they may be working in branches all over the world, the Internet, their intranet, and common file platforms (for shared files) become the grammar for their language.

Performance optimization when retrofit design is involved, is unique to each application. For instance, one might agree that steam injection can (among other benefits) boost the power developed by a gas turbine. However, the actual retrofit is designed for an existing installation. However, once a retrofit is successful, it may be incorporated by the OEM or independent designers into future installations involving the same model of gas turbine and accessory systems.

Most of the gas turbine installations in the world were built before oil soared above $70 a barrel in 2006. So retrofits for performance optimization will probably increase.

The best way to illustrate some of the potential ways gas turbine performance may be optimized is by using case studies. This is one area where they are more suitable than overall general theory, because as the reader will note, different aspects of highly specific theory may be relevant in each case.

Power generation remains and will continue as the largest industry sector in the world for a few decades, perhaps longer. It is therefore appropriate that many of the examples selected belong to that industry. Due to the relatively large real estate involved in large-scale power production, this sector also has the most physical space to add economizers, additional HRSG capacity and so forth.

In summary, the cases deal with different optimization strategies as follows:

Case 2. A Systems Approach to Hot Section Component Life Management. This case discusses extension of hot section component lives and outlines work done by a metallurgical repair facility.

Case 3. Strategies for Integration of Advanced Gas and Steam Turbines in Power Generation Applications. This case was written by an EPC contractor held responsible for the performance of a powerplant at turnover to a client. It discusses ascertaining that the OEM power generation package and components are delivered according to specification and if not, how short they fall as well as "tweaks" to the operational system that can help attain on-performance curve operation.

Case 4. A Study on the Life Cycle Impact of Steam Injection deals with the benefits of steam injection on an LM6000 (a CF6-80C2 aeroderivative). This case is written by a metallurgical repair facility. The LM6000 is different from most other aeroderivatives in that it is controlled not just by low-pressure TIT and compressor discharge temperature.

Case 5. Augmentation of Gas Turbine Power Output by Steam Injection. This case was authored by an OEM.

Case 6. Integrating Gas Turbines in Power and Cogeneration Applications. Written by an EPC contractor, this case discusses primarily "F" class gas turbines.

Case 7. An Integrated Combined-Cycle Plant Design that Provides Fast Start Capability at Base-Load. Written by an OEM, this case is of relevance to power dispatchers. (The simple cycle GT starts faster than a traditional steam turbine because of the boiler startup period.)

Case 8. Challenges in the Design of High Load Cycling Operation for Combined-Cycle Power Plants, This case deals with the same issue as the preceding case, but this is written from an EPC contractor's viewpoint.

CASE STUDY 2: A SYSTEMS APPROACH TO HOT SECTION COMPONENT LIFE MANAGEMENT*

One approach that has been successfully employed to more accurately identify the usable life of turbine blades is to perform metallurgical testing on a representative sample blade during major overhauls. This allows the extent of degradation by oxidation, corrosion, microstructural over aging, and creep occurring under the specific operating conditions of the individual engine to be characterized. On the basis of this testing, the serviceability of the balance of the blade set can be evaluated and an estimate of remaining life can be made.

In instances where unacceptable lives are obtained from blades, the information obtained by metallurgical testing can be used to identify appropriate changes to operating conditions, blade material or coating to extend life. The reparability of unserviceable blades can be assessed on the bases of the same tests.

Oxidation and Hot Corrosion

At the temperatures at which gas turbine blades operate, significant interaction between the blade alloys and the gas environment occurs. In clean gas environments, the principal reaction is with oxygen resulting in gradual oxidation of the airfoil.

However, in environments that contain contaminants such as sulfur, sodium, potassium, vanadium, or lead as contaminants in the fuel or air, a more aggressive form of attack, referred to as hot corrosion, occurs. The main factors affecting the severity of hot corrosion are also influenced by contaminant levels. Thus there can be significant difference in environmental attack in identical blades operating in a different service.

The most important effect of oxidation and hot corrosion on the turbine blade life is the change in airfoil section resulting from attack. The reduction in load bearing area and resulting increase in stress can lead to failure or creep or fatigue.

To measure the loss in section resulting from environmental attack, metallographic sections are prepared from the airfoil surfaces and examined by optical or scanning electron microscopy. The depth of attack includes the thickness of the scale, as well as internally oxidized or sulfidized zones and alloy depleted layer that may have formed.

Microstructural Degradation

Gas turbine blade alloys are primarily strengthened by precipitation of second phase particles of gamma prime (Ni3Al) and carbide phases along grain boundaries (M23C6 or M6C). During exposure at service operating temperatures, these phases are subject to aging reactions that alter their morphology.

In addition, topologically close packed phases, not present in the new alloy, can be formed in certain alloys during service. These changes alter the mechanical properties of the alloy and, thus, can significantly influence the life of the blade.

The creep strength of turbine blade alloys is strongly influenced by the morphology of the gamma prime precipitates. After heat treatment, new blades typically have either a simple structure of gamma prime precipitates of 10–50 nm in size or a duplex structure of large particles of 0.5–2 μm diameter and smaller particles of 10–50 nm size. During service, both types of structure age by an Ostwald ripening mechanism in which large particles grow at the expense of smaller particles causing a decrease in creep strength.

Metallographic examination of samples from the blade airfoil, after service exposure, allows the degree of gamma prime over aging to be determined (Figure 10–35). Because of the small particle size, electron microscopy must be used to image the gamma prime microstructure.

Mechanical Testing

Mechanical testing is frequently performed during metallurgical analysis of service exposed turbine blades to characterize the material properties. The results are largely influenced by the microstructural degradation mechanisms

* Courtesy Liburdi Engineering. Adapted extracts from R. J. Pistor, "A Systems Approach to Hot Section Component Life Management," 03-IAGT-208.

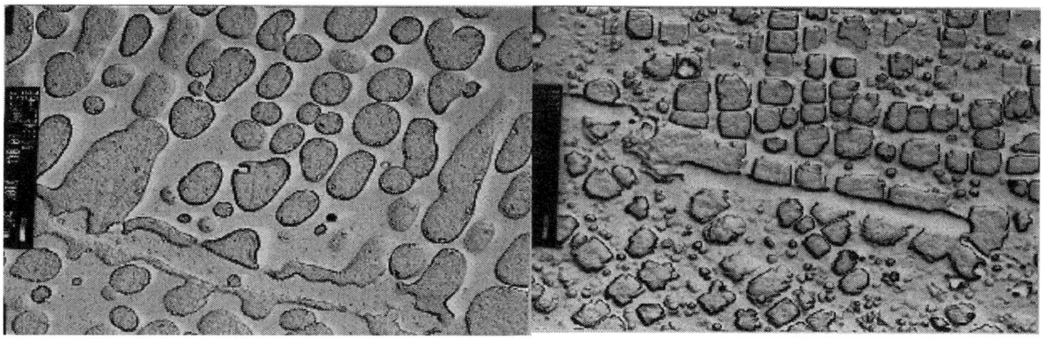

FIGURE 10–35 GTD 11 material microstructure before (left) and after (right) rejuvenation heat treatment. *(Source: Liburdi Engineering.)*

discussed above. However, mechanical test results provide quantitative measure of that degradation and, in addition, can detect some types of damage which are not microstructurally related. Typically creep rupture testing is conducted to provide an estimate of the decrease in safety margin relative to the new part and, by setting appropriate limits, can be used as a serviceability test.

Impact properties can also be of concern in instances where there is a history of FOD failures. In such cases, Charpy V-notch impact specimens are removed from the airfoil and tested at ambient or elevated temperature conditions. The results are not used predictively but are compared to a serviceability criterion arrived at by previous experience.

Metallurgical approach pros:

- The most direct and accurate way to determine remaining life in a component
- A mature O&M strategy that has demonstrated several years of success
- Does not require extensive mechanical design information

Metallurgical approach cons:

- Requires a shutdown and overhaul
- Requires destructive testing of one or more components
- Has limited ability to extrapolate future service if different from past

Mechanical Engineering Approach

In designing hot section components mechanical engineers will target a design life. For example, one may target a 24,000 hour life in a first stage turbine blade. This would mean that the designer would limit the maximum steady and unsteady stresses to that which the material can safely endure for 24,000 hours of normal operation.

Creep damage, coating life, and low cycle thermal mechanical fatigue constraints will dictate maximum allowable stresses and temperatures in the blade. Applying these mechanical constraints on the components during the design phase requires significant modeling of the component and its material system. These models form the basis for design and are used in specifying equivalent operating hour criteria.

FEA and CFD

Finite element analysis and computational fluid dynamics are very useful modeling techniques mechanical engineers employ to design hot section components.

FEA is used to predict the operating stresses and temperature that a component will experience in service. Heat transfer and fluid dynamic boundary conditions determined by CFD are applied to the FEA model. The resulting stresses and temperatures are then entered into a material model to predict the component "design life."

From an O&M perspective the overall system model can be used to evaluate remaining life. Even more useful, these models can be used to predict remaining life based on operating condition. For example, if the user were to operate their components at a different load than design, the life impact of this can be directly estimated with the model.

Long-Term Material Steady Behavior

Such behavior can be characterized using Larson-Miller curves and creep rate equations combined with safety factors to account for the unknowns in the model. The Larson-Miller curve is a generally accepted method used to determine the creep life of a component given time, temperature and stress and is represented in Figure 10–36.

Material creep strain rates are used in modeling the nonlinear plasticity experienced in hot section components. For example, the long-term load shedding of a turbine root form can be estimated to predict a more representative design stress. Material creep strain rates can be represented by the classical power law for creep known as the Bailey-Norton law. The expression for uniaxial creep strain in terms of the uniaxial stress and time is represented in the following equation:

$$\varepsilon c = C0^{*}\sigma(C1)t(C2)e(-CT/T)$$

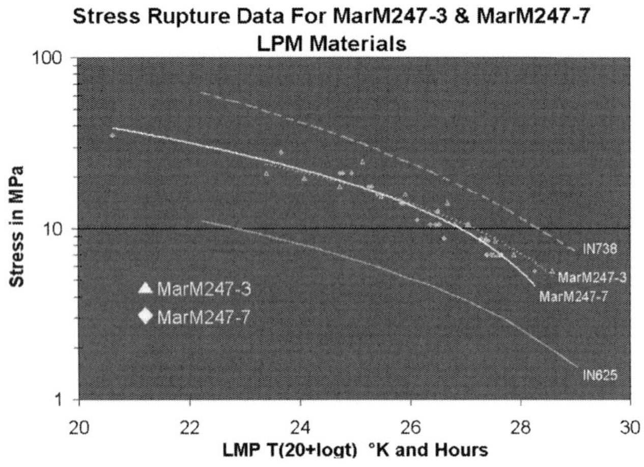

FIGURE 10–36 Larson-Miller plot. *(Source: Liburdi Engineering.)*

where

T = Temperature
σ = Stress
t = Time
C0, C1, C2, CT = Material Constants

Long-Term Material Cyclic Behavior

This behavior can be classified as either high cycle or low cycle fatigue. The classification of fatigue behavior into two types characterizes the different material model damage mechanisms. Low cycle fatigue or LCF deals primarily with 1000s of cycles.

From a practical perspective, LCF is tied with engine starts or load variations which are linking with thermal and stress cycles. High cycle fatigue or HCF results from cycles much greater than 1000s and more likely in the 10^7 range. Practically HCF is associated with cyclic stress loading tied to the engine operating speed. For example, blades rotating past a number of vane wakes will accumulate a number of stress cycles every time the blade passes by during one revolution. Both HCF and LCF are dealt with using different material models.

HCF can be modeled using a Goodman diagram as illustrated in Figure 10–37. A mechanical designer will design the HCF loading of the part to fall in a region on the Goodman diagram where safe operating stresses occur that ensure near infinite operating cycles.

LCF material properties are primarily deduced from laboratory testing. Most of the laboratory testing conducted to characterize the low cycle fatigue resistance of materials is carried out at constant temperature using simple waveforms with steady strain rates. By contrast a startup or trip cycle on a gas turbine hot gas path can generate quite a complex stress cycle with transient thermal loading.

To predict the fatigue life of a component exposed to the more complex cycles based on the lab test results, some simplifications must be made. The first simplification addresses the isothermal nature of most lab testing. The assumption is made that the damage accumulated during the thermal-mechanical fatigue cycles is equivalent to damage generated in isothermal testing at the peak temperature of the thermal-mechanical fatigue cycle.

Secondly, the differences between the complex strain cycle experienced by parts and the simpler cycles used in lab testing must be addressed. Under low cycle fatigue conditions, fatigue life is a function of the strain range (the maximum strain in the cycle less the minimum strain

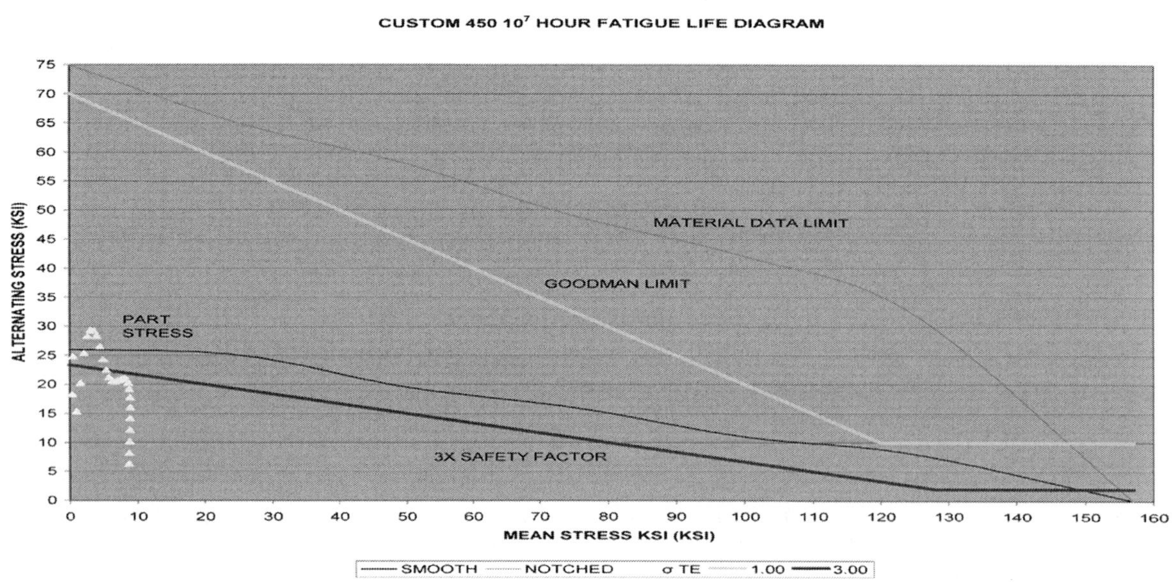

FIGURE 10–37 Goodman diagram. *(Source: Liburdi Engineering.)*

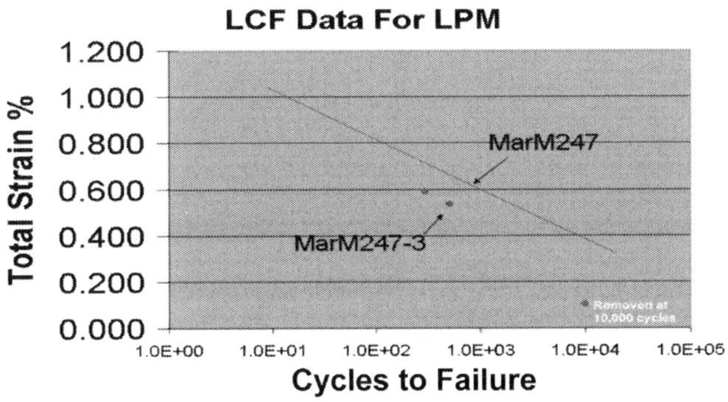

FIGURE 10–38 Typical LCF curve. *(Source: Liburdi Engineering.)*

in the cycle) and test results are most commonly presented as graphs of cyclic life to failure versus strain range.

The complicated strain cycle a component sees in service is thus characterized by the strain range, to predict low cycle fatigue life from simple lab cycles. The predicted life can then be read from the LCF curve for the material at that strain range (see Figure 10–38).

Mechanical approach pros:

- Has the capability to forecast future component life based on load profile
- Is a quantitative method that can be used to set hours and starts
- Models can be used in an optimization scheme to maximize life

Mechanical approach cons:

- Relies on specific design information of component of interest
- Relies greatly on the accuracy of the material models used to predict life
- Relies on the "state of the art" of FEM and CFD computing resources

Performance Engineering Approach

The performance engineering perspective on O&M is completely different from the metallurgical or mechanical approach. A performance engineering evaluation will let one know if the component is behaving the way it did historically in terms of its thermodynamic ability. For example, one could plot gas turbine efficiency over time and track its degradation. Non-recoverable efficiency loss due to erosion, oxidation, wear, or long-term bowing is measured.

Performance engineering diagnostics analyzes the gas turbine performance to try and determine the degradation in axial compressor flow, axial compressor efficiency, turbine throat area change, and axial turbine efficiency. Data can be gathered real-time from available plant instrumentation and entered into a performance analyzer such as GTAP™. In addition to determining component degradations, GTAP™ can be used to evaluate inter-stage gas path conditions like pressure, temperature, and flow.

Under certain O&M criteria, compressor washes or component refurbishments may be implemented once performance levels drop below what is considered acceptable.

Performance engineering pros:

- Can be implemented inexpensively to gather the most out of data that are already being logged
- Can be used to establish maintenance schedules to improve/maintain performance
- Is largely hands free and automatic requiring no machine downtime

Performance engineering cons:

- Will not tell you remaining life information about a specific blade row

Systems Engineering Approach

As demonstrated above metallurgical, mechanical and performance engineering takes a very different approach to O&M. Although different, each discipline can be tied together in one central computing server and provided as a web service to facilitate O&M.

A component tracking and management system or CTMS has been developed to track component data at the serial number level. Its features are listed as follows:

- Designed to track hot section components in and outside of the engine.
- While out of the engine CTMS records repair shop practices implemented by serial number:
    - Dimensional measurements
    - Moment weights (if applicable)
    - Repair level and type
    - Coating
    - Rejuvenation heat treatments and history
    - Life analysis reports
- While in the engine CTMS records by serial number:
    - Accumulated hours
    - Accumulated starts
    - GTAP™ gas path history
    - OEM specified equivalent operating hours
    - Metallurgical/mechanical equivalent operating hours
- Designed with a level of automation that will alarm and notify specified personnel of key indicators such as:
    - Performance degradation (fouling, wash cycles etc.)
    - Equivalent operating hours reaches a specified value

- Step changes in operating performance
- Combustion inspection due
- Blade path inspection due
- Hosted by a central server accessible through a secure web connection

Two potential CTMS application examples follow.

Example 1: Trailing Edge Thinning of Rotating Frame 7EA Row 2 Turbine Bucket. Metallurgical analysis revealed an engine axial inter-granular crack along the span of the trailing edge up to 7/8″ long. Metallographic investigation concluded that a creep related mechanism was responsible for the trailing edge cracking. Additionally oxidation and wear had reduced the thickness of the trailing edge to 76% of the new condition.

Mechanical analysis showed that the strain range experienced during a shutdown and a full load trip was between 0.003 and 0.004. The boundary conditions that resulted in this strain range were determined from a transient performance analysis which provided the gas path transient temperatures, pressures and flows. The LCF life was predicted to decrease significantly with reduced trailing edge thickness. So much that a 25% thickness reduction limit was set for the bucket's trailing edge before retirement from service.

As one can see, each analysis on its own could not establish a minimum wall thickness criterion. Metallurgical analysis revealed the type of failure. Mechanical evaluation predicted strain ranges using the result of performance engineering transient gas path analysis. By combining the results of the three disciplines, a retirement criterion has been established. To implement this in practice a systems health monitoring solution would track each bucket by serial number in a database. This database would contain a remaining life quantity for each serial number. GTAP would be added to the health monitor to track gas path conditions to calculate remaining life in the part based on the mechanical engineering models described above. This remaining life number would be updated periodically in the database and when it reaches a certain number, alarms and notifications would be given by the health monitoring system to the O&M engineer.

Example 2: Over-/Under-Firing a Frame 3 Gas Turbine. In some cases when the market conditions are right an operator may elect to over-fire their gas turbine to get additional capacity out of the unit during peak operation. The economic conditions in certain peak hours can be so advantageous that over-firing is very attractive and the consequential impact on maintenance needs to be estimated. Conversely there are cases where a reduced firing rate would substantially offset the maintenance cost to make it economically viable. These decisions can be made with a greater degree of confidence if a systems approach to O&M were implemented.

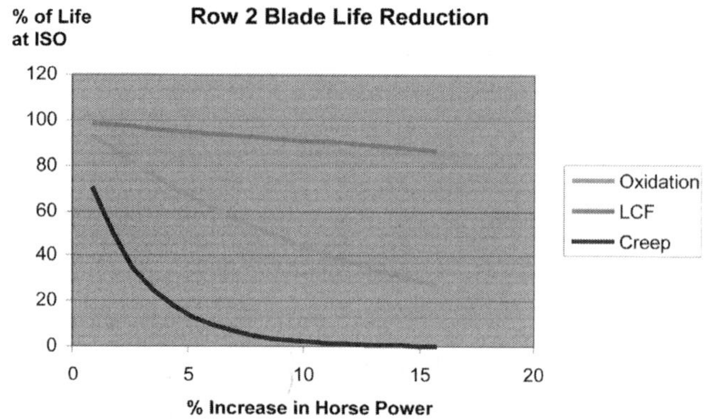

FIGURE 10–39 Impact of over-firing on the Row 2 bucket. *(Source: Liburdi Engineering.)*

By archiving and tracking component remaining life in a database, the life impact of over-/under-firing can be managed and optimized. Metallurgical and performance history will dictate the life consumed thus far and forward looking mechanical analysis can predict the impact of the load changes on the remaining life. Figure 10–39 shows the impact of over-firing on the Row 2 bucket as a function of the original life and % over fire. The integral of this curve could be taken to estimate the remaining life in the bucket and then be used to optimize load vs. economics.

The next case is another example of industry working in concert with and benefiting the practical education within academia. (See Chapter 17, Training and Education.)

CASE STUDY 3: STRATEGIES FOR INTEGRATION OF ADVANCED GAS AND STEAM TURBINES IN POWER GENERATION APPLICATIONS*

Combined cycle power plants (CCPPs) using fossil fuel generate the cleanest and most efficient form of electrical power. CCPP technologies have evolved significantly in providing better, more cost-effective products: gas turbines (GTs), steam turbines (STs), heat recovery steam generators (HRSGs), heat sinks, pollutant removal technologies, balance of plant (BOP), water treatment and fuel treatment equipment, etc. A major reason for these improvements was the introduction of G, H and now J technologies for gas turbines, in which an inseparable thermodynamic and physical link was created between the primary and secondary power generation systems by using steam instead of air, in a closed loop to perform most (or all) turbine cooling activities.

A successful and reliable operation can be achieved only when all the components listed above are harmoniously

* Extracts from "Strategies for Integration of Advanced Gas and Steam Turbines in Power Generation Applications," GT2007-27979 Courtesy of J. Zachary.

integrated. This section describes the challenges of the integration process from an engineering, procurement and construction (EPC) contractor's perspective.

Air Permit

To get a head start on the increasingly daunting and lengthy permit process for a domestic power plant, prospective plant owners seek to obtain the services of an environmental consultant well before selecting the EPC contractor. In recent years, the permitting process in the US has become increasingly complex, requiring the involvement of several entities capable of identifying and quantifying the effect of the environmental commitments on overall project success.

A good approach is to involve the EPC contractor early. The contractor brings a larger view based on proven project designs, its database of lessons learned from other advanced equipment projects, and its startup/operating experience—a perspective that could closely reflect the owner's continuing objectives over the life of the plant.

Improving the Permitting Process

Since the air permit is usually secured as quickly as possible, "ideal case" assumptions are sometimes made. The concern is that key factors—project design, fuel selection, type of operation, back-end equipment selection, and performance—may be determined upon submission of the air application even if all pertinent information and inputs are not available. While this tactic is useful, it is invaluable for projects having advanced GTs as prime movers.

The process definitely requires a team effort. The environmental consultant provides valuable input to the permitting strategy, giving the owner and EPC contractor broader project perspective. Both owner and contractor will ultimately need to consider issues such as constructability, initial compliance testing, long-term emission compliance, and operability and maintenance. Consequently, the process should be structured and comprehensive, encompassing the following:

- **Detailed characterization of the fuels.** This characterization provides the basis for sizing the gas turbine and determining the amount of supplementary firing, which ultimately define plant performance. This is followed by selecting the type and capacity of the pollution control equipment, performing steam cycle analysis, and determining heat and material balances necessary to generate pertinent emissions data for the air permit application.
- **Development of a site-specific arrangement and characterization of pollution sources.** If this arrangement is developed without a detailed design, the resulting layout is likely to represent a configuration that is unworkable from a design, construction, and operations point of view.

Examples of key site arrangement issues include:

1. Power block arrangement and dimensions
2. Main stack height location and base elevation
3. Stack parameters of emissions sources (stack exit diameter, velocity, height, temperature)
4. Size and orientation of cooling tower
   - Definition and evaluation of supplier emissions guarantees.
     The air permitting process also needs to be supported by credible emission guarantees from major equipment (e.g., GT, HRSG, auxiliary boiler) suppliers with experience and technical expertise in providing systems capable of demonstrating compliance within the proposed emission limitations. In many cases, the technical information from the suppliers is assessed using a model, which facilitates critical evaluation of the component performance by pollutant.
   - Definition of the startup/shutdown processes.
     Increasingly, air permits include implicit or explicit emission limitations for the startup and shutdown periods. Although these emissions are often exempt from emission compliance determinations, many state permit writers include specific compliance conditions for the startup and shutdown periods. In some cases, these conditions address warm and hot startups and the number of annual starts. In other cases, the emission limits cover all periods of operation (i.e., with no exclusions for startup or shutdown) or define specific startup/shutdown emission limits (lb/hr, lb/MMBTU). In the current market, where merchant power plants need to account for daily starts and stops, such restrictions could directly influence the profitability of the operation.
     Therefore, it is essential to adequately characterize these special modes of operation and consequently determine the emission and load profiles. Limitations on startup emissions will also affect the specification of major plant equipment (i.e., GT, ST, HRSG, and auxiliary boiler).
   - Assessment of the impact of monitoring uncertainty.
     In the past, measurement uncertainty was not an important issue, in part because compliance limits were large enough to make such margins less significant. The measurement methodology for emissions values was sufficiently accurate that both federal and state regulators considered stack and continuous emissions monitoring to be "presumptively accurate." However, the influence of these

protective factors has disappeared. Today's ultra-low emission limits (below 2 ppm) especially for NO_x and CO, have almost eliminated any compliance margins. Since the uncertainty is a relatively larger percentage of the overall limit, actual equipment emissions must be lower than the permit limits for the plant to remain in compliance.

Gas Turbines

Equipment Selection

Before selecting the equipment from different suppliers, a thorough investigation should be conducted to ensure that the owner's pro forma objectives for power output, heat rate/efficiency, exhaust energy, emissions, reliability, and availability are met.

The information in the database is constantly updated with results from field tests. Figure 10–40 represents a selection of 37 GT units recently tested.

When establishing the HRSG and ST design conditions, it is a good engineering practice to allow for some variability of GT exhaust flow and temperature values. This way, the design could accommodate either shortfalls or better-than-guaranteed exhaust energy (flow and temperature). Thus, more realistic and competitive values can be predicted for the CC performance.

Our experience indicates that owners/operators are well advised to engage an experienced and bankable EPC contractor in purchase or reservation agreements for GTs or power islands. As mentioned above, such collaboration becomes crucial for projects involving advanced GTs. The EPC contractor can verify that terms and conditions essential to managing project design, construction, and commissioning are adequately covered in these agreements. An EPC contractor, with direct experience with this type of GT, can also work with the equipment supplier and the customer to ensure that the scope is complete and all interfaces are well defined. Areas where the EPC contractor can add value to an owner's reservation agreement include:

- Establishing adequate coverage of performance test tolerances and measurement uncertainty
- Addressing the impact of GT performance offsets (between power output and exhaust energy) on HRSG and ST sizing
- Ensuring pollutant levels and units included in the plant permits and GT supplier guarantees are consistent
- Evaluating technology risk issues (e.g., use of first-of-a-kind technologies and impact of the continuous design improvement process)
- Assessing the effects of cycling operation on the plant components and their interaction

Gas Turbine Startup and Commissioning Issues

EPC contractors are responsible for starting up a CC plant and incur significant penalties if successful, timely operation is not achieved. Therefore, EPC contractors must evaluate the cost-effectiveness of approaches that facilitate rapid startup. The problem becomes even more critical for merchant power plants, where cycling and part-load operation are often required. Depending on specific completion requirements and previous experience with a particular type of advanced GT, the startup schedule must allow time for unforeseen events. Some examples of these include:

- **Dual-fuel operation.** The design of an advanced GT focuses on using natural gas as the primary fuel. Dual-fuel capacity with oil as an alternative fuel is an option that attracts many owners because it can produce power when natural gas is not available. Dual-fuel capability adds even more complexity to already very

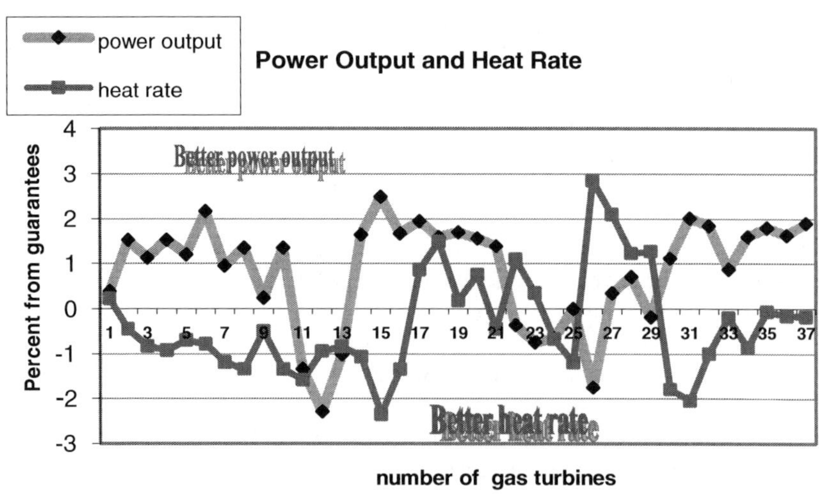

FIGURE 10–40 Comparison between guarantees and test results for GT power output and heat rate.

complicated combustion systems and controls. A difficult challenge is to achieve a switchover from one fuel type to another at a reasonably high power level. For some manufacturers, this activity took longer than expected and affected the commissioning schedule.

- **Implementation of modifications in the field.** Several critical advanced GT integration lessons learned are related to field modifications. On many occasions, unscheduled outages are used not only to correct a problem but also to implement a number of changes based on experience accumulated from other sites. Significant modifications incorporated include not only physical changes in the hardware but also control software. This process creates a "ripple effect" requiring additional changes in the complete plant control software and start sequences. Managing these changes is critical to maintaining schedule on advanced GT projects.
- **Combustion system commissioning and tuning.** To meet the strict emissions requirements, all advanced GT combustion systems operate with dry low-NO_x (DLN) combustion systems. Combustion system operation from diffusion mode at low loads to full premix mode at base load takes place in several complicated steps and stages, requiring very close control of the fuel flow and exhaust temperature. The process is sensitive to ambient conditions, combustion-associated instabilities, and even manufacturing or assembly tolerances. Currently, each GT is individually adjusted to meet the performance guarantees and emissions requirements without combustion oscillations. This practice has become a standard feature of GT commissioning.

However, this activity impacts the EPC contractor. The execution schedule is extended to perform a water wash of the compressor before the tuning process and to install and remove temporary instrumentation for full-blown performance testing of the GT. Because emissions limits must be met at all ambient conditions, adjustments made in the field might modify the performance correction curves for ambient temperature.

Additional Gas Turbine Lessons Learned

Advanced GTs with DLN combustion technology may require the ability to both heat and cool fuel gas during startup. As GTs proceed to higher outputs and efficiencies, the fuel gas supply pressure requirements typically exceed 30 barg and can sometimes exceed 35 barg for can-type combustion systems.

These supply pressures often require supplemental fuel gas compression. During startup at low fuel flow rates, operation of supplemental compressors can result in significant increases in fuel gas temperature due to compression heat. However, DLN combustors typically require "cold" fuel for initial startup. As a result, a startup or pilot cooler may be required downstream of the supplemental fuel gas compressor. Later on, as the GT ramps up in load, heated fuel would be required for full pre-mix operation and the cooler would need to be isolated or bypassed.

On projects with widely varying fuel gas supply pressures, supplemental compressors can often be bypassed, even during startup. Therefore, both a dewpoint heater and a startup/pilot cooler would be required for the fuel gas supply system to meet startup fuel gas requirements under all scenarios.

Steam Turbines

In the last 10 years, STs in CC applications have evolved significantly from small 80 MW, two-pressure, non-reheat configurations to large multiple admission pressure reheat turbines with outputs reaching the 350 MW.

Significant differences exist between STs designed for CCs and those designed for conventional Rankine Cycle (RC) applications. Firstly, feedwater heaters are not normally used in the thermal design of the bottoming cycle for a CC. Secondly, because steam is admitted from several points in the HRSG, heat extraction from the GT exhaust energy is maximized. For the same high-pressure (HP) main steam flow, the CC low-pressure (LP) exhaust steam flow, when compared with the RC LP steam flow, can be up to 35 percent higher. Thirdly, CC designs use duct firing to compensate for reduced GT output at high ambient temperatures, which coincides with maximum summer demand for power. In the US, it has also become quite common to almost double the ST output by using massive amounts of supplementary firing to capitalize on peak summer demand.

Finally, competitive market pressures have pushed suppliers to offer a compact plant layout with axial steam exhaust using only two standard cylinders, reducing the cost of manufacturing and installation as well as the erection schedule. Although all ST manufacturers typically employ a modular building block system with standardized and corresponding components, during the plant design process, it is recommended to verify that the steam blade path should be modified. If an EPC contractor is the CC developer, it might use equipment from different manufacturers, thus requiring further optimization.

Another example of optimization is operation with substantial supplementary firing. Due to currently high natural gas prices, even small blade path changes can translate into operational savings. The customization of the LP steam turbine last stage path system is an additional case in which a "standard design" might not be the most suitable for the plant heat sink, especially for 2 × 1 and 3 × 1 multi-shaft CC configurations.

Another decision needed on most cogeneration projects is whether to design the ST and associated steam cycle equipment for a case in which no steam is required for the process. In this scenario, a larger LP section and an increased downstream electrical equipment (transformer, isophase bus, etc.) capacity are required. In this decision process, economic benefits of the extra power availability and operational constraints due to possible infrequent occurrences of such conditions should be carefully considered. The selection process should be transparent and capable of evaluating not only the technology risks associated with the equipment itself but also the integration risks.

Increasing the last-stage exit area is a key factor in reducing plant capital cost and minimizing exhaust losses. For cycling plants, however, more work is necessary to demonstrate that the efficiency as well as aero-mechanical behavior during part-load operation are not much worse than in the conventional design.

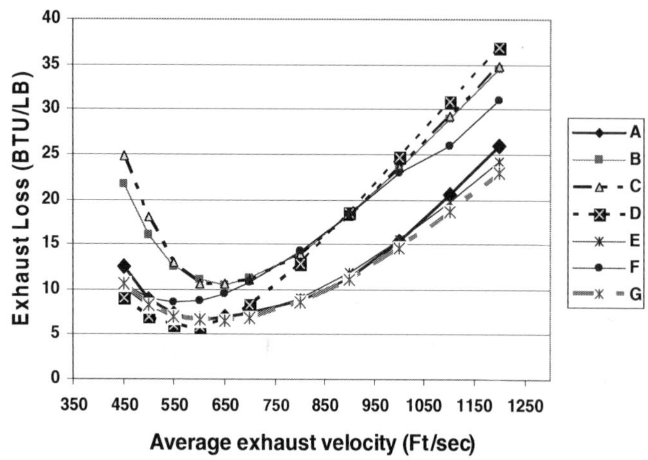

FIGURE 10–41 Typical exhaust loss curves from different manufacturers.

Exhaust Loss Curves

One of the important features of STs used to integrate other components and, in particular, optimize the heat sink is the exhaust loss curve. The ST thermal kit provided by the manufacturers to plant designers contains information on LP exhaust losses in the traditional form of an "exhaust losses curve." This curve gives the specific enthalpy loss for an exhaust average steam velocity. The steam velocity value depends on the back pressure and/or steam flow. The use of this curve requires corrections for moisture content. Since the average steam exhaust velocity is a function of the volumetric flow for a given geometry, its value depends on both back pressure and steam mass flow. Figure 10–41 provides a number of exhaust loss curves for various exhaust areas. The increased exhaust loss at lower exhaust velocities (less than 550 feet per second [fps]), occurring at low part-load conditions, is due to the formation of a reverse vortex at the blade root.

Development of new generation and often non-conventional profiles, with large variations along the blade height, has produced considerable benefits in terms of the stage's efficiency and reduced exhaust losses. Figure 10–42 presents the exhaust curve loss for blades from two manufacturers that have almost identical exhaust

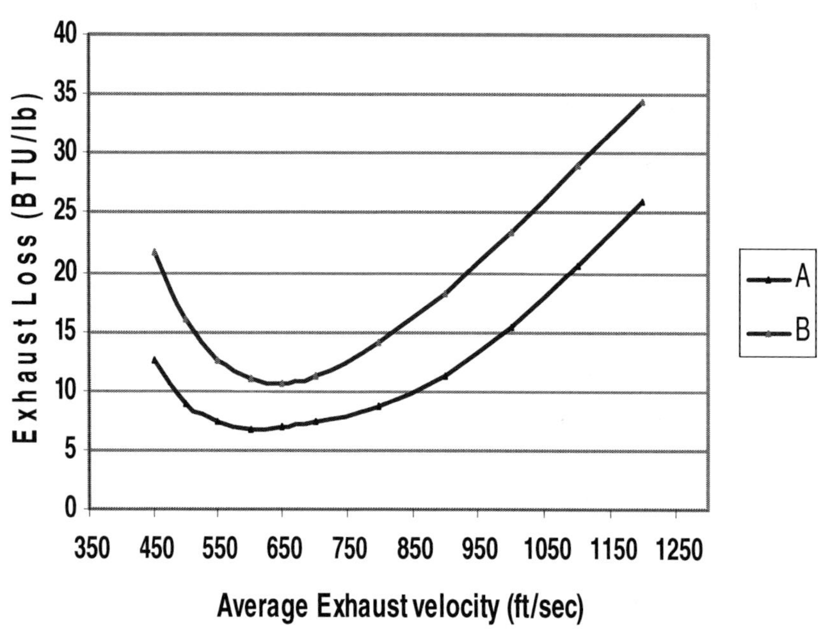

FIGURE 10–42 Exhaust loss curves from two manufacturers with very similar exhaust area and blade length.

and blade lengths. It can be concluded that a more advanced 3D profile design (Curve A) leads to a lower exhaust loss. Since the total exhaust loss curve also includes the contribution of the exhaust hood, its design should be aerodynamically effective.

Steam Turbine Startup and Commissioning Issues

ST startup flexibility and commissioning time play a significant role in startup of the entire CC plant. Due to higher fuel costs and increased electrical reserve margins, CC plants are dispatched as intermediate-duty units rather than base load units, as originally envisioned. Achieving the goal of a fast and reliable startup requires careful design and integration of the ST, GT, and BOP requirements. On one hand, the ST supplier should provide more flexible ST startup parameters (such as greater steam temperature mismatch and more relaxed steam purity) while maintaining reasonable constant life consumption, controlling low cycle fatigue by monitoring the maximum wall temperature differences and permissible ramp rates. On the other hand, an EPC contractor can employ the entire arsenal of auxiliary equipment available to assist in the process. Examples of such measures include means to improve the heat retention after shutdown, design of advanced water treatment systems capable of achieving steam purity more quickly, provisions for additional warmup lines, and use of an auxiliary steam boiler to reach desired condenser vacuum more rapidly.

Additional Steam Turbine Lessons Learned

Some GT/HRSG/ST combinations are better suited for rapid cycling than others. It is not possible to accurately predict CC power plant startup times from the typical startup curves for each piece of equipment. Each major equipment vendor will make assumptions about startup conditions that are most favorable for that vendor's individual piece of equipment. These assumptions are usually different, often conflicting, and sometimes incompatible. The overall startup integration must be considered early in the project, and consistent requirements need to be provided at the bid phase for each vendor. The drive to increase ST efficiencies results in tighter clearances to reduce leakages, but these usually require longer startup times to avoid rubbing and reduce thermal stresses. Stringent steam temperature matching requirements may also be requested. These requirements vary between ST vendors and cannot always be met by every GT/HRSG combination. The GT operating at low loads has a limited capability to maintain high exhaust gas temperature. This capability varies significantly between manufacturers.

Combining a GT with limited ability to control exhaust gas temperatures at low loads with an ST that has very strict temperature requirements at startup can lead to unacceptably long startup times (greater than 8 hours) for cycling units. The HRSG may not be capable of accommodating the GT and ST requirements simultaneously without significant bypass capability. Although this scenario might seem unlikely to occur, Bechtel has witnessed this circumstance more than once, when owners procured each piece of major equipment separately, considering only performance and price with no attention to integrated startup requirements.

Cycling Considerations

The merchant plant concept implies that electric power must be supplied to the grid only when it is commercially justifiable. Such operation requirements must typically be met on very short notice. Therefore, these plants need to be started up quickly and must have flexible operating ranges. Merchant plants are normally in cycling service, which can be considered part-load operation or daily on/off duty. Heat sink considerations for a cycling plant include the following:

- Use an auxiliary boiler to sparge the condenser hotwell to prevent the condensate from subcooling
- Maintain condenser vacuum during periods of shutdown
- Use a cooling tower bypass for low-load operation
- Use pre-coat condensate treatment (condensate polisher) with air-cooled condensers, which may have large carbon steel surface areas in contact with steam and condensate
- Include provisions to control all heat sink fans from the control room. For cogeneration plants, the quality and quantity of water returned to the power plant from the process must be considered. Depending on the quality of the returned water, further deaeration and chemical treatment may be necessary. If not all the exported water is returned, additional cycle makeup will be required. An external deaerator will be required if the total cycle makeup exceeds 3 to 5 %.

Balance of Plant

Balance of Plant Equipment Selection

BOP equipment for a CCPP includes boiler feedwater pumps, condensate pumps, circulating water pumps, closed cooling water pumps, air compressors, fuel metering equipment, etc.

Proper system integration of BOP equipment also requires that interconnecting piping/support systems be optimized to the power island (GT, ST, HRSG, heat sink). Based on the arrangement of the plant and amount of BOP equipment, the engineering evaluation will start by determining the amount of circulating water, steam, air, and fuel required to operate the plant at the expected power output. The optimization of sizes, material, layout, and cost of BOP equipment and critical piping is no less important to the performance and cost of the entire plant than the power island major equipment. One of the important lessons learned was to review and continuously validate the specifications for each of the major BOP systems (main steam, feedwater, condensate, and circulating water) to rapidly respond to new requirements and incorporate the experience gained from previous projects.

TABLE 10–8 List of Projects with Bechtel Participation

No. GTs	Type of GTs	Manufacturer	Country
18	7FA	GE	USA, Mexico
13	9FA	GE	UK, Turkey
4	GT26	Alstom	UK
6	V94.3A	Siemens	Netherlands, Egypt
4	W501G	Siemens	USA
9	W501FD	Siemens Westinghouse	Brazil, Mexico, USA
2	M701F	MHI	Egypt
56 Total			

Auxiliary Loads

A critical step in each project is to assess, with a high level of accuracy, the plant's total auxiliary load at given guaranteed conditions. Auxiliary loads, representing the amount of power consumed by each piece of equipment during normal plant operation, can vary significantly due to the type of equipment and its configuration. The theoretical estimation at the beginning of the plant design must also be confirmed during performance testing. This process requires a careful engineering evaluation of the equipment in operation and the utilization factor to be applied to each piece of equipment.

A good indication of a successful design is to achieve a close match between the auxiliary load calculated and the actual values measured during the test. By conducting upfront detailed engineering to obtain actual pump performance curves and using a database of past projects' auxiliary loads (both calculated and measured), Bechtel was able to predict with a very high degree of accuracy the auxiliary loads for the projects associated with the turbines presented in Tables 10–8 and 10–9.

Material Selection

Tailoring material selection to the specific operating conditions of the power plant reduces the cost and ensures proper plant operation. Over the years, Bechtel has developed a material selector guide that assists the design engineer in selecting the proper material and in matching its properties to the required design parameters.

One interesting lesson learned relates to the selection of a high temperature pipe material. In the initial stage, the selected material was adequate and met the operational requirements. However, a better material, which exceeded the requirements, was found to be less expensive because the pipe diameter could be smaller and a standard schedule size could be used instead of minimum wall pipe. In another example of tailored material selection, metal pipe was replaced with high-density plastic in a low-pressure and temperature application, resulting in substantial savings.

Water Chemistry

Controlling the amount of dissolved oxygen in the condensate is critical, especially during startup. The most critical operational backstop to prevent dissolved oxygen (DO) from entering the feedwater is to avoid breaking condenser vacuum overnight (or ever). Maintaining condenser vacuum and preventing the hotwell from subcooling dramatically decreases the morning startup time and almost eliminates the huge swings in chemistry associated with cold startup. Following are lessons learned on how to minimize corrosion potential due to DO:

- **Use an auxiliary boiler.** For units that shut down completely at night, an auxiliary source of steam is used to maintain steam to the ST seals, thereby sustaining condenser vacuum. This auxiliary steam source is also used to sparge the condenser hotwell to prevent the condensate from subcooling and picking up DO. Additionally, the condenser sparging system is used during part-load conditions to prevent subcooling of the hotwell. The auxiliary steam source also minimizes the amount of makeup demineralized water required to support operation of the sky valves.
- **Design the condenser vacuum system to utilize LP motive steam to power the steam jet air ejectors (SJAEs).** If using vacuum pumps, verify that the vacuum pump cooling water supply is cold enough to stay ahead of flashing due to low vacuum conditions. A hybrid system can also be used (an arrangement where vacuum pumps take suction from SJAEs).

TABLE 10–9 STs from Major Suppliers

Supplier	No. of Units	Nominal Power (MW)	Characteristics	CC Configuration
Alstom	7	250		2 × 1
Alstom	12	85/155	Unfire /duct-fired	1 × 1
Alstom	1	300		3 × 1
GE	1	300		2 × 1
GE	2	250		3 × 1
GE	1	150		1 × 1
GE	1	120	Desalinization noncondensing unit	1 × 1
GE	3	200		2 × 1
Hitachi	2	250		2 × 1
MHI	2	100	Cogen with large process steam	2 × 1
SWPC/MHI	1	120		1 × 1
SWPC	4	150		1 × 1
SWPC	1	295		3 × 1
Toshiba	1	300		3 × 1
Westinghouse	1	230		2 × 1
Total	40			

- **Consider putting the ST gland steam condenser on the closed cooling water loop.** Doing this will eliminate the need to operate the large condensate pumps during shutdown.
- **Always maintain a positive pressure in the HRSG drums during shutdown.** This is accomplished by closing the stack damper and all outlet valves to the HRSG after the GT is secured. At least approximately 8 to 16 hours are needed to cool the HP drum to saturated conditions of 400 to 500°F. After the boiler has cooled to within the range of the auxiliary steam source pressure, open the HP evaporator spargers and keep the drum as close as possible to drum soak conditions. This would constitute a warm start of the HRSG with no ramp-up restrictions. The heat migration from the HP section to the IP and LP sections of the boiler will tend to keep those drums at significant pressure. The LP drum must be vented throughout the night to prevent overpressurization.

If the HRSG is taken off line for maintenance under cold conditions, apply and monitor a nitrogen blanket above all water spaces and follow the manufacturer's recommendations for chemical dosing of the drums. As the steam space pressure decreases, align the nitrogen system to push and replace the steam, which will minimize air in-leakage.

CASE STUDY 4: A STUDY ON THE LIFE CYCLE IMPACT OF STEAM INJECTION*

It is commonly believed that steam injection into a gas stream will cause an increase in heat transfer coefficient due to increase of the heat conductivity and specific heat. The criterion that sets the maximum recommended maintenance interval in industrial heavy-duty applications is based on baseload continuous operation with natural gas and no steam or water injection. For operation that differs from the baseline, maintenance factors are established that determine the increased frequency of maintenance required. Water or steam injection is one of the key factors in determining the maintenance interval requirement.

* Source: [10-1] A Study on the Life Cycle Impact Of Steam Injection, Courtesy of Liburdi engineering. Note: *As with other cases through this book, the purpose of a case is to give the reader the highlights of specific operating and/ or design experience. If further information is required, the reader ought to then consult the original source of the case work.*

The impact of steam injection on the hot section component life of an aero-derivative LM6000PC gas turbine has been studied. Specifically, HP turbine stage 1 blade was selected as the most representative hot section component in relation to expected change in component temperature, degradation rate and serviceable life produced by the introduction and variation of the amount of steam injection.

Liburdi Turbine Services were engaged by Mighty River Power Limited of New Zealand to assess the estimated impact of increased steam injection on the LM6000PC life cycle; this section is based on that work.

Background

The industrial LM6000 gas turbine is derived from the General Electric CF6-80C2 high bypass turbofan aircraft engine. The twin spool LM6000 consists of a five-stage LP compressor, a 14-stage HP compressor, a two-stage, air-cooled HP turbine, and a five-stage LP turbine. The LM6000 does not have an aerodynamically coupled power turbine. A drive flange is available on both LP compressor and LP turbine, offering the option of either cold end or hot end drive. The LM6000PC uprated from the previous "PA/PB" model was developed in 1995. The first production unit was built in 1997. The "PC" model incorporates efficiency improvements in the LP compressor and a larger exit area in the LP turbine. The standard material for the HP turbine stage 1 blades is the Rene-142; a high strength nickel-based directionally solidified superalloy. Single crystal superalloy Rene-N5 is also used as the material upgrade. The overall pressure ratio for the LM6000PC is 29.5 to 1.

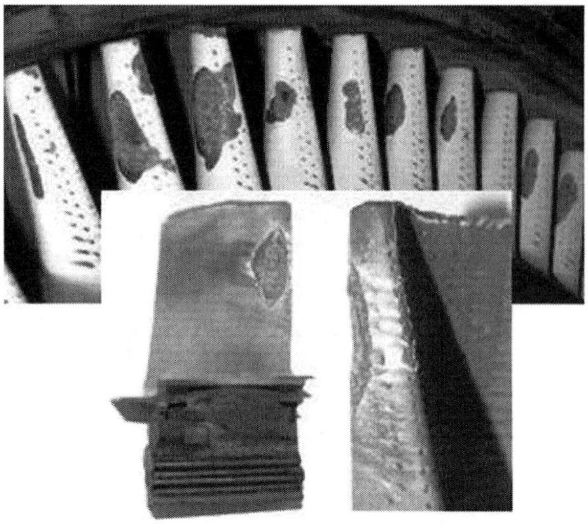

FIGURE 10–43 Coating loss and oxidation damage [10-1].

A review of repair history of the stage 1 blades revealed coating losses on the suction side leading edge at low times (Figure 10–43). Metallurgical analysis identified the most likely root cause as oxidation spallation of the thermal barrier coating (TBC).

Figure 10–44 shows the blade internal cooling design. The cooling air is introduced at the blade root and exit at the blade tip. The extracted cooling airflow rate and temperature were outputted from the engine aero-thermal program. Heat transfer correlation for fully developed turbulent internal flow was used to calculate the internal heat transfer coefficients. The internal heat transfer coefficients were found to increase with increased steam injections (Figure 10–45). This can be explained by the fact that

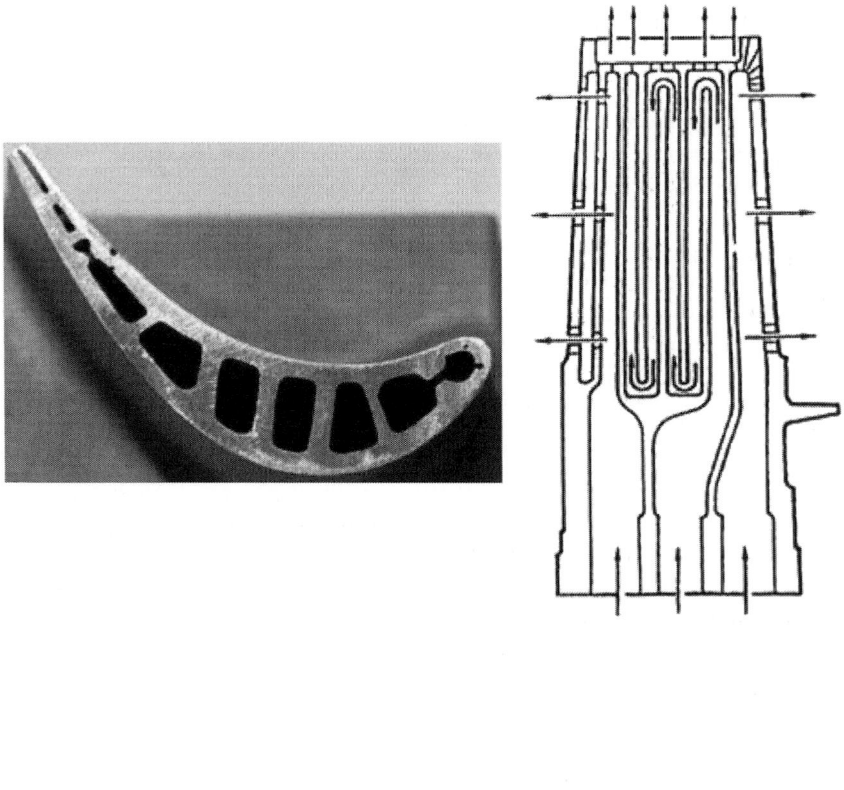

FIGURE 10–44 LM6000 HP stage 1 blade internal cooling circuits [10-1].

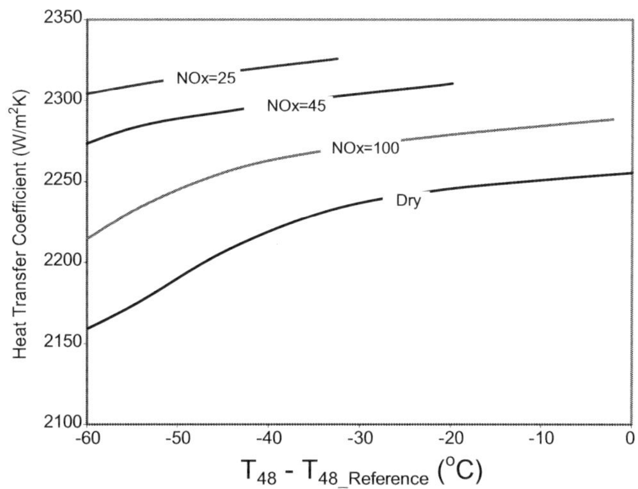

FIGURE 10–45 Cooling-side heat transfer coefficient [10-1].

steam injection increases HP compressor discharge pressure resulting in more cooling flow extraction.

Discussion

All turbines, including aeroderivatives, have "base ratings." In the case of aeroderivatives, when natural gas is used as the fuel and the engine is operated at the base-load power turbine inlet temperature control setting, its base rating corresponds to a hot-section repair interval of approximately 25,000 hours. In this analysis, the calculated blade metal temperature at the design base-load condition was used as the baseline reference. Figure 10–46 shows the stage 1 blade local metal temperature versus steam injection. It shows that for constant LP turbine inlet temperature control, the blade metal temperatures increase with increasing steam injections.

Part life impact from steam or water injection is also related to the way the engine is controlled. Most aeroderivative gas generators are controlled by the constant power turbine or LP turbine inlet temperature. Heavy-duty industrial gas turbines are in general operated with the constant firing temperature by means of a linear relationship between exhaust temperature and compressor pressure or pressure ratio. The control system on base-load application reduces firing temperature as water or steam is injected. This counters the effect of the higher heat transfer on the gas side and its impact on blade life. If the control system is designed to maintain firing temperature constant with steam injection level, this would result in additional output but the part life consumption would be accelerated. Unlike most aero-derivative gas turbines, the LM6000 is controlled by the LP turbine inlet temperature (T_{48}) as well as the HP compressor discharge temperature (T_3). With increased steam injections, engine operation would be primarily limited by the HP compressor discharge temperature. In this case, fuel flow (firing temperature) would be regulated by the engine control system to operate within the maximum allowable HP compressor discharge temperature and pressure limit.

Figure 10–47 shows the predicted stage 1 blade oxidation life change with steam injection. With the constant LP turbine inlet temperature, a 3.3% steam injection (25 ppm NO_x) increases the hot-gas thermal conductivity by 3.5%. This would cause a 6.2% increase in the gas-side heat transfer coefficients. It was found that the internal cooling-side heat transfer coefficient was also increased by 3.4%. The increased heat transfer led to an average 6.6°C (12°F) increase in the local metal temperature resulting an approximate 20% reduction in the part life.

The predicted life impact of steam injection in general agrees with the reference (Figure 10–48). In an example

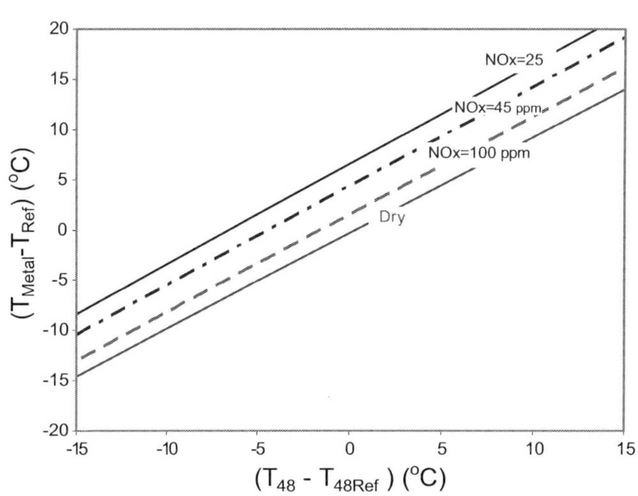

FIGURE 10–46 Predicted blade metal temperatures [10-1].

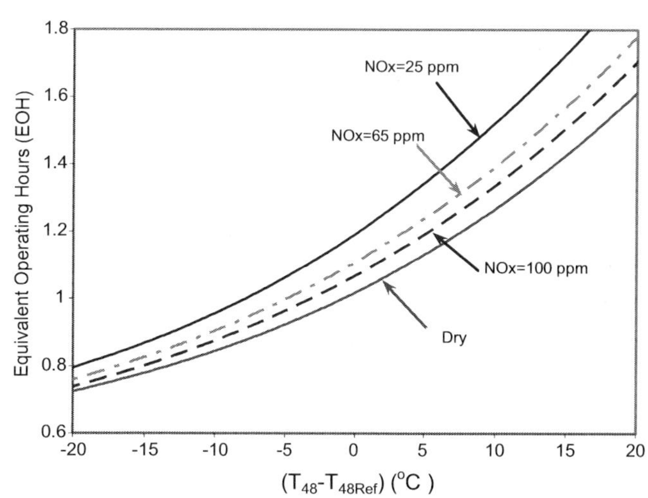

FIGURE 10–47 Predicted blade oxidation lives with steam injections [10-1].

FIGURE 10–48 Steam injection and blade life [10-1].

LM6000PC HPT1 Blade Analysis

	DRY	25PPM	DIFF
Gas Transport Properties			
k - Thermal Conductivity	0.0488	0.0505	+3.48%↑
Cp - Specific Heat	0.288	0.288	↔
μ - Viscosity	0.1208	0.1205	↔
Steam Injection			
- O2	0.154	0.144	
- CO2	0.052	0.053	
- H2O	0.043	0.079	+3.6%
- SO2	~	~	
- N2	0.74	0.71	
- AR	0.013	0.012	
Heat Transfer Coefficients			
- Gas Side	352	374	+6.16%
- Cooling Side	396	410	+3.41%
Metal Temperature			
- External Surface	T_{ref1}	T_{ref1}+6.5C	+6.5C
- Internal Surface	T_{ref2}	T_{ref2}+6.8C	+6.8C
Life Change (Oxidation)			
- EOH	1	1.195	-19.5%

For Constant T48 Control

GER 3620J

Steam/Water Injection Increases Metal Temperature of Hot-Gas-Path Components

- Water Affects Gas Transport Properties:
  - k - Thermal Conductivity ↑
  - C_p - Specific Heat ↑
  - μ - Viscosity ↔
- This Increases Heat Transfer Coefficients:
- Which Increases Metal Temperature and Decreases Bucket Life

Example (MS7001EA Stage 1 Bucket):
3% Steam (25 ppm NO_x)
H = +4% (Heat Transfer Coefficient)
T_{Metal} = +15 F (8 C)
Life = - 33%

For Constant Firing Temperature

given by the GER-3620J on the industrial Frame 7EA stage 1 bucket, for constant firing temperature, a 3% steam injection (25 ppm NO_X) would result in an 8°C (15°F) increase in blade metal temperature and 33% reduction in life. It should be noted that the engine control system for the heavy duty MS7001EA gas turbine is different in design from the LM6000PC. The MS7001EA is controlled at the constant firing temperature, while the LM6000PC is controlled at the constant LP turbine inlet temperature. Secondary, the lifing criteria used by the GER-3620J may or may not be the oxidation as identified for the LM6000PC stage 1 blade.

CASE STUDY 5: AUGMENTATION OF GAS TURBINE POWER OUTPUT BY STEAM INJECTION*

Steam injection into the combustion chambers of gas turbines (GT) or just upstream of them (around the discharge of the compressor module) increases their power output. Additionally, the thermal efficiency can be raised, if steam is generated by exhaust heat. The types of steam injected gas turbines (STIG) are distinguished by the amount of steam that can be injected. A gas turbine is called *partial* STIG if the steam turbine cannot utilize the total amount of steam that could be generated by the gas turbine exhaust heat. The limit is given by the flow capacity of the turbine. If, on the other hand, the gas turbine is sized such that the entire amount of steam producible can be utilized, it is called *full* STIG.

Three different partial STIG cooling models were selected for analysis. Compressor surge turned out to be the strongest limit for overloading the gas turbine. At the point of maximum overload—where safe operation is still guaranteed—the steam mass flow amounts to one tenth of the nominal compressor air mass flow. At this operating point, the power output can be raised by more than 30% with a simultaneous increase in efficiency (see Figure 10–49).

Based on the gas turbine configurations used for the partial STIGs, the preliminary designs of two full STIG cycles have been developed. However, for full STIG operation by injection of the total amount of steam producible, either the compressor or the turbines of the original gas turbine have to be modified. In this case, the steam flow exceeding that required for cooling has to be injected into the compressed air in front of the combustor. Depending on whether the compressor is scaled down or the turbines are scaled up, the power output of full STIGs is 30 to 135% higher than that of the original gas turbine. The gross thermal efficiency is about 50.5%.

Among the cycles that mix exhaust gas and steam, the steam injected gas turbine has a high specific power output and an intermediate efficiency. The two other types mentioned, the evaporative cycle and the chemically recuperated gas turbine, offer some advantages,

* Source: Courtesy Alstom Power. Adapted extracts from A. C. Fischer, H. U. Frutschi, H. Haselbacher, "Augmentation of Gas Turbine Power Output by Steam Injection," 2001-GT-0107 (resources/tools provided by Alstom Power, two authors from the Vienna University of Technology).

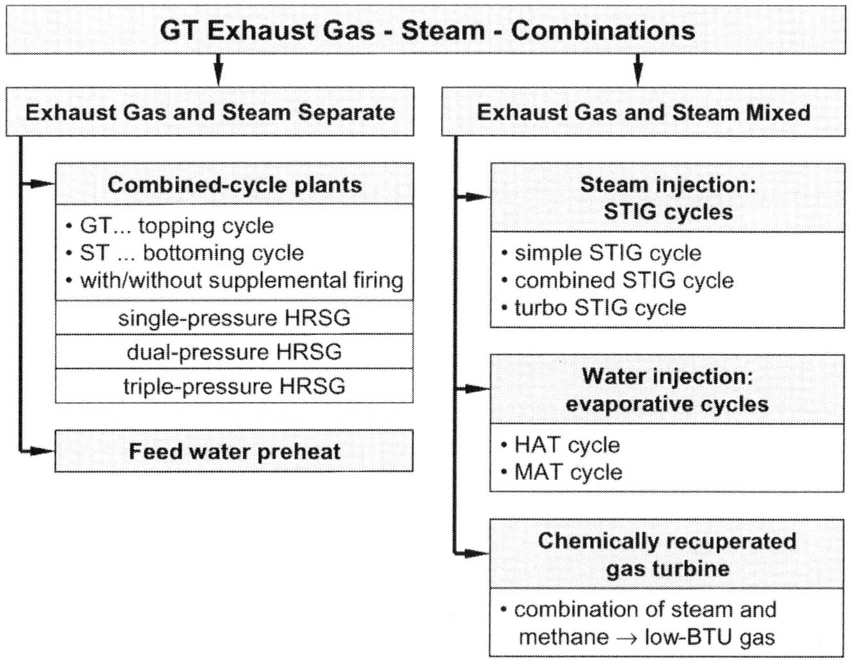

FIGURE 10–49 Configurations of gas turbine exhaust heat utilization for steam generation. *(Source: Alstom Power.)*

too. Water, instead of steam, is injected into and evaporated in the humid air turbine (HAT) and the moisture air turbine (MAT). In the chemically recuperated gas turbine, heat is transferred very effectively from the exhaust gas to the low-BTU gas which is created by mixing natural gas and steam. Therefore, the heat exchanger size is optimized.

The steam to fuel ratios range from 0.5–1.5. Steam to fuel ratios of 4 to 5 are desirable (full STIG) as that would raise the power output of gas turbines significantly, however the gas turbine needs to handle the change in composition of gases and it has its limits.

Nomenclature

$\dot{m}$	[kg/s]	mass flow rate
P	[MW]	power
TIT	[°C]	turbine inlet temperature
η	[–]	efficiency
λ	[–]	excess air ratio
Π	[–]	pressure ratio

Gas Turbine with Sequential Combustion

This study used the Alstom gas turbine technology with the SEV (sequential environmental burner). Alstom models that use this technology include the GT24 and the GT26. Basically, gaseous products of combustion from one combustion chamber section have fuel injected and are reignited. See Figure 10–50. The gas turbine is a single-shaft machine with a compressor (C) and two turbines (T), each with an annular combustion chamber (CC).

A specified portion of the compressed air is taken from the compressor at an intermediate pressure, designated low pressure (LP). This amount of air, which is cooled by water, is used to cool the low-pressure section of the gas turbine. Another amount of air, smaller than that for LP cooling, is branched off at the compressor outlet and fed into the high-pressure (HP) section after having been cooled by water.

Sequential combustion has the following features: Firstly, the specific power output is higher than that of gas turbines without reheat with the same pressure ratio (TI) and the same turbine inlet temperature (TIT).

Secondly, the exhaust temperature of a gas turbine with sequential combustion is higher than that of a gas turbine without reheat. This is important for combined-cycle applications. In the case of high-pressure ratios and a given turbine inlet temperature, the turbine outlet temperature of gas turbines without reheat may become too low for efficient steam generation.

The turbine inlet temperature cannot be increased, because it is limited by the materials used and the cooling systems employed. Therefore, the only way to reach higher turbine outlet temperatures for efficient steam generation is to re-heat the gas. Both high specific GT power output and high turbine outlet temperature improve combined-cycle power plant efficiencies.

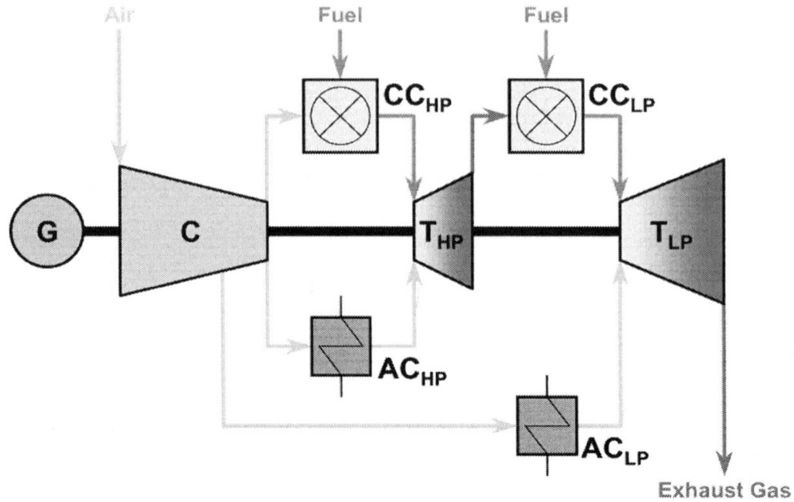

FIGURE 10–50 Gas turbine with sequential combustion. *(Source: Alstom Power.)*

Figure 10–51 shows the thermal efficiency plotted against the specific work output of gas turbine cycles with sequential combustion.

The performance map indicates that if the compressor pressure ratio is lower than 10, the thermal efficiency is decreasing with rising turbine inlet temperature. This due to the fact that the fuel input to heat up the gas to higher temperatures is higher than the work output that can be obtained from the gas during the expansion in the turbines.

On the other hand, the specific power output is rising with increasing turbine inlet temperature because the gas expands at higher temperatures.

If the compressor pressure ratio is higher than 10, the thermal efficiency and the specific power output are rising with increasing turbine inlet temperature.

The data given in Table 10–10 represent the design point of the gas turbine employed in the case study. The characteristics of several components that were used for the calculations of the off-design points with steam injection, e.g., compressor performance map, pressure losses in combustion chambers and heat exchangers, turbine performance maps, etc., are not presented here.

The nominal power output of the turbogroup without steam injection is 268 MW. The thermal efficiency at generator terminals amounts to 38.5%.

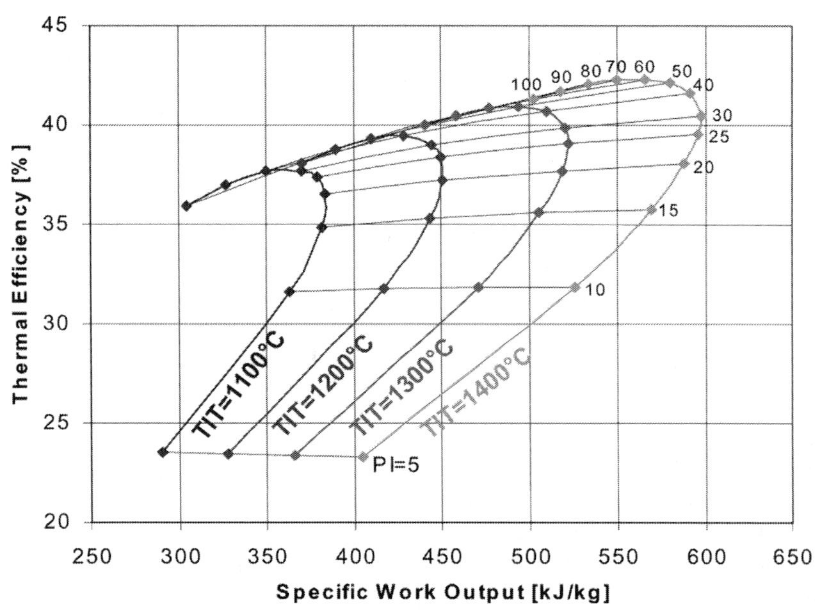

FIGURE 10–51 Performance of gas turbine cycles with sequential combustion. *(Source: Alstom Power.)*

TABLE 10–10 Data of the Turbogroup Design Point

ISO-Conditions		
Ambient pressure	1.013	bar
Ambient temperature	15.0	°C
Relative humidity	60.0	%
Compressor		
HP ratio	30.0	
LP ratio	17.0	
Compressor mass flow rate	540.0	kg/s
HP cooling air mass flow rate	86.4	kg/s
LP cooling air mass flow rate	108.0	kg/s
HP Air Cooler		
Cooling temperature	300.0	°C
HP Turbine		
Pressure ratio	1.8	
TIT	1235.0	°C
LP Air Cooler		
Cooling temperature	300.0	°C
LP Turbine		
Pressure ratio	15.53	
TIT	1295.0	°C
Mechanical efficiency	0.995	
Electrical efficiency	0.98	

(Source: Alstom Power.)

Partial STIG Cycle

The partial STIG is basically a gas turbine that is overloaded by steam injection. To raise the efficiency the steam is generated by the heat of the exhaust gas.

The increase in turbine mass flow causes additional mechanical and thermal loads on the turbine blading, the rotor, the stator, and the combustion chambers of the gas turbine.

However, frequent short-term overloading does not have such a strong negative impact on the useful life of the components. No additional cooling may be required. The steam injected is used to cool the cooling air in a heat exchanger. This type of turbine cooling is called *fixed indirect cooling* (FIC).

If the turbine is overloaded for long periods of time, additional cooling is required to avoid a reduction in the lifetime of blades and other components.

Additional cooling can be provided by surface cooling of the increased cooling airflow by water or steam, called *adjusted indirect cooling* (AIC). Second, water or steam is injected into the cooling air. This cooling system is called *adjusted direct cooling* (ADC).

Figure 10–52 shows the partial STIG with fixed indirect cooling. Feed water is evaporated in the heat recovery steam generator (HRSG) at a temperature lower than that of the cooling air, which is 300°C.

Just a portion of the exhaust gas is required for the steam generation. Therefore, the exhaust gas stream is divided. The excess exhaust gas is passed by the heat recovery steam generator and then mixed with the other stream downstream the HRSG.

To compensate the pressure drop of the gas in the heat recovery steam generator, a suction blower is installed. It is mounted at the cold end of the steam generator because compression work as well as the thermal loading on the components of the blower are smaller at lower temperatures.

In the superheater (SH) the steam is superheated while cooling the HP cooling air. Then the superheated steam is injected into the main flow in front of the HP combustor. Since the amount of steam may not be sufficient to cool the entire amount of cooling air, the HP air cooler (ACHP) is still required. The HP cooling airflow is divided according to the amount of steam available for cooling in the superheater.

If the steam flow exceeds that needed for cooling, the surplus steam is passed by the superheater and combined again downstream with the cooling steam flow.

The partial STIG with adjusted indirect cooling, depicted in Figure 10–53, is quite similar to the partial STIG with fixed indirect cooling. The significant difference is the higher amount of cooling air.

Since the LP cooling air mass flow is a fixed portion of the compressor air mass flow, an adequate portion of the steam mass flow is injected into the LP cooling system. The indirect cooling is only used in the HP cooling system.

The additional amount of HP cooling air for maintaining the cooling of the HP system is calculated by comparing the enthalpy transport capacities of air and steam. During the heat transfer from the hot gas to the cooling air the temperature drop of the hot gas and the temperature rise of the cooling air should remain constant when overloading the gas turbine. Taking these two requirements and the change of the amount of combustion air mass flow available into account, an equation can be derived to calculate the amount of cooling air necessary for a particular quantity of steam.

Another less significant difference between the fixed and the adjusted indirect cooling is the combination of the superheater and the HP air cooler. Both the series and the parallel arrangements of superheater and HP air cooler are a compromise between pressure loss and size of the heat exchangers.

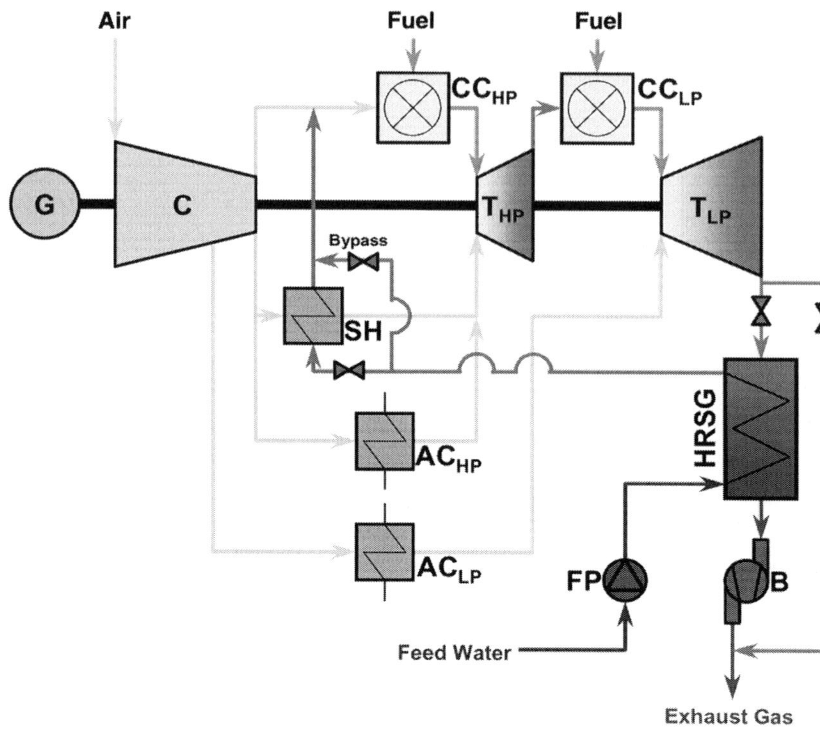

FIGURE 10–52 Partial STIG with fixed indirect cooling (FIC). *(Source: Alstom Power.)*

The parallel arrangement is preferred when the steam injection changes frequently within the entire range of the possible steam flow rates. Both apparatus are to be designed for the maximum heat transfer capacity to cool the entire amount of cooling air.

Redundancy is built into the system's heat exchanger capacity. In case one of them fails to perform the other can take over. This arrangement has a positive side-effect: If the airflow is distributed equally, the pressure loss in each heat exchanger amounts to just one fourth of that of the design point.

The series arrangement is better for constant steam injection into the gas turbine. Between superheater and HP air cooler the cooling air has an intermediate temperature

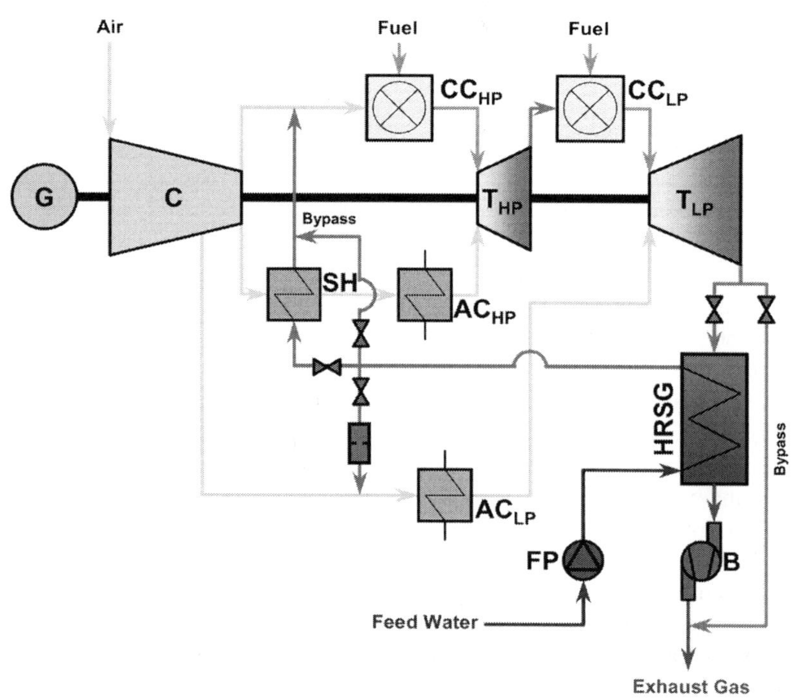

FIGURE 10–53 Partial STIG with adjusted indirect cooling (AIC). *(Source: Alstom Power.)*

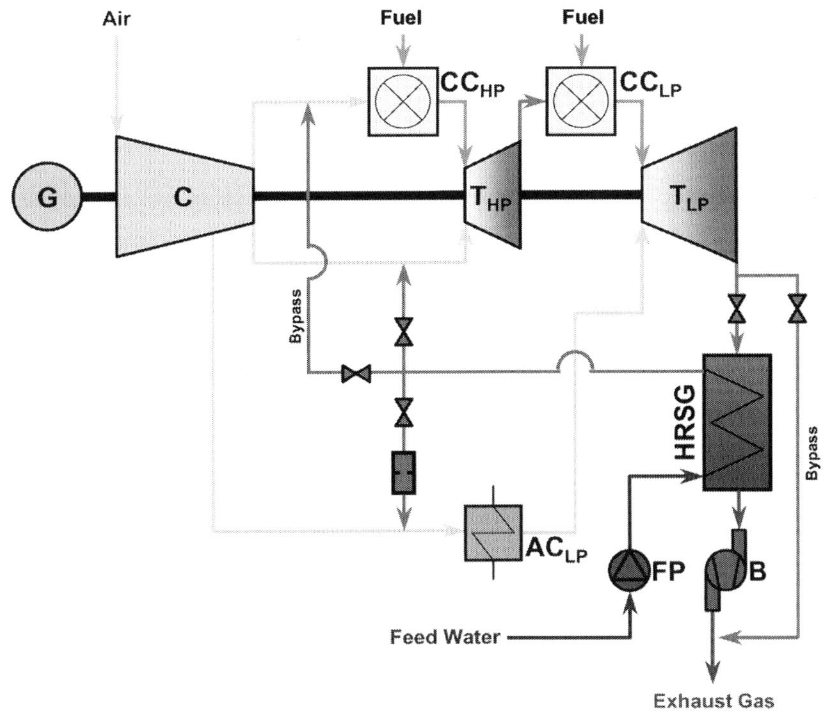

FIGURE 10–54 Partial STIG with adjusted direct cooling (ADC). *(Source: Alstom Power.)*

that depends on the quantities of heat given off from the air in each of the two heat exchangers. As air temperature decreases in the superheater (a counter-current heat exchanger), its logarithmic mean temperature rises. This results in smaller heat exchanger dimensions.

The partial STIG with adjusted direct cooling is shown in Figure 10–54. This cycle is equipped with just one air cooler for the LP cooling system. HP cooling air is cooled by water or steam injection.

For the startup of the machine, a minimum quantity of cold water has to be injected to cool the HP cooling air. During operation, the physical state of the water injected changes depending on the amount of water fed. The state of water ranges from water to wet steam to saturated steam. If water is injected into the HP cooling system, it has to be finely dispersed to avoid erosion by water droplets.

In this cycle, the cooling air is partly replaced by steam. As for the partial STIG with adjusted indirect cooling, a similar procedure using the enthalpy transport capacity of water and steam is employed to calculate the amount of air required for cooling. At a certain steam to compressor air ratio the entire HP cooling air is replaced by steam. From now on the surplus steam is injected in front of the HP combustion chamber.

The results of the parameter study of the partial STIGs are presented in Figures 10–55 to 10–60. All parameters are plotted against the steam to compressor air ratio (steam to air ratio).

The gross thermal efficiency, depicted in Figure 10–55, includes the auxiliary equipment of the plant, e.g., feed pump, suction blower, etc. Its increase with rising steam to air ratio is produced by the partly utilized exhaust heat and the increasing pressure ratios of the turbines (see also Figures 10–56 and 10–57).

The plant with fixed indirect cooling has higher efficiencies at low steam to air ratios than the machines with adjusted indirect and direct cooling. This is because the amounts of cooling air and cooling steam are lower and the entire amount of steam is injected into the HP section of the gas turbine.

At a steam to air ratio of 7.5%, the efficiency of the cycle with fixed indirect cooling has a slight bend. This bend is caused by the amount of steam exceeding that needed to cool the HP cooling air in the superheater. The surplus steam is not superheated before injection anymore. A higher quantity of fuel is required to heat the saturated steam.

The efficiencies of the plants with adjusted cooling have a similar course. The bends are at higher steam to air ratios than that of the plant with fixed indirect cooling. Since a temperature difference between air and steam is required in the superheater, the efficiency of plants with indirect cooling is slightly lower than that of plants with direct cooling.

The increase of the power output of the gas turbine is shown in Figure 10–56. A 10% increase in turbine mass flow by water injection leads to a 30% increase in power output. This is mainly caused by the higher specific heat of

FIGURE 10–55 Gross thermal efficiency of partial STIG cycles. *(Source: Alstom Power.)*

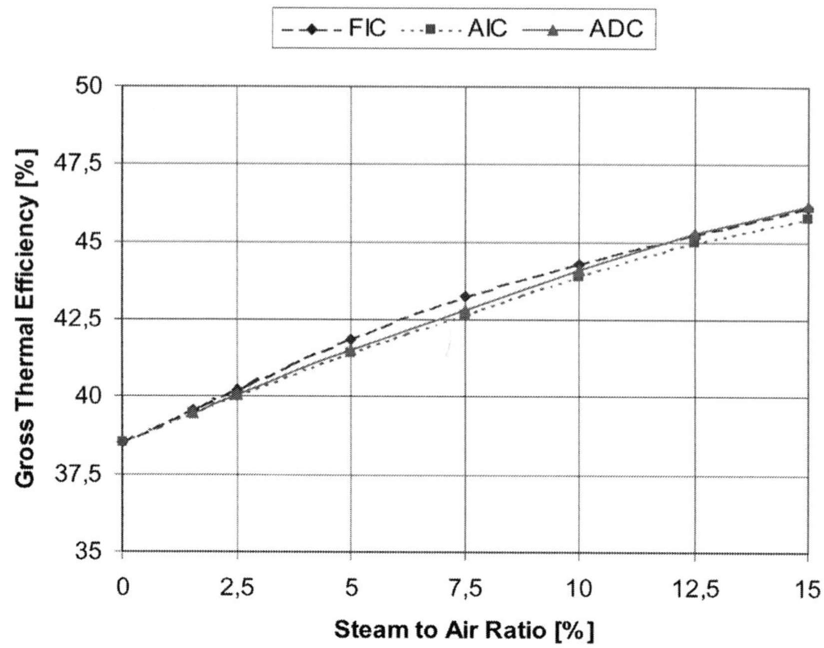

FIGURE 10–56 Power increase of partial STIG cycles. *(Source: Alstom Power.)*

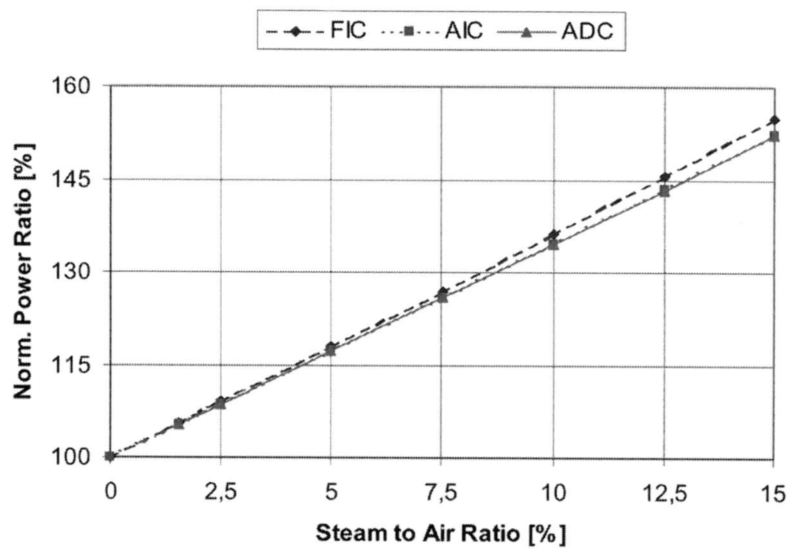

steam (double that of air) and the increasing pressure ratio due to higher turbine mass flow.

The normalized* pressure ratios of the HP turbine and the LP turbine are presented in Figures 10–57 and 10–58. To take in the increasing mass flows, the inlet pressures of the turbines and the outlet pressure of the compressor have to rise.

* The normalized parameters are all based on the values of the design point of the gas turbine (without steam injection).

While the pressure ratio of the LP turbine rises linearly, that of the HP turbine remains almost constant, although the difference between inlet and outlet pressure of the HP turbine increases. The minor increase of the HP turbine pressure ratio of the cycle with fixed cooling is caused by the fact that the total amount of steam is fed into the HP turbine, while in the cycles with adjusted cooling the steam mass flow rates for HP and LP turbines are divided like that of air.

The steam injection into existing gas turbines has a strong impact on some of their components. In this section,

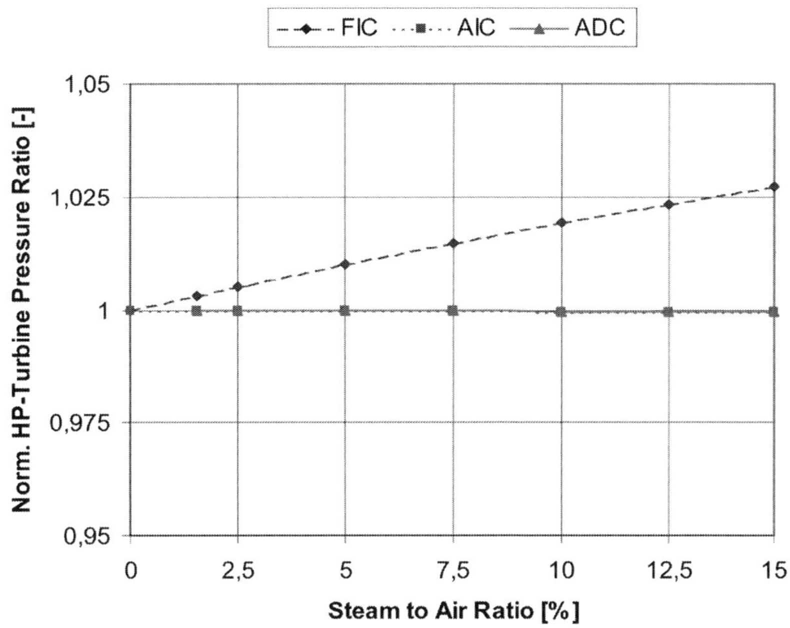

FIGURE 10–57 HP turbine pressure ratio of partial STIGs. *(Source: Alstom Power.)*

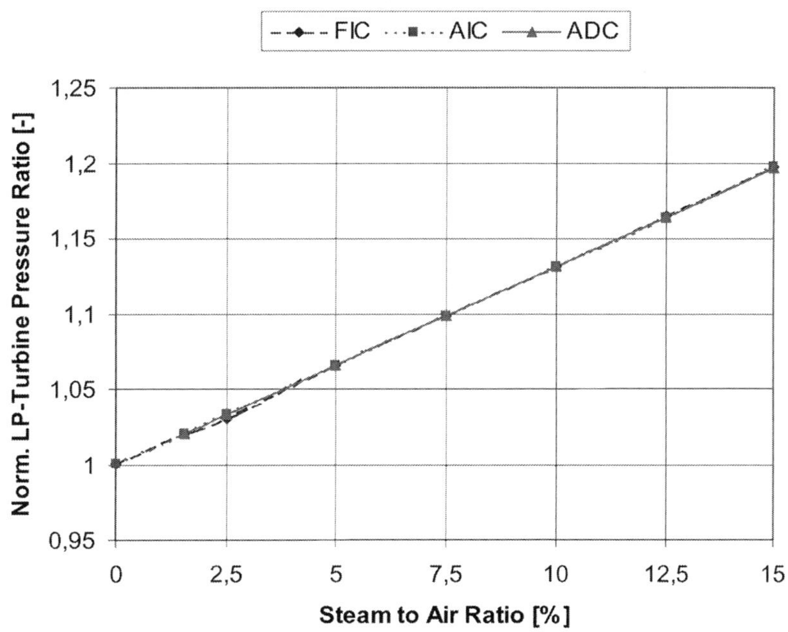

FIGURE 10–58 LP turbine pressure ratio of partial STIGs. *(Source: Alstom Power.)*

two examples are given: The compressor surge and the axial thrust of the rotor.

Figure 10–59 shows the normalized pressure ratio, the operating limit, and the surge limit of the gas turbine. The operating limit is at the mid-point between the design point and the surge limit of the gas turbine. Since the pressure ratio of the plant with fixed cooling has a greater slope than that of the machines with adjusted cooling (the whole amount of steam is fed into the HP system of the gas turbine), the operating limit is reached earlier. In the case of adjusted cooling, the steam to air ratios are about 1.5 percentage points higher than that of the cycle with fixed cooling before reaching the operating limit.

FIGURE 10–59 Compressor pressure ratio of partial STIGs. *(Source: Alstom Power.)*

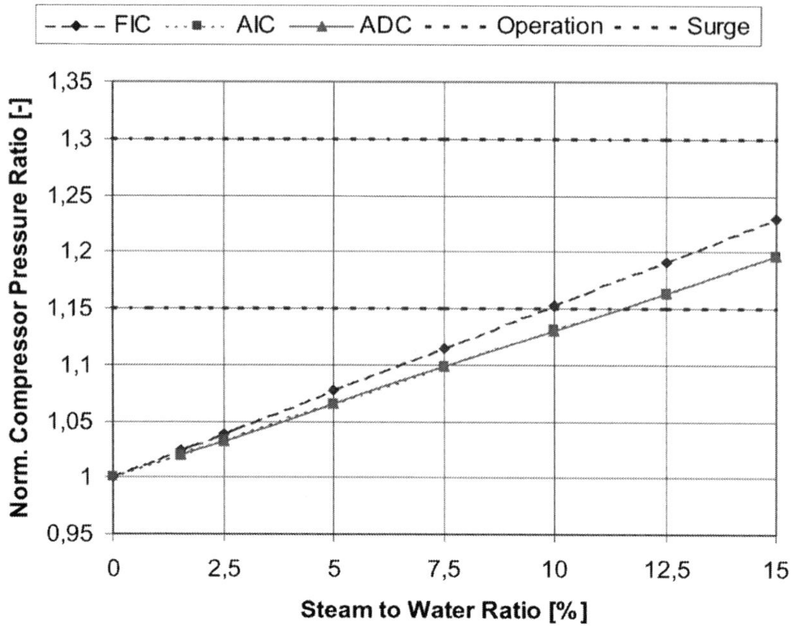

FIGURE 10–60 Axial thrust of partial STIGs. *(Source: Alstom Power.)*

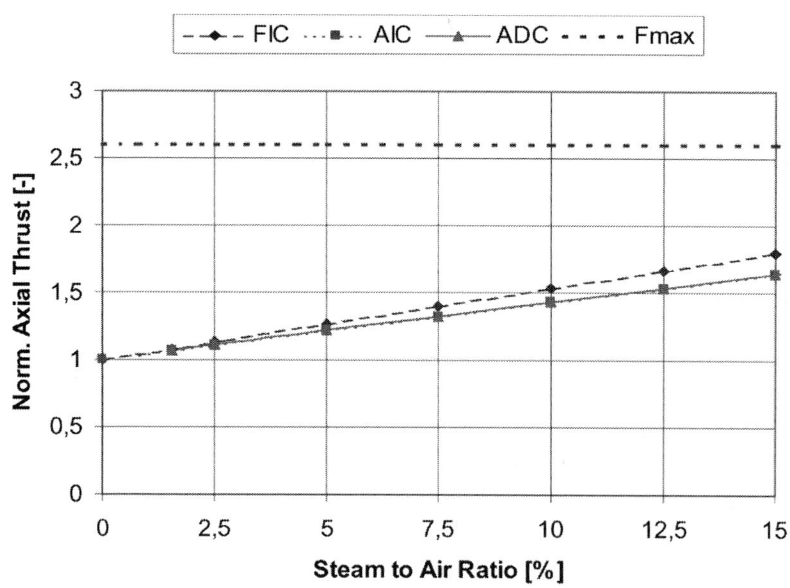

The resultant axial thrust of the gas turbine rotor is proven not to be a serious limit for steam injection into the gas turbine as depicted in Figure 10–60. Fmax represents the maximum axial thrust of the rotor that can be absorbed by the existing thrust bearing.

As to the design of the gas turbine, the following is noted: The axial thrust, pointing in the direction of the main flow, is mainly created by the static gas pressures at the shaft shoulders. The contribution of the axial thrust caused by the flows through the compressor and the turbine blading is quite small. The slight upward bend of the curves represents the increase of thrust caused by the fact that the mass flow rate through the turbines is higher than that through the compressor.

Full STIG cycles are designed to utilize the entire amount of steam that can be produced by the exhaust gas

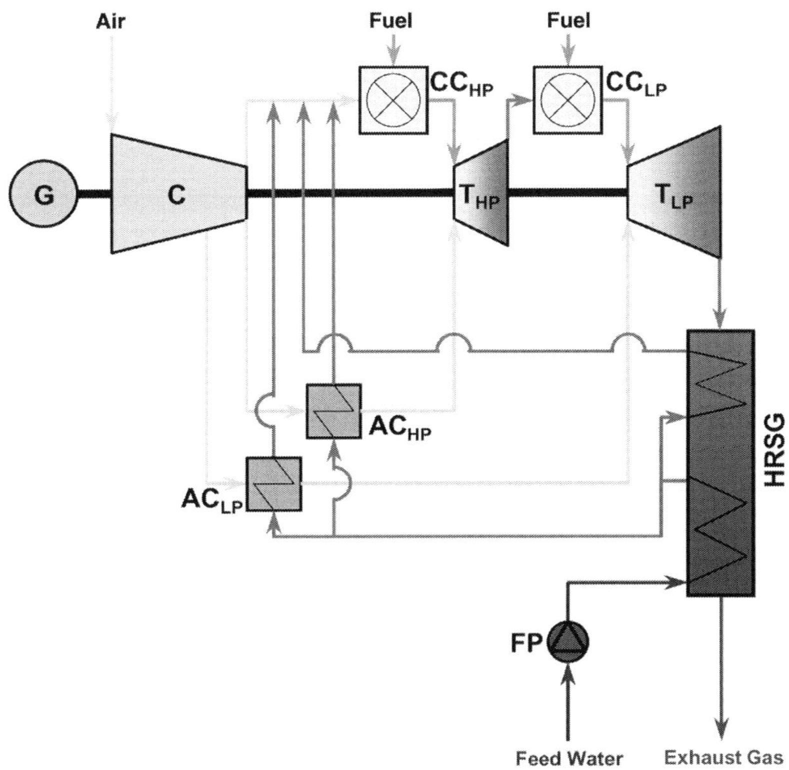

FIGURE 10–61 Full STIG with adjusted indirect cooling (AIC). *(Source: Alstom Power.)*

flow. Depending on the type of cooling, there may be other limitations to the amount of steam injected, e.g., the excess air ratio in the combustion chambers.

The full STIG cycle with adjusted indirect cooling, depicted in Figure 10–61, is the simplest of the cycles investigated. Saturated steam is used to cool the cooling air, thereby becoming superheated. The steam exceeding that needed for cooling is superheated by exhaust heat. The three superheated steam flows are injected into the main flow of the gas turbine in front of the HP combustor. The excess air coefficient must be greater than 1.1 to guarantee complete combustion.

As for the partial STIG with adjusted indirect cooling, the amount of cooling air required is calculated by employing the enthalpy transport capacities of air and steam. Since a great amount of air is passing through the cooling system, the amount of steam is reached relatively early, at which the combustor airflow is just sufficient for complete combustion.

Figure 10–62 shows the full STIG cycle with adjusted direct cooling. In this cycle, no surface heat exchangers are required anymore. The air is cooled by the water or the steam injected. Since the combustion chambers and the turbines are cooled by steam instead of air, sufficient combustion air is available. The only thermodynamic limit is the exhaust heat available.

In case the steam cooling system fails to perform, two valves, controlling the HP and the LP cooling air, are opened admitting cooling air into the cooling system during a brief period of malfunction of the steam cooling. If the steam cooling cannot be restarted, the gas turbine has to be turned off slowly to avoid increased thermal loading.

In both types of full STIG operation, the turbine flow rates are greater than the compressor flow rates. Either the size of the compressor has to be reduced when the turbines are maintained, or the dimensions of the turbines have to be increased when the compressor remains unchanged. STIGs have been designed for both cases, the parameters of which are given in Table 10–11. The turbine inlet temperatures and the pressure ratios at the design point of the four STIGs are taken from the gas turbine.

If the HPT and the LPT are used for full STIG, then the volumetric flow rates of the turbines have to be constant to maintain the pressure ratios of the gas turbine. Since the main turbine gas flow of the full STIGs contains higher quantities of steam than that of the gas turbine produced during combustion, the turbine mass flow rates of the STIGs are lower than that of the gas turbine.

The reduction in compressor airflow and the exhaust heat utilization result in a greater power output and gross thermal efficiency of the STIGs.

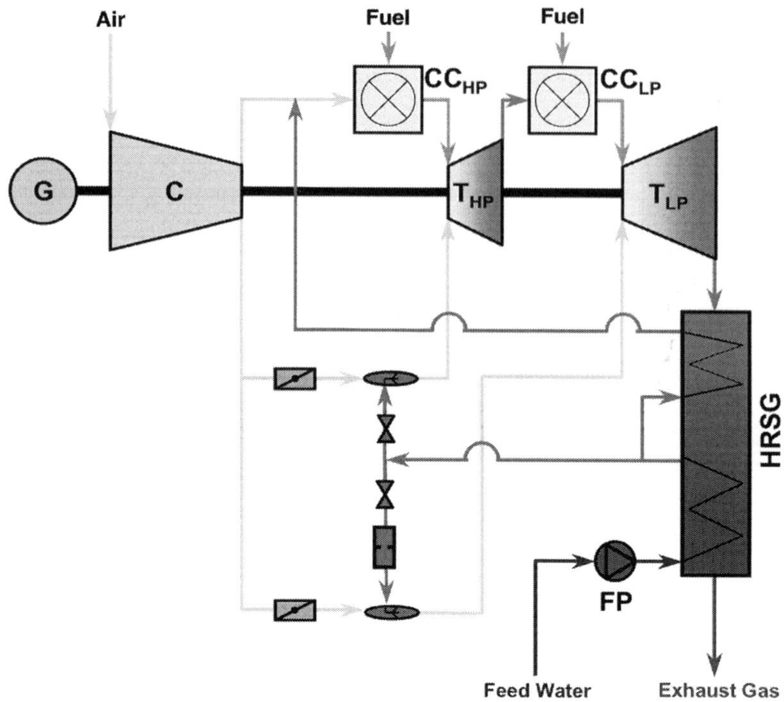

FIGURE 10–62 Full STIG with adjusted direct cooling (ADC). *(Source: Alstom Power.)*

TABLE 10–11 Comparison GT—Full STIGs

P	[MW]	268.06	345.10	360.83	508.51	628.54
mr_{Air}	[kg/s]	540.00	360.00	310.00	540.00	540.00
mr_{H_2O}	[kg/s]	0.00	91.50	124.00	135.00	216.00
T1	[%]	38.52	47.29	50.16	47.29	50.16
nC	[-]	30.00	30.00	30.00	30.00	30.00
IIHP	[-]	1.80	1.80	1.80	1.80	1.80
TITHP	[°C]	1235.00	1235.00	1235.00	1235.00	1235.00
XHP	[-]	2.40	1.10	2.08	1.10	2.08
riLP	[-]	15.53	15.53	15.53	15.53	15.53
TITLP	[°C]	1295.00	1295.00	1295.00	1295.00	1295.00
XLP	[-]	2.96	1.10	1.60	1.10	1.60

(Source: Alstom Power.)

For the cycles with constant compressor size, the dimensions of the turbines have to be scaled up. To get an idea of how much bigger the turbines of the two STIGs have to be made, one-dimensional flow calculations for the LP turbines were carried out. The results show that the blade height of the last turbine stage increases by about 33%. This results in a major design change which does not seam to be practical.

The smaller STIGs with downsized compressors and a slightly modified turbine design are favorable because they cause decisively lower expenses for development and they can be installed in small as well as in large power plants.

CASE STUDY 6: INTEGRATING GAS TURBINES IN POWER AND COGENERATION APPLICATIONS*

To meet the requirements of the highly competitive combined-cycle (CC) power plant market, steam turbines (STs) have had to follow the same evolutionary path as heavy-duty gas turbines (GTs). These two types of equipment have become even more interdependent with the introduction of the new "G" and "H" GT technology classes. These new classes have created an inseparable thermodynamic and physical link between the primary and secondary power generation systems by using steam, instead of air, in a closed loop to perform most if not all of the GT cooling duty. The previous generation of GTs, the "F" class, has also undergone extensive modifications and upgrades. In addition to producing higher output, the "F" class machines are producing substantially more exhaust energy for the bottoming steam cycle.

This case describes a contractor's experience with modern STs in all phases of a CC project: from the initial selection through construction, startup, and testing. The STs include all the major manufacturers, such as Alstom, GE, Mitsubishi, Siemens Westinghouse, and Toshiba, in several CC configurations. Several CC projects that have been completed or are in advanced stages of construction, which use the new generation of GTs and STs, are also considered.

Cycle Selection and Optimization

In the last 10 years, STs in CC applications have evolved from small 80 MW dual admission nonreheat configurations to multiple-pressure-admission reheat turbines with outputs reaching the 350 MW range. Because of this rapid evolution, many original small turbine designs have had to be modified or adjusted to compete in performance and capacity.

Differences between STs designed for CCs and conventional Rankine cycle (RC) applications are:

- Feedwater heaters are not normally used in the thermal design of the bottoming cycle for a CC; so exhaust flows are much higher.
- Steam is added to the steam path flow at several pressures from the heat recovery steam generator (HRSG) to allow the maximum amount of heat to be extracted from the GT exhaust energy. The CC low-pressure (LP) exhaust steam flow, when compared with RC LP steam flow, can be up to 35% greater than the main steam flow.
- CC designs may use supplemental or duct firing in the HRSGs to compensate for the GT's reduced output at high ambient temperatures, which coincides with maximum (summer) power demand.

In the US, it is now common to almost double the ST output by using massive supplementary firing to capitalize on peak summer demand. Finally, competitive market pressures have pushed the ST suppliers to offer a compact plant layout with axial steam exhaust using only two standard cylinders.

This reduces manufacturing and installation costs as well as length of the construction schedule.

Although a modular building block system with standardized components and turbine parts is typically employed by all ST manufacturers, note that the steam blade path is individually designed for each application to achieve high efficiency. Even small blade path improvements can translate into large operational savings.

Figure 10–63 shows the typical thermal efficiency values for three modules (high pressure [HP], intermediate pressure [IP], and LP) for several recent projects for both unfired and fired cases. As the figure indicates, turbine cylinder efficiency in the range of 94 to 96% is not uncommon. In this analysis, the power output for the fired case, with main steam pressure at 131 bara (1900 psia), was 55 to 65% higher than the unfired case, with the main steam pressure at 69 bara (1000 psia).

With sliding pressure operation, HP and IP module efficiency does not change significantly between the fired and unfired cases, indicating that ST operation at part load does not adversely affect the power plant heat rate. The LP module efficiency, however, does change significantly, since it is affected by the specific selection and sizing of the module and heat sink, which determine the ST operating backpressure.

Plant Configuration

One of the earliest tasks in the development phase is selection of plant configuration. Large contractors tend to maintain standardized plant designs to accommodate the OEM preferred packages. (Note that OEMs are frequently their own contractor and owner/operator.) For this contractor, they include:

- 1 × 1 (one GT, one HRSG, and one ST)
- 2 × 1 (two GTs, two HRSGs, and one ST)
- 3 × 1 (three GTs, three HRSGs, and one ST)

A single ST for multiple GTs is less costly than a separate ST for each GT. However, multiple 1 × 1 configuration trains offer some significant advantages, as follows:

- *Phased construction flexibility.* Owners can add units later.

* Reference: Extracts from the collected work of J. Zachary, including but not limited to the papers credited to J. Zachary, "Strategies for Integrating Advanced Gas Turbines in Power and Cogeneration Applications" and R. Narula, J. Zachary, and J. Olson, "Matching Steam Turbines with the New Generation of Gas Turbines."

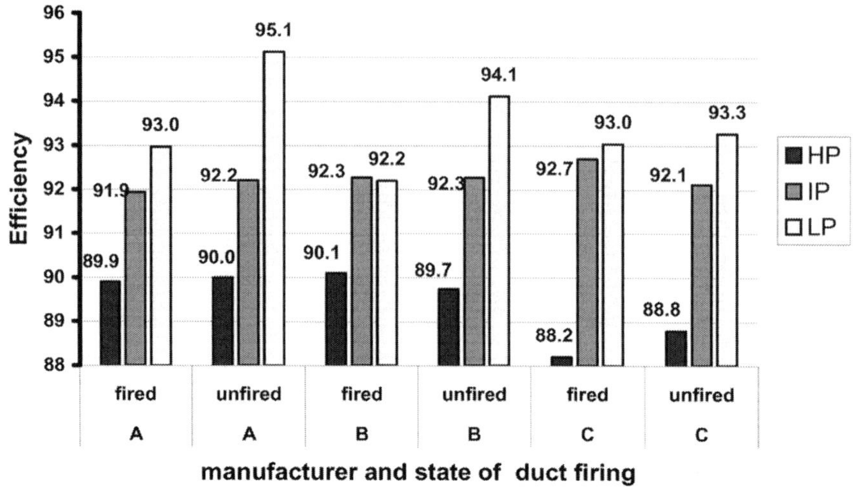

FIGURE 10–63 Steam turbine module efficiencies for duct-fired and unfired cases. *(Source: J. Zachary.)*

- *Speed to market.* As the construction duration of a 1 × 1 configuration is shorter than a 3 × 1 configuration, the first train can be commissioned, while the remaining units are still undergoing construction.
- *Closer matching with dispatch demand.* A site with multiple 1 × 1 trains using supplemental duct firing can more efficiently match dispatch requirements.
- *Greater plant redundancy.* Each unit is completely redundant.
- *Optimized spare parts inventory.* Identical components are used for all trains.

OEM Selection

Before selecting equipment from different suppliers, a thorough investigation is necessary to ensure that the plant owner's requirements for power output, heat rate, startup times, reliability and availability, etc. are met.

The process includes an independent technology assessment of the equipment's operating history and quality control for the engineering and manufacturing processes. In addition, the performance offered by the original equipment manufacturers (OEMs) for a specific project is normalized and reconciled with past performance of the same equipment in a similar configuration.

Special consideration is given to cogeneration plants where selection of cycle pressures and locations of the steam extractions is limited by the commercial availability of STs that can meet the full range of specified conditions. Most cogeneration projects require the ability to cold start the ST while the plant continues to supply full process steam requirements to the host. At the other extreme, it is also critical to determine whether to design the ST and associated steam cycle equipment for the maximum steam case when no steam is required for the process. For this instance, a larger LP section and an increased capacity for downstream electrical equipment (transformer, isophase bus, etc.) are required. Economic benefits to be realized from the availability of the extra power and operational constraints due to possibly infrequent occurrences of such conditions must be considered. Equipment selection requires the unique expertise and value-added service that only an experienced EPC contractor can provide to the customer.

Experience History

Table 10–12 lists STs from major suppliers that have been used in this contractor's projects (96 projects using 170 GTs and more than 96 STs in CC applications).

Figure 10–64 provides a comparison of the relative efficiencies of all three configurations (1 × 1 × 1, 2 × 2 × 1, 3 × 3 × 1) as a function of the amount of process steam duty. Relative efficiency $Rcog_{SEP}$ is defined as the ratio between the thermal efficiency of each configuration and a reference efficiency of a conventional cycle consisting of a combined cycle producing electricity and a boiler producing process steam. In the example, for 300 MW process steam duty, a 3 × 3 × 1 configuration is slightly more efficient ($Rcog_{SEP} = 1.1323$) than a 3 × (1 × 1 × 1) configuration, each exporting 100 MW ($Rcog_{SEP} = 1.1304$). However, for cases where the steam duty varies, three trains, each in a 1 × 1 × 1 configuration, provide substantially greater efficiency than a 3 × 3 × 1 configuration due to the increased operating flexibility.

Steam Path Optimization

While ST design is a mature technology, manufacturers continue to improve the output and efficiency of their equipment through improved blade designs and superior materials. Pressures to maintain a small footprint have

TABLE 10–12 Steam Turbines from Major Suppliers

Nominal Power Supplier	No. of Units	Power (MW)	Characteristics	CC Configuration
Alstom	7	250		2 × 1
Alstom	12	85/155	Unfired/duct-fired	1 × 1
Alstom	1	300		3 × 1
GE	1	300		2 × 1
GE	2	250		3 × 1
GE	1	150		1 × 1
GE	1	120	Desalinization noncondensing unit	1 × 1
GE	3	200		2 × 1
MHI	2	100	Cogen with large process steam	2 × 1
SWPC/MHI	1	120		1 × 1
SWPC	4	150		1 × 1
SWPC	1	295		3 × 1
Toshiba	1	300		3 × 1
Westinghouse	1	230		2 × 1
Total	38			

(Source: J. Zachary.)

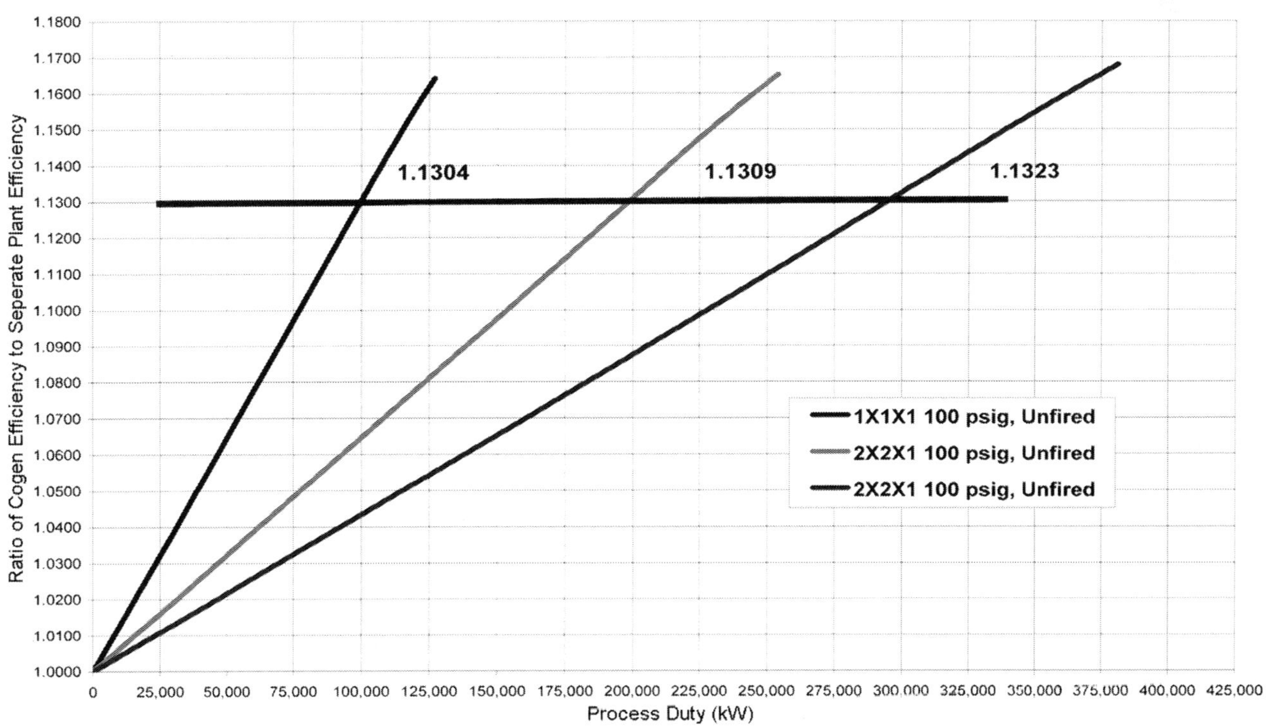

FIGURE 10–64 Comparison of cycle efficiency variation as a function of steam process duty and plant configuration. *(Source: J. Zachary.)*

improved blade path parameters and offer increased stage count, reduced blade root diameter, and optimized reaction values to minimize leakages.

Evaluation of the Exhaust System: Axial, Downward, or Lateral

With the high LP flows associated with CC STs, greater emphasis is placed on the last-stage blade (LSB) dimensions and material. One major loss in the ST is the kinetic energy of the steam as it leaves the LSBs—the lower the kinetic energy that can be achieved, the higher the resulting ST efficiency. The amount of loss is proportional to the ratio of the volumetric steam flow rate through the LSBs and the annular area of the turbine exit. To decrease this loss, a larger turbine exit annulus is required. Challenges face designers, for instance aeroelastic instability, which is one of the more challenging design problems for very large LSBs. Aeroelastic instability is a vibration induced by flow, which occurs at off-design conditions in the regions of low axial steam flow and/or at high condenser pressures. The phenomenon can lead to flow separation from the airfoil, blade stall flutter, and buffeting of the blades.

Some manufacturers use titanium as the LSB material because of its high strength/weight ratio and superior corrosion resistance. However, titanium has reduced damping properties; therefore, it is generally necessary to change blade construction from freestanding to an interlocked design for increased mechanical stiffness.

The axial exhaust design has superior thermodynamic performance through lower pressure losses, greater pressure recovery in the diffuser, and simpler and less expensive construction. Because of the compact design and lower elevation, the capital cost of the ST and the turbine building can be reduced by 40% over using a double flow LP module with a bottom exhaust arrangement.

Construction

With the pursuit of compact, more efficient designs by ST manufacturers, construction tolerances and startup requirements have become more critical. As the capacity of CC STs has continued to grow, economic factors have pushed manufacturers to try to maintain the same cylinder sizes. These designs have resulted in reduced design margins and equipment that is much more sensitive to construction variables.

Fewer bearings used to support the rotor in the turbine generator set allows for shorter overall length and lower costs. Use of fewer bearings results in higher bearing loadings and increased sensitivity to vibration, which in turn leads to additional startup delays due to the added balancing time.

With the axial configuration, the location of the outboard LP bearing is an additional concern. Normally, bearing housings are easily accessible from the outside, and the housing is maintained at atmospheric pressure or at a slightly negative pressure by the vapor extractors on the lube oil reservoir. With an axial design, this bearing is in the exhaust region and is subject to condenser backpressure conditions. In some cases, a small amount of leakage across the housing flanges or inspection openings introduces oil into the cycle. Because this leakage is difficult to detect, the resulting cleanup can be extensive.

The axial exhaust design has an additional feature that is sensitive to design tolerances—the large axial loads on the turbine during startup that are transmitted to the turbine foundations. The foundation must handle this loading with minimal deflection, and the slide plates under the turbine are designed to freely allow the cylinder to grow. This large axial load, due to both differential and thermal pressure, results in varying amounts of deflections, and the variation in slide plate friction results in further variation.

Further temperature stratification of the HP and IP cylinder due to differential warming or cooling during startup or shutdown has resulted in rotor binding if not adjusted properly. The sensitivity of the ST to these conditions results in further delays in project startup.

The conversion of standard 50 Hz designs to 60 Hz conditions has also resulted in unforeseen problems. Steam turbine sets that have had many reliable operating hours in their original 50 Hz configuration experience a variety of startup delays. The most significant is high vibrations that are transmitted through the STs and generators into the piping or foundations.

Startup

The ST startup flexibility and commissioning time play a significant role in startup of the entire CC plant. Heavy penalties for failing to bring the unit on line as scheduled mandate that predicted unit startup times be achieved.

Also, due to higher fuel costs and increased electrical reserve margins, CC plants are being dispatched as intermediate-duty units rather than base load, as originally envisioned. Achieving the goal of a fast and reliable startup requires careful integration of the ST and BOP requirements. The ST supplier should provide more flexible ST startup parameters (such as greater steam temperature mismatch and more relaxed steam purity) while maintaining reasonable constant life consumption, controlling low cycle fatigue by monitoring the maximum wall temperature differences and permissible ramp rates.

The EPC contractor can employ the entire arsenal of auxiliary equipment available to assist in providing the narrow preoperational and operating conditions that the modern ST requires.

Improving heat retention after shutdown, using advanced water treatment systems to achieve steam purity more quickly, providing additional lines for warm-up, and using an auxiliary steam boiler to reach desired condenser vacuum more rapidly are only a few examples of the measures that can be taken. However, many of these steps can significantly increase the overall cost of the project.

Thermal Performance

Contractually, the EPC contractor must conduct a performance test for the entire CC power plant to demonstrate to the owner that thermal performance guarantees are met. However, on many occasions, component performance testing for GT, HRSG, and ST is also performed concurrently with the CC plant test. ASME Performance Test Code, PTC 6.2, is for STs in CC applications, wherein the power output is the only guarantee. The new code demands specific correction curves to account for the interaction between the HRSG and ST and sets forth stringent requirements for instrumentation accuracy.

The outcome of ST tests conducted by this contractor (13 units, 5 OEMs) is in Figure 10–65. As can be seen, many of the units did not meet their guarantees, and additional work was required to improve their performance. OEMs may under competitive pressures offer aggressive performance guarantees.

Cogeneration

Factors to be considered include:

- *Supplementary firing.* A popular feature of cogeneration is the use of supplementary firing (duct firing) within the HRSG. The reduction of electric power that occurs upon an increase in process steam demand can be easily overcome by duct firing in the HRSG to generate more steam. Hence, the plant is able to meet increased steam demand without reducing power output. If an additional source of unconventional off-gas fuel is available, as is often the case in chemical plants or refineries, using it for duct firing is an ideal application.
- *Low-pressure turbine sizing.* The design team and the owner should both agree on a decision to size the ST and associated equipment based on the "zero export steam" case. An ST passing all the steam generated will have a larger low-pressure (LP) section, larger electrical equipment (generator, transformer, etc.), and obviously a higher electrical output. Economic and thermal performance considerations must be part of the decision-making process.
- *Reliability and availability.* Since the export steam is used in many cases as the heat source for a chemical or industrial process, its reliability becomes a primary consideration in equipment selection and plant design. A steam outage could have a disastrous economic impact on the host facility. Table 10–13 presents the steam export reliability requirements from one Bechtel project. The reliability values vary at different steam flows and pressure ranges.

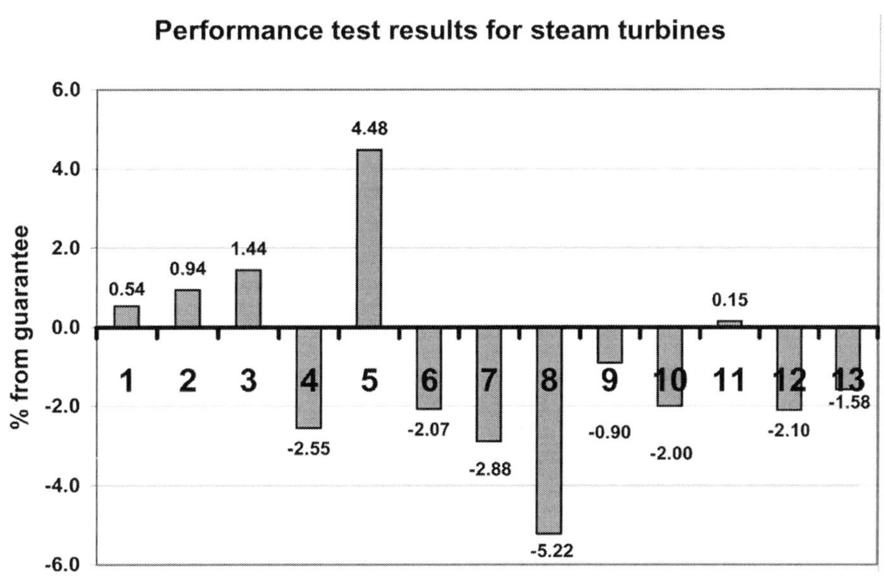

FIGURE 10–65 Steam turbine output guarantee test results. *(Source: J. Zachary.)*

TABLE 10–13 Example of Reliability Values as a Function of Flow and Pressure

Flow Requirements		Pressure Requirements	
Flow Range (lb/hr)	Reliability (%)	Pressure Range (psia)	Time Allowable
0–250,000	100	600–615	Indefinitely
		575–600	
250,000–500,000	99.9	615–6300	4 hours
500,000–1,250,000	99	Below 575	5 minutes

(Source: J. Zachary.)

CASE STUDY 7: AN INTEGRATED COMBINED-CYCLE PLANT DESIGN THAT PROVIDES FAST START CAPABILITY AT BASE-LOAD*

While efficiency is still the major driver in combined-cycle power plant design, there is an increasing need for operational flexibility along with additional operating cost reduction. Nightly and weekend shutdowns and subsequent startup capabilities should be considered during the overall evaluation of a power plant. Advancements in gas turbine technology have increased power and efficiency while decreasing emissions and life-cycle costs without sacrificing reliability.

Despite advances in gas turbine power and efficiency, when used in combined-cycle application, plant owners were traditionally forced to choose between high efficiency or operational flexibility (at the expense of efficiency). Although it has always been important to provide operational flexibility in order to enhance the value and agility of power generation assets, it has not been economical.

Gas turbines have traditionally been required to compromise their fast-loading capability to accommodate the limitations of the HRSG, steam turbine, and other components. By properly designing and integrating the plant components to allow a fast start capability while dramatically reducing startup fuel and emissions as well as water consumption, an economical solution is now available.

Changing the operational regime toward more intermediate duty significantly compromises its economics, especially when the asset is designed for base-load operation. The major challenge facing asset managers, with plants operating somewhere between intermediate and continuous-duty, is responding quickly to dispatch demands during high-profit opportunities and dispatching while revenue generation is feasible. Most combined-cycle plants recently built do not fully support these challenges.

This case will address the challenges to EPC contractors to develop improved plant concepts incorporating the latest advancements in GT technology and cycle integration that go beyond current considerations for plants designed to cycle.

This OEM's integration of advanced GT technologies into fast-start combined-cycle plants addresses the following challenges faced by operators:

- Increased operating costs due to fuel gas consumption during non-dispatched times (i.e., startup)
- Gas turbines being operated for a prolonged period of time and more frequently in higher emission operating regimes, which may encumber emissions permitting and operational flexibility
- Lowered revenue and/or permitting issues due to increased demineralized water consumption

Advanced Gas Turbines

The Siemens F- and G-class engines share several common design attributes that are found on engines dating back to the 1960s. The evolutionary design approach and product maturation philosophy employed by Siemens PG is apparent in examining these engines. Some common features of the base design include single variable inlet guide vanes (IGVs), interlocked discs, 3-D blading, a four-stage turbine, two bearings, axial exhaust, and cold end-driven generators.

Plant Design Considerations for Fast-Start/Cycling Capability

In today's market conditions it is assumed that combined-cycle power plants require the operational flexibility that allows them to operate anywhere between intermediate and continuous duty with the same efficiency as that of an advanced technology base-load continuous duty plant. It was found that the long established differentiation between hot, warm, and cold starts with the corresponding downtime assumptions of 8 hr (hot), 48 hr (warm) do not really reflect current operating scenarios. Therefore the fast-start plants are oriented along the following more challenging operating regime:

	Down Time	Starts per Year
Overnight	8–16 hr	200
Weekend	Up to 64 hr	50
Extended	>64 hr	2

* Source: Courtesy of Siemens. Adapted extracts from M. T. McManus and R. Baumgartner, "An Integrated Combined-Cycle Plant Design that Provides Fast Start Capability at Base-Load."

To increase the plant operational flexibility eight additional extended shutdowns were taken into consideration for component lifetime evaluation. While the plant is designed to accommodate the 260 starts per year with approximately 4000 hrs per year, it preserves its base-load operating capability. The flexibility of the design allows any combination of the number of overnight, weekend, and extended starts—up to 260 starts per year.

With regard to startup, the plant components restrict each other in terms of maximum allowable gradients/transient or necessary soaking time. This is an effect of interdependent systems and cannot be neglected. This OEM's strategy enables:

- The gas turbine (GT) load to be ramped up in combined-cycle applications as quickly as they are in simple-cycle applications, unrestricted by downstream components including the HRSG and ST
- The ability to vary and quantify the impact of a startup on the steam turbine maintenance schedule, enabling the asset manager flexibility in setting a business strategy considering the ST and HRSG along with GT maintenance

The fast-start plant start begins with concurrent ignition of both gas turbines (1) the gas turbine generators are synchronized at (2). After an unrestricted GT ramp-up the GTs reach rated power at (3). At this point the steam turbine is also rolled off. The ST generator synchronizes at (4) while the hot reheat and low-pressure system bypass valves are closed at (5) and the ST accepts all generated steam. At this point the plant reaches approximately 97% of its rated power output.

This optimized startup scheme makes the fast-start plant's startup approximately one-half that of its predecessor after overnight and weekend shutdowns. One of the key enablers for this improvement is Siemens' patented Benson®* Once-Through Steam Generator (Benson-OTSG), which can readily be applied with and without duct firing and SCR/CO-catalyst systems.

Benson Once-Through Steam Generator (OTSG)

The Benson technology has proven its capability and reliability over the past 70 years. This technology serves both sub- and supercritical designs in conventional steam power plants. Approximately 1000 Benson boilers with a cumulative steam output of more than 700,000 tons per hour have been built to date.

The authors' company also successfully applied the Benson evaporator principle in a 390 MW combined-cycle application at Cottam Development Center, Nottinghamshire (United Kingdom). This state-of-the-art single-shaft facility owned by Powergen Plc has been in commercial operation since 1999. During this period the Benson-OTSG has shown its direct applicability for fast startup and cycling operation, even though the Cottam plant was not explicitly designed for this duty.

Based on the increased market demand for fast startup/shutdown and cycling capabilities and the successful testing and operating experience at Cottam, the fast-start plants incorporate the Benson-OTSG in their designs. Due to industry problems encountered with once-through technology of other HRSG suppliers, the design features that distinguish the Siemens Benson-OTSG from other once-through designs will be detailed/described.

The evaporator stage is characterized by two bundles of vertical tubes, which are arranged in series in the horizontal gas-path. All tubes within an evaporator bundle are connected in parallel, creating natural circulation characteristics. The low mass flux design allows the fluid mass flow to self-adjust to the heat input in each tube row, meaning tubes with higher heat absorption will see an increased fluid mass flow. This design ensures static and dynamic flow stability throughout a wide load range. In addition it results in lower pressure losses across the evaporator compared to other once-through concepts.

The Benson-OTSG is capable of handling high temperature transients during a fast startup due to the elimination of thick-walled steam drums. The HP drum, in particular, impedes fast startup of conventional HRSGs by dictating long soaking periods and slow maximum allowable temperature ramp rates. The Benson-OTSG eliminates these restrictions by replacing the drum with a small separator vessel, which also provides a more economical arrangement.

The separator performs the function of water/steam separation during startup and shutdown. At steady-state operation, including low loads, the steam flow passes through the separator, as part of the interconnecting piping toward the superheater. The steam at the evaporator outlet is slightly superheated; consequently no separation occurs.

Another common feature of once-through technology is utilized to control the main steam temperature. In contrast to a drum-type HRSG, the Benson OTSG is able to control the main steam temperature by modulating the feedwater massflow.

This enables the system to compensate the effects of changing ambient conditions or GT loads within a certain range without placing the steam attemperators in operation. Attemperators are incorporated as interstage as well as final-stage desuperheaters for the high-pressure and reheat steam systems. This arrangement ensures compliance with the steam temperature limits, dictated by the steam turbine, during plant startup.

Simply stated, the Benson-OTSG retains all the positive features offered by the traditional drum-type HRSG,

* Benson is a registered trademark of Siemens Aktiengesellschaft.

including provisions for SCR, CO catalyst, duct firing, steam power augmentation, etc., while providing additional operating flexibility over drum-type HRSGs.

Steam Turbine

The steam turbine (ST) is started with power output in mind, which means that it is rapidly loaded with reduced and nearly constant steam temperatures until the bypass valves are closed with the GTs at base load. This start method eliminates steam vents to atmosphere at the HRSG, due to the incorporation of an auxiliary boiler. Since there are no HRSG- or ST-induced holds of the GT startup process, special care must be taken in handling the generated steam in order to meet the optimal steam temperature for fast loading of the ST.

The steam turbine is designed for full-arc admission without the use of a control stage. This design greatly reduces blade stresses during the startup process. The combination of full-arc admission with variable-pressure operation (which has become a standard for HRSGs in the industry) provides high efficiency, maximum operating flexibility, and minimized maintenance.

The application of high-capacity steam bypass systems for HP, IP, and LP steam is key to the operating flexibility and reliability/availability of any plant. The fast-start plants incorporate special provisions to route the HP bypassed steam into the cold-reheat of the HRSG to minimize thermal stressing of the HRSG reheater section during startup.

In order to provide further flexibility, the authors' company's patented Turbine Stress Controller (TSC) is supplied. The TSC consists of a stress evaluation system based on an ST startup program with an on-line life cycle counter. It calculates and controls stresses in all critical ST "thick wall" components (including the valves, HP casing and rotor body, and the IP rotor body) depending on three different ramp rates, i.e., planned, accelerated, and fast. The function of the TSC is to control the startup ramp rates in order to minimize material fatigue within the constraints of operating requirements and at the same time to calculate the cumulative fatigue of the monitored turbine parts. The TSC assigns an appropriate number of equivalent starts for each planned, accelerated, and fast start. With this knowledge of the effect of each type of start on the maintenance schedule of the steam turbine, the owner can make prudent business decisions regarding whether or not the asset should be used to satisfy transient dispatch requirements.

Steam/Water Cycle and Operational Chemistry

While the power market requirements have changed and many combined-cycle plants are being operated closer to intermediate duty with demand for fast startup/shutdown and cycling capability, asset owners still expect base-load plant efficiency levels. This is achieved not only by retaining the triple-pressure reheat steam/water cycle, but also by integration of the GT and cycle to maximize efficiency through systems such as rotor air cooling, fuel gas heating, and steam cooling of the GT transitions (in the W501G).

This cycle design has been thoroughly analyzed and as previously mentioned was first introduced at the Cottam Development Center. The Benson-OTSG is a once-through steam generator for the HP and IP sections (IP as required), while the LP section (and IP as required) retains a drum-type design. This HRSG configuration requires a combination of pH control with ammonia and oxygenated treatment. The oxygenated treatment creates and maintains a protective layer of magnetite and hematite inside the tubes.

As a further enhancement to rapidly achieve proper steam quality and to support rapid startup, the cycle incorporates a condensate polisher. The polisher provides high quality feedwater to the Benson-OTSG thus avoiding deposits in the evaporator tubes and thereby decreasing the evaporator pressure losses. The condensate is de-aerated in the condenser hotwell by vacuum, while the polisher removes the ammonia, which must then be added to the polisher effluent for proper pH adjustment.

The combined oxygen/ammonia treatment with the integrated condensate polisher of the fast-start plant has already contributed to the high reliability of Cottam Development Center and reduced the time to meet steam purity requirements to minutes.

BOP Considerations for Fast Startup/Shutdown and Cycling Operation

Noteworthy measures that are also applied in the fast-start plant are:

- Component redundancies
- Auxiliary boiler
- Vacuum pumps
- Stack damper
- Automated drain and vent valves
- Optimized steam piping warm-up concept
- High level of system and plant automation

Benefits

Since this plant can be operated equally as well (without an efficiency impact) in fast-start, intermediate-load, or base-load modes, there is no need to optimize or design for one operating regime over another. The ability to reach dispatchable load rapidly allows the owner to participate in more revenue generation opportunities. There are other advantages that can be quantified, including:

- Reduction of emissions
- Fuel savings
- Decrease in water consumption

The importance of startup emissions will be amplified as the facility tries to maximize flexibility. Without the cycle restricting the GT ramp rate, the GT reaches a low emission level rapidly producing fewer total emissions per startup. The plant thus has a significant reduction in startup emissions due to the GT reduction coupled with the benefit of any post combustion emissions controls (i.e., SCR or CO catalyst). The additional benefit, which is difficult to quantify generically, is that without these emission reductions, plant operation could be restricted or curtailed, e.g., in non-attainment areas.

Due to the lifting of cycle restrictions on the GT startup, the fast-start plants significantly reduce the fuel required for startup. This results in a savings of approximately $2.8 million a year ($23.7 M over 20 years) when the 2 × 1 W501F is operated in a nightly and weekend shutdown scenario.

Since the fast-start plant eliminates startup vents, there is also a saving in water consumption of almost 6 million gallons per year (approximately 63,000 m^3) for the 2 × 1 W501F.

CASE STUDY 8: CHALLENGES IN THE DESIGN OF HIGH LOAD CYCLING OPERATION FOR COMBINED-CYCLE POWER PLANTS*

This case concerns itself with the same issue as the OEM sourced case above.

Selling electricity profitably demands dispatch responsiveness. Merchant plants are connected to the grid only when commercially justifiable. These plants need to start up quickly and have a flexible operating range at low and high capacities in a short response time. Modifications in the balance of plant (BOP) design specifications regarding sizing, materials, and operational changes affect the operational success and life of the cycling plant.

Cycling refers to operating modes that respond to changes in system load requirements. This can be as gentle as load following to part-load operation and as abrupt as daily on/off duty. Granularity in the cycling definition is achieved by dissecting specific plant operating modes usually programmed in the distributed control system (DCS). Most power island equipment providers define an approximate cooldown criteria based on the number of hours that the equipment is shut down (Table 10–14).

TABLE 10–14 Cooldown Criteria

Mode	Combustion Turbine (hours)	HRSG (hours)	Steam Turbine (hours)
Cold start	8	48	72
Warm start	4	24	48
Hot restart	2	6	8

(Source: J. Zachary.)

Plant configuration was discussed in an earlier case in this chapter.

Cycle Pressure

Choosing and optimizing the cycle pressure rating is a difficult task. Relative to high mode cycling plants, where thermal stresses need to be minimized, the lower the pressure class, the better. Since the mid-1980s, the industry has moved from 1800–2000# pressure cycles and in some cases, 2400# cycles. The output benefits and heat rate benefits, along with the cost and startup time delay penalties, are shown in Table 10–15.

Cost differentials do not include a schedule analysis or cost associated with schedule issues.

Combustion Turbine

The basic design of a CT makes it the most suitable power train component for high mode cycling service. OEMs may provide EOH figures for starts/stops or LCA counters or call-up inspections after a given number of starts/stops.

Although the CTs are well suited for cycling operation, there are several options that can be exercised to enhance plant cycling ability:

- CT hot air bleed: This option involves piping a compressor air bleed connection to the inlet of the CT. Extracting this relatively small amount of hot air enables CT NO_x compliance down to approximately 60% of CT load. NO_x compliance increases the part-load operating range. Without this option, the minimum emission limit turndown on a heavy frame CT is about 70–80%.
- Reduced inspection interval packages: Exercising this option extends the interval between hot gas path inspections and required maintenance.
- CT fast start option: This option is not offered by all CT manufacturers. Instead of a typical ramp rate of 5 MW per minute, this option enables a higher ramp rate of about 7–9 MW per minute. The fast start option enables

* Reference: Extracts from the work of J. Zachary, including A. Chruściel, J. Zachary, and S. Keith, "Challenges in the Design of High Load Cycling Operation for Combined Cycle Power Plants."

TABLE 10–15 Output Benefits and Heat Rate Benefits

HP Nominal Throttle Pressure Rating	Output Increase (% MW)	Heat Rate Decrease (% BTU)	Net Cost Increase (% NPV Relative Dollars)	Cold Startup Time Increase (Minutes)	Warm Restart Time Increase (Minutes)	Hot Restart Time Increase (Minutes)
1800#	Base	Base	Base	Base	Base	Base
2000#	0.1	0.1	0.5	+15	+10	+5
2250#	0.2	0.3	1.0	+20	+15	+10
2400#	0.3	0.4	1.2	+35	+30	+20

(Source: J. Zachary.)

a faster start only if the HRSG and BOP systems are designed to support a rapid CT ramp to full power. However, the increased loading rate increases the CT's equivalent operating hours (EOH) for each fast start.

- Dehumidifying system options: Some manufacturers offer dehumidifying packages that protect the CT from internal corrosion when the unit is down. Depending on ambient conditions and the reliability of CT enclosure climate control, this option may not be required.

HRSG

By having a dedicated team follow the HRSG market, including failures and successes, continually updated commercial and technical specifications are assured. These specifications capture the following design experience.

Thermal Cycling

Thermal cycling is the premier offender in HRSG system failures. The cumulative fatigue damage from cycling cannot be reversed. Eliminate cyclic fatigue damage, and the HRSG's life cycle will increase substantially.

From vendor data (about combating cyclic fatigue):

- Every time the HRSG is started/stopped or cycled down to part-load, thermal fatigue damage occurs. Therefore, minimizing this operating regimen is an obvious cure for this failure. Where permitted, some operators turn down their CTs to absolute minimum load through the night, so that the HRSGs can remain online and avoid a cold or warm restart in the morning. HRSG vendors claim that fatigue damage from cold starts is 20–40 times greater than from warm starts. Keep the HRSG warm overnight or over the weekend by closing the stack damper, installing insulation up to the damper, steam sparging, or running the SCR ammonia vaporizer heaters.
- The HRSG superheaters can become supercoolers during shutdown or hot/warm restarts. The condensate flow out of a hot superheater is impressive when the CT rolls back in load, resulting in the exhaust temperature cooling the finned tubes. We have been asked to assure that the DCS automatically sequences all superheater drains open if the potential for condensate collection exists when operating at reduced CT loads (especially when purging the boiler and during shutdown). Additionally, lowering the pressure/temperature of the superheated steam by following the ramp down of the CT exhaust temperature will minimize condensate formation in the lower headers upon shutdown.
- In general, all superheater drains should be generously sized and located strategically to assure all bottom headers can be completely drained. These drains are headered together under the HRSG and headered along the side of the HRSG to the blowdown or flash tank. Assure that all lines slope in the proper direction in the hot condition.
- Automatic DCS control (with manual override) should be assured for all vents and drains required for startup and shutdown.
- Hot economizers should be protected from cold water. For warm restarts, the condensate supply must match economizer temperature within 25–50°F.

Thermal Stress

The need to understand and positively control and guide thermal expansion within the boiler is paramount. The HRSG vendors offer these design features to minimize thermal stress:

- Use lower cycle pressures and higher-grade materials to keep the high-pressure (HP) drum, HP headers, and tube connections to a minimum thickness. Consider a once-through HRSG design to eliminate the HP drum altogether. The material of choice for the superheater tubes is T91.
- In the technical specification, pay for and have the HRSG vendors perform their own fatigue analyses based on American Society of Mechanical Engineers (ASME) Section VIII and BS1113. Use the results to

justify where full penetration/full strength welds are required. The fatigue life of full penetration welds is at least 10 times greater than that of partial penetration welds. In many applications, the riser connection at the HP steam drum is identified as the critical area.

- Verify with the HRSG vendor that the tube offsets, harp support systems, liner panels, tube wash collection system, and all other components that will thermally expand are adequately designed to preclude thermal binding, galling, or other negative expansion effects.
- Heat treat all alloy tube bends to reduce residual stresses, avoid construction techniques that result in stress concentrations, use integral reinforced drum nozzles and full penetration nozzles, contour nozzle corners, and use blended tube to drum and other pressure boundary welds to reduce stress discontinuities.

Other Mitigation Steps

- Keep corrosion fatigue in check by assuring complete venting of economizers.
- Coat the inside of the main stack breaching and the inside portion of the stack up to the damper. Units that turn down at or near the acid or water dew point will experience wet corrosive surfaces in the back-end of the HRSG.
- Use a low-pressure (LP) economizer or condensate preheater bypass system or recirculation system to keep the temperature above the acid and water dew point past the last row of tubes in the HRSG. Consider upgrading the material on the last several rows of tubes (stainless steel is allowed for a condensate preheater/feedwater heater).
- Be aware that part-load operation of the HRSG may induce steaming in the intermediate pressure (IP) and LP economizer sections. This will accelerate local internal erosion. To prevent this, have a remote means of venting the offending economizer section directly to the drum or locate the drum level control valve downstream of the economizer section to assure that adequate pressure is always present in the economizer.
- Assure that the HRSG and connecting BOP systems are designed for the higher velocities, since part-load operation of the HRSG introduces low pressures and corresponding high specific volumes. Limit HP superheated steam and LP superheated steam and saturated steam to reasonable values. This will minimize erosion and noise at part-load conditions.
- Recognize that part-load operation (including sliding pressure) requires more separation equipment in the drum to prevent carryover to the superheater and eventually the ST. Typically ST manufacturers require a steam purity of 0.05–0.10 ppm total dissolved solids (TDS). This, unfortunately, will increase the size and thickness of the drum, which is counter to cycling requirements. In sliding pressure/part-load HRSG operation, the HP drum is typically sized for smooth operation down to 50–75% maximum continuous rating (MCR) throttle pressure.
- Size the drums to accommodate shrinking and swelling of the drum level as the load swings on the HRSG. The drum level control valve and feedwater system should be designed to keep up and successfully overcome a surge event. A relatively small surge margin on the boiler feedwater pump is usually all that is required to restore drum level during an upset condition. Additionally, assure the HP drum intermittent blowdown and drain system (including the blowdown system and the blowdown tank) are sized to control drum level upon an upset depressurization of the drum.
- Use T11 on LP riser bends to the LP drum. This increases erosion resistance for two phase and flashing flow.
- Follow the HRSG manufacturer's lay-up procedures for nitrogen blanketing, water levels, and residual chemicals.
- Size the blowdown system generously. This will enhance the removal of corrosion byproducts upon a boiler restart and facilitate water chemistry compliance.

SCR and CO Contribution to Extended Startup Times

The presence of a CO catalyst does not contribute significantly to the startup time required for the HRSG. Since the CO catalyst is located forward in the HRSG, the CT exhaust gases typically heat up the blocks within 30 minutes.

Conversely, the selective catalytic reduction (SCR) system requires significant startup time, especially if the unit is cold. The ammonia injection grid feed piping contains a tremendous amount of metal mass and requires significant heat to obtain operating temperature. If the HRSG uses an internal gas return system (hot gas feed from downstream of the catalyst) to vaporize the ammonia, then an additional heat source is required to enable fast HRSG startup by quickly getting the SCR catalyst and piping in the required operating temperature range of 600–800°F. Several methods exist to preheat the catalyst blocks and piping:

- For SCR systems with hot gas supply through a connection downstream of the HSRG SCR, provide a hot gas recirculation system. Channel a portion of the CT exhaust gas, through a valved piping system, from the transition duct area to the inlet of the ammonia vaporizer. This system will preheat the SCR and associated piping with sufficiently hot CT gases to enable faster NO_x compliance.

- Use an electric startup heater or permanent electric heater. With electric heat, this system can be activated before starting the CT.
- Heat trace the ammonia injecting feed piping and headers. Without an external source of heat, it takes at least 4 hours to heat the ammonia piping using a hot gas tap downstream of the catalyst.

Water Chemistry

Controlling the amount of dissolved oxygen in the condensate is critical, especially during startup. The most critical operational backstop to keep dissolved oxygen (DO) out of the feedwater is not to break condenser vacuum overnight (or ever). Maintaining condenser vacuum and preventing the hotwell from subcooling dramatically decreases the morning startup time and almost eliminates the huge swings in chemistry associated with cold startup. Successful strategies recommended to minimize corrosion potential due to DO include:

- Use an auxiliary boiler. For units that shut completely down at night, an auxiliary source of steam is used to maintain steam to the ST seals, thereby enabling condenser vacuum to be sustained. This auxiliary steam source is also used to sparge the condenser hotwell to prevent the condensate from subcooling and picking up DO. Additionally, the condenser sparging system is used during part-load conditions to assure the hotwell does not subcool. The auxiliary steam source also minimizes the amount of makeup demineralized water required to support operation of the sky valves.
- Design the condenser vacuum system to utilize LP motive steam to power the steam jet air ejectors (SJAEs). If using vacuum pumps, assure the vacuum pump cooling water supply is cold enough to stay ahead of flashing due to low vacuum conditions. A hybrid system can also be used (an arrangement where vacuum pumps take suction from SJAEs).
- Consider putting the ST gland steam condenser on the closed cooling water loop. This precludes having to operate the large condensate pumps during shutdown.
- Always maintain a positive pressure in the HRSG drums during shutdown. This is accomplished by closing the stack damper and all outlet valves to the HRSG after the CT is secured. It takes approximately 8–16 hours or longer to cool the HP drum to saturated conditions at 400–500°F. After the boiler has cooled to within the range of the auxiliary steam source pressure, open the HP evaporator spargers and keep the drum as close as possible to drum soak conditions. This would constitute a warm start of the HRSG with no ramp-up restrictions. The heat migration from the HP section to the IP and LP sections of the boiler will tend to keep those drums at significant pressure. The LP drum will have to be vented throughout the night to prevent overpressurization.
- If the HRSG is taken off line for maintenance under cold conditions, apply and monitor a nitrogen blanket above all water spaces and follow the manufacturer's recommendations for chemical dosing the drums. As the steam space pressure decreases, align the nitrogen system to push and replace the steam. This minimizes air in-leakage.

Steam Turbine

For one-on-one and two-on-one plants, the procedures are straightforward. For three-on-one plants, the startup procedure is involved, as is the hot/warm restart procedure. Direct bypasses, cascaded bypasses, auxiliary steam sources (if any), turbine backpressure, steam purity, sky valve operation, etc., all play an important part in steam turbine roll off.

A large East Coast combined-cycle facility was designed to accommodate the cycling requirements in Table 10–16 over the design life of the plant.

The areas of study included crack initiation factors, centrifugal stress, creep, and material optimization. The facility has been successfully cycling daily (two-shift operation with overnight shutdowns), with weekend shutdowns, since 1992.

Some of the lessons learned from this facility and other similar installations are:

- Rapid shutdown of the ST is preferred over forced cooling (gradually reducing load before shutdown).
- Components should be chosen that are designed for minimal centrifugal stress. Using finite element analysis, have the ST supplier verify that the design of the last stage blade does not have significant stress concentrators.

TABLE 10–16 Combined-Cycle Facility Requirements

Startups	
Hot Starts	10,000
Warm Starts	1700
Cold Starts	2000
Shutdowns	
Trip Shutdowns of CTs	500
Fast Shutdowns of CTs	1000
Normal Shutdown of CTs	Remainder

(Source: J. Zachary.)

- Materials optimized for creep resistance should be specified.
- Lower ST throttle pressure will allow reduced casing thickness.
- For temperature matching, especially in three-on-one installations, independent steam bypasses to the condenser are mandatory.
- Several operators use ST heater blankets to warm the casing to enable the ST to operate faster.
- To maximize cycle efficiency, the unit should be operated in sliding pressure mode, where the ST throttle valve is valve wide open (VWO). Assure the HRSG drum is sized to accommodate the higher specific volumes.
- Full-arc steam admission is commonly paired with sliding pressure boilers to improve minimum load operation and efficiency. Hybrid partial-arc admission is recommended for part-load performance and the ability to respond quickly to load changes.
- All major STs are provided standard with online stress analyzers and rotor stress indicators. Assure these are operational before turbine start.
- All extraction steam requirements (NO_x injection, offsite process, SJAEs, etc.) need to be cataloged to assure the ST has minimum cooling flow and can operate as designed during part-load dispatch. Steam turbines typically require at least 15% design steam flow to remove heat from any given section.

Balance of Plant Considerations

Proper system integration assures that the facility will produce output and heat rate in accordance with specifications. System integration via the BOP equipment and interconnecting/support systems is the glue that enables the CT, HRSG, and ST to generate power reliably at all design loads and configurations.

Table 10—17 is a checklist of BOP design issues.

Water Treatment System Design Features

Equipment design approaches that should be evaluated to minimize water treatment problems inherent in cycling and part-load operations include: feedwater, drum water, and steam purity requirements. Ramp-up rates may be limited by the need to avoid exceeding ASME and equipment suppliers' limits for transient drum water and steam purity.

Figure 10—66 shows a typical fluctuation of drum water quality as an 1800 psig steam generator is cycling.

As shown in Figure 10—67, the startup data graphs for an actual combined-cycle plant, a typical warm startup should take about 2½ hours, while a cold startup takes about 3½ hours. If this time is extended due to water treatment problems, peak rate revenue opportunities may be jeopardized.

Table 10—18 shows some of the inherent causes and effects that can be caused by an inability to consistently achieve and maintain the target feedwater, drum water, and steam purity requirements.

During these transient conditions, drum water characteristics, especially pH, are difficult to control and stabilize.

As a result of these problems, passivating films can be removed; corrosion products can be displaced and transported; feedwater internal treatment chemicals can be overfed or underfed; and steam purity can be compromised.

Also, when extended high-rate blowdown and corresponding makeup rates are required to achieve and maintain the drum water and steam purity requirements, costs are incurred to provide demineralized water production and storage capacity. Energy costs can also become significant.

Erosion, corrosion, deposition, hot spots, and all other associated problems are well understood. The most immediate risk to owner-operators and contractors with startup responsibility is exceeding equipment manufacturer's limits.

Longer term, these phenomena affect equipment serviceability and facility service life. Long-term effects can be monitored to provide "early warnings" with test heat exchangers and a comprehensive corrosion coupon program.

Operations-Based Design Criteria

Before any design features can be considered and evaluated, some basic criteria predicated on the anticipated pattern of operation should be established. Table 10—19 summarizes these operational requirements.

Owner-operators, major equipment suppliers, and design engineering contractors should agree on all of these operating criteria and schedule and develop a project document that clearly defines these operational requirements.

Ideally, a present worth could be developed in terms of dollars per hour of ramp-up time either from a warm start or cold start situation.

Specific Design Features for Cycling and Part-Load Operation

Figure 10—68 shows design features that should be carefully evaluated in terms of cost-effectiveness to facilitate cycling and part-load operation.

Condensate System

- Steam sparging in the condenser hotwell complements the use of auxiliary boiler and condenser vacuum pumps to degasify condensate in hotwells during short-term shutdown and to more rapidly degasify during ramp-up, ramp-down, and part-load operation.

TABLE 10—17 BOP Cycling Checklist

Plant Component	Design Feature	Comments
General	Properly characterize the plant operating modes	Specify the number of cold, warm, and hot starts; the number of shutdowns and the characterization of the shutdowns (normal, emergency, etc.).
	Define the lowest part-load operation point of the plant	For example, 1 (50%) × 1 × 0 for a 3 × 3 × 1 plant. The lowest load point usually corresponds to the minimum load on the CT while still meeting emissions.
	Assure all equipment tech specs include the requirement for cycling and part-load operations	
Heat sink	Part-load spargers in hotwell	This prevents the hotwell from subcooling and the subsequent O_2 entrainment in the condensate. This is important to prevent system corrosion. Additionally, the spargers can be used to pre-warm the hotwell upon cold start (in conjunction with the Auxiliary Boiler option).
	Condenser air removal system	Use vacuum pumps or alternatively use low-pressure SJAEs (100 psi) in conjunction with the Auxiliary Boiler option.
	Cooling tower low load operation	Use a bypass or some other option to prevent cooling tower freezing. Assure the fans can be controlled from the control room.
Main steam system	Motive steam supply to the SJAEs	If an auxiliary boiler is not available to provide seal steam, consider having two seal steam sources: HP and cold reheat (CRH). The HP source is available almost immediately on startup, thus enabling the faster availability of the condenser. When conditions permit, the lower energy source (CRH) can be used to improve plant efficiency.
	Assure pipe sizes are appropriate for the higher velocities associated with higher specific volumes during part-load cases	Follow the recommended velocity criteria for superheated for upset or emergency excursions.
	Cascade bypass	Cascaded bypass, as opposed to direct bypass, keeps the reheater section under positive pressure at all times (especially during startup, where a direct bypass system would put the reheater under vacuum for the initial 20 minutes until the intercepts open). Condensate can be readily blown out (not educted out), which enhances turbine water induction protection.
	Overboard steam line drains	Don't try to collect the condensate from any steam drain line. Route all ST upstream steam drain lines to atmosphere. Route only the downstream ST throttle valve drains to the condenser.
Feedwater system	Boiler feedwater pumps should be horizontal ring section type	Temperature gradients are OK. Pump pre-warming is not required.
	Variable speed drives for part-load operation of BFPs and other large rotating equipment	Promotes a cycle efficiency increase, especially with highly duct-fired units (>400 MMBTU/hr).
Condensate system	Assure condensate is clean and preheated prior to admission to a cold HRSG	Don't break condenser vacuum. Hotwell sparging and side stream polishers (ACC only) should provide for clean condensate.
	Properly size condenser air removal equipment	HEI standards are not sufficient at part-load conditions. As the condenser load decreases, backpressure decreases along with an increase of suction pressure to the vacuum removal system. Assure the system is designed to take the lower suction pressure and higher volumetric flow rate.
Natural gas system	Heat trace downstream of last natural gas filter	Prevents hydrocarbons from saturating in the dead leg portion of the gas line downstream of the startup heater or cycle heater (upon CT startup).
	Natural gas supply system to have high turndown capability	Assure gas line equipment can operate efficiently and satisfy the supplier gas spec at low loads. Gas scrubbers have only a 4—1 turndown.

TABLE 10–17 BOP Cycling Checklist—cont'd

Plant Component	Design Feature	Comments
Circulating water system	Auxiliary circulating pump at the cooling tower	Consider using an auxiliary circulating pump to handle overnight cooling loads, so the main circulating pumps can be isolated.
Miscellaneous	Freeze protection	Assure adequate freeze protection is provided, since overnight or weekend shutdown may occur during sub-zero conditions.
	Use on-line analyzers and automatic chemical feed systems	Where appropriate, this reduces operator interface requirements.
	Rate motors for three starts per hour	Assure electrical motors have adequate restart capability.

(Source: J. Zachary.)

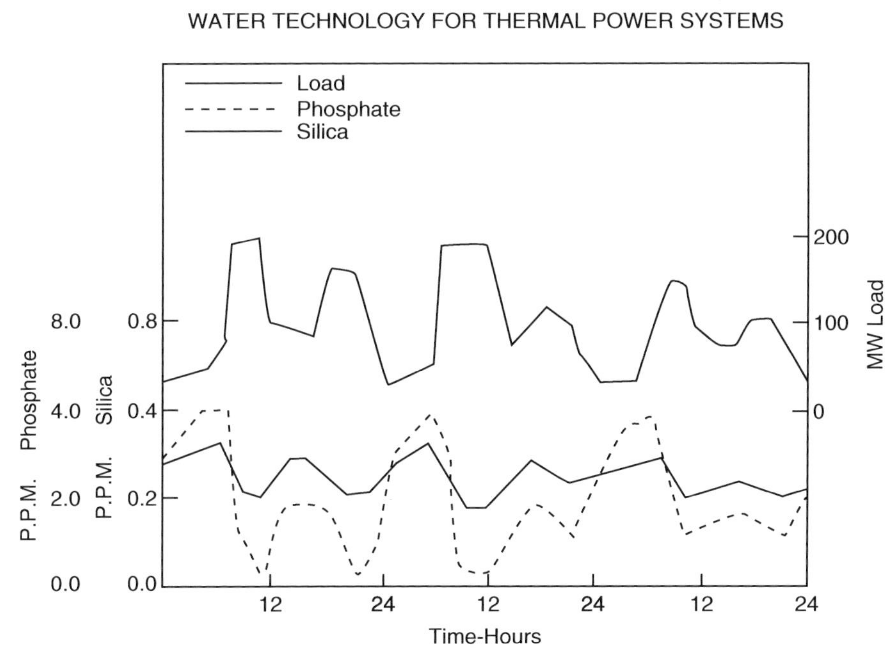

FIGURE 10–66 Variation of phosphate and silica with load cycling. *(Source: J. Zachary.)*

- Use of pre-coat condensate treatment system (using either cellulose filter medium or powdered ion exchange resin) may be justified in terms of decreased startup time and fewer short- and long-term operational problems.
- Pre-coat condensate treatment systems can operate as a crud-removal filter during startup and be switched to powdered ion exchange resin after operation is stabilized. This also provides limited protection from condensate contamination due to condenser leaks.
- Pre-coat condensate treatment and use of filming amines should especially be considered for use with air-cooled condensers, which may have large carbon steel surface areas in contact with steam and condensate.
- Though more costly, the "old fashioned" stand-alone tray type de-aerator with pegging steam and reservoir is the most effective way to de-aerate feedwater. This is especially true for cycling and part-load operations. When demineralized makeup exceeds 10–15% during sustained periods of operation (such as for cogeneration export steam), the tray type de-aerator may be required to consistently achieve the oxygen concentration level typically required for thermal-mechanical de-aeration.
- Installation of a powdered ion exchange resin polisher for desuperheat water should also be evaluated to prevent introduction of contaminants from condensate directly to steam lines.

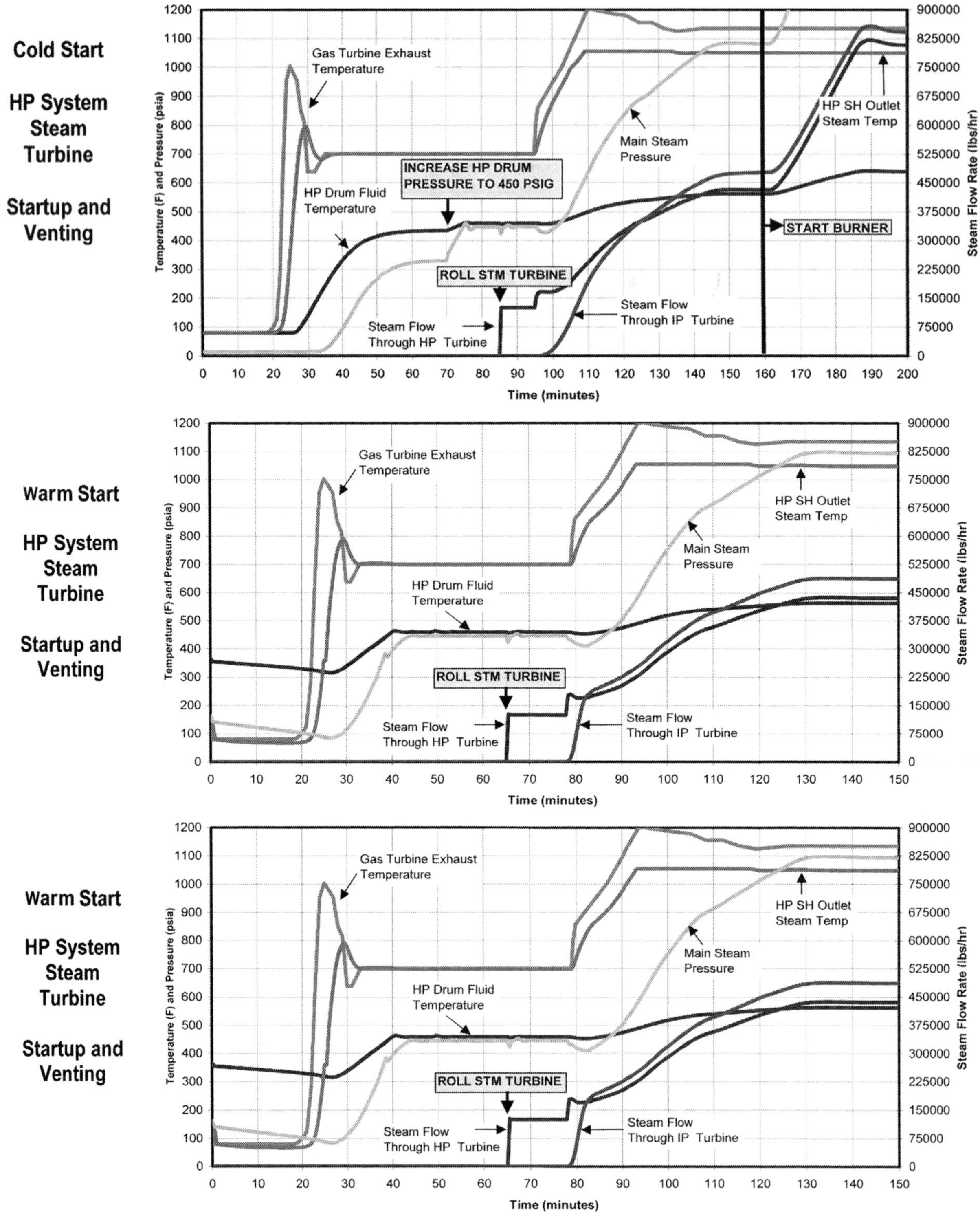

FIGURE 10–67 Startup duration graphs. *(Source: J. Zachary.)*

Feedwater Internal Chemical Treatment

- Owner-operator and design engineering contractor need to agree, as part of design criteria, on the feedwater internal treatment chemical approach that will be used.
- For cycling and part-load operation, use of equilibrium phosphate systems is increasing, as that appears to minimize the occurrence of phosphate hideout and the associated problems with stabilizing the drum water chemistry and the risk of hot spots or accelerated

TABLE 10–18 Feedwater, Drum Water, and Steam Purity: Inherent Causes of Problems; Short-Term and Long-Term Effects

Inherent Causes of Problems
Temperature excursions throughout the feedwater, condensate, and steam systems
Internal treatment chemical concentrations fluctuate throughout the feedwater, condensate, and steam systems
Circulation in drum not established; drum water levels fluctuate
Makeup as a proportion of total feedwater fluctuates
Short-term Effects
Startup time and responsiveness to dispatching
Extended high-rate blowdown and makeup
Longer-term Effects
Steam purity and associated problems
Accelerated corrosion of water side and steam side components

(Source: J. Zachary.)

TABLE 10–19 Operational Design Requirements: Schedule and Availability and Ramp-Up Rate Limitations

Schedule and Availability Requirements
Overall annual availability requirement
Planned outage and/or maintenance periods
Percentage of operation at full load and part-load
Number of cold and warm starts anticipated per year
Elapsed time between shutdowns and startups
Variation in that interval; i.e., daytime-nighttime, weekend, seasonal
Ramp-up Rate Requirements and Limitations
Ramp-up time from warm status to full load
Ramp-up time from cold status to full load
HRSG and ST manufacturers' thermal ramp-up and ramp-down rate limits

(Source: J. Zachary.)

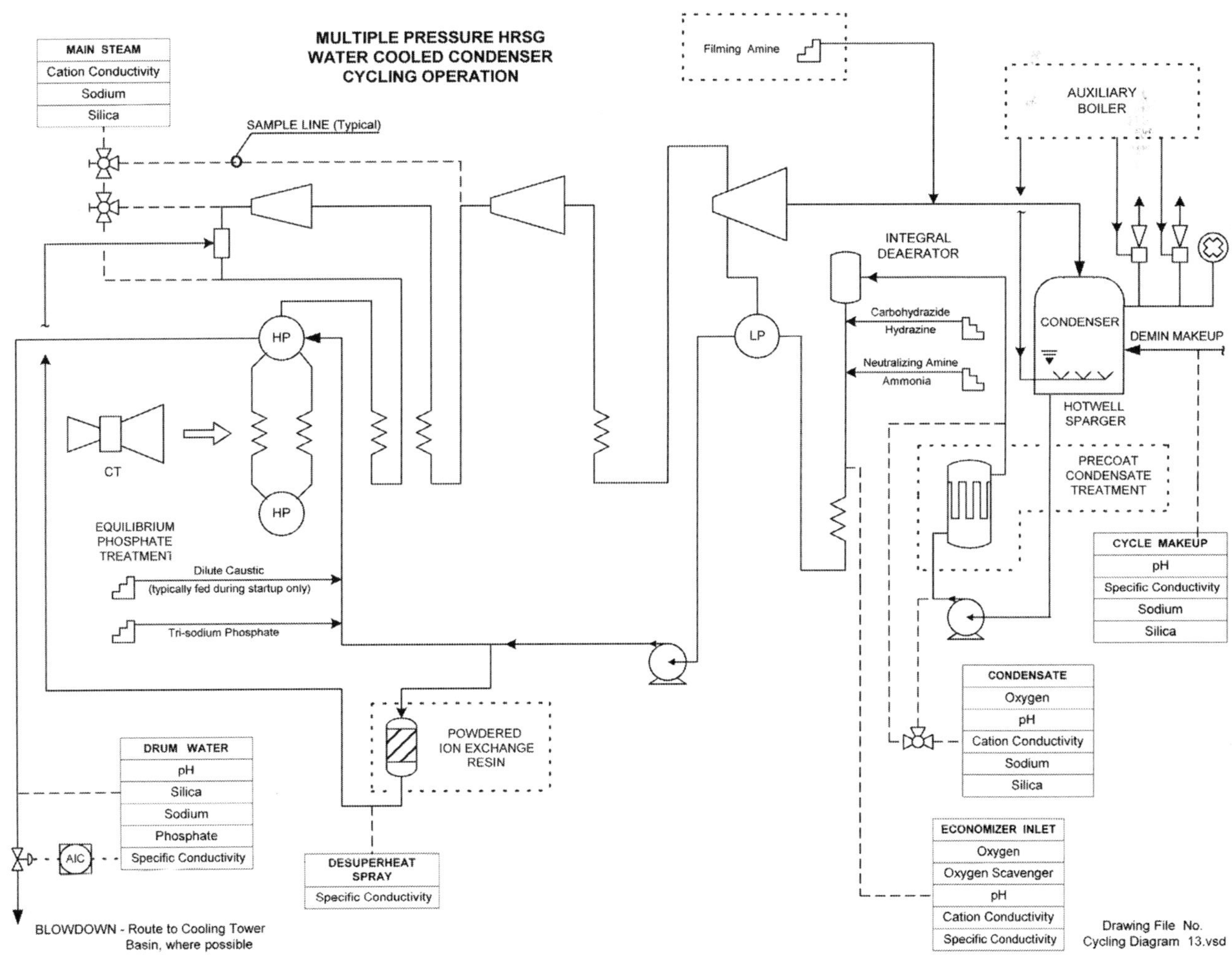

FIGURE 10–68 Design features for water-cooled condensers. *(Source: J. Zachary.)*

corrosion due to phosphate precipitation and/or scale accumulation.

- Dual feed capabilities should be considered, with one metering pump feeding from an organic amine tote and another feeding from an aqueous ammonia tank, with a common spare.
- An organic oxygen scavenger is needed that is relatively effective at low temperature but that will not create such a reducing atmosphere that it strips off protective oxidized films (typically magnetite). This must be a passivating type such as carbohydrazide. Provisions to continuously monitor excess oxygen scavenger as well as dissolved oxygen are also needed.

Sampling and Analysis

- Providing a comprehensive sampling and analysis system is probably the most cost-effective design provision to facilitate cycling and part-load operation.
- All samples need to be continuous with a centralized sample system and sink for all online analyzers. Sample system should include adequate sample conditioning, including filtration. Nonfiltered samples can quickly disable on-line analyzers. Provisions are also needed to take grab samples as close as possible to the origination of the sample line. Flush-out connections should also be provided at the origination of the sample lines.
- The capability to store and trend data for these parameters via a DCS system is also important.
- An empirical nomograph of expected values and control limits for the various parameters should be developed and available so that the chemical feed and blowdown can be manually controlled during ramp-up.

Makeup Demineralized Water System

- As previously discussed, the makeup demineralized water system design should be based on criteria that clearly describe the anticipated pattern of operation.
- This pattern determines the necessary demineralized water storage capacity to allow faster stabilization during startup when blowdown rate is high. It also can allow the makeup demineralized water system to be sized and divided into trains in a way that more closely matches the flow rates required.
- If reverse osmosis (RO) makeup water treatment is anticipated, provisions should be included to automatically recycle RO product water when the demineralized water or condensate storage tank is nearly full. This will minimize RO system cycling and shutdown.
- Providing variable frequency drives for RO feedwater pumps should also be evaluated to minimize hydraulic transients in element housings when systems are started up or as trains are brought in and out of service.

Heat Rejection with Water-Cooled Condenser and Cooling Tower

- Establishing and maintaining vacuum in the condenser is usually done with staged SJAEs (one set for hogging; one set for steady state operation). This is not necessarily effective over a wide range of part-load operations. Providing vacuum pumps should be evaluated.
- Using a cooling tower to receive HRSG blowdown is another method to be considered.
- Using cooling tower sidestream filtration makes circulating water quality easier to stabilize on startup and shutdown. It reduces consumption of chlorine and other internal treatment chemicals and will remove suspended solids that may be introduced from condensate dumps or HRSG blowdown.
- Copper-containing alloys should not be used because they are incompatible with amines, especially the high concentrations of amines and dissociated ammonia that are encountered during startups and shutdowns.
- Condenser tubes should be seal welded.
- Coupon testing with all types of metallurgy present in the circulating water system is important to monitor the cumulative effects of fluctuation in the circulating water chemistry.

Air-Cooled Condenser Heat Rejection

- ACCs sometimes are proposed with undersized air removal equipment, as HEI standards have been demonstrated to be undersized for some ACC applications, especially in cycling and part-load operation. Generously size ACC air removal equipment.
- As previously discussed, pre-coat condensate treatment and use of filming amines should especially be considered for use with air-cooled condensers, which may have large carbon steel surface areas in contact with steam and condensate.

Chapter 11

Gaseous Emissions and the Environment

"Normal people . . . believe that if it ain't broke, don't fix it. Engineers believe that if it ain't broke, it doesn't have enough features yet."
—Scott Adams, *The Dilbert Principle*

Chapter Outline

Human activity* creates a stress on the environment. Land, air, and water are inundated with the detrimental effects of industry. Air has received more attention and publicity than land and water in terms of the pollution it suffers. Besides, air pollution at some point affects land and water, either directly or indirectly. In the interests of brevity, this chapter will deal mostly with air pollution. Separate standards exist, for instance, for legislating the water effluent from power stations.

Gas turbines running at optimal emissions status tend to have TIT profiles and internal air cooling that promote the best TBOs and component longevity, particularly in the hot section. If TITs or temperatures anywhere within the gas turbine increase over optimum design settings (due to blockage or partial blockage in any of the modules), component lives and emissions are adversely affected. Thus engineers, especially those who have considerable field and overhaul experience, frequently concur that environmental emissions-lowering technologies mean better turbine component lives, as they lead to the engine operating cooler (than it would without the technologies in question for a given power setting).

As noted in Chapter 4 on gas turbine components, high temperature promotes the formation of NO_x emissions. When NO_x emissions rise, so do other emissions. When TITs rise, the potential for hot section wear, due in part to elevated temperatures, rises. Less hot section wear generally means longer TBOs and reduced costs per fired hour.

GASEOUS EMISSIONS

Traditionally, Europeans tend to be more environmentally proactive than people in the United States, except in the case of California. This is now changing as the United States realizes that the global warming "myth" is part of worldwide climate change, escalating formation of deserts and drought trends, increasingly felt by the United States. Countries like Canada, Australia, and New Zealand like to be environmentally conscious countries and work with the EU and the USA in information sharing agreements. New

* [11-1] Working case notes, Claire Soares, 1975 through 2003; AMSE IGTI panel sessions 1985 through 2003, "Engine condition monitoring systems as they relate to life extension of gas turbine engine components," chairman, Claire Soares.

partnerships are formed increasingly. The USA and China now have an agreement to work together on CSS.

Environmental consciousness is growing overseas as well. India had considerable success with its tree planting program (for CO_2 absorption purposes). This is a similar initiative to New Zealand's calculations when building the Stratford station which led to a mandate that the station's builders plant several acres of forest to absorb the increased CO_2 emissions.

Brazil has instigated a sugar cane ethanol fuel program that has made the country energy independent and helped reduce its emissions, were it to use some of the cheaper alternatives to clean oil and gas. These alternatives used by countries poor in clean fossil fuels include residual or bunker oil.

It is a physical fact that CO_2 does not proceed past the stratosphere and that it absorbs heat. One can argue "how much CO_2," but that amount is increasing with human activity and CO_2's properties include heat retention. As do methane's, the chlorofluorocarbons once used as refrigerants (they still are around in many post-Communism economies), unburned hydrocarbons from gas turbine combustion, and a variety of other gases less abundant than CO_2. Granted some of them, like natural gas (25 times worse in terms of trapping the free oxygen in stratospheric ozone), are more dangerous per unit weight, but they are vastly less abundant than CO_2.

Also CO_2 ppm by volume in the atmosphere has increased by several hundred percent in just a few decades. Trapped bubbles, representative of atmosphere at the time, in glacial cores that represent a few centuries' worth of atmosphere (and other things) prove that beyond any doubt. That rate is without precedent in the planet's history and there is no prior experience for where these trends may go.

Arctic Warming Evidence*

In Ilulissat, Greenland, massive sections of ice break off the Sermeq Kujalleq glacier and drift into the Arctic Ocean, with increasing frequency. That frequency and size of the icefalls indicate that the frozen sheet covering the world's largest island is thinning. In 2002–2003, a 6-mile-long stretch of the Sermeq Kujalleq glacier broke off near Ilulissat, 155 miles north of the Arctic Circle. Although Greenland, three times the size of Texas, is a glaring case, climate change is increasingly evident from the northward spread of spruce beetles in Canada to melting permafrost in Alaska and northern Russia.

The Inuit point out that once "we could walk on the ice in the fjord between the icebergs for a six-month period during the winter, drill holes and fish. We can only do that for a month or two now. It has become more difficult to drive dog sleds because the ice between the icebergs isn't solid anymore."

The Sirmilik glacier in southern Greenland has retreated nearly 7 miles, and the Sermeq Kujalleq glacier near Ilulissat is also shrinking. In 1967, satellite imagery measured it moving 4.3 miles per year. In 2003, the rate was 8.1 miles.

Greenland's seal hunters foresee a reduction in their trade in seals. The seals will head north to colder places.

The Arctic sea ice has decreased by about 8%, or more than 380,000 square miles, over the past 30 years. In Sisimiut, Greenland's second largest town, lakes have doubled in size in the last decade. The average ocean temperature off Greenland's west coast has risen in recent years, from 38.3°F to 40.6°F. Within a century, this trend can result in an Arctic that is ice free in the summer.

With warmer temperatures, some species of bacteria, plants, and animals disappear. For instance, polar bears depend on sea ice to breed and forage and have been filmed drowning when ice floes are too far apart for them to cross.

Sweden's Sami tribe was once a nomadic, reindeer-herding people. Increased warmth breeds new plant species that suffocate other plants that are the main food for the reindeer, the basis for the tribe's existence.

In the Yamalo-Nenets region in western Siberia, ethnic Nenet live mostly off hunting, fishing, and deer breeding. Bream fish in their river, from warmer waters, now prey upon eggs of indigenous fish.

Melting permafrost has damaged hundreds of buildings, railway lines, airport runways, and gas pipelines in Russia; reference the 2004 Arctic Climate Impact Assessment commissioned by the Arctic Council, an intergovernmental body. Similar evidence in countries that have a permafrost zone mounts.

*Reference: Associated Press article, J. Olsen, M. Danilova, J. Heintz (Moscow), Ritter (Stockholm), and B. Duff-Brown (Toronto), 2005.

As the industrial world continues to develop, storms become more severe, sea levels rise, entire species and human cultures are poised on the brink of extinction. The clearest evidence for global warming appears to be in the Arctic.

It is equally true that some areas of the world are growing colder (like parts of Labrador and northern Europe) and other areas of the world are growing ominously hotter, in some cases drier. What was permafrost beneath the supports of the Alaskan pipeline is becoming muskeg. True, engineers there can swiftly ensure that, with refrigeration coils or similar technology, the supports will not bend, that is, bend enough to contribute to the pollution from leaking oil pipelines (whose pervasive silent onslaught way exceeds the pollution committed by the *Exxon Valdes*) globally.

If you can produce power by burning less carbon per unit of power, then you are more efficient in terms of conserving fossil reserves. If you can design gas turbines with cogeneration and waste-heat-recovery systems that:

- Cool inlet air to a gas turbine compressor, thereby increasing the air mass per unit volume available for combustion given any set of atmospheric conditions that is available for combustion, and thereby increasing the power production potential of a gas turbine for a given set of atmospheric conditions,

- Inject water or steam into a gas turbine increasing the power potential of that turbine for a given set of atmospheric conditions,
- Intercool air during the gas turbine's compression process, thereby increasing the mass flow per unit volume of air exiting the compressor,
- Raise stoichiometric efficiency (reduce unburned hydrocarbons and "excess fuel" required) in gas turbine combustors,
- Reduce airfoil losses,
- Reheat turbine products of combustion during the turbine expansion process,
- Collect the products of combustion (which include some unburned hydrocarbons), add another burst of fuel, and ignite the mixture to produce still more power (that is what pilots call an *afterburner*, but some power generation engineers do it too now and use different terminology),
- Use the "waste" heat in products of combustion to help keep a town's population warm (combined heat and power) or help grow tomatoes in a greenhouse,
- Reinject the exhaust CO_2 somewhere for the purpose of making something else (like oil) rise to the surface faster (without having to spend energy on just that process),

you are then using less fuel per unit of work or energy (rate of change of work) for which you can potentially earn tax credits or charge a tariff. You are more efficient, which ultimately means less pollution, of CO_2 and everything else that fossil fuel contains.

Contemporary gas turbines are no longer limited to burning just gas or even LNG. They can burn gasified coal (either pulverized coal or coal "gas" from steam injection into a coal seam), paper production ("black liquor") waste, flue gases from steel mills, petrochemical waste, and a host of nonconventional fuels. Several of these fuels will release a larger load of CO_2 to the atmosphere when burned, so end users would be well advised to follow New Zealand's Stratford station example.

If we compare the weight of carbon in unburned coal, oil, and methane gas (input-to-power-production-system basis), we see that the values of weight of carbon in kg/GJ are as follows:*

Coal = 90 to 100
Oil = 73 to 74
Gas = 50

* Source: C. Soares, *Environmental Technology and Economics: Sustainable Development in Industry* (Boston: Butterworth-Heinemann, 1999).

If we compare the weight of carbon dioxide (output-from-power-production-system basis) resulting from combustion of coal, oil, and gas, we see that the values of carbon dioxide in kg/MW-hr are as follows:

Coal = 1000 to 1200
Oil = 750 to 850
Gas = 200 to 500

Even without commitment to treaties such as the Kyoto protocol, countries with a tradition of responsible environmental behavior continue to try to find ways to mitigate CO_2 production. This can be at pre-combustion stages (design, legislation, taxes) or post-combustion (reinjection into rock, absorption by some means), which is generally more expensive.

Statoil's work in this vein is summarized as a case at the end of this chapter. It provides interesting insight into what can be done to limit CO_2 atmospheric emissions when a company values doing that.

The entire field of environmental emissions deserves and gets its own books and separate books on each subtopic. So this book deals with the main topics in general gas turbine operation that affect environmental emissions, as follows:

- Combustion chamber (CC) and low-NO_x CC technology (see Chapter 4)
- Alternative fuels (see Chapter 7)
- Cycle modifications (see Chapters 3 and 10)
- Template for environmental emissions and permitting (this chapter)

Gas turbine operation intrinsically involves steam turbines, when in the combined-cycle mode. If supplementary fuel is involved to boost steam production, the emissions from those boiler processes are likely to be similar to those created by solo-steam turbine operation. Emissions that result will generally be higher (per unit weight of fuel) than emissions from gas turbines, regardless of what fuel the gas turbine uses. Depending on the quality of the coal used, SO_x (oxides of sulfur) can be an issue.

The reader is asked to note that environmental affecting issues and implications occur in every facet of gas turbine technology either directly or indirectly. For instance, pulsation technology (see Chapter 9 on controls, instrumentation, and diagnostics) may pick up incipient failure in combustion liners, which in turn may have affected flame pattern, temperature distribution in the CCs, and therefore emissions produced (see Tables 11−1 to 11−9).

EFFECTS OF EMISSIONS ON AIRCRAFT GAS TURBINE ENGINES

All gas turbines have to contend with the intake of emissions, solid, liquid or gaseous. The nature of these and their effect on the gas turbine will vary with the gas

TABLE 11—1 Emission Factors[a] for Nitrogen Oxides (NO_x) and Carbon Monoxide (CO) from Stationary Gas Turbines

Turbine Type	Nitrogen Oxides		Carbon Monoxide	
Natural gas-fired turbines[b]	(lb/MMBtu)[c] (fuel input)	Emission factor rating	(lb/MMBtu)[c] (fuel input)	Emission factor rating
Uncontrolled	3.2 E-01	A	8.2 E-02[d]	A
Water-steam injection	1.3 E-01	A	3.0 E-02	A
Lean-premix	9.9 E-02	D	1.5 E-02	D
Distillate oil-fired turbines[e]	(lb/MMBtu)[f] (fuel input)	Emission factor rating	(lb/MMBtu)[f] (fuel input)	Emission factor rating
Uncontrolled	8.8 E-01	C	3.3 E-03	C
Water-steam injection	2.4 E-01	B	7.6 E-02	C
Landfill gas-fired turbines[g]	(lb/MMBtu)[h] (fuel input)	Emission factor rating	(lb/MMBtu)[h] (fuel input)	Emission factor rating
Uncontrolled	1.4 E-01	A	4.4 E-01	A
Digester gas-fired turbines[j]	(lb/MMBtu)[k] (fuel input)	Emission factor rating	(lb/MMBtu)[k] (fuel input)	Emission factor rating
Uncontrolled	1.6 E-01	D	1.7 E-02	D

[a]*Factors are derived from units operating at high loads (≥80% load) only. For information on units operating at other loads, consult the background report for this chapter, available at www.epa.gov/ttn/chief.*
[b]*Source classification codes (SCCs) for natural gas-fired turbines include 2-01-002-01, 2-02-002-01, 2-02-002-03, 2-03-002-02, and 2-03-002-03. The emission factors in this table may be converted to other natural gas heating values by multiplying the given emission factor by the ratio of the specified heating value to this average heating value.*
[c]*Emission factors based on an average natural gas heating value (HHV) of 1020 Btu/scf at 60°F. To convert from (lb/MMBtu) to (lb/10^6 scf), multiply by 1020.*
[d]*It is recognized that the uncontrolled emission factor for CO is higher than the water-steam injection and lean-premix emission factors, which is contrary to expectation. The EPA could not identify the reason for this behavior, except that the data sets used for developing these factors are different.*
[e]*SCCs for distillate oil-fired turbines include 2-01-001-01, 2-02-001-01, 2-02-001-03, and 2-03-001-02.*
[f]*Emission factors based on an average distillate oil heating value of 139 MMBtu/10^3 gallons. To convert from (lb/MMBtu) to (lb/10^3 gallons), multiply by 139.*
[g]*SCC for landfill gas-fired turbines is 2-03-008-01.*
[h]*Emission factors based on an average landfill gas heating value of 400 Btu/scf at 60°F. To convert from (lb/MMBtu) to (lb/10^6 scf) multiply by 400.*
[j]*SCC for digester gas-fired turbine is 2-03-007-01.*
[k]*Emission factors based on an average digester gas heating value of 600 Btu/scf at 60°F. To convert from (lb/MMBtu) to (lb/10^6 scf) multiply by 600.*

turbine's application and physical site. Gas turbines offshore, for instance, ingest oil fumes offshore, which then provides a sticky "base" on which other airborne pollutants may stick. When offshore, for naval vessels, low flying craft (helicopters) and offshore platform gas turbines, salt ingestion can be an issue. In these applications, regular, even frequent online washing can help a great deal. Also the platform and marine vessels can also use inlet air filtration to help their problem. Gas turbines in the tropics may ingest huge hordes of insects (see the chapter on air filters) in their air filters requiring frequent filter element changes. In cold climates, air filtration has to reduce the possibility that ice may form in the gas turbine inlet.

However, aircraft gas turbines are particularly vulnerable to the increasing pollutants in the atmosphere. The extent of their problem will vary depending on their flight altitude and where in the world they fly. However, they do have a problem that requires some attention.

Consider, for instance, Iceland's E15 volcano eruption in 2010, where ash often reached 30,000 feet. While no aircraft actually flew through the ash, airlines and related industries lost $5 billion with shutdowns and rerouting. An ASME IGTI panel convened in Vancouver in 2010 reached the consensus[1]

1. From: "First IGTI Forum on Jet Engine Volcanic Ash Ingestion" by L. Langston, published in Global Gas Turbine News, Dec 2011 (p. 58).

TABLE 11–2 Emission Factors[a] (Uncontrolled) for Criteria Pollutants and Greenhouse Gases from Stationary Gas Turbines

	Natural Gas-Fired Turbines[b]		Distillate Oil-Fired Turbines[d]	
Pollutant	(lb/MMBtu)[c] (fuel input)	Emission Factor Rating	(lb/MMBtu)[e] (fuel input)	Emission Factor Rating
CO_2[f]	110	A	157	A
N_2O	0.003[g]	E	ND	NA
Lead	ND	NA	1.4 E-05	C
SO_2	0.94Sh	B	1.01S[h]	B
Methane	8.6 E-03	C	ND	NA
VOC	2.1 E-03	D	4.1 E-04[j]	E
TOC[k]	1.1 E-02	B	4.0 E-03[i]	C
PM (condensable)	4.7 E-03[i]	C	7.2 E-03[i]	C
PM (filterable)	1.9 E-03[i]	C	4.3 E-03[i]	C
PM (total)	6.6 E-03[i]	C	1.2 E-02[i]	C

[a]*Factors are derived from units operating at high loads (≥80% load) only. For information on units operating at other loads, consult the background report for this chapter, available at www.epa.gov/ttn/chief. ND = no data, NA = Not applicable.*
[b]*SCCs for natural gas-fired turbines include 2-01-002-01, 2-02-002-01 & 03, and 2-03-002-02, and 03.*
[c]*Emission factors based on an average natural gas heating value (HHV) of 1020 Btu/scf at 60°F. To convert from (lb/MMBtu) to (lb/10^6 scf), multiply by 1020. Similarly, these emission factors can be converted to other natural gas heating values.*
[d]*SCCs for distillate oil-fired turbines are 2-01-001-01, 2-02-001-01, 2-02-001-03, and 2-03-001-02.*
[e]*Emission factors based on an average distillate oil heating value of 139 MMBtu/10^3 gallons. To convert from (lb/MMBtu) to (lb/10^3 gallons), multiply by 139.*
[f]*Based on 99.5% conversion of fuel carbon to CO_2 for natural gas and 99% conversion of fuel carbon to CO_2 for distillate oil. CO_2 (natural gas) [lb/MMBtu] = (0.0036 scf/Btu)(%CON)(C)(D), where %CON = weight percent conversion of fuel carbon to CO_2, C = carbon content of fuel by weight, and D = density of fuel. For natural gas, C is assumed at 75%, and D is assumed at 4.1 E+04 lb/10^6scf. For distillate oil, CO_2 (distillate oil) [lb/MMBtu] = (26.4 gal/MMBtu) (%CON)(C)(D), where C is assumed at 87%, and the D is assumed at 6.9 lb/gallon.*
[g]*Emission factor is carried over from the previous revision to AP-42 (Supplement B, October 1996) and is based on limited source tests on a single turbine with water-steam injection (Reference 5).*
[h]*All sulfur in the fuel is assumed to be converted to SO_2. S = percent sulfur in fuel. Example, if sulfur content in the fuel is 3.4%, then S = 3.4. If S is not available, use 3.4 E-03 lb/MMBtu for natural gas turbines, and 3.3 E-02 lb/MMBtu for distillate oil turbines (the equations are more accurate).*
[i]*Emission factors are based on combustion turbines using water-steam injection.*
[j]*VOC emissions are assumed equal to the sum of organic emissions.*
[k]*Pollutant referenced as THC in the gathered emission tests. It is assumed as TOC, because it is based on EPA Test Method 25A.*

that 2 mg/cubic meter was a conservative upper limit for atmospheric volcanic ash density for safe jet aircraft flight.

Volcanoes aside, around five billion tons[2] of dust and aerosol particles are emitted into the atmosphere, mainly due to natural processes. The dust from desert regions of the world averages 1.5 billion tons, the sand from the Sahara in Africa accounts for 60% of the sand.

Studies done by this reference used data from several hundred overhauls conducted by the MTU overhaul shop in Hannover, Germany. They found "the average MCpFH (maintenance costs per flight hour) of certain selected key customers vary by approx. 20 percent below and above the overall average. Operators with high activities in erosive areas have shown the highest negative cost variation. The combination of reduced durability and increased maintenance costs can raise the average MCpFH for engine maintenance for an average operator by between 35 and 50 percent."

2. From: "Environmental Influences on Engine Performance Degradation," Wensky, Winkler and Friedrichs, GT2010-22748.

Emissions from Coal as a Fuel

The acceptance of the effect of human-activity-caused carbon dioxide emissions has been slower in some countries, even highly developed ones like the United States. There is still a school of thought that says that climate changes will happen humans or not. However, gas turbine engineers accept that if they burn less fuel, they save on

TABLE 11–3 Emission Factors[a] (Uncontrolled) for Criteria Pollutants and Greenhouse Gases from Stationary Gas Turbines

	Landfill Gas-Fired Turbines[b]		Digester Gas-Fired Turbines[d]	
Pollutants	(lb/MMBtu)[c]	Emission Factor Rating	(lb/MMBtu)[e]	Emission Factor Rating
CO_2[f]	50	D	27	C
Lead	ND	NA	<3.4 E-06[g]	D
PM-10	2.3 E-02	B	1.2 E-02	C
SO_2	4.5 E-02	C	6.5 E-03	D
VOC[h]	1.3 E-02	B	5.8 E-03	D

[a]*Factors are derived from units operating at high loads (≥80% load) only. For information on units operating at other loads, consult the background report for this chapter, available at www.epa.gov/ttn/chief. ND = no data, NA = not applicable.*
[b]*SCC for landfill gas-fired turbines is 2-03-008-01.*
[c]*Emission factors based on an average landfill gas heating value (HHV) of 400 Btu/scf at 60° F. To convert from (lb/MMBtu) to (lb/10^6 scf), multiply by 400.*
[d]*SCC for digester gas-fired turbine include 2-03-007-01.*
[e]*Emission factors based on an average digester gas heating value of 600 Btu/scf at 60° F. To convert from (lb/MMBtu) to (lb/10^6 scf), multiply by 600.*
[f]*For landfill gas and digester gas, CO_2 is presented in test data as volume percent of the exhaust stream (4.0–4.5%).*
[g]*Compound was not detected. The presented emission value is based on one half of the detection limit.*
[h]*Based on adding the formaldehyde emissions to the NMHC.*

fuel bill money and emissions. If they work in a country where emissions are taxed, they save more money. Some maintenance engineers realize that lower emissions in many cases means a cooler running engine, great component life and therefore reduced costs per fired hour. Especially in the world's largest industry (power generation), lowering emissions was therefore always a bright idea, despite a political tendency to call Kyoto threatening because it "makes us uncompetitive" in both the USA and Australia.

The current gas supply boon (from fracking) notwithstanding, coal is still the world's most plentiful fossil fuel. Renewables and smart grids (which make renewables easier to add to the power supply mix) notwithstanding, the global power demand growth is nothing short of a surge and everyone alive today will not live to see the death of fossil fuel use.

In the USA, fervent campaigns by organizations like the Sierra Club have literally forced some coal power plants to shut down. That in turn forced many coal miners out of work. The coal lobby may not be as strong as the gas and oil equivalents in Washington DC, but the loss of jobs during an already rough economy has resulted in proactivity in research areas that will benefit coal. The truth is that China, many countries in Asia, Europe, and Africa have already been proactive, within their means, for many decades already. However, the economic resources of the USA are always a great ally for the rest of the world to have.

Given that conventional coal plants are inefficient and give off twice as much CO_2 as a GT natural gas plant, countries other than the USA had installed many supercritical steam plants. Supercritical steam teachnology has been around since the 1950s. However, now the US DOE is working on metallurgy that will accept steam that is over 700°C. This then takes us to ultra-supercritical (USC) and advanced USC steam plants, with consequential increases in efficiency.

The Europeans, particularly the Norwegians and Swedes, have actively researched and practiced CO_2 sequestration. Statoil experience in this area is discussed later in this chapter. Now the USA is assigning resources to this field as well, most significantly in the FutureGen project in Illinois. One slated to employ IGCC technology, FutureGen will use oxycombustion that is expected to reduce emissions more than IGCC would. There are also plans to employ CO_2 sequestration in this project.

CARBON DIOXIDE SEQUESTRATION*

Underground sequestration of CO_2 presents technical, legal and public acceptance issues. Current demonstration project will require years of operation in order to determine the long-term impact in the injection process on the

* [11-2]: Courtesy of J. Zachery. Extracts from "CO_2 Sequestration by Conventional and Alternative Means," GT2010-22318.

TABLE 11–4 Emission Factors (Uncontrolled) for Hazardous Air Pollutants from Natural Gas-Fired Stationary Gas Turbines[a,b]

Pollutant	Emission Factor (lb/MMBtu)[c]	Emission Factor Rating
1,3-Butadiene[d]	<4.3 E-07	D
Acetaldehyde	4.0 E-05	C
Acrolein	6.4 E-06	C
Benzene[e]	1.2 E-05	A
Ethylbenzene	3.2 E-05	C
Formaldehyde[f]	7.1 E-04	A
Naphthalene	1.3 E-06	C
PAH	2.2 E-06	C
Propylene oxide[d]	<2.9 E-05	D
Toluene	1.3 E-04	C
Xylenes	6.4 E-05	C

[a]*SCC for natural gas-fired turbines include 2-01-002-01, 2-02-002-01, 2-02-002-03, 2-03-002-02, and 2-03-002-03. Hazardous air pollutants as defined in Section 112 (b) of the Clean Air Act.*
[b]*Factors are derived from units operating at high loads (≥80% load) only. For information on units operating at other loads, consult the background report for this chapter, available at www.epa.gov/ttn/chief.*
[c]*Emission factors based on an average natural gas heating value (HHV) of 1020 Btu/scf at 60° F. To convert from (lb/MMBtu) to (lb/106 scf), multiply by 1020. These emission factors can be converted to other natural gas heating values by multiplying the given emission factor by the ratio of the specified heating value to this heating value.*
[d]*Compound was not detected. The presented emission value is based on one half of the detection limit.*
[e]*Benzene with Sconox catalyst is 9.1 E-07, rating of D.*
[f]*Formaldehyde with Sconox catalyst is 2.0 E-05, rating of D.*

environment. Alternative methods are used to convert CO_2 into minerals that can be reused or at least stored in a solid form.

The only storage technology that has reached the demonstration phase is geologic sequestration. Although geological storage is considered permanent, additional CO_2 plume monitoring data are required to address the possibility of CO_2 leakage. A question also remains as to who would be liable for the CO_2 should a leak occur from a geologic storage site. Continuous research programs have been done into ways to store CO_2 in solid phase as a carbonate that is permanent and has no chance of leakage. These technologies are not so energy intensive as those requiring the compression and transport of the elemental CO_2. These technologies include a process that uses the CO_2 to produce carbonates, permanently sequestering CO_2 as environmentally benign carbonate materials.

TABLE 11–5 Emission Factors (Uncontrolled) for Hazardous Air Pollutants from Distillate Oil-Fired Stationary Gas Turbines[a,b]

Pollutant	Emission Factor (lb/MMBtu)[c]	Emission Factor Rating
1,3-Butadiene[d]	<1.6 E-05	D
Benzene	5.5 E-05	C
Formaldehyde	2.8 E-04	B
Naphthalene	3.5 E-05	C
PAH	4.0 E-05	C

[a]*SCCs for distillate oil-fired turbines include 2-01-001-01, 2-02-001-01, 2-02-001-03, and 2-03-001-02. Hazardous air pollutants as defined in Section 112 (b) of the Clean Air Act.*
[b]*Factors are derived from units operating at high loads (≥80% load) only. For information on units operating at other loads, consult the background report for this chapter, available at www.epa.gov/ttn/chief.*
[c]*Emission factors based on an average distillate oil heating value (HHV) of 139 MMBtu/103 gallons. To convert from (lb/MMBtu) to (lb/103 gallons), multiply by 139.*
[d]*Compound was not detected. The presented emission value is based on one half of the detection limit.*

TABLE 11–6 Emission Factors (Uncontrolled) for Metallic Hazardous Air Pollutants from Distillate Oil-Fired Stationary Gas Turbines[a,b]

Pollutant	Emission Factor (lb/MMBtu)[c]	Emission Factor Rating
Arsenic[d]	<1.1 E-05	D
Beryllium[d]	<3.1 E-07	D
Cadmium	4.8 E-06	D
Chromium	1.1 E-05	D
Lead	1.4 E-05	D
Manganese	7.9 E-04	D
Mercury	1.2 E-06	D
Nickel[d]	<4.6 E-06	D
Selenium[d]	<2.5 E-05	D

[a]*SCCs for distillate oil-fired turbines include 2-01-001-01, 2-02-001-01, 2-02-001-03, and 2-03-001-02. Hazardous air pollutants as defined in Section 112 (b) of the Clean Air Act.*
[b]*Factors are derived from units operating at high loads (≥80% load) only. For information on units operating at other loads, consult the background report for this chapter, available at www.epa.gov/ttn/chief.*
[c]*Emission factors based on an average distillate oil heating value (HHV) of 139 MMBtu/103 gallons. To convert from (lb/MMBtu) to (lb/10^3 gallons), multiply by 139.*
[d]*Compound was not detected. The presented emission value is based on one half of the detection limit.*

TABLE 11–7 Emission Factors (Uncontrolled) for Hazardous Air Pollutants from Landfill Gas-Fired Stationary Gas Turbines[a,b]

Pollutant	Emission Factor (lb/MMBtu)[c]	Emission Factor Rating
Acetonitrile[d]	<1.2 E-05	D
Benzene	2.1 E-05	B
Benzyl chloride[d]	<1.2 E-05	D
Carbon tetrachloride[d]	<1.8 E-06	D
Chlorobenzene[d]	<2.9 E-06	D
Chloroform[d]	<1.4 E-06	D
Methylene chloride	2.3 E-06	D
Tetrachloroethylene[d]	<2.5 E-06	D
Toluene	1.1 E-04	B
Trichloroethylene[d]	<1.9 E-06	D
Vinyl chloride[d]	<1.6 E-06	D
Xylenes	3.1 E-05	B

[a]*SCC for landfill gas-fired turbines is 2-03-008-01. Hazardous air pollutants as defined in Section 112 (b) of the Clean Air Act.*
[b]*Factors are derived from units operating at high loads (≥80% load) only. For information on units operating at other loads, consult the background report for this chapter, available at www.epa.gov/ttn/chief.*
[c]*Emission factors based on an average landfill gas heating value (HHV) of 400 Btu/scf at 60°F. To convert from (lb/MMBtu) to (lb/106 scf), multiply by 400.*
[d]*Compound was not detected. The presented emission value is based on one half of the detection limit.*

TABLE 11–8 Emission Factors (Uncontrolled) for Hazardous Air Pollutants from Digester Gas-Fired Stationary Gas Turbines[a,b]

Pollutant	Emission Factor (lb/MMBtu)[c]	Emission Factor Ratings
1,3-Butadiene[d]	<9.8 E-06	D
1,4-Dichlorobenzene[d]	<2.0 E-05	D
Acetaldehyde	5.3 E-05	D
Carbon tetrachloride[d]	<2.0 E-05	D
Chlorobenzene[d]	<1.6 E-05	D
Chloroform[d]	<1.7 E-05	D
Ethylene dichloride[d]	<1.5 E-05	D
Formaldehyde	1.9 E-04	D
Methylene chloride[d]	<1.3 E-05	D
Tetrachloroethylene[d]	<2.1 E-05	D
Trichloroethylene[d]	<1.8 E-05	D
Vinyl chloride[d]	<3.6 E-05	D
Vinylidene chloride[d]	<1.5 E-05	D

[a]*SCC for digester gas-fired turbines is 2-03-007-01. Hazardous air pollutants as defined in Section 112 (b) of the Clean Air Act.*
[b]*Factors are derived from units operating at high loads (≥80% load) only. For information on units operating at other loads, consult the background report for this chapter, available at www.epa.gov/ttn/chief.*
[c]*Emission factors based on an average digester gas heating value (HHV) of 600 Btu/scf at 60°F. To convert from (lb/MMBtu) to (lb/10^6 scf), multiply by 600.*
[d]*Compound was not detected. The presented emission value is based on one half of the detection limit.*

Post-Combustion Capture

Post-combustion capture (PCC) of CO_2 from flue gases can be done by various methods: distillation, membranes, adsorption, physical and chemical absorption. Absorption in chemical solvents, such as amine types, is a proven technology and in many applications performed consistently and reliable. It is used in natural gas sweetening and hydrogen production. The reaction between CO_2 and amines offers currently the most cost-effective solution to directly obtain high purity CO_2 The flue gases from the power plant are cooled and treated for reduction of particulates and SO_x and NO_x. Then the flue gases, boosted by a fan to overcome pressure drops in the system, pass through an absorber. A lean amine solution counter-currently interacts with the flue gases and absorbs the CO_2. The clean flue gases continue to the stack. The CO_2 rich amine solution is pumped into a stripper (regenerator) to separate the amine from the CO_2. The energy to desorb the CO_2 from the solution is provided by steam. The CO_2- rich solution at the top of the stripper is condensed for water removal and the gaseous CO_2 is sent for further drying and compression.

Post-Combustion Chilled Ammonia

Ammonia is the lowest form of amine. Like other amines, it can absorb CO_2 at atmospheric pressure, but at a slower rate than that of MEA. The chilled ammonia system uses a CO_2 absorber similar to SO_2 absorbers and is designed to operate with slurry. The process requires the flue gas to be chilled to 35°F before entering the cleanup system. The cooled flue gas flows upwards in counter current to a slurry containing a mix of dissolved and suspended ammonium carbonate (AC) and ammonium bicarbonate (ABC). More than 90% of the CO_2 from the flue gas is captured in the absorber. The CO_2-rich spent ammonia is regenerated under pressure. This reduces the CO_2 liquefaction

TABLE 11–9 Emission Factors (Uncontrolled) for Metallic Hazardous Air Pollutants from Digester Gas-Fired Stationary Gas Turbines[a,b]

Pollutant	Emission Factor (lb/MMBtu)[c]	Emission Factor Rating
Arsenic[d]	<2.3 E-06	D
Cadmium[d]	<5.8 E-07	D
Chromium[d]	<1.2 E-06	D
Lead[d]	<3.4 E-06	D
Nickel	2.0 E-06	D
Selenium	1.1 E-05	D

[a]*SCC for digester gas-fired turbines is 2-03-007-01. Hazardous air pollutants as defined in Section 112 (b) of the Clean Air Act.*
[b]*Factors are derived from units operating at high loads (≥80% load) only. For information on units operating at other loads, consult the background report for this chapter, available at www.epa.gov/ttn/chief.*
[c]*Emission factor based on an average digester gas heating value (HHV) of 600 Btu/scf at 60°F. To convert from (lb/MMBtu) to (lb/10⁶ scf), multiply by 600.*
[d]*Compound was not detected. The presented emission value is based on one half of the detection limit.*

compression energy requirement. The remaining low concentration of ammonia in the clean flue gas is captured by cold-water wash and returned to the absorber. The clean flue gas, which now contains mainly nitrogen, excess oxygen, and low concentration of CO_2, flows to the stack.

Pre-Combustion Capture IGCC

The main advantage of IGCC pre-combustion CO_2 capture is the fact that the amount of the fluid to be processed is much smaller than in the case of post-combustion for a coal-fired plant or a combined cycle. In the IGCC case only the syngas is treated, whereas in the PCC case the entire exhaust flue gas flow must be processed. For the oxygen blown IGCC, the syngas main components are hydrogen and carbon monoxide (CO) with some CO_2, steam, N_2 and traces of other elements. The raw syngas produced by the gasifier must be cleaned from contaminants including mercury, sulfur, and fluorides. The chemical processes, known commercially as Rectisol or Selexol, are capable of removing a certain amount of CO_2. However, the actual conversion of the CO into CO_2 and H_2 occurs in a water shift process. In this process steam and syngas are mixed in the presence of a catalyst to covert the CO to CO_2 in an exothermic reaction. The shift stage can be integrated into the process either before (sour shift) or after the sulfur removal (sweet shift) stage. The fuel to be burnt in the gas turbine is mainly H_2 with additives.

Oxy-Combustion

In an oxy-combustion-based power plant, oxygen rather than air is used to combust fuel resulting in a highly pure carbon dioxide (CO_2) exhaust that can be captured at relatively low cost and sequestered. Often, the oxygen is mixed with flue gas to regulate burning as well as achieve a high carbon dioxide level in the flue gas. In the case of Rankine steam cycle, the volume of flue gas leaving the boiler is considerably smaller than the conventional air-fired volume (explained by the fact that nitrogen in the air is not part of the flue gas and that the amount of flue gases is approximately 75% less for combustion with oxygen than with air) and consists primarily of carbon dioxide.

The process utilizes an air separation unit (ASU), a facility requiring high electricity consumption. To reduce the auxiliary load, new and less energy intensive oxygen separation technologies are in development, including ion transport membrane (ITM), oxygen transport membrane (OTM) and BOC's ceramic auto thermal recovery (CAR) oxygen production process.

Oxy-combustion is also associated with other promising combined cycles involving gas and steam turbines. The Graz cycle and the semi-closed oxy-combustion combined cycle are two examples, which at the present time are under theoretical investigation. This oxy-combustion concept is applicable for a variety of fuels, including methane, syngas, or biomass gasification. In the Graz cycle the working fluid following the combustion process is a mixture of steam (approx. 75%) and CO_2 (approx. 24%), with some small amounts of N_2 and O_2. The expected cycle efficiency is in the range of 50%. Pilot demonstration plants will be operational around 2015.

Post-Capture CO_2 Treatment

An important aspect of post-capture CO_2 processes is related to its compression or liquefaction in order to be transported to an underground storage place. This process is associated with large energy penalties. According to P. Baldwin ("Capturing CO_2. Gas compression vs. liquefaction," *Power Magazine*, May 2009), a typical 1000 MW coal-fired plant requires 120 MW of auxiliary power to produce the required compression needed.

CO_2 Compression Issues

A typical CO_2 processing system includes compression, dehydration, and purification/liquefaction. As mentioned above, this process is one of the major contributors to auxiliary power consumption and higher costs for the power plant. The compression process includes at least two compressors, intercoolers, water separators, dehydrators, and purifiers. The amount of impurities in the CO_2 stream has a major impact on the process. The presence of H_2O may decrease the amount of compression work, while the existence of N_2, O_2, and Ar may increase it. In the selection process, the intercooler temperature must be above the condensing temperature of the mixture. Additionally, CO_2 compression equipment requires stainless steel construction due to the presence of water vapors and potential corrosion.

A discussion of turbo-machinery for the sequestration part of the plants would not be complete without mentioning CO_2 compression technology. The major effort in this area is dedicated to identifying processes capable of reducing power consumption, which represents 40% of the auxiliary loads. In some cases it represents 8–12% of plant power output.

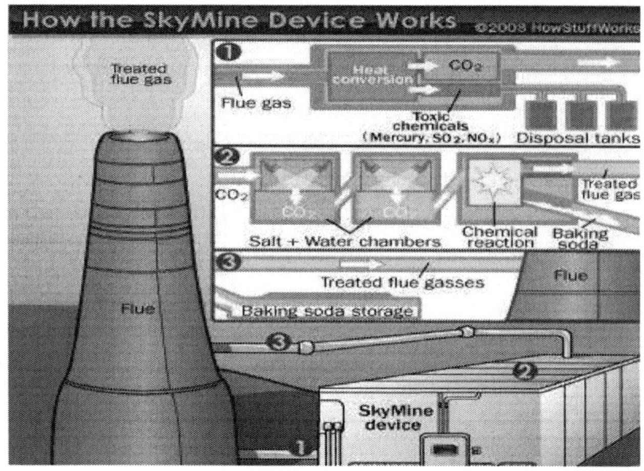

FIGURE 11–1 The SkyMine process. [11-2]

Other Methods of CO_2 Sequestration

In order to find more economic solutions beside the sequestration and storage underground, several technologies have been developed that require less energy and offer an alternative solution to geological storage of CO_2.

Brief Description of Technologies

SkyMine Process

SkyMine™ is a technology that converts CO_2 from power plant flue gas into sodium bicarbonate (baking soda) suitable for long-term landfill storage. The process uses sodium hydroxide produced on site through seawater electrolysis to react with the CO_2.

Hydrogen and chlorine are produced as byproducts of the seawater electrolysis that creates sodium hydroxide and may be sold as a revenue stream for the system. The developer claims that this process is capable of removing between 85% and 97% of the mercury, acid rain gases, and CO_2 from the flue gases that pass through the system.

Figure 11–1 shows the schematic for the process. The CO_2 enters a series of absorption chambers. Once inside, sodium hydroxide is injected (produced from electrolysis of NaCl in seawater) into the chambers. A chemical reaction takes place: $CO_2 + H_2O + Nacl \rightarrow NaHCO_3 + H_2 + Cl_2$.

The sodium bicarbonate (baking soda) is in solid (crystalline) form, and the absorption chambers dump it into a storage area. The hydrogen and chlorine gases are stored separately from the baking soda. SkyMine then returns the remaining, mostly harmless, flue gases to the power plant, where they are released into the atmosphere. The reaction's main byproduct, sodium bicarbonate, is harmless and potentially useful commercially. However, the most practical way to deal with the byproduct is landfill disposal.

The Calera Process

The Calera process was originally designed to use a base to absorb CO_2 to create calcium and magnesium carbonates as cement additive materials. One of them, seawater contains about 1290 ppm of magnesium and 410 ppm of calcium and other minerals. Calera's research work has also included using base materials extracted from other sources such as fly ash from ash ponds or ash piles to make mineral carbonates.

Fly ash contains calcium oxide and magnesium oxides that, when mixed with water, form calcium and magnesium hydroxides. Hydroxides can be used to capture CO_2 to form carbonates based on the following equation:

$$Ca(OH)_2 + CO_2 \rightarrow CaCO_3 + H_2O$$

A 10 tpd pilot plant is located at the Moss Landing plant near Los Angeles (see Figure 11–2). A number of tests have been performed to test the reactivity of calcium oxide in waste fly ash.

The Calera process can have applications in the utility industry, at power plants where high quality fly ash materials are available. Note that the reactivity of waste fly ash with CO_2 is slower than that of other reagents such as amine or ammonia, but waste fly ash is free in most cases. This process can be economical for some coal-based projects. Waste mineral piles including iron/steel slag, fly ash, and blast furnace and mine tailings containing mineral oxides that may be used to capture CO_2 have been investigated by other process developers. Mine slime—water that is left over after a mining operation—has a high concentration of calcium and magnesium that can be used for carbonation reagents.

LLNL Seawater Carbonation Process

The Lawrence Livermore National Laboratory (LLNL) has developed a novel seawater scrubbing system using

FIGURE 11–2 Moss landing plant. [11-2]

calcium carbonate to capture CO_2. This reaction will result in the production of calcium bicarbonate according to the following equation. The resulting water-soluble calcium bicarbonate is then returned to sea, where it will provide necessary nutrients for marine life.

$$CaCO_3 + CO_2 + H_2O \rightarrow Ca(HCO_3)_2$$

The process employs a reactor vessel that allows a CO_2-rich flue gas stream to flow over or through a porous bed of limestone particles that are wetted by a continuous spray or flow of water. CO_2 passes through the reactor to contact the water and wetted surfaces, forming carbonic acid, which in turn reacts with the carbonate solids (e.g., calcium carbonate) to produce bicarbonate HCO_3 in solution.

When the calcium bicarbonate is returned to the sea, it will eventually decompose back to calcium carbonate and the absorbed CO_2 will be released back to the seawater. However, due to the wide dispersion, the CO_2 released can dissolve in the seawater and may not release to the atmosphere.

Silicate Mineral Carbonation

Calcium and magnesium carbonates form in nature during a process referred to as weathering of rocks. Although this reaction is exothermic and thermodynamically favored, it occurs very slowly over geologic time scales. In this process, calcium and magnesium ions are leached from rocks in the presence of water. The ions react with CO_2, forming solid calcium and magnesium carbonates that are stable and can be disposed of as mine filler materials for long-term storage of the CO_2. In theory, the metal oxides in the earth's crust could permanently bind all CO_2 that could be produced from the combustion of all existing fossil fuel reserves. The idea behind mineral carbonation is to find a way to speed up this natural weathering process so that CO_2 from large point sources, such as fossil power plants, can be permanently stored.

Many institutions are studying ways to increase the reaction rate for mineral carbonation. Most involve some sort of pretreatment or grinding of the minerals, as well as increased pressures and temperatures in the reactor. Pretreatment alone, although it creates much faster kinetics by dissolving the mineral in solution, could result in as high as a 20% energy penalty to the plant, and grinding may have an even higher energy penalty. This penalty would be in addition to the energy penalty associated with capturing and transporting the CO_2 to the mineral carbonation site.

Although mineral carbonation, unlike geologic sequestration, does not have the issue of leakage, it is not a completely environmentally benign sequestration option. Minerals need to be mined, and environmental impacts will be similar to those of coal mining.

Figure 11–3 is a process block flow diagram showing the potential route for using calcium silicate ($CaSiO_3$) as CO_2 sequestration material. Using a carbonation reactor at 30 bar pressure, calcium silicate and CO_2 will combine to form $CaCO_3$ using acetic acid (CH_3COOH) as the reactant.

Algae Farming

Algae is the fastest-growing vegetation known. Based on recent tests at Arizona Public Service's (APS's) Redhawk station, the US Department of Energy (DOE) has estimated that a growth rate of about 70 metric tons per acre per year can be achieved. Algae cultivation methodologies under investigation include photo bioreactor, open pond, and closed loop system. Photo bioreactor employs nutrient-laden feedstock to grow algae such as carbonates or bicarbonates to supply the required CO_2 for algae photosynthesis reaction. CO_2 gas or flue gas can be sparged into the circulating water in lieu of using solid carbonates as its

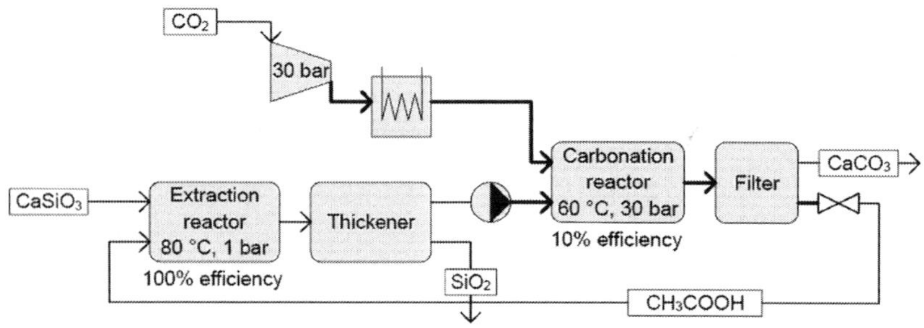

FIGURE 11–3 Calcium silicate carbonation. [11-2]

nutrient. Open pond methodology is more suitable for use in subtropical or tropical regions, where pond temperature variations are limited. The closed loop system prevents contact of air with algae-growing water and controls residual emissions from the algae farm.

When harvested, algae contain up to 50% moisture. Some algae species contain up to 40% lipid on a dry basis. Algae lipid can be processed into biodiesel, and the remaining material can be used to make methanol through fermentation. Algae can also be gasified into syngas, which can be used to make hydrocarbon liquid via the Fischer-Tropsch (FT) reactor process. The cost and land requirement for algae farming are expected to be high. DOE has awarded funds to APS to support its research and development in algae farming and processing technologies.

CASE STUDY 1: THE CAPTURE, STORAGE, AND UTILIZATION OF CARBON DIOXIDE BY STATOIL*

Climate Change

Carbon dioxide gas (CO_2) is a natural, fluctuating component of the earth's atmosphere and has been present throughout most of geological time. However, since the industrial revolution the concentration has risen by about a third (from 280–370 parts per million) and may well reach at least twice the preindustrial level by 2100. Most of this increase is attributed to the burning of carbon-rich fossil fuels—coal, natural gas, and oil—and is widely thought to be a contributory factor in trapping heat radiating from the earth's surface. This, in turn, may lead to global warming—the greenhouse effect—and stimulate climate change. To what extent this may happen is not known; some say it will lead to disastrous consequences while others foresee relatively slight but noticeable variations. Either way, something has to be done about it. The obvious answer is to increase energy efficiency and rapidly convert to alternative energy sources, such as solar and wind power. But this is easier said than done. Switching to alternative sources will be a gradual process, because about 85% of the world's present energy needs are being met by plentiful and relatively inexpensive fossil fuels. In contrast, non-fossil fuel energy sources are expensive, and onshore renewables need large land areas to produce even modest quantities of power (e.g., windmill parks).

A more pragmatic approach is to stabilize atmospheric concentrations gradually at or below 550 parts per million. But this too is an enormous challenge, requiring a 50% reduction in CO_2 emissions from projected levels by 2050. New technologies are therefore needed to lower the cost of alternative energy sources, strengthen the removal and storage of CO_2 from today's fossil-fueled industries, and replace oil and coal by less carbon-intensive natural gas. Nevertheless, this is the more attractive proposition as it promises to allow present fossil-fuel industries and fossil-fuel-rich countries to continue operating profitably while giving time for alternative energy sources to realistically come to the fore.

Statoil and Climate Policy*

For Statoil the issue is not whether the world faces a climate problem or how severe it may be, but how harmful emissions may best be overcome. The Kyoto protocol is therefore acceptable as a good basis for a rational global policy, including the introduction of a broad-based system of emission trading,[3] as long it is tied to Kyoto mechanisms.

* Source: Adapted from A. T. Buller, O. Kårstad, and G. de Koeijer. *Carbon Dioxide, Capture, Storage and Utilization, Research and Technology Memoir No. 5* (Stavanger, Norway: Statoil, April 2004). This case study has been included in the second edition because of its exhaustive cover of a field that globally is still in its infancy, and in which Statoil can accurately claim to be one of the leaders. The Statoil website needs to be consulted for current information. As an example, towards the end of this case study, I have included data drawn in 2013 on one of the sequestration sites.

* Source: Courtesy of Statoil, Norway.

3. Emission trading gives industrial countries the opportunity to meet obligatory reductions in emissions by purchasing quotas from other industrial nations or by cutting emissions for them.

Statoil also cooperates widely with other companies and authorities and is a significant player in global affairs through its membership in the World Business Council for Sustainable Development, the Energy and Biodiversity Initiative (EBI), and bodies such as the IEA Greenhouse Gas R&D Programme and the IPIECA. At home our specialists keep abreast of the latest developments in scientific knowledge about the greenhouse effect and the social, economic, and competitive impact of climate policies aimed at the petroleum industry and the energy market. The chief executive officer also regularly meets environmental and consumer organizations to discuss issues ranging from the disposal of produced water[4] to reducing greenhouse gas[5] emissions, of which CO_2 is the most important.

The company's intention is to reduce CO_2 emissions from its operating facilities by about one third by 2010. Based on the findings of a comprehensive corporate program (1997–2001), the primary measures are the injection of CO_2 into saline aquifers and reservoir rocks for long-term storage or to improve oil recovery, using hydroelectricity from the Norwegian grid to power installations presently employing on-site generation, and increasing energy efficiency.

Research History in Brief

One of our earliest engagements in CO_2 capture and storage was in the late 1980s when the Continental Shelf Institute[6] was commissioned to carry out a pilot study on environment-friendly gas power and CO_2 injection for improved oil recovery. Similar research began at the Statoil Research Centre in 1989, but it was not until the early 1990s that internal activities really began to intensify. In 1992 Statoil joined forces with Kværner Process Systems, NTNU, and SINTEF to examine whether membrane technology for capturing CO_2 from power station emissions would lead to significant weight, space, and cost reductions.

At about the same time, Statoil and partners decided that excessive amounts of CO_2 contained in natural gas from the offshore Sleipner field should be stripped off and injected into a saline aquifer situated above the hydrocarbon reservoirs. The primary goal was long-term storage to protect the natural environment. To learn as much as possible from the Sleipner case, Statoil and the IEA Greenhouse Gas R&D Programme organization set up the European Commission's SACS5 project (phases 1 and 2, 1998–2003), which led to the Sleipner experience becoming a truly multinational concern with global applications in mind. The present CO_2 Store project (2003–2005) is essentially a SACS extension, addressing long-term predictions of the aquifer's behavior and the transfer of approaches and methods to onshore and near-shore industrial sites.

The aims of the complementary, BP-coordinated CO_2 Capture Project (2001–2003) were to reduce capture costs by more than 50% at existing plants and by 75% at new ones. Emphasis was placed on the development and qualification of technology for capturing CO_2 emitted by gas turbines and power stations. The project involved eight major oil and energy companies,[7] and included three distinct regional programs run in the United States, Norway, and the European Union. Statoil headed the Norwegian "Klimatek–NorCap" contribution. And in common with the SACS initiative, the participants wished to demonstrate that CO_2 storage is safe, measurable, and verifiable.

Statoil is also looking at ways of transforming the CO_2 challenge into viable business opportunities. One area under investigation is the transport of CO_2 by ship and pipeline to mature offshore fields requiring gas-based improved oil recovery (IOR) programs. The idea is to use CO_2 instead of hydrocarbon gas as an oil-miscible component to improve sweep efficiency.

Awards

In 2002 the group received two major awards: the World Petroleum Congress's technology development prize for its pioneering efforts in underground carbon dioxide storage and a 2002 World Summit Business Award for Sustainable Development Partnerships, in association with EBI colleagues. These awards testify that Statoil's long-term efforts in environmental stewardship are paying off both in terms of industrial application and global awareness.

Options

Long-term oceanic and underground storage promises to help nature cope with excessive carbon dioxide emissions to the air—the latter being the most realistic solution, at least in the near future.

Trees and other plants use up vast quantities of CO_2 by absorbing it as they grow and retaining it throughout their lifetimes: much is also taken up by seas and oceans. However, these natural mechanisms appear to be inadequate to constrain current levels of anthropogenic

4. Produced water is natural-formation water contaminated during the hydrocarbon production process.

5. Non-CO_2 greenhouse gases include methane, N_2O, and engineering chemicals such as HFCs, PFCs, and SF_6.

6. IKU, now SINTEF Oil and Energy (Trondheim, Norway).

7. SACS is Saline Aquifer CO_2 Storage. Collaborating organizations: industry—BGS, BRGM, GEUS, IFP, NITG-TNO, and SINTEF; institutes—NERC, GECO, and the IEA Greenhouse Gas R&D Programme; national bodies—ministries and research councils in Norway, Denmark, the Netherlands, the United Kingdom, and France.

(human-made) emissions, especially with continuing denudation of the rain forests and the ravaging of fertile ground by sprawling urbanization. Clearly, there is a pressing need for new measures to be introduced, such as the disposal of CO_2 in the ocean and long-term underground storage.

The oceanic storage concept involves the bubbling of gas directly into the sea at concentrations low enough to avoid damaging surrounding ecosystems, and at sufficient depths to ensure that it stays there. Various methods have been suggested, including droplet plumes emanating either from the outlets of deep pipelines linked to onshore CO_2 pumping stations or from pipes dangled from CO_2 transport ships. Other possibilities include the injection of CO_2 from offshore pumping stations into abyssal depths to accumulate as stagnant lakes, and the dropping of solid CO_2 into the sea in the form of dry ice.

Although oceanic storage offers the greatest storage capacity, there are major uncertainties about the environmental impact and retention times. Statoil is therefore no longer actively engaged in oceanic disposal storage research but closely follows the latest scientific developments.

Long-term underground (subsurface) storage is regarded as the more reliable solution, requiring CO_2 to be injected into deeply buried geological formations. The main candidates are depleted oil and gas reservoirs, deeply buried saline aquifers and unminable coal seams.

The attraction of using depleted oil and gas reservoirs is obvious: they are proven traps; the reservoir geology is well known; and infrastructures can be readily adapted for CO_2 transport and injection. Indeed, depleted hydrocarbon gas fields and saline aquifers have long been used on a commercial basis to inject, store and withdraw natural gas according to supply and demand. At present there are 595 underground storage sites worldwide, whose collective working gas storage capacity is equivalent to 11% of the world's consumption.

There are also innumerable saline aquifers around the world that could be used for long-term CO_2 storage. In both cases—depleted reservoirs and saline aquifers—much of the injected gas will eventually dissolve in the formation water, while some may react with the minerals to form carbonate precipitates.

An important storage issue is sealing capacity; that is, the ability of the overlying (cap) rocks to stop the CO_2 from leaking out and rising back to the surface. To fulfill this criterion, cap rocks should be almost impermeable and ductile rather than brittle if natural and induced fractures are to be avoided. Onshore leakage can affect water supplies and devastate vegetation cover.

For coal seams, the theory is that injected CO_2 will be permanently locked in the coal by adsorption while enhancing methane production by preferential displacement.

Rough IEA estimates of how much CO_2 could be stored in these various geological options are >15 Gt[8] in unminable coal seams, 920 Gt in depleted oil and gas fields, and 400–10,000 Gt in deep saline aquifers. With the atmosphere today containing about 730 Gt of CO_2, saline aquifers obviously hold considerable promise. Onshore CO_2-based improved oil recovery is an established practice, which is yet to be tried offshore (see later).

Sleipner West Gas Field

The Sleipner asset notched up two world firsts in pursuit of environmental protection—large-scale offshore carbon dioxide separation and injection into a saline aquifer 1000 meters below the sea bed.

The Statoil-operated Sleipner West[9] field is one of the largest gas producers in the Norwegian sector of the North Sea, with a daily gas export capacity of 20.7 million cubic meters and a daily output of 60,000 barrels of stabilized condensate (light oil). It was discovered in 1974 close to the British/Norwegian sector divide and is linked to Sleipner East. Both fields are produced by a single operations organization.

During field development planning (1990), it was realized that the 4–9.5% CO_2 content in the natural gas would have to be reduced to less than 2.5% if it were to be fed directly into sales gas pipelines to Europe. A small team of technical experts came up with the unprecedented idea of capturing the CO_2 offshore and injecting it into a saline aquifer beneath the Sleipner installations. In this way, the Sleipner asset would minimize CO_2 emissions—the prime motive—while avoiding environmental taxes.[10] Despite its pioneering nature, this became the partner-approved solution.

Of various possibilities, the Elf-patented separation process was selected for CO_2 capture, because it was deemed cheaper to run and more compact than competing systems. One of the greatest challenges, however, was to scale down the process plant sufficiently so that it could be accommodated on a platform. Even so, the "miniaturized" version of the extraction module weighed 8200 tons—the heaviest module ever to be lifted offshore—and measured 50 m × 20 m × 35 m.

By the time the field came onstream in 1996, the Sleipner organization had notched up two world firsts: the installation of a large-scale offshore CO_2 extraction plant at the Sleipner T (Treatment) platform and the facilities for saline aquifer injection from the Sleipner East A platform.

8. One gigton or Gt is 1000 million tons.
9. The present partners are ExxonMobile, Norsk Hydro, and Total.
10. At this time the Norwegian government was discussing climate change and the possibility of introducing a national carbon tax. The latter became law in 1991 and currently stands at US $40 per ton.

Carbon Dioxide Capture and Injection

The carbon dioxide content in the natural gas can now be kept below 2.5% by increasing the amine circulation rate and total heat input.

Carbon Dioxide Capture Process

The first stage in the Sleipner CO_2 capture process entails the mixing of an amine-water solution with the natural gas in two parallel columns (absorbers A and B), both of which are kept at high pressure (100 bara[11]) and moderate temperature (60–70°C). The amine—an organic compound derived from ammonia—selectively absorbs the CO_2 by weak chemical bonding and separates out at the bottom of the columns. Thereafter it is transferred via a turbine to a 15 bara flash drum in which the coabsorbed hydrocarbons are removed. The amine is subsequently heated and depressurized to 1.2 bara in a second flash drum where the CO_2 is boiled off. By now the gas is almost (95%) pure CO_2.

As the lean liquid amine still contains residual CO_2, some 10% is subject to thermal regeneration where the CO_2 is stripped off by steam in a desorber column operating at 120°C. The remaining, even leaner amine is then mixed with the regenerated amine and pumped back to the absorbers for a new separation cycle.

Experimental Investigations

In practice it has been difficult to keep consistently within the 2.5% goal because the process has proved somewhat unstable. This led to several modifications, including new internals for the gas scrubber to reduce carryover (1997–1998), a comprehensive rebuilding of absorber A (1999), and new internals for absorber B (2000). However, faced with continuing irregularities, the Sleipner Amine Task Force asked this source [Statoil] to devise a solution.

By now this source had become familiar with the removal of CO_2 at high pressure while attempting to improve the plant's performance. Indeed, one of the most important experimental observations was that pressure has a significant effect on the absorption capacity of the amines—adsorption capacity decreases with increasing pressure. This was a cause for concern as it could impact the effectiveness of CO_2 capture in the absorber columns.

Spurred on by the task force's request, the company stepped up its engagement through a major experimental and modeling investigation aimed at better understanding and predicting high-pressure CO_2 capture mechanisms. This involved experimenting with genuine natural gas under realistic pressures and temperatures. The effects of various amine solution additives were tested under similar conditions.

However, it was not until 2003 that a major breakthrough was made. Exploiting the research center's unique laboratory facilities, the answer was found to lie in increasing the amine circulation rate and the heating energy used to separate CO_2 from the amine. Subsequent offshore tests at Sleipner resulted in a stable performance while reducing the CO_2 content to 2.25%. The new operational procedure and equipment can now be installed at Sleipner T, enabling the amine plant to meet quality specifications when operating at full capacity.

Aquifer Injection

Once the CO_2 has been captured, its pressure is boosted by four compressors to 80 bara prior to being transferred to the Sleipner East A platform for pumping into the base of the saline aquifer. Since 1996 about 1 million tons of compressed CO_2 have been injected annually.

Another requirement is that the well casing and other hardware used in the capture and injection plant have to be made of stainless steel, because even minute quantities of water mixed with CO_2 produce a weak corrosive carbonic acid (H_2CO_3).

Investment costs amounted to some US $80 million ($CO_2$ capture costs excluded). Although this was a considerable sum, the partners would otherwise have faced an annual tax bill of about US $50 million if the CO_2 had simply been vented into the air.

Geological Aquifer and Cap Rock Characterization

The storage capacity of the saline Utsira aquifer is thought to be greater than 100 times the volume of annual European carbon dioxide emissions from power plants.

The aquifer in question is the Utsira Formation, which the SACS team believes was deposited as part of a submarine turbidite fan system[12] above the Sleipner reservoir rocks. Today it is encountered some 1000 meters below the seabed,[13] and comprises an exceptionally porous and permeable sequence of poorly consolidated, fine- to medium-grained quartz-rich sandstones. Subcropping almost exclusively in the Norwegian sector of the North Sea, it is more than 200 meters thick, over 50 kilometers wide, and extends for some 500 kilometers in a sinuous strip beneath the Brage, Oseberg, Grane, and Sleipner fields and the Tampen production center to the north. The aquifer's areal coverage is thus about 26,000 square kilometers.

Delineation and mapping of the top of the formation is particularly important for defining its closure. If aquifers

11. Bara—bar absolute (not to be confused with barg—bar gauge).

12. *Turbidites* is a relatively loose term for sediments carried from shallow to deep water by gravity-induced flows and deposited at the base of slopes. Here they may form fanlike or lobate bodies, fringed by thin-bedded sand/mud couplets.

13. The Utsira Formation sandstones have porosities between 24 and 40% and permeabilities between 1 and 3 Darcy.

form large domal structures, the CO_2 will be constrained and slight structural uncertainties can be ignored. However, precise and detailed depth mapping is vital if they undulate gently, as at Sleipner, where the top of the aquifer above the injection point is relatively flat. This is because minor variations may have a major effect on CO_2 movement (migration routes), areas of accumulation, and overall storage potential.

The regional mapping was done using 2D seismic datasets, while more detailed work was carried out around the injection point using 3D seismic.[14] Petrophysical data from some 300 wells were also available for study, plus limited rock samples in the form of drill cuttings and cores. Much sedimentological, geochemical, and rock-mechanical research is still being done on the complex cap rock/overburden sequence, which at Sleipner is about 700 meters thick. A dedicated 9-meter core was cut from this interval in the summer of 2002.

Another consideration is the possible presence and continuity of faults running through the aquifer and cap rock along which CO_2 may escape to the seabed. Fortunately, no significant faults have been detected from the seismic surveys (also see later). The injection process itself could lead to local microseismicity or the opening of incipient, pressure-induced fractures, but the required injection pressures at Sleipner are sufficiently low for this to be regarded as unlikely.

Furthermore, current thinking suggests that the plasticity of the overburden is such that faults and fractures are unlikely to serve as escape conduits. In other words the sealing capacity appears to be good.

Seismic Monitoring

Seismic monitoring has revealed no carbon dioxide leakage in the overburden

Another taxing question was whether the dynamic behavior of the injected CO_2 and its potential impact on cap rock integrity could be monitored using modern geophysical techniques, especially seismic. After much discussion, it was agreed that time-lapse seismic would probably be suitable, because the velocity of sound waves should be able to differentiate between saltwater-bearing (higher velocity) and CO_2-bearing (lower velocity) sandstones. Time-lapse seismic, which is also known as 4D seismic, involves comparing the results of 3D seismic surveys repeated at considerable time intervals: differences between the survey results are attributed to fluid or pressure changes.

Four seismic surveys have been conducted so far: a pre-SACS baseline survey in 1994 prior to CO_2 injection and three monitoring surveys carried out in 1999, 2001, and 2002 during CO_2 injection. The latter have not only successfully traced the injection of the CO_2 and expansion of the "bubble" but have also yielded extremely sharp images of the aquifer's overall geometry, internal structure and flow behavior. As expected, gravitational separation is the dominant physical process because of the CO_2's buoyancy.

A particularly striking result is that the distribution and migration paths of the CO_2 are strongly controlled by intraaquifer mud rock horizons. With an extraordinary seismic detection limit of about 1 meter or less, much of the CO_2 can be seen to have migrated upwards between the Utsira Formation mud rock terminations, as witnessed by a distinct seismic chimneylike column appearing on repeated seismic surveys. What is more, it has traveled up to about 1450 meters laterally beneath individual mud rock layers after six years of injection. The lateral speeds at which the CO_2 fronts move range from 0 to about 100 meters per annum—at least in recent years.

This remarkable precision prompted the team to estimate seismically the quantity of injected CO_2—on the assumption (among others) that none has been dissolved in the saline formation water. By comparing the seismically based result with the injected volumes, it appears that all of the CO_2 is accounted for by the seismic data. This, of course, is another argument for suggesting that no significant leakage has occurred, although the lack of seismically observed CO_2 in the overburden remains the most persuasive factor. It is wise, however, to recall that there is always a margin of error associated with the seismic method—albeit relatively minor in this case.

Gravimetric Aquifer Monitoring

Time-lapse gravity can potentially be used to better determine carbon dioxide density and mass distribution.

Although gravimetry has a lower spatial resolution than its seismic counterpart, repeated high-precision micro-gravity monitoring potentially provides better constraints on CO_2 density and mass distribution. It may also give an early warning signal if considerable amounts are escaping upwards through the overburden, as well as yielding relatively inexpensive information on the long-term dissolution of the CO_2 in the formation water once injection has ceased.

The company's latest offshore time-lapse gravity surveying technique is being used, having been successfully employed at the Troll field to image and monitor changes in the (hydrocarbon) gas/water contact. Developed in association with Scripps (University of California, San Diego) and cofunded by the US Department of Energy, the state-of-the-art seafloor gravimeter contains three gravity sensors and three pressure sensors, which enable the instrument to monitor small vertical changes in the seafloor

14. In 3D surveys, seismic lines are shot so close together that the data can be represented as seismic data "cubes." In 2D surveys, seismic lines are often several kilometers apart, requiring geoscientists to interpret what goes on in between them.

as well as small gravity changes. The gravitational accuracy is about 5×10^{-9} of the earth's total gravity field.

A 7×3 kilometer baseline survey was obtained at Sleipner in August 2002, against which future surveys will be compared. So far the results have exceeded expectations: not only may it be possible to detect vertical changes in the seafloor as small as 0.5 centimeters, but the time-lapse detection threshold may also be as low as 5 μGal[15]—some 50% better than that suggested by a pre-survey modeling exercise.

However, there are other considerations to be taken into account, such as the gravitational effect of further production from the underlying Sleipner gas-condensate reservoirs. The technique also depends on lowering the gravimeter onto separate concrete blocks installed on the seafloor one at a time, although this did not prove to be a hindrance.

Aquifer Flow Modeling

Simulations suggest that the carbon dioxide "mega-bubble" may reach its ultimate size after a few hundred years, thereafter shrinking and finally disappearing within a few thousand years

Whereas geophysical surveys are designed to determine rock and fluid distributions, reservoir simulations are designed to predict how fluids will behave with time. In a case like this, it is naturally wise to make a preinjection simulation to test operational feasibility, as was done at Sleipner before the SACS project started.

The SACS team has subsequently built a detailed postinjection model to verify and improve the seismic and geological interpretation of the aquifer around the injection site; and a coarser, larger-scale model to predict CO_2 migration over a period of several thousand years. The areas covered by the models are 7 square kilometers and 128 square kilometers, respectively. In both cases, the seismically inferred mud rock distributions were imported into the reservoir models, because it is almost impossible to trace individual mudstone layers from well to well, even when they are close together. Calibration of the 3D repeated seismic data with a local reservoir model is thus a fundamental prerequisite.

The results from the larger model suggest that most of the CO_2 will eventually coalesce to form a single "mega-bubble" beneath the cap rock a few years after injection has ceased. It will also gradually spread along the top of the salty formation water according to the local topography of the cap rock seal. This, however, must be tempered by the fact that CO_2 will diffuse from the "mega-bubble" into the underlying brine column, a phenomenon that is usually ignored in standard reservoir simulations because it is extremely slow compared with other transport processes. But given time, the CO_2-enriched brine on top of the column will become denser than that beneath, resulting in a downward flow compensated by convection plumes. This, in turn, will enhance dissolution and increase the probability of the CO_2 remaining in the aquifer.

When dissolution is included in the simulations, the "mega-bubble" will probably reach its ultimate size after a few hundred years, thereafter shrinking and finally disappearing within a few thousand years.

Further Investigations

The ultimate objective is to combine chemical- and flow-oriented modeling approaches for making reliable, long-term predictions.

The main product of the SACS project is a comprehensive Best Practice Manual (2003). This contains a suggested procedure for evaluating CO_2 storage from a technical point of view, besides information aimed at satisfying authorities and the general public as to the feasibility, safety, and reliability of the storage process.

The Sleipner case, which is being used as a full-scale natural laboratory, has yielded copious information on CO_2 transport rates and geophysical properties and has gone some way towards assessing the sealing capacity of the overburden.

These are considerable shorter-term achievements, of which the seismic monitoring is the most conspicuous. However, some of the most telling challenges still lie ahead, particularly the making of reliable long-term predictions, recalling that *long-term* in this context refers to several hundred to several thousand years hence.

Ongoing investigations in the CO_2 Store program (2003–2005) include assessments of whether the free and dissolved CO_2 remain in the host aquifer or migrate elsewhere and whether the sealing capacity of the cap rock will be maintained, realizing that CO_2-rich water is slightly acidic and may lead to mineral dissolution.

Other important issues are whether and how much of the injected CO_2 can be permanently fixed by chemical reactions and in what form and whether such chemical changes will impair porosity and permeability, thereby reducing aquifer storage capacity while (possibly) improving retention. The conditions under which CO_2 might ultimately be dissolved in its entirety are also receiving attention.

In short, the CO_2 Store program aims to extend the capabilities of two-phase (gas, water) reservoir simulators to better handle extremely long-term simulations, including the migration of CO_2 in its dissolved form, and chemical-oriented modeling to predict the maximum potential for

15. Gal—a unit of gravitational acceleration equal to 1 centimeter per second.

CO_2 reaction with the Utsira Formation sediments and the cap rock.

The ultimate objective is to combine chemical modeling with the flow-oriented modeling approach of reservoir simulation to produce merged "chemical and reactive transport models" constrained by geological and geochemical understanding—a highly ambitious undertaking. Geomechanical modeling is also coming to the fore in a number of related investigations.

Snøhvit and In Salah Projects

With the addition of Snøhvit and the In Salah gas projects, the company is now involved in the world's first three carbon dioxide storage projects solely aimed at protecting the natural environment.

The Snøhvit Development in the Barents Sea

Statoil and partners are planning the second largest offshore carbon dioxide storage project at Snøhvit based on the Sleipner West experience.

Moving on from Sleipner West, Statoil and partners[16] are planning the world's second largest offshore CO_2 storage project for the Snøhvit unit in the central part of the Hammerfest basin in the Barents Sea. The production area extends across seven unitized licences, covering the Snøhvit field itself and the Albatross and Askeladd satellites. All three accumulations contain natural gas and small quantities of condensate. The Snøhvit unit is scheduled to come on stream in 2006, some 25 years after the first gas discovery was made at Askeladd in 1981.

The unitized complex will be developed entirely using subsea production installations, linked by a record-breaking 143-kilometer multiphase flow pipeline to a processing and cryogenic gas liquefaction plant located at Melkøya—a small island outside Hammerfest.

The main product, LNG (liquefied natural gas), will be shipped to the United States and continental Europe in four purpose-built vessels, each 290 meters long and capable of carrying about 140,000 cubic meters of LNG in spherical tanks. Condensate and liquefied petroleum gas (LPG) will also be produced in relatively minor quantities. The LNG plant's production capacity will be about 5.7 billion cubic meters of gas per year.

The Snøhvit LNG project is the first oil and gas development in the environmentally sensitive Barents Sea and the first LNG-based gas field development in Europe. Furthermore, it is the first Norwegian offshore development with no surface installations. With all of the production equipment residing in water depths of 250–345 meters, none will interfere with fishing activities. Operations will be remotely controlled from land.

The gas in all three accumulations contains 5–8% CO_2, which will have to be reduced to less than 50 parts per million prior to liquefaction. This means that about 700,000 tons of CO_2 will have to be captured each year. Having reviewed several disposal options, it was decided that the CO_2 will be injected into the Tubåen Formation—a deeply buried saline sandstone aquifer encountered at the Snøhvit field about 2600 meters below the seafloor and about 60 meters beneath the main natural gas reservoir in the Stø Formation.

The Tubåen Formation contains some shale intervals that are difficult to correlate from well to well. Good interconnection between the sand bodies is thus anticipated. And with a thickness of about 47–75 meters, a net-to-gross ratio[17] of 0.8–0.9, and good reservoir properties,[18] the formation should be able to cope easily with the estimated storage requirement of about 23 million tons of CO_2 during the 30-year lifetime of the Snøhvit project. The formation is sealed by shaley cap rocks of the intervening Nordmela Formation, which should be sufficient to stop the injected CO_2 from rising to contaminate the natural gas reservoirs above.

The separation process will again be amine-based and cost about US $100 million. The main difference between the Sleipner T and the Snøhvit plants is that the latter will be capable of stripping residual CO_2 from all of the lean amine solution after initial separation. And it goes without saying that Statoil's world-class expertise gained at Sleipner is being fully exploited during the planning phase. *The In Salah Gas Project, Central Algeria*[19] plans are under way to reduce greenhouse gas emissions from the jointly operated In Salah Gas project by about 60%.

The third CO_2 injection project concerns the jointly operated Sonatrach[20]-BP-Statoil In Salah dry gas project—the third largest of its kind in Algeria. The agreement covers the development of eight hydrocarbon gas discoveries in the Ahnet-Timimoum Basin in the central Saharan region of the country and proposes to deliver 9 billion cubic meters per annum.

Some of the gas streams will contain CO_2 concentrations as high as 10%, whereas export sales gas specifications require a CO_2 concentration of less than 0.3%. To achieve this target, it has been estimated that some 1.2 million tons of CO_2 will probably have to be stripped off and stored each year from mid-2004 when the gas comes on stream.

16. Petro, Total, Gaz de France, Amerada Hess, and RWE-DEA.

17. Ratio of total (net) sandstone thickness or volume to total (gross) thickness or volume.

18. Porosity 10–16%; permeability 130–890 mD.

19. Most of this section is based on BP and Sonatrach material in the public domain.

20. Sonatrach is the Algerian State Oil and Gas Company.

Prior to Statoil's entry in 2003, BP and Sonatrach began searching for suitable subsurface sites close to the planned In Salah processing facilities that would be blessed with large storage potentials, reasonably good reservoir properties, and good sealing capacities. Several possibilities emerged, ranging from storage sites at each field to a single centralized facility. Of these various options, the latter solution was favored because of the high cost and complexity associated with distributed subsurface storage.

BP and Sonatrach thereafter identified the Krechba field as the most appropriate candidate. Here the CO_2 will be pumped through two or three horizontal wells directly into the aquifer section (below the gas-oil contact) around the northern periphery of the shallow Krechba sandstone reservoir. The reservoir rocks are of carboniferous age and were deposited by a tidally influenced estuary infilling an incised valley system. Reservoir thicknesses vary from 5–24 meters within a broad, largely unfaulted, anticlinal fold. It is thought that the injected CO_2 will migrate only upwards into the producing structure of the main field once the field has been fully depleted and abandoned.

From a rigorous evaluation using reservoir simulation models, BP estimated that breakthrough of the injected CO_2 into central gas wells is unlikely to occur within the first 15 years of production. By then, reductions in CO_2 production from the other In Salah fields will have freed CO_2 handling capacity at the Krechba surface facilities. The model, which will be updated, has also been used to identify well locations that satisfy criteria such as accessing connected sandstone volumes to accommodate the predicted quantities of CO_2 and minimizing flow line distances.

The separation plant will again be amine based.

Power Plants

Past and present research is bringing the possibility of cost-effective carbon dioxide capture and storage from power plants ever closer.

Kårstø and Osaka Pilot Plants

Membrane technology promises to reduce weight and space at CO_2 separation plants, as well as costs.

Besides CO_2 capture and storage from gas fields, the company has done much research on the cleaning of flue gas emissions from coal- and gas-fired power plants. Although the power industry is the greatest contributor to global CO_2 emissions, CO_2 capture remains an elusive goal because of the low concentrations of CO_2 contained in exhaust fumes and the cost of removing them.

Initial work was largely centered on a flagship venture, which started in 1992 as a joint industry project with Kværner Process Systems, NTNU, and SINTEF. The main aim was to investigate whether membrane contactors could replace the bulky amine towers currently used for CO_2 absorption and desorption. Ongoing studies are concerned with verifying the current technology and making further improvements.

The membrane concept is simple: a gas absorption membrane serves as a contact device between a CO_2-rich flue gas flow on one side and the flow of an absorption fluid on the other. Separation occurs as the CO_2 is selectively drawn through the membrane by the attraction of the absorption fluid. The CO_2 is then removed from the liquid at elevated temperature and pressure by essentially reversing the procedure so that the membrane acts as a desorption device. The membranes are generally made from porous, hydrophobic materials, although their exact constructions are far more complex.

W L Gore and Associates (Gmbh) supplied designer membranes for installation at two major pilot plants—one in Norway (at Kårstø) and one in Japan (Osaka). The Japanese leg was in cooperation with the Kansai Electric Power Company, Mitsubishi Heavy Industries, and the Carbon Dioxide Capture Project (see later). The main difference between the two investigations is that a conventional amine was used at Kårstø, while a recently developed alternative was tested at Osaka.

The main advantage of membrane technology is that it reduces weight and space by about 50%, thus making it especially suitable for offshore CO_2 as well as onshore capture. What is more, degradation of the absorbent can be reduced and entrainment and foaming are totally avoided.

The results from both pilot plant tests are very encouraging, not only because of the weight/space advantages but also because of potential cost reductions.

Carbon Dioxide Capture Project

While this work was under way, the company decided to widen its technological net by participating in several broad-based international studies, the latest and most comprehensive being the Carbon Dioxide Capture Project (CCP).

As stated in the Introduction, the main aim of the project (2001–2003[21]) was to reduce the costs of CO_2 separation and capture by more than 50% for existing plants and by 75% for new ones. Other objectives included ways of demonstrating to authorities and the public that CO_2 storage is safe, measurable, and verifiable and to advance technologies towards a "proof of concept" stage.

With such an immense scope, it is hardly surprising that most of the conclusions point to further research. Even so, there are many significant results. For CO_2 capture, the consortium developed a broad portfolio of technologies that will serve as the basis for the next generation and has shown

21. Although the project is officially over, several results are still being received.

that all three of the main technical areas—pre-combustion, post-combustion, and oxyfuel—have considerable potential for reducing costs. Furthermore, a new set of tools has been developed for managing long-term CO_2 storage, as well as a unique risk-based approach covering both aspects.

However, it was concluded that CO_2 capture and storage from heat and power production is still too expensive in the light of current oil and gas prices and taxes. Moreover, no technology is emerging as a clear leader, although several promising areas may be on the brink of commercialization.

Perhaps the project's most resounding achievement was that industries and governments came together in an international forum to promote strong technical leadership.

The CO_2 Store Project

Another investigation is currently assessing whether the Sleipner experience can be extended to onshore industrial sites. This is being done as part of the aforementioned CO_2 Store project, in which four European locations have been earmarked for extensive feasibility studies:

1. *Denmark:* The Energi E2-operated power plant and Statoil's Kalundborg refinery, which constitute the largest, single Danish CO_2 emission point source accounting for some 6 million tons of CO_2 per year. The power plant fuel is coal and orimulsion—a fuel consisting of a bitumen-in-water emulsion.
2. *South Wales (UK):* A prospective gasification/combined-cycle power station to be developed by Progressive Energy using a mixture of anthracitic coal and green petcoke. The proposed technical solution involves pre-combustion CO_2 capture, possibly amounting to 1−2 million tons per year.
3. *Part of the Trøndelag Platform:* An area off the Norwegian coast that contains several CO_2 point sources and where others are being planned.
4. *Germany:* The Schwarze Pumpe power plant operated by Vattenfall's subsidiary VEAG (Vereinigte Energiwerke AG). The plant is fueled with lignite and each block emits about 5 million tons of CO_2 per year.

Both onshore and nearshore repositories will be investigated, including those with large structural closures, depleted fields and regional deep saline aquifers.

The intention is that research will progress to such a stage that industry and national authorities can make an informed decision as to whether injection is a practical proposition in their respective areas. This, however, is not as straightforward as it seems, because judicial, safety and geological considerations will differ significantly from place to place. The Sleipner experience will therefore have to be adapted to meet local conditions.

By the time this part of the CO_2 Store project has run its course, the participants hope to offer a tool kit of various technologies and procedures that can be matched to suit the requirements of any deep saline aquifer, wherever it may be. The SACS 2003 Best Practice Manual will be updated accordingly.

Other Projects

Although space limitations preclude extensive coverage of all but the most eye-catching projects, it is important to mention that other collaborative ventures and proprietary in-house projects are systematically examining the entire range of available technologies. Among them, Statoil is a partner in three integrated, multipartner ventures which are partly funded by the European Union; namely, ENCAP, CASTOR, and CO_2Sink. These are concerned with pre-combustion, post-combustion, and storage, respectively.

The company is also jointly involved in several basic research projects, two of them dealing with CO_2 capture and one with the use of CO_2 for improved oil recovery (discussed next). The aims are to elucidate fundamental physicochemical phenomena in the hope of making significant breakthroughs.

Carbon Dioxide Utilization

The world's consumption of brewed and carbonated drinks falls well short of using up the vast quantities of excessive anthropogenic CO_2, and the European market for so-called food grade CO_2 is currently running at only 2.7 million tons per year.

Other enticing possibilities are self-defeating in the sense that they normally involve the use of energy, thereby producing even more CO_2.

One way of overcoming this is to use CO_2 as a means of improving oil recovery. When the conditions are right, the injection of CO_2 into a petroleum reservoir results in partial storage while improving the cash flow.

At the last count, some 70 or more onshore fields in the United States use CO_2-based miscible[22] gas injection to squeeze more oil out of the reservoirs. Here there are several natural sources[23] of high-grade, high-pressure CO_2, as well as a pipeline infrastructure linked to the fields in question.

A comprehensive study[24] of 115 onshore fields concluded that the average improvement in the oil recovery factor is about 12% for sandstone reservoirs and as high as 17% for carbonates. Moreover, about 71% of the injected CO_2 (on average) remains in the reservoirs while back-produced[25] CO_2 is recovered and reinjected.

22. *Miscible* refers to the ability of CO_2 to mix with the oil.
23. In the United States, 32 million tons of CO_2 are acquired each year from natural sources and 11 million tons per year from industrial sources.
24. Statoil commissioned SINTEF to perform this study.
25. Injected CO_2 returning to the surface as part of the well stream.

The next step—although a major one—is to transfer this decade-old practice from onshore to offshore, using industrial rather than natural sources of CO_2.

The North Sea and even parts of the Norwegian Sea are regarded as potentially suitable targets, because numerous oil fields, gas processing sites, and CO_2 sources are relatively concentrated when compared with other offshore regions around the world.

But where will the CO_2 come from?

Industrial Sources and Transport

The world's industrial sources can be classified according to their CO_2 concentration, and the costs of capturing high-concentration sources are usually lower than those for low-concentration sources.

The most prolific sources are ammonia plants, hydrogen plants, gas processing centers, cement factories, and iron and steel blast furnaces.

Another source for Statoil is its own gas processing sites (e.g., Sleipner and Snøhvit), where the gas consumer has already paid for the CO_2 capture.

But how would the CO_2 be transported to the fields?

On land, much CO_2 is transported via pipelines and is thus a proven technology.[26]

This means that the adaptation of present and new pipelines systems is a feasible proposition in parts of the North and Norwegian Seas. However, another serious contender is the shipment of CO_2 in tankers that are somewhat akin to those for transporting liquefied petroleum gas. In certain cases shipping may prove to be the cheaper alternative and will certainly be the more flexible of the two solutions.

Statoil, SINTEF, Vigor AS, and the Teekay Shipping Corporation[27] have just completed a research project aimed at designing a suitable vessel. The planned ship is 177 meters long by 31 meters wide, and contains four to six tanks capable of holding 20,000 cubic meters of liquefied CO_2 at around 7 bara and a temperature of −50°C. One notable innovation is the development of equipment for directly discharging liquefied CO_2 at production platforms. The ship is also designed as a multipurpose carrier suitable for transporting LPG and similar products.

Having weighed up the relative costs of initially supplying about 10 million tons of CO_2 by pipeline or ship, Statoil believes that both alternatives could be involved either separately or in combination.

But what about the technical and economic viability of offshore CO_2-based improved oil recovery?

Carbon Dioxide-Based Improved Oil Recovery

Detailed screening studies show that several Statoil-operated fields may be suitable candidates.

Over the years, Statoil has made enormous strides in improved oil recovery processes, not least the injection of water and natural gas to sweep more oil out of reservoirs (e.g., WAG[28]) and the use of bacteria to mobilize more oil from pore surfaces (M IOR[29]). The IOR potential for Statoil-operated fields is thought to be significant, particularly in the Tampen and Halten production centers off the southern and mid-Norwegian coast.

The possibility of using CO_2 instead of natural gas has long been discussed, but it is only during the last five years that the topic has received rigorous attention. The question is not whether CO_2-based IOR is feasible—this has already been proved on land—but whether it is viable for the specific conditions encountered in the North Sea.

The general principle is broadly the same as that for natural gas-based techniques, bearing in mind that CO_2 under elevated pressure and temperature is far denser than natural gas. After natural (primary) depletion[30] and (secondary) water injection, CO_2 can be injected into reservoirs where it mixes with the remaining oil and pushes a bank of additional oil towards production wells. CO_2 injection is sometimes referred to as a tertiary recovery process.

Many of the determining factors are common to any CO_2-based IOR project whether on land or at sea; others, however, are peculiar to the offshore operating environment.

On one side of the equation is cost, where gross and net values of additional oil have to be balanced against operational and investment expenditure. Examples include the importation of vast quantities of CO_2, its mode of transport, and the distances involved. Platform modifications also come into the picture, including the construction of production and injection systems and facilities for handling back-produced CO_2.

On the other side of the equation are technical considerations; for example, well coverage and drainage areas, bearing in mind that distances between offshore wells are far greater than their onshore counterparts.

Other issues concern the ways in which the physical and chemical characteristics of the reservoir rocks and fluids may be affected by CO_2 and the influence of recovery processes and drainage strategies used earlier in the life of a field.

26. About 90 million tons/year of CO_2 was transported by a pipeline in the United States, according to an IEA Greenhouse Gas R&D Programme report in 2001.

27. Formerly the Statoil-owned Navion company.

28. WAG—water-alternating-gas.

29. MIOR—microbial improved oil recovery.

30. In the North and Norwegian Seas water injection is carried out straight away, as natural depletion is relatively ineffective.

The eventual challenge of injecting CO_2 through subsea wells (i.e., those installed on the seabed) must also be taken into account now that subsea IOR is much in focus.

And in some cases there is a possibility of CO_2 infiltrating natural gas and weak carbonic acid corroding pipe work and pipelines.

The most important aspect, however, is the potential cost-effectiveness of CO_2-based IOR compared with other recovery processes.

Taking all of this into account, it is clear that the importation of a land-based practice into the offshore arena cannot be undertaken lightly. Even so, there are certain technical advantages: CO_2 increases miscibility, which leads to improved displacement efficiency; its high density results in better sweep efficiency; and the consequent swelling of the oil improves mobility. Oil production is also generally high when the conditions are right.

Screening evaluations have been performed on numerous Statoil-operated and partner oil fields, each of which has been categorized on a scale ranging from "promising" to "inappropriate."

As expected, reservoir performance simulations show that there are considerable variations in the amounts of extra oil that could be produced. In the worst cases there is almost no benefit at all, whereas in the best cases additional production could amount to some 10% of the original oil in place.

The challenge facing Statoil's reservoir experts is to reduce the economic and technical risks by improving the precision of reservoir simulation and discuss the pros and cons of implementation with asset teams and partners.

Of those fields falling into the "promising" category, the Statoil-operated Gullfaks field is the most extensively studied candidate for a CO_2-based M WAG pilot (M = miscible).

The field is located on the western flank of the North Viking Graben (Norwegian sector, North Sea), and largely comprises highly faulted and compartmentalized marine and fluvio-deltaic sandstone reservoirs belonging to the prolific Middle Jurassic Brent Group. Production at Gullfaks began in 1986 and the field has 112 wells drilled from three production platforms.

In short, Statoil's plan is to reduce the costs of CO_2 capture, generate electricity and hydrogen from natural gas, and consolidate its position in renewables.

The case above represents the most thorough information I could find on Statoil's work in 2006, when I submitted the first edition of this book for publication. Overall, it still is.

MIT retains a log of sequestration sites worldwide and the information below* on Mongstad comes from there. MIT reference the Statoil website, so that is also provided below.

However, Statoil's website** provides more current updates on new and individual projects. An example follows.

Statoil Mongstad Fact Sheet: Carbon Dioxide Capture and Storage Project

Company/Alliance: *Phase 1*: The European CO_2 Test Centre Mongstad (TCM) owned by the Norwegian Government, Statoil, Sasol and Shell

Phase 2: Statoil and Norwegian Government

Location: Mongstad, Norway

Feedstock: Natural gas

Process: Exhaust gases from a Residue Catalytic Cracker (RCC) and Natural Gas Combined Heat and Power (CHP) plant

Size: Combined Heat and Power Facility: 350 MW heat, 280 MW electricity

Phase 1: 100,000 T/Yr CO_2

Phase 2: 1.5 MT/Yr

Capture Technology: *Phase 1:* Post-combustion: Chilled ammonia (80,000 T/Yr CO_2) and amine-based capture (20,000 T/Yr CO_2)

CO_2 Fate: Sequestration in Saline Formation

Timing: *Phase 1*: Start-up of the cogeneration facility and pilot plant (2012)

Phase 2: Awaiting Norway's Government funding decision (2016): a decision basis for the construction of the larger scale Phase 2 project is to be submitted to the authorities in 2013/2015

Motivation/Economics:

The project will be developed in two phases to reduce technical and financial risk. Phase 1 includes capturing at least 80,000 tons of CO_2 using chilled ammonia and 20,000 tons of CO_2 with amine technology.

In September 2010 an estimate was that the CCS plant at the Mongstad refinery and power station will cost about 6 billion kroner ($1.02 billion). In October 2010 the Norwegian Government announced that it was increasing CCS spending by 1/3 to Nkr 2.7 billion ($462 million). The majority of this would go to the Mongstad Project.

Comments:

The Mongstad CCS project was started in May 2012.

Phase 1: The goal of the TCM is to develop the most cost effective way to capture CO_2 with

* Source: http://sequestration.mit.edu/tools/projects/statoil_mongstad.html

** Reference: Statoil website: http://www.statoil.com/en/OurOperations/TerminalsRefining/ProdFacilitiesMongstad/Pages/EnergiverkMongstad.aspx

post-combustion technology. The TCM will test two capture technologies:

1: Aker Clean Carbon's amine-based CO_2 capture; and

2: Alstom's chilled ammonia-based technology. There is availability to test other technologies. In July 2012 Alstom successfully completed the feasibility study and has proceeded to the next step of the CO_2 capture technology qualification program for the full scale CCS plant.

Phase 2: The Norwegian oil and energy ministry said in March 2011 that they were postponing the decision to invest in Mongstad until 2016. The Norwegian Government had originally said in May 2011 that it would delay the decision to fund the CCS project at Mongstad until after 2014 when the present parliament's term is over. However, there may be a possible compromise if Statoil decreases the total size of the Mongstad. The emissions permit from the Norwegian Ministry of Environment requires that the development of the full scale CCS project proceed in parallel with the construction of the CHP plant.

APPENDIX 11A: EMISSIONS LEGISLATION

If an end user in any NIC (newly industrialized country) were to state BAT (best available technology) as his requirement, he would then probably have cleared the environmental legislation limits in his own country with a safety margin. However, frequently, legislators in NICs chose a template that they can work towards or adopt in increments, realistic to their own circumstances.

If one were to exclude CO_2 from consideration (other than it is reduced by using more efficient thermodynamic cycles), then most templates from Western countries would suffice as a guide template. Not all Western countries have adopted CO_2 taxes. Many now have NO_x and SO_x taxes. Countries like the United States trade NO_x and SO_x emissions between power plants, in each individual state.

The information on these standards is publicly available, and while there are some differences, the air quality agreement between the United States and Canada makes their standards relatively compatible. The United States has a federal Environmental Protection Agency (EPA) and each state has its own EPA, which may all have slightly different standards.

What follows are extracts from a version of the US EPA limits on stationary gas turbines (their section 3.1). The reader is referred to the source for the latest update and revisions. What follows is merely a guide and is deliberately not current. The parameters and substances discussed within are still the substances tracked by emissions monitoring; however, limits change with expanding technology. For instance, the version below does not include any NO_x ppm information on Xonon flameless combustors. It would, however, be fair to say that any standard for emissions from stationary gas turbines would involve these pollutant substances in roughly these percentages, until gas turbine technology takes the next giant leap forward. The author has left in a partial reference list for this document, so that the reader can seek updates for those sources as well.

From the US EPA's Section 3.1.1 General [1] *

Gas turbines, also called "combustion turbines," are used in a broad scope of applications including electric power generation, cogeneration, natural gas transmission, and various process applications. Gas turbines are available with power outputs ranging in size from 300 horsepower (hp) to over 268,000 hp, with an average size of 40,200 hp [2]. The primary fuels used in gas turbines are natural gas and distillate (No. 2) fuel oil [3].

From 3.1.3 Emissions *

The primary pollutants from gas turbine engines are nitrogen oxides (NOX), carbon monoxide (CO), and to a lesser extent volatile organic compounds (VOC). Particulate matter (PM) is also a primary pollutant for gas turbines using liquid fuels. Nitrogen oxide formation is strongly dependent on the high temperatures developed in the combustor. Carbon monoxide, VOC, hazardous air pollutants (HAP), and PM are primarily the result of incomplete combustion. Trace to low amounts of HAP and sulfur dioxide (SO_2) are emitted from gas turbines. Ash and metallic additives in the fuel may also contribute to PM in the exhaust. Oxides of sulfur (SOX) will only appear in a significant quantity if heavy oils are fired in the turbine. Emissions of sulfur compounds, mainly SO_2, are directly related to the sulfur content of the fuel.

Available emissions data indicate that the turbine's operating load has a considerable effect on the resulting emission levels. Gas turbines are typically operated at high loads (greater than or equal to 80% of rated capacity) to achieve maximum thermal efficiency and peak combustor zone flame temperatures. With reduced loads (lower than 80%), or during periods of frequent load changes, the combustor zone flame temperatures are expected to be lower than the high load temperatures, yielding lower thermal efficiencies and more incomplete combustion. The emission factors for this section are presented for gas turbines operating under high load conditions. Section 3.1 background information documents and emissions

* [11-3]: Federal EPA, USA, Standards on Combustion Turbines. Contact source for applicable updates or revisions.

database contain additional emissions data for gas turbines operating under various load conditions.

Gas turbines firing distillate oil may emit trace metals carried over from the metals content of the fuel. If the fuel analysis is known, the metals content of the fuel ash should be used for flue gas emission factors assuming all metals pass through the turbine.

If the HRSG is not supplementary fuel fired, the simple cycle input-specific emission factors (pounds per million British thermal units [lb/MMBtu]) will also apply to cogeneration/combined-cycle systems. If the HRSG is supplementary fired, the emissions attributable to the supplementary firing must also be considered to estimate total stack emissions.

3.1.3.1 Nitrogen Oxides

Nitrogen oxides formation occurs by three fundamentally different mechanisms. The principal mechanism with turbines firing gas or distillate fuel is thermal NO_x, which arises from the thermal dissociation and subsequent reaction of nitrogen (N_2) and oxygen (O_2) molecules in the combustion air. Most thermal NO_x is formed in high temperature stoichiometric flame pockets downstream of the fue linjectors where combustion air has mixed sufficiently with the fuel to produce the peak temperature fuel/air interface.

The second mechanism, called prompt NO_x, is formed from early reactions of nitrogen molecules in the combustion air and hydrocarbon radicals from the fuel. Prompt NO_x forms within the flame and is usually negligible when compared to the amount of thermal NO_x formed. The third mechanism, fuel NO_x, stems from the evolution and reaction of fuel-bound nitrogen compounds with oxygen. Natural gas has negligible chemically-bound fuel nitrogen (although some molecular nitrogen is present). Essentially all NO_x formed from natural gas combustion is thermal NO_x. Distillate oils have low levels of fuel-bound nitrogen. Fuel NO_x from distillate oil-fired turbines may become significant in turbines equipped with a high degree of thermal NO_x controls. Otherwise, thermal NO_x is the predominant NO_x formation mechanism in distillate oil-fired turbines.

The maximum thermal NO_x formation occurs at a slightly fuel-lean mixture because of excess oxygen available for reaction. The control of stoichiometry is critical in achieving reductions in thermal NO_x. Thermal NO_x formation also decreases rapidly as the temperature drops below the adiabatic flame temperature, for a given stoichiometry. Maximum reduction of thermal NO_x can be achieved by control of both the combustion temperature and the stoichiometry. Gas turbines operate with high overall levels of excess air, because turbines use combustion air dilution as the means to maintain the turbine inlet temperature below design limits. In older gas turbine models, where combustion is in the form of a diffusion flame, most of the dilution takes place downstream of the primary flame, which does not minimize peak temperature in the flame and suppress thermal NO_x formation.

Diffusion flames are characterized by regions of near-stoichiometric fuel/air mixtures where temperatures are very high and significant thermal NO_x is formed. Water vapor in the turbine inlet air contributes to the lowering of the peak temperature in the flame, and therefore to thermal NO_x emissions. Thermal NO_x can also be reduced in diffusion type turbines through water or steam injection. The injected water-steam acts as a heat sink lowering the combustion zone temperature, and therefore thermal NO_x. Newer model gas turbines use lean, premixed combustion where the fuel is typically premixed with more than 50% theoretical air which results in lower flame temperatures, thus suppressing thermal NO_x formation.

Ambient conditions also affect emissions and power output from turbines more than from external combustion systems. The operation at high excess air levels and at high pressures increases the influence of inlet humidity, temperature, and pressure [4]. Variations of emissions of 30% or greater have been exhibited with changes in ambient humidity and temperature. Humidity acts to absorb heat in the primary flame zone due to the conversion of the water content to steam. As heat energy is used for water to steam conversion, the temperature in the flame zone will decrease resulting in a decrease of thermal NO_x formation. For a given fuel firing rate, lower ambient temperatures lower the peak temperature in the flame, lowering thermal NO_x significantly. Similarly, the gas turbine operating loads affect NO_x emissions. Higher NO_x emissions are expected for high operating loads due to the higher peak temperature in the flame zone resulting in higher thermal NO_x.

3.1.3.2 Carbon Monoxide and Volatile Organic Compounds

CO and VOC emissions both result from incomplete combustion. CO results when there is insufficient residence time at high temperature or incomplete mixing to complete the final step in fuel carbon oxidation. The oxidation of CO to CO_2 at gas turbine temperatures is a slow reaction compared to most hydrocarbon oxidation reactions. In gas turbines, failure to achieve CO burnout may result from quenching by dilution air. With liquid fuels, this can be aggravated by carryover of larger droplets from the atomizer at the fuel injector. Carbon monoxide emissions are also dependent on the loading of the gas turbine. For example, a gas turbine operating under a full load will experience greater fuel efficiencies, which will reduce the formation of carbon monoxide. The opposite is also true, a gas turbine operating under a light to medium load will experience reduced fuel efficiencies (incomplete combustion), which will increase the formation of carbon monoxide.

The pollutants commonly classified as VOC can encompass a wide spectrum of volatile organic compounds, some of which are hazardous air pollutants. These compounds are discharged into the atmosphere when some of the fuel remains unburned or is only partially burned during the combustion process. With natural gas, some organics are carried over as unreacted, trace constituents of the gas, while others may be pyrolysis products of the heavier hydrocarbon constituents. With liquid fuels, large droplet carryover to the quench zone accounts for much of the unreacted and partially pyrolized volatile organic emissions.

Similar to CO emissions, VOC emissions are affected by the gas turbine operating load conditions. Volatile organic compounds emissions are higher for gas turbines operating at low loads as compared to similar gas turbines operating at higher loads.

3.1.3.3 Particulate Matter [13]

PM emissions from turbines primarily result from carryover of noncombustible trace constituents in the fuel. PM emissions are negligible with natural gas firing and marginally significant with distillate oil firing because of the low ash content. PM emissions can be classified as "filterable" or "condensable" PM. Filterable PM is that portion of the total PM that exists in the stack in either the solid or liquid state and can be measured on a EPA Method 5 filter. Condensable PM is that portion of the total PM that exists as a gas in the stack but condenses in the cooler ambient air to form particulate matter. Condensable PM exists as a gas in the stack, so it passes through the Method 5 filter and is typically measured by analyzing the impingers, or "back half," of the sampling train. The collection, recovery, and analysis of the impingers is described in USA EPA Method 202 of Appendix M, Part 51 of the Code of Federal Regulations. Condensable PM is composed of organic and inorganic compounds and is generally considered to be all less than 1.0 micrometers in aerodynamic diameter.

3.1.3.4 Greenhouse Gases [5–11]

Carbon dioxide (CO_2) and nitrous oxide (N_2O) emissions are all produced during natural gas and distillate oil combustion in gas turbines. Nearly all of the fuel carbon is converted to CO_2 during the combustion process. This conversion is relatively independent of firing configuration. Methane (CH_4) is also present in the exhaust gas and is thought to be unburned fuel in the case of natural gas or a product of combustion in the case of distillate fuel oil.

Although the formation of CO acts to reduce CO_2 emissions, the amount of CO produced is insignificant compared to the amount of CO_2 produced. The majority of the fuel carbon not converted to CO_2 is due to incomplete combustion.

Formation of N_2O during the combustion process is governed by a complex series of reactions and its formation is dependent upon many factors. However, the formation of N_2O is minimized when combustion temperatures are kept high (above 1475°F) and excess air is kept to a minimum (less than 1%).

3.1.3.5 HAP Emissions

Available data indicate that emission levels of HAP are lower for gas turbines than for other combustion sources. This is due to the high combustion temperatures reached during normal operation. The emissions data also indicate that formaldehyde is the most significant HAP emitted from combustion turbines. For natural gas-fired turbines, formaldehyde accounts for about two thirds of the total HAP emissions. Polycyclic aromatic hydrocarbons (PAH), benzene, toluene, xylenes, and others account for the remaining one third of HAP emissions. For No. 2 distillate oil-fired turbines, small amounts of metallic HAP are present in the turbine's exhaust in addition to the gaseous HAP identified under gas-fired turbines. These metallic HAP are carried over from the fuel constituents. The formation of carbon monoxide during the combustion process is a good indication of the expected levels of HAP emissions. Similar to CO emissions, HAP emissions increase with reduced operating loads. Typically, combustion turbines operate under full loads for greater fuel efficiency, thereby minimizing the amount of CO and HAP emissions.

*3.1.4 Control Technologies [12]**

There are three generic types of emission controls in use for gas turbines, wet controls using steam or water injection to reduce combustion temperatures for NO_x control, dry controls using advanced combustor design to suppress NO_x formation or promote CO burnout, and post-combustion catalytic control to selectively reduce NO_x or oxidize CO emission from the turbine. Other recently developed technologies promise significantly lower levels of NO_x and CO emissions from diffusion-combustion-type gas turbines. These technologies are currently being demonstrated in several installations.

Emission factors in this section have been determined from gas turbines with no add-on control devices (uncontrolled emissions). For NO_x and CO emission factors for combustion controls, such as water-steam injection and lean premix units, are presented. Additional information for controlled emissions with various add-on controls can be obtained using the section 3.1 database. Uncontrolled, lean-premix, and water injection emission factors were presented for NO_x and CO to show the effect of combustion modification on emissions.

3.1.4.1 Water Injection

Water or steam injection is a technology that has been demonstrated to effectively suppress NO_x emissions from gas turbines. The effect of steam and water injection is to increase the thermal mass by dilution and thereby reduce peak temperatures in the flame zone. With water injection, there is an additional benefit of absorbing the latent heat of vaporization from the flame zone. Water or steam is typically injected at a water-to-fuel weight ratio of less than 1.

Depending on the initial NO_x levels, such rates of injection may reduce NO_x by 60% or higher. Water or steam injection is usually accompanied by an efficiency penalty (typically 2–3%) but an increase in power output (typically 5–6%). The increased power output results from the increased mass flow required to maintain turbine inlet temperature at manufacturer's specifications. Both CO and VOC emissions are increased by water injection, with the level of CO and VOC increases dependent on the amount of water injection.

3.1.4.2 Dry Controls

Since thermal NO_x is a function of both temperature (exponentially) and time (linearly), the basis of dry controls are to either lower the combustor temperature using lean mixtures of air or fuel staging or decrease the residence time of the combustor. A combination of methods may be used to reduce NO_x emissions such as lean combustion and staged combustion (two-stage lean/lean combustion or two-stage rich/lean combustion).

Lean combustion involves increasing the air-to-fuel ratio of the mixture so that the peak and average temperatures within the combustor will be less than that of the stoichiometric mixture, thus suppressing thermal NO_x formation. Introducing excess air not only creates a leaner mixture but it also can reduce residence time at peak temperatures.

Two-stage lean/lean combustors are essentially fuel-staged, premixed combustors in which each stage burns lean. The two-stage lean/lean combustor allows the turbine to operate with an extremely lean mixture while ensuring a stable flame. A small stoichiometric pilot flame ignites the premixed gas and provides flame stability. The NO_x emissions associated with the high temperature pilot flame are insignificant. Low NO_x emission levels are achieved by this combustor design through cooler flame temperatures associated with lean combustion and avoidance of localized "hot spots" by premixing the fuel and air.

Two-stage rich/lean combustors are essentially air-staged, premixed combustors in which the primary zone is operated fuel rich and the secondary zone is operated fuel lean. The rich mixture produces lower temperatures (compared to stoichiometric) and higher concentrations of CO and H_2, because of incomplete combustion. The rich mixture also decreases the amount of oxygen available for NO_x generation. Before entering the secondary zone, the exhaust of the primary zone is quenched (to extinguish the flame) by large amounts of air and a lean mixture is created. The lean mixture is preignited and the combustion completed in the secondary zone. NO_x formation in the second stage is minimized through combustion in a fuel lean, lower temperature environment. Staged combustion is identified through a variety of names, including dry-low NO_x (DLN), dry-low emissions (DLE), or $SoLoNO_x$.

3.1.4.3 Catalytic Reduction Systems*

Selective catalytic reduction (SCR) systems selectively reduce NO_x emissions by injecting ammonium (NH_3) into the exhaust gas stream upstream of a catalyst. Nitrogen oxides, NH_3, and O_2 react on the surface of the catalyst to form N_2 and H_2O. The exhaust gas must contain a minimum amount of O_2 and be within a particular temperature range (typically 450–850°F) in order for the SCR system to operate properly.

The temperature range is dictated by the catalyst material, which is typically made from noble metals, including base metal oxides such as vanadium and titanium, or zeolite-based material. The removal efficiency of an SCR system in good working order is typically from 65–90%. Exhaust gas temperatures greater than the upper limit (850°F) cause NO_x and NH_3 to pass through the catalyst unreacted. Ammonia emissions, called NH_3 slip, may be a consideration when specifying an SCR system.

Ammonia, either in the form of liquid anhydrous ammonia or aqueous ammonia hydroxide, is stored on-site and injected into the exhaust stream upstream of the catalyst. Although an SCR system can operate alone, it is typically used in conjunction with water-steam injection systems or lean-premix system to reduce NO_x emissions to their lowest levels (less than 10 ppm at 15% oxygen for SCR and wet injection systems). The SCR system for landfill or digester gas-fired turbines requires a substantial fuel gas pretreatment to remove trace contaminants that can poison the catalyst. Therefore, SCR and other catalytic treatments may be inappropriate control technologies for landfill or digester gas-fired turbines.

The catalyst and catalyst housing used in SCR systems tend to be very large and dense (in terms of surface area to volume ratio) because of the high exhaust flow rates and long residence times required for NO_x, O_2, and NH_3, to react on the catalyst. Most catalysts are configured in a parallel-plate, "honeycomb" design to maximize the surface area-to-volume ratio of the catalyst. Some SCR installations incorporate CO catalytic oxidation modules along with the NO_x reduction catalyst for simultaneous CO/NO_x control.

Carbon monoxide oxidation catalysts are typically used on turbines to achieve control of CO emissions, especially turbines that use steam injection, which can increase the concentrations of CO and unburned hydrocarbons in the exhaust. CO catalysts are also being used to reduce VOC and organic HAPs emissions. The catalyst is usually made of a precious metal such as platinum, palladium, or rhodium. Other formulations, such as metal oxides for emission streams containing chlorinated compounds, are also used. The CO catalyst promotes the oxidation of CO and hydrocarbon compounds to carbon dioxide (CO_2) and water (H_2O) as the emission stream passes through the catalyst bed. The oxidation process takes place spontaneously, without the requirement for introducing reactants. The performance of these oxidation catalyst systems on combustion turbines results in 90+% control of CO and about 85–90% control of formaldehyde. Similar emission reductions are expected on other HAP pollutants.

3.1.4.4 Other Catalytic Systems [14,15]

New catalytic reduction technologies have been developed and are currently being commercially demonstrated for gas turbines. Such technologies include, but are not limited to, the Sconox and the Xonon systems, both of which are designed to reduce NO_x and CO emissions. The Sconox system is applicable to natural gas-fired gas turbines. It is based on a unique integration of catalytic oxidation and absorption technology. CO and NO are catalytically oxidized to CO_2 and NO_2. The NO_2 molecules are subsequently absorbed on the treated surface of the Sconox catalyst. The system manufacturer guarantees CO emissions of 1 ppm and NO_x emissions of 2 ppm. The Sconox system does not require the use of ammonia, eliminating the potential of ammonia slip conditions evident in existing SCR systems. Only limited emissions data were available for a gas turbine equipped with a Sconox system. These data reflected HAP emissions and were not sufficient to verify the manufacturer's claims.

References for Section 3.1

1. Alternative Control Techniques Document—NO_X Emissions from Stationary Gas Turbines, EPA 453/R-93—007, January 1993.
2. C. C. Shih et al. Emissions Assessment of Conventional Stationary Combustion Systems, vol. 2, Internal Combustion Sources, EPA-600/7-79-029c. Cincinnati, OH: U. S. Environmental Protection Agency, February 1979.
3. Final Report—Gas Turbine Emission Measurement Program, GASLTR787. Westbury, NY: General Applied Science Laboratories, August 1974.
4. Standards Support and Environmental Impact Statement, vol. 1, Proposed Standards of Performance For Stationary Gas Turbines, EPA-450/2-77-017a. Research Triangle Park, NC: U.S. Environmental Protection Agency, September 1977.
5. L. P. Nelson et al. Global Combustion Sources of Nitrous Oxide Emissions, Research Project 2333–4 Interim Report. Sacramento: Radian Corporation, 1991.
6. R. L. Peer et al. Characterization of Nitrous Oxide Emission Sources. Research Triangle Park, NC: U.S. Environmental Protection Agency, Office of Research and Development, 1995.
7. S. D. Piccot et al. Emissions and Cost Estimates for Globally Significant Anthropogenic Combustion Sources of NO_x, N_2O, CH_4, CO, and CO_2. Research Triangle Park, NC: U.S. Environmental Protection Agency, Office of Research and Development, 1990.
8. G. Marland and R. M. Rotty. Carbon Dioxide Emissions from Fossil Fuels: A Procedure for Estimation and Results for 1951–1981, DOE/NBB-0036 TR-003. Oak Ridge, TN: Carbon Dioxide Research Division, Office of Energy Research, U.S. Department of Energy, 1983.
9. G. Marland and R. M. Rotty. Carbon Dioxide Emissions from Fossil Fuels: A Procedure for Estimation and Results for 1950–1982, Tellus 36B:232–261, 1984.
10. Inventory of U.S. Greenhouse Gas Emissions and Sinks: 1990–1991, EPA-230-R-96-006. Washington, DC: U.S. Environmental Protection Agency, November 1995.
11. IPCC Guidelines For National Greenhouse Gas Inventories Workbook. Intergovernmental Panel on Climate Change/Organization for Economic Cooperation and Development, Paris, France, 1995.
12. L. M. Campbell and G. S. Shareef, Sourcebook: NO_x Control Technology Data, Radian Corp., EPA-600/2-91-029, Air and Energy Engineering Research Laboratory, U.S. Environmental Protection Agency, Research Triangle Park, July 1991.
13. In-Stack Condensable Particulate Matter Measurement and Permitting Issues for Maryland Power Plants. Maryland Department of Natural Resource, prepared by Versar, Inc., January 1998.
14. Catalysts for Power Generation, The Sconox System. Goal Line Environmental Technologies, www.glet.com/gl_prod_SCONOX.htm, March 1998.
15. Emission Factor Documentation of AP-42 Section 3. 1, Stationary Combustion Turbines, EPA Contract No. 68-D7-0070. Raleigh, NC: Alpha-Gamma Technologies Inc., April 2000.

Emissions Permits [11-1]

When seeking building permission, any operator of a stationary gas turbine needs to go through the permitting procedure for the country, state, or province in which it will operate. This is a complex procedure and can vary depending on where in the world one is and which governing body has jurisdiction over the project. For instance, in China, for a power plant under 200 MW, the individual province and not the federal government has authority. The licenses generally take a much shorter time.

Even in the United States, where brownouts in the late 1990s California raised the demand for power considerably, technology affects the time taken by the licensing process. At a panel session during the 1998 IGTI annual meeting, a company (now out of business) that was then in the business of making flameless combustors informed the audience that they were fitting catalytic flameless combustors that emitted only 2–3 ppmv NO_x and got their permits a few months before other stations with conventional low NO_x burners. Depending on the station size, that represents millions of revenue dollars to the power seller. Research work on flameless combustors continues today.

Figure 11–4 is an example, *deliberately not current,* of an Emissions Activity Category form (source: Ohio EPA). If the reader needs to use this form, current ones are available on the Internet (for example: http://www.epa.ohio.gov/dapc/fops/eac/eacforms.aspx. Accessed July 2014).

FOR OHIO EPA USE
FACILITY ID: ________________
—

EMISSIONS ACTIVITY CATEGORY FORM
STATIONARY INTERNAL COMBUSTION ENGINE

This form is to be completed for each stationary reciprocating or gas turbine engine. State/Federal regulations which may apply to stationary internal combustion engines are listed in the instructions. Note that there may be other regulations which apply to this emissions unit which are not included in this list.

1. Reason this form is being submitted (Check one)

   ❒ New Permit ❒ Renewal or Modification of Air Permit Number (e.g. P001)______________

2. Maximum Operating Schedule: ________hours per day; ________days per year

   If the schedule is less than 24 hours/day or 365 days/year, what limits the schedule to less than maximum? See instructions for examples. ______________________

3. Engine type: ❒ Gas turbine ❒ Reciprocating

4. Purpose of engine: ❒ Driving pump or compressor ❒ Driving electrical generator

5. Normal use of engine: ❒ Emergency only ❒ Non-emergency

6. Engine Manufacturer: ____________________ Model No: ____________________

7. Engine exhaust configuration: (for turbines only)
   ❒ simple cycle *(no heat recovery)*
   ❒ regenerative cycle *(heat recovery to preheat combustion air)*
   ❒ cogeneration cycle *(heat recovered to produce steam)*
   ❒ combined cycle *(heat recovered to produce steam which drives generator)*

8. Input capacities (million BTU/hr): Rated:__________ Maximum:__________ Normal:__________

   Supplemental burner (duct burner) input capacity, if equipped (million BTU/hr):

   Rated:__________ Maximum:__________ Normal:__________

9. Output capacities (Horsepower): Rated:__________ Maximum:__________ Normal:__________

   (Kilowatts): Rated:__________ Maximum:__________ Normal:__________

   (lbs steam/hr)*: Rated:__________ Maximum:__________ Normal:__________

   **required for cogeneration or combined cycle units only*

FIGURE 11–4 Emissions activity category form, stationary internal combustion engine [11-3]. *(Source: [11-3] Ohio EPA.)*

10. Type of ignition: ❒ non-spark (diesel) ❒ spark

11. Type of fuel fired (check all that apply):

❒ single fuel
❒ dual fuel

❒ No. 2 oil, low-sulfur
❒ No. 2 oil, high-sulfur
❒ gasoline
❒ other, explain____________________

❒ natural gas
❒ diesel
❒ propane

❒ landfill gas
❒ digester gas

12. Complete the following table for all fuels identified in question 11 that are used for the engine and any supplemental (duct) burners, if equipped:

Fuel	HeatContent (BTU/unit)	wt.% Ash	wt.% Sulfur	Fuel Usage Estimated Maximum Per Year	Normal Per Hour	Max. Per Hour
Nat. gas	BTU/cu ft		gr/scf	cu ft	cu ft	cu ft
No. 2 oil	BTU/gal			gal	gal	gal
Gasoline	BTU/gal			gal	gal	gal
Diesel	BTU/gal			gal	gal	gal
Landfill/digester gas	BTU/cu ft		ppm	cu ft	cu ft	cu ft
Other (show units)						
List supplemental (duct) burner fuel and information below (show units):						

13. Type of combustion cycle (check all that apply):

❒ 2-stroke
❒ rich-burn
❒ carbureted
❒ other, explain______________________________

❒ 4-stroke
❒ lean-burn
❒ fuel injected

14. Emissions control techniques (check all that apply):

❒ prestratified charge
❒ catalytic oxidation (CO)
❒ air/fuel ratio
❒ 2-stage rich/lean combustion
❒ water/steam injection
❒ other, explain____

❒ nonselective catalytic reduction (NSCR)
❒ selective catalytic reduction (SCR)
❒ injection timing retard (ITR)
❒ 2-stage lean/lean combustion
❒ preignition chamber combustion (PCC)

For each emissions control technique checked above, explain what pollutants are controlled by each technique: ____
____
____

FIGURE 11–4 *(Cont'd).*

INSTRUCTIONS FOR COMPLETION OF THE EMISSIONS ACTIVITY CATEGORY FORM FOR STATIONARY INTERNAL COMBUSTION ENGINES

GENERAL INSTRUCTIONS:

Provide complete responses to all applicable questions. If an item does not apply to the emissions unit, write in "Not Applicable" or "NA." If the answer is not known, write in "Not Known" or "NK." If you need assistance in understanding a question after reading the instructions below, contact your Ohio EPA District Office or Local Air Agency for assistance. Submittal of an incomplete application will delay application review and processing. In addition, the application may be returned as incomplete if all applicable questions are not answered appropriately.

APPLICABLE REGULATIONS

The following State and Federal Regulations may be applicable to stationary internal combustion engines. Note that there may be other regulations which apply to this emissions unit which are not included in this list.

Federal: 40 CFR 60, (NSPS) Subpart A (General Provisions), Subpart GG (Stationary Gas Turbines), Subparts Da, Db, and Dc (Steam Generating Units)
40 CFR 63, (NESHAPS/MACT) Subpart A (General Provisions), Subpart YYYY (Combustion Turbines), Subpart DDDDD (Steam Generators), Subpart ZZZZ (Reciprocating Internal Combustion Engines).

State: OAC rule 3745-31-02 (Permit to Install)
OAC rule 3745-35-02 (Permit to Operate)
OAC rule 3745-17-11(B)(4) - Particulate emission limits
OAC rule 3745-18-06(F) - Sulfur dioxide emission limits

If you would like a copy of these regulations, contact your Ohio EPA District Office or Local Air Agency. State regulations may also be viewed and downloaded from the Ohio EPA website at http://www.epa.state.oh.us/dapc/regs/regs.html. Federal regulations may be viewed and downloaded at http://www.epa.gov/docs/epacfr40/chapt-I.info/subch-C.htm.

CALCULATING EMISSIONS:

Manufacturers of some types of emissions units and most types of control equipment develop emissions estimates or have stack test data which you can request. Stack testing of the emissions may be done. Emissions unit sampling test data may be either for this emissions unit or a similar one located at the facility or elsewhere. You may develop your own emission factors by mass balance or other knowledge of your process, if you can quantify inputs and outputs accurately. You may be able to do this on a small scale or over a short period of time, if it is not practical during regular production. If you have control equipment, you may be able to quantify the amount of pollutants collected over a known time period or production amount. Any emission factor calculation should include a reference to the origin of the emission factor or control efficiency.

USEPA has developed emission factors for many types of emissions units and published them in a document titled "Compilation of Air Pollutant Emission Factors, AP-42", available from the following website: http://www.epa.gov/ttn/chief/ap42/index.html See Chapters 3.1, 3.2, 3.3 and 3.4 depending on the type of engine.

FIGURE 11–4 *(Cont'd).*

SPECIFIC INSTRUCTIONS:

1. Indicate whether this is an application for a new permit or an application for permit renewal. If applying for a permit renewal, provide the 4-character OEPA emissions unit identification number.

2. Provide the maximum number of hours per day and days per year the stationary internal combustion engine is expected to operate. The following are examples of why the maximum number of hours per day may be less than 24 or the maximum number of days per year may be less than 365 (this list is not all-inclusive):

   - The facility can only operate during daylight hours.
   - The process can only operate within a certain range of ambient temperatures.
   - The process is limited by another operation (i.e., a bottleneck).

5. Emergency use engines normally operate less than 500 hours per year. “Non-emergency” denotes the engine can be used at any time. OAC rule 3745-31-03 contains permit exemption provisions for emergency use generators and other small internal combustion engines.

7. For turbines only. Describe method of recovering exhaust heat, if applicable.

8. Provide the manufacturer’s rated, maximum and normal heat input capacities in million BTU (British Thermal Unit) per hour. Also specify the input capacities of any supplemental (duct) burners according to the manufacturer’s specifications. For turbines, the maximum heat input capacity should be determined at an ambient temperature of no greater than 0 degrees Farenheit.

9. Provide the output capacities in units appropriate for the application, i.e., Horsepower for pumps and compressors, Kilowatts for generators, etc. Note steam output is required for turbine cogeneration or combined cycle units only.

11. Check all types of fuel fired in the internal combustion engine and any supplemental (duct) burners. Be sure to indicate whether the unit is single or dual fuel.

12. Specify the heat content and %Ash and % Sulfur (if applicable) for all fuels used in the engine and supplemental (duct) burners. Provide estimated fuel consumption quantities based on normal and maximum operation.

13. Check the type of combustion cycle design.

14. Identify any emission control techniques employed in the engine application and indicate the pollutant or pollutants controlled by that technique.

FIGURE 11–4 *(Cont’d).*

Chapter 12

Maintenance, Repair, and Overhaul

"Learn from the mistakes of others, you'll never live long enough to make them all yourself."

—Old Pilots' saying

. . . but. . . .
"You're never too old to become younger."

—Mae West

Chapter Outline

Maintenance,* repair, and overhaul make up a major component of the cost per fired hour for a gas turbine system. The initial specification sets the tone for the end user's experience with that turbine. Systems and accessories can be specified that will help raise the turbine's TBOs and component lives (see Chapter 9, ECMS, and Chapter 10, Performance, Performance Testing, and Performance Optimization). The end user's negotiated warranty, service, and training package (see Chapter 14, The Business of Gas Turbines) also has the ability to improve matters in both the operator's and the turbine's life. Repair and overhaul are inevitable, as the turbine ages; however, this component of cost per fired hour can

* Source: [12-1] Claire Soares, working case notes 1975 through 2003 and Proceedings of AMSE IGTI panel sessions, "Engine Condition Monitoring Systems as They Relate to Life Extension of Gas Turbine Engine Components," 1985 through 2003, chair Claire Soares.

be minimized if the end user is proactive in terms of strategy in this vein.

OPERATING AND MAINTENANCE STRATEGIES

Maximized availability, or "up time," and the troubleshooting process (i.e., how easy it is to troubleshoot a machine) depend largely on the type of maintenance philosophy applied during the machine's operational life. This basic philosophy also affects the monitoring process.

Equally, the strategies and philosophy behind a turbine's selection play a key role in defining maintenance requirements. Unfortunately, a number of operators never link maintenance philosophy and troubleshooting to an appropriate extent. Therefore, they either leave themselves wide open for disastrous repair bills or spend more than they need to on maintenance.

This is poor maintenance management, as it can increase turbine system costs per fired hour. This section describes three main strategies appropriate for all rotating machinery, including gas turbine systems. A choice of these strategies should not be made by the manufacturer. It needs to fit the operator's specific application and comfort level. The manufacturer's resources and technical expertise should be utilized to support the decision process by providing relevant information.

The operator, typically, proceeds as follows. Basic goals are itemized. The highest priority is to maximize production. Optimizing production per unit of energy is part of that aim. Maximum availability and reliability (i.e., no unplanned downtime) are also critical. Operators want to minimize the maintenance, service, and repair costs.

Reactive Strategy

Too little maintenance results in unexpected failures and consequential major losses of production or customers. This approach is termed *reactive strategy* (or wait until something fails) and should be avoided on all machinery in critical applications.

Optimum maintenance strategy balances reasonable costs with maximum possible availability and reliability. The two most common main maintenance strategies employed by companies today are labeled *predictive strategy* and *preventive strategy.*

Predictive Strategy

Predictive maintenance strategies operate without a regular plan for service work or exchange of parts. A maintenance plan is set up only if there is proof of deterioration. Consequently, a company with a predictive strategy favors minimizing cost over maximizing use. Annual cost of this strategy may typically only average out to 1–2% of the prime equipment price.

With a predictive maintenance strategy, long-term plans generally involve some or all of the following.

Monitoring of operating data as follows:

- Gas path (mass flow, pressure head, efficiency)
- Water coolant (differential temperature)
- Oil analysis (water content, deterioration of anti-aging additives)
- Temperature monitoring of gas path, bearing oil.
- Incorporation of some of these parameters in a performance analysis (PA) system (see subsection in this chapter)
- Rotating speed or rpm of the different turbine rotors
- Incorporation of some of these parameters in an algorithm that is custom designed, usually by the original equipment manufacturer (OEM), to calculate life cycle usage in a life cycle assessment (LCA) system

Vibration analysis measurements as follows:

- FFT analysis (shaft, pinions), using eddy current probes at normal load and turndown
- FFT analysis (all bearing housings), at normal load

Preventive Strategy

In contrast with predictive strategy, preventive strategy aims toward maximum safety against unexpected failures. The concept here is to predict the average lifespan of a part and replace it before the end of that lifespan. Annual cost therefore is higher (perhaps 8–10% of the prime equipment price) because of the higher numbers of spare parts that need to be purchased and warehoused.

Besides the effects of choice of maintenance strategy on the troubleshooting time and effort required, the application service the unit is in also has an effect. With increasingly tough environmental legislation that, in turn, demands maximum energy usage or recovery, power recovery (cogeneration) processes are increasing in number. The deregulation of the power industry results in the increase of small power producers, or SPPs (such as process plants that can use existing process fluids to assist in energy efficiency and cogeneration). So deregulation also serves to increase the number of SPPs.

Evolving Strategy in Land Versus Air Versus Marine Applications

About three decades ago, it was common for gas turbine users to talk about the number of operating hours their unit had been running. The mid-1970s saw the beginnings of change to the concept of "cycles of operation."

The realization in aircraft engine application that sponsored that change was the acknowledgment that service factors on engines varied with mission profile and individual engine roles. The life (cycle) accumulation on the lead aircraft in aerobatic formation was up to 20 times less than that of the "followers'" engines. The pilots had to fidget with their power settings, back and forth, to stay the prerequisite distance from the leaders' wingtips.

In land-based applications, people began to realize that a start and stop with a Solar Saturn (nominal 1100 HP) engine may cost as little as the equivalent of 3–5 hours of full load operation. However, a start and stop of a GE Frame 5 (nominal 25,000 HP) might cost that engine as much as 500 hours of its operational life.

Research commenced on developing algorithms for life cycle assessment (LCA). For engines built by OEMs that use a high degree of cooling, such as Rolls Royce, the only relevant parameters in their algorithms for specific models may be rpm (excursions) versus time. For engines that did not have as much cooling, a temperature parameter in the algorithm was in order.

The service bulletins (SBs) that evolved to reveal these algorithms were as a result of end-user pressure. The end-user community thus was able to get longer lives out of their most expensive gas turbine components. LCA was done with LCAC (life cycle assessment counters). Instrumentation and condition monitoring OEMs that specialized in LCAC, such as Vosper Thornycroft, UK (also called HSDE), soon began to be approved and "certified for retrofits" by a wide range of GT OEMs.

Before LCA, preventive maintenance charts used to state a schedule of mandatory inspections, component removals, and tasks at different "total operating hours accumulated" of a gas turbine. Now those charts might call up cycle (as measured by the LCA counters) numbers.

In the case of engines whose metallurgy places their stress levels under the stress endurance curve, operation may be "on condition" (as observed with borescopes and other nonintrusive means or as observed by a limited "opening up" at an HSI, or hot section inspection).

HSIs are conducted on all gas turbine applications, land, sea, or air. There may also be unscheduled, limited or otherwise "opening up" as prompted by alarming vibration readings, borescope observations, or other monitoring means.

Due to the variable nature of aircraft engine applications, the engine external features and the aircraft are inspected visually before every flight during a pilot "walk around." Oil or hydraulic leaks may be observed at this stage. During the pilot's "run up" on the tarmac prior to takeoff, problems with excessive temperatures and pressures may be observed, sometimes necessitating a "return to the gate." ECMS are sufficiently sophisticated today that postflight analysts can tell which pilots like to "red line" their engines. Advancements in (air-to-ground) telemetry make it possible for ground staff to know which module of an aircraft engine they may want to "swap" when the aircraft next lands.

Because of their relative isolation (compared with ground applications), operators of air (say, a CF6-80C2) and marine (LM2500) gas turbines require a regular maintenance schedule that is far more exacting than their ground-based (LM2500) counterparts. This is particularly true if that land-based counterpart also runs at steady load (as in, for instance, power generation).

So aircraft engine operators (as legislated by the US FAA, the European JAA, or other international flying legislative authorities) must conduct exacting:

- Routine maintenance (after every flight, daily, or at some specified time)
- Routine inspections (such as HSIs at fixed intervals of cycles accumulated)
- Planned maintenance (if the engine is not an "on-condition" design) of varying complexity

For instance, the Pratt and Whitney JT 8D fleet that operates on a preventive maintenance philosophy (particularly with its earlier models) has to have "ESV1s" and "ESV2s" to bring it back to "zero time" condition. The list of work items, SBs both mandatory and optional, and "work, based on observed condition" for an ESV2 is far longer and more complex than for an ESV1. Other OEMs may have their own designation (other than ESV-) and stated intervals (in hours if it's an old fleet, or cycles if it has an LCA counter).

Of course, once one is talking about an "ESV1 or 2," one is now into the "repair and overhaul" (R&O) phase of that gas turbine's life. R&O is a highly specialized subdivision of overall maintenance. For the most part, it is distinguished from "regular and routine" maintenance in that many work items are not routine but "as needed" and "based on condition observed" only.

What follows is a summary of regular maintenance on an aeroengine. It provides a good template for all gas turbines, as land-based and marine gas turbines need the same basic functions checked. Exactly which instruments, features, and functions need checking and their frequency (daily, weekly, monthly, as needed) is spelled out in the maintenance manuals supplied with every gas turbine.

MAINTENANCE*,†

Maintenance covers both the work that is required to maintain the engine and its systems in an airworthy

* Source: Adapted, with permission, from Rolls Royce, *The Jet Engine*, 1986, Rolls Royce Plc: UK.

† Also applicable to most aircraft gas turbines.

condition while installed in an aircraft (on-wing or line maintenance) and the work required to return the engine to airworthy condition when removed from an aircraft (overhaul or shop maintenance). On-wing maintenance is covered in this section.

Because many aspects of maintenance are subject to the approval of a recognized authority, it should be fully understood that the information given in this section is of a general nature and is not intended as a substitute for any official instructions.

The comprehensive instructions covering the actual work to be done to support scheduled maintenance and unscheduled maintenance are contained in the aircraft maintenance manual. Both this publication, and the aircraft maintenance schedule mentioned later, are based on manufacturers' recommendations and are approved by the appropriate airworthiness authority.

The maximum time an engine can remain installed in an aircraft (engine life) is limited to a fixed period agreed between the engine manufacturer and airworthiness authority. On some engines this period is referred to as the time between overhaul (TBO) and on reaching it the engine is removed for complete overhaul.

Because the TBO is actually determined by the life of one or two assemblies within the engine, during overhaul, it is generally found that the other assemblies are mechanically sound and fit to continue in service for a much longer period. Therefore, with the introduction of modular engines and the improved inspection and monitoring techniques available, the TBO method on limiting the engine's life on-wing has been replaced by the "on-condition" method.

Basically this means that a life is not declared for the total engine but only for certain parts of the engine. On reaching their life limit, these parts are replaced and the engine continues in service, the remainder of the engine being overhauled "on condition." Modular constructed engines are particularly suited to this method, as the module containing a life-limited part can be replaced by a similar module and the engine returned to service with minimum delay. The module is then disassembled for life limited part replacement, repair, or complete overhaul as required.

On-Wing Maintenance

On-wing maintenance falls into two basic categories: scheduled maintenance and unscheduled maintenance.

Scheduled Maintenance

Scheduled maintenance embraces the periodic and recurring checks that have to be effected in accordance with the engine section of the appropriate aircraft maintenance schedule. These checks range from transit items, which do not normally entail opening cowls, to more elaborate checks within specified time limits, usually calculated in aircraft flying hours and phased with the aircraft check cycle.

Continuous "not-exceed-limit" maintenance, whereby checks are carried out progressively and as convenient within given time limits rather than at specific aircraft check periods, has been widely adopted to supersede the check cycle. With the progressive introduction of condition monitoring devices of increased efficiency and reliability, a number of traditionally accepted scheduled checks may become unnecessary. Extracts from a typical maintenance schedule are shown in Table 12–1.

Unscheduled Maintenance

Unscheduled maintenance covers work necessitated by occurrences that are not normally related to time limits, e.g., bird ingestion, a strike by lightning, a crash, or heavy landing. Unscheduled work required may also result from malfunction, troubleshooting, scheduled maintenance, and occasionally manufacturer's specific recommendations. This type of maintenance usually involves rectification adjustment or replacement.

Condition Monitoring

Condition monitoring devices must give an indication of any engine deterioration at the earliest possible stage and also enable the area or module in which deterioration is occurring to be identified. This facilitates quick diagnosis that can be followed by scheduled monitoring and subsequent programmed rectification at major bases, thereby avoiding in-flight shutdown, with resultant aircraft delay, and minimizing secondary damage. Monitoring devices and facilities can be broadly categorized as flight deck indicators, in-flight recorders, and ground indicators.

Flight Deck Indicators

Flight deck indicators are used to monitor engine parameters such as thrust or power, rpm, turbine gas temperature, oil pressure, and vibration. Other devices, however, may be used and these include:

- Accelerometers for more reliable and precise vibration monitoring
- Radiation pyrometers for direct measurement of turbine blade temperature
- Return oil temperature indicators
- Remote indicators for oil tank content
- Engine surge or stall detectors
- Rub indicators to sense eccentric running of rotating assemblies

TABLE 12–1 A Typical Maintenance Schedule (Extracts)

Item	Not Exceed Limit	Requirement
Engine oil tank	Flight termination	Check oil level. Replenish as necessary. Record amount taken
Cowls	Transit	Check the pod cowls for damage and external evidence of fuel and oil leaks
Caps and access panels	Transit	Check secure
Engine intake	Transit	Check clear. Free from damage and loose objects
Turbine and exhaust collector	Transit	Visually inspect for signs of damage and metal deposits
Engine intake	25 hours	Visually inspect front of engine through air intake for signs of damage paying particular attention to intake guide vanes and leading stage rotor blades
Turbine and exhaust collector	25 hours	Visually inspect L.P.2 turbine blades, nozzle guide vanes and mixer unit for cracking and damage by viewing from rear using a strong spot light
Fuel filter	125 hours	Drain sample and check for water contamination
Magnetic chip detector	200 hours	Remove and inspect
Igniter plugs	200 hours	Audibly check operation
Oil pressure filter	600 hours	Check and clean/renew filter element
Fuel filter	800 hours	Remove filter element, check and renew

(Adapted, with permission, from Rolls Royce, *The Jet Engine*, 1986, Rolls Royce Plc: UK.)

In-Flight Recorders

Selected engine parameters are recorded, either manually or automatically, during flight. The recordings are processed and analyzed for significant trends indicative of the commencement of failure. An in-flight recording device that may be used is the time/temperature cycle recorder. The purpose of this device is to accurately record the engine time spent operating at critical high turbine gas temperatures, thus providing a more realistic measure of "hot-end" life than that provided by total engine running hours.

Automatic systems known as aircraft integrated data systems (A.I.D.S.) are able to record parameters additional to those normally displayed, e.g., certain pressures, temperatures, and flows.

Many of the electronic components used in modern control systems have the ability to monitor their own and associated component operation. Any fault detected is recorded in its built-in memory for subsequent retrieval and rectification by the ground crew. On aircraft that feature electronic engine parameter flight deck displays certain faults are also automatically brought to the flight crew's attention.

Ground Indicators

The devices used or checked on the ground, as distinct from those used or checked in flight, may conveniently be referred to as ground indicators; this title is also taken to embrace instruments used for engine internal inspection.

Internal viewing instruments can be either flexible or rigid, designed either for end or angled viewing and, in some instances, adaptable for still or video photography that may be linked to closed circuit television. These instruments are used for examining and assessing the condition of the compressor and turbine assemblies, nozzle guide vanes (Figure 12–1), and combustion system, and can be inserted through access ports located at strategic points in the engine main casings.

The engine condition indicators include magnetic chip detectors, oil filters and certain fuel filters. These

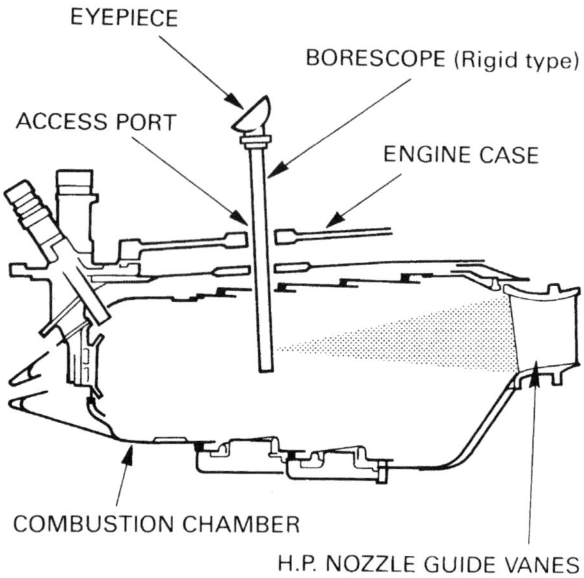

FIGURE 12–1 Inspection of H.P. nozzle guide vanes. *(Source: Rolls Royce.)*

indicators are frequently used to substantiate indications of failures shown by flight deck monitoring and in-flight recordings. For instance, inspection of the oil filters and chip detectors can reveal deposits from which experienced personnel can recognize early signs of failure. Some maintenance organizations progressively log oil filter and magnetic chip detector history and catalogue the yield of particles. Fuel filters may incorporate a silver strip indicator that detects any abnormal concentration of sulfur in the fuel.

Maintenance Precautions

During engine maintenance, it is necessary to observe certain precautions. The ignition system is potentially lethal and, therefore, before any work is done on the high energy ignition units, igniter plugs or harness, the low tension supply to the units must be disconnected and at least 1 minute allowed to elapse before disconnecting the high tension lead. Similarly, before carrying out work on units connected to the electrical system, the system must be made safe, either by switching off power or by tripping and tagging appropriate circuit breakers. With some installations, the isolation of certain associated systems may be required.

When the oil system is being replenished, care must be taken that no oil is spilled. If any oil is accidentally spilled, it should be cleaned off immediately as it is injurious to paintwork and to certain rubber compounds such as could be found in the electrical harnesses. Oil can also be toxic through absorption if allowed to come into contact with the human skin for prolonged periods. Care should be taken not to overfill the oil system; this may easily occur if the aircraft is not on level ground or if the engine has been stationary for a long period before the oil level is checked.

Before an inspection of the air intake or exhaust system is made it must be ascertained that there is no possibility of the starter system being operated or the ignition system being energized.

A final inspection of the engine, air intake and exhaust system must always be made after any repair, adjustment or component change, to ensure that no loose items, no matter how small, have been left inside. Unless specific local instructions rule otherwise, air intake and exhaust blanks or covers should be fitted when engines are not running.

Troubleshooting

The procedure for locating a fault is commonly referred to as troubleshooting, and the requirement under this procedure is for quick and accurate diagnosis with the minimum associated work and the prevention of unnecessary unit or engine removals.

The basic principle of effective troubleshooting is to clearly define and interpret the reported symptom and then proceed to a logical and systematic method of diagnosis (Figure 12–2).

The reported symptom will frequently originate from flight deck instrument readings and, unless it is apparent from supporting information that the readings are genuine, instrumentation should be checked before proceeding

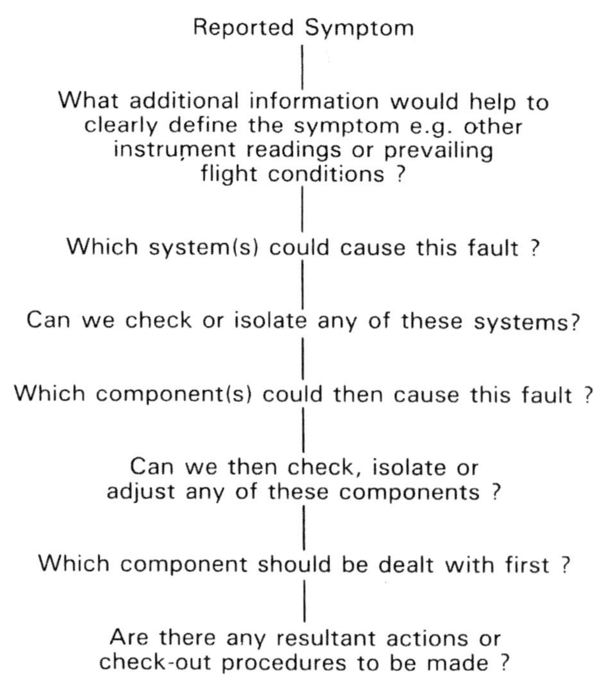

FIGURE 12–2 Troubleshooting—logical. *(Source: Rolls Royce.)*

further. Similarly, quick elimination checks should normally be undertaken before more involved tasks. The manufacturers' maintenance manual contains troubleshooting information, usually in chart form and Figure 12—3 shows a typical example.

The progressive introduction of improved and more reliable condition monitoring devices will have considerable influence on accepted troubleshooting practice, since to a large extent these devices are designed to pin-point, at an early stage, the specific system or assembly at fault. The development of suitable test sets could eventually eliminate the need for engine ground testing after troubleshooting.

Adjustments

There are usually some adjustments that can be made to the engine controlling the fuel trimming devices. Typical functions for which adjustment provision is normally made include idling and maximum rpm, acceleration and deceleration times, and compressor air bleed valve operation.

Adjustment of an engine should be made only if it is quite certain that no other fault exists that could be responsible for the particular condition. The maintenance manual instructions relative to the adjustment must be closely adhered to at all times. In many instances, subject to local instructions, a ground adjustment can be made with the engine running.

Adjusters are usually designed with some form of friction locking (Figure 12—4) that dispenses with locknuts, lock-plates, and looking wire. On some engines, provision is also made for fitting remote adjustment equipment (Figure 12—5) that permits adjustment to be made during ground test with the cowls closed, the adjustment usually being made from the flight deck.

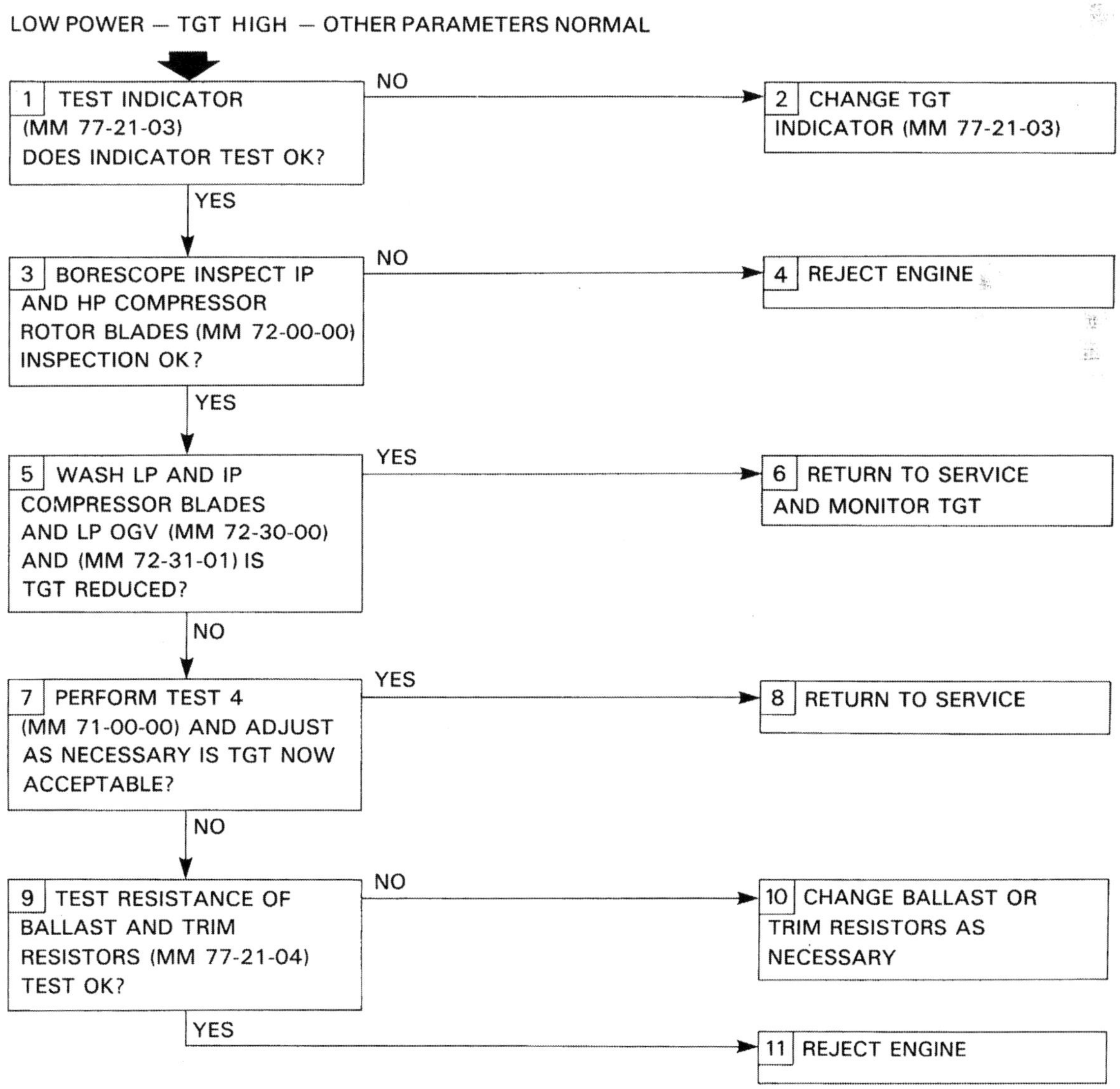

FIGURE 12—3 A typical troubleshooting chart. *(Source: Rolls Royce.)*

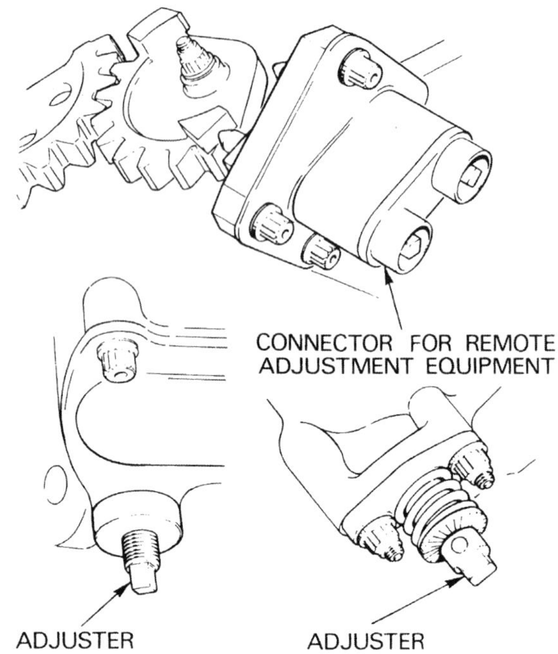

FIGURE 12–4 Typical friction locked adjusters. *(Source: Rolls Royce.)*

Ground Testing

The basic purpose of engine ground testing is to confirm performance and mechanical integrity and to check a fault or prove a rectification during troubleshooting. Ground testing is essential after engine installation, but scheduled ground testing may not normally be called for where satisfactory operation on the last flight is considered to be the authority or acceptance for the subsequent flight. In some instances, this is backed up by specific checks made in cruise or on approach and, of course, by evidence from flight deck indicators and recordings.

For economic reasons and because of the noise problem, ground testing is kept to a minimum and is usually only carried out after engine installations, during troubleshooting, or to test an aircraft system. With the improved maintenance methods and introduction of system test sets that simulate running conditions during the checking of a static engine, the need for ground testing, particularly at high power, is becoming virtually unnecessary.

Before a ground test is made, certain precautions and procedures must be observed to prevent damage to the engine or aircraft and injury to personnel.

Because of the mass of air that will be drawn into the intake and the resultant high velocity and temperature of the exhaust gases during a ground test, danger zones exist at the front and rear of the aircraft. These zones will extend for a considerable distance, and a typical example is shown in Figure 12–6. The jet efflux must be clear of buildings and other aircraft. Personnel engaged in ground testing must ensure that any easily detachable clothing is securely fastened and should wear acoustic earmuffs.

The aircraft should be headed into wind and positioned so that the air intake and exhaust are over firm concrete, or a prepared area that is free from loose material and loose objects, and clear of equipment. Where noise suppression installations are used, the aircraft should be positioned in accordance with local instructions. When vertical takeoff aircraft are being tested, protective steel plates and deflectors may be used to prevent ground erosion and engine ingestion of exhaust gases and debris. Aircraft wheels should be securely chocked and braked; with vertical takeoff aircraft, anchoring or restraining devices are also used. Adequate fire fighting equipment must be readily available and local fire regulations must be strictly enforced.

Before an engine is started, the air intake and jet pipe must be inspected to ensure that they are free from any debris or obstruction. Each operator will detail his individual pre-start inspection requirements: a typical example of this for a multi-engined aircraft is shown in Figure 12–7.

The starting drill varies between different aircraft types and a starting check procedure is normally used. Generally, all non-essential systems are switched or selected off: warning and emergency systems are checked when applicable. Finally, after ensuring that the low-pressure fuel supply is selected on, the starting cycle is initiated.

At a predetermined point during the starting cycle, the high pressure fuel shutoff valve (cock) is opened to allow fuel to pass to the fuel spray nozzles, this point varying with aircraft and engine type; on some installations the shutoff valve may be opened before the starting cycle is initiated. During the engine light-up period and subsequent acceleration to idling speed, the engine exhaust gas temperature must be carefully monitored to ensure that the maximum temperature limitation is not exceeded. If the temperature limitation appears likely to be exceeded, the shutoff valve must be closed and the starting cycle canceled; the cause and possible effect of the high temperature must then be investigated before the engine is again started.

When a turbo-propeller engine is being started, the propeller must be set to the correct starting pitch as recommended by the engine manufacturer. To provide the minimum resistance to turning and thus prevent an excessive exhaust gas temperature occurring during the starting cycle, some propellers have a special fine pitch setting.

Throttle movements should be kept to a minimum and be smooth and progressive to avoid thermal stresses associated with rapid changes in temperature. Rapid throttle movements to check the acceleration and deceleration capabilities of the engine should be made only after all other major checks have proved satisfactory and after some

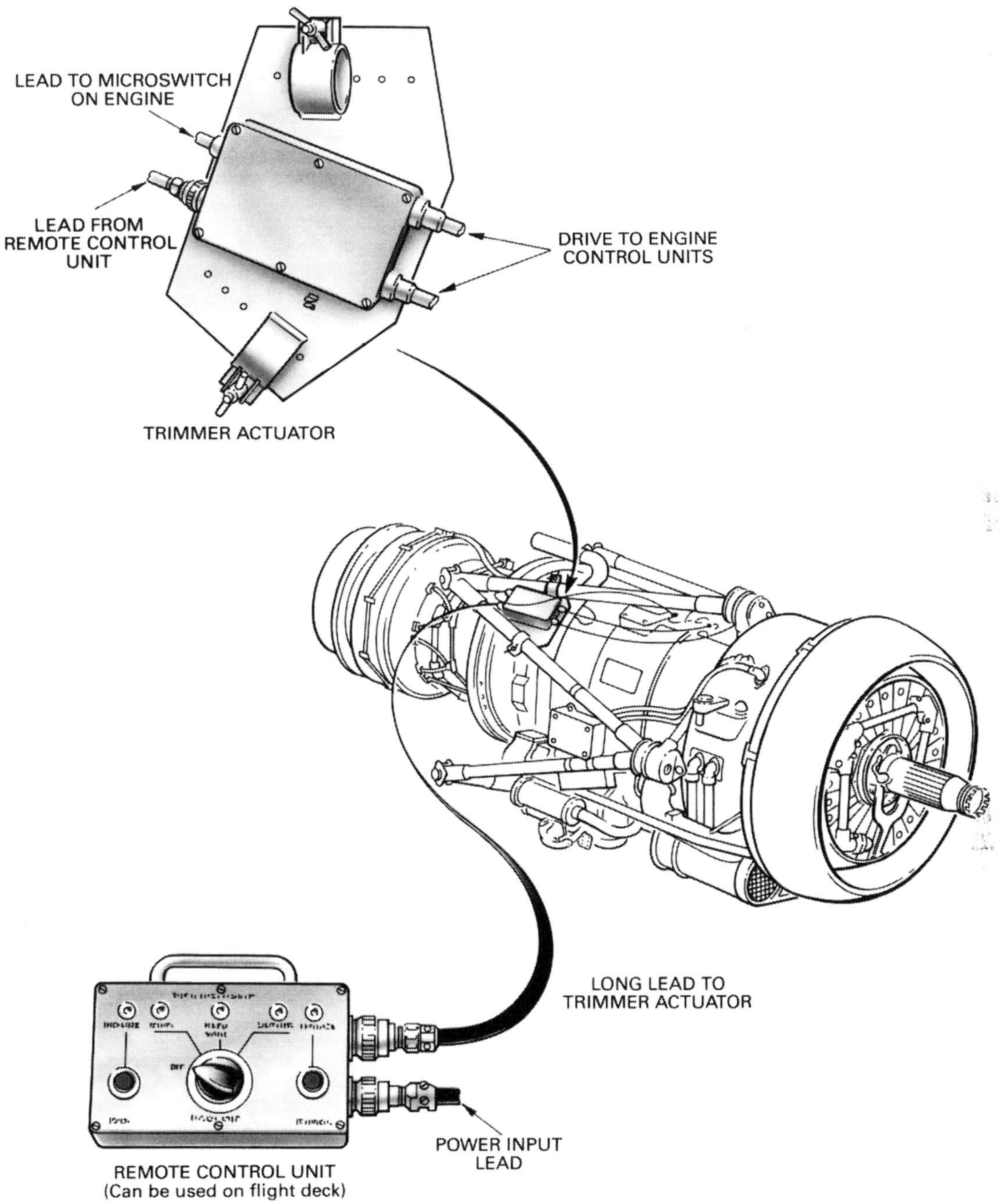

FIGURE 12–5 Remote adjustment fitted to a turbo-propeller engine. *(Source: Rolls Royce.)*

slower accelerations and decelerations have proved successful.

Before an engine is stopped, it should normally be allowed to run for a short period at idling speed to ensure gradual cooling of the turbine assembly. The only action required to stop the engine is the closing of the shutoff valve. The shutoff valve must not be re-opened during engine rundown, as the resulting supply of fuel can spontaneously ignite with consequent severe overheating of the turbine assembly. An example of turbine blades that have been subjected to overheating is shown in Figure 12–8.

The time taken for the engine to come to rest after the shutoff valve is closed is known as the "rundown time" and this can give an indication of any rubbing inside the engine. However, it should be borne in mind that variations in wind velocity and direction may affect the run-down time of an engine.

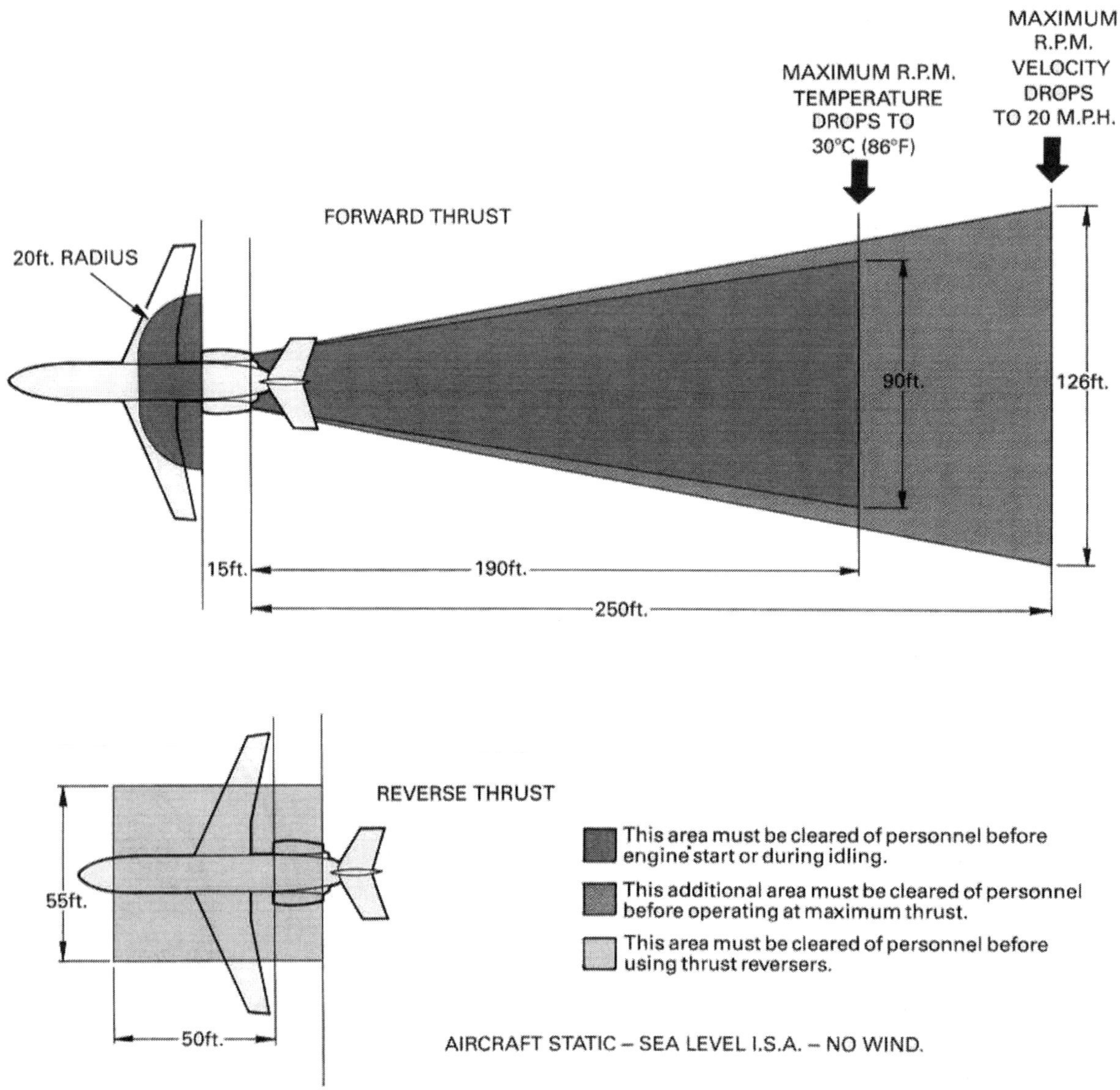

FIGURE 12–6 Ground running danger zones. *(Source: Rolls Royce.)*

MAINTENANCE INFORMATION SYSTEMS

With all gas turbine and gas turbine system applications MIS (maintenance information systems) are generally maintained by larger organizations as a way of tracking their assets and asset usage. Various information components are entered into the system, and information access and input are controlled with individual user identification numbers and passwords. Coded access limits information input to individual workers knowledge scope and specific responsibilities. Information input data may include:

- Details of specific maintenance work items, all spares, materials, and personnel hours used
- Specification of all critical machinery and spares maintained
- Changes in reorder points for spares inventory
- Changes in reorder points for materials
- Consumption of consumables

Information gleaned from MIS systems can then be used to:

- Improve equipment selections for specific applications
- Track performance of specific maintenance practices and personnel
- Provide data and planning information for plant audits and shutdowns
- Review performance of OEM spares
- Review effectiveness of overhaul, and individual repairs by both internal personnel and external vendors
- Differentiate between effectiveness of different vendors who supply generic products or repairs

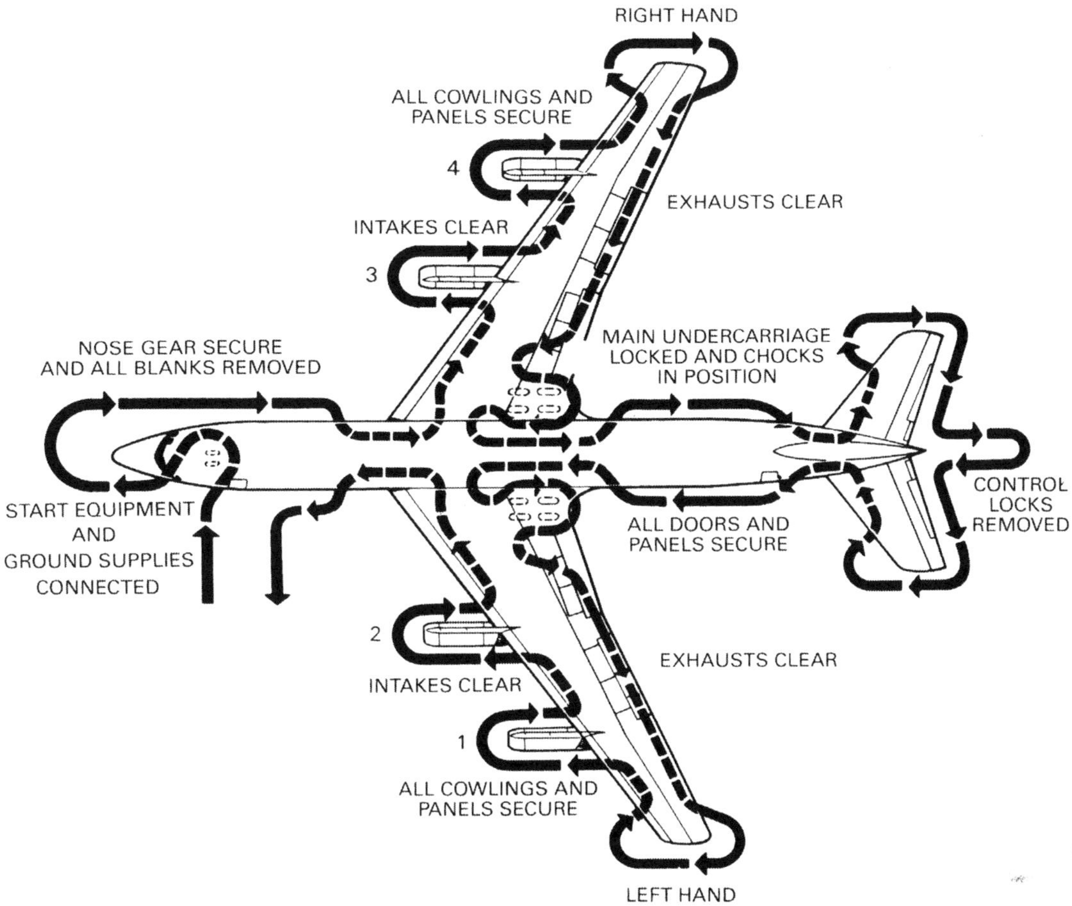

FIGURE 12–7 A pre-start inspection sequence. *(Source: Rolls Royce.)*

- Minimize spares inventory, by using a "JIT" (just-in-time) philosophy
- Track rental equipment costs so vendor billing can be monitored

End users have used MIS systems to justify major pressure on OEMs with respect to turbine component lives and campaign for, for instance, LCA algorithm(s) and retrofitted LCA counters to economize on their utilization of spares.

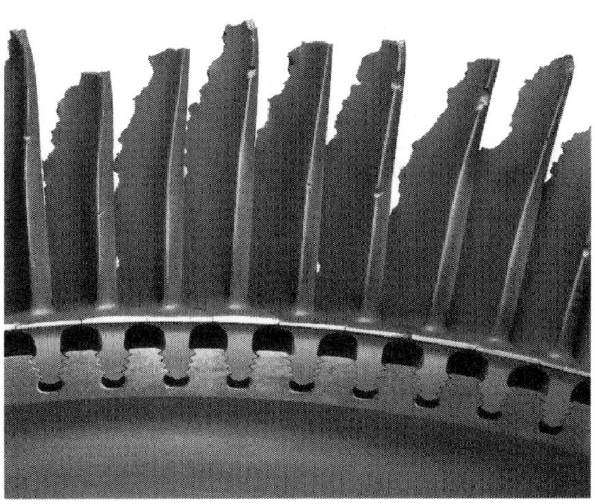

FIGURE 12–8 Overheated turbine blades. *(Source: Rolls Royce.)*

AUDITS OF AND RETROFITS WITH GT COMPONENTS AND SYSTEMS*

In a perfect world, an audit would be conducted *before* maintenance (other than that regularly scheduled to occur), repair, and overhaul (R&O) processes to assess their need. In reality, sometimes a failure occurs first, followed by a hasty R&O to get the plant back on stream, then by an audit full of "hindsight is 20–20 vision" pronouncements and sometimes some actual preventive technology for future use. So *audit* in this book need not mean something that occurs "pre-need definition" or "prefailure or -efficiency drop off." It may be well after the first horse has fled the gate. An audit could also be conducted after the introduction of new technology to assess the potential and method required

* Source: [12-2] Claire Soares, course notes, "Audits of Operating Gas Turbine Systems," 2000.

to utilize that technology in a plant. The reader therefore is advised to alter any previous mental frame that thought of "audit" as implying "before (R&O)" to either "before, during, or after or some combination thereof."

The appearance of any kind of issue or problem that requires a response on the part of an OEM, end user, or vendor of any kind generally results in an audit, under today's conditions. The stakes are too high for this not to happen. An audit generally leads to one or more of the following for a GT, its components, or system:

- Changed scope of repair, overhaul, or regular maintenance
- Newly developed or modified repair or overhaul work scope
- Modified business aspect(s) of conducting the repair(s)
- Knowledge about a mode of failure (that may be halted with a repair or not)
- Transfer of failure or troubleshooting knowledge back into the design optimization process
- Transfer of knowledge from the audit findings into the regular maintenance schedule(s)
- Retrofit(s) to reduce overall costs per fired hour, optimize maintenance, and reduce required repair(s)

In other words, repair and overhaul is now sophisticated enough that the results of new developments in those fields are often transferred back into the OEM's manufacturing processes or methods. OEM, independent repair shop, and end user frequently work or can work in concert, for TBO, MTBF, and operational optimization.

Most land- and marine-based gas turbine plants today, be they facilities for power generation, oil and gas production, or petrochemical or process in nature, generally are commissioned with as much haste as safety considerations will allow. Business considerations and communication problems dictate that a startup where adequate time is taken to check that startup parameters result in optimum performance of the plant is rare. Generally, the push to turn out product or power as soon as possible is the highest priority. Optimization comes later—time permitting—after the plant has been run in, when it is showing signs of wear, or when process conditions have changed enough that partial redesign or retrofit is required. Quite simply, an audit generally occurs because the owner/operator either wants to stay in business, maximize profit, or both. Or it may occur because of changing legislation or the threat of fines for pollution limits exceeded or other operational problems.

When the GE Frame 9F fleet was first commissioned around 1995, it underwent severe vibration problems. An audit that cost $1 billion and the labor of 400 engineers resulted in a "fix" that essentially meant no repair. Instead it meant improved quality control procedures during engine assembly.

Aeroengine plants are rigorously tested in a test cell and subject to shock loads, and deliberate FOD in the form of frozen chickens or worse. Most of the time, this means that when the aeroengine is hung on a wing, it performs as expected. On occasion, performance inconsistencies occur. One example of this was that, when the CFM-56 was originally hung on the Airbus A-321, fairly intense vibration was felt by passengers in certain seat rows. Some synchronous vibration had been excited. A similar vibration did not occur when the aircraft type was fitted with a different engine type, the V2500. When such issues occur, they may or may not prompt a repair (if there is deemed to be no danger, nothing may happen) that may or may not result in some form of design change.

The data collection, processing, assessment, and action plan formulation for this operational optimization process is generally called an *audit*. The amount of information relevant in this audit is a reflection of how well preceding design and project management activities were conducted.

An audit is basically a technical accounting process. The scope of an audit is a function of the events—design, specification, safety, operational, or maintenance in nature—that prompted it being initiated. How much ground it actually covers depends totally on the expertise of the person or persons conducting the audit.

For instance, an audit may be designated as being purely a safety audit, aimed at uncovering immediate physical hazards in a plant. However, an audit engineer who has operating expertise also may uncover items relating to performance of the machinery systems during inspection. If the audit is designated a maintenance audit, an engineer with appropriate experience might reveal areas of performance optimization. Engineers conducting a design audit of a new plant might find areas where changes would minimize required repair and overhaul.

Audits may be scheduled (for safety, design, project management, operational, maintenance, process optimization audits, and so forth) or unscheduled. The latter type occurs when precipitated by unforeseen events, such as catastrophic failures and unexpected loss or halting of production flow. When an unscheduled audit has to occur, it is appropriate that the audit person be an individual with expertise in design, specification, operations, maintenance, and performance optimization, as all these areas may need some work.

Aims of an Audit

Primarily these are:

1. *Optimization of operational costs per operating hour.* An audit of the operation and maintenance of an existing plant is primarily aimed at reducing ownership cost per operational hour. After the action items resulting from an audit are complete, the plant should

operate on a par with the best of equivalent plants. The key to optimizing audits is to consider the rotating machinery as the heart of the entire plant system, with its associated components (piping, filters, instrumentation, and so forth) as entities that contribute to the health of the machinery and the plant as a whole.

2. *Changing legislation, particularly environmental.* Another major impetus for audits today is changing legislation requiring retrofits to existing plants. Changing environmental law is probably one of the major areas of changing legislation today. In certain areas, such as Southeast Asia, which is dependent on financing from the West, the real logic prompting the audit is that Western banks demand certain environmental standards for the plants in question. Getting continued financing for other plants or expansions requires the plant operators to demonstrate compliance with relevant legislation. They must show that they have undergone an independent audit that proves they have met the requirements.
3. *Potential for increased thermal efficiency.* Frequently, a look at changing environmental legislation also highlights the potential for increased thermal and fuel efficiency. Cogeneration (waste heat recovery) system design and installation often results. A design audit of the existing plant before commencing the project is advisable, and an operational plant audit after commissioning the system is also advisable.
4. *Catching potential problems.*

Tasks

The tasks in a typical comprehensive audit of an existing plant are to:

1. Survey and correct deficiencies that curtail availability and reliability.
2. Assess current efficiency and diagnose reasons for any dropoffs in peak expected values.
3. Examine the quality control (QC) and spare parts ordering and distribution procedures for operations and maintenance, with reference to how they might improve items 1 and 2.
4. Assess condition monitoring systems and instrumentation in terms of their adequacy, potential for updating, and effectiveness.
5. Examine training and communications within the organization, as they might be improved to positively affect items 1 through 4.
6. Perform all other investigations that might result in improved performance or availability as well as reduced risk of downtime or lost efficiency.
7. Consider all the preceding with reference to the plant's maintenance history and any known history worldwide of the machinery models in question.
8. When item 6 is complete, amend as required frequencies for maintenance inspection (such as gas turbine hot section inspections) and condition monitoring as a whole (frequency, type, and position of monitoring points).
9. Using the information from items 1 to 7, review for the long term:
   - Performance optimization
   - Failure prevention
   - Failure analysis, for the entire plant

Audit Planning

A thorough operations or maintenance audit of a gas turbine system includes consideration of the following items.

Phase I. Planning the Audit Administration

1. Operations management basic procedures
2. Responsibilities required of various groups
3. Operations organization and responsibility splits
4. Operations maintenance organization
5. Operations engineering organization
6. Department interfaces
7. Philosophy for maximum communication and optimized interdepartmental interface
8. Maintenance auditing process
9. Specifications and basic practices
10. One time repairs and contracts
11. Cost estimates of regular/unscheduled maintenance
12. Cost control
13. Schedule assessment of large maintenance tasks
14. Ongoing communications for offsite maintenance
15. Formal feedback
16. Intelligence feedback
17. Performance assessment

Phase II. Activities Planned during the Audit

A. Operations Activities

1. Interruptions in production, equipment failure, and contingency plans
2. Assessment of specialists and other requirements
3. Reference books and manuals
4. Internal and contractor status work reports
5. Inquiries, bids, purchase orders, requisitions, and record keeping
6. Spare parts management
7. Internal and vendor print control
8. Expediting
9. Data sheet examples
10. Maintenance information systems

B. Maintenance Activities

1. Feedback and communication with other departments
2. Local codes, regulations, and warranty considerations
3. Turnaround schedules, tie-ins, and segmented completion
4. Heavy lifts
5. Safety and emergency procedures
6. Hot and cold work permits
7. Administering subcontracts
8. Cost control and productivity
9. Tooling
10. Cost savings potential
11. Time allocation, documentation, and follow-up for changes
12. Material delivery checks
13. Facilitator and pro-active role of the audit engineer

These are basic outlines and open to modification, adaptation, or specific omissions.

General Audit Procedures

How is this Information Gathered?

An audit primarily revolves around the following key tasks:

1. *"In the field," on-site survey of gas turbine systems.* For this item, the audit person normally works against customized checklists using some or all of the material in this section.
2. *Survey of operational procedures.* This is conducted by completing thorough interviews with
   - **a.** operational (plant operator, naval seaman, pilot),
   - **b.** maintenance,
   - **c.** managerial staff members.

Frequently problems arise due to miscommunication among departments, along the lines of "I know what the manual says, but the only way we can get this to work is by . . . we never got around to telling engineering" or "we found we could save time if we . . . but if we had told management, they would have . . ."

How Does this Information Provide Improvements to Maintenance and Operations?

3. Procedure items 1 and 2 are reviewed in the context of
   - Would different procedures have helped?
   - Would a different organizational structure help?
   - Do personnel have adequate training?
   - Are routine checks of trips and system safety shutdowns working?
   - Are these checks done regularly and as they should be?
   - Is the spare parts support system working as it should?
   - Are routine maintenance supplies available when they should be?
   - Are all reports complete and comprehensive?
   - How does the (engine) condition monitoring system (ECMS) work, that system being real time, online; not real time, online; or intermittent (readings taken on an as-needed basis)? And, if it is one of these system types, does it need to be upgraded or downgraded?

This review is generally first conducted by the audit engineer.

4. Item 3 is repeated with all involved personnel present to provide information, feedback, suggestions, and opinions that initial interviews may not have uncovered.

How Might Standard Operating Procedures and Standard Maintenance Procedures be Altered or Affected?

5a. During procedure 4, the most advisable members to make up a team that will recommend changes to standard operating procedures (SOP) and standard maintenance procedures (SMP), usually becomes self-evident. The team, however, should represent operations (plant operator personnel), maintenance, management, and long-range planning.

5b. For a complete audit, including the internal condition of the existing machinery, the machinery must be shut down and inspected. This is advisable in cases such as:

- When taking over mature units or plants from a previous owner where proper and complete records were not maintained.
- When unsure whether the condition monitoring system on the equipment is giving as complete information as you need.
- When looking for information for specific projects, such as extending the OEM's recommended time between overhauls or stated component lives.

Meetings have to be conducted beforehand to agree on the exact timing of these internal inspections and to arrange for the appropriate personnel, equipment, and tools to be available. Just before taking the machine out of service, a complete operational test should be made at different load settings—zero, half load, and full load are advisable.

5c. These tests should be conducted regardless of whether internal machinery inspections are included as part of the audit or not. The results of these tests should be compared to the manufacturers' performance charts and corrected for local atmospheric conditions. Note that these tests should also be conducted before and after a major unit is overhauled. An overspeed check should be done as well, and the reaction of the governor and tripping instrumentation noted.

Together with information in the plant's maintenance information system, the results of these runs should be considered in the light of the following.

For Gas Turbines Systems, Have Any of the Following Occurred?

- Increase or change in vibration
- Compressor discharge pressure decreased
- Change in lube oil temperatures or pressures
- Air or combustion gases blowing out the shaft seal
- Incorrectly reading thermocouples
- Change in wheel space temperatures
- Fuel oil or gas leakage
- Fuel control valve operation
- Hydraulic control pressure changes
- Does the turbine governor "hunt"?
- Are the gear boxes emitting any sound?
- Overspeed devices, do they operate properly?
- Babbitt or other material found on lubricating oil screens
- Lube oil analysis shows corrosion increase
- Pressure drop across the heat exchangers changes

5d. Before removing turbine flange bolts or disturbing the normal turbine setting, clearance readings between the wheel shroud and the last row of turbine blades should be made at both horizontal and vertical settings. The main turbine flange should be checked for spread or warpage with feeler gauges between each of the flange bolts. Elevation checks should be compared with original readings at each of the turbine supports.

5e. Some maintain that thorough and regular borescope inspections provide enough information that machinery disassembly for an audit is not required. Ultimately, it is the machinery owner/operator's choice. If facilities for accurate borescope information are provided and borescope checks are done by an individual qualified to interpret his observations, such checks can provide:

- Internal on-site visual checks without disassembly.
- Extending periods between scheduled internal inspections
- Accurate planning and scheduling of maintenance
- Monitoring condition of internal components
- Improved ability to predict required resources (spares, tools, personnel, expertise) for optimized operation and maintenance

5f. Some gas turbine system designs favor disassembly ease of hot section inspections better than others; that is, where duct flanges allow the hot section and downstream turbine sections to be unbolted and backed away from the rest of the turbine without restriction.

5g. First stage IGVs (inlet guide vanes) should be inspected closely for signs of operational problems. Liquid droplets in gaseous fuel is a particularly common problem, often called *hot corrosion* by some.

5h. Disc assemblies:

- Blade tip clearances (first and last stage of the turbine) should be checked at at least four points along the circumference, to check for blade tip rub (that may relate to excessive turbine inlet temperatures).
- Fir trees should be checked for premature cracks. Note that this is a particularly hard area to detect cracks. Discs and fir tree slots should also be checked for cracks. This also is a difficult area to detect cracks.

(Note that retrofit designs have used steam injection to add enough cooling that, for instance, discs on a GE Frame 5 were taken out of crack range. A "side" benefit of the retrofit was a substantial gain in horsepower and reduced NO_x emissions.)

5i. Journal bearings may suffer from abnormal wear in the event of excessive starts. For required repairs or new repair development, such as blade tip robotic welding, specification of "new" coatings, additional heat treatment, HIPing, and so forth, or spares requirements and changes in operation limits, see the OEM manuals and reference your own and other user experience. When these two sources do not coincide, the OEM must be negotiated with until a satisfactory compromise is reached.

CHANGING LEGISLATIVE REQUIREMENTS

Let us consider emissions standards as a case in point. In many cases, proof of an independent audit often is what investors require before approving continued financing of new plants, plant expansions, and major expenditures.

Tightening SO_x regulations have been responsible for "repowering" old steam turbine stations particularly (but not exclusively) in the Western hemisphere with gas turbine and combined-cycle plants. (Coal and lignite still may be used to supplement boiler heat provided by gas turbine exhaust or superheat the steam.) Inevitably, the progress of coal as a fuel affects the gas turbine business.

For instance, China's emission standards (effective at the beginning of 1997) require sulfur oxides (SO_x) emissions limits of 2100 or 1200 mg/cubic meter depending on whether the fuel has less than 1% sulfur or more. NO_x emissions are stated only for units larger than 300 MW units: if liquidized slag removal is conducted in a pulverized coal boiler, 1000 mg/cubic meter is the limit; for solid slag removal, the corresponding figure is 650. For flue dust (ash or particulates), the mg/cubic meter values are 200 for urban areas, 500 for the suburbs, and 600 for old units with a residual life of more than 10 years. These are not ideal standards, but they are a help if followed. Thermal plants in China turn over 7 million tons of sulfur dioxide into the atmosphere annually, so the pressure was on investors to be seen to be supporting sound environmental policy. Most power companies in coal-rich Asia now also operate combined-cycle and gas turbine (peaking or otherwise) plants. For those that use, in part, coal or lignite fuel for their power systems, environmental legislation changes are an issue.

Korea, Southeast Asia's most developed country after Japan, is also changing its emission standards in manageable increments. In 1995, local anthracite fueled plants had to meet 2000 mg/cubic meter for sulfur emissions, and imported coal-fueled plants had to meet 1430 mg/cubic meter. (Current NO_x standard is 350 mg/cubic meter.) By 1999, however, all plants must meet 770 mg/cubic meter standards.

In Taiwan in 1992, SO_2 emissions allowed were 2145 mg/cubic meter. In 1999, that was lowered to 1430 mg/cubic meter; NO_x standards stayed at 720—1025 mg/cubic meter.

Lowering emissions standards with, for instance, flue gas desulfurization (FGD) techniques is an expensive and long process, with the least expensive FGD installation currently costing on the order of $85/megawatt (MW). Components external to the turbomachinery, such as installation of new furnace burners (for lowered emissions or greater fuel diversity), FGD scrubbers, precipitators, and additional stacks all potentially cause major pressure drops in the overall plant system, which should be checked in an operational audit after commissioning.

Components internal to turbomachinery sometimes are less easy to retrofit and generally come as part of a newer model, such as low nitrous oxide (low NO_x) burners in a gas turbine. OEMs frequently supply retrofit kits if internal modification has been found to benefit the bulk of the fleet. If a modification is no longer prototype in status, an audit is probably not necessary. However, if it is accompanied by a change in process conditions, such as fuel heating value or moisture content, the operational balance of the system (including time between inspections or overhauls) may be changed, until investigated and corrected in an audit.

RETROFITS AIMED AT OPERATIONAL OPTIMIZATION

1. Cogeneration retrofits are frequently bulky (using waste heat off a gas turbine for greenhouse heat, adding waste heat recovery heat exchanges, and so forth). As with FGD units, major changes in system flows, back pressures, and heat balances result. An audit is advisable after the new system has been commissioned to check on the effects of the modifications.
2. One of the most common systems to be retrofit with operational optimization as a high priority, is performance analysis. The data gained from a PA system can also be used as inputs into a life cycle counter for life cycle analysis. Why do we need PA and LCA? PA help optimize fuel consumption and pick appropriate intervals to wash or clean compressors and gas turbines. LCA can be used to extend TBOs' and components' lives.
3. Addition of retrofits, such as water or steam injection to reduce NO_x levels that was not part of the original design package, should also be audited to check for proper operation, after commissioning.
4. I&C (instrumentation and control) retrofit packages for increased reliability or optimized performance are perhaps the most common retrofit on existing plants.

Some I&C retrofit cases follow.

Case Study 1: Brent Platform Retrofits for Extended Life*

A monumental production expansion was undertaken at Britain's largest oil and gas field, Brent. What had been primarily oil production was being altered to tap massive gas reserves and reduced oil supplies instead. This change required many changes in platform equipment and layout. It also gave rise to a requirement for optimization of the platforms' control systems. The original controls on the platforms, unreliable or inaccurate by contemporary standards, would have left margin for undesirable cost increases per unit of production flow.

Summarized production changes at Brent are as follows. The 1992 and 1993 expansion initiation was aimed at extending the field life to at least 2010 (from 1998). Basically, pressure exerted on the field was dropped so hydrocarbons vaporize into gas. Oil production, which requires higher injection pressures, was reduced. Increased recovery as a result is approximately 34 million barrels of oil and 1.5 trillion cubic feet of gas.

The changes occurred in two stages. In 1997, two of Brent's four platforms were modified, water injection stopped, and gas injection started, to begin the depressurization process. After 2000, stage 2, involving deep depressurization, begins, as water previously used for pressurization is drawn off with submersible pumps. Increased gas production and water pumping facilities are required.

Brent's daily production consequently is expected to be about 600 million cubic feet of gas and 175,000 barrels of oil, well into this millennium. Changes to the platforms are expensive. Even with the steep learning curves cutting costs, Bravo platform cost £439 million to adapt, Charlie

* [12-3] Data from HSDE (Vosper Thornycroft) UK, 1998.

£363 million, with £314 million anticipated on Delta. The Achilles heel in all this ambitious planning was the control systems. The controls are also a vital part of an anticipated 15% savings in engineering and maintenance costs until possibly well into this century.

Each of Brent's platforms once had four modules: a drill deck, production system, power system, and accommodation block. After the expansion changes, only the power system was left. Prior to expansion, the field owner, Shell, had ordered studies on reliability and new power requirements. It knew then that it had 15 years left in the field and wanted optimized, efficient running. Brent already had four Avon power plants. There were also four MW generators, two each per platform. There were Rustons on two platforms and Solar Centaurs on one platform.

In the original design, there were four Avons, since one was a backup. Part of the optimization and streamlining that optimized controls made possible was that one Avon was removed. However, now the three others had to be more reliable. Also, appropriate power management was required. One study Shell commissioned looked at the reliability of the original controls. The old control system had many relay problems, expensive in terms of their impact on production. The retrofit vendor, HSDE, supplied:

- Three governor and sequencer packages for the Avons
- One color graphics man-machine interface (MMI) for terminal-to-machine readings for the three Avons
- Two governor and sequencer packages per platform (for either the Rustons or the Solars)
- One remote color graphics MMI in the main control room
- One Brush power management system, including a fast load shedding option

Each of the five gas turbine governor and sequencer packages manages the GT system in question. It controls

- The GT itself
- Ventilation
- Lubrication
- Generation

on each of the 17.5 MW Avon or the Ruston or Solar 4 MW units.

The control package continuously calculates power "remaining" available. For instance, if 10 MW were generated, the system calculates how much net power is available for platform duty. This information is sent to the power management system.

Another task for this control feature is that, as per Rolls-Royce service bulletins, these Avons had to continuously run low-cycle fatigue (LCF) and thermal fatigue algorithms. Rolls Royce's design development on the machine components meant that considerably extended component lives were possible with certain new modifications, provided extensive calculations on operational speeds and other data were performed. Anxious to preserve the life of expensive components, some operators are "hand performing" these calculations. On the Brent platforms, the LCA system does this automatically, major cost savings in time and skilled personnel.

All GT controllers are connected to Brent's main SCADA system. Information from the potential 600 data point systems is radioed to shore, where Shell's engineers in Aberdeen can assess their offshore operation from onshore. They also can time maintenance and parts removal with better accuracy.

The Brent platform and similar retrofits frequently use the stepper motor valve and MMIs.

Stepper Motor Valve

The stepper valve is a fast response electrically operated valve (pioneered by HSDE). It is generally used in fuel systems. It provides the fast response required by aeroderivative and some industrial gas turbines. Before the stepper valve made its appearance, some users tried to use hydraulic and pneumatic actuation to provide the required response time, but this increased the overall complexity of the fuel system. As always with instances where system complexity is escalated, system cost rose while mean time between failures and availability decreased.

The design fulfills the original design aims, which included the following safety considerations:

- The liquid fuel version of the valve incorporates a pressure relief valve, protecting the system against overpressure and the fuel pump running on empty or "dead heading," caused by closure of valves downstream of the fuel valve during system operation.
- A fail freeze or fail closed option, depending on whether the operator is a pipeline (in which case, turbine shutdown on valve failure is required) or a power generation facility ("freezing" at the last power setting is then required).
- High-speed response of less than 60 mS required by aeroderivative gas turbines to prevent overspeed in block offload conditions.
- 24 volts DC is the maximum drive voltage that ensures personnel safety.
- Explosion-proof actuation to BASEEFA/CENELEC 50014 and 50018 category EEx d IIB T4 90°C ambient and CSA Class 1, Group D, Divisions 1 and 2, which allows operation in hazardous methane service.
- Resistance to fuel contaminants including tar, shale, water, and sand.
- Corrosion resistance in components exposed to wet fuel, and corrosion resistance to all parts if the service is sour gas.

Other operational objectives that shaped the design were an operator's requirements for:

- Higher mean time between failures (MTBF). A target of 50,000 hours was set and achieved.
- Low mean time to repair (LMTR). The target of 1 hour, achieved with modular design, together with the target MTBF provided an availability of 99.998%.
- Low maintenance costs, since the modular design can be repaired by an individual with relatively low expertise. Service intervals are 12 months.
- Low power consumption, since an electric motor of less than 100 W is used. This eliminates the need for additional hydraulic or pneumatic systems. Also black starting is more reliable if the fuel system is powered by the same batteries as the controller.
- Large control ratio, which allows control over the ignition to full load as well as full speed ranges to be possible with one fuel valve. Fuel pressure variation compensation is provided. The additional speed ratio type control valve found in many other industrial gas fueled installations is not required here.

MMIs

A man-machine interface (MMI) is the display unit interface with the I,C&D OEMs PLC (for example, HSDE's Digitrend unit). About 500 to 600 readings are fed into the PLC, which provides:

- Functional mimic diagrams for monitoring and control of the turbine/generator set for power generation or the turbine/driven equipment system for mechanical drive applications
- Alarm annunciation
- Trending, data storage, and retrieval

As with the packages of many OEMs that build C,I&D packages, this system was built to rival the OEM's equivalent units, such as the GE Speedtronic Mark V system. It thus can displace GE packages for LM 1600, LM 2500, GE Frame 3s, and GE Frame 5s. End user requirements include:

- Better reliability than with analog systems that may experience frequent outages
- Better availability with digital feedback start control and consistently repeatable starting under all ambient conditions
- Improved performance minus the drift seen with alteration of ambient conditions in some older systems
- Optimized maintenance, since the digital controller does not require regular maintenance or calibration Cooler starting, steadier operation, and more precise temperature control (than possible with analog systems) reduces required engine maintenance
- A modular design concept that means that the controller's processing requirements are dealt with on a platform dedicated to the performance and response requirements of the particular engine being controlled
- Integration with digital SCADA system data storage is possible with standard serial communication links and protocols
- Improved life cycle costs result, as is generally the case with digital versus analog controlled engine system
- Reduced operational risk is achieved particularly during commissioning, as digital controls provide a greater degree of influence over process fluctuations and surges that are common during startup

C,I&D vendors seek compatibility for their products with as many GT OEMs "original" systems as possible.

Case Study 2: Flotta Terminal*

The Flotta terminal is just off the Scottish coast on the Orkney Isles. It acts as a collection and shipping center for oil produced at the Claymore, Scapa, Highlander, Ivanhoe/ Rob Roy, and Tartan rigs. Elf Enterprise Caledonia, the terminal operator, decided on a retrofit that would:

- Reduce its required personnel level on the platform
- Revamp existing instrumentation to improve operational maintainability, availability, and running costs

The terminal's power is generated by six Ruston TD4000 gas turbines that drive Brush alternators. The number of units run at any time varies. Three or four normally are run, but all six may be required if the terminal is exporting. The C,I&D vendor's controls solution included:

- Replacing existing turbine controls with digital (HSDE Digicon) controllers. Communication multiplexers assist this within existing unit control panels, therefore requiring a minimum of disturbance to the station's equipment.
- Fitting a new power management system that incorporates individual autosynchronizers instead of the previous shared unit. The new control system uses fiber optic links to incorporate gas turbine and electrical generation integration control among the six gas turbines.
- Existing pneumatic fuel valves (the turbines run on gas or liquid fuel) were replaced with electrically actuated valves. The improved response time reduces wear on turbine components, which could result from liquid fuel "slugging" in what is supposed to be gaseous fuel.
- Gas fuel pressure regulators with fixed settings were fitted.

* [12-3].

- New visual display unit (VDU) workstations and interfaces in the Oil Movements Control Room (OMCR) and Power House Control Room (PHCR).

The VDUs made it possible to reduce PHCR personnel (only three to zero personnel are required now) and control it from the OMCR. They also provide far better graphics than the previous dot matrix printout and provide 20% more readings. The VDUs online information is digital, so accuracy is improved. The computers that run the VDUs can recall about 3 months worth of data, as well as recall archival data for condition monitoring and troubleshooting.

The VDUs are easy for operators to read and very user friendly. Other useful data items the VDU terminals present include:

- Power factor values, which make power control easier.
- An EGT spread diagram, which makes gas turbine "hot spots" or combustor problems graphically evident. (The system allows for one thermocouple out of two in each quadrant to fail and still provide a reading. The failed probes are removed from the average EGT calculation and nuisance trips thus are avoided.)
- Digital readings for fuel lower heating value (LHV) and volume flow rates (not part of the fiscal accounting system) are available for machinery performance monitoring.

The new control system allows for generators 4, 5, and 6 to be connected, via the export board, to the Scottish Hydro Electric PLC distribution network (for sale of any excess power the station produces) or to the station's terminal board. When power is exported, the control system provides equal power sharing between generators at a given base load. (Generators 1, 2, and 3 may be connected to the terminal board only.) Thus power is efficiently and economically distributed at all times.

Case Study 3: Forties Platform Retrofits*

Original oil production estimates at Forties were once 1.8 billion barrels recoverable over 17 years. New technology and exploration methods increased that estimate, in 1996, to 2.48 billion barrels recoverable (from 4.2 billion in situ) over a life of 30 years or more. At the end of 1996, 180 million recoverable barrels were thought to be left. That is, until Forties' partners revealed that they might be able to extract a further 400 million barrels and run till 2011. Key to achievement of these goals was an optimized control system.

The original controls were no longer state of the art by any means. The 20-year-old system was an analog one with potentiometers. Problems associated with a system of this vintage were typical. They included:

- Sporadic reliability. Unscheduled shutdowns numbered 32 a month on average.
- Start availability could not be relied on.
- System response time left a great deal to be desired.
- An antiquated surge control system took its cue from a straight line (instead of a curve) that operated 15–20% from the surge line, causing major efficiency penalties.
- During annual startup procedures, the gas turbines would keep tripping when they were switched to gas fuel after being started on diesel.
- This would also result in 2–3 days' lost production time when a fuel change over was required.
- "Overfuel" also frequently resulted when fuel was changed over. This was costly in terms of turbine hot section component lives.
- Many spares required for the original system are now obsolete.
- The diesel fuel bill was unnecessarily large due to inadequate controls' response times.

For updating the Forties NGL compressor control systems, an I&C vendor supplied the following control equipment:

- A control system (HSDE Digicon Series 2 Ace), which governs and sequences the gas turbine, NGL compressor, and associated systems. The controller also includes fault and failure monitoring for the entire turbomachinery train, including turbine, compressor, and pump units.
- NGL compressor antisurge system.
- NGL compressor load share system.
- An MMI system.
- Local operator panel.
- Liquid fuel valve and actuator.
- Gas fuel valve and actuator.

The fuel control system now continues to support gas, diesel, or combined fuel usage, only now it is far more reliable. Consumption of costly diesel fuel is minimized.

* [12-3].

Case Study 4: Al-Ain Flameout Problems*

Al-Ain, a power station in Abu Dhabi, has Westinghouse W251 generators that run Brush alternators. A C,I&D vendor was awarded the contract to replace the existing control panels for the generators. The original governor control system had been pneumatic. This system had experienced constant problems with sticking pneumatic valves during startup. This, in turn, caused flameout problems with the gas turbines. The system was both temperamental and unreliable. Only one operator understood how

* [12-3].

to fix the problem, so the station could have found itself at a standstill in this man's absence. This was an intolerable situation from the operator's standpoint.

The station ordered replacement control panels from an I&C vendor, the main features of which are:

- A control system (Digicon Series II Ace Control System, which incorporates the Digicon Overspeed Trip Unit, Input/Output Expander units, which extend potential input) and a fuel valve actuator drive unit. This unit is motor driven, which removes the unreliability problem experienced with pneumatics.
- An MMI system with a capacity for 200 inlet data points.
- CEC vibration monitors.
- 16-way annunciator with indicator lights.
- Modified pressure gauge panel adapted for reduced pneumatic control.
- Rehousing of the existing pneumatic system.
- Dual power 120/240 V supply unit.

Modifications made to the gas turbine instrumentation include replacement of pneumatically controlled gas and liquid valves with actuator-driven throttle valves. The gas turbines are controlled and monitored by the I&C vendor's (Digicon Ace) units. The controls permit independent operation of both turbines, so one can run without the other being affected in any way—a vital feature to most operators of similar stations.

Each turbine control panel also has an ABB Synchrotact 4 dual channel unit. This unit safeguards the generator synchronization, includes the vibration monitors, and has an overspeed trip detector (Digicon OTU).

Via a dual communications link, the Digicon Ace controls the interface with the MMIs in the turbine control panels. They in turn provide dual communications to the two MMIs in the control room. The MMIs provide the operators with mimics that reveal present problems or operational trends.

PERFORMANCE ANALYSIS**

What is performance analysis/monitoring? Basically it is a check of the fluid path of an item of turbomachinery (gas turbine, compressor, and so forth) to determine that it is doing what it was designed and bought to do. To perform this check, the pressures, temperatures, and flow along the machinery's gas path may be used to compare its performance curves of:

- Pressure versus flow,
- Power developed or required versus flow or other performance parameters,
- Efficiency versus power developed or required,

and so forth. The readings may then be computed for the operator to either deduce what corrective action is required or read off the required action on an automated system display.

Performance analysis systems occur in three main "formats":

1. Systems that have a calculation module based on actual instrumentation readings. Such systems may cost about $15,000 and up for the first unit, with discounts for additional units of the same model.
2. Systems that have a calculation module based on actual instrumentation readings, as well as a calculation module that uses predictive flow formulae theory and compares the two for better quality information and indication of which instrument may be malfunctioning. Such systems may cost from about $60,000, for the first unit with discounts for additional units of the same model.
3. Systems that perform the functions of 2 (above) and are part of an overall online (frequently real time as well) health monitoring system that incorporates vibration monitoring and other monitoring. Such a system uses artificial intelligence techniques to attempt automated problem solving. Such a system is generally extremely expensive. Typically, these systems cost from about $100,000 per machine train (no discounts for similar trains sometimes) to $250,000 per train.

They may frequently also be "overkill" for an application. Typically they take about 600 readings/signals as inputs to arrive at their conclusions. Frequently, their specifications also demand a requirement that "eliminates" the competition. Their manufacturers may claim its system's use absolves the operator of using his or her own system knowledge or reasoning by providing "ready" answers. Sometimes this works for certain problems, if those were programmed into the artificial intelligence logic. The customer may get the best value for money from comprehensive vibration analysis and a system that incorporates the capabilities of the system in category 2.

In this section, we discuss a system in generic category 2. This generally provides the best economic return in terms of ($ saved/$ capital cost) for a PA system and, therefore, is one of the most important items in this chapter.

We consider:

- Aims (goals) of a PA system
- Summarization of the cost, operational benefits, and return on investment of a PA system
- How such a system works, to compare it against other options and ask manufacturers the right questions
- Some advantages and additional applications of PA systems
- Use of a PA system to extend TBOs and reduce repair costs

** [12-2].

The aims of a PA system are:

1. Check detailed performance for different modules
2. Predict performance through the load range
3. Check machine in normal operation, not just when it has just been cleaned or newly overhauled
4. Check performance of CC or other complex cycles
5. Self-check and help troubleshoot
6. System must be interactive and simple to run
7. Check costs for different configurations
8. Check costs differential for different operational conditions

When successful, a PA system enhances:

1. Predictive maintenance
2. Reduced performance test cost
3. Improved operations cost prediction
4. Improved system efficiency

While setting up reference data the following parameters were varied:

1. Barometric pressure
2. Ambient temperature
3. Axial compressor flow degradation
4. Axial compressor efficiency degradation
5. Axial turbine efficiency degradation
6. Anti-icing system operation

The diagnostic module input data included:

1. Barometric pressure
2. Ambient dry bulb temperature
3. Ambient wet bulb temperature
4. Fuel consumption
5. Compressor delivery pressure
6. Compressor delivery temperature
7. Gross power
8. Exhaust gas temperature
9. Inlet guide vane settings

Any measurement errors are usually with 4, 5, 7, and 8.

Gas path analysis steps (from preceding instrument data):

1. Calculate compressor inlet flow
2. Calculate TIT from turbine swallowing capacity
3. Calculate turbine efficiency degradation
4. Calculate compressor efficiency
5. Calculate "clean" compressor flow and efficiency, then calculate flow and efficiency degradation

Experience indicates that the following ranges usually apply:

- For compressor flow degradation: 0–8%
- For compressor efficiency degradation 0–6%
- For turbine efficiency degradation 0–3%
- An appropriate program will calculate items 1 through 3.

In the predictive module, predictive values are calculated (fuel flows are interpolated). Fuel consumption is calculated for all combinations of:

1. Three barometric pressures
2. Five ambient temperatures
3. Eight degradation combinations
4. Five to 11 power levels, which gives 600–1320 data points.

A second set is required if you have an anti-icing system.

Predictive programs are useful for CC, cogeneration, and complex cycles. This set of "grid" values can be included in automatic, online, and real-time systems. A practical use of modeling is in the sale of product (gas, power) versus fuel consumption.

An end user is advised to look for the following features in a PA system:

- Analysis is aerodynamic, not just a heat balance with the machine represented by a black box.
- Predicts stage efficiency, pressure ratio, exit angle, gas properties, and swallowing capacity.
- Subroutine handles the following losses—profile, incidence, trailing edge thickness, tip clearance, shock, end wall, and secondary.
- Cooling air, flame flare, and similar flow interruptions are treated as blockages to help accurate modeling.

The advantages and uses of such a system include:

- Application in sales based on fuel consumption measured more accurately than with normal fuel flow meters.
- Ability to set up the system to determine the optimum combination in a machinery train for a certain load level.
- Cooling flow modifications can be analyzed to prevent problems such as disc cracking.
- Similarly, blade cooling modifications can be made.
- Analyzing steam and water injection effects (reduced maintenance, power increases, NO_x reduction).
- Analysis of gasifiers (GT operated to drive the compressor only) to supply pressurized gas. Being able to analyze each stage helped this development of modified running of a gas turbine (consider applications that require large quantities of pressurized gas).
- Predictive emissions monitoring.

Note that it should be simple to correct for data inputs that can cause errors due to factors such as:

1. Compressor delivery temperature being affected by heat transfer in the combustor; solution, calibrate each engine.
2. Inlet and exhaust pressure loss errors; solution, measure periodically.

3. Fuel composition; solution, get accurate samples.
4. Modifications (e.g., IGVs) are added; solution, model into solution.
5. Atmospheric condition errors; solution, do not forget barometer and rat. humidity readings daily (if not incorporated in real time into program).

Case Study 5: Extending TBOs of Gas Turbines by Preventing Premature Turbine Disc Failure in a GE Frame 5 (Old Model)*

Consider that turbine disc lives in some industrial gas turbines are limited to less than 100,000 hours. This is due to high temperature creep cracks formed when the engine is operated at full load, which, in turn, causes costly repairs and disastrous failures. This can be avoided and disc lives extended to over 100,000 hours by using performance monitoring software to analyze changes to the disc cooling. Note that, before this was done, changing the disc material was tried, but this did not work. The cracks persisted.

The cracks were at the bottom of the fir tree and difficult to see. The investigators noted that the cracks occurred in the blade root slots of the disc. Typically cracks would start along grain boundaries.

In gas path terms, at each disc root compressor air and hot gas path air accumulates. In other words, hot air accumulates where it should not. The solution in this case was to feed cool air through diaphragms. No rotating components were affected, just the diaphragms. Rows 3 and 4 had compressor delivery (discharge) air, row 5 had intermediate stage compressor air from the bleed valve (see Table 12—2). The net effect on performance was negligible:

- 70 HP (52 kW) increase
- 0.05 decrease in thermal efficiency
- 12°F (7°C) increase in TIT
- 1°C (2°F) increase exhaust temperature
- 0.75 psi (5.2 kPa) decrease in combustor shell

TABLE 12—2 Comparison of Measured and Predicted Values of Engine Parameters [12-4]

	Predicted	Actual
Increased combustion temperature	6.7—7.2	6.7
Increased exhaust temperature	2.2—3.9	1.1
Decreased compressor delivery pressure	0.9 psi	0.7 psi
	6.2 kPa	4.8 kPa
Increased Fuel Flow		
(lb/s)	0.011—0.019	0.016
(kg/s)	0.005—0.009	0.007

Case Study 6: Power Addition for GT in Cogeneration Service Using Steam Injection*

Operation of this system works best when:

- Steam is injected only when a certain power is reached
- Best efficiency is to inject all excess steam and let control system vary IGVs and fuel flow
- Keep steam lines hot with a small amount of condensate even when steam is not running

Summary: 30% more power is possible when injecting steam equivalent to 7.5% of compressor inlet flow. Note: NO_x levels were down from 83 ppm to 12 ppm.

*Integration of Detection, Assessment, and Planning in Audits***

If qualified people are conducting the audit, the detection (of faults and potential improvements), assessment (of how much of a fault or potential benefit is possible), and planning (how to implement the steps necessary) are a set of steps that blend seamlessly with each other. To be alerted to potential problems, most operators (land, sea, and air) depend on vibration analysis to be a first line of defense.

Vibration Used to Assess Gas Turbine Combustor Problems

With the development of more accurate vibration probes, it has been possible to pick up the signal noise that occurs with deterioration, such as cracks, of gas turbine combustion cans and liners. Ideally, the health of these gas path items is monitored continually, and there will be indications via vibration monitors of impending work in this area.

The inspections, work, and additional sparing can be done as part of the preparation for the actual audit.

This illustrates, as on previous occasions, that the shape nonroutine audit work takes follow, at least in part, the results of previous troubleshooting efforts. If vibration analysis does not pick up combustion section problems (they have been known to pick them up before exhaust gas temperature, EGT, monitoring with thermocouples), then

* Source: [12-4] Courtesy of Liburdi Engineering, Canada, various technical papers (consult source for full list).

** [12-2].

EGT monitoring or regular borescope inspections should detect this condition before disastrous failure occurs.

Optimizing Vibration Analysis to Extend Its Problem Detection Capability

For the most part, mature plants have displacement, velocity probes, or both. These cannot detect high-frequency problems well. All three kinds of probes should be used for maximum effectiveness.

Look for red herrings. With vibration analysis, these frequently come from outside the machine, even if their effects are read on the machine.

Sources of vibration from outside the machine include:

1. Piping stresses—static, cantilevered
2. Foundation problems, including
   - Foundation settling
   - Frost melting or permafrost problems
   - Moving soil (muskeg or other shifting soil insufficiently removed)
   - Foundation inclusions (grout problems, soft feet, and so forth)
3. Extreme climatic change causing additional stresses in piping and so forth

Anticipating Repairs Needed Based on Gas Path Analysis

When the results of vibration, EGT, and borescope monitoring are fed back into audit information collecting, the audit team may decide to:

1. Alter periodicity of hot section inspections,
2. Run certain components on condition,
3. Run the gas turbine derated,
4. Use a different fuel,
5. Buy or stock updated hot section component parts,
6. Ask OEMs for life cycle algorithms,
7. Install life cycle counters, or any other relevant work items or any combination thereof.

With respect to the fuel system, they may decide they need to:

8. Change the fuel nozzle design,
9. Change the fuel nozzles more often,
10. Install a fuel treatment system,
11. Check on the working of the fuel treatment system,
12. Change the wash cycle frequency,
13. Change the wash cycle fluid/cleaner,
14. Accurately monitor and analyze gas path performance for system optimization (this may have to be retrofitted),
15. Install additional filters, preheaters, or other components in the fuel system,
16. Install water injection,
17. Install additional cooling air flow into the hot section,
18. Prompt the start of an engineering change,

or any relevant combination thereof.

Assessing TBO and Maintenance Changes Made Necessary by Changing Fuel or Fuel Composition

Fuel selection and systems can be the biggest cause of adverse effects on TBO and maintenance schedules (see Chapter 7, Gas Turbine Fuel Systems and Fuels).

As noted in Chapter 7, there are differences in composition with common liquid fuels. A change in fuel type may require extensive changes in associated fuel systems. Maintenance intervals for gas turbines can be influenced by fuel type. Fuel selection can have a drastic effect on TBOs and maintenance intervals. Heavier fuels cause higher TITs and consequently reduce turbine component life still further.

More and more, turbine combustion systems are being designed to burn gaseous, liquid, or two-phase fuels. Audit of a mature facility may reveal a need for a change in fuel type, in the event of

1. Availability issues arising from, for instance,
   a. Decrease in production gas flow,
   b. An economic moratorium on gas,
   c. Disruptions in gas flow (for example, storms at sea in Thailand interrupt production gas flow and necessitate use of diesel fuel for prolonged periods of time).
2. Molecular weight changes, common with mixed fields, which almost always result in a change in heating value of the gas. Liquid fuels do not exhibit as much variation in heating value as gaseous fuels. Heating values of gas may vary for between 1050 BTU/cubic foot, to 300 BTU/cubic foot for process gas. This is a differential of over 300%, which means that the fuel system required to handle either extreme for the same turbine would be vastly different.
3. Cleanliness issues, nearly always accompany changing field composition conditions. Frequently gas quantities decline, and oil and water carryover in gas flows start. Water leads to oxidation in the fuel system and poor combustion.
4. This carryover, in turn, can promote corrosion, hot corrosion, and fuel nozzle spray problems as liquid carryover flashes on contact with first-stage IGVs. Solid particulates and metallic particulates that are not strained out of the fuel supply can also be contaminants. They can plug fuel nozzles. Fuel treatment systems can be used to remove harmful metal contaminants such as vanadium.

5. Deposition and fouling may occur in compression trains that handle the gaseous product.

Any audit should include a comprehensive battery of tests for the fuel being used, in addition to molecular weight:

1. Carbon residue is found by burning a sample of fuel and weighing the carbon left. This property provides an indication of a fuel's tendency to deposit carbon on the fuel nozzles and combustion liner. The value should be compared with the one from the last audit or plant startup or last significant operational date.
2. Viscosity affects the pressure loss in pipe flow.
3. Pour point is the lowest temperature at which fuel can be poured under gravitational influence. Items 2 and 3 affect a fuel's tendency to foul a fuel system and determine to what extent preheating of the fuel system is a requirement.
4. Ash content of liquid fuels helps determine the cleanliness, deposition, and corrosion characteristics of a fuel. Ash is present as either oil- or water-soluble traces of metals or solid particles (sediment). High ash fuels are highly corrosive.
5. Contaminants in the gaseous fuel that may have increased since the last analysis include tar, coke, sand, lamp black, and lube oil.

Consider a typical gaseous fuel specification (see Chapter 7, Gas Turbine Fuel Systems and Fuels). Commonly, the maximum variation tolerated on a given gas turbine fuel system of these properties is $\pm 10\%$. As design development of gas turbines continues, and partly towards contribution of OEMs environmental effort, fuel systems that handle gas of heating values of less than 100 BTU/cubic foot are being developed.

With a typical liquid fuel specification, the limits set for viscosity, water, and sediment percentages by volume are required to prevent clogging of the fuel system.

If pour point criteria are not met, fuel preheating is required. The hydrogen minimum limits fuel smoking. The sulfur maximum limits corrosion due to sulfur.

Vanadium, sodium, potassium, and lead form compounds that cause corrosion at elevated temperatures. They, as well as compounds formed with calcium, can cause deposits on blades. Alkaline sulfates of sodium and potassium, as well as liquid vanadium, contribute to hot corrosion, particularly on the first stage IGVs, as they reduce the layer of protective oxide on the metal.

The test to determine carbon content of a fuel depends on whether the fuel is light or heavy. With a heavy fuel, carbon content in 100% of the sample is determined. For a light fuel, about 90% of the sample is vaporized and the carbon content is found in the remaining amount. This test is to determine the fuel's tendency to form deposits in the combustion system.

Ash is the material left after fuel combustion has occurred. The ash contains solid particles from fuel combustion and oil- or water-soluble metallic compounds. These metallic compounds cause the fuel to be corrosive.

Changes in the fuel's viscosity may require some pumping system changes.

The specific gravity of a fuel determines parameters on a centrifugal fuel washing system.

A luminosity test measures the amount of thermal radiation potential (chemical energy content) of a fuel.

Volatility determines how heavy a fuel is. A residual fuel is fuel that is left after a distillation process is complete.

Sulfur content determines sulfidation potential. Increase in sulfur content of gases can promote hot corrosion of the first stage nozzle guide vanes.

Consider the following reactions:

- In nickel-based alloys, protective oxide films are formed as follows:

  Nickel + oxygen = nickel oxide
  Chromium + oxygen = chromium trioxide

- The nickel alloy is exposed to oxidizing gas, so a protective layer of oxide forms as per these reactions.
- Alkaline sulfates are formed as follows:

  Sodium + sulfur + oxygen = sodium sulfate

- The source of the sodium is salt (NaCl). Sulfur comes from the fuel.

Metal ions in the protective oxidizing film combine with molten alkaline sodium sulfate, which destroys the protective layer. Sulfidation products include nickel sulfide and chromium trisulfide. Chloride ions from the salt (NaCl) cause intergranular corrosion as they collect at the grain boundaries. As these reactions occur, the oxide film becomes porous and deterioration occurs at an increasingly fast rate.

Other oxides can be formed, as per these examples:

Vanadium + oxygen = vanadium pentoxide
Molybdenum + oxygen = molybdenum oxide

Catastrophic oxidation requires the presence of sodium sulfate and elements such as vanadium or molybdenum.

Crude oils are high in vanadium. Their ash contains 65% or higher vanadium pentoxide.

Corrosion is accelerated with temperature rise. At temperatures over 815°C, sulfidation takes place rapidly. Hence, the low temperature philosophy used by some gas turbine models (such as ABB's 11N2) versus their equivalent horsepower "sister models" that attain higher efficiencies but use higher TITs. At lower temperatures, vanadium pentoxide can catalyze oxidation to attain damage rates that exceed those caused by sulfidation. Hence, the fuel treatment system that is required for all designs that burn low-grade fuels.

Typical fuel treatment systems are covered in some detail in Chapter 7, Gas Turbine Fuel Systems and Fuels.

Financial Factors with Respect to Fuel

An operator should burn the most economical fuel in terms of what fuel choice does to the cost per fired hour versus the cheapest fuel. This choice may change in the life of a gas turbine. In the case of users with a "natural" supply of clean, dry, sweet natural gas, this is likely to be their optimum fuel, for the time being. "For the time being," as the price of natural gas is being kept artificially low by most of the countries where it occurs naturally, including Malaysia and North Sea producers such as England. However, the current moratorium on gas in England indicates that this may not always be the case. Part of the logic for this is the high premium natural-gas-poor countries, such as Japan and Thailand, will pay for natural gas or LNG.

If natural gas tariffs were raised to realistic values versus other fuels available, other fuels could better "compete" with natural gas even in gas-rich countries. In terms of environmentally responsible fuels, on the other hand, natural gas produces only half the carbon dioxide of oil and an even smaller fraction than produced by coal combustion. Solidification of carbon dioxide after combustion is about 20 years away from being commercialized.

All this notwithstanding, the demand for viable alternative, if not regenerable, fuels that, for instance, may be gaseous or liquid by-products of a chemical process, is increasing. Currently, users conduct what fuel treatment (changes) may be required, use technology to remove NO_x and SO_x, and tolerate the carbon dioxide emissions (or plant trees).

See also Chapter 7, Gas Turbine Fuel Systems and Fuels.

Assessing Changes Required for Turbine Cleaning Procedures

Fuel treatment and washing, if done properly, remove corrosion risk even with poor-quality fuel, such as residual fuel. However, the fuel ash and magnesium compounds deposit on turbine airfoil surfaces. For intermittent operation of 100 hours or less, the deposit does not build up, as it falls off each time the turbine is fired. If the turbine is in continuous operation, however, the first stage nozzles eventually plug at the rate of between 5 and 12% per 100 hours.

For any application where residue is built up and to extend time between shutdown intervals, mild abrasive cleaners may be used. Examples of such cleaners are walnut shells, rice, and spent catalyst. Generally, dry cleaning is initiated after a 5–10% power dropoff; typically, this removes 50% of the deposits. To get better cleaning (100% power recovery), water or liquid cleaning becomes necessary.

See also Chapter 8, Accessory Systems.

Hot Section Maintenance Assessment

Combustion chambers can be inspected directly in a hot section inspection. Borescope checks may also reveal cracks or burned areas. Many small cracks are common. The key is, Are the cracks such that they might join together and cause a loss of a part of the combustor metal? Depending on length and OEM manual limits, cracks of this kind often can be weld repaired. Burned or warped areas can be cut out and replacement metal welded in place. Extensions to OEM allowed tolerances need to be made with appropriate care and experience.

The location of all combustion chamber damage should be considered with respect to:

1. Irregularities in flame patterns
2. Blockages in air or fuel flow
3. Cross tube damage or incipient failure
4. Erosion of or dirt in fuel nozzles
5. "Hot spots" that require investigation as to the root cause, such as excessive flow
6. Excessive quick starts or overload
7. Liquid droplet contamination in gaseous fuel

Combustion chamber position should be located with a reference point; that is, an annular combustor should be put back in the same position between physical inspections. With can annular combustors, each can's location in the turbine should be referenced, each can have its own permanent number. Complete records should be maintained on all combustion sections, including repair scope and dates. The can location hardware should be checked for signs of vibration or thermal movement.

Plan Maintenance, Engineering Changes, and Retrofits Based on Operational Detection and Assessment

Obviously, there are close links between audits, troubleshooting, and life cycle assessment. Data sheet formats, as well as typical system arrangements for turbomachinery types for most API specifications need to be collected. These give the operator a list of many obvious items to check that frequently are so obvious that they are missed.

External Factors and Operations Considerations

Ground-based operators would pay particular (but not limit their) attention to:

1. Environmental emissions regulations and changes in this area that might offer opportunities for optimized operations and maintenance. For instance, Alstom GTs that uses a silo combustor design for many of its gas turbines may require water injection because only one burner has a tendency to promote NO_x formation. Now, Alstom offers a retrofit of a series of EV burners on its GT-10s and -8s instead of the single burner plus the water injection. For operators that have clean natural gas or good quality diesel, this is helpful, as water injection requires boiler feedwater quality. The operator would get no additional horsepower "caused" by the water injection, however.
2. Problem generators that have any history of coil burnout.
3. Changing fuel availability (like EGAT, Thailand, whose offshore gas supply frequently is cut off by storms at sea) or government fuel policy (for instance, the United Kingdom's current moratorium on gas).
4. Boiler feedwater pumps.
5. Cooling water discharge temperature and pollutant content.

With aeroengines, emissions is the main environmental concern.

However, by far the greatest concern to any operator is minimizing cost per operational hour. Cost of maintenance can make up a sizeable portion of that, and within that portion, the cost of spare parts is frequently a major issue with operators.

All audits should aim at moving in the direction of "on condition" operation, whether operators are upstream oil and gas, power generation, petrochemical, or anything else.

Assess Parts Pools Requirements

Smaller operators frequently resort to a communal "parts pool" to avoid warehousing and holding inventory. This has both obvious advantages and disadvantages.

Larger operators adapt their spares philosophy to match the OEM with which they have to deal. For instance, Solar gives its customers a spare engine while their own is overhauled, so these customers do not keep any spares.

Decide on Repair Development: OEMs or Independents

OEMs are reluctant to develop more repairs than necessary. It means they cannot sell as many spare parts. So, for instance, for the project involving cracked GE Frame 5 discs that we looked at previously, the operator had to get an independent to work on it. Note that this may end up being a "one-off" project, as prematurely cracking discs or other components failing prematurely may occur as a consequence of changes in individual operator's process or inconsistency in the OEMs manufacturing QC. (Another example of inconsistency was the GE 9F rotor vibration problem in 1995 and 1996. Not all the engines delivered had the same extent of disassembly and other problems.)

Generally, when an impatient operator goes to an independent metallurgical shop and gets a repair process that works, the OEMs feel pressured into developing their own or getting a subcontractor to do that process for them. Or even if they have their own process, they may subcontract "overflow."

Powder metallurgy is a good example. Both GE and Westinghouse give repair work to independent vendors. Some major OEMs can also buy the powder metallurgy technology license and equipment from the independent repair shop that developed the technology. Blade tip robotic welding is another. Pratt and Whitney bought the entire rig and technology license from a vendor that developed it for their Singapore repair shop.

There is not enough time to cover all the modern useful repair techniques on the market, which enhance life cycle usage. We will briefly cover some that have the most money and problem saving potential below.

Also, at the end of this chapter, Case 9 describes the painstaking development of repairs for -F technology by an independent facility. This gives the reader a good appreciation of the investment and lead time required to develop reliable repairs.

Powder Metallurgy*

The "powder" is a plasticine type material used to fill in damaged portions or contours on blades and vanes to have the powder blend with the original material. This method retains most of the strength of the original airfoil and has far better properties than lower strength weld fillers. This repair is suitable for high-strength nickel and cobalt superalloys, such as IN738 and Rene 80. Because it is a nonfusion method, it is useful for crack-sensitive alloys that cannot be weld repaired. The powder mix can be engineered for better strength, oxidation, or wear resistance. The repair is suitable for:

- Wide gap repairs that follow the routing of material cracks of defects
- Repair of structural areas in blades, vanes, and cases
- Dimensional buildup of worn parts
- Repair of complex geometries that are difficult to weld
- Manufacture of blade tips with better oxidation and erosion resistance

Powder metallurgy technology also, in the case of a few vendors, can be purchased under a license agreement. Costs for equipment and the license range from about $500,000 upward.

---

* [12-4].

Many OEMs have not opted to own their own technology for this process. Some have a few subsuppliers to which they contract this work. This, therefore, can be a high lead time item under these circumstances. In some cases, an OEM gives a kind of blanket approval to an independent facility, which has done a great deal of work for it and whose metallurgical integrity it has checked. This would generally need to be done by destructive testing, such as sectioning or X-ray diffraction. Once a facility has that kind of approval, it can solicit business directly from end users, particularly those whose equipment is no longer under warranty.

If the independent facility is large enough, then the OEM sometimes has its own representatives resident at that facility to approve "one-off" or "general rule" repairs for specific components, which are then generally the intellectual property of that facility.

This is an involved and complex technology as Case 10 at the end of this chapter indicates.

Blade Tip Robotic Welding*

This technology is getting better accepted all the time, its main restriction being the cost of the repair equipment, about $500,000 and up for first-time overhead costs.

The actual welding that occurs with this system is not new in itself. The advantage is that robotics provides a steady "feed" rate, which then eliminates the warpage that results from human handwork, however skilled. This then also means that the depth down the airfoil height that can be repaired is also increased, thereby extending the life of rotating components, which might have worn tips but not used up their useful metallurgical life.

This system also can be used to follow the geometry and repair of shrouds, high-pressure blade tips and shrouds, air seals, and compressor blade tips.

CNC TIG Welding

These systems use computer numerical control similar to the previous system's. Again, their advantage is that uniform feed rates eliminate warpage caused by uneven heat rates. Typically, this repair can be used to repair slots on fan cases, slots on fan blades, worn segments on turbine module cases, and casings and so forth. The rig is programmed differently for each component.

Plasma Robots

Plasma layers have been used for years by manual sprayers to build up module casing flanges and lips, so everything will fit within tolerances again. Robots remove the old disadvantage, which was that, when the layers were not put on uniformly, the plasma often tended to flake, causing expensive "recycling" in the overhaul procedure, which the novice customer might end up paying for, if it had not learned to question the item.

CNC Measuring Machines

These machines, capable of five-decimal place accuracy, are programmed for every component they measure. They are used for critical bearing and seal components and checking overhauled cases and sub-assemblies. The working limitation on these is table size. The machine with a table that will fit bearing and seal components generally is about 60% of the cost of the unit that handles midsize cases of, say, 20,000 HP machines. To handle larger engine components, a larger machine has to be bought. These machines have to be housed in climatically controlled rooms.

Coatings for Erosion and Corrosion Resistance

This is one of the fastest moving fields in repair technology, and many coating manufacturers and metallurgical facilities continue to develop new coatings with higher upper performance ceilings as measured by corrosion, oxidation (high temperatures), or erosion resistance.

One erosion-resistant coating is reactive ion coating (RIC). RIC is a plasma-assisted electron beam deposition process used to deposit wear-resistant titanium nitride coatings. Some processes (different facilities develop their own) use a deposition temperature of 450°C on a variety of component thicknesses.

Another erosion-resistant coating type is low-temperature chemical vapor deposition, which applies titanium nitride at 550–625°C.

These low-temperature applications avoid the problem associated with distortion and tempering of tool type steels by application temperatures (1000°C) in conventional chemical vapor deposition.

Several processes can be used to apply aluminide coatings to hot section components. A preferred method uses chemical vapor deposition and complex geometry, including internal cooling passages of blades and vanes so their external surfaces can be coated. This enhances oxidation and corrosion resistance.

Laser Machining

With laser machining, a part that might have taken 20 hours to machine conventionally can be finished in about half an hour. This is of considerable advantage, except for initial overhead, again about $500,000 and up. Warpage caused by conventional machining's heat and subsequent heat treatment to remove this warpage is avoided.

Laser Welding, Cladding, and Layer Removal

Laser work is always quicker and less heat prone than conventional processes. It is also extremely accurate. In

* [12-4].

layer paint removal, for instance, grit blasting was conventionally used. This caused uneven layer removal, a major disadvantage for parts that had to be recoated, and unintentional and uneven heat treatment.

Laser Drilling

Normally, laser drilling is more common in manufacture with airfoils that have cooling airflow passages such as the GE-9E's and -F's blades and vanes. The laser drilling is also useful in overhaul, however, to clear out any clogged passages. A different intensity is used.

Heat Treatment

Like chemical cleaning, there are too many processes on the market to deal with individually. Their purposes are varied—reduce stress level, restore original component profile, extend component life, reduce component stress level, and so on. It is probably the repair process most prone to divergence in quality, which is difficult to check except by destructive testing. End users normally use word-of-mouth references to select their heat treatment facilities. As do OEMs, albeit generally with more paperwork and signed guarantees.

Hot isostatic pressing (HIPing)* is one of the best regarded heat treatment processes today, noted for restoring most of the metallurgical properties of a material.

All these methods help restore dimensional integrity to the component. This in turn helps the values obtained during performance analysis, and therefore LC usage, to be within optimum limits. This also might minimize any looseness and bad fits that might precipitate vibration, with consequential costs in LC usage.

Warranty Issues**

Frequently, independent repair shops are owned and run by "runaway" ex-OEM staff members. They know their products very well. Some of them are new, but a large number have a long track record. It is up to the operator to decide how much leverage to exert to attain the goal of on condition maintenance.

In negotiations with the OEM, knowing the experience of the rest of the user base globally is of key importance.

*Planning the Next TBO or Overhaul***

During an audit is the best time to contemplate what you believe is attainable and realistic for your operation in terms of TBO, overhaul cost, and eventually desired condition status for your plant.

* [12-4].

** [12-1].

You may not be where you wanted to be in terms of bargaining success with your OEMs, but you could set targets and time goals for your next audit.

*Working out Changes in Maintenance and OEM Repair Specs***

Getting OEM Cooperation and Warranty Issues

The major item of concern here for the operator is passing on observations to the OEM regarding the condition of its parts and being taken seriously enough that the OEM takes measures to extend part life. User groups have been very successful in exerting joint pressure on the OEM. Algorithms and life cycle counters have been successful in "buying" additional component life for the operator in a number of cases.

Life cycle assessment, in some ways, is a kind of audit, as is evident from the definition that follows, only it is done continually. Life cycle assessment is the art of being able to use the following data:

1. Data about a machine, its operating history, and any failures,
2. Information about the fleet of that type and model of machine,
3. Information about changing process conditions in the plant,

to assess or extend the life of turbomachinery components or the machine itself. The data are used for one of the following:

1. Avoid catastrophic failures.
2. Improve the quality of the machine's running (lower operational temperature, promote less hot erosion).
3. Better assess the life cycles being accumulated on a machine to avoid premature overhauls or parts change.
4. A combination of all of these.

Life Cycle

A cycle of life is a unit that defines a measurable unit of life in an item of turbomachinery. In many cases, with lower inlet temperature gas turbines, one cycle is an hour. If the gas turbine always ran at base load, this might be a fair assessment. However, if the gas turbine is at a peak temperature, overload, or otherwise stressful condition, an hour of running may add up to more used-up life than if the machine were at base load. If one cycle is what is incurred by running for 1 hour at base load, then more demanding operation "costs" more.

A usage cycle may or may not embody time as a parameter necessary for definition. The OEM may define its cycle, for instance, by maximum and minimum speed values, each as a percentage of maximum speed.

OEMs might conduct LCA at the request of customers concerned about premature parts removal. The reason for removal of parts might vary from machine model to machine model or from component to component.

Algorithms for Life Cycle Use

An *algorithm to calculate life cycle usage* is just a technical term for a calculation to determine how much useful life of a component has been used up. If this is information that an operator needs to know—in other words, the OEM may or may not track it for you—then the OEM will give all owners of that machinery model a service bulletin (SB) to cover the details they need to know.

Let us consider the following hypothetical example that describes the information that an OEM would provide in such an SB.

Format of an LCA SB

1. Number and title of the SB and model number(s) applicability.
2. Statement of why the SB has been introduced: at operators' request, for improved parts life, optimized operational cost, and so forth.
3. OEM's definition of *cycle* in terms of
   - Cycle type (for instance, "fatigue cycle" if low cycle fatigue, LCF, is a factor),
   - Operational parameters that define the cycle (for instance, "a speed change from 0% gas generator speed to gas generator, GG, reference speed and back to 0%").
4. Calculations and corresponding action outlined in detail in individual service bulletins.

The OEM will specify the calculation sequence. A sequence may be laid out for how to represent all operational peaks in one cycle, which peak to apply the calculation to first and how to proceed to the next step.

When this is complete, the peaks that equate to a life cycle are totaled to give cyclic usage factor for that operational cycle only. Some machinery comes with life cycle counters to automatically add these life cycles up.

Frequently, when an OEM gets as far as designing an algorithm, it follows up with a new part modification that generally has a considerably extended part life. In some cases, the new part may even run on condition, as observed in Case 8.

Life Cycle Usage in Specific Applications

Life cycles are always used up faster if the application requires many stops and starts. The type of fuel also plays a major role in life cycle usage, as we see in the next section. Also the type of turbine involved affects how many cycles or hours are used up in a normal start. Depending on the algorithms used, these figures will vary.

However, before expecting an LCA program to solve problems, basic "commonsense" troubleshooting (as is illustrated in the following example cases) ought to be completed.

Case Study 7: Glass Bead Peening*

A gas turbine's compressor rotor in operation developed blade tensile stresses that would quickly lead to failure. The OEM specified that a glass bead peening process needed to be conducted all over the surface of the compressor blades. This peening would insert a layer of compressive stress that would reduce the net effect of the tensile stress and the rotor could be run safely. Past a certain number of cycles, the compressive stress layer would dissipate and the entire turbine would need to be overhauled again.

During one of the turbine's overhauls, the repair facility mistakenly glass bead peened only one side of the compressor blades. Two engines failed in operation before the error was discovered. It was not known with which turbines the error had been made. The only way to determine this without doubt was to destructively test the compressor blades with X-ray diffraction. The entire fleet of engines that the overhaul shop had handled had to be recalled.

The OEM went on to design new modifications and models, spurred on with pressure from the user base, and the "compensation" practice is not used any more. (This case is also discussed from a business culture perspective in Chapter 14, The Business of Gas Turbines.)

Case Study 8: First-Stage Turbine Blades†

A gas turbine's design required the first-stage turbine blades to be removed at a time halfway between the specified times for major overhauls. The entire expensive row had to be replaced with half of an overhaul cycle of life wasted.

The OEM redesigned the blades. With some newer models, the blades were of a different size, but in terms of LCA, the most important change was to the material selection, which then affected the actual life cycles used up during operation by changing the stress that the blades saw in operation. The blades were under the stress endurance curve and their life was "limitless and on condition."

---

* [12-1], [12-2].

† [12-2].

ASSESSING AUDIT FINDINGS*

The Basics

First, consider the basics for routine assessment, detection, and troubleshooting:

1. One really useful tool to have is a portable spectrum analyzer. If a vibration system already is installed, one needs to see if it is of benefit to retrofit more probes on the installation and work out how many are needed for similar future installations. A portable system is very useful if it has
   - A portable probe or probes (velocity or acceleration transducer),
   - A spectrum analyzer, including storage capacity to store successive plots,
   - A chart recorder to make a hard copy of the spectrum.
2. One should study the instrumentation, OEM supplied or otherwise, on the installation and learn about its accuracy, usefulness, and ability to have its signal fed into a retrofitted PLC (programmable logic controller) or PC. Consider what additional instrumentation, if any, might be useful.
3. Concentrate on gas path monitoring parameters, as these are the most useful. Generally, most systems, however basic, as supplied by an OEM, have enough data for you to fit a performance analysis system. This is useful for
   - Determining the health of the gas path,
   - Helping diagnose failed blades, combustion liners, crossover tubes, and so forth,
   - Determining when a module (compressor or turbine) needs to be washed,
   - Determining if premature shutdown or maintenance is required.
4. Consider what the return on investment might be if you were to get a comprehensive online (perhaps real-time) condition monitoring system. Consider also if it would ever make life trouble free for the operator.

Eliminate Obvious Problems

Before time-consuming brainstorming, eliminate obvious problems. This is determined by the accuracy of your instrumentation when it has to provide inputs for early warning signals and solving troubleshooting cases. Examples of troubleshooting problems follow.

Fouled Gas Turbine Compressor

The following observations on a compressor could confirm the existence of fouling in the compressor.

* [12-2].

Vibration: Rises
PA system data: P2/P1 drops; T2/T1 rises; compressor efficiency drops Corrective action: The compressor is washed and performance recovery looked for.
A Compressor in Surge
Vibration: Fluctuates, often wildly PA system data: P2/P1 varies; T2/T1 no change; compressor efficiency drops
Other data: Bleed chamber pressure fluctuates; temperature differential across bearing may be observed to increase; bearing pressure rises.
However, the vibration and the PA system data would be enough to diagnose the high probability of surge.

A Damaged Gas Turbine Compressor Blade

Vibration: Rises
PA system data: P2/P1 drops; T2/T1 rises; compressor efficiency drops Other data: Bleed chamber pressure fluctuates.
Again, the vibration and the PA system data would be enough to diagnose the high probability of surge.

A Gas Turbine (Compressor Module) Bearing Failure

Vibration: Rises
PA system data: No change
Other data: Temperature differential across the bearing rises; bearing pressure drops; bleed chamber pressure stays constant.
Note that just the vibration reading should be enough to detect incipient bearing failure, even though not supported (or negated) by PA data.

Gas Turbine Combustor Crossover Tube Failure

Fuel pressure: Up or down
Unevenness of flame in combustor (sound indication): No change
Exhaust temperature spread: Up considerably Exhaust temperature (average): No change

Cracked Combustion Liner

The symptoms are as for the previous problem, except there is audible unevenness (noise) in combustor. Also, vibration readings may increase.

Combustor Fouling

The symptoms are as for the previous problem, except the exhaust temperature drops and vibration levels may not indicate any change.

Incipient Bearing Failure

Differential temperature (bearing): Up Bearing pressure: Down Vibration: Up

Damaged Turbine Blades

Vibration increase, large
Exhaust temperature increase

These examples help illustrate that vibration readings and PA analysis should solve most serious problems. Whether the other data back up these two systems or not may not be essential to these diagnoses. Very often marketers of expensive "expert" systems try to indicate the additional data are vital. While it may be useful for specific problems, it may not be worth the extra initial capital outlay, as well as cost of operator or engineer training to interpret the data or consultants' fees for them to interpret the data. (For example, the fee for consultants to interpret data turned out by an expert system installed on F404 engines was about $1 million. Bear in mind that the expert system was more justifiable on a critical flight engine, despite triple redundancy in its control systems.)

When a choice has to be made among several options, the risk and weighting factors method can be used. This is best illustrated by an example.

Risk and Weighting Factors Method

This may be brought on by various factors including a gas shortage or changing policy. When faced with an array of factors embodied in different options that affect TBOs and LCA, it is advisable to do a risk and weighting analysis. The main steps of this process are as follows:

1. List all your options. Let us say the task is to choose a type of fuel for a gas turbine that will give maximum component lives and TBOs. The options would be different fuel types. Draw columns representing these options. Subdivide each of these columns into two, for the weighting (priority) and risk (probability).
2. List all the factors that are important. These may include:
   - Maximum component lives,
   - Optimum TBOs,
   - Fuel cost savings,
   - Fuel consumption savings,
   - Minimum fuel treatment system requirement,
   - Minimum water injection system requirement,
   - Comply with or allow for current and future environmental regulations (CO_2 tax),
   - Minimum expertise or training level required from operators,
   - Maximum efficiency (may be linked with fuel cost savings),
   - Design that is easy to overhaul (aimed at making the overhaul process local or in house), and so forth.

   Obviously, some of these factors will work in opposition to each other. List the factors at right angles to the columns in 1.
3. Give a weight (importance) to the factors in 2. Use a descending scale of 1 to 10.
4. Estimate likelihood of these factors occurring in the right column for each choice. Use a scale of 1 to 10.
5. Multiply the weight and the risk. Make the resulting quantity either positive or negative depending on whether the item favors choice of that option or not.
6. Sum the products in 5 and arrive at your selection.

Questions to List Potential Factors and Causes

Let us assume that a problem has occurred. Use the examples in this section for illustration. Ask these questions with reference to the occurrence:

What?
When?
How?
Why?

"Who" is a deliberate omission. Why? Then ask
What needs to be done?

When?
How?
Why?
Will this affect anything else?
What is the cost of doing nothing?
How much production will be lost meanwhile?
Can anything else be corrected while correcting this problem?

What can be Learned for Future Installations?

Overhaul and Repair*

Overhaul and repair has become a highly specialized science. Overhaul specific to an engine model is spelled out in a gas turbine's engine manual. Repairs are included as updates as they are developed: these may be designated service bulletins and further subdivided into "emergency, don't operate until you complete," "required to maintain warranty period/legal terms of, R&O agreement with OEM," "optional," or "for specific owners with engines that have (these) features, only."

As R&O is such a big business, independent shops spring up and thrive, especially in fast developing

* Source: Adapted, with permission, from Rolls Royce, *The Jet Engine*, 1986, Rolls Royce Plc: UK.

countries. These shops operate within their scope as licensed by national authorities. For instance the FAA (Federal Aviation Administration) licenses shops to be able to conduct overhaul that may occur within or outside the United States. So, JT-8D-17 engines that underwent an ESV2 (refers to level of overhaul) in the Thai Airways shop in Bangkok carry notes from the R&O history that the next FAA-approved shop will see and can also expect that the last overhaul was done to the same standards as it might be expected to provide for the same work.

Most large shops that service a high population of any OEM's engines might then have a "resident" OEM liaison person to check specific "out of limits" cases that arise. Sometimes, an independent shop's power is such that it may develop repairs that are unique to that shop and that the OEM signs off on as being "okay to use by that shop only" or "okay to use just this one time."

Standards are unique to each country. In the United States, for instance, "engineers" that have an FAA mechanics license A&P (airframe and powerplant) or P (powerplant) have sign-off authority on any set of limits or process that may occur during overhaul. The quote marks around engineers means that the individuals in question may not actually have an engineer's education or training. In the case of conservatively designed engines, such as the P&W JT-8D fleet, this may suffice as experience acquaints an "engineer" with the divergence on manual specified fits, clearances, and tolerances that still get an engine through the test cell. Test cell rejects are expensive. Just the transportation to and from an assembly plant (for convenience often situated at the nonpassenger end of an airport) to the test cell (generally situated in nonpopulated neighborhoods, because of the noise) can cost about $20,000. Fits and clearances that the engine will tolerate likely result in that engine passing vibration limits during test.

However, newer engines have tighter tolerances and standards, not all of them resulting from physical dimensions. There may be certain requirements/or damage incurred by mistakes during overhaul. One example is a heat treatment process conducted at the wrong conditions. The damage that can result is not visible to the naked eye and understanding what then needs to be rejected may require not just an engineering degree, but one with the appropriate number of metallurgical courses.

What follows is a summary of the main overhaul processes required on an aeroengine. For details on individual engines, consult the current R&O manuals for that model.

It is most important that the cost of maintaining an engine in service is considered at the design stage. All aspects of engine repairability are also considered, both to reduce the requirement for overhaul or repair and to avoid, where possible, designs that make repairs difficult to effect. Engine construction must allow the operator to complete the overhaul or repair work as quickly and cheaply as possible.

In service, the engine is inspected at routine periods based on manufacturers' recommendations and agreed between the operator and the relevant airworthiness authority. The engine is removed from the aircraft when it fails, during these inspections, to meet the specified standards. This concept is a form of "on condition" monitoring. However, regardless of condition, some engines are removed when a stipulated number of engine flying hours have been achieved; this concept is known as time between overhauls (TBO). Operators will often remove engines in order to acquire "fleet stagger," thus preventing a situation when an unacceptable number of engines require removal at the same period of time.

The length of TBO varies considerably between different engine types, being established as a result of discussions between the operator, the airworthiness authority and the manufacturer, when such considerations as the total experience gained with the particular engine series, the type of operation, the utilization, and sometimes climatic conditions are taken into account. In improving the overhaul period the airworthiness authority may take into consideration the background of the operator, his servicing facilities and the experience of his maintenance personnel.

When a new type of engine enters service, sampling (i.e., engine removal, dismantling and inspection) may be called for at a modest life. The sampling will be continued until the life at which the engine should be overhauled is indicated by the condition of the sample engines or by its reliability record in service. In some instances, the ultimate life obtained may be two, three, or even four times the original period permitted. The development of the TBO from the introduction of an engine into service, through several years of operation, is shown as an example in Figure 12—9.

Among the main factors affecting the overhaul period for an engine is the use to which it is put in service. For example, a military engine will generally have a much lower TBO than its civil counterpart, as performance capability is the operating criterion rather than economics. Due to the effect of rapid temperature changes in the hot parts of the engine, the most arduous treatment is the frequent changing of power output to which short-haul transports and fighter aircraft are subjected.

When aircraft are based in areas with exceptionally high humidity or salt content in the atmosphere, there exists the added danger of corrosion, resulting in the need for more frequent overhauls. In peace time, some military aircraft have a very low utilization; this introduces the additional problem of certain materials used in its construction deteriorating before the engine has otherwise reached a condition that would normally require an overhaul. Elapsed time, as well as flying hours, would then influence the overhaul period.

In addition to scheduled overhauls, there are problems that arise from damage and defects. A proportion of these, which are uneconomic or impractical to rectify in the

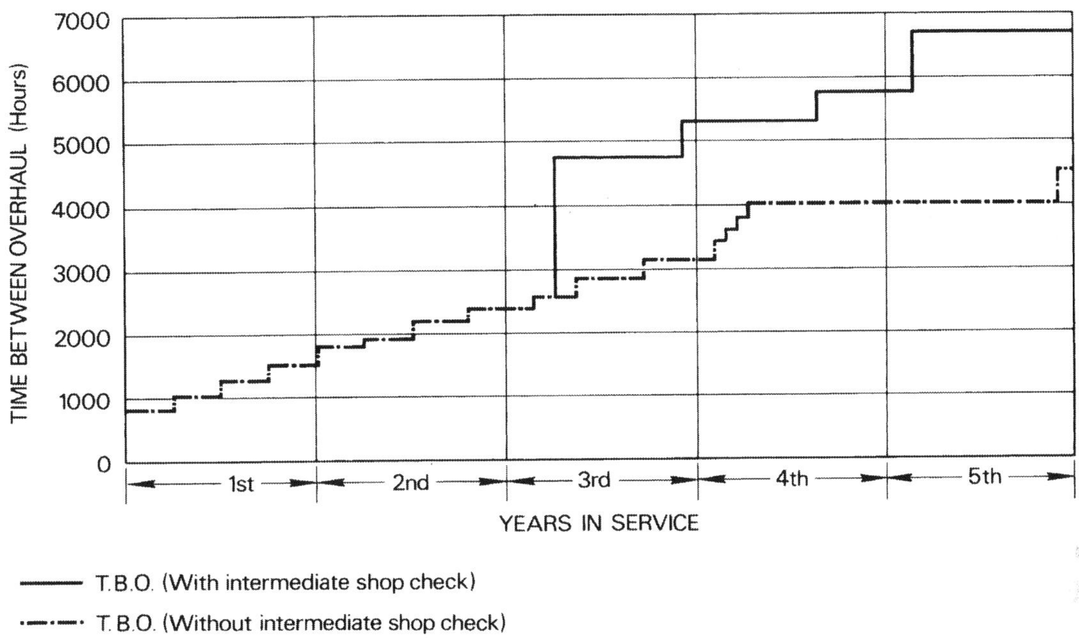

FIGURE 12–9 Example of growth of time between overhaul (TBO). *(Source: Rolls Royce.)*

aircraft, necessitate unscheduled removals and require the engine to be returned to an engineering base or an overhaul shop for partial or complete overhaul.

The purpose of overhaul is to restore an engine enabling it to complete a further life by complying with new engine performance acceptance limitations and maintaining the same reliability. This is achieved by dismantling the engine in order that parts can be inspected for condition and to determine the need for renewal or repair of those parts whose deterioration would reduce the performance, or would not remain in a serviceable condition until the next overhaul.

The design of the modular constructed engine makes it particularly suited to a different technique of overhaul/repair. This technique is based on "on condition" monitoring. This means that a life is not declared for the total engine but only certain parts of the engine. On reaching their life limit, these parts are replaced and the engine returned to service, the remainder of the engine being overhauled "on condition."

Modular construction, together with associated tooling, enables the engine to be disassembled into a number of major assemblies (modules). Modules that contain life-limited parts can be replaced by similar assemblies and the engine returned to service with minimum delay. The removed modules are disassembled into mini-modules for life limited part replacement, repair, or complete overhaul as required.

Overhaul/Repair Scope

The high cost of new engines has a considerable influence on the overhaul/repair arrangements, as the number of spare engines normally bought by the operator is kept to an absolute minimum. This means that an unserviceable engine must be quickly restored to serviceability by changing a module, or a part if the modular construction will permit it, or by careful scheduling of planned removals for overhauls at time expiry. This scheduling, through the workshop, of engines or modules that require the use of specialized equipment for repair is important, both to keep the flow of work even and to stagger removals to avoid aircraft being grounded by shortage of serviceable engines or modules.

Because the work that is to be implemented must be planned and subsequently recorded, the engine or module is received in the workshop with documents to show its modification standard and its reason for rejection from service. The planning will include a list of the modifications that can or must be incorporated to improve engine reliability or performance or to reduce operating costs.

The layout of the overhaul/repair workshop is designed to facilitate work movement through the complete range of operations, to achieve maximum utilization of floor space and to allow special equipment to be sited in positions that will suit the general flow pattern. All these considerations are aimed at achieving a quick turnaround of engines. As an example of how shop layouts may be planned, a typical arrangement is shown in Figure 12–10.

Disassembly

The engine can be disassembled in the vertical or horizontal position. When it is disassembled in the vertical position, the engine is mounted, usually front end downwards, on a

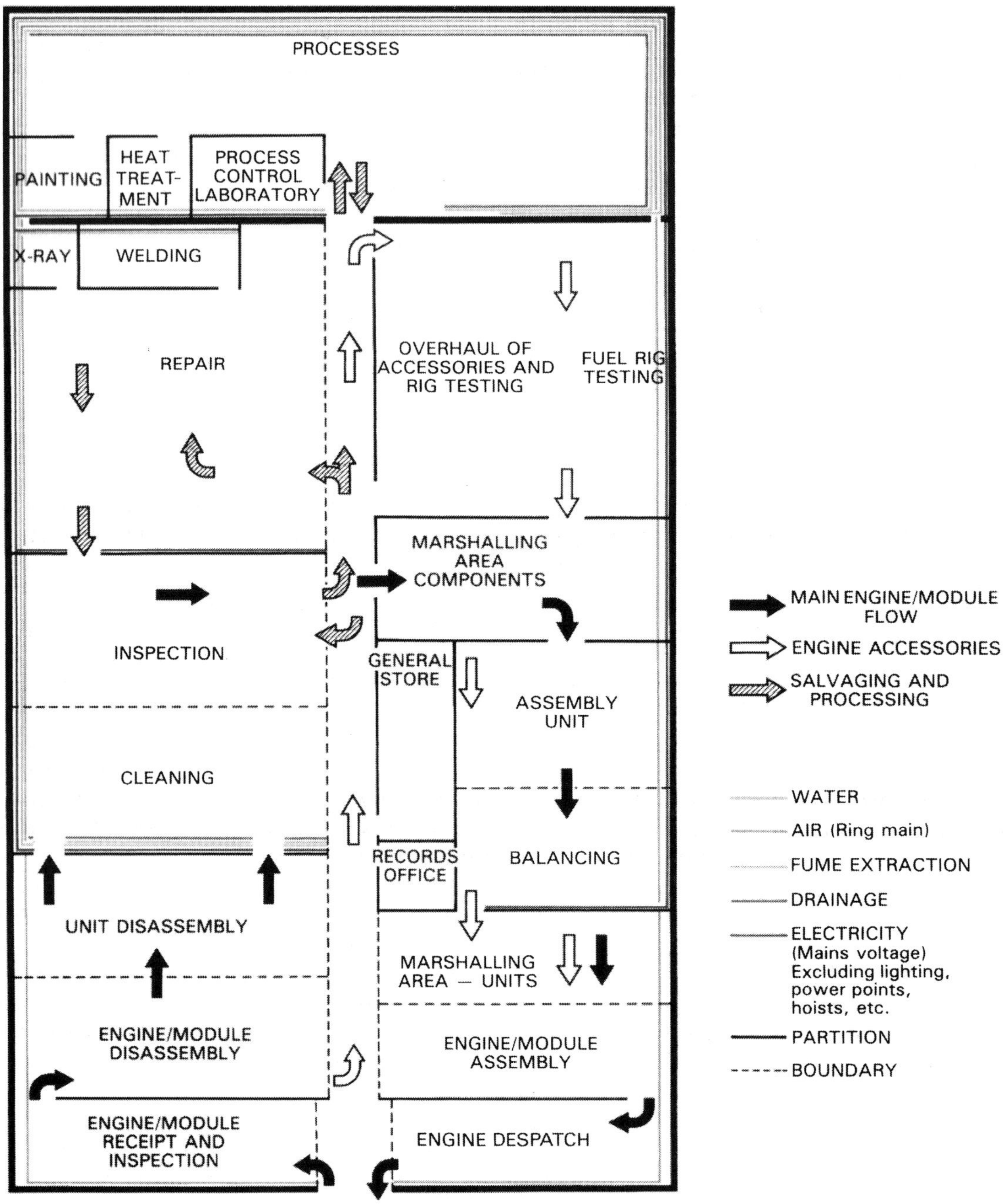

FIGURE 12–10 Typical overhaul workshop layout. *(Source: Rolls Royce.)*

floor-fixture stool or a ram-top fixture. To enable it to be disassembled horizontally, the engine is mounted in a special turnover stand.

When the floor-fixture stool is used, the personnel use a mobile work platform to raise themselves to a reasonable working position around the engine. When the ram-top fixture is used, the ram and engine are retracted into a pit, so enabling the workmen to remain at floor level.

The engine is disassembled into main sub-assemblies or modules, which are fitted in transportation stands and dispatched to the separate areas where they are further disassembled to individual parts. The individual parts are conveyed in suitable containers to a cleaning area in preparation for inspection.

Cleaning

The cleaning agents used during overhaul range from organic solvents to acid and other chemical cleaners, and extend to electrolytic cleaning solutions.

Organic solvents include kerosene for washing, trichloroethane for degreasing and paint stripping solutions that can generally be used on the majority of components for carbon and paint removal. The more restricted and sometimes rigidly controlled acid and other chemical cleaners are used for corrosion, heat scale and carbon removal from certain components. To give the highest degree of cleanliness to achieve the integrity of inspection that is considered necessary on certain major rotating parts, such as turbine discs, electrolytic cleaning solutions are often used.

Aircraft that operate at high altitudes can become contaminated with radioactive particles held in the atmosphere, this radioactivity is retained in the dirt and carbon deposits in the engine.

If contamination is suspected the radioactivity level of the engine must be determined to ensure the limitations agreed by the health authorities are not exceeded. Evidence of contamination will entail additional cleaning in a designated region, separate from the overhaul area, to safeguard the health of personnel in the workshop. Arrangements have to be made with the health authorities for disposal of the waste radioactive cleaning material.

Inspection

After cleaning, and prior to inspection, the surfaces of some parts, e.g., turbine discs, are etched. This process removes a small amount of material from the surface of the part, leaving an even matt finish that reveals surface defects that cannot normally be seen with the naked eye. The metal removal is normally achieved either by an electrolytic process in which the part forms the anode, or by immersing the part for a short time in a special acid bath. Both methods must be carefully controlled to avoid the removal of too much material.

After the components have been cleaned they are visually and, when necessary, dimensionally inspected to establish general condition and then subjected to crack inspection. This may include binocular and magnetic or penetrant inspection techniques, used either alone or consecutively, depending on the components being inspected and the degree of inspection considered necessary.

The non-dimensional inspections can be divided into visual examination for general condition and inspection for cracks. The visual examination depends on the inspector's judgment, based on experience and backed by guidance from the manufacturer. Although the visual examination of many parts of the engine conform to normal engineering practice, for some parts the acceptance standards are specialized, for example, the combustion chambers, which are subjected to very high temperatures and high speed airflows in service.

Dimensional inspection consists of measuring specific components to ensure that they are within the limits and tolerances laid down and known as "Fits and Clearances." Some of the components are measured at each overhaul, because only a small amount of wear or distortion is permissible or to enable the working clearances with mating components to be calculated. Other components are measured only when the condition found during visual inspection requires dimensional verification. The tolerances laid down for overhaul, supported by service experience, are often wider than those used during original manufacture.

The detection of cracks that are not normally visible to the naked eye is most important, particularly on major rotating parts such as turbine discs, since failure to detect them could result in crack propagation during further service and eventually lead to component failure. Various methods of accentuating these are used for inspection, the two principal techniques being penetrant inspection for nonmagnetic materials and electromagnetic inspection for those parts that can be magnetized.

Two forms of penetrant inspection in common use are known as the dye penetrant and the fluorescent test. With the dye test, a penetrating colored dye is induced to enter any cracks or pores in the surface of the part. The surface is then washed and a developer fluid containing white absorbents is applied. Dye remaining in cracks or other surface defects is drawn to the surface of the developer by capillary action and the resultant stains indicate their locations.

Fluorescent testing is based on the principle that when ultraviolet radiation falls on a chemical compound, known as fluorescent ink, it is absorbed and its energy re-emitted as visible light. If a suitable ink is allowed to penetrate surface cavities, the places where it is trapped will be revealed under the rays of an ultraviolet lamp by brilliant light emissions.

Magnetic crack testing (Figure 12–11) can only be applied to components that can be magnetized. The part is first magnetized and then sprayed with, or immersed in, a low viscosity fluid that contains ferrous particles and is known as "ink." The two walls of a crack in the magnetized part form magnetic poles and the magnetic field between these poles attracts the particles in the ink, so indicating the crack (Figure 12–12). In some instances, the ink may contain fluorescent particles that enable their build-up to be viewed under an ultraviolet lamp. A part that has been magnetically crack tested must be de-magnetized after inspection.

Chromic acid anodizing may be used as a means of crack detection on aluminum parts, e.g., compressor blades. This process, in addition to providing an oxide film that protects against corrosion, gives a surface that reveals even the smallest flaws.

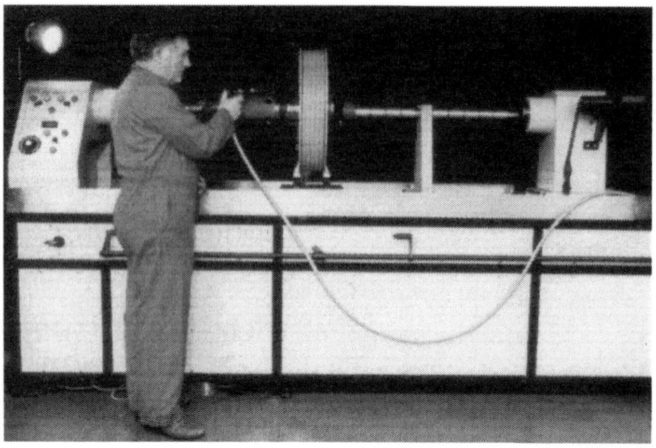

FIGURE 12–11 Magnetic crack testing. *(Source: Rolls Royce.)*

When the requirement for a detailed inspection on a component such as a turbine disc is necessary, etching of the disc surfaces would be followed by binocular inspection of the blade retention areas. The whole disc would then be subjected to magnetic crack test, followed by re-inspection of the disc including a further binocular inspection of the blade retention areas.

Repair

To ensure that costs are maintained at the lowest possible level, a wide variety of techniques are used to repair engine parts to make them suitable for further service. Welding, the fitting of interference sleeves or liners, machining and electroplating are some of the techniques employed during repair.

FIGURE 12–12 Cracks revealed by magnetic crack detection. *(Source: Rolls Royce.)*

Welding techniques are extensively used and range from welding of cracks by inert gas welding to the renewing of sections of flame tubes and jet pipes by electric resistance welding.

On some materials now being used for gas turbine engine parts, different techniques may have to be employed. An example of this is the high strength titanium alloys that suffer from brittle welds if they are allowed to become contaminated by oxygen during the cooling period. Parts made in these alloys, which have to withstand high stress loadings in service, are often welded in a bag or plastic dome that is purged by an inert gas before welding commences.

More advanced materials and constructions may have to be welded by electron-beam welding. This method not only enables dissimilar metals to be welded, but also complete sections of the more advanced fabricated constructions, e.g., a section of a fabricated rotor drum, to be replaced at a low percentage cost of a new drum.

Some repair methods, such as welding, may affect the properties of the materials and, to restore the materials to a satisfactory condition, it may be necessary to heat treat the parts to remove the stresses, reduce the hardness of the weld area or restore the strength of the material in the heat affected area. Heat treatment techniques are also used for removing distortions after welding. The parts are heated to a temperature sufficient to remove the stresses and, during the heat treatment process, fixtures are often used to ensure the parts maintain their correct configuration.

Electroplating methods are also widely used for repair purposes and these range from chromium plating, which can be used to provide a very hard surface, to thin coatings of copper or silver plating, which can be applied to such areas as bearing locations on a shaft to restore a fitting diameter that is only slightly worn.

Many repairs are effected by machining diameters and/or faces to undersize dimensions or bores to oversize dimensions and then fitting shims, liners, or metal spraying coatings of wear resistant material. The effected surfaces are then restored to their original dimensions by machining or grinding.

The inspection of parts after they have been repaired consists mainly of a penetrant or magnetic inspection. However, further inspection may be required on parts that have been extensively repaired and this may involve pressure testing or X-ray inspection of welded areas.

Re-balancing of the main rotating assembly will be necessary during overhaul, even though all the original parts may be refitted, and this is done as described next.

Balancing

Because of the high rotational speeds, any unbalance in the main rotating assembly of a gas turbine engine is capable of

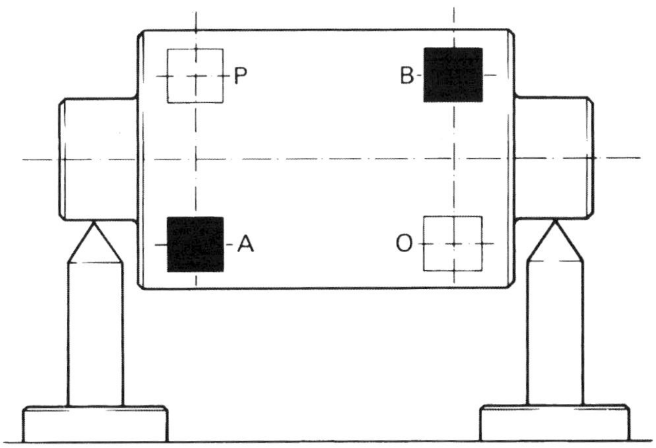

FIGURE 12–13 Unbalance couples due to centrifugal force. *(Source: Rolls Royce.)*

producing vibration and stresses that increase as the square of the rotational speed. Therefore very accurate balancing of the rotating assembly is necessary.

The two main methods of measuring and correcting unbalance are single plane (static) balancing and two-plane (dynamic) balancing. With single plane, the unbalance is only in one plane, i.e., centrally through the component at 90° to the axis. This is appropriate for components such as individual compressor or turbine discs.

For compressor and/or turbine rotor assemblies possessing appreciable axial length, unbalance may be present at many positions along the axis. In general it is not possible to correct this combination of distributed unbalance in a single plane. However, if two correction planes are chosen, usually at axially opposed ends of the assembly, it is always possible to find a combination of two unbalance weights that are equivalent for the unbalances present in the assembled rotor, hence two plane balancing.

To illustrate this point refer to Figure 12–13, the distribution of unbalance in the rotor has been reduced to an equivalent system of two unbalances "A" and "B." The rotor is already in static balance because in this example "A" and "B" are equal and opposed. However, when the part is rotating, each weight produces its own centrifugal force in opposition to the other causing unbalance couples, with the tendency to turn the part end-over-end. This action is restricted by the bearings, with resultant stresses and vibration. It will be seen, therefore, that to bring the part to a state of dynamic balance, an equal amount of weight must be removed at "A" and "B" or added at "P" and "O." When the couples set up by the centrifugal forces are equal, it is said that a part is dynamically balanced. Unbalance is expressed in units of ounce-inches, thus 1 ounce of excess weight displaced 2 inches from the axis of a rotor is 2 ounce inches of unbalance.

When balancing assemblies such as L.P. compressor rotors, the readings obtained are inconsistent due to blade scatter. Blade scatter is caused by the platform and root or retaining pin clearances allowing the blades to interlock at the platforms and assume a different radial position during each balancing run. This only occurs at the relatively low rpm used for balancing, because, during engine running, the blades will assume a consistent radial position as they are centrifuged outwards.

To obtain authentic balance results when blade scatter is present, it is necessary to record readings from several balance runs, e.g., eight runs, thereafter determining a vector mean.

A typical dynamic balancing machine for indicating the magnitude and angular position of unbalance in each plane is shown in Figure 12–14. Correction of unbalance may be achieved by one or a combination of the following basic methods: redistribution of weight, addition of weight, and removal of weight.

Redistribution of weight is possible for such assemblies as turbine and compressor discs, when blades of different weight can be interchanged and, on some engines, clamped weights are provided for positioning around the disc.

FIGURE 12–14 Dynamic balancing machine. *(Source: Rolls Royce.)*

FIGURE 12–15 Simulated engine rotor assemblies. *(Source: Rolls Royce.)*

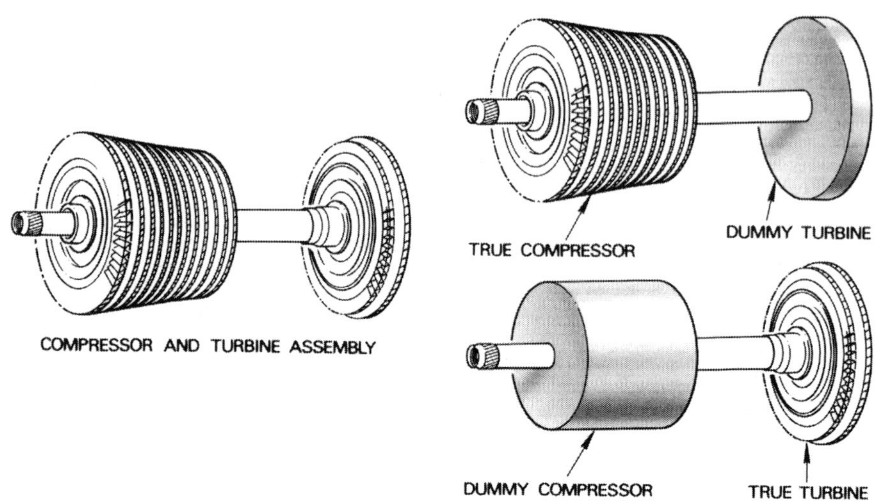

The addition of weight is probably the most common method used, certain parts of the assembly having provision for the fitting of screwed or riveted plugs, heavy wire, balancing plates, or nuts.

Removing weight by machining metal from balancing lands is the third basic method, but normally it is only employed on initial manufacture when balancing a component, e.g., a turbine shaft or a compressor shaft, that is part of a larger assembly.

Modular assembled engines demand different balancing methods that include the use of simulated engine rotors. The dummy rotors must reproduce the bearing span, weight, center of gravity and dynamic characteristics of the sub-assembly it replaces and be produced and maintained so that their own contribution to the measured unbalance is minimal. In order to obtain the correct dynamic reactions when balancing a compressor and/or turbine rotor assembly on its own, with the intention of making it an independent module, a simulated engine rotor must be used to replace the mating assembly (Figure 12–15). The compressor and/or turbine rotor assembly having then been independently balanced with the appropriate dummy rotor is thus corrected both for its own unbalance and influence due to geometric errors on any other mating assembly.

Moment Weighing of Blades

With the introduction of the large fan blade, moment weighing of blades has assumed a greater significance (Figure 12–16). This operation takes into account the mass of each blade and also the position of its center of gravity relative to the center line of the disc into which the blade is assembled. The mechanical system of blade moment weighing may be integrated with a computer, shown in Figure 12–17, which will automatically optimize the blade distribution. The moment weight of a blade in units, i.e., g.mm. or oz.in., is identical to the unbalance effect of the blade when installed into a disc. The recorded measurement of blade moment weights enables each blade to be distributed around the disc in order that these unbalances are cancelled.

Assembling

The engine can be built in the vertical or horizontal position, using the ram or stand illustrated in Figures 12–18 and 12–19, respectively. Assembling of the engine sub-assemblies or modules is done in separate areas, thus minimizing the build time on the build rams or stands.

During assembling, inspection checks are made. These checks can establish dimensions to enable axial and radial clearances to be calculated and adjustments to be made, or they can ascertain that vital fitting operations have been correctly effected. Dimensional checks are effected during disassembly to establish datums that must be repeated on subsequent re-assembly.

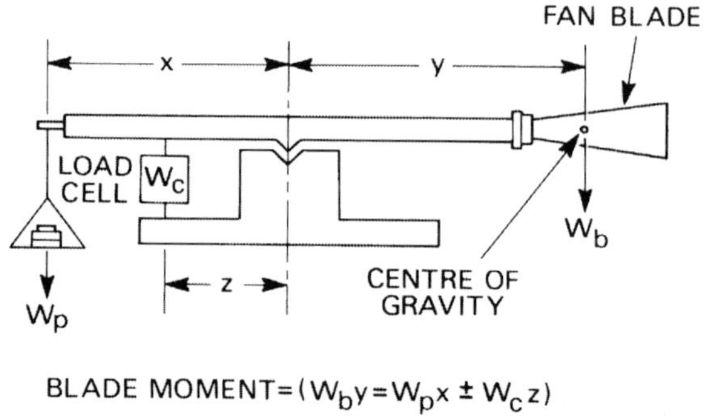

FIGURE 12–16 Principle of blade moment weighing. *(Source: Rolls Royce.)*

FIGURE 12–17 Integrated blade moment weighing. *(Source: Rolls Royce.)*

To ensure that the nuts, bolts, and setscrews throughout the engine and its accessories are uniformly tight, controlled torque tightening is applied, Figure 12–20, the torque loading figure is determined by the thread diameter and the differing coefficients of friction allied with thread finish, i.e., silver or cadmium plating and the lubricant used.

Testing

On completion of assembly, every production and/or overhauled engine must be tested in a "sea-level" test cell (Figure 12–21), i.e., a test cell in which the engine is run at ambient temperature and pressure conditions, the resultant

FIGURE 12–19 Engine assembly–horizontal. *(Source: Rolls Royce.)*

FIGURE 12–18 Engine assembly–vertical. *(Source: Rolls Royce.)*

FIGURE 12–20 Torque tightening. *(Source: Rolls Royce.)*

FIGURE 12–21 A "sea-level" test cell. *(Source: Rolls Royce.)*

FIGURE 12–22 A high altitude test cell. *(Source: Rolls Royce.)*

performance figures being corrected to International Standard Atmosphere (I.S.A.) sea-level conditions.

To ensure that the engine performance meets that guaranteed to the customer and the requirements of the government licensing and purchasing authorities, each engine is tested to an acceptance test schedule.

In addition to the "sea-level" tests, sample engines are tested to a flight evaluation test schedule. These tests cover such characteristics as anti-icing, combustion and reheat efficiencies, performance, mechanical reliability, and oil and fuel consumptions at the variety of conditions to which the engine may be subjected during its operational life. Flight evaluation testing can be effected by installing the engine in an aircraft or in an altitude test cell (Figure 12–22) to test the variations of air humidity, pressure and temperature on the engine, its accessories and the oil and fuel systems. When in an aircraft, the engine is operated at the actual atmospheric conditions specified in the schedule, but in an altitude test cell, the engine is installed in an enclosed cell and tested to the schedule in conditions that are mechanically simulated.

Mechanical simulation comprises supplying the engine inlet with an accurately controlled mass airflow at the required temperature and humidity, and adjusting the atmospheric pressure within the exhaust cell to coincide with pressure at varying altitudes.

The data that are accumulated from either "sea-level" or altitude testing are retained for future development, engine life assessment, material capabilities, or any aspect of engine history.

During the testing of turbo-jet engines there is a need to reduce the exhaust noise to within acceptable limits. This may be achieved by several different means, each involving costly equipment. However, a typical silencer would do this by expansion in the first section, damping by acoustic tubes and final diffusion by a large exit through which the hot gas would be directed upwards at a low velocity.

Preparing for Storage/Dispatch

The preparation of the engine/module for storage and/or dispatch is of major importance, since storage and transportation calls for special treatment to preserve the engine. To resist corrosion during storage, the fuel system is inhibited by special oil and all apertures are sealed off. The external and internal surfaces of the engine are also protected by special inhibiting powders or by paper impregnated with inhibiting powder and the engine is enclosed in a re-usable bag (Figure 12–23) or plastic sheeting into which a specific amount of desiccant is inserted. If transportation by rail or sea is involved, the inhibited and bagged engine may be packed in a wooden crate or metal case.

MAJOR REPAIR AND OVERHAUL CASE STUDIES*

The preceding material gives the reader the basic framework with which to develop R&O strategy. Consider also that, with modern gas turbines, the actual technology development that occurs around this framework is rather detailed as the following cases prove.

* [12-1].

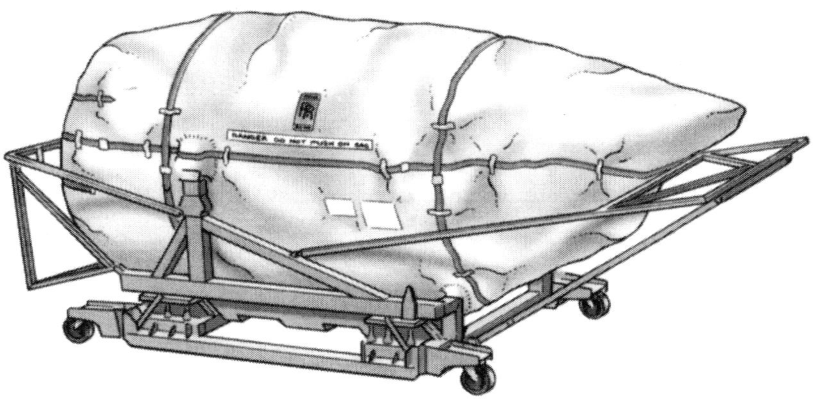

FIGURE 12–23 Transportation stand and storage bag. *(Source: Rolls Royce.)*

As occurs elsewhere in this book, certain cases lend themselves to inclusion in the chapter on maintenance, repair and overhaul (Chapter 12), as well as the chapter on performance, performance testing and optimization (Chapter 10). An interest in either chapter is better explored by reading both. For all these cases, as elsewhere in this book, the reader ought to contact the original source of the case to get more detail than I elected to include here.

Case Study 9 deals with the role of metallurgical analysis in gas turbine maintence, and is written by a metallurgical repair facilty. A self-evident connection to some operators, the case provides specifics on cases where selective metallurgical analysis has helped determine component lives or TBO.

Case Study 10 describes the painstaking development of repairs for "F" technology by an independent facility. The case includes considerable discussion on coatings that certain "F" technology machines used extensively. This gives the reader a good appreciation of the investment and the lead time required to develop reliable repairs. It also explains how good repairs can greatly extend the component lives of gas turbine engines.

Case Study 11 describes the details of some power metallurgy repairs with examples from an independent facility that developed a proprietary process for this technology. Again, this illustrates what a difference innovative repair technology can make to an end user's costs per fired hour.

Case Study 12 presents an OEM's (Siemens AG) strategy with extending the life of specific hot section components used in power generation applications.

Case Study 13, written by an EPC contractor (sometimes responsible for operation of a plant for a client), deals with "Assessing performance degradation in gas turbines for power applications."

Case Study 14, written by a metallurgical repair facility, entitled "Remaining Life Assessment of Power Turbine Disks," which discusses how component geometry and fluid flows can affect the individual lives of components that are identical by model number.

Case Study 9: Metallurgical Analysis Used in Gas Turbine Maintenance and Extending TBOs*

Metallurgical evaluation of the post-service condition of turbine blades provides information needed to make informed maintenance decisions, maximize potential cost savings, and manage risk. Original engine manufacturers (OEM) provide maintenance guidelines for service interval periods and component repair recommendations.

However, the operating conditions between engine users and specific sites differ, resulting in a considerable variation in the extent of part deterioration and post-service condition. Therefore, characterization of the blade set condition can be used to optimize maintenance decisions and provide insight towards the potential of service interval extension.

The operating conditions of a gas turbine are such that deterioration of turbine blades is inherent and blades therefore have a finite operating life. Under ideal base load conditions, degradation of hot section components is primarily a function of operating temperature. In such conditions, the component life is generally based upon (a) the depletion of the protective coatings or surface oxidation and (b) strength degradation and creep damage of the base alloy. Thermal degradation is exponentially correlated to temperature and minor changes in temperature can have large effects upon the rate of deterioration. In peaking applications, engine starts generate thermal-mechanical fatigue damage. In combination, these mechanisms are considered by OEMs to predict damage accumulation and establish criteria used to determine the component life and service interval length.

* Source: "The Role of Metallurgical Analysis in Gas Turbine Maintenance," Liburdi Turbine Services.

Non-ideal conditions such as corrosive environments, over-firing, or impact damage can result in accelerated component deterioration. These modes of degradation may not be included within the OEM service interval length guidelines and, if they occur, may have implications relevant to the maintenance and repair of the blades. Identification of these issues can be used to generate corrective actions in order to minimize risk of component failure and maintaining the component reparability after each service interval.

Conversely, many units are operated at part-load conditions under which damage will accumulate more slowly and so the OEM lifing criteria may be conservative.

As an alternative to predicting the damage expected after a service interval, destructive metallurgical analysis allows complete characterization of the actual post-service condition. Life analysis refers to a metallurgical examination that characterizes the overall condition of a blade including the material state, coating degradation, and damage within the internal passages. The data collected from the analysis can be interpreted to:

- **a.** Assess reparability of the blade set. Minimize risk of otherwise undetectable damage
- **b.** Determine repair processes required to return the set to serviceable condition
- **c.** Evaluate the performance and selection of the coatings (if applicable)
- **d.** Provide indications of abnormal or detrimental engine operating conditions
- **e.** Assess the potential of a service interval extension

This case will summarize the typical modes of deterioration found in metallurgical analysis of service exposed blades and present guidelines for using that information to make repair, maintenance, and engine operating decisions (Table 12–3).

Metallurgical Analysis Procedure

The techniques used for life analysis are well established in literature and practice. One blade, which is either representative of the set or exhibits independent, unrepairable damage (such as impact damage), is selected for destructive examination. The initial steps include review of the blade history, visual examination, non-destructive testing, and gathering of relevant surface deposits.

The blade is typically sectioned to remove samples from the lower, mid and upper portions of the airfoil, the tip or shroud and any other features of interest. As the operating temperature of the root is below that at which material degradation occurs, a section of the root is examined to provide a pre-service material benchmark. The samples are prepared for microscopic examination by conventional metallographic polishing techniques.

The evaluation includes assessment of the base alloy microstructure, the coating/external surface condition, and the internal surface condition. Modes of degradation can be identified through various forms of microscopy and microanalysis techniques. Depending on the size/design of the component, mechanical testing of the base metal may also be performed.

Base Metal Deterioration

Alloy Aging and Creep Damage

Turbine blades are manufactured from precipitation-hardened nickel-based superalloys that develop their high strength at high temperatures from gamma prime, Ni_3Al precipitates within the material. In the as-new condition, the size, density and morphology of the gamma prime precipitates are optimized for maximum strength (Figure 12–24a). However, during operation at elevated temperatures, the microstructure undergoes various changes. The strengthening gamma prime precipitates spheroidize, enlarge and join, resulting in a reduction in material strength (Figure 12–24b and c). Topologically close-packed (TCP) phases can also form resulting in a dramatic loss of ductility and reduction in creep strength (Figure 12–25a). Microstructural degradation takes place by diffusion mechanisms and the rate of degradation is therefore a function of both time and component temperature.

Due to the centrifugal loading and operating temperatures, the airfoil material also undergoes creep deformation. Creep is a diffusion-based mechanism correlated exponentially to temperature. Eventually, creep results in the formation of voids that link to cause creep cracking (Figure 12–25b). While the detection of creep voids is a reliable indicator of creep damage, in most alloys, voids only form after a large fraction of the creep life is exhausted and hence they may be absent even in significantly creep-damaged material. Prior to the formation of such voids, it is difficult to identify damage due to creep. Creep damage may be evident by reduced creep rupture life measured by mechanical testing; however, the effect is confounded with the aging effects.

Both creep and microstructural damage can be repaired through appropriate heat treatments. Knowledge of the material condition can be used to ensure the appropriate thermal processing is implemented to restore the material properties needed for continued service at minimal risk. Base material that exhibits creep damage requires hot isostatic pressing (HIP) at appropriate temperature and pressure to collapse creep voids that may have formed. Alloy degradation is corrected by heat treatments capable of fully restoring the microstructure and material strength.

TABLE 12–3 Summary of Degradation, Consequences and Maintenance Considerations

Location	Degradation Mode	Causes and Consequences	Maintenance/Repair Considerations
Base Alloy	**Base Alloy Aging**	• Material degradation due to high temperature operation. • Reduction in material strength and impact properties. • Temperature dependent.	• Life limiting mechanism for base load applications. • Undetectable by NDT methods. • Reparable with proper heat treatment. • To (a) minimize repair costs and (b) minimize risk of insufficient heat treatment, heat treatment should be based on analysis findings. • Abnormal heat patterns can indicate engine operating irregularities. • Limited aging may indicate potential for service interval extension.
	Creep Damage	• Combination of high temperature and stress causes material to creep. • Advanced stages of creep damage results in void formation. • Creep damage can lead to deformation, cracking and failure. • Occurs simultaneously with alloy aging. • Temperature and stress dependent.	
	High Cycle Fatigue	• Repetitive/cyclic loading can result in fatigue cracking. • Can lead to catastrophic failure.	• Analysis to determine the cause of crack initiation. • Blade set reparability based upon location and size of crack(s). • Detectable by NDT methods. • Heat treatment can alleviate cyclic damage accumulation (if not cracked). • Prevention requires elimination of cyclic loading or part modification.
	Thermal-Mechanical Fatigue Damage (TMF)	• Accumulation of cyclic stresses from temperature changes can result in low-cycle fatigue cracking. • Associated with number of engine starts/stops/trips. • Life limiting mechanism for peaking or high start applications. • Can result in cracking and failure.	• Life limiting mechanism for peaking and high-start engine applications. • TMF cracking can occur internally and externally. • Heat treatment can alleviate cyclic damage accumulation (if not cracked). • Location and size of TMF cracks can determine set reparability. • TMF on external surfaces only—with use NDT to assess reparability on a blade-by-blade basis. • TMF on internal - retire blade set.
External Coating/ Surface Condition	**Coating Depletion**	• Depletion of protective elements in coating. • Temperature and start/stop dependent. • Complete depletion can result in coating failure/ breaching and damage to base alloy.	• Life limiting mechanism for base load applications. • Coating can be re-applied for continued service. • Analysis to assess coating performance and selection. • If irreparable damage sustained to base alloy, consider upgrading coating. • Limited depletion may indicate potential for service interval extension.
	Coating Cracks	• Usually due to engine starts/trips or impact damage. • Localized oxidation damage to underlying base alloy.	• Requires removal and re-application of coating. • If irreparable damage sustained to base alloy, consider changing coating.

(Continued)

TABLE 12–3 Summary of Degradation, Consequences and Maintenance Considerations—cont'd

Location	Degradation Mode	Causes and Consequences	Maintenance/Repair Considerations
	TBC Coating Loss	• Coating loss due to thermally grown oxide beneath TBC or impact. • Oxide growth temperature dependent. • Coating loss results in elevated base metal operating temperature and accelerated base metal damage.	• Coating loss may result in increased base metal damage requiring adjusted repairs. • TBC is a consumable coating and only deemed inadequate if irreparable damage to the blade is sustained.
	Hot Corrosion	• Accelerated damage to coatings and base alloy. • Can result in irreparable damage or part failure. • Requires the presence of corrosion agents (Na, K with S, or V, Pb independently).	• Metallurgical assessment to determine corrosion type, corrosive agents and reparability. • Significant corrosion may be cause to retire blade set. • Elimination of corrosive agents' source(s) ideal solution. • Consider applying coating with appropriate corrosion resistance.
	Impact Damage	• Damage can range from superficial to coating damage to catastrophic failure.	• Material/objects from up-stream. • Sources can be intake through filtration, up-stream failure or material left in engine during maintenance. • Prevention based upon elimination of source.
	Deposits	• Potential for cooling hole blockage.	
Internal Surfaces	**Coating Depletion**	• Consumption of protective properties of coating. • Temperature and start/stop dependent. • Can result in coating failure/breaching and damage to base alloy.	• Coating can be re-applied during repairs. • May need to be removed depending on repair heat treatment process for base alloy.
	Internal Surface Alloy Depletion (Un-Coated)	• Depletion of strengthening elements at surface, weakening surface layer. • Temperature dependent. • Resulting in (a) lower surface fatigue resistance and (b) reduction in effective wall thickness.	• Sustained damage irreparable. • Decision to continue servicing based upon tolerating existing damage. • If depth of damage unacceptable, retire blade set. • If acceptable, monitor condition between service intervals. • Undetectable by NDT methods.
	Intergranular Oxidation (IGO) (Un-Coated)	• Oxidation damage along material grain boundaries. • Potential for IGO damage causing brittle oxides for form along grain boundaries.	
	Hot Corrosion	• Accelerated damage to coatings and base alloy. • Can result in irreparable damage or part failure. • Requires the presence of corrosion agents (Na, K with S, or V, Pb independently).	• Metallurgical assessment to determine corrosion type, corrosive agents and reparability. • Significant corrosion may be cause to retire blade set. • Corrosive agents from up-stream compressor before the cooling air off-take. • Elimination of source ideal solution.
	Internal Deposits	• Foreign material within cooling passages, potential for cooling hole blockage.	• Metallurgical assessment to determine nature of deposit. • Deposits from up-stream compressor before the cooling air off-take, usually air-intake. Can indicate filtration concerns. • Elimination of source ideal solution.

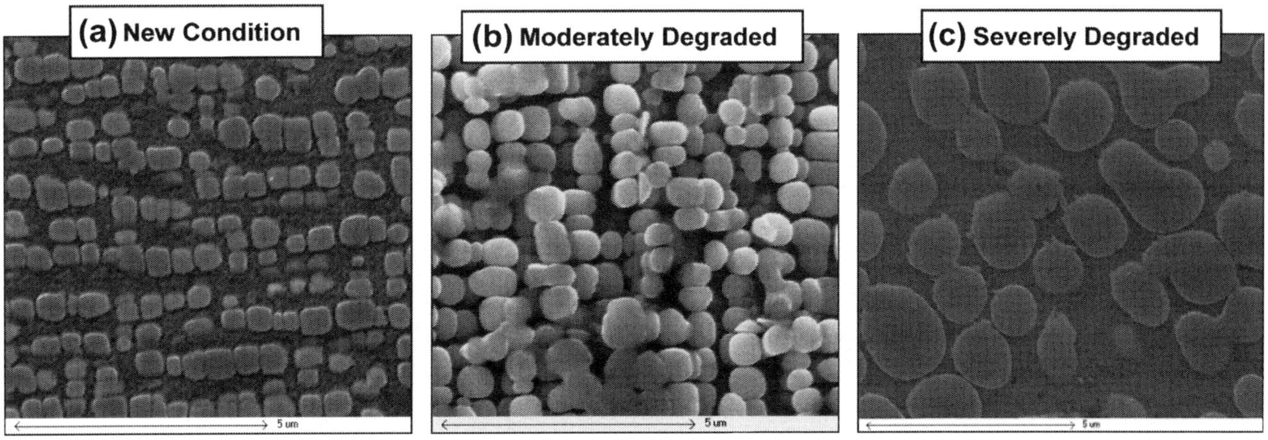

FIGURE 12–24 Scanning electron images displaying the gamma prime precipitates indicative of material condition of René 80 nickel-based superalloy in the (a) new, (b) moderately degraded, and (c) severely degraded conditions.

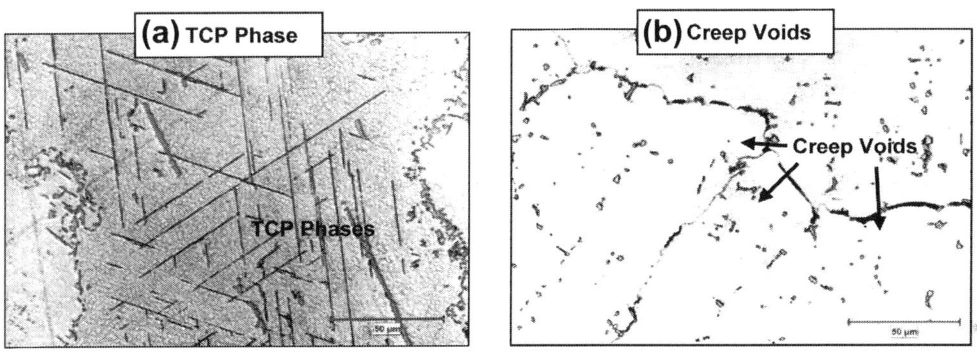

FIGURE 12–25 Optical micrographs displaying material degradation in the form of (a) TCP phase formation (dark gray, acicular phase) and (b) creep voids. As-polished condition.

As microstructural degradation occurs as a function of temperature, the condition of the base alloy provides a relative thermal map of the blade's operating temperatures. Once a typical thermal profile has been benchmarked for a given blade design, abnormal engine operation affecting the operating temperature of the blade can be detected.

Examples include short-term excessive overheating, long-term overheating, and temperature profile shift.

The base alloy condition can also provide insight into the suitability of the service interval length. Gross microstructural damage or the presence of creep voiding can indicate an inappropriate service interval at the conditions exposed. In the absence of an identifiable cause, it would be recommended to shorten the service interval in future. Conversely, if only moderate base alloy deterioration is present, service interval extension could be considered.

High Cycle Fatigue Cracking

Repetitive loading at moderate stresses can result in the formation of fatigue cracks and potentially failure. These cracks are generally attributed to vibrations, rubs, or resonant frequency events. The stages of fatigue cracking include the accumulation of material damage, crack initiation, crack propagation, and then ultimately final failure. At the common frequencies experienced by gas turbines, cycles are accumulated rapidly.

Endurance limit cycles in the order of 10^7 can occur within hours or days. Therefore, parts are designed to avoid experiencing cyclic stress intensities or events that would result in fatigue crack formation.

While NDT methods can detect cracking, metallurgical analysis is used to determine the nature of crack initiation (Figure 12.26) and to determine the reparability of the remainder of the blade set. Efforts are made to determine if crack initiation was due to an isolated cause (i.e., impact, material defect, etc.) or if the entire set was susceptible (i.e., resonant frequency). Once identified as fatigue cracking, further investigation as to the engine condition responsible for the elevated loading may be warranted.

Reparability is based upon the location and size/depth of the crack. Fatigue cracks generally initiate on the external surfaces and non-destructive techniques can be

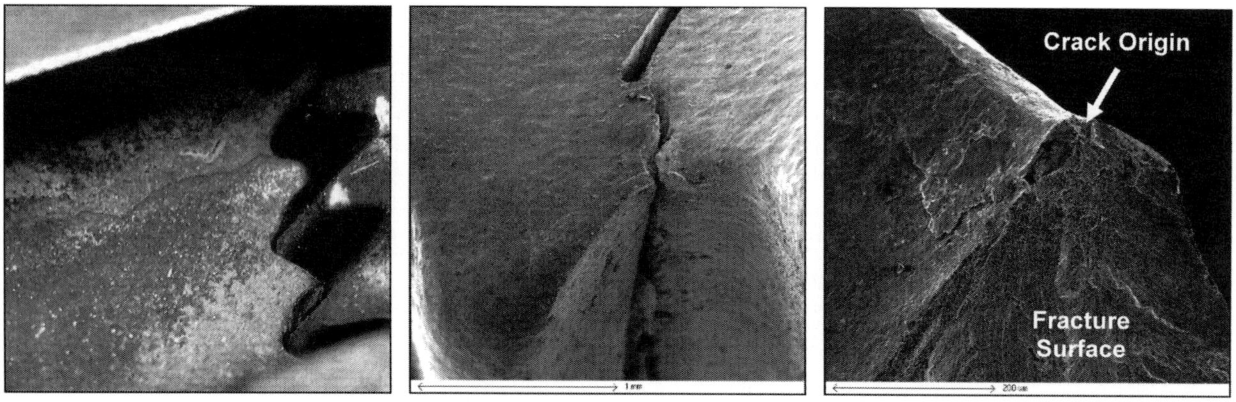

FIGURE 12–26 Photograph and SEM images displaying a high cycle fatigue crack initiating within the root.

used to determine reparability blade by blade. Cracks at locations such as the root and the majority of the airfoil would be cause to retire the blade. In the case where the entire blade set was susceptible to fatigue cracking, appropriate heat treatments should be applied to alleviate accumulated, cyclic base metal damage.

Thermal-Mechanical Fatigue Damage

Temperature gradients in blades result from both rapid changes in the operating temperatures and steady state differences in temperature resulting from cooling effects.

These in turn can result in thermal-mechanical fatigue (TMF) damage from thermal stresses associated with the gradient. Repeated cycling can result in low-cycle fatigue crack initiation and propagation (an example of TMF is provided in **Case Study 2,** Figure 12–32). Therefore, thermal-mechanical damage often is the critical life limiting factor for engines with a high number of engine starts and trips. Damage is a function of part design, operating temperature, number of cycles, hours of operation between cycles, and temperature change rate.

When thermal-mechanical cracking is present only on the external surfaces, non-destructive testing can be used to sort reparable blades from irreparable. Blades that are deemed reparable should be subjected to appropriate heat treatments in order to alleviate accumulated material damage. As there is no reliable, non-destructive method to detect internal cracking, if metallurgical analysis detects internal cracking within the airfoil, the remainder of the blade set should be retired.

External Coating Condition

Aluminide/MCrAlY Coating Deterioration in Clean Environments

Diffusion and overlay coatings provide oxidation protection for operating environments for which the base alloy material does not have acceptable resistance. Oxidation resistance of coatings is principally based on aluminum content. The high aluminum content existing as β-aluminide forms a protective aluminum oxide layer slowing down the oxidation damage process. This process is exhaustive and once the aluminum content is depleted below a critical level, the β-aluminide transforms to gamma prime. At this point, a mixed oxide is formed that does not provide the same protective nature.

Figure 12–27 displays examples of coatings after service with minimal, moderate, and complete β-aluminide depletion. In simplification, the coating protective life is based upon the amount of β-aluminide remaining.

Coating breaches can occur due to either complete β-aluminide coating depletion or thermal-mechanical coating cracks (Figure 12–28). Generally, coating failure is deemed when the base alloy sustains oxidation damage that is either irreparable and/or increases the risk of failure to unacceptable levels.

Optical examination of the coating is able to determine the amount of remaining β-aluminide in a coating, examine for cracks as well as oxidation damage to the coating and underlying base alloy. Data obtained from characterizing the coating/subsurface condition can be used to (a) assess if the coating provided adequate protection during previous service, (b) if inadequate, determine a more suitable coating based upon the damage mechanisms, and (c) assess if there is potential for service interval extension (incorporating other modes of damage).

On sets that the coating had failed to prevent damage to the base metal, the coating type, service interval, and operating conditions should be reviewed. In cases where the base alloy did not exhibit excessive alloy degradation, the coating may be defined as the primary life-limiting mechanism. In such cases, a coating upgrade may be beneficial.

Full depletion of the coating, combined with gross base alloy degradation, may suggest elevated operating temperatures or too long of a service interval at the

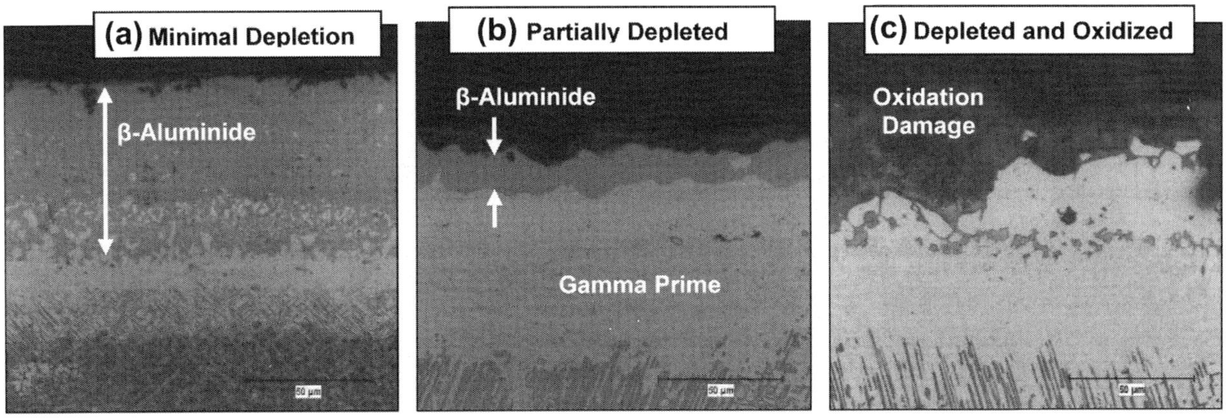

FIGURE 12–27 Optical micrographs displaying aluminide diffusion coatings after service exhibiting (a) minimal β-aluminide depletion, (b) partial depletion, and (c) fully depleted and oxidation damaged.

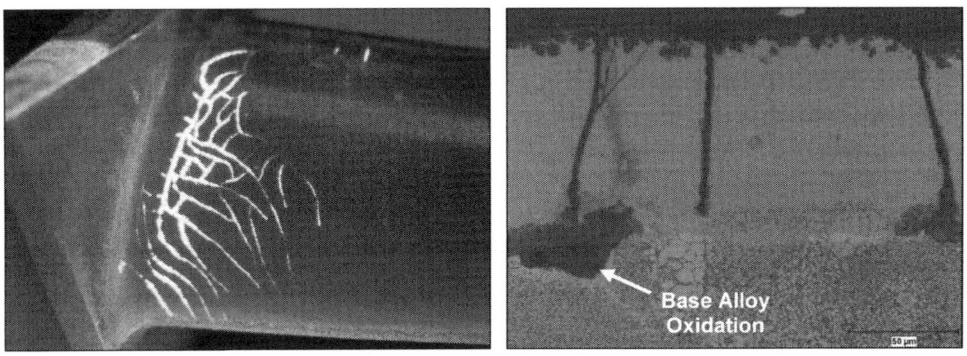

FIGURE 12–28 Photograph aided by fluorescent penetrant inspection techniques and an optical micrograph displaying coating cracks due to thermal cycling.

operating conditions applied. If neither the coating nor any other form of degradation has approached their life limit, there may be the potential to increase the service intervals.

Thermal Barrier Coating Loss

Thermal barrier coatings (TBC) are applied to internally cooled blades to insulate the base metal from the hot gas environment. Due to their porous nature, TBCs are applied overtop of a diffusion or overlay coating called a "bond coat." This bond coat provides the oxidation protection for the blade's base metal.

TBC coating spallation is generally attributed to the formation of a brittle oxide layer between the TBC and the bond coat. Spallation occurs when this thermally grown oxide (TGO) reaches a critical thickness and thermal stresses cause the oxide layer to crack and spall. Figure 12–29 displays a TGO adjacent to the spalled region of the airfoil.

Formation of this TGO is a function of the bond coat's oxidation resistance and operating temperature. Operating at higher temperatures increases the rate of TBC spallation initiation and continuation. Once TBC loss has initiated and no longer has the local insulation, neighboring coating loss is accelerated. TBC coating loss can also be initiated by impact damage. TBC coating loss by impact damage can be differentiated from TGO spallation by the sporadic nature of coating loss associated with impact damage.

Examination can determine the cause of coating loss and the extent of damage caused by the loss. Generally, TBC loss results in increased bond coat/surface and base alloy degradation due to a local elevated operating temperature. Repairs to the blade set may have to be adjusted pending the extent of base alloy damage. It should be noted that TBC coating is considered consumable and the coating performance is only considered to have been inadequate for service if the damage sustained to the blades is irreparable.

Hot Corrosion

Hot corrosion is an accelerated attack of the coating and base metal that occurs due to the presence of contaminants in the hot gas environment. In particular, elements that can cause hot corrosion include sulfur combined with either

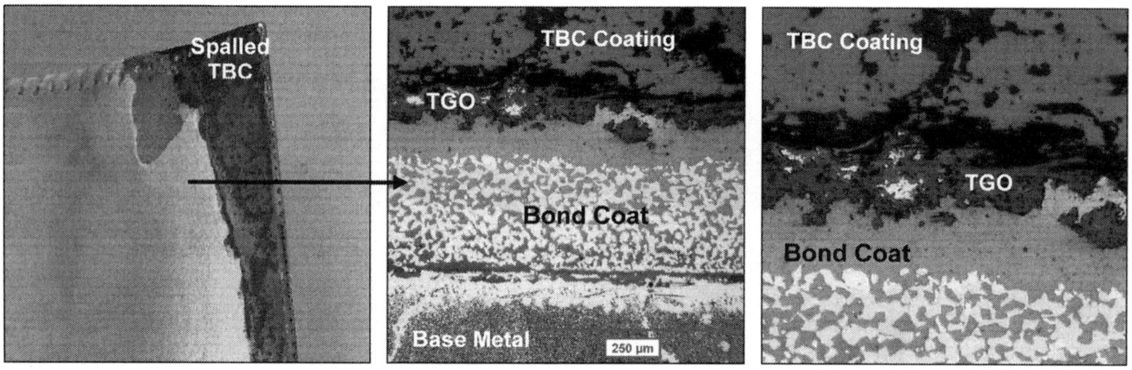

FIGURE 12–29 Photograph and micrographs displaying the TBC coating adjacent a spalled region. Spallation occurs due to the formation of a TGO layer and thermal stresses from engine starts/stops.

sodium or potassium salts or, independently, vanadium and lead. Unlike in clean environments, the presence of these elements interferes with the formation of a protective oxide layer and accelerates the rate of attack.

Two broad classes of hot corrosion have been recognized, divided by the temperature ranges at which they occur. Type I hot corrosion occurs above 871°C while Type II occurs between 649 and 871°C. Although the mechanisms differ, the causes, implications, and preventive measures are similar.

Metallurgical analysis can identify the type of corrosion and elements present causing the hot corrosion. Ideally, the prevention of hot corrosion is by the elimination of corrosion agents at the source. Corrosive elements can originate through either fuel or air intake. Seawater humidity is a constant source of these salts and hot corrosion by these salts can indicate filtration inadequacies (see Figure 12–30).

A less ideal option but at times more feasible is applying coatings that provide corrosion protection. Overlay coatings consisting of the protective elements chromium and/or cobalt can provide some measure of protection to the base alloy. Platinum aluminide coatings also exhibit corrosion protection. Determination of the type of hot corrosion, elements causing corrosion and the extent of corrosion should be taken into account when selecting coatings for corrosion protection.

Impact Damage

Impact damage is the result of object(s) impacting blade(s) at high speeds. Sustained damage can range from insignificant to coating damage to catastrophic failure. Evidence of impact damage indicates concerns upstream of the turbine blades. Impact damage can be a symptom of upstream component(s) failure or filtration inadequacies. It can also point towards poor maintenance practices such as leaving equipment/objects within the gas path during shutdowns.

Metallurgical analysis can assess the extent of damage, both surface and subsurface, if the coating has been compromised. Information can be used to determine the reparability of the blade set. Due to the sporadic nature of impact damage, inspection methods can be developed to salvage blades from those that are irreparable.

Airfoil Deposits

Deposit build-ups on airfoils can reduce efficiency or provide an indication of an upstream issue. Indirect damage to the airfoil can be sustained from cooling holes becoming

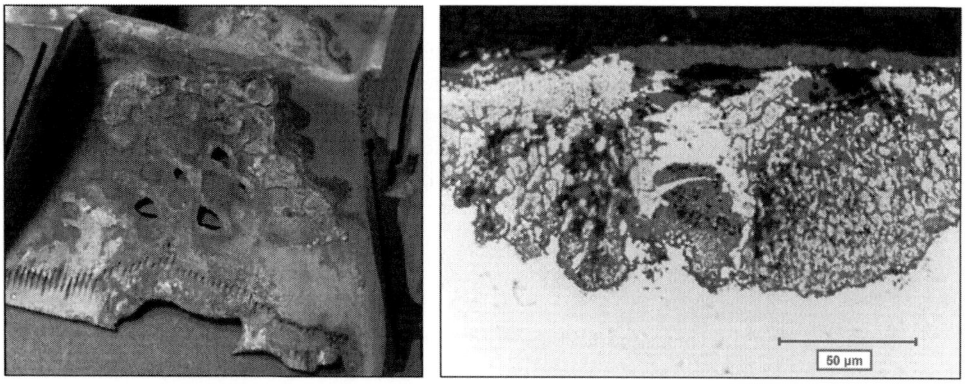

FIGURE 12–30 Photograph and micrograph displaying an example of hot corrosion.

blocked. The origins of deposits can range from ingested material, parts rubbing or upstream failures (melted and adhered metal).

Determining the composition of deposits building up on turbine blades can provide insight as to the material origins. Similar to impact damage, deposits can indicate issues with filtration, upstream component failures, or objects left within the gas path during maintenance. Adhered airfoil deposits generally do not directly affect the reparability of a blade set.

Internal Surface Condition

Due to the inaccessibility and nature of damage incurred within the internal cooling passages, internal damage to the base alloy cannot be repaired. The decision to continue servicing the parts should include an evaluation of whether the existing internal damage can be tolerated. As non-destructive testing cannot detect the majority of internal damage modes, to determine whether a set of internally cooled blades can be safely returned to service requires a metallurgical life analysis.

Uncoated Internal Surfaces

Blades that are not internally coated can exhibit multiple forms of damage. Uncoated surfaces can experience alloy depletion at the surface. Base alloy elements such as aluminum and titanium can diffuse to the surface leaving behind a surface layer void of gamma prime (Figure 12–31a). Excessive thickness of this depleted area results in both reduced fatigue strength and a decrease in effective wall thickness when considering mechanical loading of the component. Note that the depleted material is indistinguishable by ultrasonic wall measurements. Metallurgical analysis is the only method to determine the remaining, effective wall thickness capable of sustaining service loading.

Other forms of oxidation damage that uncoated, internal surfaces are susceptible to are intergranular oxidation (IGO) and stress assisted grain boundary oxidation (SAGBO).

An example of IGO is presented in Figure 12–31b. Oxidation occurs along the base alloy grain boundaries resulting in oxide spikes into the material. These oxide spikes are brittle and can lead to concerns of thermal-mechanical cracking.

Internal surfaces can sustain some damage and remain fit for continued service. Some blade designs have defined specifications as to assessing remaining life. Examination and monitoring of the internal damage is recommended until the damage is beyond tolerable limits.

Internally Coated Surfaces

Some blades are manufactured with internal diffusion coatings to protect the internal surfaces from oxidation damage. Internal coatings behave similar to the external coatings and degradation occurs due to the depletion of its aluminum content.

The only method to characterize the extent of internal coating degradation is by destructive microscopy. Assessment of whether the internal coating should be replaced prior to continued service can be made. It should be noted that due to the high temperatures of some heat treatments used to repair the base alloy that removal of the internal coating may be necessary regardless of its condition.

Hot Corrosion on Internal Surfaces

Hot corrosion within the internal cooling passages occurs by the same mechanisms as on the external surfaces. Due to the internal temperatures, hot corrosion generally occurs by Type II hot corrosion. The concern of hot corrosion is damage to the internal coatings and base alloy.

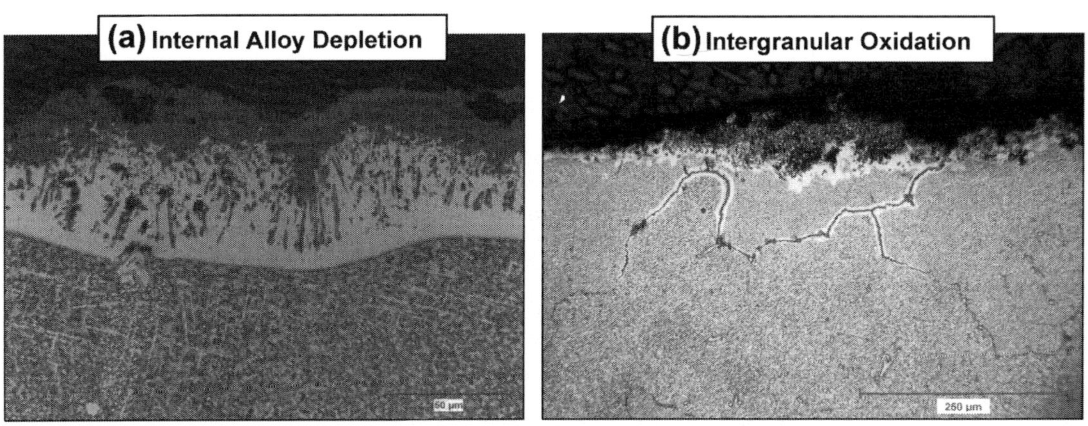

FIGURE 12–31 Optical micrographs displaying the internal surfaces exhibiting (a) alloy and oxidation damage and (b) intergranular oxidation (IGO).

The suitability of a blade set with internal hot corrosion for repair and continued service would be based upon the extent of hot corrosion. Only superficial internal hot corrosion damage would be tolerable. Any significant depth of internal hot corrosion damage would be cause for retirement.

Coatings that can be applied internally offer little corrosion resistance and therefore prevention of internal hot corrosion requires the elimination of the corrosion agent source. As the air used to cool the blades is taken from the compressor, the corrosion agents causing the hot corrosion on the internal passages are generally from the air intake. Hot corrosion on internal surfaces may indicate inadequate filtration for the engine operating environment.

Internal Deposits

Internal deposits can cause the blockage of cooling holes, disrupt cooling and accelerate local thermal degradation. As the deposit material was transferred through the cooling air, the source of the material is upstream of the compressor cooling air discharge.

Analysis of the deposits can determine the nature of the material and provide insight as to its origins. One common source is material ingested through the air intake and substantial deposits may indicate filtration inadequacies for the engine operating environment.

Use of Metallurgical Assessment Findings

Repair Decisions and Development

The information obtained through metallurgical examination can be used to (a) identify the reparability of the remaining blade set, (b) identify repair procedures necessary for continued service, (c) determine if a change in the coating type would be beneficial and (d) determine the cause of indications or conditions noted during inspections.

Metallurgical evaluation is by far the most thorough way of characterizing the sustained damage and therefore can be used to develop repair processes customized to the part and ensure the appropriate restorations are completed.

Risk Management

Risk is managed by making decisions to repair blades, how to repair the blades and ultimately to re-service blades based on a detailed understanding of the part condition.

A metallurgical assessment can provide a thorough characterization of the damage sustained, many modes of which are undetectable by non-destructive inspection techniques. This minimizes the risk of damage undetectable by other methods that may have implications upon the continued serviceability of the components.

Engine Operating Indicators

The feedback that metallurgical analysis provides to an engine manager varies over the life cycle of an engine program. Analysis of the first set of service run components is used to evaluate levels of base material degradation, coating deterioration, and assess for potential design deficiencies. This initial stage of investigation may be performed upon the engine fleet leader, new models or after-component upgrades/modifications that may alter the operating conditions.

Once a suitable data population has been obtained to adequately benchmark the typical blade deterioration rate for a fleet of engines, examination of subsequent blades can identify differences in engine operation. This can be used to detect abnormal operating conditions such as short-term over-firing, long-term over-firing, or a shift in the thermal profile within the engine. Regular monitoring of the fleet leading components will help to manage the life-limiting mechanism of those components. Continued metallurgical analysis of gas turbine components can identify conditions that may need to be addressed by the engine manager.

For example, re-occurring incidences of high cycle fatigue cracks on blade sets may warrant the review of operating records/practices to ensure that the engines are not being operated in speed zones known to result in detrimental resonant frequencies. The presence of continuous hot corrosion cases is another example the engine manager should attend to. Determination as to whether the corrosion agents responsible for the damage and the location of corrosion (internally and/or externally) will provide guidance to the engine manager to further investigate plant filtration or gas quality.

If minimal to moderate degradation was sustained, there may be potential for a service interval extension. Further work and assessment of other critical parts would be necessary to positively identify an interval extension. Engine models can be developed based on this assessment so that extended intervals could be reliably managed based on operating conditions. Needless to say significant maintenance cost savings could be realized if the service intervals could be extended.

Case example 1

A combined-cycle facility running four GE MS7001 EA engines had been uprated from a 2020°F to a 2055°F firing temperature. The engines experienced extremely rapid part deterioration and the facility decided to return to a 2020°F firing temperature. Upon returning to what was thought to be the same 2020°F firing temperature that was used before the uprating the components still experienced relatively rapid deterioration.

This facility had been conducting life analysis on turbine blades as a standard practice for over a decade. The standard deterioration rate at the original 2020°F firing temperature had been well benchmarked. Examination of components operating at the 2055°F firing temperature found a substantial increase in both the base alloy and coating deterioration and it was concluded that the parts were unlikely to survive a full 24,000 hour service interval. However, examination of the components after attempts to return to a 2020°F firing temperatures still found a relatively fast base alloy degradation rate compared to the benchmarked components. Life analysis evaluation concluded the buckets were operating at higher temperatures than the original 2020°F temperature and, therefore, the firing temperature had not been successfully returned to the preuprating condition.

In response to the life analysis results, modeling was conducted that found that the exhaust temperature control curve had underestimated the firing temperatures since the uprating. Improper control curve constants had underestimated both the 2055°F firing temperature as well as the attempts to return to the 2020°F firing temperature.

The OEM ultimately had to apply corrections to the control curves as a result of the investigation. The engine operator had concrete evidence of the incorrect firing temperature as a result of the metallurgical analysis program that he had subscribed to over the life of his components.

Case example 2

A set of Rolls-Royce RB211 24G HP turbine blades was submitted for repair. The blade had a total of 49,000 hours since new and 28,000 hours since last repair. During operation, the engine had experienced a thermal spread condition and, to avoid trips, several thermocouples had been disengaged. Inspection found several blades to exhibit thermal-mechanical fatigue cracks on the lower airfoil. To determine the extent of damage, two blades were selected for metallurgical analysis.

Optical examination of the damaged region confirmed the damage was typical of thermal-mechanical fatigue. Both blades exhibited cracks initiating on the external and internal surfaces as illustrated in Figure 12–32b.

The overall base metal degradation was considered more advanced than typical for blades of similar service

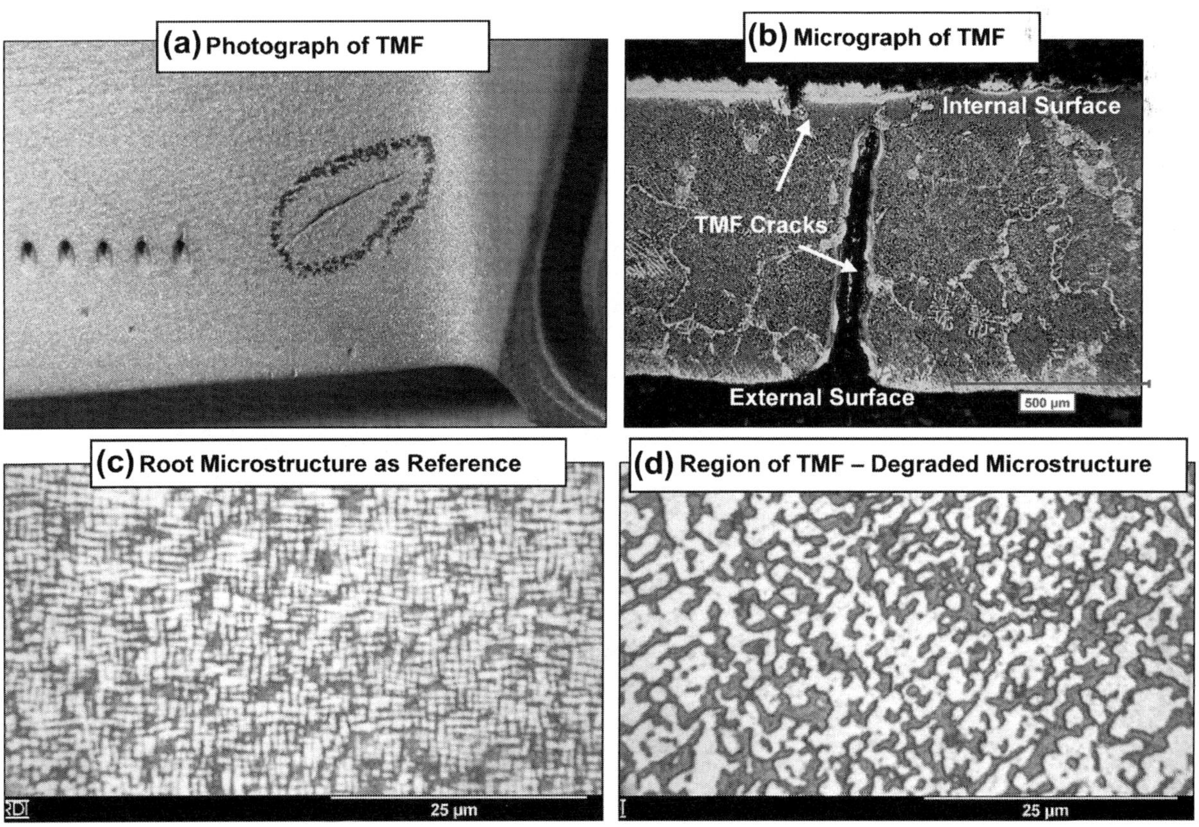

FIGURE 12–32 Optical micrographs displaying the TMF cracking and microstructural degradation relative to the root structure of RB211 24G HP Blade.

hours. The region of TMF damage exhibited significant base alloy degradation indicating this area had a relatively high operating temperature. This elevated operating temperature would have (a) resulted in large thermal changes during engine starts/stop and (b) have reduced the material strength. Both factors would have contributed to thermal cracking.

Without destructive testing and the detection of the internal TMF cracking, a repair shop could run the risk of basing reparability upon external features alone. In this example, only a portion of the blades exhibited external cracking. However, because internal TMF cracking was detected, the entirety of the blade set was recommended for retirement. Also, the engine operator was made aware that the advanced degree of base metal degradation indicated the engine had been over-firing, which had increased the blades' susceptibility to TMF.

Case Study 10: Evolution of IGT "F" Class Repair Technology**

Starting with OEM replacement philosophy based on theoretical design calculations, the gas turbine industry has evolved to a condition-based criteria for repair and life extension of expensive critical hardware. By using proper metallurgical and mechanical analyses and design criteria, it is possible to develop safe repairs capable of maximizing the life of the blades and vanes. However, significant differences exist in the execution, technology and standards used by various repair facilities, which have resulted in confusion and discrepancies in the ultimate life and reliability of repairs. The need for some form of industry standard is clearly evident, especially for "F" class component repairs, where safety margins are low and technology requirements are high. Experience has demonstrated that excellent results can be obtained in a variety of "F" class components, which should reduce the dependence on expensive long-term service agreements.

In hindsight, looking at practices employed in the 1960s and 1970s to assess the life and replacement of hot section components from such engines as the W101, W191, Frame 3, and Frame 5, we can see how conservative the OEM's were and the resultant waste in replacement parts. The basic method used by the OEMs was to destructively test sample blades at their overhaul interval and reject anything that did not meet the new material specification. Often, blades were scrapped because the stress rupture elongation fell below the specified 5% minimum and designers did not know the implications of operating with a lower arbitrary value. Most components were automatically replaced when they reached their theoretical creep life limit, which ranged from 40,000–100,000 hours on the turbines operating at 760°C (1450°F). Subsequently, these same components were rejuvenated and obtained lives in excess of 200,000 hours.

Some of the operators and OEM engineers started to question the need for automatic component replacement in the early 1970s and started to develop improved analytical techniques for examining micro-structural aging and heat treatment reactions that led to development of the rejuvenation process at Westinghouse Canada. The first paper coining the term "rejuvenation" to describe the successful reheat treatment of super alloys was presented in 1977 by Westinghouse. However, it was Chromalloy that led to the commercialization by repairing Frame 5 second stage buckets with a creep life limit.

Most of the repair processes used at the start were adapted from aero repairs from such engines as the P&W JT8D, which were conservative in nature and regulated by the FAA. Since no such restrictions were present in the industrial market, every new entrant tried to imitate, as well as push, the repair limits to try to recover more parts. Ultimately this led to varying standards for the same part repair and in some cases in failures.

The problems with the IGT market were aggravated in the 1990s with the entry of several new players and spin-off companies in the Houston area, lured by the growing installed base and flawed predictions as to their overhaul requirements. This coincided with the introduction of the "F" class turbines with their higher firing temperature and more sophisticated component designs. Operators concerned with the level of technology available in the aftermarket and the maintainability of their units sought refuge in OEM long-term service agreements.

Following the ENRON correction, there was a significant drop in demand for repair services and companies started to compete more on price than technology. In some cases, repairs have become a commodity item, with little regard to technical subtleties and long-term life. Caution must be exercised when applying relaxed standards to advanced components since it can lead to failures.

The best value is obtained by applying the appropriate level of understanding and technology to obtain a repair that provides the ultimate life for the asset in question. For example, a Frame 7 EA first stage bucket can be repaired by several methods that all appear to be alike, but may be different in significant aspects relating to how well the microstructure and the coatings are restored to a "like new" condition. For example, by following conventional practices that simplify the processing by not stripping the internals and by not doing a full high temperature solution treatment, the ultimate life of the bucket is limited to about 48,000 hours. However, by engineering a complete rejuvenation of the structure and coatings, the life for the same

** Source: Courtesy Liburdi Turbine Services, from J. Liburdi and P. Lowden, "Evolution of IGT 'F' Class Repair Technology," Turbine Forum 2005, Nice, April 20–22, 2005.

component can be extended to over 100,000 hours with resultant savings in overall replacement cost estimated in the millions.

For the "F" class components, the question of value and repair technology is even more important. By using advanced welding, powder metallurgy, coating, and rejuvenation techniques, it is possible to extend the life of these expensive components and lower the operating costs to levels significantly lower than LTS Agreement levels. This can be best illustrated by the examples of successful repairs that combine advanced aero experience with the knowledge of IGT design and metallurgical processes.

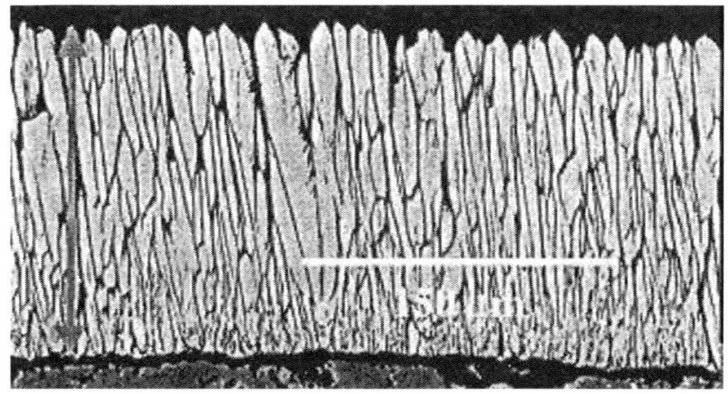

FIGURE 12–33 The F generation engines were the first to employ thermal barrier coatings on rotating components to lower metal temperature during operation. *(Source: Liburdi Engineering.)*

Technology

The high firing temperatures in F technology engines are achieved by the combination of several different material and design technologies. While the resulting performance improvement is a quantum increase over previous engines, the technologies themselves have mostly been adapted from those that were in use in older industrial and aeroengines. From a repair perspective, this means that the techniques used in repair of F class hardware have mostly been used previously on either industrial or aero hardware.

Features of F Class Hardware

Advanced Coatings The F class of engines makes more extensive use of coatings than previous generations of engines:

- F technology engines are the first industrial designs to use thermal barrier coatings (TBCs) on rotating airfoils. Previously, the use of TBC coatings in industrial designs was restricted to stator vanes and combustion hardware applied by air plasma spraying. To meet the demands of coating the more critical rotating blades, advanced techniques such as electron beam evaporation and high energy plasma spraying whose use had been restricted to aero components in the past are now used to generate more strain tolerant coatings (Figure 12–33).
- Coatings are applied to downstream stages that were uncoated in earlier designs. The coatings are the same type of MCrAlY coatings that were used in the first stages of previous engines.
- Internal coatings play a more critical role. The thin walls associated with the complex serpentine cooling passages result in higher surface temperatures on the cooling hole walls and therefore need protection against oxidation. The coating processes, however, are the same processes used in earlier industrial and aero designs.

Single Crystal and DS Materials The first use of single crystal alloys in industrial gas turbines is found in F technology engines (Figure 12–34). While such alloys have been used for many years in aero-engines, the challenge of casting larger industrial components prevented their use in industrial engines until recently. The F technology engines also see much more extensive use of directionally solidified

FIGURE 12–34 Single crystal alloys used in F technology engines have higher creep and low cycle fatigue strength then the conventional cast alloys used in older engines. *(Source: Liburdi Engineering.)*

alloys, which saw use only in the first stage blades of some of the previous generation of industrial engines.

Complex Geometry The geometry of components used in the current generation of industrial engines is more complex than in previous generations. The schemes used to cool the airfoils feature serpentine passages and surface film cooling holes similar to those used previously in aircraft parts. Also, the geometry of combustor hardware has increased in complexity due to the demands of increased cooling and lean premixed combustion to achieve low emissions at higher turbine inlet temperatures.

Repair Technologies

Stripping Since coatings were first introduced to protect turbine airfoils, stripping has played a central role in repair processing. First, coatings are often largely consumed by environmental attack in service. Chemical stripping is the first step towards replacing the spent coating. Second, for most coated airfoils, it is essential that the coating be removed prior to any solution heat treatment in order to prevent excessive diffusion and rumpling of the surface. The acid solutions used to dissolve the aluminum-rich phases in the coating are very aggressive. To prevent intergranular attack of the base material or dissolution of Ni_3Al (γ') eutectic islands in some of the newer high strength superalloys, the temperature and compositions of solutions and immersion times must be carefully controlled.

The ability to strip coating from internal passages is a requirement for both F technology components and earlier generation hardware. However, because of the higher temperatures at which the F components operate, coating removal is necessary to replace exhausted internal coatings, whereas in earlier hardware, it was typically necessary to allow full rejuvenation heat treatment. To accomplish this, a modern chemical stripping process must be able to flow the hot solution in a controlled manner through the internal cooling passages, in order to remove any internal coatings, while simultaneously stripping the thicker external coatings. Special care must be taken to prevent chemical attack on the critical blade root surfaces and inside internal cavities. An experienced technical staff is required to safely operate and maintain such advanced stripping facilities.

A new stripping need has developed in the F engines due to the higher temperatures in the cooling passages. At these temperatures, thick layers of oxides can form during service. These oxides are not dissolved by the solutions used to remove coatings and act as a barrier to both coating stripping processes and re-coating operations. On the external surfaces, this is not a problem, since the oxides can be removed by abrasive cleaning (grit blasting). However, on the internal surfaces a chemical approach must be used.

Advanced cleaning techniques are now being qualified to clean the internals and provide superior coating protection for the internal cavities.

Rejuvenation Vacuum heat treatments are traditionally used during the repair process to restore the microstructure, improve weldability, apply brazes and diffuse coatings. Significant metallurgical knowledge and experience are required to tailor the heat treatment steps to the superalloy being processed. To ensure that the creep, fatigue, and aging damage incurred during service are reversed as much as possible, and that the correct microstructure and mechanical properties are produced after all welding and coating steps have been completed, a full solution heat treatment, graphically illustrated in Figure 12–35, is essential.

The benefits of regularly performing such rejuvenation in combination of other repair steps have been well established on earlier generation engines. By restoring the microstructure through heat treatment on a routine basis, the life of the components can be extended well beyond the life at which they would otherwise need to be retired (Figure 12–36). For example, the use of rejuvenation at appropriate intervals for Frame 7EA first and second stage buckets can extend their life to more than twice that achievable through re-coating only (Figure 12–37). The economic benefit at a site operating four units was calculated to be approximately \$5 million over a 5-year period.

DS and Single Crystal Alloys The same principle of extending the life of hot section components by rejuvenation heat treatments has been applied to directionally solidified components as they have been introduced in the earlier industrial and aero engines. The heat treatments function in the same way to restore the gamma prime microstructure and related mechanical properties.

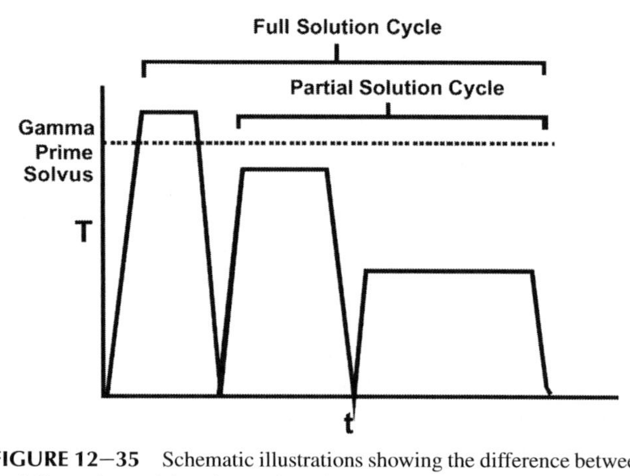

FIGURE 12–35 Schematic illustrations showing the difference between a full and partial rejuvenation heat treatment. *(Source: Liburdi Engineering.)*

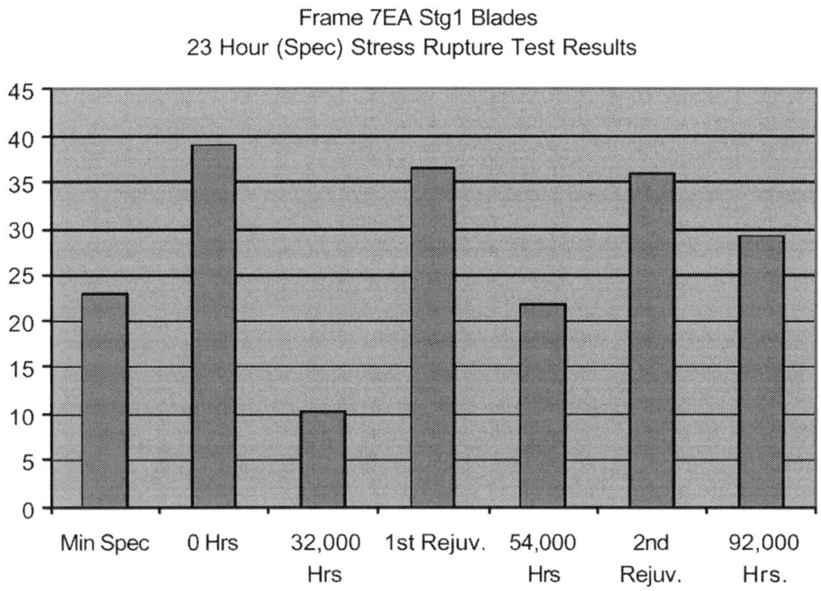

FIGURE 12–36 Stress rupture test results showing the degradation of properties and their rejuvenation. *(Source: Liburdi Engineering.)*

An additional concern for DS and single crystal alloys is the possibility that recrystallization will take place during heat treatment if the components have been subject to plastic deformation. These alloys develop their strength in part from controlling the orientation of grain boundaries or, in the case of single crystals, eliminating them all together. Figure 12–38 shows recrystallization of a single crystal alloy subjected to shot peening to introduce surface deformation. If such damage were to occur on an airfoil surface, for example, as a result of impact, the LCF and creep properties would be reduced. Rejuvenation of DS and single crystal alloys must therefore include examination of the etched airfoil surfaces for evidence of recrystallization after the heat treatments.

Repair Materials

Welding Many repairs involve the replacement of missing material or repair of cracking. Welding is the most common method used to repair, join, or replace material in hot section components. The conventional means of depositing material is by fusion welding, which has been used for repair of airfoils since the 1970s. The process can vary from manual TIG to automated plasma- or laser-based systems with the best results obtained from tightly controlled integrated equipment.

With the more readily weldable alloys used in combustors and stator vanes, extensive weld repair of these components is possible, although distortion can become a problem. However, with the higher strength alloys used for rotating blades, the tendency of the base metals to crack is a serious limitation. For this reason, repairs are typically limited to the low stressed regions at the tip of the blades. Furthermore, in more recent engines, including the F class engines, oxidation resistance has also become an important issue.

The experience with the RB211 HP blades shown in Figure 12–39 represent a good example of how the use of superior welding and materials technology can be used to produce a longer life repair, by harnessing the use of higher oxidization resistant materials that are extremely difficult to weld. The same processes were subsequently applied to successfully restore the tips on the Frame 7FA and V84.3A blades discussed below.

Selection of weld filler materials is a balance of the properties required in the joint and the weldability needed. Ideally, welds would be performed with filler metal compositions matching those of the base material, which typically have strengths approaching those of the cast alloy. This can typically be achieved with the cobalt alloys used in many stator components and weaker nickel alloys used in combustors, because the alloys are fairly readily weldable. However, for higher strength nickel-based alloys used for rotating blades, weldability becomes a serious problem and most repairs are performed with weaker alloys, such as Inconel 625, that have good weldability. By limiting repairs to low stressed regions, this approach has been used to successfully extend the life of older generation blades.

In the 1990s, as firing temperatures increased, first in aeroengines and then in the F class engines, oxidation began to become a factor in weld alloy selection. The lower strength alloys used for conventional repair have lower aluminum content and are therefore less oxidation resistant.

FIGURE 12—37 Photo of 90,000+ hour bucket. Originally from engine serial no. 13 of the 7EA fleet. *(Source: Liburdi Engineering.)*

For example, new Rolls Royce RB211-24C high-pressure turbine blades (Figure 12—39) typically lose material from the shroud edges by oxidation after one overhaul cycle (24,000 hours), necessitating weld repair. However, experience with conventional weld repairs showed that the repaired shroud edges oxidized at an accelerated rate, resulting in their removal from service for further repair after 8000 hours (Figure 12—40).

To select an appropriate weld filler material for such oxidation limited applications, candidate alloys with higher aluminum content were compared to the base material and conventional repair materials by cyclic oxidation testing. The higher aluminum alloys have significantly better oxidation resistance. However, the weldability of these alloys is significantly less than the conventional materials and cannot be reliably applied by manual welding because of cracking problems.

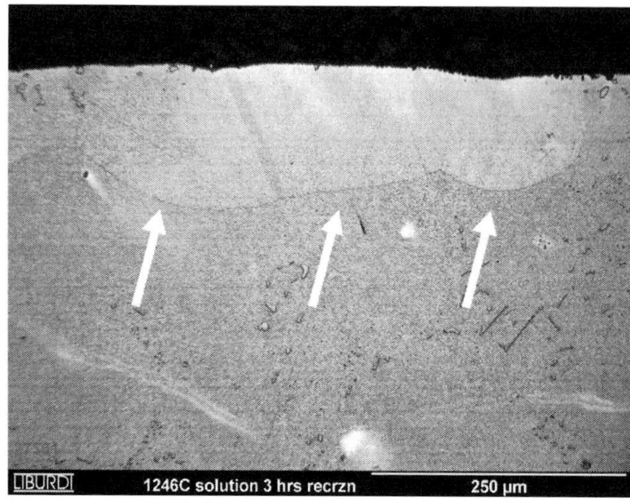

FIGURE 12—38 Micrograph showing recrystallization of a single crystal alloy subjected to shot peening to introduce surface deformation. *(Source: Liburdi Engineering.)*

It has been shown that automating welding processes for superalloys reduces their tendency to crack. The automated techniques were therefore developed for repairs with these oxidation resistant fillers. This approach was used to repair the RB211-24C high-pressure blades, with the result that the repaired blades last to the 24,000 overhaul interval in better condition than new hardware (Figure 12—41).

LPM™ Powder Metallurgy In many cases, welding is not desirable because of excessive distortion or cracking considerations. For this reason, there has been considerable effort in developing non-fusion weld repair techniques. Over the past two decades, many wide gap and modified wide gap processes have been developed and employed in the repair and manufacture of hot section gas turbine

FIGURE 12—39 Photograph showing oxidation of the shrouds on RB 211 24-C high-pressure turbine blades after 24,000 hours of operation. *(Source: Liburdi Engineering.)*

FIGURE 12–40 Oxidation of conventionally repaired RB211 24-C high-pressure turbine blade shrouds after 8000 hours of operation. *(Source: Liburdi Engineering.)*

FIGURE 12–41 Oxidation of RB211 24-C high-pressure turbine blade shrouds repaired with oxidation resistant fillers after 24,000 hours of operation. *(Source: Liburdi Engineering.)*

components; G.E.'s advanced diffusion healing (ADH), Snecma's rechargement per brasage diffusion (RBD), Howmet's effective structural repair (ESR), Chromalloy's surface reaction braze (SRB), Turbofix, Sermafill, and Liburdi Engineering's LPM™. All of these processes use high and low melting components to limit the amount of melting point depressant and to help bridge large gaps. However, the patented LPM™ process is the only modified wide gap process capable of filling large cavities resulting from the removal of structural cracks or casting core holes. In particular, the LPM™ process has been used to rebuild components that are missing large amounts of material while maintaining very low levels of melting point depressants.

The LPM™ process is a multi-layer process incorporating high and low melting metal powders to effect sintering and controlled liquid phase bonding. The powders are mixed with organic binders to produce flexible tapes or clay-like putties. The component is then heated slowly in a vacuum furnace to burn off the binders and finally heated to and held at a temperature between the melting points of the high and low melt materials to promote liquid phase sintering, transient liquid phase solidification (isothermal solidification) and diffusion bonding (Figure 12–42). The resulting deposits show excellent mechanical strength, creep strength, and ductility when compared to conventional wide gap deposits, without the need for lengthy diffusion cycles.

Because of its ability to bridge such wide gaps, the LPM™ process can be used to repair very large, structural cracks as well as to build up material on worn surfaces for dimensional or throat area restoration. Since the liquid phase consolidation takes place while the part is at a uniform temperature in the vacuum furnace, there is no lateral shrinkage or distortion. Since its introduction in the early 1990s, the process has been used to repair and upgrade numerous hot section vanes, blades, combustors and transitions.

Experience with the repair of the frame 7EA nozzles demonstrates the effectiveness of the LPM™ process. After an initial 24,000-hour service interval since new, these FSX 414 cobalt alloy nozzles typically show extensive cracking especially around the leading edges, trailing edges and between the two airfoils on the outer shroud. The cracking is primarily caused by low cycle thermomechanical fatigue. The cracking at the leading edges is branched and extended through to the internal vane cavities. Some oxidation erosion and associated cracking also occurs on the outer and inner diameter shrouds between the airfoils. Because of the degree of damage, conventional weld repairs would either crack or result in excessive distortion of the nozzles.

In preparation for LPM™, the cracks are ground out and the eroded surfaces are blended back to unaffected base alloy. This leaves large holes ranging from 25 mm × 100

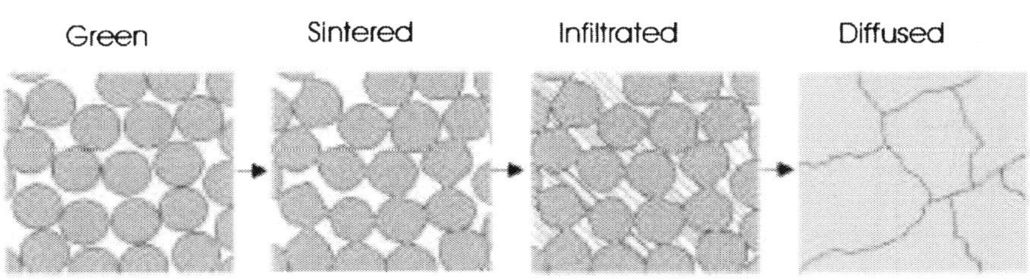

FIGURE 12–42 Schematic illustration of the bonding process that takes place at elevated temperature during LPM™ repair. *(Source: Liburdi Engineering.)*

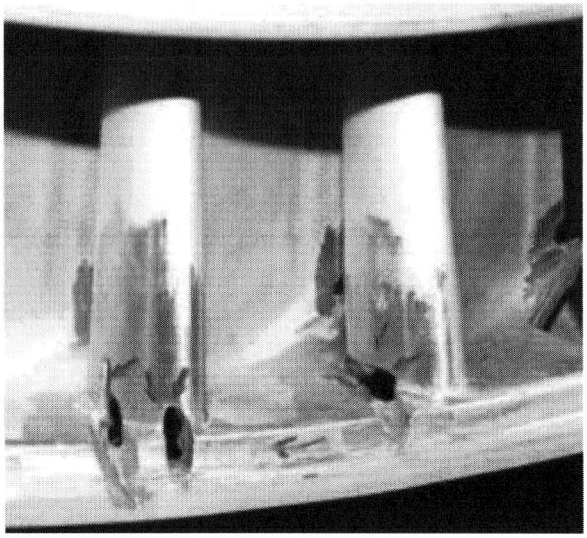

FIGURE 12–43 Frame 7EA first stage nozzle with cracks removed for LPM™ repair. *(Source: Liburdi Engineering.)*

mm in size at the leading edges of some airfoils to areas larger than 50 mm × 50 mm requiring 0.75 mm build-ups (Figure 12–43). The large holes, which are typical of vane repairs where the structural cracks are removed, were filled with LPM™ putty that was sculpted to the original contour of the nozzle. The eroded areas were covered with a 1 mm thick LPM™ tape to restore the dimensions. Small cracks and impact damage were repaired with the LPM™ putty (Figure 12–44). During repair these areas were successfully repaired using MarM247-7 LPM™.

After the second service interval of approximately 24,000 hours (~48,000 hours total operating time), the nozzles do not show the large thermomechanical fatigue cracks at the leading edges or between airfoils on the outer shroud surface formed in the new parts. This suggests that

FIGURE 12–44 Frame 7EA first stage nozzle with LPM™ putty applied to crack areas and LPM™ tape applied to eroded areas. *(Source: Liburdi Engineering.)*

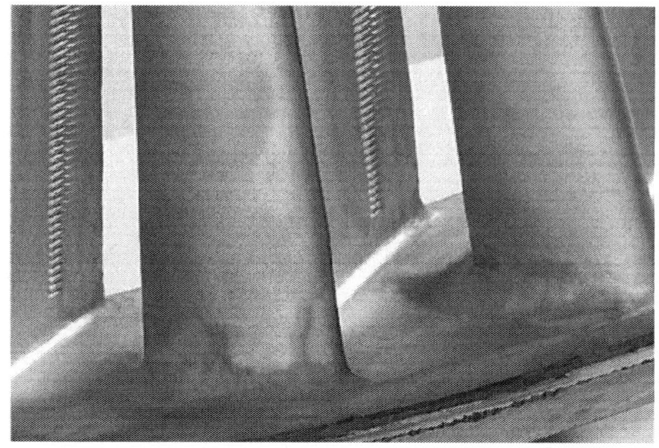

FIGURE 12–45 Photograph of nozzle after 48,000 hours and one repair cycle. The darker areas are LPM™ material. *(Source: Liburdi Engineering.)*

the LPM™ had improved the performance of these nozzles due to the improved strength and low cycle fatigue resistance of the LPM™ repair (Figure 12–45).

After 48,000 hours of service, the nozzles had small cracks on the shrouds, oxidation erosion on the internal and external surfaces, and thinned trailing edges. The thinned trailing edges were particularly of concern as the throat area between the airfoils had increased significantly due to material loss during service and repair. This increase in throat area can lead to harmonic imbalance and efficiency loss. As part of the second repair the small cracks and oxidation damage were blended in preparation for LPM™ application and the thinned trailing edges and oxidized areas were thickened by applying LPM™ tape to both surfaces. The cracks were repaired using LPM™ putty. After the second repair the nozzles were again returned to service. The nozzles have since been returned for an LPM™ repair at ~72,000 hours. After repair, the nozzles will be returned for an additional service interval.

Coatings Coatings are an integral part of any repair process. In principle, the application of coating during repair is no different than during the original manufacture. However, the overhaul cycle presents an excellent opportunity to review how well the original manufacturer's coating is performing and substitute a more appropriate coating where possible or apply coating to previously uncoated surfaces. For example, the original GT 29 coating applied to the frame 7EA first stage buckets had poor oxidation resistance, while the replacement GT 29 plus coating was too brittle and cracked in service. The repair of these buckets involved the application of a more ductile and oxidation resistant over-aluminized NiCoCrAlY by GE.

A properly engineered repair will specify a proven-McrAIY coating appropriate to the engine operating environment, followed by an internal and external aluminide coating. Where needed, a thermal barrier coating is applied over the MCrAlY layer to provide an added level of insulation.

F Class Repairs Just as the technology used in the manufacture of the F class components evolved from those used in the previous generation of turbines, the repair of these components use the same technologies that have been developed and proven over the last 20 years. The typical repair workscope still involves the following basic steps; incoming inspection, removal of coatings, repair of missing material by welding or LPM™, rejuvenation heat treatment, and application of appropriate coatings. However, as in all repair applications, it is essential that the damaged mechanisms are well understood so that appropriate repair techniques and materials are chosen.

The following examples illustrate the application of repair to the latest generation of turbine hardware.

GE Frame 7F

First Stage Buckets The first stage buckets in the GE frame 7F engine are manufactured from directionally solidified GTD 1 1 1 alloy. They feature a complex serpentine cooling scheme and are coated with an over-aluminized NiCoCrAlY coating on the external surfaces and a simple aluminide coating on the internal passages. The most typical damage requiring repair is oxidation and cracking of the tip (Figure 12–46) and general consumption of the coating life.

The repair workscope typically involves the following steps:

- Incoming inspection
- Strip coatings (internal and external)
- Fluorescent penetrant inspection

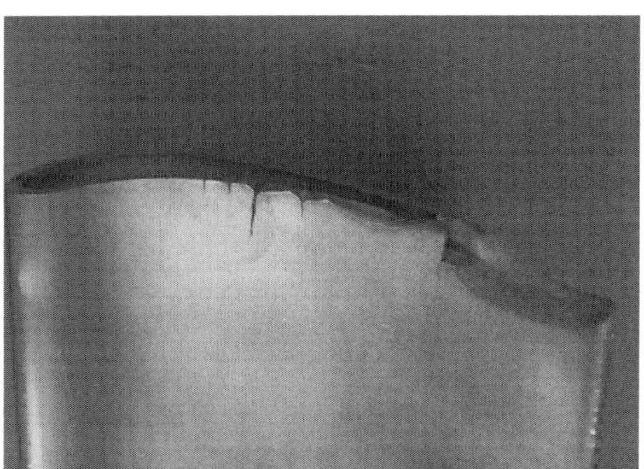

FIGURE 12–46 Cracking and oxidation damage at the tip of a frame 7F first stage bucket. *(Source: Liburdi Engineering.)*

- Tip weld repair with oxidation resistant filler (Figure 12–47)
- Finish to contour
- Rejuvenation heat treatment
- Internal and external coating
- Final inspection

Second Stage Buckets The second stage buckets on the frame 7F suffer from a similar problem as the 7EA buckets. In both engines, the shrouds lift due to the centrifugal forces acting on them, resulting in misalignment of the Z notch contact faces. The repair developed for this type of damage involves applying material by the LPM™ process to realign the contact faces and restoring the contacts surfaces by hard face welding (Figure 12–48). During this repair, it is possible to realign the shroud to lower the stresses and reduce the tendency for future shroud lift.

FIGURE 12–47 Weld repair of tip damage on a frame 7F first stage bucket. *(Source: Liburdi Engineering.)*

The typical workscope for repair of these buckets would include:

- Incoming inspection
- Strip coating
- Fluorescent penetrant inspection
- LPM™ reinforcement
- Weld hard facing
- Rejuvenation heat treatment
- Finish to contour
- Reapply coating
- Final inspections

Nozzles While the geometry of the nozzles in the frame 7F turbine are different than those in the EA model, the materials and construction and general features are very similar. The repair of these components, therefore, is similar to that described for the 7EA components described above. The differences in geometry, of course, require different tooling (Figure 12–49).

The typical workscope for repair of 7F nozzles includes:

- Incoming inspection
- Remove cooling inserts
- Strip coatings
- Fluorescent penetrant inspection
- Welded and LPM™ repair
- Finish to contour

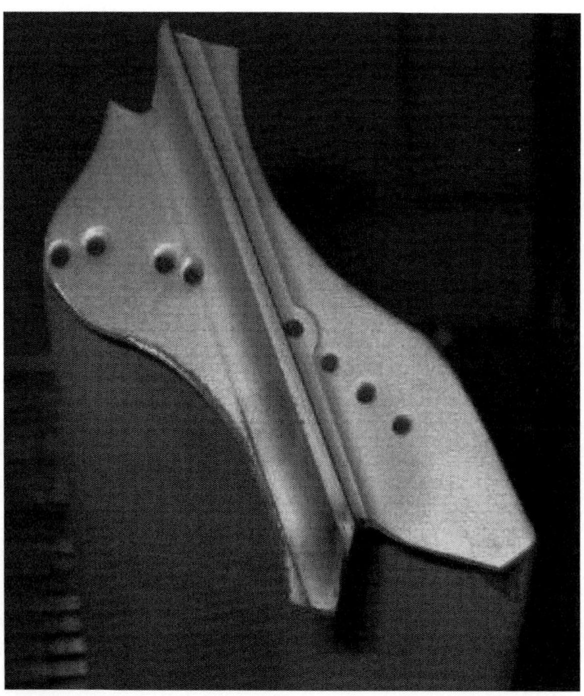

FIGURE 12–48 Photograph of frame 7F second stage bucket after hard face buildup. *(Source: Liburdi Engineering.)*

FIGURE 12–49 Frame 7F nozzles being machined to final diameter. *(Source: Liburdi Engineering.)*

- Rejuvenation heat treatment
- Coating
- Final inspection

Combustors Since the materials used in the manufacture of combustors are readily weldable, repairs are typically based on welding cracks and missing material. However, there are many applications where distortion may be an issue, such as when surfaces need to be built up over a large area, and in these instances, the LPM™ technique is more suitable. The principal difference between the F class combustors and earlier models is the increasing complexity in geometry. For this reason, having appropriate gauging and tooling is increasingly important (Figure 12–50).

FIGURE 12–50 Frame 7F transition pieces been checked in an alignment fixture. *(Source: Liburdi Engineering.)*

A typical combustor workscope would entail:

- Incoming inspection
- Disassemble
- Strip coatings
- Fluorescent penetrant inspection
- Welded and LPM™ repair
- Finish to contour
- Rejuvenation heat treatment
- Coating
- Final inspection

Siemens V.84.3

Rotor Blades The first two stages of rotating blades in the Siemens V.84.3 are manufactured from a single crystal alloy PWA 1483 and are coated with an MCrAlY coating on the external services and an aluminide on the internal surfaces. The remaining rows of blades are made from conventional cast alloys. All of the blades are of an unshrouded design, with the first stages employing a serpentine cooling scheme.

The rotor blades in the Siemens V.84.3 exhibit many of the same types of damage observed in the GE frame 7F first stage buckets, including tip cracking and oxidation. In addition, these blades suffer from short cracks along the trail edges. The repairs therefore are very similar in scope to those employed on the 7F buckets. Because of their location, the cracking the along the trailing edges can only be repaired by blending away the damaged material.

The repair workscope for the Siemens V 84.3 rotor blades involves the following steps:

- Incoming inspection
- Strip coatings (internal and external)
- Fluorescent penetrant inspection
- Tip weld repair with oxidation resistant filler
- Finish to contour
- Rejuvenation heat treatment
- Internal and external coating
- Final inspection

Stator Vanes The stator vanes in the V.84.3 design differ from those in the GE design in several ways. In addition to the obvious geometry differences, these vanes are manufactured from high strength nickel-based superalloys used only for rotating components in GE designs.

From our repair perspective, this means that repair by welding techniques will not be practicable. While welding is possible, the tendency for cracking to occur within the base material will limit the yield achievable by a welding process. By contrast, the LPM™ process avoids the stresses associated with the welding process making full yield repairs possible. Figure 12–51 shows trailing edge cracking in a V.84.3 second stage vane during the LPM repair process.

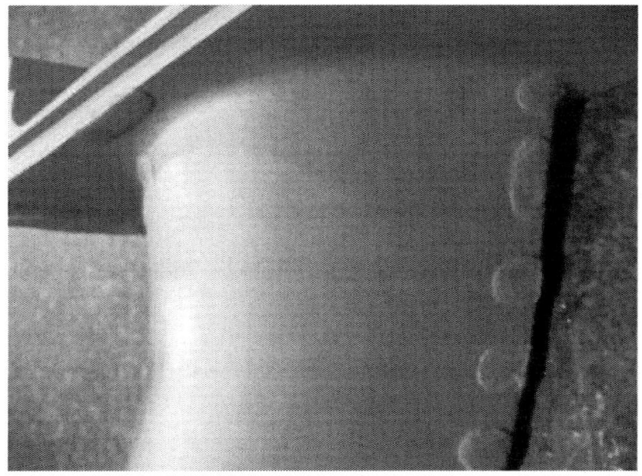

FIGURE 12–51 Photograph of LPM™ deposits in previously cracked areas after consolidation in treatment. *(Source: Liburdi Engineering.)*

The typical workscope for repair of V.84.3 stator airfoils includes:

- Incoming inspection
- Remove cooling inserts (as required)
- Fluorescent penetrant inspection
- LPM™ repair
- Finish to contour
- Rejuvenation heat treatment
- Coating
- Final inspection

Alstom 11N2

Rotor and Stator Airfoils The Alstom 11N2 has a lower firing temperature than either the GE or Siemens design and does not employ single crystal or directionally solidified materials in the hot section. Nonetheless, it employs the same coating systems, namely MCrAlY coatings on rotating blades and thermal barrier coatings on the stationary airfoils. It is also subject to the same types of degradation and the repairs are therefore similar in scope.

The repair of the first stage blades from these engines is a good example of using repair as an opportunity to improve the components. Unlike the Siemens and GE hardware, the first stage blades as originally supplied by Alstom do not have a coating on the internal cooling passages. As a consequence, significant oxidation typically occurs during service. Ideally, it would be beneficial to provide an internal coating at repair to extend the life of the component. However, the oxide layer acts as a barrier to the formation of the aluminide coating.

To remove the oxide layer, a unique stripping process was developed. The process effectively removes the oxide layer so that aluminum can come in contact with the substrate to form an aluminide coating during the slurry coating process. This aluminide layer will prevent oxidation from becoming a life-limiting factor in continued operation and thus will extend the total life of the blades.

Combustor Hardware Because the combustor in the 11N2 engine is of a silo type, the components appear to have little similarity to those in a GE engine (Figure 12–52). However, in both instances, the components are manufactured from readily weldable materials, so the repair workscope is actually very similar. The components typically suffer from wear and cracking, which can be repaired by welding. Surfaces that are protected by a thermal barrier coating will have the coating removed and reapplied as part of the repair process.

The typical workscope for repair of 7F nozzles includes:

- Incoming inspection
- Strip coatings [as applicable]
- Fluorescent penetrant inspection
- Welded and LPM™ repair
- Finish to contour
- Rejuvenation heat treatment
- Coating
- Final inspection

Conclusions

The F class of engines operates at significantly higher firing temperatures than previous generations. To achieve this they use a combination of advanced materials, coatings, and component design features. However, the earlier industrial gas turbine designs incorporated many of these features individually and significant experience had also been accumulated in aeroengines. For this reason, the designs can be seen as evolutionary rather than revolutionary.

By the same token, the development of repairs for the F class of engines has been an evolutionary process. The techniques used to repair the previous generation of components have largely been applicable to the F components. By employing these techniques, it is possible not only to economically extend the life of the components, but, in some instances, to improve the components through the use of more oxidation resistant materials and coatings.

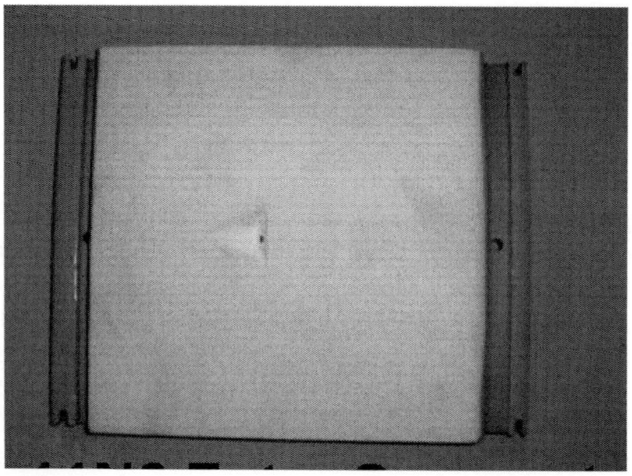

FIGURE 12–52 Typical tile components from an 11N2 silo type combustor. *(Source: Liburdi Engineering.)*

Case Study 11: Liburdi Powder Metallurgy, Applications for Manufacture and Repair of Gas Turbine Components*

The Liburdi Powder Metallurgy (LPM™) joining process has been improved through the addition of new superalloy compositions and the development of hard particle-based compositions. Two new compositions based on MarM247 have been developed, tested and applied to hot section gas turbine components. These new compositions show improved microstructures and excellent mechanical properties. In addition to the conventional superalloy LPM™ materials, a hard facing LPM™ material based on wear resistant and abrading compositions incorporating hard particles in an oxidation resistant matrix have also been developed. This section discusses the mechanical and physical properties of the different materials developed and presents examples of different applications for these materials.

Over the past two decades many wide gap and modified wide gap processes have been developed and employed in the repair and manufacture of hot section gas turbine components; GE's Advanced diffusion healing ADH, Snecma's rechargement per brasage diffusion RBD, Howmet's effective structural repair ESR, Chromalloy's surface reaction braze SRB, Turbofix, Sermafill, and Liburdi Engineering's LPM™. All of these processes involve the use of high and low melting components to help limit the amount of melting point depressant and to help bridge large gaps. The patented LPM™ process is the only modified wide gap process capable of filling large cavities resulting from the removal of structural cracks or casting core holes. In particular the LPM™ process has been used to rebuild components that are missing large amounts of material while maintaining very low levels of melting point depressants. Additionally, the LPM™ process has been extended to applying abrading and wear resistant materials to change the surface properties of hot section components.

* Source: Courtesy Liburdi Engineering, from R. Sparling and J. Liburdi, "Liburdi Powder Metallurgy, Applications for Manufacture and Repair of Gas Turbine Components."

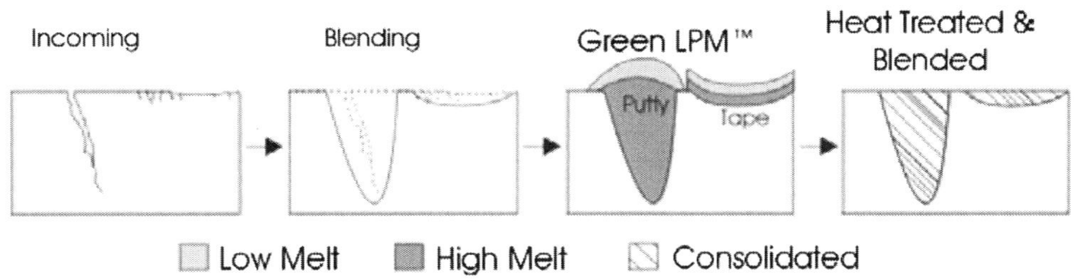

FIGURE 12–53 Sketch showing the steps in an LPM™ repair process. *(Source: Liburdi Engineering.)*

The patented LPM™ process is a multi-layer process incorporating high and low melting metal powders to effect sintering and controlled liquid phase bonding. The powders are mixed with organic binders to produce flexible tapes or clay like putties. During application of the tape or putty to the new component or repair area: the area is prepared for LPM™, the low melt material is applied on top of the high melting material, and the ratio of the two materials is carefully controlled (Figure 12–53). The component is then heated slowly in a vacuum furnace to burn off the organic binders and finally heated to and held at a temperature between the melting points of the high and low melt materials. During the heating process the high melting point material sinters to form a rigid structure. Once the melting point of the low melting material is exceeded the rigid structure is infiltrated by the low melting material at which point liquid phase sintering, transient liquid phase solidification (isothermal solidification) and diffusion bonding occur. The resulting deposits show excellent mechanical strength, creep strength, and ductility when compared to conventional wide gap deposits without the need for lengthy diffusion cycles.

The major advantage of using the LPM™ process is that it does not result in the distortion or cracking commonly associated with welding. This allows for large surface restorations that would otherwise distort or crack the component beyond repairable limits. It also makes it possible to repair cast superalloys components that would crack beyond acceptable limits during welding.

LPM™ has been used in the repair of nozzles, vanes, blades, and buckets from GE Frame 3, 5, 6B, 7B, 7E, 7EA, 7FA, LM1600 and LM2500, Siemens Westinghouse W191, W251 B8, B10, B12, and W501D5, Rolls Royce T56, RB211, Avon and Spey, Alstom Tornado and TB5400, Dresser Rand DJ270, and others. LPM has also been used in the new manufacture of W501G Row 1 and Row 2 blades and some Solar turbine blades. LPM™ has proven an effective tool in the manufacture and repair of hot section turbine components. The following sections discuss the properties of the newer MarM247-based, abrading, and wear resistant LPM™ compositions and present some typical LPM uses.

Superalloy LPM™

The initial LPM™ composition developed was based on IN738. This composition showed good properties but it was felt that improved properties could be developed using other alloys. A matrix of various high and low melt components was developed and various alloy combinations were evaluated for suitability in the LPM™ process. Both cobalt and nickel-based compositions were evaluated. The results showed that compositions based on MarM247 had a very desirable microstructure. Two new MarM247 based compositions were further tested and evaluated MarM247-3 and MarM247-7.

Microstructure

The microstructure of the original IN738 LPM™ material contained elongated borides. The 100 μm to 500 μm borides were inter-granular and had length to width ratios of greater than 10:1. This microstructure resulted in lower ductility but tended to perform well in service.

Both of the newer MarM247 LPM™ compositions have small cuboidal boride particles. The 10 μm to 50 μm cuboidal particles are found intra-granularly and have a length to width ratio of about 1:1. As a result, these two new compositions show improved ductility and better fracture properties than the IN738 LPM™ material. The MarM247-7 composition has a slightly higher volume fraction of borides than the MarM247-3 but both have lower boride volume fractions than the IN738-based material. The borides in the MarM274-3 structure tend to be smaller than those in the MarM247-7 structure.

All LPM™ compositions show a good interface with the base alloy. LPM™ materials have been applied to the following alloy systems: GTD111, GTD222, Rene N4, IN738, IN939, IN625, MarM002, MarM247, Rene 80, Rene 125, Rene 142, Hastelloy X, CMSX 4, C1023, U500, U520, U700, U720, 304 stainless steel, 321 stainless steel, 347 stainless steel, ECY768, X40, X45, FSX414 and others. No needle-like topologically close packed phases (TCP), microcracks, or abnormal phases were found at the interface between LPM™ and any of the aforementioned alloys. The interface between LPM™ is generally a 50 μm interdiffusion zone that contains a coherent interface with the base alloy.

Microscopic examinations of large cross-sections of the original IN738 LPM™ and the newer MarM247-based compositions have shown that any pores tend to be smaller than 125 μm in size and total porosity levels are lower than 1%. As the LPM™ material is produced from fine powders, the resulting microstructure has a very fine grain size.

Composition

Due to the multi-layer nature of the LPM™ process the resulting deposits have a very low weight percent of boron, the melting point depressant used in the low melt component (Table 12–4). As boron is a very light element having an atomic mass of 11 compared to 59 for nickel, small weight percent changes in boron result in very large changes in the atom percentages. Most conventional wide gap processes contain 1.5 weight percent boron or more. The LPM™ materials contain less than 0.9 weight percent boron due to the deposits containing less than 30% low melting material. The small weight percent and subsequent atomic percent of boron in the LPM™ materials leads to low volume fractions of boride and better ductility than other brazing processes. The less than 30% low melting material also results in high levels of γ' forming elements and subsequently excellent creep properties as is evident in the relatively large amounts of Al and Ti in the LPM™ material when compared to conventional wide gap materials.

Test Bar Manufacture

Most of the studies done on conventional and modified wide gap processes have concentrated on joint widths not larger than about 1.5 mm. These small joints widths allow for large amounts of boron to diffuse out of the joint into the base material. Often the joints are given long diffusion cycles of up to 24 hours to allow for the diffusion of the boron into the base alloy. Thus the data generated from these test bars is not representative of the mechanical properties of overlays or relatively thick repair joints formed on actual component repairs. It was felt that for this testing the butt joint should be 6.3 mm wide such that diffusion into the base alloy did not play a role in the joint properties at the centerline of the joint. The test bars with a 6.3 mm joint width should be representative of the bulk LPM™ properties and should also test the interfacial strength.

6.3 mm × 6.3 mm slots were cut down the center of several 50.8 m × 50.8 mm × 9.5 mm MarM247 blocks. These slots were filled with the MarM247 LPM™ materials and heat treated for not more than 2 hours at the maximum temperature reached during the alloy solution cycle. The blocks were then aged for 2 hours at 1120°C and 16 hours at 843°C to produce the desired γ' precipitate structure. Stress rupture and tensile bars were cut from the blocks such that a 6.3 mm region in the center of the gauge was LPM™ (Figures 12–54 and 12–55). The bars were X-ray and fluorescent penetrant inspected. Test bars with defects larger than 500 μm were rejected.

TABLE 12–4 Chart Showing the Compositions of the Various LPM™ Materials, a Typical 50:50 MarM247 Wide Gap Material, and the Nominal Composition of MarM247

Element	MarM247	247-3	247-7	IN738 LPM™	MarM247 Wide Gap
Ni	59.90	58.13	65.26	63.32	66.95
Cr	8.50	11.95	10.11	15.4	11.00
Co	10.00	13.00	7.52	8.95	9.75
Mo	0.60	0.42	0.45	1.23	0.30
W	10.00	7.00	7.52	1.82	5.00
Ta	3.00	3.00	2.26	1.98	1.50
Hf	1.50	1.05	1.13	0.00	0.75
Al	5.50	3.85	4.14	3.43	2.75
Ti	1.0	0.70	0.75	2.38	0.5
Nb	N/A	N/A	N/A	0.63	N/A
B	0.01	0.90	0.87	0.81	1.5
Atom% B	0.01	4.7	4.5	4.2	7.5

(Source: Liburdi Engineering.)

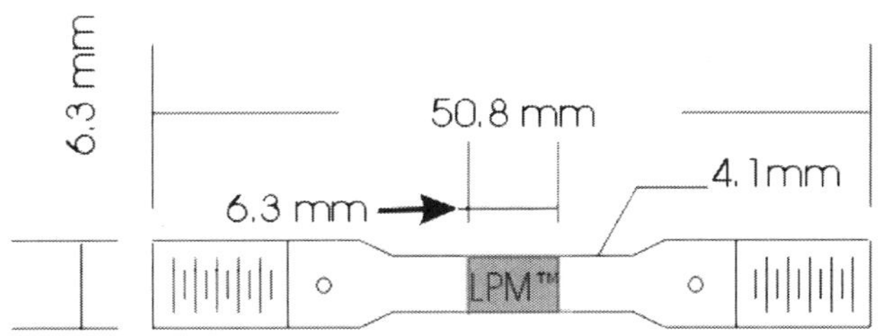

FIGURE 12–54 Sketch of the test bar geometry. *(Source: Liburdi Engineering.)*

Tensile Properties

Tensile test bars were tested at a variety of temperatures. Table 12–5 shows the ultimate tensile strength and reduction in area for each of the test bars. The MarM247-3 material shows slightly better ductility due to the small boride particles and the smaller volume percent of borides. The ductility of the MarM247 based LPM™ materials is only slightly lower than that for common nickel-based superalloys and is considerably higher than the levels for conventional wide gap brazing processes that generally show little or no ductility. None of the bars failed at the interface and all bars failed intergranularly as would be expected. The modulus for both materials was approximately 170 GPa.

TABLE 12–5 Tensile Properties of the MarM247-3 and MarM247-7 LPM™ Materials

Material	Temperature	UTS in MPa	Reduction in Area%
MarM247-7	21°C	640	0.8
MarM247-7	549°C	680	0.8
MarM247-7	760°C	590	0.4
MarM247-7	871°C	510	0.6
MarM247-3	21°C	770	1.6
MarM247-3	649°C	730	1.4
MarM247-3	704°C	490	1.5
MarM247-3	927°C	340	2.4

(Source: Liburdi Engineering.)

Creep Properties

Stress rupture tests were performed using the bars described previously. The tests were conducted to ASTM E139. The stress time and temperature were varied to generate a broad range of Larson Miller parameter values. The temperature varied from 787–982°C and the stress varied from 5–35 MPa. The Larson Miller plot is shown in Figure 12–56. The average reduction in area was 4.2% for the MarM247-7 LPM™ and 7.3% for the MarM247-3 LPM™. These values are similar to those for cast MarM247.

Due to the fine-grained nature of LPM™ materials, they tend to have lower creep strengths when compared to large grained cast nickel-based superalloys. The LPM™ materials have creep strengths higher than conventional welding alloys such as IN625 for nickel-based superalloys and L605 for cobalt-based superalloys.

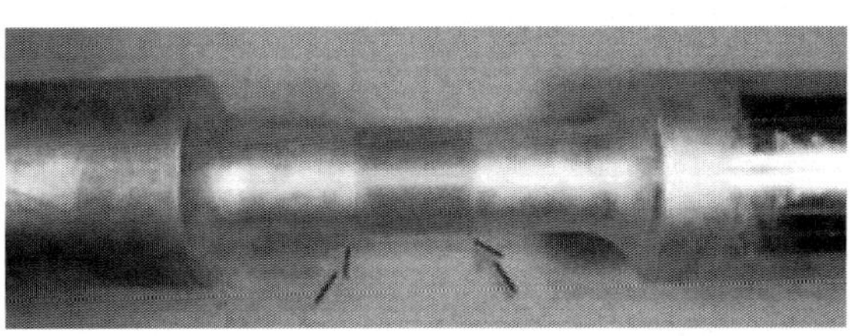

FIGURE 12–55 Photograph of typical test bar in rough machined state. *(Source: Liburdi Engineering.)*

The creep strength compares more closely to fine grained wrought superalloys and cobalt-based superalloys. During the studies it was also found that the creep properties of LPM™ varied little with the composition. It has been hypothesized that the creep properties are highly dependent on the grain size. Studies to this effect will be conducted in the future by increasing the grain size through long diffusion cycles.

Low Cycle Fatigue

Low cycle fatigue test bars were produced as described previously and were inspected for surface defects larger than 0.50 mm. The testing was conducted at 850°C using a trapezoidal waveform at an R ratio of −1.0. The specimens

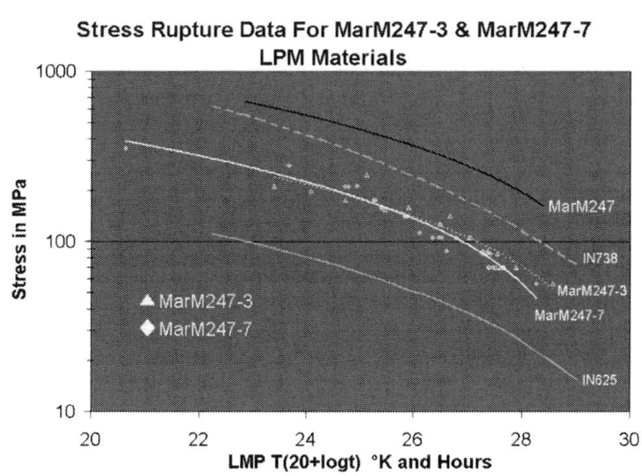

FIGURE 12–56 Graph showing the stress rupture properties of the MarM247-7 and MarM247-3 LPM™ materials. *(Source: Liburdi Engineering.)*

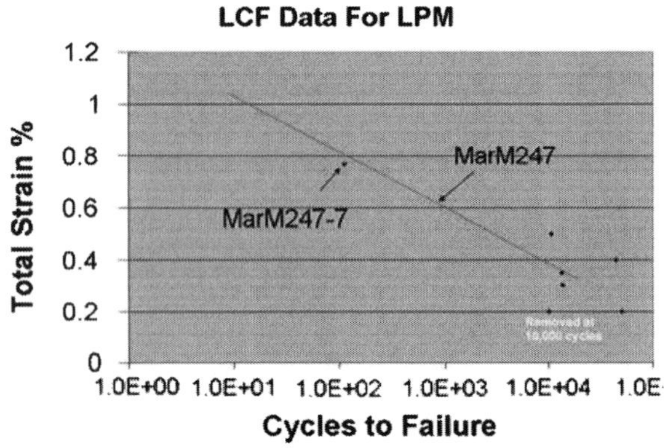

FIGURE 12–57 Graph of LCF data for MarM247-7. *(Source: Liburdi Engineering.)*

were held at the maximum compressive strain for 2 minutes during each cycle. Failure was defined as the point at which the maximum stress amplitude decreased to 50% of the value at 100 cycles. It should be noted that the butt joint in the middle of the test bars produces an effective gauge length of only 6.3 mm and the strain is measured over the total gauge length. During testing the majority of the elongation occurs in the 6.3 mm joint. This means the strain in the joint area is much higher than that measured. The data reported in Figures 12–57 and 12–58 are for the total strain over the gauge length of the test bar. Testing indicated the fatigue properties are similar to that of cast MarM247 for MarM247-7 and likely slightly lower properties for the MarM247-3.

Superalloy LPM™ Example #1: General Electric MS7001EA Row 1 Nozzles

After the initial 24,000 hour service interval, these FSX 414 cobalt alloy nozzles show extensive cracking especially around the leading edges, trailing edges, and between the two airfoils on the outer shroud. The cracking was primarily caused by low cycle thermomechanical fatigue. The cracking at the leading edges was branched and extended

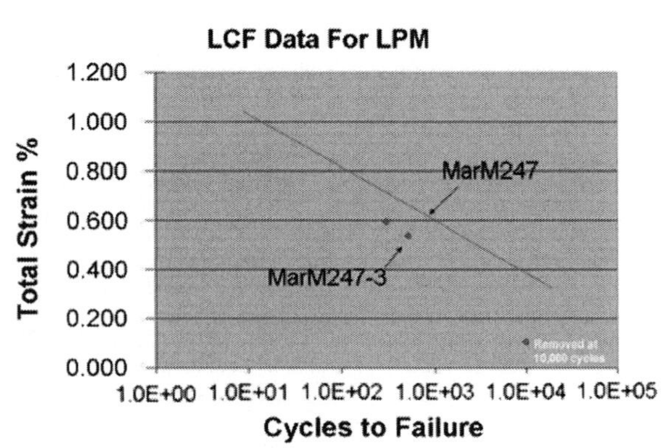

FIGURE 12–58 Graph of LCF data for MarM247-3. *(Source: Liburdi Engineering.)*

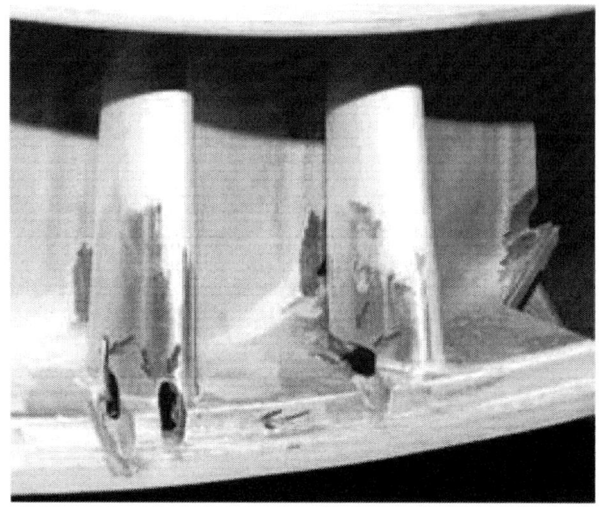

FIGURE 12–59 Photograph showing vane segment prepared for LPM™. *(Source: Liburdi Engineering.)*

through to the internal vane cavities. Some oxidation erosion and associated cracking had also occurred on the outer and inner diameter shrouds between the airfoils. Based on the degree of damage it was felt that conventional weld repairs would either crack or result in excessive distortion of the nozzles. It was decided the LPM™ was the best approach for repair.

In preparation for LPM™ the cracks were ground out and the eroded surfaces were blended back to unaffected base alloy. This left large holes approximately 25 mm × 100 mm in size at the leading edges of the airfoils of several nozzles and areas larger than 50 mm × 50 mm requiring 0.75 mm build-ups (Figure 12–59). The large holes, which are typical of vane repairs where the structural cracks are removed, were filled with LPM™ putty that was sculpted to the original contour of the nozzle. The eroded areas were covered with a 1 mm thick LPM™ tape to restore the dimensions. Small cracks and impact damage were repaired with the LPM™ putty. During repair these areas were successfully repaired using MarM247-7 LPM™.

After the second service interval of approximately 24,000 hours (~48,000 hours total operating time), the nozzles did not show the large thermomechanical fatigue cracks at the leading edges or between airfoils on the outer shroud surface. This suggested the LPM™ had improved the performance of these nozzles due to the improved strength and low cycle fatigue resistance of the LPM™ repair (Figure 12–60).

After 48,000 hours of service, the nozzles had small cracks on the shrouds, oxidation erosion on the internal and external surfaces and thinned trailing edges. The thinned trailing edges were particularly of concern as the throat area between the airfoils had increased significantly due to material loss during service and repair. This increase in throat area can lead to harmonic imbalance and efficiency

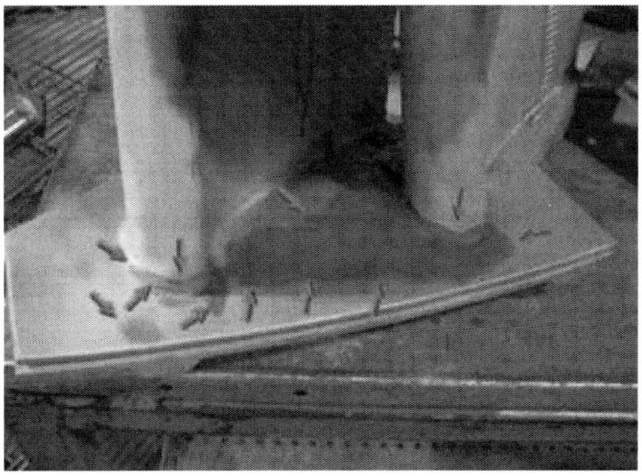

FIGURE 12–60 Photograph of nozzle after 48,000 hours and one repair cycle. The darker areas are LPM™ material. *(Source: Liburdi Engineering.)*

loss. As part of the second repair the small cracks and oxidation damage were blended in preparation for LPM™ application and the thinned trailing edges and oxidized areas were thickened by applying LPM™ tape to both surfaces. The cracks were repaired using LPM™ putty. After the second repair the nozzles were again returned to service. The nozzles are expected to return for LPM™ repair at ~72,000 hours in early 2004.

Superalloy LPM™ Example #2: Siemens Westinghouse W501G Row 1 and 2 Core Closure

Liburdi Engineering's LPM™ is specified for closure of the core holes on Siemens Westinghouse 501G row 1 and row 2 directionally solidified MarM247 turbine blades. The casting core holes in the 501G blades are necessary to retain the core during manufacture of the blades. These large core holes are difficult to close via welding due to cracking of the directionally solidified grain boundaries. Conventional wide gap brazing of the 25.4 mm × 6.3 mm casting holes is also not practical (Figure 12–61). LPM™ was selected to close the blade tips as it has excellent creep properties, very low porosity, and can close the casting holes without generating cracks.

LPM™ is applied as putty into the core holes such that each hole is slightly overfilled. The blades are then heat treated to consolidate the LPM™. After finishing to contour and machining the tip to nominal dimensions it is not possible to visually determine where the casting holes were (Figure 12–62). To date over 49 sets of row 1 and row 2 blades have been closed using this process. Figure 12–63 shows a blade tip after more than 8000 hours of service.

Hard Particle LPM™ Materials

Recently LPM™ compositions based on hard particles have been developed for use on hot section turbine components. These materials use hard particles such as chrome carbide, tungsten carbide and other materials to locally change the surface properties of hot section components. This allows designers and repair engineers to achieve extended life on hot section components.

Wear Resistant LPM™

Wear resistant LPM™ (known as LZN™) compositions based on chrome carbide have been developed and used in the manufacture and repair of hot section gas turbine components. These materials are composites of hard particles in an oxidation resistant matrix. The wear resistant LPM™ was designed to replace plasma, high velocity oxyfuel (HVOF) and weld hard facing materials.

The major drawback of plasma and HVOF chrome carbide coatings is their poor mechanical bond to the base alloy. The LZN™ compositions form a metallurgical bond with the base alloy. This metallurgical bond consistently results in failure of the adhesive in the ASTM C633 adhesion test for flame sprayed coatings. The other drawback of plasma spray and HVOF coatings is the

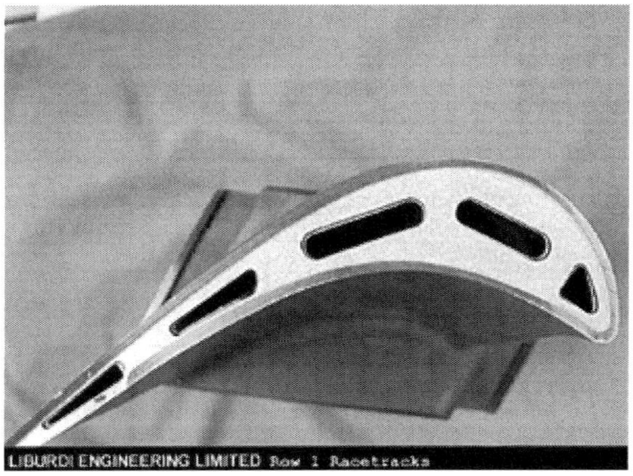

FIGURE 12–61 W501G Row 1 blade prior to LPM™ closure. *(Source: Liburdi Engineering.)*

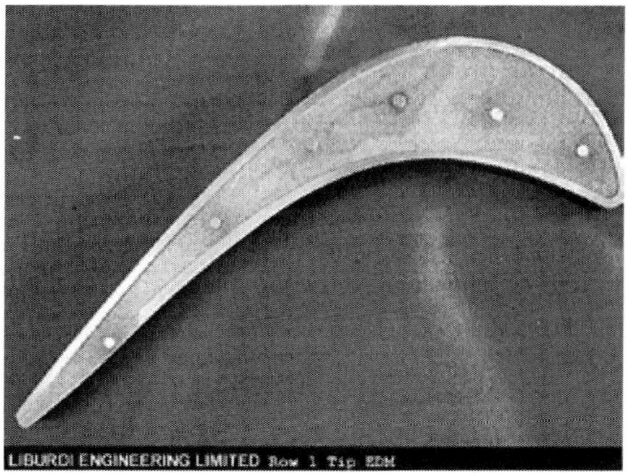

FIGURE 12–62 W501G Row 1 blade tip after LPM closure. *(Source: Liburdi Engineering.)*

FIGURE 12–63 W501G Row 1 blade tip after 8000 hours of service. *(Source: Liburdi Engineering.)*

necessity for 90% spray angles near 90% to the surface. The LZN can be applied into slots on opposing surfaces and on surfaces that have no direct line of site for a plasma or HVOF gun.

In comparison to weld hard facing, the LZN™ material does not result in base alloy cracking and shows improved hardness. Weld hard facing materials often result in cracking at grain boundaries intersecting the face to be welded. In some components welding results in cracks at or beyond acceptable limits resulting in high levels of deviated or scrapped material. Due to the nature of the LPM™ process LZN™ does not result in cracking of the base alloy. During weld hard facing, the weld pool mixes with the base alloy reducing the hardness of the deposit reducing hardness values to be below 50 Rc (Rockwell "C"). The LZN™ materials have higher volume percents of carbide particles resulting in hardness values of 60 to 65 Rc. The volume percent of carbide is more than 55% of the deposit. Voids are typically smaller than 25 μm and the porosity levels are below 5%.

Because the LZN™ material can be used for hard-facing of shroud contact faces, transition piece outlets, floating seals, etc., it must be oxidation resistant. The composition of the material is shown in Table 12–6. The chrome carbide particles are expected to start decomposing above 815°C and it is felt that the oxidation resistant matrix should be stable to temperatures above 980°C. Oxidation testing at 870°C for 1000 hours showed a native chrome oxide scale 2 μm thick had formed on the deposit and no significant attack of the deposit had occurred.

TABLE 12–6 Composition of LZN

Element	Weight%
Ni	33
Cr	60
Si	4
C	3

(Source: Liburdi Engineering.)

The LZN material has been used to hard face the shroud contact faces on CMSX 4 single crystal turbine blades. The blades in question had no direct line of site for plasma or HVOF hard facing and the geometry of the contact face did not allow access for a gas tungsten arc weld hard face. As a result LZN was chosen to replace the conventional hard facing techniques. The LZN material is applied to the contact faces as an LPM™ tape material and is given appropriate heat treatments to consolidate it. The LZN material showed no abnormal phases or unusual microstructures at the interface with the CMSX 4 base alloy. The material was machined to contour using conventional aluminum oxide grinding wheels. After machining the contact faces did not show any cracking or spalling and the LZN had successfully hard faced the z-notch on these blades with a very difficult geometry.

Abrading LPM™

Abrading LPM™ compositions based on large angular hard particles have been developed for use on the tips of unshrouded turbine blades. Unshrouded turbine blade tips run against shroud blocks to ensure a tight air seal and subsequently high engine efficiency. Unfortunately the turbine blades are often worn by the shroud blocks as opposed to the shroud blocks being worn by the turbine blades. This results in the loss of material from the tips of the difficult to repair turbine blades. It also results in the removal of any protective coatings from the blade tips. A need was recognized for a harder material that would machine into the shroud blocks, preventing the blade tips from being worn away and prevent the blade tips from oxidizing.

The abrading LPM™ material is a mixture of large hard particles in an MCrAlY-based matrix. This material is made into an LPM™ tape and applied to the blade tip. The tape is contoured to the blade tip in the green state and is given an appropriate heat treatment for the base alloy. After heat treatment the abrading tip is contoured and ground to the appropriate tip height.

In summary, Superalloy LPM™ compositions based on IN738 and MarM247 have proven extremely successful for the repair and manufacture of hot section gas turbine components. These materials show excellent creep, tensile, and stress rupture properties for a brazing process. The materials have proven successful during service on a variety of components from different engines.

Hard particle LPM™ compositions used for wear resistance and abrading have also been developed and tested. The wear resistant LZN™ material shows very good hardness and can be used in wear applications where welding, plasma spray and HVOF hard facing methods do not work. The abrading LPM™ material can be used on unshrouded turbine blades where tight tolerances and wear of the tip shrouds into the blade tips is an issue.

Case Study 12: Hot-Gas-Path Life Extension Options for the V94.2 Gas Turbine*

Gas turbine and combined-cycle power plants are typically designed for a service life of over 30 years. If operated at base load in continuous duty, the gas turbine hot-gas-path components, for example in a combined-cycle power plant, need repair and replacement according to the maintenance program several times during plant life. Most of the hot components would reach the end of their service life, e.g., 100,000 equivalent operating hours (EOH), after 10–12 years. As this is well before the end of the overall plant service life defined in the power plant concept, such plant applications therefore necessitate life extension measures enabling to continue operation beyond 100,000 EOH. This case presents strategic options for hot-gas-path component life extension.

While repair techniques are considered only briefly, the focus of this section is on the strategic aspects for upgraded operation and maintenance. Salient items are:

- The use of upgraded hot-gas-path components with higher durability to reduce maintenance outage time by extended inspection intervals and reduced scope of overhaul and repair
- The implications of residual life assessment after one hot-gas-path inspection interval or after 100,000 EOH of operation
- The driving role of upgraded hot components to achieve higher gas turbine performance during plant life extension after 100,000 EOH

Because of the operating experience involved, the considerations are based on the hot-gas-path components of the V94.2 gas turbine of the authors' company. This model operates on 50 Hz. To a large extent the statements are also valid for the V84.2 gas turbine, which is the 60 Hz derivative of the V94.2.

The first V94 gas turbine (including the earlier V94.0 and V94.1 variants) was connected to the grid in 1974 with an output of approximately 90 MW and a turbine inlet temperature of 850°C according to ISO definition (Figure 12–64). In various development steps, the current design has achieved an output of close to 160 MW at a turbine inlet temperature of 1060°C (ISO definition). During this evolution the gas turbine efficiency was raised from 30.3–34.6%. To date, 112 turbines are in service, and an additional 22 are currently being manufactured or commissioned. Collectively, the V94.2 fleet has up to now logged 3.8 million hours of operation at base load, 5.5 million equivalent operating hours and 77,000 starts. The majority of these gas turbines are operated in the base and intermediate load ranges in combined-cycle application.

* Source: Courtesy of Siemens, extracts from G. Bohrenkamper, H. Bals, U. Wrede, and R. Umlauft, "Hot-Gas-Path Life Extension Options for the V94.2 Gas Turbine," in Proceedings of the ASME Turboexpo 2000: Munich, May 8–11, 2000, 2000-GT-0178.

Nomenclature

EOH: Equivalent operating hours are determined from base load operating hours, the start/stop cycles and also operations such as turbine trips, with corresponding weighting factors. For example, 1 hour of operation at base load is defined as 1 EOH and 1 start with normal loading gradient is weighted 10 EOH.

LCF: Low-cycle fatigue

HCF: High-cycle fatigue

TIT: Turbine inlet temperature according to ISO definition

HGP: Hot-gas-path

HGPI: Hot-gas-path inspection

Maintenance Program and Upgraded Spare Parts

Not only the turbine inlet temperature has been increased but also the hot-gas-path (HGP) inspection interval has been extended from initially 16,000 equivalent operating hours (EOH) to 25,000 EOH and then to 33,000 EOH in 1994 (Figure 12–64). The application of the last step eliminates two hot-gas-path inspections within a plant's life of for example 200,000 EOH. The table in Figure 12–64 shows the predicted service life data of hot components according to the design life under certain basic assumptions (natural gas firing, continuous duty etc.). Apart from the 33,000 to 66,000 EOH replacement interval for turbine rotor blades stage 1 and 2 the other hot components are scheduled for 100,000 EOH of operation. To allow this, the refurbishment intervals, also shown in this table, have to be met for these components. Service life extension involves the following cases:

- Components that are in good condition during overhauls and that have remaining useful life require no action to be taken and these components can remain in service.
- The physical condition of the component requires life extending repair measures, such as refurbishing of turbine blades, to enable further use of the component.
- The component is replaced with one of equal value.

FIGURE 12–64 Maintenance program for hot-gas-path components, V94.2. *(Source: Siemens.)*

Component	Repair / Refurbishing interval(EOH)[1]	Replacement interval (EOH)
Flame tube ceramic tiles-	–	according to findings during MI
Mixing casing	33,000	100,000
Inner casing	33,000	100,000
Rotor blades stage1	–	3,000 to 66,000[2]
Rotor blades stage2	–	3,000 to 66,000[2]
Rotor blades stage3	33,000	100,000
Rotor blades stage4	–	100,000
Stator blades stage1	33,000	100,000
Stator blades stage2	33,000	100,000
Stator blades stage3	Depending on version	100,000
Stator blades stage4	–	100,000

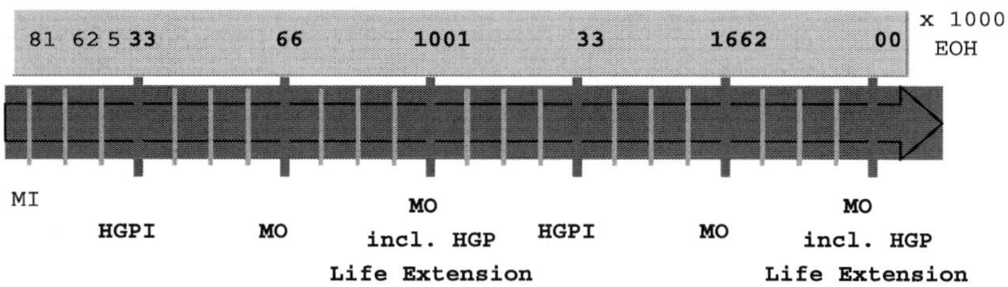

1) according to findings are placement of single parts is expected
2) depending on version

MI = Minor inspection; HGPI = Hot-Gas-Path Inspection;
MO = Major Overhaul; EOH = Equivalent Operating Hours

Minor inspection interval = 4000-8000 EOH depending on operating mode

- The component is replaced with one of higher value with focus on higher robustness, increased output, and efficiency.

As replacement options, hot component upgrades that have been developed during the design evolution of the V94.2 over the last two decades are available (Table 12–7). The decision depends on the condition of the component and on the assessment of possible remaining usable life. Evaluation of the cost and benefits of these possibilities have a significant influence on the choice.

A life assessment of all V94.2 hot-gas-path components is reported below. By detailed material investigations on turbine blading of the front stages after 33,000 EOH of operation, valuable experience has been gained. Several hot components that we have monitored and investigated have proven potential for further extension of the HGP inspection interval to longer assessment of the other hot components after 100,000 EOH of operation is reported.

Hot Component Life Assessment Procedure

The hot component life assessment procedure has to be based on the design criteria for these parts.

Design Criteria

Turbine blading is designed, based on creep strength for a period of 100,000 equivalent operating hours and, in the V94.2 turbine, is made of nickel-based alloys.

Surface coatings are applied to the front stages to protect the base material from high-temperature corrosion and oxidation. These coatings are consumed and are then renewed, e.g., after one hot-gas-path inspection interval of 33,000 EOH.

Depending on the blade row, thousands of starts with normal loading gradients or hundreds of turbine trips are taken into account. The resulting loads are caused by changes in mechanical forces such as centrifugal force as well as restrained thermal expansion and are partially in the plastic range. This type of loading is characterized by low-cycle fatigue (LCF) strength. The number of load cycles, which can be achieved, is mainly dependent on the extent of plastic strain and component temperature. Furthermore the rotor blades are designed for withstanding high-cycle fatigue (HCF). This means that the effects of vibration caused by rotor dynamics, aerodynamic, and combustion effects are below a certain

TABLE 12–7 Upgrading V94.2 Hot-Gas-Path Components

Component	Original	Upgrade	Latest Upgrade
Mixing casings, Inner Casing	austenitic steel	Incoloy 800 H with higher strength, improved cooling	Inconel 617 with higher strength, improved cooling
Rotor blade stage 1	IN738LC, serpentine cooling, chromized	improved cooling by additional turbolators	separate cooling channel at leading edge, VPS coating SICOAT 2231
Rotor blade stage 2	IN738LC, uncooled, chromized	cooling with 6 radial bores	improved cooling with 12 radial bores, VPS coating SICOAT 2231
Rotor blade stage 3	Nimonic 90	U520 or U720 with higher strength, chromized	IN738LC, new design, aerodynamic improved
Rotor blade stage 4	Nimonic 80A, damping pins	Nimonic 90 with higher strength, free standing blade without damping pins is more efficient	U520 with higher strength, new design, aerodynamic improved
Stator blade No. 1	IN738LC, chromized	improved cooling	improved cooling, VPS coating SICOAT 2231
Stator blade stage 2	U500, chromized	IN738LC with higher strength	VPS coating SICOAT 2231
Stator blade stage 3	U500, chromized	IN738LC, chromized	IN939 uncoated
Stator blade stage 4	austenitic steel, segments	N155 with higher strength, single airfoils	IN939 with higher strength

(Source: Siemens.)

stress value so that the component can be exposed to essentially unlimited numbers of stress cycles occurring at high frequency.

The mixing casings and the inner casing are designed similarly to the turbine blading based on creep strength and low-cycle fatigue strength. As casings in the hot-gas-path are exposed to significant low-cycle fatigue loading, these components require regenerative heat treatment and weld repair during hot-gas-path inspections.

Design criteria are established according to the different loads in the operated gas turbine and calibrated on measurements with gas turbines in test beds and in operation. The sophisticated and validated philosophy and criteria for design, manufacturing, repair, and quality assurance is fully available only at the original manufacturer.

Life Assessment

Based on the design philosophy described above, the residual life assessment encompasses stress analysis, operation, and repair history, quality control as well as material investigations of the service-exposed parts.

A preliminary step for plant life extension involves the compilation of all operating data such as the number of starts, hours of operation, and rapid temperature changes (e.g., due to load rejection or turbine trip, Figure 12–65).

The history of maintenance measures and the trends of findings of the most recent fact-finding on minor and major inspections must also be accounted for.

Following selection of the components to be considered, the corresponding original design criteria and corresponding analysis are used and updated if necessary.

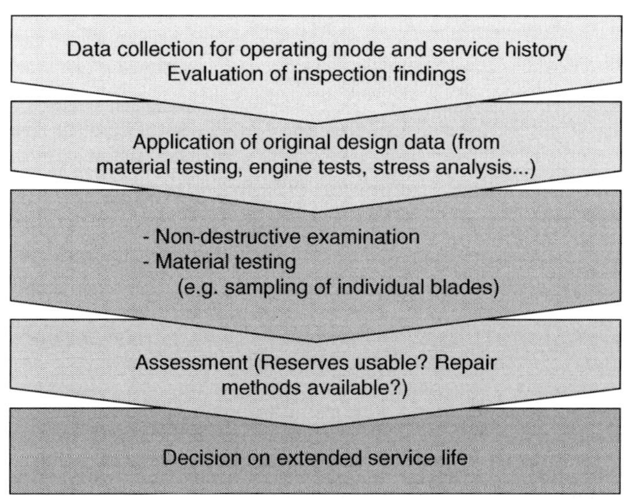

FIGURE 12–65 Hot component life assessment procedure. *(Source: Siemens.)*

Material investigation corresponding to the design criteria includes both non-destructive evaluation and destructive testing of components:

- Visual examination is first required to determine whether a blade has sustained oxidation/corrosion. This is followed by surface crack examination. Selected areas are then subjected to metallographic examination to evaluate the surfaces with regard to consumption of the protective coating, the depth of attack in the base material and for assessment of the microstructure. Investigations with the scanning electron microscope of the γ'-phase may be included if useful.
- Creep tests are performed on selected components, the samples being taken from the blade airfoil out of areas of the highest loading. Test time is reduced by performance at test temperatures that exceed the metal temperature in operation and by increased test stress levels above the operating stress. The same tests are performed for comparison purposes on samples from the blade root, which is not subjected to creep.
- Corresponding low-cycle fatigue tests are required to enable conclusions regarding low-cycle fatigue strength, the material property used for evaluation of service life expenditure caused by starts/stops, and rapid temperature changes. These tests are performed at room temperature. Samples from both the blade airfoil and root are also tested here. Results can then be directly compared with the low-cycle fatigue strength design curve, the lower bounding curve.

Crack examination yields preliminary indications of exceeding high-cycle fatigue strength. In the event that cracks occur due to cyclic loading, these can be assigned to loading based on the crack initiation site and crack propagation. If the blade is free of cracks, the design strength and the high-cyclic loading limits are used for further evaluation. An example of such investigations is described below.

Based on these material investigations, the original design and mode of operation of the gas turbine, and reusability of the component(s) in question are evaluated from a technical standpoint. Following consideration of the costs and benefits of measures such as requisite repairs, the final decision is made regarding continued use or replacement.

Hot Component Life Extension

The rotor blades and stator blades of stage 1 and 2 are highly temperature loaded, internally-cooled blades, and rotor blades stage 1 and 2 are designed with low wall thicknesses. Stage 1 and stage 2 are coated with a highly efficient coating on the outer surface. Compared to stage 1 and 2 the stages 3 and 4 are moderately temperature loaded. Apart from the moderately cooled stator blade stage 3 the other rows are uncooled solid airfoils so that life limiters such as internal oxidation or low wall thickness do not play a role here.

All turbine blade stages showed excellent creep strength during operation. No failures, negligible creep deflection, and excellent microstructures of investigated blades have been found.

Stage 1 and 2 of the Turbine Blading

Rotor Blade Stage 1 (Tla1) The original design of the Tla1 was developed in the late seventies. This blade already uses—compared to the state of the art in those years—an advanced serpentine type convective cooling channel. Turbolators have been added to the cooling configuration over the years. The cooling holes at the trailing edge had to be electrochemically manufactured (ECM) at that time. Since then, several steps of increasing turbine inlet temperature have been carried out. The original design revealed several findings during the refurbishment after 33,000 EOH of operation. This led to the decision to limit the service lifetime to one hot-gas-path inspection interval of 33,000 EOH. The findings were mainly:

- Considerable loss of wall thickness due to oxidation at the outer surface for the chromized type of this blade
- Significant internal oxidation especially at the trailing edge that caused certain cracking at the cooling holes in combination with cyclic operation due to LCF
- Occasional LCF cracking of the leading edge due to cyclic operation

Therefore, about 6 years ago this blade was redesigned with several features. Due to the progress in casting technology it was feasible to cast-in cooling outlet slots instead of the ECM manufactured holes. This measure resulted in a reduced concentration factor, thus significantly improving the low cycle fatigue strength. The outlet slots were provided with six rows of cylindrical pins to connect the suction side and the pressure side. These pins increased turbolance effects and enlarged the cooling surface for a more effective heat transfer.

The cooling of the leading edge was improved by a separate cooling channel with turbulators and a cooling air outlet at the blade tip. Additionally this feature made the airfoil more resistant against potential foreign object damage.

Additionally this blade was protected with the MCrAlY type vacuum plasma spray coating SICOAT 2231 on the outer surface.

A systematic life assessment program was carried out on service exposed rotor blades stage 1 after 33,000 EOH of operation to verify these improvements.

Samples from base load units revealed excellent coating behavior even at the leading edge, which is the position of the highest material temperature. It retains still usable coating and therefore potential for longer refurbishment intervals.

A unit being operated at intermediate duty with numerous start-stop cycles over one HGP inspection interval showed comparable excellent coating behavior. Only at the leading edge was the coating nearly consumed while other regions looked as good as new.

The leading edge of the upgraded design, opposite to the original design, showed no crack indications after comparable operating conditions. Also no notable internal oxidation was found with the upgraded design. This is of special interest for the life of the trailing edge. The trailing edge of the upgraded design was found crack free in all cases.

In addition a microstructural temperature evaluation by investigating the coarsening of γ'-phase was performed for the leading edge. In accordance with the computational analysis this investigation showed that the material temperature is about 50°C lower compared to the original design (Figure 12–66).

With the upgraded design the usable life of the Tla1 has been extended by 100% from one to two HGP inspection intervals including the refurbishment after 33,000 EOH.

Rotor Blade Stage 2 (Tla2) The MCrAlY type SICOAT 2231 protective coating was implemented to the outer surface of rotor blade stage 2 as well. On samples taken from such blades after service, excellent coating and base material conditions have been investigated after one HGP inspection interval of 33,000 EOH in every case. One example in Figure 12–67 shows a still usable coating thickness of >150 m and no base material attack, as was the case with the chromized type. Therefore, with this enhanced coating, comparable to the Tla1 the usable life of the Tla2 has been extended by 100% from one to two HGP inspection intervals including the refurbishment after 33,000 EOH. The enhanced coating is not only applied on new blades but also available for the blades of the original design.

Stator Blades Stage 1 and 2 These rows are also supplied with the above-mentioned SICOAT 2231 coating to meet the refurbishment intervals of 33,000 EOH. The refurbishment includes cleaning of the oxidized outer surface by the SICLEAN® method, recoating of the outer surface and potential repair of thermally caused cracks at the trailing edge. During refurbishment the internal cooling passages also require nondestructive surface testing. With increasing operating time, degradation of the internal cooling channels accumulates and also repair measures might not be allowed several times on the same blade. Therefore the scrap rate increases with the number of refurbishment cycles and refurbishment after reaching the 100,000 EOH milestone is considered to be not realistic for these blades.

Stage 3 and 4 of the Turbine Blading and Hot-Gas-Path Casings

Stator blades in stages 3 and 4 undergo residual life assessment when reaching the 100,000 EOH milestone, as they have potential for continued service. Highly influenced by the operating and repair history and also depending on the material and design variant, the inner casing and the mixing casings are assessed for their residual life after 100,000 EOH.

In summary, life extension options for hot components during regular HGP inspections and after long operating periods 100,000 EOH were studied.

Repair methods and protective coatings have been continuously improved for those hot components that need regular life extension measures during hot-gas-path inspections.

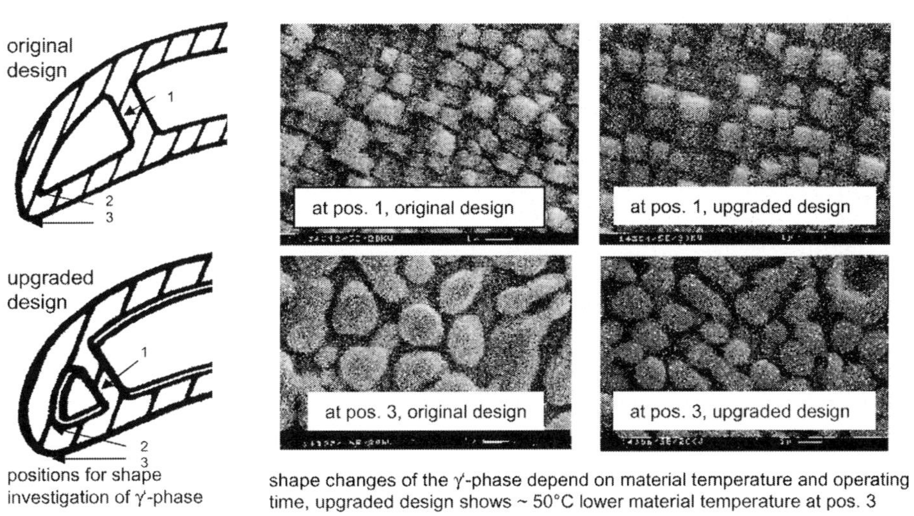

FIGURE 12–66 Tia1, V94.2, material temperature evaluation after ≈33,000 EOH. *(Source: Siemens.)*

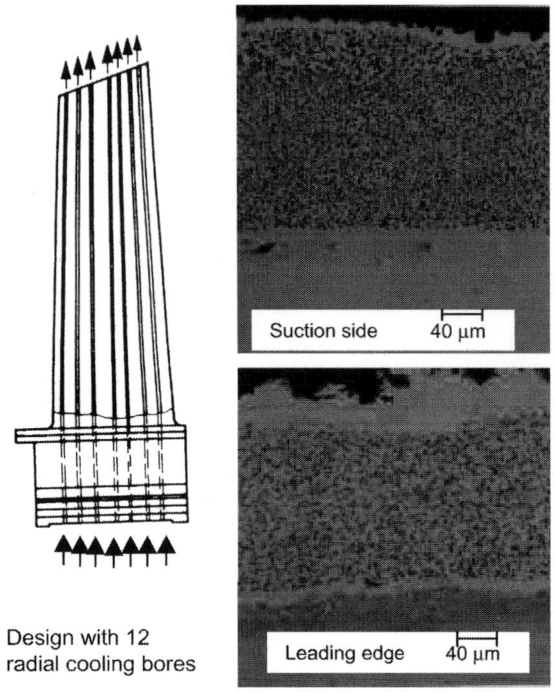

FIGURE 12–67 Upgraded rotor blade row no. 2, V94.2, coating condition after ≈33,000 EOH. *(Source: Siemens.)*

The investigations for life extension of hot-gas-path components after the long operating period of 100,000 EOH have revealed that several hot-gas-path components have definitely reached the end of their usable life, while others, e.g., rear rows of the turbine blades, may be usable for one more 33,000 EOH period or even longer.

Plant life extension including hot-gas-path life extension of V94.2 gas turbines has been applied on 11 gas turbines V94.2. Also a large number of additional plant life extensions are due to take place in the coming years. While the basic statements for life assessment and life extension are valid for the whole V94.2 fleet, it must be emphasized that each gas turbine needs to be evaluated individually because of its individual operating conditions. Up to now, life extension projects involved the early versions of V94.2 with relatively low turbine inlet temperature (TIT_{ISO}) × 1000°C and with the original hot component design, while many of the future projects will involve a higher $TIT_{ISO} > 1000°C$ for base load operation with different hot component design.

The most profitable plant life extension projects:

- include not only residual life assessment of the hot-gas path components,
- but also focus on strategic maintenance aspects to use upgraded hot components for higher durability, longer inspection intervals, lower scope of inspection and overhaul work and subsequently higher plant availability,
- and also care for modernization opportunities to improve the performance and competitiveness of the operating plant.

The best way is to evaluate these items in a joint project involving the electric power producer and the gas turbine manufacturer.

Case Study 13: Assessing Performance Degradation in Gas Turbines for Power Applications*

In a combined-cycle configuration, due to lengthy commissioning activities of various systems, acceptance performance tests are conducted only after a significant number of hours in operation have been accumulated.

The degradation of gas turbine (GT) thermal performance is playing a significant role in the economic viability of the highly competitive merchant power plant market. In combined-cycle applications, the degradation impacts not only to the traditional performance guarantees for power output and heat rate, but also to GT exhaust flow and exhaust temperature.

This case addresses the impact on plant performance guarantees due to qualifications applied to degradation process terms such as "new and clean condition," "fired and equivalent operating hours," and "bench-mark testing." In addition, the document will touch on how degradation evaluation may affect the long-term service agreements (LTSA) between owners and equipment manufacturers.

Since a gas turbine has several major components (compressor, combustor, turbine), interconnected mechanically and thermodynamically, any change in the behavior of one affects the others. It is therefore important, when examining degradation impact, to treat the GT as a "complete system."

Compressor Degradation

Compressor degradation accounts for 70–80% of GT performance losses. Consequently, manufacturers focused their efforts on analyzing its degradation mechanism and effective means of performance recovery. For a large industrial turbine, a drop in compressor isentropic efficiency by 1% causes a reduction in output of 1.2%. A combined effect of 1% decrease in airflow and efficiency yields a 1.8% power output reduction and an increase of 0.8% in the GT heat rate. There are two major contributors to loss of output and efficiency:

- Compressor airflow reduction
- Compressor increased aerodynamic losses operation in adverse pressure gradient

Erosion, corrosion, and fouling change the blades surface roughness and shape and affect both compressor capacity

* Courtesy of J. Zachary, Extracts from "Assessing Performance Degradation in Gas Turbines for Power Applications," GT2007-27976.

and optimal aerodynamic behavior. Compressor capacity is further reduced due to increased parasitic airflow caused by larger gaps in seals and clearances. The reduction in capacity is typically 1.6 times the drop in compressor efficiency. It should be noted that local degradation has a chain effect.

If a particular compressor stage has a higher loss, the stage exit pressure falls, the temperature increases, and all other stages operate at mismatched conditions.

Fouling

Fouling is caused by adherence of particulates, usually smaller than 3 μm in diameter, to the surface of airfoils and air passages. The water mist from the inlet air and oil leaks are the main bonding elements for these ingested particles. Lubrication oil is drawn into the flow path, due to the low-pressure upstream of the first rotor. While the oil acts only as glue for the first few rows, the increased temperature at the subsequent stages of the compressor causes the oil to form coke deposits.

Modern transonic compressors, operating near choking conditions, are particularly sensitive to fouling. Material build-up and thicker boundary layers, due to surface roughness, accentuate the airflow reduction. Heavy-duty industrial gas turbines operating at constant speed experience a more severe loss of capacity, since they cannot compensate for the loss by rotating at a higher speed. The fouling of the first stage will reduce the mass flow (flow coefficient) and the performance of the following stages will be affected. The operating point of the first stage will move from the design point to a higher pressure, higher density at the inlet of the second stage and a further flow coefficient reduction. The phenomenon of propagation in the following stages could eventually trigger a surge.

Erosion

Large particles, typically 20–30 μm in diameter, passing through the flow path have an abrasive impact on the airfoils. The result is removal of material and increased surface roughness. For industrial gas turbines, this problem is less severe because of extensive utilization of inlet air filtering systems. The recent use of inlet fogging or wet compression systems has the potential to cause damage if large droplets of water hit compressor blades. Blade erosion degrades the aerodynamic performance and could also generate structural problems. Due to several problems with the first rotating blade of the compressor, some of the manufacturers insist on using only their own design for an online water wash system (rather than a generic one) and imposed limitations in the use of inlet fogging for extended periods of time.

Clearances

During GT operation, the clearances between stationary and rotating parts have a predisposition to widen, resulting in substantial increases of parasitic flows (leakages). Of particular significance are the tip clearances between the rotating blades and the stationary casings. The increased leakage flow bypasses the stage and mixes lower pressure air with the mainstream air. The result is a lower pressure head at the stage exit. An increase of rotor tip clearance from 1% of blade chord to 3.5% of blade cord reduces the pressure ratio of a stage by up to 15%.

It is important to note that a reduction in pressure rise negatively affects the compressor surge margin and consequently the operational range. Another source of increased airflow leakages is reduced inter-stage labyrinth seals' tightness. Obviously the mode of GT operation affects the pace of degradation. During the initial "engine break-in," axial or blade tip rubs occur and result in increased clearances, which quickly reach nominal values. After this initial period, the clearances will remain relative constant over time. Fast loading and unloading of the gas turbine, trips, and load rejection events accelerate the widening of the clearances and cause more rapid degradation.

Turbine Degradation

The typical causes of turbine degradation are associated with alteration in airfoil geometry (leading and training edge thickness, surface roughness, and profile changes) and increase in parasitic losses (tip clearances, secondary and cooling air).

Degradation can alter the amount and distribution of the secondary cooling air, which might create severe overheating of metal surfaces and low efficiency/high blockage of the flow path. In the absence of cooling air, the component damage could be extensive, beyond salvage. Worn seals, enlarged clearances, and damaged cooling flow passages have a substantial negative effect on performance. Larger amounts of cooling air either bypass the blading without producing work or enter the main exhaust gas stream at a wrong location, spoiling the flow high momentum. A poorer turbine section efficiency of 1% has a powerful impact on the GT performance, causing a 0.8% power output drop and 0.5% worse heat rate.

Combustion Degradation

While the combustion process efficiency is unlikely to decrease, the degradation impact on the combustion system is felt in several ways:

- Change in the burner geometry—This affects the combustion process and increases emissions due to either hot spots (higher NO_x) or cold spots (higher CO). In particular, the dry low NO_x (DLN) burners, which have very lean premix flames and require precise tuning, are extremely sensitive to even minor burner and combustion chamber alterations.

- Change in the exhaust temperature distribution. A distorted combustion profile may alter a predetermined control temperature setting, either under-firing or over-firing the GT. In each case, the results are not desirable, causing loss of power (under-firing) or reducing components' life (over-firing).
- Change in pressure drop across the combustor—This affects the gas flow through the turbine, since the combustor exit pressure is proportional to P turbine inlet plane.
- Change in the secondary cooling flows—Combustor metal liners are affected by erosion and corrosion, which increase the cooling air holes and thus the cooling air consumption. On the other hand, fuels containing ash might partially block the cooling holes, creating high temperature spots on the liner surface.

External Factors Affecting Degradation

Gaseous Fuel

In general, gaseous fuels are considered more suitable than liquid fuels for continuous base-load operation. Fuel quality has an important role in performance degradation of gas turbines. The ideal fuel is a clean, dry gas, with a low sulfur content, which can originate from different sources: pipeline quality natural gas, vaporized LNG or LPG, and process synthesis gas.

Fuels containing traces of metal create deposits on the nozzles and blades, impacting GT integrity and performance. In one particular application, Ni deposits concentrated on the tip of the burner forced a burner redesign. In many other cases, the deposits plug the blades' cooling holes and create areas of overheating, which eventually lead to cracks in the material. Special coatings on the blades are often used to minimize the corrosion. Another negative effect of deposits, as mentioned above, is accelerated thermal performance degradation. In the current market, with gas prices near record levels, increased demand, and strained supplies, there is little incentive for the producers to maintain a consistent gas quality. Traditionally, gas liquids, such as propane and butane, were removed and sold as chemical feed-stocks because their value was higher than in the fuel gas market. Not any more.

Some experts estimate that as much as one third of the gas produced in the US is no longer processed to remove liquid hydrocarbons. Therefore, the physical and chemical characteristics of pipeline gas are significantly different from the "original design fuel." This "out of specification fuel" can create mechanical problems and accelerate the degradation process.

Liquid Fuel

Heavy-duty gas turbines are capable of burning a variety of liquid fuels, ranging from light petroleum distillates to heavy residuals. The combustion process of light distillates such as naphtha, kerosene, and No. 2 fuel oil is well understood and has been proven through millions of hours of operation. The degradation rates on these types of liquid fuels are predicted to be 20% higher than on natural gas fuel. The real challenges for GT design and operation come from the use of heavier ash-forming fuels. The combustion of these fuels can be extremely detrimental to GT life, due to corrosion, erosion, and fouling of the blades. Hot corrosion, caused by traces of elements such as sodium (Na), potassium (K), lead (Pb), and particularly vanadium (V) in the fuel, has a major effect on GT performance and the service life of the hot gas path. The number of equivalent operating hours (EOH), which serves as the basis for degradation and maintenance intervals, increases more rapidly based on fuel contaminant concentration, accounting for a factor of 1.5 for each operating hour (by certain estimates).

Water

The effect of water on GT degradation depends on its use and point of entry. At the inlet of the compressor, droplets of water generated by the fogging system may cause erosion on the blades and corrosion of the inlet ducting and plenum. Water and steam injection in the combustion system is used for NO_x reduction. On one hand, the increased hot gas mass flow causes a rise in the pressure ratio and power output. Thus, the mechanical loading of the blades is higher. On the other hand, the higher moisture content in the hot gas affects the heat transfer properties and consequently increasing the metal temperature of the blades. Finally, exposure to gases with higher moisture content reduces the wear resistance of the turbine blades' coatings. Therefore, a weighting factor on the number of operating hours should be applied as a function of the amount of water injected.

Marine and Offshore Environment

The airborne salt in different forms (aerosol, spray or crystal) is impacting the fouling of the compressor. High environmental relative humidity causes the salt to remain as supersaturated droplets. The database concerning the amount and form of the salt content indicates considerable variation and dependency on the wind speed on the level of contamination.

Means of Reducing Degradation

Recoverable and Non-Recoverable Degradation

The process of degradation cannot be avoided, but its impact on performance and hardware life can be mitigated by undertaking a number of preventive actions, starting with conceptual design, and continuing through the project execution and operation and maintenance activities. Here it is appropriate to introduce the definitions of recoverable and non-recoverable degradation. The recoverable degradation is considered the portion of the GT performance, which can be restored by a series of maintenance activities without opening the machine or replacing major parts. This degradation recovery is commonly achieved by conducting compressor washing and minor engine adjustments (such as resetting the inlet guide vanes). The remaining degradation is regarded as non-recoverable. Equipment manufacturers predict the non-recoverable degradation for the equipment in the form of degradation curves.

Figure 12–68 provides an example of recoverable degradation by online and offline compressor water wash. It should be noted that these notions of recoverable and non-recoverable degradation could be confusing, since after major overhauls, the GT performance can be restored to almost the level achieved in "new and clean" conditions.

Limits on allowable degradation, recoverable and non-recoverable from base load conditions, should be established with the equipment manufacturers. Operating below these limits increases the risk of compressor surge. In the project development phase, the impact of the recoverable and non-recoverable degradation, scheduled outages, etc., should be duly accounted for in the project pro forma analysis.

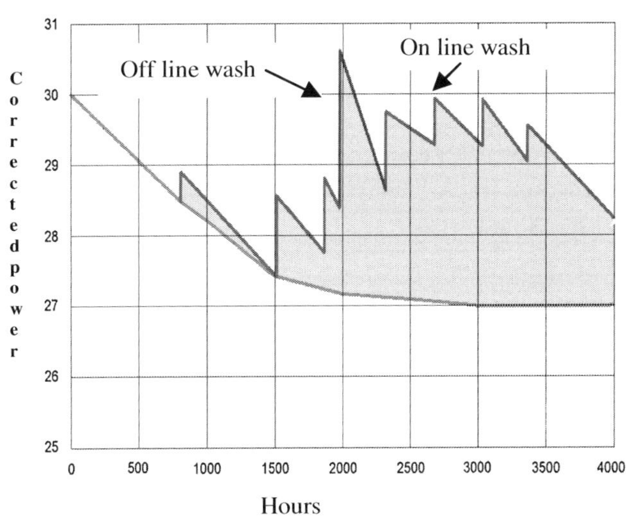

FIGURE 12–68 Recoverable degradation. *(Source: Courtesy of Turbotect Ltd.)*

Fuel Treatment

During the initial design phase, actual specifications should be provided for fuel quality entering the gas turbine. If GT manufacturer fuel requirements cannot be met, a comprehensive precombustion fuel treatment must be developed and implemented, aimed at reducing the concentration of contaminants. It is highly recommended to implement for gaseous fuels an online, real-time, continuous composition monitoring system, based on gas chromatography. The pretreatment process for liquid fuels should involve chemical reactions and/or mechanical systems. Na and K salts are soluble in water and can be removed by the on-site treatment known as fuel washing. Other methods of fuel treatment include the use of high-temperature corrosion additives. To reduce the amount of expensive additives, a preliminary step for improving fuel quality entails using of commercially available mechanical fuel treatment skids composed of centrifugal separators, pumps, and sludge tanks. The devices can reduce the concentration of harmful contaminants (Na, K, S, and V). A typical two-stage system can reduce the Na content from 150 ppm to 1 ppm for heavy salts.

Air Filtering System

The air filtering system must have the capability to eliminate 99.95% of all particulates above 20 μm, since these particles are the main source for compressor erosion. Installations located in the desert, in coastal areas or near dust emitting sources (coal dust, cement dust, fly ash, and pollen) should be equipped with self-cleaning systems that have an effective dust removal system. The ambient air for many combined cycles located close to refineries contains air-borne heavy hydrocarbon matter that eventually could sinter on compressor blades and cause non-recoverable degradation. In the absence of an efficient and proper designed inlet system, the ingestion of humid and salty air at a seaside location may lead to erosion and corrosion in the compressor as well as in the turbine section of the gas turbine. Finally, it is recommended that an ambient air quality analysis be conducted during initial site development, as extremely useful information.

Degradation Curves

To cover the gap from the "new and clean" performance guaranteed by the manufacturer to the performance measured during the test, equipment degradation curves have been created. The predicted degradation curve, in logarithmic shape, is a function of operational hours. This trend represents the non-recoverable degradation, which cannot be restored without opening of engine covers to refurbish the components back to a new and clean condition. Such events occur during hot gas path inspection or major overhaul. This cumulative degradation curve represents the contributions of each individual engine components. Since degradation, as described above, is not simply affected by the number of fired hours, some GT manufacturers created a methodology where degradation is a function of EOH, a weighted number of hours accounting for different operating conditions: type of fuel, number of starts, stops, trips, sudden changes in load, and use of diluents. For example, a normal start adds 10 EOH and each hour operating with water injection is considered to add 1.5 EOH. An hour of peak load operation at increased firing temperature equals 3 EOH. Some manufacturers indicate that each trip could represent between 25 and 400 EOH, depending on the GT load at the time of the trip. The determination of the weighting factor for each of the operational conditions on the EOH value is done solely by the equipment manufacturers. It is assumed that these factors are determined on the basis of the fleet performance experience and inhouse modeling.

Operations and LTSA (Long-Term Service Agreement)

Assessing the GT degradation is important not only for acceptance performance tests but also for day-to-day operations. Asset managers/operators should carefully plan the operation and scheduled maintenance of the gas turbine, to keep the number of EOH as low as possible. However, in the current opportunistic power market, decisions on the way to operate the equipment are guided by commercial considerations rather than technical. To mitigate the uncertainty associated with maintenance costs, many owners enter in a maintenance contract often called a long-term service agreement (LTSA). An LTSA typically commits the OEM or non-OEM service contractor to provide, on a relative, fixed price basis, parts and maintenance services for gas or steam turbines.

Commercially, an LTSA should offer many advantages to owners, including the predictability of relatively fixed long-term costs, while keeping the OEM motivated through performance bonuses to provide support and ensure a predetermined level of performance (see Thomson, R., and J. Yost. "Avoiding Pitfalls in Service Contracts." *Turbo Machinery*).

An LTSA must be clearly written and discussed, so that the parties involved do not have false expectations about the extent of its coverage. In many LTSA the validation of the actual GT degradation is an important element of the agreement. The suggestions and recommendations concerning the evaluation of performance degradation made above should be implemented in the LTSA documents and carried out in the field.

Case Study 14: Remaining Life Assessment of Power Turbine Disks*

This case presents the results of a study conducted to determine the life expectancy of a power turbine disk. The purpose of the study is to revisit the original design calculations with current numerical computing techniques. By determining the state of stress and temperature in the vicinity of the stress concentrations, combined with material properties, the life expectancy of power turbine disks can be established.

Most industrial power turbines commissioned in the early 1970s have accumulated approximate 200,000 operating hours, well beyond the minimum design life reported by some OEMs. Operators running into this situation have to make decisions on whether or not to replace the most expensive rotors and disks. Some operators are willing to take certain calculated risks, in return to gain value through reliable life extensions. Their decision making would be largely based upon the real life expectancy of the components and the risk of failure.

Power turbine rotor disks are subject to high loads and moderate temperatures. Metallurgically, the remaining life of power turbine disks is in general limited by the remaining creep or fatigue life. When a failure occurs in a disk, it is usually due to the crack initiation in the serrations where the stress concentrations are the greatest. The design life limits established by OEMs on those power turbines were based on the technologies in the 1970s, e.g., analytical approach, past experience, and understanding of material response to the expected operating envelope. The objective of the remaining life analysis is to revisit the original design calculations with current numerical computing techniques and determine the real life expectancy of power turbine disks.

Background

Turbine rotor disks are highly stressed in the rim area where the blade root attachment occurs and in the hub of bored disks where high burst strength is required. Temperatures are moderately low in the power turbine disk hub but can be high in the rim depending on the blade and cooling design. Rotor disks are also subject to high cycle fatigue loads from blade vibration and low cycle fatigue loads from start/stop cycles.

A combination of high tensile strength, fatigue resistance and moderate creep resistance is required for disk materials (see Schilke, P.W., "Advanced Gas Turbine

* Source: [12-5]: Copyright: Remaining Life Assessment of Power Turbine Disks by H Jin, P Lowden, Proceedings of ASME Turbo Expo 2008: Power for Land, Sea and Air, Paper number GT2008-51010, copyright 2008, Courtesy ASME

Materials and Coatings," GER-3569F). The materials used for industrial turbines rotor disks in the early 1970s were high strength steels (Cr-Mo-V), 12Cr stainless steels, and Fe-Ni base super-alloys (e.g., A-286). Table 12—8 lists the chemical composition of common industrial turbine rotor disk materials.

During service, turbine rotor disk materials undergo various types of time and cycle dependent degradations due to exposure in the operating environment. Data are rarely available for metallurgical characterization of gas turbine rotor disk materials after long-term service exposure. One reference reported material testing results of samples taken from a high-pressure steam turbine rotor that had been operated for 167,500 hours. The test results indicated that the rotor steel was slightly softened and temper embrittled as a result of service exposure after 26 years.

Metallurgically, the remaining life of power turbine rotor disks is in general limited by the remaining creep or low cycle fatigue life. High cycle fatigue (HCF) is not considered since HCF damage can be accumulated very rapidly under non-design conditions. As such, turbine rotors are not designed with an HCF life—they should have infinite life under normal operating conditions.

Technical Approach

Temperature, stress, and material are the three major factors determining life consumption. To estimate the temperature distribution across the rotor disk, an engine aero-thermal model, a blade external and internal heat transfer model, and an overall disk and blade assembly heat transfer model have to be developed. The engine aero-thermal model calculates gas path temperature, pressure, and flow. The calculated gas path aero-thermal values are then used as boundary conditions for the blade/disk assembly heat transfer analysis.

TABLE 12—8 Chemical Composition of Disk Materials (% wt) [12-5]

Element	ASTM-A470 Class 8	AEI T-G	A286 Spec. MS-618A/D/E
Ni	0.75 max.	3.0—3.75	24.0-27.0
Cr	0.90—1.50	0.8—1.6	13.5-16.0
Si	0.15—0.35	0.30 max.	0.40 max.
Mn	1.00 max.	0.25—0.55	0.50 max.
C	0.25—0.35	0.25—0.35	0.08 max.
Mo	1.00—1.50	0.40—0.60	1.0-1.5
V	0.20—0.30	0.10—0.15	0.10-0.50
S	0.018 max.	0.025 max.	0.025 max.
P	0.015 max.	0.025 max.	0.025 max.

The blade root and disk interaction is quite complex. The blade root and disk groove have complicated geometries with numerous grooves where stress concentrations develop. Elastic-plastic deformation occurs, especially in the regions of the stress concentrations. The load applied to the blade root is transmitted to the disk via contact surfaces of the lands. The geometry of the blade root and disk can be measured using a computer-controlled coordinate measurement machine (CMM) or by a precisely calibrated optical comparator where the blade root and the disk root form casting are magnified by ~10 times.

A plane strain/stress analysis of the blade root and disk interaction can be conducted using a finite element computer program to calculate stress concentrations or stress intensity factors. Temperature-dependent material properties have to be developed for the FEA model. When failure occurs in a disk, it is typically due to cracks that develop in the disk grooves where the stress concentrations are the greatest. With determining the state of stress and temperature in the vicinity of those stress concentrations, combined with the time dependent be estimated non-destructively. In this analysis, the disk was operated in continuous operation, so low cycle fatigue was not considered to be a life-limiting factor.

Geometry

The power turbine studied is a two-stage design with cooling air bleeding from the gas generator compressor to the first stage disk rim. The power turbine has accumulated 183,400 operating hours since installation in the late 1970s. The power turbine inlet temperature at ISO base load according to the published information is 1240°F and the exhaust temperature is 849°F. The same power turbine model is also packaged with gas generators with higher exhaust temperatures. The life expectancies of the power turbine disks under the elevated turbine inlet temperature conditions were also evaluated (Figure 12—69).

Aero-Thermal Analysis

Engine Modeling

To simulate the blade- and disk-operating environment, an aero-thermal model for the power turbine was developed. The aero-thermal model is a two-stage turbine mean-line program with the gas path geometry constructed from reverse engineering actual hardware. The aero-thermal program calculates hot gas path temperature, pressure, and flow across the turbine stages using the published power turbine performance data at the rated conditions. The calculated aero-thermal results relative to the stage 1 blades were then used as the boundary conditions for the blade heat transfer analysis.

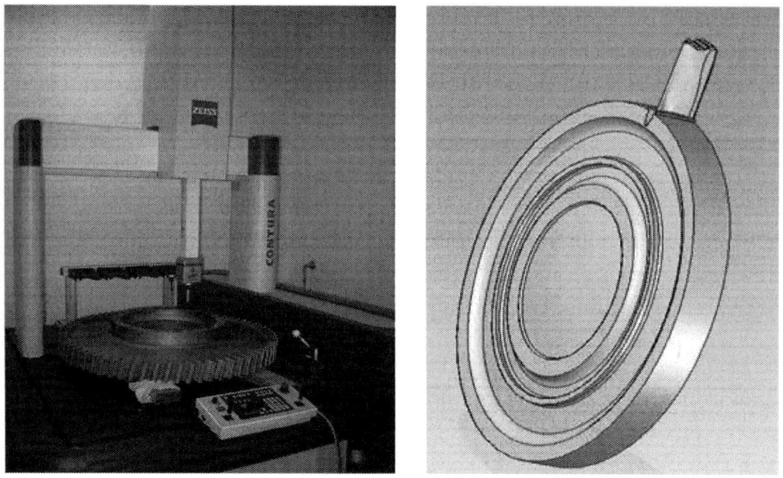

FIGURE 12–69 Dimensional measurement. [12-5]

Disk Cooling

The operating temperature of the first stage power turbine disk is strongly dependent on the cooling air temperature inside the front disk cavity. The power turbine rim cooling system is a series of pipes, connectors, flexible hoses, drilled nozzle vanes and temperature probes. Cooling air is bled from the gas generator intermediate compressor stage to the power turbine first stage disk rim and nozzles. It is designed to protect the first stage turbine disk from the extreme heat of gas generator exhaust gases.

From the power turbine historical operating data, it was found that the measured disk cavity temperature is a function of power turbine inlet temperature (Figure 12–70). The average measured disk cavity temperature based on the typical power turbine load conditions was used in the analysis.

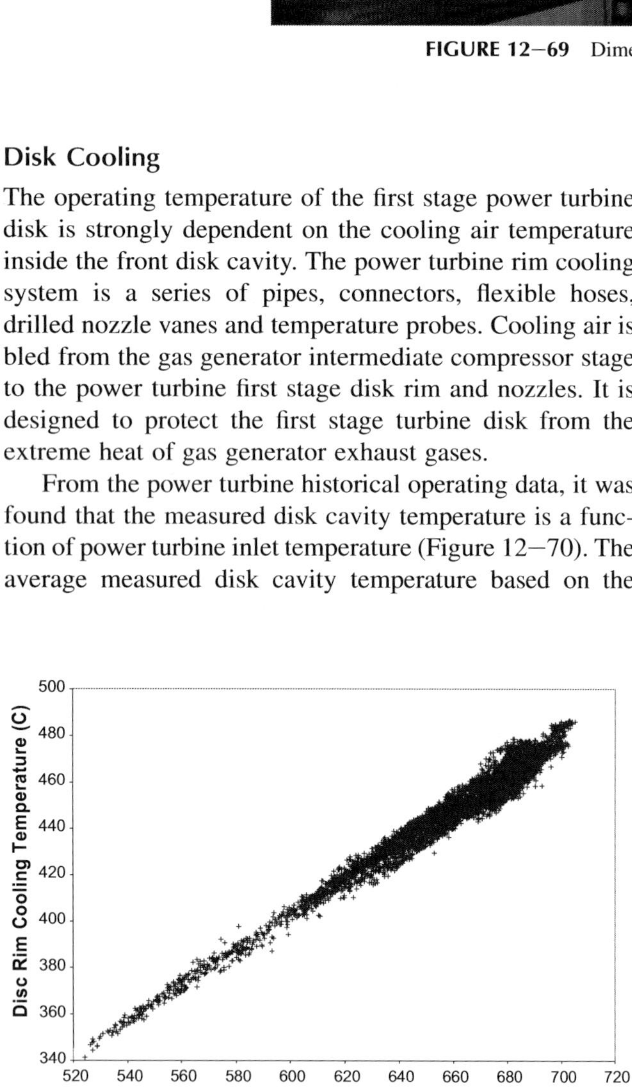

FIGURE 12–70 Disk cavity temperature. [12-5]

The case goes on to detail work that involved blade/disk heat transfer calculations, peak stress, stress relaxation, creep rupture, probabilistic modeling, and then life expectancy calculations. Comparisons can then be made between different turbine disks. The calculations are too lengthy to repeat in this case.

In summary, an engineering approach has been developed to analyze real life expectancy of aged power turbine disks. The approach integrates mechanical design, analytical calculation, aero-thermal analysis, finite element modeling and probabilistic modeling of material properties. Applying the approach to a power turbine disk with 183,400 service hours, it was found that the disk could run for an infinite life assuming continued operation at the current conditions.

This conclusion is consistent with the result of a nondestructive examination conducted prior to the analysis. The dimensional inspection indicated no measurable plastic deformation per engineering drawings, while fluorescent penetrant testing found no cracking. The good dimensional stability after 27 years of operation suggests that it is unlikely the disk will fail due to primary stress related ductile bursting in the near future.

Since the power turbine inlet temperature and cooling air system have a significant impact on the disk metal temperature, and hence the life, operators should implement a fleet-wide monitoring program that would include the disk cavity temperature measurement and periodic inspections for any unforeseen damage mechanisms to ensure safe long-term operation of their power turbine disk fleet.

Chapter 13

Installation

"Throw your dreams into space like a kite, and you do not know what it will bring back, a new life, a new friend, a new love, a new country."

—Anais Nin

Chapter Outline

INSTALLATION OF AIRCRAFT ENGINES*

When a gas turbine engine is installed in an aircraft it usually requires a number of accessories fitting to it and connections made to various aircraft systems. The engine, jet pipe, and accessories, and in some installations a thrust reverser, must be suitably cowled and an air intake must be provided for the compressor, the complete installation forming the aircraft power plant.

Power Plant Location

The power plant location and aircraft configuration are of an integrated design and this depends upon the duties that the aircraft has to perform. Turbo-jet engine power plants may be in the form of pod installations that are attached to the wings by pylons (Figure 13—1), or attached to the sides of the rear fuselage by short stub wings (Figure 13—2), or they may be buried in the fuselage or wings. Some aircraft have a combination of rear fuselage and tail-mounted power plants, others, as shown in Figure 13—3, have wing-mounted pod installations with a third engine buried in the tail structure. Turbo-propeller engines, however, are normally limited to installation in the wings or nose of an aircraft.

The position of the power plant must not affect the efficiency of the air intake, and the exhaust gases must be discharged clear of the aircraft and its control surfaces. Any installation must also be such that it produces the minimum drag effect.

Power plant installations are numbered from left to right when viewed from the rear of the aircraft.

Supersonic aircraft usually have the power plants buried in the aircraft for aerodynamic reasons. Vertical lift aircraft can use either the buried installation or the podded power plant, or in some instances both types may be combined in one aircraft.

Air Intakes

The main requirement of an air intake is that, under all operating conditions, delivery of the air to the engine is achieved with the minimum loss of energy occurring through the duct. To enable the compressor to operate satisfactorily, the air must reach the compressor at a uniform pressure distributed evenly across the whole inlet area.

The ideal air intake for a turbo-jet engine fitted to an aircraft flying at subsonic or low supersonic speeds is a short, pitot-type circular intake (Figure 13—4). This type of

* Source: Adapted, with permission, from Rolls Royce. *The Jet Engine,* 1986, Rolls Royce Plc: UK.

FIGURE 13–1 Wing-mounted pod installation. *(Source: Rolls Royce.)*

FIGURE 13–2 Fuselage mounted pod installation. *(Source: Rolls Royce.)*

FIGURE 13–3 Tail- and wing-mounted pod installation. *(Source: Rolls Royce.)*

FIGURE 13–4 Pitot-type intake. *(Source: Rolls Royce.)*

intake makes the fullest use of the ram effect on the air due to forward speed, and suffers the minimum loss of ram pressure with changes of aircraft attitude. However, as sonic speed is approached, the efficiency of this type of air intake begins to fall because of the formation of a shock-wave at the intake lip.

The pitot-type intake can be used for engines that are mounted in pods or in the wings, although the latter sometimes require a departure from the circular cross-section because of the wing thickness (Figure 13–5).

Single engined aircraft sometimes use a pitot-type intake; however, because this generally involves the use of a long duct ahead of the compressor, a divided type of intake on each side of the fuselage is often used (Figure 13–6).

The disadvantage of the divided type of air intake is that when the aircraft yaws, a loss of ram pressure occurs on one side of the intake, as shown in Figure 13–7, causing an uneven distribution of airflow into the compressor.

FIGURE 13–5 Wing leading edge intakes. *(Source: Rolls Royce.)*

FIGURE 13–6 Single engined aircraft with fuselage intakes. *(Source: Rolls Royce.)*

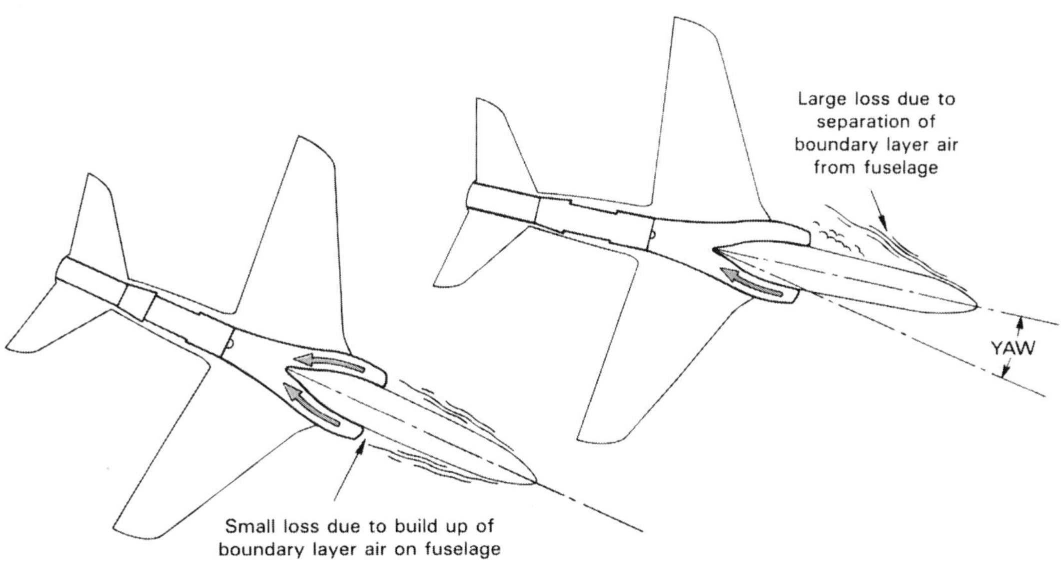

FIGURE 13–7 Loss of ram pressure in divided intakes. *(Source: Rolls Royce.)*

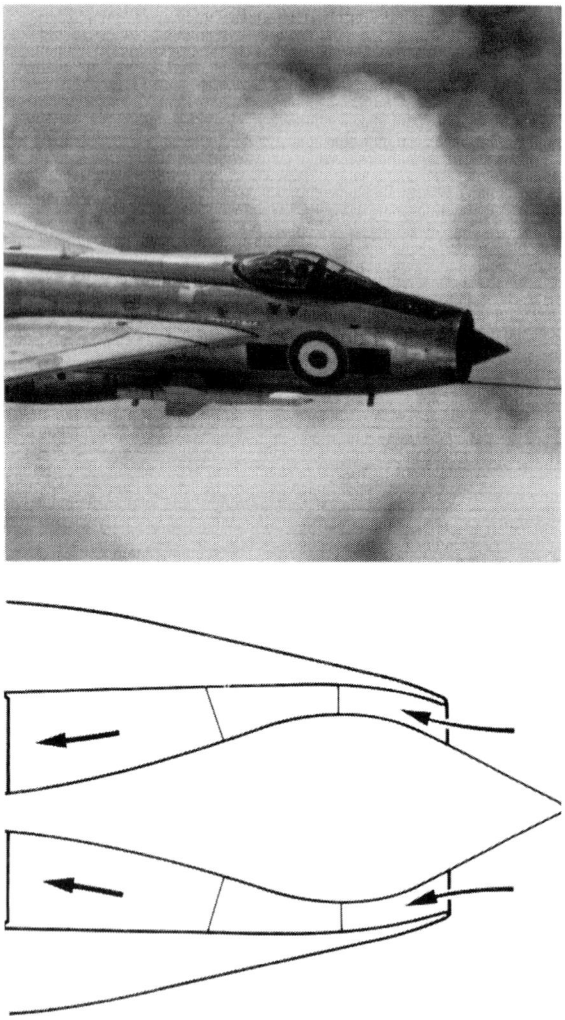

FIGURE 13–8 External/internal compression intake. *(Source: Rolls Royce.)*

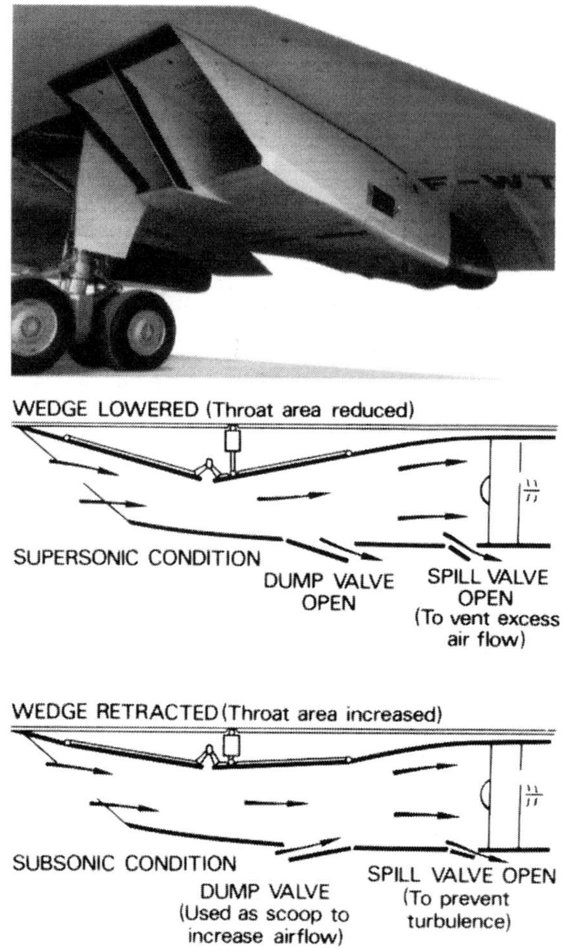

FIGURE 13–9 Variable throat area intake. *(Source: Rolls Royce.)*

At higher supersonic speeds, the pitot type of air intake is unsuitable due to the severity of the shockwave that forms and progressively reduces the intake efficiency as speed increases. A more suitable type of intake for these higher speeds is known as the external/internal compression intake (Figure 13–8). This type of intake produces a series of mild shockwaves without excessively reducing the intake efficiency.

As aircraft speed increases still further, so also does the intake compression ratio and, at high Mach numbers, it is necessary to have an air intake that has a variable throat area and spill valves to accommodate and control the changing volumes of air (Figure 13–9). The airflow velocities encountered in the higher speed range of the aircraft are much higher than the engine can efficiently use; therefore, the air velocity must be decreased between the intake and the engine air inlet. The angle of the variable throat area intake automatically varies with aircraft speed and positions the shockwave to decrease the air velocity at the engine inlet and maintain maximum pressure recovery within the inlet duct. However, continued development enables this to be achieved by careful design of the intake and ducting. This, coupled with auxiliary air doors to permit extra air to be taken in under certain engine operating conditions, allows the airflow to be controlled without the use of variable geometry intakes. The fuselage intakes shown in Figure 13–10 are of the variable throat area type.

Engine and Jet Pipe Mountings

The engine is mounted in the aircraft in a manner that allows the thrust forces developed by the engine to be transmitted to the aircraft main structure, in addition to supporting the engine weight and carrying any flight loads. Because of the wide variations in the temperature of the engine casings, the engine is mounted so that the casings can expand freely in both a longitudinal and a radial direction. Types of engine mountings, however, vary to suit the particular installation

FIGURE 13–10 Fuselage intakes. *(Source: Rolls Royce.)*

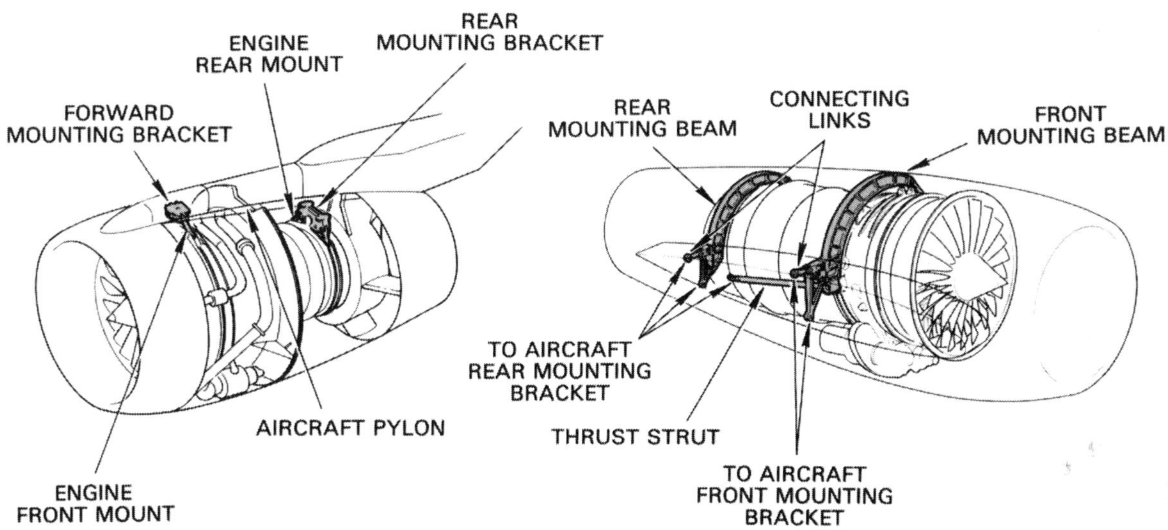

FIGURE 13–11 Typical turbo-jet engine mountings. *(Source: Rolls Royce.)*

requirement. Turbo-jet engines are usually either side mounted or underslung as illustrated in Figure 13–11. Turbo-propeller engines are mounted forward on a tubular framework as illustrated in Figure 13–12.

The jet pipe is normally attached to the rear of the engine and supported by the engine mountings. In some installations, particularly where long jet pipes are employed, an additional mounting is provided, usually in the form of small rollers attached to each side of the jet pipe. The rollers locate in airframe-mounted channels and support the weight of the jet pipe, while still allowing it to freely expand in a longitudinal direction.

Accessories

An aircraft power plant installation generally includes a number of accessories that are electrically operated, mechanically driven, or driven by high-pressure air.

Electrically operated accessories such as engine control actuators, amplifiers, air control valves, and solenoids, are supplied with power from the aircraft electrical system or an engine-driven dedicated electrical generator.

Mechanically driven units, such as generators, constant speed drive units, hydraulic pumps, low- and high-pressure fuel pumps, and engine speed signaling, measuring, or governing units are driven from the engine through internal and external gearboxes.

Air-driven accessories, such as the air starter and possibly the thrust reverser, afterburner, and water injection pumps, are driven by air tapped from the engine compressor. Air conditioning and cabin pressurization units may have a separate air-driven compressor or use air direct from the engine compressor. The amount of air that is taken for all accessories and services must always be a very small percentage of the total airflow, as it represents a thrust or power loss and an increase in specific fuel consumption.

Cowlings

Access to an engine mounted in the wing or fuselage is by hinged doors; on pod and turbo-propeller installations the main cowlings are hinged. Access for minor servicing is by small detachable or hinged panels. All fasteners are of the quick-release type.

A turbo-propeller engine, or a turbo-jet engine mounted in a pod, is usually far more accessible than a buried engine because of the larger area of hinged cowling that can be provided. The accessibility of a podded turbo-fan engine is shown in Figure 13—13 and that of a turbo-propeller engine is shown in Figure 13—12.

FIGURE 13—12 Engine accessibility, turbo-propeller engine. *(Source: Rolls Royce.)*

FIGURE 13—13 Engine accessibility, turbo-fan engine. *(Source: Rolls Royce.)*

The fan cowl doors provide a continuous external aerodynamic surface for the nacelle while allowing easy access to all the engine fan case counted accessories. This access is achieved by having the fan cowl doors hinged to the aircraft pylon.

The top half of each cowl door is fire proof as the volume underneath them is a designated fire zone containing the fuel pumps and fuel lines. The airflow under the fan cowl doors must not be obstructed, so that the engine accessories are cooled and ventilated.

Fan cowl doors are typically made from composite materials with a number of access panels for maintenance. Some large fan cowl doors might have powered opening devices; all have "hold open" rods that have to ensure safe opening on the ground in winds of up to 110 kmh (60 knots).

Configuration is a short, near-circular, pitot-type intake. This design is highly efficient for subsonic operation, as low levels of pressure loss are achieved under all operating conditions. Civil pitot intakes are suitable for wing and rear fuselage mounted nacelles. For tri-jet configurations, s-duct intakes are a design option when the engine is buried in the rear fuselage.

Civil Nacelles

A nacelle is a streamlined enclosure that fits around a dressed engine and interfaces with the aircraft structure.

The primary objectives of a nacelle are to

- provide low drag, achieved through aerodynamic design of the nacelle itself and its interaction with the fuselage and smooth surfaces
- ensure good engine performance throughout the aircraft flight and ground envelopes
- reduce engine noise with acoustic treatment of nacelle structure
- prevent ice impact damaging fan blades.

The nacelle must also be manufactured cost effectively, and be easy to install and remove. This must be achieved at the lowest possible weight, while remaining durable and repairable in service.

Nacelles are composed of an air intake, fan cowl doors, nozzle and tail cones, and, optionally, a thrust reverser. Typical civil turbofan nacelles are fitted under the wing on a pylon or fitted to the rear fuselage via a stub wing (Figure 13—14).

Two further nacelle options are the long (mixed exhaust) nacelle—the Trent 700—and the short (separate core and bypass jets) nacelle—Trent 800 and 500. Long nacelles can give a performance gain for some engines due to the mixing of the exhaust, and also have a greater acoustic treatment area, at the cost of extra weight and drag (Figures 13—15 and 13—16).

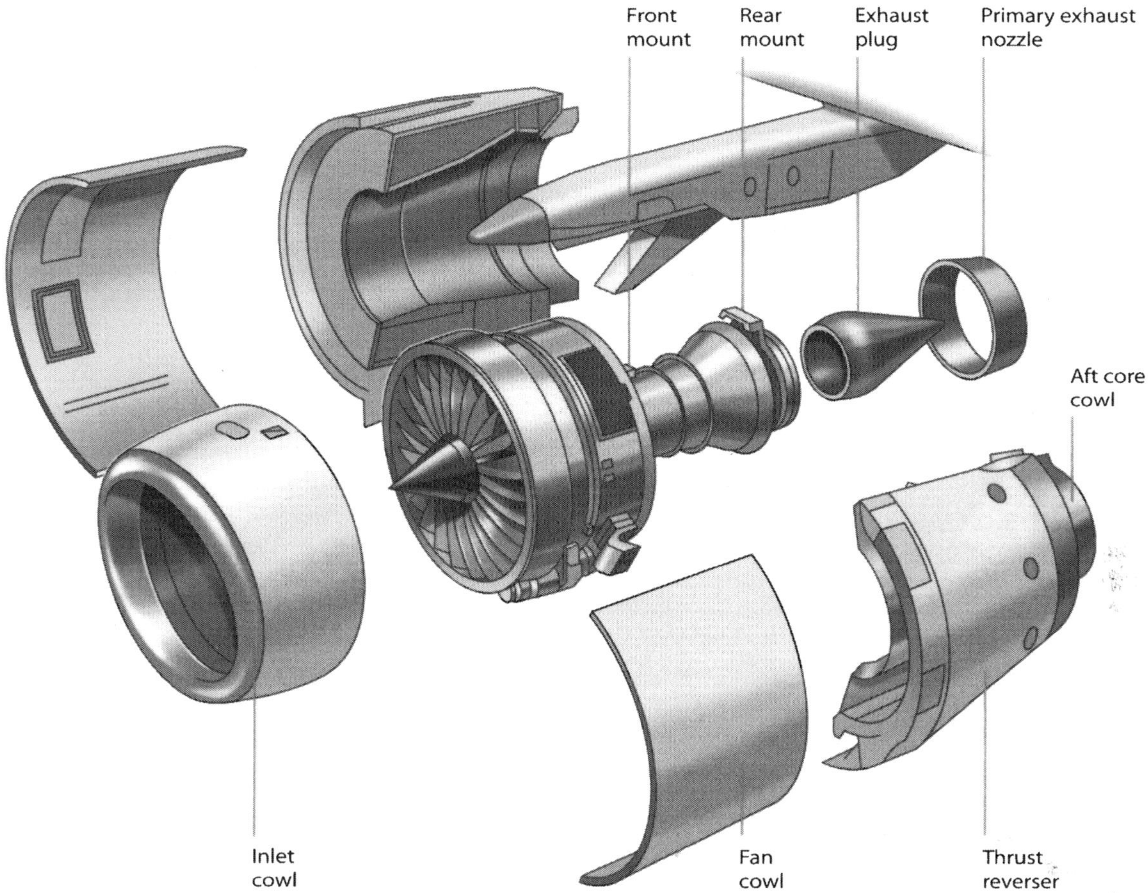

FIGURE 13–14 Typical nacelle elements as on the Trent 500.

FIGURE 13–15 The fuselage-mounted BR710 on the Gulfstream GV.

FIGURE 13–16 The underwing-mounted Trent 700, showing the long nacelle option for mixing bypass and core jets before exhausting to the atmosphere.

The Air Intake

The purpose of the intake of civil turbofan engines is to ensure that, under all operating conditions, the engine is supplied with the correct quantity of air, and that the air has sufficient flow uniformity to allow efficient and stable engine operation. The intake design is integrated into the engine nacelle to obtain the lowest level of drag at the operating design point, typically cruise.

For civil turbofans, the optimum intake configuration is a short, near-circular, pitot-type intake. This design is highly efficient for subsonic operation, as low levels of pressure loss are achieved under all operating conditions. Civil pitot intakes are suitable for wing and rear fuselage-mounted nacelles. For tri-jet configurations, s-duct intakes are a design option when the engine is buried in the rear fuselage (Figures 13–17 and 13–18).

A pitot intake consists of two geometric regions, the "lip" and the "diffuser." The forward section, the lip, is similar in section to an aerofoil and is shaped to guide the airflow into the engine under all operating conditions; the lip is also optimized to prevent flow separation under cross-wind and incidence operation. If flow separation occurs, this produces significant asymmetry in the total pressure within the intake increasing fan blade stresses and, in severe cases, may result in engine surge.

The internal surface of the lip is heated with hot air to ensure that there is no ice accretion, which could shed and damage the engine fan blades.

The lip contracts to a minimum area, sized for the engine flow requirements, known as the throat. Aft of the throat, the airflow passes into the diffuser where the flow area is increased up to the fan entry plane. The diffuser acts as a settling length to improve the uniformity of the airflow entering the fan. The diffuser section is lined with sound-absorbing, acoustic panels to reduce noise emissions.

Civil intake aerodynamic surface shapes are generated as mathematically defined 3D surfaces using CAD tools. The intake surface designs are evaluated and optimized using CFD codes, and the final design is validated by wind tunnel testing.

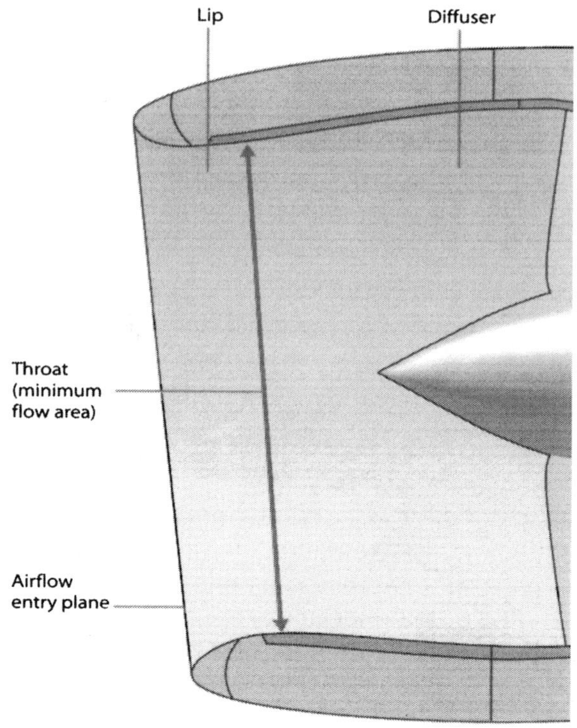

FIGURE 13–17 Civil pitot intake geometric features.

Before flight testing, to demonstrate that the intake and engine are fully compatible, further testing is conducted over a full range of flow conditions using a machine to simulate high cross-wind speeds.

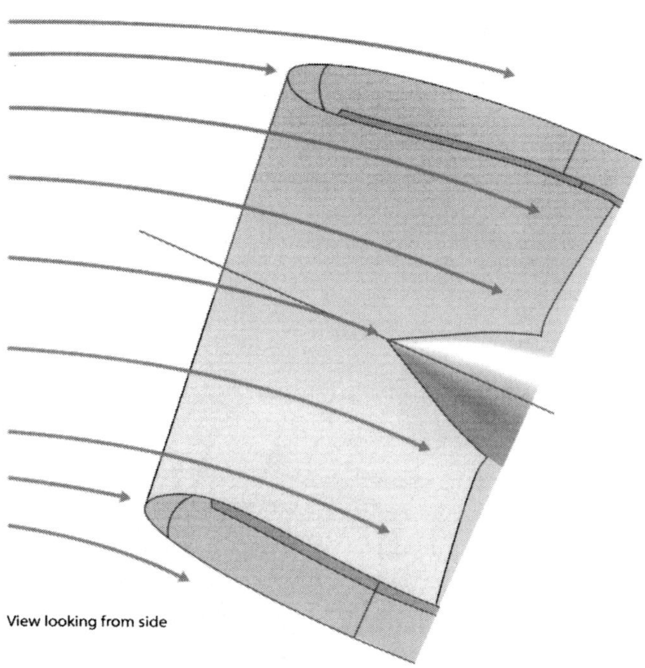

FIGURE 13–18 High incidence climb intake turning airflow onto engine axis.

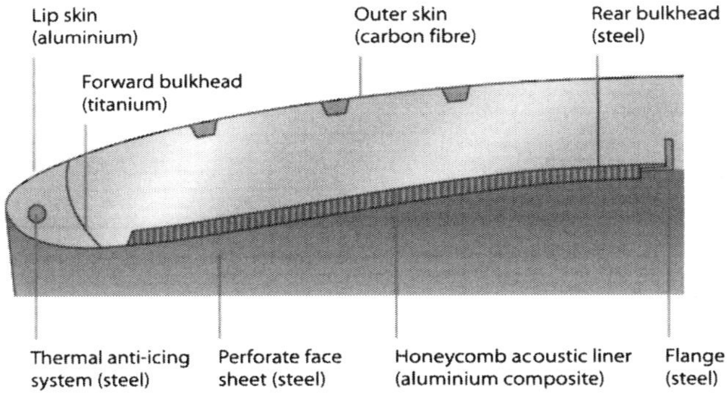

FIGURE 13–19 Typical intake construction materials.

Intake construction is, typically, an aluminium intake lip for durability and compatibility with the ice protection system, a composite outer skin, an acoustic honeycomb inner barrel, and metal structure bulkheads. The intake is normally bolted onto the engine fan case. Some air intakes might not be circular due to ground clearance constraints and non-uniform accessory distribution around the fan case (Figure 13–19).

Following the retirement of Concorde, no immediate replacement existed for civil supersonic air travel. However, any future civil supersonic engines would probably use an external/internal compression intake with variable geometry similar to that used on Concorde and current military aircraft; such an intake provides higher efficiency at supersonic speed.

Thrust Reverser

The thrust reverser unit (TRU) has three key functions: to provide a continuous external aerodynamic surface for the nacelle; to provide a fan flowpath for the engine in forward thrust mode; and, of course, to reverse the exhaust flow after the aircraft touches down to assist with aircraft deceleration.

Generally, pilots and airlines want TRUs on jet engines. They can reduce aircraft landing distance, especially on wet and icy runways, while also reducing brake and tyre wear. They improve ground handling on wet and icy runways and taxiways—and improve rejected take-off margins in similar conditions. For military applications, TRUs provide the possibility of operating from bases with shorter runways giving greater operational flexibility.

In civil applications, no certification credit on landing distances is given for TRU fitment; an aircraft landing distance will be determined by the use of anti-lock brakes and aerodynamic drag devices such as flaps, airbrakes, and parachutes. TRUs are not essential for safe landing; however, they do provide increased safety.

There are four main types of TRU in use today:

- Translating sleeve and pivot door (fan air systems). These are used for large turbofan engines as the majority of the thrust is generated by the fan.
- Target door and pivot door (mixed stream systems). These are used for small, low bypass ratio engines as the split of fan and core thrust is more equal.

Most large fan engines have "C" duct TRUs that are split into two halves and hinged to the aircraft pylon, providing access to the engine core components. The TRU also provides a discrete fire zone containing fuel pipes, fuel nozzles, and combustion chambers.

Part of the thrust reverser optimization process includes ensuring that the hot air/gases neither impinge on the aircraft wing or fuselage nor are re-ingested into the engine intake, which could cause engine surge. It is also important to minimize any lift component from the thrust reverser in order to maximize braking efficiency.

Typically, most TRUs are constructed from carbon composite panels, aluminum structural beams, a metal firewall bulkhead, and a suitable thermal blanket (usually stainless steel) on the inner wall to ensure the epoxy in the carbon composite can withstand the combustion and turbine case temperatures.

Actuation of the translating sleeves or pivot doors is either hydraulic or electrical and three separate locks are provided to ensure there is no TRU deployment in flight. One of these locks will be separately operated, while the other two will be operated and controlled by the engine. For a short nacelle, the TRU also forms the cold, bypass air nozzle.

Nozzles and Tail Cones

There are two types of nacelle nozzle: the combined cold fan and hot gas nozzle, as seen, for example on the long nacelle of the Trent 700; and the hot gas nozzle, seen, for example, on the short nacelles of the Trent 500 and 800. Tail cones are standard and vary only in their length, cone angle, and whether or not they are acoustically treated. A combined nozzle assembly can reduce engine noise emissions by the fitting of acoustic honeycomb panels. Both combined and hot gas nozzles are fitted to the engine LP turbine flange. The tail cone is fitted to the turbine bearing housing at the engine center and provides a smooth increase in the hot gas exit transitional area (Figure 13–20).

Military Fuselage Intakes

Where, as in most military installations, the engine or engines are accommodated within the aircraft, the intakes are incorporated into either the fuselage or wing roots and become a much more integrated part of the aircraft design. As with the civil nacelle intake, the main requirement for such intakes is to supply air to the engine with the minimum loss of pressure and the least increase in aircraft drag. Similarly, for the compressor to operate efficiently and stably, the air delivered to the engine face must be of an acceptable quality in terms of

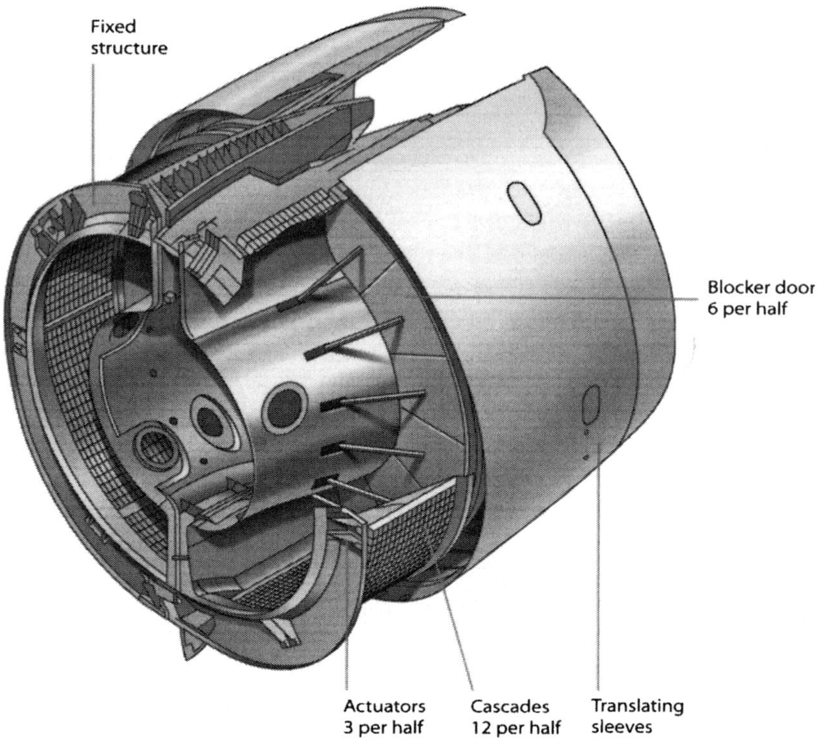

FIGURE 13–20 The translating sleeve thrust reverser unit on the Trent 500.

velocity, angularity, and total pressure uniformity. On this form of intake, it is often this last requirement that becomes the most difficult to satisfy because of the physical constraints imposed by the more highly integrated installation and the need for the intake to operate over a wider range of flight speeds and aircraft attitudes.

A further, military-specific, requirement that is becoming increasingly important is for the intake to conform with the aircraft structure and obscure line-of-sight views of the engine face in such a way as to reduce the aircraft's observability by radar and infra-red detectors. This can lead to highly convoluted intake ducts and unusual intake opening shapes and locations.

Where the operational flight speed range is mainly subsonic, a pitot intake will normally offer the most efficient solution in terms of pressure recovery and drag. In a single-engined aircraft, this usually involves the use of a divided or bifurcated type of intake set on each side of the fuselage.

One disadvantage of the side-mounted type of intake is that, when the aircraft yaws, a loss of ram pressure occurs on one side of the intake (Figures 13–21 and 13–22).

This, together with a tendency for the flow over the intake lip to separate when the aircraft is flying at a nose-up incidence, will cause deterioration in both the pressure recovery and uniformity of the air presented to the engine face. The potential influence on engine performance (known as the intake/engine compatibility) demands understanding and attention.

A further example of a fuselage intake where ensuring intake engine compatibility is particularly challenging is the intake for the LiftFan® in the Joint Strike Fighter (F-35 JSF).

In this installation, the LiftFan is required to operate efficiently and stably behind an extremely short pitot intake at flight speeds of up to 460 kmh (250 knots) where the free-stream air is traveling at 90 degrees to the axis of the aircraft. This contrasts with a more normal installation, where conventional, much longer, civil and military intakes operate inside a 55 kmh (30 knot) crosswind limitation.

Despite careful intake design, including the use of a rear-mounted intake door to help turn the flow, the non-uniformity of the pressure and the angularity of the air presented to the fan face is more severe than that normally encountered in more conventional intake/engine installations (Figure 13–23).

To meet this challenge and ensure compatibility between LiftFan® and intake, novel techniques have had to be developed to replicate during ground test conditions seen in flight.

The pitot type of air intake becomes increasingly unsuitable at higher supersonic speeds due to the severity of the compression shockwave that forms and which progressively reduces the intake pressure recovery. A more suitable type of intake for these higher speeds is known as the external/internal compression intake, which improves the achievable pressure recovery by allowing the air to compress via a series of weaker shockwaves (Figure 13–24).

FIGURE 13–21 Exhaust nozzle and tail cone of a separate jets nacelle.

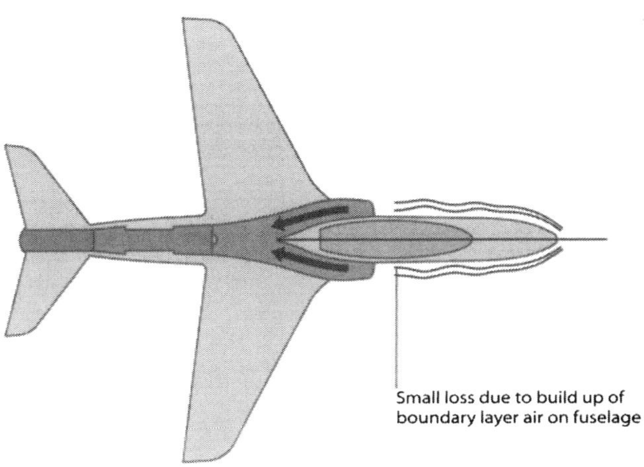

FIGURE 13–22 Loss of ram pressure in divided intakes.

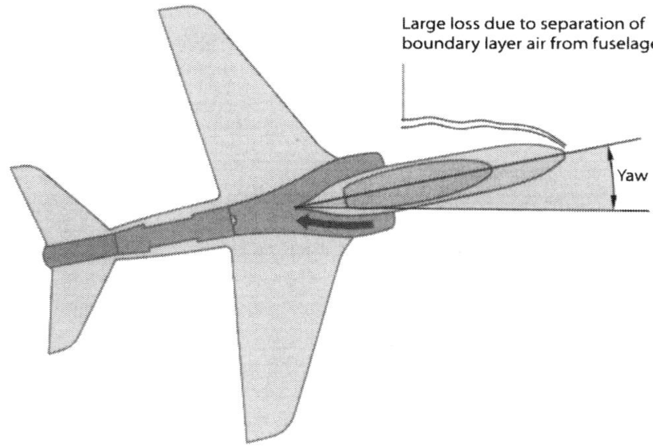

FIGURE 13–23 The effect of aircraft yaw on side-mounted intakes.

As aircraft speed increases still further, so also does the intake compression ratio, and it is necessary, at high Mach numbers, to have an air intake that includes variable geometry features to accommodate and control the changing volumes of air. The airflow velocities encountered in the higher speed range of the aircraft are much greater than the engine can efficiently use; therefore, the air velocity must be decreased between the intake and engine face. The angle of the variable throat area intake automatically varies with aircraft speed, and positions the shockwave to decrease the air velocity at the engine face and maintain maximum pressure recovery. This, coupled with auxiliary air doors used at static and very low speed conditions, allows an efficient match of the intake airflow to the engine demand over the full range of flight conditions.

The Panavia Tornado has a twin-engined installation that retains the side mounting but uses an external/internal compression intake combined with a variable-throat area and auxiliary inlets (Figure 13–25).

The Typhoon, a more agile aircraft, has a twin-engined installation with the intakes mounted under the fuselage (Figure 13–26). As well as avoiding the problems of fuselage shielding in yaw, this arrangement has the added advantage of using the under-fuselage surface to turn the air into the intake when maneuvering in a nose-up attitude. This not only off-loads the intake lips to avoid separation at high incidence, but also, at high flight speeds, pre-compresses the inlet air so improving pressure recovery. This intake also has a variable geometry bottom lip, which is used both to improve performance at high incidence and to achieve a better match of the intake capture area to the flight speed—avoiding the need for auxiliary inlets.

Stealth Aircraft

Enhanced survivability is an important emerging requirement for military aircraft. One way of meeting

FIGURE 13–24 The JSF engine installation, showing the two-stage vertical LiftFan®, the two rolls of post ducts, and the three bearing swivel module.

this requirement is through the reduction of aircraft "signatures" so the aircraft is less easily detected or traced by potential threats. This type of aircraft is known as a low observable, or "stealthy," aircraft.

Reducing signature has considerable implications for the engine installation. The major signatures of interest are the radar cross-section (RCS), infra-red emissions—and, to a lesser extent, visibility and noise. The engine air inlet and exhaust duct designs of such stealthy aircraft are driven primarily by the need to achieve the requirements for a minimum signature without too great an impact on aerodynamic performance, weight, and cost. This results in designs that are visibly very different from those of conventional, non-stealthy aircraft.

FIGURE 13–25 Side-mounted intakes for supersonic flow on the twin RB199-engined Tornado.

Cavities, of which the engine air intake and exhaust duct are the largest, are a potentially large source of RCS emissions from a stealthy aircraft. The stealthy engine air inlet duct design obscures the engine fan face using either an inlet entry grill (as on the F-117 Nighthawk), or a convoluted air inlet duct (for example, the F-35 JSF), or blocker vanes immediately upstream of the engine (F/A-18 E/F Super Hornet). These additional surfaces are typically treated with radar absorbent material.

In addition, the engine air intake may itself be located so that it is shielded by the airframe from potential threat radars, and the inlet lips angled to align with the wing leading edges—for example, the X-45 Unmanned Combat Air

FIGURE 13–26 The under-fuselage-mounted supersonic intakes for the twin EJ200-engined Typhoon.

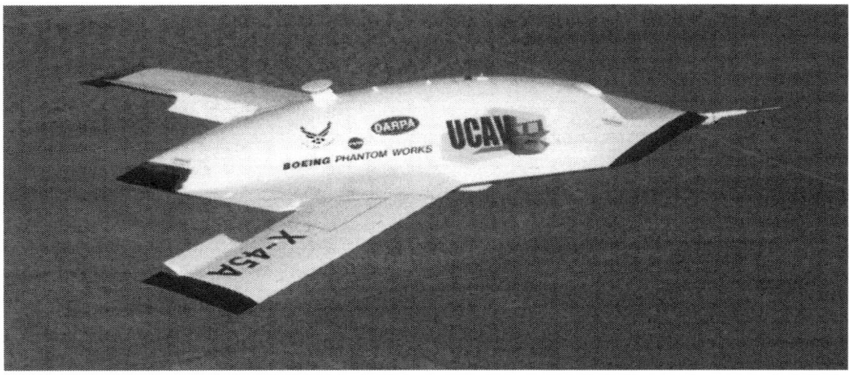

FIGURE 13–27 The X-45 UCAV has its intake shielded by the fuselage to reduce detection from the ground.

Vehicle. The inlet duct is also designed to be free of any steps or gaps that may also contribute to the RCS (Figure 13–27).

Similar design features (with the exception of grills) may be used to control the RCS of the engine exhaust system, although the high temperature of the exhaust plume makes this more difficult.

One immediate consequence of the need to geometrically integrate the exhaust nozzle with the trailing edge of the airframe is a trend towards high aspect ratio rectangular nozzles—for example, the F/A-22 Raptor.

The engine exhaust system components and plume, being the hottest parts of the aircraft, dominate its infra-red signature. Consequently, an engine exhaust system designed for a stealthy aircraft is heavily compromised by the need both to shield the view into the hottest parts from the ground and to cool the exhaust plume by mixing it rapidly with the surrounding atmosphere—this will also help reduce jet exhaust noise. In addition, the exhaust system may employ aggressive cooling of the exhaust system components and the application of controlled emissivity materials, which make hot surfaces appear cooler than they actually are.

Flying Test Beds

A flying test bed (FTB) is usually a production aircraft converted to test a new engine type before the first flight of a new aircraft type. An FTB is fitted with data acquisition equipment and also has a number of simulated systems. A flying test bed requires a new test pylon or strut adaptor for the specific test engine; it may also need structural modifications (Figure 13–28).

Historically, before the development of altitude test facilities, a flying test bed was the only means of altitude testing a new engine. There are still specific tests that cannot be done using a ground-based altitude test facility: for example, various nacelle tests and g-load engine tests.

FIGURE 13–28 A Boeing 747 being used as a flying test bed for the Rolls Royce Trent 800 engine.

Airframe manufacturers and test pilots normally insist on FTBs for all engine programs, to evaluate engine operability with representative loads and inlet conditions before the first flight of the prototype aircraft.

ENERGY, (POWER GENERATION), AND MARINE INSTALLATIONS

Energy and marine engine packages are generally supplied with all engine auxiliaries in place, leaving the builder of the application to provide starter power and fuel, water, and electrical connections.

Intake System

The intake system has to provide protection against snow, rain, and foreign object damage. Marine intakes are corrosion-resistant, often made of composite materials; industrial intakes require dust filters. The intake's large flow area reduces filter pressure loss and avoids ingesting snow or rain.

Cold, humid environments may require heating of the intake to prevent ice formation. Silencing is provided by flow splitters, consisting of sound absorbent material covered in a perforated sheet.

Enclosure

The enclosure provides weather protection (where appropriate), fire protection, and silencing. Ventilation is required to maintain a cooling flow past the engine. Engine accessories are often mounted within the enclosure but off the engine to allow quicker access and maintenance. Access is a key consideration for energy and marine installations to minimize

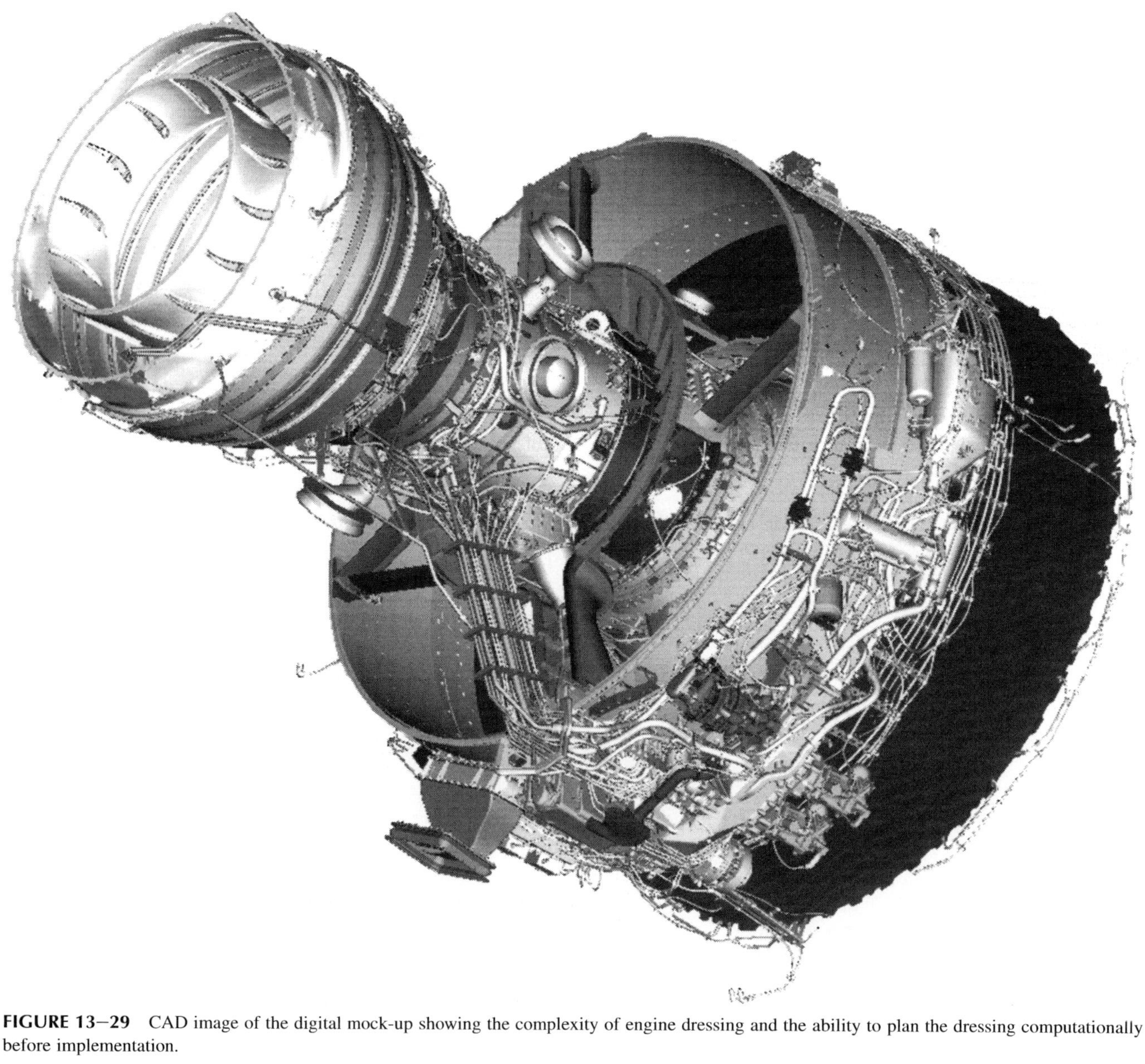

FIGURE 13–29 CAD image of the digital mock-up showing the complexity of engine dressing and the ability to plan the dressing computationally before implementation.

TABLE 13–1 SGT-100 Turbine for Power Generation (ISO) 4.35/4.70/5.05/5.25MW(e)

Weights and Dimensions SGT-100 Generator Set	
Diagrams, Weights, and Dimensions are for Typical Standard Equipment	
Length (with package mounted controls)	10.0 m (394 ins)
Length (without package mounted controls)	8.0 m (315 ins)
Width	2.40 m (94 ins)
Height (to top of enclosure)	3.2 m (126 ins)
Weight	35,460 kg (78,175 lbs)

(Source: Siemens).

downtime for maintenance. All enclosures have access panels. Some marine installations, like the WR-21, also allow engine removal via the intake. The design must consider issues such as safe working practices (for example, access control without entrapment), achieving a noise level below 80 dB at 1 m, and avoiding dangerous hot surfaces.

Although marine and industrial installations are not concerned with the flight speed aspects so important to the aero engine, size remains an issue. The WR-21 enclosure was designed so that it would fit in the footprint of existing marine engines.

Externals and the Engine Build Unit

Engine externals are all the elements on a fully dressed engine that connect the engine accessories and controls:

- fuel, oil, and pneumatic pipes
- brackets and attachments
- wiring looms and attachments

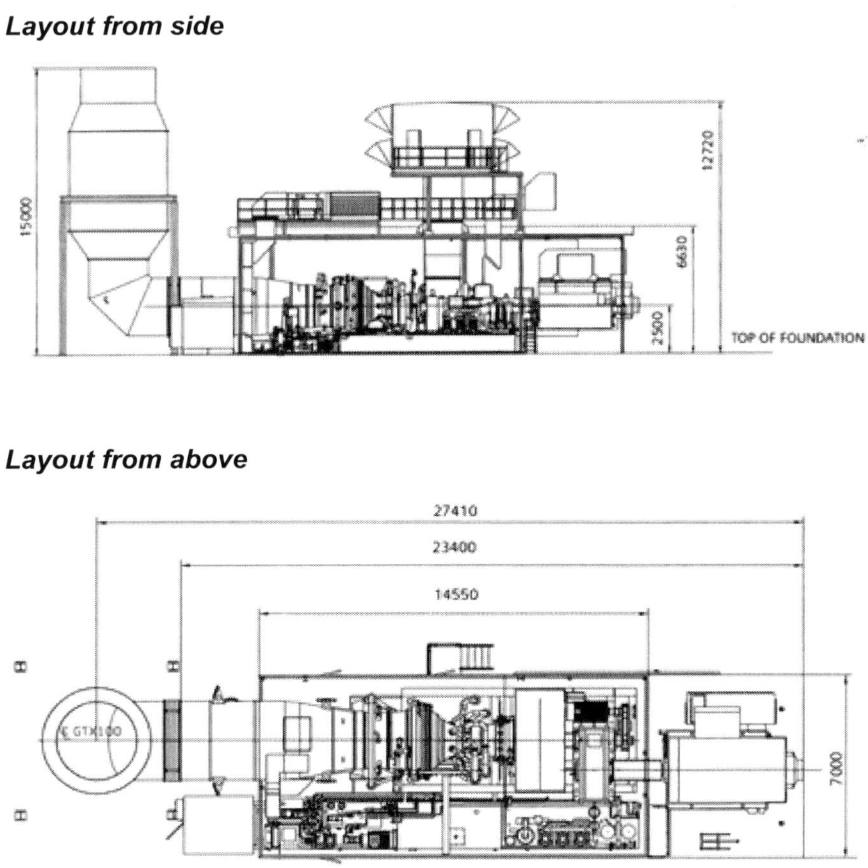

FIGURE 13–30 SGT-800 industrial gas turbine—45 MW Layout *(Source: Siemens).*

The placement and routing of the engine externals is defined using a digital mock-up, which provides both static clash detection to assist positioning the externals and also dynamic clash detection to help demonstrate maintainability (Figure 13–29).

The engine build unit comprises the dressed engine with externals along with all the interfaces that need to be connected between the dressed engine and the airframe or nacelle:

- cabin air ducts
- engine mounts and struts
- electrical and hydraulic feeds

INSTALLATION OF LAND-BASED AND MARINE ENGINES

The installation of land-based and marine engines is different from that of aircraft engines in that they generally come with a base or skid on which the entire system is mounted. The system may be mechanical drive or power generation, on a ship, ferry, or offshore platform. The essential difference for offshore packages with the same gas turbine (as an equivalent land-based system) is that weight is curtailed as much as possible on offshore platforms.

On land applications, weight is not an issue. Generally, the base is grouted (filled with grout through grouting holes left in the skid surface). Care is taken to not leave any voids in the grout to lessen the chance of "soft feet" below a critical gearbox or bearing mount.

How much of an issue weight is on marine applications depends on the particular application and size of the marine vessel.

The main issue on land and at sea is space. The end user needs to ask a few questions:

- How much space does the total installation/skid take up? (See Table 13–1 and Figure 13–30.)
- If weight is critical, what are the weights of all the system components?
- Is it likely that future additional propulsion requirements will need another gas turbine package?
- Can the skid be situated so that all plant components, such as inlet and outlet piping to the driven equipment in mechanical drive applications be routed so as to not cause excessive or cantilevered loads on the machinery flanges?

Other siting issues include items such as, does the gas turbine package need a "house" or will the system skid sit on the plant floor "as is?" Is the site appropriate for the potential needs of safety systems, such as risk posed by fluid moved by the driven equipment? My book *Process Engineers Equipment Handbook* (New York: McGraw-Hill, 2001) has a section on the risk of explosion on power generation gas turbine trains. The exception to permanent siting requirements is the mobile gas turbine package, which is frequently used by merchant power producers.

Chapter 14

The Business of Gas Turbines

"Hear reason, or she'll make you feel her."

—Benjamin Franklin

Chapter Outline

Business*,† decisions in any sector involve much subjective thought, most of it rooted in the decision maker's personality, culture, and life exposure to that point. Despite any changes that prolonged war, the increasing awareness of climate change, the impending emissions economy or expansion of same (in some countries), the essence of how gas turbine business flows or stagnates hasn't changed much since the first edition of this book.

Both personal experience and statistics offer many opportunities for brilliance or failure. Also, what is "right" one year is "wrong" 10 years from now, as technology, corporate, and national alliances change.

This chapter includes some excellent studies that detail the economics of specific cases. However, technical papers tend to leave out all but the bare technical and economic bones. Any business is far more complex than that. It involves cultures, alliances, rivalries, combatants, both old and new, seen and unseen.

The culture within the USA with respect to acknowledging the fact of human effects on climate change and

* [14-1] Claire Soares, "Separating the Elements of Plants, Process, and Personnel," 1994, Turbomachinery Congress, Bangkok, Thailand.

† [14-2] Claire Soares, working notes 1975 through present; and Proceedings ASME IGTI panel sessions, "Engine Conditioning Monitoring Systems as They Relate to Life Extension of Gas Turbine Components," 1985 through 2003. Chair, Claire Soares.

increasing greenhouse gas emissions has been evolving. Several factors contribute. Organizations like the Sierra Club have been attacking coal, resulting in some coal plants and mines being closed. Some power plants across the country have converted their coal-fired steam turbine plants to GT CC plants, raising fuel efficiency and curtailing emissions. Some have converted coal-fired steam turbine plants to natural gas-fired plants, which while it curtails emissions to some degree, does not raise efficiency, which is the main "gain" factor in reducing emissions.

Meanwhile, the politics within the USA will not let coal die, despite its renewed gas supplies. The FutureGen project will feature oxycombustion of coal (instead of the originally planned IGCC). The US DOE and EPRI continue to work on USC (ultra-supercritical steam) and advanced USC steam, which plan steam temperatures of about 760°C.

SC steam has been around since the 1950s and because of its coal-rich resources, China made some great strides in this area. EU OEMs (original equipment manufacturers), like Siemens, supplied some of their SC and USC steam turbines, but China also has her own steam turbine manufacturers.

One of the main features of FutureGen will be the CSS (carbon sequestration and storage system) that will be part of the power generation system being built and relatively "new" to the USA. That said, corporations like Statoil (Norway) and Vatenfall (Sweden) have been practicing CSS for decades. As Scandinavia practices a more environmentally conscious culture, people there did not flinch at SO_X and NO_X taxes, even when they were among the first in the world to levy them.

There seems little doubt at this point that the world is headed towards a carbon emissions trading facet in the global economy. The details have yet to be ironed out. Meanwhile organizations like IEA collect and feature the work of pilot projects like FutureGen that occur elsewhere in the world. Now China and the USA have a cooperative agreement for working together on certain CSS work.

All this does not mean that renewables have been abandoned, although the development funds for them were decidedly diminished over bad economic years and they still are not at the top of any major economy's list (except for countries like Denmark of course). That said, smart grids may do a great deal to revive interest in these small and intermittent producers of power. Power production conglomerates think increasingly in terms of system integration (STs, SC STs, GT CCs, GTs, and renewables in concert). There remains the question of the archaic and crumbling transmission line infrastructure that traverses so much of the USA, but fortunately that's a question that this book can avoid.

The "macro" business picture can periodically be dictated by wars (for instance large new orders of turbomachinery slated for the United States rebuilding program in Iraq), trading bloc protocols (EU, ASEAN, NAFTA), or in the case of the United States, a national penchant for moving toward countries with cheap labor (migration of manufacturing jobs to China and India). The "intermediate" business scene is painted by larger corporations taking on smaller ones (for instance, GE buying Rotoflow and Bentley Nevada to name a few), large corporate role expansions (like OEMs becoming IPPs, oil companies becoming IPPs), and expanding global presence (like the major OEMs' growing presence in NICs, like China and India).

However, the detailed ("micro") business scene is dictated by the working engineers, regardless of their function: design, operations and maintenance, and so forth. This, of course, is to some extent limited or promoted by circumstances beyond our control. The field of operations (including maintenance, R&O, and troubleshooting/failure analysis), however, is where the most visible and controversial examples of engineers affecting corporate business occur.

For most turbomachinery engineers, some of their work, particularly if they are in operations, has direct, immediate (or almost immediate), and major impact on their customers' net operating profits. Selected at random, some of the cases in point stretch from a purge oil system that threatened to shut down a 120,000 oil barrels a day plant, to an aeroengine failure that threatened to ground a world fleet, to small(er) issues like an aeroengine that had failed the post-overhaul test six times. Solving each one involves far more than just technical facts. So, before the case studies in this chapter, a summary of what that case paper's title means in the context of work in the power generation, oil and gas, and aeroengine sectors is probably useful. Note that the experience and observations that follow are based on my working notes and experience [14-2]. The reader's experience may differ. However, the points where a reader differs are just as helpful as those that concur. The overall subject of business is highly subjective. Disagreement and controversy broaden one's frame of reference and are useful educators.

The gas turbine engineer needs to maintain a broad base of reference. People everywhere tend to keep their gas turbines until they literally have nothing more than they can move or operate with them, and even then, they may try to resell them for "spares." Engineers who work in performance optimization and troubleshooting may find themselves developing a repair process for a superalloy component one day. The next day, they may be designing steam injection for gas turbines made from materials that are 1960s in composition, that operate on very low TITs, just to get another 20% power out of an ancient fleet and not need to install a new machine for additional power, in a crowded plant.

CONTEMPORARY BUSINESS CLIMATE

The world is a tumultuous place some years into a new century. With an item such as the gas turbine, the fuel selected for the gas turbine system is key to the gas turbine's success and competitiveness.

Fuels

With the first edition of this book, the gas turbine fuels of choice were gas, clean liquid fuels, residual or process waste, such as paper liquor (with many changes to the fuel system). With coal gasification making the strides that it has, the gas turbine can square off with the steam turbine even in coal rich countries. Just one indicator of this fact is the following news clip: "Siemens[1] Energy[2] is adding a new high-performance model, the SFG-850, to its range of entrained-flow gasifiers. The SFG-850 is designed to enhance the profitability of future gasification plants by reducing specific plant costs, along with the associated production costs of synthesis gas. An SFG-850 gasifier can convert around 3,000 metric tonnes of coal per day into more than 5 million standard cubic metres (Nm3) of high-quality synthesis gas.... In addition, Siemens can offer an optimised configuration of gasifiers and latest-generation Siemens gas turbines with the new SFG-850 gasifier. This will make power generation in IGCC (Integrated Gasification Combined Cycle) plants more economically appealing and, hence, more competitive than current solutions."

Further material on the gas turbine's increasing adaptability to a widening range of fuels can be found through the rest of this book, particularly in chapter 7 (fuels).

War

At the ASME IGTI TurboExpo 2002, in the year prior to the second US-Iraq war and the year after the 2001 US-Afghanistan war, the large OEMs admitted to a major downturn in orders for the following two years. Major unrest in the Middle East, Afghanistan, India, and Pakistan continues. Iran's national pride in accumulating centrifuges to make power-generation-grade uranium is evident. Estimates as to when they will have enough centrifuges to get that up to weapons grade vary. Meanwhile, a section of Pakistan has reverted to Taliban life philosophies and North Korea is testing nuclear missiles. Central Europe remains uneasy and relatively poor. Russia is booming, despite pockets of civil unrest. The unofficial unspoken reaction to all of these happenings by some has been "if it takes war[s] to make up for our lost orders of a few years ago, then bring on war."

In some cases, such as in Iraq, the need to develop some infrastructure hastily is so great that there are orders for old models. Some of these may have been "lying around" in some "excess equipment yard" waiting to turn a larger profit margin than possible if the buyer were a poorer country or one not being funded by the United States.

Development and Delivery Lead Times

The gas turbine business demands a massive investment for new or expanded installations. The lead times with gas turbines and gas turbine systems are much larger than they are with turbines for "renewables" such as wind (partly because the unit power sizes are on a ratio of anywhere from 1—300 compared with a typical wind turbine). Also, unlike energy sources like wind turbines, one cannot just buy major components from other manufacturers, assemble them (partially on site), and start producing power. With gas turbines, the extent of integrated assembly of controls and hardware, as well as systems testing, which has to occur in the OEMs facility, prior to shipping, places the gas turbine at a disadvantage in an unstable and fragmented world. Such will remain the case, even if US gas prices drop.

Conferences as Harbingers

ASME's IGTI annual conference frequently heralds shifts in the gas turbine business. As of the year 2000, the reduced participation of large gas turbine OEMs in the exhibit section of the conference became increasingly marked. The IGTI annual conference in 2002 saw an exhibit with relatively small participation from large OEMs (it had been shrinking for several years), and when they did appear, it was with a niche-market developed component, such as Pratt and Whitney exhibiting new brush seals it developed. Smaller firms with cutting edge products that give OEMs an edge, such as optical pyrometers, and new twists to online condition monitoring systems fared well.

PowerGen[3] conferences continue to be a hive of activity in the US and other venues. The conferences certainly host large exhibit sections and attendee groups that number in the thousands.

At one point, in the 1980s, the ASME IGTI conference also hosted about 5000 engineers annually, when it was a meeting that catered more to end users than

1. http://www.iea-coal.org.uk/site/2010/news-section/news-items/siemens-launches-new-gasifier? (extracted on May 19, 2014)
2. http://www.gasworld.com/news/regions/north-pacific/siemens-launches-new-gasifier/2003763.article (extracted on May 19, 2014)

3. A trademark of Penwell publishing. PowerGen conferences feature non-juried papers where commercialism tends to be less of an issue at juried paper conferences, such as IGTI.

researchers and academics. Today, OEMs, individual firms and researchers that work on the cutting edge of their fields still patronize the conference, but its attendance numbers have shrunk, since it began to feature more research and academic work. The exhibit area has shrunk correspondingly.

At the 2002 IGTI conference, and in subsequent years the shift in emphasis became increasingly marked. Even in the applications area, the presenters with an academic background formed an unusually large part of the overall content. OEM and end-user originated papers made up a smaller percentage of the overall technical paper roster. The end-user and OEM papers tended to be on newer technology, such as hybrid systems. Fuel cell technology, in combination with gas turbines, is favored by legislative trends toward more environmentally friendly technology. Indeed legislation dictates that the energy mix in the United States should aim for 20% renewables by 2020. Other countries will follow suit. Some countries will exceed this target appreciably: consider Denmark and its 50% wind energy power mix by 2020.

Around 2004, the IGTI gave up trying to hold its design and end-user communities together in one meeting. Deciding that end users preferred a more practical meeting, IGTI started end-user annual meetings, generally in the same location as a larger meeting, such as Penwell's PowerGen. The "design types" IGTI meeting is held separately. The significance of this to the engineering world is that IGTI is one of the few engineering associations that present juried papers, each with three preprint reviews. This fact served the gas turbine community (and IGTI) particularly well in the 1980s and 1990s. It provided an optimum climate for "mixed" industry sector attendees (land-based, mechanical drive and powergen, as well as marine, and aviation) at the panel session[4] that ran for 19 years (1985 through 2003): ECMS as GT ECMS as They Relate to Life Extension of Gas Turbine Components. Attendees were keen to learn from other sectors and "not reinvent the wheel" in their own.

OEM Acquistion strategy, National and International trends

Larger OEMs have made several acquisitions of smaller or medium sized support system manufacturing companies. This may be to maintain a competitive advantage with the ownership of support systems that they used to purchase. (Consider GE's purchase of Bentley Nevada and Altair Filters, UK.)

Some OEMs may be keen to absorb as much in the way of high-tech niche products that will save them engineering R&D time, when the market comes back. And come back it always does, if often in unsettlingly short spurts that do not always match the lead times required to turn out a gas turbine, especially ones over 200 MW. That said, some OEMs are scaling down, selling off divisions of well-engineered gas turbines to other OEMs. The location of OEM divisions/management can shift and with that, corporate culture may also change.

The balance of overall engineer populations continues to shift. China turned out five times the number of engineers than the United States in 2005. That figure rises to 10 times if you adjust the US number for foreign students that cannot get US work visas in wartime. India graduated about half the engineers China did. Especially with what were once US jobs in manufacturing (of items that include gas turbine systems) reaching their shores, it is likely that the Asian engineers will come to know what the insides of turbomachinery look like sooner than some US counterparts.

In terms of the mechanical-drive gas turbines used in gas and oil production, the overall global gas turbine market is likely to stay stable to growing for at least the next decade. Business gurus promise that the US consumer will feel "pressured enough when gas goes from $3 to $4 and $5. That's when you'll see a massive change in the US consumer." No marked trend was visible when gas prices did rise and when they fell again, with the advent of fracking, the behavior of the US consumer did not seem markedly different.

The recent increase in natural gas supply from fracking has both consumers and OEMs optimistic. The exceptions are environmental groups who note the problems that small contractors (who hold themselves perhaps to a lesser standard than a large oil company would) can cause with their practice of the fracking process.

The small (smaller in MW) gas turbine market for ferry, naval ships, small power generation applications (including offshore platform and merchant power "portable skids"), and mechanical drive applications grows, as the "news items" sections in the trade journals indicate.

The military aircraft engine market was stable to growing till about 2008 in wartime-induced sales, but that does not do much for improvements in the optimized design of aircraft engines as a whole. Not that long ago, the flow of technology went from NASA via military aircraft engines to commercial aircraft engines to aeroderivatives to industrial models. At some point in the last decade, commercial aircraft engine development got ahead of military engine development and recent wars have not changed that on a macro scale. Defense budgets are being cut notably.

4. "Gas turbine engine condition monitoring systems as they relate to life extension of gas turbine components", Chair: Claire Soares, on behalf of the Aircraft Engine, Controls Diagnostics and Instrumentaion, Applications (Pipelines), Marine, Materials and Metallurgy and Utilities (Large PowerGeneration) committees; ViceChair for several years: Dr. James Hartsel.

As of 2014, the US military cut the number of Osprey aircraft it ordered to half.[5]

Where is this taking the world? Inside of the next three decades, probably to the age of smaller distributed land-based power, and ultimately the personal turbine (PeT). PeTs, at some point, will become as prevalent as PCs now are. The rest of the global power mix will consist of some large plants, particularly in the newly developing countries, such as India and certain countries in Latin America, where the power wastage with faulty and badly designed transmission systems is currently as high as 10–30%.

The gas turbine, in all sizes and models, will continue, wars notwithstanding, to flourish because of their application potential in all sizes of aircraft and marine vessels. The ability to take an aeroengine model and adapt that for ground-based application will continue with a shorter time span for designing the transition. There may be some grassroots land-based designs that can be adapted for fitting into an aeroengine nacelle. However, more about all this in the chapter on the future of gas turbines.

CULTURE

Culture is an extremely complex factor that can rule business and technology outcomes. Industry subcultures vary as does the industry's treatment of and response to, gas turbines. The subculture varies within each industry. Within the military, the coastguard and the army have different procedures and preferences regarding what may operationally be similar engines. The powergeneration sector varies from the mechanical drive gas turbine industries (which include oil and gas, refining, process). Within for instance oil and gas, the "conventional oil and gas" operators think differently from those who make synthetic crude for instance. These cultures blend with national cultures to form their own unique stamp.

International conglomerates, within and outside of OEM conglomerates, form very different schools of thought. The consortium that makes the IAE V2500 engine have different philiosophies from the one that makes the rival CFM 56 engine. When OEMs buy a division in another OEM, the resulting "sub-company" may find it takes years for its new culture to be established. This can happen even if the companies are from the same country. An end-user who owns a gas turbine system that he bought as a European Gas Turbine model or later as an Alstom acquired model may find both advantages and disadvantages with Siemens acquisition of that model, in terms of component updates, potential for complex repairs, servicing.

5. Berard, Yamil. "Bell to lay off 325 workers as V-22 orders decline" *Fort Worth Star-Telegram*, 5 May 2014. Accessed: 19 May 2014.

In part because gas turbines are expensive items to manufacture, operate, and overhaul, many divisions of, or entire companies of gas turbine manufacturers are bought out by or merge with larger manufacturers. Rolls Royce bought Allison. Siemens merged with Westinghouse. Brown Boveri became Asea Brown Boveri (ABB), which merged with European Gas Turbine (which once was Ruston), which then merged with Alstom to form ABB Alstom, which then changed its name to Alstom, the Swedish division of which was bought by Siemens. Profit margins are hard won. Costs per pound of thrust (or horsepower or kilowatt) with manufacturers, costs per fired hour with operators, and percentage net profit per overhauled engine with repair shops are watched closely by eagle-eyed "bean counters." As frequently happens in a highly technical business, the value of the guidance financial gurus provide is directly proportional to their technical knowledge. With gas turbines, engineers, particularly those with a broad background that includes design, repair, and operations, make the best marketing or business guides.

Historically, European and Japanese firms drew their business analysts and sales force from those with engineering backgrounds, the broader the better. US firms have been more inclined to employ business or finance "purists," often to their detriment. Europeans are now trending more toward the US management model, which may not help increase long-term profits.

Repair and Overhaul Shop Culture

Gas turbine overhaul companies are constantly reorganizing, changing ownership, "going under," or all three. It is possible to have a "small" repair shop. The shop may deal with only one or two repair processes, such as heat treatment or welding, and not have to accumulate the capital overhead an OEM needs for a comprehensive shop. When the shop gets successful, particularly if it developed some unique process, a larger shop or an OEM may buy the shop to extend its potential in-house work scope.

The other important factor in the business of gas turbines is one's network. The gas turbine world, for all of the gas turbine industry's growth, is basically small and, in any one specific subsector, for instance, aircraft engine repair and overhaul, all the main players know each other. Each subsector has its own unique culture. Some subsectors are friendlier than others. In aircraft engine overhaul, for instance, I always found that rival shops would sell each other a spare part that the other needed desperately (and expect the favor to be remembered and returned). In the oil patch, however, I recall a rival company flatly refusing to sell my company a couple of land based mechanical drive gas turbine packages for which they had no use and had left sitting in the desert, even for a most reasonable sum. (The two oil companies have since merged, which makes the tale all the more ironic.)

In this contemporary global village, recognizing culture is crucial to business success. The 1980s and 1990s saw the turbulent demise and ascendant of many overhaul facilities, particularly in the aviation business. Those shops, managed by staff who had prospered through the 1960s and 1970s, some of them getting away with a "local" client base (say, US domestic only) and bullying their small, foreign client base ("behave yourself or I won't fix your engine"), may have been bewildered by what the 1980s brought. If they were and did not react quickly, they could have found themselves out of business in a decade or two.[6] Excess equipment facilities made it possible for small repair shops and "fly by nights" (short lived) to turn a profit and exert "guerilla warfare" tactics in some cases. They might dent the work order load of one of the large shop's business centers, say, the plating shop or the heat treatment department, or they might form an alliance to get work items that the large shop could not perform more profitably.

I recall the massive aeroengine overhaul shop I worked for in the late 1980s, subcontracting some EDM drilling work to a tiny local shop. We had the equipment in-house, we just took too long and used too much time and personnel to get it to the right work stations in our shop.

In the US business climate of the 1980s, it became critical to observe the cultural preferences of international clients during all phases of the relationship, to remember that European and Asian clients ask technical questions that may be beyond the capability of a nonengineer sales force. Rival European, Asian, and Canadian facilities knew this and had adapted. It was essential to know that the equivalent of US "blue-collar" workers in Europe were "craftsmen," who were probably in the same union as their engineers. It was bad business to not realize that 1980's US mechanics might often be, veterans of the Vietnam war, who wanted, their management to properly motivate as well as train and compensate them. Unlike the traditional postwar Germans and English, they would not shift into "autopilot" if they disrespected their management, just so everyone could keep their jobs. Japanese executives roll up their sleeves, get on the floor with their people, and start suggestion boxes toward which they demonstrate a commitment. The US workforce is poorer in this category of executive. However, when they have a "people's person chief," it shows. As was evident to all who watched Herb Kelleher of SW Airlines offer his shareholders a three for two split in 2001, even as rival airlines, overhaul shops, and support businesses were forced to close their doors.

In these days of mega-corporations, portfolio diversification, takeovers, and mergers are common. Sometimes, successful large corporations are tempted into an acquisition or set of acquisitions that gets away from their core competence. As the stock exchange "graveyard" and "just treading water" lists reveal, acquisition is not always an exercise in prudence. Some such ambitious ventures did not get as far as mergers or joint ventures with OEMs. Instead, companies like Ryder concentrated on acquiring aircraft engine overhaul shops in the 1980s. After all, a gas turbine overhaul shop is really just part of the transportation business right? Wrong! Besides, the "products" from aeroengine overhaul shops are vastly different from the orderly sameness of a manufacturing facility or the routine of truck or school bus rentals. This feature is amplified in the case of an independent overhaul shop that has a dog's breakfast of small customers who may own just a few engines (sometimes, only one plane) and a few large airline customers who sometimes grace them with their "overflow." Then, you not only have a different culture "on the shop floor" but every customer, from Chile to Thailand, adds its own cultural profile to daily operations.

Military versus civilian mandates make different cultures, too. One of the engines from the "old" Airforce One arrived on my former nightshift beat also came with an NCO (non-commissioned officer) who watched it faithfully and mopped up a grease spill if he saw one. On the other hand, a freight hauler like Federal Express might not send a rep into an independent shop to steer it through overhaul and may want to only "make the 500 operating cycles that'll take us past the Christmas season."

Degree of knowledge can create major cultural barriers. I was once berated for scrapping a critical turbine disc that someone had forgotten in the acid bath. It should have been out of there in about half an hour, but it was not. I didn't like my hydrogen embrittlement odds. Another engineer said "it looks just fine" and took it off scrap status. He did not want further dialogue. The disc looked fine—on the outside. The decision makers were not degreed engineers, and those that were may not have had to study metallurgy. The good news is that, if hydrogen embrittlement does not "get you" in fairly short order, the "heat treatment" from just the gas turbine hot section operating temperatures will cause the troublesome ions to lose their problem potential over time. I never heard about that particular engine failing, so that customer, and the shop, were lucky that time.

It is always a customer's responsibility to "beware" as a buyer; in other words, never assume your seller has the same values as you. The customer may be someone who "sticks to the work scope (or contract)" but his seller may

6. Aviall, a Ryder acquisition in the 1980s was bought once in the 1990s and once about a decade later. Stock values fell from $24 a share to less than $7 in one particularly bad year in the 1990s.

be of the school that "sells what [it] can get away with." As a case in point, I once suggested we ask for a work scope extension on some used engines that had been bought by a customer, after I had seen a couple of them opened up. The seller of said engines had claimed it had completed an "ESV2" overhaul on all of them. ESV2 was Pratt and Whitney shorthand for a very thorough overhaul. So those engines ought to have looked, if not new, then very clean. However, when we opened them, we found cracks in the diffuser case, missing internal turbine air seals, and a bunch of other transgressions.

If that was an ESV2, the claimants of same were, no doubt, tooth fairies. The unsuspecting customer had specified a quick "look see" work scope, which demonstrated great trust and a complete lack of knowledge of the seller. My overhaul shop employer could comply with that work scope and be, technically at any rate, within his rights to not inform them of the mess. At that time, I was sole engineering support to an entire second shift of about 250 mechanics. We worked when all the other engineers, customers, and other staff had left for the day (except for about a two-hour common timeslot). My own "culture" could not let what a couple of floor supervisors politely called "ratshit" slide. Despite the flimsy work scope that essentially asked me to do nothing, I provided many graphic photographs that the customer's contact engineer could include in his final report. I also stuck a copy of the pictures in the turnover log, as my position did not liaise directly with the customer or write final reports. Whether the customer had enough technical background to read the glaring evidence in the photographs or removed the engines from service henceforth and shut up to save face, I do not know.

Sense of community within each company varies globally. An overhaul shop in Scotland does not think or react like one in Texas. If acquired by the same owner, the executives may see themselves as jockeying for position. Frequently they do.

When Ryder (known for their school buses and trucks), decided to buy Aviall in the 1980's, (as distinguished from the Aviall of today, which is more a parts supplier and the small remainder of what was once Ryder Aviation Services), then the largest global independent overhaul facility for aircraft engines. Prior to 1986, the firm serviced industrial engines as well. This presumably was to extend Ryder's reach within the global transportation sector. All the cultural implications may have been news to some of chairman Tony Burns' key decision makers. When he added what was the former British Caledonian Airways facility in Scotland, to Aviall's assets, much of this potential melee might still have been a mystery to key personnel. So it was that an overhaul facility with a combined shop capacity of about 1000 engines a year and a stable share value of about $24 at the end of 1991, was divorced by its Ryder parent in the face of abysmal losses a few years later. The "new" corporate stock that fetched about $14 on initial offering would have difficulty raising between $5 and $7 just a year later. A relieved and helpless management sold out about then. The new Miami company management struggled with what was residual debt load, probably some inherited customer bad will, a different worker culture, or all three. It, too, sold out. The new owner, General Electric, had other shops to farm the work to, if it pleased. The aircraft engine overhaul world rumbled, raised eyebrows, found alternative shops, or used more than one shop; then all was "business as usual" again. In May 2005 or thereabouts, GE closed that shop's doors forever, keeping just the engine test cells. Assets that supported their CFM line overhaul were probably transferred to their other shops. The tooling for the V2500 engine line (I had commissioned that program in 1991 with the first engine input from Indian Air), a rival to the CFM56, was sold to Pratt and Whitney.

End User or Operator Culture

Can such a stew of cultures be managed and turned to advantage? Southwest Airlines is living proof that it can. Whether you're an overhaul shop, an end user/operator, or OEM, the answer is emphatically, "Yes." The proviso is, however, that you must have the leadership of a Herb Kelleher or a Robert ("he's a son-of-a-bitch, but he's *our* son-of-a-bitch") Crandall to make it happen.

With end user/operators, commercial size and global business affect the worker's life and therefore the corporate success and synergy. Before conventional oil prices hit the floor in the early 1980s, I enjoyed working for them. Even in the more stringent tax environment in Canada, money spent on maintenance came off pretax gross revenues, and as long as oil prices stayed high, we never had to stay in less expensive hotels or rent the smallest cars. Even for the environmentally friendly types, car size was a consideration if you were driving out to fields over a mix of snow and black ice and you wanted a car that would "hold" the road.

It behooves employers to recognize synergy potential with its individual employees and their working circumstances. Employees need to return the favor. Some people thrive on pioneer type field projects. The more difficult the circumstances the better. Syncrude's first synthetic crude (oil sands) 170,000 barrels-a-day facility in northern Alberta, Canada, worked through eight-month winters with average temperatures of −30°C to get commissioned on schedule in 1977/78. The average age in what was then a 35,000-people strong boom town was 15. A boy scout troop made its money for airfares across the country by collecting empty beer bottles that littered the highway between towns and the weekend playground (all things are relative) that was Edmonton. I went to the local pub only if I had many friends for company. They had shooting matches in there, not all of them friendly. The workforce stuck together through years of 16-hour days and went cross-country skiing or curling on Sundays. We were rock solid as a community, in and out of the work place.

The plant cost about $4 billion in 1970s currency and most of the critical machinery was a prototype in some way. Many of us had some of our most valuable working years on that project. The learning intensity on subsequent projects, under more "structured" environments, curiously, often is reduced.

Those of us involved with rotating machinery at the Syncrude of the late 1970s found ourselves rewriting the Exxon, Shell, and Gulf bibles for prototype machinery. That machinery is no longer "different" today. Faced with the "oil down pipeline" date, as well as OEM puzzlement and caution in the face of unanticipated obstacles that would require machinery design changes, we made swift, sweeping changes to machinery and plant systems. That may have cost us heavily, had they not worked.

Today, many oil companies in the Ft. McMurray area are building other plants, but the early 1980s saw a temporary setback in oil sands and all nonconventional energy development because of the global oil prices crisis. Had this not happened, the use of nonconventional fuels, such as orimulsion (the Venezuelan equivalent of synthetic crude) in gas turbines, would be much further along.

Money Makes Attitudes

As a propulsion systems manager for the Transport Directorate in Air Command for the Canadian Air Force, I had six aging helicopter engine fleets and four NCOs to supervise. Later in my three-year term, there would be a budding engine program newcomer (the Pratt and Whitney PW100) to be part of the Kiowa replacement program, as well as studies into the Sea King replacement program. That notwithstanding, we still had to make old chopper engines operate safely. In that, we found the US Coast Guard's operations (whose budget does not match those of other US military branches) repaired most engine problems. Canadian military choppers at the time were overhauled to commercial standards, so they could all do grueling pipeline service, if necessary. This was appropriate, as the Canadian military in peacetime performs like a US Coast Guard, chasing smugglers and illegal aliens, as well as performing search and rescue. In other words, there is no room for error during overhaul of engines that must then be put through mission profiles as demanding as those imposed by wartime service.

To illustrate the difference financial resources make to operations: once, I asked a US naval lieutenant colonel how they had handled a bearing failure on one of those aging fleets (the T58 engine). He informed me, laughing, that they did not bother. "We just remove the engine, leave it in the Arizona desert and put a new one in." This was then the mid-1980s I hasten to add. Many things have changed since then.

OEM (Manufacturer) Culture

"Customers Buy Whole Engines Not Components"

Operator response to two engines I learned to admire in my military years has been consistently positive. The engines are General Electric's F404 on the F-18 fighter/bomber jet and the T-700 helicopter engine. The latter was one of the candidates for a power plant in the Canadian Forces Sea King replacement study, so I had occasion to study it thoroughly. I read everything I could on the F404's condition monitoring system (CMS) and triple redundancy controls architecture, all of which was to prove handy for some power generation project work years later. Triple redundancy controls are increasingly observed on critical land-based installations. The lesson here is that technology developments in one sector eventually visit the other sectors.

As mentioned elsewhere in this book, I ran an annual condition monitoring session that five committees in ASME IGTI's annual TurboExpo cosponsored, from 1985 through 2003. The session aimed at pooling the best brains in condition monitoring (land, sea, and air; OEM; end user; repair shop; whatever) to link CMS systems that monitor an engine's condition to changes in gas turbine operating hardware effective lives. One of the members of the design team of both the T-700 and the F-404 engines also became my vice chair at those panel sessions.

The additional cooling that gives the F-404 and the T-700 engine part of their edge in service was a contribution this colleague insisted on making, in the face of opposition from the "efficiency freaks." Given the war being continuously waged to better one another's efficiency, many enthusiastic designers shave the safety margin on cooling air very close to the bone. When adverse circumstances arise in unexpected combination, a pilot may be on the quick road to oblivion. This gentleman told his team that he would rather have just a little less efficiency and less of a chance of "meeting someone's angry widow. . . . I realized a long time ago that people don't buy turbine sections. They buy engines. Whole engines. And not just an efficiency figure."

The Ruggedness School

Operators and end users like rugged. If you are a pilot facing possible death, you can get downright passionate about it. The pilot of an early V2500 (Airbus A321) flight over Calcutta airport's approach path may not have realized one of his engine's had just ingested a 24-pound vulture. The approach takes aircraft over a garbage dump, so the vulture was where he or she belonged. Its weight was confirmed by the size of a few leftover feathers. The engine just broke two of its wide chord Rolls Royce-designed fan blades and kept steadily on. I have seen one of the V2500's rivals (equivalent thrust class) "corncobbed" by a two-pound pigeon. This has much to do with the effectiveness of the wide chord fan blade design (which some years later "appeared" on other engines, such as the GE 90).

However, ruggedness is a school of thought, a culture. Its followers are generally those who add items like an extra margin of cooling air or oil to their equations, over and above those dictated by accepted insurance-worthy calculations. The V2500 has four (originally five) quite excellent design partners, but a sigh of relief went up among overhaul shops and savvy pilots when we knew that the oil system was Rolls Royce designed. Rolls does not stint on its cooling. Its algorithms for calculating engine cycle usage (in the late 1990s) on engine models like the Spey and Olympus engines confirm that.

So did some of my "overhaul witness" work. When witnessing teardown of the Rolls Avon engines that drove Cooper RBB barrel compressors (the first prototype application machines in high-pressure, high-volume natural gas reinjection service in northern Alberta), I expected to see a noteworthy amount of wear and tear. The load on the barrels varied to some extent as with all mechanical drive applications and I did not expect things to look as clean as with a similar unit in power generation. I expected to develop a substantial "to-do" list. With those Avons, I felt about as useful as another star in a tropical sky. The engines were almost embarrassingly clean.

So was MTU's (Motoren Turbinen Union's) low-pressure turbine section on the repeatedly shock tested (accelerated cycles accumulation) V2500 that was bared in 1990 in Munich. There were many engineers at that gathering, and all of us just wandered and wandered around in circles trying to find a nick, a scratch, anything. If the objective for that meeting had begun as "let's see what flaws may occur early with this V2500," it ended as "let's have a good dinner so the trip's not wasted."

Interestingly enough, those turbine manufacturers who prefer gaining efficiency points to providing extra cooling may end up with hot sections that look every bit as clean during overhaul, *provided* nothing tips the scales. If there is no unanticipated flashback with sophisticated burner design having quality control problems, or no partial blockages for oil to the bearing closest to the turbine inlet temperatures, or no upstream compressor blockage caused by seasonal bugs hardened on compressor airfoil passages, the scales stay even. If, if, if. Operators and end users would, in general, be unequivocal about getting the "extra" cooling—if that choice was theirs to make.

Market Entry Monopoly

Conventional gas and oil production was busy in the 1950s and 1960s. Production fields might be large or small, with many years of supply or just a few. The gas molecular weight was prone to change, as these were mixed fields, with the onset of heavier-molecular-weight components an unknown variable. End users/operators needed a set of small compressors with adaptable staging and gas turbine drivers that could tolerate a wide range of loads. Efficiency was not a key requirement, stable operation within load ranges and easily available rental units were essential. The manufacturer who filled this bill, and stood relatively unchallenged at the time, was Solar turbines.

While not high on efficiency ratings by current standards, much of that original Solar fleet is likely to be around after most readers of this first edition of this book have expired. End users like the machinery they are familiar with, even if it has faults. Justification of new machinery may just not be there, given declining field economics.

Aware they had a giant grasp of the 1000- to 3500-hp mechanical drive gas turbine market, Solar was, for some years, from my operator's perspective (advising field offices that asked me to get certain questions relating to their machinery answered, which they had not had an answer to) not the best ear for its customers to approach. The original Centaur was about 3300 hp. Uprate kits took it to about 3500 and then 3800 hp. Solar told its customers their warranties were void unless they adopted the additional horsepower, regardless of whether their application needed it or not. Everyone understands that spare parts are expensive to stock and service, but when the uprate kits cost upwards of 25% the cost of a new machine, customers squeal—and with the help of end-user lobbies, such as CIPs (component improvement programs), roar. The once noteworthy Gas Turbine Users Association (GTUA) is no more, at least for the time being, but it gave the late 1970s user base a good sounding board.

Its early monopoly over, Solar immersed itself in much progressive development, DOE programs, quality control award programs, and other successful endeavors. Its sophisticated current R&D work outshines its late 1970s to early 1980s image, where it faced an aggressive end-user community bent on optimizing their own costs per fired hour. Both OEM and end users benefited. Solar won the Malcolm Baldrige award in 1998. Today Solar is a highly effective company in its own niche market of smaller gas turbines and compressors, with a high level proactivity in terms of optimizing its designs.

Conglomerate and Joint Venture Cultures

When giant OEMs absorb smaller OEMs, OEM divisions, or support systems/accessories manufacturers, it is inevitable that cultures will change. Whose culture governs is both an obvious and subtle answer. The larger company officially has the edge. The actual dynamics depend on the personalities on a per-department or per-division basis.

The same is true of joint ventures. For instance, the companies involved in the V2500 success are very different. The Japanese Aeroengine Consortium (JAEC) is a consortium of three large Japanese manufacturers with aeroengine manufacturing potential, making the V2500 low-pressure compressor module, complete with wide chord fan blades, a Rolls Royce design. Rolls Royce makes the high-pressure compressor; Pratt and Whitney, the hot section; MTU, the low-pressure turbine. Every other engine assembled is put together in Rolls Royce, Derby, England, the alternating engine(s) in the Pratt and Whitney facilities in the United States. Four more-different partners it would be hard to find, and yet the synergy works.

Selecting tooling for a joint-venture engine can be interesting. At module interfaces and for some other functions, both OEMs involved at that interface would design tooling for the same functions. I recall a German-designed balancing tool (to be accurate, it was designed by a German balance machine manufacturer, not MTU) that cost $55,000. The Japanese equivalent cost $17,000. I never did get to try the Japanese version, as for a variety of reasons, we had opted to buy the German manufacturer's balance machine package as a whole, for the Dallas-based V2500 overhaul program. However, both tools worked.

In summary, when synergy is on your side, even the most unusual international conglomerates work wonderfully. It is the same with overseas subsidiaries.

All the major manufacturers—GE, Pratt and Whitney, Siemens Westinghouse, and Alstom—have low-labor-cost subsidiaries overseas. Many are in China and India. Many, much to some people's surprise, have gained ground by directly employing or promoting the education and training of formerly illiterate women. These subsidiaries have put "home" (US and European-based) labor out of work and face much political opposition, but they continue.

There are a growing number of international joint ventures in power generation. An example is the one between Siemens Westinghouse and Malaysia's YTL (an IPP company operating in Malaysia). The two companies actually continue to own and operate these assets, versus a BOOT (build, own, operate [for the years required to train the locals], and transfer) project.

Most of the major OEMs were quick to set up components manufacture in NICs (newly industrialized countries). Siemens and GE are among those that provided a great deal of local training in China, India, and a host of other countries, so their locals can make everything from blades to composites to relays. However, when they form these manufacturing joint ventures, then also train local people to operate their own power infrastructure, that is an even more valuable contribution.

Educators and Training

There is a separate chapter on training, so I will not dwell on this much here, except for a few facts. The first is that economic deprivation breeds an almost frightening level of ambition. Sometimes, that ambition seizes responsibility at what others may call a premature point in history.

A good example is the Chinese aerospace industry workforce, which covered in 10 years what it took the Western world 60 years to cover.

The key factor in training is always the students themselves. I have faced students on both sides of the Pacific and Atlantic who did not understand why, in the course of a week's training, I could not give them the content of all my years of experience. And they did not like "homework" for that week either. This is a common frustration with trainers.

I have met others who were angry that I would attempt to give them a snapshot of performance analysis systems when they did not want to be bothered to learn more than the basics of Process Plant Machinery (or whatever the course was called). I met still others who were angry that I would present the basics of a gas turbine (and I firmly believe that there can be no comprehensive engineering career in power generation, energy production, or process engineering, which makes up 80% of global activity, without knowing about gas turbines) because they currently owned only a few steam turbines. Again, as we say in Texas, go figure.

Integration with Environmental Technology Culture

With the wisdom that comes from a good bottle of wine, a colleague and I once reasoned that the Scandinavians were "so much better at all this environmental stuff" since they were Vikings at heart and used to not wasting anything on long sea voyages. Much of their economy with resources stemmed from their early roots. You rarely have to sell a Scandinavian on emissions taxes.

A holistic perspective of gas turbine operations must acknowledge that cooler TITs mean longer TBOs, not just lower emissions. Some things are hard to collect enough data to prove conclusively. Nevertheless, we must make our best judgments in the absence of all the facts. To this point at any rate, I believe a design that best favors optimal environmental performance from a gas turbine (discounting other factors like design development costs and other one-time only costs) also favors optimal costs per fired hour and TBO for a given engine.

So, environmental performance ought not to be considered as an item separate from overall performance but more as an ally in attaining optimum design and operation of any engine. Promoting an environmentally admirable design is part of optimizing gas turbine performance (see the chapter on optimizing gas turbine performance), even though it may seem that you are "throwing" away a few efficiency decimal points.

RISK

Selection and Specification Process for Gas Turbines and Gas Turbine Systems

In global engine population terms, the main reason any gas turbine (GT) OEM is successful during any major bid process is not an engine's overall design attributes, but the finance packages offered. General Electric makes fine engines, but one reason it sells so many more engines than its competitors, who also make fine engines, is the strength of its financial services corporate arm.

During a bid where the V2500 and the CFM56 were squaring off for an order with a large Chinese customer, the customer noted the V2500 model had 4% greater efficiency than its current rival's model. Fuel is an issue with China. The only fossil fuel China had was its own large supply of coal, and the only other one it could access easily due to a global glut is residual oil. Today natural gas from fracking[7] shale may change all this, but that was not a factor at the time of this bid. For that bid, the CFM team paid the Chinese the difference in fuel efficiency costs over the stated official life of the engines. It essentially gave away the initial capital cost of the gas turbine engines free to the customer. The losers sniffed their disappointment and remarked that the less rugged (in terms of FOD), lesser cooling air (in terms of cooling air per pound of engine thrust) of the winner would soon recoup their outlay in spare parts costs. They may have had a point, but the lesson is clear. Particularly in the developing world, he with the deepest pockets wins, regardless of any risk inherent in a customer's financially driven choice.

The Luxury of Choice, Maybe

Large oil companies are among the few gas turbine customers who can make their gas turbine selections independent of financial factor considerations. However, very often, if the warranty package offers good enough security to result in the effective cost per fired hour they need, they opt for either the best financial choice or one driven by deliverable dates. Deliverable dates count, because all manufacturers of power generation machinery can generally sell every machine they produce. They are heavily backlogged, and that will remain a factor in the future of the gas turbine business for the foreseeable future, although the extent may diminish in a bad recession. The aircraft engine and naval gas turbine market can be more competitive. I say "can," because if the commercial aviation business is in a recession-goaded slump, the fortunes of commercial aviation gas turbine manufacturers follows the curve. If the year was 2001 or 2003, war was once again the best business to be in. Military engine (aviation and naval) programs were hastily taken out of mothballs and given massive infusions of cash. A California comedy show host on NBC's the *Tonight Show* pointed out sometime in April 2006 that the Secretary of Defense need not worry about his generals hitting him with their collective lack of confidence with respect to his performance in the Iraq War. "He's still got lots of friends: General Dynamics, General Electric . . ."

The times when gas turbine selections are made on design merit alone are therefore rare. Reliability and end-user comfort level are two main deciding factors, if the end user does not need OEM financing. I recall power-plant selection for power production in Esso's Norman Wells proposed project. I chose the GE Frame 5 (old model) although it was vastly oversized for the project's needs. The year was 1980 and I did not want to tangle with the then-new Frame 3's rash of cross tube and other problems. I did not want a still relatively new Solar Mars. I needed a stolid, reliable design that had worked well in another sub-Arctic site (the Syncrude project) I had experienced. For the $500,000,000 (1980s dollars) proposed Judy Creek expansion,* I chose Solar Centaur-driven seawater pump packages. Esso was the largest Solar turbines operator in Canada and the operators knew the machines well. Spare rental machines would not be a problem.

In terms of gas turbine system (such as the engine condition monitoring system) selection, whether new or in retrofitted/reengineered applications, the selection and risk factors are varied and several. They are discussed at some length in this book's Chapter 9, "Controls, Instrumentation, and Diagnostics."

Risk Factors and Their Mitigation in Gas Turbine Design and Operation

Risk in gas turbine design and operation is defined differently depending on the audience in question. Nonetheless, three key components of risk to profit margin, when a

7. See: "Can China Join the Fracking Game?" A. Sreekumar, extracted on June 30 2013; also at http://www.dailyfinance.com/2013/06/30/a-first-step-for-chinese-shale-gas-development/, extracted July 24, 2013.

* Both the Judy Creek and the Norman Wells expansions were put on hold in the face of the oil price crisis in the early 1980s, but both were revived later in better oil market times.

system is being analyzed by potential financiers and end users are:

- Unavailability (as may be caused by unscheduled shutdowns) and inconsistency of supply (as may be caused by "brownouts")
- Fuel costs (which may be affected by political and market factors)
- The costs of spare parts (which are affected by items such as design complexity, operational component lives, design reliability, vulnerability to untrained operators, and so forth)

An increasingly competitive market demands that every iota of potential increased efficiency be squeezed out of a power plant, particularly a new project that seeks financing. The return on investment (ROI) figures ultimately look much better to bankers this way. Operations engineers sometimes groan when they work with the hard reality of the turbine outages that result. Their main job is to keep the plants running constantly and they know that some of the design methods used to raise efficiency are going to make their maintenance lives very difficult. This dichotomy poses several interesting questions, depending on the turbine design philosophy in question. To some extent, it also depends on the consistency of the quality control program enforced by the manufacturer.

The Higher Turbine Inlet Temperature School

The greatest risk in raising efficiency is the risk of unexpected outages. The way better efficiencies are primarily achieved is by increasing the turbine inlet temperature (TIT). This is increasingly evident as manufacturers introduce their G-, H- and now J-series gas turbine technology, which then gains incremental design-calculated efficiency points over their last closest-equivalent designs, by raising TITs. Cases I have included in this book on testing G, H and J machines include some from OEMs who are very thorough about making sure every aspect of their design works. They also tend to be the OEMs who raise their TITs in careful increments, testing on-load performance at each step. Not all OEMs are that cautious. There have been cases when the steam cooling system required to get a gas turbine to the TITs corresponding to this class of gas turbine have leaked, with certain OEMs. It is the end-users responsibility to check with the user community to decide which OEM would be their best choice.

Some progress has been made in increasing the cooling-system efficiency of turbine blades and vanes, using elaborate blade design, and steam cooling. If we consider General Electric's aeroderivatives such as the LM series, the technology for design features like active clearance controls (ACC, a cooling design feature) was already developed by the aircraft engine CF6-80C2 team. Now MHI has its version of ACC using steam cooling. MHI uses both a closed loop and an open loop, where the steam is reinjected back into the gas path for a power boost. The closed loop was a tricky feat to pull off, and it does MHI considerable credit. (See the case study on MHI's steam cooling technology in Chapter 4.)

The exposure of hot gas parts to the higher turbine inlet temperatures will naturally shorten the lifespan of the exposed parts. This effect is limited by the use of aviation engine metallurgy and manufacturing processes. Both Siemens and MHI have pointed out that they use aero-engine developments to their advantage.

The targets for combined cycle efficiency of percentage points—60% plus by the turn of the century—as laid out for suggested targets in the 1990s by the US DOE (Department of Energy) proved to be attainable. All the OEM "majors" "have new blade designs and cooling schemes, but these efficiency levels are far from easy to achieve with consistent high availability for every OEM. Certain OEMs have tested their steam cooling systems on their -H machines, and they did not leak then or later in operation, under full load. It is always the end user's responsibility to check technical claims made by bidder OEMs as they might apply to the application in question.

Manufacturers' (OEMs') Guarantees

Increasingly, manufacturers are taking over the maintenance of their own machines, usually with power-by-the-hour contracts. Some, like Siemens and General Electric are also buying large stakes in power generation stations globally. Overseas, where training is an issue that can directly affect unscheduled outages and brownouts, manufacturers may contract some of their involvement in BOO (build, own, and operate), BOT (build, operate, and transfer), and BOOT (build, own, operate, and transfer) projects.

Both for overseas projects in NICs and in industrialized areas, such as the United States and United Kingdom, manufacturers frequently base their contracts on availability and outage figures. For instance, for the first year, the OEM may base contract prices on an availability of, say, 87% with a fixed number of unscheduled shutdowns, say, 10–15. If that target is exceeded, the OEM gets bonus points, if the reverse is true it pays penalty points. If all goes well, the next year's contract may be based on perhaps 91% availability and 7–9 unscheduled shutdowns. Thus, the OEM and its customers negotiate the share of increased risk with newer designs.

Supply of New Machines

In a rapidly burgeoning deregulation climate, most customers cannot always base purchase selections on an analysis of potential OEM candidates. There simply are too few machines being made to fill global demand growth.

California is one of the best examples of a US state that no longer has enough power. It also has industries that can be crippled by power outage, as in Silicon Valley; loss of power means more than lost production time. It could also mean losses of huge amounts of data. Therefore, customers have to buy from the OEMs that can supply them faster than anyone else and manage their risk with contract clauses, power-by-the-hour contracts, peak purchase agreements, emergency power purchase from MPPs (merchant power producers) as well as spares and spare machine strategy.

Effect of Design Factors

When the market is thriving, it is tough for OEMs to keep up with new orders, even with improvements in manufacturing technology. That notwithstanding, the design philosophies adopted by various manufacturers do affect availability of existing machines. Exactly how much depends on human factors that involve operators, as well as the level of service available and demanded by, a certain design.

If one considers ABB GT 24 and 26 technology for instance: These designs can operate at lower TITs than some of their competitors' designs. This means less NO_x emissions in pounds per megawatt (lbs/MW) even though ppm (parts per million) NO_x levels might sound not that different from competitors. The SEV (sequential environmental) burner design, in essence, can be said to have two hot sections (and two TITs) because it uses the reheat concept. Ultimately, it needs less airflow for both cooling and reaction with fuel, which also means fewer greenhouse gas emissions.

That being said, the increased number of parts involved may affect repair and availability adversely. The reduced number of discs in ABB's first −11N2s meant much higher bearing loads with resultant availability (reduced) potential consequences.

Today the 200+ MW machines pose the main questions with respect to all the three major aspects of risk—availability, spare parts costs, actual fuel costs. There is a larger amount of end-user information available on mature machines in the 100−130 MW size and still more on small machines that make up the 10−35 MW range.

However, as temperature is arguably the single highest promoter of hot section deterioration, in the bigger machine sizes, Alstom, for instance, can be considered to have managed risk well from the operator's perspective, with specific reference to its cooler-running engines. This permits the use of cheap residual fuel in the case of certain models and allows a wide variety of low BTU fuels. Bankers and lenders look at fuel bill calculations that fall with increased theoretical efficiency and TIT values. However, they may weigh that third risk factor (cost of and fluctuations in the cost of fuel) just as heavily as risks from unscheduled power outages and higher spare parts requirements, in the ultimate derived figure of cost per fired hour.

Another contributing factor to availability and reduced component life (or increased spare parts costs) is the low NO_x combustion burner. With some of these burner designs, low NO_x combustion is achieved without steam or water injection. The disadvantage of this technology, however, can be high pulsation levels in the combustion section. The vibrations of the parts exposed to the high inlet temperature accelerate wear significantly and contribute to early component failure.

Machines using the single-burner technology combined with water injection to achieve the NO_x requirements have lower pulsations, which affects the parts' lifetimes positively. There are far fewer failures with these designs. Alstom has been proactive in this regard, with water injection options designed into its −13E (and formerly −13D, now −11N2 with a gearbox to service the 50 Hz market) designs. The alternative is to have a series of burners as Siemens-Westinghouse does with its V94.3 designs, and as ABB has designed a retrofit for, with many of their smaller models, like the GT-8. These designs can then be run without water injection. (From an operations perspective, the worst problem with water injection is that it requires boiler feedwater quality, which then means an additional system that can be prone to problems).

Critical attention to quality control details by the OEM is also a major factor. When this is not done, small inconsistencies in well-proven designs can add up to operational unavailability.

In other words, sacrificing 1 or 2% efficiency and choosing a proven burner concept with low combustion pulsations and reduced or no tendency for backflow does pay off in the long run with high availability of the machine. In "peaker" applications starting reliability and availability are crucial. Increasingly, CC (combined-cycle) power plants are stopped and started more frequently. Reliability and availability, for startup and continued operation, are increasingly important for CC plants.

Life Cycle Assessment

It says a great deal about an OEM's faith in the cooling features of its turbine hot section if its algorithms that tally the life cycle count only measure rpm (revolutions per minute) against time. For instance, in Rolls Royce's case, the development of these algorithms and the retrofit of life cycle assessment (LCA) counters was outlined in a service bulletin (SB) for its Spey and Olympus. It was instigated by end-user pressure and high costs of replacement parts. The SB has been successful in extending hot section component lives by factors of up to 200% from the previous standards (which had specified a fixed number of operating hours for each component).

Most LCA algorithms take into account temperatures (whether peak, base load, or both), rpm, and time. The development of LCA is another item first pioneered with aviation engines and slowly handed down to industrial and aeroderivative gas turbines. LCA counters are being refined constantly, as are materials and instrumentation for temperature measuring systems, in the constant fight to bring down the cost of spare parts.

For more information, see Chapter 9, Controls, Instrumentation, and Diagnostics.

Performance Assessment

Performance assessment (PA) systems basically monitor the gas path of a turbine in terms of flow versus pressure. Temperature readings are also taken. Some systems compute parameters derived from measurements versus parameters calculated from predictive formulas and end conditions. These PA systems then compare the two and can troubleshoot problems that may affect both reliability and availability.

In competent hands, PA systems can be used to design retrofit modifications that can raise availability. One example is retrofitted steam injection (for cooling) that took critical GE Frame 5 (old version) turbine discs out of the crack range, for several thousand more cycles. PA systems can also be used to make new design changes that can extend component lives and thus reduce the cost of spare parts.

PA systems can be used to monitor emissions and changes in emissions resulting from component deterioration. Ultimately, they measure aerodynamic performance and indicate the source of losses that can be corrected. About a 0.5% difference in turbine section efficiency on one GE Frame 7 (end-user Alyeska) made an $800,000 difference in an annual fuel bill (based on 1995 Alaskan gas prices).

PA systems are then a major tool in risk assessment and limitation. Although vibration analysis (VA) can be considered part of the reliability and safety aspects of machine operation (versus availability), VA is often incorporated into an overall condition monitoring system by either the OEM or a contractor hired by the end user. Especially when they work well (which largely depends on how suitable the package components are for the application), reliability and safety systems contribute indirectly to availability and the cost of spare parts.

For more information see Chapter 9, Controls, Instrumentation, and Diagnostics.

The Push from War, Emergencies, Politics, and Policies Design Compromise

War sometimes breeds design compromise. One of the six gas turbine fleets assigned to me, in the Canadian Air Force, was the Kiowa engine, which had been so active in Vietnam. The compressor section of what was the C-18 model of Allison's 250 series could not, in the "raw metal" state, withstand the centrifugal loads during operation independently. To get the tensile stress to within allowable limits, a layer of compressive stress was imposed by glass bead peening on the rotor blades. The rotor had to be turned and peened again, so that both sides of the compressor blading received benefit of the compressive stress layer. This engine might not have entered service with this design feature had the aircraft not been needed that badly when it was introduced. When two engines in the Canadian fleet failed (the pilot safely autorotated in both cases), alarm bells began in the global Kiowa community. Well past the Vietnam days, the fleet was still active commercially in many countries. When a third engine failed and this time its aircraft crashed, killing pilot and passenger, the fleet was held within a hair's breath of being grounded, pending the results of the investigation I was sent to complete.

The situation was worsened by the fact that it had been discovered that some personnel at the overhaul facility we used did not realize that the rotor blades had to be glass bead peened on both sides. We had no way of knowing how many "bad" rotors were out there. We did not know if a failed engine had fatally disoriented the pilot for vital seconds during his practice exercises (night landings using the tail lights of a jeep). The world fleet was taken off "potential grounding" status after I was able to prove that the engine, at any rate, was operating as per specification when it hit the ground. Later, X-ray diffraction performed by the manufacturer revealed that this rotor was one of the "good" (both sides peened) ones. Nevertheless, an immediate recall was initiated for our fleet and every rotor "redone." As the only way, at the time, that we could be sure stress levels were what they should be was destructive X-ray diffraction, we had no choice but to recall the entire fleet.

The design compromise that left a fleet open to deficiencies at the overhaul stage is never one that makes OEMs comfortable. It may be one that they come up with, based on the needs at the time (war, politics), however. As soon as possible, they upgrade the design, try to pressure the end-user base into upgraded models (consider the T 55 fleet changes discussed earlier) and move on. Allison's later model C-30 series, and for that matter their earlier (post C-18) C-20B engines (in the Jet Ranger fleet), did not require the peening "safety umbrella."

In today's economic climate, most operators in the Western hemisphere operate newer models with superior technology. However, in the working lifetime of anyone reading this book, these older models with their flaws and compromises are out there, some of them in the developing, struggling world, some of them closer than affluent society would like to have. Without the appropriate knowledge or records, telling "good" from "risky" is not possible.

When the Canadian Department of Defence called for a Kiowa replacement program, the designated power plant manufacturer was the Canadian-based small engine division of Pratt and Whitney. Political convention required that a Canadian manufacturer be chosen. Pratt and Whitney designed the highly successful Pratt and Whitney PW100. The core was then used to scale up to other equally successful models, like the PW300, an example of political influence buying positive results and products for the gas turbine community.

Operational Compromise

The Falklands War revealed that the Rolls Royce "Peggy" (Pegasus) could withstand the temporary (limited hours) operation on heavy fuel that her British designers might not have anticipated. Then again they might, having seen two World Wars fought on their home turf.

In the mid-1980s, I was collecting information on NATO studies of gas turbines using "off spec" fuels. The Canadian navy had a fleet of LM2500s on its vessels. The TITs in the LM2500 were not conducive to using fuel with high vanadium content. So fuel treatment (which is expensive) was a must. Then, too, the gas turbine must be run at temperatures low enough (derated) to not have the compound that results from fuel treatment destroy the turbine blades. Regular washing is also a must (see the chapter on fuels).

The engine fleet extension studies I was tasked with in the mid-1980s are another example of the requirement for contingency measures. The six aging engine fleets in my care were being asked for "another potential 20–25 years." The overhaul shop promptly returned their verdict of "yes we have enough spares." I submitted this statement to the management, with a warning, referencing the T-55-Cs that the OEM did not want to continue servicing the -C model. Why would other OEMs not eventually (and much sooner than 20–25 years) take the same position with old models, I asked. Flatly calling the fleet extension programs "pipe-dreams" due to logistical problems was not something my rank could pull off. When the penny finally dropped, the fleet together with the Chinook helicopters they powered were sold off to a north African nation, so then it became their problem. Other military users quickly migrated to the newer variants like the ALF 502.

The Secondhand Merchants

There are many "good" secondhand merchants out there. There are many gas turbine packages, both power generation and mechanical drive, that have gathered dust but never been commissioned. Others have been used, but especially since they were well maintained and all critical component records on them carefully preserved, they have useful life in them. There are many excess commercial aircraft on sale, some of them "cheap." But, "not all that still gleams" is worthwhile. If the rates are "fire sale" that is likely for a reason. Entire 747s for $0.75 million each very likely have "no useful cycle" JT9s in them. What will it cost to zero time them? Hush kit them if they are to be used where noise regulations matter? Who will provide future spares and service? As always, "let the buyer beware."

Some news sources have claimed that well over $1–2 billion worth of bogus and "useful-life-ended" parts are in global commercially operating flight engines. The figures are hard to prove. That it happens is unquestioned. "How" is simple. Instead of permanently scoring the part surface when it is deemed scrap during overhaul (as a bartender would do to a bottle of name brand hooch), the parts are left some place where the nocturnal "elves" might carry them away.

Aircraft are not the only recipients. One of those no-longer-serviced-by-the-OEM engines I mentioned earlier ended up in a power boat. When cruising along the water's surface, one of the turbine wheels came flying through the side of the hull.

Again, many secondhand merchants are reputable. Excess or partially used power generation machinery from the Western hemisphere is being installed and commissioned in growing areas like Latin America and Asia. Aside from knowing with whom you are dealing, technical knowledge obviously helps steer these dicey waters. Whether the seller will also agree to provide some kind of warranty, parts and service as well as commissioning service for its wares (via subcontractors it will back) is the acid test of a vendor's reliability. It's also hard to find.

Technical Risk Mitigation

What follows are short paraphrased extracts from EPRI work on technical risk mitigation.* The fleet studied is in a land-based power generation application, but the methodology used for calculations is statistical, thorough, and worth consideration (with any required modifications) for all applications of gas turbine.

In this EPRI work, the major areas of uncertainty related to technical risk are addressed: scheduled maintenance frequency, unplanned maintenance frequency, costs, and outage duration.

Key unplanned maintenance-dependent variables (outage hours and maintenance costs) can be represented as statistical

* Reference: D. Grace. EPRI "Technical Risks and Mitigation Measures in Combustion Turbine (Design) Project Development," IGTI 2001-GT-0472.

distributions determined from the failure rate (i.e., events per time period), outage hours, and cost distributions. Unplanned outage hours and unplanned maintenance costs are calculated from three levels of event statistics (i.e., minor, major, and catastrophic) for the selected duty cycle by summing as follows (for 8760 hours per year):

$$\text{Unplanned outage hours} = \sum\Big[(\text{Failure rate})_{\text{Type}} \times \text{Service factor} \times 8760 \times (\text{Outage hours per event})_{\text{Type}}\Big]$$

$$\text{Unplanned maintenance cost} = \sum\Big[(\text{Failure rate})_{\text{Type}} \times \text{Service factor} \times 8760 \times (\text{Cost per event})_{\text{Type}}\Big]$$

Note that, in this EPRI study,
Minor is $800 per event.
Major is $10,000 per event.
Catastrophic is $240,000 per event.

Technology risk primarily affects scheduled maintenance and unplanned maintenance. Variability in these areas can affect plant net revenue and profits considerably. The Combustion Turbine Project Risk Analyzer results, combined with an overall plant financial analysis, can represent data on variables of interest. For example, variations in plant profitability can be represented as a function of the variability in maintenance frequency.

"SHIFTING TARGET" DATA DURING PROJECT DEVELOPMENT, NEGOTIATION, AND NEW MODEL INTRODUCTION

Promising gas turbine powered projects are negotiated only to die later. They may be reborn a while later. Depending on how opportunistic the culture is, this cycle might be swift indeed. In an article I wrote for a power generation periodical that specialized in the Chinese market, I mentioned the demise of a specific project. The cancelled project was owned by a specific partnership at the time of cancellation. Cancellation was due to a variety of circumstances: permitting, financing, and ROI negotiations (the exact mix I no longer recall and it is not germane to the issue at hand here). The information I had was current at the time of going to print as confirmed by a well-known, reputable Hong Kong source. About a week later, a new IPP Hong Kong firm called the editor I had written the article for, complaining wrathfully. The project was going ahead they said, they now owned it. They neglected to add how recent their acquisition was. As much publicity as they could find for their new "baby" would raise the success of their new public offering on it, so their wrath was financially driven, if not entirely justified.

Aircraft engine programs have long been the victims of "on again, off again" financial romances. Their success often depends to a large extent on the potential for or presence of war. The Osprey tilting rotor helicopter program's fortunes dipped in the climate that cost the Dallas—Fort Worth area well over 50,000 jobs in the early 1990s, as what was then General Dynamics, a USAF base, and Bell helicopter slashed budgets and workforce. Does some of the lost momentum or staffing changes that such projects get put through translate into small design deficiencies and oversights later? Much depends on how well documentation was kept. In the case of changing partners for specific power plant joint ventures, just philosophical design differences between differing OEM partners on issues such as damage made by tooling limit levels that trigger component replacement and can initiate massive warranty cases later. On the matter of the Osprey itself, to cut a very long back-and-forth-with-funding story short, on 28 September 2005, the Pentagon formally approved full-rate production for the V-22.[8] In 2013 the USMC leveraged the unique abilities of the V-22 Osprey to form an intercontinental response force, the Special Purpose Marine Air-Ground Task Force for Crisis Response.[9]

Potentially differing quality control methods among licensees can also cause warranty problems. During GE's 9F problem in the mid-1990s (the model was newly introduced on the market), end users debated long and hard about whether −9F problems were greater with US assembled or GEC licensee assembled models. It took $1 billion, the involvement of about 400 GE engineers, and new quality control methods in rotor assembly that applied to all GE manufacturing and licensee facilities, to stop set the rash of −9F vibration problems.

Risk in Negotiating IPP Projects

As of 2010, the United States was the largest consumer of power globally. China (very close to USA's consumption in 2010) and India have yet to fulfill their growth potential. Asia remains the largest target for IPPs. As it is home to many of the world's oldest and complex cultures, which then have to be the recipient of the swift, sometimes brash cultures of the occidental hemisphere, conflict and confusion can result. The gas turbine business reveals and is revealed by the circumstances of IPP growth.

8. From: "Osprey OK'd." *Defense Tech,* 28 September 2005.
9. From: "Marines, Army form quick-strike forces for Africa." *USA Today.* 15 June 2013. Retrieved 23 June, 2013.

International Negotiation

International negotiations in the gas turbine (or combined-cycle) business are fraught with many conflicting factors:

- The end-users' financial circumstances, which then dictate the terms of financing, joint venture percentages, controlling interests, potential for countertrade, and so forth.
- Level of technical expertise of the end user. This will dictate the extent to which the end user can opt for specific turbine features, refuse others with conviction, know exactly what options contribute to his application and which should stay with other users.
- Bargaining power of the end user. If the two preceding factors are ones in which the end user is poor, then he may find himself paying more for certain items. For instance, a study I was doing on costs per fired hour and the cost of spare parts on a specific OEM's turbines revealed that lesser educated clients were charged more (as they were prone to operate turbines erroneously then claim "warranty" when failures occurred).
- Joint venturing potential of the end user. If an end user has enough financial strength or bartering power to be "almost" equal partners with an OEM in project development, then as in the case of the Siemens Westinghouse YTL project (also see next chapter), a joint venture company with almost equal share partners can be formed. The end user can get training from the OEM as well as eventual transfer of all operational duties (see BOOT projects in the next chapter) as part of the deal.
- Political climate offered by the end user.
- Financial exchange facilities offered by the end user. Can the IPP, OEM, or IPP OEM receive payment in some international currency? Countertrade?
- Infrastructure offered by the end user. The fact that China's current infrastructure is superior to that in many areas of India makes China a preferred choice despite its politics and difficult exchange problems.
- What, if any, are the main cultural barriers in the end-user's culture that also present legal hurdles to the developer? For instance, if middlemen negotiators in the Middle East are used to "baksheesh" what does a US businessperson faced with the Corrupt Practices Act do?

This is a short list of a mesmerizing array of possibilities. Only upon successful negotiation of all these factors does a gas turbine or a gas turbine package (combined cycle or other option) get revealed in its most advantageous light, in terms of efficiency and costs per fired hour.

At this level, environmental standards maintained by the OEM are not an issue. Most end users, including many of those in the Western hemisphere, specify "emissions to be compatible with BAT or better, but not to exceed [some easily attainable figure given contemporary technology]." Environmental standards maintained by the end user, however, do make a major difference to project operating costs. Does the OEM anticipate NO_x and SO_x taxes? Does it anticipate a CO_2 tax? Does it care about CO_2 sequestering? Does its plant offer potential for the gas turbine system to have:

- Compressor inlet cooling?
- Compressor intercooling?
- Reheat within the turbine module?
- Additional waste heat recovery?
- A land-based "afterburner" (see the chapter on performance optimization)?

Is the end user ambitious enough or sufficiently trained and supported to work with any combination of these? Or is he potentially an increased warranty risk to the OEM, if the OEM supplies the option.

MARKET ASSESSMENT RISK

Market assessment of various factors, of which market entry timing is particularly critical, can make or break the progress of a gas turbine. If that turbine's core is applicable to many growth models, then market factors can make or cripple an entire family of engines.

The gas turbine's potential has changed dramatically as it crossed first the 200 MW barrier and, in 2002, the 450 MW (in CC GT systems) barrier. The excuse to stay with "steam only" diminishes drastically with these entries on the power train stage. And yet ironically, steam remains a player, because of the technology refinements that come with combined cycles, steam reinjection for power augmentation and NO_x reduction, supercritical steam production for efficiency increments, cogeneration applications, and a variety of other embellishments. Now USC (ultra-supercritical) and A USC (advanced ultra-supercritical) steam turbines are being developed. (See Appendix A on steam turbines in Chapter 3.)

The GE LM series, heralded by the entry of the LM 2500 series, was well served by the success of commercial aviation's CF6-80C2. The aeroderivative LM series model is essentially the CF6's child sitting on a base with an oil console. That CF6 model was, in turn, well timed and well served by rivalry with the Pratt and Whitney PW4000 and the Rolls Royce RB211 series, which pushed it to all levels of excellence in the scrap for every last decimal point of incremental efficiency.

Solar, back in the 1950s, could not have picked a more optimal market entry time for its initial Saturn and Centaur 3300 offerings, which would, in part because they filled a much needed market hole that no one else came close to, lead to its prime positioning today. The finances for the development of its vastly more modern Mars and Titan technology came on the backs on these more humdrum, humble workhorses that made no claims on breaking any efficiency barriers.

The Rolls Royce Avon was well served by being "on time" for being the power plant on the Alyeska pipeline. The prototype land-based aeroderivative of a successful aeroengine power plant called the RB211 picked its aero-derivative market entry timing to coincide with Trans-Canada pipeline's needs for an aeroderivative gas turbine fleet.

PLANT SITING

Determining the site of a plant involves several factors. Among the most critical in a complex infrastructure, like the United States', are environmental standards, consequential permitting requirements, and emissions trading potential. This is a subject that almost merits a separate book. However, a few links are provided for the reader's perusal. Government or government-funded organizations such as EPRI (Electric Power Research Institute) can have pertinent information available on their websites:

http://www.energy.ca.gov/sitingcases/jargon.html

http://www.perf.org/pdf/MMiller.pdf (EPRI)

Each state in the United States has its own air quality/permitting (for gas turbines) site. One example is http://www.maine.gov/dep/air/licensing/gasturbine.htm.

DESIGN DEVELOPMENT AND OPERATIONAL ASSESSMENT BY BOTH OEMS AND END USERS

Aeroengine design precedent has much to offer land-based gas turbine design, whether the engine is "officially" an aeroderivative or not. Development costs of gas turbine design improvements are massive, and OEMs strive to avoid "reinventing the wheel," as a normal business practice. This is illustrated next by extracts from Case Study 1 on combined cycle gas turbine power generation plants. The reader may note that aeroengine precedent has played a critical role in the development of this OEM's industrial F, G, and H technology models. The full-load testing mentioned in this case is vital, especially with larger gas turbine models.

It is sound business practice for the customer to attend the full-load test of any machinery package that he will end up owning and operating, if at all possible. A full-load test gives one the opportunity to observe the performance of cooling systems such as any steam cooling (closed or open loop) and ensures that the system does not leak, especially on the massive H and J technology machines.

Case Study 2 is of interest in that it presents an EPC contractor's perspective. Depending on the contractor's work scope, his concerns may be equivalent to an end user's. The case describes checks and tests that an operator (EPC contractor or otherwise) or end user can conduct to compare how an installed plant fares against the equipment's measured output in the manufacturer's facility.

Case Study 3 is an OEM's work on plant profitability options using PC, AFBC, CC, IGCC, ACFBC, SPFBC, and CPFBC technologies (see the case study for a description of these acronyms). It needs to be noted that an OEM may also be its own plant contractor and its own operator or end user, especially in BOOT projects. Some OEM projects are BOOs, where the project is never transferred and the OEM retains part ownership (such as the YTL-Siemens IPP power plants in Malaysia). Further, as per the author's description in the introductory material of this book, the cases presented are potentially of interest for both their technical and historical comparison value.

A few years after this source material was written, the prices of natural gas escalated steeply, which could change many of the comparisons made here and in similar cases throughout this book. Several years later, thanks to increased natural gas supply from fracking, gas prices dropped again. In other words, if you don't think a case applies to your market conditions for any reason, just wait some years for the supply-demand pendulum to swing in the opposite direction.

CASE STUDY 1: ENHANCING RELIABILITY AND REDUCING O&M EXPENDITURES IN ADVANCED COMBINED CYCLE GAS TURBINE POWER PLANTS*

Explosive growth in power plants in the United States and elsewhere is occurring through combined-cycle advanced gas turbine technology. Significant fuel expenditure savings are possible through such technology. Much of the recent advances have been possible through the transfer of aeroengine design tools, elevated temperature materials, coatings, and sealing technologies. A balanced approach that melds the aeroengine technologies and longstanding field experience from large industrial frame gas turbines is key to ensuring reliability, availability, maintainability, and durability (RAM-D) design objectives are met. Future costs associated with power generation and distribution with these advanced gas turbine combined-cycle power plants will undoubtedly be driven by major decisions of OEMs pertaining to design and validation approach, material

* Source: [14-3] Courtesy of Mitsubishi Power Systems (MPS). Extracts from the MPS collection of academic papers, with specific reference to V. Kallianpur, E. Akita, and Y. Tsukuda. "Enhancing Reliability and Reducing O&M Expenditures in Advanced Combined Cycle Gas Turbine Power Plants," IGTI 2001-GT-0499.

selection, reparability, design criteria, and design features for quicker maintainability.

Figure 14—1 shows the evolution of Mitsubishi Heavy Industry (MHI) gas turbines during the past two decades. The introduction of F-class gas turbines resulted in an increase in firing temperature of 2000 through the use of newer materials and advanced blade cooling concepts. With firing temperatures around 1300—14,000, the combined cycle efficiency of the F-class combined-cycle power plants is 56—57% (LHV). In 1997, MHI introduced the G-class gas turbines at 15,000 firing temperature. G-class gas turbines apply steam cooling to combustor baskets and achieve a combined-cycle efficiency of 58% (LHV). The field experience with steam cooling application was further expanded to include other hot gas path components in MHI's H-class gas turbine. MHI's H-class gas turbine applies closed circuit steam cooling to combustor baskets, rows 1 and 2 turbine vanes and blades, and turbine rings to achieve thermal efficiency of 60%.

Combustor technology is shifting toward meeting single-digit NO_x emissions. The changes occurring under deregulated generation and dispatch also necessitate that the NO_x emission levels are met at a broader load range. Other emissions, such as CO, VOC, and particulates, also have to be minimized. The combustion control to stabilize the flame at a lean fuel-to-air ratio is one of the most important challenges for all gas turbine manufacturers.

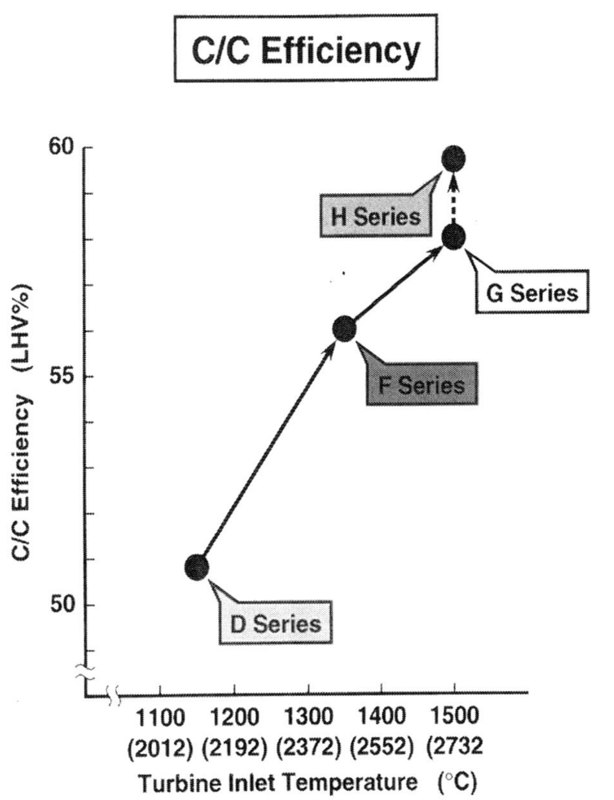

FIGURE 14—1 Evolution of MHI industrial gas turbines. *(Source: Mitsubishi Heavy Industries.)*

The thermal efficiency of today's combined-cycle power plants are much higher than conventional steam power plants. For example, a combined-cycle power plant with the F-class gas turbine having firing temperatures between 1300 and 1400°C attains thermal efficiency of 55—57%. The combined-cycle efficiency has been further improved by raising the firing temperature, as shown in Figure 14—1, and it has even reached 58—60% for the G- and H-class gas turbine.

In order to use valuable clean energy efficiently, several key technologies for efficient and environmentally benign combined-cycle power generation have been applied:

1. Dry, low NO_x combustion technology
2. Blade cooling technology
3. Heat resistant materials and casting technology
4. Aerodynamic design methodology
5. Numerical flow simulation (computational fluid dynamics)
6. Seal and clearance control technology

It has been demonstrated that the development of these key technologies for large frame gas turbines can be made economical by using the knowledge and experience acquired from aeroengine developments. However, in many cases, there are limitations in extrapolating the aeroengine technologies directly to large frame gas turbines due to size and duty cycle differences. For example, casting of large directionally solidified and single-crystal blades is much more difficult than smaller aeroengine blades. Alloy composition and heat treatment are generally altered for improving castability and production yields of the larger industrial castings. Those chemistry alterations also affect the long-term material behavior aspects. In aeroengines, the hold time exposure at peak transient stress and temperature are significantly shorter, typically around two minutes. By comparison, the LCF hold time exposure at peak stress and temperatures are several hours long in industrial applications, for peaking and base load operation. Therefore, the design approach for achieving high reliability in new high-efficiency industrial gas turbines requires retaining the proven designs features and field experience from the frame machines, while transferring aeroengine design technologies and field experience for optimizing stage efficiencies, heat rates, and output.

Design Decisions Influencing Reliability

Key features that are important from the standpoint of reliability are use of proven design structural features, maintaining same scaling procedures, use of proven materials, and extensive verification testing.

Use of Proven Design Structural Features

Figure 14—2 compares rotor cross-sections of the Mitsubishi M501D, M501F, M501G, and M501H gas turbines. There is a striking resemblance between the rotors. Except for the spool-type compressor in the M501D, the compressor and turbine sections in the F, G, and H gas turbines even maintain the exact same bolt count: 12. By retaining the proven design features in the product evolution (see Table 14—1), the design engineers continue to work within the field experience window relative to temperatures and stresses.

MHI rotor designs make use of "positive torque" features for enhancing torque transmission without slippage. In fact, field experience has proven that the use of radial pins in the compressor and curvic couplings in the turbine enhance the overall resistance to interstage slippage, which is particularly important from a startup/shutdown consideration. Another necessary design feature in advanced design compressors is the provision to replace foreign object damage of compressor blades quickly on site without transporting the rotor to a service overhaul repair facility.

Maintaining Same Scaling Procedures

The scaling procedure from 60 Hz designs to 50 Hz designs has been steadfastly maintained throughout MHI product lines. For example, the first two rows of blades and vanes of the turbine are common across the 60 Hz (M501 series) and 50 Hz (M701 series) versions. Likewise, the combustors are common between the two frame sizes. The field experience and design calibration with this scaling approach is retained across MHI's D, F, and G classes of gas turbines. By contrast, others have deviated significantly in scaling method, rotor structure, materials, and coatings.

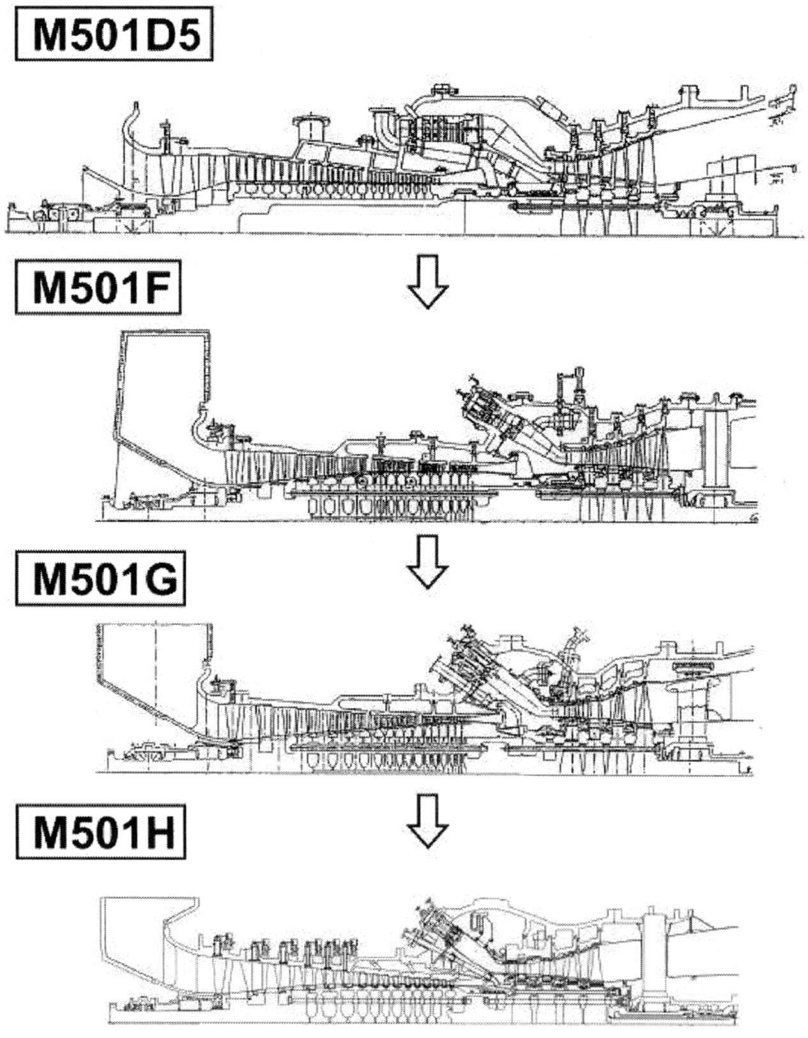

FIGURE 14—2 Similar rotor cross-sections retained. *(Source: Mitsubishi Heavy Industries.)*

TABLE 14–1 Proven Design Retained

Feature	501D5	501F	501G	501H
Cold end generator drive	✓	✓	✓	✓
2-Bearing rotor	✓	✓	✓	✓
4-Stage turbine	✓	✓	✓	✓
Individual combustors	✓	✓	✓	
Horizontal split casing	✓	✓	✓	✓
Single row 1 vanes	✓	✓	✓	✓
Cooled & filtered rotor air	✓	✓	✓	(Steam Cooling)

(Source: Mitsubishi Heavy Industries.)

Use of Proven Materials

Table 14–2 compares the materials used in MHI's F-, G-, and H-class machines. The commonality in materials simplifies transfer of field repair experience and practices. Also, the use of fewer standardized materials makes it possible to concentrate on long-term material properties characterization.

By retaining materials and field proven structural design features, MHI's design focus has been on improving stage efficiencies in the compressor and turbine through state-of-the-art 3D aerocomputational techniques. Figure 14–3 is a picture of the advanced 3D designed compressor blades in the H machines. The use of such advanced tools has made it possible to achieve a 25:1 pressure ratio, as compared to 20:1 with the G machine. More importantly, the higher pressure ratio is achieved with two fewer stages than the G. Thus, similar rotor spans are maintained for the G and H machines, thereby optimizing on design operating experience and reliability aspects.

The M501 gas turbine has a 16-can dry premix low NO_x combustor to satisfy the worldwide NO_x emission regulations. The established concept of the premix combustor is retained, such as the two-stage burner assembly and the bypass valve system. The feature of the combustor is the application of a multiple fuel nozzle system to improve flame stability in lean burning. The combustor has a pilot nozzle in the center and eight surrounding main nozzles supplying the premixed air and fuel. Proper adjustment of fuel flow rates through the pilot and the main nozzles allows the NO_x emission to be maintained at a lower level than that of its predecessor in the entire load range. The combustor wall and the transition piece are cooled using efficient double-wall cooling structures developed by Mitsubishi, named MTFIN™. The structures efficiently provide more air to the combustion region for the lean, low-NO_x burning. The F-class premix combustor is shown in Figure 14–4.

TABLE 14–2 Materials Commonality Retained

Materials used in F, G, & H	
MGA 1400	Turbine blades rows 1 to 4
MGA 2400	Turbine vanes rows 1 to 3
Tomilloy	Combustor transition piece
Low alloy steel	Rotor

(Source: Mitsubishi Heavy Industries.)

Extensive Verification Testing

Throughout MHI's almost 40 years as an industrial gas turbine OEM, new technologies have been made

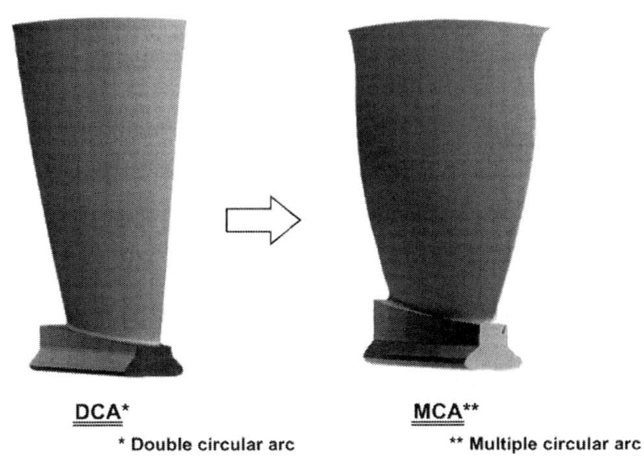

FIGURE 14–3 Advanced 3D designed compressor blades. *(Source: Mitsubishi Heavy Industries.)*

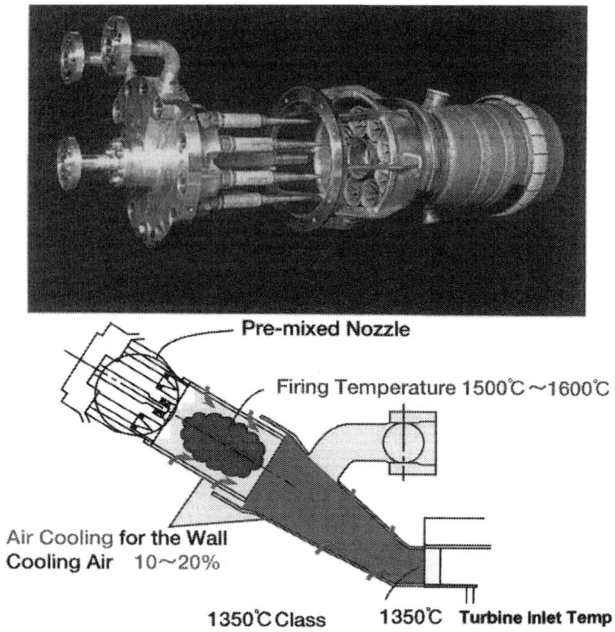

FIGURE 14–4 MHI F-class premix combustor. *(Source: Mitsubishi Heavy Industries.)*

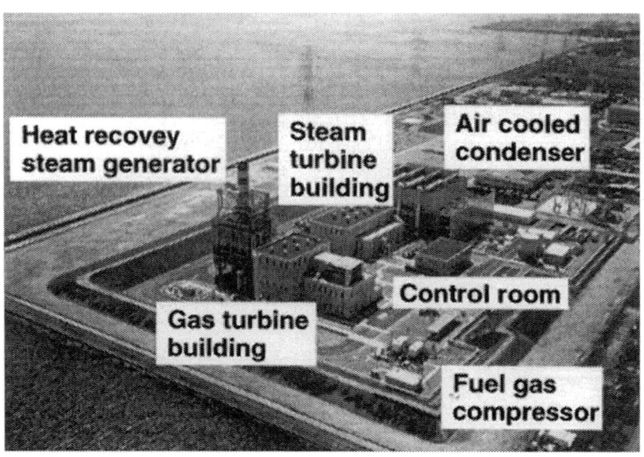

FIGURE 14–5 MHI's T-Point combined cycle power plant. *(Source: Mitsubishi Heavy Industries.)*

commercially available only after full-scale full-load testing at MHI's internal facilities. Figure 14–5 is a picture of a full-scale MHI combined-cycle power plant at Takasago, called T-Point facility. The facility is operated usually under daily start-stop operation, especially during the summer peak season, and power is transmitted to a local utility. Thus reliability and availability aspects have to be met according to stringent commercial contractual expectations. Thorough inspections following the peak dispatch season enable continuous validation and modifications for enhancing component durability, power plant reliability, and availability. For example, minor design adjustments made during early stages of machine verification and operation resulted in enhancements to cooling circuits, resulting in superior thermal distributions and improved material/coating performance and component durability prior to commercial introduction of G technology. MHI followed a similar approach before commercial introduction of F technology.

Figures 14–6 through 14–8 show the excellent condition of hardware at the fourth inspection on the M501G machine at T-Point following 10,185 hrs and 561 starts. The facility is also utilized for validating the H machine. For the previously mentioned reason, the common rotor span between the G and H machines makes it possible to use the T-point facility to conduct design validation tests at the facility.

FIGURE 14–6 Service exposed M501G combustor from T-Point. *(Source: Mitsubishi Heavy Industries.)*

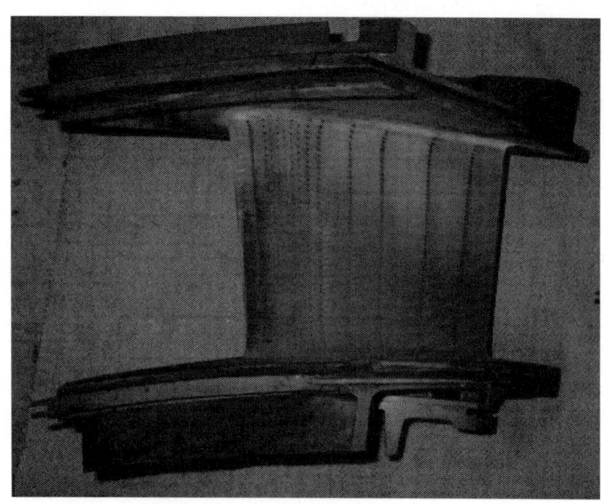

FIGURE 14–7 Service exposed M501 G row 1 vane from T-Point. *(Source: Mitsubishi Heavy Industries.)*

FIGURE 14–8 Service exposed M501 G row 1 turbine blades from T-Point. *(Source: Mitsubishi Heavy Industries.)*

CASE STUDY 2: HOW CLOSE IS THE MEASURED PERFORMANCE TO THE TRUE OUTPUT AND HEAT RATE? THE PROOF IS IN THE TESTING!*

The huge increase in electrical power demand continues worldwide and especially in the United States. Therefore, an increasing number of existing power plants plan thermal performance tests to check their power output range (not necessarily the same as when they were commissioned), before they commit to any further power deliveries.

A thermal performance test on a new plant is a test where the end user checks that his newly commissioned plant meets the power output guarantees made by manufacturers and contractors. The proper conductance of the test, data collection, and correction to reference conditions have many contractual, technical, and commercial implications for all parties.

* Reference: [14-4] J. Zachary, "How Close is the Measured Performance to the True Output and Heat Rate? The Proof is in the Testing!" from the collective papers of J. Zachary, including work presented at PowerGen International 2001.

The practical aspects of testing, from contract negotiations to field implementation, in which the thermal evaluation of a power plant facility and its major components is carried out, are described as follows.

ASME Performance Test Codes

The American Society of Mechanical Engineers (ASME) performance test codes (PTCs) serve as the guiding documents for all aspects of performance testing. The ASME test codes have achieved worldwide acceptance and are applied in Asia, Australia, South America, and Europe.

In the last six years, to keep up with the rapid technological changes, most of the PTCs have been updated and reissued. PTC-46, Power Plant Performance Test Code (1996), is routinely implemented in combined-cycle facilities. PTC-22, Gas Turbine Performance Test Code, will expand its scope to cover the testing and calculations for exhaust temperature and exhaust flow, in addition to power output and heat rate. This new version is because the industry needs to provide guarantees for exhaust energy and exhaust temperature in combined-cycle applications. The methodology will be based on a complete mass and energy balance around the combustion turbine. A new version of PTC-6, Steam Turbine, was issued in 1996. A committee was formed in 2000 to develop a new PTC-6.2 dealing exclusively with steam turbine performance in combined cycles. A new and comprehensive Measurement Uncertainty Code, PTC-19.1, was issued in 1998. A draft for a new PTC-4.4 for heat recovery steam generators (HRSGs) was released for industry review. PTC-4.4 (HRSGs) and PTC-22 (gas turbines) are being revised in close cooperation so that they use the same methodology for calculating combustion turbine exhaust heat (input source for the HRSG). New codes in the PTC-19 series dealing with measurement techniques are close to publication, of particular interest being PTC-19.5, which covers flow measurement.

Combined-Cycle Guaranteed Values

The testing of combined-cycle power plants, including combustion turbines, HRSG, and steam turbines, requires two measurements:

- The electrical power output on the high side of the transformers
- The fuel heat input—flow and composition—at the facility fence

Accurate evaluation of the thermal performance in a power plant is challenging. Errors in measurement of power output and heat rate could translate to millions of dollars in liquidated damages (LDs). An example of the commercial risks for missing the guarantees even by very small

TABLE 14–3 Potential Commercial Liquidated Damages—Typical Case [14-4]

	Guarantee	Error	Error (%)	LDs	LDs
Power	8000,000 kW	500 kW	0.0625	$1000/kW	$500,000
Heat rate	6000 BTU/kWh	4 kW	0.06667	$150,000/(BTU/kWh)	$600,000

amounts is presented in Table 14–3. A typical 800 MW power plant with a 6000 BTU/kWh heat rate could pay LDs in excess of $1 million for shortages of 0.07% in performance guarantees.

Preparations for test should start at the beginning of a project, when considering conditions and configuration. All of the parameters that might influence the test results need to be properly identified. As part of the contract definitions, all the ambient conditions at the inlet of the combustion turbine and in the vicinity of the heat sink must be stated. One must also define the location of the electrical power metering, auxiliary power measurement, power factor range, fuel quality and heating value, plant configuration during test, makeup water, and blowdown valve positions.

Test Significance

While test codes offer the basic framework, additional practical issues must be resolved to implement code recommendations in real-life situations. Most of the commercial contracts specify that test procedures must be mutually agreed on. Many end users do not realize the true aim of the thermal performance tests and combine other plant acceptance criteria with the thermal evaluation of the plant.

- *The thermal performance test is not a reliability or endurance test.* Very often, owners and end users request that a lengthy performance test be conducted, far beyond the recommendations of the codes. Requesting a test to be carried out for an excessive number of hours (for example, 8 or 10 hours for a combined cycle) might actually have a negative impact on the results. During these continuous runs, a wide variation in ambient temperature could result in the plant operating at conditions further away from guarantee conditions and thus cause errors.
- *Thermal performance tests are used to verify the correction curves.* This controversial issue is common with performance testing. In the following paragraph, the issue of correction curves will be discussed. However, plant performance guarantees are usually made for a single point. Often, the test program requires conducting several tests at different ambient conditions and plant configurations. This does not relate to the performance test but is intended to verify if the hardware behaves as correction curves predict. This is a separate task with a different objective than the performance of the equipment. So if the parties involved agree, it should be conducted before the performance test, results should be mutually accepted to by all, and correction curves should be redrawn accordingly.
- *Test repeatability is not an acceptance criterion.* A facility thermal performance test is a demonstration of the output and heat rate. Conducting repeated tests and comparing the corrected results might not indicate a poor test. Correction curves might be slightly different at different ambient conditions, or random error values could be higher. Verifying the instrumentation and proofing the conversion methodology are part of the pretest preparations and the mutually agreed-on test procedure, not of the test itself.
- *Conformance with important design criteria of equipment.* One prerequisite of the performance test procedure is to operate the power plant at base load. In most cases, there is not a clear definition of what *base load* means. It should be mentioned that all parties expect the equipment to perform in accordance with manufacturer's design criteria for "safe continuous" operation. One of the most contested and difficult issues between owners, bank engineers, contractors, and manufacturers is the definition of the combustion turbine control algorithm, which is considered proprietary by the manufacturer. All of the parties need to ensure that the control algorithm used during the thermal performance test will not be further modified, which could invalidate the results.
- *Conductance of the thermal performance test should not be linked to the punch list items.* Arguments often arise between the parties because of the component status during the test. For example, a test can be carried out with valves or subsystems operating in manual rather than automatic mode. The performance test scope is to confirm the power output and heat rate guarantees only. As long as the safety and integrity of the equipment are maintained, the actual status of components or system, which could be fixed later as part of the punch list, should not nullify the test.

Measurement Uncertainty and Test Tolerance

The determination of the power output and the heat rate of a power plant is associated with errors in the measuring devices, corrections to guarantees, and inadequate test conductance. A rigorous measurement uncertainty analysis is now required by all PTCs to assess statistically the potential gap between the test results and the "actual true" power output and heat rate of the plant. The analysis, fully described and commented on in PIC-19.1, basically identifies systematic error (an offset from the actual value obtained by nonstatistical methods) and precision errors (random errors derived by statistical methods) of each primary variable, from the sensor to the data acquisition system.

These values are multiplied by a sensitivity coefficient quantifying how each variable influences the result of either power output or heat rate. The square root sum of squares for all parameters is calculated to determine a ± tolerance band. The area, defined by the band applied around the measured test results, creates a domain where there is a 95% confidence level that the "true power output or heat rate" could be found. An example is presented in Table 14–4, where the power output uncertainty is 0.5% and the heat rate uncertainty is 0.87%. For the same example of an 800 MW power plant, the uncertainty for power equals ±4 MW valued at $4 million, using the same LD criteria as in Table 14–3. Respectively, for a heat rate value of 6000 BTU/kWh, a 0.87% uncertainty equals ±52 BTU/kWh and has a value of $7.8 million. Thus, the commercial value of measurement uncertainty equals close to $12 million or 2.8% of the entire typical project cost.

Correction Curves

Power plants are rarely tested at the guaranteed conditions (which usually includes ambient conditions, generator power factor, and other specific plant configuration requirements, such as blowdown). To demonstrate if a test has achieved the guaranteed power output and heat rate, a well-established methodology of correction curves is applied. A correction curve for a specific parameter, for example, ambient temperature, is developed using a computer simulation program, where all of the other variables are kept constant at their guaranteed conditions. Only the parameter in question is allowed to vary, and the impact on power output and heat rate is recorded. A curve is drawn, showing the variation of the power output versus ambient temperature, as shown in the example in Figure 14–9. The curve is then converted to a polynomial equation and further used with other correction curves in a large spreadsheet. The entire process is tedious, time consuming, and expensive.

There are many significant problems with this methodology, which could very significantly skew the test results.

As can be seen from Table 14–5, the use of correction curves started with the simple process of a single homogeneous fluid expansion, which could be depicted quite accurately. As the concept is expanded to more than one thermodynamic process, the correction curves become two and three dimensional.

An alternative solution to the problem is to apply a complete computer simulation program for the plant behavior. Thus, all variables can be adjusted simultaneously and all the basic laws for mass, momentum, and energy balance are truly maintained. The correction curve methodology cannot account for secondary effects on the turbomachinery because of the coupled effect of changes in more than one variable. Ambient temperature and relative humidity are the most common examples.

Degradation Curves

The final thermal performance tests of a combined-cycle plant are conducted after all of the components—combustion turbine, HRSG, and steam turbine—are commissioned and fine-tuned, a process that requires a significant number of hours of operation for each of the components. To cover the gap from the "new and clean" performance guaranteed by the manufacturer to the performance measured during the test, equipment degradation curves have been used.

The most influential degradation curve is associated with nonrecoverable degradation of the combustion turbine performance. The combustion turbine manufacturer produces these curves based on the results of the performance fleet and in-house evaluation. Great significance is attached to the shape of the curves for obvious impact on plant performance.

Particularly for new models where fleet experience is not totally applicable, conservative degradation curves are used. Attempts to develop specific curves based on actual tests are extremely difficult because changes in performance for small numbers of operating hours are the same order of magnitude as the measurement uncertainty.

Inlet Cooling Devices

Inlet cooling devices are a safe and profitable means to achieve better performance at low risk and low cost. Evaporative cooling, inlet fogging, inlet air refrigeration using mechanical compression, and absorption systems are popular options. This has created new challenges for testing staff. The performance of the combustion turbine, and the Rankine cycle components, depends not on ambient conditions at the fitter house but on those at the inlet belmouth of the compressor. The methodology to assess the conditions at this particular plane is far more

TABLE 14–4 Typical Measurement Uncertainty for a Combined-Cycle Plant [14-4]

Parameter	Sensitivity	Systematic Uncertainty	Systematic Uncertainty Contribution	Random Uncertainty	Random Uncertainty Contribution
Power Output Measurement Uncertainty					
Ambient temperature	0.26051 %/°F	1.33°F	0.34648°F	0.20780°F	0.05413
Facility net output	1.00000 %/%	0.258	0.25800	0.15800	0.15800
Barometric pressure	4.23000 %/psia	0.039 psia	0.16497 psia	0.02000 psia	0.08460
Ambient RH	−0.00156 %/%RH	2.0040	−0.00313	0.50000	−0.00078
Circ. water temp.	0.01428 %/°F	4.170°F	0.05955°F	1.00000°F	0.01428
CTG shaft speed	0.02549 %/rpm	0.271 rpm	0.00691 rpm	0.50000 rpm	0.01275
CTG gross output	0.01052 %/%	0.318	0.00335	0.20000	0.00210
STG gross output	0.00389 %/%	0.3260	0.00127	0.20000	0.00078
CTG power factor	0.01875 %/%	1.030	0.01931	0.00800	0.00015
STG power factor	0.00490 %/%	1.0300	0.00504	0.00800	0.00004
RSS			0.46674		0.18821
Total uncertainty				0.5033	
Heat Rate Measurement Uncertainty					
Gas fuel flow	1 %/%	0.801°F	0.80100	0.05000°F	0.05000
Gas heating value	1 %/%	0.150°F	0.15000	0.04000	0.04000
Ambient temperature	0.0056 %/°F	0.562°F	0.00315	0.20780°F	0.00116
Facility net output	1 %/%	0.258	0.25800	0.15800	0.15800
Barometric pressure	0.38 %/psia	0.0039 psia	0.00148	0.02000	0.00760
Ambient RH	0.0023 %/%RH	2.0000	0.00460	0.50000	0.40115
Circ. water temp.	0.012 %/°F	4.170°F	0.05004	1.00000	0.01200
CTG shaft speed	0.0024662 %/rpm	0.271	0.00067	0.50000	0.00123
CTG gross output	0.011 %/%	0.318	0.00350	0.20000	0.00220
STG gross output	0.0039 %/%	0.3260	0.00127	0.20000	0.00078
CTG power factor	0.019 %/%	1.030	0.01957	0.00800	0.00015
STG power factor	0.0049 %/%	1.0300	0.00505	0.00800	0.00004
RSS			0.85657		0.17110
Total uncertainty				0.87344	

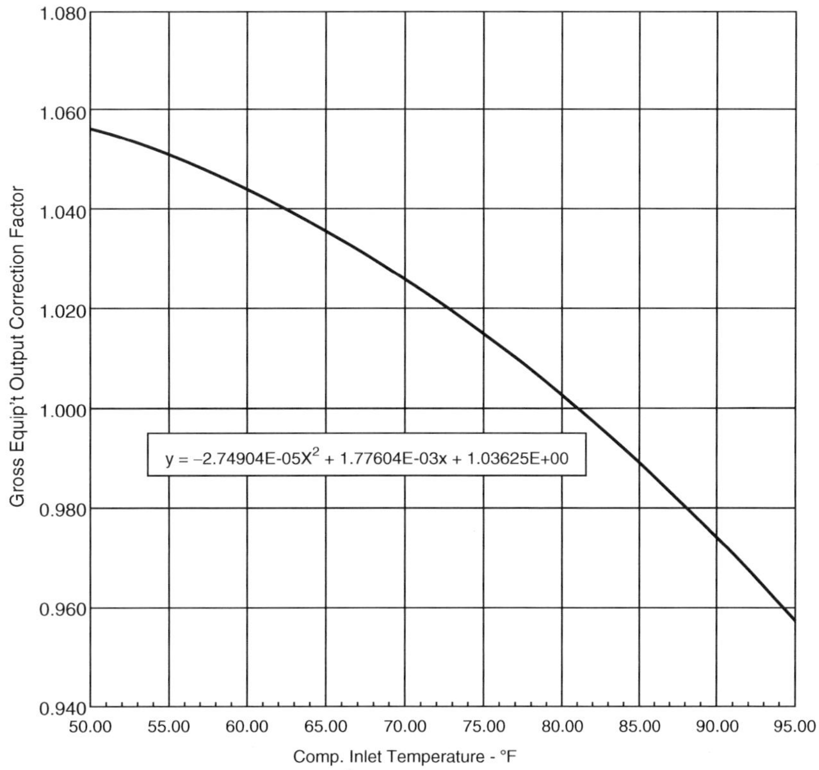

FIGURE 14–9 Example of correction curves [14-4].

difficult and can create errors for the overall power output and heat rate values.

So is the inlet cooling device an integral part of the power plant? With mechanical chillers, should we include the entire refrigeration unit as part of the power plant?

Manufacturers have created a system of interstage cooling for aeroderivatives. How should we deal with performance testing and corrections under these conditions? Of particular concern is the use of inlet foggers with a water injection amount beyond the 100% relative humidity, commercially called *wet compression*.

TABLE 14–5 Thermodynamic Processes [14-4]

Component to be Tested	Type of Thermodynamic Process	Comments
Steam turbine	Adiabatic expansion	Single homogeneous fluid
Combustion turbine	Adiabatic compression	Air
	Combustion	Fuel and air
	Adiabatic expansion	Exhaust gases
HRSG	Heat transfer	Steam and exhaust gases
	Vaporization	
Heat sink	Condensation	Steam
	Various heat removal methods using phase change	Water and air
Combined cycles	All of the above	
	All of the above	
Integrated gasification combined cycle (IGCC)	Gasifier Air separation unit	

So some recommend that one exclude all the inlet or interstage cooling devices from the applicable control volume of the power plant. Particularly for combustion turbine tests, verification of the guaranteed exhaust flow and exhaust temperature is almost impossible when the inlet-cooling device is running. However, one can conduct tests with the inlet cooling on and then off for comparison purposes.

CASE STUDY 3: COMPARATIVE EVALUATION OF POWER PLANTS WITH REGARD TO TECHNICAL, ECOLOGICAL AND ECONOMICAL ASPECTS*

The final evaluation of power generation technologies is based mainly on its electricity generation costs or life cycle cost. Considering today's fuel prices and price development projections over the plant lifetimes, the economical analysis performed can serve as a framework for project and investment decisions.

More stringent environmental regulations raise production cost. Long-term planning in power plant capacities has become uncertain as never before. Changing fuel prices are the third parameter that has to be considered before investing in power plants. According to forecasts of the world energy conference and others, the fossil fuels, like coal, oil, and natural gas, will be also in the near future our main energy source for power generation (Figure 14—10).

Therefore it is very important to use these nonrenewable energies with great care and burn them only in highly efficient plants. This saves not only our limited reserves but also cuts emissions in order to protect life, environment, and climatic stability.

An overview follows of state-of-the art and future power plant technologies, environmental issues, cost, and economic considerations. We employ the following abbreviations:

PC = Pulverized coal steam power plant
AFBC = Atmospheric fluidized bed combustion
CC = Combined-cycle plant
IGCC = Integrated gasification combined cycle
ACFBC = Atmospheric circulating fluidized combustion bed
SPFBC = Stationary pressurized fluidized combustion bed
CPFBC = Circulating pressurized fluidized combustion bed

* Source: [14-5] Courtesy of Siemens. Extracts from A. Lezuo and R. Taud. "Comparative Evaluation of Power Plants with Regard to Technical, Ecological and Economical Aspects," 2001-GT-0504.

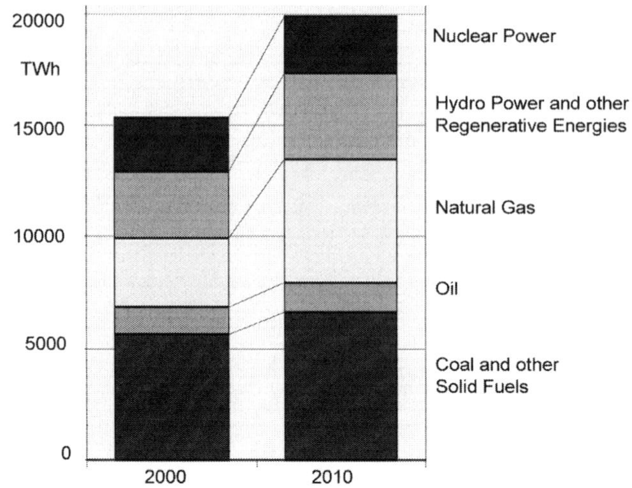

FIGURE 14—10 Worldwide electricity generation by fuels [14-5].

Available Technologies

Considering the time frame up to the year 2005, no revolutionary new power plant technology will be commercially available. The improvement will be found mainly in higher efficiencies due to enhancement of components and an increase in the upper process temperatures of the well-known processes. For all plant designs, the same boundary conditions have been assumed unless otherwise stated, which are in Table 14—6.

Coal-Fired Power Plants

Due to large reserves and the worldwide availability of coal, both lignite and hard coal, this energy carrier also will play an important role for electricity generation in the future. Although the increase in consumption will be lower than that of natural gas, the relative contribution to the total electricity production will still amount to approximately 33—34%. Therefore, investment in the development of coal technologies also will be of great importance in the future.

TABLE 14—6 Assumed Boundary Conditions [14-5]

Ambient temperature	°C/°F	15/59
Ambient air moisture	%	60
Ambient air pressure	bar/psi	1013/14.69
Condenser pressure	mbar/psi	50/0.725
NO_x emission	mg/m^3 STP	150
SO_x emission	mg/m^3 STP	100
Dust emission	mg/m^3 STP	50

STP = standard temperature pressure

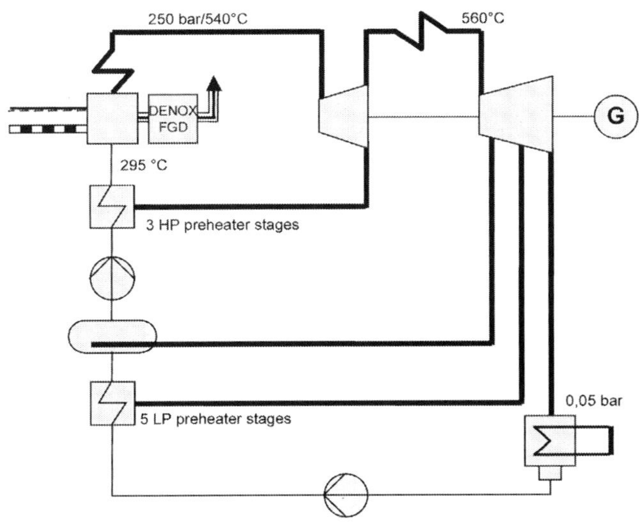

FIGURE 14–11 Pulverized-coal-fired steam power plant [14-5].

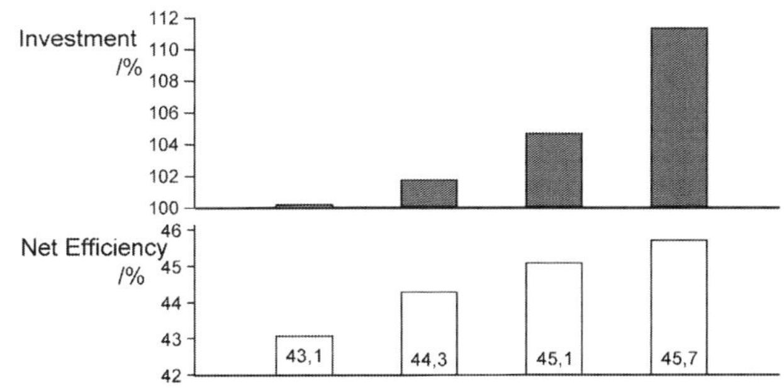

Main steam pressure	250	270	285	285	bar
temperature	540	580	600	600	°C
reheat temperature 1	560	600	620	620	°C
reheat temperature 2	-	-	-	620	°C

FIGURE 14–12 Influence of steam condition of pulverized coal power plants on efficiency and investment [14-5].

Pulverized Coal Power Plants

Today's atmospheric pulverized coal power plants (Figure 14–11) are characterized by high efficiencies, high availability, and high reliability. They serve a power range from 20–1000 MW per block, whereas the lower capacities up to 300 MW are usually built as atmospheric fluidized bed combustion systems. For large grids, large-sized double units are preferred for cost reasons.

The design points out that net efficiencies for a modern power plant with wet cooling towers and secondary measures for flue gas cleaning can be well above 40% LHV. Due to startup losses and partial load operation, the mean annual net efficiency can be lowered, however, by up to 3 percentage points. For these investigations, only the design points have been considered.

In industrialized countries, newly built large power plants usually have steam conditions of 250 bar and 540°C and are equipped with reheat. In a few cases, power plants with steam conditions of 285 bar and 600°C are under construction. The influence of the steam parameters on the efficiency and investment is shown in Figure 14–12. From this diagram, it can be seen that the investment is increasing more than proportionally compared to the efficiency increase.

With the help of desulfurization plants (FGD), DENOX, and dust filters, all the emission limits set up by any regulation authority can be met easily. The limits applied in this study (see Table 14–6) are more stringent than any current limits in the world, but they are expected to be relevant in future power plants.

Pressurized Fluidized Bed Power Plant

The power plant with stationary pressurized fluidized bed combustion (PFBC) was introduced into the market in the 1990s. Since then, seven plants have been built and a cumulated operating experience of more than 100,000 hours could be gained. While the first plants had a capacity of only 70 MW, in Japan the first plant with an electrical output of 350 MW is in the test phase.

The advantage of this type of plant is its possibility to reduce the sulfur and nitrogen emissions only by primary measures, thus saving investment for secondary measures. In addition, this plant can be fed a very wide range of coal types with different compositions. Due to the low operating temperature in the steam generator of approximately 850°C, the formation of thermal NO_x can be kept at a minimum. The reduction of SO_x is being managed by adding limestone directly into the fuel, where it is reacting with the sulfur and bonded in the ash. The degree of reduction can be adjusted by the sulfur/limestone ratio. As long as today's emission standards are applied, the limit values can be maintained, but for lower values, secondary measures, at least for NO_x, have to be taken, such as the SCR installed with the 350 MW Karita plant in Japan. This, however, reduces the cost advantage of this concept.

By means of expansion of the exhaust of the supercharged steam generator (12–18 bar) in a bottoming gas turbine, the high temperature also can be used for power generation and hence reducing exergetic losses. The gas turbine drives the compressor for the combustion air and in addition an electric generator, which produces about 20% of the total power output. In this way, a net efficiency of approximately 42% LHV can be reached.

An improvement of the efficiency by increasing the process temperature is not possible because of the ash melting point. Higher temperatures would make it impossible to reduce the ash content in the flue gas by cyclones, which are needed to protect the bottoming gas turbine against erosion.

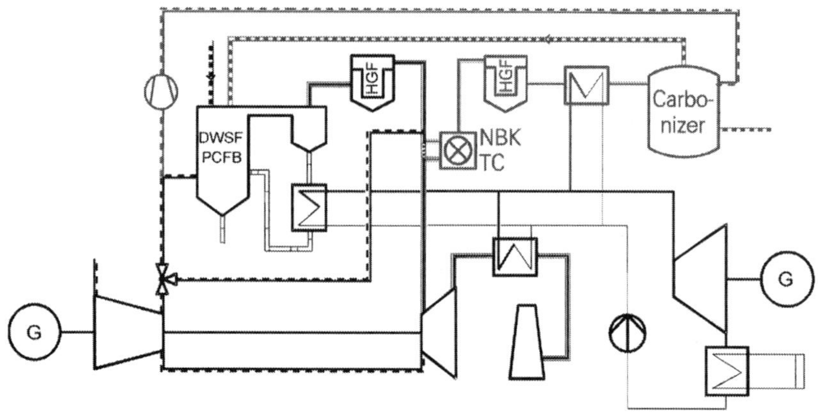

FIGURE 14–13 Circulating pressurized fluidized bed combustion [14-5].

Because of the limitation of temperature, this concept cannot take advantage of the highly efficient modern gas turbines with high inlet temperatures. Therefore, the so-called second generation circulating pressurized fluidized bed combustion (Figure 14–13) is being developed.

Its higher thermal power density enables very high efficiencies of above 50% LHV. This concept consists of a circulating pressurized fluidized bed combustor (CPFBC) with a pressurized partial gasifier in parallel that generates a syngas. The gasifier converts only 60–80% of the carbon, and the rest is burned completely in the CPFBC. After cleaning the flue gas from particles in two filter stages, its temperature can be raised to gas turbine inlet temperature by means of the syngas of the gasifier. The exhaust gas of the gas turbine can be utilized in a steam process in the same way as it is done in a conventional combined-cycle plant.

To maintain today's emission standards, an adequate sulfur bonding can be reached by admixing additives into the fuel. Because of the high process temperature, an SCR plant for NO_x reduction will be necessary.

This concept is still under development, and it has to be demonstrated that the particles in the flue gas can be kept below the permissible values of the gas turbine.

Integrated Coal Gasification Combined Cycle

The concept of an integrated gasification combined-cycle plant incorporates an oxygen or air-blown gasifier operating at high pressure and producing raw gas, which is cleaned of most pollutants and burned in the combustion chamber of the gas turbine (Figure 14–14). The sensible heat of the raw gas and the hot exhaust gas of the gas turbine are used to generate steam, which is expanded in a steam turbine. The mechanical energy of the gas turbine as well as the steam turbine is used to produce electric power.

While the power plant systems are well known and commercially available, the different gasification processes, including their auxiliary systems, are still at the beginning of commercial-scale production.

The potential of a high net efficiency of more than 50% LHV and very low sulfur and dust emissions—they are removed before passing the combustion process—make it an attractive option for future power generation.

Until 1998, five demonstration plants for coal gasification with outputs of between 100 and 300 MW have been put into operation worldwide.

Gas-Fired Power Plants

Because of the quick gas turbine development, driven by strong competition, during the past 10 years, the conventional gas-fired steam plants are no more of interest. This is not only due to their relatively low efficiency but also due to their relatively high investment.

The outstanding performance of natural gas-fired combined-cycle (CC) plants (Figure 14–15) has led to the situation in which many of the old conventional power plants have been shut down in favor of new CC plants.

These CC plants are available in a wide power range from 50 MW up to 800 MW per block. The large units reach a net efficiency level of approximately 56–58% today. Even the small plants rated at about 100 MW are still well above 50%. New plants, which are already under development and in construction, are supposed to reach the target value of 60% by the year 2002.

Besides high efficiencies, combined-cycle plants are characterized by low investment, low service cost, and outstanding environmental behavior. The only negative aspect in the row of benefits is the availability of the gas and the gas price, which is very sensitive to market alterations.

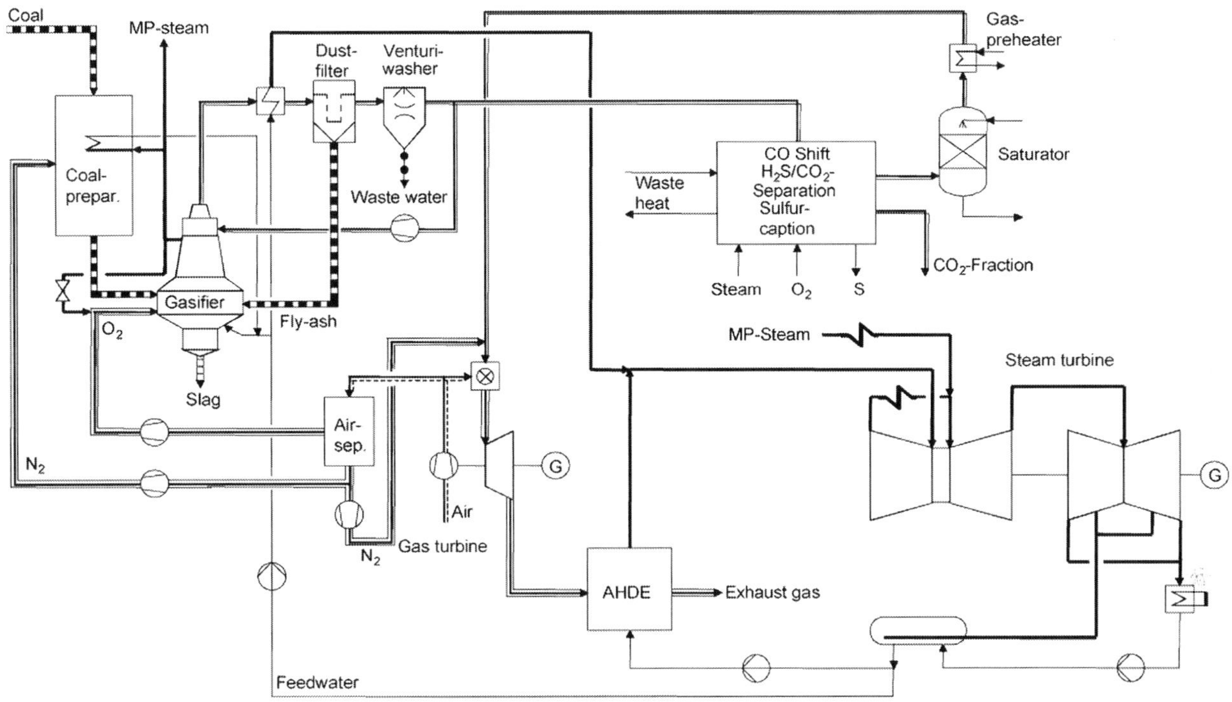

FIGURE 14–14 Integrated gasification combined cycle [14-5].

Efficiencies

The comparison of efficiencies (Figure 14–16) shows that the whole range of modern base-load power plants is able to reach a net efficiency of more than 40%. The outstanding performance of the combined-cycle plants is not only due to its technology but also a result of the use of a high-valued clean fuel.

The conventional coal-fired steam power plants have only a limited potential in increasing its efficiency by raising the steam parameters or adding a second reheat stage.

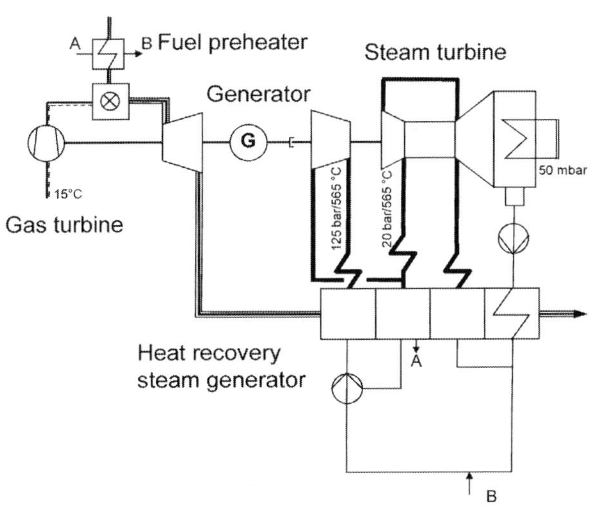

FIGURE 14–15 Single-shaft combined cycle power plant [14-5].

Only the new technologies like IGCC and CPFBC are able to overcome the limitations that restrict conventional steam turbine cycles. These technologies, such as the natural-gas-fired CC plants, use a combination of gas turbine and steam turbine and hence are able to use the entire temperature gradient from the combustion chamber.

These technologies are very attractive since they run on almost any type of coal, which is the longest lasting fossil fuel source.

Emissions

The consciousness of a benign environment is more and more important to the people all over the world. Power plants are besides the traffic the branch with the highest output of gaseous emissions. Several ecological anomalies, like acid rain, the greenhouse effect, and overheating of rivers, can be attributed to power plants.

In the meantime, government and industry in many countries have been responding with emission regulations and the positive results can be seen already in many areas. Basically all emissions are reduced by increasing plant efficiency. That means less fuel is required to generate the same amount of electricity and thus also less pollution is efficient. But this is not enough, and other measures have to be taken to limit the amount of pollutants.

Dust filters are now standard equipment in many of the power plants all over the world, admittedly with differing

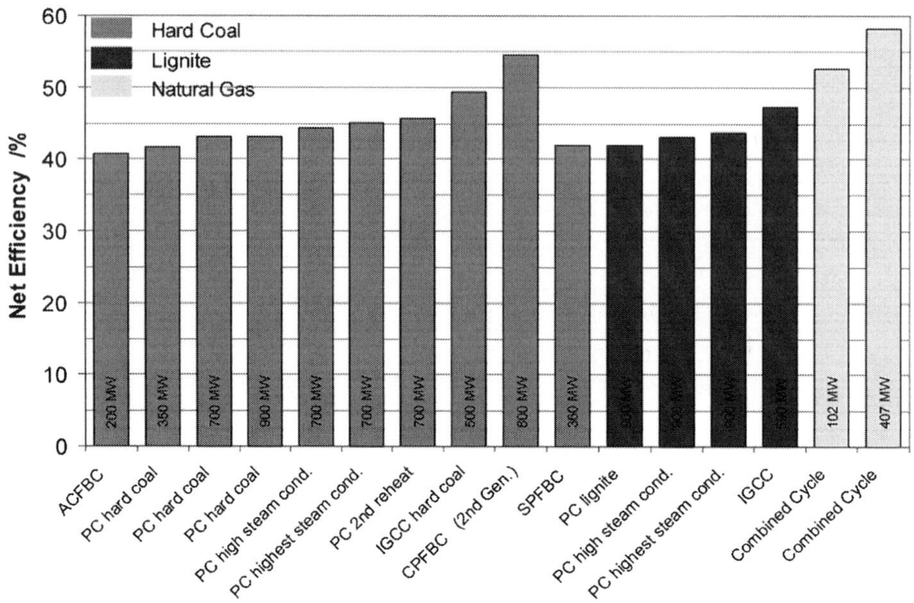

FIGURE 14–16 Comparison of efficiencies [14-5].

limit values. Certainly, a lot more can be done for a cleaner environment.

The reduction of SO_x and NO_x is technically somewhat more complicated and more costly. Therefore, only industrialized countries have been ready so far to invest in these technologies. Flue gas desulfurization plants and $DeNO_x$ plants are state of the art and enough experience exists for reliable operation without too much influence on the power plant performance. A reduction down to a concentration of 100 mg of SO_x or NO_x per m^3 of flue gas with secondary measures is easily achievable.

Fluidized bed combustion systems are able to reduce SO_x by primary measures only down to approximately 150 mg/m^3 STP by means of adding dolomite directly into the coal feed where it reacts with the sulfur to form $CaSO_4$ (gypsum), with the result that no further bonding agent is required for the ash product to set when water is added. Due to the low combustion temperature of approximately 850°C in FBC systems, the formation of thermal NO_x can be avoided right on the spot. This leads to NO_x emissions of only approximately 150 mg/m^3 STP, eliminating the need for secondary equipment. Only the CPFBC needs an SCR system, because it makes use of the advanced gas turbines with high inlet temperatures.

The desulfurization process in gasification systems is the most effective one, because the sulfur is captured already in the raw gas of the gasifier, where elemental sulfur, a saleable product, can be produced.

The use of the clean fuel "natural gas" in the combined cycle results neither in SO_2 output nor in any output of waste.

To make the different power plant technologies comparable regarding their main pollutants and wastes, shown in Figures 14–17 and 14–18, the various pollutants are related to 1 kWh of electricity output.

The emission of the greenhouse gas CO_2 from power plants is dependent on the type of fuel and the efficiency (Figure 14–19). The capture of CO_2 is technically feasible and there are even solutions for the long-term deposit of the CO_2, such as in the oceans or in depleted oil fields, but economical reasons make a realization of these ideas today impossible.

Power Plant Investment

Figure 14–20 shows 18 power plant concepts classified according to fuel type. For the pulverized-coal-fired power

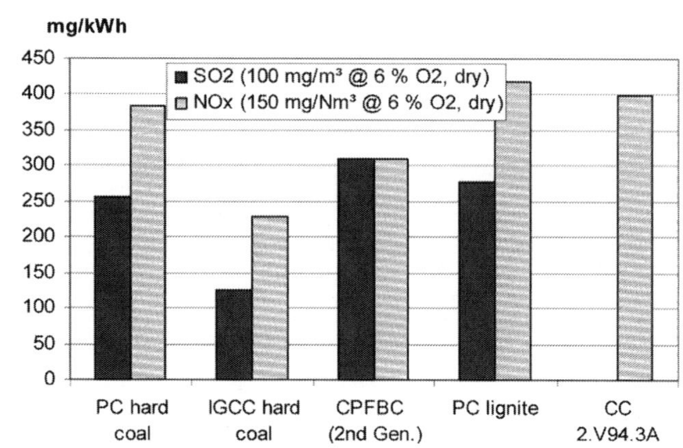

FIGURE 14–17 Flue gas pollutants [14-5].

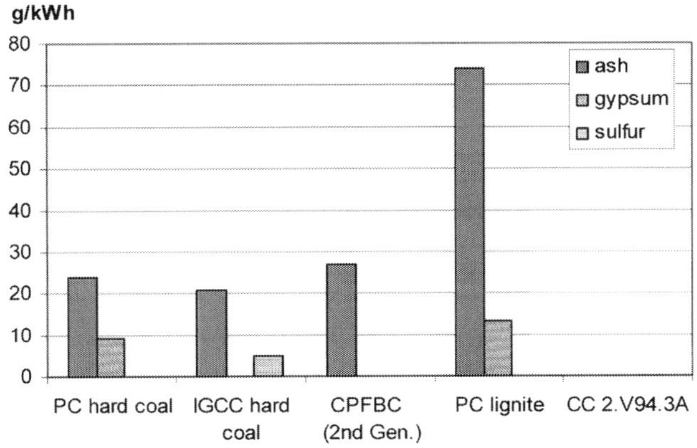

FIGURE 14–18 Power plant wastes [14-5].

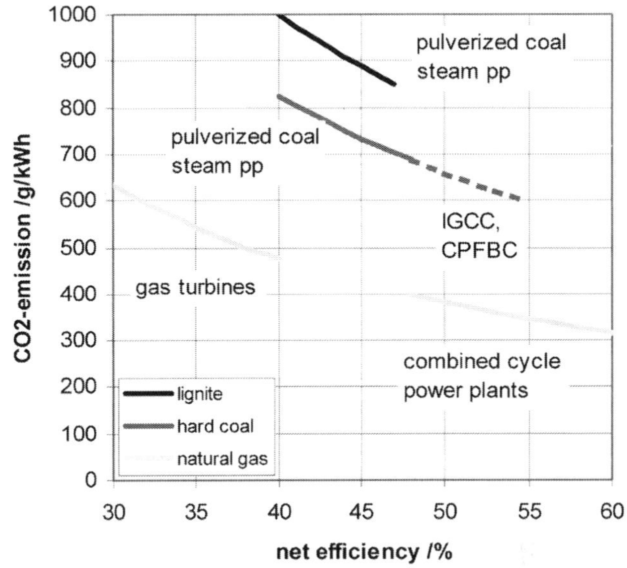

FIGURE 14–19 Specific CO_2 emissions of power plants [14-5].

plants both the influence on capacity size and the additional cost for higher steam parameters are given. All costs include the necessary environmental equipment to reach the standard as specified in Table 14–6.

The cost data specified for the emerging technologies have to be understood as target costs for the coming years, when enough design experience exists to cut costs considerably. Their cost accuracy can be assumed to be within 15–20%.

The investment for lignite-fired power plants is approximately 20 percentage points above that of hard-coal-fired power plants because of the larger steam generator and the higher expenditure for coal feed and ash removal systems. Because of the low BTU fuel, which would cause very high transportation cost, these plants are usually built as mine-mouth plants.

The natural-gas-fired combined-cycle plants are attractive not only because of their outstanding high efficiency but also by their very low investment, which is only about 40% of a hard-coal-fired power plant. But one never should forget that it consumes only high-priced, noble fuels, such as natural gas or light fuel oil with limited reserves.

Economic Evaluation

Besides the plant investment, the fuel price is the most influential portion of electricity generation cost. Furthermore, the fuel price can be subject to relatively large

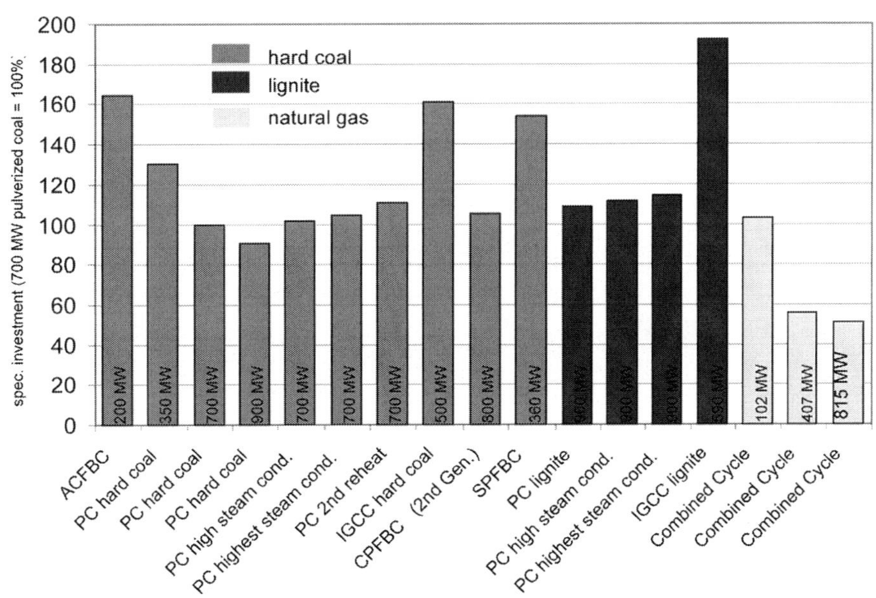

FIGURE 14–20 Comparison of investment cost [14-5].

TABLE 14–7 Assumed Fuel Prices [14-5]

Prices as of Year	Fuel Price	Price Increase	
	EUR/GJ	\$/MMBTU	%/a (real)
Hard coal	1.3	1.3	0.5
Lignite	1.5	1.5	0.0
Natural gas	3.1	3.2	1.0

TABLE 14–8 Economic Boundary Conditions [14-5]

Price basis	Year	2000
Start of operation	Year	2005
Interest rate	%/a	8
Depreciation period	a	15
Full-load operating hours	h/a	7000

changes over time. Therefore it is the most critical factor for long-term planning.

For this study the fuel prices according to Table 14–7 have been assumed, which represents an estimation of average prices in Europe.

Selections of additional important boundary conditions that have been used for the economic evaluation are given in Table 14–8. Operation and maintenance costs have been assumed individually for each plant.

The resulting power generation costs show clearly the advantage of the natural gas-driven combined-cycle plants. Only the pulverized-coal-fired plant of high capacity can compete with them. The "second generation" circulating pressurized FBC plant is in the same cost range, but this plant has to prove whether the assumed low investment can be reached after successful market introduction. All these plants come out with lower generation costs than the pulverized-coal-fired reference plant of the 600 MW class and moderate steam conditions (Figure 14–21). Evidently, on the basis of the low coal price, advanced steam conditions do not justify their higher investment. Also lignite-fired

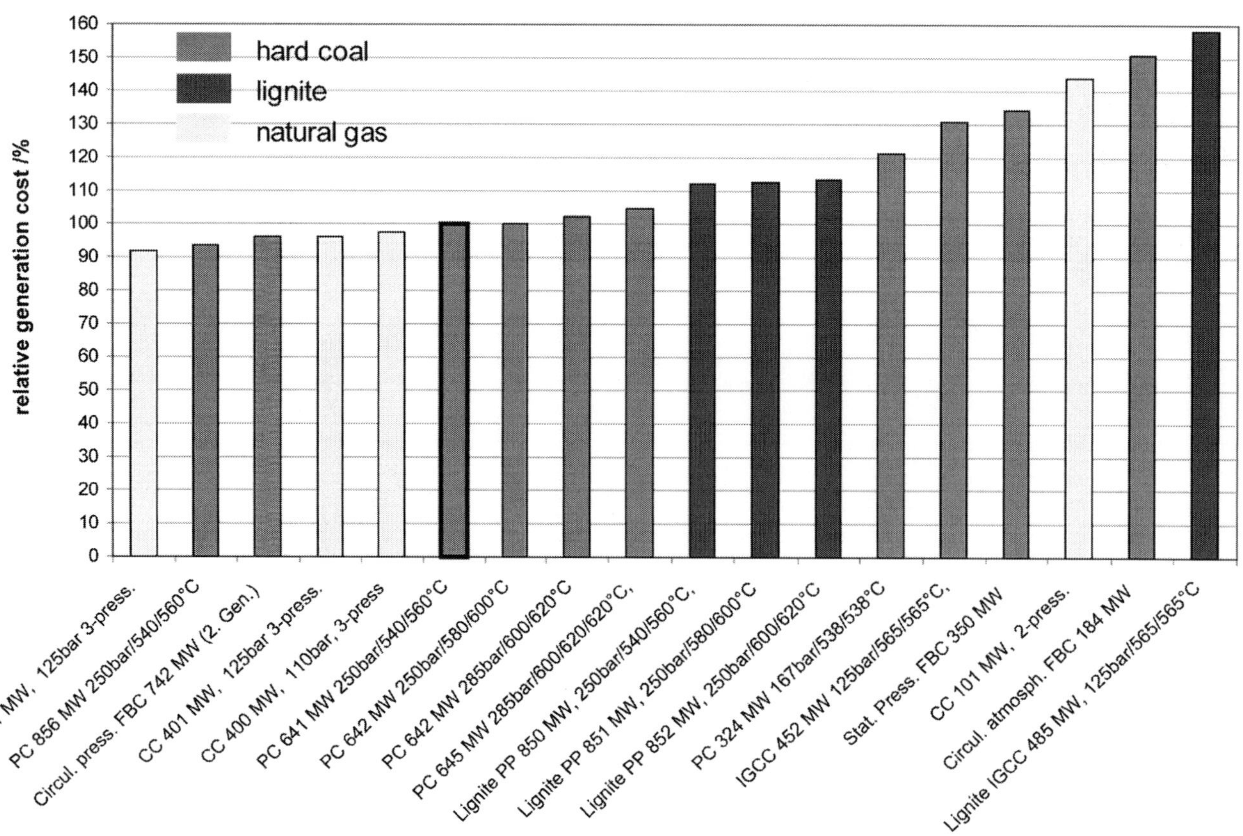

FIGURE 14–21 Relative power generation costs in percent [14-5].

power plants, generally of higher investment costs compared to hard coal power plants, come out with 12% higher generation cost than the reference plant. A price increase of natural gas of only 13% equals the generation costs of the reference coal power plant and the CC plant.

IGCC plants as well as pressurized FBC plants with their relatively high efficiency do not appear economical at today's low coal prices.

Power generation in small capacity power plants (100–200 MW) result in approximately 40–50% higher generation costs.

Chapter 15

Manufacturing, Materials, and Metallurgy

"There are old pilots, and there are bold pilots, but there are no old, bold pilots."

—Old Pilots' saying

Chapter Outline

Manufacturing methods,* materials, and metallurgy (MMM) in gas turbines, like other technical elements of turbine technology, are dictated by the global business climate. The summary that follows describes the main elements of this climate as it affects MMM. The importance of each element at any given time depends on global economics as dictated by currencies (consider the Asian monetary crises of the late 1990s), politics, and wars (see also Chapter 14).

Just 30 or so years ago, *industrial gas turbine engine* could mean a heavy unsophisticated engine compared to its aviation counterpart. Industrial engines are still relatively heavier (with their large flat base and sometimes a "noise protection house" around them). However, their metallurgical sophistication may be quite comparable to aeroderivative and even aviation engines. The increasing sophistication in an OEM's F, G, and H technology is not that different from that in its aeroderivatives and flight engines. Rolls Royce, for instance, has consistently used its aircraft engine metallurgical technology in its land-based models, which it calls by the same generic stem: Avon, RB211, Trent, and so forth. Siemens Westinghouse is not officially an aircraft engine manufacturer, neither are Mitsubishi Power Systems (MPS) and Alstom (although ABB Stal, which was absorbed under the Alstom umbrella in 2000, had developed the aeroderivative GT35). However, their metallurgy is every bit as sophisticated as that which their rivals, who also make aircraft engines, employ.

In today's age of workers who can travel or change jobs internationally or get different work visas within a week, scarcely anything is sacred in design technology. The wide chord fan blade that proved such an asset to Rolls Royce aeroengine designs showed up on the GE 90 some years after its original design, with different "insides." The example list is endless and travels in both directions across oceans.

* Working case notes 1975 through 2007, Claire Soares.

To keep costs down, every major manufacturer in the western hemisphere contracts the manufacture of specific components to countries with low labor rates. Korea used to be a favorite, but then Singapore, Malaysia, and Thailand became hotbeds of electronic component manufacture. China and India followed, with contracts acquired for every type of component: mechanical (such as blade and vane airfoils) or electrical (such as switchgear) and electronic (controls, PLCs, PCs, and so forth). The OEMs "follow" the lowest labor rates available and train the local workforce.

The contemporary power generation industry also sees the OEMs hard at work selling, installing, and running power plants in these newly industrializing countries (NIC). Typically, they then train the local workforce to operate and eventually take over these plants, which therefore are build, own, operate, transfer (BOOT) projects. The initial "operate" phase then positions the OEM well for a "power by the hour" contract (see Chapter 12, Maintenance, Repair, and Overhaul). All this activity tends to promote the eventual transfer of repair and overhaul (R&O) functions to local repair shops. The OEM's workforce may stay approximately the same globally: numbers that are hired in Asia and Latin America equate to layoffs in countries with higher labor costs.

The large manufacturers such as General Electric have licensees, such as GEC and Nuovo Pignone, that assemble entire engines for them. These licensees, in turn, are large enough that they may establish factories in NICs.

Acquisitions, mergers, and joint venture programs also move technology into different OEM firms. Ruston became European Gas Turbine (EGT), which developed new grassroots models, like the Cyclone, Tempest, and Typhoon that differed considerably from the old Ruston's more traditional, conservative fare. Then EGT became part of ABB, which in 1999 became ABB Alstom, and Alstom Power in 2000. Later, Siemens bought portions of Alstom Power and its engine lines. The combined product range with respect to gas turbines is both broad and varied, the range of manufacturing methods and design development strategy even more so. Also, before this, Siemens and Westinghouse merged. Rolls Royce gained a US partner in Allison. Both Rolls and Siemens can now bid for US DOE grants along with US companies.

Reorganizations internally can shift centers of special knowledge, domestically and abroad. To conserve on overhead, Pratt and Whitney closed its West Palm Beach military engine facility and sent that workforce to join the commercial aircraft engine group in Connecticut.

One example of a joint venture gas turbine is the V2500 engine (with the three Japanese manufacturers that make up the Japanese Aeroengine consortium making the low-pressure compressor, including the wide chord bladed fan, to a Rolls Royce design). Rolls Royce designs and makes the high-pressure compressor and the oil system. Pratt and Whitney designs and makes the hot high-pressure section and MTU (Motoren Turbinen Union) designs and makes the low-pressure turbine. Alternate serial numbers are assembled at Rolls Royce, Derby, England, and Pratt and Whitney, Connecticut. Another joint venture example is GE/Snecma's CFM 56.

Perhaps to gain competitive advantage, some OEMs also buy the manufacturers that make some of their accessories. GE bought either entire or partial ownership in several aircraft engine overhaul shops, firms that provide risk insurance for some of its engines, and a manufacturer that makes condition-monitoring equipment (Bentley Nevada).

Other OEMs have systems or components made under license, then stamp them with their logo before including them in their engine assembly packages. Turbine wash systems are a common "farmed-out" item. Other OEMs farm out increasingly complex manufactured items. An entire components industry has sprung up around firms that make only one of either combustion liners, blades and vanes, fuel nozzles, exhaust cases, gears, or accessory drives. To stay competitive, these firms try to minimize overhead. In the process, they may develop increasingly narrower product ranges that require greater expertise in fewer components.

Some accessory manufacturers have been non-OEM for much longer, such as the firms that design and make wash systems. Depending on the year of the order, the actual design of a wash system, for instance, on a given large engine fleet, may be quite different as a consequence of being made by two entirely different manufacturers' contracted suppliers, as may be the case with an inlet air fogging system or inlet air filtration system.

There are specialist plating shops, heat treatment shops, casting foundries, forging facilities, machine shops that may perform just one function, such as laser drilling of air cooling passages in blades and vanes—the list is endless. Basically, an OEM tries to keep in house as much work as it can profitably maintain, or as much as might cause it to lose a great deal in warranty claims or specialized transportation, if farmed out.

The bottom line in maintaining healthy profit margins in the manufacturing business is spare parts sales. An OEM makes far more money on new spare parts than on overhauled, repaired, or refurbished ones. Very often OEMs will not tell certain customers (who do not know better) that they ought to ask about potential repairs. Some of these repairs may be developed by the OEM to keep its more knowledgeable customers happy or developed for "internal use only" by an independent shop (see Chapter 12, Maintenance, Repair, and Overhaul).

Sometimes, the OEM develops its own repairs, other times it licenses a trustworthy external facility. Some "renegade" ex-OEM staff members start their own firms. Not taking designs with them does not stop them. They know enough to "reverse engineer" new components without OEM blueprints. Frequently, the customer base of such "guerilla" shops is confined to a client base with machinery well past its warranty period. Generally, these former employees also are shrewd enough to check most of the reverse engineering themselves, thus minimizing their risk profile (see Chapter 14, The Business of Gas Turbines). And, they may have the partial security of feedback from an end-user lobby group (with which they can compare notes) regardless of whether those operators have a "power by the hour" contract with the OEM. However, once they are "power[ed] by the hour," the need for end users to maintain an engineering staff clever enough to bargain with OEMs and find "good deals" outside of OEM dealings goes away. Power by the hour contracts come at a huge price. How huge may depend on the country of the application. If an OEM feels that a customer, in Pakistan, for instance, would be more apt to yell "warranty" in cases where the fault was clearly caused by the operators, the OEM might raise the "power by the hour" or even the cost of basic spare parts to cover these losses.

As the business chapter pointed out, however, increasingly, especially in the power generation business, the turbine buyer is the OEM as it may also be the IPP. It is a situation of relative autonomy for the OEM, as it then can write its own spare parts (to be purchased) lists. It also can test prototype components in real-life operating environments with the minimum of negotiation with or concessions to the end user. It can conduct realistic full-load tests on new models, an undertaking that ordinarily is prohibitively expensive. It can be legitimate regular members of end-user lobby groups where, otherwise, it could attend only sessions where "OEMs are invited."

The newer, but no less powerful, IPPs on the scene are the oil companies, which then write the contracts to sell their own fuel to themselves.

The section that follows deals with the elements of basic manufacture. The source (Rolls Royce) wrote the original of this section to describe aircraft engine manufacture. However, with today's technology, the section is appropriate for describing the manufacture of aeroderivative (used in land, offshore, and naval applications) and contemporary industrial engines.

BASIC MANUFACTURE*

During the design stages of the aircraft gas turbine engine, close liaison is maintained between design, manufacturing, development, and product support to ensure that the final design is a match between the engineering specification and the manufacturing process capability.

The functioning of this type of engine, with its high power-to-weight ratio, demands the highest possible performance from each component. Consistent with this requirement, each component must be manufactured at the lowest possible weight and cost and also provide mechanical integrity through a long service life. Consequently, the methods used during manufacture are diverse and are usually determined by the duties each component has to fulfill.

No manufacturing technique or process that in any way offers an advantage is ignored and most available engineering methods and processes are employed in the manufacture of these engines. In some instances, the technique or process may appear by some standards to be elaborate, time consuming, and expensive, but is only adopted after confirmation that it does produce maximized component lives comparable with rig test achievements.

Engine components are produced from a variety of high tensile steel and high temperature nickel and cobalt alloy forgings. A proportion of components are cast using the investment casting process. While fabrications, which form an increasing content, are produced from materials such as stainless steel, titanium and nickel alloys using modern joining techniques i.e., tungsten inert gas welding, resistance welding, electron beam welding, and high temperature brazing in vacuum furnaces.

The methods of machining engine components include grinding, turning, drilling, boring, and broaching whenever possible, with the more difficult materials and configurations being machined by electro-discharge, electro-chemical, laser hole drilling, and chemical size reduction.

Structural components, i.e., cold spoiler, location rings, and by-pass ducts, benefit by considerable weight saving when using composite material.

In addition to the many manufacturing methods, chemical and thermal processes are used on part-finished and finished components. These include heat treatment, electroplating, chromate sealing, chemical treatments, anodizing to prevent corrosion, chemical and mechanical cleaning, wet and dry abrasive blasting, polishing, plasma spraying, electrolytic etching, and polishing to reveal metallurgical defects. Also a variety of barreling techniques for removal of burrs and surface improvement. Most processes are concerned with surface changes, some give resistance to corrosion while others can be used to release unwanted stress.

The main structure of an aero gas turbine engine is formed by a number of circular casings (see Figure 15—1) that are assembled and secured together by flanged joints and couplings located with dowels and tenons. These engines use curvic and hurth couplings to enable accurate concentricity and mating assemblies that, in turn, assist an airline operator when maintenance is required.

Manufacturing Strategy

Manufacturing is changing and will continue to change to meet the increasing demands of aeroengine components for fuel efficiency, cost and weight reductions, and being able to process the materials required to meet these demands.

* Source: Adapted, with permission, from Rolls Royce, *The Jet Engine*, 1986, Rolls Royce Plc: UK.

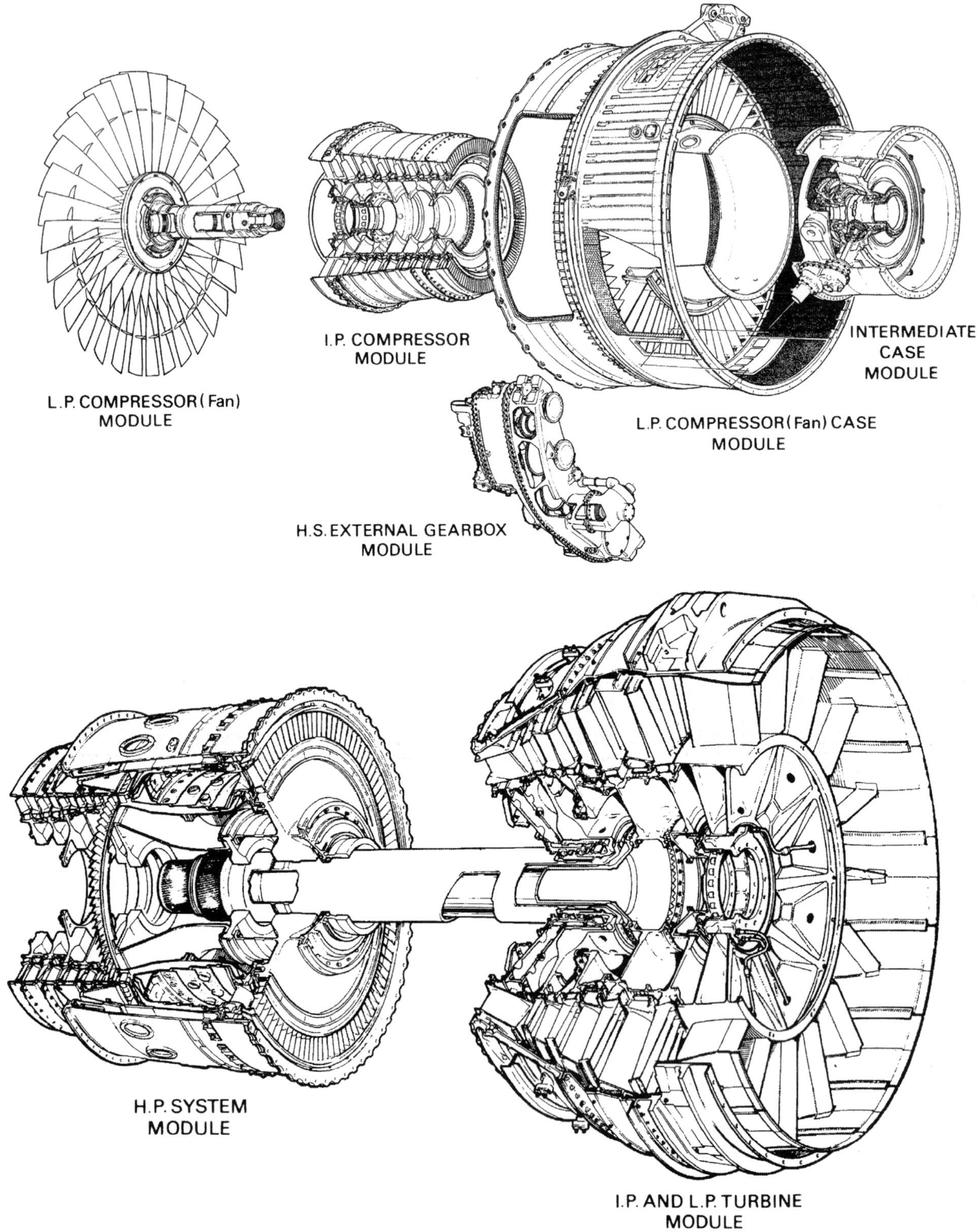

FIGURE 15–1 Arrangements of a triple-spool turbo-jet engine. *(Source: Rolls Royce.)*

With the advent of microprocessors and extending the use of the computer, full automation of components considered for in-house manufacture are implemented in line with supply groups manufacturing strategy, all other components being resourced within the worldwide supplier network.

This automation is already applied in the manufacture of cast turbine blades with the seven cell and computer numerical controlled (C.N.C.) grinding centers, laser hard facing, and film cooling hole drilling by electro-discharge machining (E.D.M.). Families of turbine and compressor discs are produced in flexible manufacturing cells, employing automated guided vehicles delivering palletized components from computerized storage to C.N.C. machining cells that all use batch of one technique. The smaller blades, with very thin airfoil sections, are produced by integrated broaching and 360° electro-chemical machining (E.C.M.) while inspection and processing are being automated using the computer.

Tolerances between design and manufacturing are much closer when the design specification is matched by the manufacturing proven capability.

Computer aided design (C.A.D.) and computer aided manufacture (C.A.M.) provides an equivalent link when engine components designed by C.A.D. can be used for the preparation of manufacturing drawings, programs for numerically controlled machines, tool layouts, tool designs, operation sequence, estimating, and scheduling. Computer simulation allows potential cell and flow line manufacture to be proven before physical machine purchase and operation, thus preventing equipment not fulfilling their intended purpose.

Each casing is manufactured from the lightest material commensurate with the stress and temperatures to which it is subjected in service. For example, magnesium alloy composites and materials of sandwich construction are used for air intake casings, fan casings, and low-pressure compressor casings, since these are the coolest parts of the engine. Alloy steels are used for the turbine and nozzle casings where the temperatures are high and because these casings usually incorporate the engine rear mounting features. For casings subjected to intermediate temperatures, i.e., by-pass duct and combustion outer casings, aluminum alloys and titanium alloys are used.

Forging

The engine drive shafts, compressor discs, turbine discs, and gear trains are forged to as near optimum shape as is practicable commensurate with non-destructive testing, i.e., ultrasonic, magnetic particle, and penetrant inspection. With turbine and compressor blades, the accurately produced thin airfoil sections with varying degrees of camber and twist, in a variety of alloys, entails a high standard of precision forging (see Figure 15–2). Nevertheless precision forging of these blades is a recognized practice and enables one to be produced from a shaped die with the minimum of further work.

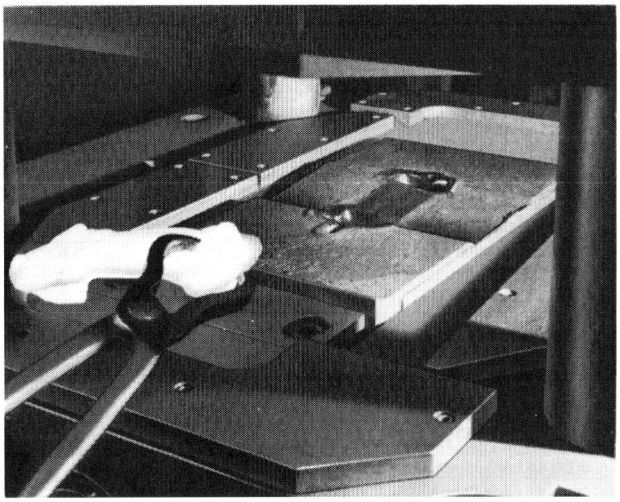

FIGURE 15–2 Precision forging. *(Source: Rolls Royce.)*

The high operating temperatures at which the turbine discs must operate necessitate the use of nickel-based alloys. The compressor discs at the rear end of the compressor are produced from creep-resisting steels, or even nickel-based alloys, because of the high temperatures to which they are subjected. The compressor discs at the front end of the compressor are produced from titanium. The higher strength of titanium at the moderate operating temperatures at the front end of the compressor, together with its lower weight, provides a considerable advantage over steel.

Forging calls for a very close control of the temperature during the various operations. An exceptionally high standard of furnace control equipment, careful maintenance and cleanliness of the forging hammers, presses, and dies are essential.

Annular combustion rings can be cold forged to exacting tolerances and surfaces, which alleviates the need for further machining before welding together to produce the combustion casing.

H.P. compressor casings of the gas turbine engine are forged as rings or half rings which, when assembled together, form the rigid structure of the engine. They are produced in various materials, i.e., stainless steel, titanium, and nickel alloys.

Casting

An increasing percentage of the gas turbine engine is produced from cast components using sand casting (Figure 15–3), die casting, and investment casting techniques; the latter becoming the foremost in use because of

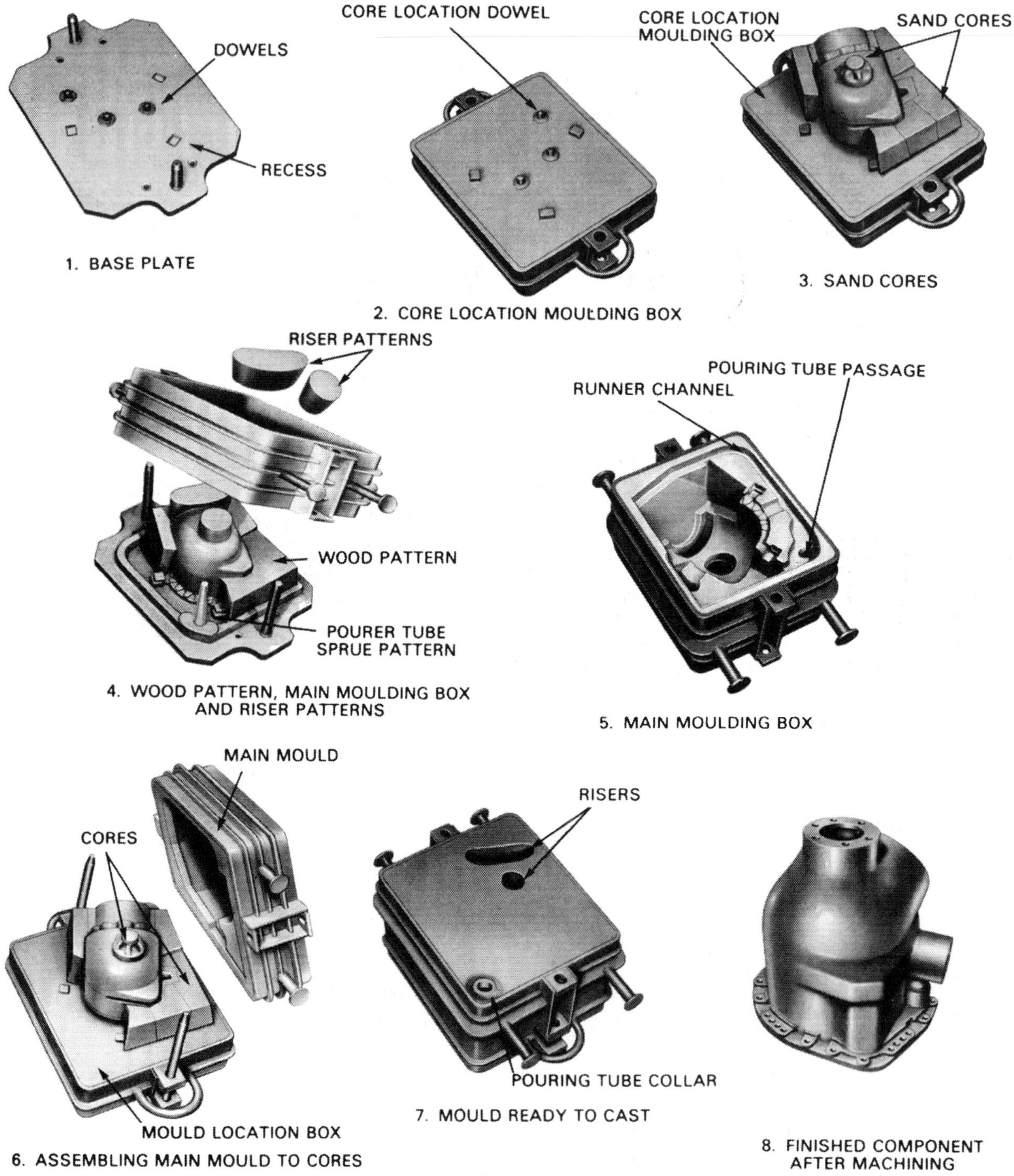

FIGURE 15–3 Method of producing an engine component by sand casting. *(Source: Rolls Royce.)*

its capability to produce components with surfaces that require no further machining. It is essential that all castings are defect free by the disciplines of cleanliness during the casting process otherwise they could cause component failure.

All casting techniques depend upon care with methods of inspection such as correct chemical composition, testing of mechanical properties, radiological and microscopic examination, tensile strength, and creep tests.

The complexity of configurations together with accurate tolerances in size and surface finish is totally dependent upon close liaison with design, manufacturing, metallurgist, chemist, die maker, furnace operator, and final casting.

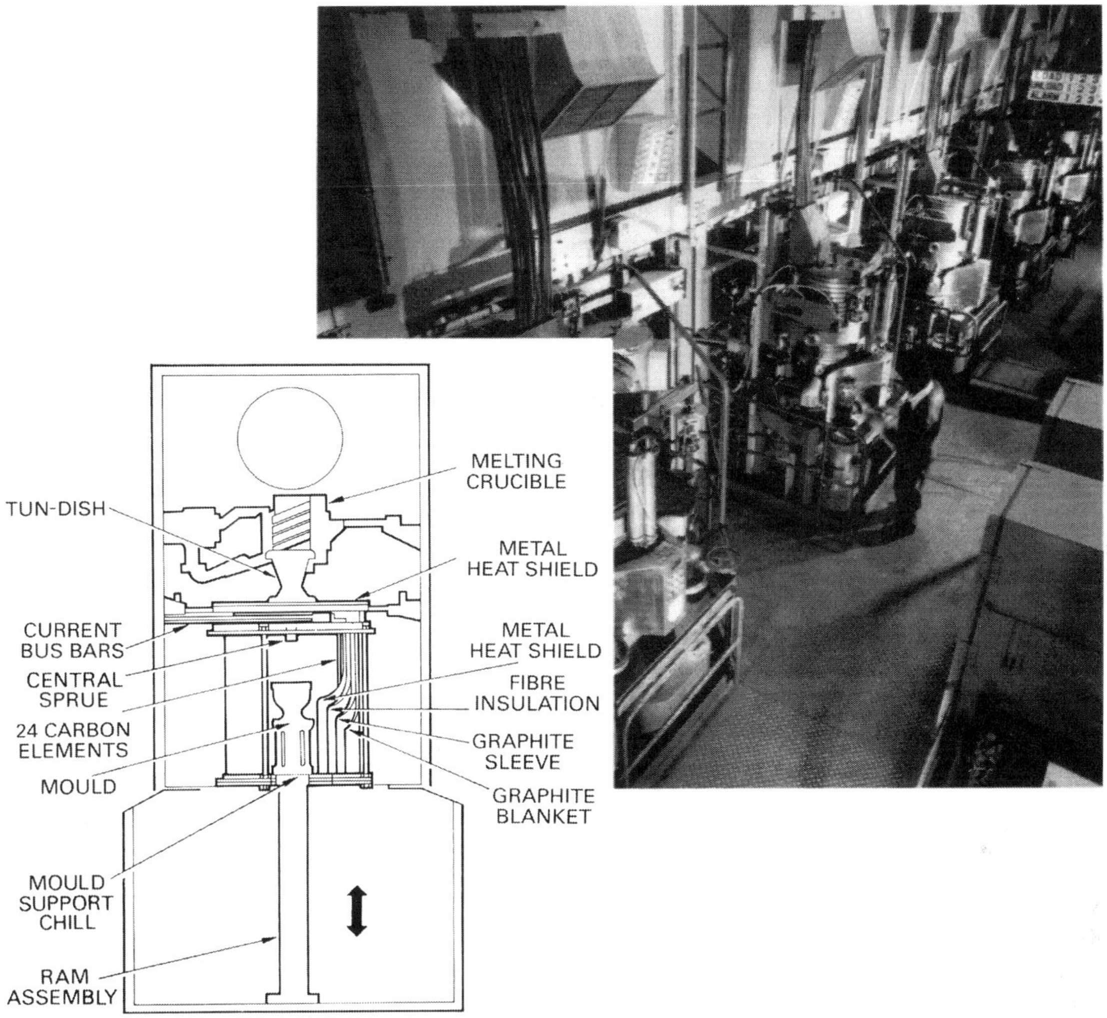

FIGURE 15–4 Automatic investment casting. *(Source: Rolls Royce.)*

In the pursuit of ever increasing performance, turbine blades are produced from high temperature nickel alloys that are cast by the investment casting or "lost wax" technique. Directionally solidified and single crystal turbine blades are cast using this technique in order to extend their cyclic lives.

Figure 15–4 illustrates automatic casting used in the production of equi-axed, directional solidified, and single crystal turbine blades. The lost wax process is unparalleled in its ability to provide the highest standards of surface finish, repeatable accuracy, and surface detail in a cast component. The increasing demands of the engine have manifested in the need to limit grain boundaries and provide complex internal passages. The molds used for directional solidified and single crystal castings differ from conventional molds in that they are open at both ends, the base of a mold forms a socketed bayonet fitting into which a chill plate is located during casting. Metal is introduced from the central sprue into the mold cavities via a ceramic filter. These and orientated seed crystals, if required, are assembled with the patterns prior to investment. Extensive automation is possible to ensure the wax patterns are coated with the shell material consistently by using robots. The final casting can also have their rises removed using elastic cut-off wheels driven from robot arms (see Figure 15–5).

Fabrication

Major components of the gas turbine engine, i.e., bearing housings, combustion and turbine casings, exhaust units, jet pipes, by-pass mixer units, and low-pressure compressor casings, can be produced as fabricated assemblies using sheet materials such as stainless steel, titanium, and varying types of nickel alloys.

Other fabrication techniques for the manufacture of the low pressure compressor wide chord fan blade comprise rolled titanium side panels assembled in dies, hot twisted in a furnace, and finally hot creep formed to achieve the necessary configuration. Chemical milling is used to recess the center of each panel which sandwiches a honeycomb

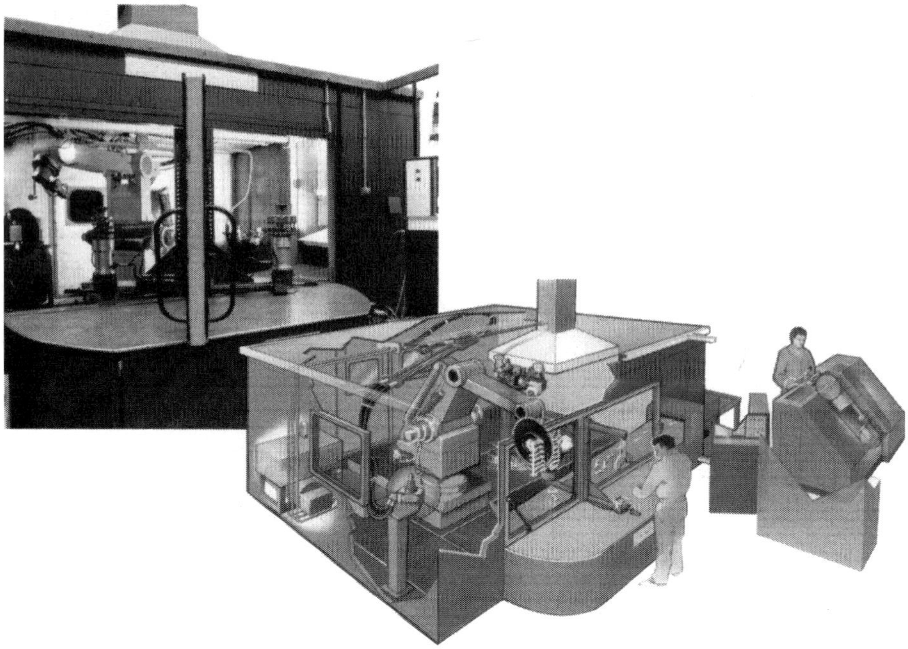

FIGURE 15–5 Robot cut-off. *(Source: Rolls Royce.)*

core; both panels and the honeycomb are finally joined together using automated furnaces where an activated diffusion bonding process takes place (see Figure 15–6).

Welding

Welding processes are used extensively in the fabrication of gas turbine engine components, i.e., resistance welding by spot and seam, tungsten inert gas, and electron beam are amongst the most widely used today. Care has to be taken to limit the distortion and shrinkage associated with these techniques.

Tungsten Inert Gas (T.I.G.) Welding

The most common form of tungsten inert gas welding, Figure 15–7, in use is the direct current straight polarity, i.e., electrode negative pole. This is widely used and the most economical method of producing high quality welds for the range of high strength/high temperature materials used in gas turbine engines. For this class of work, high purity argon shielding gas is fed to both sides of the weld and the welding torch nozzle is fitted with a gas lens to ensure maximum efficiency for shielding gas coverage. A consumable 4% thoriated tungsten electrode, together with a suitable non-contact method of arc starting, is used and the weld current is reduced in a controlled manner at the end of each weld to prevent the formation of finishing cracks. All welds are visually and penetrant inspected and, in addition, welds associated with rotating parts, i.e., compressor and/or turbine, are radiologically examined to quality acceptance standards. During welding operations and to aid in the control of distortion and shrinkage the use of an expanding fixture is recommended and, whenever possible, mechanized welding employed together with the pulsed arc technique is preferred. A typical T.I.G. welding operation is illustrated in Figure 15–8.

Electron Beam Welding (E.B.W.)

This system, which can use either low or high voltage, uses a high power density beam of electrons to join a wide range of different materials and of varying thickness. The welding machine (Figure 15–9) comprises an electron gun, optical viewing system, work chamber and handling equipment, vacuum pumping system, high or low voltage power supply, and operating controls. Many major rotating assemblies for gas turbine engines are manufactured as single items in steel, titanium, and nickel alloys and joined together, i.e., intermediate and high-pressure compressor drums. This technique allows design flexibility in that distortion and shrinkage are reduced and dissimilar materials, to serve quite different functions, can be homogeneously joined together. For example, the H.P. turbine stub shafts requiring a stable bearing steel welded to a material that can expand with the mating turbine disc. Automation has been enhanced by the application of computer numerical control (C.N.C.) to the work handling and manipulation. Seam tracking to ensure that the joint is accurately followed and close loop under bead control to guarantee the full depth of material thickness is welded.

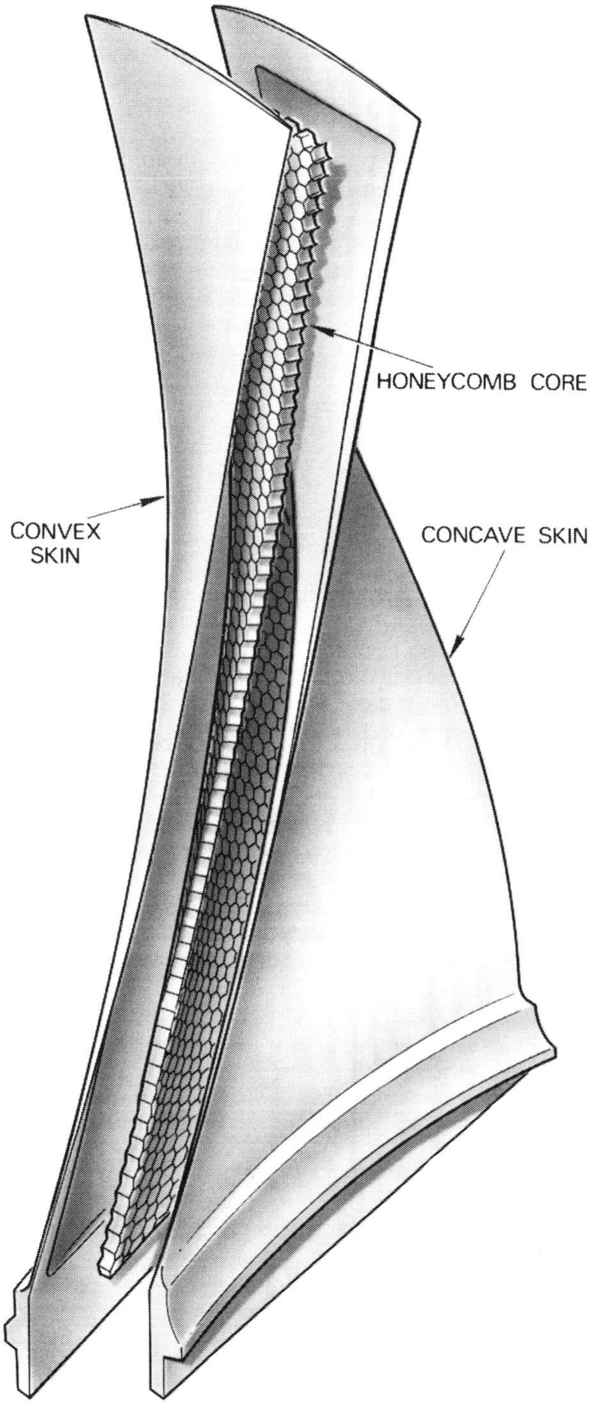

FIGURE 15–6 Wide chord fan blade construction. *(Source: Rolls Royce.)*

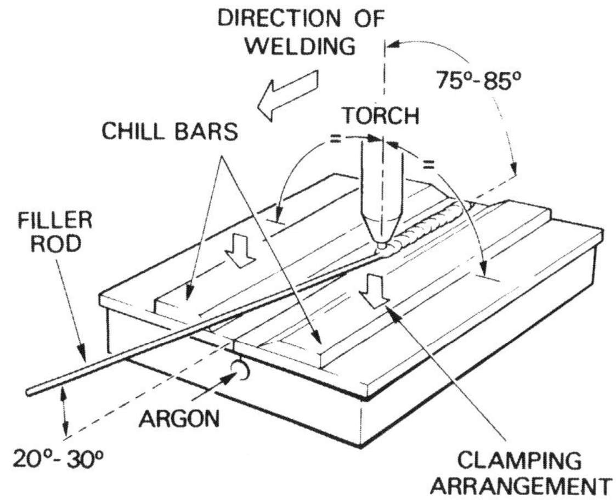

FIGURE 15–7 Typical tungsten inert gas welding details. *(Source: Rolls Royce.)*

FIGURE 15–8 Tungsten inert gas welding. *(Source: Rolls Royce.)*

Focus of the beam is controlled by digital voltmeters. See Figure 15–10 for weld examples.

Electro-Chemical Machining (E.C.M.)

This type of machining employs both electrical and chemical effects in the removal of metal. Chemical forming, electrochemical drilling, and electrolytic grinding are techniques of electro-chemical machining employed in the production of gas turbine engine components.

The principle of the process is that when a current flows between the electrodes immersed in a solution of salts, chemical reactions occur in which metallic ions are transported from one electrode to another (Figure 15–11). Faraday's law of electrolysis explains that the amount of chemical reaction produced by a current is proportional to the quantity of electricity passed.

In chemical forming (Figure 15–11), the tool electrode (the cathode) and the workpiece (the anode) are connected into a direct current circuit. Electrolytic solution passes, under pressure, through the tool electrode and metal is removed from the work gap by electrolytic action. A

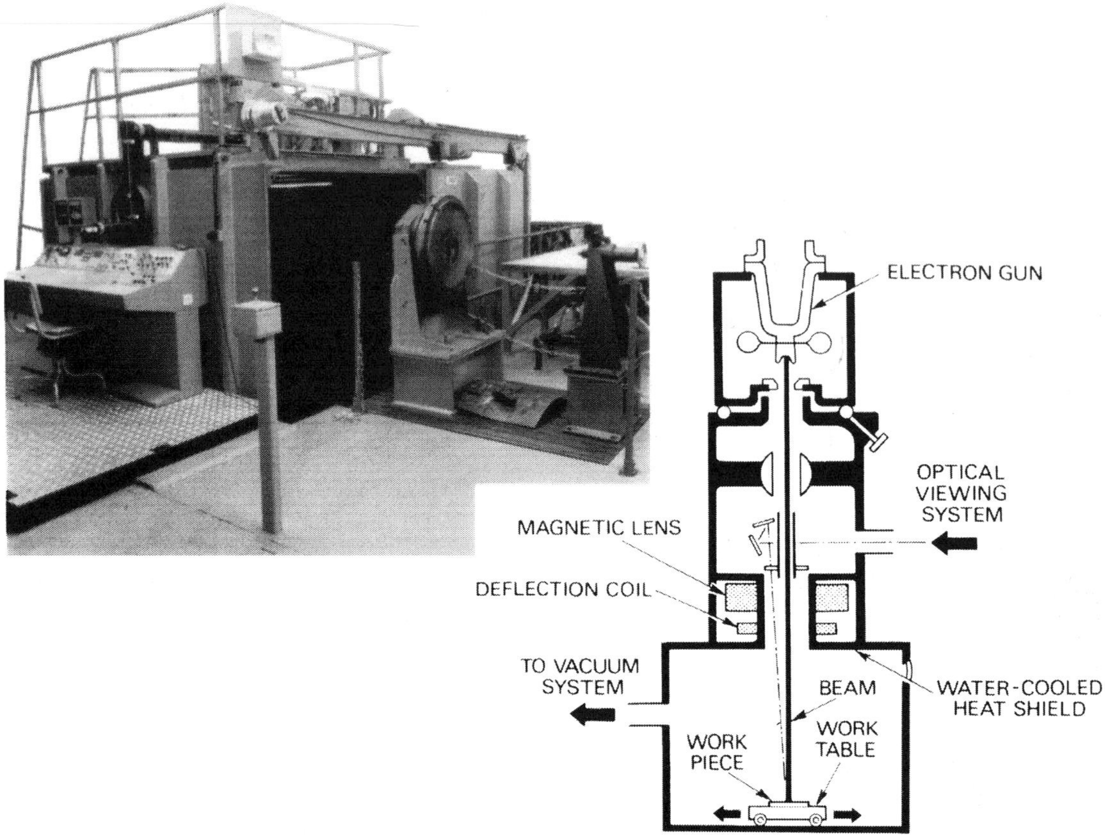

FIGURE 15–9 Electron beam welding. *(Source: Rolls Royce.)*

hydraulic ram advances the tool electrodes into the workpiece to form the desired passage.

Electrolytic grinding employs a conductive wheel impregnated with abrasive particles. The wheel is rotated close to the surface of the workpiece in such a way that the actual metal removal is achieved by electro-chemical means. The by-products, which would inhibit the process, are removed by the sharp particles embodied in the wheel.

Stem drilling and capillary drilling techniques are used principally in the drilling of small holes, usually cooling holes, such as required when producing turbine blades.

Stem Drilling

This process consists of tubes (cathode) produced from titanium and suitably insulated to ensure a reaction at the tip. A 20% solution of nitric acid is fed under pressure onto the blade producing holes generally in the region of 0.026 in. diameter. The process is faster in operation than electro-discharge machining and is capable of drilling holes up to a depth 200 times the diameter of the tube in use.

Capillary Drilling

This process is similar to stem drilling but uses tubes produced from glass incorporating a core of platinum wire (cathode). A 20% nitric acid solution is passed through the tube onto the workpiece and is capable of producing holes as small as 0.009 in. diameter. Depth of the hole is up to 40 times greater than the tube in use and therefore determined by tube diameter.

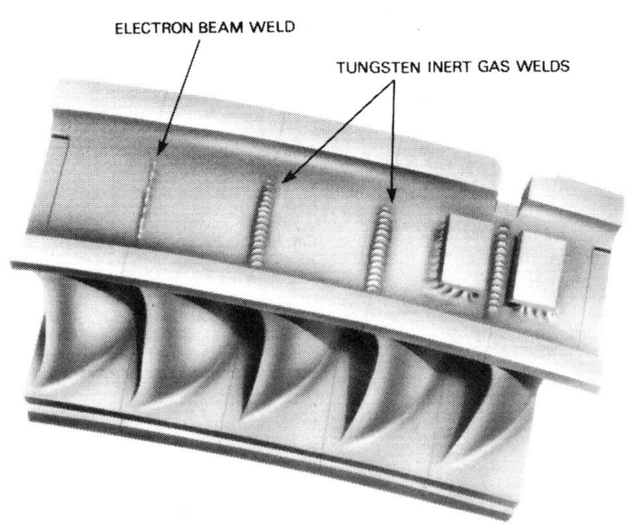

FIGURE 15–10 Examples of T.I.G. and E.B. welds. *(Source: Rolls Royce.)*

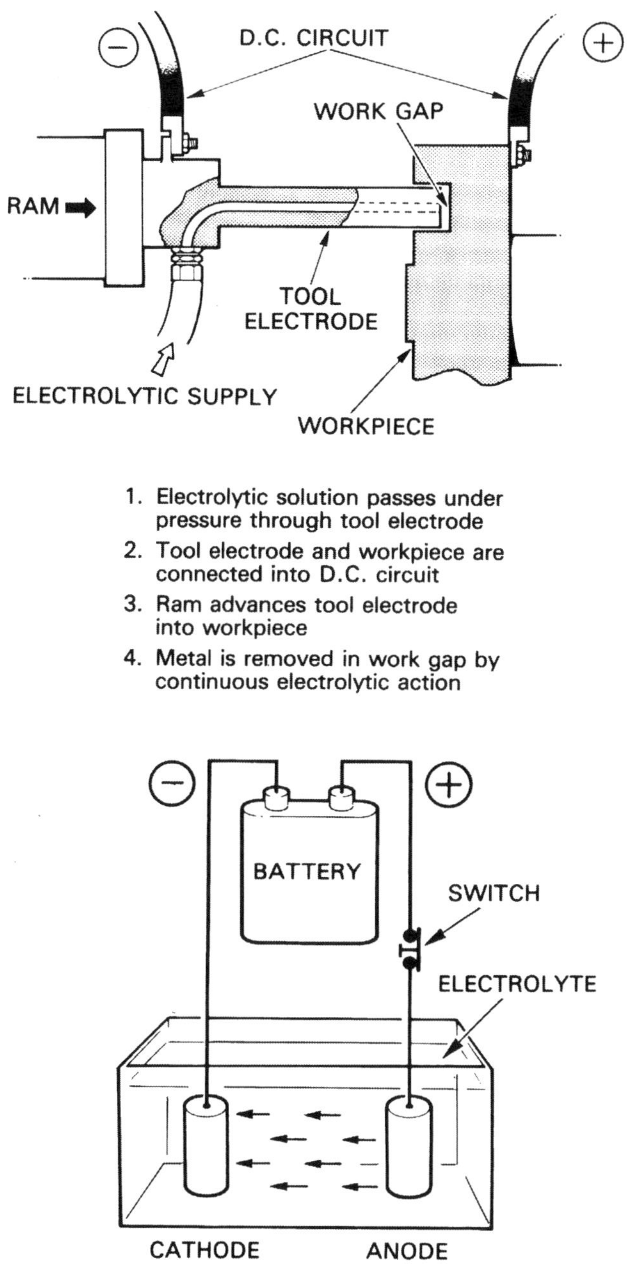

FIGURE 15–11 Electro-chemical machining. *(Source: Rolls Royce.)*

Automation has also been added to the process of electro-chemical machining (E.C.M.) with the introduction of 360° E.C. machining of small compressor blades, see Figure 15–12. For some blades of shorter length airfoil, this technique is more cost effective than the finished shaped airfoil when using precision forging techniques. Blades produced by E.C.M. employ integrated vertical broaching machines which take pre-cut lengths of bar material, produce the blade root feature, such as a fir-tree, and then by using this as the location, fully E.C.M. from both sides to produce the thin airfoil section in one operation.

Electro-Discharge Machining (E.D.M.)

This type of machining removes metal from the workpiece by converting the kinetic energy of electric sparks into heat as the sparks strike the workpiece.

An electric spark results when an electric potential between two conducting surfaces reaches the point at which the accumulation of electrons has acquired sufficient energy to bridge the gap between the two surfaces and complete the circuit. At this point, electrons break through the dielectric medium between the conducting surfaces and, moving from negative (the tool electrode) to positive (the workpiece), strike the latter surface with great energy; Figure 15–13 illustrates a typical spark erosion circuit.

When the sparks strike the workpiece, the heat is so intense that the metal to be removed is instantaneously vaporized with explosive results. Away from the actual center of the explosion, the metal is torn into fragments, which may themselves be melted by the intense heat. The dielectric medium, usually paraffin oil, pumped into the gap between the tool electrode and the workpiece, has the tendency to quench the explosion and to sweep away metallic vapor and molten particles.

The amount of work that can be effected in the system is a function of the energy of the individual sparks and the frequency at which they occur.

The shape of the tool electrode is a mirror image of the passage to be machined in the workpiece and, to maintain a constant work gap, the electrode is fed into the workpiece as erosion is effected.

Composite Materials and Sandwich Casings

High power to weight ratio and low component costs are very important considerations in the design of any aircraft gas turbine engine, but when the function of such an engine is to support a vertical take-off aircraft during transition, or as an auxiliary power unit, then the power to weight ratio becomes extremely critical.

In such engines, the advantage of composite materials allows the designer to produce structures in which directional strengths can be varied by directional lay-up of fibers according to the applied loads.

Composite materials have and will continue to replace casings, which, in previous engines, would have been produced in steels or titanium. By-pass duct assemblies comprising three casings are currently being produced up to 4ft 7in. in diameter and 2ft 0in. in length using pre-cured composite materials for the casing fabric. Flanges and mounting bosses are added during the manufacturing process, which are then drilled for both location and machined for peripheral feature attachment on C.N.C. machining centers, which at one component load, completely machine all required features. Examples of

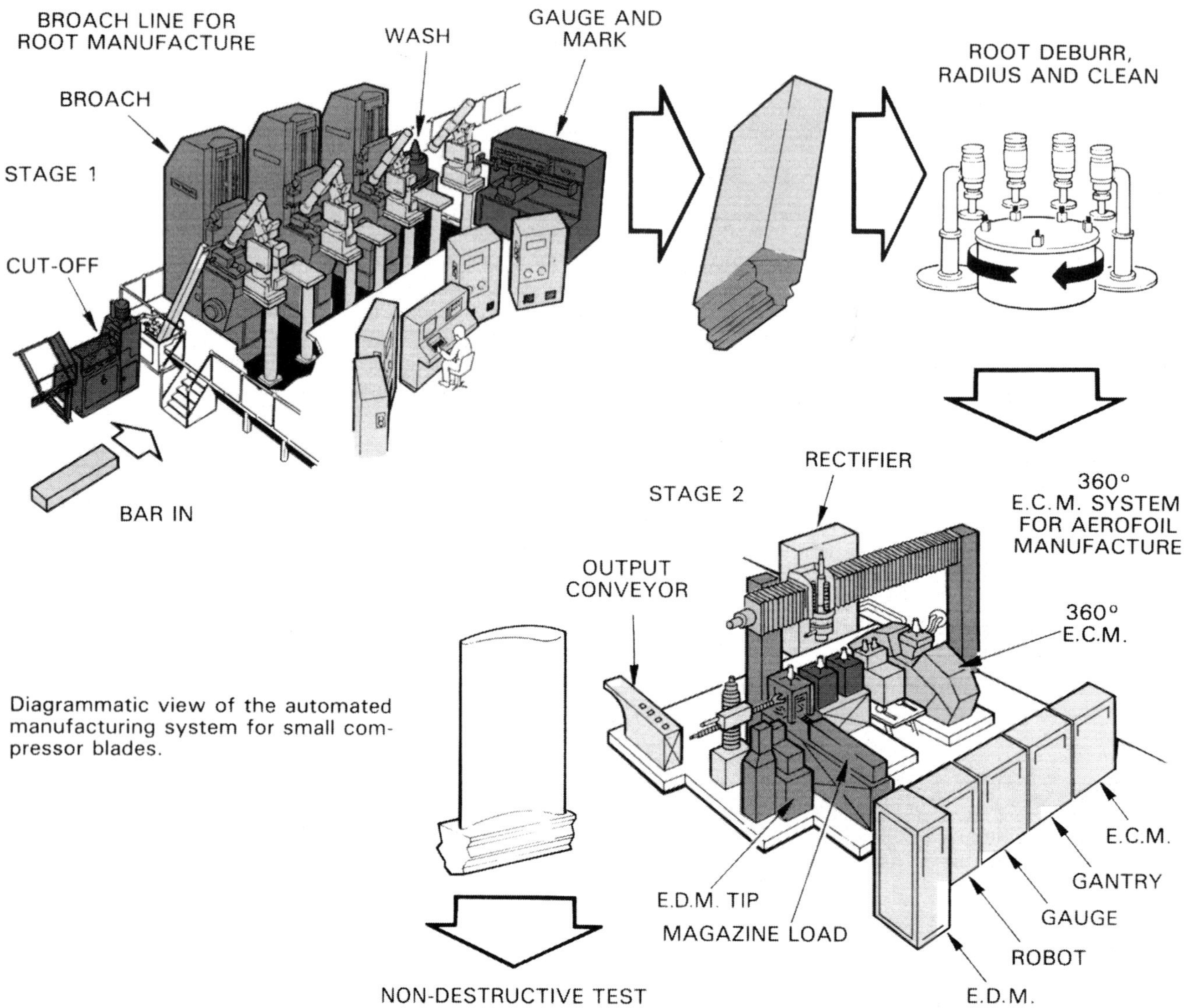

FIGURE 15–12 Typical automated manufacture of compressor blades. *(Source: Rolls Royce.)*

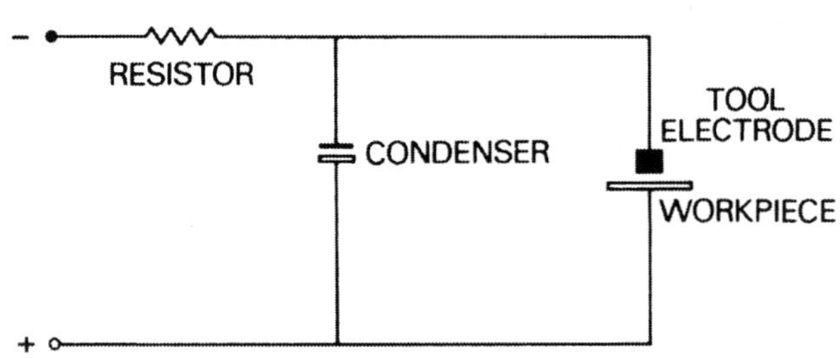

FIGURE 15–13 Electro-discharge machining circuit. *(Source: Rolls Royce.)*

composite material applications are illustrated in Figure 15–14.

Conventional cast and fabricated casings and cowlings are also being replaced by casings of sandwich construction that provide strength allied with lightness and also act as a noise suppression medium. Sandwich construction casings comprise a honeycomb structure of aluminum or stainless steel interposed between layers of dissimilar material. The materials employed depend upon the environment in which they are used.

INSPECTION

During the process of manufacture, component parts need to be inspected to ensure defect-free engines are produced. Using automated machinery and automated inspection, dimensional accuracy is maintained by using multi-directional applied probes that record sizes and position of features. The C.N.C. inspection machine can inspect families of components at pre-determined allotted intervals without further operator intervention. In the chip

FIGURE 15–14 Some composite material applications. *(Source: Rolls Royce.)*

machining (i.e., turning, boring, milling etc.) and metal forming processes C.N.C. machine tools enable consistency of manufacture that can be statistically inspected, i.e., one in ten. Component integrity is achieved by use of ultrasonic, radiological, magnetic particle and penetrant inspection techniques, as well as electrolytic and acid etching to ensure all material properties are maintained to both laboratory and quality acceptance standards.

Smart Materials

Today, development work on smart materials continues. These are materials that can change their shape in response to their environment. Clearly, when further advances are made, this could revolutionize the face of manufacture.

Smart materials are designed materials that have one or more properties that can be significantly changed in a controlled fashion by external changes such as pressure or temperature. The application potential in the gas turbine industry is obvious but in the earlier stages of research. An abstract from published work relating to just one example from the research in this field, follows below. This is a rapidly growing field and holds great potential for several industries. Other keywords related to smart material are such as shape memory material (SMM) and shape memory technology (SMT).

This Small Business Innovation Research Phase I* project focuses on developing a novel thermographic phosphor thermometry technique, combined with a thermal barrier coating (TBC) material, for monitoring gas turbine surface temperature and TBC condition (i.e., to monitor the coating for spallation and deterioration). Design of efficient and low-emission power generation gas turbines requires detailed knowledge of the temperatures, heat fluxes and flow regimes experienced by hot-gas path (HGP) components under realistic operating conditions. Hence, there is an urgent need for innovative and non-intrusive experimental techniques suitable for accurate measurement of surface temperatures in excess of 2500 K for HGP components in high-temperature regions of the gas turbine, where conventional techniques cannot be applied. In this project, a unique upconversion phosphor thermometry technology will be developed which will overcome the issues confronting widely-used down-conversion thermometry techniques. The proposed temperature sensing mechanism is independent of variations in the excitation power and environmental conditions, and hence is especially suitable for hazardous gas turbine applications. The broader impact/commercial potential of this project will be significant, because gas turbines are used to power aircraft, trains, ships, generators, and ground vehicles. Almost all manufacturers in the aerospace, transportation, and energy generation fields would benefit from the technology developed in this project. If successful, this Phase I project will directly result in an enabling technique for non-intrusive monitoring of temperature and surface conditions of materials experiencing extremely high-temperature environments. The data gathered by such a technique will also add to the scientific understanding of high-temperature combustion and fluid flow processes.

* Source: Abstract from http://www.sbir.gov/sbirsearch/detail/2545, extracted May 19, 2014.

Reverse Engineering

This is a process that is normally conducted by an independent shop, not the OEMs. That said, some of these independent shops are owned and operated by people who have retired from the OEM's ranks, so they may know the engine in question very well.

Particularly in the case of legacy engines, spare parts can be hard to find. However, end-users tend to keep their gas turbines, even if there are new models available that have far better efficiency and operating parameters. If the owner wants to keep his old engine(s) running and can't find parts, he then gets components reengineered. Turbine blades are among the more common candidates for reengineering.

Dimensional data are captured with 3D (usually laser) scanning. Many commercial scanners are accurate to 0.001". Higher precision machines measure within 0.0003".[1]

Point cloud data and reverse engineering software produces a 3D shape that is then submitted for finite element analysis. The results are then transferred to a CAD package the results from which are then compared by a CMM (coordinate measuring machine) with the actual part. Verification processes are then done. The part then proceeds to

[1] Reference: "Today's Reverse Engineering Cycle for Turbine Blades." S. Shock, *Global Gas Turbine News*, Dec 2011 (p. 60).

fabrication stages, which vary considerably, depending on the performance and metallurgy required in the application.

Reengineering by Overhaul Facilities

Highly specialized metallurgical facilities develop varying processes that recover component lives and extend the time between overhauls for OEM components. Examples include powder metallurgy (using a kind of metal putty to fill in damage voids in aerfoils), hot isostatic pressing (HIPing) where a heat treatment process recovers the metallurgical grain structure of an aerfoil, different brazing techniques, coatings for erosion or corrosion resistance.

For the first five years or so after a new model is released, the OEM will not allow such an independent facility to work on their components. This makes sense, as the OEM wants to ensure his prototype has no major flaws. If none surface in five years, the OEM may himself go the independent facility to develop some repair processes that improve their image with customers. The independent facility can generally develop repairs cheaper than a large OEM. Besides they have a wealth of specialized knowledge from their customer base, which includes many whose GT warranty period has run out.

For this reason, end-users are advised to read this chapter in conjunction with Chapter 12, Maintenance, Repair and Overhaul. In fact, end-users selecting new GT packages ought to study an OEM's track record of having their components of their older engines overhauled by independent facilities. Ones that offer a range of opportunities in this area mean potentially less downtime in the case of plant malfunction or machinery failure.

Optimization of Component Manufacture by OEMs

To optimize their cost effectiveness, manufacturers build on, extrapolate from and selectively reengineer the cores of their already successful engines. A case history in point follows. It describes how Mitsubishi developed its core experience with TITs of 1150 to 1350 and then to a ceiling of 1500°C (for 50 and 60 Hz gas turbines in power-generation combined-cycle applications).

OPTIMIZING GAS TURBINES WITH MANUFACTURING TECHNOLOGY

Case Study 1: Upgrading the Core Engine*

The next generation G class gas turbine, with turbine inlet gas temperature in the 1500°C range, has been developed by Mitsubishi Heavy Industries Ltd. (MHI). Many advanced technologies, including a high efficiency compressor, a steam cooled low NO_x combustor, a high temperature and high efficiency turbine, etc., are employed to achieve high combined-cycle performance. Actually, MHI has been accumulating the operating experiences of the M501 G (60 Hz machine), a combined-cycle verification plant in MHI Takasago, Japan, achieving high performance and reliability. Also, M701G (50 Hz machine) has been accumulating the operating experience in Higashi Niigata Thermal Power Station of Tohoku Electric Power Co., Inc. in Japan.

What follows describes the technical features of M501 G/M701 G, and up-to-date operating status of the combined-cycle power plant in MHI Takasago, Japan.

MHI developed a turbine inlet temperature of 1150°C class M701D gas turbine and 1350°C class M501F/M701F gas turbines. Their high efficiency and reliability for combined-cycle applications have been proven in field operations. Based on the fact that the combined-cycle efficiency is highly dependent on the gas firing temperature, MHI has newly developed the 1500°C class M501G/M701G, both "G" series gas turbines.

Table 15–1 shows the main introductory performance specification of the M501G, in comparison with the M501F. The M501G is a heavy-duty gas turbine designed to serve the 60 Hz power generating utility. The output of the M501G is 254 MW at a turbine inlet temperature of 1°C on natural gas fuel.

Development of the "G" series gas turbine was started in the early 1990s and the design was based on the following concepts, which give excellent performance and high reliability to the "G" series.

Firstly, proven features from the "F" series are kept. Secondly, the most advanced technologies in aerodynamic design, heat transfer design, and new materials for the "G" series are introduced. Thirdly, our conventional design criteria in industrial gas turbines are applied. Fourthly, the turbine inlet temperature is increased to the extent where the NO_x and the metal temperature can be maintained at the same level as the "F" series. Finally, verification tests are carried out before production of the M501G gas turbine to secure their high reliability.

TABLE 15–1 Performance of G01 g Gas Turbine [15-1]

	M501G	M501F
Speed (rpm)	3600	3600
G/T Output (MW)	254	185.4
G/T Thermal efficiency (%-LHV)	38.7	37.0
Pressure ratio	20	16

* Source: [15-1] Courtesy of Mitsubishi Power Systems. Extracts from Y. Tsukuda, E. Akita, H. Arimura, Y. Tomita, M. Kuwabara, and T. Koga. "The Operating Experience of the Next Generation M501 G/M701 G Gas Turbine," Mitsubishi Heavy Industries Ltd. (MHI) 2001-GT-0546.

The trial operation of the M501G was conducted in 1997 in the combined-cycle verification plant in MHI Takasago, Japan. It was successful and the M501G commercial operation was started sequentially in June 1997. Since then, four inspections were conducted and even though some operational events were observed, the condition of each component was quite sound. Also, the availability, which defined as the actual power supply hours divided by demanded power supply hours (excluding scheduled shutdown such as periodical inspection, during the operation) was 99.2%, which indicates the high reliability of the M501G.

Design Features

The rotor of the M501G is shown in Figure 15–15.

The "G" series inherits the proven NMI technology of the "F" series and employs the advanced technology shown in Figure 15–16.

FIGURE 15–15 The M501G rotor [15-1].

Overall Structure

The basic structure of the "G" series is the same as the "F" series gas turbines. Several basic design concepts such as the two bearing single shaft structure, cold end generator drive, and axial exhaust are evident.

The single shaft rotor is made of compressor and turbine disks that are bolted with spindle bolts. The rotor is supported by two bearings, which are two-element tilting pad bearings. The thrust bearing is a double-acting type using the direct lubrication system on the compressor side.

The air inlet system delivers air to the compressor by a plenum bell mouth and covers the bearings. The compressor is an axial flow design with seventeen stages. A four-stage axial turbine is applied to maintain the optimum aerodynamic loading even at the increased firing temperature and pressure ratio.

All gas turbine casings can be horizontally split to facilitate maintenance without rotor removal. Eight radial struts support the front bearing housing and six tangential struts support the rear bearing housing. Airfoil-shaped covers protect the tangential struts from the blade path gas and support the inner and outer diffuser cones.

The individual inner turbine casings that are called blade rings are used for the stationary part of each turbine stage and can be easily removed and replaced without rotor removal. The similar structure of blade rings is applied to the compressor rear stages. Another feature of these blade rings is that they have high thermal response being independent of the outer casing and can be aligned concentric to the rotor to minimize the blade tip clearances.

Cooling circuits for the turbine section are similar to those of the "F" series. They consist of a rotor cooling circuit and

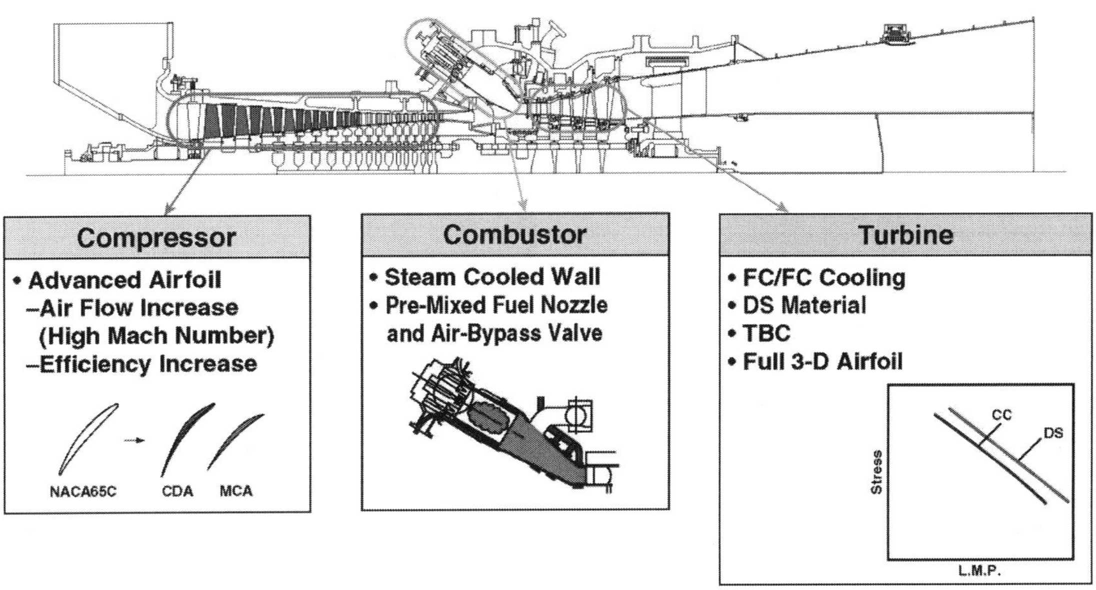

FIGURE 15–16 Advanced technologies applied to the M501G [15-1].

four stationary vane-cooling circuits. Rotor cooling air is provided from compressor discharge air extracted from the combustor shell. This air is supplied as the cooling and seal air of the turbine disks and rotating blades after being externally cooled and filtered. Direct compressor discharge air is used to cool the first stage vane while the compressor bleed air from intermediate stages is used as the cooling air for turbine blade ring cavities and second, third, and fourth stage vanes.

Advanced Compressor

The compressor has highly efficient axial flow design. The "G" series compressor requires a larger flow rate and higher efficiency than the "F" series. The result is that this compressor has a higher Mach number. To meet this requirement, the M501G incorporates multiple circular arc (MCA) airfoils in the forward four stages of the rotating blades. Also, controlled diffusion airfoils (CDA) are applied for the rest of the stages of the rotating blades, namely from the fourth to the seventeenth stage, and all stages of the stationary vanes. The latest three-dimensional compressor blade design is employed.

The aerodynamic performance is verified with both the cascade test and model compressor test. Model compressor test results show that primary parameters of mass flow and efficiency meet the design range as shown in Figure 15—17. Also, aerodynamic performance is proven with the MF221 gas turbine, which has a half scale compressor of the M501G.

The compressor is also equipped with variable inlet guide vanes, which improve the compressor surge characteristics during start-up and are used in the combined-cycle applications to improve part-load performance.

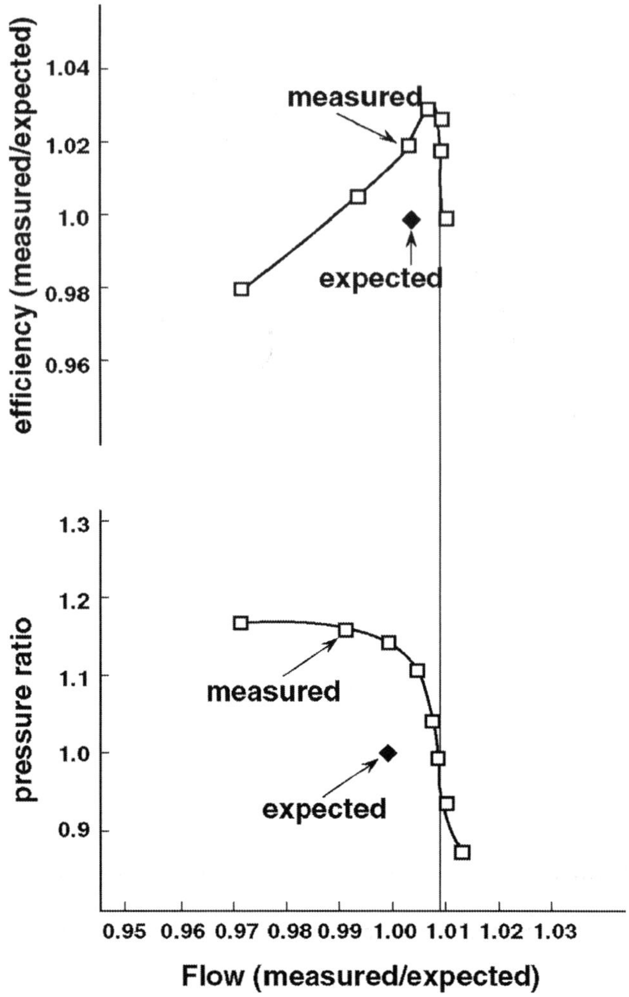

FIGURE 15—17 Result of M501G compressor model test [15-1].

Dry Low NO_x Steam-Cooled Combustor

The combustor design is based on the successful can-annular dry low NO_x combustor developed for the "F" series. This combustor currently operates at less than 25 ppm NO_x level at 1400°C turbine inlet temperature (TIT) defined as the averaged combustor outlet gas temperature.

In order to keep the same NO_x level as the "F" series at 1500°C TIT, all the air discharged from the compressor should be supplied for combustion. For this reason, a premixed combustor with closed circuit steam cooling system was applied for the "G" series combustor.

To optimize the swirl strength and the configuration of fuel ejection holes, a cold flow test was conducted. Also, atmospheric and high-pressurized combustion tests were conducted using a full-scale combustor and a fuel nozzle as shown in Figure 15—18. The test results showed 25 ppm NO_x level and satisfactory wall temperatures without cooling steam leakage. The field verification test of this steam-cooled combustor has already been conducted in the M701F gas turbine to verify its durability.

High Temperature Turbine

It is essential to keep the metal temperature below the allowable limit to secure the same level of reliability at 1500°C TIT with minimum cooling airflow. To achieve a lot of advanced technologies like full coverage film cooling (FCFC), thermal barrier coating (TBC), new heat resistant material, and directionally solidification (DS) casting technology are introduced.

Several heat transfer tests like measurement of film cooling effectiveness around the airfoil and heat transfer characteristics of serpentine cooling passage with turbulence promoters under rotating conditions were conducted. Based on the results obtained from these fundamental tests, the advanced cooled turbine airfoils were designed. Figure 15—19 shows the first stage blade of the turbine.

The actual turbine blades and vanes were first verified in the hot cascade test facility shown in Figure 15—20. In the facility, the test was conducted with the maximum average inlet gas temperature up to 1550°C in order to check the

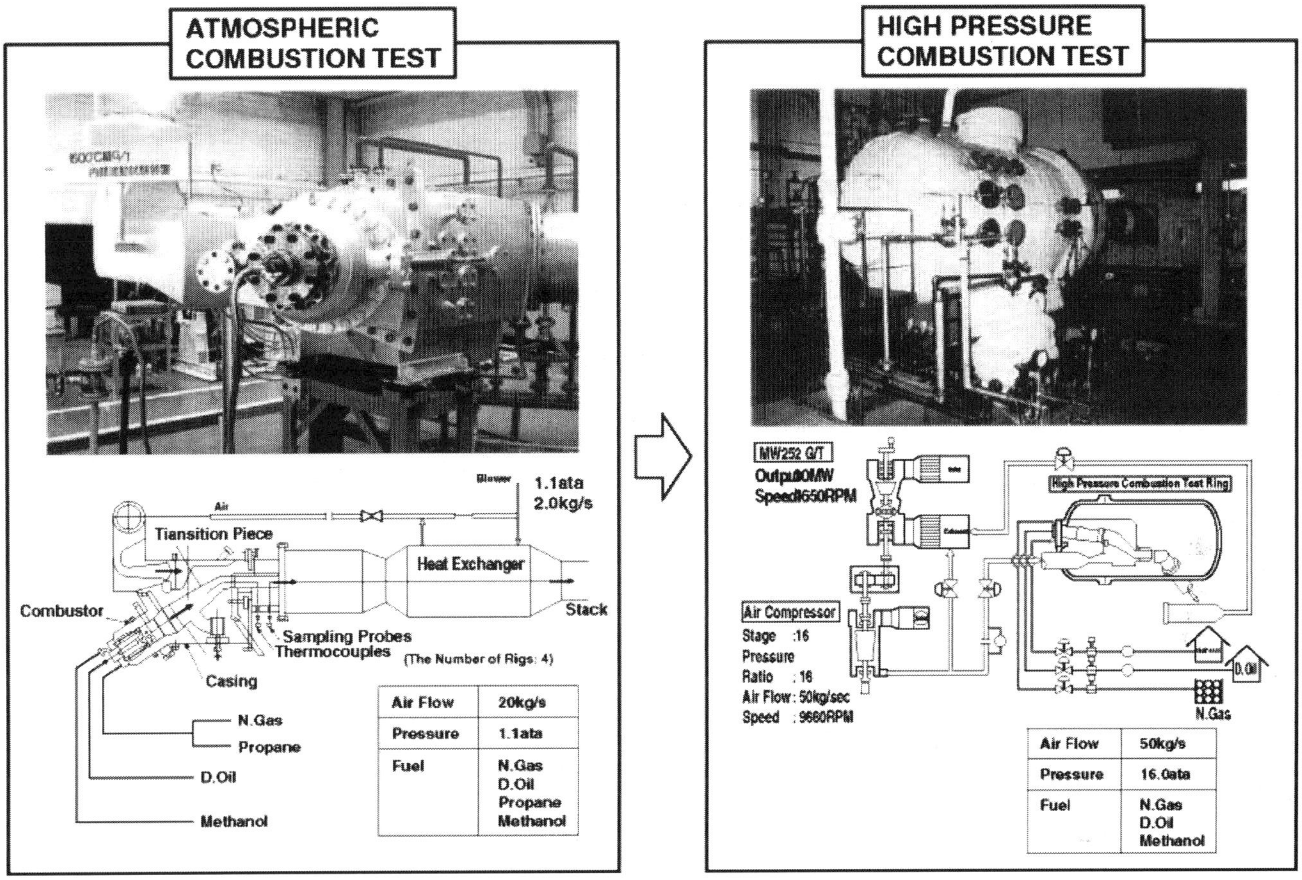

FIGURE 15–18 Combustor component tests [15-1].

margin. Gas temperature distribution at the cascade inlet is measured with total temperature probes and the metal temperature of the vane and the blade is measured with the embedded thermocouples. With the hot cascade tests, the airfoil metal temperature distribution was obtained and the cooling effectiveness was verified.

These features and technologies were finally verified through a high temperature demonstration unit (HTDU) by testing the airfoils under certain conditions. The HDTU shown in Figure 15–21 is a special core turbine with 0.6 scale turbine blades and vanes of the M501G for demonstrating the key technologies.

Materials

The advanced gas turbine requires heat resistant materials as well as advanced cooling technologies. For this purpose, some advanced materials for turbine blades and vanes, and thermal barrier coatings are developed. For the rotating blade material, MGA1400 alloy was developed. MGA1400 is a nickel-base super alloy and can be used for DS casting as well as for conventional casting. Compared with the IN738LC conventionally cast blade, the creep strength of the MGA1400DS blade is 50°C higher, and that of the conventionally cast MGA1400 is 30°C higher as shown in Figure 15–22. Both DS and CC blades are applied in the "G" series gas turbine.

For the stationary vane material, MGA2400 alloy was developed. MGA2400 is also a nickel-base super alloy that has excellent resistance against thermal fatigue, oxidation, and hot corrosion as well as high creep strength. It also has good weldability for settlement of accessory parts and repair. Table 15–2 summarizes the materials for turbine vanes and blades.

Long-Term Operating Experience

NMI constructed a long-term verification test facility with the M501G as a complete combined-power plant in its Takasago Machinery Works. NMI started the trial operation of the M501G in January 1997 to verify the performance and reliability. During the trial operation, more than 1800 instrumentation probes were installed in the gas turbine. Figure 15–23 shows the special measurement items. The flow path characteristics, metal temperatures, pressures,

FIGURE 15–19 The turbine row 1 blade [15-1].

strains, sound pressure levels, exhaust emissions, etc., were measured over the full range of operating conditions.

All important characteristics, including the cooling characteristics and component reliability, were verified. As an example, measured vibratory stresses of the blades and vanes are shown in Figure 15–24. They were measured for the first stage to the fourth stage blades, the first stage to the eighth stage and the seventeenth stage vanes of the compressor and the first stage to the fourth stage blades of the turbine during starting period, rated speed, and up to 110% of the rated speed. These items were measured by a non-contact method using optical fibers for compressor blades and by strain gauges for compressor vanes and turbine blades (with telemeter). The results show that the vibratory stresses are low enough compared with the allowable values.

At the end of June 1997, the combined-cycle verification plant passed the Ministry of International Trade and Industry (MITI)'s qualification test as an industrial power plant, which consists of one M501G gas turbine, one HRSG and one steam turbine and started commercial operation as the long-term verification test. The plant layout is shown in Figure 15–25. The electricity is sent to the grid of a domestic utility company. As this plant operates on demand of the utility company, it has mainly been operated with Daily Start and Stop (DSS) mode. Accumulated total operating hours/start-and-stop cycles are 10,898 hours/612 cycles at the end of October 2000. It is still accumulating operating experience successfully.

The following is the detailed explanation of each operation and inspection. The result shows the high reliability of the M501G.

June–October 1997

The M501G started its first verificational operation for summer peaking duty. In October, summer peaking duty

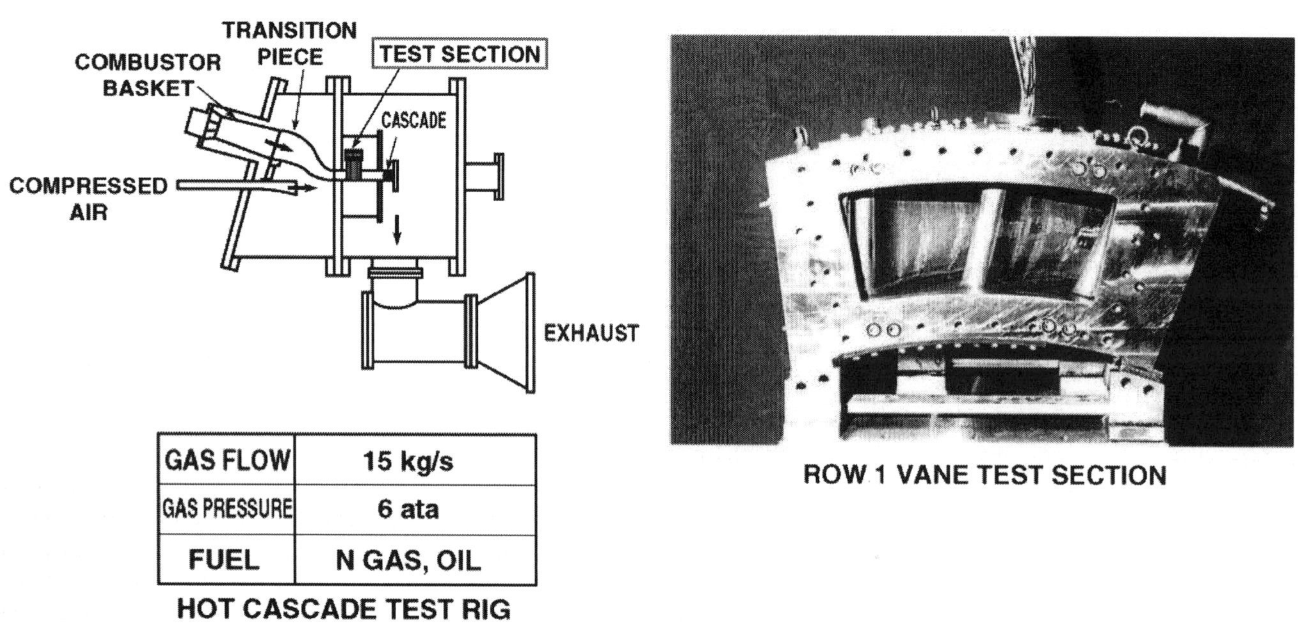

FIGURE 15–20 The hot cascade test ring for turbine cooling performance verification [15-1].

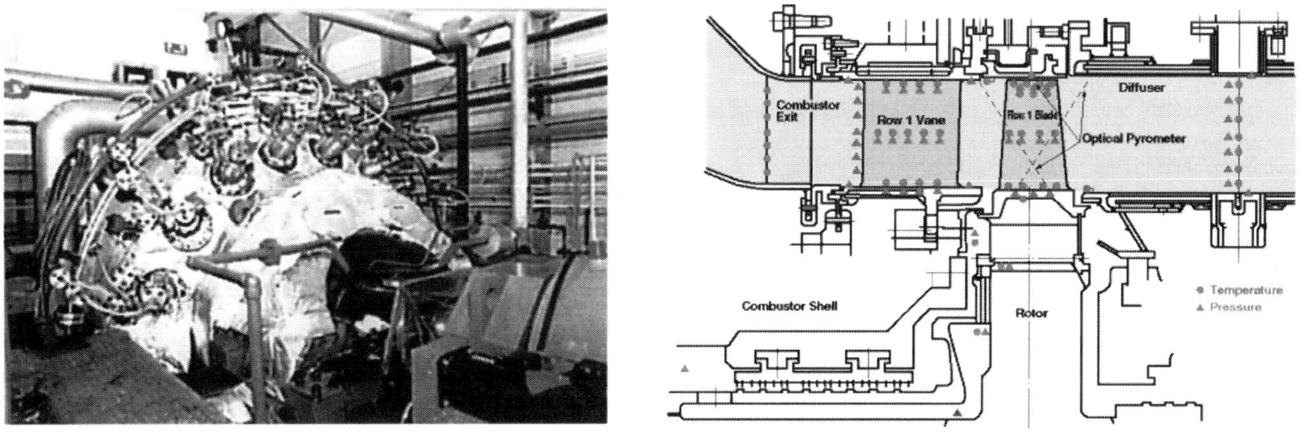

FIGURE 15–21 The HTDU facility [15-1].

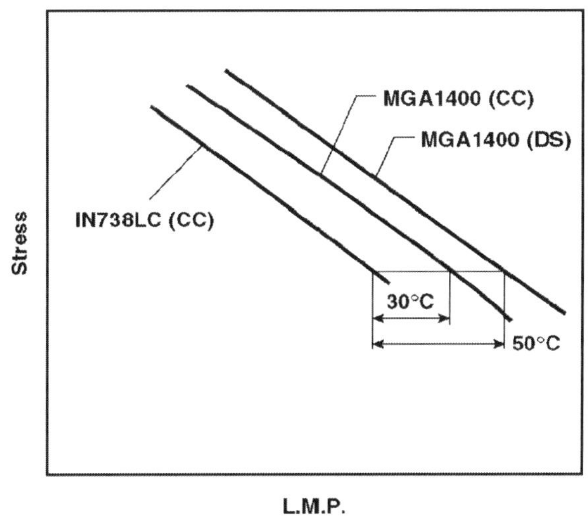

FIGURE 15–22 Creep rupture strength of the turbine blade material [15-1].

TABLE 15–2 The M501G Turbine Blade Material [15-1]

	M501G	M501F
Vane		
Row 1	MGA2400	EC768
Row 2	MGA2400	EC768
Row 3	MGA2400	X-45
Row 4	MGA2400	X-45
Blade		
Row 1	MGA1400 (DS)	IN738LC
Row 2	MGA1400 (DS)	IN738LC
Row 3	MGA1400	IN738LC
Row 4	MGA1400	U520

TBC is also important for its heat shield effect. The durability and the heat shield effect have been confirmed through many long-term field operational experiences of MF61, MF111, MF221, M7011, M501F, M701F, etc.

was completed and the first inspection of the M501G started. The hot gas path components like steam-cooled combustor, advanced cooled turbine vanes, and blades were found to be sound except minor cracks of steam cooled combustor transition piece exhaust mouth comers.

December 1997–March 1998

After the first inspection, the M501G was back in service in December 1997. The load demand in winter tends to be lower, but to confirm the soundness in preparation for the next summer peaking duty, the second inspection was conducted in March 1998. The inspection proved the hot gas path components to be sound.

June 1998–November 1998

The M501G was back in service. The result of the inspection in November 1998 showed no severe problems and everything was in sound condition. Until the next operation started in July 1999, the operating test of the M501H, which has steam cooled first and second stage vanes and blades as well as the steam-cooled combustor, was conducted and the results were successful.

July 1999–March 2000

The M501G was in service again for summer peaking duty. Accumulated operating hours/start-and-stop cycles were 8633 hours/514 cycles. The result of the inspection in March 2000 was very good as shown in Figure 15–26 even though partial peeling of TBC at row 1 blade was observed.

(A) PERFORMANCE

1 AIR FLOW
2 INLET TEMP. & PRESS
3 EXHAUST TEMP. & PRESS
4 FUEL FLOW
5 GENERATOR OUTPUT
6 COMPRESSOR
(SURGE MARGIN)
(STAGE EFF.)
7 TURBINE
(STAGE EFF.)
(DIFFUSER EFF.)

(B) METAL TEMP.

8 COMBUSTOR TRANSITION PIECE
9 TURB. ROW 1-4 BLADE
10 TURB. ROW 1-4 VANE
11 BEARING METAL
12 OUTER CASING
13 INNER CASING
14 EXHAUST CYLINDER

(C) STRESS/VIBRATION

15 COMPRESSOR BLADE
16 COMPRESSOR VANE
17 COMBUSTOR TRANSITION PIECE
18 TURBINE BLADE
19 ROTOR VIB.
20 CASING VIB.
21 ROTOR TORSIONAL VIB.

(D) OTHERS

22 COOLING AIR NETWORK
• FLOW
• TEMPERATURE
• PRESSURE
23 THRUST LOAD
24 EXHAUST EMISSION
25 ROTOR/CASING EXPANSION
26 NOISE
27 LUBE OIL TEMP.

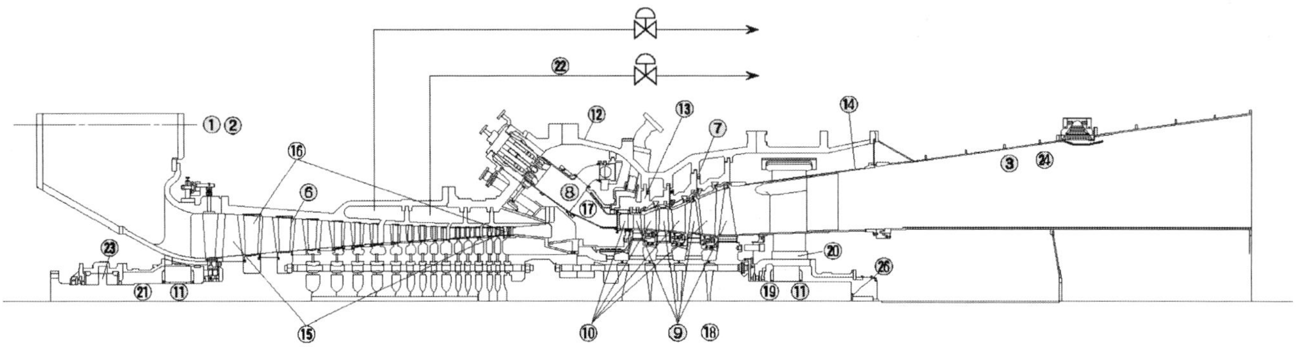

FIGURE 15–23 Special measurement items [15-1].

May–October 2000

After the fourth inspection, the M501G has returned to service and is successfully accumulating operating experience. The result of the inspection in October 2000 was fine by this inspection, and accumulated operating hours/start-and-stop cycles are 10,898 hours/612 cycles. The availability defined as the actual power supply hours over the demanded power supply hours reaches 99.2%. Also, the M701G, the "G" series gas turbine of 50 Hz, started a trial operation in October 1998. The plant started the commercial operation for the domestic utility company in July 1999 and has been successfully accumulating the operating hours and start-and-stop cycles of 9842 hours/63 times for the No. 1 gas turbine and 9350 hours/52 times for the No. 2 gas turbine. Since the gas turbine has never stopped except for the scheduled outage, availability defined as the actual power supply hours over the demanded power supply hours is 100%.

Because the high reliability of the M501G/M701G is verified through these operations, MHI has already received the order of 17 M501Gs and 7 M701Gs by the end of October 2000 and further potential orders are expected especially for the M501G. This would be a great help for energy savings and environmental conservation.

In summary, the M501G full load test with special measurement, which was conducted during the trial operation in February 1997, showed its high efficiency and high reliability and low environmental impact. At the end of June 1997, the combined-cycle plant with the M501G was put into long-term verificational operation after receiving MITI's certification.

By the end of October 2000, the plant has been successfully accumulating the 10,898 operating hours to keep up with the electricity demand from the utility company and start-and-stop cycles reached 612 cycles. The availability of 98.6% is kept. The M501G experienced five inspections and the condition of each component was quite good. These inspection results showed the high reliability of the M501G. The M501G would make a great contribution to energy savings and good global environmental issues.

Raising Serviceability Ceilings

With TIT upper limits being pushed upward, designers of different components within the gas turbine are all forced to seek new materials or manufacture methods or both. The different OEMs employ different technologies and strategies, several of which are discussed in cases in several of this book's chapters.

One summarized example here is a study made of the spray forming of a high(er)-temperature casing (participants were DERA, UK, and Rolls Royce, UK). One of the key parameters of concern with any such work is creep behavior.

A summary of this work's conclusions, to the point reported in a paper, is as follows.*

* Source: [15-2] Courtesy of Rolls Royce. A. Partridge, M. B. Henderson, D. G. Cole, and P. Andrews. "The Implementation of Spray Forming for High Temperature Casing Applications," in *Proceedings of the ASME Turbo Expo 2001,* June 4–7, 2001, New Orleans, 2001-GT-0581.

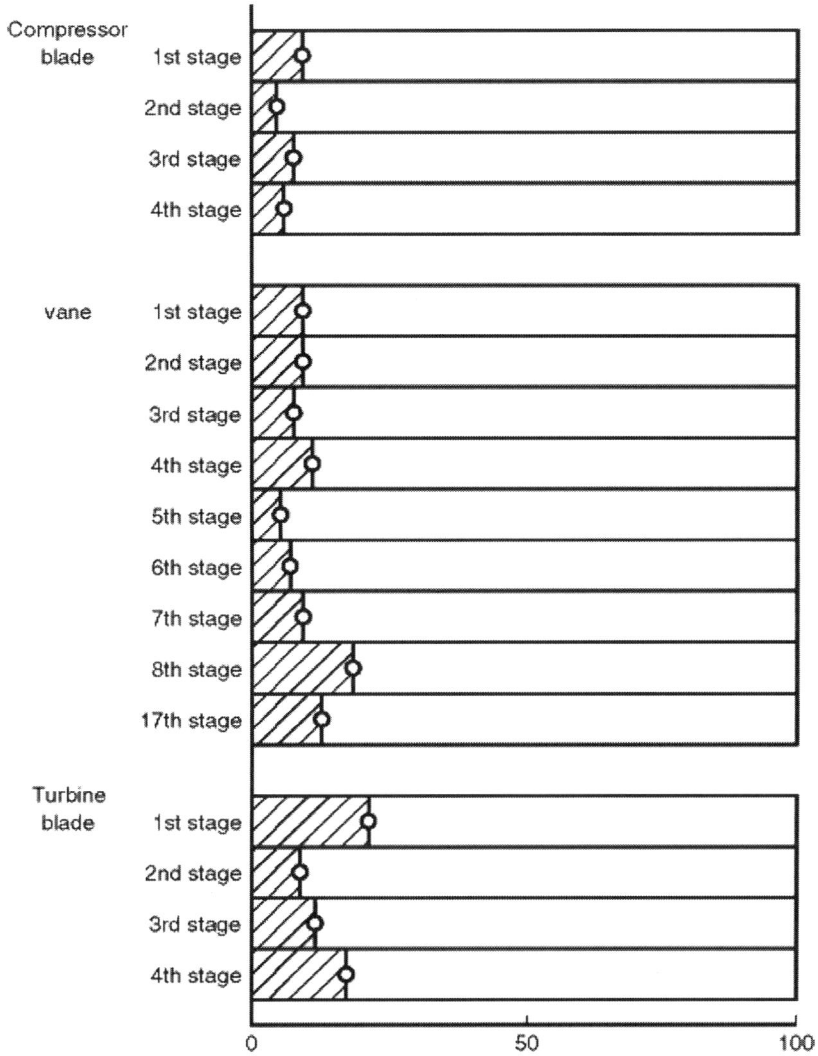

FIGURE 15–24 Summary of vibratory stress [15-1].

FIGURE 15–25 Plant layout of M501G combined-cycle verification plant of Takasago machinery works [15-1].

Many current gas turbine casings are manufactured using conventional wrought processing routes. Although well established this approach often requires numerous and complex processing steps. This can result in relatively long component lead times and high part costs. In an attempt to reduce lead times and cost, the production of parts using the spray-forming process is under consideration.

In modern gas turbine designs structural casings provide a gas tight pressure vessel, structural support for internal and external fixtures, and in the event of a catastrophic failure component containment. These casings are generally nickel, steel, or titanium depending upon the usage temperature and specific application. However, in all cases the casings are manufactured using a wrought processing route.

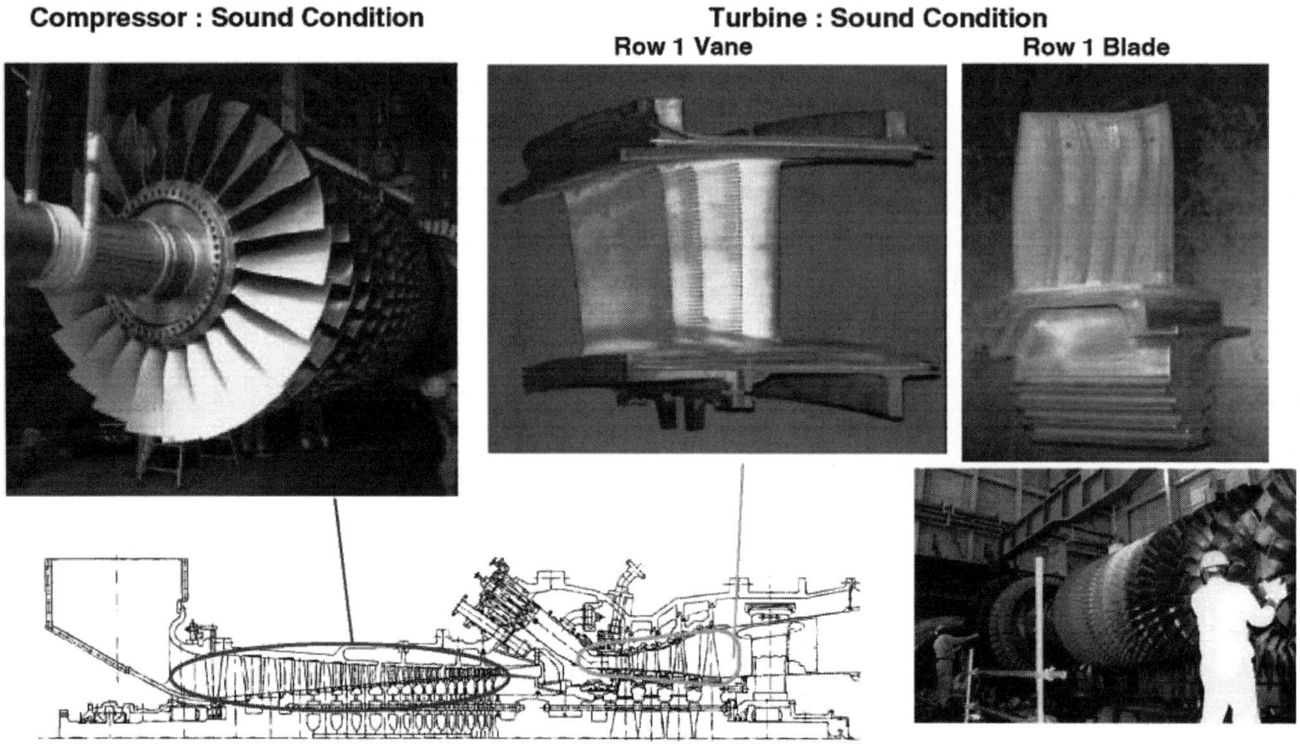

FIGURE 15–26 Inspection results in March 2000 [15-1].

Generally casing structures are fabricated from ring rolled flanges and wrought sheet that are welded together to produce the final component geometry. A typical process route for the production of such a casing would be to take a cast ingot, forge to size, pierce, and then ring roll the pierced component to produce a flange of the required diameter and thickness. The wrought sheet would then be formed to shape and welded to the ring-rolled flanges. This process involves many manufacturing steps, is costly, time consuming, and requires the selected alloy to exhibit good forming and welding characteristics.

For alloys that are produced in large quantities the production of casings via the wrought route can be carried out reasonably cost effectively. However, due to the requirement for increased combustion temperatures and higher pressures many of the current materials utilized as combustion and turbine casings are rapidly reaching the extremes of their temperature capability. In the case of the combustion casing support structure, the front of the combustion casing sees wall temperatures approaching 650°C. In this case alloys, such as the nickel-base alloy W718, which is widely used as a combustor support casing, are becoming severely creep limited. Therefore, new higher temperature nickel-base alloys, such as RS5, W939, CM247LC, and MarM002, are under consideration to replace W718, since they offer improved creep capability.

Much development effort has been expended in developing and assessing these new alloys for engine applications and a significant proportion of the required work has now been completed. However, as a general rule an alloy that exhibits improved creep resistance (i.e., improved resistance to deformation at elevated temperature) is also more difficult to produce via conventional wrought processing routes, such as forging and rolling. Furthermore, ease of weldability can be an issue with these higher performance alloys, as can the availability and cost of wrought product. Consequently, the introduction of these elevated temperature materials may demand the application of alternative processing techniques. One alternative technique under consideration is spray forming.

Spray Forming

The spray forming of components requires the atomization and spraying of molten metal onto a suitably shaped mandrel. Generally the material is melted by induction heating, and then atomized using an inert gas (usually Ar). The atomized metal is sprayed into a deposition chamber held under a reduced pressure and deposited onto a rotating mandrel. Computer control of the spray deposition process means that profiled "near net shaped" casings can be produced. Following deposition the porous pre-form is consolidated by hot isostatic pressing, ring rolled (if

TABLE 15–3 Typical Chemical Compositions (wt%) of Alloys IN718 and RS5 [15-2]

Element	IN718	RS5	Element	IN718	RS5
C	0.05	0.08	B	0.003	0.006
Cr	18.5	16.0	Zr	0.006	0.005
Co	0.1	10.0	Fe	18.0	—
Mo	3.0	4.8	Mn	0.04	0.05
W	—	2.0	Si	0.07	0.05
Nb	5.0	4.8	S	0.003	0.003
Ti	0.8	2.7	P	0.007	0.005
Ta	—	1.5	Ni	Bal.	Bal.
Al	0.5	1.0			

necessary) to the final size, and then machined to the final component geometry. The large reduction in process steps that this approach enables makes significant cost reductions feasible for many alloys. However, for some of the higher performance alloys that are difficult to wrought process, spray forming may prove to be the only viable production route. In addition to cost spray forming also has the advantage that a fine grain sized and chemically homogeneous product is produced, which again offers benefits over conventional process routes.

Casing Fabrication

In this section the fabrication of a combustion chamber support casing via spray forming is considered. The material selected for study is the high performance nickel-base alloy RS5, which has been identified as a potential replacement alloy for 1N718 for a number of casing applications. The composition of RS5 compared to 1N718 is shown in Table 15–3. The microstructure and mechanical properties of a hot isostatic pressed (HIP) and heat-treated pre-form was evaluated.

Microstructure of Processed and Heat Treated RS5

The microstructure of the spray-formed RS5 after HIPping, ring-rolling and heat treatment is shown in Figure 15–27. As can be seen from this figure the microstructure is primarily equi-axed, with a grain size ranging from 50–350 microns. The mean grain size was determined as being 100 microns. Electron back-scattered diffraction patterns showed that the microstructure did not contain any evidence of texture, as would be expected from a spray-formed deposit.

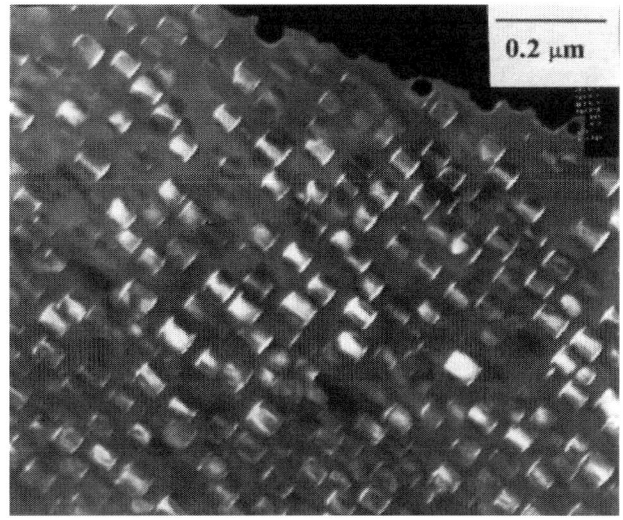

FIGURE 15–27 Microstructure of spray-formed, HIPped, ring-rolled, and heat-treated RS5 [15-2].

Mechanical Properties of RS5

A full mechanical property evaluation of the spray-formed RS5 was carried out to assess the suitability of the material for casing applications.

Tensile: The tensile strength of spray-cast, HIPped and ring rolled RS5 is comparable to, or better than, typical values for wrought 1N718 over the temperature range evaluated; with the strength of RS5 being retained to at least 60°C higher than 1N718. The elongation of RS5 is comparable to 1N718 in the range 20–600°C but at temperatures greater than 750°C it is significantly lower.

Creep: Comparison of the creep behavior of RS5 and 1N718 again indicates that RS5 exhibits a temperature advantage of around 60°C over 1N718 at elevated temperatures.

Fatigue: In terms of the fatigue properties, RS5 was found to be approximately comparable to typical values for wrought.

In summary, the deposition and consolidation of spray-formed RS5 results in a chemically homogeneous and equiaxed product, with little evidence of porosity or voiding. Mechanical testing of the spray-formed and processed material has clearly demonstrated that the material is broadly comparable to IN718, but exhibits a significant improvement in its properties at elevated temperature. An analysis of the data indicates that RS5 has a temperature advantage of around 60°C over IN718, in terms of tensile strength and creep resistance.

However, tensile, creep, and fatigue testing have all shown that the material is highly susceptible to intergranular failure at elevated temperatures. It is clear though that considerable scope exists to modify either the composition or processing routes in an attempt to improve the behavior of the material at elevated temperatures.

TABLE 15—4 Compositions of Heat-Resistant Austenitic Stainless Alloys Processed into Foils (wt%) [15-3]

Alloy/Vendor	Fe	Cr	Ni	Mo	Nb	C	Si	Ti	Al	Others
347 steel (Allegheny-Ludlum)	68.7	18.3	11.2	0.3	0.64	0.03	0.6	0.001	0.003	0.2 Co
Modified 803 (Special Metals, developmental)	40	25	35	n.a.	n.a.	0.05	n.a.	n.a.	n.a.	n.a.
Thermie-alloy (Special Metals)	2.0	24	48	0.5	2.0	0.1	0.5	2.0	0.8	20 Co
Alloy 120 (Haynes International)	33	25	32.3	2.5 max	0.7	0.05	0.6	0.1	0.1	3 Co max, 3 W max, 0.2 N
Alloy 230 (Haynes International)	3 max	22	52.7	2	—	0.1	0.4	—	0.3	5 Co max, 14 W, + trace La
Alloy 214 (Haynes International)	3.0	16	76.5	—	—	—	—	—	4.5	+ minor Y
Alloy 625 (Special Metals)	3.2	22.2	61.2	9.1	3.6	0.02	0.2	0.23	0.16	

n.a.—not available

It is clear that RS5 does offer an increased temperature capability compared to IN718. However, at present the cost of processing RS5 via the spray-forming route is prohibitive. The major advantage offered by the spray-forming route is the ability to process high strength alloys that are not susceptible to wrought processing. As higher combustion temperatures are realized the need for spray-formed product is likely to increase considerably.

Creep Resistance Advancements

Creep Resistance of Materials for Microturbine Recuperators

Microturbines evolved from automotive and small aerospace turbine applications. They have caught the US Department of Energy's (DOE's) attention enough that they are among DOE's closely pursued optimization projects. They have also caught public attention to the point that it will not be long before the "PT" (personal turbine) is talked about as the "PC" (personal computer) now is. (Also see Chapter 16 on microturbines and hybrids.)

Microturbines run faster than conventional gas turbines and the heat balance of a microturbine system is such that a recuperator is essential for optimum efficiencies. With the recuperator, as with many other components on gas turbine engines, the limiting design parameter is creep, as a development study outlines. That study,* funded by Oak Ridge National Laboratories, concludes: "A group of heat-resistant and oxidation/corrosion resistant austenitic stainless alloys have been processed into 1.3 mm foils and creep-rupture tested at 750°C and 100 MPa." See Table 15—4.

Ceramic Components

Considered a distant, esoteric prospect only a scant few decades ago, the ceramic turbine is fast approaching reality. Gas turbines that use ceramic coatings and selected components that are ceramic or partially ceramic are already part of mainstream technology. Improvement of this technology continues, sponsored in the United States, in part by the US DOE. The following cases provide details of design, testing, and operational strategies that involve ceramics components and point the way for work in this discipline to continue.

Case Study 2: Ceramic Vanes for a Model 501-K Industrial Turbine Demonstration*

The approach for this project is to design and demonstrate ceramic first-stage vanes in a Rolls Royce Allison Model 501-K turbine. These vanes operate at similar (and somewhat higher) temperatures as the second-stage vanes of advanced turbines. Consequently, a successful demonstration of first-stage vanes in the current generation Model 501-K turbine could provide a stepping stone to using ceramic vanes in second-stage and downstream rows of future generation advanced turbines.

* Source: [15-3] Courtesy of US DOE. P. J. Maziasz and R. W. Swindeman. "Selecting and Developing Advanced Alloys for Creep-Resistance for Microturbine Recuperator Applications," in *Proceedings of the ASME Turbo Expo 2001*, June 4—7, 2001, New Orleans, 2001-GT-0541.

* Source: [15-4] Courtesy Rolls Royce. Extracts from R. A. Wenglarz and KI. Kouns, "Ceramic Vanes for a Model 501-K Industrial Turbine Demonstration," in Proceedings of the ASME Turbo Expo 2000, May 8-11, 2000, Munich, 2000-GT-731.

Project Summary

Activities of the DOE/Rolls Royce Allison ceramic vane project include the following:

- Design, analyses, and fabrication of ceramic first-stage vanes and their metal mounts for retrofit into Model 501-K turbines
- Thermal shock proof tests of the ceramic vanes
- Engine validation test of the ceramic vanes and their metal mounts
- Phase 1 demonstration of the first-stage ceramic vanes/mounts in a Model 501-K turbine at a commercial site to assess design integration and materials compatibility
- Phase 2 demonstration with an optimized vane/mount redesign

These activities have been completed through the Phase 1 demonstration. Operation of an Exxon turbine for a total of 815 hr (793 hr at an Exxon plant and 22 hr proof test) was achieved. The Phase 2 demonstration was originally planned to accommodate any redesign needs of the metal mounts for the ceramic vanes, because past experience has shown the challenges of integrating structural ceramic components with metallic support structures. However, the metallic mounts performed successfully in the 501-KB5 Phase 1 demonstration of first-stage ceramic vanes. Since the Phase 1 demonstration identified a significant issue of ceramic oxidation, the Phase 2 demonstration will address the resolution of the oxidation issue.

Ceramic Vane/Mount Design and Analyses

Vanes

The AlliedSignal AS 800 silicon nitride ceramic was used for the first-stage ceramic vanes. The ceramic vanes were designed to facilitate manufacturability and relatively low cost. The vane design involves a single airfoil and simple platforms. The mounting arrangement (described in the next section) enables minimal final machining of the vane to reduce costs. The ceramic vane has been designed with a thinner airfoil than for the Model 501 turbine first metal vane. Analyses showed that the thinner vane was necessary for acceptable thermal shock stresses during emergency shutdown of the turbine (the highest stress condition).

Thermal and stress analyses indicated a maximum emergency shutdown stress of 207 MPa (30.0 ksi) in the vane mid span trailing edge region. A maximum steady-state stress of 192 MPa (27.9 ksi) was calculated at maximum continuous power for the turbine. The location of the maximum steady-state stress was also in the mid span trailing edge region. The calculated fast fracture probability of survival for the 60-vane engine set exceeds 99.99% for all of the emergency shutdown and maximum continuous power conditions that were analyzed.

Stress rupture tests of the AS 800 silicon nitride material used for the ceramic vane with durations up to 10,000 hr have been conducted at Oak Ridge National Laboratory (ORNL) and the University of Dayton at significantly higher temperatures and somewhat lower stresses than calculated for the vane at maximum continuous power conditions. No failures were experienced in 4500 hr by any of the specimens from a properly processed batch of the AS 800 ceramic. Retained strengths of specimens measured after surviving the 4500 hr stress rupture test exceeded 490 MPa (71 ksi), which is substantially higher than the highest calculated emergency shutdown stress of 207 MPa (30.0 ksi) for the Model 501-K ceramic vanes. The results of these tests provided a basis for conducting a long-term field test of the AS 800 ceramic vanes.

Mounts

Figure 15—28 illustrates the mounting arrangement for the ceramic vanes. To minimize mount loads and stresses, the vanes are not gripped or otherwise hard mounted but instead are seated in circumferential grooves of the inner and outer mount segments. Ceramic cloth is used at the interfaces between the metal mount segments and the outer surfaces of the vane platforms. This compliant material reduces ceramic contact loads and stresses from the mounts and also insulates the ceramic/metal interface. Insulation allows the mounts to operate at lower temperatures in their alloy working range and reduces the thermal gradient in the ceramic vane, which is the primary source of steady-state stress.

Ceramic Vane/Mount Assembly

Figure 15—29 shows the first-stage ceramic vane/mount assembly that replaces a similar metal vane assembly used for conventional Model 501-K turbines. The inner vane

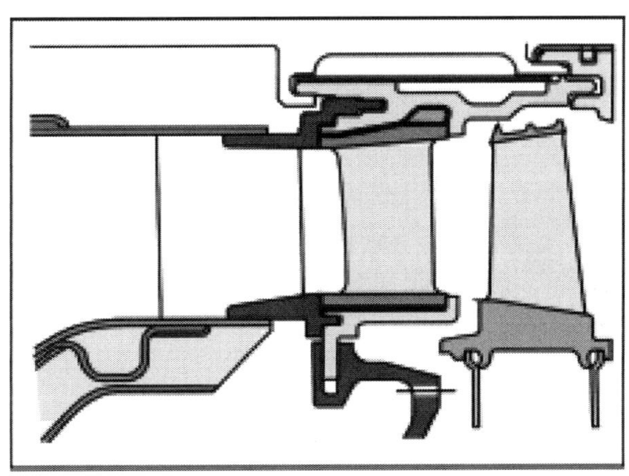

FIGURE 15—28 Ceramic vane mounting arrangement [15-4].

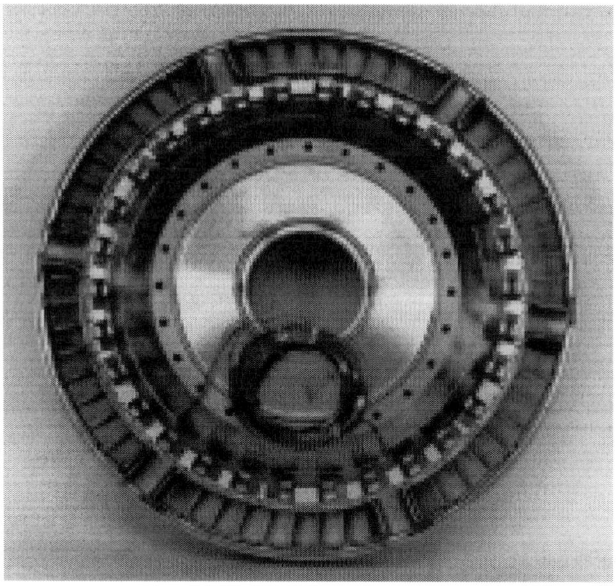

FIGURE 15–29 Ceramic vane/mount assembly [15-4].

support and an outer sheet metal band that holds the assembly together are common to both the ceramic and metal vane assemblies. All of the other parts of Figure 15–29 were designed for retrofit of the first-stage ceramic vanes into Model 501-K turbines. The Figure 15–29 front view also shows the saddles that position the transition sections from the six can combustors of the Model 501-K turbines. Consequently, 10 of the 60 first-stage ceramic vanes are located at the exit of each of the transition sections from the six combustor cans.

Ceramic Vane Thermal Shock Tests

Trailing edge thickness and other measurements were taken on each of the 87 AS 800 ceramic vanes received from AlliedSignal. The vanes were analyzed using fluorescent penetrant and x-ray analyses. The vanes were then exposed in an atmospheric combustor rig to thermal shock tests. Visual, fluorescent, and x-ray analyses conducted after the thermal shock tests revealed that only one vane had evidence of a possible crack and was questionable for operation in the turbine.

Ceramic Vane/Mount Proof Tests

Turbine proof tests of the ceramic vanes and metal mounts were conducted prior to operation at a commercial site. Objectives of these tests were to verify that (i) the metal mounts do not transmit excessive contact loads to the ceramic vanes, (ii) turbine power and performance are not adversely affected by the vane/mount designs, and (iii) the uncooled ceramic vanes do not result in excessive temperatures in the adjacent metal support structures. Turbine power and performance were considered potential issues because of the challenges of maintaining tight gaps, clearances, and airfoil positioning due to a difference of more than a factor five between the thermal expansion coefficients for the metal mount materials and the ceramic vane materials. The temperature of the metal support adjacent to the ceramic vanes was also a potential issue because the mount parts common to both the metal vane and ceramic vane assemblies do not benefit from the metal vane cooling when operated with ceramic vanes.

The approach to accomplish the above objectives consisted of a baseline performance test of the Exxon 501-KB5 turbine with metal vanes, performance tests of the same turbine except with the metal first-stage vane assembly replaced with the ceramic first-stage vane assembly, and thermocouple measurements of the temperatures of the metal vane support during ceramic vane runs. The performance of the turbine with the metal and ceramic vane assemblies was compared. The turbine was disassembled and inspected after the final ceramic vane test to determine whether the metal mounts had transmitted excessive loads to the ceramic vanes.

The total operation time of the turbine proof test with the ceramic vane/mount assembly was 22 hr. Post-test visual and fluorescent penetrant analyses revealed no fractured or cracked vanes. However, two vanes had platform chips, one of which was suspected to have occurred during disassembly. Evaluations of the other chipped vane and the few vanes that experienced chips during the later demonstration led to the conclusion that the ceramic vane/mount design provided stress relief of inevitable localized contact loads at platform edges associated with both ceramic and metal part tolerances. Oxide scale formation observed on chipped platform surfaces during the later demonstration inspections indicated that the vanes operated many hours after chipping and cracks had not propagated from the chipped surfaces.

All six performance calibration point measurements referenced to standard day conditions for the Exxon turbine with the ceramic vane assembly indicated greater power than for the same turbine operated with the metal vane assembly. The average calculated standard day power for the six ceramic vane performance runs exceeded that for the metal vane run by 1.7%. Some of the improved performance for the ceramic vanes probably resulted from a thinner airfoil than the metal vanes without the flow disruption of trailing edge cooling airflow discharge. Also, the ceramic vane was designed with more advanced aerodynamic analyses approaches than used for the standard metal vane.

Thermocouple measurements of the metal vane support temperatures adjacent to the ceramic vanes showed temperatures well below the working limits of the vane support alloy.

FIGURE 15—30 Exxon 501-KB5 turbine at site [15-4].

Following reassembly of the ceramic vane assembly after inspection, the Exxon turbine used in the engine proof tests was sent to the Exxon Mobile Bay facility for the field demonstration.

Ceramic Vane/Metal Mount Demonstration

The first-stage ceramic vane Phase 1 field demonstration occurred at an Exxon natural gas processing plant near Mobile, Alabama. Three identical Rolls Royce Allison Model 501-KB5 turbine skids provided power and process steam at this Exxon plant. Figure 15—30 shows one of the turbines in its skid enclosure at Exxon. The turbine with the ceramic first-stage vanes was installed and operated under normal commercial service conditions at maximum continuous power, except for preplanned teardowns and ceramic vane assembly inspections with removal of vanes for analyses. These inspections were originally scheduled at 200, 500, and 1000 hr. The turbine sustained power and performance throughout its operation with ceramic vanes. The turbine at Exxon experienced an emergency shutdown (the highest vane stress condition) at 525 hr and a full power water wash of its compressor with no detectable adverse effects on further operation or the ceramic vanes. The cause of the emergency shutdown was a malfunctioning vibration sensor and was unrelated to the ceramic vane assembly. There was no need to shut the turbine down associated with the ceramic vanes except for the planned inspections. None of the ceramic vanes failed or cracked and the metal mount design was shown to perform successfully. As mentioned earlier, a few vanes experienced chipping at platform edges, which appeared to alleviate local contact loads and enable further operation for many hours without propagation of cracks or failure.

FIGURE 15—31 Ceramic vanes before and after 815 hr operation in the turbine [15-4].

Analyses of the vanes removed during the periodic inspections showed unexpected rates of material loss due to oxidation of the silicon nitride ceramic in the turbine operation environment. These rates of materials loss were determined to be excessive for industrial turbine applications, which require part lifetimes of 20,000 hr and longer. Analyses of the vanes showed the need for additional measures (e.g., environmental coatings) to achieve industrial turbine lifetimes. Since the Phase 1 demonstration had achieved its objective of identifying redesign needs, this demonstration phase was discontinued after a total operation time of 815 hr (22 hr proof test and 793 hr demonstration) for the Exxon turbine with ceramic vanes.

Analyses of Ceramic Vanes

Visual evidence of ceramic oxidation was apparent during the turbine inspections. As illustrated by Figure 15—31, the

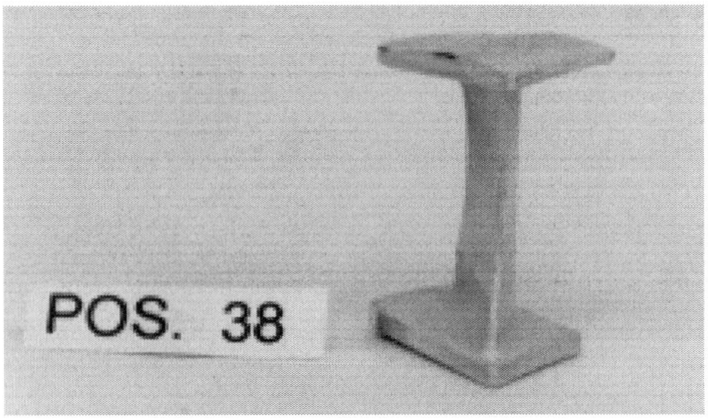

FIGURE 15–32 Ceramic material loss on concave surface [15-4].

FIGURE 15–33 Ceramic material loss on convex surface [15-4].

color of the vanes had changed from tan/gray (vane on right) to white (vane on left). The ceramic vanes were covered by a white powder. The metal rotor blades and vanes in the two rows downstream of the ceramic vanes also had a dusting of white powder. The white powder was analyzed by Oak Ridge National Laboratory (ORNL) and found to be an oxide of silicon and lanthanum from the intergranular phase of the ceramic. Figures 15–32 and Figure 15–33 illustrate oxidation material losses visible to the unaided eye on vane concave and convex surfaces, respectively. These vanes had experienced 815 hr operation in the turbine.

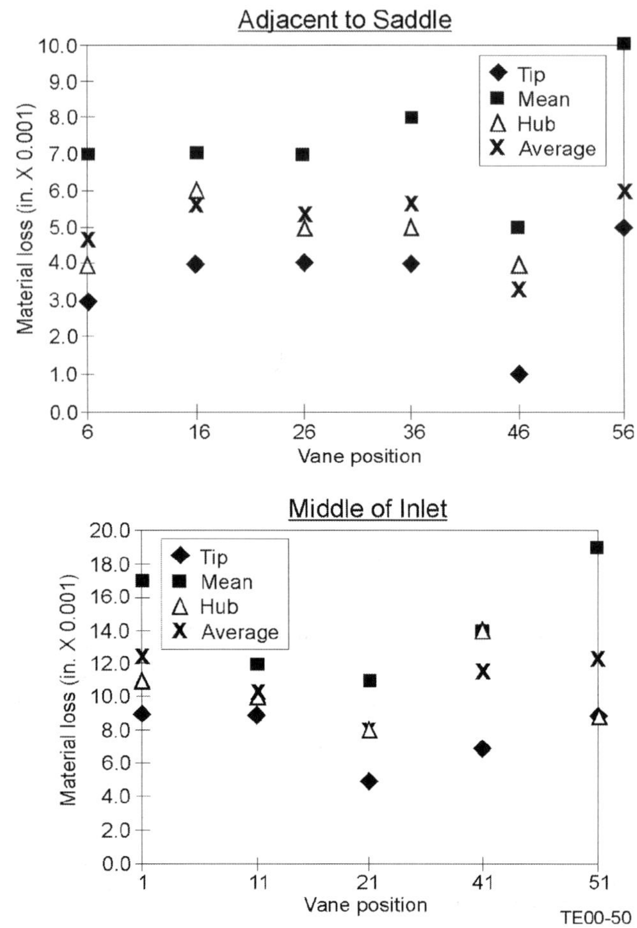

FIGURE 15–34 Trailing edge thickness loss at 815 hr [15-4].

Oxidation Versus Position in the Combustor Pattern

The trailing edge thickness was measured for each vane removed during inspections and after the demonstration to determine the oxidation loss of material at that location. Figure 15–34 shows trailing edge thickness losses at 815 hr versus radial location on the vanes and circumferential position in the six combustor cans of the turbine. The plot on the top shows thickness losses for vanes located at the same position at the exit of the combustor cans in the saddle region near the outside of the annular transition section from the combustor cans. The plot on the bottom shows thickness losses for vanes at the same location at the exit of the combustor cans near the center of the annular transition from the combustors. Both plots show losses at the hub, mid span, and tip trailing edge regions. These plots indicate that the greatest trailing edge losses were measured at the vane mid span, corresponding to the hottest and highest velocity location in the radial direction. The trailing edge losses were also higher at circumferential positions nearer to the center of the combustor outlet transition sections, again corresponding to higher temperatures and velocities. Similar circumferential patterns of trailing edge loss versus circumferential vane location were observed for the six combustors, but the overall level of loss varied by as much as a factor of 2 between combustors.

Oxidation Versus Time

Figure 15–35 shows the measured trailing edge losses at mid span (the trailing edge radial location of greatest loss) versus vane operation time in the turbine. The data suggest insignificant trailing edge thickness losses for times less

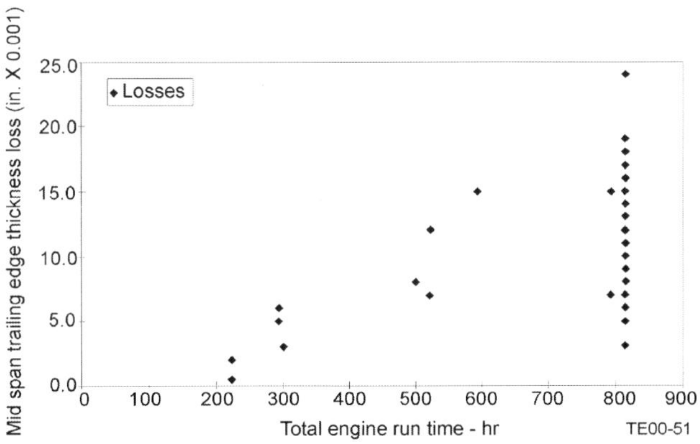

FIGURE 15–35 Mid-span trailing edge thickness loss versus time [15-4].

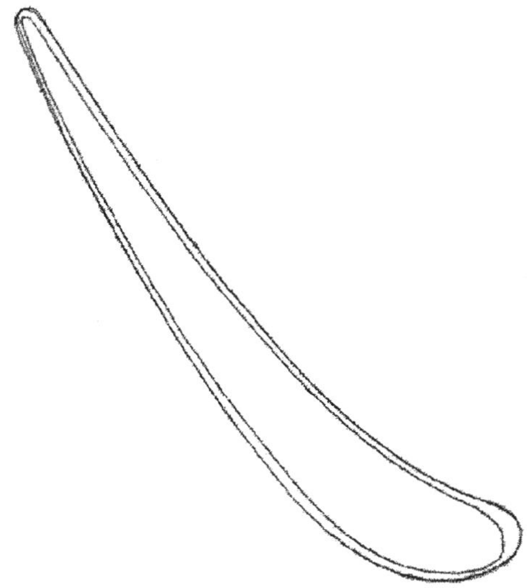

FIGURE 15–36 Mid-span profile loss [15-4].

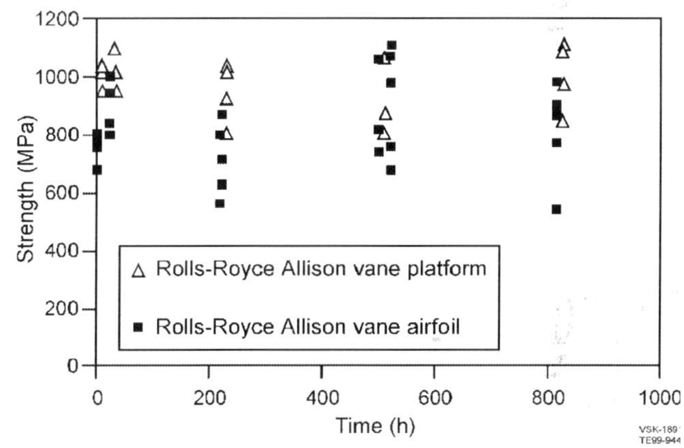

FIGURE 15–37 Strength versus time for specimens from vanes [15-4].

than 200 hr. Analyses by ORNL identified an initial oxide scale on the new ceramic vanes resulting from processing. Apparently this scale was much more adherent and impermeable to further oxidant penetration to the underlying substrate than the oxide scale formed in the turbine environment. Figure 15–35 indicates that the two vanes removed after 222 hr of operation had very little measurable trailing edge material losses when compared to the pretest measurements. However, the vanes removed after 522 hr had measured trailing edge material losses up to 0.32 mm (0.0125 in.). Since the test was discontinued after the 815 hr inspection, all 60 vanes removed from the assembly were measured for trailing edge thickness losses. Except for vanes with low oxidation levels because of their locations in untypical flow and temperature regions behind the saddles, the decreases in trailing edge mid span thickness at 815 hr range from 0.13 mm (0.005 in.) to a maximum loss of 0.61 mm (0.024 in.). Percentage decreases in trailing edge thickness range from 12% to 50% of original values. The trailing edge thickness measurements indicated that the rate of oxidation might have been decreasing with time.

Oxidation Versus Position on Airfoil Surface

AlliedSignal took profile measurements of several of the ceramic vanes that had operated 815 hr in the turbine and compared those to profiles measured on the same vanes after fabrication. Figure 15–36 gives the mid span profile comparisons for one of the vanes that had the greatest material loss. Although the gas velocity near the airfoil surface varies from near zero at the nose stagnation point to about 573 m/s (1880 ft/sec) at the trailing edge, the loss does not greatly differ over the airfoil surface. This suggests that gas velocity is a secondary factor in the rates of oxidation and that relatively expensive high velocity rigs (e.g., cascades) might not be needed for ceramic materials oxidation tests. However, such ceramic materials tests do need to replicate the water vapor content and pressures of the turbine flow path environments.

Retained Strength of Ceramic Vanes

Several of the ceramic vanes removed at various operation intervals from the demonstration turbine were sent to ORNL for retained strength measurements. Small disks (about 6 mm diameter and about 0.6 mm thick) were cut from the vanes and then fractured using a load at their center. Figure 15–37 shows results of the ORNL retained strength measurements versus operation time. The retained strength after 22 hr exposure to the turbine environment (in the engine proof test) increased over the as-processed strengths (zero time exposure). The average retained strength of specimens did not greatly change from initial values for increasing exposure times up to 815 hr, although the scatter in the strength measurements did increase with time. The lowest retained strength measurement in

TABLE 15–5 Porous Ceramic and Metal Filter Technology Development [15-4]

Monolithic Matrices	Fiber-Reinforced Composites	Metals/Intermetallics
Coors P-100A-1 alumina/mullite	McDermott oxide-based CFCC	US Filter Haynes 214, 556, 188, 230
Pall clay-bonded silicon carbide (442T, 326, 181)	Techniweave oxide-based CFCC	Inconel 600, Hastelloy X, 3105 (US Filter, Pall, Mott Corp.)
Schumacher clay-bonded silicon carbide (F40, FT20)	3M oxide-based CFCC	Iron aluminide (US Filter, Pall)
GTE cordierite and cordierite-silicon nitride	Americon oxide-based CFCC 3M CVI-SiC	FeCrAlY (US Filter, Technetics, others)
AiResearch reaction bonded and sintered silicon nitride	DuPont SiC-SiC	
Enstomullite-bonded alumina	Textron nonoxide-Based CFCC	
Blasch mullite-bonded alumina	**Filament wound**	
Specific surface cordierite	Honeywell PRD-66	
IF&P recrystallized silicon carbide		
Vacuum infiltrated chopped fibers		
IF&P fibrosics™		
Foseco aluminosilicate		
Reticulated foam		
Ultramet CVI-SiC		
Selee oxide-based foam		

Figure 15–37 of about 500 MPa is approximately 2.4 times greater than the maximum calculated vane stress of 207 MPa at the highest stress vane location and worst (emergency shutdown) condition (see earlier discussion of stress analyses). Consequently, the ORNL data suggest that the vane oxidation is a surface phenomenon that does not substantially affect the strength of the underlying substrate material. Much longer vane lifetimes appear possible if the loss of surface material by oxidation can be inhibited (e.g., through use of environmental coatings).

In summary, a first-stage ceramic vane and metal mount design has been developed and demonstrated for 815 hr under commercial industrial turbine conditions. This design appears to have successfully addressed many of the challenges of integrating stationary structural ceramics with metallics in an industrial turbine. However, the ability to operate a turbine for extended durations afforded by the ceramic vane/mount design has brought to light the issue of excessive long-term ceramic oxidation in the industrial turbine flow path environment. Additional measures are needed to improve materials durability so that commercial lifetimes of at least 20,000 hr can be achieved for industrial turbine ceramic parts. Measures to inhibit surface oxidation should advance structural ceramics toward this life goal since the strength of the ceramic structure underlying the oxidizing and recessing surfaces exposed to the flow stream was not observed to be markedly decreasing.

Case Study 3 describes the assessment of ceramic versus metal media filters for gas turbine power systems (conducted by Siemens Westinghouse).

Case Study 3: Assessment of Ceramic and Metal Media Filters in Advanced Power Systems*

Siemens Westinghouse Power Corporation (SWPC) has been involved in hot gas filter materials technology development since 1988. Emphasis was initially focused on the development and use of oxide- and nonoxide-based monolithic filter materials in bench-scale test programs and field applications (Table 15–5). With thermal fatigue issues being encountered by the oxide-based monoliths, and the oxidation and/or high temperature creep issues resulting in the nonoxide-based filter matrices during pressurized fluidized-bed combustion (PFBC) and/or pressurized circulating fluidized-bed combustion (PCFBC) operation, development of fracture toughened, continuous fiber reinforced ceramic composites (CFCC) and intermetallic filter elements was undertaken in 1994. Several issues were identified with respect to the long-term durability, response, and performance of the CFCC filter elements during extended service life. These included manufacturing and structural integrity, load bearing capability, and ease of fixturing within the filter system.

Numerous advancements continued to be made since initiating development and manufacture of the monolithic

* Source: [15-5] Courtesy of Siemens. Extracts from M. A. Alvin, "Assessment of Ceramic and Metal Media Filters in Advanced Power Systems," in *Proceedings of the ASME Turbo Expo 2001*, June 4-7, 2001, New Orleans, 2001-GT-0574.

TABLE 15—6 Field and Extended Life Testing [15-4]

	Maximum Operating Time, hrs		Equivalent Exposure Time, hrs
Filter Supplier & Matrix	PFBC—AEP	PCFBC—FW Karhula	Field & Accelerated Testing
Schumacher F40	5855	227(1)	
Schumacher FT20	1705	2201(2)	11,080; 11,080
Pall Vitropore 442T	1705	1341(1)	
Pall 326		2201(2)	9662; 11,080
Coors P-100A-l alumina/mullite	2815	716(1)	
		2201(2)	8485; 12,211
		3311(3)	13,356
3M CVI-SIC	1705	627(4)	
DuPont PRD-66	1705	581(5)	7517(6)
3M oxide CFCC			
McDermott oxide CFCC		581(5)	10,626
Techniweave oxide CFCC			
Blasch mullite-bonded alumina		581(5)	10,626
Ensto			
IF&P REECER		581(5)	5611
Other			5030(6)

(1) 1992—1994 test campaign.
(2) 1995—1997 test campaign.
(3) 2201 hrs of operation under PCFBC conditions and 1110 hrs of operation under PFBC conditions.
(4) 1995—1996 test campaign.
(5) 1997 test campaign.
(6) New element without field exposure.

and composite ceramic filters in 1988 and 1994, respectively. These included:

- Development and use of high temperature, creep resistant binders in the clay-bonded silicon carbide filter materials
- Efforts by alternate domestic and foreign suppliers (i.e., Blasch, Ensto, etc.) to bring to the market potentially viable filter materials and/or filter elements manufactured by alternate processes and production techniques
- Development of alternate filter concepts (i.e., sheet filter; inverted candle filter; etc.).
- Development of oxidation resistance coatings for nonoxide matrices
- Incorporation of a membrane along the o.d./i.d. surfaces of the filter elements
- Enhanced flange strength and surface abrasion resistance

As issues arose with respect to the long-term viability of monolithic and CFCC filter elements, efforts were directed in 1998 to the development of a more robust, metal-based filter element.

Note: Table 15—6 provides a summary of the operating hours accumulated for each candle filter type tested in the various programs.

Ceramic Filter Elements—Summary Comments

Table 15—7 identifies the manufacturing, as well as thermal, mechanical, and chemical degradation, issues that have been identified during the initial and extended stages of life of the oxide and nonoxide-based, monolithic ceramic, fiber reinforced, and filament wound filter materials. Although microstructural changes occur in the porous ceramic filter matrices as a result of exposure to process operating conditions, the residual strength of all materials tends to stabilize after an initial conditioning period. An area that remains to be addressed is whether microstructural or phase changes will continue to occur within the various filter matrices with extended aging of

TABLE 15–7 Matrix/Component Response to Process Operation—PFBC/PCFBC [15-4]

Initial Life			
Matrix Operating Time	**Plant Commissioning**	**1 Year**	**Extended Life $>$ 3 Years**
Monoliths			
Oxides	Failure resulting from process thermal transients	Thermal fatigue Matrix phase changes; component conditioning	Full matrix crystallization Microcrack formation Plant process control required for material/ component stability and performance
Nonoxides	Failure resulting from extreme process thermal transients Initial matrix oxidation $\geq$700°C	Matrix phase changes; component conditioning Impact of reformulation of binder phase limiting high temperature creep TBD Oxidation $\geq$700°C	Crystallization of binder/ligaments Oxidation of SiC grains creep TBD Limited to $<$700–750°C applications
Fiber Reinforced			
Oxides		Potential stability and performance issues related to manufacturing, homogeneity, and reproducibility of elements; lower strength; lower load bearing capabilities; and fixturing need to be addressed Oxidation $>$700°C Fiber/SiC bonding and embrittlement; reduced matrix fracture toughness	Fiber embrittlement TBD
Nonoxides	Initial oxidation $>$700°C Fiber/SiC bonding and embrittlement	Potential stability and performance issues related to manufacturing, homogeneity, and reproducibility of elements; lower strength; lower load bearing capabilities; and fixturing need to be addressed Oxidation $>$700°C Fiber/SiC bonding and embrittlement; reduced matrix fracture toughness	Not recommended for high temperature oxidizing applications
Filament Wound			
Oxides		Potential stability and performance issues related to manufacturing, homogeneity, and reproducibility of elements; lower strength; lower load bearing capabilities; fixturing and membrane stability (devoting, matrix pull-out) need to be addressed	TBD

elements during field operation, and whether continued phase change will impact the residual strength, thermal and/or mechanical properties, and filtration characteristics of the various filter materials. Continued grain growth and secondary phase formations are projected for some materials during extended field operation, which may have a significant impact on the ultimate viability of these filter elements during commercial operation.

Metal Filter Media Development

In 1998, SWPC initiated a program with DOE/NETL to develop and evaluate the use of porous, high temperature, metal media candle filters in PFBC systems. Working with US Filter/Fluid Dynamics, Pall Advanced Separation Systems, Mott Corporation, Fairey Microfiltrex, Technetics, and Ultramet, SWPC produced composite filter elements for exposure to a simulated 1200°F (650°C), 1400°F (760°C), and 1550°F (840°C) PFBC process gas environment containing 200 ppmv S02/S03. The advanced alloys selected by SWPC for enhanced corrosion resistance in this environment were Haynes 230, 214, 188, and 556, and FeCrAIY. Elements constructed from 310S, Inconel 600, Hastelloy X, and FeAI were included as commercially available materials for comparison of performance in this program.

The filtration media of the advanced alloy filter elements principally included metal fibers, powders, wire laminates, and open-cell, metal-based, reticulated foams. Typically the US Filter/Fluid Dynamics, Fairey Microfiltrex, and Technetics sinter bonded fiber-containing metal media is formed into an ~1 mm thick filtration mat layer which is supported by an underlying, dense, perforated, cylindrical metal support structure. An open mesh outer containment layer is included either along the outer surface of the porous filter body, encapsulating the filtration mat layer, or is embedded within, as well as sintered bonded to one surface of the porous filtration mat layer. The outer confinement mesh and filtration mat are generally longitudinally welded together, forming a tight fit around the underlying support structure. In contrast, filter elements fabricated from sinter bonded metal media powder or particulates typically are seamless (with the exception of Pall Hastelloy X), are generally thicker walled in comparison to fiber-containing metal filter elements, and do not have an outer confinement layer or underlying metal support structure.

The as-manufactured, high temperature, tensile strength of the commercially available and developmental porous advanced metal filter media are shown in Figure 15–38. Post-test exposure strength of the composite elements after 250 and 500 hours of simulated PFBC testing at 1200°F (650°C) is also shown. In general, the tensile strength of the as-manufactured US Filter/Fluid Dynamics filter elements at 1200°F (650°C) is between 10,000 and 17,000 psi. The strength of the fiber-containing US Filter/Fluid Dynamics filter media generally appeared to increase during the initial

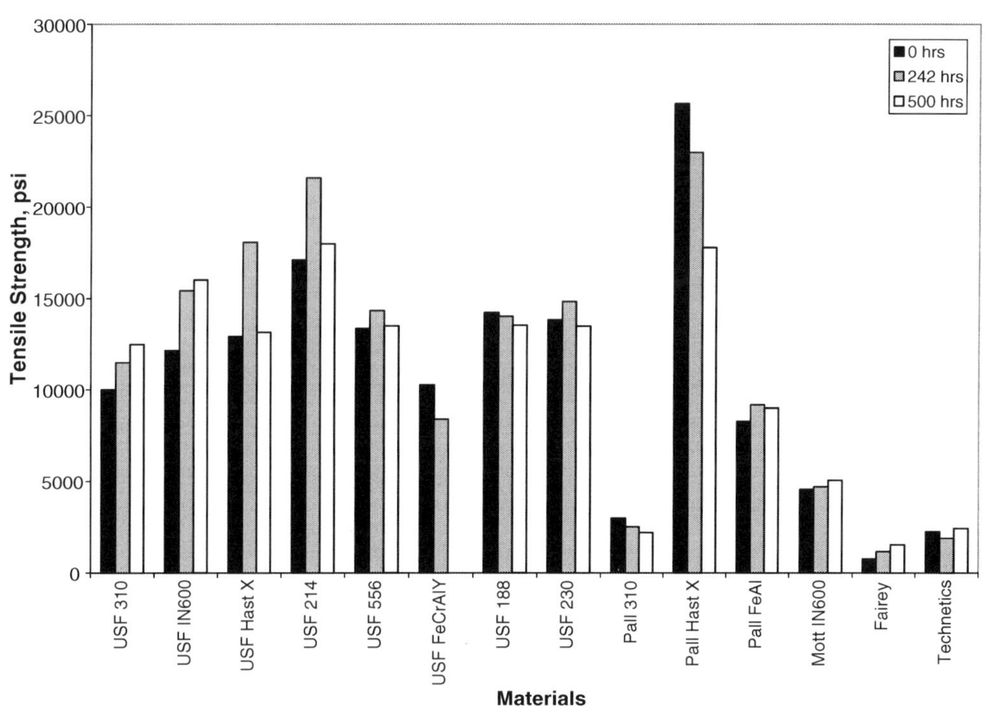

FIGURE 15–38 Tensile strength of the porous metal filter media at 1200°F (650°C) as a function of exposure time to simulated PFBC process operating conditions [15-5].

250 hours of exposure to the simulated PFBC operating conditions. After 500 hours of exposure, a minor reduction in the 1200°F (650°C) tensile strength resulted for the advanced US Filter/Fluid Dynamics filter media, as well as Hastelloy X, while a continued strengthening of the commercially available 310S and Inconel 600 filter media resulted. Minor strengthening of the Mott Inconel 600 and Pall FeAI powder-containing media, and Fairey Microfiltrex fiber-containing media resulted during initial operation in the 1200°F (650°C) simulated PFBC environment. In contrast, the 1200°F (650°C) tensile strength of the powder-containing Pall 310 and Hastelloy X porous filter media decreased as a function of operating time. Due to the variation in the use of coarse vs. fine outer containment mesh, trends in the Technetics residual tensile strength were not readily apparent for the porous FeCrAIY filter media. Currently testing is ongoing at SWPC STC to demonstrate the potential viability of the various porous filter media during 1400°F (760°C) simulated PFBC operation.

Chapter 16

Microturbines, Fuel Cells, and Hybrid Systems*

"Faith makes all things possible. Love makes all things easy. Hope makes all things work."

—Unknown

Chapter Outline

A microturbine*, as the name suggests, is a small gas turbine. Research on microturbines predicts that affordable commercially available models eventually will shrink to thumbnail size and become PTs, or personal turbines, much the way large room-sized computers became PCs.

The first edition of this book was released in 2008, about the time that the US recession became entrenched. The rest of the world has suffered similarly or worse. Research funding has been scarce on many programs and the ones that are given priority mainly relate to emissions reduction potential. In this economic climate, microturbines, fuel cells and hybrids in distributed power generation have not reached the potential they had hoped to, prior to the latter half of the last decade. There have been gains made in the transportation sector, however, with hybrids powering buses, small trains and other vehicles. Fuel cells, as we see later in this chapter, now are installed on German naval submarines.

The margin for gain with research and design development is best with larger GTs, so projects like FutureGen and its associated CSS have priority. The US DOE concentrates much effort on developing metallurgy for USC and AUSC STs. A couple of OEMs have rolled out their GT-J models, which offer incrementally higher efficiency than the -H and -G models did. Smart grids have gained attention because of their potential to add renewables to the main transmission power mix. Fuel cells can count as part of the renewables mix. One OEM that has invested and continues to invest in fuel cells is Siemens. This chapter contains summarized material on their work with microbial fuel cells that work with wastewater systems. Siemens have also had success with tests of arrays of fuel cells on submarines. This is a good application location for fuel cells as submarine have other operational conerns that rank higher than cost per kW.

Some other turbomachinery fields have suffered financially. Turboexpanders is one field that has not fared well financially in the last decade. The microturbine world is holding its own with steady sales, but there have not been monumentally *different* technical developments in the past seven years, although there will always be incremental (Table 16—1) developments in recuperator and alternative fuels technology. Further, the commercialization (in terms of greater affordability) of related technologies, for instance the cost of a fuel cell per kW of power produced, has not progressed appreciably. Had this happened, microturbine-fuel cell hybrids would have expanded their global market base.

For the time being "affordable, commercially available" microturbines are about the size of a household refrigerator and run on a variety of waste (like wood waste,

* Source: [16-1] Unless otherwise specified, Claire M. Soares, *Microturbines and Fuel Cells* (Boston: Elsevier).

TABLE 16—1 Microturbine Overview [16-1]

Commercially available	Yes (limited)
Size range	25—500 kW
Fuel	Natural gas, hydrogen, propane, diesel
Efficiency	20—30% (recuperated)
Environmental	Low (<9—50 ppm) NO_x
Other features	Cogen (50—80°C water)
Commercial status	Small volume production, commercial prototypes now

biomass, or methane rising off landfill) and low BTU gases. Microturbines require a recuperator to develop anywhere close to an efficiency (about 40%) that makes them commercially viable. Commercial microturbines deliver from 50—500 kW most commonly and can power small schools, remote process plants, and small industry. A tie in to the local grid, if grid access is available, for backup power can be provided.

Hybrid is the name given to a gas turbine in conjunction with a fuel cell. Fuel cells are given some space in this chapter because of their ability for application in concert with gas turbines. In contemporary work, the gas turbine in a hybrid system is most commonly a microturbine. The hybrid then results from the partnership of a heat engine and a nonheat engine (two different cycles). Because the term *combined cycle* is reserved for the combination of two different heat cycles (gas and steam turbine cycles), the gas turbine-fuel cell is called a *hybrid*. The recent work on hybrid systems has been intense.

Most of the US work on hybrid systems has been funded by the US Department of Energy (DOE). The DOE gives grants to US manufacturers, so when Siemens became Siemens Westinghouse and Rolls Royce merged with Allison, the Europeans could then get US DOE grants, too. Due to the lack of emissions and the potential for future power distribution for large consumers as well as single households, DOE's sponsorship is well founded. Minus the greenhouse gas emissions of conventional fossil fueled plants and the hybrid's potential to eventually make dinosaurs of power lines and massive transmission systems, the hybrid's future may be bright indeed. It may be a few years before its price is brought down from current levels of $900/kW to $1500/kW uninstalled. In fact some initial commercial systems may cost in excess of $3000/kW. A few tested systems today boast lower figures than that value; however, the reader needs to check all data that may affect system-related costs in "local" conditions to arrive at an accurate "cost per kW of this system." When target price objectives are reached however, it is anticipated that hybrids will become competitive and then drop in price below conventional power generation unit costs.

MICROTURBINES

Microturbines** produce between 25 kW and 500 kW of power. Microturbines were derived from turbocharger technologies found in large trucks or the turbines in aircraft auxiliary power units (APUs). Most microturbines are single-stage, radial flow devices with high rotating speeds of 90,000 to 120,000 revolutions per minute. However, a few manufacturers have developed alternative systems with multiple stages and/or lower rotation speeds.

Microturbines are currently at partly commercial status. Capstone, for example, has delivered over 2400 microturbines to customers as of 2003. However, many of the microturbine installations are still undergoing field tests or are part of large-scale demonstrations.

Microturbine generators can be divided in two general classes:

- Recuperated microturbines, which recover the heat from the exhaust gas to boost the temperature of combustion and increase the efficiency, and
- Unrecuperated (or simple cycle) microturbines, which have lower efficiencies, but also lower capital costs.

While some early product introductions have featured unrecuperated designs, the bulk of developers' efforts are focused on recuperated systems. The recuperator recovers heat from the exhaust gas in order to boost the temperature of the air stream supplied to the combustor. Further exhaust heat recovery can be used in a cogeneration configuration. Figure 16—1 illustrates a recuperated microturbine system.

FUEL CELLS

We now look at the basics of fuel cell technology, specifically solid oxide fuel cell (SOFC) technology.

Many† gas turbine OEMs now have divisions refining microturbines, fuel cells (SOFC or otherwise), and hybrids. We will look at Siemens SOFC design. Siemens Power Generation developed tubular SOFC technology with the support of the US Department of Energy's advanced fuel cell research program and the German Ministry of Economics and Labor (BMWA). More recently, Siemens have used PEM (polymer electrolyte membrane) type fuel cells in naval submarines. The following figure points out

** Source: [16-2] Courtesy of the US DOE.

† Source: [16-3] Courtesy Siemens Power Generation. Adapted with permission.

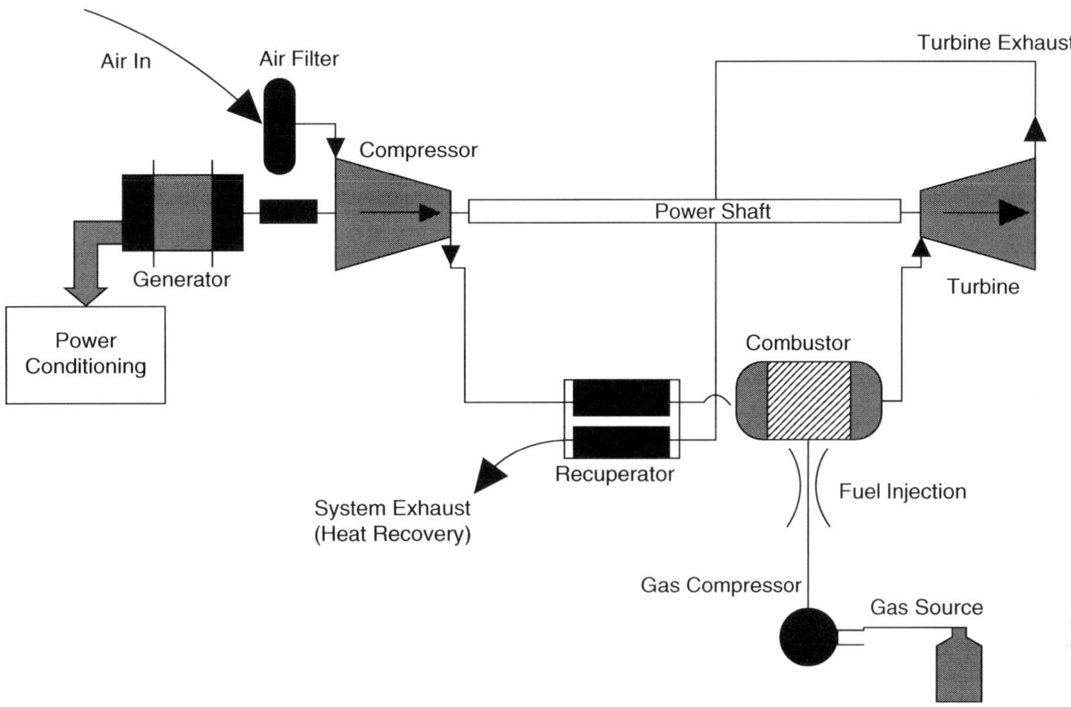

FIGURE 16–1 A recuperated microturbine system [16-2].

applications for both the PEM and the SOFC basic designs (Figures 16–2 to 16–5).

The PEM-type design[1] turns on and off quickly, has low voltage degradation characteritics, a low operating temperature (80°C), can be overloaded and has no liquid corrosive electrolyte. The basic operation and components are simple, as illustrated in the next three figures.

Power Generation Tubular SOFC Technology

Siemens Power Generation's Stationary Fuel Cells (SFC) division makes an SOFC type design. The Siemens Power Generation solid oxide fuel cell is made up of an electrolyte and two electrode layers in a unique tubular design. This design eliminates the need for seals required by other types of fuel cells, and also allows for thermal expansion. In a tubular SOFC design, air flows through the interior of the cell, and fuel flows on the outside of the cell. At elevated temperatures, the oxygen in the air ionizes and the resulting ions flow through the electrolyte and combine with the fuel on the cell's exterior. This is an electrochemical reaction, so electrons are released. With proper connections, they can flow through an external circuit as electricity.

The design of the stack is simple. It is cooled using process air, and during normal operation consumes no external water. It also has integrated thermally and hydraulically within its structure a natural gas reformer that produces the hydrogen and carbon monoxide utilized by the cell. Also, except during startup, no external heat source is needed.

Siemens Power Generation's first fuel cell product was the SFC-200 (Figure 16–6). This is a 125 kW SOFC cogeneration system, operating on natural gas at atmospheric pressure, with electrical efficiency of 44–47% at full load. An overall system energy efficiency of >80% is expected, assuming steam/hot water or other cogeneration.

Hybrids

If an SOFC is pressurized, an increased voltage results, thus leading to improved performance. For example, operation at 3 atmospheres increases the power output by ~10%. However, this improved performance alone may not justify the expense of pressurization, but what may is the ability to integrate the SOFC with a gas turbine, which needs a hot pressurized gas flow to operate. Since the SOFC stack operates at 1000°C it produces a high temperature exhaust gas. If operated at an elevated pressure, the exhaust becomes a hot pressurized gas flow that can be used to drive a turbine. If an SOFC is pressurized and integrated with a gas turbine, the pressurized air needed by the SOFC can be provided by the gas turbine's compressor, the SOFC can act as the system combustor, and the exhaust from the SOFC can drive the compressor and a separate generator. This yields a dry (no steam) hybrid-cycle power system that

1. Courtesy Siemens Power Generation, www.siemens.com/marine. Adapted with permission.

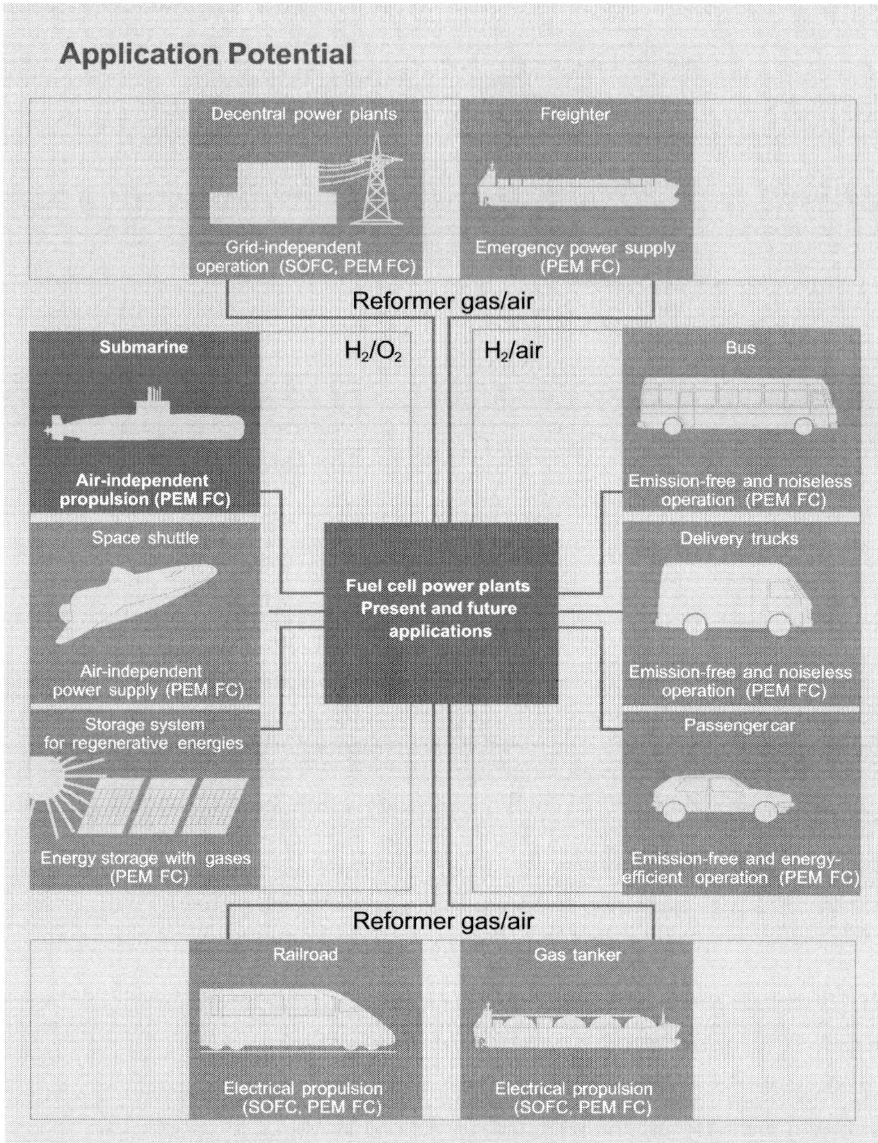

FIGURE 16–2 Possible applications for fuel cell power plants [16-3].

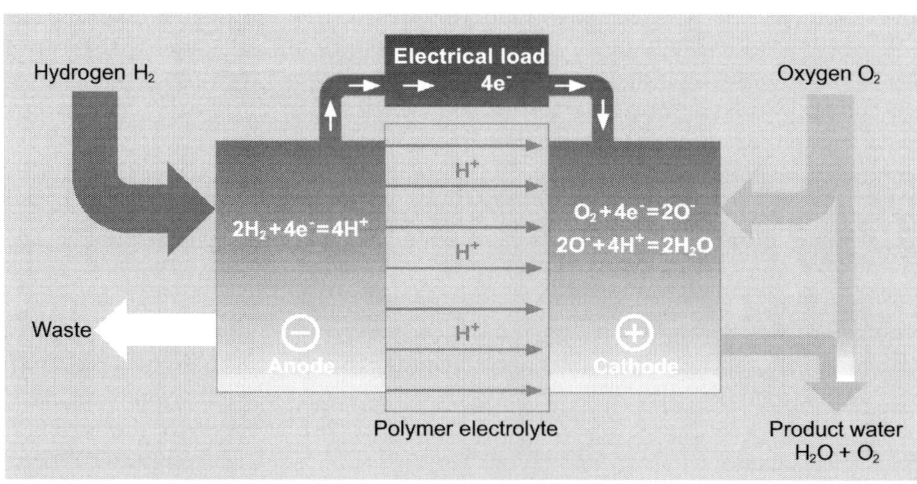

FIGURE 16–3 Functional principle [16-3].

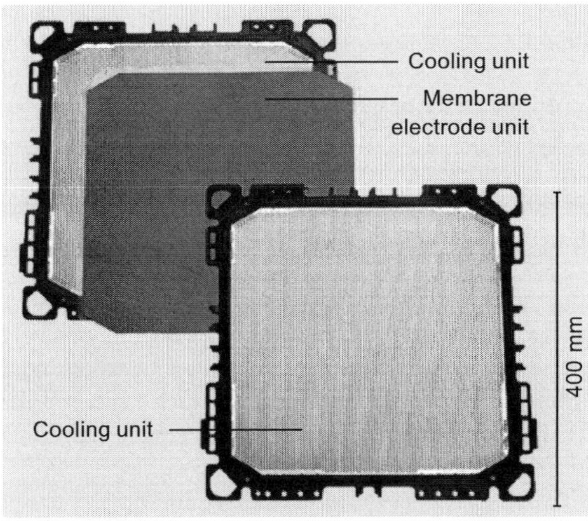

FIGURE 16–4 Components of cell [16-3].

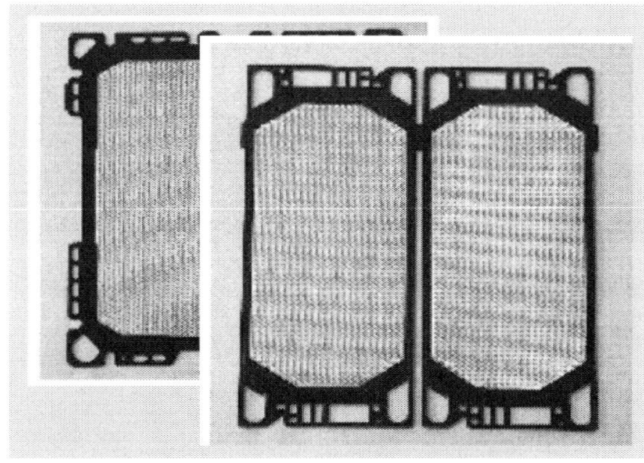

FIGURE 16–5 Comparison of cells [16-3].

promotes unprecedented electrical generation efficiency (Figure 16–7).

During normal operation of the pressurized SOFC hybrid, air enters the compressor and is compressed to −3 atmospheres. This compressed air passes through the recuperator where it is preheated and then enters the SOFC. Pressurized fuel from the fuel pump also enters the SOFC and the electrochemical reactions take place along the cells (Figure 16–8). The hot pressurized exhaust leaves the SOFC and goes directly to the expander section of the gas turbine, which drives both the compressor and the generator. The gases from the expander pass into the recuperator and then are exhausted. At −200°C the exhaust is hot enough to make hot water. Electric power is thus generated by the SOFC (dc) and the generator (ac) using the same fuel/air flow. Analysis indicates that with such SOFC/GT hybrids an electrical efficiency of 55% can be achieved at power plant capacities as low as 250 kW, and −60% as low as 1 MW using small gas turbines. At the 2–3 MW capacity level with larger, more sophisticated gas turbines, analysis indicates that electrical efficiencies of up to 70% are possible.

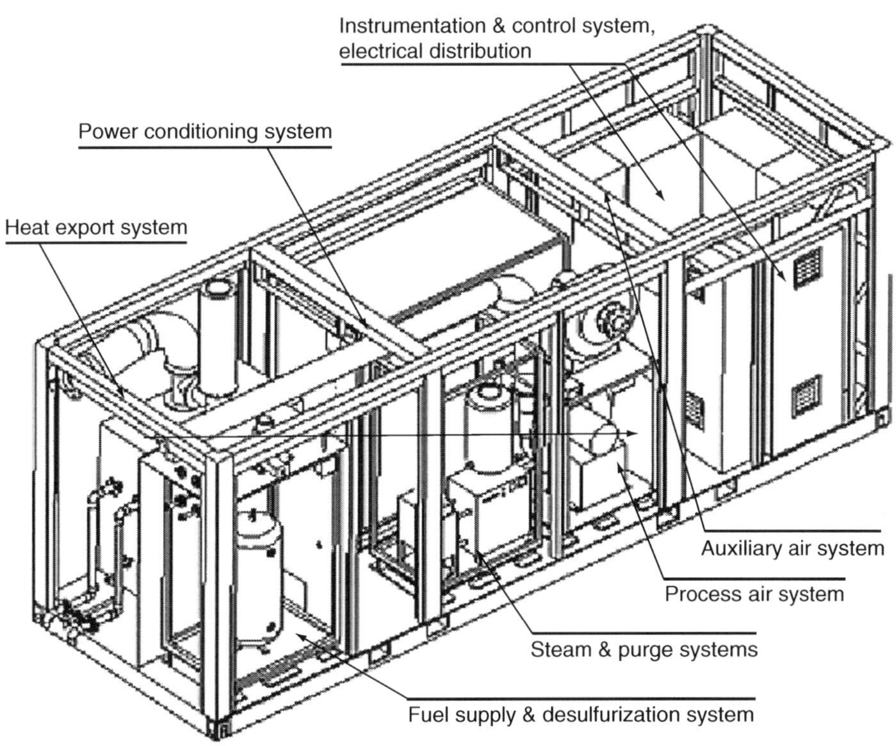

FIGURE 16–6 The SFC-200 will be the building block for systems up to 500 kW [16-3].

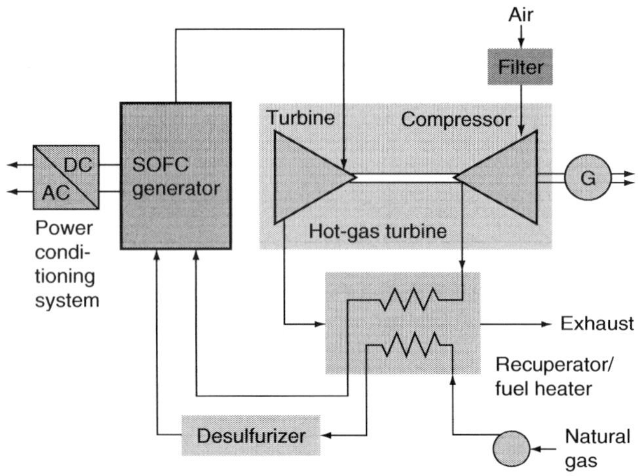

FIGURE 16–7 SOFC/gas turbine hybrid cycle diagram [16-3].

Future of SOFC/Gas Turbine Hybrids

Numerous alternative SOFC/GT hybrid configurations under a Cooperative Agreement with the US Department of Energy, National Energy Technology Laboratory have been studied. These studies have included hybrids with pressurized and atmospheric pressure fuel cell modules, turbines heated only by SOFC exhaust or also with supplemental duct heating and multiple, staged (cascaded) fuel cell/turbine reheat cycles. These studies have shown that while many different cycles offer considerable promise efficiency is maximized in a pressurized SOFC hybrid without supplemental firing, where the SOFC acts as the combustor for the gas turbine.

For large-scale applications ($\sim$20 MWe) staged reheat cycles have indicated that electrical generating efficiencies could reach as high as $\sim$70%. These studies led to demonstrations of pressurized SOFC/GT hybrids in pursuit of these high electrical efficiencies. Through testing of a 220 kW and a 300 kW pressurized hybrid Siemens have verified the feasibility of the pressurized SOFC/GT hybrid concept. With these demonstrations Siemens have also demonstrated that the close-coupled pressurized hybrids are quite costly, but because of their feasibility and potential further development is warranted.

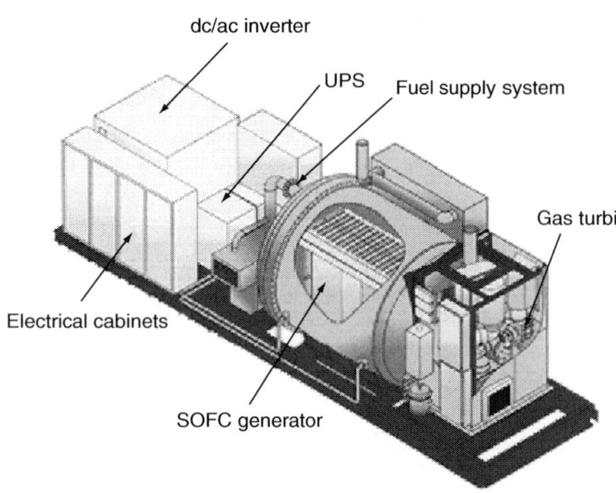

FIGURE 16–8 SOFC/Gas Turbine Hybrid System [16-3].

Studies of hybrid cycles indicated that atmospheric cycles, while offering somewhat lower efficiency than pressurized cycles, would be less complex to develop and quicker to implement. Because they would require less integration of the SOFC and gas turbine, they have the potential also to be less expensive and could accommodate a wider variety of gas turbines. Further study of other SOFC/gas turbine hybrid cycles is planned to assess their potential.

Further efforts are also needed on the part of government and industry to ensure the availability of appropriately sized small gas turbines. For pressurized hybrids especially, suitable mass flow, pressure ratio, and other characteristics are necessary.

APPLICATIONS AND CASE STUDIES*

Grassroots design development is expensive and this field of small gas turbine systems providing distributed power lends itself to several OEM joint ventures and component supplier arrangements. Capstone turbine, for instance, uses a recuperator developed by Solar Turbines under the US DOE funding umbrella. Back in 2002, Ingersoll-Rand (IR) had formed an alliance with a division of Siemens to integrate the IR 70-kilowatt and 250-kilowatt PowerWorks microturbine products to create Microturbine Energy Systems. More recently, that division of IR has changed hands.[2] IR's microturbine technology functions as an on-site, micropower generation plant that can complement or supplement the user's other sources of power. These systems utilize natural gas or other fossil fuels to produce small-scale electricity in the ranges of 70 kilowatts for a single unit, to 3 megawatts of capacity with multiple units. Exhaust heat from the microturbine can also be used to produce hot water or steam in a highly efficient combined heat and power (CHP) plant.

CASE STUDY 1: MICROBIAL FUEL CELLS (MFC)

This work is still in developmental stages, but it is simple and promising.

---

* [16-1]

2. A news release in Irvine, CA, January 18, 2011 announced that Flex-Energy, a cleantech company completed the acquisition of the Energy Systems business of Ingersoll-Rand.

Recent advances[3] in MFC technology expanded the basic concept to production of fuels and chemicals such as hydrogen, methane, and ethanol, as well as using MFC for desalination.

MFC technology is gaining more attention as companies push for sustainability. However, there are many challenges to scaling these. Most designs have been tested only in the laboratory, and many tests are conducted in fed batch mode with synthetic wastewater. The viability of scaling up MFCs could open up new ways of generating renewable energy while treating wastewater, contributing to a more sustainable society.

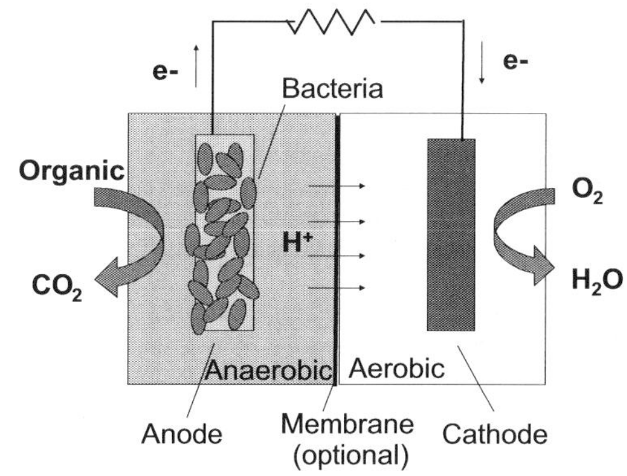

FIGURE 16–9[3]

How the MFC Works[3]

MFC is a bio-electrochemical system using microorganisms on one or both electrode(s) to catalyze the oxidation or reduction reactions, and thus can be deployed for general water treatment applications. In the simple configuration depicted in Figure 16–9, the system could produce net electrical power.

$$\text{Organics} + O_2 \rightarrow \text{Water} + CO_2 + \text{electricity}$$

(Organics in wastewater) (O_2 from the air)

Oxidation reaction at anode:

Glucose is oxidized by microorganisms, generating electrons and protons and CO_2. $C_{12}H_{22}O_{11} + 13H_2O \rightarrow 12CO_2 + 48H^+ + 48e^-$

Reduction reaction at cathode:

Electrons and protons are consumed in the cathode compartment, combining with oxygen to form water.

$$O_2 + 4H^+ + 4e^- \rightarrow 2H_2O$$

Objective

To investigate the issues related to scale-up of MFC for wastewater treatment and explore practical applications.

Methodology

The performances of MFC under different test conditions were evaluated in terms of chemical oxygen demand (COD) removal efficiency, power generation and coulombic efficiency (CE). CE is defined as the percent of electrons recovered as current versus that in the starting organic matter.

Results and Discussion

Flow-through MFC showed a consistent power output throughout the operation while the power output of batch-mode MFC decreased with the depletion of COD in wastewater feed (Figure 16–10a–d).

Performance of Scaled-Up MFC

Effect of Wastewater pH (Figure 16–11)

The COD removal efficiency decreased ~10% as the pH fell below 7. The efficiency of power generation and CE were not affected significantly.

Effect of Wastewater Conductivity (Figure 16–12)

Power density increased significantly with increasing wastewater conductivity. The conductivity of wastewater did not affect the COD removal and CE.

Effect of COD in Wastewater (Figure 16–13)

Spiking the domestic wastewater with acetate (~700 ppm COD) immediately increased performance; whereas increasing conductivity (from 1 – 7 mS/cm) did not show significant improvement.

Synthetic versus Real Wastewater (Figure 16–14)

Real food and beverage wastewater performed similarly to synthetic wastewater.

Conclusions

Performance of up-scaled continuous mode MFC was comparable to that of small batch-mode MFC, confirming that the MFC design is scalable.

Performance of MFC was maintained when using real wastewater. Industrial wastewater with COD > 500 ppm is suitable for MFC technology.

3. Courtesy of Siemens. For further information, see www.siemens.com/siww. (Accessed October 14, 2013).

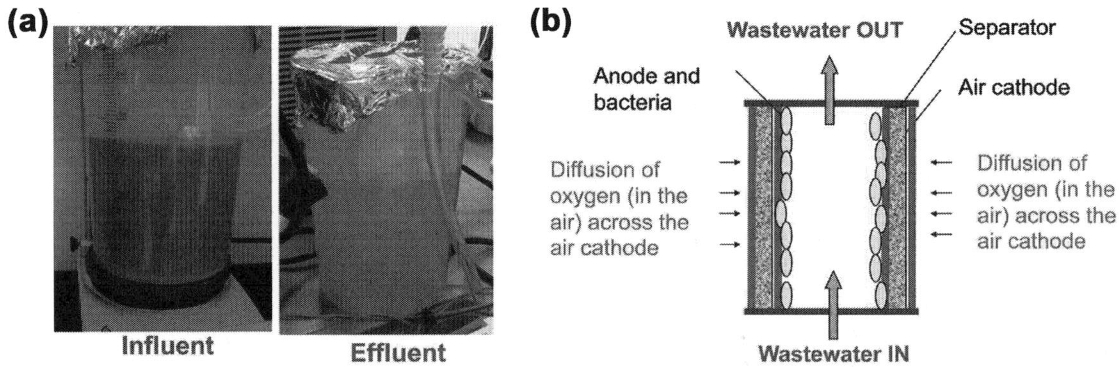

FIGURE 16–10a&b Schematic of a Flow-through MFC Module (480 mL).[3]

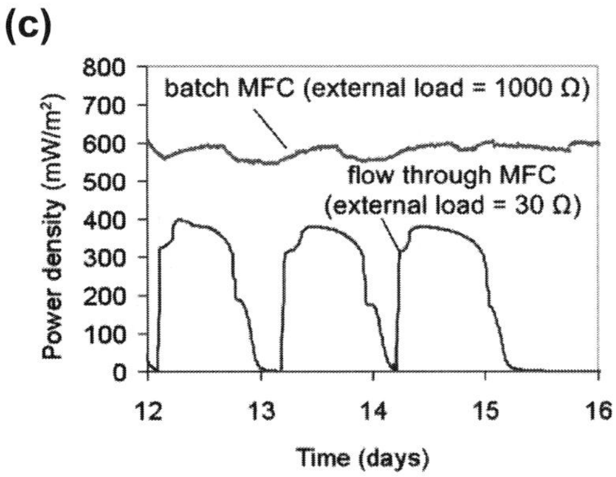

(d)

	Batch MFC	Flow-through MFC
Feed	Acetate (1 g/L)	Acetate (1g/L)
External load during operation (Ω)	1000	30
Maximum potential output (V)	~ 540 at 1000 Ω	~ 290 at 30 Ω
Maximum power density (mW/m²)	761	947
Average COD removal	3.8 % per hour	6.3 % per hour
Average Coulombic efficiency (%)	25	59

FIGURE 16–10c&d Batch versus Flow-through MFC.[3]

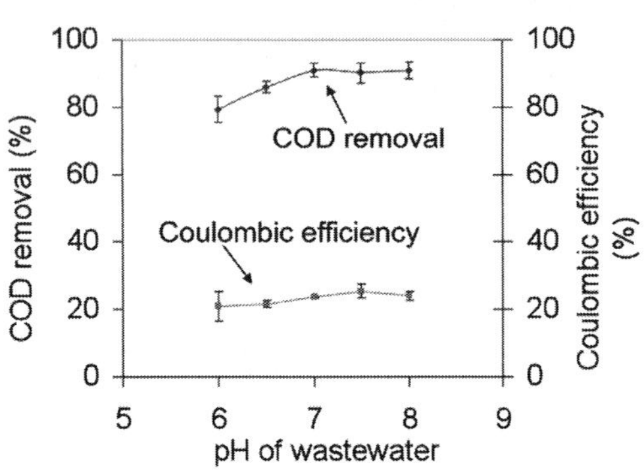

FIGURE 16–11 Effect of wastewater pH.[3]

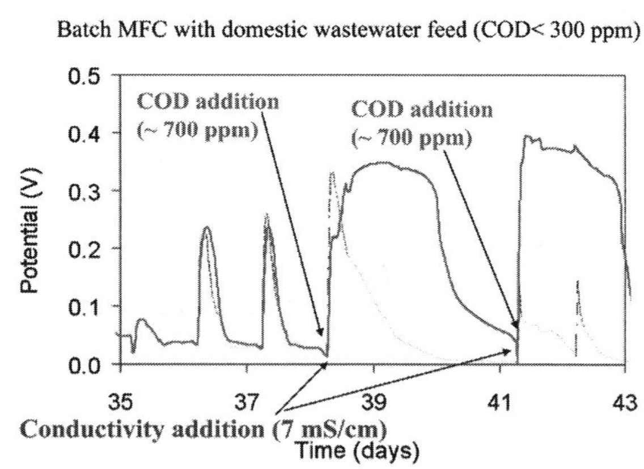

FIGURE 16–13 Effect of COD in wastewater.[3]

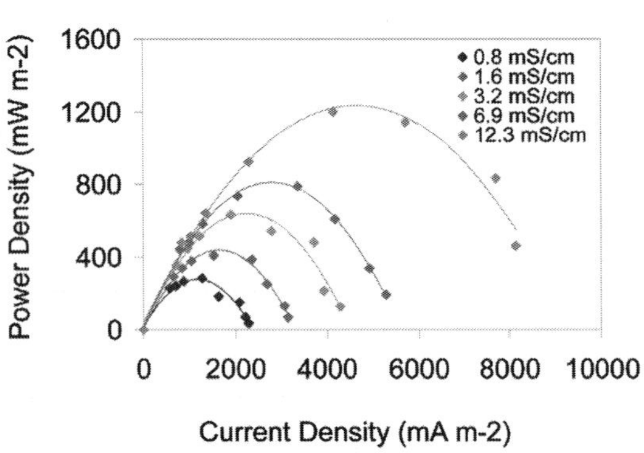

FIGURE 16–12 Effect of wastewater conductivity.[3]

	Synthetic Wastewater (Acetate)	Real Food & Beverage Wastewater
Reactor	Flow-through 480 mL	
Influent COD (ppm)	1300 – 1600	600 – 800
Maximum power density (mW/m²)	992 at 6 Ω	1042 at 6 Ω
Average COD removal	30% at 6 Ω	50% at 6 Ω
Average coulombic efficiency (%)	65 – 75% at 6 Ω	85% at 6 Ω

FIGURE 16–14 Synthetic versus real wastewater.[3]

Development of pilot scale MFC with real industrial wastewater is under way.

CASE STUDY 2: PEM FUEL CELLS (FCs) ON NAVAL SUBMARINES[4]

Figures illustrating voltage degradation, module output versus load current and efficiency for the two models Siemens makes (34 kW and 120 kW, respectively) and installs on submarines, follow (Figures 16–3, 16–4, and 16–5).

The following figures illustrate the peripheral systems (in Siemens application cases on naval submarines discussed earlier) with which the PEM FCs interface and a bank of FCs in a rack (Figures 16–15 and 16–16).

The data table (Table 16–2) on the two PEM fuel cells does not point out one important feature of the fuel cells in submarine application: they are noiseless.[5]

CASE STUDY 3: MICROTURBINE IN A CHP APPLICATION**

A CHP application example (Oregon) follows. In 2004, Siemens contracted to design, build, operate, and maintain a Combined Heat and Power Plant (CHP) to serve Oregon Health and Science University's (OHSU) new River Campus in Portland, Oregon.

The CHP provides OHSU with a clean, flexible, reliable and efficient source of heat and electric power. It features five natural gas-fired microturbines, which produce all of the heat required by the facility, as well as 34% of its electric power.

The CHP serves Building One, a 400,000 square-foot facility that houses medical offices, outpatient surgery, research laboratories, a wellness center, an imaging center, conference center, retail outlets, and parking. It is the first building being built in the university's expansion along the Willamette River in the new South Waterfront District.

According to the US Department of Energy, a CHP design will reduce the amount of fuel consumed by almost 40% when compared to a traditional fossil fuel-fired utility power plant and customer-owned boilers. In addition, OHSU estimates that CHP will reduce its CO_2 emissions by roughly 9 million pounds per year.

A chilled water production plant will serve the cooling needs of the building. Both the CHP and chilled-water plants will be integrated and controlled as a coordinated central utility plant.

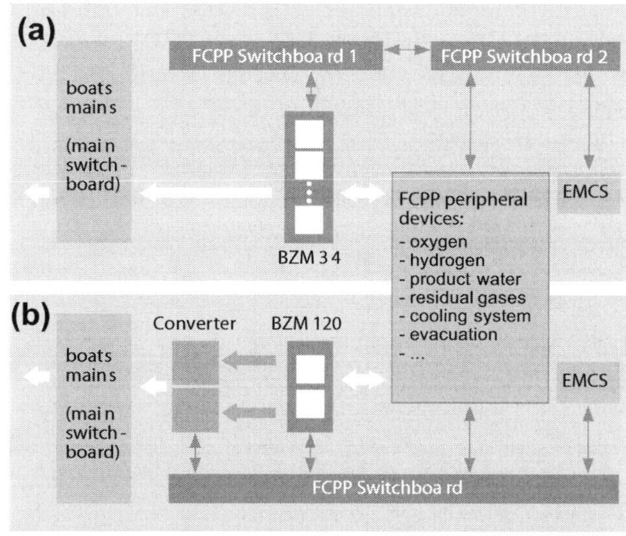

FIGURE 16–15 Two types of fuel cell power plants.[4]

FIGURE 16–16 PEM fuel cell modules assembled in a test rack.[4]

TABLE 16–2 Technical Data[4]

	BZM 34	BZM 120
Rated power	34 kW	120 kW
Voltage range	50–55 V	208–243 V
Efficiency at rated load, approx.	59%	58%
Efficiency at 20% load, approx.	69%	68%
Operating temperature	80°C	
H_2 pressure	2.3 bar abs.	
O_2 pressure	2.6 bar abs.	
Dimensions	H = 48cm	H = 50cm
	W = 48 cm	W = 53 cm
	L = 145 cm	L = 176 cm
Weight (without module electronics)	650 kg	900 kg

4. Courtesy of Siemens. Adapted with permission www.siemens/marine. (Accessed October 14, 2013).
5. For further details, see http://www.industry.siemens.com/verticals/global/en/marine/products-and-solutions/pgd/pem_full_cell/Pages/Default.aspx. (Accessed October 14, 2013).
** Source: [16-4] Siemens Power Generation.

OHSU and its design and construction team are utilizing the US Green Building Council's LEED (Leadership in Energy and Environmental Design) rating system to gauge the sustainability of the project. OHSU's goal is to achieve a gold LEED rating. Innovative features of the new OHSU facility, such as using rainwater to flush public fixtures and a internal bioremediator to treat waste, might just earn it LEED's highest platinum rating.

Siemens is helping OHSU to secure financial incentives for the CHP. State tax credits from the Oregon Department of Energy, combined with financial incentives from an Energy Trust of Oregon pilot project, contribute to making the project an effective choice for OHSU.

CASE STUDY 4: A FUEL CELL APPLICATION*

The target this DOE-sponsored partnership's effort (Rolls Royce Allison and stack manufacturer) was aimed at producing between 3 and 30 MW units at under $1500/kW uninstalled. Interim progress is discussed. Work in this field continues.

Rolls Royce is working with a molten carbonate fuel cell manufacturer to accelerate commercialization of a pressurized power generation package. The intent of the participants is to introduce stack and reformer improvements coupled with a new system integration approach to achieve a product with commercially competitive qualities. Cost-effectiveness and durability must be brought to fuel cell systems that already achieve high efficiency and low emissions.

A key element of the program will be endurance testing in base load mode for extended periods. Stack performance will be tracked in order to assess the maximum useful working life of this high investment component.

Load following generally requires careful management in fuel cell systems, typically because the stack/reformer system should be kept in a near-steady state environment. This will be demonstrated.

Existing systems have usually been designed somewhat similarly to a chemical process plant, using off-the-shelf components. This approach leads to a highly predictable but excessively complex and costly system. System simplification and cost reduction efforts have been initiated, with a goal of a 3:1 reduction in the cost of the balance-of-plant.

Preliminary results of this work are not available due to a delay in the start of the program. The cost reduction goal is to bring the plant uninstalled cost to less than $1500/kW. The projected costs of early commercial systems as currently estimated are almost twice this goal.

* Source: [16-5] Courtesy of Rolls Royce. Extracts from S. A. Ali and R. Moritz. "A Prototype for the First Commercial Pressurized Fuel Cell System," in *Proceedings of the ASME Turbo Expo 2000,* May 8-11, 2000, Munich, 2000-GT-551.

Project Description

The baseline system to be tested in the evaluation program for performance and durability consists of three distinct modules: the power module, the mechanical module, and the electrical module. This is the configuration selected for most exploratory fuel cell plants. The primary purpose of this first system is to demonstrate stack life, performance, and operational characteristics of the new stack and reformer. The prototype system test bed configuration represents a more flexible but more expensive system than that required for a commercial product.

A simplified schematic of the pressurized molten carbonate fuel cell (MCFC) test system is presented in Figure 16—17. The power module contains the fuel cell stack assembly, reformer, manifolds, and the anode (fuel side) recirculation ejector. The mechanical module includes all other major air and fuel gas management equipment including the turbo-machinery, gas desulfurization system, cathode recycle blower, valves, and interconnect piping. The third module, the electrical module, provides power conditioning, electronics, and system controls.

High Temperature Fuel Cell Plants

The three fuel cell systems studied by Rolls Royce Allison under the DOE High Efficiency Fossil Power Plants (HEFPP) conceptualization program have the following major components in common:

- Fuel cell stack
- Turbogenerator
- Power conditioning
- Electronic controls
- Fuel desulfurization

In pressurized systems, the fuel cells act in place of a combustor for the turbogenerator. In the unpressurized system, the fuel cell is placed in the turbogenerator exhaust stream but also supplies heat and residual fuel to a heat exchanger acting as an indirect combustor.

Additional major components for individual systems are:

- 1100°F heat exchanger (pressurized SOFC)
- 1600°F heat exchanger (1 atmosphere MCFC)
- 1500°F exchanger/reformer (pressurized MCFC)

Though each system has distinctive merits and challenges, a common problem is that each module is of high cost.

The power module uses raw materials that are expensive and heavy, whether they are ceramic or nickel based. At present, each kilowatt requires 10—20 lbs. of active fuel cell and 10—20 lbs. of associated structure, or more than 15 tons of stack per megawatt. This ratio is considerably higher than for a simple cycle gas. The fuel cell active surfaces can provide a range of current density in which the

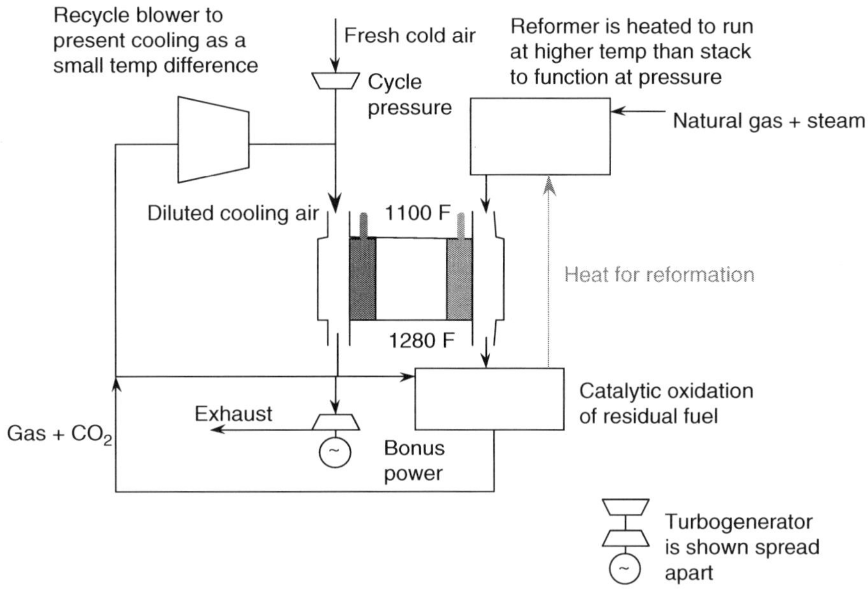

FIGURE 16–17 Simplified pressurized MCFC [16-5].

delivered voltage (efficiency) drops as current rises but there is little scope for increasing the current per unit area at the high efficiency. It follows that the required route is to increase the active surface area in a given stack volume, a requirement very similar to that of producing compact high effectiveness heat exchangers. Figure 16–18 gives an indication of the comparative compactness of available fuel cells and heat exchangers.

Stack pressurization is required in order to provide enough fuel and oxidizer gas to these compact active surfaces with low parasitic pumping loss. This is the primary reason for the belief that pressurized systems are the long-term valid solution. Though more heat is released per unit volume of stack, it is carried away by proportionally more gas, thus avoiding excessive temperature. Pressurization also offers important secondary gains, both increasing the power that may be drawn at a given level of stack efficiency and making it easier to convert cycle heat into electricity in associated turbomachinery (accounting for 10–20% of total system power), without using more fuel.

A preliminary comparison of performance characteristics and rough order of magnitude cost of solid oxide and molten carbonate systems was conducted. The results indicate that near term (before 2010), the two offer a similar

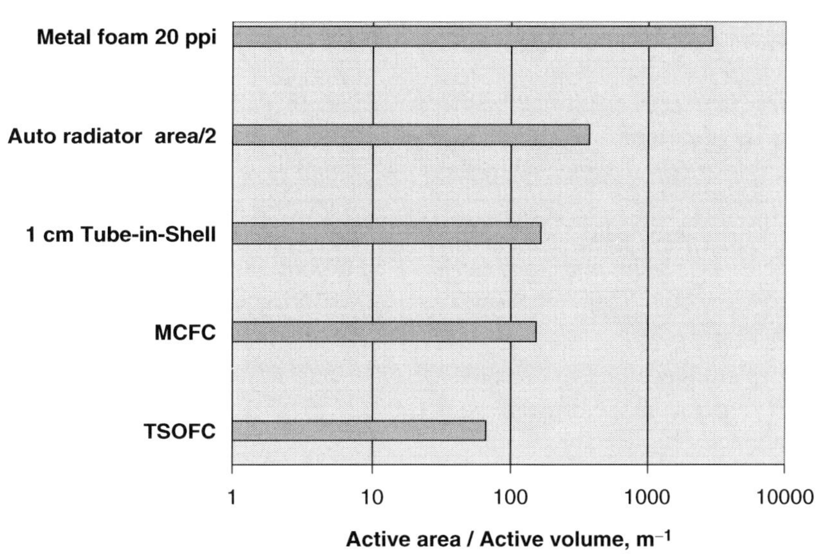

FIGURE 16–18 Fuel cell vs. heat exchangers comparison [16-5].

level of performance and technical challenges. The somewhat simpler system advantage of the solid oxide is offset by its often higher operating temperature. The lower cost of the balance-of-plant system for the SOFC tends to be offset by higher stack cost.

The MCFC system's pressurized reformer has to be supplied with an excess of steam in order to suppress carbon formation. This phenomenon poses a significant complication, and requires the continuous addition of treated water to the cycle.

The higher operating temperature capability and the flexibility provided by a planar solid oxide stack arrangement indicates that the solid oxide system will dominate in the longer term. Over the long term, the SOFC stack offers the following potential advantages:

- More durable chemistry
- Compatibility of internal reforming with potentially high pressures at the higher temperature
- Eliminating the need for excess steam

An opportunity has been made available to partner with M-C Power in the pressurized MCFC programs. This opportunity represents the earliest possibility to launch a fuel cell hybrid product. Pressurized MCFC and SOFC systems will enable moving towards longer-term aspirations of launching higher power density products. The intent is to introduce into operation the first in-service pressurized MCFC system. The plan is to participate in designing and developing the prototype systems, currently funded by the DOE MCFC Product Design and Improvement (PDI) program.

Pressurized MCFC System Design

M-C Power gained considerable insight into the design and functioning of pressurized MCFC systems. This valuable experience was gained by running the plant installed at the Miramar Naval Air Station (test units ranging in size from 75–250 kW). This system is not a full hybrid. Although turbomachinery is used to turbocharge the system to about 3 atmospheres, no attempt is made to generate bonus electrical power with it.

A plant design for the PDI program was developed by the M-C Power partnership based on the operating experience gained at the Miramar station. The PDI program incorporates a number of improvements:

- Power increased to 450 kW (improved stack mechanical and thermal design, and materials modifications)
- Temperature stable reformer/catalytic combustor
- Pressurization increased
- Turbogenerator used for about 10% power bonus
- Once-through steam generator (unattended)
- Reduced footprint

The proposed PDI system analysis made it apparent that the cost of the system was very high as designed. The first reaction is, of course, that the stack and reformer are determined to be expensive because they are complex parts produced on a "once-off handmade" basis. This was expected. There is a coherent plan to decrease their cost by an order of magnitude. The electrical module cost was based on off-shelf power conditioning and power electronics components. The cost of this module represented a minor portion of the overall system cost. This is partly because good electronic control systems are very competitively priced and inverters are continuing to decrease in cost.

What came as a surprise, however, was that the collection of pipes and valves in the mechanical module, specified to service the system, is the major cost challenge. Unlike the stack and reformer, the mechanical module is an assembly of standard, volume produced, components that are not likely to reduce cost. The PDI plant is configured in a fairly straightforward "chemical plant" style, with pipe work and valves linking and controlling the major processes. The mechanical module assembly, manufactured as prototypes with soft tools and jigs, requires a large number of pipes and valves. The components of the mechanical module alone cost more than the intended early production cost of the whole plant. This high system cost problem is being attacked to resolve it by a different approach. The plan is to modify the overall system operational control, and greatly simplify system integration. The following integration changes have been proposed:

- Put all critical pipe connections inside the pressure vessel that houses the stack and reformer.
- Air delivered from the intake compressor is discharged into the pressure vessel.
- Turbine is driven by the exhaust gases as they emanate and then pass to the external exhaust heat recovery unit.
- All other flow manipulation is accomplished inside the pressure vessel, and can therefore be handled in lightweight ducts (with small pressure differences across the ducts).
- Small leaks are acceptable and this opens the way to using slip joints if necessary.
- Eliminate motorized air/gas valves. Select and use the turbomachinery and its electrical loading to master flow control, probably aided by a waste gate system.
- Provide an integrated turbomachinery package. Replace the present large low speed industrial hot recycle blower and its variable frequency drive and drive motor by a small high speed fan. The fan is driven by a free power turbine using the same hot gas stream as the turbogenerator turbine.

The resulting system package, which is closely integrated, is sketched in Figure 16–19.

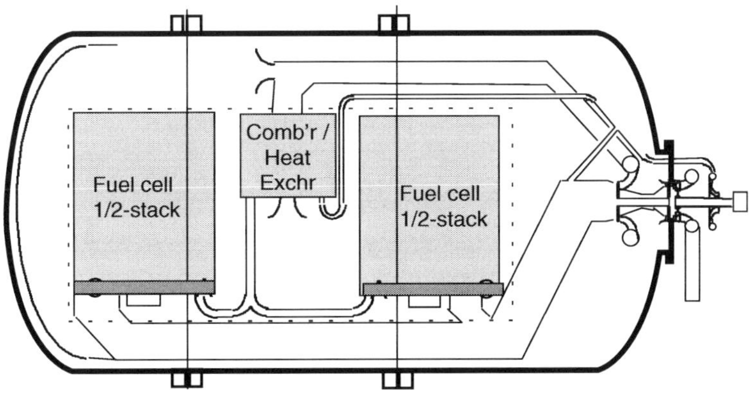

FIGURE 16–19 Simplified MCFC system [16-5].

Revised system cost estimates indicate that these changes decrease the total initial production plant cost (uninstalled) by 30% and the long-term plant cost by 45%.

Wide Application Fuel Cell Turbomachinery*

The DOE is urging the development of high efficiency hybrid fuel cell/gas turbine systems generating electric power in the range of 20–30 MW. The long-range plan is to enable these systems to serve the 21st century power generation needs in the US and globally. Fuel cell manufacturers, working on the High Efficiency Fossil Power Plant (HEFPP) program, examined a variety of fuel cell cycle arrangements reaching the level of efficiency specified. From this work, it is possible to draw a number of conclusions:

1. The attainment of 70% efficiency is challenging yet attainable, requiring improvement in stack efficiency, fuel utilization, and supporting system efficiency.
2. Turbomachinery is mostly configured as though for a recuperated cycle (compressor discharge flow supplied to stack and returned to turbine) with generator.
3. The required pressure ratio may be less than that available in high-pressure ratio gas turbines.
4. Cycles may require intercooled compressors.
5. Supporting machinery flow size is relatively low, usually less than 50 lbs./sec of air being required for a 20 MW hybrid system.
6. Each stack module power output is much less than 20 MW, so plant may divide the total flow between multiple small turbines, perhaps 5 lbs./sec each, depending on modular concept selected. Suitable gas turbines are not available at this size.
7. Approximately 40,000 hours of base load operation poses a major challenge to small turbomachinery. To achieve this life, special bearings may be needed.
8. Turbine entry conditions are mostly limited by stack capability below level demanding blade cooling.

* [16-2]

What emerges from these preliminary system definitions above is the need for a turbogenerator with a minimum specification quite close to that of a "fuel cell flexible" (FCF) turbine. These turbines will be designed to a pressure ratio of 4–8, turbine temperature of 1700°F, recuperated and fitted with a generator. The flow size of emerging microturbines is limited to about 2 lbs./sec, but larger versions may follow. Using FCF turbines offers a big cost benefit because they are to be mass produced.

Currently available microturbines fall short of the pressure ratio and mass flows indicated for maximum efficiency pressurized fuel cell systems. An optimum pressure ratio of 7 appears to be suitable for both SOFC and MCFC systems. This is a coincidental outcome, because the cycles are different due to diverse characteristics of active components of the molten carbonate and solid oxide fuel cells.

Therefore, there is the possibility that a dedicated style of gas turbine should be developed, which could serve several types of fuel cell systems. Its specification is likely to be:

Pressure ratio	7
Turbine entry temp	1700°F max
Flow size	5–20 lbs./s
Stall margin	Generous
Combustor	Start only
Compressor	One-stage radial
Intercooler	None
Ducting	As if recuperated
Turbine	Two-stage axial
Bearings	Non-contact
Adaptability	Scalable

This unit has the potential to serve as a high efficiency small turbogenerator if recuperated. By combining fuel cell and stand-alone turbogenerator market opportunities, it

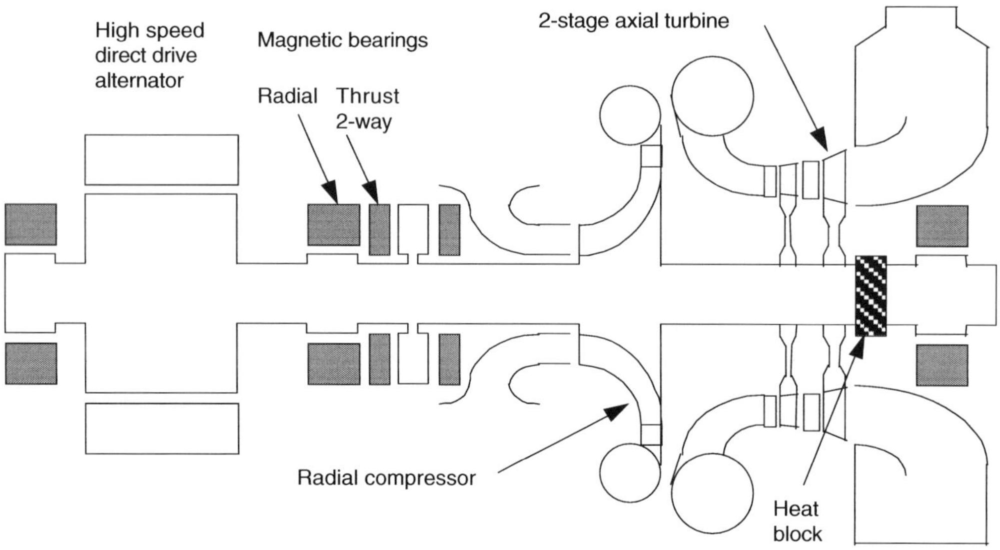

FIGURE 16–20 Fuel cell flexible turbogenerator integration for low cost [16-2].

may be possible to provide a generous return on the initial investment, and minimize the fuel cell system cost.

A schematic indicating the configuration of this gas turbine is presented in Figure 16–20. The design proposed in this figure represents an approach of a fuel cell flexible or fuel cell neutral turbomachinery.

The design approach represented for hybrid fuel cell systems is usually analogous to gasification systems and pressurized fluidized bed combustion systems. These approaches have tended to lead to expensive system configurations.

At the bench-top level of fuel cell experiments, it was naturally convenient to use readily available components to build the pressurized systems. For pressurization, the most convenient source of mildly compressed air is a radial compressor driven by a synchronous electric motor. This type of pressurization requires no direct control of the overall system because the user can add motorized valves to select pressures and flows as required.

Incorporating a fuel cell flexible gas turbine that is specifically designed to serve the needs of pressurized fuel cell systems would enable achieving these key benefits:

- A fuel cell flexible gas turbine would help simplify operational control by minimizing high temperature ducting and a vast number of valves.
- A fuel cell flexible gas turbine would also enhance operational stability of a fuel cell hybrid by controlling the generator loading, therefore spool speed in a single shaft machine.
- It would also help improve the system load following capability.
- The turbine would help minimize the fuel cell stack pressure and temperature fluctuations, by maintaining mass flow uniformity, resulting in uniform pressure and temperature distributions within the stack.
- The operational stability of the fuel cell stack (facilitated by the FCF gas turbine) is necessary to achieve hybrid system life in the range of 40,000 hours under steady state base load operating conditions.
- A gas turbine designed specifically to be compatible with fuel cell operating requirements would facilitate overall design simplification.
- With system design simplification, both in the mechanical module and the power module, it would be easier to meet the system cost reduction goal specified earlier.
- A FCF gas turbine would serve the needs of the various high temperature pressurized hybrid systems. Meeting this goal is important in order to reduce the cost of turbomachinery, since these are likely to be produced initially in comparatively small quantities.
- A FCF turbine would impact favorably the cost of the mechanical module.
- Another important potential benefit of having a gas turbine designed specifically for fuel cells is the ability to ensure power quality of the system due to the fact that the entire system would have a tendency to operate with great stability. An operationally stable fuel cell hybrid system would allow the meeting high power quality requirements on a sustained basis.
- A pressurized fuel cell hybrid system utilizing a performance optimized compatible gas turbine would derive other benefits such as the system's ability to simplify the grid interconnect.
- The system would be able to access grid interconnect directly to an AC power line. The Institute of Electronics and Electrical Engineers (IEEE) is developing

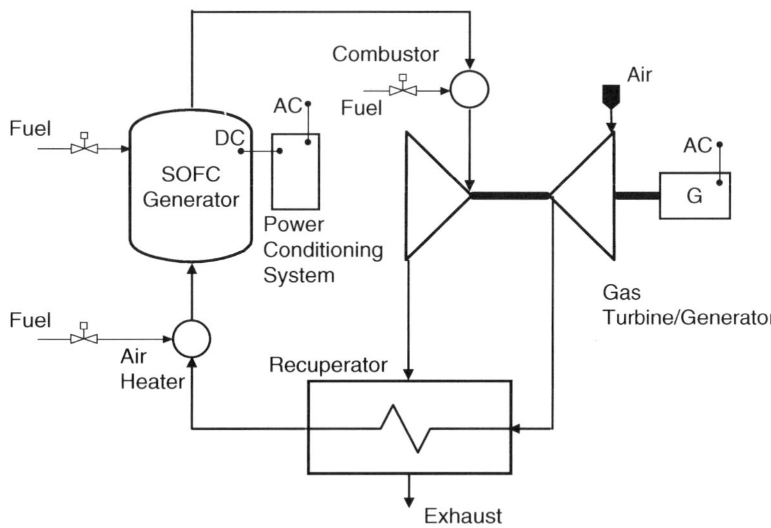

FIGURE 16–21 PSOFC/GT hybrid power system cycle—simplified schematic [16-6].

an interconnect standards under the 519 code. It would be possible to couple the system to the grid through a properly designed gas turbine and a correspondingly compatible electrical module. Such an approach would eliminate use of a DC load bank and a more complicated grid interconnect.

- The system incorporating several of the cost reduction techniques would also have the capability to serve in the combined heat and power (CHP) mode.
- It is necessary for these distributed resources systems to offer not only the electrical output, but also heat and cogeneration capability.

The ultimate objective of the cost reduction and design simplification program is to reduce stack cost by at least 40%, and overall system cost by approximately 45%. These numbers represent the type of cost reduction from current levels required to achieve system costs of <$1500/kW (uninstalled). Preliminary economics evaluations indicate that at the target costs, fuel cell hybrid systems will be competitive on the basis of cost of electricity with competing energy sources.

CASE STUDY 5: TUBULAR SOLID OXIDE FUEL CELL/GAS TURBINE HYBRID CYCLE POWER SYSTEMS*

This case involved a partnership between Siemens Westinghouse and SOFC Power Generation. Funding was provided by the US DOE, Ontario Hydro Technologies and their Canadian funding partners, and the Institute for Advanced Energy of Japan.

A pressurized SOFC/GT hybrid cycle power system is depicted schematically simplified in Figure 16–21.

The SOFC generator is pressurized, operating on recuperatively heated process air supplied by the compressor. The power system based upon this cycle achieves increased power output and higher efficiencies due to the utilization by the gas turbine generator of thermal energy in the pressurized SOFC exhaust stream. Power system performance is also enhanced by SOFC generator operation at elevated pressure. For a given cell operating current, cell voltage, cell power output, and efficiency increase logarithmically with pressure. The gas turbine combustor and the air heater are fired during system startup operations, and they could be fired during peak-power or turndown periods. For maximum system efficiency during steady-state operation at system rating, neither combustor is fired.

A recent study concluded that high cycle efficiencies are achieved by directly expanding SOFC exhaust gas at the gas turbine without firing supplemental fuel at the GT combustor, highlighting the synergy of SOFC and gas turbine integration. The projected performance of small-capacity SOFC/GT hybrid power systems based on specific small gas turbines is discussed. In addition, performance estimates for the first constructed SOFC/GT hybrid power system are provided.

Siemens Westinghouse Program Status

Focus has been on the operation of a 100 kWe atmospheric-pressure SOFC-CHP demonstration power system in the Netherlands, and on the design and fabrication of a

* Source: [16-6] Courtesy of Siemens. Extracts from S. E. Veyo et al. "Tubular Solid Oxide Fuel Cell/Gas Turbine Hybrid Cycle Power Systems—Status," in *Proceedings of the ASME Turbo Expo 2000*, May 8–11, 2000, Munich, 2000-GT-550.

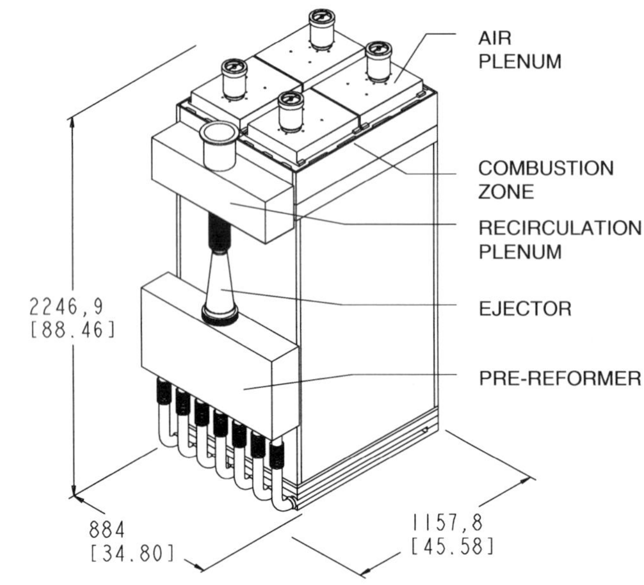

FIGURE 16–22 Basic building block (576-cell substack) [16-6].

220 kWe PSOFC/GT hybrid cycle power system. The 100 kWe system, sponsored by EDB/ELSAM, a consortium of Dutch and Danish energy generating and distribution companies, is installed on a utility site near Arnhem, and has logged over 6000 operating hours. It generates approximately 110 kWe net AC power at 46% efficiency (net AC/LHV) for the utility grid, and hot water for the local district heating system. The demonstrated energy efficiency is nearly 75%.

The SOFC generator for this system uses a stack of 1152 tubular cells, each cell having a diameter of 22 mm and an active length of 1500 mm. The cell stack can be viewed as composed of two 576-cell substacks, each having its own ejector, ducting, and reformer assembly for the recirculation of moisture-containing depleted fuel and the reformation of incoming fresh fuel, and a zone above the cells for the combustion of electrochemically-unreacted fuel. A substack, pictured in Figure 16–22, is the building block for the SOFC generators to be discussed later.

The 220 kWe PSOFC/GT hybrid cycle power system, the world's first such hybrid and is sponsored by Southern California Edison (SCE), was designed and built by Siemens Westinghouse, and is scheduled for installation and operation at the University of California, Irvine, in early 2000. It is a proof-of-concept system, and a major objective of the program is to demonstrate the integration of the SOFC and gas turbine technologies, and hybrid system operation. The system is based on the cycle depicted in Figure 16–23.

The gas turbine, a recuperated, two-shaft machine, was supplied by Northern Research and Engineering Company (NREC), a division of Ingersoll-Rand. The engine is a modified version of the 75 kW gas turbine that NREC is developing and commercializing for power generation, mechanical drive, and CHP applications. Figure 16–24 provides a pictorial view of the skid-mounted hybrid cycle power system. Approximate overall dimensions for the system are 7.4 m (l), 2.8 m (d), and 3.9 m (h). The SOFC generator consists of a cell stack that is housed in the vertical cylindrical pressure vessel. The stack in this particular power system is composed of two 576-cell substacks, and is of the design used in the 100 kWe atmospheric-pressure SOFC-CHP system. No significant modifications were needed to adapt the stack design for pressurized operation. The gas turbine receives hot SOFC exhaust at the gas turbine combustor inlet, and it sends recuperatively-heated process air to the SOFC generator inlet. The hot gas is expanded partially across the gasifier turbine, which rotates at approximately 70,000 rpm, and drives the compressor. The expansion of the hot gas is completed across the power turbine, where the derived shaft power turns the generator. A typical power turbine shaft speed is 44,000 rpm. Initially, the electric power produced by both the gas turbine and the SOFC generator will be dissipated.

As indicated in Figure 16–23, the SOFC generator is valved to permit the flow of air around the cell stack. The gas turbine can therefore be started before the SOFC generator, by firing the gas turbine combustor, and air can be directed around and through the SOFC generator in desired proportions.

With the gas turbine started, exhaust heat recovered at the recuperator is available to aid the SOFC heat-up process. The air heater is fired during this process, but the flow of fuel to the heater and to the GT combustor are turned back to zero as steady-state operation at the peak-efficiency rating point is achieved.

Power system performance estimates are presented in Table 16–3. For these estimates the efficiency is calculated assuming all power is exported to the AC grid, and reasonable power conditioning and turbine generator efficiencies are applied.

Commercial Product Design Studies

Based upon market study, the conclusion is that PSOFC/GT hybrid cycle power systems will have competitive advantage, due to their high efficiency and low installed cost, in the distributed generation market in the 200 kWe to 10 MWe capacity range. To begin defining potential products in this range, conceptual design studies have been undertaken to project system configuration and appearance, and to estimate performance and cost. Aspects of two studies

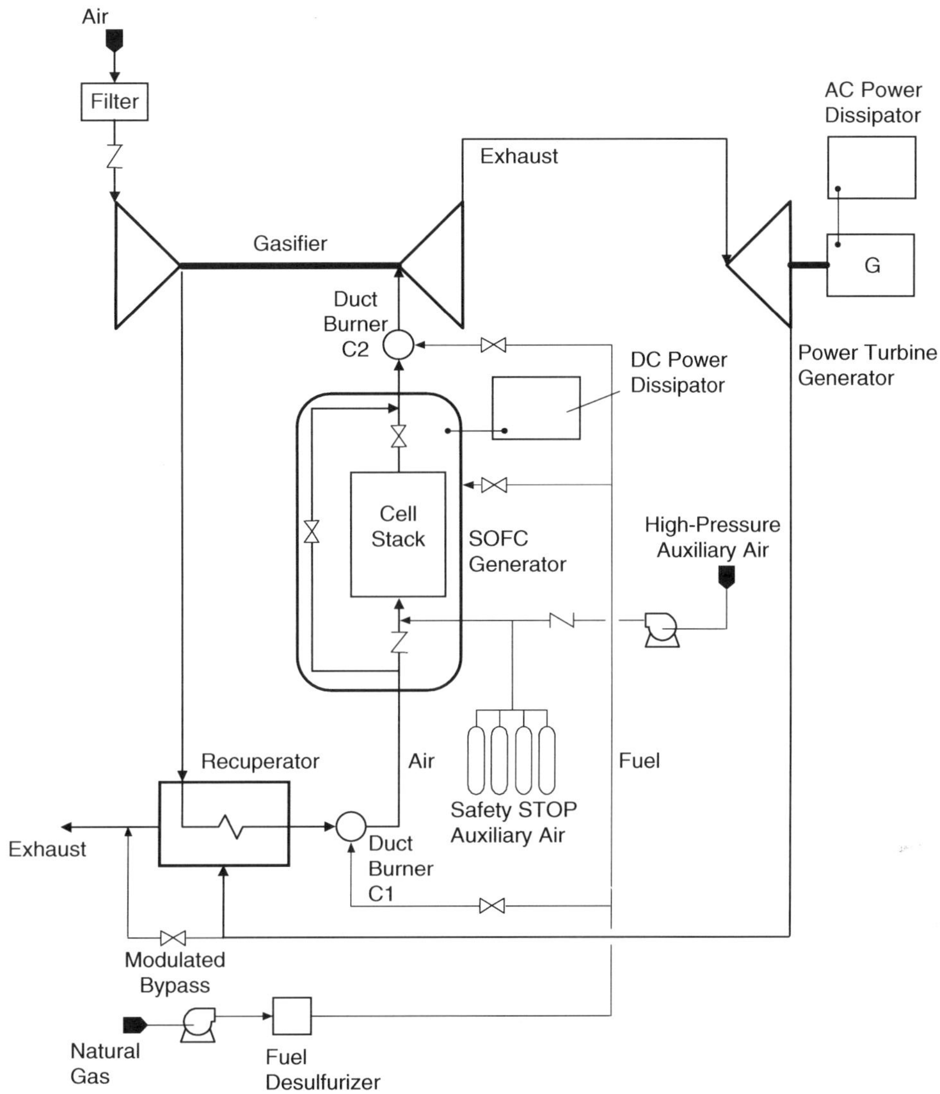

FIGURE 16–23 220 kWe PSOFC/GT power system cycle [16-6].

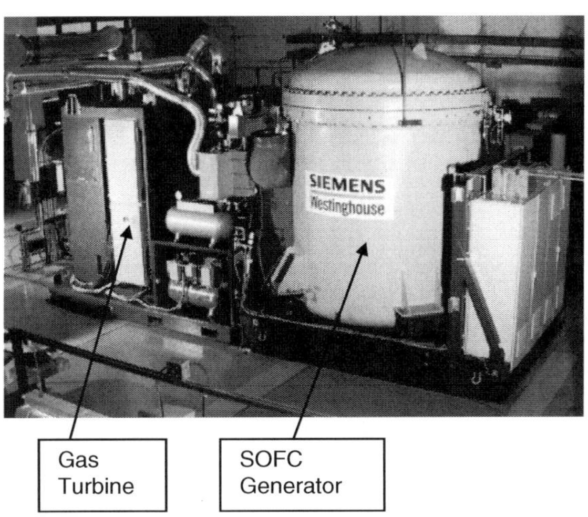

FIGURE 16–24 220 kWe PSOFC/GT power system [16-6].

TABLE 16–3 220 kWe PSOFC/GT Power System Performance Estimates [16-6]

Cell current	250 amps
Cell voltage	0.610 V
Compressor pressure ratio	2.9:1
Air intake rate	0.58 kg/s
Turbine inlet temperature	780°C
SOFC DC power	187 kW
SOFC gross AC power	176 kW
Gas turbine AC power	47 kW
System net AC power	217 kW
Efficiency (net AC/LHV)	57%

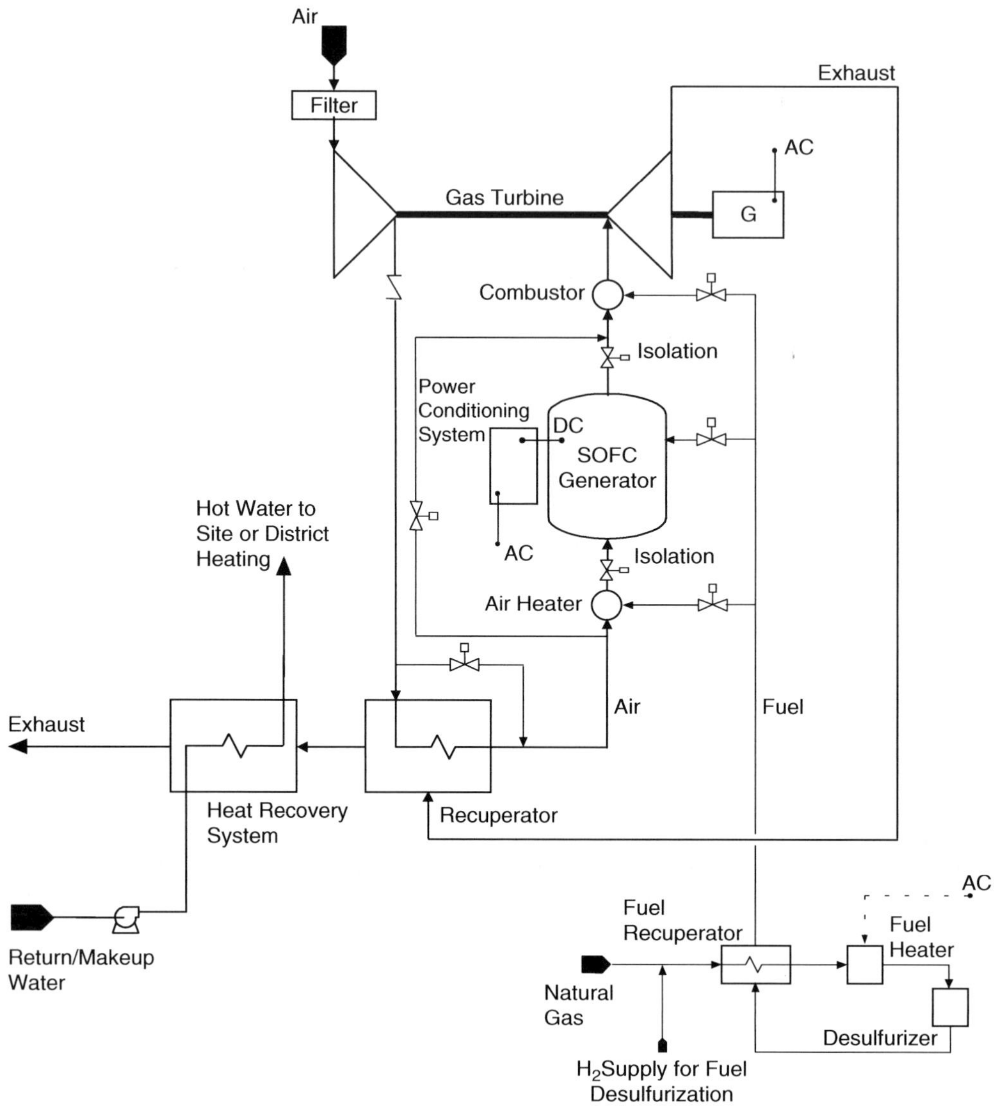

FIGURE 16–25 320 kWe PSOFC/GT power system cycle [16-6].

are discussed. In one, the Allied-Signal Parallon 75 turbogenerator was specified, and the SOFC generator was sized to match the turbine. The result is a power system concept that would be rated at approximately 300 kWe, with an efficiency of 57%.

The objective of the second study was to conceptualize a hybrid power system with rating of at least 1 MWe. Gas turbines under development by Solar Turbines and NREC could be deployed in the system concept that resulted, and a net AC efficiency of near 60% is projected. The performance estimates presented herein for both power systems are pre-commercial values, which are expected to be achieved in demonstration systems. Higher power outputs and efficiencies are projected for the mature-product SOFC technology, which will be available by 2010 or sooner.

300 kWe-Class Hybrid Power System

The cycle for this system is depicted in Figure 16–25. It is shown in CHP mode, with a heater for the preparation of hot water for site or district using heat recovered from the system exhaust. The Allied-Signal turbogenerator is a single-shaft recuperated gas turbine that powers a high-speed, direct-drive alternator. An important function of the turbine is to supply a steady flow of air to the SOFC generator. The ability to modulate the GT electric load to maintain shaft speed at set point, and the ability of the turbine to operate across a range of turbine inlet temperatures, enables steady air flow over a range of flows, an important consideration for SOFC thermal management.

The SOFC generator was sized to match the turbine flow characteristics, assuming no firing of the gas turbine

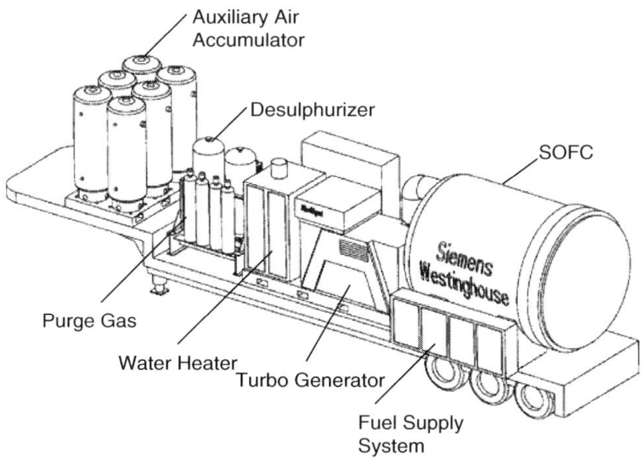

FIGURE 16–26 320 kWe PSOFC/GT power system [16-6].

combustor. The resulting cell stack, consisting of three 576-cell substacks, is best arranged in a horizontal monolithic configuration, which in turn is well suited for installation in a horizontal cylindrical pressure vessel. A system pictorial view is presented in Figure 16–26. It is envisioned that the power system would be factory-assembled on a single truck transporter. The skid would then be jacked into place at the installation site, or the transporter could be designed with a removable wheel and axle assembly, and the bed containing the system would simply be left at the site. System performance estimates are provided in Table 16–4.

TABLE 16–4 300 kWe-Class PSOFC/GT-CHP System Performance Estimates (Pre-Commercial) [16-6]

Cell current	250 amps
Cell voltage	0.635 V
Compressor pressure ratio	3.5:1
Air intake rate	0.64 kg/s
Turbine inlet temperature	1145 K
Combustor and air heater fuel flow rate	0
SOFC DC power	270 kW
SOFC gross AC power	251 kW
Gas turbine AC power	67 kW
System net AC power	307 kW
Efficiency (net AC/LHV)	57%
Gas temperature at water heater inlet	477 K
Hot-water heat recovery rate	95 kWt
Hot-water temperature	393 K
Hot-water flow rate	0.3 kg/s
Heat recovery efficiency (kWt/LHV)	18%
Energy efficiency [(net AC + kWt)/LHV]	75%
Exhaust flue temperature	330 K

MWe-Class Hybrid Power System

The simplified cycle diagram for a MWe-class PSOFC/GT power system is presented in Figure 16–27. It is similar to that for the 300 kWe system, with the exception that the MWe-class studies did not consider hot water heat recovery. As in the 300 kWe-class system, the gas turbine is a recuperated, single-shaft machine, with the alternator load modulated to control shaft speed and air flow to the SOFC generator.

The SOFC generator for this system uses 10 576-cell substacks in a horizontal pressure vessel. The power system is configured as two factory-assembled skids; the SOFC and gas turbine equipment is installed on one skid, and balance-of-plant hardware is installed on the second skid. Each skid will be transported to the installation site in a single truck shipment. At the site, trailer wheel assemblies are removed, and the transporter skids are lowered into place on prepared pads. Other installation activity at the site will involve, in addition to the site preparation work, the interconnection of the skids, and the completion of the system/site interfaces. Performance estimates for the MWe-class power system are presented in Table 16–5. For these particular estimates the gas turbine was modeled using information on a developmental 300 kW Solar Turbines engine.

CASE STUDY 6: A TURBOGENERATOR FOR A FUEL CELL/GAS TURBINE HYBRID POWER PLANT*

This work was also funded by the US DOE. When Rolls Royce merged with Allison, its gas turbine range then included very small gas turbines that powered helicopters. Fuel cell manufacturers have been concerned that the small gas turbines that are available are not reliable enough to power commercially viable hybrid systems. So DOE is sponsoring the development of turbogenerators for use with hybrids.

Special Purpose Gas Turbine for Fuel Cells

Stack designs have to simultaneously achieve good fuel and air distribution, uniform heat dispersal, minimized

* Source: [16-7] Courtesy of Rolls Royce. Extracts from S. S. Ali and R. R. Moritz, "A Turbogenerator for Fuel Cell/Gas Turbine Hybrid Power Plant," in *Proceedings of the ASME Turbo Expo 2001*, June 4–7, 2001, New Orleans, 2001-GT-0524.

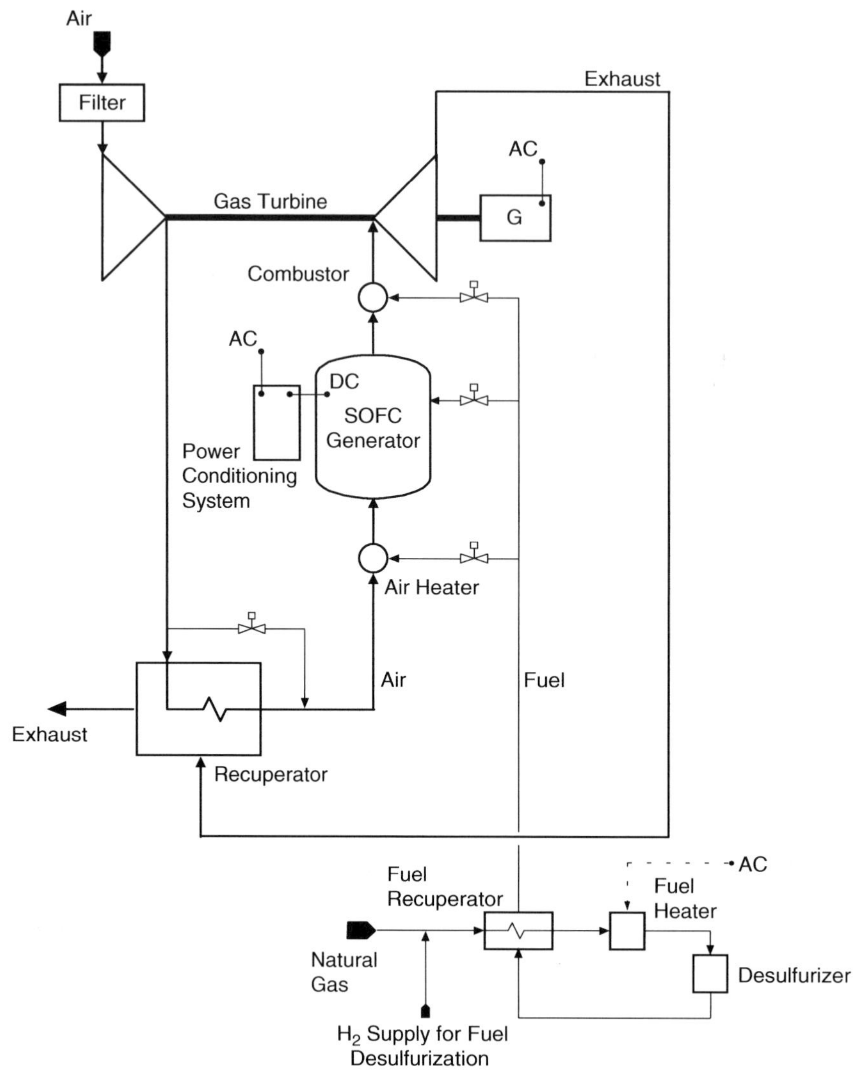

FIGURE 16–27 1220 kWe PSOFC/GT power system cycle [16-6].

thermal gradients in stacks, comply with mechanical and thermal stress limitations, and provide consistent cell to cell electrical continuity. In several cases, these features are achieved in two adjacent types of zone, the reforming zone and the active cell zone. The complexity of meeting these requirements increases with stack power level. Currently, the three major "full-size" technologies, solid oxide, molten carbonate, and direct (molten carbonate), have all been limited to a maximum single stack assembly power level of about 0.3 MW. An installation of such low power level would inevitably result in very high plant capital cost per unit power. Manufacturers, realizing a limited market potential for such units, have proceeded to build multi-stack plants. A 1 MW hybrid is often seen as the smallest plant size likely to have substantial cost competitive, environmental and commercial appeal.

It is prudent to ship fuel cell systems largely assembled and this implies limiting module dimensions to those that can be trucked economically, i.e., as indicated for 1–2 MW hybrids. Design proposals by three manufacturers for very high efficiency 20 MW plant, arising from feasibility studies funded by the US Department of Energy, concluded that these systems would be based on multiples of the same stack modules. Large plant systems were evaluated under a DOE High Efficiency Fuel Cell Power Plant feasibility study program in 1999–2000. This program showed that fuel cell plant efficiency can be raised from about 55% to >70%.

A relatively simple, low-cost turbogenerator will be the best option for fuel cell plant of up to approximately 10 MW hybrid capacity, supporting efficiencies >65%. The same gas turbine can serve as the core of a larger capacity turbogenerator for up to a 30 MW fuel cell plant.

TABLE 16–5 MWe-Class PSOFC/GT Power System Performance Estimates (Pre-Commercial) [16-6]

Cell current	239 amps
Cell voltage	0.643 V
Compressor pressure ratio	3.3:1
Air intake rate	2.0 kg/s
Turbine inlet temperature	1144 K
Combustor and air heater fuel flow rate	0
SOFC DC power	879 kWe
SOFC gross AC power	818 kWe
Gas turbine AC power	218 kWe
System net AC power	1014 kWe
Efficiency (net AC/LHV)	59%
Exhaust flue temperature	515 K

The turbogenerator to be designed on this project will be matched initially to serve the needs of a 1–3 MW hybrid plant. The unit will combine the following features:

- Oil-free
- Active magnetic bearings
- Direct drive alternator
- Low parts count
- High efficiency
- Zero emissions under steady state
- Very high reliability and durability

A stand-alone gas turbine will also result as an important by-product of this design. By using recuperation and appropriate ceramic components, this stand-alone gas turbine could achieve 40% thermal efficiency.

Turbogenerator Operating Characteristics

The initial mass flow and pressure ratio requirements are based on the latest concepts of the fuel cell manufacturers for their hybrid systems. Some fuel cells start with larger flow requirements, both the solid oxide and molten carbonate fuel cells need 1 kg/sec/MW generated. Stack unit power is a consistent factor.

Four major hybrid configurations are presented in Table 16–6. The data given are representative and will be reviewed with the fuel cell manufacturers to confirm their latest standards. The last cycle shown in Table 16–6 represents a more complex higher efficiency solid oxide system that requires a minimum of one module-pair (stacks operating in two different pressure zones).

The solid oxide ambient pressure hybrid was not included in this assessment, because the system becomes economically unattractive in MW class range. Such a system would require a very high temperature recuperator. To date, fuel cell manufacturers have expressed limited interest in a SOFC ambient pressure hybrid system.

The table contains the total airflow required for the various types of fuel cells, which is roughly the same for each system. Even more significant is that the higher flow requirements are also compatible with higher operating pressure ratio. This tabulation supports the notion that a properly designed single turbogenerator would satisfy all the high temperature fuel cell systems. The last cycle configuration points out that turbine reheat may be needed for this hybrid system. Using this preliminary data, a turbogenerator is being configured to provide 2.8 kg/s

TABLE 16–6 Preliminary Cycle Parameters (Fuel Cell/Gas Turbine Hybrids) [16-7]

Cell Type	Cycle	Total Power, MW	Turbine Press, Bar	Stack T in, °C	Stack T out, °C	Total Airflow, Kg/s	Plant Efficiency, %
Solid oxide	Pressurized stack, recuperated	1.3	4–6	600	870	2	60
Molten carbonate	Atmospheric stack, recuperated	3.1	9	580	660	2.8	65
Molten carbonate	Pressurized stack, recirculated	2.1	8	590	690	2.5	65
Solid oxide	Series modules, pressurized and recuperated	2.5 2nd 3	1st >6700	600 and	870	2.3	65

1. Baseline solid oxide hybrid system. Pressurized stack. Heat recovery by recuperation.
2. Molten carbonate hybrid system. Pressurized stack. Heat recovery by recirculation.
3. Molten carbonate hybrid system. Ambient pressure stack. Heat recovery by recuperation plus indirect residual fuel combustion.
4. Alternate solid oxide hybrid system with series modules. Pressurized stack. Recuperated.

airflow, work at a pressure ratio of 9 and accept 1000°C at turbine entry. Such a turbogenerator would serve all systems. It would be operated at part speed to lower its pressure ratio for the solid oxide systems. If used as a stand-alone module, the initial unit would develop <1 MW power.

Turbogenerator Design Parameters

Industrial gas turbines generating above 5 MW capacity have a maintenance-free life expectancy of 30,000 hours when operated at the lower temperatures required in fuel cell hybrid systems. This long life, however, is not achieved by small gas turbines. Gears and rotating element bearings are examples of components that have finite fatigue life associated with operating high speed, which results in Hertzian stresses. As power is proportional to length squared, a 200 kW machine theoretically would have about 1/5th the life of a 10 MW machine. This would suggest 40,000/5 = 8000 hours working life for a small gas turbine generator. This is consistent with experience.

The turbogenerator being designed on this project utilizes the following key components:

- Active magnetic bearings
- Direct drive alternator
- Redesigned existing gas turbine core

Active Magnetic Bearings

For the design being analyzed, it was decided to eliminate all gears and rotating element bearings. The company has extensive experience in the application of active magnetic bearings (AMB) in <100 kW units. Such a generator was designed, developed and tested in early 1990. This prototype performed successfully in automobiles as demonstration vehicles. The unit demonstrated great advantage of AMBs. Commercially available systems have inadequate sensing/control systems. These systems were completely redesigned in-house and gave very satisfactory results. Some 300 hours of engine running experience confirmed that the system was functionally fully effective.

The advantages of AMBs include:

- Very long system life
- Elimination of most parasitic losses
- Elimination of ancillary mechanical systems (e.g., oil pumps)
- Relaxation of shaft balancing requirements

Direct Drive Alternator

Parasitic losses normally become a major penalty as engine size decreases (presumably oil viscous shear stress increases because velocity differences remain the same but the clearances as a function of film thickness decrease). Conventional gears and bearings produce over 10% loss in a small machine compared to approximately 2% in a MW class machine.

A high-frequency direct drive alternator (DDA) is a necessity if gears are to be eliminated. Microturbine manufacturers have shown that DDA can be used effectively, but this class of generator is not a mature technology. Widely different designs are being developed and offered. The goal for this project is to achieve alternator efficiency >95%. The combination of AMB and DDA yields a turbogenerator that uses no oil. The DDA requires inverters to produce 50 or 60 Hz AC output. But this is a minor addition compared to the fuel cell requirements for larger inverters to convert their DC power to 50 or 60 Hz AC output.

Turbine Core Design

Previous cycle analysis showed that both atmospheric and pressurized MCFC cycles were fairly insensitive to the turbogenerator pressure ratio level selected. A range between 6 and 15 PR seems acceptable. The SOFC cycle as shown in Table 16–5 is more constrained in maximum pressure ratio by the requirement to heat the stack inlet flow by heat exchange with the turbine exhaust.

The determining parameters are the stack minimum inlet temperature and the turbine efficiency. With nominal stack and an efficient turbine, the maximum workable pressure ratio is approximately 4½. If the turbine efficiency were 5% worse, implying less heat removal, the acceptable pressure ratio would rise to 6. If the turbine remained efficient but the minimum acceptable stack inlet decreased 100°C (200°C has been indicated as a target), then the acceptable pressure ratio would rise to 8.

Selecting a turbogenerator design pressure ratio of 9 thus appears to provide immediate support to MCFC systems. This turbogenerator would also permit operation of early versions of standard SOFC hybrid plants at part speed and a pressure ratio of up to 6. This approach enables retention of the flexibility for likely development of SOFC stacks of higher temperatures. Maintaining high turbine high efficiency over the range 4½ to 9 pressure ratio would be accomplished by allowing a low cost compressor diffuser vane sets to be compatible with high or low PR applications.

The Model 250-C40 radial compressor provides a high efficiency basis for this machine at the correct flow size. Without the constraints of helicopter installation, the design can be adapted to still higher efficiency. A conservative maximum efficiency turbine for 9 pressure ratio would have three stages, but the high speed of the 250 impeller helps to enable a two stage version with reasonable loading and acceptable efficiency.

TABLE 16–7 Performance of the Proposed Turbogenerator and Derivatives [16-7]

Configuration	Airflow, kg/s	Press. Ratio	Turbine Inlet, °C	Recup. Inlet, °C	Stack Temp., °C	Power, MW	Thermal Efficiency
Turbogenerator design pt	2.80	9.00	900		474	0.476	0.253
Operation with PSOFC	1.86	6.00	900		540	0.313	
Operation with MCFC	2.82	8.15	650		345	0.201	
Stand alone at 1000 TIT	2.73	9.22	1000		540	0.587	0.268
As above + recuperator	2.70	9.30	1000	550	339	0.525	0.354
As above + reheat	2.70	9.29	1000	700	351	0.508	0.493
Recup. + cooled turbine	2.69	9.32	1100	594	343	0.587	0.365

Stand-alone unrecuperated efficiency of this unit is approximately the same as typical existing 3 MW industrial gas turbines. It appears to have potential for similar, i.e., combined heat and power (CHP) applications. If recuperated, its efficiency becomes about 35%. This level is good enough to compete with some diesel engines and to serve in the distributed generation role without resorting to CHP or other energy recovery systems. The low operating temperature significantly reduces greenhouse gas emissions. This feature in combination with the elimination of oil makes this turbogenerator a very "green" and cost effective alternate to a diesel. With very long life expectancy, it also promises to be more suitable for base-load service than a diesel.

The combination of relatively high-pressure ratio and reduced parasitic losses helps to improve the potential of this unit as a stand-alone unit. Table 16–7 indicates the performance of the unit and its possible derivatives.

This machine will differ from any available unit by combining the following features:

- Oil-free
- Active magnetic bearings
- Direct drive alternator
- Low parts count but high efficiency (cost saving)
- Zero emissions burden (fired only for startup of plant)
- Very high durability and reliability

Improvements in Fuel Cell Hybrids

Joint studies with fuel cell manufacturers showed that an approach for fuel cell hybrid system design by merely assembling off-the-shelf components in a simplistic "chemical plant" manner, fails to achieve a marketable product resulting in a complex and expensive system. This experience suggests that, when a higher efficiency power plant is designed, requiring several of the available fuel cell stack modules, it will be preferable to use several of the same integrated modules rather than a farm of fuel cell stack vessels connected to a single turbogenerator and balance of plant. Five of the basic 2 MW modules can be packaged into such installations, to provide plant capability of up to 10 MW size.

Fuel cell manufacturers are working to increase power density. It is anticipated that individual truckable modules will grow in power output (just as any other power plant does). This improvement would cause the turbogenerator to grow also. The baseline turbogenerator will be operating at part speed for SOFC plant initially (to avoid exceeding the best cycle pressure ratio). The likely second generation SOFC plant temperature flexibility and acceptable pressure ratio should also rise. The turbogenerator can expand its capability simply through speeding the unit up to design speed. This feature will probably accommodate 30–50% power increase.

It is anticipated that the dynamics of the turbogenerator rotor with only two radial bearings will not be amenable to the addition of a "zero stage." So further growth of this configuration will probably be achieved by geometrical scaling. The relative merits of designing to accommodate a zero-stage (possibly adding a third radial bearing) versus accepting the change through scaling will be evaluated.

Performance Comparison with Available Turbogenerators

There are several aero-derivative turbomachinery spools of approximately the same mass flow and pressure ratio as proposed. However, these fall short of the requirement specified earlier. The undesirable features of these aero-derivatives that need to be modified are: (1) contact bearings and rotating assemblies with many separate elements,

TABLE 16–8 A Comparison of Performance of the Proposed Turbogenerator with Existing Units [16-7]

Manufacturer/Model	Airflow, kg/s	Press. Ratio	Turbine Inlet, °C	Recup. Inlet, °C	Stack Temp., °C	Power, MW	Thermal Efficiency
Proposed							
Baseline simple cycle unit	2.73	9.22	1000		540	0.587	0.268
Unit + recuperator	2.70	9.30	1000	550	339	0.525	0.354
Simple cycle existing							
ASE8-1000*	3.58	10.6	935		496	0.548	0.221
Volvo VT600*	3.58	8.9	977		538	0.670	0.235
P&W ST6L-795*	3.22	7.4	1000~		589	0.678	0.247
Recuperated existing							
Capstone microturbine(1)	0.27~	3.25~	880~	600~	271(1)	0.028(1)	0.259(1)
GM404 Patriot (ISO day)	1.82	4.3	1020	700**	313	0.240	0.315

**Data taken from Gas Turbine World 1998–1999 Performance specs.*
***Used a metal regenerator rather than a recuperator.*
(1)From Capstone website.
~Approximate.

(2) gear-driven generators, (3) gas paths to be directly compatible with connection to the fuel cell system in place of the usual combustor, (4) expensive aviation hardware which offsets cost reductions that may accrue from volume production, and (5) an older style design, using a combination of axial stages and a radial impeller to achieve the same duty served by an impeller alone.

The proposed generator extends the range of high-speed direct drive alternator products, from the 30–100 kW range (now being introduced in microturbines) to >0.5 MW. This is a significant technology enhancement step and is expected to be an important development in a continuing evolution to higher direct drive generation, which is paced economically by progressive decreases in power conditioning system prices.

The improvements for fuel cell hybrids provided by the proposed system (see Table 16–8) are:

- Much improved durability and reliability for the turbogenerator
- High pressure ratio (relative to micro-turbines) improves fuel cell hybrid efficiency
- Lower first cost, particularly as the turbogenerator production volume increases
- Easier integration into the fuel cell hybrid plant gas path
- Better environmental quality, both by eliminating oil usage and by maximizing plant efficiency
- Direct drive gives better cycle control during load following

Market for the Turbogenerator

Based on experience in working with prospective fuel cell manufacturers, we have seen that the microturbines being developed for one sector of the distributed generation market are an enabling technology for very small fuel cell hybrid systems. But these are not big enough to serve a commercially attractive plant. The availability of the correct size turbogenerator with promises of good durability has been of great importance to early fuel cell hybrid designs. Some fuel cell hybrids have actually been designed to fit the available microturbines.

With flexibility to serve a variety of the planned fuel cell hybrids, it is believed that an assured US fuel cell hybrid market exists, arising with the success of fuel cells. There is promise that these small turbogenerators will support plant in sizes from 0.8 MW up to 12 MW through 2010 and beyond.

The merits of a nominal 1 MW class stand-alone generator of direct drive design include local power quality improvement. Very low emissions (uncooled turbine will be run at emissions-favorable temperature), low weight, good efficiency by recuperation or CHP (suitable exhaust temperature), and low noise and vibration, will make this unit a good candidate to displace many diesel generator sets. This would result in a large production volume.

As fuel cell stack technology improves, and the power generated in a transportable modular unit increases, a need will develop for a turbogenerator having a higher unit flow capacity and power than the machine proposed. With this in mind, the turbogenerator design will be laid out, as far as is possible, for literal geometrical scalability, to facilitate low cost derivation. One product of the program will be the understanding of the extent to which this is feasible.

Preliminary configuration studies are being conducted under contract DE-FC26-00NT40914 for the US DOE. These incorporate cell manufacturer and DOE inputs to selection of the best duty cycle, and determine a mechanical design with acceptable dynamic and durability characteristics. Particular attention will be given to non-contact bearing selection and direct drive alternator design.

There is a need to provide a new high-quality, low-cost, long-life turbogenerator to be compatible with hybrid fuel cell systems for the following reasons:

1. There is a good possibility that a single machine can have wide applicability to the various US manufactured fuel cell system modules anticipated.
2. Though individual module power may only be 1–3 MW, larger plant installations will probably use multiples of the base module, up to 10 MW.
3. The proposed turbogenerator will support some system power growth, but design evaluation is required to determine whether scaling is required to support growth beyond 50% power increase in fuel cells.

Chapter 17

Training and Education

"I was gratified to be able to answer promptly. I said I don't know."

—Mark Twain

Chapter Outline

With† the gas turbine developing at the rate that it does, education and teaching methods struggle to keep pace. This chapter provides specific examples and case studies of training and education in both the industrial and academic worlds. All cases are detailed with regard to program structure and content. However, in the hands of creative in-house trainers in industry or professors, they could suggest potential for creating one's own tailor-made program.

INDUSTRY TRAINING

All OEMs offer some kind of training to their customers. These presentations can vary from "sales-type" commercial overviews to those that are considerably more comprehensive. The latter are generally given to customers who bought the OEM's equipment, and the training is included in the price, in other words, "free."

In addition to theoretical courses, the OEMs offer practical training at the customer's plant or in their own plants, with a very "hands-on" approach. This type of training is generally given to key mechanics, operators, and service crew.

Independent firms, which frequently are owned and staffed by ex-OEM senior people, offer similar training. They point out that they offer a more objective perspective than an OEM might.

The job titles given to engineers in field operations or field operations consulting are varied: *Field service representative, field service engineer, applications engineer, field engineer*, or simply *OEM's rep* are common. *Project application engineers* is the term used by the OEM authors of Case 1. In some cases, that engineer is one that the OEM dispatches to solve field problems that an end user may have. In others, the engineer may be one who follows an individual project within the OEM's walls, through all phases, including but not limited to initial customer specification, manufacture, testing, commissioning, and even optimization (also called *streamlining or debottlenecking*). These individuals are well equipped to then also serve the OEM as field troubleshooters.

Depending on the corporate culture (OEM or otherwise) involved, a successful field service or project applications engineer may be moved to "technical sales" or marketing. In certain corporations, the senior officers have been promoted from these "nuts and bolts" beginnings.

The following extracts were written by senior staff within an OEM's organization. This OEM makes small- and medium-sized gas turbines, as well as associated

† [17-1] Working case notes, Claire Soares, 1975 through present.

packages for mechanical drive and power generation service. However, the template they provide is a sound one for all gas turbine OEMs and, in fact, all OEMs.

CASE STUDY 1: OEM PROJECT APPLICATION ENGINEERS TRAINING*

Project application engineers (PAE) are a critical link between the customer and the OEM. His or her knowledge of both internal capabilities and the customer's business make the PAE an important contributor in developing new projects. Despite the fact that this position can be an important and satisfying career path for engineers, its existence is widely unknown among engineering graduates. Therefore, a description of the tasks, duties, and challenges for project application engineers is provided.

We (i.e., Solar Turbines) cover the training and education of application engineers for packaged industrial gas turbines. However, most of the concepts are applicable to application engineers for most engineered products.

Packaged industrial gas turbines, especially in the oil and gas industry, consist of pre-engineered and custom-engineered components. The gas turbine and power turbine themselves are usually standardized products. The package components may, depending on the manufacturer's philosophy and the project requirements, be either mostly pre-engineered with some customized components or mostly custom designed. The driven equipment may be standardized (such as a generator), or as for most compressors and pumps, either customized using pre-engineered components or entirely custom designed.

The Job Description

The application engineer is the interface between in-house technical, commercial, and legal disciplines, on the one hand, and the sales force and ultimately the customer, on the other hand. The exact boundaries vary from company to company and also depend on the extent of project related special engineering (Table 17−1).

In-house technical disciplines include engineers specialized in equipment design, manufacturing and testing of the equipment, or the design of ancillaries and balance of plant items. They also include legal and commercial specialists.

However, this should not be construed as the PAE being just a transmission vehicle. Rather, his knowledge of both internal capabilities and knowledge and the customer's business make him an important contributor in developing new projects. Again, depending on company strategy, the distribution of tasks between the sales engineer and the project application engineer can vary. Ideally, they work in close cooperation as a team. Table 17−2 shows the typical steps in the project cycle. The PAE supports primarily the selling process with his technical expertise and product knowledge. In this role, the PAE helps define the scope of the project, as well as the type and the features of the product offered. An important milestone in a project is the handover process from the business development organization (which sells and defines the product) to the business management organization (which is responsible to manage all necessary steps to execute the project). The project application engineer tends to stay involved during this time to make sure that the project is fully defined.

Custom-engineered products, such as centrifugal gas compressors, require more involvement by the engineering organization, while standardized or pre-engineered products may be handled entirely by applications engineers, equipped with product knowledge and application guidelines.

Typical projects include land-based power generation with simple-cycle or combined-cycle gas turbines, gas compression stations for pipelines, power generation, gas compression, and pump applications in offshore applications (Figures 17−1 through 17−3).

In many cases, the scope includes not only the turbomachinery but also the ancillary equipment; balance of plant items may be included and, often, complex testing requirements as well as legal and commercial aspects have to be covered.

To meet this rather broad requirement, the application engineer has to be able to communicate both with internal organizations as well as the sales force and with the customer. This, in turn, requires that he or she understands the tasks, options, capabilities, and preferences of the internal organization. This is more than just "product knowledge" but rather the capability to think as if a member of a variety of internal organizations. This is the reason why a large number of application engineers have worked in one or more of the (usually technical- or commercial-oriented) internal organizations.

Another important task lies in the understanding of the customer's business. Since turbomachinery products are bought by the industry in order to achieve commercial success, a good application engineer has also to understand the customer's business (which is why sometimes application engineers are former employees of gas turbine users). In a sense, the task of acting as an interface between internal organizations and customers exists in a similar fashion for the project manager, who is

* Source: [17-2] Courtesy of Rainier Kurz. Extracts from R. Kurz and H. Andrade, "The Training and Education of Application Engineers," 2005-GT-68097. (Both authors work for Solar Turbines.)

TABLE 17—1 Training Module List [17-2]

No.	Module Name	Priority	Name of PAE	%
1	INTRODUCTION	1	Manager	100
2	DEPARTMENT CKECK IN	2	Admin./Mentor PAE	100
3	OVERVIEW/PAE PROC.	3	Mentor	100
4	EQUIP. PERFORMANCE/TEST	4	Video/Mentor PAE	100
5	PROPOSAL PROCEDURES	5	Designated Trainee	100
6	FUEL SUITABILITY	6	Mentor PAE	100
7	REVIEW OF PRODUCTS	12	Mentor	100
8	RESOURCES FOR PAE	7	Mentor	100
9	CUSTOMER SERVICES	14	Mentor	70
10	CONSTRUCTION SERVICES	15	Mentor	80
11	EQUIPMENT SUPPLIERS	13	Various	70
12	CONTRACTS	9	Legal department	70
13	FINANCIAL EVALUATION	10	Mentor	70
14	INCOTERMS TERMS	11	Mentor	75
15	API REQUIREMENTS	16	Mentor	75
16	QUALITY ASSURANCE	17	Mentor	50
17	BOOKING A PROJECT	18	Mentor	67
18	PROJECT WORK	8	Mentor	100
19	PROGRESS REVIEW	STATUS	Manager	90
	% COMPLETED			85

responsible for the project execution in the order fulfillment process.

Finding the Right People

Obviously, the wide spectrum of knowledge and experience just described is not achievable without training and education. There is—to our knowledge—no program in any US or European university dedicated to the professional development of application engineers. Nor would such a course be feasible without a close interaction between academia and industry. Even general descriptions of careers in the gas turbine industry sometimes omit the existence of application engineers. On the other hand, successful application engineers often come from the ranks of the technical or commercial (such as business management) disciplines within the companies whose products they are supposed to support. Another successful route tends to be people who worked operating or specifying the equipment in question. The "right people" are therefore not only defined by a predetermined education but rather by a certain mindset, including good communication skills, the capability to learn the basics in new fields fast, the capability to work in a team and with limited supervision, the capability to work multiple tasks and prioritize, and a wide educational and professional background.

Because most of the business in the gas turbine industry is international, it is of significant advantage for any application engineering organization to retain employees with a variety of language capabilities and cultural backgrounds. In this context, differences in the educational and business structure in different countries have to be understood.

Development

The ideal application engineer in an ideal application engineering organization is a person with working knowledge in all facets of topics required to represent the product and in-depth knowledge in one or more of these topics. The

TABLE 17–2 Typical Training Module [17-2]

Module 8	
Resources for Application Engineers	
This is one of the most important sections of the training for the application engineer	
2 Hour	Review the contents of the application engineer support on the LOCAL drive including: scope and status of each application engineer tool: - API comments, - API datasheets, - Typical drawings, - ...
1 Hr	Review the application engineer Web page Location of contacts list
4 Hr	Review of Solar intranet: Controls Homepage, Technical information department, Engineering specifications, On-line drawings, Customer service, ...
2 Hr	Review of local drive "S:\drive" and library's video tapes:
15 Min	Identify the on-line locations of the following references: Product Information Letters (PIL) Product Information Bulletins (PIB) Customer Services Service Bulletins (SB) Management Decisions (MD) Sign up to marketing's CBR reports distribution list Engineering specifications Online pictures

ideal department has engineers such that at least one expert is available for any topic required. In many instances, the screening process comes as a result of months of internal networking searches for future candidates. Together with the networking efforts, managers can seek aid from the human resource personnel to develop a solid interview

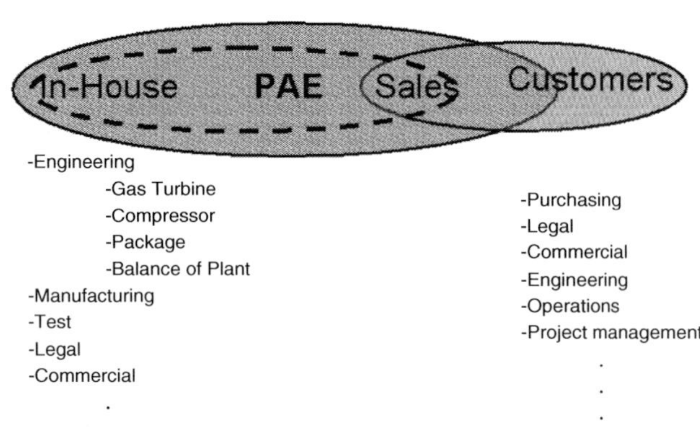

FIGURE 17–1 The project application engineer (PAE) as the interface between the manufacturer's internal organization, the sales organization, and the customer's organization [17-2].

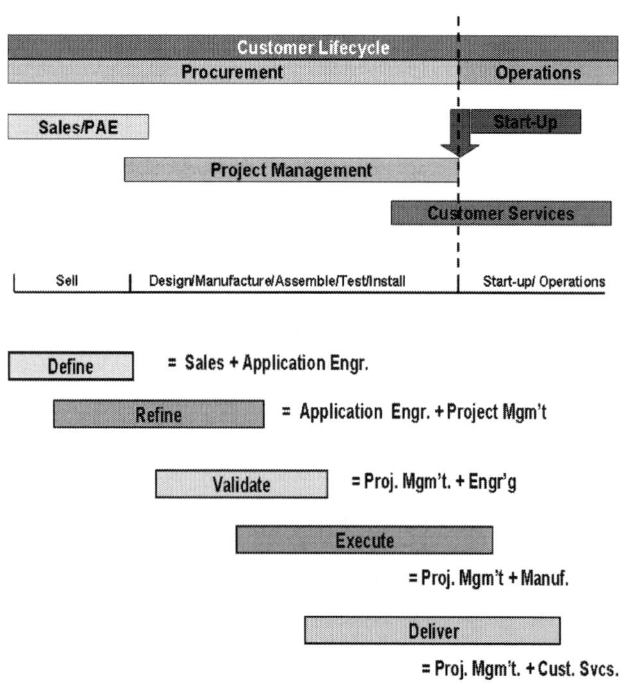

FIGURE 17–2 Project cycle (top) and steps (bottom) [17-2].

FIGURE 17–3 Gas turbine application on an offshore platform [17-2].

FIGURE 17–4 Multiple gas-turbine-driven gas compressors in a pipeline compressor station [17-2].

process. Areas of consideration for the development of a good interview process should be discussed and agreed upon. Points of reference follow:

1. Review of previous performance appraisal and interview supervisors and peers to determine the candidate's track record
2. Request for a writing sample
3. Panel interview to target technical background and gauge interpersonal skills
4. Development of a scheme to measure applicant's problem-solving skills

Training Program

The complete self-paced training program must consist of individual modules with specific objectives targeting business applications. Still, for the training program to work, managers are recommended to promote initiative and assertiveness from all participants.

The components of the training program follow and are described next:

1. Self-paced modules
2. Hands-on training
3. Formal training
4. Project work
5. Progress review

Self-Paced Modules

If not already available, management must develop a list of applicable themes related to the industry and accountabilities related to the position. As an example, Figure 17–4 provides commercial and technical themes applicable to the turbomachinery industry.

Each of these modules should be structured with a self-study outline covering related topics. An illustration detailing a typical module is provided in Figure 17–5. As noted in Figure 17–5, the module provides a list of the reference material to be reviewed and highlights the main points under study. In addition, the module topics can incorporate estimated times for completion.

FIGURE 17–5 Power generation on an FPSO (floating production, storage, and offloading) vessel [17-2].

Introductory Modules (1–3 and 8)

These first modules should familiarize newcomers with the organization and must be written to allow a smooth transition. Typical items covered in these sections should encompass a checklist to work logistics (credit card, phone card, cubicle location, etc), PC hardware and software, passwords (networks, programs, etc.), and provide direction to coordinate software application training.

Technical Modules (4–7 and 9–11)/Commercial/ Legal Modules (12–14)

As a minimum, the technical aspects should encompass areas related to product packaging, turbine and compressor theory, fuel suitability analysis, etc. This portion should be matched with specific areas of the business (i.e., customer services, balance of plant equipment, engineering, etc.). Other examples of technical areas to consider are listed next:

- Fuel suitability analysis
- Turbine and compressor performance guarantees
- Packaging and controls theory
- API and industry specification review and explanations
- Auxiliary equipment

In turn, the commercial modules must provide a full understanding of company's policies from a legal and contractual perspective. Examples of commercial/legal topics to consider are denoted here:

- Review of company's policy of terms and conditions of sale exchange policy
- Trade terms (incoterms)
- Commercial tools (i.e., bank guarantees, bonds, letters of credit, etc.)
- Responsibility matrix
- Credit rating policy
- Department internal policies
- Pricing policies

At the end of the self-paced training, an application engineer must be able to develop an understanding for technical considerations from a commercial/legal perspective. In turn, the training process will allow for good decision making while managing risk.

Hands-on Training

A newly hired application engineer should spend several days in the workshop and test cells. The idea is to provide new application engineers with a better understanding of the product and present a global picture of the business. Typical areas for review during this portion of the training program are:

1. Testing capabilities (i.e., ASME PTC 10, ASME PTC 22, American Petroleum Institute test requirements, etc.)
2. Manufacturing capabilities (product lines, shop lineup, lead-time factors, etc.)
3. Understanding of demand management functions
4. Review of bill of materials, package maintenance, controls, etc.

Formal Training

Application engineering management must balance the preceding training with the company's formal training program. For example, some of the areas useful to application engineers can be tied to human resources courses related to interpersonal skills:

- Time management/organization skills
- Negotiation skills
- Communication skills
- Technical writing course
- Business management
- Managing professional growth

Project Work

Depending on training progress, management must bring application engineers in direct contact with customer inquiry work from day 1. Application engineers can be exposed to these areas in a step fashion to start building their accountabilities. For example, a trainee can help a senior application engineer with budgetary proposals, help solve technical queries, and obtain exposure as an observer during technical clarification meetings. These tasks help trainees develop the required confidence to cope with the assigned work responsibilities.

Progress Review

Finally, a monthly status report must be completed to monitor the progress of trainees and obtain feedback on the effectiveness of the training program. The monthly status report should include a signed copy of the sections that have been completed during the given month.

TRAINING PROGRAMS WITHIN ACADEMIA

When it comes to undergraduate engineering courses, standards vary all over the global map. However, the ease of communications and especially the Internet is slowly closing gaps between curricula in different countries.[†]

At one point in the 1960s, universities (all universities) offered little more than a few ideal and theoretical cycle diagrams with equally theoretical heat balance calculations. If the university was fortunate enough to have well-funded labs, the practical equipment may have included a steam turbine, a diesel turbine, perhaps some car engines for the students to strip and reassemble.

The first university/OEM joint educational accredited course venture in the United States may have been the

[†] [17-1].

University of Cincinnati/General Electric collaboration back in the 1920s. Their example was not one other US universities were swift to follow but follow they eventually did.

The early 1970s saw the start of a landmark program in the United Kingdom. In 1971, Dr. D.B. Spalding, then professor of Heat and Mass Transfer at Imperial College, London, England, conceived COBALT (computer-based learning and teaching). He wanted experimental rigs constructed or modified to demonstrate the efficacy of prediction flow formulas. Experimental readings on the rigs were used in nontheoretical calculations and compared with the computer program results that iterated flow parameter values, starting with boundary layer conditions. The program would thereby introduce masters' level program material to second-year undergraduates. Some of those predictive flow formulas are still used in performance analysis (PA) systems, such as are described (in Chapter 10 on Performance Optimization) and designed 20 years later, so that earlier work is still current. That summer in 1971, I was one of two students hired to realize Spalding's plan, and after the other became unavailable, the only one. The initial work scope I had been tasked with was finished in the summer of 1972.

OEM/university joint ventures are still not quite as common as they could be. When they are in place, however, they benefit students greatly. University students may be able to complete some experiments that the OEM cannot perform more inexpensively in its own facility. This can and ought to be organized throughout the undergraduate degree program and, if possible, still earlier.

With most gas turbines, simulators, flight or otherwise, are used to train plant operators, naval seamen, and pilots or "refresh" their own skills periodically. Schools and universities often cannot afford the investment that good simulators or real gas turbines (even small ones) require. They depend on donations and other help from industry (see the examples and cases that follow).

Today, students are lured into fields seen as more lucrative, such as law or business. Some engineers may soon realize that, if they want to make more money, they need to go back to school for an MBA. Of the few who stay loyal to the basic profession, still fewer may be attracted to the subdisciplines within the engineering field that need new blood. On the other hand, if somehow an engineering student gets a detailed enough look at the reality of operating turbomachinery when a student, his interest is often anchored for life. Employers find the "catch-up" time that fresh graduates take to come up to speed is then reduced to a substantial extent.

Engineering student recruiting numbers stand at low to dwindling. Robert Herbold (former COO, Microsoft) compiled the data in Table 17–3.

TABLE 17–3 Engineering Degrees*

	B.S./B.A. Degrees (000s)	B.S. Engineering (000s)	B.S. Engineering (%)
United States	1253.0	59.5	5%
China	567.9	219.6	39%
South Korea	209.7	56.5	27%
Taiwan	117.4	26.6	23%
Japan	542.3	104.5	19%

Source: http://www.vcconfidential.com/2006/04/leisure_class.html.

TABLE 17–4 Doctoral Degrees

	1987	2001	Growth
US citizens getting Ph.D.s	4700	4400	(6.4%)
Asian Citizens	5600	24,900	345%

Note: Parentheses indicate negative growth.

All four Asian countries have four to eight times the per capita number of students going into engineering and China and Japan have four times and twice, respectively, the annual number of B.S. degrees in absolute terms compared to the United States.

Consider also the numbers for students receiving Ph.D.s in physical science or engineering, shown in Table 17–4.

Data sources indicate that the United States in 2005 graduated only 10% in US-resident engineer numbers that China did and 20% the number that India graduated. Many of those engineers will work for US and European corporation branch offices without ever leaving their home country. Also, the total US graduate engineer population is effectively halved because half of them are foreign and cannot get a work permit since the start of the US-Afghanistan war in September 2001.

Given the current US policy of exporting technical jobs to countries with cheap labor, the figures are not likely to change much in the near future. Many US students argue that there is no future in engineering.

The cultures in these countries that rival the United States for the globally available engineering jobs are more apt to place an engineer in a top management position than a lawyer or an accountant. For many years, Europe was similar in this regard. The "MBA-rules" perspective has taken over with some European companies, but it is still not as widespread as in the United States.

To demonstrate the effectiveness as well as the "nuts and bolts" of specific academic programs, I have included some examples and case studies of academia's efforts to bring the outside world and gas turbines into their classrooms. They provide excellent templates for ambitious professors and the potential for endless variations.

The first example involves an industry (BMW-Rolls Royce and MTU Munich)-academia (Berlin University of Technology) joint venture in Germany.

CASE STUDY 2: INDUSTRY SUPPORTED MULTIMEDIA AEROENGINE DESIGN CASE*

The Jet Propulsion Laboratory at the Berlin University of Technology introduced an aeroengine design project, Project Jet Propulsion. The seminar, in which the German aeroengine industry is actively involved, is directed at students who have successfully finished their basic engineering training course.

The objective of the course is to provide experience in all stages of the complete design process of an engine component in a genuine industrial working environment. The component is selected to ensure that a wide range of design requirements, including customer requirements, aero-thermodynamic and mechanical design, pricing and certification, must be considered. An experiment supporting the design completes the course.

The course lasts for one year. Minutes are kept and action lists checked. The students are encouraged to work on their own.

Industry participates with telephone conferences, giving their comments on the work of the group and answering questions, a video conference sometimes, and in-house OEM progress reviews at intermediate and final design stage.

The incentive to start this program was prompted by the realization that a basic engineering education tends to produce engineers who know a little about many different subjects, while postgraduate and Ph.D. education tends to produce engineers who know a lot about a few subjects. Industry [used to] promotes this trend developing generalists who in the long run know nothing about everything and specialists who know everything about nothing. . . .

Classical teaching has not changed significantly over the past 50 years. Most of the reforms introduced in this period of time have been concerned with the opening of the universities to a larger number of young people rather than with a quality improvement. The universities have little flexibility to adapt to a changing environment.

The Germans are trying to remedy matters.

The objective of the Jet Propulsion Project is to lead the students through the interdisciplinary process required to define and develop a gas turbine component or system.

* [17-3] Extracts from J. Hourmouziadis, N. Schroeder, K. Beigi, J. Schmidt, S. Servaty, and W. Gärtner. "A Multimedia Aeroengine Design Course with Industry Support," 2000-GT-585.

Several theory courses provide the background basics, including:

- Thermodynamic cycle design
- Component characteristics (intake, compressor, combustor, turbine, nozzle)
- Engine performance and partial-load operation
- Engine systems (air, oil, control system)
- Measurement techniques (laboratory, engine)
- Experimental facilities
- Certification requirements and procedures
- Safety and reliability aspects

A course in turbomachinery aerodynamics presents:

- Short review of fundamental gas dynamics
- 1D axisymmetric hub-to-tip and 2D blade-to-blade design (quasi-3D design system)
- Overview of CFD techniques and applications
- Boundary layer theory and application to profile design
- Secondary flows
- Transsonic and supersonic cascade flow

A course on mechanical design and cooling teaches:

- Engine architecture alternatives
- Hot and cold annulus diagram
- Bearings and oil system design
- Air system components (restrictors, seals) requirements
- Design principles and analysis
- Heat transfer and thermal control of engine parts
- Thermal transients and clearance effects
- Cooling methods and design principles

The Jet Propulsion Project is an integrating course. One of the objectives is getting used to preparing a complete solution to a technical problem and presenting the results in a team. The participants have to communicate data, technical, and time scale information and learn how to behave in a team to achieve a common goal within schedule constraints.

The objectives of the first phase, which lasts about one third to one half the course, are as follows:

- Provide the necessary background information, like requirements, performance data, design constraints, etc.
- Preliminary study of alternative solutions, evaluation, and selection of realistic configurations; one may be selected after a review
- Preparation of a plan and schedule for the second phase

The second design phase includes:

- Mechanical design
- Failure analysis
- Material selection, stressing, and lifing
- Development and certification program
- Costing of engineering work and costing of product

TABLE 17–5 Comparison between the Model Plant and the Result of the Students, Major Components [17-4]

Item	Neubrandenburg	Student Result
Gas turbine	2 × 23.7 MW_{el} (ABB GT 10B)	2 × 25 MW (ABB GT 10)
Steam turbine	1 × 27 MW_{el} (ABB V40AHH)	1 × 24 MW (ABB VAX HP 16)
Pressure levels	2 levels	70 bars, 7 bars
Waste heat boilers	2	2
Supplementary firing	10–100%	Up to 100%
Bypass	Yes	Possibility to add
Transformer	2 × 40 MVA for 110 kV	For 3 transformers for feeding the electric grid

The last meeting is a final design review at a supporting company. The course is finished with a final report and an evaluation of the course by the students.

Two problems treated by the Jet Propulsion Project were the design of an HP-Compressor bleed system for a two-spool turbofan (1997/98) and the design of an LP shaft failure protection system for a turbofan engine (1998/99). The results from this project included student-designed items that were "industry ready." One is described next.

Two electronic systems for an emergency fuel shutoff were investigated. The first uses two shaft speed gauges, one on the fan end of the shaft and the other on the LP turbine end. The signal difference is used for actuation. The second system uses the LP turbine speed gauge and responds to the acceleration rate of the rotor. To determine if the signal is sufficiently fast, transient performance of the engine was analyzed. It was found that acceleration rates after shaft failure is several orders of magnitude greater than in normal operation. Both systems require a speed gauge in the "hot" part of the engine.

System accuracy ensures that no undesirable fuel shutoffs occur.

The system using the rotor acceleration rate appeared preferable, because it uses only one speed measurement and is therefore more reliable. To reduce response time, the students came up with digital electronics for a faster evaluation of the signal than in current systems. This proved to be of major importance for the response time. To support this evaluation, experiments were performed in the laboratory with three speed gauges and the signals were analyzed in the time domain to define the requirements on the electronic treatment.

The investigations ended with a final assessment of the sequence of events after engine failure and the resulting turbine overspeed. This required an analysis of the dynamics of the system and the engine during and after fuel shutoff, when the HP rotor feeds energy back into the core mass flow. Three out of four systems satisfied the FAA requirements, so no extra overspeed capability has to be designed into the LP turbine disc design.

Since all the systems investigated were new, tests were needed as part of the compliance with the FAA certification requirements. The students produced an engine development plan lasting two years, which used eight engines.

CASE STUDY 3: THEORETICAL CALCULATIONS COMPARED WITH ACTUAL COGENERATION PLANT†

In another project, the Royal Institute of Technology in Sweden ran a program, assisted by ABB Stal. The students chose an optimum design configuration for a cogeneration application and then were able to visit the actual plant and compare their own work with ABB Stal's choices. Table 17–5 indicates how close they came.

The following two cases* are courtesy of K.W. van Treuren, Baylor University. As with most gas turbine academics, he is keen to spread the word on gas turbines and he has presented his work at different venues, including IGTI conferences.

The first case involves using commercial software and spreadsheets for gas turbine design that accommodates a flight mission profile. The second is work on a land-based gas turbine that the university was given to work on in its laboratory. The starting point for the first case was the US Air Force Academy (USAFA). The USAFA program was adapted for use at Baylor University in Texas.

† [17-4] Courtesy of Alstom Power. Reference S. Svensdotter, P. Almqvist, and T.H. Fransson. "Introduction of Project Based Learning for Designing and Heat and Power Plant into the Last Year Curriculum," 2000-GT-583.

* Source: [17-5] The work of K.W. van Treuren, including extracts from K.W. van Treuren and B.A. Haven. "Undergraduate Gas Turbine Engine Design Using Spreadsheets and Commercial Software," IGTI 2000-GT-587.

CASE STUDY 4: UNDERGRADUATE ENGINE DESIGN PROGRAM*,[1]

A unique, three-part undergraduate gas turbine engine design project was developed to acquaint students, working in teams of two or three, with the process of engine cycle selection. The design application is a low-flying, close air support (CAS) aircraft using a separate exhaust turbofan engine. Both spreadsheets and commercial software are used. The commercial software is included with the course textbook, *Elements of Gas Turbine Propulsion*, by Dr. Jack D. Mattingly. Using commercial software, reinforced by classroom lectures, allows the students to focus on the design decisions.

The first part of the project is mission analysis, which introduces the student teams to the design problem. A spreadsheet template is given to each student team that includes aircraft and mission profile specifications. The students must complete the spreadsheet and develop the relationships for lift, drag, thrust required, and fuel burn to calculate a usable fuel remaining at the end to the mission. The spreadsheet allows the students to obtain an average specific fuel consumption that results in 1500 lb_m of fuel remaining at the end of the mission. This target value is used in the second part of the design process, on-design parametric cycle analysis (PCA), as a basis for engine cycle selection. Parametric cycle analysis is accomplished using the program PARA.EXE. PARA.EXE generates a carpet plot of possible engine design choices by varying the compressor pressure ratio, bypass ratio, and fan pressure ratio. From these carpet plots the students must identify three possible engine cycles that meet the target value for specific fuel consumption found during the mission analysis.

Trade-offs between thrust and fuel consumption are discussed, and the students are required to justify their choices for the engine cycle. The last part of the project is the off-design engine performance analysis (EPA) using the program PERF.EXE. The chosen engines must fly the mission and meet the required performance and mission constraints. Based on the overall mission performance, the students narrow the field of three possible engine cycles to one. Each student team then does a sensitivity study to determine if there is an additional benefit for slight changes in the design choices. The result of this sensitivity study is the students' final engine cycle. With this cycle, an additive drag calculation is made using the program DADD.EXE to account for losses (off-design) and these losses are then factored back into the performance spreadsheet to check the engine's capabilities for completing the mission. The iterative nature of the design process is emphasized throughout, but only one pass through the process is accomplished. Units are given in English Engineering, as that is what is required for the project. Both SI and English Engineering units are taught in the course.

Nomenclature

C_D	=	Drag coefficient
C_L	=	Lift coefficient
D_{add}	=	Additive drag, lb_f
F	=	Uninstalled thrust, lb_f
$F/\dot{m}$	=	Specific thrust, $lb_f/(lb_m/sec)$
k	=	Drag due to lift factor for given wing platform
$\dot{m}$	=	Mass flow rate of air, lb_m/sec
$\dot{m}_f$	=	Mass flow rate of fuel, lb_m/sec
S	=	Uninstalled specific fuel consumption, $(lb_m/hr)/lb_f$, $(\dot{m}_f/F)$
T	=	Installed thrust, lb_f, $(F-D_{add})$
TO	=	Takeoff
TSFC	=	Installed specific fuel consumption, $(lb_m/hr)/lb_f$, $(\dot{m}_f/T)$

Greek

π_f	=	Fan pressure ratio
π_c	=	Compressor pressure ratio
α	=	Bypass ratio
φ	=	Installation loss coefficient (D_{add}/F)

The need for design in an engineering curriculum is further emphasized in the recent engineering criteria (EC) 2000 developed by the Accreditation Board of Engineering and Technology (ABET). The design described next is not a capstone design experience but is meant to emphasize the conceptual design portion of the process. For the engineer, this is often thought of as creative problem solving. The problem can be defined in great detail or may be something very ill-defined. Whatever the detail, the engineer must develop an acceptable solution that satisfies the requirements. This process of arriving at an acceptable solution uses the cyclical nature of the design process as seen in Figure 17—6. Here, engineers use all their mental faculties to be creative with ideas, analyze the ideas, and then decide on their relative merit, as is shown in Figure 17—7. This often leads to more ideas to be evaluated

1. For readers with additional interest, references for Case 4:

Oates, Gordon C., 1988. *Aerothermodynamics of Gas Turbine and Rocket Propulsion*. Washington, DC: American Institute of Aeronautics and Astronautics, Inc.

Mattingly, Jack D., 1996. *Elements of Gas Turbine Propulsion*. New York: McGraw-Hill, Inc.

Mattingly, J.D., Heiser, W.H., and Daley, D.H., 1987. *Aircraft Engine Design*. New York: American Institute of Aeronautics and Astronautics, Inc.

Brandt, S.A., Stiles, R.J., Bertin, J.J., and Whitford, R., 1997. *Introduction to Aeronautics: A Design Perspective*. Reston, VA: American Institute of Aeronautics and Astronautics, Inc.

Raymer, D.P., 1992. *Aircraft Design: A Conceptual Approach*. Washington, DC: American Institute of Aeronautics and Astronautics, Inc.

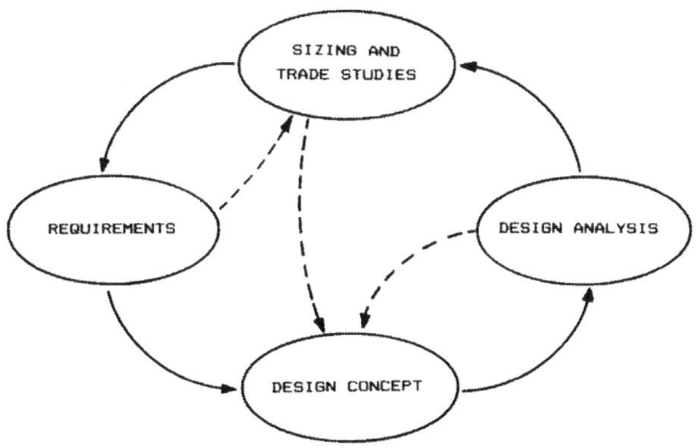

FIGURE 17–6 The design wheel (Raymer, 1992) [17-5].

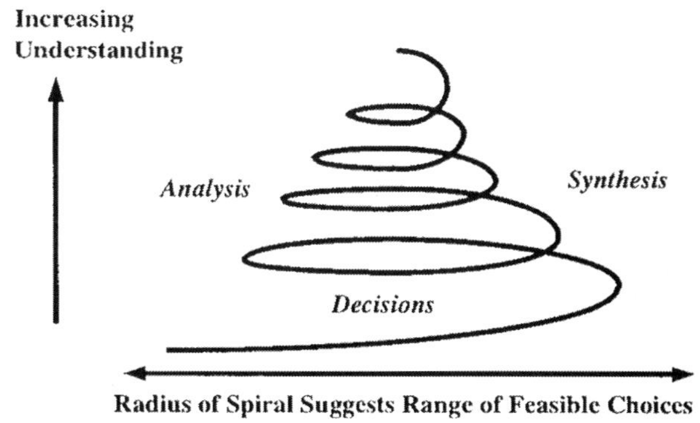

FIGURE 17–8 Design spiral (Brandt, 1997) [17-5].

and further decisions that must be made. Eventually, the engineer repeats this cycle many times until an acceptable solution is determined, as shown in Figure 17–8.

The project described in this section acquaints the students, working in teams of two or three, with the process of engine cycle selection and reinforces the design decisions necessary for the conceptual design of a gas turbine engine. Cycle analysis, according to Oates (1988), is the process to obtain estimates of the performance parameters (primarily thrust and specific fuel consumption) in terms of design limitations (such as maximum allowable turbine temperature), flight conditions (the ambient pressure and temperature and the Mach number), and design choices (such as compressor pressure ratio, fan pressure ratio, and bypass ratio). With these goals in mind, the initial development of the project was accomplished at the US Air Force Academy by Lt. Col. Brenda Haven and has been subsequently adapted for use at Baylor University in an elective course on propulsion systems for mechanical engineers. The degree program at Baylor University is a more traditional undergraduate B.S. in Engineering with a mechanical or electrical emphasis. Baylor does not have a graduate engineering program, but approximately 13% of the students take additional study at other institutions and receive advanced degrees. Approximately 10 students take this course in the fall of their senior year. Mechanical engineering students often have a keen interest in aerospace topics. This project reinforces that interest and presents the complexities of designing a gas turbine engine. Prerequisites for this course are fluid mechanics and an advanced course in thermodynamics. Heat transfer is taken concurrently. Propulsion systems are presented as an application of these fields.

The project has great value in light of ABET EC 2000. Teamwork is strongly emphasized as the students work in teams of two or three. Student teams are thought of as industry competitors. They are allowed to work with their partner and the professor but not allowed to talk with other students about the details of their project or use material from previous semesters. At the end of the course the student teams present their design and discuss its performance. After each team presents its results, the class decides on the "winner" of the competition. A final, written report is also required, thus exposing the students to both written and oral presentation. Higher-order thinking skills are developed, as the students must conduct multiple paper studies of various engines and make engineering trade-offs. While there are many solutions, the desire is to optimize the results for the "best" solution possible. In the process of selecting the engine, the students are using design tools such as spreadsheets and performance analysis programs.

The design project is accomplished in three parts. Basic physical concepts and analysis techniques required by the design process are emphasized throughout the course and provide the basis for these parts. The first part is a mission analysis, which introduces the student teams to the design

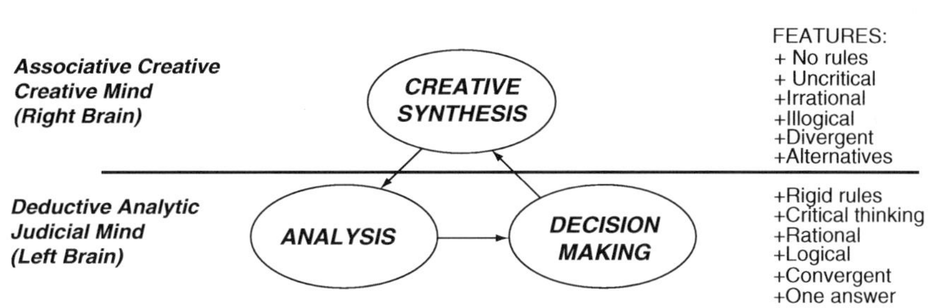

FIGURE 17–7 Mental activity in the design cycle (Brandt, 1997) [17-5].

problem. The request for proposal (RFP) calls for an aircraft to fly the close air support mission. The students are given the requirements for the mission and need to determine lift, drag, thrust required, and fuel burn at each mission point. The next part is the parametric cycle analysis, where the students are required to select the design choices for the engine cycle after comparing potential engine performance at the chosen engine on-design point. The last part of the design process is the engine performance analysis, where the chosen engine is "flown" off-design to see if it satisfies the mission requirements. After the engine is shown to meet these design requirements, an additive drag calculation is accomplished and the engine run again to determine if the engine still satisfies the design requirements. This last check of performance emphasizes the iterative nature of the design process.

MISSION ANALYSIS

The first part of the project, mission analysis, introduces the student teams to the design problem. The students are placed in the scenario of being young propulsion design engineers working for a contractor. Their task is determining a suitable engine cycle for the USAF's next generation two-engine, thrust vectoring, close air support aircraft. This part of the design process is the conceptual design phase, where engineers, logisticians, and technical support team that includes aircraft and mission profile specifications (see Table 17—6 for completed spreadsheet).

The students must complete the spreadsheet and develop the relationships for lift, drag, thrust required, and fuel burn to calculate a usable fuel remaining at the end of the mission. The design constraint for this mission is the aircraft must land with 1500—1510 lb_m of usable fuel remaining after engines are shut down. The students are led through an algorithm for this analysis (see Figure 17—9). Not all necessary information is available at this time, and the students must use an average specific fuel consumption that results in required fuel remaining at the end of the mission. Class time is used to review the RFP to ensure the students know what is required. The mission presented is a simple mission as given in Figure 17—10.

The mission profile consists of the following mission legs: 0—1 Warmup/Taxi, 1—2 Takeoff, 2—3 Climb, 3—4 High Cruise, 4—5 Descent, 5—6 Combat, 6—7 Low Cruise, and 7—8 Landing/Taxi. Guidelines for fuel usage are given in the RFP. Standard percentages for fuel used are given for all legs except the High Cruise, Combat, and Low Cruise legs. These are the legs where the team must determine the amount of thrust required for the aircraft. The aircraft is expected to combine some of the best features of the A-10 and AH-64 platforms, including survivability, lethality, efficiency, and maneuverability, while being cheaper, lighter, and more versatile in extreme weather. Thrust vectoring only occurs at the combat condition and the vector angle is 45°. The aircraft loiters for 100 minutes at the combat conditions and then expends its ordinance. The aircraft specifications are given in Table 17—6. These values are programmed into the spreadsheet template that each student team is given. Students then develop the drag model and ultimately calculate the uninstalled thrust. A separate exhaust turbofan has been selected as compressor pressure ratio, fan pressure ratio, and the bypass ratio. The designers are required to calculate the uninstalled thrust, F, and the uninstalled specific fuel consumption, S. To relate these quantities to their installed counterparts, T and TSFC, the teams use a first "guess" for the installation loss coefficient, φ or PHI, as given in Table 17—6.

Fuel-used calculations for cruise legs are made with a user input value of S. The spreadsheet is developed in such a way as to allow an update of the engine performance S values at the off-design condition, when this analysis is accomplished later in the design process. A sample spreadsheet with example values is given to the students so they can validate their spreadsheet formulas for accuracy. However, the numbers given do not represent a viable engine solution.

To satisfy the first phase of the conceptual design process, Design Project I, the team must deliver a written report containing:

1. A working spreadsheet
2. Information on current technology medium/high-bypass turbofan engines (i.e., provide typical values for specific thrust, specific fuel consumption, engine diameter, π_c, π_f, α) for attack aircraft
3. A discussion of the purpose of doing a mission analysis
4. The complete mission description
5. Full and complete discussions of the equations used in the spreadsheet and their corresponding assumptions
6. The average S needed for the mission and what drives the selection of average S
7. The effect of installation losses on engine thrust required and specific fuel consumption
8. A selection and discussion of the engine "design point"

Several skills are used in this part of the exercise. The students use spreadsheets for calculation. While a template is supplied into which equations for lift, drag, and uninstalled T and TSFC are coded, the students still must develop and validate their equations in the spreadsheet. The spreadsheet equations also make use of basic principles learned in fluid mechanics and aerodynamics, reinforcing the concepts of lift, drag, thrust, and fuel consumption. Next, the students must research current technology engines of this size. They are shown the importance of historical data from which to make projections for future technology capabilities.

TABLE 17–6 Design Project—Part 1: Mission Analysis [17-5]

Aircraft Specifications	
Number of Engines:	2
Ordinance (lb_f):	16,000
Dry A/C Weight (lb_f):	22,000
Usable Fuel (lb_m):	9500
Parasite C_D:	0.023
k:	0.045
Wing Area (ft^2):	400
C_{Lmax}:	2.8
TO distance (ft):	3000

Mission Specs:	Altitude ft	Density lb_m/ft^3	Temp R	Mach No.	Velocity (ft/s)	Time (minutes)	PHI-
0-1 Warmup/Taxi	0	0.07647	518.7	-	-	-	-
1-2 Take-Off	0	0.07647	518.7	-	-	-	0.17
2-3 Climb	n/a	-	-	-	-	-	-
3-4 High Cruise	2000	0.0721	511.6	0.65	721	20	0.02
4-5 Descent	n/a	-	-	-	-	-	-
5-6 Combat	200	0.07609	518.0	0.40	446	100	0.03
6-7 Low Cruise	200	0.07609	518.0	0.57	636	20	0.02
7-8 Landing/Taxi	0	0.07647	518.7	-	-	-	-

Mission Particulars:

Leg	Fuel Used (lb_f)	A/C Weight @ Start of Leg (lb_f)	New A/C Weight (lb_f)	C_L	C_D	T Required (lb_f) per Engine	TSFC (lb_m/hr)/lb_f	F Required (lb_f)per Engine	Average S: (lb_m/hr)/lb_f
Warmup/Taxi	38	47,500	47,462	-	-	-	-	-	-
Take-Off	475	47,462	46,987	-	-	6312	-	7605	-
Climb	470	46,987	46,518	-	-	-	-	-	-
High Cruise	1266	46,518	45,251	0.1998	0.0248	2886	0.6582	2945	0.645
Descent	5	45,251	45,247	-	-	-	-	-	-
Combat	4731	45,247	24,516	0.4483	0.0320	2134	0.6649	2200	0.645
Low Cruise	996	24,516	23,520	0.1282	0.0237	2270	0.6582	2317	0.645
Landing/Taxi	12	23,520	23,507	-	-	-	-	-	-

Parametric Cycle Analysis

The second part of the design project is to accomplish the on-design parametric cycle analysis. Mattingly et al. (1987) state three reasons for PCA:

Firstly, off-design analysis cannot begin until the design point and the size of the engine have been chosen by some means. On-design analysis is much less tedious and time consuming than off-design, often providing mathematical optima that can be directly exploited. Thirdly, and most important, identifying the combinations of design choices that provide the best performance at each mission flight condition reveals trends that illuminate the way to the best solution.

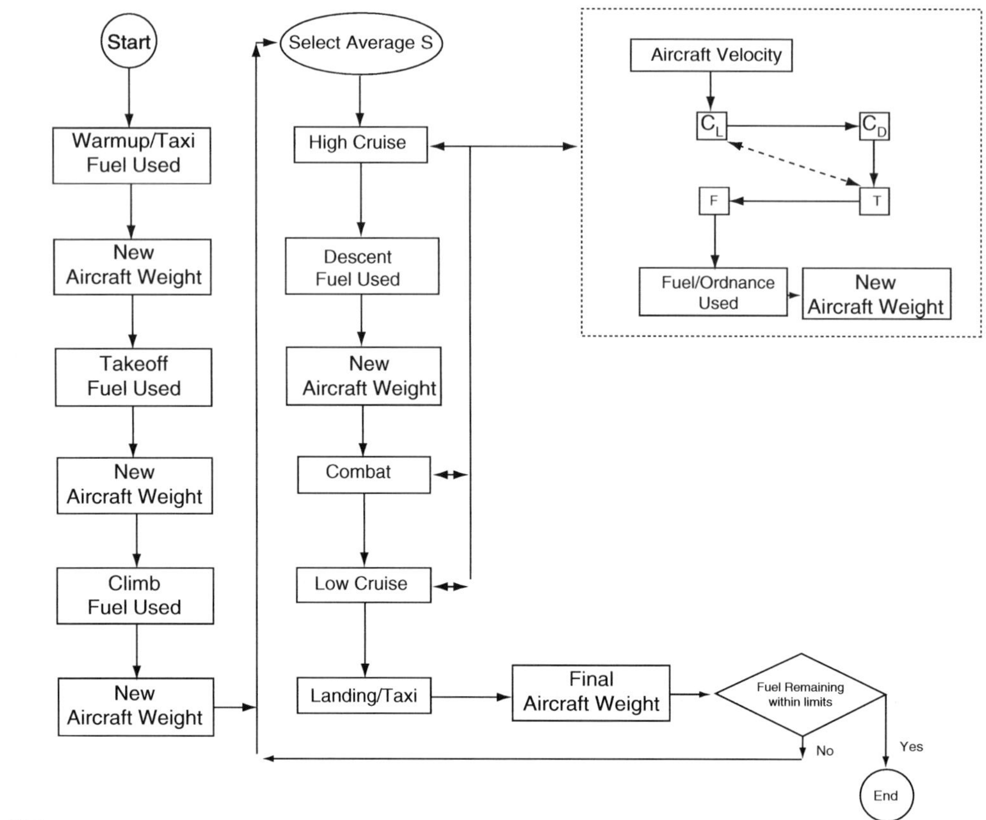

Notes:

1. ⟷ means execute the dashed-boxed routine and return.

2. ⇠⇢ means you must iterate on a solution for the *combat* phase.

FIGURE 17–9 Algorithm for mission analysis [17-5].

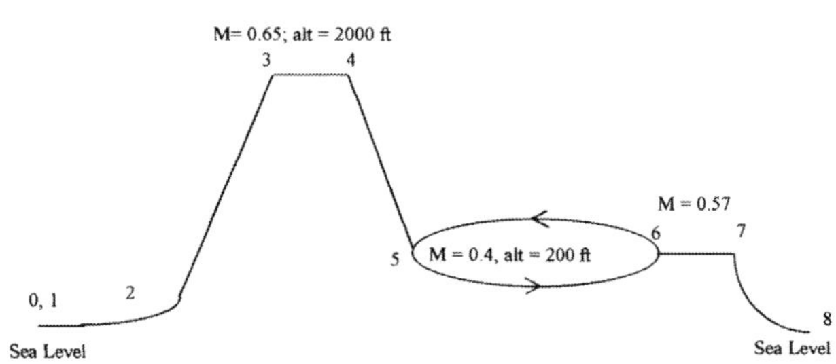

FIGURE 17–10 Mission profile [17-5].

Computer labs introduce the students to each of the commercial programs used: PARA.EXE and PERF.EXE. In the lab, the students are given a self-paced tutorial worksheet to familiarize themselves with the operation of the program. This has worked remarkably well as PARA.EXE is a Windows-based program and is menu driven. The program calculates the on-design cycle performance that was previously discussed in class. In the Design Project I handout, the students are given a discussion of how to generate data and how to present data using a carpet plot. A second page explains how to import the data into a spreadsheet and then create the plot. PARA.EXE calculates engine performance and generates the data. The carpet plot displays possible engine design choices that result from varying the compressor pressure ratio, bypass ratio, and fan pressure ratio (see Figure 17–11). This is a parametric study of the three design variables available to the turbofan designer; π_f, π_c, and α, and observation of the trends in specific thrust, $F/\dot{m}$, and specific fuel consumption, S, as the design variables are changed. The program PARA.EXE, developed by Dr. Jack Mattingly, is used to input the on-design conditions and output the $F/\dot{m}$. and S for variations in the design choices. The results allow the design teams to narrow the field of possible engine designs. The students are given a range of values for the design choices that must be explored, as reported in Table 17–7.

A carpet plot shows how $F/\dot{m}$ and S vary for a given π_f, and variable π_c and α. Plots are generated for the range of π_f. For each π_f value, the design team must determine combinations of π_c and α that might satisfy the maximum average S requirement determined from the previous Design Project I exercise. Drawing a horizontal line on the carpet plots corresponding to the maximum allowable

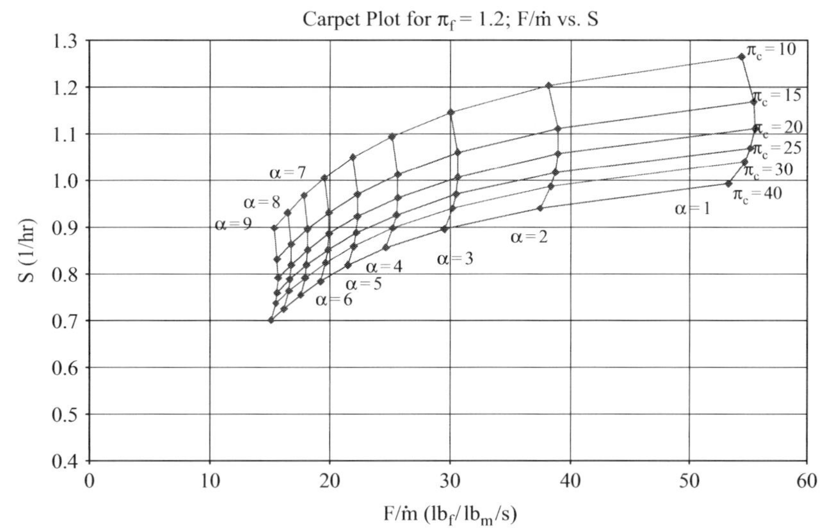

FIGURE 17–11 Typical carpet plot [17-5].

TABLE 17–7 PCA Range of Design Choices [17-5]

Design Choice	Limits	Increment
π_c	10–40	10
α	1–9	1
π_f	1.4–2.3	0.3

average S does this. Based on these plots, the teams have now identified many different engines that have the same value of S at the design point (albeit different values of π_c, π_f, and α), with this value equaling the average S calculated from Design Project I. Each possible engine design might meet the mission requirements, although it is still too early in the design process to tell for sure, as the analysis is incomplete without an off-design calculation. The students must choose the final engine cycle after carefully considering the trade-offs required. The students must also justify their answer by looking at factors such as pressure rise per stage, number of stages, cost, weight, size, etc.

After generating the carpet plots, the students must identify three possible engines that will satisfy the mission constraint of 1500–1510 lb_f fuel remaining at the end of the mission. Trade-offs between thrust and fuel consumption are discussed, and the students are required to justify their choices for the engine cycle based on their knowledge of technology, performance, cost, etc. Class time is devoted to going over the Design Project II handout and parametric cycle analysis. During this part of the design process, the teams begin to explore the vast number of possible turbofan designs that might be encouraged to use the combat loiter flight condition (M = 0.4, altitude = 200 ft) as the on-design point. Since the aircraft will spend about 70% of the mission time at this condition, this is a logical candidate.

The requirements for Design Project II are as follows:

1. Justification for doing a PCA. What does it reveal, in general?
2. The significance of picking an on-design point. Why is an on-design point needed?
3. Inclusion of Design I mission analysis and how it relates to PCA.
4. Physical explanations of the trends in S and F/ṁ when each of the three design variables are changed.
5. Selection of three engine designs (unique combination of π_c, π_f, and α) that meet the maximum average S requirement. Also required is a discussion of the justification with regard to size, potential cost, unique technology requirements, etc.

Design Project I is intended to be an introduction to Design Project II as well as part of the final report (Design Project III). Previous spreadsheet errors should be corrected by now. Students are encouraged to begin collecting a list of figures, nomenclature page, references or documentation, etc. as they go. They also should include a closing statement/paragraph regarding how this PCA effort will provide the foundation for the next step in the design process, engine performance analysis.

Engine Performance Analysis

The last part of the project is the off-design engine performance analysis accomplished on the three engines chosen in Design Project II. The chosen engines must "fly"

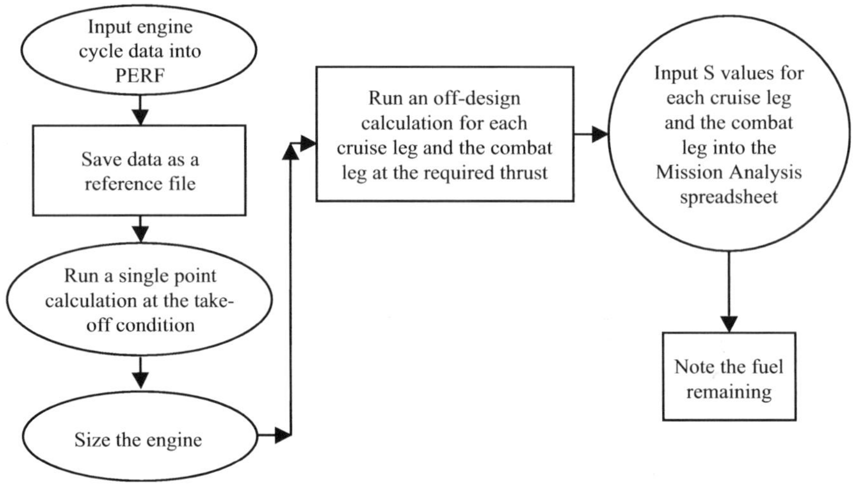

FIGURE 17–12 Algorithm for EPA [17-5].

the mission and meet the required performance and the mission constraints. The EPA is conducted using commercial software PERF.EXE. The on-design file from PARA.EXE is input into PERF.EXE and Figure 17–12 shows the algorithm for each cruise condition. For the chosen engines, the off-design values of the uninstalled specific fuel consumption, S, at the two cruise legs and the combat leg will be input into the mission analysis spreadsheet written in Design Project I. A final selection of the best engine and a justification of this choice is required.

By comparing the different fuel-remaining values generated for each engine, along with relative knowledge of engine technology, cost, reliability, weight, materials, and size, the student teams refine the engine design for the aircraft and decide on a final engine configuration described by a unique set of π_c, π_f, and α values. The output of the spreadsheet program, as previously discussed, will be a final fuel-remaining value for the engine/airframe combination. Using this selected engine, the students do a sensitivity study to determine if there is an additional benefit for slight changes in the design choices. Five additional engines are run to determine if the selected design choices are optimum (see Figure 17–13). From the fuel remaining vs. bypass ratio plot and relative knowledge of cost, technology, etc., the design team chooses one engine (a combination of π_c, π_f, and α) to satisfy the RFP. This engine must lie on one of the two π_c lines from the sensitivity analysis, it must meet the fuel remaining limit (no more, no less than 1500 lb_f), and it will most likely require linear interpolation to determine the appropriate. The design team must then take this new engine, size it, and run the mission spreadsheet with the new S values to ensure the final design will satisfy the requirements.

As a final step, a calculation is made to determine the actual additive drag of the chosen engine at the takeoff, high cruise, combat, and low cruise mission legs. To do

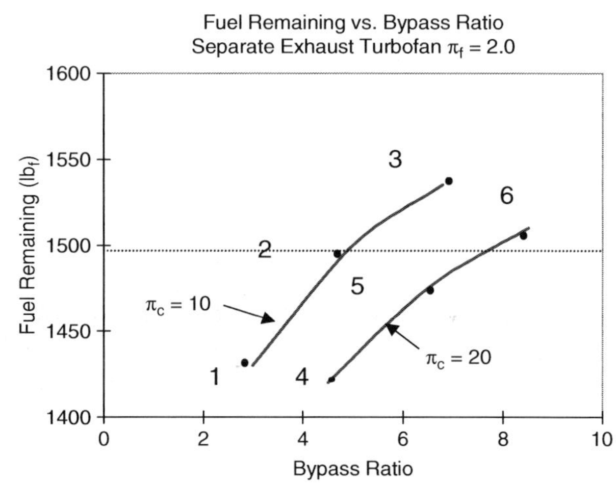

FIGURE 17–13 Sensitivity study [17-5].

this, the program DADD.EXE is used. This program uses the PERF.EXE results at the different mission legs. The students also perform one hand calculation at the takeoff condition for verification purposes. The program DADD.EXE will ask for on-design and off-design values of the Mach number, altitude, and mass flow. Once the additive drag is calculated, the new values for φ are input into the mission spreadsheet and the fuel remaining is recalculated. Students discuss how good the original assumptions were for additive drag and what the impact on the final design is. The next step, iteration, is discussed but not required for this project. The iterative nature of the design process is emphasized throughout the course but only one pass through the process is accomplished. The students must deliver a final design report incorporating Design Projects I and II with the following new information added:

1. A general discussion which answers the "What is it?" and "How is it done?" of EPA.

2. A summary table listing the results of the EPA for the chosen engine. Include values of π_c, π_f, α, S (at each cruise/combat condition), and the amount of fuel remaining at the completion of the mission.
3. A plot of fuel remaining vs. bypass ratio, along with a complete explanation of the observable trends in fuel remaining as functions of π_c and α.
4. A table for the additive drag. Comment on how close the estimated values are to the calculated values. How do the calculated values affect the fuel remaining? If the fuel remaining decreases due to the higher additive drag, what might the designer do?
5. Strong justification for why the chosen engine should be produced.
6. Computer/spreadsheet output that provides the supporting information for all plots, tables, figures, calculations, and numerical results.

Again, a variety of skills are used to arrive at the finished product. Written skills are present with the final design report. Students had the benefit of feedback on their writing skills for Design Projects I and II. An oral presentation is given at the close of the project, giving important performance parameters and a justification for why their engine is the best. After all students have presented their engines, the class determines which engine should be declared the "winner" of the competition. The students used a variety of tools, including spreadsheets and commercial software, to aid in their design analysis. They made difficult choices concerning the trade-offs associated with design of gas turbine engines. A sense of accomplishment is evident in their confidence when they complete the course.

CASE STUDY 5: GAS TURBINE UNIVERSITY LABORATORY STUDY*

An important part of teaching a gas turbine course is exposing students to the practical applications of the gas turbine. This laboratory proposes an opportunity for students to view an operating gas turbine engine in an aircraft propulsion application and to model the engine performance. A Pratt and Whitney PT6A-20 turboprop was run at a local airfield and engine parameters typical of cockpit instrumentation were taken.

The students, in teams of two, then modeled the system using the software PARA and PERF in an attempt to match the manufacturer's specifications. This laboratory required students to research the parameters necessary to model this engine that were not part of the data set provided by the manufacturer. The research and modeling encompassed areas such as technology level, efficiencies, fuel consumption, and performance.

The end result was a two-page report containing the students' calculations comparing the actual performance of the engine with the manufacturer's specifications. Supporting graphs and figures were included as appendices. The same type laboratory could be adapted for cogeneration gas turbines. Over 121 colleges and universities have cogeneration facilities on campus and that presents a unique opportunity for the students to observe the operation of a land-based gas turbine used for power generation.

A 5 MW TB5000 manufactured by Ruston (Alstom) Gas Engines is available on the Baylor University campus and is highlighted as an example. Potential problems encountered with using the Baylor University gas turbine are discussed, which include lack of appropriate engine instrumentation.

Much emphasis is being placed by ABET EC2000 (www.abet.org) on topics such as the students having an ability to design and conduct experiments, as well as analyze and interpret data; however, with budget constraints, purchasing commercial equipment for many types of practical laboratories is prohibitive due to cost. Floor space at many universities is also at a premium.

If a school has the equipment for student laboratories, leaving the equipment set up year round is often not an option. At Baylor University's mechanical engineering laboratory, space available for use by undergraduates is very limited. The single fluids/thermal laboratory must serve as classroom, laboratory, and storeroom for any experimental apparatus.

Much of the equipment is used for demonstration purposes only once or twice each year. Due to cost and space limitations, the addition of desirable labs involving large-scale hardware is restricted. The possibility exists for virtual laboratories or computer exercises to fill some of this void. However, students need exposure to actual "hands-on" experiments with hardware to solidify much of what is learned in the classroom. There is no substitute for practical experience.

In thermodynamics, the basics of the first and second law are discussed and an introduction to the Brayton cycle is accomplished. The students learn about the individual components, such as the compressor, combustor, and turbine and link these components in a cycle toward the end of the course.

A senior elective course at Baylor University looks at the gas turbine engine as a propulsion system. The course has students design an engine cycle for an aircraft application. This includes choosing the appropriate cycle compressor pressure ratio, fan pressure ratio, and bypass ratio. After selecting these design choices, the student looks at the engine cycle's off-design performance. As part of

* Source: [17-6] The work of K.W. van Treuren, including extracts from K.W. van Treuren. "An Application Oriented Gas Turbine Laboratory Experience," 2004-GT-5376.

the course, two lessons each on rotating machinery, combustors, and inlets/nozzles are included. One lesson shows the history of the component with current technology and future trends. Without hardware, appropriate pictures of components are included for the students to see. The additional lesson develops the design methodology and shows design considerations for the components. Included in these lessons are short discussions on operational issues such as stall, surge, flame instability, and emissions. After studying these topics it would be a natural extension of the class to look at the performance of an actual gas turbine engine.

Nomenclature

C_{po}	=	Specific heat
C	=	Work output coefficient
C_{prop}	=	Propeller work output coefficient
C_{tot}	=	Total work output coefficient
ESHP	=	Equivalent shaft horsepower
f	=	Fuel air ratio
$\dot{m}_o$	=	Mass flow rate of air
rpm	=	Revolutions per minute
shp	=	Shaft horsepower
T_0	=	Atmospheric temperature

Aircraft Gas Turbine Engine Experiment

Several options exist to give the students an experimental laboratory experience with gas turbines when such a facility does not currently exist on campus. The simplest option, and perhaps the most costly, would be the purchase of a commercial gas turbine test system. Prices range from \$30,000 to well over \$100,000.

Looking at the capabilities, these machines are very versatile and offer the greatest opportunities for students to learn about gas turbines in a laboratory setting. Most engine test systems commercially offered have the capability to perform basic cycle analysis in addition to detailed experiments on component performance. The test systems also have the capability of being integrated with computers for control of the experiment/engine and for data acquisition.

However, the cost is prohibitive for most universities. At best, these high-dollar items must be budgeted several years in advance and integrating their use into the curriculum during the appropriate courses to the maximum extent possible is absolutely necessary to justify their cost.

Another option would be to design and build a gas turbine test system similar to the commercial systems. Suitable engines exist, such as those currently being used in R/C model aircraft applications. While the design process itself is beneficial, the cost is still high (about \$5000 to \$10,000) and the time involved with the development cycle can be long.

As part of the gas turbine propulsion elective course, the class visited the Baylor University Department of Aviation Science test facility at a local airport and was able to take data from an operating turboprop engine. Not every university has a Department of Aviation Science with such a facility. Other options to run a turboprop engine might be available from a local community college or aviation maintenance school.

Another avenue for exposure to gas turbine engine operation is a cogeneration plant, such as the one located on the Baylor University campus. This work explores the use of the Baylor University Department of Aviation Science test facility and suggests the use of the Baylor University cogeneration plant to augment Brayton cycle instruction in the classroom.

In preparation for the visit to the airfield, two lessons on turboprops are given as a precursor to the turboprop laboratory. At this point in the course the students have studied cycle analysis, component performance, and engine cycle off-design performance. Students understand performance parameters and what figures of merit are used to characterize gas turbine operation. They understand efficiency, specific fuel consumption, and specific thrust. The turboprop lessons introduce them to turboprop operation, including work coefficient. The first lecture develops the equations of performance to include the core and power turbine work coefficient. Since the course is a propulsion system design course, propeller efficiency is also discussed.

The second lecture looks specifically at the engine to be tested. The engine is a Pratt and Whitney PT6A-20 turboprop with the specifications given in Table 17—8. A cross-section diagram of the engine gas path is discussed as well as prominent features of the engine (see Figure 17—14). The students are to run the engine, collect typical cockpit data, and then model the performance of the engine for comparison to the manufacturer's data. The laboratory requirement includes a two-page report with supporting graphs and figures as appendices.

Baylor University is fortunate to have this particular engine available through its Department of Aviation Sciences. The department is focused on the development and qualification of alternative fuels. As part of their program, they have a PT6A20 mounted on a truck bed (Figures 17—15 and 17—16). The engine runs regular aviation fuel in addition to fuels such as ethanol and bio-fuels. The airfield is approximately 10 minutes from campus and is easily accessed by the students. Extra time must be allocated for this laboratory above the normal class time. The students were given one class period (approximately 1 hour and 20 minutes) for compensation but the overall laboratory takes approximately 2 hours including transit time. Upon arrival, the students were given a tour of the engine and facilities. Components, such as the inlet, starter-generator, compressor, etc., were identified by the

TABLE 17–8 Manufacturer's Data in Cockpit Instrumentation Units [17-6]

	ESHP	SHP	Prop RPM	Jet Thrust (lbs)	Fuel Consumption (lb/ESHP/hr)	Fuel Consumption (lb/hr)
Takeoff	579	550	2200	72	0.649	376
Max Cont.	579	550	2200	72	0.649	376
Max Climb	566	538	2200	70	0.653	370
Max Cruise	550	495	2200	68	0.067	369

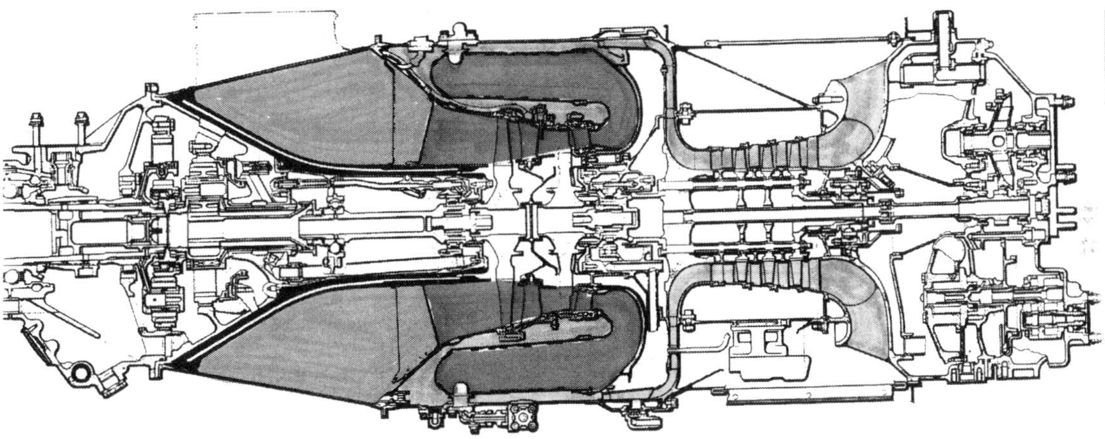

FIGURE 17–14 PT6A-20 Cross-section diagram [17-6].

FIGURE 17–15 Baylor University PT6A-20 gas turbine [17-6].

FIGURE 17–16 Students examining the engine [17-6].

students. The students also examined the propeller connected to the power turbine as shown in Figure 17–18. Eventually the engine technician explained the starting procedure for the engine prior to initiating the sequence. The control panel was equipped with the standard engine instruments found in any cockpit (see Figure 17–17). Table 17–9 displays the experimental data. The engine was run at five power settings and is allowed to stabilize prior to taking data. These data formed the basis for the comparison with the manufacturer data and the theoretical engine simulation.

Colocated at the same airfield with Baylor University's Department of Aviation Science is the Texas State Technical College (TSTC), which also possesses a PT6A-20 engine on a test stand. Figure 17–18 shows this test

FIGURE 17–17 Control panel [17-6].

TABLE 17–9 Typical PT6A-20 Cockpit Engine Data (Fall 2002) [17-6]

Turb inlet (°R)	1410	1437	1482	1590	1626
Torque (ft-lbf)	100	290	500	850	950
Nl RPM (%)	51	70	80	90	93
Fuel	132.5	125.0	220.0	295.0	325.0
Flow (lb/hr)					
Power (ft-lb/s)	11,750	4676	92,153	176,243	20,353
Power (hp)	18	89	160	241	327

stand. It is used for students to become familiar with running an engine and test engines after their disassembly and reassembly. The engine test stand contains all cockpit controls as shown in Figure 17–19. The same cockpit data can be taken with this engine as compared to the Baylor engine. An added bonus with visiting the TSTC campus is the cutaway engines (Figures 17–20 and 17–21) and the spare parts available for viewing (Figure 17–22).

When this on-design engine was run off-design at sea level, several problems were encountered. Firstly, it was not clear where the quoted mass flow rate from the manufacturer was taken. If one assumes the mass flow rate and compressor pressure ratio stated by the manufacturer were given as the sea-level values, then the on-design mass flow and compressor pressure ratio had to be adjusted.

When this was accomplished, the original engine would not run at sea level and input values were changed to determine a combination that would work. After much iteration, the mass flow and compressor pressure ratio were correct for sea level, but the power distributions between the core and the propeller had been changed significantly. More research must be done to find the proper engine data to provide a more accurate model. Calculations were made with engine data, as shown in Table 17–9, but the engine was not able to be run at maximum power on the ground.

As stated previously, some problems were encountered finding parameters, such as efficiencies, and the calculated output did not always closely match the manufacturer's data or the experimental results. The exercise was valuable, as the students learned to do a sensitivity analysis for the various input parameters to decide which parameter might be in error and by how much.

FIGURE 17–18 Turboprop at Texas State Technical College [17-6].

FIGURE 17–19 Control panel for Texas State Technical College [17-6].

FIGURE 17–20 Early turbojet engine [17-6].

FIGURE 17–21 Allison 250 cutaway engine [17-6].

FIGURE 17–22 Miscellaneous engine parts [17-6].

Cogeneration Gas Turbines

Another alternative to the aviation gas turbine is the cogeneration plants. Baylor University is one of over 121 colleges and universities across the United States that has such facilities on campus. Baylor's cogeneration plant is located in the center of the campus and generates approximately 4.1 MW of electrical power 24 hours a day for seven days a week.

FIGURE 17–23 Cogeneration power facility [17-6].

See Figures 17–23 and 17–24 for pictures of the facility. The gas turbine is a 5 MW TB5000 manufactured by Ruston (Alstom) Gas Engines installed in 1989. It uses exhaust gases to heat water and also includes a mister for the inlet to increase performance during the hot Texas summers. The thermodynamics and freshman orientation classes have been visiting the facility as a field trip. The purpose of these trips is to expose the students to power generation as an illustration of the cycles discussed in class.

Of all the colleges and universities listed with cogeneration facilities, only one has made academic use of the facility. The University of Florida had a laboratory/control room simulator with "read-only" computer output. At this time the website is no longer functional, indicating the university has suspended this program. At Baylor University, not enough parameters are being monitored by the facility to enable calculations of performance. The energy complex at Baylor University is only concerned with electrical power output and not with efficiency. More work could be done to look at the possibility of instrumenting the engine to take engine health parameters. Discussions are in progress to explore possibilities. Any data would be taken from existing monitoring screens as shown in Figure 17–25.

In summary, this work investigated several alternatives to purchasing commercial gas turbine demonstrators. Using an aviation gas turbine on a dedicated test stand is the option discussed in the paper, however, facilities at either a local community college, aviation maintenance school, or at a fixed-base operator at the local airfield could provide a suitable alternative.

As an alternative to the turboprop laboratory, cogeneration facilities exist at numerous colleges and universities.

FIGURE 17–24 Closeup of the gas turbine and generator housing [17-6].

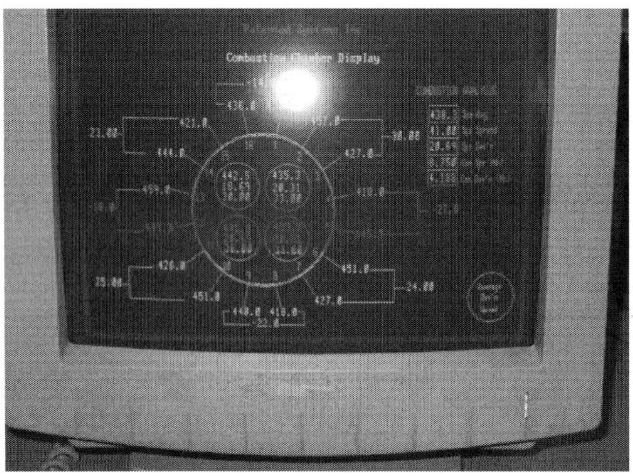

FIGURE 17–25 Closeup of the control computer screen [17-6].

These facilities are underutilized and could be developed for academic purposes.

CASE STUDY 6: OEM WORKING WITH SEVERAL UNIVERSITIES ON GAS TURBINE PROTOTYPE DEVELOPMENT*

The Siemens H-class system (Figure 17–26) is the result of an intensive R&D program to evolve a competitive, efficient, and flexible fully air-cooled engine (Figure 17–26). It is also the first new frame to be developed since the merger of Siemens KWU and Westinghouse aiming to combine the best features of both companies' existing product lines with advanced technology.

With less complexity in the engine and plant leading to greater operational flexibility, the fully air-cooled engine offers a much quicker startup time than a steam-cooled system, as shown in Figure 17–27, e.g., a hot restart to base load within less than 1 hour.

Extensive component, subcomponent and material testing were performed in the development phase of the SGT5-8000H. Materials testing focused mainly on thermal and mechanical properties. Subcomponent tests involved testing of key features of individual components, for example blade cooling holes. Component testing involved testing of actual or scaled parts under conditions that simulated actual engine conditions as closely as possible. Most of the key gas turbine components (including the turbine and secondary air system as presented below) have already been pre-validated by a number of tests. For example, testing programs were conducted or are in process at the following sites:

- Texas A&M University (College Station, Texas), high Reynolds number turbine airfoil, advanced trailing edge for film and heat transfer, labyrinth seals
- Northeastern University (Boston, Massachusetts), high Reynolds number turbulated channel augmentation and Impingement Leading Edge
- Forschungszentrum Karlsruhe GmbH (Karlsruhe, Germany), high Reynolds number turbulated serpentine channel and pressure drop
- Virginia Tech (Blacksburg, Virginia), impact of turbulence on turbine airfoil film and heat transfer
- University of Duisburg (Duisburg, Germany), pre-swirl particle separation
- Siemens Mülheim Testing Lab (Mülheim, Germany), pre-swirler effectiveness/aerodynamics, advanced sealing concepts
- University of Central Florida (Orlando, Florida), Florida Turbine Technologies, aerodynamics and heat transfer
- Siemens Berlin Test Facility (Berlin, Germany), turbine component testing

The validation effort is culminating in an 18-month prototype test program that began in December of 2007 at E.ON Energie's Irsching, Germany, power plant. Figure 17–28 shows an external view of the SGT5-8000H.

Figure 17–29 shows the SGT5-8000H prototype power plant.

Nomenclature

LCC	Life Cycle Cost
PDP	Siemens Product Development Process
TBC	Thermal Barrier Coating
T/C	Thermocouple
TI	Turbulence Intensity

* Source: [17–7] Courtesy of Siemens Energy, Inc., Adapted with permission from GT 2009-60137, "Innovative Design Features of the SGT5-8000H Turbine and Secondary Air System" 2009.

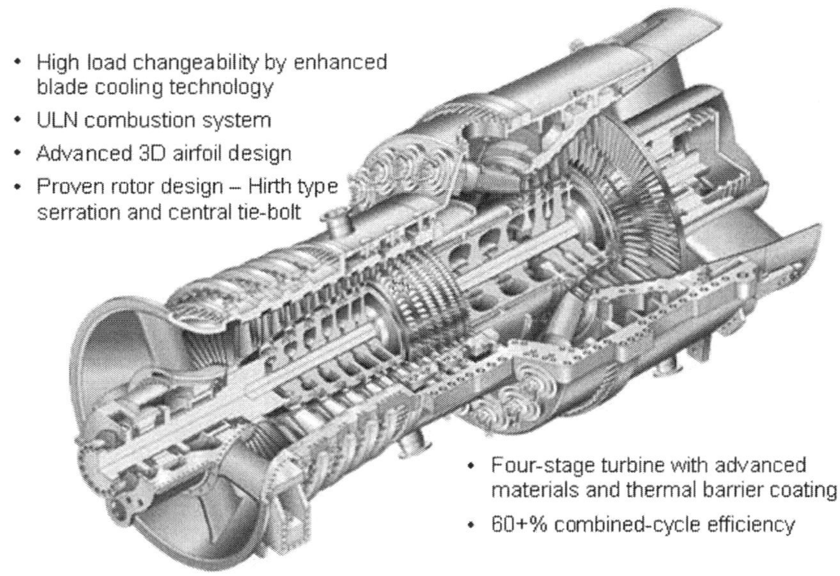

FIGURE 17–26 The new Siemens 340+ and MW SGT5-8000H gas turbine [17–7].

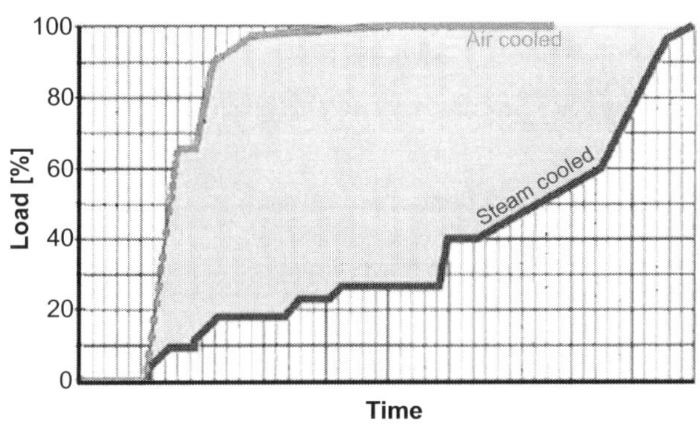

FIGURE 17–27 Expected startup times for an exclusively air-cooled turbine compared to a steam-cooled H-class system [17–7].

FIGURE 17–29 The SGT5-8000H prototype power plant at Irsching [17-7].

FIGURE 17–28 The SGT5-8000H gas turbine in final assembly [17-7].

Turbine Validation

As previously mentioned a number of universities and institutions were selected early on to conduct generic research studies to support the down selection of technologies that would be incorporated into the final design phase. After several airfoil designs had reached a sufficient level of maturity component specific test programs were initiated to support the overall risk mitigation and validation strategy. Two of the more comprehensive test programs concentrated on the first stage highest customer value including performance and operational availability.

Figure 17–30 provides a general overview of the wind tunnel at Virginia Tech where the first vane heat transfer

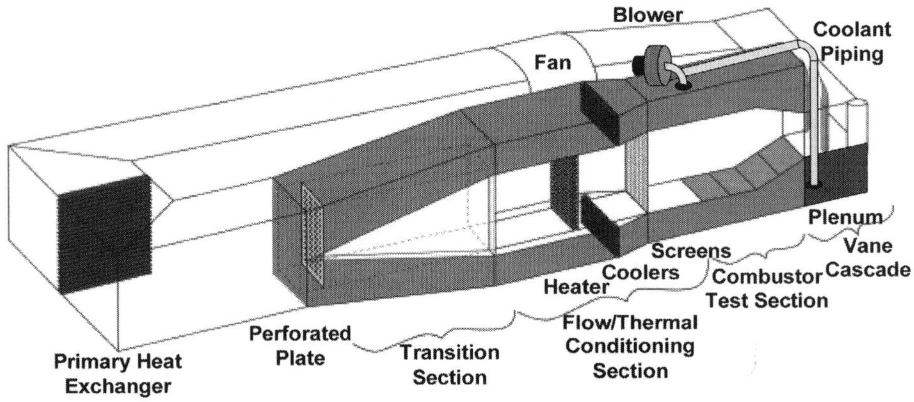

FIGURE 17–30 Overview of the Virginia Tech wind tunnel [17-7].

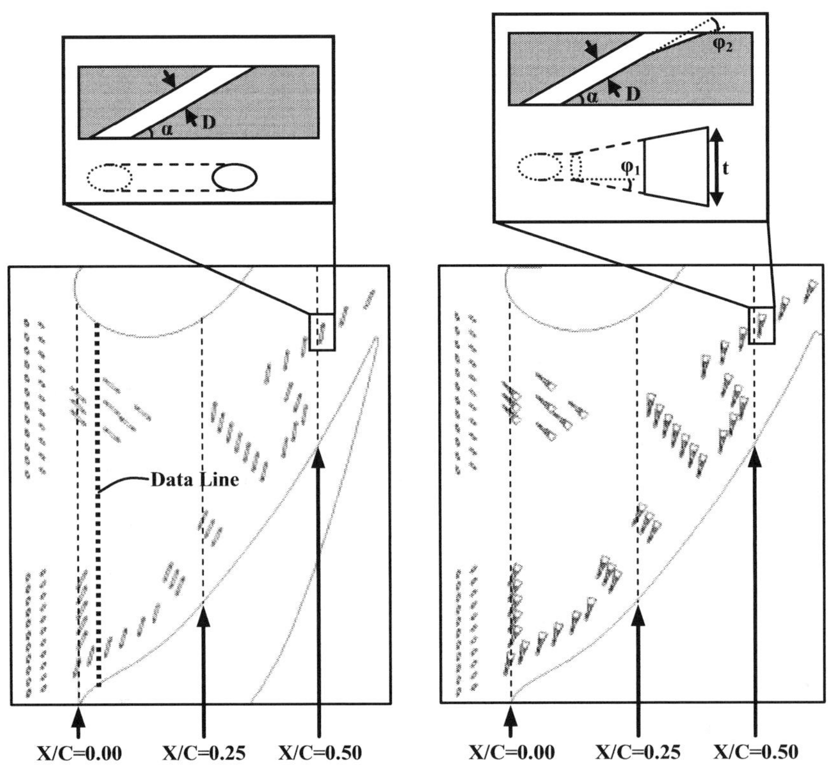

FIGURE 17–31 Film cooling hole configurations tested in the Virginia Tech wind tunnel [17-7].

and film cooling tests were conducted on both the airfoil and endwalls (see Figures 17–31 and 17–32). Although there are numerous papers that have investigated the impact of shaped holes on film effectiveness on flat plates, there is little information found in literature on the impact on airfoil and endwall surfaces. Due to the additional cost of incorporating shaped geometries for cooling holes, additional detailed information was required to verify that the increased cost for manufacturing was offset by a reduction in cooling flow.

The CASCADE test that was conducted at Texas A&M followed a similar philosophy as the Virginia Tech test program, in that the results needed to prove that their incorporation produced the best overall LCC for this particular component. In order to improve data relevance, the CASCADE testing facility was upgraded to achieve near full engine design inlet Mach numbers. Figure 17–33 provides a general overview of the CASCADE test facility.

Turbine Testing Validation

The final assembly of the SGT5-8000H was conducted at the Siemens Berlin manufacturing plant. (Consult the source for details of the testing methods and other details.)

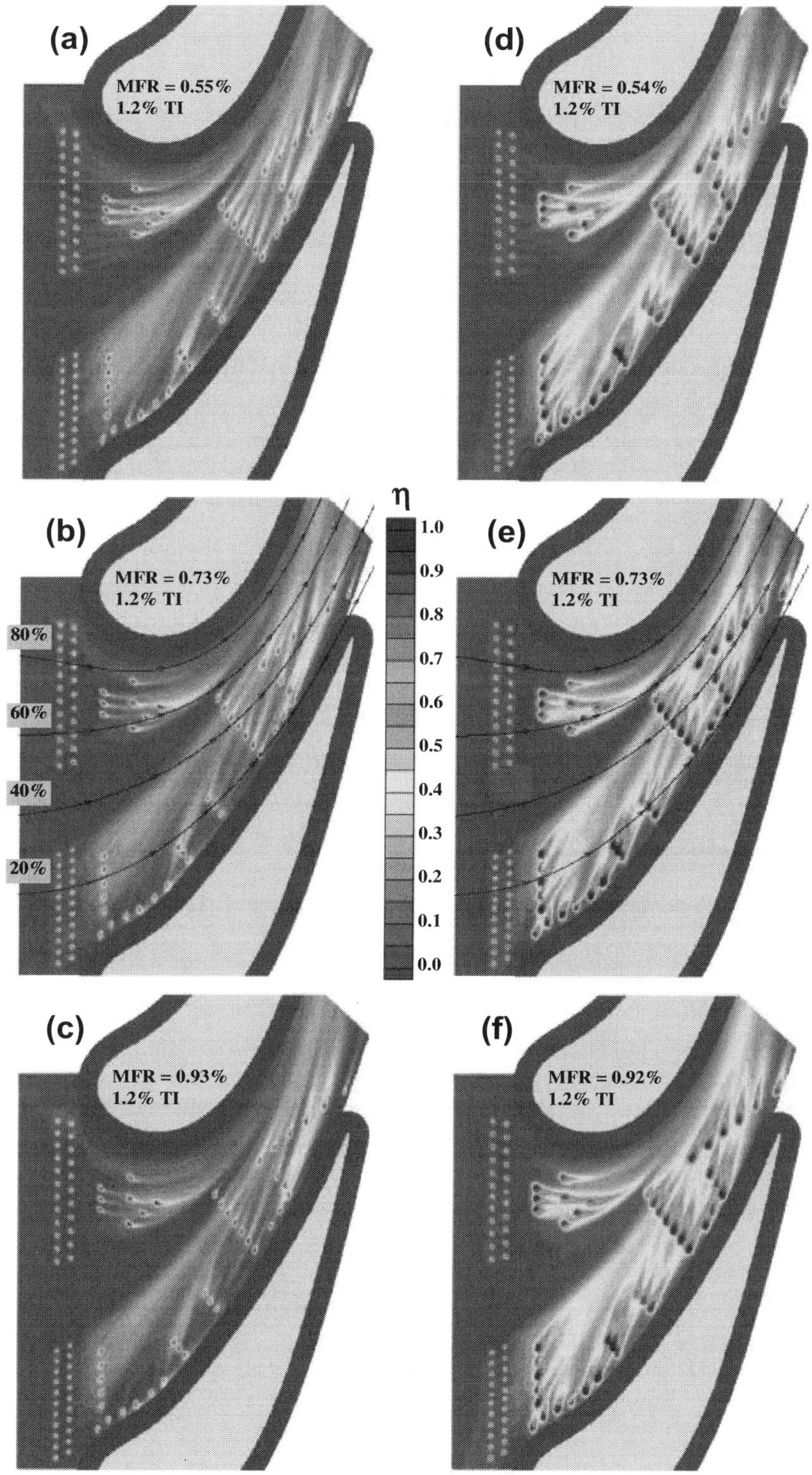

FIGURE 17–32 Turbine vane 1 film cooling comparison of cylindrical and shaped holes on endwalls [17-7].

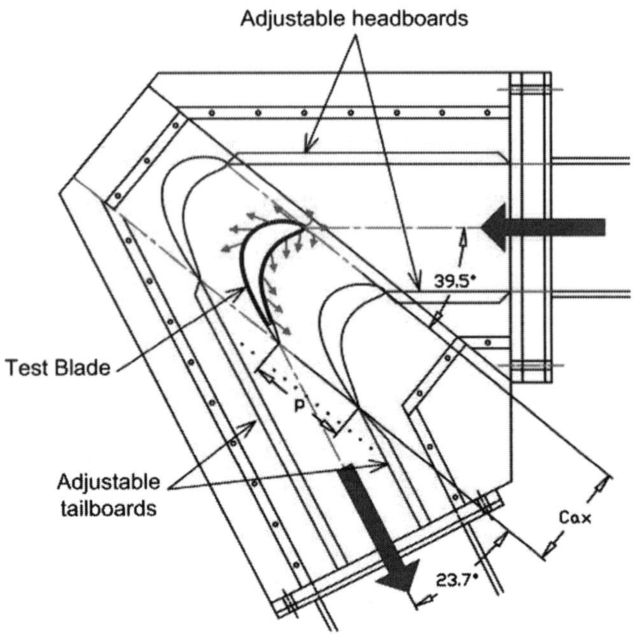

FIGURE 17–33 The SGT5-8000H 1st stage blade CASCADE test facility [17-7].

Chapter 18

Future Trends in the Gas Turbine Industry

"I am enough of an artist to draw freely on my imagination. Imagination is more important that knowledge. Knowledge is limited. Imagination encircles."

—Albert Einstein

Chapter Outline

The future[†] of gas turbine systems design development and the gas turbine business is steered by several factors. Business and political factors are a far greater influence on technology than the average engineer feels comfortable acknowledging. The reader is advised to read this chapter in conjunction with Chapter 14 on business.

What will steer the global scene in the gas turbine industry today?

There may be many answers to that question, however perhaps the most significant change in the gas turbine and gas turbine systems industries over the past several years, since I wrote and edited the first edition, has been the changes in turbine fuels strategy. In 2007 I wrote "A quick look at the world in 2007 indicates that oil and gas will continue to fuel the engine of global growth." That referred primarily to power generation, still the world's largest industry by far. In 2014, that is still true, but the details inherent within the fuel use chart, in specific countries such as the United States, have altered. In the power generation and land-based turbine sector, coal has lost its "number 1" place in the United States, due mostly to the advent of natural gas fracking exploration and production. Coal still remains number 1 in countries like China and much of Eastern Europe, because of those countries huge coal reserves.

SOME NEWER TECHNOLOGIES

Coal Fuel Combustion Related

This is an appropriate point at which to mention two areas of newer development with respect to coal being used as a fuel in gas turbines: oxy-combustion and hydrogen turbines.

Illinois' new FutureGen project will now feature oxy-combustion (not IGCC as originally planned).

Oxy-fuel combustion potentially can be used in plants based on both conventional and advanced technology. Studies have shown that plants equipped with oxy-fuel

[†] [18-1] Working course notes, Claire Soares, 1975 through present; proceedings ASME IGTI panel sessions 1985 to 2003, "ECMS as they relate to life expectancy of GT components," Chair, Claire Soares.

systems could reach nominal efficiencies in the 30% range with today's steam turbines when fueled with natural gas and when capturing the CO_2. With anticipated advances in gasification, oxygen separation, and steam turbine technology, plants using oxy-fuel systems are expected to achieve efficiencies in the mid-40% range, with near-100% CO_2 capture and near-zero NO_x emissions.

In the near-term, efforts are focused on the development of oxy-fuel turbine and combustor technologies for highly efficient (50–60%), near-zero emissions, coal-based power systems.

Oxy-Combustion

The mission of the U.S. Department of Energy/National Energy Technology Laboratory (DOE/NETL) Existing Plants, Emissions & Capture (EPEC) Research & Development (R&D) Program is to develop innovative environmental control technologies to enable full use of the nation's vast coal reserves, while at the same time allowing the current fleet of coal-fired power plants to comply with existing and emerging environmental regulations. The EPEC R&D Program portfolio of post- and oxy-combustion carbon dioxide (CO_2) emissions control technologies and CO_2 compression is focused on advancing technological options for the existing fleet of coal-fired power plants in the event of carbon constraints.

Oxy-combustion, or burning of fuel in an atmosphere of nearly pure oxygen (O_2) to generate flue gas consisting primarily of CO_2 and water, has been established as a credible means to facilitate CO_2 capture from coal power plants. The economics of conventional oxy-combustion processes are currently limited by the parasitic power required for cryogenic O_2 production in conventional air separation units. A further limitation of oxy-combustion is the requirement that a portion of the CO_2 in the exhaust must be cooled and recycled in order to maintain the temperature in the combustion chamber within practical limits.

The CANMET Energy Technology Centre-Ottawa (CETC-O) has developed a world class pilot-scale research facility to perform advanced oxy-combustion research. The CanmetENERGY CO_2 R&D Consortium (Consortium), comprising national and international industry and government partners, was formed to pursue fundamental and pre-competitive research exploring the capture of CO_2 from fossil fuel power plants using O_2 enhanced combustion technologies. DOE provides funding to the Consortium through an agreement with the International Energy Agency Green-house Gas Program.

The CanmetENERGY CO_2 R&D Consortium program started in 1994 with develop-ment of an advanced oxy-combustion pilot-scale research facility—the Vertical Combustion Research Facility (VCRF)—and has undergone eight successive phases of R&D to help bring oxy-combustion, advanced power cycles, integrated multi-pollutant, and CO_2 capture tech-nologies to a level of commercial acceptance. The Consortium has reached Phase 9, development of the first-of-its-kind CO_2 capture and compression unit (CO_2CCU). This unit is capable of separating and compressing CO_2 from combustion flue gas streams for pipeline transport and storage. Part of this work involves studying CO_2 phase changes and the impact of impurities in the gas stream on the capture process. Other ongoing R&D activities include the modeling of advanced near-zero-emissions cycles (including super-critical and ultra-supercritical oxy-combustion plants with CO_2 capture), the development and testing of multi-pollutant control strategies, and testing and optimization of a novel multi-function oxy-fuel/steam burner.

The project goal is to conduct research to further the development of oxy-combustion for retrofit to coal-fired power plants using a 0.3 megawatt thermal (MWth; 1 million Btu per hour) modular pilot-scale facility.

The objectives of the Consortium Phase 9 program are to evaluate an expanded array of new technologies and process options for application in new, clean fossil power plants with CO_2 capture, including:

- Modeling and analysis of an advanced supercritical (ASC) oxy-combustion plant. The work also includes a cost analysis, impact of O_2 purity and O_2 partial enrichment, and overall process performance analysis.
- Performance testing of a pilot-scale CO_2CCU.
- Experimental investigation of CO_2 phase change at liquid and supercritical states in gas mixtures resulting from the oxy-combustion of fossil fuels.
- Testing and performance optimization of a novel multi-function oxy-combustion/steam burner.
- Development of an efficient mercury (Hg) removal process and analysis of multi-pollutant control strategies for oxy-combustion fired power plants.
- Development of an advanced gas turbine cycle (100 MW class) for near-zero-emission combustion systems.
- Development of tools and processes for near-zero-emission advanced gas turbine cycle integrated with a solid-oxide fuel cell (SOFC).

From: National Energy Technical Laboratory, ***www.netl.gov***

Hydrogen Turbines

With hydrogen turbines, turbine systems and components targeted for improvement include combustor technology, materials research, enhanced cooling technology, and coatings development. These technologies are considered key components of FutureGen, which will use technology developed from the core R&D program, including Advanced Turbines, to build and operate the world's first near-zero emissions power plant that will produce electricity and hydrogen from coal while capturing and storing CO_2 through sequestration. It is intended that the successful prototype would be a model for other near-zero emission coal plants that can produce electricity at no more than a 10% increase in cost over a conventional plant, and produce hydrogen at $4/million Btu (wholesale).

Fracking

Natural gas fracking as a process incorporates what could be technically sound technology. Whether it actually is practiced as such, is another matter however. It is often cheaper for oil companies to employ small contractors to conduct elements of the fracking process: the drilling, the explosions to create fissures in the gas-carrying rock, the water injection, the collection of waste water and the dumping of waste water. Small contractors may be less concerned about their corporate image or liability profile than major oil companies. Therefore, working the Marcellus shale in Pennsylvania, for instance, may yield waste water containing poisonous toxins, that some unnamed operator may haul to and dump in New York state, at such times as New York state law does not restrict them from doing so (and Pennsylvania's does).

Media footage that shows water "catching on fire"[1] has attracted the attention of the general public. One of the filmmakers (Josh Fox) who featured that type of event in his film admitted later that he knew the water could have been set alight with or without the neighboring fracking drill probe. In other words, the earth is a seismic place and natural gas occurs not just within rock formation but also at the junction of rock seams. Seismic disturbances, man-made or natural, can result in the natural gas migrating to a point where the gas dissolves partially in ground water.

The truth about fracking "done responsibly and well" or fracking "that causes a public hazard," must therefore be decided on an individual basis. With oil companies chasing themselves to pay (unseen horizontal) drill holes through land, that may include housing developments, the time to assess the "after drilling" results (other than meter the gas produced) may not always happen.

1. www.fb.com/FrackNation (Accessed October 12, 2013)

Drilling sagas notwithstanding, natural gas from fracking burns in a gas turbine, much as conventionally produced natural gas does. The emissions do not alter appreciably in terms of the gas undergoing combustion in a gas turbine.

Improvement of Established Technologies

Steam turbines are significant in the gas turbine industry for two main reasons. First, development of higher efficiency steam turbines can, in some cases, discourage the use of gas turbines that can employ coal fuel, but using newer technologies, such as those discussed above. Second, higher temperature steam pushes the advancement of boiler materials. Even if the boiler is a relatively large one for a steam-only plant, this will benefit the boilers used in gas turbine CC operation.

Supercritical steam has been around since the 1950s, but USC (ultrasupercritical steam) has become the "norm" for many coal fuel users. Now researchers and manufacturers are exploring AUSC (advanced ultrasupercritical) steam.

Ultrasupercritical Steam

OEMs and research are taking the steam turbine towards ultrsupercritical steam (steam in excess of 700 degrees Celsius). Metallurgy is developed, often by government agencies, such as the US DOE or OEM's given a grant (by the US DOE in the United States or the EU in Europe). See the appendix on steam turbines in Chapter 4.

Supercritical Steam

Higher temperature steam gives a power facility more heat to extract than lower temperature steam. Supercritical steam plants have been able to attain efficiencies of about 42%, based on LHV of the coal fuel used versus about 31% with some conventional steam and single-cycle gas turbine stations in existence. Condensing turbines extract the most possible heat from the steam. Reheat, single cycle or otherwise, is used to help attain supercritical steam conditions.

For instance, Shidongkou, China's first supercritical steam plant has 2600 MW units and uses single-cycle reheat steam and condensing steam turbines to attain 42% thermal efficiency. This is a steam-turbine-only plant, but it illustrates that coal derived steam would work as well with a combined-cycle steam turbine. Put another way, for the same coal input, if the heating surfaces required in a subcritical boiler total 24,100 square meters, the figure for a supercritical boiler would be 15,810 square meters. (Also, if the boilers were the same 12 meters in diameter, the subcritical one would be about 10% taller.) These figures directly affect efficiency and thus fuel consumption and emissions in both combined-cycle and steam-turbine-only applications.

The cost of power production not counting environmental equipment is typically 5% lower for a supercritical steam plant versus an only-steam-turbine plant. With environmental equipment factored in, this percentage rises, depending on the environmental equipment required.

Efficient Steam Turbine Condensing

Because a condenser can seriously affect the back pressure on a turbine, its efficiency can considerably alter a plant's overall efficiency (and therefore its fuel consumption). For instance, for a condenser backpressure of 45 mbar where, in a specific plant 3 MW electrical output power is "gained," the corresponding "gain" for 80 mbar is 1 MW and for 115 mbar, 3 MW are lost.

Note also that it is important to the environment that cooling water be "returned to nature" as cool as possible, so condensers also have a direct effect on the environment.

Removal of Emissions After They have been Produced

After the fuel (coal, oil, distillate, or gas) has been burned, there are SO_x, fly ash, and particulates, NO_x, as well as dioxins and heavy metals, to be removed. Separate systems can handle each of these or combination solutions can be sought by a power company, depending on the plant configuration.

CO_2 is vastly more expensive to remove from solution (after it has been produced) than to limit its production in the first place. The latter is done by raising cycle efficiencies to reduce the fuel burned.

Fuel System Variables and Versatility

Regardless of the emissions reduction method, the variables involved include type and grade of fuel and external atmospheric conditions.

As coal is gas's greatest fuel rival, progress with coal fuel and steam turbines directly affect the gas turbine business. Fuel quality affects how "good" the coal is. Consider, for instance, the following grades of boiler coal used in China:

Source/Grade/LHV (lower heating value) in MJ/kg/ash % by weight.

Jiangxi/Anthracite/24/19

Datong/HV (high volatile) bituminous/28/13

Kailuan/HV bituminous/13/41

Hebi/LV (low volatile) bituminous/28/16

Shulan/brown/12/30

Although all these coals contain roughly between 1% and 2% of sulfur, the furnace, boiler, and associated environmental equipment for each obviously differ greatly, as does the thermal efficiency (amount of fuel required per unit of power).

Original equipment manufacturers use a combination of techniques for NO_x reduction, each with their own trade name. Alstom's, for instance, include a technique they call *tangential firing* (TF), which uses dedicated fuel and air compartments; *TF with overfire air* (OA), which subdivides the combustion air to reduce the excess oxygen provided and thus retard NO_x formation; and *low NO_x concentric firing system* (LNCFS). The latter is an extension of TF with OA as it delays the secondary air from mixing with the fuel.

Several OEMs now make trifuel burners, which are important with the fuel versatility required especially in newly industrializing countries. Mitusi Babcock's trifuel burner, specifications for which claim below 10 ppm NO_x formation minimum, have been retrofitted in Castle Peak, Hong Kong, a station owned by China Light and Power.

Carbon Capture

Carbon capture is a challenging and expensive prospect and increasingly countries are pooling their resources to improve current technology. China and the U.S.,[2] which together burn more than 40% of the world's coal, have agreed to jointly develop technology to capture carbon dioxide from power plants and take other steps to combat climate change. The agreement came during the U.S.-China Strategic and Economic Dialogue that started today in Washington. Talks are hosted by Treasury Secretary Jacob J. Lew and Secretary of State John Kerry and include counterparts Vice Premier Wang Yang and State Councilor Yang Jiechi.The two nations will implement "large-scale, integrated" demonstration projects aimed at capturing, utilizing or storing carbon dioxide, according to a statement released by the U.S. State Department. "These demonstrations will engage companies in both countries and allow for enhanced trade and commerce." The countries will also work together to lower emissions from heavy-duty vehicles, increase energy efficiency in buildings and improve greenhouse gas data collection, according to the statement. Implementation plans for those targets will be ready by October, according to the statement.

Technologies that Reduce Required Infrastructure

Offshore platforms and underwater pipelines continue to be the primary means of tapping undersea petroleum based resources. That could change in the future. In 2017[3], the world's largest floating gas factory for producing liquefied

2. http://www.bloomberg.com/news/2013-07-10/china-to-join-u-s-in-carbon-capture-projects.html (Accessed July 24, 2013).
3. http://business.financialpost.com/2013/12/03/record-breaking-lng-ship-launched-bigger-one-planned/?__lsa=a3b7-943e (Accessed May 24, 2014).

natural gas will go to work, north west of Australia. The "ship" does not power itself, it has to be towed, then moored. However, unlike offshore platforms, the *Prelude* can be moved after it has worked the Prelude field for an estimated 25 years. The project is a joint brainchild of Shell (who owns the Prelude field being tapped), Samsung and Technip.

Distributed energy is an effective way to reduce otherwise required infrastructure. The microturbine market, which can accept a wide range of fuels, including biomass, without the need for power transmission lines, is a good example (see Chapter 16).

One of the newer candidates being developed produces gas(Aeous fuel from waste[4] (process belongs to Green and Smart Sdn. Bhd.) using an anaerobic digester to burn in gas engines made by GE. Local bamboo and wood will be the feedstocks. Not all gaseous fuel engines are gas turbines (as microturbines are): some of them are piston engines. However, when machinery increases a country's distributed energy potential, that will affect the gas turbine market.

Resultant Energy/Fuels Usage Chart

The U.S. Energy Information Administration (EIA)[5] has released its International Energy Outlook 2013. It projects that world energy use will climb 56% by 2040. That rise will be driven by growth in the developing world, including increased prosperity in China, according to the report. World net electricity generation is expected to increase by 93% by 2040, increasing from 20.2 trillion kilowatt hours in 2010 to 39 trillion kWh in 2040.

Renewable energy sources will be the fastest growing sources of electricity generation, increasing by 2.8% per year from 2010 to 2040. Natural gas and nuclear power are expected to be second-fastest growing sources of generation, increasing by 2.5% per year.

Coal-fired power generation is expected to remain the largest source of world power generation through 2040 despite increasing by an annual average of 1.8% per year, although the EIA notes the outlook for coal could be changed considerably by any future national policies or international agreements aimed at reducing or limiting the growth of greenhouse gas emissions. The EIA projects total world coal consumption to rise at an average rate of 1.3% per year.

The EIA notes that China, the U.S. and India are expected to account for around 75% of global coal usage in 2040, an increase from the 70% the three countries use in 2010. China is expected to account for around 55% of global coal usage in 2040, while the U.S. is expected to use around 9.2% and India around 10.2%.

About 80% of the increase in renewable power generation is expected to come from hydropower and wind power. Of the 5.4 trillion kWh expected to be added by 2040, 52% is expected to come from hydropower. 28% is expected to come from wind power projects.

The EIA also looks at world carbon dioxide emissions, projecting that total emissions will increase from 31.2 billion metric tons in 2010 to 36.4 billion metric tons in 2020 and 45.5 billion metric tons in 2040 for a total increase of 46%. Coal is expected to account for 20.6 billion metric tons, or around 45.3%, of CO_2 emissions in 2040, while natural gas is expected to account for 10.1 billion metric tons, or around 22.2%, and liquid fuels are expected to account for 14.7 billion metric tons, or around 32.3%.

Europe tends to be more proactive in renewables technology if there is funding to support their efforts. Renewables have already started to have an effect on small sections of the global gas turbine market as the following example indicates.

E.ON[6] announced idling of its 430 MW gas fired Malzenice plant in Slovakia, citing market conditions, in July 2013. Recntly EnBW announced the closure of four of its plants for similar reasons. E.ON SE said the move was prompted in light of low electricity prices. The Malzenice combined-cycle gas turbine was only commissioned less than three years ago at a cost of $522m. The German utility blamed the proliferation of renewable energies across vast parts of Europe for the decision to shut down the power plant as well as muted energy demand in an economy that is stagnating at best. E.ON said that the Malzenice plant operated for only around 5,600 hours, adding the plant is designed to produce power at least 4,000 hours to 5,000 hours per year.

Smart Grid

The power grid in the United States is about 100 years old. It delivers 1 million megawatts of generating capacity, along 200,000 miles of high voltage transmission lines and 5.5 million miles of distribution lines.[7] It is however, an old outdated system that is not very clean environmentally.

In 2007, power plants in the United States emitted 2,500 million metric tons of carbon dioxide; total CO_2 emissions

4. http://www.genewscenter.com/Press-Releases/GE-Signs-Agreement-with-Green-and-Smart-Sdn-Bhd-to-Jointly-Develop-Biogas-to-Power-Solution-for-Mala-45c0.aspx (Accessed May 24, 2014).
5. From: http://www.iea-coal.org.uk/site/2010/news-section/news-items/usa-eia-predicts-world-energy-use-to-climb-56-percent-by-2040? (Accessed July 29, 2013).
6. From: http://www.powerengineeringint.com/articles/2013/07/renewable-surge-sees-another-eon-plant-enter-storage.html (Accessed July 29, 2013).
7. EPRI, The Green Grid: Energy Savings and Carbon Emission Reductions Enabled by a Smart Grid, Technical Update, June 2008. (Accessed July 29, 2013).

nationwide were 6,022 million metric tons, 75.9 million more than in 2006.[8]

The United States's electricity consumption is estimated to increase by 41% by 2030. All forms of fuel, renewable and fossil will be used to fill this gap. Renwables have the disadvantage that they do not produce power at a consistent constant rate.

Recent years have seen the initial construction in the United States of what it will take to finally place renewable energies in a competitive position with respect to the rest of the fossil fuel family: the smart grid. Basically this transmission grid offers high voltage transmission, ensuring that power generated, sometimes intermittently, by power generation sources like wind and solar, will be added to the "power available" mix. The customer can still be charged a price that reflects if the power he has chosen includes more expensive renewables.

This is very good news for the green crowd in the United States. I say "in the United States" because smart grids are not new and other countries are already comfortable with them. One might expect that with sophisticated European countries like Denmark, with their 50% overall wind generation mix, but those countries also include China.

There are enough renewable energies in the world that it could free itself of fossil fuel dependence in several decades, any time it chose to initiate that progress. However, using renewable, particularly in the United States, requires the development of several budding technologies, and large oil interests not hiding the successful ones they already bought out. The technological tone for the developing world is set by the United States for the most part, and the United States policy is set by lobbyists, who are frequently powerful enough they can threaten senators and congressional representatives with nonelection "the next time."

Transmission and Distribution Technology Improvements

Some of the easier items to address in the drive for greater efficiency involve retrofitting modern transmission and distribution equipment, including:

- Modern voltage regulators
- Replacing old conductors
- Reactive power compensation equipment, such as static var compensators
- Dynamic power compensation equipment
- Infrastructure review

This requires administrative involvement mainly on a national level and less of the international technical involvement that revamping of the power plant machinery often needs. Action items here, among others, are:

- Improvement of training procedures to reduce outages and inefficiencies not due to equipment
- Review of the incentive system for SPPs
- Incentives added for SPPs, as these can take up the bulk of a demand curve's peaking load in a mature economy
- Review of the transmission system available to take an SPP's contribution to the national grid. Countries like Thailand offer SPPs incentive but their contribution is limited by transmission line size.
- Consideration of the location and performance of substations, although this is best done at design time. For instance, the IPP plant at Kuala Langat, Malaysia, is situated close to one of the national power company's substations as well as next door to the mill that receives some of its steam

Fuels and their Emissions

The gas turbine is now a rather mature animal. Changes that result from its use in land, sea, and air applications result from growth of core engine technology (to produce larger power plants or fill a customer's requirement), innovative repair developments, improved metallurgy and materials strategies and optimized electronic controls, instrumentation, and diagnostics to some extent.

The area of change that may produce more development than the other areas is that of fuel selection and emissions control. Both those two players, fuel and emissions, are frequently run by politics and economics. Consider the following figure.

Competition among fuels for power generation driven by changes in fuel prices.[9]

Technology aside, China, India, and Indonesia (collectively more than half the world's population) have large reserves of coal and access to cheap residual oil that the Middle East is happy to sell. Based on the needs of their swelling population and economic growth, it is likely these giants, all their Asian neighbors, and the countries of South and Central America will change their overall required-fuels mix.

That said, in fairness, China has shown a great eagerness to keep up with the frontrunners in new technology. Even though she might continue to use her old coal plants for instance, China is extremely advanced and proactive in areas that include IGCC, for instance. China's boiler technology (for GT CC or otherwise) is the best because she has worked hard on ultra supercritical steam with her own steam turbine manufacturers.

8. From: EIA, http://www.eia.doe.gov/oiaf/1605/ggrpt/pdf/0573(2007).pdf (Accessed July 29, 2013).

9. http://www.eia.gov/todayinenergy/detail.cfm?id=7090 (Accessed July 13, 2013).

Fossil fuels will continue to reign supreme for probably the next 30–40 years, with renewables making a small but growing dent in the fuels pie. Certain proactive countries, like Brazil, justifiably concerned about their rain forest and its people, have developed an alternative to "big oil:" ethanol made from homegrown sugar cane. They are now energy independent. Note that the sugar cane ethanol produces about 8:1 times the energy required to produce it. With U.S. corn ethanol production, the ratio is 1:1.

The corn lobby in the United States is a strong one, however. The corn growers make more money with the ethanol train than they do selling the corn as food. Less food in the pipeline generally also mean some poor people somewhere go hungry.

Whatever happens, the gas turbine is going to be around for a very long time and will continue to improve efficiency, either on its own or in concert with cogeneration adaptations. It will continue as large conventional utility models or distributed energy size microturbines. It will burn an increasing number of fuels.

Those fuels, as we saw earlier, include gasified coal or syngas (IGCC); and the coal oxy-combustion process. Finally, the world has reached consensus that a greenhouse gas like carbon dioxide that collects, like a giant spherical sock just below the stratosphere, will warm up our planet. When that happens, the warming is not even, so climate change results: "We have more rain and didn't need it," "we have worse storms, who needs those," "Our desert is growing maybe we ought to go to war for water next." The United States held out on signing Kyoto, but now is doing all kinds of things, some of them strange (like burning natural gas instead of coal in ancient steam turbines), to bring down its level of carbon dioxide emissions.

Carbon storage and sequestration therefore now takes a prominent position in the United States, as it already had in countries like Norway (in their North Sea fields). This is particularly evident in projects such as Future Gen (which will be built in Illinois).

California is more proactive than the rest of the United States with respect to the environment, and is struggling to allow renewables to take over from fossil fuels in the state. If any state could pull this off in the United States, it is likely to be California. When California succeeds, this could create a model for some states, but much depends on a state's individual culture. The study at the end of this chapter (authored by the California Environmental Council) has been shortened in this edition, (Case Study 1 in the previous edition), as some of the parameters the study used have altered, however elements of the study may still be of interest to the reader.

The oil and gas lobby is *very* strong in the United States. The oil and gas "fuel share" in the gas turbine sector, may fluctuate now and then, but may not alter appreciably in our lifetimes. Although 2013 saw an increase in natural gas prices in the USA, the fracking trend may outweigh any gains made by the coal gasification lobby in the USA.

FUTURE BUSINESS TRENDS

The gas turbine's technological progress depends on money just as much as it does human intellect. Without money no turbines are bought, no installations completed. Without actual sales and project investment, there is no working experience to affirm designs, point the way to improvements, or provide a "real-conditions test run" for validating a new feature. However, in turn, money does not "happen" without the right politics, international or domestic infrastructure, or trust among international original equipment manufacturers (OEMs). Therefore part of this chapter is devoted to explaining the business and financial climate in which gas turbines attempt to progress and prosper while maintaining legislated environmental requirements.

Those basic factors remain essentially unchanged in this second edition. Business acumen can add credibility to the text of "request(s) for funding." A large part of that today is political correctness, which today includes "green awareness" in the United States. There is currently a demoncrat in office.

The gas turbine's future is also dictated by the size and growth of the different industrial sectors that use gas turbines. Today, in order, those are power generation, energy production (including oil and gas), aviation (commercial, military, and commuter), marine (ferries, frigates, and other naval ships; cruise liners), and small (domestic, cars) use.

The order in which relevant factors affect the business and their relative importance depend on the demographic circumstances and how they work in concert.

The following factors are the main ones that affect the future of gas turbines in the short (up to 5 years), interim (the next 15 years), and long term (over 15 years).

1. Changing tides (financial, political, legislative, or technological) in the world, including:
   (a) Advantages afforded certain practices because of political favors. The gas industry lobby may be more astute than many with respect to legal positioning. Consider one clause in the U.S. Federal Energy Policy Act of 2005.[10] This bill exempts fluids used in the natural gas extraction process of Hydraulic fracturing from protections under the Clean Air Act, Clean Water Act, Safe Drinking Water Act, and CERCLA. It creates a loophole that exempts companies drilling for natural gas from disclosing the chemicals involved in fracking

10. From: "Natural gas in the U.S. Electric Power Sector," Appendix B, May 2012, published by the Center for Climate and Energy Solutions and the University of Texas's Energy Institute and the Energy Management and Innovation Center.

operations, normally required under federal clean water laws. The loophole is commonly known as the"Halliburton loophole" since former Halliburton CEO Dick Cheney was reportedly instrumental in its passage. The proposed Fracturing Responsibility and Awareness of Chemicals Act would ordinarily repeal these exemptions.

(b) The recession that started in the United States around 2008 brought about the demise of several development and research programs. Programs like FutureGen, now being constructed, were in holding pattern for years. The EEC did not escape lightly either. Siemens was awarded the EEC grant to work on higher temperature steam turbines, only to have their funding placed on hold.

(c) The change in priorities caused by wars. Although the Iraq war is over and plans are underway for the USA to pull its troops out of Afghanistan, war creates business alliances and dependencies that can last for decades after military conflict ends. Things get scarce in war-torn countries. The Iraq rebuilding efforts saw a surge in the market for "excess" and "mothballed" equipment lying around.

Companies that service the military look forward to better business when unrest in the Middle East or elsewhere erupts.

(d) The constant drive for profits. Many gas turbine and gas turbine service industries have undergone reductions in work force, with many of the casualties being engineers who work on gas turbine system optimization or related items.

(e) Global deregulation within the power generation business, the changing mix of IPPs (independent power producers), SPPs (small power producers), and MPPs (merchant power producers).

(f) The escalation of technology for distributed power and small power plants that will run a small farm, factory, and even a household.

This refers in part to microturbines in microturbines in hybrid application (with fuel cells or otherwise). The recession that came to roost at the end of the George W. Bush era and persists through 2013 is responsible for little progress being made in terms of further commercialization of these technologies. For instance, the cost per kilowatt for a fuel cell has not dropped appreciably.

Small companies that sell mini-wind turbines or photovoltaic generation systems maintain reasonable sales and survive, if their overhead is not too high. Legislation with local or state governments can help, if for instance, laws make it illegal for neighborhood associations to disallow installation of solar panels on roofs. This has happened in certain states within the United States. Also tax rebates, federal and state have been stepped up in some cases, to reward the installation of distributed renewables with households and businesses.

(g) Global business trends including international trading blocs, resource conservation, and financial crises. The United States' recent recessions (the one prior to 2006 and the one after 2008) has cost many small companies dearly. Some promising technologies have been significantly compromised. One such player has been the turboexpander industry, where the few remaining players either joined forces or have gone under.

(h) The progress of large rival (rivals of fossil fuel) power technologies (including nuclear) and growing rivals (like wind and tidal power).

(i) The progress of support technologies to the small (or large) growing rivals, such as the development and construction of smart grids.

(j) Natural disasters in major financial center areas, like potentially a mega tsunami on the United States east coast or a crippling earthquake in Tokyo or on the U.S. west coast. Tokyo's recent nuclear crisis caused Japan some severe hardships.

2. Environment related factors, including environmental legislation and national and international caucuses (such as Kyoto, Montreal, and Rio) and the technologies and priorities they promote.

3. OEM growth and diversification, particularly:
- OEM acquisitions, joint ventures, licensees, and resulting technological transfer
- OEM business affairs, including production backlogs and vendor alliances
- OEM technological research and development as well as support by and for government programs
- Significant (OEM or government agency funded) technological breakthroughs that will revolutionize the state of the industry
- Performance optimization trends in system or component design

4. OEM financial failures: sometimes promising new technologies are delayed in their progress due to conditions that include the recent economic crisis in the United States. A company called Catalytica that made flameless, ultra-low NO_x combustors was one player that did not survive the onslaught.

5. The effect of powerful gas turbine or end-user lobby groups. For instance:
- Sugar cane ethanol (Brazil's variety) is better for the environment than corn ethanol (which the United States plans to make), by an energy return for energy input ratio difference of 8:1. However, if the corn

lobby is stronger than the sugar lobby in the USA, corn ethanol will still be made.
- What's the use of developing heat pumps that use solar energy trapped in the ground (free) if you lack the muscle to force legislative subsidies that would make that product an affordable option?
- Why would you develop alternative energy technologies that could curtail much of the dependence on fossil fuels if you have no lobby to promote them? Actually, there is a good answer to that one. You develop it and then wait for an oil company to offer you millions to buy it, so it can stay "off the shelves."

6. E-trading of energy (fuel), power (kW), emissions credits, and gas turbine spares.
7. New and unconventional fuel resources.
8. Distributed power: How large does a power plant need to be?

The short- and medium-term future of gas turbines will be largely dictated by the success of the OEMs in terms of their technology. Their business strengths will be dictated by all these factors and can change their finances over decades or in the short term. Further discussion on how the above eight factors can alter the course of the gas turbine world follows.

The Agents of Change

Politics

The lead time for a new gas turbine after production begins, is frequently upward of two years, so the beating the market takes when political unrest on a global scale occurs, is evident. That the major OEMs will survive the fray and be ready to meet the surge in orders, which reinstates itself after power generation development lags growth in GDP, is a given. Many small systems suppliers who survive from monthly order to monthly order may not be that fortunate. The small businesses that will thrive in all circumstances are those that the OEM needs for day-to-day business, such as repair shops that complete piecework the OEM does not want to spend liquid capital on support "in house." Niche suppliers of specialized instrumentation, such as optical pyrometers, that are required for design development and operational investigation probably always will have enough business to maintain their overhead.

War will see some resurgence of orders. Rebuilding Iraq was one example. Current unrest in the Middle East, suggest that history may repeat itself. This may not always mean new gas turbine orders, though, due to the lead times involved. A service industries acquaintance mentioned several gas turbines ordered for Iraq that were "existing, old." Several of these packages are all over the world, collecting dust because the customer "cancelled the order." The already industrialized countries would not likely buy older models if they can afford to wait for the latest ones. However, a nation now officially out of Iraq while keen to be seen doing everything it can to rebuild that county's damaged infrastructure is likely to buy "what still works." In many cases, these old gas turbine packages were never installed, so they are operationally "new" (zero time).

Aviation as Affected by Politics

Commercial flying has seen a turndown due the events in global politics. The result has been felt all through the chain: airline to aircraft manufacturer to aircraft engine OEM to the OEM's employees (layoffs and early retirement packages).

The military engine market saw more activity as recent politics revived some hitherto temporarily neglected military aviation programs. Not that this would necessarily help the progress of gas turbine technology as a whole, however. In the last 20 years, the level of technology with commercial aircraft engines crept ahead of their military engine counterparts. In short, at the turn of the century, war and political unrest essentially do little or nothing to promote gas turbine technology (as different from the second World War, which did much to pave market entry for the gas turbine).

Marine Sector

Gas turbines in commercial marine applications is a relatively new field and activity here has stayed steady. Installations on cruise ships and ferries increase.

Gas turbines in military marine applications (like the U.S. Navy's LM2500 fleet) continues to stay an active business. GE, the OEM on this model, is one of the US military's largest contractors. The very size of the LM2500 fleet promotes further sales, including applications in the non-military sector. The LM2500 fleet in North Sea offshore operations is a large one.

Land-Based Applications

Earlier in this chapter, I discussed the growing profile of coal in the gas turbine world. Technology's direction with coal gasification advances and CSS will promote that.

Any niche-technology system that saves money was given a gracious hearing as oil soared above $70 a barrel in 2006. Not all the new players that this produced still survive today. The commercial industry still scrutinizes expenditures and overhead, cutting back on attendance at trade shows, perks, and even the corporate commuter jet fleet.

Meanwhile the government-sponsored programs for fuel cells, hybrid systems, the oil-free large gas turbine engine program and microturbine CHP systems continue steadily. Government sponsorship, via agencies such as the U.S. DOE continues, even if private industry contributions diminish.

Legislation, Financing, and Business Models

Global Deregulation in Power Generation

The largest economic sector in the world today is power generation. Oil companies now realize that they are prudent to get into the business as independent power producers and become their own best customer for their oil, gas, as well as nonconventional energy and fuel production. Smaller process and petrochemical plants decide to join in the fray, particularly if their incentives include:

- A potential fuel source that ordinarily would be a waste by-product in their process
- Tax incentives to become power producers
- A major utility grid that will buy the power it makes
- Some combination of those factors

Gas turbine manufacturers such as Alstom stabilize their portfolios by buying major shares in mega-sized power companies. Alstom's share in the Midlothian power plant was a case in point.

Government regulations everywhere are changing in terms of deregulation of the power business, environmental regulations, fair competition laws, and trading bloc requirements.

Smaller "guerilla" business units, termed *merchant power producers,* rent or buy stationary or portable turbine packages to supply power for what may be a limited period of time. They generally are short on permanent assets and technical expertise, but they nip persistently at the heels of large IPPs and national utilities, generally for quick profits before they vanish altogether or go to a different location.

IPP candidates have no shortage of countries that would be willing to receive their investment dollars. The catch then is to estimate gains both monetary and otherwise and the work involved beyond that of designing, building, and operating a power plant. What follows could never be exhaustive from a perspective of global IPP activity, but it does illustrate the main features of the IPP market, with particular reference to the continent of long-term maximum potential power demand growth—Asia.

Who and Where Are the IPPs?

The answer to this question is becoming infinitely more complex than it was only 10 years ago. The acronym IPP used to refer to firms that existed for the sole purpose for investing in and building power plants then selling the power to a national governing body. It still does, except the IPP ranks are swelling to include "small" power producers (SPPs).

Small producers include large industrial entities, such as refineries and manufacturing plants, that buy their own power production machinery (generally to avoid expensive brownouts or outages) and make their own power. In some countries, they sell their excess power back to the national grid. The limits of this sale generally are set by the size of the distribution lines available. This small power producer generally gets less of a tariff for its power than it pays for national grid supplied power. As such power producers increase, they lessen demand growth and therefore the required size of new large power plants.

IPP ranks are further being swelled by IPP joint venture companies, which can have as a major or controlling interest partner one of the turbine manufacturers, such as Alstom, Siemens, or GE. Interesting variations on a theme of joint ventures can be arranged contractually with OEMs. Consider the case of Malaysia. ABB (prior to being part of Alstom) made a turnkey delivery and transfer (BOT: build, operate, and transfer) arrangement on the Kuala Langat plant with the Genting Corporation and the Lumut plant with Segari Ventures. GE (General Electric) had a turnkey arrangement with the power plant owned by the Port Dickson group. The 51–49% joint venture that Siemens Westinghouse has with the YTL Corporation has Siemens the controlling partner. In the latter project, the number of Siemens and other expatriate staff will slowly dwindle to zero six years after commissioning. The national goodwill created by firms that assist newly industrialized countries (NICs) with attaining financial independence is immense and hard to measure.

Options for IPPs

- *Build, operate, and own (BOO).* The developer builds, operates, and owns the plant for its entire life. This owner generally secures a PPA (power purchase agreement) and a fuel purchase agreement before commencing construction.
- *Build, operate, and transfer (BOT).* The developer builds and operates the plant for a contractual number of years. Then, the plant is sold to another authority, often a government department or agency after that time.
- *Repower (replace steam with gas or combined-cycle turbines)/refurbish, operate, and transfer (ROT).* In these contracts, the developer may operate the plant for as long as it takes to recover its investment plus a designated profit margin. Another firm or government agency retains ownership of the plant.
- *International purchase.* Especially when caught in the crisis of unexpected growth, countries resort to the

emergency strategy of buying power from a neighbor. Malaysia, in its early 1990s shortage crunch, bought power from Singapore. Malaysia sells some power to Thailand. Whom the power contract is with may have something to do with desired national fuel resource conservation policy. For instance, both Thailand and Singapore would like to buy a great deal more gas than Malaysia currently sells them. Why do some gas-rich countries "hoard" their gas reserves? Possibly because they are aware that any artificially low prices for natural gas will give way to larger profit margins in some years (enough to outdistance the diminishing value of unit currencies).

SE Asian Currency 1990s Crises and Subsequent Recovery

Currency problems in the late 1990s sent SE Asian currencies tumbling. Recent economic woes in the United States have allowed many currencies in and beyond SE Asia to recover much of that lost ground. However, the bottom line is that IPP investments generally yield 18–20% long term.

Overall Asian Investment

The number of IPP opportunities in Asia is immense—bigger than anywhere else in the world. Even once the current economic hassle sorts itself out, the catch is getting financing, and that depends on where the project is and what incentives or drawbacks go with the location.

Any global IPP scene summation must include China, the highest power demand growth area in the world, with India and Indonesia as close seconds.

China has potential investors wary for many reasons, which go beyond any current economic crisis. The Chinese frequently offer 16% ROI (return on investment). It took much gut wrenching to get it up from 13%. Other countries offer between 18 and 20%. Attractive, if their currency is stable.

Older problems in China include a non-globally traded currency, although moves have begun to slowly change that. The currency also carries some devaluation uncertainties. Tariff pricing issues are a further stumbling block.

Major drawbacks with development in India have been the lack of infrastructure and government corrupt practices. Consider though that "corrupt" is often in the eyes of the beholder. "Baksheesh" is a way of doing business in many countries and frequently Anglo-Saxon cultures (other than those restricted by the U.S. Corrupt Practices Act) just pay the "expediting fees" and do business.

All the main turbomachinery players—Alstom Power, GE, and Siemens Westinghouse—are in India, each with a long-term agenda in the power generation industry. GE arrived late, behind its traditional rivals, but its agenda is ambitious. As are Alstom's (separately ABB and GEC Alstom when they first entered India) and Siemens Westinghouse's (Siemens when they first entered India). GE picked a variety of local firms with relevant expertise to joint venture with for market entry. Siemens came in through Siemens India, and ABB came in alone.

India does not smile on 100% foreign ownership or joint ventures in non-high priority industries. Political waves have meant a great deal of stalled legislation and project approvals, even in critical industries such as power. As of 1995, however, the way was cleared for the first few "fast track" power projects. Support industries, such as telecommunications, have been opened to privatization, which helps add overall economic momentum.

Latin (South and Central) America

Global unrest due to situations and conjecture in Israel, Iraq, North Korea, and Afghanistan notwithstanding, some Latin American countries always experience their share of internal unrest. The responsible factors include politics, military coups, vandalism by guerrillas, drug cartels, poverty, and in Columbia for instance, "a culture of nonpayment among our clients." Also someone like Venezuela's Hugo Chavez can affect the United States adversely by tightening his country's oil exports.

Other factors include slow progress of the privatization process and waste of power in an ineffective transmission system (as high as 29% at one power company in Columbia).

The Latin American countries are struggling to rebuild their infrastructure despite all handicaps. Large OEMs have the deep pockets to wait out the period it takes to get recoup their investment.

Africa

The situation is similar to that in Latin America, exacerbated by worse poverty and disease. Nonetheless, development of industry and power generation continues, particularly in countries such as Nigeria that have oil resources. Certain countries have proved to be models of relative stability, such as Ghana and Tanzania. South Africa.

Gas turbine-related development, however, is unlikely to reach the per capita levels currently prevalent in India and China in any foreseeable timeframe.

Global Financing Agencies

The capital available for and invested in a project ultimately gives gas turbine systems and technology the proving ground they require. With the world's drastically changing economy bringing turmoil and political disarray, some of this finance is in short supply. The basic methods of raising it essentially are the same and deserve short mention. This subsection will make constant mention of China and India. This is logical, as the countries collectively hold about half the world's people. Neither country is exactly devoid of unrest. However, both populations demonstrate an intellect and capacity for the study of technology. That bodes well for investing companies to find the trainable labor that gas turbine-related projects require. As an indicator, China took 10 years in the field of aviation technology to cover what had taken the Western world 60 years to develop.

Granting loans (by private banks or multilateral agencies, such as the World Bank) in most contemporary instances depends on the project supporting "sustainable development," which in turn requires sound environmental practices. Approved projects have been cancelled in midstream on this basis alone. For instance ,in July 2013, the World Bank (WB)[11] agreed to limit financing of coal-fired power plants to "rare circumstances" as part of a new energy strategy aimed at tackling climate change. The WB said in its "Energy Sector Directions Paper," updated every 10 years, that it will amend its lending policies for new coal-fired power projects, restricting financial support to countries that have "no feasible alternatives" to coal, as it seeks to balance environmental efforts with the energy needs of poor countries.

The WB has traditionally argued that funding coal-fired power plants is sometimes necessary to bring energy to the world's poorest nations and to help them eradicate poverty, so while a tougher stance has been adopted, there still appears to be wriggle room in order to facilitate countries with no alternative in terms of energy security and the combating of poverty. Analysts say coal is often the cheapest energy source in places like Kosovo, where the WB is mulling whether to support the country's plans for a coal-fired power plant. In its latest paper (post 2000), the WB also backed increased support for hydroelectric power, reversing its decision to abandon those projects in the 1990s under pressure from aid groups that warned they would displace people.

A quick scan of the headlines in IEA's regular email indicates that even as the USA is taking steps to shut down coal plants, the rest of the world is increasingly developing its coal-fired power generation. This is in part due to the fact that the US' generation infrastructure has many rather old steam turbine facilities that are expensive to "refurbish." The Sierra Club and other environmental agencies have also hurt the coal industry in the USA. At the same time however, projects like FutureGen are going ahead. Whether this project will herald more like it in the USA remains to be seen. The answer depends to some extent on the progress made with certain technologies, such as coal gasification.

The concerns of project developers follow those of their lenders. The same factors affect a project's profitability. The factors, such as emissions taxes, may alter over time.

11. http://www.powerengineeringint.com/articles/2013/07/world-bank-agrees-to-limit-financing-of-coal-fired-power.html (Accessed July 29, 2013).

Political Incentives, Infrastructure, and SPPs

Political infrastructure does not always see fit to give incentives to SPP formation. Singapore Power used to be quite adamant about not buying excess power made by SPPs, such as oil and petrochemical firms. For instance, until the late 1990s, Singapore Power had been able to supply all its consumers profitably. It now allows IPP agreements and is likely to relent further in terms of purchases from SPPs. As a consequence of the initial policy, firms such as Petrochemical Corporation of Singapore (PCS) were careful to generate power that was just below their own requirement of about 25 MW, after they had installed their in-house Alstom GT 10, so there was no question of them supplying any to the national grid.

Other SE Asian countries had future SPP development stalled for a while as IPPs have long-tenured power purchase agreements in place. The government is obliged to buy the power they produce. For the most part, these IPPs have been running at base load since being commissioned around the late 1990s. Tenaga Nasional Berhad (TNB), Malaysia's national power company, therefore has not seen much incentive for encouraging its own plants to retrofit cogeneration facilities, let alone encourage SPPs.

Thailand, on the other hand, will take everything it can get. It had created an incentive structure for receiving power from SPPs by the mid 1990s. This is excellent planning, particularly in its recent predicament with devalued currency and a sudden dearth of willing international IPP investors. The Electrical Generating Authority of Thailand (EGAT) is the authority that buys SPPs' excess power. Thailand provides good incentives for SPPs, as is illustrated by the Esso refinery in Sriracha, Thailand. It has two ABB GT35s for its power needs and sells the excess to the national grid. Unfortunately, this is limited by the transmission lines available for the purpose, only 15 kV rated, as in the case in some areas of Thailand.

Thailand's first trash burning plant contract (20 MW) was let to Kvaerner in the 1990s. Thailand is likely to see far more of these plants; many of them will probably be IPP facilities or SPPs who need steam produced from the trash fuel for their other processes.

Thailand also is progressing on biomass burning in around 20 MW increments, using yet another fuel (rice husks, bark, and other plant material) of which it is not short. Other SE Asian countries, such as Vietnam, have used biomass on a small family-sized scale for centuries before their oil and gas resources, as well as the potential for power generation revenues, brought the major OEMs to their countries.

Merchant Power Producers

The latest trend in the fast moving business of independent power producers is that of merchant power producers. An MPP is an IPP that has no long-term contract for sales of most of the power it produces; in other words, no measurable financial security at the time of making the decision to build the plant. MPPs' business and negotiating skills therefore have to be superior, as does their credibility, or they will make little or no profit. New MPPs can compete with established IPPs whose contracts are running out, on level turf.

MPPs mark a trend that is barely five years old. It was fostered by the fact that deregulation creates intense global competition in this industry. Deregulation in turn mutually fosters decentralization, which causes rural industry and power demand to grow. A good example is Brazil, which managed to get inflation below 10%, down from 2500% annually. The net effective spending power of its people rose accordingly, creating a middle class with appliance and gadget hunger to match. This also contributes to small rural industries, which now have enough stable capital to invest in industries such as small mining rural plants.

These plants need power in quantities that make a great deal of sense for small- to medium-sized power plants owned by IPPs or MPPs. The plants owned by national companies generally are much larger and more centrally located. Power purchased from them often is an inefficient proposition, because of tariff structures, transmission losses, or both.

MPP contracts are being signed on an almost daily basis now. MPPs build small- to medium-sized facilities that require less of an outlay than some of the massive IPP plants built (which had a guaranteed power purchaser or purchasers).

IPP strategy, in face of competition from MPPs, is for raised efficiency, optimized operation, automation that reduces required personnel, and anything else that cuts cost per fired hour. Some of the optimization stems from the growing trend in tougher environmental legislation. In England, an IPP facility at Dagenham (an English-Canadian joint venture) purchased a retrofitted engine condition monitoring system that helps optimize CO_2 emissions taxes, in the 1990s. Plants in Scandinavia and other European countries were ahead of the curve and already equipped to cut costs.

The Oil Company Model Thus Far

Some 30 years or so ago, the only consideration for an oil company to produce its own power was the size of its power requirement or the remoteness of its facility. Commissioned about 20 years ago, the first 170,000 barrels-a-day Syncrude oil sands plant in northern Alberta needed to run machinery in a $3.5 billion (in 1976 dollars) plant, as well as support the massive startup load of its mining draglines. It required its own power plant, as did Esso's Norman Wells oil production operations in Canada's remote north. Efficiency was less of a power production machinery selection criteria than availability. No one talked about combined cycles or cogeneration much. Also, if the government tariff for power it bought back from the operator was not advantageous, that was not considered a major issue. Power companies felt enough in control of turning healthy profits to discourage potential SPPs.

This is changing drastically but at varying rates throughout the world. The burden of emissions regulations and, in progressive countries such as Sweden, high taxes on weight of emissions, has made burgeoning industries of retrofit and environmental engineering. Faced with increased difficulties in turning a profit, many formerly smug national power companies are allowing "independents" (power producers that make only power) as well as SPPs (that may be refineries, process plants, and mills) to enter the fray.

Some nonremote oil company facilities produce what power it takes to help them get their products delivered. Take Elf, U.K., for instance, which produces all the power it requires at its Flotta terminal and sells any excess back to the national grid. Offshore platforms always have made their own power. Occasionally, as with the BP Forties field in the North Sea, if platforms are close enough together, the oil company may lay down underground cables for platforms to share power produced by one of them. However, underground cables are expensive to lay. Also, changes in production flow could occur, as is the case with Forties and Brent after their fields were found to be larger than thought. Power to platforms could get interrupted if the "main" platform were to shut down its generation turbines for any reason.

Today, there is a proactive movement by oil companies to get involved with the power generation business internationally. It does provide an end market for the fuels they provide. If and when this is done by an operating plant for its own power consumption (with excess being sold to the national grid), such as Syncrude, it also provides consolidation of specific resources, such as specialist staff services and repair facilities. A number of contemporary projects involve oil companies in joint venture plans, with governments and other power companies, for actual ownership in power plants.

TABLE 18—1* Exxon Mobil Generating Facilities in China

	Capacity	Fuel	Ownership (%)
Castle Peak A	1400	Coal/fuel oil	60
(Hong Kong)			
Castle Peak B	2708	Coal/fuel oil/gas	60
Black Point	2500	Gas	60
Penny's Bay	300	Diesel	60
Guangdong	600	Pumped storage	51

**Reference: Black Paint Station management.*

BP Gas, not always the most assertive of the oil companies in terms of power generation, nevertheless, admits that it currently is considering several power project joint ventures, including ones in Vietnam and Columbia.

Shell International acquired 50% of InterGen, a joint venture between Bechtel Enterprises Ltd. and Shell Generating Ltd. One of InterGen's ventures, the 770 MW station at Rocksavage, England, was opened on July 31, 1998. Only recently did Shell officially recognize the need to use major increases in power generation capability to make its 1996-initiated "fuels to power" strategy effective. Another proactive example of oil companies providing a market for their fuel with power plant investments was thus set.

Exxon Mobil's publicly stated strategy on power generation, particularly in lucrative SE Asian markets, is assertive and focused. The former Exxon's statement read:

Pursue attractive power generation investment opportunities worldwide to grow earnings while capitalizing on synergies with other Exxon businesses.

Maximize the long-term value of current power interests in Hong Kong.

Exxon had noted the global trend toward deregulation and therefore the emergence of power investment opportunities in many countries. The first area of major interest selected was Hong Kong. Exxon Energy Limited (EEL) holds 60% ownership of three power stations in Hong Kong. China Light and Power (CLP) operates the stations and owns the remaining 40%. CLP also owns all transmission and distribution facilities in Kowloon and the New Territories. EEL also owns 51% the company, which has off-take rights to half the capacity of a pumped storage station in Guangdong Province, China.

Exxon Mobil's generating facilities in China are listed in Table 18—1.

POSITIONING WITH RESPECT TO TECHNOLOGY

The Gas Turbine's Main Rivals

Nuclear Power

Nuclear power's pitfalls can (as many recent crises in that industry segment demonstrate) include (but are not confined to):

- Inadequate or complacent management in some firms (the now defunct Ontario Hydro is a case in point)
- Waste fuel management technology
- Politics revolving around waste fuel management (like the unpopularity of Yucca Mountain with locals in the United States)

The management of these issues can make the difference between whether a nuclear power plant provides economic power or not. The best illustrations of this fact are acquired by looking at some of the recent events in the nuclear industry in some of the countries attempting to sell their nuclear technology to newly industrialized countries in SE Asia: Canada, Finland, Japan, the United Kingdom, and the United States.

The Ontario Hydro debacle or the far worse messes of Three-Mile Island and Chernobyl will not stop the growth of nuclear power, as sector estimates for different countries' growth indicates. It is worth pointing out, however, that, of all forms of power generation, nuclear power by fission is one that leaves the public at considerable potential risk. Unlike the case with fossil plants, nuclear plants may not provide a gradual warning of future problems.

Nuclear reactors that use the fusion process avoid the waste problem that fission reactors have. Some current estimates to get fusion reactors into commercial operation state 2030 as an approximate realistic target date.

Fossil fuel poor countries will continue to advocate nuclear fission power. Japan is still the world's second largest economy, with power needs to match. It imports all but 0.4% of its oil, 5.6% of its coal, and 4% of its gas; hence, its emphasis on nuclear power. Its recent nuclear disaster was a setback, but not enough for a national moratorium on nuclear.

Japan needs to import the fuel but otherwise can consider nuclear power a semi-domestic industry. It consumes about 140 GW of power and 30% of that is nuclear provided. By 2010, nuclear's sector share climbed to 42%.

Public Opinion of Nuclear Power

The global domestic sector currently is about 20% of overall global activity. One might be tempted to think that isolated individual consumers, such as households or small farms that install mini-hydro or windmill facilities, might not affect the large power producer's territory. This is totally untrue, as illustrated by the two nuclear plants in Sweden that were cancelled when 200,000 households installed individual geothermal heat pumps. Not everyone has the Swedes' characteristic environmental verve, however.

Some Lessons Learned with Large Nuclear Power Plants

One lesson learned from the severe ice storms suffered by Canada and the United States early in 1998 is that smaller IPP installations might prove less of an Achilles heel to overall power demand than a few large national power plants.

Problem-riddled national nuclear industries in Canada, the United States, and Japan are testament to overly optimistic life prognoses of nuclear fission reactors. They have and will continue to be decommissioned. This can result in several smaller IPPs taking up the slack.

Coal-Fired Thermal Units (Gas Turbines and Steam Turbines)

Energy estimates in years of fuel left in global reserves vary according to their source. In the late 1990s, the world was thought to have about 70 years worth of natural gas supply left. New deep-sea exploration moved that figure up to 90 years. Then fracking technology appeared on the scene and few people would swear to the new figure. That said, many countries have a great deal of coal and mining it (or turning it into coal gas with steam) is well known and predictable. Developments that use coal syngas are on the rise everywhere. Much as legislation may point out that natural gas causes only half the greenhouse gases and global warming that coal does, countries that have coal want to use their coal.

Besides, countries that have an abundance of natural gas, such as Malaysia, are thinking in terms of its export more than its domestic consumption. At a power generation conference in 1996, Malaysia's Prime Minister, Dr. Mahathir Mohammed, stated that he would like to see less "dependence on gas turbines." Translated, that is likely to have meant, "we can burn cheap coal, imported or otherwise and sell our natural gas at a profit to Japan, Thailand, and so forth . . . also if we put off selling it, the prices are sure to go up."

Frequently, in countries like the United States, fuel-burning facilities, such as power plants, "share" or exchange their pollution allocations to meet a legislated area average.

After-the-fact retrofits to help environmental emissions cost 300% what they cost to put them in at initial installation stage, so they are not popular with plant owners.

Now new technologies being reinforced with projects that that the EEC began long before FutureGen have appeared and make coal a viable gas turbine fuel, if still not a preferred one.

The Economics of Coal

Coal, particularly in developing areas, is the fuel of choice. Apart from the size of known global reserves, there are other economic reasons. To look at these in more detail, it is worth considering SE Asia, where more than half the world's population lives. Coal is their most common fuel.

Gas is cleaner but expensive if you do not own it. Poor in natural resources, Japan currently maintains most of its stringent environmental standards by burning expensive LNG and continuing to develop nuclear power.

The Philippines, despite the Camago Malampaya gas field discovery, nevertheless has a long-term energy policy that places coal uppermost in its priorities. The offshore field has gas reserves enough to keep 3000 MW (1 MW = 1000 kilowatts) of electrical generation capacity running for 15–25 years. Nonetheless, the 25 GW (1 GW = 1000 MW) of anticipated power generation new installation before 2010 breaks down into 60% of coal-fueled power, 10% hydro power, and the rest gas. Coal takes the lead because it is still the cheapest and most abundant fuel that the country can use.

Back in 1995, Indonesia started a national policy to reduce its dependence on oil. Instead, it exports its oil and will base power growth mainly on coal and hydro resources. Currently Indonesia's 10 GW are fueled thus: 25% gas, 20% hydro, 20% diesel, 35% coal, and 5% others, including geothermal. It is intended that, by 2020, 50% or higher will be coal and 40% oil and gas. Indonesia, in the early 1990s, was gifted with Finnish technology and joint venture capital aimed at developing its vast natural peat resources as a fuel. Therefore, Indonesia has yet another economic reason to curtail the burning of natural gas and oil, its cleaner fuels, and concentrate more on coal (and peat on a smaller scale).

New trifuel (coal/oil/gas) burner technology (for boiler and steam plant designs) can limit NO_x (or oxides of nitrogen that combine with water to make acid rain) levels to below 20 ppm (parts per million, by volume). This standard is good enough for the most stringent environmental NO_x standards in the world.

Despite an increase in China's natural gas and liquid fueled facilities, it is expected that 75% of that power will be, as it now is, coal fired. China is the world's largest market for coal-fired stations, and power is its largest industry sector. China releases over 7 million tons of SO_2 (sulfur dioxide, which combined with water also adds to the acid rain problem) to the atmosphere annually. Coal burning, of course, is the main culprit.

As a consequence of fast changing environmental standards, the wealthier nations of SE Asia are retrofitting scrubbers (to clean up coal emissions) as fast as they can. Countries with a higher standard of living, like Taiwan and Korea, have more stringent standards. Those standards are progressively tightened every five to seven years to produce still cleaner emissions.

Coal Gasification and Coal Gas

Coal gasification technology (to use coal gas instead of natural gas in gas turbines) is on the upswing. Gas turbines then will grow in popularity.

This is because, despite the world's tendency to promote steam units due to its desire to use coal, once coal gasification is widely commercialized, the gas turbine population will increase and help make coal-burning steam plants redundant. Users also want to take advantage of the higher efficiencies offered by the gas turbine and gas turbine CC systems. A higher efficiency by 20% (IGCC versus conventional steam turbine) means one uses 20% less fuel.

Coal gas now also can be produced by underground coal gasification. Basically, the process is simple. Two boreholes from the surface are drilled. One supplies oxygen and water; the other removes the gas produced. The former is generally drilled vertically; the other follows the curvature of the coal seam. The process can "work" coal seams up to 1 km deep. What the process essentially does is avoid all the cost and dangers of underground mining and results in a gaseous coal product suitable for power generation, with the ash and several other unwanted constituents in the ground.

The prevalent tendency is to think "steam" and "steam turbine" or "thermal unit" at the mention of the words *coal fuel.* Since coal fields in the middle to late twenty-first century may look more like oil fields than current mining establishments, the vision is probably outdated.

Commercially, Siberia and Uzbekistan are the only countries that have used this process. However, recently, a European Union (EU) project in Spain demonstrated the feasibility of this process. The EU, like the rest of the world, is looking for a hedge against natural gas prices soaring when escalating requirements and security of supply raise gas prices.

The process can mine deep coal seams that would be uneconomic to mine using conventional methods. Some of these coal seams, interestingly enough, have been found when companies were drilling for oil and gas.

Now this in-situ conversion of coal into gas gives the gas turbine world a potentially plentiful source of fuel. The acronym UCG (underground coal gasification) has been coined. The Spanish project's coal gas had a heating value of 11 MJ/cubic meter. Optimization could yield 16 MJ/cubic meter.

"Renewables"

These include turbines driven by wind and tidal power. Their profile is growing globally.

Other regenerables include wave energy and biomass energy. Wave energy is in its infancy. Biomass, including peat, are not yet widespread on a large scale.

USING TECHNOLOGY TO ADVANTAGE

Fuel Technologies

The gas turbine continues to burn a widening range of fuels including but not limited to:

- Residual fuel
- Biomass
- Waste fluids from petrochemical processes
- Waste liquor from paper production
- "Syngas"
- Pulverized coal

Designers continue to improve the gas turbine's technology with respect to burner residence time, flame temperature, cooling, reading hot section temperatures, and so on.

Repowering

Repowering is a major activity in Europe and the United States. The incentives are adherence to Kyoto objectives, but the higher efficiencies available with gas turbine options can make IPPs far more competitive. Although emissions taxes are not yet reality in Asia, they are in Europe, and the trends will eventually spread. They should, particularly if we look at the economic gains in an example such as the Peterhead station in Scotland. The two boiler, two GE 115 MW Frame 9E station had been designed to operate on heavy fuel oil, LNG, sour gas, and natural gas. In 1998, the decision was made to increase plant capacity with three Siemens Westinghouse V94.3 combined cycle units. The V94.3 is a scaled-up version of the V84.3, which can run at both 60 and 50 cycles. The economics of the situation are heavily influenced by fuel sources now made available by the U.K.'s gas supplies. It is important to note, however, that station efficiency will jump from 38% to between 50 and 55%. NO_x emissions will be reduced by 85%.

Another major reason for the repowering trend is that what was thought to be 60 years worth of natural gas left in global supply terms was "updated" to 70 years plus recently. Evidence from ongoing exploration indicates that this figure will climb. Despite China's anxiety to use its coal and the Middle East's desire to use its residual oil, the trend toward gas turbines burning cleaner fuels will continue as lending agencies increasingly tie up their loans with environmental standards as conditions. As gas turbines get better at burning atomized coal, residual fuel, and other erosive or corrosive fuels, use of the gas turbine will increase. As will use of these fuels—at least until emissions taxes are felt as heavily as they are in countries like Sweden. NO_x and SO_x taxes will become the norm fairly early in the new millennium, as they are a source of tax revenue from power companies. Deregulation will tend to favor establishing these taxes worldwide. A CO_2 tax will soon follow in their wake and, with it, the operational economics that may make more-expensive natural gas start to approach competitiveness with residual fuel and coal. Operational economics consider overall costs per running hour, which includes factors such as TBO (time between overhauls, which extends with cleaner fuels) and longevity of components. The gas turbine then is in the ascendant, the remaining question is, whose turbine?

The primary reason that gas turbine packages are bought is good financial purchase packages (GE, for instance, excels in the financial arena), not technology. However, good service features (which technology dictates), fuel price, flexibility in terms of choices, and parts longevity are high on that list, too.

Desalination

A complacent world is getting alarmingly short of fresh drinkable water at a pace that accelerates in proportion to global industrial activity. I ought to say, *some countries* instead of *world,* although it may as well be the world. The larger powers have more of all basic resources, including water, and wake up calls, such as part of California's ground level sinking as it drains its usable water, go relatively unsung in terms of widespread public alarm. On the other hand, gains in fossil fuel reserves, whether by exploration or some new technology, get greater press.

We always take for granted what we have in abundance. So it is with the Middle East and her fuel reserves. Always water hungry, however, a new technology that may as yet be underestimated in terms of global reach, now gives the fossil fuel rich moguls a reason to care about efficiency. No doubt, this technology will spread.

The Middle East has fuel resources in abundance—natural gas, clean oil, and residual oil—and technology available that will help it burn each of its options with what it terms "acceptable" (not optimum values in most Western countries) efficiency. This technology also helps it achieve its major operational objective, which is to extend time between overhauls and therefore parts life (as defined by life cycle analysis, LCA) over the previous generation of power-producing machinery it operated. Saving fuel for its own sake is not as critical to this market as it is to areas that do not have natural gas and oil in abundance. However, a new reason that the Middle East has for caring about efficiency has emerged in the last decade: desalination. The Middle East is very short of fresh water.

How Desalination Works

There are many different techniques for desalination. Among the main ones are the multistage flash (MSF) process and reverse osmosis. Qatar's technology showpieces include the Ras Abu Fontas B power and desalination plant, which cost $1 billion to build. Dubai, in the United Arab Emirates (UAE), has a 60-million-gallon per day desalination plant at Jebel Ali. The plant's eight MSF units are part of a cogeneration power facility, and the fresh water would have been at least as much incentive in the Middle East as the increased total thermal efficiency. The desalination equipment in both cases was supplied by Weir Westgarth. Weir Westgarth designed the MSF process. The principle of the system is simple: Water and steam in a closed system can be made to boil at temperatures lower than at standard temperature and pressure by reduction of the system pressure. MSF plants contain a series of closed chambers—as many as 20, each held at a lower pressure than the preceding one.

Heated saltwater is passed through the overall system. In each chamber, some of the saltwater vaporizes into steam. Moisture droplet separators remove saltwater droplets. The steam condenses to fresh water when faced with cold tubes and is collected for storage. The last chamber's brine is quite cool, and it is used as the coolant fluid. Then it starts to pick up the latent heat of condensation and increases in temperature. Only a small amount of additional heat investment is required to prepare this stream for entry into the first flash chamber. One source of this steam, of course, is low-pressure steam from a power station.

The key to the reverse osmosis (RO) process is a suitable semipermeable membrane. Improvements in membrane technology now mean that the process can apply to industrial-scale plants. Common contemporary membrane selections are made of a cellulose-based polymer or a polyamide layer applied to a microporous polymer film. This membrane is bonded to a porous polyester sheet for structural stiffness. This composite is rolled into a spiral. Spun hollow fine fibers are the finished product. The semipermeable layer is on the outside of the fibers. The total thickness of the composite is about 24 microns. The outside diameter of the tube is about 95 microns, making for a large surface area for rejecting salt. The fibers are made into bundles that are sealed with epoxy in a fiberglass pressure container.

The Global Drive for Desalination

The main forces behind the optimized commercialization of desalination technology are ironies in the light of conditions in the Middle East. Concern regarding seawater contamination of fresh water aquifers in more freshwater-rich countries was one. (Seawater takes the place of fresh water as the aquifer pressure drops with water extraction.)

The need to give nuclear energy a better image (by using its waste heat to produce fresh water) was another.

With respect to the first concern, desalination alleviates use of underground fresh water aquifers, which then reduces the risk of seawater seeping in to make up the balance of fresh water pumped out of the ground. Japan is gas and oil poor. A major user of nuclear power, it has been instrumental in promoting the use of desalination processes in conjunction with its nuclear facilities. The water is produced only in enough quantities to be used by the plants themselves. However, the Nuclear Power Technology Development section of the International Atomic Energy Agency (IAEA) has been seeking to promote large-scale desalination in conjunction with nuclear power production since 1989. The IAEA no doubt hopes that a fresh water supply would overcome the general public's reluctance to be situated close to a nuclear station: Water rises in cost if it has to be transported further.

Interestingly, despite the IAEA agreeing that MSF is promising, it does not contemplate using it for any of its large-scale attempts at fresh water production because of its inflexibility at partial load operation. MSF's tendency to corrosion and scaling versus other desalination techniques does not help either.

Metallurgical Selections

The amount of seawater handled for a given membrane selection and seawater temperature varies directly as the applied pressure. Gulf seawater typically contains 19,000 ppm of salt. Typical temperature gradients are 20–30°C. Metallic corrosion at 30° is four times what it would be at 20° (double the rate of corrosion for each 5°C rise in temperature). This then requires corrosion-resistant alloy selections for the pumps.

To design for a 25-year pump life for a Danish nuclear plant, Alfa Laval used copper nickel alloys in the evaporator and titanium for the heat transfer tubing.

Of course, desalinated water requires fewer chemical additives for water treatment. This is an advantage in terms of overall system cost.

ENVIRONMENTAL INTERNATIONAL CAUCUSES

The effects of international caucuses such as Kyoto, Rio and Montreal, include additionally stringent environmental legislation. The massive acid rain damage noted in the 1970s and 1980s made NO_x and SO_x emissions the first major emissions priority. Scandinavian coutnries, such as Sweden were quick to assess taxes per unit weight of NO_x or SO_x.

Carbon dioxide emissions were very much part of Kyoto, but only in early 2006 did the mainstream U.S. media (*Time* magazine) began the cry of "be worried . . . be very worried" with respect to global warming. Acknowledged U.S. presidential concern began towards the end of the Goerge W. Bush era. The overall effects of growing concern are evident with the revival of projects such as FutureGen and the growing list of new technologies or existing ones being optimized. They are covered in some detail in Chapters 7 and 11 of this book, also indirectly, to some extent in Chapters 10 and 12. For further details on carbon dioxide removal and sequestering, see Chapter 11, Environmental Technology.

With matters relating to the cost of environmental controls, the European countries have always been more proactive than the United States. However, things are changing as is evident with an increase in international joint ventures, such as the one between China and the United States, aimed at sharing technology on carbon capture (see earlier in this chapter).

OEM CHANGING FORTUNES

OEM Acquisitions, Joint Ventures, and Licensees

Which "Parent" Philosophy Governs Which Model?

"Parent" companies also are changing rapidly. As we saw in the introductory to the chapter, due to acquisitions, mergers, joint ventures, and engineers who work for more than one OEM, it gets hard to recognize certain gas turbine models from their insides.

Certain key developments are well known; and it is easy to identify where, when, and by whom they might have been originally designed. For instance, the SEV (sequential environmental) burner was an ABB (Asea Brown Boveri) design, designed when the company was ABB, not Brown Boveri (as it once was), and not ABB Alstom, later Alstom, as it later became. The wide chord fan blade was a Rolls Royce original. Steam cooling in closed-cycle loops to support H technology, which was a success both in design and full load test, was probably a Mitsubishi first. This may be debated by some, but note that "first" refers to no leakage on a full load test. The Cyclone, Tempest, Tornado, and Typhoon were grassroots designs engineered by European Gas Turbine (before ABB bought it) when it was EGT, not in its formerly Ruston days. Now these models are owned by Siemens. Sooner or later, every OEM has a variation on a theme or a better alternative for which it may not have done the grassroots development.

Grassroots model development is expensive. The very fact that the Cyclone was developed in the first place indicates that the manufacturers foresaw potential market share in a power bracket that would also equate to a Solar Mars. At this power range, the compressors partnered with mechanical drive gas turbines tend to require adherence to API (American Petroleum Institute) specification. It could have been anticipated that this would put the OEM on equal footing with manufacturers like Solar, who currently has the lion's portion of the installed mechanical drive market share in sectors such as SE Asia's oil and gas production market. It similarly dominates in the installed-machinery oil and gas production market in the United States and Canada. Solar's historical advantage was based partly on market entry timing and partly on the fact that their smaller compressors did not always need API specification adherence to win contracts.

Licensee Growth and Development

When a licensee does well, the company that licenses it to build gas turbines also gains strength.

General Electric is a major conglomerate with a large stable of licensees for different gas turbine product lines. They include Nuovo Pignone, GEC, Stewart and Stevenson, and Kvaerner.

Consider the growth of Kvaerner. Kvaerner, originally founded in 1853, made steel products. This was extended to include high-head Francis and Pelton type turbines: current unit sizes exceed 400 MW. The 1956 agreement signed with General Electric licensed Kvaerner to build marine steam turbines and later industrial and aero-derivative gas turbines.

Kvaerner's client base is largest in Scandinavia but they now market aggressively in SE Asia, as an individual company and also in joint venture agreements. Kvaerner bought John Brown, another licensee of GE, in April 1996. John Brown was already seasoned in Asia and its plants included one of the earliest combined-cycle plants in the world, built in Hong Kong and commissioned in 1972. John Brown has been renamed Kvaerner Energy Limited. Its consortium with Kvaerner is called Kvaerner Energy Thermal Power.

Asian projects include a combined-cycle plant in China. The 330 MW station in Zhenai, Zhejiang province, China, is heavy-oil fueled and then requires technology similar to that employed by the plant in Shunde, Guangdong province, that runs two ABB (now Alstom) 13D-3s in combined-cycle mode. In the case of the 13Ds, the design already incorporates the low turbine inlet temperatures (TITs) required to make the turbine run successfully on residual or bunker oil.

In Kvaerner's case, the GE-9E combustors are adapted for burning residual. The fuel is washed and chemically treated. It is also heated, with steam from the steam turbine, to lower its viscosity so that it flows well enough for fuel injection. As with all residual fuel applications, expensive high-quality distillate is required on startup and shutdown to avoid the fuel nozzles clogging. Deposits build up on turbine blades and have to be washed off regularly or the effect gas path area will eventually clog. The frequency of the washes depends on the turbine design and fuel quality. In this plant's case, washing is about once weekly.

The plant cost over $100 million to build. When necessary, it can be operated in simple-cycle mode. The plant is run 16 to 18 hours a day for at least 300 days a year, adding up to 5000 to 6000 hours a year.

In 1995, Kvaerner opened an office in Kuala Lumpur, Malaysia. Its first contract was for one LM2500 gas turbine driver for an Ebara compressor, which will be used in Shell's Central Luconia project in Sarawak.

A Kvaerner-built foil cat that serves the route between Hong Kong and Macau runs on Kvaerner supplied gas turbines. Kvaerner services some of the Indonesian navy's LM2500s in its overhaul and repair facility in Agotnes, Norway.

OEM Business Strategy Including Production Backlogs and Vendor Alliances

The state of all manufacturers' backlogged orders is proof that the power business in booming with unprecedented vigor. Although manufacturers are strict about not making predictions regarding future order numbers, indications are that the backlog condition will escalate in the next few years. Many customers, new and old, have stopped looking at individual design features that can directly affect their costs per fired hour, even if they know how to assess them. They just take whichever machine they can get delivered the quickest. If that machine has problems due to ironing out new design updates, the customers need to compensate for this by writing "guarantee points" into the contract (the manufacturer promises a certain availability for the first year, collects bonus points for meeting targets, demerit points for not doing so, and raises the availability figure with increasing experience with any one model).

Even large customers who have their own engineering staff are opting for "power by the hour" contracts with the OEM and leaving all parts replacement decisions to the OEM. This may prove to be expensive, as OEMs make the bulk of their profits with aftersale spares and service. Yet, increasingly U.S. customers expect their OEMs to also do their service, rather than staff up to do their own or even their own assessment of the effectiveness of overhauls.

The power producer mix is shifting to include OEMS, oil companies that are also becoming IPPs, SPPs, and hungry MPPs anxious for a quick return. The demand for gas turbines in all sizes and system formats (simple or combined cycles) keeps climbing.

The strategy of major OEMs has also been to acquire all the turbine systems accessories and support companies they can. GE is a good illustration of this. By staying abreast of these developments, customers can turn their OEM's diversification and growth in their favor.

Another noteworthy trend among the OEMs is the development of power options that use renewable resources for fuel. It is a considerable goodwill factor in their annual statements bottomline, something that is increasingly acknowledged within the United States. This sometimes can be rewarded with public awards for performance and service. To what extent gas turbine OEMs allow their renewables portfolio to flourish depends on which of their lines (and profitability) will decline if they do. Their overall publicity campaign (e.g., "ecoimagination" ads) support their decisions regarding profitability mix, government contracts sought, and gains in political alliances. OEMs have continued to buy companies, stakes in new technology development, and e-trading businesses, all of which will affect the turbine market, in the United States and globally.

OEM Development Programs with the U.S. Government (DOE)

This has been well illustrated in this book with several case studies of the work done by the OEMs under U.S. DOE (Department of Energy) sponsorship.

Oil and Gas Production Portfolios Help OEMs

Examples of this include: GE Energy Rentals bought Jenbacher, an Austrian natural gas power generation reciprocating engines. GE Power Systems purchased Nuovo Pignone (which makes compressors, gas and steam turbines, pumps, and an assortment of process and power accessories), Gemini (which makes reciprocating compressors for natural gas applications), and Thermodyn (which makes small compressors and steam turbines for oil and gas as well as power generation).

The larger OEMs' philosophy seems now to be to maintain a presence with both the fuel supplier and the customer base for their own machines.

OEMs Communications Infrastructure Grows Portfolio

A case in point is this: GE bought Young Generators in April 2000, a company that is a leading provider of power generation rental equipment and services for the motion picture, broadcast, and events businesses. This increases GE's reach with respect to temporary distributed power solutions. "Outage Optimizer" is another Internet service provided by GE. It helps turbine owners plan outages for their machines, evaluate service bids, and purchase them online.

Service Company Acquisitions

Most OEMs have secured technology transfer agreements with highly sophisticated companies that developed specialized repairs: coatings, blade tip robotic welding, powder metallurgy, online wash systems, and so forth. Despite items such as online trading in used or reconditioned parts, sometimes an OEM may make these available to a specific customer only if that client knows enough to ask for the option. (Not all these developments end up as general service bulletins.) To a smaller company, this "current options" information can save a great deal of money. It is always worth a customer's while to investigate the OEM's current state of technology share agreements to pursue every last advantage possible in the fight to drive down costs.

Accuracy in Reading the Target Market

The end customer must consider an interesting twist to the market equation. It is a consideration tied irrevocably to the changing manufacturing partners, as this affects the largest questions on the purchaser's mind: How much will good service cost me? Who will provide it? What are their service and engineering philosophies—old, new, negotiated compromise?

Consider that what was ABB Stal's (now Siemens by acquisition) grass-roots designed GTX-100, 43 MW has not yet made a dent in LM6000 territory. Reasons may include GE's financial services competence with structuring deals or perhaps a preference for an aeroderivative's modular design over an industrial one. Will the GTX eventually take a sizable market share because of its ability to withstand dirtier fuels and a wider range of heating values?

Reading Renewables

Smart grid development and legislation will really help the progress of renewables in the United States. The rest of the world, including the developing world, like China, has already been proactive in this field for years.

Staying current with environmental legislation helps direct where OEMs must eventually head. OEMs are aware that the century of renewable energy fuels is anywhere from three to five decades ahead, and they are quietly getting ready. Siemens, for instance, is working hard on fuel cell (and many other energy products) development even as it develops different coal gasification technologies. Mitsubishi develops wind turbines (and other energy products) even as it develops its gas and steam turbines. While not in a hurry to abandon their conventional technologies, all of them want to be able to sell their client base fuel cells and tidal turbines when those markets turn profitable. An informed customer will use this to tailor contemporary demand curves within his control to best advantage when in negotiations with the OEM.

END-USER ASSOCIATIONS

End-user associations exist for practically every model in the gas turbine world. For instance, the Alstom-11N population has their own meeting, the GE-7E people meet every year, and every military helicopter engine has its own CIP (component improvement program) meeting. One needs to find one's own crowd and stay abreast of their global community's problems.

The paradigm "You get what you negotiate" is true with overhaul contracts as much as anything else. When the contracts cross international boundaries, this may be even more accurate. Increasingly, customers need to investigate contract term options with the worldwide client base of their particular original equipment manufacturer.

Overhaul contracts are negotiated on a variety of bases. The end user must pick what best befits its application. Power by the hour is one of the most common bases for contracts.

Others specify hot section overhaul only, with everything else on an as-needed basis. The term used by some for this kind of contract is *hot gas path inspection protection plan.* A monthly charge applies per machine.

Some power producers turn over overhaul and repair entirely to the OEM or an OEM company.

The end user's aim ought to be to get the best cost per fired hour. The end users with sufficient experience with repair and overhaul achieve this with a combination of OEM-bought parts and maintenance, independent repair vendor-bought parts and maintenance, some maintenance from in-house end-user staff, and constant experience comparisons with other end users in the business via end-user associations.

Two of the most valuable items on which end users can compare notes are independent contractors and spare parts price/power by the hour price variations.

The Independent Contractor Factor

End-user associations force OEMs to accommodate machinery owner priorities and preferences by sheer weight of numbers. Another powerful and growing asset in the quest to reduce overhaul costs is the independent overhaul contractor workforce.

Frequently, the main executives at independent overhaul contractors are ex-OEM employees who may have designed or pioneered the technology they now are "going independent on." Such contractors have an edge over others with "reverse engineering," the term used to describe working out the steps in the manufacture of a component just by having one physically to work with, without benefit of the OEM's drawings or designs.

Some such contractors will also confidently tackle "re-engineering," which is actually engineering the manufacture of a component again. Metallurgy, heat treatment, coatings, and so forth may differ from the original in a re-engineered component. Typically, a component is re-engineered to save money, improve its operation or life, or all of these. It requires extreme engineering competence and confidence to do this. It also requires considerable technical competence on the part of the end user to assess such a contractor's ability to match its claims. When it can, however, the end user stands to save a great deal of money. Therefore, as lost revenue is always a source of ire to the OEM, a further bargaining asset for the end user is available.

Such independent overhaul contractors have made considerable strides in the United States and western Europe. This is starkly contrasted against their minimal potential in countries that are, or recently were, communist. An entrepreneur in this sector once mentioned his frustration when visiting Vietnam, when he invited end users to a free seminar on his firm's range of services. Some of the end users, used in their domestic industrial culture to being totally dependent on the OEM, expected to be paid for listening to the seminar.

In newly industrialized countries (NICs), such contractors have been viewed with interest for some time. Growing "trial usage" of their services is underway. These firms are anxious to break into territory traditionally occupied exclusively by the major OEMs. See Chapter 12, Maintenance, Repair, and Overhaul, for further details.

Varying Spare Parts and Service Prices

Spare parts, whether paid for singly or as part of a power-by-the-hour contract, are a major component of overhaul costs.

End users sometime stumble upon major differences in prices internationally. A U.S. IPP representative investigating joint venture potential for a company in Pakistan once found that spare parts prices may vary by as much as 300%. To the consumer-power culture in the United States, this seems a large differential. However, the OEM may have built in allowance for "expediting payments" in the chain of customs command in the country in question, which are totally acceptable in the local culture. Also, if the end user is not yet seasoned in the use of the models in question, its expectations of what the warranty may normally exclude (such as component failures resulting from faulty operation) might not be the same as everywhere else. Some OEMs understandably mark up prices for that user to accommodate this difference.

DISTRIBUTED POWER: HOW LARGE DOES A POWER PLANT NEED TO BE?

The Age of the Personal Turbine

The age of the personal turbine (PT) will probably arrive in a few decades. LG, a Korean company, and the German Fraunhofer Institute for Solar Energy Systems (ISE) just launched a product that is a fuel cell system integrated into a laptop computer. A mini fuel cell, hydrogen fuel tanks, and electronics have replaced the computer battery. This may be one a series of steps to the day when we carry our personal turbine/computer (PT/PC) to our car, plug it in, drive to and from work, come home, plug it in, and have our own personal heating system incorporated in this PT, the size of a CD player. Massive transmission line systems may become redundant dinosaurs in many locations.

For more information, see Chapter 16, Microturbines, Fuel cells, and Hybrids.

THE POWER MIX

The mix (of fuel types) and targets will vary with changing politics and governments, particularly in countries like the USA. Nevertheless, even when under a Republican government, the USA's target(s) of "renwables in the mix" continue to rise. Politics does not always coincide with technological data. When President George W. Bush was in office, he gained one ally in Australia, inexplicably, given Australia's previous reputation for caring about their environment. Australia's prime minister opted to "toe the Bush [U.S.] line" claiming "thousands of Australian jobs will be lost" if his country meets the Kyoto protocol. As it turns out, Australia is closer to the Kyoto targets it negotiated than most countries. The implication to its people of meeting Kyoto is the cost of abating 20 million tons of CO_2. This works out to an insignificant $1.95 (2002 prices) rise in their annual power bill: an all-time storm in a kangaroo's teacup.

The target by 2020 for both the USA and the EU is currently 20% renewables in the mix. How a country gets there varies. Denmark plans to have 50% wind power in her mix by 2020 and to have all her energy from renewables by 2050.

Within a country like the USA, the strategy and policy of individual states will also vary. Consider the case of California. "Established in 2002 under Senate Bill 1078, accelerated in 2006 under Senate Bill 107 and expanded in 2011 under Senate Bill 2, California's Renewables Portfolio Standard (RPS) is one of the most ambitious renewable energy standards in the country. The RPS program requires investor-owned utilities, electric service providers, and community choice aggregators to increase procurement from eligible renewable energy resources to 33% of total procurement by 2020. The California Public Utilities Commission (CPUC) and the California Energy Commission jointly implement the RPS program.[12]

California's history with respect to renewable is an interesting one and the state has stuck to its stated values overall, despite being poorer in resources such as clean natural gas or oil than states like Texas, for instance. Some of the milestones in California's arrival at it's stuff-of-dreams "33% by 2020 target" follow.[13]

In 2002, California established its Renewables Portfolio Standard (RPS) Program, with the goal of increasing the percentage of renewable energy in the state's electricity mix to 20 percent of retail sales by 2017. The *2003 Integrated Energy Policy Report* recommended accelerating that goal to 20 percent by 2010, and the *2004 Energy Report Update* further recommended increasing the target to 33 percent by 2020. The state's **Energy Action Plan** supported this goal. In 2006 under Senate Bill 107, California's 20 percent by 2010 RPS goal was codified. The legislation required retail sellers of electricity to increase renewable energy purchases by at least 1 percent per year with a target of 20 percent renewables by 2010. Publicly owned utilities set their own RPS goals recognizing the intent of the legislature to attain the 20 percent by 2010 target.

On November 17, 2008, Governor Arnold Schwarzenegger signed Executive Order S-14-08 requiring that "...[a]ll retail sellers of electricity shall serve 33 percent of their load with renewable energy by 2020." The following year, Executive Order S-21-09 directed the California Air Resources Board, under its AB 32 authority, to enact regulations to achieve the goal of 33 percent renewables by 2020.

In the ongoing effort to codify the ambitious 33 percent by 2020 goal, SBX1-2 was signed by Governor Edmund G. Brown, Jr., in April 2011. In his signing comments, Governor Brown noted that "This bill will bring many important benefits to California, including stimulating investment in green technologies in the state, creating tens of thousands of new jobs, improving local air quality, promoting energy independence, and reducing greenhouse gas emissions."

This new RPS preempts the California Air Resources Board's 33 percent Renewable Electricity Standard and applies to all electricity retailers in the state including publicly owned utilities, investor-owned utilities, electricity service providers, and community choice aggregators. All of these entities must adopt the new RPS goals of 20 percent of retails sales from renewables by the end of

12. Source: http://www.cpuc.ca.gov/PUC/energy/Renewables/overview.htm (Accessed May 24, 2014).

13. Source: http://www.energy.ca.gov/renewables/ (Accessed May 24, 2014).

2013, 25 percent by the end of 2016, and the 33 percent requirement being met by the end of 2020.

Timeline of California's Renewables Portfolio Standard

- 2002: Senate Bill 1078 establishes the RPS program, requiring 20% of retail sales from renewable energy by 2017.
- 2003: Energy Action Plan I accelerated the 20% deadline to 2010.
- 2005: Energy Action Plan II recommends a further goal of 33% by 2020.
- 2006: Senate Bill 107 codified the accelerated 20% by 2010 deadline into law.
- 2008: Governor Schwarzenegger issues Executive Order S-14-08 requiring 33% renewables by 2020.
- 2009: Governor Schwarzenegger issues Executive Order S-21-09 directing the California Air Resources Board, under its AB 32 authority, to adopt regulations by July 31, 2010, consistent with the 33% renewable energy target established in Executive Order S-14-08.
- 2011: Senate Bill X1-2, signed by Gov. Edmund G. Brown, Jr., codifies 33% by 2020 RPS.

Brief extracts from a study entitled Does Califorinia Need Liquefied Natural Gas (LNG)? (Excerpt from 2006 study).*

(Author's note: The following excerpts from the study named 'Does California Need Liquiefied Natural Gas (LNG)?' are included in its entirety in Edition 1 of Gas Turbines as Case 1).

Our analysis shows that renewables and energy efficiency could produce 133% to 381% of the projected additional gas demand in California by 2016.

This report looks critically at natural gas supply and consumption projections and concludes that California's energy efficiency and renewable energy mandates could readily meet expected additional natural gas demand and, therefore, eliminate the need for LNG import terminals along our coast (Figures 18–1 and 18–2 are included from Edition 1 of Gas Turbines, however, the reader is cautioned to check the current levels of fluctuation in natural gas vs. coal/coal gas demand).

California's Renewable Energy Potential

Table 18–2 depicts the energy production under the mandated, or likely to be mandated, renewable energy goals in California.

Is California's Renewable Energy Market Viable?

The following sections discuss estimates of California's renewable energy potential, from the Energy Commission and other reliable sources. Table 18–3 summarizes the potential from the various renewable resources in California.

Wind

In 2004, California generated 4258 GWh of electricity using wind power, 1.5% of the gross system power and equivalent to about 7 of a large LNG import terminal.

Expanding wind power capacity from the 2,096 MW of capacity in 2004 to about 8,540 MW in 2017, as is expected by the Energy Commission, would produce 19,760 GWh per year, equivalent to 33% of a large LNG import terminal. The technical potential for wind power is of course much larger—the Energy Commission recently estimated 127,000 MW of potential in the state.

Solar

Solar energy is another renewable resource that is easily accessible in many parts of California with significant expansion potential. The technical potential for photovoltaic and concentrating solar power systems in California exceeds 17 *million* MW. The state currently has about 60,000 MW of generation capacity.

Geothermal

Though wind and solar resources have perhaps the largest potential in California, geothermal, biomass, and small hydroelectric facilities currently contribute more to California's total renewable energy resource base. Geothermal power contributed 13,571 GWh, or about 4.9% of the gross system power in 2004.

Biomass and Waste to Energy

Biomass energy is generated from organic wastes such as woody agricultural wastes and forest thinnings. Biomass power plants provided 5997 GWh of electricity in California in 2004—about 2.2% of the gross system power. In its 2005 updated biomass assessment, the Energy Commission found an additional technical potential of 4700 MW of biomass power by 2017, using current technologies.

Small Hydroelectric

Small hydroelectric plants (30 MW capacity or less) are considered renewable due to the relatively small amount of water required for their operation and consequent minimal environmental impacts when compared to large hydroelectric projects. In 2004, about 1.7% of the electricity generated in California was produced by small hydroelectric plants. Small hydroelectric power potential is estimated at 2280 GWh from new facilities,[14] plus 667 GWh from water pipelines among municipal water utilities and irrigation districts.

Ocean Power

The ocean is also a viable resource for energy production, especially in California. Wave power along the coast—from surface wave energy conversion alone—has a technical potential of 18,912 GWh, *at primary sites* only.

Additional Supplies of Natural Gas in North America

(Author's note: This study was made without consideration of the now prevalent fracking "gold rush" that has captured the United States as a whole, if not California to the same degree. The reader needs to consider the following in that light).

It is certainly possible that the state will not meet its renewable energy mandates by 2010, let alone the likely new

(*Continued*)

Cont'd

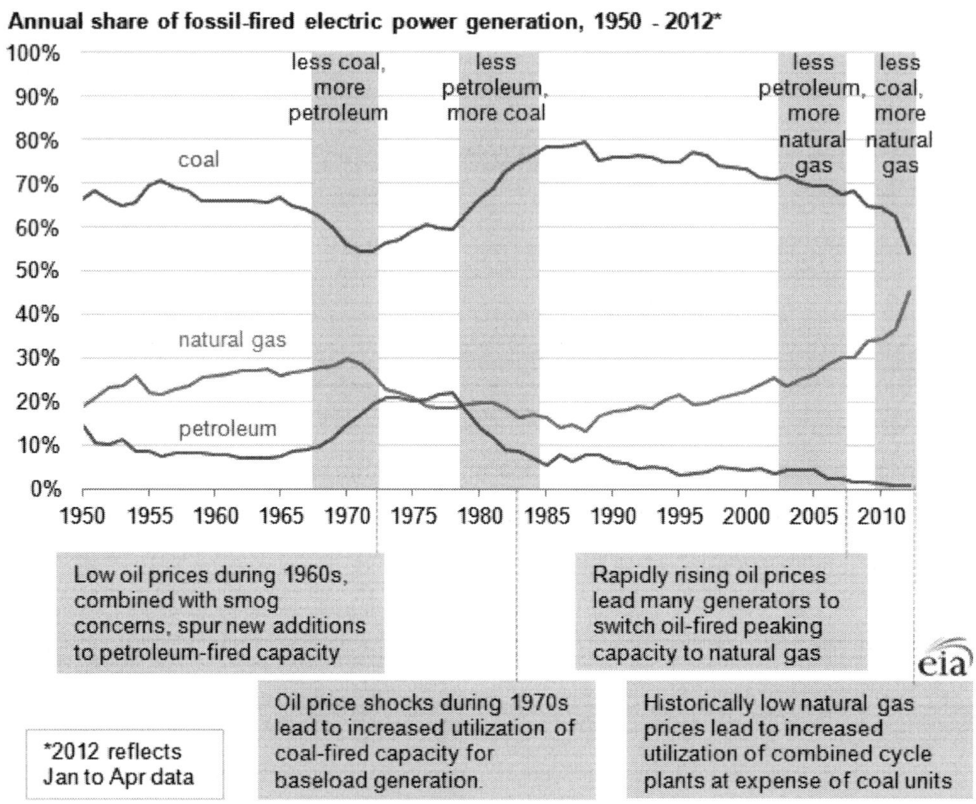

FIGURE 18–1 Annual share of fossil-fired electric power generation 1950–2012. *(Sources: U.S. Energy Information Administration, Annual Energy Review and Electric Power Monthly.)* Notes: Shares are calculated using generation supplied by all sectors. Data for 2012 include January through April. Start and end years for the grey boxes are approximate. A very small percentage age (generally less than 0.5%) of fossil-fired generation is from other gases, not shown.

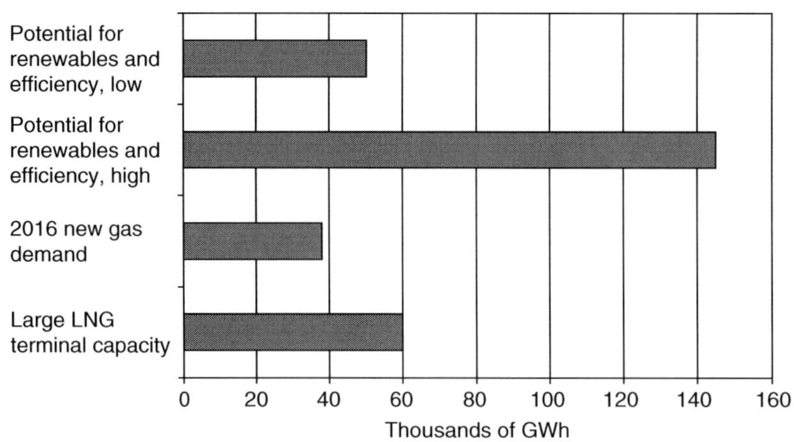

FIGURE 18–2 Energy efficiency and renewable energy potential vs. natural gas demand [18-2].

TABLE 18–2 California's Current and Prospective Renewable Energy Mandates [18-2]

Mandate	Generation (GWh)	2016 New Gas Demand	Equivalent Large LNG Terminals
20% by 2010	29,533[a]	(Already included)	49%
30% by 2016	32,781[b]	87%	55%
33% by 2020	47,323[c]	125%	79%

[a] *20% of projected 2010 electricity demand.*
[b] *This figure does not include the 20% by 2010 amount, so represents new generation above the 20% RPS goal*
[c] *This figure represents the total new generation by 2020, not already accounted for in the 2010 RPS goal.*

TABLE 18–3 California's Renewable Energy Potential [18-2]

Resource	Estimated Generation by 2017[a] (GWh)	2016 New Gas Demand	Equivalent Large LNG Import Terminals
Wind	19,760[b]	52%	33%
Solar PV	4,139[c]	11%	7%
Solar thermal	NA[d]	NA	NA
Concentrating solar	18,615[e]	49%	31%
Geothermal	22,654[f]	60%	38%
Biomass/landfill gas	35,000[g]	93%	58%
Small hydroelectric	2,947[h]	8%	5%
Ocean power	4,728[i]	12%	8%
Totals	107,843	285%	180%

[a] *We project new generation through 2017, instead of 2016, which is the time frame for the Energy Commission's natural gas assessment, because state agencies have done a considerable amount of work looking at potential by 2017, which was the original date for achieving the 20% RPS. Due to the large leeway our numbers provide in meeting future natural gas demands from renewable energy, this one year discrepancy should not be that important.*
[b] *California Energy Commission,* Renewable Resources Development Report, *2003 (RRDR), p. 98 (estimating 6644 MW of new wind projects throughout California by 2017). The CPUC report,* Achieving a 33% Renewable Energy Target *(Nov. 2005), finds a larger potential of 6960 MW in California (it also includes consideration of nearby out-of-state resources as a separate figure), plus a separate figure for potential from repowering wind turbines at Altamont Pass (p. 39). We use the lower estimate for present purposes because the CPUC report does not state that this full potential could be developed by 2016, the timeframe for our analysis.*
[c] *Based on the California Solar Initiative's stated goal, with rebates likely to be approved by the California Public Utilities Commission in January 2006, of 3000 MW of new solar, in addition to the approximately 150 MW already installed in California. This production figure assumes a 15% capacity factor. The CPUC report* Achieving a 33% Renewable Energy Target *finds a 5000 MW capacity for solar PV (p. 41). We find the lower figure more realistic at this time, but are optimistic that 5000 MW or more will be installed, given the rapid pace of innovation in this technology.*
[d] *The California Energy Commission does not currently provide an estimate of solar thermal potential in California, either in its 2005 draft solar power assessment or its 2003 renewable resources development report. We hope this oversight will be corrected soon, as solar thermal technologies have vast potential for displacing natural gas and electricity demand, as evidenced by the fact that the largest source of renewable energy in the world (other than large hydroelectric, which is not considered renewable in California) is solar thermal, due largely to the numerous installations in China.*
[e] *The California Energy Commission estimates 1 million MW of concentrating solar potential in its 2005 draft staff paper on California solar resources, but we don't believe, for obvious reasons, that anywhere near this amount of CSP will be built by 2017 or even 2027. We have assumed that 10 systems like the 850 MW CSP facility recently approved for SCE by the PUC could be built by 2017, leading to a total of 8,500 MW of CSP by 2017, and 18,615 GWh per year by 2017, at a 25% capacity factor. The CPUC report* Achieving a 33% Renewable Energy Target *finds a similar potential, at 10,200 MW (p. 41).*
[f] *California Energy Commission,* California Geothermal Resources, *CEC-500-2005-070, April 2005, p. 7. The GWh total assumes, as the CEC does, a 90% capacity factor for 2862 MW. The CPUC report* Achieving a 33% Renewable Energy Target *(Nov. 2005) finds a similar figure of 2565 MW of potential in California (p. 40).*
[g] *California Energy Commission,* Biomass Resources in California: Preliminary 2005 Assessment *(April 2005), p. v. The figure cited is the low estimate in that report. The CPUC report* Achieving a 33% Renewable Energy Target *finds a much lower potential, at 1,775 MW, equivalent to 13,994 GWh (p. 40). The report explained this discrepancy: "Biomass power in general has favorable economics. But the development potential of biomass is contingent on securing long-term fuel supplies, with each project requiring a narrow range of fuel specification. Biomass projects tend to be of modest scale and linked geographically to local fuel sources. For these reasons, biomass was only projected to supply 10% of the renewable energy needs" (p. 42). We use the higher estimate due to our expectation that reliable baseload capacity will be developed where it is economically feasible, making biomass generation very attractive where feedstock is available, and due to the more detailed (and more convincing) analysis in the CEC's Biomass Resources report.*
[h] *California Energy Commission, draft report, "California Small Hydropower and Ocean Energy Resources" (May 2005), p. 4.*
[i] *Ibid., p. 16. We ignore the secondary sites identified in this report and assume that only 25% of the primary site potential identified by the Energy Commission will be developed by 2017.*

(*Continued*)

Cont'd

TABLE 18–4 Summary of Energy Efficiency and Renewable Energy Mandates [18-2]

	Low Estimate	High Estimate
Energy efficiency mandates (GWh)	12,881	12,881
Renewable energy mandates (GWh)	29,533	64,781[a]
Total (GWh) percent of a large LNG import terminal	42,414	77,662
	71%	129%

[a]*This figure represents 32,781 GWh of new generation by 2016 plus 32,000 GWh already included in the Energy Commissions' natural assessment, representing new renewable generation by 2010.*

TABLE 18–5 Energy Efficiency and Renewable Energy Total Potential [18-2]

	Low	High
Energy efficiency potential (GWh)	17,497	68,305
Renewable energy potential (GWh)	32,781	75,843
Total (GWh)	50,278	144,148
Percent of projected demand by 2016	133%	381%
Percent of a large LNG import terminal	84%	240%

mandate of 3% by 2020. It is even possible that the CPUC's ambitious and funded energy efficiency programs with the investor-owned utilities will not produce expected savings.

However, even if the state slips in meeting its own mandates, California need not be overly concerned about natural gas supplies, as significant additional supplies will come online in North America in the next decade from a number of sources:

- Domestic U.S. natural gas production is expected to increase over the next decade, while Canadian imports are projected to decrease. However, the decrease from Canada is projected to be more than offset by increases in U.S. production.[14]
- A consortium of oil companies has proposed a natural gas pipeline from Alaska and Canada to the contiguous United States. This project will provide 1.5–2.0 trillion cubic feet of natural gas per year and should be completed by 2016. If completed, this pipeline would forestall the apparent peak in North American natural gas production by a number of years because it would provide access to otherwise stranded natural gas resources.
- An additional pipeline, from the MacKenzie region of Canada's Yukon, is expected to be online by 2013. **(Authors Note: As Canadian pipelines come online that forestalls by the number of years the anticipated declines in Canadian production. Also check http://www.capp.ca/canadaIndustry/oil/Pages/PipelineMap.aspx for current updates)**.

Tables 18–4 and 18–5 summarize the energy efficiency and renewable energy potential in California.

* Source: [18-2] Courtesy of the Community Environmental Council, adapted extracts from T. Hunt, A. Chan, J. Phillips, and S. Wright. *Does California Need Liquefied Matured Gas? The Potential for Energy Efficiency and Renewable Energy to Replace Future Natural Gas Demand* (Santa Barbara, CA: Community Environmental Council, 2006).

14. California Energy Commission, *California Small Hydropower and Ocean Wave Energy Resources,* CEC-500-2005-074, April 2005, p. 4.

Chapter 19

Basic Design Theory

"Never mistake motion for action."

—Ernest Hemingway

Chapter Outline

At some point, a gas turbine operations or repair and overhaul engineer has to consider the design of the gas turbine package in his or her care. This may be for:

- Specification of a new package or system
- An engineering a retrofit package
- Arguing a warranty case
- Troubleshooting or failure analysis

Regardless of the reason for a look at the gas turbine system design, it is useful to know the terminology and basic theory behind the OEM's design. What follows is a summary of an OEM's basic theory and terminology in the design areas of:

- Operational envelope
- Properties and charts for dry air, combustion products, and other working fluids
- "Design point" definitions and terminology

After these sections, there are two case studies and a Mach number/altitude chart developed by an aircraft engine designer.

The very basic thermodynamic laws and cycle diagrams used by operational engineers with little or no design (retrofit or otherwise) or specification responsibility are in the chapter on gas turbine cycles.

OPERATIONAL ENVELOPE*

The performance—thrust or power, fuel consumption, temperatures, shaft speeds, etc.—of a gas turbine engine is

* Source: Courtesy Rolls Royce, *Gas Turbine Performance*, Walsh and Fletcher, Blackwell Science, 1998. Adapted with permission. The reader ought to note that this section, while covering the basics required for this book, is abbreviated considerably from the original work. Gas turbine "pure design" engineers ought therefore to consider getting the source book for design calculations.

crucially dependent upon its inlet and exit conditions. The most important items are pressure and temperature, which are determined by the combination of ambient values and any changes due to flight speed, or pressure loss imposed by the installation.

The full range of inlet conditions that a given gas turbine engine application could encounter is encompassed in the operational envelope. This comprises:

- An environmental envelope, defining ambient pressure, temperature, and humidity
- Installation pressure losses
- A flight envelope for aircraft engines

The Environmental Envelope

The environmental envelope for an engine defines the range of ambient pressure (or pressure altitude), ambient temperature, and humidity throughout which it must operate satisfactorily. These atmospheric conditions local to the engine have a considerable effect upon its performance.

International Standards

The International Standard Atmosphere (ISA) defines standard day ambient temperature and pressure up to an altitude of 30,500 m (100,066 ft). The term ISA conditions alone would imply zero relative humidity.

US Military Standard 210 (MIL 210) is the most commonly used standard for defining likely extremes of ambient temperature versus altitude. This is primarily an aerospace standard, and is also widely used for land-based applications though with the hot and cold day temperature ranges extended. Table 19−1 shows the ambient pressure and temperature relationships of MIL 210 and ISA.

For land-based engines, performance data are frequently quoted at the single point ISO conditions, as stipulated by the International Organization for Standardization (ISO). These are:

- 101.325 kPa (14.696 psia), sea level, ambient pressure
- 15°C ambient temperature
- 60% relative humidity
- Zero installation pressure losses

Ambient Pressure and Pressure Altitude

Pressure altitude, or geo-potential altitude, at a point in the atmosphere is defined by the level of ambient pressure, as per the International Standard Atmosphere. Pressure altitude is therefore not set by the elevation of the point in question above sea level. For example, due to prevailing weather conditions a ship at sea may encounter a low ambient pressure of, say, 97.8 kPa, and hence its pressure altitude would be 300 m.

Table 19−1 includes the ISA definition of pressure altitude versus ambient pressure, and Figure 19−1 shows the relationship graphically. It will be observed that pressure falls exponentially from its sea level value of 101.325 kPa (14.696 psia) to 1.08 kPa (0.16 psia) at 30,500 m (100,066 ft).

The highest value of ambient pressure for which an engine would be designed is 108 kPa (15.7 psia). This would be due to local conditions and is commensurate with a pressure altitude of −600 m (−1968 ft).

Ambient Temperature

Table 19−1 also presents the ISA standard day ambient temperature, together with MIL 210 cold and hot day temperature, versus pressure altitude. Figure 19−2 shows these three lines of ambient temperature plotted versus pressure altitude.

Standard day temperature falls at the rate of approximately 6°C per 1000 m (2°C for 1000 ft) until a pressure altitude of 11,000 m (36,089 ft), after which it stays constant until 25,000 m (82,000 ft). This altitude of 11,000 m is referred to as the tropopause; the region below this is the troposphere, and that above it is the stratosphere. Above 25,000 m standard day temperature rises again.

The minimum MIL 210 cold day temperature of 185.9 K (−87.3°C) occurs between 15,545 m (51,000 ft) and 18,595 m (61,000 ft). The maximum MIL 210 hot day temperature is 312.6 K (39.5°C) at sea level.

Relative Density and the Speed of Sound

Relative density is the atmospheric density divided by that for an ISA standard day at sea level. Table 19−1 includes relative density, the square root of relative density, and the speed of sound for cold, hot and standard days. Figures 19−3 through 19−5 present this data graphically. These parameters are important in understanding the interrelationships between the different definitions of flight speed.

Density falls with pressure altitude such that at 30,500 m (100,066 ft) it is only 1.3% of its sISA sea level value. The maximum speed of sound of 689.0 kt (1276 km/h, 792.8 mph) occurs on a hot day at sea level. The minimum value is 531.6 kt (984.3 km/h, 611.6 mph), occurring between 15,545 m (51,000 ft) and 18,595 m (61,000 ft).

Specific and Relative Humidity

Atmospheric *specific humidity* is variously defined either as:

1. the ratio of water vapor to *dry air* by mass
2. the ratio of water vapor to *moist air* by mass

The former definition is used exclusively herein; for most practical purposes the difference is small anyway. *Relative*

TABLE 19–1 Ambient Conditions Versus Pressure Altitude

		MIL STD 210A Cold Atmosphere				Standard Atmosphere				MIL STD 210A Hot Atmosphere			
Pressure Altitude (m)	Pressure (kPa)	Temp (K)	Relative Density	√Relative Density	Speed of Sound (kt)	Temp (K)	Relative Density	√Relative Density	Speed of Sound (kt)	Temp (K)	Relative Density	√Relative Density	Speed of Sound (kt)
(a) SI Units: 0–15,000 m													
0	101.325	222.1	1.298	1.139	581.0	288.2	1.000	1.000	661.7	312.6	0.922	0.960	689.0
250	98.362	228.2	1.226	1.107	589.0	286.6	0.976	0.988	659.8	310.9	0.900	0.949	687.1
500	95.460	234.4	1.158	1.076	596.9	284.9	0.953	0.976	658.0	309.1	0.878	0.937	685.2
750	92.631	240.6	1.095	1.046	604.7	283.3	0.930	0.964	656.1	307.4	0.857	0.926	683.3
1000	89.873	245.3	1.042	1.021	610.7	281.7	0.907	0.953	654.2	305.6	0.836	0.914	681.3
1250	87.180	247.1	1.004	1.002	612.8	280.0	0.885	0.941	652.3	303.8	0.816	0.903	679.3
1500	84.558	247.1	0.973	0.987	612.8	278.4	0.864	0.929	650.4	302.1	0.796	0.892	677.3
1750	81.994	247.1	0.944	0.972	612.8	276.8	0.842	0.918	648.5	300.3	0.777	0.881	675.4
2000	79.496	247.1	0.915	0.957	612.8	275.2	0.822	0.906	646.6	298.5	0.757	0.870	673.3
2250	77.060	247.1	0.887	0.942	612.8	273.5	0.801	0.895	644.7	296.7	0.739	0.859	671.3
2500	74.683	247.1	0.860	0.927	612.8	271.9	0.781	0.884	642.8	294.8	0.720	0.849	669.2
2750	72.367	247.1	0.833	0.913	612.8	270.3	0.761	0.873	640.9	293.0	0.702	0.838	667.2
3000	70.106	247.1	0.807	0.898	612.8	268.6	0.742	0.861	639.0	291.2	0.685	0.827	665.2
3250	67.905	246.8	0.783	0.885	612.4	267.0	0.723	0.850	637.1	289.5	0.667	0.817	663.2
3500	65.761	245.7	0.761	0.872	611.1	265.4	0.705	0.839	635.1	287.8	0.650	0.806	661.3
3750	63.673	244.2	0.741	0.861	609.3	263.8	0.686	0.829	633.2	286.1	0.633	0.796	659.3
4000	61.640	242.7	0.722	0.850	607.4	262.1	0.669	0.818	631.2	284.4	0.616	0.785	657.3
4250	59.657	241.2	0.703	0.839	605.5	260.6	0.651	0.807	629.3	282.6	0.600	0.775	655.3
4500	57.731	239.7	0.685	0.828	603.6	258.9	0.634	0.796	627.3	280.8	0.585	0.765	653.2
4750	55.852	238.2	0.667	0.817	601.7	257.3	0.617	0.786	625.3	279.1	0.569	0.754	651.2
5000	54.022	236.6	0.649	0.806	599.7	255.7	0.601	0.775	623.4	277.3	0.554	0.744	649.1
5250	52.242	235.1	0.632	0.795	597.8	254.1	0.585	0.765	621.4	275.5	0.539	0.734	647.0
5500	50.507	233.5	0.615	0.784	595.8	252.4	0.569	0.754	619.4	273.7	0.525	0.724	644.9
5750	48.820	231.9	0.599	0.774	593.8	250.8	0.554	0.744	617.4	271.9	0.511	0.715	642.8

(Continued)

TABLE 19–1 Ambient Conditions Versus Pressure Altitude—cont'd

		MIL STD 210A Cold Atmosphere				Standard Atmosphere				MIL STD 210A Hot Atmosphere			
Pressure Altitude (m)	Pressure (kPa)	Temp (K)	Relative Density	√Relative Density	Speed of Sound (kt)	Temp (K)	Relative Density	√Relative Density	Speed of Sound (kt)	Temp (K)	Relative Density	√Relative Density	Speed of Sound (kt)
6000	47.178	230.4	0.582	0.763	591.8	249.2	0.538	0.734	615.4	270.2	0.497	0.705	640.8
6250	45.584	228.8	0.567	0.753	589.7	247.5	0.524	0.724	613.4	268.5	0.483	0.695	638.8
6500	44.033	227.2	0.551	0.742	587.7	245.9	0.509	0.714	611.4	266.8	0.469	0.685	636.8
6750	42.525	225.6	0.536	0.732	585.5	244.3	0.495	0.704	609.3	265.1	0.456	0.675	634.7
7000	41.063	224.0	0.521	0.722	583.5	242.7	0.481	0.694	607.4	263.4	0.443	0.666	632.7
7250	39.638	222.3	0.507	0.712	581.3	241.0	0.468	0.684	605.3	261.7	0.431	0.656	630.6
7500	38.254	220.7	0.493	0.702	579.2	239.4	0.454	0.674	603.2	259.9	0.419	0.647	628.5
7750	36.909	219.0	0.479	0.692	577.0	237.8	0.441	0.664	601.2	258.2	0.407	0.638	626.5
8000	35.601	217.4	0.466	0.682	574.9	236.2	0.429	0.655	599.2	256.5	0.395	0.628	624.3
8250	34.330	215.8	0.453	0.673	572.7	234.5	0.416	0.645	597.1	254.7	0.383	0.619	622.2
8500	33.096	214.1	0.440	0.663	570.4	232.9	0.404	0.636	595.0	252.9	0.372	0.610	620.0
8750	31.899	212.3	0.427	0.654	568.1	231.3	0.392	0.626	592.9	251.2	0.361	0.601	617.9
9000	30.740	210.6	0.415	0.644	565.9	229.7	0.381	0.617	590.9	249.5	0.350	0.592	615.8
9250	29.616	209.1	0.403	0.635	563.8	228.0	0.369	0.608	588.7	247.9	0.340	0.583	613.8
9500	28.523	208.0	0.390	0.624	562.3	226.4	0.358	0.599	586.6	246.2	0.329	0.574	611.8
9750	27.463	208.0	0.375	0.613	562.3	224.8	0.347	0.589	584.6	244.6	0.319	0.565	609.7
10,000	26.435	208.0	0.361	0.601	562.3	223.2	0.337	0.580	582.4	242.9	0.310	0.556	607.6
10,250	25.441	208.0	0.348	0.590	562.3	221.5	0.327	0.571	580.3	241.2	0.300	0.548	605.6
10,500	24.475	208.0	0.335	0.578	562.3	219.9	0.317	0.563	578.2	239.7	0.290	0.539	603.6
10,750	23.540	208.0	0.322	0.567	562.3	218.3	0.307	0.554	576.0	238.2	0.281	0.530	601.7
11,000	22.628	208.0	0.309	0.556	562.3	216.7	0.297	0.545	573.9	236.7	0.272	0.521	599.8
11,250	21.758	208.0	0.297	0.545	562.3	216.7	0.286	0.534	573.9	235.2	0.263	0.513	597.9
11,500	20.914	208.0	0.286	0.535	562.3	216.7	0.275	0.524	573.9	233.6	0.255	0.505	595.9
11,750	20.106	208.0	0.275	0.524	562.3	216.7	0.264	0.514	573.9	232.1	0.246	0.496	593.9
12,000	19.331	208.0	0.264	0.514	562.3	216.7	0.254	0.504	573.9	231.0	0.238	0.488	592.5

12,250	18.583	208.0	0.254	0.504	562.3	216.7	0.244	0.494	573.9	230.6	0.229	0.479	592.0
12,500	17.862	208.0	0.244	0.494	562.3	216.7	0.234	0.484	573.9	230.8	0.220	0.469	592.3
12,750	17.176	208.0	0.235	0.485	562.3	216.7	0.225	0.475	573.9	231.0	0.211	0.460	592.5
13,000	16.512	207.0	0.227	0.476	560.9	216.7	0.217	0.466	573.9	231.1	0.203	0.451	592.8
13,250	15.872	205.1	0.220	0.469	558.3	216.7	0.208	0.456	573.9	231.3	0.195	0.442	593.0
13,500	15.257	202.7	0.214	0.463	555.1	216.7	0.200	0.447	573.9	231.5	0.187	0.433	593.2
13,750	14.669	200.3	0.208	0.456	551.8	216.7	0.193	0.439	573.9	231.8	0.180	0.424	593.5
14,000	14.105	197.9	0.203	0.450	548.4	216.7	0.185	0.430	573.9	232.0	0.173	0.416	593.8
14,250	13.558	195.4	0.197	0.444	545.0	216.7	0.178	0.422	573.9	232.2	0.166	0.408	594.1
14,500	13.034	193.0	0.192	0.438	541.6	216.7	0.171	0.414	573.9	232.4	0.160	0.399	594.3
14,750	12.530	190.7	0.187	0.432	538.4	216.7	0.164	0.406	573.9	232.6	0.153	0.391	594.6
15,000	12.045	188.7	0.182	0.426	535.6	216.7	0.158	0.398	573.9	232.8	0.147	0.384	594.9
(b) SI units: 15,250–30,500 m													
15,250	11.579	187.1	0.176	0.420	533.2	216.7	0.152	0.390	573.9	233.1	0.141	0.376	595.2
15,500	11.131	186.1	0.170	0.412	531.9	216.7	0.146	0.382	573.9	233.2	0.136	0.368	595.4
15,750	10.702	185.9	0.164	0.405	531.6	216.7	0.140	0.375	573.9	233.3	0.130	0.361	595.5
16,000	10.287	185.9	0.157	0.397	531.6	216.7	0.135	0.367	573.9	233.4	0.125	0.354	595.6
16,250	9.889	185.9	0.151	0.389	531.6	216.7	0.130	0.360	573.9	233.5	0.120	0.347	595.7
16,500	9.509	185.9	0.145	0.381	531.6	216.7	0.125	0.353	573.9	233.6	0.116	0.340	595.9
16,750	9.142	185.9	0.140	0.374	531.6	216.7	0.120	0.346	573.9	233.7	0.111	0.334	596.0
17,000	8.789	185.9	0.134	0.367	531.6	216.7	0.115	0.340	573.9	233.7	0.107	0.327	596.0
17,250	8.446	185.9	0.129	0.359	531.6	216.7	0.111	0.333	573.9	233.8	0.103	0.321	596.1
17,500	8.118	185.9	0.124	0.352	531.6	216.7	0.107	0.326	573.9	233.9	0.099	0.314	596.2
17,750	7.806	185.9	0.119	0.346	531.6	216.7	0.102	0.320	573.9	234.0	0.095	0.308	596.4
18,000	7.502	185.9	0.115	0.339	531.6	216.7	0.098	0.314	573.9	234.1	0.091	0.302	596.5
18,250	7.213	185.9	0.110	0.332	531.6	216.7	0.095	0.308	573.9	234.2	0.088	0.296	596.6
18,500	6.936	185.9	0.106	0.326	531.6	216.7	0.091	0.302	573.9	234.2	0.084	0.290	596.7
18,750	6.668	186.8	0.102	0.319	532.8	216.7	0.088	0.296	573.9	234.3	0.081	0.284	596.8
19,000	6.410	188.1	0.097	0.311	534.8	216.7	0.084	0.290	573.9	234.4	0.078	0.279	596.9

(Continued)

TABLE 19–1 Ambient Conditions Versus Pressure Altitude—cont'd

		MIL STD 210A Cold Atmosphere				Standard Atmosphere				MIL STD 210A Hot Atmosphere			
Pressure Altitude (m)	Pressure (kPa)	Temp (K)	Relative Density	√Relative Density	Speed of Sound (kt)	Temp (K)	Relative Density	√Relative Density	Speed of Sound (kt)	Temp (K)	Relative Density	√Relative Density	Speed of Sound (kt)
19,250	6.162	189.5	0.092	0.304	536.7	216.7	0.081	0.284	573.9	234.5	0.075	0.273	597.1
19,500	5.924	190.9	0.088	0.297	538.7	216.7	0.078	0.279	573.9	234.6	0.072	0.268	597.2
19,750	5.695	192.2	0.084	0.290	540.5	216.7	0.075	0.273	573.9	234.7	0.069	0.263	597.3
20,000	5.475	193.5	0.080	0.284	542.3	216.7	0.072	0.268	573.9	234.8	0.066	0.258	597.4
20,250	5.263	194.7	0.077	0.277	544.0	216.7	0.069	0.263	573.9	234.9	0.064	0.252	597.5
20,500	5.060	195.9	0.073	0.271	545.6	216.7	0.066	0.258	573.9	235.1	0.061	0.247	597.8
20,750	4.864	197.0	0.070	0.265	547.2	216.7	0.064	0.253	573.9	235.4	0.059	0.242	598.1
21,000	4.676	198.1	0.067	0.259	548.8	216.7	0.061	0.248	573.9	235.6	0.056	0.238	598.5

Notes: The source for this chart provides data for up to 100,200 m pressure altitude.
To convert kt to m/s multiply by 0.5144.
To convert kt to km/h multiply by 1.8520.
To convert K to °C subtract 273.15.
To convert K to °R multiply by 1.8.
To convert K to °F multiply by 1.8 and subtract 459.67.
Density at ISA sea level static = 1.2250 kg/m^3.
Standard practice is to interpolate linearly between altitudes listed.
(Source: Rolls Royce.)

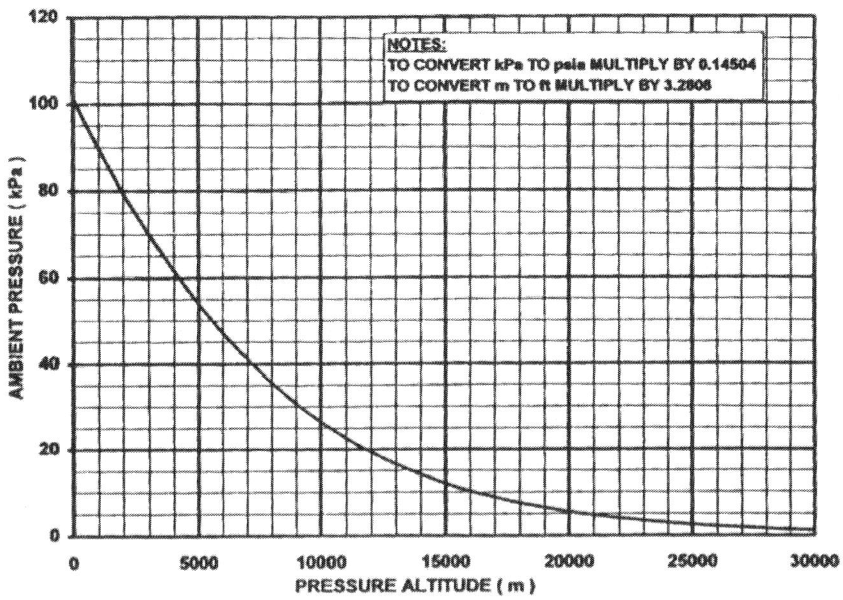

FIGURE 19–1 Ambient pressure versus pressure altitude. *(Source: Rolls Royce.)*

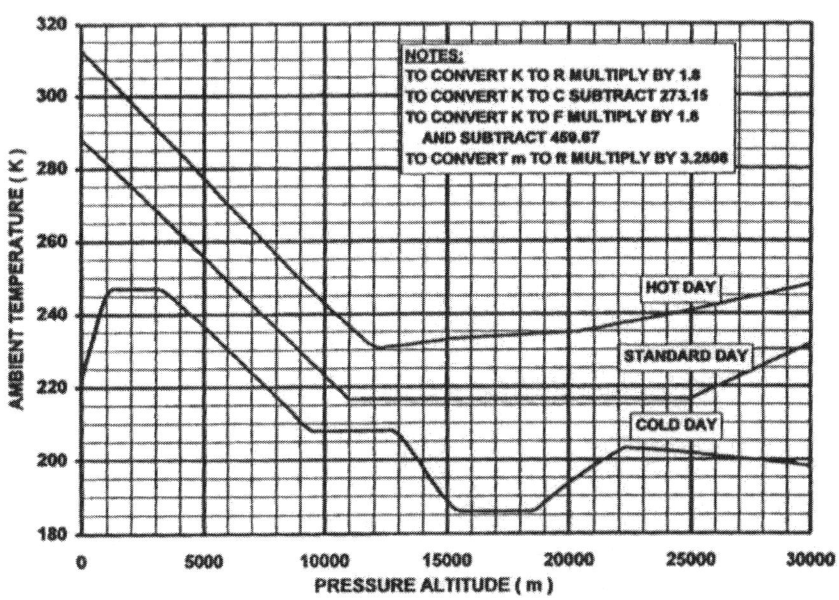

FIGURE 19–2 Ambient temperature versus pressure altitude. *(Source: Rolls Royce.)*

humidity is specific humidity divided by the saturated value for the prevailing ambient pressure and temperature.

Humidity has the least powerful effect upon engine performance of the three ambient parameters. Its effect is not negligible, however, in that it changes the inlet air's molecular weight, and hence basic properties of specific heat and gas constant. In addition, condensation may occasionally have gross effects on temperature. Wherever possible humidity effects should be considered, particularly for hot days with high levels of relative humidity.

For most gas turbine performance purposes, specific humidity is negligible below 0°C, and also above 40°C. The latter is because the highest temperatures only occur in desert conditions, where water is scarce. MIL 210 gives 35°C as the highest ambient temperature at which to consider 100% relative humidity.

Figure 19–6 presents specific humidity for 100% relative humidity versus pressure altitude for cold, standard, and hot days. For MIL 210 cold days specific humidity is almost zero for all altitudes. The maximum specific humidity will never exceed 4.8%, which would

FIGURE 19–3 Relative density versus pressure altitude. *(Source: Rolls Royce.)*

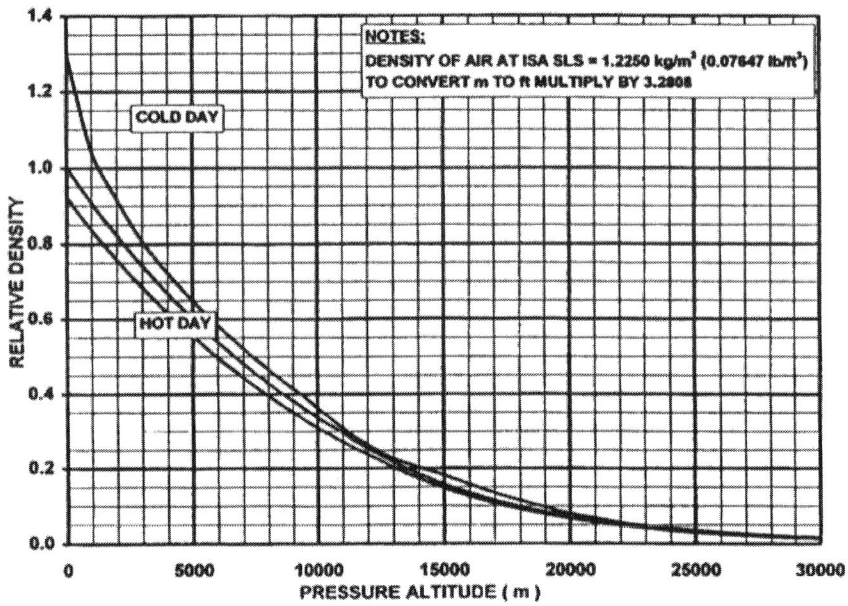

FIGURE 19–4 Square root of relative density versus pressure altitude. *(Source: Rolls Royce.)*

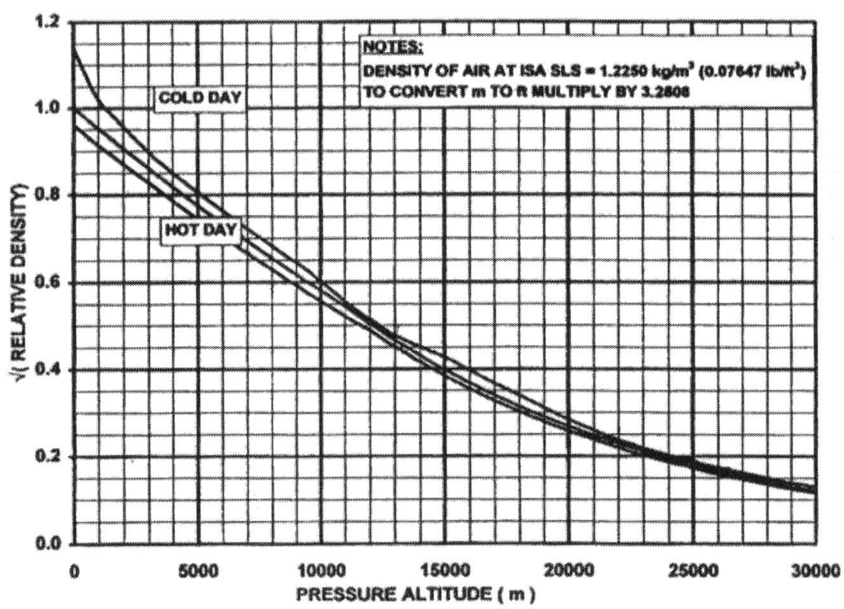

occur on a MIL 210 hot day at sea level. In the troposphere (i.e., below 11,000 m) specific humidity for 100% relative humidity falls with pressure altitude, due to the falling ambient temperature. Above that, in the stratosphere, water vapor content is negligible, almost all having condensed out at the colder temperatures below.

Figures 19–7 and 19–8 facilitate conversion of specific and relative humidities. Figure 19–7 presents specific humidity versus ambient temperature and relative humidity at sea level. For other altitudes Figure 19–8 presents factors to be applied to the specific humidity obtained from Figure 19–7. For a given relative humidity specific humidity is higher at altitude because whereas water vapor pressure is dependent only on temperature, air pressure is significantly lower.

Industrial Gas Turbines

The environmental envelope for industrial gas turbines, both for power generation and mechanical drive

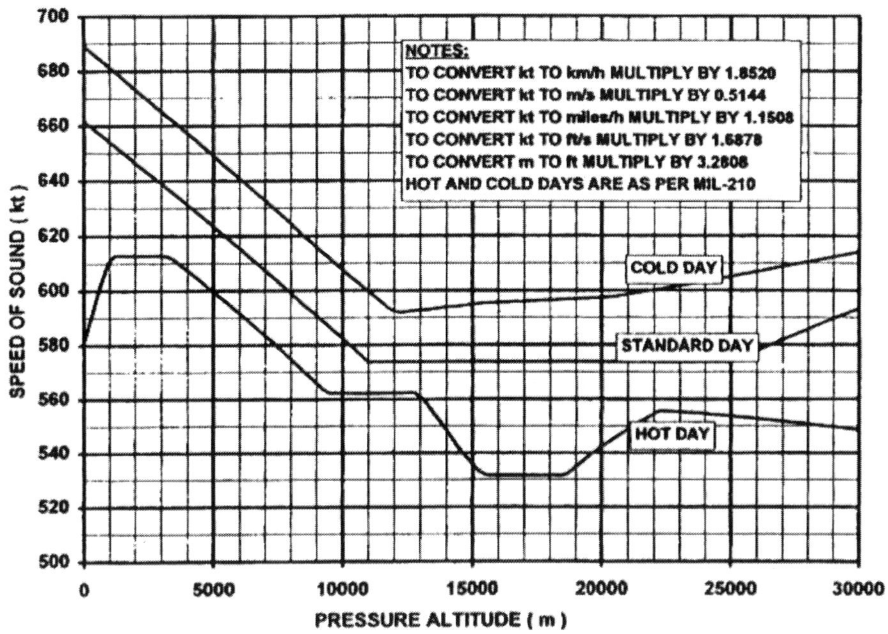

FIGURE 19–5 Speed of sound versus pressure altitude. *(Source: Rolls Royce.)*

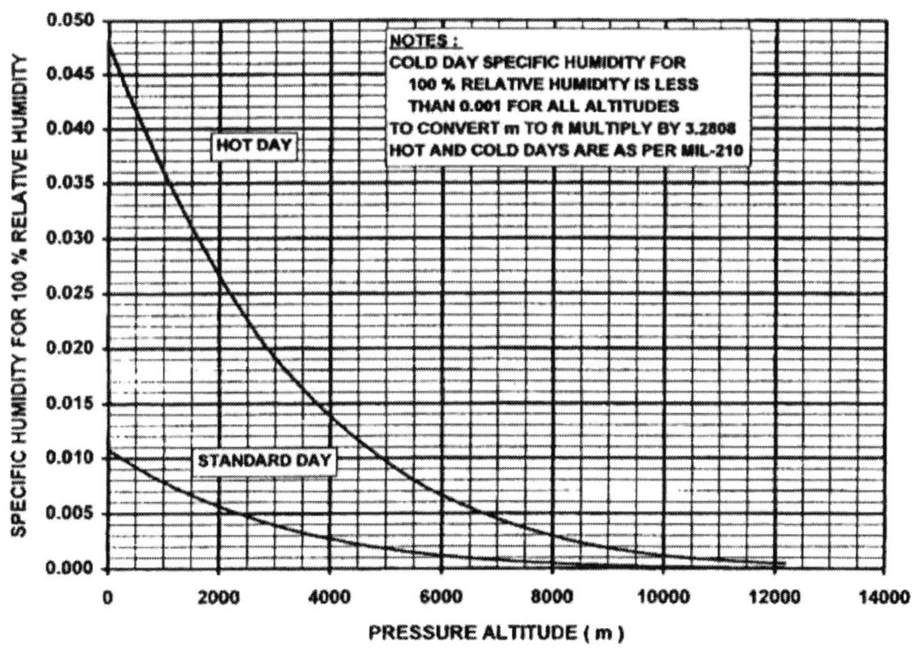

FIGURE 19–6 Specific humidity versus pressure altitude for 100% relative humidity. *(Source: Rolls Royce.)*

applications, is normally taken from Table 19–1 up to a pressure altitude of around 4500 m (or 15,000 ft). Hot and cold day ambient temperatures beyond those of MIL 210 are often used, ±50°C being typical at sea level. For specific fixed locations, altitude is known, and *S curves* are available defining the annual distribution of ambient temperature: these allow lifing assessments and rating selection. (The name derives from the characteristic shape of the curve, which plots the percentage of time for which a particular temperature level would be exceeded.)

The range of specific humidities for an industrial gas turbine would be commensurate with 0–100% relative humidity over most of the ambient temperature range, with some alleviation at the hot and cold extremes.

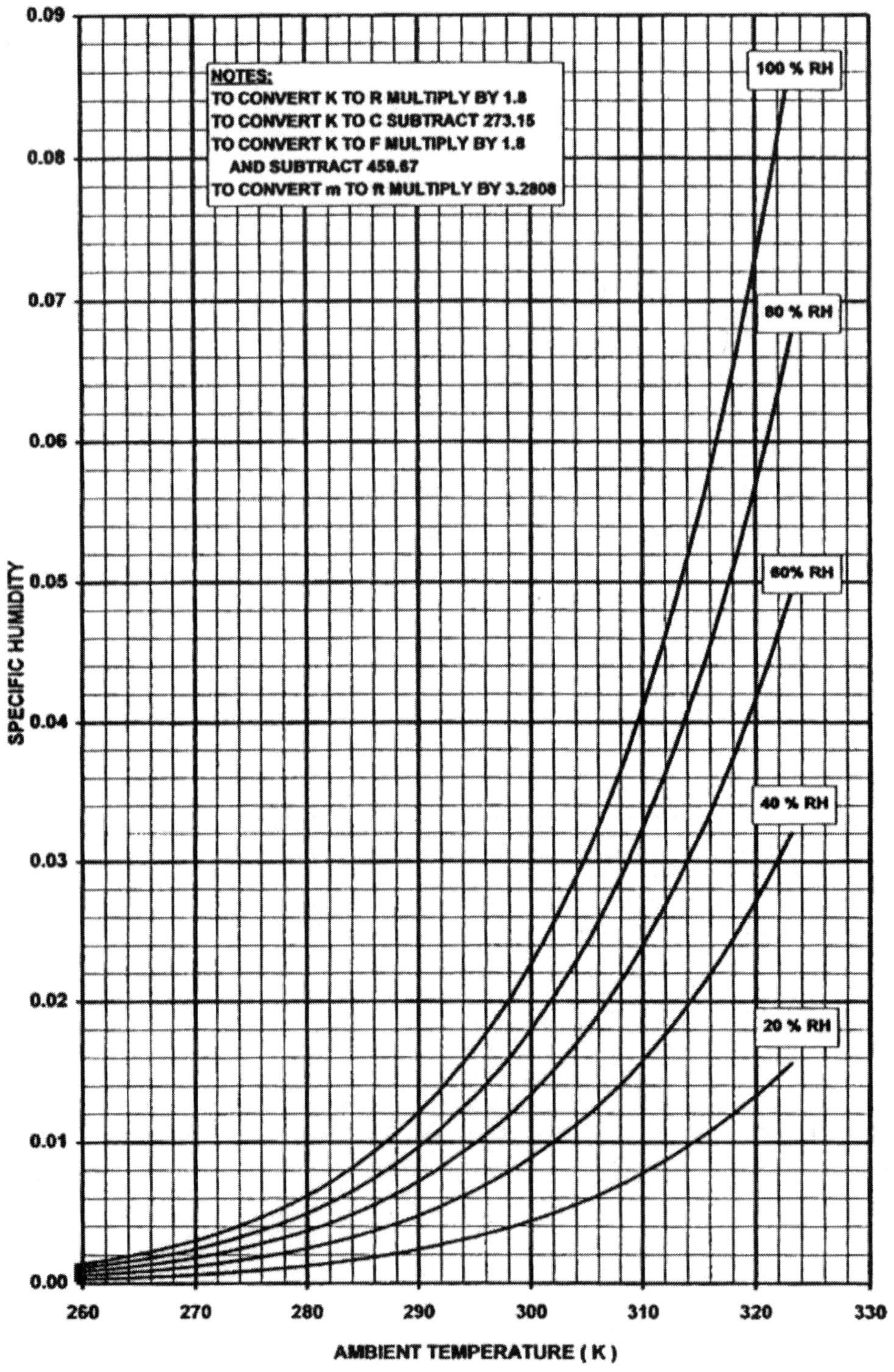

FIGURE 19–7 Specific humidity versus relative humidity and ambient temperature at sea level. *(Source: Rolls Royce.)*

Automotive Gas Turbines

Most comments are as per industrial engines, except that narrowing down the range of ambient conditions for a specific application based on a fixed location is not appropriate.

Marine Gas Turbines

The range of pressure altitudes at sea is governed by weather conditions only, as the element of elevation that significantly affects all other gas turbine types is absent. The practice of the US Navy, the most prolific user of marine gas

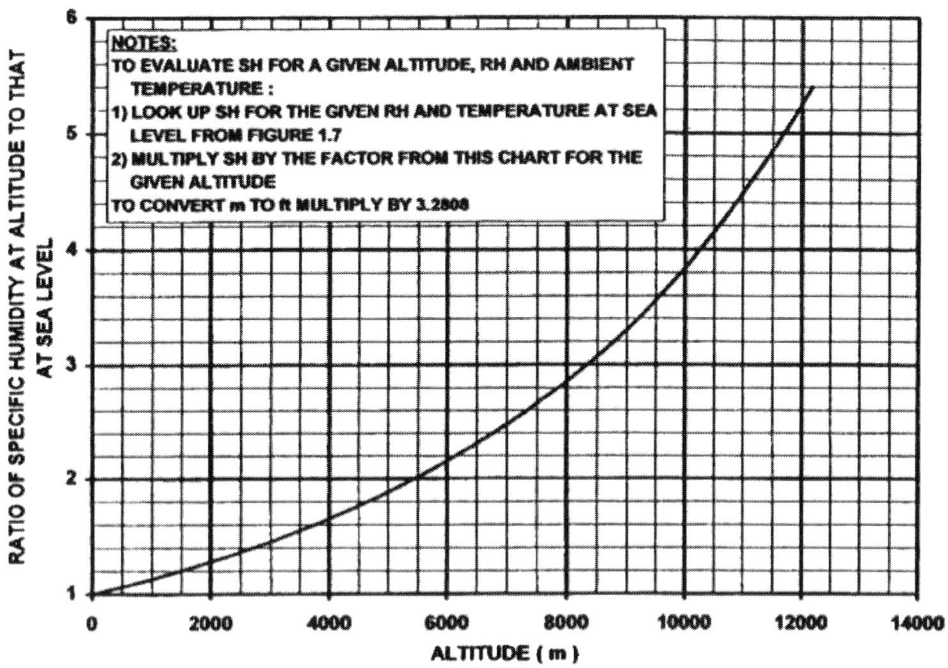

FIGURE 19–8 Ratio of specific humidity at altitude to that at sea level. *(Source: Rolls Royce.)*

turbines, is to take the likely range of ambient pressure as 87–108 kPa (12.6–15.7 psia). This corresponds to a pressure altitude variation of −600–1800 m (−1968–5905 ft).

At sea, free stream air temperature (i.e., that not affected by solar heating of the ship's decks) matches sea surface temperature, day or night. Owing to the vast thermal inertia of the sea there is a significant reduction in the range of ambient temperature that marine gas turbines encounter when on the open sea relative to land based or aircraft gas turbines. However, ships must also be able to operate close to land, including polar ice fields and the Persian Gulf. Consequently for operability (if not lifing) purposes, a wide range of ambient temperature would normally be considered for a marine engine. The most commonly used ambient temperature range is that of the US Navy, which is −40–50°C. US Navy ratings are proven at 38°C, giving some margin on engine life.

The relative humidity range encountered by a marine gas turbine would be unlikely to include zero, due to the proximity of water. In practice values above 80% are typical. The upper limit would be commensurate with 100% relative humidity over most of the ambient temperature range, again with some alleviation at the hot and cold extremes.

Aircraft Engines

The environmental envelope for aircraft engines is normally taken from Table 19–1 up to the altitude ceiling for the aircraft. The specific humidity range is that corresponding to zero to 100% relative humidity as per Figure 19–6.

Installation Pressure Losses

Engine performance levels quoted at ISO conditions do not include installation ducting pressure losses. This level of performance is termed *uninstalled* and would normally be between inlet and exit planes consistent with the engine manufacturer's supply. Examples might include from the flange at entry to the first compressor casing to the engine exhaust duct exit flange, or to the propelling nozzle exit plane for thrust engines. When installation pressure losses, together with other installation effects, are included the resultant level of performance is termed *installed.*

For industrial, automotive, and marine engines installation pressure losses are normally imposed by plant intake and exhaust ducting. For aircraft engines there is usually a flight intake upstream of the engine inlet flange, which is an integral part of the airframe as opposed to the engine; however, for high bypass ratio turbofans there is not normally an installation exhaust duct. An additional item for aircraft engines is intake ram recovery factor. This is the fraction of the free stream dynamic pressure recovered by the installation or flight intake as total pressure at the engine intake front face.

Pressure losses due to installation ducting should *never* be approximated as a change of pressure altitude reflecting the lower inlet pressure at the engine intake flange. While intake losses do indeed lower inlet pressure, exhaust losses *raise* engine exhaust plane pressure. Artificially changing ambient pressure clearly cannot simulate both effects at once.

For industrial, automotive, and marine engines installation pressure losses are most commonly expressed as mm H_2O, where 100 mm H_2O is approximately 1% total pressure loss at sea level (0.981 kPa, 0.142 psi). For aircraft applications installation losses are more usually expressed as a percentage loss in total pressure (%ΔP/P).

Industrial Engines

Overall installation inlet pressure loss due to physical ducting, filters and silencers is typically 100 mm H_2O at high power. Installation exhaust loss is typically 100–300 mm H_2O (0.981 kPa, 0.142 psi to 2.942 kPa, 0.427 psi); the higher values occur where there is a steam plant downstream of the gas turbine.

Automotive Engines

In this instance both installation inlet and exhaust loss are typically 100 mm H_2O (0.981 kPa, 0.142 psi).

Marine Engines

Installation intake and exhaust loss values at rated power may be up to 300 mm H_2O (2.942 kPa, 0.427 psi) and 500 mm H_2O (4.904 kPa, 0.711 psi), respectively, dependent upon ship design. Standard values used by the US Navy are 100 mm H_2O (0.981 kPa, 0.142 psi) and 150 mm H_2O (1.471 kPa, 0.213 psi).

Aircraft Engines

For a pod mounted turbofan cruising at 0.8 Mach number, the total pressure loss from free stream to the flight intake/engine intake interface due to incomplete ram recovery and the installation intake may be as low as 0.5% ΔP/P, whereas for a ramjet operating at Mach 3 the loss may be nearer 15%. For a helicopter engine buried behind filters the installation intake total pressure loss may be up to 2%, and there may also be an installation exhaust pressure loss due to exhaust signature suppression devices.

The Flight Envelope

Typical Flight Envelopes for Major Aircraft Types

Aircraft engines must operate at a range of forward speeds in addition to the environmental envelope. The range of flight Mach numbers for a given altitude is defined by the flight envelope. Figure 19–9 presents typical flight envelopes for the seven major types of aircraft.

For each flight envelope the minimum and maximum free stream temperatures and pressures the engine would experience are shown, together with basic reasons for the shape of the envelope. Where auxiliary power units are employed the same free stream conditions are experienced as for the propulsion unit. The intake ram recovery is often lower, however, due both to placement at the rear of the fuselage and drag constraints on the intake design.

Free Stream Total Pressure and Temperature

The free stream total pressure (P0) is a function of both pressure altitude and flight Mach number. Free stream total temperature (T0) is a function also of ambient temperature and flight Mach number. Both inlet pressure and temperature are fundamental to engine performance. They are often used to refer engine parameters to ISA sea level static conditions, via quasi dimensionless parameter groups. To do this the following ratios are defined:

$$\text{DELTA}(\delta) = \text{P0}/101.325\ \text{kPa}$$

$$\text{THETA}(\theta) = \text{T0}/288.15\ \text{K}$$

Definitions of Flight Speed

Traditionally aircraft speed has been measured using a pitot-static head located on a long tube projecting forward from the wing or fuselage nose. The difference between total and static pressure is used to evaluate velocity, which is shown on a visual display unit or gauge in the cockpit. The device is normally calibrated at sea level, which has given rise to a number of definitions of flight speed:

- *Indicated air speed* (VIAS) is the speed indicated in the cockpit based upon the above calibration.
- *Calibrated air speed* (VCAS) is approximately equal to VIAS with the only difference being a small adjustment to allow for aircraft disturbance of the static pressure field around the pitot-static probe.
- *Equivalent air speed* (VEAS) results from correcting VCAS for the lower ambient pressure at altitude versus that embedded in the probe calibration conducted at sea level, i.e., for a given Mach number the dynamic head is smaller at altitude. When at sea level VEAS is equal to VCAS.
- *True air speed* (VTAS) is the actual speed of the aircraft relative to the air. It is evaluated by multiplying VEAS by the square root of relative density as presented in Figure 19–4. This correction is due to the fact that the density of air at sea level is embedded in the probe calibration that provides VIAS. Both density and velocity make up dynamic pressure.

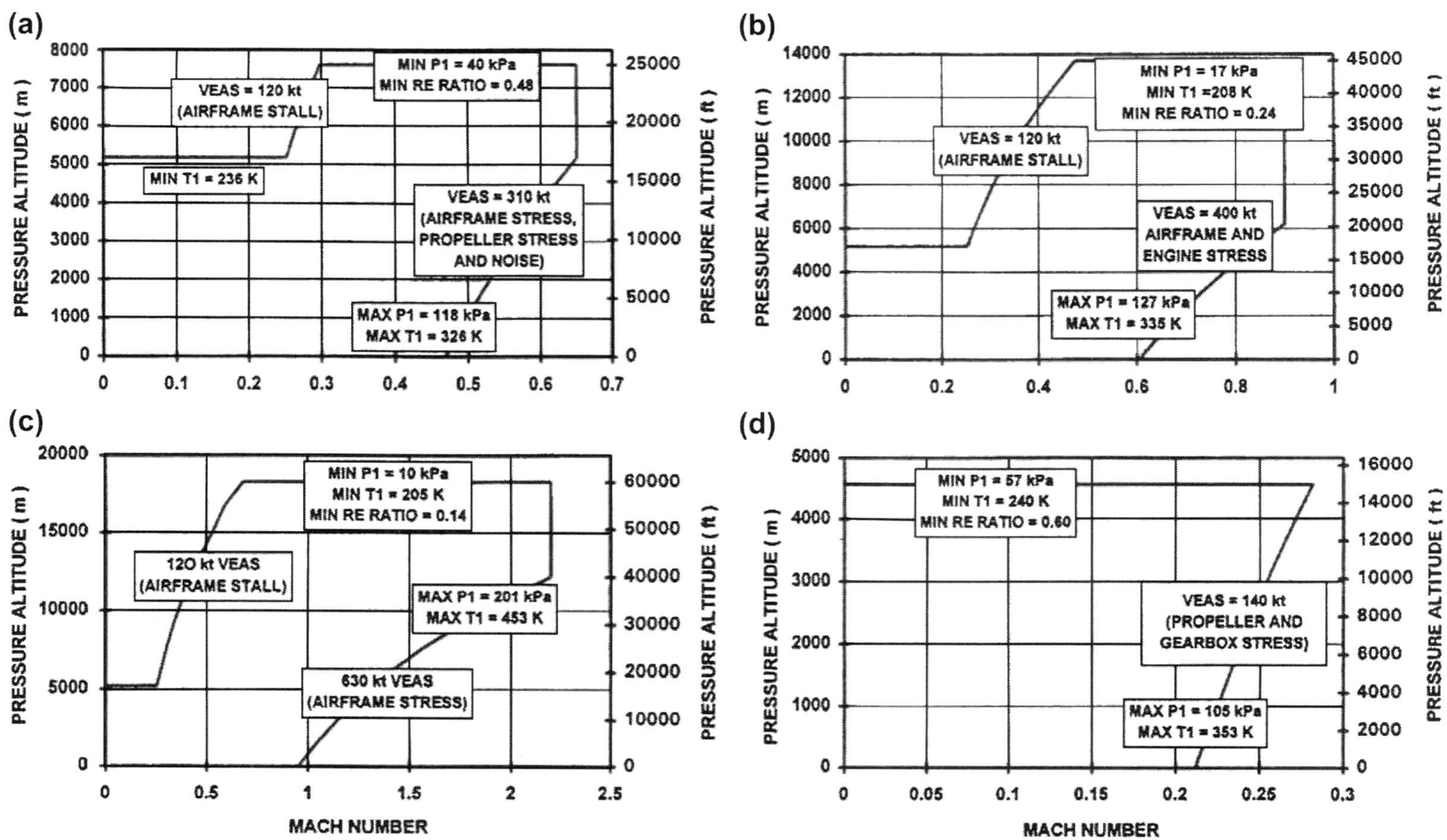

FIGURE 19–9 Flight envelopes for the major aircraft types. (a) Conventional civil transport turboprop. (b) Subsonic civil transport turbofan. (c) Supersonic civil transport. (d) Helicopter. (e) Subsonic airbreathing missile, drone, or RPV. (f) Supersonic airbreathing missile. (g) Advanced military fighter. *(Source: Rolls Royce.)*

Notes: APUs have lower intake RAM recovery than propulsion engines.
Pressures shown are free stream, i.e., 100% RAM recovery.
To convert temperatures in K to R multiply by 1.8.
To convert temperatures in K to C subtract 273.15.
To convert temperatures in K to F multiply by 1.8 and subtract 459.67.
To convert pressures in kPa to psia multiply by 0.145038
To convert speeds in kt to km/h m/s miles/h ft/s multiply by 1.8520, 0.5144, 1.1508, 1.6878.
Maximum temperatures shown are for MIL STD 210 Hot day.
Minimum temperatures shown are for MIL STD 210 Cold day.
Minimum Reynolds' Number ratios shown are for MIL STD 210 Hot day.
Unducted fans would have similar flight envelope to commercial turbofans.
"RPV"= Remotely Piloted Vehicle.
All numbers shown are indicative, for guidance only.

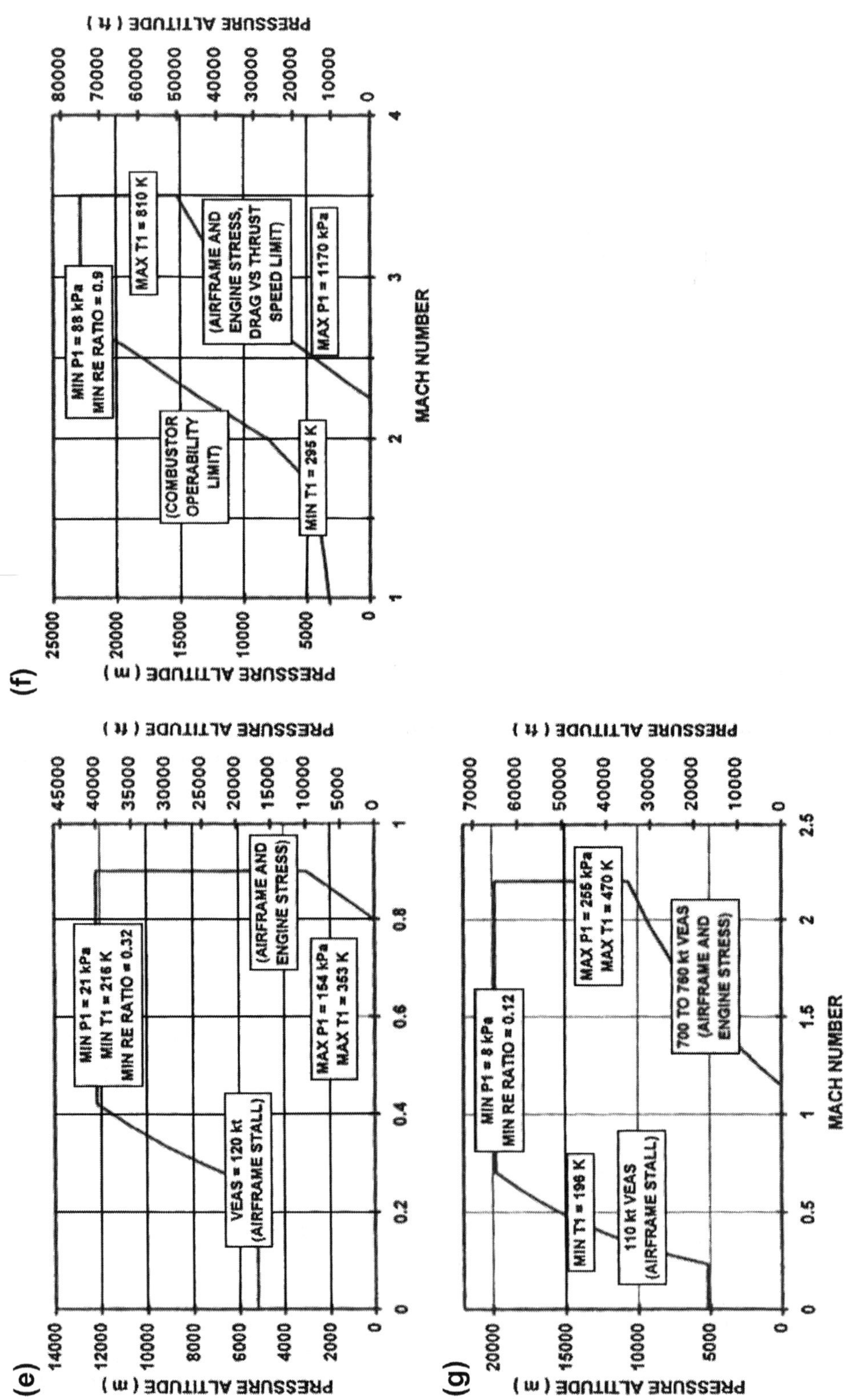

FIGURE 19–9 *(Cont'd).*

- *Mach number* (M) is the ratio of true air speed to the local speed of sound.
- *Ground speed* is VTAS adjusted for wind speed.

VEAS and VCAS are functions of pressure altitude and Mach number only. The difference between VEAS and VCAS is termed the *scale altitude effect* (SAE), and is independent of ambient temperature. Conversely, VTAS is a function of ambient temperature and Mach number, which is independent of pressure altitude.

To a pilot both Mach number and ground air speed are important. The former dictates critical aircraft aerodynamic conditions such as shock or stall, whereas the latter is vital for navigation. For gas turbine engineers Mach number is of paramount importance in determining inlet total conditions from ambient static. Often, however, when analyzing engine performance data from flight tests only VCAS or VEAS are available.

PROPERTIES AND CHARTS FOR DRY AIR, COMBUSTION PRODUCTS, AND OTHER WORKING FLUIDS*

The properties of the working fluid in a gas turbine engine have a powerful impact upon its performance. It is essential that these gas properties are accounted rigorously in calculations, or that any inaccuracy due to simplifying assumptions is quantified and understood. This chapter describes at an engineering level the fundamental gas properties of concern, and their various interrelationships. It also provides a comprehensive database for use in calculations for:

- Dry air
- Combustion products for kerosene or diesel fuel
- Combustion products for natural gas fuel
- Helium, the working fluid often employed in closed cycles

Description of Fundamental Gas Properties

Equation of State for a Perfect Gas

A *perfect gas* adheres to Formula F19.1. All gases employed as the working fluid in gas turbine engines, except for water vapor, may be considered as perfect gases without compromising calculation accuracy. When the mass fraction of water vapor is less than 10%, which is usually the case when it results from the combination of ambient humidity and products of combustion, then for performance calculations the gas mixture may still be considered perfect. When water vapor content exceeds 10% the assumption of a perfect gas is no longer valid and for rigorous calculations steam tables must be employed in parallel, for that fraction of the mixture.

A physical description of a perfect gas is that its enthalpy is only a function of temperature and *not* pressure, as there are no intermolecular forces to absorb or release energy when density changes.

Molecular Weight and the Mole

The molecular weight for a pure gas is defined in the Periodic Table. For mixtures of gases, such as air, the molecular weight may be found by averaging the constituents on a *molar* (volumetric) basis. This is because a mole contains a fixed number of molecules, as described.

A mole is the quantity of a substance such that the mass is equal to the molecular weight in grams. For any perfect gas 1 mole occupies a volume of 22.4 liters at 0°C, 101.325 kPa. A mole contains the Avogadro's number of molecules, 6.023×10^{23}.

Specific Heat at Constant Pressure (CP) and at Constant Volume (CV)

These are the amounts of energy required to raise the temperature of 1 kilogram of the gas by 1°C, at constant pressure and volume, respectively. For gas turbine engines, with a steady flow of gas (as opposed to piston engines where it is intermittent) only the specific heat at constant pressure, CP, is used directly. This is referred to hereafter simply as specific heat.

For the gases of interest specific heat is a function of only gas composition and static temperature. For performance calculations total temperature can normally be used up to Mach numbers of 0.4 with negligible loss in accuracy, since dynamic temperature remains a low proportion of the total.

Gas Constant (R)

The gas constant appears extensively in formulae relating pressure and temperature changes, and is numerically equal to the difference between CP and CV. The gas constant for an individual gas is the universal gas constant divided by the molecular weight, and has units of J/kg K. The universal gas constant has a value of 8314.3 J/mol K.

Ratio of Specific Heats, Gamma (γ) (Formulae F19.6–F19.8)

This is the ratio of the specific heat at constant pressure to that at constant volume. Again, it is a function of gas composition and static temperature, but total temperature may be used when the Mach number is less than 0.4. Gamma appears extensively in the "perfect gas" formulae relating pressure and temperature changes and component efficiencies.

* Source: Courtesy Rolls Royce. *Gas Turbine Performance*, Walsh and Fletcher, Blackwell Science, 1998. Adapted with permission.

Dynamic Viscosity (VIS) and Reynolds Number (RE) (Formulae F19.9)

Dynamic viscosity is used to calculate the Reynolds number, which reflects the ratio of momentum to viscous forces present in a fluid. The Reynolds number is used in many performance calculations, such as for disc windage, and has a second-order effect on component efficiencies. Dynamic viscosity is a measure of the viscous forces and is a function of gas composition and static temperature. As viscosity has only a second-order effect on an engine cycle, total temperature may be used up to a Mach number of 0.6. The effect of fuel air ratio (gas composition) is negligible for practical purposes.

The units of viscosity of N s/m^2 are derived from N/(m/s)/m; force per unit gradient of velocity. Gas velocity varies in a direction perpendicular to the flow in the boundary layers on all gas washed surfaces.

Description of Key Thermodynamic Parameters

The key thermodynamic parameters most widely used in gas turbine performance calculations are described below. Their interrelationships are dependent upon the values of the fundamental gas properties described above.

Total or Stagnation Temperature (T) (Formula F19.10)

Total temperature is the temperature resulting from bringing a gas stream to rest with no work or heat transfer. Note that here "at rest" means relative to the engine, which may have a flight velocity relative to the Earth. The difference between the total and static temperatures at a given point is called the dynamic temperature. The ratio of total to static temperature is a function of only gamma and Mach number, as per Formula F19.10.

In general for gas turbine performance calculations total temperature is used through the engine, evaluated at engine entry from the ambient static temperature and any ram effect. At locations between engine components total temperature is a valid measure of energy changes. In addition, this aids comparison between predictions and test data, as it is only practical to measure total temperature. For most component design purposes, however, static conditions are also relevant, as for example the Mach number is often high (1.0 and greater) at entry to a compressor stator or turbine rotor blade.

Total temperature is constant for flow along ducts where there is no work or heat transfer, such as intake and exhaust systems. Total and static temperature diverge much less rapidly versus Mach number than do total and static pressure, as described below.

Total or Stagnation Pressure (P) (Formulae F19.11, F19.12, and F19.13)

Total pressure is that which would result from bringing a gas stream to rest without any work or heat transfer, and without any change in entropy. Total pressure is therefore an idealized property.

The difference between total and static pressure at a point is called either the dynamic pressure, dynamic head, or velocity head (Formulae F19.12 and F19.13). The term head relates back to hydraulic engineering. The ratio of total to static pressure, as for temperature, is a function of only gamma and Mach number. Most performance calculations are conducted using total pressure, that at engine inlet again resulting from ambient static plus intake ram recovery.

Total pressure is not constant for flow through ducts, being reduced by wall friction and changes in flow direction, which produce turbulent losses. Both these effects act on the dynamic head; the pressure loss in a duct of given geometry and inlet swirl angle is almost always a fixed number of inlet dynamic heads. For this reason for performance calculations both the total and static pressure must often be evaluated at entry to ducts. Again, for component design purposes both the total and static values are of interest.

Total and static pressure diverge much more rapidly versus Mach number than do total and static temperature. Calculation of pressure ratio from temperature ratio is far more sensitive to errors in the assumption of the mean gamma than the reverse calculation.

Specific Enthalpy (H) (Formulae F19.14, F19.15)

This is the energy per kilogram of gas relative to a stipulated zero datum. Changes in enthalpy, rather than absolute values, are important for gas turbine performance. Total or static enthalpy may be calculated, depending on which of the respective temperatures is used. Total enthalpy, like total temperature, is most common in performance calculations.

Specific Entropy (S)

Traditionally the property entropy has been shrouded in mystery, primarily due to being less tangible than the other properties discussed in this section. Entropy relates to other thermodynamic properties relevant to gas turbine performance, and thereby helps overcome these difficulties.

During compression or expansion the increase in entropy is a measure of the thermal energy lost to friction, which becomes unavailable as useful work.

Composition of Dry Air and Combustion Products

Dry Air

Dry air comprises the elements in Table 19—2.

TABLE 19–2 Composition of Dry Air

	By Mole or Volume (%)	By Mass (%)
Nitrogen (N_2)	78.08	75.52
Oxygen (O_2)	20.95	23.14
Argon (Ar)	0.93	1.28
Carbon dioxide	0.03	0.05
Neon	0.002	0.001

(Source: Rolls Royce.)

There are also trace amounts of helium, methane, krypton, hydrogen, nitrous oxide, and xenon. These are negligible for gas turbine performance purposes.

Combustion Products

When a hydrocarbon fuel is burned in air, combustion products change the composition significantly. Atmospheric oxygen is consumed to *oxidize* the hydrogen and carbon, creating water and carbon dioxide, respectively. The degree of change in air composition depends both on *fuel air ratio* and *fuel chemistry*. The fuel air ratio such that all the oxygen is consumed is termed *stoichiometric*.

Distilled liquid fuels such as kerosene or diesel each have relatively fixed chemistry. Properties of their combustion products can be evaluated versus fuel air ratio and temperature using unique formulae, with the fuel chemistry inbuilt. In contrast, the chemistry of natural gas varies considerably. All natural gases have a high proportion of light hydrocarbons, often with other gases such as nitrogen, carbon dioxide, or hydrogen. Because the composition of natural gas combustion products varies, along with the fuel chemistry, unique formulae for their gas properties do not exist, hence the calculation is more complex.

The Use of CP and Gamma, or Specific Enthalpy and Entropy, in Calculations

Either CP and gamma, *or* specific enthalpy and entropy, are used extensively in performance calculations. The manner of their use is described below in order of increasing accuracy and calculation complexity. This list covers all gas turbine components except for the combustor.

Constant, Standard Values for CP and Gamma

This normally uses the following approximations:

- Cold end gas properties: CP = 1004.7J/kg K, gamma = 1.4
- Hot end gas properties: CP = 1156.9J/kg K, gamma = 1.33
- Component performance: Formulae use values of CP and gamma as above

This is the least accurate method, giving errors of up to 5% in leading performance parameters. It should only be used in illustrative calculations for teaching purposes, or for crude "ballpark" estimates.

Values for CP and Gamma Based on Mean Temperature

For formulae using CP and gamma it is most accurate to base these values on the mean temperature within each component, i.e., the arithmetic mean of the inlet and exit values. It is less accurate to evaluate CP and gamma at inlet and exit, and then take a mean value for each.

For dry air and combustion products of kerosene or diesel the formulae given for CP as a function of temperature and fuel air ratio give accuracies of within 1.5% for leading performance parameters. The largest errors occur at the highest pressure ratios.

Specific Enthalpy and Entropy—Dry Air, and Diesel or Kerosene

For fully rigorous calculations changes in enthalpy and entropy across components must be accurately evaluated. This improves accuracy to be within 0.25% for leading parameters at all pressure ratios. Here polynomials of specific enthalpy and entropy are utilized, obtained by integration of the standard polynomials for specific heat. In these methods, a formula for specific heat is therefore still required.

The use of specific enthalpy and entropy for performance calculations is now almost mandatory for computer "library" routines in large companies.

Specific Enthalpy and Entropy—Natural Gas

For the combustion products of natural gas it is logical to use specific enthalpy and entropy only if CP is evaluated accurately.

Database for Fundamental and Thermodynamic Gas Properties

Molecular Weight and Gas Constant

Data for gases of interest are listed in Table 19–3.

Figure 19–10 shows the gas constant resulting from the combustion of leading fuel types in air plotted versus fuel air ratio. It is not possible to provide all encompassing data for natural gas due to the wide variety of blends that occur. For indicative purposes combustion of a sample natural gas has been used. The following are apparent:

- For kerosene, molecular weight and gas constant are not changed noticeably from the values for dry air up to stoichiometric fuel to air ratio.
- For diesel molecular weight and hence gas constant change minimally, in a linear fashion versus fuel to air ratio. For performance calculations there is negligible loss in accuracy by ignoring these small changes and using data for kerosene.

TABLE 19–3 Molecular Weight and Gas Constants

	Molecular Weight	Gas Constant (J/kg K)
Dry air	28.964	287.05
Oxygen	31.999	259.83
Water	18.015	461.51
Carbon dioxide	44.010	188.92
Nitrogen	28.013	296.80
Argon	39.948	208.13
Hydrogen	2.016	4124.16
Neon	20.183	411.95
Helium	4.003	2077.02

Note: The universal gas constant is 8314.3 J/mol K.
(Source: Rolls Royce.)

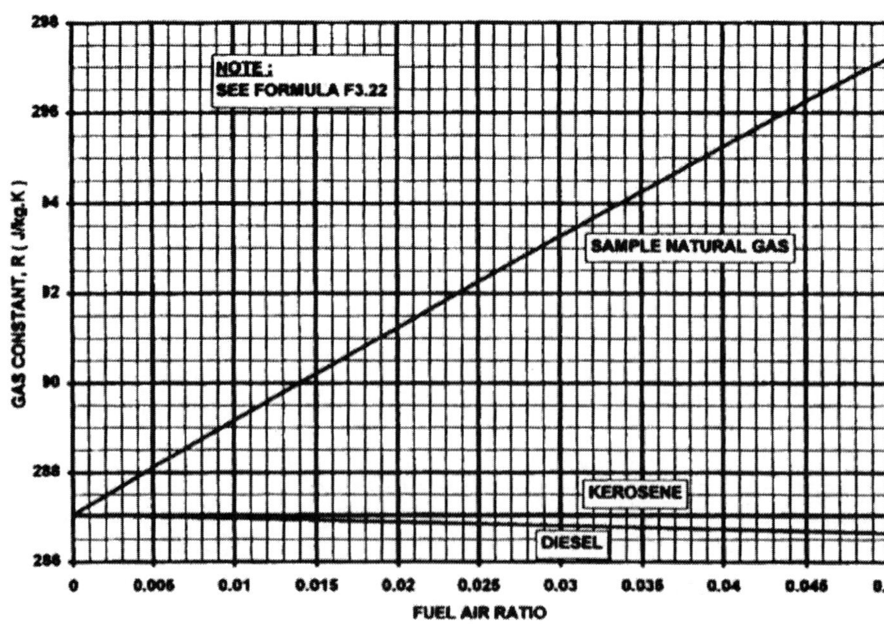

FIGURE 19–10 Gas constant, R, for combustion products of kerosene, diesel, and natural gas versus fuel air ratio. *(Source: Rolls Royce.)*

- For the sample natural gas molecular weight and gas constant vary linearly with fuel air ratio from the values for dry air to 27.975 and 297.15 J/kg K, respectively, at a fuel to air ratio of 0.05. A significant loss of accuracy will occur if this change is not accounted.

The effect of gas fuel is more powerful than that of liquid fuels, primarily due to the constituent hydrocarbons being lighter (i.e., containing less carbon and more hydrogen); this results in a higher proportion of water vapor after combustion, which has a significantly lower molecular weight than the other constituents.

Specific Heat and Gamma

Figures 19—11 and 19—12 present specific heat and gamma respectively for dry air and combustion products versus static temperature and fuel air ratio for kerosene or diesel fuels.

Figure 19—13 shows the ratio of specific heat following the combustion of the sample natural gas to that for kerosene, versus fuel to air ratio. This plot is sensibly independent of temperature. As stated earlier, specific heat is noticeably higher following the combustion of natural gas due to the higher resultant water content, which significantly impacts engine performance.

Figure 19—14 shows specific heat respectively versus temperature for the individual gases present in air and combustion products. The higher value for water vapor is immediately apparent. For inert gases such as helium, argon, and neon, specific heat and gamma do not change with temperature.

Temperature Entropy Diagram for Dry Air

Most heat engine cycles are taught at university level via schematic illustration on a temperature—entropy (T—S) diagram. This approach becomes laborious to extend to "real" engine effects such as internal bleeds and cooling flows, but remains a useful indication of the overall thermodynamics of a known engine cycle. Figure 19—15 presents an actual temperature—entropy diagram for dry air, complete with numbers, showing lines of constant pressure. Such a diagram is rare in the open literature.

The following are important:

- Raising temperature at constant pressure (e.g., by adding heat in a combustor) raises entropy.
- Reducing temperature at constant pressure (e.g., by removing heat in an intercooler) lowers entropy.
- Compression from a lower to a higher constant pressure line (i.e., by adding work) produces minimum change in temperature (i.e., requires minimum energy input) if entropy does not increase. Isentropic compression is an idealized process.
- In reality entropy does increase during compression, hence extra energy must be provided, beyond the ideal work required for the pressure change. This extra energy is converted to heat.
- Expansion from a higher to a lower constant pressure line produces maximum change in temperature (i.e., produces maximum work) if entropy does not increase. Isentropic expansion is also an idealized process.
- In reality entropy does increase during expansion, hence less work output is obtained than the ideal work produced pressure change. This "lost" energy is retained as heat.

Entropy may be defined as *thermal energy not available for doing work.* In real compressors and turbines some energy goes into raising entropy, as some pressure is lost to real effects such as friction. The ideal work would be required or produced if entropy did not change, i.e., the process were *isentropic. Isentropic efficiency* is defined as the appropriate ratio of actual and ideal work, and is always less than 100%. (The term *adiabatic efficiency* is also commonly used, but is strictly incorrect. It only excludes heat transfer but not friction, and an isentropic process would have neither.)

Gas turbine cycles utilize the above processes, and rely on one other vital, fundamental thermodynamic effect:

Work input, approximately proportional to temperature rise, for a given compression ratio from low temperature is significantly lower than the work output from the same expansion ratio from higher temperature.

This is because on the T—S diagram lines of constant pressure diverge with increasing temperature and entropy.

This can be seen by considering a sample compression, heating, and expansion between two lines of constant pressure, using Figure 19—15. At an entropy value of 1.5 kJ/kg K the temperature rise required to go from 100 to 5100 kPa is 500 K. If fuel is now burned at this pressure level such that entropy increases to 2.75 kJ/kg K, and temperature to 1850 K, an expansion back to 100 kPa will achieve a temperature drop of around 1000 K. This clearly illustrates the rationale behind the Brayton cycle.

Schematic T—S Diagrams for Major Engine Cycles

Figures 19—16 through 19—21 show the key cycles of interest to gas turbine engineers.

Figure 19—16 shows the Carnot cycle. This is the most efficient cycle theoretically possible between two temperature levels. Gas turbine engines necessarily do not use the Carnot cycle, as unlike steam cycles they cannot add or reject heat at constant temperature.

Figure 19—17 shows the Brayton cycle. This is the basic cycle utilized by all gas turbine engines where heat is input at constant pressure. The effect of component inefficiency is shown by the non-vertical compression and expansion lines, a further difference from the ideal Carnot cycle. The

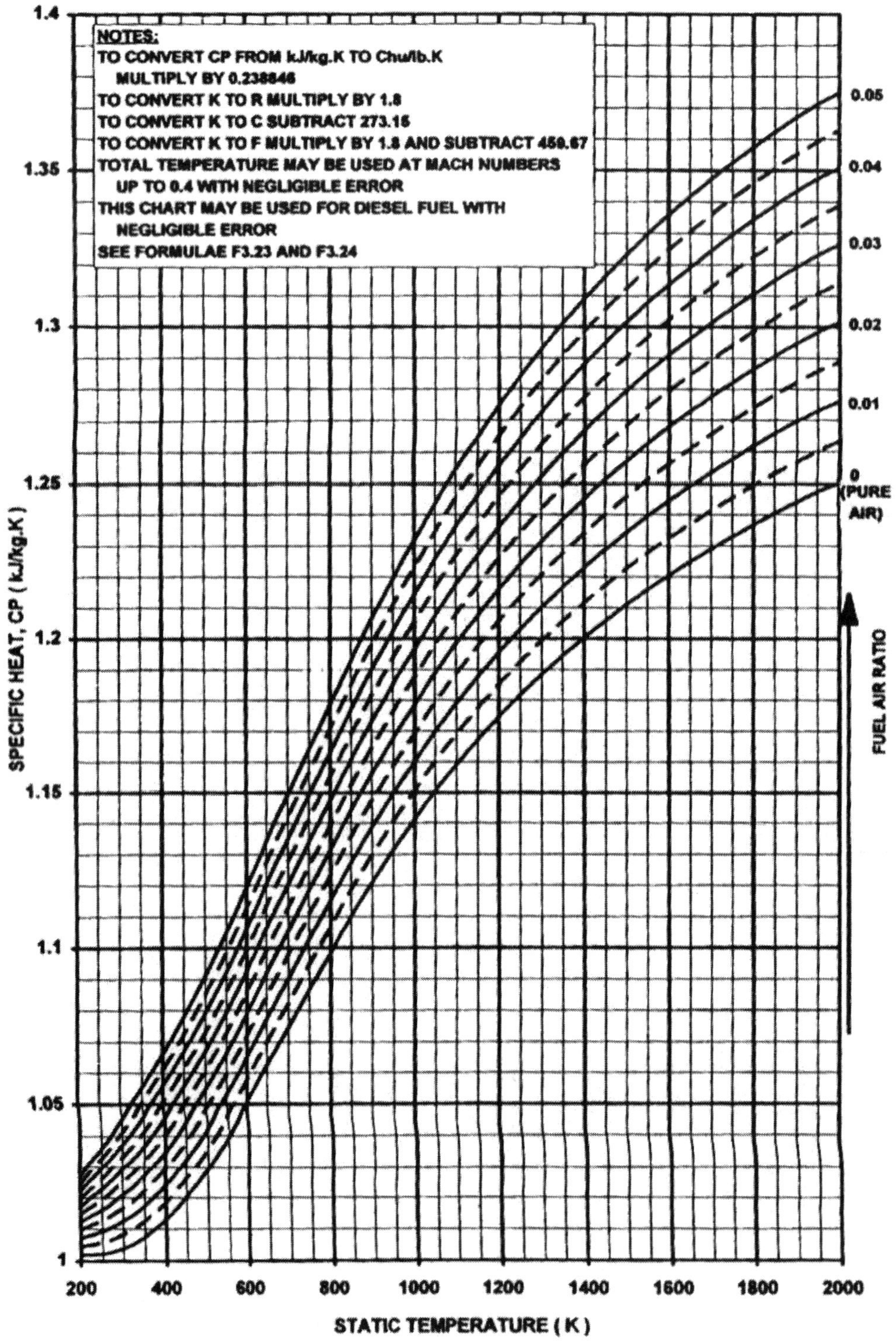

FIGURE 19–11 Specific heat, CP, for kerosene combustion products versus temperature and fuel air ratio. *(Source: Rolls Royce.)*

form of the Brayton cycle is modified for heat exchangers and bypass flows.

Figure 19–18 presents the cycle for a turbofan. The bypass stream only undergoes partial compression, and no heating before expansion back to ambient pressure.

Figure 19–19 shows a heat exchanged cycle. Waste heat from exhaust gases is used to heat air from compressor delivery prior to combustion, thereby reducing the required fuel flow.

Figure 19–20 presents an intercooled cycle, where heat is extracted downstream of an initial compressor. This

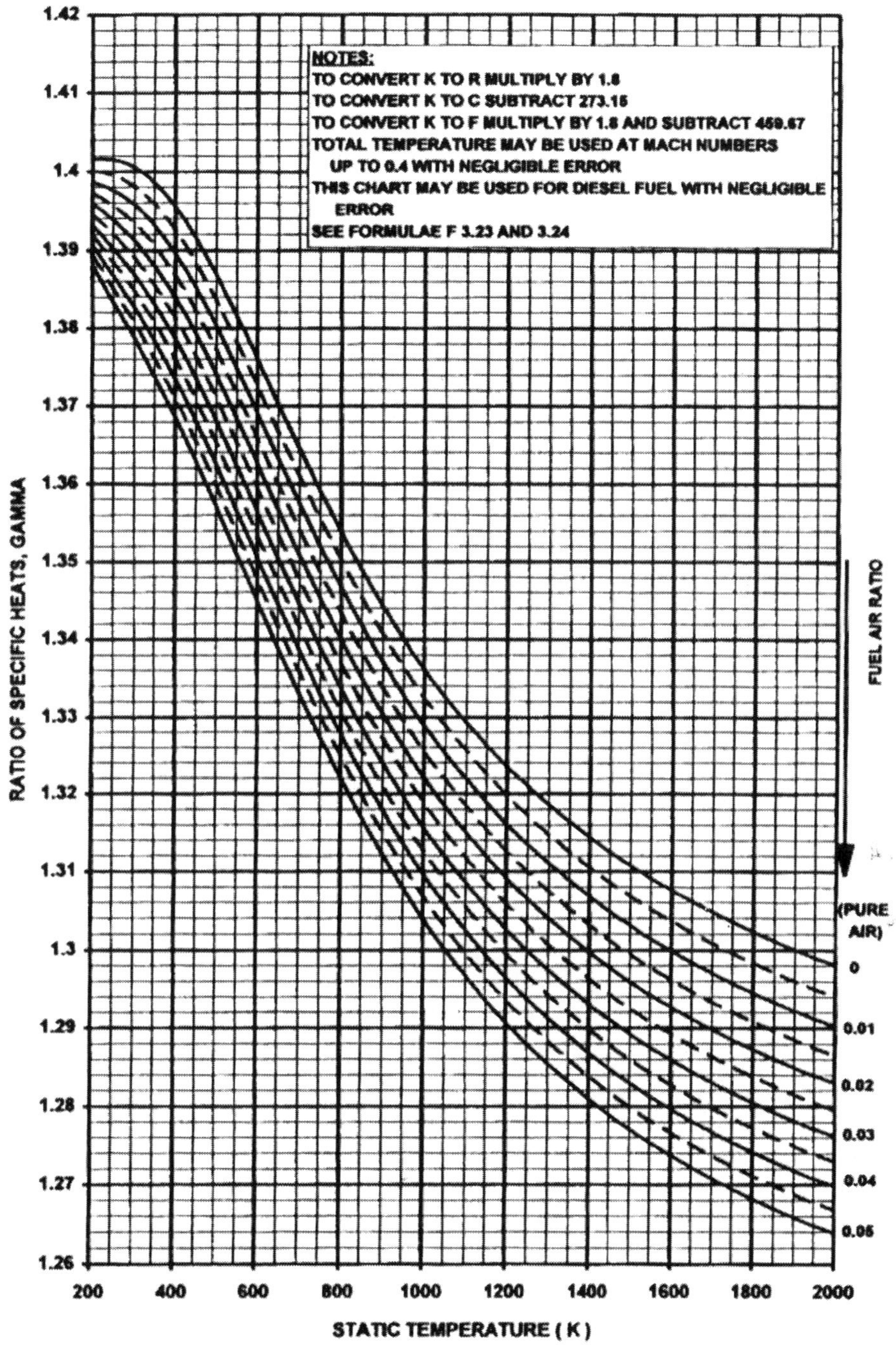

FIGURE 19–12 Gamma for kerosene combustion products versus temperature and fuel air ratio. *(Source: Rolls Royce.)*

reduces the work required to drive a second compressor, and thereby increases power output.

Figure 19–21 shows a Rankine cycle with superheat. This is used in combined-cycle applications, with the gas turbine exhaust gases providing heat to raise steam. Where heat is added at constant temperature during evaporation a close approximation to the Carnot cycle is achieved, the main deviation being the non-ideal component efficiencies.

Formulae

F19.1. Equation of state for perfect gas

$$RHO = PS/(R \times TS)$$

F19.2. Specific heat at constant pressure (J/kg K) = fn(specific enthalpy (J/kg), static temperature (K))

$$CP = dH/dTS$$

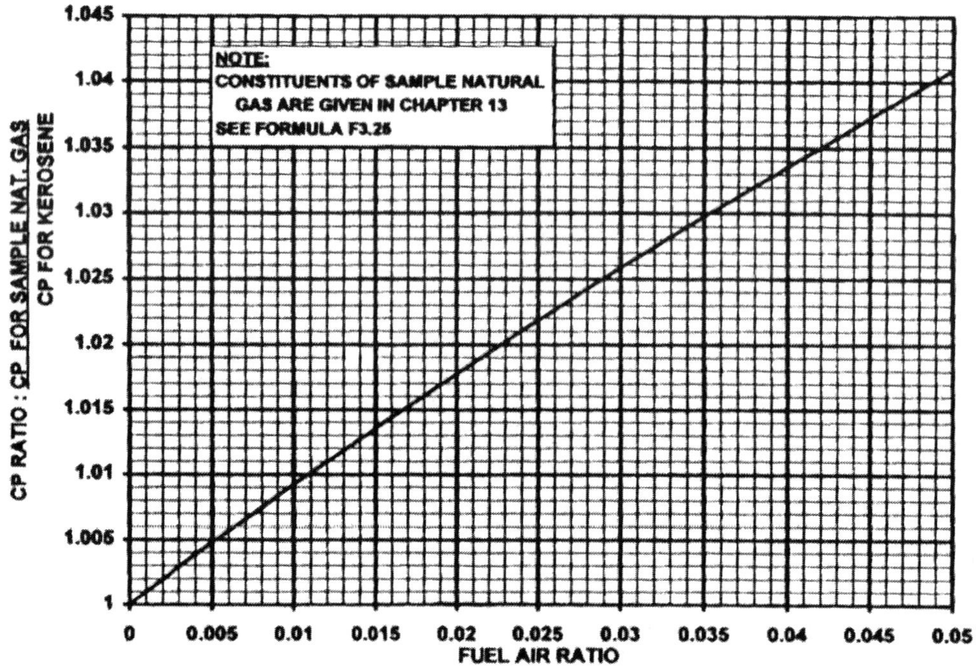

FIGURE 19–13 Specific heat, CP, for typical natural gas combustion products relative to kerosene versus fuel air ratio. *(Source: Rolls Royce.)*

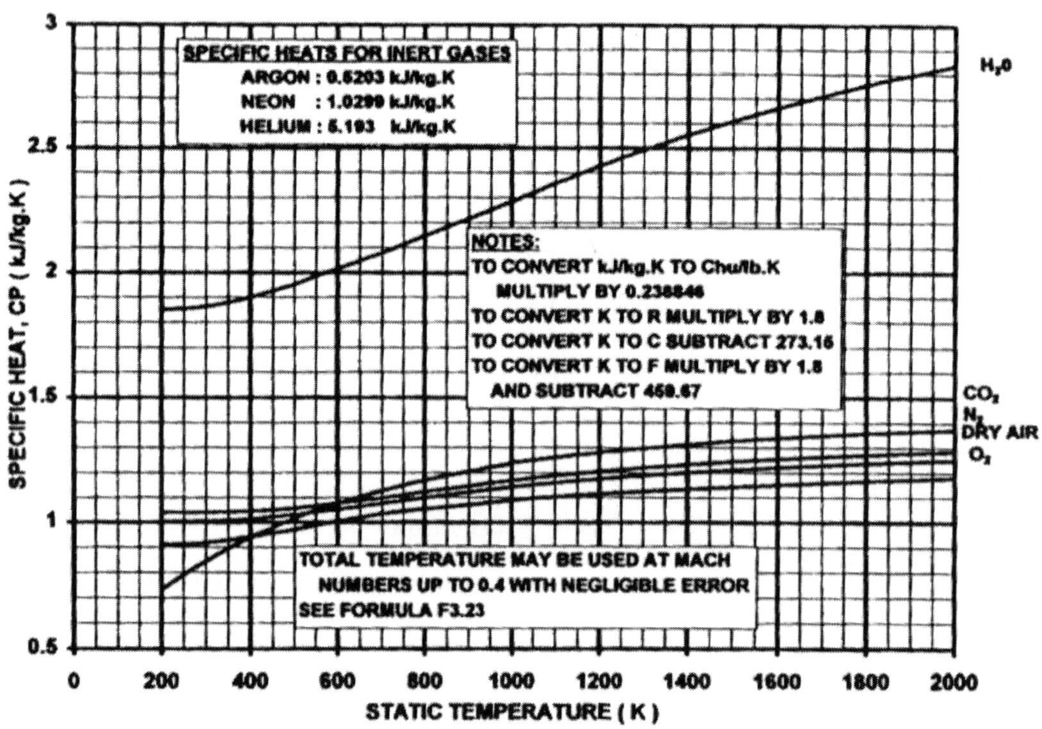

FIGURE 19–14 Specific heat, CP, for the constituents of air and combustion products versus temperature. *(Source: Rolls Royce.)*

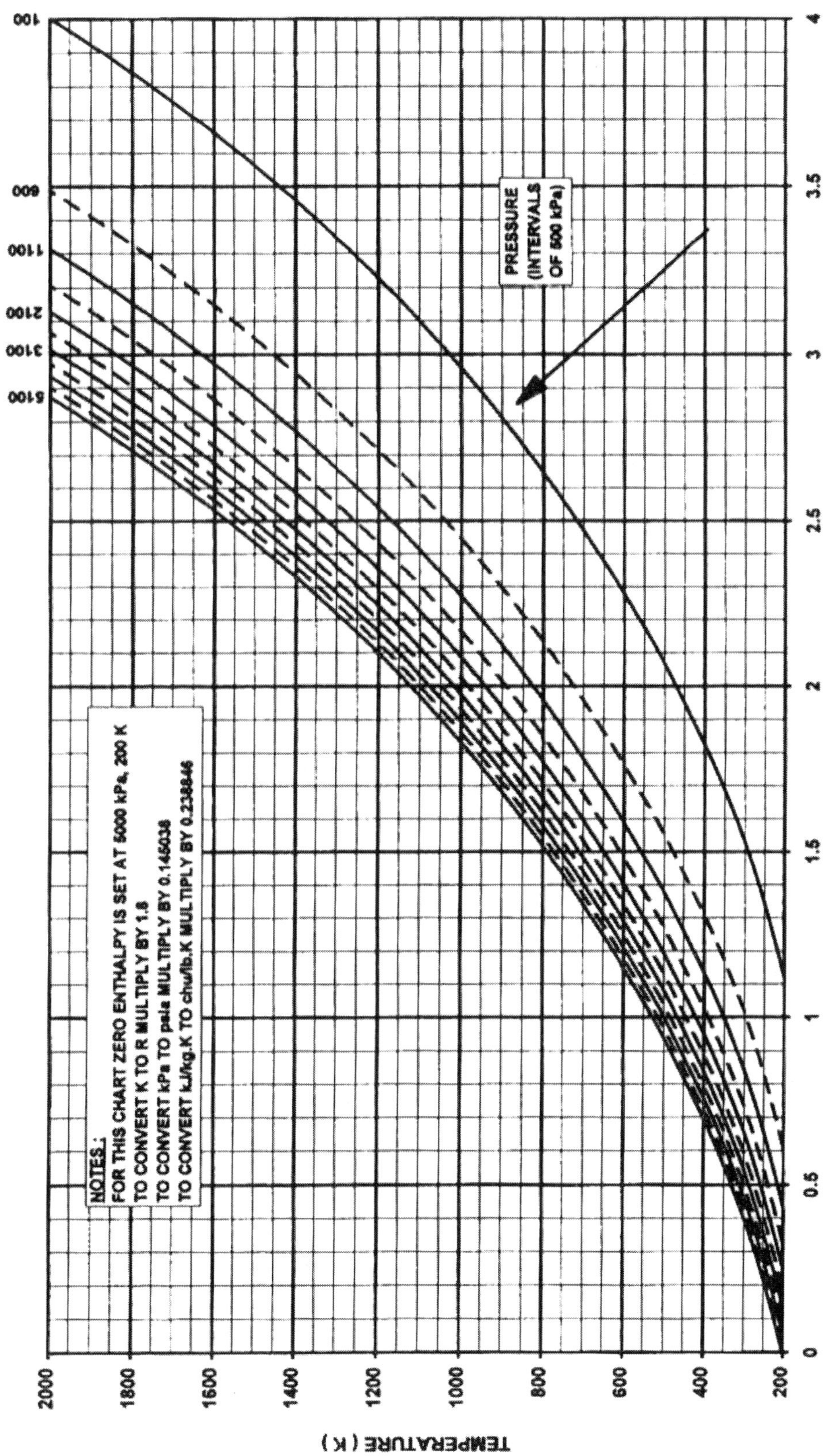

FIGURE 19–15 Temperature–entropy (T–S) diagram for dry air. *(Source: Rolls Royce.)*

FIGURE 19–16 Ideal Carnot cycle. *(Source: Rolls Royce.)*

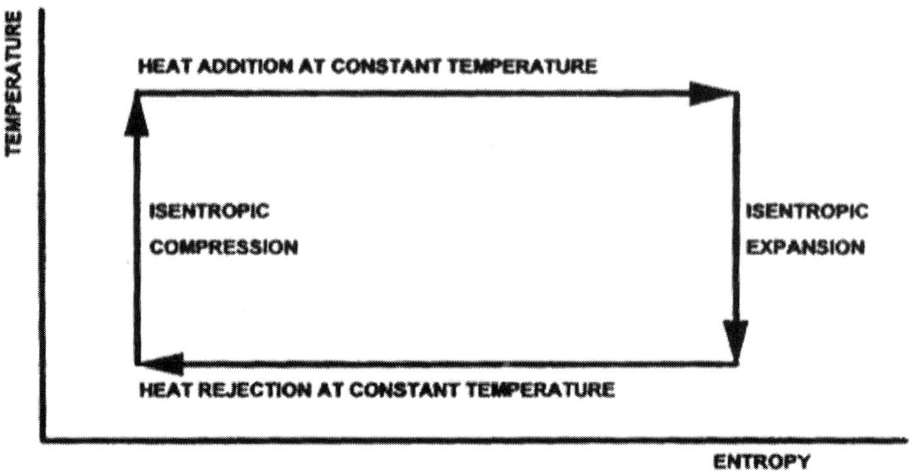

FIGURE 19–17 Brayton cycle for turboshaft, turboprop, turbojet, or ramjet. *(Source: Rolls Royce.)*

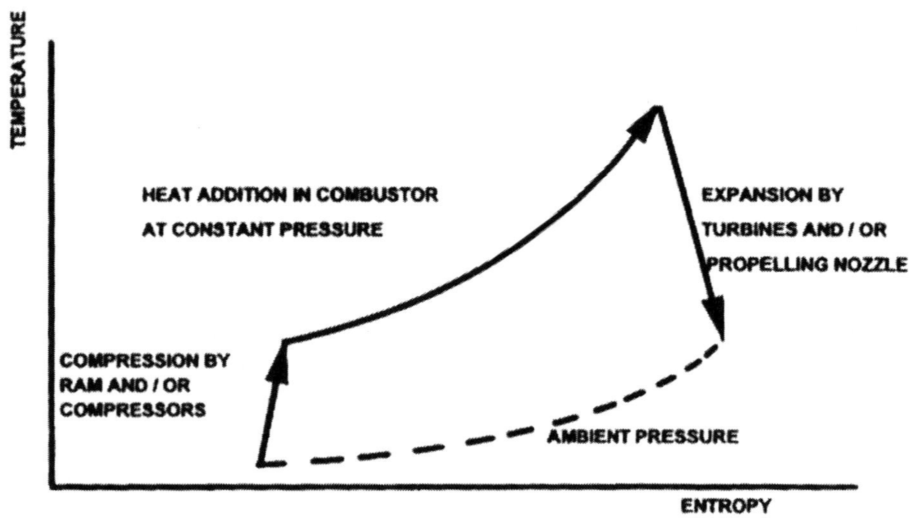

FIGURE 19–18 Cycle for turbofan. *(Source: Rolls Royce.)*

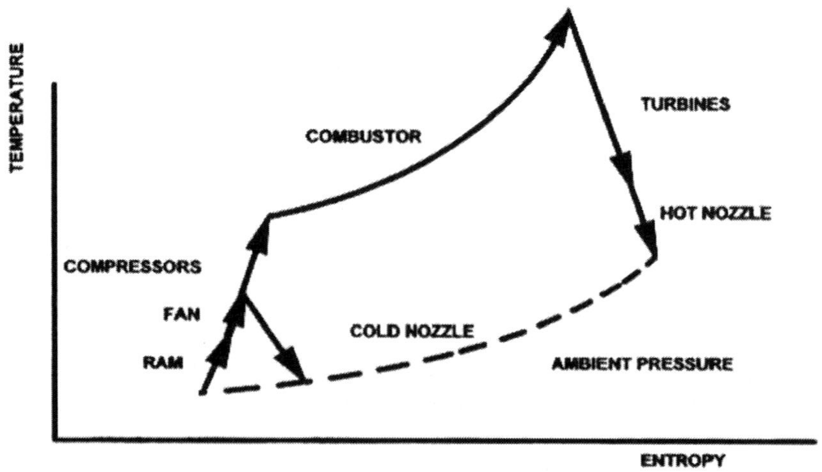

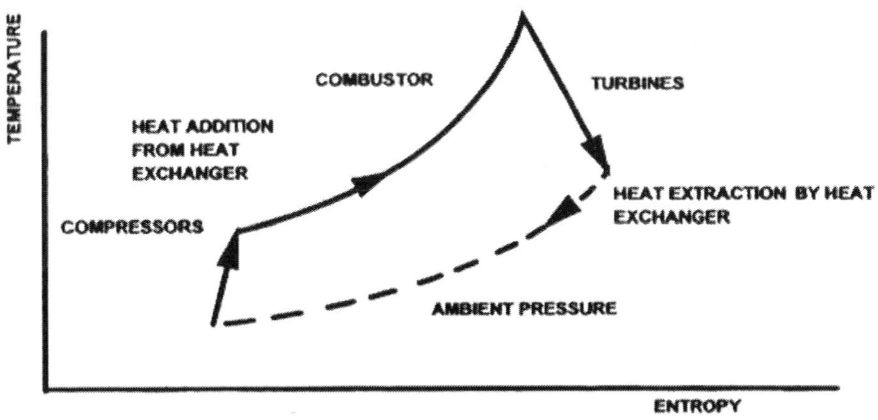

FIGURE 19–19 Cycle with heat recovery for shaft power applications. *(Source: Rolls Royce.)*

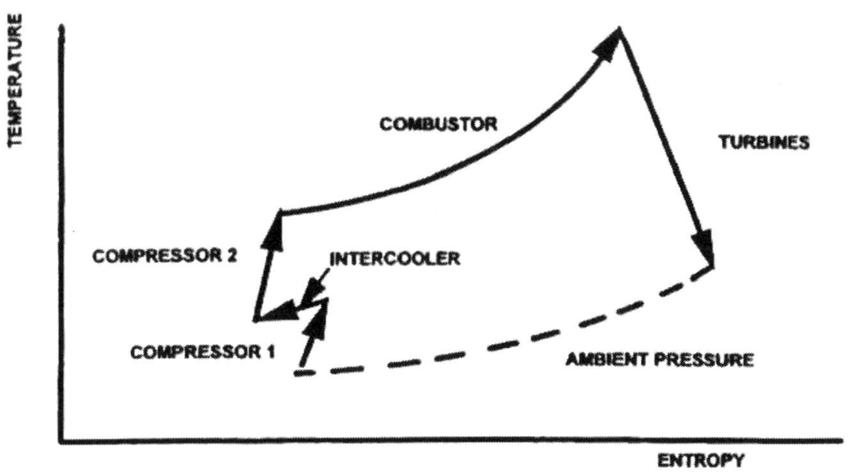

FIGURE 19–20 Intercooled cycle for shaft power applications. *(Source: Rolls Royce.)*

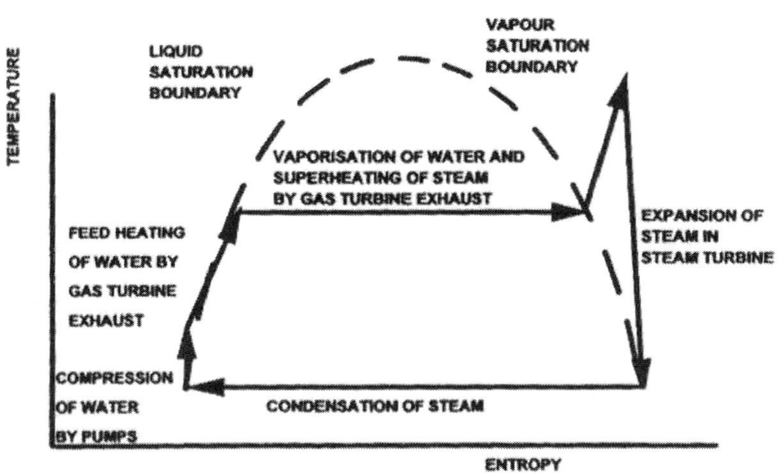

FIGURE 19–21 Rankine cycle with superheat; typical steam cycle used with a gas turbine for combined-cycle power generation. *(Source: Rolls Royce.)*

F19.3. Specific heat at constant volume (J/kg K) = fn(specific internal energy (J/kg), static temperature (K))

$$CV = dU/dTS$$

F19.4. Gas constant (J/kg K) = fn(universal gas constant (J/kg K), molecular weight)

$$R = Runiversal/MW$$

where the universal gas constant Runiversal = 8314.3 J/mol K.

F19.5. Gas constant (J/kg K) = fn(CP (J/kg K), CV (J/kg K))

$$R = CP - CV$$

F19.6. Gamma = fn(CP (J/kg K), CV (J/kg K))

$$\gamma = CP/CV$$

F19.7. Gamma = fn(gas constant (J/kg K), CP (J/kg K))

$$\gamma = CP/(CP - R)$$

F19.8. The gamma exponent (γ − 1)/γ = n(gas constant (J/kg K), CP (J/kg K))

$$(\gamma - 1)/\gamma = R/CP$$

F19.9. Dynamic viscosity of dry air (N s/m^2) = fn(shear stress (N/m^2), velocity gradient (m/s m), static temperature (K))

$$VIS = Fshear/(dV/dy)$$

1. Fshear is the shear stress in the fluid.
2. V is the velocity in the direction of the shear stress.
3. dV/dy is the velocity gradient perpendicular to the shear stress.

F19.10. Total temperature (K) = fn(static temp (K), gas velocity (m/s), CP (J/kg K))

$$T = TS + V\hat{}2/(2\times CP)$$

F19.11. Total pressure (kPa) = fn(total to static temperature ratio, gamma)

$$DP = PS \times (T/TS\hat{})(\gamma/(\gamma - 1))$$

Note: This is the definition of total pressure.

F19.12. Dynamic head (kPa) = fn(total pressure (kPa), static pressure (kPa))

$$VH = P - PS$$

F19.13. Dynamic head (kPa) = fn(density (kg/m^3), velocity (m/s), Mach number)

$$VH = 0.5\times RHO \times V\hat{}2((1 + 0.5 \times (\gamma - 1)\times M\hat{}2) - 1)\times 2/(\gamma \times M\hat{}2)$$

For incompressible flow, such as that of liquids, it is sufficient to only use the first term—this is the well-known Bernoulli equation.

F19.14. Specific enthalpy (kJ/kg) = fn(temperature (K), CP (kJ/kg K))

$$H = H0 + \int CP\, dT$$

H0 is an arbitrarily defined datum. The datum is unimportant in gas turbine performance as it is changes in enthalpy that are of interest.

F19.15. Change in enthalpy (kJ/kg) = fn(temperature (K), CP (kJ/kg K))

$$DH = CP \times (T2 - T1)$$

"Design Point" Engine Design, Definitions, and Terminology*

The Engine Design Point and Design Point Performance

For initial definition work, the operating condition where an engine will spend most time has been traditionally chosen as the engine design point. For an industrial unit this would normally be ISO base load, or for an aeroengine cruise at altitude on an ISA day. Alternatively some important high power condition may be chosen. Either way, at the design point the engine configuration, component design, and cycle parameters are optimized. The method used is the design point performance calculation. Each time input parameters are changed and this calculation procedure is repeated, the resulting change to the engine design requires a different engine geometry, at the fixed operating condition.

For the concept design phase described here the component design points are usually at the same operating condition as the engine design point. In a detailed design phase, however, this may not be true. For example, in the detailed design phase an aeroengine fan may be designed at the top of climb, the highest referred speed and flow, whereas the engine design point would be cruise. The term design point refers to the engine design point in the concept design phase, which is taken to be coincident with the component design points.

Design Point Performance Parameters, Definitions

Engine Performance Parameters

A number of key parameters that define overall engine performance are utilized to assess the suitability of a given

* Source: Courtesy Rolls Royce, *Gas Turbine Performance*, Walsh and Fletcher, Blackwell Science, 1998. Adapted with permission.

engine design to the application, or compare several possible engine designs. These engine performance parameters are described as follows:

- Output power or nett thrust (PW, FN). The required output power or nett thrust is almost always the fundamental goal for the engine design. It is evaluated via the overall cycle calculation. The term effective or equivalent power is used for turboprops and turboshafts, where any residual thrust in the exhaust is converted to a power value and added to the shaft output power.
- Exhaust gas power. For a turboshaft engine core this is the output power that would be produced by a power turbine of 100% efficiency. It is of interest when engine cores are tested or supplied without their free power turbine, which may remain with the installation or be supplied by another collaborating company.
- Specific power or thrust (SPW, SFN). This is the amount of output power or thrust per unit of mass flow entering the engine. It provides a good first-order indication of the engine weight, frontal area, and volume. It is particularly important to maximize specific power or thrust in applications where engine weight or volume are crucial, or for aircraft that fly at high Mach numbers where the drag per unit frontal area is high. For turboprops and turboshafts and effective specific power may be evaluated, based on the effective power value.
- Specific fuel consumption (SFC). This is the mass of fuel burned per unit time per unit of output power or thrust. It is important to minimize SFC for applications where the weight and/or cost of the fuel is significant versus the penalties of doing so. When quoting SFC values it is imperative to state the calorific value of the fuel, and whether it is the higher or lower heating value, to ensure valid "back to back" comparisons. Again, for turboprops and turboshafts an effective SFC may be evaluated.
- Thermal efficiency for shaft power engines (ETATH). This is the engine power output divided by the rate of fuel energy input, usually expressed as a percentage. It is effectively the reciprocal of SFC, but is independent of fuel calorific value. However, when quoting thermal efficiencies it is still important to state whether the values are based upon the higher or lower fuel heating value. For combined-cycle applications the terms gross and nett thermal efficiency are used. Gross thermal efficiency does not deduct the power required to drive the steam plant auxiliaries, whereas nett values do.
- Thermal efficiency is usually quoted for industrial gas turbines, and SFC for aircraft turboshafts and turboprops. For marine and automotive gas turbines both terms are commonly used. This difference reflects the higher importance of fuel weight in aero applications, and the lower likely variation of fuel calorific values.
- Heat rate for shaft power cycles (HRATE). Heat rate is a parameter used only in the power generation industry, and is the rate of fuel energy input divided by the useful power output. Hence it is comparable to SFC but is independent of fuel calorific value. Again it is important to state whether higher or lower calorific value has been assumed in calculating fuel energy input, and for combined-cycle applications whether the values are gross or net.
- Exhaust temperature (T6). For engines used in combined cycle for industrial power generation, high exhaust temperature is vital in maximizing overall efficiency. For combined heat and power the optimum value depends on the relative demand of heat versus power. In both cases there is a limit to the allowable exhaust temperature due to mechanical integrity considerations in the steam plant.
- For military aircraft applications low exhaust gas temperature is important to reduce the infrared signature presented to heat seeking missiles.
- Exhaust mass flow (W6). For engines used in combined-cycle or combined heat and power applications the exhaust mass flow is important in indicating the heat available in the gas turbine exhaust, and hence the overall plant thermal efficiency.

For aircraft thrust engines there are a number of secondary performance parameters. These do not in themselves describe overall engine performance, but do help the engine designer understand the variation of the primary performance parameters across a number of designs. These parameters are as follows:

- Thermal efficiency. Thermal efficiency for aircraft thrust engines is defined as the rate of addition of kinetic energy to the air divided by the rate of fuel energy supplied, usually expressed as a percentage. The energy in the jet is proportional to the difference in the squares of jet and flight velocities. Generally thermal efficiency increases as pressure ratio and SOT increase together, as this results in a higher jet velocity for a given energy input.
- Propulsive efficiency (ETAPROP). Propulsive efficiency for aircraft thrust engines is defined as the useful propulsive power produced by the engine divided by the rate of kinetic energy addition to the air, again usually expressed as a percentage. The nett thrust is proportional to the difference in the jet and flight velocities. Since power is force times velocity, propulsive power is proportional to the flight speed times the difference in the jet and flight velocities.
- Propulsive efficiency is improved by low jet velocities, due to lower energy wastage as jet kinetic energy. This requires high-pressure ratio and low SOT. However, low jet

velocities produce lower thrust output, hence to achieve high propulsive efficiency, as well as a required thrust, high engine mass flow must be coupled with low jet velocities. This leads to engines of low specific thrust, which are large and heavy. Turbofan engines are based upon this principle.

Thrust SFC is directly proportional to flight speed, and inversely proportional to both thermal and propulsive efficiencies. The dependency on the efficiencies is intuitively obvious. The choice of pressure ratio and SOT for minimum SFC is a compromise between maximizing propulsive and thermal efficiency.

The dependency of thrust SFC on flight speed arises because fuel flow relates to power with the thermal and propulsive efficiencies fixed, and propulsive power is directly proportional to flight speed for a given thrust.

Cycle Design Parameters

The fundamental thermodynamics of gas turbine cycles in relation to temperature entropy diagrams show that the changes in pressure and temperature that the working fluid experiences strongly affect the engine performance parameters. The degree of change of pressure and temperature are reflected via the following cycle design parameters.

Overall Pressure Ratio

This is total pressure at compressor delivery divided by that at the engine inlet.

Stator Outlet Temperature (SOT)

This is the temperature of the gas able to do work at entry to the first turbine rotor. Other terms are also used to reflect maximum temperatures in a cycle:

- Rotor inlet temperature (RIT): this term is sometimes used in North America, and means the same as SOT.
- Combustor outlet temperature (COT): this is the temperature at the first turbine nozzle guide vane leading edge.
- Turbine entry temperature (TET): this can have either of the above meanings.

The standard definition for SOT, used herein, is:

The fully mixed out temperature resulting from combustion delivery gas mixing with all cooling air that enters upstream of the first turbine rotor, and is able to do work due to having momentum comparable to the nozzle guide vane flow.

Hence, in this definition, nozzle guide vane or platform cooling air entering upstream of the throat, or trailing edge cooling air ejected with the same momentum and direction as the main flow, would be included. However, front disc face cooling airflow would not be considered in evaluating SOT since it will not do work in the turbine rotor. For most gas turbine engine types it is desirable to raise SOT to as high a level as possible within mechanical design constraints.

In addition for turbofan engines two further cycle design parameters occur due to the parallel gas paths.

Fan Pressure Ratio

This is the ratio of fan delivery total pressure to that at fan inlet, and is usually lower for the core stream than the bypass due to lower blade speed.

Bypass Ratio

This is the ratio of mass flow rate for the cold stream to that for thc hot strcam.

Component Performance Parameters

A plethora of parameters define component performance in terms of efficiency, flow capacity, pressure loss, etc. As the level of component performance parameters improves, at fixed values of cycle design parameters, then the design point engine performance parameters also improve.

Changes in component performance parameters have a secondary effect on the optimum values of engine cycle design parameters.

Mechanical Design Parameters

For a given performance design point to be practical the mechanical design parameters must be kept within the limits of the materials, manufacturing, and production technology available. These are mechanical design restraints in relation to gas turbine engines:

- Creep as a function of material type, metal temperatures, stress level, or AN^2
- Oxidation as a function of material and coating type, and metal temperatures
- Cyclic life (low cycle fatigue) as a function of material type and metal temperatures
- Disc and blade tensile stress as a function of rim speed or AN^2
- Casing rupture as a function of compressor delivery pressure
- Choke or stall flutter as a function of fan or compressor referred speed
- Vibration (high cycle fatigue) of rotating components as a function of rotational speed and excitation parameters such as upstream blade numbers and pressure levels
- Shaft critical speeds

Life Parameters

The two major life parameters are:

1. Time between overhauls (TBO).
2. Cyclic life (also called low cycle fatigue life): this is the number of times the engine is started, accelerated to full power, and eventually shut down, between overhauls.

The TBO is governed mainly by creep and oxidation life, while cyclic life is dictated by thermal stress levels. Typical life requirements for the major gas turbine applications are as shown in Table 19—4.

Fuel Type

Kerosene is the standard aviation fuel while marine engines burn diesel and most industrial applications use natural gas. The highly distilled forms of diesel used make little difference to performance compared with kerosene, but natural gas gives performance improvements because of the higher resulting specific heat of the combustion products.

Design Point Diagrams

A design point diagram is created by plotting engine performance parameters versus the cycle parameters. To produce the diagram the design point calculations are repeated varying each cycle parameter in turn through the range of interest; every point is a different engine geometry. These diagrams are useful in the engine design process for selecting the optimum cycle parameters, or for comparing the performance of different gas turbine types. Initially design point diagrams are prepared using constant component performance parameter levels for all combinations of cycle parameters. Later in the engine design process different individual component performance levels are used for each combination of cycle parameters, based on aerothermal component design work.

Referred Parameters

The design point charts may be applied to any altitude if the referred form of specific power or thrust, SFC, etc. is used. In this way the datum values for the charts are adjusted. A change in flight Mach number does change engine matching, however, unless all nozzles are choked, and also changes ram drag. The charts are not directly adaptable to other flight Mach numbers.

Linearly Scaling Components and Engines

During the concept design process the scaling of existing components, or occasionally a complete engine, will be considered wherever possible. If viable this will significantly reduce program cost and development risk.

Design Point Exchange Rates

Design point exchange rates show the impact of a small change, typically 1%, in leading component performance levels on engine design point performance parameters. All other component efficiencies and cycle parameters are held constant as the parameter in question is varied. In practice this would require many other components to be redesigned to change flow sizes. Tables of design point exchange rates are extremely useful in the engine design process in highlighting the most sensitive component design areas, and should always be produced.

TABLE 19—4 Typical Life Requirements

Application	TBO (hours)	Cycles
Power generation—base load	25,000—50,000	3000
Power generation—standby/peak lopping	25,000	10,000
Gas and oil pumping	25,000—100,000	5000—10,000
Automotive—family saloon	5000	10,000
Automotive—truck	10,000	5000—10,000
Marine—military	5000—20,000	2000—3000
Marine—fast ferry	5000—10,000	3000
Aeroengine—civil	15,000	3000
Aeroengine—military fighter*	25—3000	25—3000

There is a large difference between past achievements and future targets as the emphasis has moved from performance to cost of ownership.
(Source: Rolls Royce.)

Synthesis exchange rates are quite different, and are utilized for off-design analysis, whereas the component in question is modified no other components are redesigned. All components move or rematch to a different non-dimensional operating point. In consequence component performance parameter levels change, as well as cycle parameters.

Open Shaft Power Cycles

With one exception all gas turbine engine configurations are open cycle. Air that enters the front of the engine is not recirculated, but is exhausted back into the atmosphere.

Simple Cycle

This is the basic shaft power cycle. First, air is compressed, and then it is heated by burning fuel in the combustor. Next, expansion through one or more turbines produces power in excess of that required to drive the compressors, which is available as output. Whether the engine is single spool or free power turbine is only reflected in a design point diagram by any small change in assessed overall turbine efficiency:

- Thermal efficiency and specific power generally increase with SOT. Though less convenient, combustor temperature rise is a better gauge of the cycle's ability to deliver power. When this is low, excess power is low and the losses due to component inefficiencies dominate, reducing both thermal efficiency and specific power.
- The optimum pressure ratio for thermal efficiency increases with SOT, being 12:1 at 1100 K and rising to over 40:1 at 1800 K. The fact that there is an optimum is because as well as high combustion temperature rise, high thermal efficiency also requires low exhaust temperature to minimize energy wastage. Maximum thermal efficiency occurs at the minimum value of the ratio of combustor temperature rise to exhaust temperature, reflecting the ratio of heat input to heat wastage. Achieving this mainly requires low exhaust temperature, hence the optimum pressure ratios are relatively high. At too high a pressure ratio the low combustor temperature rise offsets the low exhaust temperature, which reduces thermal efficiency.
- The optimum pressure ratio for specific power also increases with SOT but is only around half that for thermal efficiency, being only 7:1 at 1100 K rising to 20:1 at 1800 K. Maximum specific power occurs at the maximum difference between the combustor temperature rise and the exhaust temperature, which reflects work output. Achieving this is less dominated by exhaust temperature and more by the need to reduce compressor work.
- For mechanical integrity the combustor entry temperature must be limited to between 850 and 950 K, depending on the technology level.
- If the engine is to be used in combined cycle or combined heat and power then, exhaust temperature must be limited to between 800 and 900 K, depending on the technology level of the heat recovery system.

Recuperated Cycle

This is as per the simple cycle except that a heat exchanger transfers some of the heat in the exhaust to the compressor delivery air. If the gas and air streams do not mix the heat exchanger is known as a recuperator; if they do mix, as with the automotive ceramic rotating matrix variety, it is known as a regenerator:

- Thermal efficiency and specific power generally increase with SOT.
- At optimum pressure ratios the thermal efficiency is around 10% better than that for the simple cycle, due to the heat recovery reducing the fuel requirement. This difference reduces as SOT increases as the corresponding simple cycle becomes more efficient.
- The optimum pressure ratio for thermal efficiency is comparatively low since the difference between the exhaust and compressor delivery temperature is high, hence more heat may be recovered. This is the dominant effect of pressure ratio.
- The optimum pressure ratio for specific power is 7.5:1 at 1100 K and 23:1 at 1800 K, and is independent of effectiveness. It is very similar to that for simple cycle, being only slightly reduced by the additional pressure losses in the recuperator.
- Effectiveness is one of the few component design parameters that has a strong effect upon the optimum pressure ratio for thermal efficiency. As it is increased, the optimum pressure ratio decreases, since increased heat recovery more than offsets a poor simple cycle efficiency.
- Combustor inlet temperature rises with falling pressure ratio for a given SOT. This is because of the reduced power requirement to drive the compressor, which reduces the temperature drop in the turbines.
- For mechanical integrity the recuperator and combustor entry temperatures must be limited to between 850 and 950 K, depending on the technology level.

Intercooled Cycle

In this configuration the temperature of the air is reduced part way through the compression process using a heat exchanger and an external medium such as water. This increases power output via reduced compressor work, which is directly proportional to compressor inlet temperature. Generally thermal efficiency reduces as more fuel is required to reach a given SOT:

- Thermal efficiency and specific power generally increase with SOT.
- Thermal efficiency is marginally worse than for the simple cycle below the optimum simple cycle pressure ratio and significantly better above it. This is driven by the relative magnitudes of the extra power output and the extra fuel flow required to raise the lower compressor delivery temperature to the given SOT.
- The optimum pressure ratio for thermal efficiency is 18:1 at 1100 K and over 40:1 at 1800 K. This is higher than for simple cycle, as the intercooler gives most benefit with a comparatively high-pressure ratio.
- As the intercooler position is moved, thermal efficiency peaks with it 30% through the compression ratio, such that the first compressor pressure ratio would be the overall value raised to an exponent of 0.30. (The design point diagrams are drawn for the intercooler placed 50% through the compression such that the compressors have equal pressure ratios.)
- Specific power is approximately 20–40% higher than for simple cycle due to the reduction in the power to drive the compressor stages after the intercooler. This drive power is directly proportional to compressor inlet temperature for a given pressure ratio.
- The optimum pressure ratio for specific power is 12:1 at 1100 K and over 40:1 at 1800 K. Again this is higher than for simple cycle, and for the same reason.
- As the intercooler position is moved, specific power peaks with the intercooler 50% through the compression ratio, such that the compressors would have equal pressure ratios. This corresponds to rejecting more heat than for the case of optimum thermal efficiency, and clearly requires more fuel energy input.

Intercooled and Recuperated Cycle

This is the combination of intercooling and recuperation. The thermal efficiency loss due to intercooling is offset by increased exhaust heat recovery due to the lower recuperator air-side entry temperature:

- Thermal efficiency and specific power generally increase with SOT.
- Thermal efficiency is around 20% higher than for simple cycle for a given SOT at the respective optimum pressure ratios, as in concert both the intercooler and recuperator reduce fuel consumption.
- The optimum pressure ratio for thermal efficiency is 7:1 at 1100 K and 20:1 at 1800 K. However, the cycle curves are flatter versus pressure ratio than for the other configurations above.
- Again, increasing recuperator effectiveness significantly reduces the optimum pressure ratio for thermal efficiency.
- Specific power is around 10% lower than for an intercooled only cycle due to the recuperator pressure losses, but still significantly higher than for simple cycle.
- The optimum pressure ratio for specific power is 13:1 at 1100 K and over 30:1 at 1800 K. Again, the cycle curves are flatter versus pressure ratio than for the other configurations above.
- For mechanical integrity the recuperator and combustor entry temperatures must be limited to between 850 and 950 K, depending on the technology level.
- As the intercooler position is moved, thermal efficiency peaks with it 40% through the pressure rise, and specific power peaks with it at 50%.

Combined Cycle

This is where a steam (Rankine) cycle also generates power, using exhaust heat from a simple cycle gas turbine. In addition, supplementary firing may be employed whereby a separate boiler supplements the gas turbine heat output when necessary. The curves are for a triple pressure reheated steam plant:

- Thermal efficiency and specific power generally increase with SOT.
- Thermal efficiency exceeds that of all other configurations by around 20–30%.
- The optimum pressure ratio for thermal efficiency is 4:1 at 1100 K and around 21:1 at 1800 K. These values are lower than for simple cycle, as here increasing pressure ratio at a given SOT reduces the exhaust heat available for the steam cycle, reducing its efficiency. At the highest SOT levels the thermal efficiency curves are flatter versus pressure ratio than for all other configurations above; the steam cycle trends offset those of the gas turbine, benefiting from the higher exhaust temperature when gas turbine efficiency falls.
- As a rule of thumb, the pressure ratio that gives the best combined-cycle thermal efficiency, while retaining an acceptable gas turbine exit temperature, is the optimum value for simple cycle specific power.
- The optimum pressure ratio for specific power is less than 4:1 at 1100 K and 7:1 at 1800 K. These values are even lower than those for simple cycle due to the effect of high gas turbine exit temperature on steam plant power.
- The exhaust temperature limits of 800–900 K discussed for simple cycle engines must be considered here.

Combined Heat and Power

This configuration differs from those above in that the engine exhaust heat is also utilized in the application, either directly in an industrial process or to raise steam for space heating. A varying amount of exhaust heat may be utilized or wasted, hence *heat to power ratio* is an important parameter in cycle comparisons. Again, supplementary firing may be employed:

- Thermal efficiency increases with SOT.
- Thermal efficiency increases almost linearly with heat to power ratio, as would be expected.
- Increasing pressure ratio increases thermal efficiency, though with diminishing returns above a value of 12:1. This reflects the corresponding simple cycle trends.
- Relatively high exhaust temperature, as given by a high specific power gas turbine, is often beneficial as high-grade heat is more useful for a variety of processes. The exhaust temperature limits of 800–900 K discussed for simple cycle engines must be considered here, however.

Closed Cycles

Here the working fluid is recirculated through the cycle. Heat exchangers are employed in place of the intake/exhaust and combustion processes, passing heat to and from external media. Maximum temperatures are limited by heat exchanger mechanical integrity, and unlike for open cycles, compressor entry pressure may be controlled to vary power level. Applications involve energy sources unsuited to direct combustion within a gas turbine engine, such as nuclear reactors or alternative fuels. The working fluid is normally helium, due to its high specific heat:

- Thermal efficiency and "power ratio" generally increase with SOT.
- The "power ratio" as presented is specific power multiplied by the ratio of the density of helium at the inlet temperature and pressure to that of air at ISO conditions. This reflects the impact on the required engine size.
- The optimum pressure ratio for thermal efficiency is around 4:1.
- Reducing inlet temperature raises thermal efficiency and power ratio, via increased non-dimensional SOT, i.e., SOT/T1.
- Raising inlet pressure increases power ratio, via increased fluid density.
- The optimum pressure ratio for power ratio is around 6:1.
- Practical SOT levels are low, due to heat exchanger integrity limits.

Aircraft Engine Shaft Power Cycles

Here a simple cycle configuration produces shaft power, along with a residual exhaust thrust. SFC is plotted rather than thermal efficiency:

- SFC and specific power generally improve with SOT.
- At ISA SLS the optimum pressure ratio for SFC is 12:1 at 1100 K SOT and over 30:1 at 1700 K. At the typical turboprop cruise flight condition of 6000 m 0.5 M it is 17:1 at 1100 K SOT, and over 30:1 at 1700 K. The small increase in optimum pressure ratio is due to the increased cycle temperature ratio at the cooler inlet conditions at altitude, despite the change in matching produced by the flight Mach number.
- At both conditions the optimum pressure ratio for specific power is around 7:1 at 1100 K SOT and 19:1 at 1700 K.

Aircraft Engine Thrust Cycles

Turbojets

Here a simple cycle configuration produces thrust via high exhaust gas velocities.

Turbofans

Here air from the first compressor of what is otherwise a turbojet engine bypasses the remaining compressor and turbine stages. This air produces thrust directly, additional to that from the hot exhaust gases.

Ramjets

Here fuel is burned in air compressed solely by the intake ram effect; there is no turbomachinery. Engine performance parameters are therefore a function only of SOT and flight Mach number.

The Engine Concept Design Process

This section presents a simplified overview of the complex concept design process. To lead an engine concept design effectively requires a full understanding of gas turbine performance and all the associated design processes and their interrelationship. Equally, others involved in the process must have some knowledge of performance if they are to contribute to it beyond their own subject areas.

Statement of Requirements

A statement of requirements or specification for a new engine may be presented by a customer organization, or may be specified within the cycle designer's company to meet a perceived market need. Realism is required in terms of technical challenge as well as potential development and unit production costs. The document must not just reflect what "Marketing" or "the Customer" desires but also what "Engineering" can achieve. It should be written jointly.

With respect to performance the statement of requirements must contain a minimum of:

- Performance targets both at design point and other key operating conditions. The relationship to average or minimum engine and in service deterioration must also be specified, as well as future growth potential.
- Full definition of operating conditions for the above, and also the entire operational envelope, including ambient temperature, pressure or pressure altitude, humidity, flight Mach number, and installation losses.
- The starting envelope, as well as starting and above idle transient response times.
- Fuel type and emissions requirements.
- Engine diameter, length, and weight.
- Time between overhauls and cyclic life.
- Design and development program duration and cost, as well as unit production cost.
- Any derivative engines that should be considered.

"First Cut" Design Points

Generic design point diagrams will show what engine configurations, together with the levels of cycle parameters, are likely to meet the statement of requirements. Next, initial component performance targets are set. For the narrow range of cycle parameters of interest, refined design point diagrams are then produced. Usually component efficiency variation with engine cycle parameters is unknown at this stage. As far as possible the cycle designer takes account of mechanical design limits such as temperature levels and considers the use of existing or scaled components.

From these design point diagrams a first cut design point for each engine configuration under consideration is formally issued to component designers. An example of multiple configurations is that both simple cycle and recuperated engines may be competitive for a shaft power application.

For clarity all design points should have an identifying version number, as many more updates will follow in what is a highly iterative design process.

"First Cut" Aero-Thermal Component Design

The component specialists now reassess the component performance inputs assumed in the design points, and perform initial component sizing. Their analysis uses component design computer codes and empirical data. Mechanical design parameters are monitored and where there is doubt regarding limits mechanical technologists must be consulted.

Proper choice of rotational speed is crucial. It is key to achieving the target compressor and turbine efficiency levels as well as number of stages and diameter, and the compressor and turbine designers must agree upon it. The practicality or otherwise of using the number of spools that the cycle designer had in mind when setting the initial design points will soon be apparent, and weight and cost implications will become clear.

Component performance levels are now bid back to the performance engineer for each design point.

Design Point Calculations and Aero-Thermal Component Design Iterations

By repeating the design point calculations using the new bid component performance levels the concept design team can decide which engine configuration is the most suitable. If a bid is changed even for only one component, it will change the design point requirements for all others.

The process is repeated a number of times for the chosen configuration so that a number of cycle parameter combinations, each with specific component performance levels, can be compared.

Engine Layout

It is important to commence engine layout drawings as early as possible to highlight potential difficulties. For example, for a turbofan there would normally be a rigid engine diameter constraint. The core diameter required by the compressor and turbine designs, combined with the bypass duct annulus area to keep Mach number to a level where pressure loss is acceptable, may exceed this limit. In this instance the cycle designer may have to reduce bypass ratio, or the component designers increase rotational speed.

Off-Design Performance

Once a good candidate design point is emerging from the above iterations the cycle designer must then freeze the engine geometry and issue off-design performance. This should cover other key operating points as well as all corners of the operational envelope. Often at this juncture the component designers will not have created off-design maps for the "frozen" component designs so maps from existing components of similar design must be used, with factors and deltas applied to align them to the design point. The performance engineer must decide how the engine will be controlled for key ratings before running the off-design cases.

It is good practice at this stage for the performance engineer to highlight the operational cases where the maximum and minimum value of all key parameters occur, and any margins that need to be applied for say operating temperatures for a minimum engine.

Performance, Aero-Thermal, and Mechanical Design Investigation

The off-design performance data must be examined in detail by all parties. The cycle designer must check whether the engine design meets other performance requirements besides that at the design point. The aero-thermal and mechanical component designers must ensure that the components are satisfactory throughout the operational envelope with respect to component performance levels, stress, and vibration.

Basic Starting and above Idle Transient Performance Assessment

Engine operability is assessed by ensuring that surge and weak extinction margins are commensurate with the required accel, decel, and start times. A judgment is made with respect to the need for variable stator vanes or blow off valves (BOVs); these also impact compressor design, unless the BOVs are at compressor delivery, and may be inherited anyway if an existing compressor design is utilized. Unless there are unusual or severe requirements, modeling of transient performance or starting would not normally be conducted in the concept design phase. However, when such requirements are present or if a novel engine configuration is employed, then transient performance modeling is essential as transient considerations can significantly impact the fundamental engine design in terms of cycle, number of spools, etc.

Iterations

A number of iterations through the whole process concludes the concept design phase. The resulting engine design is then compared with the statement of requirements and a judgment made as to whether to proceed into a detailed design and development program.

Margins Required When Specifying Target Performance Levels

It is important that no ambiguities are present when performance targets for an engine design are specified, and that certain less obvious issues are not forgotten; some of the latter are outlined below.

Minimum Engine

Owing to manufacture and build tolerances there will be significant engine-to-engine variation in production. Usually the engine designers deal with average engine data, that is, the average performance of all production engines. However, if the customer has been guaranteed a given performance level then even the minimum engine must achieve this standard. Hence a design margin must be built into the average engine performance targets. The shortfall in power or thrust and SFC of a minimum engine in a production run will be 1–3% versus the average engine if run at constant temperature, depending on build quality and engine complexity. If the customer has been guaranteed levels of power or thrust then this minimum engine will run up to 20 K hotter than an average engine to achieve power or thrust. This must be allowed for in the mechanical design.

Design and Development Program Shortfall

At the outset of a program the risk of component performance levels falling short of the values used in any preliminary design point analysis must be assessed. Ideally some engine performance risk analysis should be employed to quantify the potential shortfall versus confidence level. Suitable margins should then be applied to the performance target levels.

However, at the outset of a program this level of risk analysis may not be commensurate with time or resource constraints and the available component performance database. In this instance judgment must be used. The level of the margins should reflect the confidence of component performance levels employed in any preliminary analysis. Each case must be considered individually, but typical margins are presented in Table 19–5 for three confidence levels that the components will meet their design targets.

When in competitive tender for an engine application the luxury of SFC margins may not be allowed due to commercial constraints. However, for shaft power engines building in power margin is often practical by sizing the engine larger. In this instance, should the component performance levels employed in the preliminary analysis be achieved then almost invariably the customer will accept the extra power. In aircraft applications it may be more difficult to provide margin for thrust due to tight restrictions on weight and frontal area.

Growth Potential

Almost any engine must be capable of adaptation to higher power or thrust levels. In aircraft applications the airframe usually requires it, intentionally or otherwise. In addition, addressing a wider market with development costs spread over a family of engines is attractive economically.

TABLE 19–5 Typical Performance Margins, by Confidence Level

	High	Medium	Low
Power or thrust	+2.5%	+5%	+7.5%
SFC	−2%	−4%	−6%

(Source: Rolls Royce.)

Examples for turbofans include the Rolls Royce RB211/ Trent family (thrust growth of 180 to 400+ kN in 25 years), the Rolls Royce civil and military Spey family, and the General Electric CF6 range. Margin is required on levels of temperature and speed, and there must be a practical route to increasing engine mass flow.

Engine Deterioration

At the concept design stage it is unusual to specify performance requirements after deterioration over a number of hours in service. However, if this is the case then a margin must be allowed for from the outset. The amount required depends on whether in service the engine is governed to fixed levels of temperature, speed or thrust/ power. Depending on engine complexity typical margins at 10,000 hours, expressed at constant SOT, are: +3 to +6% SFC; −5 to −15% power.

Performance degradation is usually exponential with time, and after 10,000 hours there is minimal further deterioration. Practices vary regarding the accounting or not of running in effects caused by initial seal cutting as part of deterioration. The aim of running in is to improve performance levels by ensuring a gentle first contact on key seals, which removes less material than would a first in-service rub.

One further effect is compressor fouling where accretion of airborne particles on aerofoil surfaces reduces flow and efficiency, raising engine temperature levels. This applies mainly to ground based engines and is recoverable via compressor washing, for which appropriate fluid injection nozzles must be provided. The temperature increase depends entirely on site location, filter effectiveness, and washing frequency.

Installation Losses

Usually performance targets are stipulated as uninstalled, with the engine performance quoted from the engine intake flange to the engine exhaust or propelling nozzle flange. If the performance targets are installed then the magnitude of all installation effects must be stated, as follows:

- Plant or airframe intake pressure loss
- Plant exhaust or airframe jet pipe pressure loss
- Customer auxiliary power offtake (gas turbine accessory power requirements should be accounted even for uninstalled performance)
- Customer bleed offtake
- Whether shaft power output and thermal efficiency are at turbine, gearbox, or alternator output
- Whether the thrust and SFC include any pod drag utilize

Two case histories and one Mach number-altitude chart follow. Although the cases involve aircraft engine design, the principles are applicable to all gas turbines. The cases were written in 1972 and 1988, respectively, but they still draw interest today. Because the 1972 case is now hard to access, it is reprinted almost entirely.

CASE STUDY 1: PREDICTION EFFECTS OF MASS-TRANSFER COOLING ON THE BLADE-ROW EFFICIENCY OF TURBINE AIRFOILS*

Two analytical models for calculating the effect of injected coolant on cascade efficiency are presented. Both methods assume that the performance losses that accompany coolant injection are due wholly to the mixing of coolant and mainstream gases and may be superimposed on the basic boundary layer loss. One model employs the "influence coefficient" method, assuming that the differential form of the one-dimensional equation for compressible flow with injection may be applied to finite injection rates. This method, although approximate, does agree well with more exact calculations using one-dimensional constant static pressure mixing. The approximation also reveals the significant mixing loss parameters in a relatively simple form, showing that coolant injection effects are primarily dependent on mainstream Mach number, coolant fraction, coolant injection angle, and the ratios of coolant-to-mainstream total temperature and velocity. The second analytical model, primarily suited for numerical computation, employs the concept of one-dimensional mixing layers to calculate losses due to surface injection (film- or transpiration-cooling) but employs a two-dimensional solution for the flow downstream of the trailing edge in order to include the effects of airfoil boundary layers and trailing-edge coolant bleed. The numerical method has been applied to several stator-cooling configurations and good agreement with available data is obtained.

Nomenclature

c_p = Specific heat at constant pressure
M = Mach number
P = Pressure
s = Height of trailing-edge coolant jet when expanded to pressure P_1
$s^* = s/(t \cos \alpha_1)$
t = Blade-row pitch
t_e = Trailing-edge thickness
T = Temperature
V = Velocity
W = Mass flowrate
α = Cascade flow exit angle
γ = Ratio of specific heats

* Source: [19-1] Courtesy of J.E. Hartsel from J.E. Hartsel. "Prediction Effects of Mass-Transfer Cooling on the Blade-Row Efficiency of Turbine Airfoil," AIAA 72−11, 1972.

δ^* = Total boundary layer displacement thickness at the trailing edge based on average freestream (Station 1) conditions
$\delta^{**} = \delta^*/t \cos \alpha_1)$
θ = Total momentum thickness at the trailing edge based on average mainstream (Station 1) conditions
$\theta^* = \theta/(t \cos \alpha_1)$
$\delta_{TE} = t_e/(t \cos \alpha_1)$
η = Cascade thermodynamic efficiency
ξ = Ratio of coolant-to-mainstream mass flowrates
ρ = Density
ϕ_c = Angle of coolant injection measured from local mainstream direction

Subscripts

0 = Inlet
1 = Average mainstream condition at the trailing-edge plane
2 = Fully-mixed equilibrium downstream state
c = Coolant
f = Local condition downstream of injection
g = Local mainstream condition
t = Total or stagnation condition

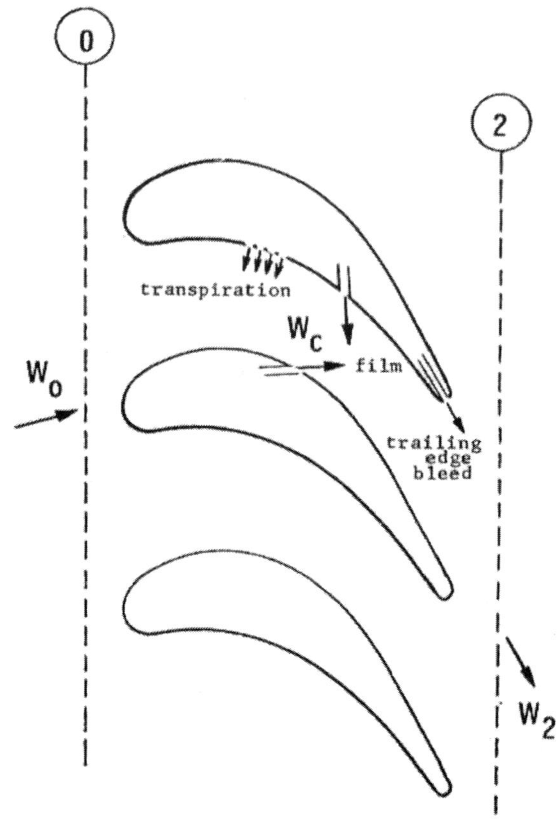

FIGURE 19—22 General cooled cascade nomenclature [19-1].

The increasing use of mass-transfer cooling in turbines has allowed significant increases in turbine inlet temperatures and overall engine performance. The problem of predicting the aerodynamic effects of injecting coolant air has grown in importance as mass-transfer cooling has increased in both extent of application and amount of cooling air employed. The rising costs of experimental testing have encouraged the development of analytical methods to predict the effects on blade-row performance of film-, transpiration-cooling, and the introduction of spent cooling air by methods such as airfoil trailing-edge bleed. The present case examines the purely aerodynamic loss, often called the "mixing loss," due to coolant injection within a turbine airfoil blade row and its subsequent mixing with the mainstream flow. The general methods developed, however, may be applied to any portion of the turbine where coolant is injected.

Consider the turbine airfoil blade row of Figure 19—22, which represents a rotor or stator employing mass-transfer cooling. The thermodynamic efficiency for the cascade shown is defined as

$$\eta \equiv \frac{\text{actual exit kinetic energy}}{\text{ideal exit kinetic energy}}$$

$$= \frac{\left(1+\varepsilon\frac{T_{t_c}}{T_{t_0}}\right)\left[1-\left(\frac{P_2}{P_{t_2}}\right)^{\frac{\gamma-1}{\gamma}}\right]}{1+\varepsilon\frac{T_{t_c}}{T_{t_0}}-\left(\frac{P_2}{P_{t_0}}\right)^{\frac{\gamma-1}{\gamma}}-\varepsilon\frac{T_{t_c}}{T_{t_0}}\left(\frac{P_2}{P_{t_c}}\right)^{\frac{\gamma-1}{\gamma}}} \tag{19.1}$$

where Equation 19.1 is written for like, perfect gases and assumes that the ideal available kinetic energy of the exit flow is equal to the sum of the ideal kinetic energies of the coolant and inlet flows when expanded isentropically (without mixing) to the exit static pressure. For the particular case where $P_{t_c} = P_{t_0}$, the efficiency expression reduces to the classical cascade efficiency,

$$\eta = \frac{1-\left(\frac{P_2}{P_{t_2}}\right)^{\frac{\gamma-1}{\gamma}}}{1-\left(\frac{P_2}{P_{t_0}}\right)^{\frac{\gamma-1}{\gamma}}} \tag{19.2}$$

In engine applications P_{t_c} is an external variable, usually dictated by the cycle coolant air availability, thus the airfoil cascade efficiency is determined by the average exit total pressure P_{t_2} for a fixed coolant supply pressure. The prediction of cooled blade-row efficiency is thus equivalent to predicting the total pressure change $\Delta P_t = P_{t_c} - P_{t_o}$ caused by viscous effects, trailing-edge blockage, and coolant-primary flow mixing. The methods of this case are based on the assumption that these losses are additive and may be determined independently. This implies that the injected coolant mixes with the mainstream flow rather than remaining entrained in the boundary layer. Although the true situation is undoubtedly something between the two extremes of complete entrainment and complete mixing, there does exist experimental evidence to substantiate the present simple model.

One-Dimensional Methods

The simplest model for coolant-mainstream mixing is the one-dimensional slot shown in Figure 19—23. This may be viewed as an appropriate idealization of a single row of film-cooling holes or, if the imagination is willing, an "average" of a complex cooling geometry. Two solutions are readily obtained for the one-dimensional case.

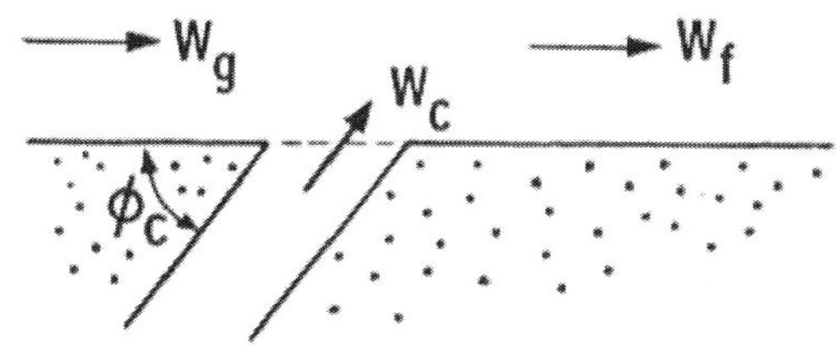

FIGURE 19—23 The one-dimensional mixing model [19-1].

Constant Static Pressure Mixing

If the upper boundary of the flow shown in Figure 19—23 is assumed to be flexible, then coolant and mainstream flows may be mixed under the assumption of a constant static pressure. Applying the equations of continuity, streamwise momentum, energy, and assuming thermally and calorically perfect gases for both mainstream and coolant, the result

$$\frac{\Delta P_t}{P_{t_g}} = \left(\frac{1+\frac{\gamma-1}{2}M_f^2}{1+\frac{\gamma-1}{2}M_g^2}\right)^{\frac{\gamma-1}{\gamma}} - 1 \quad (19.3)$$

$$M_f^2 = \left[\frac{c_P(\gamma-1)(1+\varepsilon)(T_{t_g}+\varepsilon T_{t_c})}{(V_g+\varepsilon V_c\cos\phi)^2} - \frac{(\gamma-1)}{2}\right]^{-1} \quad (19.4)$$

is obtained.

A Linear Approximation

The one-dimensional mixing of Figure 19—23 may also be approximated by adopting the "influence coefficient" method of Shapiro in the following manner.

For the compressible flow of a perfect gas having constant external and frictional forces, the differential variation of total pressure may be expressed as

$$dP_t = \left(\frac{\partial P_t}{\partial A}\right)dA + \left(\frac{\partial P_t}{\partial T_t}\right)dT_t + \left(\frac{\partial P_t}{\partial W}\right)dW$$

with partial derivatives of

$$\frac{\partial P_t}{\partial A} = 0$$

$$\frac{\partial P_t}{\partial T_t} = -\frac{\gamma M^2 P_t}{2T_t}$$

$$\frac{\partial P_t}{\partial W} = -\frac{\gamma M^2 P_t}{W}\left(1-\frac{V_c}{V}\cos\phi_c\right)\frac{dW}{W}$$

as obtained from the derivation by Shapiro. Rearranging gives

$$\frac{dP_t}{P_t} = \frac{-\gamma}{2}M^2\frac{dT_t}{T_t} - \gamma M^2\left(1-\frac{V_c}{V_g}\cos\phi_c\right)\frac{dW}{W} \quad (19.5)$$

which assumes that the injected fluid enters the mainstream at the local static pressure. Exact solution of Equation 19.5 requires integration of the functions, but, assuming that relatively small variations of T_t and W occur, the differentials may be considered Δ's. The term "linear approximation" is thus derived from the fact that the equation is thus linear in the independent variables dT_t/T_t and dW/W. The term dT_t/T_t may be simplified by assuming equal mainstream and coolant specific heats and writing the energy equation for a perfect gas in differential form as

$$T_{t_g}W_g + T_{t_c}dW_g = (W_g + dW_g)(T_{t_g} + dT_{t_g})$$

which, when expanded and second-order terms are dropped, becomes

$$\frac{dT_{t_g}}{T_{t_g}} = \frac{dW_g}{W_g}\left(\frac{T_{t_c}}{T_{t_g}} - 1\right)$$

Since dW/W = ξ, Equation 19.5 may then be written in the more convenient form

$$\frac{\Delta P_t}{P_{t_g}} = \frac{-\gamma}{2}M_g^2\xi\left[1+\frac{T_{t_c}}{T_{t_g}} - 2\frac{V_c}{V_g}\cos\phi_c\right] \quad (19.6)$$

A further simplification results when $P_{t_c} = P_{t_g}$, in which

$$\frac{V_c}{V_g} = \sqrt{\frac{T_{t_c}}{T_{t_g}}}$$

since $\gamma_c = \gamma_g = \gamma$ has been assumed.

The linear approximation method is valuable, not merely because it allows rapid hand calculation, but also since it reveals the primary mixing-loss parameters in their simplest form. This is seen by comparing Equations 19.3 and 19.4 with Equation 19.6. The latter relationship shows that gas properties affect the mixing result only weakly through the specific heat ratio, while mainstream Mach number and coolant fraction are far more significant, as are the coolant-mainstream temperature ratio, velocity ratio, and the coolant injection angle. As would be expected from a one-dimensional analysis, the losses predicted for injection normal to the mainstream flow ($\phi_c = 90°$) are independent of coolant velocity since the streamwise component of coolant momentum is zero for all coolant velocities.

Figures 19–24 and 19–25 present a comparison of calculated losses using both of the one-dimensional methods. A constant coolant area for each temperature ratio has been chosen so that coolant and mainstream total pressures are equal at a coolant mass fraction $\xi = .03$. Coolant velocity for high coolant pressures was limited to the sonic value for these calculations. Aside from showing the close agreement of the linear approximation with constant static pressure mixing, the effects of total temperature ratio and coolant injection angle are quite apparent.

The close agreement of the two calculation methods is not surprising since they are, of course, identical in the limit of zero coolant addition. Figure 19–26 presents the linear approximation in terms of a dimensionless loss parameter in order to show the effects of temperature ratio and coolant injection angle for the ideal case in which coolant and mainstream gases are identical in composition, have equal total pressures, but differ in total temperature. Figure 19–26 is merely a graphic portrayal of the bracketed term in Equation 19.6. The result shows that not only do losses decrease with decreasing angle of injection, but it also reveals that as injection angle decreases, the effect of coolant-to-mainstream temperature ratio becomes less pronounced for higher temperature ratios and more evident for lower ratios. As coolant total temperature approaches extremely low values, the intuition that injection angle becomes less important is confirmed. For total temperature ratios in the general vicinity of ½, however, the effect of coolant injection angle is quite significant.

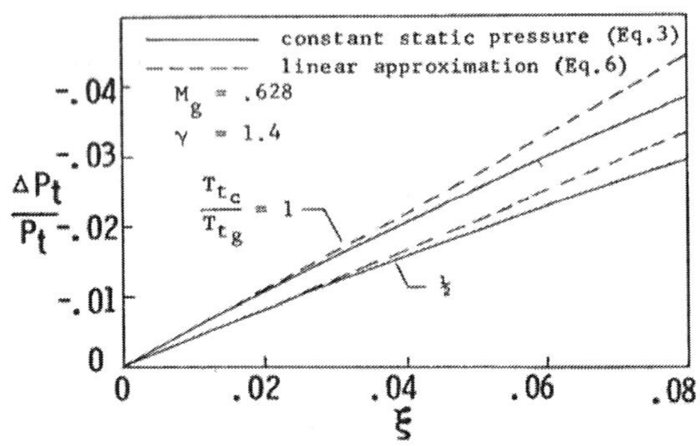

FIGURE 19–24 One-dimensional mixing results for coolant injection normal to the mainstream flow direction [19-1].

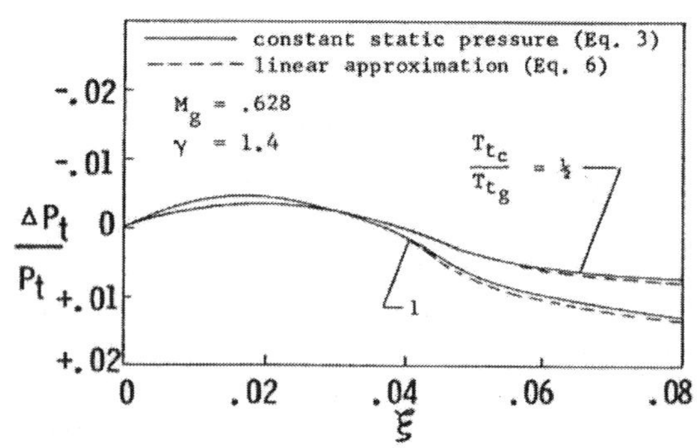

FIGURE 19–25 One-dimensional mixing results for coolant injection inclined 30° from the mainstream flow direction [19-1].

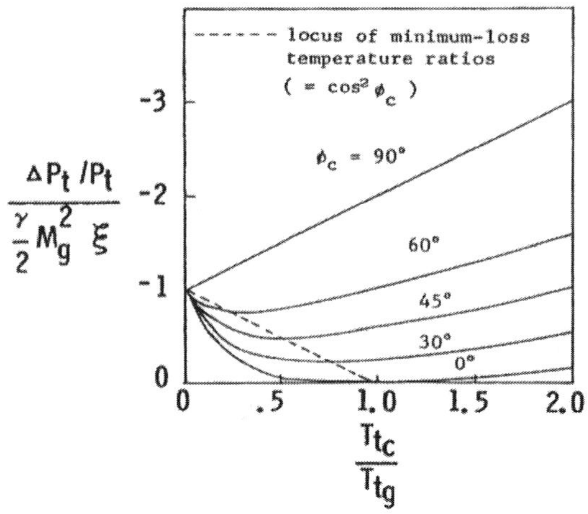

FIGURE 19–26 One-dimensional mixing loss for ideal coolant and mainstream gases ($\gamma_c = \gamma_g$ and $P_{t_c} = P_{t_g}$) [19-1].

It is also interesting to note that the one-dimensional methods discussed above are in close agreement with one-dimensional constant area mixing methods, which are far more difficult to apply, particularly for transonic Mach numbers.

The TOTLOS Method

The one-dimensional methods discussed above are based on the concept of coolant injection at a single location into a uniform mainstream. A more practical problem is that posed by complex cooling geometries in which injection occurs at a variety of locations having differing mainstream and coolant conditions. It is possible to apply the one-dimensional techniques by assuming coolant and mainstream parameters which are "typical" of the lumped cooling system, but the method of selecting the average parameters for such a simplification is not fore-known. The following method, called the TOTLOS method, allows the calculation of the total blade-row ΔP_t due to viscous boundary layers, trailing-edge blockage, and coolant injection. The method, as with the one-dimensional ones just discussed, is based on the assumption that losses due to coolant-mainstream mixing may be superimposed on those due to boundary layers and other causes unrelated to cooling per se, such as airfoil base pressure drag and end-wall losses.

Total pressure loss due to film or transpiration-cooling of the airfoil surface is found by applying the one-dimensional constant static pressure mixing method within a "mixing layer" over each surface, allowing coolant and mainstream flow to be mixed at successive "nodes" along the surface. These nodes may correspond to either rows of injection holes or arbitrary subdivisions of a porous region. The local static pressure of each node is equal to that on the airfoil surface at that particular station. As shown in Figure 19–27, the unaffected mainstream flow is then mixed with the contents of both surface-mixing layers (again using the one-dimensional constant static pressure mixing method) at a static pressure equal to the average mainstream value in the trailing edge plane (P_1).

No theoretical basis for selecting the thickness of the mixing layers has been formulated to date, but numerical computations using the TOTLOS method show repeatedly that the final result is not significantly dependent on this parameter. As more experience is gained using the method as a comparison with experimental data, a dependence of mixing-layer size on other parameters may be discerned. Good candidates for such parameters would be cooling geometry, airfoil pressure distribution, mainstream turbulence, etc. For the present calculations, each mixing layer height at Station 0 has been arbitrarily set at 5% of blade-row pitch, and variation of this initial height from 3–30% of inlet pitch shows little effect on the final calculated loss.

The inclusion of viscous losses, trailing-edge blockage effects, and the effect of coolant injection from the trailing edge are accomplished using the two-dimensional model shown in Figure 19–28. The method used is based directly on that of Stewart, with some modification to include a coolant jet in the airfoil base. The desired result of the two-dimensional calculation between Stations 1 and 2 is the fully-mixed, uniform downstream state, particularly the total pressure and flow direction. The equations of continuity, tangential momentum, and axial momentum for the control volume shown are written as

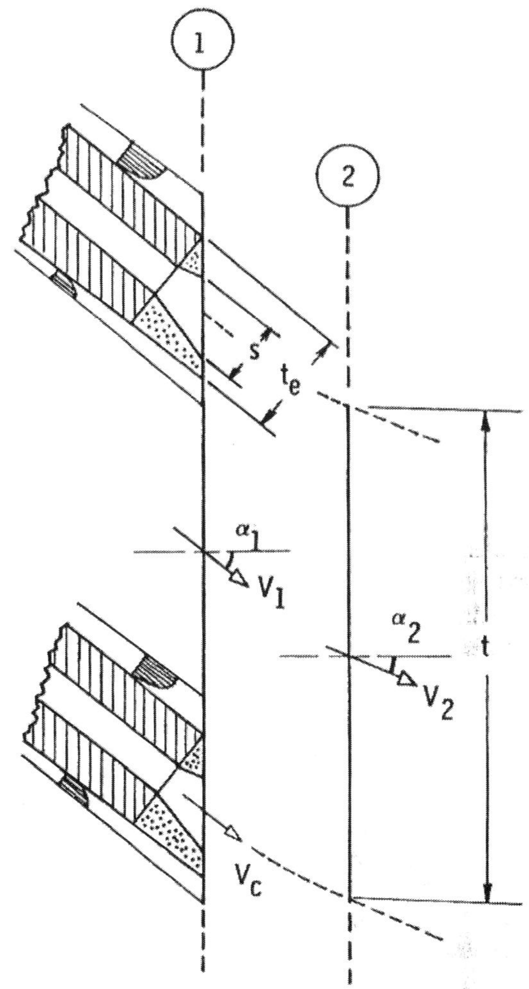

FIGURE 19–28 Schematic of the TOTLOS trailing-edge mixing control volume [19-1].

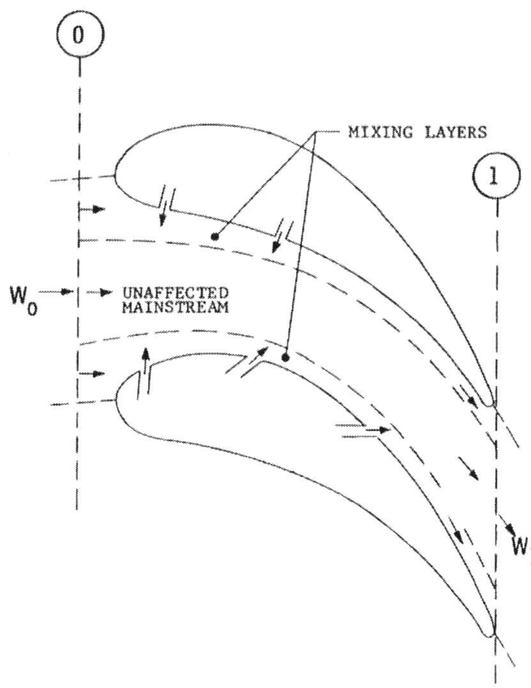

FIGURE 19–27 Schematic of the TOTLOS mixing-layer model for calculating losses due to surface coolant injection [19-1].

$$
\begin{aligned}
C &= \cos\alpha_2(\rho V)_2 \\
&= \cos\alpha_1(\rho V)_1(1-\delta_{TE}-\delta^{**}) + (s/t)(\rho V)_c \\
Q &= \cos\alpha_2 \sin\alpha_2 (\rho V^2)_2 \\
&= \cos\alpha_1 \sin\alpha_1 \left[(\rho V^2)_1 (1-\delta_{TE}-\delta^{**}-\theta^*) + s^*(\rho V)_c\right. \\
A &= \cos^2\alpha_2 (\rho V^2)_2 + P_2 \\
&= \cos^2\alpha_1 \left[(1-\delta_{TE}-\delta^{**}-\theta^*)(\rho V^2)_1 + s^*(\rho V^2)_c\right] + P_1
\end{aligned}
$$

By introducing the energy and state equations, the system may be solved. The solution for the case of a perfect gas is a direct one, obtaining the downstream density as

$$\rho_2 = -\frac{A\left(\frac{\gamma}{\gamma-1}\right)}{\frac{Q^2}{C^2} - 2c_pT_{t_2}} + \sqrt{\left[\frac{A\left(\frac{\gamma}{\gamma-1}\right)}{\frac{Q^2}{C^2} - 2c_pT_{t_2}}\right]^2 - \left[\frac{C^2\left(1 - \frac{2\gamma}{\gamma-1}\right)}{\frac{Q^2}{C^2} - 2c_pT_{t_2}}\right]}$$

from which the complete flow condition at Station 2 may then be found using standard relationships.

Two required inputs for the inclusion of boundary layer losses in the final result are the displacement and momentum thicknesses at the trailing edge. The program of McDonel has been adopted for this purpose, but any comparable method could be used. The TOTLOS calculation scheme can become quite laborious if more than a few coolant-injection locations are specified, thus the method has been programmed for numerical computations, allowing the inclusion of real gas properties. Cooling geometries having up to 70 injection locations have been analyzed using the programmed version.

The TOTLOS calculation method has been applied to a series of cooled stator designs tested by The Lewis Research Center of NASA. Figures 19–29 through 19–31 present the variation of stator thermodynamic efficiency, both measured and predicted, for the three vane cooling configurations tested. These designs included film-cooling through holes drilled normal to the surface, transpiration-cooling employing porous wire-mesh surface segments, and trailing-edge injection through rectangular slots. The film- and transpiration-cooled designs incorporated coolant injection over the entire vane surface and all data were obtained in an annular stator at a coolant-to-mainstream total temperature ratio of 1.

Since the TOTLOS method is two-dimensional, a "pitch-line" analysis at a nominal cascade pressure ratio of 1.5 was used to calculate the predicted stator efficiencies shown here. The calculations were carried out for each vane

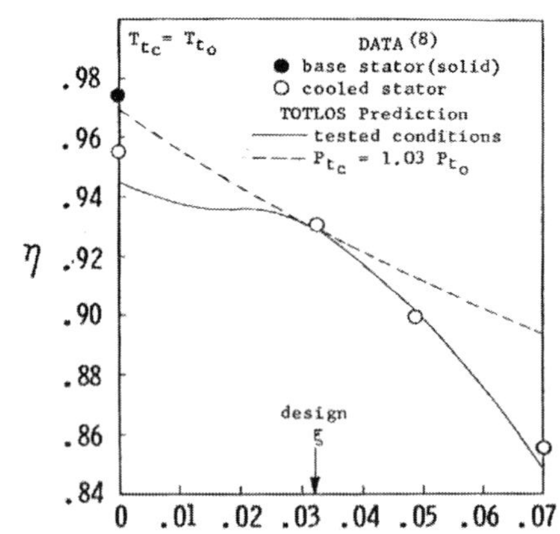

FIGURE 19–29 Comparison of measured and predicted stator efficiency for film-cooled vanes ($\varphi_c = 90°$, discrete coolant holes) [19-1].

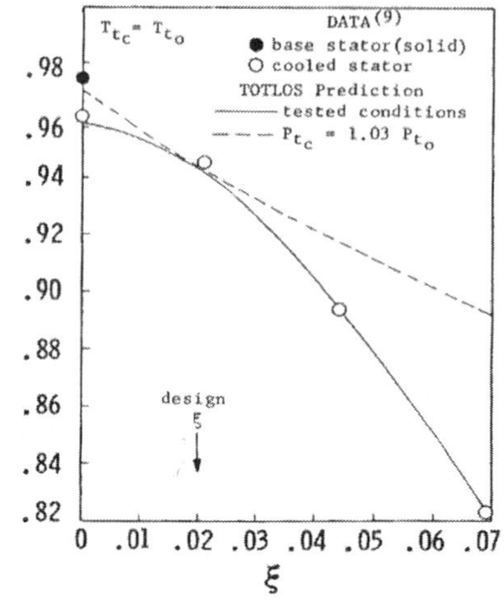

FIGURE 19–30 Comparison of measured and predicted stator efficiency for transpiration-cooled vanes ($\varphi_c = 90°$, wire-mesh surface) [19-1].

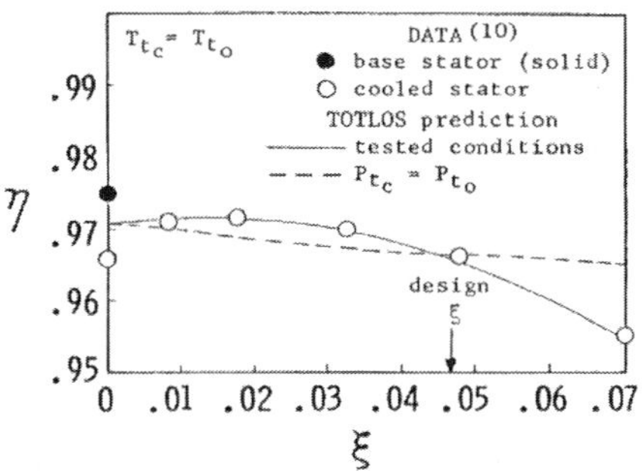

FIGURE 19–31 Comparison of measured and predicted stator efficiency for vanes with trailing-edge injection [19-1].

design for two distinctly different cases. One case, shown as the solid line in each figure, was for the tested conditions, i.e., coolant flow variation was achieved by varying the coolant supply pressure. This case obviously allows recirculation of mainstream flow within the vane coolant cavity for the film- and transpiration-cooled vanes at coolant total pressures below the stator inlet value.

To calculate a "net" coolant injection distribution over the surface in this range, the surface pressure distribution and the known coolant fraction were employed in an iterative solution to find the coolant supply pressure and the distribution of inflow or outflow over the vane surface. Once this surface flow distribution was found, the mainstream flow into the vane was neglected and only surface outflow was used in the loss calculation.

A second case, calculated for each vane design and shown as a dashed line in each figure, was the "design" case in which the coolant supply pressure was held constant at the design value, approximately equal to the inlet total pressure. The predicted efficiencies agree well with the data for coolant pressures equal to or above the design value, the region in which coolant flow distribution could be most accurately predicted. The large differences between efficiencies calculated for the "tested conditions" case and the "design" case at large coolant flows is due primarily to the fact that the ideal available kinetic energy of the coolant is charged to the stator at a total pressure considerably greater than inlet total pressure. As would be expected, trailing-edge coolant injection causes far less loss of efficiency than either normally-injected film- or transpiration-cooling for both the "test conditions" and "design" cases.

In summary, the effects of coolant injection on the flow through an airfoil cascade are admittedly complex and require analyses far more sophisticated than those given here if a detailed description of the resultant flowfield is desired. To the turbine designer, however, the methods presented here allow a relatively rapid determination of the effect of coolant injection on blade-row efficiency by predicting the state of the fully-mixed exit flow, a quantity of great interest to the engineer. Comparisons of predicted efficiency variations with data, such as those presented here, have shown the methods to be reliable and consistent. While their most useful application is in the evaluation of the aerodynamic performance of various cooling designs, reducing the amount of expensive testing required, the one-dimensional model also reveals the relative roles of mainstream Mach number, coolant-to-mainstream temperature and velocity ratios, and coolant injection angle in the generation of observed losses.

CASE STUDY 2: ADVANCED TECHNOLOGY ENGINE SUPPORTABILITY: PRELIMINARY DESIGNER'S CHALLENGE*

The United States Government is currently sponsoring, through industry, an ambitious, high-risk engine initiative. One goal is to double fighter engine thrust-to-weight by the next century. This initiative is called the "Integrated High Performance Turbine Engine Technology," or IHPTET. Parts of the initiative are currently under contract to the major military aircraft engine manufacturers in the United States. Major advances are planned and under development in engine materials, design innovations, and performance. All design disciplines are involved including supportability.

In the IHPTET engine design program, conventional thinking is out; revolutionary but fundamentally sound thinking is in vogue. It's back to basics, both in design and supportability. One of the problems in the past was the mistaken and simplistic notion that all that was needed during the preliminary or conceptual design to ensure supportability was *simplicity*. While simplicity is a virtue, it does not, by itself, ensure that something is inherently reliable, maintainable, or repairable. However, coupled with innovative thinking and a thorough understanding of the mechanisms and relative importance of failure, repair, accessibility, cost, etc., the preliminary designer can make decisions at the outset of a design which will allow an engine to be very supportable; but he must be armed with the proper training, motivation, attitude, and *then* he can make it simple and supportable.

General Electric Aircraft Engine (GEAE) Preliminary Design and Supportability have been cooperating, utilizing advanced new materials and design innovations, in an effort to conceptualize an easily supported engine which can be maintained with standard and special tools which fit in a mechanic's standard tool box.

Historical Trends/Recent GEAE Experience

One of the misconceptions that persists in the aviation industry is "that things were better in the old days." This holds true for supportability and one may still hear tales of aircraft and engines that "never had to be touched." The comparison must always be tempered by judging the performance of the machines. It is the development of high energy cores that has allowed the dramatic increases in engine thrust-to-weight ratio and specific thrust and which offers the aircraft designer the option of trading range for payload or lets the engine designer trade fan pressure ratio, T_{41}, and bypass ratio to help the system optimization. Thirty or 40 years ago, the engine power-to-weight ratio was a constant, and the world was trying to design bigger and better superchargers to get more performance out of the engine.

The jet age started in the United States with the GE I-A in 1941 with a thrust-to-weight ratio of 1.4. Now fighters are

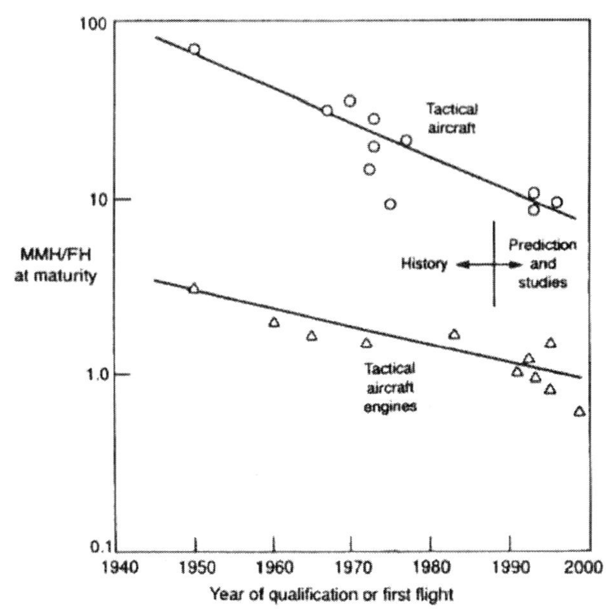

FIGURE 19–32 Tactical aircraft and engine maintenance trends [19-2].

* Source: [19-2] Courtesy of J.E. Hartsel. J.J. Ciokajlo and J.E. Hartsel. Extracts from "Advanced Technology Engine Supportability: Preliminary Designer's Challenge," AIAA-88-2796, 1988.

powered by engines with ratios over 7, with ratios of 10, 15, and possibly even 20 foreseen within the next two decades.

How has supportability fared? Figure 19–32 shows a few data points from the authors' personal collection of Maintenance Man Hours per Flight Hour values. Things have been getting better, at least for tactical aircraft and engines, and current study aircraft and engines seem to fall on the same improving historical trend lines.

On the surface, the philosophy of testing, analyzing, and fixing appears to be the framework upon which the engine development process is built. Avionics developers have identified the need for more than one pass through the development cycle as a key to reliability improvements. This strategy is referred to as Maturational Development.

In the case of engine development, the Maturational Development approach is totally out of the question, based upon the long lead times involved. A key to supportable engine designs is designing supportability into the components during the earliest phases of component research and development and improving on the supportability throughout the whole technology development process. Care must be taken when integrating these components into a final propulsion system design so that all is not lost in the end product.

A recent supportability success story from GE Aircraft Engines is the T700. In this case, durable/reliable components were integrated into a final engine designed for ease of access and overall maintenance.

The IHPTET initiative demands major performance improvements in our future generations of engines. The challenge facing today's engine designers is how to reach these performance goals while still improving supportability. The only way to make this happen is for supportability to be a design criterion from the earliest component research and development projects through preliminary design, demonstration, and production. This case outlines an approach to the design of an easily supportable engine.

Advanced Materials

Advanced materials are the most important element in the IHPTET initiative. The supportability community, like the design community, must recognize this fact. Metal matrix composites, ceramic matrix composites, intermetallic compounds, high temperature titanium alloys, sapphire fiber reinforced nickel aluminides, and carbon-carbon will be a way of life. Material costs will dictate net shape manufacturing techniques that are still to be defined. Repairability of these materials is still a question at present. Most of the advanced metallic materials will be powders in their initial manufacturing states. It is envisioned that intricate parts, which are currently castings fabricated from molten metals, will initially be injection molded powder slurries—dried, hardened, and sintered into metallic form. Welding will be replaced by applying a putty-like intermetallic compound and then sintering locally to cure the "weld." Fiber reinforced matrices will be manufactured by either a metallic spray over ceramic fibers or by layers of metallic foil interspersed with layers of ceramic fibers. In either approach, the final step will be fusion by the Hot Isostatic Press (H.I.P.) process.

Proper attachment methods (bolts, flanges, etc.) are currently being defined or developed. Material strength, ductility, and rupture properties may be different in all three directions. It may be possible to repair "weld" a component in the longitudinal direction but not in the transverse direction. The title "metallurgist" may become obsolete; "materials engineer" will probably be a more apt description.

Engine Preliminary Design

General Approach

Attaining the high thrust-to-weight goal will be accomplished by integrating all engineering disciplines. It is estimated that 50% of the goal will come from new materials, 35% from design innovations, and 15% from engine performance (aerodynamic and thermodynamic) advances. Improved engine performance will be helped by higher efficiencies and turbine inlet temperatures. Design innovations will include multifunctional components, high temperature lubricated bearings, 360° static structures, and design simplicity. Fiber-reinforced metal matrix composites (silicon carbide and sapphire fiber), carbon-carbon (C-C), ceramic matrix composite (CMC), intermetallics, and high temperature titanium alloys will be the new materials for future, very advanced, lightweight fighter engines.

The conceptual engine design process begins with the clear enunciation of the requirements. Before the first cycle calculation has been made, the Design Management and Design Team must understand the requirements for performance, weight, cost, and supportability. Requirements for reliability, maintainability, and repairability must be emphasized and given "teeth" by the Design Management.

Engine considerations are initiated with the engine stick flowpath. The "stick" flowpath is a sketch the preliminary engine designer receives from the aerodynamic designers. The engine preliminary designer must know all engine design disciplines and be able to modify the engine flowpath without violating the design intent or engineering laws. The engine preliminary designer at this point must (1) produce a concise compact fan, compressor, combustor, turbine, augmentor, and exhaust nozzle flowpath, (2) mentally and physically determine component mechanical configuration, and (3) determine module configuration. At this point, the preliminary designer must be permitted to struggle alone as he addresses the engine design goals of performance, cost, weight, and supportability.

Once the designer sets design direction involving all design disciplines, the engine's initial design configuration begins. The skill with which the preliminary designer understands and balances the often-conflicting requirements as he sets the design direction can have a major influence on the supportability characteristics of the final design. If supportability isn't considered here, it can conceivably be locked out of the final product.

Specific Approach

The crucial first step in the preliminary mechanical design of an engine is the configuration of the sumps and bearings supporting the rotors. Grossly miscalculate initial rotor dynamics and the preliminary design is time wasted. Current engines have multipiece sump, seal, and bearing configurations requiring manual build-up at engine assembly. Future engines will be lubricated by high temperature lubricants with design simplicity as a top priority. Sumps will be one-piece cartridge bearings and seal configurations for easy assembly, maintainability, and removal. The quick replaceability of the cartridge bearing is an example of design simplicity and supportability interaction.

Once the sumps and shafts are in place, remaining engine components can be wrapped around them, starting with the combustor and rotors. Fan, compressor, and turbine rotors are located relative to the bearings as dictated by preliminary rotor dynamic analysis. Advanced compression rotor systems are simple, integrated, lightweight disks and blades called blisks. Compressor blade replaceability is a maintainability tool eliminated by blisks; but to compensate, the rotors will be tolerant of substantial imbalance and will have rugged, damage tolerant, low aspect ratio, high-pressure rise airfoils, and a minimum number of compression stages. Finally, the casings, shells, augmentor, and exhaust nozzle are added to form an initial engine configuration.

Initial Layouts

Once the responsible designer has received and understands the requirements, has initial cycle data in hand, and has held an initial consultation with all design disciplines and management, a series of configuration options should be created by the designer, the designer-draftsperson, and a small team. Modularity and simplicity should be the guidelines. Upon completion, each configuration is critiqued by the full design group and management. These initial layouts will merely be first attempts to incorporate past experience and design practices, but they should also be allowed to contain the new and creative ideas that have been waiting for a home. During this phase of the design process, the drafting room population should be kept to a minimum except when critiques are invited. Supportability interests and considerations, such as modularity, multifunctional components, ruggedness, accessibility, etc., must be given the same emphasis as the other "traditional" requirements during this critique and iteration phase.

Specific Examples

Overall Engine

Supportability must be designed into the engine in the preliminary design phase, preferably with little or no increase in engine weight or cost. There must be a game plan for assembly, disassembly, and module selection. Figure 19—33 illustrates a schematic of an engine on an

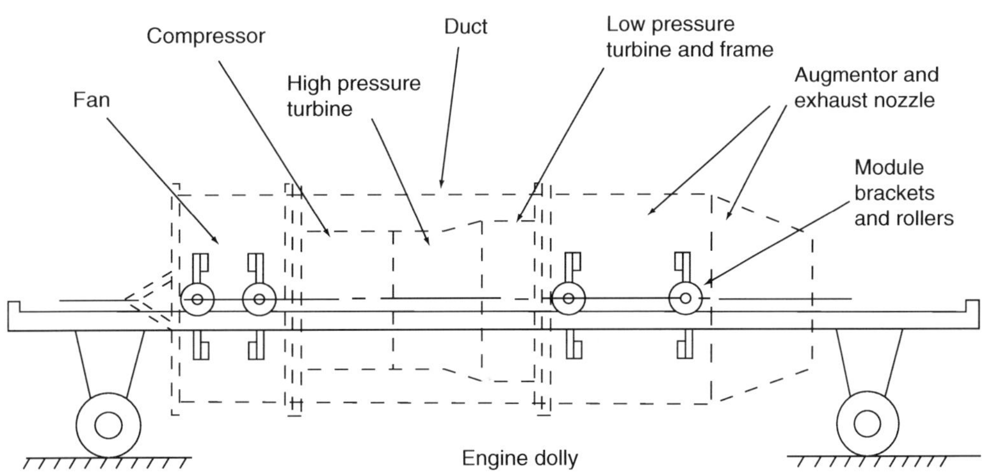

FIGURE 19—33 Modular engine and dolly arrangement [19-2].

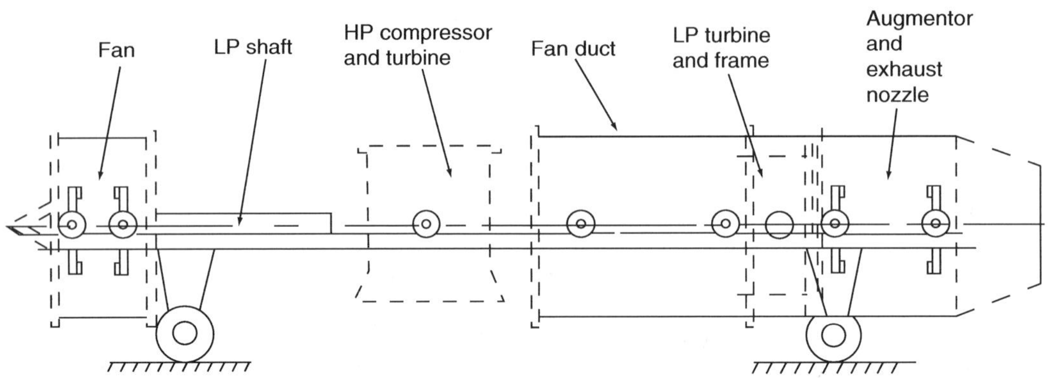

FIGURE 19—34 Modular engine disassembly and core access [19-2].

assembly dolly. Here major module assemblies such as the augmentor and exhaust nozzles, fan duct, low pressure turbine, and fan and fan frame are mounted and attached to mini dollies so that after disconnecting electrical and service lines, engine axial disassembly can be attained as illustrated in Figure 19—34. Rapid access to the high-pressure turbine and combustor module is attained.

This type of disassembly/assembly is considered highly desirable in many military scenarios.

The engine core or high pressure section of a turbofan engine, normally a small volume portion of the engine, can be rotated vertically in the dolly for further access to either the high pressure turbine or compressor modules and the compressor rear frame assembly. This engine disassembly approach is nothing new—it has been with us for years. This principle was in General Electric's J-85 engine design, which now has seen 30 years of service.

Component Design

Figures 19—35 and 19—36 show specific examples of how the maintainability aspect of supportability can be incorporated in the engine in the preliminary design phase. The advent of new composite materials allows the designer to custom tailor materials for specific maintainability problems. Figure 19—35 illustrates a conventional shaft spline detail complete with interference fit pilots. Conventional design practice requires pilot interference fits of 0.0005 inches per inch of pilot radius and

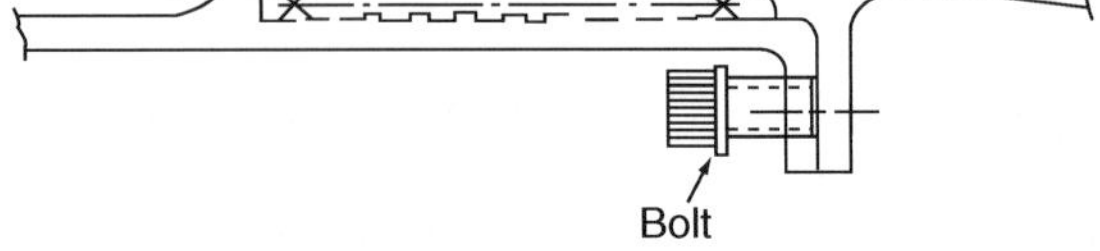

FIGURE 19—35 Conventional spline with backoff bolts [19-2].

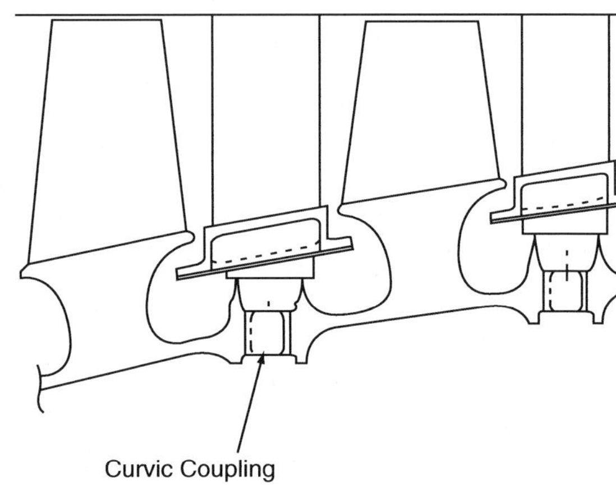

FIGURE 19—36 Curvic coupling detail [19-2].

special tooling to assemble/disassemble splined components (see Figure 19—37).

This brute force assembly technique has been effective for past rotor designs using monolithic, metal rotor materials, but requires simultaneous heating and chilling or large special tools. However, this technique is antiquated when applied to advanced design rotors of different materials operating at elevated temperatures. By selecting the material of the respective shafts, the thermal growth coefficients can be matched so a loose fit or snug close tolerance pilot fit is obtained at assembly room temperature and a very tight interference fit, caused by differential radial thermal growths, is achieved at pilot/spline operating temperature. This simple, old, but effective technique, where applicable, can be applied throughout the engine, thus eliminating the use of large and heavy special tooling and allowing the use of simple techniques such as "backoff" bolts, as illustrated in Figure 19—35. These simple tools can either remain on the engine or fit in a mechanic's standard toolbox.

Controlled shaft heating is another feature available when splined shafts have mismatched thermal growths. Selective heating of the inner shaft by a portable electric heater causes inner and outer shaft relative axial growth, unseating the locking nut for easy nut removal. This simple

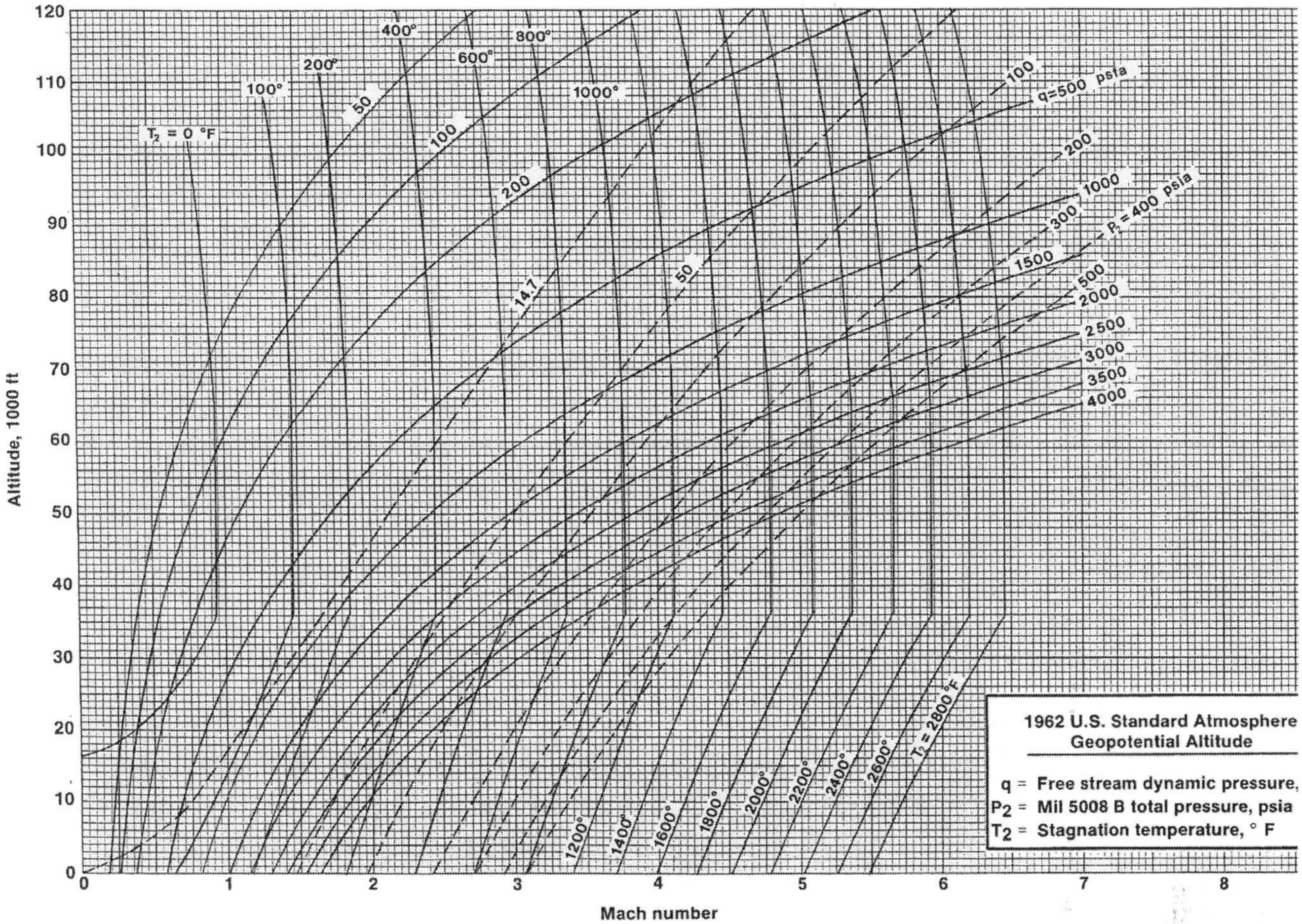

FIGURE 19—37 Altitude-mach number chart [19-2].

concept, if properly controlled, eliminates the need of heavy load cells or torque multiplier wrenches for rotor assembly/disassembly.

A second simple, but effective, design assembly/disassembly feature is the old technique of "lead-ons" to protect critical and fragile engine sump seals and other components. For decades engine sumps have been designed as an assortment of small components, hand assembled at engine assembly and requiring interim special dimensional checks. More recent engine sump designs have eliminated special assembly dimensional checks but still require delicate assembly procedures. Why not design the bearings, seals, oil jets, and other related components as a one piece mini module? This principle has been accomplished in electronic design—it's called the circuit board, used extensively in television sets.

If special lead on procedures for components are required, "lead on" rods are a simple and effective method that (1) if left on the engine add little weight, and (2) if removed can fit in a mechanic's tool box. Long rods or pins, attached via flanges to the engine casing, allow a "lead on" to protect the sumps, shrouds, or other critical components at assembly/disassembly. A circumferential series of rods or pins, with long protrusions, are rigidly attached to one component and engage with and align a second component prior to engagement of the respective component sump seals, rotor shrouds, or other critical parts. A design such as this requires reasonable, but not special, care at assembly.

Another design detail that can be utilized in the preliminary design phase is the old design friend, the curvic coupling. The use of engine compressor blisks (integral one piece disks and blades) reduces compressor rotor weight but can add difficulties to rotor blade repairability; and, in some instances, impairs compressor rotor accessibility by eliminating the compressor split casing. This one piece casing unit can further reduce engine weight. Therefore, another method must be devised for compressor rotor accessibility, and curvic couplings are an example of this.

As stated in Specific Examples—Overall Engine, the engine core module of a turbofan is a small volume and could be rotated vertically in the engine dolly. By machining curvic couplings into the compressor blisk ends, the compressor rotor can be held together as one piece by a single tie bolt. The T700 compressor rotor is a blisk design held together in this manner. Figure 12–36 illustrates the rotor blisk curvic coupling feature. With the engine core module in the vertical position, heat can be applied to the core rotor tie bolt shaft, inducing shaft axial growth and unloading the nut. With the axial load reduced or eliminated, the coupling nut can be removed.

A final example to aid engine supportability in the preliminary design phase is the design of engine static components. Static components in advanced engines account for approximately 60% of the weight. Engine supportability can best be served by minimizing potential problems. There can be no frame problems if there are no frames. Designing 360°, one-piece, multi-functional components that perform functions previously required by two or three units not only saves engine weight but also minimizes supportability problems by component elimination.

In summary, the US military services are participating with the aircraft engine industry in an ambitious program to develop high performance engine technologies for advanced future military applications. This radical program, starting from a "clean sheet of paper," is a golden opportunity to integrate many radically advanced but fundamentally correct engineering disciplines, including supportability, into future, very high thrust-to-weight fighter engines. The thrust-to-weight ratio goal is double that of current military engines under development and almost triple the thrust-to-weight ratio of fighter engines currently flying. This accomplishment, scheduled for completion in 15 years, will require the total cooperation and understanding between management, supportability, and engineering. Engine supportability techniques may change but fundamental basics will apply. New materials, not yet commercially available, must be understood. The preliminary design engineer will always be a radically thinking, analytically oriented person with his/her head in the clouds; and the supportability engineer will have both feet on the ground beset with day-to-day problems. Some technical crossbreeding between both will be necessary to meet the supportability challenges of the future.

Chapter 20

Additional References and Appendix for Unit Conversion

"The only true wisdom is in knowing you know nothing."

—Socrates

Chapter Outline

ADDITIONAL GENERAL REFERENCES*

1. Websites

   A great deal more information is available on the websites of the following sources:

   Siemens Power Generation
   Rolls Royce
   Alstom Power
   Mitsubishi Power Systems
   The US Department of Energy (DOE)
   The US Environmental Protection Agency (EPA)
   The European Turbine Network (ETN)
   The IEA Clean Coal Center
   EPRI

Actual web addresses are not included in the location as these can change. The reader is advised to do a Google search as these organizations may have more than one website or subsite.

Also for additional (and in some cases, the same in different words) theory on gas turbines: http://www.netl.doe.gov/technologies/coalpower/turbines/refshelf/handbook/TableofContents.html

2. Emailed newsletters

   Most major corporations and even smaller companies now direct their information flow with mass email. Getting on a distribution list is very easy, so one needs to pick the vital companies or industries with care. The selection of "those who've got the best information" may also vary, so I will not outline a specific list, but suggest that keywords in this subsection can point the way. At this point in the gas turbine's and the power generation industry's history, there may be no more important development than what is likely to evolve into a carbon economy in many countries and eventually a global carbon economy. One of the key information sources for keeping up with coal and coal related projects, are the sites run by the IEA and the ETN. This agency has branches all over the world, including the USA and they all share information. Special attention needs to be paid to the change in the gas turbine fuels industry. In the first edition of this book, I did point out that the field was widening with gas turbines able to burn fuels like residual oil, biomass, paper liquor, flue gas from making steel and a host of other materials (with appropriate design modifications). Now, despite the fracking activity globally which has driven down the price of natural gas, the biggest player globally, especially in coal rich countries, may continue to be coal. In the energy field, there has been considerably more research done on coal gasification since the first edition

* Publishing is a communications mode for engineering technology, not the technology itself. Publishers are also subject to worsening financial conditions in their industry. Therefore, a great deal of still current and useful material has been removed from publishers' stocking shelves. Several good technical journals are no longer in business. That said, I will acknowledge these books and technical journals where I quoted or referenced extracts of previous work. There sources are available from used book Internet-shop outlets and many engineers keep copies of their magazine subscriptions. The reader may be well advised to pursue secondary sources to obtain these and similar references if possible. This material may not always be available in publishers' latest offerings or on web-published sites.

of this book was published in 2007. Different grades of coal will have different impurities that have to be "cleared for use" or removed before the coal gas can be used as gas turbine fuel. If the impurities are dealt with appropriately, then the gas burns cleanly, as methane would. The other area that has received considerably more attention in the gas turbine field is that of carbon sequestration and storage (CSS). Formerly an item that was given attention by the Scandinavians (the Statoil projects have been left in this edition) because of their emissions taxes, it is now an item of interest globally. The USA is proceeding with the construction of FutureGen and this bodes well for the CSS industry in the USA. The Chinese are working with several partners, including the USA on CSS.

3. Other firms whose material is sourced or referenced can also be found using the Google search engine, for instance:
   Liburdi Engineering, Canada
   Westfalia Separators, Germany
   Rotadata, United Kingdom
   Statoil, Norway

This way the reader can get current information on new branches or country representatives.

4. CD Roms (Index for proceedings generally available on organiation websites)
   All of the firms mentioned in 1 and 2 preceding publish their works in conference proceedings such as ASME's (American Society of Mechanical Engineers) IGTI (International Gas Turbine Institute) and Pennwell PowerGen conferences. Conference proceedings can be purchased from the conference offices, in per-paper format or as a conference CD Rom.
5. Books (many available on CD Roms from publisher; check publishers websites)
   Soares, C. *Environmental Technology and Economics.* Boston: Butterworth-Heinemann, 1999.
   Soares, C. *Process Engineers Equipment Handbook.* New York: McGraw Hill, 2001.
   Bloch, H., and C. Soares. *Turbo Expanders.* Houston: Gulf, 2000.
   Bloch, H., and C. Soares. *Process Plant Machinery,* 2nd edn. Boston: Butterworth-Heinemann, 1998.
6. Technical Journals (abstracts if not entire paper generally available on organization websites)
   Throughout this book, I used relevant extracts of articles I have written for various technical journals, including but not limited to the following:
   Asian Electricity[++], 1997 through 1999.
   Middle East Electricity[++], 1997 through 1999.
   The Petroleum Economist, 1998.
   Modern Power Systems, 1998.
   Pollution Engineering, 2001.
   International Power Generation (IPG)[++], 1997 through 2006.
   European Power News[++], 2006

[++] These publications are not in print or circulation at this time, so individual article titles are not included here.

SOME SPECIFIC REFERENCES

Chapter 5, Cooling and Load Bearing Systems

Lubrication engineers: http://www.stle.org/cls/directory.cfm

Oil analysis: http://www.oilanalysis.com/learning_center/default.asp?sectionlink=LubricantSelection

Turbine Oils

Association of Iron and Steel Engineers. *The Lubrication Engineers Manual,* 2nd edn. Pittsburgh: AISE, 1996.

Bloch, H. P. *Practical Lubrication for Industrial Facilities.* Lithburn, GA: Fairmont Press, 2000.

Exxon Mobil Corporation. *Turbine Inspection Manual.* Fairfax, VA: Exxon.

Swift, S. T., D. K. Butler, and W. Dewald. *Turbine Oil Quality and Field Applications Requirements: Turbine Lubrication in the 21st Century.* ASTM STP 1407. West Conshohocken, PA: ASTM, 2001.

ASTM. *Standard Practice for In-Service Monitoring of Mineral Turbine Oils for Steam and Gas Turbines.* ASTM D4378-97, Annual Book of ASTM Standards, vol. 05.01. ASTM, 1997.

Chapter 9, Controls, Instrumentation, and Diagnostics

Control valves: http://www.ccivalve.com/presspower.shtml

Electrostatic charge monitoring: http://scitation.aip.org/getabs/servlet/GetabsServlet?prog=normal&id=JOTRE900012400000200028800000l&idtype=cvips&gifs=yes

Failure prevention: http://www.mfpt.org/tcproc_a_z.htm

Resensors/transducers: http://sensors-transducers.globalspec.com/Industrial-Directory/analysis_csi_vibration http://www.harcolabs.com/new-sensors.html

Temperature-sensitive paint: http://www.aiaa.org/content.cfm?pageid=231&masterid=62 http://www1.mengr.tamu.edu/thtl/projects.html

Ultrasound diagnosis: http://www.swantech.com/technology.html

Chapter 19, Basic Design Theory

Walsh and Fletcher. *Gas Turbine Performance.* Blackwell Science, 1998.

Unit Conversions

This section presents, in alphabetical order, all unit conversions likely to be required for gas turbine performance calculations. For all tables, except that covering pressure or stress, a quantity in given units is multiplied by the value in the next column to the right to convert it to the units in the column to the right of the conversion factor. The conversions presented may be combined, for example for acceleration mile/hs may be converted to ft^2/s by multiplying by 0.447 and then by 3.28084.

From	Multiply By	To/From This	Multiply By	To
Acceleration				
ft^2/s	0.3048	m/s^2	3.28084	re's
km/hs	0.27778	m/s^2	3.6	km/hs
mile/hs	1.609344	km/hs	0.621371	mile/hs
mile/hs	0.447	in/s^2	2.23714	mile/hs
Area				
in^2	645.16	M_{M2}	0.00155	in^2
in^2	6.4516^{-04}	m^2	1550.0	in^2
ft^2	0.092903	m^2	10.7639	ft^2
Density				
lb/ft^3	16.0185	kg/m^3	0.062428	lb/ft^3
lb/in^3	27.6799	kg/m^3	0.0361273	lb/in^3
lb/UKgal	0.0997763	kg/m^3	10.0224	lb/UKgal
lb/USgal	0.0830807	kg/m^3	12.0365	lb/USgal
Energy				
Btu	0.555558	Chu	1.8	Btu
Btu	778.169	ft lbf	0.28507^{-03}	Btu
Calorie	1.05506	kJ	0.947817	Btu
Calorie	401868	J	0.238846	Calorie
Calorie	0.0022046	Chu	453.597	Calorie
Chu	1899.105	J	0.0005265	Chu
Chu	1400.7	ft lbf	7.1393^{-04}	Chu
hp h	1.98+6	ft lbf	5.0505^{-05}	hp h
hp h	2.68452	MJ	0.372506	hp h
kWh	2.65522^{+6}	ft lbf	3.76617^{-05}	kWh
kWh	3.600	MJ	0.277778	kWh
Force				
kgf	9.80665	N	0.0101972	kgf
lhf	0.4535924	kgf	2.20462	lhf
lhf	3.2174	lb/ft s^2 p(11)	0.031081	lhf
lhf	4A4822	N	0.224809	lhf
tonf (Imperial)	9964.02	N	1.00361^{-04}	tonf

(Continued)

—cont'd

From	Multiply By	To/From This	Multiply By	To
Fuel consumption				
mile/UKgal	0.354006	km/liter	2.82481	mile/UKgal
mile/UKgal	0.83267	mile/USgal	1.20096	mile/UKgal
mile/USgal	0.29477	km/liter	3.39248	mile/USgal
Length				
ft	0.3048	m	3.28084	ft
in	0.0254	m	39.3701	in
in	25.4	mm	0.0393701	in
mile	1.609344	km	0.621371	mile
nautical mile*	1.852	km	0.539957	nautical mile
yard	0.9144	m	1.09361	yard
Mass				
lb	0.45359237	kg	2.20462	lb
lb	0.031056	slug	32.2174	lb
ounce	28.3495	g	0.035274	ounce
tonne	1000	kg	0.001	tonne
UK ton	1016.05	kg	9.84207^{-04}	UK ton
UK ton	1.01605	tonne	0.984207	UK ton
Moment of inertia				
lb ft^2	0.0421401	kg m^2	23.7304	lb ft^2
in ft^2	2.9264^{-04}	kg m^2	3417.17	kg m^2
Momentum—angular				
lb ft^2/s	0.0421401	kg m^2/s	23.7304	lb ft^2/s
Momentum—linear				
lb ft/s	0.138255	kg m/s	7.23301	lb ft/s
Power				
Btu/s	0.555558	Chu/s	1.799992	Btu/s
Btu/s	778.169	ft lbf/s	1.28507^{-03}	Btu/s
Btu/s	1.05506	kW	0.947817	Btu/s
Chu/s	2.54674	hp	0.39266	Chu/s
Chu/s	1.899105	kW	0.5265	Chu/s
ft lbf/s	1.35582	W	0.737562	ft lbf/s
hp	550	ft M/s	1.81818^{-03}	hp
hp	0.7457	kW	1.34102	hp
PS**	0.98632	hp	1.01387	PS
PS	75	kgf m/s	0.0133333	PS
PS	735.499	W	1359.62^{-06}	PS

—cont'd

From	Multiply By	To/From This	Multiply By	To
Specific energy				
Btu/lb	2.326	kJ/kg	0.429923	Btu/lb
Chu/lb	45066.1	ft^2/s^2	2.219^{-05}	Chu/lb
Chu/lb	4.1868	kJ/kg	0.238846	Chu/lb
ft lbf/lb	2.98907	J/kg	0.334553	ft lbf/lb
Specific fuel consumption (SFC)				
kg/kW h	0.735499	kg/PS h	1.35962	kg/kW h
lb/lbf h	0.10197	kg/N h	9.80665	lb/lbf h
lb/lbf h	1.0197	kg/daN h	0.980681	lb/lbf h
lb/hp h	0.60828	kg/kW h	1.64399	lb/hp h
lb/hp h	0.447387	kg/PS h	2.2352	lb/hp h
Specific heat				
Chu/lb K	1	131u/lb R	1	Chu/lb K
Chu/lb K	4186.8	J/kg K	2.38846^{-04}	Chu/lb K
ft lbf/lb R	5.38032	J/kg K	0.185863	ft lbf/lb R
HPs/lb K	1643.99	J/kg K	6.08277^{-04}	HPs/lb K
Specific thrust				
lbf s/lb	9.80665	N s/kg	0.1019716	lbf s/lb
Torque				
114 ft	0.138255	kgfm	7.23301	lbf ft
Kit	1.35582	N m	0.737562	lb ft
lbf in	0.112985	N m	8.85075	lbf in
Velocity—angular				
deg/s	0.0174533	rad/s	57.2958	deg/s
rev/min (rpm)	0.104720	rad/s	9.54930	rev/min (rpm)
rev/s	6.28319	rad/s	0.159155	rev/s
Velocity—linear				
ft/s	0.59248	kt	1.68782	ft/s
kt[†]	1.852	km/h	0.539957	kt
kt	0.514444	m/s	1.94384	kt
mile/h	1.46667	ft/s	0.681818	mile/h
mile/h	1.609344	km/h	0.621371	mile/h
mile/h	0.86896	kt	1.1508	mile/h
mile/h	0.44704	m/s	2.23694	mile/h
Viscosity—dynamic				
lb/ft s	1.48816	kg/m s	0.671969	lb/ft s
lb/in s	17.858	kg/m s	0.055997	lb/in s
$lbfb/ft^2$	0.172369	$MN s/m^2$	5.80151	$lbf h/ft^2$

(Continued)

—cont'd

From	Multiply By	To/From This	Multiply By	To
lbfs/ft^2	47.8803	kg/ms	0.0208854	lbfs/ft^2
Pa s (kg/m s)	1000	cP	0.001	kg/m s
Pa s	1.0	N OW	1.0	Pa s
Viscosity				
cSt	106	m^2/s	106	cSt
ft^2/s	0.092903	m^2/s	10,7639	ft^2/s
in^2/s	6.4516	cm^2/s	0.155	in^2/s
Volume				
in^3	16.3871	cm^3	0.0610237	in^3
ft^3	28.3168	liter	0.0353147	ft^3
UKgal	4.54609	liter	0.219969	UKgal
UKgal	1.20095	USgal	0.832674	UKgal
USgal	3.785	liter	0.2642	USgal
yard^3	0.764555	m^3	1.30795	yard^3

**This is the international nautical mile; the UK nautical mile is obsolete.*
***The PS is also called a metric.*
†This is for the international knot; the UK nautical mile is obsolete.

PRESSURE AND STRESS

See the following table for pressure and stress conversions.

Unit Conversions

	atm	bar	in Hg	in H_2O	kgf:cm^2	mm Hg	mm H_2O	lbf/in^2 (psi)	kPa
atm		1.01325	29.9213	406.782	1.03323	760.0	10,332.3	14.6959	101.325
bar	0.986923		29.53	401.463	1.01972	750.062	10,197.2	14.5038	100
in Hg	0.0334211	0.0338639		13.5951	0.0345316	25.4	345.316	0.491154	3.38639
in H_2O	0.0024583	0.002491	0.073556		0.00254	1.86832	25.4	0.036127	0.249089
kgf/cm^2	0.96784/	0.980665	28.959	393.701		735.559	10,000	14.2233	98.0665
mm Hg	0.0013158	0.0013332	0.03937	0.53524	0.0013595		13.5951	0.0193368	0.133322
mm H_2O	0.0000978	0.0000981	0.002896	0.0393701	0.0001	0.073556		0.0014223	0.009807
lbf/in^2 (psi)	0.068046	0.0689476	2.03602	27.68	0.070307	51.7149	703.07		6.89476
kPa	0.D098692	0.01	0.2953	4.01463	0.0101972	7.50062	101.972	0.145038	

Notes: To convert a value in the units in the left-hand column to those in the top row multiply by the number at the junction of the row and column. The conversion factors for columns of H_2O are for water at a uniform density of 1000 kg/m^3 under the standard gravity of 9.80665 m/s^2. The conversion factors for columns of Hg are for mercury at a uniform density of 13.590 kg/m^3 under the standard gravity of 9.80665 m/s^2. 1 kPa is equivalent to 1 kN/m^2.

TEMPERATURE

Unit Conversions

Conversion shown is to convert a quantity *from* units after the = sign *to* units before it.

$$C = K - 273.15$$

$$F = 1.8 \times K - 459.67$$

$$R = 1.8 \times K$$

$$K = C + 273.15$$

$$C = (R - 491.67)/1.8$$

$$F = R - 459.67$$

$$R = 1.8 \times (C + 273.15)$$

$$K = (F + 459.67)/1.8$$

$$C(F - 32)/1.8$$

$$F = 1.8 \times C + 32$$

$$R = F + 459.67$$

$$K = R/I.8$$

C Celsius
F Fahrenheit
K Kelvin
R Rankine

Index

Note: Page numbers with "b," "f," and "t" denote boxes; figures; and tables.

B

F

J

K

L

M

P

Q

R

S

T

U

V

W

Edwards Brothers Malloy
Ann Arbor MI. USA
July 25, 2017